MEMPHIS LIBRARIES
P9-CDJ-444

STOP
NOT TO BE
TAKEN FROM
THE
LIBRARY!

Comprehensive Natural Products Chemistry

Comprehensive Natural Products Chemistry

Editors-in-Chief
Sir Derek Barton†
Texas A&M University, USA
Koji Nakanishi
Columbia University, USA

Executive Editor
Otto Meth-Cohn
University of Sunderland, UK

Volume 1

POLYKETIDES AND OTHER SECONDARY METABOLITES INCLUDING FATTY ACIDS AND THEIR DERIVATIVES

Volume Editor
Ushio Sankawa
Toyama Medical and Pharmaceutical University, Japan

1999
ELSEVIER
AMSTERDAM – LAUSANNE – NEW YORK – OXFORD – SHANNON – SINGAPORE – TOKYO

Elsevier Science Ltd., The Boulevard, Langford Lane, Kidlington, Oxford, OX5 1GB, UK

First edition 1999

Library of Congress Cataloging-in-Publication Data
Comprehensive natural products chemistry / editors-in-chief, Sir Derek Barton, Koji Nakanishi ; executive editor, Otto Meth-Cohn. -- 1st ed.
p. cm.
Includes index.
Contents: v. 1. Polyketides and other secondary metabolites including fatty acids and their derivatives / volume editor Ushio Sankawa
1. Natural products. I. Barton, Derek, Sir, 1918-1998. II. Nakanishi, Koji, 1925- . III. Meth-Cohn, Otto.
QD415.C63 1999
547.7--dc21 98-15249

British Library Cataloguing in Publication Data
Comprehensive natural products chemistry
1. Organic compounds
I. Barton, Sir Derek, 1918-1998 II. Nakanishi Koji III. Meth-Cohn Otto
572.5

ISBN 0-08-042709-X (set : alk. paper)
ISBN 0-08-043153-4 (Volume 1 : alk. paper)

∞™ The paper used in this publication meets the minimum requirements of the American National Standard for Information Sciences—Permanence of Paper for Printed Library Materials, ANSI Z39.48–1984.

Typeset by BPC Digital Data Ltd., Glasgow, UK.
Printed and bound in Great Britain by BPC Wheatons Ltd., Exeter, UK.

Contents

The chapter *Cyclosporin: The Biosynthetic Path to a Lipopeptide* by H. Von Döhren and H. Kleinkauf was originally planned to be included in Volume 4. However, due to unforseen circumstances during the preparation of that volume it now appears as Chapter 20 in Volume 1.

Introduction

For many decades, Natural Products Chemistry has been the principal driving force for progress in Organic Chemistry.

In the past, the determination of structure was arduous and difficult. As soon as computing became easy, the application of X-ray crystallography to structural determination quickly surpassed all other methods. Supplemented by the equally remarkable progress made more recently by Nuclear Magnetic Resonance techniques, determination of structure has become a routine exercise. This is even true for enzymes and other molecules of a similar size. Not to be forgotten remains the progress in mass spectrometry which permits another approach to structure and, in particular, to the precise determination of molecular weight.

There have not been such revolutionary changes in the partial or total synthesis of Natural Products. This still requires effort, imagination and time. But remarkable syntheses have been accomplished and great progress has been made in stereoselective synthesis. However, the one hundred percent yield problem is only solved in certain steps in certain industrial processes. Thus there remains a great divide between the reactions carried out in living organisms and those that synthetic chemists attain in the laboratory. Of course Nature edits the accuracy of DNA, RNA, and protein synthesis in a way that does not apply to a multi-step Organic Synthesis.

Organic Synthesis has already a significant component that uses enzymes to carry out specific reactions. This applies particularly to lipases and to oxidation enzymes. We have therefore, given serious attention to enzymatic reactions.

No longer standing in the wings, but already on-stage, are the wonderful tools of Molecular Biology. It is now clear that multi-step syntheses can be carried out in one vessel using multiple cloned enzymes. Thus, Molecular Biology and Organic Synthesis will come together to make economically important Natural Products.

From these preliminary comments it is clear that Natural Products Chemistry continues to evolve in different directions interacting with physical methods, Biochemistry, and Molecular Biology all at the same time.

This new Comprehensive Series has been conceived with the common theme of "How does Nature make all these molecules of life?" The principal idea was to organize the multitude of facts in terms of Biosynthesis rather than structure. The work is not intended to be a comprehensive listing of natural products, nor is it intended that there should be any detail about biological activity. These kinds of information can be found elsewhere.

The work has been planned for eight volumes with one more volume for Indexes. As far as we are aware, a broad treatment of the whole of Natural Products Chemistry has never been attempted before. We trust that our efforts will be useful and informative to all scientific disciplines where Natural Products play a role.

D. H. R. Barton† K. Nakanishi O. Meth-Cohn

Preface

It is surprising indeed that this work is the first attempt to produce a "comprehensive" overview of Natural Products beyond the student text level. However, the awe-inspiring breadth of the topic, which in many respects is still only developing, is such as to make the job daunting to anyone in the field. Fools rush in where angels fear to tread and the particular fool in this case was myself, a lifelong enthusiast and reader of the subject but with no research base whatever in the field!

Having been involved in several of the *Comprehensive* works produced by Pergamon Press, this omission intrigued me and over a period of gestation I put together a rough outline of how such a work could be written and presented it to Pergamon. To my delight they agreed that the project was worthwhile and in short measure Derek Barton was approached and took on the challenge of fleshing out this framework with alacrity. He also brought his long-standing friend and outstanding contributor to the field, Koji Nakanishi, into the team. With Derek's knowledge of the whole field, the subject was broken down into eight volumes and an outstanding team of internationally recognised Volume Editors was appointed.

We used Derek's 80th birthday as a target for finalising the work. Sadly he died just a few months before reaching this milestone. This work therefore is dedicated to the memory of Sir Derek Barton, Natural Products being the area which he loved best of all.

OTTO METH-COHN
Executive Editor

SIR DEREK BARTON

Sir Derek Barton, who was Distinguished Professor of Chemistry at Texas A&M University and holder of the Dow Chair of Chemical Invention died on March 16, 1998 in College Station, Texas of heart failure. He was 79 years old and had been Chairman of the Executive Board of Editors for Tetrahedron Publications since 1979.

Barton was considered to be one of the greatest organic chemists of the twentieth century whose work continues to have a major influence on contemporary science and will continue to do so for future generations of chemists.

Derek Harold Richard Barton was born on September 8, 1918 in Gravesend, Kent, UK and graduated from Imperial College, London with the degrees of B.Sc. (1940) and Ph.D. (1942). He carried out work on military intelligence during World War II and after a brief period in industry, joined the faculty at Imperial College. It was an early indication of the breadth and depth of his chemical knowledge that his lectureship was in physical chemistry. This research led him into the mechanism of elimination reactions and to the concept of molecular rotation difference to correlate the configurations of steroid isomers. During a sabbatical leave at Harvard in 1949–1950 he published a paper on the "Conformation of the Steroid Nucleus" (*Experientia*, 1950, **6**, 316) which was to bring him the Nobel Prize in Chemistry in 1969, shared with the Norwegian chemist, Odd Hassel. This key paper (only four pages long) altered the way in which chemists thought about the shape and reactivity of molecules, since it showed how the reactivity of functional groups in steroids depends on their axial or equatorial positions in a given conformation. Returning to the UK he held Chairs of Chemistry at Birkbeck College and Glasgow University before returning to Imperial College in 1957, where he developed a remarkable synthesis of the steroid hormone, aldosterone, by a photochemical reaction known as the Barton Reaction (nitrite photolysis). In 1978 he retired from Imperial College and became Director of the Natural Products Institute (CNRS) at Gif-sur-Yvette in France where he studied the invention of new chemical reactions, especially the chemistry of radicals, which opened up a whole new area of organic synthesis involving Gif chemistry. In 1986 he moved to a third career at Texas A&M University as Distinguished Professor of Chemistry and continued to work on novel reactions involving radical chemistry and the oxidation of hydrocarbons, which has become of great industrial importance. In a research career spanning more than five decades, Barton's contributions to organic chemistry included major discoveries which have profoundly altered our way of thinking about chemical structure and reactivity. His chemistry has provided models for the biochemical synthesis of natural products including alkaloids, antibiotics, carbohydrates, and DNA. Most recently his discoveries led to models for enzymes which oxidize hydrocarbons, including methane monooxygenase.

The following are selected highlights from his published work:

The 1950 paper which launched Conformational Analysis was recognized by the Nobel Prize Committee as the key contribution whereby the third dimension was added to chemistry. This work alone transformed our thinking about the connection between stereochemistry and reactivity, and was later adapted from small molecules to macromolecules e.g., DNA, and to inorganic complexes.

Barton's breadth and influence is illustrated in "Biogenetic Aspects of Phenol Oxidation" (*Festschr. Arthur Stoll*, 1957, 117). This theoretical work led to many later experiments on alkaloid biosynthesis and to a set of rules for *ortho-para*-phenolic oxidative coupling which allowed the predication of new natural product systems before they were actually discovered and to the correction of several erroneous structures.

In 1960, his paper on the remarkably short synthesis of the steroid hormone aldosterone (*J. Am. Chem. Soc.*, 1960, **82**, 2641) disclosed the first of many inventions of new reactions—in this case nitrite photolysis—to achieve short, high yielding processes, many of which have been patented and are used worldwide in the pharmaceutical industry.

Moving to 1975, by which time some 500 papers had been published, yet another "Barton reaction" was born—"The Deoxygenation of Secondary Alcohols" (*J. Chem. Soc. Perkin Trans. 1*, 1975, 1574), which has been very widely applied due to its tolerance of quite hostile and complex local environments in carbohydrate and nucleoside chemistry. This reaction is the chemical counterpart to ribonucleotide→ deoxyribonucleotide reductase in biochemistry and, until the arrival of the Barton reaction, was virtually impossible to achieve.

In 1985, "Invention of a New Radical Chain Reaction" involved the generation of carbon radicals from carboxylic acids (*Tetrahedron*, 1985, **41**, 3901). The method is of great synthetic utility and has been used many times by others in the burgeoning area of radicals in organic synthesis.

These recent advances in synthetic methodology were remarkable since his chemistry had virtually no precedent in the work of others. The radical methodology was especially timely in light of the significant recent increase in applications for fine chemical syntheses, and Barton gave the organic community an entrée into what will prove to be one of the most important methods of the twenty-first century. He often said how proud he was, at age 71, to receive the ACS Award for Creativity in Organic Synthesis for work published in the preceding five years.

Much of Barton's more recent work is summarized in the articles "The Invention of Chemical Reactions—The Last 5 Years" (*Tetrahedron*, 1992, **48**, 2529) and "Recent Developments in Gif Chemistry" (*Pure Appl. Chem.*, 1997, **69**, 1941).

Working 12 hours a day, Barton's stamina and creativity remained undiminished to the day of his death. The author of more than 1000 papers in chemical journals, Barton also held many successful patents. In addition to the Nobel Prize he received many honors and awards including the Davy, Copley, and Royal medals of the Royal Society of London, and the Roger Adams and Priestley Medals of the American Chemical Society. He held honorary degrees from 34 universities. He was a Fellow of the Royal Societies of London and Edinburgh, Foreign Associate of the National Academy of Sciences (USA), and Foreign Member of the Russian and Chinese Academies of Sciences. He was knighted by Queen Elizabeth in 1972, received the Légion d'Honneur (Chevalier 1972; Officier 1985) from France, and the Order of the Rising Sun from the Emperor of Japan. In his long career, Sir Derek trained over 300 students and postdoctoral fellows, many of whom now hold major positions throughout the world and include some of today's most distinguished organic chemists.

For those of us who were fortunate to know Sir Derek personally there is no doubt that his genius and work ethic were unique. He gave generously of his time to students and colleagues wherever he traveled and engendered such great respect and loyalty in his students and co-workers, that major symposia accompanied his birthdays every five years beginning with the 60th, ending this year with two celebrations just before his 80th birthday.

With the death of Sir Derek Barton, the world of science has lost a major figure, who together with Sir Robert Robinson and Robert B. Woodward, the cofounders of *Tetrahedron*, changed the face of organic chemistry in the twentieth century.

Professor Barton is survived by his wife, Judy, and by a son, William from his first marriage, and three grandchildren.

A. I. SCOTT
Texas A&M University

Reprinted from *Tetrahedron*, 1998, **54**, 8847

Contributors to Volume 1

Dr. C. Abell
University Chemical Laboratory, University of Cambridge, Lensfield Road, Cambridge, CB2 1EW, UK

Dr. I. D. G. Campuzano
Division of Biochemistry and Molecular Biology, School of Biological Sciences, University of Southampton, Basset Crescent East, Southampton, SO16 7PX, UK

Ms. H. Daiyasu
Department of Bioinformatics, Biomolecular Engineering Research Institute, 6-2-3 Furuedai, Suita, Osaka 565-0874, Japan

Dr. L. B. Davin
Institute of Biological Chemistry, Washington State University, PO Box 646340, Pullman, WA 99164-6340, USA

Dr. R. A. Dixon
Plant Biology Division, Samuel Roberts Noble Foundation, PO Box 2180, Ardmore, OK 73402, USA

Professor H. G. Floss
Department of Chemistry, University of Washington, Box 351700, Seattle, WA 98195, USA

Professor G. Forkmann
Lehrstuhl für Zierpflanzenbau, Technische Universität München, D-85350 Freising, Germany

Dr I. Fujii
Faculty of Pharmaceutical Sciences, The University of Tokyo, 7-3-1 Hongo, Bunkyo ku, Tokyo 113, Japan

Professor W. H. Gerwick
College of Pharmacy, Oregon State University, Corvallis, OR 97331-3507, USA

Dr. F. Greulich
Department of Bioscience and Chemistry, Faculty of Agriculture, Hokkaido University, Kita-ku, Kita 9, Nishi 9, Sapporo 060-0809, Japan

Dr. N. Hamanaka
Research Institute, Ono Pharmaceutical Co. Ltd., 3-1-1 Sakurai, Shimamoto-cho, Mishima-gun, Osaka 618, Japan

Professor A. Hatanaka
Department of Life Science, Graduate School of Integrated Science and Art, University of East Asia, 2-1 Gakuen Ichinomiya, Shimonoseki 751, Japan

Dr. W. Heller
GSF Forschungszentrum für Umwelt und Gesundheit, Institut für Biochemische Pflanzenpathologie, D-85764 Oberschleissheim, Germany

Professor M. A. Hughes
Department of Biochemistry and Genetics, The Medical School, University of Newcastle upon Tyne, Newcastle upon Tyne, NE2 4HH, UK

Professor A. Ichihara
Department of Bioscience and Chemistry, Faculty of Agriculture, Hokkaido University, Kita-9-jo, Nishi-9-chome, Kita-ku, Sapporo 060, Japan

Dr. A. Iwamoto-Kihara
Department of Life Sciences, Graduate School of Arts and Sciences, The University of Tokyo, Tokyo 153-8902, Japan

Professor S. Iwasaki
Institute of Molecular and Cellular Biosciences, The University of Tokyo, 1-1-1 Yayoi, Bunkyo-ku, Tokyo 113, Japan

Professor A. Kawaguchi
Department of Life Sciences, Graduate School of Arts and Sciences, The University of Tokyo, Tokyo 153-8902, Japan

Professor C. Khosla
Department of Chemical Engineering, Chemistry and Biochemistry, Stanford University, Stanford, CA 94305-5025, USA

Mr. R. Kikuno
Department of Bioinformatics, Biomolecular Engineering Research Institute, 6-2-3 Furuedai, Suita, Osaka 565-0874, Japan

Dr. H. Kleinkauf
Institute of Biochemistry and Molecular Biology, Technische Universität Berlin, Franklinstrasse 29, D-10587 Berlin, Germany

Dr. D. Kreusch
Institut für Pharmazeutische Biologie, Philipps-Universität Marburg, Deutschhausstrasse 17A, D-35037 Marburg, Germany

Professor E. Leistner
Institut für Pharmazeutische Biologie, Rheinische Friedrich-Wilhelms-Universität Bonn, Nussallee 6, D-53115 Bonn, Germany

Professor N. G. Lewis
Director, Institute of Biological Chemistry, Washington State University, PO Box 646340, Pullman, WA 99164-6340, USA

Dr. B. J. J. Lugtenberg
Institute of Molecular Plant Sciences, Leiden University, Clusius Laboratory, Wassenaarseweg 64, NL-2333 AL Leiden, The Netherlands

Dr. P. Lüer
Institut für Pharmazeutische Biologie, Philipps-Universität Marburg, Deutschhausstrasse 17A, D-35037 Marburg, Germany

Professor U. Matern
Institut für Pharmazeutische Biologie, Philipps-Universität Marburg, Deutschhausstrasse 17A, D-35037 Marburg, Germany

Dr. R. E. Minto
Department of Chemistry, The Johns Hopkins University, Remsen Hall, 3400 North Charles Street, Baltimore, MD 21218, USA

Dr. B. S. Moore
Department of Chemistry, University of Washington, Box 351700, Seattle, WA 98195, USA

Professor A. Murai
Division of Chemistry, Graduate School of Science, Hokkaido University, Kita-10-jo, Nishi-8-chome, Kita-ku, Sapporo 060, Japan

Dr. H. Oikawa
Department of Bioscience and Chemistry, Faculty of Agriculture, Hokkaido University, Sapporo 060, Japan

Professor R. J. Parry
Department of Chemistry, Rice University, MS60, 6100 South Main, Houston, TX 77005-1892, USA

Dr. M. Richardson
Department of Chemical Engineering, Chemistry and Biochemistry, Stanford University, Stanford, CA, 94305-5025, USA

Dr.T. Ritsema
Institute of Molecular Plant Sciences, Leiden University, Clusius Laboratory, Wassenaarseweg 64, NL-2333 AL Leiden, The Netherlands

Dr. S. Sakuda
Department of Applied Biological Chemistry, The University of Tokyo, 1-1-1 Yayoi, Bunkyo-ku, Tokyo 113, Japan

Professor U. Sankawa
Faculty of Pharmaceutical Sciences, Toyama Medical and Pharmaceutical University, 2630 Sigitani, Toyama 930-01, Japan

Dr. N. Sato
School of Life Science, Tokyo University of Pharmacy and Life Science, Horinouchi, Tokyo 192-0392, Japan

Professor J. Schröder
Institute für Biologie II, Universität Freiburg, Schänzlestrasse 1, D-79104 Freiburg, Germany

Professor H. Seto
Institute of Molecular and Cellular Biosciences, The University of Tokyo, 1-1-1 Yayoi, Bunkyo-ku, Tokyo 113-0032, Japan

Professor P. M. Shoolingin-Jordan
Division of Biochemistry and Molecular Biology, School of Biological Sciences, University of Southampton, Basset Crescent East, Southampton, SO16 7PX, UK

Professor H. P. Spaink
Institute of Molecular Plant Sciences, Leiden University, Clusius Laboratory, Wassenaarseweg 64, NL-2333 AL Leiden, The Netherlands

Dr. J. Staunton
University Chemical Laboratory, University of Cambridge, Lensfield Road, Cambridge, CB2 1EW, UK

Dr. T. Sugiura
Faculty of Pharmaceutical Sciences, Teikyo University, Sagamiko, Kanagawa 199-01, Japan

Dr. H. Toh
Department of Bioinformatics, Biomolecular Engineering Research Institute, 6-2-3 Furuedai, Suita, Osaka 565-0874, Japan

Professor C. A. Townsend
Department of Chemistry, The Johns Hopkins University, Remsen Hall, 3400 North Charles Street, Baltimore, MD 21218, USA

Dr. H. von Döhren
Institute of Biochemistry and Molecular Biology, Technische Universität Berlin, Franklinstrasse 29, D-10587 Berlin, Germany

Professor K. Waku
Faculty of Pharmaceutical Sciences, Teikyo University, Sagamiko, Kanagawa 199-01, Japan

Dr. B. Wilkinson
Bioprocessing Research Unit, GlaxoWellcome Research and Development, Medicines Research Centre, Gunnels Wood Road, Stevenage, SG1 2NY

Dr. Y. Yamada
Department of Biotechnology, Faculty of Engineering, Osaka University, 2-1 Yamada-oka, Suita-shi, Osaka 565, Japan

Professor S. Yamamoto
Department of Biochemistry, Tokushima University School of Medicine, 3-18-15 Kuramoto-cho, Tokushima 770, Japan

Professor T. Yoshihara
Department of Bioscience and Chemistry, Faculty of Agriculture, Hokkaido University, Kita-ku, Kita 9, Nishi-9, Sapporo 060-0809, Japan

Abbreviations

The most commonly used abbreviations in *Comprehensive Natural Products Chemistry* are listed below. Please note that in some instances these may differ from those used in other branches of chemistry

A	adenine
ABA	abscisic acid
Ac	acetyl
ACAC	acetylacetonate
ACTH	adrenocorticotropic hormone
ADP	adenosine 5'-diphosphate
AIBN	2,2'-azobisisobutyronitrile
Ala	alanine
AMP	adenosine 5'-monophosphate
APS	adenosine 5'-phosphosulfate
Ar	aryl
Arg	arginine
ATP	adenosine 5'-triphosphate
B	nucleoside base (adenine, cylosine, guanine, thymine or uracil)
9-BBN	9-borabicyclo[3.3.1]nonane
BOC	*t*-butoxycarbonyl (or carbo-*t*-butoxy)
BSA	*N*,*O*-bis(trimethylsilyl)acetamide
BSTFA	*N*,*O*-bis(trimethylsilyl)trifluoroacetamide
Bu	butyl
Bu^n	*n*-butyl
Bu^i	isobutyl
Bu^s	*s*-butyl
Bu^t	*t*-butyl
Bz	benzoyl
CAN	ceric ammonium nitrate
CD	cyclodextrin
CDP	cytidine 5'-diphosphate
CMP	cytidine 5'-monophosphate
CoA	coenzyme A
COD	cyclooctadiene
COT	cyclooctatetraene
Cp	η^5-cyclopentadiene
Cp*	pentamethylcyclopentadiene
12-Crown-4	1,4,7,10-tetraoxacyclododecane
15-Crown-5	1,4,7,10,13-pentaoxacyclopentadecane
18-Crown-6	1,4,7,10,13,16-hexaoxacyclooctadecane
CSA	camphorsulfonic acid
CSI	chlorosulfonyl isocyanate
CTP	cytidine 5'-triphosphate
cyclic AMP	adenosine 3',5'-cyclic monophosphoric acid
CySH	cysteine
DABCO	1,4-diazabicyclo[2.2.2]octane
DBA	dibenz[*a*,*h*]anthracene
DBN	1,5-diazabicyclo[4.3.0]non-5-ene

DBU	1,8-diazabicyclo[5.4.0]undec-7-ene
DCC	dicyclohexylcarbodiimide
DEAC	diethylaluminum chloride
DEAD	diethyl azodicarboxylate
DET	diethyl tartrate (+ or -)
DHET	dihydroergotoxine
DIBAH	diisobutylaluminum hydride
Diglyme	diethylene glycol dimethyl ether (or bis(2-methoxyethyl)ether)
DiHPhe	2,5-dihydroxyphenylalanine
Dimsyl Na	sodium methylsulfinylmethide
DIOP	2,3-*O*-isopropylidene-2,3-dihydroxy-1,4-bis(diphenylphosphino)butane
dipt	diisopropyl tartrate (+ or -)
DMA	dimethylacetamide
DMAD	dimethyl acetylenedicarboxylate
DMAP	4-dimethylaminopyridine
DME	1,2-dimethoxyethane (glyme)
DMF	dimethylformamide
DMF-DMA	dimethylformamide dimethyl acetal
DMI	1,3-dimethyl-2-imidazalidinone
DMSO	dimethyl sulfoxide
DMTSF	dimethyl(methylthio)sulfonium fluoroborate
DNA	deoxyribonucleic acid
DOCA	deoxycorticosterone acetate
EADC	ethylaluminum dichloride
EDTA	ethylenediaminetetraacetic acid
EEDQ	*N*-ethoxycarbonyl-2-ethoxy-1,2-dihydroquinoline
Et	ethyl
EVK	ethyl vinyl ketone
FAD	flavin adenine dinucleotide
FI	flavin
FMN	flavin mononucleotide
G	guanine
GABA	4-aminobutyric acid
GDP	guanosine 5'-diphosphate
GLDH	glutamate dehydrogenase
gln	glutamine
Glu	glutamic acid
Gly	glycine
GMP	guanosine 5'-monophosphate
GOD	glucose oxidase
G-6-P	glucose-6-phosphate
GTP	guanosine 5'-triphosphate
Hb	hemoglobin
His	histidine
HMPA	hexamethylphosphoramide (or hexamethylphosphorous triamide)
Ile	isoleucine
INAH	isonicotinic acid hydrazide
IpcBH	isopinocampheylborane
Ipc_2BH	diisopinocampheylborane
KAPA	potassium 3-aminopropylamide
K-Slectride	potassium tri-*s*-butylborohydride

LAH	lithium aluminum hydride
LAP	leucine aminopeptidase
LDA	lithium diisopropylamide
LDH	lactic dehydrogenase
Leu	leucine
LICA	lithium isopropylcyclohexylamide
L-Selectride	lithium tri-*s*-butylborohydride
LTA	lead tetraacetate
Lys	lysine
MCPBA	*m*-chloroperoxybenzoic acid
Me	methyl
MEM	methoxyethoxymethyl
MEM-Cl	ß-methoxyethoxymethyl chloride
Met	methionine
MMA	methyl methacrylate
MMC	methyl magnesium carbonate
MOM	methoxymethyl
Ms	mesyl (or methanesulfonyl)
MSA	methanesulfonic acid
MsCl	methanesulfonyl chloride
MVK	methyl vinyl ketone
NAAD	nicotinic acid adenine dinucleotide
NAD	nicotinamide adenine dinucleotide
NADH	nicotinamide adenine dinucleotide phosphate, reduced
NBS	*N*-bromosuccinimider
NMO	*N*-methylmorpholine *N*-oxide monohydrate
NMP	*N*-methylpyrrolidone
PCBA	*p*-chlorobenzoic acid
PCBC	*p*-chlorobenzyl chloride
PCBN	*p*-chlorobenzonitrile
PCBTF	*p*-chlorobenzotrifluoride
PCC	pyridinium chlorochromate
PDC	pyridinium dichromate
PG	prostaglandin
Ph	phenyl
Phe	phenylalanine
Phth	phthaloyl
PPA	polyphosphoric acid
PPE	polyphosphate ester (or ethyl *m*-phosphate)
Pr	propyl
Pr^i	isopropyl
Pro	proline
Py	pyridine
RNA	ribonucleic acid
Rnase	ribonuclease
Ser	serine
Sia_2BH	disiamylborane
TAS	tris(diethylamino)sulfonium
TBAF	tetra-*n*-butylammonium fluoroborate
TBDMS	*t*-butyldimethylsilyl
TBDMS-Cl	*t*-butyldimethylsilyl chloride
TBDPS	*t*-butyldiphenylsilyl
TCNE	tetracyanoethene

TES	triethylsilyl
TFA	trifluoracetic acid
TFAA	trifluoroacetic anhydride
THF	tetrahydrofuran
THF	tetrahydrofolic acid
THP	tetrahydropyran (or tetrahydropyranyl)
Thr	threonine
TMEDA	*N,N,N',N'*,tetramethylethylenediamine[1,2-bis(dimethylamino)ethane]
TMS	trimethylsilyl
TMS-Cl	trimethylsilyl chloride
TMS-CN	trimethylsilyl cyanide
Tol	toluene
TosMIC	tosylmethyl isocyanide
TPP	tetraphenylporphyrin
Tr	trityl (or triphenylmethyl)
Trp	tryptophan
Ts	tosyl (or *p*-toluenesulfonyl)
TTFA	thallium trifluoroacetate
TTN	thallium(III) nitrate
Tyr	tyrosine
Tyr-OMe	tyrosine methyl ester
U	uridine
UDP	uridine 5'-diphosphate
UMP	uridine 5'-monophosphate

Contents of All Volumes

An Historical Perspective of Natural Products Chemistry

KOJI NAKANISHI

Columbia University, New York, USA

To give an account of the rich history of natural products chemistry in a short essay is a daunting task. This brief outline begins with a description of ancient folk medicine and continues with an outline of some of the major conceptual and experimental advances that have been made from the early nineteenth century through to about 1960, the start of the modern era of natural products chemistry. Achievements of living chemists are noted only minimally, usually in the context of related topics within the text. More recent developments are reviewed within the individual chapters of the present volumes, written by experts in each field. The subheadings follow, in part, the sequence of topics presented in Volumes 1–8.

1. ETHNOBOTANY AND "NATURAL PRODUCTS CHEMISTRY"

Except for minerals and synthetic materials our surroundings consist entirely of organic natural products, either of prebiotic organic origins or from microbial, plant, or animal sources. These materials include polyketides, terpenoids, amino acids, proteins, carbohydrates, lipids, nucleic acid bases, RNA and DNA, etc. Natural products chemistry can be thought of as originating from mankind's curiosity about odor, taste, color, and cures for diseases. Folk interest in treatments for pain, for food-poisoning and other maladies, and in hallucinogens appears to go back to the dawn of humanity

For centuries China has led the world in the use of natural products for healing. One of the earliest health science anthologies in China is the Nei Ching, whose authorship is attributed to the legendary Yellow Emperor (thirtieth century BC), although it is said that the dates were backdated from the third century by compilers. Excavation of a Han Dynasty (206 BC–AD 220) tomb in Hunan Province in 1974 unearthed decayed books, written on silk, bamboo, and wood, which filled a critical gap between the dawn of medicine up to the classic Nei Ching; Book 5 of these excavated documents lists 151 medical materials of plant origin. Generally regarded as the oldest compilation of Chinese herbs is Shen Nung Pen Ts'ao Ching (Catalog of Herbs by Shen Nung), which is believed to have been revised during the Han Dynasty; it lists 365 materials. Numerous revisions and enlargements of Pen Ts'ao were undertaken by physicians in subsequent dynasties, the ultimate being the Pen Ts'ao Kang Mu (General Catalog of Herbs) written by Li Shih-Chen over a period of 27 years during the Ming Dynasty (1573–1620), which records 1898 herbal drugs and 8160 prescriptions. This was circulated in Japan around 1620 and translated, and has made a huge impact on subsequent herbal studies in Japan; however, it has not been translated into English. The number of medicinal herbs used in 1979 in China numbered 5267. One of the most famous of the Chinese folk herbs is the ginseng root *Panax ginseng*, used for health maintenance and treatment of various diseases. The active principles were thought to be the saponins called ginsenosides but this is now doubtful; the effects could well be synergistic between saponins, flavonoids, etc. Another popular folk drug, the extract of the Ginkgo tree, *Ginkgo biloba* L., the only surviving species of the Paleozoic era (250 million years ago) family which became extinct during the last few million years, is mentioned in the Chinese Materia Medica to have an effect in improving memory and sharpening mental alertness. The main constituents responsible for this are now understood to be ginkgolides and flavonoids, but again not much else is known. Clarifying the active constituents and mode of (synergistic) bioactivity of Chinese herbs is a challenging task that has yet to be fully addressed.

The Assyrians left 660 clay tablets describing 1000 medicinal plants used around 1900–400 BC, but the best insight into ancient pharmacy is provided by the two scripts left by the ancient Egyptians, who were masters of human anatomy and surgery because of their extensive mummification practices. The Edwin Smith Surgical Papyrus purchased by Smith in 1862 in Luxor (now in the New York Academy of Sciences collection), is one of the most important medicinal documents of the ancient Nile Valley, and describes the healer's involvement in surgery, prescription, and healing practices using plants, animals, and minerals. The Ebers Papyrus, also purchased by Edwin Smith in 1862, and then acquired by Egyptologist George Ebers in 1872, describes 800 remedies using plants, animals, minerals, and magic. Indian medicine also has a long history, possibly dating back to the second millennium BC. The Indian materia medica consisted mainly of vegetable drugs prepared from plants but also used animals, bones, and minerals such as sulfur, arsenic, lead, copper sulfate, and gold. Ancient Greece inherited much from Egypt, India, and China, and underwent a gradual transition from magic to science. Pythagoras (580–500 BC) influenced the medical thinkers of his time, including Aristotle (384–322 BC), who in turn affected the medical practices of another influential Greek physician Galen (129–216). The Iranian physician Avicenna (980–1037) is noted for his contributions to Aristotelian philosophy and medicine, while the German-Swiss physician and alchemist Paracelsus (1493–1541) was an early champion who established the role of chemistry in medicine.

The rainforests in Central and South America and Africa are known to be particularly abundant in various organisms of interest to our lives because of their rich biodiversity, intense competition, and the necessity for self-defense. However, since folk-treatments are transmitted verbally to the next generation via shamans who naturally have a tendency to keep their plant and animal sources confidential, the recipes tend to get lost, particularly with destruction of rainforests and the encroachment of "civilization." Studies on folk medicine, hallucinogens, and shamanism of the Central and South American Indians conducted by Richard Schultes (Harvard Botanical Museum, emeritus) have led to renewed activity by ethnobotanists, recording the knowledge of shamans, assembling herbaria, and transmitting the record of learning to the village.

Extracts of toxic plants and animals have been used throughout the world for thousands of years for hunting and murder. These include the various arrow poisons used all over the world. *Strychnos* and *Chondrodendron* (containing strychnine, etc.) were used in South America and called "curare," *Strophanthus* (strophantidine, etc.) was used in Africa, the latex of the upas tree *Antiaris toxicaria* (cardiac glycosides) was used in Java, while *Aconitum napellus*, which appears in Greek mythology (aconitine) was used in medieval Europe and Hokkaido (by the Ainus). The Colombian arrow poison is from frogs (batrachotoxins; 200 toxins have been isolated from frogs by B. Witkop and J. Daly at NIH). Extracts of *Hyoscyamus niger* and *Atropa belladonna* contain the toxic tropane alkaloids, for example hyoscyamine, belladonnine, and atropine. The belladonna berry juice (atropine) which dilates the eye pupils was used during the Renaissance by ladies to produce doe-like eyes (belladona means beautiful woman). The Efik people in Calabar, southeastern Nigeria, used extracts of the calabar bean known as esere (physostigmine) for unmasking witches. The ancient Egyptians and Chinese knew of the toxic effect of the puffer fish, fugu, which contains the neurotoxin tetrodotoxin (Y. Hirata, K. Tsuda, R. B. Woodward).

When rye is infected by the fungus *Claviceps purpurea*, the toxin ergotamine and a number of ergot alkaloids are produced. These cause ergotism or the "devil's curse," "St. Anthony's fire," which leads to convulsions, miscarriages, loss of arms and legs, dry gangrene, and death. Epidemics of ergotism occurred in medieval times in villages throughout Europe, killing tens of thousands of people and livestock; Julius Caesar's legions were destroyed by ergotism during a campaign in Gaul, while in AD 994 an estimated 50,000 people died in an epidemic in France. As recently as 1926, a total of 11,000 cases of ergotism were reported in a region close to the Urals. It has been suggested that the witch hysteria that occurred in Salem, Massachusetts, might have been due to a mild outbreak of ergotism. Lysergic acid diethylamide (LSD) was first prepared by A. Hofmann, Sandoz Laboratories, Basel, in 1943 during efforts to improve the physiological effects of the ergot alkaloids when he accidentally inhaled it. "On Friday afternoon, April 16, 1943," he wrote, "I was seized by a sensation of restlessness... ." He went home from the laboratory and "perceived an uninterrupted stream of fantastic dreams ... ." (*Helvetica Chimica Acta*).

Numerous psychedelic plants have been used since ancient times, producing visions, mystical fantasies (cats and tigers also seem to have fantasies?, see nepetalactone below), sensations of flying, glorious feelings in warriors before battle, etc. The ethnobotanists Wasson and Schultes identified "ololiqui," an important Aztec concoction, as the seeds of the morning glory *Rivea corymbosa* and gave the seeds to Hofmann who found that they contained lysergic acid amides similar to but less potent than LSD. Iboga, a powerful hallucinogen from the root of the African shrub *Tabernanthe iboga*, is used by the Bwiti cult in Central Africa who chew the roots to obtain relief from fatigue and hunger; it contains the alkaloid ibogamine. The powerful hallucinogen used for thousands of years by the American Indians, the peyote cactus, contains mescaline and other alkaloids. The Indian hemp plant, *Cannabis sativa*, has been used for making rope since 3000 BC, but when it is used for its pleasure-giving effects it is called cannabis and has been known in central Asia, China, India, and the Near East since ancient times. Marijuana, hashish (named after the Persian founder of the Assassins of the eleventh century, Hasan-e Sabbah), charas, ghanja, bhang, kef, and dagga are names given to various preparations of the hemp plant. The constituent responsible for the mind-altering effect is 1-tetrahydrocannabinol (also referred to as 9-THC) contained in 1%. R. Mechoulam (1930–, Hebrew University) has been the principal worker in the cannabinoids, including structure determination and synthesis of 9-THC (1964 to present); the Israeli police have also made a contribution by providing Mechoulam with a constant supply of marijuana. Opium (morphine) is another ancient drug used for a variety of pain-relievers and it is documented that the Sumerians used poppy as early as 4000 BC; the narcotic effect is present only in seeds before they are fully formed. The irritating secretion of the blister beetles, for example *Mylabris* and the European species *Lytta vesicatoria*, commonly called Spanish fly, was used medically as a topical skin irritant to remove warts but was also a major ingredient in so-called love potions (constituent is cantharidin, stereospecific synthesis in 1951, G. Stork, 1921–; prep. scale high-pressure Diels–Alder synthesis in 1985, W. G. Dauben, 1919–1996).

Plants have been used for centuries for the treatment of heart problems, the most important being the foxgloves *Digitalis purpurea* and *D. lanata* (digitalin, diginin) and *Strophanthus gratus* (ouabain). The bark of cinchona *Cinchona officinalis* (called quina-quina by the Indians) has been used widely among the Indians in the Andes against malaria, which is still one of the major infectious diseases; its most important alkaloid is quinine. The British protected themselves against malaria during the occupation of India through gin and tonic (quinine!). The stimulant coca, used by the Incas around the tenth century, was introduced into Europe by the conquistadors; coca beans are also commonly chewed in West Africa. Wine making was already practiced in the Middle East 6000–8000 years ago; Moors made date wines, the Japanese rice wine, the Vikings honey mead, the Incas maize chicha. It is said that the Babylonians made beer using yeast 5000–6000 years ago. As shown above in parentheses, alkaloids are the major constituents of the herbal plants and extracts used for centuries, but it was not until the early nineteenth century that the active principles were isolated in pure form, for example morphine (1816), strychnine (1817), atropine (1819), quinine (1820), and colchicine (1820). It was a century later that the structures of these compounds were finally elucidated.

2. DAWN OF ORGANIC CHEMISTRY, EARLY STRUCTURAL STUDIES, MODERN METHODOLOGY

The term "organic compound" to define compounds made by and isolated from living organisms was coined in 1807 by the Swedish chemist Jons Jacob Berzelius (1779–1848), a founder of today's chemistry, who developed the modern system of symbols and formulas in chemistry, made a remarkably accurate table of atomic weights and analyzed many chemicals. At that time it was considered that organic compounds could not be synthesized from inorganic materials *in vitro*. However, Friedrich Wöhler (1800–1882), a medical doctor from Heidelberg who was starting his chemical career at a technical school in Berlin, attempted in 1828 to make "ammonium cyanate," which had been assigned a wrong structure, by heating the two inorganic salts potassium cyanate and ammonium sulfate; this led to the unexpected isolation of white crystals which were identical to the urea from urine, a typical organic compound. This well-known incident marked the beginning of organic chemistry. With the preparation of acetic acid from inorganic material in 1845 by Hermann Kolbe (1818–1884) at Leipzig, the myth surrounding organic compounds, in which they were associated with some vitalism was brought to an end and organic chemistry became the chemistry of carbon compounds. The same Kolbe was involved in the development of aspirin, one of the earliest and most important success stories in natural products chemistry. Salicylic acid from the leaf of the wintergreen plant had long been used as a pain reliever, especially in treating arthritis and gout. The inexpensive synthesis of salicylic acid from sodium phenolate and carbon dioxide by Kolbe in 1859 led to the industrial production in 1893 by the Bayer Company of acetylsalicylic acid "aspirin," still one of the most popular drugs. Aspirin is less acidic than salicylic acid and therefore causes less irritation in the mouth, throat, and stomach. The remarkable mechanism of the anti-inflammatory effect of aspirin was clarified in 1974 by John Vane (1927–) who showed that it inhibits the biosynthesis of prostaglandins by irreversibly acetylating a serine residue in prostaglandin synthase. Vane shared the 1982 Nobel Prize with Bergström and Samuelsson who determined the structure of prostaglandins (see below).

In the early days, natural products chemistry was focused on isolating the more readily available plant and animal constituents and determining their structures. The course of structure determination in the 1940s was a complex, indirect process, combining evidence from many types of experiments. The first effort was to crystallize the unknown compound or make derivatives such as esters or 2,4-dinitrophenylhydrazones, and to repeat recrystallization until the highest and sharp melting point was reached, since prior to the advent of isolation and purification methods now taken for granted, there was no simple criterion for purity. The only chromatography was through special grade alumina (first used by M. Tswett in 1906, then reintroduced by R. Willstätter). Molecular weight estimation by the Rast method which depended on melting point depression of a sample/camphor mixture, coupled with Pregl elemental microanalysis (see below) gave the molecular formula. Functionalities such as hydroxyl, amino, and carbonyl groups were recognized on the basis of specific derivatization and crystallization, followed by redetermination of molecular formula; the change in molecular composition led to identification of the functionality. Thus, sterically hindered carbonyls, for example the 11-keto group of cortisone, or tertiary hydroxyls, were very difficult to pinpoint, and often had to depend on more searching experiments. Therefore, an entire paper describing the recognition of a single hydroxyl group in a complex natural product would occasionally appear in the literature. An oxygen function suggested from the molecular formula but left unaccounted for would usually be assigned to an ether.

Determination of C-methyl groups depended on Kuhn–Roth oxidation which is performed by drastic oxidation with chromic acid/sulfuric acid, reduction of excess oxidant with hydrazine, neutralization with alkali, addition of phosphoric acid, distillation of the acetic acid originating from the C-methyls, and finally its titration with alkali. However, the results were only approximate, since *gem*-dimethyl groups only yield one equivalent of acetic acid, while primary, secondary, and tertiary methyl groups all give different yields of acetic acid. The skeletal structure of polycyclic compounds were frequently deduced on the basis of dehydrogenation reactions. It is therefore not surprising that the original steroid skeleton put forth by Wieland and Windaus in 1928, which depended a great deal on the production of chrysene upon Pd/C dehydrogenation, had to be revised in 1932 after several discrepancies were found (they received the Nobel prizes in 1927 and 1928 for this "extraordinarily difficult structure determination," see below).

In the following are listed some of the Nobel prizes awarded for the development of methodologies which have contributed critically to the progress in isolation protocols and structure determination. The year in which each prize was awarded is preceded by "Np."

Fritz Pregl, 1869–1930, Graz University, Np 1923. Invention of carbon and hydrogen microanalysis. Improvement of Kuhlmann's microbalance enabled weighing at an accuracy of 1 μg over a 20 g range, and refinement of carbon and hydrogen analytical methods made it possible to perform analysis with 3–4 mg of sample. His microbalance and the monograph *Quantitative Organic Microanalysis* (1916) profoundly influenced subsequent developments in practically all fields of chemistry and medicine.

The Svedberg, 1884–1971, Uppsala, Np 1926. Uppsala was a center for quantitative work on colloids for which the prize was awarded. His extensive study on ultracentrifugation, the first paper of which was published in the year of the award, evolved from a spring visit in 1922 to the University of Wisconsin. The ultracentrifuge together with the electrophoresis technique developed by his student Tiselius, have profoundly influenced subsequent progress in molecular biology and biochemistry.

Arne Tiselius, 1902–1971, Ph.D. Uppsala (T. Svedberg), Uppsala, Np 1948. Assisted by a grant from the Rockefeller Foundation, Tiselius was able to use his early electrophoresis instrument to show four bands in horse blood serum, alpha, beta and gamma globulins in addition to albumin; the first paper published in 1937 brought immediate positive responses.

Archer Martin, 1910–, Ph.D. Cambridge; Medical Research Council, Mill Hill, and Richard Synge, 1914–1994, Ph.D. Cambridge; Rowett Research Institute, Food Research Institute, Np 1952. They developed chromatography using two immiscible phases, gas–liquid, liquid–liquid, and paper chromatography, all of which have profoundly influenced all phases of chemistry.

Frederick Sanger, 1918–, Ph.D. Cambridge (A. Neuberger), Medical Research Council, Cambridge, Np 1958 and 1980. His confrontation with challenging structural problems in proteins and nucleic acids led to the development of two general analytical methods, 1,2,4-fluorodinitrobenzene (DNP) for tagging free amino groups (1945) in connection with insulin sequencing studies, and the dideoxynucleotide method for sequencing DNA (1977) in connection with recombinant DNA. For the latter he received his second Np in chemistry in 1980, which was shared with Paul Berg (1926–, Stanford University) and Walter Gilbert (1932–, Harvard University) for their contributions, respectively, in recombinant DNA and chemical sequencing of DNA. The studies of insulin involved usage of DNP for tagging disulfide bonds as cysteic acid residues (1949), and paper chromatography introduced by Martin and Synge 1944. That it was the first elucidation of any protein structure lowered the barrier for future structure studies of proteins.

Stanford Moore, 1913–1982, Ph.D. Wisconsin (K. P. Link), Rockefeller, Np 1972; and William Stein, 1911–1980, Ph.D. Columbia (E. G. Miller); Rockefeller, Np 1972. Moore and Stein cooperatively developed methods for the rapid quantification of protein hydrolysates by combining partition chromatography, ninhydrin coloration, and drop-counting fraction collector, i.e., the basis for commercial amino acid analyzers, and applied them to analysis of the ribonuclease structure.

Bruce Merrifield, 1921–, Ph.D. UCLA (M. Dunn), Rockefeller, Np 1984. The concept of solid-phase peptide synthesis using porous beads, chromatographic columns, and sequential elongation of peptides and other chains revolutionized the synthesis of biopolymers.

High-performance liquid chromatography (HPLC), introduced around the mid-1960s and now coupled on-line to many analytical instruments, for example UV, FTIR, and MS, is an indispensable daily tool found in all natural products chemistry laboratories.

3. STRUCTURES OF ORGANIC COMPOUNDS, NINETEENTH CENTURY

The discoveries made from 1848 to 1874 by Pasteur, Kekulé, van't Hoff, Le Bel, and others led to a revolution in structural organic chemistry. Louis Pasteur (1822–1895) was puzzled about why the potassium salt of tartaric acid (deposited on wine casks during fermentation) was dextrorotatory while the sodium ammonium salt of racemic acid (also deposited on wine casks) was optically inactive although both tartaric acid and "racemic" acid had identical chemical compositions. In 1848, the 25 year old Pasteur examined the racemic acid salt under the microscope and found two kinds of crystals exhibiting a left- and right-hand relation. Upon separation of the left-handed and right-handed crystals, he found that they rotated the plane of polarized light in opposite directions. He had thus performed his famous resolution of a racemic mixture, and had demonstrated the phenomenon of chirality. Pasteur went on to show that the racemic acid formed two kinds of salts with optically active bases such as quinine; this was the first demonstration of diastereomeric resolution. From this work Pasteur concluded that tartaric acid must have an element of asymmetry within the molecule itself. However, a three-dimensional understanding of the enantiomeric pair was only solved 25 years later (see below). Pasteur's own interest shifted to microbiology where he made the crucial discovery of the involvement of "germs" or microorganisms in various processes and proved that yeast induces alcoholic fermentation, while other microorganisms lead to diseases; he thus saved the wine industries of France, originated the process known as "pasteurization," and later developed vaccines for rabies. He was a genius who made many fundamental discoveries in chemistry and in microbiology.

The structures of organic compounds were still totally mysterious. Although Wöhler had synthesized urea, an isomer of ammonium cyanate, in 1828, the structural difference between these isomers was not known In 1858 August Kekulé (1829–1896; studied with André Dumas and C. A. Wurtz in Paris, taught at Ghent, Heidelberg, and Bonn) published his famous paper in Liebig's *Annalen der Chemie* on the structure of carbon, in which he proposed that carbon atoms could form C–C bonds with hydrogen and other atoms linked to them; his dream on the top deck of a London bus led him to this concept. It was Butlerov who introduced the term "structure theory" in 1861. Further, in 1865 Kekulé conceived the cyclo-hexa-1:3:5-triene structure for benzene (C_6H_6) from a dream of a snake biting its own tail. In 1874, two young chemists, van't Hoff (1852–1911, Np 1901) in Utrecht, and Le Bel (1847–1930) in Paris, who had met in 1874 as students of C. A. Wurtz, published the revolutionary three-dimensional (3D) structure of the tetrahedral carbon Cabcd to explain the enantiomeric behavior of Pasteur's salts. The model was welcomed by J. Wislicenus (1835–1902, Zürich, Würzburg, Leipzig) who in 1863 had demonstrated the enantiomeric nature of the two lactic acids found by Scheele in sour milk (1780) and by Berzelius in muscle tissue (1807). This model, however, was criticized by Hermann Kolbe (1818–1884, Leipzig) as an "ingenious but in reality trivial and senseless natural philosophy." After 10 years of heated controversy, the idea of tetrahedral carbon was fully accepted, Kolbe had died and Wislicenus succeeded him in Leipzig.

Emil Fischer (1852–1919, Np 1902) was the next to make a critical contribution to stereochemistry. From the work of van't Hoff and Le Bel he reasoned that glucose should have 16 stereoisomers. Fischer's doctorate work on hydrazines under Baeyer (1835–1917, Np 1905) at Strasbourg had led to studies of osazones which culminated in the brilliant establishment, including configurations, of the Fischer sugar tree starting from D-(+)-glyceraldehyde all the way up to the aldohexoses, allose, altrose, glucose, mannose, gulose, idose, galactose, and talose (from 1884 to 1890). Unfortunately Fischer suffered from the toxic effects of phenylhydrazine for 12 years. The arbitrarily but luckily chosen absolute configuration of D-(+)-glyceraldehyde was shown to be correct sixty years later in 1951 (Johannes-Martin Bijvoet, 1892–1980). Fischer's brilliant correlation of the sugars comprising the Fischer sugar tree was performed using the Kiliani (1855–1945)–Fischer method via cyanohydrin intermediates for elongating sugars. Fischer also made remarkable contributions to the chemistry of amino acids and to nucleic acid bases (see below).

4. STRUCTURES OF ORGANIC COMPOUNDS, TWENTIETH CENTURY

The early concept of covalent bonds was provided with a sound theoretical basis by Linus Pauling (1901–1994, Np 1954), one of the greatest intellects of the twentieth century. Pauling's totally interdisciplinary research interests, including proteins and DNA is responsible for our present understanding of molecular structures. His books *Introduction to Quantum Mechanics* (with graduate student E. B. Wilson, 1935) and *The Nature of the Chemical Bond* (1939) have had a profound effect on our understanding of all of chemistry.

The actual 3D shapes of organic molecules which were still unclear in the late 1940s were then brilliantly clarified by Odd Hassel (1897–1981, Oslo University, Np 1969) and Derek Barton (1918–1998, Np 1969). Hassel, an X-ray crystallographer and physical chemist, demonstrated by electron diffraction that cyclohexane adopted the chair form in the gas phase and that it had two kinds of bonds, "standing (axial)" and "reclining (equatorial)" (1943). Because of the German occupation of Norway in 1940, instead of publishing the result in German journals, he published it in a Norwegian journal which was not abstracted in English until 1945. During his 1949 stay at Harvard, Barton attended a seminar by Louis Fieser on steric effects in steroids and showed Fieser that interpretations could be simplified if the shapes ("conformations") of cyclohexane rings were taken into consideration; Barton made these comments because he was familiar with Hassel's study on *cis*- and *trans*-decalins. Following Fieser's suggestion Barton published these ideas in a four-page *Experientia* paper (1950). This led to the joint Nobel prize with Hassel (1969), and established the concept of conformational analysis, which has exerted a profound effect in every field involving organic molecules.

Using conformational analysis, Barton determined the structures of many key terpenoids such as ß-amyrin, cycloartenone, and cycloartenol (Birkbeck College). At Glasgow University (from 1955) he collaborated in a number of cases with Monteath Robertson (1900–1989) and established many challenging structures: limonin, glauconic acid, byssochlamic acid, and nonadrides. Barton was also associated with the Research Institute for Medicine and Chemistry (RIMAC), Cambridge, USA founded by the Schering company, where with J. M. Beaton, he produced 60 g of aldosterone at a time when the world supply of this important hormone was in mg quantities. Aldosterone synthesis ("a good problem") was achieved in 1961 by Beaton ("a good experimentalist") through a nitrite photolysis, which came to be known as the Barton reaction ("a good idea") (quotes from his 1991 autobiography published by the American Chemical Society). From Glasgow, Barton went on to Imperial College, and a year before retirement, in 1977 he moved to France to direct the research at ICSN at Gif-sur-Yvette where he explored the oxidation reaction selectivity for unactivated C–H. After retiring from ICSN he made a further move to Texas A&M University in 1986, and continued his energetic activities, including chairman of the *Tetrahedron* publications. He felt weak during work one evening and died soon after, on March 16, 1998. He was fond of the phrase "gap jumping" by which he meant seeking generalizations between facts that do not seem to be related: "In the conformational analysis story, one had to jump the gap between steroids and chemical physics" (from his autobiography). According to Barton, the three most important qualities for a scientist are "intelligence, motivation, and honesty." His routine at Texas A&M was to wake around 4 a.m., read the literature, go to the office at 7 a.m. and stay there until 7 p.m.; when asked in 1997 whether this was still the routine, his response was that he wanted to wake up earlier because sleep was a waste of time—a remark which characterized this active scientist approaching 80!

Robert B. Woodward (1917–1979, Np 1965), who died prematurely, is regarded by many as the preeminent organic chemist of the twentieth century. He made landmark achievements in spectroscopy, synthesis, structure determination, biogenesis, as well as in theory. His solo papers published in 1941–1942 on empirical rules for estimating the absorption maxima of enones and dienes made the general organic chemical community realize that UV could be used for structural studies, thus launching the beginning of the spectroscopic revolution which soon brought on the applications of IR, NMR, MS, etc. He determined the structures of the following compounds: penicillin in 1945 (through joint UK–USA collaboration, see Hodgkin), strychnine in 1948, patulin in 1949, terramycin, aureomycin, and ferrocene (with G. Wilkinson, Np 1973—shared with E. O. Fischer for sandwich compounds) in 1952, cevine in 1954 (with Barton Np 1966, Jeger and Prelog, Np 1975), magnamycin in 1956, gliotoxin in 1958, oleandomycin in 1960, streptonigrin in 1963, and tetrodotoxin in 1964. He synthesized patulin in 1950, cortisone and cholesterol in 1951, lanosterol, lysergic acid (with Eli Lilly), and strychnine in 1954, reserpine in 1956, chlorophyll in 1960, a tetracycline (with Pfizer) in 1962, cephalosporin in 1965, and vitamin B_{12} in 1972 (with A. Eschenmoser, 1925–, ETH Zürich). He derived biogenetic schemes for steroids in 1953 (with K. Bloch, see below), and for macrolides in 1956, while the Woodward–Hoffmann orbital symmetry rules in 1965 brought order to a large class of seemingly random cyclization reactions.

Another central figure in stereochemistry is Vladimir Prelog (1906–1998, Np 1975), who succeeded Leopold Ruzicka at the ETH Zürich, and continued to build this institution into one of the most active and lively research and discussion centers in the world. The core group of intellectual leaders consisted of P. Plattner (1904–1975), O. Jeger, A. Eschenmoser, J. Dunitz, D. Arigoni, and A. Dreiding (from Zürich University). After completing extensive research on alkaloids, Prelog determined the structures of nonactin, boromycin, ferrioxamins, and rifamycins. His seminal studies in the synthesis and properties of 8–12 membered rings led him into unexplored areas of stereochemisty and chirality. Together with Robert Cahn (1899–1981, London Chemical Society) and Christopher Ingold (1893–1970, University College, London; pioneering mechanistic interpretation of organic reactions), he developed the Cahn–Ingold–Prelog (CIP) sequence rules for the unambiguous specification of stereoisomers. Prelog was an excellent story teller, always had jokes to tell, and was respected and loved by all who knew him.

4.1 Polyketides and Fatty Acids

Arthur Birch (1915–1995) from Sydney University, Ph.D. with Robert Robinson (Oxford University), then professor at Manchester University and Australian National University, was one of the earliest chemists to perform biosynthetic studies using radiolabels; starting with polyketides he studied the biosynthesis of a variety of natural products such as the C_6–C_3–C_6 backbone of plant phenolics, polyene macrolides, terpenoids, and alkaloids. He is especially known for the Birch reduction of aromatic rings, metal–ammonia reductions leading to 19-norsteroid hormones and other important products (1942–) which were of industrial importance. Feodor Lynen (1911–1979, Np 1964) performed studies on the intermediary metabolism of the living cell that led him to the demonstration of the first step in a chain of reactions resulting in the biosynthesis of sterols and fatty acids.

Prostaglandins, a family of 20-carbon, lipid-derived acids discovered in seminal fluids and accessory genital glands of man and sheep by von Euler (1934), have attracted great interest because of their extremely diverse biological activities. They were isolated and their structures elucidated from 1963 by S. Bergström (1916–, Np 1982) and B. Samuelsson (1934–, Np 1982) at the Karolinska Institute, Stockholm. Many syntheses of the natural prostaglandins and their nonnatural analogues have been published.

Tetsuo Nozoe (1902–1996) who studied at Tohoku University, Sendai, with Riko Majima (1874–1962, see below) went to Taiwan where he stayed until 1948 before returning to Tohoku University. At National Taiwan University he isolated hinokitiol from the essential oil of *taiwanhinoki*. Remembering the resonance concept put forward by Pauling just before World War II, he arrived at the seven-membered nonbenzenoid aromatic structure for hinokitiol in 1941, the first of the troponoids. This highly original work remained unknown to the rest of the world until 1951. In the meantime, during 1945–1948, nonbenzenoid aromatic structures had been assigned to stipitatic acid (isolated by H. Raistrick) by Michael J. S. Dewar (1918–) and to the thujaplicins by Holger Erdtman (1902–1989); the term tropolones was coined by Dewar in 1945. Nozoe continued to work on and discuss troponoids, up to the night before his death, without knowing that he had cancer. He was a remarkably focused and warm scientist, working unremittingly. Erdtman (Royal Institute of Technology, Stockholm) was the central figure in Swedish natural products chemistry who, with his wife Gunhild Aulin Erdtman (dynamic General Secretary of the Swedish Chemistry Society), worked in the area of plant phenolics.

As mentioned in the following and in the concluding sections, classical biosynthetic studies using radioactive isotopes for determining the distribution of isotopes has now largely been replaced by the use of various stable isotopes coupled with NMR and MS. The main effort has now shifted to the identification and cloning of genes, or where possible the gene clusters, involved in the biosynthesis of the natural product. In the case of polyketides (acyclic, cyclic, and aromatic), the focus is on the polyketide synthases.

4.2 Isoprenoids, Steroids, and Carotenoids

During his time as an assistant to Kekulé at Bonn, Otto Wallach (1847–1931, Np 1910) had to familiarize himself with the essential oils from plants; many of the components of these oils were compounds for which no structure was known. In 1891 he clarified the relations between 12 different monoterpenes related to pinene. This was summarized together with other terpene chemistry in book form in 1909, and led him to propose the "isoprene rule." These achievements laid the foundation for the future development of terpenoid chemistry and brought order from chaos.

The next period up to around 1950 saw phenomenal advances in natural products chemistry centered on isoprenoids. Many of the best natural products chemists in Europe, including Wieland, Windaus, Karrer, Kuhn, Butenandt, and Ruzicka contributed to this breathtaking pace. Heinrich Wieland (1877–1957) worked on the bile acid structure, which had been studied over a period of 100 years and considered to be one of the most difficult to attack; he received the Nobel Prize in 1927 for these studies. His friend Adolph Windaus (1876–1959) worked on the structure of cholesterol for which he also received the Nobel Prize in 1928. Unfortunately, there were chemical discrepancies in the proposed steroidal skeletal structure, which had a five-membered ring B attached to C-7 and C-9. J. D. Bernal, Mineralogical Museums, Cambridge University, who was examining the X-ray patterns of ergosterol (1932) noted that the dimensions were inconsistent with the Wieland–Windaus formula. A reinterpretation of the production of chrysene from sterols by Pd/C dehydrogenation reported by Diels (see below) in 1927 eventually led Rosenheim and King and Wieland and Dane to deduce the correct structure in 1932. Wieland also worked on the structures of morphine/strychnine alkaloids, phalloidin/amanitin cyclopeptides of toxic mushroom *Amanita phalloides*, and pteridines, the important fluorescent pigments of butterfly wings. Windaus determined the structure of ergosterol and continued structural studies of its irradiation product which exhibited antirachitic activity "vitamin D." The mechanistically complex photochemistry of ergosterol leading to the vitamin D group has been investigated in detail by Egbert Havinga (1927–1988, Leiden University), a leading photochemist and excellent tennis player.

Paul Karrer (1889–1971, Np 1937), established the foundations of carotenoid chemistry through structural determinations of lycopene, carotene, vitamin A, etc. and the synthesis of squalene, carotenoids, and others. George Wald (1906–1997, Np 1967) showed that vitamin A was the key compound in vision during his stay in Karrer's laboratory. Vitamin K (K from "Koagulation"), discovered by Henrik Dam (1895–1976, Polytechnic Institute, Copenhagen, Np 1943) and structurally studied by Edward Doisy (1893–1986, St. Louis University, Np 1943), was also synthesized by Karrer. In addition, Karrer synthesized riboflavin (vitamin B_2) and determined the structure and role of nicotinamide adenine dinucleotide phosphate ($NADP^+$) with Otto Warburg. The research on carotenoids and vitamins of Karrer who was at Zürich University overlapped with that of Richard Kuhn (1900–1967, Np 1938) at the ETH Zürich, and the two were frequently rivals. Richard Kuhn, one of the pioneers in using UV-vis spectroscopy for structural studies, introduced the concept of "atropisomerism" in diphenyls, and studied the spectra of a series of diphenyl polyenes. He determined the structures of many natural carotenoids, proved the structure of riboflavin-5-phosphate (flavin-adenine-dinucleotide-5-phosphate) and showed that the combination of NAD-5-phosphate with the carrier protein yielded the yellow oxidation enzyme, thus providing an understanding of the role of a prosthetic group. He also determined the structures of vitamin B complexes, i.e., pyridoxine, *p*-aminobenzoic acid, pantothenic acid. After World War II he went on to structural studies of nitrogen-containing oligosaccharides in human milk that provide immunity for infants, and brain gangliosides. Carotenoid studies in Switzerland were later taken up by Otto Isler (1910–1993), a Ruzicka student at Hoffmann-La Roche, and Conrad Hans Eugster (1921–), a Karrer student at Zürich University.

Adolf Butenandt (1903–1998, Np 1939) initiated and essentially completed isolation and structural studies of the human sex hormones, the insect molting hormone (ecdysone), and the first pheromone, bombykol. With help from industry he was able to obtain large supplies of urine from pregnant women for estrone, sow ovaries for progesterone, and 4,000 gallons of male urine for androsterone (50 mg, crystals). He isolated and determined the structures of two female sex hormones, estrone and progesterone, and the male hormone androsterone all during the period 1934–1939 (!) and was awarded the Nobel prize in 1939. Keen intuition and use of UV data and Pregl's microanalysis all played important roles. He was appointed to a professorship in Danzig at the age of 30. With Peter Karlson he isolated from 500 kg of silkworm larvae 25 mg of α-ecdysone, the prohormone of insect and crustacean molting hormone, and determined its structure as a polyhydroxysteroid (1965); 20-hydroxylation gives the insect and crustacean molting hormone or ß-ecdysone (20-hydroxyecdysteroid). He was also the first to isolate an insect pheromone, bombykol, from female silkworm moths (with E. Hecker). As president of the Max Planck Foundation, he strongly influenced the postwar rebuilding of German science.

The successor to Kuhn, who left ETH Zürich for Heidelberg, was Leopold Ruzicka (1887–1967, Np 1939) who established a close relationship with the Swiss pharmaceutical industry. His synthesis of the 17- and 15-membered macrocyclic ketones, civetone and muscone (the constituents of musk) showed that contrary to Baeyer's prediction, large alicyclic rings could be strainless. He reintroduced and refined the isoprene rule proposed by Wallach (1887) and determined the basic structures of many sesqui-, di-, and triterpenes, as well as the structure of lanosterol, the key intermediate in cholesterol biosynthesis. The "biogenetic isoprene rule" of the ETH group, Albert Eschenmoser, Leopold Ruzicka, Oskar Jeger, and Duilio Arigoni, contributed to a concept of terpenoid cyclization (1955), which was consistent with the mechanistic considerations put forward by Stork as early as 1950. Besides making the ETH group into a center of natural products chemistry, Ruzicka bought many seventeenth century Dutch paintings with royalties accumulated during the war from his Swiss and American patents, and donated them to the Zürich Kunsthaus.

Studies in the isolation, structures, and activities of the antiarthritic hormone, cortisone and related compounds from the adrenal cortex were performed in the mid- to late 1940s during World War II by Edward Kendall (1886–1972, Mayo Clinic, Rochester, Np 1950), Tadeus Reichstein (1897–1996, Basel University, Np 1950), Philip Hench (1896–1965, Mayo Clinic, Rochester, Np 1950), Oskar Wintersteiner (1898–1971, Columbia University, Squibb) and others initiated interest as an adjunct to military medicine as well as to supplement the meager supply from beef adrenal glands by synthesis. Lewis Sarett (1917–, Merck & Co., later president) and co-workers completed the cortisone synthesis in 28 steps, one of the first two totally stereocontrolled syntheses of a natural product; the other was cantharidin (Stork 1951) (see above). The multistep cortisone synthesis was put on the production line by Max Tishler (1906–1989, Merck & Co., later president) who made contributions to the synthesis of a number of drugs, including riboflavin. Besides working on steroid reactions/synthesis and antimalarial agents, Louis F. Fieser (1899–1977) and Mary Fieser (1909–1997) of Harvard University made huge contributions to the chemical community through their outstanding books *Natural Products related to Phenanthrene* (1949), *Steroids* (1959), *Advanced Organic Chemistry* (1961), and *Topics in Organic Chemistry* (1963), as well as their textbooks and an important series of books on Organic Reagents. Carl Djerassi (1923–, Stanford University), a prolific chemist, industrialist, and more recently a novelist, started to work at the Syntex laboratories in Mexico City where he directed the work leading to the first oral contraceptive ("the pill") for women.

Takashi Kubota (1909–, Osaka City University), with Teruo Matsuura (1924–, Kyoto University), determined the structure of the furanoid sesquiterpene, ipomeamarone, from the black rotted portion of spoiled sweet potatoes; this research constitutes the first characterization of a phytoallexin, defense substances produced by plants in response to attack by fungi or physical damage. Damaging a plant and characterizing the defense substances produced may lead to new bioactive compounds. The mechanism of induced biosynthesis of phytoallexins, which is not fully understood, is an interesting biological mechanistic topic that deserves further investigation. Another center of high activity in terpenoids and nucleic acids was headed by Frantisek Sorm (1913–1980, Institute of Organic and Biochemistry, Prague), who determined the structures of many sesquiterpenoids and other natural products; he was not only active scientifically but also was a central figure who helped to guide the careers of many Czech chemists.

The key compound in terpenoid biosynthesis is mevalonic acid (MVA) derived from acetyl-CoA, which was discovered fortuitously in 1957 by the Merck team in Rahway, NJ headed by Karl Folkers (1906–1998). They soon realized and proved that this C_6 acid was the precursor of the C_5 isoprenoid unit isopentenyl diphosphate (IPP) that ultimately leads to the biosynthesis of cholesterol. In 1952 Konrad Bloch (1912–, Harvard, Np 1964) with R. B. Woodward published a paper suggesting a mechanism of the cyclization of squalene to lanosterol and the subsequent steps to cholesterol, which turned out to be essentially correct. This biosynthetic path from MVA to cholesterol was experimentally clarified in stereochemical detail by John Cornforth (1917–, Np 1975) and George Popják. In 1932, Harold Urey (1893–1981, Np 1934) of Columbia University discovered heavy hydrogen. Urey showed, contrary to common expectation, that isotope separation could be achieved with deuterium in the form of deuterium oxide by fractional electrolysis of water. Urey's separation of the stable isotope deuterium led to the isotopic tracer methodology that revolutionized the protocols for elucidating biosynthetic processes and reaction mechanisms, as exemplified beautifully by the cholesterol studies. Using MVA labeled chirally with isotopes, including chiral methyl, i.e., -CHDT, Cornforth and Popják clarified the key steps in the intricate biosynthetic conversion of mevalonate to cholesterol in stereochemical detail. The chiral methyl group was also prepared independently by Duilio Arigoni (1928– , ETH, Zürich). Cornforth has had great difficulty in hearing and speech since childhood but has been helped expertly by his chemist wife Rita; he is an excellent tennis and chess player, and is renowned for his speed in composing occasional witty limericks.

Although MVA has long been assumed to be the only natural precursor for IPP, a non-MVA pathway in which IPP is formed via the glyceraldehyde phosphate-pyruvate pathway has been discovered (1995–1996) in the ancient bacteriohopanoids by Michel Rohmer, who started working on them with Guy Ourisson (1926–, University of Strasbourg, terpenoid studies, including prebiotic), and by Duilio Arigoni in the ginkgolides, which are present in the ancient *Ginkgo biloba* tree. It is possible that many other terpenoids are biosynthesized via the non-MVA route. In classical biosynthetic experiments, ^{14}C-labeled acetic acid was incorporated into the microbial or plant product, and location or distribution of the ^{14}C label was deduced by oxidation or degradation to specific fragments including acetic acid; therefore, it was not possible or extremely difficult to map the distribution of all radioactive carbons. The progress in ^{13}C NMR made it possible to incorporate ^{13}C-labeled acetic acid and locate all labeled carbons. This led to the discovery of the nonmevalonate pathway leading to the IPP units. Similarly, NMR and MS have made it possible to use the stable isotopes, e.g., ^{18}O, ^{2}H, ^{15}N, etc., in biosynthetic studies. The current trend of biosynthesis has now shifted to genomic approaches for cloning the genes of various enzyme synthases involved in the biosynthesis.

4.3 Carbohydrates and Cellulose

The most important advance in carbohydrate structures following those made by Emil Fischer was the change from acyclic to the current cyclic structure introduced by Walter Haworth (1883–1937). He noticed the presence of α- and ß-anomers, and determined the structures of important disaccharides including cellobiose, maltose, and lactose. He also determined the basic structural aspects of starch, cellulose, inulin, and other polysaccharides, and accomplished the structure determination and synthesis of vitamin C, a sample of which he had received from Albert von Szent-Györgyi (1893–1986, Np 1937). This first synthesis of a vitamin was significant since it showed that a vitamin could be synthesized in the same way as any other organic compound. There was strong belief among leading scientists in the 1910s that cellulose, starch, protein, and rubber were colloidal aggregates of small molecules. However, Hermann Staudinger (1881–1965, Np 1953) who succeeded R. Willstätter and H. Wieland at the ETH Zürich and Freiburg, respectively, showed through viscosity measurements and various molecular weight measurements that macromolecules do exist, and developed the principles of macromolecular chemistry.

In more modern times, Raymond Lemieux (1920–, Universities of Ottawa and Alberta) has been a leader in carbohydrate research. He introduced the concept of *endo*- and *exo*-anomeric effects, accomplished the challenging synthesis of sucrose (1953), pioneered in the use of NMR coupling constants in configuration studies, and most importantly, starting with syntheses of oligosaccharides responsible for human blood group determinants, he prepared antibodies and clarified fundamental aspects of the binding of oligosaccharides by lectins and antibodies. The periodate–potassium permanganate cleavage of double bonds at room temperature (1955) is called the Lemieux reaction.

4.4 Amino Acids, Peptides, Porphyrins, and Alkaloids

It is fortunate that we have China's record and practice of herbal medicine over the centuries, which is providing us with an indispensable source of knowledge. China is rapidly catching up in terms of infrastructure and equipment in organic and bioorganic chemistry, and work on isolation, structure determination, and synthesis stemming from these valuable sources has picked up momentum. However, as mentioned above, clarification of the active principles and mode of action of these plant extracts will be quite a challenge since in many cases synergistic action is expected. Wang Yu (1910–1997) who headed the well-equipped Shanghai Institute of Organic Chemistry surprised the world with the total synthesis of bovine insulin performed by his group in 1965; the human insulin was synthesized around the same time by P. G. Katsoyannis, A. Tometsko, and C. Zaut of the Brookhaven National Laboratory (1966).

One of the giants in natural products chemistry during the first half of this century was Robert Robinson (1886–1975, Np 1947) at Oxford University. His synthesis of tropinone, a bicyclic amino ketone related to cocaine, from succindialdehyde, methylamine, and acetone dicarboxylic acid under Mannich reaction conditions was the first biomimetic synthesis (1917). It reduced Willstätter's 1903 13-step synthesis starting with suberone into a single step. This achievement demonstrated Robinson's analytical prowess. He was able to dissect complex molecular structures into simple biosynthetic building blocks, which allowed him to propose the biogenesis of all types of alkaloids and other natural products. His laboratory at Oxford, where he developed the well-known Robinson annulation reaction (1937) in connection with his work on the synthesis of steroids became a world center for natural products study. Robinson was a pioneer in the so-called electronic theory of organic reactions, and introduced the use of curly arrows to show the movements of electrons. His analytical power is exemplified in the structural studies of strychnine and brucine around 1946–1952. Barton clarified the biosynthetic route to the morphine alkaloids, which he saw as an extension of his biomimetic synthesis of usnic acid through a one-electron oxidation; this was later extended to a general phenolate coupling scheme. Morphine total synthesis was brilliantly achieved by Marshall Gates (1915–, University of Rochester) in 1952.

The yield of the Robinson tropinone synthesis was low but Clemens Schöpf (1899–1970) , Ph.D. Munich (Wieland), Universität Darmstadt, improved it to 90% by carrying out the reaction in buffer; he also worked on the stereochemistry of morphine and determined the structure of the steroidal alkaloid salamandarine (1961), the toxin secreted from glands behind the eyes of the salamander.

Roger Adams (1889–1971, University of Illinois), was the central figure in organic chemistry in the USA and is credited with contributing to the rapid development of its chemistry in the late 1930s and 1940s, including training of graduate students for both academe and industry. After earning a Ph.D. in 1912 at Harvard University he did postdoctoral studies with Otto Diels (see below) and Richard Willstätter (see below) in 1913; he once said that around those years in Germany he could cover all *Journal of the American Chemical Society* papers published in a year in a single night. His important work include determination of the structures of tetrahydrocannabinol in marijuana, the toxic gossypol in cottonseed oil, chaulmoogric acid used in treatment of leprosy, and the Senecio alkaloids with Nelson Leonard (1916–, University of Illinois, now at Caltech). He also contributed to many fundamental organic reactions and syntheses. The famous Adams platinum catalyst is not only important for reducing double bonds in industry and in the laboratory, but was central for determining the number of double bonds in a structure. He was also one of the founders of the *Organic Synthesis* (started in 1921) and the *Organic Reactions* series. Nelson Leonard switched interests to bioorganic chemistry and biochemistry, where he has worked with nucleic acid bases and nucleotides, coenzymes, dimensional probes, and fluorescent modifications such as ethenoguanine.

The complicated structures of the medieval plant poisons aconitine (from *Aconitum*) and delphinine (from *Delphinium*) were finally characterized in 1959–1960 by Karel Wiesner (1919–1986, University of New Brunswick), Leo Marion (1899–1979, National Research Council, Ottawa), George Büchi (1921–, mycotoxins, aflatoxin/DNA adduct, synthesis of terpenoids and nitrogen-containing bioactive compounds, photochemistry), and Maria Przybylska (1923–, X-ray).

The complex chlorophyll structure was elucidated by Richard Willstätter (1872–1942, Np 1915). Although he could not join Baeyer's group at Munich because the latter had ceased taking students, a close relation developed between the two. During his chlorophyll studies, Willstätter reintroduced the important technique of column chromatography published in Russian by Michael Tswett (1906). Willstätter further demonstrated that magnesium was an integral part of chlorophyll, clarified the relation between chlorophyll and the blood pigment hemin, and found the wide distribution of carotenoids in tomato, egg yolk, and bovine corpus luteum. Willstätter also synthesized cyclooctatetraene and showed its properties to be wholly unlike benzene but close to those of acyclic polyenes (around 1913). He succeeded Baeyer at Munich in 1915, synthesized the anesthetic cocaine, retired early in protest of anti-Semitism, but remained active until the Hitler era, and in 1938 emigrated to Switzerland.

The hemin structure was determined by another German chemist of the same era, Hans Fischer (1881–1945, Np 1930), who succeeded Windaus at Innsbruck and at Munich. He worked on the structure of hemin from the blood pigment hemoglobin, and completed its synthesis in 1929. He continued Willstätter's structural studies of chlorophyll, and further synthesized bilirubin in 1944. Destruction of his institute at Technische Hochschule München, during World War II led him to take his life in March 1945. The biosynthesis of hemin was elucidated largely by David Shemin (1911–1991).

In the mid 1930s the Department of Biochemistry at Columbia Medical School, which had accepted many refugees from the Third Reich, including Erwin Chargaff, Rudolf Schoenheimer, and others on the faculty, and Konrad Bloch (see above) and David Shemin as graduate students, was a great center of research activity. In 1940, Shemin ingested 66 g of ^{15}N-labeled glycine over a period of 66 hours in order to determine the half-life of erythrocytes. David Rittenberg's analysis of the heme moiety with his home-made mass spectrometer showed all four pyrrole nitrogens came from glycine. Using ^{14}C (that had just become available) as a second isotope (see next paragraph), doubly labeled glycine $^{15}NH_2{}^{14}CH_2COOH$ and other precursors, Shemin showed that glycine and succinic acid condensed to yield δ-aminolevulinate, thus elegantly demonstrating the novel biosynthesis of the porphyrin ring (around 1950). At this time, Bloch was working on the other side of the bench.

Melvin Calvin (1911–1997, Np 1961) at University of California, Berkeley, elucidated the complex photosynthetic pathway in which plants reduce carbon dioxide to carbohydrates. The critical $^{14}CO_2$ had just been made available at Berkeley Lawrence Radiation Laboratory as a result of the pioneering research of Martin Kamen (1913–), while paper chromatography also played crucial roles. Kamen produced ^{14}C with Sam Ruben (1940), used ^{18}O to show that oxygen in photosynthesis comes from water and not from carbon dioxide, participated in the *Manhattan* project, testified before the House UnAmerican Activities Committee (1947), won compensatory damages from the US Department of State, and helped build the University of California, La Jolla (1957). The entire structure of the photosynthetic reaction center (>10 000 atoms) from the purple bacterium *Rhodopseudomonas viridis* has been established by X-ray crystallography in the landmark studies performed by Johann Deisenhofer (1943–), Robert Huber (1937–), and Hartmut Michel (1948–) in 1989; this was the first membrane protein structure determined by X-ray, for which they shared the 1988 Nobel prize. The information gained from the full structure of this first membrane protein has been especially rewarding.

The studies on vitamin B_{12}, the structure of which was established by crystallographic studies performed by Dorothy Hodgkin (1910–1994, Np 1964), are fascinating. Hodgkin also determined the structure of penicillin (in a joint effort between UK and US scientists during World War II) and insulin. The formidable total synthesis of vitamin B_{12} was completed in 1972 through collaborative efforts between Woodward and Eschenmoser, involving 100 postdoctoral fellows and extending over 10 years. The biosynthesis of fascinating complexity is almost completely solved through studies performed by Alan Battersby (1925–, Cambridge University), Duilio Arigoni, and Ian Scott (1928–, Texas A&M University) and collaborators where advanced NMR techniques and synthesis of labeled precursors is elegantly combined with cloning of enzymes controlling each biosynthetic step. This work provides a beautiful demonstration of the power of the combination of bioorganic chemistry, spectroscopy and molecular biology, a future direction which will become increasingly important for the creation of new "unnatural" natural products.

4.5 Enzymes and Proteins

In the early days of natural products chemistry, enzymes and viruses were very poorly understood. Thus, the 1926 paper by James Sumner (1887–1955) at Cornell University on crystalline urease was received with ignorance or skepticism, especially by Willstätter who believed that enzymes were small molecules and not proteins. John Northrop (1891–1987) and co-workers at the Rockefeller Institute went on to crystallize pepsin, trypsin, chymotrypsin, ribonuclease, deoyribonuclease, carboxypeptidase, and other enzymes between 1930 and 1935. Despite this, for many years biochemists did not recognize the significance of these findings, and considered enzymes as being low molecular weight compounds adsorbed onto proteins or colloids. Using Northrop's method for crystalline enzyme preparations, Wendell Stanley (1904–1971) at Princeton obtained tobacco mosaic virus as needles from one ton of tobacco leaves (1935). Sumner, Northrop, and Stanley shared the 1946 Nobel prize in chemistry. All these studies opened a new era for biochemistry.

Meanwhile, Linus Pauling, who in mid-1930 became interested in the magnetic properties of hemoglobin, investigated the configurations of proteins and the effects of hydrogen bonds. In 1949 he showed that sickle cell anemia was due to a mutation of a single amino acid in the hemoglobin molecule, the first correlation of a change in molecular structure with a genetic disease. Starting in 1951 he and colleagues published a series of papers describing the alpha helix structure of proteins; a paper published in the early 1950s with R. B. Corey on the structure of DNA played an important role in leading Francis Crick and James Watson to the double helix structure (Np 1962).

A further important achievement in the peptide field was that of Vincent Du Vigneaud (1901–1978, Np 1955), Cornell Medical School, who isolated and determined the structure of oxytocin, a posterior pituitary gland hormone, for which a structure involving a disulfide bond was proposed. He synthesized oxytocin in 1953, thereby completing the first synthesis of a natural peptide hormone.

Progress in isolation, purification, crystallization methods, computers, and instrumentation, including cyclotrons, have made X-ray crystallography the major tool in structural. Numerous structures including those of ligand/receptor complexes are being published at an extremely rapid rate. Some of the past major achievements in protein structures are the following. Max Perutz (1914, Np 1962) and John Kendrew (1914–1997, Np 1962), both at the Laboratory of Molecular Biology, Cambridge University, determined the structures of hemoglobin and myoglobin, respectively. William Lipscomb (1919–, Np 1976), Harvard University, who has trained many of the world's leaders in protein X-ray crystallography has been involved in the structure determination of many enzymes including carboxypeptidase A (1967); in 1965 he determined the structure of the anticancer bisindole alkaloid, vinblastine. Folding of proteins, an important but still enigmatic phenomenon, is attracting increasing attention. Christian Anfinsen (1916–1995, Np 1972), NIH, one of the pioneers in this area, showed that the amino acid residues in ribonuclease interact in an energetically most favorable manner to produce the unique 3D structure of the protein.

4.6 Nucleic Acid Bases, RNA, and DNA

The "Fischer indole synthesis" was first performed in 1886 by Emil Fischer. During the period 1881–1914, he determined the structures of and synthesized uric acid, caffeine, theobromine, xanthine, guanine, hypoxanthine, adenine, guanine, and made theophylline-D-glucoside phosphoric acid, the first synthetic nucleotide. In 1903, he made 5,5-diethylbarbituric acid or Barbital, Dorminal, Veronal, etc. (sedative), and in 1912, phenobarbital or Barbipil, Luminal, Phenobal, etc. (sedative). Many of his syntheses formed the basis of German industrial production of purine bases. In 1912 he showed that tannins are gallates of sugars such as maltose and glucose. Starting in 1899, he synthesized many of the 13 α-amino acids known at that time, including the L- and D-forms, which were separated through fractional crystallization of their salts with optically active bases. He also developed a method for synthesizing fragments of proteins, namely peptides, and made an 18-amino acid peptide. He lost his two sons in World War I, lost his wealth due to postwar inflation, believed he had terminal cancer (a misdiagnosis), and killed himself in July 1919. Fischer was a skilled experimentalist, so that even today, many of the reactions performed by him and his students are so delicately controlled that they are not easy to reproduce. As a result of his suffering by inhaling diethylmercury, and of the poisonous effect of phenylhydrazine, he was one of the first to design fume hoods. He was a superb teacher and was also influential in establishing the Kaiser Wilhelm Institute, which later became the Max Planck Institute. The number and quality of his accomplishments and contributions are hard to believe; he was truly a genius.

Alexander Todd (1907–1997, Np 1957) made critical contributions to the basic chemistry and synthesis of nucleotides. His early experience consisted of an extremely fruitful stay at Oxford in the Robinson group, where he completed the syntheses of many representative anthocyanins, and then at Edinburgh where he worked on the synthesis of vitamin B_1. He also prepared the hexacarboxylate of vitamin B_{12} (1954), which was used by D. Hodgkin's group for their X-ray elucidation of this vitamin (1956). M. Wiewiorowski (1918–), Institute for Bioorganic Chemistry, in Poznan, has headed a famous group in nucleic acid chemistry, and his colleagues are now distributed worldwide.

4.7 Antibiotics, Pigments, and Marine Natural Products

The concept of one microorganism killing another was introduced by Pasteur who coined the term antibiosis in 1877, but it was much later that this concept was realized in the form of an actual antibiotic. The bacteriologist Alexander Fleming (1881–1955, University of London, Np 1945) noticed that an airborne mold, a *Penicillium* strain, contaminated cultures of *Staphylococci* left on the open bench and formed a transparent circle around its colony due to lysis of *Staphylococci*. He published these results in 1929. The discovery did not attract much interest but the work was continued by Fleming until it was taken up further at Oxford University by pathologist Howard Florey (1898–1968, Np 1945) and biochemist Ernst Chain (1906–1979, Np 1945). The bioactivities of purified "penicillin," the first antibiotic, attracted serious interest in the early 1940s in the midst of World War II. A UK/USA team was formed during the war between academe and industry with Oxford University, Harvard University, ICI, Glaxo, Burroughs Wellcome, Merck, Shell, Squibb, and Pfizer as members. This project resulted in the large scale production of penicillin and determination of its structure (finally by X-ray, D. Hodgkin). John Sheehan (1915–1992) at MIT synthesized 6-aminopenicillanic acid in 1959, which opened the route for the synthesis of a number of analogues. Besides being the first antibiotic to be discovered, penicillin is also the first member of a large number of important antibiotics containing the ß-lactam ring, for example cephalosporins, carbapenems, monobactams, and nocardicins. The strained ß-lactam ring of these antibiotics inactivates the transpeptidase by acylating its serine residue at the active site, thus preventing the enzyme from forming the link between the pentaglycine chain and the D-Ala-D-Ala peptide, the essential link in bacterial cell walls. The overuse of ß-lactam antibiotics, which has given rise to the disturbing appearance of microbial resistant strains, is leading to active research in the design of synthetic ß-lactam analogues to counteract these strains. The complex nature of the important penicillin biosynthesis is being elucidated through efforts combining genetic engineering, expression of biosynthetic genes as well as feeding of synthetic precursors, etc. by Jack Baldwin (1938–, Oxford University), José Luengo (Universidad de León, Spain) and many other groups from industry and academe.

Shortly after the penicillin discovery, Selman Waksman (1888–1973, Rutgers University, Np 1952) discovered streptomycin, the second antibiotic and the first active against the dreaded disease tuberculosis. The discovery and development of new antibiotics continued throughout the world at pharmaceutical companies in Europe, Japan, and the USA from soil and various odd sources: cephalosporin from sewage in Sardinia, cyclosporin from Wisconsin and Norway soil which was carried back to Switzerland, avermectin from the soil near a golf course in Shizuoka Prefecture. People involved in antibiotic discovery used to collect soil samples from various sources during their trips but this has now become severely restricted to protect a country's right to its soil. M. M. Shemyakin (1908–1970, Institute of Chemistry of Natural Products, Moscow) was a grand master of Russian natural products who worked on antibiotics, especially of the tetracycline class; he also worked on cyclic antibiotics composed of alternating sequences of amides and esters and coined the term depsipeptide for these in 1953. He died in 1970 of a sudden heart attack in the midst of the 7th IUPAC Natural Products

Symposium held in Riga, Latvia, which he had organized. The Institute he headed was renamed the Shemyakin Institute.

Indigo, an important vat dye known in ancient Asia, Egypt, Greece, Rome, Britain, and Peru, is probably the oldest known coloring material of plant origin, Indigofera and Isatis. The structure was determined in 1883 and a commercially feasible synthesis was performed in 1883 by Adolf von Baeyer (see above, 1835–1917, Np 1905), who founded the German Chemical Society in 1867 following the precedent of the Chemistry Society of London. In 1872 Baeyer was appointed a professor at Strasbourg where E. Fischer was his student, and in 1875 he succeeded J. Liebig in Munich. Tyrian (or Phoenician) purple, the dibromo derivative of indigo which is obtained from the purple snail Murex bundaris, was used as a royal emblem in connection with religious ceremonies because of its rarity; because of the availability of other cheaper dyes with similar color, it has no commercial value today. K. Venkataraman (1901–1981, University of Bombay then National Chemical Laboratory) who worked with R. Robinson on the synthesis of chromones in his early career, continued to study natural and synthetic coloring matters, including synthetic anthraquinone vat dyes, natural quinonoid pigments, etc. T. R. Seshadri (1900–1975) is another Indian natural products chemist who worked mainly in natural pigments, dyes, drugs, insecticides, and especially in polyphenols. He also studied with Robinson, and with Pregl at Graz, and taught at Delhi University. Seshadri and Venkataraman had a huge impact on Indian chemistry. After a 40 year involvement, Toshio Goto (1929–1990) finally succeeded in solving the mysterious identity of commelinin, the deep-blue flower petal pigment of the Commelina communis isolated by Kozo Hayashi (1958) and protocyanin, isolated from the blue cornflower Centaurea cyanus by E. Bayer (1957). His group elucidated the remarkable structure in its entirety which consisted of six unstable anthocyanins, six flavones and two metals, the molecular weight approaching 10 000; complex stacking and hydrogen bonds were also involved. Thus the pigmentation of petals turned out to be far more complex than the theories put forth by Willstätter (1913) and Robinson (1931). Goto suffered a fatal heart attack while inspecting the first X-ray structure of commelinin; commelinin represents a pinnacle of current natural products isolation and structure determination in terms of subtlety in isolation and complexity of structure.

The study of marine natural products is understandably far behind that of compounds of terrestrial origin due to the difficulty in collection and identification of marine organisms. However, it is an area which has great potentialities for new discoveries from every conceivable source. One pioneer in modern marine chemistry is Paul Scheuer (1915–, University of Hawaii) who started his work with quinones of marine origin and has since characterized a very large number of bioactive compounds from mollusks and other sources. Luigi Minale (1936–1997, Napoli) started a strong group working on marine natural products, concentrating mainly on complex saponins. He was a leading natural products chemist who died prematurely. A. Gonzalez Gonzalez (1917–) who headed the Organic Natural Products Institute at the University of La Laguna, Tenerife, was the first to isolate and study polyhalogenated sesquiterpenoids from marine sources. His group has also carried out extensive studies on terrestrial terpenoids from the Canary Islands and South America. Carotenoids are widely distributed in nature and are of importance as food coloring material and as antioxidants (the detailed mechanisms of which still have to be worked out); new carotenoids continue to be discovered from marine sources, for example by the group of Synnove Liaaen-Jensen, Norwegian Institute of Technology). Yoshimasa Hirata (1915–), who started research at Nagoya University, is a champion in the isolation of nontrivial natural products. He characterized the bioluminescent luciferin from the marine ostracod *Cypridina hilgendorfii* in 1966 (with his students, Toshio Goto, Yoshito Kishi, and Osamu Shimomura); tetrodotoxin from the fugu fish in 1964 (with Goto and Kishi and co-workers), the structure of which was announced simultaneously by the group of Kyosuke Tsuda (1907–, tetrodotoxin, matrine) and Woodward; and the very complex palytoxin, $C_{129}H_{223}N_3O_{54}$ in 1981–1987 (with Daisuke Uemura and Kishi). Richard E. Moore, University of Hawaii, also announced the structure of palytoxin independently. Jon Clardy (1943–, Cornell University) has determined the X-ray structures of many unique marine natural products, including brevetoxin B (1981), the first of the group of toxins with contiguous *trans*-fused ether rings constituting a stiff ladder-like skeleton. Maitotoxin, $C_{164}H_{256}O_{68}S_2Na_2$, MW 3422, produced by the dinoflagellate *Gambierdiscus toxicus* is the largest and most toxic of the nonbiopolymeric toxins known; it has 32 alicyclic 6- to 8-membered ethereal rings and acyclic chains. Its isolation (1994) and complete structure determination was accomplished jointly by the groups of Takeshi Yasumoto (Tohoku University), Kazuo Tachibana and Michio Murata (Tokyo University) in 1996. Kishi, Harvard University, also deduced the full structure in 1996.

The well-known excitatory agent for the cat family contained in the volatile oil of catnip, *Nepeta cataria*, is the monoterpene nepetalactone, isolated by S. M. McElvain (1943) and structure determined by Jerrold Meinwald (1954); cats, tigers, and lions start purring and roll on their backs in response to this lactone. Takeo Sakan (1912–1993) investigated the series of monoterpenes neomatatabiols, etc.

from Actinidia, some of which are male lacewing attractants. As little as 1 fg of neomatatabiol attracts lacewings.

The first insect pheromone to be isolated and characterized was bombykol, the sex attractant for the male silkworm, *Bombyx mori* (by Butenandt and co-workers, see above). Numerous pheromones have been isolated, characterized, synthesized, and are playing central roles in insect control and in chemical ecology. The group at Cornell University have long been active in this field: Tom Eisner (1929–, behavior), Jerrold Meinwald (1927–, chemistry), Wendell Roeloff (1939–, electrophysiology, chemistry). Since the available sample is usually minuscule, full structure determination of a pheromone often requires total synthesis; Kenji Mori (1935–, Tokyo University) has been particularly active in this field. Progress in the techniques for handling volatile compounds, including collection, isolation, GC/MS, etc., has started to disclose the extreme complexity of chemical ecology which plays an important role in the lives of all living organisms. In this context, natural products chemistry will be play an increasingly important role in our grasp of the significance of biodiversity.

5. SYNTHESIS

Synthesis has been mentioned often in the preceding sections of this essay. In the following, synthetic methods of more general nature are described. The Grignard reaction of Victor Grignard (1871–1935, Np 1912) and then the Diels–Alder reaction by Otto Diels (1876–1954, Np 1950) and Kurt Alder (1902–1956, Np 1950) are extremely versatile reactions. The Diels–Alder reaction can account for the biosynthesis of several natural products with complex structures, and now an enzyme, a Diels–Alderase involved in biosynthesis has been isolated by Akitami Ichihara, Hokkaido University (1997).

The hydroboration reactions of Herbert Brown (1912–, Purdue University, Np 1979) and the Wittig reactions of Georg Wittig (1897–1987, Np 1979) are extremely versatile synthetic reactions. William S. Johnson (1913–1995, University of Wisconsin, Stanford University) developed efficient methods for the cyclization of acyclic polyolefinic compounds for the synthesis of corticoid and other steroids, while Gilbert Stork (1921–, Columbia University) introduced enamine alkylation, regiospecific enolate formation from enones and their kinetic trapping (called "three component coupling" in some cases), and radical cyclization in regio- and stereospecific constructions. Elias J. Corey (1928–, Harvard University, Np 1990) introduced the concept of retrosynthetic analysis and developed many key synthetic reactions and reagents during his synthesis of bioactive compounds, including prostaglandins and gingkolides. A recent development is the ever-expanding supramolecular chemistry stemming from 1967 studies on crown ethers by Charles Pedersen (1904–1989), 1968 studies on cryptates by Jean-Marie Lehn (1939–), and 1973 studies on host–guest chemistry by Donald Cram (1919–); they shared the chemistry Nobel prize in 1987.

6. NATURAL PRODUCTS STUDIES IN JAPAN

Since the background of natural products study in Japan is quite different from that in other countries, a brief history is given here. Natural products is one of the strongest areas of chemical research in Japan with probably the world's largest number of chemists pursuing structural studies; these are joined by a healthy number of synthetic and bioorganic chemists. An important Symposium on Natural Products was held in 1957 in Nagoya as a joint event between the faculties of science, pharmacy, and agriculture. This was the beginning of a series of annual symposia held in various cities, which has grown into a three-day event with about 50 talks and numerous papers; practically all achievements in this area are presented at this symposium. Japan adopted the early twentieth century German or European academic system where continuity of research can be assured through a permanent staff in addition to the professor, a system which is suited for natural products research which involves isolation and assay, as well as structure determination, all steps requiring delicate skills and much expertise.

The history of Japanese chemistry is short because the country was closed to the outside world up to 1868. This is when the Tokugawa shogunate which had ruled Japan for 264 years was overthrown and the Meiji era (1868–1912) began. Two of the first Japanese organic chemists sent abroad were Shokei Shibata and Nagayoshi Nagai, who joined the laboratory of A. W. von Hoffmann in Berlin. Upon return to Japan, Shibata (Chinese herbs) started a line of distinguished chemists, Keita and Yuji Shibata (flavones) and Shoji Shibata (1915–, lichens, fungal bisanthraquinonoid pigments, ginsenosides); Nagai returned to Tokyo Science University in 1884, studied ephedrine, and left a big mark in the embryonic era of organic chemistry. Modern natural products chemistry really began when three extraordinary organic chemists returned from Europe in the 1910s and started teaching and research at their respective faculties:

Riko Majima, 1874–1962, C. D. Harries (Kiel University); R. Willstätter (Zürich): Faculty of Science, Tohoku University; studied urushiol, the catecholic mixture of poison ivy irritant.

Yasuhiko Asahina, 1881–1975, R. Willstätter: Faculty of pharmacy, Tokyo University; lichens and Chinese herb.

Umetaro Suzuki, 1874–1943, E. Fischer: Faculty of agriculture, Tokyo University; vitamin B_1(thiamine).

Because these three pioneers started research in three different faculties (i.e., science, pharmacy, and agriculture), and because little interfaculty personnel exchange occurred in subsequent years, natural products chemistry in Japan was pursued independently within these three academic domains; the situation has changed now. The three pioneers started lines of first-class successors, but the establishment of a strong infrastructure takes many years, and it was only after the mid-1960s that the general level of science became comparable to that in the rest of the world; the 3rd IUPAC Symposium on the Chemistry of Natural Products, presided over by Munio Kotake (1894–1976, bufotoxins, see below), held in 1964 in Kyoto, was a clear turning point in Japan's role in this area.

Some of the outstanding Japanese chemists not already quoted are the following. Shibasaburo Kitazato (1852–1931), worked with Robert Koch (Np 1905, tuberculosis) and von Behring, antitoxins of diphtheria and tetanus which opened the new field of serology, isolation of microorganism causing dysentery, founder of Kitazato Institute; Chika Kuroda (1884–1968), first female Ph.D., structure of the complex carthamin, important dye in safflower (1930) which was revised in 1979 by Obara *et al.*, although the absolute configuration is still unknown (1998); Munio Kotake (1894–1976), bufotoxins, tryptophan metabolites, nupharidine; Harusada Suginome (1892–1972), aconite alkaloids; Teijiro Yabuta (1888–1977), kojic acid, gibberrelins; Eiji Ochiai (1898–1974), aconite alkaloids; Toshio Hoshino (1899–1979), abrine and other alkaloids; Yusuke Sumiki (1901–1974), gibberrelins; Sankichi Takei (1896–1982), rotenone; Shiro Akabori (1900–1992), peptides, C-terminal hydrazinolysis of amino acid ; Hamao Umezawa (1914–1986), kanamycin, bleomycin, numerous antibiotics; Shojiro Uyeo (1909–1988), lycorine; Tsunematsu Takemoto (1913–1989), inokosterone, kainic acid, domoic acid, quisqualic acid; Tomihide Shimizu (1889–1958), bile acids; Kenichi Takeda (1907–1991), Chinese herbs, sesquiterpenes; Yoshio Ban (1921–1994), alkaloid synthesis; Wataru Nagata (1922–1993), stereocontrolled hydrocyanation.

7. CURRENT AND FUTURE TRENDS IN NATURAL PRODUCTS CHEMISTRY

Spectroscopy and X-ray crystallography has totally changed the process of structure determination, which used to generate the excitement of solving a mystery. The first introduction of spectroscopy to the general organic community was Woodward's 1942–1943 empirical rules for estimating the UV maxima of dienes, trienes, and enones, which were extended by Fieser (1959). However, Butenandt had used UV for correctly determining the structures of the sex hormones as early as the early 1930s, while Karrer and Kuhn also used UV very early in their structural studies of the carotenoids. The Beckman DU instruments were an important factor which made UV spectroscopy a common tool for organic chemists and biochemists. With the availability of commercial instruments in 1950, IR spectroscopy became the next physical tool, making the 1950 Colthup IR correlation chart and the 1954 Bellamy monograph indispensable. The IR fingerprint region was analyzed in detail in attempts to gain as much structural information as possible from the molecular stretching and bending vibrations. Introduction of NMR spectroscopy into organic chemistry, first for protons and then for carbons, has totally changed the picture of structure determination, so that now IR is used much less frequently; however, in biopolymer studies, the techniques of difference FTIR and resonance Raman spectroscopy are indispensable.

The dramatic and rapid advancements in mass spectrometry are now drastically changing the protocol of biomacromolecular structural studies performed in biochemistry and molecular biology. Herbert Hauptman (mathematician, 1917–, Medical Foundation, Buffalo, Np 1985) and Jerome Karle (1918–, US Naval Research Laboratory, Washington, DC, Np 1985) developed direct methods for the determination of crystal structures devoid of disproportionately heavy atoms. The direct method together with modern computers revolutionized the X-ray analysis of molecular structures, which has become routine for crystalline compounds, large as well as small. Fred McLafferty (1923–, Cornell University) and Klaus Biemann (1926–, MIT) have made important contributions in the development of organic and bioorganic mass spectrometry. The development of cyclotron-based facilities for crystallographic biology studies has led to further dramatic advances enabling some protein structures to be determined in a single day, while cryoscopic electron micrography developed in 1975 by Richard Henderson and Nigel Unwin has also become a powerful tool for 3D structural determinations of membrane proteins such as bacteriorhodopsin (25 kd) and the nicotinic acetylcholine receptor (270 kd).

Circular dichroism (c.d.), which was used by French scientists Jean B. Biot (1774–1862) and Aimé Cotton during the nineteenth century "deteriorated" into monochromatic measurements at 589 nm after R.W. Bunsen (1811–1899, Heidelberg) introduced the Bunsen burner into the laboratory which readily emitted a 589 nm light characteristic of sodium. The 589 nm $[\alpha]_D$ values, remote from most chromophoric maxima, simply represent the summation of the low-intensity readings of the decreasing end of multiple Cotton effects. It is therefore very difficult or impossible to deduce structural information from $[\alpha]_D$ readings. Chiroptical spectroscopy was reintroduced to organic chemistry in the 1950s by C. Djerassi at Wayne State University (and later at Stanford University) as optical rotatory dispersion (ORD) and by L. Velluz and M. Legrand at Roussel-Uclaf as c.d. Günther Snatzke (1928–1992, Bonn then Ruhr University Bochum) was a major force in developing the theory and application of organic chiroptical spectroscopy. He investigated the chiroptical properties of a wide variety of natural products, including constituents of indigenous plants collected throughout the world, and established semiempirical sector rules for absolute configurational studies. He also established close collaborations with scientists of the former Eastern bloc countries and had a major impact in increasing the interest in c.d. there.

Chiroptical spectroscopy, nevertheless, remains one of the most underutilized physical measurements. Most organic chemists regard c.d. (more popular than ORD because interpretation is usually less ambiguous) simply as a tool for assigning absolute configurations, and since there are only two possibilities in absolute configurations, c.d. is apparently regarded as not as crucial compared to other spectroscopic methods. Moreover, many of the c.d. correlations with absolute configuration are empirical. For such reasons, chiroptical spectroscopy, with its immense potentialities, is grossly underused. However, c.d. curves can now be calculated nonempirically. Moreover, through-space coupling between the electric transition moments of two or more chromophores gives rise to intense Cotton effects split into opposite signs, exciton-coupled c.d.; fluorescence-detected c.d. further enhances the sensitivity by 50- to 100-fold. This leads to a highly versatile nonempirical microscale solution method for determining absolute configurations, etc.

With the rapid advances in spectroscopy and isolation techniques, most structure determinations in natural products chemistry have become quite routine, shifting the trend gradually towards activity-monitored isolation and structural studies of biologically active principles available only in microgram or submicrogram quantities. This in turn has made it possible for organic chemists to direct their attention towards clarifying the mechanistic and structural aspects of the ligand/biopolymeric receptor interactions on a more well-defined molecular structural basis. Until the 1990s, it was inconceivable and impossible to perform such studies.

Why does sugar taste sweet? This is an extremely challenging problem which at present cannot be answered even with major multidisciplinary efforts. Structural characterization of sweet compounds and elucidation of the amino acid sequences in the receptors are only the starting point. We are confronted with a long list of problems such as cloning of the receptors to produce them in sufficient quantities to investigate the physical fit between the active factor (sugar) and receptor by biophysical methods, and the time-resolved change in this physical contact and subsequent activation of G-protein and enzymes. This would then be followed by neurophysiological and ultimately physiological and psychological studies of sensation. How do the hundreds of taste receptors differ in their structures and their physical contact with molecules, and how do we differentiate the various taste sensations? The same applies to vision and to olfactory processes. What are the functions of the numerous glutamate receptor subtypes in our brain? We are at the starting point of a new field which is filled with exciting possibilities.

Familiarity with molecular biology is becoming essential for natural products chemists to plan research directed towards an understanding of natural products biosynthesis, mechanisms of bioactivity triggered by ligand–receptor interactions, etc. Numerous genes encoding enzymes have been cloned and expressed by the cDNA and/or genomic DNA-polymerase chain reaction protocols. This then leads to the possible production of new molecules by gene shuffling and recombinant biosynthetic techniques. Monoclonal catalytic antibodies using haptens possessing a structure similar to a high-energy intermediate of a proposed reaction are also contributing to the elucidation of biochemical mechanisms and the design of efficient syntheses. The technique of photoaffinity labeling, brilliantly invented by Frank Westheimer (1912–, Harvard University), assisted especially by advances in mass spectrometry, will clearly be playing an increasingly important role in studies of ligand–receptor interactions including enzyme–substrate reactions. The combined and sophisticated use of various spectroscopic means, including difference spectroscopy and fast time-resolved spectroscopy, will also become increasingly central in future studies of ligand–receptor studies.

Organic chemists, especially those involved in structural studies have the techniques, imagination, and knowledge to use these approaches. But it is difficult for organic chemists to identify an exciting and worthwhile topic. In contrast, the biochemists, biologists, and medical doctors are daily facing exciting life-related phenomena, frequently without realizing that the phenomena could be understood or at least clarified on a chemical basis. Broad individual expertise and knowledge coupled with multidisciplinary research collaboration thus becomes essential to investigate many of the more important future targets successfully. This approach may be termed "dynamic," as opposed to a "static" approach, exemplified by isolation and structure determination of a single natural product. Fortunately for scientists, nature is extremely complex and hence all the more challenging. Natural products chemistry will be playing an absolutely indispensable role for the future. Conservation of the alarming number of disappearing species, utilization of biodiversity, and understanding of the intricacies of biodiversity are further difficult, but urgent, problems confronting us.

That natural medicines are attracting renewed attention is encouraging from both practical and scientific viewpoints; their efficacy has often been proven over the centuries. However, to understand the mode of action of folk herbs and related products from nature is even more complex than mechanistic clarification of a single bioactive factor. This is because unfractionated or partly fractionated extracts are used, often containing mixtures of materials, and in many cases synergism is most likely playing an important role. Clarification of the active constituents and their modes of action will be difficult. This is nevertheless a worthwhile subject for serious investigations.

Dedicated to Sir Derek Barton whose amazing insight helped tremendously in the planning of this series, but who passed away just before its completion. It is a pity that he was unable to write this introduction as originally envisaged, since he would have had a masterful overview of the content he wanted, based on his vast experience. I have tried to fulfill his task, but this introduction cannot do justice to his original intention.

ACKNOWLEDGMENT

I am grateful to current research group members for letting me take quite a time off in order to undertake this difficult writing assignment with hardly any preparation. I am grateful to Drs. Nina Berova, Reimar Bruening, Jerrold Meinwald, Yoko Naya, and Tetsuo Shiba for their many suggestions.

8. BIBLIOGRAPHY

"A 100 Year History of Japanese Chemistry," Chemical Society of Japan, Tokyo Kagaku Dojin, 1978.
K. Bloch, *FASEB J.*, 1996, **10**, 802.
"Britannica Online," 1994–1998.
Bull. Oriental Healing Arts Inst. USA, 1980, **5**(7).
L. F. Fieser and M. Fieser, "Advanced Organic Chemistry," Reinhold, New York, 1961.
L. F. Fieser and M. Fieser, "Natural Products Related to Phenanthrene," Reinhold, New York, 1949.
M. Goodman and F. Morehouse, "Organic Molecules in Action," Gordon & Breach, New York, 1973.
L. K. James (ed.), "Nobel Laureates in Chemistry," American Chemical Society and Chemistry Heritage Foundation, 1994.
J. Mann, "Murder, Magic and Medicine," Oxford University Press, New York, 1992.
R. M. Roberts, "Serendipity, Accidental Discoveries in Science," Wiley, New York, 1989.
D. S. Tarbell and T. Tarbell, "The History of Organic Chemistry in the United States, 1875–1955," Folio, Nashville, TN, 1986.

1.01 Overview

USHIO SANKAWA
Toyama Medical and Pharmaceutical University, Japan

1.01.1 INTRODUCTION

The term "polyketides" is not clearly defined, although the textbooks of chemistry and/or biosynthesis of natural products describe fatty acids as a subclass of polyketides.[1–3] The word polyketides is used in different senses. One covers a wide range of compounds including fatty acids while the other is limited to non-fatty acid polyketides. Sometimes polyketides is applied to polyketomethylene intermediates of aromatic polyketides. We use the word polyketides instinctively for a limited range of compounds without considering the definition that the term polyketides includes fatty acids. This is due to the historical background of natural products chemistry. Common fatty acids are primary metabolites and have been regarded as compounds for study by biochemists. Non-fatty acid polyketides are secondary metabolites and they are the particular focus for natural products chemists.

Recent investigations on the molecular biology of polyketide biosynthesis have demonstrated that the genes of fatty acid synthases (FAS) and microbial polyketide synthases (PKS) have a significant homology and they are recognized to have evolved from the same prototype gene. The chain elongation mechanism in the biosynthesis of reduced polyketides such as erythromycin and rapamycin is basically the same as that with FAS. The distinction between primary and secondary metabolites has disappeared and many chapters of this volume describe the multidisciplinary approach on the studies of biosynthesis of corresponding classes of compounds, indicating that this approach is essential for studies of the biosynthesis of natural products. It is reasonable therefore that fatty acids are included in polyketides when their biosynthesis is discussed at the enzyme and gene levels.

The most remarkable feature of natural products is the diversity of chemical structures, which reflects the diversity of enzymes and genes responsible for the biosynthesis of complex structures. Molecular diversity is the reason why natural products have been the target of new drug discovery. The advantages of screening natural products to find new lead compounds for drug development are the *unpredictability* and *unexpectedness* of the chemical structures found. Many examples of complex and diverse structures of microbial metabolites are evident in the following sections. Toxic fungal metabolites aflatoxin B_1, patulin, and penicillic acid are good examples of fungal polyketide mycotoxins. They are synthesized by extensive modifications of rather simple compounds, orsellinic acid, 6-methylsalicylic acid (6MS), and norsolorinic acid, respectively. They are good examples of biosynthesis which epitomize the diversity of natural product structures. Fatty acid derived compounds are also extensively modified for a variety of reactions. Prostaglandins (PG) and jasmonic acid (JA) are well-known compounds deriving from fatty acids, linolenic and arachidonic acids (AA), respectively. The structures of PG and JA look similar, however, the ring formation reaction is completely different. Microbial polyketides produced by actinomycetes and filamentous fungi have been used for medicinal purposes as antibiotics (erythromycin and tetracycline), antitumor agents (anthracyclinones), cholesterol-lowering drugs (pravastin and mevinolin), and immunosuppressors (FK 506, rapamycin). They are synthesized from the basic building units by chain elongation enzymes (PKS) and following modification reactions such as Diels–Alder cyclization. The polypeptide chain elongation reaction of cyclic polypeptide cyclosporin is very similar to that of macrolide biosynthesis. The shikimate pathway is well developed in higher plants, and phenylalanine, which is supplied from shikimate pathways, is the precursor of all C_6–C_3 and C_6–C_3–C_6 compounds. Phenylalanine ammonia lyase (PAL) is a characteristic enzyme of higher plants and responsible for the formation of cinnamic acid, which is the precursor of most abundant plant constituents, lignin, lignans, coumarins (C_6–C_3), and flavonoids (C_6–C_3–C_6). PAL is a gateway to secondary metabolism from primary metabolism. The other categories of compounds are rather miscellaneous and the biosyntheses of compounds containing sulfur, C—P bonds and CN groups have been described. The mechanism of *de novo* C—S bond formation is different for each compound, reflecting the diversity of biosynthesis. This volume covers a wide range of compounds; however, it is intended to focus on the multidisciplinary approaches to elucidate complex and diverse reactions in their biosynthesis.

1.01.2 FATTY ACIDS AND POLYKETIDES

From the early days of organic chemistry, chemists noticed that straight-chain fatty acids occurring in nature consisted of an even number of carbon atoms. This led to the hypothesis that fatty acids are formed by the head-to-tail condensation of acetate. At the end of the nineteenth century Collie proposed a hypothesis that ketene ($CH_2{=}C{=}O$) or its equivalent was the building unit of aromatic polyketides. His hypothesis was demonstrated by the synthesis of phenolic compounds from so-called polyacetates.[4] The concept of a polyketide hypothesis was further elaborated by Robert Robinson and he compiled a textbook entitled *The Structural Relationships of Natural Products* in 1955.[5] Two years before Robinson's textbook was published, Birch wrote his first paper on the acetate hypothesis in 1953 when he was a professor at Sydney University.[6] His concept of the acetate hypothesis is based on the detailed analysis of the structures of phenolic natural products, particularly lichen substances, depsides, and depsidones, as well as flavonoids and stilbenes. The participation of acetate in fatty acid biosynthesis was proved by Lynen in the late 1940s and this undoubtedly had an influence on the concept of the acetate hypothesis. At that time it was believed that in the biosynthesis of fatty acids, head-to-tail condensation of acetate yielded β-ketoacyl intermediates and the carbonyl was then reduced to methylene by successive reactions. The acyl group was further condensed with another acetate and repeated reaction cycles yielded fatty acid. On the other hand, if head-to-tail acetate condensation proceeded without reductive removal of oxygen, all the carbonyl groups were retained and thus formed polyketomethylene intermediates which were cyclized by aldol or Claisen condensation to give resorcinol or phloroglucinol type phenolics. This is the theoretical background of the acetate hypothesis. Scientists engaged in the structural elucidation of lichen substances noticed that resorcyclic acid homologues, consisting of depsides and depsidones, had an odd number of carbon side chains. This was reasonably explained by the acetate hypothesis. Shoji Shibata, who studied in the laboratory of Asahina, said reminiscently, "We should have thought seriously how depsides and depsidones were synthesized in lichens. Everybody knew that the side chains of depsides and depsidones consisted of odd numbers of carbons". Birch mentioned in his lecture that he could do nothing but use his brain because of the lack of staff and facilities when he moved to Sydney from Oxford. This situation gave him a time to see the obvious acetate hypothesis, as indicated by the title of his autobiography.[7]

1.01.2.1 Early Studies of Polyketide Biosynthesis

The first practical evidence to verify the acetate hypothesis was an incorporation and degradation experiment on 6MS with ^{14}C-labeled acetate to prove the labeling at expected carbons.[8] The advent of ^{14}C in the biosynthetic investigations made it possible to obtain definite evidence which changed the hypothesis to reality. Extensive studies by Birch and his collaborators on many fungal polyketides rigorously established that acetate was the building unit of phenolic polyketides. The presence of polyketomethylene intermediates was proved by the isolation of tetraacetic acid lactones from the cultures of a tropolone producing fungus *Penicillium stipitatum* when it was cultured in the presence of ethionine, an inhibitor of methionine.[9] The intermediate en route to stipitatonic acid, a tropolone, was shown to be methylorsellinic acid, formed by the introduction of a methyl group from methionine to the polyketomethylene intermediate (Figure 1). Cyclization afforded methylorsellinic acid, which underwent ring expansion and oxidation to give the seven-membered tropolone ring. Tetraacetic acid lactone afforded orsellinic acid under mild alkaline conditions, thus leading to the acetate hypothesis being proved by biochemical and biomimetic experiments in the 1960s.

In contrast, fatty acid biosynthesis was studied by biochemists using a biochemical approach and the characterization of enzymes involved in FAS was the mainstream of research. The basic building unit of fatty acid biosynthesis was soon shown to be not acetic acid but acetyl CoA. This is an activated form of acetate in all living organisms and serves as the precursor of fatty acids, phenolic polyketides, and isoprenoids. During the investigations of FAS with crude enzyme extracts, in addition to acetyl CoA and NADPH, other cofactors were found to be essential for the reaction. They were soon identified as biotin and carbonate. This led to the finding that acetyl CoA was converted into malonyl CoA by acetyl CoA carboxylase and is the true C_2 chain elongation unit. The scheme of fatty acid biosynthesis in vertebrate FAS is illustrated in Figure 2 and in more detail in Chapter 1.02. The participation of malonyl CoA was soon proved in polyketide biosynthesis by feeding experiments with ^{14}C-labeled malonate and it acted as a chain elongation unit. In some cases ^{14}C-labeled diethylmalonate gave higher incorporation ratios because of its better permeability through the cell membrane. As it appears in Figure 2, acetyl CoA is a "starter," the starting unit, and malonyl CoA is the chain elongation unit, the "extender." Ketoacyl synthase (KS) mediates decarboxylative condensation between Sp-acyl and Sc-malonyl groups which are transferred from acetyl and malonyl CoAs by acyl transferase (AT) to give β-ketothioester. Reduction of the β-keto group by ketoreductase (KR) followed by dehydration (DH) and further enoyl reduction (ER) gives saturated acyl thioester. Repetition of the cycle leads to the formation of C_{16} or C_{18} saturated acyl thioesters, which are cleaved by thioesterase (TE) to afford free fatty acids.

In contrast to the ample data on FAS, from bacteria to mammals, up until recently 6-methylsalicylic acid synthase (6MSase) was the sole PKS characterized and purified.[10,11] 6MSase is a single polypeptide of high molecular weight (ca. 270 kDa), containing domains of KS, AT, DH, KR, and acyl carrier protein (ACP). The steps of reduction and dehydration of one carbonyl group are involved in the reaction and it requires NADPH. When NADPH was omitted from the reaction mixture an incomplete derailed product was released from the enzyme as triacetic acid lactone (TAL). This led to the definition of the timing of carbonyl reduction at the C_6 stage; the scheme of 6MS biosynthesis is illustrated in Figure 3. Two aromatic hydrogen atoms of 6MS are derived from malonyl CoA. Investigations with chiral malonate revealed that one of the two hydrogens derived from the proR and the other from the proS precursor (Chapter 1.14). The modes of biosynthesis of FAS and PKS clearly show a marked similarity.

The structural diversity of microbial polyketides were attractive targets for biosynthetic studies by natural product chemists. Radioisotopes, ^{14}C and ^{3}H, were the only available tools for investigations on biosynthesis in the early days. In late 1960s, tracer studies with ^{13}C and its NMR detection were developed into highly sophisticated methodology. In particular, incorporation experiments with ^{13}C–^{13}C double-labeled acetate unambiguously clarified the labeling patterns (Figure 4). The first experiment on mollisin biosynthesis by Cary, Seto, and Tanabe clearly demonstrated the labeling pattern of intact acetate units.[12] The advantage of double-labeled acetate was not only in

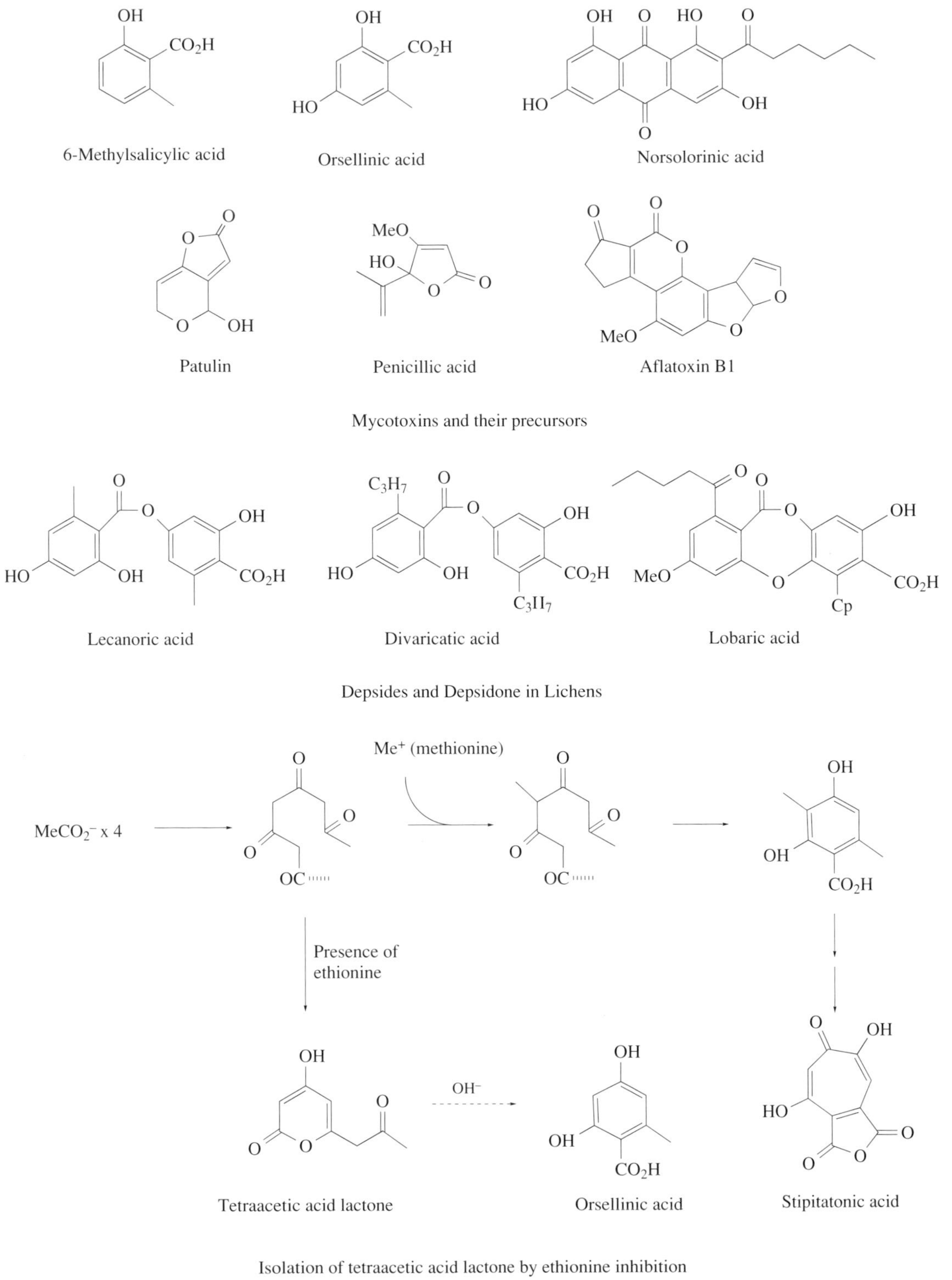

6-Methylsalicylic acid

Orsellinic acid

Norsolorinic acid

Patulin

Penicillic acid

Aflatoxin B1

Mycotoxins and their precursors

Lecanoric acid

Divaricatic acid

Lobaric acid

Depsides and Depsidone in Lichens

$MeCO_2^-$ x 4

Me^+ (methionine)

Presence of ethionine

OH^-

Tetraacetic acid lactone

Orsellinic acid

Stipitatonic acid

Isolation of tetraacetic acid lactone by ethionine inhibition

Figure 1 Mycotoxins and their precursors. Production of tetraacetic acid lactone by ethionine inhibition in tropolone producing fungus, *Penicillium stipitatum*, and biomimetic cyclization affording orsellinic acid.

determining the labeling patterns of intact acetate units but also detecting ^{13}C signals even if the incorporation of acetate was very poor because ^{13}C–^{13}C coupled signals appeared as satellites not hindered by the 1.1% ^{13}C natural abundance. Double-labeled acetate was extensively used in polyketide biosynthesis. For example, the labeling patterns of rubrofusarin and norsolorinic acid were assigned and the folding patterns of polyketo-chains could not be clarified without using

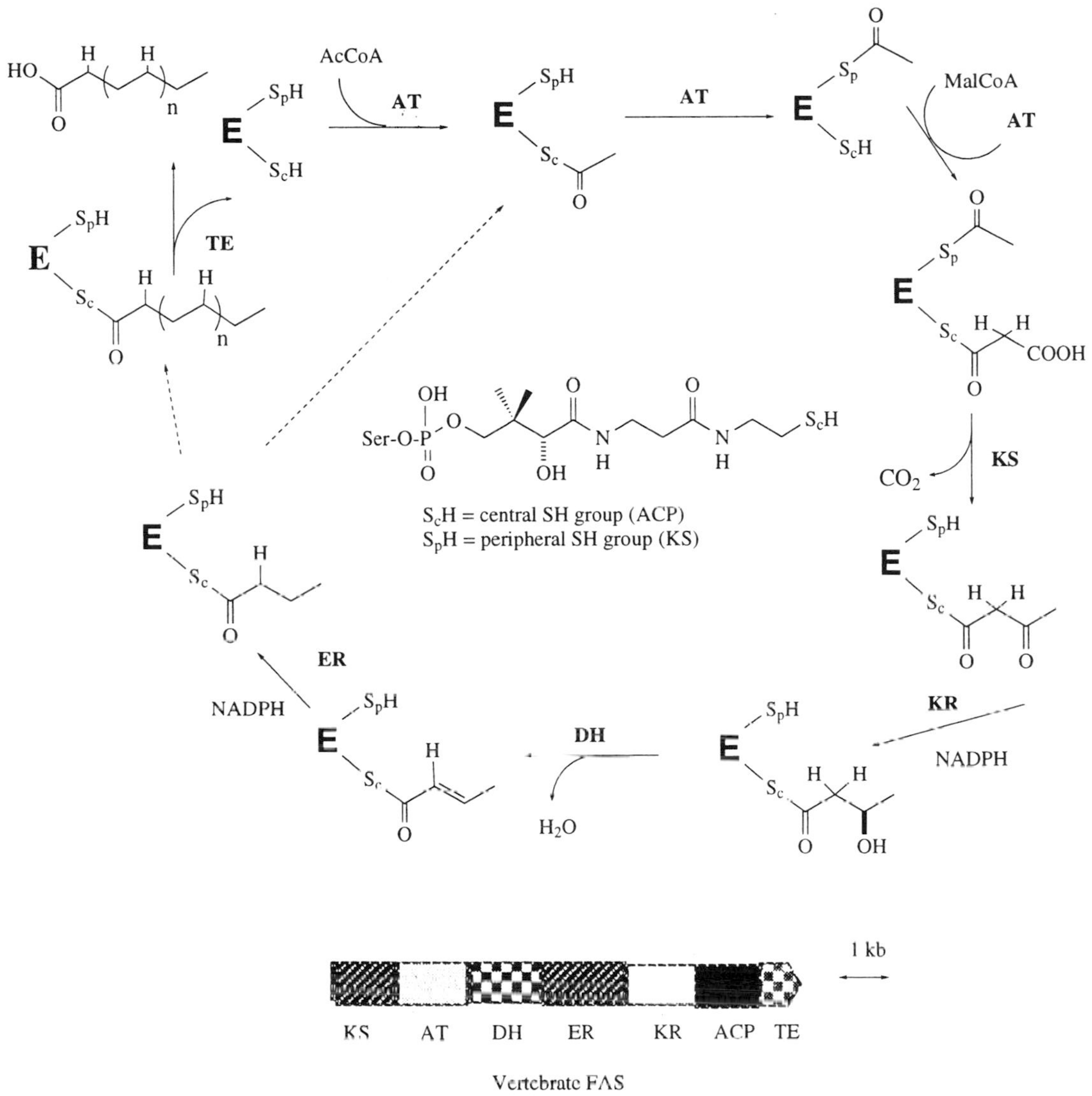

Figure 2 Catalytic cycle of vertebrate FAS and its gene construction.

incorporation experiments with double-labeled acetate.[13] Incorporation of double-labeled acetate into scytalone revealed a 1 : 1 ratio of two different labeling patterns, indicating that a symmetrical intermediate, tetrahydroxynaphthalene, was reduced in either of the two benzene rings in equal proportion.[14] Splitting NMR signals by ^{13}C–^{13}C coupling observed in the compounds labeled with double labeled acetate indicated the present of ^{13}C at adjacent carbons. This means information on adjacent atoms can be obtained by measuring the ^{13}C NMR of compounds labeled with double-labeled acetate. The method was soon extended to the combination of ^{13}C with ^{2}H[15] and ^{18}O,[16] and the ^{13}C NMR spectrum gave coupled and/or shifted signals and the site of labeling could be easily detected.

Enzymes involved in polyketide biosynthesis were particularly difficult to characterize; this is clear from the fact that 6MSase was the only aromatic polyketide synthetase purified and characterized. This problem was overcome by the advent of gene cloning and functional expression techniques in this area; however the characterization of enzymes of fungal polyketide synthases is still rather limited.

1.01.2.2 Fatty Acid Biosynthesis

The structure of FAS has been well characterized by gene cloning and enzymatic studies. Fatty acids are constituents of the cell membranes of all living organisms except for some archaebacteria,

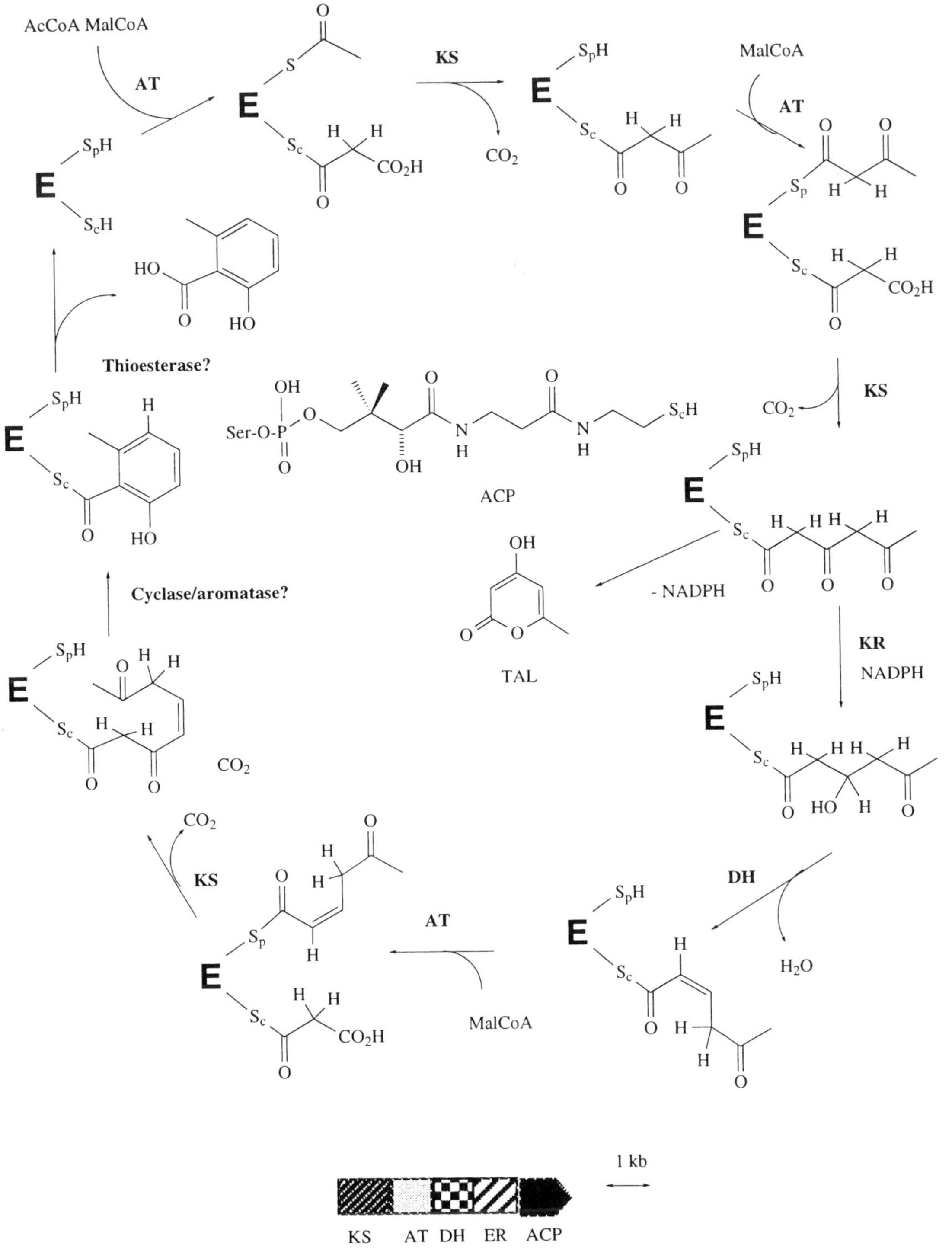

KS AT DH ER ACP

Fungal 6MS PKS

Figure 3 Reaction of 6MSase and its gene construction.

and FAS is present in all living organisms from bacteria to human beings. FASs are divided into two major classes, type I and type II, according to their protein constructions. FASs of bacteria and plants belong to type II in which each enzyme is an independent protein as it appears in the gene cluster map of *E. coli* (Figure 5). Type II FAS enzyme proteins form a multienzyme complex to catalyze fatty acid synthesis in living cells. In contrast, vertebrate type I FAS consists of a single polypeptide of 2500 amino acids with a molecular weight of 270 kDa. Its native and functional state

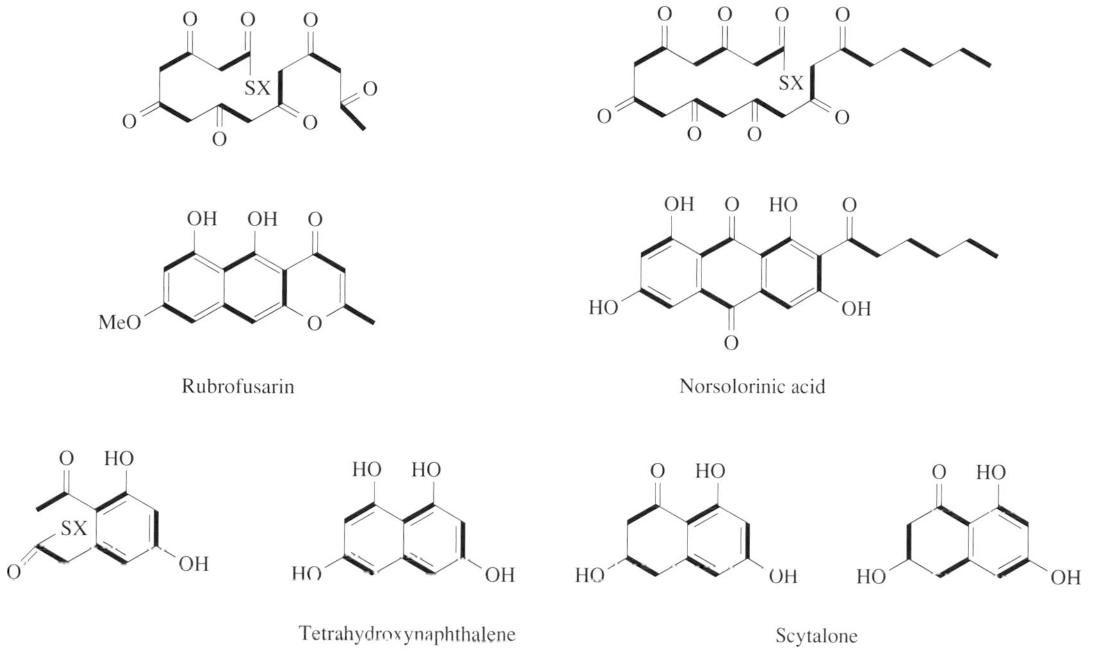

Rubrofusarin

Norsolorinic acid

Tetrahydroxynaphthalene

Scytalone

Figure 4 Labeling and cyclization patterns in fungal polyketides.

is an α_2 dimer. Yeast and fungal FAS are also type I; however, they consist of two proteins and form a dodecamer $\alpha_6\beta_6$ when they function as active FAS. Each enzyme and catalytic domain was identified by the comparison of amino acid sequences with known motifs of FAS proteins. The amino acid sequences of corresponding catalytic sites are retained from bacteria to vertebrates and the whole amino acid sequences of the regions were assigned to corresponding functional domains from limited catalytic site sequences. Acyl carrier proteins (ACP) possess 4′-phosphopantetheine moiety at a serine residue which is the key SH group, acting as a swinging arm to carry growing acyl groups. Gene constructions of type I and type II FASs and PKSs are summarized in Figure 5. Fatty acids with nonstraight chains are biosynthesized either by using nonacetyl CoA starter units or modification of double bonds to cyclopropane rings. ω-Cyclopentyl, cyclohexyl, and cycloheptyl fatty acids are biosynthesized with corresponding starters, although the biosynthesis of these cyclic acyl CoAs are unique processes (see Chapter 1.03).

1.01.2.3 Unsaturated Fatty Acids and Related Compounds

Excess fatty acids produced or taken from foods are stored as triglyceride. Saturated fatty acids are degraded by β-oxidation to yield acetyl CoA which is used as an energy source in mitochondria. Unsaturated fatty acids are formed by the dehydrogenation of saturated fatty acids in various organisms and the major unsaturated fatty acids in plants that we take as foods are oleic, linoleic, and linolenic acids. These unsaturated fatty acids are not only constituents of phospholipid but are also the precursors of a wide variety of bioactive natural products. Leguminous plants have been used by farmers as biological fertilizers. This is because the legume plants accumulate nitrogen in nodules, formed on roots. Nodule formation is a complex process and the soil bacteria *Risobium* is responsible for nodule formation. Flavonoids are chemotactic-factors for *Risobium* and lipochitins are essential in the formation of symbiotic nodules. *Nod* genes, associated with the nodule formation of *Rizobium*, have a homology with *Streptomyces* PKS genes. *NodE* and *NodF* are now clarified to be genes for the biosynthesis of a unique *trans*-unsaturated fatty acid of lipooligosaccharide, lipochitin. This is a typical example of natural products, exchanging signals between microorganism and plant. This means that "conversation" among organisms in nature is at the chemical level (see Chapter 1.13). The simple aldehyde responsible for the "green odor" of plant leaves is derived from linoleic and linolenic acids by peroxidation with lipoxygenase followed by lyase reaction. Jasmonic acid (JA) and its methyl ester are known as the fragrance of jasmine (see Figure 6). JA has been recognized as responsible for tuber formation in potatoes, yams, and onions. It also causes growth inhibition, senescence, and leaf abscission and also acts as a signal transmitter in elicitor treated

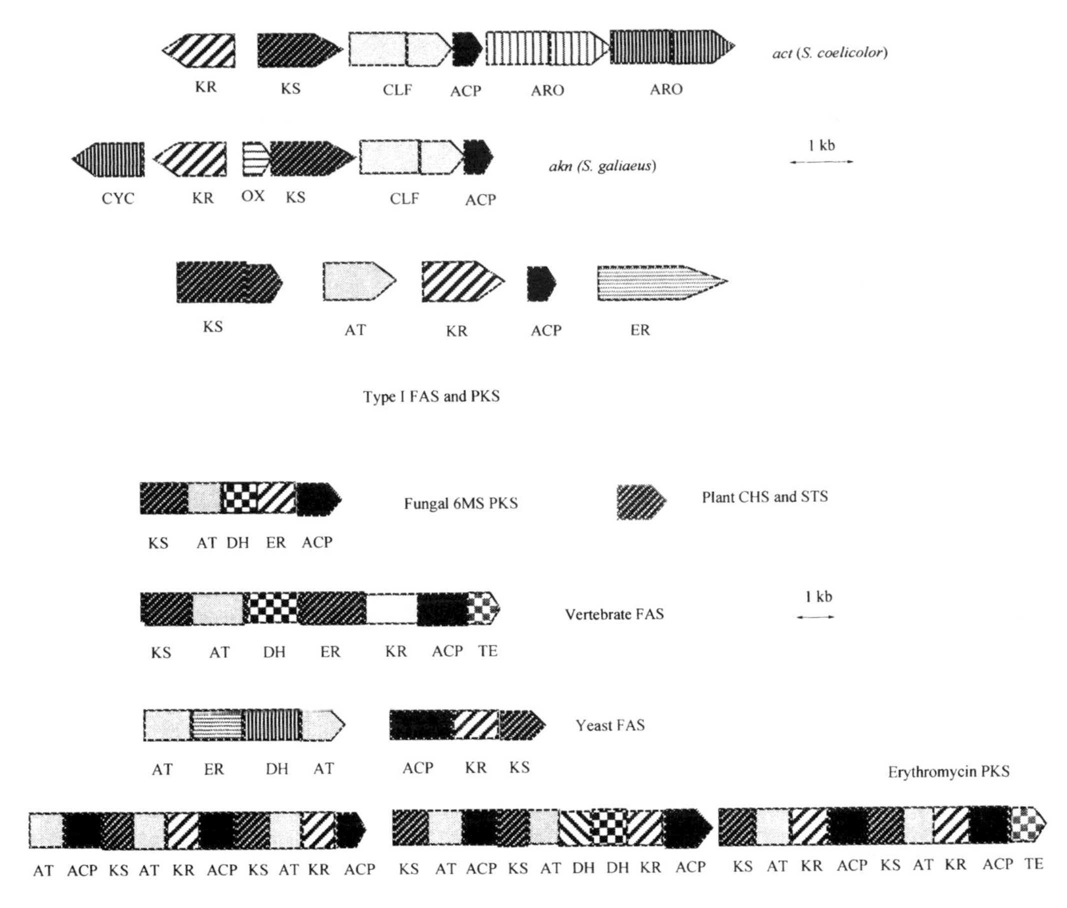

Figure 5 Gene constructions of type I and type II FASs and PKSs.

cells inducing the production of phytoalexins. JA is now characterized as the seventh plane hormone, the others being auxin, cytokinin, gibberellin, ethylene, abscicic acid, and brasinolide. The cyclopentane structure of JA seemed to indicate that its synthesis is analogous to prostaglandins; however, the reaction mechanism is completely different. Peroxidation of linolenic acid followed by dehydration affords allene epoxide, which generates the oxypentadienyl cation and cyclizes to give the oxocyclopentene skeleton (see Chapter 1.05). Prostaglandins of marine organisms are produced via the oxypentadienyl cation. A considerable number of compounds are derived from unsaturated fatty acids by oxygenation and bromocation-induced cyclization (see Chapters 1.08 and 1.12). A polyene(yne) alkene derived from unsaturated fatty acids is the precursor of a cyclic bromoether of marine algae. Laurencin is formed from hydroxyacetylenic alkene by bromocation-induced cyclization catalyzed by haloperoxidase. A unique heptadiene ring of ectocarpene is formed by lipoxygenase-induced radical reaction followed by electrocyclic reaction. Ectocarpene is a signal substance in the sexual cycle of algae and acts as an attractant pheromone of gametes.[17] Another example of a signal substance is A factor which is formed by aldol condensation between branched short fatty acid and glycerol derivatives and induces antibiotic production in *Streptomyces*.

1.01.2.4 Arachidonate Cascade

Among the compounds derived from fatty acids, prostaglandins (PGs) and other eiconsanoids are among the most important findings of the twentieth century. The history of PGs can be traced back to the 1930s and after a long dark period of no progress, the chemistry and biology of PGs flourished in the 1960s. The physiological functions of PGs were clarified using chemically synthesized PGs, since the amount of PGs obtained from tissues is minuscule. Aspirin had been used as an antipyretic, analgesic, and antiinflammatory drug from the end of nineteenth century without

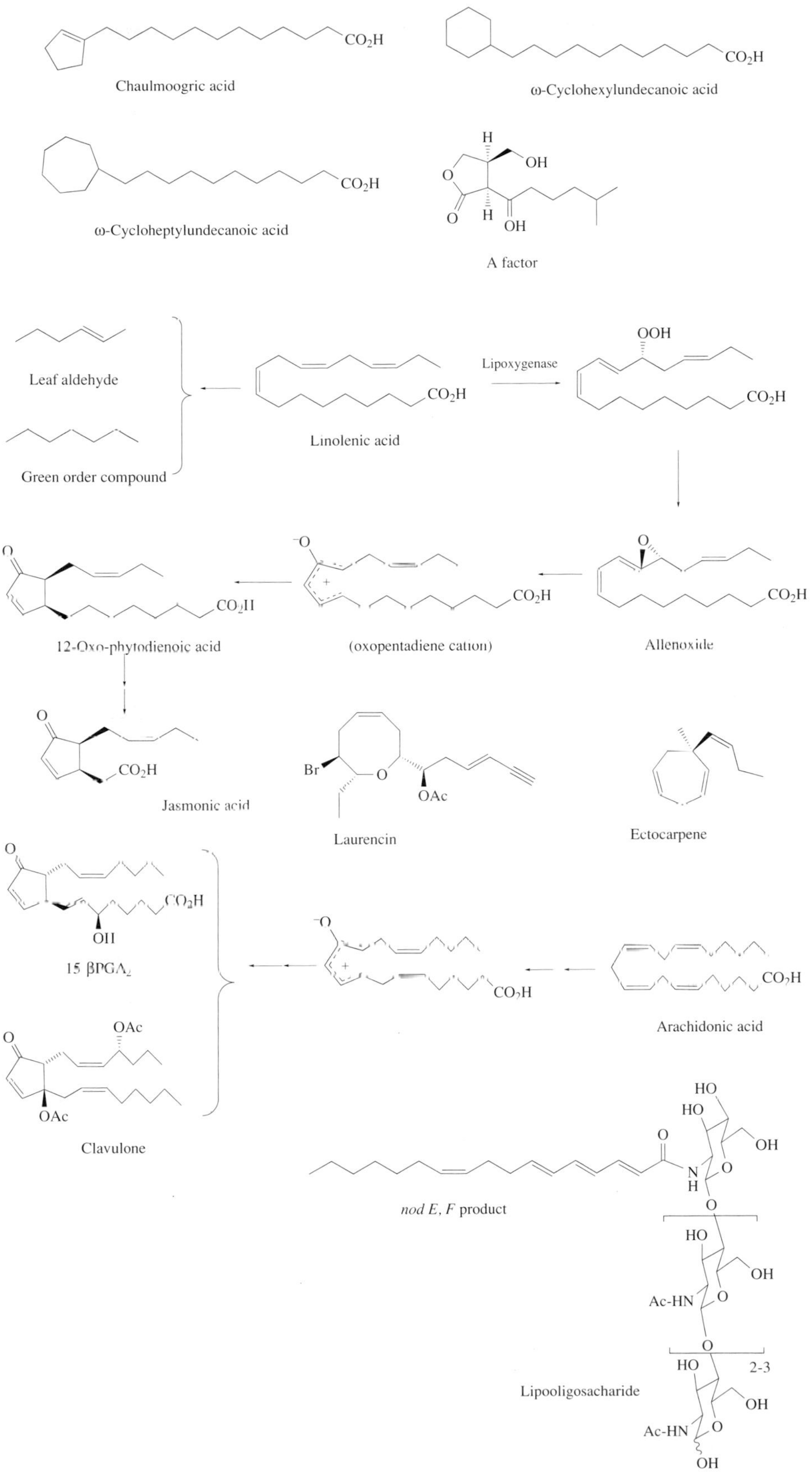

Chaulmoogric acid

ω-Cyclohexylundecanoic acid

ω-Cycloheptylundecanoic acid

A factor

Leaf aldehyde

Green order compound

Linolenic acid

Lipoxygenase

12-Oxo-phytodienoic acid

(oxopentadiene cation)

Allenoxide

Jasmonic acid

Laurencin

Ectocarpene

15 βPGA$_2$

Arachidonic acid

Clavulone

nod E, F product

Lipooligosacharide

Figure 6 Fatty acid related compounds and biosynthesis of leaf aldehyde, jasmonic acid, and marine prostaglandins.

knowing what the target of aspirin was. In 1971 Vane discovered that nonsteroidal antiinflammatory drugs such as aspirin and indomethacin exhibited their action by inhibiting prostaglandin biosynthesis in cyclooxygenase (COX) reaction. Aspirin is a unique drug in its inhibitory activity of COX. Aspirin inhibits COX irreversibly by acetylating a serine residue near catalytic center. The X-ray crystal structure of COX clearly demonstrated the mechanism of action of nonsteroidal anti-inflammatory drugs.[18] The discovery of thromboxane (TX) and prostacyclin (PGI_2) from platelet and blood vessel endothelial cells is another achievement in prostaglandin studies. Thromboxane A_2 acts to induce platelet coagulation, while prostacyclin inhibits platelet coagulation. The opposite activities of TX and PGI_2 answer the simple question as to why blood does not coagulate in blood vessels. TX and PGI_2 are synthesized by the corresponding synthase from PGH_2 produced by COX in platelets and endothelial cells. The sensitivities of COX are different in both cells and platelet COX is more sensitive to aspirin. Moreover, aspirin inhibits COX irreversibly and TX is not synthesized *de novo* in platelets since platelets have no nucleus, whereas in endothelial cells COX can be synthesized leading to the formation of PGI_2. This selectivity led to the administration of low-dose aspirin to patients who suffer from myocardial infarction to prevent platelet coagulation which triggers thrombosis. More recently inducible COX was found in various cells and was found to be responsible for inflammation. The new COX was named as COX2 and the previously known COX as COX1. COX1 is present in almost all tissues and plays important physiological roles. Selective inhibition of COX2 is the target in the development of new antiinflammatory drugs (see Chapters 1.07 and 1.09).

Arachidonic acid also serves as a precursor of leukotrienes (LT), originally found as slow reacting substances of anaphylaxis (SRSA). After a long blank as with PG studies, SRSA-like compounds were detected in leukocytes and their structures finally determined by total synthesis. Since these eiconsanoids were found in leukocytes, they are called leukotrienes. The first enzyme of PG, TX, and PGI_2 synthesis is COX, while arachidonate 5-lipoxygenase is the first enzyme reaction leading to the LT group. Metabolism from arachidonate is called the arachidonate cascade, which is named after metabolic flow like a cascade falling with several separation streams (see Figures 7 and 8). Studies on PG and LT have recently focused on their receptors. Multiple types of receptors, specific to various tissues and cells, are now the targets for drug development. The molecular evolution of arachidonate cascade enzymes and PG and LT receptors are summarized in Chapter 1.10. The structure of platelet activating factor (PAF) resembles those of phospholipids, however the ether linkage is unique to PAF. Its biosynthesis looks simple, but the reactions taking place in PAF biosynthesis are interesting and reasonable from the point of view of organic chemistry (see Chapter 1.11).

1.01.3 POLYKETIDE BIOSYNTHESIS IN ACTINOMYCETES AND FUNGI

Microorganisms produce natural products with diverse structures. Following the discovery of penicillin and streptomycin, antibiotics have been used as medicines and tools to study biological phenomena. The molecular diversity of microbial metabolites posed a challenge to organic chemists to determine their complex structures (see Figure 9). Their structural diversity and complexity have been challenging problems in biosynthetic investigations for many years and various sophisticated methods have been developed during investigations on their biosynthesis.

1.01.3.1 Type II Aromatic Polyketide Synthetases in Actinomycetes

The wave of progress in molecular biology overwhelmed existing methodologies and gene cloning has become the most common and essential technique for the investigation of the biosynthesis of polyketides. The methodology of polyketide biosynthesis in *Streptomycetes* at the gene level has been established by the great contribution of David Hopwood. His longstanding study on the genetics of *Actinomycetes* led to the fruitful development of molecular biology in the 1980s. Cloning and identification of genes involved in the biosynthesis of actinorhodin had a tremendous impact on the biosynthesis of polyketides (see Figure 10). Although actinorhodin itself is not used for medicinal purposes, it is the first *Actinomycetes* polyketide studied at the gene level. The approach developed by Hopwood is based on the preparation of mutants in which actinorhodin biosynthesis was blocked in *Streptomyces coelicolar* and a restoration of actinorhodin production by complementation with shuttle vectors containing random fragments of genes from wild type *S. coelicolar*.

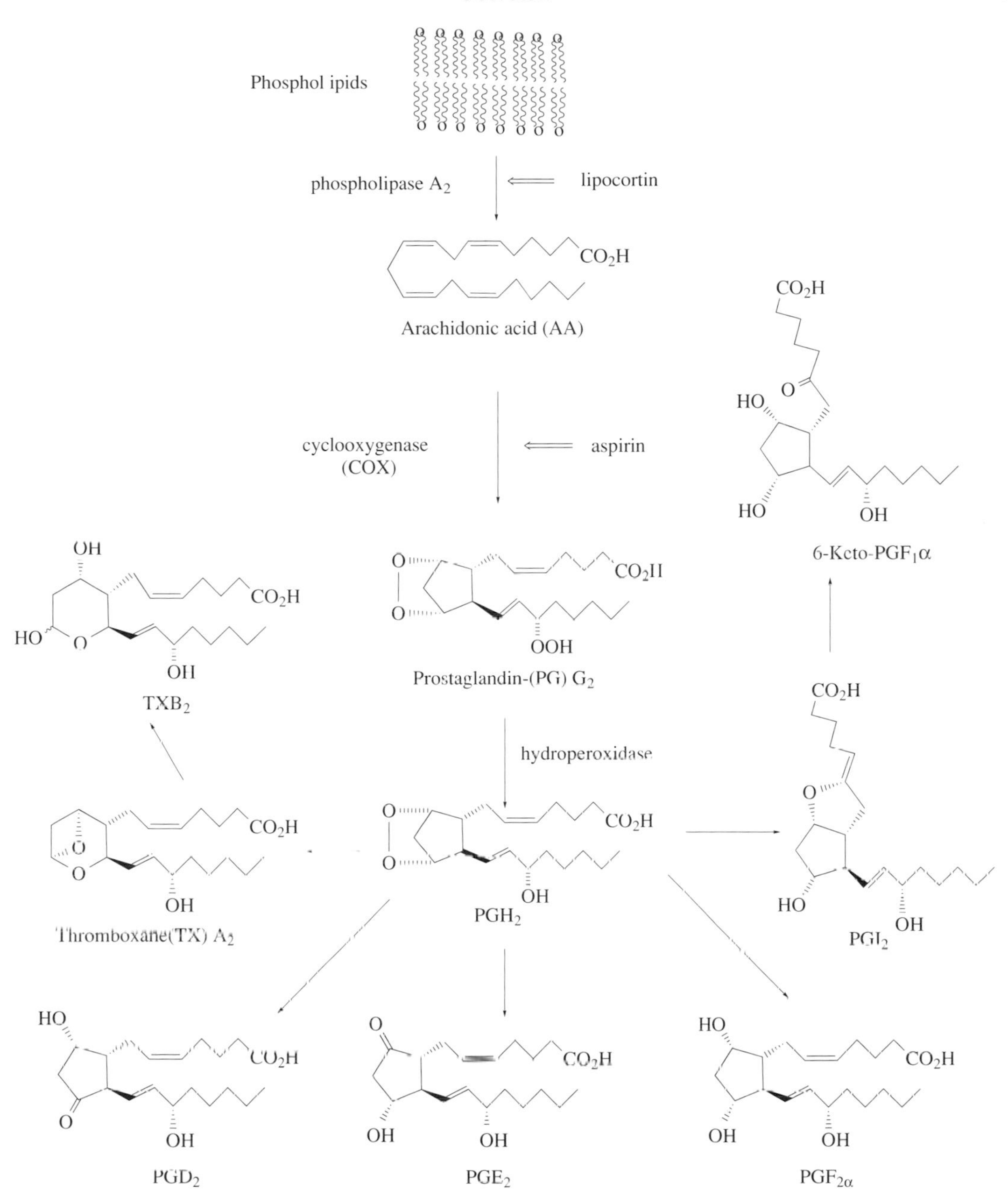

Figure 7 Arachidonate cascade; cyclooxygenase (COX) pathway.

Complementation of the mutated gene with a genomic DNA fragment was easily identified by the intense color of actinorhodin on exposure to ammonia vapor. This simple chemical reaction brought about a great leap in the molecular biology of *Streptomyces*. An important finding was that biosynthesis genes form a cluster and one or more self-resistant genes were found in the cluster. The scheme for actinorhodin biosynthesis, elucidated by Hopwood's group, appears in Figure 10. *Act I*, *II*, *III*, etc. are the names of mutants as well as the genes encoding corresponding enzymes. For example, *act III* is KR, with an amino acid sequence homologous to NAD(P)H dependent reductase. KS, chain length determining factor (CLF), and ACP are the minimal and essential genes of polyketide biosynthesis, encoded in a region of *act I*. The hypothetical pathway is deduced from the derailed products, produced by mutants or recombinant strains.[19] Many of the derailed or artificially produced compounds were assumed to possess γ-pyrone structures. α-Pyrone is normally much more stable than γ-pyrone and structures proposed as γ-pyrone should be reinvestigated (see Figure 14). More than 20 PKS gene clusters have been cloned by complementation, resistance, or *act* probe hybridization. They include oxytetracycline (*S. rimosus*), tetracenomycin (*S. glaucescens*),[20]

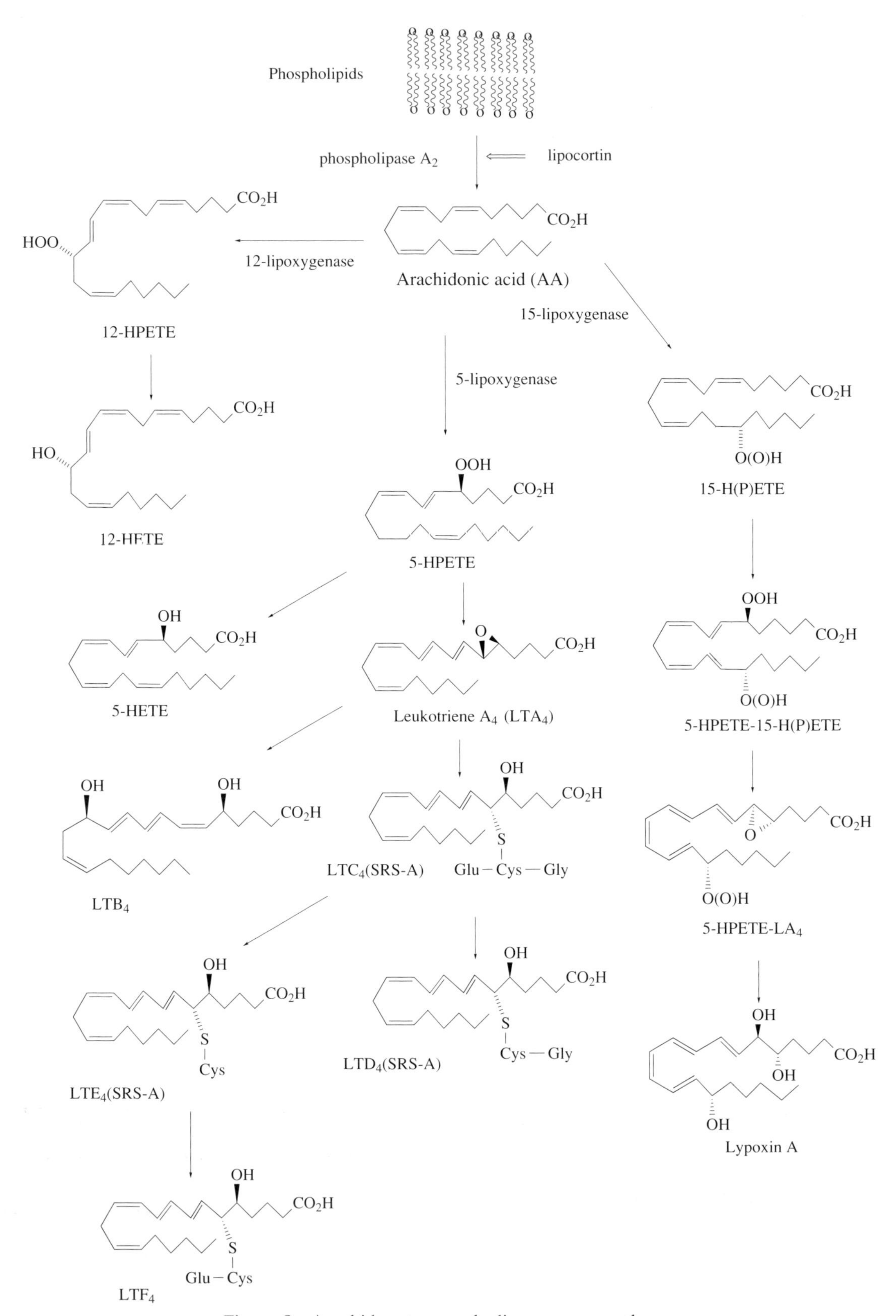

Figure 8 Arachidonate cascade; lipoxygenase pathway.

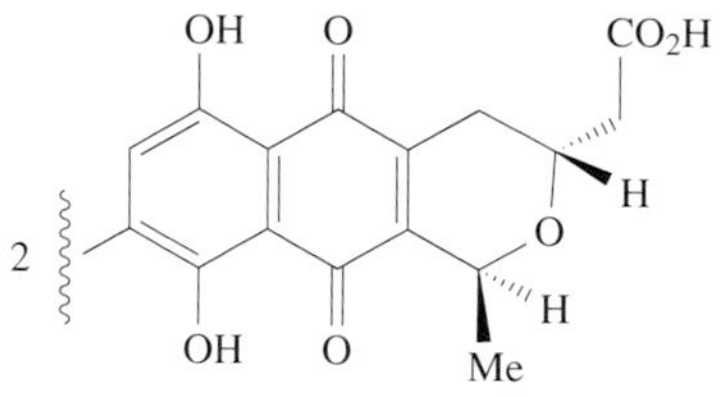

Actinorhodin

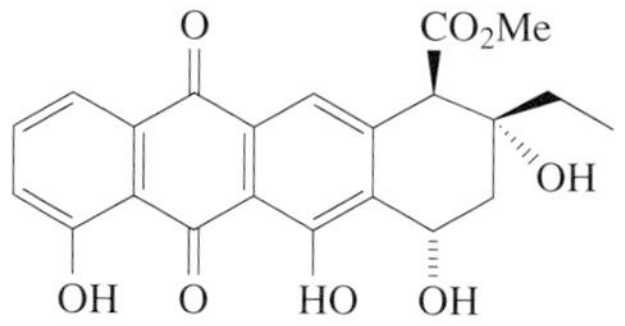

Aclavinone

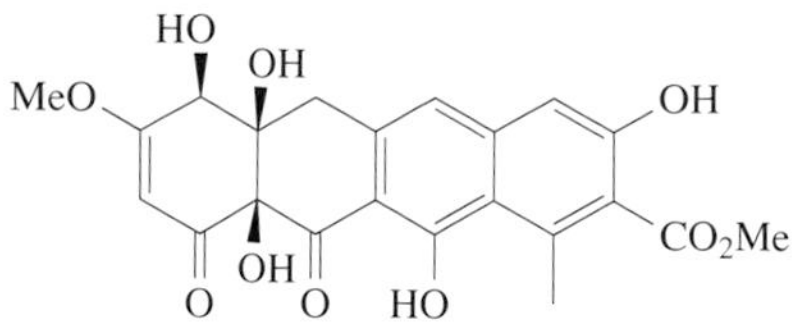

Tetracenomycin C

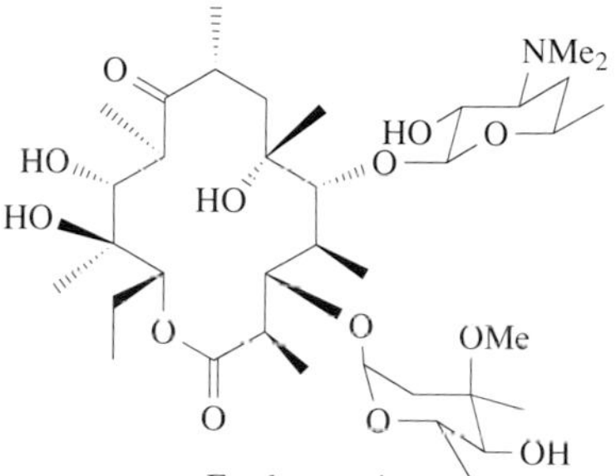

Erythromycin

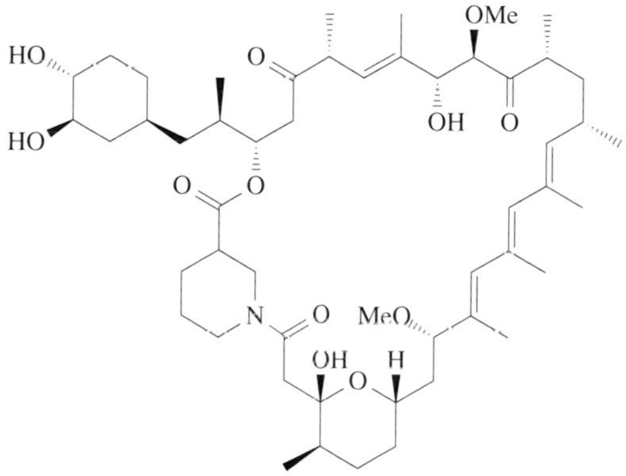

Rapamycin

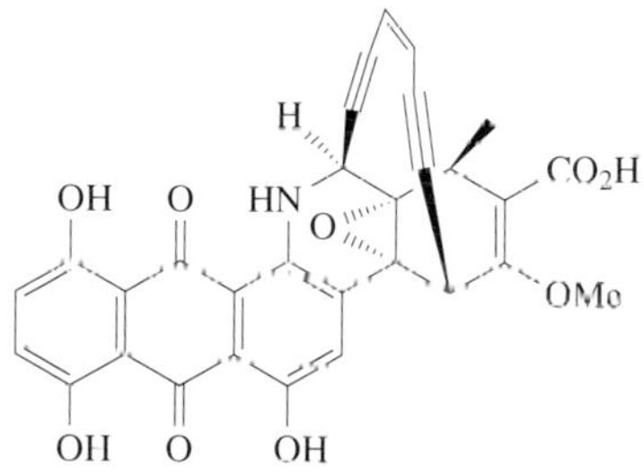

Dynemicin A

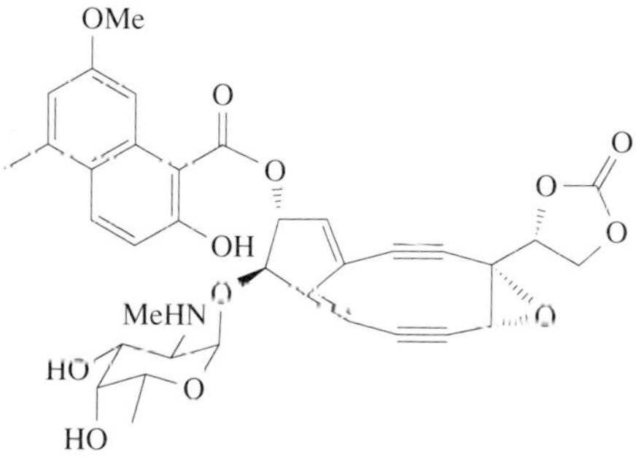

Neocarzinostatin chromophore

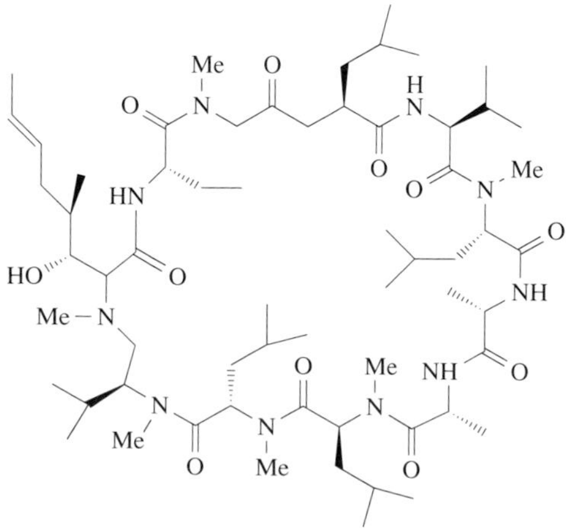

Cyclosporin

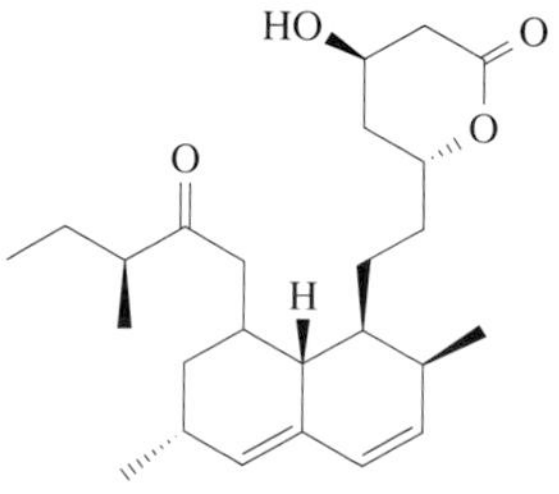

Mevinolin

Figure 9 Polyketides from actinomycetes and fungi.

frenolicin (*S. roseofluvus*), and aclavinone (*S. galiaeus*).[21] The small gene *akn X* between KR and KS in the *akn* gene of aklavinone producing *S. galiaeus* is the oxygenation enzyme of an anthrone intermediate and requires no cofactor in the reaction. The reaction mechanism of an unusual oxygenase was investigated by site-directed mutagenesis, however, no definite conclusion was obtained.[21] In contrast, fungal anthrone oxygenase required nonheme ferric iron as a cofactor.[22] The engineering of *Actinomycetes* aromatic polyketide synthases to produce novel compounds by combinatorial biosynthesis is discussed in Chapter 1.18.

1.01.3.2 Type I Polyketide Synthase in Actinomycetes

The participation of propionate in erythromycin biosynthesis was correctly predicted by Woodward when its structure was elucidated and later verified by feeding experiments, that the erythromycin skeleton was built from seven propionate units. In the reactions of the erythromycin-synthesizing enzyme, one propionyl CoA was a starter and 6-methylmalonyl CoA was an extender to elongate the polyketide chain. Starting from resistant gene *ermE*, both sides were sequenced to find PKS associated with erythronolide synthesis. Erythronolide PKS was far away from *erm E*, surrounding by modification enzyme genes open reading frames (ORF). 6-Deoxyerythronolide PKS consists of three large ORFs, *ery AI*, *AII*, and *AIII*, each 10 kb with more than 300 amino acids. Sequencing and functional expression uncovered the unique nature of the three deoxyerythronolide synthase (DEBS) enzymes. They consist of modules of FAS-like architecture, programmed to generate a polyketide chain with the right structure for each unit. The synthase for the units possessing hydroxyl contain ACP, KS, KR, but lack DH and ER, and that for carbonyl also lacks KR. The unit for the synthesis of the saturated moiety possesses complete sets of enzymes like FAS. Macrocyclic bacterial antibiotics, rapamycin FK 506, avermectin,[23] and the ansamycin family have similar modular PKS genes, programmed to afford the polyketide chain of the correct structure. The modular structures are accurately constructed to give the correct final products. This must be achieved by molecular evolution, although it is just the results of successive random mutation from the ancient FAS genes (see Chapter 1.19).

Endiyne antibiotics are extremely potent antitumor compounds with unique molecular structures. Dynemycin, esperamycin, and neocarcinostatin chromophores are typical examples. The chemistry of endiynes has been extensively studied to give an account of their antitumor activity and benzene diradicals formed by the cyclization of endiynes are active forms for antitumor activity. To date, biosynthesis of endiynes has been investigated only by incorporation experiments and no studies at the gene level have been reported. The structures of enediynes suggest that their biosynthesis enzymes are probably type I (see Chapter 1.21).

1.01.3.3 Polyketide and Polypeptide Biosynthesis in Fungi

The biosynthesis genes of fungal polyketides have not been investigated as extensively as in bacteria, partly due to the difficulty of purifying the corresponding enzymes as well as due to the large size of the genes. For example, the PKS of simple structure compound 6-methylsalycilic acid, 6MSase, consists of ca. 1800 amino acids and it was a sole example of fungal PKS whose enzyme protein was purified and characterized. Another difficulty was the lack of suitable expression vectors such as that used in *Streptomyces* PKS studies. pTAex vector, an expression vector under the control of α-amylase promoter of *Asparagus oryzae*, can be used in *A. oryzae* as well as in *A. nidlans*. The 6MSase gene, cloned from *A. terreus* where it was a silent gene, was successfully expressed by pTAex. The *wA* gene encoding a spore color producing protein in *Aspergillus nidlans* was expressed with pTAwA; the inserted *wA* gene originally cloned from *A. nidlans* produced an isocoumarin derivative of C_{14} carbon atoms although the expected naphthopyrone product was not detected. Reinvestigation of C-terminal revealed that there was a mistake in the original sequencing and missing one base had led to a shorter ORF. The correct ORF was 2157 aa, 170 aa additional length, and revealed the presence of a TE-like domain in the missing region. At the terminal of erythromycin synthesis gene *ery AIII* was provided a TE domain which serves as thioesterase as well as a lactone-forming enzyme (see Chapter 1.19). TE in fungal PKS catalyzes thioester cleavage as well as a Claisen type condensation to form the second ring of the *wA* naphthopyrone pigment. In addition to the *wA* gene for naphthopyrone synthatase, PKS1 (*Cocliobolus longenarium*) for tetrahydroxynaphthalene synthase, STCA (*Aspergillus nidlans*), and PKSA (*Aspergillus parasiticus*) for

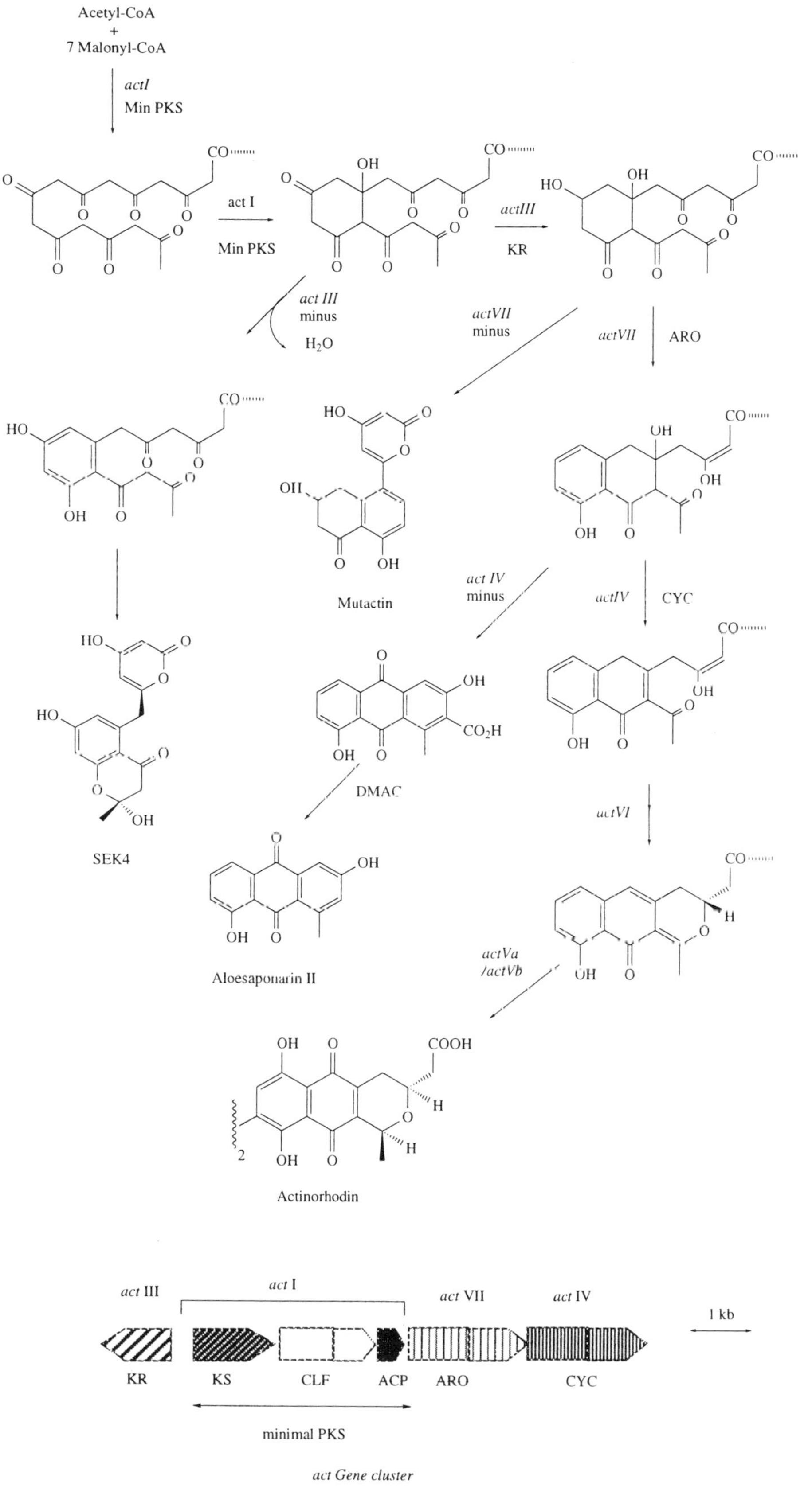

Figure 10 Biosynthesis of antinorhodin and gene clusters.

norsolorinic acid synthase genes showed a similar construction, KS-AT-ACP-(ACP)-TE, while other known PKSs, 6MS, lobastatin, and T-toxin synthases lack the terminal TE. This suggests the important function of the TE-like gene product in fungal PKS which uses a Claisen condensation to form the final rings. Amino acid sequences of fungal PKSs are not module structures and are similar to 6MSase and type I FAS, indicating that polyketide chain elongation is basically the same as 6MSase and type I FAS where chain elongation takes place by a shuttle between central SH in ACP and peripheral SH in KS. The reason why there are two ACP in some fungal PKS is unknown at present and this problem has to be solved in the future (see Chapter 1.16). The biosynthesis of the highly toxic fungal metabolites aflatoxins has been studied at various levels. It is impossible to deduce the first polyketide for aflatoxin biosynthesis from its structure. The biosynthetic pathway of aflatoxin biosynthesis has been proposed from studies on co-metabolite analysis, precursor feeding, and model chemical reaction experiments. Starting from an anthraquinone norsolorinic acid, biosynthesis proceeds via averufin (anthraquinone), averantin (anthraquinone), and sterigmatocystin (xanthone) and finally to aflatoxin B_1. The diversity of reactions involved in the pathway is the reason why aflatoxin biosynthesis has attracted the attention of many scientists. A study at the gene level has revealed that genes for aflatoxin biosynthesis form a cluster as in actinomycetes PKS and the reaction mechanisms in the pathway are likely to be clarified in the near future (see Chapter 1.17).

The Diels–Alder reaction was regarded as typical organic reaction for synthesis. The structural analysis of fungal metabolites from a biosynthetic viewpoint suggested that the biosynthesis of a significant number of compounds could be accounted for by the Diels–Alder reaction. Among the candidates studied by incorporation experiments with potential intermediates for biological Diels–Alder reactions, betaenones, chaetoglobosins, solanopyrones, and macrophomic acid were singled out for further study. Solanopyrone is the sole example whose biological Diels–Alder reaction was investigated at the enzyme level. Enzyme reactions were carried out with crude preparations and inevitably accompanied by chemical reactions yielding racemic products. Ample evidence has been accumulated by chemical level studies and further support from the characterization of enzymes and genes is expected (see Chapter 1.15).

Cyclosporin is an immunosupressive compound in medical use for organ transplantation as FK 506. The size of the cyclosporin-synthesizing enzyme gene termed *simA* is the largest among the known peptide synthetases. It resembles erythromycin biosynthesis modular gene *eryA* due to its structure for programmed biosynthesis of the peptide chain. Starting from the N-terminal region, each amino acid is linked successively with the aid of a condensing enzyme carrying the swinging arm of a 4′-phosphopantetheine residue. Instead of acyl CoA in polyketide synthase, constituent amino acids are converted into adenylate by adenylate-forming enzymes in each domain, which reacts with amino acids on 4′-phosphopantetheine to give an elongated peptide chain. When peptide chain transfer to the next module occurs, the peptide chain increases in length by one amino acid and finally a condensing enzyme cyclizes the linear peptide. The whole construction of the enzyme is shown in Figure 11.

1.01.4 BIOSYNTHESIS OF NATURAL PRODUCTS INVOLVING THE SHIKIMATE PATHWAY

The first investigation of the shikimate pathway was undertaken by Davis and Sprinson in the 1950s to clarify the biosynthesis of aromatic amino acids, phenylalanine, tyrosine, and tryptophan. Humans lack the ability to synthesize these aromatic amino acids and phenylalanine and tryptophan are essential amino acids supplied by foods. Shikimic acid was identified as an intermediate in the blocked mutants of *E. coli* and *Neurospora crassa* and it is the origin of the name of the pathway. Shikimic acid itself was isolated from the fruits of the Japanese shikimi tree (*Illicium religiosum*) by Eykmann at the end of the nineteenth century. Eykmann was a Dutch professor who taught chemistry at that time in the Pharmacy School of the University of Tokyo. The structure of shikimic acid was not determined until much later due to its complexity. The name shikimate pathway was given after isolation of the first cycle intermediate in the pathway (see Figure 12).[24]

1.01.4.1 Enzymology and Molecular Biology of the Shikimate Pathway

Classical work by Davis has developed to modern study at the gene level and it is now possible to investigate the enzyme reaction in detail by using functional expression of the enzyme produced

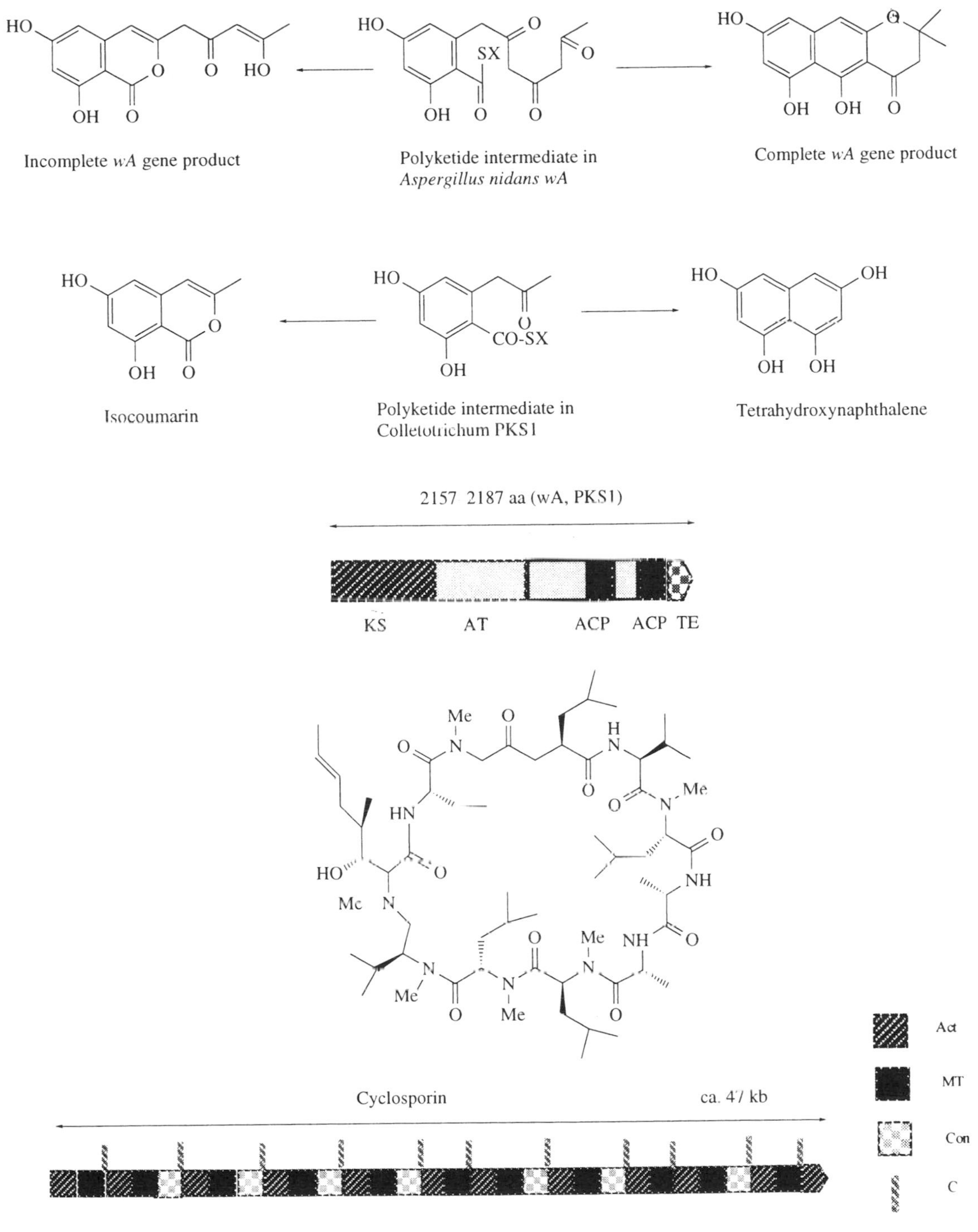

Incomplete *wA* gene product

Polyketide intermediate in *Aspergillus nidans wA*

Complete *wA* gene product

Isocoumarin

Polyketide intermediate in Colletotrichum PKS1

Tetrahydroxynaphthalene

2157 2187 aa (wA, PKS1)

KS AT ACP ACP TE

Cyclosporin

ca. 47 kb

Act MT Con C

Cyclosporin synthesis gene; Act-adenylate formine domain, MT-*N*-methyltransferase, Con-Condensing domain, C-carrier

Figure 11 Biosynthesis of fungal PKS, cyclosporin and their biosynthesis genes.

in transformed microorganisms. This study is a fine example of the combination of organic chemistry and molecular biology and offers a model for contemporary biosynthetic study. The shikimate pathway branches at chorismate into two directions. One is the path to prephenate and the other is to isochorismate. Isochorismate is the important intermediate to *o*-succinylbenzoate, which is further converted into menaquinone and other natural products.

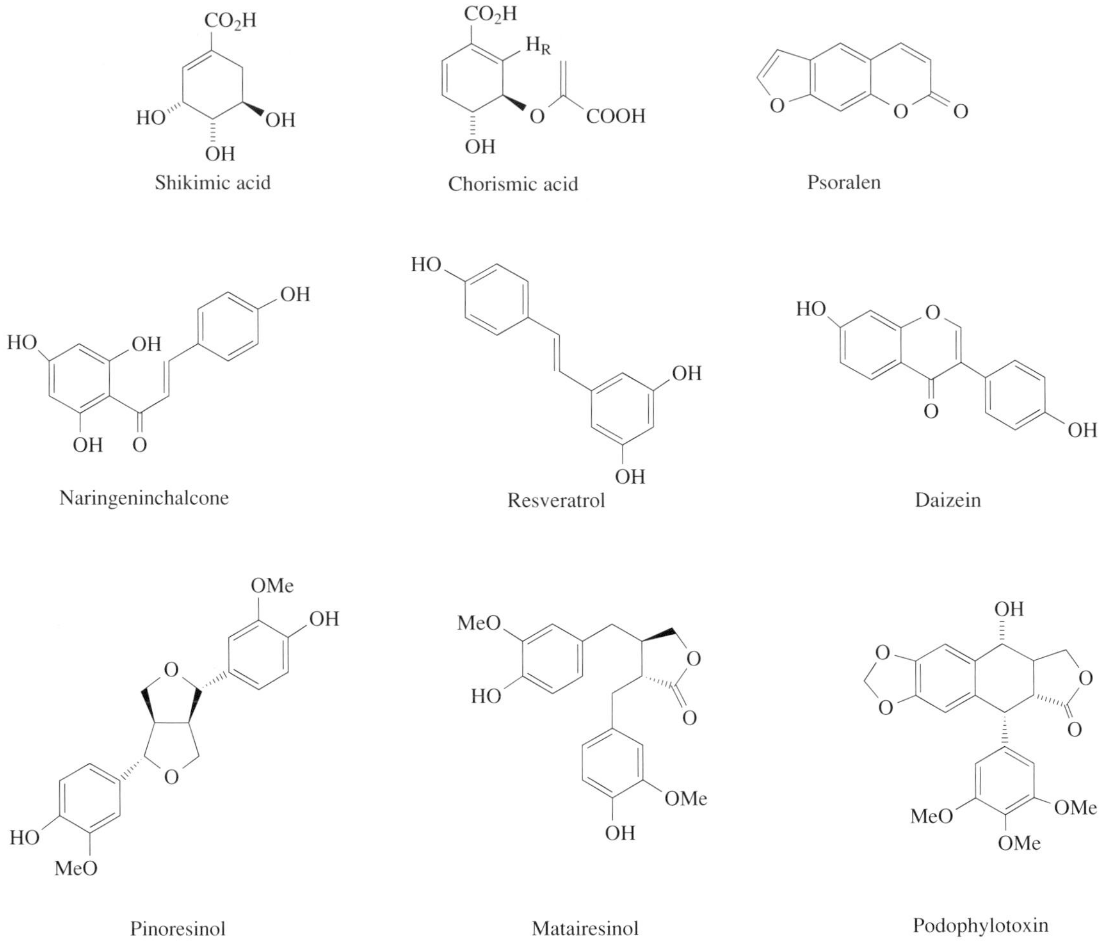

Shikimic acid Chorismic acid Psoralen

Naringeninchalcone Resveratrol Daizein

Pinoresinol Matairesinol Podophylotoxin

Figure 12 Shikimic acid pathway derived compounds.

1.01.4.2 Biosynthesis of C_6–C_3 Compounds: Coumarins and Lignans

Plants produce compounds with C_6–C_3 structures, phenylpropanoids, which are derived from the amino acid phenylalanine. Phenylalanine is a shikimate pathway metabolite and the high production of phenylalanine in plants is closely associated with photosynthesis. Metabolic pressure generated by photosynthesis causes the flow of photosynthetic products running into the shikimate pathway and this leads to the production of phenylalanine. Phenylalanine ammonia lyase (PAL) is a key enzyme of the phenylpropanoid pathway for supplying the basic material for producing $(C_6–C_3)_n$ compounds, including lignin. The phenylpropanoid pathway is present in all plants and cinnamic acid is converted into *p*-coumaric acid, caffeic acid, ferulic acid, feruloyl CoA, and coniferyl alcohol. This core reaction of phenylpropanoid pathways results in the synthesis of lignin, an essential compound for fortifying the plant cell wall so that trees are able to stand higher than 20 m. Lignin is a polymer of coniferyl and cinapyl alcohols and polymerization is a radical reaction catalyzed by peroxidase. The polymerization reaction is nonstereo- and nonregiospecific and as a result lignin is optically inactive. In normal plant growth lignin is formed according to synthesis of the cell wall; however, lignin formation is induced upon injury or phytopathogenic microorganism infection (see Chapter 1.25). In principle, lignans are the dimers of phenylpropanoids formed by oxidative or other dimerization reactions. Radical coupling reaction of coniferyl alcohol by oxidase yielded racemic pinoresinol, a bisphenylbistetrahydrofuran type lignan. The biosynthesis of optically active pinoresinol requires laccase-like oxidase and an additional protein called a dirigent protein. The role of the dirigent protein in optically active dimerization was investigated in detail by using cloned and expressed protein. It was unexpected that a bisbenzylbutyrolactone type lignan (matairesinol) was formed from pinoresinol by successive ring opening by NADPH-dependent reductase.

Coumarins are cinnamate-derived C_6–C_3 compounds widely distributed in the plant kingdom. The conversion of *p*-coumaric acid into umbelliferone looks simple; however, the exact mechanism of isomerization of the *trans*-double bond to *cis* is obscure. Prenylation at the benzene ring is a frequently occurring modification as is further transformation into isopropyldihydrofuranocourmarin and furanocoumarin. The formation of furanocoumarin psoralen from (+)-marmesin by a P450-catalyzed reaction proved to be a radical reaction that is initiated by hydrogen abstraction followed by C—C bond cleavage to yield the isopropanol radical which is hydroxylated by a rebound reaction. Thus-formed acetone ketal was trapped chemically and its stoichiometric formation in parallel with psoralen formation was demonstrated (see Chapter 1.24).

1.01.4.3 Biosynthesis of Flavonoids

Flavonoids form a large family in higher plants and their distribution among plants is ubiquitous. All flavonoids are formed from one molecule of *p*-coumaroyl CoA and three molecules of malonyl CoA and have a basic structure of C_6–C_3–C_6 (see Figure 13). They are classified into subclasses according to their skeletons: chalcone, flavanone, flavone, flavonol, flavan-3,4-diol, anthocyanin, and cathechin. Isoflavone has a unique structure whose phenyl group attaches to C-2 of the pyrone ring. The ubiquitous occurrence of flavonoids suggests their physiological roles in plants. For example, flavone and flavonol are contained in all plant leaves and strongly absorb UV light, thus preventing DNA damage by UV irradiation. This solves the conflict between the requirement for light for photosynthesis and damage by UV irradiation. Anthocyanins are responsible for red to blue flower colors and they attract insects for pollination. Flavonoids induce chemotaxis of *Rizobium* for nodule formation in legume roots and isoflavonoids are phytoalexins induced upon elicitation by outer stress (see Chapter 1.28).

The first enzyme reaction in flavonoid biosynthesis is catalyzed by chalcone synthase (CHS), which forms chalcone from *p*-coumaroyl CoA and malonyl CoA. The reaction is typical for PKS; however, CHSs have no homology with FAS and microbial PKSs. The molecular evolution of CHS is therefore independent of other PKSs. Two different modes of cyclization are possible for polyketo intermediates. Claisen condensation yields chalcone and decarboxylative aldol condensation gives stilbene. CHSs have been cloned in more than 40 plants, while the cloning of stilbene synthase (STS) is limited to several plants. STS and CHS are very similar in size, ca. 4.3 kDa, and their DNA sequences also show great homology. Acridone synthase (AS) which produces acridone from *N*-methylanthranoyl CoA and malonyl CoA was cloned from *Ruta graveolens*. It showed great homology with CHS and STS, indicating that CHS superfamily genes are present in the plant kingdom and catalyze various plant PKS reactions (see Chapter 1.27). Flavonoids in legumes are mainly of the deoxy type and polyketide reductases (PKR) were cloned from several legume plants. CHS yielded deoxy type chalcones in the presence of PKR and NADPH. This is similar to the type II FAS reactions; however, PKR showed no homology with KR in FAS. A new type of CHS-like cDNA has been cloned from *Hydrangea macrophylla var. thunbergii* to produce *p*-coumaroyltriacetic acid lactone (CTAL).[25] Detailed investigation of CHS reaction revealed that a significant amount of CTAL was formed along with other by-products, bisnoryangonin and TAL. As shown in Figure 14, CHS produces a small but significant amount of stilbene and STS produces chalcone.[26] To date it is not clear whether these are cross-reactions catalyzed by enzymes or just chemical reactions. Further investigation of the reactions of CHS, STS, and CTA synthase (CTAS) are required to investigate their relation to the reactions of the CHS superfamily of enzymes (see Chapter 1.27).

Most of the biosynthetic studies on isoflavonoids were carried out with tissue cultures of leguminous plants which had induced isoflavonoid production via elicitors. Elicited cells produced isoflavonoid phytoalexins belonging to pterocarpan via isoflavone daizein. Deoxy-type chalcone could be converted into the flavanone, liquiritigenin by chalcone-flavanone isomerase (CHI). The branching reaction to isoflavonoids is the migration of the phenyl ring from C-2 of isoliquiritigenin to C-3 of daizein. The enzyme was characterized as a P450. Experiments using $^{18}O_2$ clarify the reaction mechanism of phenyl migration catalyzed by P450. Abstraction of hydrogen from C-3 followed by phenyl migration affords a radical at C-2, which is hydroxylated by P450 by a rebound reaction, a normal hydroxylation by P450. Thus formed 2-hydroxyisoflavanone is dehydrated by dehydratase (DH), which was purified to apparent homogeneity and characterized.[27] It is amazing that a simple compound such as daizein required four steps and six enzymes, CHS, PKR, CHI, P450, cytochrome reductase, and DH (see Chapter 1.28).

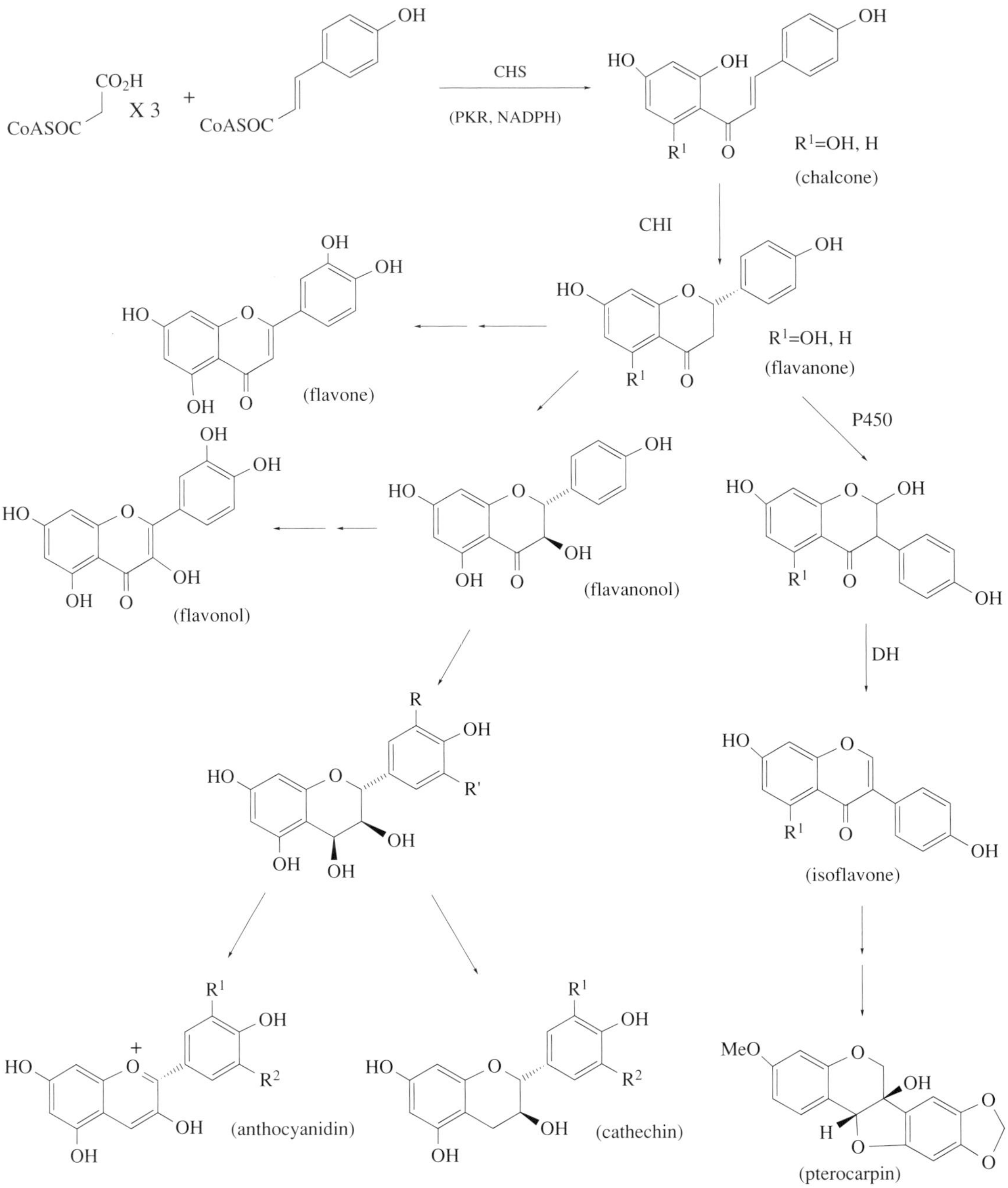

Figure 13 Flavonoid biosynthesis.

1.01.5 BIOSYNTHESIS OF COMPOUNDS CONTAINING SULFUR, A C—P BOND OR A CN GROUP

Chapters 30 to 31 deal with the natural products containing sulfur, CN, and C phosphate (see Figure 15). Biotin and lipic acid are well-known co-enzymes and in despite of extensive investigations the mechanism of C—S bond formation is not completely clear. Numerous examples of sulfur containing compounds indicate that their biosynthesis are all different. As for compounds with a C—P bond, the biosynthesis of bialaphos and fosfomycin have been extensively investigated at the enzyme and gene levels. In both cases, C—P bond formation is the migration of phosphate in phosphoenolpyruvate (PEP) or reaction of phosphonoformic acid with PEP. Cyanogenic glucosides cause cyanogenesis which means the release of hydrogen cyanide from damaged plants. Hydrogen cyanide is liberated from cyanogenic glycoside by hydrolysis with β-glucosidase.

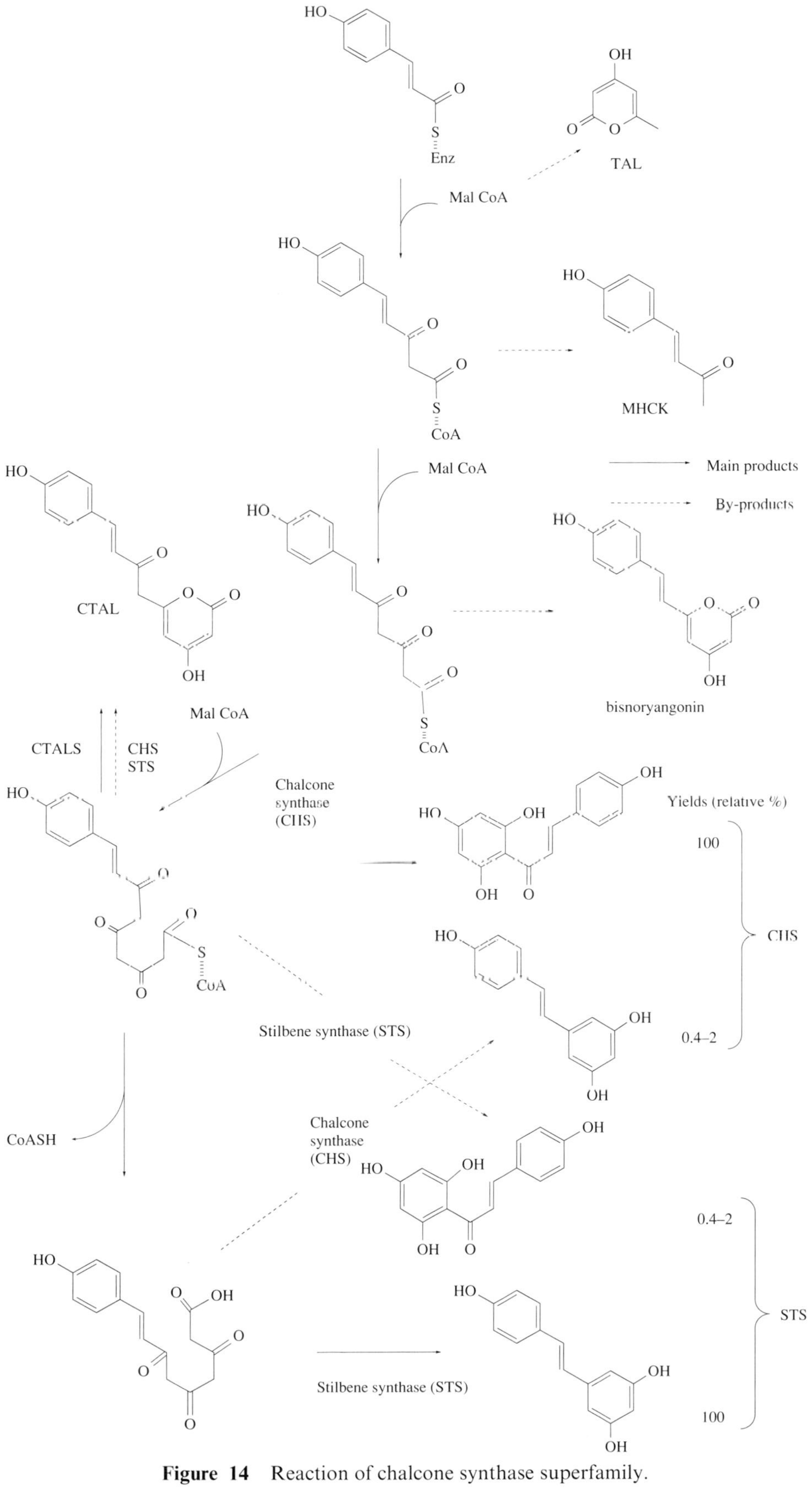

Figure 14 Reaction of chalcone synthase superfamily.

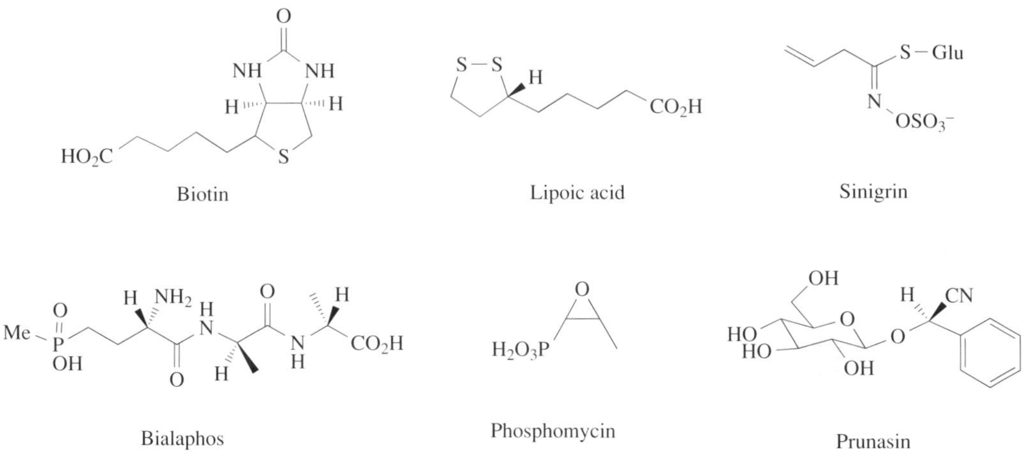

Biotin Lipoic acid Sinigrin

Bialaphos Phosphomycin Prunasin

Figure 15 Structure of compounds containing sulfur, CN, and CP.

1.01.6 REFERENCES

1. P. Manitto, "Biosynthesis of Natural Products," Ellis Horwood, New York, 1981.
2. K. B. G. Torssell, "Natural Product Chemistry: A Mechanistic and Biosynthetic Approach to Secondary Metabolism," Wiley, Chichester, 1983.
3. D. O'Hagan, "The Polyketide Metabolites," Ellis Horwood, New York, 1991.
4. J. N. Collie, *J. Chem. Soc.*, 1893, **63**, 329.
5. R. Robinson, "The Structural Relationships of Natural Products," Oxford University Press, Oxford, 1955.
6. A. J. Birch and F. W. Donovan, *Austr. J. Chem.*, 1953, **6**, 360.
7. A. J. Birch, "To See The Obvious," American Chemical Society, Washington DC, 1995.
8. A. J. Birch, R. A. Massey-Westropp, and C. P. Moye, *Austr. J. Chem.*, 1955, **8**, 539.
9. R. Bentley and P. M. Zwitkowits, *J. Am. Chem. Soc.*, 1967, **56**, 681.
10. F. Lynen and M. Tada, *Angew. Chem.*, 1961, **73**, 513.
11. J. Beck, S. Rpka, A. Signer, E. Schiltz, and E. Schweizer, *Eur. J. Biochem.*, 1990, **192**, 487.
12. L. J. Cary, H. Seto, and M. Tanabe, *J. Chem. Soc., Chem. Commun.*, 1973, 867.
13. F. J. Leeper and J. Staunton, *J. Chem. Soc., Perkin Trans. 1*, 1984, 2919.
14. U. Sankawa, T. Sato, T. Kinoshita, and K. Yamasaki, *Chem. Pharm. Bull.*, 1981, **29**, 3586.
15. M. J. Garson and J. Staunton, *Chem. Soc. Rev.*, 1979, 539.
16. J. C. Vederas, *J. Am. Chem. Soc.*, 1980, **102**, 374.
17. L. Jaenicke and W. Boland, *Angew. Chem. Int. Ed. Engl.*, 1982, **9**, 643.
18. D. Picot, P. J. Loll, and R. M. Garavito, *Nature*, 1994, **365**, 243.
19. D. A. Hopwood, *Chem. Rev.*, 1997, **97**, 2465.
20. C. R. Hutchinson, *Chem. Rev.*, 1997, **97**, 2525.
21. I. Fujii and Y. Ebizuka, *Chem. Rev.*, 1997, **97**, 2511.
22. Z. G. Chen, I. Fujii, Y. Ebizuka, and U. Sankawa, *Phytochemistry*, 1955, **38**, 299.
23. H. Ikeda and S. Omura, *Chem. Rev.*, 1997, **97**, 2591.
24. E. Haslam, "Shikimic Acid, Metabolism and Metabolites," Wiley, Chichester, 1993.
25. T. Akiyama, M. Shibuya, and Y. Ebizuka, personal communication.
26. M. Nishioka, M. Shibuya, Y. Ebizuka, and U. Sankawa, unpublished results.
27. T. Hakamatsuka, K. Mori, S. Ishia, Y. Ebizuka, and U. Sankawa, *Phytochemistry*, 1998, **49**, 497.

1.02 Biosynthesis and Degradation of Fatty Acids

AKIHIKO KAWAGUCHI, ATSUKO IWAMOTO-KIHARA, and NORIHIRO SATO
The University of Tokyo, Japan

1.02.1 INTRODUCTION

Fatty acids have four major physiological roles. First, they are major building blocks of phospholipids and glycolipids, which are important components of biological membranes. Second, they are lipophilic modifiers of proteins. Many proteins are acylated through covalent attachment of fatty acids, which targets them to membrane locations. Third, fatty acids are fuel molecules. They are stored as triacylglycerols (neutral fats or triglycerides) and over 90% of the stored energy of triacylglycerols resides in their fatty acid components. Fourth, fatty acids are precursors of eicosanoids which act as hormones and intracellular messengers. The fatty acids required for these lipids can be derived either from dietary sources or through *de novo* synthesis. In general, saturated and monoenoic fatty acids can be derived from either source but polyenoic fatty acids can only be derived from dietary linoleic or α-linolenic acids which have been derived directly or indirectly from a plant source. Since these fatty acids cannot be synthesized by animal tissue, they are often referred to as essential fatty acids.

Fatty acids in biological systems usually contain an even number of carbon atoms, typically between 14 and 24. C_{16} and C_{18} fatty acids are most common. Fatty acid carbon atoms are usually numbered from the carboxy terminus. Carbon atoms 2 and 3 are often referred to as the α and β carbons, respectively. The carbon atom at the methyl-terminal end of the chain is called the ω carbon. The hydrocarbon chain may be saturated or it may contain one or more double bonds. The configuration of the double bonds in most unsaturated fatty acids is *cis*. The double bonds in polyenoic fatty acids are separated by at least one methylene group. The position of a double bond is represented by the symbol Δ followed by a superscript number. Alternatively, the position of a double bond can be denoted by counting from the methyl terminus with the ω carbon as number 1. Table 1 lists the more common fatty acids with their trivial names and double-bond positions.

Table 1 Some naturally occurring fatty acids in animals. The number immediately before the colon (:) refers to the number of carbon atoms in the fatty acid chain, while the number after the colon refers to the number of double bonds; thus 16:1 indicates a fatty acid of 16 carbon atoms with one double bond.

Trivial name	*Shorthand notation*	*Systematic name*
Palmitic acid	16:0	*n*-hexadecanoic acid
Palmitoleic acid	Δ^{9}-16:1	*cis*-9-hexadecanoic acid
Stearic acid	18:0	*n*-octadecanoic acid
Oleic acid	Δ^{9}-18:1	*cis*-9-octadecanoic acid
cis-Vaccenic acid	Δ^{11}-18:1	*cis*-11-octadecanoic acid
Linoleic acid	$\Delta^{9,12}$-18:2	*cis, cis*-9,12-octadecadienoic acid
Linolenic acid	$\Delta^{9,12,15}$-18:3	*all-cis*-9,12,15-octadecatrienoic acid
Arachidonic acid	$\Delta^{5,8,11,14}$-20:4	*all-cis*-5,8,11,14-eicosatetraenoic acid
Eicosapentaenoic acid	$\Delta^{5,8,11,14,17}$-20:5	*all-cis*-5,8,11,14,17-eicosapentaenoic acid

In this chapter the biosynthesis of saturated and unsaturated fatty acids is described, and then the oxidative degradation of fatty acids is mentioned. When fatty acid oxidation was found to occur through oxidative removal of successive acetyl groups in the form of acetyl-CoA, it was assumed that the biosynthesis of fatty acids would prove to proceed through reverse of the enzymatic steps for the oxidation. It turned out, however, that the biosynthesis occurs through a different pathway, is catalyzed by a different set of enzymes, and takes place in a different part of the cell.

1.02.2 BIOSYNTHESIS OF SATURATED FATTY ACIDS

The biosynthesis of fatty acids may be considered as a complex series of integrated reactions. The key enzymes concerned in the *de novo* synthesis of fatty acids, i.e., the conversion of acetates to fatty acids, are acetyl-CoA carboxylase and fatty acid synthase (Figure 1). The metabolic fate of either dietary or *de novo* synthesized fatty acids depends on the balance of a series of enzymatic reactions, which are as follows: chain elongation, desaturation, incorporation into complex lipids, and β-oxidation. The product of each of the above reactions then becomes a potential substrate for one of the other pathways shown in Figure 1(a). The preferred route of any particular fatty acid will depend not only on the structure of the fatty acid in relation to the specificity of each of the participating enzymes, but also on various factors which are capable of influencing metabolism. An example of the effects of enzyme specificity and the reaction rate is shown in Figure 1(b), i.e., the metabolism of palmitic acid, the end-product of fatty acid synthase of animal tissues.[1] Palmitic acid synthesized in the cytosol is transported to the endoplasmic reticulum and then converted to a CoA ester. In the endoplasmic reticulum each substrate potentially competes for desaturation, chain elongation, or incorporation into complex lipids. The rates of elongation are considerably greater than those of desaturation. The preferred chain length of a substrate for Δ^9-desaturation is that of stearoyl-CoA. Therefore, the major fatty acids found in the *de novo* synthesis system are stearic, oleic, and palmitic acids.

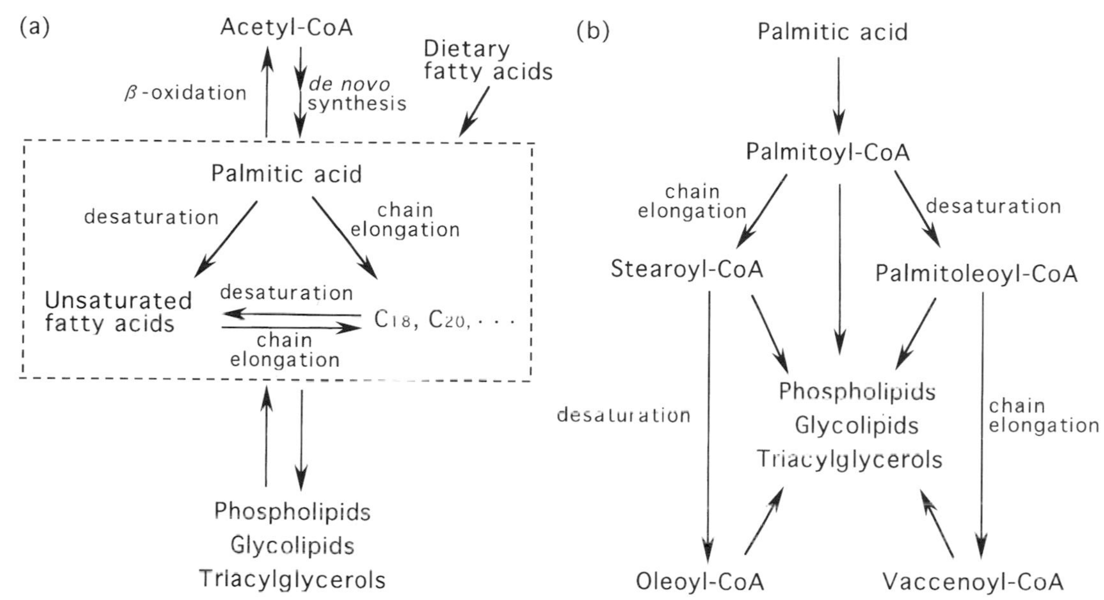

Figure 1 Synthesis and metabolic fate of fatty acids (a), and metabolic fate of palmitic acid (b).

1.02.2.1 *De novo* Synthesis

Fatty acids are synthesized through the concerted actions of two enzymes, acetyl-CoA carboxylase and fatty acid synthase. The stereochemical courses of the reactions catalyzed by the two enzymes have been reviewed elsewhere.[2] The major product of fatty acid synthase is palmitic acid. Further elongation and the introduction of double bonds are carried out by totally different enzyme systems.

1.02.2.1.1 Acetyl-CoA carboxylase

Fatty acid synthesis starts with the carboxylation of acetyl-CoA to malonyl-CoA, as shown in Equation (1). This reaction is catalyzed by acetyl-CoA carboxylase, which contains a biotin prosthetic group. The carboxyl group of biotin is covalently bound to the ε-amino group of a specific lysine residue. Acetyl-CoA is carboxylated through two partial reactions. First, a carboxybiotin intermediate is formed at the expense of ATP (Equation (2)). The activated CO_2 group in this intermediate is then transferred to acetyl-CoA to give malonyl-CoA (Equation (3)).

$$CH_3COSCoA + HCO_3^- + ATP \rightarrow {}^-OOCCH_2COSCoA + ADP + Pi + H^+ \quad (1)$$

$$HCO_3^- + ATP + \text{Biotin-Enzyme} \rightarrow CO_2^-\text{-Biotin-Enzyme} + ADP + Pi \quad (2)$$

$$CO_2^-\text{-Biotin-Enzyme} + CH_3COSCoA \rightarrow {}^-OOCCH_2COSCoA + \text{Biotin-Enzyme} \quad (3)$$

Acetyl-CoA carboxylase consists of three different functional components, biotin carboxylase (BC), biotin-carboxyl-carrier protein (BCCP), and carboxyl transferase (CT). In fact, the acetyl-CoA carboxylase of *Escherichia coli* consists of three corresponding dissociable components. Protein fractionation of the *E. coli* carboxylase initially gives rise to two protein fractions, E_a and E_b, which catalyze Equations (2) and (3), respectively.[3,4] Purified E_a can be separated further into two proteins by gel electrophoresis. Therefore, the bacterial carboxylase dissociates into three components: BC (molecular weight, 102 kDa), BCCP (22.5 kDa), and CT (130 kDa). They may function as an enzyme complex in the cell.

A single subunit of eukaryotic acetyl-CoA carboxylase exhibits the functions of BC, BCCP, and CT. The eukaryotic enzyme has a highly integrated structure, representing a multifunctional protein (polypeptide).[5] Figure 2 shows a schematic model of the eukaryotic enzyme. The biotinyl lysine has a long arm attached to a ureido ring. The length and flexibility of the link between biotin and its carrier protein enable the activated carboxyl group to move between the active domains of BC and CT. Therefore, eukaryotic and prokaryotic acetyl-CoA carboxylases have different structural organizations. The former are composed of multiple monofunctional polypeptides, whereas the latter consist of a single, integrated multifunctional polypeptide.

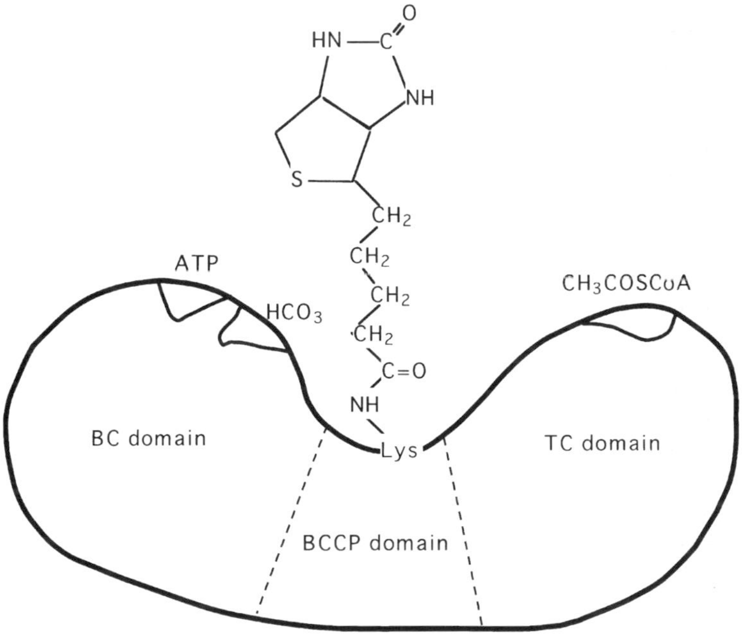

Figure 2 Schematic model of eukaryotic acetyl-CoA carboxylase.

The biotin-dependent carboxylases, which include acetyl-CoA carboxylase, propionyl-CoA carboxylase, oxalacetate decarboxylase, pyruvate carboxylase, and transcarboxylase, share common catalytic mechanisms. Biotin-dependent carboxylases other than oxalacetate decarboxylase and transcarboxylase, which lack biotin carboxylase, exert their catalytic activities through the three functional domains.[5] The three functional domains of the enzymes are encoded on the genome in various ways (Figure 3). In some enzymes (prokaryotic acetyl-CoA carboxylase and transcarboxylase), the three components are encoded by three different genes. In other enzymes, two functional components are encoded by a single gene, the other component being encoded by a different gene. In eukaryotic acetyl-CoA carboxylase and pyruvate carboxylase, all components are encoded by a single gene. Even when two or three components are encoded by a single gene, both the order and the combination of functional domains in the primary structure differ from enzyme to enzyme.

The amino acid sequence of BCCP is similar to those of the lipoyl domains of various enzymes. The lipoyl domain also contains a specific lysine residue, at which a lipoic acid is covalently bound to form a lipoamide.[6] The lipoyl domain is found in the subunits of enzyme complexes such as pyruvate dehydrogenase, 2-oxoglutamate dehydrogenase, and branched-chain α-keto acid dehydrogenase.[7] The lipoyl domain is also found in a subunit of the glycine cleavage enzyme.[8] The lipoamide transfers acetyl, succcinyl, or aminomethyl moiety between two different active sites in the enzyme complex.

The amino acid sequence of BC is similar to those of the duplicated domains of carbamoyl-phosphate synthetases, which are involved in the metabolic pathway for arginine and pyrimidine biosynthesis.[6,9,10] The N-terminal half of carbamoyl-phosphate synthetases is homologous to the

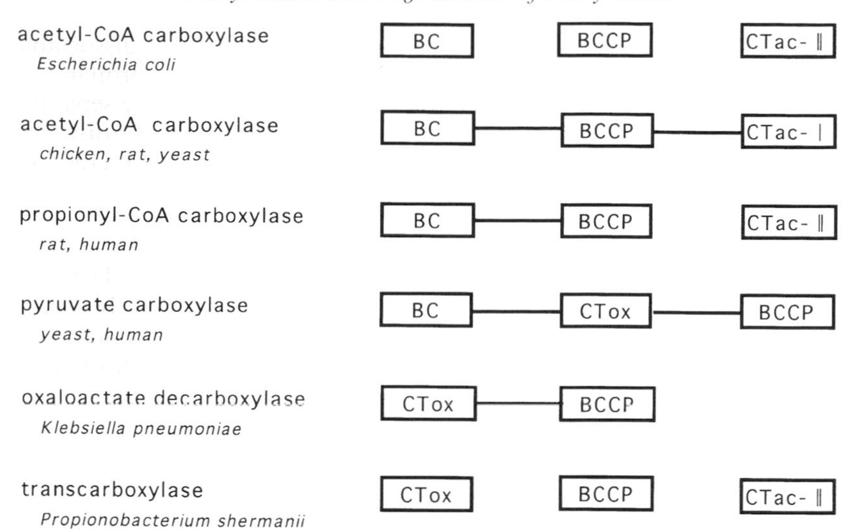

Figure 3 Schematic diagram of the domains and/or subunit structures of various biotin-dependent carboxylases. Boxes indicate functional domains and bars connecting the boxes indicate that the functional domains are fused to be encoded by a gene. The order of the connected boxes corresponds to the order of the functional domains in the gene. CTac or CTox show whether CT uses acyl-CoA or 2-oxo acid as its acceptor molecule (after Toh *et al.*[5]).

C-terminal half. The repeats in carbamoyl-phosphate synthetases catalyze similar but distinct reactions,[11] which are also similar to the reactions catalyzed by BC.

There are two types of CT. One of them uses acyl-CoA (acetyl-CoA, propionyl-CoA, or 3-methylcrotonyl-CoA) as the acceptor of CO_2, and the other uses 2-oxo acid. In Figure 3, the former is referred to as CTac, and the latter as CTox. The amino-acid sequences of CTac do not show any similarity to that of CTox.[5] The CTac domains of eukaryotic acetyl-CoA carboxylases are considered to be carried on the C-terminal regions. However, significant sequence similarity has not been detected between these regions and *E. coli* CT. It is difficult, on amino-acid sequence comparison, to determine whether prokaryotic and eukaryotic carboxyl transferases share a common ancestral gene or have different origins.

1.02.2.1.2 Fatty acid synthase

Fatty acid synthase catalyzes various reactions to produce long-chain fatty acids from acetyl-CoA, malonyl-CoA, and NADPH (and/or NADH). The overall reaction is summarized in Equation (4). Acetyl-CoA forms the methyl end of the fatty acid and malonyl-CoA contributes the remaining carbon atoms of the molecule.

$$\text{Acetyl-CoA} + 7\,\text{malonyl-CoA} + 14\,\text{NADPH} + 14\,\text{H}^+ \rightarrow \text{palmitate} + 7\,\text{CO}_2 + 14\,\text{NADP}^+ + 8\,\text{CoA} + 6\,\text{H}_2\text{O} \quad (4)$$

All the intermediates in fatty acid synthesis are linked to the sulfhydryl terminus of 4′-phosphopantetheine, which is attached as a phosphodiester to the hydroxyl group of a specific serine residue of fatty acid synthase (Figure 4).[12,13] The priming reaction is catalyzed by acetyl transacylase. The acetyl group is transferred from the sufhydryl group of the 4′-phosphopantetheine moiety of coenzyme A to the sulfhydryl group of 4′-phosphopantetheine (pantetheine-SH, ScH in Figure 4) covalently associated with the enzyme. This acetyl group is then transferred to a specific cysteine-SH (SpH in Figure 4) of 3-oxoacyl synthase (condensing enzyme). The above process is generally accepted for the priming reaction. As discussed later, however, it is not clear whether the acetyl group is transferred directly to the cysteine-SH or via the pantetheine-SH. The liberated pantetheine-SH group on the enzyme is available to accept a malonyl group from malonyl-CoA in the reaction catalyzed by malonyl transacylase forming acetyl-malonyl-enzyme. Acetyl transacylase can transfer an acyl group other than the acetyl group. The synthesis of fatty acids with odd numbers of carbon atoms starts with propionyl-CoA. However, the reaction rate decreases with increasing chain length of the "priming" acyl-CoA derivative. The malonyl transacylase of most fatty acid synthases, on the other hand, shows a high degree of specificity to malonyl-CoA. Exceptionally, the fatty acid synthase isolated from Harderian gland can utilize both malonyl-CoA and methylmalonyl-CoA.[14]

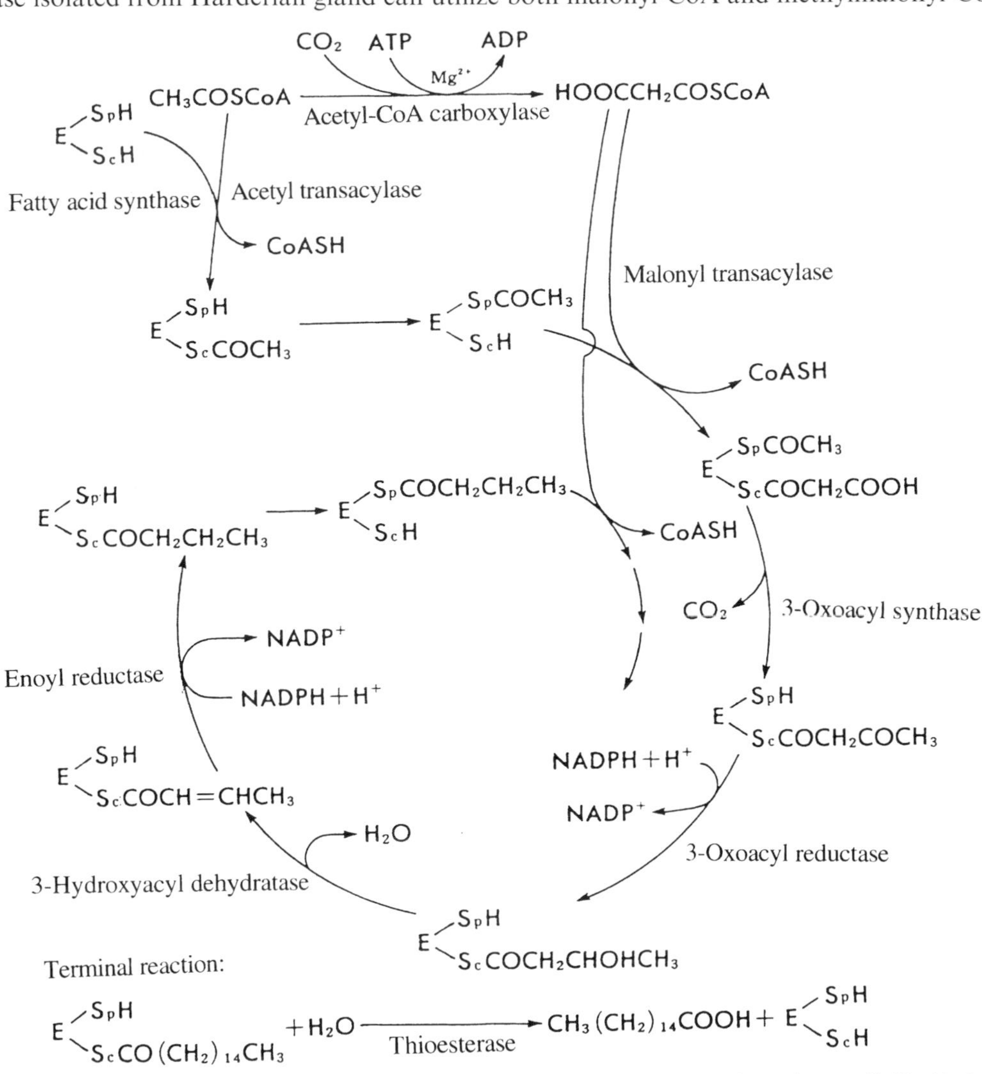

Figure 4 Schematic representation of the reactions catalyzed by fatty acid synthases. ScH: 4′-phosphopantetheine group bound to a serine residue of the enzyme. SpH: a cysteine residue of the enzyme.

The acetyl and malonyl groups bound to the enzyme react to form acetoacetyl-enzyme. In this condensation reaction, a four-carbon unit is formed from a two-carbon unit and a three-carbon one with the concomitant release of CO_2. Why isn't the four-carbon unit formed from two carbon-units? The answer is that the equilibrium for the synthesis of acetoacetyl-enzyme from two acetyl moieties is unfavorable. Since ATP is used to carboxylate acetyl-CoA to malonyl-CoA, the condensation reaction is indirectly driven by ATP. The free energy stored in malonyl-CoA is released on decarboxylation during the formation of the new C–C bond. Although HCO_3^- is required for fatty acid synthesis, its carbon atom does not appear in the fatty acid produced. Therefore, all the carbon atoms of fatty acids containing even numbers are derived from acetyl-CoA.

The next three steps reduce the 3-oxo group to a methylene group. First, 3-oxoacyl reductase catalyzes the NADPH-dependent reduction of acetoacetyl-enzyme to 3*R*-3-hydroxybutyryl-enzyme. Then this is dehydrated to crotonyl-enzyme by 3-hydroxyacyl dehyratase. The resulting crotonyl-enzyme is normally reduced through a second NADPH-dependent reaction by enoyl reductase yielding butyryl-enzyme. These last three reactions—reduction, dehydration, and second reduction—complete the first elongation cycle. Fatty acid synthases from animal sources show specificity for NADPH. NADH is either poorly utilized or not utilized at all. Dual reduced pyridine nucleotide requirements have been observed for synthases isolated from bacteria. It is generally acknowledged for bacterial systems that 3-oxoacyl reductase is specific for NADPH and enoyl reductase for NADH (and NADPH).[12] During 3-oxoacyl reduction, the enzyme shows specificity for the pro-4*S* hydrogen of NADPH, and the hydride originating from NADPH attacks from the *si* face of the substrate regardless of the source of the enzyme.[2] On the contrary, the several stereochemically distinct pathways for enoyl reduction produce saturated acyl chains: pro-4*R* and pro-4*S* of a reduced pyridine nucleotide, attacks of hydrides and protons on one or other stereoheterotopic face of the *trans*-enoyl substrate.[2,15–25]

In the second round of fatty acid synthesis, a butyryl group is transferred to the cysteine-SH of 3-oxoacyl synthase, and then this butyryl group condenses with a malonyl group bound to the enzyme and the reaction sequence continues as above. The elongation cycles continue until the C_{16}-enzyme is formed and this intermediate is not a substrate for the 3-oxoacyl synthase. Free palmitic acid formed through the hydrolysis of palmitoyl-enzyme is the typical final product of animal fatty acid synthases. The hydrolytic cleavage of the fatty acid chain is carried out by a thioesterase, an integral part of the multifunctional enzyme complex.[13]

The last step of the elongation cycles in yeast and some bacterial fatty acid synthases involves transacylation with CoA rather than hydrolysis of the free fatty acid, as found for the animal enzymes. Palmitoyl transacylase catalyzes the chain termination instead of thioesterase.[12] It is not clear as yet why the products of animal synthases are free fatty acids, and those of bacterial and yeast enzymes CoA esters.

In most bacteria and plants, as discussed later, the constituent enzymes of fatty acid synthases and acyl carrier protein (ACP) become dissociated when cells are disrupted. The detailed mechanism of fatty acid synthesis in bacteria and plants is slightly different from that shown in Figure 4. The first condensation between acetyl and malonyl groups is catalyzed by acetoacetyl-ACP synthase (synthase III). This synthase purified from *E. coli* and *Spinacia oleracea* condenses malonyl-ACP and acetyl-CoA rather than with acetyl-ACP.[26,27] Several lines of evidence suggest that (i) butyryl-ACP is formed principally through condensation of malonyl-ACP and acetyl-CoA, and (ii) acetyl-ACP is a minor participant in fatty acid biosynthesis. *E. coli* contains three 3-oxoacyl-ACP synthases. 3-Oxoacyl-ACP synthase I is primarily responsible for the elongation of butyryl-ACP to C_{16}-ACP. 3-Oxoacyl-ACP synthase II is required for convering palmitoleoyl-ACP to *cis*-vaccenoyl-ACP. The antibiotic, cerulenin, irreversibly inhibits both synthases I and II, while synthase III is resistant to cerulenin.[28,29] On the other hand, the antibiotic thiolactomycin, inhibits synthetase III, but not I or II.

(i) Architecture of fatty acid synthases

Major variations of the fatty acid biosynthesis pathway occurring in different organisms do not lie in the individual reactions of the pathway, but rather in the architecture of the fatty acid synthases. There are two major classes of fatty acid synthase found in nature (Table 2). A type I synthase is a multifunctional enzyme with several active sites on each polypeptide. Type I synthases can be divided into three subclasses: IA, IB, and IC. The IA enzymes are found as homodimers (α_2) in animal tissues, and the IC ones are obtained as homooligomers (α_6) from some bacteria (Corynebacteriaceae and Mycobacteriaceae). The type IB synthases are heterooligomers ($\alpha_6\beta_6$) found in fungi such as yeast. These enzymes catalyze the synthesis of free fatty acids (type IA) or acyl-CoAs (type IB and IC) from acetyl-CoA, malonyl-CoA, and NADPH (and/or NADH). In contrast, plants and most bacteria have a type II synthase, each reaction of the fatty acid synthase being catalyzed by an individual enzyme.

All the intermediates of type II fatty acid synthases are bound as thioesters of ACP. ACP was first demonstrated in an *E. coli* system, and a little later the requirement of ACP in plant fatty acid synthesis was also recognized.[63,64] *E. coli* ACP contains 77 amino acids and the substrate binding site of ACP is 4′-phosphopantetheine, which is linked as a phosphodiester to the hydroxyl group of the serine at residue 36 of the protein.[65] In spinach and barley ACPs, the amino acid sequence adjacent to the serine residue is (-G-A-D-S*-L-D-), which is the same as that of the ACP from *E. coli*, although the sequence of the peripheral amino acids is quite different.[66] All the proteins are rich in glutamic and aspartic acid residues, thus accounting for their acidic nature, and have a typical globular structure.[67] This consensus sequence is also observed in the ACP domains of various type I synthases (Figure 5). In type I synthases, such a prosthetic group is found as an integral part of the multifunctional enzyme complex, and it is impossible to separate it as an individual protein.

Type II fatty acid synthases, monofunctional enzymes, can be isolated separately and perform individual partial reactions *in vitro*, but the rate of fatty acid production is far slower than that *in*

Table 2 Classification of fatty acid synthases.

Type	*Source*	*Major Product*	*Ref.*
IA: Multifunctional proteins homodimer (α_2)	Mammalian liver	Palmitic acid	30–35
	Avian liver	Palmitic acid	30,36
	Adipose tissue	Palmitic acid	31,37
	Mammary gland	Palmitic acid	38–41
	Harderian gland	Branched fatty acids	42
	Insect	Palmitic acid	43
IB: Multifunctional proteins heterooligomer ($\alpha_6\beta_6$)	Fungi (yeast, *Aspergillus*, *Neurospora*, *Penicillium*, *Cephalosporium*)	Acyl-CoAs	44–48
IC: Multifunctional protein homooligomer (α_6)	Some bacteria (*Brevibacterium*, *Corynebacterium*, *Mycobacterium*)	Acyl-CoAs	49–51
II: Individual enzymes + ACP	Bacteria (*E. coli*, *Bacillus*, *Clostridium*, *Pseudomonas*) Plants and algae	Acyl-ACPs	52–62

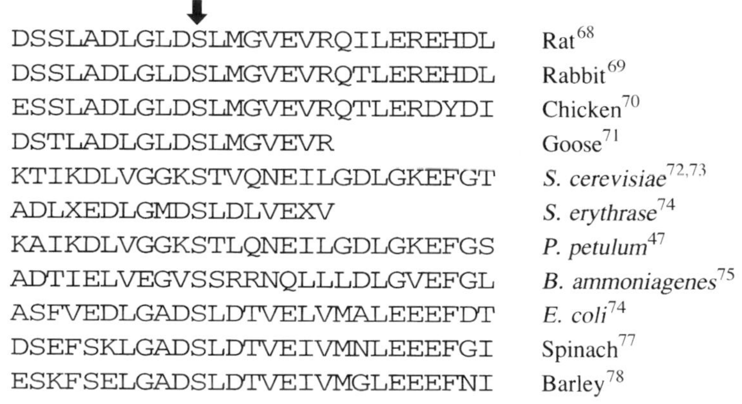

DSSLADLGLDSLMGVEVRQILEREHDL Rat[68]
DSSLADLGLDSLMGVEVRQTLEREHDL Rabbit[69]
ESSLADLGLDSLMGVEVRQTLERDYDI Chicken[70]
DSTLADLGLDSLMGVEVR Goose[71]
KTIKDLVGGKSTVQNEILGDLGKEFGT S. cerevisiae[72,73]
ADLXEDLGMDSLDLVEXV S. erythrase[74]
KAIKDLVGGKSTLQNEILGDLGKEFGS P. petulum[47]
ADTIELVEGVSSRRNQLLLDLGVEFGL B. ammoniagenes[75]
ASFVEDLGADSLDTVELVMALEEEFDT E. coli[74]
DSEFSKLGADSLDTVEIVMNLEEEFGI Spinach[77]
ESKFSELGADSLDTVEIVMGLEEEFNI Barley[78]

Figure 5 Amino acid sequence conservation in the acyl carrier protein.

vivo,[79] so they are also considered to be associated in a certain order *in vivo*. It has been reported that the ACP in *E. coli* appears to be somewhat loosely associated with the inner surface of the plasma membrane of the cell.[80]

As a model for type I fatty acid synthases, Lynen proposed a multienzyme complex comprising seven different proteins after he had analyzed the fatty acid synthase from yeast.[81,82] This is the most famous structure described in various textbooks on biochemistry and was considered to be the common structure of all fatty acid synthases. This concept was built on several findings: seven different N-terminal amino acids were detected in the yeast enzyme;[82] and a peptide containing 4′-phosphopantetheine was isolated from a guanidinium chloride-treated synthase.[83] Similar results were also reported for mammalian enzymes like that from pigeon liver.[84] According to this model, seven different enzyme proteins (subunits) are arranged in a regular sequence around a central-SH (pantetheine-SH), and the intermediates covalently bound to the central-SH would be in the vicinity of the active site of each enzyme. In other words, acyl intermediates bound to the central-SH will move sequentially between the related enzymes arranged as a complex.

But in the early 1970s some investigators pointed out several inconsistencies in the above model. The fatty acid synthases from yeast and animals are composed of multifunctional subunits rather than multienzyme complexes.[85,86] The previous finding of seven or more peptides was probably the result of nonspecific proteolysis during the isolation of the enzyme. At present the enzyme from yeast is thought to be a heterooligomer and the mammalian enzymes to each be a homodimer.

The fatty acid synthases from animal tissues each have a molecular weight of around 500 kDa and consist of two identical subunits. One mole of pantetheine is found per mole of subunit in the cases of the chicken and rat enzymes, indicating that the two polypeptide subunits are identical.[87] The sedimentation coefficient of the type IA synthase isolated from guinea pig Harderian gland suggested a nonglobular structure for this enzyme.[42] The electron-micrographic image of the enzymically active dimer is ellipsoid in shape with major and minor axes of approximately 22 nm and

15 nm, respectively (Figure 6(a)). Two monomers lie together in an antiparallel fashion. Each subunit has a rodlike structure of about 22 nm long and 5 nm wide. There is a groove between the two subunits, giving the appearance of the Greek letter θ. The two subunits are in strong contact at their centers. The active dimer is dissociated into inactive monomers at low temperature or on hypotonic treatment. The dissociated monomers reaggregate into the active dimer with an increase in temperature of the solution or on dialysis against a high-ionic-strength buffer. Neither an active monomer nor an inactive dimer is observed in the inactivation and reactivation processes.[42,88] The cofactor, NADPH, reduces the velocity of inactivation caused by hypotonic treatment or exposure to low temperature, indicating that the binding of the two subunits is related to the binding site of NADPH. According to the results of the study of Tsukamoto *et al.*[89] on the chicken liver enzyme, 3-oxoacyl reductase and enoyl reductase reside in the central part of each subunit. Since the active enzyme is a homodimer, an entire active site should exist in each monomer. However, dissociated monomers cannot synthesize fatty acids. The formation of a dimer seems necessary for the activity. Considering the long and slender subunit, a "head-to-tail" arrangement of the two subunits is attractive to explain the active form of the enzyme.[90] It becomes possible to locate all the active domains near the pantetheine-SH, to which all the acyl intermediates bind (Figure 6(a)). The flexibility and 20 nm maximal length of the phosphopantetheinyl moiety are critical for the function of the enzyme. The enzyme subunit does not need to undergo large structural rearrangements to interact with a substrate. The substrate is on a long, flexible arm, which can reach each of the various active sites. The organization of fatty acid synthase enhances the efficiency of the overall process because all the intermediates are directly transferred from one active site to the next. Another advantage of the enzyme is that covalently bound intermediates are protected from competing side reactions.

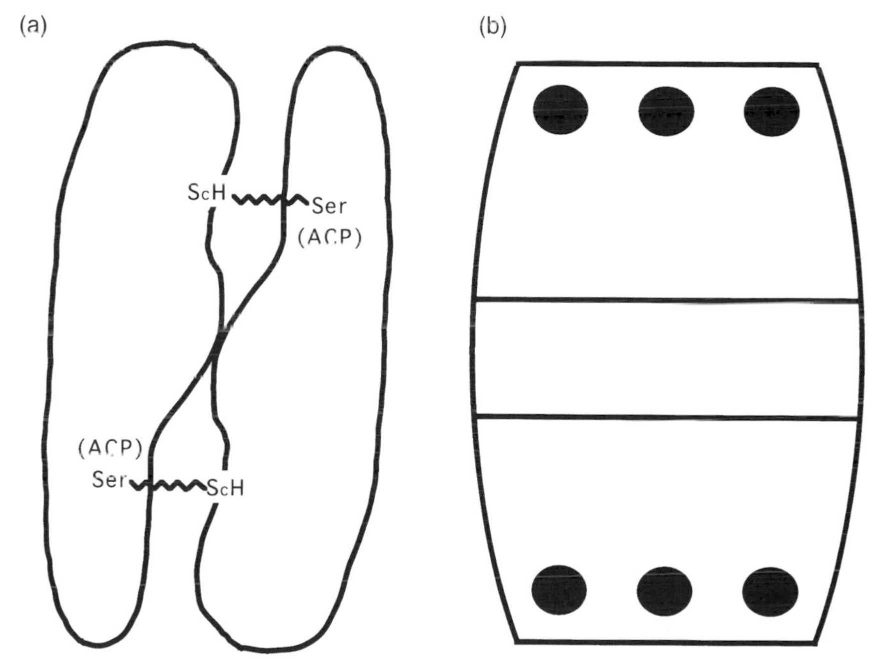

Figure 6 Schematic model of fatty acid synthases. (a) Type IA synthase. Two subunits are drawn in a head-to-tail arrangement, and the wavy line represents the 4′-phosphopantetheine prosthetic group. (b) Type IC synthase. The shaded circle indicates the 4′-phosphopantetheine site in the hexameric fatty acid synthase.

Further support for this head-to-tail arrangement was obtained in hybridization experiments.[91] In these experiments, the pantetheine-SH group was chemically modified with chloroacetyl-CoA and the cysteine-SH group with iodoacetoamide. The resulting enzyme variants were dissociated and hybridized. The hybrid containing an intact pantetheine-SH group on one subunit and a cysteine-SH group on the other subunit actively produced palmitic acid. These results strongly suggest that the active cysteine-SH group on one subunit is opposite to the pantetheine-SH on the other subunit, and that condensation of acetyl and malonyl groups proceeds between the two subunits. Since the dimeric synthase possesses two sets of each catalytic activity involved in fatty acid synthesis, the model presented in Figure 6(a) predicts the presence of two active centers for fatty acid synthesis. As to whether there is half- or full-site reactivity, some have proposed a half-site reactivity mechanism for the lactating rat mammary gland enzyme,[92,93] but others have reported a full-site one for the chicken liver fatty acid synthase.[94]

The estimated molecular weight of yeast fatty acid synthase (type IB) suggest that the native yeast synthase is an $\alpha_6\beta_6$ complex. According to the results of electron microscopic studies, Stoops *et al.*[36] proposed a model with an oval structure that has on its short axis a plate-like protein structure (six α subunits) on which six arch-like proteins (β subunits) are distributed, three on either side.

The fatty acid synthase from *Brevibacterium ammoniagenes* is a hexameric protein, and the electron micrographic image of the negatively stained enzyme is a barrel with major and minor axes of approximately 27 nm and 18 nm, respectively (Figure 6(b)).[49] A thin line can be seen on the minor axis, and white spots, which were not penetrated by the stain solution, can also be clearly seen. These features resemble those of the yeast enzyme,[36] which is quite different from the *B. ammoniagenes* enzyme with respect to subunit structure. The pantetheine-SH sites are in the peripheral region of the barrel-shaped molecule far from the central bar structure.

At present there is no reasonable explanation for why an $\alpha_6\beta_6$ (type IB) or α_6-(type IC) structure is necessary for the fatty acid synthase activity.

(ii) Distribution of active centers on subunits of type I fatty acid synthases

Multifunctional proteins catalyze a series of reactions in the same metabolic sequence. It is assumed that the close proximity of constituent catalytic sites favors the kinetics of the overall process. A particular topology of the catalytic domains within an enzyme molecule may be instrumental for both the reaction mechanism and the reaction kineitcs. The multifunctional fatty acid synthase genes of various organisms have been isolated and sequenced. From the results, we can deduce the orders of the catalytic domains in the genes (Table 3).

Table 3 Order of catalytic domains of fatty acid synthases.

Type	*Source*	*Order of catalytic domains*	*Comment*
IA	chicken liver	OS-AT/MT-DH-OR-ER-ACP-TE	MW 267,288
	rat liver	OS-AT/MT-DH-ER-OR-ACP-TE	MW 272,340 2505 amino acids
IB	*S. cerevisiae*		
	FAS 1 (β subunit)	AT-ER-DH-MT/PT	MW 220,077 1980 amino acids
	FAS 2 (α subunit)	OR-OS-ACP	MW 207,863 1894 amino acids
	Y. lipolytica FAS 1	AT-ER-DH-MT/PT	MW 229,980 2076 amino acids
	P. natulum FAS 2	OR-OS-ACP	MW 204,500 1857 amino acids
IC	*B. ammoniagenes*	AT-ER-DH-MT/PT-OR-OS-ACP	MW 327,466 3104 amino acids

Abbreviations: AT, acetyl transacylase; MT, malonyl transacylase; OS, 3-oxoacyl synthase; OR, 3-oxoacyl reductase; DH, 3-hydroxyacyl dehydratase; ER, enoyl reductase; TE, thioesterase; PT, palmitoyl transacylase, ACP, acyl carrier protein.

The order of domains of type IA synthases is 3-oxoacyl synthase–acetyl- and malonyl transacylases–3-hydroxyacyl dehydratase–3-oxoacyl- and enoyl reductases–acyl carrier protein–thioesterase. Limited tryptic digestion of chicken liver fatty acid synthase gives three major functional domains.[95] Polypeptide I (molecular weight 127 kDa) carries the acetyl and malonyl transacylase as well as the 3-oxoacyl synthase activities. Polypeptide II (107 kDa) carries the 3-hydroxyacyl dehydratase, 3-oxoacyl reductase, and enoyl reductase activities as well as that of ACP; and polypeptide III (33 kDa) carries the thiesterase activity. Polypeptide I catalyzes the entry of the acetyl and malonyl moieties into the process of palmitate synthesis and the condensation of the two moieties. Both transacylases transfer the acyl groups to an active serine-OH residue on the protein (Figure 7). Since the acetyl and malonyl groups both bind to a common active serine, they competitively inhibit each other's binding to the protein. Both groups are bound as an *O*-ester prior to their transfer to an appropriate thiol (pantetheine-SH or cysteine-SH). It is not clear whether the acetyl group is transferred directly from the serine *O*-ester to the cysteine-SH or whether it is first transferred to the pantetheine-SH and then to the cysteine-SH of 3-oxoacyl synthase as shown in Figure 4. In the case of type I synthases it has not been possible as yet to determine which of the two pathways the acetyl group follows. The sequence of the amino acids of a peptide containing this serine is highly conserved and retains the sequence motif of -G-X-S-G- that characterizes enzymes with a reactive serine, such as acetyl and malonyl/palmitoyl transacylases among type IB and IC synthases, thioesterases among type IA synthases, lipases, lipoprotein lipases, trypsin, and carboxylase.

```
           ⬇
MGLKPDGIIGHSLGEVACGYADGCLSQRE     AMT. Rat[68]
           SLGEVA                 AMT. Rabbit[69]
AGLQPDGILGHSVGELACGYADNSLSHEE     AMT. Chicken[70]
MGLRPDGIIGHSLGEVARAYYNGRISQEE     AMT. Goat[96]
QPEGPVRVAGYSPGACVAFEMCSQLQQGP     TE. Rat[68]
QPEGPVRIAGYSPGACVAFEMCSQLQQNA     TE. Chicken[70]
LIPADATFAGHSLGEYAALASLADVMSIE     MPT. S. cerevisiae[97]
LVPVDATFAGHSLGEYSALASLGDVMPIE     MPT. Y. lipolytica[98]
VKSQDTYTAGHSVGEYNALAAYAQVLSLE     MPT. B. ammoniagenes[75]
LRSYLKGATGHSQGLVTAVAIAETDSWES     AT. S. cerevisiae[97]
VRDNLKGATGHSQGLITAIATSASDSWDE     AT. Y. lipolytica[98]
    PSGATGHSQG                    AT. C. caerulens[99]
DPADAVAHIGHSQGALATYISSGRAQAAE     AT. B. ammoniagenes[75]
```

Figure 7 Amino acid sequence conservation in acyl transacylases. The abbreviations for partial activities used are: AMT, acetyl/malonyl transacylase; TE, thioesterase; MPT, malonyl/palmitoyl transacylase; AT, acetyl transacylase.

Genetic complementation studies revealed the presence of two fatty acid synthase gene loci (FAS 1 and FAS 2) on the *S. cerevisiae* genome.[100] FAS 1 and FAS 2 encode a pentafunctional polypeptide (β subunit) and a trifunctional polypeptide (α subunit), respectively. The order of domains in the pentafunctional polypeptide is acetyl transacylase enoyl reductase–3-hydroxyacyl dehyratase–malonyl/pamitoyl transacylase, and that in the trifunctional polypeptide is 3-oxoacyl reductase–3-oxoacyl synthase–acyl carrier protein. The sequences of the amino acids around the malonyl- and palmitoyl binding serine residues of the respective transacylases are identical. However, the amino acid sequences around the acetyl-binding peptides are different. From these results it is suggested that the malonyl and palmitoyl transacylase domains share a common active serine-OH on the protein. The acetyl/malonyl transacylases and thioesterase among type IA synthases, and malonyl/palmitoyl and acetyl transacylases among type IB synthases have strikingly homologous amino acid sequences around their substrate-binding serine residues (Figure 7). It appears likely that type IA acetyl/malonyl transacylase and type IB malonyl/palmitoyl transacylase on one side, and type IA thioesterase and type IB acetyl transacylase on the other side have evolved from common ancestral enzymes. The generation of the acetyl transacylase site and the site responsible for the termination at mutually inverted positions may be explained by these homologies.

The order of the catalytic domains of the fatty acid synthase gene of *B. ammoniagenes* is colinear, with hypothetical head-to-tail fusion of the two yeast fatty acid synthase genes, FAS 1 and FAS 2.[75] These findings confirm and extend protein-chemical results on the relative locations of active domains in the *B. ammoniagenes* synthase protein.[101] Comparison of the *B. ammoniagenes* synthase sequence to those of other synthetases revealed a particularly high degree of similarity to the products of the two yeast genes (30% identical and 46% identical plus closely related amino acids). This similarity extends over the entire lengths of the genes and involves not only the primary sequences of individual component enzymes but also their sequential order within the multi-functional proteins.

The *B. ammoniagenes* synthase has the unique feature of producing not only saturated fatty acids (palmitic and stearic acids) but also a monoenoic fatty acid (oleic acid).[102] To synthesize oleic acid it is assumed that this synthetase carries an additional component enzyme, 3-hydroxydecanoyl β,γ-dehydratase. Therefore, this protein is a nonfunctional polypeptide. The catalytic domain of 3-hydroxydecanoyl β,γ-dehydratase has not been assigned yet. The 3-hydroxyacyl α,β-dehydratase domain was tentatively assigned, although the degree of sequence similarity to yeast synthetase is not very high in this region. Also, the large enoyl reductase-dehydratase interdomain region may represent such a functionally unassigned DNA segment. The exact locations of both α,β- and β,γ-dehydratase will have to be confirmed by additional, genetic studies involving *B. ammoniagenes fas*-mutants.

1.02.2.2 Chain Elongation

The *de novo* fatty acid synthesis in animals and yeast occurs in the cytosol. Longer fatty acids are formed through elongation reactions catalyzed by the enzymes on the cytosolic face of the endoplasmic reticulum membrane in eukaryotic cells. For studies on these elongation reactions, the membrane is fractionated into vesicles called microsomes. Microsomes add two-carbon units sequentially to the carboxyl ends of both saturated and unsaturated fatty acids. Malonyl-CoA is the two-carbon donor involved in the elongation of fatty acyl-CoAs. Microsomal fatty acid synthases are considered to function primarily in the chain elongation systems for saturated fatty acids produced through the *de novo* synthesis and for dietary essential unsaturated fatty acids.[103] Therefore, chain elongation of linoleic acid in microsomes appears to be the major pathway in the formation of arachidonic and other polyenoic fatty acids.[104] The microsomal chain elongation of fatty acids and the cytosolic *de novo* fatty acid synthesis follow similar pathways involving the following steps: (i) condensation of a fatty acyl-CoA with malonyl-CoA, (ii) reduction of the resulting 3-oxoacyl derivative to a secondary alcohol, (iii) dehydration of the alcohol to a *trans*-2-enoyl intermediate, and (iv) reduction of the enoyl intermediate to give an acid of two carbon atoms longer than the primer.[103,105,106] However, there are several fundamental differences between the cytoplasmic and microsomal reactions: (i) the intermediates in microsomal chain elongation are CoA thioesters; (ii) these reactions have different requirements for pyridine nucleotides, the *de novo* fatty acid synthesis primarily utilizes NADPH, whereas the microsomal elongation reaction utilizes NADPH and NADH equally well; and (iii) cytochrome b_5 is involved in the transfer of reducing equivalents to the microsomal elongation system.

In plants *de novo* fatty acid synthesis proceeds in chloroplasts to produce palmitoyl-ACP. The chain elongation to stearoyl-ACP is carried out by an enzyme system different from that for the *de novo* fatty acid synthesis.[107]

Prior to the discovery of what is now known as fatty acid synthase in liver cytosol, it was suggested that fatty acid synthesis might be the reverse of mitochondrial β-oxidation. Subsequently, the function of the mitochondrial elongation system was thought to be restricted to the chain elongation of palmitoyl-CoA to stearoyl-CoA, and the elongation of polyenoic fatty acids. It now appears that these more limited functions are performed by microsomal elongation systems and not mitochondrial systems. However, the elongation of short- and medium-chain fatty acids can occur in the mitochondrial matrix with the three β-oxidation enzymes in reverse order: acetyl-CoA acetyltransferase (thiolase), 3-hydroxyacyl-CoA dehydrogenase, and enoyl-CoA hydratase. The fourth β-oxidation enzyme (acyl-CoA dehydrogenase) is replaced by an enoyl-CoA reductase.[108] There are two major differences between mitochondrial elongation and microsomal elongation: (i) mitrochondrial elongation reactions utilize fatty acyl-CoA plus acetyl-CoA, and (ii) the maximal rate of mitochondrial elongation is only attained when both NADH and NADPH are present.

1.02.3 BIOSYNTHESIS OF UNSATURATED FATTY ACIDS

A large variety of naturally occurring unsaturated fatty acids can be distinguished by differences in the position and number of double bonds and by variations in chain length. There are two distinct biochemical mechanisms for the introduction of *cis* double bonds. In many bacteria, a *cis*-3 double bond is introduced into a medium chain length fatty acid (usually C_{10}) through the dehydration of a *de novo* fatty acid synthesis intermediate, the 3-hydroxyacyl thioester derivative of ACP, by a specific enzyme, 3-hydroxydecanoyl thioester dehydratase.[109] Chain elongation of the *cis*-3 derivative gives rise to long-chain monoenoic fatty acids. This so-called "anaerobic" pathway is identical (except for the formation of the β,γ-unsaturated intermediate) to the pathway for saturated fatty acid biosynthesis and is not considered in this chapter.

A second mechanism for *cis* double bond formation involves the direct, oxygen-dependent desaturation of long-chain fatty acids. Although certain bacteria and presumably all eukaryotic organisms utilize molecular oxygen to produce unsaturated fatty acids, there are differences in the nature and availability of substrates, cofactor requirements, and specificities of the enzyme systems themselves among various groups of organisms.

1.02.3.1 Biochemical Characterization of the Desaturation Reaction

The desaturation reaction of fatty acids is catalyzed by protein components constituting the electron transport system from a primary reductant, NADH or NADPH, to molecular oxygen. Generally, microsomes of animals and plants possess an NADH-dependent desaturation system, whereas plastids of plants have an NADPH-dependent one. The terminal oxygenases which introduce double bonds into fatty acids are designated as desaturases, being classified into three groups, i.e., acyl-CoA, acyl-ACP, and acyl-lipid desaturases, on the basis of their substrate specificities. Although the stereospecificity of hydrogen atoms which are removed from acyl chains during the desaturation reaction seems indistinguishable for the various types of desaturases so far investigated, the mechanism determining the positions of carbon atoms for the introduction of a double bond depends on the type of desaturase.

1.02.3.1.1 Desaturation in animals

In animal cells, essential fatty acids such as $\Delta^{9,12}$-18:2 and $\Delta^{9,12,15}$-18:3, and an endogenously synthesized saturated one, 18:0, are converted into more highly unsaturated fatty acids through the chain elongation and the desaturation at the C-9, C-6, C-5, and C-4 positions (Figure 8(a)). Tetraenoic and pentaenoic fatty acids are not only assembled into membrane phospholipids such as phosphatidylcholine (PC) and phosphatidylethanolamine (PE), but also serve as precursors for signaling molecules such as prostaglandins, thromboxanes, and leukotrienes.

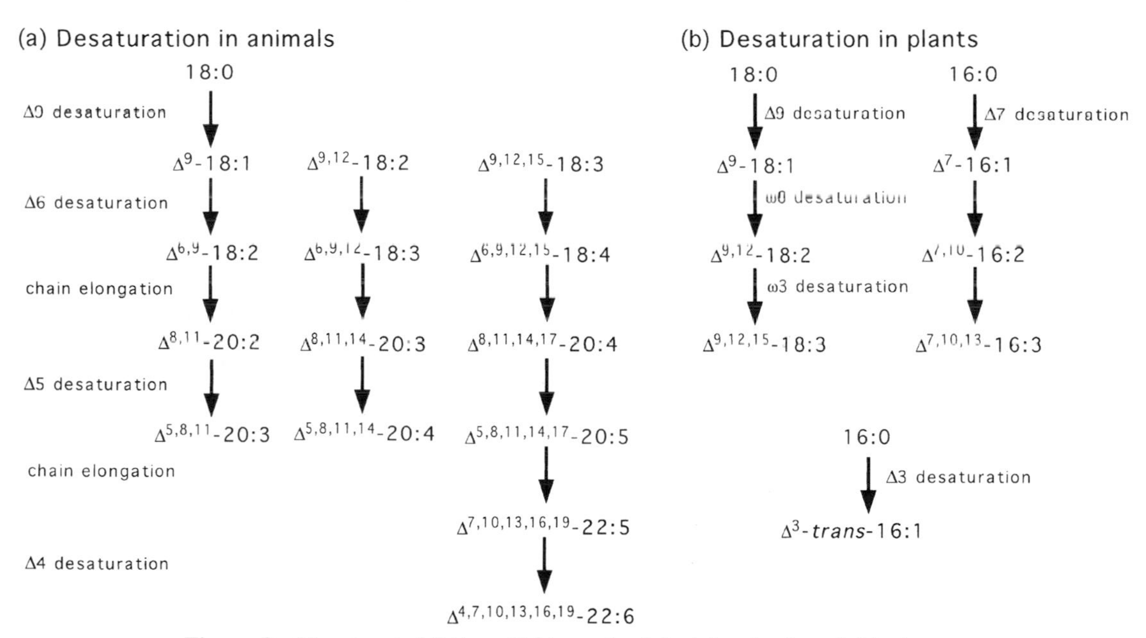

Figure 8 Unsaturated fatty acid biosynthesis in (a) animals and (b) plants.

Δ9 desaturase, which utilizes 18:0-CoA as a substrate, was purified as a single polypeptide of 41.4 kDa from microsomal membranes of rat liver,[110] and the desaturation reaction was reconstituted with the desaturase, and cytochrome b_5 and NADH-cytochrome b_5 oxidoreductase. This indicates that the desaturation process comprises sequential electron transport, as shown in Figure 9. The reaction is catalyzed by a complex of three membrane-bound enzymes: NADH-cytochrome b_5 oxidoreductase, cytochrome b_5, and desaturase. First, electrons are transferred from NADH to the FAD moiety of NADH-cytochrome b_5 oxidoreductase, and then the heme iron atom of cytochrome b_5 is converted to the Fe^{2+} state. The nonheme iron of the desaturase is subsequently reduced to the Fe^{2+} state. This ferrous form interacts with oxygen and the saturated acyl-CoA to introduce a double bond, with the concomitant formation of two molecules of H_2O. Two electrons come from NADH and two from the single bond of saturated acyl-CoA.

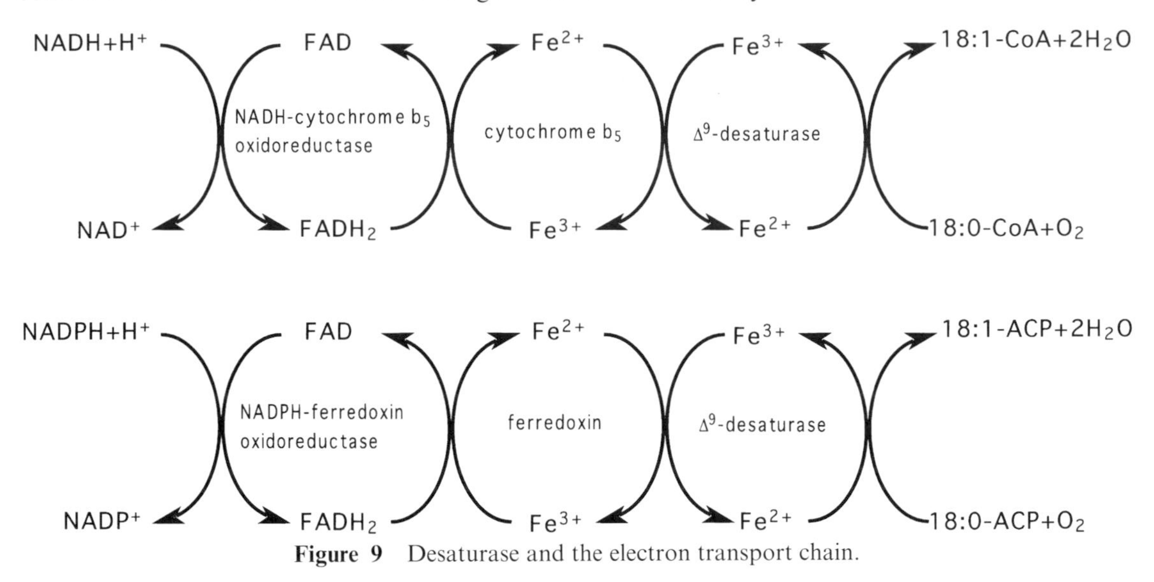

Figure 9 Desaturase and the electron transport chain.

Both NADH-cytochrome b_5 and cytochrome b_5 consist of a globular hydrophilic region exposed to the cytoplasm and anchored to the microsomal membrane via a hydrophobic tail. The hydrophobicity of the desaturase is very similar to that of the hydrophobic tail of the cytochrome b_5 molecule. Since the active center of the desaturase must be accessible to electrons from cytochrome b_5, which is exposed to the cytosol, it is supposed that most of the desaturase is located in the membrane with only a small part exposed to the cytosol (Figure 10).

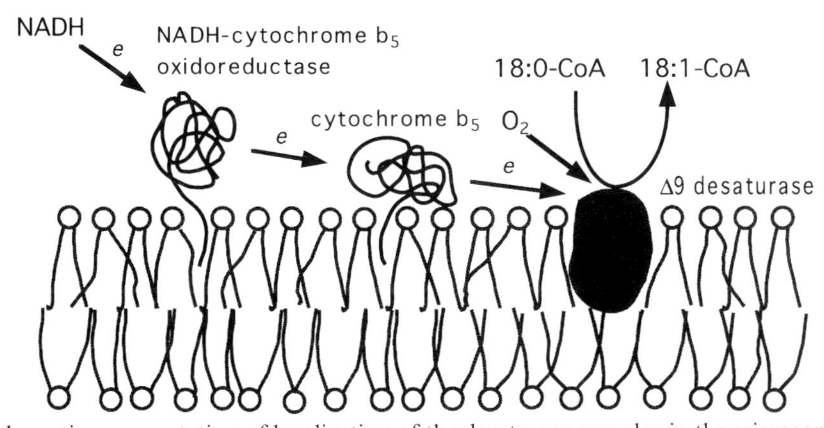

Figure 10 Schematic representation of localization of the desaturase complex in the microsomal membrane.

Δ6 desaturase, which acts on acyl-CoAs, was purified from rat liver microsomal membranes as a single polypeptide of 66 kDa, a similar electron transport chain to that of Δ9 18:0-CoA desaturase being revealed through reconstitution experiments.[111] Desaturation from $\Delta^{8,11,14}$-20:3 to $\Delta^{5,8,11,14}$-20:4, i.e., Δ5 desaturation, was shown to proceed on both phospholipids (PC and PE) and CoA-esters for microsomal membranes of rat liver, although nothing is known about the components of the desaturation system.[112]

1.02.3.1.2 Desaturation in cyanobacteria and plants

Acyl-ACP desaturases are present in the stroma of plant plastids. Acyl-lipid desaturases are bound to the endoplasmic reticulum, the chloroplast membrane in plant cells, and the thylakoid membrane in cyanobacterial cells.[113] The acyl-lipid desaturases can be further classified into two subgroups according to their electron transport systems. One subgroup, which is present in the endoplasmic reticulum of plant cells, uses cytochrome b_5 as the electron donor. The other, which is present in chloroplasts in plant cells and in cyanobacterial cells, uses ferredoxin as the electron donor.

Cyanobacterial strains can be classified into four groups in terms of fatty acid desaturation.[113] Group 1 is characterized by the presence of only saturated and monoenoic fatty acids, while groups 2, 3, and 4 contain polyenoic fatty acids. In the strains in group 1, a double bond is introduced only at the C-9 position of fatty acids esterified at either the *sn*-1 or *sn*-2 position of the glycerol moiety. The strains in group 2 can introduce double bonds at the C-9, C-12, and C-15 positions of *sn*-1 C_{18} fatty acids, and at the C-9 and C-12 positions of *sn*-2 C_{16} fatty acids (Figure 11). Group 2 is most similar to the chloroplasts of plants in terms of the desaturation of fatty acids. The strains in group 3 can introduce double bonds at the C-6, C-9, and C-12 positions of *sn*-1 C_{18} fatty acids, and those in group 4 can introduce ones at the C-6, C-9, C-12, and C-15 positions of *sn*-1 C_{18} fatty acids.

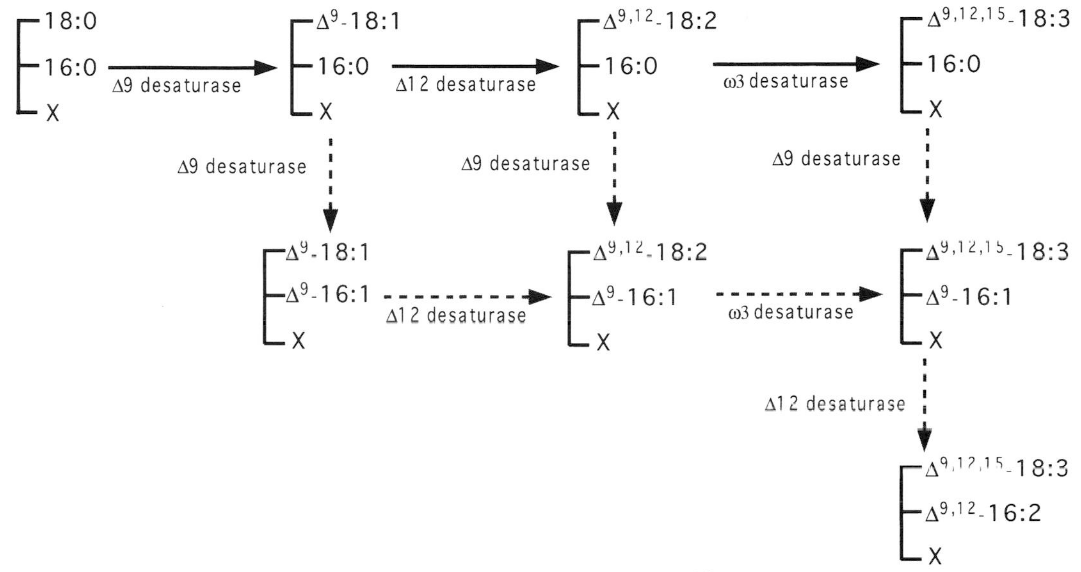

Figure 11 The acyl-lipid desaturation in cyanobacteria of group 2.[113] Black arrows show the desaturation occuring in MGDG, SQDG and PG. Dashed arrows show that occurring only in MGDG.

Plastids are the major site for fatty acid synthesis in plants, mainly producing 16:0 and 18:0 as the ACP derivative. Plants possess several desaturation systems, converting 16:0 into Δ^3-*trans*-16:1 and/or $\Delta^{7,10,13}$-16:3, and 18:0 into $\Delta^{9,12,15}$-18:3 (Figure 8(b)). The desaturation from 18:0 to Δ^9-18:1 occurs in plastids with 18:0-ACP as the substrate, while subsequent desaturation from Δ^9-18:1-ACP and 16:0-ACP proceeds either in plastids or the endoplasmic reticulum after incorporation into glycerolipids.[114] In plastids, 16:0 is desaturated to Δ^3-*trans*-16:1 on phosphatidylglycerol (PG), and, in a higher plant group designated as 16:3-plants, to $\Delta^{7,10,13}$-16:3 on monogalactosyldiacylglycerol (MGDG), whereas Δ^9-18:1 is desaturated to $\Delta^{9,12,15}$-18:3 on chloroplast lipids, MGDG, digalactosyldiacylglycerol (DGDG), sulfoquinobosyldiacylglycerol (SQDG), and PG. On the other hand, in the endoplasmic reticulum, Δ^9-18:1 is converted into $\Delta^{9,12,15}$-18:3 on extrachloroplast lipids such as PC and PE.

Δ9 18:0-ACP desaturase is present in plastid stroma as a soluble protein, and has been partially purified from safflower seeds as a homodimer of a 36 kDa polypeptide.[115] The Δ9 desaturases of both a green alga and a higher plant utilize an electron transport chain composed of NADPH-ferredoxin oxidoreductase and ferredoxin (Figure 9).[116] In contrast, fatty acid desaturases utilizing glycerolipids as substrates have been little characterized biochemically owing to their being membrane proteins. However, the substrate specificities of these individual desaturases have been deduced from the lipid phenotypes of mutants of green plants defective in the corresponding desaturases. Mutants defective in the desaturation from 16:0 to Δ^3-*trans*-16:1 on PG, and that from 16:0 to Δ^7-16:1 on MGDG, respectively, have little effect on all the other desaturation reactions, indicating that Δ3 desaturase is specific to 16:0-PG and Δ7 desaturase to 16:0-MGDG.[117–119] In contrast, mutants defective in the chloroplast desaturation from Δ^7-16:1 to $\Delta^{7,10}$-16:2 on MGDG are also impaired in the desaturation from Δ^9-18:1 to $\Delta^{9,12}$-18:2 on MGDG, DGDG, SQDG, and PG, indicating that neither the polar head groups of chloroplast lipids nor carbon chain length is distinguished by the desaturase to introduce the second double bond.[120,121] A similar property is postulated for the chloroplast ω3 desaturase required for the conversion of dienoic into trienoic fatty acids, through the analysis of a mutant impaired in the corresponding desaturase.[122] Among membrane-bound acyl-lipid desaturases, plastid $\omega 6$ desaturase was solely purified from chloroplast envelope membranes of spinach as a single polypeptide of 37 kDa, being found to require ferredoxin and NADPH-ferredoxin oxidereductase as an electron transport system, as for $\Delta 9$ 18:0-ACP desaturase (Figure 9).[123,124] Interestingly, this spinach desaturase acts on not only chloroplast lipids, but also free acids such as Δ^9-18:1 and Δ^{13}-22:1, and lysophosphatidate containing Δ^9-18:1, inferring that this desaturase also accepts nonesterified monoenoic fatty acids and lipid classes other than chloroplast lipids, but that the actual substrates *in vivo* are the chloroplast lipids present in substantial amounts.

A mutant devoid of microsomal desaturation from Δ^9-18:1 to $\Delta^{9,12}$-18:2 on PC, PE, and phosphatidylinositol (PI) has been isolated, suggesting that this microsomal $\omega 6$ desaturase acts on all phospholipids in the endoplasmic reticulum.[125] A mutant deficient in the microsomal desaturation from $\Delta^{9,12}$-18:2 to $\Delta^{9,12,15}$-18:3, i.e., $\omega 3$ desaturation, suggests that the microsomal $\omega 3$ desaturase also makes little distinction as to substrates regarding phospholipid classes in the endoplasmic reticulum.[126] The $\omega 6$ desaturation of safflower microsomes occurs with NADH-dependent reduction of cytochrome b_5, implying that the electron transport system for this desaturation is composed of NADH-cytochrome b_5 oxidoreductase and cytochrome b_5,[127] as for the $\Delta 9$ desaturation of animal cells (Figure 9(a)). The involvement of cytochrome b_5 in the election transport system is also suggested for the microsomal $\omega 6$ desaturation by the inhibition of the desaturation by antibodies raised against cytochrome b_5.[128]

1.02.3.1.3 Desaturation in a marine bacterium

Shewanella sp. strain SCRC-2738, which was isolated from the intestinal contents of Pacific mackerel, produces a large amount of eicosapentaenoic acid (20:5).[129] This bacterium exhibits a novel feature of fatty acid biosynthesis.[130] The *de novo* synthesis is carried out by the type II fatty acid synthase to produce 16:0-ACP and Δ^9-16:1-ACP. Δ^9-16:1-ACP is synthesized through the anaerobic pathway and this monoenoic fatty acid is unable to be the precursor of 20:5. C_{20}-fatty acids are supplied through the chain elongation system involving 16:0-ACP as a substrate. *Shewanella* sp. does not exhibit any acyl-lipid desaturation, but shows acyl-ACP desaturase activity. This desaturation system is specific for acyl-ACP, and requires ferredoxin and NADPH ferredoxin oxidoreductase.

1.02.3.1.4 Stereospecificity of hydrogen removal during desaturation

Two hydrogen atoms are removed from a fatty acid on the desaturation reaction. The stereospecificity of the removal of hydrogen atoms was first shown for oleate synthesis from stearate in *Corynebacterium diphtheria*.[131] Incubation of *C. diphtheria* independently with four stereospecific stearate isoforms, i.e., *R*-[9-^{3}H]-, *S*-[9-^{3}H]-, *R*-[10-^{3}H]-, and *S*-[10-^{3}H]stearate, results in the synthesis of [^{3}H]oleate not from *R*-[9-^{3}H]- or *R*-[10-^{3}H]stearate, but from *S*-[9-^{3}H]- and *S*-[10-^{3}H]stearate. Thus, the removal of two hydrogens for stearate desaturation is specific to the pro-9*R* and pro-10*R* hydrogen atoms. This dehydrogenation is simply considered as *syn* elimination from stearate. Removal of pro-*R* hydrogen atoms has been reported also for the desaturation from stearate to oleate in *Chlorella vulgaris*,[132,133] and for the desaturation from oleate to linoleate in *C. vulgaris*[132,133] and castor bean tissues.[134] The collective evidence at this stage suggests that all desaturases, irrespective of their origin, exhibit the same stereospecificity.[2]

1.02.3.1.5 How do desaturases determine the positions to introduce double bonds?

The mechanism to determine the desaturation positions have been studied by identification of the position at which a double bond is introduced after incubation of cells with precursor fatty acids of various carbon chain lengths. When *C. vulgaris* was incubated with C_{14} to C_{18} saturated fatty acids, the synthesis of $\Delta 9$ monoenoic fatty acids of corresponding chain length was observed.[135] *Synechochystis* PCC6803 supplemented with a saturated C_7 acid likewise produced novel $\Delta 9$ monoenoic fatty acids, Δ^9-17:1 and Δ^9-19:1, in addition to a normal monoenoic fatty acid, Δ^9-18:1. These results indicate that the unusual products of fatty acid synthesis, 17:0 and 19:0, are desaturated at the C-9 position.[136] These results also prove that the desaturases which introduce the first double bond recognize the position from the carboxyl end, and thus are designated as Δ9 desaturases.

However, desaturases which act after the Δ9 desaturases recognize desaturation positions through distinct mechanisms from those in the cases of the Δ9 desaturases. When *C. vulgaris* and the seeds of a higher plant, *Ricinus communis*, were incubated with monoenoic C_{16} to C_{19} acids with double bonds at the Δ9 or ω9 position, they produced dienoic acids, the second double bonds of which were introduced on the side of the methyl end, and separated by methylene from the first one. These results indicate that the desaturases which introduce the second double bond determine the position from the distance from the first double bond.[137] On the other hand, in *Synechocystis* PCC6803 incubated with a C_7 saturated acid, novel odd-numbered tetraenoic fatty acids, $\Delta^{6,9,12,14}$-17:4 and $\Delta^{6,9,12,16}$-19:4, were detected in addition to a normal one, $\Delta^{6,9,12,15}$-18:4, showing that the desaturase originally responsible for the introduction of the double bond at the Δ15 position of a C_{18} acid counted the position from the methyl end.[136]

1.02.3.2 Genes for Fatty Acid Desaturases

The purification of desaturases, and the cloning of the corresponding genes and cDNAs are indispensable for characterization of the reaction and regulation mechanisms for fatty acid desaturation at the biochemical and molecular levels. So far, purification has been successfully performed for soluble Δ9 18:0-ACP desaturases from higher plants, membrane-bound Δ9 18:0-CoA desaturases from animals, and a chloroplast ω6 desaturase from a higher plant, all of which were used to isolate the respective cDNAs for elucidation of the primary sequences of the desaturases. Other desaturases have not yet been purified, mainly due to their being integral membrane proteins. However, the genes and cDNAs for these desaturases have been cloned, especially ones from higher plants and cyanobacteria, using mutants, which are defective in various desaturation steps, and various techniques. The latter include complementation through shotgun introduction of genomic DNA, identification of a tagged chromosomal DNA produced by insertional mutagenesis, mapping of a damaged genetic locus, and cloning with the use of nucleotide and amino acid sequences already known for some desaturases for heterologous probes and the design of PCR primers.

1.02.3.2.1 Δ9 desaturases in animals and fungi

On the basis of the observation that Δ9 18:0-CoA desaturase activity was induced in the livers of rats fed a fat-free diet followed by starvation, a cDNA clone expressed in the induced ones, but not in uninduced rats, was isolated as a candidate for the desaturase cDNA.[138] The clone was hybridized with total poly(A+) RNA from induced rats for recovery of the corresponding mRNA, which was then translated *in vitro*. The polypeptide product could be immunoprecipitated with antiserum against the purified Δ9 18:0-CoA desaturase of rats, proving that the cDNA clone possessed the nucleotide sequence of the desaturase.[138] The cDNA clone, covering only a partial nucleotide sequence of the desaturase, was then compiled to full-length by primer extension, the resultant complete sequence containing an open reading frame for a polypeptide of 358 amino acids.[139] The amino acid composition was similar to that of the purified desaturase protein, and hydrophobic amino acids accounted for 62% of the total, which is consistent with the enzyme being a membrane protein.[110] cDNA and/or genes later cloned from the animal kingdom were identified as coding for Δ9 18:0-CoA desaturases on the basis of the primary sequence homology of 39–92% to that of the rat Δ9 18:0-CoA desaturase.[140–143]

The Δ9 18:0-CoA desaturase gene was also isolated from a yeast, *Saccharomyces cerevisiae*, as a genomic clone complementing the Δ^9-18:1-requiring phenotype of the *OLE1* mutant. The clone contained an open reading frame (ORF) encoding a protein of 510 amino acids, the sequence of which is 36% identical to that of rat Δ9 18:0-CoA desaturase.[144] Interestingly, the yeast desaturase, as compared with the rat counterpart, exhibited an N-terminal extension of 113 amino acids which is homologous to the heme-binding domain and the electron-transfer motif of cytochrome b_5.[145] Truncation or disruption of the cytochrome b_5-like domain brought about unsaturated fatty acid auxotrophy in the yeast, while disruption of the cytochrome b_5 gene resulted in no requirement for unsaturated fatty acids, indicating that the cytochrome b_5-like domain, but not cytochrome b_5 itself, is essential for the functioning of the Δ9 18:0-CoA desaturase in the yeast.[145] A lethal phenotype was complemented by expression of the Δ9 18:0-CoA desaturase cDNA of rat for the *OLE1*-disrupted yeast mutant, but not for the double mutant with disruption of both the cytochrome b_5- and *OLE1* genes, which demonstrate that the electron transport system for the desaturation in yeast is available also for the rat desaturase, and that cytochrome b_5 is an indispensable component of the system.[145]

1.02.3.2.2 ω3 desaturase in Caenorhabditis elegance

A database search for the expressed sequence tag (EST) of *C. elegance* identified some EST clones originating from a single gene as showing high scoring to membrane-bound ω3 and ω6 desaturases of microsomes and chloropasts of higher plants, which utilize acyl-lipids as substrates.[146] A full-length cDNA corresponding to these cDNAs was isolated, which was shown to encode a protein of 402 amino acids, the sequence of which was 32% and 35% identical to those of microsomal ω3 and ω6 desaturases of *Arabidopsis*, respectively, and to possess a C-terminal motif of the retention signal in endoplasmic reticulum membranes, i.e., KAKAK. The expression of this cDNA in *Arabidopsis* brought about a 90% increase in $\Delta^{9,12,5}$-18:3 synthesis from $\Delta^{9,12}$-18:2 in *in vitro* grown roots, and also the novel synthesis of $\Delta^{5,8,11,14,17}$-20:5 and $\Delta^{8,11,14,17}$-20:4 in leaves when sprayed with $\Delta^{5,8,11,14}$-20:4 and $\Delta^{8,11,14}$-20:3, respectively. These results indicate that the *Caenorhabditis* cDNA encodes a ω3 desaturase with substrate specificity towards both C_{18} and C_{20} fatty acids, possibly esterified to lipids, in endoplamic reticulum membranes.

1.02.3.2.3 Δ9 desaturases in cyanobacteria and plants

Δ9 18:0-lipid and Δ9 18:0-ACP desaturases are the enzymes in cyanobacteria and higher plants, respectively, that introduce the first double bonds into saturated fatty acids. An ORF showing a deduced amino acid sequence similar to those of membrane-bound Δ9 18:0-CoA desaturases was found in *Anabaena variabilis* during the genome sequencing of the 5′-upstream region of the Δ12 desaturase gene, and was used to probe the counterpart from *Synechochystis* PCC6803.[147] The amino acid sequences, comprising 272 and 318 residues in *A. variabilis* and *Synechochystis* PCC 6803, respectively, showed 62% homology to each other, 29–31% homology to those of the 18:0-CoA desaturases of rat, mouse, and yeast, but no significant homology to those of other membrane-bound desaturases or soluble 18:0-ACP desaturases of higher plants. *E. coli* cells transformed with the *Anabaena* gene produced Δ^{9}-18:1, which was not originally involved in the *E. coli* cells.

The cDNAs of Δ9 18:0-ACP desaturases, which are soluble proteins in stroma, were cloned from castor beans, cucumber, and safflower.[148,149] The primary sequences deduced from the cDNAs, which are respectively composed of 396 amino acids, are unrelated to those of Δ9 18:0-CoA and other membrane-bound desaturases. The N-terminal sequence of the safflower desaturase, in comparison to that of the purified enzyme, had an extension of 33 amino acids which is compatible with a transit peptide for plastid stroma. Cell extracts of *E. coli*, in which the 18:0-ACP desaturase of safflower was overexpressed, demonstrated a remarkable increase in the desaturation activity with exogenously added ferredoxin. Thus far, 18:0-ACP desaturase cDNAs have been isolated from several species of higher plants, exhibiting more than 80% identity in these amino acid sequences.

1.02.3.2.4 Δ12 and ω6 desaturases in cyanobacteria and plants

The Δ12 desaturases of cyanobacteria, and chloroplast and microsomal ω6 desaturases of green plants are membrane-bound proteins that introduce the second double bonds into monoenoic fatty acids esterified to glycerolipids. The gene for Δ12 desaturase of *Synechochystis* PCC6803, which is responsible for the synthesis of $\Delta^{9,12}$-18:2 from Δ^{9}-18:1, was first cloned as an acyl-lipid desaturase gene.[150] This gene was obtained as a genomic DNA clone which could complement a mutant of *Synechochystis* PCC6803 defective in the desaturation from Δ^{9}-18:1 to $\Delta^{9,12}$-18:2[150], and used as a heterologous probe for isolation of the counterparts from other cyanobacterial species.[151] The respective cyanobacterial Δ12 desaturase genes contained ORFs for proteins of 347–351 amino acids, the sequences of which showed more than 57% homology.[151]

Chloroplast ω6 desaturases in higher plants and green algae catalyze the desaturation from Δ^{7}-

16:1 to $\Delta^{7,10}$-16:2 and/or from Δ^{9}-18:1 to $\Delta^{9,12}$-18:2 in choloroplasts.[152,121] The ω6 desaturase cDNAs have been cloned from several higher plants such as spinach, *Glycine max*, *Brassica napus*, and *Arabidopsis thaliana*.[153–155] The chloroplast ω6 desaturase cDNA was also isolated from green alga, *Chlamydomonas reinhardtii*.[156] These cDNAs with ORFs for proteins of 418 to 447 amino acids have been functionally identified as encoding chloroplast ω6 desaturases from the observation that expression of the cDNAs or the corresponding genomic DNA clones rescued the mutants of green plants from a lesion in the chloroplast ω6 desaturation,[155,156] or conferred a novel ω6 desaturation activity on a cyanobacterial strain.[154] The latter result indicates that chloroplast ω6 desaturases can also function in cyanobacteria. The amino acid sequences of the desaturases of higher plants are 60–70% identical to one another, and 50% to that of *C. reinhardtii*, whereas green plant sequences show 50% homology to those of cyanobacterial Δ12 desaturases.[156] The N-terminal sequence deduced from spinach cDNA, compared with that of the purified enzyme, had an extension of 65 amino acids with some features of chloroplast transit peptides.[153] Transit peptides for targetting to chloroplasts are also presumed for the N-terminal sequences of other chloroplast ω6 desaturases deduced from the cDNAs.[153,154,156]

cDNA and genomic DNA for microsomal ω6 desaturase, which synthesizes $\Delta^{9,12}$-18:2 from Δ^{9}-18:1, were probed from *Arabidopsis* with tagged chromosomal DNA of an *Arabidopsis* mutant defective in the desaturase activity, which was produced by insertional mutagenesis with *Agrobacterium tumefaciens* T-DNA.[157] The cDNA was expressed to complement the desaturation lesion in the mutant, and thus was functionally identified as encoding the microsomal ω6 desaturase.[157] With the *Arabidopsis* cDNA as a heterologous probe, two cDNAs for isoforms of microsomal ω6 desaturase were cloned from *G. max*.[158] These three cDNAs encode proteins of 383 to 387 amino acids, showing more than 68% identity in sequence, but only limited identity (below 30%) to chloroplast ω6 desaturases of green plants and Δ12 desaturases of cyanobacteria.

1.02.3.2.5 ω3 desaturases in cyanobacteria and plants

The ω3 desaturases of cyanobacteria and higher plants are membrane proteins which catalyze the conversion of dienoic into trienoic fatty acids. The gene of ω3 desaturase, which produces $\Delta^{9,12,15}$-18:3 from $\Delta^{9,12}$-18:2, was probed from *Synechochystis* PCC6803 with a partial fragment of the structural gene for its Δ12 desaturase.[159] This gene contains an ORF for a protein of 359 amino acids, the sequence of which is 28% homologous to that of the Δ12 desaturase, and 49–50% to those of chloroplast ω6 desaturases of higher plants. The introduction of this gene into *Synechococcus* PCC7942, together with the Δ12 desaturase gene, conferred the ability to synthesize $\Delta^{9,12,15}$-18:3 on this strain, which contain Δ^{9}-18:1 as a sole C_{18} unsaturated fatty acid. The result proved that this gene is responsible for the conversion of $\Delta^{9,12}$-18:2 to $\Delta^{9,12,15}$-18:3.

The cDNAs for two isoforms of chloroplast ω3 desaturases, which synthesize $\Delta^{9,12,15}$-18:3 and $\Delta^{7,10,13}$-16:3 from $\Delta^{9,12}$-18:2 and $\Delta^{7,10}$-16:2, respectively, have been cloned from *A. thaliana*.[160–162] These *Arabidopsis* cDNAs were identified as encoding the desaturase from their ability to complement a defective mutation in *Arabidopsis* as to chloroplast ω3 desaturation. Chloroplast ω3 desaturase cDNAs were then probed from *B. napus* and *G. max* with one of the *Arabidopsis* cDNAs. These chloroplast ω3 desaturase cDNAs contain ORFs for proteins of 404–457 amino acids with sequence identity of more than 69% to one another, and of more than 66% to the microsomal ω3 desaturases. N-terminal extensions of 60–70 amino acid residues were identified for chloroplast ω3 desaturases in comparison to the microsomal ω3 desaturases, showing characteristic features of transit peptides of chloroplast proteins encoded by the nuclear genome.

The cDNA for microsomal ω3 desaturase responsible for $\Delta^{9,12,15}$-18:3 synthesis from $\Delta^{9,12}$-18:2 was screened for in *Brassica* with a yeast artificial chromosome clone of wild-type *Arabidopsis* as a probe, which was found to contain the genetic locus damaged in an EMS (ethyl methanesulfonate)-induced *Arabidopsis* mutant defective in the desaturation,[163] while it was screened for in *Arabidopsis* with the use of a tagged genomic DNA as a probe, which was obtained from another, but phenotypically similar *Arabidopsis* mutant produced through T-DNA insertion.[160] These cDNAs were expressed to complement the respective *Arabidopsis* mutants. cDNAs with ORFs for proteins homologous to the microsomal ω3 desaturase were also cloned from *G.* max and *B napus* with the *Arabidopsis* cDNA as a heterologous probe. These cDNAs contain ORFs for proteins of 377–383 amino acid residues with sequence identities of more than 68%. Besides, the cDNAs of *Brassica* and *Arabidopsis* possess lysyl residues at the third and fifth residues from the C-terminus, which is sufficient for membrane proteins of animals to stay on the endoplasmic reticulum.

1.02.3.2.6 Unusual desaturases

The cDNAs for soluble desaturases responsible for the syntheses of uncommon fatty acids have been cloned from some plants. A cDNA for Δ6 18:0-ACP desaturase isolated from coriander, the seed oil of which mainly comprises Δ^6-18:1, contains an ORF for a protein of 385 amino acids with 70% identity in sequence to castor bean Δ9 18:0-ACP desaturase.[164] Expression of this coriander cDNA in tobacco resulted in the production of a novel fatty acid, Δ^6-18:1. cDNA of *Thunbergia alata*, the seed oil of which is mainly composed of Δ^6-16:1, was isolated as a candidate that codes for Δ6 16:0-ACP desaturase through PCR with degenerate oligonucleotide primers based on primary sequences conserved in Δ9 18:0-ACP and Δ6 18:0-ACP desaturases.[165] The deduced amino acid sequence composed of 387 residues with 66% and 57% identity to those of the castor bean Δ9 18:0-ACP and the coriander Δ6 18:0-ACP desaturases, respectively, and was confirmed to be Δ6 16:0-ACP desaturase cDNA by the observation that extracts of *E. coli* cells transformed with this cDNA catalyzed the Δ6 desaturation of 16:0-ACP. A cDNA for Δ9 14:0-ACP desaturase was isolated from *Pelargonium xhortorum*, which produces Δ^9-14:1 as a precursor of the Δ^7-22:1 and Δ^{19}-24:1 anacardic acids involved in pest resistance.[166] This cDNA contains an ORF for a protein of 368 amino acids, the sequence of which is 79% similar to that of the castor bean Δ9 18:0-ACP desaturase, and was functionally identified as Δ9 14:0-ACP desaturase cDNA from the appearance of Δ9 desaturation activity toward 14:0-ACP in extracts of *E. coli* overexpressing this cDNA. These three novel desaturases, as compared with mature Δ9 desaturases, had an N-terminal extension possibly corresponding to a transit peptide for plastids, suggesting their localization in plastid stroma as Δ9 18:0-ACP desaturase.[164–166]

As to a membrane-bound desaturase, a gene and a cDNA for Δ6 desaturase were isolated from a cyanobacterium and a higher plant, respectively. *Synechochystis* PCC6803 possesses Δ6 desaturation activity to produce $\Delta^{6,9,12}$-18:3 and $\Delta^{6,9,12,15}$-18:4 from $\Delta^{9,12}$-18:2 and $\Delta^{9,12,15}$-18:3, respectively, using acyl-lipid as substrates. A gene isolated from its cosmid genomic DNA library was found to confer Δ6 desaturation activity on another cyanobacterial strain, *Anabaena* PCC7120, originally with no Δ6 desaturation activity. The primary sequence deduced from the Δ6 desaturase gene is composed of 359 amino acids with only limited identity to those of other membrane-bound desaturases.[167] On the other hand, cDNA was cloned from borage, the seeds of which contain as much as 20–25% of $\Delta^{6,9,12}$-18:3, through PCR with degenerated primers for amino acid sequences conserved in membrane-bound ω3 an ω6 desaturases of higher plants.[168] The cDNA contained an ORF for a protein of 448 amino acid residues with limited identity in sequence (less than 22%) to other membrane-bound desaturases including Δ6 desaturase of *Synechocystis* PCC6803. Expression of the cDNA in tobacco resulted in the production of $\Delta^{6,9,12}$-18:3 and $\Delta^{6,9,2,15}$-18:4 from $\Delta^{9,12}$-18:2 and $\Delta^{9,12,15}$-18:3, respectively. The borage Δ6 desaturase is similar to yeast Δ9 desaturase in possessing a region with some identity to cytochrome b_5,[167] but, in contrast to the yeast Δ9 desaturase, the region is located not at the N-terminus, but at the C-terminus.[168]

1.02.3.3 Reaction Center of Desaturases

Soluble and membrane-bound desaturase reactions show common features such as requirements for molecular oxygen and iron,[169] and the stereospecificity of hydrogen atom removal,[131] but amino acid sequence homology is limited, especially among membrane-bound desaturases with distinct catalytic characteristics, and is unrelated between soluble and membrane-bound desaturases. However, X-ray resolution of a soluble desaturase and primary sequence alignment of desaturases, and of desaturases with proteins of other functions, have provided important clues for elucidation of the mechanism of the desaturase reaction.

1.02.3.3.1 Stearoyl-ACP desaturase

Fox *et al.*[170,171] utilized the stearoyl-ACP desaturase of *R. communis* overproduced in *E. coli* for atomic absorption spectroscopy to determine that the desaturase contains four iron atoms per homodimer. They also showed that the iron ions constitute a diiron-cluster through optical, Mossbauer, and resonance Raman spectroscopies, and proposed possible amino acid ligands involved in the cluster on the basis of primary sequence alignment of the stearoyl-ACP desaturases of various species of higher plants with class II diiron-oxo proteins, i.e., bacterial hydroxylases such as ribonucleotide reductase and methane monooxygenase, the amino acid ligands of which had been elucidated by X-ray crystallography.

The X-ray structure of the stearoyl-ACP desaturase was then solved, showing a diiron cluster in the interior of the desaturase, the structure of which is consistent with that indicated by alignment with the class II diiron-oxo proteins.[172] The oxygen atoms of the carboxyl groups of E143 and E229 are bridging ligands for the two iron ions, while those of E105 and E196 are bidentate ligands for one and the other iron ion, respectively (Figure 12(a)). A nitrogen atom, Nδ1, in H146 and H232 are also ligands for the respective iron ions. The X-ray crystallography also revealed that a channel extends deeply inside from the surface of the desaturase, and passes via the vicinity of the diiron center on the same side of the proposed oxygen binding site. Modeling of a stearate to this channel made the carbon atom at the Δ9 position close to one of the iron ions, and also to a small pocket. The authors suggested that this pocket is probably occupied by an oxygen molecule bound to the diiron center, and that a peroxide radical is produced to remove the hydrogen atoms at the C-9 position.

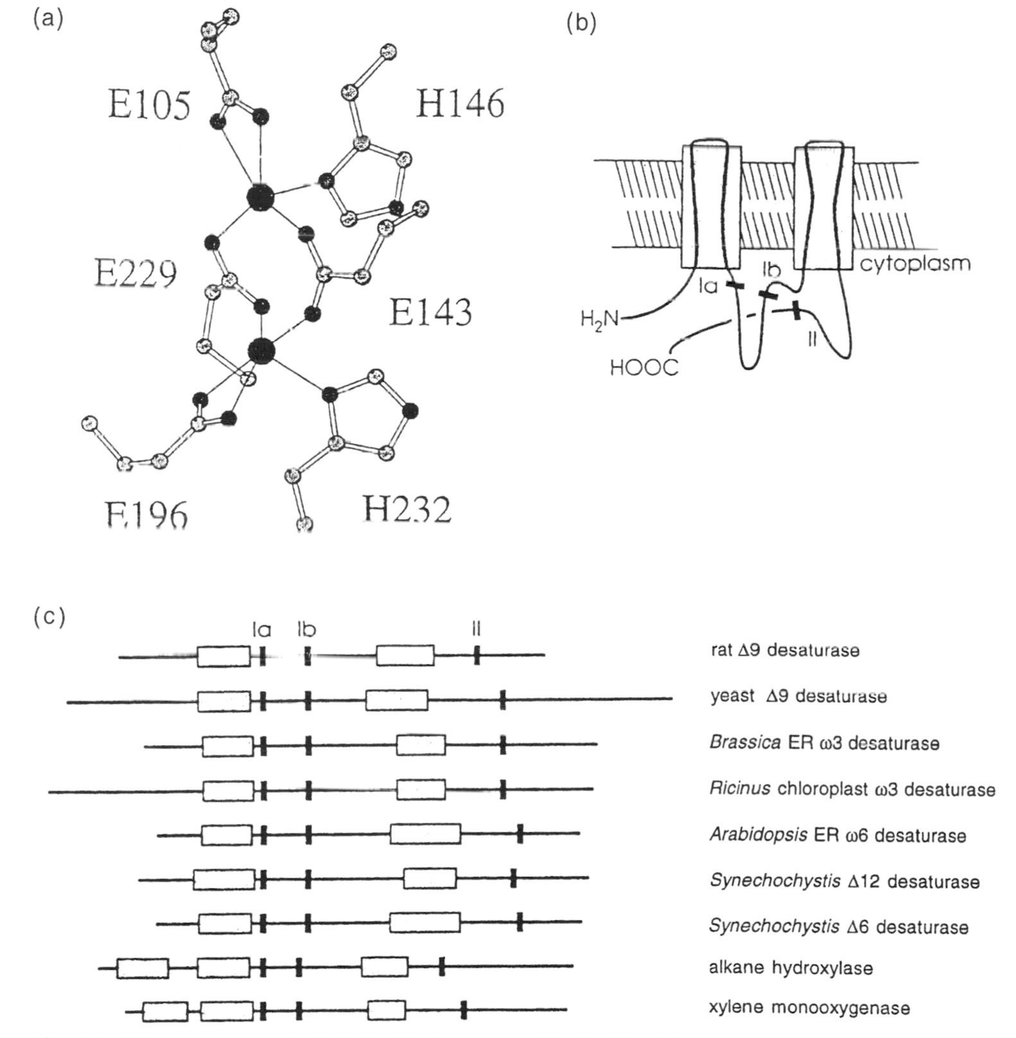

Figure 12 (a) Diiron center of 18:0-ACP desaturase.[172] (b) Model of the topology of the desaturase sequences.[176] (c) Hydrophobic domain structure of the membrane desaturases and hydroxylases.[176] Open boxes represent hydrophobic domains containing greater than 40 amino acid residues, capable of spanning the membrane twice. The locations of His-containing regions Ia, Ib, and II are indicated by solid boxes.

1.02.3.3.2 Membrane-bound desaturases

So far, only two types of membrane-bound desaturases have been purified, i.e., stearoyl-CoA[139,110] and chloroplast ω6 desaturases.[123] The stearoyl-CoA desaturase of rat liver showed an absorption spectrum characteristic of an oxo-bridged diiron cluster.[110,173] However, the iron-containing active sites have not been further characterized as to the properties of an integral membrane protein. His-motifs responsible for ligands to iron atoms were first proposed for rat and yeast stearoyl-CoA desaturases on comparison of the deduced primary sequences,[174] in view of the fact that closely spaced His residues are characteristic of chelate transition metals with high affinity.[175] Alignment of the primary sequences of acyl-CoA desaturases together with those of other membrane-bound desaturases acting on lipid substrates, and bacterial membrane hydroxylases, i.e., alkane hydroxylase and xylene monooxygenase, showed the consensus sequences, HX(3 or 4)H, HX(2 or 3)HH, and HX(2 or 3)HH.[176] The requirement of the His residues for desaturase reactions was shown for the rat stearoyl-CoA desaturase and *Synechochystis* Δ12 desaturase by site-directed mutagenesis.[176,177] Hydropathy analysis of the membrane desaturases and hydroxylases predicted that the His-motifs are located in hydrophilic regions, and the distance from the preceding region spanning the membrane twice to each His-motif is similar for these enzymes (Figure 12(b), (c)). Since electron donors for the membrane enzymes, cytochrome b_5, ferredoxin, rubredoxin, and xylA, are peripheral membrane or soluble proteins, it was proposed that the His-motifs are localized at or near the membrane surfaces.

1.02.3.4 Physiological Significance of Fatty Acid Desaturation

Fatty acid desaturation is the process that supplies unsaturated fatty acid constituents of glycerolipids for biomembranes, and also for lipid bodies which are utilized as an energy reservoir. The role of fatty acid desaturation in organelle biogenesis and tissue development has been indicated by analyses of desaturation mutants, and of the special expression pattern for desaturase genes. Some desaturation processes are dependent on external factors such as temperature and nutrition, which reflects the importance of the desaturation for the organisms to adapt to changing environments. The unsaturation levels of membrane lipids are increased by a downward shift of the growth temperature in order to maintain the membrane fluidity in several organisms. Other than their roles in the production of components of the cellular architecture such as membranes, fatty acid desaturation provides animals and higher plants with unsaturated fatty acids, which become substrates for bioactive compounds such as eicosanoids and jasmonates responsible for the physiological responses to, for example, injury.

1.02.3.4.1 Organelle biogenesis and tissue development

Some mutants defective in fatty acid desaturation revealed lesions in the biogenesis of organelle. A high-temperature sensitive mutant of *S. cerevisiae* owing to damage to the structural gene of *OLE1* encoding Δ9 18:0-CoA desaturase showed abnormal aggregation of mitochondrial vesicles and an inability to transfer mitochondria to growing buds at a nonpermissive temperature.[178] The mutational phenotypes were complemented by the addition of Δ^9-18:1 to the cell culture, indicating an essential role of the *OLE1* gene in mitochondrial morphology and distribution, and eventually in the viability of the yeast cell. In higher plants, chloroplast development during leaf cell maturation is accompanied by increases in the contents of highly unsaturated fatty acids, $\Delta^{7,10,13}$-16:3 and/or $\Delta^{9,12,15}$-18:3, of chloroplast lipids, which suggests the involvement of fatty acid desaturation in the biogenesis of the chloroplast membrane.[179] A mutant of *A. thaliana* defective in the *FAD7* gene coding for chloroplast ω3 desaturase showed a 45% reduction in the cross-sectional area of chloroplasts, and a 30–40% reduction in grana and stroma thylakoid membranes. However, this mutant is similar to the wild-type in photosynthetic characteristics and growth rates probably owing to the compensation of the loss of photosynthetic membranes per chloroplast by the 45% increase in chloroplast number per cell. These observations indicate that the *FAD7* gene is required for normal biogenesis by chloroplasts.[180]

Tissue-specific expression of desaturase genes has been observed in mice and higher plants, suggesting some particular roles of the respective desaturases in tissue development. Mice possess two isoforms of 18:0-CoA desaturase encoded by SCD1 and SCD2, respectively, which show distinct tissue-specific expression patterns at the transcript level: high expression levels in adipose tissue for the SCD1 gene, and in brain for the SCD2 gene, low expression levels in lung for the SCD1 gene, and in lung, spleen, kidney, and adipose for the SCD2 gene, and undetectable levels in brain, heart, spleen, kidney, and liver for the SCD1 gene, and in heart and liver for the SCD2 gene.[140,141] *G. max* contains two isoforms of microsomal ω6 desaturase coded by the *FAD2-1* and *FAD2-2* genes, the transcripts of which were detected specifically in developing seeds, and in both developing seeds and vegetative tissues, respectively.[181] Judging from the fatty acid compositions of transgenic plants, in which the *FAD2-1* or *FAD2-2* gene was repressed by expression of the corresponding antisense RNA, the authors suggested that the *FAD2-1* gene contributes to $\Delta^{9,12}$-18:2 synthesis for storage lipids of seeds, while the *FAD2-2* gene contributes to the synthesis for membrane lipids of both seeds and vegetative tissues. *A. thaliana* possessed the transcript of chloroplast ω3 desaturase encoded by the *FAD7* gene specifically in photosynthetic active tissues, and analysis of *FAD7* promoter::*GUS* fusion gene expression in a tobocco transformant revealed the involvement of this promoter in the transcriptional regulation of the tissue-specific expression of the *FAD7* gene.[182]

1.02.3.4.2 Adaptation to temperature

The unsaturation levels of membrane lipids are altered in various organisms exposed to changes in environmental temperature. The increases in the unsaturation levels of constituent fatty acids of membrane lipids, reported for a wide range of organisms shifted to lower temperatures, are interpreted as a means of lowering the transition temperature of the membranes for maintenance of the membrane fluidity. Three possible mechanisms have been proposed for the increases in unsaturation levels which depend on the balance between fatty acid synthetic and desaturation activities providing saturated and unsaturated fatty acids, respectively.

The first is activation of the desaturase pre-existing in the membranes at low temperatures. The activities of Δ12 and Δ15 desaturation were enhanced in *Tetrahymena* cells shifted to a lower temperature, but not in ones that simultaneously incorporated polyenoic fatty acids into membrane lipids and thus with increased membrane fluidity.[183] These results suggested that the desaturation activities were regulated by membrane fluidity, and not by temperature *per se*. The role of membrane fluidity in the control of desaturation activity was also inferred for *Candida lipolytica*, the Δ12 desaturation activity of which became twofold greater after catalytic hydrogenation of microsomal membranes leading to a decrease in the unsaturation level.[184] The second is elevation of the ratio of the activities of desaturation and fatty acid synthesis at low temperatures owing to distinct activation energies between these two reactions, which was suggested for an increase in ω6 desaturation in developing seeds.[185]

Accumulation of information on fatty acid desaturases at the molecular level has led to successes in the cloning of desaturase genes, and strongly supported the third mechanism, i.e., up-regulation of desaturase gene expression at low temperatures. The transcript levels of desaturase genes have been shown to increase with a downward shift of the growth temperature, as in the cases of Δ6, Δ12, and ω3 desaturases of *Synechochystis* PCC6803,[186] Δ12, and ω3 desaturases of *Synechococcus* PCC7002,[187] one chloroplast ω3 desaturase isozyme of *A. thaulina*,[162] and Δ9 18:0-CoA desaturase of *Tetrahymena thermophila*.[143] These results indicate that several desaturation genes are up-regulated at least at the tanscript level. The elevation of both the transcription rates and transcript stability contributes to the increased transcript levels of Δ6, Δ12, and ω3 desaturases of *Synechochystis* PCC6803, and the Δ9 18:0-CoA desaturase of *T. thermophila*,[186,143] it being also shown to correlate to increased levels of the desaturase proteins in *Synechochystis* PCC6803.[186] The results indicate that these desaturases are both transcriptionally and post-transcriptionally regulated.

In contrast, some species of higher plants exposed to elevated temperatures exhibit decreases in the unsaturation levels of thylakoid membrane lipids concomitantly with enhancement of heat-tolerance of photosystem II (PSII) activity, one of the crucial functions of the membranes, inferring some role of the lowered unsaturation levels in adaptation of the plants to high temperatures.[188,189] The protective role of lowered unsaturation levels against high temperature was also demonstrated by enhancement of the heat-tolerance of PSII activity in mutants as to chloroplast ω6 desaturases of *C. reinhardtii*[190] and *A. thaliana*,[152] and chloroplast ω9 desaturase of *A. thaliana*[191] with decreased unsaturation levels of chloroplast lipids, as compared with the wild-types. Introduction of the chloroplast ω6 desaturase gene into a *Chlamydomonas* mutant resulted in a decrease in the heat-tolerance of PSII activity as well as recovery from the desaturation defect, directly indicating the involvement of the unsaturation level in the tolerance to high temperature.[156] The reduced tendency of nonbilayer forming lipid, MGDG, to phase-separate owing to the lowered unsaturation levels may increase the stability of PSII activity at high temperatures, as was suggested by the observation that the thylakoid membrane ultrastructure together with PSII activity were stabilized in pea chloroplasts at high temperatures after saturation of the thylakoid membranes by catalytic hydrogenation.[192] The molecular mechanism underlying the decrease in the unsaturation levels at high temperature remains to be investigated.

1.02.3.4.3 Adaptation to nutritional conditions

In animals, in response to dietary conditions, Δ9 18:0-CoA desaturase activity changes. The starvation of rats and mice decreased Δ9 18:0-CoA desaturase activity,[193,194] whereas cycles of starvation and refeeding with a fat-free diet highly enhanced the desaturase activity.[195] Diabetic rats with lower Δ9 18:0-CoA desaturase activity than normal rats showed an increase in the desaturase activity when fed with insulin or fructose, and the feeding of normal rats with sucrose or fructose stimulated the desaturation activity, suggesting that both insulin and intermediates of carbohydrate metabolism are regulatory factors for stearoyl-CoA desaturation.[196]

Changes in the transcript levels due to dietary control were observed with two isoforms of mouse Δ9 acyl-CoA desaturase encoded by the SCD1 and SCD2 genes, respectively.[140,141] The SCD1 transcript level is higher in liver, kidney, and lung of mice fed a fat-free diet after starvation, than in mice fed a normal diet, whereas regardless of the dietary conditions, the transcript levels are almost equal in adipose tissue. On the other hand, the SCD2 transcript level is increased in adipose tissue, kidney, and lung with the starvation/refeeding cycle, while it is constitutively expressed in brain, heart, and spleen. These results indicate that SCD1 and SCD2 are distinctly regulated by the dietary conditions in the respective tissues. The authors suggested that the constitutive expression of the SCD2 gene in brain is for the supply of unsaturated fatty acids to synthesize membrane phospholipids for myelination, and that constitutive expression of the SCD1 gene in adipose tissue, and induction of the SCD2 gene in adipose tissue, kidney, and lung with the starvation/refeeding cycle reflect adipose storage in the monoenoic form and membrane phospholipid synthesis.

The availability of fatty acids in the environment controls the expression of the *OEL1* gene of *S. cerevisiae*, which encodes the Δ9 18:0-CoA desaturase responsible for the synthesis of Δ^9-16:1 and Δ^9-18:1. The Δ9 desaturase activity was increased 46% to 75% in cells supplemented with saturated fatty acids such as 16:0 and 18:0, whereas it was decreased to an undetectable level in cells fed with both Δ^9-16:1 and Δ^9-18:1, and with $\Delta^{9,12}$-18:2.[197] The decrease in the enzyme activity was accompanied by a 10-fold reduction in the level of the *OLE1* transcript, indicating that the decrease in desaturase activity can be accounted for at least partially by the down-regulation of the *OLE1* transcript level. Various unsaturated fatty acids originally present or absent in *S. cerevisiae* were examined as to their specificity toward the repression of the *OLE1* gene expression through promoter analysis of *OLE1* promoter::*lacZ* fusion gene.[198] Expression of the reporter gene activity was repressed by mono-, di-, and trienoic fatty acids, respectively, containing a double bond at the Δ9 position, indicating that expression of the *OLE1* gene is, at least partially, transcriptionally controlled by these fatty acids. Some unusual fatty acids, such as $\Delta^{9,12}$-18:2 and $\Delta^{9,12,5}$-18:3, were incorporated into membrane lipids and/or supported the growth of an *OLE1*-disrupted strain, which indicates that these fatty acids are structurally and functionally complementary to the original ones in *S. cerevisiae*. Thus, the down-regulation of *OLE1* gene expression can be regarded as a mechanism that represses the metabolic pathway becoming less necessary on supplementation with unsaturated fatty acids. Despite its abilities of incorporation into membrane lipids and sustaining *OLE1*-disruptant growth, Δ^{10}-17:1 had little effect on the reporter gene expression, but reduced the transcript level of the endogenous *OLE1* gene, indicating that the *OLE1* gene is regulated post-transcriptionally also.

On detection analysis of the *OLE1* promoter region of fusion reporter genes, a 111 bp fatty acid-regulated (FAR) sequence was identified as that responsible for repression of the transcription by unsaturated fatty acids, and also for activation of the transcription.[199] In view of the fact that fatty acids are converted to CoA-esters for incorporation into membrane lipids, the authors investigated the effects of disruption of the genes for acyl-CoA synthetases and acyl-CoA binding protein on the FAR-mediated regulation. The disruption of two genes for acyl-CoA synthetases, *FAA1* and *FAA4*, resulted in a loss of the down-regulation of the FAR-mediated reporter gene expression by unsaturated fatty acids, whereas disruption of the gene coding for acyl-CoA binding protein enhanced both the FAR-mediated activated level of transcription of the reporter gene and the transcript level of the endogenous *OLE1* gene. These results indicated the crucial role of acyl-CoA metabolism in the regulation of *OLE1* gene expression.

Post-transcriptional regulation of *OLE1* gene expression was also studied in detail with *S. cerevisiae*. The half-life of the *OLE1* transcript decreased from 10 min to less than 2.5 min on the addition of $\Delta^{9,12}$-18:2 to the cell culture.[200] The chimeric *GAL1::OLE1* construct, in which the *GAL1* promoter replaced the *OLE1* promoter in the *OLE1* gene, still exhibited an unsaturated fatty acid-induced decrease in the half-life of the transcript. However, deletion of the 5′ untranslated region of the *OLE1* gene from the chimeric construct resulted in no response to unsaturated fatty acids, indicating that the 5′ untranslated region of the *OLE1* transcript contributes to its destabilization in response to exogenously added unsaturated fatty acids.

1.02.4 DEGRADATION OF FATTY ACIDS

Fatty acids are essential components of membranes and are important sources of metabolic energy in all organisms. The contraction of cardiac muscles physiologically depends on energy derived from fatty acids. Germinating plant seeds and growing bacteria degrades fatty acids to obtain the metabolic energy and precursors required for macromolecular biosynthesis.

β-Oxidation is an essential metabolic pathway for the degradation of fatty acids. The β-oxidation system is located in mitochondria and peroxisomes in animal cells, peroxisomes (glyoxisomes) in plants, peroxisomes (microsomes) in yeasts and fungi, and the cells of bacteria. At least five enzyme activities are involved in the oxidation of fatty acids to acetyl-CoAs (Figure 13). Fatty acids are first activated to acyl-CoA by acyl-CoA synthetase coupled with ATP hydrolysis. The resulting acyl-CoA is oxidized through sequential reactions with acyl-CoA dehydrogenase (or acyl-CoA oxidase), 2-enoyl-CoA hydratase, 3-hydroxyacyl-CoA dehydrogenase, and 3-oxoacyl-CoA thiolase, and acetyl-CoA being finally liberated. Most fatty acids in nature, which have even numbers of carbon atoms, are degraded completely to acetyl-CoA by these enzymes.

1.02.4.1 Acyl-CoA Synthetase

Fatty acids imported from outside of a cell are activated by acyl-CoA synthetase prior to β-oxidation. This enzyme forms the thioester bond between the carboxyl group of a fatty acid and CoA coupled with hydrolysis of ATP to AMP and PPi. This reaction is a two-step process. In the first step, an acyl-adenylate intermediate is formed with the concomitant release of inorganic phosphate (Equation (5)), and then the acyl-adenylate is converted to acyl-CoA with the concomitant release of AMP (Equation (6)).

$$\text{RCO}_2\text{H} + \text{ATP} \rightarrow \text{RCO-AMP} + \text{PPi} \tag{5}$$

$$\text{RCO-AMP} + \text{CoASH} \rightarrow \text{RCO-SCoA} + \text{AMP} \tag{6}$$

In the mitochondrial matrix, the activities of short-chain and medium-chain acyl-CoA synthetases are found.[201–203] Long-chain acyl-CoA synthetases are located in the mitochondrial outer membranes and peroxisomes.[204] Bacterial cells have an acyl-CoA synthetase preferring long-chain acyl-CoA as a substrate.[205] Rat liver peroxisomes contain another very long-chain acyl-CoA specific enzyme.[206] The amino acid sequence deduced from the cDNA was not similar to that of other acyl-CoA synthetases except for the domains involved in adenylation and thioester formation.[207]

Nucleotide sequences coding acyl-CoA synthetases have been published so far for rat (ACS1, ACS2, ACS3, and ACS4),[208–211] mouse,[212] man,[213–216] *S. cerevisiae* (*FAA1*, *FAA2*, *FAA3*, and *FAA4*),[217–219] *E. coli* (*fadD*),[220] *Pseudomonas oleovarans* (*alkK*),[221] and *B. napus*.[222] Acyl-CoA synthetase is a member of an enzyme family that includes luciferase from firefly,[223] 4-coumarate-CoA ligase from parsley,[224] *Methanothrix soehngenii* acetyl-CoA synthetase,[225] *E. coli* 2-acylglycerophosphoethanolamine acyltransferase/acyl-ACP synthetase,[226] *Bacillus brevis* gramicidine S synthetase 1,[227] and *Penicillium chrysogenum* δ-(L-α-aminoadipyl)-L-cysteinyl-D-valine synthetase.[228,229] All of these enzymes require ATP as a cofactor, and catalyze the adenylation and thioester formation of their substrates. The amino acid sequences show that these proteins consist of two regions which are significantly conserved among the proteins in this family (Figure 14). The conserved sequence of 18 amino acids of *E. coli* acyl-CoA synthetase (FadD), between Asn-431 and Lys-455 (431-NGWLHTGDIAVMDEEGFLRIVRKK-455), was analyzed by site-directed mutagenesis.[230] The substitution of Val-451, Asp-452, Lys-454, and Lys-455 with alanine increased the activity with decanoate; however, it decreased the activity with oleate. Two enzymes, yeast Faa2p (a protein coded by *FAA2*) and *P. oleovarans* acyl-CoA synthetase, which bind to medium-chain fatty acids

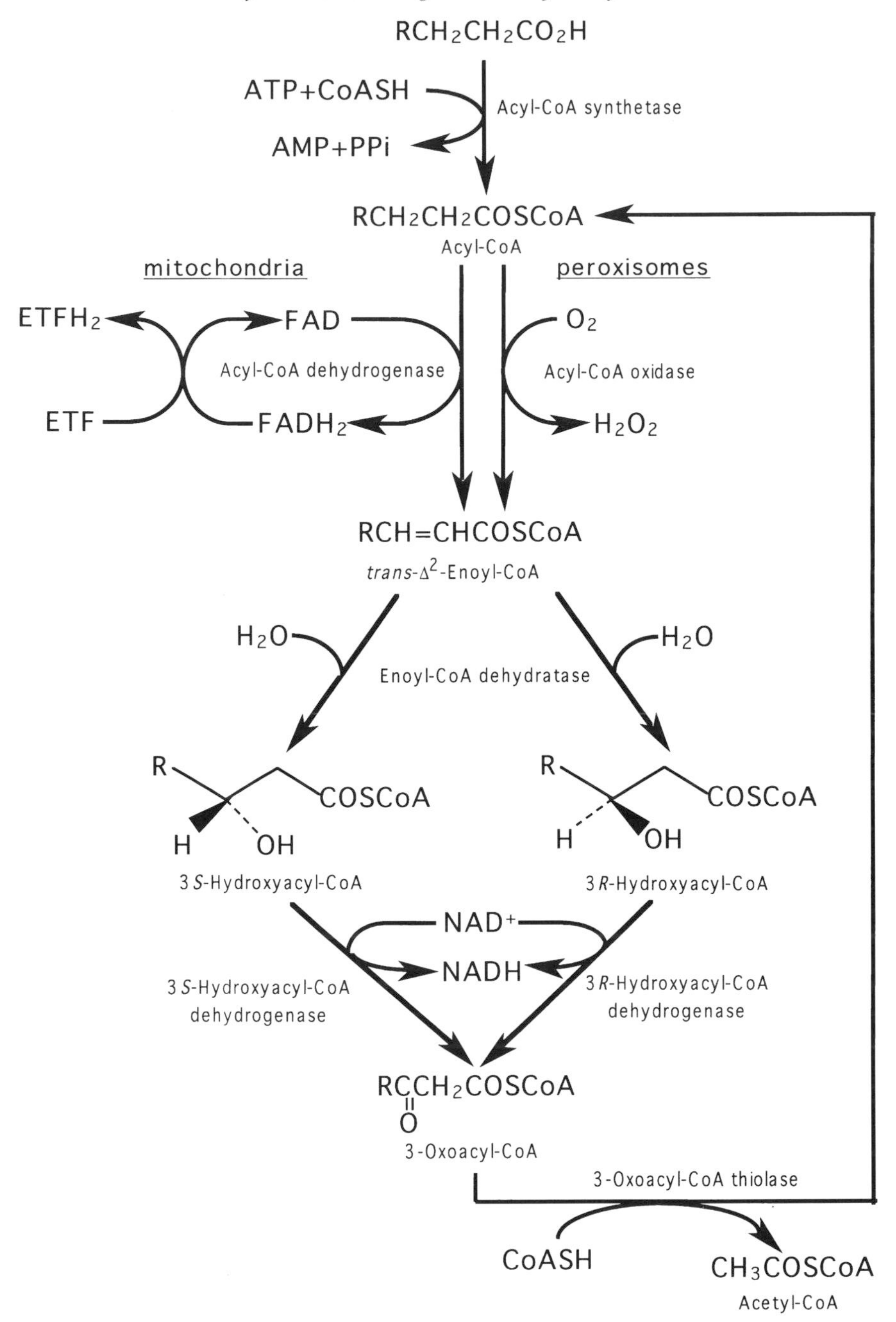

Figure 13 Reaction mechanism of the β-oxidation of fatty acids. β-Oxidation has been considered to proceed via 3*S*-hydroxyacyl-CoA, but another pathway via 3*R*-hydroxyacyl-CoA has been observed in some organisms. ETF: electron transferring flavoprotein.

as substrates, lack the conserved Lys/Arg residues corresponding to Lys-454 in FadD, suggesting Lys-454 may be involved in the acyl chain specificity of substrate binding.

S. cerevisiae has at least four acyl-CoA synthetases including Faa2p. Because a disruptant lacking the four genes of acyl-CoA synthetases still grows on a long-chain fatty acid as a sole carbon source,[219] a fifth gene may exist in *S. cerevisiae*. The properties of the four enzymes, Faa1p, Faa2p, Faa3p, and Faa4p, were demonstrated by the purified proteins.[231,232] All four proteins were expressed in an *E. coli* strain defective in *fadD*, with carboxyl-terminal histidine tags, and the purified enzymes were assayed with C_3–C_{24} fatty acids. Faa1p preferred 12:0–16:0, with highest activities for myristic (14:0) and pentadecanoic (15:0) acids. Faa2p could accommodate a wider range of acyl chain lengths, 9:0–13:0 being the most active substrates. However, Faa3p exhibited lower activities with 9:0–18:0 substrates compared with the other Faaps, and only Faa3p was able to activate 22:0 and

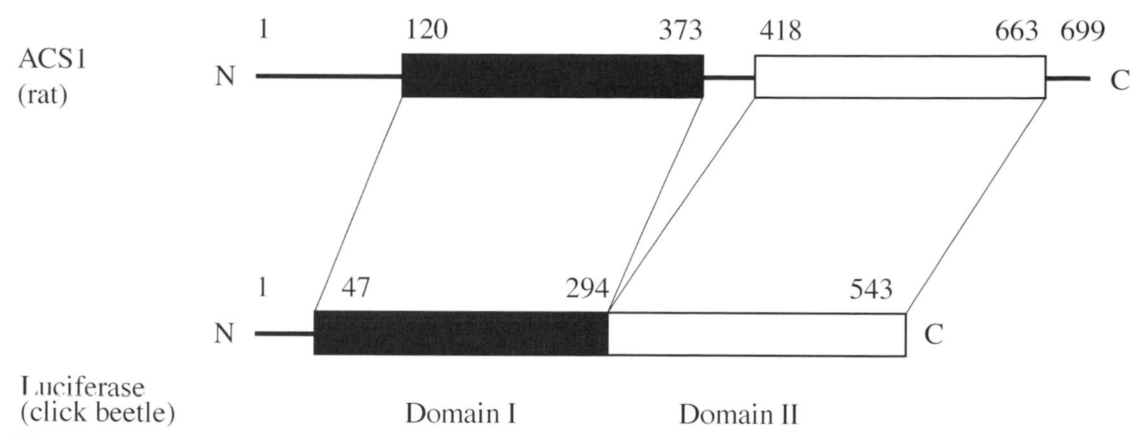

Figure 14 The amino acids in rat ACS1 compared with those in click beetle luciferase. Domains I and II (filled and opened boxes) are conserved between ACS1 and luciferase.

24:0.[231] Faa4p showed higher activities with palmitoleic (16:1) and oleic (18:1) acids.[232] These substrate specificities *in vitro* suggest that each of the four enzymes has a different role in fatty acid metabolism, not simple redundancy of the genes. Hemagglutinin-tagged Faa2p was located in peroxisomes which were fractionated by equilibrium density centrifugation.[233] The resistance to protease degradation and observation by immunoelectron microscopy showed that molecules of hemagglutinin-tagged Faa2p were located on the matrix side of the peroxisomal membranes.[233] Faa2p may activate fatty acids transported in peroxisomes through a medium-chain fatty acid transporter that is not known yet. Genetic analysis also suggested that the substrates of Faa2p are endogenous fatty acids.[219] In contrast, Faa1p and Faa4p are responsible for the activation of exogenous long-chain fatty acids. Supplementation of long-chain fatty acids to the culture medium and either Faa1p or Faa4p were required for cell viability when fatty acid synthetase was blocked with inhibitors such as celurenin.[219,234] Two acyl-CoA transport proteins in yeast peroxisomes were found to be long- (very long-) acyl CoAs, Pxa1p (Pat2p) and Pxa2p (Pat1p).[233,235] These belong to the ATP binding cassette (ABC) transporter super family,[236] possibly transporting acyl-CoA activated in other organelle by Faa1p, Faa3p, or Faa4p. Human peroxisome membranes also have homologous transporters adenoleukodystrophy protein (ALDP),[237] PMP70,[238,239] and ALDP-related protein.[240] Mutations in the ALDP gene cause X-linked adenoleukodystrophy (X-ALD),[241] which is a peroxisomal disorder involving impaired β-oxidation of very long-chain fatty acids and which leads to very severe neurological disability. Mutations in the PMP70 gene have been identified in two patients with Zellweger syndrome,[242] which is an inborn error of peroxisome biogenesis. However, their contribution to this disease remains unclear.

Acyl-CoA synthetase of *Pseudomonas fragi* was purified in the authors' laboratory.[243] The gene was cloned from a *P. fragi* genomic library, which was designated as *faoC*. The deduced amino acid sequence shows the highest similarity to that of the *E. coli* enzyme among the sequences published so far. The enzyme expressed in *E. coli* cell was active on long-chain fatty acids as substrates. Interestingly, another ORF was found upstream of *faoC*, named *faoD*. Although the sequence was remarkably similar to that of *faoC*, a polypeptide expressed in *E. coli* cells was inactive with long-straight-chain fatty acids. The expression and function of this putative protein are unknown. The gene of *fadD* is the exclusive acyl-CoA synthetase gene in *E. coli* because *fadD* mutants are not able to grow on minimal medium containing a fatty acid as a sole carbon source.

1.02.4.1.1 Carnitine carries acyl-CoA into the mitochondrial matrix

Fatty acids are activated on the outer mitochondrial membrane, whereas they are oxidized in the mitochondrial matrix. Since acyl-CoA molecules do not traverse the inner mitochondrial membrane, a special transport system is necessary. Acyl-CoAs are carried across the inner mitochondrial membrane with the aid of carnitine (Figure 15).[244] The acyl group is transferred from the sulfur atom of CoA to the hydroxyl group of carnitine to form acyl carnitine. This reaction is catalyzed by carnitine acyltransferase I.[245] Acyl carnitine is then shuttled across the inner membrane by a carnitine translocase to the matrix side of mitochondria.[246,247] The acyl group is transferred back to CoA in the mitochondrial matrix. This reaction is catalyzed by carnitine acyltransferase II.[248]

Finally, carnitine is returned to the outside of the inner membrane by the translocase. Because the *O*-ester bond of acylcarnitine is unique in having free energy similar to that of the thioester bond of acyl-CoA, this process is thermodynamically feasible.

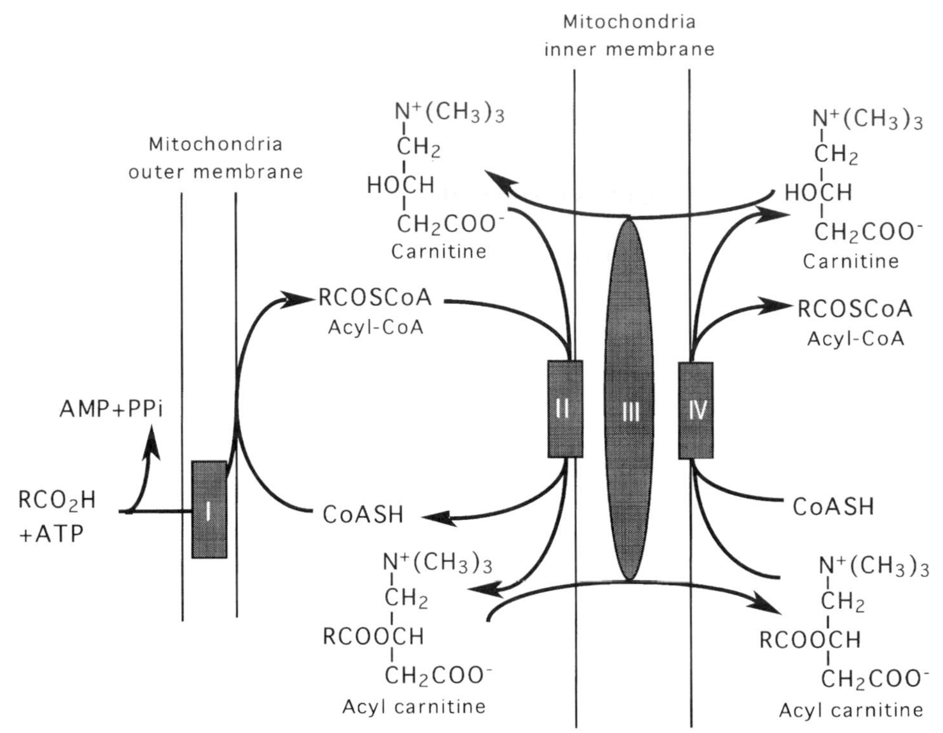

Figure 15 The entry of acyl carnitine into the mitochondrial matrix. I: Acyl-CoA synthetase; II: carnitine acyltransferase I; III: carnitine-acylcarnitine translocase; IV: carnitine acyltransferase II.

1.02.4.2 Acyl-CoA Dehydrogenase and Acyl-CoA Oxidase

The first reaction in the β-oxidation of straight-chain fatty acids is the introduction of the double bond to acyl-CoAs resulting in 2-*trans*-enoyl-CoAs. This step is catalyzed by acyl-CoA dehydrogenase in animal mitochondria and bacterial cells, but by acyl-CoA oxidase in peroxisomes of fungi and animals (Figure 13). All of these enzymes bind to FAD molecules tightly. Acyl-CoA dehydrogenase uses an electron transfer flavoprotein-quinone oxidoreductase (ETF) as an electron receptor; on the other hand, acyl-CoA oxidase donates electrons directly to molecular oxygen, thereby producing H_2O_2. A deficiency of medium-chain acyl-CoA dehydrogenase is the most common human mitochondrial β-oxidation defect.

In mitochondria, multiple acyl-CoA dehydrogenases have been demonstrated, which have different chain-length specificities. Short- (C_4–C_8), medium- (C_4–C_{16}), and long- (C_6–C_{16}) chain acyl-CoA specific proteins consisting of a homotetramer of 41–45 kDa subunits form an enzyme family located in the mitochondrial matrix. Very long-chain (C_{12}–C_{24}) acyl-CoA dehydrogenase is a homodimer of 71 kDa subunits and is located in mitochondrial inner membranes.[249] Its amino acid sequence shows the highest similarity (38%) with that of the short-chain specific enzyme.

Medium-chain acyl-CoA dehydrogenase (MCADH) is the most characterized acyl-CoA dehydrogenase. Residue Glu376 of the pig kidney enzyme (Glu401 in the precurser protein) was suggested to abstract a proton of acyl-CoA during the dehydrogenation reaction because 2-octynoyl-CoA bound to the residue covalently.[250] Site-directed mutagenesis of Glu376 → Gln indicated that the carboxyl moiety of the glutamic acid played an important role during the initial step of catalysis.[251] These findings were confirmed by the crystal structure of MCADH from pig liver. X-ray analysis at 2.4 Å resolution, with or without fatty acyl-CoA ligands, showed that the alkyl chain of the bound fatty acyl-CoA was deeply buried inside the enzyme.[252] This view of the C_α—C_β bond of acyl-CoA located between the carboxyl moiety of Glu376 and the flavin ring of FAD supported the role of Glu376 as a base abstracting the proton in the dehydrogenenation reaction. Alignment of acyl-

CoA dehydrogenases from many sources showed that residue Glu376 is conserved in not only medium-chain specific enzymes, but also short-chain acyl-CoA dehydrogenase (SCADH) and very long-chain acyl-CoA dehydrogenase (VLCADH) (Figure 16). However, long-chain acyl-CoA dehydrogenase (LCADH) had no glutamate residue at the corresponding position.[253] A molecular model of the human LCADH based on atomic coordinates of the pig MCADH suggested the functioning of Glu261 as the catalytic base in the active site.[254] The mutant enzyme having the Glu261 → Gln substitution exhibited less than 0.02% of the wild-type enzyme activity. Isovaleryl-CoA dehydrogenase (IVD), a branched-chain specific acyl-CoA dehydrogenase in amino acid metabolism, has a glutamate at 254 (261 in the sequence of LCADH). All mutant IVDs (Glu254 → Gly, Glu254 → Asp, Glu254 → Gln, and Glu254 → Gly/Ala375 → Glu) showed no significant activity except for the Asp mutant,[255] suggesting that the carboxylate of Glu254 was essential for the catalysis. A mutant of the human MCADH carrying Glu376 → Gly/Thr255 → Glu double mutations was constructed.[256,257] In the mutant enzyme, a glutamate residue moved from position 376 to position 255, which corresponds to the position of Glu261 in LCADH (Figure 16). The chain-length specificity and catalytic mechanism of long-chain acyl-CoA dehydrogenase were examined by X-ray analysis of the mutant enzyme.[257]

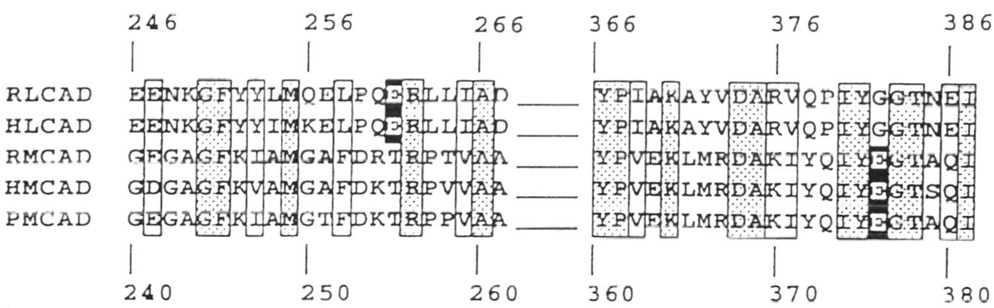

Figure 16 A portion of the sequence alignment of LCADs and MCADs.[254] The numbers above the sequences correspond to LCADs, and those below the sequences correspond to the mature forms of the MCAD enzymes. Identical residues are indicated by stippled boxes; unstippled boxes indicate similar ones; and the catalytic residues are labeled with white letters on a black background. R: rat; H: human; P: pig.

The *E. coli* genome carries a putative acyl-CoA dehydrogenase gene which has an upstream sequence motif responsible for transcriptional regulation of β-oxidation enzymes.[258,259] There is a glutamate residue at the homologous position to Glu376 of animal MCADHs.

Acyl-CoA oxidase proteins are distributed in eukaryotic peroxisomes.[260,261] The amino acid sequences show slight similarity to those of acyl-CoA dehydrogenases.[253,262,263] In animal and plant peroxisomes, there are multiple enzymes with different acyl-chain specificities, similar to those of acyl-CoA dehydrogenases. Three enzymes have been isolated from rat liver peroxisomes, i.e., palmitoyl-CoA oxidase, pristanoyl-CoA oxidase, and trihydroxycoprostanoyl-CoA oxidase.[264,265] On the other hand, human peroxisomes contain palmitoyl-CoA oxidase and blanched-chain oxidase.[266] Short-, medium-, and long-chain specific enzymes have been isolated from cucumber and maize.[267,268]

1.02.4.3 Multifunctional Protein with the Activities of 2-enoyl-CoA Hydratase, 3-hydroxyacyl-CoA Dehydrogenase, and 3-oxoacyl-CoA Thiolase

2-Enoyl-CoA, the product with acyl-CoA dehydrogenase or acyl-CoA oxidase, is hydrated to 3-hydroxyacyl-CoA by enoyl-CoA hydratase (Figure 13). Sequentially, 3-hydroxyacyl-CoA is oxidized to 3-oxoacyl-CoA by 3-hydroxyacyl-CoA dehydrogenase, and then thiolytically cleaved by oxoacyl-CoA thiolase, resulting in acetyl-CoA and acyl-CoA shortened by two carbon atoms. The resulting acyl-CoA is degraded through the β-oxidation cycle again.

In bacteria, such as *E. coli* and *P. fragi*, these three metabolic reactions are catalyzed by an enzyme complex that is called HDT or multifunctional protein (MFP).[269,270] This complex also has Δ^3,Δ^2-enoyl-CoA isomerase activity, which catalyzes the isomerization of Δ^3-*cis*-enoyl-CoA to Δ^2-*trans*-enoyl-CoA. This activity is necessary for the β-oxidation of unsaturated fatty acids. The enzyme complex consists of an α subunit ($\approx$70 kDa) and a β subunit ($\approx$40 kDa) in the ratio of

$\alpha_2\beta_2$. Animal mitochondria contain similar multifunctional proteins bound to inner membranes, which consist of $\alpha_4\beta_4$.[271] Animal peroxisomes (microsomes) and plant glyoxisomes have monomeric multifunctional (bifunctional) proteins which catalyze the hydration of 2-enoyl-CoA and the oxidation of 3*S*-hydroxyacyl-CoA, but not the thiolytic cleavage of 3-oxoacyl-CoA.[272–274] Amino acid sequences are conserved between the bacterial α subunit, mitochondrial α subunit, peroxisomal enzyme, and glyoxisomal enzyme, suggesting a common ancestry. In addition, the amino-terminal halves of these proteins show significant similarities to that of monofunctional 2-enoyl-CoA hydratase, which is located in the mitochondrial matrix.[275] On the other hand, the amino acid sequences of the carboxyl-terminal halves show similarities to the sequence of 3*S*-hydroxylacyl-CoA dehydrogenase in the mitochondrial matrix.[276,277] The carboxyl-terminal half of glyoxisomal MFP of cucumber represents a 3*S*-hydroxyacyl-CoA dehydrogenase domain (Figure 17).[278] An important residue for catalysis was suggested by site-directed mutagenesis. Residue Glu139 of the *E. coli* α subunit is conserved in the multifunctional enzymes and the monofunctional 2-enoyl-CoA hydratases. The mutation of Glu139 → Gln remarkably reduced the enoyl-CoA hydratase activity without decreasing the 3-hydroxyacyl-CoA dehydrogenase and 3-oxoacyl-CoA thiolase activities.[279]

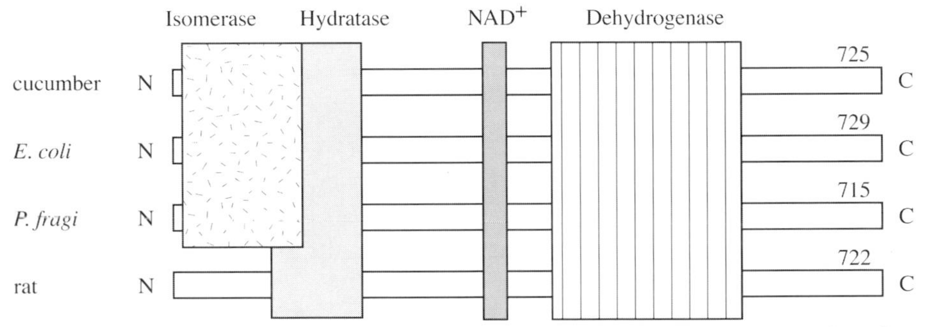

Figure 17 Schematic diagram of HDTs and MEPs indicating the regions of single enzyme domains. Dehydrogenase, 3-hydroxyacyl-CoA dehydrogenase; hydratase, 2-enoyl-CoA hydratase; isomerase, Δ^3,Δ^2-enoyl-CoA isomerase.

S. cerevisiae, *Neurospora crassa*, and *Candida tropicalis* also have a bifunctional enzyme with the activities of 2-enoyl-CoA hydratase and 3*R*-hydroxylacyl-CoA dehydrogenase.[280–282] Molecular analyses of the enzymes revealed amino acid sequences are conserved in them (42–45% identical), while they exhibit no significant homology with other multifunctional proteins described above. The first third (5–856) and second third (947–1767) of the gene for the bifunctional protein of *C. tropicalis* are 59% identical.[283] The genes of *S. cerevisiae* and *N. crassa* have the same feature, that indicates partial duplication occurred during evolution. The enzyme of *S. cerevisiae* without the one-third from the carboxyl terminus (271 amino acids) lacked hydratase activity even though it retained dehydrogenase activity, which suggested the carboxyl terminal region is responsible for the 2-enoyl-CoA hydratase activity (Figure 18).[282]

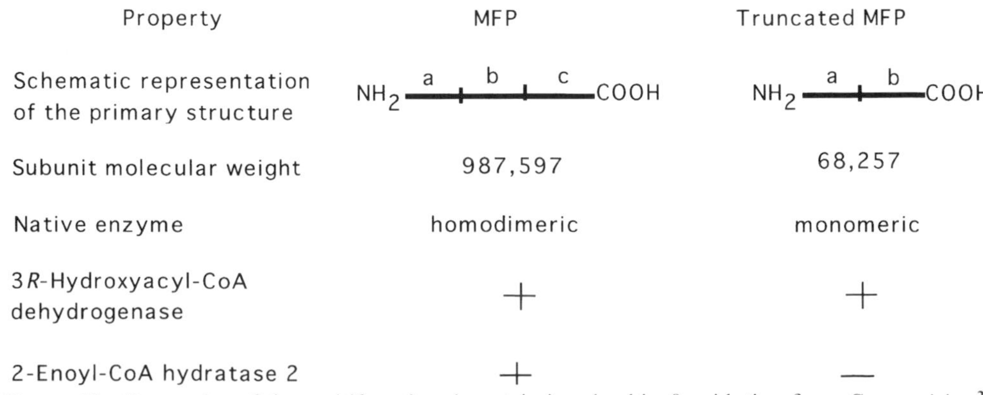

Figure 18 Properties of the multifunctional protein involved in β-oxidation from *S. cerevisiae*.[282]

Novel multifunctional proteins have been purified from animal peroxisomes.[284,285] The enzymes from man and rat were thought to be dimers of 77 kDa subunits, which carry the activity of 3*R*-

hydroxyacyl-CoA dehydrogenase but not that of 3*S*-hydroxyacyl-CoA dehydrogenase. The cloning and sequencing of the cDNAs from rat,[286] man,[287] and guinea pig[288] revealed that these proteins were homologous to fungal bifunctional proteins. The human enzyme was abundantly expressed in the peroxisomes of cultured fibroblasts and HepG2 cells.[289] Analysis of the catalytic activities toward various substrates suggested that the physiological roles of the enzymes were the oxidation of 2-methyl-blanched fatty acids and the side-chain shortening of cholesterol for bile acid formation.[289,290]

The multifunctional protein of *P. fragi* (HDT) was purified in the authors' laboratory.[291] The native complex was estimated to be close to 240 kDa, consisting of two α subunits (73 kDa) and two β subunits (42 kDa). The enzyme complex carried 2-enoyl-CoA hydratase, 3-hydroxyacyl-CoA dehydrogenase, 3-oxoacyl-CoA thiolase, Δ^3,Δ^2-enoyl-CoA isomerase, and 3*S*-hydroxyacyl-CoA epimerase activities. A β-oxidation system reconstituted *in vitro* with the purified multifunctional enzyme together with acyl-CoA synthetase and acyl-CoA oxidase completely oxidized saturated fatty acids of C_4–C_{18} chain length. The reconstituted system also degraded unsaturated fatty acids having *cis* double bonds extending from odd-numbered carbon atoms. Fatty acids having *cis* double bonds extending from even-numbered carbons were completely oxidized to acetyl-CoA when 2,4-dienoyl-CoA reductase was added to the system. These results suggest that the weak 3-hydroxyacyl-CoA epimerase activity of the HDT is not able to degrade polyenoic fatty acids. The genes encoding the α subunit (*faoA*) and β subunit (*faoB*) of HDT were isolated using antiserum raised against the purified enzyme.[291] The genes of *faoA* and *faoB* formed a single operon (*faoAB*). The deduced amino acid sequences of the α and β subunits were significantly similar to those of the same subunits of the *E. coli* enzyme.

Ishikawa *et al.*[292] overexpressed the α and β subunits individually in *E. coli* cells by means of the T7 promoter system. Each subunit formed a homodimeric complex. The α_2 complex had the activities of 2-enoyl-CoA hydratase and 3*S*-hydroxyacyl-CoA dehydrogenase, whereas the β_2 complex did not have 3-oxoacyl-CoA thiolase activity. Reconstitution of the $\alpha_2\beta_2$ complex from the β_2 complex together with the α_2 complex restored the 3-oxoacyl-CoA thiolase activity, indicating that the interface between the α_2 and β_2 complexes should be crucial for the activity of 3-oxoacyl-CoA thiolase.

The HDT activity of *P. fragi* was induced by straight-chain fatty acids in the culture medium. Long-chain fatty acids of C_{13}–C_{15} induced acyl-CoA dehydrogenase in a similar manner,[293] suggesting a common regulation mechanism for β-oxidation enzymes. A promoter assay using *lacZ* as a reporter suggested that the upstream region of *faoA* contained the transcriptional regulatory element that might bind the putative repressor. Deletion of 136 bp containing the transcriptional initiation site of the *faoAB* operon caused constitutive expression of β-galactosidase without a supply of fatty acids in the medium, suggesting that a regulatory element was located in the 136 bp fragment.[294] DiRusso and co-workers investigated a common regulatory element of genes involved in *E. coli* β-oxidation.[259,295–297] The consensus sequence, 5′-ANCTGGTCNGANCT/AGTN-3′, was found upstream of *fadBA* (multifunctional enzyme gene), *fadD* (acyl-CoA synthetase gene), and *fadL* (fatty acids transport protein gene), and was bound by a repressor protein, FadR. The inhibition of transcription was relieved by releasing the FadR protein which was bound to an acyl-CoA ligand. Because there is no FadR binding sequence in the upstream region of *faoAB* of *P. fragi*, the repressor of *P. fragi* genes may be different from in the case of FadR. The authors have cloned the acyl-CoA synthetase gene (*faoC*), as described previously, and found consensus sequences in the upstream regions of *faoAB* and *faoC*.[243] These sequences may be the elements responsible for transcriptional regulation of the β-oxidation enzymes of *P. fragi*.

1.02.5 CONCLUSIONS

The biosynthesis and oxidative degradation of fatty acids comprise complex series of integrated reactions. Many enzymes are involved in these reactions. Some of the enzymes are highly integrated to form multifunctional proteins in which different enzymes are linked covalently. Various types and different combinations of multifunctional proteins involved in both the biosynthesis and degradation of fatty acids have been observed. An advantage of this arrangement is that the syntheses of different enzymes are coordinated. Multifunctional proteins consisting of covalently organized enzymes may be more stable than ones formed through noncovalent attraction. Furthermore, intermediates of the reactions can efficiently move from one active site to another without leaving the enzyme molecules. The time for diffusion is markedly reduced and side reactions are minimized.

Fatty acid synthesis is not simply a reversal of the degradation pathway. Some important features of the pathway for the *de novo* biosynthesis of fatty acids can be summarized as follows:

(i) Synthesis takes place in the cytosol (yeast and animal) or choloroplasts (plant), in contrast with degradation, which occurs in the mitochondrial matrix or peroxisomes (glyoxisomes).

(ii) Intermediates in fatty acid synthesis are covalently bound to the sulfhydryl groups of an ACP, while the intermediates in fatty acid oxidation are linked to coenzyme A.

(iii) The growing fatty acid chain is elongated through the sequential addition of two-carbon units derived from acetyl-CoA. The activated donor of the two-carbon units in the elongation step is malonyl-ACP and the elongation reaction proceeds through the release of CO_2.

Elongation by fatty acid synthase stops upon the formation of palmitic acid. The longer fatty acids are synthesized by other enzyme systems with fatty acyl-CoAs as the substrates. Malonyl-CoA is the two-carbon donor in the elongation of acyl-CoAs. The condensation is also driven by the decarboxylation of malonyl-CoA.

The roles of the desaturases fall very clearly into two categories. The $\Delta 9$ desaturases are primarily responsible for the conversion of saturated fatty acids into monoenoic fatty acids. The other desaturases act primarily on unsaturated fatty acids and in conjunction with the elongation system supply various polyenoic fatty acids (Figure 8). The desaturases themselves are separate and distinct proteins, but they share common electron-transport proteins. The exact nature of the interaction of desaturases with the electron-transport proteins is unknown. Is cytochrome b_5, for example, free to move about in the lipid bilayer of the membrane transitorily interacting with desaturases or does each desaturase exist as a complex bound to its own electron-transport proteins?

Mammals carry acyl-CoA desaturases but not acyl-lipid desaturases. The mammalian acyl-CoA desaturases have no ability to introduce double bonds at carbon atoms beyond C-9 in the fatty acid chain. Therefore, mammals cannot synthesize linoleic and linolenic acids. These acids must be supplied in the diet and are the starting points for the synthesis of various other unsaturated fatty acids.

Studies on fatty acid oxidation have a long history. Franz Knoop made a critical contribution to elucidation of the mechanism of fatty acid oxidation at the beginning of the twentieth century. He fed dogs straight-chain fatty acids in which a phenyl group was bound to the ω carbon atom. The urine of these dogs contained a derivative of phenylacetic acid when they were fed phenylbutyrate. In contrast, a derivative of benzoic acid was formed when they were fed phenylpropionate. Phenylacetic acid was produced whenever a fatty acid containing an even number of carbon atoms was fed, whereas benzoic acid was formed whenever a fatty acid containing an odd number was fed. Knoop deduced from these findings that fatty acids are degraded through oxidation at the β carbon. These experiments were a landmark in biochemistry because they were the first to use a synthetic label (the phenyl group) to elucidate a reaction mechanism. Isotopes became available in the biochemical field several decades later. Despite the long history of the studies, the β-oxidation of fatty acids is still an exciting field of research. The stereochemical courses of the reactions depend on enzyme sources, and the organization and architecture of the enzymes concerned are different in different organisms.

1.02.6 REFERENCES

1. R. Jeffcoat, in "Essays in Biochemistry," Academic Press, London, 1979, vol. 15, p. 1.
2. A. Kawaguchi and Y. Seyama, in "Pyridine Nucleotide Coenzymes Part A," eds. D. Dolphin, R. Poulson, and O. Avramovic, Wiley, New York, 1987, vol. 2, p. 367.
3. H. G. Wood and R. E. Barden, *Annu. Rev. Biochem.*, 1977, **46**, 385.
4. S. Numa and T. Tanabe, in "Fatty Acid Metabolism and Its Regulation," ed. S. Numa, Elsevier, Amsterdam, 1984, p. 1.
5. H. Toh, H. Kondo, and T. Tanabe, *Eur. J. Biochem.*, 1993, **215**, 687.
6. F. Lim, C. P. Morris, F. Occhiodoro, and J. C. Wallace, *J. Biol. Chem.*, 1988, **263**, 11 493.
7. R. N. Perham, *Biochemistry*, 1991, **30**, 8501.
8. Y. Motokawa and G. Kikuchi, *Arch. Biochem. Biophys.*, 1974, **164**, 634.
9. T. Takai, C. Yokoyama, K. Wada, and T. Tanabe, *J. Biol. Chem.*, 1988, **263**, 2651.
10. F. Lopez-Casillas, D.-H. Bai, X. C. Luo, I.-S. Kong, M. A. Hermodoson, and K. H. Kim, *Proc. Natl. Acad. Sci. USA*, 1988, **85**, 5784.
11. L. E. Post, D. J. Post, and F. M. Raushel, *J. Biol. Chem.*, 1990, **265**, 7742.
12. Y. Seyama and A. Kawaguchi, in "Pyridine Nucleotide Coenzymes Part B," eds. D. Dolphin, R. Poulson, and O. Avramovic, Wiley, New York, 1987, vol. 2, p. 381.
13. S. J. Wakil, *Biochemistry*, 1989, **28**, 4523.
14. Y. Seyama, A. Kawaguchi, S. Okuda, and T. Yamakawa, *J. Biochem.*, 1978, **84**, 1309.
15. Y. Seyama, T. Kasama, T. Yamakawa, A. Kawaguchi, and S. Okuda, *J. Biochem.*, 1977, **81**, 1167.

16. Y. Seyama, T. Kasama, T. Yamakawa, A. Kawaguchi, K. Saito, and S. Okuda, *J. Biochem.*, 1977, **82**, 1325.
17. Y. Seyama, A. Kawaguchi, T. Kasama, K. Sasaki, K. Arai, S. Okuda, and T. Yamakawa, *Biomed. Mass Spectrom.*, 1978, **5**, 357.
18. K. Saito, A. Kawaguchi, S. Okuda, Y. Seyama, and T. Yamakawa, *Biochim. Biophys. Acta*, 1980, **618**, 202.
19. R. H. White, *Biochemistry*, 1980, **19**, 9.
20. B. Sedgwick and C. Morris, *Chem. Commun.*, **1980**, 96.
21. A. Kawaguchi, T. Yoshimura, K. Saito, Y. Seyama, T. Yamakawa, and S. Okuda, *J. Biochem.*, 1980, **88**, 1.
22. K. Saito, A. Kawaguchi, Y. Seyama, T. Yamakawa, and S. Okuda, *J. Biochem.*, 1981, **90**, 1697.
23. K. Saito, A. Kawaguchi, Y. Seyama, T. Yamakawa, and S. Okuda, *Eur. J. Biochem.*, 1981, **116**, 581.
24. A. G. McInnes, J. A. Walter, and J. L. C. Wright, *Tetrahedron*, 1983, **39**, 3515.
25. V. E. Anderson and G. G. Hammes, *Biochemistry*, 1984, **23**, 2088.
26. S. Jackowski and C. O. Rock, *J. Biol. Chem.*, 1987, **262**, 7927.
27. R. C. Clough, A. L. Matthis, S. R. Barnum, and J. G. Jaworski, *J. Biol. Chem.*, 1992, **267**, 20 992.
28. S. Jackowski, C. M. Murphy, J. E. Cronan, Jr., and C. O. Rock, *J. Biol. Chem.*, 1989, **264**, 7624.
29. J. G. Jaworski, R. C. Clough, and S. R. Barnum, *Plant Physiol.*, 1989, **90**, 41.
30. D. N. Burton, A. G. Haavik, and J. W. Porter, *Arch. Biochem. Biophys.*, 1968, **126**, 141.
31. J. K. Stoops, P. Ross, M. J. Arslanian, K. C. Aune, S. J. Wakil, and R. M. Oliver, *J. Biol. Chem.*, 1979, **254**, 7418.
32. H. Dutler, M. J. Coon, A. Kull, H. Vogel, G. Waldvogel, and V. Prelog, *Eur. J. Biochem.*, 1971, **22**, 203.
33. I. C. Kim, C. J. Unkefer, and W. C. Deal, Jr., *Arch. Biochem. Biophys.*, 1977, **178**, 475.
34. D. A. Roncari, *J. Biol. Chem.*, 1974, **249**, 7035.
35. D. A. Roncari, *Can. J. Biochem.*, 1974, **52**, 221.
36. J. K. Stoops, M. J. Arslanian, K. C. Aune, and S. J. Wakil, *Arch. Biochem. Biophys.*, 1978, **188**, 348.
37. D. B. Martin, M. G. Horning, and P. R. Vagelos, *J. Biol. Chem.*, 1961, **236**, 663.
38. S. Smith and S. Abraham, *J. Biol. Chem.*, 1970, **245**, 3209.
39. C. R. Strong, *Int. J. Biochem.*, 1972, **3**, 369.
40. J. E. Kinsella, D. Bruns, and J. P. Infante, *Lipids*, 1975, **10**, 227.
41. I. Grunnet and J. Knudsen, *Biochem. J.*, 1978, **173**, 929.
42. T. Kitamoto, M. Nishigai, A. Ikai, K. Ohashi, and Y. Seyama, *Biochim. Biophys. Acta*, 1985, **827**, 164.
43. A. M. Municio, M. A. Lizarbe, E. Relano, and J. A. Ramos, *Biochim. Biophys. Acta*, 1977, **487**, 175.
44. J. K. Stoops, E. S. Awad, M. J. Arslanian, S. Gunsberg, S. J. Wakil, and R. M. Oliver, *J. Biol. Chem.*, 1978, **253**, 4464.
45. N. M. Packter and A. Alam, *Biochim. Biophys. Acta*, 1980, **615**, 497.
46. J. Elovson, *J. Bacteriol.*, 1975, **124**, 524.
47. P. Wiesner, J. Beck, K.-F. Beck, S. Ripka, G. Muller, S. Lucke, and E. Schweizer, *Eur. J. Biochem.*, 1988, **177**, 69.
48. H. Tomoda, A. Kawaguchi, S. Omura, and S. Okuda, *J. Biochem.*, 1984, **95**, 1705.
49. N. Morishima, A. Ikai, H. Noda, and A. Kawaguchi, *Biochim. Biophys. Acta*, 1982, **708**, 305.
50. H. W. Knoche and K. E. Koths, *J. Biol. Chem.*, 1973, **248**, 3517.
51. W. I. Wood, D. O. Peterson, and K. Bloch, *J. Biol. Chem.*, 1978, **253**, 2650.
52. W. J. Lennarz, R. J. Light, and K. Bloch, *Proc. Natl. Acad. Sci. USA*, 1962, **48**, 840.
53. P. Goldman, A. W. Alberts, and P. R. Vagelos, *J. Biol. Chem.*, 1963, **238**, 1255.
54. E. L. Pugh, F. Sauer, M. Waite, R. E. Toomey, and S. J. Wakil, *J. Biol. Chem.*, 1966, **241**, 2635.
55. P. Goldman and P. R. Vagelos, *Biochem. Biophys. Res. Commun.*, 1962, **7**, 414.
56. P. H. Butterworth and K. Bloch, *Eur. J. Biochem.*, 1970, **12**, 496.
57. M. I. Ernst-Fonberg and K. Bloch, *Arch. Biochem. Biophys.*, 1971, **143**, 392.
58. R. Sirevag and R. P. Levine, *J. Biol. Chem.*, 1972, **247**, 2586.
59. J. L. Brooks and P. K. Stumpf, *Arch. Biochem. Biophys.*, 1966, **116**, 108.
60. K. Saito, A. Kawaguchi, S. Okuda, Y. Seyama, T. Yamakawa, Y. Nakamura, and M. Yamada, *Plant Cell Physiol.*, 1980, 21, 9.
61. J. L. Brooks and P. K. Stumpf, *Biochim. Biophys. Acta.*, 1965, 98, 213.
62. T. Shimakata and P. K. Stumpf, *Arch. Biochem. Biophys.*, 1982, 217, 144.
63. P. Goldman, A. W. Alberts, and P. R. Vagelos, *Biochem. Biophys. Res. Commun.*, 1961, **5**, 280.
64. P. Overath and P. K. Stumpf, *J. Biol. Chem.*, 1964, **239**, 4103.
65. P. W. Majerus, A. W. Alberts, and P. R. Vagelos, *Proc. Natl. Acad. Sci. USA*, 1965, **53**, 410.
66. S. Matsumura and P. K. Stumpf, *Arch. Biochem. Biophys.*, 1968, **125**, 932.
67. T. Takagi and C. Tanford, *J. Biol. Chem.*, 1968, **243**, 6432.
68. C. M. Amy, A. Witkowski, J. Naggert, B. Williams, Z. Randhawa, and S. Smith, *Proc. Natl. Acad. Sci. USA*, 1989, **86**, 3114.
69. A. D. McCarthy, A. Aitken, and D. G. Hardie, *Eur. J. Biochem.*, 1983, **136**, 501.
70. S. S. Chirala, R. Kasturi, M. Pazirandeh, D. T. Stolow, W.-Y. Huang, and S. J. Wakil, *J. Biol. Chem.*, 1989, **264**, 3750.
71. A. J. Poulose, R. F. Bonsall, and P. E. Kolattukudy, *Arch. Biochem. Biophys.*, 1984, **230**, 117.
72. E. Schweizer, G. Muller, L. M. Roberts, M. Schweizer, J. Rosch, P. Wiesner, J. Beck, D. Stratmann, and I. Zauner, *Fett Wiss. Technol.*, 1987, **89**, 570.
73. A. H. Mohamed, S. S. Chirala, N. H. Mody, W.-Y. Huang, and S. J. Wakil, *J. Biol. Chem.*, 1988, **263**, 12 315.
74. R. S. Hale, K. N. Jordan, and P. F. Leadlay, *FEBS Lett.*, 1987, **224**, 133.
75. G. Meurer, G. Biermann, A. Schutz, S. Harth, and E. Schweizer, *Mol. Gen. Genet.*, 1992, **232**, 106.
76. T. C. Vanaman, S. J. Wakil, and R. L. Hill, *J. Biol. Chem.*, 1986, **243**, 6420.
77. T. M. Kuo and J. B. Ohlrogge, *Arch. Biochem. Biophys.*, 1984, **234**, 290.
78. P. B. Hoj and I. Svendsen, *Carlsberg Res. Commun.*, 1983, **48**, 285.
79. J. E. Cronan, Jr. and E. P. Gelmann, *Bacteriol. Rev.*, 1975, **39**, 232.
80. H. van den Bosch, J. R. Williamson, and P. R. Vagelos, *Nature*, 1970, **28**, 338.
81. M. Ilton, A. W. Jevans, E. D. McCarthy, D. Vance, H. B. White, III, and K. Bloch, *Proc. Natl. Acad. Sci. USA*, 1971, **68**, 87.
82. F. Lynen, *BBA Libr.*, 1964, **4**, 132.
83. K. Willecke, E. Ritter, and F. Lynen, *Eur. J. Biochem.*, 1969, **8**, 503.

84. P. C. Yang, P. H. Butterworth, R. M. Bock, and J. W. Porter, *J. Biol. Chem.*, 1967, **242**, 3501.
85. E. Schweizer, B. Kniep, H. Castorph, and U. Holzner, *Eur. J. Biochem.*, 1973, **39**, 353.
86. J. K. Stoops, M. J. Arslanian, Y. H. Oh, K. C. Aune, T. C. Vanaman, and S. J. Wakil, *Proc. Natl. Acad. Sci. USA*, 1975, **72**, 1940.
87. M. J. Arslanian, J. K. Stoops, Y. H. Oh, and S. J. Wakil, *J. Biol. Chem.*, 1976, **251**, 3194.
88. S. Smith, *Biochim. Biophys. Acta*, 1971, **251**, 477.
89. Y. Tsukamoto, H. Wong, J. S. Mattick, and S. J. Wakil, *J. Biol. Chem.*, 1983, **258**, 15 312.
90. J. K. Stoops and S. J. Wakil, *J. Biol. Chem.*, 1981, **256**, 5128.
91. Y. S. Wang, W. X. Tian, and R. Y. Hsu, *J. Biol. Chem.*, 1984, **259**, 13 644.
92. P. R. Clements, R. E. Barden, P. M. Ahmad, and F. Ahmad, *Biochem. Biophys. Res. Commun.*, 1979, **86**, 278.
93. L. J. Libertini and S. Smith, *Arch. Biochem. Biophys.*, 1979, **192**, 47.
94. N. Singh, S. J. Wakil, and J. K. Stoops, *J. Biol. Chem.*, 1984, **259**, 3605.
95. Y. Tsukamoto and S. J. Wakil, *J. Biol. Chem.*, 1988, **263**, 16 225.
96. J. Mikkelsen, P. Hojrup, M. M. Rasmussen, P. Roepstorff, and J. Knudsen, *Biochem. J.*, 1985, **227**, 21.
97. M. Schweizer, L. M. Roberts, H. J. Holtke, K. Takabayashi, E. Hollerer, B. Hoffmann, G. Muller, H. Kottig, and E. Schweizer, *Mol. Gen. Genet.*, 1986, **203**, 479.
98. H. Kottig, G. Rottner, K. F. Beck, M. Schweizer, and E. Schweizer, *Mol. Gen. Genet.*, 1991, **226**, 310.
99. H. Tomoda, A. Kawaguchi, T. Yasuhara, T. Nakajima, S. Omura, and S. Okuda, *J. Biochem.*, 1984, **95**, 1712.
100. M. Schweizer, in "Fatty Acid Metabolism and Its Regulation," ed. S. Numa, Elsevier, Amsterdam, 1984, p. 59.
101. N. Morishima and A. Ikai, *J. Biochem.*, 1987, **102**, 1451.
102. A. Kawaguchi and S. Okuda, *Proc. Natl. Acad. Sci. USA*, 1977, **74**, 3180.
103. D. H. Nugteren, *Biochim. Biophys. Acta*, 1965, **106**, 280.
104. J. T. Bernert, Jr. and H. Sprecher, *J. Biol. Chem.*, 1977, **252**, 6736.
105. S. R. Keyes, J. A. Alfano, I. Jansson, and D. L. Cinti, *J. Biol. Chem.*, 1979, **254**, 7778.
106. S. R. Keyes and D. L. Cinti, *J. Biol. Chem.*, 1980, **255**, 11 357.
107. J. G. Jaworski, E. E. Goldschmidt, and P. K. Stumpf, *Arch. Biochem. Biophys.*, 1974, **163**, 769.
108. W. Hinsch, C. Klages, and W. Seubert, *Eur. J. Biochem.*, 1976, **64**, 45.
109. L. R. Kass, D. J. Brock, and K. Bloch, *J. Biol. Chem.*, 1967, **242**, 4418.
110. P. Strittmatter, L. Spatz, D. Corcoran, M. J. Rogers, B. Setlow, and R. Redline, *Proc. Natl. Acad. Sci. USA*, 1974, **71**, 4565.
111. T. Okayasu, M. Nagao, T. Ishibashi, and Y. Imai, *Arch. Biochem. Biophys.*, 1981, **206**, 21.
112. E. L. Pugh and M. Kates, *J. Biol. Chem.*, 1977, **252**, 68.
113. N. Murata and H. Wada, *Biochem. J.*, 1995, **308**, 1.
114. P. G. Roughan and C. R. Slack, *Annu. Rev. Plant Physiol.*, 1982, **33**, 97.
115. T. A. McKeon and P. K. Stump, *J. Biol. Chem.*, 1982, **257**, 12 141.
116. J. B. Ohlrogge, D. N. Kuhn, and P. K. Stump, *Proc. Natl. Acad. Sci. USA*, 1979, **76**, 1194.
117. J. Browse, P. McCourt, and C. R. Somerville, *Science*, 1985, **227**, 763.
118. J. Maroc, A. Tremolieres, J. Garnier, and D. Guyon, *Biochim. Biophys. Acta*, 1987, **893**, 91.
119. L. Kunst, J. Browse, and C. R. Somerville, *Plant Physiol.*, 1989, **90**, 943.
120. J. Browse, L. Kunst, S. Anderson, S. Hugly, and C. R. Somerville, *Plant Physiol.*, 1989, **90**, 522.
121. N. Sato, M. Tsuzuki, Y. Matsuda, T. Ehara, T. Osafune, and A. Kawaguchi, *Eur. J. Biochem.*, 1995, **230**, 987.
122. J. Browse, P. McCourt, and C. R. Somerville, *Plant Physiol.*, 1986, **81**, 859.
123. H. Scmidt, T. Dresselhaus, F. Buck, and E. Heinz, *Plant Mol. Biol.*, 1994, **26**, 631.
124. H. Schmidt and E. Heinz, *Biochem. J.*, 1993, **289**, 777.
125. M. Miquel and J. Browse, *J. Biol. Chem.*, 1992, **267**, 1502.
126. J. Browse, M. McConn, D. James, Jr., and M. Miquel, *J. Biol. Chem.*, 1993, **268**, 16 345.
127. M. A. Smith, A. R. Cross, O. T. Jones, W. T. Griffiths, S. Stymne, and K. Stobart, *Biochem. J.*, 1990, **272**, 23.
128. E. V. Kearns, S. Hugly, and C. R. Somerville, *Arch. Biochem. Biophys.*, 1991, **284**, 431.
129. K. Watanabe, C. Ishikawa, K. Yazawa, K. Kondo, and A. Kawaguchi, *J. Mar. Biotechnol.*, 1996, **4**, 104.
130. K. Watanabe, K. Yazawa, K. Kondo, and A. Kawaguchi, *J. Biochem.*, 1997, **122**, 467.
131. G. J. Schroepher, Jr. and K. Bloch, *J. Biol. Chem.*, 1965, **240**, 54.
132. L. J. Morris, R. V. Harris, W. Kelly, and A. T. James, *Biochem. Biophys. Res. Commun.*, 1967, **28**, 904.
133. L. J. Morris, R. V. Harris, W. Kelly, and A. T. James, *J. Biochem.*, 1968, **109**, 673.
134. L. J. Morris, *Biochem. Biophys. Res. Commun.*, 1967, **29**, 311.
135. D. Howling, L. J. Morris, and A. T. James, *Biochim. Biophys. Acta*, 1968, **152**, 224.
136. S. Higashi and N. Murata, *Plant Physiol.*, 1993, **102**, 1275.
137. D. Howling, L. J. Morris, M. I. Gurr, and A. T. James, *Biochim. Biophys. Acta*, 1972, **260**, 10.
138. M. A. Thiede and P. Strittmatter, *J. Biol. Chem.*, 1985, **260**, 14 459.
139. M. A. Thiede, J. Ozols, and P. Strittmatter, *J. Biol. Chem.*, 1986, **261**, 13 230.
140. J. M. Ntambi, S. A. Buhrow, K. H. Kaestner, R. J. Christy, E. Sibley, T. J. Kelly, Jr., and M. D. Lane, *J. Biol. Chem.*, 1988, **263**, 17 291.
141. K. Kaestner, J. M. Ntambi, T. J. Kelly, Jr., and M. D. Lane, *J. Biol. Chem.*, 1989, **264**, 14 755.
142. K. Mihara, *J. Biochem.*, 1990, **108**, 1022.
143. S. Nakashima, Y. Zhao, and Y. Nozawa, *Biochem. J.*, 1996, **317**, 29.
144. J. E. Stukey, V. M. McDonough, and C. E. Martin, *J. Biol. Chem.*, 1990, **265**, 20 144.
145. A. G. Mitchell and C. E. Martin, *J. Biol. Chem.*, 1995, **270**, 29 766.
146. J. P. Spychalla, A. J. Kinney, and J. Browse, *Proc. Natl. Acad. Sci. USA*, 1997, **94**, 1142.
147. T. Sakamoto, H. Wada, I. Nishida, M. Ohmori, and N. Murata, *J. Biol. Chem.*, 1994, **269**, 25 576.
148. J. Shanklin and C. Somerville, *Proc. Natl. Acad. USA*, 1991, **88**, 2510.
149. G. A. Thompson, D. E. Scherer, S. F.-V. Aken, J. W. Kenny, H. L. Young, D. K. Shintani, J. C. Kridl, and V. C. Knauf, *Proc. Natl. Acad Sci. USA*, 1991, **88**, 2578.
150. H. Wada, Z. Gombos, and N. Murata, *Nature*, 1990, **347**, 200.
151. T. Sakamoto, H. Wada, I. Nishida, M. Ohmori, and N. Murata, *Plant Mol. Biol.*, 1994, **24**, 643.

152. S. Hugly, J. Kunst, J. Browse, and C. R. Somerville, *Plant Physiol.*, 1989, **90**, 1134.
153. H. Schmidt, T. Dresselhaus, F. Buck, and E. Heinz, *Plant Mol. Biol.*, 1994, **26**, 631.
154. W. D. Hitz, T. J. Carlson, J. R. Booth, Jr., A. J. Kinney, K. L. Stecca, and N. S. Yadav, *Plant Physiol.*, 1994, **105**, 635.
155. D. L. Falcone, S. Gibson, B. Lemieux, and C. R. Somerville, *Plant Physiol.*, 1994, **106**, 1453.
156. N. Sato, S. Fujiwara, A. Kawaguchi, and M. Tsuzuki, *J. Biochem.*, 1997, **122**, 1224.
157. J. Okuley, J. Lightner, K. Feldmann, N. Yadav, E. Lark, and J. Browse, *Plant Cell*, 1994, **6**, 147.
158. E. P. Heppard, A. J. Kinney, K. L. Stecca, and G. H. Miao, *Plant Physiol.*, 1996, **110**, 311.
159. T. Sakamoto, D. A. Los, S. Higashi, H. Wada, I. Nishida, M. Ohmori, and N. Murata, *Plant Mol. Biol.*, 1994, **26**, 249.
160. N. S. Yadav, A. Wierzbicki, M. Aegerter, C. S. Caster, L. Perez-Grau, A. J. Kinney, W. D. Hitz WD, J. R. Booth, Jr., B. Schweiger, and K. L. Stecca, *Plant Physiol.*, 1993, **103**, 467.
161. K. Iba, S. Gibson, T. Nishiuchi, T. Fuse, M. Nishimura, V. Arondel, S. Hugly, and C. Somerville, *J. Biol. Chem.*, 1993, **268**, 24 099.
162. S. Gibson, V. Arondel, K. Iba, and C. Somerville, *Plant Physiol.*, 1994, **106**, 1615.
163. V. Arondel, B. Lemieux, I. Hwang, S. Gibson, H. M. Goodman, and C. R. Somerville, *Science*, 1992, **258**, 1353.
164. E. B. Cahoon, J. Shanklin, and J. B Ohlrogee, *Proc. Natl. Acad. Sci. USA*, 1992, **89**, 11 184.
165. E. B. Cahoon, A. M. Cranmer, J. Shanklin, and J. B. Ohlrogge, *J. Biol. Chem.*, 1994, **269**, 27 519.
166. D. J. Schultz, E. B. Cahoon, J. Shanklin, R. Craig, D. L. Cox-Foster, R. O. Mumma, and J. I. Medford, *Proc. Natl. Acad. Sci. USA*, 1996, **93**, 8771.
167. A. S. Reddy, M. L. Nuccio, L. M. Gross, and T. L. Thomas, *Plant Mol. Biol.*, 1993, **22**, 293
168. O. Sayanova, M. A. Smith, P. Lapinskas, A. K. Stobart, G. Dobson, W. W. Christie, P. R. Shewry, and J. A. Napier, *Proc. Natl. Acad. Sci. USA*, 1997, **94**, 4211.
169. J. Nagai and K. Bloch, *J. Biol. Chem.*, 1968, **243**, 4626.
170. B. G. Fox, J. Shanklin, C. Somerville, and E. Munck, *Proc. Natl. Acad. Sci. USA*, 1993, **90**, 2486.
171. B. G. Fox, J. Shanklin, J. Ai, T. M. Loehr, and J. Sanders-Loehr, *Biochemistry*, 1994, **33**, 12 776.
172. Y. Lindqvist, W. Huang, G. Schneider, and J. Shanklin, *EMBO J*, 1996, **15**, 4081.
173. J. Sanders-Loehr, in "Ion Carriers and Ion Proteins," ed. T. M. Loehr, VCH, New York, 1989, p. 375.
174. J. E. Stukey, V. M. McDonough, and C. E. Martin, *J. Biol. Chem.*, 1990, **265**, 20 144.
175. F. H. Arnold and B. L. Haymore, *Science*, 1991, **252**, 1796.
176. J. Shanklin, E. Whittle, and B. G. Fox, *Biochemistry*, 1994, **33**, 12 787.
177. M. H. Avelange-Macherel, D. Macherel, H. Wada, and N. Murata, *FEBS Lett.*, 1995, **361**, 111.
178. L. C. Stewart and M. P. Yaffe, *J. Cell Biol.*, 1991, **115**, 1249.
179. R. M. Leech, M. G. Rumsby, and W. W. Thomson, *Plant Physiol.*, 1973, **52**, 240.
180. P. McCourt, L. Kunst, J. Browse, and C. R. Somerville, *Plant Physiol.*, 1987, **84**, 353.
181. P. H. Elmer, A. J. Kinney, K. L. Stecca, and G. H. Miao, *Plant Physiol.*, 1996, **110**, 311.
182. T. Nishiuchi, T. Nakamura, T. Abe, H. Kodama, M. Nishimura, and K. Iba, *Plant Mol. Biol.*, 1995, **29**, 599.
183. C. E. Martin, K. Hiramatsu, Y. Kitajima, Y. Nozawa, L. Skriver, and G. A. Thompson, Jr., *Biochemistry*, 1976, **15**, 5218.
184. I. Horvath, Z. Torok, L. Vigh, and M. Kates, *Biochim. Biophys. Acta*, 1991, **1085**, 126.
185. J. Browse and C. R. Slack, *Biochim. Biophys. Acta*, 1983, **753**, 145.
186. D. A. Los, M. K. Ray, and N. Murata, *Mol. Microbiol.*, 1997, **25**, 1167.
187. T. Sakamoto and D. A. Bryant, *Mol. Microbiol.*, 1997, **23**, 1281.
188. R. Pearcy, *Plant Physiol.*, 1978, **61**, 484.
189. J. X. Raison, J. K. M. Roberts, and J. A. Berry, *Biochim. Biophys. Acta*, 1978, **688**, 218.
190. N. Sato, K. Sonoike, A. Kawaguchi, and M. Tsuzuki, *J. Photochem. Photobiol.*, 1996, **36**, 333.
191. L. Kunst, J. Browse, and C. Somerville, *Plant Physiol.*, 1989, **91**, 401.
192. P. G. Thomas, P. J. Domony, L. Vigh, A. R. Mansourian, P. J. Quinn, and W. P. Williams, *Biochim. Biophys. Acta*, 1986, **849**, 131.
193. N. Oshino, Y. Imai, and R. Sato, *J. Biochem.*, 1971, **69**, 155.
194. O. Mercuri, R. O. Peluffo, and M. E. De Tomas, *Biochim. Biophys. Acta*, 1974, **369**, 264.
195. N. Oshino and R. Sato, *Arch. Biochem. Biophysiol.*, 1972, **149**, 369.
196. M. R. Prasad and V. C. Joshi, *J. Biol. Chem.*, 1979, **254**, 997.
197. M. A. Bossie and C. E. Martin, *J. Bacteriol.*, 1989, **171**, 6409.
198. V. M. McDonough, J. E. Stukey, and C. E. Martin, *J. Biol. Chem.*, 1992, **267**, 5931.
199. J.-Y. Choi, J. Stukey, S. Y. Hwang, and C. E. Martin, *J. Biol. Chem.*, 1996, **271**, 3581.
200. C. I. Gonzalez and C. E. Martin, *J. Biol. Chem.*, 1996, **271**, 25 801.
201. E. Bergman, R. Reid, M. Murray, J. Brockway, and F. Whitelaw, *Biochem. J.*, 1965, **97**, 53.
202. H. Mahler, S. J. Wakil, and R. Bock, *J. Biol. Chem.*, 1953, **204**, 453.
203. P. G. Killenberg, E. D. Davidson, and L. T. Webster, Jr., *Mol. Pharmacol.*, 1971, **7**, 260.
204. S. Miyazawa, T. Hashimoto, and S. Yokota, *J. Biochem.*, 1985, **98**, 723.
205. K. Kameda, L. K. Suzuki, and Y. Imai, *Biochim. Biophys. Acta*, 1985, **840**, 29.
206. Y. Uchida, N. Kondo, T. Orii, and T. Hashimoto, *J. Biochem.*, 1996, **119**, 565.
207. A. Uchiyama, T. Aoyama, K. Kamijo, Y. Uchida, N. Kondo, T. Orii, and T. Hashimoto, *J. Biol. Chem.*, 1996, **271**, 30 360.
208. H. Suzuki, Y. Kawarabayashi, J. Kondo, T. Abe, K. Nishikawa, S. Kimura, T. Hashimoto, and T. Yamamoto, *J. Biol. Chem.*, 1990, **265**, 8681.
209. T. Fujino and T. Yamamoto, *J. Biochem.*, 1992, **111**, 197.
210. T. Fujino, M. J. Kang, H. Suzuki, H. Iijima, and T. Yamamoto, *J. Biol. Chem.*, 1996, **271**, 16 748.
211. M. J. Kang, T. Fujino, H. Sasano, H. Minekura, N. Yabuki, H. Nagura, H. Iijima, and T. Yamamoto, *Proc. Natl. Acad. Sci. USA*, 1997, **94**, 2880.
212. J. Berger, C. Truppe, H. Neumann, and S. Fross-Petter, *FEBS Lett.*, 1998, **425**, 305.
213. T. Abe, T. Fujino, R. Fukuyama, S. Minoshima, N. Shimizu, H. Toh, H. Suzuki, and T. Yamamoto, *J. Biochem.*, 1992, **111**, 123.
214. B. Ghosh, E. Barbosa, and I. Singh, *Mol. Cell. Biochem.*, 1995, **151**, 77.

215. M. Piccini, F. Vitelli, M. Bruttini, B. R. Pober, J. J. Jonsson, M. Villanova, M. Zollo, G. Borsani, A. Ballabio, and A. Renieri, *Genomics*, 1998, **47**, 350.
216. Y. Cao, E. Traer, G. A. Zimmerman, T. M. McIntyre, and S. M. Prescott, *Genomics*, 1998, **49**, 327.
217. R. J. Duronio, L. J. Knoll, and J. I. Gordon, *J. Cell Biol.*, 1992, **117**, 515.
218. D. R. Johnson, L. J. Knoll, N. Rowley, and J. I. Gordon, *J. Biol. Chem.*, 1994, **269**, 18 037.
219. D. R. Johnson, L. J. Knoll, D. E. Levin, and J. I. Gordon, *J. Cell Biol.*, 1994, **127**, 751.
220. P. N. Black, C. C. DiRusso, A. K. Metzger, and T. L. Heimert, *J. Biol. Chem.*, 1992, **267**, 25 513.
221. J. B. van Beilen, G. Eggink, H. Enequist, R. Bos, and B. Witholt, *Mol. Microbiol.*, 1992, **6**, 3121.
222. M. Fulda, E. Heinz, and F. P. Wolter, *Plant Mol. Biol.*, 1997, **33**, 911.
223. J. R. de Wet, K. V. Wood, M. DeLuca, D. R. Helinski, and S. Subramani, *Mol. Cell. Biol.*, 1987, **7**, 725.
224. E. Lozoya, H. Hoffmann, C. Douglas, W. Schulz, D. Scheel, and K. Hahlbrock, *Eur. J. Biochem.*, 1988, **176**, 661.
225. R. I. Eggen, A. C. Geerling, A. B. Boshoven, and W. M. de Vos, *J. Bacteriol.*, 1991, **173**, 6383.
226. S. Jackowski, P. D. Jackson, and C. O. Rock, *J. Biol. Chem.*, 1994, **269**, 2921.
227. K. Hori, Y. Yamamoto, T. Minetoki, T. Kurotsu, M. Kanda, S. Miura, K. Okamura, J. Furuyama, and Y. Saito, *J. Biochem.*, 1989, **106**, 639.
228. D. J. Smith, A. J. Earl, and G. Turner, *EMBO J.*, 1990, **9**, 2743.
229. B. Diez, S. Gutierrez, J. L. Barredo, P. van Solingen, L. H. van der Voort, and J. F. Martin *J. Biol. Chem.*, 1990, **265**, 16 358.
230. P. N. Black, Q. Zhang, J. D. Weimar, and C. C. DiRusso, *J. Biol. Chem.*, 1997, **272**, 4896.
231. L. J. Knoll, D. R. Johnson, and J. I. Gordon, *J. Biol. Chem.*, 1994, **269**, 16 348.
232. L. J. Knoll, O. F. Schall, I. Suzuki, G. W. Gokel, and J. I. Gordon, *J. Biol. Chem.*, 1995, **270**, 20 090.
233. E. H. Hettema, C. W. T. van Roermund, B. Distel, M. van den Berg, C. Vilela, C. Rodrigues-Pousada, R. J. A. Wanders, and H. F. Tabak, *EMBO J.*, 1996, **15**, 3813.
234. L. J. Knoll, D. R. Johnson, and J. I. Gordon, *J. Biol. Chem.*, 1995, **270**, 10 861.
235. N. Shani and Valle, D. *Proc. Natl. Acad. Sci. USA*, 1996, **93**, 11 901.
236. C. F. Higgins, *Annu. Rev. Cell Biol.*, 1992, **8**, 67.
237. J. Mosser, A. Douar, C. Sarde, P. Kioschis, R. Feil, H. Moser, A. M. Poustka, J. L. Mandel, and P. Aubourg, *Nature*, 1993, **361**, 726.
238. K. Kamijo, T. Kamijo, I. Ueno, T. Osumi, and T. Hashimoto, *Biochim. Biophys. Acta*, 1992, **1129**, 323.
239. K. Kamijo, S. Taketani, S. Yokota, T. Osumi, and T. Hashimoto, *J. Biol. Chem.*, 1990, **265**, 4534.
240. G. Lombard-Platet, S. Savary, C. O. Sarde, J. J. Mandel, and G. Chimini, *Proc. Natl. Acad. Sci. USA*, 1996, **93**, 1265.
241. J. F. Lu, A. M. Lawler, P. A. Watkins, J. M. Powers, A. B. Moser, H. W. Moser, and K. D. Smith, *Proc. Natl. Acad. Sci. USA*, 1997, **94**, 9366.
242. J. Gärtner, H. Moser, and D. Valle, *Nat. Genet.*, 1992, **1**, 16.
243. K. Yamaguchi, A. Iwamoto-Kihara, and A. Kawaguchi, in preparation.
244. J. Bremer and H. Osmundsen, in "Fatty Acid Metabolism and Its Regulation," ed. S. Numa, Elsevier, Amsterdam, 1984, p. 113.
245. K. R. Norum and J. Bremer, *J. Biol. Chem.*, 1967, **242**, 407.
246. S. V. Pande, *Proc. Natl. Acad. Sci. USA*, 1975, **72**, 883.
247. R. R. Ramsay and P. K. Tubbs, *FEBS Lett.*, 1975, **54**, 21.
248. J. F. Chase and P. K. Tubbs, *Biochem. J.*, 1972, **129**, 55.
249. K. Izai, Y. Uchida, T. Orii, S. Yamamoto, and T. Hashimoto, *J. Biol. Chem.*, 1992, **267**, 1027.
250. P. J. Powell and C. Thorpe, *Biochemistry*, 1988, **27**, 8022.
251. P. Bross, S. Engst, A. W. Strauss, D. P. Kelly, I. Rasched, and S. Ghisla, *J. Biol. Chem.*, 1990, **265**, 7116.
252. J. J. Kim, M. Wang, and R. Paschke, *Proc. Natl. Acad. Sci. USA*, 1993, **90**, 7523.
253. Y. Matsubara, Y. Indo, E. Naito, H. Ozasa, R. Glassberg, J. Vockley, Y. Ikeda, J. Kraus, and K. Tanaka, *J. Biol. Chem.*, 1989, **264**, 16 321.
254. S. Djordjevic, Y. Dong, R. Paschke, F. E. Frerman, A. W. Strauss, and J. J. Kim, *Biochemistry*, 1994, **33**, 4258.
255. A. W. Mohsen and J. Vockley, *Biochemistry*, 1995, **34**, 10 146.
256. A. Nandy, V. Kieweg, F. G. Krautle, P. Vock, B. Kuchler, P. Bross, J. J. Kim, I. Rasched, and S. Ghisla, *Biochemistry*, 1996, **35**, 12 402.
257. H. J. Lee, M. Wang, R. Paschke, A. Nandy, S, Ghisla, and J. J. Kim, *Biochemistry*, 1996, **35**, 12 412.
258. A. Farewell, A. A. Diez, C. C. DiRusso, and T. Nyström, *J. Bacteriol.*, 1996, **178**, 6443.
259. P. N. Black and C. C. DiRusso, *Biochim. Biophys. Acta*, 1994, **1210**, 123.
260. T. Takahashi, *Ann. NY Acad. Sci.*, 1996, **804**, 86.
261. G. P. Mannaerts and P. P. van Veldhoven, *Ann. NY Acad. Sci.*, 1996, **804**, 99.
262. S. Miyazawa, H. Hayashi, M. Hijikata, N. Ishii, S. Furuta, H. Kagamiyama, T. Osumi, and T. Hashimoto, *J. Biol. Chem.*, 1987, **262**, 8131.
263. J. I. Pedersen, G. Eggertsen, U. Hellman, U. Andersson, and I. Bjorkhem, *J. Biol. Chem.*, 1997, **272**, 18 481.
264. T. Osumi, T. Hashimoto, and N. Ui, *J. Biochem.*, 1980, **87**, 1735.
265. P. P. van Veldhoven, G. Vanhove, S. Assselberghs, H. J. Eyssen, and G. P. Mannaerts, *J. Biol. Chem.*, 1992, **267**, 20 065.
266. G. F. Vanhove, P. P. van Veldhoven, M. Fransen, S. Denis, H. J. Eyssen, R. J. Wanders, and G. P. Mannaerts, *J. Biol. Chem.*, 1993, **268**, 10 335.
267. T. Kirsch, H. G. Loffler, and H. Kindl, *J. Biol. Chem.*, 1986, **261**, 8570.
268. M. A. Hooks, K. Bode, and I. Couée, *Biochem. J.*, 1996, **320**, 607.
269. S. Pawar and H. Schulz, *J. Biol. Chem.*, 1981, **256**, 3894.
270. S. Imamura, S. Ueda, M. Mizugaki, and A. Kawaguchi, *J. Biochem.*, 1990, **107**, 184.
271. Y. Uchida, K. Izai, T. Orii, and T. Hashimoto, *J. Biol. Chem.*, 1992, **267**, 1034.
272. T. Osumi and T. Hashimoto, *Biochem. Biophys. Res. Commun.*, 1979, **89**, 580.
273. S. Furuta, S. Miyazawa, T. Osumi, T. Hashimoto, and N. Ui, *J. Biochem.*, 1980, **88**, 1059.
274. W. Behrends, K. Engeland, and H. Kindl, *Arch. Biochem. Biophys.*, 1988, **263**, 161.
275. N. Minami-Ishii, S. Taketani, T. Osumi, and T. Hashimoto, *Eur. J. Biochem.*, 1989, **185**, 73.

276. K. G. Bitar, A. Perez-Aranda, and R. A. Bradshaw, *FEBS Lett.*, 1980, **116**, 196.
277. N. Ishii, M. Hijikata, T. Osumi, and T. Hashimoto, *J. Biol. Chem.*, 1987, **262**, 8144.
278. R. Preisig-Müller, K. Gühnemann-Schäfer, and H. Kindl, *J. Biol. Chem.*, 1994, **269**, 20 475.
279. S. Yang, X. Y. He, and H. Schulz, *Biochemistry*, 1995, **34**, 6441.
280. M. de la Garza, U. Schultz-Borchard, J. W. Crabb, and W. H. Kunau, *Eur. J. Biochem.*, 1985, **148**, 285.
281. R. Thieringer and W. H. Kunau, *J. Biol. Chem.*, 1991, **266**, 13 110.
282. J. K. Hiltunen, B. Wenzel, A. Beyer, R. Erdmann, A. Fosså, and W. H. Kunau, *J. Biol. Chem.*, 1992, **267**, 6646.
283. W. M. Nuttley, J. D. Aitchison, and R. A. Rachubinski, *Gene*, 1988, **69**, 171.
284. L. L. Jiang, A. Kobayashi, H. Matsuura, H. Fukushima, and T. Hashimoto, *J. Biochem.*, 1996, **120**, 624.
285. L. L. Jiang, S. Miyazawa, and T. Hashimoto, *J. Biochem.*, 1996, **120**, 633.
286. Y. M. Qin, M. H. Poutanen, H. M. Helander, A. P. Kvist, K. M. Siivari, W. Schmitz, E. Conzelmann, U. Hellman, and J. K. Hiltunen, *Biochem. J.*, 1997, **321**, 21.
287. J. Adamski, T. Normand, F. Leenders, D. Monte, A. Begue, D. Stehelin, P. W. Jungblut, and Y. de Launoit, *Biochem. J.*, 1995, **311**, 437.
288. F. Caira, M. C. Clémencet, M. Cherkaoui-Malki, M. Dieuaide-Noubhani, C. Pacot, P. P. van Veldhoven, and N. Latruffe, *Biochem. J.*, 1998, **330**, 1361.
289. L. L. Jiang, T. Kurosawa, M. Sato, Y. Suzuki, and T. Hashimoto, *J. Biochem.*, 1997, **121**, 506.
290. M. Dieuaide-Noubhani, S. Asselberghs, G. P. Mannaerts, and P. P. van Veldhoven, *Biochem. J.*, 1997, **325**, 367.
291. S. Sato, M. Hayashi, S. Imamura, Y. Ozeki, and A. Kawaguchi, *J. Biochem.*, 1992, **111**, 8.
292. M. Ishikawa, Y. Mikami, J. Usukura, H. Iwasaki, H. Shinagawa, and K. Morikawa, *Biochem. J.*, 1997, **328**, 815.
293. S. Sato, S. Imamura, Y. Ozeki, and A. Kawaguchi, *J. Biochem.*, 1992, **111**, 16.
294. S. Sato, Y. Ozeki, and A. Kawaguchi, *J. Biochem.*, 1994, **115**, 286.
295. C. C. DiRusso, T. L. Heimert, and A. K. Metzger, *J. Biol. Chem.*, 1992, **267**, 8685.
296. N. Raman and C. C. DiRusso, *J. Biol. Chem.*, 1995, **270**, 1092.
297. N. Raman, P. N. Black, and C. C. DiRusso, *J. Biol. Chem.*, 1997, **272**, 30 645.

1.03
Biosynthesis of Cyclic Fatty Acids Containing Cyclopropyl-, Cyclopentyl-, Cyclohexyl-, and Cycloheptyl-rings

BRADLEY S. MOORE and HEINZ G. FLOSS
University of Washington, Seattle, WA, USA

1.03.1 INTRODUCTION

The occurrence of carbocyclic fatty acids is very significant in specific genera of bacteria and plants. Their apparent functional difference, along with branched-chain fatty acids, from straight-chain fatty acids is their effect on membrane fluidity. Membrane lipids containing straight-chain fatty acids are adjusted to their appropriate fluidity by unsaturated fatty acids, whereas the percentage of cyclic and/or branched-chain fatty acids governs the membrane fluidity in other systems. In some cases, cyclic fatty acids are essential for cell survival, as they provide a dense membrane structure that enables certain bacteria to thrive under extreme environmental conditions.

ω-Alicyclic fatty acids such as ω-cyclopentenyl, ω-cyclohexyl, and ω-cycloheptyl fatty acids are biosynthesized during log growth from cyclic starter units corresponding to their alicyclic carbonyl-CoA thioesters. The mechanism for chain extension involves malonyl-CoA in essentially the same manner as for branched- and straight-chain fatty acids. The only difference between the pathways thus involves the different primers. For the most part, ω-alicyclic fatty acid synthetases are not very specific and accept other cycloalkyl carboxylic acid CoA thioesters as well as branched short chain acyl-CoA thioesters, but not acetyl CoA, as primer units.[1] Detailed biosynthetic pathways for some of these cyclic carboxylate primers have been examined and are reviewed in this chapter.

Conversely, cyclopropyl fatty acids are vastly different from the ω-alicyclic fatty acids in structure and in biosynthesis. The three-membered carbocycle is typically positioned mid-chain, rather than at the ω-position, 9–11 carbon atoms removed from the carbonyl group. They are biosynthesized by a postsynthetic modification of unsaturated fatty acids that are already incorporated into the membrane-localized lipids during the onset of stationary growth. Cyclopropanation involves the addition of a methylene group from the activated methyl of *S*-adenosyl-L-methionine (AdoMet) to the *cis* double bond of the unsaturated fatty acid.

In this chapter, we will review cycloalkyl fatty acid biosynthesis. Each section concentrates on a cyclic fatty acid class and commences with a brief overview on their occurrence and, if appropriate, function in their producing organisms. Reviews on the distribution and biosynthesis of plant cyclopropyl and cyclopentyl fatty acids[2] and *Alicyclobacillus* ω-cycloalkyl fatty acids[3] have appeared.

1.03.2 CYCLOPROPYL AND CYCLOPROPENYL FATTY ACIDS

1.03.2.1 Distribution and Function

Cyclopropane fatty acids are the most prevalent of the cyclic fatty acids and are widely distributed among both gram-positive and gram-negative bacteria, including Lactobacilli, Streptococci, Clostridia, Enterobacteria, and Brucellaceae.[4] *cis*-11,12-Methyleneoctadecanoic acid (lactobacillic acid) (**1**) and *cis*-9,10-methylenehexadecanoic acid (**2**) are the most commonly found cyclopropanoid fatty acids in bacteria where they are components of the phospholipids. These fatty acids are dead-end metabolites and thus predominate at the end of the growth cycle when up to 80% of the unsaturated fatty acids are cyclopropanated.

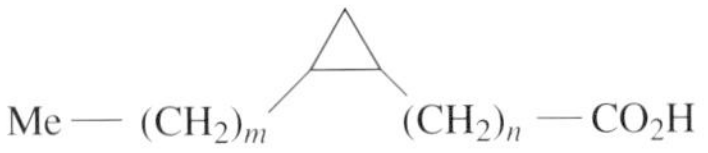

$Me-(CH_2)_m-$[cyclopropane]$-(CH_2)_n-CO_2H$

	n	*m*	common name
(**1**)	9	5	lactobacillic acid
(**2**)	7	5	
(**3**)	7	7	dihydrosterculic acid
(**4**)	6	7	dihydromalvalic acid

Cyclopropanoid as well as cyclopropenoid fatty acids are also characteristic lipid components present in many plant tissues of the families Sterculiaceae, Malvaceae, Bombacaceae, and Tiliaceae.[2] The largest proportions of these fatty acids are detected in the seeds where they are principally present as components of the triglycerides. The saturated dihydrosterculic (**3**) and dihydromalvalic (**4**) acids are typical higher plant cyclopropanoid fatty acids and are often accompanied by larger amounts of the cyclopropenoid fatty acids sterculic (**5**) and malvalic (**6**) acid.[2] To a lesser extent, the cyclopropenoids sterculynic (**7**) and 2-hyroxysterculic (**8**) acid have been reported.

$Me-(CH_2)_m-$[cyclopropene]$-(CH_2)_n-CO_2H$

	n	*m*	common name
(**5**)	7	7	sterculic acid
(**6**)	6	7	malvalic acid

CO₂H
H

(7)

OH
CO₂H

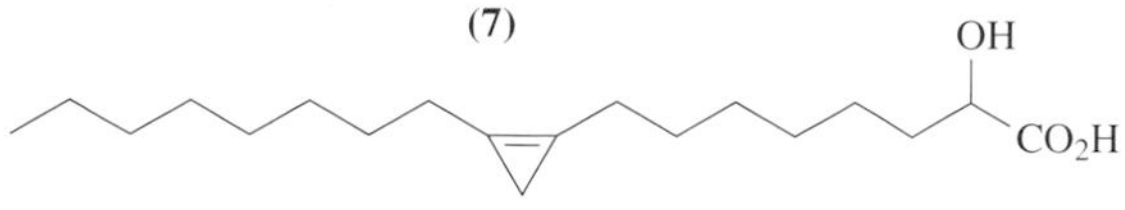

(8)

Structurally unique fatty acids containing cyclopropane rings have also been detected in marine isolates. The digestive gland of the sea hare *Bursatalla leachii* contains the C_{20} diunsaturated cyclopropanoid fatty acid (**9**), accounting for over 75% of the fatty acid content.[5] Sea hares typically accumulate terpenoidal natural products in their digestive glands from their algal diet, yet neither (**9**) nor an appropriate biosynthetic intermediate have been identified from alga.[6] Cladocroic acid (**10**) from the deep-water New Caledonian sponge *Cladocroce incurvata* is a straight-chain fatty acid which uniquely contains a cyclopropane ring adjacent to the carboxylic acid functionality as well as a terminal enyne.[7]

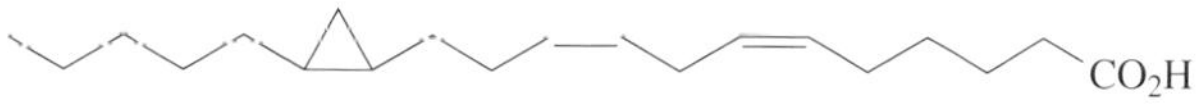

CO₂H

(9)

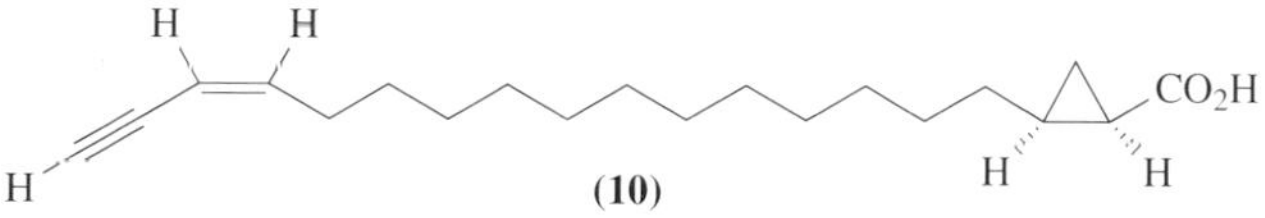

H H
CO₂H
H H H

(10)

The biological function of cyclopropane fatty acid-containing lipids has been most extensively examined in *Escherichia coli* and studies indicate that, though not essential, they contribute to the increased structural integrity of the cell wall. *E. coli* mutants deficient in cyclopropane fatty acids[8] and *E. coli* cyclopropane fatty acid overproducers[9] are phenotypically identical to the wild type under most growth conditions. Cyclopropane fatty acid-deficient mutants are, however, more sensitive to repeated freeze–thaw manipulations.[8] The occurrence of cyclopropane fatty acid-containing membrane lipids dramatically increases as cultures enter stationary growth phase, suggesting that these fatty acids offer protection during stationary growth.[10]

In the halophilic eubacterium *Pseudomonas halosaccharolytica*, cyclopropane fatty acids increase in proportion with increased growth salinity.[11] Likewise, the expression of *Mycobacterium tuberculosis* cyclopropane mycolic acid synthase genes in the nonproducing *Mycobacterium smegmatis* results in a higher phase transition temperature.[12] These studies further imply that the functional role of cyclopropanation lies in the overall structural integrity of the cell membrane.

Cyclopropyl rings are not just limited to certain fatty acid lipids. A number of other classes of natural products have been found to contain cyclopropyl rings, including the quinquecyclopropane-containing cholesteryl ester transfer protein inhibitor U-106305 (**11**) from *Streptomyces* sp. UC 11136[13,14] and the dinoflagellate sterol gorgosterol (**12**).[15] The modes of cyclopropanation in these metabolites, as well as that for the cyclopropanoid fatty acids, are related and involve the net addition of a methylene group from AdoMet across a double bond. A comprehensive review on marine sterol side-chain biosynthesis, including the formation of cyclopropane rings, has been published.[16]

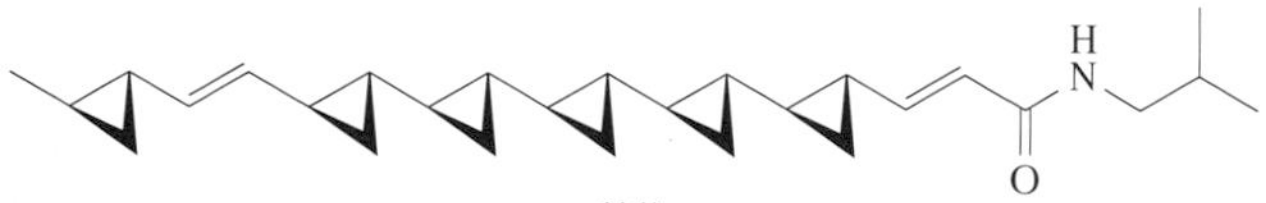

H
N
O

(11)

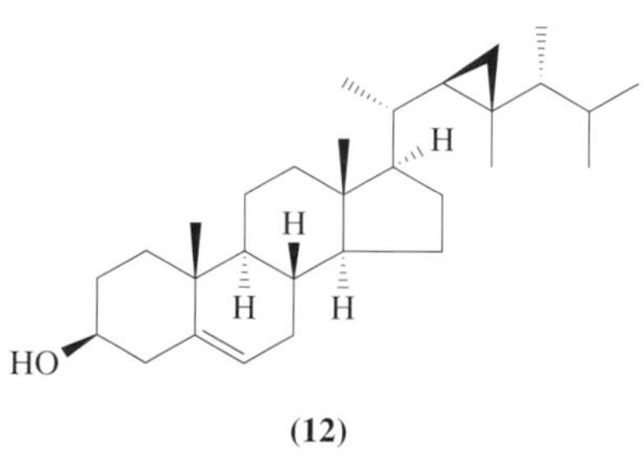

H
H
H H
HO

(12)

1.03.2.2 Biosynthesis

1.03.2.2.1 Cyclopropanation

The biosynthesis of cyclopropanoid fatty acids has been examined in a host of organisms and involves the addition of a methylene group from the activated methyl of AdoMet to the *cis* double bond of an unsaturated fatty acid already esterified into the membrane-localized lipids.[17] The reaction is thus a postsynthetic modification of the lipid bilayer. Two of the three methyl hydrogens, the *cis* geometry of the double bond, and the vinyl hydrogens of the unsaturated fatty acid are retained in the product. This reaction is a variant of the alkene methylation/proton elimination/ene reduction reaction series involved in unsaturated fatty acid methylations[18] and in sterol side-chain methylations[16] (Figure 1).

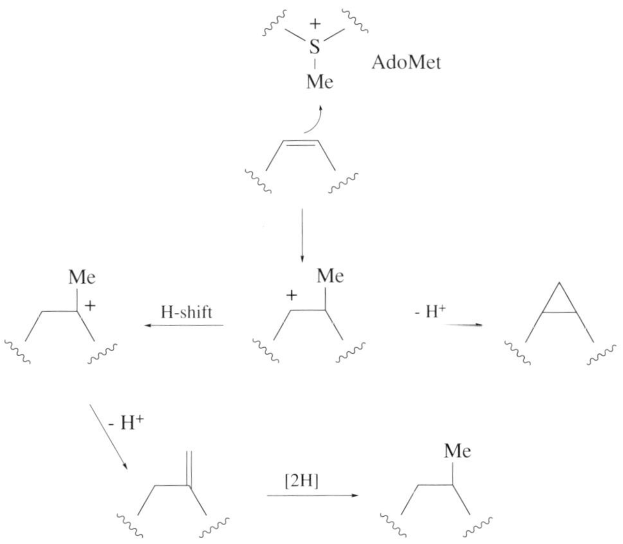

Figure 1 Biomethylation reactions catalyzed by *S*-adenosyl-L-methionine (AdoMet) dependent proteins.

The mechanism of fatty acid cyclopropanation in *Lactobacillus plantarum* has been examined in greatest detail, mainly by the groups of Arigoni and Buist, and is highlighted in this section. Cyclopropanation of endogenous *cis*-11-octadecanoic (*cis*-vaccenic) acid (**13**) and non-endogenous *cis*-9-octadecanoic (oleic) acid (**14**) by *L. plantarum* results in the formation of (**1**) and (**3**), respectively (Figure 2).[19] Both of the olefinic hydrogens are retained in the cyclopropanoid product as demonstrated through a feeding experiment with [9,10-$^2H_2$]-(**14**) (Figure 3).[19,20]

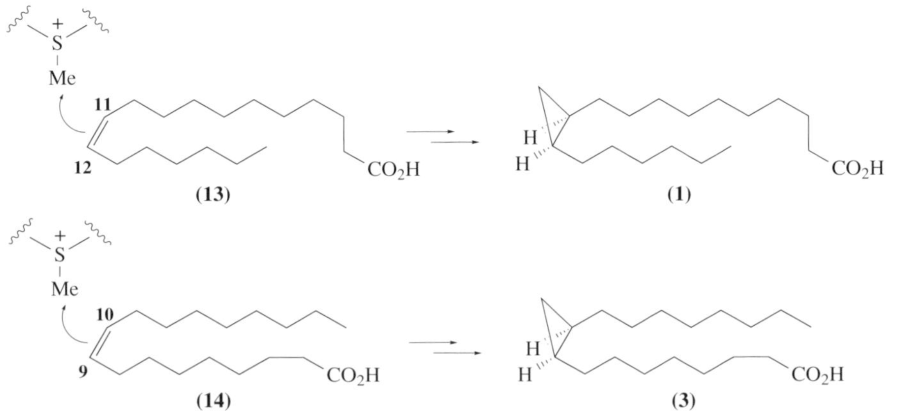

Figure 2 Cyclopropanation of *cis*-vaccenic (**13**) and oleic (**14**) acids in *L. plantarum* to (11*R*,12*S*)-lactobacillic (**1**) and (9*S*,10*R*)-dihydrosterculic (**3**) acids, respectively.

D CO_2H *L. plantarum* AdoMet D CO_2H

D D

[9,10-$^2H_2$] (**14**) [9,10-$^2H_2$] (**3**)

Figure 3 Biosynthetic rate of [9,10-$^2H_2$]oleic acid (**14**) in *L. plantarum*.

Arigoni and Rásonyi have determined the absolute configurations of the cyclopropanoid products based on the procedure of Tocanne[21,22] and found them to be opposites.[23] Comparison of the optical rotations of α-cyclopropylketones resulting from the chromic acid oxidation of *L. plantarum* (**1**) and (**3**) methyl esters with known values from *Brucella melitensis* samples[21] indicated that (11*R*,12*S*)-(**1**) is biosynthesized from (**13**), whereas (9*S*,10*R*)-(**3**) is generated from (**14**).[23] These results suggest that precursors (**13**) and (**14**) bind in the enzyme active site of cyclopropane synthase with different orientations resulting in the presentation of opposite olefinic faces towards AdoMet (Figure 2).

Complementary findings were established by Buist and co-workers through fluorine substituent effects, and these results are presented in Table 1.[24,25] Fluorine substitution at the 12-position of (**14**) had a greater rate-retarding effect on cyclopropanation than substitution at the 7-position.[24] The exact opposite effect was measured for the (**13**) series, in which fluorine substitution at the 9-position had a greater rate-retarding effect on biomethylation than that at the 14-position.[25] These results also indicate that both substrates bind to the same enzyme active site with opposite orientations. The location of the carbocation formed during the biomethylation reactions could not be determined from these studies, as it was not known at which point the pathway was affected by the fluorine-substituent. The problem is complicated by the fact that the deprotonation step is reversible.[26]

Table 1 Percent biomethylation of fluorinated olefinic fatty acids as a function of double bond position in *L. plantarum*.[25]

Fatty acid	*% Biomethylation*
Oleic acid (**14**)	78.0
7-Fluorooleic acid	25.0
12-Fluorooleic acid	3.8
cis-Vaccenic acid (**13**)	82.1
cis-9-Fluorovaccenic acid	2.2
cis-14-Fluorovaccenic acid	29.7

Arigoni and co-workers have determined the stereochemical course of the cyclopropanation of (**13**) and (**14**) through feeding experiments with chiral-methyl methionines and have thus deduced that the location of the carbocation in each series differs (Figure 4).[23,27] Methionine with an *S* methyl group was fed to *L. plantarum* and the resultant (**1**) was analyzed and found to have the *R*-configuration of the doubly labeled methylene group in the cyclopropane ring.[27] An S_N2-displacement of the *S*-methyl group from SAM by the alkene of (**13**) must occur at C-11 to give (**15**) with inversion of stereochemistry. Deprotonation from the side of the methyl terminus of the chain would thus result in the formation of the cyclopropane ring of (**1**) with the *R* configuration. The converse was true for the cyclopropanation of (**14**).[23] Incubation with methionine carrying an *R*-methyl group resulted in the formation of (**3**) also with the *R*-configuration. Deprotonation in this series must take place from the side of the carboxy terminus of the chain. This apparent contradiction is in full agreement with the model presented in Figure 2.

1.03.2.2.2 Enzymes and encoding genes

Cyclopropane fatty acid (CFA) synthase was discovered in 1963 by Zalkin *et al.*[28] The enzyme from *E. coli* has since been extensively studied by Cronan and co-workers. CFA synthase is a soluble enzyme located in the cell cytoplasm whose substrates are the soluble AdoMet and the insoluble phospholipid bilayer containing unsaturated fatty acids.[29] The extreme lability of the protein has

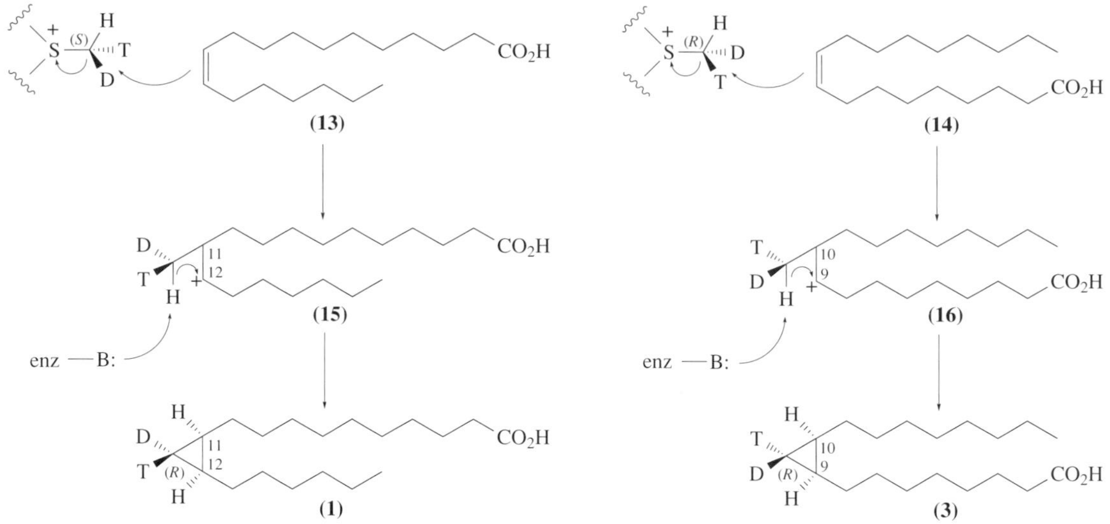

Figure 4 Stereochemistry of the cyclopropanation of (**13**) and (**14**) in *L. plantarum*.

precluded its purification to homogeneity in its native form. Isolation of the *cfa* gene[8,9] has since allowed for its overproduction and subsequent purification to homogeneity.[30]

Sequence analysis indicated that the *cfa* gene encodes a protein of 382 residues with a calculated molecular weight of 43 913 Da, which showed minor sequence homology to other AdoMet-dependent enzymes.[30] Three conserved overlapping motifs between residues 171 and 182 are believed to be the AdoMet binding site. Interestingly, the protein lacks a long hydrophobic region common to membrane proteins, thus raising the question of how CFA synthase accesses the inner and outer membranes. Cronan and co-workers speculate that the active site is located within the protein structure, rather than on the enzyme surface, and is exposed to the acyl chain double bond upon a conformational change during lipid binding.[30]

Inhibition studies with long chain alkylmaleimides (*N*-ethyl to *N*-heptyl) demonstrated that the rate of inactivation increased with extended inhibitor chain lengths, thus indicating that an essential sulfhydryl group from cysteine is located in a hydrophobic environment.[30] The essential sulfhydryl group was protected by the addition of phospholipids. Of the eight cysteine residues in CFA synthase, cysteine-354 appears to be the active residue as it is located within a predicted hydrophobic region and its deletion among the last 50 residues of CFA synthase resulted in an inactive protein.

Even though the formation of cyclopropyl fatty acids takes place as cultures enter the stationary phase of growth, the level of CFA synthase only increases about two- to threefold.[29] Transcriptional analysis of *cfa* indicated the presence of two promoters of apparently equal strength.[10] The more upstream promoter is active throughout the growth cycle, whereas the proximal promoter is only expressed during the log-to-stationary phase transition. The increase of CFA synthase activity during the stationary phase and concurrent cessation of phospholipid synthesis are believed to result in the observed time course of synthesis of cyclopropane fatty acids.

Two *E. coli cfa* related genes (*cma1* and *cma2*) have been identified by Barry and co-workers in *M. tuberculosis*, which produces α-mycolates, such as (**21**), that contain two *cis*-cyclopropane rings on the long (mero) chain (Figure 5).[12,31,32] Mycolic acids are long chain (approximately 60–80 carbons) α-alkyl-β-hydroxy fatty acids unique to mycobacteria. The proposed biosynthetic pathway for the formation of (**21**) involves a Claisen condensation between the malonate (**20**) and the long chain meromycolic acid (**19**) (Figure 5).[31]

The two cyclopropanations at the distal and proximal positions in (**21**) are catalyzed by the protein products encoded by the genes *cma1*[31] and *cma2*,[12] respectively. At the amino acid level, cyclopropane mycolic acid synthase (CMAS-1) is 34% identical to the *E. coli* CFA synthase.[31] Two regions of important homology include the SAM binding motif and the conserved cysteine-290, which has been implicated in catalysis in the *E. coli* system. The gene *cma2* is 52% identical to *cma1* and 73% identical to the *Mycobacterium leprae* putative cyclopropane synthase identified from the *M. leprae* genome sequencing project.[12] Heterologous expression of *cma1* and/or *cma2* in *M. smegmatis*, which does not cyclopropanate its mycolic acids, resulted in the formation of mycolic acids cyclopropanated at either or both of the distal and proximal positions.

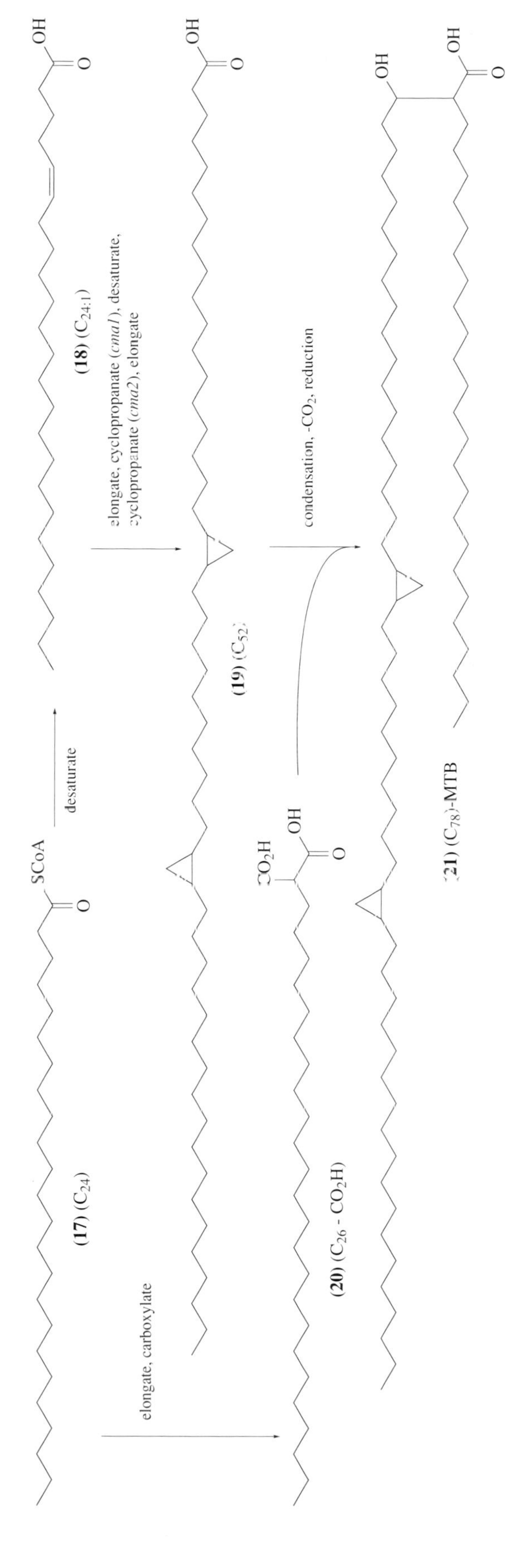

Figure 5 Proposed biosynthesis of the mycolic acid **(21)** in *M. tuberculosis* H37Ra.

1.03.3 ω-CYCLOPENTYL AND ω-CYCLOPENTENYL FATTY ACIDS

1.03.3.1 Distribution

Plants belonging to the tribes *Oncobeae*, *Pangieae*, and *Flacourtieae* within the family Flacourtiaceae contain considerable amounts of ω-cyclopent-2-enyl fatty acids (up to 90%), predominantly in their seeds and as minor constituents in leaves, chloroplasts, and cell cultures.[2] They are mainly detected in triacylglycerols, but also occur in phospholipids, glycolipids, and as free fatty acids. These plant fatty acids may function as an energy reserve that is utilized during germination, as they are largely found in the seeds. Cyclopentenyl fatty acids and their derivatives have been used in the chemotherapy of leprosy since ancient times.

A homologous series of even-numbered ω-cyclopent-2-enyl fatty acids ranging from C_6 to C_{20}, which are saturated **(22)**–**(29)** and monounsaturated **(30)**–**(32)** in the aliphatic chain, have been identified.[2] Hydnocarpic **(27)** and chaulmoogric **(28)** acids are the most common of the saturated chain ω-cyclopent-2-enyl fatty acids. In addition to the diunsaturated cyclopentenyl fatty acids, monoaic **(30)** and gorlic **(31)** acids that are unsaturated in the chain at the Δ^6-position, and Δ^4- and Δ^9-isomers of these cyclic fatty acids have also been detected.[33,34] The *R* configuration of **(28)** has been deduced from synthesis via (+)-(1*S*)-cyclopent-2-ene-1-acetic acid and ethyl hydrogen brassylate followed by saponification.[35] The same *R* configuration is presumed for the other homologues, as they are also dextrorotatory and isolated from the same plant family.

H
—$(CH_2)_n$–CO_2H

	n	common name
(22)	0	aleprolic acid
(23)	2	alepraic acid
(24)	4	aleprestic acid
(25)	6	aleprylic acid
(26)	8	alepric acid
(27)	10	hydnocarpic acid
(28)	12	chaulmoogric acid
(29)	14	hormelic acid

H
—$(CH_2)_m$CH=CH–$(CH_2)_n$$CO_2H$

	n	*m*	common name
(30)	4	4	manoaic acid
(31)	4	6	gorlic acid
(32)	6	6	oncobic acid

Trace amounts of the saturated ω-cyclopentyl fatty acid dihydrohydnocarpic acid **(33)** have been identified within some species belonging to the Flacourtiaceae, including *Hydnocarpus anthelminthica* and *Taraktogenus kurzii*, in addition to large concentrations of the unsaturated ω-cyclopent-2-enyl fatty acids.[3,4] ω-Cyclopentyl fatty acids **(33)** and dihydrochaulmoogric **(34)** have additionally been identified in red algae of the Solieriaceae, where they, but no ω-cyclopent-2-enyl fatty acids, are present in high percentages (6–25%).[36]

—$(CH_2)_n$– CO_2H

	n	common name
(33)	10	dihydrohydnocarpic acid
(34)	12	dihydrochaulmoogric acid

1.03.3.2 Biosynthesis

1.03.3.2.1 ω-Cyclopent-2-enyl fatty acids

Spener and co-workers have examined ω-cyclopent-2-enyl fatty acid biosynthesis in a variety of tissue samples from Flacourtiaceae.[2] Oxidative ring closure of polyunsaturated straight-chain fatty acids, analogous to that in prostaglandin biosynthesis, was immediately ruled out as a possible pathway, as (i) no structural relationships between isolated unsaturated straight-chain and ω-cyclopent-2-enyl fatty acids were found[33] and (ii) [1-^{14}C]acetate was incorporated only into the aliphatic chain.[37,38] Schmidt degradation of [1-^{14}C]acetate-enriched hydnocarpic acid showed that 20% of the ^{14}C resided at the carboxyl carbon, indicating that a non-C_1-acetate derived C_6-primer is chain-extended by five malonyl-CoA additions. ω-Cyclopent-2-enyl fatty acids are biosynthesized from the (2*R*)-cyclopent-2-enylcarboxylic acid (aleprolic acid (**22**)) primer unit, most likely as its CoA thioester (Figure 6).[37] Its role was examined in a variety of tissues in Flacourtiaceae through feeding studies with [1-^{14}C]-(**22**); for instance, ω-cyclopent-2-enyl fatty acids in the seeds of *H. anthelminthica* and *Caloncoba echinata* are almost exclusively labeled by [1-^{14}C]-(**22**).

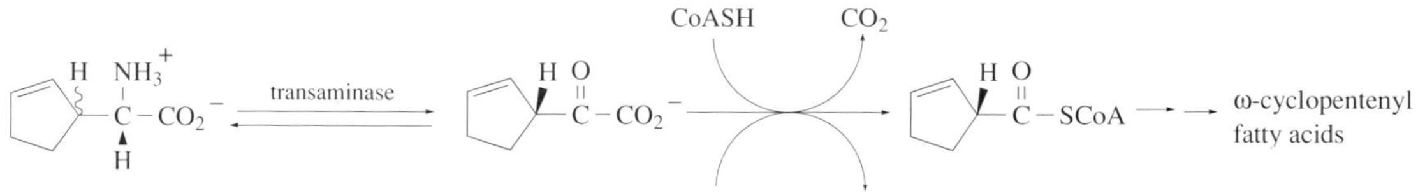

Figure 6 Biosynthesis of ω-cyclopent-2-enyl fatty acids.

The starter unit itself is derived from the nonproteinogenic amino acid 2-(2′-cyclopentenyl)glycine (**35**) via transamination and subsequent oxidative decarboxylation (Figure 6).[38] Cyclopentenyl [2-^{14}C]glycine was efficiently incorporated into the cyclic fatty acids in various tissues of *H. anthelminthica*. Competition experiments in embryonic tissue of maturing *H. anthelminthica* seeds with excess unlabeled (**22**) decreased the incorporation of [2-^{14}C]-(**35**), establishing that (**35**) is a precursor of (**22**). This pathway has precedence in the biosynthesis of iso and anteiso branched-chain fatty acids that are formed from the amino acids valine, leucine, and isoleucine.[1] Naturally occurring 2-(2′-cyclopentenyl)glycine is a constituent of the free amino acid pool in Flacourtiaceae up to the extent of 2% and exists as a pair of diastereomeric L-amino acids epimeric at C-1′.[39] The *S* configuration of the α-carbon was deduced through treatment with L- and D-amino acid oxidases, whereas the *R* and *S* configurations at C-1′ were determined by NMR spectroscopy.

Spener and co-workers explored several different hypotheses in the biosynthesis of cyclopentenylglycine, including (i) its formation via shikimic acid pathway intermediates, (ii) the condensation of aspartate and pyruvate to α,ε-diaminopimelate, an intermediate in lysine biosynthesis in higher plants, followed by transamination and cyclization (C_4+C_3 pathway), and (iii) two successive C_1 chain elongations of α-ketoglutarate (**37**) to α-ketopimelate (**39**) followed by cyclization of the reduced semialdehyde (**40**) (C_5+2C_1 pathway) (Figure 7). ^{14}C-Labeled shikimic acid and glucose were not incorporated into (**35**), disproving hypothesis (i).[2,40] [U-^{14}C]glucose was rather efficiently incorporated into the glycerol moiety of the triacylglycerols and [U-^{14}C]shikimate labeled only the aromatic amino acids.

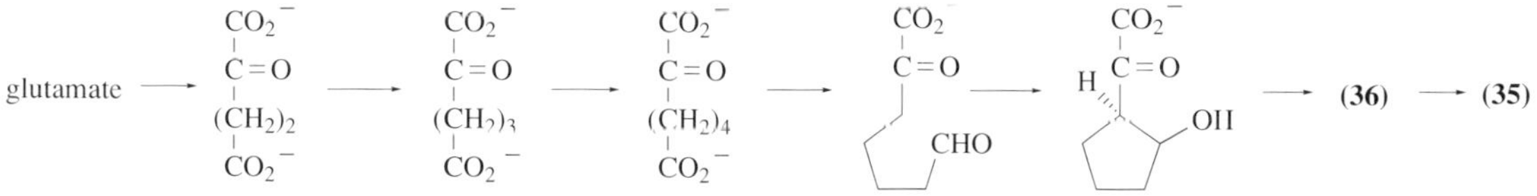

Figure 7 Proposed biosynthesis of 2-(2′-cyclopentenyl)glycine (**35**) via α-ketopimelate (**39**) (C_5+2C_1 pathway).

In feeding experiments in *C. echinata* leaves and chloroplasts with [U-^{14}C]aspartate plus acetate, α,ε-[1,7-$^{14}C_2$]diaminopimelate and [U-^{14}C]lysine, these intermediates of the C_4+C_3 pathway were modestly incorporated into cyclopentenylglycine.[39] This contrasts with the results of incorporation experiments with C_5+2C_1 pathway precursors which were more efficiently incorporated. Incubation of *C. echinata* leaves and chloroplasts with [U-^{14}C]glutamate plus acetate and α-[6-^{14}C]aminoadipate plus acetate resulted in up to 42% of the ^{14}C incorporated into the free amino acids present in cyclopentenylglycine.[39] Similarly, α-[1,2-$^{14}C_2$]ketopimelate was a highly efficient intermediate with enrichments of up to 63% in (**35**) from *I. polycarpa* callus cultures, providing further evidence for hypothesis (iii), the C_5+2C_1 pathway (Figure 7).[40] Spener has proposed that the (**35**) biosynthetic pathway involves two successive C_1 chain elongations of (**37**) to (**39**). Reduction of the ε-carboxyl group to α-ketopimelate semialdehyde (**40**) followed by an intramolecular aldol condensation gives

2′-hydroxycyclopentylglyoxylate (**41**). Dehydration to (**36**) and a subsequent transamination would then give (**35**). The one cause of concern about this pathway proposal entails the dehydration step which involves removal of one of the nonactivated C-3′ methylene hydrogens in preference to the activated C-1′ methine hydrogen.

ω-Cyclopent-2-enyl fatty acids are pure enantiomers with an *R* configuration, indicating that only (1′*R*,2*S*)-2-(2′-cyclopentenyl)glycine serves as the precursor. We speculate that cyclopentenylglyoxylate (**36**) is labile and that non-enzymatic epimerization at C-1′ takes place at this stage and not at the level of the amino acid. Transamination of racemic (**36**) to the diastereomeric (**35**) must equally involve either enantiomer, whereas the oxidative decarboxylation of (**36**) to the primer unit (**22**) is apparently a stereospecific process only involving the *R* isomer (Figure 6).

Both L-2-(2′-cyclopentenyl)glycine diastereomers are apparently utilized in the biosynthesis of cyanogenic glycosides containing a cyclopentenoid aglycone (**42**)–(**47**).[41,42] These metabolites are also synthesized in some tribes of Flacourtiaceae, as well as other closely related plant families within the order Violales. The cyclopentenoids occur as mixtures of stereoisomers epimeric at the carbon bearing the cyanohydrin functionality. This contrasts with the classical cyanogenic glycosides which are single isomers derived from the common amino acids phenylalanine, tyrosine, valine, isoleucine, or leucine and present in all major groups of flowering plants. Natural and unnatural cyclopentyl cyanohydrin glycosides are synthesized in *Passiflora morifolia* from racemic 2-cyclopentenecarbonitrile and cyclopentanecarbonitrile, respectively.[43]

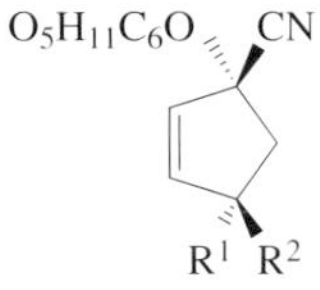

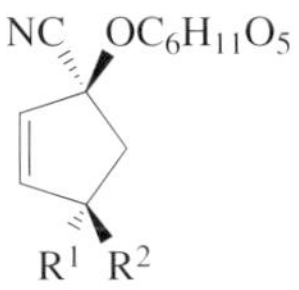

$O_5H_{11}C_6O$ CN   NC $OC_6H_{11}O_5$

R^1 R^2 R^1 R^2

(**42**) $R^1 = R^2 = H$ deidaclin
(**43**) $R^1 = H$, $R^2 = OH$ volkenin
(**44**) $R^1 = OH$, $R^2 = H$ taraktophyllin

(**45**) $R^1 = R^2 = H$ tetraphyllin A
(**46**) $R^1 = H$, $R^2 = OH$ epivolkenin
(**47**) $R^1 = OH$, $R^2 = H$ tetraphyllin B

1.03.3.2.2 ω-Cyclopentyl fatty acids

The biosynthesis of the saturated ω-cyclopentyl fatty acids (**33**) and (**34**) has not been examined, but in Flacourtiaceae, it may involve the saturated starter unit cyclopentanecarboxylic acid (**49**), which may arise from a double bond isomerization of (**22**) to the Δ^1-position (**48**) followed by reduction (Figure 8). This pathway has precedence in the cyclohexane- and cycloheptanecarboxylic acid biosynthetic pathways (see Figures 10 and 15). Conversely, the ω-cyclopentyl fatty acids (**33**) and (**34**) in Solieriaceae were proposed to be formed from the direct end-cyclization of the unusual Δ-ω5 monounsaturated fatty acids 16:1ω5 (**50**) and 18:1ω5 (**51**), respectively, which are also prominent in these algae.[36] This hypothesis was proposed because no cyclopent-2-enyl metabolites were detected in the algae.

(**22**) —isomerase→ (**48**) C(=O)–SCoA —enoyl reductase→ (**49**) C(=O)–SCoA → → ω-cyclopentyl fatty acids

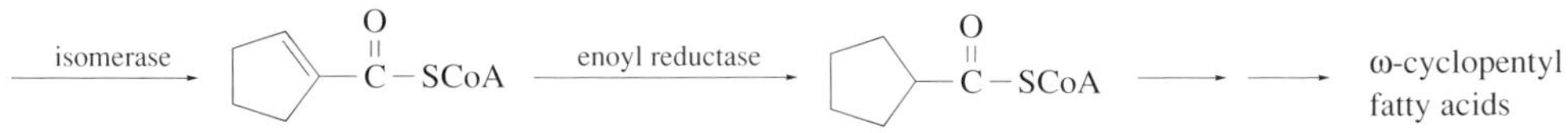

Figure 8 Hypothetical biosynthesis of ω-cyclopentyl fatty acids.

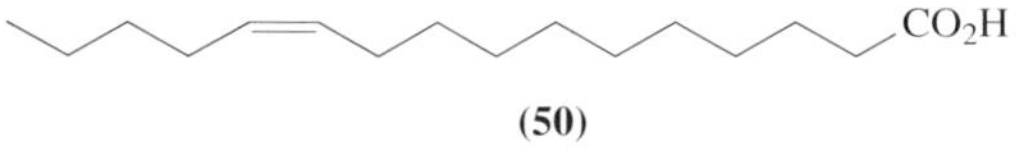

CO_2H

(**50**)

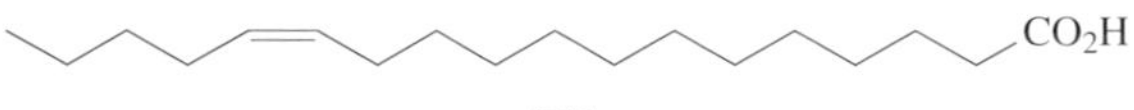

CO_2H

(**51**)

1.03.4 ω-CYCLOHEXYL FATTY ACIDS

1.03.4.1 Distribution and Function

The cell membrane of the thermoacidophilic bacterium *Alicyclobacillus acidocaldarius* (formerly *Bacillus acidocaldarius*) is uncommonly composed of ω-cyclohexyl fatty acid-containing lipids and hopanes (pentacyclic triterpenes).[44] *A. acidocaldarius* was originally isolated in 1971 by Darland and Brock from acidic thermal springs (44–72 °C, pH 2.5–3.3) in Yellowstone National Park and from acidic fumarole soil in the Hawaiian Volcano National Park.[45] ω-Cyclohexylundecanoic acid (**52**) and ω-cyclohexyltridecanoic acid are the main fatty acid components in this organism (70–90%) and are suggested to have special physiological importance for the cells at high temperature and low pH. Their dense packing properties provide a unique membrane structure which plays a key role in enabling the organism to thrive under such extreme conditions. Consequently, mutants of *A. acidocaldarius* deficient in ω-cyclohexyl fatty acid biosynthesis grow poorly at high temperatures and low pH.[46] Model membranes consisting of lipids containing ω-cyclohexyl fatty acids are relatively dense even beyond the phase transition temperature.[47]

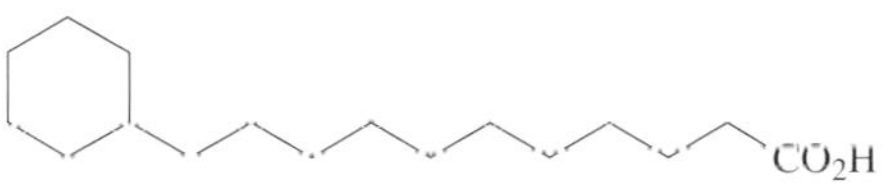

ω-cyclohexylundecanoic acid (**52**)

ω-Cyclohexyl fatty acids have also been isolated from another thermoacidophile, *Alicyclobacillus acidoterrestis*,[48] and the mesophile *Curtobacterium pusillum*.[49] When these organisms are grown at pH 4, the percentage of ω-cyclohexyl fatty acids increases as the growth temperature is raised.[49,50]

This physiological adaptation may also be operative in the thermoacidophile *Alicyclobacillus cycloheptanicus*, where ω-cycloheptyl fatty acids occur in high percentage (see Section 1.03.5).[51] The three ω-alicyclic fatty acid-containing *Bacillus* species have been placed in the new genus *Alicyclobacillus* where they form a distinct phylogenetic group supported by 16S rRNA sequence data.[52]

The occurrence of a fully saturated, monosubstituted cyclohexane ring is rare but has also been observed in long-chain ω-cyclohexylalkanes isolated from soil extracts[53] and from shoots of *Achyranthes aspera*[54] and in several *Streptomyces* antibiotics, including ansatrienin A[55] (mycotrienin I)[56,57] (**53**) from *Streptomyces collinus*. ω-Cyclohexyl fatty acids have additionally been detected in *S. collinus* in minor amounts (0.1%) that are dramatically increased to 20% upon the addition of cyclohexanecarboxylic acid to the fermentation.[38] *Streptomyces collinus* is capable of generating other ω-cycloalkyl fatty acids as well upon the addition of cycloalkylcarboxylic acids, including ω-cyclobutyl, ω-cyclopentyl, and ω-cycloheptyl fatty acids; however, the percentages decrease if the growth temperature is increased or the pH is decreased. Thus, ω-cyclohexyl fatty acids do not appear to play a similar membrane-stabilizing role as in *A. acidocaldarius*. *Streptomyces antibioticus*, which is not known to produce secondary metabolites containing cyclohexanecarboxylate, only synthesizes ω-cyclohexyl fatty acids upon the addition of cyclohexanecarboxylic acid.[58]

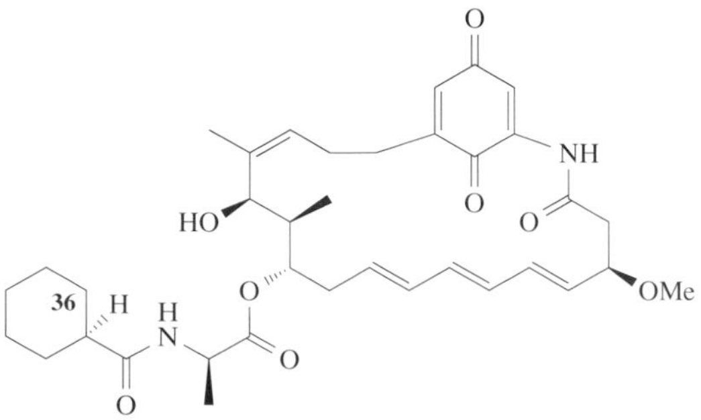

ansatrienin A (**53**)

1.03.4.2 Biosynthesis

1.03.4.2.1 Formation of the cyclohexanecarboxylic acid starter unit

Several groups have determined that the cyclohexanecarboxylic acid starter unit in ω-cyclohexyl fatty acid synthesis is derived from shikimic acid.[59–61] Shikimate is a general precursor of aromatic metabolites, including the aromatic amino acids, *p*-aminobenzoic acid, salicylic acid, and vitamin K.[62] De Rosa *et al.* reported in 1974 that the conversion of shikimic acid to cyclohexanecarboxylic acid in *A. acidocaldarius* proceeds through 1-cyclohexenecarboxylic acid,[59] yet the metabolic route by which shikimate was "deoxygenated" largely remained unexplored until the pathway was examined in detail in the ansatrienin A producer *S. collinus*.

As with the ω-cyclohexyl fatty acids, the cyclohexanecarboxylic acid moiety of ansatrienin A was found to arise intact from the seven carbon atoms of shikimate via 1-cyclohexenecarboxylic acid.[63] ^{13}C- and ^{2}H-labeled samples of shikimic acid were used to probe the stereochemistry of processing the cyclohexane ring of shikimic acid and to establish the fate of all the precursor hydrogens in this transformation.[64,65] A sample of [2-^{13}C]shikimic acid was fed to *S. collinus*, and ^{13}C in the resulting ansatrienin was found to reside exclusively at the pro-*S* position of the cyclohexanecarboxylate moiety, i.e., C-36.[64] The 1-cyclohexenecarboxylic acid (**55**) accompanying cyclohexanecarboxylic acid (**56**) in a minor amount from the hydrolysis of the biosynthetic sample of ansatrienin carried the ^{13}C label not at C-2 but at C-6. Deuterated samples of shikimic acid (**54**) were fed to *S. collinus*, and deuterium from C-2, C-3, C-4, and C-5 was effectively incorporated and occupied the 36*R* (axial), 35*R* (equatorial), 34*E* (equatorial), and 33*R* (axial) positions, respectively, in the resulting (**53**) (Figure 9).[65] However, absolutely no deuterium from C-6 of shikimic acid was retained in the product. The transformation of (**54**) into (**56**) must therefore involve two proton eliminations at C-6 including both of the stereochemically opposite hydrogens.

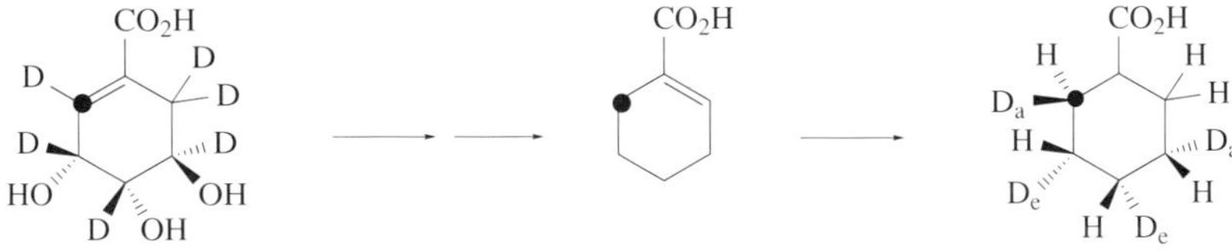

(**54**) (**55**) (**56**)

Figure 9 Summary of the labeling pattern of the 1-cyclohexene- and cyclohexanecarboxylic acid moieties of ansatrienin A (**53**) after feeding labeled shikimic acid (**54**) samples to *S. collinus*.

Based on these results, a series of feeding experiments in *S. collinus* with ^{13}C- and ^{2}H-labeled potential intermediates established that the proposed pathway involves a series of alternating dehydrations and double bond reductions arranged such that the ring system never becomes aromatic (Figure 10).[65] A key result was the finding that (**60**) deuterated at C-5 was efficiently incorporated and labeled an axial (pro-33*R*) rather than an equatorial hydrogen (pro-35*R*) in (**53**). This experiment identified the labeled position in the precursor (**60**) as corresponding to C-5, and not C-3, of (**54**). The 4-hydroxy isomer of (**60**), carrying deuterium at C-4, was also incorporated, but since it labeled the 34*Z* hydrogen, rather than the 34*E* position as did [4-^{2}H]-(**54**), it cannot be a normal pathway intermediate.

The formation of the cyclohexanecarboxylic acid starter unit of ω-cyclohexyl fatty acids from shikimic acid in *A. acidocaldarius* and two blocked mutants was similarly examined and found to follow the same biosynthetic course.[66] The combined knowledge of the stereochemical fate of the carbon-bound hydrogens of shikimate in *S. collinus* with the stereochemical information determined in the *A. acidocaldarius* blocked mutants was crucial to the ultimate delineation of the stereochemistry of each reaction in the formation of cyclohexanecarboxylic acid.

A. acidocaldarius blocked mutant 2 is autotrophic for cyclohexanecarboxylic acid and accumulates (1*S*,3*S*)-3-hydroxycyclohexanecarboxylic acid (**61**) in its spent media.[66] The absolute stereochemistry of (**61**) was determined using Mosher's method. This finding indicated that 5-hydroxycyclohex-1-enecarboxylic acid (**60**) is not dehydrated to the cross-conjugated diene (**63**) prior to double bond reduction to 2-cyclohexenecarboxylic acid (**62**) as originally proposed (Figure 11).[67] The pathway via (**63**) was initially favored in view of the efficient incorporation of [7-^{13}C]-(**63**) into (**53**)[65] and the reduction to (**56**) at the CoA thioester level by cell-free extracts of *S. collinus*.[68] Rather, the double bond of (**60**) is first reduced to give the hydroxy acid (**61**), which is accumulated in mutant 2, before

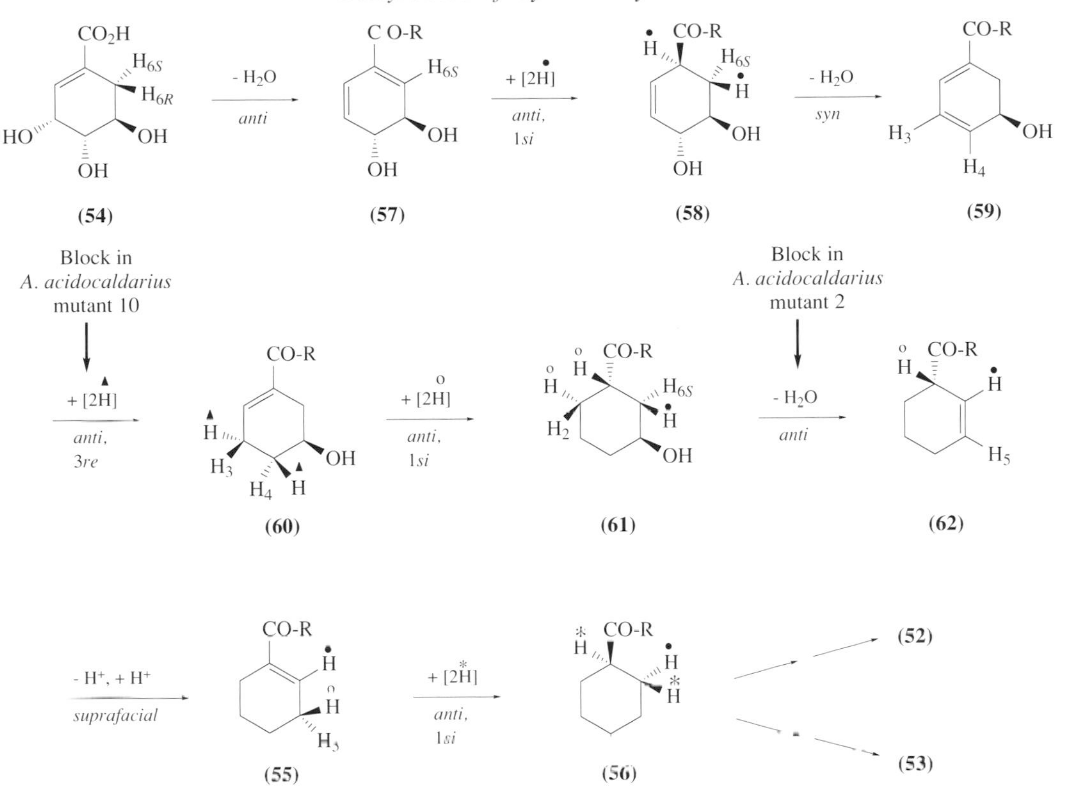

Figure 10 Biosynthetic pathway of cyclohexanecarboxylic acid (**56**) formation in *A. acidocaldarius* and *S. collinus*. The stereochemical fate of the carbon-bound hydrogens of shikimic acid (**54**) is depicted. Each hydrogen is denoted with a numerical subscript referring to its carbon of origin in (**54**), R = -OH or -SCoA.

dehydration to (**62**). Feeding experiments with racemic [3-^{2}H]-(**61**) in *S. collinus* and *A. acidocaldarius* verified this scenario. This mutant also accumulated smaller amounts of the 3-epimer of (**61**) as a result of epimerization of (**60**) by oxidation to the ketone and subsequent reduction with opposite stereochemistry.

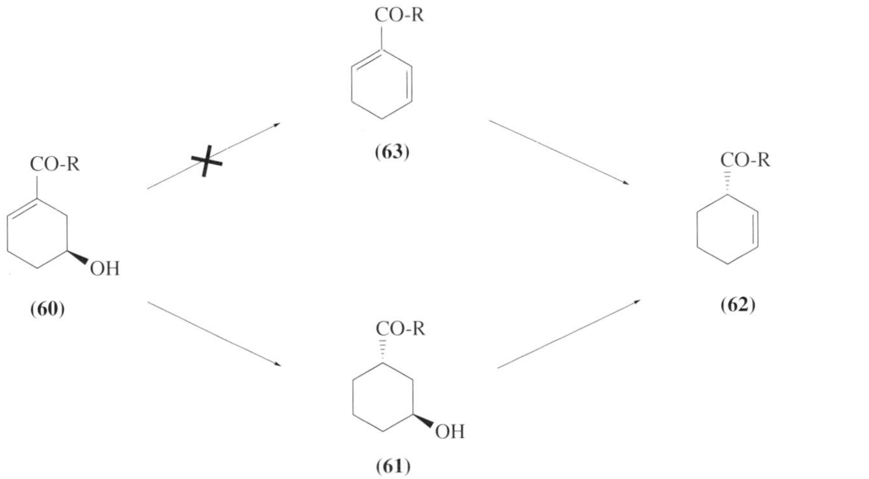

Figure 11 Sequence of reduction and dehydration steps in the conversion of (**60**) to (**62**).

The stereochemistries of several of the pathway reactions were elucidated based on the analysis of labeling patterns in the accumulated product (**61**), as it allowed for the dissection of the pathway at a stage when one C-6 hydrogen of (**54**) was still present. Stereospecifically deuterated (6*R*)-[6-^{2}H$_1$]shikimic acid was administered to *A. acidocaldarius* blocked mutant 2 and labeled the resultant (**61**) with deuterium at the pro-2*S* position; in contrast, deuterium from (6*S*)-[6-^{2}H$_1$]shikimic acid was not retained.[69] The 1,4-conjugate elimination of water from shikimic acid to (**57**) thus occurs in an *anti* fashion.

This conjugate elimination is analogous to the conversion of 5-enolpyruvylshikimate-3-phosphate to chorismate, which involves the removal of the pro-6*S* hydrogen and loss of phosphate with overall *anti* stereochemistry.[70,71] In fact, chorismic acid undergoes cleavage of its enolpyruvyl side chain to produce (**57**) in *Klebsiella pneumoniae*.[72] Feeding experiments with [2,6,10,10-$^2H_4$]chorismate in *S. collinus* and *A. acidocaldarius*, and inhibitor experiments with glyphosate in *A. acidocaldarius*, have shown that the biosynthesis of cyclohexanecarboxylic acid branches off from the shikimate pathway at a point prior to enolpyruvylshikimate-3-phosphate, either at shikimate or shikimate-3-phosphate.[73]

Deuterium from (2*S*)-[2-^{2}H]-(**61**), which was generated from [6-^{2}H]shikimic acid in *A. acidocaldarius* mutant 2, was not retained in the resultant ω-cyclohexyl fatty acids when administered to *A. acidocaldarius* mutant 10.[66] This mutant blocks the reduction of (**59**) to (**60**), resulting in the accumulation of the decomposition products benzoic and 3-hydroxybenzoic acid. The dehydration of (**61**) to (**62**) consequently involves the overall removal of the nonacidic pro-2*S* hydrogen and the C-3 hydroxyl group in an *anti* fashion. This reaction is in contrast with the majority of biological dehydration and similar elimination reactions which involve the removal of a proton α to a carbonyl or some other activating group and a leaving group β to the carbonyl. The dehydration involves an *anti* 1,2-elimination which may be facilitated by the protonation or phosphorylation of the hydroxy group. The fact that the stereochemistry of the reaction is consistent with a concerted, ionic mechanism may only be fortuitous. The dehydration of (**61**) is directly analogous to the radical dehydration of 4-hydroxybutyryl-CoA to vinylacetyl-CoA in *Clostridium aminobutyricum* by 4-hydroxybutyryl-CoA dehydratase.[74]

The final two steps in the pathway involve isomerization of (**62**) to (**55**) followed by double bond reduction to the saturated (**56**). This series of reactions was observed in cell-free extracts of *S. collinus* at the level of the CoA esters.[68] The isomerization of 2-cyclohexenylcarbonyl-CoA to 1-cyclohexenylcarbonyl-CoA proceeds via a suprafacial 1,3-allylic hydrogen shift with a stereochemical preference for the 1*S* isomer of 2-cyclohexenylcarbonyl-CoA.[75]

The stereochemistry of the double bond reduction of (**55**) was probed by feeding [2,6,6-$^2H_3$]-(**55**) to the ansatrienin A producer and found to proceed by an *anti* addition of hydrogen.[65] This result is in agreement with the enzymatic model which showed that the reduction of 1-cyclohexenylcarbonyl-CoA occurs by the addition of the pro-4*S* hydrogen of NADPH to the *si* face at C-2 of the cyclohexene ring and addition of a solvent proton at C-1.[76] In fact, the absolute stereochemistry of all three Δ^1-double bond pathway reductions is identical, suggesting that the same enoyl-CoA reductase may be responsible for all three reactions.[66]

1.03.4.2.2 Enzymes and encoding genes

Reynolds *et al.* have purified this enoyl-CoA reductase, designated 1-cyclohexenylcarbonyl-CoA reductase, to homogeneity from *S. collinus*.[77] The enzyme is a homodimer with a 36 000 Da subunit molecular weight which showed an absolute requirement for NADPH as a cofactor and the coenzyme A thioester of its substrate; neither the free acid nor the *N*-acetylcysteamine thioester were active substrates. The purified protein is not only able to catalyze the reduction of 1-cyclohexenylcarbonyl-CoA as originally reported,[77] but can reduce (**57**) and (**60**) with similar kinetic constants.[78]

The gene encoding 1-cyclohexenylcarbonyl-CoA reductase was cloned by reverse genetics from a genomic library of *S. collinus* Tü 1892,[78] using sequence information from two internal cleavage peptides generated from the homogenous protein.[77] Sequence analysis indicated that the gene encodes a 280 amino acid protein (calculated molecular weight of 29.7 kDa), which showed homology not to other enoyl-CoA reductases but rather to members of the short chain alcohol dehydrogenase superfamily. Overexpression in *E. coli* gave a protein with identical characteristics to the native enzyme. Deletion of the 1-cyclohexenylcarbonyl-CoA reductase gene in *S. collinus* resulted in the loss of ansatrienin production and the inability to reduce the Δ^1-double bond of (**55**) and (**60**). Ansatrienin A production, however, could not be restored upon supplementation with cyclohexanecarboxylic acid, suggesting that the gene deletion affected the expression of the ansatrienin biosynthetic gene cluster. 1-Cyclohexenylcarbonyl-CoA reductase deleted mutants were able to produce ω-cyclohexyl fatty acids only after supplementation with cyclohexanecarboxylic acid, demonstrating that the 1-cyclohexenylcarbonyl-CoA gene is indeed involved in cyclohexanecarboxylic acid biosynthesis.

1.03.4.2.3 Related biosynthetic pathways

The streptomycete immunosuppressants rapamycin (**64**),[79,80] ascomycin (FK520) (**65**),[81,82] and FK506 (**66**)[83–85] each contain a more oxygenated cyclohexanecarboxylic acid moiety derived from (1*R*,3*R*,4*R*)-3,4-dihydroxycarboxylic acid (**69**) which serves as the polyketide starter unit in the biosynthesis of these metabolites. Their polyketide nature has been confirmed through feeding experiments[86,87] and, in the case of rapamycin, by extensive genetic studies.[88–90] The latter have resulted in the cloning and complete sequence analysis of the (**64**) biosynthetic gene cluster, which encodes a modular polyketide synthase (PKS) as well as additional processing enzymes.

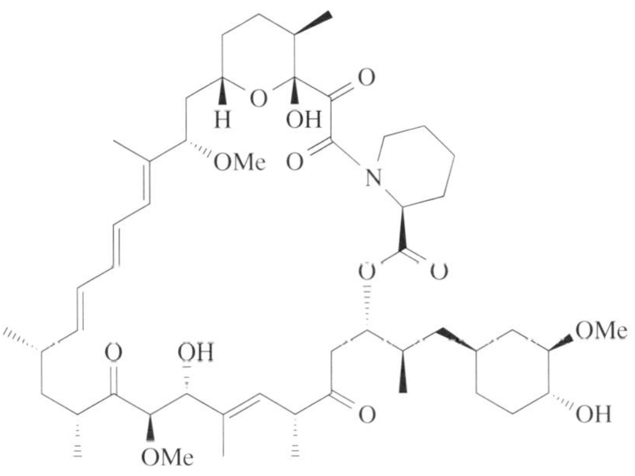

rapamycin (**64**)

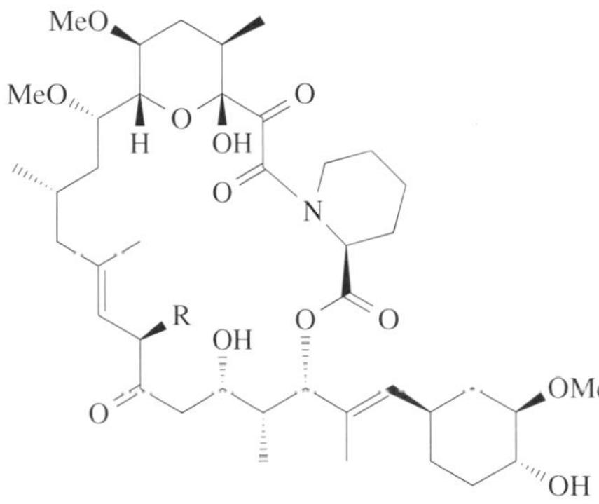

FK 520 (**65**), R = CH_2Me
FK 506 (**66**), R = $CH_2CH{=}CH_2$

The dihydroxycyclohexane moiety (**69**) in (**64**) and (**65**) is derived from all seven carbon atoms of shikimic acid,[87,91] as is the case for (**56**) (Figure 10). Detailed feeding experiments in the (**65**)-producer indicated that the pathway proceeds from (**54**) to the presumed, and in the case of (**64**) confirmed,[92] starter unit (**69**) (Figure 12).[93] The pathway shares the first step with the cyclohexanecarboxylic acid biosynthetic pathway, *anti* elimination of H_{6R} from (**54**),[94,95] but diverges at the stage of (**57**). Both (**57**) and (**68**) were incorporated into (**65**), implicating (**67a**) or its C-1 epimer (**67b**) as a likely intermediate. However, deuterium labeling revealed that the enoyl reductions in this pathway proceed with a different stereochemistry from those in the pathway from (**54**) to (**56**) (Figure 10). Since the C-1 configuration of the inferred intermediate (**67**) is not known, two alternative stereochemical pathways, one proceeding through (**67a**) and the other through its C-1 epimer (**67b**), could not be distinguished (Figure 12).[93] Consistent with the difference in enoyl reduction stereochemistry, the analysis of the (**64**) biosynthetic gene cluster suggests that the enoyl reductase which is part of the first PKS module is responsible for the last double bond reduction in the starter unit.[90] In contrast, the 1-cyclohexenylcarbonyl-CoA reductase of the ansatrienin pathway is a distinct, separate enzyme with no homology to enoyl reductases.[77]

1.03.5 ω-CYCLOHEPTYL FATTY ACIDS

1.03.5.1 Distribution and Function

The fatty acid mixture from the lipids of the thermoacidophilic soil bacterium *Alicyclobacillus cycloheptanicus* is dominated by ω-cycloheptyl fatty acids, which are unique to this organism.[51,96,97] The optimal temperature and pH for growth are 48 °C and 3.5–4.5, respectively. ω-Cycloheptylundecanoate (**72**), -tridecanoate (**74**), and -α-hydroxyundecanoate (**73**) comprise nearly 80% of the fatty acids obtained upon saponification of *A. cycloheptanicus* total lipids. Three additional minor ω-cycloheptyl fatty acids have been identified as ω-cycloheptylnonanoate (**70**), -decanoate (**71**), and -α-hydroxytridecanoate (**75**).[98] The remaining fatty acids from this organism are predominantly a mixture of branched-chain saturated fatty acids.

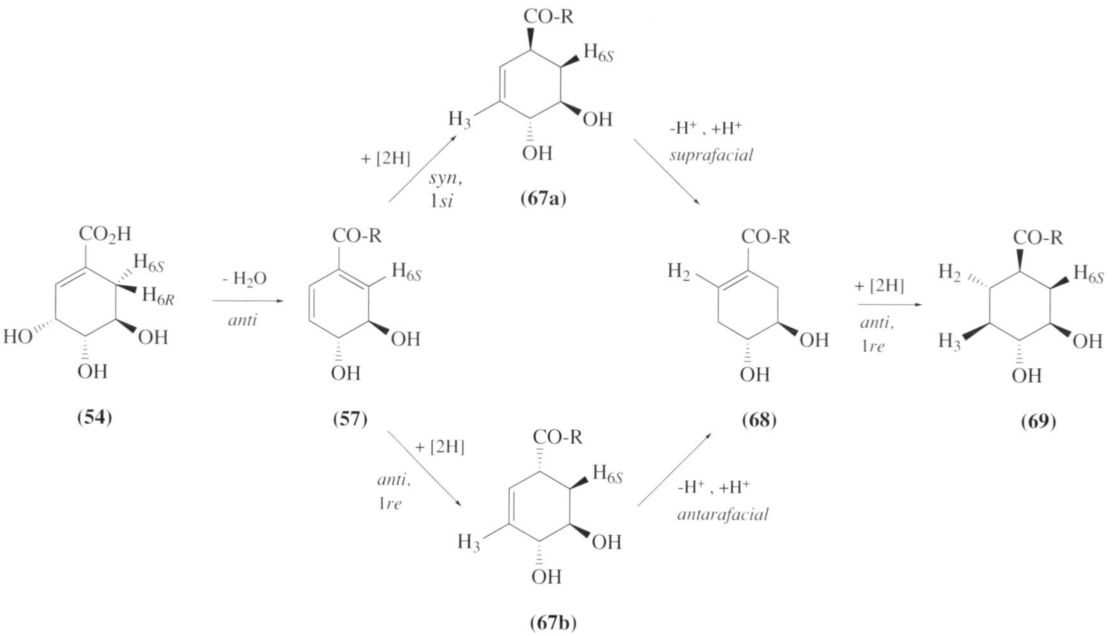

Figure 12 Biosynthetic pathway of the (1*R*,3*R*,4*R*)-3,4-dihydroxycyclohexanecarboxylic acid starter unit of FK520 (**65**) in *S. hygroscopicus*. R = -OH or -SCoA.

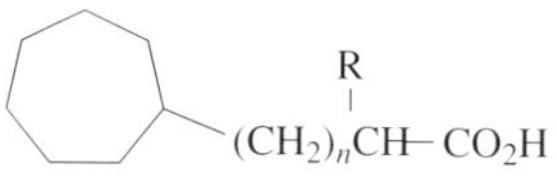

$(CH_2)_nCH(R)-CO_2H$

	n	R
(70)	7	H
(71)	8	H
(72)	9	H
(73)	9	OH
(74)	11	H
(75)	11	OH

ω-Cycloheptyl fatty acids are presumed to enable the organism to grow in acidic, hot media by providing a more dense cell membrane as do the homologous *Alicyclobacillus* ω-cyclohexyl fatty acids. This raises the intriguing question as to whether the functional equivalence of these cyclic fatty acids is a result of a divergent or convergent biosynthesis. For instance, are the ω-cycloheptyl fatty acids produced by a modification of the ω-cyclohexyl fatty acid biosynthetic pathway or is the cycloheptane ring generated in a completely different fashion?

1.03.5.2 Biosynthesis

1.03.5.2.1 Formation of the cycloheptanecarboxylic acid starter unit

Several different hypotheses for the biosynthesis of ω-cycloheptyl fatty acids were tested in specific feeding experiments,[99] including (i) ω-oxidation of a straight-chain fatty acid followed by cyclization to a terminal cycloheptane ring, (ii) addition of a one-carbon unit to an intermediate of the cyclohexanecarboxylic acid pathway followed by ring expansion, and (iii) formation of cycloheptanecarboxylic acid (**76**) from a hypothetical seven-membered ring homologue of shikimic acid, which could arise from the lipopolysaccharide constituent 3-deoxy-D-*manno*-2-octulosonate 8-phosphate (KDO). All of these hypotheses were, however, disproved through feeding experiments (Figure 13).

Since acetate labeled only C-1, though C-10 of (**72**), the main component of the fatty acids, and [8-^{13}C]cycloheptanecarboxylic acid were efficiently incorporated and exclusively labeled C-11

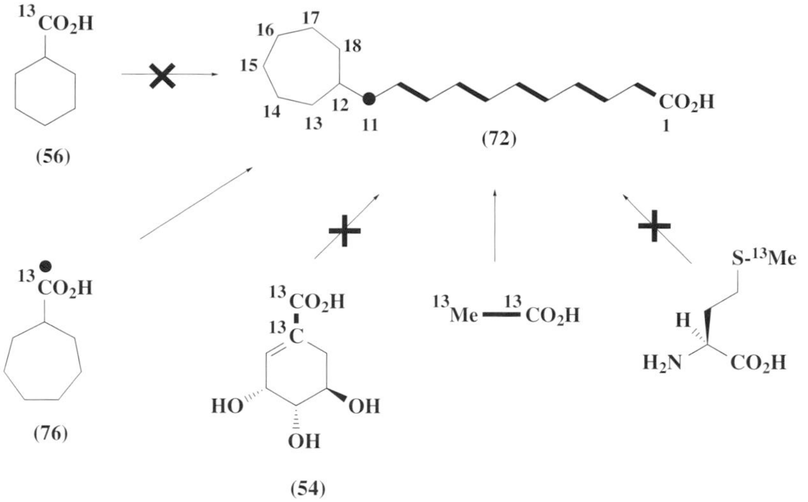

Figure 13 Precursors of ω-cycloheptylundecanoic acid (**72**).

of (**72**), hypothesis (i) was eliminated. Hypothesis (ii) was ruled out from a number of experiments. Neither the methyl group of methionine nor an acetate carbon was incorporated into the (**76**)-derived carbons of (**72**), and no ^{13}C ^{13}C coupling between C-11 and C-12 was observed in a feeding experiment with [1,7-$^{13}C_2$]shikimic acid. [7-^{13}C]Cyclohexanecarboxylic acid was also not incorporated, although it gave rise to the formation of the corresponding ω-cyclohexyl fatty acids in *A. cycloheptanicus* without dilution of the isotope by endogenous material. The possible involvement of a hypothetical "homoshikimic acid" intermediate, hypothesis (iii), was probed in feeding experiments with uniformly and positionally ^{13}C-labeled glucose samples. Analysis of the resultant labeling and coupling patterns in the cycloheptyl moiety of (**72**) was clearly not compatible with this hypothesis, but rather suggested a shikimate pathway origin of cycloheptanecarboxylic acid via aromatic amino acids (Figure 14).[99]

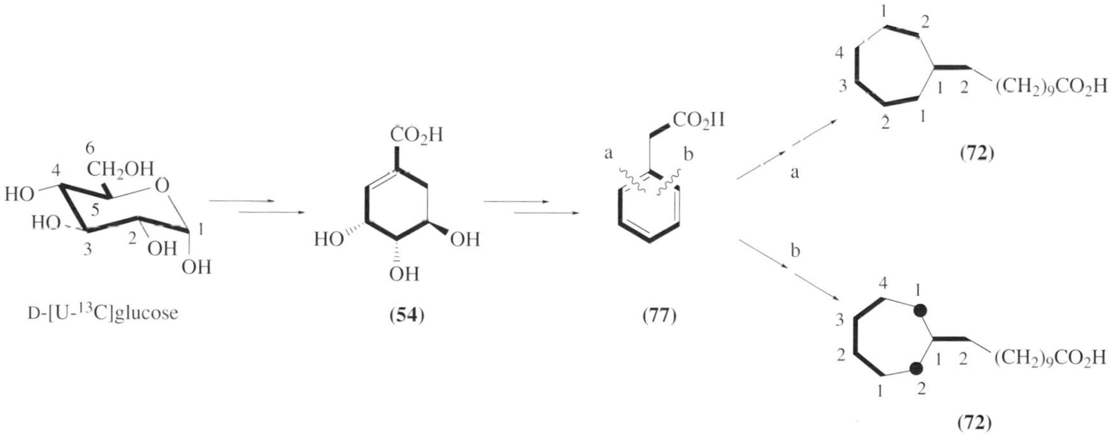

Figure 14 ^{13}C-Labeling and ^{13}C–^{13}C coupling patterns in ω-cycloheptylundecanoic acid (**72**) biosynthesized from [U-^{13}C]glucose in *A. cycloheptanicus*.

The ^{13}C–^{13}C coupling pattern of (**72**) derived from [U-$^{13}C_6$]glucose was complex due to the symmetry of the cycloheptane ring in (**72**) (Figure 14).[99,100] The labeling pattern was identical to that reported by Cane *et al.* for the biosynthesis of thiotropocin (**78**), a *Pseudomonas* metabolite containing a carbon framework identical to that of cycloheptanecarboxylic acid.[101] As Cane *et al.* had recognized, this coupling pattern arises from the ring expansion of phenylacetic acid (**77**) formed plausibly from (**54**) via L-phenylalanine (**79**). They confirmed this interpretation by demonstrating incorporation of [1,2-$^{13}C_2$]-(**77**) to give (**78**) with the expected labeling pattern.[101] Similar feeding experiments with [1,2-$^{13}C_2$]-(**77**) to *A. cycloheptanicus* gave ω-cycloheptylundecanoic acid showing ^{13}C enrichment in C-11 and C-12 (20–27%) and strong one-bond coupling between these two nuclei. The suspected origin of (**77**) from (**79**) was verified by a number of feeding experiments with deuterium-labeled samples of (**79**).

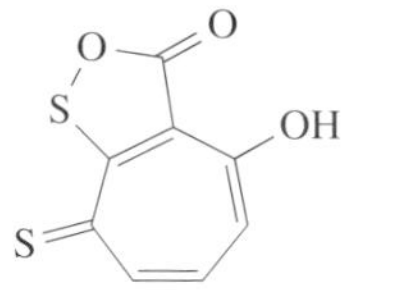

thiotropocin (**78**)

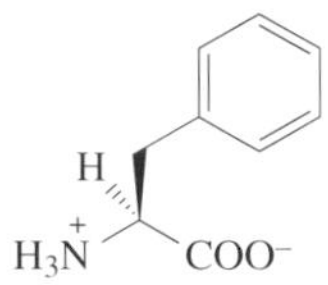

L-Phenylalanine (**79**)

To gain more information on the transformation of (**77**) to (**76**), the fates of the aromatic and benzylic hydrogens of phenylacetic acid were traced by deuterium labeling. The results are summarized in Figure 15, which also shows the hypothetical pathway for the conversion of (**77**) into (**76**) based on the available data.[99] Of the two methylene hydrogens of (**77**), H_R is completely eliminated, suggesting initiation of the reaction by an oxidative attack at the benzylic carbon of (**77**). Mandelic acid, the C-2 hydroxylation product of (**77**), containing a deuterium atom at the benzylic position, however, was not incorporated into (**72**). The other benzylic hydrogen of (**77**), H_S, migrates to C-14 of the cycloheptane ring where it becomes equivalent to one of the two *meta* hydrogens (H_m) from the precursor. The two *ortho* hydrogens (H_o), the *para* hydrogen (H_p) and the other *meta* hydrogen of (**77**) appear in the expected positions in the ring of (**72**), at C-13, C-17, C-15, and C-16, respectively. The data also revealed that the biosynthesis must proceed through a symmetrical intermediate, since the labeling patterns on the two sides of the ring have been equilibrated. For example, 50% of H_o at C-13 is found in the equatorial position and the other 50% in the axial position, presumably at C-18. Likewise, H_p is located 50% in the equatorial position at C-15 and 50% in the axial position, presumably at C-16. This is true for every labeled hydrogen on the cycloheptane ring. Finally, half of H_S and half of H_m, with which it has become equivalent, are eliminated. Whether this elimination of the H_S and H_m hydrogen is the result of "symmetrization" of the ring or whether it reflects a nonstereospecific hydrogen elimination, as suggested in Figure 15, is not clear.

The deuterium labeling data provide important boundary conditions for the formulation of a biosynthetic pathway, but they are insufficient to define the pathway completely. The hypothetical route shown in Figure 15 satisfies most of the boundary conditions but is entirely hypothetical. Evidence is available to support some of the later steps in the pathway.[99] Compound (**80**) and its double bond isomer (**84**) were tested as precursors of (**72**) (Figure 16). Compound (**84**) was incorporated to the extent of 4% vs. 20% for (**78**), and with retention of only 10% of the deuterium from C-1, as opposed to 50% of H_S from (**78**). Most importantly, the deuterium in the product, although it had undergone the 1,3-migration, was located exclusively in the axial position of C-14 of the cycloheptane ring. This makes it unlikely that the conversion of (**84**) is part of the natural biosynthetic pathway to (**72**). The data for (**80**), 8% incorporation with complete retention of deuterium at C-14, are more consistent with the intermediacy of this compound in (**72**) biosynthesis. The location of the incorporated deuterium, half in the axial and half in the equatorial position, is also consistent with the pattern of incorporation of (**77**). However, since (**80**) was deuterated nonstereospecifically, this distribution would also have resulted from a nonphysiological conversion.

The intermediacy of the diene (**81**) and monoenes (**82**) and (**83**) was supported by the efficient incorporation of their carboxy ^{13}C-labeled samples.[99] Furthermore, the enzymatic reduction of the two monoene CoA thioesters was demonstrated in cell-free extracts of *A. cycloheptanicus*.[102] Feeding experiments with cycloheptyl mimics showed that neither compound (**60**) nor compound (**63**), both efficient precursors of ω-cyclohexyl fatty acids in *A. acidocaldarius*, were converted into ω-cyclohexyl fatty acids when fed to *A. cycloheptanicus* (Figure 17). However, diene (**85**), a mimic of (**81**), which is not converted into ω-cyclohexyl fatty acids in *A. acidocaldarius*, but rather gives rise to the 3-cyclohexenyl analogues, was efficiently reduced in *A. cycloheptanicus* and converted into ω-cyclohexyl fatty acids. Likewise, the two cyclohexenyl monoenes (**55**) and (**62**) were converted into ω-cyclohexyl fatty acids in *A. cycloheptanicus*. However, despite these initial results, substantially more work, particularly the synthesis and feeding of the other postulated intermediates, will be necessary to evaluate the working hypothesis for the pathway from (**77**) to (**72**).

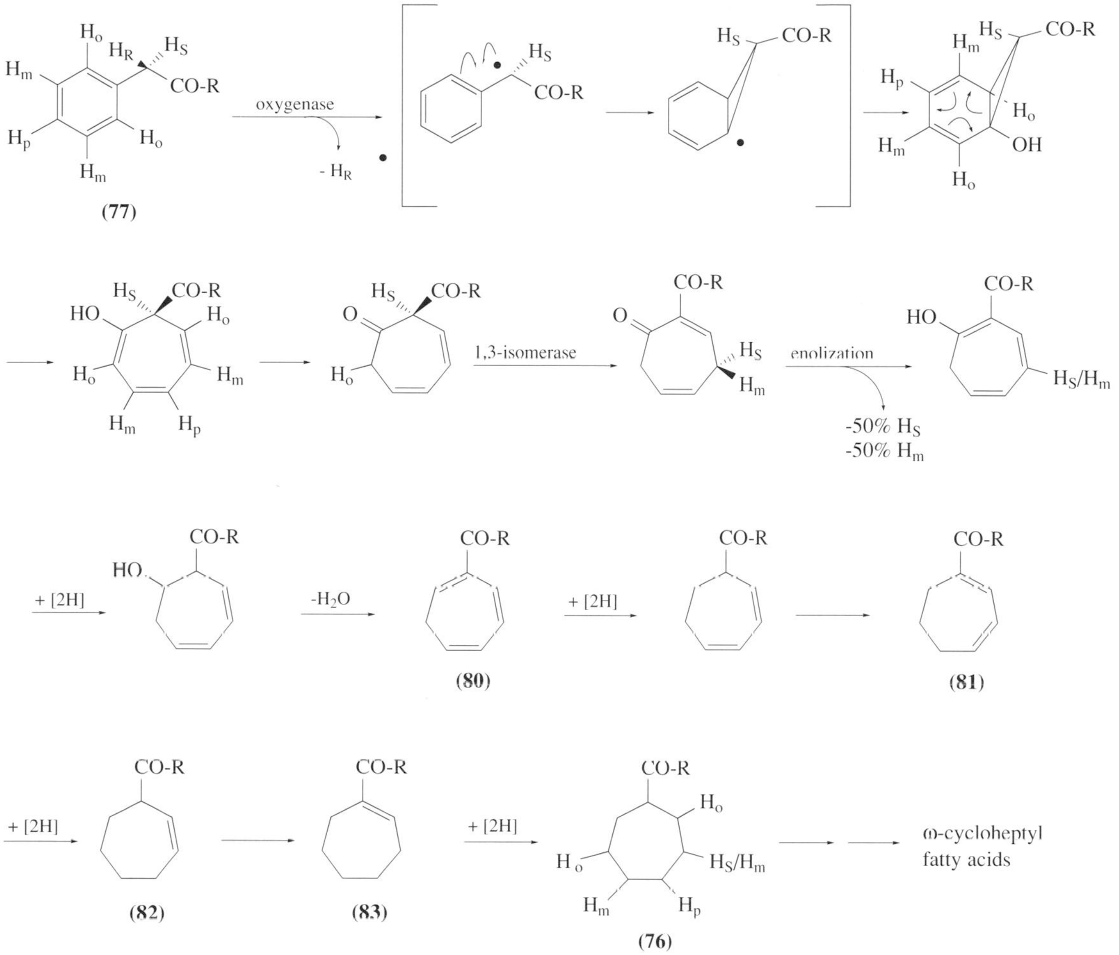

Figure 15 Proposed biosynthetic pathway for the conversion of (78) into (76) and the resultant deuterium distribution in (72) from deuterated (78). R = -OH or -SCoA.

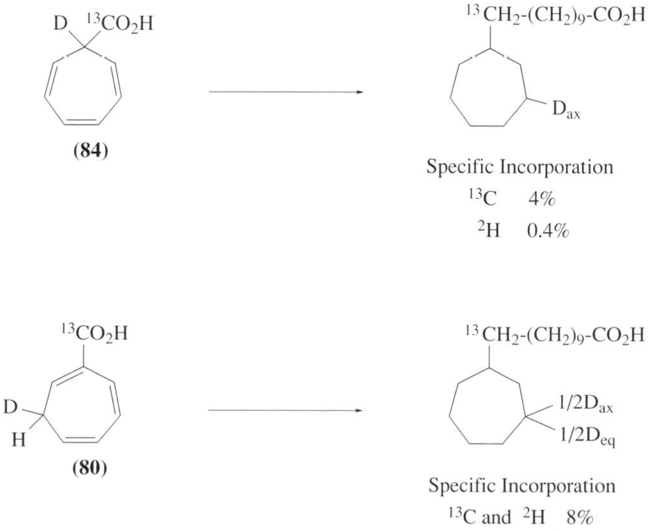

Figure 16 Incorporation of labeled cycloheptatrienecarboxylic acids into (72).

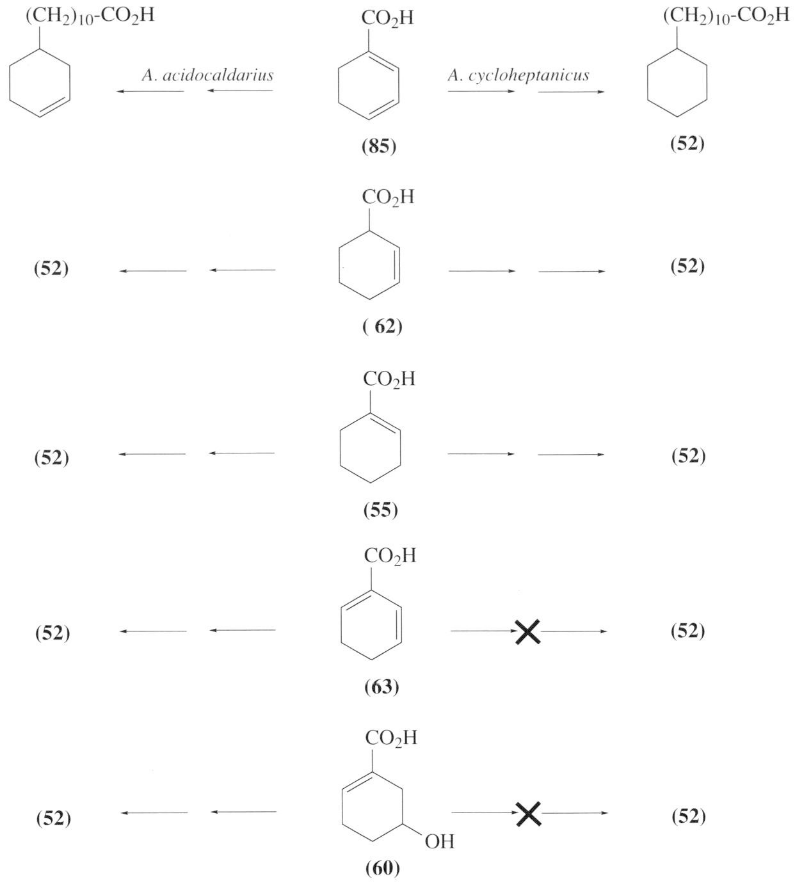

Figure 17 Conversion of cyclohexylcarboxylic acid derivatives into ω-cyclohexyl fatty acids in *A. acidocaldarius* and *A. cycloheptanicus*.

1.03.5.2.2 Biosynthetic interrelationships

The biosynthetic interrelationships of the fatty acids were examined by feeding ^{13}C-labeled ω-cycloheptylundecanoate (**72**) and -α-hydroxyundecanoate (**73**), which had been prepared by a feeding experiment with [8-^{13}C]cycloheptylcarboxylic acid.[98] Both of the ^{13}C-labeled acids were converted to ω-cycloheptyldecanoate (**71**), which is one carbon shorter in length, presumably through α-dehydrogenation and oxidative decarboxylation. α-Hydroxylation of (**72**) was also observed, but not chain extension to ω-cycloheptyltridecanoate (**74**). The results also suggest that these transformations take place at the level of the free acids which are not in complete equilibrium with the corresponding fatty acid moieties in the lipids.

ACKNOWLEDGMENTS

Research on the biosynthesis of *Alicyclobacillus* ω-cycloalkyl fatty acids and *Streptomyces* antibiotics containing cyclohexyl moieties was supported by the National Institutes of Health through grant AI 20264.

1.03.6 REFERENCES

1. T. Kaneda, *Microbiol. Rev.*, 1991, **55,** 288.
2. H. K. Mangold and F. Spener, in "The Biochemistry of Plants," eds. P. K. Stumpf and E. E. Conn, Academic Press, New York, 1980, vol. 4, pp. 647–663.
3. G. Deinhard and K. Poralla, *Biospektrum*, 1996, **2,** 40.
4. H. Goldfine, *Adv. Microbiol. Physiol.*, 1972, **8,** 1.

5. W. Fenical, H. L. Sleeper, V. J. Paul, M. O. Stallard, and H. H. Sun, *Pure Appl. Chem.*, 1979, **51**, 1865.
6. W. H. Gerwick, *Chem. Rev.*, 1993, **93**, 1807.
7. M. V. D'Auria, L. G. Paloma, L. Minale, R. Riccio, and A. Zampella, *J. Nat. Prod.*, 1993, **56**, 418.
8. D. W. Grogan and J. E. Cronan Jr., *J. Bacteriol.*, 1986, **166**, 872.
9. D. W. Grogan and J. E. Cronan Jr., *J. Bacteriol.*, 1984, **158**, 286.
10. A.-Y. Wang and J. E. Cronan Jr., *Mol. Microbiol.*, 1994, **11**, 1009.
11. M. Monteoliva-Sanchez, A. Ramos-Cormenzana, and N. J. Russell, *J. Gen. Microbiol.*, 1993, **139**, 1877.
12. K. M. George, Y. Yuan, D. R. Sherman, and C. E. Barry III, *J. Biol. Chem.*, 1995, **270**, 27 292.
13. M. S. Kuo, R. J. Zielinski, J. I. Cialdella, C. K. Marschke, M. J. Dupuis, G. P. Li, D. A. Koosterman, C. H. Spilman, and V. P. Marshall, *J. Am. Chem. Soc.*, 1995, **117**, 10 629.
14. A. G. M. Barrett, D. Hamprecht, A. J. P. White, and D. J. Williams, *J. Am. Chem. Soc.*, 1996, **118**, 7863.
15. N. C. Ling, R. L. Hale, and C. Djerassi, *J. Am. Chem. Soc.*, 1970, **92**, 5281.
16. J.-L. Giner, *Chem. Rev.*, 1993, **93**, 1735.
17. J. H. Law, *Acc. Chem. Res.*, 1971, **4**, 199.
18. G. Jaureguiberry, M. Lenfant, R. Toubiana, R. Azerad, and E. Lederer, *J. Chem. Soc., Chem. Comm.*, 1966, 855.
19. J. W. Polacheck, B. E. Tropp, J. H. Law, and J. A. McCloskey, *J. Biol. Chem.*, 1966, **241**, 3362.
20. P. H. Buist and D. B. MacLean, *Can. J. Chem.*, 1981, **59**, 828.
21. J. F. Tocanne, *Tetrahedron*, 1972, **28**, 363.
22. J. F. Tocanne and R. G. Bergmann, *Tetrahedron*, 1972, **28**, 373.
23. S. Rásonyi, Ph.D. Dissertation, ETH, Zürich, 1995.
24. P. H. Buist, J. M. Findlay, G. Leger, and R. A. Pon, *Tetrahedron Lett.*, 1987, **28**, 3891.
25. P. H. Buist and R. A. Pon, *J. Org. Chem.*, 1990, **55**, 6240.
26. P. H. Buist and J. M. Findlay, *Can. J. Chem.*, 1985, **63**, 971.
27. J.-P. Obrecht, Ph.D. Dissertation, ETH, Zürich, 1982.
28. H. Zalkin, J. H. Law, and H. Goldfine, *J. Biol. Chem.*, 1963, **238**, 1242.
29. F. R. Taylor and J. E. Cronan Jr., *Biochemistry*, 1979, **18**, 3292.
30. A.-Y. Wang, D. W. Grogan, and J. E. Cronan Jr., *Biochemistry*, 1992, **31**, 11 020.
31. Y. Yuan, R. E. Lee, G. E. Besra, J. T. Belisle, and C. E. Barry III, *Proc. Natl. Acad. Sci. USA*, 1995, **92**, 6630.
32. Y. Yuan and C. E. Barry III, *Proc. Natl. Acad. Sci. USA*, 1996, **93**, 12 828.
33. F. Spener and H. K. Mangold, *Biochemistry*, 1974, **13**, 2241.
34. W. W. Christie, E. Y. Brechany, and V. K. S. Shukla, *Lipids*, 1989, **24**, 116.
35. K. Mislow and I. V. Steinberg, *J. Am. Chem. Soc.*, 1955, **77**, 3807.
36. J. Miralles, M. Aknin, L. Micouin, E. M. Gaydou, and J. M. Kornprobst, *Phytochemistry*, 1990, **29**, 2161.
37. U. Cramer and F. Spener, *Biochim. Biophys. Acta*, 1976, **450**, 261.
38. U. Cramer and F. Spener, *Eur. J. Biochem.*, 1977, **74**, 495.
39. U. Cramer, A. G. Rehfeldt, and F. Spener, *Biochemistry*, 1980, **19**, 3074.
40. I. Tober and F. Spener, *Plant Cell Rep.*, 1982, **1**, 193.
41. I. Tober and E. E. Conn, *Phytochemistry*, 1985, **24**, 1215.
42. E. S. Olafsdottier, L. B. Jorgensen, and J. W. Jaroszewski, *Phytochemistry*, 1992, **31**, 4129.
43. J. W. Jaroszewski, A. B. Rasmussen, H. B. Rasmussen, C. E. Olsen, and L. B. Jorgensen, *Phytochemistry*, 1996, **42**, 649.
44. B. Hippchen, A. Roell, and K. Poralla, *Arch. Microbiol.*, 1981, **129**, 53.
45. G. Darland and T. D. Brock, *J. Gen. Microbiol.*, 1971, **67**, 9.
46. W. Krischke and K. Poralla, *Arch. Microbiol.*, 1990, **153**, 463.
47. E. Kannenberg, A. Blume, and K. Poralla, *FEBS Lett.*, 1984, **172**, 331.
48. G. Deinhard, P. Blanz, K. Poralla, and E. Altan, *Syst. Appl. Microbiol.*, 1987, **10**, 47.
49. K. Suzuki, K. Saito, A. Kawaguchi, S. Okuda, and K. Komagata, *J. Gen. Appl. Microbiol.*, 1981, **27**, 261.
50. M. De Rosa, A. Gambacorta, and J. D. Bu'lock, *J. Bacteriol.*, 1974, **117**, 212.
51. G. Deinhard, J. Saar, W. Krischke, and K. Poralla, *Syst. Appl. Microbiol.*, 1987, **10**, 68.
52. J. D. Wisotzkey, P. Jurtshuk, G. E. Fox, G. Deinhard, and K. Poralla, *Int. J. Sys. Bacteriol.*, 1992, **42**, 263.
53. M. Schnitzer, C. A. Hindle, and M. Meglic, *Soil Sci. Soc. Am. J.*, 1986, **50**, 913.
54. T. N. Misra, R. S. Singh, H. S. Pandey, C. Prasad, and B. P. Singh, *Phytochemistry*, 1993, **33**, 221.
55. M. Damberg, P. Russ, and A. Zeeck, *Tetrahedron Lett.*, 1982, 59.
56. M. Sugita, Y. Natori, T. Sasaki, K. Furihata, A. Shimazu, H. Seto, and N. Otake, *J. Antibiot.*, 1982, **35**, 1460.
57. M. Sugita, T. Sasaki, K. Furihata, H. Seto, and N. Otake, *J. Antibiot.*, 1982, **35**, 1467.
58. Y. Hu, S. Handa, and H. G. Floss, unpublished results.
59. M. De Rosa, A. Gambacorta, and J. D. Bu'lock, *Phytochemistry*, 1974, **13**, 1793.
60. M. Oshima and T. Ariga, *J. Biol. Chem.*, 1975, **250**, 6963.
61. J. Furukawa, T. Tsuyuki, N. Morisaki, N. Uemura, Y. Koiso, B. Umezawa, A. Kawaguchi, S. Iwasaki, and S. Okuda, *Chem. Pharm. Bull.*, 1986, **34**, 5176.
62. E. Haslam, "Shikimic Acid: Metabolism and Metabolites," Wiley, Chichester, 1993.
63. T. S. Wu, J. Duncan, S. W. Tsao, C. J. Chang, P. J. Keller, and H. G. Floss, *J. Nat. Prod.*, 1987, **50**, 108.
64. R. Casati, J. M. Beale, and H. G. Floss, *J. Am. Chem. Soc.*, 1987, **109**, 8102.
65. B. S. Moore, H. Cho, R. Casati, E. Kennedy, K. A. Reynolds, J. M. Beale, U. Mocek, and H. G. Floss, *J. Am. Chem. Soc.*, 1993, **115**, 5254.
66. B. S. Moore, K. Poralla, and H. G. Floss, *J. Am. Chem. Soc.*, 1993, **115**, 5267.
67. H. G. Floss, H. Cho, K. A. Reynolds, E. Kennedy, B. S. Moore, J. M. Beale, U. Mocek, and K. Poralla, in "Environmental Science Research," eds. R. J. Petroski and S. P. McCormick, Plenum Press, New York, 1992, p. 77.
68. K. A. Reynolds, P. Wang, K. M. Fox, and H. G. Floss, *J. Antibiot.*, 1992, **45**, 411.
69. S. Handa and H. G. Floss, *J. Chem. Soc., Chem. Commun.*, 1997, 153.
70. R. K. Hill and G. R. Newkome, *J. Am. Chem. Soc.*, 1969, **91**, 5893.
71. D. K. Onderka and H. G. Floss, *J. Am. Chem. Soc.*, 1969, **91**, 5894.
72. I. G. Young and F. Gibson, *Biochim. Biophys. Acta*, 1969, **177**, 182.

73. B. S. Moore and H. G. Floss, *J. Nat. Prod.*, 1994, **57**, 382.
74. P. Willadsen and W. Buckel, *FEMS Microbiol. Lett.*, 1990, **70**, 187.
75. K. A. Reynolds, N. Seaton, K. M. Fox, K. Warner, and P. Wang, *J. Nat. Prod.*, 1993, **56**, 825.
76. K. A. Reynolds, K. M. Fox, Z. Yuan, and Y. Lam, *J. Am. Chem. Soc.*, 1991, **113**, 4339.
77. K. A. Reynolds, P. Wang, K. M. Fox, M. K. Speedie, Y. Lam, and H. G. Floss, *J. Bacteriol.*, 1992, **174**, 3850.
78. P. Wang, C. D. Denoya, M. R. Morgenstern, D. D. Skinner, K. K. Wallace, R. Digate, S. Patton, N. Banavali, G. Schuler, M. K. Speedie, and K. A. Reynolds, *J. Bacteriol.* 1996, **178**, 6873.
79. C. Vezina, A. Kudelski, and S. N. Sehgal, *J. Antibiot.*, 1975, **28**, 721.
80. S. N. Sehgal, H. Baker, and C. Vezina, *J. Antibiot.*, 1975, **28**, 727.
81. H. Hatanaka, M. Iwami, T. Kino, T. Goto, and M. Okuhara, *J. Antibiot.*, 1988, **41**, 1586.
82. H. Hatanaka, T. Kino, S. Miyata, N. Imunara, A. Kuroda, T. Goto, H. Tanaka, and M. Okuhara, *J. Antibiot.*, 1988, **41**, 1592.
83. T. Kino, H. Hatanaka, M. Hashimoto, M. Nishiyama, T. Goto, M. Okuhara, M. Kohsaka, H. Aoki, and H. Imanaka, *J. Antibiot.*, 1987, **40**, 1249.
84. T. Kino, H. Hatanaka, S. Miyata, N. Inamura, M. Nishiyama, T. Yajima, T. Goto, M. Okuhara, M. Kohsaka, H. Aoki, and T. Ochiai, *J. Antibiot.*, 1987, **40**, 1256.
85. H. Tanaka, A. Kuroda, H. Murusawa, H. Hatanaka, T. Kino, T. Goto, M. Hashimoto, and S. T. Taga, *J. Am. Chem. Soc.*, 1987, **109**, 5031.
86. N. L. Paiva, A. L. Demain, and M. F. Roberts, *J. Nat. Prod.*, 1991, **54**, 167.
87. K. Byrne, A. Shaffiee, J. B. Nielsen, B. Arison, R. L. Monaghan, and L. Kaplan, *Dev. Ind. Microbiol.*, 1993, **32**, 29.
88. T. Schwecke, J. F. Aparicio, I. Molnar, A. König, L. E. Khaw, S. F. Haydock, M. Oliynyk, P. Caffrey, J. Cortes, J. B. Lester, G. A. Böhm, J. Staunton, and P. F. Leadlay, *Proc. Natl. Acad. Sci. USA*, 1995, **92**, 7839.
89. I. Molnar, J. F. Aparicio, S. F. Haydock, L. E. Khaw, T. Schwecke, A. König, J. Staunton, and P. F. Leadlay, *Gene*, 1996, **169**, 1.
90. J. F. Aparicio, I. Molnar, T. Schwecke, A. König, S. F. Haydock, L. E. Khaw, J. Staunton, and P. F. Leadlay, *Gene*, 1996, **169**, 9.
91. N. L. Paiva, M. F. Roberts, and A. L. Demain, *J. Ind. Microbiol.*, 1993, **12**, 423.
92. P. A. S. Lowden, G. Böhm, J. Staunton, and P. F. Leadlay, *Angew. Chem. Int. Ed. Engl.*, 1996, **35**, 2249.
93. K. K. Wallace, K. A. Reynolds, K. Koch, H. A. I. McArthur, M. S. Brown, R. G. Wax, and B. S. Moore, *J. Am. Chem. Soc.*, 1994, **116**, 11 600.
94. K. A. Reynolds, S. Handa, K. Wallace, M. S. Brown, H. A. I. McArthur, and H. G. Floss, *J. Antibiot.*, 1997, **50**, 701.
95. P. S. Lowden, S. Handa, J. Staunton, P. F. Leadley, and H. G. Floss, unpublished results.
96. H. Allgaier, K. Poralla, and G. Jung, *Liebigs Ann. Chem.*, 1985, 378.
97. K. Poralla and W. A. König, *FEMS Microbiol. Lett.*, 1983, **16**, 303.
98. B. S. Moore, K. Poralla, and H. G. Floss, *J. Nat. Prod.*, 1995, **58**, 590.
99. B. S. Moore, K. Walker, I. Tornus, S. Handa, K. Poralla, and H. G. Floss, *J. Org. Chem.*, 1997, **62**, 2173.
100. T. Pratum and B. S. Moore, *J. Mag. Reson., Series B*, 1994, **102**, 91.
101. D. E. Cane, Z. Wu, and J. E. Van Epp, *J. Am. Chem. Soc.*, 1992, **114**, 8479.
102. I. Tornus and H. G. Floss, unpublished results.

1.04
Biosynthesis of So-called "Green Odor" Emitted by Green Leaves

AKIKAZU HATANAKA
University of East Asia, Shimonoseki, Japan

1.04.1 INTRODUCTION

The so-called "fresh green odor" components emitted by green leaves involve eight volatile C_6 aldehydes and C_6 alcohols, including so-called leaf aldehyde, (*E*)-2-hexenal, and leaf alcohol (*Z*)-3-hexenol (Figure 1). These volatile components are the major contributors to the characteristic fresh green odor of green leaves of various classes of plants[1–3] and have been the subject of several reviews.[4–9] In response to various environmental stimuli, green leaves emit physiologically significant green odors whose characteristics are dependent on the concentrations of the eight volatile C_6 compounds. The subtle differences in the composition of each of the eight green odor components are thought to be used by plants to communicate with or attack other species, and also to attract or repel insects: allelopathy.[10] In addition, plants can kill certain bacteria such as *Dermatophytes* and *Staphylococcus* species by using the green odor components (main component: leaf aldehyde) at various concentrations which act as a phytonocide. Certain ants take green odor compounds into their bodies by consuming green leaves and then use them as pheromones for communication, alarm and attack, etc.[11,12] Humans also find the green odor emitted by green leaves in forests refreshing. This is also an example of pheromones. Following the pioneering studies on the green odor of green leaves in 1881 by Reinke,[13,14] a botanist at the University of Göttingen, leaf aldehyde, 2-hexenal, was first isolated from the green leaves of bushes in 1912 by Curtius,[15,16] an organic chemist at the University of Heidelberg. Leaf alcohol, 3-hexenol, was found in black tea during fermentation[17] and also in fresh tea (*Thea sinensis*) leaves.[18] The study of leaf alcohol was continued from 1933 to 1942 by Takei at Kyoto University.[18–28] Since 1957, the author of this chapter has studied green odor in plants by using a multidisciplinary approach involving synthetic chemistry, natural product chemistry, flavor chemistry, plant biochemistry, molecular biology, and plant physiology.[6,29–43] 2-Hexenal and 3-hexenol were determined as the (*E*)-form for the former and as the (*Z*)-form for the latter by comparison with synthetic specimens.[44–46] Leaf alcohol and leaf aldehyde are synthesized industrially on a fairly large scale, $\sim 2.5 \times 10^5$ kg yr^{-1} for the former and $\sim 5 \times 10^4$ kg yr^{-1} for the latter, and they are widely utilized in the perfume and food technology fields.

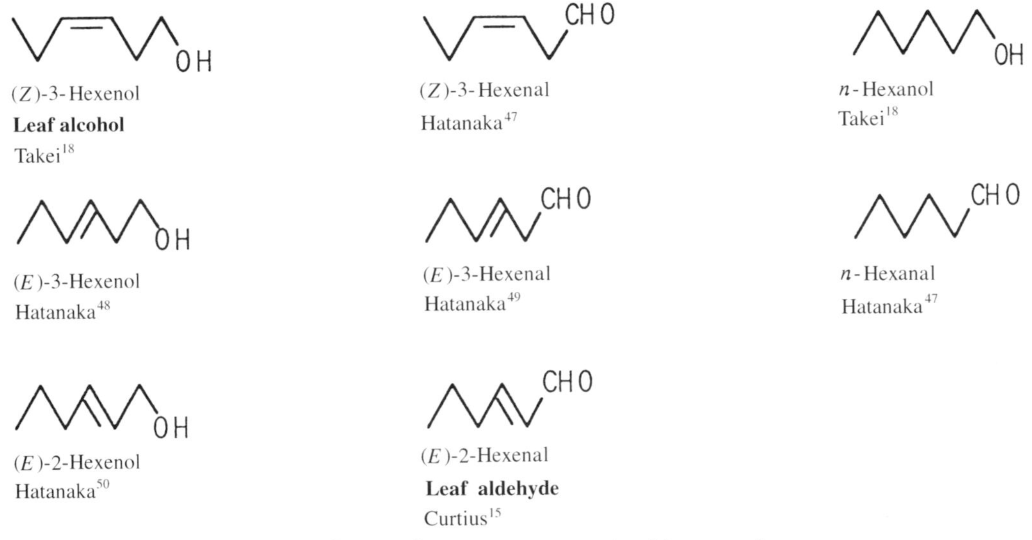

Figure 1 Green odor components emitted by green leaves.

In addition to leaf alcohol and leaf aldehyde, their isomers, (*E*)-3-hexenol, (*E*)-2-hexenol, (*Z*)-3-hexenal, (*E*)-3-hexenal, *n*-hexanol, and *n*-hexanal, were found to be present in *T. sinensis* (*Camellia japonica*) in studies from 1960 to 1981.[47–54] (*Z*)-3-Hexenal, *n*-hexanal, and (*Z*)-3-hexenol in fresh tomato volatiles were quantitatively determined by an excellent method using Tenax trapping and $CaCl_2$ enzyme deactivation.[55] Through these studies, it was found that the "green odor" components emitted by green leaves consist of eight volatile compounds, C_6 aldehydes and C_6 alcohols (Figure 1). In 1973, using tea chloroplasts, (*Z*)-3-hexenal and (*E*)-2-hexenal were found to be produced enzymatically from α-linoleic acid.[47] α-Linolenic and linoleic acids were reported to be possible precursors in fruit tissues of hexenals and *n*-hexanal, respectively.[56,57] This was confirmed in 1975 by labeling experiments involving the incubation of [U-^{14}C]-α-linolenic acid with tea chloroplasts.[58–60] 13-Hydroperoxylinoleic acid was found to be a reaction intermediate in the formation of *n*-hexanal from linoleic acid in chloroplasts.[61]

The results of these studies suggested that the biogeneration of green odor follows the route shown in Scheme 1: due to various environmental stimuli, lipolytic acyl hydrolase (lipase) first acts to form α-linolenic or linoleic acids by hydrolysis of neutral fats or phospholipids in chloroplasts. Subsequently, hydroperoxygenation of these fatty acids by lipoxygenase forms the corresponding 13-(*S*)-hydroperoxides and cleavage of the double bond between C-12 and C-13 in these hydroperoxides (HPOs) by HPO lyase gives (*Z*)-3-hexenal or *n*-hexanal. These aldehydes are converted into other C_6 aldehydes and C_6 alcohols by alcohol dehydrogenase (ADH) and/or an isomerization enzyme (isomerization factor, IF). Green odor components are formed by the action of an enzyme system involving four enzymes, one of which is bound to the chloroplast membrane. This chapter describes: (i) the synthesis of positional and geometric isomers including leaf alcohol, (*Z*)-3-hexenol, and leaf aldehyde, (*E*)-2-hexenal; (ii) relationships between the organoleptic properties of green odor and the structure of *n*-hexenals and *n*-hexenols; (iii) the chemistry of the formation of "Lipton black tea aroma" (aromatic compounds) from leaf alcohol homologues (aliphatic compounds), under simple and general reaction conditions; (iv) the formation of α-linolenic or linoleic acids from neutral fats and phospholipids by lipase; (v) the establishment of the biosynthetic pathway of green odor components; (vi) the distribution of green odor-forming enzymes in plant species; (vii) the biochemistry of lipoxygenase and HPO lyase of the enzyme system producing green odor; and (viii) plant physiology and environmental stimuli: the changes in enzyme activities throughout the year with temperature, solar radiation, and photosynthesis conditions, growing season and seed development, with ambient temperature and dark–light conditions, etc.

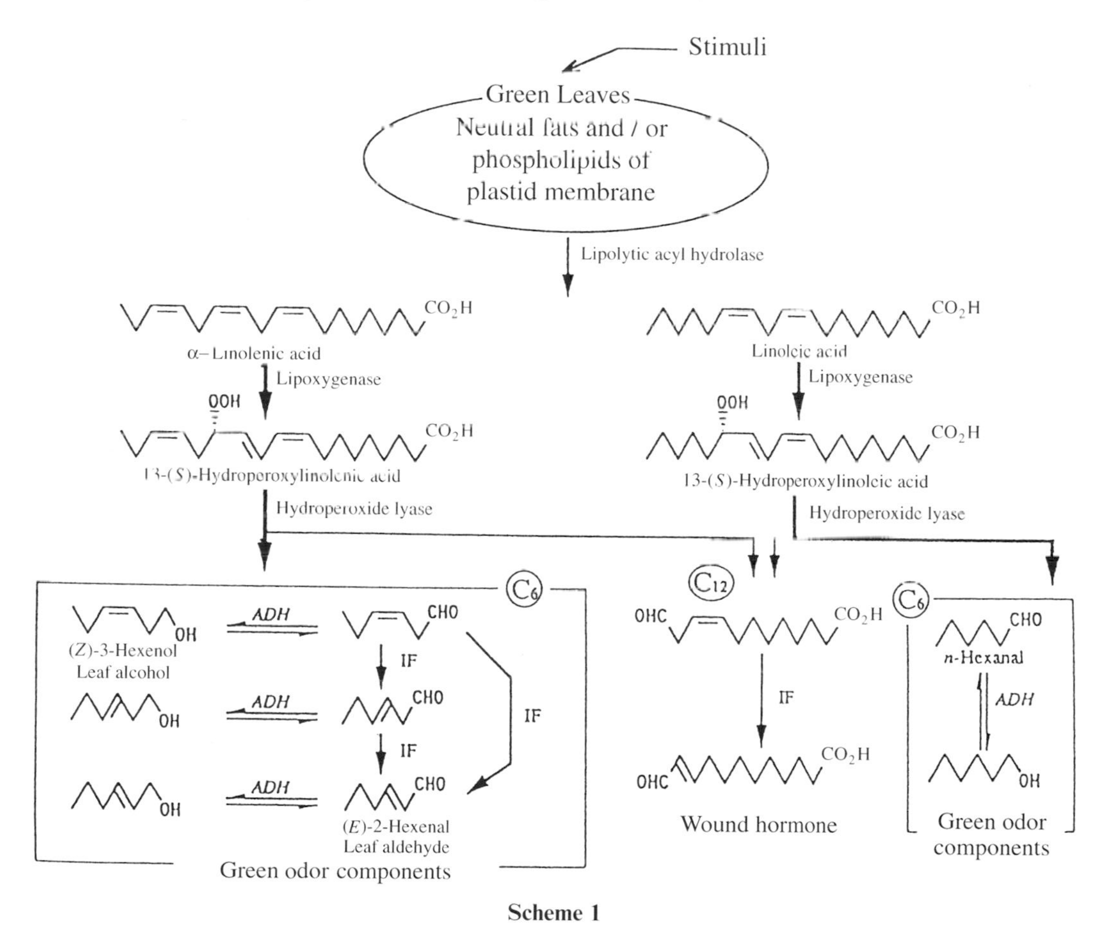

Scheme 1

1.04.2 SYNTHESIS OF THE SERIES OF POSITIONAL AND GEOMETRIC ISOMERS OF UNSATURATED C_6 ALCOHOLS AND C_6 ALDEHYDES

Leaf alcohol, (*Z*)-3-hexenol, has been synthesized through a three-step reaction with carbon chain elongation from sodium acetylide in liquid ammonia with over 98% geometric purity and at an overall yield of 40%.[29] Leaf aldehyde, (*E*)-2-hexenal, has been synthesized through a four-step reaction from butyric acid chloride and acetylene with an overall yield of 50%.[44] By modifying these synthetic methods, the entire series of positional and geometric isomers of leaf alcohol, i.e., (*E*)-2-, (*Z*)-2-, (*E*)-3-, (*E*)-4-, (*Z*)-4-, and 5-hexenols, were systematically obtained in good yields.[29] These compounds were effective for identifying green odor components: the isomers of leaf alcohol, (*E*)-3- and (*E*)-2-hexenols, and (*Z*)-3- and (*E*)-3-hexenals, in *T. sinensis* leaves.[47–54] The detection and identification of (*Z*)-3-hexenal in tea chloroplasts helped to elucidate the biosynthetic pathway of green odor released from green leaves as shown in Scheme 1 (see Section 1.04.5), as did a follow-up series of studies.[50,51] To examine the relationships between chemical structure and the characteristics of green odor, the corresponding aldehydes with the above seven hexenols, including (*Z*)-3-hexenol, were synthesized in high purity by established oxidation reactions.[62,63]

1.04.3 CHEMICAL STRUCTURE–ODOR CHARACTERISTICS RELATIONSHIPS IN *n*-HEXENOLS AND *n*-HEXENALS

The relationships between chemical structure and the characteristics of green odor were examined in a series of highly purified *n*-hexenals.[29,44,62,63] The threshold values of odor and taste for these compounds are summarized in Table 1. It appears from this table that the effect of geometry on the threshold values of odor and taste was much less than that of the double bond position. The values for hexenols were 10–1000 times higher than those for the corresponding hexenals having the same double bond position and (*E*)/(*Z*) geometry. The olfactory characteristics of the *n*-hexenols and *n*-hexenals are represented using radial charts in Figure 2. Although the absolute values for *n*-hexenals in the chart could not be directly compared with those for *n*-hexenols because of differences in dilution ratios, close resemblances were found between the *n*-hexenols and *n*-hexenals having the same regio- and stereo-double bonds. For 2-hexenols and 2-hexenals, fruity, sweet, and fresh flavors were dominant. The double bond at the ω-end leads to the odor profiles having oily–fatty and herbal components. However, a critical effect of double bond isomerism was not clearly revealed by these charts. Data from sensory evaluation were analyzed using principal component analysis[64] to characterize the odor and flavor profiles of the compounds tested. The analysis was carried out with all 14 compounds using 10 explanatory variables (i.e., descriptive terms). Table 2 shows the resulting eigenvalues, eigenvectors, and the proportions of the correlation matrices with six principal components. Two major principal components contributed over 70% to the cumulative proportion (Figure 2) and the contributions below the third principal component were small, therefore the analysis was performed on the first and second principal components to evaluate the odor images. Figure 3 shows a representation of the score of each compound in coordinates which visualize the vector patterns where the x-axis represents the first and the y-axis the second principal component. Fresh, fruity, and sweet flavors are substantially correlated with the x-axis, whereas green and spicy are correlated with the y-axis. Aldehydes having a double bond at the C-2 position are located on the right side of the diagram, which indicates that they have a fruity and sweet character. The corresponding alcohols are located in negative regions of the y-axis and show a sweeter character, but less green, than the corresponding aldehydes. (*Z*)-3-Hexenol shows the highest values for both types of component, which means that this isomer has a high green and fresh note. The compounds having a double bond at the C-4 or C-5 position, especially 5-hexenol and 5-hexenal, are located on the left side of the diagram, indicating that they have strong oily–fatty, insect-like green and herbal

Table 1 Threshold values of *n*-hexenols and *n*-hexenals for odor and taste according to double bond position and (*E*)/(*Z*) geometry.

Double bond position	*C-2*		*C-3*		*C-4*		*C-5*
Geometry C_6 compounds	(E)	(Z)	(E)	(Z)	(E)	(Z)	
Odor							
n-Hexenol	10	10	1	1	0.1	0.1	1
n-Hexenal	0.01	0.01	0.001	0.001	0.001	0.001	0.001
Taste							
n-Hexenol	0.6	2	1	0.03	0.3	0.09	2
n-Hexenal	0.06	0.08	0.006	0.0008	0.008	0.002	0.002

Table 2 Principal component analyses for the odor of *n*-hexenols and *n*-hexenals.

Odor description	*PC1*[a]	*PC2*	*PC3*	*PC4*	*PC5*	*PC6*
Leafy green	0.6473	0.6558	0.0226	−0.2643	−0.0111	0.2352
Grassy green	0.3818	0.7667	0.4528	0.1582	−0.1058	−0.1086
Insect-like green	−0.8935	0.2881	0.0149	0.0999	0.0348	−0.2155
Vegetable-like green	−0.7932	0.4983	−0.1188	0.1201	−0.1982	−0.0404
Fruity	0.9281	0.1573	0.1485	0.1830	0.0457	0.0523
Sweet	0.7808	−0.3105	−0.1854	0.4850	0.0349	0.0385
Fresh	0.9243	0.2753	0.0076	0.0292	0.0788	−0.1020
Spicy	−0.2476	0.8235	−0.4590	0.1385	0.0269	0.0991
Oily–fatty	−0.8475	−0.2111	0.2697	0.2315	−0.1675	0.2715
Herbal	−0.8223	0.2401	0.1868	0.0584	0.4618	0.0791
Eigenvalue	5.773	2.326	0.595	0.467	0.306	0.220
Proportion	57.732	23.265	5.947	4.666	3.063	2.196
Cumulative proportion	57.732	80.997	86.944	91.610	94.673	96.869

[a]PC = principal component.

odors. 4-Hexenals are located in the upper left part which indicates a strong impression of spicy and vegetable-like green odor in contrast to the corresponding alcohols located in the lower left part. (*Z*)-3-Hexenal, located in the upper part of the diagram near the *y*-axis, has a strong spicy and grassy green odor and can be distinguished from (*Z*)-3-hexenol, which is located on the right side, by its fruity and fresh odor. (*E*)-3-Hexenal was weaker in grassy green and spicy characteristics than (*Z*)-3-hexenal and is located in the lower part of the diagram. It is not possible from this analysis to establish whether the position of a double bond is closely related to the scores of the first principal component, whereas the type of functional group (alcohol/aldehyde) is clearly related to the second principal component affecting an odor. Only (*Z*)-3-hexenol does not fit this pattern and will require additional study to clarify its structure–activity relationships. This work is being extended by using the entire series of positional and geometric isomers of C_{7-9} monoenols and C_9 dienols involving the (2*E*,6*Z*)-nonadienol (cucumber alcohol) system.[65,66]

1.04.4 AROMATIZATION OF LEAF ALCOHOL

In 1935, a compound with a boiling point of 240 °C was obtained by refluxing leaf alcohol with sodium metal at 160–170 °C for 8 h. It had a citrus fruit or Lipton black tea-like fine odor.[20] This compound was identified as 2-propyl-5-ethylbenzyl alcohol by degradation methods,[67,68] and its identity was confirmed by comparison with a specimen synthesized independently.[69] The starting materials used for the aromatization reaction were extended to $\alpha\beta$-,$\beta\gamma$-unsaturated (*Z* and *E*)-C_4 to -C_6 enols and the corresponding β-enals.[68,70,71] This aromatization reaction has become known as the "leaf alcohol reaction" (Scheme 2).[67] A mechanism of formation has been proposed based on confirmation of the intermediates involved[70,72–79] and by a tracer study.[80] It is interesting that the aromatization reaction involved Michael, aldol, and Cannizzaro reactions (Scheme 3). As leaf alcohol has been found in black tea during fermentation,[17] in 1966, 1000 kg of black tea was steam distilled and extracted with diethyl ether in a search for this aromatic compound, but it could not be found.

1.04.5 BIOSYNTHETIC PATHWAY OF GREEN ODOR

1.04.5.1 Changes in Fatty Acid Content and C_6 Alcohols and C_6 Aldehydes During the Blending of Tea Leaves

In preliminary investigations on the biosynthesis of leaf alcohol and leaf aldehyde,[81–87] α-linolenic and linoleic acids were found to decrease markedly during the blending of tea leaves (Table 3).[50,88] Without blending, 95% of the total α-linolenic acid and 77% of the total linoleic acid were present in the neutral fat. The ratio of α-linolenic acid to linoleic acid in the total lipids of summer leaves was ~3:1 but in winter leaves the ratio increased to ~5:1. After blending the summer leaves for 3 min, half of the α-linolenic and linoleic acids disappeared from the neutral fat, whereas these acids

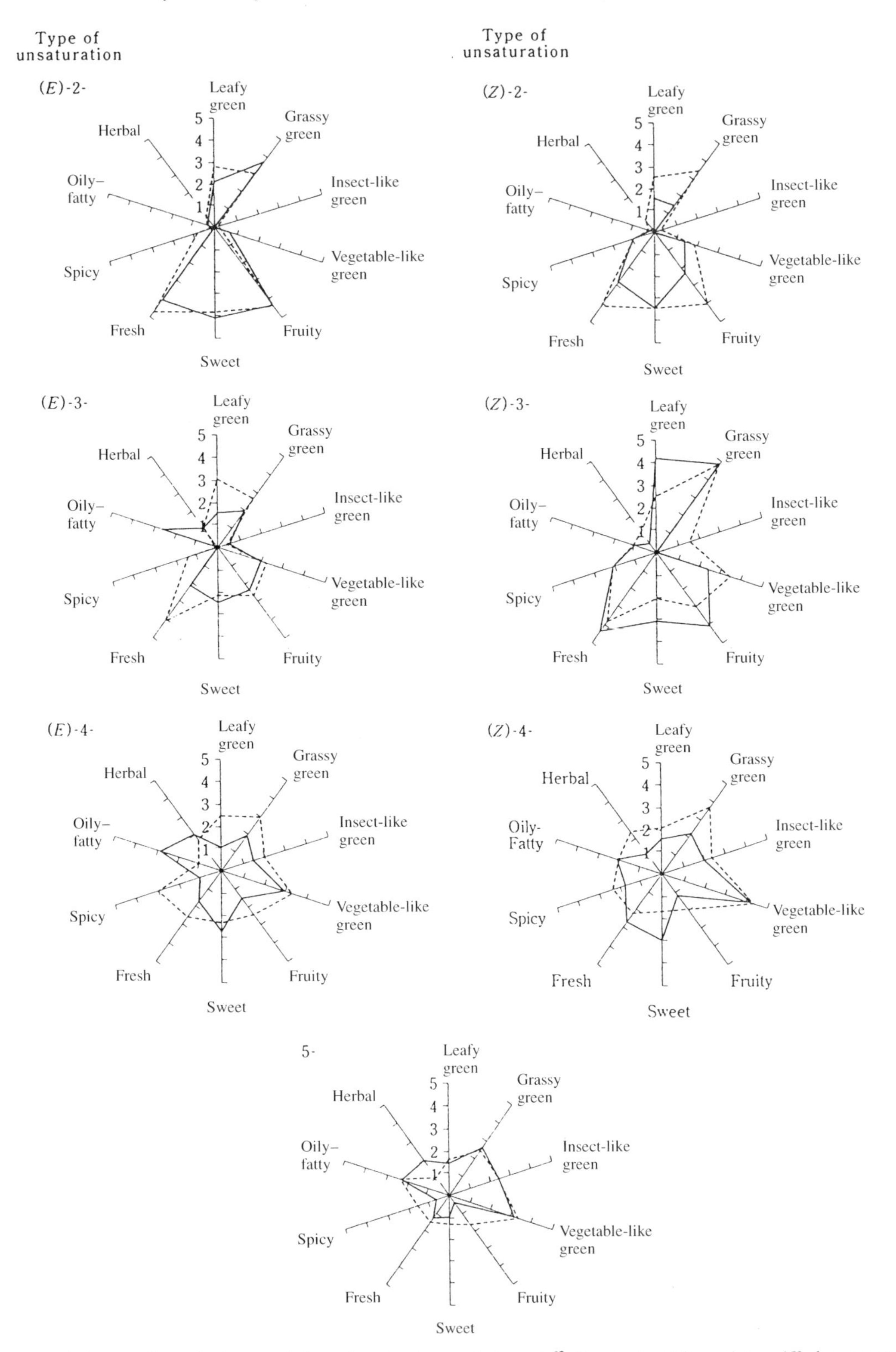

Figure 2 Odor profiles of *n*-hexenols (- - - -) and *n*-hexenals (——).[62] Five trained flavorists sniffed paper strips dipped in ethanolic solutions of hexenols or hexanals. They were asked to describe the sensory characteristics using seven sensory descriptors (the term "green odor" (1) was further classified into four descriptive terms). 1-1, Leafy green: an image of tree leaves; 1-2, grassy green: an image of grass; 1-3, insect-like green; 1-4, vegetable-like green: an image of vegetables, e.g., tomatoes, green peppers, cabbages, etc.; 2, fruity: an image of fruits, e.g., apples, berries, pears, etc.; 3, sweet: degree of sweetness; 4, fresh: degree of freshness; 5, spicy: an image of pepper, nutmeg, cinnamon, etc.; 6, oily–fatty, waxy, rancid; and 7, herbal: an image reminiscent of the bitterness of crude drugs. Score sheets with a six-point scale ranging from threshold to very intense (0,

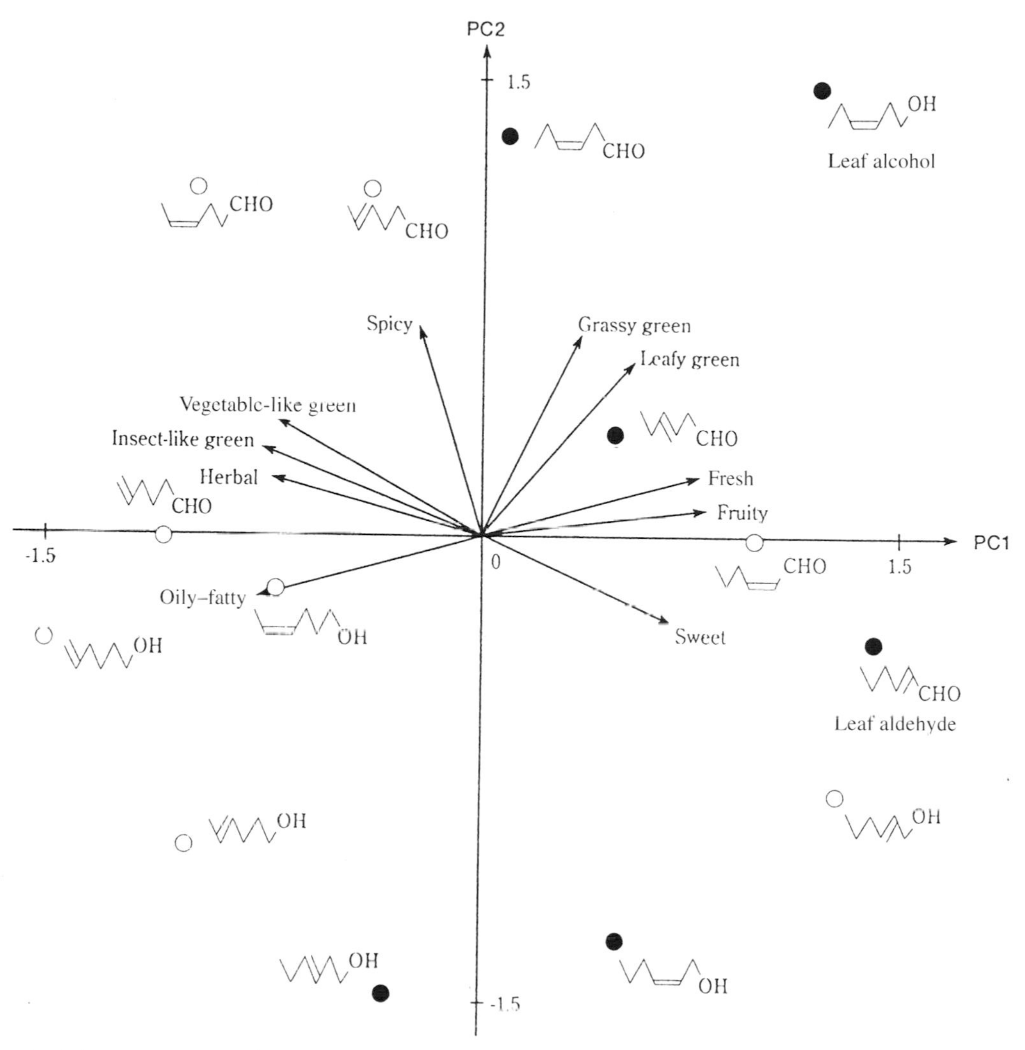

Figure 3 The score plots and the vectors of the eigenvalues on the plane of the first principal component vs. the second principal component in *n*-hexenols and *n*-hexenals. ●, green odor components by green leaves; ○, synthetic analogues.

decreased only slightly in the phospholipids of summer leaves, and free fatty acids remained constant in low concentrations. At the same time, a strong green odor was also apparent in the homogenate of the tea leaves. This indicated that α-linolenic and linoleic acids were converted into green odoriferous compounds during blending. The latter were thought to be C_6 compounds, such as (*E*)-2-hexenal and (*Z*)-3-hexenol. In an experiment on blending, (*E*)-2-hexenal and (*Z*)-3-hexenol increased as the α-linolenic acid content declined (Figure 4).[50,88] In this experiment, which employed steam distillation, (*Z*)-3-hexenal was not detected because isomerization of the aldehyde to (*E*)-2-hexenal is inevitable under these experimental conditions. In order to detect (*Z*)-3-hexenal, headspace vapor analysis was necessary.[50,89]

1.04.5.2 The Biosynthetic Pathway for C_6 Aldehyde Formation in Leaves

When [U-^{14}C]-α-linolenic acid was incubated with tea chloroplasts, two radioactive peaks were detected by radio gas chromatography of the headspace vapor at the positions corresponding to

threshold; 1, very weak; 2, weak; 3, medium; 4, intense; and 5, very intense) were used for evaluation of test samples. The average scores of the five panelists were adopted as the odor strength of the sensory attributes. These scores were subjected to principal component analysis using "ANALYST" (analyser programs for statistical data, Fujitsu, Tokyo, Japan) on a FACOM M-380 computer (Fujitsu, Tokyo, Japan).[64]

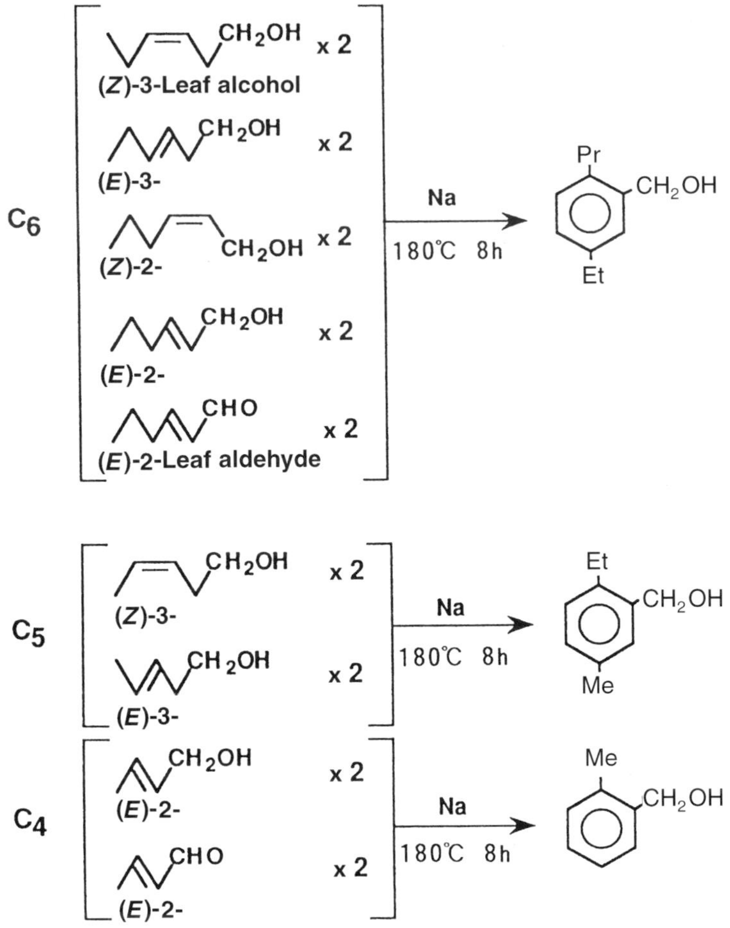

C_6

(*Z*)-3-Leaf alcohol CH_2OH x 2

(*E*)-3- CH_2OH x 2

(*Z*)-2- CH_2OH x 2

(*E*)-2- CH_2OH x 2

(*E*)-2-Leaf aldehyde CHO x 2

Na, 180℃ 8h → Pr, CH_2OH, Et

C_5

(*Z*)-3- CH_2OH x 2

(*E*)-3- CH_2OH x 2

Na, 180℃ 8h → Et, CH_2OH, Me

C_4

(*E*)-2- CH_2OH x 2

(*E*)-2- CHO x 2

Na, 180℃ 8h → Me, CH_2OH

Scheme 2

CHO, OHC — Michael, base → CHO, OHC — Aldol → CHO, Na, H, H, H

Na ↓

H, C–O–Na, H — -NaH → CHO — Cannizzaro → CH_2OH

[CO_2H]

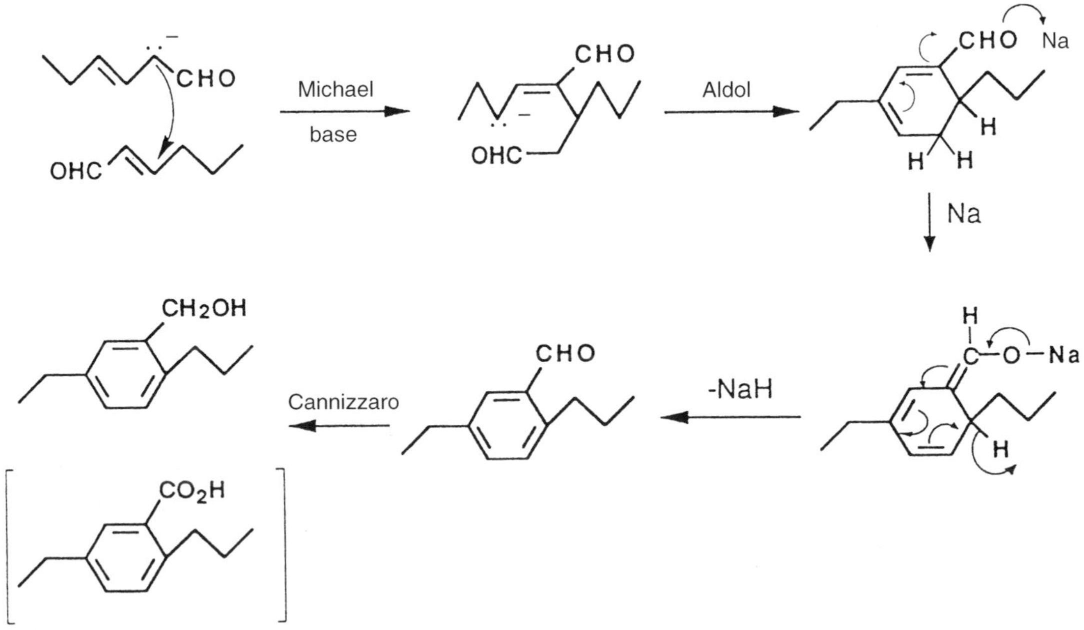

Scheme 3

Table 3 Changes in fatty acid content of tea leaves during blending.

	Blending time			
	α-Linolenic acid[a]		*Linoleic acid*[a]	
Fraction	*0 min*[b]	*3 min*	*0 min*[b]	*3 min*
Free acid	trace (trace)	trace (trace)	trace (trace)	trace (trace)
Neutral fat	218.3 (164.0)	99.3 (145.8)	55.6 (18.5)	23.5 (15.8)
Phospholipid	11.5 (8.4)	10.5 (1.4)	16.8 (13.8)	13.3 (3.5)
Total	229.8 (172.4)	109.8 (147.2)	72.4 (32.4)	36.8 (19.3)

[a]mg 100 g^{-1} fresh tea leaves. Summer leaves were harvested on June 6, winter leaves, shown in parentheses, on November 26. [b]Zero blending time: fractions prepared from leaves inactivated by heating at 80 °C for 15 min.

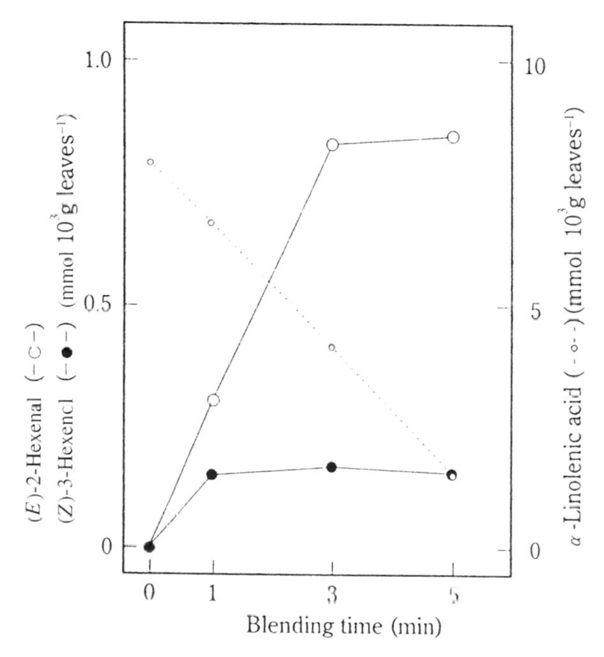

Figure 4 Relationship between decrease in α-linolenic acid and increase in (*E*)-2-hexenal and (*Z*)-3-hexenol during the blending of summer tea leaves.

(*Z*)-3-hexenal and (*E*)-2-hexenal, as shown by the spiked trace in Figure 5(a).[58] When the incubation times were prolonged, the radioactivity in (*Z*)-3-hexenal decreased and that in (*E*)-2-hexenal increased. [1-^{14}C]-α-Linolenic acid gave five radioactive peaks as shown in Figure 5(b).[59] In order from the far right of this figure, α-linolenic acid, an unknown compound, 11-formyl-(*E*)-10- and 11-formyl-(*Z*)-9-undecenoic acid, and 8-formyloctanoic acid were detected. With prolonged incubation, 11-formyl-(*E*)-10-undecenoic acid increased with a decrease in the (*Z*)-9-form.[59,60] These results indicate that (*Z*)-3-hexenal and 11-formyl-(*Z*)-9-undecenoic acid were formed from α-linolenic acid and underwent isomerization to the corresponding (*E*)-10 isomers. From (i) the finding of (*E*)-3- and (*E*)-2-hexenol and (*Z*)-3- and (*E*)-3-hexenal in addition to leaf aldehyde and leaf alcohol in tea leaves;[47–54] (ii) the enzymatic production of (*Z*)-3- and (*E*)-2-hexenal from α-linolenic acid during maceration;[50] (iii) the experiments on tea chloroplasts with [U-^{14}C]- and [1-^{14}C]-α-linolenic acid;[58,59] and (iv) confirmation of 13-HPO involvement in the formation of *n*-hexanal from linoleic acid by tea chloroplasts,[61] a plausible biosynthetic pathway for green odor components emerged.

The pathway to the fresh green odor components in green leaves involves the following steps: hydrolysis of natural fats and phospholipids in the chloroplast membrane by lipase;[88] hydroperoxygenation of the C-13 position of α-linolenic and linoleic acids by lipoxygenase;[58] cleavage between C-12 and C-13 in the HPOs by HPO lyase;[58,88] reduction of aldehydes to alcohols by ADH;[84–87,90] and isomerization of (*Z*)-3-hexanal to (*E*)-2-hexenal via (*E*)-3-hexenal or by a direct isomerization enzyme.[37] (*Z*)-3-Hexenal, (*E*)-2-hexenal, and 1-undecene were first found in *Farfugium japonicum* Kitamura (Japanese silver) leaves. Macerating the leaves in the presence of oxygen

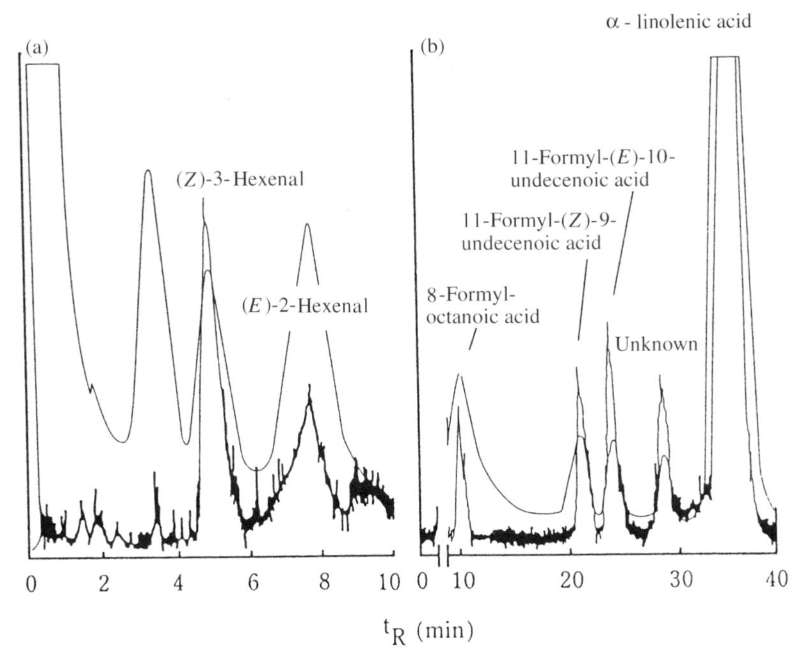

Figure 5 Radio gas chromatograms of the aldehydes and oxo acids formed from (a) [U-^{14}C]-α-linolenic acid and (b) [1-^{14}C]-α-linolenic acid in incubation with chloroplasts. The GC traces show the radioactivity (lower trace) and mass (top trace).

generated (*E*)-2-hexenal via a similar biosynthetic pathway to that for *T. sinensis* leaves, i.e., via (*Z*)-3-hexenal from α-linolenic acid. However, neither (*Z*)-3-hexenol (leaf alcohol) nor (*E*)-2-hexenol was found in *F. japonicum* leaves, in contrast to *T. sinensis* leaves.[91]

1.04.6 ENZYME SYSTEMS IN GREEN ODOR BIOSYNTHESIS

1.04.6.1 Enzymes Producing Green Odor Components

The enzyme system for green odor formation involves the following steps (see Scheme 1): (i) in response to various environmental stimuli, lipolytic acyl hydrolase is first activated to form α-linolenic and linoleic acids from galactolipids, phospholipids, and triglycerides in chloroplasts;[88,89,92,93] (ii) these free fatty acids are oxygenated to the corresponding 13-(*S*)-HPOs by lipoxygenase, the peroxygenation-catalyzing enzyme; (iii) the fatty acid chains of the HPOs are cleaved between C-12 and C-13 by fatty acid HPO lyase to form C_6 aldehyde ((*Z*)-3-hexenal or *n*-hexenal) and C_{12}-oxoacid (11-formyl-(*Z*)-9-undecenoic acid) (Scheme 1);[59,94] (iv) (*Z*)-3-hexenal or *n*-hexanal is converted into other C_6 aldehydes and C_6 alcohols by ADH[84,86,87] and IF;[37] and (v) the geometric and/or positional structures of the double bond are converted into those suitable to produce green odor components by IF. This subsection focuses on lipoxygenase and HPO lyase of the enzyme system which is bound to chloroplasts. Although the biosynthetic pathway and the participating enzymes have been fully elucidated,[95–98] the exact physiological role(s) of this enzyme system is not clear.

1.04.6.2 Distribution of Enzyme Systems Producing Green Odor

About 90% of the enzyme activities producing green odor in plants are localized in green leaves[3,99] (Table 4). When the enzyme activities in tomato leaves were compared with those in tomato fruits, which are reported to contain the C_6 aldehyde-forming enzyme system,[100,101] a higher C_6 aldehyde-forming activity was found in the green leaves than in the fruits.[2] During ripening, HPO lyase activity per gram fresh weight (fr. wt.) of fruit decreased with decrease in the chlorophyll content, but it was sufficient to produce a considerable amount of the C_6 aldehydes. The C_6 aldehyde-forming activity in tomato fruit was very low, even with increasing lipoxygenase activity and sufficient HPO lyase activity.[2] This may be explained by the fact that the lipoxygenase in tomato fruits mediates 9-

HPO formation, as reported by Galliard and Matthew.[100] Thus, the possibility cannot be excluded that the lipoxygenase in other plant species mediates 9-HPO formation (see Section 1.04.6.3.2), and that even apparently high lipoxygenase activity does not always result in a high activity for C_6 aldehyde formation. The product specificity of lipoxygenase and the substrate specificity of HPO lyase must be considered when studying the relationships between C_6 aldehyde formation and the individual enzyme activities. It is not yet known why green leaves contain HPO lyase. Possibly it is a result of the degradation or detoxification of peroxidized fatty acids formed in chloroplasts by photooxidation. Enzyme activities responsible for volatile C_6 aldehyde formation were accompanied by lipoxygenase and HPO lyase in the green leaves of 28 plant species (Table 5).[2] Most species tested showed C_6 aldehyde-forming activity, but its value varied with the plant considered. All the plants had both lipoxygenase and HPO lyase activities. This is evidence that, in general, the C_6 aldehyde-forming enzyme system in green leaves consists of lipoxygenase and HPO lyase, for example, in tea,[102] watermelon,[103] tomato,[100] and pear.[104] The amount of lipoxygenase activity varies with the plant species, as has been reported by Pinsky *et al.*[104] Cabbage and lettuce, which are used as raw food, had low lipoxygenase activities (Table 5).[1] HPO lyase was widespread in green leaves in relatively high concentrations (Table 5),[1] but leaves with low chlorophyll contents had low levels of HPO lyase activity (Figure 6). When the correlation coefficient (r) was calculated between HPO lyase activity and chlorophyll content from the values ($n = 37$) in Table 4 and for those of the green leaves of *Phaseolus vulgaris*, a value of 0.712 was obtained (Figure 6).[2] There was no correlation found between lipoxygenase activity and chlorophyll content. This suggests the possibility that the degree of HPO lyase activity depends on some function of the chloroplasts. A linear regression equation, $y = 2.85x + 0.39$, was obtained by the least-squares method for HPO lyase activity (y) and the chlorophyll content (x) (Figure 6). When x was extrapolated to zero in the equation, y was 0.39, which indicates the possible presence of a chlorophyll-independent HPO lyase(s). Thus, HPO lyase probably exists in multiple forms, a chlorophyll-dependent form in chloroplasts and a chlorophyll-independent form in nonphotosynthetic organelles and membranes, since HPO lyase activity has been found in various nongreen tissues.[100,103,104–108] As shown in Table 5, the C_6 aldehyde-forming activity that resulted from the sequential actions of lipoxygenase and HPO lyase varied widely. Gingko (*Gymnospermae*) seeds (edible) and mulberry belong to a group with low C_6 aldehyde-forming activity. The mulberry leaf homogenate was very viscous and turned brown immediately, probably because of the presence of a large amount of polyphenols which might be oxidized with peroxidase. The leaves in this group showed low lipoxygenase activity rather than low HPO lyase activity—evidence that lipoxygenase activity affects the formation of C_6 aldehydes. A similar phenomenon has been observed in tea plants; seasonal changes in C_6 aldehyde formation are caused by changes in lipoxygenase activity, not by changes in HPO lyase activity.[102] Among ~40 plants investigated, the green leaves of many dicotyledonous plants show high activities for this enzyme system, which forms C_6 aldehydes from α-linolenic and linoleic acids. However, edible leafy vegetables, fruits, and monocotyledonous plants have low activities.[1] Lower plants, e.g., *Marchantia polymorpha* (a liverwort),[109,110] *Chlorella pyrenoidosa* (a green alga),[111] and *Oscillatoria* sp. (a cyanobacterium),[111] were reported to possess this enzyme system.

Table 4 Localization of the enzyme activity producing green odor by plant organ.

Organ	n-*Hexanal*[a] Lipoxygenase and HPO lyase[b]	n-*Hexanal*[a] HPO lyase[c]	Chlorophyll[d]
Leaf	1.21 (89.6)[e]	1.78 (73.9)	255.8 (73.4)
Stem with buds	0.09 (6.7)	0.33 (13.7)	60.7 (17.4)
Cotyledon	0.01 (0.7)	0.11 (4.6)	21.3 (6.1)
Root	0.04 (0.3)	0.19 (7.9)	10.6 (3.0)

[a]μmol g^{-1} fresh leaves. [b]Substrate: Linoleic acid. [c]Substrate: 13-Hydroperoxylinoleic acid. [d]μg g^{-1} fresh leaves. [e]Values in parentheses are percentages.

1.04.6.3 Properties of Enzymes

1.04.6.3.1 Lipoxygenase

Two types of lipoxygenase exist in plants: one is present in soybean seeds and tea leaves, etc., and oxygenates specifically at C-13 of α-linolenic or linoleic acids, and the other is present in potato

Table 5 Lipoxygenase, HPO lyase and C_6 aldehyde-forming activities in the green leaves of various plants.

Plant	*Lipoxygenase* (μmol O_2 min^{-1} 0.5 g^{-1} fr. wt.)	*HPO lyase* (μmol 0.5 g^{-1} fr. wt.)	*C_6 aldehyde-forming activity* (μmol 0.5 g^{-1} fr. wt.)	*Chlorophyll* (mg 0.5 g^{-1} fr. wt.)	*Date harvested*
Ginkgo (*Ginkgo biloba*)	0.10	0.31	0.01	0.474	July
Pumpkin (*Cucurbita maxima*)	0.24	2.08	0.01	0.499	July
Watermelon (*Citrullus vulgaris*)	2.75	4.89	5.21	0.904	July
Cabbage (*Brassica oleracea* var. *capitata*)	0.04	0.18	Trace	0.020	September
Chinese cabbage (*Brassica napus*)	0.14	0.17	Trace	0.030	September
Tea (*Thea sinensis*)	0.30	2.12	1.08	0.343	June
Camellia (*Camellia japonica*)	0.50	1.98	0.83	0.197	June
Sasanqua (*Camellia sasanqua*)	0.20	0.39	0.04	0.160	June
Japanese persimmon (*Diospyros kaki*)	0.10	1.50	0.26	0.343	July
False acacia (*Robinia pseudoacacia*)	0.40	1.71	0.36	0.913	July
Alfalfa (*Medicago sativa*)	0.53	3.07	1.66	0.712	June
White clover (*Trifolium repens*)	1.70	3.00	3.00	1.287	July
Soybean (*Glycine max*)	2.70	2.06	1.08	0.540	July
Kidney bean (*Phaseolus vulgaris*)	0.38	5.01	0.81	1.250	June
Holly (*Ilexintegra thunb*)	0.20	0.89	0.37	0.579	July
Mulberry (*Morus bombycis*)	1.92	0.92	0.03	0.060	July
Sweet gum (*Liquidambar styraciflua*)	0.58	1.55	0.79	0.356	July
Spinach (*Spinacia oleracea*)	0.25	4.25	Trace	0.420	September
Potato (*Solanum tuberosum*)	0.17	2.42	0.03	0.713	June
Sweet potato (*Ipomoea batatas*)	0.40	1.92	0.47	0.364	July
Tobacco (*Nicotiana tabacum*)	1.31	0.72	0.91	0.720	August
Eggplant (*Solanum melongena*)	0.91	1.96	0.19	0.741	June
Tomato (*Lycopersicon esculentum*)	0.37	1.84	0.48	0.630	June
Sunflower (*Helianthus annuus*)	0.05	1.58	0.60	0.770	July
Burdock (*Arctium lappa*)	1.20	1.92	0.48	0.740	July
Lettuce (*Lactuca sativa*)	0.08	1.19	Trace	0.170	September
Rice (*Oryza sativa*)	0.13	0.40	0.22	0.598	July
Corn (*Zea mays*)	0.55	1.10	0.43	0.830	June

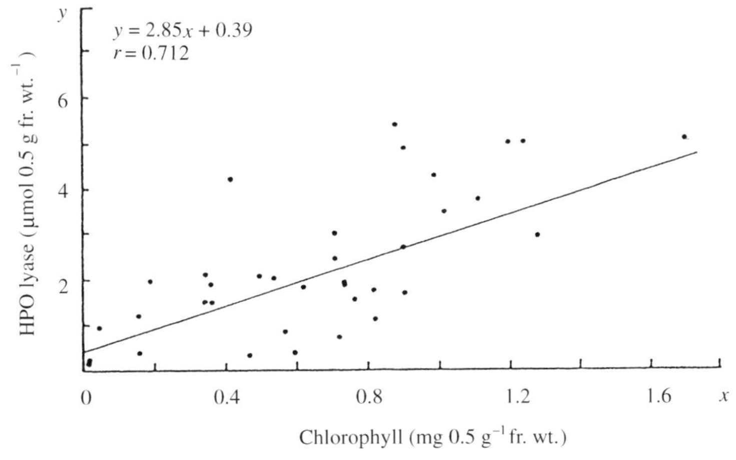

Figure 6 Correlation between HPO lyase activity and the chlorophyll content of green leaves.

tuber, etc., and oxygenates at C-9.[112] Lipoxygenase in legume seeds or cereal grains has been extensively investigated, but research indicates that it exists in a wide variety of plant species including lower plants, such as green algae and cyanobacteria.[113,114] Furthermore, it has been detected in most plant tissues, including cotyledons, leaves, roots, stems, and fruits. In some cases it is present in a soluble form and in others in a membrane-associated form.[115] In fresh tea leaves, the majority of the lipoxygenase activity is associated with a membrane fraction.[89] Lipoxygenase in tea leaves shows the highest activity at pH 6.3 with linoleic acid and α- and γ-linolenic acids as substrates. A second optimal pH of 4.5 is also observed, although the origin of this second peak has not yet been elucidated. Because tea leaves are rich in polyphenols, attempts to solubilize lipoxygenase from the tea leaf membrane have failed. Addition of lipolytic acyl hydrolase to the membrane caused inactivation of lipoxygenase activity, but not of the fatty acid HPO lyase activity. These observations indicate that lipoxygenase in tea leaves requires a specific protein–lipid interaction to exert its oxygenation activity.[93]

1.04.6.3.2 HPO lyase

Fatty acid HPO lyases can be grouped into two types that cleave either 9-(*R*)-HPOs or 13-(*S*)-HPOs. These lyases give either two C_9 fragments, or a C_6 and a C_{12} fragment, respectively (Scheme 4; the C_6 and C_9 groups are shown in Figures 1 and 7, respectively). The lyases that cleave 9-(*R*)-HPOs occur in pear fruit[104,117] and those that cleave 13-(*S*)-HPOs occur in watermelon seedlings,[103] tea leaves,[119,120] cultured tobacco cells,[121] tomato fruits,[122] green bell pepper fruits,[123] alfalfa seedlings,[124] and soybean seeds.[125] In cucumber cotyledons, both C_6 and C_9 aldehydes are formed by HPO lyase.[126–128] Because it had not been elucidated whether these activities are attributable to one enzyme which cleaves both 13- and 9-HPOs, or to two or more enzymes, each of which specifically cleaves 13- or 9-HPO, an attempt to separate HPO lyase activity by ion-exchange chromatography was made. The activity was resolved into two fractions. One cleaved specifically linoleic acid 13-HPO, whereas the other cleaved specifically the 9-isomer.[128] The specific activity of 13-HPO was most active at pH 8.0 and that of the 9-HPO at pH 6.5. Sulfhydryl reagents inhibited both of the lyases, but to different extents. HPO lyase was first purified to homogeneity from the membrane fraction of tea leaves.[129,130] Table 6 shows the purification of HPO lyase from tea leaves. Fresh tea leaves were macerated with phosphate buffer containing polyvinylpyrrolidone along with reducing agents, glutathione and ascorbic acid, in order to avoid nonspecific aggregation of protein by polyphenols and then the crude fraction was obtained. This fraction was centrifuged at 100 *g* for 10 min and the supernatant was centrifuged at 1×10^5 *g* for 60 min to obtain the membrane fraction. The HPO lyase activity was separated by hydroxyapatite gel chromatography (Figure 8) (Table 6)[42] and its M_r was found to be 5.5×10^4 by SDS–PAGE (Figure 9).[129] Nordihydroguaiaretic acid and butylated hydroxyanisole are potent inhibitors of lipoxygenase,[43] which inhibit HPO lyase. Tea leaf HPO lyase was rapidly and irreversibly inactivated by linoleic acid 13-HPO, its natural substrate.[130] Fruit fatty acid HPO lyase was also purified to apparent homogeneity from immature fruits of green bell pepper (*Capsicum annuum* L.) by differential centrifugation, ion-exchange chromatography, hydroxyapatite chromatography, and gel filtration.[131,132] The enzymatic activity was separated into two fractions (HPO lyases I and II) by hydroxyapatite. Both iso-forms were deduced to be trimers of M_r 5.5×10^4 subunits and to have similar enzymatic properties. Peptide maps revealed only slight differences between them. Immunoblot analysis showed that an antibody raised against HPO lyase I reacted with HPO lyase II as strongly as with the original antigen. These results indicate that there is only limited heterogeneity in amino acid sequence and/or posttranslational modification. The activities of both HPO lyases were significantly inhibited by lipophilic antioxidants, such as nordihydroguaiaretic acid and α-tocopherol. The activities of HPO lyases against 13-hydroperoxy-(9*Z*,11*E*,15*Z*)-octadecatrienoic acid were about 12 times higher than those against 13-hydroperoxy-(9*Z*,11*E*)-octadecadienoic acid. In contrast, no significant activity was detectable against 13-hydroperoxy-(9*E*,11*E*)-octadecadienoic acid, a geometric isomer, or against 9-hydroperoxy-(10*E*,12*Z*)-octadecadienoic acid, a positional isomer. Tissue print immunoblot analyses using antiserum against HPO lyase indicated that HPO lyase was most abundant in the outer parenchymal cells of the pericarp. Fatty acid HPO lyase is an enzyme that cleaves HPOs of polyunsaturated fatty acids to form short-chain aldehydes and ω-oxo acids. Spectrophotometric analyses of HPO lyase highly purified from green bell pepper fruits indicated that it is a heme protein. The heme species was revealed to be heme b (proto-heme IX) from the absorption spectrum of the pyridine hemochromogen. Although the spectrum closely resembles that of allene oxide synthase, a plant cytochrome P450 from flaxseed, CO treatment of the reduced enzyme did not result in a peak at 450 nm, which is an essential diagnostic feature of a cytochrome P450. Internal amino acid sequences determined with peptide fragments obtained from the lyase showed no homology with any reported sequences.[131]

Vegetable fruit green odor (3*Z*,6*Z*)-nonadienal (melon-like odor),[118] is formed from α-linolenic acid via 9-(*R*)-hydroperoxide through a similar pathway to Scheme 4[117] and is then converted into the corresponding alcohol (watermelon odor) by ADH, or to (2*E*,6*Z*)-nonadienal (cucumber odor) by an IF. This aldehyde is reduced to the corresponding alcohol (cucumber bitter odor) by ADH.[117,133] (*Z*)-3- and (*E*)-2-Nonenols and their corresponding aldehydes are also formed from

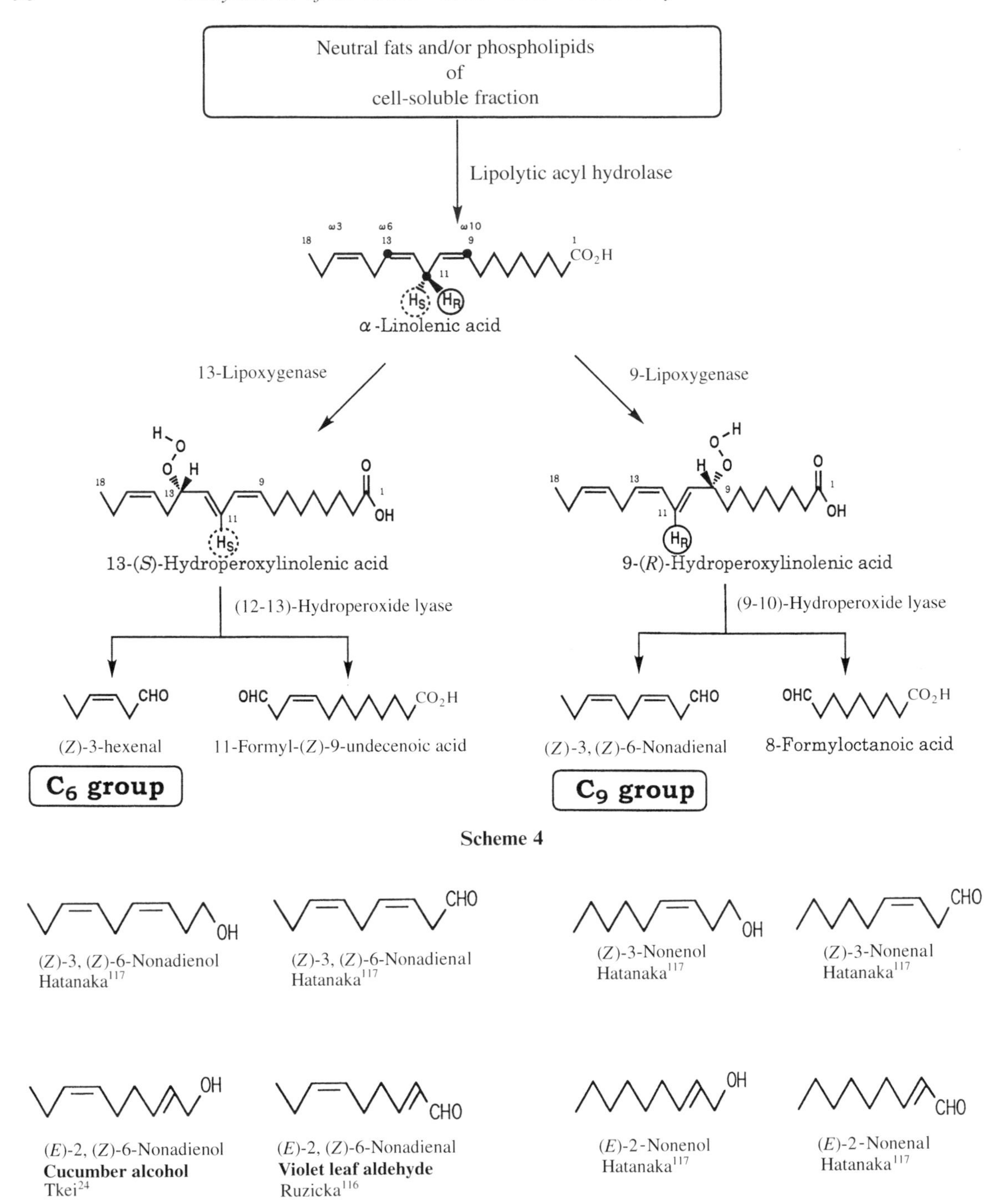

Scheme 4

Figure 7 The C_9 aldehydes and C_9 alcohols of fresh green odor emitted by fruits.

linoleic acid. The young and fresh green odor of fruits results from a mixture of not only the eight volatile C_6 compounds, but also eight C_9 compounds consisting of (3*Z*,6*Z*), (2*E*,6*Z*)-nonadienols, and (3*Z*,2*E*)-nonenols, and their corresponding aldehydes (Scheme 4,[117] Figure 7).[118,134] The ratio of C_9 to C_6 is characteristic for each fruit species, e.g., in cucumber fruits 90% are C_9 and 10% are C_6 compounds[117] and in banana fruits the ratio changes during ripening.[135,136]

1.04.6.4 Substrate Specificities of Lipoxygenase and HPO Lyase

This section describes the substrate specificity of two key enzymes, lipoxygenase and HPO lyase, which catalyze the biosynthesis of C_6 aldehydes.

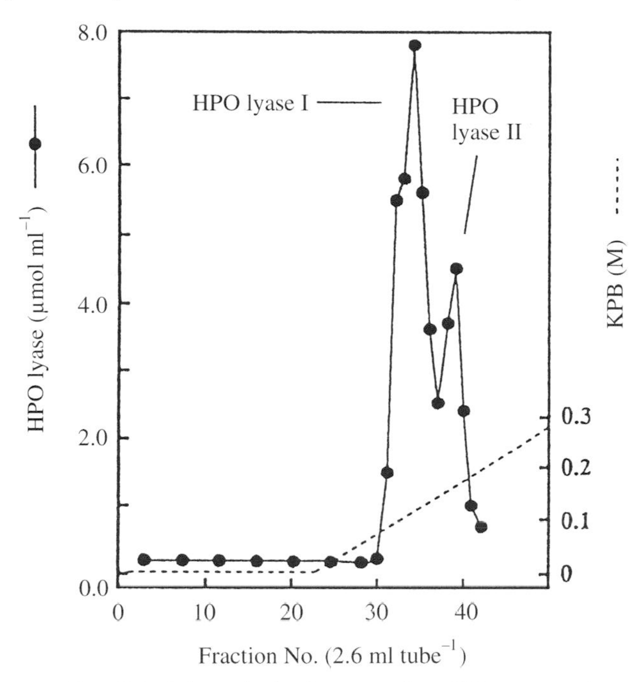

Figure 8 Elution profile of HPO lyase by hydroxyapatite. KPB = potassium phosphate buffer.

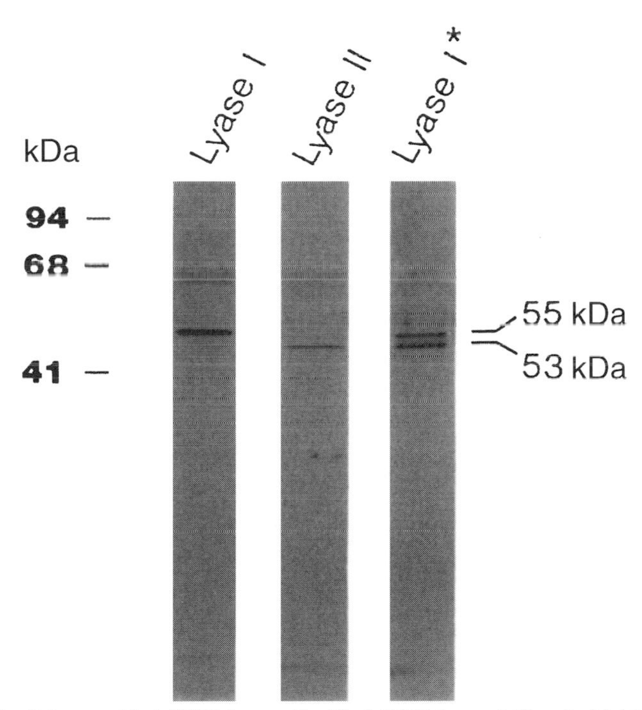

Figure 9 SDS–PAGE of the purified HPO lyase. Purified HPO lyase I (line 2; 55 kDa), II (line 3; 53 kDA) and I* (I kept at 4°C for one week) were analyzed with SDS–PAGE (12.5% gel). The M_r standards used (line 1) were soybean lipoxygenase-1 (94 kDa), bovine serum albumin (68 kDa), and yeast ADH (41 kDa).

1.04.6.4.1 Substrate and product specificities of lipoxygenase

The activity of lipoxygenase was evaluated on the basis of oxygen uptake,[137] and/or spectrophotometrically by following the formation of the HPOs (formation of the 1-hydroperoxy-(2*E*,4*Z*)-pentadiene system from the (1*Z*,4*Z*)-system in the UV region at 234 nm).[138] It has been widely accepted that lipoxygenase recognizes the (1*Z*,4*Z*)-pentadiene moiety. However, unsaturated fatty acids with C_{8-12} chains could not act as substrates for tea chloroplast lipoxygenase even though they

Table 6 Purification of HPO lyase from tea leaves.

	Total protein (mg)	*Total activity* (μmol)	*Specific activity* (μmol mg^{-1})	*Yield* (%)	*Purification* (fold)
Crude	9515	1444	0.15	100	1
Membrane	1332	720	0.54	49.9	3.61
Solubilized	925	1383	1.5	95.8	9.97
PEG 6000	73.5	799	10.9	55.4	72.5
DEAE-Cellulofine	2.49	366	147	25.4	984
DEAE-Toyopearl	nd[a]	160		11.1	
Hydroxyapatite					
fraction I	0.055	44.7	872	3.10	5816
fraction II	0.010	9.84	879	0.68	5860

[a]Not determined. PEG, polyethylene glycol; DEAE, diethyl amino ethyl.

possessed the (1*Z*,4*Z*)-pentadiene moiety in their structure.[120,131] α-Linolenic and linoleic acids are natural substrates, and α-linolenic acid is a better substrate than linoleic acid (Table 7)[139] for the enzyme system which forms C_6 aldehydes in plant tissues: this includes both lipoxygenase and HPO lyase activities. γ-Linolenic and arachidonic acids are also good substrates. All the C_{18} fatty acids acting as substrates had a (1*Z*,4*Z*)-pentadiene moiety between the C-9 and C-12 positions, which correspond to the ninth and sixth positions from the end methyl group: ω6 and ω9. The geometric isomers of linoleic acid, (9*E*,12*E*)-, (9*E*,12*Z*)-, and (9*Z*,12*E*)-octadecadienoic acids, did not act as substrates (Table 7).[139]

Table 7 Substrate specificities of lipoxygenase and C_6 aldehyde-forming activities.

Substrate	*Lipoxygenase*[a] (%)	*C_6 aldehydes* (%)
C_{18}-acid		
(3*Z*,6*Z*)-dienoic	30	0
(4*Z*,7*Z*)-dienoic	38	0
(5*Z*,8*Z*)-dienoic	35	0
(6*Z*,9*Z*)-dienoic	27	0
(7*Z*,10*Z*)-dienoic	25	0
(8*Z*,11*Z*)-dienoic	56	0
(9*Z*,12*Z*)-(linoleic acid)	100	100 hexanal
(9*Z*,12*Z*)-(methyl linoleate)	75[b]	14 hexanal
(9*Z*,12*Z*)-(linoleyl alcohol)	84[b]	16 hexanal
(9*Z*,12*Z*)-(linoleyl aldehyde)	68[b]	8 hexanal
(9*Z*,12*E*)-dienoic	nd[c]	0
(9*É*,12*Z*)-dienoic	nd[c]	0
(9*E*,12*E*)-dienoic	0	0
(10*Z*,13*Z*)-dienoic	44	0
(11*Z*,14*Z*)-dienoic	60	0
(12*Z*,15*Z*)-dienoic	31	0
(13*Z*,16*Z*)-dienoic	29	0
(9*Z*,12*Z*,15*Z*)-trienoic (α-linolenic acid)	181	114 hexenals (3*Z*,2*E*)
(6*Z*,9*Z*,12*Z*)-trienoic (γ-linolenic acid)	147	18 hexanal
C_{20}-acid		
(5*Z*,8*Z*,11*Z*,14*Z*)-tetraenoic (arachidonic acid)	108	0

[a]Oxygen uptake to linoleic acid defined as 100%. [b]Assayed at pH 4.5. [c]Not determined.

Ten positional isomers of the diene system in linoleic acid from (3*Z*,6*Z*)- to (13*Z*,16*Z*)-octadecadienoic acid, which do not occur naturally except for the (9*Z*,12*Z*)-acid, linoleic acid, were synthesized and used as substrates for tea chloroplast lipoxygenase.[139] The rate of oxygen uptake was ~30–60% of that with linoleic acid (Table 7). Only certain isomers formed 13-HPOs and gave hexanals by the tea chloroplast HPO lyase reaction (Table 7).[39] From the reactivities of the synthetic isomers, it is apparent that the configurational and structural features required in the substrate for lipoxygenase activity are (1*Z*,4*Z*)-pentadienes with ω6 and ω9 positions. The substrate specificity of the hydrophilic pocket of tea chloroplast lipoxygenase was examined using linoleic acid, methyl linoleate, and linoleyl alcohol. From these experiments, it became clear that the carboxy group at C-1 is important, but not essential.[140,141] The major product from linoleic acid was 13-HPO (96%), predominantly 13-(*S*)-hydroperoxy-(9*Z*,11*E*)-octadecadienoic acid, and methyl linoleate behaved similarly (Table 8).[140] When linoleyl alcohol was used as a substrate, the stereospecificity of the formation of the 13-HPO decreased to 48/52 in the (*R*)/(*S*) configuration (Table 8).[140] Hence the functional group at C-1 of a substrate plays a key role in determining the stereochemistry of the product. When linoleic acid was used as a substrate, the product specificity in the soybean lipoxygenase-1 reaction was similar to that in the tea chloroplast lipoxygenase reaction,[140] but artificial substrates such as the methyl ester and alcohol demonstrated differences between tea chloroplast and soybean lipoxygenases.[139] Soybean lipoxygenase-1 produced 13-(*S*)-HPO from linoleyl alcohol stereoselectively, whereas the racemic 13-HPO was formed from the corresponding methyl ester.[140] These results suggest that there are some differences in the enzyme–substrate interaction at C-1 as shown in the lipoxygenase of soybean and tea chloroplasts.

Table 8 Compositions of positional geometric, and optical isomers of linoleic acid, methyl linoleate, and linoleyl alcohol HPOs formed by tea chloroplast and soybean lipoxygenases from substrates.

	Optical		*Positional*	*Geometric*	
	13-HPO R:S	*9-HPO* R:S	*HPO* 13:9	*13-HPO* Z-9,E-11 : E-9,E-11	*9-HPO* E-10,Z-12 : E-10,E-12
LA[a]	20:75	3:2	96:4	80:16	3:1
	(8:82)[d]	(5:5)	(88:12)	(83:5)	(7:5)
LM[b]	9:87	2:2	97:3	96:1	1:2
	(18:48)	(5:19)	(66:34)	(42.24)	(16:18)
LAL[c]	48:52	nd[e]	87:13	68:19	9:4
	(22:78)	nd	(91:9)	(82:9)	(5.4)

[a]LA, linoleic acid. [b]LM, methyl linoleate. [c]LAL, linoleyl alcohol. [d]Soybean lipoxygenase values are in parentheses. [e]nd = Not determined.

1.04.6.4.2 Enzyme specificity observed by using a systematically synthesized series of substrates

In order to clarify a recognition mechanism by the hydrophilic pocket of lipoxygenase, an entire series of ($\omega 6Z,\omega 9Z$)-C_{12-24} dienoic (A group)[142] and ($\omega 3Z,\omega 6Z,\omega 9Z$)-$C_{12-24}$ trienoic (B group) acids[143] were synthesized as substrates (Figure 10). They have a fixed carbon chain from $\omega 1$ to $\omega 10$ incorporating a ($\omega 6Z,\omega 9Z$)-diene or ($\omega 3Z,\omega 6Z,\omega 9Z$)-triene structure together with an elongated carbon chain of various lengths from $\omega 11$ toward the terminal carboxy group. They are analogues of the natural fatty acids, α-linolenic, linoleic, and γ-linolenic acids, which have a common structure of (1*Z*,4*Z*)-pentadiene between the $\omega 6$ and $\omega 10$ carbon positions (Figure 10).[42] In order to examine the environment of the hydrophobic pocket of this enzyme, (9*Z*,12*Z*)-C_{14-24} dienoic acids (C group) (Figure 10), with a fixed (9*Z*,12*Z*)-C_{13} diene carboxy moiety and successively elongated carbon chains from C_{14} to C_{24}, were synthesized.[42] These substrates were synthesized by Jones oxidation of the corresponding alcohols, prepared via Wittig or Grignard coupling reactions of two counterparts, using acetylene chemistry.[142,143] The soybean lipoxygenase-1 (EC 1.13.11.12) was purified to homogeneity from the soluble protein fraction of soybean seeds using an established procedure with slight modifications (purification at 20-fold, specific activity at 134.5 U mg^{-1}, M_r 94 kDa by SDS–PAGE, without isozyme).[42,144]

(i) Substrate specificity of lipoxygenase

Lipoxygenase-1 activities in substrate specificity investigations were determined spectrophotometrically at 25 °C by following the formation of the HPOs at 234 nm (of conjugate systems for $\omega 7$-E,$\omega 9$-Z) at pH 9.0. On the other hand, product specificities were determined by the formation of corresponding $\omega 6$-(*S*)-HPOs (A′, B′, and C′ groups in Figure 11) from A, B, and C groups of synthetic fatty acids in Figure 10. The reactions were carried out through dispersion with 0.2% Tween 20, addition of purified soybean lipoxygenase-1, and stirring for 9 h at 5 °C under O_2 atmosphere. Positional and geometric analyses were carried out by HPLC and GC–MS, and optical analyses by GC of MTPA derivatives (α-methoxy-α-trifluoromethylphenyl acetate).[42] Substrate specificities of lipoxygenase-1 for the A group were compared with the relative activity of linoleic acid to form $\omega 6$-(*S*)-hydroperoxy-($\omega 7E,\omega 9Z$)-C_{18}-dienoic acid defined as 100% and the B group also with that of α-linolenic acid form $\omega 6$-(*S*)-hydroperoxy-($\omega 3Z,\omega 7E,\omega 9Z$)-$C_{18}$-trienoic acid (Fig-

Natural fatty acids : C_{18}

$\omega 6$ $\omega 10$

CO_2H α-Linolenic acid

CO_2H Linoleic acid

CO_2H γ-Linolenic acid

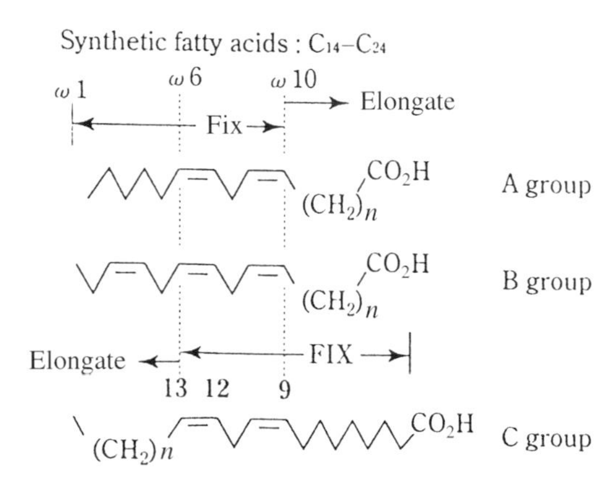

Figure 10 Synthetic fatty acids used as substrate. $n = 7$: A group, linoleic acid; B group, α-linolenic acid; $n = 4$: C group, linoleic acid.

ure 11).[42] Lipoxygenase-1 showed broad substrate specificities for compounds of the A and B groups with increasing activity from C_{16} to C_{20} and then decreasing from the maximum at C_{20} to C_{24}. Appreciable activity was not detected with C_{14} and C_{15} (Figure 11, A′ and B′ groups). The C′ group showed little activity except for C_{18}, linoleic acid (Figure 11, C′ group).[42] Maximum activities were 120% and 135% at C_{20} for A and B groups and 100% at C_{18} for the C group (Figure 11, A′–C′ groups).[42] These results indicate that the substrate requirement for the hydrophilic pocket of lipoxygenase is fairly broad. In contrast, that for the hydrophobic pocket is strictly defined. Lipoxygenase has a clearly defined pocket groove for the hydrophobic pocket (Figure 12).[42] It is supported by the tertiary structure[145,146] and also by modification of the arginine residue on activity.[147] It is also supported by the tertiary structure of the iron-binding domain in soybean lipoxygenase-1.[148,149] Lipoxygenases were highly purified from soybean seed, cucumber cotyledons, and wheat seed. The substrate specificities of these lipoxygenases were studied by using an entire series of ($\omega 6Z$,$\omega 9Z$)-C_{14-24}-dienoic acids, A group, as shown in Figure 13. Soybean lipoxygenase-1 and cucumber lipoxygenase showed broad specificities for the substrates whereas wheat lipoxygenase showed narrow specificities.[150] The position of dioxygenation for each substrate was analyzed by HPLC. After a synthetic substrate had been dispersed with 0.2% Tween 20, it was suspended in borate buffer (pH 9.0) and purified soybean lipoxygenase-1 was added and the mixture was stirred for 9 h at 5 °C under an O_2 atmosphere. With soybean lipoxygenase-1, elongation of the distance between the terminal carboxy group and the site of hydrogen removal in a substrate decreased the positional specificity of dioxygenation, whereas with cucumber lipoxygenase, shortening the distance decreased the specificity. It was suggested that cucumber lipoxygenase and soybean lipoxygenase-1 recognized the terminal carboxy group of a substrate to arrange it in only one orientation at the reaction center. In the case of wheat lipoxygenase, recognition of the carboxy group was crucial and essential to secure the activity.[150]

(ii) Positional, geometric, and optical isomers of the HPOs

The positional and geometric isomers of the HPOs prepared using tea chloroplasts were analyzed after reduction of the HPO function in the form of the corresponding hydroxymethyl esters by using

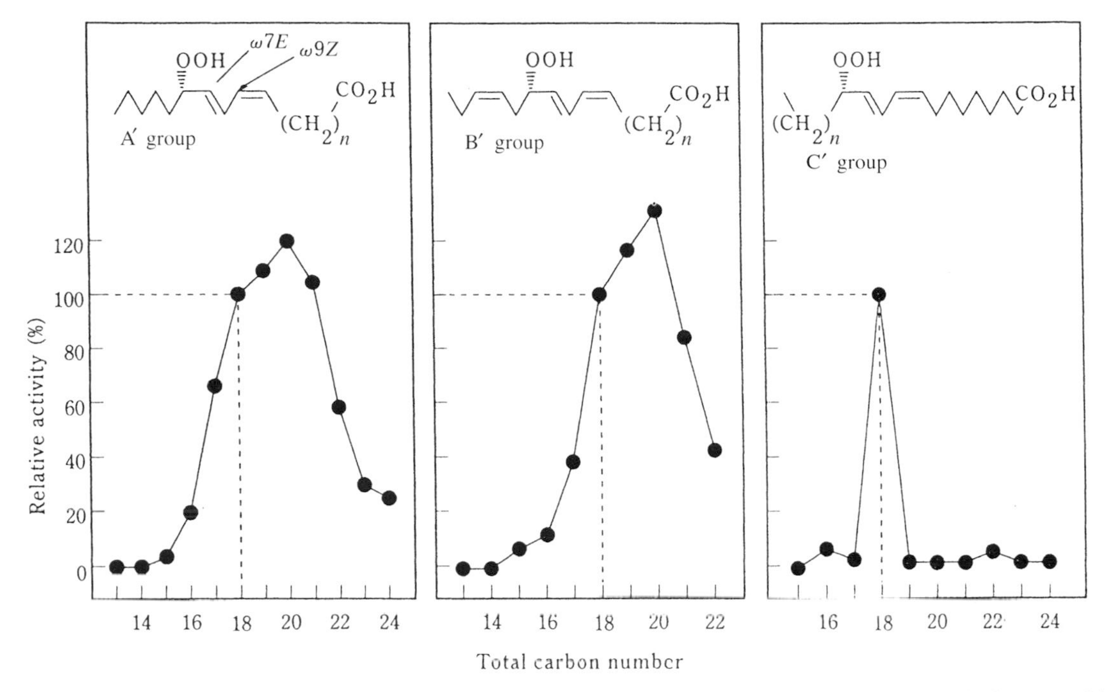

Figure 11 Substrate specificity of soybean lipoxygenase-1. $n = 7$: A group and B group; $n = 4$: C group. %: Relative activity of formation for ω6-(*S*)-HPO-C_{18} defined as 100%.

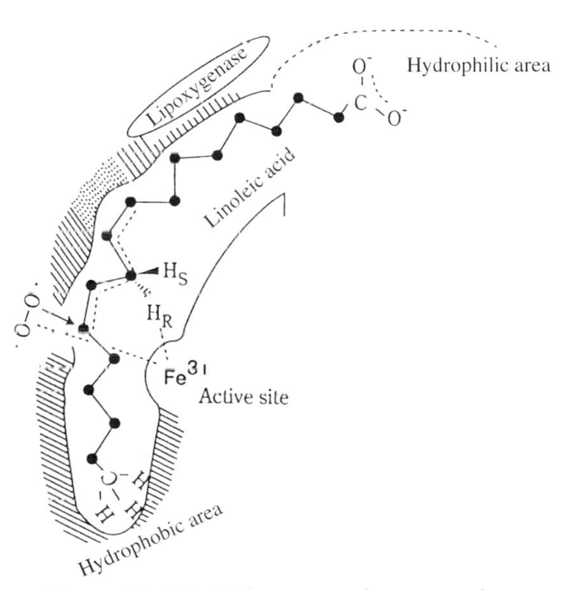

Figure 12 Model for enzymatic oxygenation.

normal phase HPLC and GC–MS (Table 9).[151] The configuration at the ω6-carbon of the HPO was determined by GC using the corresponding MTPA ester derivatives (Table 9). The positional and geometric specificities of HPOs prepared by soybean seed, cucumber cotyledon, and wheat seed lipoxygenase are given in Table 10.[36,114,150,152] The (ω6*Z*,ω9*Z*)-dienols could also act as substrates (Figure 14).[142] Soybean lipoxygenase shows the highest activity for $C_{15:2}OH$ ((6*Z*,9*Z*)-pentadecadienol) among alcohol analogues and this activity reached 76% of that for linoleic acid (Table 7). All the substrates used here have two possible oxygenation sites, at ω6 and ω10, and the regioisomer ratios of lipoxygenase products varied significantly depending on the substrates used. The oxygenation activity at the ω10 was the highest for $C_{15:2}OH$ and the activity at ω6 was the highest for $C_{14:2}OH$ (Figure 14). Higher activity at ω10 than at ω6 was observed only for $C_{15:2}OH$; in the other dienols the activity at ω6 was lower than that at ω10. Table 11 indicates ω6/ω10 ratios for regional isomers with optical purity, (*R*)/(*S*) ratio, at each oxygenation position.

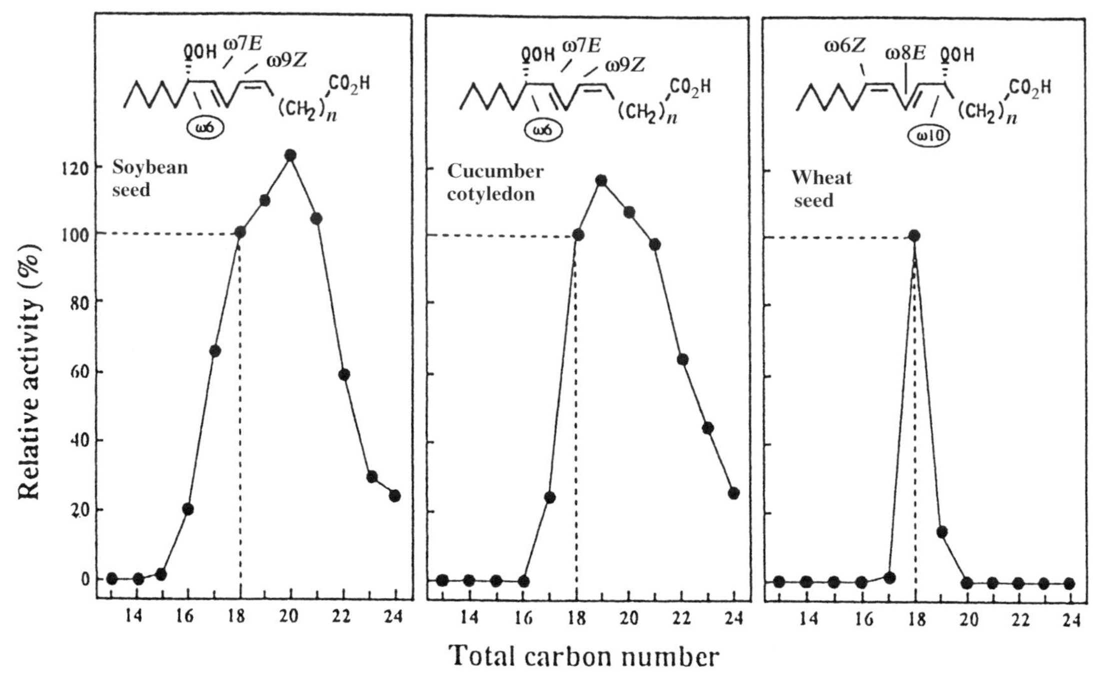

Figure 13 Substrate specificities of soybean, cucumber, and wheat lipoxygenases for the synthetic substrates.

Table 9 Geometric and optical purity of ω6-(*S*)-HPOs.

	ω6-(S)-*hydroperoxy*-(ω3*Z*,ω7*E*,ω9*Z*)-*trienoic acids*	
Total carbon number	*Geometric purity* (%)	*Optical purity* (% ee)
14	85.3 (86.8)[a]	91.6 (60.0)
15	88.6 (94.4)	98.0 (70.8)
16	87.2 (99.3)	98.0 (51.4)
17	89.1 (95.3)	83.2 (89.6)
18	99.7 (96.3)	72.4 (84.0)
19	88.2 (95.0)	75.2 (90.2)
20	92.7 (93.6)	98.0 (93.0)
21	91.3 (90.2)	88.8 (86.6)
22	97.1 (92.6)	85.4 (85.0)
23	78.4 (86.7)	84.2 (82.0)
24	91.7 (75.6)	72.6 (71.0)

[a]All values in parentheses are for (ω7*E*,ω9*Z*)-dienoic acids.

1.04.6.4.3 Substrate and product specificities of HPO lyase

13-HPOs of α-linolenic and linoleic acids are the best substrates for membrane-bound HPO lyase in tea leaves, but their 9-isomers do not act as substrates[151] (Table 12). When 13-hydroperoxy-(6*Z*,9*Z*,11*E*)-octadecatrienoic acid, the 13-HPO of γ-linolenic acid, was used as substrate, the reactivity was 22% of that for 13-hydroperoxy-(9*Z*,11*E*)-octadecadienoic acid, whereas 15-hydroperoxy-(5*Z*,8*Z*,11*Z*,13*E*)-eicosatetraenoic acid (15-hydroperoxyarachidonic acid) did not act as a substrate for *n*-hexanal formation, although lipoxygenase was capable of producing HPO. Conversion of the carboxy group into a methyl ester or alcohol at pH 6.3 markedly reduced the reactivity to 27% and 53%, respectively. 12-Ketohydroxy- and 13-hyroxylinoleic acids did not act as substrates (Table 12).[119,144] As for the configuration of the hydroperoxy group in the 13-HPOs of linoleic acid, the (*S*)-configuration is favored by the tea chloroplast enzyme.[153] The substrates for tea chloroplast HPO lyase require the following structural and stereochemical features: a C_{18} straight-chain fatty acid with a free carboxy group; a hydroperoxy group at ω6 with (*S*)-configuration; and a (*Z*)-double bond at ω9 and (*E*)-double bond at ω7. Further introduction of a (*Z*)-double bond at C-15 in

Table 10 Positional and geometric specificity of soybean seed, cucumber cotyledons, and wheat seed lipoxygenases.

Total carbon number	*ω6-OOH*		*ω10-OOH*	
	(ω7E,ω9Z)	(ω7E,ω9E)	(ω6Z,ω8E)	(ω6Z,ω8E)
Soybean seed (9.0)[a]				
15	94.4	1.5	3.1	1.0
16	99.3	trace	0.7	trace
17	95.3	1.7	1.0	2.0
18	96.3	0.3	3.0	0.4
19	95.0	1.9	1.5	1.6
20	93.6	3.2	3.2	trace
21	90.2	5.3	2.1	2.4
22	92.6	3.4	4.0	trace
23	86.2	3.4	9.9	trace
24	75.6	2.3	22.1	trace
Cucumber cotyledons (6.3)[a,b]				
17	68.7	4.0	19.4	7.9
18	79.2	2.9	17.4	4.1
19	72.5	4.9	16.3	6.3
20	90.0	3.4	4.4	2.2
21	92.4	4.0	2.8	0.8
22	93.1	5.2	1.1	0.6
23	90.4	7.3	1.8	0.5
24	69.2	2.5	27.6	0.7
Wheat seed (6.9)[a,c]				
17	10.8	2.8	84.1	2.4
18	7.1	2.8	87.7	2.4
19	16.1	5.3	74.7	3.9

[a]Values in parentheses are optimum pHs. [b]From ref. 128. [c]From ref. 150.

13-hydroperoxy-(9*Z*,11*E*)-octadecadienoic acid increased the reactivity by 2%, and that at C-6 decreased it by 78%, as judged by headspace analysis (Table 12).[119] In order to learn more about the recognition mechanism for the hydrophilic pocket of HPO lyase, fatty acids, i.e., those of the A and B groups (Figure 10), were systematically synthesized and converted into their HPOs, A′ and B′ groups, using soybean lipoxygenase (Figure 15). A spectrophotometric assay to follow the decrease in absorbance at 234 nm was used for determination of substrate specificity, and product specificity was identified by comparison with an authentic specimen of the hydrazone derivative of *n*-hexanal for the A′ group and that of (*Z*)-3-hexenal for the B′ group using GC–MS and HPLC analyses. (*Z*)-3-Hexenal, but not (*E*)-2-hexenal, was formed from the B′ group and 12-oxo-(*Z*)-9-dodecenoic acid, but not the (*E*)-10-isomer, from each of the A′ and B′ groups.[42] These results indicate that HPO lyase retains the (*Z*)-configuration of substrates and isomerization to the (*E*)-configuration occurs after the cleavage reaction brought about by HPO lyase. Because the natural substrates, HPOs of linoleic, α-linolenic, and γ-linolenic acids prepared with soybean lipoxygenase-1, contained less than 3% geometric and positional isomers, these reaction products were used without further purification as substrates for HPO lyase. From calculation of the initial velocity, the relative activities of the purified HPO lyase with hydroperoxylinoleic acid, hydroperoxy-α-linolenic acid, and hydroperoxy-γ-linolenic acid were 100%, 920%, and 7%, respectively (Figure 15). This indicates that HPO lyase in tea leaves has strict substrate specificity for natural fatty acid HPOs. The entire series of substrates, ω6-(*S*)-HPOs, was prepared with soybean lipoxygenase-1 by oxygenation of the corresponding fatty acids. Tables 9 and 10 show the regio- and enantiospecificity of the reaction products.[138] Some variations in the specificity were inevitable and further purification of the HPOs was not successful. With both the dienoic and trienoic acid HPOs, the substrate specificity for tea leaf HPO lyase was broad (Figure 15).[42,138] Elongation from a C_{18} carbon chain (natural substrate) to C_{22} between the terminal carboxy group and the hydroperoxy group caused enhancement of the activity in the reaction of tea leaf HPO lyase. However, elongation beyond C_{22} decreased the activity. The reactivities of the trienoic acid hydroperoxides were always 4–10 times higher than those of the dienoic acid HPOs with the same carbon number. Introduction of a double bond between the ω3 and ω4 is very effective in increasing activity. It is assumed that the rotation of the more compact ω-terminal end containing a double bond facilitates recognition by HPO lyase.

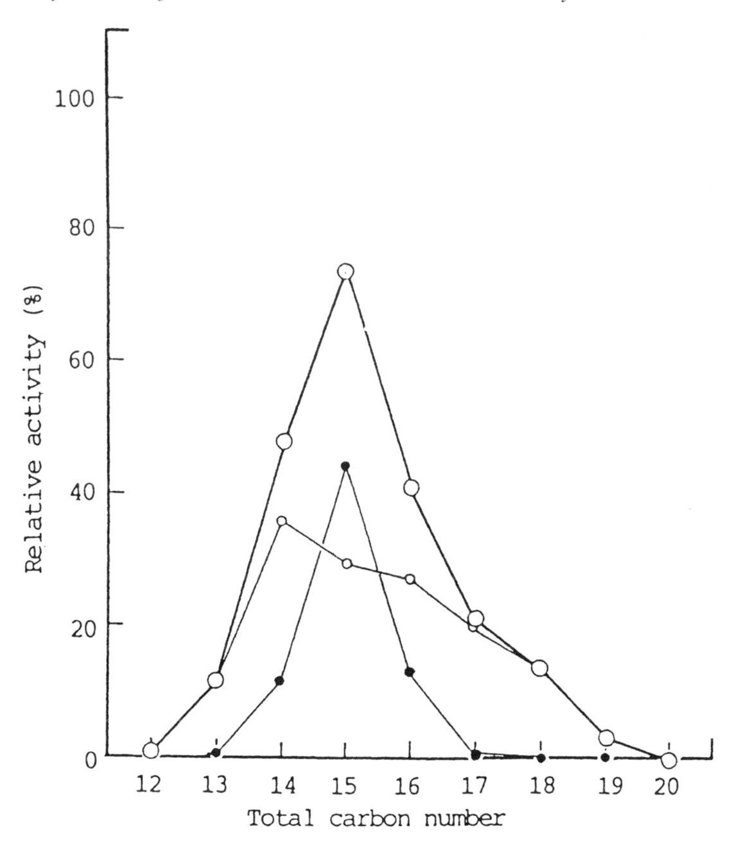

Figure 14 Relative activity of lipoxygenase-1 for synthetic substrates plotted against total carbon number of (*Z*-ω6,*Z*-ω9)-dienols. Relative ω6 and ω10 oxygenation activities (○), relative ω6 oxygenation activity (○), and that of ω10 (●) according to HPLC analyses of the products derived from the respective substrates.

Table 11 Product specificity of lipoxygenase-1 for optical and positional isomers of (ω6*Z*,ω10*Z*)-pentadecadienol.

Product	*Ratio* (%)	*Regioisomer* (%)
OOH, OH ω6(*S*)-HPO	24.4	33.0
OOH, OH ω6(*R*)-HPO	8.6	
OOH, OH ω10(*S*)-HPO	40.2	67.0
OOH, OH ω10(*R*)-HPO	26.8	

γ-Linolenic acid 13-HPO was catalyzed at a rate of only ~2% of that of α-linolenic acid 13-HPO (Figure 15).[42,138] In summary, recognition of the chain length ranging from ω10 to the terminal carboxy group was not strict for tea leaf HPO lyase, particularly when the chain length was longer than those in linoleic or α-linolenic acid. However, introduction of a double bond into this segment decreased the activity substantially. The pattern of product specificities was also very similar to that of substrate specificities.[142]

Table 12 Substrate specificity of tea chloroplast HPO lyase.

Substrate	*Relative activity* (%)	*Product*
13-Hydroperoxy-(9*Z*-11*E*)-octadecadienoic acid	100	hexanal
-methyl ester	27[a]	hexanal
-alcohol	53[a]	hexanal
13-Hydroxy-(9*Z*,11*E*)-octadecadienoic acid	0	
9-Hydroperoxy-(10*E*,12*Z*)-octadecadienoic acid	0	
13-Hydroperoxy-(9*Z*,11*E*,15*Z*)-octadecatrienoic acid	102	hexenals (3*Z*,2*E*)
13-Hydroperoxy-(6*Z*,9*Z*,11*E*)-octadecatrienoic acid	22	hexanal
9-Hydroperoxy-(10*E*,12*Z*,15*Z*)-octadecatrienoic acid	0	
15-Hydroperoxy-(5*Z*,8*Z*,11*Z*,13*E*)-eicosatetraenoic acid	0	
12-Oxo-13-hydroxy-(9*Z*)-octadecenoic acid (α-ketol)	0	

[a]Assayed at pH 6.3.

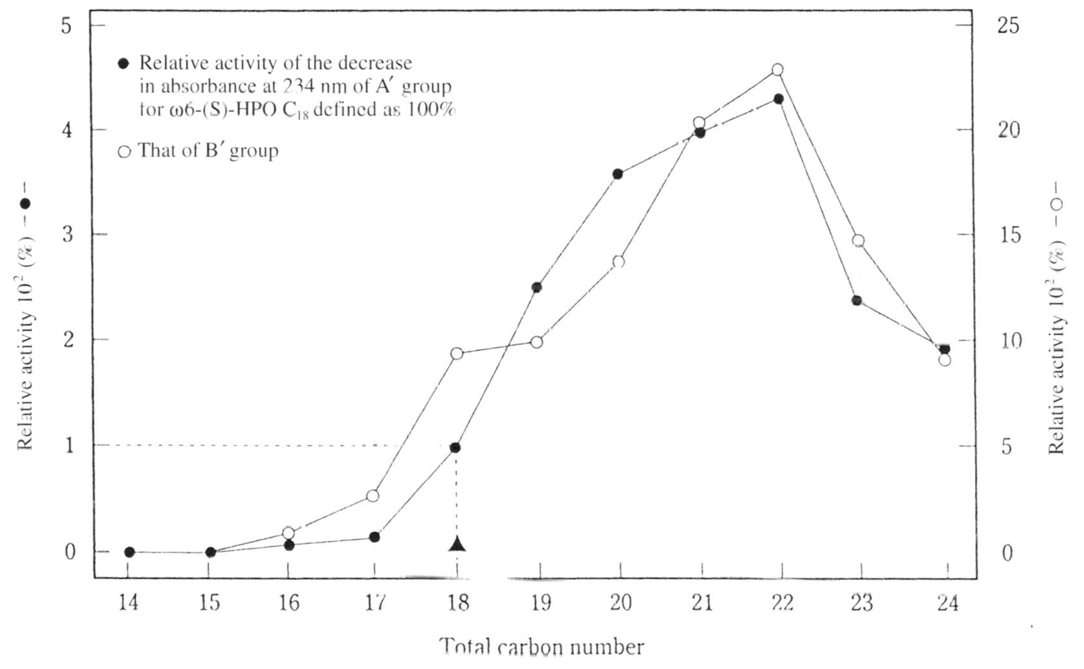

Figure 15 Substrate specificity of HPO lyase.

1.04.6.5 Enzymatic Peroxygenation in Linoleic Acid

Oxidation of linoleic acid with either tea chloroplast or soybean lipoxygenase gave the same product, i.e., 13-(*S*)-hydroperoxy-(9*Z*,11*E*)-octadecadienoic acid. In the oxidation of the C-13 of linoleic acid, the formation of a radical is expected. Figure 16 shows the signals of the ESR spectrum of the spin adduct of 2-methyl-2-nitrosopropane with the free radical formed during incubation of linoleic acid and tea chloroplasts. The hyperfine constants of 15.25 G and 2.00 G indicate the presence of a hydrogen atom at the β-position (β-hydrogen in Figure 16).[154] This supports the view that enzymatic oxygenation involves the formation of a free radical at the C-13 of linoleic acid. From this result and from other findings,[154–161] the mechanism of the chloroplast hydroperoxidase-catalyzed hydroperoxygenation is considered to be similar to that in soybean lipoxygenase (Figure 17).[162,163] In the initial step the pro-(*S*) hydrogen is abstracted stereospecifically from the methylene group at C-11. The double bond ((*Z*)-form) at C-12 is delocalized by resonance to C-11 ((*E*)-form), producing a free radical at C-13. This radical may be stabilized by the formation of a fatty acid–lipoxygenase complex. As shown in Figure 17, activated oxygen attacks the free radical at the C-13 position from the *si*-face specifically, to produce 13-(*S*)-hydroperoxy-(9*Z*,11*E*)-octadecadienoic acid.[6] This has been clearly established by the crystallographic determination of the active site iron and its ligand (tertiary structure of iron-binding domain) in soybean lipoxygenase-1.[148,149] The surfaces of the cavities are superimposed on the α-carbon traces of the enzyme. The tertiary structure of soybean lipoxygenase-1 was also clarified by Boyington[146] in 1993 and the resolved structure indicated the cavity accommodating oxygen and another cavity the substrate, and the position of iron.

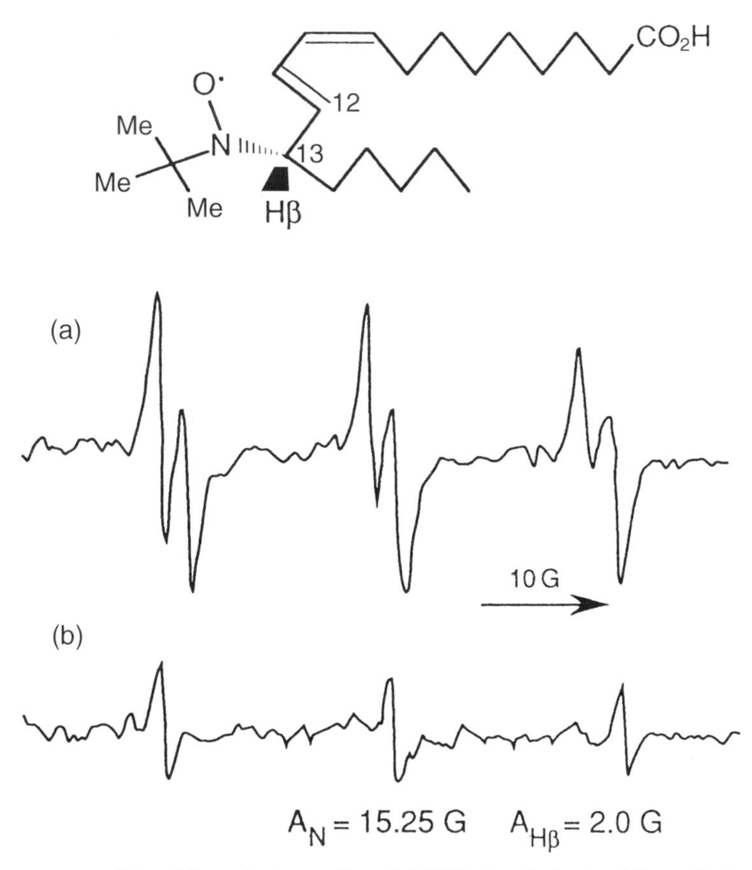

Figure 16 ESR spectrum of tea chloroplasts reaction. (a) ESR signal obtained from the incubation of linoleic acid, chloroplasts of tea leaves, and 2-methyl-2-nitrosopropane for 25 min. (b) ESR signal obtained from the incubation of linoleic acid, heated chloroplasts of tea leaves, and 2-methyl-2-nitrosopropane for 26 min.

1.04.6.6 Enzymatic Cleavage of 13-(*S*)-Hydroperoxylinoleic Acid

13-HPO-specific HPO lyase in tea leaves, and also in soybean seed, cleaves the (*S*)-enantiomer of 13-hydroperoxylinoleic acid to C_6 aldehydes and C_{12} oxo acid.[143,164] 13-(*S*)-Hydroperoxyoctadecadienol, an analogue of 13-(*S*)-hydroperoxylinoleic acid, is also cleaved stereospecifically to *n*-hexanal (C_6) and 12-oxo-(*Z*)-9-dodecenol (C_{12}) by the HPO lyase of tea leaves.[32,42,94] To clarify the mechanism of the cleavage reaction, both oxygen atoms of the hydroperoxy group of 13-(*S*)-hydroperoxyoctadecadienol were labeled with ^{18}O and this was incubated with tea chloroplasts. In order to reduce the rapidly exchangeable carbonyl ^{18}O-labeled oxygen of the cleaved product with the oxygen of water in the reaction medium,[165,166] the cleaved products, *n*-hexanal and 12-oxo-(*Z*)-9-dodecenol, were reduced to the corresponding alcohols with sodium borohydride immediately after incubation. These alcohols were analyzed by GC–MS. The ^{18}O of the HPO was not incorporated into *n*-hexanol but into 1,12-(*Z*)-9-dodecenediol.[32] From these findings, the reaction mechanism appeared to be similar to that of the acid-catalyzed rearrangement of the 13-HPO in aprotic solvents.[167] In the first step, HPO lyase catalyzes cyclization of the protonated HPO to a 12,13-epoxycarbonium ion with loss of a molecule of water. This leads to an allylic ether cation with charge located at C-13 adjacent to oxygen. Addition of hydroxy to the carbonium ion and subsequent rearrangement ultimately yield the C_6 and C_{12} compounds (Figure 18).[32] Vinyl ethers formed by potato tuber extracts are formed by a mechanism similar to that proposed for lyase.[168–170]

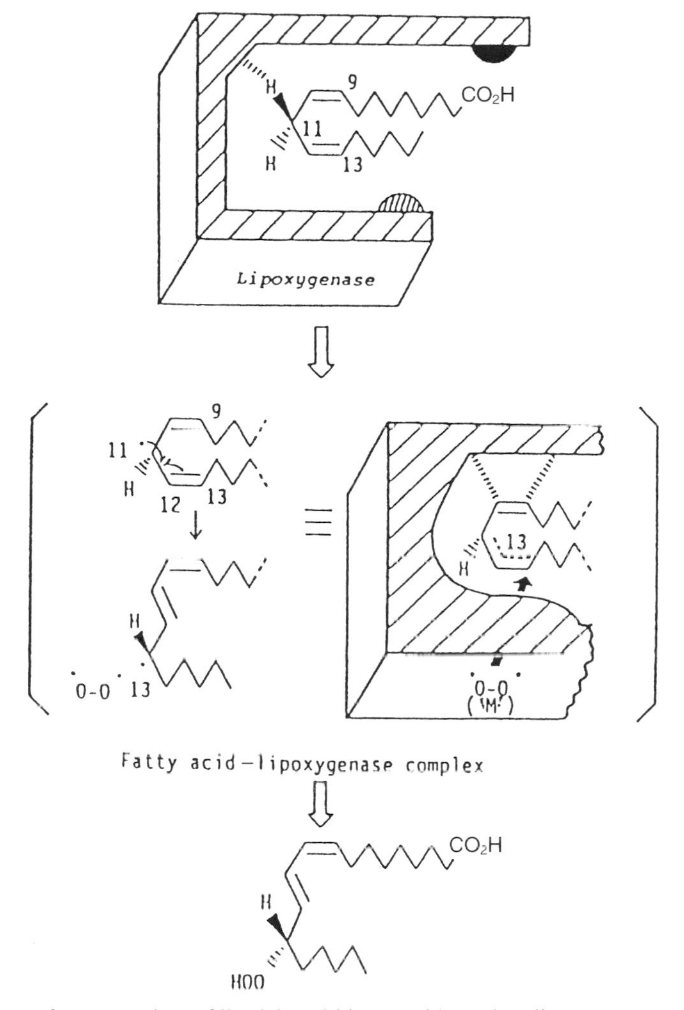

Figure 17 Mechanism of oxygenation of linoleic acid by tea chloroplast lipoxygenase. M = iron; two stereoscopic views of the active iron center of soybean lipoxygenase-1. The iron atom is represented by a single three-dimensional contour (at five times the rms density) from the anomalous difference map. The five protein groups that act as ligands are represented: His-499, 504, and -690, Asn-694, and C-terminal carboxylate of Ile-839.[145,148,149]

1.04.7 RELATIONSHIP BETWEEN ENVIRONMENTAL STIMULI AND ENZYME SYSTEM ACTIVITIES

Each of the volatile C_6 aldehydes and alcohols in green leaves has its own distinct green odor. The quantitative changes in the eight volatile compounds of green leaves produce a green odor characteristic of the plant species and of the seasonal changes in each plant.

1.04.7.1 Seasonal Changes in C_6 Aldehyde Formation in Tea Leaves

The activities of enzymes involved in (*Z*)-3-hexenal and *n*-hexanal formation have been shown to increase from April to July and decrease in the autumn in a homogenate of fresh young tea leaves (Figure 19). Relatively high activities were found from May to October and activities were maximum around July and August. No significant activities were detected in the winter. Similar results were observed for enzyme activities in chloroplasts isolated from tea leaves.[171] The seasonal changes in enzyme activities are closely related to temperature and incident solar radiation (Figure 19). When the minimum temperature dropped below 10 °C, the C_6 aldehyde-forming activities almost disappeared. In contrast, the enzyme activity reached its maximum in July and August when the temperature and solar radiation were highest.[171] Low lipoxygenase activities against α-linolenic and

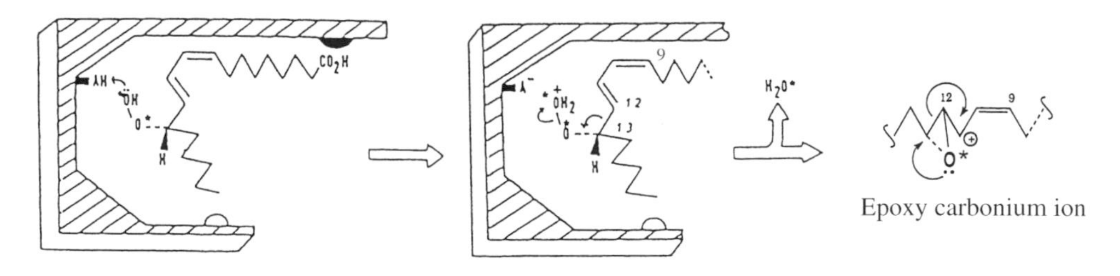

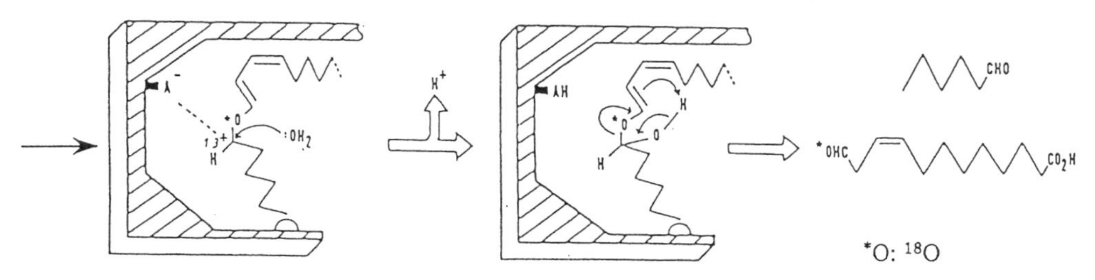

Figure 18 Mechanism of the cleavage reaction of HPO to *n*-hexanal and 12-oxo acid by tea chloroplast HPO lyase.

linoleic acids were detected in winter leaves. These activities began to increase in spring, reached their maxima in July, and then decreased in the autumn. The maximum activity was 3–4 U g^{-1} fr. wt. in summer leaves (Figure 20(a)). Expressed on a chlorophyll basis, lipoxygenase activities showed similar patterns.[172] However, high HPO lyase activity was found throughout the year (Figure 20(b)).[172] The HPO lyase activity in summer leaves although less than in winter leaves was still high enough to produce C_6 aldehydes. The overall C_6 aldehyde-forming activity, which is a sequential reaction involving lipoxygenase and HPO lyase, showed a seasonal change similar to that of lipoxygenase. This means that the step determining the seasonal changes is that catalyzed by lipoxygenase rather than HPO lyase.

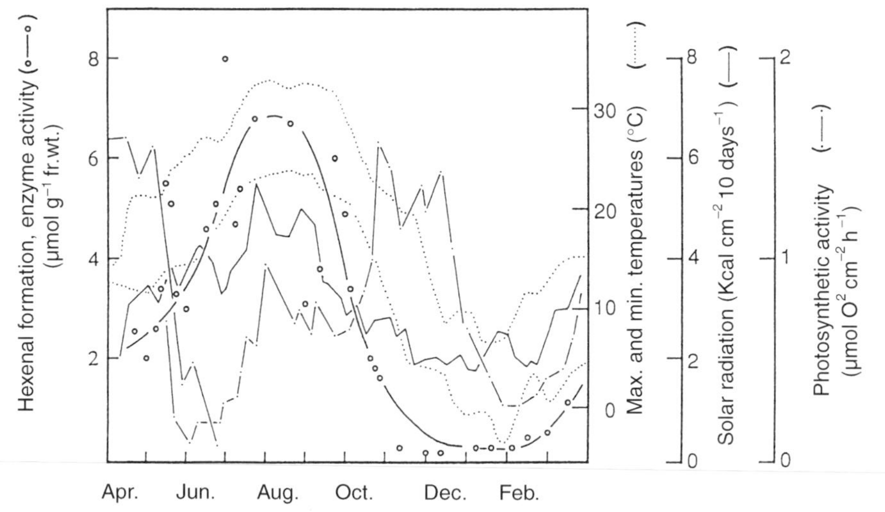

Figure 19 Seasonal changes in the activities of the enzyme system producing (*Z*)-3-hexenal and (*E*)-2-hexenal in homogenate of fresh young tea leaves; 1 cal = 4.186 J.

1.04.7.2 Relationship of Lipoxygenase and HPO Lyase Activities to Various Environmental Stimuli

The relationships between temperature and enzyme activities that produce C_6 aldehyde by pot growth of *T. sinensis* leaves are shown in Figure 21.[173] Lipoxygenase activity is proportional to

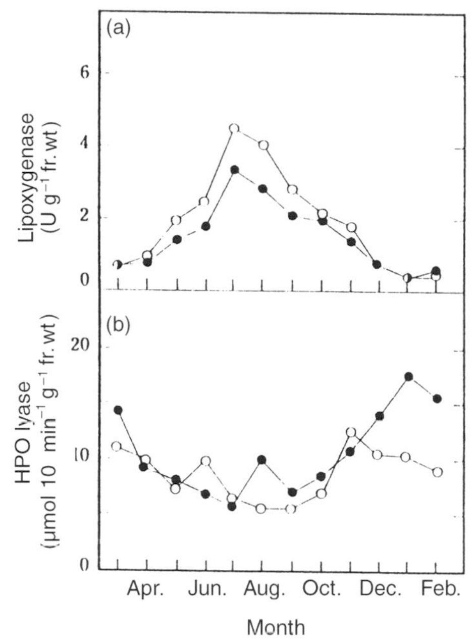

Figure 20 Seasonal changes in lipoxygenase and HPO lyase activities of tea leaves homogenate. (a) Lipoxygenase activity with α-linolenic acid (○) and linoleic acid (●). (b) HPO lyase activity with 1.2 mM linolenic acid HPO (○) and 1.2 mM linoleic acid HPO (●).

temperature, but the activity of HPO lyase is independent of temperature.[173] C_6 Aldehydes from intact tea leaves were quantitatively analyzed. Emission of the aldehydes increased in mid-May, when enzyme activities involved in aldehyde formation from lipid began to increase. The amounts of C_6 aldehydes accumulated in tea leaves also increased. However, the composition of the accumulated C_6 aldehydes did not always coincide with those of the emitted aldehydes.[41]

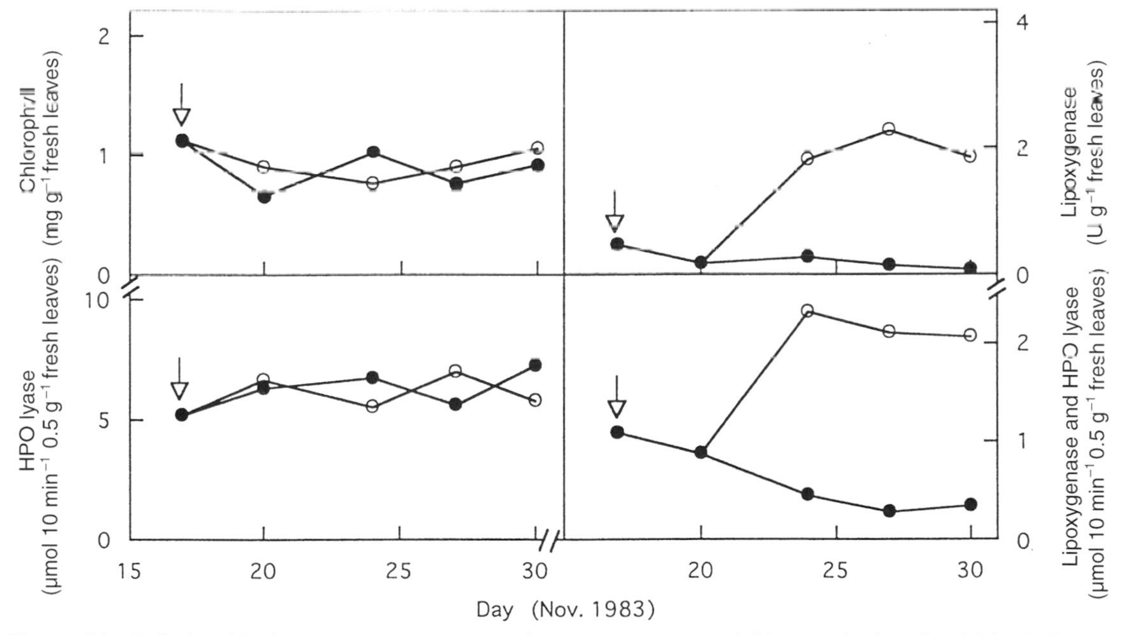

Figure 21 Relationship between temperature and enzyme system activities producing C_6 aldehyde by pot growth of *T. sinensis* leaves. Outside pot growth at 7 °C (●), inside growth chamber at 25 °C (○), change point from 7 °C to 25 °C indicated by arrows.

In tomato fruit, the highest lipoxygenase activity was found between the skin and the flesh, although HPO lyase was found ubiquitously in the fruit. Lipoxygenase specifically formed 9-(Z,E)-hydroperoxylinoleic acid from linoleic acid, whereas HPO lyase specifically cleaved the 13-(E,Z)-

hydroperoxide.[40] Although a low level of n-hexanal was detected in intact tomato fruit (0.36 ± 0.069 nmol g^{-1} fr. wt.), HPOs were not detected. When tomato fruit was injured by cutting into eight fragments and incubating at 25 °C, the n-hexanal content increased to 1.642 nmol g^{-1} fr. wt. after 30 min. On homogenizing at pH 6.3, n-hexanal increased to 21.1 nmol g^{-1} fr. wt. after 30 min of incubation. UV irradiation of tomato fruit also increased the formation of n-hexanal. From these results, lipoxygenase and HPO lyase were considered to exist in latent form and to express their activity upon injury.[40]

The effect of illumination on the enzyme activities relating to C_6 aldehyde formation in cultured alfalfa cells has been examined. Alfalfa green cells cultured for 40 days in the light were transferred to dark conditions at time zero (Figure 22). The cells in the dark were again transferred to light 72 h after the first transfer. The chlorophyll content and enzyme activities were determined at the times indicated.[174] When the green cells were transferred to the dark from light conditions, lipoxygenase and the C_6 aldehyde-forming activity decreased almost to zero, and low levels of activity were maintained during the following dark period. In contrast, HPO lyase activity decreased temporarily to 50% of the original level after dark incubation for 24 h, but subsequently returned to the original level even in the dark. When the cells were transferred to the light again, lipoxygenase and the C_6 aldehyde-forming activity were restored to their original levels. Therefore, the lipoxygenase activity of the cultured green cells of alfalfa is light-dependent, but HPO lyase is only slightly light-dependent. Consequently, the level of the C_6 aldehyde-forming activity of the cultured green cells is dependent on the illumination. Kidney bean plants (*P. vulgaris*) were found to have the capability to produce C_6 aldehydes (n-hexanal and n-hexenals) from linoleic and linolenic acid. The various plant parts tested had lipoxygenase and HPO lyase activities responsible for the C_6 aldehyde formation. Young

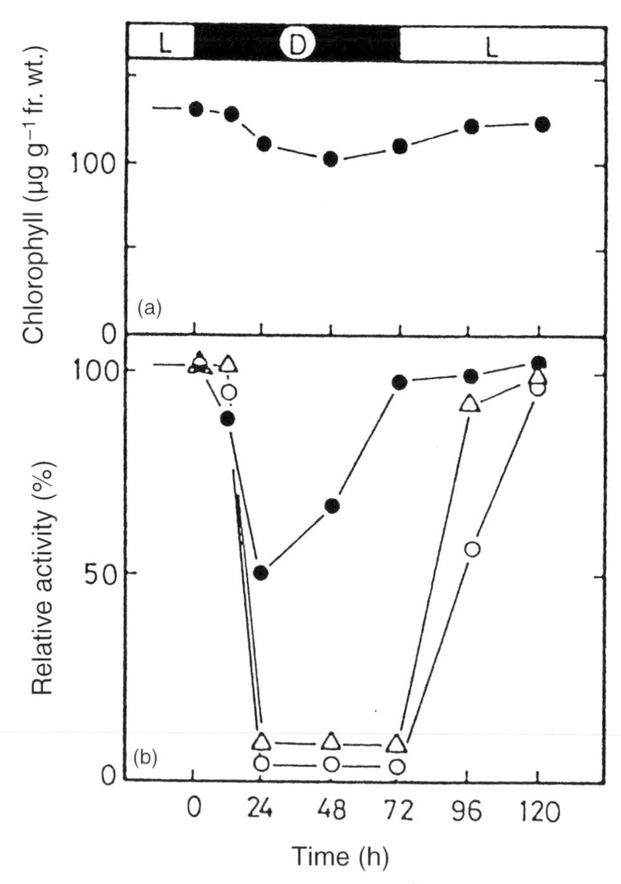

Figure 22 Effect of illumination on the enzyme activities involved in C_6 aldehyde formation in cultured alfalfa cells. Alfalfa green cells cultured for 40 days in the light were transferred to dark conditions at time zero. The cells in the dark were again transferred to the light 72 h after the first transfer. The chlorophyll content (a) and the enzyme activities (b) were determined at the times indicated. (△), lipoxygenase activity; (●), HPO lyase activity; (○), n-hexanal-forming activity from linoleic acid.

leaves showed relatively high activities, and activities decreased gradually with leaf development. Seedlings and seeds with cotyledons showed low activities for C_6 aldehyde formation because of the occurrence of an inhibitory factor in the cotyledons. The substrate specificity of the enzymes was essentially the same among the various developmental stages of leaves tested (Table 13).[175] Changes in the activities of lipoxygenase and HPO lyase during seed development of soybean were also examined.[176]

It was found that the activity for (*Z*)-3-hexenal and *n*-hexanal formation from α-linolenic and linoleic acids was also maintained in alfalfa cells cultured *in vitro*.[177] The green cells contained C_6 aldehydes and C_6 alcohols in the essential oil but white cells contained much lower concentrations. These results suggest that a biosynthetic pathway of C_6 aldehydes and C_6 alcohols similar to that of tea leaves is involved in this culture.

1.04.8 PERSPECTIVE

In the twenty-first century, the study of the relationships between chemical structure and green odor characteristics will be extended to research on plant and human relationships, to establish why people are refreshed by the fresh green odor emitted by green leaves in forests. In order to demonstrate this theme, the study should begin with the transfer affinity and nervous mechanism of green odor on human olfactory bulb nerves and the function of sense of smell, activation of neurosecretion and of immune systems by green odor and, in addition, on the reduction of stress and schizoid tendencies, etc.[178,179]

Research on the biosynthesis of volatile compounds in terrestrial plants has been extended to a study of the biogeneration of sex pheromones in marine brown algae. These are acyclic or cyclic hydrocarbons ($C_{11}H_{14}$, $C_{11}H_{16}$, and $C_{11}H_{18}$).[180] It is presumed that the pheromones are biosynthesized in female gametes from polyunsaturated C_{20} fatty acids (eicosapentaenoic and arachidonic acids) by a sequence of oxygenation and cleavage reactions which are catalyzed by lipoxygenase and fatty acid HPO lyase. It has been found that oxygenation activity was present in gametes, although the cleavage enzyme is unknown. Gametes secrete not only the species-specific pheromone but also a complex mixture of related compounds.[178] The composition of the pheromone bouquet depends on the specificity of the enzymes involved in the biosynthesis. Future studies are required to determine the site of pheromone biosynthesis, the cellular pathways leading to secretion, precise structure–activity relationships, and the nature of the chemoreceptors for pheromones.

In the plant enzyme system that produces C_6 aldehydes and/or alcohols, the first step, the lipolytic acyl hydrolase-catalyzed liberation of free fatty acids, is a key step which regulates the whole sequence. This is because: (i) the substrates of acyl hydrolase lipids are abundant in plant cells; (ii) the concentrations of free fatty acids and their HPOs in plant cells are very low; and (iii) C_6 aldehydes and C_6 alcohols are formed very rapidly on homogenizing plant tissues.[181] A similar regulatory system exists in an arachidonic cascade in mammalian cells, although calcium ion has no significance in the regulatory system in plants. There must be a novel and plant-specific regulation mechanism in the aldehyde/alcohol formation system of plants. It seems likely that external stimuli such as wounding or pest invasion are amplified by a special signal transduction pathway to trigger a response. A regulation mechanism for free fatty acid liberation must be included in such a signal transduction pathway.[181]

Furthermore, it has also been shown that nonheme iron can be removed from soybean lipoxygenase with little effect on the tertiary structure of the enzyme. Finally, more detailed information on the precise structure and formation mechanism of the green odor-generating enzyme complex may be obtained by NMR spectroscopy through dynamic analysis of lipoxygenase-[U-^{13}C]-13-hydroxy-(9*Z*,11*E*)-octadecadienoic acid (a potent competitive inhibitor).[182]

ACKNOWLEDGMENTS

I should like to dedicate this chapter to my teachers, the late Professors S. Takei and M. Ohno. I also gratefully acknowledge the contributions of my co-workers, Drs. T. Kajiwara, J. Sekiya, and K. Matsui. Research by the author was carried out at the Institute for Chemical Research, Kyoto University from 1957 to 1968 and at Yamaguchi University from 1968 to 1996. The unpublished data cited at the end of the chapter comes from work by K. Matsui, I. Shimada (Department of

Table 13 Changes in the activities of lipoxygenase, HPO lyase, and *n*-hexanal formation from linoleic acid in kidney bean plants at different developmental stages.

Days after planting	*Organ*	*Fresh weight* (g plant^{-1})	*Lipoxygenase*[a] (μmol O_2 min^{-1})	*HPO lyase*[a] (μmol)	*Hexanal-forming activity from linoleic acid*[b] (μmol)	*Chlorophyll* (mg)	*Protein* (mg)
0	Dry whole seed	0.50 (0.50)	1.48 (1.80)	0.36 (0.06)	0.14 (0.10)	0 (0)	167 (173)
	Dry hypocotyl	0.005	5.80	2.74	0.78	0	258
	Dry cotyledon	0.50	1.44	0.30	0.12	0	167
1	Whole seedling	1.00	2.16	0.84	0.12	0	78.0
4	Shoot	1.29 (1.29)	63.0 (122)	3.62 (1.14)	0.72 (0.14)	tr (tr)	60.0 (60.0)
	Primary leaf	0.10 (0.10)	96.0 (191)	6.52 (2.92)	3.72 (0.64)	tr (tr)	56.0 (56.0)
	Cotyledon	0.89 (0.89)	52.0 (101)	1.20 (0.54)	0.22 (0.06)	tr (tr)	67.0 (67.0)
	Hypocotyl	0.30 (0.30)	132 (264)	4.18 (1.62)	1.70 (0.58)	tr (tr)	26.0 (26.0)
7	Primary leaf	0.50	23.0	10.2	3.66	3.40	45.0
	Stem and cotyledon	1.95	6.50	2.10	0.16	0.20	19.9
	Root	0.32	9.16	3.20	1.12	0	9.8
13	Primary leaf	1.00 (1.00)	1.24 (4.00)	10.9 (4.40)	2.32 (2.28)	1.76 (2.34)	30.3 (33.6)
23	Primary leaf	1.60 (1.60)	0.40 (0.36)	3.52 (3.36)	0.68 (0.90)	1.64 (1.64)	19.7 (19.7)
	First trifoliate leaf	0.60 (0.60)	0.40 (0.70)	5.42 (3.14)	0.84 (0.40)	1.82 (1.82)	22.5 (22.5)
	Second trifoliate leaf	0.30 (0.30)	0.40 (0.90)	8.60 (3.36)	1.28 (0.72)	2.01 (2.01)	28.2 (28.2)
	Stem	1.70 (1.70)	0.38 (1.30)	1.14 (0.76)	1.18 (0.60)	0.20 (0.20)	6.5 (6.5)
29	Primary leaf	1.70 (1.70)	0.14 (0.34)	3.66 (3.60)	0.52 (0.14)	1.27 (1.27)	17.4 (17.4)
	First trifoliate leaf	0.90 (0.90)	0.36 (0.54)	7.02 (3.36)	0.94 (0.28)	2.05 (2.05)	21.9 (21.9)
	Second trifoliate leaf	0.50 (0.50)	0.42 (0.80)	7.54 (3.62)	1.16 (0.20)	2.25 (2.25)	25.2 (25.2)
	Third trifoliate leaf	0.20 (0.20)	0.76 (1.86)	10.0 (3.64)	1.62 (0.68)	2.50 (2.50)	34.1 (34.1)
	Stem	2.10 (2.10)	1.00 (2.40)	1.38 (0.60)	1.44 (0.50)	0.11 (0.11)	7.5 (7.5)
50	Immature seed and pod[c]	0.70	54.0	8.58	1.86	0.42	22.5

[a]The substrates for lipoxygenase and HPO lyase were represented by oxygen uptake to linoleic acid and by the decrease in absorbance at 234 nm with disappearance of the conjugate system 9*Z*,11*E* in linoleic acid hydroperoxide, respectively. [b]Values in parentheses are for α-linolenic acid hydroperoxide. The enzyme activities, chlorophyll, and protein are given as the values per gram fresh weight. [c]4–5 cm long.

Pharmacy, Tokyo University), and A. Hatanaka, supported by a Grant-in Aid for Scientific Research on Priority Areas No. 07229103 from the Ministry of Education, Science and Culture, Japan (1994–1995).

1.04.9 REFERENCES

1. A. Hatanaka, T. Kajiwara, and J. Sekiya, *Phytochemistry*, 1978, **17**, 869.
2. J. Sekiya, T. Kajiwara, T. Munechika, and A. Hatanaka, *Phytochemistry*, 1983, **22**, 1867.
3. A. Hatanaka, K. Munechika, and M. Imoto, Proceedings of the *Agricultural Biology and Chemistry* Annual Meeting, Fukuoka, 1980, p. 471.
4. A. Hatanaka, *Bull. Fac. Agric., Yamaguchi Univ.*, 1968, **19**, 1107.
5. A. Hatanaka, *Bull. Inst. Chem. Res. Kyoto Univ.*, 1983, **61**, 180.
6. A. Hatanaka, T. Kajiwara, and J. Sekiya, *Chem. Phys. Lipids*, 1987, **44**, 341.
7. A. Hatanaka, *Phytochemistry*, 1993, **34**, 1201.
8. A. Hatanaka, T. Kajiwara, and J. Sekiya, *Z. Naturforsch.*, 1995, **50c**, 467.
9. A. Hatanaka, *Food Res. Int.*, 1996, **12**, 303.
10. E. L. Rice, "Allelopathy," 2nd edn., Academic Press, New York, 1984.
11. L. M. Riddiford and M. C. Williams, *Science*, 1967, **155**, 89.
12. L. M. L. Riddiford, *Science*, 1967, **158**, 140.
13. I. Reinke, *Ber. d. d. Chem. Ges.*, 1881, **14**, 2144.
14. I. Reinke and L. Kraetschmar, *Untersuchungen, a. d, botan. Lab. d. Univ. Gettingen*, 1883, **IV**, 18.
15. T. Curtius and H. Franzen, *Justus Liebigs Ann. Chem.*, 1912, **390**, 89.
16. T. Curtius and H. Franzen, *Justus Liebigs Ann. Chem.*, 1914, **404**, 93.
17. P. von Romburgh, *Chem. Zentralbl.*, 1920, **1**, 83.
18. S. Takei and Y. Sakato, *Bull. Inst. Phys. Chem. Res. Tokyo*, 1933, **12**, 13.
19. S. Takei, Y. Sakato, and M. Ohno, *Bull. Inst. Phys. Chem. Res. Tokyo*, 1934, **13**, 128.
20. S. Takei, Y. Sakato, and M. Ohno, *Bull. Inst. Phys. Chem. Res. Tokyo*, 1935, **14**, 303.
21. S. Takei, T. Imaki, and Y. Tada, *Bull. Inst. Phys. Chem. Res. Tokyo*, 1935, **14**, 507.
22. S. Takei, Y. Sakato, M. Ohno, and Y. Kuroiwa, *Bull. Agric. Chem. Soc. Jpn.*, 1938, **14**, 709.
23. S. Takei, M. Ohno, Y. Kuroiwa, T. Takahata, and T. Sima, *Bull. Agric. Chem. Soc. Jpn.*, 1938, **14**, 717.
24. S. Takei and M. Ohno, *Bull. Agric. Chem. Soc. Jpn.*, 1939, **15**, 193.
25. S. Takei, M. Ohno, and K. Sinosaki, *Bull. Agric. Chem. Soc. Jpn.*, 1940, **16**, 772.
26. S. Takei and M. Ohno, *Bull. Agric. Chem. Soc. Jpn.*, 1942, **18**, 19.
27. S. Takei, T. Imaki, and Y. Tada, *Chem. Ber.*, 1935, **68**, 953.
28. S. Takei, M. Ohno, and K. Shinosaki, *Chem. Ber.*, 1940, **73**, 950.
29. A. Hatanaka, M. Hamada, and M. Ohno, *Bull. Agric. Chem. Soc. Jpn.*, 1960, **34**, 115.
30. E. Honkanen, T. Moisio, M. Ohno, and A. Hatanaka, *Acta Chem. Scand.*, 1963, **17**, 2051.
31. A. Hatanaka, T. Kajiwara, and J. Sekiya, Proceedings of the ACS Symposium, eds. T. H. Parliment and R. Croteau, American Chemical Society, Washington, DC, 1986, no. 317, **13**, 167.
32. A. Hatanaka, T. Kajiwara, J. Sekiya, and H. Toyota, *Z. Naturforsch.*, 1986, **41c**, 359.
33. J. Sekiya, H. Koiso, A. Morita, and A. Hatanaka, in "Proceedings of the 8th International Symposium on Plant Lipids," eds. P. K. Stampf, J. B. Mudd, and W. D. Nes, Plenum, New York, 1987, p. 377.
34. A. Hatanaka, T. Kajiwara, and J. Sekiya, in "Proceedings of the 8th International Symposium on Plant Lipids," eds. P. K. Stampf, J. B. Mudd, and W. D. Nes, Plenum, New York, 1987, p. 391.
35. A. Hatanaka, T. Kajiwara, and K. Matsui, *Z. Naturforsch.*, 1988, **43c**, 308
36. K. Matsui, T. Kajiwara, K. Hayasi, and A. Hatanaka, *Agric. Biol. Chem.*, 1988, **52**, 3219.
37. A. Hatanaka, T. Kajiwara, K. Matsui, and T. Matsunaga, *Z. Naturforsch.*, 1989, **44c**, 161.
38. K. Matsui, T. Kajiwara, and A. Hatanaka, in "Proceedings of the 9th International Symposium on Plant Lipids," eds. P. J. Quinn and J. L. Harwood, Portland Press, London, 1990, p. 292.
39. A. Hatanaka, T. Kajiwara, and K. Matsui, in "Proceedings of the 9th International Symposium on Plant Lipids," eds. P. J. Quinn and J. L. Harwood, Portland Press, London, 1990, p. 295.
40. A. Hatanaka, T. Kajiwara, K. Matsui, and A. Kitamura, *Z. Naturforsch.*, 1992, **47c**, 369.
41. A. Kitamura, K. Matsui, T. Kajiwara, and A. Hatanaka, *Plant Cell Physiol.*, 1992, **33**, 493.
42. A. Hatanaka, T. Kajiwara, and K. Matsui, in "Proceedings of the International Conference in Flavour Precursor Studies," eds. P. Schreier and P. Winterhalter, *Perfumer and Flavorist*, Wurzburg, 1993, p. 151.
43. K. Matsui, T. Kajiwara, and A. Hatanaka, in "Proceedings of the International Conference in Flavour Precursor Studies," eds. P. Schreier and P. Winterhalter, *Perfumer and Flavorist*, Wurzburg, 1993, p. 171.
44. A. Hatanaka and M. Ohno, *Agric. Biol. Chem.*, 1961, **25**, 7.
45. L. Crombie and H. Harper, *J. Chem. Soc.*, 1950, 873.
46. F. Sondheimer, *J. Chem. Soc.*, 1950, 887.
47. A. Hatanaka and T. Harada, *Phytochemistry*, 1973, **12**, 2341.
48. A. Hatanaka and M. Ohno, *Z. Naturforsch.*, 1960, **15b**, 415.
49. A. Hatanaka and T. Kajiwara, *Z. Naturforsch.*, 1981, **36b**, 755.
50. A. Hatanaka and M. Ohno, *Agric. Biol. Chem.*, 1971, **35**, 1044.
51. A. Hatanaka and M. Ohno, *Bull. Agric. Chem.*, 1960, **24**, 614.
52. T. Kajiwara, T. Harada, and A. Hatanaka, *Agric. Biol. Chem.*, 1975, **39**, 243.
53. A. Hatanaka and T. Ohgi, *Agric. Biol. Chem.*, 1972, **36**, 1263.
54. A. Hatanaka, *Bull. Inst. Chem. Res. Kyoto*, 1972, **50**, 135.
55. R. G. Buttery, R. Teranishi, and L. C. Ling, *J. Agric. Chem.*, 1987, **35**, 540.
56. F. Drawert, W. Heimann, R. Emberger, and R. Tressel, *Justus Liebigs Ann. Chem.*, 1966, **694**, 200.

57. T. Galliard and J. A. Matthew, *Phytochemistry*, 1977, **16**, 339.
58. A. Hatanaka, T. Kajiwara, and J. Sekiya, *Phytochemistry*, 1976, **15**, 1125.
59. A. Hatanaka, T. Kajiwara, J. Sekiya, and Y. Kido, *Phytochemistry*, 1977, **16**, 1828.
60. T. Kajiwara, J. Sekiya, Y. Kido, and A. Hatanaka, *Agric. Biol. Chem.*, 1977, **41**, 1793.
61. A. Hatanaka, T. Kajiwara, J. Sekiya, and K. Fujimura, *Agric. Biol. Chem.*, 1979, **43**, 175.
62. A. Hatanaka, T. Kajiwara, H. Horino, and K. Inokuchi, *Z. Naturforsch.*, 1992, **47c**, 183.
63. H. Horino, T. Kajiwara, and A. Hatanaka, *Chem. Express*, 1992, **7**, 925.
64. J. Suzuki, N. Ichimura, and T. Eto, *Food Rev. Int.*, 1990, **6**, 537.
65. Y. Sakoda, K. Matsui, T. Kajiwara, and A. Hatanaka, *Z. Naturforsch.*, 1995, **50c**, 757.
66. Y. Sakoda, K. Matsui, Y. Akakabe, J. Suzuki, T. Kajiwara, and A. Hatanaka, *Z. Naturforsch*,, 1996, **51c**, 841.
67. A. Hatanaka, M. Ohno, and Y. Inouye, *Angew. Chem.*, 1962, **74**, 291.
68. M. Ohno, A. Hatanaka, and Y. Inouye, *Agric. Biol. Chem.*, 1962, **26**, 460.
69. A. Hatanaka, T. Kajiwara, and M. Ohno, *Agric. Biol. Chem.*, 1965, **29**, 662.
70. A. Hatanaka, T. Kajiwara, and M. Ohno, *Agric. Biol. Chem.*, 1967, **31**, 964.
71. M. Ohno and A. Hatanaka, *Bull. Inst. Chem. Res. Kyoto Univ.*, 1962, **40**, 322.
72. M. Ohno and A. Hatanaka, *Agric. Biol. Chem.*, 1964, **28**, 908.
73. A. Hatanaka and M. Ohno, *Agric. Biol. Chem.*, 1964, **28**, 910.
74. M. Ohno and A. Hatanaka, *Bull. Inst. Chem. Res. Kyoto Univ.*, 1964, **42**, 232.
75. M. Ohno and A. Hatanaka, *Bull. Inst. Chem. Res. Kyoto Univ.*, 1964, **42**, 227.
76. M. Ohno, Y. Inouye, A. Hatanaka, and T. Kajiwara, *Bull. Inst. Chem. Res. Kyoto Univ.*, 1965, **43**, 231.
77. A. Hatanaka, T. Kajiwara, and M. Ohno, *Agric. Biol. Chem.*, 1967, **31**, 969.
78. M. Ohno, A. Hatanaka, T. Kajiwara, and H. Miyawaki, *Bull. Inst. Chem. Res. Kyoto Univ.*, 1967, **45**, 184.
79. T. Kajiwara, A. Hatanaka, Y. Inouye, and M. Ohno, *Agric. Biol. Chem.*, 1969, **33**, 409.
80. M. Hamada, Y. Nagata, and A. Hatanaka, *Agric. Biol. Chem.*, 1972, **36**, 324.
81. A. Hatanaka, O. Adachi, and M. Ameyama, *Agric. Biol. Chem.*, 1970, **34**, 1574.
82. A. Hatanaka, O. Adachi, T. Chiyonobu, and M. Ameyama, *Agric. Biol. Chem.*, 1971, **35**, 1142.
83. A. Hatanaka, O. Adachi, T. Chiyonobu, and M. Ameyama, *Agric. Biol. Chem.*, 1971, **35**, 1304.
84. A. Hatanaka and T. Harada, *Agric. Biol. Chem.*, 1972, **36**, 2033.
85. A. Hatanaka, T. Kajiwara, and S. Tomohiro, *Agric. Biol. Chem.*, 1974, **38**, 1819.
86. A. Hatanaka, T. Kajiwara, S. Tomohiro, and H. Yamashita, *Agric. Biol. Chem.*, 1974, **38**, 1835.
87. J. Sekiya, W. Kawasaki, T. Kajiwara, and A. Hatanaka, *Agric. Biol. Chem.*, 1975, **39**, 1677.
88. A. Hatanaka, T. Kajiwara, and J. Sekiya, *Phytochemistry*, 1976, **15**, 1889.
89. J. Sekiya, S. Numa, T. Kajiwara, and A. Hatanaka, *Agric. Biol. Chem.*, 1976, **40**, 185.
90. A. Hatanaka and T. Ohgi, *Agric. Biol. Chem.*, 1972, **36**, 1263.
91. A. Hatanaka, T. Kajiwara, J. Sekiya, and H. Hirata, *Agric. Biol. Chem.*, 1976, **40**, 2177.
92. T. Galliard, *Eur. J. Biochem.*, 1971, **21**, 90.
93. J. Sekiya, T. Kajiwara, M. Imoto, S. Inouye, and A. Hatanaka, *J. Agric. Food Chem.*, 1982, **30**, 183.
94. T. Kajiwara, J. Sejiya, Y. Kido, and A. Hatanaka, *Agric. Biol. Chem.*, 1977, **41**, 1793.
95. A. Hatanaka, J. Sekiya, and T. Kajiwara, *Plant Cell Physiol.*, 1977, **16**, 107.
96. J. Sekiya and A. Hatanaka, *Plant Sci. Letters*, 1977, **10**, 165.
97. A. Hatanaka, A. Kajiwara, J. Sekiya, and T. Kido, *Phytochemistry*, 1978, **17**, 548.
98. J. Sekiya, T. Kajiwara, and A. Hatanaka, *Plant Cell Physiol.*, 1978, **19**, 553.
99. A. Hatanaka, J. Sekiya, T. Kajiwara, and K. Munechika, *Agric. Biol. Chem.*, 1982, **46**, 2705.
100. T. Galliard and J. A. Matthew, *Phytochemistry*, 1977, **16**, 339.
101. S. J. Kazeniac and R. M. Hall, *J. Food Sci.*, 1970, **35**, 519.
102. A. Hatanaka, T. Kajiwara, J. Sekiya, M. Imoto, and S. Inouye, *Plant Cell Physiol.*, 1982, **23**, 91.
103. B. A. Vick and D. C. Zimmerman, *Plant Physiol.*, 1976, **57**, 780.
104. A. Pinsky, S. Grossman, and M. Trop, *J. Food Sci.*, 1971, **36**, 571.
105. P. Schreier and G. Lorenz, *Z. Naturforsch.*, 1982, **37c**, 165.
106. J. Sekiya, H. Kamiuchi, and A. Hatanaka, *Plant Cell Physiol.*, 1982, **23**, 631.
107. J. Sekiya, T. Kajiwara, and A. Hatanaka, *Agric. Biol. Chem.*, 1979, **43**, 969.
108. K. Matsui, H. Narahara, T. Kajiwara, and A. Hatanaka, *Phytochemistry*, 1991, **30**, 1499.
109. A. B. Vick and C. D. Zimmerman, *Plant Physiol.*, 1989, **90**, 125.
110. I. S. Kim and W. Grosch, *J. Agric. Food Chem.*, 1981, **29**, 1220.
111. R. H. Andrianarison, J. L. Beneytout, and M. Tixier, *Plant Physiol.*, 1989, **91**, 1280.
112. J. Sekiya, H. Aoshima, T. Kajiwara, T. Togo, and A. Hatanaka, *Agric. Biol. Chem.*, 1977, **41**, 827.
113. J. F. G. Vliegenthart and G. A. Veldink, in "Free Radicals in Biology," ed. W. Pryor, Academic Press, New York, 1982, vol. 5, p. 29.
114. A. B. Vick and C. D. Zimmerman, in "The Biochemistry of Plants," ed. P. K. Stampf, Academic Press, New York, 1987, vol. 9, p. 53.
115. W. H. Gardner, *Biochim. Biophys. Acta*, 1991, **1084**, 221.
116. L. Ruzica, H. Schinz, and B.-P. Susz, *Helv. Chim. Acta*, 1944, **27**, 1561.
117. A. Hatanaka, T. Kajiwara, and T. Harada, *Phytochemistry*, 1975, **14**, 2589.
118. T. Kajiwara, J. Sekiya, Y. Odake, and A. Hatanaka, *Agric. Biol. Chem.*, 1977, **41**, 1481.
119. A. Hatanaka, T. Kajiwara, J. Sekiya, and S. Inouye, *Phytochemistry*, 1982, **21**, 13.
120. A. Hatanaka, T. Kajiwara, and T. Koda, *Agric. Biol. Chem.*, 1979, **43**, 2115.
121. J. Sekiya, S. Tanigawa, T. Kajiwara, and A. Hatanaka, *Phytochemistry*, 1984, **23**, 2439.
122. A. J. Matthew and T. Galliard, *Phytochemistry*, 1978, **17**, 1043.
123. Y. Shibata, K. Matsui, T. Kajiwara, and A. Hatanaka, *Plant Cell Physiol.*, 1995, **36**, 147.
124. J. Sekiya, T. Kajiwara, and A. Hatanaka, *Agric. Biol. Chem.*, 1979, **43**, 969.
125. T. Matoba, H. Hidaka, H. Narita, K. Kitamura, N. Kaizuma, and M. Kito, *J. Agric. Food Chem.*, 1985, **33**, 852.
126. T. Galliard, D. R. Phillips, and J. Reynols, *Biochim. Biophys. Acta*, 1976, **441**, 181.
127. T. Galliard and D. R. Phillips, *Biochim. Biophys. Acta*, 1976, **431**, 278.

128. K. Matsui, Y. Shibata, T. Kajiwara, and A. Hatanaka, *Z. Naturforsch.*, 1989, **44c**, 883.
129. K. Matsui, H. Toyota, T. Kajiwara, and A. Hatanaka, *Phytochemistry*, 1991, **30**, 2109.
130. K. Matsui, T. Kajiwara, and A. Hatanaka, *J. Agric. Food Chem.*, 1992, **40**, 175.
131. Y. Shibata, K. Matsui, T. Kajiwara, and A. Hatanaka, *Biochem. Biophys. Res. Commun.*, 1995, **207**, 438.
132. Y. Shibata, K. Matsui, T. Kajiwara, and A. Hatanaka, *Plant Cell Physiol.*, 1995, **36**, 147.
133. J. Sekiya, T. Kajiwara, and A. Hatanaka, *Phytochemistry*, 1977, **16**, 1043.
134. T. Kajiwara, J. Sekiya, Y. Odake, and A. Hatanaka, *Agric. Biol. Chem.*, 1975, **39**, 1617.
135. A. Hatanaka, J. Sekiya, and T. Kajiwara, *Agric. Biol. Chem.*, Annual Joint Meeting of Kansai and Nishinippon Branches, Okayama, 1976, p. 5.
136. A. Hatanaka, *Koryo*, 1977, **117**, 25.
137. A. Hatanaka, T. Kajiwara, J. Sekiya, M. Imoto, and S. Inouye, *Plant Cell Physiol.*, 1982, **23**, 91.
138. A. Hatanaka, T. Kajiwara, K. Matsui, and H. Toyota, *Z. Naturforsch.*, 1992, **47c**, 677.
139. T. Kajiwara, T. Koda, and A. Hatanaka, *Agric. Biol. Chem.*, 1979, **43**, 1781.
140. A. Hatanaka, T. Kajiwara, J. Sekiya, and M. Asano, *Z. Naturforsch.*, 1984, **39c**, 171.
141. T. Kajiwara, N. Nagata, A. Hatanaka, and Y. Naoshima, *Agric. Biol. Chem.*, 1980, **44**, 437.
142. A. Hatanaka, T. Kajiwara, K. Matsui, and M. Yamaguchi, *Z. Naturforsch.*, 1989, **44c**, 64.
143. A. Hatanaka, T. Kajiwara, K. Matsui, and M. Ogura, *Z. Naturforsch.*, 1990, **45c**, 1161.
144. A. Hatanaka and T. Kajiwara, *Nippon Kagaku Kaishi*, 1981, **5**, 684.
145. B. Axelrod, M. T. Cheesbrough, and S. Laakso, *Methods Enzymol.*, 1981, **71**, 441.
146. J. C. Boyuington, B. J. Gaffney, and L. M. Amzel, *Science*, 1993, **260**, 1482.
147. K. Matsui, H. Shibata, T. Kajiwara, and A. Hatanaka, *Z. Naturforsch.*, 1995, **50c**, 37.
148. J. Steczko, G. P. Donoho, J. C. Clemens, J. E. Dixon, and B. Axelrod, *Biochemistry*, 1992, **31**, 4055.
149. W. Minor, J. Steckzko, J. T. Bolin, Z. Otwinoski, and B. Axelrod, *Biochemistry*, 1993, **32**, 6320.
150. K. Matsui, H. Shinta, H. Toyota, T. Kajiwara, and A. Hatanaka, *Z. Naturforsch.*, 1992, **47c**, 85.
151. A. Hatanaka, T. Kajiwara, K. Matsui, and H. Toyota, *Z. Naturforsch.*, 1992, **47c**, 677.
152. J. M. Wallace and E. L. Wheeler, *J. Food Chem.*, 1975, **23**, 146.
153. T. Kajiwara, J. Sekiya, M. Asano, and A. Hatanaka, *Agric. Biol. Chem.*, 1982, **46**, 3087.
154. H. Aoshima, J. Sekiya, T. Kajiwara, and A. Hatanaka, *Agric. Biol. Chem.*, 1977, **41**, 1787.
155. A. Hatanaka, J. Sekiya, T. Kajiwara, and T. Miura, *Agric. Biol. Chem.*, 1979, **43**, 735.
156. H. Aoshima, T. Kajiwara, A. Hatanaka, H. Nakatani, and K. Hiromi, *Int. J. Peptide Protein Res.*, 1977, **10**, 219.
157. H. Aoshima, T. Kajiwara, A. Hatanaka, and H. Hatano, *J. Biochem.*, 1977, **82**, 1559.
158. H. Aoshima, T. Kajiwara, A. Hatanaka, H. Nakatani, and K. Hiromi, *Biochim. Biophys. Acta*, 1977, **486**, 121.
159. J. Sekiya, H. Aoshima, T. Kajiwara, T. Togo, and A. Hatanaka, *Agric. Biol. Chem.*, 1977, **41**, 827.
160. H. Aoshima, T. Kajiwara, A. Hatanaka, and H. Nakatani, *Agric. Biol. Chem.*, 1979, **43**, 167.
161. H. Aoshima, T. Kajiwara, and A. Hatanaka, *Agric. Biol. Chem.*, 1981, **45**, 2245.
162. M. R. Egmond, J. F. G. Vliegenthart, and J. Boldingh, *Biochem. Biophys. Res. Commun.*, 1972, **48**, 1055.
163. J. J. M. C. De Groot, G. A. Veldink, J. F. G. Vliegenthart, J. Boldingh, R. Wever, and B. F. Gelder, *Biochim. Biophys. Acta*, 1975, **377**, 71.
164. T. Matoba, H. Hidaka, H. Narita, K. Kitamura, N. Kaizuma, and M. Kito, *J. Agric. Food Chem.*, 1985, **33**, 852.
165. M. Wurzenberger and W. Grosch, *Biochim. Biophys. Acta*, 1984, **794**, 18.
166. M. Byrn and M. Calvin, *J. Am. Chem. Soc.*, 1966, **88**, 1916.
167. H. W. Gardner and R. D. Plattner, *Lipids*, 1984, **19**, 294.
168. L. Crombie, D. O. Morgan, and E. H. Smith, *J. Chem. Soc., Chem. Commun.*, 1987, 502.
169. L. Crombie and D. O. Morgan, *J. Chem. Soc., Chem. Commun.*, 1987, 503.
170. L. Crombie, O. Morgan, and E. H. Smith, *J. Chem. Soc.*, 1991, 567.
171. J. Sekiya, T. Kajiwara, and A. Hatanaka, *Plant Cell Physiol.*, 1977, **18**, 283.
172. J. Sekiya, T. Kajiwara, and A. Hatanaka, *Plant Cell Physiol.*, 1984, **25**, 269.
173. A. Hatanaka, T. Kajiwara, J. Sekiya, T. Takeo, Y. Saijiyo, and T. Tusida, The experiment for elucidation of biosynthetic route of young tea odor (leaf alcohol) by green tea, Special Grant in Aid Research. 1981–1982, Report from the Ministry of Agriculture, 1982, p. 21.
174. J. Sekiya, K. Munechika, and A. Hatanaka, *Agric. Biol. Chem.*, 1983, **47**, 1647.
175. J. Sekiya, H. Kamiuchi, and A. Hatanaka, *Plant Cell Physiol.*, 1982, **23**, 631.
176. J. Sekiya, T. Monma, T. Kajiwara, and A. Hatanaka, *Agric. Biol. Chem.*, 1986, **50**, 521.
177. J. Sekiya and A. Hatanaka, *Plant Sci. Lett.*, 1977, **10**, 165.
178. H. Sugano, M. Uchida, N. Sato, A. Hatanaka, and K. Sano, *Jpn. Assoc. Study of Taste and Smell Res.*, 1996, **3**, 672.
179. T. Sawada, N. Tokuda, T. Fukumoto, A. Hatanaka, and K. Sano, unpublished results.
180. T. Kajiwara, A. Hatanaka, and K. Matsui, Proceedings of the ACS Symposium series, eds. R. Teranishi, P. G. Buttery, and H. Sugisawa, American Chemical Society, Washington, DC, 1993, 525, p. 103.
181. K. Matsui, T. Kajiwara, and A. Hatanaka, in "Proceedings of the XV International Botanical Congress at Yokohama, Japan, 1993," ed. M. Furuya, Organization Committee, Tokyo, p. 368.
182. K. Matsui, I. Shimada, and A. Hatanaka, unpublished results.

1.05
Biosynthesis of Jasmonoids and Their Functions

TERUHIKO YOSHIHARA and FRIEDEMANN GREULICH
Hokkaido University, Sapporo, Japan

1.05.1 INTRODUCTION

(−)-Jasmonic acid methyl ester (JA-Me) (**2b**) was isolated as a fragrant constituent of the essential oil of jasmine (*Jasminum grandiflorum* L.).[1] The free acid, (−)-jasmonic acid ((−)-(JA)) (**1b**), was isolated as a plant growth inhibitor from culture filtrates of the fungus *Lasiodiplodia theobromae* (synonym *Botryodiplodia theobromae* Pat.).[2] This was the first report indicating that jasmonic acid (JA) has a regulatory effect on plant growth. Using different methods such as GC, GC–MS, HPLC, radioimmunoassay, and bioassays, JA (**1**) and JA-Me (**2**) were found to be widespread in the plant kingdom.[3] A conjugate of JA with an amino acid was first isolated from the metabolites of *Gibberella fujikuroi*.[4]

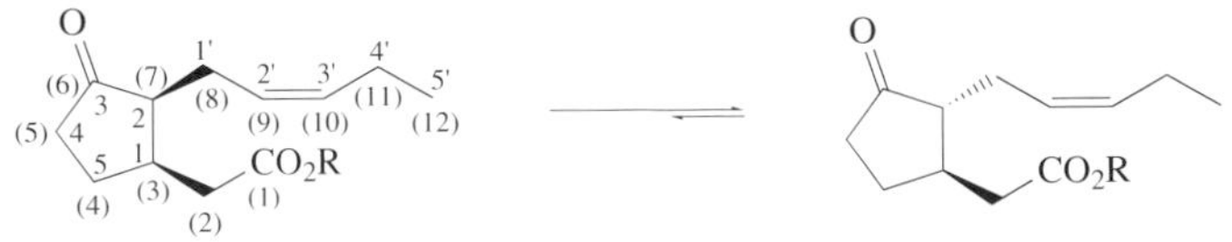

R = H	(+)-Epijasmonic acid	R = H	(–)-Jasmonic acid
(**1a**)	*cis*-Jasmonic acid	(**1b**)	*trans*-Jasmonic acid
	2-Epijasmonic acid		(–)-JA
	2-Isojasmonic acid		
	(+)-7-Isojasmonic acid		
	(+)-JA		
R = Me	Methyl ester	R = Me	Methyl ester
(**2a**)	(+)-JA-Me	(**2b**)	(–)-JA-Me

Cucurbic acid (**3**) together with its glucoside and the glucoside of its methyl ester were isolated as plant growth inhibitors from cucumber seeds (*Cucurbita pepo* L.) guided by a gibberellin bioassay using dwarf rice (*Oryza sativa* L. cv. Tan-ginbozu) seedlings.[5,6] Epicucurbic acid lactone was isolated from jasmin absolute.[7]

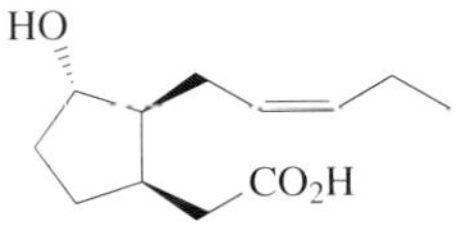

Cucurbic acid
(**3**)

Two types of hydroxy-JA are known. 11-Hydroxy-JA (**4**) was isolated from *B. theobromae*.[8] 12-Hydroxy-JA (**5**) was isolated as a lactone from jasmine (*J. grandiflorum* L.).[9] The glucoside was isolated as a potato tuber-inducing substance from potato leaves and the aglycone was named tuberonic acid.[10,11] *N*-Acetyl jasmonylphenylalanine was isolated from *Praxelis clematidea* (Gardn.) K. et R.[12]

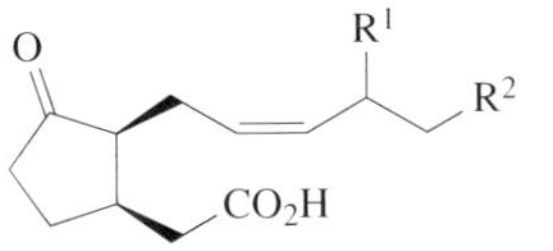

R^1 = OH, R^2 = H 11-hydroxy-JA
(**4**)
R^1 = H, R^2 = OH Tuberonic acid
(**5**)

Dihydro-JA (**6**) was isolated as the conjugate with isoleucine from the metabolites of *G. fujikuroi*.[4]

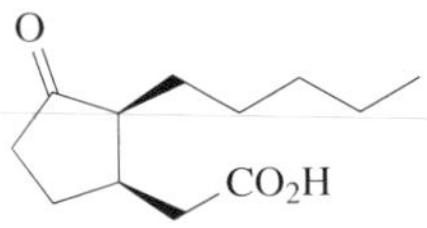

Dihydro-JA
(**6**)

The growing number of jasmonates with their various plant growth regulating and signal-mediating activities appears to be a distinct group of plant hormones and signal transmitters. In this text we will use the term "jasmonates" (besides its chemically correct interpretation as esters or salts of JA) for the physiological group of native JA derivatives of plants, but "jasmonoids" as the more general term, including all structurally and functionally related compounds.

1.05.2 BIOSYNTHESIS OF JASMONIC ACID

1.05.2.1 Biosynthetic Pathway

Jasmonic acid is called plant prostaglandin because of similarities in structure, biosynthesis, and function. Prostaglandin arises from arachidonic acid with C-20:4 via β-oxidation by lipoxygenase and is involved in inflammatory events and pain responses of animals. JA (**1**) arises through the oxidative cyclization of α-linolenic acid (**7**) with C-18:3. (The complete reaction sequence established at the time of writing is summarized in Scheme 1, later in the chapter.) Both galactolipids in the plastid membrane and phospholipids in the plasma membrane contain large amounts of linolenic acid (**7**), which can be liberated from these lipids either by lipid acyl hydrolase or by phospholipase A_2. During the first steps of the following pathway the 18 carbons of linolenic acid are conserved, which led to the name "octadecanoid pathway."

1.05.2.1.1 Pathway to 12-oxo-PDA (8) from α-linolenic acid (7)

The key intermediate product in the biosynthetic pathway is 8-[2-(*cis*-pent-2′-enyl)-3-oxo-*cis*-cyclopent-4-enyl]octanoic acid (12-oxo-*cis*-10,15-phytodienoic acid; 12-oxo-PDA) (**8**), which was formed by incubation of α-linolenic acid (**7**) with an extract of flaxseed (*Linum usitatissimum* L.) acetone powder in 0.05 M potassium phosphate buffer, pH 7.0, incubated for 90 min at 25 °C (Equation 1).[13] The chemical structure of 12-oxo-PDA (**8**) was determined and the name phytonoic acid was proposed for the parent compound, an 18 carbon fatty acid containing a five-membered ring between C-9 and C-13.[13] Substrate specificity studies of the flaxseed extracts showed that n-3,6,9 unsaturation was an absolute requirement for the conversion of polyunsaturated fatty acids into analogous products containing a cyclopentenone ring.[14] Fatty acids with 18, 20, or 22 carbons that satisfied this requirement were effective substrates. 12-oxo-PDA (**8**) was also one of the products when α-linolenic acid (**7**) reacted with a crude extract of pericarp of *Vicia faba* L. cv. Windsor.[15] 12-oxo-PDA (**8**) has the *cis* configuration of carbon chains with respect to the cyclopentenone ring. Treatment with acid, base, or heat (190 °C) isomerized 12-oxo-PDA to the *trans* isomer.[16]

CO_2H Linolenic acid (**7**) → O, CO_2H 12-oxo-PDA (**8**) (1)

The initial reaction of the pathway in which α-linolenic acid (**7**) is converted to 12-oxo-PDA (**8**) was catalyzed by soybean lipoxygenase in the presence of O_2 and the resulting hydroperoxide products were (13*S*)-hydroperoxylinolenic acid ((13*S*)-hydroperoxy-*cis*-9,15-*trans*-11-octadecatrienoic acid; (13*S*)-HPLA) (**9**) and 9-HPLA as a minor product.[14] A regiospecificity study with partially purified lipoxygenase from *V. faba* pericarp showed that 92% of the product was (13*S*)-HPLA (**9**) and 8% was the 9-hydroperoxide isomer.[15]

CO_2H, OOH

(**9**) (13*S*)-HPLA

Incubation of (13*S*)-HPLA (**9**) with hydroperoxide dehydrase from homogenized defatted corn (*Zea mays* L.) germ meal was found to catalyze the conversion into an unstable allene oxide derivative, (12*S*, 13*S*)-epoxylinolenic acid ((12*S*, 13*S*)-EOD) (**10**).[17] The half-life of this compound was estimated to be 32–34 s at 0 °C[17] and 9 s at 10 °C.[18] In water it suffered nucleophilic attack at C-13 followed by the opening of the epoxide ring and ketonization to afford the α-ketol, 12-keto-13-hydroxy-(9*Z*)-octadecenoic acid, as the main stable end product.[17] 12-oxo-PDA and a γ-ketol were found as minor by-products of this reaction.[18]

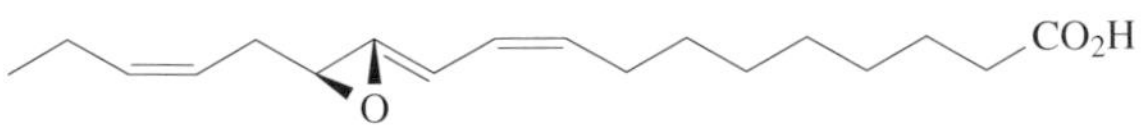

CO_2H

O

(10) (12*S*, 13*S*)-EOD

The enzyme, allene oxide synthase, was found in the microsomal fraction of fresh tissues. It was purified from flaxseeds by ammonium sulfate fractionation (0–45%), washing on octyl Sepharose (CL-4B column), anion exchange chromatography (Mono-Q column), and finally chromatofocusing (Mono-P column). Characterization of the dehydrase enzyme revealed that it is a 55 kDa hemoprotein. The spectral characteristics of this dehydrase revealed it to be a cytochrome P450. It operates with the remarkable activity of $\geqslant$1000 turnovers per second.[18] This enzyme's International Union of Biochemistry and Molecular Biology name is hydroperoxide dehydratase (EC 4.2.1.92), and it has also been known as hydroperoxide isomerase, hydroperoxide cyclase, and fatty acid hydroperoxide dehydrase.

The allene oxide cyclase, a novel enzyme in the metabolism of oxygenated fatty acids, was discovered from corn homogenates fractionated by differential centrifugation. While the concentration of the allene oxide synthase was high in the 105 000 *g* particle fraction, the cyclase activity could only be found after resuspension with the supernatant.

The enzymatic conversion of the allene oxide (12*S*,13*S*)-EOD (**10**) yielded optically pure 12-oxo-PDA (**8**). The enzyme was partially characterized and found to be a soluble protein with an apparent molecular weight of about 4.5×10^4 that specifically catalyzed the conversion of allene oxide (**10**) into 12-oxo-PDA (**8**).[19]

In two other studies the incubation of (13*S*)-HPLA (**9**) with ammonium sulfate precipitate of defatted corn germ[20] or with acetone powder extract of flaxseed[21] yielded 12-oxo-PDA (**8**) not optically pure but as a mixture of enantiomers in a ratio of 82:18[20] or as a racemic mixture together with the α-ketol.[21] These results are probably due to low or missing allene oxide cyclase activity in the preparations used.

1.05.2.1.2 Pathway to jasmonic acid from 12-oxo-PDA

For the pathway from 12-oxo-PDA (**8**) to JA (**1**), it was proposed that first the ring double bond of 12-oxo-PDA (**8**) is saturated, then β-oxidation enzymes remove six carbons from the carboxyl side chain of the ring.[15]

When 12-oxo-PDA (**8**) preparations with different combinations of isotope labels ([U-^{14}C], [^{18}O], [^{16}O]) were incubated with thin transverse sections of *V. faba* pericarp tissue, four radioactive metabolites were detected. The predominant product was 3-oxo-2-(2-pentenyl)cyclopentaneoctanoic acid (OPC-8:0) (**11**) and one of the minor products was JA (**1**) (Equation (2)). This result showed that 12-oxo-PDA (**8**) is a precursor in the jasmonic acid (**1**) biosynthesis via OPC-8:0 (**11**).[15]

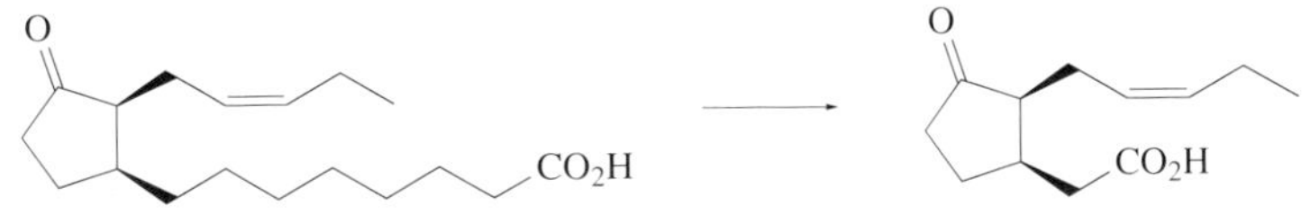

OPC-8:0 **(11)** → (+)-JA **(1a)** (2)

12-oxo-PDA reductase (12-oxo-phytodienoate 10,11-reductase, EC 1.3.1.42), which catalyzes the reduction of the double bond in the cyclopentanone ring of 12-oxo-PDA (**8**), has been characterized from the kernel and seedling of corn.[22] The molecular weight of the enzyme, estimated by gel filtration, was 5.4×10^4. Optimum enzyme activity was observed over a broad pH range, from pH 6.8 to 9.0. The enzyme had a K_m of 190 μM for its substrate, 12-oxo-PDA (**8**). The preferred reductant was NADPH, for which the enzyme exhibited a K_m of 13 μM, contrasting with 4.2 mM for NADH. Reductase activity was initially low in the corn kernel but increased five-fold by the fifth day after germination and then gradually declined.

12-oxo-PDA reductase was purified to apparent homogeneity from a cell suspension culture of *Corydalis sempervirens*.[23] The 1.0×10^5 *g* supernatant of a protein extract from late-log-phase cultures (6 days' postinoculation) served as the source of crude enzyme. The activity precipitated between 30% and 60% saturation with ammonium sulfate. The 40–53% saturation cut was used further, representing a 61% yield of total enzyme activity enriched 4.7-fold. The purification consisted of two steps of anion-exchange chromatography, followed by two sequential native gel electrophoretic separations. The enzyme is soluble and a monomer with an apparent molecular mass of 41 kDa which prefers NADPH over NADH to reduce the 10,11-double bond of 12-oxo-PDA (**8**). The enzyme accepts both the *cis* and the *trans* isomers of 12-oxo-PDA (**8**), with a preference for the *cis* isomer (6:1). 12-oxo-PDA reductase also converted the synthetic substrate 2-cyclohexenone to cyclohexanone, but the enzyme did not reduce some other cyclic α,β-unsaturated ketones such as the plant hormone abscisic acid or the steroids testosterone and progesterone. When subcellular preparations were fractionated on sucrose gradients, the activity appeared exclusively in the fraction of soluble proteins and was not found associated with any of the organelles including intact chloroplasts and peroxisomes in which early and late steps in the biosynthetic pathway to jasmonates are localized. Hence, 12-oxo-PDA reductase is most likely a cytosolic enzyme.[23]

Using partial amino acid sequence information for 12-oxo-PDA reductase from *C. sempervirens*, a homologous enzyme was cloned from a cDNA library of *Arabidopsis thaliana*, race Columbia, shoots. The open reading frame of the cDNA encodes a polypeptide of 372 amino acids ($M_r = 41\ 165$). The enzyme was functionally expressed from its cDNA in *Escherichia coli* and thus proven to encode 12-oxo-PDA reductase.[24] Four plant species metabolized ^{18}O-labeled 12-oxo-PDA (**8**) to short chain cyclic fatty acids.[25] The plant species were corn (*Zea mays* L.), eggplant (*Solanum melongena* L.), flax (*L. usitatissimum* L.), and wheat (*Triticum aestivum* L.). The products were OPC-8:0 (**11**), OPC-4:0 (**13**), and JA (**1**). Observation of the presence of the intermediate product OPC-6:0 (**12**) was not possible; however, this metabolite was detected in the experiments with *V. faba* pericarp.[15]

The presence of enzymes which convert 12-oxo-PDA (**8**) to JA (**1**) in several plant species indicates that this may be a general metabolic pathway in plants. None of the tissues tested showed conversion of 12-oxo-PDA (**8**) to fatty acids with an odd number of carbons, demonstrating that the chain-shortening is a result of β-oxidation and not α-oxidation reactions. Naturally occurring cyclopentane fatty acids have an even number of carbons in the carboxyl side chain with the single exception of a compound from the culture filtrate of the fungus *B. theobromae*.[26]

12-oxo-PDA (**8**), biosynthesized by plant enzyme preparations, and (+)-epi-JA (**1a**), produced by *B. theobromae*, were converted into a common derivative, methyl 3-hydroxy-2-pentylcyclo-pentane-1-octanoate. The two products had the same retention time by GLC analysis. Therefore, the stereochemistry of 12-oxo-PDA (**8**) was confirmed to be the (9*S*,13*S*) configuration which is identical to the configurations of the corresponding carbons of (+)-epi-JA (**1a**). Thus it has been established that 12-oxo-PDA (**8**) serves as the precursor of (+)-epi-JA (**1a**) in the biosynthetic pathway of plants (Scheme 1).[27]

1.05.2.2 Green Algae

In addition to the higher plants, other organisms are also known to possess the JA pathway. The green alga *Chlorella pyrenoidosa* was examined for its ability to metabolize (13*S*)-HPLA (**9**). The study showed that *Chlorella* extracts possessed hydroperoxide dehydrase and other enzymes of the JA pathway. However, under normal laboratory conditions for culture growth, neither JA (**1**) nor metabolites of the JA pathway were present in *Chlorella*.[28]

1.05.2.3 Jasmonoid Production by Microorganisms

At least two fungi are known to produce JA and other jasmonoids. They are *B. theobromae*[2] and *G. fujikuroi*.[4] There is not much known about jasmonoid biosynthesis by these fungi or any endogenous role of these metabolites. However, since oxylipid-derived pathways to endogenous signal compounds are conserved in the plant and animal kingdom (prostaglandins)[29] these pathways seem to be ancient and might have been developed by divergent evolution from common ancestral organisms.

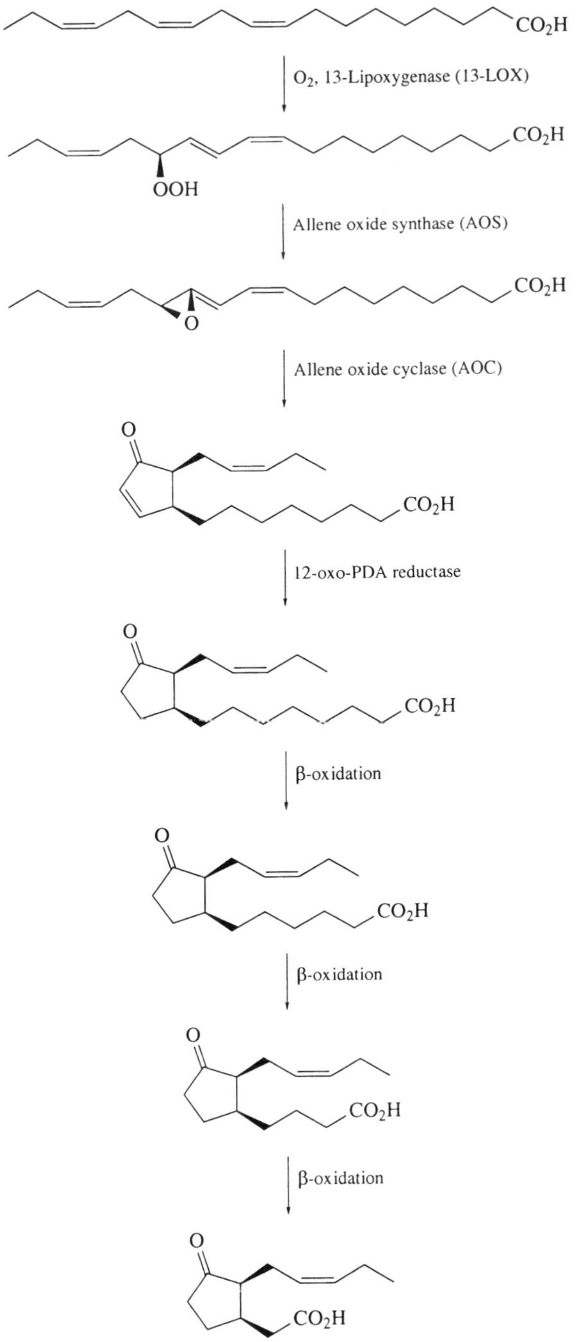

CO_2H (7) Linolenic acid

O_2, 13-Lipoxygenase (13-LOX)

CO_2H OOH (9) (13*S*)-HPLA

Allene oxide synthase (AOS)

CO_2H O (10) (12*S*,13*S*)-EOD

Allene oxide cyclase (AOC)

O CO_2H (8) 12-oxo-PDA

12-oxo-PDA reductase

O CO_2H (11) OPC-8:0

β-oxidation

O CO_2H (12) OPC-6:0

β-oxidation

O CO_2H (13) OPC-4:0

β-oxidation

O CO_2H (1a) (+)-JA

Scheme 1

The role of the fungal metabolites in the plant–microbe interaction also deserves further elucidation.

One group of bacteria consisting of several pathovars of *Pseudomonas syringae* is able to produce coronatine (**14**), a structural analogue of the amino acid conjugates of JA with strong jasmonate-like activity.[30,31] The two moieties are biosynthetically derived from the polyketide pathway and from isoleucine, respectively. The related genes were identified and characterized.[32] An asymmetric total synthesis of the chiral coronatine is available.[33]

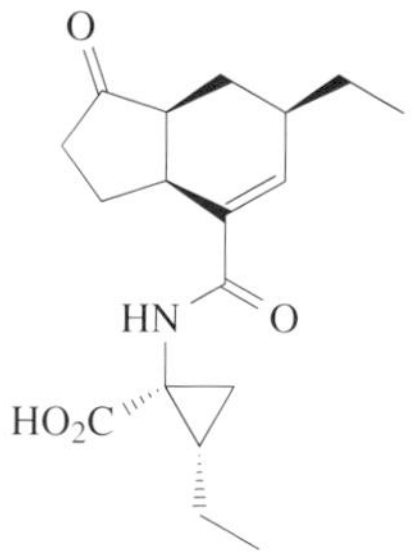

(14) Coronatine

1.05.2.4 Inhibitors of the Octadecanoid Pathway

Several inhibitors which block the octadecanoid pathway at different steps have been reported. Quinacrine (**15**) is known to be a phospholipase A_2 inhibitor and was used successfully to prevent a jasmonate-dependent elicitation signal in rice.[34]

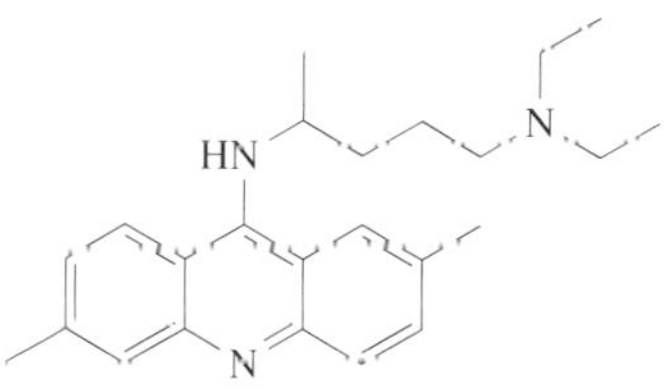

(15) Quinacrine

Lipoxygenase, the first enzyme in the JA biosynthetic pathway, can be blocked by several inhibitors, e.g., nordihydroguaiaretic acid (**16**),[34] salicylhydroxamic acid (*N*,2-dihydroxybenzamide) (**17**),[35] and ibuprofen (α-methyl-4-(2-methylpropyl)benezeneacetic acid) (**18**).[35]

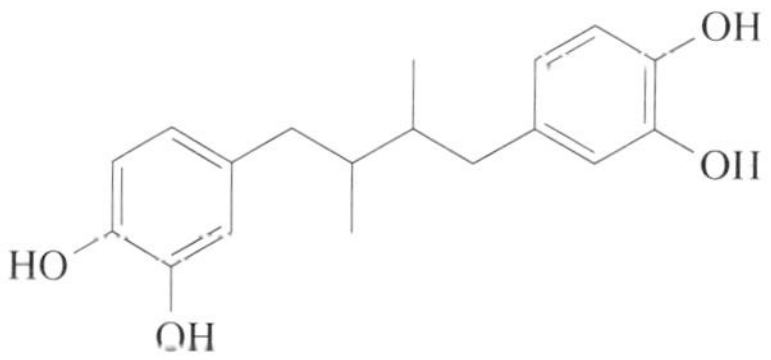

(16) Nordihydroguaiaretic acid

(17) Salicylhydroxamic acid

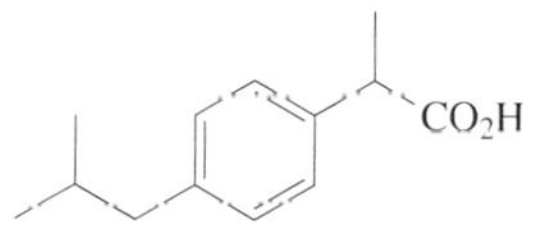

(18) Ibuprofen

Aspirin (acetylsalicylic acid) (**19**), *n*-propyl gallate (**20**), and DIECA (sodium diethyldithiocarbamate) (**21**) are inhibitors of cyclooxygenase, the putative enzyme complex of AOS and AOC. Quantitation of endogenous levels of JA showed that aspirin blocks the increase of this phytohormone normally observed as a result of wounding. Linolenic acid (**7**) and (13*S*)-HPLA (**9**) did not induce the expression of Pin2, cathepsin D inhibitor, and threonine deaminase in the presence of aspirin. However, 12-oxo-PDA and JA were able to overcome the inhibitory effect of this substance. These results strongly indicate that aspirin prevents wound-induced gene activation by inhibiting the hydroxyperoxide-dehydrase activity that mediates the conversion of (13*S*)-HPLA to 12-oxo-PDA.[36]

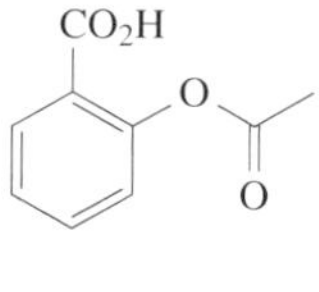

(19) Aspirin

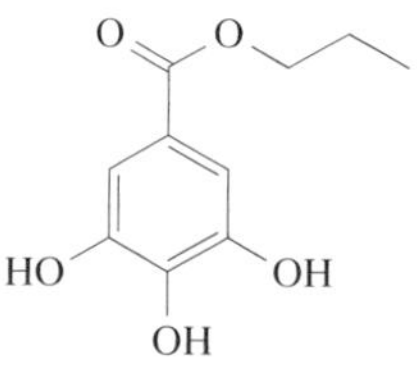

(20) *n*-propyl gallate

DIECA (**21**) is a strong reducing agent and inhibits the octadecanoid pathway by efficiently converting (13*S*)-HPLA (**9**) to 13-hydroxylinolenic acid, which is not a signaling intermediate, thereby shunting the pathway into a dead end.[37]

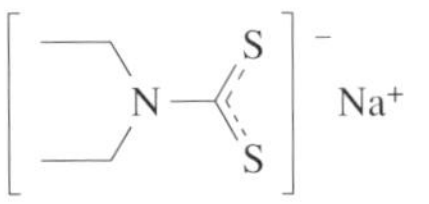

S
N
Na+
S

(**21**) DIECA

1.05.2.5 Spatial Distribution and Accumulation

The natural levels of JA in plant tissues vary as a function of tissue type and developmental stage. The highest levels of jasmonates are reported from flowers and reproductive tissues, whereas much lower levels are found in roots and mature leaves. Extracts of untreated *Nicotiana sylvestris* Speg et Comes plants contained free JA (**1**) at concentrations of more than 20 ng g^{-1} fresh mass in the root part and less than 10 ng g^{-1} fresh mass in the shoot part. The overall conjugate pool was estimated to be only about 10% of the free JA pool. The biggest amount of JA could be released from the extracted conjugates by hydrolysis with β-glucosidase. Minor amounts were obtained by hydrolysis with pronase, α-amylase, or 3 N HCl.[38]

In soybean (*Glycine max* L.), the endogenous levels of JA (**1**) were studied in the pericarp (vascular bundles and remaining part) and seed parts (hilum, testa, cotyledons, embryo axis) at four different stages of fruit development.[39] The highest levels were found in the vascular bundles of the pericarp at all stages investigated. Somewhat smaller amounts were found in the hilum and testa of the seed. Lowest JA concentrations were found in the cotyledons and embryo axes during the whole development of fruits.

In dormant apple (*Malus domestica*) seeds, free JA (**1**) concentrations of about 10 ng g^{-1} fresh weight and slightly higher levels of unidentified JA conjugates were detected. During stratification JA accumulated transiently to more than 1 µg g^{-1} fresh weight. This was paralleled by unidentified JA conjugates; however, the amount was six times higher than the free acid.[40]

Several environmental stresses like osmotic stress, burning, UV light, mechanical wounding, herbivores, or pathogens can lead to a local or systemic rise of endogenous jasmonates (see Section 1.05.3.2).

Endogenous jasmonates accumulated rapidly and transiently (starting at about 20 min) after treatment of plant cell suspension cultures of *Rauvolfia canescens* and *Eschscholtzia californica* with a yeast cell wall elicitor.[41]

Wounded soybean seedling stems also rapidly accumulated jasmonate.[42] The tissue of soybean seedling stems was wounded by chopping it into 1–3 mm sections. Jasmonate levels rapidly increased within 2 h. Levels of jasmonate increased further by 8 h after wounding and remained elevated for at least 24 h.

1.05.2.6 Metabolism and Transport

The biotransformation of (+)-9,10-dihydro-JA (**6**) was studied in 6 day-old barley seedlings.[43] Both [2-^{14}C] and [U-^{3}H]dihydro-JA (**6**) were fed to excised shoots and the formed metabolites analyzed after 72 h. Dihydro-JA (**6**) was converted into two major and some minor metabolites. The major metabolites were (−)-9,10-dihydro-11-hydroxy-JA (**22**) and its (11*O*)-β-D-glucopyranoside (**23**) (Scheme 2).[43] Minor metabolites from the same source were valine, leucine, or isoleucine conjugates of either 9,10-dihydro-JA or its 11-hydroxy, 12-hydroxy, or 3-hydroxy derivatives.[44]

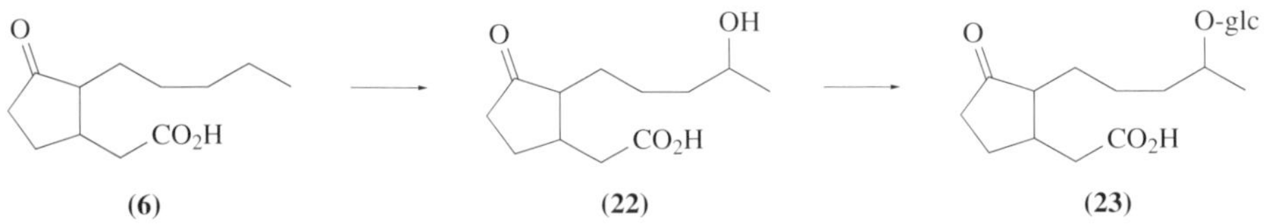

O
CO_2H
(**6**)

O OH
CO_2H
(**22**)

O O-glc
CO_2H
(**23**)

Scheme 2

Cell-free extracts from cell suspension cultures of *Lycopersicon peruvianum* were found to catalyze the glucosylation of [U-^{3}H]dihydro-JA (**6**) in the presence of UDP-glucose. The products of enzymatic reactions were identified as glucosyl esters of dihydro-JA (**6**).[45]

The metabolism of JA (**1**) itself was investigated using suspension cultures of *E. californica*.[46] From the cells incubated with racemic JA, the major metabolite was isolated and the chemical structure was determined to be the β-D-glucoside (**24**) of 11-hydroxy-JA (**4**). The configuration at C-11 was determined as *R* by the Horeau-Brooks method.

Metabolism and transport of JA (**1**) were studied in potato plants.[47] The [2-^{14}C]JA (**1**) applied on the leaves was metabolized to a tuberonic acid (**5**)-like substance and its β-D-glucoside (**25**) and distributed throughout the whole plant (Scheme 3). Besides the treated leaf, the upper leaves contained most of the total radioactivity, followed by the stems and lower leaves.

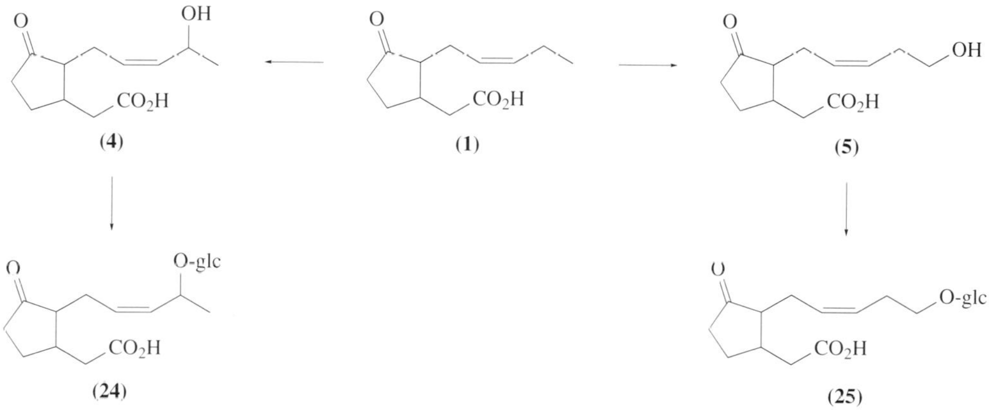

Scheme 3

In young *N. sylvestris* plants [2-^{14}C]JA (**1**), applied to the youngest fully expanded leaf, was rapidly translocated as the free acid mainly into the roots and also into untreated young leaves, but not into mature leaves, thus indicating phloem transport. More-polar JA metabolites were found only in small amounts. Interestingly, JA was immobilized to a great extent into less-polar metabolites after translocation.[38] Plant cell suspension cultures were used to investigate the release of JA from the cells into the culture medium. After stimulation of the JA biosynthesis by addition of a yeast cell wall elicitor, the endogenous and exogenous concentrations of JA and 12-oxo-PDA were observed for 4 h. Cells of *Phaseolus vulgaris* and *Rauwolfia serpentina* were able to secrete JA into the medium up to concentrations 2–4 times higher than the intracellular levels. In contrast, *Tinospora cordifolia*, *Mahonia nervosa*, and *E. californica* did not accumulate any JA in the culture medium in spite of high internal concentrations. *Agrostis tenius* cells exhibited a transient intra- and extracellular increase of JA, indicating a rapid reuptake and metabolism. However, 12-oxo-PDA was always found to be exclusively cell associated.[48]

The data show that one major metabolic pathway of both JA (**1**) and dihydro-JA (**6**) is hydroxylation with subsequent glucosylation. It is still unknown whether this metabolism leads to inactivation of these physiologically highly potent molecules and/or to more polar transport forms. Moreover, the formation of amino acid conjugates and their physiological role is not fully established.

Other metabolites can be expected to be discovered from different species and/or specialized plant tissues. Forms for transport, or for storage and subsequent release, or for inactivation and recycling, as well as specific signal derivatives could be expected.

1.05.3 FUNCTIONS

1.05.3.1 Plant Growth and Development

Jasmonates are known to influence a wide variety of physiological processes in plants (Table 1).[49,50] Bioassays are essential for the isolation of the active compounds and are briefly outlined in the following sections where suitable.

Table 1 Examples of the multiplicity of jasmonate effects in plants.

Inhibition	*Induction*	*Promotion*
Growth (seedlings, roots, cell or tissue cultures)	tuber formation	senescence (pericarp, leaves)
Germination of pollen and seed	microtubuli disruption	abscission of leaves
Flower bud formation	tendril coiling	stomata closure
Embryogenesis	alkaloid synthesis	chlorophyll degradation
Pigment formation	PAL activity	respiration
Photosynthetic activities	phytoalexin synthesis	
Rubisco synthesis	seed storage proteins	
Translation/transcription of "normal" cell proteins	functionally unknown leaf proteins	
	leaf proteinase inhibitor I and II	

Source: Sembdner and Parthier[49] and Gross and Parthier.[50]
PAL, phenylalanine ammonia-lyase.

1.05.3.1.1 Growth inhibition

During the course of investigation of gibberellins using the bioassay with rice seedlings, JA (**1**) was isolated from the immature seeds of *P. vulgaris* and *Dolichos lablab* L. as well as from leaves of *Castanea crenata* Sieb. et Zucc. as a growth inhibitor.[51] However, this (antigibberellin) growth inhibiting effect requires relatively high jasmonate concentrations and does not seem to be very specific.

1.05.3.1.2 Senescence

The leaf senescence response to jasmonates includes a loss of chlorophyll, degradation of chloroplast proteins such as ribulose bisphosphate carboxylase, and expression of specific new proteins.

JA-Me (**2**) was isolated from wormwood shoots (*Artemisia absinthium* L.) as a senescence-promoting factor[52] using the loss of the chlorophyll in oat (*Avena sativa* L.) leaf segments as a bioassay system (anticytokinin). The senescence-promoting effect of JA-Me was much stronger than that of abscisic acid.

Similarly, barley (*Hordeum vulgare* L.) primary leaf segments responded to the application of JA-Me with accelerated senescence, as detected by the loss of chlorophyll and the rapid decrease in activity and immunoreactive protein content of ribulose-1,5-bisphosphate carboxylase/oxygenase (RuBP carboxylase, EC 4.1.1.39). The senescence-promoting action of JA-Me was found to differ in light and in darkness; e.g., the initial rates of chlorophyll and RuBP carboxylase breakdown were markedly higher in light than in darkness in the presence of 4×10^{-5} M JA-Me. Cytokinin (benzyladenine) stopped the loss of chlorophyll and RuBP carboxylase during senescence; however, the rapid drop induced by JA-Me in the early phase of leaf segment senescence could not be prevented by concomitant or previous addition of benzyladenine. However, benzyladenine added 24 h after JA-Me application resulted in a recovery of chlorophyll and RuBP carboxylase at the later stages, indicating a possible rapid inactivation of JA-Me in the tissues. The activities of a number of other chloroplastic and cytosolic enzymes were not significantly altered in JA-Me-treated leaf segments compared with controls floated on water. Time-dependent chlorophyll decrease in isolated chloroplasts did not change upon JA-Me addition to the isolated organelles. It is suggested that JA-Me acts on chloroplast senescence by promoting cytoplasmic events which eventually bring about the degradation of chloroplast constituents.[53]

An example of a bioassay will now be described[52] in which oat (*A. sativa* L. cv. Victory) leaves were used as test objects. Seeds were germinated in Vermiculite and seedlings were grown under continuous white fluorescent light of about 5000 lux at 25 °C for 7 days. The upper 3 cm segments of the first leaves were excised for the assay. Ten such leaf segments were placed on two layers of filter paper moistened with 5 ml of the test solutions in a sterile Petri dish. The test substances were dissolved in water containing 500 mg ml^{-1} of the surfactants Tween 80:Span 80 (7:3 w/w) with or without 2 μg ml^{-1} kinetin. Additional controls were treated with the same medium without kinetin. The Petri dishes were kept in a dark room at 25 °C. The leaf segments were sampled at 2 or 4 days after treatment and their chlorophyll contents were determined photometrically after extraction with boiling ethanol (80%).

1.05.3.1.3 Leaf abscission

Jasmonates have often been compared with abscisic acid regarding their senescence-promoting effect on detached leaves,[52] but only a few reports have focused on the role of jasmonates in the process of leaf abscission.

JA promoted the abscission of bean petiole explants via the degradation of cell wall polysaccharides in the abscission zone.[54]

In intact *Kalanchoe blossfeldiana* Poelln. plants JA-Me stimulated ethylene production and promoted leaf abscission. Spraying of the plants with JA-Me at a concentration of 500 mg l^{-1} caused a complete loss of all leaves, and lower concentrations also stimulated the process. Treatments of the plants with exogenous ethylene at high concentrations (1000 mg l^{-1}) as well as abscisic acid alone (100 mg l^{-1}) did not induce leaf abscission.[55]

1.05.3.1.4 Formation of storage organs

The tuberization of potato plants is controlled by photoperiod. Short days promote tuberization whereas long days inhibit the process. With grafting experiments, the occurrence of a tuberization stimulus was demonstrated.[56] Using cultures of single-node segments of potato stems as a bioassay system (*in vitro* tuberization test), the active compound was isolated and the chemical structure was determined to be 3-oxo-2-(5′-β-D-glucopyranosyl-2′-*cis*-pentenyl)cyclopentane-1-acetic acid (12-hydroxy-JA glucoside, tuberonic acid glucoside).[57–59] Exogenous free JA (**1**) also induced tuberization of potato stolons *in vitro*.[60]

The tuberization of Jerusalem artichoke plants (*Helianthus tuberosus* L.) is also induced by short days. From the leaves of this species, JA (**1**) and the methyl ester of the tuberonic acid glucoside were isolated, again using the *in vitro* tuberization test.[61] Similarly, JA was isolated from leaves of the monocot yam plant (*Dioscorea batatas* Decne.cv. Hontokkuri), guided by a modified *in vitro* tuberization test using yam stem segments.[62]

[2-^{14}C]JA (**1**) applied on potato leaves was metabolized to a tuberonic acid glucoside-like compound and transferred to the stolons.[47]

The cells of potato tuber tissue expanded in response to JA as a consequence of water uptake, not growth.[63] At an early stage of potato tuberization, radial cell expansion of stolons occurs as well.[64] The direction of cell expansion depends to a considerable extent on the orientation of cellulose microfibrils in the cell wall. Their orientation is considered to be controlled by cortical microtubules. Similarly, bulb formation of onions (*Allium cepa* L.) is known to be accompanied by the disruption of cortical microtubules in the cells of leaf sheaths. In addition, microtubule-disrupting compounds, such as colchicine and cremat, are known to cause lateral expansion of cells in onion leaf sheaths.[65]

Microtubule-disrupting activity by JA-Me (**2**) was examined using tobacco BY-2 cells. A small number of the cultured cells at the logarithmic growth phase were synchronized by treatment with aphidicolin, an inhibitor of DNA polymerase-α. Disruption of microtubules was found exclusively in cells at the S phase of the cell cycle.[66] Accordingly, the influences of JA-Me on microtubules in potato cells and formation of potato tubers were studied.[67] JA-Me exhibited microtubule-disrupting activity in this system as well.

Bulb development in onion plants is considered to be regulated by bulbing and antibulbing hormones. Since bulbing involves the disruption of microtubules, jasmonates are candidates for the bulbing hormone. The amount of JA (**1**) per plant in leaf sheaths of bulbing onion plants was about three times higher than that of nonbulbing onion plants, although the difference in levels of JA (**1**) in leaf blades between bulbing and nonbulbing onion plants was quite small.[68]

An example of a bioassay will now be described.[57] Single-node stem segments (ca. 2 cm long) including an axillary bud were prepared from etiolated potato shoots. The segments were sterilized with 1% sodium hypochlorite solution for 1 h, washed five times with sterile water and then trimmed by 5 mm at each end. Three such stem segments were placed horizontally with the axillary bud upward in an 100 ml Erlenmeyer flask containing 10 ml White's basal agar medium supplemented with the sample to be tested. Five replicates were prepared. The medium was adjusted to pH 5.6 and solidified with 0.6% Bacto-agar (Difco) before autoclaving for 7 min. The cultures were maintained at 25 °C in darkness for 3 weeks. Thereafter, the number of microtubers was counted and the rate of tuberization was calculated. The criterion for a valid microtuber was a distinct visible swelling at the apex or at lateral buds of the stolons grown from the original axillary bud.

1.05.3.1.5 Pollen germination

Flowers were the first source of jasmonates and can contain large amounts of (−)-JA-Me (**2b**).[1] Three different conjugates of JA (**1**) from flowers are known: *N*-[(−)-jasmonoyl]-(*S*)-tyrosine,[69] *N*-[(−)-jasmonoyl]-(*S*)-tryptophan,[70] and *N*-[(+)-cucurbinoyl]-(*S*)-tryptophan,[70] all from *Vicia faba*.

N-[(−)-jasmonoyl]-(*S*)-tyramine has been isolated from pollen of *Petunia hybrida*, the first example of a jasmonate amine conjugate from plants. It can be discussed as a metabolite of the known corresponding amino acid conjugate. In the same pollen material the free JA was quantified to more than 600 ng per gram pollen.[71]

Two other jasmonates, *N*-[(−)-jasmonoyl]-(*S*)-isoleucine and *N*-(7-isocucurbinoyl)-(*S*)-isoleucine, were identified from mature pollen of *Pinus mugo*, whereas the corresponding free acids could not be detected.[72] The isoleucine conjugate of JA (JA-Ile) inhibited the pollen germination in *P. mugo*, whereas free JA was neither inhibiting nor stimulating.[73]

JA and JA-Me have been isolated from pollen and anthers of tea plants (*Camellia sinensis* L.). Interestingly, in this species JA, but not its methyl ester, inhibited pollen germination.[74]

An *Arabidopsis* mutant lacking unsaturated fatty acids, and as a consequence lacking jasmonates, shows male sterility.[75]

These findings indicate an important role of jasmonates in the maturation and/or germination of pollen and possibly in the regulation of the many complex physiological processes linked to flowering and pollination.

1.05.3.1.6 Seed germination and dormancy

Application of JA or JA-Me initiates dormancy breakage in seeds of apple (*M. domestica*)[40] and Douglas fir (*Pseudotsuga menziesii* (Mirb.) Franco).[76] However, in the nondormant seeds of sunflower (*Helianthus annuus* L.) JA inhibited germination.[77]

Similarly, in nondormant lettuce (*Lactuca sativa* L. cv. Great Lakes "Topmark") seeds the germination was partially inhibited and delayed in the presence of 10–100 ppm JA.[51] During the natural dormancy-breaking period in apple seeds a transient increase of endogenous free JA as well as of unknown JA conjugate(s) was observed by GC–MS/SIM and ELISA, respectively. Surprisingly, exogenously applied JA-Ile completely inhibited seed germination, whereas JA and JA-Me each stimulated seed germination.[40]

In Douglas fir seeds, the expression of specific stratification responsive genes was induced by treatment with JA-Me. JA-Me at a concentration of 45 μmol l^{-1} promoted seed germination to percentages equivalent to the effect of 3 weeks of chilling. However, JA-Me-treated seeds generated seedlings with discolored and swollen roots which failed to establish once transplanted into a soil-based medium.[76]

1.05.3.1.7 Tendril coiling

A coiling-inducing factor was isolated from tendrils of *Bryonia dioica* Jacq. and identified as α-linolenic acid (**7**).[78] When applied to detached tendrils, exogenous α-linolenic acid, but not linoleic acid or oleic acid, induced tendril coiling. Further investigations showed that metabolites of α-linolenic acid, JA (**1**), and JA-Me (**2**) are highly effective inducers of tendril coiling in *B. dioica*. JA-Me was most active when administered by air and, in atmospheric concentrations as low as 40–80 nM, induced a full free-coiling response with kinetics similar to mechanical stimulation. Even at atmospheric levels as low as 4–5 nM, JA-Me was still found to be significantly active. JA-Me could be one of the endogenous chemical signals produced in mechanically stimulated parts of a tendril and, being highly volatile, act as a diffusible gaseous mediator spreading through the intracellular spaces to trigger free coiling of the tendril.

Tendril coiling is induced by mechanical stimulation, and roles for auxin and ethylene have been proposed. Although jasmonate can stimulate ethylene biosynthesis by inducing activity of the ethylene-forming enzyme, studies on tendril coiling showed that jasmonate can mediate this response in the presence of an inhibitor of ethylene biosynthesis.[79]

An example of a bioassay will now be described.[78] The test compounds were dissolved in 50 mM potassium phosphate buffer (pH 6.5) with 0.02% (v/v) Tween 20. The test solution (20 ml) was delivered into glass Petri dishes (9 cm diameter). Apical segments of 6–8 cm were cut from unstimulated tendrils of 12–16 cm length and immediately transferred into the test solution (three per dish) avoiding direct contact. The appearance of the tendrils was recorded at regular intervals for up to 24 h. Each assay was repeated at least four more times on different days. Whenever possible, experiments were carried out in the phytotron chambers where the plants had been grown.

To determine the physiological activity of airborne JA-Me, aliquots of a methanolic solution of JA-Me were applied to a cotton plug inside a 13 l glass chamber with a water-saturated atmosphere into which freshly cut shoots of *B. dioica* with one or two ~12–14 cm long tendrils had been introduced in such a way as to avoid any possibility of mechanical stimulation during the experiment. The shoots were kept in a solution of 0.5% (w/v) glucose and 0.1 mM KCl. Controls were incubated in a separate chamber under otherwise identical conditions (the cotton received the appropriate amount of solvent). The average temperature during incubations was 26 °C.

1.05.3.2 Responses to Environmental Factors and Related Signal Transduction

1.05.3.2.1 Gene expression and jasmonate-induced proteins

Apparently, all of the different plant responses to jasmonates, whether applied externally or released internally, appear to be correlated with alterations in gene expression followed by diverse metabolic and physiological changes. However, many of the observed reactions seem to be secondary or tertiary steps embedded in complex cascades of events.

It is not known whether the jasmonate signals are transduced by a receptor and second messenger system, or by interaction with regulatory enzymes or any other molecular interaction.

At least two genes are known which are expressed to the transcriptional level after jasmonate treatment even in the presence of cycloheximide. That means that they are expressed without the *de novo* synthesis of mediating proteins.

The transcription of a novel cytochrome P450 gene (*CYP93A1*) was induced specifically by 30 μM JA-Me in suspension cultured soybean cells in the presence of cycloheximide.[80] Low homology of this gene with other plant P450s reported so far indicates that it belongs to a novel cytochrome P450 family in plants. The corresponding enzyme could be involved in the synthesis or regulation of signal chain components, or alternatively in the biosynthesis of secondary metabolites such as flavonoids, terpenoids, or alkaloids. The biosynthesis of some defensive secondary metabolites known to be induced by jasmonates also involves the activity of cytochrome P450 enzymes (see Section 1.05.3.2.2).

A mRNA from tobacco (*Nicotiana tabacum* cv. Bright Yellow) cell cultures, which is rapidly inducible by 100 μM JA-Me in the presence of cycloheximide, codes for a putative glycosyltransferase.[81] The corresponding enzyme is homologous to UDP-sugar glycosyltransferases previously characterized from several plant species and was named jasmonate-inducible glycosyltransferase (JIGT). Besides JA-Me, as little as 10^{-9} M coronatine (**14**) induced JIGT mRNA. A sequence highly homologous to JIGT is present as a single copy in the genomes of *Nicotiana sylvestris* and *Nicotiana tomentosiformis*. Plant glycosyltransferases are known to glycosylate a wide variety of secondary metabolites. They are involved in the biosynthesis of flavonoids, glycoalkaloids, cyanogenic glucosides, and glucosinolates. They are also active in the sugar conjugation of phytohormones and their metabolites, such as salicylic acid, indoleacetic acid, and gibberellins.[73]

The name "jasmonate-induced proteins" (JIPs) is a term which has been used for proteins, often with unknown functions, induced after jasmonate treatment in a specific test system. In leaves of barley treated with JA (**1**) or JA-Me (**2**), the composition of polypeptides was changed as shown by SDS-polyacrylamide gel electrophoresis.[53] The accumulated proteins were named jasmonate-induced proteins (JIPs). Many JIPs have been reported (Table 2).[49,82]

During vegetative growth, soybean leaves and immature organs accumulate three polypeptides of approximately 27 (vsp27 or vspα), 29 (vsp29 or vspβ), and 94 (vsp94) kDa. These three polypeptides constitute a group of proteins termed the vegetative storage proteins (VSPs).[93]

VSPs respond to nitrogen status and are believed to be involved in the temporary storage of nitrogen. JA (**1**) increased the level of expression and accumulation of the VSPs in soybean suspension cultures and seedlings.[93] Low levels of atmospheric JA-Me (**2**) induce the accumulation of soybean VSPs.[94]

The VSPs, vsp27 and vsp29 are glycosylated and consist of two related polypeptides which form heterodimers and homodimers. The third, vsp94, is a lipoxygenase that is localized in paraveinal mesophyll cell vacuoles.[90,99]

Table 2 Examples of JIPs.

Function	*Source*	*Ref.*
Inhibitors		
PIs	tomato (*Lycopersicon esculentum* Mill.)	83,84
	potato (*Solanum tuberosum* L.)	85
Cathepsin D inhibitors	tomato (*Lycopersicon esculentum* Mill.)	85
Enzymes		
Leucine aminopeptidase	tomato (*Lycopersicon esculentum* Mill.)	86
Threonine deaminase	tomato (*Lycopersicon esculentum* Mill.)	87
Alkaloid-synthesizing enzyme	*Catharanthus roseus*	88
Lipoxygenases	*Arabidopsis thaliana*	89
	soybean (*Glycine max* L.)	90
	barley (*Hordeum vulgare* L.)	91
Storage proteins		
Napin	*Brassica napus*	92
Cruciferin	*Brassica napus*	92
VSPs	soybean (*Glycine max* L.)	93,94
Amino acid-rich proteins		
Hydroxyproline-rich proteins	soybean (*Glycine max* L.)	42
Others		
Ribosome-inacting proteins	barley (*Hordeum vulgare* L.)	95
Thioninivat	barley (*Hordeum vulgare* L.)	96
Functionally unknown proteins	cotton (*Gossypium hirsutum* L.)	97
	barley (*Hordeum vulgare* L.)	53
	tomato (*Lycopersicon esculentum* Mill.)	98
Late embryogenesis abundant proteins	Douglas fir (*Pseudotsuga menziesii* (Mirb.) Franco)	76

Addition of JA-Me (**2**) to soybean suspension cultures increased mRNA levels for three wound-responsive genes (coding for chalcone synthase, VSP, and proline-rich cell wall protein).[42] This suggests a role for jasmonate in the mediation of several changes in gene expression associated with the plant's response to wounding.

1.05.3.2.2 Induction of secondary metabolites

Jasmonates are important signal transducers in plant secondary metabolism. Cell suspension cultures of 36 plant species could be elicited with respect to the accumulation of specific secondary metabolites in the absence of elicitors by exogenously supplied JA-Me (**2**). For example, *Crotolaria cobalticola* produced isobavachalcone, *Glycine max* genisteine, *Lactuca sativa* lettucenin A, *Rauvolfia canescens* raucaffricine, *Rubia tinctorum* rubiadin, and *Ruta chalepensis* rutacridone.[100] The content of these metabolites in the medium increased 10–30 times. Addition of JA-Me was shown to initiate the *de novo* transcription of related genes, such as phenylalanine ammonia lyase.

Glucosinolates are a group of thioglucosides found in all cruciferous plants. When tissues are damaged, they are hydrolyzed by a thioglucosidase enzyme (myrosinase, EC 3.2.3.1) to release various products, some of which contribute to the plant's defense against microorganisms and pests. Glucosinolates are constitutive to oilseed rape (*Brassica napus*) tissues, but they also accumulate in response to fungal infection, actual and simulated insect damage, and other forms of stress. Oilseed rape plants sprayed with a solution of JA-Me (**2**) or exposed to JA-Me vapor accumulated indolyl glucosinolates in their leaves in amounts that depended on the concentrations applied.[101] The predominant compounds of the response were 3-indolylmethyl- and 1-methoxy-3-indolylmethyl-glucosinolates.

A biosynthetically active microsomal enzyme system catalyzing the conversion of phenylalanine into phenylacetaldoxime was identified in seedlings of *Tropaeolum majus* L.[102] By exposure of the seedlings to the combined treatment of JA (**1**), ethanol, and light, the enzyme activity was stimulated nine-fold, compared with that of untreated, dark-grown seedlings. Furthermore, this enzyme system responsible for the oxime production in *T. majus* resembles the system involved in the biosynthesis of *p*-hydroxybenzylglucosinolate in *Sinapis alba* L. in that they are both dependent on cytochrome P450.

JA was found to be a key signal transducer between the recognition of an elicitor (*N*-acetylchitoheptaose) and the production of a phytoalexin, momilactone A, in rice cells. In suspension-cultured rice cells (*Oryza sativa* L.), treatment with *N*-acetylchitoheptaose induced the production of phytoalexins.[103]

Exogenously applied JA (**1**) clearly induced the production of momilactone A, a major phytoalexin, in suspension-cultured rice cells.[104] However, in rice cells treated with *N*-acetylchitoheptaose, endogenous JA was rapidly and transiently accumulated prior to accumulation of momilactone A. Treatment with ibuprofen, an inhibitor of JA biosynthesis, induced production of momilactone A in the cells treated with *N*-acetylchitoheptaose, but the addition of JA increased the production of momilactone A to levels higher than those in the control of elicited rice cells. These results strongly suggest that JA functions as a signal transducer in the induction of the biosynthesis of momilactone A by *N*-acetylchitoheptaose in suspension-cultured rice cells.

In rice plants, $CuCl_2$ is a potent abiotic elicitor. The endogenous level of JA (**1**) increased rapidly in $CuCl_2$-elicited rice leaves, and exogenously applied JA caused a large amount of phytoalexins, sakuranetin, and momilactone A, to accumulate in rice leaves.[105]

A phytoalexin of opium poppy (*Papaver somniferum* L.) is the alkaloid sanguinarine. Treatment of *P. somniferum* L. with JA-Me (**2**) resulted in an induction of tyrosine/DOPA decarboxylase (TYDC, EC 4.1.1.25), which catalyzes the conversion of L-tyrosine and L-β-3,4-dihydroxyphenylalanine (L-DOPA) to tyramine and dopamine, respectively, the first steps in sanguinarine biosynthesis.[106]

1.05.3.2.3 Osmotic stress

Sorbitol and other osmotically active substances provoked a marked increase of endogenous jasmonates (JA and amino acid conjugates) in barley leaf tissue (*Hordeum vulgare* L. cv. Salome). After treatment of the leaf segments with sorbitol, mannitol, or polyethylene glycol, an increase of JA (**1**) could be observed. In contrast, salt stress had absolutely no stimulating effect on the levels of endogenous jasmonates. From barley leaf segments exposed to sorbitol (1 M) for 24 h, JA was identified as the major accumulating compound.[107] Similarly, jasmonates were induced in response to stress of desiccation.[108]

1.05.3.2.4 Wounding and defense

Wounding is a serious challenge for every plant. The mechanical damage requires an immediate adjustment of the metabolic activities to stop the loss of water, turgor, assimilation, and to compensate for the loss of structural compartmentalization. New cell division and/or necrotization of affected tissue parts may close the wound.

Simultaneously, the plant prepares for possible invading microorganisms. Even in the case of abiotic damage, the systemic induction of basic defense mechanisms against herbivorous animals seems to complete the typical wound response to avoid further damage. However, the specific perception of the presence of feeding insects at the wounding site can lead to a modification of the wound response and to the activation of additional defense mechanisms. Similarly, the specific detection of pathogenic microorganisms may lead to a more specific antimicrobial activity and subsequent resistance.

JA (**1**) appears to be an integral part of a general signal transduction system regulating inducible defense genes in plants in responses to wounding. JA and its fatty acid precursors rapidly accumulate in cases of wounding.[42,109] External application of jasmonates induces the wound response without wounding.[82,87,98] The inhibition of different enzymes of the octadecanoid pathway prevented the accumulation of wound-induced proteins,[35,36] thus showing jasmonate biosynthesis to be essential for the general wound response. However, an additional wound-induced pathway controlling the expression of specific genes independent of JA was detected in the JA-insensitive coi1 mutant of *Arabidopsis thaliana*.[110]

Wound-induced proteinase inhibitors (PIs) are powerful inhibitors of major intestinal proteinases of animals, which make the plant less palatable to feeding insects.[83] The relationship between wounding and PIs was revealed as follows.[111] Local wounding of the leaves of potato or tomato plants by adult Colorado potato beetles or their larvae induced a rapid accumulation of a PI throughout the aerial part of the plant. This effect of insect damage could be simulated by mechanically wounding the leaves.

The accumulation of the wound-inducible PI proteins was found to be induced by an endogenous polypeptide named systemin, isolated from tomato leaves.[83] The polypeptide, consisting of 18 amino acids, was able at very low concentrations to induce the synthesis of two wound-inducible PI proteins when applied to young tomato plants. Systemin is transported out of wounds to distal tissues.

JA-Me (**2**) induced the synthesis of defensive PI proteins in treated plants and also in nearby plants when applied to the surface of tomato leaves.[84] The presence of JA-Me in the atmosphere of chambers containing plants resulted in the accumulation of PIs in their leaves. When sagebrush (*Artemisia tridentata*), shown to possess JA-Me in leaf surface structures, was incubated in chambers with tomato plants, PI accumulation was induced in the tomato leaves. That is to say, interplant communication occurred from the leaves of one plant species to the leaves of another species to activate the expression of defensive genes.

When octadecanoid precursors of JA (**1**), i.e., linolenic acid (**7**), (13*S*)-HPLA (**9**), and 12-oxo-PDA (**8**), were applied to the surface of tomato leaves, these compounds also served as powerful inducers of PI I and II synthesis, mimicking a wound response.[112]

However, the octadecanoid pathway may be involved in some of the more special, not necessarily wound-inducible, reactions as well. Some defense mechanisms against pathogenic microorganisms also require *de novo* synthesis of JA as demonstrated for the elicitor-induced synthesis of phytoalexins[41] (see Section 1.05.3.2.2).

Several plants respond to herbivore feeding by additionally emitting diverse blends of volatiles from their leaves that in turn attract carnivorous enemies of the herbivores.[113,114] Just as there are elicitors derived from microbial pathogens that induce plant signal transduction pathways, there are also specific elicitors of plant responses associated with herbivorous insects.[115] The transduction of the signal proceeds via compounds from the octadecanoid pathway and ends with the transcription of genes and the *de novo* synthesis of enzymes and their products.[116] Most of these responses can be activated by JA, but conjugates of JA with aliphatic amino acids like isoleucine and leucine can act as specific signals.[117]

Environmental stimuli inducing JA biosynthesis do not always trigger all the responses which are obtained in cases of exogenous application of JA and vice versa. Though both the wound- as well as the elicitor-induced gene expression require the activation of the octadecanoid pathway, the results do not strictly correlate with the effects of JA or α-linolenic acid.[118]

A reversible protein phosphorylation step has been reported to be involved in the wound signal transduction and assumed to act downstream of JA.[119] A staurosporine-sensitive protein kinase negatively regulated the pathway and a protein phosphatase, most probably of type 2A, activated the JA-responsive gene expression in *A. thaliana* plantlets. Activation via this pathway was found to be blocked by exogenously applied auxins as well as the JA-insensitive mutants jin1, jin4, and coi1. Linear signal transduction models can no longer cope with the complexity of plant life. It seems likely that the different initial molecular events involved in the perception of exogenous or endogenous stimuli, as well as commonly initiating the octadecanoid pathway, can differentially modulate the jasmonate signal transduction. This may involve the formation of JA metabolites with specific activities and/or the modification of jasmonate-perceiving components in addition to the downregulation of the pathway.

1.05.3.2.5 Disease resistance

Besides several constitutive barriers against microorganisms, plants have developed numerous antimicrobial defense mechanisms which are activated locally in cases of infections, for example the hypersensitive death of affected cells, deposition of callose and lignin in cell walls, cross-linking of hydroxyproline-rich cell wall glycoproteins, stimulation of oxidative enzymes such as peroxidase and polyphenoloxidase (PPO), and the biosynthesis of phytoalexins.[120]

Phytoalexins are low molecular weight antimicrobial compounds that are synthesized and accumulate in plants after their exposure to pathogenic microorganisms. Plant defense reactions against pathogens include the induced synthesis of phytoalexins. Biotic elicitors that are derived from the cell surfaces of both pathogenic microbes as well as host plants trigger the defense response. It has been hypothesized that an elicitor molecule combines with a plant membrane receptor and that the complex activates a series of specific genes, resulting in the synthesis of phytoalexins. JA

(**1**) rapidly accumulates during elicitation,[100] and often was found to be a key signal transducer for the induction of phytoalexin production (see Section 1.05.3.2.2).

In many higher plants an initial inoculation or natural infection by a necrotizing microorganism can induce a defense status that protects the plant against subsequent infection. This systemic acquired resistance (SAR) correlates at least partially with the induction of the hypersensitive reaction and is associated with the local and systemic appearance of at least five families of pathogenesis-related (PR) proteins.[121] Tissue localization, timing of appearance, and known functions of some PR proteins (e.g., chitinases and glucanases) suggest their involvement in the constitution of SAR. They are often used as SAR indicators. PR proteins were originally identified in plants upon viral infection,[122] but inducers as well as targets of the SAR belong to viruses, bacteria, and fungi.[123] A key compound in the local establishment of the SAR status is salicylic acid, but the mobile systemic signal is still unknown.[124]

The SAR signal is believed to be independent of JA because wounding or JA alone do not lead to accumulation of salicylic acid and to SAR.[125] However, JA is involved in the direct interaction with the pathogens (elicitation and phytoalexin response). Moreover, salicylic acid derivatives are known to inhibit the octadecanoid pathway[35,37] and ethylene biosynthesis,[126,127] while both of them are required for the general wound response.

The complex interactions of these important signal molecules, their hierarchy, compartmentalization, and their fine-tuned chronological counterregulation are far from understood.

1.05.3.3 Interaction with Other Plant Hormones

Interactions of the jasmonate signal with several plant growth regulating compounds were reported from different plant species and biotest systems. Some prominent examples include abscisic acid (ABA) and ethylene.

JA (**1**) or JA-Me (**2**) and ABA often show similar effects on physiological processes or act in an interdependent way. One of the few exeptions is their effect on dormant seeds, where application of ABA inhibits while JA or JA-Me stimulates seed germination.[76] Phytohormones such as ABA and ethylene are involved in the wound-induced activation of the *pin2* gene. Local wounding of potato or tomato plants resulted in the accumulation of PIs I and II throughout the aerial part of the plant. In contrast to wild-type plants, ABA-deficient mutants of potato (droopy) and tomato (sit) showed a drastically reduced induction of these genes in response to plant wounding.[128] High levels of PI II gene expression were obtained in mutant and wild-type plants upon exogenous application of ABA. Experiments with wild-type plants showed that wounding results in systemically increased levels of this phytohormone in wounded and nonwounded leaves.

Potato and tomato wild-type leaves also showed increased endogenous levels of ABA and JA after systemin treatment.[129] Similar to wounding, the peptide did not affect the endogenous concentration of ABA or JA in ABA-deficient plants. Exogenous ABA promoted an increase of endogenous JA levels in both wild-type and ABA-deficient plants. Conversely, potato or tomato leaves treated with JA did not show any change of internal ABA levels.[87] This demonstrates that the site of action of JA is located downstream of the site of action of ABA in the wound response.

Wounding induces ethylene biosynthesis as well.[130] Systemin[131] and JA[31] (**1**) have both been shown to induce ethylene in suspension-cultured cells of tomato. Ethylene is required in the transduction pathway of the wound signal, and ethylene and jasmonates act together to regulate *pin* gene expression during the wound response.[132]

JA-Me applied to *Dendrobium* and *Petunia* flowers as an aqueous solution through the cut stem or as a vapor accelerated senescence and increased the production of ethylene and its precursor, 1-aminocyclopropane-1-carboxylic acid (ACC), in proportion to the dose of JA-Me. Inhibitors of ethylene biosynthesis or of the action of ethylene completely inhibited the senescence effect of JA-Me, which indicates that JA-Me enhanced the flower senescence via the promotion of ACC and ethylene production.[133] Ethylene had an additive effect on the jasmonate-induced loss of chlorophyll from cucumber cotyledons.[134]

However, tulip leaf senescence induced by JA-Me did not affect ethylene production, ACC oxidase activity, or ACC content.[135] During fruit ripening, very complex patterns of interaction between jasmonates and ethylene were observed. For instance, in preclimacteric apple fruits JA-Me led to a strong stimulation of all parts of ethylene biosynthesis, but inhibited ethylene production in climacteric and postclimacteric apples.[136]

1.05.4 INITIATION OF JASMONATE BIOSYNTHESIS

Systemin represents one of the initial steps in the signal transduction pathway, regulating the wound response via the octadecanoid pathway.[137] However, the actual mode of action of systemin is still unknown. It is possible that there are other inducers of jasmonate biosynthesis, independent of systemin.

Since exogenous application of precursors of JA can effectively cause jasmonate effects in some systems,[79,112] it seems that all of the enzymes responsible for the synthesis of jasmonates from linolenic acid are constitutionally expressed. Thus, a key event in the jasmonate signaling would be the release of linolenic acid by the activation, liberation, or synthesis of a lipase. In a hypothetical model, the activation of a membrane-based lipase by a supposed protein kinase was suggested to initiate the octadecanoid pathway.[112] In general, post-translational modification via protein phosphorylation by protein kinases or dephosphorylation by phosphoprotein phosphatases appears to be the principal mechanism for the rapid activation or deactivation of specific regulatory enzymes. In mammals a mitogen-activated protein (MAP) kinase activates the cytoplasmic phospholipase A_2 ($cPLA_2$), a cell surface protein which cleaves phospholipids to release arachidonic acid, the precursor of the prostaglandins[138] (see Section 1.05.2.1).

A complementary DNA encoding a 46 kDa MAP kinase homologue has been isolated from tobacco plants. Transcripts of the corresponding gene were not observed in healthy tobacco leaves but began to accumulate as early as 1 min after mechanical wounding, reaching a maximum level within 1 h and rapidly declining thereafter. In tobacco plants transformed with the cloned complementary DNA, *trans*-inactivation of the endogenous homologous gene occurred, and both production of wound-induced JA and accumulation of wound-inducible gene transcripts were inhibited. The corresponding enzyme was designated WIPK (wound-induced protein kinase).[139]

Similar wound-inducible, fast-responding protein kinases, all with a molecular mass of 46 kDa, were detected in leaves of a wide variety of plant species including dicotyledonous and monocotyledonous plants. The name PMSAP (plant multisignal-activated protein) kinase was proposed, because this enzyme was also activated by various signals other than wounding.[140] In this study the relationship of the new protein kinases with the initiation of jasmonate biosynthesis was not investigated, but all known properties of these enzymes are consistent with the WIPK of the parallel-published paper.[139]

Moreover, similar patterns of 46 kDa protein kinase activation were found without wounding in roots and tissues containing the shoot apical meristem and flower buds as well as after application of elicitors.[140] This seems to correlate with sites of high jasmonate activity.

When tobacco leaf protein extracts containing the active kinase were treated with serine/threonine-specific or tyrosine-specific protein phosphatase, the kinase activity was abolished. These results suggest that the active form of the kinase is phosphorylated at both serine/threonine and tyrosine residues. It seems likely that it can be activated by dual phosphorylation.

Using tobacco leaf disks the 46 kDa protein kinase was shown to be desensitized immediately after the first activation and was unable to respond to a second cutting for at least 1 h, but recovered later. This rapid inactivation could be blocked in the presence of cycloheximide, indicating that the *de novo* synthesis of another protein, for instance a protein phosphatase, is required for this negative regulation.[140]

The details of the transformation of the mechanical wounding process into the specific activation of these protein kinases are still unknown. Similar but separate mechanisms for the various environmental factors which stimulate the octadecanoid pathway can be expected.

1.05.5 STRUCTURE–ACTIVITY RELATIONSHIPS

JA (**1**) has two asymmetric carbons and therefore there are four diastereomers. The two naturally occurring isomers are shown in Section 1.05.1. The absolute configuration of the native JA, (+)-epi-JA (**1a**), and its methyl ester (**2a**) is 1*R*,2*S*. Of the four diastereomers, only (**2a**) has the typical odor.[141] It is known that (**2a**) is readily isomerized into an equilibrium mixture of about 94–95%(−)-JA-Me (**2b**) and 5–6% (+)-epi-JA-Me (**2a**).[142] It is probable that most physiological studies have been completed with the synthetic racemate containing all four diastereomers. Such mixtures reportedly contain less than 2.5%(+)-epi-JA.[142] When the chiral purity of a sample was not specified, it has to be assumed to be the racemate.

Ethyl *trans*-cinnamate, (+)-epi-JA-Me (**2a**), (−)-JA-Me (**2b**) and (*R*)-(−)-mellein were identified from the hairpencils of male oriental fruit moths (*Grapholitha molesta* Busck). The blend of these compounds attracted sex pheromone-releasing females from several centimeters away.[143] Ethyl *trans*-cinnamate proved to be active by itself and (**2a**), not (**2b**), acted as a synergist.

In a study by Ueda *et al.* (**2a**) had a stronger inhibitory effect on kinetin-induced retardation of oat leaf senescence than (**2b**).[144]

JA-related compounds were synthesized, and their inhibitory activities on rice seedling growth were investigated.[145] Three functional groups, $—CH_2CO_2H$ or $—CH_2CO_2Me$ at C-1, $—CH_2CH{=}CHCH_2Me$ or $—(CH_2)_4Me$ at C-2, and $=O$ or $—OH$ at C-3, were essential for exhibiting inhibitory activity in this series of compounds. The structural requirements for growth-inhibiting activity on rice seedlings[145] are very similar to those for microtuber formation of potato,[146] and the induction of the expression of the two PI genes of potato.[85]

Coronatine (**14**) is a bacterial blight toxin produced by several pathovars of *Pseudomonas syringae* von Hall which induces chlorosis, increased ethylene production, and accelerated senescence. These effects are not limited to the host plants of the pathogens. The structure,[147] including the absolute configuration of the cyclopentane portion, is very similar to the isoleucine conjugate of the native (+)-JA: *N*-((+)-jasmonoyl)-(*S*)-isoleucine (JA-Ile) (**26**). Coronatine and jasmonates produced similar effects in biotests using the ethylene production of tomato cell cultures,[31] the coiling response of *B. dioica* tendrils,[148] and potato microtuber formation,[149] indicating that coronatine is a mimic of jasmonates.[30]

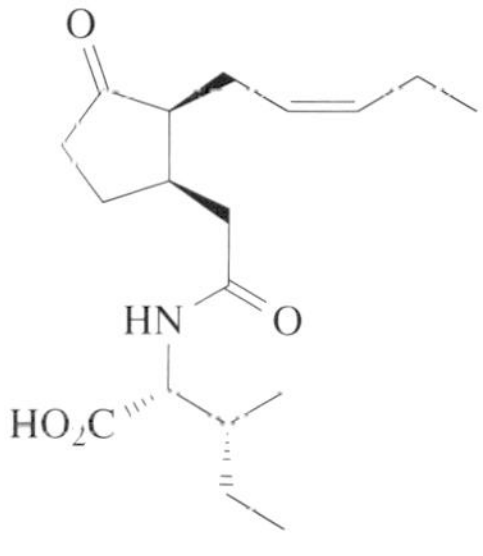

(**26**) *N*-((+)-Jasmonoyl)-(*S*)-isoleucine

The opposite effects of JA and JA-Ile regarding seed germination[40] indicate that the latter has its own activity, not related to its transformation into free JA or activity as a structural analogue.

Further evidence for the specific activity of the JA conjugates comes from experiments investigating the emission of volatiles from Lima bean (*Phaseolus lunatus*). Coronatine (**14**), as well as the conjugate of synthetic 1-oxo-indan-4-carboxylic acid (**27**) with isoleucine (In-Ile), mimic the action of JA (**1**) or JA-Ile (**26**). Crucially, the component parts of In-Ile are completely inactive, attributing the biological activity exclusively to the conjugate. These findings are independently confirmed by the lack of biological activity for derivatives of JA which are not able to form conjugates with amino acids, like jasmonoyl aldehyde or a nitrojasmonate with a nitro group in place of the carboxylate.[117]

Tests recording the up- or downregulation of various genes in tomato leaves gave the following results: (i) 12-oxo-PDA and other intermediates of the octadecanoid pathway have to be β-oxidized to give a JA response; (ii) octadecanoids which can not be β-oxidized are inactive; (iii) JA, its methyl ester (JA-Me), and its amino acid conjugates are the most active signals in tomato leaves leading to upregulation of mainly wound-inducible genes and downregulation of mainly "housekeeping" genes; and (iv) some compounds carrying a JA/JA-Me or JA amino acid conjugate-like structure induce/repress only a subset of genes, suggesting diversity of jasmonate signaling.[150]

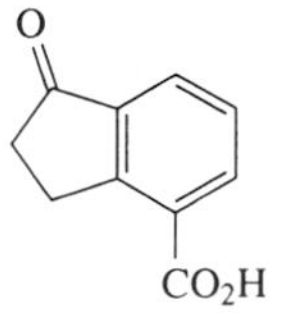

(**27**) 1-Oxo-indan-4-carboxylic acid

1.05.6 REFERENCES

1. E. Demole, E. Lederer, and D. Mercier, *Helv. Chim. Acta*, 1962, **45**, 675.
2. D. C. Aldridge, S. Galt, D. Giles, and W. B. Turner, *J. Chem. Soc. C*, 1971, 1623.

3. A. Meyer, O. Miersch, C. Butter, W. Dathe, and G. Sembdner, *J. Plant Growth Regul.*, 1984, **3**, 1.
4. B. E. Cross and G. R. B. Webster, *J. Chem. Soc. C*, 1970, 1839.
5. H. Fukui, K. Koshimizu, S. Usuda, and Y. Yamazaki, *Agric. Biol. Chem.*, 1977, **41**, 175.
6. H. Fukui, K. Koshimizu, Y. Yamazaki, and S. Usuda, *Agric. Biol. Chem.*, 1977, **41**, 189.
7. R. Kaiser and D. Lamparsky, *Tetrahedron Lett.*, 1974, **38**, 3413.
8. O. Miersch, G. Schneider, and G. Sembdner, *Phytochemistry*, 1991, **30**, 4049.
9. E. Demole, B. Wilhalm, and M. Stoll, *Helv. Chim. Acta*, 1964, **47**, 1152.
10. Y. Koda, E. A. Omer, T. Yoshihara, H. Shibata, S. Sakamura, and Y. Okazawa, *Plant Cell Physiol.*, 1988, **29**, 1047.
11. T. Yoshihara, E. A. Omer, H. Koshino, S. Sakamura, Y. Kikuta, and Y. Koda, *Agric. Biol. Chem.*, 1989, **53**, 2835.
12. F. Bohlmann, P. Wegner, J. Jakupovic, and R. M. King, *Tetrahedron*, 1984, **40**, 2537.
13. D. C. Zimmerman and P. Feng, *Lipids*, 1978, **13**, 313.
14. B. A. Vick and D. C. Zimmerman, *Plant Physiol.*, 1979, **63**, 490.
15. B. A. Vick and D. C. Zimmerman, *Biochem. Biophys. Res. Commun.*, 1983, **111**, 470.
16. B. A. Vick and D. C. Zimmerman, *Lipids*, 1979, **14**, 734.
17. M. Hamberg, *Biochim. Biophys. Acta*, 1987, **920**, 76.
18. W.-C. Song and A. R. Brash, *Science*, 1991, **253**, 781.
19. M. Hamberg, *Biochem. Biophys. Res. Commun.*, 1988, **156**, 543.
20. M. Hamberg and M. A. Hughes, *Lipids*, 1988, **23**, 469.
21. S. W. Baertschi, C. D. Ingram, T. M. Harris, and A. R. Brash, *Biochemistry*, 1988, **27**, 18.
22. B. A. Vick and D. C. Zimmerman, *Plant Physiol.*, 1986, **80**, 202.
23. F. Schaller and E. W. Weiler, *Eur. J. Biochem.*, 1997, **245**, 294.
24. F. Schaller and E. W. Weiler, *J. Biol. Chem.*, 1997, **272**, 28 066.
25. B. A. Vick and D. C. Zimmerman, *Plant Physiol.*, 1984, **75**, 458.
26. O. Miersch, J. Schmidt, G. Sembdner, and K. Schreiber, *Phytochemistry*, 1989, **28**, 4049.
27. M. Hamberg, O. Miersch, and G. Sembdner, *Lipids*, 1988, **23**, 521.
28. B. A. Vick and D. C. Zimmerman, *Plant Physiol.*, 1989, **90**, 125.
29. M. Hamberg and H. W. Gardner, *Biochim. Biophys. Acta*, 1992, **1165**, 1.
30. F. Greulich, T. Yoshihara, H. Toshima, and A. Ichihara, in "XV International Botanical Congress, Abstracts, Yokohama (Japan), 1993," p. 388.
31. F. Greulich, T. Yoshihara, and A. Ichihara, *J. Plant Physiol.*, 1995, **147**, 359.
32. C. L. Bender, H. Liyanage, D. Palmer, M. Ullrich, S. Young, and R. Mitchell, *Gene*, 1993, **133**, 31.
33. S. Nara, H. Toshima, and A. Ichihara, *Tetrahedron*, 1997, **53**, 9509.
34. O. Kodama, *Mycotoxins*, 1996, **42**, 7.
35. P. E. Staswick, J.-F. Huang, and Y. Rhee, *Plant Physiol.*, 1991, **96**, 130.
36. H. Pena-Cortes, T. Albrecht, S. Prat, E. W. Weiler, and L. Willmitzer, *Planta*, 1993, **191**, 123.
37. E. E. Farmer, D. Caldelari, G. Pearce, M. K. Walker-Simmons, and C. A. Ryan, *Plant Physiol.*, 1994, **106**, 337.
38. Z. P. Zhang and I. T. Baldwin, *Planta*, 1997, **203**, 436.
39. R. Lopez, W. Dathe, C. Brückner, O. Miersch, and G. Sembdner, *Biochem. Physiol. Pflanz.*, 1987, **182**, 195.
40. R. Ranjan, O. Miersch, G. Sembdner, and S. Lewak, *Physiol. Plant.*, 1994, **90**, 548.
41. H. Gundlach, M. J. Müller, T. M. Kutchan, and M. H. Zenk, *Proc. Natl. Acad. Sci. USA*, 1992, **89**, 2389.
42. R. A. Creelman, M. L. Tierney, and J. E. Mullet, *Proc. Natl. Acad. Sci. USA*, 1992, **89**, 4938.
43. A. Meyer, D. Gross, S. Vorkefeld, M. Kummer, J. Schmidt, G. Sembdner, and K. Schreiber, *Phytochemistry*, 1989, **28**, 1007.
44. A. Meyer, J. Schmidt, D. Gross, E. Jensen, A. Rudolph, S. Vorkefeld, and G. Sembdner, *J. Plant Growth Regul.*, 1990, **10**, 17.
45. E. Schwarzkopf and O. Miersch, *Biochem. Physiol. Pflanz.*, 1992, **188**, 57.
46. Z.-Q. Xia and M. H. Zenk, *Planta Med.*, 1993, **59**, 575.
47. T. Yoshihara, M. Amanuma, T. Tsutsumi, Y. Okumura, H. Matsuura, and A. Ichihara, *Plant Cell Physiol.*, 1996, **37**, 586.
48. S. Parchmann, H. Gundlach, and M. J. Müller, *Plant Physiol.*, 1997, **115**, 1057.
49. G. Sembdner and B. Parthier, *Annu. Rev. Plant Physiol. Plant Mol. Biol.*, 1993, **44**, 569.
50. D. Gross and B. Parthier, *J. Plant Growth Regul.*, 1994, **13**, 93.
51. H. Yamane, H. Takagi, H. Abe, T. Yokota, and N. Takahashi, *Plant Cell Physiol.*, 1981, **22**, 689.
52. J. Ueda and J. Kato, *Plant Physiol.*, 1980, **66**, 246.
53. R. A. Weidhase, J. Lehmann, H. Kramell, G. Sembdner, and B. Parthier, *Physiol. Plant.*, 1987, **69**, 161.
54. J. Ueda, K. Miyamoto, and M. Hashimoto, *Plant Cell Physiol.*, 1993, **34** (Suppl.), 102.
55. M. Saniewski and E. Wegrzynowicz-Lesiak, *Acta Hortic.*, 1995, **394**, 315.
56. L. Gregory, *Am. J. Bot.*, 1956, **43**, 281.
57. Y. Koda, Y. Okazawa, T. Yoshihara, H. Shibata, S. Sakamura, and Y. Okazawa, *Plant Cell Physiol.*, 1988, **29**, 969.
58. Y. Koda and E. A. Omer, *Plant Cell Physiol.*, 1988, **29**, 1047.
59. T. Yoshihara, E. A. Omer, H. Koshino, S. Sakamura, Y. Kikuta, and Y. Koda, *Agric. Biol. Chem.*, 1989, **53**, 2835.
60. A. M. Pelacho and A. M. Mingo-Castel, *Plant Physiol.*, 1991, **97**, 1253.
61. H. Matsuura, T. Yoshihara, A. Ichihara, Y. Kikuta, and Y. Koda, *Biosci., Biotechnol., Biochem.*, 1993, **57**, 1253.
62. Y. Koda and Y. Kikuta, *Plant Cell Physiol.*, 1991, **32**, 629.
63. K. Takahashi, K. Fujino, Y. Kikuta, and Y. Koda, *Plant Sci. (Limerick, Irel.)*, 1994, **100**, 3.
64. Y. Koda and Y. Okazawa, *Jpn. J. Crop Sci.*, 1983, **52**, 592.
65. T. Mita and H. Shibaoka, *Plant Cell Physiol.*, 1983, **24**, 109.
66. M. Abe, H. Shibaoka, H. Yamane, and N. Takahashi, *Protoplasma*, 1990, **156**, 1.
67. T. Matsuki, H. Tazaki, T. Fujimori, and T. Hogetsu, *Biosci., Biotechnol., Biochem.*, 1992, **56**, 1329.
68. H. Nojiri, H. Yamane, H. Seto, I. Yamaguchi, N. Murofushi, T. Yoshihara, and H. Shibaoka, *Plant Cell Physiol.* 1992, **33**, 1225.
69. C. Brückner, R. Kramell, G. Schneider, H.-D. Knöfel, G. Sembdner, and K. Schreiber, *Phytochemistry*, 1986, **25**, 2236.
70. C. Brückner, R. Kramell, G. Schneider, J. Schmidt, A. Preiss, G. Sembdner, and K. Schreiber, *Phytochemistry*, 1988, **27**, 275.

71. O. Miersch, H.-D. Knöfel, J. Schmidt, R. Kramell, and B. Parthier, *Phytochemistry*, 1998, **47**, 327.
72. H.-D. Knöfel and G. Sembdner, *Phytochemistry*, 1995, **38**, 569.
73. G. Sembdner, R. Atzorn, and G. Schneider, *Plant Mol. Biol.*, 1994, **26**, 1459.
74. H. Yamane, H. Abe, and N. Takahashi, *Plant Cell Physiol.* 1982, **23**, 1125.
75. M. McConn and J. Browse, *Plant Cell*, 1996, **8**, 403.
76. S. B. Jarvis, M. A. Taylor, J. Bianco, F. Corbineau, and H. V. Davies, *J. Plant Physiol.*, 1997, **151**, 457.
77. F. Corbineau, R. M. Rudnicki, and D. Come, *Plant Growth Regul.*, 1988, **7**, 157.
78. E. Falkenstein, B. Groth, A. Mithofer, and E. W. Weiler, *Planta*, 1991, **85**, 316.
79. E. W. Weiler, T. Albrecht, B. Groth, Z.-Q. Xia, M. Luxem, H. Lib, L. Andert, and P. Spengler, *Phytochemistry*, 1993, **32**, 591.
80. G. Suzuki, H. Ohta, T. Kato, T. Igarashi, F. Sakai, D. Shibata, A. Takano, T. Masuda, Y. Shioi, and K. I. Takamiya, *FEBS Lett.*, 1996, **383**, 83.
81. S. Imanishi, K. Hashizume, H. Kojima, A. Ichihara, and K. Nakamura, *Plant Cell Physiol.*, 1998, **39**, 202.
82. S. Reinbothe, B. Mollenhauer, and C. Reinbothe, *Plant Cell*, 1994, **6**, 1197.
83. G. Pearce, D. Strydom, S. Johnson, and C. A. Ryan, *Science*, 1991, **253**, 895.
84. E. E. Farmer and C. A. Ryan, *Proc. Natl. Acad. Sci. USA*, 1990, **87**, 7713.
85. A. Ishikawa, T. Yoshihara, and K. Nakamura, *Biosci., Biotechnol., Biochem.*, 1994, **58**, 544.
86. K. Herbers, S. Prat, and L. Willmitzer, *Planta*, 1994, **194**, 230.
87. T. Hildmann, M. Ebneth, H. Pena-Cortes, J. J. Sanchez-Serrano, L. Willmitzer, and S. Prat, *Plant Cell*, 1992, **4**, 1157.
88. R. J. Aerts, D. Gisi, E. De Carolis, V. De Luca, and T. W. Baumann, *Plant J.*, 1994, **5**, 635.
89. E. Bell and J. E. Mullet, *Plant Physiol.*, 1993, **103**, 1133.
90. T. J. Tranbarger, V. R. Franceschi, D. F. Hildebrand, and H. D. Grimes, *Plant Cell*, 1991, **3**, 973.
91. I. Feussner, B. Hause, K. Voeroes, B. Parthier, and C. Wasternack, *Plant J.*, 1995, **7**, 949.
92. R. W. Wilen, G. J. H. van Rooijen, D. W. Pearce, R. P. Pharis, L. A. Holbrook, and M. M. Moloney, *Plant Physiol.*, 1991, **95**, 399.
93. J. M. Anderson, *J. Plant Growth Regul.*, 1988, **7**, 203.
94. V. R. Franceschi and H. D. Grimes, *Proc. Natl. Acad. Sci. USA*, 1991, **88**, 6745.
95. B. Chaudhry, F. Müller-Uri, V. Cameron-Mills, S. Gough, D. Simpson, K. Skriver, and J. Mundy, *Plant J.*, 1994, **6**, 815.
96. I. Andresen, W. Becker, K. Schlüter, J. Burges, B. Parthier, and K. Apel, *Plant Mol. Biol.*, 1992, **19**, 193.
97. S. Reinbothe, A. Machmudowa, C. Wasternack, C. Reinbothe, and B. Parthier, *J. Plant Growth Regul.*, 1992, **11**, 7.
98. E. E. Farmer, R. R. Johnson, and C. A. Ryan, *Plant Physiol.*, 1992, **98**, 995.
99. H. D. Grimes, D. S. Koetje, and V. R. Franceschi, *Plant. Physiol.*, 1992, **100**, 433.
100. H. Gundlach, M. J. Müller, T. M. Kutchan, and M. H. Zenk, *Proc. Natl. Acad. Sci. USA*, 1992, **89**, 2389.
101. K. J. Doughty, G. A. Kiddle, B. J. Pye, R. M. Wallsgrove, and J. A. Pickett, *Phytochemistry*, 1995, **38**, 347.
102. L. Du and B. A. Halkier, *Plant Physiol.*, 1996, **111**, 831.
103. A. Yamada, N. Shibuya, O. Kodama, and T. Akatsuka, *Biosci., Biotechnol., Biochem.*, 1993, **57**, 405.
104. H. Nojiri, M. Sugimori, H. Yamane, Y. Nishimura, A. Yamada, N. Shibuya, O. Kodama, N. Murofushi, and T. Omori, *Plant Physiol.* 1996, **110**, 387.
105. R. Rakwal, S. Tamogami, and O. Kodama, *Biosci., Biotechnol., Biochem.*, 1996, **60**, 1046.
106. P. J. Facchini, A. G. Johnson, J. Poupart, and V. De Luca, *Plant Physiol.*, 1996, **111**, 687.
107. R. Kramell, R. Atzorn, G. Schneider, O. Miersch, C. Brückner, J. Schmidt, G. Sembdner, and B. Parthier, *J. Plant Growth Regul.*, 1995, **14**, 29.
108. S. Reinbothe, C. Reinbothe, J. Lehmann, and B. Parthier, *Physiol. Plant.*, 1992, **86**, 49.
109. R. A. Creelman, M. L. Tierney, and J. E. Mullet, *Proc. Natl. Acad. Sci. USA*, 1992, **89**, 4938.
110. E. Titarenko, E. Rojo, J. Leon, and J. J. Sanchez-Serrano, *Plant Physiol.*, 1997, **115**, 817.
111. T. R. Green and C. A. Ryan, *Science*, 1972, **175**, 776.
112. E. E. Farmer and C. A. Ryan, *Plant Cell*, 1992, **4**, 129.
113. T. C. J. Turlings, J. H. Loughrin, P. J. McCall, U. S. R. Rose, W. J. Lewis, and J. H. Tumlinson, *Proc. Natl. Acad. Sci. USA*, 1995, **92**, 4169.
114. J. Takabayashi and M. Dicke, *Trends Plant Sci.*, 1996, **1**, 109.
115. K. L. Korth and R. A. Dixon, *Plant Physiol.*, 1997, **115**, 1299.
116. W. Boland, J. Hopke, J. Donath, J. Nüske, and F. Bublitz, *Angew. Chem.*, 1995, **107**, 1715.
117. T. Krumm, K. Bandemer, and W. Boland, *FEBS Lett.*, 1995, **377**, 523.
118. M. Ellard-Ivey and C. J. Douglas, *Plant Physiol.*, 1996, **112**, 183.
119. E. Rojo, E. Titarenko, J. Leon, S. Berger, G. Vancanneyt, and J. J. Sanchez-Serrano, *Plant J.*, 1998, **13**, 153.
120. R. Hammerschmidt and J. C. Schultz, in "Phytochemical Diversity and Redundancy in Ecological Interactions," eds. J. T. Romeo, J. A. Saunders, and P. Barbosa, Plenum, New York, 1996, p. 121.
121. A. J. Enyedi, N. Yalpani, P. Silverman, and I. Raskin, *Cell*, 1992, **70**, 879.
122. L. C. van Loon, Y. A. Gerristsen, and C. E. Ritter, *Plant Mol. Biol.* 1987, **9**, 593.
123. J. Ryals, S. Uknes, and E. Ward, *Plant Physiol.*, 1994, **104**, 1109.
124. B. Vernooij, L. Friedrich, A. Morse, R. Reist, R. Kolditz-Jawhar, E. Ward, S. Uknes, H. Kessmann, and J. Ryals, *Plant Cell*, 1994, **6**, 959.
125. N. Yalpani, P. Silverman, T. M. A. Wilson, D. A. Kleier, and I. Raskin, *Plant Cell*, 1991, **3**, 809.
126. C. A. Leslie and R. J. Romani, *Plant Cell Rep.*, 1986, **5**, 144.
127. S. Pennazio and P. Roggero, *Biol. Plant.*, 1991, **33**, 58.
128. H. Pena-Cortes, J. J. Sanchez-Serrano, R. Mertens, L. Willmitzer, and S. Prat, *Proc. Natl. Acad. Sci. USA*, 1989, **86**, 9851.
129. H. Pena-Cortes, J. Fisahn, and L. Willmitzer, *Proc. Natl. Acad. Sci. USA*, 1995, **92**, 4106.
130. H. Kende, *Annu. Rev. Plant Physiol. Plant Mol. Biol.*, 1993, **44**, 283.
131. G. Felix and T. Boller, *Plant J.*, 1995, **7**, 381.
132. P. J. O'Donnell, C. Calvert, R. Atzorn, C. Wasternack, H. M. O. Leyser, and D. J. Bowles, *Science*, 1996, **274**, 1914.
133. R. Porat and A. H. Halevy, *Plant Growth Regul.*, 1993, **13**, 297.

134. F. B. Abeles, W. L. Hershberger, and L. J. Dunn, *Plant Physiol.*, 1989, **89**, 664.
135. J. Puchalski, P. Klim, M. Saniewski, and J. Nowacki, *Acta Hortic.*, 1989, **251**, 107.
136. M. Saniewski, *Acta Hortic.*, 1995, **394**, 85.
137. C. P. Constabel, D. R. Bergey, and C. A. Ryan, *Proc. Natl. Acad. Sci. USA*, 1995, **92**, 407.
138. L. L. Lin, M. Wartmann, A. Y. Lin, J. L. Knopf, A. Seth, and R. J. Davis, *Cell*, 1993, **72**, 269.
139. S. Seo, M. Okamoto, H. Seto, K. Ishizuka, H. Sano, and Y. Ohashi, *Science*, 1995, **270**, 1988.
140. S. Usami, H. Banno, Y. Ito, R. Nishihama, and Y. Machida, *Proc. Natl. Acad. Sci. USA*, 1995, **92**, 8660.
141. T. E. Acree, R. Nishida, and H. Fukami, *J. Agric. Food Chem.*, 1985, **33**, 425.
142. R. Nishida, T. E. Acree, and H. Fukami, *Agric. Biol. Chem.*, 1985, **49**, 769.
143. T. C. Baker, R. Nishida, and W. L. Roelofs, *Science*, 1981, **214**, 1359.
144. J. Ueda, J. Kato, H. Yamane, and N. Takahashi, *Physiol. Plant.* 1981, **52**, 305.
145. H. Yamane, J. Sugawara, Y. Suzuki, E. Shimanuma, and N. Takahashi, *Agric. Biol. Chem.*, 1980, **44**, 2857.
146. Y. Koda, Y. Kikuta, H. Tazaki, Y. Tsujino, S. Sakamura, and T. Yoshihara, *Phytochemistry*, 1991, **30**, 1435.
147. A. Ichihara, K. Shiraishi, H. Sato, S. Sakamura, K. Nishiyama, R. Sakai, A. Furusaki, and T. Matsumoto, *J. Am. Chem. Soc.*, 1977, **99**, 636.
148. E. W. Weiler, T. M. Kutchan, T. Gorba, W. Brodschelm, U. Niesel, and F. Bublitz, *FEBS Lett.*, 1994, **345**, 9.
149. Y. Koda, K. Takahashi, Y. Kikuta, F. Greulich, H. Toshima, and A. Ichihara, *Phytochemistry*, 1996, **41**, 93.
150. C. Wasternack, B. Ortel, O. Miersch, R. Kramell, M. Beale, F. Greulich, I. Feussner, B. Hause, T. Krumm, W. Boland, and B. Parthier, *J. Plant Physiol.*, 1998, **152**, 345.

1.06 Biosynthesis of Butyrolactone and Cyclopentanoid Skeletons Formed by Aldol Condensation

SHOHEI SAKUDA
The University of Tokyo, Japan

and

YASUHIRO YAMADA
Osaka University, Japan

1.06.1 INTRODUCTION

The biosynthesis of two bioactive natural products are described in this chapter. The first is that of a signal molecule from *Streptomyces*, virginiae butanolide A (VB A). The other is that of the chitinase inhibitor allosamidin. Each of them has interesting biological activities and unique structural features. Although their structures and biological activities have no notable features in common, these compounds do possess a five-membered lactone or cyclopentane component. Both compounds are produced by *Streptomyces*. The butyrolactone and cyclopentanoid skeletons present in molecules of VB A and allosamidin, respectively, are unique among natural products and are key moieties for the expression of the biological activities of these compounds. To investigate their biosynthetic pathways, feeding experiments with a variety of ^{13}C, ^{15}N, or ^{2}H-labeled precursors were carried out, and the mechanisms of the formation of butyrolactone and cyclopentanoid skeletons were studied by conversion experiments with synthetic intermediates in a cell-free system or feeding experiments with stereospecifically ^{2}H-labeled precursors. As a result, it has become clear that C—C bond formation by the aldol reaction is a key step in the biosynthetic pathways to the butyrolactone and cyclopentanoid skeletons. This chapter provides an overview of the biosynthesis of VB A and allosamidin as well as related compounds. The chemistry, biochemistry, and mechanism of action of signal molecules in microorganisms are reviewed in detail in Volume 8 of this series.

1.06.2 BIOSYNTHESIS OF VIRGINIAE BUTANOLIDE A—A BUTYROLACTONE AUTOREGULATOR FROM *STREPTOMYCES*

Streptomyces is one of the most important producers of useful bioactive compounds, such as antibiotics, enzyme inhibitors, or anticancer agents. In this microbe, a number of signal molecules or autoregulators which regulate secondary metabolite production or cytodifferentiation are known. Three types of endogenous signal molecules have been isolated from *Streptomyces*, and their structures have been characterized. The first characterized compound was A factor (**1**), found by Khokhlov and co-workers, which induces production of streptomycin and formation of aerial mycelium and spores in *Streptomyces griseus*.[1] The next were virginiae butanolides (VBs) A–E (**2**), (**3**), (**4**), (**5**), and (**6**), which induce the production of virginiamycin in *Streptomyces virginiae*.[2,3] The third was IM-2 (**7**), which induces the production of a blue pigment and nucleoside antibiotics in *Streptomyces* sp. FRI-5.[4,5] Besides these endogeneous molecules, exogenous factors were also searched for in metabolites of *Streptomyces*, and several molecules were found by Gräfe *et al*. They first isolated factor 1 (**8**) from a culture broth of *Streptomyces viridochromogenes* as an inducer of the formation of aerial mycelia and leukaemomycin in *S. griseus*.[6] Then they found factors (**2**), (**9**), and (**10**) in a culture broth of *Streptomyces bikiniensis* and *Streptomyces cyaneofuscatus*, which induce the production of anthracycline in *S. griseus*.[7]

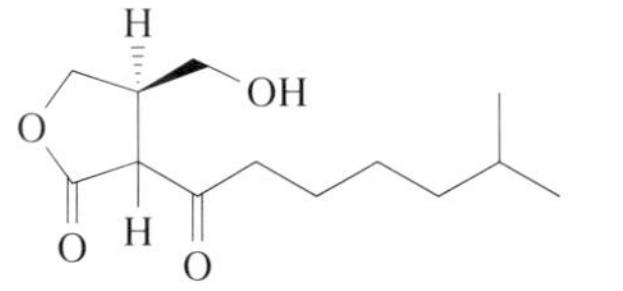

(**1**)

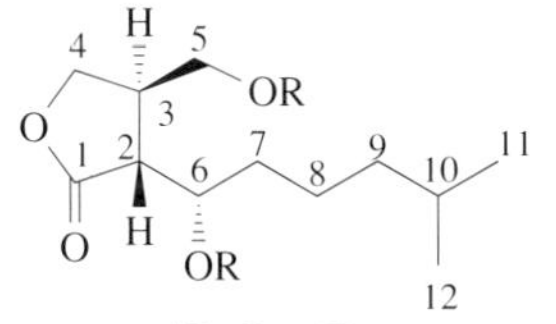

(**2**) R = H
(**2a**) R = Bz

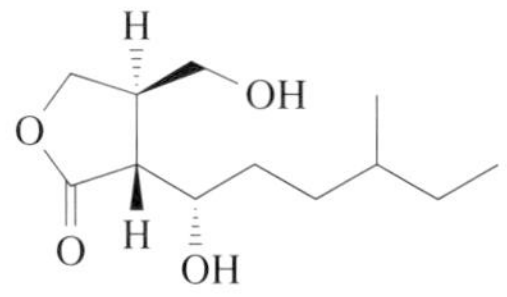

(**3**)

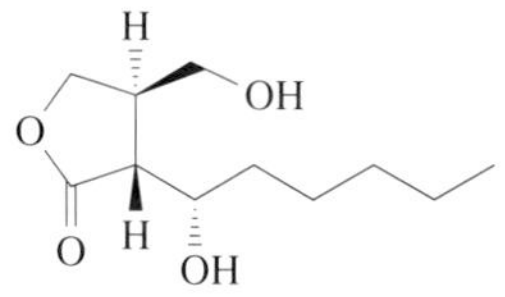

(**4**)

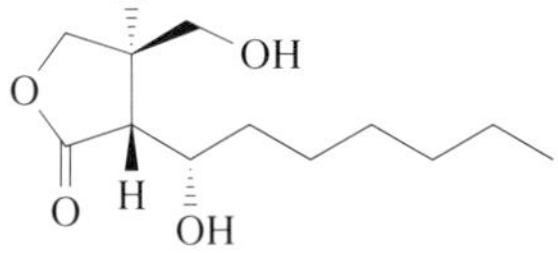

(**5**)

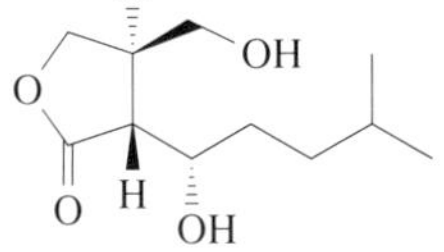

(**6**)

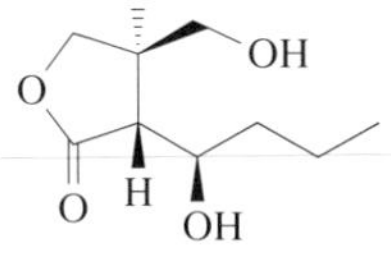

(**7**)

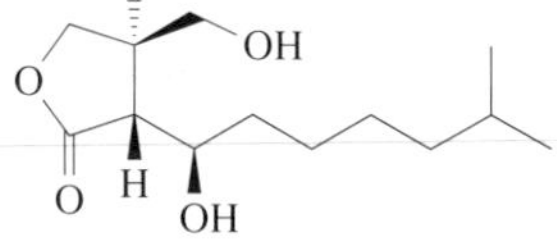

(**8**)

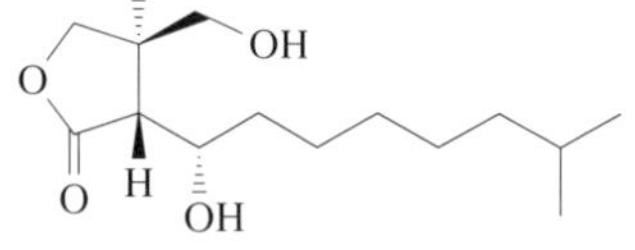

(**9**)

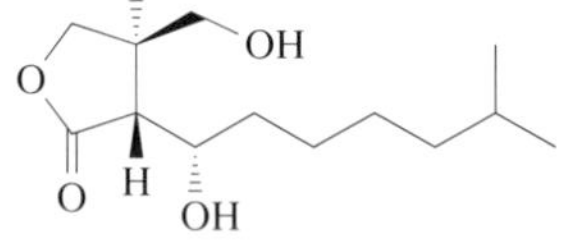

(**10**)

All of these endo- and exogenous signal molecules have common structural features. They possess a 2,3-disubstituted butanolide skeleton but differ in the C-2 side chain containing functional groups, such as 6-hydroxy or 6-keto groups, and in the length or branching of the alkyl chain. All autoregulators which contain a C-6 hydroxyl group have a 2,3-*trans* configuration, but the stereochemistry at C-6 is different between the VB-type and IM-2-type molecules.[8] The absolute configurations of A factor, VB A, VB B, VB C, and IM-2 have been assigned to (**1**), (**2**), (**3**), (**4**), and (**7**), respectively, with their chiral synthesis being accomplished by Mori *et al.*[9,10] and Mizuno *et al.*[11] All of them have the (3*R*) configuration.

This unique butanolide skeleton is known only in metabolites of *Streptomyces*. A bioassay to detect the activity of VB, A factor, or IM-2 was used to investigate the distribution of such molecules in *Streptomyces*. In spite of their slight structural differences, they show very low cross-activity with each other. From the results of the bioassay using culture broths of randomly selected strains it was shown that such molecules are distributed widely in *Streptomyces*, occurring in at least 60%. These molecules are active at extremely low concentrations, and this suggests that specific receptor proteins are involved in the expression of their activity, as in the cases of mammalian hormones.[12–14] Owing to the significant nature of these signal molecules, biosynthetic studies are required not only from the standpoint of a new approach to the understanding of the mechanism of secondary metabolite production in *Streptomyces* but also for application to the production of physiologically useful compounds in *Streptomyces*. However, it has been very difficult to study the biosynthesis of these molecules by the usual feeding experiments with labeled precursors because they are produced only in trace amounts in culture broths. For example, only a few micrograms of VB A (**2**) were obtained from 1 L of *S. virginiae* broth. However, the authors' group found a strain of *Streptomyces antibioticus* which produces a few milligrams of VB A (**2**) per liter of culture broth during the work on the distribution of metabolites having VB activity mentioned above.[15] This finding has made it possible to elucidate the biosynthesis of VB A (**2**) by incorporation experiments with ^{13}C-labeled precursors. In the following section, the biosynthesis of VB A (**2**) by *S. antibioticus* is described.

1.06.2.1 Origin of the Carbon Atoms of Virginiae Butanolide A

Before the incorporation experiments of labeled precursors could take place, culture conditions for *S. antibioticus* for VB A (**2**) production and a convenient isolation procedure for VB A (**2**) from the culture broth had to be investigated. One strain of *S. antibioticus* IFO 12838 was selected as a high producer of VB A (**2**) by single-cell isolation, and was used for the biosynthetic studies on VB A (**2**). *S. antibioticus* constantly produced more than 1 mg of VB A (**2**) per liter under cultivation in a 500 ml Sakaguchi flask containing 100 ml of medium on a reciprocating shaker. Under these culture conditions, VB A (**2**) production started after 24 h of cultivation, and reached a maximum after a total of 96 h of cultivation. The effect of sodium acetate on the yield of VB A (**2**) was investigated before the feeding experiments with ^{13}C-labeled acetate. Since the addition of acetate increased the yield of VB A (**2**) by about twofold, sodium acetate was added twice to the culture, at the 24th and 48th hour of cultivation. After 96 h of cultivation, VB A (**2**) was isolated as its dibenzoate (**2a**) from the culture broth. This conversion of VB A (**2**) to its dibenzoate (**2a**) facilitates the detection of VB A (**2**) by HPLC using a UV detector.

Since the structure of the carbon skeleton of VB A (**2**), especially that of the C-2 side chain, suggested that acetate may be involved in its biosynthesis, incorporation experiments with labeled acetate were carried out first. Before the feeding experiments with [^{13}C]acetate, sodium [2-$^{2}H_3$] acetate was used to establish the experimental conditions for feeding. Based on the results of MS spectral analysis of the ^{2}H-labeled (**2a**) obtained, a mixture of sodium [1-^{13}C]- or [2-^{13}C]acetate and unlabeled sodium acetate was administered twice to the culture. The ^{13}C NMR spectrum of the ^{13}C-labeled (**2a**) obtained showed enrichment at C-1 and C-6, and C-2 and C-7, respectively, indicating that each C_2 unit of C-1/C-2 and C-6/C-7 was derived from an intact acetate molecule.

Next, to clarify the origin of the five carbons atoms C-8 to C-12, sodium [1-^{13}C]isovalerate was administered to the culture. Since the addition of a large amount of sodium isovalerate at one time caused strong growth inhibition, 2.5 mg of labeled isovalerate was added to 100 ml of culture broth eight times at 2 h intervals. The VB A benzoate (**2a**) obtained showed only one enriched peak at C-8 in its ^{13}C NMR spectrum. This indicated that C-8 to C-12 of VB A (**2a**) derive from isovaleric acid.

Finally, in order to clarify the origin of the remaining three carbons, C-3 to C-5, the incorporation of [1,3-$^{13}C_2$]glycerol[16] was investigated. In this case, potato starch was used as the carbon source for the cultivation in place of glycerol to avoid dilution of the ^{13}C-labelled glycerol. Accordingly, the yield of (**2a**) significantly decreased. However, the ^{13}C NMR spectrum of (**2a**) (Figure 1), which was measured using 62 μg of the labeled sample obtained from 500 ml of culture broth, clearly showed enriched peaks at C-4 and C-5 as expected. In this spectrum, the C-2 and C-7 signals were also enriched due to the metabolism of [1,3-$^{13}C_2$]glycerol to [2-^{13}C]acetic acid and reincorporation into (**2**). Unfortunately, under the experimental conditions used for ^{13}C NMR, an expected two-bond coupling between C-4 and C-5 could not be observed because of its small value. Thus, the incorporation of the intact glycerol molecule into (**2**) was not verified by NMR. However, the chemical ionization (CI)–MS spectrum of labeled (**2a**) indicated that the increased ratio of dilabeled molecule calibrated by the MS spectrum was approximately consistent with the percentage incorporation of C-4 or C-5 estimated by NMR. Since the only possible site of incorporation of the intact glycerol molecule is the C-3 to C-5 moiety of the VB A (**2**) molecule, it was concluded that glycerol was incorporated in an intact form. From the results of the incorporation experiments mentioned above, the origin of all the carbon atoms in the VB A (**2a**) molecule was established (Scheme 1).[17]

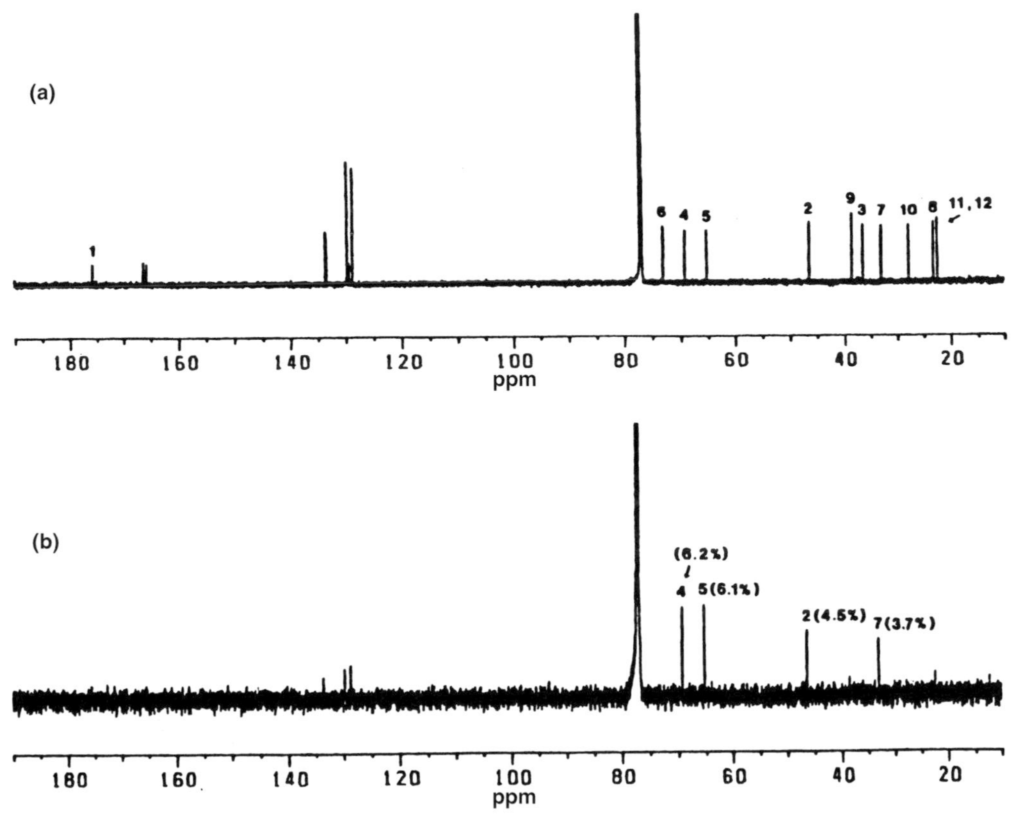

Figure 1 ^{13}C NMR spectra of (**2a**): (a) natural abundance (3.0 mg in 0.6 ml of $CDCl_3$, 1515 scans) and (b) derived from [1,3-$^{13}C_2$]glycerol (0.062 mg in 0.6 ml of $CDCl_3$, 18211 scans).

1.06.2.2 *In Vivo* Studies on the Biosynthetic Pathway of Virginiae Butanolide A

The results obtained in the previous section made it clear that the VB A (**2**) molecule is an assembly of two acetate, one isovalerate, and one glycerol moieties. The most probable biosynthesis of VB A (**2**) is a coupling between a β-keto acid derivative and a dihydroxyacetone-type C_3 unit from glycerol, as shown in Scheme 1. In order to verify this pathway, the incorporation of [2,3-$^{13}C_2$]-3-oxo-7-methyloctanoic acid *N*-acetylcysteamine thioester (**11**) was undertaken. This thioester is a mimic of β-keto acyl CoA,[18–20] which is the plausible key precursor of VB A (**2**). The labeled thioester (**11**) was prepared from the ^{13}C-dilabeled β-keto acid methyl ester (**12**), which was prepared by coupling between methyl [2-^{13}C]bromoacetate and [1-^{13}C]5-methylhexanenitrile using a modified Blaise reaction (Scheme 2).[21,22] Incorporation of the β-keto acid (**13**) itself and its methyl ester (**12**) was unsuccessful, but the thioester (**11**) was successfully incorporated into VB A (**2**). In the ^{13}C

NMR spectrum of the resulting (**2a**) (Figure 2), the enriched signals at C-2 and C-6, which are coupled to each other with a coupling constant of 39.7 Hz, were observed. This fact indicated that the *β*-keto acid moiety of (**11**) was incorporated in an intact form without cleavage of the C—C bond between C-2 and C-3, suggesting that a *β*-keto acid derivative is involved in the biosynthetic pathway of VB A (**2**).

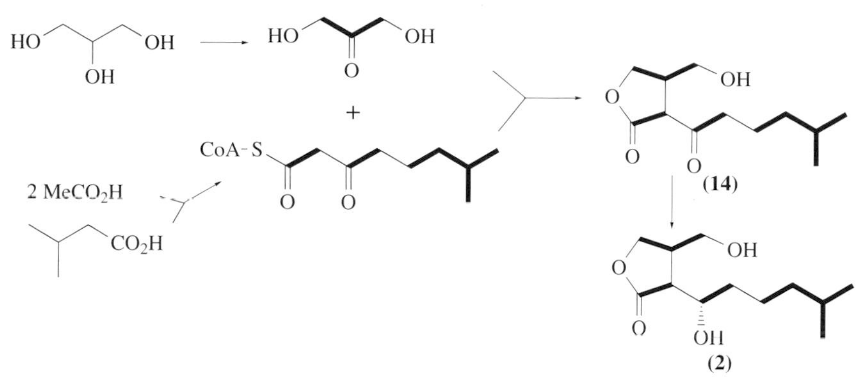

Scheme 1

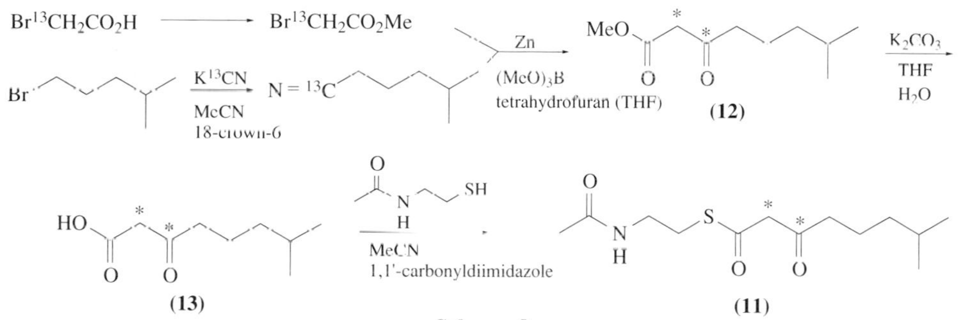

Scheme 2

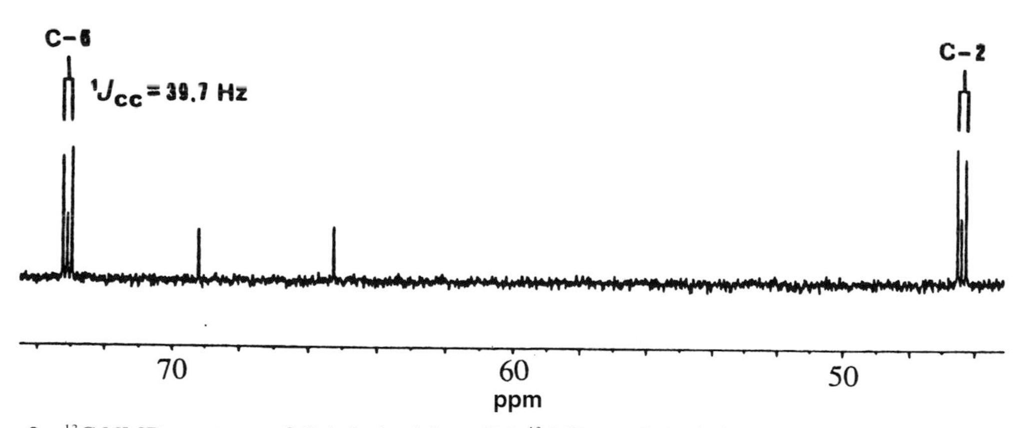

Figure 2 ^{13}C NMR spectrum of (**2a**) derived from [2,3-$^{13}C_2$]3-oxo-7-methyloctanoic acid *N*-acetylcysteamine thioester (**11**) (0.37 mg in 0.6 ml of $CDCl_3$, 20043 scans).

To ascertain the nature of the C_3 unit derived from glycerol in the biosynthetic pathway, an incorporation experiment using [$^2H_5$]glycerol was carried out. Cultivation on a 25 L scale using 250 flasks of 500 ml capacity was necessary to obtain a large enough amount of labeled (**2a**) for the ^2H NMR measurement. The ^2H NMR spectrum of the (**2a**) obtained shows that no ^2H signal was observed on C-3 when it was compared with the ^1H NMR spectrum of (**2a**). This indicated that the ^2H on C-2 of the glycerol molecule was lost during the incorporation of [$^2H_5$]glycerol, suggesting that a C_3 unit having a C-2 oxo group, such as dihydroxyacetone or a derivative, may be the precursor. Finally, in order to verify the reduction step from 6-dehydro-VB A (**14**) to (**2**), incorporation of [4,5-$^2H_3$]-6-dehydro-VB A, which was synthesized in a racemic form, was performed. By analysis of the CI-MS and c.d. spectra of the labeled (**2a**) obtained, it was revealed that 6-dehydro-VB A (**14**) was converted to (**2**) stereospecifically.

These *in vivo* incorporation experiments with a labeled β-keto acid derivative, glycerol, and 6-dehydro-VB A (**14**) afforded proof of the biosynthetic pathway of VB A (**2**), as shown in Scheme 1. Based on this pathway, a plausible reaction mechanism for the formation of the VB A (**2**) skeleton, was proposed (Scheme 3).[23] In this route, glycerol is oxidized to a dihydroxyacetone or a derivative, which is then acylated by β-keto acyl CoA via an acyltransferase to form the β-keto ester. The ketone group of the dihydroxyacetone moiety of the β-keto ester then undergoes an intramolecular aldol condensation on the C-2 methylene of the β-keto acid moiety to afford the butyrolactone skeleton. Dehydration and reduction leads to the C-6 oxo skeleton. Finally, the C-6 oxo group is reduced to a hydroxyl group by an alcohol dehydrogenase to give the VB A (**2**) molecule. In this pathway, if it is assumed that the acylation step to form the ester resembles that in glycerolipid biosynthesis,[24] phosphate groups of hypothetical intermediates may be present as shown in Scheme 3.

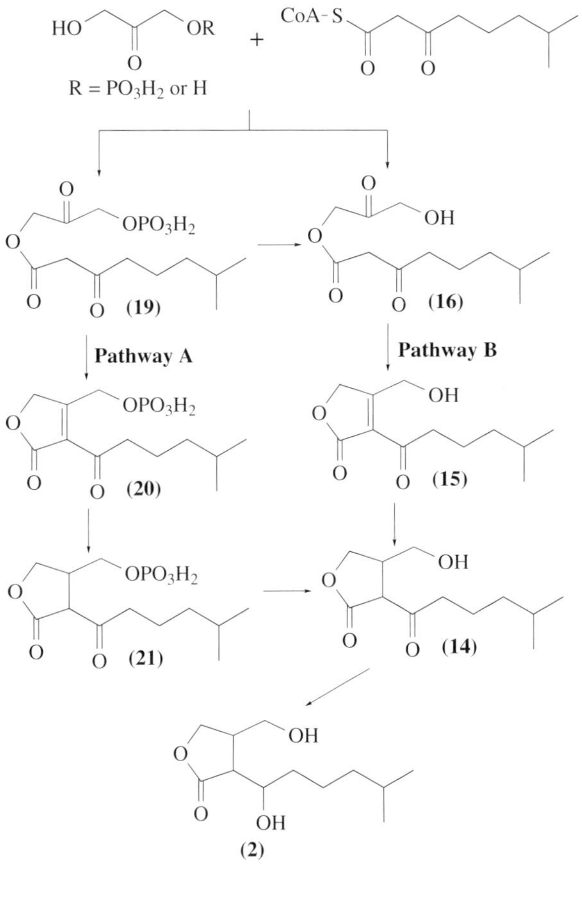

Scheme 3

1.06.2.3 *In Vitro* Studies on the Biosynthetic Pathway of Virginiae Butanolide A

To investigate the biosynthetic pathway of VB A (**2**) in an *in vitro* system, possible biosynthetic intermediates were synthesized, and conversion experiments with them in a cell-free system were carried out. All compounds numbered in Scheme 3, except for compound (**15**), could be prepared. Because of the instability of the β-keto ester (**16**), a rapid preparation method was developed to obtain it, as shown in Scheme 4. The trityl ether of dihydroxyacetone was coupled with the β-keto acid to prepare the ester (**17**).[25] The pure ester (**17**) was obtained by purification with reverse-phase HPLC. In chromatography on a silica gel column, only the butenolide (**18**) was obtained. This indicates that cyclization by intramolecular aldol condensation and subsequent dehydration occurred during the chromatography. Deprotection of the trityl group afforded the ester (**16**). The ester (**16**) formed was not recovered from a silica gel column, as in the case of (**17**). Moreover, (**16**) could not be purified by reverse-phase HPLC because of the instability of (**16**) in aqueous solution (half-life in 0.1 M phosphate buffer pH 7.0 at 28 °C is ~15 min). Only rapid treatment of the reaction mixture on a Sep-Pak silica cartridge afforded pure (**16**). To prepare the butenolide (**15**), cleavage of the trityl group of (**18**) was also attempted, but this reaction gave no significant product, indicating that the butenolide (**15**) was too labile to be isolated. The phosphate of the β-keto ester (**16**), (**19**), the butenolide phosphate (**20**), 6-dehydro-VB A phosphate (**21**), and 6-dehydro-VB A (**14**) were prepared according to conventional methods.[26,27] The butenolide phosphate (**20**) was also unstable, but could be prepared from (**19**).

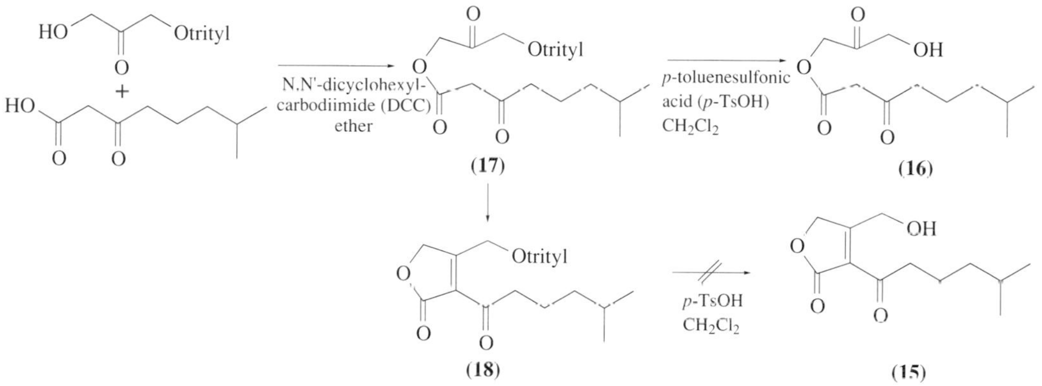

HO Otrityl O + HO O O
N,N'-dicyclohexylcarbodiimide (DCC) ether
O Otrityl O O O (**17**)
p-toluenesulfonic acid (*p*-TsOH) CH_2Cl_2
O OH O O O (**16**)
O Otrityl O O O (**18**)
p-TsOH CH_2Cl_2
OH O O O (**15**)

Scheme 4

Using the compounds synthesized above, conversion experiments were then performed. A cell-free system for the experiments was constructed with the mycelial extract of *S. antibioticus*. Since cells cultivated for 28 h showed the highest 6-dehydro-VB A (**15**) to VB A (**2**) transformation activity, cells were collected by filtration at that time and disrupted by sonication to give a cell-free system. After the conversion experiments with the cell-free system, the VB A (**2**) formed was converted into its dibenzoate (**2a**) and quantified by analysis using reverse-phase HPLC, in which racemic VB D (**5**) was used as an internal standard.

Conversion experiments with 6-dehydro-VB A (**14**) were carried out first. 6-Dehydro-VB A (**14**) has been proved to be a biosynthetic precursor of (**2**) by the incorporation experiment with ^{2}H labeled (**14**) in an *in vivo* system, described above. In the cell-free system for this conversion experiment, enzyme solution, which was roughly purified by ammonium sulfate precipitation, was used. From the results of the conversion experiments with racemic 6-dehydro-VB A (**14**), it was shown that (**14**) was effectively converted into VB A (**2**) in the presence of NADPH or NADH. Because (**14**) was more effectively converted into VB A (**2**) in the presence of NADPH than NADH, it seemed that NADPH was mainly used in the reduction step of the 6-keto group of (**14**). As the optical rotation value of (**2a**) obtained by this conversion was about the same as that of (**2a**) derived from natural VB A (**2**), it was suggested that this reduction step occurred stereospecifically, which had been suggested by the *in vivo* experiment mentioned previously. To confirm this in an *in vitro* system, conversion experiments with optically active (3*R*)- or (3*S*)-(**14**), which were prepared by enzymatic resolution of racemic (**14**) with lipase,[11] were performed. When (3*R*)-(**14**) (97.2% *ee*), which has the same configuration as that of natural VB A (**2**), was used in the experiment, the amount of VB A (**2**) formed became about two times higher than that formed from racemic (**14**) mentioned previously. On the other hand, in the case of the conversion experiment with (3*S*)-(**14**) (94.4% *ee*), the amount of VB A (**2**) formed was much reduced, to about 10% of that from racemic (**14**). These results strongly suggested that the enzyme involved in the reduction step has a high stereospecificity. Further experiments using a purified enzyme are necessary to prove this unequivocally.

Next, transformation experiments with the ester (**16**) were performed. Since the ester (**16**) was unstable in aqueous solution mentioned previously, it was added to the cell-free system five times at 30 min intervals. The amount of VB A (**2**) formed by the conversion of (**16**) was much smaller than that from 6-dehydro-VB A. However, the conversion from ester (**16**) to VB A (**2**) was clearly observed when both NADPH and NADH were added to the reaction solution. Because the addition of NADPH to the solution only (which was needed to convert (**14**) to VB A (**2**) mentioned previously) was less effective for the conversion of (**16**) to VB A (**2**), NADH was preferentially used in the reduction step of the butenolide (**15**), which was a putative intermediate during the biosynthetic process from ester (**16**) to 6-dehydro-VB A (**14**).

To investigate the mechanism of the conversion process from ester (**16**) to VB A (**2**) with a cell-free system, a conversion experiment with ester (**16**) in D_2O solution was carried out. The cells of *S. antibioticus* were disrupted in a buffer prepared with D_2O. Ester (**16**) and 6-dehydro-VB A (**15**) were both converted to (**2**) in the cell-free system in the presence of NADPH and NADH. Figure 3 shows the ^1H NMR spectra of the resulting (**2a**). In the spectra, a reduced area of each ^1H signal, which is shown as a percentage in Figure 3, indicates the rate of deuterium incorporation in each signal during the conversion process. It was observed that deuterium was extensively incorporated at C-2 in both of the conversion experiments due to the enolization of the β-keto system. Incorporation of deuterium at C-6 suggested that NADPD (NADD) was produced in the reaction solution. In the conversion experiment with ester (**16**), a high incorporation of deuterium was observed at C-4. The observed area of both 4-H_R and 4-H_S signals became less than half of the natural value, but the rate of deuterium incorporation on 4-H_R was a little smaller than that of 4-H_S. This result suggested that the hydrogen atoms on C-4 were being rapidly exchanged for deuterium in the solution during the conversion process, and at that time 4-H_S was preferentially lost compared with 4-H_R. Low incorporation of deuterium was observed at C-3 in the case of ester (**16**), suggesting that the hydrogen on C-3 came from NADH in the reduction process.

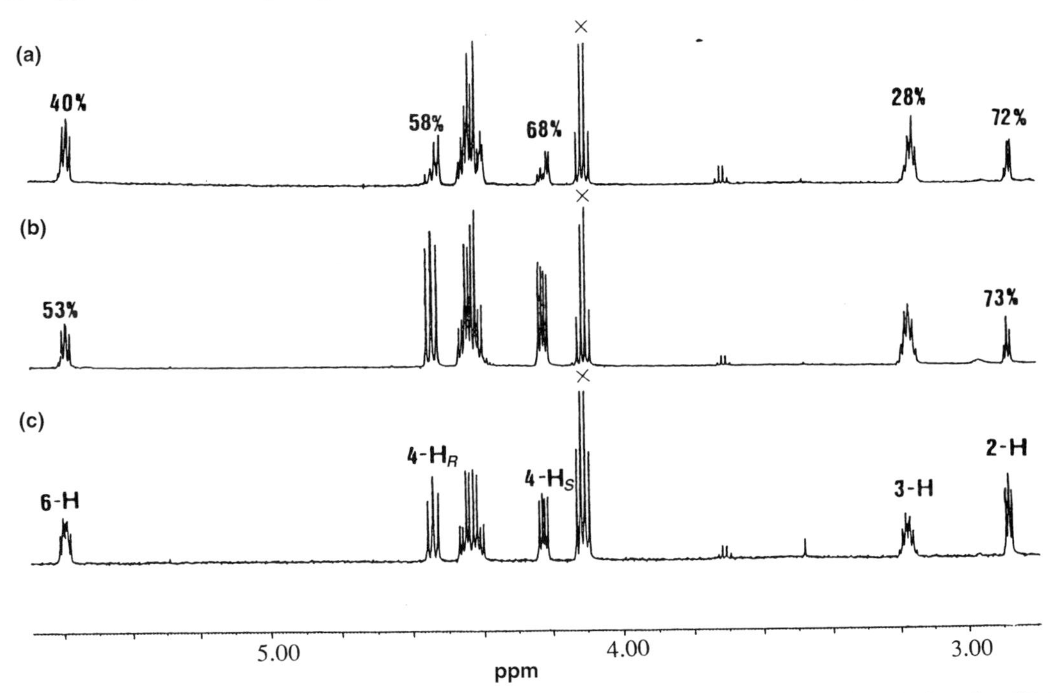

Figure 3 ^1H NMR spectra of (**2a**): (a) converted from the β-keto ester (**16**) in D_2O solution (in $CDCl_3$), (b) converted from 6-dehydro-VB A (**14**) in D_2O solution (in $CDCl_3$), and (c) natural abundance (in $CDCl_3$). The percentages indicate the rate of deuterium incorporation on each carbon position. x, contamination by ethyl acetate.

Significant conversion of the butanolide phosphate (**21**) into VB A (**2**) was observed. However, the amount of VB A (**2**) formed by the conversion was much smaller than that from 6-dehydro-VB A (**14**). This indicated that 6-dehydro-VB A (**14**) was produced in the conversion experiment with (**21**) by the action of a nonspecific or specific phosphatase. Transformation of the ester phosphate (**19**) to VB A (**2**) was less effective than that of ester (**16**) to (**2**), and conversion of the butenolide phosphate (**20**) to (**2**) was not detected under the tested conditions. These results may indicate the ester phosphate (**19**) was first converted into the ester (**16**) by a phosphatase to produce (**2**). However, to confirm whether pathway A in Scheme 3 is present or not, a conversion experiment investigating the conversion of (**19**) to (**20**) with a highly purified enzyme is necessary.

From the results of the conversion experiments with possible biosynthetic intermediates of VB A (**2**) in a cell-free system mentioned previously, it was proved that biosynthetic pathway B from the ester (**16**) to VB A (**2**) in Scheme 3 is unambiguously present in *S. antibioticus*, and a more detailed possible mechanism for the formation of the VB A (**2**) skeleton from ester (**16**) has been proposed (Scheme 5).[27] After intramolecular aldol condensation, dehydration occurred, to give the unsaturated butanolide intermediate as a keto–enol tautomeric mixture. The subsequent reduction with NADH of the intermediate led to the 6-oxo skeleton. Finally, the 6-oxo group was reduced to a hydroxyl group with a NADPH-dependent dehydrogenase to give the VB A (**2**) molecule.

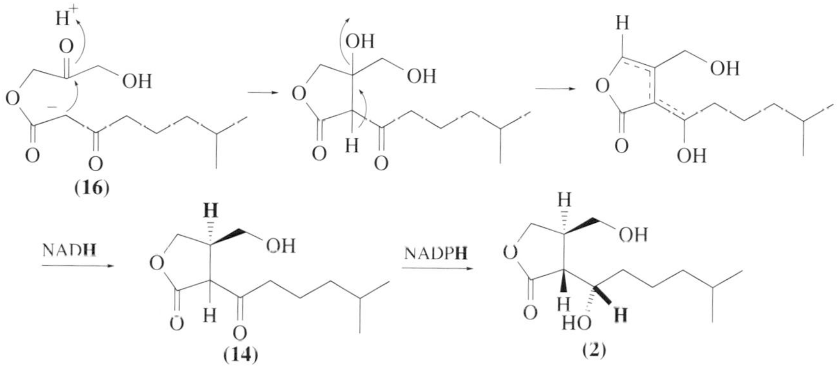

Scheme 5

1.06.2.4 Biosynthesis of Compounds Related to Virginiae Butanolide A

The biosynthetic pathway of VB A (**2**) in *S. antibioticus* has been established, as shown in Scheme 3 described in the previous sections. It is believed that the pathway is common to all species of *Streptomyces* for the biosynthesis of butyrolactone autoregulators. In the pathway, β-keto acyl CoA is a key precursor, and couples with a C_3 unit from a glycerol molecule. In the case of VB A (**2**) biosynthesis, the β-keto acyl CoA on VB A (**2**) is thought to be synthesized from isovaleryl CoA as a starter molecule and two malonyl CoAs derived from two acetate molecules, as in polyketide biosynthesis. Thus, the variety of starter molecules and the number of malonyl CoA molecules may determine the length and branching of the C-2 side chains among the autoregulators. For example, in A factor (**1**) biosynthesis, isobutyryl CoA would be the starter and three malonyl CoAs would be involved, and acetyl CoA as the starter and two malonyl CoAs would be involved in IM-2 (**7**) biosynthesis, as shown in Scheme 6.

Since the acylation of a glycerol derivative such as dihydroxyacetone 3-phosphate or glycerol 3-phosphate is well known in the biosynthesis of glycerolipids,[24] formation of a β-keto ester between a β-keto acid derivative and a dihydroxyacetone-type C_3 unit is not an unusual process. The cyclization of the β-keto ester to form a butenolide skeleton occurs easily, and this has been confirmed by the results of the synthetic studies mentioned above. Subsequent reduction can produce an A factor-type molecule having a C-6 oxo skeleton. By further reduction of the C-6 oxo skeleton, a VB- or IM-2-type molecule having a C-6 hydroxyl group is produced. Reduction enzymes of the C-6 keto group are probably classified into two groups, depending on the orientation of the hydroxyl group produced on C-6. A gene, named *afsA*, has been identified by Horinouchi *et al.*, with the biosynthesis of A factor in *S. grieus*,[28] but it is still unknown which reaction in the biosynthetic pathway is catalyzed by the *afsA* protein.

Other butyrolactone derivatives, named NFX-2 (**22**) and NFX-4 (**23**), were isolated from the culture broth of *S. antibioticus*, and their absolute configurations were determined by their chiral synthesis.[29,30] They have weak virginiamycin-inducing activities against *S. virginiae*. If we assume that the biosynthetic pathway of NFX resembles that of VB A (**2**), the pathway shown in Scheme 7 may be possible. In this pathway, a β-keto acid derivative might also be the key intermediate, and a C—C bond may be formed between C-1 of a glyceraldehyde-type C_3 unit and the C-2 methylene of β-keto acid. The biosynthetic pathway of the compound (**24**),[31] which was isolated from a marine actinomycete and has a similar butanolide skeleton to NFX, might be basically the same as NFX,

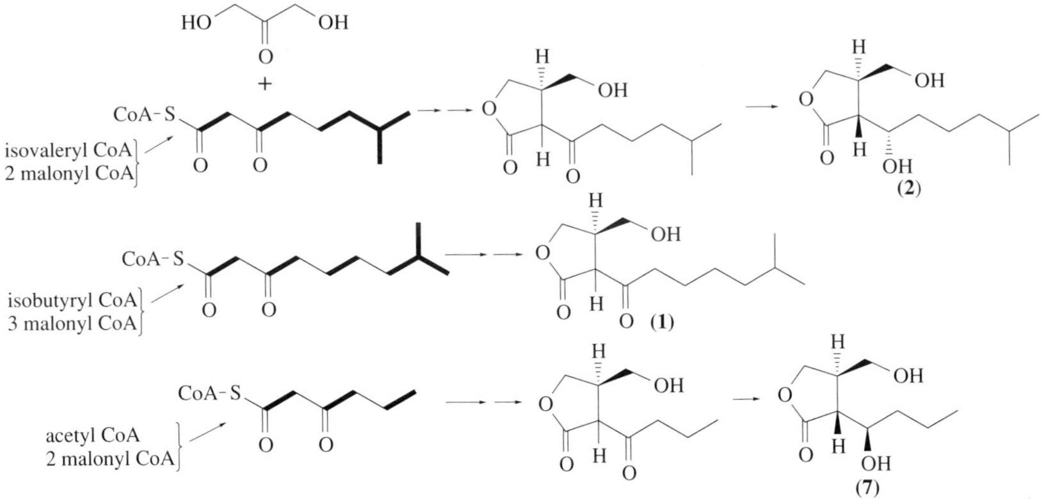

Scheme 6

as shown in Scheme 7. Sylingolide (**25**), an elicitor of the bacterial plant pathogen *Pseudomonas syringae* pv. tomato,[32–34] is thought to be biosynthesized by a similar pathway to VB A (**2**), in which the β-keto acyl xylulose ester (**26**) is a putative key intermediate like the ester (**16**) in the biosynthesis of VB A (**2**), as shown in Scheme 8.

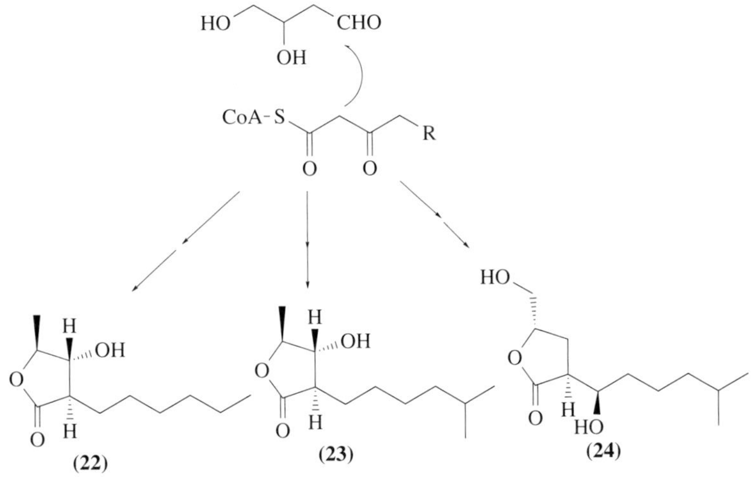

Scheme 7

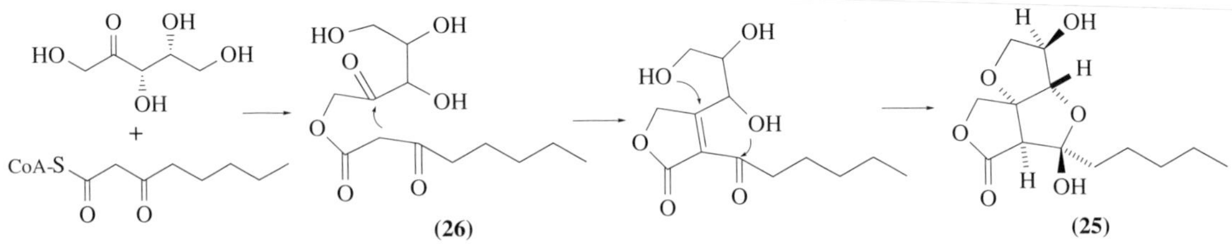

Scheme 8

1.06.3 BIOSYNTHESIS OF THE CHITINASE INHIBITOR ALLOSAMIDIN

In chitin-containing organisms, such as insects or fungi, the turnover of chitin plays an important role in their growth.[35,36] Among the chitin metabolite enzymes, chitin synthase is essential for the processes of insect ecdysis or fungal cell wall formation, as has been shown by experiments using its specific inhibitors, such as polyoxins[37] or nikkomycins.[38] However, chitinase activity has also been detected during the growth of organisms. The role of chitinase in insects and fungi has not been clarified, mainly owing to lack of a specific inhibitor of chitinase. Chitinase inhibitors are also of interest as possible candidates as a new type of insect growth regulator or fungicide. Chitinase inhibitors are therefore being searched for, and a novel compound, allosamidin (**27**), has been isolated from the metabolites of *Streptomyces* sp. as the first chitinase inhibitor.[39]

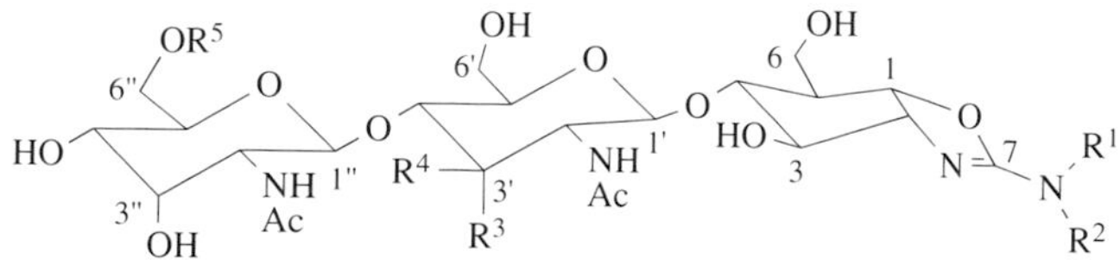

		R^1	R^2	R^3	R^4	R^5
Allosamidin	(**27**)	Me	Me	OH	H	H
Demethylallosamidin	(**29**)	Me	H	OH	H	H
Didemethylallosamidin	(**30**)	H	H	OH	H	H
Methylallosamidin	(**31**)	Me	Me	OH	H	Me
Methyl-*N*-demethylallosamidin	(**32**)	Me	H	OH	H	Me
Glucoallosamidin A	(**33**)	Me	Me	H	OH	Me
Glucoallosamidin B	(**34**)	Me	H	H	OH	Me

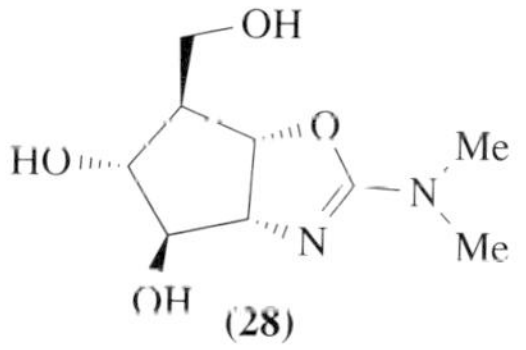

(**28**)

Allosamidin has a unique pseudotrisaccharide structure consisting of two units of *N*-acetyl-D-allosamine and one unit of a novel aminocyclitol derivative, allosamizoline (**28**).[40–42] Allosamizoline (**28**) has a cyclopentanoid structure, which is highly oxygenated and fused with a dimethyl-aminooxazoline ring. After discovering allosamidin (**27**), six allosamidin derivatives (**29**)–(**34**) were isolated as natural products from *Streptomyces* sp.[43–45] They have a similar structure, but differ in the number of *N*-methyl groups, the stereochemistry of the hydroxyl group or the presence of *O*-methyl group. The structure of allosamidin (**27**) is a mimic of chitin, and its allosamizoline (**28**) moiety interacts with the active center of a chitinase molecule, which has been shown by X-ray analysis of the complex between chitinase and allosamidin.[46] Since a structure having a six-membered ring is common to pseudosaccharide moieties contained in known glycosidase inhibitors, the cyclo-pentane ring of allosamizoline is the first case of a five-membered ring which interacts with the active center of an enzyme. Many synthetic studies on allosamidins,[47–52] allosamizoline,[53–57] and their analogues[58–60] have been reported.

Allosamidin shows interesting biological activities against chitin-containing organisms. For example, it inhibits ecdysis of lepidopteran insects,[39] which first demonstrated that insect chitinase has an essential role during ecdysis stage,[61] and its inhibitors are possible insect growth regulators. When demethylallosamidin (**29**), a specific inhibitor of yeast chitinase, was added to the culture of *Saccharomyces cerevisiae*, cell groups in a clustered form were observed.[62] This phenomenon showed the role of chitinase in cell separation[63] for the first time. A similar abnormal morphology was observed in the growth of *Candida albicans* and *Geotricum candidum*[64] by addition of demethyl-allosamidin (**29**). It is known that all chitinases grouped in family 18 from various sources, such as insects, fungi, bacteria, parasites, humans or shrimps, are inhibited by allosamidin (**27**).[65–78] Allosamidins having an *N*-monomethyl group show much stronger activities against yeast chitinases than those having an *N*-dimethyl group.[44] A series of synthetically prepared *N*-monoalkyl derivatives of allosamidin, such as *N*-monoethyl or *N*-monopropyl allosamidin, showed much weaker activities compared with allosamidin.[79]

Because of the novel structure and biological activities, allosamidin (**27**) or allosamizoline (**28**) is expected to be a lead compound in the design of new bioactive compounds. The unique structure of each of the two components in the allosamidin molecule also prompted us to study its biosynthetic pathway. *N*-Acetyl-D-allosamine is a C-3 epimer of *N*-acetyl-D-glucosamine hitherto unknown in nature. Allosamizoline (**28**) contains a cyclopentane ring in its molecule, and as a cyclopentanoid structure of carbohydrate origin is relatively rare in natural products the mechanism of its formation has not been well studied compared with that of compounds containing cyclohexane rings. The biosynthetic origin of the dimethylaminooxazoline moiety is also interesting. In the following sections, our studies on the biosynthesis of allosamidin (**27**) and the mechanism of the cyclopentane ring formation of allosamizoline (**28**) are described.

1.06.3.1 Origin of the Carbon and Nitrogen Atoms of Allosamidin

To investigate the origin of the carbon and nitrogen atoms of allosamidin (**27**), incorporation experiments with a variety of labeled precursors were performed. *Streptomyces* sp. AJ 9463 was selected as a high producer of allosamidin (**27**), and used throughout this biosynthetic work. After cultivation, allosamidin (**27**) was isolated from the aqueous methanol extract of mycelia. Since the carbohydrate pool may be the biosynthetic origin for each moiety of D-allosamine and the cyclopentane ring of allosamizoline (**28**), feeding experiments with labeled glucose were carried out first. To avoid a high level of dilution of the labeled glucose with the nonlabeled form contained in the medium, the usual medium was exchanged with a replacement medium containing little glucose when the labeled glucose was aded to the culture. Since a high level of incorporation was observed in the experiment with [^{14}C]glucose, [1-^{13}C]- or [6-^{13}C] D-glucose was added to the culture under the replacement conditions. The ^{13}C NMR spectrum of ^{13}C-labeled allosamidin (**27**) showed enrichment at C-1, C-1′, and C-1″, and C-6, C-6′, and C-6″, indicating that the carbon skeleton of D-allosamine and the cyclopentane ring of allosamizoline were derived from D-glucose.

Next, labeled glucosamine was evaluated as a precursor, since the nitrogen atom on C-2 of allosamine or allosamizoline (**28**) strongly suggested that glucose may be incorporated via glucosamine. In the feeding experiment with [1-^{14}C]-glucosamine, a high degree of incorporation into each moiety of allosamine and allosamizoline (**28**) was observed without the use of replacement conditions. In order to verify the incorporation of the nitrogen of D-glucosamine into each nitrogen atom on C-2, C-2′, and C-2″ of allosamidin, a feeding experiment with doubly labeled D-[1-^{13}C, 2-^{15}N]glucosamine[80] was undertaken. The ^{13}C NMR spectrum of the resulting sample showed enriched peaks at C-1, C-1′, and C-1″, but an expected clear two-bond coupling with ^{15}N on C-2, C-2′, or C-2″ could not be observed at any of the enriched carbon signals, due to its small value.[81] Further analysis of the CI–MS spectra of labeled allosaminitol peracetate (the corresponding alditol acetate of D-allosamine) and the triacetate of allosamizoline, which were derivatized from the labeled allosamidin (**27**), indicated that, in both cases, the increased ratios of monolabeled molecules were negligible and those of dilabeled molecules were approximately consistent with the increased ^{13}C percentage estimated by NMR spectroscopy for C-1′ or C-2″, and C-1. These facts demonstrated that the doubly labeled glucosamine was incorporated into each moiety of D-allosamine and allosamizoline (**28**) without cleavage of the ^{13}C—C—^{15}N bonds. Considering the results obtained with the labeled glucose mentioned above, it was concluded that the carbon skeleton and nitrogen atom of D-glucosamine were incorporated into each moiety of D-allosamine and allosamizoline (**28**) in an intact form.

The biosynthetic origin of the dimethylaminooxazoline moiety of allosamizoline (**28**) was next elucidated. Since the skeleton of the moiety partly resembles that of the methylaminooxazolinone moiety of indolmycin (**35**), in which the C-2 of the ring is derived from C-2 highlighted in the guanidino carbon of arginine (Equation (1)),[82] labeled arginine was evaluated as a precursor. A high level of incorporation of L-[guanidino-^{14}C]arginine into allosamizoline (**28**) was observed in a feeding experiment. Then, multiply labeled L-[guanidino-^{13}C,$^{15}N_2$]arginine[83] was fed to the culture in order to confirm the carbon position incorporated and investigate the origin of the nitrogen atom of the dimethylamino group. In the ^{13}C NMR spectrum of the labeled allosamidin (**27**) obtained, the enriched C-7 showed doublet signals. This indicated that the ^{13}C and one of the ^{15}N atoms of the guanidino group of labeled arginine were incorporated together into allosamizoline (**28**) without cleavage of the bond. The position of the labeled nitrogen should be the dimethylamino group because it had already become clear that the nitrogen atom on C-2 originated from glucosamine, as mentioned above. The origin of the two *N*-methyl carbons was confirmed as methionine by a feeding experiment with [methyl-^{13}C]methionine. The basic building blocks of allosamidin obtained above are summarized in Scheme 9.[84]

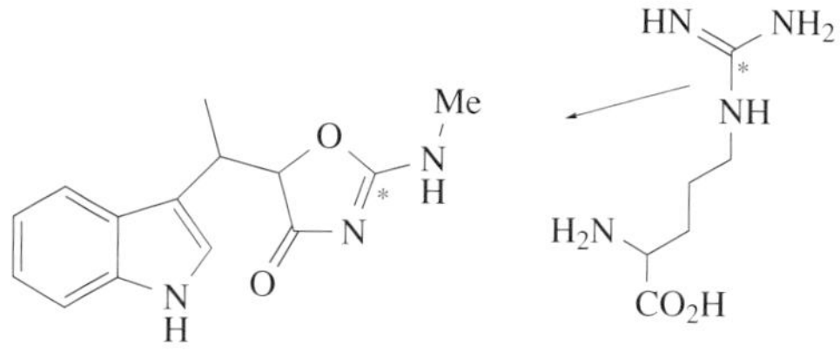

(35) L-arginine (1)

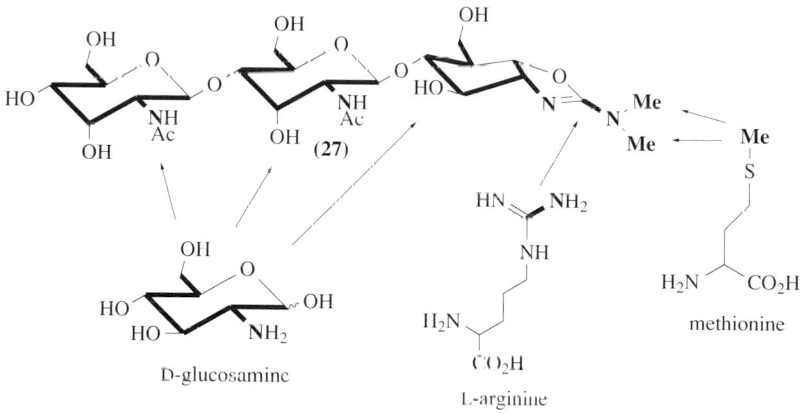

(27) D-glucosamine L-arginine methionine

Scheme 9

1.06.3.2 Conversion Experiments with Labeled Allosamidin Analogues

Streptomyces sp. AJ 9463 produces three allosamidins, allosamidin (**27**), demethylallosamidin (**29**), and didemethylallosamidin (**30**), each having a different number of *N*-methyl groups. The *N*-methylation steps involved in the biosynthesis of (**27**) are very important because the number of methyl groups on the aminooxazoline moiety strongly affects the biological activities of allosamidins, as mentioned above. Since both (**29**) and (**30**) are likely intermediates in the *N*-methylation steps, conversion experiments with labeled (**27**), (**29**), and (**30**) were attempted to verify the biosynthetic steps.

As the quaternary carbon atom of the aminooxazoline moiety of (**27**) originated from the guanidino group of arginine (Scheme 9), [7-^{14}C]-(**27**) and -(**29**) were prepared by feeding experiments with [guanidino-^{14}C]arginine. Because (**30**) is produced only in a trace amount in the culture broth, a sufficient quantity of labeled (**30**) was prepared synthetically according to the reactions in Scheme 10. It is known that the reaction of (**27**) or (**29**) with ammonia water affords (**30**) together with (**36**) as a by-product at high temperatures in a sealed tube.[79] Therefore, reaction of [7-^{14}C]-(**27**) or -(**29**) with ammonia water afforded [7-^{14}C]-(**30**). By this reaction, [7-^{14}C]-(**36**) and, -(**37**) and -(**38**) were also prepared from [7-^{14}C]-(**27**) and -(**29**), respectively.

Conversion experiments on these ^{14}C-labeled compounds were performed using an *in vivo* system. In all cases with compounds (**27**), (**29**), (**30**), (**36**), (**37**), and (**38**), much radioactive isotope was incorporated into the cells, which was very important because allosamidins are present in the mycelia. After cultivation, the fraction containing all allosamidins and compounds (**36**), (**37**), and

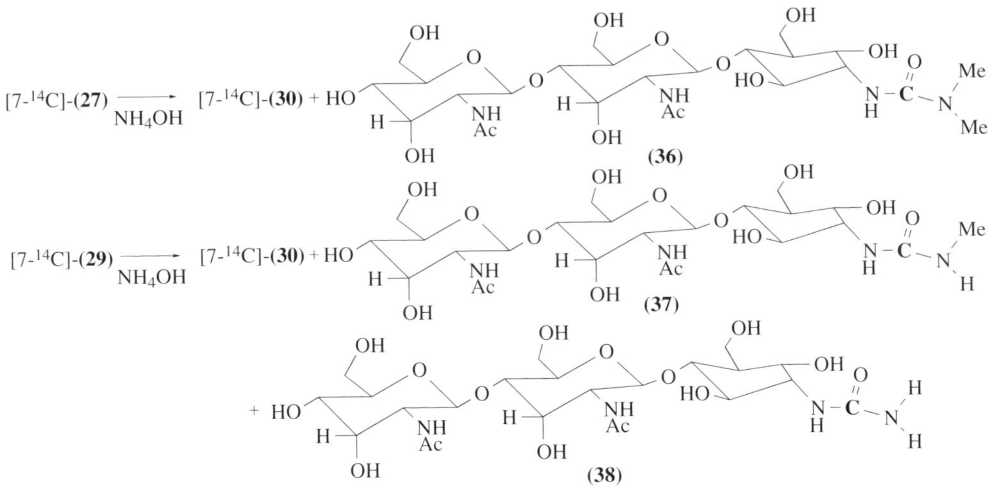

$[7\text{-}^{14}C]$-(**27**) $\xrightarrow{NH_4OH}$ $[7\text{-}^{14}C]$-(**30**) + (**36**)

$[7\text{-}^{14}C]$-(**29**) $\xrightarrow{NH_4OH}$ $[7\text{-}^{14}C]$-(**30**) + (**37**) + (**38**)

Scheme 10

(**38**) was analyzed by HPLC apparatus equiped with UV and radio isotope (RI) detectors. When $[7\text{-}^{14}C]$-(**29**) was fed to the culture, 75% of the radioactivity in the cells was associated with the peak of (**27**), indicating that (**29**) was efficiently converted to (**27**). On the other hand, in the case of $[7\text{-}^{14}C]$-(**27**), more than 90% of the radioactivity in the cells was retained in the peak of (**27**). Unexpectedly, conversion of $[7\text{-}^{14}C]$-(**30**) to (**27**) or (**29**) was not observed at all. These results strongly indicated that (**29**) was a biosynthetic intermediate of (**27**), but (**30**) was not.

Since a compound having a ureido group at C-2 was presumed to be a candidate for a precursor just before the formation of the oxazoline ring, conversion experiments with $[7\text{-}^{14}C]$-(**36**), -(**37**), and -(**38**) were also done. But none of these compounds were converted to (**27**), (**29**), or (**30**) in spite of the high incorporation of radioactivity into the cells being detected. However, it was found that the amount of (**27**) produced in the culture was reduced by the addition of (**36**) by 71% and 93% at a concentration of 10 $\mu g\ ml^{-1}$ and 100 $\mu g\ ml^{-1}$ of (**36**), respectively. This inhibitory activity of (**36**) suggested that a compound structurally similar to (**36**) might be present as a biosynthetic precursor of (**27**).

The results of the conversion experiments mentioned previously are summarized in Scheme 11.[45] The second *N*-methylation step from (**29**) to (**27**) was proved to be the final step of the biosynthesis of (**27**). Since (**30**) was not converted to (**27**) or (**29**), the first *N*-methyl group of (**27**) may be introduced before the cyclization to the aminooxazoline ring. For example, it is assumed that an intermediate having a guanidino group at C-2 could accept a methyl group and then cyclize, leading to (**29**). In the case of the biosynthesis of (**30**), an oxazoline ring would be formed before the methylation.

There is no information about the assembly steps of the two allosamine moieties and an allosamizoline moiety at present. In a trial, it was examined whether allosamine or allosamizoline (**28**) could be a possible precursor by means of a feeding experiment with $[3\text{-}^{2}H]$allosamine or $[7\text{-}^{14}C]$allosamizoline. However, no incorporation into (**27**) was observed in either case, indicating that allosamine or allosamizoline (**28**) cannot be used as a precursor in the biosynthesis of (**27**).

1.06.3.3 Mechanism of the Cyclopentane Ring Formation of Allosamizoline

It is a most interesting point in the biosynthesis of (**27**) how the cyclopentane ring of allosamizoline (**28**) is formed. A cyclopentanoid skeleton biosynthesized from carbohydrate is uncommon in natural products compared with a cyclohexanoid skeleton such as inositol, shikimic acid, or their derivatives. Besides allosamidins, pactamycin (**39**),[85] bacteriohopane (**40**),[86] aristeromycin (**41**),[87] and trehazoline (**42**)[88] are known to have a cyclopentanoid skeleton to which a hydroxymethyl group is attached, (the numbers 1 and 6 in (**39**)–(**42**) indicate the positions originating from C-1 and C-6 of glucose, respectively). The cyclopentanoid ring of pactamycin, bacteriohopane, and

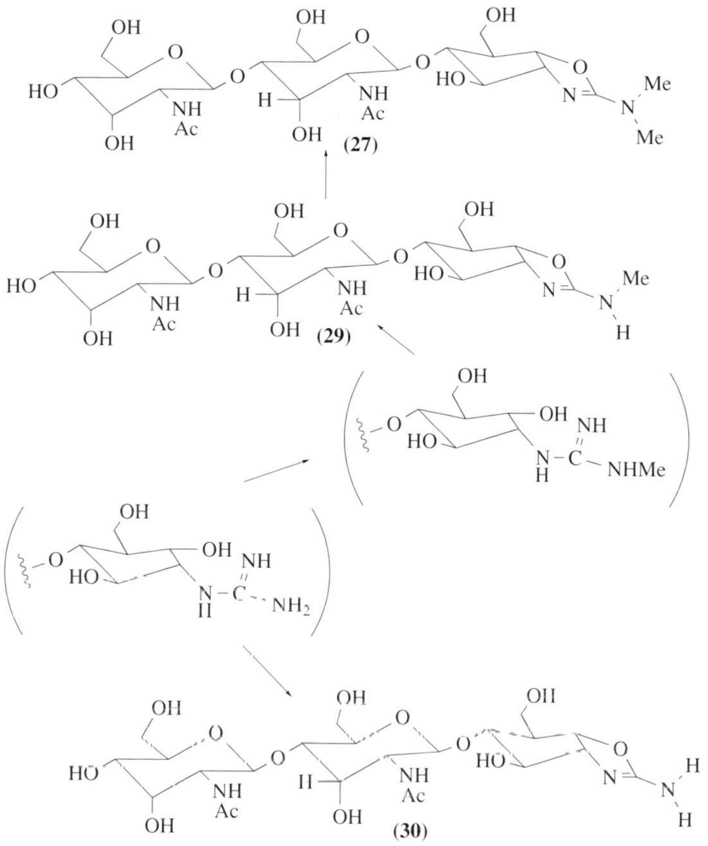

(27)

(29)

(30)

Scheme 11

aristeromycin has proved to be biosynthesized from glucose. The cyclopentane ring of allosamizoline (**28**) is also formed from glucose, but it is the first known example of a natural product in which glucosamine is its close precursor. In the biosynthesis of pactamycin and bacteriohopane, the hydroxymethyl carbon originates from C-6 of glucose, and C—C bond formation occurs between C-5 and C-1 of glucose. Since the hydroxymethyl carbon of allosamizoline (**28**) originates from C-6 of glucose as mentioned above, the cyclopentane ring of allosamizoline (**28**) can be classified into this type. On the other hand, the C-1 carbon of glucose is the origin of the hydroxymethyl carbon in the case of aristeromycin, in which a C—C bond formation occurs between C-2 and C-6 of glucose. In the latter case, the mechanism of the cyclopentane ring formation has been studied by feeding experiments with specifically ^{3}H-labeled glucose.[87] However, in the former case the mechanism has not been studied yet.

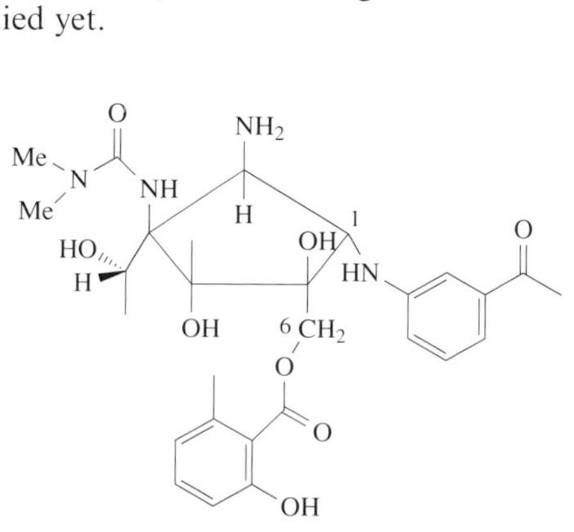

(39)

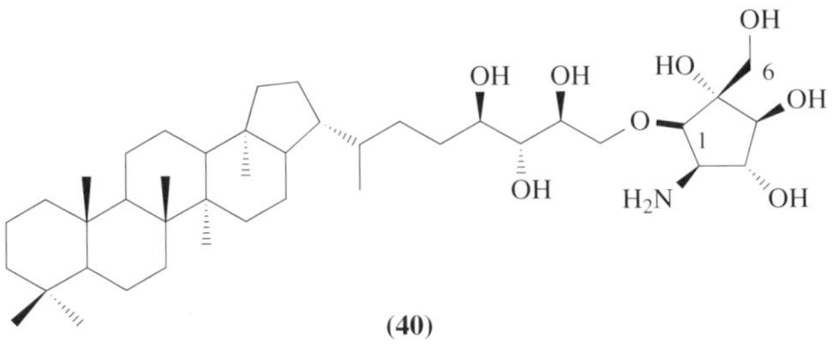

(40)

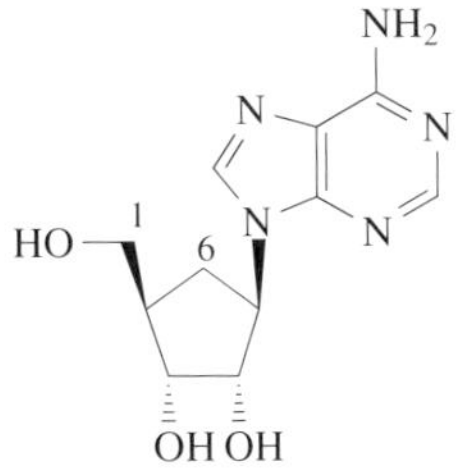

(41) 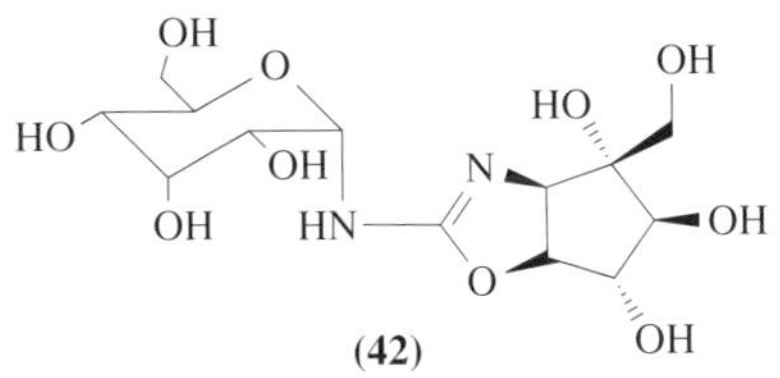 **(42)**

The mechanisms of formation of cyclohexane rings observed in the biosynthetic pathways of inositol or shikimic acid are well known.[89] By analogy to the mechanism of the biosyntheses of *myo*-inositol 1-phosphate (**43**) and the dehydroquinate (**44**), involved in the biosynthetic pathway of shikimic acid (Scheme 12), it is presumed that the cyclization forming the cyclopentane ring of allosamizoline (**28**) proceeds via a 4-keto or 6-aldehyde (or their enol equivalents) glucosamine derivative, which would undergo an aldol condensation of C-5 with C-1.[89] Therefore, three pathways to form the cyclopentane ring of allosamizoline are possible, as shown in Scheme 13. Pathways A and B are analogous mechanisms of cyclization during inositol biosynthesis. On the other hand, pathway C has an analogy to the mechanism of cyclization during shikimic acid biosynthesis.

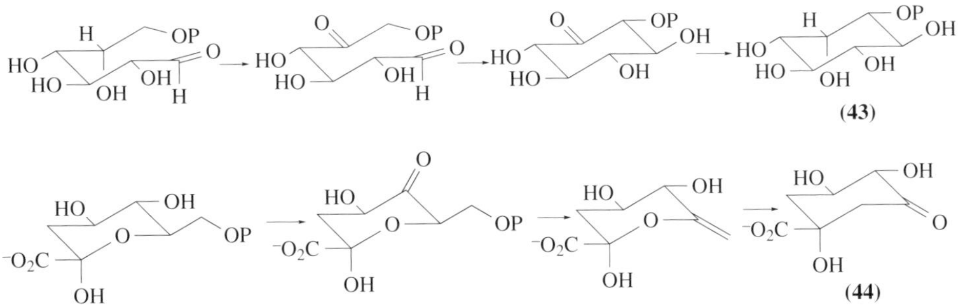

(43)

(44)

Scheme 12

To elucidate by which pathway the cyclopentane ring of allosamizoline (**28**) is formed, feeding experiments with specifically ^{2}H-labeled glucosamines were carried out. Labeled glucosamine was added to an *in vivo* system. After cultivation, the labeled allosamidin (**27**) obtained was hydrolyzed with acid to afford labeled allosamine and allosamizoline (**28**), which were investigated by ^{2}H NMR in order to confirm the position of the incorporated deuterium. Deuterium enrichment in labeled allosamine and allosamizoline (**28**) was evaluated by CI–MS analysis of labeled allosaminitol peracetate and the triacetate of allosamizoline derivatized from each of them. Since glucosamine is a common precursor of both allosamine and allosamizoline (**28**), as mentioned already, the comparison of deuterium incorporation into allosamizoline (**28**) with that into allosamine was very useful to evaluate whether a deuterium loss from labeled glucosamine had specifically occurred during the biosynthesis of the cyclopentane ring of allosamizoline (**28**) or not.

Four labeled glucosamines, [3-^{2}H],[90] [4-^{2}H],[90] [5-^{2}H], and [6-$^2H_2$]-D-glucosamine, were prepared,[91–93] and feeding experiments with each of them were carried out. The results of the experiments are summarized in Equation (2). In the feeding experiment with [4-^{2}H]-D-glucosamine, deuterium was incorporated into each C-4 of allosamine and allosamizoline from the allosamidin (**27**) obtained.

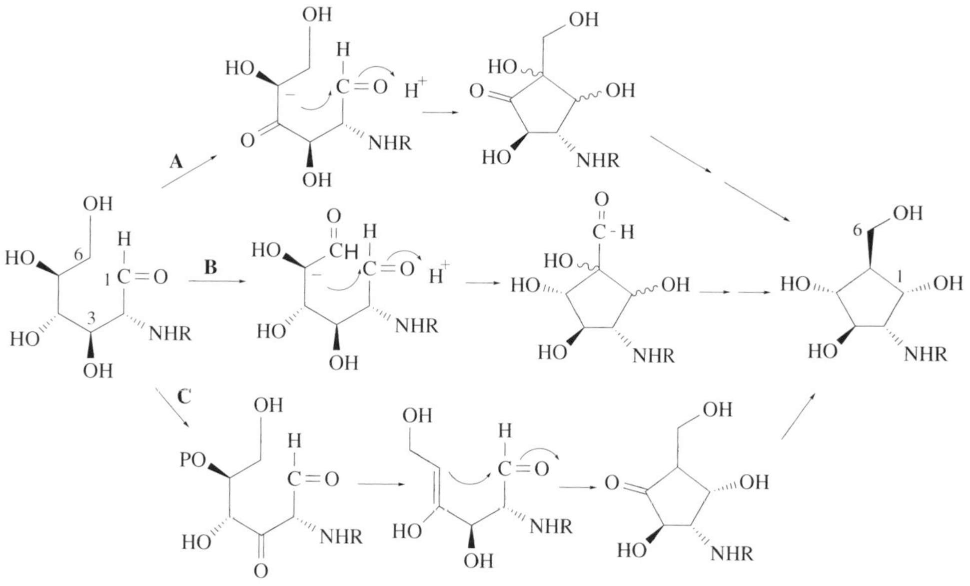

Scheme 13

Deuterium enrichment was observed at C-3 of allosamizoline (**28**), and also of allosamine from the (**27**) obtained in the experiment with [3-^{2}H] D-glucosamine. This deuterium incorporation at C-3 of allosamine suggested that the epimerization of a hydroxy group at C-3 occurred with retention of the deuterium on C-3 of glucosamine. Such a hydrogen retention during a process of epimerization has been observed in, for example, the reaction of UDP-D-glucose-4-epimerase.[94] In the feeding experiment with [5-^{2}H]-D-glucosamine, deuterium incorporation was observed at C-5 of allosamine from the (**27**) obtained, but no incorporation into allosamizoline (**28**) was detected. This indicated that deuterium on C-5 of glucosamine was lost during the formation of the cyclopentane ring. In the ^{2}H NMR spectrum of labeled allosamine prepared from (**27**) obtained from the feeding experiment with [6-^{2}H$_2$]-D-glucosamine, two deuterium signals, whose chemical shifts corresponded to those of the methylene protons on C-6 of allosamine, were observed (Figure 4). On the other hand, only one deuterium signal, which had the same chemical shift as that of one of the two protons on C-6 of allosamizoline (**28**), was observed in the spectrum of labeled allosamizoline (**28**) (Figure 4). Furthermore, the CI–MS spectrum of labeled allosaminitol peracetate indicated that the dilabeled molecules increased mainly, but an increase only in monolabeled molecules was observed in the spectrum of the triacetate of the labeled allosamizoline. These facts showed that one of the two deuterium atoms on C-6 of glucosamine was lost stereospecifically during the formation of the cyclopentane ring.

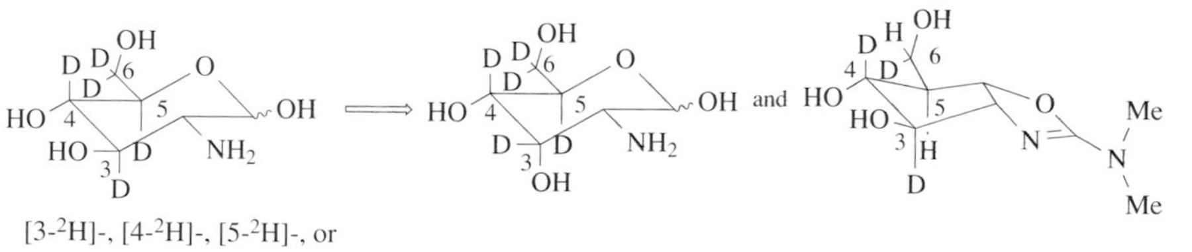

[3-^{2}H]-, [4-^{2}H]-, [5-^{2}H]-, or [6-^{2}H$_2$]-D-glucosamine (2)

The results from the feeding experiments with ^{2}H-labeled glucosamines mentioned above strongly suggest that the cyclopentane ring of allosamizoline (**28**) is probably formed through pathway B in Scheme 13.[95] In the pathway, the C-6 position of the glucosamine derivative is once oxidized to an aldehyde group, which facilitate C—C bond formation between C-5 and C-1 by an aldol condensation. After the cyclopentane ring formation, C-5 and C-6 are reduced to methine and hydroxymethyl groups, respectively, to create the allosamizoline skeleton. The reduction of the 6-aldehyde

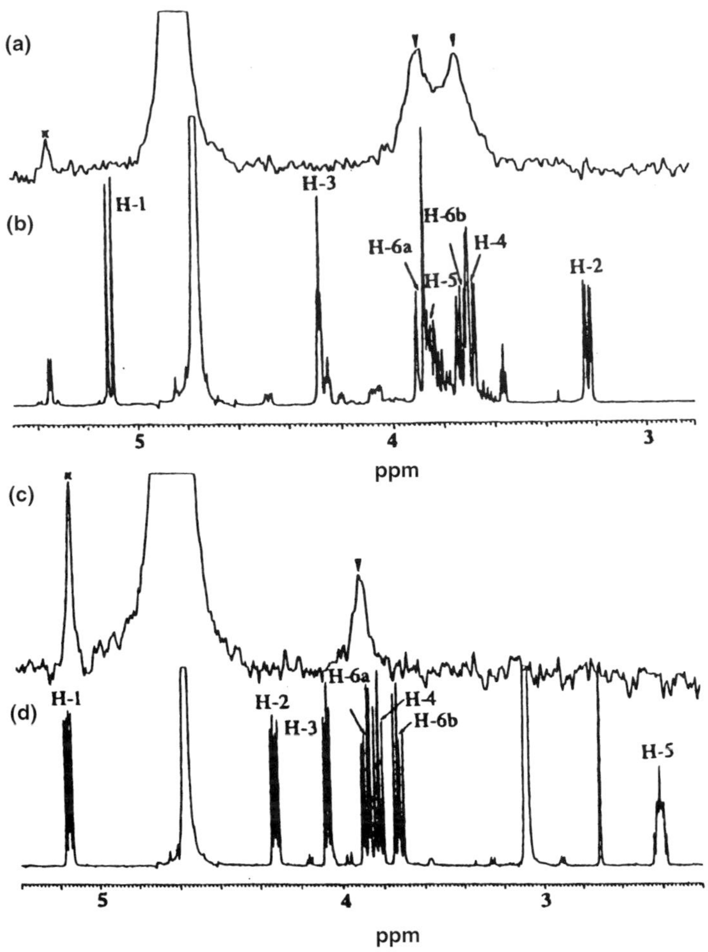

Figure 4 (a) ^{2}H NMR spectrum of allosamine derived from [6-$^2H_2$]-D-glucosamine (in D_2O). (b) ^{1}H NMR spectrum of natural allosamine (in D_2O). Signals of the β anomer are mainly observed. (c) ^{2}H NMR spectrum of allosamizoline (**28**) derived from [6-$^2H_2$]-D-glucosamine (in D_2O). (d) ^{1}H NMR spectrum of natural allosamizoline (**28**) (in D_2O).

group should occur stereospecifically because only one deuterium signal was observed in the ^{2}H NMR spectrum of allosamizoline from [6-$^2H_2$]-D-glucosamine. The stereochemistry of this reduction step as well as that of the first oxidation step of the 6-hydroxymethyl group to the 6-aldehyde group remains to be investigated, which is interesting from the viewpoint of comparison with stereochemistry observed in reactions to form cyclohexane rings such as inositol[96] or shikimic acid.[97] A reduction mechanism to remove a hydroxyl group from C-5 is not clear yet. A radical mechanism, which is known in the C—O bond cleavage reaction catalyzed by CDP-6-$\Delta^{3,4}$-glucoseen reductase,[98] might be possible, as opposed to a dehydration to produce a double bond between C-1 and C-5 followed by a subsequent reduction process. An enzyme system to catalyze this cyclization is now under investigation. Such a system would provide a convenient *in vitro* method to form a cyclopentanoid skeleton from a simple carbohydrate.

1.06.4 REFERENCES

1. E. M. Kleiner, S. A. Pliner, V. S. Soifer, V. V. Onoprienko, T. A. Balashova, B. V. Rosynov, and A. S. Khokhlov, *Bioorg. Khim.*, 1976, **2**, 1142.
2. Y. Yamada, K. Sugamura, K. Kondo, M. Yanagimoto, and H. Okada, *J. Antibiot.*, 1987, **40**, 496.
3. K. Kondo, Y. Higuchi, S. Sakuda, T. Nihira, and Y. Yamada, *J. Antibiot.*, 1989, **42**, 1873.
4. K. Sato, T. Nihira, S. Sakuda, M. Yanagimoto, and Y. Yamada, *J. Ferment. Bioeng.*, 1989, **68**, 170.
5. K. Hashimoto, T. Nihira, S. Sakuda, and Y. Yamada, *J. Ferment. Bioeng.*, 1992, **73**, 449.
6. U. Gräfe, W. Schade, I. Eritt, W. F. Fleck, and L. Radics, *J. Antibiot.*, 1982, **35**, 1722.
7. U. Gräfe, G. Reinhardt, W. Schade, I. Eritt, W. F. Fleck, and L. Radics, *Biotechnol. Lett.*, 1983, **5**, 591.

8. S. Sakuda and Y. Yamada, *Tetrahedron Lett.*, 1991, **32**, 1817.
9. K. Mori, *Tetrahedron*, 1983, **39**, 3107.
10. K. Mori and N. Chiba, *Liebigs Ann. Chem.*, 1990, 31.
11. K. Mizuno, S. Sakuda, T. Nihira, and Y. Yamada, *Tetrahedron*, 1994, **50**, 10 849.
12. S. Okamoto, K. Nakamura, T. Nihira, and Y. Yamada, *J. Biol. Chem.*, 1995, **270**, 12 319.
13. M. Ruengjtchatchawalya, T. Nihira, and Y. Yamada, *J. Bacteriol.*, 1995, **177**, 551.
14. H. Onaka, N. Ando, T. Nihira, Y. Yamada, T. Beppu, and H. Horinouchi, *J. Bacteriol.*, 1995, **177**, 6083.
15. H. Ohashi, Y.-H. Zheng, T. Nihira, and Y. Yamada, *J. Antibiot.*, 1989, **42**, 1191.
16. T. S. Chen, C. Chang, and H. G. Floss, *J. Am. Chem. Soc.*, 1981, **103**, 4568.
17. S. Sakuda, A. Higashi, T. Nihira, and Y. Yamada, *J. Am. Chem. Soc.*, 1990, **112**, 898.
18. S. Yue, J. S. Duncan, Y. Yamamoto, and C. R. Hutchinson, *J. Am. Chem. Soc.*, 1987, **109**, 1253.
19. D. E. Cane and C.-C. Yang, *J. Am. Chem. Soc.*, 1987, **109**, 1255.
20. D. E. Cane and W. R. Ott, *J. Am. Chem. Soc.*, 1988, **110**, 4840.
21. J. Cason, K. L. Rinehart, Jr., and S. D. Thornton, Jr., *J. Org. Chem.*, 1953, **18**, 1594.
22. M. W. Rathke and A. Lindert, *J. Org. Chem.*, 1970, **35**, 3966.
23. S. Sakuda, A. Higashi, S. Tanaka, T. Nihira, and Y. Yamada, *J. Am. Chem. Soc.*, 1992, **114**, 663.
24. H. V. D. Bosch, *Annu. Rev. Biochem.*, 1974, **43**, 243.
25. R. W. Rickards, in "Proceedings of the 6th Asian Symposium on Medicinal Plants and Spices, 1989," p. 91.
26. A. K. Hajra, T. V. Saraswathi, and A. K. Das, *Chem. Phys. Lipids*, 1983, **33**, 179.
27. S. Sakuda, S. Tanaka, K. Mizuno, O. Sukcharoen, T. Nihira, and Y. Yamada, *J. Chem. Soc., Perkin Trans. 1*, 1993, 2309.
28. S. Horinouchi, H. Suzuki, M. Nishizawa, and T. Beppu, *J. Bacteriol.*, 1989, **171**, 1206.
29. W. Li, T. Nihira, S. Sakuda, T. Nishida, and Y. Yamada, *J. Ferment. Bioeng.*, 1992, **74**, 214.
30. T. Nishida, T. Nihira, and Y. Yamada, *Tetrahedron*, 1991, **47**, 6623.
31. C. Pathirana, R. Dwight, P. R. Jensen, W. Fenical, A. Delgado, L. S. Brinen, and J. Clardy, *Tetrahedron Lett.*, 1991, **32**, 7004.
32. M. J. Smith, E. P. Mazzola, J. J. Sims, S. L. Midland, N. T. Keen, V. Burton, and M. M. Stayton, *Tetrahedron Lett.*, 1993, **34**, 223.
33. S. Kuwahara, M. Moriguchi, K. Miyagawa, M. Konno, and O. Kodama, *Tetrahedron*, 1995, **51**, 8809.
34. J. P. Henschke and R. W. Rickards, *Tetrahedron Lett.*, 1996, **37**, 3557.
35. E. Cabib, *Adv. Enzymol. Rel. Areas Molec. Biol.*, 1987, **59**, 59.
36. G. W. Gooday, *Progr. Indust. Microbiol.*, 1989, **27**, 139.
37. K. Isono and S. Suzuki, *Heterocycles*, 1979, **13**, 333.
38. U. Dahn, H. H. Hagenmaier, W. A. König, G. Wolf, and H. Zähner, *Arch. Microbiol.*, 1976, **107**, 143.
39. S. Sakuda, A. Isogai, S. Matsumoto, and A. Suzuki, *J. Antibiot.*, 1987, **40**, 296.
40. S. Sakuda, A. Isogai, S. Matsumoto, A. Suzuki, and K. Koseki, *Tetrahedron Lett.*, 1986, **27**, 2472.
41. S. Sakuda, A. Isogai, T. Makita, S. Matsumoto, K. Koseki, H. Kodama, and A. Suzuki, *Agric. Biol. Chem.*, 1987, **51**, 3251.
42. S. Sakuda, A. Isogai, S. Matsumoto, A. Suzuki, K. Koseki, H. Kodama, and Y. Yamada, *Agric. Biol. Chem.*, 1988, **52**, 1615.
43. A. Isogai, M. Sato, S. Sakuda, J. Nakayama, and A. Suzuki, *Agric. Biol. Chem.*, 1989, **53**, 2825.
44. Y. Nishimoto, S. Sakuda, S. Takayama, and Y. Yamada, *J. Antibiot.*, 1991, **44**, 716.
45. Z.-Y. Zhou, S. Sakuda, M. Kinoshita, and Y. Yamada, *J. Antibiot.*, 1993, **46**, 1582.
46. A. C. Terwisscha van Scheltinga, S. Armand, K. H. Kalk, A. Isogai, B. Henrissat, and B. W. Dijkstra, *Biochemistry*, 1995, **34**, 15 619.
47. D. A. Griffith and S. J. Danishefsky, *J. Am. Chem. Soc.*, 1991, **113**, 5863.
48. J. L. Maloisei, A. Vasella, B. M. Trost, and D. L. Van Vranken, *J. Chem. Soc., Chem. Commun.*, 1991, 1099.
49. J. L. Maloisei, A. Vasella, B. M. Trost, and D. L. Van Vranken, *Helv. Chim. Acta*, 1992, **75**, 1515.
50. S. Takahashi, H. Terayama, and H. Kuzuhara, *Tetrahedron Lett.*, 1992, **33**, 7565.
51. S. Takahashi, H. Terayama, and H. Kuzuhara, *Tetrahedron Lett.*, 1994, **35**, 4149.
52. D. A. Griffith and S. J. Danishefsky, *J. Am. Chem. Soc.*, 1996, **118**, 9526.
53. B. M. Trost and D. L. Van Vranken, *J. Am. Chem. Soc.*, 1993, **115**, 444.
54. M. Nakata, S. Akazawa, S. Kitamura, and K. Tatsuta, *Tetrahedron Lett.*, 1991, **32**, 5363.
55. S. Takahashi, H. Terayama, and H. Kuzuhara, *Tetrahedron Lett.*, 1991, **32**, 5123.
56. N. S. Simpkins, S. Stokes, and A. J. Whittle, *Tetrahedron Lett.*, 1992, **33**, 793.
57. T. Kitahara, N. Suzuki, K. Koseki, and K. Mori, *Biosci. Biotechnol. Biochem.*, 1993, **57**, 1906.
58. H. Terayama, H. Kuzuhara, S. Takahashi, S. Sakuda, and Y. Yamada, *Biosci. Biotechnol. Biochem.*, 1993, **57**, 2067.
59. S. Takahashi, H. Terayama, H. Kuzuhara, S. Sakuda, and Y. Yamada, *Biosci. Biotechnol. Biochem.*, 1994, **58**, 2301.
60. W. D. Shrader and B. Imperiali, *Tetrahedron Lett.*, 1996, **37**, 599.
61. K. J. Kramer and D. Koga, *Insect Biochem.*, 1986, **16**, 851.
62. S. Sakuda, Y. Nishimoto, M. Ohi, M. Watanabe, S. Takayama, A. Isogai, and Y. Yamada, *Agric. Biol. Chem.*, 1990, **54**, 1333.
63. N. Elango, J. U. Correa, and E. Cabib, *J. Biol. Chem.*, 1982, **57**, 1398.
64. S. Yamanaka, N. Tsuyoshi, R. Kikuchi, S. Takayama, S. Sakuda, and Y. Yamada, *J. Gen. Appl. Microbiol.*, 1994, **40**, 171.
65. D. Koga, A. Isogai, S. Sakuda, S. Matsumoto, A. Suzuki, S. Kimura, and A. Ide, *Agric. Biol. Chem.*, 1987, **51**, 471.
66. G. W. Gooday, L. J. Brydon, and L. H. Chappell, *Mol. Biochem. Parasitol.*, 1988, **29**, 223.
67. J. C. Villagomez-Castro, C. Calvo-Mendez, and E. Lopez-Romero, *Mol. Biochem. Parasitol.*, 1992, **52**, 53.
68. M. Shahabuddin, T. Toyashima, M. Aikawa, and D. C. Kaslow, *Proc. Natl. Acad. Sci. USA*, 1993, **90**, 4266.
69. K. Dickinson, V. Keer, C. A. Hitchcock, and D. J. Adams, *Biochem. Biophys. Acta*, 1991, **1073**, 177.
70. S. Milewski, P. W. O'Donnell, and G. W. Gooday, *J. Gen. Microbiol.*, 1992, **38**, 2545.
71. R. McNab and L. A. Glover, *FEMS Microbiol. Lett.*, 1991, **82**, 79.
72. M. Pedraza-Reyers and E. Lopez-Romero, *Antonie van Leewenhoek*, 1991, **59**, 183.

73. A. R. Butler, R. W. O'Donnell, V. J. Martin, G. W. Gooday, and M. J. R. Stark, *Eur. J. Biochem.*, 1991, **199**, 438.
74. E. Cabib, S. J. Silverman, and J. A. Shaw, *J. Gen. Microbiol.*, 1992, **38**, 97.
75. H. Blaak, J. Schnellmann, S. Walter, B. Henrissaat, and H. Schremph, *Eur. J. Biochem.*, 1993, **214**, 659.
76. Q. Wang, Z.-Y. Zhou, S. Sakuda, and Y. Yamada, *Biosci. Biotechnol. Biochem.*, 1993, **57**, 467.
77. K. D. Spindler and M. Spindler-Barth, *Pesticide Sci.*, 1994, **40**, 113.
78. D. Koga, K. Miziki, A. Ide, M. Kono, T. Matsui, and C. Shimizu, *Agric. Biol. Chem.*, 1990, **54**, 2505.
79. M. Kinoshita, S. Sakuda, and Y. Yamada, *Biosci. Biotech. Biochem.*, 1993, **57**, 1699.
80. M. Taniguchi, R. F. Nystrom, and K. L. Rinehart, Jr., *Carbohydr. Res.*, 1982, **109**, 161.
81. G. C. Levy and R. L. Lichter, "Nitrogen-15 Nuclear Magnetic Resonance Spectroscopy," Wiley, New York, 1979.
82. U. Hornemann, L. H. Hurley, M. K. Speedie, and H. G. Floss, *J. Am. Chem. Soc.*, 1971, **93**, 3028.
83. K. J. Martinkus, C.-H. Tann, and S. J. Gould, *Tetrahedron*, 1983, **39**, 3439.
84. Z.-Y. Zhou, S. Sakuda, and Y. Yamada, *J. Chem. Soc., Perkin Trans. 1*, 1992, 1649.
85. D. D. Weller and K. L. Rinehart, Jr., *J. Am. Chem. Soc.*, 1978, **100**, 6757.
86. G. Flesch and M. Rohmer, *Eur. J. Biochem.*, 1988, **175**, 405.
87. R. J. Parry, V. Bornemann, and R. Subramanian, *J. Am. Chem. Soc.*, 1989, **111**, 5819.
88. O. Ando, H. Satake, K. Itii, A. Sato, M. Nakajima, S. Takahashi, H. Haruyama, Y. Ohkuma, T. Kinoshita, and R. Enokita, *J. Antibiot.*, 1991, **44**, 1165.
89. G. N. Jenkins and N. J. Turner, *Chem. Soc. Rev.*, 1995, 169.
90. D. R. Bundle, H. J. Jennings, and I. C. R. Smith, *Can. J. Chem.*, 1973, **51**, 3812.
91. Y. Nishida, H. Hori, H. Ohrui, and H. Meguro, *Carbohydr. Res.*, 1987, **170**, 106.
92. W. Mackie and A. S. Perlin, *Can. J. Chem.*, 1965, **43**, 2645.
93. G. Moss, *Arch. Biochem. Biophys.*, 1960, **90**, 111.
94. R. D. Bevill, J. H. Nordin, F. Smith, and S. Kirkwood, *Biochem. Biophys. Res. Commun.*, 1963, **12**, 152.
95. S. Sakuda, Z.-Y. Zhou, H. Takao, and Y. Yamada, *Tetrahedron Lett.*, 1996, **37**, 5711.
96. M. W. Loewus, F. A. Loewus, G.-U. Brillinger, H. Otsuka, and H. G. Floss, *J. Biol. Chem.*, 1980, **255**, 11 710.
97. T. S. Widlanski, S. L. Bender, and J. R. Knowles, *J. Am. Chem. Soc.*, 1987, **109**, 1873.
98. V. P. Miller, J. S. Thorson, O. Ploux, S. F. Lo, and H.-W. Liu, *Biochemistry*, 1993, **32**, 11 934.

1.07
Eicosanoids in Mammals

NOBUYUKI HAMANAKA
Ono Pharmaceutical Co. Ltd., Osaka, Japan

1.07.1 INTRODUCTION

1.07.1.1 Definition of Eicosanoids

Eicosanoids is a general name for the metabolites that are biosynthesized by cycloxygenase, lipoxygenase, or P450 from arachidonic acid or similar polyunsaturated fatty acids as precursors. At first, since only the prostaglandins were known, they were named generically "prostanoids." Later, after a class of leukotrienes was also found to be metabolized from arachidonic acid, they were together named the arachidonic acid cascade. However, PGE_1, an important naturally occurring substance, was not included in this family, and thereafter the term "eicosanoids" was introduced as a general name for the metabolites derived from C-20 fatty acid.[1]

1.07.1.2 History

In 1930, Kurzrok and Leib first observed that lipid fractions isolated from human semen induced contraction of the human uterus. They believed that this substance was a low-molecular-weight substance.[2] This result was confirmed by Goldblatt[3] and von Euler.[4] They confirmed independently that there existed some substances in alcohol extracts which induced lowering of the blood pressure and contraction of the smooth muscle. The term "prostaglandins" (PGs) was coined in 1934 by von Euler in the belief that the newly discovered biologically active substances were produced in the prostate gland. Since the concentration of PG in human semen was very low and the analytical techniques at that time could not deal with these unstable substances, their structures were not determined; the only information gained was that they were acidic substances.

In the latter half of the 1950s, it was clear that PG in the semen was produced in the seminal vesicle. Large amounts of seminal vesicles of sheep were collected, and PGE_1 and $PGF_{1\alpha}$ were isolated from the seminal vesicle and their structures determined.[5] Thereafter, over the next 10 years, the isolation and elucidation of many substances resulted in the discovery of the diversity of their biological activities, and so a vast new field of chemically, biologically, and clinically important eicosanoids was opened up, ranking them among the most potent substances found in nature.

In contrast, in 1938 Feldberg and Kellaway injected cobra venom into guinea-pig perfused lungs and observed the release of a substance into the perfusate which differed from histamine and caused a slow contraction, of long duration, of guinea-pig jejunum.[6,7] The agent responsible was referred to as "slow reacting substance" (SRS). They demonstrated that a similar substance was produced in the effluent of guinea-pig perfused lung following challenge with an appropriate antigen. This immunologically produced mediator was later termed SRS-A—slow reacting substance of anaphylaxis. Although SRS-A was considered to be an important mediator of asthma from much investigation, the structural studies did not proceed very fast due to the stability of the substances and their extremely low concentration in mammalian tissues.

In 1979, Samuelsson *et al.*[54] found that arachidonic acid was oxidized by lipoxygenase and then reacted with peptide to afford a peptide-lipid which showed SRS activity. This breakthrough allowed the investigation of a new class of substances to begin.[8] Since the new class of compounds was obtained by metabolism using polymorphonuclear leukocytes (from the peritoneal cavity of rabbits) and possessed a conjugated triene, Samuelsson proposed the name "leukotriene", and showed that SRS from different sources contained varying amounts of different leukotrienes with SRS activity.

As described above, arachidonic acid which has a simple structure is transformed by two different pathways to produce an important class of biologically active substances. Since then, keen competition for the discovery of new substances has spread throughout the world.

1.07.2 BIOSYNTHESIS, STRUCTURAL ELUCIDATION, AND CHEMISTRY OF EICOSANOIDS

1.07.2.1 Oxidation Mechanism of Lipoxygenase

The enzymes associated with eicosanoids are discussed in detail in Chapter 1.08. In this section, the structures of compounds biosynthesized from polyunsaturated fatty acids are described. Polyunsaturated fatty acids are stored in mammalian tissues in ester form at the 2-position of phospholipid and are components of organism membranes. Usually they do not exist as free acids, and they are released from the phospholipid ester into the eicosanoid biosynthetic pathway by either the enzyme phospholipase A_2 or the combined action of phospholipase C and a diglyceride lipase. These unstable fatty acids include the divinyl methane partial structure. The systematic reaction pathway as shown in Figure 1 can be presumed.[8,9]

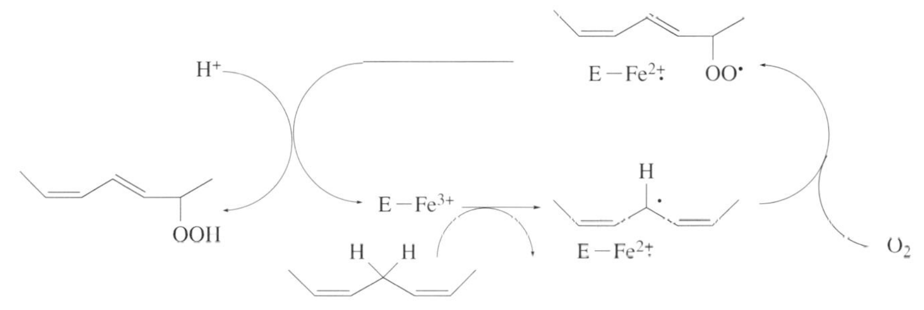

Figure 1 Plausible mechanism for lipoxygenase reaction.

The enzyme in the Fe^{3+} state, probably complexed with an alkylhydroperoxide, abstracts a hydrogen atom from the substrate, giving a bisallylic radical and leaving the iron in the Fe^{2+} state. Oxygen associates with the Fe^{2+} enzyme and oxygenates the enzyme-bound substrate radical. Subsequent electron transfer produces the product hydroperoxide and regenerates the Fe^{3+} state of the enzyme. This is the biological mechanism of lipoxygenase activity.

As for cyclooxygenase, it is possible that the hydroperoxide reacts with other intramolecular double bonds to form the endoperoxide.

In a divinyl methane system, there are two positions to which the hydroperoxide can be introduced. Since polyunsaturated fatty acids have several divinyl methane systems, a vast number of metabolites can be produced involving optical isomers. Furthermore, if the compounds which are oxidized in a multiple manner are significant substances in mammalian tissues, their complexity is incalculable. This complex eicosanoid biosystem is called a cascade.

1.07.2.2 Nomenclature of Prostaglandins and Thromboxane (Cyclooxygenase Pathway)

PG and thromboxane (TX) belong to this group. The substrates for lipoxygenase are dihomo-γ-linolenic acid, arachidonic acid, and eicosapentenoic acid. PGs consist of a cyclopentane ring and carboxylic acid containing a seven- and alkyl containing an eight-carbon side chain named a and w, respectively, with *trans* stereochemistry. Alphabetic designations (A, B, C, . . ., J) in PGs refer to the different functions in the cyclopentane moiety. Numerical subscripts represent the number of olefinic bonds in the side chains. The subscript α in $PGF_{2\alpha}$ refers to the orientation of the C-9 hydroxyl group. Prostacyclin (PGI_2) also belongs to this classification, but has an ether linkage between the C-9 oxygen and C-6 vinyl carbon. Thromboxane A_2 (TXA_2) is characterized by a tetrahydropyran nucleus derived from further skeletal rearrangement (Figure 2).

1.07.2.3 Structure and Reaction of Prostaglandins and Thromboxane

1.07.2.3.1 Classical PGs

PGE_1 and $PGF_{1\alpha}$, derived from sheep seminal vesicles by extensive chromatography, were first obtained in pure crystalline form in 1957 (see Figure 3).[10] Application of the gas chromatography/mass spectrometry (GS/MS) technique then being developed by Ryhage[11] was the key

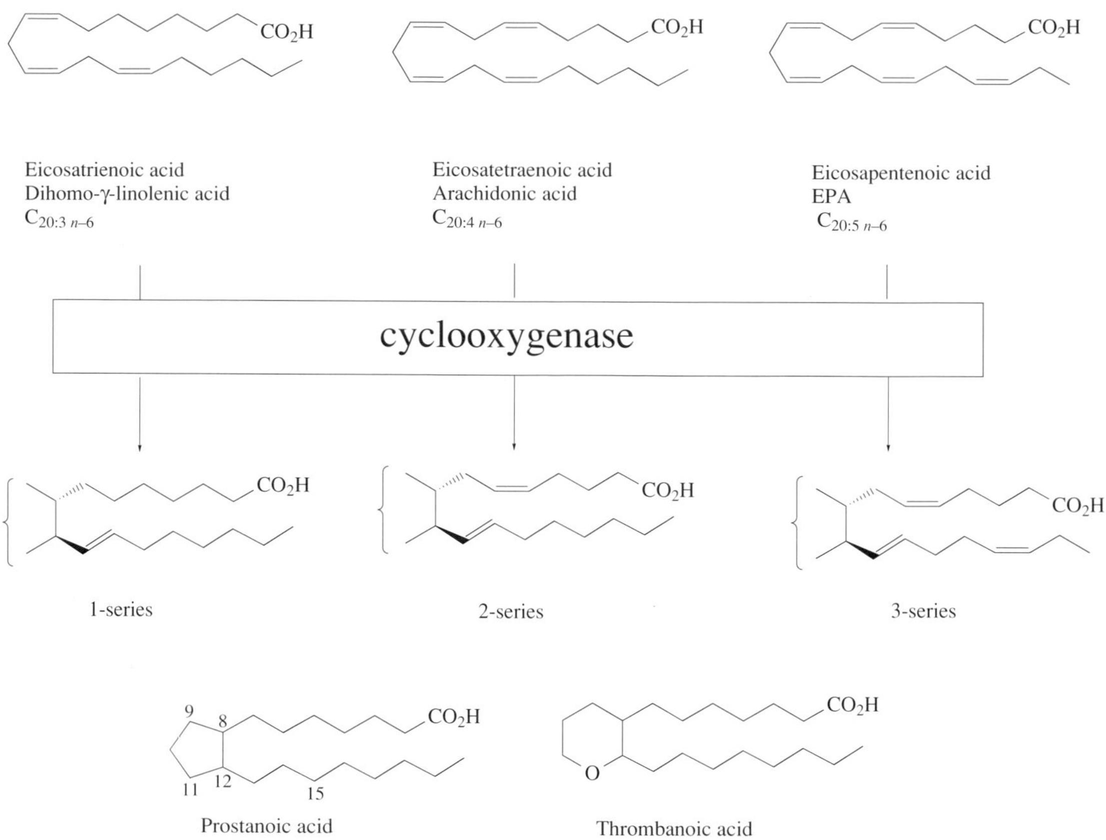

CO_2H CO_2H CO_2H

Eicosatrienoic acid
Dihomo-γ-linolenic acid
$C_{20:3\ n-6}$

Eicosatetraenoic acid
Arachidonic acid
$C_{20:4\ n-6}$

Eicosapentenoic acid
EPA
$C_{20:5\ n-6}$

cyclooxygenase

CO_2H CO_2H CO_2H

1-series

2-series

3-series

9 8 CO_2H 11 12 15

Prostanoic acid

CO_2H O

Thrombanoic acid

Figure 2 Precursor, numerical subscript, and parent structure.

to their structure elucidation; the data for the ozonolysis products derived from the basic dehydration product of dihydro-PGE was the best single proof.[12,13] The relationship between PGE_1 and $PGF_{1\alpha}$ was established by borohydride reduction,[14] which afforded a nearly equal amount of $PGF_{1\beta}$. The structural assignments were confirmed by X-ray crystallographic studies on a heavy-atom derivative of $PGF_{1\beta}$.[15] These studies also interrelated PGE_1 and its two dehydration isomers, PGA_1 and PGB_1. The skeleton and carbonyl location of PGE_1 were also confirmed by the total synthesis of a degradation product of 13,14-dihydro-PGE_1.[16] This early work and related studies on metabolites have been summarized by Samuelsson.[17] The absolute configuration was determined by identification of 2*S*-hydroxyheptanoic acid with the ozonolysis product of PGB_1.[18]

The more highly unsaturated 2-series PGs were also isolated during this period,[19] and hydrogenation studies and ozonolysis established their relationship with PGE_1 and the locations of the additional double bonds.[17] A number of 2-series PGs were produced mainly by biosynthesis in mammals; however, 1-series PGs were isolated first since they were crystalline substances which were readily purified. The biosynthetic map from arachidonic acid is shown in Figure 4.

PGF is very stable under various conditions compared with other PGs. However, isomerization occurs at the C-15 position along with the allyl rearrangement under strong acidic conditions (see Figure 5).[20]

The thermodynamic stability of the PGE-type is crucial since the α side chain is alpha to the ketone group; dehydration proceeds under isomerization conditions and it is difficult to determine the position of equilibrium. However, it was reported that PGEs and their 8-isomers exist in a 85 : 15 ratio under weakly basic conditions (see Figure 6).[21]

PGB-type substances show strong UV absorption (see Figure 7) and this can be applied to the quantitative analysis of PGs.[22] This transformation was thoroughly investigated, but PGB-type substances were converted into the polymer PGBx[23] under certain chemical conditions.

C-type PGs were not reported until the early 1970s when they were observed as further transformation products of PGAs.[24–26] The PGC structure is a logical intermediate between PGA and PGB during base-catalyzed isomerization. PGD structures were first noted as metabolites in biosynthetic incubation studies.[27–29]

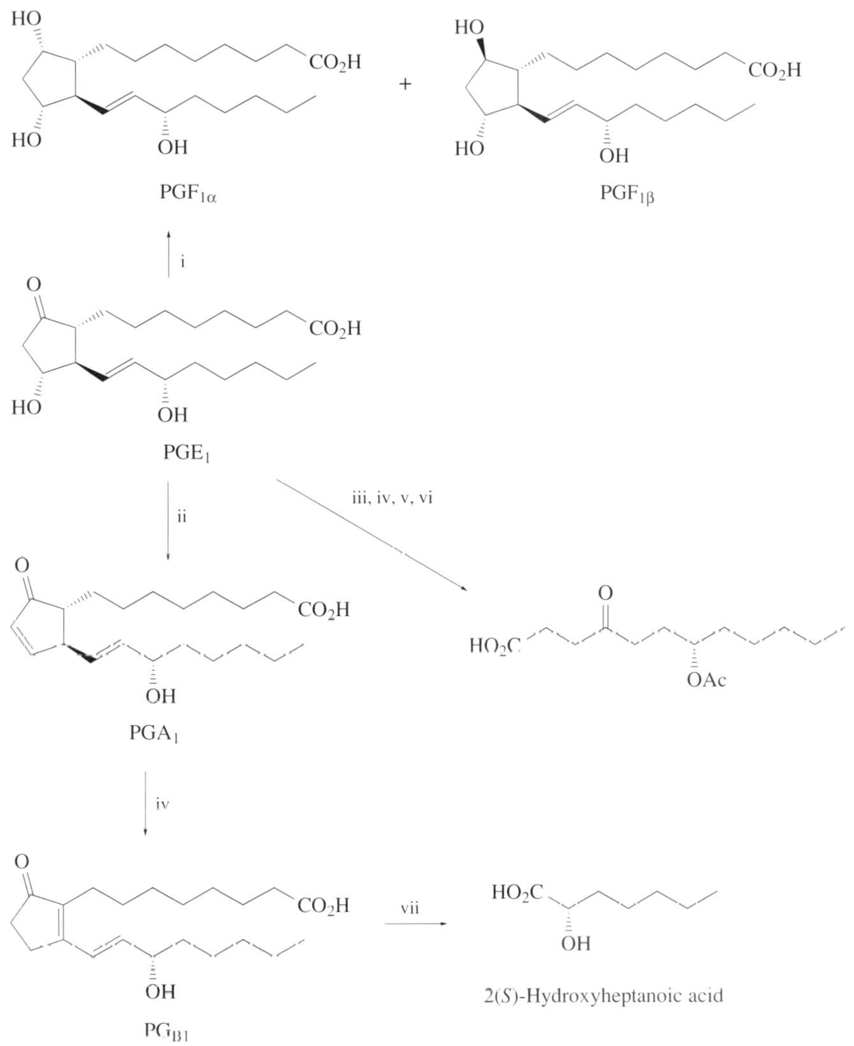

i. $NaBH_4$, ii. acid, iii. H_2/Pt, iv. base v. CH_2N_2, vi. Ac_2O, vii. O_3

Figure 3 Structural determination of PGE_1.

The 11-keto structure was supported by $NaBH_4$ reduction, which afforded $PGF_{1\alpha}$,[29] and by mass spectroscopy of the methyl ester methoxyamine derivatives.[27] PGD_2 is more unstable than PGE_2 and readily converted to PGJ_2 by a dehydration reaction (Figure 8).[29,30]

PGA, PGB, PGC, and PGJ-type substances are the decomposition products that are produced during the extraction process. PGs (PGA to PGE) isolated at an early stage are generally named the primary PGs and also called the classical PGs involving PGD. They are distinguished from the other PGs described below.

1.07.2.3.2 PGG_2 and PGH_2

In the mid-1960s, the groups of Van Dorp and Bergström simultaneously reported that prostaglandins were biosynthesized from arachidonic acid.[31,32] Studies with isotopic oxygen indicated that a labile endoperoxide intermediate was involved in this lipoxygenase ("prostaglandin synthetase") reaction.[33] In the early 1970s, three groups isolated these endoperoxides and studied their biological activities, principally induction of platelet aggregation and contraction of isolated vascular tissue.[34–37] Eventually, endoperoxides with a hydroperoxy group at C-15 were designated as PGG, while those with a C-15 hydroxyl group were designated as PGH.[36] This class of the substances is very unstable, and PGH_2 is converted into $PGF_{2\alpha}$ under reducing conditions. In addition, malon-

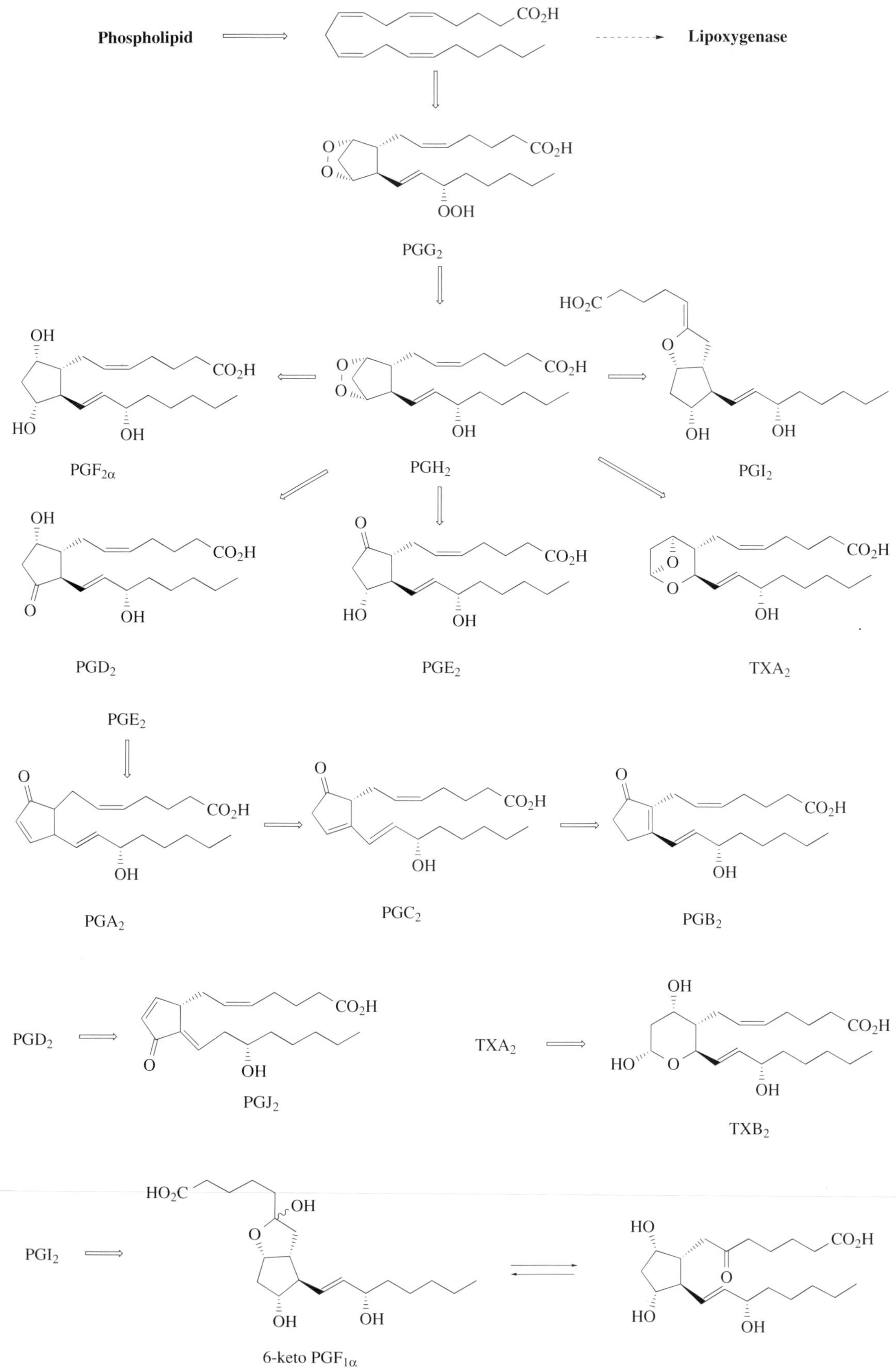

Phospholipid ⟹ → **Lipoxygenase**

PGG_2

$PGF_{2\alpha}$ PGH_2 PGI_2

PGD_2 PGE_2 TXA_2

PGE_2

PGA_2 PGC_2 PGB_2

PGD_2 ⟹ PGJ_2

TXA_2 ⟹ TXB_2

PGI_2 ⟹ 6-keto $PGF_{1\alpha}$

Figure 4 Cycloxygenase pathway (arachidonic acid cascade).

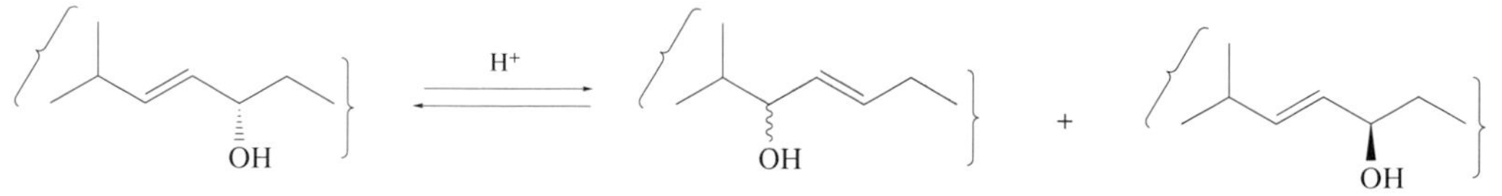
H+
OH
OH
+
OH

Figure 5 Epimerization and rearrangement at C-15.

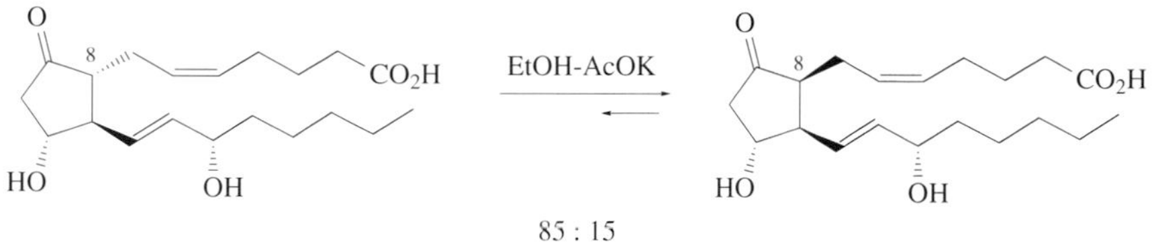
O
8
CO_2H
HO
OH
EtOH-AcOK
O
8
CO_2H
HO
OH
85 : 15

Figure 6 Isomerization of PGE_2 at C-8.

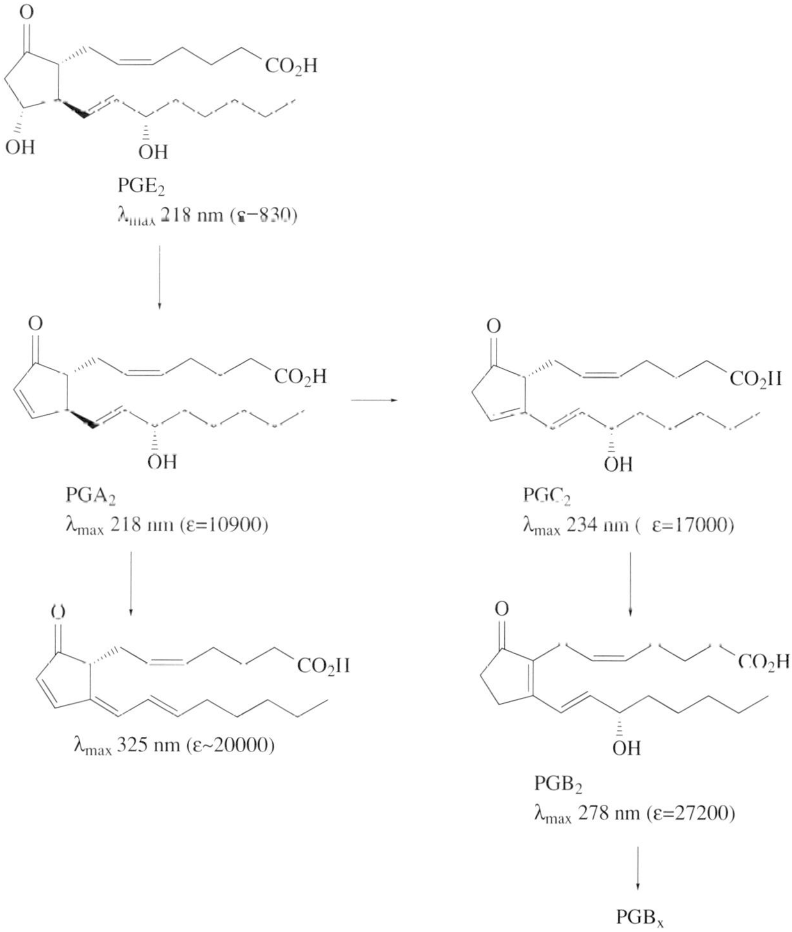
PGE2
λmax 218 nm (ε=830)
PGA2
λmax 218 nm (ε=10900)
PGC2
λmax 234 nm (ε=17000)
λmax 325 nm (ε~20000)
PGB2
λmax 278 nm (ε=27200)
PGBx

Figure 7 UV absorption of PGA_2, PGB_2, PGC_2, and derivatives.

dialdehyde was readily removed to afford 12(*S*)-hydroxy-5-*cis*,8-*trans*,10-*trans*-heptadecatrienoic acid (HHT) (Figure 9).[37]

Interesting reactivities were observed under various reaction conditions. When they were allowed to stand in a solution of DMSO–H_2O, the rearrangement reaction resulted in ring opening of the five-membered ring; the substances named Levuglandin E_2 and D_2 were produced as if PGE_2 and PGD_2 were subjected to the retro-aldol reaction (Figure 1).[38] Furthermore, the rearrangement reactions by metal catalysts afforded many kinds of derivatives.[39,40] However, such rearrangement reactions did not give PGI_2 and TXA_2 (Figure 11).

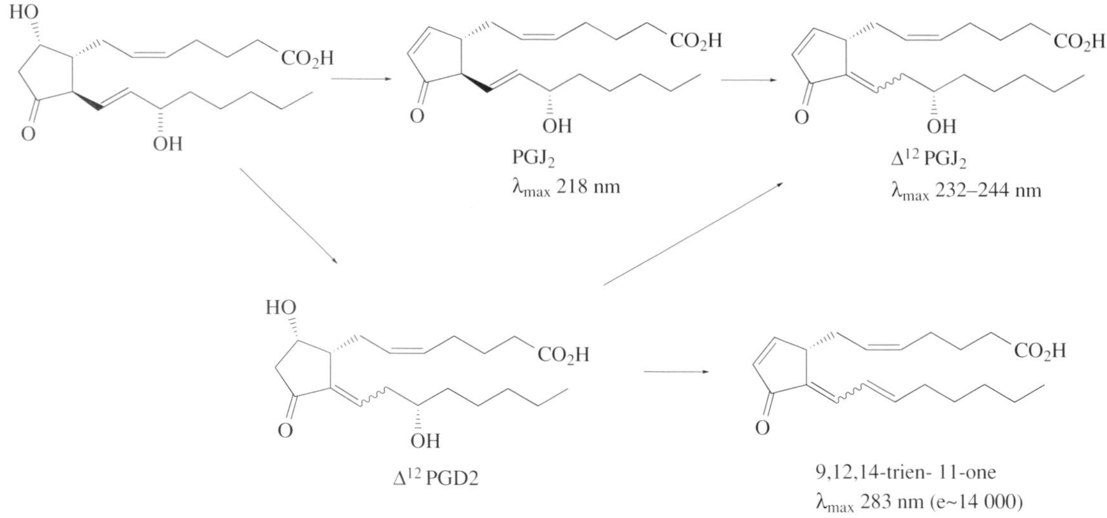

PGJ$_2$ λ_{max} 218 nm

Δ^{12} PGJ$_2$ λ_{max} 232–244 nm

Δ^{12} PGD2

9,12,14-trien- 11-one λ_{max} 283 nm (e~14 000)

Figure 8 Degradation products of PGD$_2$ and their UV absorbance.

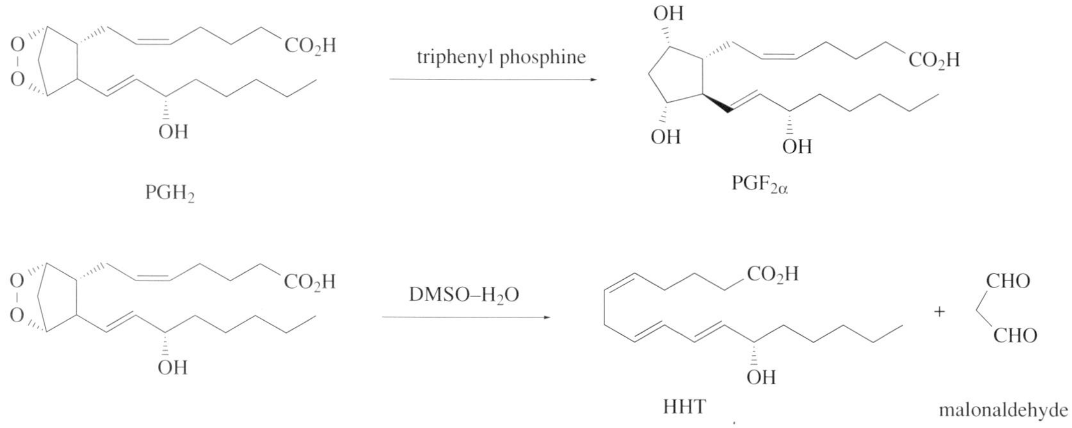

triphenyl phosphine

PGH$_2$

PGF$_{2\alpha}$

DMSO–H$_2$O

HHT

malonaldehyde

Figure 9 Reduction of PGH$_2$ and formation of HHT.

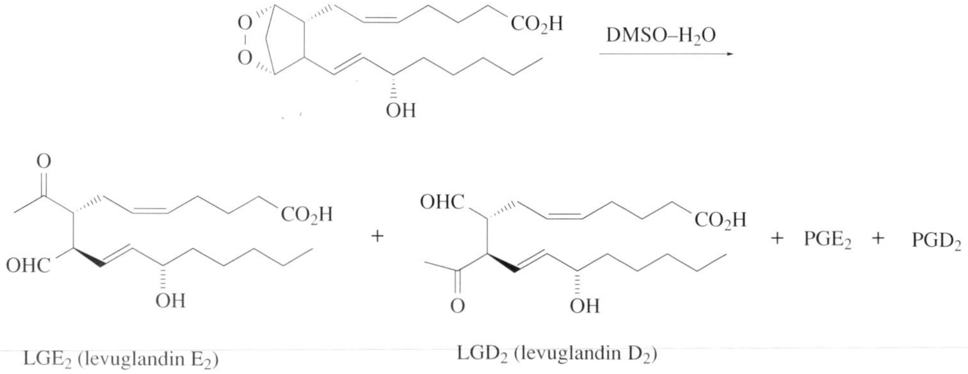

DMSO–H$_2$O

+ PGE$_2$ + PGD$_2$

LGE$_2$ (levuglandin E$_2$)

LGD$_2$ (levuglandin D$_2$)

Figure 10 Formation of levuglandin.

1.07.2.3.3 PGI$_2$ (prostacyclin)

In the mid-1970s, Moncada *et al.* described a labile prostaglandin (temporarily designated PGX) formed from the dienoic endoperoxides by a vascular microsomal enzyme.[41] This compound has very high antiplatelet aggregation activity and blood smooth muscle contraction activity. It was shown that this substance was an intermediate in the formation of 6-keto PGF$_{1\alpha}$ via a pathway described several years earlier by Pace-Asciak and Wolfe.[42,43] Upon determination of its unique

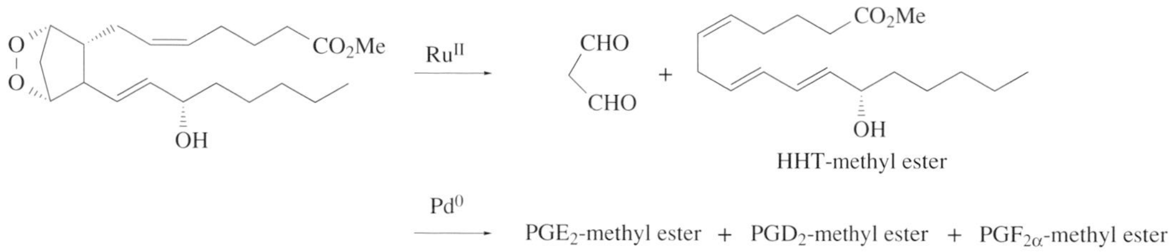

$\xrightarrow{Ru^{II}}$ CHO CHO + HHT-methyl ester

$\xrightarrow{Pd^0}$ PGE_2-methyl ester + PGD_2-methyl ester + $PGF_{2\alpha}$-methyl ester

Figure 11 Degradation of PGH_2 by metal catalyst.

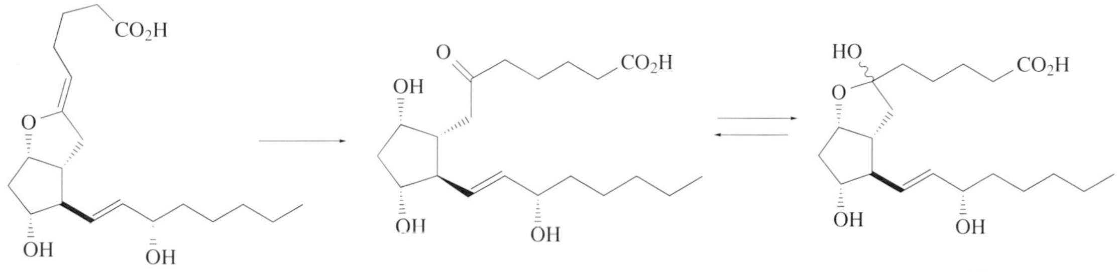

6-keto $PGF_{1\alpha}$ acetal form

Figure 12 Degradation of PGI_2.

bicyclic structure, it was termed "prostacyclin," and later renamed as PGI_2.[44] PGI_2 was readily hydrolyzed to 6-keto $PGF_{1\alpha}$ even in neutral aqueous solution and converted to a mixture of intramolecular hemiacetals (Figure 12).

6-Keto $PGF_{1\alpha}$ was oxidized enzymatically to 6-keto PGE_1, which exhibited very highly PGI_2-like biological activities (Figure 13).[45]

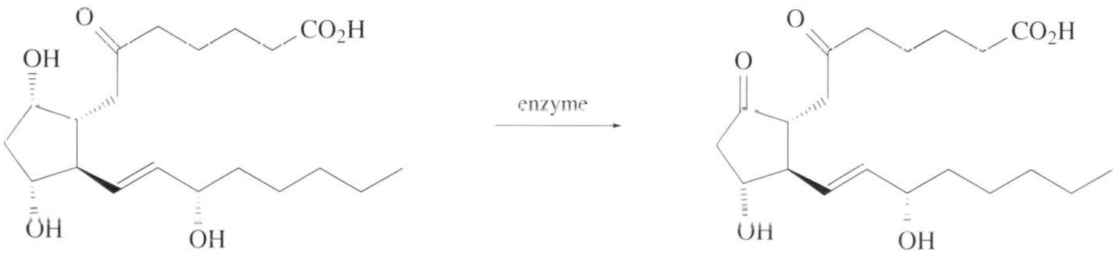

6-keto $PGF_{1\alpha}$ $\xrightarrow{enzyme}$ 6-keto PGE_1

Figure 13 Enzymatic oxidation of 6-keto $PGF_{1\alpha}$ to 6-keto PGE_1.

1.07.2.3.4 TXA_2

In 1975, Samuelsson and his colleagues discovered a new family of hydroxylated cyclized fatty acids that lacked the prostanoic acid backbone of prostaglandins. These were designated thromboxanes because of their formation and probable role in aggregating platelets and because of their characteristic oxane ring structure.[46] A-type thromboxanes are bicyclic and are derived enzymatically from prostaglandin endoperoxides by thromboxane synthetase.[47] A-type thromboxanes are highly unstable. Their structure was proposed by chemical trapping experiments that result in characteristic methoxy or azide derivatives when incubates containing TXA_2 were mixed with methanol or sodium azide, respectively (Figure 14).[46,48]

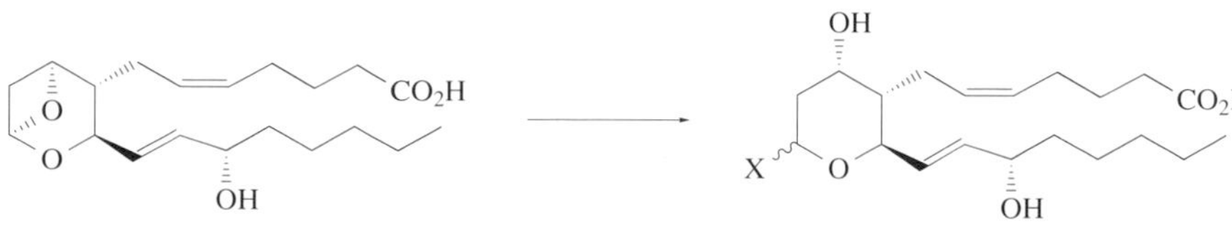

TXA_2 → X = OH (TXB_2), N_3, OMe

Figure 14 Degradation of TXA_2.

Thus, one of the pathways of the arachidonic acid cascade from PGG_2 was clear. It is reasonable that each metabolite has bioactivities; however, PGA, which was presumed to be a chemical decomposition product and a biological product, also has highly incommensurable bioactivities compared with those in other biosystems.

1.07.2.4 Lipoxygenase Pathway

Slow reacting substances of anaphylaxis (SRS-A) behave as potent mediators of airway constriction and bronchial spasm in human asthma as well as allergic hypersensitivity.[49] Studies of structural and biological properties carried out during the 1970s demonstrated that SRS-A was a polar lipid,[50–52] possibly containing the sulfur atom and having a strong UV absorption.

Parker *et al.*[53] showed that arachidonic acid could stimulate the release of SRS-A from rat basophilic leukemia cells; arachidonic acid radiolabeled with ^{13}C and ^{3}H was incorporated into purified SRS-A. Further evidence supporting arachidonic acid as a precursor of SRS was obtained from findings that the calcium ionophore A23187 stimulated the production of SRS as well as 5-HETE and LTB_4 and that the exogenous arachidonic acid considerably enhanced the response to A23187.[54] Morris *et al.* showed that pure material had an absorbance at 280 nm.[55]

On the basis of these facts, Samuelsson *et al.* conceived the idea that the unstable epoxy acid triene LTA_4 might be an intermediate in the formation of SRS. They incubated tritiated arachidonic acid with mastocytoma cells in the presence of ^{14}C-labeled cysteine and the ionophore A23187.[56] From this experiment, a very low yield of material containing both ^{14}C and tritium was separated from a mixture by HPLC purification. The isolated material showed the same UV absorption ($\lambda = 280$ nm) and reactivity with soybean lipoxygenase as SRS. Because of the incorporation of the radiocarbon label from cysteine, Samuelsson's group[57] suggested that the cysteinyl substituent was present in leukotriene C and C-6 in view of the UV data and the observed reactivity toward soybean lipoxygenase. Having these results on hand, Samuelsson proposed that LTC_4 was a 5-hydroxy-7,9,11,14-eicosatetraenoic acid containing a C-6 peptide linkage. The final geometry of the triene was undetermined, although the C-11 and C-14 double bonds were presumably *cis*.

At this point, Corey *et al.*[56] accomplished the conversion of synthetic (DL) leukotriene A_4 methyl ester to the cysteine conjugate as the first step in a comparison of synthetic and natural substances. However, the synthetic compound was not identical to the naturally occurring SRS-A. In the action of the compound, it was evident that there was a problem not in the lipid part but in the amino acid unit. Reexamination revealed that SRS-A contained glutathione and not cysteine (Figure 15).

The chemical structure of the SRS-A molecule was therefore determined as 5-hydroxy-6-*S*-glutathionyl-7,9,11,14-eicosatetraenoic acid. This compound was named leukotriene as SRS-A is formed by leukocytes and because the molecule contained a conjugated triene. The letter C and a subscript 4 were added to indicate the structure of the peptide substituent and the total number of double bonds (four) in a molecule.

Leukotriene C_4 has four asymmetrical carbon atoms and four double bonds. To determine the stereochemistry at these centers, Corey developed methods for the chemical synthesis of several stereoisomers of leukotriene C_4.[58]

By using rat basophilic leukemia cells instead of murine mastocytoma cells, LTD_4 was obtained.[59,60] A compound that was identical to the rat basophilic leukemia cell SRS in terms of chemical, physiochemical, and biological properties was obtained from leukotriene C_4 by treatment with γ-glutamyl transpeptidase from porcine kidney. Leukotriene D_4 has been identified as a major component of SRS-A from several sources.[61–63]

1.07.2.5 Other Lipoxygenase and P450 Pathways

It is a surprising fact that the important compounds for a living body are biosynthesized from simple lipids by cyclooxygenase and 5-lipoxygenase, because in organic chemistry it is common sense that the substances which involve the divinylmethane function in a molecule are very readily oxidized by air to provide lipid peroxides (auto-oxidation). It is naturally anticipated that the interesting substances will be biosynthesized by the other pathway. In fact, a number of investigations were conducted from this point of view.

As shown in Figure 16, oxygen is introduced into positions C-5, C-8, C-9, C-11, and C-12 in a molecule, which is presumed to be the mechanism discussed in Section 1.07.2.1.

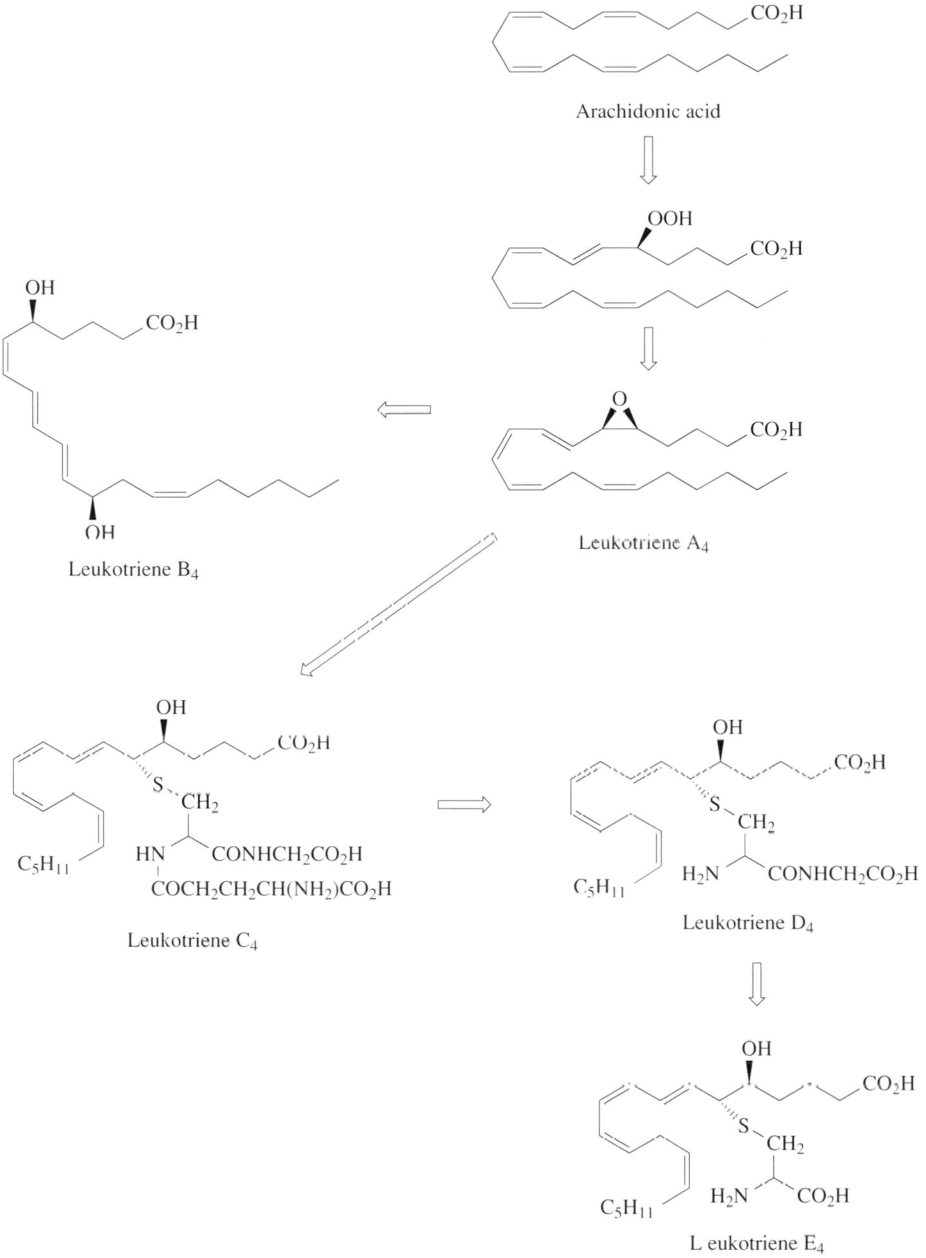

Figure 15 Lipoxygenase pathway (arachidonic acid cascade).

Actually, the first enzyme recognized was 12-lipoxygenase. Hamberg and Samuelsson discovered the production of a 12(*S*)-hydroxy-5-*cis*,8-*cis*,10-*trans*,14-*cis*-eicosatetraenoic acid (12-HETE) upon incubation of human platelets with arachidonic acid.[64] 12-HETE production suggested the presence of a lipoxygenase enzyme which oxygenated C-12 of arachidonic acid. Furthermore, Nugteren detected an enzyme with properties of lipoxygenase in the supernatant of broken bovine platelets, which produced 5(*S*)-hydroxy acid with at least two *cis*-double bonds at the C-8 and C-11 positions.[65] 5-Lipoxygenase catalyzes the pathway to LTs, and 11-lipoxygenase catalyzes the pathway to PG-TX as described above.

15-Lipoxygenase was found in rabbit reticulocytes[66] and leukocytes.[67] 8-Lipoxygenase[68] was found in mouse epidermis. However, 9-lipoxygenase has not yet been discovered. References 69–73 are review articles dealing with mammalian lipoxygenases.

As an example, the 12-lipoxygenase pathway is shown in Figure 17. 12-Lipoxygenase has a multifunctional nature, even when it is purified. Namely, the oxygen function is introduced not only at the C-12 position but also at the C-15 position of arachidonic acid: two kinds of compounds are

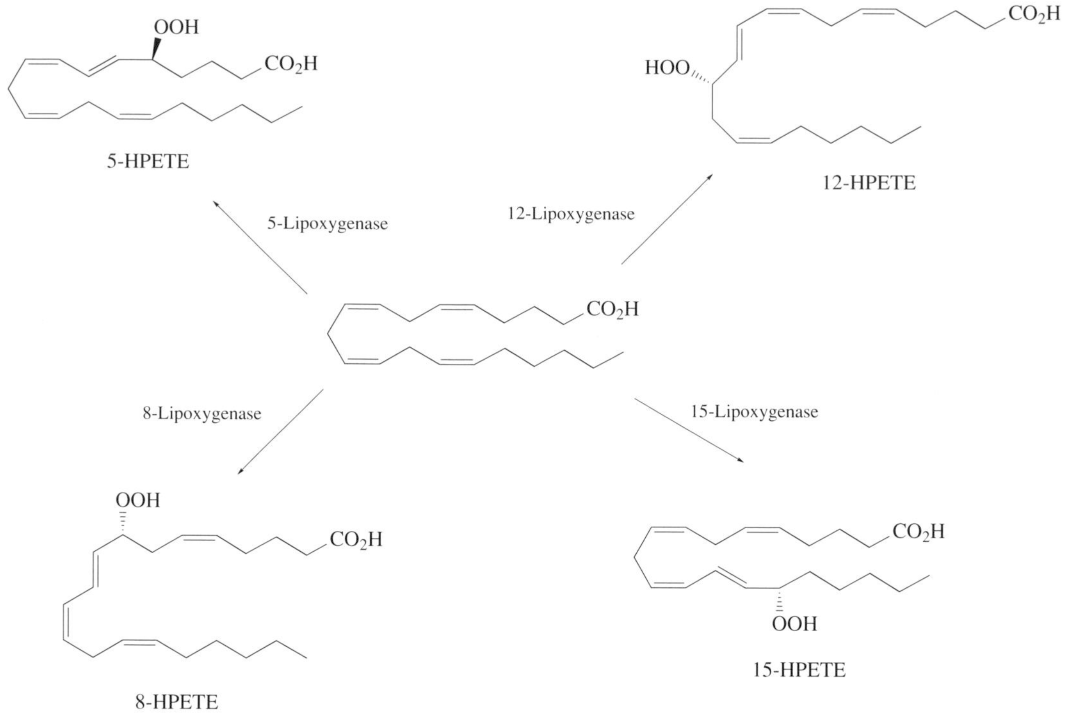

OOH CO$_2$H

5-HPETE

HOO CO$_2$H

12-HPETE

5-Lipoxygenase

12-Lipoxygenase

CO$_2$H

8-Lipoxygenase

15-Lipoxygenase

OOH CO$_2$H

8-HPETE

CO$_2$H OOH

15-HPETE

Figure 16 Mammalian lipoxygenase pathway.

biosynthesized by a single enzyme. In addition, the substrate specification of these enzymes is so low that further oxygen function is introduced into the other positions. The murine 12-lipoxygenase of leukocyte type showed a 12/15-HPETE ratio of 3 : 1, and the murine enzyme of platelet-type produced exclusively 12-HPETE.

There is a further complication in that 12-lipoxygenase transformed 15-HPETEs as substrates to a mixture of various dihydroperoxy and dihydroxy acids with a conjugated triene. These HPETEs were further converted by the other enzymes to a complicated mixture through 14,15-epoxide (14,15-LTA$_4$) (Figure 18).[74–76]

Herein, the biosynthesis of hypoxilin, trioxilin,[77,78] and lipoxin[79] is shown. There are a number of isomers, which will not be discussed here. It is generally considered that the physiological and pathological role of lipoxygenase products without PGs and LTs has not been clearly established.[80]

The things that make the lipoxygenase system complex are not only oxidation as described above but also the lack of substrate specificity of the enzyme and oxidation with P450. The multiple catalytic activities of plant lipoxygenase have been reviewed in detail.[81]

P450 is able to act as a monooxygenase and catalyzes the hydroxylation or epoxidation of a variety of hydrophobic substrates. It has also been presumed that it is related to the metabolism of PGs by ω-oxidation of fatty acids; however, it is clear that P450 forms epoxy eicosanoids by the direct addition of oxygen to the double bond of unsaturated fatty acids (Figure 19).[82,83]

Capdevila *et al.*[84] reported that treatment of arachidonic acid with rat liver microsome that was pretreated with phenobarbiturate gave epoxy eicosanoids which were a mixture of *cis*-5,6-, 8,9-, 11,12-, and 14,15-epoxyeicosatrienoic acids, which showed different properties from those of LT because they did not have triene function. Di-HETEs are produced by the hydrolysis of these substances (Figure 20).

Furthermore, P450 forms, as well as the other lipoxygenases, substances in which oxygen is introduced into positions C-5, -8, -9, -11, -12, and -15. Since these substances are not distinguished from the substances derived via lipoxygenase, there are some doubts about the physiological role of each of the HETEs.[85]

Thus, a number of eicosanoids are biosynthesized by a combination of very complex enzyme systems and nonenzymatic hydration and rearrangement reactions. It is very difficult to draw their metabolic maps. It is not until the enzymatic reactions are analyzed by the isolation and purification

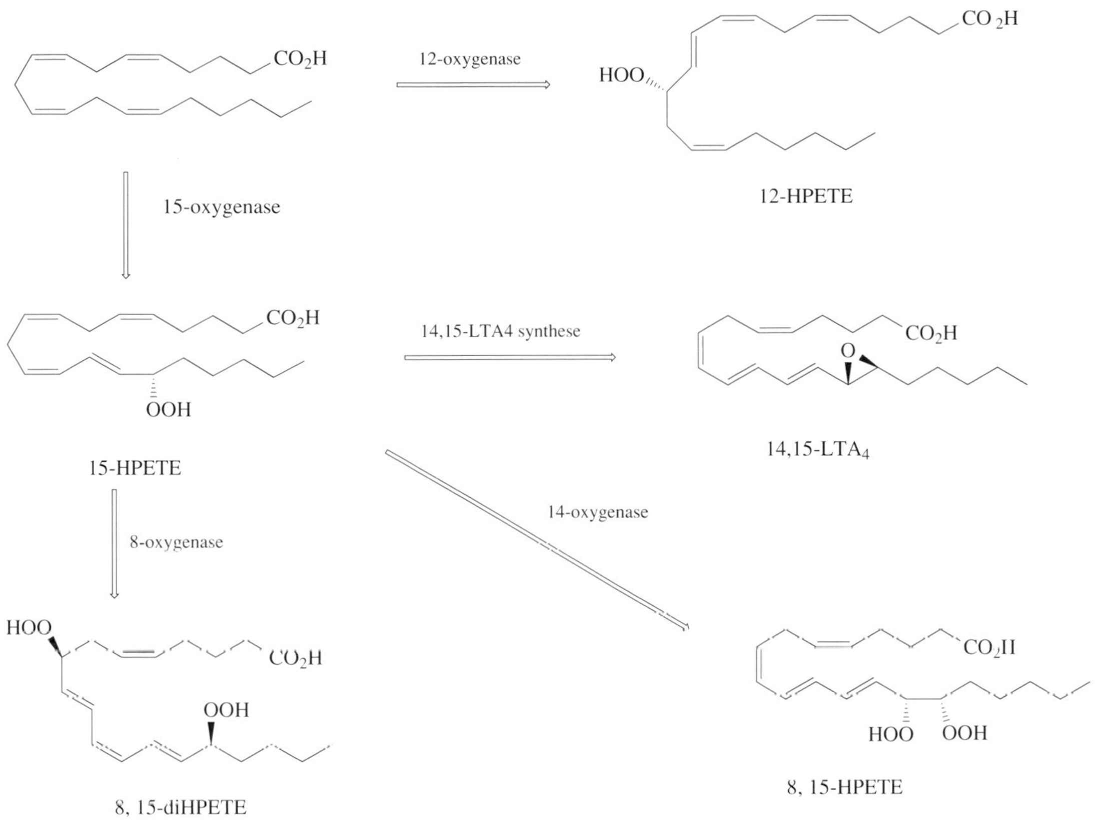

CO_2H 12-oxygenase HOO CO_2H

12-HPETE

15-oxygenase

CO_2H 14,15-LTA4 synthese O CO_2H

OOH

15-HPETE

$14,15\text{-}LTA_4$

14-oxygenase

8-oxygenase

HOO CO_2H OOH

CO_2H

HOO OOH

8, 15-diHPETE

8, 15-HPETE

Figure 17 Multifunctional nature of 12-lipoxygenases.

of each enzyme, the chemical equivalent calculated, and after consideration of many newly formed optical centers, that new developments in this field are possible. Herein the substances derived via 12-lipoxygenase are shown.[86]

1.07.2.6 Free Radical Pathway

In 1990, novel PG-like substances were isolated and elucidated, and they were named F_2-isoprostanes. These substances were 8-iso-PGF and its regioisomer and existed as racemic forms.[87,88] They are produced *in vivo* in humans by the action of free radicals and contain the same skeletal structure as PGs. However, they are not produced *in vivo* by lipoxygenase. It is very interesting that lipoxygenase which is essential for formation of the carbon framework of PGs does not produce such substances. As shown in Figure 21, they are produced by the reduction of PGH_2-like substances which are biosynthesized by the addition of an oxygen molecule to arachidonic acid followed by the cyclization reaction. In contrast with the formation of PGs derived lipoxygenase free, F_2-isoprostanes are formed *in situ* by esterification in phospholipid, and probably released by phospholipase. PGD_2 and E_2-like substances are also isolated in the same way.[89]

PGs do not appear to be stored free in tissues, but instead are biosynthesized and released on demand, but F_2-isoprostanes are proposed to be stored in tissues. Interestingly, some of these substances act as strong renal and pulmonary vasoconstrictors.[90] Furthermore, these bioactivities are inhibited by TX receptor antagonists. However, they are reported to show TX antagonistic activity in blood platelets.[91] Although PGH-like substances which are proposed to be biosynthetic intermediates have not yet been isolated, they are isomerized nonenzymatically to PGE_2/PGD_2-like substances. They not only show TX-like bioactivity as well as F_2-isoprostanes but are also utilized as markers to determine the formation of the lipid peroxide *in vitro* which is greatly related to human diseases.[92]

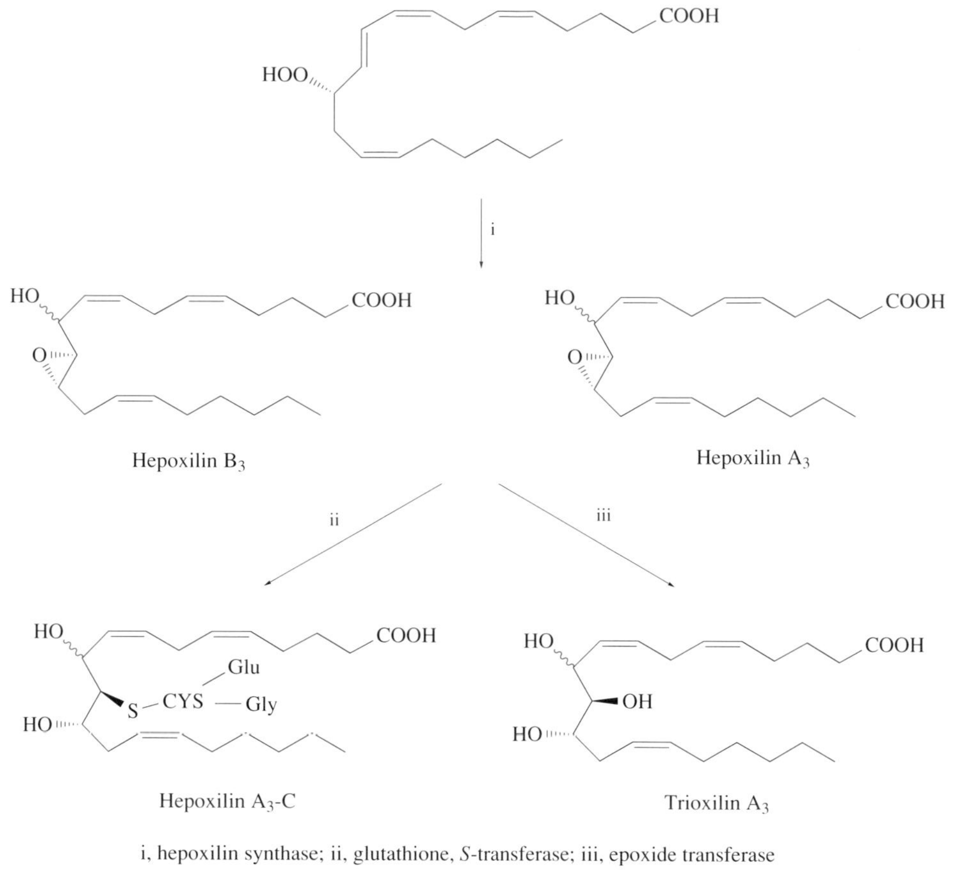

Hepoxilin B_3

Hepoxilin A_3

Hepoxilin A_3-C

Trioxilin A_3

i, hepoxilin synthase; ii, glutathione, *S*-transferase; iii, epoxide transferase

Figure 18 Formation, structures, and metabolism of hepoxilins.

1.07.3 BIOLOGICAL ACTIVITIES OF EICOSANOIDS IN MAMMALS

Eicosanoids display a highly potent and diverse spectrum of biological activities in mammalian tissues as shown in Table 1. These biological activities are very complex, and the same substances show entirely contrary activities according to the mammalian biorhythm, and so we cannot in general terms explain their biological activities. In addition, since the biosynthesized substances show entirely contrary activities to each other, in many cases in the same biosynthetic processes, it is very difficult to understand the reactions in the living body. This class of substances are biosynthesized and released on demand, and they are immediately metabolized to their inactive forms after carrying out their biological activities. Homeostasis of the living body can be maintained by such a mechanism. Although it is important that any biologically active substances must be analyzed externally and their roles in the living body must be observed, it is highly likely that there is a big difference between their biological and pathological activities due to their biological instability.

1.07.3.1 Classical Prostaglandins

PGEs and PGAs are potent vasodilators in most species in most vascular beds. Responses to $PGF_{2\alpha}$ show species variation, but vasodilatation has been observed following injection of $PGF_{2\alpha}$ and PGs A_1, B_1, E_1, and E_2 into the human brachial artery.[93] Dilatation in response to prostaglandins seemingly involves arterioles, precapillaries, sphincters, and postcapillary venules.

PGEs are not universally vasodilatory; constriction effects have been noted at selected sites. Superficial veins of the hand are contracted by $PGF_{2\alpha}$, but not by PGEs. The behavior of other large capacity veins in various animals is similar.[94] Cardiac output is generally increased by PGs E, F, and A. Weak, direct isotropic effects (increased cardiac output) have been noted in various isolated preparations. In the intact animal, however, increased force of contraction as well as increased heart rate is largely a reflex consequence of fall in total peripheral resistance. Systemic blood pressure generally falls in response to PGs E and A, and blood flow to most organs, including

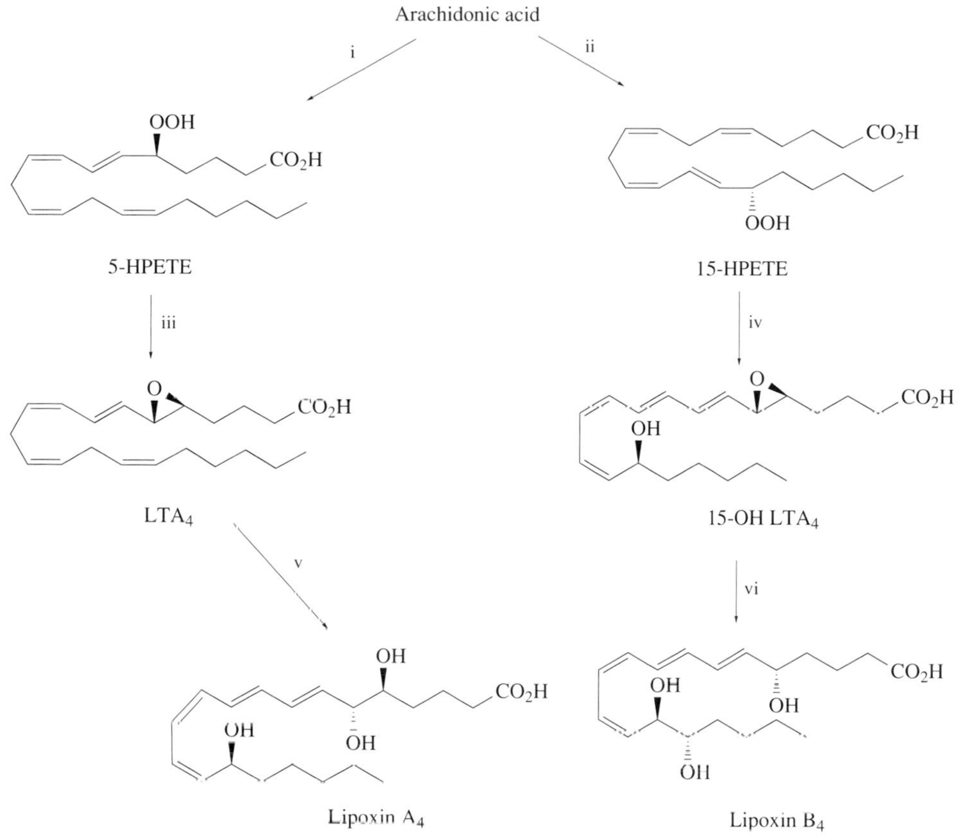

Arachidonic acid

5-HPETE

15-HPETE

LTA_4

15-OH LTA_4

Lipoxin A_4

Lipoxin B_4

i, 5-lipoxygenase; ii, 15-lipoxygenase; iii, 5-lipoxygenase; iv , 5-lipoxygenase; v, 12-lipoxygenase as lipoxin synthase; vi, epoxide hydrase

Figure 19 Structures and biosynthesis of lipoxins.

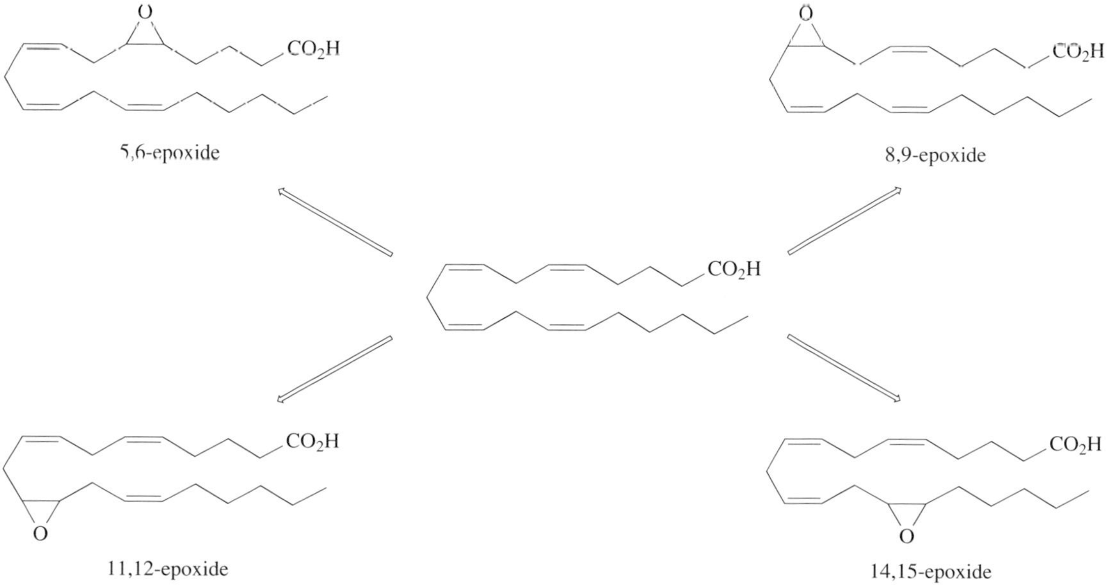

5,6-epoxide

8,9-epoxide

11,12-epoxide

14,15-epoxide

Figure 20 Epoxidation of arachidonic acid by the P450 pathway.

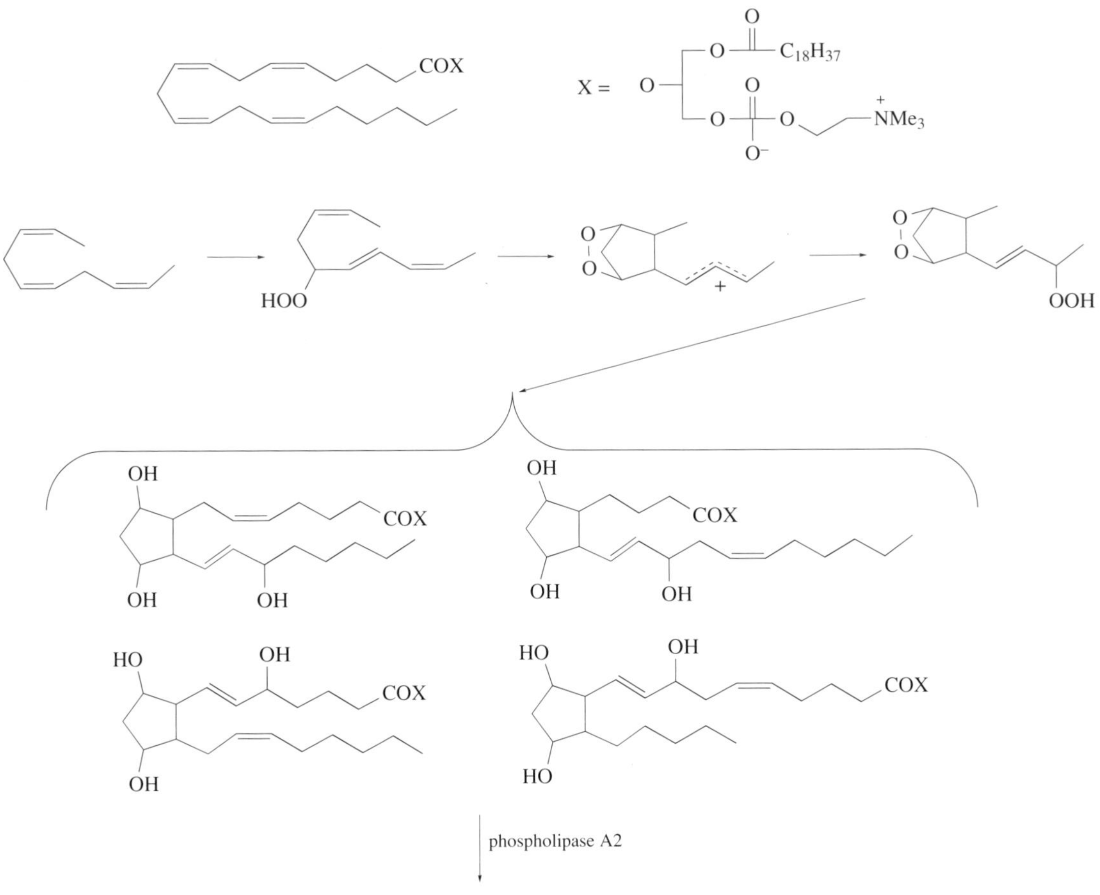

phospholipase A2

free acid (mixture of the steroisomers)

Figure 21 Free radical pathway (isoprostanes).

the heart and kidney, is increased. These effects are particularly striking in some patients with hypertensive disease.[95]

The prostaglandins exert powerful actions on platelets. Some of them, like PGE_1 and PGD_2, are inhibitors of the aggregation of human platelets *in vitro* at low concentrations. PGE_2 exerts variable effects on platelets; it is a potentiator of some forms of aggregation at low concentrations and an inhibitor at higher concentrations.[96]

In general, PGFs contract and PGEs relax bronchial and tracheal muscle from various species. Asthmatic individuals are particularly sensitive, and $PGF_{2\alpha}$ can cause intense bronchospasm. In contrast, both PGE_1 and PGE_2 are potent bronchodilators when given to such patients by aerosol.[97]

Strips of nonpregnant human uterus are contracted by PGFs but relaxed by PGs E, A, and B. Uterine strips from pregnant women are uniformly contracted by PGFs and by low concentrations of PGE_2; high concentrations of PGE_2 induce relaxation. The intravenous infusion of PGE_2 or $PGF_{2\alpha}$ to pregnant human females produces a dose-dependent increase in the frequency and intensity of uterine contraction. The roles of prostaglandins in reproductive processes have been reviewed.[98]

PGEs and PGAs inhibit gastric acid secretion stimulated by histamine or gastrin. Volume of secretion, acidity, and content of pepsin are all reduced. In addition, these prostaglandins are vasodilators in the gastric mucosa. The role of prostaglandins as regulators of gastrointestinal function has been reviewed.[99]

Infusions of PGE_2 directly into the renal arteries of dogs increase renal blood flow and provoke dieresis, natriuresis, and kaliuresis. PGEs inhibit water reabsorption induced by antidiuretic hormone in the toad bladder and in rabbit collecting tubules. In addition, PGE_2 and PGD_2 can cause the release of rennin from the renal cortex.[100]

Inflammation is one of the few conditions in which PGE_2 is a major product of cyclooxygenase and high levels of PGE_2 have been detected in many human inflammatory diseases.[101,102] It is the predominant eicosanoid detected in inflammatory conditions ranging from experimental acute edema and sunburn through to chronic arthritis in man. Prostaglandin E_2 also acts synergistically

Table 1 Physiological, pathological, and pharmacological activities of the ecosanoid family.

System	*Action and acting site*	*Related compound*
Cardiovascular	Periferal and systemic circulation Blood pressure Cardiac function	PGI_2, PGG_2, PGH_2, PGE_2, PGE_1, PGD_1 PGD_2, $PGF_{2\alpha}$, TXA_2
Respiratory	Pulmonary vascular bed Bronchoconstriction Bronchodilation	PGI_2, PGG_2, PGH_2, PGE_2, PGD_2, $PGF_{2\alpha}$ TXA_2, LTC_4, LTD_4
Gastrointestinal	Gastric juice secretion Absorption Gastrointestinal motility	PGI_2, PGE_2, $PGF_{2\alpha}$
Urogenital	Renal homeostasis Regulation of renin–angiotensin system Regulation of electrolyte	PGI_2, PGE_2, $PGF_{2\alpha}$
Endocrinological	Hormone production and secretion Regulation of hypothalamus–pituitary gland system Regulation of blood glucose	PGI_2, PGE_2, $PGF_{2\alpha}$
Reproductive	Follicle, Corpus luteum Ovulation Labor induction	PGI_2, PGE_2, $PGF_{2\alpha}$
Central nervous	Regulation of body temperature Brain function Behavior	PGI_2, PGE_2, $PGF_{2\alpha}$
Periferal nervous	Regulation of neurotransmitter Release in autonomic nervous system	PGI_2, PGE_2, PGE_1
Platelet	Adhesion, aggregation, secretion	PGI_2, TXA_2, PGE_1, $PGF_{2\alpha}$, PGD_2, PGH_2
Connective tissue	Bone absorption Matrix compound formation Osteoclast	PGE_1, PGE_2, PGI_2, $PGF_{2\alpha}$, PHD_2
Sensory	Ocular pressure	PGE_1, PGE_2, $PG_{1\alpha}$
Plasmamembrane	Permeability and fluidity	PGE_2, PGD_2, $PGF_{2\alpha}$
Adenylate cyclase	cAMP production and suppression	PGE_1, PGE_2, PGD_2, PGI_2, PGG_2, PGH_2
Inflammation	Stimulation of vascular permeability Chemotaxis	PGE_1, PGE_2, PGI_2, PGG_2, TXA_2, HPETE, LTB_4, LTC, LTD
Cancer	Regulation of proliferation Tumor invasion	PGE_1, PGE_2, PGD_2, PGI_2, PGJ_2
Immunity	T lymphocyte, B lymphocyte, macrophage LTD_4 production and secretion Anaphylaxis	PGE_2, PGI_2, TXA_2, LTC
Production and differentiation	Fetal growth Malformation DNA synthesis	PGE_2, $PGF_{2\alpha}$, TXA_2
Maturation	Aging and its control	PGI_2, TXA_2, PGE_2

with other mediators to produce inflammatory pain. Without having any direct pain-producing activity, PGE_2 sensitizes receptors on afferent nerve endings to the actions of bradykinin and histamine. PGE_2 is a potent pyretic agent and its production in bacterial and viral infections contributes to the fever associated with these diseases.[103]

1.07.3.2 PGI_2

PGI_2 is the main product of arachidonic acid in all vascular tissues. The cultures of cells from vessel walls show that endothelial cells have the greatest capacity to produce PGI_2.[104] PGI_2 relaxes isolated vascular strips and is a strong hypotensive substance.[105]

PGI_2 is the most potent endogenous inhibitor of platelet aggregation. PGI_2 inhibits platelet aggregation by stimulating adenylate cyclase, leading to an increase in cAMP level in the platelets. In this respect, PGI_2 is much more potent than either PGE_1 or PGD_2. A balance with the thromboxane system exerts a similar homeostatic control of cell behavior. The PGI_2/TXA_2 system may have wider biological significance in cell regulation. PGI_2 inhibits platelet aggregation (platelet–

platelet interaction) at much lower concentrations than those needed to inhibit adhesion. A number of diseases are associated with an imbalance in the PGI_2/TXA_2 system.[106–108]

PGI_2 interacts with endothelium-derived relaxing factor (EDRF, nitric oxide, NO), the labile humoral molecule released by vascular endothelium which is responsible for the vascular relaxant properties of some vasodilators.[109]

1.07.3.3 Thromboxane A_2

Thromboxane A_2, a powerful vasoconstrictor and promoter of platelet aggregation,[110] is released from platelets during aggregation and from guinea-pig lungs perused with arachidonic acid. Its stable degradation product, TXB_2, has considerably diminished biological activity.

TXA_2 also contracts the mesenteric artery, umbilical artery, trachea, and coronary artery and has a coronary vasospastic action in the isolated heart. In addition, it contracts helically cut strips prepared from bovine cerebral conductance arteries and isolated segments of human basila artery. Elevated TXA_2 production has been demonstrated in a number of diseases in which there is a tendency for thrombosis to develop.[107,108,111] Platelets from patients with arterial thrombosis, deep-vein thrombosis, or recurrent venous thrombosis produce more prostaglandin endoperoxides and TXA_2 than normal and have a shortened survival time.[108]

TXA_2 was also found to be a potent contractile agent in the airway.[112] This effect is demonstrable both *in vitro* and *in vivo*, and this discovery led to the hypothesis that TXA_2 might be a mediator of bronchoconstriction in asthma. The possible roles of TXA_2 in the respiratory system, as well as their interactions with the more potent LT have been outlined,[113] and reviews of the biological properties of TXA_2 have been published.[111,112]

In addition to opposing effects of PGI_2 and TXA_2 on platelet function and the cardiovascular system, PGI_2 and thromboxane A_2 have counteracting effects on the bronchopulmonary system and the stomach. In the gastric mucosa, TXA_2 is a potent ulcerogen, whereas PGI_2 can inhibit gastric damage. Likewise, in myocardial and hepatic tissue, TXA_2 can exert a cytolytic action, whereas PGI_2 can protect these tissues from damage, perhaps reflecting again the opposite poles of the same biological mechanisms, in this case cellular integrity. The possible interactions between TXA_2 and PGI_2 in the modulation of smooth muscle tone and motility in the gastrointestinal and reproductive tracts are less clear, since in general both prostanoids are only weak spasmogens on nonvascular tissue.

1.07.3.4 Biological Properties of Lipoxygenase Products

The activity of slow reacting substance of anaphylaxis (SRS-A, a mixture of leukotrienes C_4, D_4, and E_4) became a focus of research interest as it was believed to be an important mediator of human asthma. SRS-A is released concomitantly with prostaglandins and TXA_2 during anaphylaxis.[114,115]

The peptidoleukotrienes are two to three orders of magnitude more potent than histamine as bronchoconstrictors. Leukotrienes contract isolated preparations of trachea, bronchial, and parenchymal smooth muscle, but experiments *in vivo* indicate that they have a selective action on small airways. LTC_4 and LTD_4 induce a preferential reduction in lung compliance and are relatively less effective at reducing specific airway conductance.[116] Furthermore, peptidoleukotrienes have been detected in the sputum from asthmatics[117] and in nasal washes from allergic patients following antigen challenge.[118] Leukotrienes are important mediators of respiratory pathology. The first indication that lipoxygenase activation occurs in inflammation came from the observation that 12-HETE was present in the involved epidermis of patients with psoriasis.[119] LTB_4 has been detected in fluid from involved tissue in rheumatoid arthritis, gout, psoriasis, and ulcerative colitis.[101,120]

The major contribution of LTB_4 and 12-HETE to inflammation is through an effect on leukocytes. The platelet lipoxygenase products, 5-HETE, 12-HETE, and LTB_4, have chemotactic activity for PMNs.[121] LTB_4 has the most potent activity.[122,123] It seems, therefore, that 5-lipoxygenase activity in migrating leukocytes represents a local control mechanism to amplify the recruitment of inflammatory cells to damaged tissues.

1.07.4 SYNTHESIS OF EICOSANOIDS

1.07.4.1 Synthesis of PGs

When PGs were first isolated and their structure elucidated, it was only known that they had a cyclopentane ring with two side chains. The first total synthesis of PGs was reported by Just and Simonovitch.[113] At first, this synthetic work was recognized as a landmark of PG chemistry; however, the chemists of Upjohn Company improved this synthetic method as the Just–Upjohn method and this new route enabled PGs to be produced on a large and industrial scale by combination with Corey's method which is described below.[114,115]

Great interest in a highly potent and diverse spectrum of biological activities of PGs was shown by many scientists in the latter half of the 1960s. PGs are naturally occurring substances widely distributed in mammalian tissues at very low concentrations, and it is very difficult to isolate them from mammalian tissues or enzyme synthesis. Only chemical synthesis can supply adequate amounts for pharmacological and medical studies. Since 1970, much attention has been focused on synthetic chemistry, not only by scientists working in the pharmaceutical and chemical industries but also by academic researchers. In spite of the simple C_{20} organic acids, PGs have several asymmetric carbons, *E*-, *Z*-double bonds, and, furthermore, β-hydroxyl ketone. These structures resulted in a need for a different synthetic method from the usual natural products chemistry used thus far.

1.07.4.1.1 Synthesis of PGs via Corey's lactone

Corey made great efforts in the synthesis of PGs, and succeeded in controlling their stereochemistry, discovering a novel synthetic method for construction of the five-membered ring, introducing new protecting groups, and also developing new reagents. Corey raised PG synthesis to a novel, important, and elegant tool in the field of natural products chemistry. Many reactions developed in PG synthesis and in the construction of the logical synthetic scheme had a great effect on organic chemistry throughout the world.[116–125]

Although PGs are synthesized by a variety of methods and numerous synthetic routes have been reported, their evaluations are carried out not only in academic synthetic organic chemistry but also from industrial viewpoints. At present, the most useful synthetic routes to PGs are Corey's method and the 1,4-addition method.

The synthetic methods for the primary PGs are divided into two fundamental categories related to formation of the carbon framework as follows:

(i) Method 1. Construction of the five-membered ring followed by the introduction of two side chains

(ii) Method 2. The first preparation of the side chains followed by formation of the carbon framework by intramolecular cyclization.

Method 2 is not particularly good for PG synthesis. Method 1 comprises several types and it has reached high levels of achievement of PG synthesis by applying the merits of the following synthetic methods:

(i) Through the Corey lactone (Corey's method).
(ii) 1,4-Addition reaction to the cyclopentane ring.
(iii) Akylation to the cyclopentane ring.
(iv) Ring opening reaction of the cyclopropane ring.

Corey *et al.* reported many synthetic routes to PGs. The synthetic route called Corey's method is generally the synthetic route through the Corey five-membered lactone (**10**) (Figures 22 and 23).

In Figure 22, the steps in which stereocontrol is conducted are as follows:

(a) In step (ii), construction of the *trans* configuration of the two side chains at C-8 and C-12 by Diels–Alder reaction of the substituted cyclopentadiene.

(b) In step (iii), by Baeyer–Villiger reaction the hydroxyl group at C-11 is introduced in *trans* form to the substituent group at C-12.

(c) In step (iv), optical resolution is carried out.

(d) In step (v), iodolactonization made the hydroxyl group at C-9 is introduced in the *cis* manner to the substituent group at C-8.

The only nonstereoselective reaction of this synthesis was the reduction of C-15 ketone. This step was solved by using diisobutylaluminum-2,6-di-*t*-butyl-4-methyl phenoxide[126] or (*S*)-2,2′-dihydroxy-1,1′-binaphthyl-ethanol-lithium aluminium hydride reagent (BINAL-H).[127] Thus, the efficient stereoselective synthesis of PGs was achieved. As minor disadvantage points, this synthesis is i, $ClCH_2OCH_2Ph$; ii, a (2-chloroacrylonitrile); b, KOH; iii, H_2O_2; iv, a, NaOH; b, resolution;

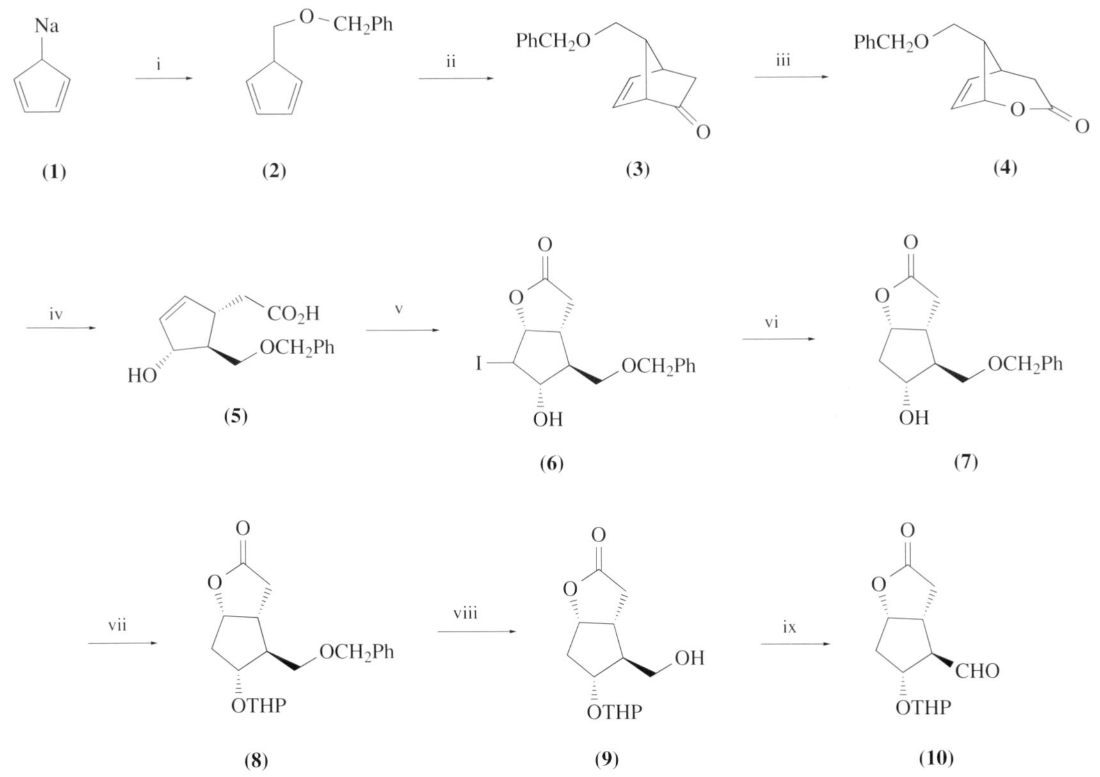

v, KI_3; vi, Bu_3SnH; vii, DHP; viii, H_2-Pd/C; ix, CrO_3-py

Figure 22 Synthesis of Corey's lactone.

straightforward but involves rather long steps, the *cis* selectivity of the Wittig reaction (introduction of the α-chain) is not perfect, and a small amount of *trans* isomer is produced as a by-product. As recognition of the effectiveness of the Corey method, numerous synthetic methods via the Corey lactone have been reported throughout the world (see Figure 24).

1.07.4.1.2 Synthesis of PGs via the 1,4-addition process

The most convenient and efficient synthetic method for the synthesis of PGs is the one in which the α- and ω-side chains are introduced simultaneously into the five-membered ring as outlined in Figure 25. Namely, the organocopper-aided conjugate 1,4-addition of an ω-side chain unit (**33**) to an *O*-protected 4-hydroxy-2-cyclopentenone (**32**), followed by electrophilic trapping of the enolate intermediate (**34**) by an α-side chain organic halide (**35**) was expected to produce the entire PG framework (**36**).

The historical synthetic method for the preparation of PGE_1 by Sih *et al.* using the 1,4-addition reaction is shown in Figure 26.[153,154] The optically active cyclopentenone derivative (**37**) and the ω-side chain (**38**) were independently prepared and then 1,4-addition with *trans*-vinyl copper reagent afforded PGE_1 methyl ester. In this 1,4-addition reaction, the ω-chain was introduced to the five-membered ring selectively in the *trans* relative configuration to the protected hydroxyl group. Furthermore, the α-side chain was introduced in the thermodynamically stable *trans* relative configuration to the ω-side chain and so the substituted functions at C-2,3 and C-3,4 of the cyclopentanone ring were in the *trans* relative configuration to each other. The methyl ester group was enzymatically hydrolyzed to PGE_1.

After publication of Sih's report, many trials on this method were reported; however, the reaction described above had a problem of reproducibility even with the use of nonsubstituted cycloi, $(MeO)_2POCH_2COC_5H_{11}$-NaH; ii, reduction–separation; iii, DHP, iv, DIBAH; v, Ph_3–$CH(CH_2)_3CO_2^-$; vi, oxidation; vii, H^+

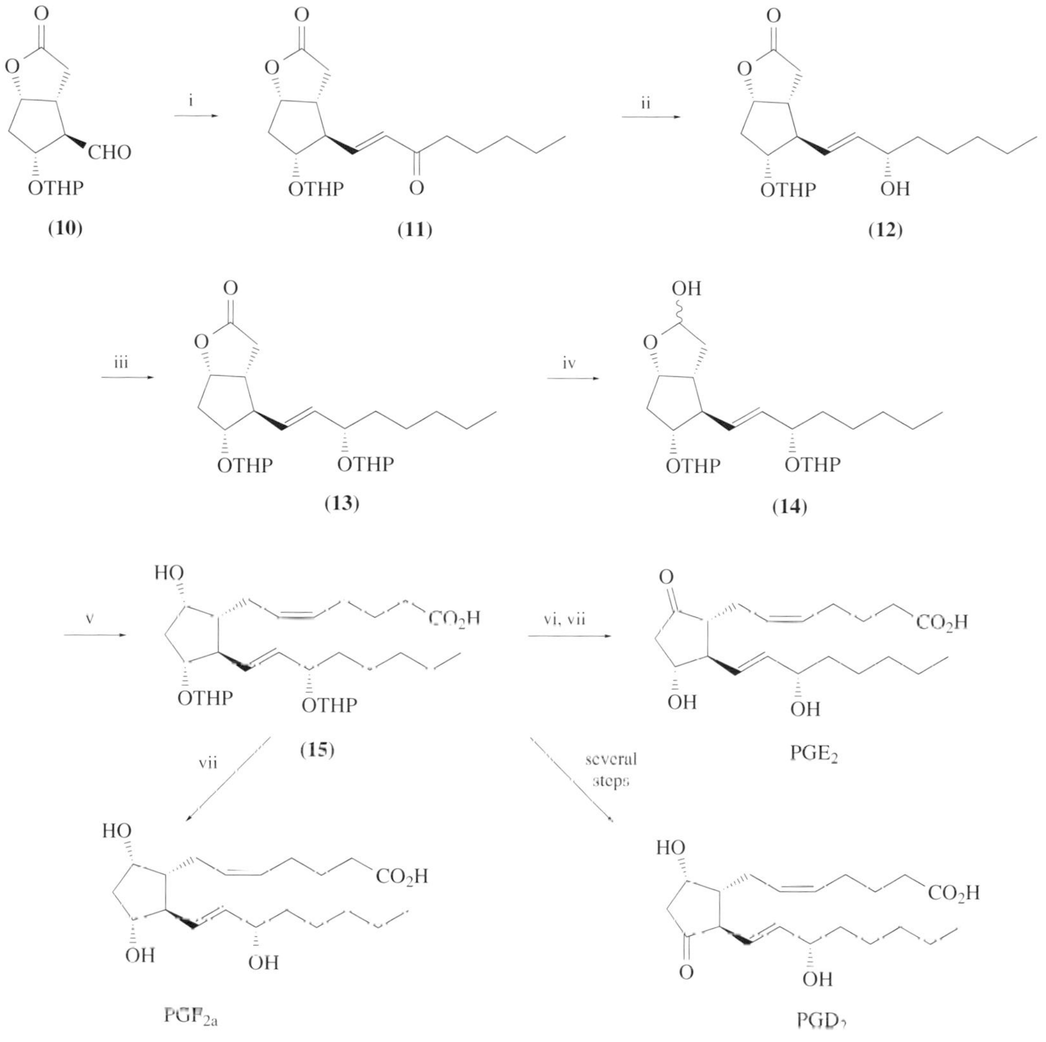

Figure 23 Synthesis of classical prostaglandins by Corey's method.

pentenone.[155–157] The copper-generated enolates lack sufficient reactivity to organic halides and are prone to result in side reactions. Therefore, the method performed was one in which a suitable function was introduced using the electrophilic reagent trapped easily by the enolate followed by the formation of an α-side chain.

Stork and Kobe obtained compound (**39**) by reaction of the cyclopentenone derivative (**32**) with the organocopper reagent followed by trapping the generated enolate with formaldehyde. Compound (**39**) was converted to the exomethylene compound (**40**), which was converted to $PGF_{1\alpha}$ via compound (**42**) by a second 1,4-addition reaction with the *cis*-vinyl copper reagent (**40**) (Figure 27).[158,159]

In contrast, Stork and Fried *et al.*[160–162] examined the yield of the addition reaction in a system similar to that of Sih.[154] To improve this step, they used the *cis*-vinyl copper reagent (**42**). The *cis*-allylic alcohol (**43**) obtained was subjected to a [2.3]-sigmatropic reaction to lead to a PG carbon framework (Figure 28). Furthermore, interestingly, kinetic resolution occurred when the racemic cyclopentenone and the excess of optically active copper reagent were utilized in this reaction to selectively furnish the compound possessing the desired stereochemistry. Additionally, treatment of the optically active cyclopentenone derivative with the excess of the copper reagent (**42**) formed exclusively the compound possessing the desired stereochemistry. The *trans* copper reagent did not cause such a phenomenon. The two-component coupling synthesis has been improved, but this method does not have wide application to the construction of versatile PG frameworks.

Thus, the limitation of the electrophilic reagent trapped by the enolate became an obstacle to

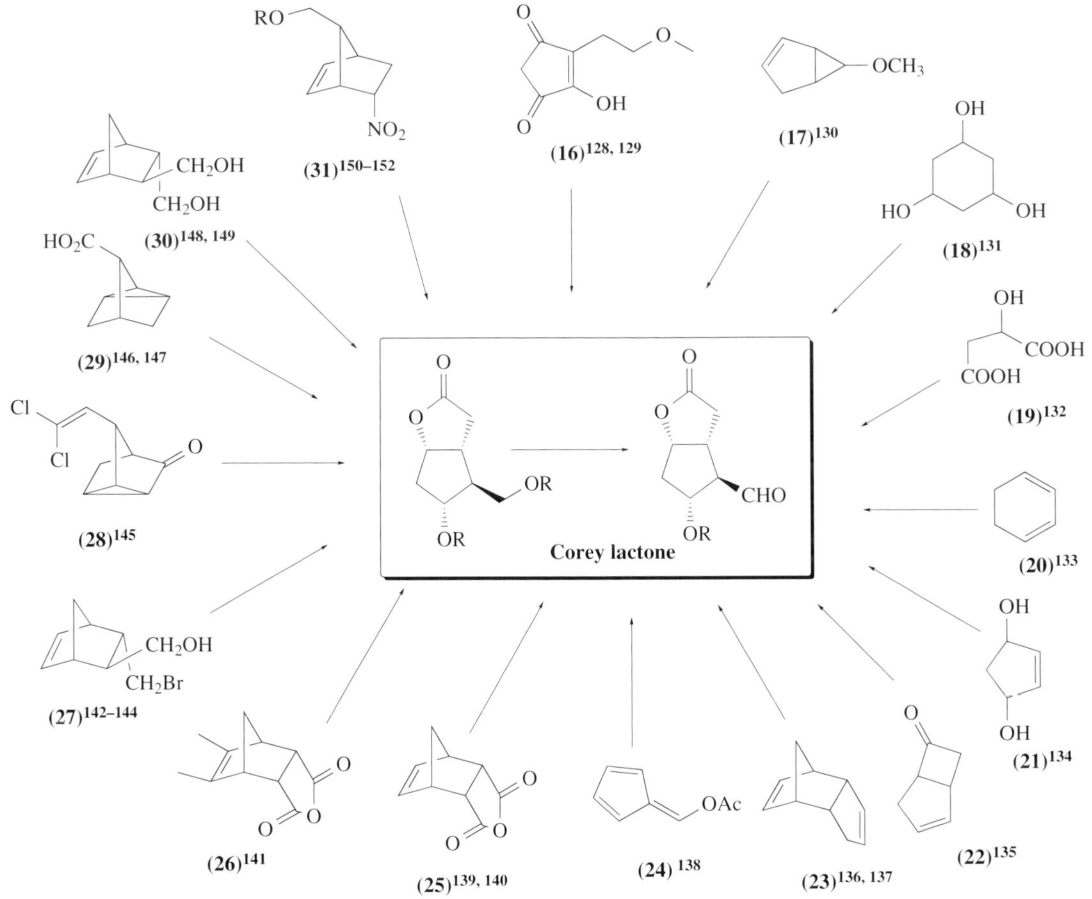

RO
NO_2
(31)[150–152]
O
O
OH
O
(16)[128, 129]
OCH_3
(17)[130]
OH
HO
OH
(18)[131]
CH_2OH
CH_2OH
(30)[148, 149]
HO_2C
(29)[146, 147]
OH
COOH
COOH
(19)[132]
Cl
Cl
O
(28)[145]
O
O
OR
OR
O
O
CHO
OR
Corey lactone
(20)[133]
CH_2OH
CH_2Br
(27)[142–144]
OH
OH
(21)[134]
O
O
O
(26)[141]
O
O
O
(25)[139, 140]
OAc
(24)[138]
(23)[136, 137]
O
(22)[135]

Figure 24 Synthesis of Corey's lactone from various precursors.

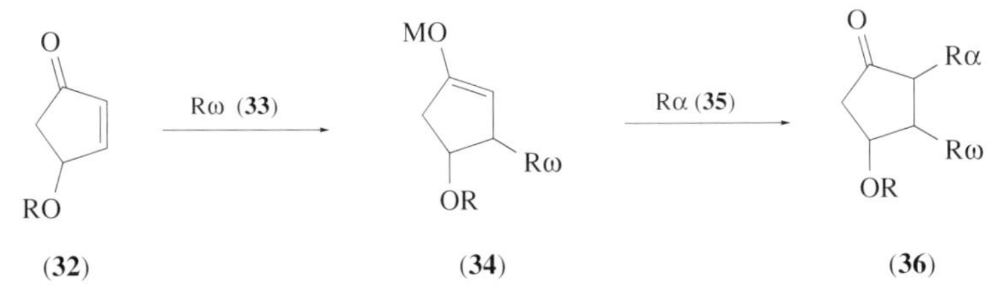

O
RO
(32)
Rω (33)
MO
Rω
OR
(34)
Rα (35)
O
Rα
Rω
OR
(36)

Figure 25 1,4-Addition method.

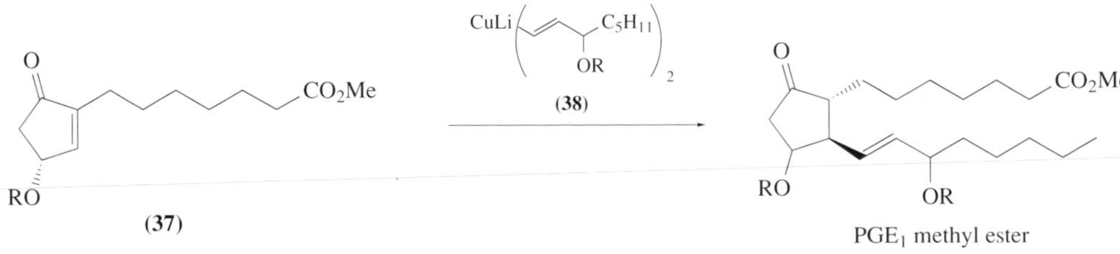

O
CO_2Me
RO
(37)
CuLi
C_5H_{11}
OR
2
(38)
O
CO_2Me
RO
OR
PGE_1 methyl ester

Figure 26 Sih's first synthesis via 1,4-addition method.

developing this synthetic method. However, Noyori and Suzuki examined the 1,4-addition reaction and trapping of the enolate which resulted in the successful and efficient formation of the enolate by use of the following two strategies:[163,164]

(i) Particularly important is the efficient conjugate transfer of sp^2 hybridized carbon to enones, which involves a 1 : 1 enone/side chain stoichiometry. The stereo- and chemoselective transfer of the

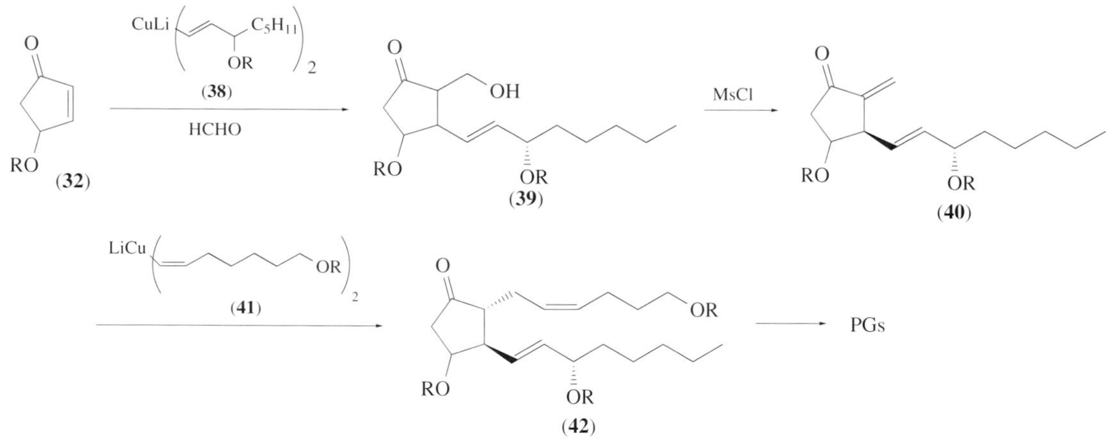

Figure 27 Enolate trapping with formaldehyde.

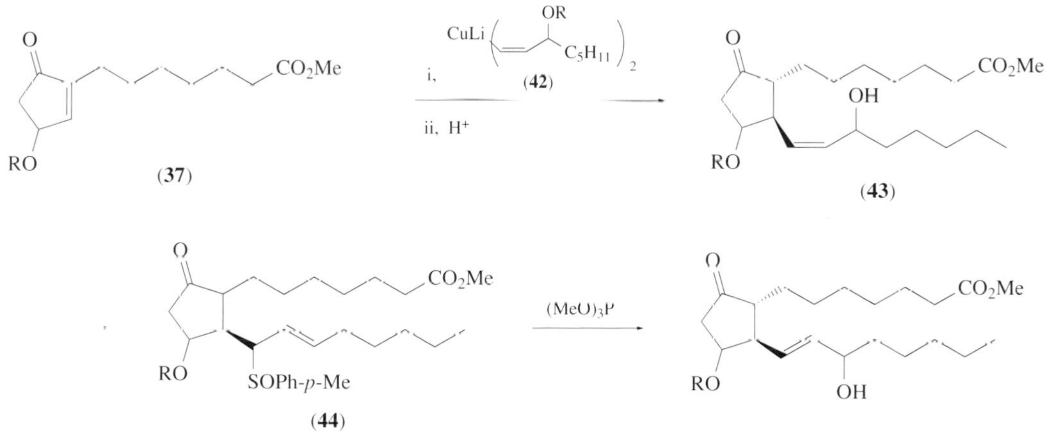

Figure 28 Addition of *cis*-vinyl copper reagent (Stork).

ω-side chain unit to the substituted cyclopentenone is possible in high yield simply by using 0.1 equivalent of the reagent.

(ii) Conjugate addition using a stoichiometric (not excess) quantity of organometallic reagent to the cyclopentenone unit is also possible by using a copper reagent formed *in situ* by mixing organolithium, copper(I) iodide, and tributylphosphine in a 1 : 1 : 2–3 mole ratio in ether or THF. They obtained the desired products in a high yield by aldol reaction or Michael-type trapping of the enolate,[165] acetylenic aldehyde, and nitroolefin.[166] The acetylenic product (**46**) is a common intermediate for the general synthesis of naturally occurring PGs of series 1 and 2 (Figure 29).

The direct alkylative trapping of the enolate intermediate cannot be achieved in this form, but is possible with the aid of organotin compounds. Thus, a one-pot, sequential treatment of the organocopper reagent with the enone, triphenyltin chloride, and α-side chain propargyl iodide led to the desired condensation product in high yield. Similarly, when the three-component coupling was performed with a *Z*-allyl iodide, a protected PGE_2 was obtained in high yield.[166–168]

Thus, total synthesis of PGE_2 from cyclopentenone was reduced to only three steps. The total chemical synthesis of PGs is reaching completion.

1.07.4.2 Synthesis of PGI_2

Syntheses of PGI_2, including structural identification, were reported simultaneously by several research groups. Most of the methods for the synthesis of PGI_2 were conducted in a similar manner in which the unstable enol ether linkage was constructed in the last step. They involve intermolecular ether cyclization with the electrophilic reagent and the elimination reaction; only the reagents vary among the different investigators.

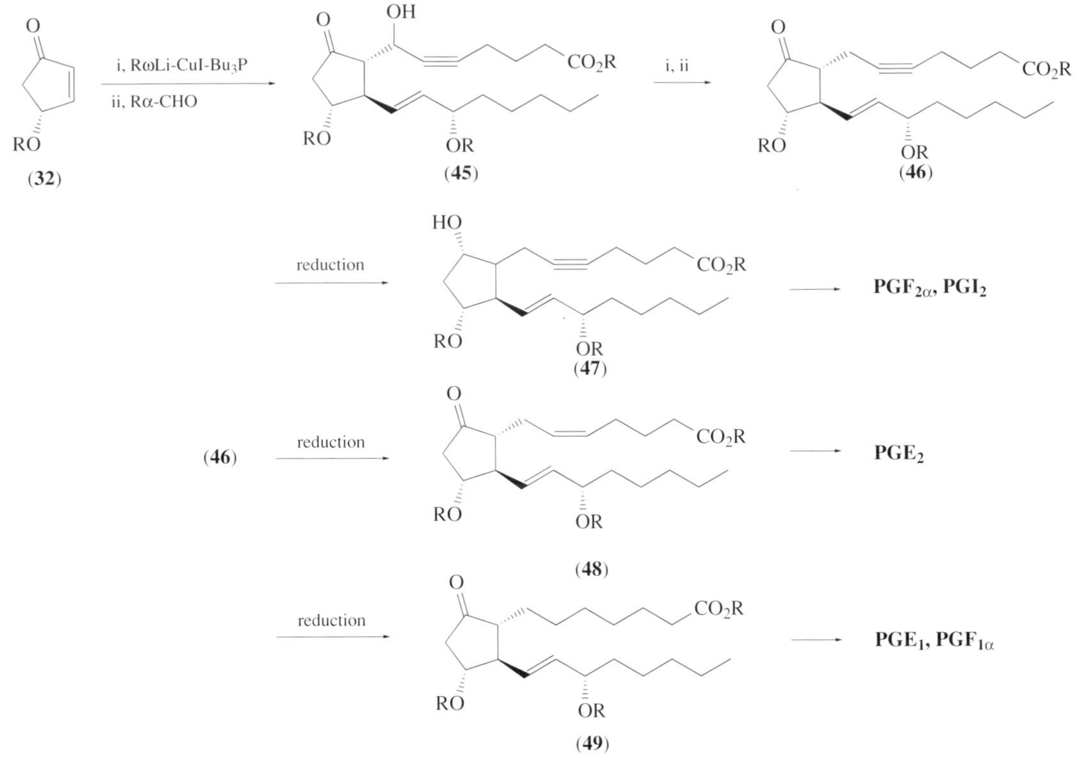

i, PhCSCl-DMAP; ii, $(Bu^n)_3SnH-(Bu^tO)_2$

Figure 29 Noyori's PG synthesis via 1,4-addition route.

The first total synthesis of PGI_2 was reported by Corey *et al.*,[169,170] including confirmation of the stereochemistry of the enol ether linkage. It was known that treatment of the $PGF_{2\alpha}$ derivative with a reagent containing a halogen such as *N*-bromosuccinimide (NBS) or iodine afforded the halo ether (**50** , **51**) between C-9 alcohol and the C-5,6 double bond. Corey *et al.* synthesized PGI_2 by a dehydrohalogenation reaction with base from this halo ether (Figure 30). Many similar synthetic routes to PGI_2 have been reported in the literature.[171–177]

Noyori and co-workers[178] obtained a PGI_2-Na salt stereoselectively by the use of oxymercurylation with mercury(II) trifluoromethyl acetate from 5,6-dehydro-$PGF_{2\alpha}$ methyl ester (**46**) as a starting material which was derived by the asymmetric three-component coupling method, followed by reductive demercurylation and hydrolysis. Noyori's method is an excellent synthetic route for obtaining PGI_2 in short steps, but it proved difficult to remove the mercury from the product. In order to overcome this problem, an improved method of cyclization was required, and his group succeeded by using a palladium complex to obtain compound (**52**) and depalladation gave PGI_2 methyl stereoselectively.[179]

1.07.4.3 Synthesis of PGH_2 and PGG_2

PG endoperoxides are the first metabolites in the cycloxygenase PG biosynthetic system. There are two kinds of PG endoperoxides, namely PGG_2 possessing 15(*S*)-hydroperoxide and PGH_2 with a 15(*S*)-hydroxyl group. Platelet aggregation and contraction of the smooth muscle activities of PGG_2 and PGH_2 are much higher than those of the primary PGs. However, both of them are very unstable in aqueous solution and their half-lives are only about 5 min. The cause of this instability derives from the carboxylic function in the molecule. The synthesis of PGH_2 methyl ester was reported by Johnson *et al.*[180] Porter *et al.* improved Johnson's synthetic method: compound (**53**) was transformed into PGH_2 methyl ester in 21% yield by treatment with anhydrous hydrogen peroxide in the presence of silver trifluoromethyl acetate in diethyl ether.[181] Afterwards, Porter *et al.* further improved and established the synthetic method for the preparation of endoperoxides.[182,183] First, the methyl ester (**53**) was hydrolyzed exclusively by the use of lipase, and then the resulting

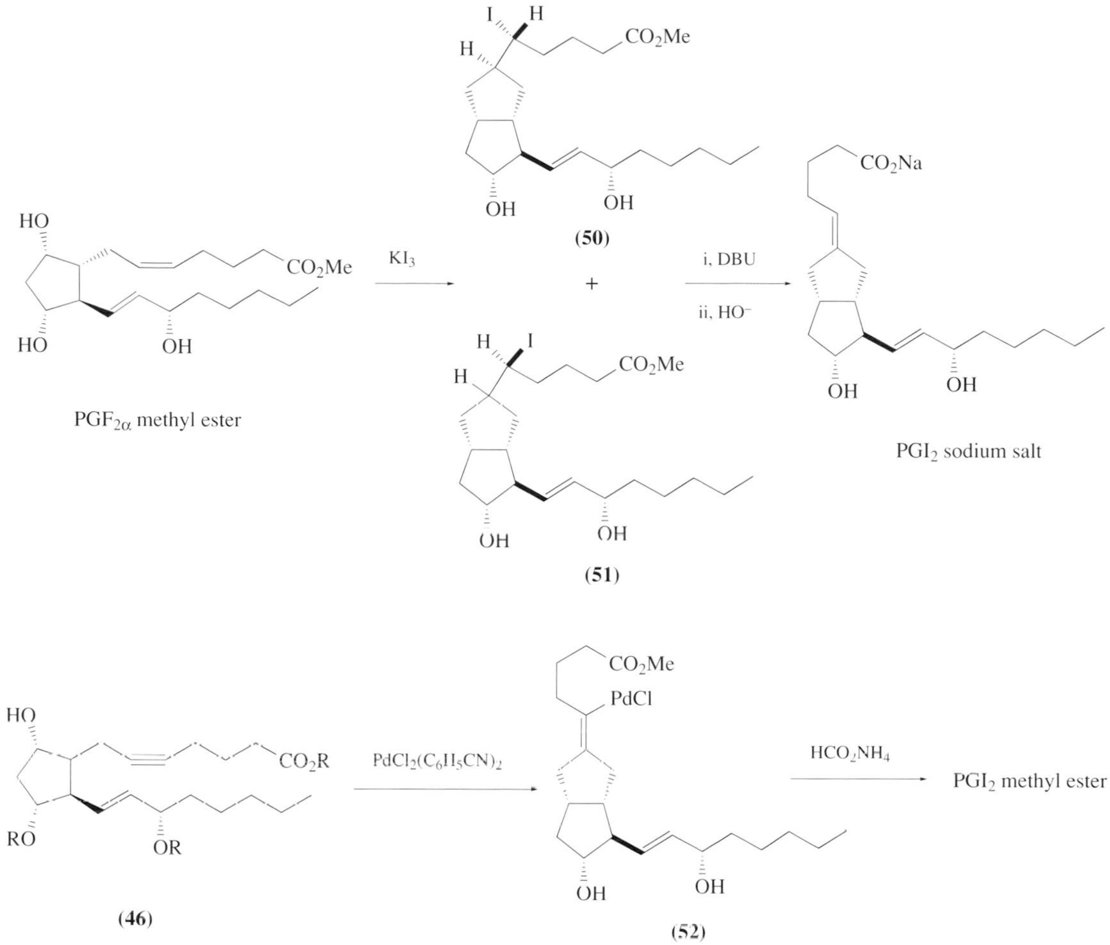

KI_3

PGF$_{2\alpha}$ methyl ester

(50)

+

(51)

i, DBU

ii, HO^-

PGI$_2$ sodium salt

(46)

$PdCl_2(C_6H_5CN)_2$

(52)

HCO_2NH_4

PGI$_2$ methyl ester

Figure 30 Synthesis of PGI$_2$ sodium salt.

compound was subjected to the endoperoxide formation reaction to give PGH$_2$. Second, the 15(S)-hydroxyl group in (**54**) was converted to 15(R)-chloro compound (**55**) by use of Mukaiyama's reagent[183] and then lipase. Treatment of (**55**) with silver trifluoromethyl acetate–anhydrous hydrogen peroxide furnished PGG$_2$ (Figure 31).[184]

PGG$_2$ and PGH$_2$ thus prepared were not unstable as reported earlier, and PGH$_2$ methyl ester could be stored at $-89\,^{\circ}C$ for several months. The constant supply of PGH$_2$ and PGG$_2$ enables studies of the chemical transformation of endoperoxide.

1.07.4.4 Synthesis of TXA$_2$

TXA$_2$ is one of the most difficult synthetic target molecules in synthetic organic chemistry for two reasons. First, the 2,6-dioxabicyclic[3.1.1]heptane ring has an unknown carbon framework, and second, the instability of TXA$_2$ whose half-life is only 30 s as reported by Samuelsson. Even if the synthesis of TXA$_2$ is achieved, its isolation and purification are extremely difficult, and furthermore there is no authentic sample in nature and also no physical data.

After some serious trials, Still *et al.* succeeded in the synthesis of TXA$_2$, starting from TXB$_2$-C-1, C-15 lactone (**57**).[185,186] In this method, the C-10 radical which is the intermediate in the reductive dehalogenation cannot attack the C-13, C-14 double bond, thus the reaction points are separated efficiently. After the C-11 hydroxyl group was dehydrated to afford the unstable enol ether (**58**), treatment with NBS provided the monobromide (**59**). Compound (**59**) was subjected to Mitsunobu reaction to produce the oxetane (**60**), which was stable enough to be purified by the usual silica gel column chromatography. The bromo group at C-10 was removed by photoirradiation with tin i, [3-ethyl-2-chloro-oxazolo[4,5-b]pyridinium BF$_4^-$] Et_4NBr ; ii, lipase ; iii, H_2O_2–AgOAc; iv, [3-ethyl-2-chloro-oxazolo[4,5-b]pyridinium BF$_4^-$] Et_4NCl

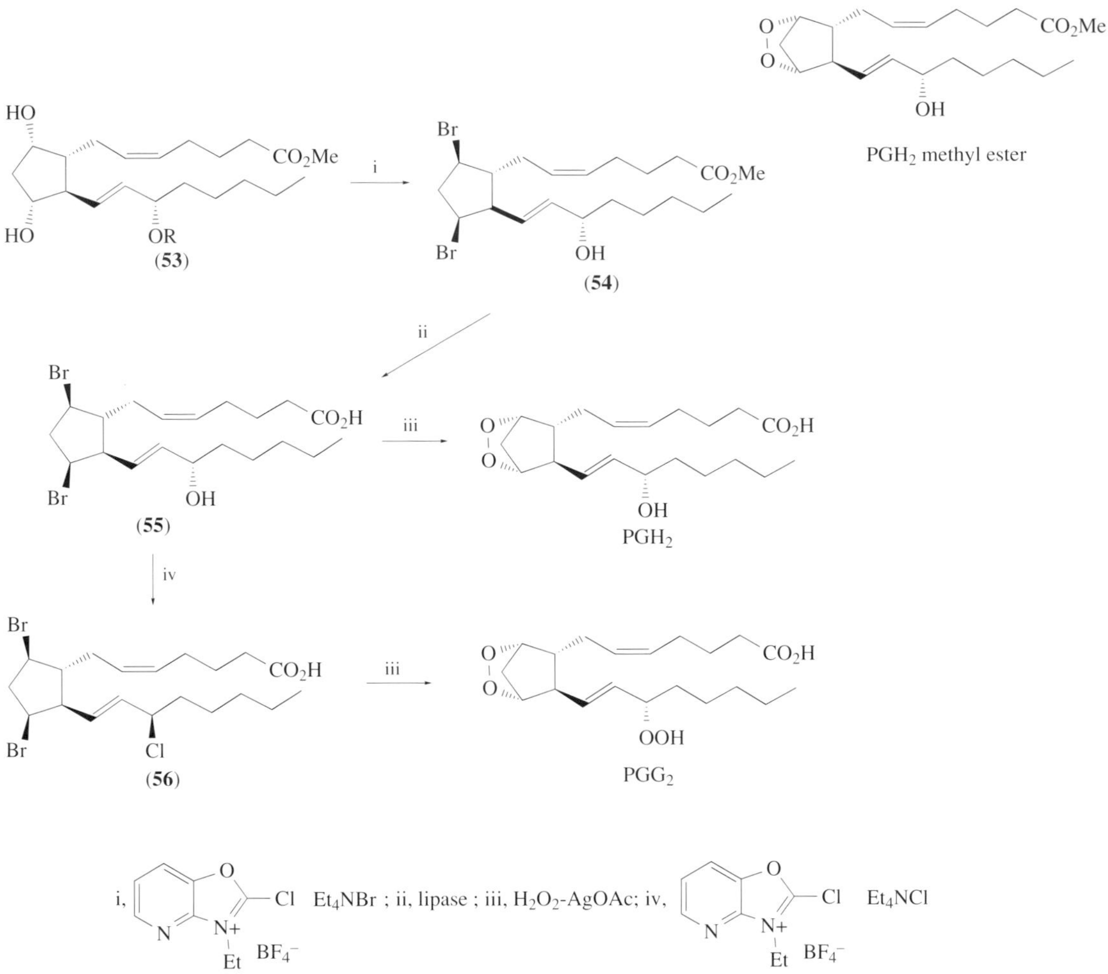

Figure 31 Synthesis of PGH_2 and PGG_2.

hydride supported with resin in the presence of AIBN to give TXA_2-C-1, C-15 lactone (**61**) as crystals (Figure 32). This lactone did not show TXA_2-like activity. Saponification of this lactone with NaOH–methanol–water provided the TXA_2–Na salt, which has the same TXA_2 activity as the one derived by enzyme.

This sodium salt was stable enough at $-20\,^\circ C$ to be stored for more than a week. The synthesis of TXA_2 was thus accomplished. However, the toxicity *in vivo* was not as high as anticipated by the *in vitro* activity.[187] This did not come from the nature of synthesized TXA_2 itself but from the decomposition mechanism *in vivo*.

1.07.4.5 Synthesis of Leukotrienes (LTs)

Borgeat and Samuelsson[188] and Corey *et al.*[189] proposed LTA_4 possessing the epoxy function as the intermediate for the biosynthesis of LTB_4 and LTC_4 since the production of both of them was in parallel. Furthermore, it was reasonable that 5-HPETE is a precursor in the biosynthesis of LTs from arachidonic acid. On the basis of this hypothesis, Corey *et al.* began the total synthesis of LT-related substances. In the structural identification studies, they synthesized all the possible analogues in the geometry of the double bonds and the configuration of the hydroxyl group, and then compared them with the natural substances.

The strategy for the synthesis of lipoxygenase metabolites is different from PGs with a ring system and is not versatile. As a basic strategy, stereoselective introduction of the hydroxyl group and the epoxy function, and geometrical introduction of the double bond are important. The strategy outline is as follows:

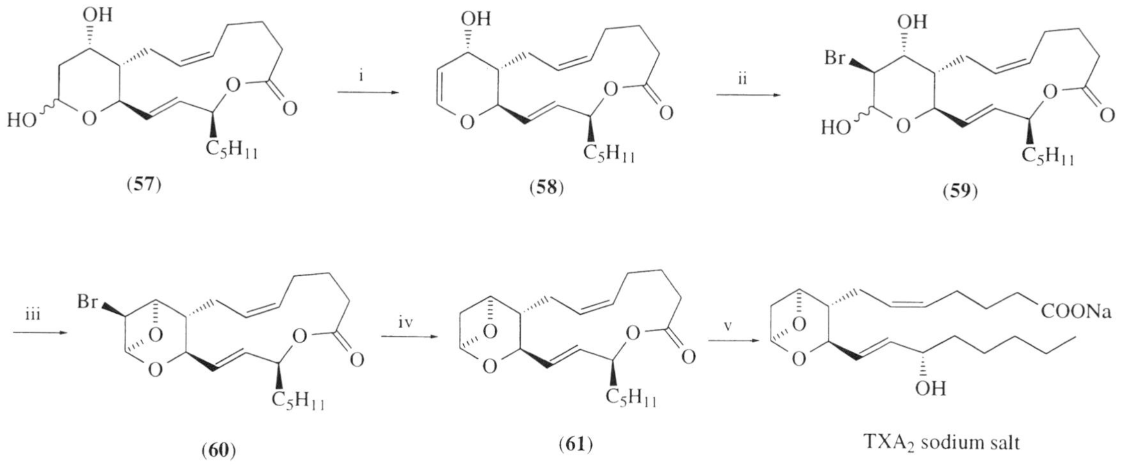

i, 2-chloro-1-methylpyridinium BF_4^-, Et_3N; ii, NBS; iii, $(MeO)_3P$/DEAD; iv, polymer supported tin hydride, AIBN, $h\nu$; v, NaOH

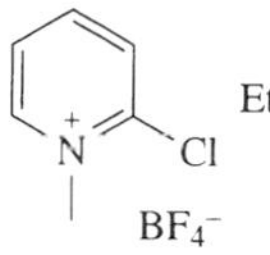

Figure 32 Synthesis of TXA_2.

(i) Synthesis of the alcohol and epoxide:
 (a) optical resolution
 (b) utilization of the sugar and the amino acids
 (c) application of the Sharpless epoxidation
 (d) utilization of the asymmetric reduction
(ii) Formation of the double bond
 (a) the Wittig reaction
 (b) reduction of the triple bond.

These are general synthetic methods for the preparation of LTs, but they are not without complications. For example, with respect to the introduction of the hydroxyl group, when sugars are utilized, there are many combinations in the selection of protecting groups to remove unnecessary parts. As for the double bond, the Wittig reaction which is reported to form the *cis* double bond produced a certain percentage of the *trans* isomer. An effective reaction to yield the isolated *trans* double bond has not yet been reported. Construction of the triple bond is rather difficult. In order to overcome these weak points, a variety of devices are used and finally the pure compounds are obtained by HPLC purification.

1.07.4.5.1 Synthesis of LTA, LTC, and LTD

In order to determine the stereochemistry of the epoxide of LTA_4, Corey *et al.* selected tribenzoyl (−)-ribose as a starting material. As shown in Figures 33 and 34, the epoxides **(63)** and **(64)** were obtained, whose stereochemistry is known. Compounds **(63)** and **(64)** are the most important intermediates for the synthesis of LTs, and are target molecules in many synthetic studies for LTs. Compound **(64)** was oxidized to the aldehyde **(65)**, and then dienal **(66)** was produced by C-4-homologation. The Wittig reaction of **(66)** with the C-9 segment gave LTA_4 methyl ester in which the stereochemistry was the same as that of natural compounds. Treatment of this LTA_4 methyl ester with the protected glutathione and other amino acids followed by deprotection afforded LTC_4 and other LTs.[190–195] Stereoisomers of LTA_4 are also obtained from epoxides **(67)**, **(68)**, and **(69)** in a similar way.[196,197] Many other syntheses of LTC_4 have been reported.[198–202]

1.07.4.5.2 Synthesis of LTB_4

LTB_4 is synthesized by hydrolysis (S_N2' reaction) of the epoxy group of LTA_4 and by introduction of hydroxyl groups. When SRS-A was isolated and elucidated, little attention was paid to LTB_4.

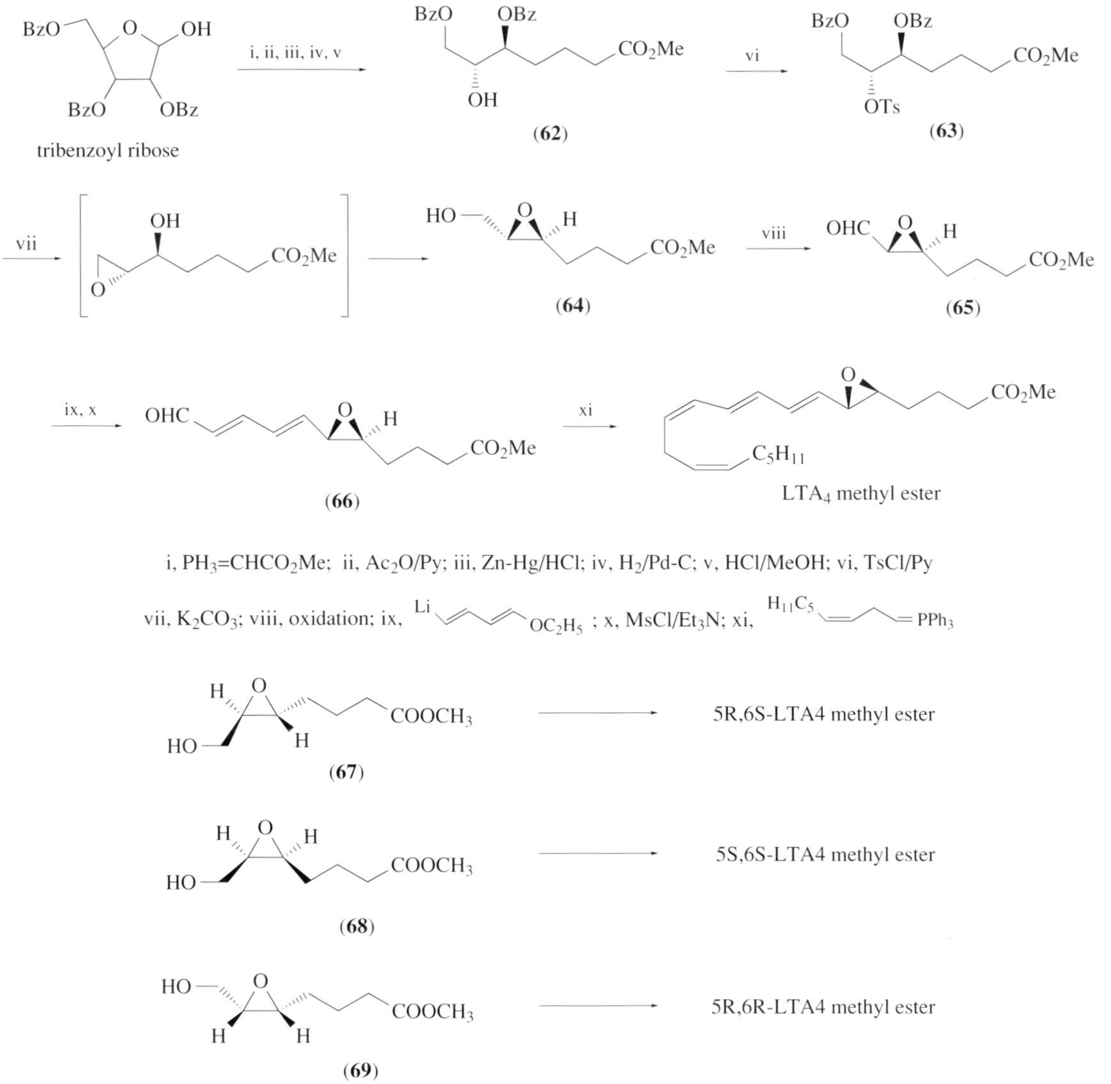

tribenzoyl ribose → (62) → (63) → (64) → (65) → (66) → LTA_4 methyl ester

i, PH_3=$CHCO_2Me$; ii, Ac_2O/Py; iii, Zn-Hg/HCl; iv, H_2/Pd-C; v, HCl/MeOH; vi, TsCl/Py

vii, K_2CO_3; viii, oxidation; ix, Li–CH=CH–CH=CH–OC_2H_5 ; x, MsCl/Et_3N; xi, $H_{11}C_5$–CH=CH–CH_2CH_2=PPh_3

(67) → 5R,6S-LTA4 methyl ester

(68) → 5S,6S-LTA4 methyl ester

(69) → 5R,6R-LTA4 methyl ester

Figure 33 Synthesis of LTA_4 methyl ester.

Afterwards it was reported that LTB_4, formed by enzymatic hydrolysis of LTA_4, was chemotactic for macrophages and neutrophils at low concentration and that it acts as a mediator in various inflammations. Since then, much attention has been focused on LTB_4. Samuelsson *et al.*[203,204] presumed that two hydroxyl groups existed at C-5(*S*) and C-12(*R*) positions and that the conjugated triene consisted of a *cis* and two *trans* double bonds by UV absorption spectrometry; however, they could not determine the exact structure. Corey *et al.*,[205] prior to the actual synthesis, considering the transition state from LTA_4 to LTB_4 and the repulsion in a molecule, started to synthesize the structural molecule of LTB_4 (Figure 35). The triene compound which was obtained by Wittig reaction of the aldehyde **(73)** with the phosphonium salt **(74)** was identified as the same as the natural compound. Many other syntheses of LTB_4 have been reported.[202–204,206–208]

1.07.5 SYNTHESIS OF AGONISTS AND ANTAGONISTS

1.07.5.1 Synthesis of Analogues of Classical PGs

Since PGs are characterized by a highly potent and diverse spectrum of biological activities, it is a matter of course that these biological activities can be applied to medicines. However, the administration of these medicines, which play a fundamental role in cellular metabolism and cell

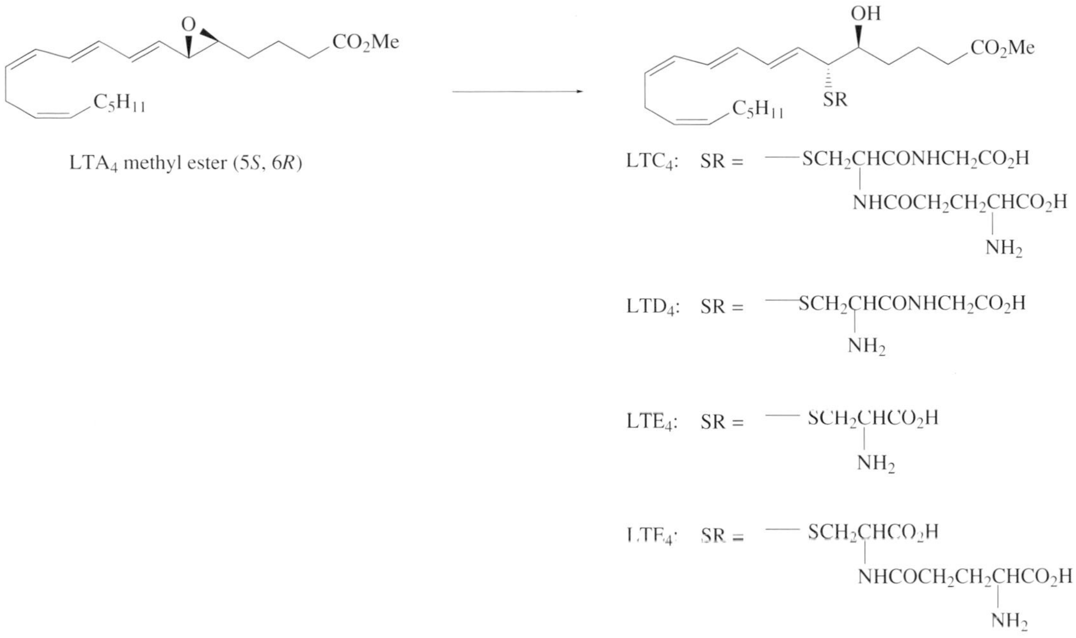

LTA$_4$ methyl ester (5*S*, 6*R*)

LTC$_4$: SR = $-SCH_2CH(NHCOCH_2CH_2CH(NH_2)CO_2H)CONHCH_2CO_2H$

LTD$_4$: SR = $-SCH_2CH(NH_2)CONHCH_2CO_2H$

LTE$_4$: SR = $-SCH_2CH(NH_2)CO_2H$

LTF$_4$: SR = $-SCH_2CH(NHCOCH_2CH_2CH(NH_2)CO_2H)CO_2H$

Figure 34 Synthesis of LTC$_4$, LTD$_4$, LTE$_4$, and LTF$_4$.

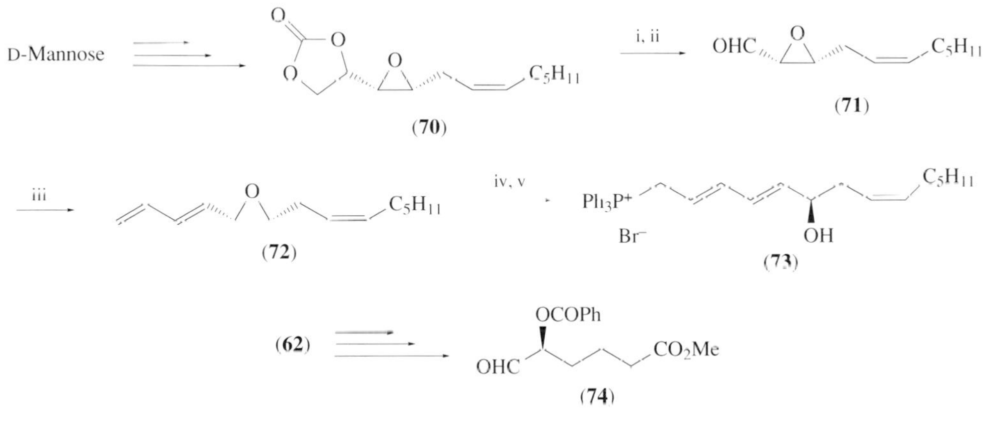

D-Mannose → (70) —i, ii→ (71) —iii→ (72) —iv, v→ (73)

(62) → (74)

i, base; ii, Pb(OAc)$_4$; iii, Wittig reaction; iv, HBr; v, Ph$_3$P

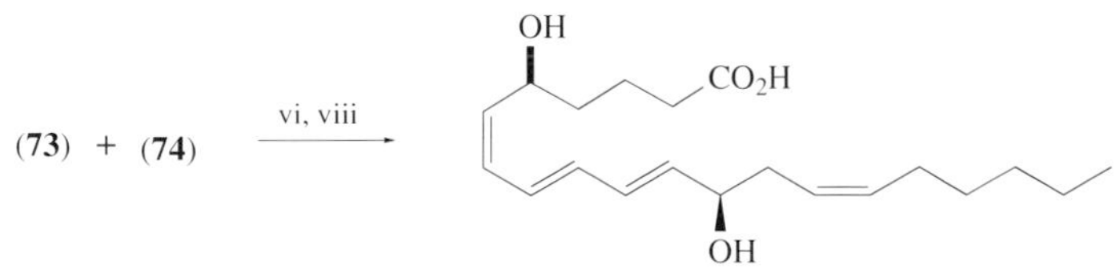

(73) + (74) —vi, viii→

vi, Bu^{n}Li; vii, base

Figure 35 Synthesis of LTB$_4$.

function, will have a great influence on the whole body. In addition, the instability of the substances results in the shortness of their biological activities. In order to overcome these difficulties, many chemists took part in the serious competition of the development of these substances.

With respect to the primary PGs, structural modification of the carbon framework was not performed, but modification of the side chains was carried out. This is due to the following:

(i) the bioactivities of PGs themselves were applied to the medicines;

(ii) conformation of the structures of PGs having a five-membered ring with two simple side chains was not clearly determined; and

(iii) as the result of (ii), the molecular design was difficult.

A tremendous number of PG analogues have been synthesized and screened. Structurally modified PGEs described in patents and papers are shown in Figure 36. Basically they are examples of α- and/or ω-chain modifications. Among them, the analogues which showed better biological profiles than natural PGs in the field of obstetrics and circulation have been investigated and developed as medicines.

However, after the chemically unstable PGH_2, PGI_2, TXA_2, etc. were elucidated and they were found to have very high biological activities, trials to synthesize the stable analogues were performed all over the world. At first, chemically interesting analogues were synthesized and they showed very high agonistic and antagonistic activities. Thereafter, pharmacological studies utilizing these PG analogues stimulated great progress in the field of PG studies.

1.07.5.2 Synthesis of PGH_2 and TXA_2 Analogues

PGH_2 has a bicyclo[2.2.1] structure and is easily synthesized; therefore many analogues have been prepared, for example **(75)**–**(85)**.[209,210] However, with a few exceptions, almost none of the synthesized analogues showed PGH_2-like biological activities. The exceptional analogue was U-46619 (**75**) and it is used at present as a TXA_2 agonist. However, the regioisomer (**76**) in the epoxymethano part had low biological activity.[211] Disulfide analogue (**77**) possessed PGH_2-like activity.[212]

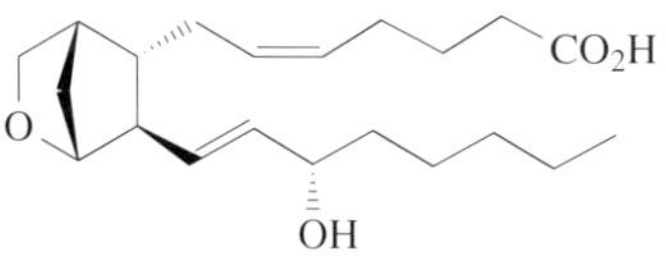

O CO₂H OH (**75**)

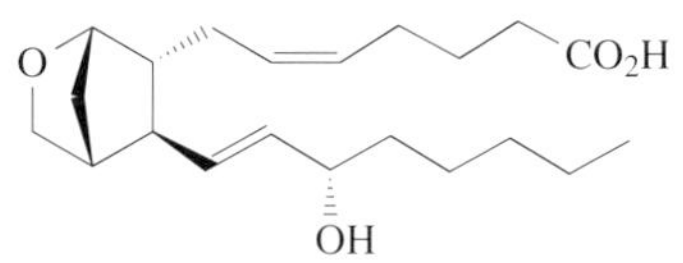

O CO₂H OH (**76**)

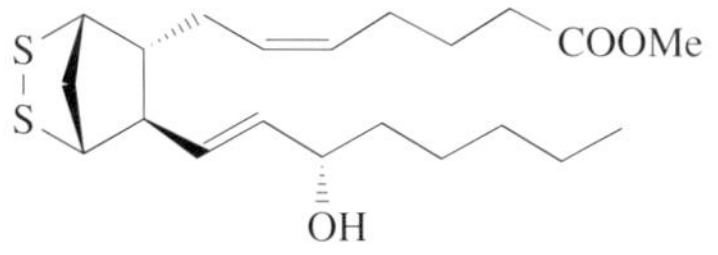

S S COOMe OH (**77**)

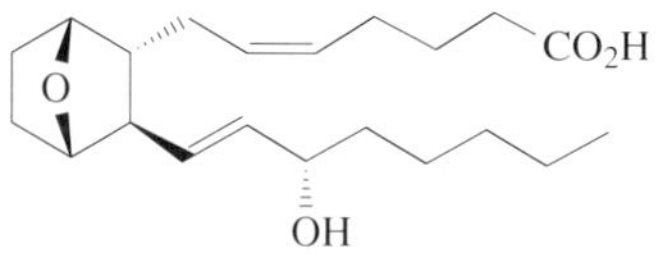

O CO₂H OH (**78**)

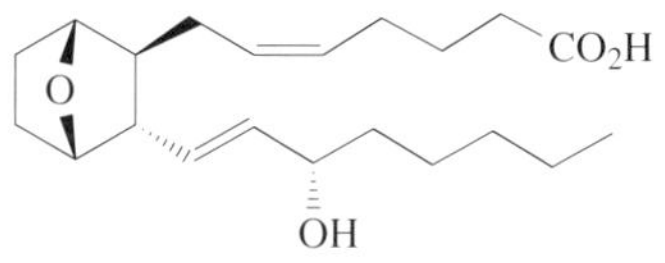

O CO₂H OH (**79**)

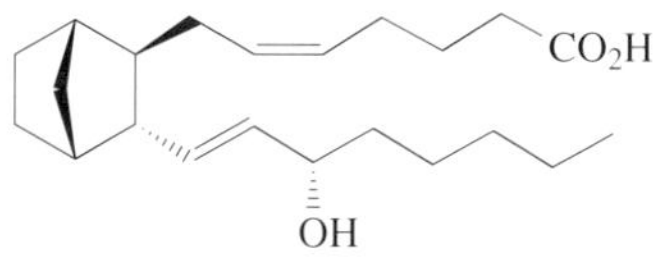

CO₂H OH (**80**)[217]

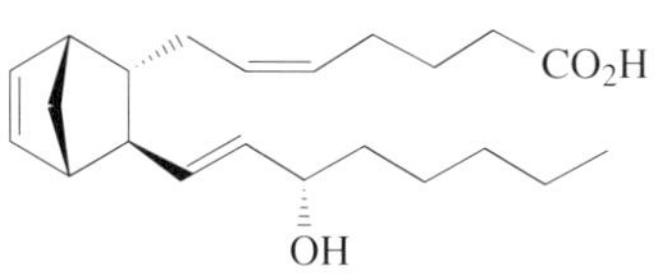

CO₂H OH (**81**)[218]

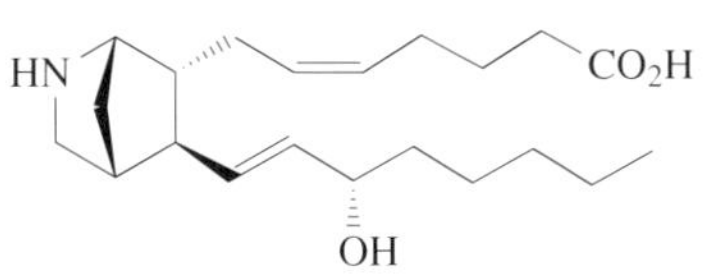

HN CO₂H OH (**82**)[219]

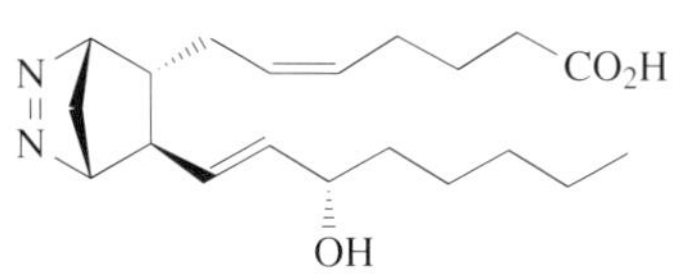

N=N CO₂H OH (**83**)[220]

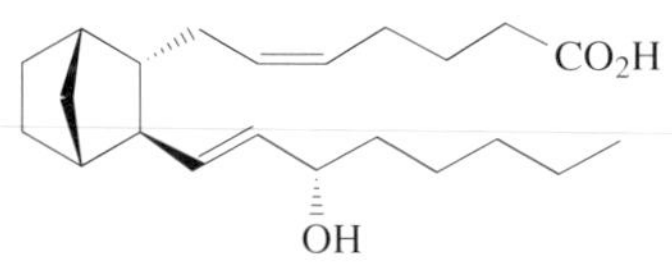

CO₂H OH (**85**)[221]

It was reported that PGH_2 and TXA_2 bound to the same receptor and disclosed their activities.[213] Compounds (**78**) and (**79**), which have the same plane structure, according to their stereo-configuration, showed agonistic and antagonistic activity, but the relationship between structure and activity was not clear.[214,215] Many other analogues, (**80**)–(**85**), have been reported.[216–220]

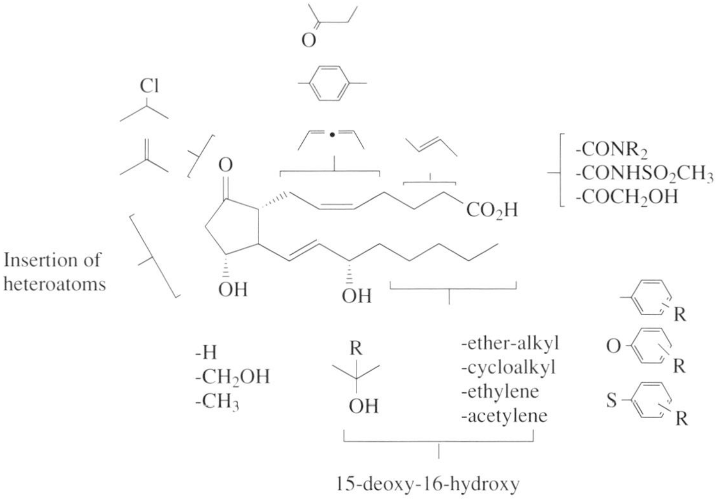

Cl
O
CO_2H
$-CONR_2$
$-CONHSO_2CH_3$
$-COCH_2OH$
Insertion of heteroatoms
OH
OH
-H
$-CH_2OH$
$-CH_3$
R
OH
-ether-alkyl
-cycloalkyl
-ethylene
-acetylene
R
O R
S R
15-deoxy-16-hydroxy

Figure 36 Structural transformation of PGE_2.

TXA_2 possesses the unstable oxetane function leaving an unsolved problem in synthetic organic chemistry, and the synthesis of TXA_2 analogues was one of the most fascinating targets. However, even when Still *et al.* accomplished the total synthesis of TXA_2 and it was applied to pharmacological studies, the stability of TXA_2 did not change and complete information was not obtained from the pharmacological experiments. The pharmacological study was possible only by the use of stable analogues.

In the case of TXA_2, modification of the unstable oxane–oxetane function was mainly conducted. The first analogue reported was compound (**86**) (PTA_2) which was synthesized using pinene, the naturally occurring substance, as a starting material.[221] This analogue showed weak TXA_2 antagonistic activity. Compound (**87**) (CTA_2) exhibited high TXA_2 agonistic activity in the blood smooth muscle and high antagonistic activity against compound (**75**) in blood platelets.[222] Furthermore, the analogue in which the two ethereal oxygens of TXA_2 were replaced by the other atoms in a molecule was reported, and 9,11-thia analogue (**88**) (STA_2) showed high TXA_2 agonistic activity,[223] but compounds (**89**)–(**93**) were reported to have no activity.[224–228] The biological activity of TXA_2 is considered to be related to the seriousness of diseases, and inhibition of activity is of medical importance. Synthesis of the agonist displayed an ability for solving the mechanism of diseases. Development of the TXA_2 antagonists is very important and significant. Many analogues that exhibited antagonistic activities were found, in which the heteroatom was introduced into the side chain of the carbon framework of the agonists.

On the other hand, BM-13505 (**94**), which does not have the PG carbon framework, and amide containing TX analogue (**95**) were reported to be TXA_2 antagonists.[229,230] Considering the structural similarity between these compounds and PGs, introduction of the sulfonamide function into the PG carbon framework resulted in high antagonistic activity being exhibited, (**96**)–(**99**),[231–234] and serious development competition in this research field spread throughout the world.

1.07.5.3 Synthesis of PGI_2 Analogues

Since PGI_2 itself has the possibility of being a medicine in the circulatory organ system, stabilization of the unstable enol–ether linkage was the biggest theme in the synthesis of its analogues. Introduction of the substituent group which contributed to stabilization of the unstable enol–ether linkage or replacement of the ethereal oxygen by other atoms were the basis for structural modification of PGI_2.[235,236]

The structural stability retaining the biological activities was successful by virtue of conjugation with the electron withdrawing group (**100**);[237] however, the profile of essential biological activities did not change and the development of structural modification has not been continued. Replacement of the enol ether part by other functions was effective in stabilization and biological activities (**101**).[238] Thia-PGI_2 (**102**)[239] is more stable than PGI_2 and its biological activity was approximately one-tenth that of PGI_2 in the platelet aggregation inhibiting property. Aza-PGI_2 (**103**)[240] possesses the enamine or the imine function instead of the enol–ether linkage, and retained the platelet aggregation inhibitory activity.

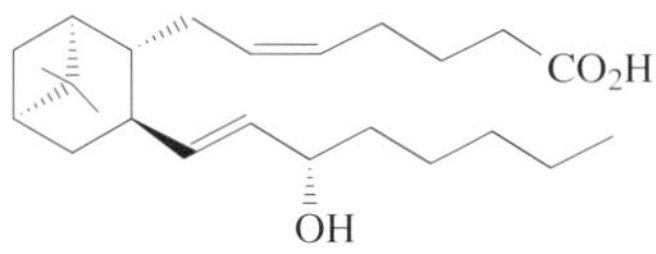

(86) 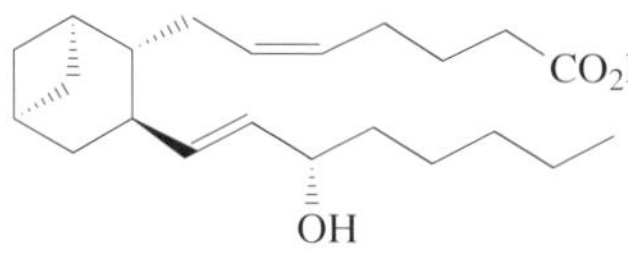 **(87)** 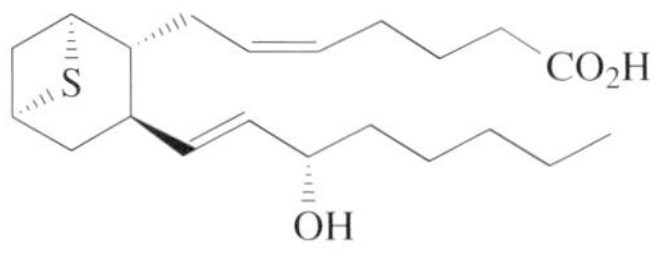 **(88)**

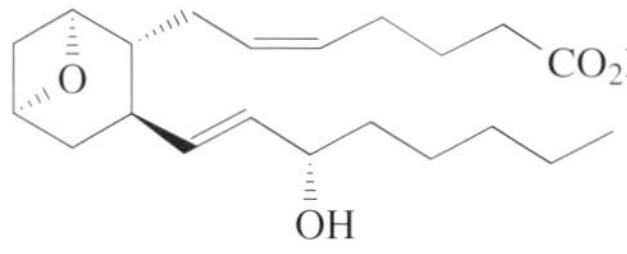

(89) 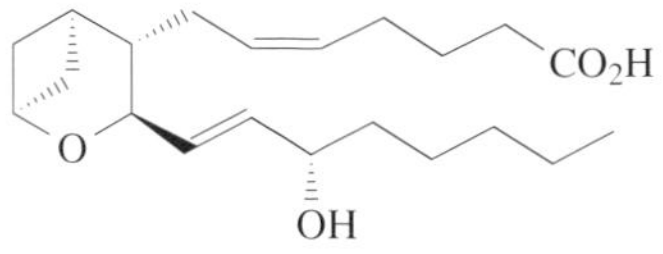 **(90)** 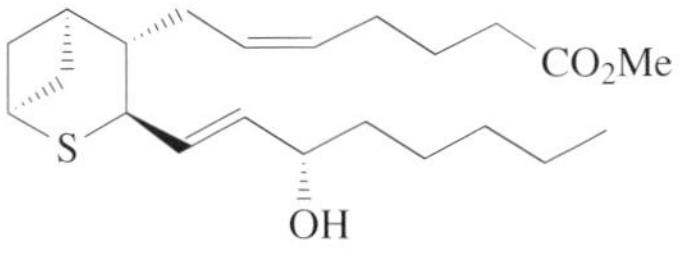 **(91)**[221]

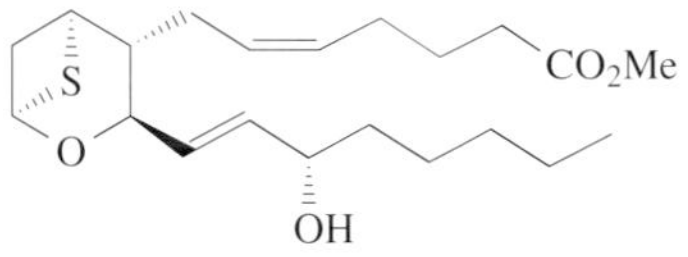

(92)[222] 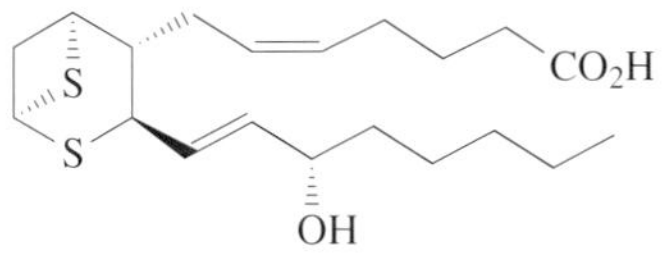 **(93)**[223]

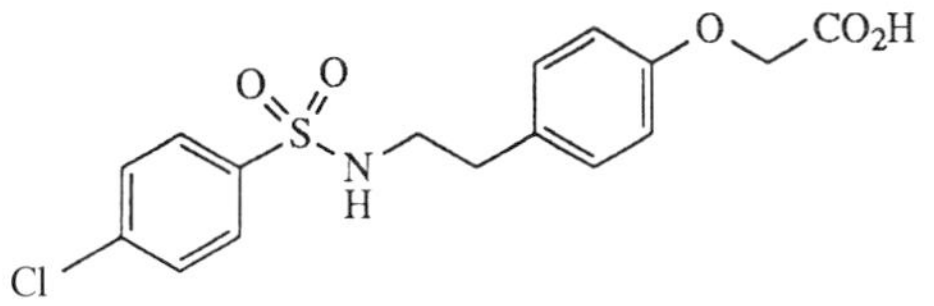

(94)[229] 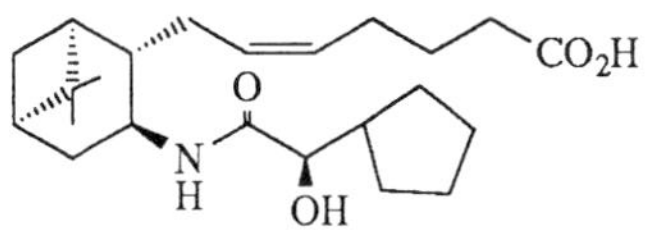 **(95)**[230]

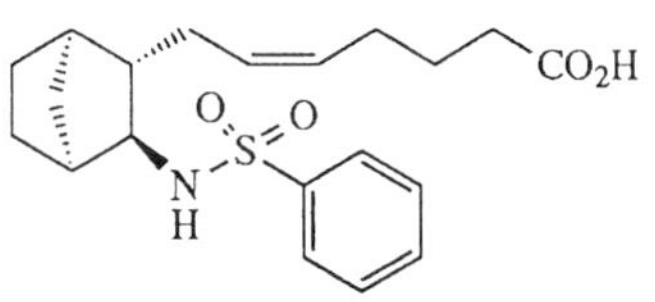

(96)[231] 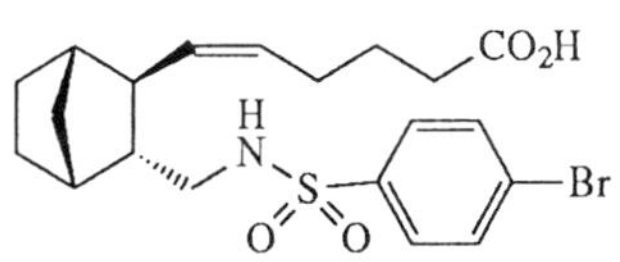 **(97)**[232]

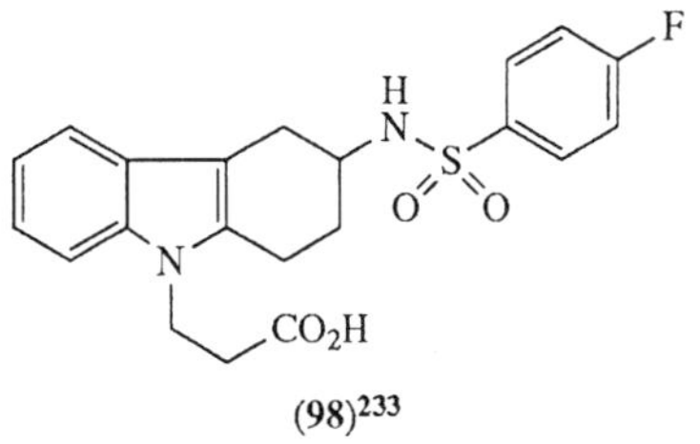

(98)[233] 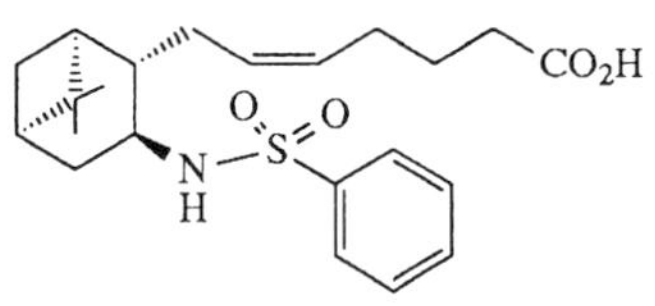 **(99)**[234]

The analogue in which the oxygen at the enol–ether linkage was replaced by the carbon is called carbacyclin (**104**)[241] and has the natural type side chain. Although its biological activity is 1/30 that of PGI_2, the mechanism of its biological activity is very similar to that of PGI_2. It is chemically stable, but its rate of metabolism is faster than that of PGI_2. Structural modification was performed as well as that of PGs, and many analogues were supplied for clinical trials. Compound (**105**) (iloprost)[242] has potent PGI_2 agonistic activity. In a series of these analogues, those which have the *Z*-configuration double bond in a molecule were potent and biologically active, but the *E*-isomers were inactive. Furthermore, the analogue which has the double bond at the 6,9α position was named isocarbacyclin (**106**), which is more potent than carbacyclin in platelet aggregation inhibitory activity, but its 6,7-double bond isomer (**107**) is inactive.[243]

Biologically active analogues were found in the interphenylene analogues in which the enol–ether linkage was replaced by the phenyl ether linkage.[244] The activity changed according to the position of the α-side chain at the phenyl ring (**108**) beraprost,[245] (**109**).[246] The effect suggests that the regiometrical configuration of both the α- and ω-side chains is very important for exhibiting PG-like activities, which is very critical for the molecular design of analogue synthesis.

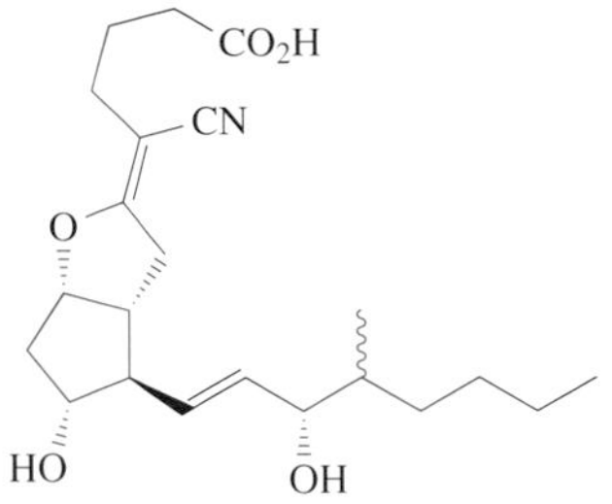

(**100**) 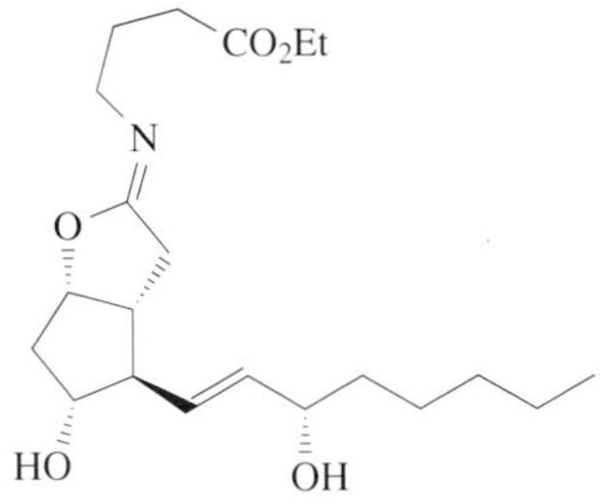 (**101**) 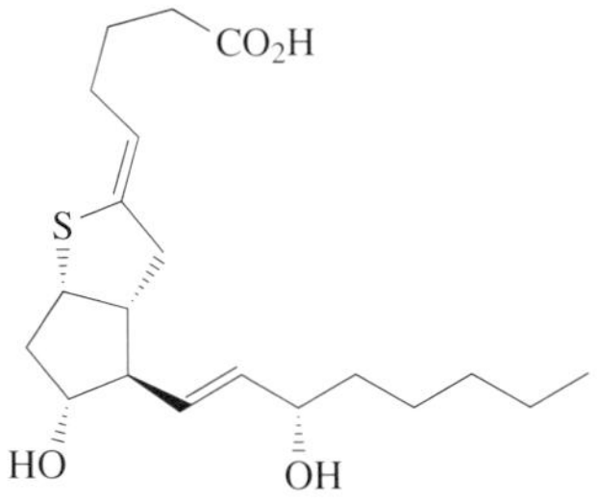 (**102**)

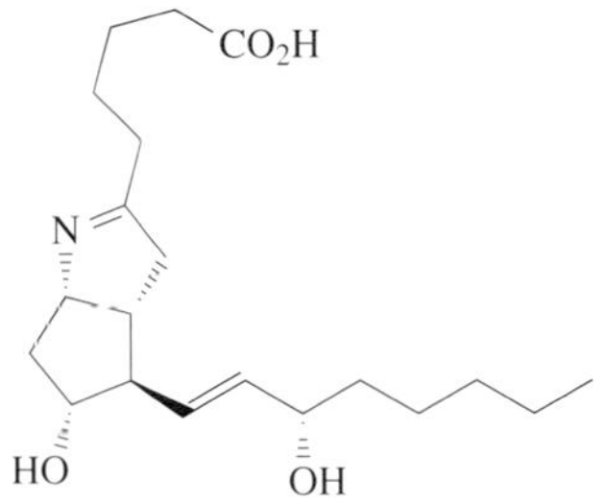

(**103**) 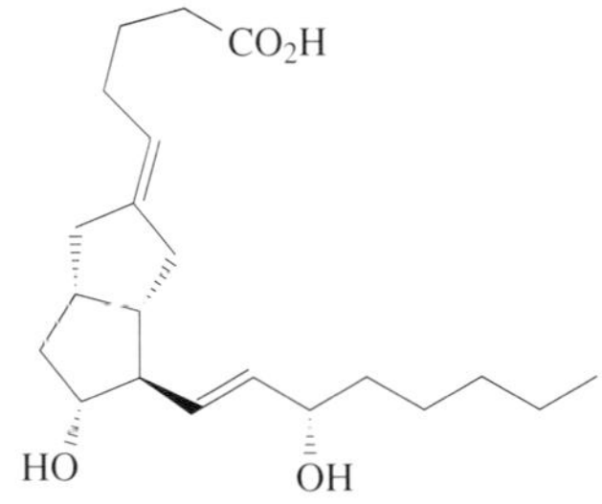 (**104**) 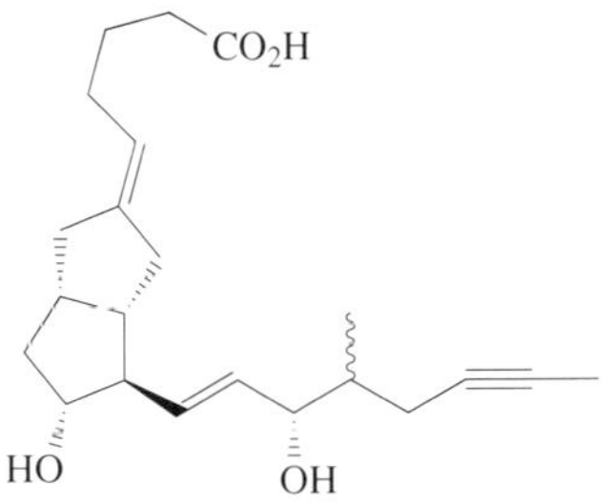 (**105**)

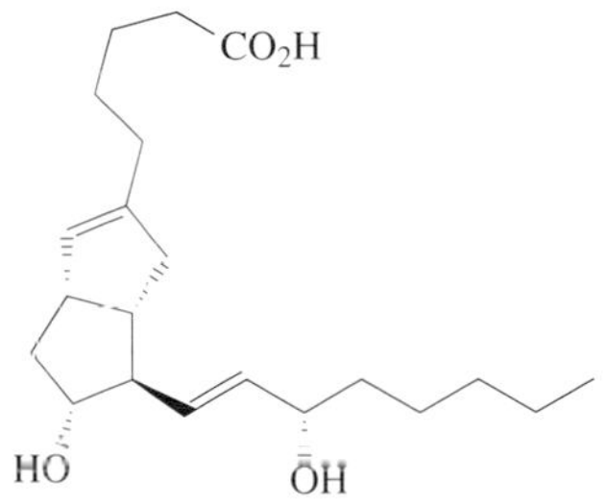

(**106**) 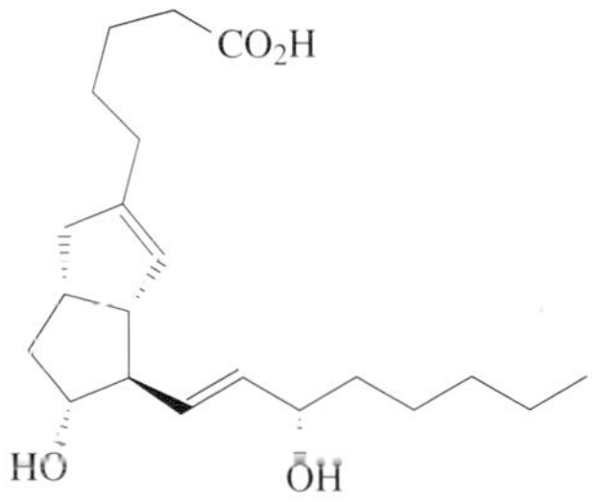 (**107**) 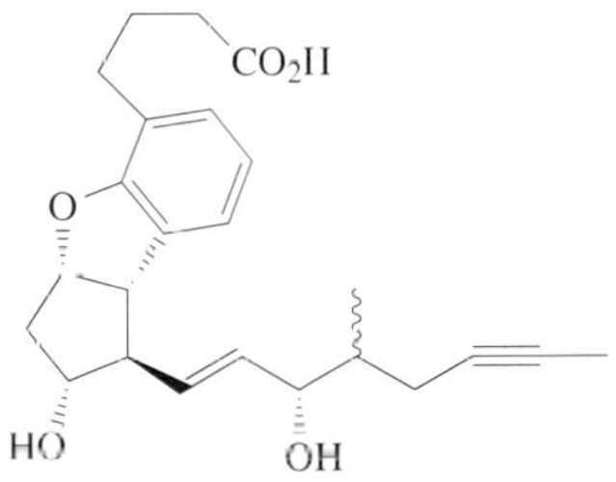 (**108**)

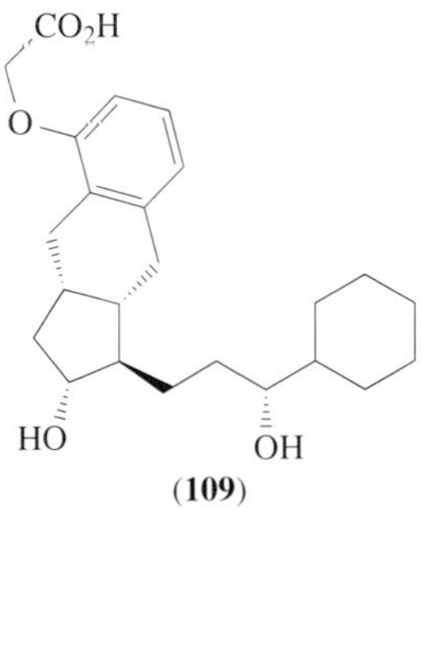

(**109**)

1.07.5.4 LTD$_4$ and LTB$_4$ Agonists and Antagonists[247]

Since LTs leave the straight C-20 carbon chain, presumption of the conformation is far more difficult than for PGs. With regard to LTD$_4$, structural modification at the alkyl part and the amino acid moiety was energetically conducted, but compounds which possessed agonistic or antagonistic properties were not found. LTD$_4$ antagonist was found not from analogue synthesis but from the random screening. FPL-55712 (**110**)[248] was first discovered as an SRS-A antagonist. Referring to this structure and the carboxylic acid function of LTs, a number of compounds were synthesized, and many antagonists were found, as described in Section 1.07.6.

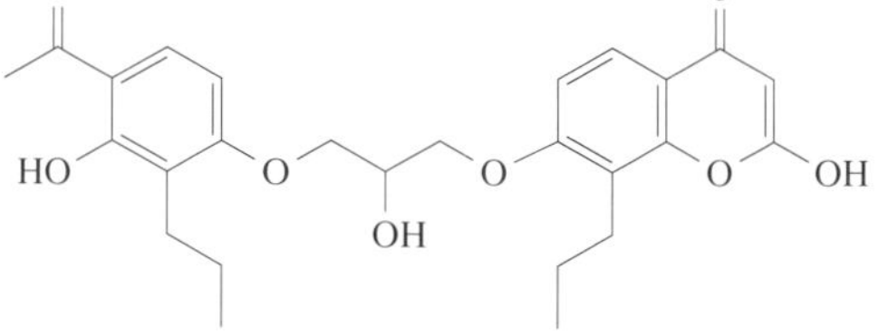

(**110**) FPL-55712 (SRS-A antagonist)

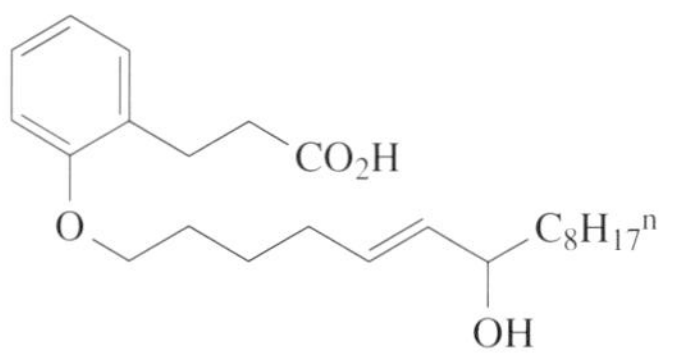

(**111**) LTB_4 agonist

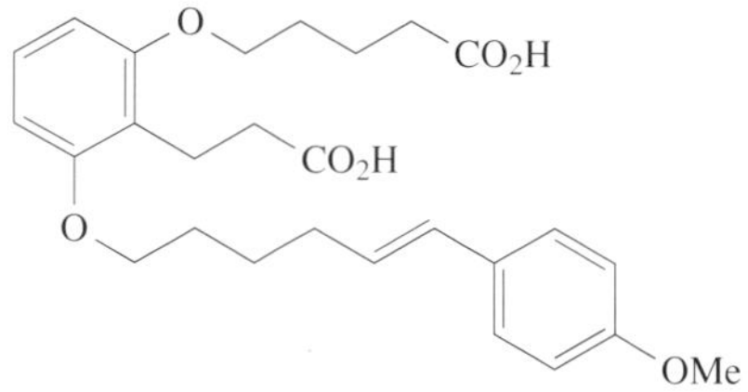

(**112**) LTB_4 antagonist

As for LTB_4, since some analogues in which the triene part was replaced by the benzene ring had LTB_4-like activities, there are many reports on modification of the carbon skeleton. Hamanaka *et al.* found that the synthesized compounds (**111**)[249] and (**112**)[250] were LTB_4 agonist and antagonist, respectively. They reported the conformational similarity of the agonist and antagonist. They presumed the conformation of LTB_4 based on these two compounds.

1.07.5.5 PG-like Substances not Possessing the PG Skeleton

Since isolation of the PG receptor, and the properties of PG-relating compounds prepared so far being defined clearly, it was clear that the pharmacological classification is not always correct. For example, Iloprost had been believed to be a pure agonist of PGI_2. However, it bound not only to the receptor of PGI_2 but also to the receptor of PGE_2. This result posed a question as to whether PGI_2 truly binds to only the PGI_2 receptor. With respect to the TX antagonists synthesized so far, they bound not only to the TX receptor but also to the EP receptor and all the binding experiments were re-examined. New trials for the development and synthesis of new analogues have been carried out. It is expected that among these analogues a novel functioning medicine which would bind to a new receptor would be discovered.

Moreover, after the properties of the compounds which had even a weak biological activity could be distinguished, this work made a marked advance resulting from isolation of the receptor. Isolation of the receptor was successful due to the existence of stable and potent antagonists, and it is possible to design a novel molecule using this receptor; isolation of the receptor and the design of a novel molecule have a complementary relationship.

It had been believed that the five-membered ring, the substituent group on the ring, and the allylic alcohol are essential for deducing PG activities. It became necessary to destroy this fallacy in order to make progress in the chemistry of PGs and for creating the novel carbon skeleton. The biological activities of PGs are far more potent than those of the other known biologically active compounds, and it was difficult to recognize compounds possessing weak biological activities as PG-like substances in the pharmacological experiments. Only compound (**94**) as a TX antagonist, phenyl substituted heterocyclic compounds (**113**),[251], (**114**),[252] tetrahydronaphthalene or dihydronaphthalene derivatives (**115**) as PGI_2 agonist,[253,254] and the dibenzoxazepine derivative (**120**) as PGE_2 agonist[255] were recognized. The synthesis and biological evaluation are now developing using these compounds as lead compounds.

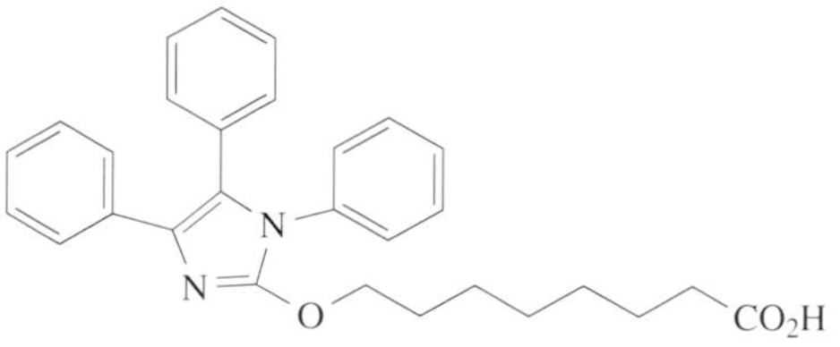

(**113**)

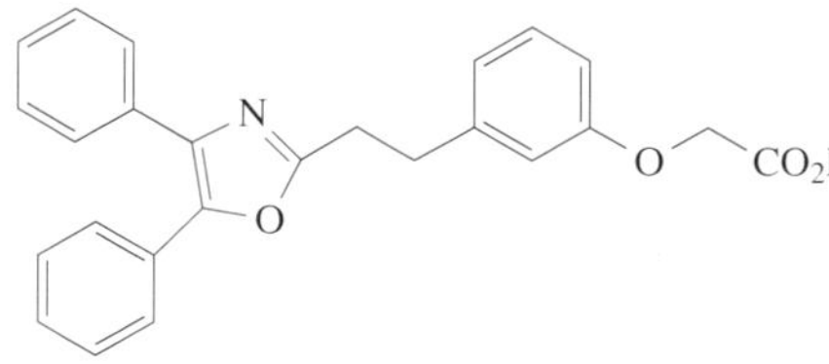

(**114**)

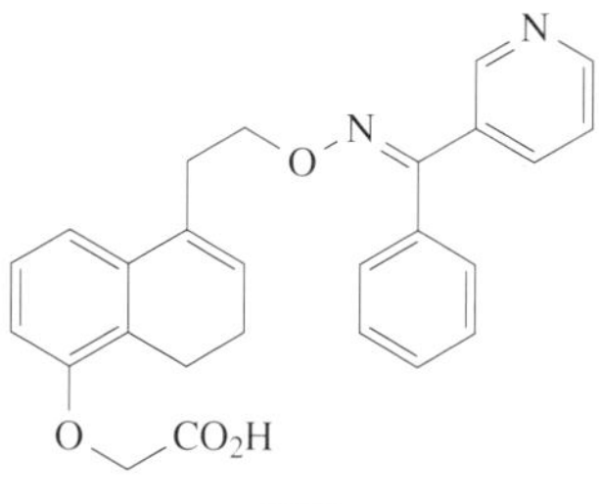

(**115**)

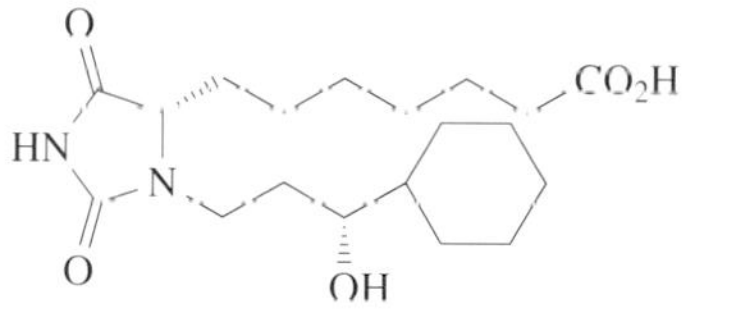

(**116**) (PGD_2 agonist)

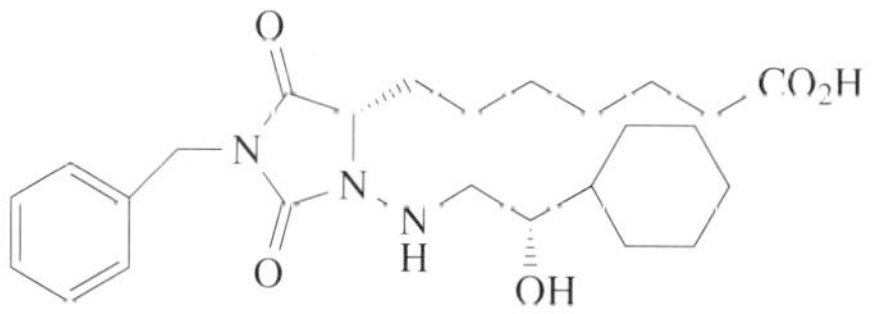

(**117**) (PGD_2 antagonist)

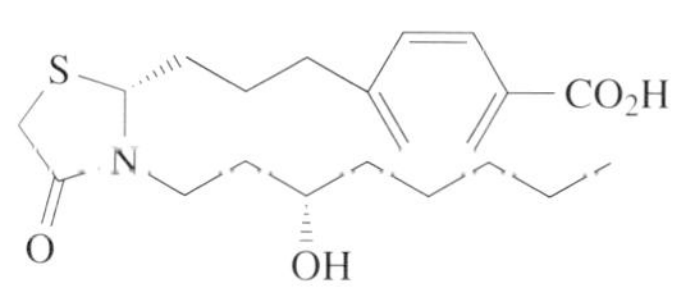

(**118**) (PGD_2 antagonist)

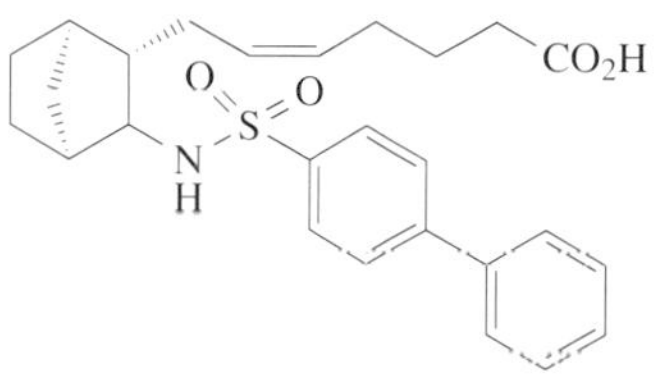

(**119**) (PGD_2 antagonist)

PGD_2 is relatively unstable and was reported to show an increasing effect on sleeping times. Synthetic studies were carried out on non-PG-like skeletons. Among them, the structures of (**116**)[256] and (**117**)[257] were constructed by replacement of the five-membered ring of PGs by the other heterocyclic ring, and each of them showed agonistic and antagonistic activities, respectively. A similar compound, (**118**),[258] and sulfonyl amide-containing compound (**119**)[259] also showed potent antagonistic activity.

1.07.6 APPLICATION TO MEDICAL USE

Since their discovery, eicosanoids have been expected to be used as medicines because of their potent biological activities. Many scientists throughout the world are involved in investigation of the biological activity of eicosanoids. The developing methods for medicines with regard to eicosanoids are divided into three categories as follows:

(i) utilizing biological activities of eicosanoids themselves or their agonistic analogues;
(ii) utilizing receptor antagonists to inhibit the pathophysiological activities of eicosanoids;
(iii) utilizing specific inhibitors of the production of eicosanoids.

With respect to (iii), it is well known that nonsteroidal antiinflammatory drugs, e.g., aspirin and indomethacine, are widely used to inhibit the action of cyclooxygenase, both cyclooxygenase types I and II. Thus, they inhibit eicosanoid production by COX-I which is an important enzyme for the production of eicosanoids to maintain physiological homeostasis. Specific inhibitors and stimulators of PG isomerases have yet to be discovered, except for TXA_2 synthase inhibitors.

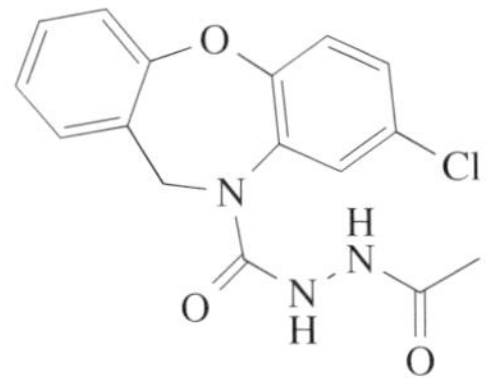

SC-19220 EP_1 antagonist
(**120**)[255]

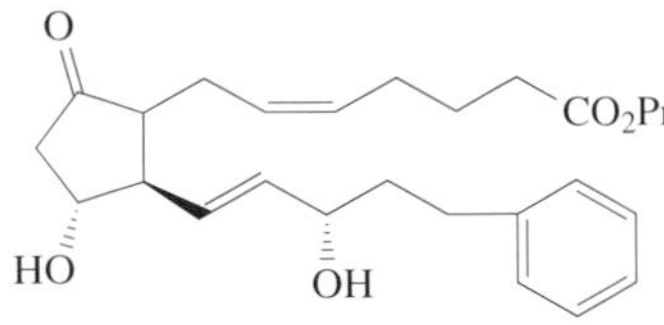

17-Phenyl PGE_2 EP_1 agonist
(**121**)[260]

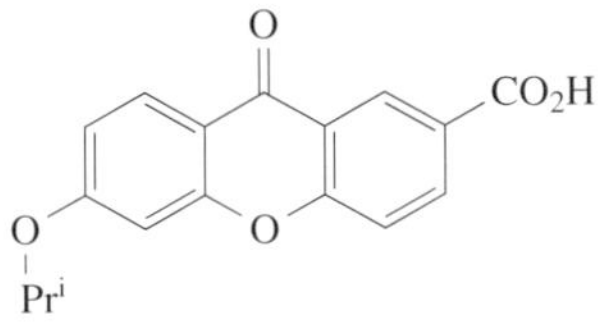

AH-6809 EP_1 antagonist
(**122**)[261]

17-Phenyl PGE_2
SC-19220
AH-13205
GR 63799
M & B28767
AH-6809

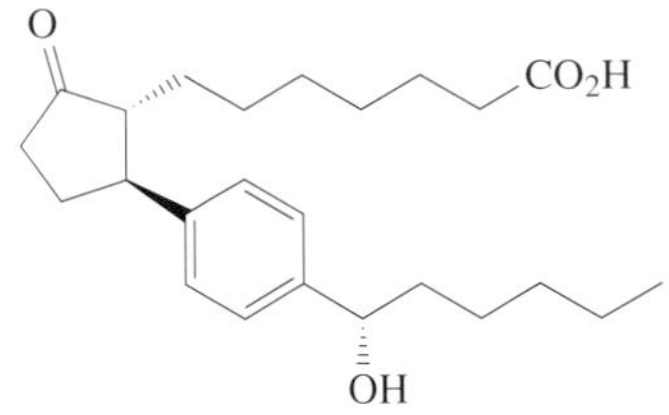

AH-13205 EP_2 agonist
(**123**)[262]

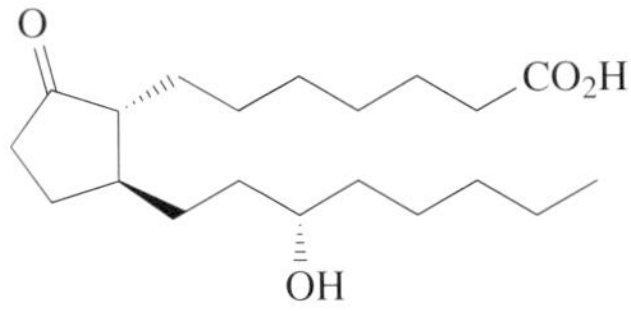

AY-23626 EP_2 agonist
(**124**)[263]

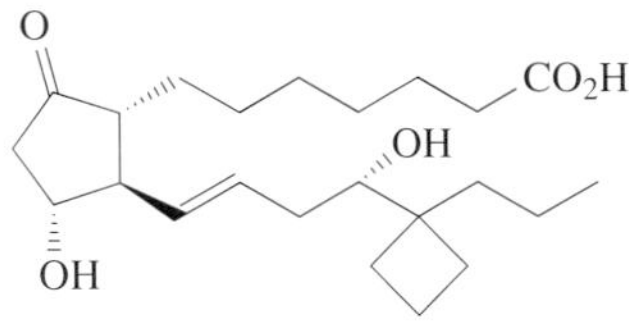

Butaprost EP_2 agonist
(**125**)[264]

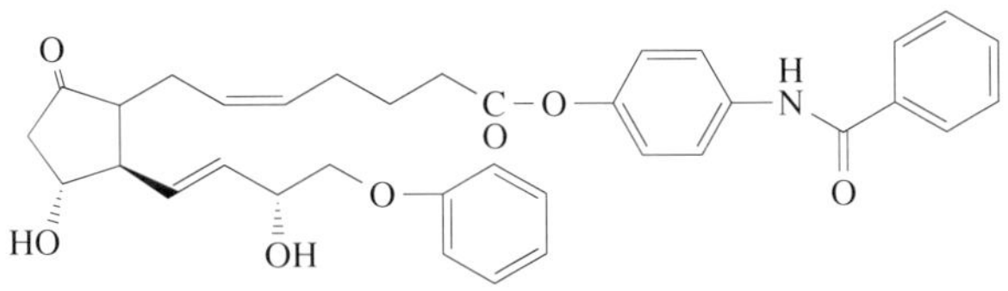

GR 63799 EP_3 agonist
(**126**)[265]

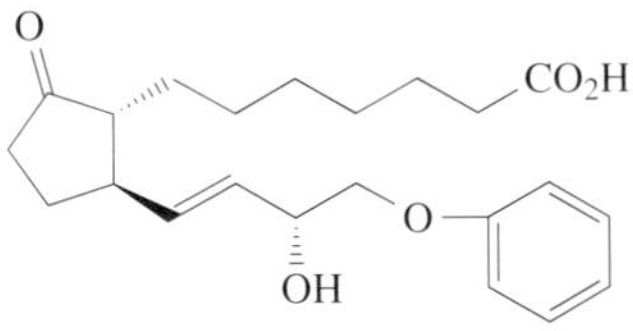

M & B28767 EP_3 agonist
(**127**)[266]

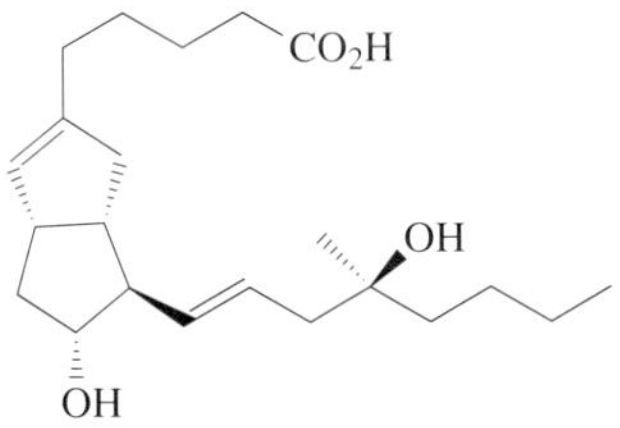

TEI-3356 EP_3 agonist
(**128**)[267]

So far, the development of medicines has been conducted in a variety of fields of diseases, but it has not always been successful. In the following, the region and the compounds which are proven medicines are discussed.

1.07.6.1 Obstetrics and Gynecology

In the second half of the 1960s, the induction of labor by intravenous infusion of $PGF_{2\alpha}$ was attempted. Since then the uterine contractile activity of PGs has been applied not only to the induction of labor and suppression of labor pain but also for induced abortion.

$PGF_{2\alpha}$ and PGE_2 were utilized in this field in the 1970s. As derivatives of natural substances, Sulprostone (**129**), Gemeprost (**130**), and Carboprost (**131**) are now used clinically.

Sulprostone is highly selective for the uterus. Rapid metabolism of the ω-chain can be blocked by replacement of the terminal alkyl group of PGs with the phenoxy moiety.[268] Gemeprost can be administered as a vaginal suppository, while $PGF_{2\alpha}$ and PGE_2 are administered intravenously and orally, respectively.[269] Carboprost is an improved analogue compared with $PGF_{2\alpha}$ in the potency of uterine contractile activity and the duration of action. It does not cause topical irritation during intramuscular administration, while $PGF_{2\alpha}$ does.[270]

Obstetrics and Gynecology

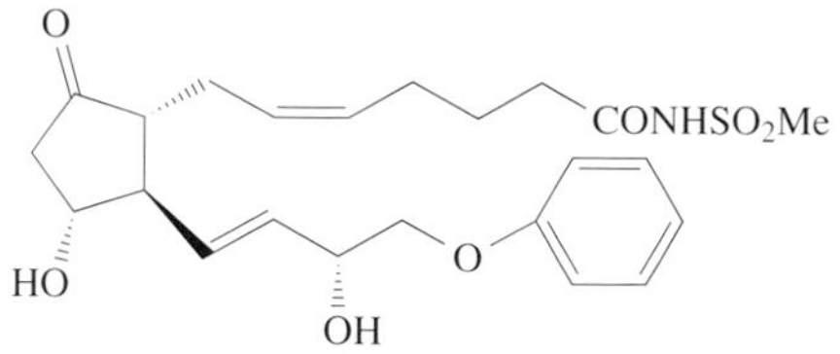

Sulprostone (**129**)

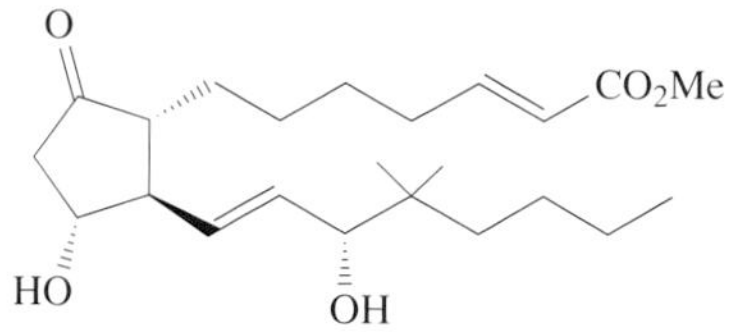

Gemeprost (**130**)

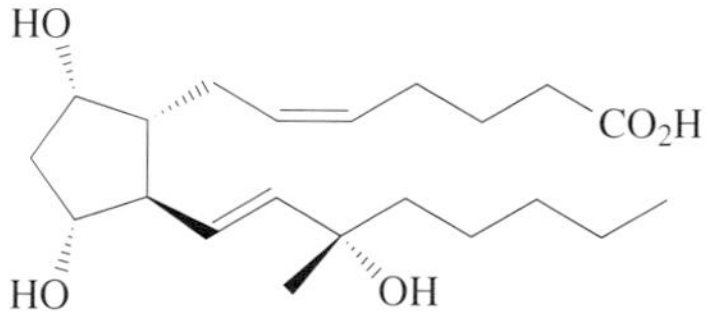

Carboprost(**131**)

Digestive System

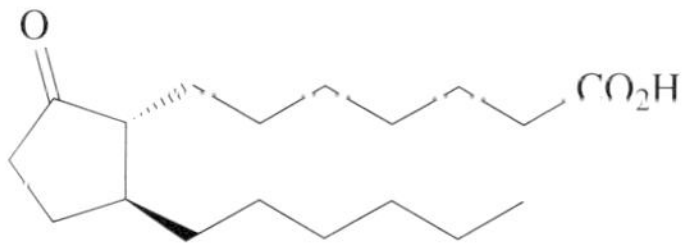

Rosaprostol (**132**)

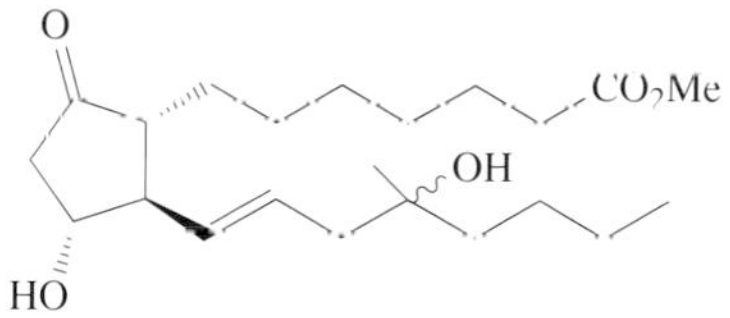

Misoprostol (**133**)

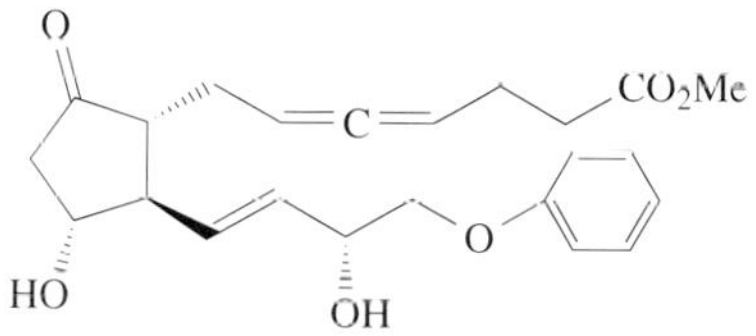

Enprostil (**134**)

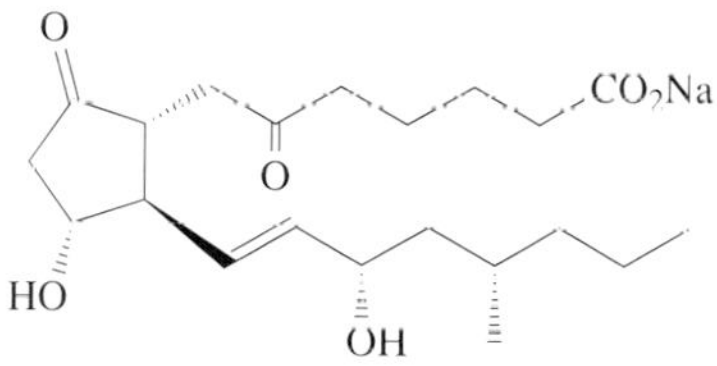

Ornoprostil (**135**)

Cardiovascular System

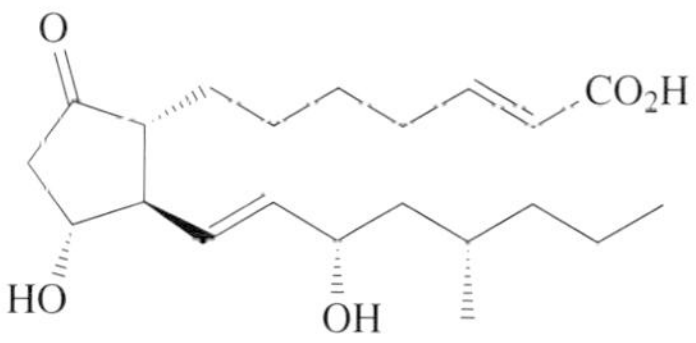

Limaprost (**136**)

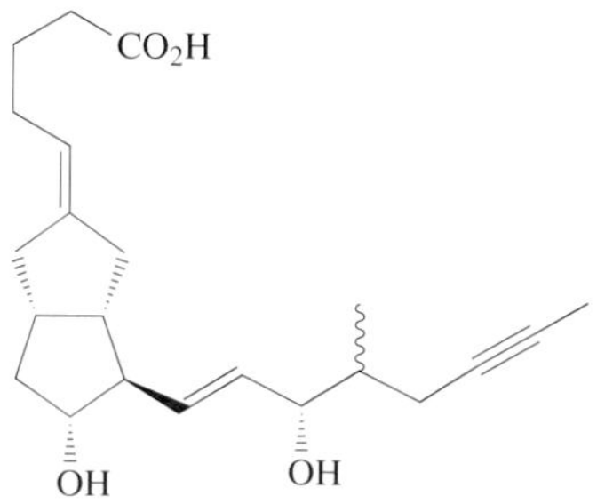

Iloprost (**105**)

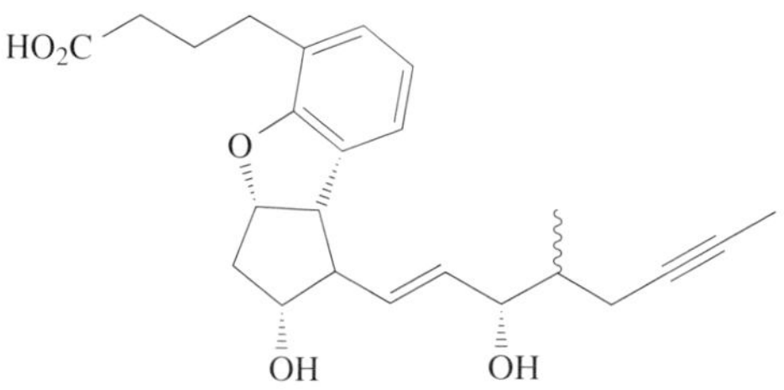

Beraprost (**108**)

1.07.6.2 Digestive System

In 1979, Robert *et al.* proposed a concept of cytoprotection of PGs by disclosing that administration of a trace amount of natural PGs which did not cause inhibition of gastric secretion could block mucosal injury of the stomach caused by a necrosis-induced substance.[271] Much attention had been paid to this concept because $PGF_{2\alpha}$ without having an inhibitory activity on gastric acid secretion was demonstrated to have an antiulcerative effect, and it has begun to be considered that PGs are enhancers of defensive factors. The mechanism of action leading to cytoprotection has been suggested to be increased blood flow in the mucous membrane, and an enhancement of secretion of muciparous and mucous HCO_3^-. In this field, Rosaprostol (**132**), Misoprostol (**133**), Enprostil (**134**), and Ornoprostil (**135**) are used.

Rosaprostol was launched as a medicine for stomach and duodenal ulcers. This compound has a C-18 framework, lacks a functional group, and is not an eicosanoid. Its biological activity is very low as well as its side effects.[272] Misoprostol has an inhibitory activity on gastric acid secretion and cytoprotective activity. It is a racemic mixture consisting of 16(*S*) and 16(*R*) isomers.[273] Enprostil is an antiulcerative agent possessing an inhibitory activity on gastric acid and pepsin secretions, a cytoprotective activity, and a lowering effect on gastrin levels in serum. It is a racemic mixture consisting of two stereoisomers of allenes.[274] Ornoprostil has an inhibitory activity on gastric acid secretion, an enhancement of mucus secretion, and an antiulcerative effect. This compound is a derivative of 6-keto PGE_1, and it has PGI_2-like activity.[275]

1.07.6.3 Cardiovascular System

A natural PGE_1 formulation possessing vasodilative activity and an inhibitory activity on platelet aggregation has been used for the treatment of peripheral circulatory failures including arteriosclerosis obliterans, Burger's disease, and Reynold's syndrome. Currently PG indications have been used in the peripheral circulation; however, it is expected that novel PGE_1 and PGI_2 analogues will be used not only for peripheral circulatory dysfunctions but also for cardiac and cerebral infarction in the future. In this field, many clinical trials of the synthesized compounds have been performed and Limaprost (**136**), Iloprost (**105**), and Beraprost (**108**) have been used.

Limaprost is an analogue of PGE_1 possessing metabolic resistance against 15-hydroxy-PG dehydrogenase. The α-cyclodextrin clathrate of Limaprost has been used clinically for the treatment of ulcers, pain, and frigidity accompanied by thromboangiitis obliterates.[276] Iloprost is a stable PGI_2 analogue and was registered as an injective formula. It has been used for treatment of ischemic heart failure and peripheral circulatory failure.[277] Beraprost is an orally active PGI_2 analogue in which the drawbacks of PGI_2 have been improved upon. It was demonstrated to have highly potent antiplatelet and vasodilatative activities. This molecule is a stabilized PGI_2 analogue possessing a phenyl ether moiety instead of vinyl ether moiety in the natural PGI_2, which could reduce the side effects of PGs such as hypotension and diarrhea.[278]

1.07.6.4 Ophthalmology

Ophthalmology is a new field in which the clinical use of PGs is expected. Localizations and some of the roles of PGs in the eye have been known for a long time. It had been discovered that PGs were topically produced and released in the eye to cause congestion, miosis, destruction of the blood–aqueous humor junction, and intraocular pressure rise. Thus, PGs have been considered to be mediators of intraocular inflammation and ocular pressure raising substances. Application of $PGF_{2\alpha}$ caused a sustained decrease in the intraocular pressure for 24 h without intraprotomerite inflammation and a change in the pupillary diameter. However, side effects such as conjunctiva congestion, ophthalmalgia, and headache were also observed. A structural feature of the PG analogues developed in this field is to possess an inactive metabolite-like substance in the ω-chain. It seems to be a product of a flexible way of thinking about the molecular design of PG analogues. At present, Unoprostone (**137**) and Latanoprost (**138**) are used clinically.

Unoprostone isopropyl ester, which has an intraocular pressure lowering effect, was launched as an antiglaucoma agent. Its basic structure is an inactive metabolite of $PGF_{2\alpha}$, 13,14-dihydro-15-keto $PGF_{2\alpha}$. The duration of its action is long when applied topically to the eye and it shows few side effects.[279] Latanoprost has an ocular pressure lowering effect. This compound is free from side effects except for slight conjunctiva congestion.[280]

Ophthalmology

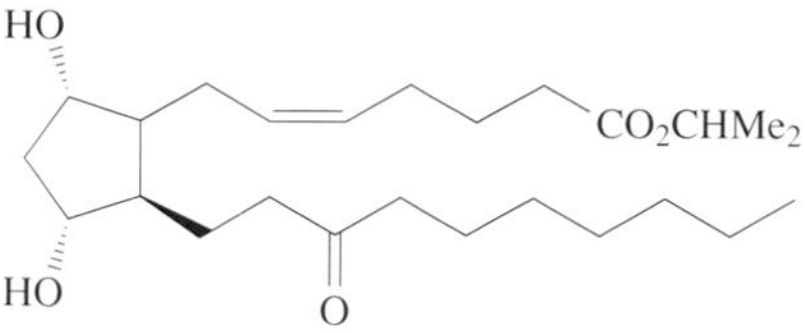

Unoprostone isopropyl ester (**137**)

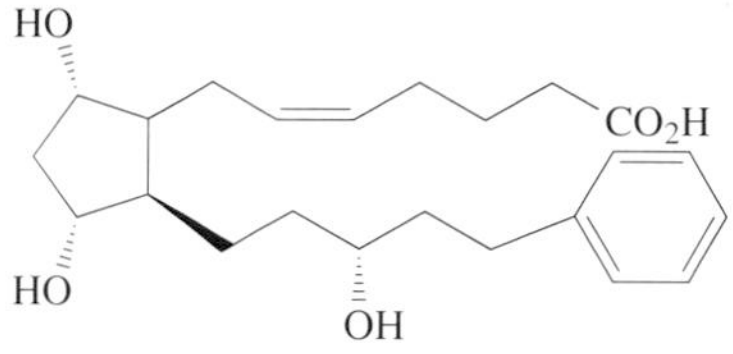

Latanoprost (**138**)

Respiratory system

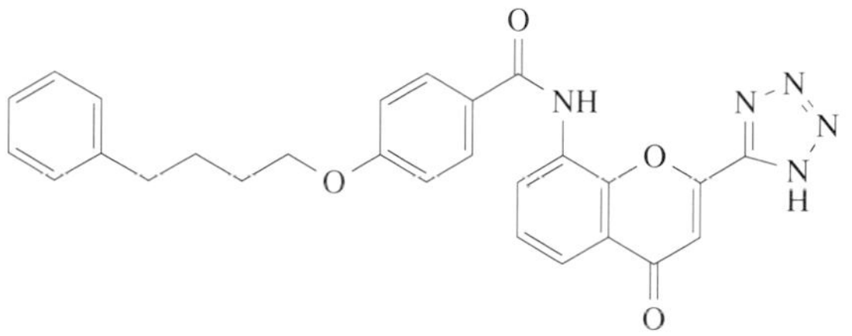

Pranrukast (**139**)

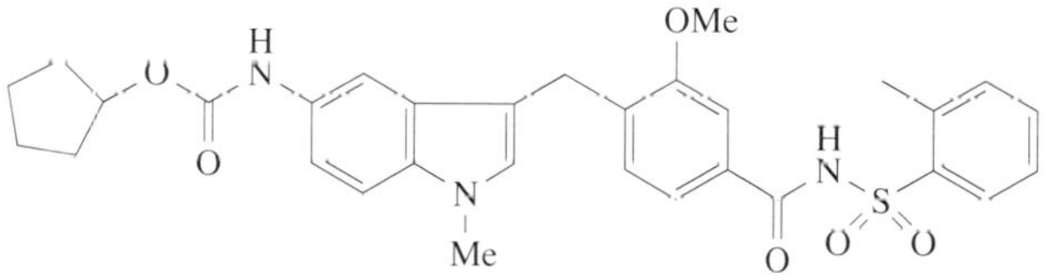

Zafirlukast (**140**)

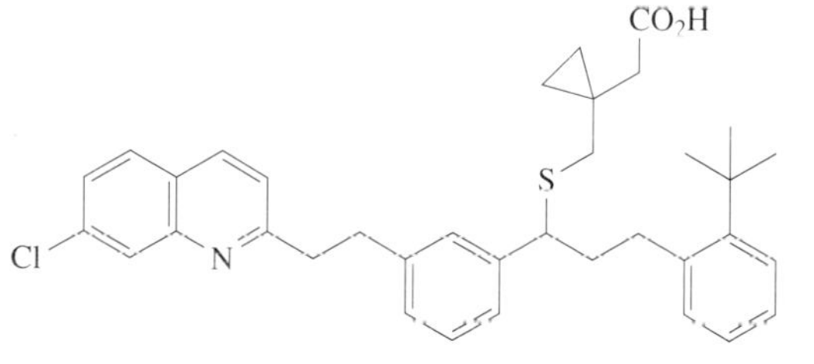

Montelukast (**141**)

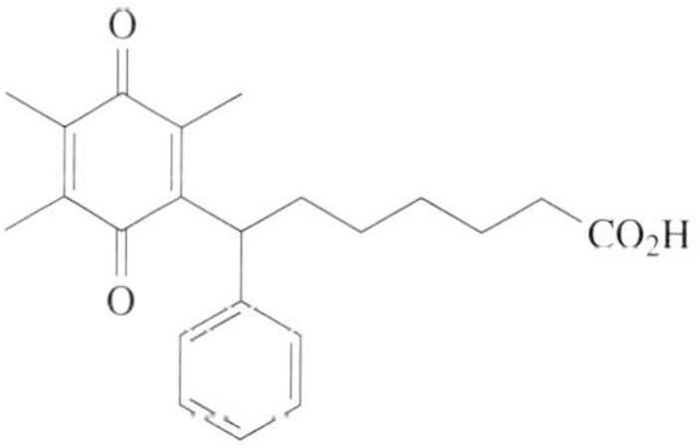

Seratorodast (**142**)

1.07.6.5 Respiratory System

The peptide LTs (LTC_4, LTD_4, and LTE_4) are endogenous biological mediators derived from arachidonic acid via the 5-lipoxygenase pathway. These compounds are potent bronchoconstrictors which are believed to play a role in the pathophysiology of several disorders, especially human allergic diseases. Elucidation of the leukotriene (LT) structures in 1979 prompted many pharmaceutical companies to explore the development of LT antagonists. Those effects resulted in the discovery of many different classes of LT antagonists. At present Pranlukast (**139**), Zafirlukast (**140**), and Montelukast (**141**) are on the market as anti-asthma medicines.

Pranlukast was launched in 1995 as an anti-asthmatic. Pranlukast was found to be a specific LT receptor antagonist in *in vitro* and *in vivo* experiments in guinea-pigs and human bronchial provocation tests. This compound significantly reduced symptoms such as wheezing and dyspnea in asthmatic patients. The effect of Pranlukast on allergic responses of the immediate type was investigated in patients with bronchial asthma in double-blind tests.[281,282] Zafirlukast was also launched in 1996 as an anti-asthmatic drug. Zafirlukast is a potent, selective, long-acting, and orally effective antagonist of LTs. *In vivo*, Zafirlukast was effective at blocking aerosolized LTD_4-induced dyspnea in conscious guinea-pigs following oral, intravenous, or aerosol administration.[283,284] Montelukast was launched in 1997. Montelukast sodium is also a potent inhibitor of LTD_4 that inhibits the binding of (^{3}H)-LTD_4 against human receptors. In many clinical trials the high intrinsic potency, oral bioavailability, and long duration of effect seen indicated that Montelukast would be clinically useful in the treatment of asthma.[285,286]

TXA_2 is also a highly potent bronchoconstrictor, and the TXA_2 antagonist is now in a stage of clinical testing as an anti-asthma agent. Seratrodast (**142**) was launched as an anti-asthma medicine, but this compound is not specific to TXA_2, possessing lipoxygenase inhibitory activity, LTD_4 antagonistic activity, and PAF antagonistic activity, and inhibiting the chemical mediator which is related to asthma.[287]

1.07.7 EICOSANOID RECEPTORS

1.07.7.1 Classification by Functional Studies

As the chemical synthesis of PGs developed, eicosanoids and related compounds were added to pharmacological studies, and biological studies were also expanded. By the mid-1970s, it was clear that PGs were capable of causing a diverse range of actions; however, the cause of the diversity was not clear. Studies on the receptor were not well advanced.

In 1948, the pioneering work of classification of the hormone was begun by Ahlquist.[288] He classified the receptors by use of the biological actions of the catecholamines, adrenaline and noradrenaline. The outcome of these studies was the classification of adrenoceptors into α and β subtypes, a classification scheme that has stood to the present day. This work was subsequently extended by Lands *et al.* in 1967 who, using the same approach, demonstrated that, although the classification proposed by Ahlquist was essentially correct, it was an oversimplification and one of Ahlquist's receptors, the β-adrenoceptor, could be further divided into two subtypes, termed $\beta 1$ and $\beta 2$.[289] This work is revolutionary not only in the field of biological studies but also in the field of medicinal chemistry and such work is continuing on biologically active substances.

In 1967, Pickles *et al.* reported on the initial work on PG receptors; however, further study was not pursued. By 1980, some reports on the receptors had been presented,[290–294] but a unified interpretation was not performed because there existed no suitable agonists and antagonists. In 1982, Kennedy *et al.* tried to classify systematically the receptors by use of the functional data of the naturally occurring PGs, PG analogues, and a few synthetic antagonists.[295,296]

Their classification of receptors into DP, EP, FP, IP, and TP recognized the fact that receptors exist that are specific for each of the five naturally occurring prostanoids, PGs, D_2, E_2, $F_{2\alpha}$, I_2, and TXA_2, respectively.[295] It was clear that at each of these receptors one of the natural prostanoids was at least one order of magnitude more potent than any of the other four. There is now evidence for the existence of four subtypes of EP receptors, termed arbitrarily EP_1, EP_2, EP_3, and EP_4. These EP receptor subtypes are all PGE_2 physiological agonists, but the reactions to a variety of PGE_2-like substances are different.

The cloning and expression of receptors for the prostanoids has not only confirmed the existence of at least four of the five classes of prostanoid receptor, EP, FP, IP, and TP, but has also supported the subdivision of EP receptors into at least four subtypes, corresponding to EP_1, EP_2, EP_3, and EP_4. The current classification and nomenclature of prostanoid receptors is summarized in Table 2. By these classifications, the details of synthetic agonists have been resolved which were not distinct before. For example, Iloprost, a well-known PGI_2 agonist, displaces [^{3}H]PGE_2 binding to the EP_1 receptor as potently as PGE_2.[296] This observation agrees with previous findings that several PGI_2 agonists including PGI_2 and Iloprost induce smooth muscle contraction via the EP_1 receptor in some tissues.

On the other hand, PGE_1 displaces [^{3}H]Iloprost binding to the IP receptor more potently than carbacyclin, a PGI_2 agonist. It is therefore important, to take such cross-reactivity into account when examining the actions of prostanoids in tissues containing several types of receptors.[296]

1.07.7.2 Structure of Eicosanoid Receptors

In spite of the progress of molecular biology, studies on eicosanoid receptors were delayed.[297–299] This was due to the lack of suitable antagonists using affinity column chromatography in spite of a

Table 2 Ligand binding properties of the cloned prostanoid receptors.

Type	*Ligand*	*Rank order of binding affinity*
TP	[^{3}H](**96**)	(**96**) > (**95**) > (**88**) > PGD_2 > PGE_2, $PGF_{2\alpha}$
EP_1	[^{3}H]PGE_2	PGE_2 > (**105**) > PGE_1 > $PGF_{2\alpha}$ > PGD_2
EP_2	[^{3}H]PGE_2	PGE_2 = PGE_1 ≫ PGD_2, $PGF_{2\alpha}$
EP_3	[^{3}H]PGE_2	PGE_2 = PGE_1 ≫ (**105**) > PGD_2, $PGF_{2\alpha}$, (**127**)
EP_4	[^{3}H]PGE_2	PGE_2 = PGE_1 ≫ PGD_2, $PGF_{2\alpha}$, 11-deoxy-PGE_1
FP	[^{3}H]$PGF_{2\alpha}$	$PGF_{2\alpha}$ > $PGF_{1\alpha}$ > PGD_2 > (**88**) > PGE_2 > (**105**)
IP	[^{3}H]Iloprost	(**105**) > PGE_1 > (**104**) > PGD_2, PGE_2, (**88**) > $PGF_{2\alpha}$
DP	[^{3}H]PGD_2	PGD_2 > (**116**) > (**117**) > (**88**) > PGE_2 > (**105**) > $PGF_{2\alpha}$

number of reports on antagonists. Biochemical studies have clarified the ligand binding properties of some of the receptors, and have indicated that their actions are mediated by G-proteins. However, none of the receptors had been isolated and cloned until the TXA receptor was purified from human blood platelets.[300] Narumiya *et al.* purified the TX receptor as a single substance using S-145 by affinity column chromatography.[301] Thereafter in 1991, Hirata *et al.* succeeded in the cloning of human cDNA.[302]

These studies revealed that the TXA_2 receptor was a G-protein-coupled, rhodopsin-type receptor with seven transmembrane domains. By homology screening in mouse cDNA libraries, the structures of seven types and subtypes of mouse prostanoid receptors have been identified. Based on these studies, the homologues of these receptors in other species have subsequently been cloned.[303–308]

Thereafter, a variety of receptors were isolated from various mammals. Considering a matter of species, there are a large number of receptors.[309] There exist seven hydrophobic functions which construct the transmembrane domains, without reference to receptors and species. The hydrophobicity analysis and homology to the other proteins suggested that they are rhodopsin type proteins with seven putative transmembrane domains. Transmembrane segments consist mainly of hydrophobic amino acids, and the three-dimensional structure model based on the sequence suggests that these hydrophobic amino acids form a hydrophobic pocket for the structure of its ligand.[310,311] These studies have clearly shown that all of the prostanoid receptors belong to the G-protein-coupled, rhodopsin-type receptor superfamily, and that they constitute a new subfamily of receptors. The membrane topology model is shown in Figure 37.

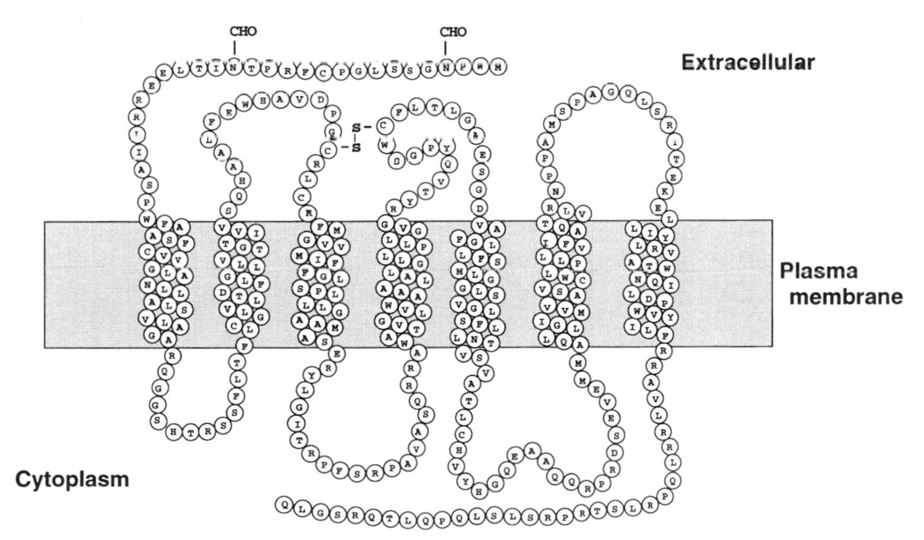

Figure 37 Membrane topology model of the human TX receptor (after S. Narumija).

On the other hand, there are two kinds of receptors in SRS-A, classified as cys-LT_1 and cys-LT_2. It was reported that cys-LT_1 was a receptor of LTD_4 and that cys-LT_2 bound strongly to LTC_4.[312] However, they have not yet been isolated.

The LTB_4 receptor was cloned by the HL cell. Its structure was identified as a seven hydrophobic-type receptor, which formed the transmembrane domains. However, this receptor has a low homology not only with PG receptors but also with the known receptors, which results in a new class of family.[313]

1.07.7.3 Molecular Biology

Previous pharmacological and biochemical studies have found that prostanoid receptors are expressed in many tissues in the body. The exact distribution of each receptor and the patterns of cellular localization, however, remain unknown because of relatively low expression levels of these receptors and the expression of multiple receptors in a single tissue. Molecular biology has provided a new approach to the study of this problem. Technique such as Northern blot analysis and *in situ* hybridization are revealing further information about prostanoid receptor distribution.[314,315]

The tissue distribution of mouse and human prostanoid receptors, as examined by a Northern blot analysis of mRNA expression, is summarized in Table 3. These analyses have shown that each receptor is specifically distributed in the body, and that expression levels are variable between tissues. Some of these findings are in good agreement with previous studies, and others are very novel, and they have facilitated further examination of the actions of prostanoids in various tissues. For example, the TP receptor was expressed most highly in the thymus, where little is known about its actions.[316]

Table 3 Distribution of prostanoid receptor mRNA in various mouse and human organs.

Organ	*mDP*	*mTP*	*mIP*	*hIP*	mEP_1	hEP_2	mEP_3	hEP_3	mEP_4	hEP_4	*mFP*
Brain	+/−	+	−	−	−	−	+/−	−	−	+/−	−
Heart	−	+	+	+	−	−	−	+	+	+	−
Lung	+	++	+	++	+	+	+	−	++	++	−
Thymus	−	+++	+++	−	−	n.d.	−	−	++	+	−
Liver	−	−	−	++	−	n.d.	+/−	−	−	−	−
Stomach	+	−	−	n.d.	−	n.d.	+++		+	n.d.	−
Spleen	−	++	++	+/−	−	n.d.	+/−	−	++	+	−
Kidney	−	+	+	++	++	+/−	+++	+++	−	++	++
Ileum	+++	−	−	+	−	n.d.	−	−	+++	++	−
Testis	+/−	−	−	n.d.	−	n.d.	−	−	−	+	−
Uterus	+	+	−	n.d.	−	n.d.	++	n.d.	+	+	−
Ovary	n.d.	n.d.	n.d.	n.d.	n.d.	n.d.	n.d.	n.d.	n.d.	n.d.	+++

In situ hybridization studies have been carried out for some of the cloned receptors and revealed the more detailed expression patterns of the prostanoid receptors within organs and tissues, which explain the cellular basis of some of the known actions of the prostanoids. For example, *in situ* hybridization studies of PGE receptors in kidney[317,318] have revealed that the EP_3 receptor is expressed mainly by tubular epithels in the medulla, the EP_1 receptor in the collecting tubulus of the papilla, and the EP_4 receptor in the glomerulus. These distribution patterns appear to correspond with the PGE_2-mediated regulation of ion transport, water reabsorption, and glomerular filtration, respectively. Many similar analyses of the other receptors have been reported.[315,316,319,320]

Although these studies have revealed detailed or novel distribution patterns of the receptors, they have limitations in the receptor localization within the cells. In order to solve the direct role of PGs in a living body an animal lacking the PG receptor has been developed. In mice lacking IP, the animals grew up normally, reproduction and blood pressure were not changed, but altered pain perception and inflammatory response were observed to be abnormal.[321]

It was shown that in mice lacking FP, failure parturition was observed,[322] and EP_4 triggered remodeling of the cardiovascular system at birth.[323]

1.07.8 CONCLUSION

Especially in the field of natural product chemistry, the studies on eicosanoids place a limit on improvement of the method of instrumental analysis when using ultramicro amounts of substances. Also in the field of synthetic chemistry, the substances which require a new developing method after the discovery of TX have not been found, and studies in this field have stopped at present. Although the studies on eicosanoids seemed to be finished at the end of the twentieth century, success in the isolation and purification of new receptors, along with recent progress in molecular biology has resulted in a new course of research. Namely, the biological activities which were expressed as diverse biological activities have begun to be solved at the molecular level. The necessity of molecular design of novel biologically active substances is a subject of research for the twenty-first century. It will take much time to structurally analyze the receptor, which is a macromolecule, investigate the receptor and ligand interaction, and investigate the signal transduction in a life phenomenon. The development of medicines related to eicosanoids has not been particularly successful, but it is expected to improve in the near future.

ACKNOWLEDGMENTS

I gratefully acknowledge enlightening discussions with Professors S. Narumiya and F. Ushikubi, and also thank them for providing figures and tables.

1.07.9 REFERENCES

1. E. J. Corey, H. Niwa, J. R. Falck, C. Mioskowski, Y. Arai, and A. Marfat, *Adv. Prostaglandin Thromboxane Res.*, 1980, **6**, 19.
2. R. Kurzrok and C. C. Leib, *Proc. Soc. Exp. Biol. Med.*, 1930, **28**, 268.
3. M. E. Goldblatt, *J. Soc. Chem. Ind. London*, 1933, **52**, 1056.
4. U. S. von Euler, *J. Physiol.*, 1937, **88**, 213.
5. S. Bergström and Sjövall, *Acta Chem. Scand.*, 1957, **11**, 1086
6. W. Feldberg and C. H. Kellaway, *J. Physiol.*, 1930, **94**, 187.
7. C. H. Kellaway and E. R. Q. Trethewie, *J. Exp. Physiol.*, 1940, **30**, 121.
8. H. Kuhn, T. Schewe, and S. M. Rapport, *Adv. Enzymol.*, 1986, **58**, 273.
9. T. Shewe, S. M. Rapport, and H. Kuhn, *Adv. Enzymol.*, 1986, **58**, 191.
10. S. Bergström and J. Sjövall, *Acta Chem. Scand.*, 1957, **11**, 1086.
11. R. Ryhage, *Ark. Kemi.*, 1959, **13**, 475.
12. S. Bergström, R. Ryhage, B. Samuelsson, and J. Sjovall, *Acta Chem. Scand.*, 1962, **16**, 501.
13. S. Bergström, R. Ryhage, B. Samuelsson, and J. Sjövall, *J. Biol. Chem.*, 1963, **238**, 3555.
14. S. Bergström, L. Krabisch, B. Samuelsson, and J. Sjövall, *Acta Chem. Scand.*, 1962, **16**, 969.
15. S. Abrahamsson, *Acta Crystallogr.*, 1963, **16**, 409.
16. B. Samuelsson and G. Stallberg, *Acta Chem. Scand.*, 1963, **17**, 810.
17. B. Samuelsson, *Angew. Chem., Int. Ed. Engl.*, 1964, **4**, 410.
18. D. H. Nugteren, D. A. Van Dorp, S. Bergström, M. Hamberg, and B. Samuelsson, *Nature*, 1966, **212**, 38.
19. S. Bergström, F. Dressler, R. Ryhage, B. Samuelsson, and J. Sjövall, *Ark. Kemi*, 1962, **19**, 563.
20. D. C. Monkhouse, L. Campen, and A. J. Aguiar, *J. Pharm. Sci.*, 1973, **62**, 576.
21. E. Daniels, W. Krueger, F. Kupiecki, J. Pike, and W. Schneider, *J. Am. Chem. Soc.*, 1968, **90**, 5894.
22. N. H. Andersen, *J. Lipid Res.*, 1969, **10**, 316.
23. T. J. Roseman and S. Yalkowsky, *J. Pharm. Sci.*, 1973, **62**, 1680.
24. R. L. Jones, *Biochem. J.*, 1970, **119**, 64.
25. R. L. Jones, *J. Lipid Res.*, 1972, **13**, 511.
26. R. L. Jones, *Prostaglandins*, 1974, **5**, 283.
27. E. Granström, W. E. M. Lands, and B. Samuelsson, *J. Biol. Chem.*, 1968, **243**, 4104.
28. M. Hamberg and B. Samuelsson, *J. Am. Chem. Soc.*, 1966, **88**, 2349.
29. D. H. Nugteren, D. H. Beerthuis, and D. A. Van Dorp, *Recl. Trav. Chim. Pays-Bas*, 1966, **85**, 405.
30. I. Mahmud, R. Alvarez, F. Miller, J. T. Nelson, D. Cho, L. G. Tokes, D. L. Smith, M. A. Whyte, and A. L. Willis, *Fed. Proc.*, 1984, **43**, 980.
31. D. A. Van Dorp, R. K. Beerthuis, D. H. Nugteren, and H. Vonkeman, *Biochim. Biophys. Acta*, 1964, **90**, 204.
32. S. Bergström, H. Danielsson, and B. Samuelsson, *Biochim. Biophys. Acta*, 1964, **90**, 207.
33. B. Samuelsson, *J. Am. Chem. Soc.*, 1965, **87**, 3011.
34. M. Hamberg and B. Samuelsson, *Proc. Natl. Acad. Sci. USA*, 1973, **70**, 899.
35. D. H. Nugteren and E. Hazelhof, *Biochim. Biophys. Acta*, 1973, **236**, 448.
36. M. Hamberg, J. Svensson, T. Wakabayashi, and B. Samuelsson, *Proc. Natl. Acad. Sci. USA*, 1974, **71**, 345.
37. A. L. Willis, F. M. Vane, D. C. Kuhn, C. G. Scott, and M. Petrin, *Prostaglandins*, 1974, **8**, 453.
38. B. S. Levision, D. B. Moller, and R. G. Salomon, *Tetrahedron Lett.*, 1984, **25**, 4633.
39. M. Suzuki, R. Noyori, and N. Hamanaka, *J. Am. Chem. Soc.*, 1981, **103**, 5607.
40. M. Suzuki, R. Noyori, and N. Hamanaka, *J. Am. Chem. Soc.*, 1982, **104**, 2024.
41. S. Moncada, R. Gryglewski, S. Bunting, and J. R. Vane, *Nature (London)*, 1976, **263**, 663.
42. C. Pace-Asciak and I. S. Wolfe, *Biochemistry*, 1971, **10**, 3657.
43. C. Pace-Asciak, *J. Am. Chem. Soc.*, 1976, **98**, 2348.
44. R. A. Johnson, D. R. Morton, J. H. Kinner, R. R. Clorman, J. C. McGuire, F. F. Sun, N. Whittaker, S. Bunting, J. Salmon, S. Moncada, and J. R. Vane, *Prostaglandins*, 1976, **12**, 915.

45. P. Y.-K. Won, W. H. Lee, P. H.-W. Chao, R. F. Reiss, and J. C. McGiff, *J. Biol. Chem.*, 1980, **255**, 9021.
46. M. Hamberg, J. Svensson, and B. Samuelsson, *Proc. Natl. Acad. Sci. USA*, 1975, **72**, 2994.
47. P. Wlodawer and S. Hammarström, *Biochem. Biophys. Res. Commun.*, 1978, **80**, 525.
48. J. B. Smith, C. Ingerman, and M. J. Silver, *J. Clin. Invest.*, 1976, **58**, 1119.
49. R. P. Orange and K. F. Austen, *J. Immunol.*, 1969, **100**, 105.
50. R. P. Orange, R. C. Murphy, R. C. Karnovsky, and K. F. Austen, *J. Immunol.*, 1973, **112**, 760.
51. H. R. Morris, G. W. Taylor, P. J. Piper, P. Sirois, and J. R. Pippen, *FEBS Lett.*, 1978, **87**, 203.
52. K. Strandberg and B. Uvnäs, *Acta Physiol. Scand.*, 1971, **82**, 358.
53. B. A. Jakshik, S. Falkenheim, and S. Parker, *Proc. Natl. Acad. Sci. USA*, 1977, **74**, 4577.
54. M. Hamberg and B. Samuelsson, *Proc. Natl. Acad. Sci. USA*, 1979, **76**, 2148.
55. H. R. Morris, G. W. Taylor, P. J. Piper, P. Sirois, and J. R. Pippen, *FEBS Lett.*, 1978, **87**, 203.
56. S. Hammerström, R. C. Murphy, B. Samuelsson, D. A. Clark, C. Mioskowski, and E. J. Corey, *Biochem. Biophys. Res. Commun.*, 1979, **91**, 1266.
57. R. C. Murphy, S. Hammerström, and B. Samuelsson, *Proc. Natl. Acad. Sci. USA*, 1979, **76**, 4279.
58. E. J. Corey, D. A. Clark, G. Goto, A. Marfat, C. Mioskowski, B. Samuelsson, and S. Hammerström, *J. Am. Chem. Soc.*, 1980, **102**, 1436.
59. R. L. Maas, J. Turk, J. A. Oates, and A. R. Brash, *J. Biol. Chem.*, 1982, **257**, 7056.
60. H. R. Morris, G. W. Tayor, P. J. Piper, M. N. Samhoun, and J. R. Tippins, *Prostaglandins*, 1980, **19**, 185.
61. M. K. Bach, J. R. Brashier, S. Hammarström, and B. Samuelsson, *Biochem. Biophys. Res. Commun.*, 1980, **93**, 1121.
62. J. Houglum, J.-K. Pai, V. Atrache, D.-E. Sok, and C. J. Sih, *Proc. Natl. Acad. Sci. USA*, 1980, **77**, 5688.
63. R. A. Lewis, K. F. Austen, J. M. Drazen, D. A. Clark, A. Marfat, and E. J. Corey, *Proc. Natl. Acad. Sci. USA*, 1980, **77**, 3710.
64. M. Hamberg and B. Samuelsson, *Proc. Natl. Acad. Sci. USA*, 1974, **71**, 3400.
65. D. H. Nugteren, *Biochim. Biophys. Acta*, 1975, **380**, 299.
66. R. W. Bryant, J. M. Bailey, T. Schewe, and S. M. Rapoport, *J. Biol. Chem.*, 1982, **257**, 6050.
67. S. Narumiya, J. A. Salmon, F. H. Cottee, B. C. Weatherley, and R. J. Flower, *J. Biol. Chem.*, 1981, **256**, 9582.
68. G. Furstenberger, H. Hagedorn, T. Jacobi, E. Besemfelder, M. Stephan, W. D. Lehmann, and F. Marks, *J. Biol. Chem.*, 1991, **266**, 15 738.
69. J. M. Schewe, S. M. Rapoport, and H. Kuhn, *Adv. Enzymol.*, 1986, **58**, 191.
70. H. Kuhn, S. M. Rapoport, and J. M. Schewe, *Adv. Enzymol.*, 1986, **58**, 273.
71. S. Yamamoto, *Prostaglandins Leukotrienes Essent. Fatty Acids*, 1989, **35**, 219.
72. A. W. Ford-Hutchinson, M. Gresser, and R. N. Young, *Annu. Rev. Biochem.*, 1994, **63**, 383.
73. C. D. Funk, *Prog. Nucl. Acid Res. Mol. Biol.*, 1993, **45**, 67.
74. C. Yokoyama, F. Shinjo, T. Yoshimoto, S. Yamamoto, J. A. Oates, and A. R. Brash, *J. Biol. Chem.*, 1986, **261**, 16 714.
75. R. L. Maas and A. R. Brash, *Proc. Natl. Acad. Sci. USA*, 1983, **80**, 2884.
76. W. C. Glasgow, T. M. Harris, and A. R. Brash, *J. Biol. Chem.*, 1986, **261**, 200.
77. C. R. Pace-Asciak, *Biochim. Biophys. Acta*, 1984, **793**, 485.
78. C. R. Pace-Asciak, *J. Biol. Chem.*, 1984, **259**, 8332.
79. B. Samuelsson, S.-E. Dahlen, J. A. Lindgren, C. A. Ruzer, and C. N. Serhan, *Science*, 1987, **237**, 1171.
80. A. R. Brash, *Circulation*, 1985, **72**, 702.
81. J. F. G. Vliegenhart and G. A. Veldink, *Free Radicals Biol.*, 1982, **5**, 29.
82. J. Capdevila, G. Snyder, and J. R. Falck, in "Microsomes and Drug Oxidation," eds. A. R. Booboobis, J. Caldwell, F. DeMatteis, and C. R. Elcombe, Taylor and Francis, London, 1985, p. 84.
83. M. Schwartzman, M. A. Carroll, N. G. Ibraham, N. R. Ferreri, E. Songu-Mize, and J. C. McGiff, *Hypertension, Suppl.*, 1985, **1**, 136.
84. J. Capdevila, L. J. Marrnett, N. Chacos, R. A. Prough, and R. W. Estabrook, *Proc. Natl. Acad. Sci. USA*, 1982, **79**, 767.
85. J. Capdevila, P. Yadagiri, S. Manna, and J. F. Falck, *Biochem. Biophys. Res. Commun.*, 1986, **141**, 1007.
86. S. Yamamoto, H. Suzuki, and N. Ueda, *Prog. Lipid Res.*, 1997, **36**, 23.
87. J. D. Morrow, K. E. Hill, R. E. Burk, T. M. Nammour, K. F. Dadr, and L. J. Roberts II, *Proc. Natl. Acad. Sci. USA*, 1990, **87**, 9383.
88. J. D. Morrow, J. A. Awad, H. J. Boss, I. A. Blair, and L. J. Roberts II, *Proc. Natl. Acad. Sci. USA*, 1992, **89**, 10 721.
89. S. W. Hhang, M. Adiaman, S. Khanapure, L. Schio, and J. Rokach, *J. Am. Chem. Soc.*, 1994, **116**, 10 829.
90. K. Takahashi, T. M. Nammour, J. Elbert, J. D. Morrow, L. J. Roberts II, and K. F. Badr, *J. Clin. Invest.*, 1992, **90**, 135.
91. M. Fukunaga, N. Makita, L. J. Roberts II, J. D. Morrow, K. Takahashi, and K. F. Badr, *Am. J. Physiol.*, 1993, **264** (*Cell Physiol.* 33) C-1619.
92. J. D. Morrow, T. A. Minton, C. R. Mukunden, M. D. Campbell, W. E. Zackert, V. C. Daniel, K. F. Badr, I. A. Blair, and L. J. Roberts II, *J. Biol. Chem.*, 1993, **269**, 4317.
93. B. F. Robinson, J. G. Collier, S. M. M. Karim, and K. Somers, *Clin. Sci.*, 1973, **44**, 367.
94. J. Nakano, in "The Prostaglandins: Pharmacological and Therapeutic Advances," ed. M. F. Cuthbert, Lippincott, Philadelphia, PA, 1973, p. 23.
95. J. B. Lee, *Arch. Intern. Med.*, 1974, **133**, 56.
96. H. Rasmussen and W. Lake, in "Prostaglandins in Hematology," eds. M. Silver, B. J. Smith, and J. J. Kocsis, Spectrum Publications, New York, 1977, p. 187.
97. M. F. Cuthbert, in "The Prostaglandins: Pharmacological and Therapeutic Advances," ed. M. F. Cuthbert, Lippincott, Philadelphia, PA, 1973, p. 253.
98. V. J. Goldberg and P. W. Ramwell, *Physiol. Rev.*, 1975, **55**, 325.
99. B. J. R. Whittle and J. R. Vane, in "Physiology of the Gastrointestinal Tract," 2nd edn., ed. L. R. Johnson, Raven Press, New York, 1987, vol. 1, p. 143.
100. S. Moncada, R. J. Flower, and J. R. Vane, in "The Pharmacological Basis of Therapeutics," 7th edn., eds. A. G. Gilman, L. S. Goodman, T. W. Rall, and F. Murad, Macmillan, New York, 1985, p. 660.
101. G. A. Higgs, S. Moncada, and J. R. Vane, *Ann. Clin. Res.*, 1984, **16**, 287.

102. D. S. Rampton and C. J. H. Wkey, *Gut*, 1984, **25**, 1399.
103. H. A. Bernheim, T. M. Gilbert, and J. T. Stitt, *J. Physiol.*, 1980, **301**, 69.
104. S. Moncada, in "Prostaglandins: Research and Clinical Update," eds. G. J. Longenecker and S. W. Schaffer, Alpha Editions, Minneapolis, MN, 1985, p. 1.
105. B. J. R. Whittle, in "Gastrointestinal Mucosal Blood Flow," Churchill Livingstone, Edinburgh, 1980, p. 180.
106. E. A. Higgs, S. Moncada, J. R. Vane, J. P. Caen, H. Michel, and G. Tobelem, *Prostaglandins*, 1978, **16**, 17.
107. S. Moncada, in "Pharmacological Control of Hyperlipidaemia," Prous, Barcelona, 1986, p. 439.
108. S. Moncada and E. A. Higgs, *Clin. Haematol.*, 1986, **15**, 273.
109. S. Moncada, R. M. J. Palmer, and E. A. Higgs, in "Thrombosis and Haemostasis," eds. M. Verstraete, J. Vermylen, H. R. Lijnen, and J. Arnout, Leuven University Press, Belgium, 1987, p. 597.
110. M. Hamberg, J. Svensson, and B. Samuelsson, *Proc. Natl. Acad. Sci. USA*, 1975, **72**, 2994.
111. S. Moncada and J. R. Vane, *Pharmacol. Rev.*, 1979, **30**, 293.
112. J. Svensson, K. Strandberg, T. Tubeno, and M. Hamberg, *Prostaglandins*, 1977, **14**, 425.
113. G. Just and C. Simonovitch, *Tetrahedron Lett.*, 1967, 2093.
114. R. C. Kelly, V. Van Reenen, I. Schletter, and M. D. Pillai, *J. Am. Chem. Soc.*, 1973, **95**, 2746.
115. Van Reenen, R. C. Kelly, and D. Y. Cha, *Tetrahedron Lett.*, 1976, 1973.
116. J. S. Bindra and R. Bindra, "Prostaglandin Synthesis," Academic Press, New York, 1978.
117. A. Mitra, "The Synthesis of Prostaglandins," John Wiliey & Sons, New York, 1977.
118. A. Garcia, L. A. Maldonado, and P. Crabbe, "Prostaglandin Research," Academic Press, New York, 1977, chap. 6.
119. M. P. L. Caton, *Tetrahedron*, 1979, **35**, 2705.
120. K. C. Nicolaou, G. P. Gasic, and W. E. Barnett, *Angew. Chem., Int. Ed. Engl.*, 1978, **17**, 293.
121. S. M. Roberts and F. Scheinmann, "Chemistry, Biochemistry, and Pharmacological Activity of Prostanoids," Pergamon, New York, 1979.
122. R. F. Newton and S. M. Roberts, *Tetrahedron*, 1980, **36**, 2163.
123. R. F. Newton and S. M. Roberts, "Prostaglandins and Thromboxanes," Butterworth Scientific, London, 1982.
124. M. P. L. Caton and T. W. Hart, *Adv. Prostagl. Thrombox. Res.*, 1980, **14**, 73.
125. S. M. Roberts and F. Scheinmann, "New Synthetic Routesto Prostaglandins and Thromboxanes," Academic Press, New York, 1982.
126. S. Iguchi, H. Nakai, M. Hayashi, H. Yamamoto, and K. Maruoka, *Bull. Chem. Soc. Jpn.*, 1981, **54**, 3033.
127. R. Noyori, I. Tomio, and M. Nishizawa, *J. Am. Chem. Soc.*, 1979, **101**, 5843.
128. P. De Clercq, D. Van Hoof, and M. Vanderwalle, *Tetrahedron*, 1976, **32**, 2747.
129. R. Coen, P. De Clercq, D. Van Harver, and M. Vanderwalle, *Bull. Soc. Chim. Belg.*, 1975, **84**, 203.
130. E. J. Corey, Z. Arnold, and J. Hutten, *Tetrahedron Lett.*, 1970, 307.
131. R. B. Woodward, J. Gosteli, I. Ernest, R. J. Friary, G. Nestler, H. Raman, R. Sitrin, Ch. Suter, and J. K. Whitesell, *J. Am. Chem. Soc.*, 1973, **95**, 6853.
132. K. G. Paul, F. Johnson, and D. Farava, *J. Am. Chem. Soc.*, 1976, **98**, 1285.
133. E. J. Corey and R. B. Sneider, *Tetrahedron Lett.*, 1973, 3091.
134. E. J. Corey and P. L. Fuchs, *J. Am. Chem. Soc.*, 1972, **94**, 4014.
135. G. Kovacs, I. Szekely, V. Simonidesz, I. Tomoskozi, and L. Gruber, *Tetrahedron Lett.*, 1976, 4639.
136. D. Brewster, M. Myers, J. Ormerod, P. Otter, A. C. B. Smith, M. E. Spinner, and S. Turner, *J. Chem. Soc., Perkin Trans. 1*, 1973, 2796.
137. D. Brewster, M. Meyers, J. Ormerod, M. E. Spinner, S. Turner, and A. C. B. Smith, *J. Chem. Soc., Chem. Commun.*, 1972, 1235.
138. E. D. Brown and T. L. Lilley, *J. Chem. Soc., Chem. Commun.*, 1975, 39.
139. S. Raganathan, D. Raganathan, and A. K. Mehrotra, *J. Am. Chem. Soc.*, 1974, **96**, 5261.
140. S. Raganathan, D. Raganathan, and A. K. Mehrotra, *Tetrahedron Lett.*, 1975, 1215.
141. S. Raganathan, D. Raganathan, and R. Iyengar, *Tetrahedron Lett.*, 1976, **32**, 961.
142. G. Jones, R. A. Raphel, and S. Wright, *J. Chem. Soc., Chem. Commun.*, 1972, 609.
143. G. Jones, R. A. Raphel, and S. Wright, *J. Chem. Soc., Perkin Trans. 1*, 1974, 1676.
144. J. S. Bindra, A. Grodski, K. F. Schaaf, and E. J. Corey, *J. Am. Chem. Soc.*, 1973, **95**, 7522.
145. R. Peel and J. Sutherland, *J. Chem. Soc., Chem. Commun.*, 1974, 151.
146. S. Takano, N. Kubodera, and K. Ogasawara, *J. Org. Chem.*, 1977, **42**, 786.
147. J. Katsube, H. Shimomura, and M. Matsui, *Agric. Biol. Chem.*, 1971, **35**, 1828.
148. J. Katsube, H. Shimomura, and M. Matsui, *Agric. Biol. Chem.*, 1972, **36**, 1997.
149. H. Shimomura, J. Katsube, and M. Matsui, *Agric. Biol. Chem.*, 1975, **39**, 657.
150. A. Fischli, M. Kraus, H. Meyer, P. Schonholzer, and R. Ruegg, *Helv. Chim. Acta*, 1975, **58**, 564.
151. C. J. Sih, P. Price, R. Sood, R. G. Salomon, G. Peruzzoti, and M. Casey, *J. Am. Chem. Soc.*, 1972, **94**, 3642.
152. A. F. Kluge, K. G. Untch, and J. H. Fried, *J. Am. Chem. Soc.*, 1972, **96**, 6774.
153. E. J. Corey, I. Vlattas, and K. Harding, *J. Am. Chem. Soc.*, 1969, **91**, 5675.
154. C. J. Sih, P. Price, R. Sood, R. G. Salomon, G. Peruzzotti, and M. Casey, *J. Am. Chem. Soc.*, 1972, **94**, 3643.
155. G. Stork and T. Takahashi, *J. Am. Chem. Soc.*, 1977, **99**, 1275.
156. J. W. Patterson, Jr. and J. H. Fried, *J. Org. Chem.*, 1974, **39**, 2506.
157. R. Davis and K. G. Untch, *J. Org. Chem.*, 1979, **44**, 3755.
158. G. Stork and M. Isobe, *J. Am. Chem. Soc.*, 1975, **97**, 4745.
159. G. Stork and M. Isobe, *J. Am. Chem. Soc.*, 1975, **97**, 6260.
160. A. F. Kluge, K. G. Untch, and J. H. Fried, *J. Am. Chem. Soc.*, 1972, **94**, 7828.
161. A. F. Kluge, K. G. Untch, and J. H. Fried, *J. Am. Chem. Soc.*, 1972, **94**, 9256.
162. J. G. Miller, W. K. Karl, K. G. Untch, and G. Stork, *J. Am. Chem. Soc.*, 1974, **96**, 6774.
163. R. Noyori and M. Suzuki, *Chemtracts: Org. Chem.*, 1990, 173.
164. R. Noyori and M. Suzuki, *Angew. Chem., Int. Ed. Engl.*, 1984, **23**, 847.
165. M. Suzuki, T. Kawagishi, T. Suzuki, and R. Noyori, *Tetrahedron Lett.*, 1982, **23**, 4057.
166. T. Tanaka, T. Toru, Okamura, A. Hazato, S. Sugiura, K. Manabe, S. Kurozumi, M. Suzuki, T. Kawaguchi, and R. Noyori, *Tetrahedron Lett.*, 1983, **24**, 5563.

167. H. Nishiyama, K. Sakuta, and K. Itoh, *Tetrahedron Lett.*, 1984, **25**, 223.
168. H. Nishiyama, K. Sakuta, and K. Itho, *Tetrahedron Lett.*, 1984, **25**, 2487.
169. E. J. Corey, G. E. Keck, and I. Szèkely, *J. Am. Chem. Soc.*, 1977, **99**, 2006.
170. E. J. Corey, H. L. Pearce, I. Szèkely, and M. Ishiguro, *Tetrahedron Lett.*, 1978, 1023.
171. J. Fried and J. Barton, *Proc. Natl. Acad. Sci. USA*, 1977, **74**, 2199.
172. R. A. Johnson, F. H. Lincoln, J. L. Thompson, E. J. Nidy, S. A. Muzsak, and U. Axen, *J. Am. Chem. Soc.*, 1977, **99**, 4182.
173. R. A. Johnson, F. H. Lincoln, E. J. Nidy, W. P. Schneider, J. L. Thompson, and U. Axen, *J. Am. Chem. Soc.*, 1978, **100**, 7690.
174. E. J. Nidy and R. A. Johnson, *Tetrahedron Lett.*, 1978, 2375.
175. N. A. Nelson, *J. Am. Chem. Soc.*, 1977, **99**, 7362.
176. I. Tömösközi, G. Galambos, V. Simonidesz, and G. Kovács, *Tetrahedron Lett.*, 1977, 2627.
177. I. Tömösközi, G. Galambos, G. Kovács, and L. Radics, *Tetrahedron Lett.*, 1978, 581.
178. M. Suzuki, A. Yanagisawa, and R. Noyori, *Tetrahedron Lett.*, 1983, **24**, 1187.
179. M. Suzuki, A. Yanagisawa, and R. Noyori, *J. Am. Chem. Soc.*, 1988, **110**, 4718.
180. R. A. Johnson, E. J. Nidy, L. Baczynskyj, and R. R. Gorman, *J. Am. Chem. Soc.*, 1977, **99**, 7738.
181. N. A. Porter, J. D. Byers, R. C. Mebane, D. W. Gilmore, and J. R. Nixon, *J. Org. Chem.*, 1978, **43**, 2088.
182. N. A. Porter, J. D. Byers, K. M. Holden, and D. B. Menzel, *J. Am. Chem. Soc.*, 1979, **101**, 4319.
183. T. Mukaiyama, S. Shoda, and Y. Watanabe, *Chem. Lett.*, 1977, 383.
184. N. A. Porter, J. D. Byers, A. E. Ali, and T. E. Eling, *J. Am. Chem. Soc.*, 1980, **102**, 1183.
185. S. S. Bhagwat, P. R. Hamann, and W. C. Still, *Tetrahedron Lett.*, 1985, **26**, 1955.
186. S. S. Bhagwat, P. R. Hamann, and W. C. Still, *J. Am. Chem. Soc.*, 1985, **107**, 6372.
187. H. H. Holzgrafe, L. V. Buchanan, and S. Bunting, *Circ. Res.*, 1987, **60**, 290.
188. P. Borgeat and B. Samuelsson, *Proc. Natl. Acad. Sci. USA*, 1979, **76**, 3213.
189. S. Hammerström, R. C. Murphy, B. Samuelsson, D. A. Clark, C. Mioskowski, and E. J. Corey, *Biochem. Biophys. Res. Commun.*, 1979, **91**, 1266.
190. E. J. Corey, J. O. Albright, A. E. Barton, and S. Hashimoto, *J. Am. Chem. Soc.*, 1980, **102**, 1435.
191. E. J. Corey, D. A. Clark, G. Goto, A. Marfat, C. Mioskowski, B. Samuelsson, and S. Hammerström, *J. Am. Chem. Soc.*, 1980, **102**, 1436.
192. E. J. Corey, D. A. Clark, G. Goto, A. Marfat, C. Mioskowski, B. Samuelsson, and S. Hammerström, *J. Am. Chem. Soc.*, 1980, **102**, 3663.
193. E. J. Corey, D. A. Clark, G. Goto, A. Marfat, C. Mioskowski, B. Samuelsson, and S. Hammerström, *J. Am. Chem. Soc.*, 1980, **102**, 1436.
194. E. J. Corey and D. A. Clark, *Tetrahedron Lett.*, 1980, **21**, 3547.
195. R. A. Lewis, J. M. Drazen, K. F. Austen, D. A. Clark, and E. J. Corey, *Biophys. Res. Commun.*, 1980, **96**, 271.
196. J. Rokach, R. Zamboni, C. K. Lau, and Y. Guindon, *Tetrahedron Lett.*, 1981, **22**, 2759.
197. J. Rokach, R. Zamboni, C. K. Lau, and Y. Guindon, *Tetrahedron Lett.*, 1981, **22**, 2763.
1989. B. Samuelsson, S. Hammerström, M. Hamberg, and C. N. Serhan, *Adv. Prostaglandin Thromboxane Res.*, 1980, **14**, 45.
199. A. Marfat and E. J. Corey, *Adv. Prostaglandin Thromboxane Res.*, 1980, **14**, 155.
200. F. Scheinmann and J. Ackroyd, "Leukotriene Syntheses: A New Class of Biologically Active Compounds Including SRS-A," Raven Press, New York, 1984.
201. J. K. Atkins and L. Rokach, "Handbook of Eicosanoids: Prostaglandins and Related Lipids," ed. A. L. Wills, CRC Press, Boca Raton, FL, 1987, vol. II, p. 175.
202. J. Rokach, "Leukotrienes and Lipoxygenases," Elsevier, New York, 1989.
203. P. Borgeat and B. Samuelsson, *J. Biol. Chem.*, 1979, **254**, 2643.
204. P. Borgeat and B. Samuelsson, *Proc. Natl. Acad. Sci. USA*, 1979, **76**, 2148.
205. E. J. Corey, A. Marfat, G. Goto, and F. Brion, *J. Am. Chem. Soc.*, 1980, **102**, 7984.
206. E. J. Corey, A. E. Barton, and D. A. Clark, *J. Am. Chem. Soc.*, 1980, **102**, 4278.
207. Y. Guindon, R. Zamboni, C. K. Lau, and J. Rokach, *Tetrahedron Lett.*, 1982, **23**, 739.
208. R. Zamboni and J. Rokach, *Tetrahedron Lett.*, 1982, **23**, 2631.
209. A. L. Willis and K. J. Stone, "Handbook of Eicosanoids: Prostaglandins and Related Lipids," ed. A. L. Wills, CRC Press, Boca Raton, FL, 1987, vol. I.
210. N. H. Wilson and R. L. Jones, *Adv. Prostaglandin, Thromboxane, Leukotriene Res.*, 1985, **14**, 393.
211. G. L. Bundy, *Tetrahedron lett.*, 1975, 1957.
212. H. Miyake, S. Iguchi, H. Itoh, and M. Hayashi, *J. Am. Chem. Soc.*, 1977, **99**, 3536.
213. P. V. Halushka, D. E. Mais, P. R. Mayeux, and T. A. Morinelli, *Eur. J. Pharmacol.*, 1986, **131**, 49.
214. T. A. Egglete, H. de Köning, and H. O. Huisman, *J. Chem. Soc., Perkin Trans. 1*, 1978, 980.
215. P. W. Sprague, J. E. Heikes, D. N. Harris, and R. Greenberg, *Adv. Prostaglandin Thromboxane Res.*, 1980, **6**, 493.
216. P. Barraclough, *Tetrahedron Lett.*, 1980, **21**, 1897.
217. T. J. Leeney, P. R. Marsham, G. A. Ritchie, and M. W. Senior, *Prostaglandins*, 1976, **11**, 953.
218. E. J. Corey, H. Niwa, M. Bloom, and P. W. Ramwell, *Tetrahedron Lett.*, 1979, 671.
219. E. J. Corey, K. C. Nicolaou, Y. Machida, C. L. Malmsten, and B. Samuelsson, *Proc. Natl. Acad. Sci. USA*, 1975, **72**, 3355.
220. T. J. Leeney, P. R. Marsham, G. A. Ritchie, and M. W. Senior, *Prostaglandins*, **1976**, 11, 953.
221. K. C. Nicolaou, R. L. Magolda, J. B. Smith, D. Aharony, E. F. Smith, and A. M. Lefer, *Proc. Natl. Acad. Sci. USA*, 1979, **76**, 2566.
222. S. Ohuchida, N. Hamanaka, and M. Hayashi, *Tetrahedron Lett.*, 1979, 3661.
223. S. Ohuchida, N. Hamanaka, and M. Hayashi, *Tetrahedron*, 1983, **39**, 4263.
224. K. M. Maxey and G. L. Bundy, *Tetrahedron Lett.*, 1980, **21**, 445.
225. E. J. Corey, J. W. Ponder, and P. Ulrich, *Tetrahedron Lett.*, 1980, **21**, 137.
226. S. Kosuge, N. Hamanaka, and M. Hayashi, *Tetrahedron*, 1983, 1345.
227. S. Ohuchida, N. Hamanaka, S. Hashimoto, and M. Hayashi, *Tetrahedron Lett.*, 1982, **23**, 2883.

228. S. Ohuchida, N. Hamanaka, and M. Hayashi, *J. Am. Chem. Soc.*, 1981, **103**, 4597.
229. P. Lumley, B. P. White, and P. P. A. Humphery, *Br. J. Pharmacol.*, 1989, **97**, 783.
230. H. Suga, N. Hamanaka, K. Kondo, H. Miyake, S. Ohuchida, Y. Arai, and A. Kawasaki, *Adv. Prostaglandin, Thromboxane, Leukotriene Res.*, 1987, **17**, 799.
231. T. Hanasaki and H. Arita, *Thromb. Res.*, 1988, **50**, 365.
232. N. Hamanaka, T. Seko, T. Miyazaki, M. Naka, K. Furuta, and H. Yamamoto, *Tetrahedron Lett.*, 1989, **30**, 2399.
233. M. G. McKenniff, P. Norman, N. J. Cuthbert, and P. J. Gardiner, *Br. J. Pharmacol.*, 1991, **104**, 585.
234. D. E. Mais, D. Knapp, P. V. Halushka, K. Ballard, and N. Hamanaka, *Tetrahedron Lett.*, 1984, **25**, 4207.
235. P. A. Aristoff, *Adv. Prostaglandin, Thromboxane, Leukotriene Res.*, 1985, **14**, 309.
236. J. M. Muchowski, in "Handbook of Eicosanoids: Prostaglandins and Related Lipids," ed. A. L. Wills, CRC Press, Boca Raton, FL, 1987, vol. II, p. 19.
237. Nileprost, *Drugs Fut.*, 1982, **7**, 643.
238. B. Radüchel, *Tetrahedron Lett.*, 1983, 3229.
239. K. C. Nicolaou, W. E. Barnette, G. P. Gasic, and R. L. Magolda, *J. Am. Chem. Soc.*, 1977, **99**, 7736.
240. G. L. Bundy and J. M. Baldwin, *Tetrahedron Lett.*, 1978, 1371.
241. *Drugs Fut.*, 1981, **6**, 753.
242. W. Skuballa and H. Vörbryggen, *Angew. Chem., Int. Ed. Engl.*, 1981, **20**, 1046.
243. M. Shibasaki, Y. Torisawa, and S. Ikegami, *Tetrahedron Lett.*, 1983, 3493.
244. L. Föhle, H. Bohlke, E. Frankas, S. M. A. Kim, Lintz, G. Loschen, B. Müller, J. Schneider, U. Seipp, W. Vollenberg, and K. Wilsman, *Arzneim Forsch. Drug. Res.*, 1983, **33**, 1240.
245. K. Ohno, H. Nagase, K. Matsumoto, H. Nishiyama, and S. Nishio, *Adv. Prostaglandin, Thromboxane, Leukotriene Res.*, 1985, **15**, 279.
246. P. A. Aristoff, P. D. Johnson, and A. W. Harrison, *J. Am. Chem. Soc.*, 1985, **107**, 7967.
247. M. Toda, Y. Arai, and M. Hayashi, *Adv. Prostaglandin, Thromboxane, Leukotriene Res.*, 1985, **14**, 427.
248. J. Augstein, J. B. Farmer, T. B. Lee, P. Sheard, and M. L. Tattersall, *Nature, New. Biol.*, 1973, **245**, 17.
249. M. Konno, T. Nakae, S. Sakuyama, M. Nishizaki, Y. Odagaki, H. Nakai, and N. Hamanaka, *Bioorg. Med. Chem.*, 1997, **5**, 1621.
250. M. Konno, T. Nakae, S. Sakuyama, Y. Odagaki, H. Nakai, and N. Hamanaka, *Bioorg. Med. Chem.*, 1997, **5**, 1649.
251. J. F. Merritt, T. J. Hallam, A. M. Brown, I. Boyfield, D. G. Cooper, D. M. B. Hickey,, A. A. Jaxa-Chamiec, A. J. Kaumann, M. Keen, E. Kelly, U. Kozlowski, J. A. Lynham, K. E. Moores, K. J. Murray, J. MacDermot, and T. J. Rink, *Br. J. Pharmacol.*, 1991, **102**, 251.
252. N. A. Meanwell, M. J. Rosenfeld, A. K. Trehan, J. L. Romine, J. J. K. Wright, C. L. Brassard, J. O. Buchanan, M. F. Federica, J. S. Fleming, M. Gamberdalla, G. B. Zavoico, and S. M. Seiler, *J. Med. Chem.*, 1992, **35**, 3498.
253. N. Hamanaka, K. Takahashi, Y. Nagao, K. Torisu, H. Tokumoto, and K. Kondo, *Bioorg. Med. Chem. Lett.*, 1995, 1083.
254. N. Hamanaka, K. Takahashi, Y. Nagao, K. Torisu, S. Shigeoka, H. Hamada, H. Kato, H. Tokumoto, and K. Kondo, *Bioorg. Med. Chem. Lett.*, 1995, **85**, 273.
255. R. A. Coleman, L. H. Denyer, and R. L. G. Scheldrick, *Br. J. Pharmacol.*, 1985, **86**, 203.
256. M.-H. Town, J. Casals-Stenzel, and E. Schillenger, *Prostaglandins*, 1983, **25**, 13.
257. D. G. Trist, B. A. Collins, J. Wood, M. G. Kelly, and A. D. Robertson, *Br. J. Pharmacol.*, 1989, **96**, 301.
258. D. H. Wright, Kathleen M. Metters, M. Abramovitz, and A. W. Ford-Hutchinson, *Br. J. Pharmacol.*, 1998, **123**, 1317.
259. T. Tsuri, T. Honma, Y. Hiramatsu, T. Okada, H. Hashizumi, S. Mitsumori, M. Inagaki, A. Arimura, K. Yasui, F. Asanuma, J. Kishino, and M. Ohtani, *J. Med. Chem.*, 1997, **40**, 3504.
260. R. A. Laurence, R. L. Jones, and N. H. Wilson, *Br. J. Pharmacol.*, 1992, **105**, 271.
261. R. A. Coleman, I. Kennedy, and R. L. G. Scheldrick, *Br. J. Pharmacol.*, 1985, **85**, 273.
262. A. Nials, C. J. Vardey, L. H. Denyer, M. Thomas, S. J. Sparrow, C. D. Shepherd, and R. A. Coleman, *Drug Rev.*, 1993, **11**, 165.
263. J. J. Reeves, K. T. Bunce, R. L. G. Scheldrick, and R. Stables, *Br. J. Pharmacol.*, 1988, **95**, 805.
264. P. J. Gardiner, *Br. J. Pharmacol.*, 1986, **87**, 45.
265. K. T. Bunce, N. M. Clayton, R. A. Coleman, E. W. Collington, H. Finch, J. M. Humphrey, P. P. A. Humphrey, J. J. Reeves, R. L. G. Scheldrick, and R. Stables, *Adv. Prostaglandin, Thromboxane, Leukotriene Res.*, 1990, **21**, 379.
266. R. A. Laurence and R. L. Jones, *Br. J. Pharmacol.*, 1992, **105**, 817.
267. F. S. F. Tam, K.-M. Chan, J.-P. Bourreau, and R. L. Jones, *Br. J. Pharmacol.*, 1967, **121**, 1413.
268. T. K. Schaaf, J. S. Bindra, J. F. Eggler, J. J. Plattner, A. J. Nelson, M. R. Johnson, J. W. Constantine, H.-J. Hess, and W. Elger, *J. Med. Chem.*, 1981, **24**, 1353.
269. D. A. Van Dorp, *Ann. NY Acad. Sci.*, 1971, **180**, 181.
270. G. Hansson and E. Grastrom, *Biochem. Med.*, 1977, **18**, 420.
271. A. Robert, J. E. Nezamis, C. Lancaster, and J. H. Alexander, *Gastroenterology*, 1979, **77**, 433.
272. J. F. Paoletto, K. F. Bernady, D. Kupfer, R. Partridge, and M. J. Weiss, *J. Med. Chem.*, 1975, **18**, 359.
273. P. W. Collins, E. Z. Dajani, D. R. Driskill, M. S. Bruhn, C. J. Jung, and R. Pappo, *J. Med. Chem.*, 1977, **20**, 1152.
274. P. Baret, E. Barreiro, A. E. Greene, J.-L. Luche, M.-A. Teixeira, and P. Crabbe, *Tetrahedron*, 1979, **35**, 2931.
275. *Drugs Fut.*, 1978, **3**, 536.
276. *Drugs Fut.*, 1982, **7**, 116.
277. W. Skuballa and H. Vörbryggen, *Angew. Chem., Int. Ed. Engl.*, 1981, **20**, 1046.
278. *Drug Fut.*, 1996, **21**, 1176.
279. M. Sakurai, M. Araie, T. Oshika, M. Mori, K. Masuda, K. Ueno, and M. Takase, *Jpn. J. Ophthalmol.*, 1991, **35**, 156.
280. B. Fristrom and S. E. G. Nilsson, *Arch. Ophthalmol.*, 1993, **111**, 662.
281. H. Fujiwara, N. Kurihara, K. Ohta, H. Hirata, H. Matsushita, H. Kanazawa, and T. Takeda, *Prostaglandin Leukotriene Essent. Fatty Acids*, 1993, **48**, 241.
282. H. Nakai, M. Konno, S. Kosuge, S. Sakuyama, M. Toda, Y. Arai, T. Obata, N. Katsube, T. Miyamoto, T. Okegawa, and A. Kawasaki, *J. Med. Chem.*, 1988, **31**, 84.
283. R. D. Krell, D. Aharony, and C. K. Buckner, *Am. Rev. Resp. Dis.*, 1990, **141**, 978.
284. V. G. Matassa, T. P. Maduskuie, Jr., H. S. Shapiro, B. Hesp, D. W. Snyder, D. Aharony, R. D. Krell, and R. A. Keith, *J. Med. Chem.*, 1990, **33**, 1781.

285. M. Labelle, M. Belley, and Y. Gareau, *Bioorg. Med. Chem. Lett.*, 1995, **5**, 283.
286. A. W. Ford-Hutchinson, *Adv. Prostaglandin, Thromboxane, Leukotriene Res.*, 1994, **69**, 74.
287. S. Fukumoto, M. Shiraishi, Z. Terashita, Y. Ashida, and Y. Inada, *J. Med. Chem.*, 1992, **35**, 2202.
288. R. P. Ahlquist, *Am. J. Physiol.*, 1948, **153**, 586.
289. A. M. Lands, A. Arnold, J. P. McAuliff, F. P. Luduena, and T. G. Brown, *Nature (London)*, 1967, **241**, 597.
290. V. R. Pickles, in "Nobel Symposium, Vol. 2: Prostaglandins," eds. S. Bergström and B. Samuelsson, 1967, p. 79.
291. N. H. Andersen and P. W. Ramwell, *Arch. Intern. Med.*, 1974, **133**, 30.
292. N. H. Andersen, T. L. Eggerman, L. A. Harker, C. H. Wilson, and B. De, *Prostaglandins*, 1980, **19**, 711.
293. P. J. Gardiner and H. O. J. Collier, *Prostaglandins*, 1980, **19**, 819.
294. E. E. Horton, *Br. Med. Bull.*, 1979, **35**, 295.
295. I. Kennedy, R. A. Coleman, P. P. A. Hunphery, C. P. Levy, and P. Lumley, *Prostaglandins*, 1982, **24**, 667.
296. R. A. Coleman, P. P. A. Hunphery, I. Kennedy, and P. Lumley, *Trends Pharmacol. Sci.*, 1984, **5**, 303.
297. R. A. Coleman, I. Kennedy, P. P. A. Hunphery, K. Bunce, and P. Lumley, in "Comprehensive Medicinal Chemistry," ed. J. C. Emmett, Pergamon, Oxford, 1990, vol. 3, p. 643.
298. A. Watabe, Y. Sugimoto, A. Honda, A. Irie, T. Namba, M. Negishi, S. Ito, S. Narumiya, and A. Ichikawa, *J. Biol. Chem.*, 1993, **268**, 20 175.
299. R. A. Armstrong, R. A. Lawrence, R. J. Jones, N. H. Wilson, and A. Collier, *Br. J. Pharmacol.*, 1989, **97**, 657.
300. P. V. Halushka, D. E. Mais, P. R. Mayeux, and T. A. Morinelli, *Annu. Rev. Pharmacol. Toxicol.*, 1989, **10**, 213.
301. F. Usikubi, M. Nakajima, M. Hirata, M. Okuma, M. Fujiwara, and S. Narumiya, *J. Biol. Chem.*, 1989, **264**, 16 496.
302. M. Hirata, Y. Hayashi, F. Usikubi, Y. Yokota, R. Kageyama, S. Nakanishi, and S. Narumiya, *Nature (London)*, 1991, **349**, 617.
303. T. Namba, Y. Sugimoto, M. Hirata, Y. Hayashi, A. Honda, A. Watabe, M. Negishi, A. Ichikawa, and S. Narumiya, *Biochem. Biophys. Res. Commun.*, 1992, **184**, 1197.
304. Y. Sugimoto, T. Namba, A. Honda, Y. Hayashi, M. Negishi, A. Ichikawa, and S. Narumiya, *J. Biol. Chem.*, 1992, **267**, 6463.
305. A. Honda, Y. Sugimoto, T. Namba, A. Watabe, A. Irie, M. Negishi, S. Narumiya, and A. Ichikawa, *J. Biol. Chem.*, 1993, **268**, 7759.
306. A. Watabe, Y. Sugimoto, A. Honda, A. Irie, T. Namba, M. Negishi, S. Ito, S. Narumiya, and A. Ichikawa, *J. Biol. Chem.*, 1993, **268**, 20 175.
307. Y. Sugimoto, T. Namba, Y. Sugimoto, M. Negishi, A. Ichikawa, and S. Narumiya, *Am. J. Physiol.*, 1994, **268**, 2712.
308. H. Namba, Y. Oida, A. Sugimoto, M. Kakizuka, A. Negishi, A. Ichikawa, and S. Narumiya, *J. Biol. Chem.*, 1994, **269**, 9986.
309. R. A. Coleman, W. L. Smith, and S. Narumiya, *Am. Soc. Pharmacol., Exp. Ther.*, 1994, **46**, 27.
310. S. Narumiya, M. Hirata, T. Namba, Y. Hayashi, F. Ushikubi, Y. Sugimoto, M. Negishi, and A. Ichikawa, *J. Lipid Mediators*, 1993, **6**, 155.
311. Y. Yamamoto, K. Kamiya, and S. Terao, *J. Med. Chem.*, 1993, **36**, 820.
312. S. P. H. Alexander and J. A. Peter, *Trends in Pharmacol. Sci.*, *Receptor and Ion Channel Nomenclature*, 1997, suppl. 50.
313. T. Yokomizo, T. Izumi, K. Chang, Y. Takuwa, and T. Shimizu, *Nature*, 1997, **387**, 620.
314. F. Ushikubi, M. Hirata, and S. Narumiya, *J. Lipid Mediators Cell Signalling*, 1995, **12**, 343.
315. S. Narumiya, N. Hirata, T. Namba, Y. Hayashi, F. Ushikubi, Y. Sugimoto, M. Negishi, and A. Ichikawa, *J. Lipid Mediators*, 1993, **6**, 155.
316. F. Ushikubi, Y. Aiba, K. Nakamura, T. Namba, M. Hirata, O. Mazda, Y. Katsura, and S. Narumiya, *J. Exp. Med.*, 1993, **178**, 1825.
317. Y. Sugimoto, T. Namba, R. Shigemoto, M. Negishi, A. Ichikawa, and S. Narumiya, *Am. J. Physiol.*, 1994, **266**, 823.
318. M. D. Breyer, H. R. Jacobson, L. S. Davis, and R. M. Breyer, *Kidney Int.*, 1993, **43**, 1372.
319. T. Namba, H. Oida, Y. Sugimoto, A. Kakizuka, M. Negishi, A. Ichikawa, and S. Narumiya, *J. Biol. Chem.*, 1994, **269**, 9986.
320. Y. Sugimoto, K. Hasumoto, T. Namba, A. Ine, M. Katsuyama, M. Negishi, A. Kakizuka, S. Narumiya, and A. Ichikawa, *J. Biol. Chem.*, 1994, **269**, 1356.
331. T. Murata, F. Ushikubi, T. Matsuoka, M. Hirata, A. Yamasaki, Y. Sugimoto, A. Ichikawa, Y. Aze, T. Tanaka, N. Yoshida, A. Ueno, S. Oh-ishi, and S. Narumiya, *Nature*, 1997, **388**, 878.
322. Y. Sugimoto, A. Yamasaki, E. Segi, K. Tsubi, Y. Aze, T. Nishimura, H. Oide, N. Yoshida, T. Tanaka, M. Hirata, F. Ushikubi, M. Negishi, A. Ichikawa, and S. Narumiya, *Science*, 1997, **277**, 681.
323. M. T. Nguen, T. Camenisch, J. N. Snouwaert, E. Hicks, T. M. Coffman, P. A. W. Anderson, N. N. Malouf, and B. H. Koller, *Nature*, 1997, **390**, 78.

1.08
Eicosanoids in Nonmammals

WILLIAM H. GERWICK
Oregon State University, Corvallis, OR, USA

1.08.1 INTRODUCTION

While eicosanoids and related compounds are best known for their important roles in mammalian physiology and disease, they in fact occur in such diverse creatures as bacteria, marine algae, and invertebrate animals as well as in terrestrial higher plants. These eicosanoid-like structures vary in chain length, oxidation pattern, cyclization pattern, and even the type of incorporated heteroatoms.

Because of the nature of eicosanoid-like compounds found in nonmammals, a somewhat expanded definition of "eicosanoid" will be used in this chapter. Specifically, polyunsaturated fatty acids

(PUFAs) of any chain length which are metabolized to distinct structures will be discussed. This expansion in the scope of material covered takes into consideration that many groups of organisms metabolize PUFAs of various chain lengths by routes similar or identical to those employed for metabolizing C_{20} acids, and as such, should be classed within the same review subject. For example, the term "oxylipin" has been introduced into the literature as a way of easily referring to PUFAs of any chain length that become enzymatically oxidized.[1]

The organization of this chapter is strictly along taxonomic lines, in contrast to several excellent reviews that have appeared in this area, which are arranged by class of metabolite or enzymatic reaction leading to their production (see, for example, Gardner[2]). The chapter begins with the most primitive group, the bacteria, including cyanobacteria (blue-green algae). This is followed by the primitive eukaryotes, fungi and yeasts, and then algae, predominately the macrophytic marine algae. Discussion of the oxylipins of higher plants ensues, and is arranged along familial lines. A review of animal oxylipins follows and is arranged sequentially, beginning with fish, then marine invertebrates, insects, amphibians, arachnids, and finally, worms.

For each group of organism, an introductory paragraph provides a brief taxonomic/biological identification of the group and an overview perspective on the type of eicosanoid metabolism found in that group. Representative chemical structures and their biosyntheses are then presented. However, for space limitation reasons, usually only a single representative structure is given per metabolic trend. Sadly, many fascinating accounts of unusual nonmammalian oxylipins have not been included. In making these difficult decisions, discussion of newer examples of oxylipins is given in favor of old, oxylipins illustrating trends over those representing exceptions, and structurally complex oxylipins in favor of those of more straightforward structure. When biosynthetic investigation of a particular oxylipin structure class has advanced notably using a particular species, then this species has been chosen as the representative example (e.g., *Glycine max*—soybean). A figure is presented in cases where the biogenesis of the class is explained by a particularly intriguing metabolic transformation. Biological activity data and chemical ecology is kept to a minimum in favor of giving expanded coverage of biosynthetic considerations.

There have been several general reviews of eicosanoids and related compounds from nonmammalian sources. These have focused on the prostaglandins in aquatic organisms,[3] a broad survey of life forms containing oxylipins presented with a synthetic organic chemical perspective,[4] the biological significance of oxylipins in invertebrates,[5] the chemistry and biology of oxylipins in insects,[6] oxylipins from marine algae[7] and marine invertebrates,[8] and oxylipins from higher plants.[2]

1.08.2 EICOSANOIDS FROM BACTERIA AND MICROORGANISMS

1.08.2.1 Eubacteria

There are only a few reports of eicosanoid-like compounds from bacteria with the majority of these coming from the cyanobacteria or blue-green algae (see Section 1.08.2.2). For example, from the fermentation broth of an unnamed actinomycetes, a potent inhibitor of sterol biosynthesis was isolated and characterized by synthesis to an absolute stereostructure.[9] The metabolite, known as L-660,631 (**1**), is a 16-carbon fatty acid with alcohol-derived functionalities at C-8, C-9, and C-10. The cyclic carbonate functionality between alcohols at C-8 and C-9 has only rarely been observed in natural products.[10] The three sequential acetylenic bonds between C-11 and C-16 represent an additional unusual modification. Conceivably, the oxygen functionalities in L-660,631 could derive in large part from an epoxy alcohol that would in turn come from a diene hydroperoxide, perhaps through action of an $\omega 9$ (C-8 oxygenation) lipoxygenase.

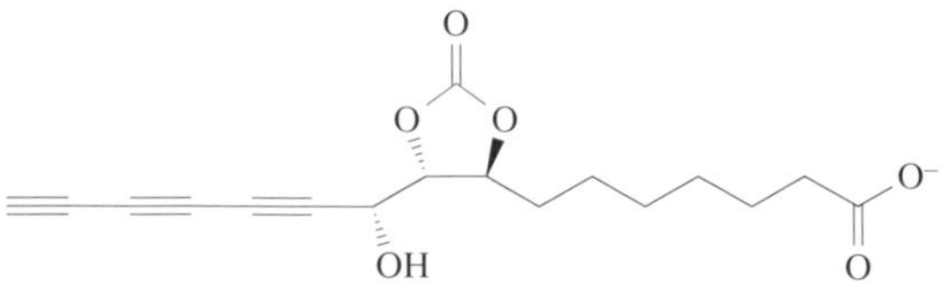

O O O OH O O⁻

(**1**)

1.08.2.2 Cyanobacteria

From the cultured freshwater cyanobacterium, *Oscillatoria* sp., a lipoxygenase (LO) activity was discovered which converts linoleic acid into a 52/48 mixture of 13*S*-hydroperoxide (HPOD) and 9*S*-HPOD at its pH optimum of 8.8.[11] The new LO has an estimated molecular weight of 124 kDa, considerably larger than other plant LOs, and the plant flavanoid esculetin was found to be an effective inhibitor of the enzyme. A hydroperoxide lyase (M_r 56 kDa and pH optimum 6.4) was subsequently reported from this same organism which accepted 13*S*-HPOD as substrate, but not 9*S*-HPOD, and produced 13-oxo-9,11-tridecadienoic acid and 1-pentanol as products.[12] Hence, oxylipin metabolism in *Oscillatoria* sp. is similar but not identical to that in *Chlorella pyrenoidosa* (1-pentanol vs. *cis*-2-pentene is produced, see Section 1.08.3.2.1(i)). From another freshwater cyanobacterium, *Anabaena flos-aquae f. flos-aquae*, two new hydroxy acids were obtained and characterized through extensive NMR spectroscopy and MS on the natural products and derivatives. The new compounds, 9*R*-hydroxy-10*E*,12*Z*,15*Z*-octadecatrienoic acid (**2**) and 9*R*-hydroxy-10*E*,12*Z*-octadecadienoic acid, likely derive from action of an ω10*R*-LO acting on either α-linolenic acid or linoleic acid.[13]

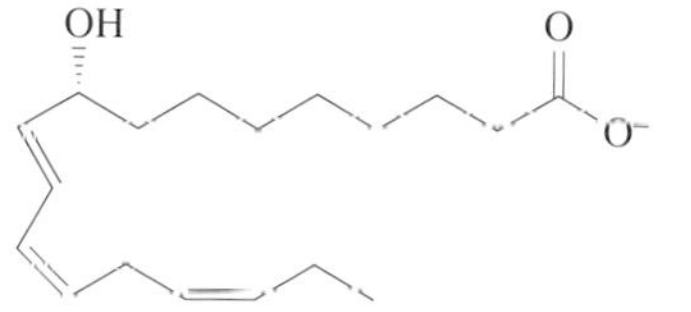

OH O O⁻

(**2**)

The marine cyanobacterium (blue-green alga) *Lyngbya majuscula* is a source of the hepoxillin-type hydrolysis product 9*S*,12*R*,13*S*-trihydroxy-10*E*,15*Z*-octadecadienoic acid (malyngic acid, (**3**)),[14] a structure related to the homologous trioxillin of mammalian occurrence. The C_{20} homologues of these triols have been isolated from various mammalian tissues and have been shown to derive from hepoxillin epoxide hydrolase opening of hepoxillin A_3 (Scheme 1). In this study of *L. majuscula* chemistry,[14] a nonacidic extraction and workup were used which favors the triol metabolite as the true natural product. Inherent in this biogenetic proposal for the formation of (**3**) is the intermediacy of 12*S*,13*S*-epoxy-9*S*-hydroxy-10*E*,15*Z*-octadecadienoic acid, itself formed via the rearrangement of 13*S*-hydroperoxy-9*Z*,11*E*,15*Z*-octadecatrienoic acid (Scheme 1). In turn, the hydroperoxide likely originates from action of an ω6-LO (C-13 oxygenation) on α-linolenic acid.

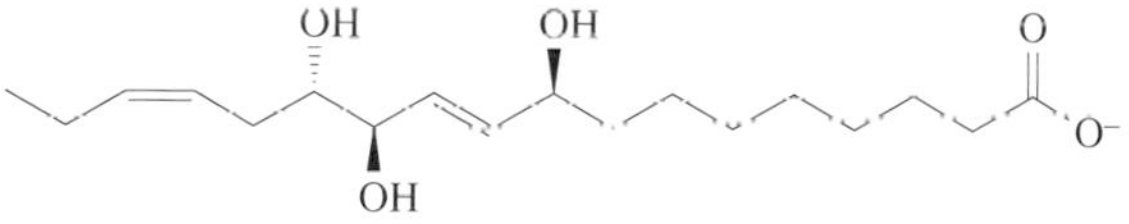

OH OH O O⁻ OH

(**3**)

1.08.3 FUNGI, ALGAE, AND PLANTS

1.08.3.1 Fungi

The fungi are a taxonomic group with a well developed metabolism of fatty acids to eicosanoids as well as other classes of oxylipins. The fatty acid chemistry of fungi has been previously reviewed in the context of fungal lipids[15] and their biological activity.[16] The taxonomic order of description of the fungi is arbitrary, and taxonomic positions are identified parenthetically as there are an insufficient number of examples within any one taxonomic grouping for an arrangement based on systematics to be useful. Taxonomic positions of those genera and species described below are given as outlined by Hawksworth *et al.*[17] In general, the fatty acids of fungi are highly polyunsaturated and are abundant in C_{18} chain lengths with C_{16} and C_{20} being common but less prevalent. In a few cases, such as the filamentous fungus *Mortierella alpina* (Zygomycota, Zygomycetes, Mucorales, Mortierellaceae), arachidonic acid is the major fatty acid, accounting for more than 70% of the total fatty acids.[18] Fatty acids with ω3 double bonds are notable in this group; from *Saprolegnia parasitica* a C-20 specific ω3 desaturase has been isolated and characterized.[19]

The functions of oxylipins in fungi appear to be many and diverse. For example, LO inhibitors added to *Lagenidium giganteum* (Oomycota, Pythiales, Pythiaceae) cultures resulted in the stage-

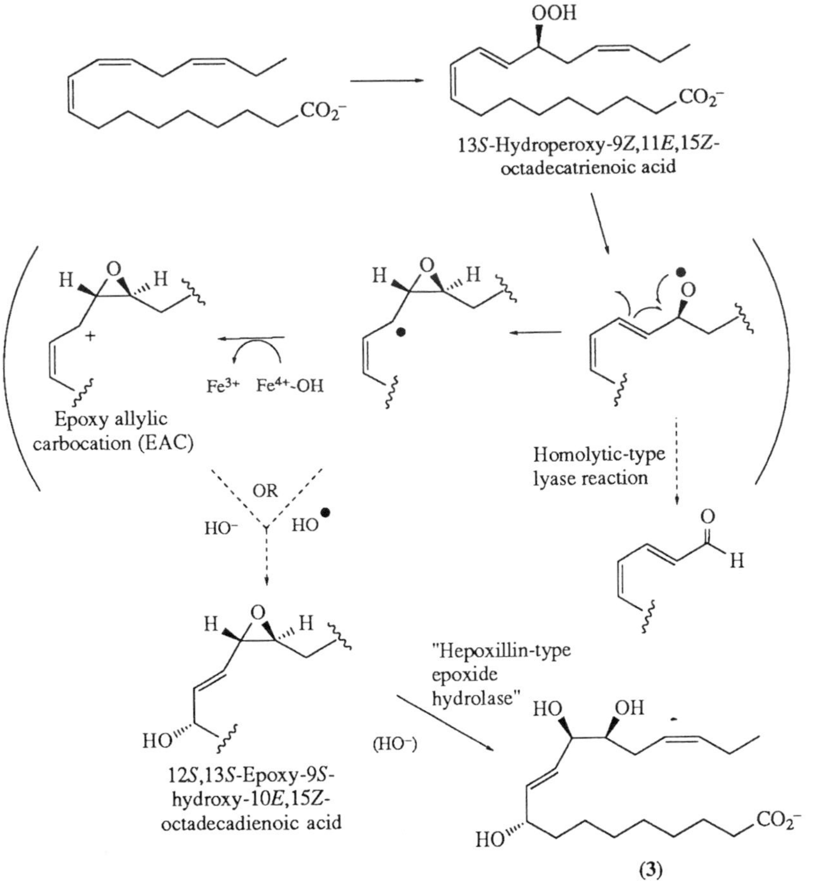

13*S*-Hydroperoxy-9*Z*,11*E*,15*Z*-octadecatrienoic acid

Epoxy allylic carbocation (EAC)

Fe^{3+} Fe^{4+}-OH

Homolytic-type lyase reaction

OR

HO^- $HO^{\bullet}$

12*S*,13*S*-Epoxy-9*S*-hydroxy-10*E*,15*Z*-octadecadienoic acid

"Hepoxillin-type epoxide hydrolase"

(HO^-)

(3)

Scheme 1

specific inhibition of various developmental and sexual processes, including oosporogenesis.[20] Application of prostaglandins have been shown to reverse this inhibition.[21] In other cases, such as *Laetisaria arvalis*,[22] oxylipins are apparently used as allelopathic agents to inhibit the growth of competing fungal species.

In the fungi, PUFA precursors are oxidatively metabolized in a variety of ways to both very simple epoxy- and hydroxy-fatty acids as well as to more complex and carbocyclic substances.[15,23] Mirroring their fatty acid composition, the majority of these derive from C_{18} fatty acids although in a few cases metabolism of C_{14}, C_{16}, and C_{20} acids occurs. In the C_{18} class, metabolism occurs in the various species by oxidative enzymes with specificity for C-8, C-9, C-10, C-12, and C-17. The enzyme systems involved in the initial oxidation of these fatty acid substrates are of both the cytochrome P450 and LO class. Various wax esters with additional sites of oxidation have also been reported from fungi.[24]

One of the more prevalent themes in fungal metabolism is the cytochrome P450 oxidation of unsaturated fatty acid precursors to epoxides, such as by *Puccia framinis tritici* (Wheat stem rust, Basidiomycota, Teliomycetes, Uredinales, Pucciniaceae)[25] and *Claviceps* spp. (Ascomycota, Hypocreales, Clavicipitaceae).[26] Apparently, oleic acid is the fatty acid most commonly epoxidized in this manner to form the (−)-*cis*-9,10-epoxystearic acid (**4**). Molecular oxygen has been shown to label the resultant epoxy fatty acid.[25] This epoxide has been reported from a number of other fungi, higher plants, and a gram negative bacterium.[26] From most, it is frequently acted upon by an epoxide hydrolase to form (+)-*threo*-9,10-dihydroxystearic acid (**5**).[25]

The oxidative enzymes of the fungus *Fusarium oxysporum* (Ascomycota, Hypocreales, Hypocreaceae) have been studied over a considerable period of time, and have yielded both an LO as well as several cytochrome P450s. The *F. oxysporum* LO has been isolated, crystallized, and partially characterized.[27] This putative LO is unique in that it has a high pH optimum (>9), is stabilized by Co^{2+}, has a relatively small size (10–30 kDa), and utilizes linoleic acid most readily.[28] The products of the fungal LO metabolism of linoleic acid at pH 9 were later characterized as a 70:30 ratio of the 9- (**6**) and 13-hydroperoxyoctadecadienoic acids (**7**). Oxygen was incorporated into these hydro-

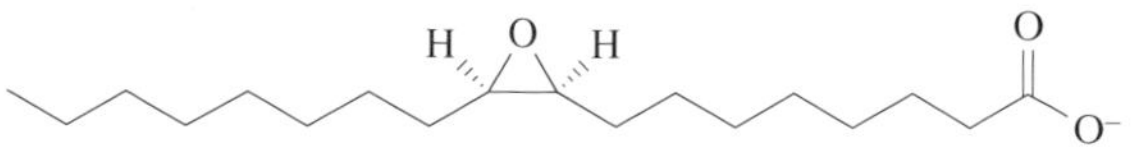

(4)

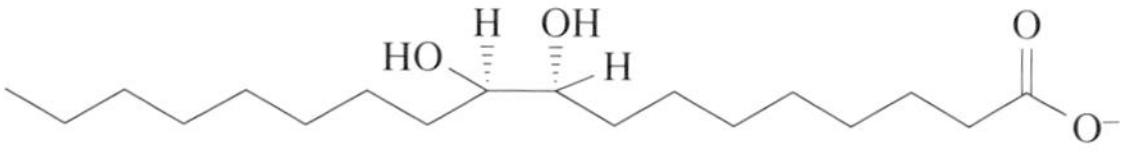

(5)

peroxy acids from $^{18}O_2$. The chirality of the process has not been investigated. Other more polar products are formed in the reaction of this LO with linoleic acid; however, they remain uncharacterized.[27] The fatty acid metabolizing cytochrome P450 from this fungus (P450oxy), which optimally oxidizes lauric acid ($C_{12:0}$) at the $\omega 2$ and $\omega 3$ positions to hydroxy-fatty acids, is also unique in that it is a catalytically self-sufficient enzyme containing both P450 and reductase domains.[29] This duality of activities as well as its molecular size (M_r = 118 kDa), high catalytic turnover, and its substrate regioselectivity are nearly identical to that of a bacterial P450 from *Bacillus megaterium*, suggesting the possibility of a close evolutionary relationship between these enzymes.[30]

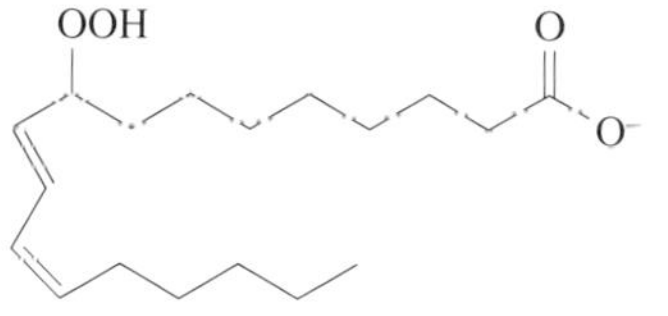

(6) 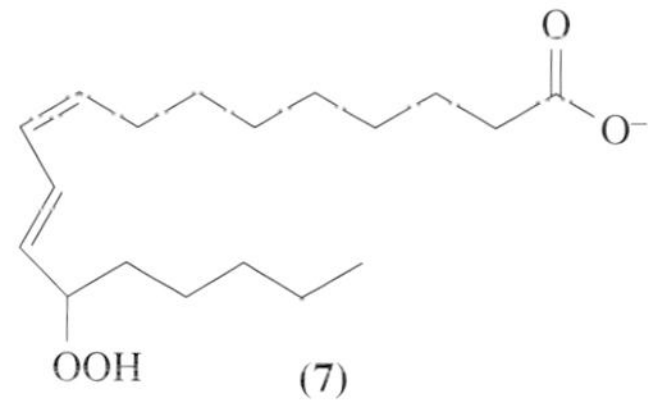 (7)

A part of the characteristic odor of mushrooms is due to the oxylipin-derived fragment 1,5*Z*-octadien-3*R*-ol (**8**).[31–33] Formation of this compound by a protein fraction from *Psalliota bispora* (Basidiomycetes, Basidiomycetes, Agaricaceae) has been studied in some detail (Scheme 2). The pathway is initiated by an unique dioxygenase which converts linoleic or linolenic acids into the corresponding 8*E*,10-hydroperoxide, presumably the 10*S*-stereoisomer.[33] This reaction is distinct from other dioxygenases in that a conjugated diene is not formed in consequence of oxidation. The 10*S*-hydroperoxyoctadeca-8*E*,12*Z*-dienoic acid is substrate for a presumed "homolytic-type"[a] hydroperoxide lyase which fragments the C-10—C-11 bond and transfers the distal oxygen of the hydroperoxide to C-14, possibly in a concerted manner.[33] The C-1—C-10 fragment, 10-oxo-8*E*-decenoic acid (**9**) has also been characterized as a product of this metabolism by *P. bispora*.[33] Curiously, the enantiomer 1,5*Z*-octadien-3*S*-ol (**10**) has been reported from the red alga *Chondrococcus hornemannni*.[34]

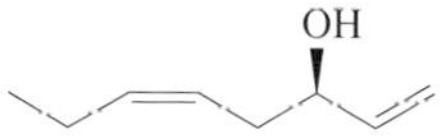

(8)

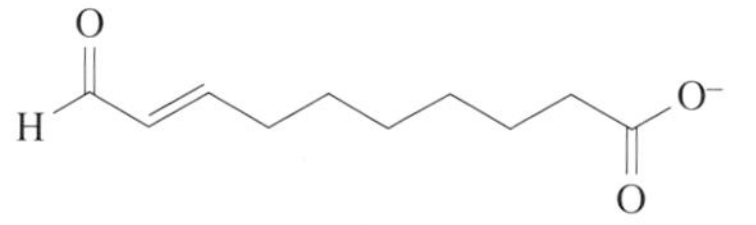

(9) 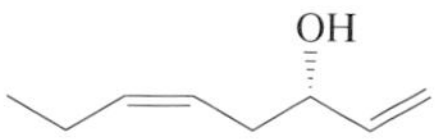 (10)

The wheat parasitic fungus *Gaeumannomyces graminis* (Ascomycota, Magnaporthaceae) possesses both a dioxygenase[35,36] and omega hydroxylase[37] and produces a variety of unusually oxidized products from linoleic acid. The major products of the dioxygenase are 8*R*-hydroxylinoleic acid (**11**) and 7*S*,8*S*-dihydroxylinoleic acid (**12**)[35] while the omega hydroxylase primarily oxidizes the $\omega 2$ and $\omega 3$ positions of a variety of fatty acid substrates. This latter enzyme, presumably a cytochrome P450, has not been studied in detail from this source, but is possibly related to the comparable system described from other fungi (e.g., *F. oxysporium*).[29] The dioxygenase in *G. graminis*, as well as that of another fungus, *L. arvalis* (Basidiomycota, Basidiomycetes, Stereales, Corticiaceae),[38] is

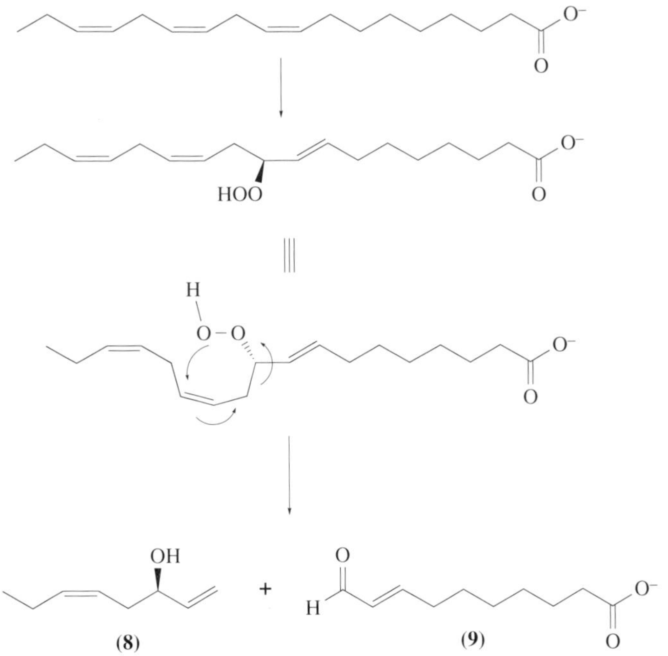

(8) **(9)**

Scheme 2

distinct from LO dioxygenases in that only a 9*Z* double bond and a saturated carboxyl side chain are required in the substrate, and molecular oxygen is added to C-8 without alteration of the position or geometry of the adjacent double bond.[35,38] For both fungi, the dioxygenase produces the 8*R*-hydroperoxide, and in both, this is reduced to form the hydroxyacid, 8*R*-hydroxylinoleic acid (**11**). In *L. arvalis*, this unique acid, named laetisaric acid (**11**), was identified as an allelopathic agent that rapidly induces hyphal lysis in potentially competing fungi.[22] *G. graminis* converts 8*R*-hydroperoxylinoleic acid into 7*S*,8*S*-dihydroxylinoleic acid (**12**) via a hydroperoxide isomerase-catalyzed intramolecular rearrangement.[35] The stereochemistries of the hydrogen abstractions in the dioxygenase and hydroperoxide isomerase reactions have been characterized; the dioxygenase removes the *pro*-8*S* hydrogen and antarafacially inserts dioxygen for a net inversion of configuration at C-8, whereas the hydroperoxide isomerase produces a *threo* diol by abstraction of the *pro*-7*S* and suprafacial intramolecular transfer of the distal hydroperoxide oxygen for net retention of configuration at C-7.[36] A similar intramolecular rearrangement of a hydroperoxide into a vicinal diol has been characterized in the red alga *Gracilariopsis lemaneiformis*, although in this case an *erythro* diol is produced.[1]

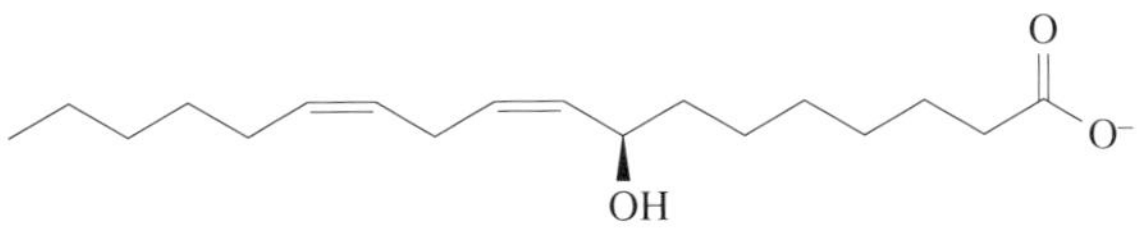

(11)

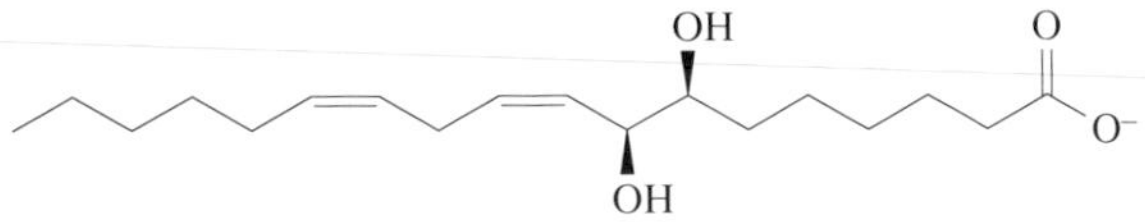

(12)

S. parasitica (Oomycota, Saprolegniales, Saprolegniaceae) is a rich source of eicosapentaenoic acid (EPA) and a methyl-directed desaturation of C_{20} fatty acids (e.g., arachidonic acid (AA)) to EPA has been described.[19] Subsequently, metabolism of AA was reported in *S. parasitica* with the isolation and complete structural description of two new epoxy alcohols, 11*S*,12*R*-epoxy-15*S*-

hydroxy-5*Z*,8*Z*,13*E*-eicosatrienoic acid (**13**) and 13*R*,14*R*-epoxy-15*S*-hydroxy-5*Z*,8*Z*,11*Z*-eicosatrienoic acid (**14**).[39,40] Some endogenous metabolism of C_{18} fatty acids to analogous epoxyalcohols was also recorded.[39] The C_{20} metabolites were shown to derive through the intermediacy of 15*S*-hydroperoxy-5,8,11,13-eicosatetraenoic acid (15*S*-hydroperoxyeicosatetraenoic acid (HPETE)), and the epoxide oxygen in each case was shown to derive from intramolecular transfer of the distal oxygen of the hydroperoxide (Scheme 3).[41] The proposed pathway therefore involves the sequential action of a 15*S*-LO and hydroperoxide isomerase. These activities were partially purified from the fungus, and based on co-chromatography on several support systems and similarity in several other properties, proposed to reside on the same 145–150 kDa polypeptide.[22] Further examination of the pathway using a variety of C_{18} substrates revealed that the hydroperoxide isomerase always introduces the epoxide with the same absolute stereochemistry, regardless of the regiochemistry or stereochemistry of the hydroperoxide (however, different epoxide stereochemistries derive from epoxidation of *cis* or *trans* double bonds) (see Scheme 3).[33]

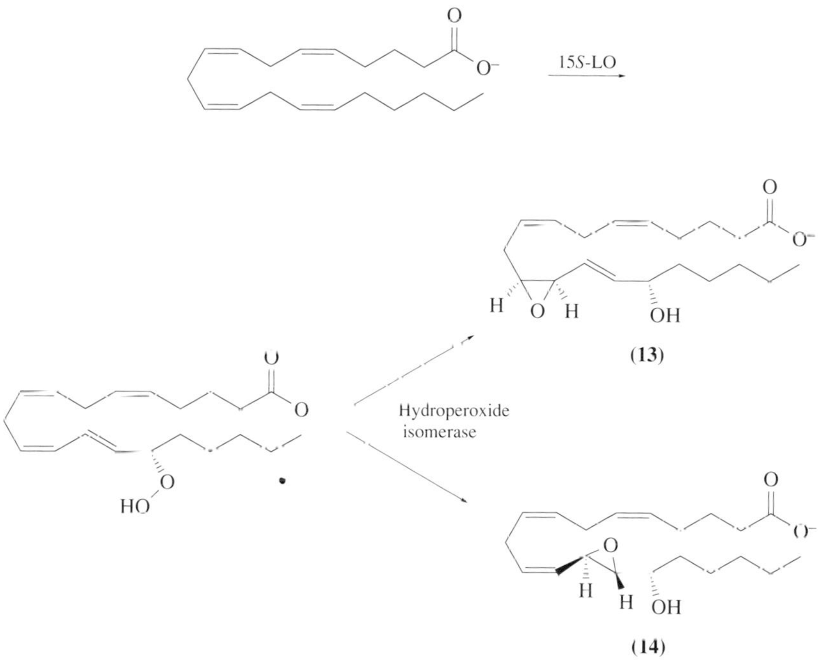

15*S*-LO

Hydroperoxide isomerase

(**13**)

(**14**)

Scheme 3

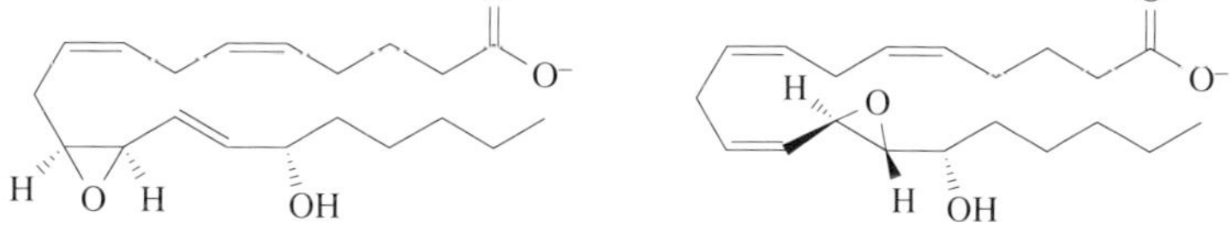

(**13**) (**14**)

1.08.3.2 Algae

The chemistry and biochemistry of algal oxylipins has been previously reviewed. In some cases, these reviews have been specifically focused (12-LO activity in red algae,[42] patterns in the positions of oxygenation[43] and in the carbocyclic nature of algal oxylipins,[44] centrality of an epoxy allylic carbocation (EAC) in algal oxylipin biosynthetic pathways[45]), whereas in other cases, these reviews have attempted to provide a more or less comprehensive coverage of algal oxylipin chemistry and biochemistry.[7,46]

1.08.3.2.1 Chlorophyta (green algae)

The major trends in oxylipin metabolism in the Chlorophyta are unique in that the common substrates that become oxidized are often of C_{18} length and of the ω3 class (e.g., stearidonic and α-linolenic acids. From the limited examination of Chlorophyta oxylipins described to date, enzymatic oxidation appears most commonly at the ω6 and ω9 positions via a lipoxygenase-type reaction (oxidation at C-9 and C-12), although in one series (*Acrosiphonia coalita*) ω3 oxidation is also apparent. In most of the product isolation studies performed to date with these algae, there is evidence for a widespread lyase metabolism of the putative hydroperoxides.

(i) Chlorella pyrenoidosa

The earliest work on a Chlorophyceaen alga, *C. pyrenoidosa*, described an unstable LO activity in which 9-HPOD (20%, (**6**)) and 13-HPOD (80%, (**7**)) were the sole products of linoleic acid metabolism.[47] Further studies in which [^{14}C]-13-HPOD was provided to a partially purified enzyme preparation from this alga gave three series of products; the peroxidase reduction product 13-hydroxyoctadecadienoic acid (13-HOD) (**15**), the lyase products 13-oxo-9*E*,11*E*-tridecadienoic acid and *cis*-2-pentene, and the novel metabolite 12,13-*trans*-epoxy-9-oxo-10*E*,15*Z*-octadecadienoic acid (**16**).[48] This latter metabolite, which was shown by use of appropriate control experiments to be a true enzymatic product, may derive from rearrangement of 13-HPOD to an EAC species[45] to which "OH" is added back, possibly via an intramolecular mechanism. Final oxidation of the resultant allylic alcohol to form the 9-oxo functionality would complete the conceptual biogenesis of this metabolite.

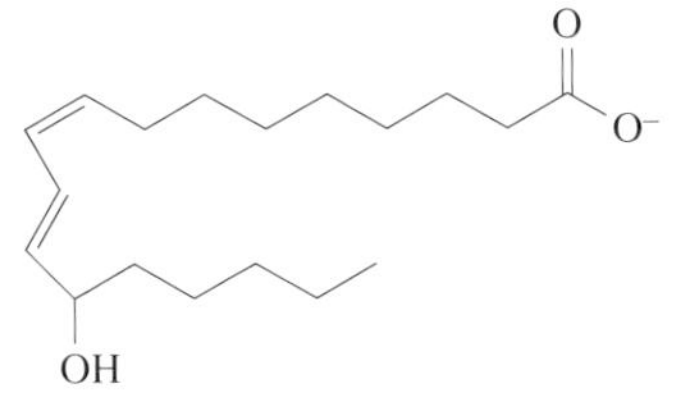

(**15**) 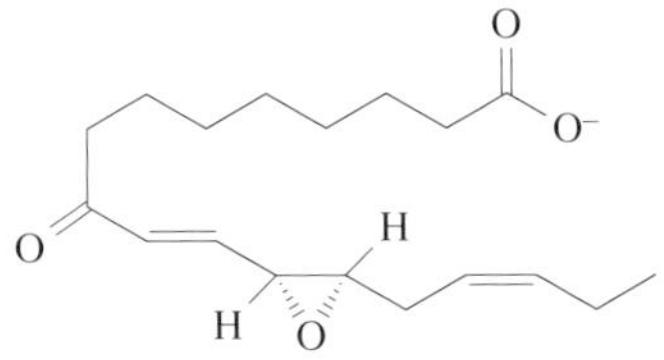 (**16**)

(ii) Acrosiphonia coalita

Phytochemical investigation of the Oregon green alga *A. coalita* led to the isolation of several C_{18} oxylipins containing an intriguing array of functional groups as well as two presumed lyase products of shorter chain length (C_{10}).[49] Several of these C_{10} and C_{18} compounds possessed similar conjugated trienone functionalities (e.g., (**17**) and (**18**)). As exemplified by compound (**17**), the C_{10} compounds were isomeric *E,E,E*- and *E,Z,E*-trienals (**17**) with an 8*S*-hydroxy group. The *E,E,E*-isomer was moderately antifungal to *Candida albicans* (100 μg ml^{-1}). Essentially the same relationship of functional groups was observed in the C_{18} metabolites, however, with a ketone rather than an aldehyde (compound (**18**)). One exception to this was a C_{18} metabolite with three apparent termini; carboxylate, high field methyl group in the ^{1}H NMR spectrum, and an aldehyde. Metabolite (**19**) was defined from detailed NMR analysis as a chain-rearranged analogue of (**18**) in which one carbon (C-9) is excised from the chain to form the aldehyde-containing branch. Two series of C_{18} regioisomeric epoxy alcohols were also isolated and fully defined as part of this study (e.g., (**20**) and (**21**)). Overall, the profile of metabolism in *A. coalita* appears to be by a combined oxidation by ω10- and ω3-LOs followed by either C-8—C-9 chain cleavage to form (**17**), dehydration of the C-9 hydroperoxide and reduction of the C-16 hydroperoxide to form (**18**), or rearrangement of the C-9 hydroperoxide and reduction of C-16 to form branched chain trienal (**19**). Alternatively, either ω10- or ω6-LO metabolism by themselves followed by rearrangement to an EAC and then addition of "HO$^-$" would yield the observed *A. coalita* epoxy alcohols (Scheme 4).[45]

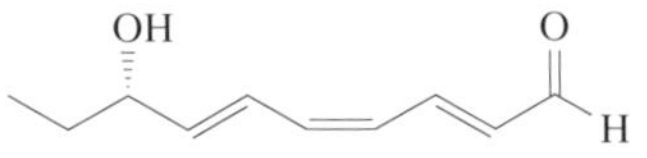

(**17**)

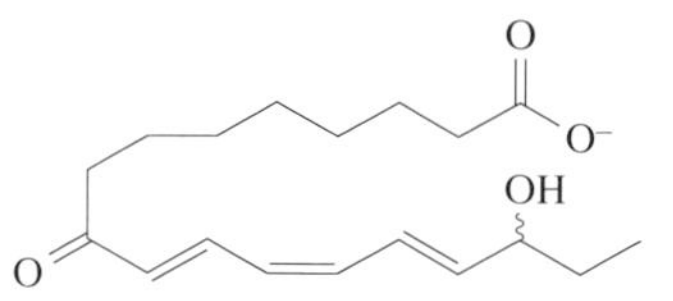

(18) 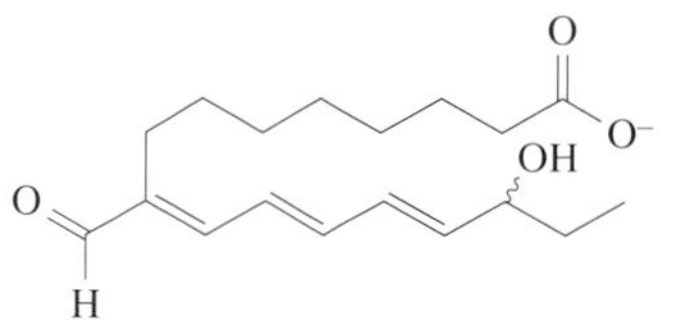 (19)

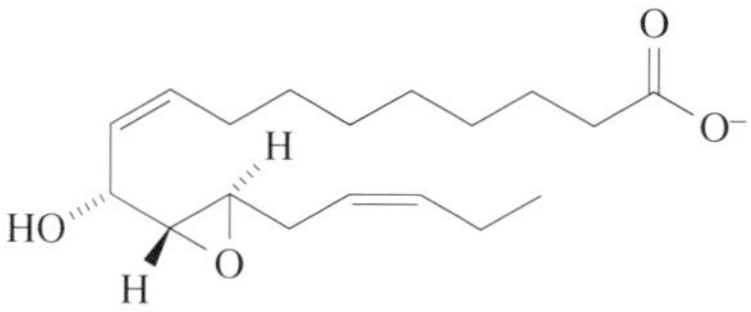

(20) 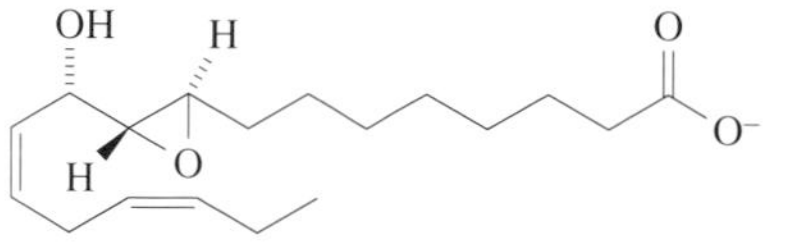 (21)

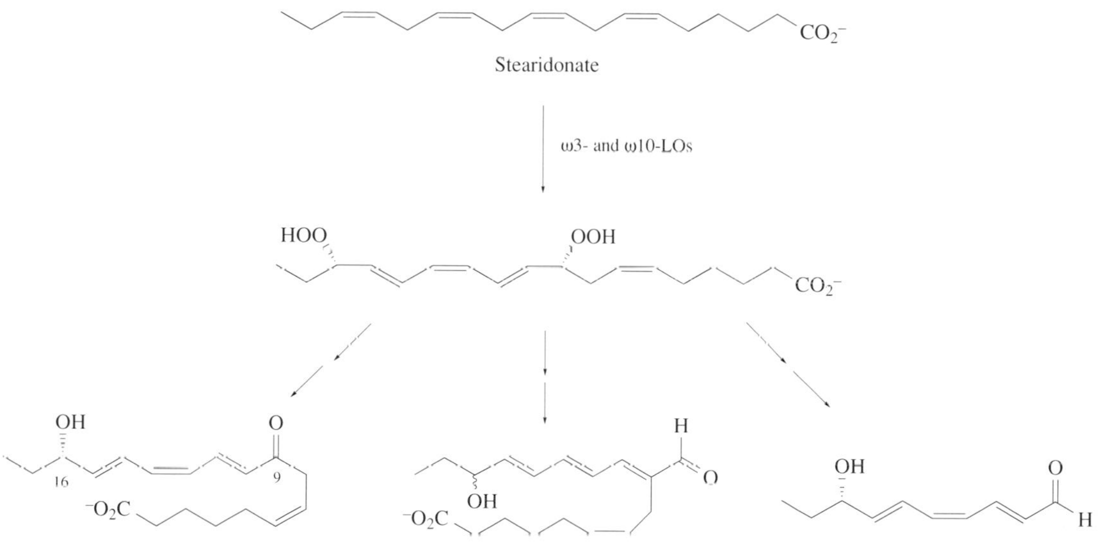

Stearidonate

ω3- and ω10-LOs

(18) (19) (17)

Scheme 4

(iii) Dictyosphaeria sericea

Dictyosphaerin (**22**) is a novel bicyclic oxylipin which is perhaps the most intriguing metabolite of this structure class isolated from a marine chlorophyte to date.[50] The alga *Dictyosphaeria sericea* was obtained from several locations along the Southern temperate coast of Australia and chromatography in each case led to the isolation of a single unstable lipid metabolite. Its structure was assembled from MS and NMR data obtained for the natural product as well as several key derivatives. NMR analysis was enhanced through partial shifting of several resonances by addition of 0.03 equivalents of $Eu(fod)_3$ shift reagent. Neither relative nor absolute stereochemistry of (**22**) was determined. This C_{22} lipid is unique in its possession of fused six-membered and five-membered rings. The hydroxyl group at C-6 in dictyosphaerin would imply a $\Delta 6$ lipid, an unsaturation family typical of C_{18} fatty acids. Hence, one possible biogenetic dissection of dictyosphaerin (**22**) is to separate the four carbons of the cyclohexene ring (C-12–C-15), with their being added from a nonfatty acid origin by a possible Diels–Alder biological equivalent. The remaining cyclopentadiene-containing octadecanoid might in turn arise from a novel carbocyclization of 6*Z*,9*Z*,12*Z*-octadecatrienoic acid (γ-linolenic acid).

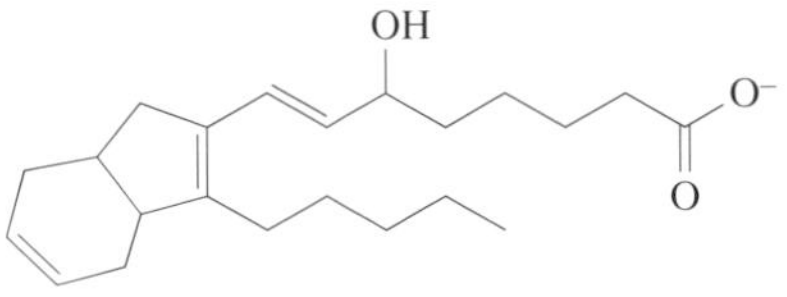

OH

O⁻

O

(22)

1.08.3.2.2 Phaeophyceae (brown algae)

Phaeophyceae (brown algae) have quickly emerged as a fertile source of structurally unique oxylipins. In most cases to date, simple or acyclic oxylipins have been isolated as cometabolites along with structurally more complex carbocyclic oxylipins. Coisolation of these acyclic and cyclic metabolites has provided considerable insight into the processes involved in the formation of the latter class. It appears that metabolism of both C_{18} and C_{20} fatty acid precursors is common, with many being of the ω3 structure class. Further, a consistent profile of oxidative metabolism at the ω6 position in both the C_{18} and C_{20} classes is evident in this array of metabolites.

(i) Giffordia mitchellae

One of the more fascinating stories to emerge in brown algal oxylipin chemistry is the discovery that a LO pathway leads to the formation of a variety of polyene hydrocarbons, specific of which are used as gamete attractants by various brown algae.[51] For example, in the brown alga *Giffordia mitchellae* (Ectocarpales), EPA appears to be the ultimate precursor to the simple C_{11}-metabolite giffordene (**23**).[52] This apparently occurs via initial metabolism by a 9-LO to intermediately form 9-hydroperoxyeicosapentaenoic acid (9-HPEPE), which in turn is subject to chain fragmentation and rearrangement yielding (**23**) (Scheme 5).

(23)

(ii) Ecklonia stolonifera *and* Egregia menziesii

The related brown algae *Ecklonia stolonifera* and *Egregia menziesii* from Japan and Oregon, USA, respectively, metabolize C_{18} and C_{20} PUFAs by similar routes to produce several of the same metabolites as well as some unique to each organism. The first report in this area was with the Japanese alga as part of a study to identify the abalone antifeedant chemistry in this organism.[53] Two novel compounds were isolated, eckloniallactones A (**24**) and B; however, they were only weakly active as antifeedants. Their structures were assembled from spectrochemical data and confirmed by X-ray diffraction analysis of crystalline eckloniallactone A (**24**). Absolute stereochemistry in this series was established from studies with (**24**) as isolated from the Oregon alga *E. menziesii* through selective opening of the epoxide at C-13, formation of the corresponding bis(*p*-bromobenzoate), and characterization by circular dichroism spectroscopy.[54] The 12,13-diols were subsequently isolated as natural products of *E. stolonifera*.[55] From the Oregon alga, an obviously related series of compounds was discovered—the egregiachlorides (e.g., (**25**))—in which a chloro functionality rather than the attachment of the lactone is found at the ω3 position.[56] Finally, an ω2 olefin analogue of the egregiachlorides (**26**) was also characterized in this latter study. The array of metabolites from these two algae is biogenetically unified by a proposed metabolism of various PUFA precursors, stearidonic acid for example, by a lipoxygenase showing ω6-specificity. The resultant hydroperoxide has been proposed to subsequently rearrange to an EAC that leads to cyclopentyl ring formation and an ω3 carbocation.[45] The metabolic fates of this carbocation, as

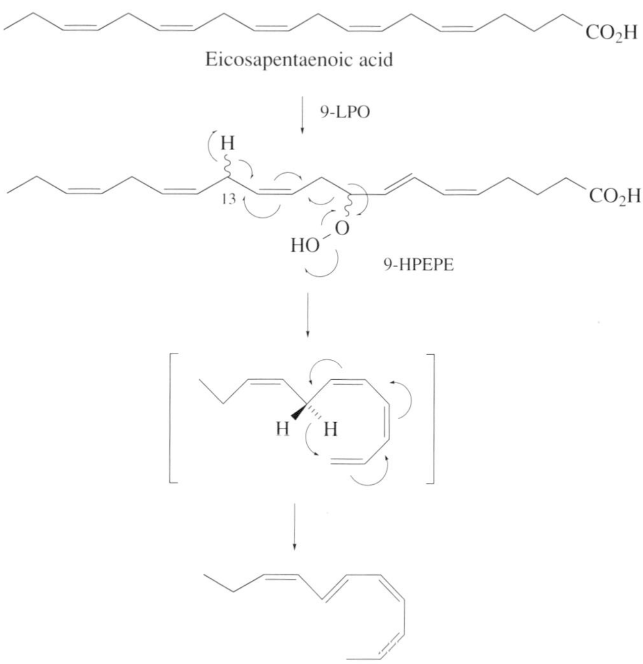

Eicosapentaenoic acid

9-LPO

9-HPEPE

(23)

Scheme 5

observed through these natural product isolations, include trapping with the carboxylate to form the macrolactone of the ecklonialactones (e.g., (**24**)), trapping with chloride to form the egregiachlorides (e.g., (**25**)), or loss of an ω2 proton to make the ω2 olefin (**26**) (Scheme 6).

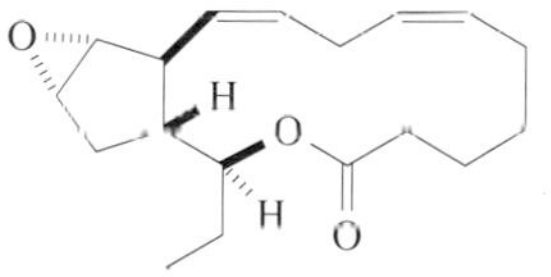

(24)

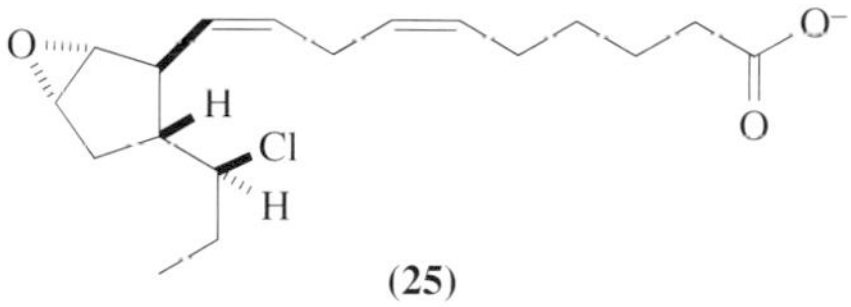

(25)

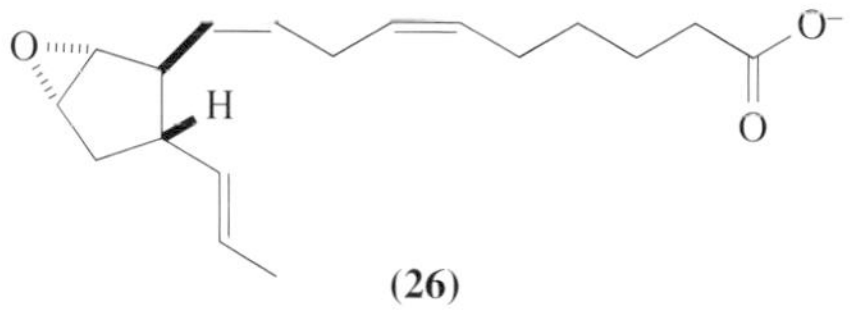

(26)

(iii) Notheia anomala

The Australian brown alga *Notheia anomala* has been a rich source of biogenetically related oxylipins which exemplify yet another metabolic theme in this group of algae. More than a dozen oxylipins, plus several of the likely precursor fatty acid-derived hydrocarbons, have been characterized from this epiphytic/parasitic alga.[57,58] The spectrum of reported oxylipins included a simple epoxide (**27**) as well as two diepoxides (e.g., (**28**)), a vicinal diol (**29**), and eight related tetrahydrofuran derivatives that varied in the position and chain length of an esterified fatty acid residue (e.g., (**30**)). Structures in this latter class were deduced via spectroscopic methods and confirmed via X-ray crystallography, which also gave relative stereochemistry.[57] Absolute stereochemistry of the C_7

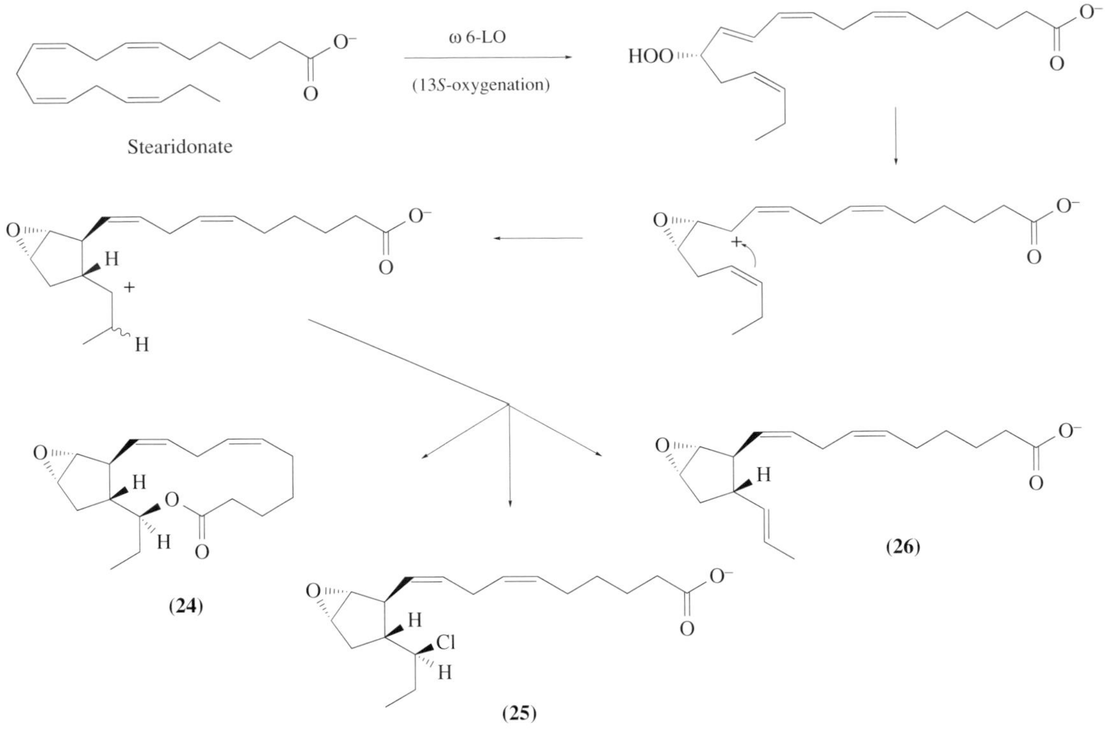

ω 6-LO

(13*S*-oxygenation)

Stearidonate

(24)

(25)

(26)

Scheme 6

alcohol was proposed on the basis of Horeau's method. The presumed precursor hydrocarbons 6*Z*,9*Z*,18-nonadecatriene (**31**) and 3*Z*,6*Z*,9*Z*,18-nonadecatetraene (**32**), were also characterized from *N. anomala*, themselves likely deriving from 11*Z*,14*Z*-eicosadienoic acid and 11*Z*,14*Z*,17*Z*-eicosatrienoic acid, respectively.[58] A biosynthetic proposal which interrelates many of these metabolites has been advanced,[58] a key aspect of which is the operation of a cytochrome P450-like epoxidation of the above hydrocarbon precursors to mono- and di-epoxides. In turn, these epoxides are subject to metabolism to diols or tetrahydrofuran derivatives.

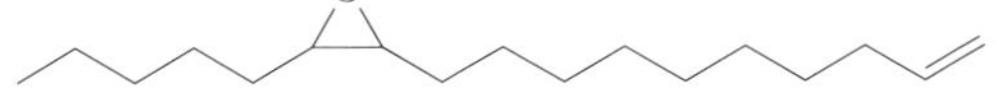

(27)

(28)

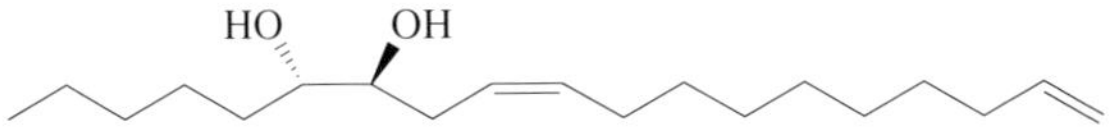

(29)

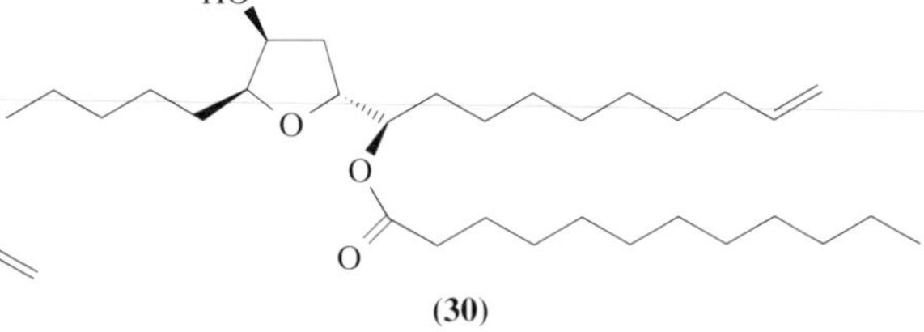

(30)

(31)

(32)

1.08.3.2.3 Rhodophyta (red algae)

The Rhodophyta, or red algae, have been the most prolific algal source for structurally unique oxylipins. The principal theme among oxylipins in these algae is of C_{20} fatty acids, mainly arachidonate and eicosapentaenoate, metabolized via 12-LO initiated pathways. What is remarkable in this algal group is the huge variety of unique structures that result from reaction or rearrangement of the resultant dienyl hydroperoxide. In a number of cases, simple 12-hydroxy eicosanoids or epoxy alcohols are the end product, while in others, complex multicyclic and multifunctional metabolites such as hybridalactone (**33**) or constanolactone A (**34**) are the final result. In other red algae, metabolism of predominately the same C_{20} precursors is via putative 5-LO, 8-LO, or 9-LO pathways with similar rearrangements occurring to the intermediate dienyl hydroperoxide to form a diversity of final products. An additional and distinct theme in oxylipin metabolism in the red algae is the production of conjugated triene and tetraene fatty acids, in some cases, via clearly demonstrated oxygen-dependent processes. Because of the relatively large number of red algae which have been reported to contain oxylipin metabolites, coverage of this group is necessarily selective. Previous reviews provide comprehensive albeit not as current coverage of this group.[42,46]

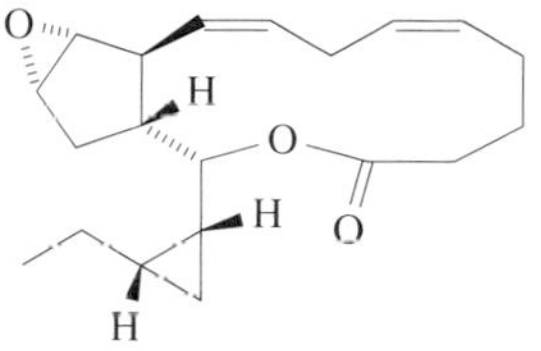

(33) 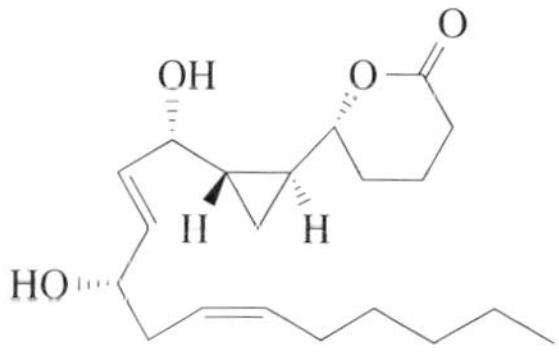 **(34)**

(i) Simple hydroxy- and epoxyhydroxy-acids from various red algae

Simple hydroxy alcohols, mostly of C_{20} chain length and oxidized at the C-12-position by a putative 12-LO, have been reported from a wide variety of red algae. It is common for these to co-occur with more complex substances, as described below. Nevertheless, 12*S*-hydroxyeicosatetraenoic acid (12*S*-HETE) (**35**) has been isolated from the following red algae: *Laurencia hybrida*,[59] *Murrayella periclados*,[60] *Laurencia spectabilis*,[61] *Constantinea simplex*,[62] *Platysiphonia miniata*,[63] and *G. lemaneiformis*,[64] while 12*S*-HEPE (**36**) has been isolated from *M. periclados*,[60] *C. simplex*,[62] and *G. lemaneiformis*.[64]

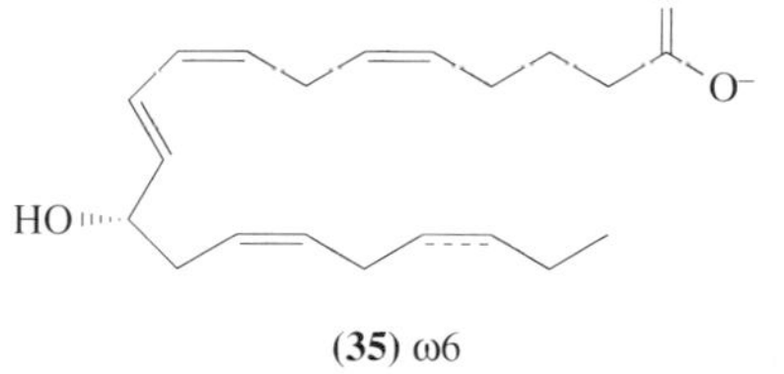

(35) ω6
(36) ω3

The occurrence of epoxyalcohols, such as hepoxillin B_3, in the lipid extracts of marine organisms was first reported from the tropical red algae, *P. miniata* and *Cottoniella filamentosa* (**33**).[65] Subsequently, both the *erythro* and *threo* isomers of hepoxillin B_3 ((**37**), (**38**)) and hepoxillin B_4 were isolated from another tropical red alga, *M. periclados* (**34**).[66] Absolute stereochemistries in two of these latter isolates were determined by derivatization and cleavage to form (−)-menthoxycarbonyl (MC) derivatives which were compared with authentic standards by GC-MS.[67] Epoxyalcohols such as these could conceptually arise via an intermediate EAC, which in turn would derive from a dienyl hydroperoxide. The resulting EACs could be trapped with water to form the corresponding epoxy alcohols. Alternatively, it is possible that the carbon centered radical formed en route to an EAC (see Scheme 1) is trapped directly through an "oxygen rebound-type" mechanism to form these epoxyalcohols.[45]

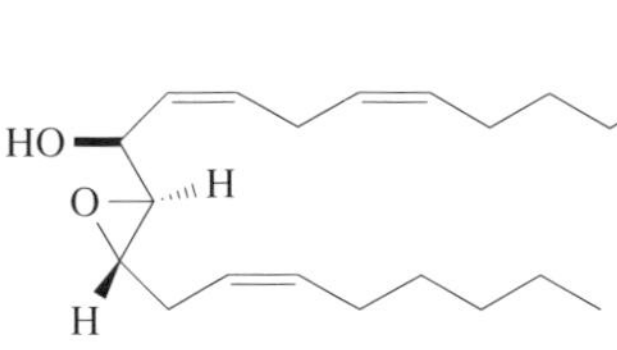

(37)

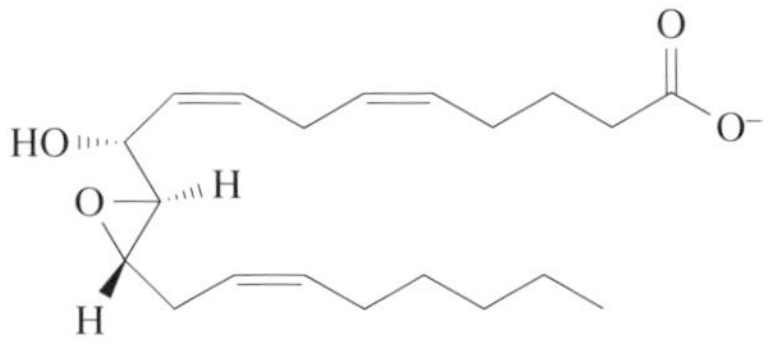

(38)

(ii) Vicinal diol biosynthesis in Gracilariopsis lemaneiformis

G. lemaneiformis is a common red alga from the Oregon coast which has been a rich source of both known as well as new oxylipins.[64] These consist of chain-shortened hydroxy acids (**39**), C_{18} and C_{20} hydroxy and dihydroxy acids which contain diene chromophores (e.g., (**40**), (**41**)), a C_{18} keto-diene (**42**) and a C_{20} keto-diene containing hydroxy groups as well (**43**),[64] and a number of diacyl monogalactosyl glycerols and diacyl digalactosyl glycerols which incorporate oxidized acids as acyl groups (e.g., (**44**)).[68] This was one of the very first marine algae in which detailed mechanistic studies of oxylipin biosynthesis was undertaken. While initial studies on vicinal diol biosynthesis utilized acetone powder preparations,[1] more detailed studies employed homogenates of fresh-frozen tissue.[69] The earlier work had shown that 12*S*-HPETE was formed first as a free intermediate by action of a 12*S*-LO. This was substrate for a hydroperoxide isomerase which intramolecularly rearranged the distal oxygen of the hydroperoxide to form a vicinal diol. The stereochemistry with which this new hydroxyl group was introduced was shown to be enzymatically predetermined and irrespective of the substrate hydroperoxide stereochemistry. Gel filtration chromatography was used to partially purify the 12-lipoxygenase activity into a protein band eluting with a molecular weight of 84–89 kDa while the hydroperoxide isomerase eluted as two peaks, one greater than 220 kDa and the

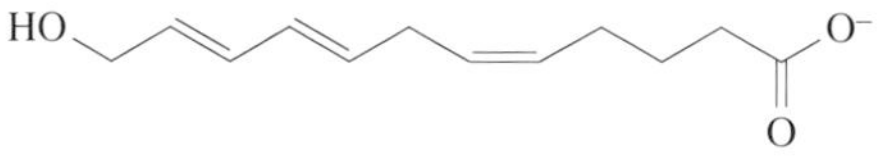

(39)

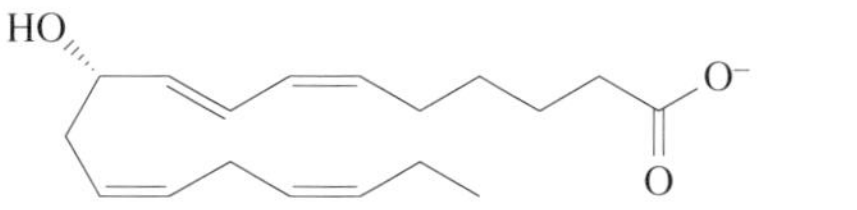

(40)

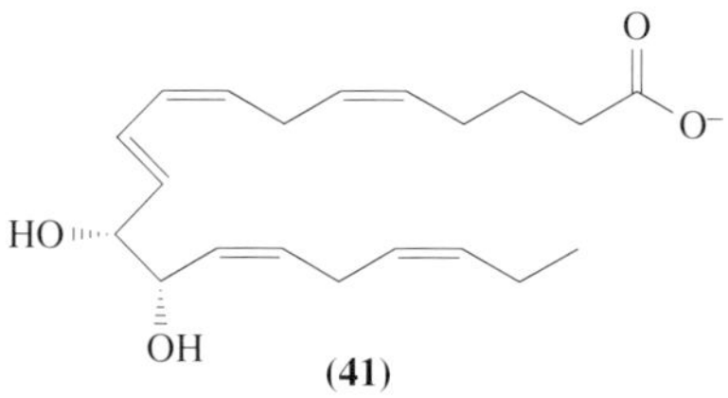

(41)

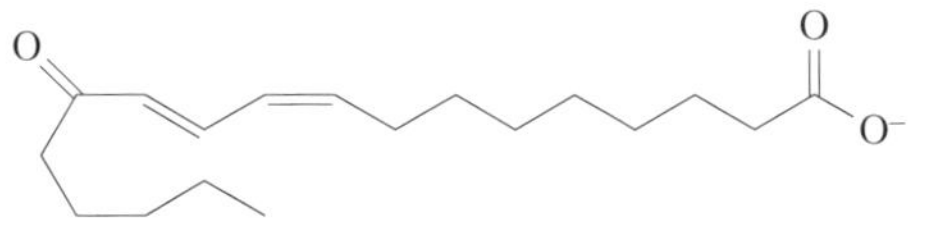

(42)

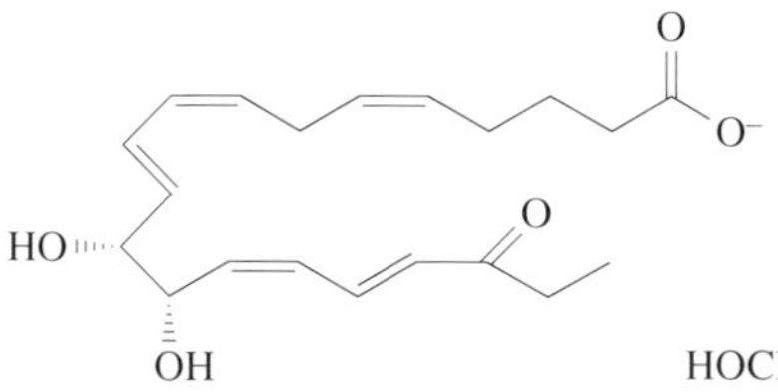

(43)

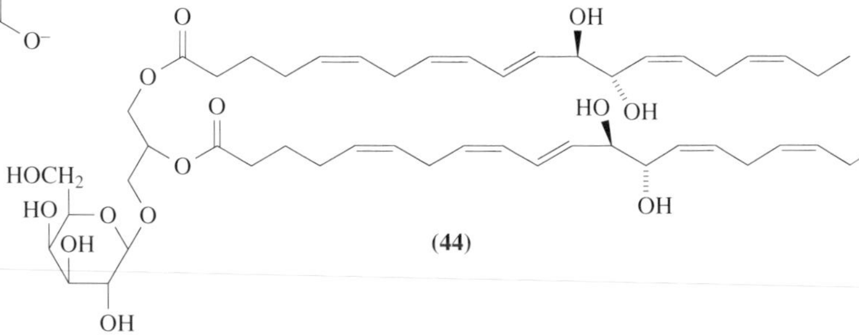

(44)

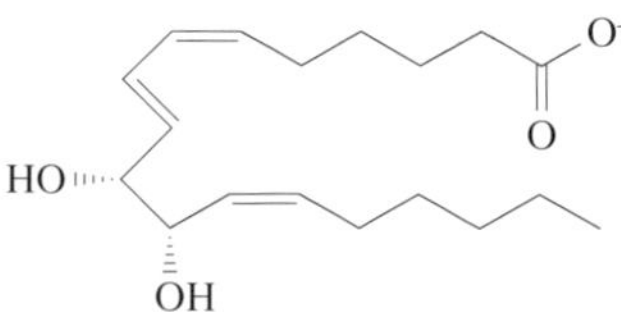

(45)

other at 40–45 kDa. A 20-fold increase in lipoxygenase activity resulted from addition of 0.8–1.0 M NaCl to fractions containing LO-activity. Interestingly, Na^+ was found to be specifically responsible for this stimulation. Of a number of other cations evaluated, only Li^+ and Mg^{2+} showed any ability to stimulate the lipoxygenase besides sodium ion. The purified LO was sensitive to inhibition by 5,8,11,14-eicosatetraynoic acid and nordihydroguaiaretic acid, two well-known LO inhibitors. No effective inhibitors of the hydroperoxide isomerase were found. The stereospecificity of hydrogen removals during vicinal diol formation by the *G. lemaneiformis* system was examined by separate incubations of 8*R*-,11*R*-, and 11*S*-deutero-6,9,12-octadecatrienoic acids. It was shown that the algal system converted these fatty acid precursors into a novel product, 10*R*,11*S*-diHOT (**45**). Formation of this diol from these various ^{2}H-labeled precursors was accompanied by loss of the majority of the [8*R*-^{2}H] and [11*S*-^{2}H] labels, but retention of the [11*R*-^{2}H] label. Loss of the 8*R* label is consistent with the stereospecificity of human platelet 12-LO.[70]

(iii) Polyneura latissima

Two new epoxyalcohols were obtained from the lipid extract of the Oregon red alga *Polyneura latissima* which illustrate the functioning of an arachidonate 9-LO (**37**).[71] These structures are unique in that the epoxide is located at C-8—C-9 and the alcohol (of both stereochemistries) is at C-7 (8*S**, 9*S**-epoxy-7*S**-hydroxy-5*Z*,11*Z*,14*Z*-eicosatrienoic acid (**46**) and 8*S**,9*S**-epoxy-7*R**-hydroxy-5*Z*,11*Z*,14*Z*-eicosatrienoic acid (**47**)). Another unusual metabolite of *P. latissima* was the divinyl ether polyneuric acid (**48**), the structure of which was assembled from detailed NMR study. As all of these metabolites co-occurred with 9*S*-HETE (**49**), it is reasonable to propose the activity of a 9-LO in this seaweed, which results in a 9-hydroperoxy intermediate. This is either reduced (**49**), rearranged to epimeric epoxyalcohols ((**46**) and (**47**)), or rearranged to the novel divinyl ether polyneuric acid (**48**).[71]

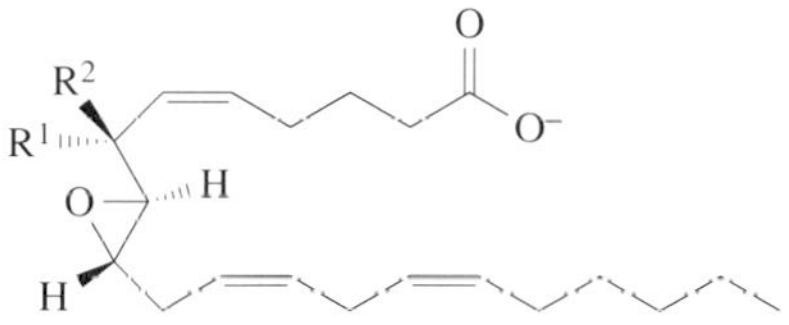

(**46**) R^1 = OH, R^2 = H
(**47**) R^2 = H, R^2 = OH

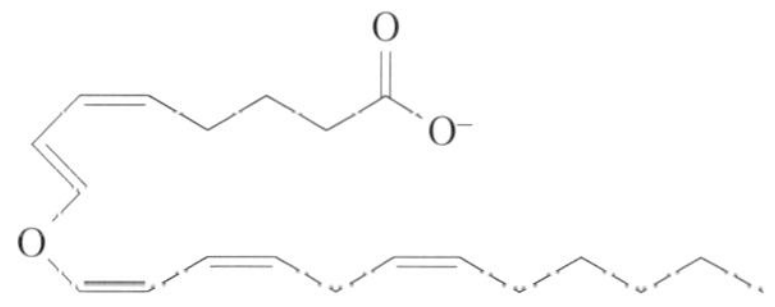

(**48**)

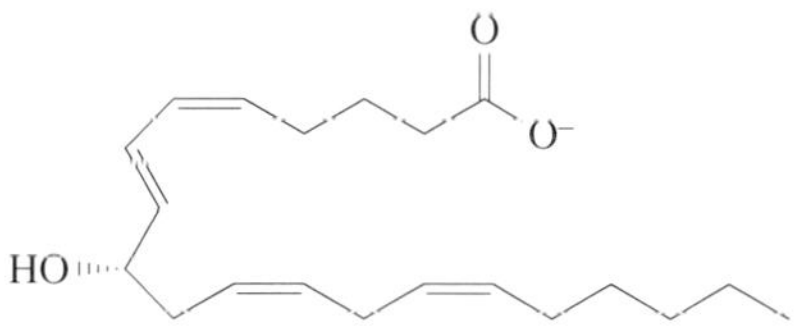

(**49**)

(iv) Agardhiella subulata

The Atlantic red alga *Agardhiella subulata* has been described to produce novel oxylipin metabolites.[72] One such metabolite, agardhilactone (**50**), which was difficult to isolate due to its inherent chemical instability, is unique in its highly functionalized skeleton which contains alcohol and diene functionalities as well as cyclopentyl, epoxide, and lactone rings. The location of the cyclopentyl ring is unique among reported oxylipins, spanning C-6 and C-10. The presence and location of the epoxide, cyclopentyl, and lactone rings suggest a biogenesis for agardhilactone which involves the intermediacy of an EAC. In this case, initial LO oxidation is thought to occur at C-8, followed by a "hydroperoxide dehydrase"-type rearrangement to the 8,9-epoxide with C-10 as an allylic carbocation. Second, a flow of electrons from the 5,6-olefin forms the cyclopentyl ring and from the carboxylate the lactone ring. Isomerization of the C-11—C-12 olefin likely occurs at an early stage in the process. Finally, a presumed 18-LO forms the diene and, following reduction, ω3 alcohol (Scheme 7).[45,72]

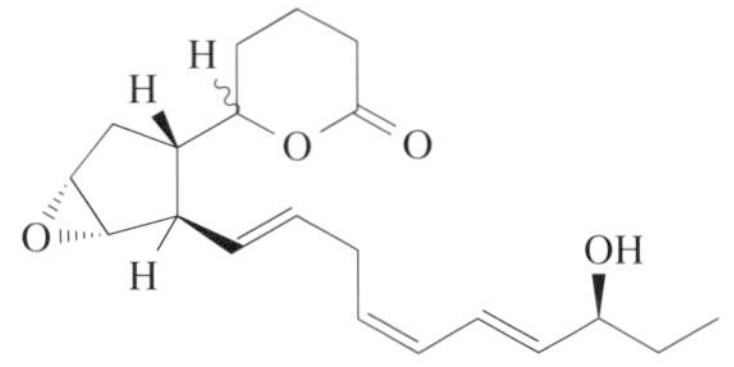

(50)

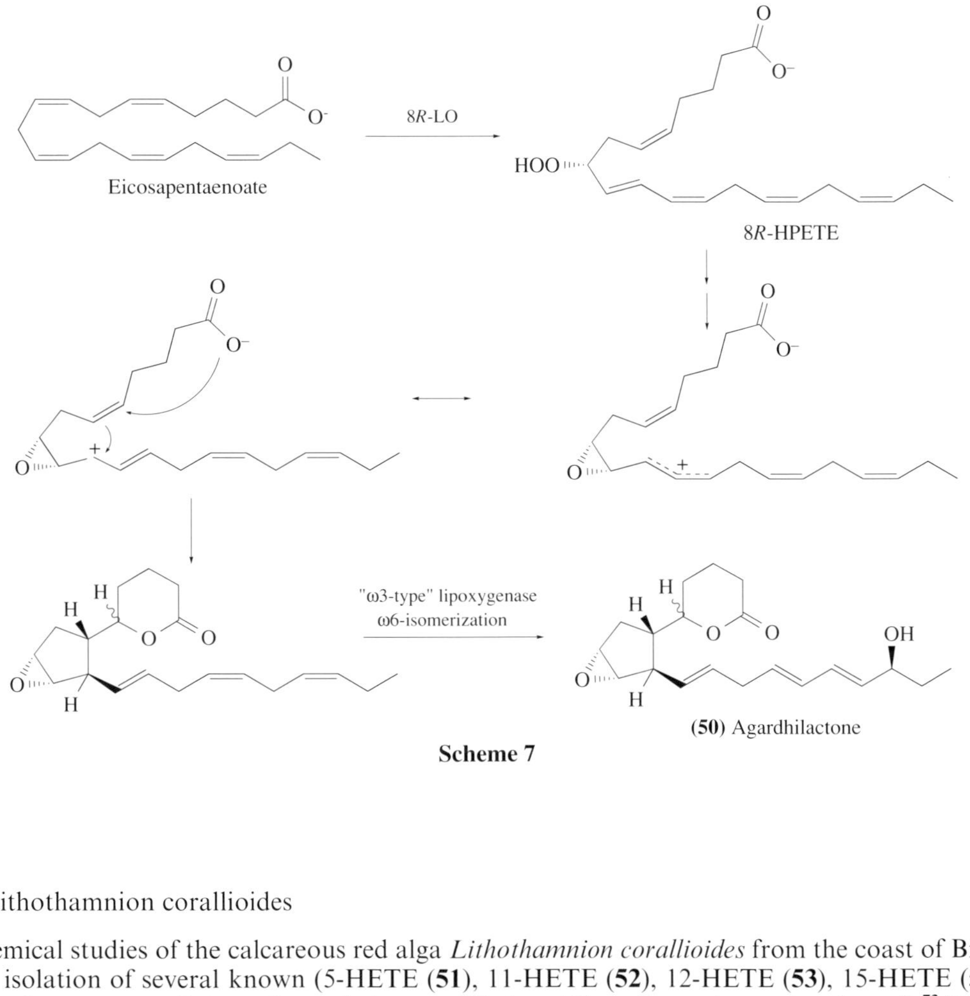

Eicosapentaenoate

8*R*-LO

8*R*-HPETE

"ω3-type" lipoxygenase
ω6-isomerization

(50) Agardhilactone

Scheme 7

(v) Lithothamnion corallioides

Chemical studies of the calcareous red alga *Lithothamnion corallioides* from the coast of Brittany led to isolation of several known (5-HETE (**51**), 11-HETE (**52**), 12-HETE (**53**), 15-HETE (**54**)) as well as three new oxylipins, all as ethyl esters (likely artifacts from storage in ethanol).[73] The new compounds, 13-hydroxyarachidonic acid (13-HAA, (**55**)), 13-hydroxyeicosapentaenoic acid, and 8-hydroxy-13-keto-5*Z*,9*E*,11*E*,14*E*-eicosatetraenoate (**56**) occurring as co-metabolites with the above well recognized LO products suggested the presence of novel pathways of oxylipin metabolism in this alga. These findings stimulated detailed studies of these biosynthetic pathways employing exogenously supplied C_{20}[74] and C_{18} fatty acids to cell free homogenates of the alga.[75,76]

Exploration of oxylipin metabolism in *L. corallioides* using [1-^{14}C]arachidonic acid as substrate yielded a resource of [1-^{14}C]13-HAA (**55**) for stereochemical studies,[74] as the original work had not established the stereochemistry of the hydroxyl group in 13-HAA (**55**).[73] Partial hydrogenation, conversion to MC derivatives, oxidative cleavage with O_3, and GC-MS analysis gave the C-9 to C-14 and C-12 to C-20 hydroxyacid fragments. These were compared with derivatives of standards and homologues of these fragments, and defined the 13-hydroxy group of (**55**) as *R*.

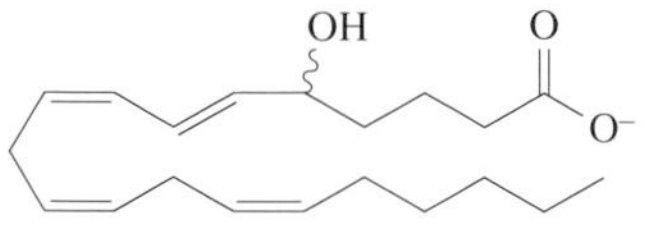

(51)

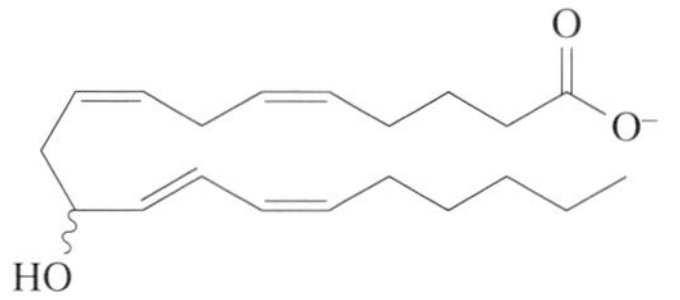

(52)

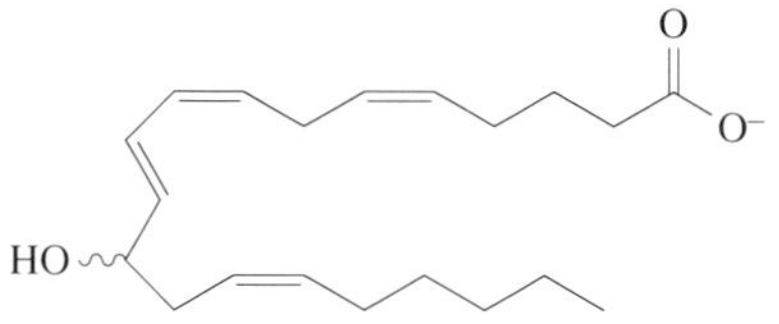

(53)

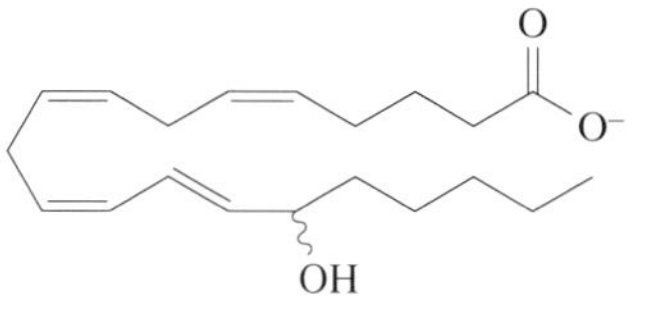

(54)

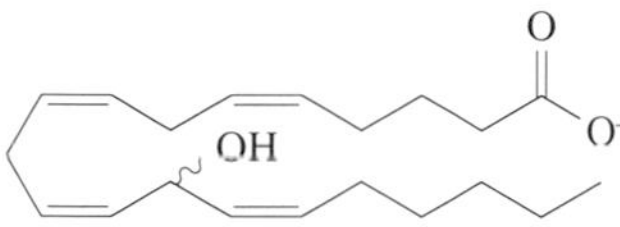

(55)

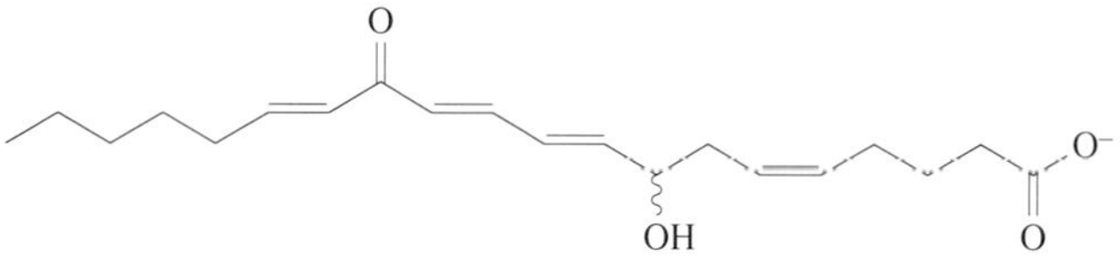

(56)

Cell free preparations of *L. corallioides* incubated in buffer enriched in $H_2{}^{18}O$ gave 13-HAA (**55**) containing appreciable ^{18}O in the 13-hydroxyl group.[74] However, this oxidation was nevertheless shown to have an absolute dependence on O_2. It may be that the enzyme catalyzing this transformation is an oxidase-type enzyme, which uses molecular oxygen as the ultimate electron acceptor. It was also found as a part of this study that 13-HAA has a chemical propensity to rearrange to 11-HETE (**52**) and 15-HETE (**54**) in the presence of acidified water.[74]

Additional study of oxylipin metabolism in *L. corallioides* utilized C_{18} substrates.[75,76] Incubation of [1-^{14}C]linoleic acid with a partially purified enzyme preparation led to the formation of 71% 11*R*-hydroxy-9*Z*,12*Z*-octadecadienoic acid (**57**), 5% 9-hydroxy-10*E*,12*Z*-octadecadienoic acid (75% 9*R*, (**58**)), 5% 13-hydroxy-9*Z*,11*E*-octadecadienoic acid (76% 13*S*, (**59**)), and 13% 11-keto-9*Z*,12*Z*-octadecadienoic acid (**60**).[75] Incubations with $^{18}O_2$ or $H_2{}^{18}O$ gave essentially the same results as obtained above with arachidonic acid for both the 11-hydroxy- and 11-keto-octadecadienoic acids, and the process was again absolutely dependent on the presence of molecular oxygen. Surprising, however, was the finding that the 9-HOD (**58**) and 13-HOD (**59**) formed by this system incorporated oxygen to an appreciable degree only from $H_2{}^{18}O$ and not from $^{18}O_2$. As steric analysis showed that only about 50% of these HOD products were formed by rearrangement of the 11-hydroxy compound (**57**), the other 50% must be formed via enzymatic oxidation with oxygen incorporation from H_2O. Together, these results suggest the operation of an unique non LO oxidation to form the chiral HOD products.

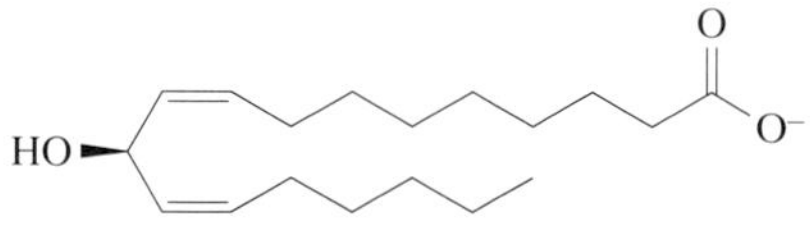

(57)

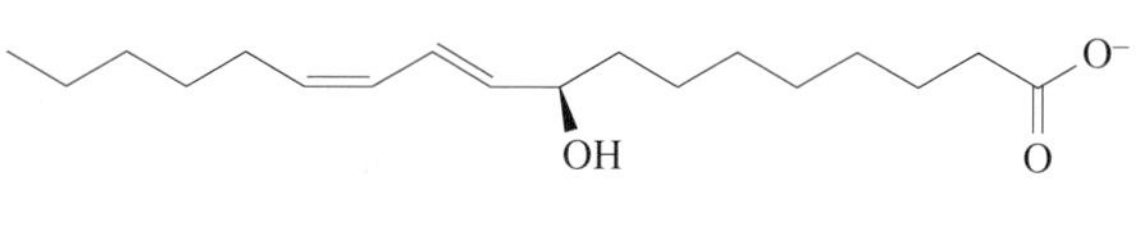

(58)

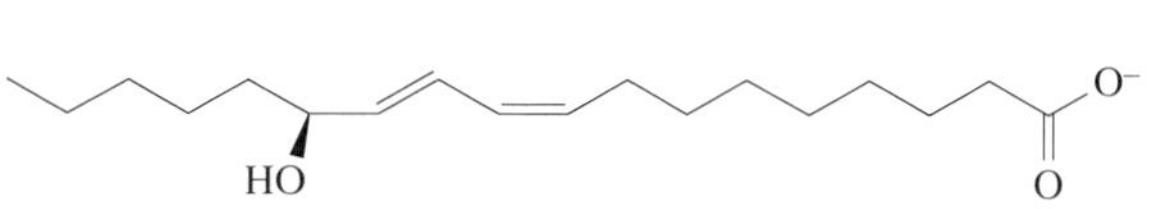

(59)

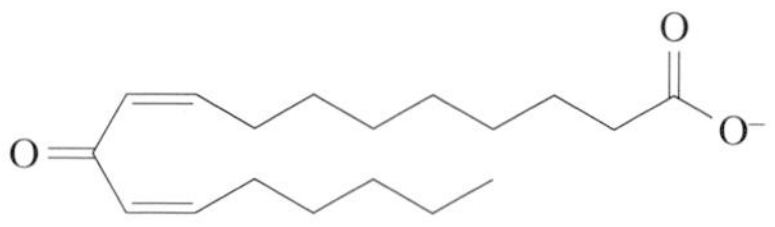

(60)

Another product obtained in the cell free incubations with *L. corallioides* was identified as 5*Z*,8*Z*,10*E*,12*E*,14*Z*-eicosapentaenoic acid (**61**) from GC-MS data obtained on the methyl ester derivative.[74] This conjugated tetraene-containing fatty acid was previously reported as a natural product from another red alga, *Bossiella orbigiana*[77] and has subsequently been isolated from a green alga, *Anadyomene stellata*, as well.[78] The mechanism of formation of conjugated tetraene products in *L. corallioides* has been examined in detail.[76] Cell free homogenates of this alga utilized 6*Z*,9*Z*,12*Z*-octadecatrienoic acid (γ-linolenic acid) as substrate to form 11*R*-hydroxy-6*Z*,9*Z*,12*Z*-octadecatrienoic acid ((**62**), 47% of the characterized products), 9-HOTω6 ((**63**), 3%), 13-HOTω6 ((**64**), 3%), and 6*Z*,8*E*,10*E*,12*Z*-octadecatetraenoic acid ((**65**), 47%). Reincubation of [1-^{14}C]-labeled (**62**) and (**65**) with the enzyme preparation showed that these two species do not interconvert. Formation of both was shown to be dependent on the presence of molecular oxygen. However, incubation studies with various stereospecifically ^{2}H-labeled octadecatrienoic acids revealed that the biosynthesis of the tetranene (**65**) occurs with loss of the *pro*-8*S* and *pro*-11*R* hydrogens while the hydroxy compound (**62**) is formed with loss of the *pro*-11*S* hydrogen. Further differentiation in the formation of these two products was shown by different sensitivities to several inhibitors and by the fact that under anaerobic conditions, formation of the tetraene (**65**) could be supported by a variety of artificial electron acceptors, while formation of the hydroxy compound (**62**) could not. Production of the tetraene was also shown to evolve a nearly stoichiometric amount of H_2O_2. Clearly, separate enzyme activities are responsible for formation of 11*R*-hydroxy-6*Z*,9*Z*,12*Z*-octadecatrienoic acid (**62**) and 6*Z*,8*E*,10*E*,12*Z*-octadecatetraenoic acid (**65**); the latter hypothesized to be a fatty acid oxidase, possibly related to D-amino acid oxidase (Scheme 8).[76]

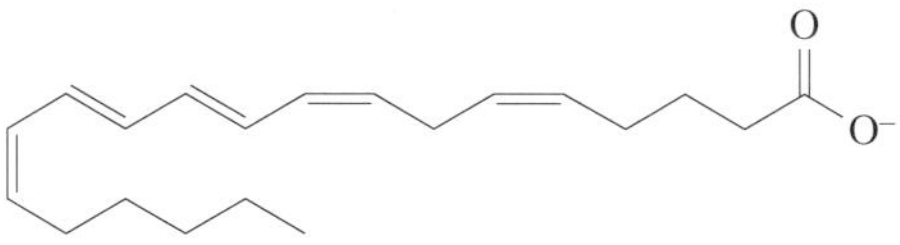

(61)

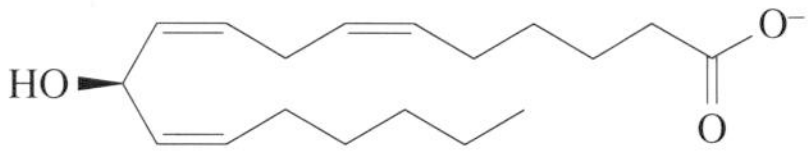

(62)

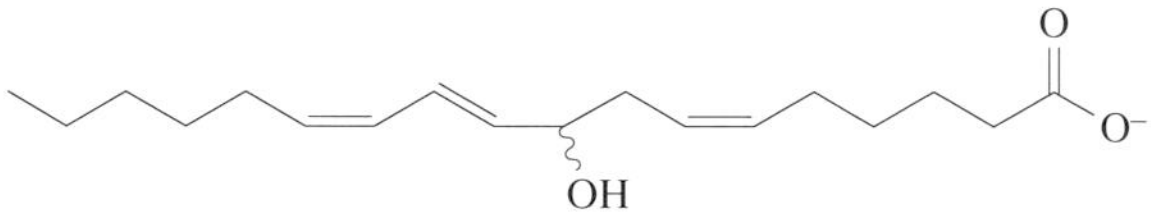

(63)

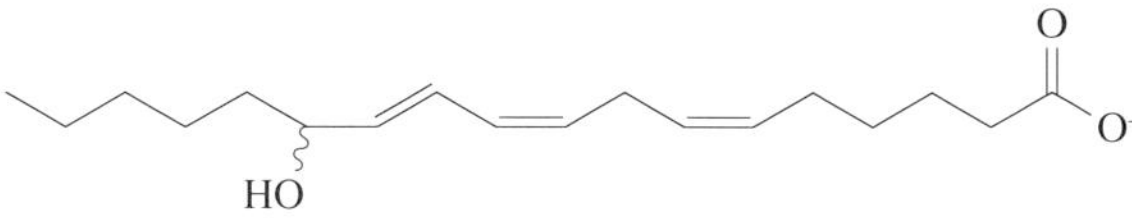

(64)

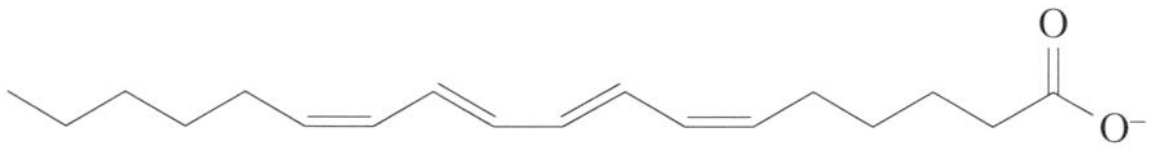

(65)

(vi) Ptilota filicina

The red alga *Ptilota filicina* from the West Coast of the USA has been the focus of detailed biochemical studies following the discovery of two distinct types of oxylipins in this organism, 5*Z*,7*E*,9*E*,14*Z*,17*Z*-eicosapentaenoic acid (**66**)[79] and ptilodene (**67**).[80] In each case, their structures were assembled principally from detailed NMR analysis. A crude tissue homogenate of freshly collected or frozen *P. filicina* was found to have a robust isomerase activity which cleanly converted exogenous EPA to the corresponding conjugated triene (**66**).[81] This activity was isolated by a standard assortment of protein purification techniques to a single band by silver-stained sodium dodecyl sulfate-polyacrylamide gel electrophoresis (SDS-PAGE), and it was able to catalyze the full conversion of EPA into the conjugated triene metabolite (**66**). A subunit molecular mass M_r of 58 119 Da was established by matrix-assisted laser desorption ionization mass spectrometry (MALDI-MS), whereas sedimentation equilibrium ultracentrifugation established the native enzyme M_r to be 125 kDa, leading to the conclusion that the enzyme is likely homodimeric.[81,82] Both flavin and iron cofactors were implicated from partial amino acid sequence analysis, absorbance spectroscopy, inductively coupled plasma mass spectrometry, and neutron activation analysis. Molecular oxygen does not appear to be required in this isomerase reaction.

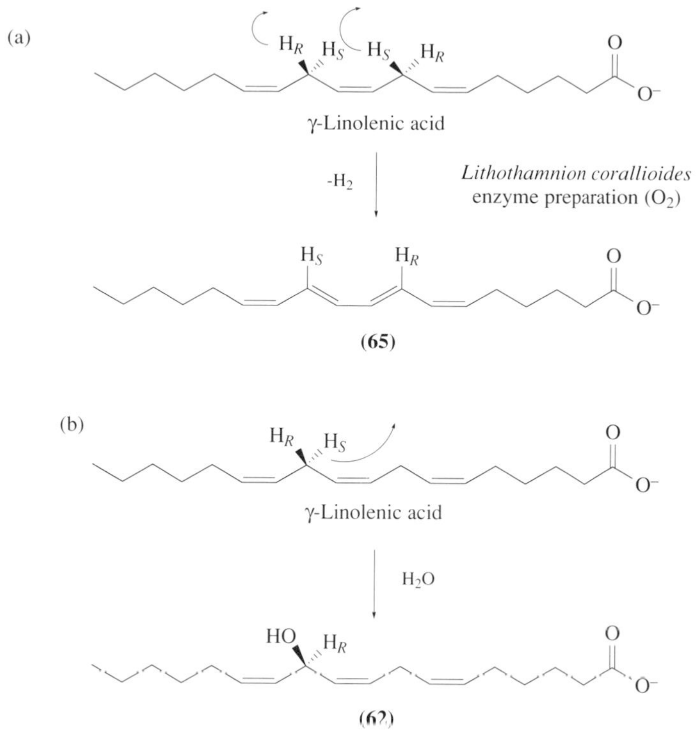

(a)

γ-Linolenic acid

$-H_2$ *Lithothamnion corallioides* enzyme preparation (O_2)

(65)

(b)

γ-Linolenic acid

H_2O

(62)

Scheme 8

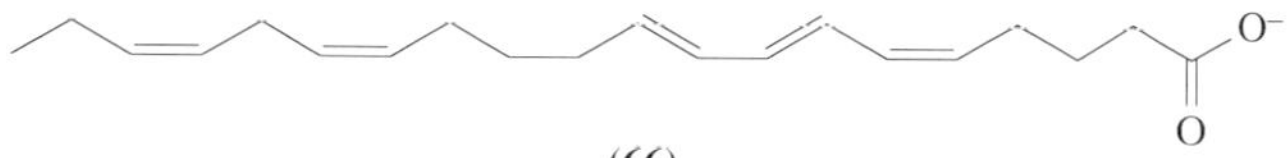

(66)

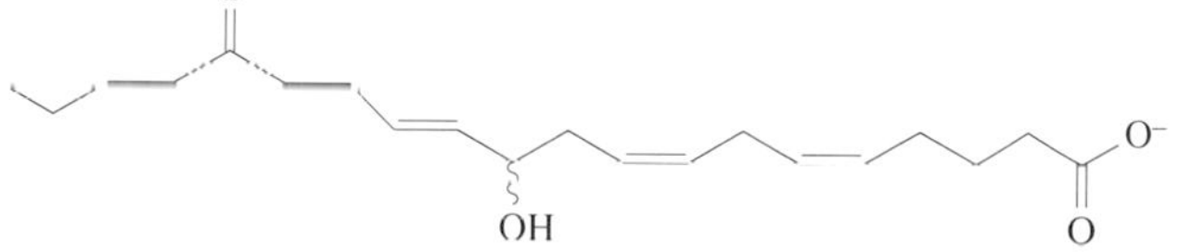

(67)

EPA was found to be the most rapidly turned over substrate, followed by AA, and then by a number of C_{22} and C_{18} substrates.[83] Analysis of kinetic parameters for these various substrates demonstrated that those with $\omega 3$ double bonds had the lowest K_m values while EPA and AA had the largest V_{max} values (6.0 and 2.8 μmol min^{-1} mg^{-1}, respectively). Investigations of the pH-dependency of polyunsaturated fatty acid isomerase (PFI) with AA, EPA, and anandamide (arachidonic acid ethanolamide, a nonionizable substrate) suggested that the enzyme may preferentially bind the protonated form of the substrate. The structures of products formed from a variety of alternate substrates, in combination with kinetic parameters for some of these, strongly suggests that PFI preferentially orients the substrate in the catalytic site relative to the methyl terminus. No effective inhibitors of the enzyme were found.

The pathway of conjugated triene production in *P. filicina* was studied using AA or a variety of ^{2}H-labeled linolenic acids.[81] With AA as substrate in a D_2O buffer solution, one of the two protons at C-11 was highly labeled with ^{2}H, indicating its protonation from a residue in the enzyme that is in equilibrium with solvent. The enzyme was sequentially investigated for its metabolism of (11*S*)-γ-[11-^{2}H]linolenate, (11*R*)-γ-[11-^{2}H]linolenate, (8*S*)-γ-[8-^{2}H]linolenate, and (8*R*)-γ-[8-^{2}H]linolenate, with both products and unreacted starting materials analyzed by GC-MS for deuterium. These experiments identified that the *pro-R* hydrogen at position C-8 in γ-linolenate is lost to the solvent, whereas the *pro-S* hydrogen at C-11 is intramolecularly transferred to the C-13 position (Scheme 9). However, the electronic nature of these hydrogen abstractions and rearrangements has not been elucidated.

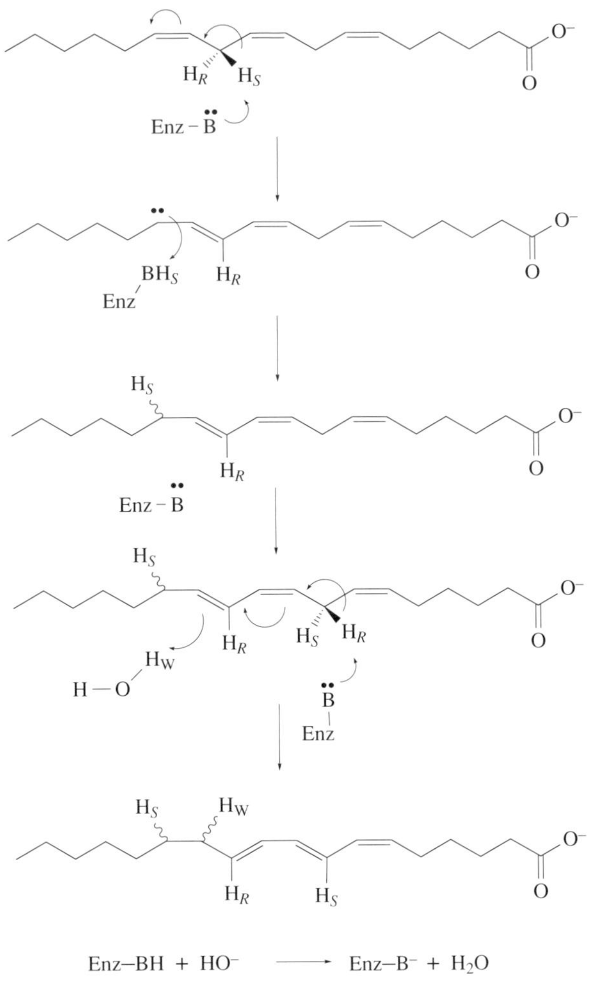

$$\text{Enz–BH} + \text{HO}^- \longrightarrow \text{Enz–B}^- + \text{H}_2\text{O}$$

Scheme 9

1.08.3.2.4 Bacillariophyceae (diatoms)

(i) Isochrysis *sp. and* Phaeodactylum tricornutum

While first detected in fish and other invertebrate sources such as the soft coral *Sarcophyton glaucum*,[84] furan fatty acids such as (**68**) have subsequently been shown to be produced only in plants and are incorporated into animals, including man, by nutritional routes.[85] Their biosynthetic pathway has been partially elucidated in studies with the diatoms *Isochrysis* sp. and *Phaeodactylum tricornutum*, and appears to be initiated by a LO with $\omega6$ specificity (C-13 oxidation) in linoleic acid. Interestingly, there is some evidence that the diatom LO locates its positional specificity in C_{18} and C_{16} fatty acids relative to the carboxyl terminus. The resulting dienyl hydroperoxides are believed to be cyclized with subsequent double bond rearrangements to yield furan fatty acids. These have been shown to be methylated by *S*-adenosyl methionine to give final products such as (**68**) (Scheme 10). Consistent with this hypothesis, the furan oxygen has been shown to derive from atmospheric O_2. The most interesting proposed reaction in this sequence is the conversion of the hydroperoxide into a cyclic species. Formally, this step is a dehydration, and as such, bears remarkable similarity to the hydroperoxide dehydrase (allene oxide synthase) reaction of higher plants, which produces an allene oxide intermediate during phytodienoic acid and jasmonic acid biosynthesis (see Sections 1.08.3.3.2(ii), (iv), and (v)).[2] This is a potentially rich area for further mechanistic and enzymological investigation.

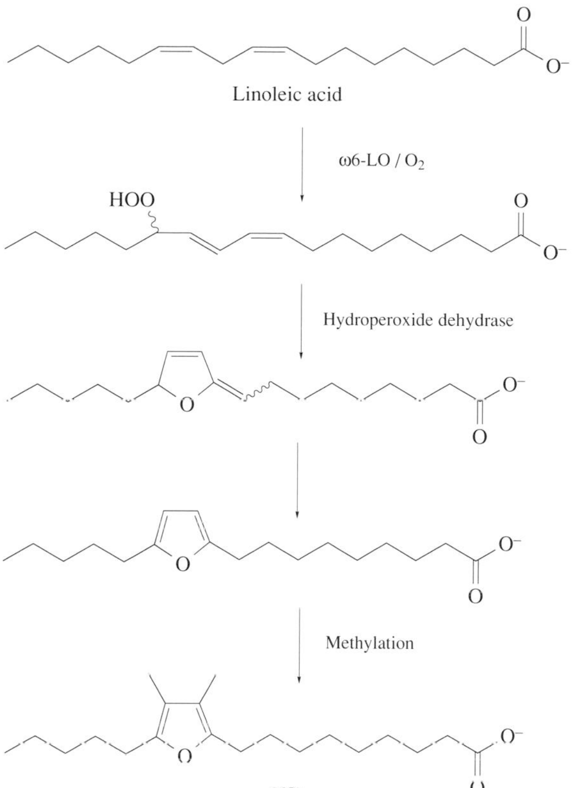

Linoleic acid

ω6-LO / O_2

HOO

Hydroperoxide dehydrase

Methylation

(68)

Scheme 10

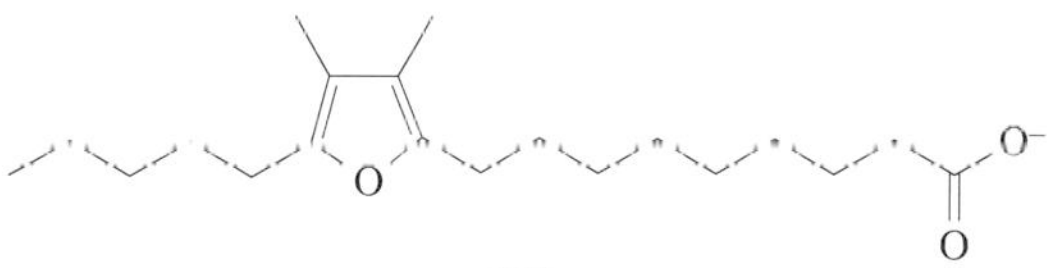

(68)

(ii) Nitzschia pungens

Two of the more interesting diatom-derived oxylipins are the bacillariolides (I **(69)**, II **(70)**), novel cyclopentane-containing eicosanoids from the marine bacillariophyte *Nitzschia pungens*.[86] This diatom species has been implicated as the causative organism in intoxications known as "amnesic shellfish poisoning" (ASP) in Eastern Canada, with domoic acid, a glutamate receptor mimic, as the presumed toxin.[87] However, as domoic acid has been used as an anthelmintic in Japan for years without problems of the type reported in Canada, it has been speculated that other constituents of the diatom may contribute to the toxicity. Field collected diatoms were extracted with methanol and the new compounds isolated by a combination of normal and reverse phased chromatography. The overall structures of the new compounds were assembled from spectroscopic data, principally NMR, whereas relative and absolute stereochemistry was indicated from an X-ray crystallographic study of a chiral derivative.[88] These eicosanoids are particularly unique in the location of the cyclopentane ring which connects C-2 and C-6. A biogenesis for these metabolites has been proposed which is initiated by a 5-LO and possibly leads, via an EAC intermediate,[45] to an epoxy alcohol[88] or an epoxy lactone intermediate.[45] Carbocyclization is envisioned to occur through formation of a C-2 carbanion and its subsequent nucleophilic attack at C-6 of the C-5—C-6 epoxide.

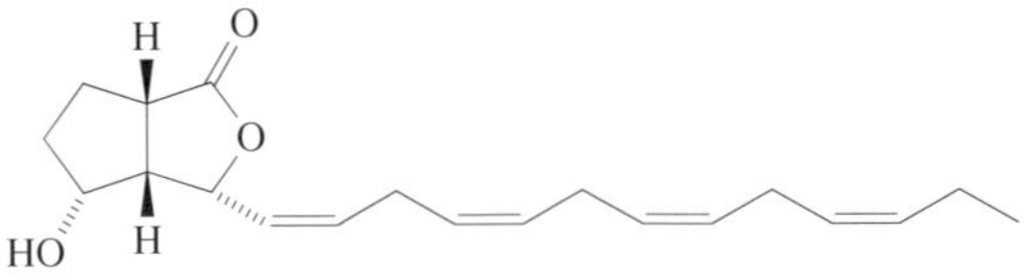

(69)

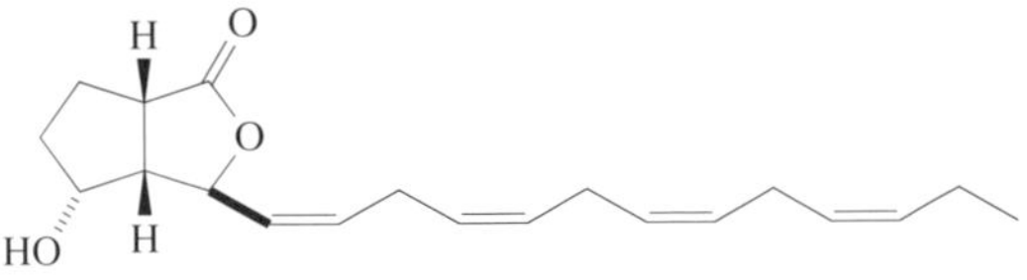

(70)

1.08.3.3 Plants

Several excellent reviews have appeared which cover various features of higher plant oxylipin metabolism, such as their LO pathways[2,89,90] and the mechanism, distribution, and biological significance of jasmonic acid formation.[91–96] In particular, the review by Gardner[2] provides an excellent account of the pathways and mechanisms of formation for most of the significant LO-derived oxylipins found in higher plants. While in that review the pathways of oxylipin metabolism are grouped by enzymatic or mechanistic type, in this chapter the coverage is organized along taxonomic lines.

1.08.3.3.1 Lower plants

(i) Ferns

There are few reports of oxylipins from ferns, with the most notable example being 8*S*-hydroxyhexadecanoic acid (**71**). While this had been isolated in a phytochemical study in 1965, it was later described as an endogenous inhibitor of spore germination in the producing species *Lygodium japonicum*.[97] Total synthesis was used to describe the absolute stereochemistry of this hydroxyacid.[98]

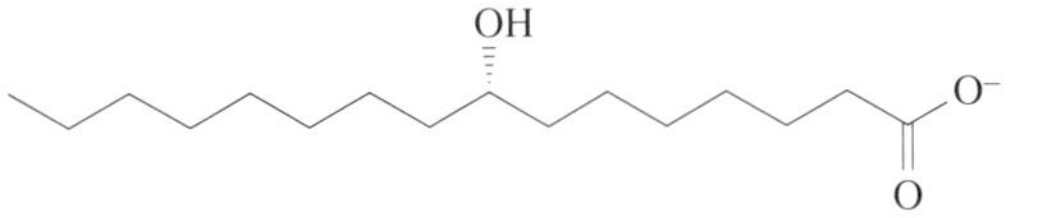

(71)

(ii) Mosses and liverworts

Mosses have been studied rather extensively for their unique oxylipins. The polymerized lipids of some mosses such as *Dicranum elongatum* are quite resistent to decay, and are rich in oxidized lipids including various hydroxy acids and dicarboxylic acids.[99] Mosses possess an abundance of AA and EPA, as well as a number of unusual acetylenic fatty acids of broad taxonomic occurrence, particularly in the Dicranales.[100,101] Of greater significance to this chapter is the finding that some of these unusual fatty acids are metabolized to hydroxy acids such as (**72**) and to unique cyclopentenoids such as dicraneone A (**73**) in the moss *Dicranum scoporium*.[102] Several stereochemical features of these cyclopentenoids have not yet been determined. Dicranenone A (**73**) and B were isolated as antifungal agents in tests using the rice-blast disease causing organism *Piricularia oryzae*. Based on the profile of metabolites obtained from this moss, as well as the precedence of other pathways to cyclopentanoids in higher plants such as flax (see Section 1.08.3.3.2(iv)), a pathway was proposed to dicranenone A which first involves lipoxygenation at C-13 and then conversion to a cyclopentenoid, presumably via an allene oxide. A 10-hydroxy "ene-yne" metabolite (**74**), as well as the corresponding ketone, has been characterized, the former without stereochemistry, from the New Zealand liverwort *Monoclea forsteri*.[103]

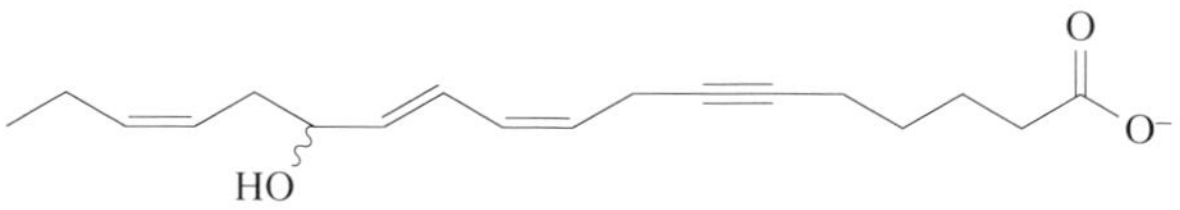

(72)

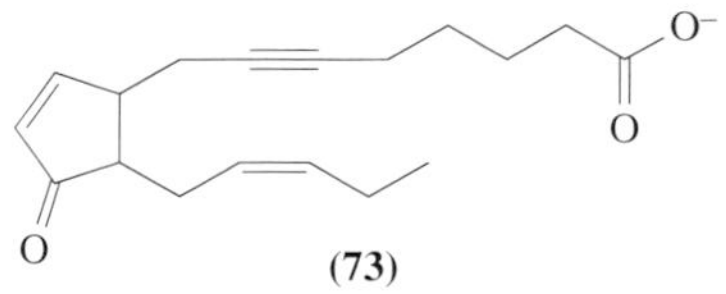

(73)

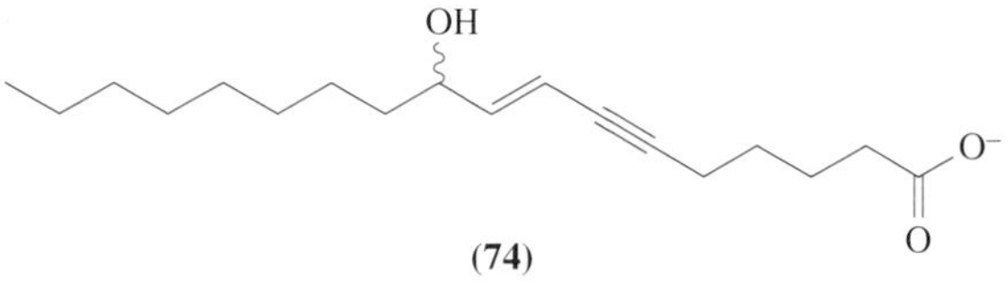

(74)

1.08.3.3.2 Higher plants

(i) Cyperaceae

(a) Eleocharis microcarpa—*sedge*. The sedge *Eleocharis microcarpa* has been studied for its competitive relationship with blue-green algae (cyanobacteria) and a number of hydroxy and hydroxy keto acids were implicated as allelopathic agents.[104] One of these **(75)** was characterized as to its planar structure by a variety of spectral techniques. Its novel C_{20} structure is exceptional for the location of a cyclopentyl ring between C-15 and C-19. This new substance also possessed three hydroxy groups at C-11, C-16, and C-18, the latter two perhaps indicating a cyclooxygenase-like mechanistic origin. A *trans* olefinic bond and an acetylenic bond were also components in this new sedge metabolite. Two additional oxidized fatty acids were subsequently reported as contributing to the anticyanobacterial chemistry of *E. microcarpa* (e.g., **(76)**).[105] These were carefully characterized as to overall structure; however, stereochemistry was not investigated. The new compounds show a close relationship to 12-oxo-phytodienoic acid (12-oxo-PDA) **(77)**, albeit with the cyclopentyl ring occurring between C-10 and C-14. A biogenesis of these metabolites has not been proposed; it is enticing, though perhaps premature, to propose the intermediacy of an allene oxide. Further studies on the chemistry and biosynthesis of unusual oxylipins from sedge are clearly warranted.

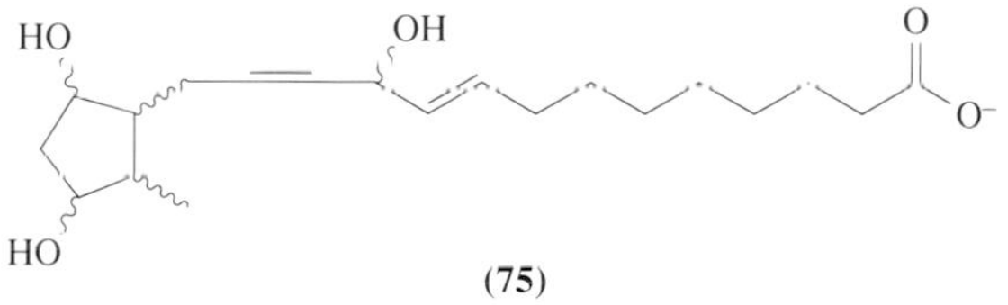

(75)

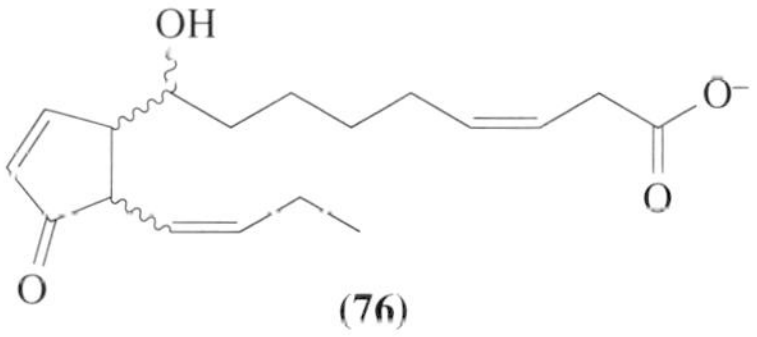

(76)

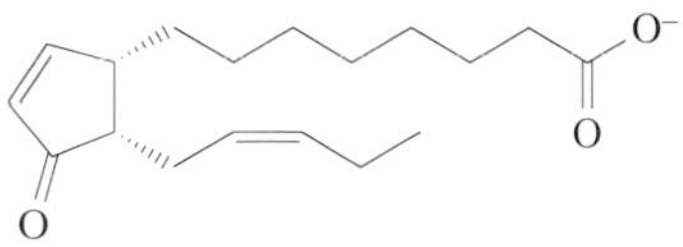

(77)

(ii) Gramineae

Three members of the Gramineae have been studied extensively for their oxylipin chemistry and biochemistry; oats (*Avena sativa*), rice (*Oryza sativa*), and corn (*Zea mays*). Two dominant biochemical themes that emerge from these studies are (i) the production of epoxy alcohols as well as corresponding triols in pathways involving the sequential functioning of LO and peroxygenase enzymes and (ii) the production of allene oxide-derived oxylipins including α-ketols, macrolactones, and carbocyclic products such as phytodienoic and jasmonic acids. While several different LOs are responsible for initiating these various pathways, the predominant LO appears to have ω6 specificity on C_{18} precursors and produces 13*S*-HPOD.

(a) Avena sativa—*oats*. Oats have been best characterized for their production of a range of hydroxy, epoxy, hydroxyepoxy, and trihydroxy metabolites of linoleic acid. Because of a complex precursor and product relationship of these oat metabolites, clarification of this pathway required the development of new analytical methods for trihydroxy compounds.[106] A soluble series of enzymes were subsequently characterized to transform linoleic acid into this mixture of products (e.g., **(78)**) through the sequential action of LO, peroxygenase, and epoxide hydrolase activities (Scheme 11).[107] Peroxygenase, a hydroperoxide-dependent oxidase producing either alcohols or epoxides, was first characterized in pea seeds.[108] In addition to these metabolites, on-going efforts to fully describe the oxylipin chemistry of *A. sativa* have resulted in the isolation and complete characterization of 15*R*-hydroxylinoleic acid **(79)**, a novel substance found in abundance in seeds of this species.[109]

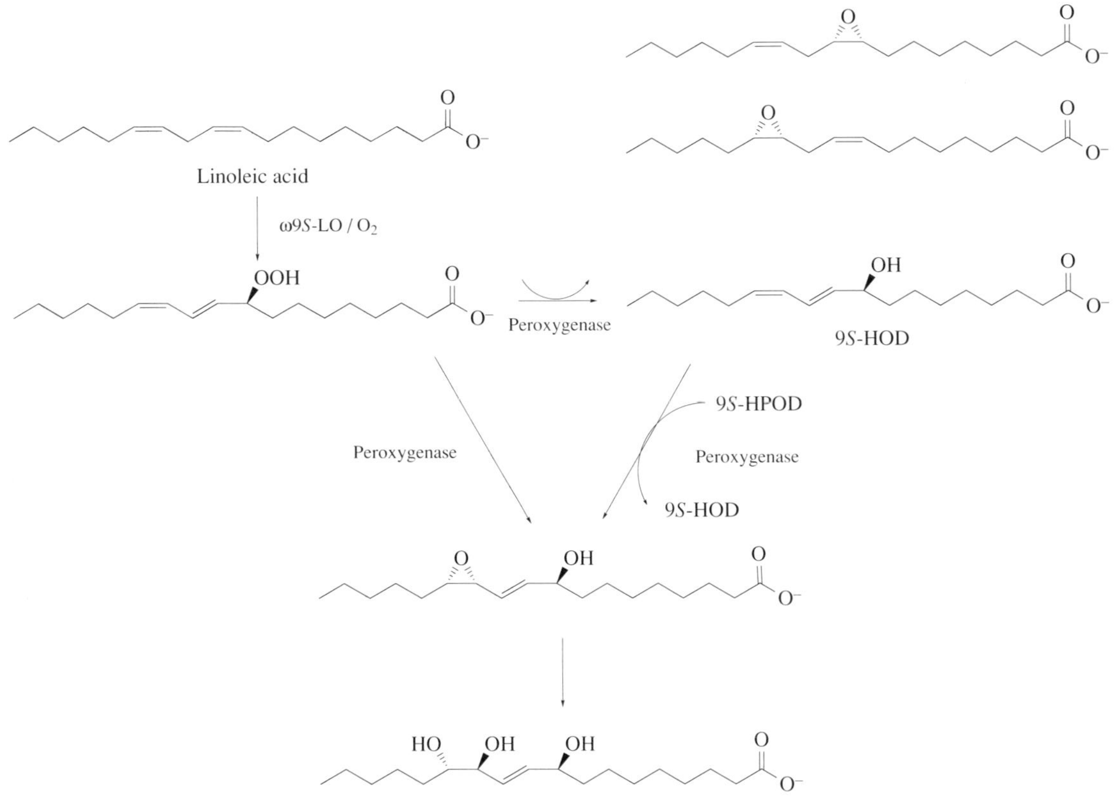

Linoleic acid

ω9*S*-LO / O_2

Peroxygenase

9*S*-HOD

Peroxygenase

9*S*-HPOD

Peroxygenase

9*S*-HOD

(78)

Scheme 11

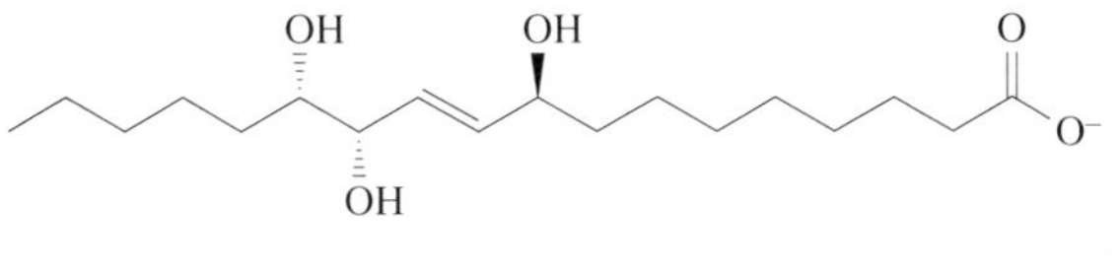

(78)

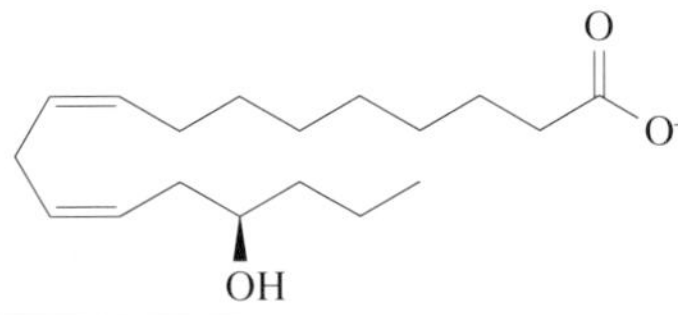

(79)

(b) Zea mays—*corn*. While homogenates of corn seedlings have been characterized to metabolize exogenous AA to a mixture of HETEs (5-HETE, 12-HETE, 15-HETE),[110] the seeds have an ω10*S*-LO (C-9 oxygenation)[111] acting on endogenous C_{18} precursors whereas leaves have an ω6*S*-LO (C-13 oxygenation)[112] acting on the same substrates. Most work with this plant species has focused on its ability to convert 13*S*-HPOD or 13*S*-hydroperoxyoctadecatrienoic acid (HPOT) into unstable allene oxides[113] with subsequent conversion by enzymatic processes into cyclopentanoids or by nonenzymatic processes into an array of products.[114] For example, the allene oxide formed from

this substrate via action of hydroperoxide dehydrase rapidly added either water or a trapping agent (methanol) to form α-ketols. Two new macrolactones were also characterized as spontaneous decomposition products of the unstable allene oxide, as was a racemic cyclopentanoid.[114] However, under direction of an allene oxide cyclase, found to have broad distribution in higher plants,[115] the 13*S*-HPOT derived allene oxide (**80**) was shown to be converted into 9*S*,13*S*-12-oxo-phytodienoic acid (**77**), an important precursor to the phytohormone 7-iso-jasmonic acid (**81**).[116]

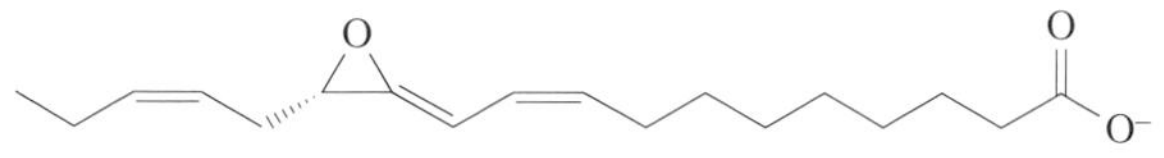

(**80**) 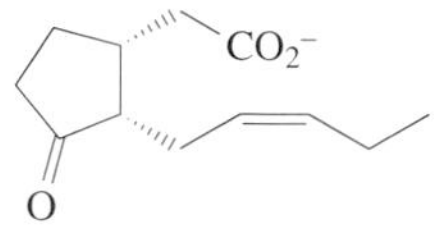 (**81**)

(iii) Lemaceae

(a) Lemna trisulca *and* Lemna minor. Investigations of the two species of *Lemna*, *L. trisulca* and *L. minor* have revealed a general capacity in these plants to metabolize C_{16} fatty acids, such as 8*Z*,11*Z*,14*Z*-hexatrienoic acid, to a variety of hydroxy acids (e.g., (**82**)) and cyclopentanoids (e.g., (**83**) and (**84**)).[117,118] In this regard, it appears that the related genera *Lemna* and *Eleocharis* have similar biosynthetic capacities (see Section 1.08.3.3.2(i)). The structure of hydroxy acid (**82**) from *L. trisulca* was assigned principally on the basis of NMR data; however, correlates of optical rotation with homologous known compounds were used to assign the 12S stereochemistry. The structure, including relative stereochemistry, of the *L. trisulca* cyclopentanoid derivative (**83**) was similarly reasoned from NMR data, and absolute stereochemistry suggested from ORD measurements.[117] From *L. minor*, two related cyclopentanoids were obtained with the cyclopentane ring being located between C-10 and C-14 in a C_{16} framework (e.g., (**84**)).[118] A similar relationship of functional groups was positioned on this carbon backbone as found in the above described compound (**83**) from *L. trisulca*. It was suggested for the metabolites from both species of *Lemna* that the hydroxy acids likely derive from LO metabolism while the carbocyclic metabolites likely come from cyclooxygenase (COX) activity as found in mammalian species. However, it seems equally likely that these cyclopentanoids are formed via a variant of the LO → hydroperoxide dehydrase (= allene oxide synthase) → cyclase pathway described in other higher plant species as the route to 12-oxo-phytodienoic acid and *iso*-jasmonic acid. In this latter regard, it is conceivable that these pathways in *Lemna* could be initiated by an ω2 LO acting on 8*Z*,11*Z*,14*Z*-hexatrienoic acid (path to (**83**)) or an ω3-LO acting on 7*Z*,10*Z*,13*Z*-hexatrienoic acid (path to (**84**)). Both of these fatty acid precursors have been described in *Lemna* spp. The additional oxidation in the acyclic portions of metabolites (**83**) (C-8) and (**84**) (C-7) might be introduced via a separate dioxygenase event early in the pathway.

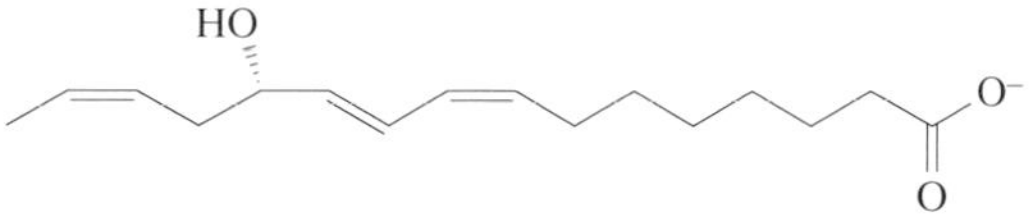

(**82**)

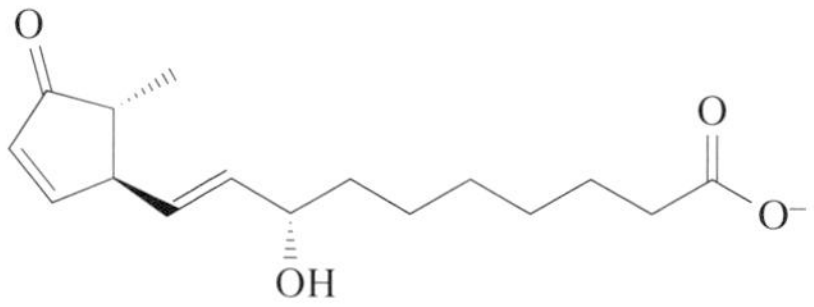

(**83**) 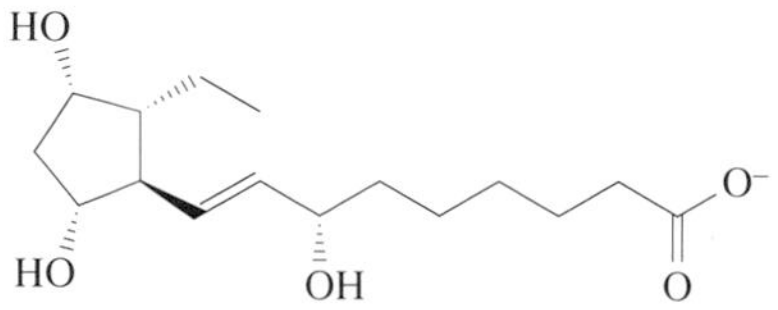 (**84**)

(iv) Linaceae

(a) Linum usitatissimum—*flax*. Flax seed (*Linum usitatissimum*) acetone powders or homogenates have played a central role in studies of oxylipin metabolism in higher plants. Indeed, 12-oxo-phytodienoic acid (**77**) was first obtained from *in vitro* metabolism studies using a flax seed

acetone powder and α-linolenic acid.[119] Subsequent detailed mechanistic studies using flax seed acetone powders by Crombie[120] showed that an allene oxide intermediate (**80**) was involved in the formation of phytodienoic acid as well as the α-ketol (**85**) and γ-ketol (**86**) products formed by spontaneous hydrolysis of (**80**). The acid- and base-catalyzed hydrolyses of 13-HPOT have been studied in some detail and appear to follow two distinct pathways.[121] Subsequently, Brash's group[122] reported the use of a flax seed preparation in the conversion of 13*S*-hydroperoxyoctadeca-9*Z*,12*Z*,15*Z*-trienoic acid into an allene oxide which was sufficiently stabilized in cold aprotic solvents to be isolated and characterized spectroscopically. Comparable studies in corn (*Z. mays*) and two species of coral (*Plexaura homomalla* and *Clavularia viridis*) (see Sections 1.08.3.3.2(ii) and 1.08.4.1.2(i), (ii), and (iii)) were nearly simultaneously giving very similar results. By providing fatty acid hydroperoxides which contained additional sites of oxidation (e.g., C-15-hydroxy in a EPA-derived substrate) to flax seed preparations, Brash's group was able to demonstrate the biosynthetic production of prostaglandin A (PGA) analogues (e.g, (**87**)) via an allene oxide pathway. This result provided a conceptual precedent for PGA_2 biosynthesis via an analogous route in the coral *P. homomalla*.[123] The enzyme, allene oxide synthase, which catalyzes this transformation was subsequently isolated by Brash's group and shown to be a member of the cytochrome P450 family of hemoproteins.[124] Structure–function aspects of this enzyme have been described.[125] Insight that the conversion of hydroperoxide to allene oxide might involve homolytic cleavage of the hydroperoxide to an alkoxyl radical was obtained through isolation of epoxyalcohol derivatives (e.g, (**88**)) as minor products when substrates containing two sites of oxygen were provided to flax seed-derived allene oxide synthase (Scheme 12).[123,126] Products produced by the allene oxide synthase appear to inhibit the functioning of flax seed LO.[127]

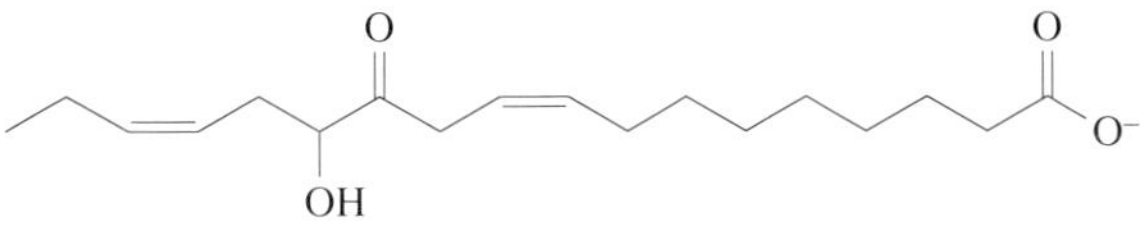

(85)

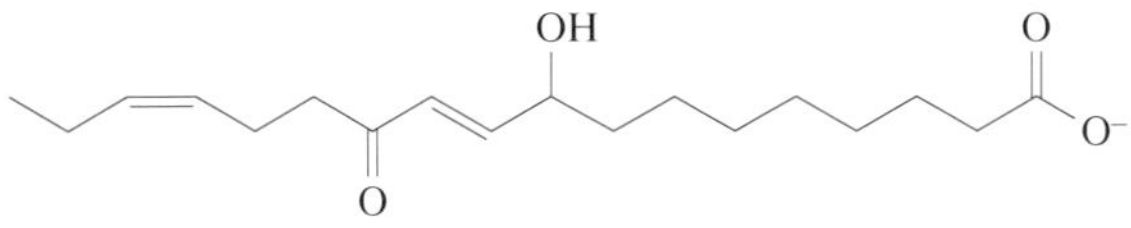

(86)

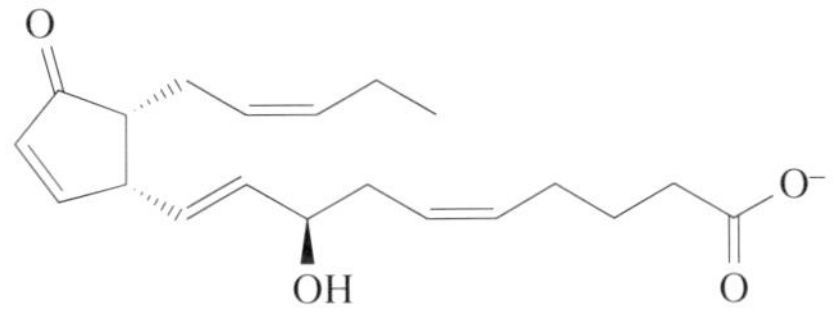

(87)

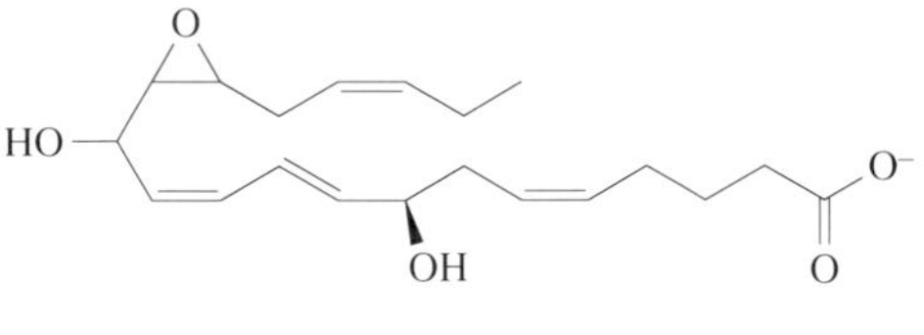

(88)

(v) Leguminosae

(a) Glycine max—*soybean*. LO metabolism in plants is perhaps most commonly recognized from members of the Leguminosae, which include the soybean (*Glycine max*). Indeed, mechanistic features of this enzyme reaction have been most thoroughly studied using soybean LO as a model. Gardner[2] provides a superb review of mechanistic features of soybean LO, which also provides numerous citations to the many previous excellent reviews on this subject. Whereas several isozymes of soybean LO have been described, soybean LO Type 1 has been used to the greatest degree in mechanistic and enzymological studies.

Soybean LO-1 has been characterized as a nonheme iron containing enzyme of M_r 94 038 Da, a size comparable to other plant LOs. The complete structures of several plant LOs are known from cloning and sequencing, and they show intriguing homology in the substrate binding site and active site, giving possible insight into the evolution and function of LOs. The enzyme is functional as a monomer which contains a single iron molecule. A high resolution (1.4 Å) crystal structure has

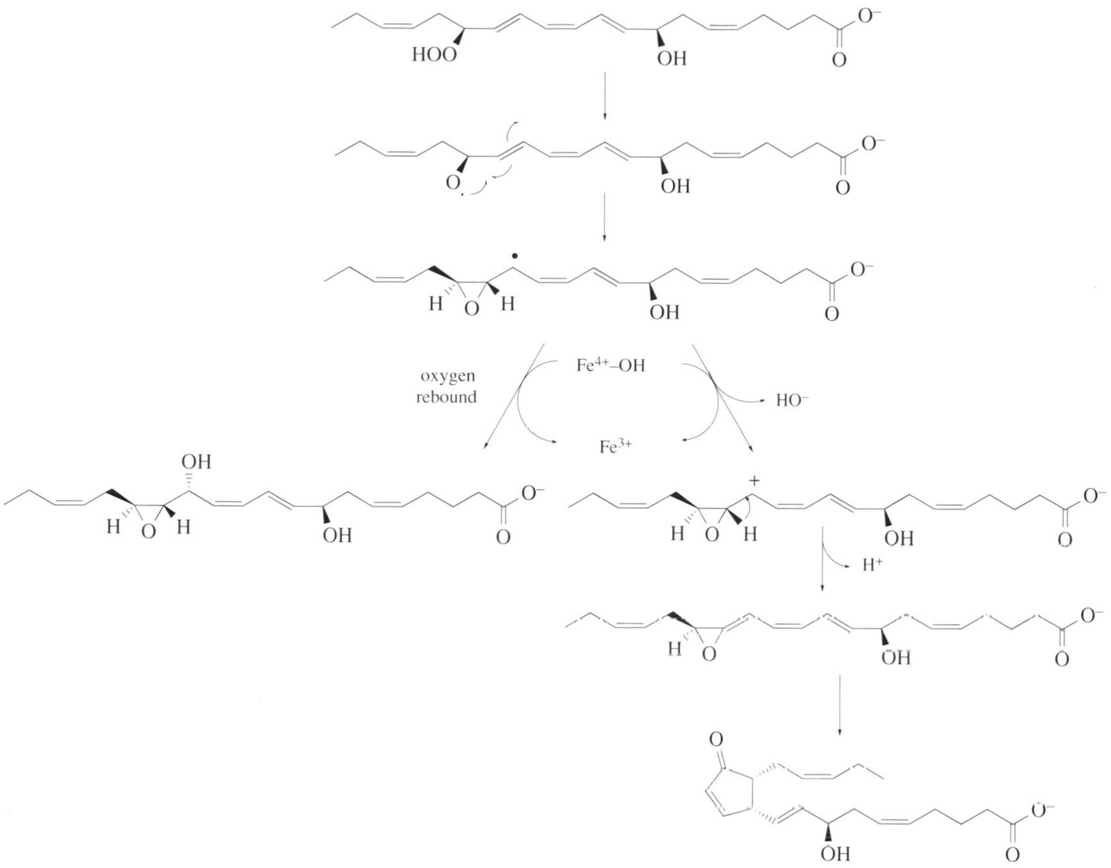

(87)

Scheme 12

clarified several aspects of the iron octahedral coordination sphere.[128] Four nitrogen ligands are afforded by three His residues and one Asn, the latter is a weak ligand which is simultaneously involved in several polar interactions with other protein groups. A fifth ligand is provided by the carboxyl function of the terminal Ile residue; it is monodentate. The sixth ligand is newly identified as a water molecule. A direct covalent link between the iron atom and the substrate fatty acid is precluded from geometrical constraints in the active site. While previous lower resolution X ray studies of this enzyme have indicated two channels to the active site, tentatively proposed as providing separate access for the fatty acid and the molecular oxygen substrates,[129] the higher resolution data raises concerns and provides several new alternatives.

The natural substrates for soybean, as for almost all plant LOs, are linoleic acid and α-linolenic acid. For soybean LO-1, optimal alignment of the dienoic substrate at high pH is with the ω6 unsaturation pattern, with enzyme recognition of substrate occurring relative to the ω terminus. The enzyme stereospecifically removes the *pro-S* hydrogen from the ω8 position and introduces molecular oxygen from the opposite face at C-13, to produce 13*S*-HPOD. On the other hand, protonation of the fatty acid substrate, which occurs at anomalously high pH,[2] allows substrate binding in a reverse orientation such that the *pro-R* hydrogen is abstracted from C-11 and 9*S*-HPOD is produced instead. It is interesting to consider that this pH dependency on orientation of substrate binding is another independent measure of the anomalous pK_a values for long chain fatty acids at physiologically relevant concentrations.[130]

The LO reaction mechanism has been actively studied with the soybean LO-1 model (Scheme 13).[2] While this mechanism has not been fully elucidated despite considerable effort (and controversy), the generally accepted aerobic mechanism begins with resting state enzyme, LO–Fe^{2+} (high-spin ferrous). Peroxide—exogenously added H_2O_2 or fatty acid autoxidation products—is required to convert the enzyme to its reactive LO–Fe^{3+} (ferric) form; hence, the lag times commonly associated with kinetic analyses of this enzyme. The activated enzyme then homolytically abstracts the *pro-S* hydrogen from C-11, probably with the C-8 to C-14 region of the substrate forming a kinked and delocalized radical structure.[128] The iron–hydroxide species most likely acts as the active site base, becoming reduced to LO–Fe^{2+} in the process. This abstraction, the rate limiting step of the LO reaction, shows the largest published intrinsic primary H/D isotope effect for an enzymatic reaction (extrapolated value = 80). However, features other than this hydrogen abstraction, including substrate binding and a H_2O/D_2O sensitive step, contribute to the overall limitation and variability in reaction rate.[130] This hydrogen abstraction has been shown to precede the binding or reaction of molecular oxygen.[131] It appears that the molecular oxygen interacts directly with the delocalized radical of the fatty acid substrate in the LO active site, forming a peroxy radical, a species for which experimental evidence exists. This peroxy radical is believed to subsequently interact with the LO–Fe^{2+}, becoming oxidized to the peroxy anion which then regenerates the oxidized and active LO–Fe^{3+} form of the enzyme. Protonation of the peroxy anion from media water yields the familiar hydroperoxide product of LO reactions.

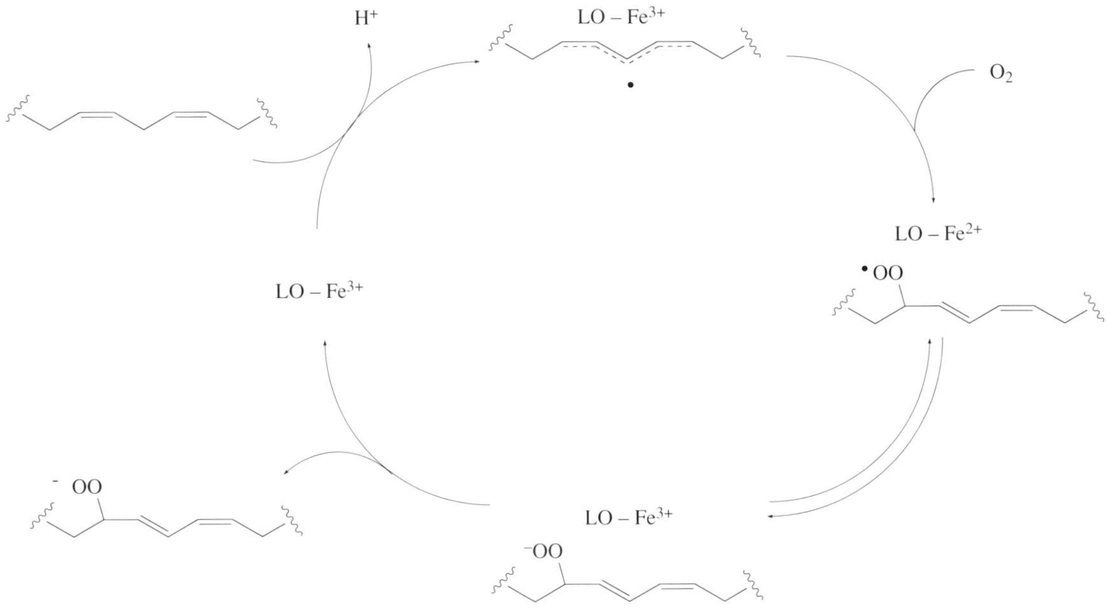

H^+ LO – Fe^{3+} O_2 LO – Fe^{2+} •OO LO – Fe^{3+} ⁻OO LO – Fe^{3+} ⁻OO

Scheme 13

As is the case with other classes of higher plants, a rich diversity of products is formed by further reaction of the LO-produced dienyl hydroperoxides in soybeans.[2] These include dihydroperoxides formed by a double dioxygenation, a process favored at low pH with AA as substrate to produce such compounds as 8*S*,15*S*-dihydroperoxy-6*E*,8*Z*,11*Z*,13*E*-eicosatetraenoic acid. Other reactions of fatty acid hydroperoxides observed in soybean involves formation of lipoxin-like compounds.

(b) Vicia faba—*broad bean*. The broad bean *Vicia faba* has played an important role in the development of our understanding of oxylipin metabolism in plants, particularly in the pathway of biosynthesis of jasmonic acid. It was in this plant that the metabolic relationship of 12-oxo-phytodienoic acid (12-oxo-PDA; (**77**)) and jasmonic acid (**89**) was first uncovered by Vick and Zimmerman.[132] Provision of labeled 12-oxo-PDA (either [U-^{14}C]- or [U-^{14}C]-12-[^{18}O]-oxo-) not only revealed the precursor–product relationship of 12-oxo-PDA and jasmonic acid, but also the nature and sequence of several key intermediates. In this pathway, the C-10—C-11 cyclopentenone double bond of 12-oxo-PDA is first reduced and then this product is subjected to three rounds of β-oxidation to yield jasmonic acid.

Subsequently, the broad bean has been studied for its hydroperoxide-dependent epoxidation of unsaturated fatty acids[133] and oxygenation of fatty acid fragments such as 3*Z*-nonenal to 4-hydroxy-2*E*-nonenal (**90**).[134] In the former work, an epoxygenase pathway was described whereby the distal oxygen of 13*S*-HPOD (**91**) is transferred to linoleic acid substrates via an epoxygenase reaction to form a mixture of enantiomeric *cis*-9,10-epoxides (**92**) and *cis*-12,13-epoxides (**93**). Minor products of this reaction included epoxyalcohols (e.g, (**94**)) arising from epoxygenase reaction with 13*S*-HOD (**59**) or 13*S*-HPOD (**91**). This pathway appears related to that involving "peroxygenase" in pea seeds[108,135] and oats (see Section 1.08.3.3.2(ii)).[107]

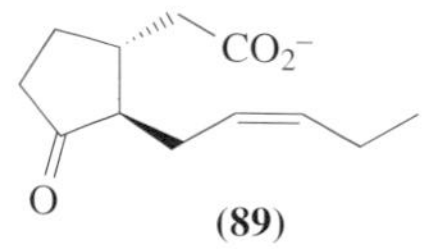

(89) 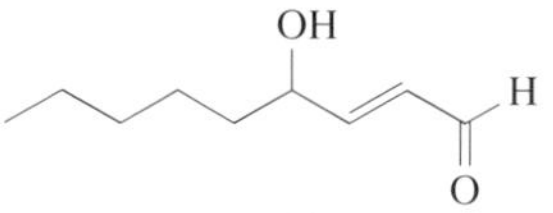 **(90)** 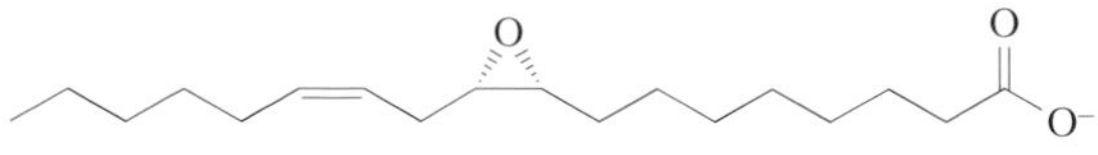 **(92)**

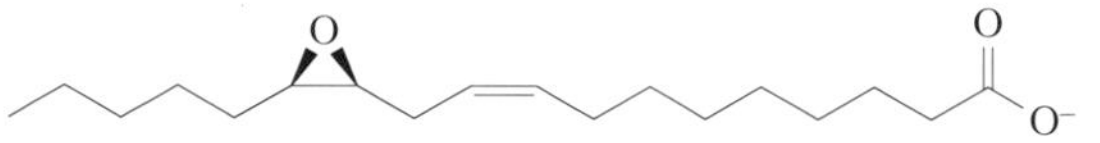

(93) 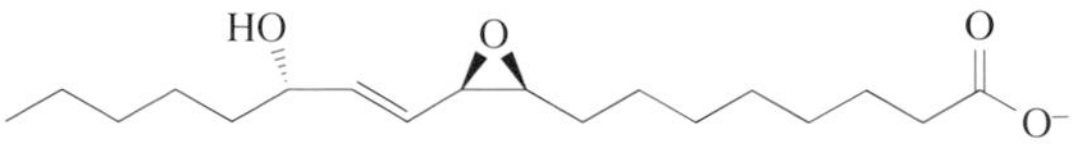 **(94)**

The mechanism of formation of 4-hydroxy-2*E*-nonenal (**90**) was found to be surprisingly complex, and arises by two distinct pathways, in some ways reminiscent of epoxyalcohol formation in oats (see Section 1.08.3.3.2(ii)). While it had been speculated that this hydroxy group might be introduced via a lipoxygenase reaction recognizing the 3*Z*-nonenal substrate as a 1,4-diene equivalent, this was shown not to be the case. A new enzyme activity was discovered in the broad bean, a 3*Z*-alkenal oxygenase, which introduces molecular oxygen at this position to form 4-hydroperoxy-2*E*-nonenal.[134] This product is then capable of transferring the distal hydroperoxide oxygen to a second molecule of 3*Z*-nonenal via an epoxygenase reaction to form two molecules of 4-hydroxy-2*E*-nonenal (**90**).

(vi) Solanaceae

(a) Solanum tuberosum—*potato*. Potato tubers were found to metabolize linoleic and linolenic acids via a LO pathway[136] into an analogous series of divinyl ether compounds termed colneleic acid (**95**) and colnelenic acid (**96**), respectively.[137,138] Both a crude potato enzyme preparation and a model solution of Fe^{2+}, each showing optimal activity at pH 5.0–5.5, were effective at catalyzing the breakdown of colneleic acid into shorter chain carbonyl compounds, and hence, this was generally identified as a possible pathway for the production of volatile plant products. While the potato LO-derived product 9*S*-hydroperoxyoctadecadienoic acid (9*S*-HPOD) was a substrate for divinyl ether biosynthesis, 13*S*-HPOD was not.[139] Whereas from linoleic acid, potato tuber LO was shown to actually produce a 92:8 mixture of the 9- and 13-hydroperoxides at pH 5.5 with the oxygen deriving from molecular oxygen as expected,[140] the major dioxygenation product from LO metabolism of exogenous AA by hairy root cultures of potato was 11-HPETE (5-HPETE was the predominant product of AA metabolism by tubers, see below).[141] $^{18}O_2$-Labeled 9*S*-HPOD was utilized to convincingly demonstrate that the divinyl ether oxygen of colneleic acid derived from molecular oxygen.[140] An EAC was suggested as a common intermediate in colneleic acid (**95**) and epoxy alcohol (**97**) biosynthesis at either basic (pH 9.0) or acidic pH, respectively. As the breakdown products of divinyl ethers were shown to be 3*Z*-nonenal and 9-oxo-nonanoic acid, divinyl ethers represent potential intermediates in some hydroperoxide lyase-like reactions. Deuterium incorporation from [9,10,12,13-$^2H_4$]linoleic acid as substrate was also consistent with a 9-hydroperoxy intermediate being transformed into an epoxy allylic carbocation which then lost a proton from C-8 with resultant cleavage of the epoxide carbon–carbon bond giving colneleic acid.[142] Hydrogen removal from C-8 of 9*S*-HPOD was shown to be stereospecific for the *pro-R* hydrogen in the biosynthesis of colneleic acid (**95**).[143]

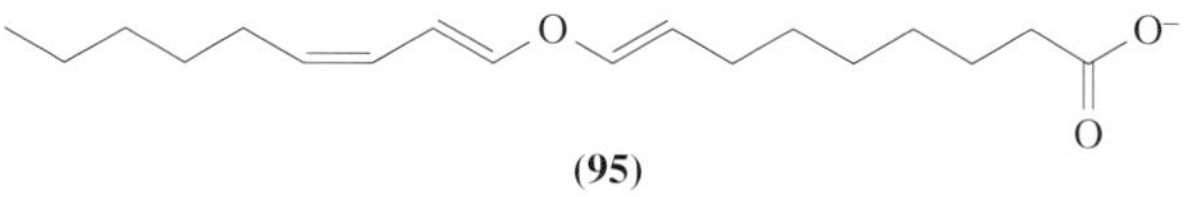

(95)

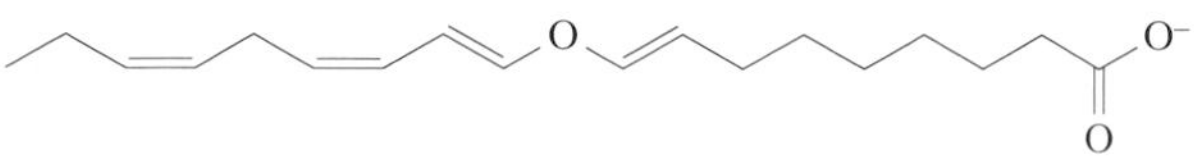

(96)

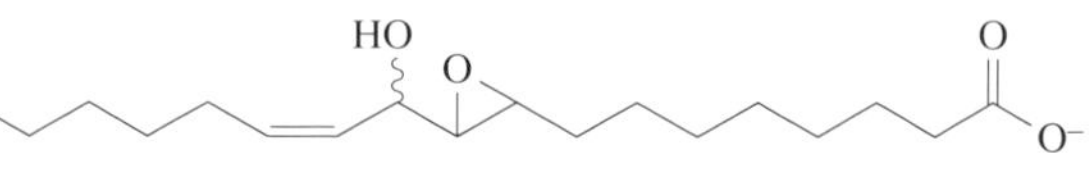

(97)

From the perspective of regulating responses to external stimuli, there are two distinct pathways of oxylipin metabolism in the potato.[144] These are the formation of methyl jasmonate from α-linolenic acid, which occurs during wounding, or the lipoxygenase metabolism of AA provided by invading fungal pathogens. These appear to upregulate separate 3-hydroxy-3-methylglutaryl-coenzyme A reductase (HMGCoA reductase) genes (*hmg1* and *hmg2*). Selective increases in transcription of these HMGCoA reductases leads in turn to the production of steroids and glycoalkaloids or to terpene-derived phytoalexins, respectively.

Products of intriguing structure have been obtained from metabolism studies using potato tubers and AA, a nonendogenous substrate but one that may be contributed by invading fungi, as indicated above, and hence, represent significant signal molecules in phytoalexin responses. These include products such as 5,6-leukotriene A_4, 5,12-diHETE, and lipoxins A and B.[2] Potato LO recognizes its substrate from the carboxyl end, first forming 5*S*-HPETE, which in turn is proposed to form 5,6-leukotriene A_4, and then is hydrolyzed into diHETE products. Alternatively, double dioxygenation of fatty acid substrates is another fate for potato hydroperoxides. For example, potato tuber metabolism (as well as soybean) of α-linolenic acid gives products which likely derive from both a 9,10-epoxy compound, such as 9*S*,16(*R*,*S*)-dihydroxy-10*E*,12*E*,14*E*-octadecatrienoic acid, as well as the double dioxygenation pathway, such as 9*S*,16(*R*,*S*)-dihydroxy-10*E*,12*Z*,14*E*-octadecatrienoic acid. However, the double dioxygenation pathway is substantially more prominent in soybean than in potato.[145]

1.08.4 ANIMALS

The oxylipins of nonmammalian animals have been reviewed several times previously, in some cases providing semicomprehensive coverage as of the date of writing,[4,146,147] and in other cases having a restricted focus, such as on aquatic fauna,[3] physiological functions of oxylipins in invertebrates,[148] marine invertebrates,[8] or mechanistic features of their biosynthesis.[149]

1.08.4.1 Aquatic Animals

1.08.4.1.1 Fish

There are four dominant themes in the oxylipin chemistry of fish: (i) the production of PGE_2 **(98)** and $PGF_{2\alpha}$ **(99)** in gills and other nonreproductive tissues for putative physiological roles (studied mainly in trout, salmon, turbot, and plaice), (ii) the production of these same two prostaglandins in reproductive tissues for mediating ovulation and spawning behaviors (studied mainly in goldfish), (iii) production of lipoxin and trioxillin-like compounds via LO pathways (studied mainly in salmon, trout, and rockfish), and (iv) production of small aldehydes and ketones via sequential functioning of LO and lyase enzymes (studied in emerald shiners, trout, and sardines). No attempt has been made to arrange the coverage of fish along taxonomic lines as there are too few examples to make such a structure meaningful. A general review by Ruggeri and Thoroughgood has summarized some of these efforts with fish as well as other aquatic fauna.[3]

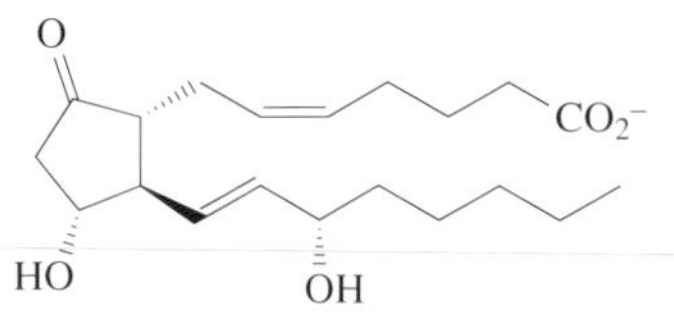

O HO CO_2^- HO OH **(98)**

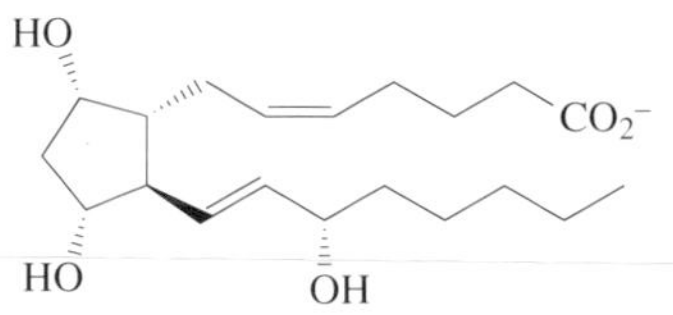

HO CO_2^- HO OH **(99)**

(i) Fish prostaglandins in nonreproductive tissues

The prostaglandin content of various osmoregulatory tissues was surveyed in the trout (*Oncorhynchus mykiss*) and the European Eel (*Anguilla anguilla*), in each case adapted to either fresh water or saltwater.[150] Kidney, gills, and the urinary bladder of both species were found to contain PGE_2 (**98**), PGE_1, $PGF_{2\alpha}$ (**99**), $PGF_{1\alpha}$, PGD_2 (**100**), and the stable metabolite of prostacyclin, 6-keto $PGF_{1\alpha}$, with PGD_2 (**100**) and 6-keto $PGF_{1\alpha}$ predominating in most organs under the two conditions of adaptation. In the gills and kidneys of eels, levels of these two decreased dramatically in seawater-adapted animals, whereas in trout the levels of PGD_2 (**100**) increased substantially in the bladders of seawater-adapted animals. While it is speculated that prostaglandins play osmoregulatory roles in the tissues of fish, the complex pattern observed upon fresh or saltwater adaptation makes it difficult to identify a clear function at the present time.

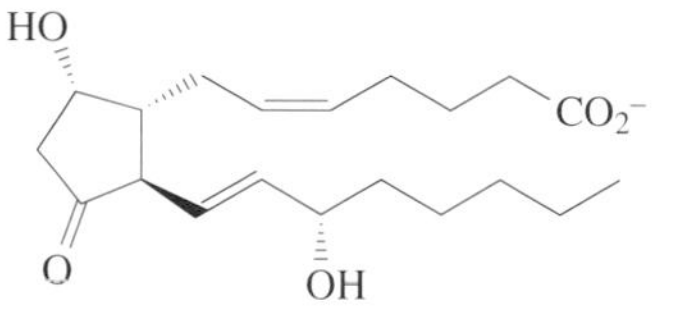

HO CO_2^- O OH

(**100**)

PGE_2 (**98**) biosynthesis has also been detected in the skin of the plaice *Pleuronectes platessa*.[151] AA was preferred over EPA as substrate for PG biosynthesis, and EPA actually caused a dose-dependent inhibition of PGE_2 synthesis from AA. PG biosynthesis was localized to a microsomal fraction and was effectively inhibited by indomethacin (1 mM). Prostaglandins with various degrees of unsaturation have been reported following modification of fatty acid precursors in turbot brain astroglial cells (*Scophthalmus maximus*)[152] and Atlantic salmon gill and kidney (*Salmo salar*).[153]

(ii) Fish prostaglandins involved in reproduction

Prostaglandins in combination with steroids play profound roles in the reproductive biology and behavior of the goldfish *Carassius auratus*, various species of carp, minnow, Japanese loach, the brook trout *Salvelinus fontinalis*,[154] and likely many other fish.[155] While $PGF_{2\alpha}$ (**99**) apparently plays a paracrine role in ovary and follicular development, circulating $PGF_{2\alpha}$ elicits female sexual behavior in the goldfish. There is some evidence that PG metabolites, such as 15-keto $PGF_{2\alpha}$, are excreted by females into the surrounding water and elicit sexual behavior among males with great potency (ca. 10^{-12} M). In the brook trout and the goldfish, production of PGs is effectively blocked by indomethacin,[155] although it has been hypothesized that LO-derived products contribute to the suite of AA metabolites involved in fish reproduction.[154]

(iii) Fish lipoxin, trioxillin, and other lipoxygenase products

Fresh-water trout have been studied extensively for their oxylipins. Early reports tentatively identified two new trihydroxy metabolites deriving from C_{20} and C_{22} precursors, 8,11,12-trihydroxy-5*Z*,9*E*,14*Z*,17*Z*-eicosatetraenoic acid (**101**) and 10,13,14-trihydroxy-4*Z*,7*Z*,11*E*,16*Z*,19*Z*-docosapentaenoic acid (**102**), respectively.[156] The cytosolic fraction of gill tissue was found to produce 12*S*-HETE (**35**), 12*S*-HEPE (**36**), and 13-hydroxydocosahexaenoic acid (**103**) from AA, EPA, and DHA precursors, a process that was strongly inhibited by several well characterized LO inhibitors.[157] Isolation of a diHETE metabolite, 8*R*,15*S*-diHETE (**104**) from trout gill led to the identification of both 12- and 15-LO activities in this tissue. The diHETE metabolite was shown to derive from the sequential action of 15-LO and 12-LO enzymes on AA substrate.[158] Subsequently, rainbow trout macrophages were identified as a rich source of HETE, leukotrienes of the B series, and lipoxins of the A series.[159] The major products were lipoxin A_4 (LXA_4) (**105**), 11-*trans*-LXA_4 (**106**), leukotriene B_4 (LTB_4) (**107**), LTB_5 (**108**), and 12-HETE (**53**) following stimulation by the calcium ionophore A23187. Other positional and stereochemical isomers were present in smaller quantity. However, it was found that the proportion of products varied depending on the macrophage stimulus.

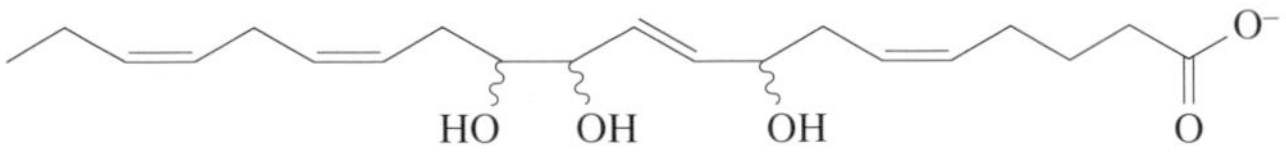

(101)

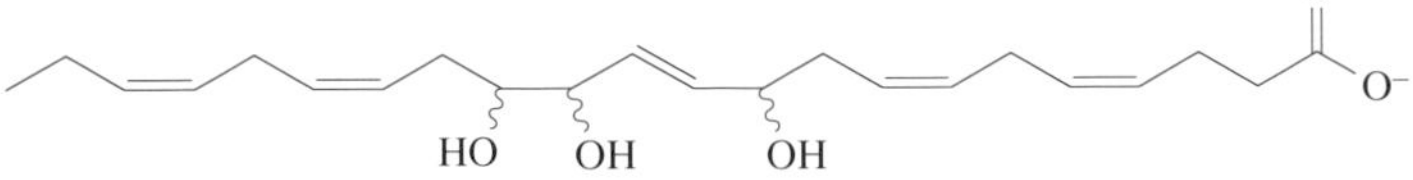

(102)

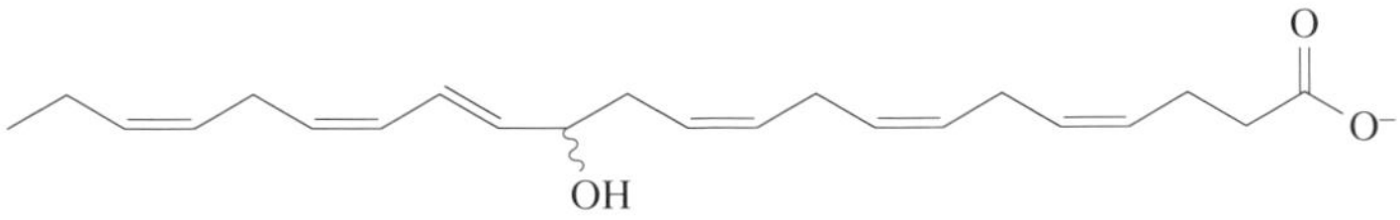

(103)

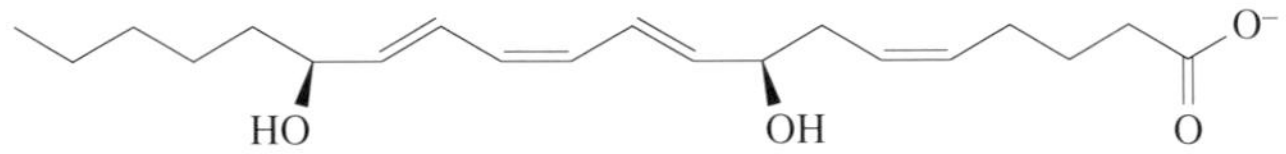

(104)

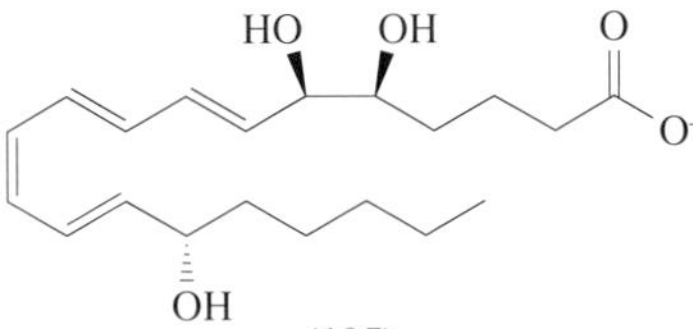

(105) 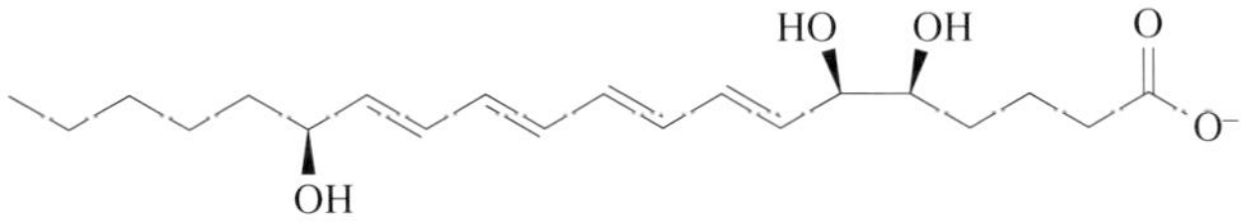 **(106)**

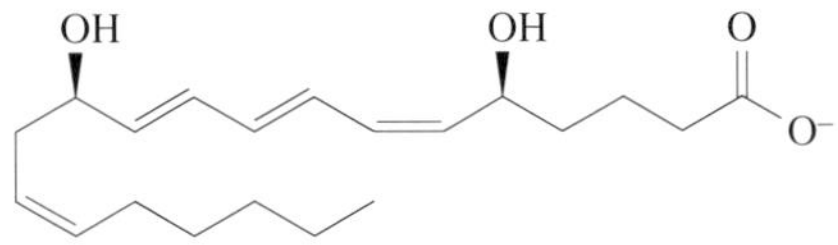

(107) 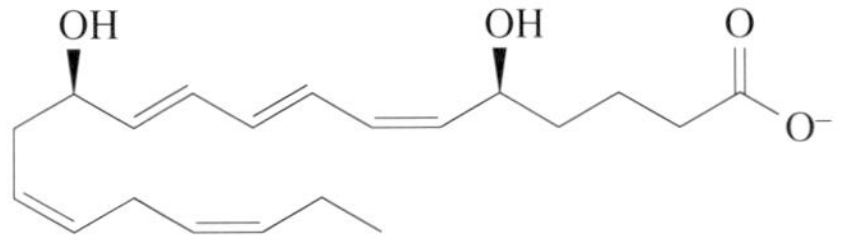 **(108)**

Lipoxins have been characterized as potent chemotactic/chemokinetic agents in the rainbow trout.[159] Subsequently, a survey of head kidney-derived macrophage preparations from five species of fish for their lipoxin-forming ability identified only two, the Atlantic salmon (*S. salar*) and mirror carp (*Cyprinus carpio*), as having this specific metabolic capacity. Moreover, the salmon preparation synthesized mainly lipoxin A isomers (e.g., LXA_4 (**105**) and 11-*trans*-LXA_4 (**106**)) whereas carp macrophages synthesized a mixture of lipoxin A_4 (**105**) and B_4 (**109**).[160] While tilapia (*Oreochromis niloticus*) did not synthesize lipoxins, appreciable amounts of LTB_4 (**107**), LTB_5 (**108**), 12-HETE (**53**), and 15-HETE (**54**) were measured. The other two species, rudd (*Scardinius erythrophthalmus*) and catfish (*Ictalusus punctatus*), did not produce appreciable quantities of any of the oxylipins analyzed in this study. Clearly, oxylipin biosynthesis in fish is highly species specific.

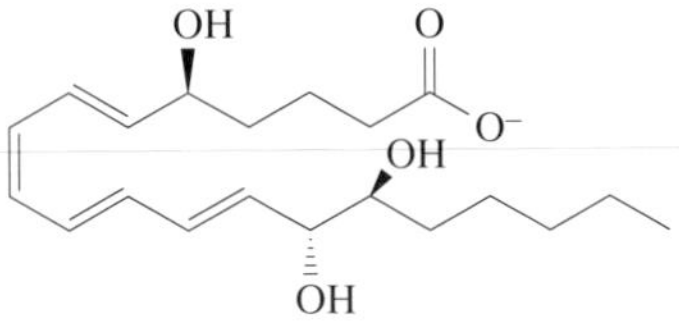

(109)

Detailed studies on the mechanism of formation of LXA_4 (**105**) in rainbow trout macrophages revealed the presence of 5-, 12-, and 15-LO activities.[161] Of a wide variety of potential intermediates in LXA_4 biosynthesis that were provided to trout macrophages, only 5-HPETE led to an increase in LXA_4 production. These results, in combination with alcohol trapping experiments, implicated a 5(6)-epoxy tetraene, possibly 5,6-epoxy-15-hydroxyeicosa-7*E*,9*E*,11*Z*,13*E*-tetraenoic acid (**110**), as a key intermediate in LXA_4 biosynthesis. This is in contrast to the route to LXA_4 production in mammalian granulocytes and mononuclear phagocytes which involve the intermediacy of 15-HPETE, a species not converted to LXA_4 by the fish cells.

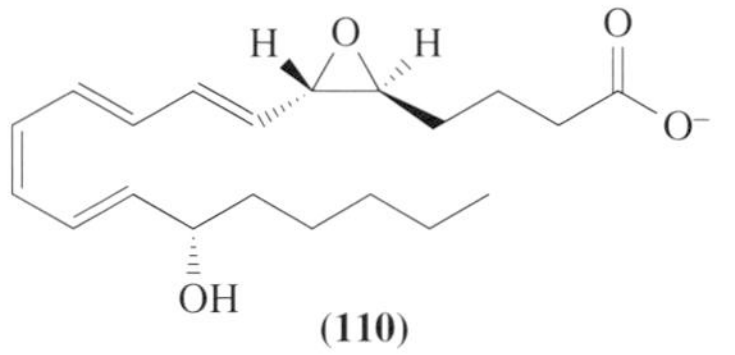

(110)

(iv) Oxylipin fragments as odorous compounds in fish

The fourth oxylipin metabolic theme in fish is their capacity to metabolize endogenous PUFAs (linoleic, α-linolenic, γ-linolenic, AA, EPA, DHA) via sequential action of LO and hydroperoxide lyase pathways to produce small odorous ketones, aldehydes, and unsaturated alcohols. This feature has been studied in several fish species, including trout (*Salmo gairdneri*),[162] emerald shiner (*Notropis atherinoides*),[163] and sardine (*Sardinops melanosticta*).[164] Emerald shiners were exhaustively investigated for their volatile aroma compounds, and many of those characterized are well-known products of hydroperoxide lyase acting on dienyl hydroperoxide precursors, such as 1,5*Z*-octadien-3-ol (**111**).[163] Based on the profile of volatiles characterized, a metabolism of EPA was proposed by putative 11-, 12-, and 15-LO enzymes with subsequent rearrangements or hydroperoxide lyase reactions to produce the observed assortment of odorous molecules.

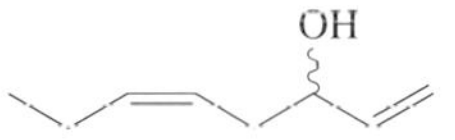

(111)

1.08.4.1.2 Cnidaria (corals)

(i) Plexaura homomalla

Prostaglandins were first discovered in a marine life form in the Caribbean gorgonian *Plexaura homomalla* forma *homomalla*.[165] These soft-bodied colonial animals were found to contain enormous quantities (a variable yield of 1–3.5% of the wet weight of the coral has been reported) of 15*R*-PGA_2 (**112**) and methyl 15*R*-acetoxy-PGA_2 (**113**). The C-15 hydroxyl group of these prostaglandins is epimeric to that found in mammalian systems. The planar structures of these metabolites were characterized by extensive chemical degradation to structurally informative fragments, and subsequent studies established their absolute stereochemistry.[166]

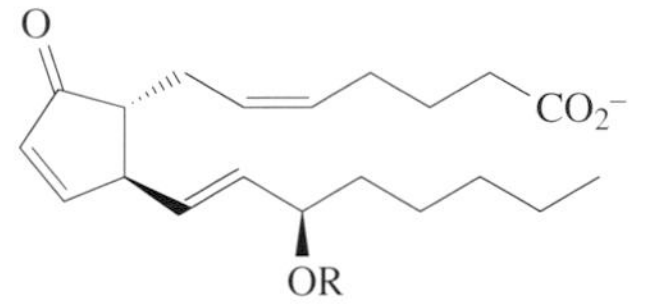

(112) R = H
(113) R = COMe

Following this initial discovery of large quantities of PGA_2 in *P. homomalla*, this coral became the subject of intense chemical,[4] biological,[167] biochemical,[8] and ecological[168] investigation as it was considered a potential industrial source of these bioactive molecules. As a result, a number of analogues of PGA and PGB, both of 15*R* and 15*S* stereochemistry (as well as derivatives of PGE and PGF), have been isolated from the coral and structurally characterized.[8] It appears that the coral's ability to make 15*R* or 15*S* prostanoids correlates with geographical region (15*R* from Florida, 15*S* from Cayman Islands). However, there are reports of individual colonies that produce both C-15 stereoisomers (e.g., both 15*R*-PGB_2 (**114**) and 15*S*-PGB_2 (**115**)).[169] It was also shown that in the coral these prostanoids largely exist in esterified form (methyl, acetoxyl). The hydrolysis of these relatively labile esters is accelerated in the presence of a coral esterase.

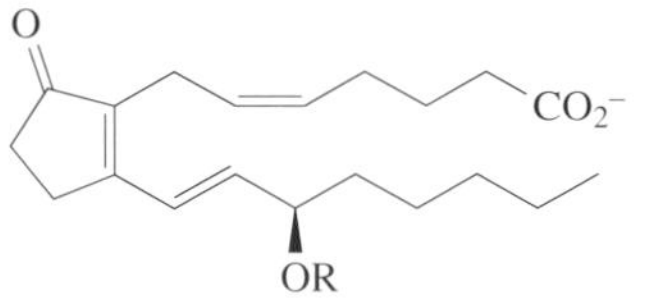

(**114**) 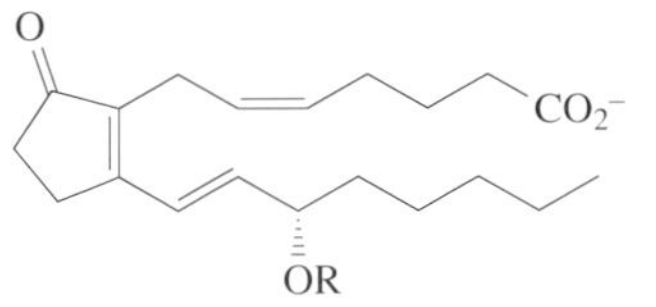 (**115**)

The role of PGA_2 ester production by *P. homomalla* appears to be involved in the chemical defense of these soft bodied organisms against predation by fish, and perhaps invertebrates as well. A series of investigations by Gerhart[170] have convincingly showed that predacious fish are deterred from feeding on these corals because the coral-derived prostaglandins (both 15*R*-PGA_2 (**112**) and 15*S*-PGA_2 (**116**)) are effective emetic agents which subsequently cause a learned aversion in the fish.

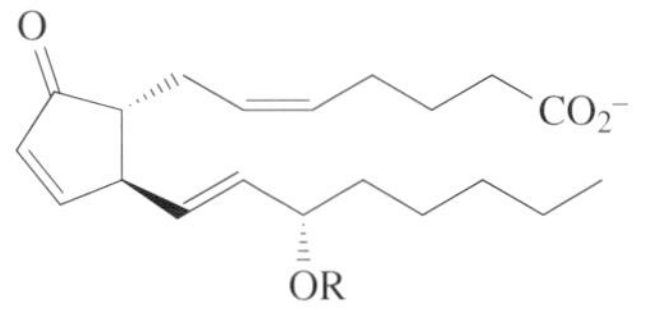

(**116**)

(ii) Prostaglandin biosynthesis in Plexaura homomalla *and* Pseudoplexaura porosa

The biosynthesis of prostaglandin derivatives by *P. homomalla* has been the subject of intense investigation.[8] As these corals contain an appreciable biomass of symbiotic microalgae in their tissues, it was questioned whether the animal or plant cells were responsible for this metabolism. Isolation of the zooxanthellae and demonstration of their inability to metabolize exogenous labeled substrate, as well as the absence of any endogenous prostanoids in these cells, indicated that they are not the site of biosynthesis or storage of prostaglandins. However, it was proposed that the algae could contribute PUFA precursor (AA) into the coral prostaglandin biosynthetic pathway.[171]

Numerous investigations of the biosynthetic pathway of these coral prostanoids, principally by the Corey group at Harvard and the Brash group at Vanderbilt, have occurred since the 1970s; however, we still do not have a complete understanding of the prostaglandin biosynthetic pathway in *P. homomalla*.[8] Early investigations illustrated that the coral utilizes a route other than the prostaglandin H synthase (cyclooxygenase) pathway typical of mammalian prostaglandin biosynthesis.[172] Subsequently, in work with *P. homomalla*[173] as well as through parallel investigations with several other species of coral (notably, *Pseudoplexaura porosa* and *Clavularia viridis*),[174] a pathway has emerged beginning with AA and its initial conversion to 8*R*-HPETE by an 8*R*-LO, dehydration to an allene oxide, and then cyclization to carbocyclic oxylipins (Scheme 14). This is clearly related to the pathway in higher plants which leads to the phytohormones phytodienoic acid (**77**) and jasmonic acid (**89**).[91] However, the exact nature of the substrate (i.e., whether it possesses any preexisting oxidation) which feeds into this coral "allene oxide" pathway is unknown.[8,173] A variety of other new prostanoids have subsequently been isolated from biosynthetic experiments with soft corals.[175] These biosynthetic investigations have been hampered, however, by the inability to develop a coral preparation which is able to perform the entire biosynthetic sequence beginning with arachidonic acid and finishing with chiral PGA_2. An addition to this story is the purification, molecular cloning, and structure of the 8*R*-lipoxygenase from *P. homomalla*.[176] The coral LO shows considerable structural relatedness to the *S*-lipoxygenases of higher plants and mammals in the location of key iron-binding residues (histidine residues) as well as overall size and amino acid composition. From these sequence data, it was concluded that *S*- and *R*-lipoxygenases are of the same family of enzymes, and that their stereospecificity is a result of the different fit of the substrate in comparable binding sites with resultant oxygenation occurring with opposite stereospecificity.[176]

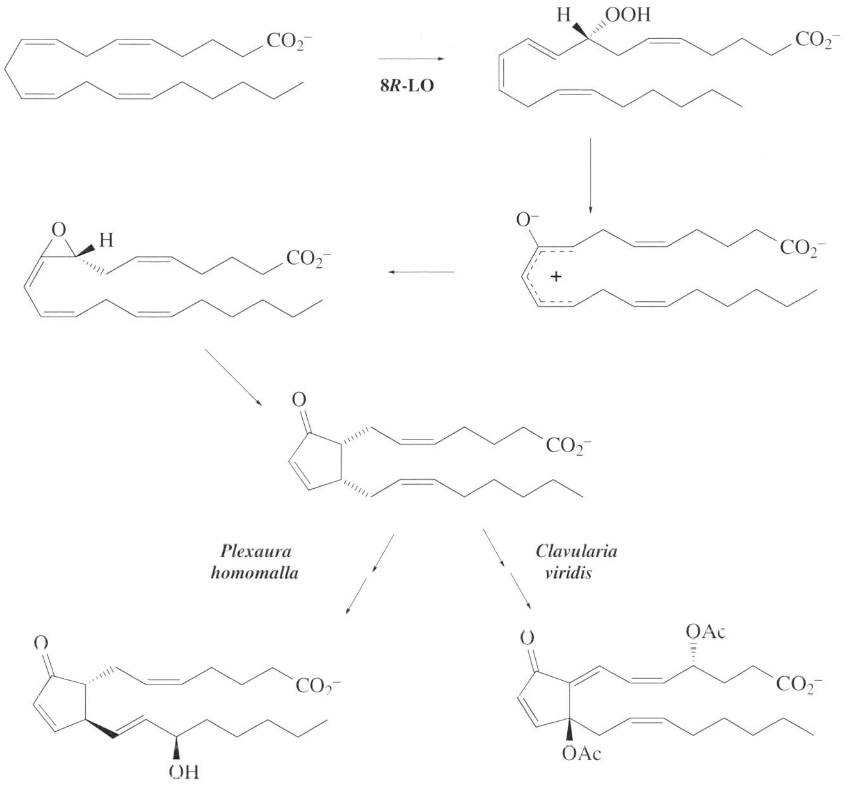

$8R$-LO

Plexaura homomalla

Clavularia viridis

(112)

(117)

Scheme 14

(iii) Clavularia viridis

The Okinawan soft coral *Clavularia viridis* has been a rich source of structurally novel and biologically active oxylipins. Reported simultaneously by two groups in 1982 were a mixture of four geometrical isomers, clavulones I–III (e.g., (**117**) and claviridenone-a.[177,178] The absolute stereochemistry of these prostanoids was deduced using degradative and chiroptical techniques by both groups the following year, and several stereospecific syntheses of the clavulones have confirmed their structures and stereochemistry.[8] A number of derivatives of these clavulone natural products have subsequently been isolated from *C. viridis* and defined using similar techniques, including the 20-acetoxy derivatives of clavulones I–III, C-10 chlorine-containing metabolites chlorovulones I–IV (e.g., (**118**)) the bromo- (**119**) and iodo- (**120**) analogues of chlorovulone I, and the 10,11-epoxide analogue of chlorovulone I.[8,179]

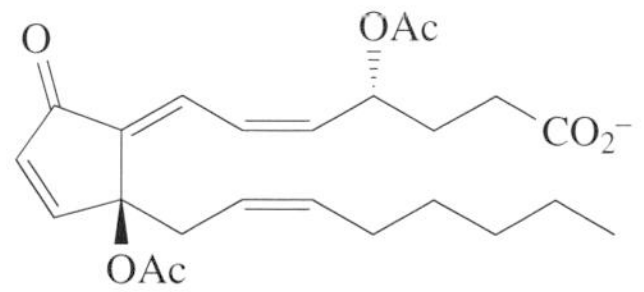

(117)

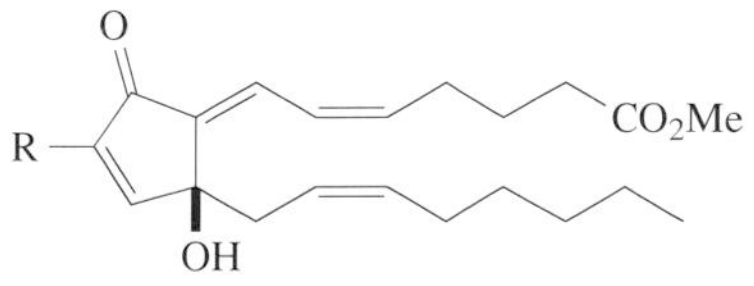

(**118**) R = Cl
(**119**) R = Br
(**120**) R = I

As noted above, the biosynthesis of these *C. viridis* prostanoids has been the source of intense interest and experimentation. Various cell free preparations of *C. viridis* (as well as other corals, e.g., *P. porosa*)[180] were shown to transform exogenous arachidonic acid to $8R$-hydroperoxy-eicosatetraenoic acid ($8R$-HPETE, (**121**))[174] This presumed 8-LO product was then shown to be converted by coral enzymes into a cyclopentanoid product, termed "preclavulone-A" (**122**), and its planar structure and relative stereochemistry were established from extensive chemical degradations and synthesis. The close structural similarity of preclavulone-A to 12-oxo-phytodienoic acid (**77**) and *iso*-jasmonic acid (**81**)[91] suggested it might be formed via analogous transformations.[181] Hence, in similarity to PG production in *P. homomalla*, it was envisioned (see Scheme 14) that lipoxygenase-derived 8*R*-HPETE (**121**) is first transformed to an allene oxide, then opened to an oxidopentadienyl cation intermediate, which finally cyclizes to preclavulone-A (**122**). While preclavulone-A possesses an obviously close structural relationship to the naturally occurring clavulones, it has subsequently been found that the intermediate allene oxide undergoes spontaneous nonenzymatic rearrangement yielding racemic preclavulone-A.[182] Thus, it remains uncertain what relationship the *in vitro* synthesis of racemic preclavulone-A has to the *in vivo* biosynthesis of chiral clavulones. The clavulones and their halogenated derivatives, along with the structurally related punaglandins, have shown remarkable cytotoxicity and antiproliferative activity against several transformed cell lines, as well as *in vivo* antitumor effects.[183]

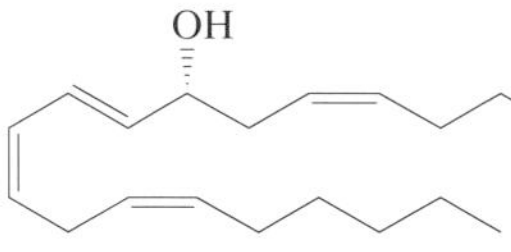

(121) 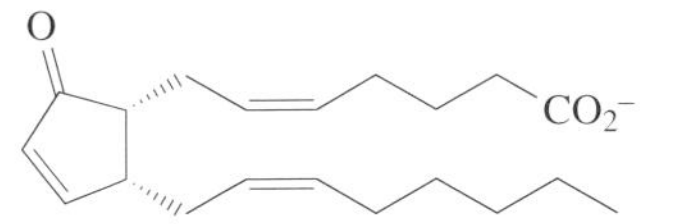 **(122)**

(iv) Gersemia fruticosa

The White Sea soft coral *Gersemia fruticosa* represents an exception to the study of coral prostaglandin biosynthesis.[184] Whereas in all other species studied it has been impossible to demonstrate the *in vitro* biosynthesis of prostaglandins, an acetone powder preparation of *G. fruticosa* was readily developed which made chiral PGD_2 (**100**), PGE_2 (**98**), $PGF_{2\alpha}$ (**99**), and 15-keto-$PGF_{2\alpha}$ (**123**) from exogenous radioactive AA. Analysis of the endogenous coral metabolites showed a substantial amount of 8-HETE (**124**) and 11*R*-HETE (**125**), as well as of the same four prostaglandins obtained above, all present in the coral as free acids. Indomethacin was found to be ineffective as an inhibitor of coral PG biosynthesis. Subsequently, the prostaglandin endoperoxide PGG_2 (**126**) was identified as an unstable product of the coral acetone powder preparation, and this was found to be converted by nonenzymatic processes to the above recorded optically active PGs.[185] Based on the lack of activity of indomethacin and the lack of any detectable prostaglandin-hydroperoxidase activity ($PGG_2 \rightarrow PGH_2$), it was speculated that either a novel COX-isozyme or a novel LO pathway is responsible for PG biosynthesis in *G. fruticosa*.

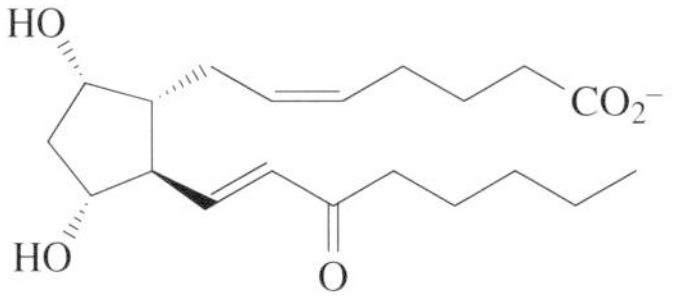

(123) 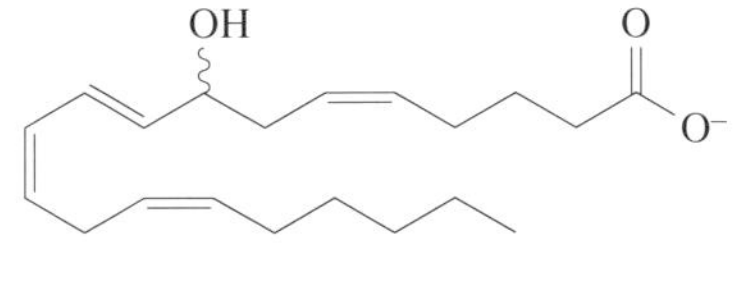 **(124)**

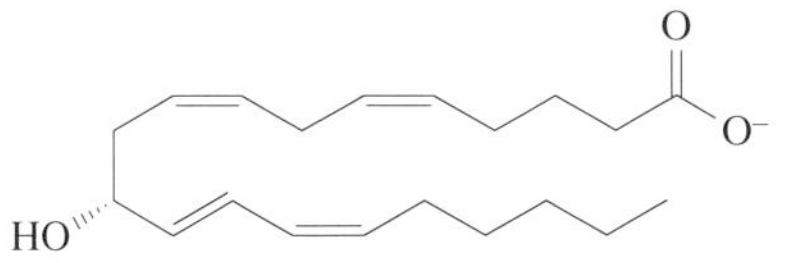

(125) 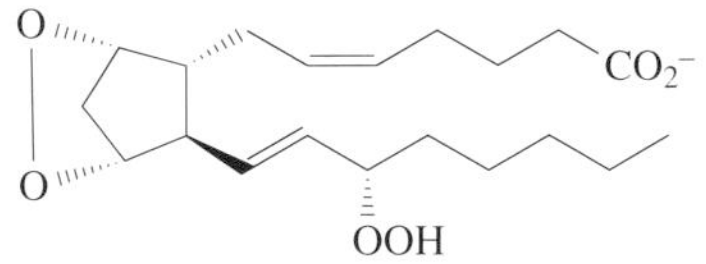 **(126)**

(v) Hydra vulgaris *and* Hydratinia echinata

Both freshwater and marine species of hydrozoans have been found to metabolize PUFAs via LO-pathways to simple hydroxyacids which exert potent activities on tentacle regeneration (in *Hydra vulgaris*),[186] control of metamorphosis (in *Hydratinia echinata*),[187] and feeding responses (in *H. vulgaris*).[188] In the freshwater hydroid *H. vulgaris*, the best studied case, the endogenous substrate for hydroxyacid formation is phospholipid-derived α-linolenic acid (9*Z*,12*Z*,15*Z*-octadecatrienoic acid) which is released by a hydrozoan phospholipase A_2 (PLA_2).[189] This is subject to oxidation by a putative ω10-LO to produce 9*R*-α-HPOTE (**127**), and following reduction or dehydration, 9*R*-α-

HOTE and 9-KOTE. A fraction of these metabolites becomes reesterified to *H. vulgaris* phospholipids, and may act as a reservoir of these oxidized PUFA metabolites.[190] While the biological activity of these α-linolenic acid-derived metabolites has not been characterized, due to their lack of availability, other hydroxyacids of a related nature (e.g., 11*R*-HETE and 9*S*-HOD) were characterized as relatively potent in enhancing the average number of tentacles produced during tentacle regeneration.[186] Metamorphosis in the marine hydroid *H. echinata* is inhibited by either of the well-known LO inhibitors nordihydroguaiaretic acid (NDGA) or eicosatetraynoic acid (ETYA), but not by indomethacin or acetylsalicylic acid.[187] Two HETE products were characterized as endogenous AA products; 8*R*-HETE (**124**) and 12-HETE (**53**). A cytosolic fraction of the hydrozoan was found to contain the 8*R*-HETE-forming ability, and this was also inhibitable by LO inhibitors but not COX inhibitors.

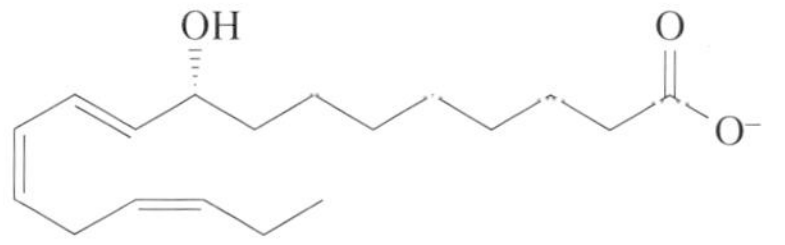

OH O O⁻

(**127**)

1.08.4.1.3 *Urochordata (tunicates)*

(i) Didemnum candidum *and* Didemnum mosleyi

Two species of the tunicate *Didemnum*, *D. candidum* and *D. mosleyi*, have been a source of closely related oxylipin metabolites. Following clarification of the structures of one series, the ascidiatrienolides (e.g., (**128**)) from *D. candidum* collected in Florida,[191,192] and confirmation of the related didemnilactone structures from the other species by synthesis,[193] a consistent metabolic theme emerges from these related animals which involves metabolism of AA by an 8*R*-LO. It has been proposed that the resulting hydroperoxide is converted to an 8,9-LTA_4-like intermediate which is subsequently attacked by the carboxylate at C-9 to form a 10-membered lactone ring.[8]

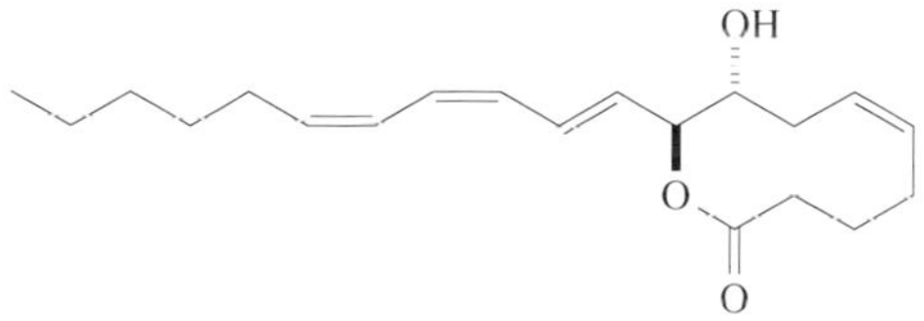

OH O O

(**128**)

1.08.4.1.4 *Porifera (sponges)*

(i) Halichondria okadai

The sponge *Halichondria okadai* was a source of cyclopropyl ring-containing oxylipins (**129,130**)[194] which showed the same relationship of functional groups within the fatty acid chain as seen for the constanolactones (**34**).[195] However, the positions of these alcohol, olefin, cyclopropyl ring, and lactone groups were shifted to C-8—C-15 vs. C-5—C-12 in the constanolactones. The planar structures of halicholactone (**129**) and neohalicholactone (**130**) were originally formulated based on spectroscopic data and partial degradation, and relative stereochemistry provided by X-ray diffraction analysis of neohalicholactone (**130**).[196] While the relative stereochemistry is identical in these metabolites to that found in the constanolactones (**34**), the absolute stereochemistry at C-15 was shown to be *R* by degradation to 1,2*R*-diacetoxyheptane and comparison with authentic materials. It is possible that the halicholactones originate from transformations analogous to those of the constanolactones, albeit initiated through a 15-LO introduced hydroperoxide.

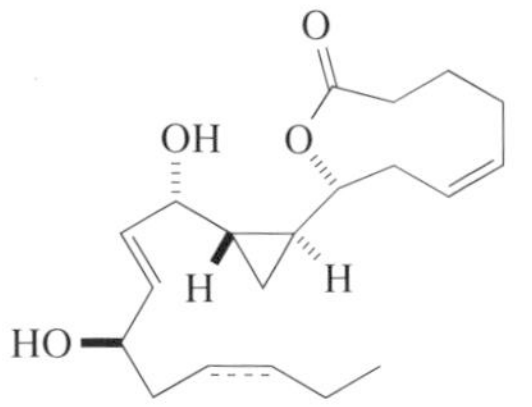

(129) ω7
(130) ω3

A potential complication in the structure of halicholactone and neohalicholactone was introduced by isolation and structural description of "neohalicholactone" from the Pacific brown alga *Laminaria sinclairii* (Phaeophyceae).[197] However, clarification of this has come from synthetic studies of these metabolites,[198] and indicates that the stereochemistry at C-15 in sponge-derived neohalicholactone is correct as originally formulated.[199]

1.08.4.1.5 Mollusca

(i) General

PGE_2 (**98**) and PGA_2 (**116**) have been detected in a variety of molluscs by bioassay and radioimmunoassay techniques, including the bivalve *Modiolus demissus*, mussels (*Mytilis edulis*), scallops (*Patinopecten yessoensis*), and oysters (*Crassostrea gigas*)[8]. These have been isolated in most cases from the gill or reproductive tissues of these various molluscan species, and in the latter case, appears to have a role in reproductive maturation and egg release. A similar role in egg spawning has been ascribed to the putative production of PGs in the red abalone *Haliotis rufescens*.[200] The gill tissue of a freshwater snail *Ligumia subrostrata* has been found to contain both a LO activity leading to production of 5-HETE (**51**) and 12-HETE (**53**) and a cyclooxygenase activity leading to production of PGE_2 (**98**).[201] Inhibition of these pathways by well described LO or COX inhibitors confirmed in each case their metabolic origins. It is speculated that the LO metabolites may be involved in ionic regulation.

(ii) Aplysia kurodai

A highly unusual dimeric oxylipin was isolated from the sea hare *Aplysia kurodai*, aplydilactone (**131**), which possessed weak phospholipase A_2 activating activity.[202] Aplydilactone is unique among oxylipin natural products in its dimeric nature, with the two units likely both deriving from EPA. The monomeric units are highly similar in structure to those of the constanolactones (**34**),[195] cyclopropane-lactones from a red alga.[195] As sea hares in general, and *Aplysia* spp. in particular, are well known to assimilate and sequester the unique secondary metabolites of their algal diets, presumably as a way of enhancing their own defense against predation, it is likely that aplydilactone or its monomeric precursors also originate in an algal source. Dimerization could be occurring by enzymatic processes either in the alga or the sea hare, or conceivably, could occur via nonenzymatic processes in the sea hare digestive gland (Scheme 15).

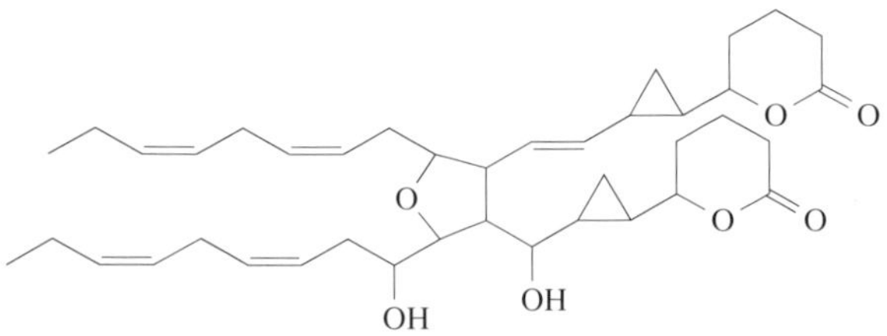

(131)

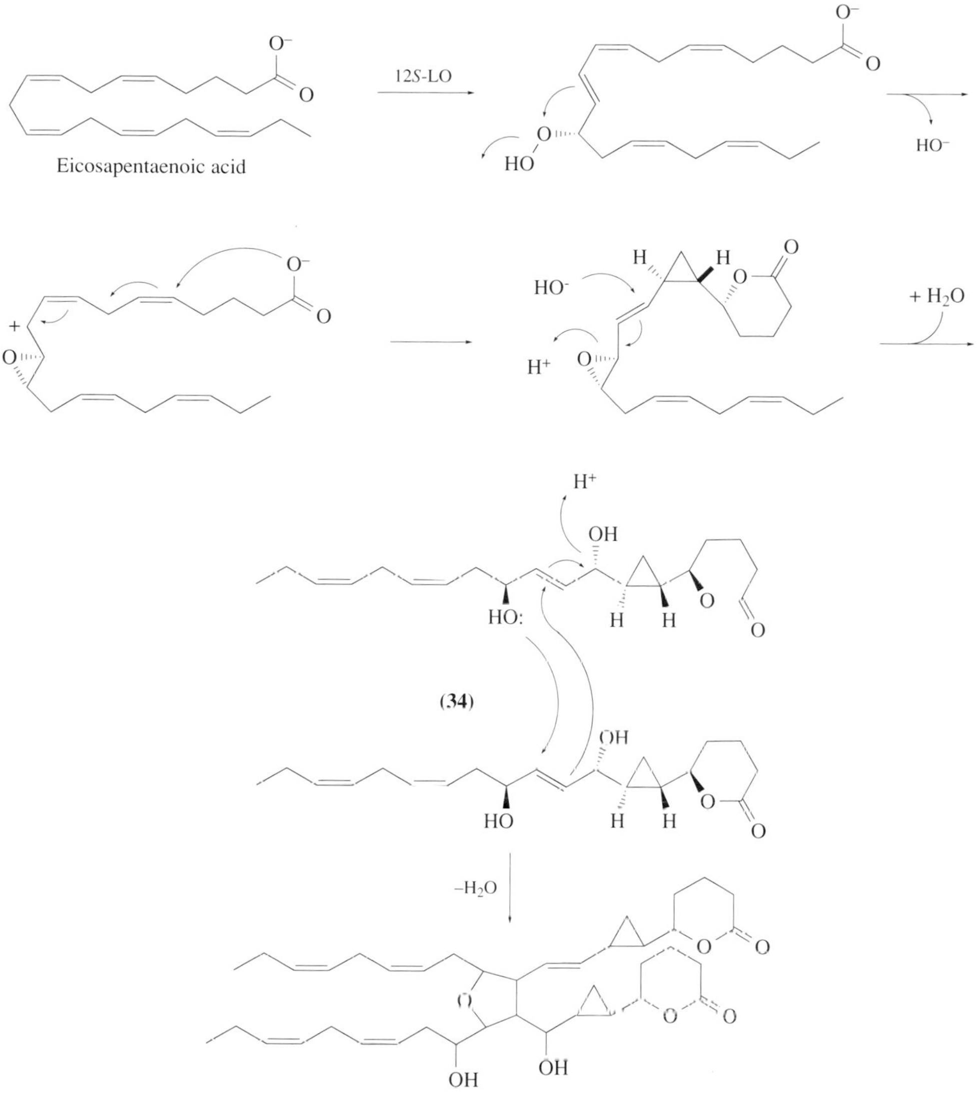

Scheme 15

(iii) Aplysia californica

The role of AA metabolites in neurochemical signal transduction in the sea hare *Aplysia californica* has been studied extensively.[8,148,203] Both COX and LO pathways appear operative in *Aplysia* neurons as exogenous arachidonic acid is converted into both PGE_2 (**98**) and $PGF_{2\alpha}$ (**99**), and 5-HETE (**51**), 12-HETE (**53**), 12-KETE (**132**), and hepoxillin A_3 (**133**). There is considerable evidence that these 12-LO metabolites function in *Aplysia* as first and/or second messengers. The pattern of putative 12-LO metabolites in *Aplysia* suggests conversion of AA into 12-HPETE and subsequent peroxidase-type reduction to 12-HETE (**53**), dehydration to 12-KETE (**132**), or rearrangement to hepoxillin A_3 (**133**). While only 12*S*-HPETE is naturally produced in *Aplysia* neurons, products from both stereoisomers have potent effects on the sea hare neurochemical signalling pathways.[204]

(iv) Tethys fimbria

The mantles of the nudibranch *Tethys fimbria* were shown to contain a rich diversity of prostanoids, including PGE_2 (**98**) and PGE_3 (**134**), the corresponding lactones, PGE_3-1,15-lactone (**135**)

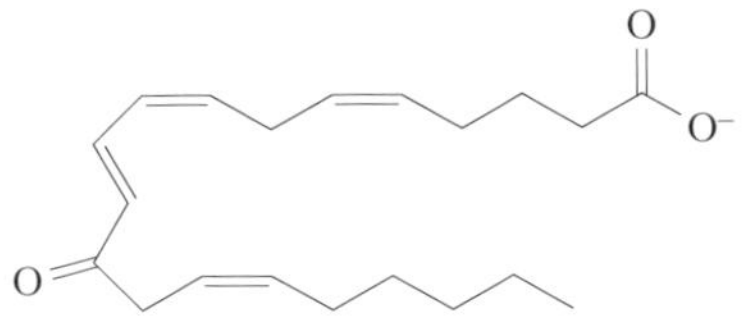

(132) 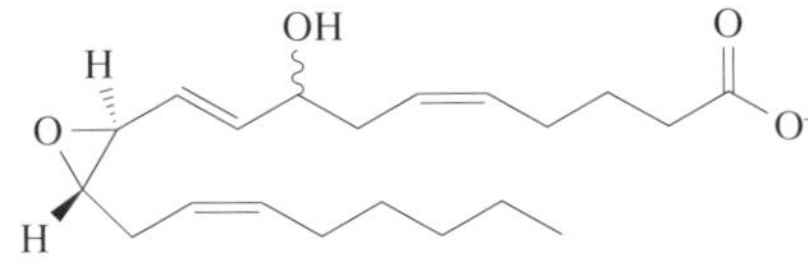 (133)

and PGE_3-1,15-lactone-11-acetate, as well as PGA_2-1,15-lactone (**136**) and PGA_3-1,15-lactone. These were identified on the basis of spectroscopic analyses and comparison with synthetically produced derivatives. Similar 1,15-lactones in the PGF-series were also characterized. It seems that these PGE derivatives serve several roles in the nudibranch; they appear to (i) be involved in oocyte maturation or hatching, (ii) function as ichthyotoxic agents in defense of fish predation, and (iii) mediate contraction of the dorsal appendage and induce mucous secretion. Biosynthetic studies of prostaglandin production in *T. fimbria* have painted a complicated picture in which the original site of production is the dorsal mantle, particularly during sexual maturation. The prostanoids are then transported to the ovotestis, the site of ultimate storage and utilization.[205,206] However, it is unknown whether cyclooxygenase or lipoxygenase metabolism is responsible for construction of the prostanoid skeleton in *T. fimbria*.

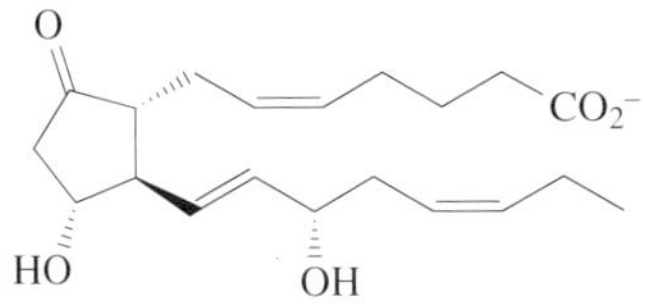

(134) 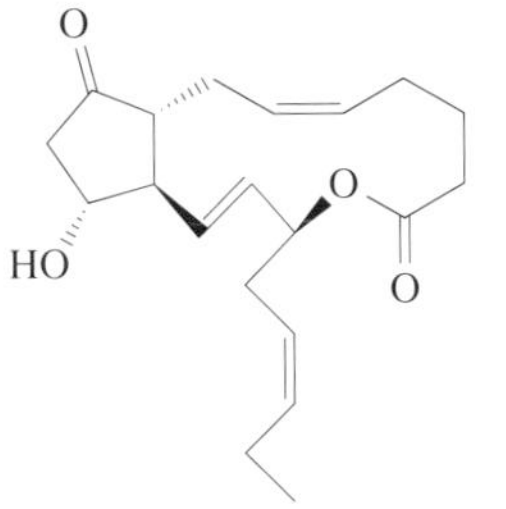 (135) 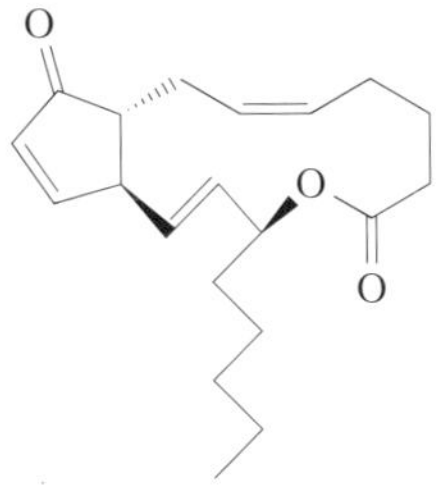 (136)

1.08.4.1.6 Arthropods

(i) Rhithropanopeus harrisii *and* Carcinus maenas

Oxylipin metabolites of several sorts have either been isolated from various Crustacea or have been shown to have potent biological effects in their tissues.[8] For example, externally applied PGF_2 and PGF_2 have both been reported to induce abdominal flexion in the estuarine crab *Rhithropanopeus harrisii*, a behavior observed during both copulation and egg laying, at concentrations as low as 10^{-12} M.[207] On the other hand, the putative hydroperoxide lyase products 1-octen-3-ol and 1,5*Z*-octadien-3-ol (**111**) have been isolated from several species of prawns and one species of lobster, and are reported to impart metallic flavors when added to crustacean tissue.[208] However, a detailed study of the oxylipins produced following stimulation of crab blood cells by the calcium ionophore A23187 has shown a substantial capacity to produce various HETE products with 8*R*-HETE (**121**) predominating.[209] In these studies with the shore crab *Carcinus maenas* both 8*R*-HETE (**121**) and 8*R*-HEPE (**137**) predominated while 5*R*,*S*-, 9*S*- (**138**), 11*R*- (**125**), 12*S*- (**35**), and 15*R*,*S*-HETE were produced in smaller quantities along with small amounts of prostaglandins (e.g., PGE_2 (**98**), TXB_2 (**139**)). When these same cells were both stimulated and provided an exogenous source of AA, 8*R*-HETE (**121**) was the major product. The responses of this system to several inhibitors was consistent with 8-HETE being formed via a LO pathway.

(137) 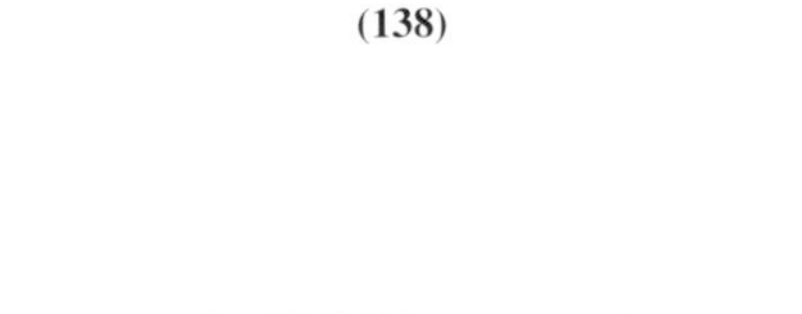 (138)  (139)

(ii) Limulus polyphemus

Blood cells (granular amebocytes) of the horseshoe crab *Limulus polyphemus*, a primitive arthropod of the subphylum Chelicerata, have been found to produce a variety of HETE products following stimulation by the calcium ionophore A23187.[210] In this study, HPLC was used in conjunction with electrospray ionization mass spectrometry to characterize and quantitate oxylipin production without derivatization. The predominating product was 8-HETE (**124**) with small quantities of other hydroxyacids also being recorded. It has been speculated that 8-HETE may function as a chemotactic agent for amebocyte migration during inflammatory events.

(iii) Balanus balanoides *and* Elminius modestus

The fertilized eggs of barnacles (e.g., *Balanus balanoides* and *Elminius modestus*) are maintained within the adult mantle cavity until a hatching factor elicits their release as stage I nauplii larvae. Because the egg masses are in direct contact with seawater, the hatching factor released from nonparental barnacles can cause the hatching in brooding adults. The chemical nature of this factor has been difficult to ascertain due to its small concentration in seawater and the production of multiple oxylipin products from *in vitro* experiments. However, it appears that barnacles have an 8*R*-LO, in similarity to other arthropods, which results in the production of 8*R*-HETE (**121**) and 8*R*-HEPE (**137**) under *in vitro* experimental conditions. Interestingly, the enantiomeric substance 8*S*-HEPE does not have barnacle egg hatching activity.[211] Confirmation of the structure and biological properties of the putative hatching factor as 8*R*-HEPE (**137**) has been accomplished through total synthesis.[212] However, a second series of more polar oxylipins, regioisomeric trihydroxy fatty acids such as 10,11,12-trihydroxy-5,8,14,17-eicosatetraenoic acid (**140**), have also been implicated as barnacle hatching factors. Extraction of water from undisturbed adult barnacles that had a demonstrated presence of hatching factor led to the conclusion that these trihydroxy fatty acids are of greater *in vivo* relevance to the process of barnacle egg hatching than are the simple hydroxy-acids.[213]

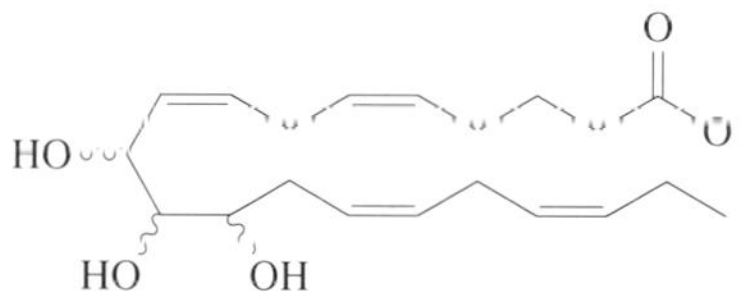

O
HO
O
HO OH

(**140**)

1.08.4.1.7 Echinoderms

Starfish and sea urchins have been studied in some detail for their oxylipin metabolites, the pathways of their biosynthesis, and their roles in echinoderm reproductive biology. Studies with several genera of starfish have shown their capacity to produce a variety of HETE products, including 8*R*-HETE (**121**), 9-HETE (**49**), 12-HETE (**53**), and 15-HETE (**54**).[214–216] However, it is 8*R*-HETE (**121**) which is of apparent biological relevance, being a very potent and selective inducer of oocyte maturation.[216] Additional isolation efforts identified trioxillin A_4 (8*R*,11*S*,12*R*-trihydroxy-eicosa-5*Z*,9*E*,14*Z*,17*Z*-tetraenoic acid (**141**)) in several species.[217] It is intriguing to note that this same metabolite, although of uncertain stereochemistry, was isolated from the trout (see Section 1.08.4.1.1(iii)) (**101**).[156] Biosynthetic studies of oxylipins in the starfish *Evasterias troschelii* showed the formation of 8*R*-HPETE via a putative LO reaction.[218] Though not revealed by the pattern of metabolites detected from whole animal isolation studies, subsequent biosynthetic investigations showed that this intermediate 8*R*-HPETE was not only reduced to the corresponding 8*R*-HETE (**121**) product, but that it was also converted by an allene oxide pathway to cyclopentanone, α-ketol and γ-ketol products, and by a hydroperoxide lyase pathway to C_7 and C_{13} products (e.g., cleavage of C-7—C-8). These latter products were not active in promoting oocyte maturation.[218]

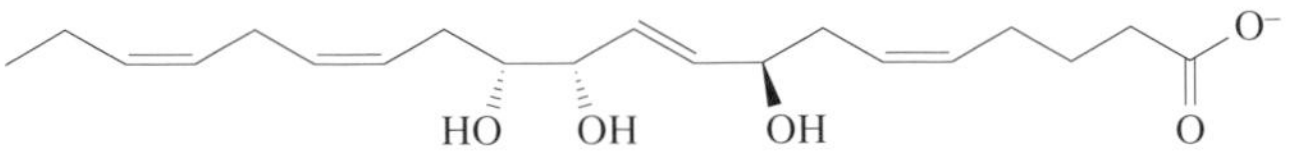

HO OH OH O O⁻

(141)

Sea urchins have been known for some time to either produce or assimilate PG- and HETE-derivatives[8]. Structure work with HETEs produced from incubating either AA or EPA with egg homogenates led to the characterization of 11*R*- (**125**) and 12*R*-HETE (**142**), and 11*R*- and 12*R*-HEPE.[219] These were subsequently shown to derive via peroxidase-type reduction of the corresponding hydroperoxides.[220] While a definitive role for these oxylipins in sea urchin biology is not yet available, there is some evidence that they have a role in preventing polyspermic fertilization.[148]

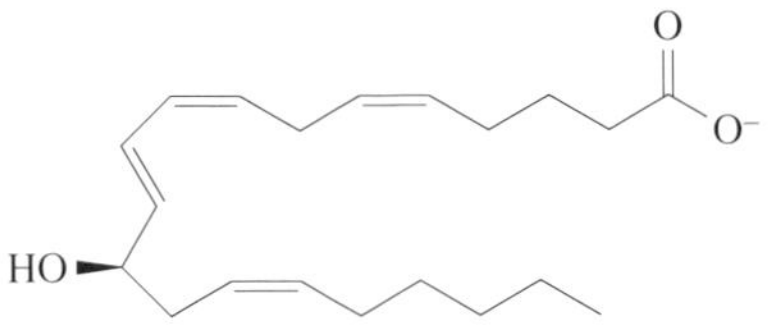

O O⁻ HO

(142)

1.08.4.2 Terrestrial Animals

1.08.4.2.1 Insects

Insects are one of the better studied groups of lower animal life forms for the occurrence and biological function of oxylipins, and several excellent reviews on this subject have appeared.[5,6,221,222] Most of these studies have focused on the biological role of prostaglandins and related AA-derived eicosanoids in insects. On the other hand, the chemistry and biochemistry of insect oxylipins is less well studied. For example, involvement of oxylipins has largely been implicated through application of specific inhibitors, and in some cases, simultaneous replacement of candidate eicosanoids with resulting restoration of the physiological process under study (rescue experiments). Most specific structural identifications of oxylipins from insects have been based on HPLC or TLC retention times. With the wealth of data on the biological roles of oxylipins in insects, the area is ripe for a detailed exploration of the chemical structures of the involved compounds and the biochemical pathways of their production.

The putative biological roles of prostaglandins in insects, as illustrated in the examples below, are in categories with direct parallel to their functions in mammalian systems. These include regulation of reproduction and development, ion and water transport, hemocytic and other immune defense functions, and temperature homeostasis. A series of experiments involving application of various PGs provided convincing evidence that PGE_2 upregulates egg-laying behavior in various cricket species.[223] In other insect species, PGs do not influence egg-laying, but nevertheless have been observed to quantitatively increase following mating.[224] $PGF_{2\alpha}$ has been implicated in the ability of mosquitos to successfully emerge from pupal molt and develop flight abilities.[225] Dexamethasone inhibition of PLA_2 in tobacco hornworm larvae was found to be highly effective at inhibiting the larvae's ability to eliminate bacteria following artificial infection, resulting in increased larval mortality. Application of exogenous AA to the inhibitor-compromised larvae reversed this sensitivity. However, the nature of the putative eicosanoid mediators was not investigated.[226] Sweating set points in desert cicadas, a thermoregulating adaptation to desert temperatures, have been shown to be modified with application of eicosanoid biosynthesis inhibitors.[227] PGs have also been implicated in the regulation of basal fluid secretion rates in the Malpighian tubules of the mosquito *Aedes aegypti*.[228]

1.08.4.2.2 Amphibians

To date, the major group of amphibians to be studied for their oxylipin-producing abilities is that of frogs and toads. Here, arachidonic acid appears to be the most common fatty acid precursor and, depending on the species, is metabolized by either a 5-LO or 12-LO route. For example, in the aquatic toad *Xenopus laevis*, 12-HETE (stereochemistry not determined, (**53**)) is the major monohydroxy product[229] while LTB_4 (**107**) is the major dihydroxy product.[230] A new leukotriene, 5*S*,12*R*-dihydroxy-6,10-*trans*-8,14-*cis*-eicosatetraenoic acid (**143**) has also been reported.[230] In contrast, from the bullfrog *Rana catebeiana*, 12-HETE and a variety of sulfidopeptides, such as LTD_4 (**144**), are produced.[231] Interestingly, when this latter organism was challenged by an experimentally induced septicemia, the products were 5-HETE (stereochemistry not determined, (**51**)) and LTB_4 (**107**).[232] Based on these few reports, the major pathways of oxylipin metabolism in amphibians appear to be the 5-LO and 12-LO conversion of AA to 5- and 12-HETE, LTB_4 and related compounds, and the sulfidopeptide leukotrienes.

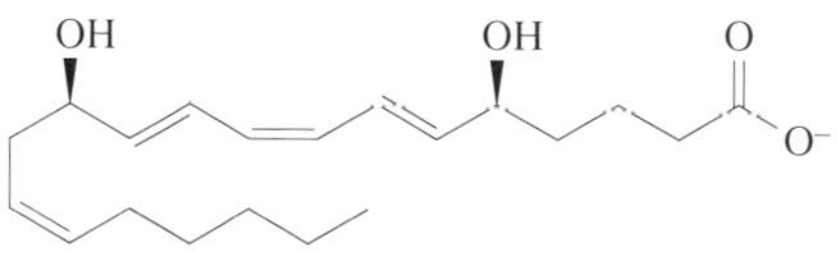

(**143**) 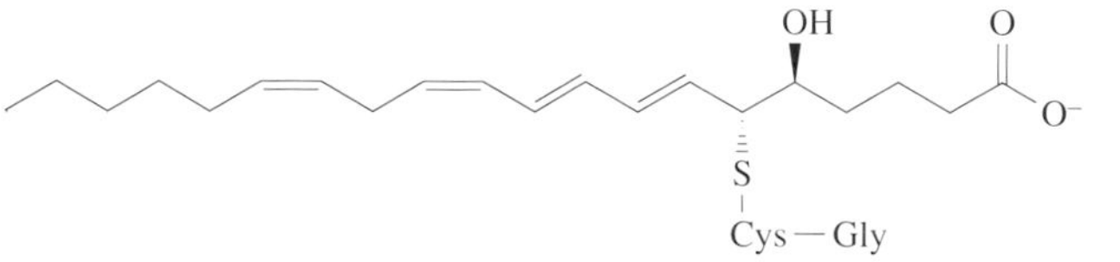 (**144**)

1.08.4.2.3 *Arachnids*

Ixodid ticks have been shown by a variety of techniques, including GC-MS, to produce and secrete from their salivary glands a variety of putative cyclooxygenase products.[233] Interestingly, the ultimate source of the AA for prostaglandin synthesis by the tick is the mammalian host. Free AA is preferentially taken up by tick salivary gland tissue and incorporated into phosphatidylethanolamine and phosphatidylcholine phospholipids. Upon initiation of feeding by the tick, a PLA_2 is stimulated to release AA in the salivary gland where it is mainly converted to PGE_2, $PGF_{2\alpha}$, and PGD_2 (**100**). There is some evidence that the tick COX is substantially different than mammalian COX, and may represent a unique target for future antitick therapies. The released prostaglandins are believed to facilitate feeding by the tick through modulation of the host physiology, such as suppression of the host immune and inflammatory responses and stimulation of hyperemia at the feeding lesion.

1.08.4.2.4 *Annelids and other worms*

A number of studies have explored the production of oxylipins by parasitic organisms that have been provided suitable PUFA precursors.[234] In general, while helminths are not themselves capable of *de novo* biosynthesis of PUFAs, they efficiently incorporate the free fatty acids of their hosts into their phospholipids as well as elongate and further desaturate these to species such as AA.[235] Metabolism of AA has been reported for a number of species in the Protozoa, Trematodes, Cestodes, and Nematodes.[235] For the most part, the structural identities of the oxylipins produced have depended on HPLC retention times sometimes in concert with radio immuno assay (RIA). Hence, there is still some uncertainty concerning the identities of the metabolites produced, and this has given rise to conflicting reports in the literature.[236] The best data available to date for Trematodes implicates the production of a "15-HETE-like" species by a LO activity obtained from *Schistosoma mansoni*.[234] The LO was found to be calcium-dependent and inhibited by several mammalian and plant LO inhibitors. Antisera to different mammalian LOs were used to characterize two immunoreactive bands from *S. mansoni* with molecular weights of ~78 kDa and 100 kDa. Other LO products, including 5-HETE (**51**) and LTB_4 (**107**), have been reported from *S. mansoni*.[235] There is GC-MS evidence of PGD_2 (**100**) production during heartworm (*Dirofilaria immitis*) infection of dogs.[237] In other groups of parasites, prostaglandins (e.g., PGE_2, PGI_2, PGD_2) are the principal products and have been detected by mostly HPLC/RIA methods.[235]

The role of oxylipin metabolism in parasites is incompletely characterized; however, an attractive proposal has emerged in which the parasite produces immunologically active oxylipins so as to inhibit the host immune response, particularly T- and B-cell functions. Alternatively, parasite success may benefit in some cases from a host inflammatory response, in part aided by parasite-derived proinflammatory oxylipins. It is still not clear if parasites use oxylipins in the regulation of their own physiological processes,[235] although there has been at least one report that suggests this may be the case.[238]

1.08.5 CONCLUSION

It is hoped that, for the dedicated reader, this chapter has been a useful review, not only in identifying some perhaps not previously recognized trends in the metabolism of PUFAs by nonmammalian species, and for the first time reviewing this class of chemistry in a truly comprehensive taxonomic sense in one place, but also in identifying some areas for further exploration. It is the author's firm conviction that much novel mechanistic biochemistry and enzymology in this chemical class remains to be discovered from nonmammalian species. This seems richest in the primitive plants, such as algae and primitive aquatic plants, and in the marine invertebrates. Several precedents exist which support the notion that such studies will not only be of value in understanding these pathways and their roles in plants and animals of economic importance, but that they will provide the knowledge to recognize that the same or related pathways are operational in mammalian tissues. Ultimately, knowledge of the biosynthetic pathways of these unusual oxylipins is of fundamental value as it identifies new enzymatic transformations, a necessary first step in identifying interesting genetic material for use by the techniques of molecular biology and engineering. Knowledge of these genes, including their molecular organization and regulation, is critical to their ultimate utilization in a variety of ways, for example in pharmaceutical drug development via molecular biological approaches.

1.08.6 REFERENCES

1. W. H. Gerwick, M. Moghaddam, and M. Hamberg, *Arch. Biochem. Biophys.*, 1991, **290**, 436.
2. H. W. Gardner, *Biochim. Biophys. Acta*, 1991, **1084**, 221.
3. B. Ruggeri and C. A. Thoroughgood, *Mar. Ecol. Prog. Ser.*, 1985, **23**, 301.
4. G. L. Bundy, *Adv. Prostaglandin Thromboxane Leukotrine Res.*, 1985, **14**, 229.
5. D. W. Stanley-Samuelson, *Am. Zool.*, 1994, **34**, 589.
6. D. W. Stanley-Samuelson and V. K. Pedibhotla, *Insect Biochem. Mol. Biol.*, 1996, **26**, 223.
7. W. H. Gerwick, *Biochim. Biophys. Acta*, 1994, **1211**, 243.
8. W. H. Gerwick, D. G. Nagle, and P. J. Proteau, *Top. Curr. Chem.*, 1993, **167**, 117.
9. M. D. Lewis and R. Menes, *Tetrahedron Lett.*, 1987, **43**, 5129.
10. J. S. Todd and W. H. Gerwick, *J. Nat. Prod.*, 1995, **58**, 586.
11. J.-L. Beneytout, R. H. Andrianarison, S. Rakotoarisoa, and M. Tixier, *Plant Physiol.*, 1989, **91**, 367.
12. R. H. Andrianarison, J.-L. Beneytout, and M. Tixier, *Plant Physiol.*, 1989, **91**, 1280.
13. N. Murakami, H. Shirahashi, A. Nagatsu, and J. Sakakibara, *Lipids*, 1992, **27**, 776.
14. J. H. Cardellina, II and R. E. Moore, *Tetrahedron*, 1980, **36**, 993.
15. D. M. Losel and M. Sancholle, in "Lipids of Pathogenic Fungi," eds. R. Prassad and M. A. Ghannoum, CRC Press, Boca Raton, FL, 1996, p. 27.
16. J. L. Kerwin, in "Ecology and Metabolism of Plant Lipids," eds. G. Fuller and W. D. Nes, American Chemical Society, 1987, p. 331.
17. D. L. Hawksworth, P. M. Kirk, B. C. Sutton, and D. N. Pegler, "Ainsworth & Bisby's Dictionary of the Fung," 8th edn., International Mycological Institute, CAB International, 1995.
18. N. Totani and K. Oba, *Lipids*, 1987, **22**, 1060.
19. J. L. Gellerman and H. Schlenk, *Biochim. Biophys. Acta*, 1979, **573**, 23.
20. J. L. Kerwin, *Prostaglandins Leukotrines Med.*, 1986, **23**, 173.
21. R. P. Herman and C. A. Herman, *Prostaglandins*, 1985, **29**, 819.
22. W. S. Bowers, H. C. Hoch, P. H. Evans, and M. Katayama, *Science*, 1986, **232**, 105.
23. L. J. Morris, *Biochem. J.*, 1970, **118**, 681.
24. W. D. Nes, *Lipids*, 1988, **23**, 9.
25. H. W. Knoche, *Lipids*, 1971, **6**, 581.
26. L. J. Morris, *Lipids*, 1967, **3**, 260.
27. Y. Matsuda, T. Beppu, and K. Arima, *Biochim. Biophys. Acta*, 1978, **530**, 439.
28. T. Satoh, Y. Matsuda, M. Takashio, K. Satoh, T. Beppu, and K. Arima, *Agric. Biol. Chem.*, 1976, ***40***, 953.
29. N. Nakayama, A. Takemae, and H. Shoun, *J. Biochem.*, 1996, **119**, 435.
30. L. O. Narhi, and A. J. Fulco, *J. Biol. Chem.*, 1986, **261**, 7160.
31. S. M. Picardi and P. Issenberg, *J. Agric. Food Chem.*, 1971, **21**, 959.
32. R. Tressl, D. Bahri, and K.-H. Engel, *J. Agric. Food Chem.*, 1982, **30**, 89.
33. M. Wurzenberger and W. Grosch, *Biochim. Biophys. Acta*, 1984, **795**, 163.

34. F. X. Wollard, B. J. Burreson, and R. E. Moore, *J. Chem. Soc., Chem. Commun.*, 1975, 486.
35. I. D. Brodowsky, M. Hamberg, and E. H. Oliw, *J. Biol. Chem.*, 1992, **267**, 14 738.
36. M. Hamberg, L.-Y. Zhang, I. D. Brodowsky, and E. H. Oliw, *Arch. Biochem. Biophys.*, 1994, **309**, 77.
37. I. D. Brodowsky and E. H. Oliw, *Biochim. Biophys. Acta*, 1992, **1124**, 59.
38. I. D. Brodowsky and E. H. Oliw, *Biochim. Biophys. Acta*, 1993, **1168**, 68.
39. M. Hamberg, *Biochim. Biophys. Acta*, 1986, **876**, 688.
40. M. Hamberg, R. P. Herman, and U. Jacobsson, *Biochim. Biophys. Acta*, 1986, **879**, 410.
41. M. Hamberg, C. A. Herman, and R. P. Herman, *Biochim. Biophys. Acta*, 1986, **877**, 447.
42. W. H. Gerwick, M. W. Bernart, M. F. Moghaddam, Z. D. Jiang, M. L. Solem, and D. G. Nagle, *Hydrobiologia*, 1990, **204/205**, 621.
43. W. H. Gerwick, M. W. Bernart, Z. D. Jiang, M. F. Moghaddam, D. G. Nagle, P. J. Proteau, M. L. Wise, and M. Hamberg, *Dev. Ind. Micro.*, 1993, **1**, 369.
44. W. H. Gerwick, *Chem. Rev.*, 1993, **93**, 1807.
45. W. H. Gerwick, *Lipids*, 1996, **31**, 1215.
46. W. H. Gerwick and M. W. Bernart, in "Marine Biotechnology, Volume 1: Pharmaceutical and Bioactive Natural Products," eds. D. H. Attaway and O. R. Zaborsky, Plenum, New York, 1993, p. 101.
47. D. C. Zimmerman and B. A. Vick, *Lipids*, 1973, **8**, 264.
48. B. A. Vick and D. C. Zimmerman, *Plant Physiol.*, 1989, **90**, 125.
49. M. W. Bernart, G. G. Whatley, and W. H. Gerwick, *J. Nat. Prod.*, 1993, **56**, 245.
50. S. J. Rochfort, R. Watson, and R. J. Capon, *J. Nat. Prod.*, 1996, **59**, 1154.
51. W. Boland, *Proc. Natl. Acad. Sci. USA*, 1995, **92**, 37.
52. K. Stratmann, W. Boland, and D. G. Muller, *Tetrahedron*, 1993, **49**, 3755.
53. K. Kurata, K. Taniguchi, K. Shiraishi, N. Hayama, I. Tanaka, and M. Suzuki, *Chem. Lett.*, 1989, 267.
54. J. S. Todd, P. J. Proteau, and W. H. Gerwick, *J. Nat. Prod.*, 1994, **57**, 171.
55. K. Kurata, K. Tanaguchi, K. Shiraishi, and M. Suzuki, *Phytochemistry*, 1993, **33**, 155.
56. J. S. Todd, P. J. Proteau, and W. H. Gerwick, *Tetrahedron Lett.*, 1993, **34**, 7689.
57. R. G. Warren, R. J. Wells, and J. F. Blount, *Aust. J. Chem.*, 1980, **33**, 891.
58. R. A. Barrow and R. J. Capon, *Aust. J. Chem.*, 1990, **43**, 895.
59. M. D. Higgs and L. J. Mulheirn, *Tetrahedron*, 1981, **37**, 4259.
60. M. W. Bernart and W. H. Gerwick, *Tetrahedron Lett.*, 1988, **29**, 2015.
61. M. W. Bernart, W. H. Gerwick, E. F. Corcoran, A. Y. Lee, and J. Clardy, *Phytochemistry*, 1992, **31**, 1273.
62. D. G. Nagle and W. H. Gerwick, *Tetrahedron Lett.*, 1990, **31**, 2995.
63. M. F. Moghaddam, W. H. Gerwick, and D. L. Ballantine, *Prostaglandins*, 1989, **37**, 303.
64. Z. D. Jiang, and W. H. Gerwick, *Phytochemistry*, 1991, **30**, 1187.
65. M. F. Moghaddam, W. H. Gerwick, and D. L. Ballantine, *J. Biol. Chem.*, 1990, **265**, 6126.
66. M. B. Bernart and W. H. Gerwick, *Phytochemistry*, 1994, **36**, 1233.
67. M. Hamberg, *Lipids*, 1992, **27**, 1042.
68. Z. D. Jiang and W. H. Gerwick, *Phytochemistry*, 1990, **29**, 1433.
69. M. Hamberg and W. H. Gerwick, *Arch. Biochem. Biophys.*, 1993, **305**, 115.
70. T. A. Dix and L. J. Marnett, *J. Biol. Chem.*, 1985, **260**, 5351.
71. Z. D. Jiang and W. H. Gerwick, *Lipids*, 1997, **32**, 231.
72. M. A. Graber, W. H. Gerwick, and D. P. Cheney, *Tetrahedron Lett.*, 1996, **37**, 4635.
73. A. Guerriero, M. D'Ambrosio, and F. Pietra, *Helv. Chim. Acta*, 1990, **73**, 2183.
74. W. H. Gerwick, P. Asen, and M. Hamberg, *Phytochemistry*, 1993, **34**, 1029.
75. M. Hamberg, W. H. Gerwick, and P. Asen, *Lipids*, 1992, **27**, 487.
76. M. Hamberg, *Biochem. Biophys. Res. Commun.*, 1992, **188**, 1220.
77. J. R. Burgess, R. I. de la Rosa, R. S. Jacobs, and A. Butler, *Lipids*, 1991, **26**, 162.
78. M. V. Mikhailova, D. L. Bemis, M. L. Wise, W. H. Gerwick, J. N. Norris, and R. S. Jacobs, *Lipids*, 1995, **30**, 583.
79. A. Lopez and W. H. Gerwick, *Lipids*, 1987, **22**, 190.
80. A. Lopez and W. H. Gerwick, *Tetrahedron Lett.*, 1988, **29**, 1505.
81. M. L. Wise, M. Hamberg, and W. H. Gerwick, *Biochemistry*, 1994, **33**, 14 223.
82. M. L. Wise, Ph.D. Thesis, Oregon State University, 1995.
83. M. L. Wise and W. H. Gerwick, *Biochemistry* 1997, **36**, 2985.
84. A. Groweiss and Y. Kashman, *Experientia*, 1978, 299.
85. A. Batna, J. Scheinkonig, and G. Spiteller, *Biochim. Biophys. Acta*, 1993, **1166**, 171.
86. R. Wang and Y. Shimizu, *J. Chem. Soc., Chem. Commun.*, 1990, 413.
87. Atlantic Research Laboratory Technical Report 57, NRCC 29086, Atlantic Research Laboratory, NRC, Halifax, Canada, 1988.
88. R. Wang, Y. Shimizu, J. R. Steiner, and J. Clardy, *J. Chem. Soc., Chem. Commun.*, 1993, 379.
89. M. Hamberg, *J. Lipid Mediators*, 1993, **6**, 375.
90. B. A. Vick and D. C. Zimmerman, in "The Biochemistry of Plants, A Comprehensive Treatise, Vol. 9, Lipids: Structure and Function," ed. P.K. Stumpf, Academic Press, Orlando, FL, 1987, p. 53.
91. M. Hamberg and H. W. Gardner, *Biochim. Biophys. Acta*, 1992, **1165**, 1.
92. H. Gundlack, M. J. Muller, T. M. Kutchan, and M. H. Zenk, *Proc. Natl. Acad. Sci. USA*, 1992, **89**, 2389.
93. A. J. Enyedi, N. Yalpani, P. Silverman, and I. Raskin, *Cell*, 1992, **70**, 879.
94. A. M. Jones, *Science*, 1993, **263**, 183.
95. S. H. Doares, T. Syrovets, E. W. Weiler, and C. A. Ryan, *Proc. Natl. Acad. Sci. USA*, 1995, **92**, 4095.
96. S. Blechert, W. Brodschelm, S. Holder, L. Kammerer, T. M. Kutchan, M. J. Mueller, Z.-Q. Xia, and M. H. Zenk, *Proc. Natl. Acad. Sci. USA*, 1995, **92**, 4099.
97. H. Yamane, Y. Sato, N. Takahashi, K. Takeno, and M. Furuya, *Agric. Biol. Chem.*, 1980, **44**, 1697.
98. Y. Masaoka, M. Sakakibara, and K. Mori, *Agric. Biol. Chem.*, 1982, **46**, 2319.
99. P. Karunen, L. Heikkila, and E. Kalviainen, *Phytochemistry*, 1987, **26**, 1723.
100. Y. Shinmen, K. Katoh, S. Shimizu, S. Jareonkitmongkol, and H. Yamada, *Phytochemistry*, 1991, **30**, 3255.

101. G. Kohn, S. Demmerle, O. Vandekerkhove, E. Hartmann, and P. Beutelmann, *Phytochemistry*, 1987, **26**, 2271.
102. T. Ichikawa, M. Namikawa, K. Yamada, K. Sakai, and K. Kondo, *Tetrahedron Lett.*, 1983, **32**, 3337.
103. M. Toyota, F. Nagashima, and Y. Asakawa, *Phytochemistry*, 1988, **27**, 2603.
104. R. T. Van Aller, L. R. Clark, G. F. Pessoney, and V. A. Rogers, *Lipids*, 1983, **18**, 617.
105. R. T. Van Aller, G. F. Pessoney, V. A. Rogers, E. J. Watkins, and H. G. Leggett, In "The Chemistry of Allelopathy, ACS Symposium Series 268," ed. A. C. Thompson, ACS, Washington, DC, 1985, p. 387.
106. M. Hamberg, *Lipids*, 1991, **26**, 407.
107. M. Hamberg and G. Hamberg, *Plant Physiol.*, 1996, **110**, 807.
108. A. Ishimaru and I. Yamazaki, *J. Biol. Chem.*, 1977, **252**, 6118.
109. M. Hamberg and G. Hamberg, *Phytochemistry*, 1996, **42**, 729.
110. B. Janistyn, *Phytochemistry*, 1990, **29**, 2453.
111. M. Hamberg, *Anal. Biochem.*, 1971, **43**, 515.
112. M. Hamberg, *Adv. Prostaglandin Thromboxane Leukotrine Res.*, 1990, **21**, 117.
113. M. Hamberg, *Biochim. Biophys. Acta*, 1987, **920**, 76.
114. M. Hamberg and M. A. Hughes, *Adv. Prostaglandin Thromboxane Leukotrine Res.*, 1989, **19**, 64.
115. M. Hamberg and P. Fahlstadius, *Arch. Biochem. Biophys.*, 1989, **276**, 518.
116. B. A. Vick and D. C. Zimmerman, *Biochem. Biophys. Res. Commun.*, 1984, **111**, 470.
117. P. Monaco and L. Previtera, *Phytochemistry*, 1987, **26**, 745.
118. L. Previtera and P. Monaco, *J. Nat. Prod.*, 1987, **50**, 807.
119. D. C. Zimmerman and P. Feng, *Lipids*, 1978, **13**, 313.
120. L. Crombie and D. O. Morgan, *J. Chem. Soc.*, 1988, 558.
121. A. N. Grechkin, R. A. Kuramshin, E. Y. Safonova, S. K. Latypov, and A. V. Ilyasov, *Biochim. Biophys. Acta*, 1991, **1086**, 317.
122. A. R. Brash, S. W. Baertschi, C. D. Ingram, and T. M. Harris, *Proc. Natl. Acad. Sci. USA*, 1988, **85**, 3382.
123. A. R. Brash, S. W. Baertschi, and T. M. Harris, *J. Biol. Chem.*, 1990, **265**, 6705.
124. W.-C. Song and A. R. Brash, *Science*, 1991, **253**, 781.
125. A. R. Brash and W.-C. Song, *J. Lipid. Mediators Cell Signalling*, 1995, **12**, 275.
126. W. C. Song, S. W. Baertschi, W. E. Boeglin, T. M. Harris, and A. R. Brash, *J. Biol. Chem.*, 1993, **268**, 6293.
127. H. Rabinovitch-Chable, J. Cook-Moreau, J. C. Breton, and M. Rigaud, *Biochem. Biophys. Res. Commun.*, 1992, **188**, 858.
128. W. Minor, J. Steczko, B. Stec, Z. Otwinowski, J. T. Bolin, R. Walter, and B. Axelrod, *Biochem.*, 1996, **35**, 10 687.
129. J. C. Boyington, B. J. Gaffney, and L. M. Amzel, *Science*, 1993, **260**, 1482.
130. M. H. Glickman and J. P. Klinman, *Biochemistry*, 1995, **34**, 14 077.
131. M. H. Glickman and J. P. Klinman, *Biochemistry*, 1996, **35**, 12 882.
132. B. A. Vick and D. C. Zimmerman, *Biochem. Biophys. Res. Commun.*, 1983, **111**, 470.
133. M. Hamberg and G. Hamberg, *Arch. Biochem. Biophys.*, 1990, **283**, 4.
134. H. W. Gardner and M. Hamberg, *J. Biol. Chem.*, 1993, **268**, 6971.
135. E. Blee and F. Durst, *Arch. Biochem. Biophys.*, 1987, **254**, 43.
136. T. Galliard and D. R. Phillips, *Biochem. J.*, 1971, **124**, 431.
137. T. Galliard and D. R. Phillips, *Biochem. J.*, 1972, **129**, 743.
138. T. Galliard, D. R. Phillips, and D. J. Frost, *Chem. Phys. Lipids*, 1973, **11**, 173.
139. T. Galliard and J. A. Matthew, *Biochim. Biophys. Acta*, 1975, **398**, 1.
140. L. Crombie, D. O. Morgan, and E. H. Smith, *J. Chem. Soc., Chem. Commun.*, 1987, 502.
141. G. R. Reddy, P. Reddanna, C. C. Reddy, and W. R. Curtis, *Biochem. Biophys. Res. Commun.*, 1992, **189**, 1349.
142. L. Crombie and D. O. Morgan, *J. Chem. Soc., Chem. Commun.*, 1987, 503.
143. P. Fahlstadius and M. Hamberg, *J. Chem. Soc., Perkin Trans. 1*, 1990, 2027.
144. D. Choi, R. M. Bostock, S. Avdiushko, and D. F. Hildebrand, *Proc. Natl. Acad. Sci. USA*, 1994, **91**, 2329.
145. D.-E. Sok and M. R. Kim, *J. Agric. Food Chem.*, 1994, **42**, 2703.
146. T. Nomura and H. Ogata, *Biochim. Biophys. Acta*, 1976, **431**, 127.
147. E. J. Christ and D. A. Van Dorp, *Biochim. Biophys. Acta*, 1972, **270**, 537.
148. D. W. Stanley-Samuelson, *Am. J. Physiol.*, 1991, R849.
149. E. J. Corey, *Pure Appl. Chem.*, 1987, **59**, 269.
150. J. A. Brown, C. J. Gray, G. Hattersley, and J. Robinson, *Gen. Comp. Endocrinol.*, 1991, **84**, 328.
151. A. A. Anderson, T. C. Fletcher, and G. M. Smith, *Comp. Biochem. Physiol.*, 1981, **70C**, 195.
152. J. G. Bell, D. R. Tocher, and J. R. Sargent, *Biochim. Biophys. Acta*, 1994, **1211**, 335.
153. J. G. Bell, B. M. Farndale, J. R. Dick, and J. R. Sargent, *Lipids*, 1996, **31**, 1163.
154. F. W. Goetz, P. Duman, M. Ranjan, and C. A. Herman, *J. Exp. Zool.*, 1989, **250**, 196.
155. P. W. Sorensen and F. W. Goetz, *J. Lipid Mediators*, 1993, **6**, 385.
156. B. German, G. Bruckner, and J. Kinsella, *Prostaglandins*, 1983, **26**, 207.
157. J. B. German, G. G. Brukner, and J. E. Kinsella, *Biochim. Biophys. Acta*, 1986, **875**, 12.
158. J. B. German and R. Berger, *Lipids*, 1990, **25**, 849.
159. T. R. Pettitt, A. F. Rowley, S. E. Barrow, A. I. Mallet, and C. J. Secombes, *J. Biol. Chem.*, 1991, **266**, 8720.
160. A. F. Rowley, *Biochim. Biophys. Acta*, 1991, **1084**, 303.
161. A. F. Rowley, P. Lloyd-Evans, S. E. Barrow, and C. N. Serhan, *Biochemistry*, 1994, **33**, 856.
162. J. B. German, R. G. Berger, and F. Drawert, *Chem. Mikrobiol. Technol. Lebensm.*, 1991, **13**, 19.
163. D. B. Josephson, R. C. Lindsay, and D. A. Stuiber, *J. Agric. Food Chem.*, 1984, **32**, 1347.
164. S. Mohri, S.-Y. Cho, Y. Endo, and K. Fujimoto, *Agric. Biol. Chem.*, 1990, **54**, 1889.
165. A. J. Weinheimer and R. L. Spraggins, *Tetrahedron Lett.*, 1969, 5185.
166. F. M. Bayer and A. J. Weinheimer, "Prostaglandins from Plexaura homomalla: Ecology, utilization and conservation of a major medical marine resource. A symposium," University of Miami Press, FL, 1974, 165 pp.
167. W. P. Schneider, R. D. Hamilton, and L. E. Rhuland, *J. Am. Chem. Soc.*, 1972, **94**, 2122.
168. J. C. Coll, *Chem. Rev.*, 1992, **92**, 613.
169. W. P. Schneider, G. L. Bundy, F. H. Lincoln, E. G. Daniels, and J. E. Pike, *J. Am. Chem. Soc.*, 1977, **99**, 1222.

170. D. J. Gerhart, *Am. J. Physiol.*, 1991, **260**, R839.
171. E. J. Corey and W. N. Washburn, *J. Am. Chem. Soc.*, 1974, **96**, 934.
172. E. J. Corey, H. E. Ensley, M. Hamberg, and B. Samuelsson, *J. Chem. Soc., Chem. Commun.*, 1975, 277.
173. W. C. Song and A. R. Brash, *Arch. Biochem. Biophys.*, 1991, **290**, 427.
174. E. J. Corey, M. D'Alarcao, S. P. T. Matsuda, P. T. Lansbury, Jr., and Y. Yamada, *J. Am. Chem. Soc.*, 1987, **109**, 289.
175. E. J. Corey, S. P. T. Matsuda, R. Nagata, and M. B. Cleaver, *Tetrahedron Lett.*, 1988, **29**, 2555.
176. A. R. Brash, W. E. Boeglin, M. S. Chang, and B.-H. Shieh, *J. Biol. Chem.*, 1996, **271**, 20 949.
177. H. Kikuchi, Y. Tsukitani, K. Iguchi, and Y. Yamada, *Tetrahedron Lett.*, 1982, **23**, 5171.
178. M. Kobayashi, T. Yasuzawa, M. Yoshihara, H. Akutsu, Y. Kyogoku, and I. Kitagawa, *Tetrahedron Lett.*, 1982, **23**, 5331.
179. K. Iguchi, S. Kaneta, K. Mori, and Y. Yamada, *Chem. Pharm. Bull.*, 1987, **35**, 4375.
180. E. J. Corey and S. P. T. Matsuda, *Tetrahedron Lett.*, 1987, **28**, 4247.
181. B. A. Vick and D. C. Zimmerman, *Biochem. Biophys. Res. Commun.*, 1983, **111**, 470.
182. A. R. Brash, *J. Am. Chem. Soc.*, 1989, **111**, 1891.
183. M. Fukushima, T. Kato, Y. Yamada, I. Kitagawa, S. Kurozumi, and P. J. Scheuer, *Proc. Am. Assoc. Cancer Res.*, 1985, **26**, 980.
184. K. Varvas, I. Jarving, R. Koljak, A. Vahemets, T. Pehk, A.-M. Muurisepp, U. Lille, and N. Samel, *Tetrahedron Lett.*, 1993, **34**, 3643.
185. K. Varvas, R. Koljak, I. Jarving, T. Pehk, and N. Samel, *Tetrahedron Lett.*, 1994, **35**, 8267.
186. V. Di Marzo, C. Gianfrani, L. De Petrocellis, A. Milone, and G. Cimino, *Biochem. J.*, 1994, **300**, 501.
187. T. Leitz, H. Beck, M. Stephan, W. D. Lehmann, L. De Petrocellis, and V. Di Marzo, *J. Exp. Zool.*, 1994, **269**, 422.
188. P. Pierobon, L. De Petrocellis, R. Minei, and V. Di Marzo, *Cell Mol. Life Sci.*, 1997, **53**, 61.
189. G. Gianfrani, V. Di Marzo, L. De Petrocellis, and G. Cimino, *Experientia*, 1995, **51**, 48.
190. V. Di Marzo, R. R. Vardaro, L. De Petrocellis, and G. Cimino, *Experientia*, 1996, **52**, 120.
191. N. Lindquist and W. Fenical, *Tetrahedron Lett.*, 1989, **30**, 2735.
192. M. S. Congreve, A. B. Holmes, A. B. Hughes, and M. G. Looney, *J. Am. Chem. Soc.*, 1993, **115**, 5815.
193. H. Niwa, M. Watanabe, H. Inagaki, and K. Yamada, *Tetrahedron*, 1994, **50**, 7385.
194. H. Niwa, K. Wakamatsu, and K. Yamada, *Tetrahedron Lett.*, 1989, **30**, 4543.
195. D. G. Nagle and W. H. Gerwick, *J. Org. Chem.*, 1994, **59**, 7227.
196. H. Kigoshi, H. Niwa, K. Yamada, T. J. Stout, and J. Clardy, *Tetrahedron Lett.*, 1991, **32**, 2427.
197. P. J. Proteau, J. V. Rossi, and W. H. Gerwick, *J. Nat. Prod.*, 1994, **57**, 1717.
198. D. J. Critcher, S. Connolly, and M. Wills, *Tetrahedron Lett.*, 1995, **36**, 3763.
199. M. Wills, University of Bath, personal communication.
200. D. E. Morse, H. Duncan, N. Hooker, and A. Morse, *Science*, 1977, **196**, 298.
201. A. F. Hagar, D. H. Hwang and T. H. Dietz, *Biochim. Biophys. Acta*, 1989, **1005**, 162.
202. M. Ojika, Y. Yoshida, Y. Nakayama, and K. Yamada, *Tetrahedron Lett.*, 1990, **31**, 4907.
203. D. Piomelli, *Am. J. Physiol.*, 1991, R844.
204. M. Abe, M. Klein, D. J. Steel, A. Thekkuveettil, E. Shapiro, J. H. Schwartz, and S. J. Feinmark, *Brain Res.*, 1995, **683**, 200.
205. V. Di Marzo, G. Cimino, A. Crispino, C. Minardi, G. Sodano, and A. Spinella, *Biochem. J.*, 1991, **273**, 593.
206. V. Di Marzo, C. Minardi, R. R. Vardaro, E. Mollo, and G. Cimino, *Comp. Biochem. Physiol.*, 1992, **101B**, 99.
207. D. J. Gerhart, A. S. Clare, K. Eisenman, D. Rittschof, and R. B. Forward, Jr., *Prog. Comp. Endocrinol.*, 1990, 598.
208. F. B. Whitfield, D. J. Freeman, J. H. Last, P. A. Bannister, and B. H. Kennett, *Aust. J. Chem.*, 1982, **35**, 373.
209. A. J. Hampson, A. F. Rowley, S. E. Barrow, and R. Steadman, *Biochim. Biophys. Acta*, 1992, **1124**, 143.
210. J. C. MacPherson, J. G. Pavlovich, and R. S. Jacobs, *Biochim. Biophys. Acta*, 1996, **1303**, 127.
211. E. M. Hill and D. L. Holland, *Proc. R. Soc. London B*, 1992, **247**, 41.
212. T. K. Shing, K. H. Gibson, J. R. Wiley, and C. I. F. Watt, *Tetrahedron Lett.*, 1994, **35**, 1067.
213. W.-C. Song, D. L. Holland, K. H. Gibson, E. Clayton, and A. Oldfield, *Biochim. Biophys. Acta*, 1990, **1047**, 239.
214. L. Meijer, J. Maclouf, and R. W. Bryant, *Dev. Biol.*, 1986, **114**, 22.
215. M. V. D'Auria, L. Minale, R. Riccio, and E. Uriarte, *Experientia*, 1988, **44**, 719.
216. L. Meijer, A. R. Brash, R. W. Bryant, K. Ng, J. Maclouf, and H. Sprecher, *J. Biol. Chem.*, 1986, **261**, 17 040.
217. I. Bruno, M. V. D'Auria, M. Iorizzi, L. Minale, and R. Riccio, *Experientia*, 1992, **48**, 114.
218. A. R. Brash, M. A. Hughes, D. J. Hawkins, W. E. Boeglin, W.-C. Song, and L. Meijer, *J. Biol. Chem.*, 1991, **266**, 22 926.
219. D. J. Hawkins and A. R. Brash, *J. Biol. Chem.*, 1987, **262**, 7629.
220. D. J. Hawkins and A. R. Brash, *FEBS Lett.*, 1989, **247**, 9.
221. D. H. Petzel, in "Insect Lipids: Chemistry, Biochemistry and Biology," eds. D. W. Stanley-Samuelson and D. R. Nelson, University of Nebraska Press, Lincoln, NE, 1993, p. 139.
222. D. W. Stanley-Samuelson, *J. Insect Physiol.*, 1994, **40**, 3.
223. W. Loher, I. Ganjian, I. Kubo, D. Stanley-Samuelson, and S. S. Tobe, *Proc. Natl. Acad. Sci. U.S.A.*, 1981, **78**, 7835.
224. D. W. Stanley-Samuelson and W. Loher, *Ann. Entomol. Soc. Am.*, 1986, **79**, 841.
225. R. H. Dadd and J. E. Kleinjan, *J. Insect. Physiol.*, 1988, **34**, 779.
226. J. S. Miller, T. Nguyen, and D. W. Stanley-Samuelson, *Proc. Natl. Acad. Sci. U.S.A.*, 1994, **91**, 12 418.
227. E. C. Toolson, P. D. Ashby, R. W. Howard, and D. W. Stanley-Samuelson, *J. Comp. Physiol.*, 1994, **164**, 278.
228. D. H. Petzel and D. W. Stanley-Samuelson, *J. Insect Physiol.*, 1992, **38**, 1.
229. F. A. Green, C. A. Herman, R. P. Herman, H.-E. Claesson, and M. Hamberg, *J. Exp. Zool.*, 1987, **243**, 211.
230. F. Stromberg, M. Hamberg, U. Rosenqvist, S.-E. Dahlen, and J. Z. Haeffstrom, *Eur. J. Biochem.*, 1996, **238**, 599.
231. K. Gronert, S. M. Virk, and C. A. Herman, *Biochim. Biophys. Acta*, 1995, **1255**, 311.
232. F. A. Green, *Biochem. Biophys. Res. Commun.*, 1987, **148**, 1533.
233. J. R. Sauer, A. S. Bowman, M. M. Shipley, C. L. Gengler, M. R. Surdick, J. L. McSwain, C. Luo, R. C. Essenberg, and J. W. Dillwith, "Insect Lipids: Chemistry, Biochemistry and Biology," eds. D. W. Stanley-Samuelson and D. R. Nelson, University of Nebraska Press, Lincoln, NE, 1993, p. 99.
234. H. A. Baset, G. P. O'Neill, and A. W. Ford-Hutchinson, *Mol. Biochem. Parasitol.*, 1995, **73**, 31.

235. A. Belley and K. Chadee, *Parasitol. Today*, 1995, **11**, 327.
236. I. Hara, S. Hara, B. Salafsky, and T. Shibuya, *Exp. Parasitol.*, 1993, **77**, 484.
237. L. Kaiser, V. L. Lamb, P. K. Tithof, D. A. Gage, B. A. Chamberlin, J. T. Watson, and J. F. Williams, *Exp. Parasitol.*, 1992, **75**, 159.
238. M. Jett, S. K. Martin, R. L. Fine, and I. Schneider, "Prostaglandins, Leukotrienes, Lipoxins and PAF, XIth Washington International Spring Symposium," ed. J. M. Bailey, 1991, Abstr. 85.

1.09
Biosynthesis and Metabolism of Eicosanoids

SHOZO YAMAMOTO
Tokushima University School of Medicine, Japan

1.09.1 ARACHIDONATE CASCADE

A variety of compounds are biosynthesized from polyunsaturated fatty acids, especially eicosapolyenoic acids like arachidonic acid (see Chapters 1.07 and 1.08). Some of these metabolites are biologically active; namely, prostaglandin (PG), thromboxane (TX), leukotriene (LT), hepoxilin, and lipoxin are bound to their specific receptors which are coupled to specific intracellular signal transduction pathways. The biosynthetic pathway of these bioactive eicosanoids is generally referred to as the arachidonate cascade, which is composed of several branches each with several enzymatic steps. The cascade is initiated by the release of arachidonic acid from position 2 of glycerophospholipids by the action of phospholipase A_2 enzyme. Then, arachidonic acid is oxygenated to its endoperoxy or hydroperoxy derivatives by cyclooxygenase and lipoxygenase enzymes. The peroxide moiety is further metabolized to produce various bioactive oxyeicosanoids. Further enzymatic modifications by dehydrogenases or cytochrome P450 enzymes result in the loss of their biological activities.

The phospholipase A_2 is a triggering enzyme releasing arachidonic acid, which is otherwise unavailable as substrate within the cell. The metabolism of arachidonic acid to PGs is regulated at the cyclooxygenase step. Thus, PGs and other bioactive oxyeicosanoids are enzymatically produced when they are necessary, and inactivated by further metabolism when their functions are over.

1.09.2 PHOSPHOLIPASE A_2

A regulatory function of phospholipase A_2 for PG release was proposed much earlier[1] by the combination of four early findings: (i) the PGs derive from unsaturated fatty acids; (ii) phospholipase A is capable of releasing PGs from homogenates and perfused intact tissues; (iii) PGs are not stored as preformed agents but are generated prior to release; and (iv) the precursor acids are known not to exist to a major extent in tissues as free acids; they are mostly constituents of complex lipids, mainly phospholipids.[2]

As shown in Scheme 1, phospholipase A_2 hydrolyzes the *sn*-2 fatty acyl ester on glycerophospholipids, which are generally unsaturated fatty acids including arachidonic acid. Although a variety of phospholipases have been found in various animal tissues,[3] two types of phospholipase A_2 are generally considered to be involved in the arachidonate release in mammalian tissues. One is nonpancreatic secretory phospholipase A_2 ($sPLA_2$) with a low molecular weight of about 1.4×10^4. The enzyme is secreted by platelets, mast cells, hepatocytes, smooth muscle cells, and other cells, and the extracellular enzyme activity is detected in inflamed tissues. The other type is cytosolic phospholipase A_2 ($cPLA_2$) with a high molecular weight of about 8.5×10^4. The enzyme is activated by free calcium ion at submicromolar concentration, and hydrolyzes preferentially phospholipids containing arachidonic acid at the *sn*-2 position. For details of these types of phospholipase A_2 readers are advised to refer to other studies.[3–6]

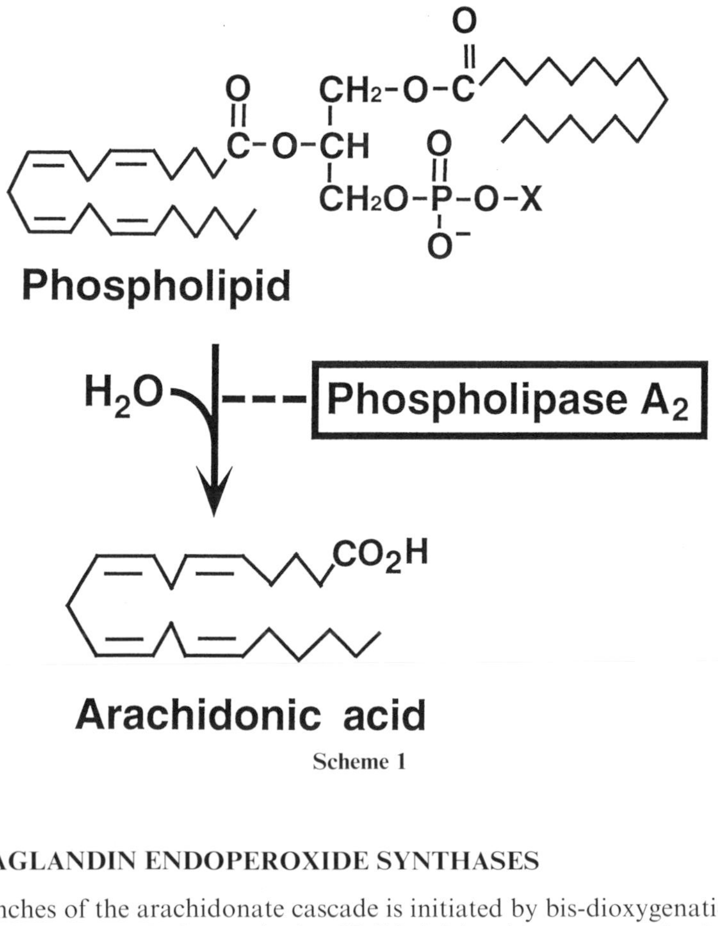

Phospholipid

H_2O — Phospholipase A_2

Arachidonic acid

Scheme 1

1.09.3 PROSTAGLANDIN ENDOPEROXIDE SYNTHASES

One of the branches of the arachidonate cascade is initiated by bis-dioxygenation of arachidonic acid. As illustrated in Scheme 2, the synthesis of PG is initiated by a stereoselective removal of *pro-*

S hydrogen at position 13 of arachidonic acid, followed by shift of double bonds, incorporation of two oxygen molecules at positions 11 and 15, and cyclization of the carbon chain to form a cyclopentane ring. The product is PGG_2, a prostaglandin with 9,11-endoperoxide and 15-hydroperoxide. The enzyme catalyzing the conversion of arachidonic acid to PGG_2 is referred to as fatty acid cyclooxygenase (COX) (Scheme 2). Then, PGG_2 is transformed to PGH_2 by reduction of 15-hydroperoxide in the presence of a reducing agent. This is a peroxidase reaction, and the enzyme is referred to as PG hydroperoxidase. Both the COX and hydroperoxidase activities are attributed to a single enzyme protein on the basis of enzymological findings.[7] Furthermore, when the cDNA for ovine seminal vesicle enzyme was cloned and expressed, both the COX and hydroperoxidase activities were found in the transfected cells.[8] The enzyme with the COX and hydroperoxidase activities is designated officially as PG endoperoxide synthase (EC 1.14.99.1). However, the enzyme with the two activities is usually referred to simply as COX or PGH synthase.

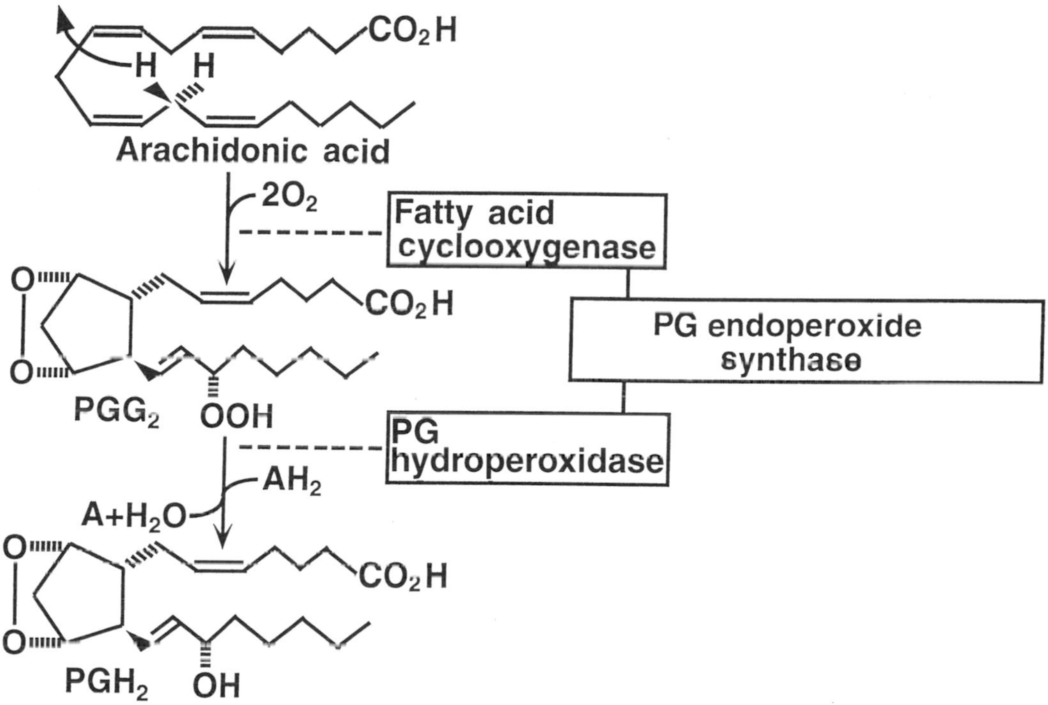

Scheme 2

Both the COX and hydroperoxidase activities require heme, and various compounds including tryptophan and monophenol are utilized as reducing agents for the hydroperoxidase reaction. Properties of the enzyme purified from bovine and ovine seminal vesicles are described in detail elsewhere.[7–9] The COX activity (but not hydroperoxidase activity) is selectively inhibited by a variety of nonsteroidal anti-inflammatory drugs.[7] In particular, aspirin (acetyl salicylate) is known to inhibit the COX activity by acetylating Ser506 of the enzyme of ovine seminal vesicle.[8]

The above-mentioned catalytic properties have been revealed with the COX preparation of bovine and ovine seminal vesicles, and a similar enzyme was also found in platelets, kidney, and stomach. However, the studies since 1991 led to the discovery of a COX isozyme. The previously known enzyme is referred to as COX-1 and the new isozyme as COX-2 (Table 1). COX-1 is a constitutive enzyme which is present constantly in tissues like seminal vesicle, platelets, stomach, and kidney, and functions as a housekeeping enzyme. In contrast, COX-2 is a product of an immediate early gene, and is an inducible enzyme. The enzyme is induced rapidly and transiently by various inducers (hormones, growth factors, and cytokines) in various types of cell. In particular, the COX-2 induction by pro-inflammatory lipopolysaccharide and cytokines in monocytes is related to inflammation. The COX-2 induction is markedly suppressed by anti-inflammatory glucocorticoid, and the enzyme activity is inhibited by most nonsteroidal anti-inflammatory drugs. For detailed descriptions of COX-2 in comparison with COX-1, readers are advised to refer to other studies.[10–19]

The discovery of COX-2 provided a new approach for the studies on inflammation and for the development of anti-inflammatory drugs. As stated by Vane,[20] "because COX-2 is induced in migratory and other cells by inflammatory stimuli and by cytokines, it is attractive to suggest that

Table 1 Two isozymes of fatty acid cyclooxygenases (COX-1 and COX-2).

	COX-1	*COX-2*
Gene	housekeeping gene no TATA box 11 exons in human and mouse chromosome 2 (mouse), 9q32–q33.3 (human)	immediate early gene with TATA box 10 exons in mouse chromosome 1 (mouse), 1q25.2–q25.3 (human)
Gene expression	constitutive ca. 3 kb mRNA several-fold increase slowly during cell differentiation	inducible ca. 4 kb mRNA 10–100-fold increase rapidly and transiently by growth factors, cytokines, hormones, and phorbol esters
Anti-inflammatory drugs		
Glucocorticoid	gene expression almost unaffected	gene expression suppressed
Aspirin	inhibited	(15*R*)-lipoxygenase activity
Indomethacin	inhibited	inhibited
NS-398, SC-558, RS57067, etc.	unaffected	inhibited
Enzyme distribution	seminal vesicle, platelets, kidney, stomach, and most other tissues	monocytes, synoviocytes, fibroblasts, osteoblasts, ovary granulosa cells, colorectal cancer cells

the anti-inflammatory actions of aspirin and its fellows are due to the inhibition of COX-2, whereas the unwanted side-effects (such as irritation of the stomach lining and toxic effects on the kidney) are due to inhibition of the constitutive enzyme, COX-1." Thus, a selective inhibitor of COX-2 as an anti-inflammatory drug without side effects is a subject of active investigations and developments.[21]

The inhibitory effects of anti-inflammatory drugs on COX-1 and COX-2 suggest pro-inflammatory functions of some PGs, which would be further elucidated by gene disruption. Knockout mice of COX-1 gene[22] and COX-2 gene[23,24] were reported. However, as reviewed and commented on by DeWitt and Smith,[25] the results were somewhat unexpected. The COX-1-deficient mice had no spontaneous stomach ulceration, and showed a reduced inflammatory response to arachidonic acid in ear swelling. The COX-2-deficient mice and normal mice responded similarly to arachidonic acid.

In connection with the drug actions, knowledge of the three-dimensional structure of COX-1 and COX-2 would be useful to understand drug action and the design of new drugs. The X-ray crystal structures of COX-1[26,27] and COX-2[28,29] have been revealed, and discussed in terms of the binding of anti-inflammatory drugs to the enzyme active sites.[28–31] According to Garavito's group, COX-1 exists as a symmetric dimer, and the enzyme protein can be divided into three distinct folding units.[26,27] The N-terminal region is an epidermal growth factor (EGF)-like domain. The second membrane-binding domain consists of four amphipathic α-helices, and inserts the enzyme into the lipid bilayer. The catalytic domain has oxygenase and peroxidase sites in the form of a long hydrophobic channel extending from the membrane-binding domain to the center of the molecule. This channel includes Tyr385 abstracting hydrogen at position 13 and Ser506 to be acetylated by aspirin. The overall structure of COX-1 is highly conserved in COX-2 which also has the three domains mentioned above.[29] It has been pointed out that a bulky isoleucine of COX-1 is replaced by Val523 and a pocket branching off from the main channel is more accessible in COX-2 by the isoleucine–valine substitution resulting in an easier binding of a selective inhibitor to COX-2 rather than COX-1.[28,29]

Epidemiological data have shown the reduced occurrence or progression of colorectal cancer and polyps by chronic administration of aspirin and other nonsteroidal anti-inflammatory drugs,[32] and this observation was supported by animal experiments (reviewed by Marnett[33]). In 1996, a correlation of colon cancer with the COX-2 induction was reported with Apc knockout mice as a model of human familial adenomatous polyposis.[34] In particular, the study on Apc Δ^{716} mice with COX-2 gene disruption suggested that COX-2 plays a key role in tumorigenesis and suggested a therapeutic efficacy of COX-2-selective inhibitors.[35]

1.09.4 PROSTAGLANDIN ENDOPEROXIDE METABOLIZING ENZYMES

Earlier papers on the synthases of PGD_2, PGE_2, $PGF_{2\alpha}$, PGI_2, and TXA_2 will not be discussed in this chapter; instead the reader is advised to refer to the original papers cited in other studies.[7,9] Only papers from the 1980s and 1990s will be cited here (Table 2).

Table 2 PG endoperoxide metabolizing enzymes.

Enzyme source	*Reductant requirement*	*Ref.*	K_m *for* PGH_2 (μM)	*Ref.*	*Molecular weight*	*Ref.*	*Amino acid number*	*Ref.*	*No. of exons*	*Ref.*	*Genomic DNA chromosome*	*Ref.*
PGD synthase												
Lipocalin type												
Human cerebrospinal fluid	sulfhydryls	35	4	35	27 000	35						
Human brain					21 016	43	190	43	7	52	9q34.2–q34.3	53
Rat brain	sulfhydryls	36	14	36	ca. 27 000	36	188 or 183	48	7	53		
Hematopoietic type												
Rat spleen	Glutathione	37	200	37	29 297	44	199	44				
PGE synthase												
Rat deferent duct	Glutathione	38	25.7	38								
Rat heart			82.5	38								
Human brain (peaks 1 and 2)	Glutathione	39	147; 308	39	42 000; 44 000	39						
PGF synthase												
Bovine lung	NADPH	40	10	40	36 666	45	323	45				
Bovine liver	NADPH	41	14.3	41	ca 36 000	41						
PGI synthase												
Rabbit aorta			20	42								
Bovine aorta					ca 52 000	46	500	42				
Rat							501	49				
Human							501	50	10	54	20q13.11–q13.13	55
TXA synthase												
Human					60 487	47	533	47	13	55,56	7q33–q34	56
Porcine							534	51				

1.09.4.1 Prostaglandin D Synthase

Enzymatic isomerization of 9,11-endoperoxide of PGH_2 produces PGD_2 with 9α-hydroxy and 11-keto groups (Scheme 3). The enzyme responsible for this transformation (PGD synthase, PGH_2 D-isomerase, EC 5.3.99.2) was classified into two types by earlier studies: the brain type and spleen type.[7] The "glutathione-independent" PGD synthase has been purified from rat brain, and the enzyme with a molecular weight of about 2.7×10^4 was stimulated nonspecifically by various sulfhydryl compounds, including dithiothreitol, glutathione, β-mercaptoethanol, cysteine and cysteamine and almost saturated around 0.5 mM.[36] The apparent K_m for PGH_2 was 14 μM, and the enzyme showed a suicide-type inactivation upon reaction with PGH_2.[37] The enzyme belongs to the lipocalin family, which is a group of secretory proteins involved in the transport of small lipophilic ligands.[58] Surprisingly, PGD synthase was found to be identical with β-trace protein, which is a major protein component of human cerebrospinal fluid.[36,59] The "brain-type" or "lipocalin type" of PGD synthase was localized predominantly in leptomeninges, choroid plexus, and oligodendrocytes of adult rat brain as examined by *in situ* hybridization, immunohistochemical staining, and enzyme activity assay. The enzyme as β-trace protein may be secreted into cerebrospinal fluid from these sites in the brain.[60]

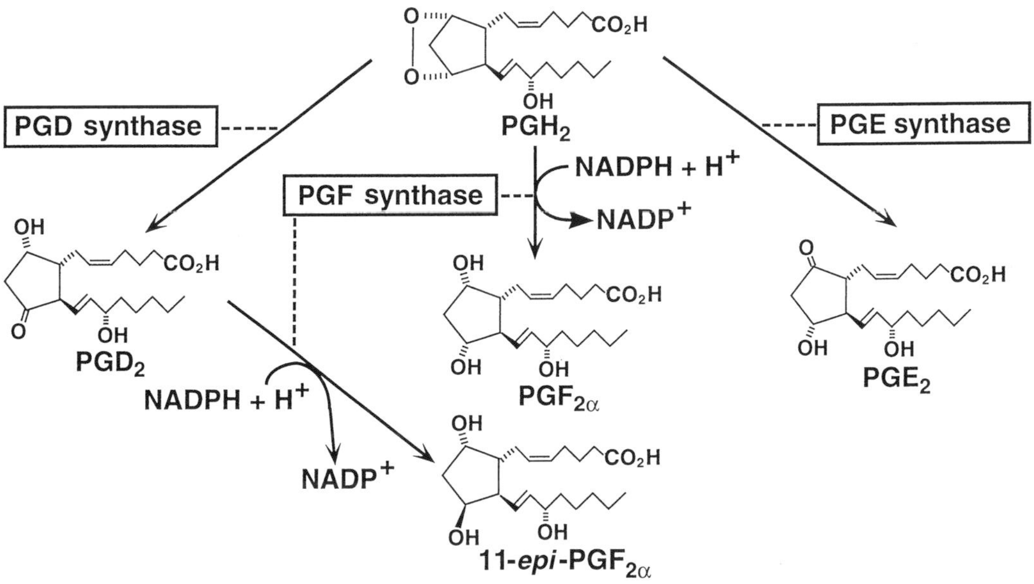

Scheme 3

The other type ("spleen type" or "hematopoietic type") of PGD synthase is characteristic of its specific requirement for glutathione. The enzyme (molecular weight of about 2.6×10^4) was purified from rat spleen,[38] and its K_m values for PGH_2 and glutathione were about 200 μM and 300 μM, respectively. The enzyme also showed the glutathione *S*-transferase activity with glutathione (K_m = 300 μM) and 1-chloro-2,4-dinitrobenzene (K_m = 5 mM), the latter of which inhibited the hematopoietic type of PGD synthase.[38] A similar enzyme was isolated from peritoneal mast cells of adult rats.[61] cDNA for rat spleen enzyme was cloned and expressed in *E. coli*. The recombinant enzyme was crystallized and its three-dimensional structure was determined at 2.3 Å resolution.[45]

1.09.4.2 Prostaglandin E Synthase

PGE synthase (PGH_2 E-isomerase, EC 5.3.99.3) isomerizes 9,11-endoperoxide of PGH_2 to 9-keto and 11α-hydroxy groups and produces PGE_2 (see Scheme 3). Apart from earlier reports with crude enzyme preparations,[7] there was little information on the properties of glutathione-requring PGE synthase. However, in 1996, sigma-class glutathione *S*-transferase purified from *Ascaridia galli* was shown to convert PGH_2 specifically to PGE_2.[62] When PGE synthesis activity was screened in various rat tissues, the glutathione-dependent PGE synthase activity was found in genital accessory organs (especially deferent duct) and kidney, whereas the glutathione-independant enzyme activity was detected in heart, spleen, uterus, and other organs. The two types of PGE synthase were localized in the microsomal fraction.[39] An enzyme was purified from the cytosol of human brain, and identified as anionic forms of glutathione *S*-transferase.[40] Elucidation of catalytic properties awaits further purification of the enzyme and its cDNA cloning.

1.09.4.3 Prostaglandin F Synthase

Three possible $PGF_{2\alpha}$ biosynthesis pathways have been considered; 9-keto reduction of PGE_2, 11-keto reduction of PGD_2, and endoperoxide reduction of PGH_2.[7] When PGD 11-keto reductase was purified from bovine lung, the enzyme of molecular weight of $\sim 3 \times 10^4$ was also active with PGH_2 producing $PGF_{2\alpha}$.[41] The product from PGD_2 was later found to be 11-*epi*-$PGF_{2\alpha}$ rather than $PGF_{2\alpha}$.[63] The enzyme reduced not only PGD_2 and PGH_2 but also various carbonyl compounds, and all the enzyme activities with the three types of substrate were attributed to the same enzyme protein, referred to as PGF synthase (EC 1.1.1.88) (see Scheme 3). Two catalytic sites were presumed for the enzyme; one for PGH_2 and the other for PGD_2 and other carbonyl compounds.[41] cDNA for the enzyme was cloned, and its nucleotide sequence revealed that the enzyme was a 323 amino acid peptide of molecular weight 36 666. In terms of amino acid sequence the PGF synthase showed high homologies with human liver aldehyde reductase and ε-crystallin of European common frog lens.[46] Similar enzymes with PGF synthase activity were purified from bovine[42] and human[64] liver.

1.09.4.4 Prostaglandin I Synthase

PGI synthase (EC 5.3.99.4) transforms PGH_2 to PGI_2 (prostacyclin) which is an unstable 6,9α-epoxy compound and is nonenzymatically decomposed to 6-keto-$PGF_{1\alpha}$ (Scheme 4). Earlier studies on the enzyme are described elsewhere.[7,9] The enzyme was solubilized from bovine aortic microsomes with Triton X-100, and purified to a single protein (molecular weight = 5.2×10^4) by immunoaffinity chromatography. The purified enzyme had a specific activity of about 1 μmol min^{-1} mg^{-1} protein at 24 °C with 50 μM PGH_2 as substrate, and showed a heme absorption peak at 418–420 nm. The reduced enzyme treated with carbon monoxide gave a difference spectrum around 440 nm.[47] The bovine aortic enzyme was also purified by affinity chromatography using a PGH_2 analogue as a ligand.[65] PGG_2, PGH_1, and PGH_3 were as active as PGH_2. The endoperoxide of PGH_1 was fragmented into malondialdehyde (MDA) and C_{17} hydroxy acid ((12*S*)-hydroxy-(*E*-8,*E*-10)-heptadecadienoic acid or HHD).[65]

Ullrich and co-workers suggested the P450 nature of PGI synthase and TXA synthase,[66] and their further investigations led to a proposal of a mechanism which explains how PGI_2 and TXA_2 are transformed from PGH_2 mediated by heme prosthetic group.[65,67] They used two PG endoperoxide analogues (9,11-epoxymethano-$PGF_{2\alpha}$ and 11,9-epoxymethano-$PGF_{2\alpha}$) and recorded difference spectra. As shown in Scheme 4, the results suggested the interaction of the ferric iron of PGI synthase with the C-11 endoperoxide oxygen and that of TXA synthase with the C-9 endoperoxide oxygen.[65,67]

cDNA of PGI synthase was cloned from bovine endothelial cells, and its nucleotide sequence analysis revealed a 1500 bp open reading frame encoding a 500 amino acid protein with a molecular weight of 56 628. The cDNA was expressed in COS-7 cells which showed the enzyme activity producing 6-keto-$PGF_{1\alpha}$ from PGH_2.[43] cDNA of human PGI synthase was cloned, and was found to contain a 1500 bp open reading frame encoding 500 amino acids. When the tissue distribution was examined by Northern blotting, the human PGI synthase mRNA was abundant in ovary, heart, skeletal muscle, lung, and prostate.[51] The cDNA of rat PGI synthase was also cloned, and had a 1503 bp open reading frame encoding 501 amino acids. According to Northern blotting with various rat tissues, the PGI synthase mRNA was highly expressed in aorta, uterus, stomach, lung, heart, testis, liver, and skeletal muscle.[50] Human PGI synthase gene was found to consist of 10 exons, and was localized on chromosome 20q13.11–q13.13.[55]

1.09.4.5 Thromboxane A Synthase

TXA synthase (EC 5.3.99.5) transforms PGH_2 to TXA_2 which is unstable and nonenzymatically degraded to TXB_2. Concomitantly, the endoperoxide moiety of a nearly equimolar amount of PGH_2

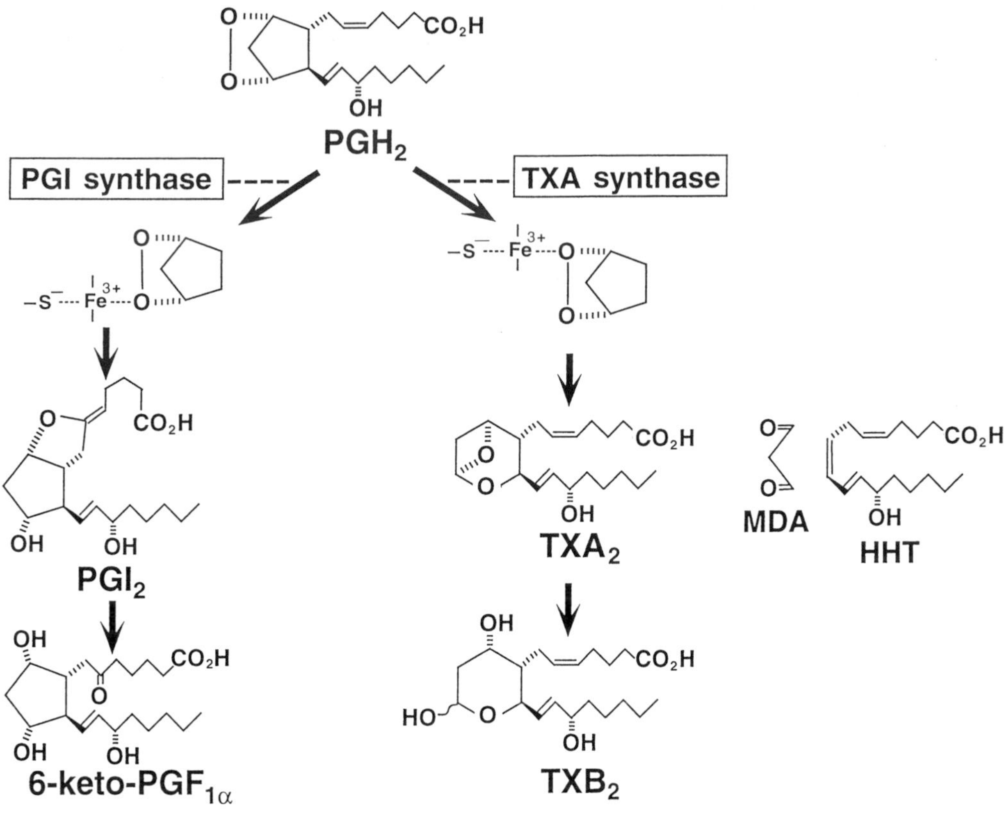

Scheme 4

is subjected to fragmentation producing MDA and (12*S*)-hydroxy-(*Z*-5,*E*-8,*E*-10)-heptadecatrienoic acid (HHT) (Scheme 4).

The enzyme was purified to homogeneity and a specific activity of 24.1 μmol min^{-1} mg^{-1} protein from the microsomes of human platelets. The molecular weight of the enzyme was estimated to be 58 800, and one heme was found per mol of enzyme.[68] The cytochrome P450 nature of the enzyme was established by a spectroscopic study utilizing two PG endoperoxide analogues (9,11-epoxymethano-$PGF_{2\alpha}$ and 11,9-epoxymethano-$PGF_{2\alpha}$). As shown in Scheme 4, the interaction of the ferric iron of the enzyme-bound heme with the C-9 endoperoxide oxygen was suggested.[65,67] Attempts were made to purify the enzyme by immunoaffinity chromatography,[69,70] but the major part of the purified enzyme was obtained as the inactive P420 form.[70] PGG_2 and PGH_3 were nearly as active as PGH_2, and converted to corresponding TXB derivatives and C_{17} acids. PGH_1 was transformed only to HHD.[65] In view of the proaggregatory and vasoconstrictive activities of TXA_2, the inhibitors of TXA synthase have been developed for clinical use.[71]

cDNA for TXA synthase was amplified from human platelets by the polymerase chain reaction technique, and the cloned cDNA encoded a 533 amino acid enzyme with a molecular weight of 60 487.[48] The cDNA for porcine TXA synthase encoding a 534 amino acid enzyme was cloned and highly expressed in Sf9 cells with the aid of baculovirus.[52] Human gene-encoding TXA synthase was isolated and found to consist of 13 exons and 12 introns, and the 5′-flanking region had potential binding sites for various transcription factors.[56,57] The human gene was localized to q33–q34 of the long arm of chromosome 7 according to fluorescence *in situ* hybridization.[55] According to RNA blotting by the use of cDNA probe for human TXA synthase, the enzyme mRNA is expressed in various human tissues, abundantly in leukocytes, spleen, lung, and liver, and at low but significant levels in kidney, placenta, and thymus.[56]

1.09.5 PROSTAGLANDIN METABOLISM

PGs and thromboxanes lose their biological activities by their further metabolism. The enzymes involved in their metabolism have been described in detail elsewhere.[7,9] NAD^+-dependent

15-hydroxyprostaglandin dehydrogenase (EC 1.1.1.141) oxidizes the 15-hydroxy group of most PGs in the presence of NAD^+ as a hydrogen acceptor, and the 15-keto metabolites produced have much lower biological activities. cDNA for the human placental enzyme was cloned, which encoded a 266 amino acid protein of molecular weight 28 975.[72] Enzymological and molecular biological studies on this and other related dehydrogenases have been reviewed by Ensor and Tai.[73]

1.09.6 LIPOXYGENASES

Lipoxygenase is a dioxygenase which recognizes the 1-*cis*, 4-*cis*-pentadiene structure of polyunsaturated fatty acids, and produces hydroperoxy acids with a conjugated diene. In mammalian tissues there are several different lipoxygenases distinguished by the oxygenation site in the unsaturated fatty acid molecules. With arachidonic acid (C_{20} fatty acid) as substrate, the enzymes oxygenate positions 5, 8, 12, and 15, and are referred to as arachidonate 5-, 8-, 12-, and 15-lipoxygenase, respectively (Scheme 5). Their primary reaction products are corresponding hydroperoxy acids: (5*S*)-hydroperoxy-6,8,11,14-eicosatetraenoic acid (5-HPETE), (8*S*)-hydroperoxy-5,9,11,14-eicosatetraenoic acid (8-HPETE), (12*S*)-hydroperoxy-5,8,10,14-eicosatetraenoic acid (12-HPETE), and (15*S*)-hydroperoxy-5,8,11,13-eicosatetraenoic acid (15-HPETE). With crude enzyme preparations such as whole cells and tissue homogenates, these hydroperoxy acids are reduced, and corresponding hydroxy acids (HETE) are found. As will be described individually, lipoxygenases are multifunctional enzymes, and these primary hydroperoxy products are further metabolized by the same enzymes. In particular, 5-lipoxygenase produces leukotriene A_4 (LTA_4) with 5,6-epoxide from arachidonic acid via 5-HPETE, and the LTA_4 is further metabolized to bioactive leukotrienes. Another catalytic property characteristic of lipoxygenases is the suicide inactivation, namely, the enzyme reaction slows down soon after its start due to a mechanism-based inactivation. The reader is referred to general review articles on mammalian lipoxygenases.[74–82]

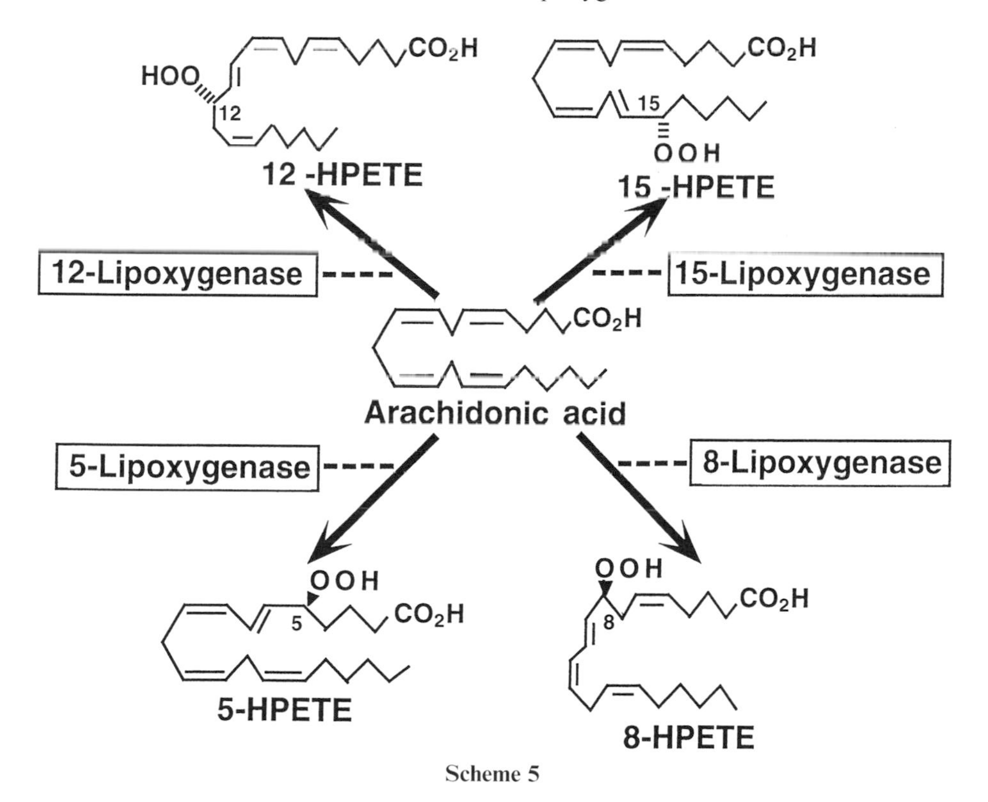

Scheme 5

1.09.6.1 Arachidonate 5-Lipoxygenase

Arachidonate 5-lipoxygenase (EC 1.13.11.34) initiates the biosynthesis of leukotrienes, which are chemical mediators of inflammation and anaphylaxis.[83,84] The enzyme is specific for eicosapolyenoic acids,[80] and oxygenates position 5 of arachidonic acid producing 5-HPETE. The 5-lipoxygenase is a bifunctional enzyme, and in addition to 5-oxygenase activity the enzyme has LTA_4 synthase activity which transforms 5-HPETE to LTA_4 with 5,6-epoxide and a conjugated triene (Scheme 6). Both 5-oxygenase and LTA_4 synthase activities are attributed to a single enzyme protein.[85] The enzyme contains an equimolar amount of nonheme iron, which is considered to be involved in the catalytic activity.[85] Site-directed mutation experiments proposed three histidine residues and the C-terminal isoleucine as the ligands of the iron.[81,82,86] The enzyme activity requires calcium ion, and is stimulated by the addition of ATP. The mechanism of their stimulatory effects is unclear.[80,85] 5-Lipoxygenase is a typical suicide enzyme.[87] The presence of a 18 kDa protein FLAP (5-lipoxygenase-activating protein) is required for the arachidonate metabolism through the 5-lipoxygenase pathway. FLAP is localized in the nuclear envelope, and facilitates the transfer of arachidonic acid to 5-lipoxygenase which migrates from the cytosol to the nuclear envelope.[85,88]

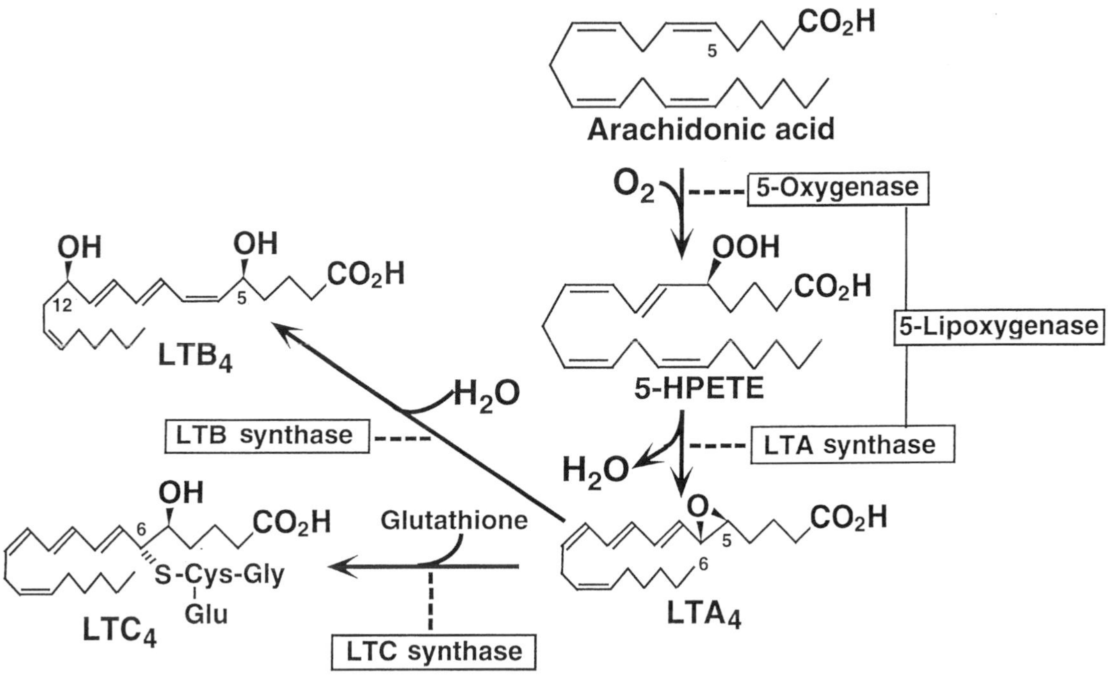

Scheme 6

1.09.6.2 Arachidonate 12-Lipoxygenase

Arachidonate 12-lipoxygenase (EC 1.13.11.31) oxygenates position 12 of arachidonic acid and produces 12-HPETE as a major product.[89,90] Unlike 5-lipoxygenase, the produced 12-HPETE is not metabolized by 12-lipoxygenase. However, the multifunctional nature of 12-lipoxygenase is observed with 15-HPETE, which is produced as a minor oxygenation product together with the major 12-HPETE. 12-Lipoxygenase either oxygenates positions 8 and 14 of 15-HPETE producing 8,15- and 14,15-dihydroperoxy acids or transforms 15-HPETE to a 14,15-epoxy acid with a conjugated triene (Scheme 7). These catalytic properties were demonstrated with leukocytes and purified enzyme.[89,90] 12-Lipoxygenase is also found in platelets, and the platelet enzyme is distinguished from the leukocyte enzyme in terms of catalytic properties. As listed in Table 3, platelet 12-lipoxygenase is active selectively with eicosapolyenoic acids, and almost inactive with linoleic and linolenic acids. Moreover, the platelet enzyme produces much less 15-HPETE and is much less active with 15-HPETE than the leukocyte enzyme. This catalytic property is related to a marked suicide inactivation of the leukocyte 12-lipoxygenase, but not of the platelet 12-lipoxygenase.[91] When the amino acid sequences are deduced from the nucleotide sequences of their cDNAs, the leukocyte enzyme is closer to 15-lipoxygenases, rather than to the platelet 12-lipoxygenase in terms of the homology of their primary structures.[89,90] As listed in Table 3, the leukocyte-type 12-lipoxygenase is found not only in leukocytes but also in various other types of cells, and the platelet-type 12-lipoxygenase is also localized in skin. In view of the tissue distribution of 12- and 15-lipoxygenases, the leukocyte-type 12-lipoxygenase was considered to be equivalent to the leukocyte 15-lipoxygenase.[90] However, despite the presence of 15-lipoxygenase in rabbit reticulocytes, the occurrence of a leukocyte-type 12-lipoxygenase in rabbit monocytes has been reported, and its cDNA cloned.[92] In murine tissues the platelet and leukocyte types of 12-lipoxygenase had been found, but a cDNA of the third type of 12-lipoxygenase was cloned from mouse epidermis. The expressed enzyme showed only 60% identity with the other 12-lipoxygenases, and linoleic and linolenic acids were not appreciably metabolized by the enzyme.[93] Enzymological and molecular biological findings with the two 12-lipoxygenase isozymes have been reviewed elsewhere.[76,78,80–82,89,90] Various biological activities have been reported for 12-HPETE and its reduced product (12-hydroxy-5,8,10,14-eicosatetraenoic acid, 12-HETE).[80,90]

Table 3 Isozymes of arachidonate 12-lipoxygenases.

	Platelet-type	*Leukocyte-type*
Substrate specificity		
Linoleic, Linolenic (free)	almost inactive	active
Arachidonic (free)	active	active
Linoleic, Arachidonic (esterified)	less active	active
Suicide inactivation	almost negative	positive
Immunoreactivity with antibody		
for leukocyte 12-lipoxygenase	inactive	active
for platelet 12-lipoxygenase	active	inactive
Amino acid homology with 15-lipoxygenase	lower	higher
Exon–intron structure compared with 15-lipoxygenase	different	similar
Distribution		
Human	platelet, skin	adrenal
Porcine		leukocyte, pituitary
Bovine	platelet	leukocyte, trachea, cornea
Canine	platelet	leukocyte, brain
Rat	platelet	leukocyte, pineal gland, aorta, lung, pancreas, spleen
Mouse	platelet, skin	leukocyte, pituitary, kidney, pineal gland

1.09.6.3 Arachidonate 15-Lipoxygenase

Arachidonate 15-lipoxygenase (EC 1.13.11.33) oxygenates position 15 of arachidonic acid producing 15-HPETE, which is further metabolized by the same enzyme either by oxygenation at positions 5 and 8 or by anaerobic transformation of 15-hydroperoxy acid to 14,15-epoxy acid with a conjugated triene (Scheme 8). The enzyme was purified from rabbit reticulocytes, and has been extensively studied by enzymological and molecular biological approaches.[74,75,94] 15-Lipoxygenase has a broad substrate specificity reacting with linoleic and linolenic acids as well as arachidonic acid,[80] and has a high homology of the primary structure with 12-lipoxygenase of leukocyte type as described above.[80] The enzyme is also active with esterified arachidonic and linoleic acids in phospholipid,[95] membrane,[96] and lipoprotein,[97] implicating its role in the pathogenesis of arteriosclerosis.[98] A new 15-lipoxygenase has been discovered in human skin, and its cDNA encoding 676 amino acids has been cloned from hair roots. The enzyme oxygenates arachidonic acid mainly at C-15 and partly at C-12. Linoleic acid was less active than arachidonic acid. The primary structure of the enzyme has only about 40% identity to the previously studied 5-, 12-, and 15-lipoxygenases but 78% identity to murine 8-lipoxygenase, which is described below.[99]

1.09.6.4 Arachidonate 8-Lipoxygenase

Arachidonate 8-lipoxygenase (EC 1.13.11.40) producing 8-HPETE has been described in several papers.[100–102] The enzyme activity was induced in mouse skin treated with phorbol ester, but was

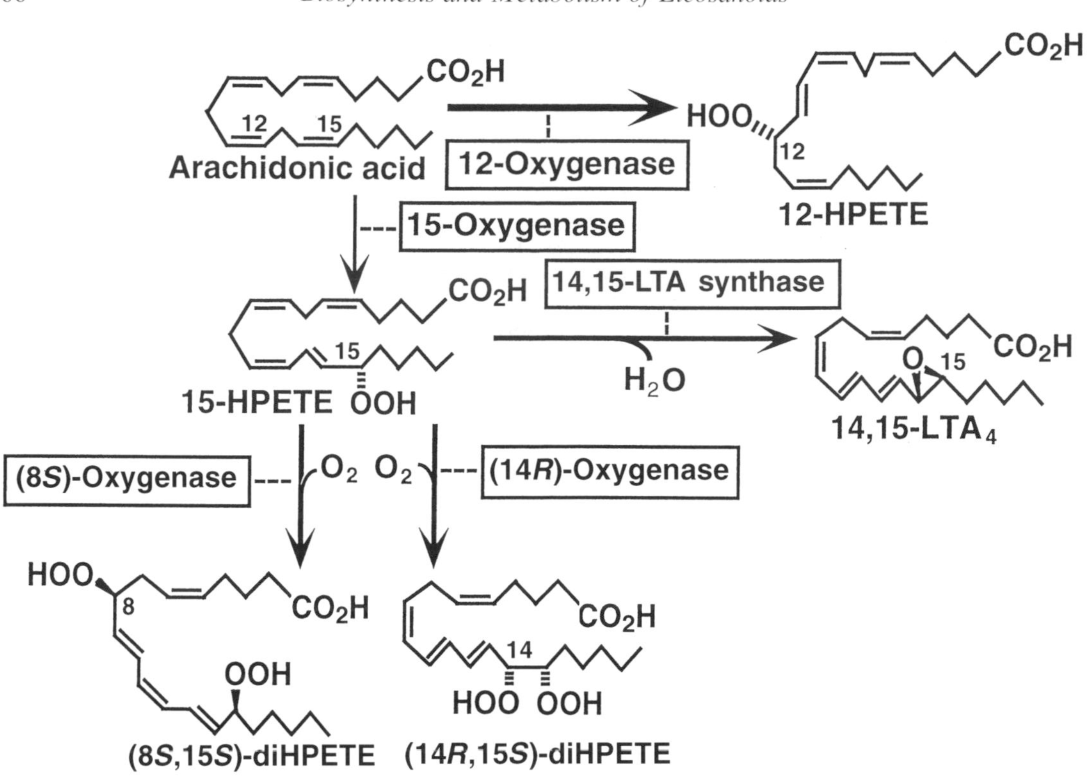

Scheme 7

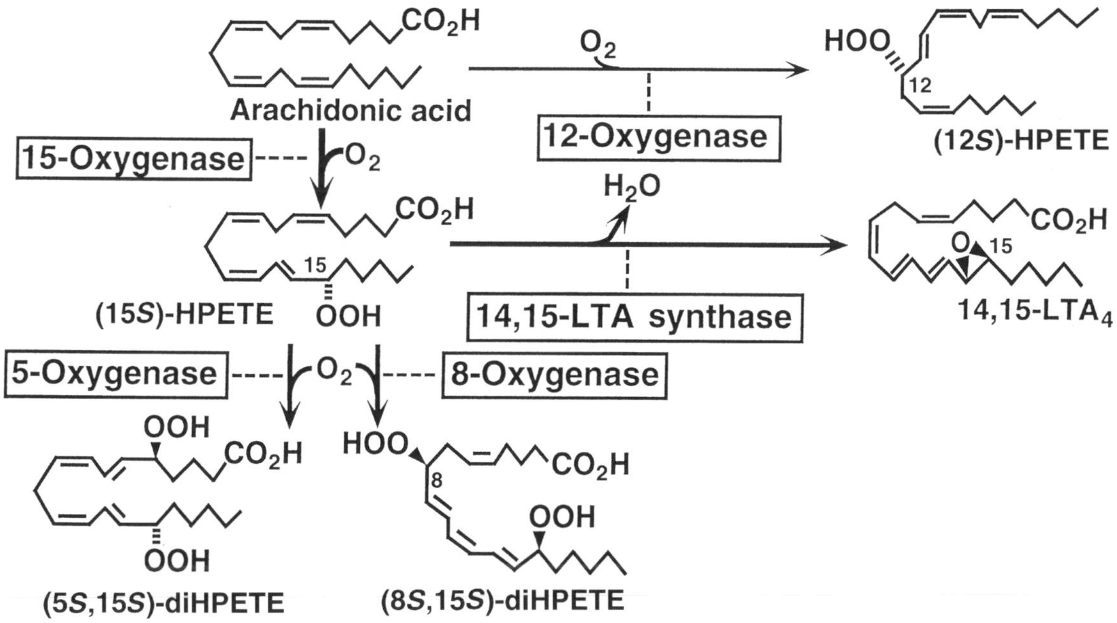

Scheme 8

not found in normal skin. The product was identified as (8*S*)-hydroxy enantiomer.[103] cDNA encoding a 677 amino acid enzyme with a 76 kDa molecular weight was cloned from mouse epidermis, and expressed in Hela cells. The enzyme oxygenates arachidonic acid exclusively to (8*S*)-HPETE, and linoleic acid to (9*S*)-hydroperoxy-octadecadienoic acid at a 2–3-fold lower rate.[104] (8*R*)-Lipoxygenase (not (8*S*)-lipoxygenase) was purified from coral, and its cDNA was cloned and sequenced.[105]

1.09.7 HYDROPEROXY- AND EPOXY-EICOSANOID METABOLIZING ENZYMES

1.09.7.1 Lipoxin Synthesis

Lipoxins are trihydroxyeicosapolyenoic acids with a conjugated tetraene with structures illustrated in Scheme 9, and arachidonic acid is a precursor for lipoxins A_4 and B_4. Two pathways for lipoxin production have been proposed. The unicellular biosynthesis is a series of 15-lipoxygenase, 5-lipoxygenase, and epoxide hydrolase reactions. The transcellular biosynthesis is LTA_4 synthesis by leukocyte 5-lipoxygenase, followed by its transformation to lipoxins by platelet 12-lipoxygenase. Their chemical structures, biosynthesis, and biological functions, especially as immunologic and hemodynamic regulators, have been reviewed by Serhan[106] and by Brady and Serhan.[107]

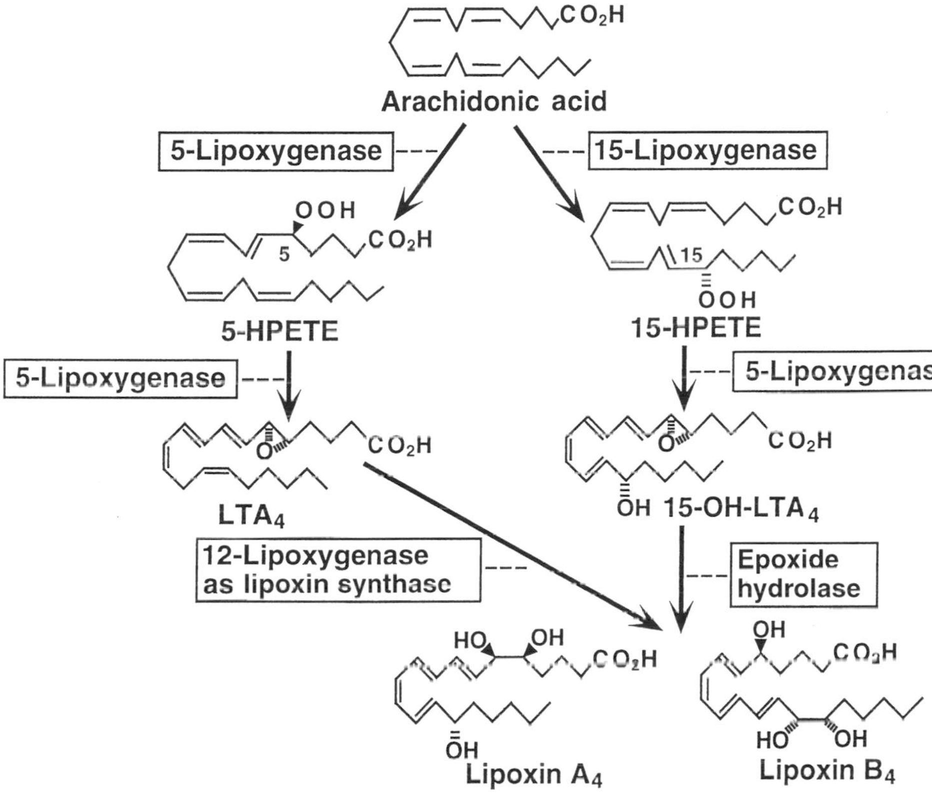

Scheme 9

1.09.7.2 Hepoxilins

Hepoxilins A_3 and B_3 are hydroxy–epoxy derivatives of arachidonic acid, and are produced by transformation of 12-HPETE (Scheme 10). The enzymatic nature of hepoxilin production from 12-HPETE was shown with rat pineal gland.[108] Various biological activities have been reported for hepoxilins.[109,110]

1.09.7.3 Leukotriene A_4 Hydrolase

The enzyme LTA_4 hydrolase is responsible for the synthesis of chemotactic LTB_4 from LTA_4. As shown in Scheme 6, 5,6-epoxide of LTA_4 is cleaved, followed by double bond migration and incorporation of a hydroxy group at position 12. Thus, LTB_4 ((5*S*),(12*R*)-dihydroxy-(*Z*-6,*E*-8,*E*-

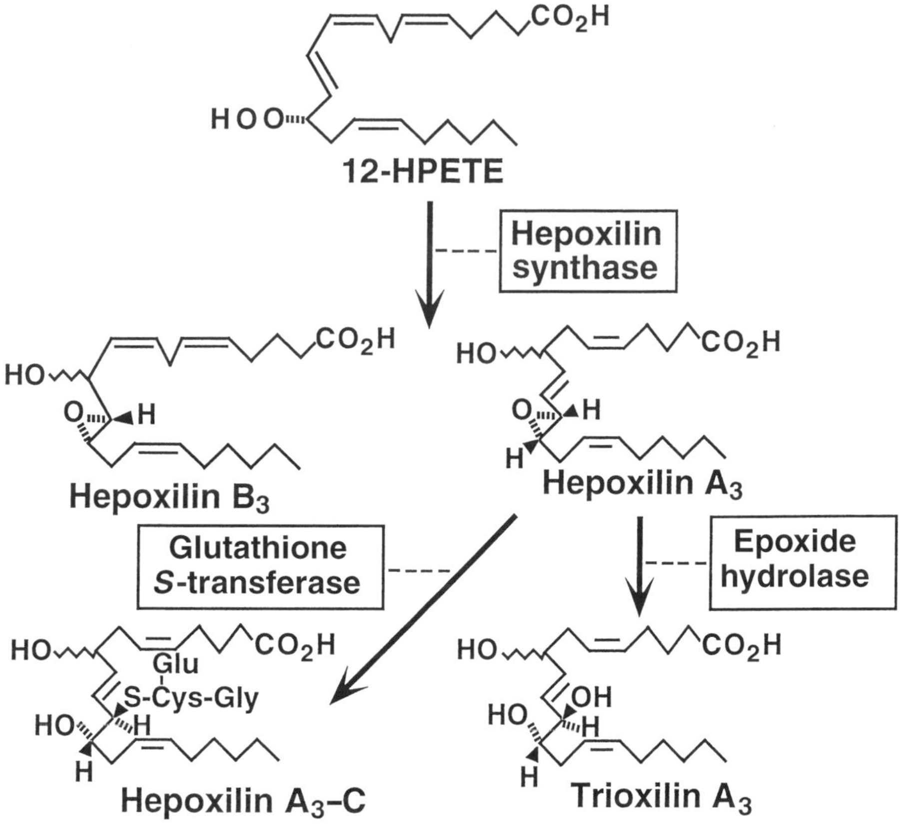

12-HPETE
Hepoxilin synthase
Hepoxilin B_3
Hepoxilin A_3
Glutathione *S*-transferase
Epoxide hydrolase
Hepoxilin A_3-C
Trioxilin A_3

Scheme 10

10,*Z*-14)-eicosatetraenoic acid) is produced by an apparent hydrolysis of LTA_4, and the enzyme is referred to as LTA_4 hydrolase (EC 3.3.2.6). Its cDNA cloning and nucleotide sequencing suggested that LTA_4 hydrolase was a member of the aminopeptidase N family, and the enzyme was found to contain an equimolar amount of Zn^{2+} and to have an aminopeptidase activity. These findings with LTA_4 hydrolase have been reviewed by Yokomizo *et al.*[111]

1.09.7.4 LTC_4 Synthase

LTC_4 and its metabolite LTD_4 are potent chemical mediators to constrict bronchi and to increase vascular permeability. LTA_4 and glutathione are the substrates for LTC_4 synthase (EC 2.5.1.37). The 5,6-epoxide is opened concomitant with the addition of glutathione to form a thioether bond (Scheme 6). As reviewed by Lam *et al.*,[112] the enzyme is localized in the microsomes of the cells of bone marrow origin, reacts specifically with LTA_4 (not with xenobiotics), and is distinguished from cytosolic and microsomal glutathione *S*-transferases. Molecular cloning of cDNA for human enzyme deduced the sequence of a 150 amino acid protein with a molecular weight of 16 567. The gene for the enzyme contains five exons and four introns, and is localized on chromosome 5q35.[113]

1.09.8 CYTOCHROME P450 IN EICOSANOID METABOLISM

The cytochrome P450 nature of TXA and PGI synthases were described earlier (see Sections 1.09.4.5 and 1.09.4.4, respectively). In addition, other roles of cytochrome P450 have been reported for the arachidonate and eicosanoid metabolism. Arachidonic acid is hydroxylated or epoxidized by monooxygenation. P450 is also involved in ω- or (ω-1)-hydroxylation of PGs and leukotrienes. This area of research has been the subject of several reviews.[114–116]

1.09.9 REFERENCES

1. W. Vogt, T. Suzuki, and S. Babilli, *Mem. Soc. Endocrinol.*, 1966, **14**, 137.
2. W. Vogt, *Adv. Prostaglandin Thromboxane Res.*, 1978, **3**, 89.
3. E. A. Dennis, *J. Biol. Chem.*, 1994, **269**, 13 057.
4. I. Kudo, M. Murakami, S. Hara, and K. Inoue, *Biochim. Biophys. Acta*, 1993, **117**, 217.
5. J. D. Clark, A. R. Schievella, E. A. Nalefski, and L.-L. Lin, *J. Lipid Mediators Cell Signalling*, 1995, **12**, 83.
6. M. Murakami, I. Kudo, and K. Inoue, *J. Lipid Mediators Cell Signalling*, 1995, **12**, 119.
7. S. Yamamoto, in "Prostaglandins and Related Substances," eds. C. Pace-Asciak and E. Granström, Elsevier, Amsterdam, 1983, p. 171.
8. W. L. Smith and L. J. Marnett, *Biochim. Biophys. Acta*, 1991, **1083**, 1.
9. C. R. Pace-Asciak and W. L. Smith, *Enzymes*, 1983, **16**, 544.
10. H. R. Herschman, *Cancer Metastasis Rev.*, 1994, **13**, 241.
11. M. Goppelt-Struebe, *Prostagl. Leukotrienes Essential Fatty Acids*, 1995, **52**, 213.
12. H. R. Herschman, W. Xie, and S. Reddy, *BioEssays*, 1995, **17**, 1031.
13. J. C. Otto and W. L. Smith, *J. Lipid Mediators Cell Signalling*, 1995, **12**, 139.
14. H. R. Herschman, *Biochim. Biophys. Acta*, 1996, **1299**, 125.
15. C. S. Williams and R. N. DuBois, *Am. J. Physiol.*, 1996, **270**, G393.
16. D. E. Griswold and J. L. Adams, *Medicinal Res. Rev.*, 1996, **16**, 181.
17. M. Pairet and G. Engelhardt, *Fundam. Clin. Pharmacol.*, 1996, **10**, 1.
18. K. K. Wu, *J. Lab. Clin. Med.*, 1996, **128**, 242.
19. W. L. Smith, R. M. Garavito, and D. L. DeWitt, *J. Biol. Chem.*, 1996, **271**, 33 157.
20. J. Vane, *Nature*, 1994, **367**, 215.
21. J. R. Vane and R. M. Botting, *Lung Biology*, 1998, **114**, 1.
22. R. Langenbach, S. G. Morham, H. F. Tiano, C. D. Loftin, B. I. Ghanayem, P. C. Chulada, J. F. Mahler, C. A. Lee, E. H. Goulding, K. D. Kluckman, H. S. Kim, and O. Smithies, *Cell*, 1995, **83**, 483.
23. S. G. Morham, R. Langenbach, C. D. Loftin, H. F. Tiano, N. Vouloumanos, J. C. Jennette, J. F. Mahler, K. D. Kluckman, A. Ledford, C. A. Lee, and O. Smithies, *Cell*, 1995, **83**, 473.
24. J. E. Dinchuk, B. D. Car, R. J. Focht, J. J. Johnston, B. D. Jaffee, M. B. Covington, N. R. Contel, V. M. Eng, R. J. Collins, P. M. Czerniak, S. A. Gorry, and J. M. Trzaskos, *Nature*, 1995, **378**, 406.
25. D. DeWitt and W. L. Smith, *Cell*, 1995, **83**, 345.
26. D. Picot, P. J. Loll, and R. M. Garavito, *Nature*, 1994, **367**, 243.
27. D. Picot and R. M. Garavito, *FEBS Lett.*, 1994, **346**, 21.
28. C. Luong, A. Miller, J. Barnett, J. Chow, C. Ramesha, and M. F. Browner, *Nature Struct. Biol.*, 1996, **3**, 927.
29. R. G. Kurumbail, A. M. Stevens, J. K. Gierse, J. J. McDonald, R. A. Stegeman, J. Y. Pak, D. Gildehaus, J. M. Miyashiro, T. D. Penning, K. Seibert, P. C. Isakson, and W. C. Stallings, *Nature*, 1996, **384**, 644.
30. P. J. Loll, D. Picot, and R. M. Garavito, *Nature Struct. Biol.*, 1995, **2**, 63.
31. R. M. Garavito, *Nature Struct. Biol.*, 1996, **3**, 897.
32. E. Giovannucci, K. M. Egan, D. J. Hunter, M. J. Stampfer, G. A Golditz, W. C. Willett, and F. E. Speizer, *New Engl. J. Med.*, 1995, **333**, 609.
33. L. J. Marnett, *Cancer Res.*, 1992, **52**, 5575.
34. S. M. Prescott and R. L. White, *Cell*, 1996, **87**, 783.
35. M. Oshima, J. E. Dinchuk, S. L. Kargman, H. Oshima, B. Hancock, E. Kwong, J. M. Trzaskos, J. F. Evans, and M. M. Taketo, *Cell*, 1996, **87**, 803.
36. K. Watanabe, Y. Urade, M. Mäder, C. Murphy, and O. Hayaishi, *Biochem. Biophys. Res. Commun.*, 1994, **203**, 1110.
37. Y. Urade, N. Fujimoto, and O. Hayaishi, *J. Biol. Chem.*, 1985, **260**, 12 410.
38. Y. Urade, N. Fujimoto, M. Ujihara, and O. Hayaishi, *J. Biol. Chem.*, 1987, **262**, 3820.
39. K. Watanabe, K. Kurihara, Y. Tokunaga, and O. Hayaishi, *Biochem. Biophys. Res. Commun.*, 1997, **235**, 148.
40. T. Ogorochi, M. Ujihara, and S. Narumiya, *J. Neurochem.*, 1987, **48**, 900.
41. K. Watanabe, R. Yoshida, T. Shimizu, and O. Hayaishi, *J. Biol. Chem.*, 1985, **260**, 7035.
42. L.-Y. Chen, K. Watanabe, and O. Hayaishi, *Arch. Biochem. Biophys.*, 1992, **296**, 17.
43. S. Hara, A. Miyata, C. Yokoyama, H. Inoue, R. Brugger, F. Lottspeich, V. Ullrich, and T. Tanabe, *J. Biol. Chem.*, 1994, **269**, 19 897.
44. A. Nagata, Y. Suzuki, M. Igarashi, N. Eguchi, H. Toh, Y. Urade, and O. Hayaishi, *Proc. Natl. Acad. Sci. USA*, 1991, **88**, 4020.
45. Y. Kanaoka, H. Ago, E. Inagaki, T. Nanayama, M. Miyano, R. Kikuno, Y. Fujii, N. Eguchi, H. Toh, Y. Urade, and O. Hayaishi, *Cell*, 1997, **90**, 1085.
46. K. Watanabe, Y. Fujii, K. Nakayama, H. Ohkubo, S. Kuramitsu, H. Kagamiyama, S. Nakanishi, and O. Hayaishi, *Proc. Natl. Acad. Sci. USA*, 1988, **85**, 11.
47. D. L. DeWitt and W. L. Smith, *J. Biol. Chem.*, 1983, **258**, 3285.
48. C. Yokoyama, A. Miyata, H. Ihara, V. Ullrich, and T. Tanabe, *Biochem. Biophys. Res. Commun.*, 1991, **178**, 1479.
49. Y. Urade, A. Nagata, Y. Suzuki, Y. Fujii, and O. Hayaishi, *J. Biol. Chem.*, 1989, **264**, 1041.
50. Y. Tone, H. Inoue, S. Hara, C. Yokoyama, T. Hatae, H. Oida, S. Narumiya, R. Shigemoto, S. Yukawa, and T. Tanabe, *Eur. J. Cell Biol.*, 1997, **72**, 268.
51. A. Miyata, S. Hara, C. Yokoyama, H. Inoue, V. Ullrich, and T. Tanabe, *Biochem. Biophys. Res. Commun.*, 1994, **200**, 1728.
52. R.-F. Shen, L. Zhang, S. J. Baek, H.-H. Tai, and K.-D. Lee, *Gene*, 1994, **140**, 261.

53. D. M. White, D. D. Mikol, R. Espinosa, B. Weimer, M. M. LeBeau, and K. Stefansson, *J. Biol. Chem.*, 1992, **267**, 23 202.
54. M. Igarashi, A. Nagata, H. Toh, Y. Urade, and O. Hayaishi, *Proc. Natl. Acad. Sci. USA*, 1992, **89**, 5376.
55. C. Yokoyama, T. Yabuki, H. Inoue, Y. Tone, S. Hara, T. Hatae, M. Nagata, E. Takahashi, and T. Tanabe, *Genomics*, 1996, **36**, 296.
56. A. Miyata, C. Yokoyama, H. Ihara, S. Bandoh, O. Takeda, E. Takahashi, and T. Tanabe, *Eur. J. Biochem.*, 1994, **224**, 273.
57. S. J. Baek, K.-D. Lee, and R.-F. Shen, *Gene*, 1996, **173**, 251.
58. H. Toh, H. Kubodera, N. Nakajima, T. Sekiya, N. Eguchi, T. Tanaka, Y. Urade, and O. Hayaishi, *Protein Eng.*, 1996, **9**, 1067.
59. A. Hoffmann, H. S. Conradt, G. Gross, M. Nimtz, F. Lottspeich, and U. Wurster, *J. Neurochem.*, 1993, **61**, 451.
60. Y. Urade, K. Kitahama, H. Ohishi, T. Kaneko, N. Mizuno, and O. Hayaishi, *Proc. Natl. Acad. Sci. USA*, 1993, **90**, 9070.
61. Y. Urade, M. Ujihara, Y. Horiguchi, M. Igarashi, A. Nagata, K. Ikai, and O. Hayaishi, *J. Biol. Chem.*, 1990, **265**, 371.
62. D. J. Meyer, R. Muimo, M. Thomas, D. Coates, and R. E. Isaac, *Biochem. J.*, 1996, **313**, 223.
63. K. Watanabe, Y. Iguchi, S. Iguchi, Y. Arai, O. Hayaishi, and L. J. Roberts II, *Proc. Natl. Acad. Sci. USA*, 1986, **83**, 1583.
64. H. Hayashi, Y. Fujii, K. Watanabe, Y. Urade, and O. Hayaishi, *J. Biol. Chem.*, 1989, **264**, 1036.
65. M. Hecker and V. Ullrich, *J. Biol. Chem.*, 1989, **264**, 141.
66. V. Ullrich, L. Castle, and P. Weber, *Biochem. Pharmacol.*, 1981, **30**, 2033.
67. V. Ullrich and R. Brugger, *Angew. Chem. Int. Ed. Engl.*, 1994, **33**, 1911.
68. M. Haurand and V. Ullrich, *J. Biol. Chem.*, 1985, **260**, 15 059.
69. R.-F. Shen and H.-H. Tai, *J. Biol. Chem.*, 1986, **261**, 11 592.
70. R. Nüsing, S. Schneider-Voss, and V. Ullrich, *Arch. Biochem. Biophys.*, 1990, **280**, 325.
71. S. Yamamoto, in "Prostaglandins and Cardiovascular Diseases," eds. T. Ozawa, K. Yamada, and S. Yamamoto, Japan Scientific Societies Press, Tokyo, 1986, p. 71.
72. C. M. Ensor, J.-Y. Yang, R. T. Okita, and H.-H. Tai, *J. Biol. Chem.*, 1990, **265**, 14 888.
73. C. M. Ensor and H.-H. Tai, *J. Lipid Mediators Cell Signalling*, 1995, **12**, 313.
74. T. Schewe, S. M. Rapoport, and H. Kühn, *Adv. Enzymol.*, 1986, **58**, 191.
75. H. Kühn, T. Schewe, and S. M. Rapoport, *Adv. Enzymol.*, 1986, **58**, 273.
76. S. Yamamoto, *Prostagl., Leukotrienes Essential Fatty Acids*, 1989, **35**, 219.
77. T. Shimizu and L. S. Wolfe, *J. Neurochem.*, 1990, **55**, 1.
78. S. Yamamoto, *Free Radical Biol. Med.*, 1991, **10**, 149.
79. E. Sigal, *Am. J. Physiol.*, 1991, **260**, L13.
80. S. Yamamoto, *Biochim. Biophys. Acta*, 1992, **1128**, 117.
81. C. D. Funk, *Prog. Nuc. Acid Res. Mol. Biol.*, 1993, **45**, 67.
82. C. D. Funk, *Biochim. Biophys. Acta*, 1996, **1304**, 65.
83. B. Samuelsson, *Science*, 1983, **220**, 568.
84. S. Hammarström, *Annu. Rev. Biochem.*, 1983, **52**, 355.
85. A. W. Ford-Hutchinson, M. Gresser, and R. N. Young, *Annu. Rev. Biochem.*, 1994, **63**, 383.
86. O. Rådmark, *J. Lipid Mediators Cell Signalling*, 1995, **12**, 171.
87. R. A. Lepley and F. A. Fitzpatrick, *J. Biol. Chem.*, 1994, **269**, 2627.
88. P. J. Vickers, *J. Lipid Mediators Cell Signalling*, 1995, **12**, 185.
89. T. Yoshimoto and S. Yamamoto, *J. Lipid Mediators Cell Signalling*, 1995, **12**, 195.
90. S. Yamamoto, H. Suzuki, and N. Ueda, *Prog. Lipid Res.*, 1997, **36**, 23.
91. K. Kishimoto, M. Nakamura, H. Suzuki, T. Yoshimoto, S. Yamamoto, T. Takao, Y. Shimonishi, and T. Tanabe, *Biochim. Biophys. Acta*, 1996, **1300**, 56.
92. B.-J. Thiele, M. Berger, H. Thiele, K. Schwarz, S. Borngräber, H. Kühn, I. Reimann, and A. Huth, "5th International Conference on Eicosanoids and Other Bioactive Lipids in Cancer, Inflammation and Related Diseases," La Jolla, CA, September 17–20, 1997, Abstract no. 23, eds. K. V. Honn, L. J. Marnett, S. Nigam, and E. Dennis, 1997.
93. C. D. Funk, D. S. Keeney, E. H. Oliw, W. E. Boeglin, and A. R. Brash, *J. Biol. Chem.*, 1996, **271**, 23 338.
94. H. Kühn and B.-J. J. Thiele, *Lipid Mediators Cell Signalling*, 1995, **12**, 157.
95. J. J. Murray and A. R. Brash, *Arch. Biochem. Biophys.*, 1988, **265**, 514.
96. T. Schewe, W. Halangk, Ch. Hiebsch, and S. M. Rapoport, *FEBS Lett.*, 1975, **60**, 149.
97. S. Ylä-Herttuala, M. E. Rosenfeld, S. Parthasarathy, C. K. Glass, E. Sigal, J. L. Witztum, and D. Steinberg, *Proc. Natl. Acad. Sci. USA*, 1990, **87**, 6959.
98. D. Harats, M. A. Mulkins, and E. Sigal, *Trends Cardiovasc. Med.*, 1995, **5**, 29.
99. A. R. Brash, W. E. Boeglin, and M. S. Chang, *Proc. Natl. Acad. Sci. USA*, 1997, **94**, 6148.
100. M. Gschwendt, G. Fürstenberger, W. Kittstein, E. Besemfelder, W. E. Hull, H. Hagedorn, H. J. Opferkuch, and F. Marks, *Carcinogenesis*, 1986, **7**, 449.
101. S. M. Fischer, J. K. Baldwin, D. W. Jasheway, K. E. Patrick, and G. S. Cameron, *Cancer Res.*, 1988, **48**, 658.
102. G. Fürstenberger, H. Hagedorn, T. Jacobi, E. Besemfelder, M. Stephan, W.-D. Lehmann, and F. Marks, *J. Biol. Chem.*, 1991, **266**, 15 738.
103. M. A. Hughes and A. R. Brash, *Biochim. Biophys. Acta*, 1991, **1081**, 347.
104. M. Jisaka, R. B. Kim, W. E. Boeglin, L. B. Nanney, and A. R. Brash, *J. Biol. Chem.*, 1997, **272**, 24 410.
105. A. R. Brash, W. E. Boeglin, M. S. Chang, and B.-H. Shieh, *J. Biol. Chem.*, 1996, **271**, 20 949.
106. C. N. Serhan, *Biochim. Biophys. Acta*, 1994, **1212**, 1.
107. H. R. Brady and C. N. Serhan, *Curr. Opin. Nephrol. Hypertension*, 1996, **5**, 20.
108. C. R. Pace-Asciak, D. Reynaud, and P. Demin, *Biochem. Biophys. Res. Commun.*, 1993, **197**, 869.
109. C. R. Pace-Asciak, *Biochim. Biophys. Acta*, 1994, **1215**, 1.
110. C. R. Pace-Asciak, D. Reynaud, and P. M. Demin, *Lipids*, 1995, **30**, 107.
111. T. Yokomizo, N. Uozumi, T. Takahashi, K. Kume, T. Izumi, and T. Shimizu, *J. Lipid Mediators Cell Signalling*, 1995, **12**, 321.

112. B. K. Lam, J. F. Penrose, K. Xu, and K. F. Austen, *J. Lipid Mediators Cell Signalling*, 1995, **12**, 333.
113. J. F. Penrose, J. Spector, M. Baldasaro, K. Xu, J. Boyce, J. P. Arm, K. F. Austen, and B. K. Lam, *J. Biol. Chem.*, 1996, **271**, 11 356.
114. J. C. McGiff, *Annu. Rev. Pharmacol. Toxicol.*, 1991, **31**, 339.
115. J. H. Capdevila, J. R. Falck, and R. W. Estabrook, *FASEB J.*, 1992, **6**, 731.
116. J. C. McGiff, M. Steinberg, and J. Quilley, *Trends Cardiovasc. Med.*, 1996, **6**, 4.

1.10
Molecular Evolution of Proteins Involved in the Arachidonic Acid Cascade

REIKO KIKUNO, HIROMI DAIYASU, and HIROYUKI TOH
Biomolecular Engineering Research Institute, Osaka, Japan

1.10.1 INTRODUCTION

Progress in the investigation of the arachidonic acid cascade has occurred quite rapidly in the 1990s. Major advances have been achieved by the cloning of the cDNAs and/or the genes for the proteins involved in the metabolic pathway. The nucleotide sequences of the cloned cDNAs and/or the genes have been determined, and have enabled us to investigate the arachidonic acid cascade from the molecular evolutionary viewpoint. In this chapter, we will discuss the molecular evolution of the enzymes, the receptors, and a transporter involved in the metabolic pathway. However, the molecular phylogeny of phospholipase A_2 is not included in the current discussion. There are many evolutionary unrelated phospholipases, the inclusion of which would be far beyond the scope of this analysis.

To construct the molecular phylogeny, the amino acid sequences of a subject protein and its relatives were first aligned. Then, the genetic distance of each aligned pair was calculated as a sequence difference with a Poisson correction. Using the set of genetic distances thus obtained, an unrooted phylogenetic tree was constructed by the neighbor-joining method.[1] However, this was drawn as a rooted tree to facilitate recognition of the tree topology. Therefore, the root was introduced to clarify the evolutionary position of the subject protein in the family. The scale bar in each figure in this chapter represents a branch length of 0.1 substitution of an amino acid residue per site. The sequence data used in this paper were taken from several sequence databases. The identification code for a protein in a database is written after or below the protein name in the phylogenetic tree and the database is indicated by an abbreviation in parentheses (gb, Genbank; pir, PIR; prf, PRF; sp, Swiss Prot).

1.10.2 FATTY ACID CYCLOOXYGENASES-1 AND -2

Cyclooxygenase[2–4] is an enzyme of about 600 amino acid residues in length, which is involved in the rate-limiting step of prostaglandin and thromboxane (TX) biosynthesis. The enzyme associates with the membranes of the endoplasmic reticulum and the nuclear envelope. Cyclooxygenase is classified into two isoforms, cyclooxygenases-1 and -2. The former is regarded as a constitutive or housekeeping enzyme, while the latter is inducibly expressed. Both isoforms are similar to each other, and the sequence identity between them is about 70%. The enzyme is a so-called mosaic protein, which consists of three domains, an EGF-like domain, a membrane-binding domain, and a catalytic domain. The EGF-like domain is about 40 amino acid residues in length, and is located in the N-terminal region of the enzyme. However, the functional role of this domain remains unknown. The membrane-binding domain is about 50 amino acid residues in length. The catalytic domain is about 500 amino acid residues in length, and shows similarity in both its primary and tertiary structures to the members of the peroxidase family.[5,6] This family includes myeloperoxidase, eosinophil peroxidase, thyroid peroxidase, lactoperoxidase, and other peroxidases. The members of the peroxidase family were first identified in mammals. Subsequently, homologues have been found in fish, *Drosophila*, squids, and *Caenorhabditis elegans*. The catalytic domains of four cyclooxygenases-1 and seven cyclooxygenases-2 were aligned with 12 representative members of the peroxidase family. The phylogenetic tree is shown in Figure 1. The tree topology suggests that the divergence of cyclooxygenase from other peroxidases is quite ancient, and occurred at least before the divergence between mammals and *C. elegans*. After this divergence, the ancestral gene for cyclooxygenase was subjected to exon shuffling, and encoded a mosaic protein with three domains. The cyclooxygenases-1 and -2 form distinct clusters. The Y and W nodes are considered to correspond to mammalian divergence, while the Z node represents avian and mammalian divergence. A duplication of an ancestral cyclooxygenase gene, which corresponds to the X node, occurred before avian and mammalian divergence, and thus generated the genes for the two isoforms.

1.10.3 PROSTAGLANDIN D_2 SYNTHASE

Prostaglandin D_2 (PGD) synthase catalyzes the isomerization of prostaglandin H_2 (PGH) to produce PGD. PGD synthase is classified into two types, brain-type PGD synthase[6–8] and hematopoietic PGD synthase.[6,7,9] The two types are evolutionarily unrelated. The former belongs to the lipocalin family, while the latter is a member of the glutathione *S*-transferase family.

1.10.3.1 Brain-type PGD Synthase

Brain-type PGD synthase is about 190 amino acid residues in length. The enzyme shows weak, but significant, sequence similarity to the lipocalins.[10] The lipocalins are a group of small secretory proteins whose genes are expressed in various secretory tissues. The gene products are secreted into body fluids, where they are involved in binding and transport of small lipophilic ligands such as retinal, retinoic acid, bilin, and some types of odorants and pheromones. Except for one enzyme, PGD synthase, the lipocalins are nonenzymatic proteins, and include α_1-microglobulin, β-lactoglobulin, α_1-acid glycoprotein, retinol-binding protein, and bilin-binding protein among others. To construct the phylogenetic tree, the amino acid sequences of 10 PGD synthases were aligned with

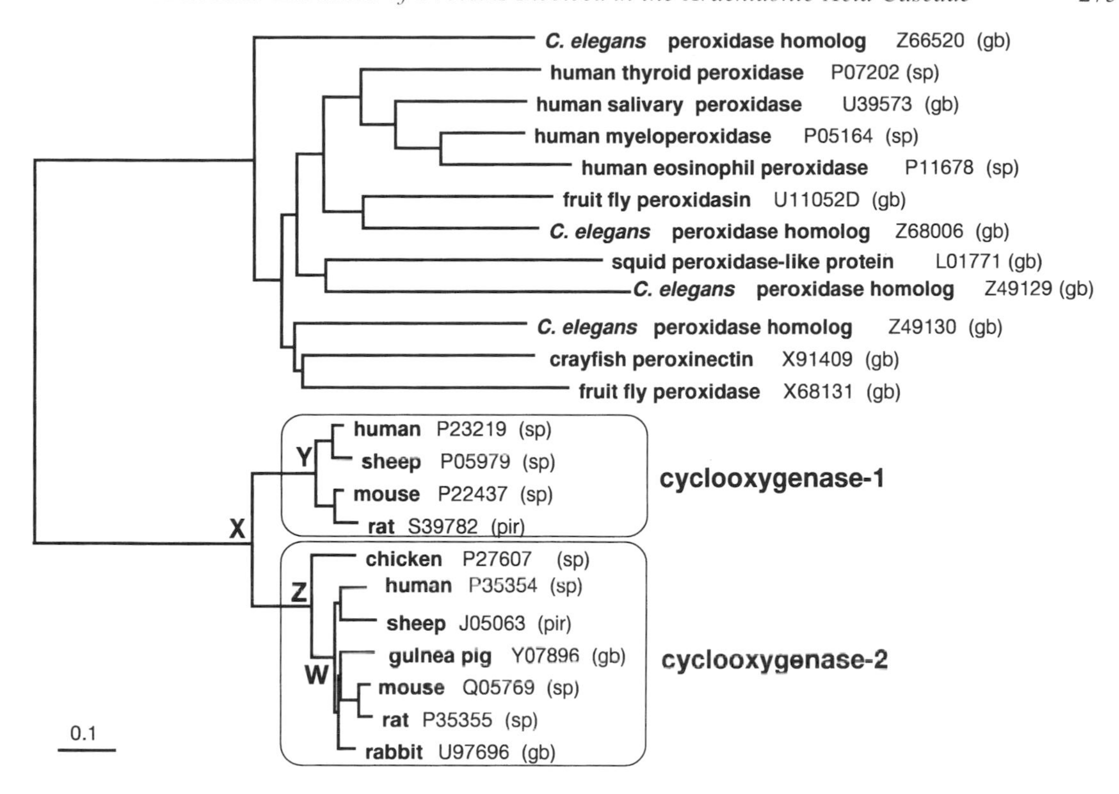

Figure 1 Molecular phylogeny of peroxidases including catalytic domains of cyclooxygenases.

16 representative lipocalins. The phylogenetic tree is shown in Figure 2. PGD synthases from mammals form a cluster in the tree, as do the amphibian counterparts of the enzymes. The topology suggests that the Z node corresponds to mammalian divergence, while the Y node represents the divergence between mammals and amphibians. The tree also suggests that the PGD synthases are distantly related to the other lipocalins. At the X node, PGD synthase diverged from the other lipocalins. The neutrophil gelatinase-associated lipocalin is most closely related to the enzymes within the family, although the overall amino acid sequence identity between the lipocalins and the PGD synthases is only 36%. It is thought that PGD synthase acquired its catalytic activity in the branch connecting the X and Y nodes.

1.10.3.2 Hematopoietic PGD Synthase

In contrast to the brain-type PGD synthase, the hematopoietic PGD synthase is expressed in peripheral tissues. In particular, the enzyme is involved in the production of the D and the J series of prostanoids in the immune system and mast cells. Hematopoietic PGD synthase requires a glutathione molecule for its catalytic activity.

Only the amino acid sequence of the rat hematopoietic PGD synthase has been obtained. The enzyme is 199 amino acid residues in length, and shows sequence similarity to glutathione *S*-transferases. The glutathione *S*-transferase family is divided into five classes, α, μ, π, σ, and θ.[11] To investigate the evolutionary position of the enzyme in the family, the amino acid sequences of representative members of four of the classes and rat PGD synthase were aligned. The θ class was not analyzed, due to the high sequence divergence. Figure 3 shows the phylogenetic tree of the glutathione *S*-transferase family, including PGD synthase. The tree suggests that PGD synthase belongs to the σ class. Previously, σ class glutathione *S*-transferases had been identified only in invertebrates, such as insects, cephalopods, flukes, and nematodes. Thus, the rat hematopoietic PGD synthase is the first vertebrate homologue of the σ class. However, the sequence similarities of the rat enzyme to the other members of the σ class are very low. The glutathione *S*-transferase from the housefly (*Musca domestica*) shows the highest sequence similarity to the rat enzyme. However, the sequence identity between them is only 40%.

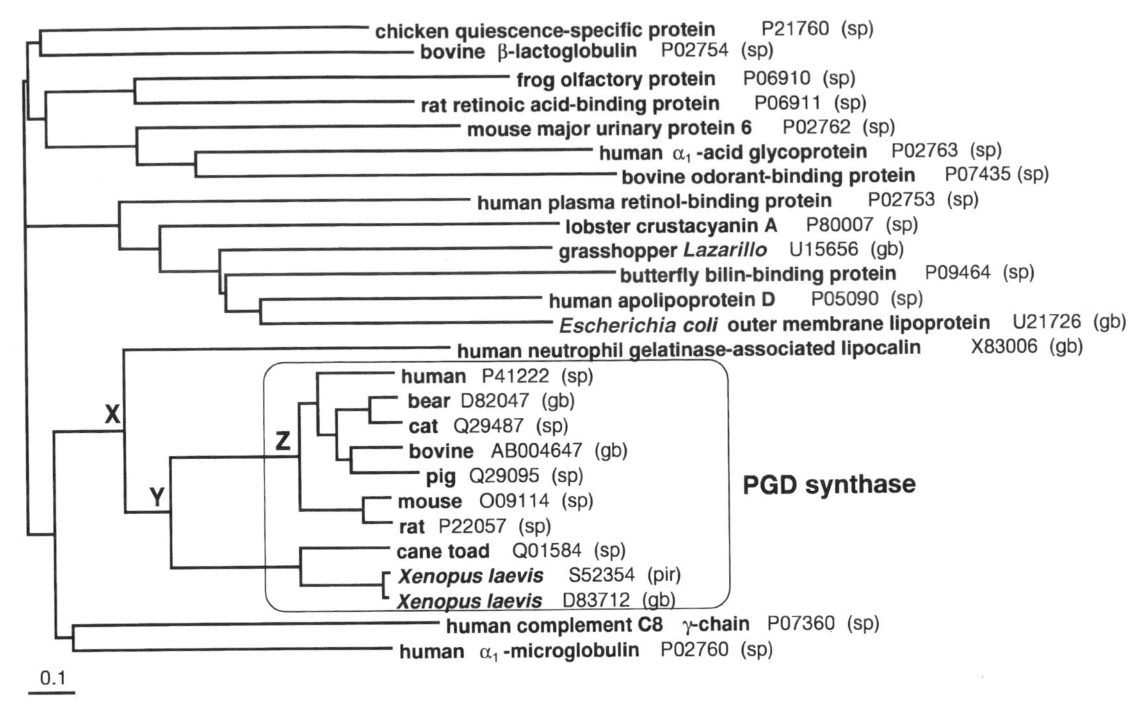

Figure 2 Molecular phylogeny of lipoalins including brain-type PGD synthases.

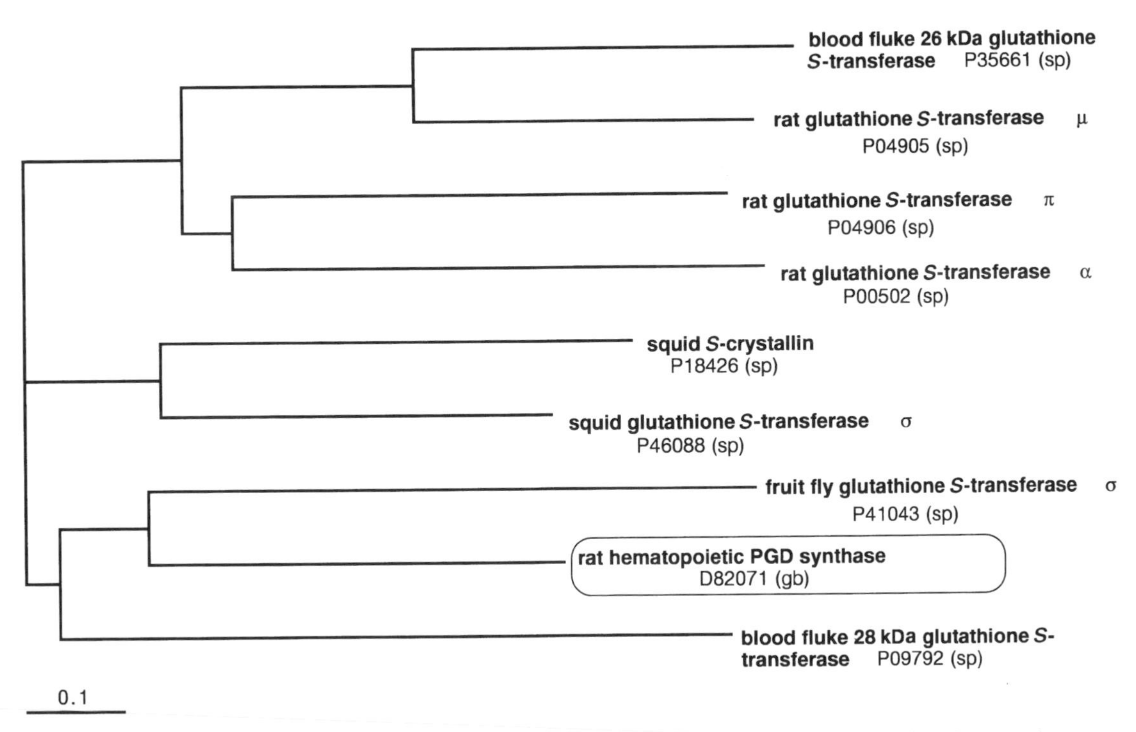

Figure 3 Molecular phylogeny of glutathione *S*-transferases including hematopoietic PGD synthase.

1.10.4 PROSTAGLANDIN $F_{2\alpha}$ SYNTHASE

The production of 11-*epi*-prostaglandin $F_{2\alpha}$ (11-*epi*-PGF) and PGF is catalyzed by PGF synthase.[7] The former is converted from PGH, while the latter is derived from PGD. Although both of the enzyme activities require NADPH, the active sites for the reactions are different from each other. The enzyme, which consists of about 300 amino acid residues, belongs to the aldo–keto reductase family.[6,7] Like other members of the protein family, PGF synthase can catalyze the reduction of several carbonyl compounds. In addition, human liver aldehyde reductase, a member of the aldo–keto reductase family, is also able to catalyze the reduction of PGH, although the enzyme cannot use PGD as the substrate. To investigate the evolutionary position of the enzyme within the aldo–keto reductase family, the amino acid sequences of three bovine PGF synthases were aligned with those of 18 representative members of the family. The phylogenetic tree is shown in Figure 4. The PGF synthases are present in a cluster of mammalian dehydrogenases. Within the cluster, the human enzymes, *trans*-1,2-dehydrobenzene-1,2-diol dehydrogenase and chlordecon reductase, are most closely related to the PGF synthases. The sequence identities between them are about 70%. It is interesting to note that frog ρ crystallin is near the cluster of mammalian dehydrogenases, including the PGF synthases.[6,7] The amino acid sequence identities between the PGF synthases and frog ρ crystallin are about 60%. Frog ρ crystallin has NADPH-binding activity, and weakly catalyzes the conversion of PGH to PGF. However, frog ρ crystallin lacks the activity to convert PGD to 11-*epi*-PGF.

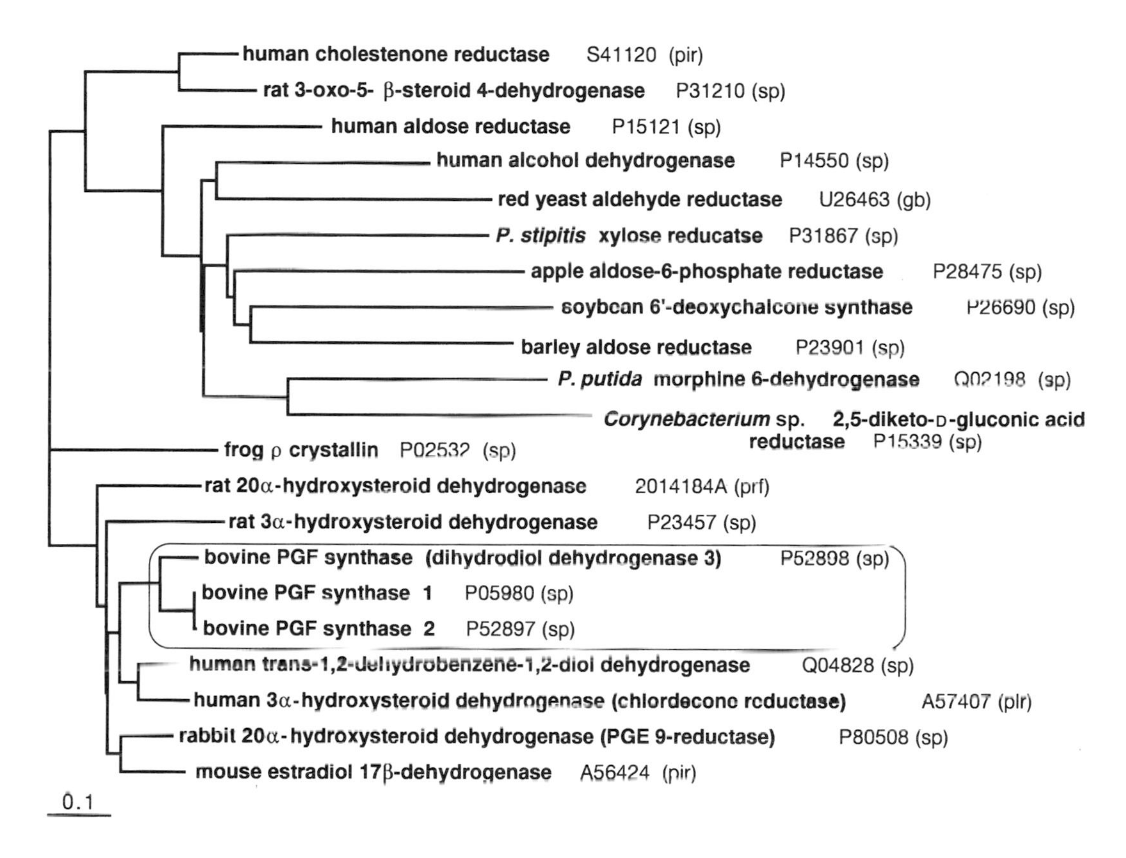

Figure 4 Molecular phylogeny of aldo–keto reductases including PGF synthases.

1.10.5 THROMBOXANE SYNTHASE AND PROSTAGLANDIN I_2 SYNTHASE

Several members of the P450 family are involved in the metabolism of eicosanoids, such as the oxygenation of arachidonic acid, and the ω-hydroxylation of leukotrienes and prostaglandins. Both TX synthase[6,12] and prostaglandin I_2 (PGI) synthase[12] belong to this family. The former catalyzes the conversion from PGH to TX, while the reaction from PGH to PGI is catalyzed by the latter. Currently, only the amino acid sequences of four TX synthases and those of three PGI synthases have been determined. The members of the P450 family are about 400–500 amino acid residues in length. An enormous amount of P450 sequence data are available. Therefore, 12 members that are relatively close to the TX and PGI synthases were selected to examine the evolutionary position of the enzyme. The phylogenetic tree is shown in Figure 5. P450 III and IV are most closely related to TX synthase, although the sequence identities between them are only 30%. On the other hand, cholesterol 7α-monooxygenase is most closely related to the PGI synthases. However, the sequence identities are 30%. As shown in the figure, the TX synthases are distantly related to the PGI synthases. That is, these two enzymes appeared independently during the course of molecular evolution. The X and Y nodes are considered to correspond to mammalian divergence.

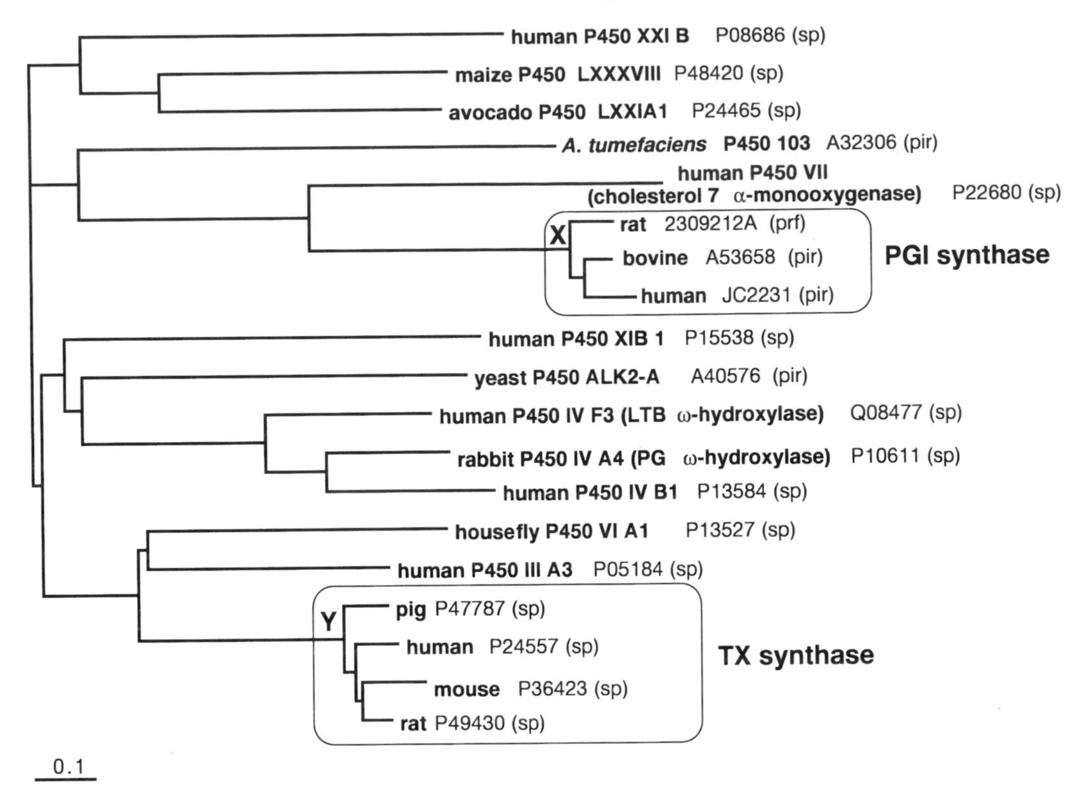

Figure 5 Molecular phylogeny of P450s including TX and PGI synthases.

1.10.6 LIPOXYGENASE

Mammalian lipoxygenases,[13–15] such as cyclooxygenase and cytochrome P450, catalyze the oxygenation of arachidonic acid. The enzymes, which are about 700 amino acid residues in length, are classified into four types, the (5*S*)-, (8*S*)-, (12*S*)-, and (15*S*)-lipoxygenases, according to the position of oxygenation of arachidonic acid. The amino acid sequences of the lipoxygenases were aligned to construct the phylogenetic tree. Various plants also have lipoxygenases which are distantly related to their mammalian counterparts. Therefore, 18 representative plant enzymes were included in the alignment as the outgroup for the mammalian lipoxygenases. The lipoxygenases derived from mammalian and plant sources form fatty acid hydroperoxides with the (*S*) stereoconfiguration, while (*R*) configuration-specific lipoxygenases have been identified in several species of invertebrates. The cDNA for an (8*R*)-lipoxygenase from a prostaglandin-containing coral, *Plexaura homomalla*, has also been sequenced, and was shown to encode a polypeptide of 715 amino acid residues. This amino acid sequence was also included in the alignment for the phylogenetic analysis.

The phylogenetic tree of the lipoxygenases is shown in Figure 6. At the A node, the gene for the coral (8*R*)-lipoxygenase and an ancestral gene for other mammalian (*S*)-lipoxygenases diverge. The tree topology suggests that an ancestral enzyme for animal and plant lipoxygenases formed fatty acid hydroperoxides with the (*S*) configuration. At the B node, the ancestral gene was duplicated again, and two copies of the (*S*)-lipoxygenase gene were generated. The descendants from one copy include the (5*S*)-, (8*S*)-, and (15*S*)-lipoxygenases, while the (12*S*)- and (15*S*)-lipoxygenases were generated from another copy. At the C node, gene duplication occurred again. One of the duplicated genes encoded an ancestral enzyme of the mammalian (5*S*)-lipoxygenase. The H node is considered to correspond to mammalian divergence, where the orthologous (5*S*)-lipoxygenases were generated by species divergence. The (8*S*)- and (15*S*)-lipoxygenases have evolved from another copy generated at the C node, and functionally diverged at the G node. The tree topology suggests that an ancestral enzyme corresponding to the D node carried the (12*S*)-lipoxygenase activity. That is, two genes generated at the D node first encoded the (12*S*)-lipoxygenase activity. The I node is considered to correspond to mammalian divergence. A gene duplication occurred at the E node. Both of the duplicated genes encoded the (12*S*)-lipoxygenases. One of them was duplicated at the F node to yield two copies of the (12*S*)-lipoxygenase genes. The J node is considered to correspond to mammalian divergence. Another (12*S*)-lipoxygenase gene, formed at the F node, diverged further at the K node. The (12*S*)- and (15*S*)-lipoxygenases that diverged at the K node may be orthologous, although their function has differentiated. If so, the K node also corresponds to mammalian divergence. The tree topology suggests that four types of mammalian (*S*)-lipoxygenases already existed before mammalian divergence. In addition, the (15*S*)-lipoxygenases were independently generated at least twice. At this stage, it is difficult to infer the function of the ancestral enzymes corresponding to the A and B nodes.

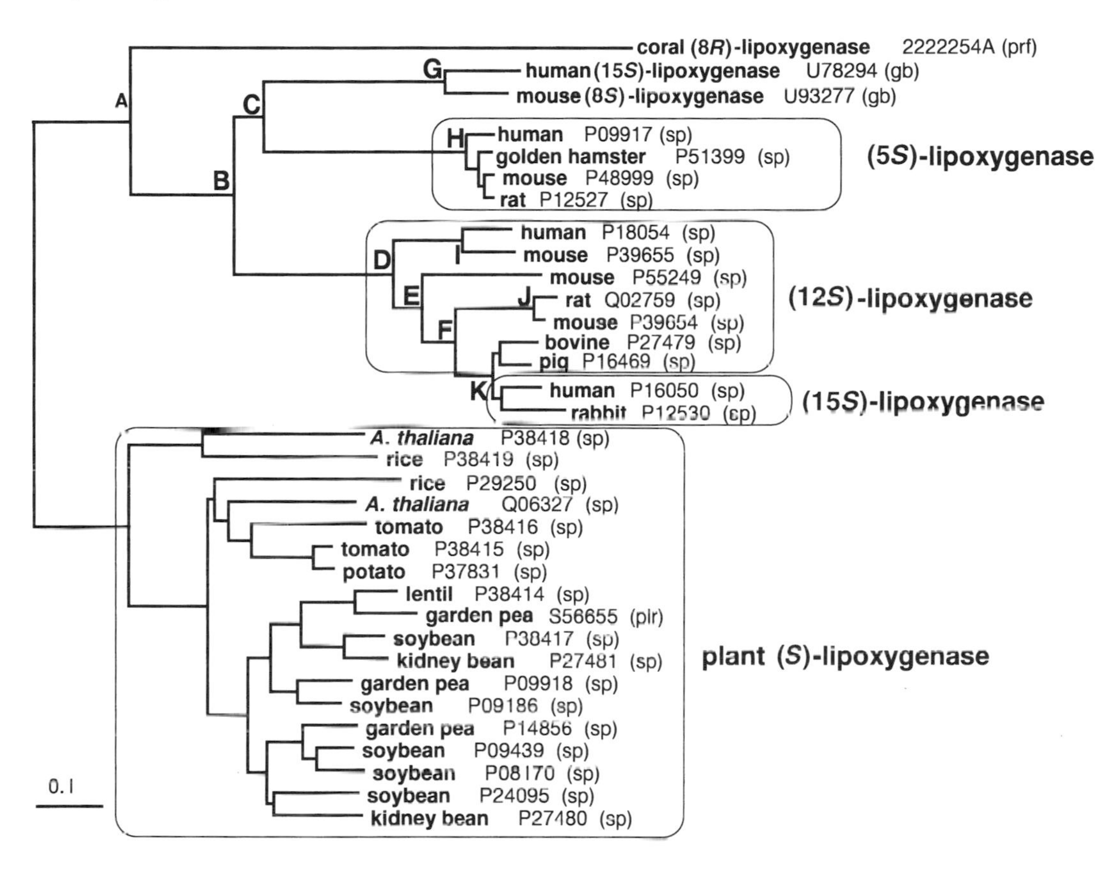

Figure 6 Molecular phylogeny of lipoxygenases.

1.10.7 LEUKOTRIENE A_4 HYDROLASE

Leukotriene A_4 (LTA) hydrolase[16–18] catalyzes the formation of leukotriene B_4 (LTB). The amino acid sequence of the enzyme shows weak, but significant, similarity to those of aminopeptidases. The substrate of the enzyme, LTA, is the product of 5-lipoxygenase. However, the expression of the gene encoding the enzyme is observed even in cells where 5-lipoxygenase is not expressed. LTA hydrolase also shows peptidase activity toward synthetic peptide substrates, such as alanine-4-nitroanilide and leucine-4-nitroanilide. This suggests that the enzyme may act as a peptidase in some cells. The amino peptidases homologous to the LTA hydrolase are known as zinc metalloproteinases.[6] Likewise, LTA hydrolase contains a zinc ion, which is essential for both the hydrolase activity and the peptidase activity.

To determine the evolutionary position of LTA hydrolase within the aminopeptidase family, the amino acid sequences of four hydrolases were aligned with those of 13 representative aminopeptidases. The phylogenetic tree is shown in Figure 7. The X node is considered to correspond to mammalian divergence. The protein derived from *Saccharomyces cerevisiae* is the most closely related to the hydrolases. The sequence identities between them are about 40%. However, the function of the protein has not been identified yet. Aminopeptidase N from *Escherichia coli* is the next closest to the hydrolases, although the sequence identities between them were only about 20%. That is, the divergence of the hydrolases from the other peptidases is considered to be ancient.

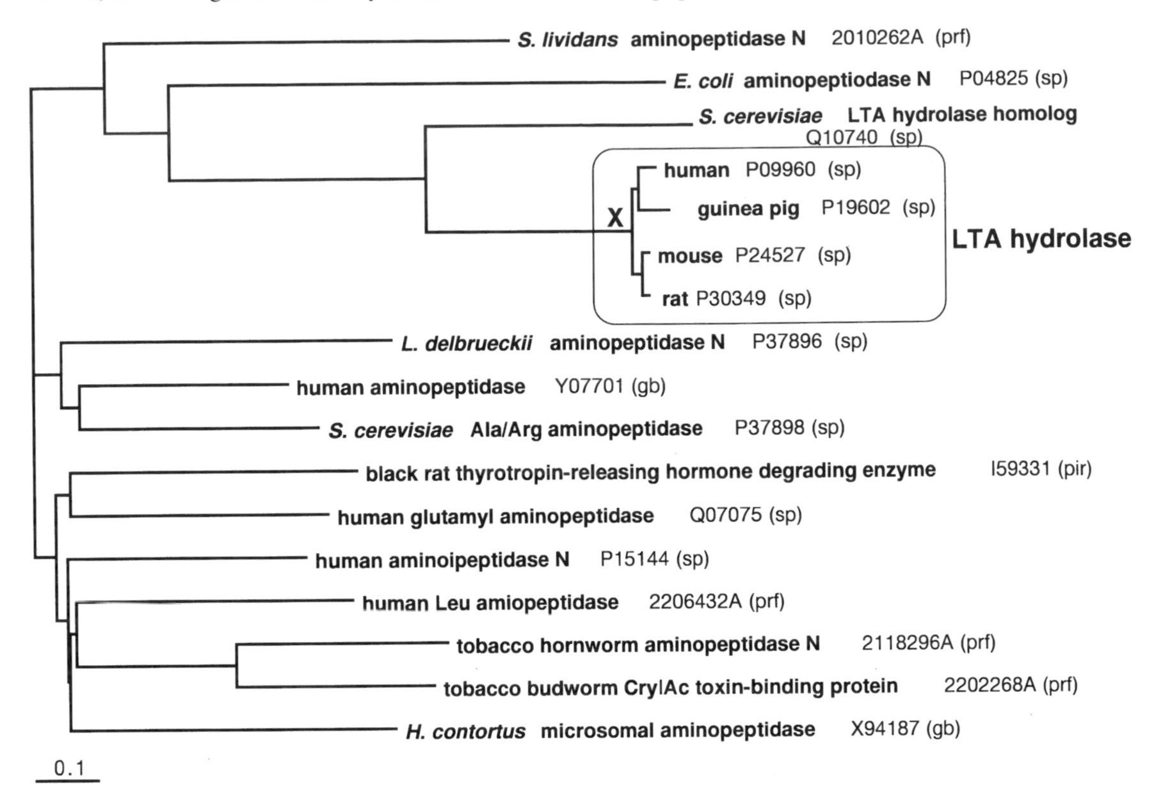

Figure 7 Molecular phylogeny of aminopeptidases including LTA hydrolases.

1.10.8 LEUKOTRIENE C_4 HYDROLASE AND 5-LIPOXYGENASE-ACTIVATING PROTEIN

The amino acid sequences of 5-lipoxygenase-activating protein (FLAP)[19] and leukotriene C_4 (LTC) synthase[20] show similarity to that of microsomal glutathione *S*-transferase II. The glutathione *S*-transferase is evolutionarily unrelated to the enzymes homologous to hematopoietic PGD synthase. Both LTC synthase and microsomal glutathione *S*-transferase II catalyze the conjugation of reduced glutathione with LTA to form LTC. In contrast, no catalytic activity has been demonstrated for FLAP. The protein binds to arachidonic acid to increase the efficiency of 5-lipoxygenase catalysis. However, MK-886, a FLAP inhibitor, can inhibit LTC synthase. They are integral membrane proteins of about 150 amino acid residues.

The phylogenetic tree of the protein family is shown in Figure 8. Microsomal glutathione *S*-transferase II is relatively close to the LTC synthase, although the sequence identity is about 40%. On the other hand, the FLAPs show about 30% sequence identity to the LTC synthases and microsomal glutathione *S*-transferase II. The tree suggests that the divergence of these proteins is ancient, and that FLAP and LTC synthase have been independently integrated into the arachidonic acid cascade after divergence. The X and Y nodes are considered to correspond to mammalian divergence.

1.10.9 EICOSANOID RECEPTORS

The physiological actions of various eicosanoids are mediated through the corresponding receptors.[21–24] The receptors belong to the rhodopsin superfamily or the family of G-protein-coupled

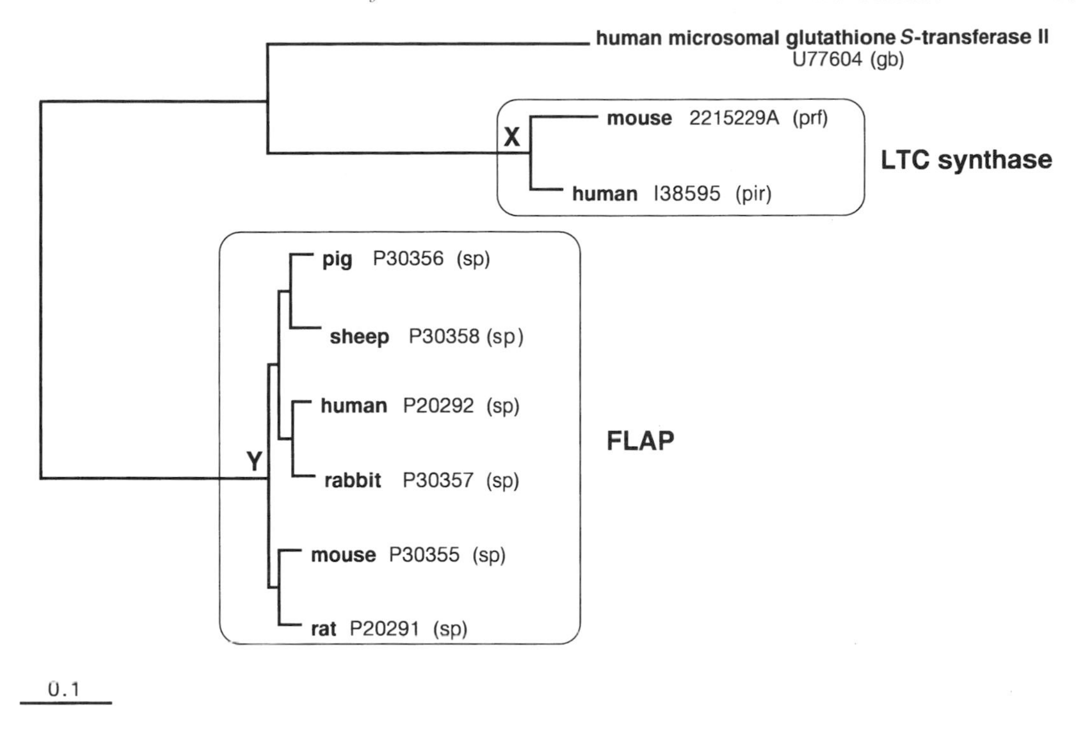

Figure 8 Molecular phylogeny of LTC hydrolases and FLAPs.

receptors with seven membrane-spanning regions. Currently, the nucleotide sequences of the cDNAs and/or genes for the receptors for prostaglandins (PGD, PGE, PGF, and PGI), TX, lipoxin A_4 (LX), and LTB have been determined. Psychoactive cannabinoid compounds also bind to the corresponding G-protein-coupled receptor in the brain.[25,26] Arachidonylethanolamide (anadamide) is an arachidonic acid derivative found in porcine brain that has been identified as the endogeneous ligand for the cannabinoid receptor. 2-Arachidonylglycerol was also found as an endogeneous ligand. Platelet-activating factor (PAF) is a lipid mediator, which also binds to the corresponding G-protein-coupled receptor.[27] The amino acid sequences of these receptors and their close relatives in the family were aligned for the phylogenetic analysis.

The phylogenetic tree is shown in Figures 9 and 10. The receptors for prostaglandins and TX constitute a distinct cluster in the tree (Figure 9), while the receptors for LX and LTB are present in the cluster of peptide receptors (Figure 10). The receptors for PAF are also found in the cluster of peptide receptors (Figure 10). In contrast, the cannabinoid receptors are distantly related to the receptors for other lipid-derived mediators (Figure 10). The cluster of receptors for prostaglandins and TX is further divided into three subclusters, I, II, and III (Figure 9). The PGE receptors are classified into four subtypes, EP_1, EP_2, EP_3, and EP_4. Subcluster I contains EP_3. In contrast, subcluster II includes EP_1, PGF receptors, and TX receptors. Subcluster III is composed of EP_2, EP_4, and PGI receptors. Thus, all of the subclusters include PGE receptors, which suggests that the ancestral receptor corresponding to the A node was a PGE receptor. EP_3 is associated with inhibition of adenylate cyclase, while EP_2 and EP_4 stimulate adenylate cyclase. Therefore, the ancestral PGE receptor may have been involved in cAMP metabolism. In contrast, EP_1 is involved in PI turnover and calcium mobilization. The EP_1 and EP_3 subtypes diverged at the B node. The functional coupling of the EP_1 subtype with PI turnover and calcium mobilization occurred after the divergence at the B node. The TX receptor then diverged from the EP_1 subtype at the C node. Subsequently, the PGF receptor diverged from EP_1 at the D node. At the E node, two PGE subtypes, EP_2 and EP_4, diverged. Then, the PGI receptor diverged from the subtype EP_2 at the F node. The G, H, I, J, K, and L nodes are all considered to correspond to mammalian divergence.

As shown in Figure 10, the cluster of peptide receptors is divided into two subclusters, one with the PAF receptors, and the other with the LX and LTB receptors. In contrast to the case of the prostaglandin and TX receptors, the receptors for LX and LTB do not form a single cluster. That is, these receptors independently diverged from the peptide receptors.

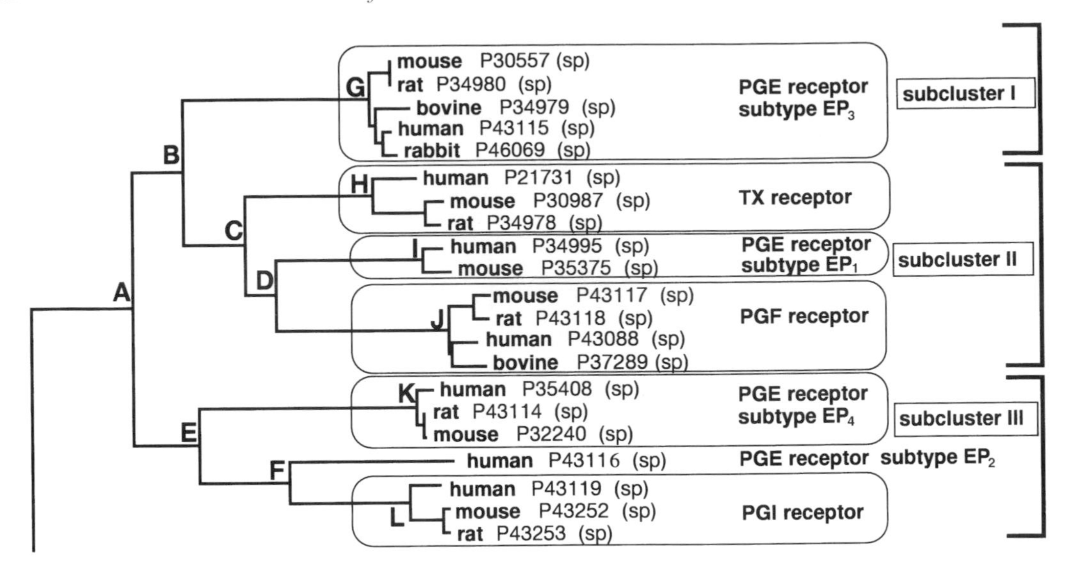

Figure 9 Molecular phylogeny of receptors for prostaglandins and TXs. This figure is continued in Figure 10.

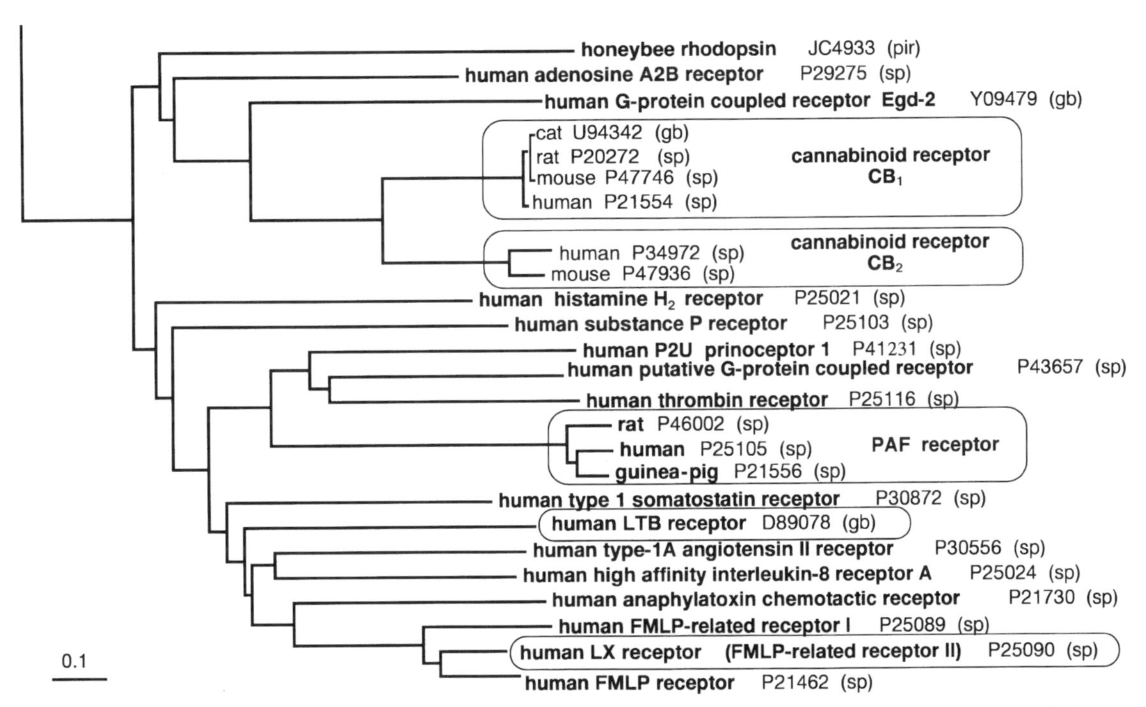

Figure 10 Molecular phylogeny of receptors for peptide ligands, lipoxin, PAF, and cannabinoids. This figure is continued from Figure 9.

1.10.10 PROSTAGLANDIN TRANSPORTER

Prostaglandins are considered to be transported across cell membranes by diffusion. The diffusion is limited by the negative charge of the prostaglandins. However, transport is augmented by a carrier in some circumstances, known as a selective prostaglandin transporter.[28] The cDNAs of the prostaglandin transporters from humans and rats have been sequenced. The cDNAs encode polypeptides of about 640 amino acid residues in length, which are involved in the specific transport of PGE_1, PGE_2, and $PGF_{2\alpha}$. The prostaglandin transporter is thought to mediate the release of newly synthesized prostaglandins from cells, the transepithelial transport of prostaglandins, and the clearance of prostaglandins from the circulation.

The amino acid sequence of the transporter shows similarity to those of mammalian Na^+-dependent organic anion transporters and their homologues from *C. elegans*. The organic anion transporters are about 700 amino acid residues in length, and contain 12 putative transmembrane segments in their primary structures. The proteins transport organic anions, such as sulfobromophthalein and bilirubin. The amino acid sequence of the prostaglandin transporter was aligned with those of six representative proteins. The phylogenetic tree is shown in Figure 11. The rat prostaglandin transporter is closely related to the anion transporters from rats and humans. The sequence identities between them are about 30%. The A and B nodes are considered to correspond to mammalian divergence. That is, the functional divergence of the prostaglandin transporter from the other anion transporters occurred long before mammalian divergence.

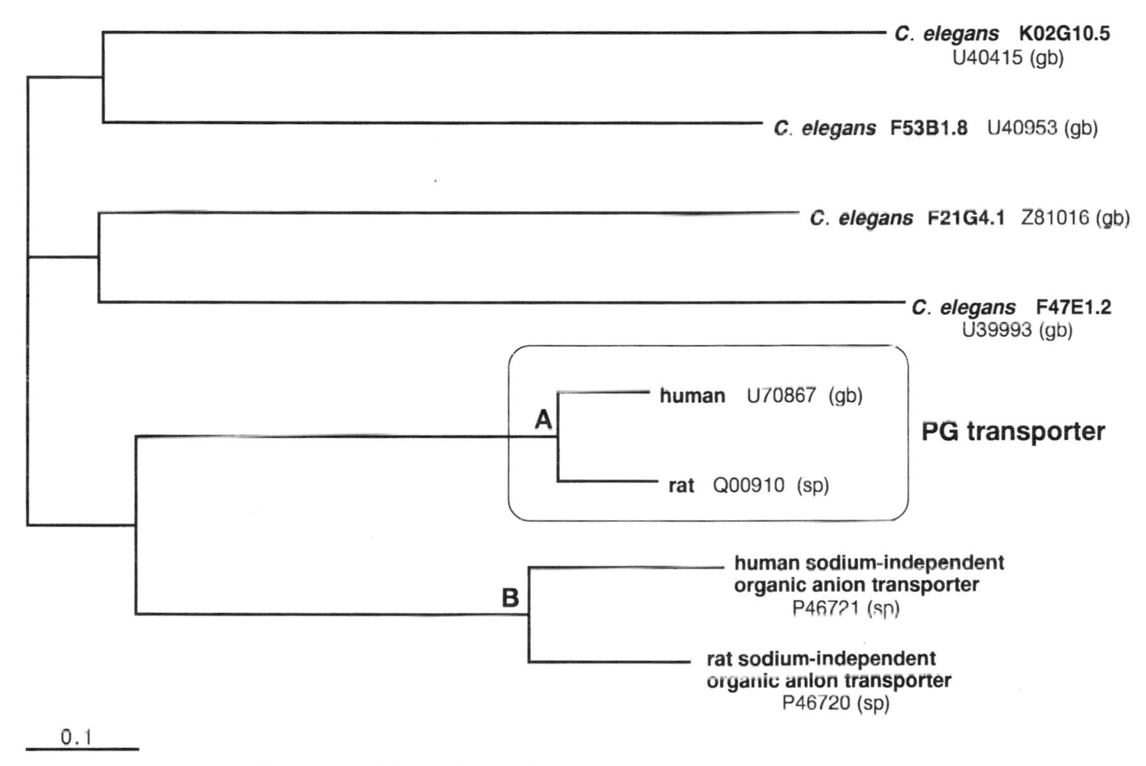

Figure 11 Molecular phylogeny of prostaglandin transporter.

1.10.11 CONCLUSION

As described above, investigation of the arachidonic acid cascade has progressed quite rapidly. In spite of this, the sequence and/or structural data are still insufficient to reveal the evolutionary mechanism of the metabolic pathway. Further accumulation of relevant sequence and structural data would be helpful to describe how and when the arachidonic acid cascade was established during the course of evolution.

1.10.12 REFERENCES

1. N. Saito and M. Nei, *Mol. Evol. Biol.*, 1987, **4**, 406.
2. H. R. Herschman, *Biochim. Biophys. Acta*, 1996, **1299**, 125.
3. W. L. Smith, R. M. Garavito, and D. L. DeWitt, *J. Biol. Chem.*, 1996, **271**, 33 157.
4. R. M. Garavito, *Nat. Str. Biol.*, 1996, **3**, 897.
5. H. Toh, C. Yokoyama, T. Tanabe, T. Yoshimoto, and S. Yamamoto, *Prostaglandins*, 1992, **44**, 291.
6. H. Toh, Y. Urade, and T. Tanabe, *Mediators Inflammation*, 1992, **1**, 223.

7. Y. Urade, K. Watanabe, and O. Hayaishi, *J. Lipid Mediators Cell Signalling*, 1995, **12**, 257.
8. H. Toh, H. Kubodera, N. Nakajima, T. Sekiya, N. Eguchi, T. Tanaka, Y. Urade, and O. Hayaishi, *Protein Eng.*, 1996, **9**, 1067.
9. Y. Kanaoka, H. Ago, E. Inagaki, T. Nakayama, M. Miyano, R. Kikuno, Y. Fujii, N. Eguchi, H. Toh, Y. Urade, and O. Hayaishi, *Cell*, 1997, **90**, 1085.
10. S. Pervaiz and K. Brew, *FASEB J.*, 1987, **1**, 209.
11. S. Tsuchida and K. Sato, *Crit. Rev. Biochem. Mol. Biol.*, 1992, **27**, 337.
12. T. Tanabe and V. Ullrich, *J. Lipid Mediators Cell Signalling*, 1995, **12**, 243.
13. S. Yamamoto, *Biochim. Biophys. Acta*, 1992, **1128**, 117.
14. H. Kühn and B.-J. Thiele, *J. Lipid Mediators Cell Signalling*, 1995, **12**, 157.
15. C. D. Funk, *Biochim. Biophys. Acta*, 1996, **1304**, 65.
16. O. Radmark and J. Haeggstrom, *Adv. Prostaglandin Thromboxane Leukotriene, Res.*, 1990, **20**, 35.
17. T. Yokomizo, N. Uozumi, T. Takahashi, K. Kume, T. Izumi, and T. Shimizu, *J. Lipid Mediators Cell Signalling*, 1995, **12**, 321.
18. A. Wetterholm, M. J. Mueller, M. Blomster, B. Samuelsson, and J. Z. Haeggstrom, *Adv. Exp. Med. Biol.*, 1997, **407**, 1.
19. P. J. Vickers, *J. Lipid Mediators Cell Signalling*, 1995, **12**, 185.
20. B. K. Lam, *Front. Biosci.*, 1997, **2**, D380.
21. H. Toh, A. Ichikawa, and S. Narumiya, *FEBS Lett.*, 1995, **361**, 17.
22. F. Ushikubi, M. Hirata, and S. Narumiya, *J. Lipid Mediators Cell Signalling*, 1995, **12**, 343.
23. S. Narumiya, *Prog. Brain Res.*, 1996, **113**, 231.
24. T. Yokomizo, T. Izumi, K. Chang, Y. Takuwa, and T. Shimizu, *Nature*, 1997, **387**, 620.
25. A. R. Schatz, M. Lee, R. B. Condie, J. T. Pulaski, and N. E. Kaminski, *Toxicol. Appl. Pharmacol.*, 1997, **142**, 278.
26. L. A. Matsuda, *Crit. Rev. Neurobiol.*, 1997, **11**, 143.
27. T. Shimizu, H. Mutoh, and S. Kato, *Adv. Exp. Med. Biol.*, 1996, **416**, 79.
28. N. Kanai, R. Lu, J. A. Satriano, Y. Bao, A. W. Wolkoff, and V. L. Schuster, *Science*, 1995, **12**, 866.

1.11 Biosynthesis of Platelet-activating Factor and Structurally Related Bioactive Lipids

TAKAYUKI SUGIURA and KEIZO WAKU
Teikyo University, Kanagawa, Japan

1.11.1 INTRODUCTION

Platelet-activating factor (PAF) was first described by Benveniste *et al.*[1] in 1972 as one of the chemical mediators released from immunoglobulin E (IgE)-sensitized rabbit basophils. In 1979, its chemical structure was proposed to be that of a unique ether-linked phospholipid, 1-*O*-alkyl-2-acetyl-*sn*-glycero-3-phosphocholine (**1**) by three separate groups,[2–4] and proved to be so in 1980.[5] Surprisingly, very low concentrations (10^{-11} to 10^{-10} M) of PAF are capable of inducing the aggregation of rabbit platelets. PAF is one of the most potent stimulants for platelets. The elucidation of the chemical structure of PAF as that of an alkyl ether-linked phospholipid shed light on the physiological significance of ether phospholipids in mammalian tissues. Indeed, the 1-acyl analogue of PAF was 200 times less active than PAF, indicating that the presence of an alkyl ether bond in the molecule is essential. The presence of an acetyl moiety at the *sn*-2 position is also crucially important; the deacetylated analogue, lysoPAF, does not exhibit any appreciable biological activity.

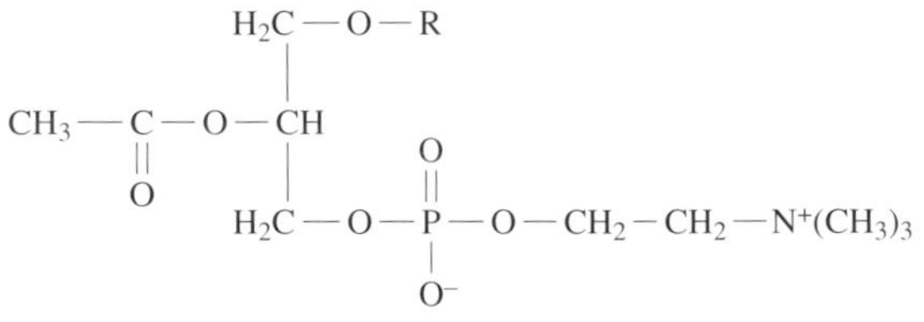

$$
\begin{array}{l}
\qquad\qquad\qquad H_2C-O-R \\
\qquad\qquad\qquad\quad | \\
CH_3-\underset{\underset{O}{\|}}{C}-O-CH \\
\qquad\qquad\qquad\quad | \\
\qquad\qquad\qquad H_2C-O-\overset{\overset{O}{\|}}{\underset{\underset{O^-}{|}}{P}}-O-CH_2-CH_2-N^+(CH_3)_3
\end{array}
$$

R = fatty chain (C_{16}–C_{18})

(1)

In addition to the activation of platelets, PAF has been shown to exhibit a variety of biological activities *in vitro* and *in vivo* through a specific PAF receptor expressed on the cell surface.[6] For example, PAF induces aggregation and degranulation of neutrophils, infiltration of eosinophils in the airway, hypotension, smooth muscle contraction, increased vascular permeability, constriction of coronary arteries in isolated heart, and glycogenolysis in perfused liver.[7–11] PAF is now assumed to be one of the important mediators of anaphylactic shock and septic shock as well as other inflammatory and allergic reactions. Interestingly, PAF is present not only in vertebrates but also in various invertebrate species, such as sea cucumbers,[12] slugs,[13] earthworms,[14] and a protozoan *Tetrahymena pyriformis*,[15] suggesting that PAF is an evolutionally conserved molecule and plays some physiological or pathophysiological role even in these lower animals. The presence of PAF has also been reported in microorganisms such as yeast.[16]

PAF is known to be released from several types of blood cells, such as neutrophils, eosinophils, macrophages, and monocytes upon stimulation.[7–11] PAF is also known to be synthesized in other tissues and cells, such as endothelial cells, kidney, intestine, heart, amnionic membranes, uterus, stomach, and brain.[7–11] Apparently, the tissue level of PAF must be strictly regulated under physiological conditions, because PAF is known to exert profound biological effects on a variety of cells and tissues at very low concentrations. The level of PAF can be controlled through either the rate of synthesis or the rate of degradation. This chapter focuses mainly on the biosynthesis of PAF. The mechanism of PAF production as well as the properties and regulation of the enzyme activities involved are described in detail. The formation of PAF-related bioactive molecules, such as short-chain fatty acid-containing phosphatidylcholines (PAF analogues) are also discussed.

1.11.2 BIOSYNTHESIS OF ETHER-LINKED PHOSPHOLIPIDS

First, the biosynthetic route for ether phospholipids is outlined (Scheme 1). A number of studies have been performed on the metabolism of ether-linked phospholipids.[17,18] The origin of the glycerol backbone of ether phospholipids is dihydroxyacetone phosphate (DHAP), which is supplied through glycolysis. DHAP is then acylated to acyldihydroxyacetone phosphate, which is further metabolized to alkyldihydroxyacetone phosphate. On the other hand, some of acyldihydroxyacetone phosphate is known to be converted by NADPH:acyldihydroxyacetone phosphate oxidoreductase to 1-acyl-*sn*-glycero-3-phosphate (lysophosphatidic acid), a common precursor molecule for the *de novo* synthesis of diacyl phospholipids. Thus, the metabolism of acyldihydroxyacetone phosphate is a branch point in the formation of ether phospholipids and diacyl phospholipids through the DHAP pathway.

1.11.2.1 Formation of Acyldihydroxyacetone Phosphate

Acyl-CoA:dihydroxyacetone phosphate acyltransferase (see Scheme 1, step i) is known to be located mainly in peroxisomes.[19] In fibroblasts obtained from Zellweger syndrome cases, in which the formation of peroxisomes is known to be insufficient, the activity of this enzyme is only 6% of the control level.[20] This may explain, at least in part, why the content of ether phospholipids is very

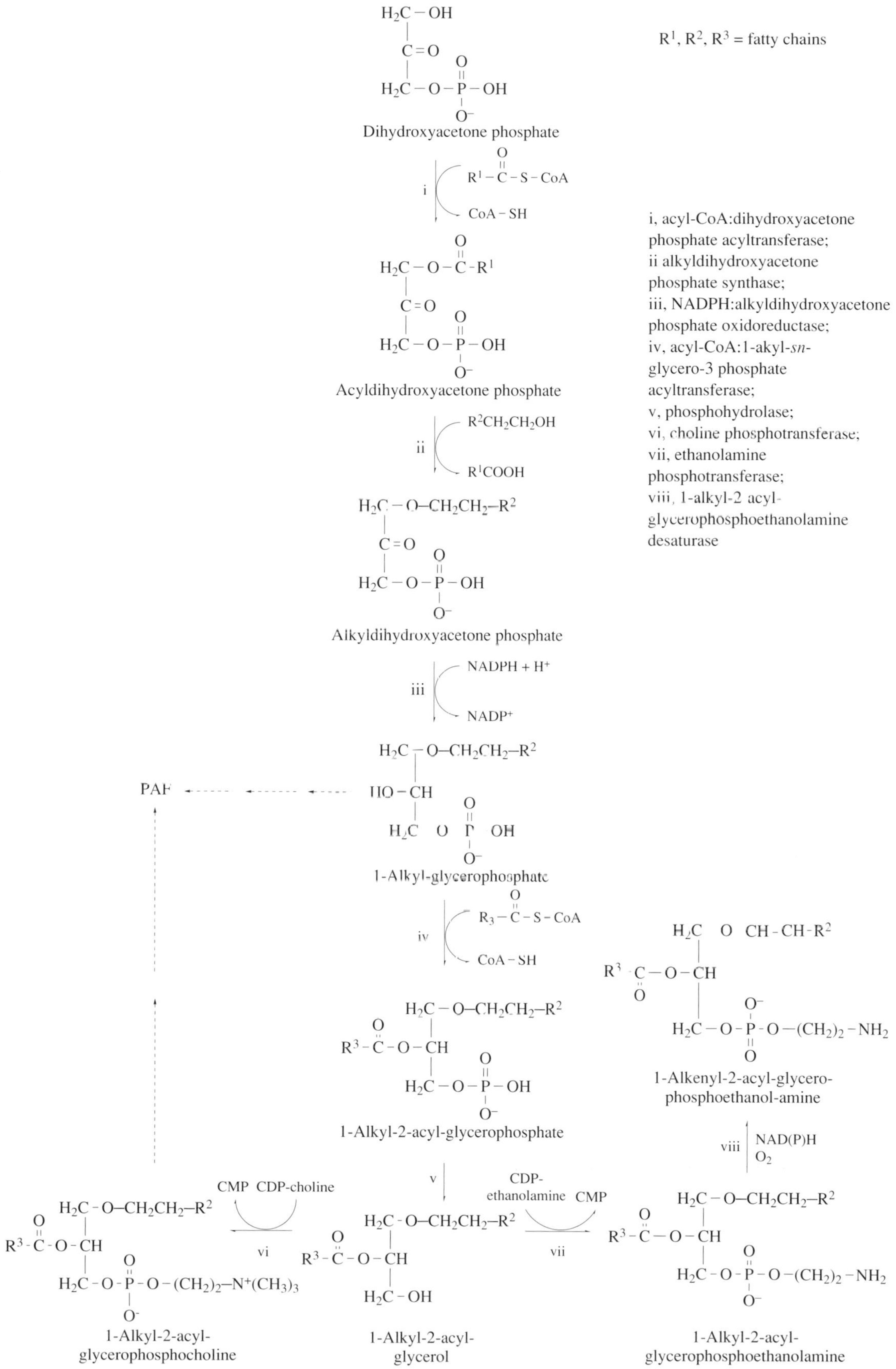

R^1, R^2, R^3 = fatty chains

Dihydroxyacetone phosphate

i: $R^1-C(=O)-S-CoA$ → CoA-SH

Acyldihydroxyacetone phosphate

ii: $R^2CH_2CH_2OH$ → R^1COOH

Alkyldihydroxyacetone phosphate

iii: NADPH + H^+ → $NADP^+$

1-Alkyl-glycerophosphate → PAF

iv: $R_3-C(=O)-S-CoA$ → CoA-SH

1-Alkyl-2-acyl-glycerophosphate

v

1-Alkyl-2-acyl-glycerol

vi: CDP-choline → CMP

1-Alkyl-2-acyl-glycerophosphocholine → PAF

vii: CDP-ethanolamine → CMP

1-Alkyl-2-acyl-glycerophosphoethanolamine

viii: NAD(P)H, O_2

1-Alkenyl-2-acyl-glycero-phosphoethanol-amine

i, acyl-CoA:dihydroxyacetone phosphate acyltransferase;
ii alkyldihydroxyacetone phosphate synthase;
iii, NADPH:alkyldihydroxyacetone phosphate oxidoreductase;
iv, acyl-CoA:1-akyl-*sn*-glycero-3 phosphate acyltransferase;
v, phosphohydrolase;
vi, choline phosphotransferase;
vii, ethanolamine phosphotransferase;
viii, 1-alkyl-2 acyl-glycerophosphoethanolamine desaturase

Scheme 1

low in such cases. Acyl-CoA:dihydroxyacetone phosphate acyltransferase was purified (3250-fold) by Webber and Hajra[21] from guinea pig liver. The molecular mass determined by sodium dodecyl sulfate polyacrylamide gel electrophoresis (SDS-PAGE) was 69 kDa and that determined by gel filtration was 90 kDa.

1.11.2.2 Formation of Alkyl Ether Bonds

Alkyldihydroxyacetone phosphate synthase (see Scheme 1, step ii) catalyzes the formation of alkyldihydroxyacetone phosphate from acyldihydroxyacetone phosphate and a long chain fatty alcohol through the displacement of the fatty acyl moiety of acyldihydroxyacetone phosphate by the long chain fatty alcohol.[22] This unique enzyme activity was found in the microsomal and/or mitochondrial fractions of several tissues. In guinea pig liver, this enzyme activity is located in peroxisomes rather than mitochondria.[19] A decrease in V_{max} and an increase in K_m of the enzyme activity in fibroblasts obtained from Zellweger syndrome cases have been observed.[20] This enzyme exclusively utilizes acyldihydroxyacetone phosphate as an acceptor of a long chain fatty alcohol. Various types of long chain (C_{10}–C_{22}) fatty alcohols were shown to be incorporated enzymatically.[23,24] Alkyldihydroxyacetone phosphate synthase was purified (13 000-fold) by Zomer *et al.*[25] from guinea pig liver. The molecular mass determined by SDS gel electrophoresis was 65 kDa.

1.11.2.3 Formation of 1-Alkyl-*sn*-glycero-3-phosphate

Alkyldihydroxyacetone phosphate is converted to 1-alkyl-*sn*-glycero-3-phosphate by NADPH: alkyldihydroxyacetone phosphate oxidoreductase (see Scheme 1, step iii). This oxidoreductase is assumed to be the same as NADPH:acyldihydroxyacetone phosphate oxidoreductase. This enzyme activity is found in the microsomal and mitochondrial fractions of several tissues.[26,27] In guinea pig liver, the activity was found in peroxisomes.[19] 1-Alkyl-*sn*-glycero-3-phosphate can also be formed through the phosphorylation of 1-alkyl-*sn*-glycerol.[28] Alternatively, 1-alkyl-*sn*-glycero-3-phosphate can be formed from 1-alkyl-*sn*-glycero-3-phosphocholine (1-alkyl-GPC, lysoPAF) or 1-alkyl-*sn*-glycero-3-phosphoethanolamine (1-alkyl-GPE) through the action of ether lysophospholipid-specific lysophospholipase D.[29] Notably, 1-alkyl-*sn*-glycero-3-phosphate (an alkyl analogue of lysophosphatidic acid) as well as 1-acyl-*sn*-glycero-3-phosphate (lysophosphatidic acid) are novel types of bioactive lipids.[30,31] 1-Alkyl-*sn*-glycero-3-phosphate has been shown to activate human and cat platelets.[31,32] Interestingly, its activity toward platelets is considerably higher than that of the corresponding acyl analogue, similar to the case of PAF.[2] 1-Alkyl (or acyl)-*sn*-glycero-3-phosphate is assumed to bind to its own putative receptor site on platelets, which is different from the PAF receptor, to elicit biological responses.[31,32]

1.11.2.4 Formation of 1-Alkyl-2-acyl-*sn*-glycero-3-phosphate

1-Alkyl-*sn*-glycero-3-phosphate is acylated by acyl-CoA:1-alkyl-*sn*-glycero-3-phosphate acyltransferase to 1-alkyl-2-acyl-*sn*-glycero-3-phosphate[33] (see Scheme 1, step iv). The fatty acid specificity of acyl-CoA:1-alkyl-*sn*-glycero-3-phosphate acyltransferase depends on the concentration of the substrate, 1-alkyl-*sn*-glycero-3-phosphate.[34] The enzyme in the brain microsomal fraction preferentially utilizes polyunsaturated fatty acyl-CoA when the concentration of 1-alkyl-*sn*-glycero-3-phosphate is low, which is somewhat different from the case of acyl-CoA:1-acyl-*sn*-glycero-3-phosphate acyltransferase. There is a possibility that acyl-CoA:1-alkyl-*sn*-glycero-3-phosphate acyltransferase and acyl-CoA:1-acyl-*sn*-glycero-3-phosphate acyltransferase are different enzymes.

1.11.2.5 Formation of 1-Alkyl-2-acyl-*sn*-glycerol

1-Alkyl-*sn*-glycero-3-phosphate is dephosphorylated by a phosphohydrolase to 1-alkyl-2-acyl-*sn*-glycerol[35] (Scheme 1, step v). It is not known whether this phosphohydrolase is the same as phosphatidic acid phosphohydrolase.

1.11.2.6 Formation of 1-Alkyl-2-acyl-glycerophosphocholine or -glycerophosphoethanolamine by Choline or Ethanolamine Phosphotransferases

1-Alkyl-2-acyl-*sn*-glycerol is converted to 1-alkyl-2-acyl-GPC or GPE by choline or ethanolamine phosphotransferases (Scheme 1, steps vi and vii). The enzymes involved in the synthesis of alkyl ether-linked phospholipids are assumed to be the same as those catalyzing the formation of the corresponding diacyl phospholipids. Ethanolamine phosphotransferase preferentially utilizes docosahexaenoic acid ($C_{22:6}$)-containing species.[36] In the brain, endogenous 1-alkyl-2-acyl-*sn*-glycerol is preferentially utilized by ethanolamine phosphotransferase rather than by choline phosphotransferase.[37] This leads to the accumulation of ether phospholipids in the ethanolamine glycerophospholipid (EGP) fraction of this tissue. On the other hand, several types of inflammatory cells, such as macrophages, exhibit a high choline phosphotransferase activity. It was postulated that endogenous 1-alkyl-2-acyl-*sn*-glycerol is utilized not only by ethanolamine phosphotransferase, but also by choline phosphotransferase to yield 1-alkyl-2-acyl-GPC in these cells. This may account, at least in part, for the abundance of 1-alkyl-2-acyl-GPC in this type of cell.[38]

1.11.2.7 Biosynthesis of Alkenyl Ether-linked Phospholipids (Plasmalogens)

1-Alkyl-2-acyl-GPE is further metabolized by 1-alkyl-2-acyl-GPE desaturase to 1-alkenyl-2-acyl-GPE (ethanolamine plasmalogen) (Scheme 1, step viii). This enzyme contains cytochrome B_5, and requires the presence of oxygen and NAD(P)H. The reaction resembles that of fatty acyl-CoA desaturases such as stearoyl-CoA desaturase. The enzyme in Fischer R-3259 sarcoma cells is stimulated by catalase, while those in pig spleen and kidney are not stimulated by catalase but are activated by cytosolic stimulating proteins.[39] 1-Alkyl-2-acyl-GPE desaturase preferentially utilizes $C_{22:6}$-containing species, similar to ethanolamine phosphotransferase.[40] Interestingly, 1-alkyl-2-acyl-GPC does not act as a substrate for this type of enzyme reaction. So far, the direct conversion of 1-alkyl-2-acyl-GPC to 1-alkenyl-2-acyl-GPC has not been reported in mammalian tissues. This seems to be in agreement with the observation that 1-alkenyl-2-acyl-GPE and 1-alkyl-2-acyl-GPC, but not 1-alkenyl-2-acyl-GPC, are accumulated in various types of inflammatory cells. 1-Alkenyl-2-acyl-GPC, which is abundant in the heart, is assumed to be formed from 1-alkenyl-2-acyl-*sn*-glycerol, directly or indirectly derived from 1-alkenyl-2-acyl-GPE, and CDP-choline through the action of choline phosphotransferase[41,42] or from 1-alkenyl-2-acyl-GPE and choline through a direct base exchange reaction.[41]

1.11.3 BIOSYNTHESIS OF PLATELET-ACTIVATING FACTOR

Two biosynthetic pathways have been proposed for PAF.[43,44] One comprises the remodeling of pre-existing 1-alkyl-2-acyl (long chain)-GPC through the sequential actions of phospholipase A_2 or CoA-independent transacylase and acetyl-CoA:lysoPAF acetyltransferase (Scheme 2 (a)). The other comprises the *de novo* synthesis from 1-alkyl-2-acetyl-*sn*-glycerol and CDP-choline through the action of CDP-choline:1-alkyl-2-acetyl-*sn*-glycerol choline phosphotransferase (Scheme 2 (b)). In this subsection, several properties of the enzyme activities involved in these two pathways are described in detail.

1.11.3.1 Hydrolysis of 1-Alkyl-2-acyl-glycerophosphocholine by Phospholipase A_2

The first step in the biosynthesis of PAF via the remodeling pathway is the hydrolysis of 1-alkyl-2-acyl (long chain)-GPC to lysoPAF, a direct precursor of PAF. Inflammatory cells such as macrophages,[45,46] neutrophils,[46–48] and eosinophils[49] are known to contain large amounts of 1-alkyl-2-acyl-GPC (16–76% of choline glycerophospholipids (CGP)), which is almost absent in many mammalian tissues, such as brain and liver.[38] The abundance of 1-alkyl-2-acyl-GPC in these cells appears to be favorable for the production of lysoPAF (1-alkyl-GPC), a direct precursor for PAF synthesis. In fact, various types of inflammatory cells have been shown to produce large amounts of lysoPAF when subjected to various stimuli, such as the calcium ionophore, A23187, and opsonized zymosan.[50,51] For example, human monocytes generate 17.7 pmol of lysoPAF and 17.8 pmol of PAF when stimulated with A23187 (2 μM) for 1 h. In human neutrophils, the generation of lysoPAF was reported to be a transient event.[52] This may be due to the rapid acylation of lysoPAF with long

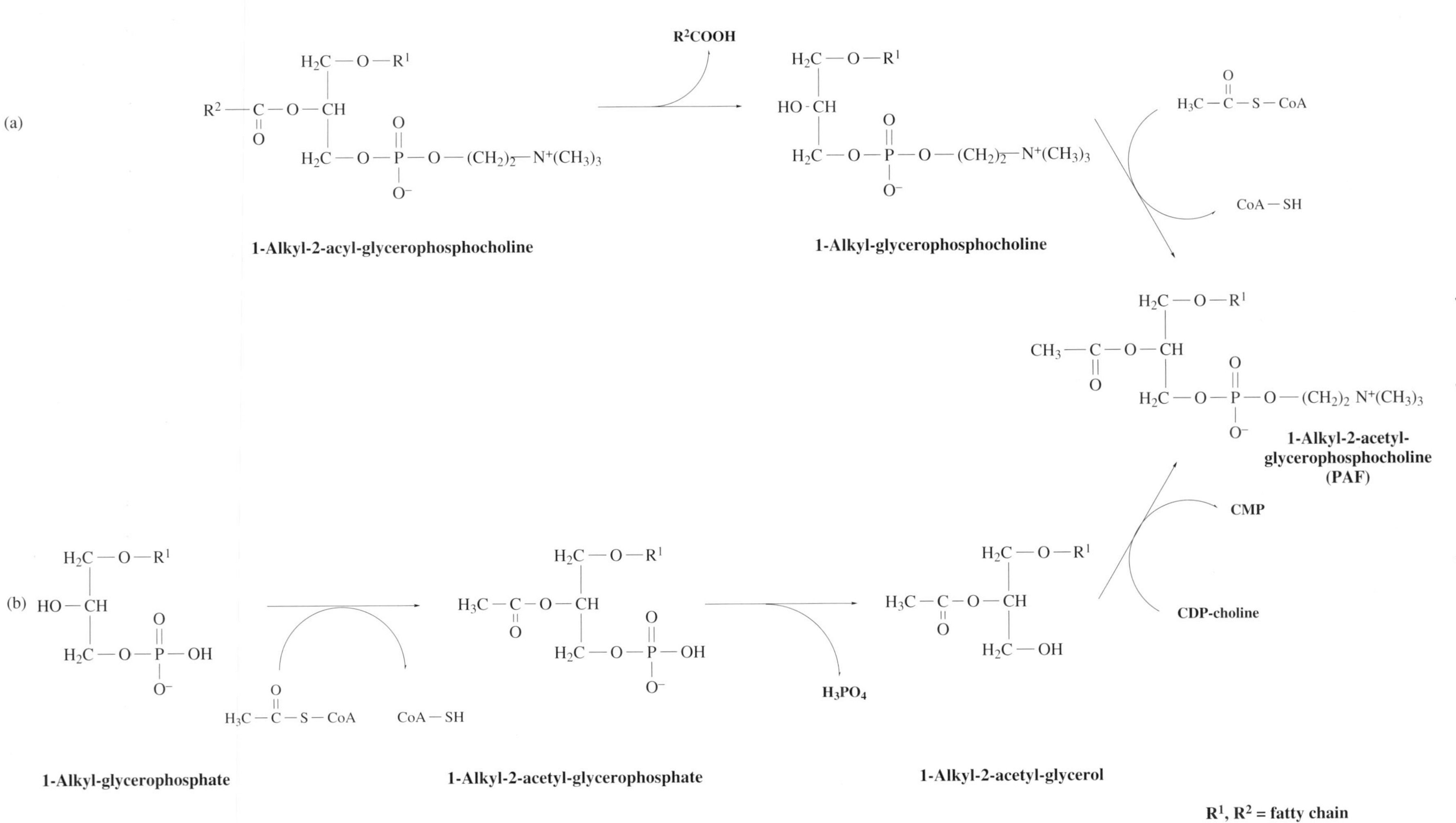
(a)
R²COOH
1-Alkyl-2-acyl-glycerophosphocholine
1-Alkyl-glycerophosphocholine
H₃C–C(=O)–S–CoA
CoA – SH
1-Alkyl-2-acetyl-glycerophosphocholine (PAF)
CMP
CDP-choline
(b)
H₃C–C(=O)–S–CoA
CoA – SH
H₃PO₄
1-Alkyl-glycerophosphate
1-Alkyl-2-acetyl-glycerophosphate
1-Alkyl-2-acetyl-glycerol
R¹, R² = fatty chain

Scheme 2

chain fatty acids in these cells. About 1/138 and 1/12 of pre-existing 1-alkyl-2-acyl-GPC was calculated to be converted to PAF in human neutrophils on stimulation with A23187 (2.5 μM) for 15 min,[53] and with A23187 (2 μM) for 1 h,[52] respectively. Thus, it appears that some of the lysoPAF formed upon the stimulation of cells is converted to PAF by acetyltransferase, and the remainder is rapidly converted to 1-alkyl-2-acyl-GPC, presumably via the CoA-independent transacylation reaction in this type of cell.

Importantly, 1-alkyl-2-acyl-GPC present in macrophages,[45] neutrophils,[48] and eosinophils[49] contains a large amount of arachidonic acid ($C_{20:4}$) at the *sn*-2 position. It was strongly suggested, therefore, that both lysoPAF, a direct precursor of PAF, and $C_{20:4}$, a precursor of eicosanoids, are released simultaneously upon the stimulation of these cells. In fact, $[^3H]C_{20:4}$ was shown to be released from the 1-alkyl-2-acyl-GPC fraction of neutrophils and macrophages on stimulation with A23187.[54–56] The release of $C_{20:4}$ from the 1-alkyl-2-acyl-GPC fraction upon stimulation with A23187 was also observed in $[^3H]$lysoPAF-labeled human neutrophils.[51] Furthermore, human hypodense eosinophils generate large amounts of both PAF[57] and leukotriene C_4[58] upon stimulation. The generation of PAF and eicosanoids should be tightly coupled events in these inflammatory cells.

The finding that two structurally different groups of lipid mediators are generated from a common precursor molecule strongly suggests a close relationship between these lipid mediators in their metabolism as well as their actions in this type of cell. In fact, Chilton *et al.*[59] reported that the action of PAF in neutrophils is mediated in part by lipoxygenase products. Furthermore, PAF is known to be rapidly metabolized to $C_{20:4}$-containing 1-alkyl-2-acyl-GPC in neutrophils[60] and platelets.[61] Also, there is evidence that $C_{20:4}$ and/or its metabolites are involved in the generation of PAF. Billah *et al.*[62] reported that the generation of either PAF or $C_{20:4}$ by A23187 (0.5 μM)-stimulated neutrophils was markedly increased in the presence of 5-hydroxyeicosatetraenoic acid (5-HETE). They also demonstrated that the addition of nordihydroguaiaretic acid (5 μM) blocked the generation of both PAF and $C_{20:4}$, and that such inhibition was relieved on the addition of 5-HETE (1 μM), 5-hydroperoxyeicosatetraenoic acid (5-HPETE) (1 μM), or leukotriene B_4 (1 μM). These results suggest that PAF synthesis is regulated by lipoxygenase products at the step of hydrolysis of 1-alkyl-2-acyl-GPC. Similar results have been reported for rat peritoneal macrophages.[63] Ramesha and Pickett[64] also stressed the significance of $C_{20:4}$-containing 1-alkyl-2-acyl-GPC in the generation of PAF, showing that PAF synthesis was decreased in neutrophils obtained from rats fed a fat-free diet and that the supplementation of $C_{20:4}$ restored the capacity for PAF synthesis. A similar reduction of PAF synthesis was observed in human monocytes obtained from fish oil-administered (6 weeks) subjects,[65] in human monocytes cultured in the presence of eicosapentaenoic acid ($C_{20:5}$),[65] and in eosinophilic leukemia cell line EOL-1 cells cultured with docosahexaenoic acid ($C_{22:6}$).[66] Although the mechanism underlying such inhibition is not yet fully understood, it seems very likely that the step of hydrolysis of 1-alkyl-2-acyl-GPC is attenuated on the treatment of cells with these polyunsaturated fatty acids. In fact, Shikano *et al.*[67] showed that a $C_{22:6}$-containing species of 1,2-diacyl-GPE interferes with the hydrolysis of a $C_{20:4}$-containing species using a purified 85 kDa cytosolic phospholipase A_2.

Several investigators have examined the phospholipase A_2 involved in the hydrolysis of 1-alkyl-2-acyl-GPC. Alonso *et al.*[68] demonstrated that phospholipase A_2 activity catalyzing the hydrolysis of 1-alkyl-2-acyl-GPC was found mainly in the cytosolic fraction of human neutrophils. The activity was Ca^{2+}-dependent and the optimal pH was 8.0. 1-Alkyl-2-acyl-GPC and 1,2-diacyl-GPC were hydrolyzed at similar rates. From the results of substrate dilution experiments, the same enzyme protein was assumed to be involved in the reaction. Kramer *et al.*[69] also demonstrated that the phospholipase A_2 partially purified from a human platelet lysate hydrolyzes both 1-alkyl ($C_{16:0}$)-2-acyl ($C_{20:4}$)-GPC and 1,2-diacyl ($C_{16:0}$, $C_{20:4}$)-GPC to similar extents. The K_m values for 1-alkyl-2-acyl-GPC and 1,2-diacyl-GPC were 0.9 μM and 1.0 μM, respectively, and the V_{max} values were 10.8 nmol min^{-1} mg^{-1} protein and 11.1 nmol min^{-1} mg^{-1} protein, respectively. The enzyme activity was Ca^{2+}-dependent and the optimal pH was around 9.0–10.0. They found that the addition of 1,2-diacyl-*sn*-glycerol augmented the enzyme activity fivefold. Similar nonselective or nonpreferential hydrolysis of 1-alkyl-2-acyl-GPC, compared with the hydrolysis of the 1,2-diacyl counterpart, has been reported by several investigators.[70,71] On the other hand, Ban *et al.*[72] demonstrated that the phospholipase A_2 activities toward 1-alkyl ($C_{16:0}$)-2-acyl ($C_{18:1}$)-GPC and diacyl ($C_{16:0}$, $C_{20:4}$)-GPC were somewhat different in their requirements of Ca^{2+} in amnionic membranes. Furthermore, Angle *et al.*[73] reported the preferential hydrolysis of 1-alkyl-2-acyl-GPC over 1,2-diacyl-GPC by phospholipase A_2 in rabbit lung. The occurrence of phospholipase A_2, which preferentially hydrolyzes 1-alkyl-2-acyl-GPC, was also found in the particulate fraction of guinea pig 1 epidermis.[74] Gross and co-workers[75,76] demonstrated the occurrence of a phospholipase A_2 that specifically hydrolyzes 1-alkenyl-2-acyl-GPC (choline plasmalogen) in canine heart and a sheep platelet lysate.

The enzyme in canine heart does not require the presence of Ca^{2+} for its activity, while that in sheep platelets requires the presence of a physiological concentration (200–800 nM) of Ca^{2+}. Plasmalogen-selective phospholipase A_2 has been purified from canine myocardial cytosol (145 000-fold)[77] and from sheep platelet cytosol (3 500-fold).[78] Notably, canine myocardial Ca^{2+}-independent cytosolic phospholipase A_2 is able to catalyze the degradation of 1-alkyl-2-acyl-GPC as well, though 1-alkenyl-2-acyl-GPC (choline plasmalogen) is a much preferred substrate. In any case, detailed studies are still required for a better understanding of the phospholipase A_2 activity involved in the hydrolysis of 1-alkyl-2-acyl-GPC in various tissues and cells. The mechanism of regulation of the hydrolysis of 1-alkyl-2-acyl-GPC by phospholipase A_2 is also an important issue to be clarified in the future.

1.11.3.2 Hydrolysis of 1-Alkyl-2-acyl-GPC by CoA-independent Transacylase

An alternative pathway providing lysoPAF is the degradation of 1-alkyl-2-acyl-GPC through the CoA-independent transacylation reaction. CoA-independent transacylase was first described by Kramer and Deykin[79] in 1983, and was then explored by several investigators.[80–86] CoA-independent transacylase is assumed to play an important role in the gradual transfer of $C_{20:4}$ as well as $C_{22:6}$ from 1,2-diacyl-GPC to 1-alkyl-2-acyl-GPC and 1-alkenyl-2-acyl-GPE observed in several types of cells, such as alveolar macrophages,[87] neutrophils,[88] and platelets.[89] Such a gradual transfer of polyunsaturated fatty acids may account for the accumulation of polyunsaturated fatty acids in ether-linked phospholipids. In 1990, Sugiura *et al.*[90] demonstrated that human neutrophil membranes contain an enzyme activity that catalyzes the transfer of $C_{20:4}$ not only from 1,2-diacyl-GPC but also from 1-alkyl-2-acyl-GPC to 1-alkenyl-GPE without the participation of any cofactors. The products of this enzyme reaction are 1-alkyl-GPC (lysoPAF) or 1-acyl-GPC and 1-alkenyl-2-acyl-GPE. It was strongly suggested, therefore, that CoA-independent transacylase is able to generate lysoPAF from 1-alkyl-2-acyl-GPC in living cells under certain conditions. They provided the evidence that the addition of 1-alkenyl-GPE to intact human neutrophils triggers the formation of lysoPAF and PAF.[90] Similar results have been obtained for HL-60 cells[91] and human neutrophils.[92] There is a possibility that CoA-independent transacylase together with phospholipase A_2 play important roles in the generation of lysoPAF in inflammatory cells. It is apparent, however, that further studies are needed to determine whether or not CoA-independent transacylase is actually involved in the formation of lysoPAF and PAF in cells subjected to more physiological stimuli. It is also necessary to elucidate the mechanism of the CoA-independent transacylation reaction using a purified enzyme protein, and to compare the enzyme's molecular properties with those of various types of phospholipase A_2.

1.11.3.3 Formation of PAF from Acetyl-CoA and LysoPAF through the Action of Acetyl-CoA:lysoPAF Acetyltransferase

The enzyme activity catalyzing the transfer of the acetyl moiety of acetyl-CoA to lysoPAF to form PAF was first described by Wykle *et al.*,[93] and then by Ninio *et al.*[94] Acetyl-CoA:lysoPAF acetyltransferase is widely distributed in various mammalian tissues, such as spleen, lung, lymph node, thymus, kidney medulla, bone marrow, and kidney cortex.[93] The enzyme activity was also detected in brain.[95] The enzyme activity was shown to be high in various types of white blood cells such as neutrophils and macrophages, but not in small lymphocytes. Obviously, acetyltransferase is a key enzyme in the production of PAF in inflammatory cells. Why small lymphocytes do not produce PAF is attributed mainly to the impairment of acetyltransferase in these cells.[52] Similar to in the case of small lymphocytes, the activity of acetyl-CoA:lysoPAF acetyltransferase in thioglycolate-induced mouse peritoneal macrophages was found to be very low, and these cells do not generate a large amount of PAF.[96,97]

Acetyltransferase activity is high in the microsomal fraction, but is low in the cytosolic and mitochondrial fractions of rat spleen.[93] The activity in rat spleen microsomes was estimated to be 10 nmol min^{-1} mg^{-1} protein when 100 μM acetyl-CoA and 30 μM lysoPAF were employed as substrates.[93] In human neutrophils, acetyl-CoA:lysoPAF acetyltransferase was found in the microsomal fraction rather than in the plasma membrane.[98] Mollinedo *et al.*[99] demonstrated that the tertiary granule fraction also contained substantial acetyl-CoA:lysoPAF acetyltransferase activity.

Several types of choline-containing lysophospholipids, such as 1-acyl-GPC and 1-alkyl-GPC, were shown to act as substrates for acetyltransferase.[93,100,101] In agreement with this, several investigators have shown that considerable amounts of the 1-acyl analogue of PAF[102–104] and the 1-alkenyl analogue of PAF[105] were generated besides PAF upon stimulation. On the other hand, the activity was decreased when the head group was changed from choline to dimethylethanolamine, monomethylethanolamine, or ethanolamine in that order.[100] The activities toward other lysophospholipids, such as 1-acyl-*sn*-glycero-3-phosphoserine, 1-acyl-*sn*-glycero-3-phosphoinositol and 1-acyl-*sn*-glycero-3-phosphate, were very low. On the other hand, Tessner and Wykle[106] found that human neutrophil sonicates contain an enzyme activity that catalyzes the transfer of the acetyl moiety of acetyl-CoA to 1-alkenyl-GPE to form 1-alkenyl-2-acetyl-GPE. The activity was considerably lower compared with when lysoPAF was used as the substrate. They demonstrated that a small amount of 1-alkenyl-2-acetyl-GPE was generated besides PAF by human neutrophils upon stimulation.[106] It is not clear, however, whether the enzyme activity catalyzing the formation of 1-alkenyl-2-acetyl-GPE is different from that of acetyl-CoA:lysoPAF acetyltransferase.

High concentrations ($>30\ \mu M$) of acceptor lysophospholipids such as 1-acyl-GPC and 1-alkyl-GPC reduced the acetyltransferase activity.[93] Such inhibition is probably due to the detergent effects of these lysophospholipids. In fact, it has been reported that relatively low concentrations of several types of detergents, such as Triton X-100, NP-40, and palmitoyl-CoA, strongly inhibit the enzyme reaction.[93,94] Acetyl-CoA:lysoPAF acetyltransferase appears to be very sensitive to detergents. The enzyme activity was also inhibited by unsaturated fatty acids such as arachidonic acid and oleic acid.[107,108] As for the fatty chain at the *sn*-1 position, the enzyme activity estimated in the presence of 1-octadecyl-GPC was higher than that estimated with 1-hexadecyl-GPC.[100] Also, the enzyme activity estimated in the presence of unsaturated 1-alkyl-GPC was somewhat higher than that estimated in the presence of a saturated species of 1-alkyl-GPC of the same chain length.[100]

The enzyme activity increased with increasing concentrations of acetyl-CoA. The apparent K_m value for acetyl-CoA was reported to be $67\ \mu M$[93] or $196\ \mu M$[101] in rat spleen, $172\ \mu M$[94] in rat peritoneal macrophages, and $320\ \mu M$[109] in rabbit alveolar macrophages. The enzyme utilizes not only acetyl-CoA but also several types of short chain fatty acyl-CoA such as propionyl-CoA.[94,100] The enzyme activity, however, decreased with increasing acyl-CoA chain length.

Because the sensitivity of acetyl-CoA:lysoPAF acetyltransferase to detergents was different from that of long chain fatty acyl-CoA:lysophospholipid acyltransferase, and the fact that the addition of acetyl-CoA did not affect the activity of long chain fatty acyl-CoA acyltransferase,[93] it is apparent that acetyltransferase and long chain fatty acyl-CoA acyltransferase are separate enzymes. This was further supported by the Ca^{2+}-dependencies of these two enzymes. Both enzyme activities did not require the addition of Ca^{2+}. However, the addition of EDTA or ethylene glycol bis(2-aminoethylether)tetraacetic acid (EGTA) totally abolished the activity of acetyltransferase, but not that of long chain fatty acyl-CoA acyltransferase.[93,94,101,110,111] The inhibition of acetyl-CoA:lysoPAF acetyltransferase by EDTA or EGTA was reversed on the addition of Ca^{2+}.[94,110,111] Taken together, it is obvious that the presence of a trace amount of Ca^{2+} is essential for the acetyl-CoA:lysoPAF acetyltransferase activity. Various types of acyltransferases other than acetyl-CoA:lysoPAF acetyltransferase usually do not require the presence of Ca^{2+}; the requirement of Ca^{2+} is thus a characteristic feature of acetyl-CoA:lysoPAF acetyltransferase in mammalian tissues. Unlike acetyltransferase in mammalian tissues, however, the acetyltransferase activity in slugs was not inhibited by EDTA.[14]

Acetyl-CoA:lysoPAF acetyltransferase is also inhibited by several sulfhydryl (SH) reagents, such as *p*-chloromercuribenzenesulfonic acid (PCMBS, 0.5 mM),[101] *N*-ethylmaleimide (NEM, 1 mM),[101] and 5,5′-dithio-bis(2-nitrobenzoic acid) (DTNB, 1 mM),[94] and by *p*-bromophenacylbromide (0.1 mM)[101] and diisopropylfluorophosphate (DFP, 10 mM).[93] As for DFP and phenylmethanesulfonyl fluoride (PMSF), Seyama and Ishibashi[101] reported that DFP (1 mM) and PMSF (5 mM) did not affect the enzyme activity. The difference may be due to the different experimental conditions.

The development of drugs that specifically inhibit PAF synthesis by inhibiting acetyl-CoA:lysoPAF acetyltransferase would be of great therapeutic value, because PAF is known to be closely involved in various allergic diseases. Shen *et al.*[112] reported that 2-[*N*-palmitoylamino]propylphosphocholine (2PPPC) and 3-[*N*-palmitoylamino]propylphosphocholine (3PPPC) inhibit acetyl-CoA:lysoPAF acetyltransferase activity. The IC_{50} values were both $5\ \mu M$. They reported that when these compounds were added (final concentration $25\ \mu M$) to A23187-stimulated mouse peritoneal macrophages, PAF production decreased to 44–70% of the control level. It is not clear, however, whether such a reduction in PAF synthesis is due only to the inhibition of acetyl-CoA:lysoPAF acetyltransferase. Sulfasalazine[113] and other anti-inflammatory drugs, such as diflunisal, benoxaprofen, and salicylate[114] have also been shown to inhibit acetyl-CoA:lysoPAF acetyltransferase in a cell-free system, although the specificities of their actions remain to be determined.

It is well known that PAF is not present in large amounts in unstimulated tissues and cells, but is produced when the tissues and cells are stimulated. Alonso *et al.*[115] and Lee *et al.*[116] demonstrated that acetyl-CoA:lysoPAF acetyltransferase activity was dramatically increased in the homogenate of human neutrophils stimulated with opsonized zymosan or A23187, compared with that in control cells. This activation of acetyltransferase was observed soon after the stimulation. The optimal dose of opsonized zymosan was $2\,mg\,ml^{-1}$. The activation reached a peak at around 10 min, the enzyme activity gradually decreasing thereafter. Such changes in the activity of acetyl-CoA:lysoPAF acetyltransferase paralleled the changes in PAF synthesis as either a function of time or a function of the dose of opsonized zymosan. Considering the fact that the activity of CDP-choline:1-alkyl-2-acetyl-*sn*-glycerol choline phosphotransferase described below was not affected in stimulated neutrophils, it was strongly suggested that the activation of acetyl-CoA:lysoPAF acetyltransferase is a key event in the induction of PAF synthesis in this type of cell. The activation of acetyl-CoA:lysoPAF acetyltransferase in cell homogenates or membrane fractions was also observed with PAF- or A23187-stimulated human neutrophils,[117] A23187- or zymosan-stimulated rat peritoneal macrophages,[97,118,119] fMLP-, ECF-A- or A23187-stimulated human eosinophils,[120] thrombin- or A23187-stimulated human platelets,[121] carbamoylcholine-stimulated guinea pig parotid glands,[111] antigen-stimulated mouse mast cells,[122] thrombin-stimulated human endothelial cells,[123] and A23187-stimulated or warmed rabbit alveolar macrophages.[109,124] Several groups have confirmed that this elevation of the enzyme activity is mainly due to an increase in V_{max}. In contrast to such increases in activity, Ihara *et al.*[125] demonstrated that dexamethasone treatment decreases the acetyl-CoA:lysoPAF acetyltransferase activity in rat spleen and liver. The details of the mechanism underlying such a reduction in enzyme activity are not clear.

Gomez-Cambronero *et al.*[110] demonstrated that the activity of acetyl-CoA:lysoPAF acetyltransferase increases in the presence of 0.1–0.2 μM Ca^{2+}, with a decreased K_m value for acetyl-CoA. They suggested that the acetyl-CoA:lysoPAF acetyltransferase activity is regulated in part by intracellular Ca^{2+}. On the other hand, Domenech *et al.*[111] reported that the concentration of Ca^{2+} required for the maximal enzyme activity (0.1 μM) in guinea pig parotid glands is not so different from the concentration of free Ca^{2+} in the same tissue, and claimed that changes in the level of free Ca^{2+} would not be the direct cause of the activation of acetyl-CoA:lysoPAF acetyltransferase. The possible regulation of acetyl-CoA:lysoPAF acetyltransferase by phosphorylation was first described by Lenihan and Lee.[126] They showed that acetyltransferase activity was markedly increased when rat spleen microsomes were incubated with the soluble fraction, ATP, Mg^{2+}, phosphatidylserine, diacylglycerol, and Ca^{2+}. The addition of *E. coli* alkaline phosphatase to the assay mixture prevented the activation of acetyl-CoA:lysoPAF acetyltransferase. These results indicate that phosphorylation of the enzyme protein is important for the activation of acetyl-CoA:lysoPAF acetyltransferase. Similar activation of acetyl-CoA:lysoPAF acetyltransferase through phosphorylation was reported for rat spleen microsomes,[127] guinea pig parotid glands,[111] human neutrophils,[117,128] mouse mast cells,[122] and rabbit alveolar macrophages.[109] The activation of the enzyme by phosphorylation is due to an increase in V_{max}, as observed in the case of the activation in stimulated intact cells.

As for the mechanism of phosphorylation, Lenihan and Lee[126] suggested the possible involvement of protein kinase C. Leyravaud *et al.*[129] also suggested the involvement of protein kinase C in the activation of acetyltransferase through phosphorylation. On the other hand, Domenech *et al.*[111] demonstrated that protein kinase C failed to activate acetyl-CoA:lysoPAF acetyltransferase in guinea pig parotid glands and concluded that the activation was mediated mainly by other kinases in this tissue. They found that isoproterenol, which is known to stimulate protein kinase A, failed to activate acetyl-CoA:lysoPAF acetyltransferase. In contrast, Ca^{2+}/calmodulin-dependent protein kinase did not further stimulate acetyl-CoA:lysoPAF acetyltransferase activity in the microsomal fraction obtained from carbacol-stimulated parotid glands. These results suggest that Ca^{2+}/calmodulin-dependent protein kinase is responsible for the activation of acetyl-CoA:lysoPAF acetyltransferase. Thus, the mechanism may differ with the tissues and cells. In any case, further studies are required to determine which enzyme system is actually involved in the phosphorylation of acetyl-CoA:lysoPAF acetyltransferase in various tissues and cells.

In addition to modification of the enzyme protein and divalent cations, the availability of a substrate is an important factor determining the enzyme activity. Mencia-Huerta *et al.*[130] found that the addition of sodium acetate or acetyl-CoA augmented PAF synthesis in rat peritoneal macrophages by 1.6–2.4-fold. They pointed out that the pool size of acetyl-CoA within the cells is a factor regulating the amount of PAF synthesized. Sugiura *et al.*[131] also demonstrated that the addition of ketone bodies such as acetoacetic acid and β-hydroxybutyric acid augmented the PAF production by stimulated human polymorphonuclear leukocytes. These ketone bodies are known to act as transporters of the acetyl unit in blood; it seems very likely that ketone bodies increase the intracellular level of acetyl-CoA, thereby augmenting the PAF production. There is a possibility, therefore, that the production of PAF is enhanced in diabetic patients, in which ketosis is very often observed. Whether or not PAF plays some role in vascular dysfunctions in diabetic patients is still to be determined.

Apparently, the activation of acetyl-CoA:lysoPAF acetyltransferase is important for the modulation of PAF synthesis. However, the activation of acetyl-CoA:lysoPAF acetyltransferase itself appears not to be sufficient for the induction of PAF production. Sugiura *et al.*[124] found that acetyl-CoA:lysoPAF acetyltransferase is transiently activated on the warming of ice-cold cells to 37°C, similar to in the case of stimulation with A23187, while these cells did not produce PAF upon the treatment. Probably, the generation of lysoPAF is essential for the induction of PAF synthesis. In fact, they demonstrated that the addition of lysoPAF to human polymorphonuclear leukocytes *per se* triggers the formation of PAF without the activation of acetyl-CoA:lysoPAF acetyltransferase.[90] The level of PAF formed in the presence of 5 μM lysoPAF was almost comparable to that formed in the presence of 4 mg ml^{-1} of opsonized zymosan. Thus, it appears that the increased availability of lysoPAF within the cells is a crucially important event for the triggering of PAF synthesis in this type of cell. The augmentation of PAF synthesis by lysoPAF in stimulated human neutrophils has also been reported by Jouvin-Marche *et al.*[52]

Obviously, the purification of acetyl-CoA:lysoPAF acetyltransferase is an important step leading to a better understanding of the enzyme reaction. Gomez-Cambronero *et al.*[132,133] reported the partial purification of acetyl-CoA:lysoPAF acetyltransferase. They solubilized acetyl-CoA:lysoPAF acetyltransferase from rat spleen microsomes using 0.4% deoxycholate. The solubilized fraction was then subjected to ammonium sulfate precipitation and sequential column chromatography. The enzyme activity was increased by 1500-fold with a 1.6% yield on partial purification. The specific activity of the partially purified enzyme was 0.317 μmol $min^{-1}mg^{-1}$ protein and the K_m value for acetyl-CoA was 137 μM. They demonstrated that the M_r 29–30 kDa band detected on SDS gel electrophoresis can be labeled with exogenously added [^{3}H]lysoPAF or [^{3}H]acetyl-CoA, and that this band material can also be by the phosphorylated by the exogenously added catalytic subunit of protein kinase A and ATP–Mg^{2+}. Based on these observations, they concluded that this protein is acetyl-CoA:lysoPAF acetyltransferase itself or one of its subunits. Further studies are required for complete purification of acetyl-CoA:lysoPAF acetyltransferase molecule and cloning of its cDNA. These would be of great value for elucidation of the mechanism of regulation of PAF synthesis under various pathophysiological conditions.

1.11.3.4 Formation of PAF from 1-Alkyl-2-acetyl-*sn*-glycerol and CDP-choline through the Action of CDP-choline:1-alkyl-2-acetyl-*sn*-glycerol Choline Phosphotransferase

Another synthetic pathway for PAF is the *de novo* synthesis through the action of CDP-choline: 1-alkyl-2-acetyl-*sn*-glycerol choline phosphotransferase described first by Snyder and co-workers.[134] In 1981, Renooij and Snyder[134] found an enzyme activity that catalyzes the formation of PAF from 1-alkyl-2-acetyl-*sn*-glycerol and CDP-choline. The enzyme activity is widely distributed in various mammalian tissues, such as spleen, lung, liver, kidney, and heart.[134] The enzyme activity was also detected in brain.[95,135–137] The enzyme activity in rat spleen was mainly found in the microsomal fraction. The optimal pH was 8.0, which was somewhat different from the optimal pH of the choline phosphotransferase utilizing 1,2-diacyl(long chain)-*sn*-glycerol as a substrate (pH 8.5). Interestingly, the sensitivities of these two types of choline phosphotransferase to dithiothreitol were considerably different from each other.[134] CDP-choline:1-alkyl-2-acetyl-*sn*-glycerol choline phosphotransferase activity was rather stabilized in the presence of 5 mM dithiothreitol, while CDP-choline:1,2-diacyl (long chain)-*sn*-glycerol choline phosphotransferase activity was markedly inhibited by 5 mM dithiothreitol. They further explored this point in detail using rat kidney medulla microsomes.[138] Besides the differences in the optimal pH and sensitivity to dithiothreitol, differences were also found in their sensitivities to temperature, deoxycholate, and ethanol. These results strongly suggest that these two reactions are catalyzed by different enzyme proteins. In rat kidney medulla, the activity of acetyl-CoA:lysoPAF acetyltransferase is very low compared with that of CDP-choline:1-alkyl-2-acetyl-*sn*-glycerol choline phosphotransferase (about 1/120). Furthermore, a considerable amount of PAF (3.3 pmol $min^{-1}mg^{-1}$ protein) was formed when kidney medulla microsomes were incubated with CDP-choline, suggesting that microsomes contain a substantial amount of endogenous 1-alkyl-2-acetyl-*sn*-glycerol. Snyder and co-workers[138,139] concluded that the biosynthesis of PAF in rat kidney medulla proceeds mainly via the *de novo* pathway catalyzed by dithiothreitol-insensitive CDP-choline:1-alkyl-2-acetyl-*sn*-glycerol choline phosphotransferase, rather than via the remodeling pathway catalyzed by acetyl-CoA:lysoPAF acetyltransferase, and that PAF continuously synthesized in kidney medulla via the *de novo* pathway may play physiologically important roles such as in the regulation of blood pressure. They demonstrated that PAF newly formed via the *de novo* pathway was rapidly translocated from its intracellular site of enzymatic synthesis to the plasma membrane.[140] On the other hand, Lianos and Zanglis,[141] Pirotzky *et al.*,[142] and Wang *et al.*[143] reported that PAF can be synthesized not only via the *de novo* pathway but also via the remodeling pathway in kidney glomerular cells. Hence, the biosynthetic routes for PAF may differ in different regions even in the same tissue, and may differ with the type of stimulus.

How, then, is 1-alkyl-2-acetyl-*sn*-glycerol, the direct precursor for the *de novo* synthesis of PAF, provided in living tissues? Lee *et al.*[144] demonstrated that rat spleen microsomes contain an acetyltransferase activity that catalyzes the formation of 1-alkyl-2-acetyl-*sn*-glycero-3-phosphate from 1-alkyl-*sn*-glycero-3-phosphate and acetyl-CoA. The apparent K_m for acetyl-CoA was 226 μM, and the optimal concentration of 1-alkyl-*sn*-glycero-3-phosphate ranged between 16 μM and 25 μM. They also found the occurrence of a phosphohydrolase activity that catalyzes the formation of 1-alkyl-2-acetyl-*sn*-glycerol from 1-alkyl-2-acetyl-*sn*-glycero-3-phosphate in rat spleen microsomes. These results clearly indicate that 1-alkyl-2-acetyl-*sn*-glycerol can actually be supplied through stepwise enzyme reactions in living tissues. They showed that the thermal stability of acetyltransferase that catalyzes the formation of 1-alkyl-2-acetyl-*sn*-glycerol from 1-alkyl-*sn*-glycero-3-phosphate and acetyl-CoA is considerably different from that of acetyl-CoA:lysoPAF acetyltransferase.[144] They also found that the addition of lysoPAF did not affect the activity of acetyl-CoA:1-alkyl-*sn*-glycero-3-phosphate acetyltransferase.[144] Furthermore, Baker and Chang[145] demonstrated that the sensitivity of acetyl-CoA:1-alkyl-*sn*-glycero-3-phosphate acetyltransferase to ATP-Mg^{2+} is greater than that of acetyl-CoA:lysoPAF acetyltransferase, and that the distributions of these two enzyme activities in brain subcellular fractions are somewhat different from each other. These results suggest that these two reactions are catalyzed by separate enzyme proteins. Lee and co-workers[146] further demonstrated that the substrate specificity, optimal pH, effects of detergents, ethanol, and cations, and thermal stability of 1-alkyl-2-acetyl-*sn*-glycero-3-phosphate phosphohydrolase are different from those of the phosphohydrolase that acts on 1,2-diacyl (long chain)-*sn*-glycero-3-phosphate and 1-alkyl-2-acyl (long chain)-*sn*-glycero-3-phosphate, suggesting that they are different enzymes. Overall, it is evident that PAF can be synthesized through the sequential actions of specific enzymes for PAF *de novo* synthesis in several tissues. Snyder and co-workers[43,44] pointed out that the rate-limiting step of this pathway is the acetylation of 1-alkyl-*sn*-glycero-3-phosphate; the activity of acetyl-CoA:1-alkyl-*sn*-glycero-3-phosphate acetyltransferase is low compared with those of other enzyme activities involved in this pathway.

As for the regulation of the enzyme activities involved in the *de novo* synthesis of PAF, the details remain unclear. Heller *et al.*[147] demonstrated that the activities of both acetyl-CoA:1-alkyl-*sn*-glycero-3-phosphate acetyltransferase and CDP-choline:1-alkyl-2-acetyl-*sn*-glycerol choline phosphotransferase are elevated in human endothelial cells stimulated with 12-*O*-tetradecanoylphorbol 13-acetate (TPA), which is known to activate protein kinase C. Baker and Chang[148] also showed that acetyl-CoA:1-alkyl-*sn*-glycero-3-phosphate acetyltransferase activity increased 2.4 times on preincubation with ATP and $MgCl_2$, suggesting that phosphorylation of the enzyme protein augmented the enzyme activity similar to in the case of acetyl-CoA:lysoPAF acetyltransferase. It is not known, however, which protein kinase is involved in the phosphorylation of the enzyme protein. On the other hand, Lee and co-workers[149] reported that the treatment of saponin-permeabilized rabbit platelets with 0.2 mM sodium oleate enhanced the production of PAF from 1-alkyl-2-acetyl-*sn*-glycerol by fivefold. They suggested the possibility that CTP:phosphocholine cytidyltransferase is activated through oleic acid-induced translocation of the enzyme protein from the cytosol to the membrane, thereby providing a larger amount of CDP-choline for PAF synthesis. This would be reasonable because CTP:phosphocholine cytidyltransferase is known to be a rate-limiting enzyme in the *de novo* synthesis of long chain fatty acid-containing phosphatidylcholine, and its intracellular localization is regulated by fatty acids.[150] Thus, the availability of CDP-choline is an important factor determining the rate of *de novo* synthesis of PAF. In any case, the regulation of these enzyme activities involved in the *de novo* synthesis of PAF under various conditions as well as possible intracellular trafficking of PAF precursors remain to be further clarified.

Despite the accumulation of knowledge concerning individual enzyme activities, not much direct evidence has been obtained concerning the physiological or pathophysiological significance of the *de novo* synthesis of PAF in various tissues and cells, except in several cases, compared with that of the remodeling pathway. In order to resolve this, it is essential, for example, to determine the exact tissue levels of individual enzyme substrates. However, information concerning the levels of endogenous substrates for the *de novo* synthesis of PAF is quite limited. It was assumed that the sustained decrease in blood pressure induced by the intravenous injection of 1-alkyl-2-acetyl-*sn*-glycerol is due to PAF formed from 1-alkyl-2-acetyl-*sn*-glycerol through the action of CDP-choline:1-alkyl-2-acetyl-*sn*-glycerol choline phosphotransferase.[151]. Satouchi *et al.*[152] also reported that the addition of 1-alkyl-2-acetyl-*sn*-glycerol induced the aggregation of rabbit platelets and suggested that 1-alkyl-2-acetyl-*sn*-glycerol was converted to PAF, thereby inducing the aggregation of platelets. A similar result was reported by Blank *et al.*[153] Thus, this pathway operates if a sufficient amount of the substrate, 1-alkyl-2-acetyl-*sn*-glycerol, is provided in living tissues.

Previously, Bussolino *et al.*[154] reported that PAF was formed in chick retina stimulated with dopamine and acetylcholine, and that the activity of dithiothreitol-insensitive CDP-choline:1-alkyl-2-acetyl-*sn*-glycerol choline phosphotransferase was increased sixfold, whereas the activity of acetyl-CoA:lysoPAF acetyltransferase remained unchanged under these experimental conditions. It is possible that the generation of PAF in nervous tissues upon stimulation proceeds, at least in part, via the *de novo* synthetic pathway. Goracci and Francescangeli,[135,136] and Baker and Chang[145] suggested that both the *de novo* synthetic pathway and the remodeling pathway are important in the production of PAF in brain. They pointed out that the biosynthetic route for PAF may differ with the type of stimulus and the region of the brain, as in the case of kidney. Nieto *et al.*[155] reported that PAF is formed through the *de novo* pathway in neutrophils stimulated with TPA, which is different from in the case of stimulation with A23187 and opsonized zymosan, where the production proceeds mainly via the remodeling pathway. TPA-induced PAF formation through the *de novo* pathway has also been reported for human endothelial cells.[147] The possible involvement of the *de novo* pathway in addition to the remodeling pathway in the production of PAF in fetal rabbit lung has also been reported by Hoffman *et al.*[156] On the other hand, Fernadez-Gallardo *et al.*[157] suggested that PAF synthesis in rat glandular gastric mucosa proceeds mainly via the *de novo* pathway. Furthermore, Appleyard and Hillier[158] provided evidence that PAF in inflamed human colon mucosa is produced *de novo*. In addition, PAF present in *Tetrahymena pyriformis*[159] and in sea cucumber intestine[160] were suggested to be produced exclusively via the *de novo* synthetic pathway rather than via the remodeling pathway. These invertebrate cells and tissue lack appreciable acetyl-CoA:lysoPAF acetyltransferase activity. Apparently, the *de novo* synthesis of PAF would be an important pathway providing PAF in several tissues and cells in some circumstances. However, further detailed studies are required to clarify the relative importance of the *de novo* and remodeling pathways in the production of PAF in various tissues and cells under various physiological and pathophysiological conditions. The purification and cloning of cDNAs of enzymes involved in these two synthetic pathways should also be helpful for a thorough understanding of the molecular mechanism of the enzymatic synthesis of this potent bioactive lipid molecule.

1.11.4 FORMATION OF PAF-LIKE LIPIDS

Various types of cells and tissues are known to generate the 1-acyl analogue of PAF besides PAF upon stimulation.[102–104] Perfused guinea pig heart has also been shown to contain the 1-alkenyl analogue of PAF besides PAF.[105] These types of PAF analogues are assumed to be produced in a similar way to the biosynthesis of PAF, as mentioned before. The platelet-stimulating ability of the 1-acyl analogue of PAF is two hundred times lower than that of PAF,[2] and the activity of the 1-alkenyl analogue of PAF is about one-fifth that of PAF.[161] Their physiological or pathophysiological significance has not yet been well elucidated. Pinckard and co-workers[162] pointed out the possibility that the 1-acyl analogue of PAF plays an important role in the priming of neutrophils.

Sugiura *et al.*[14] detected a significant amount of the 2-propionyl analogue of PAF in earthworms. This molecule is also assumed to be produced through biosynthetic routes similar to those for PAF in mammals even in this lower animal.[14] The biological activity of the 2-propionyl analogue of PAF toward rabbit platelets is about one-third that of PAF. However, the physiological significance of this novel type of PAF analogue found in earthworms so far remains unclear.

New insight into the pathophysiological significance of PAF-like lipids was gained later in the 1980s. Tokumura and co-workers[163–166] found that bovine brain contains large amounts of various novel species of 1-long-chain acyl-2-short-chain acyl-*sn*-glycero-3-phosphocholine besides PAF. The fatty chain at the *sn*-1 position was exclusively a long-chain fatty acid such as palmitic acid and stearic acid. The short-chain acyl moiety at the *sn*-2 position comprises a mixture of a number of species of hydroxycarboxylic acids, monocarboxylic acids, and dicarboxylic acids. They detected 3-hydroxypropionyl, 4-hydroxybutyryl, 5-hydroxyvaleroyl, 6-hydroxycaproyl, 8-hydroxycapryloyl, and 9-hydroxypelargonoyl moieties as hydroxycarboxylyl residues, acetyl, propionyl, acryloyl, butyryl, valeryl, caproyl, and heptanoyl moieties as monocarboxylyl residues, and succinyl, glutaryl, adipoyl, pimeloyl, suberoyl, and azelaoyl moieties as dicarboxylyl residues. Importantly, these short-chain fatty acids other than acetic acid can be formed through the peroxidation of unsaturated fatty acids. There is a possibility, therefore, that most of these novel short-chain fatty acid-containing phosphatidylcholines are formed through the direct peroxidation of brain membrane phospholipids, especially polyunsaturated fatty acid-containing phosphatidylcholines.

Such a possibility was examined by several investigators. Tanaka *et al.*[167] demonstrated that four kinds of phosphatidylcholine having a short-chain monocarboxylate, dicarboxylate, semialdehyde, or ω-hydroxylate moiety at the *sn*-2 position were formed from synthetic polyunsaturated fatty acid-containing phosphatidylcholines and bovine brain phosphatidylcholine upon peroxidation with Fe^{2+}/ascorbate/EDTA. They further demonstrated that various species of short-chain fatty acid-containing phosphatidylcholines are present in the Cu^{2+}-oxidized lipoproteins in egg yolk.[168] On the other hand, Itabe *et al.*[169] detected phosphatidylcholine with an *sn*-2-azelaoyl group as a fragmented product formed during the peroxidation of linoleic acid-containing phosphatidylcholines by oxyhemoglobin. Stremler *et al.*[170] also found that phosphatidylcholine with a 5-oxovaleroyl group at the *sn*-2 position was formed from arachidonic acid-containing phosphatidylcholine upon lipoxygenase-catalyzed peroxidation. In addition, Kamido *et al.*[171] isolated various species of short-chain aldehyde-containing phosphatidylcholines from the copper-catalyzed peroxidation product of human plasma lipoproteins. These observations support the idea that lipid peroxidation is an important pathway for the generation of short-chain fatty acid (and/or fatty aldehyde)-containing phosphatidylcholines, i.e., PAF analogues.

The question was raised as to whether these short-chain fatty acid-containing phosphatidylcholines possess biological activities like those of PAF. Smiley *et al.*[172] demonstrated that oxidatively fragmented phosphatidylcholines such as 1-acyl-2-(5-oxovaleroyl)-GPC are able to activate human neutrophils through the receptor for PAF. They provided evidence that such an oxidatively fragmented phosphatidylcholine with leukocyte-stimulating activity was released by endothelial cells when they were exposed to peroxide.[173] They further demonstrated that copper-oxidized low density lipoprotein (LDL) stimulates the growth of smooth muscle cells through a PAF receptor-dependent mechanism.[174] In addition to their actions on leukocytes and smooth muscle cells, Tanaka *et al.*[175] reported that various species of short-chain fatty acid-containing phosphatidylcholines, derived from polyunsaturated fatty acid-containing phosphatidylcholines through Fe^{2+}/ascorbate/EDTA-induced oxidation, are capable of stimulating rabbit platelets. These results raise the possibility that short-chain fatty acid-containing phosphatidylcholines formed from polyunsaturated fatty acid-containing phosphatidylcholines through peroxidation play some important pathophysiological roles in vascular diseases such as atherosclerosis. In relation to this, it should be mentioned that the accumulation of oxidized LDL in atherosclerotic lesions has already been described by several investigators.[176,177]

Interestingly, short-chain fatty acid-containing phosphatidylcholines are good substrates for PAF acetylhydrolase, a PAF-degrading enzyme.[170,178] This suggests that PAF acetylhydrolase plays an important role in the elimination of not only PAF but also short-chain fatty acid-containing phosphatidylcholines, which may stimulate leukocytes, platelets, and smooth muscle cells, and thereby have deleterious effects on the vascular system, from the circulation. In fact, significant amounts of short-chain fatty acid-containing phosphatidylcholines were detected in Cu^{2+}-oxidized LDL obtained from human plasma when plasma PAF acetylhydrolase had been blocked by pre-treatment with DFP.[168] Such short-chain fatty acid-containing phosphatidylcholines were not detected in intact LDL. Liapikos *et al.*[179] also reported that PAF was found in oxidized human LDL when plasma PAF acetylhydrolase was inhibited by PMSF. Considering the fact that oxidized LDL plays a crucial role in the pathogenesis of atherosclerosis,[180] and that oxidized phosphatidylcholines were actually detected in atherosclerotic lesions,[176,177] along with the fact that oxidized phosphatidylcholines exhibit proinflammatory activities, it is tempting to speculate that short-chain fatty acid (and/or fatty aldehyde)-containing phosphatidylcholines, i.e., PAF analogues, formed through lipid peroxidation are involved in the induction of atherosclerosis.

1.11.5 CONCLUDING REMARKS

Since the elucidation of its chemical structure, numerous studies have been performed on PAF. Evidence is accumulating concerning the pathophysiological roles of PAF in various allergic and inflammatory diseases in mammals. For example, based on the results with PAF receptor antagonists, it has been strongly suggested that PAF is involved in anaphylactic shock and septic shock. The possible involvement of PAF in bronchial asthma has also been suggested by a number of investigators. Hence, the control of PAF synthesis under pathological conditions is of potential therapeutic importance.

As described in this chapter, the outline of the biosynthesis of PAF has been elucidated through the studies already conducted. Nevertheless, details of the mechanism of regulation of the PAF synthesis remain unknown. Furthermore, the detailed molecular properties of individual enzymes involved in PAF synthesis remain unclear, because none of them has yet been purified. Also, not much is known concerning the intracellular processing of the PAF molecule as well as any possible intracellular binding site(s) for PAF. Thus, further studies are needed to clarify these important issues. Such efforts should be helpful for a better understanding of PAF under physiological or pathophysiological conditions, and for the development of drugs which attenuate PAF production.

It should be emphasized that the discovery of PAF-like lipids such as short-chain fatty acid (and/or fatty aldehyde)-containing phosphatidylcholines has added a new area to the field of PAF research. These PAF-like lipids, together with other bioactive lipids, may play crucial roles in the pathogeneses of vascular diseases such as atherosclerosis. At least, some of their actions are assumed to be mediated through the PAF receptor. Because not much is known concerning PAF-like lipids, additional studies are required to determine their structures, distributions, synthesis, actions, and catabolism in more detail. Such studies are essential for establishing their pathophysiological significance in human diseases.

1.11.6 REFERENCES

1. J. Benveniste, P. M. Henson, and C. G. Cochrane, *J. Exp. Med.*, 1972, **136**, 1356.
2. C. A. Demopoulos, R. N. Pinckard, and D. J. Hanahan *J. Biol. Chem.*, 1979, **254**, 9355.
3. J. Benveniste, M. Tence, P. Varenne, J. Bidault, C. Boullet, and J. Polonsky, *C.R. Acad. Sci. (Paris)*, 1979, **289D**, 1037.
4. M. L. Blank, F. Snyder, L. W. Byers, B. Brooks, and E. E. Muirhead, *Biochem. Biophys. Res. Commun.*, 1979, **90**, 1194.
5. D. J. Hanahan, C. A. Demopoulos, J. Liehr, and R. N. Pinckard, *J. Biol. Chem.*, 1980, **255**, 5514.
6. T. Izumi and T. Shimizu, *Biochim. Biophys. Acta*, 1995, **1259**, 317.
7. D. J. Hanahan, *Annu. Rev. Biochem.*, 1986, **55**, 483.
8. F. Snyder (ed.), "Platelet-Activating Factor and Related Lipid Mediators," Plenum, New York, 1987.
9. S. M. Prescott, G. A. Zimmerman, and T. M. McIntyre, *J. Biol. Chem.*, 1990, **265**, 17 381.
10. P. Braquet, L. Touqui, T. Y. Shen, and B. B. Vargaftig, *Pharmacol. Rev.*, 1987, **39**, 97.
11. P. J. Barnes, C. P. Page, and P. M. Henson (eds.), "Platelet Activating Factor and Human Disease," Blackwell, Oxford, 1989.
12. T. Sugiura, T. Fukuda, T. Miyamoto, and K. Waku, *Biochim. Biophys. Acta*, 1992, **1126**, 298.
13. T. Sugiura, T. Ojima, T. Fukuda, K. Satouchi, K. Saito, and K. Waku, *J. Lipid Res.* 1991, **32**, 1795.
14. T. Sugiura, A. Yamashita, N. Kudo, T. Fukuda, T. Miyamoto, N.-N. Cheng, S. Kishimoto, K. Waku, T. Tanaka, H. Tsukatani, and A. Tokumura, *Biochim. Biophys. Acta*, 1995, **1258**, 19.
15. M. Lekka, A. D. Tselepis, and D. Tsoukatos, *FEBS Lett.*, 1986, **208**, 52.
16. R. Nakayama, H. Kumagai, and K. Saito, *Biochim. Biophys. Acta*, 1994, **1199**, 137.
17. R. L. Wykle and F. Snyder, in "The Enzymes of Biological Membranes," ed. A. Martonosi, Plenum, New York, 1976, vol.2, p. 87.
18. L. A. Horrocks and M. Sharma, in "Phospholipids," eds. J. N. Hawthorne and G. B. Ansell, Elsevier, Amsterdam, 1982, p. 51.
19. A. K. Hajra and J. E. Bishop, *Ann. N.Y. Acad. Sci.*, 1982, **386**, 170.
20. K. O. Webber, N. S. Datta, and A. K. Hajra, *Arch. Biochem. Biophys.*, 1987, **254**, 611.
21. K. O. Webber and A. K. Hajra, *Arch. Biochem. Biophys.*, 1993, **300**, 88.
22. A. K. Hajra, *Biochem. Biophys. Res. Commun.*, 1969, **39**, 1037.
23. Z. L. Bandi, E. Aaes-Jørgensen, and H. K. Mangold, *Biochim. Biophys. Acta*, 1971, **239**, 357.
24. F. Snyder, M. Clark, and C. Piantadosi, *Biochem. Biophys. Res. Commun.*, 1973, **53**, 350.
25. A. W. M. Zomer, W. F. C. de Weerd, J. Langeveld, and H. van den Bosch, *Biochim. Biophys. Acta*, 1993, **1170**, 189.
26. E. F. LaBelle, Jr. and A. K. Hajra, *J. Biol. Chem.*, 1972, **247**, 5825.
27. K. Chae, C. Piantadosi, and F. Snyder, *J. Biol. Chem.*, 1973, **248**, 6718.
28. C. O. Rock and F. Snyder, *J. Biol. Chem.*, 1974, **249**, 5382.
29. R. L. Wykle and J. M. Schremmer, *J. Biol. Chem.*, 1974, **249**, 1742.
30. W. H. Moolenaar, *J. Biol. Chem.*, 1995, **270**, 12 949.
31. A. Tokumura, *Prog. Lipid Res.*, 1995, **34**, 151.
32. T. Sugiura, A. Tokumura, L. Gregory, T. Nouchi, S. T. Weintraub, and D. J. Hanahan, *Arch. Biochem. Biophys.*, 1994, **311**, 358.
33. R. L. Wykle and F. Snyder, *J. Biol. Chem.*, 1970, **245**, 3047.
34. P. J. Fleming and A. K. Hajra, *J. Biol. Chem.*, 1977, **252**, 1663.
35. F. Snyder, M. L. Blank, and B. Malone, *J. Biol. Chem.*, 1970, **245**, 4016.
36. Y. Nakagawa and K. Waku, *Eur. J. Biochem.*, 1985, **152**, 569.

37. L. Freysz, L. A. Horrocks, and P. Mandel, *J. Neurochem.*, 1980, **34**, 963.
38. T. Sugiura and K. Waku, in "Platelet-Activating Factor and Related Lipid Mediators," ed. F. Snyder, Plenum, New York, 1987, p. 55.
39. F. Paltauf, *Eur. J. Biochem.*, 1978, **85**, 263.
40. Y. Masuzawa, T. Sugiura, Y. Ishima, and K. Waku, *J. Neurochem.*, 1984, **42**, 961.
41. M. L. Blank, V. Fitzgerald, T.-C. Lee, and F. Snyder, *Biochim. Biophys. Acta*, 1993, **1166**, 309.
42. J. C. Strum and L. W. Daniel, *J. Biol. Chem.*, 1993, **268**, 25 500.
43. F. Snyder, *Biochim. Biophys. Acta*, 1995, **1254**, 231.
44. F. Snyder, *Biochem. J.*, 1995, **305**, 689.
45. T. Sugiura, M. Nakajima, N. Sekiguchi, Y. Nakagawa, and K. Waku, *Lipids*, 1983, **18**, 125.
46. T. Sugiura, Y. Onuma, N. Sekiguchi, and K. Waku, *Biochim. Biophys. Acta*, 1982, **712**, 515.
47. H. W. Mueller, J. T. O'Flaherty, and R. L. Wykle, *Lipids*, 1982, **17**, 72.
48. H. W. Mueller, J. T. O'Flaherty, D. G. Greene, M. P. Samuel, and R. L. Wykle, *J. Lipid Res.*, 1984, **25**, 383.
49. A. Ojima-Uchiyama, Y. Masuzawa, T. Sugiura, K. Waku, H. Saito, Y. Yui, H. Tomioka, *Lipids*, 1988, **23**, 815.
50. D. H. Albert and F. Snyder, *J. Biol. Chem.*, 1983, **258**, 97.
51. F. H. Chilton, J. M. Ellis, S. C. Olson, and R. L. Wykle, *J. Biol. Chem.*, 1984, **259**, 12 014.
52. E. Jouvin-Marche, E. Ninio, G. Beaurain, M. Tence, P. Niaudet, and J. Benveniste, *J. Immunol.*, 1984, **133**, 892.
53. M. Oda, K. Satouchi, K. Yasunaga, and K. Saito, *J. Immunol.*, 1985, **134**, 1090.
54. C. L. Swendsen, J. M. Ellis, F. H. Chilton, J. T. O'Flaherty, and R. L. Wykle, *Biochem. Biophys. Res. Commun.*, 1983, **113**, 72.
55. D. H. Albert and F. Snyder, *Biochim. Biophys. Acta*, 1984, **796**, 92.
56. Y. Nakagawa, K. Kurihara, T. Sugiura, and K. Waku, *Biochim. Biophys. Acta*, 1986, **876**, 601.
57. A. Ojima-Uchiyama, Y. Masuzawa, T. Sugiura, K. Waku, T. Fukuda, and S. Makino, *Lipids*, 1991, **26**, 1200.
58. T. Kajita, Y. Yui, H. Mita, N. Taniguchi, H. Saito, T. Mishima, and T. Shida, *Int. Arch. Allergy Appl. Immunol.*, 1985, **78**, 406.
59. F. H. Chilton, J. T. O'Flaherty, C. E. Walsh, M. J. Thomas, R. L. Wykle, R. R. DeChatelet, and B. M. Waite, *J. Biol. Chem.*, 1982, **257**, 5402.
60. F. H. Chilton, J. T. O'Flaherty, J. M. Ellis, C. L. Swendsen, and R. L. Wykle, *J. Biol. Chem.*, 1983, **258**, 7268.
61. B. Malone, T.-C. Lee, and F. Snyder, *J. Biol. Chem.*, 1985, **260**, 1531.
62. M. M. Billah, R. W. Bryant, and M. I. Siegel, *J. Biol. Chem.*, 1985, **260**, 6899.
63. H. Saito, A. Hirai, Y. Tamura, and S. Yoshida, *Prostaglandins Leukotrienes Med.*, 1985, **18**, 271.
64. C. S. Ramesha and W. C. Pickett, *J. Biol. Chem.*, 1986, **261**, 7592.
65. R. I. Sperling, J. L. Robin, K. A. Kylander, T. H. Lee, R. A. Lewis, and K. F. Austen, *J. Immunol.*, 1987, **139**, 4186.
66. M. Shikano, Y. Masuzawa, and K. Yazawa, *J. Immunol.*, 1993, **150**, 3525.
67. M. Shikano, Y. Masuzawa, K. Yazawa, K. Takayama, I. Kudo, and K. Inoue, *Biochim. Biophys. Acta*, 1994, **1212**, 211.
68. F. Alonso, P. M. Henson, and C. C. Lesile, *Biochim. Biophys. Acta*, 1986, **878**, 273.
69. R. M. Kramer, J. A. Jakubowski, and D. Deykin, *Biochim. Biophys. Acta*, 1988, **959**, 269.
70. J. Wijkander and R. Sundler, *Eur. J. Biochem.*, 1991, **202**, 873.
71. E. Diez, F. H. Chilton, G. Stroup, R. J. Mayer, J. D. Winkler, and A. N. Fonteh, *Biochem. J.*, 1994, **301**, 721.
72. C. Ban, M. M. Billah, C. T. Truong, and J. M. Johnston, *Arch. Biochem. Biophys.*, 1986, **246**, 9.
73. M. J. Angle, F. Paltauf, and J. M. Johnston, *Biochim. Biophys. Acta*, 1988, **962**, 234.
74. B. Wong, W. Tang, and V. A. Ziboh, *FEBS Lett.*, 1992, **305**, 213.
75. R. A. Wolf and R. W. Gross, *J. Biol. Chem.*, 1985, **260**, 7295.
76. L. A. Loeb and R. W. Gross, *J. Biol. Chem.*, 1986, **261**, 10 467.
77. S. L. Hazen, R. J. Stuppy, and R. W. Gross, *J. Biol. Chem.*, 1990, **265**, 10 622.
78. S. L. Hazen, L. A. Loeb, and R. W. Gross, *Methods Enzymol.*, 1991, **197**, 400.
79. R. M. Kramer and D. Deykin, *J. Biol. Chem.*, 1983, **258**, 13 806.
80. T. Sugiura and K. Waku, *Biochem. Biophys. Res. Commun.*, 1985, **127**, 384.
81. M. Robinson, M. L. Blank, and F. Snyder, *J. Biol. Chem.*, 1985, **260**, 7889.
82. T. Sugiura, Y. Masuzawa, and K. Waku, *Biochem. Biophys. Res. Commun.*, 1985, **133**, 574.
83. P. V. Reddy and H. H. O. Schmid, *Biochem. Biophys. Res. Commun.*, 1985, **129**, 381.
84. T. Sugiura, Y. Masuzawa, Y. Nakagawa, and K. Waku, *J. Biol. Chem.*, 1987, **262**, 1199.
85. J. D. Winkler, C.-M. Sung, C. F. Bennett, and F. H. Chilton, *Biochim. Biophys. Acta*, 1991, **1081**, 339.
86. J. I. S. MacDonald and H. Sprecher, *Biochim. Biophys. Acta*, 1991, **1084**, 105.
87. T. Sugiura, O. Katayama, J. Fukui, Y. Nakagawa, and K. Waku, *FEBS Lett.*, 1984, **165**, 273.
88. F. H. Chilton and R. C. Murphy, *J. Biol. Chem.*, 1986, **261**, 7771.
89. O. Colard, M. Breton, and G. Bereziat, *Biochem. J.*, 1984, **222**, 657.
90. T. Sugiura, T. Fukuda, Y. Masuzawa, and K. Waku, *Biochim. Biophys. Acta*, 1990, **1047**, 223.
91. Y. Uemura, T.-C. Lee, and F. Snyder, *J. Biol. Chem.*, 1991, **266**, 8268.
92. Nieto, M. L., M. E. Venable, S. A. Bauldry, D. G. Greene, M. Kennedy, D. A. Bass, and R. L. Wykle, *J. Biol. Chem.*, 1991, **266**, 18 699.
93. R. L. Wykle, B. Malone, and F. Snyder, *J. Biol. Chem.*, 1980, **255**, 10 256.
94. E. Ninio, J. M. Mencia-Huerta, F. Heymans, and J. Benveniste, *Biochim. Biophys. Acta*, 1982, **710**, 23.
95. E. Francescangeli, K. Domanska-Janik, and G. Goracci, *J. Lipid Mediators Cell Signalling*, 1996, **14**, 89.
96. R. Roubin, J. M. Mencia-Huerta, A. Landes, and J. Benveniste, *J. Immunol.*, 1982, **129**, 809.
97. R. Roubin, A. Dulioust, I. Haye-Legrand, E. Ninio, and J. Benveniste, *J. Immunol.*, 1986, **136**, 1796.
98. G. Ribbes, E. Ninio, P. Fontan, M. Record, H. Chap, J. Benveniste, and L. Douste-Blazy, *FEBS Lett.*, 1985, **191**, 195.
99. F. Mollinedo, J. Gomez-Cambronero, E. Cano, and M. Sanchez-Crespo, *Biochem. Biophys. Res. Commun.*, 1988, **154**, 1232.
100. T.-C., Lee, *J. Biol. Chem.*, 1985, **260**, 10 952.
101. K. Seyama and T. Ishibashi, *Lipids*, 1987, **22**, 185.
102. K. Satouchi, M. Oda, K. Yasunaga, and K. Saito, *Biochem. Biophys. Res. Commun.*, 1985, **128**, 1409.

103. M. Triggiani, W. C. Hubbard, and F. H. Chilton, *J. Immunol.*, 1990, **144**, 4773.
104. Y. Nakagawa, M. Sugai, K. Karasawa, A. Tokumura, H. Tsukatani, M. Setaka, and S. Nojima, *Biochim. Biophys. Acta*, 1992, **1126**, 277.
105. R. Nakayama and K. Saito, *J. Biochem.*, 1989, **105**, 494.
106. T. G. Tessner and R. L. Wykle, *J. Biol. Chem.*, 1987, **262**, 12 660.
107. E. Remy, G. Lenoir, A. Houben, C. Vandesteene, and J. Remacle, *Biochim. Biophys. Acta*, 1989, **1005**, 87.
108. M. C. Garcia, S. Fernadez-Gallardo, M. A. Gijon, C. Garcia, M. L. Nieto, and M. Sanchez-Crespo, *Biochem. J.*, 1990, **268**, 91.
109. T. Sugiura, A. Ojima-Uchiyama, Y. Masuzawa, M. Fujita, Y. Nakagawa, and K. Waku, *Lipids*, 1991, **26**, 974.
110. J. Gomez-Cambronero, M. L. Nieto, J. M. Mato, and M. Sanchez-Crespo, *Biochim. Biophys. Acta*, 1985, **845**, 511.
111. C. Domenech, E. Machado-DeDomenech, and H. D. Söling, *J. Biol. Chem.*, 1987, **262**, 5671.
112. T. Y. Shen, S.-B. Hwang, T. W. Doebber, and J. C. Robbins, in "Platelet-Activating Factor and Related Lipid Mediators," ed. F. Snyder, Plenum, New York, 1987, p. 153.
113. L. D. Faison and H. L. White, *Prostaglandins*, 1992, **44**, 245.
114. H. L. White and L. D. Faison, *Prostaglandins*, 1988, **35**, 939.
115. F. Alonso, M. G. Gil, M. Sanchez-Crespo, and J. M. Mato, *J. Biol. Chem.*, 1982, **257**, 3376.
116. T.-C. Lee, B. Malone, S. I. Wasserman, V. Fitzgerald, and F. Snyder, *Biochem. Biophys. Res. Commun.*, 1982, **105**, 1303.
117. T. W. Doebber and M. S. Wu, *Proc. Natl. Acad. Sci. USA*, 1987, **84**, 7557.
118. E. Ninio, J. M. Mencia-Huerta, and J. Benveniste, *Biochim. Biophys. Acta*, 1983, **751**, 298.
119. J. Gomez-Cambronero, P. Inarrea, F. Alonso, and M. Sanchez-Crespo, *Biochem. J.*, 1984, **219**, 419.
120. T.-C., Lee, D. J. Lenihan, B. Malone, L. L. Roddy, and S. I. Wasserman, *J. Biol. Chem.*, 1984, **259**, 5526.
121. I. Alam and M. J. Silver, *Biochim. Biophys. Acta*, 1986, **884**, 67.
122. E. Ninio, F. Joly, C. Hieblot, G. Bessou, J. M. Mencia-Huerta, and J. Benveniste, *J. Immunol.*, 1987, **139**, 154.
123. M. Hirafuji, J. M. Mencia-Huerta, and J. Benveniste, *Biochim. Biophys. Acta*, 1987, **930**, 359.
124. T. Sugiura, T. Fukuda, N.-N. Cheng, and K. Waku, *Lipids*, 1991, **26**, 861.
125. Y. Ihara, R. A. Frenkel, and J. M. Johnston, *Arch. Biochem. Biophys.*, 1993, **304**, 503.
126. D. J. Lenihan and T.-C. Lee, *Biochem. Biophys. Res. Commun.*, 1984, **120**, 834.
127. J. Gomez-Cambronero, S. Velasco, J. M. Mato, and M. Sanchez-Crespo, *Biochim. Biophys. Acta*, 1985, **845**, 516.
128. M. L. Nieto, S. Velasco, and M. Sanchez-Crespo, *J. Biol. Chem.*, 1988, **263**, 4607.
129. S. Leyravaud, M. J. Bossant, F. Joly, G. Bessou, J. Benveniste, and E. Ninio, *J. Immunol.*, 1989, **143**, 245.
130. J. M. Mencia-Huerta, R. Roubin, J. L. Morgat, and J. Benveniste, *J. Immunol.*, 1982, **129**, 804.
131. T. Sugiura, A. Ojima-Uchiyama, Y. Masuzawa, and K. Waku, *FEBS Lett.*, 1989, **258**, 351.
132. J. Gomez-Cambronero, S. Velasco, M. Sancez-Crespo, F. Vivanco, and J. M. Mato, *Biochem. J.*, 1986, **237**, 439.
133. J. Gomez-Cambronero, J. M. Mato, F. Vivanco, and M. Sanchez-Crespo, *Biochem. J.*, 1987, **245**, 893.
134. W. Renooij and F. Snyder, *Biochim. Biophys. Acta*, 1981, **663**, 545.
135. E. Francescangeli and G. Goracci, *Biochem. Biophys. Res. Commun.*, 1989, **161**, 107.
136. G. Goracci and E. Francescangeli, *Lipids*, 1991, **26**, 986.
137. R. R. Baker and H.-Y. Chang, *Biochim. Biophys. Acta*, 1993, **1170**, 157.
138. D. S. Woodard, T.-C. Lee, and F. Snyder, *J. Biol. Chem.*, 1987, **262**, 2520.
139. T.-C. Lee, B. Malone, D. Woodard, and F. Snyder, *Biochem. Biophys. Res. Commun.*, 1989, **163**, 1002.
140. D. S. Vallari, M. Record, and F. Snyder, *Arch. Biochem. Biophys.*, 1990, **276**, 538.
141. E. A. Llanos and A. Zanglis, *J. Biol. Chem.*, 1987, **262**, 8990.
142. E. Pirotzky, E. Ninio, J. Bidault, A. Pfister, and J. Benveniste, *Lab. Invest.*, 1984, **51**, 567.
143. J. Wang, M. Kester, and M. J. Dunn, *Biochim. Biophys. Acta*, 1988, **969**, 217.
144. T.-C. Lee, B. Malone, and F. Snyder, *J. Biol. Chem.*, 1986, **261**, 5373.
145. R. R. Baker and H.-Y. Chang, *Biochim. Biophys. Acta*, 1996, **1302**, 257.
146. T.-C. Lee, B. Malone, and F. Snyder, *J. Biol. Chem.*, 1988, **263**, 1755.
147. R. Heller, F. Bussolino, D. Ghigo, G. Garbarino, G. Pescarmona, U. Till, and A. Bosia, *J. Biol. Chem.*, 1991, **266**, 21 358.
148. R. R. Baker and H.-Y. Chang, *Biochim. Biophys. Acta*, 1994, **1213**, 27.
149. T.-C. Lee, B. Malone, M. L. Blank, V. Fitzgerald, and F. Snyder, *J. Biol. Chem.*, 1990, **265**, 9181.
150. D. Vance, in "Phosphatidylcholine Metabolism," ed. D. Vance, CRC Press, Boca Raton, FL, 1989, p. 33.
151. M. L. Blank, E. A. Cress, and F. Snyder, *Biochem. Biophys. Res. Commun.*, 1984, **118**, 344.
152. K. Satouchi, M. Oda, K. Saito, and D. J. Hanahan, *Arch. Biochem. Biophys.*, 1984, **234**, 318.
153. M. L. Blank, T.-C. Lee, E. A. Cress, B. Malone, V. Fitzgerald, and F. Snyder, *Biochem. Biophys. Res. Commun.*, 1984, **124**, 156.
154. F. Bussolino, F. Gremo, C. Tetta, G. P. Pescarmona, and G. Camussi, *J. Biol. Chem.*, 1986, **261**, 16 502.
155. M. L. Nieto, S. Velasco, and M. Sanchez-Crespo, *J. Biol. Chem.*, 1988, **263**, 2217.
156. D. R. Hoffman, M. K. Bateman, and J. M. Johnston, *Lipids*, 1988, **23**, 96.
157. S. Fernandez-Gallardo, M. A. Gijon, M. C. Garcia, E. Cano, and M. Sanchez-Crespo, *Biochem. J.*, 1988, **254**, 707.
158. C. B. Appleyard and K. Hillier, *Clin. Sci.*, 1995, **88**, 713.
159. D. C. Tsoukatos, A. D. Tselepis, and M. E. Lekka, *Biochim. Biophys. Acta*, 1993, **1170**, 258.
160. T. Sugiura, unpublished results, 1993.
161. R. Nakayama, K. Yasuda, K. Satouchi, and K. Saito, *Biochem. Biophys. Res. Commun.*, 1988, **151**, 1256.
162. R. N. Pinckard, H. J. Showell, R. Castillo, C. Lear, R. Breslow, L. M. McManus, D. S. Woodard, and J. C. Ludwig, *J. Immunol.*, 1992, **148**, 3528.
163. A. Tokumura, K. Kamiyasu, K. Takauchi, and H. Tsukatani, *Biochem. Biophys. Res. Commun.*, 1987, **145**, 415.
164. A. Tokumura, T. Asai, K. Takauchi, K. Kamiyasu, T. Ogawa, and H. Tsukatani, *Biochem. Biophys. Acta*, 1988, **155**, 863.
165. A. Tokumura, K. Takauchi, T. Asai, K. Kamiyasu, T. Ogawa, and H. Tsukatani, *J. Lipid Res.*, 1989, **30**, 219.
166. A. Tokumura, T. Tanaka, T. Yotsumoto, and H. Tsukatani, *Biochem. Biophys. Res. Commun.*, 1991, **177**, 466.
167. T. Tanaka, H. Minamino, S. Unezaki, H. Tsukatani, and A. Tokumura, *Biochim. Biophys. Acta*, 1993, **1166**, 264.

168. A. Tokumura, M. Toujima, Y. Yoshioka, and K. Fukuzawa, *Lipids*, 1996, **31**, 1251.
169. H. Itabe, Y. Kushi, S. Handa, and K. Inoue, *Biochim. Biophys. Acta*, 1988, **962**, 8.
170. K. E. Stremler, D. M. Stafforini, S. M. Prescott, G. A. Zimmerman, and T. M. McIntyre, *J. Biol. Chem.*, 1989, **264**, 5331.
171. H. Kamido, A. Kuksis, L. Marai, and J. J. Myher, *J. Lipid Res.*, 1995, **36**, 1876.
172. P. L. Smiley, K. E. Stremler, S. M. Prescott, G. A. Zimmerman, and T. M. McIntyre, *J. Biol. Chem.*, 1991, **266**, 11 104.
173. K. D. Patel, G. A. Zimmerman, S. M. Prescott, and T. M. McIntyre, *J. Biol. Chem.*, 1992, **267**, 15 168.
174. J. M. Heery, M. Kozak, D. M. Stafforini, D. A. Jones, G. A. Zimmerman, T. M. McIntyre, and S. M. Prescott, *J. Clin. Invest.*, 1995, **96**, 2322.
175. T. Tanaka, M. Iimori, H. Tsukatani, and A. Tokumura, *Biochim. Biophys. Acta*, 1994, **1210**, 202.
176. S. Ylä-Herttuala, W. Palinski, M. E. Rosenfeld, and S. D. Steinberg, *J. Clin. Invest.*, 1989, **84**, 1086.
177. H. Itabe, E. Takeshima, H. Iwasaki, J. Kimura, Y. Yoshida, T. Imanaka, and T. Takano, *J. Biol. Chem.*, 1994, **269**, 15 274.
178. U. P. Steinbrecher and P. H. Pritchard, *J. Lipid Res.*, 1989, **30**, 305.
179. T. A. Liapikos, S. Antonopoulou, S. A. P. Karabina, D. C. Tsoukatos, C. A. Demopoulos, and A. D. Tselepis, *Biochim. Biophys. Acta*, 1994, **1212**, 353.
180. M. T. Quinn, S. Parthasarathy, L. G. Fong, and D. Steinberg, *Proc. Natl. Acad. Sci. USA*, 1987, **84**, 2995.

1.12
Biosynthesis of Cyclic Bromoethers from Red Algae

AKIO MURAI
Hokkaido University, Sapporo, Japan

1.12.1 INTRODUCTION

1.12.1.1 Cyclic Bromoethers from *Laurencia* Species

The red algae, *Laurencia* species, belonging to the Rhodomelaceae family in the Ceramiaceae order, are distributed widely in the world and produce a host of nonterpenoid C_{15} compounds, which arise from fatty acid metabolism. Chemical studies on the constituents of *Laurencia* commenced with the pioneering isolation of laurencin (**1**) from *Laurencia glandulifera* in 1965 by Irie *et al.*[1,2] Since that time, many researchers have joined in studies on the isolation of new C_{15} acetogenins related to (**1**) from *Laurencia* sp.[3–6] The representative cyclic compounds are revealed to be as follows: obtusin (**2**)[7,8] from *L. obtusa*, laureepoxide (**3**),[9] (3*Z*)-(**4**) and (3*E*)-kumausynes (**5**),[10] (3*Z*)-(**6**) and (3*E*)-deacetylkumausynes (**7**),[10] kumausallene (**8**),[11] and laureoxolane (**9**)[12] from *L. nipponica* with oxolane skeletons, dactylyne (**10**),[13] isodactylyne (**11**),[14] and (3*Z*)-(**12**)[15] and (3*E*)-dactomelynes (**13**)[15] having oxane systems from *Aplysia dactylomela* considered to be of algal origin, and isolaurepinnacin (**14**)[16] from *L. pinnata* and (3*Z*)-(**15**)[17] and (3*E*)-isoprelaurefucins (**16**)[17] from *L. nipponica* with seven-membered cyclic ethers. As the eight-membered systems, (**1**), deacetyllaurencin (**17**),[18,19] laurefucin (**18**),[20,21] and laureoxanyne (**19**)[22] have been isolated from *L. nipponica*. These four oxocane compounds are characteristic in respect of their C-6*R* and C-7*R* configurations. We have designated these as "lauthisan" compounds. On the other hand, another eight-membered series having *S* and *S*-configurations at the requisite positions have also been isolated from *L. nipponica*, such as laureatin (**20**),[23] isolaureatin (**21**),[23] prelaureatin (**22**),[24] and laurallene (**23**).[25] These compounds have been named "laurenan" compounds. Furthermore, (3*Z*)-(**24**)[26–29] and (3*E*)-obtusenynes (**25**)[15] from *L. obtusa* and *A. dactylomela*, isolaurallene (**26**)[30] from *L. nipponica*, and obtusallene (**27**)[31] from *L. obtusa* have also been known. These C_{15} acetogenins display the following structural features: (i) cyclic ether rings of various sizes, (ii) an enyne or allenic side-chain, and (iii) at least one bromine atom and, generally, an additional halogen (bromine or chlorine) atom.

Apparently, it has generally been accepted with no experimental evidence that these metabolites arise from C_{15} linear hydroxy, halohydroxy, or epoxypolyenynes that, in turn, trace their origin to hexadeca-4,7,10,13-tetraenoic acid (**28**)[32] (Scheme 1). The isolation of (3*Z*)-(**29**) and (3*E*)-laurediols (**30**) and their respective acetates ((**31**) and (**32**)) by Irie's group in 1972[18] and of (3*Z* and 3*E*,12*Z*)-laurediols ((**33**) and (**34**)) and (3*Z* and 3*E*)-12,13-dihydrolaurediols ((**35**) and (**36**)) by the author's group in 1993[19] from *L. nipponica* provides some support for this postulate. Yamada's group has also succeeded in isolation of (3*Z*)-(**37**) and (3*E*)-laurencenynes (**38**) and (3*Z*)-(**39**) and (3*E*)-neolaurencenynes (**40**) from *L. okamurai*.[33,34] Detection of the four intermediates ((**33**)–(**36**)) suggests the primary existence of hexadeca-4,7,10-trienoic acid (**41**) and enables us to propose that neolaurencenynes ((**39**) and (**40**)) would be converted into laurencenynes ((**37**) and (**38**)) by enzymatic dehydrogenation at the 12,13-position, and then (*Z*)-6,7-epoxidation of (**37**) and (**38**) followed by the epoxide-opening reaction would give the *threo* diols ((**33**) and (**34**)), which would finally be isomerized into (12*E*)-laurediols ((**29**) and (**30**)) (Scheme 1). Otherwise, an alternative pathway from (**39**) and (**40**) through (**35**) and (**36**) may proceed independently. Therefore, all of the four compounds ((**33**)–(**36**)) isolated by the author's group could be the direct precursors of laurediols.

1.12.1.2 Proposed Biogenesis for Cyclic Bromoethers

As described in Section 1.12.1.1, a variety of halogenated C_{15} acetogenins have been isolated and structurally clarified. Furthermore, the structural relationships between these compounds have been gradually clarified and the reasonable biogenetic routes have been illustrated to some extent.[3–6] When we started the biosynthetic studies, there had been no experimental report on the chemical mechanisms of the introduction of halogen atoms into the second metabolites from red algae. Two suggestive routes were proposed initially for introduction of the bromine atom into a straight-chain C_{15} skeleton. In 1975, Gonzalez[35] reported that the ring closure of the polyene system would proceed involving participation of a positive bromine atom in the biogenetic pathway of bromine-containing terpenoids (Scheme 2). This proposal has been accepted to be reasonable without any proof since that time.[3–6,36]

On the other hand, in 1979, Kurosawa[37] proposed alternatively in the field of bromoterpenoids that the starting polyene would be oxidized biosynthetically to give the terminal epoxide (Scheme 3). The epoxide could be attacked with a negative bromine atom to afford the corresponding bromohydrin, which would be cyclized internally with acid to provide a brominated ring.[37] Concerning introduction of a chlorine atom, it had generally been proposed that the chlorine atom might be introduced in the form of either a negative chlorine[36–38] or positive chlorine.[39]

Although any attempts to establish laurediols or their original polyenes as the real biosynthetic precursors have not been reported, two biogenetic routes would be possible if these proposals are applied to laurencin (**1**) (Scheme 4). Starting from the presumable (3*E*,12*E*)-laurencenyne (**42**), the stereoselective oxidation at C-6 and C-7 could provide (6*R*,7*R*)-laurediol (**30**) that, on the intramolecular cyclization reaction with a positive bromine, would give rise to laurencin (**1**). Alternatively, if the oxidation of natural (3*E*,12*Z*)-laurencenyne[33,34] only at the terminal 12,13-olefin would afford the 12,13-epoxy derivative, it would provide laurencin (**1**) by way of the oxidative ring contraction of the oxirane ring with a negative bromine. Now, which path is more plausible in nature? Is the bromine atom introduced to polyenynes as a positive bromine or a negative bromine? According to biogenetical consideration, it was supposed that construction of cyclic bromoether systems in the algae might be attributable to the enzymatic reaction of laurediols ((**29**) and (**30**)) with an enzyme-bound bromonium ion[40] *in vivo*. Commencement of a study of these enzymatic reactions was triggered by the establishment of the mechanism of the peroxidase reaction carried out by Yamazaki and co-workers.[41–47] They reported that, during the peroxidative oxidation, only a negative iodine atom could be changed into a positive iodine atom on treatment with hog thyroid

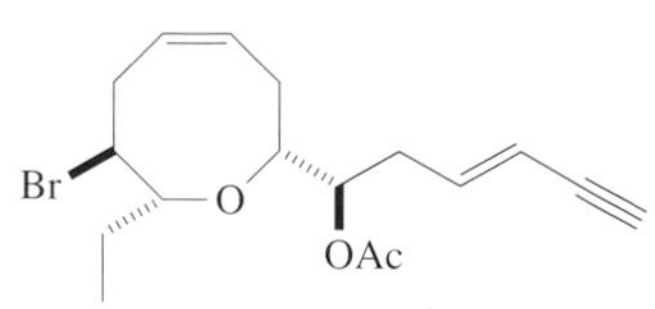

Br O OAc
(1)

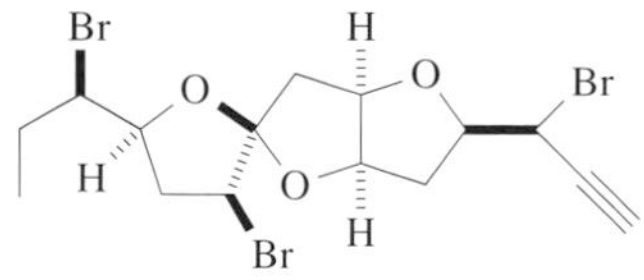

Br H O O O H Br H Br
(2)

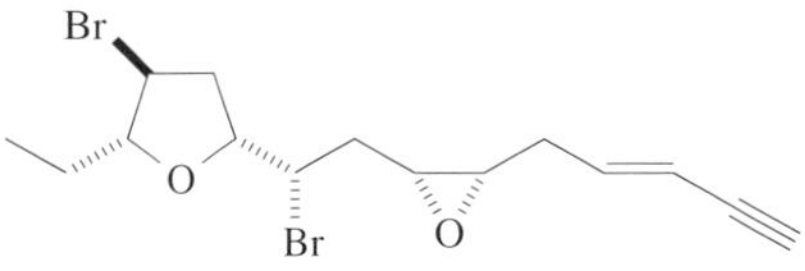

Br O Br O
(3)

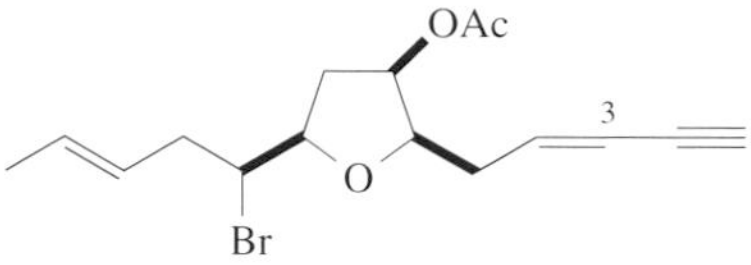

OAc O 3 Br
(4) *Z*-3 **(5)** *E*-3

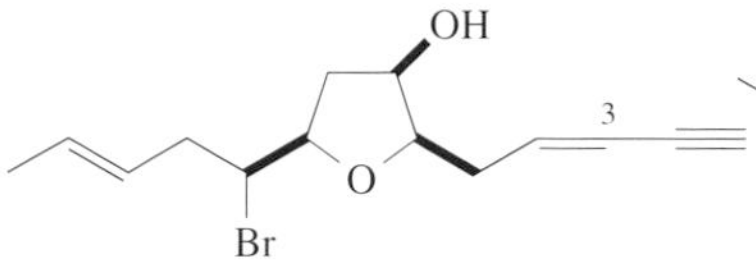

OH O 3 Br
(6) *Z*-3 **(7)** *E*-3

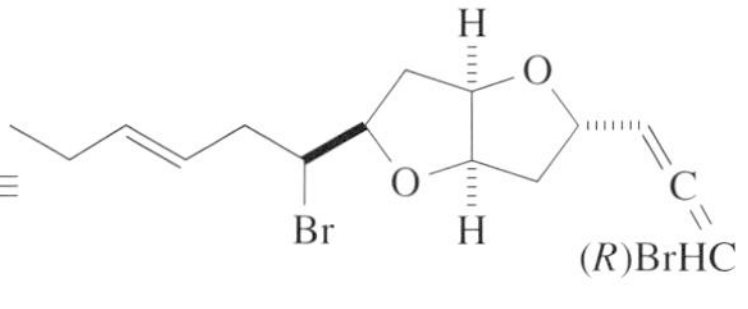

H O O Br H C (*R*)BrHC
(8)

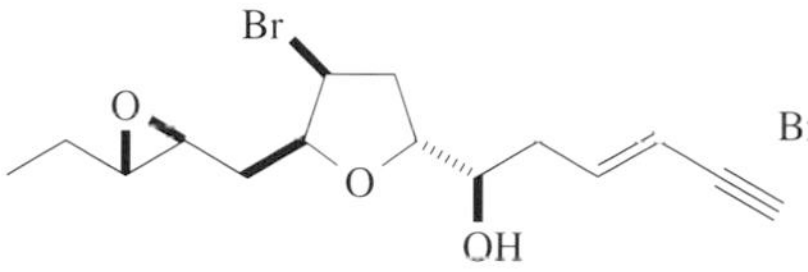

Br O O OH
(9)

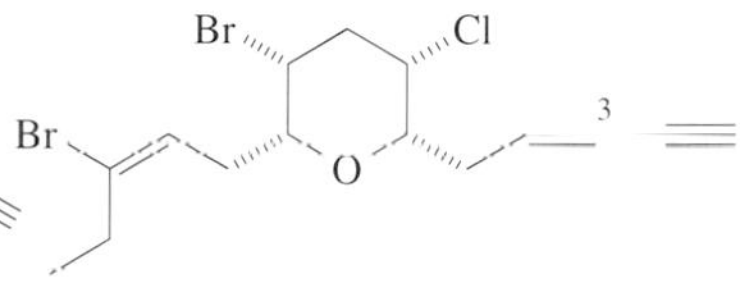

Br Cl Br O 3
(10) *Z*-3 **(11)** *E*-3

H O Cl Br H O 3 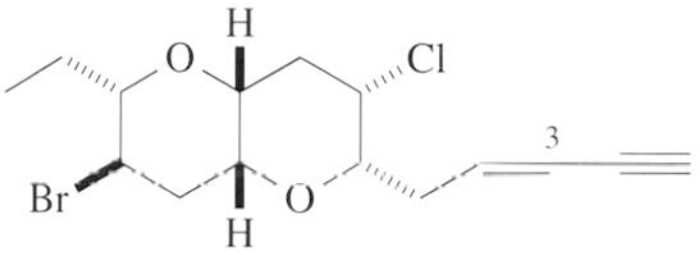
(12) *Z* 3 **(13)** *E* 3

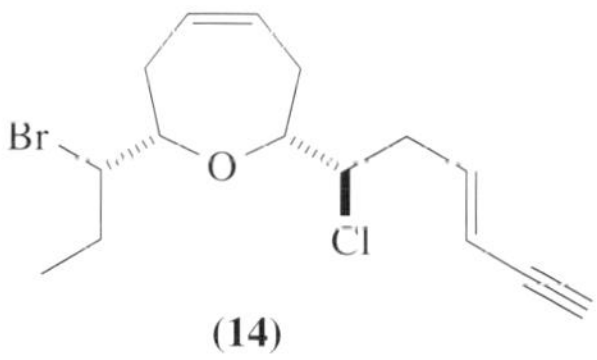

Br O Cl
(14)

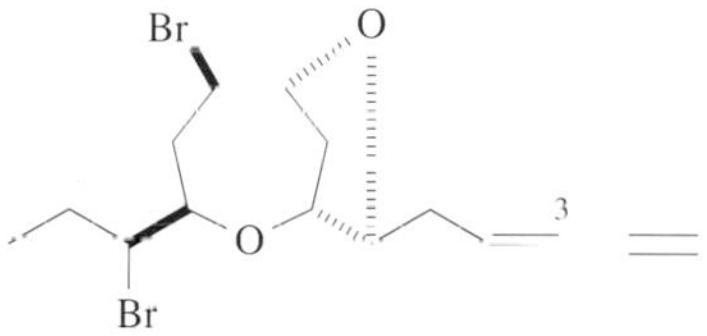

Br O Br O 3 Br
(15) *Z*-3 **(16)** *E*-3

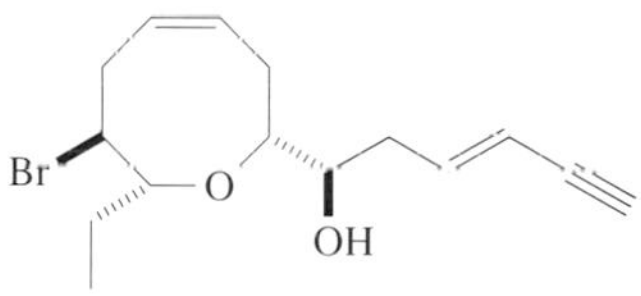

Br O OH
(17)

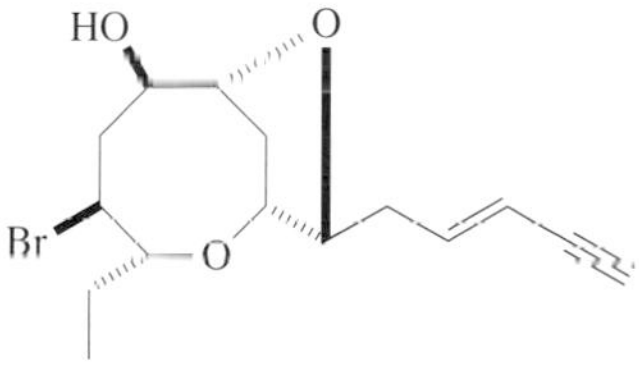

HO O Br O
(18)

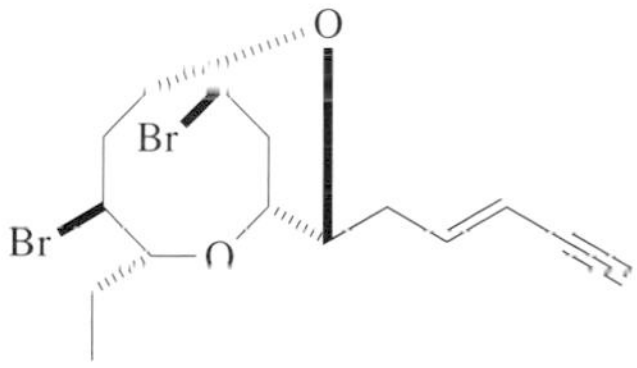

O Br Br O
(19)

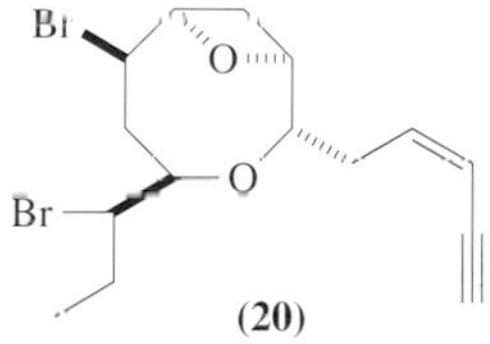

Br O O Br
(20)

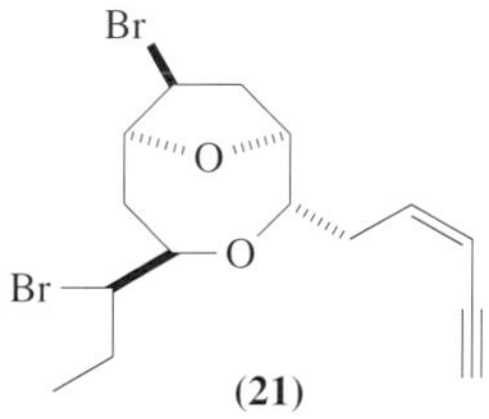

Br O O Br
(21)

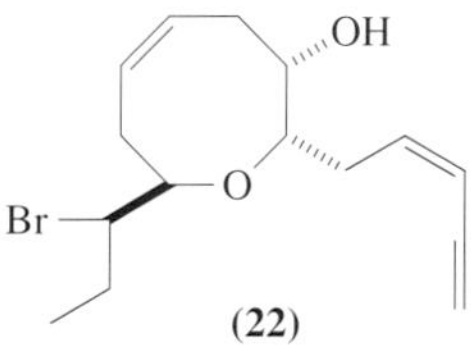

OH Br O
(22)

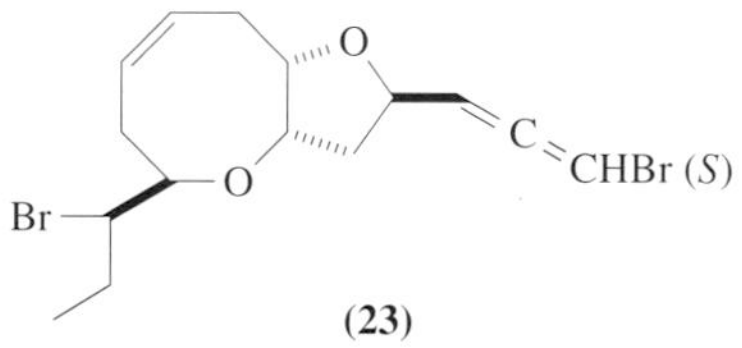

O C CHBr (*S*) Br O
(23)

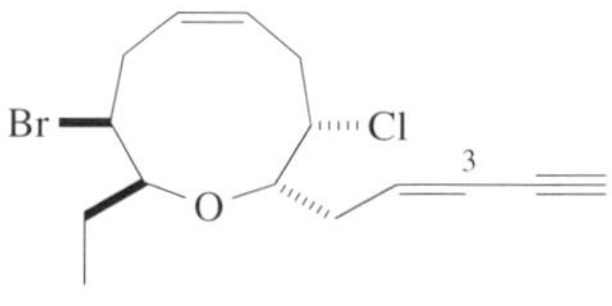

Br Cl O 3
(24) *Z*-3 **(25)** *E*-3

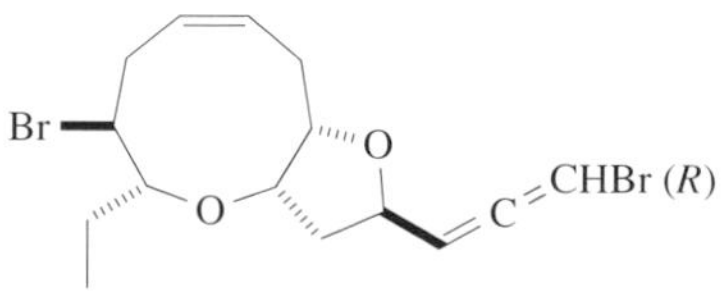

Br O O C CHBr (*R*)
(26)

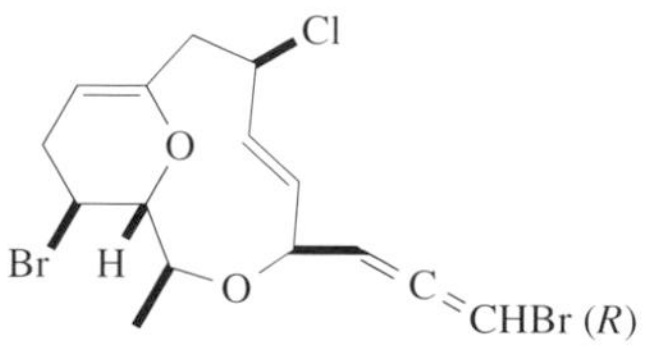

Cl O Br H O C CHBr (*R*)
(27)

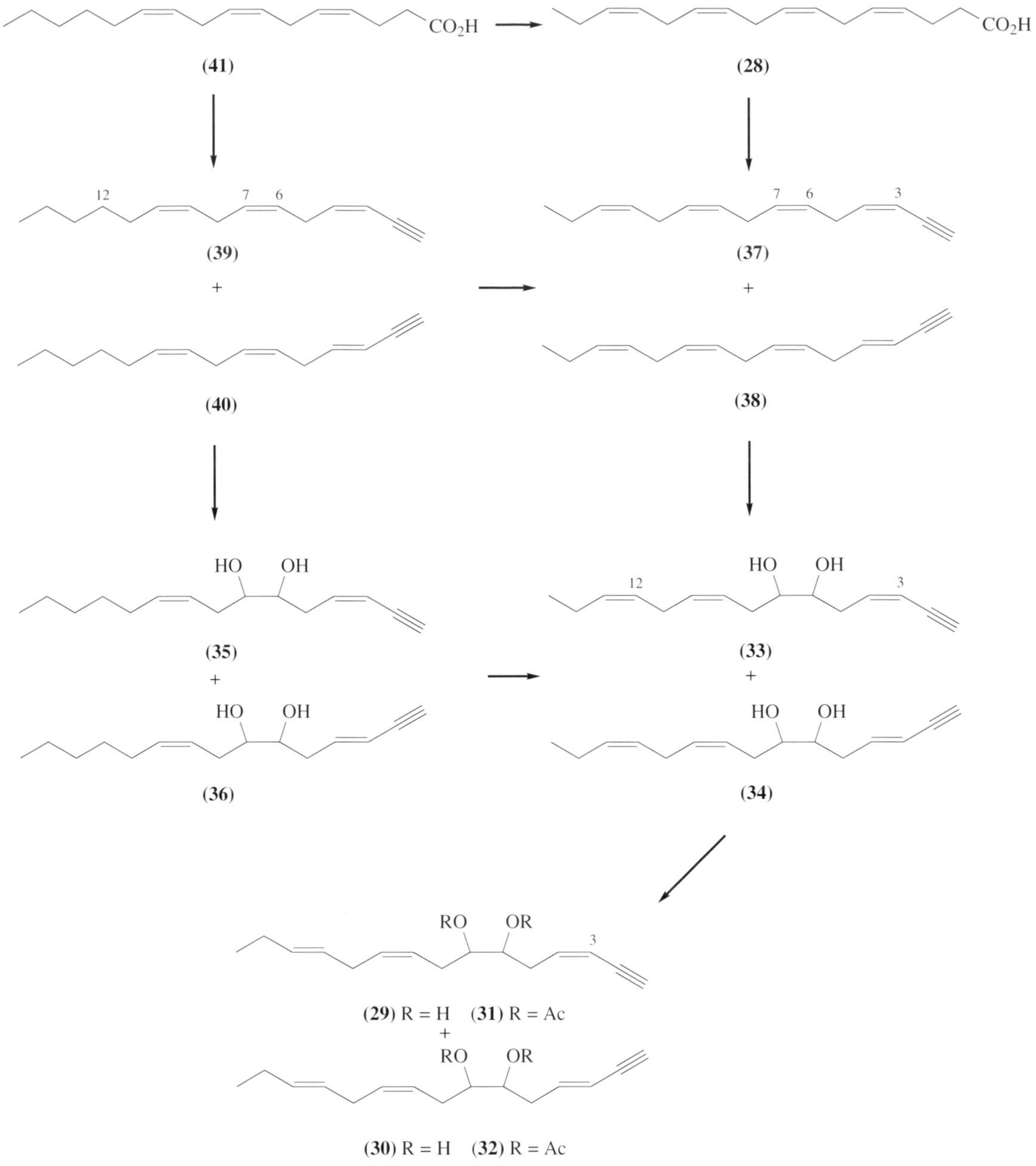

CO$_2$H (41) → CO$_2$H (28)

12 7 6 (39) + (40) → 7 6 3 (37) + (38)

HO OH (35) + HO OH (36) → HO OH 12 3 (33) + HO OH (34)

RO OR 3

(29) R = H (31) R = Ac

+

RO OR

(30) R = H (32) R = Ac

Scheme 1

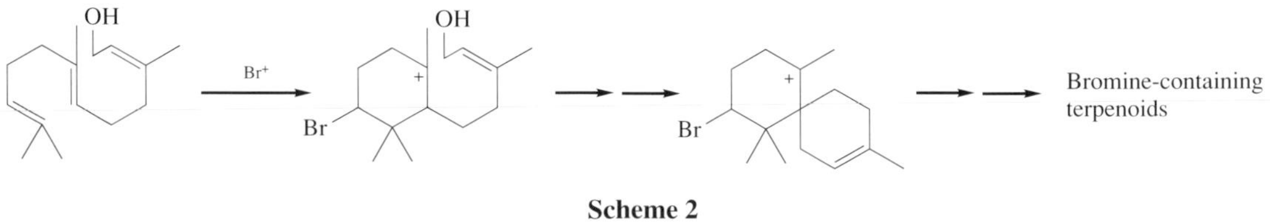

OH → Br$^+$ → OH + Br → → + Br → → Bromine-containing terpenoids

Scheme 2

peroxidase and hydrogen peroxide (H_2O_2) by way of a two-electron oxidation, but not a one-electron oxidation, although peroxidase enzyme had generally been known as an initiator of one-electron oxidation.[41–47] Bromoperoxidase (abbreviated as BPO)[40] was proposed as the plausible enzyme for bromoetherification, because it was expected that the peroxidase enzyme included in

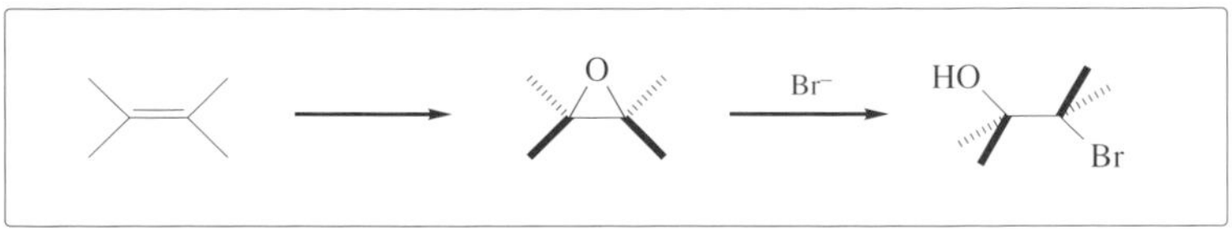

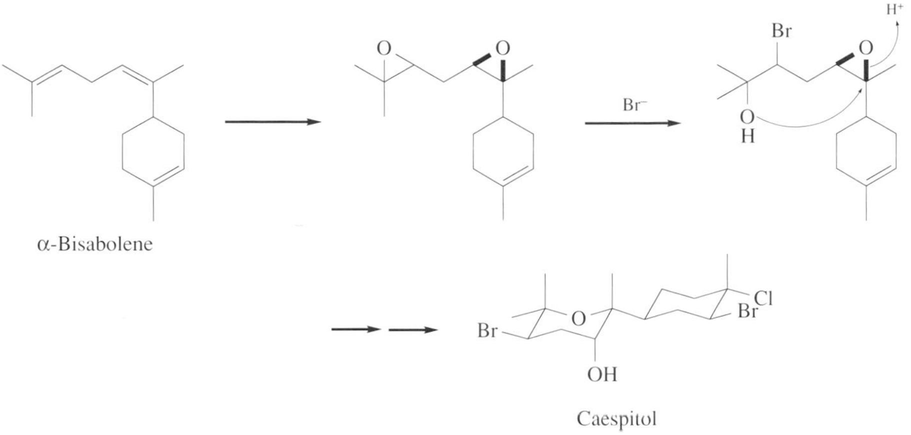

α-Bisabolene

Caespitol

Scheme 3

algae might be the generator of a positive bromine on reaction with H_2O_2 and a negative bromine. When the authors' group started to study such enzymatic reactions, there had been no report on the existence of the peroxidase enzyme concerning two electron oxidation of the bromide ion in living bodies.

Bromoperoxidase had not yet been isolated from *Laurencia* species. Accordingly, commercially available lactoperoxidase (abbreviated as LPO) was chosen first as an enzyme related in character to BPO. LPO can be prepared from cow's milk, and has been well characterized. Its relative molecular mass is 78 500. LPO contains an iron-porphyrin thiol as the prosthetic group. In 1987, the structure of LPO heme (**43**) was clarified by Australian biochemists.[48] They proposed that the unusual strength of binding of the prosthetic group to the apoprotein is due to formation of a disulfide bond from a cysteine residue to the porphyrin thiol. As shown in Scheme 5, the iron atom (oxidation value of Fe part is 3) in the porphyrin skeleton can be oxidized in contact with H_2O_2 to give compound I (oxidation value of Fe part changes to 5). The conversion of the native enzyme to compound I is very fast as shown by the rate constant (1.2×10^7 M^{-1} s^{-1}). Compound I is allowed to react with a bromide ion to afford enzyme-oxygen-bromine (EOB), which is tentatively named. The rate constant was estimated as 2.2×10^6 M^{-1} s^{-1}. The results revealed that the bromide ion was oxidized by the enzyme by way of two-electron transfer (not a bromine radical). The reaction species generated by iron porphyrin in LPO has been established to be an enzyme-bound bromonium ion (porphyrin-Fe^{III}—O—$Br^{\delta+}$, i.e. EOB). When some substrate is treated with EOB, the bromide atom of EOB would attack the substrate as a positive bromine. After completion of the reaction, the ferric form would be converted to the native enzyme.

1.12.2 BIOSYNTHESIS WITH LACTOPEROXIDASE

1.12.2.1 Model Study

First of all, the author's group carried out the preliminary experiments using pentadecenediols ((**44**), (**45**), and (**46**)) as the second substrates, all of which could be laurediol mimics[49] (Scheme 6). (6*S*,7*S*,9*E*)-9-Pentadecene-6,7-diol ((**44**), 18 mg, 74 nmol, 3.7 mM) as the substrate was dissolved in DMSO (0.2 ml) and the solution was injected into a phosphate buffer (pH 5.5, 50 mM, 20 ml) containing NaBr (60 nmol, 3 mM). To the mixture was added an aliquot of each solution of H_2O_2 (16 nmol, 0.80 mM) and LPO (32 nM) divided into 12 portions during 2 h. The mixture was stirred in the dark at 23 °C for 24 h and extracted twice with EtOAc. The extracts were washed with distilled

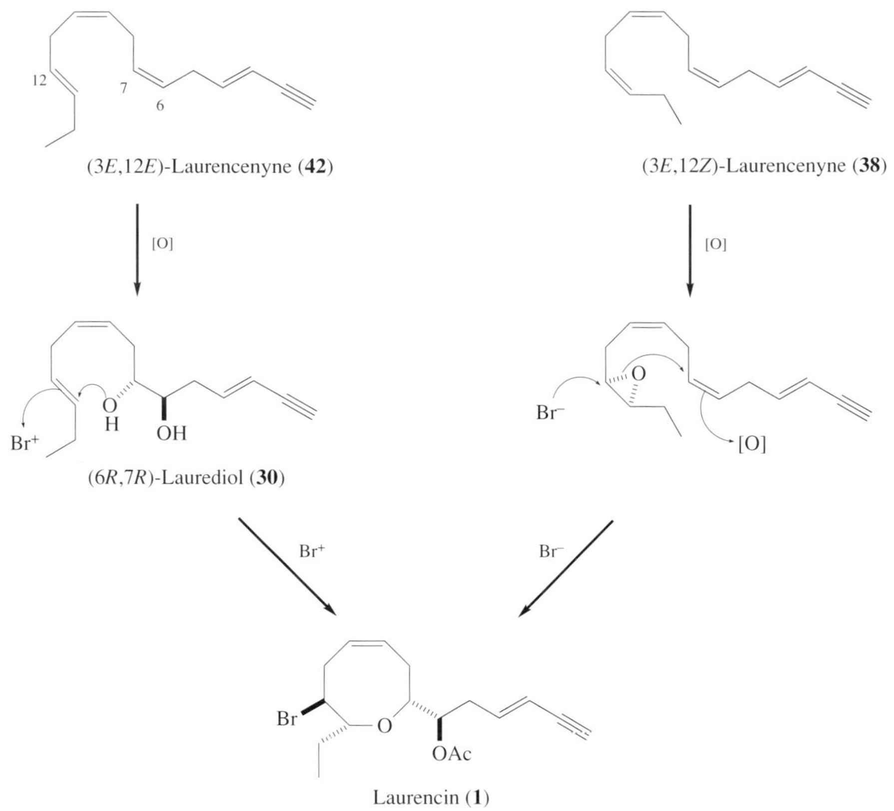
12
7
6
(3*E*,12*E*)-Laurencenyne (**42**)
(3*E*,12*Z*)-Laurencenyne (**38**)
[O]
[O]
O
H
OH
Br+
(6*R*,7*R*)-Laurediol (**30**)
O
Br−
[O]
Br+
Br−
Br
O
OAc
Laurencin (**1**)

Scheme 4

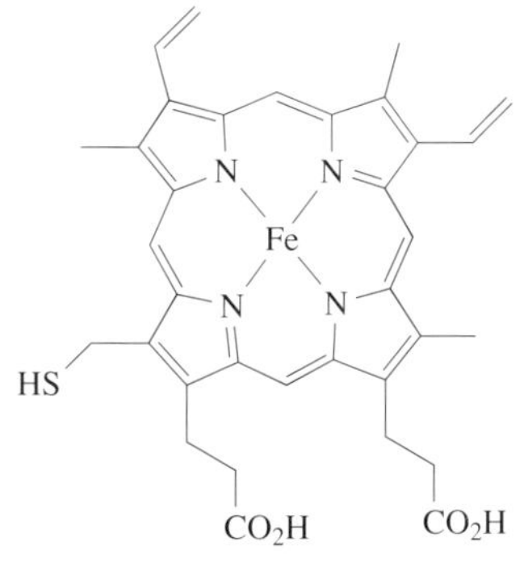
N
N
Fe
N
N
HS
CO2H
CO2H

(43)

water, dried, and concentrated *in vacuo*. The residue was then subjected to chromatography over SiO_2 and purified by HPLC to give the bromoether ((**47**), 2.1 mg, 8.8%), the recovered starting material (13 mg, 72%), and a mixture of bromohydrins (3.1 mg, 12.3%), which was not further characterized. Next, the reaction was attempted under the same conditions as above without LPO and it was observed that no reaction proceeded. The above reaction involving formation of (**47**) could be regarded as a straightforward enzymatic reaction. Furthermore, a mixture of the above bromohydrins, when repeatedly subjected to the same reaction conditions, gave no bromoethers. The results eliminate the possibility of bromoether cyclization via a bromohydrin pathway.[37] On the other hand, when the (9*Z*)-isomer ((**45**), 17.0 mg, 35 nmol, 3.5 mM) was treated under the same conditions, two bromoethers ((**48**), 0.53 mg, 2.35% and (**49**), 0.27 mg, 1.2%), were obtained along with bromohydrins (4.5 mg, 18.6%) and the recovered starting material (11.0 mg, 64.7%).

Furthermore, the substrate (**46**), (6*R*,7*R*,12*E*)-12-pentadecene-6,7-diol (16.1 mg) was subjected to the enzymatic reaction under the same conditions as mentioned above to afford the desired eight-

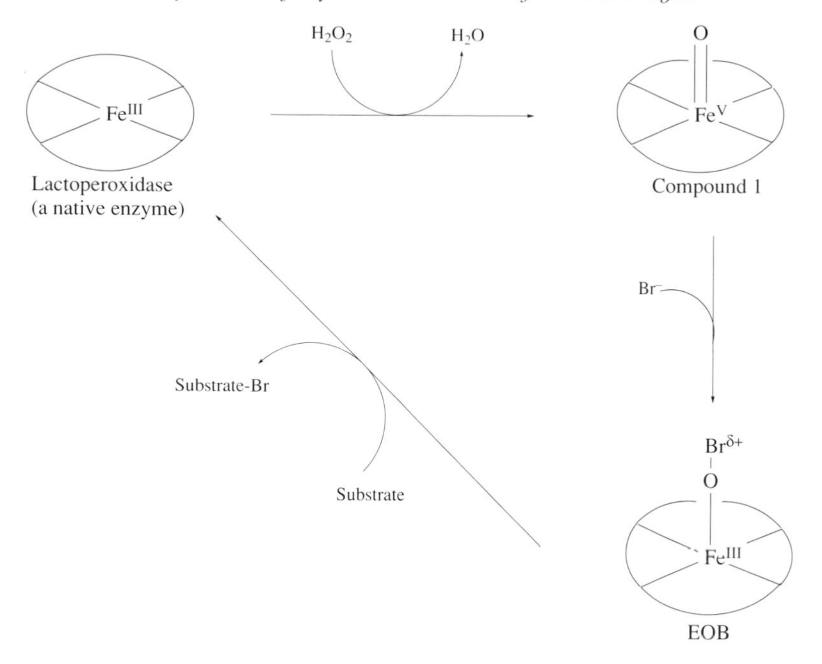

Scheme 5

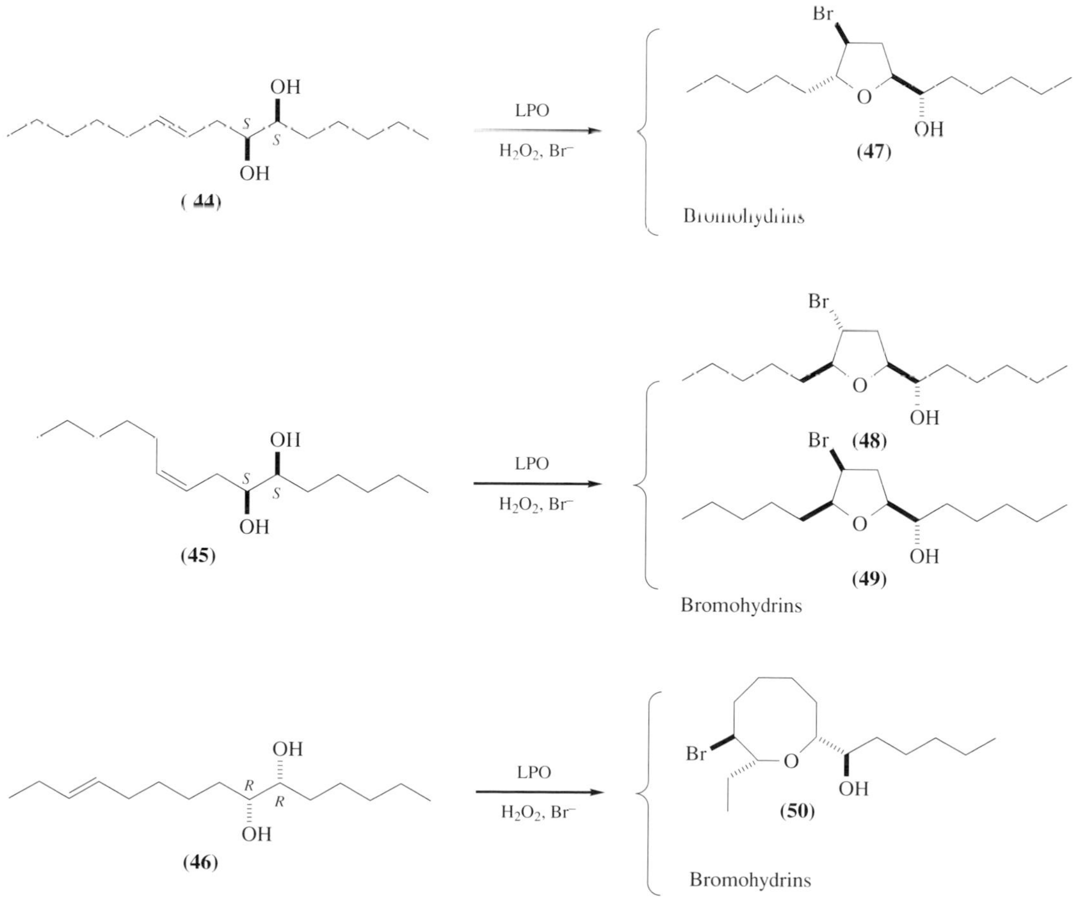

Scheme 6

membered bromoether ((**50**), 0.3 mg, 1.4%) and bromohydrins (3.9 mg, 17.5%) as the major products with the recovered starting material (8.7 mg, 54.0%). Compound (**50**) was identical in all respects with deacetyloctahydrolaurencin prepared from natural laurencin (**1**).

In view of the chemical reaction mechanisms, bromocationic etherification of the substrates ((**44**) and (**45**)) via the corresponding bridged bromonium intermediates would possibly give rise to four cyclic bromoether products, respectively. However, the enzymatic reactions produced only endocyclic bromooxolanes from these olefin diols, respectively. It is to be noted that (*E*)-olefin alcohol cyclizes exclusively in an *anti*-addition manner, while the (*Z*)-isomer cyclizes both in *anti*- and in *syn*-addition manner. The results indicate that LPO has a high regio- but a low stereoselectivity for these enzymatic reactions. Conversion of (**46**) into the eight-membered bromoether (**50**) reveals that there should be a conformationally implicit role of LPO in the enzymatic reaction, since the internal bromocyclization of (**46**) never proceeds with any chemical reagents generating a positive bromine (i.e. *N*-bromosuccinimide or 2,4,4,6-tetrabromocyclohexane-2,5-dien-1-one). Consequently, it is concluded that LPO clearly recognized the olefin alcohols ((**44**), (**45**), and (**46**)) as analogues of laurediols.

1.12.2.2 Lauthisan Series

The various eight-membered cyclic bromoether compounds have been produced from *Laurencia* sp. These are characterized by possessing either a (6*R*,7*R*)- or (6*S*,7*S*)-configuration with one exceptional example, i.e. laureepoxide (**3**) having (6*S*,7*R*)-oxirane.[9] The former compounds are designated as "lauthisan" and the latter as "laurenan" compounds. On the basis of the preliminary experiments mentioned in Section 1.12.2.1, the corresponding enzymatic reactions were attempted with highly unstable laurediol (**34**) aiming at the enzymatic synthesis of lauthisan compounds. Herein the enzymatic synthesis of deacetyllaurencin (**17**) starting from (3*E*,6*R*,7*R*)-laurediol (**51**) is described.[50]

It is impossible to provide (3*E*,6*R*,7*R*)- (**51**) and (3*E*,6*S*,7*S*)-laurediols (**52**) in pure state from natural sources. Masamune and co-workers reported the total synthesis of (3*E*,6*S*,7*S*)-laurediol (**52**) starting from (2*R*,3*R*)-(+)-tartaric acid[51] (**53**) (Scheme 7). The tartaric acid was converted into (2*S*,3*S*)-1-benzyloxy-3,4-epoxy-2-butanol (**54**) in 51% yield by a modification of the known procedure.[52] Elongation of the two-carbon unit (C-9–C-10) was achieved by treatment of (**54**) with a lithium acetylide–ethylenediamine complex, giving acetylene glycol, which easily formed its acetonide (**55**). Further treatment of (**55**) with LDA and then with (*E*)-2-pentenyl bromide effected formation of undecenyne (**56**). Hydrogenation of (**56**) over a Lindlar catalyst afforded the corresponding undecadiene, which underwent the Birch reduction to give (5*Z*,8*E*)-undecadiene-1,2,3-triol 2,3-acetonide (**57**) (52% from (**54**)), constituting the C-5–C-15 unit of (**52**). Transformation of (**57**) into (**52**) was commenced by formation of epoxy alcohol (**58**), which was smoothly performed by mesylation of (**57**) and subsequent treatment with an acid and then with a base. Compound (**58**) was treated with magnesium cyanide[53] to give glycol nitrile, which again formed its acenonide (**59**). Hydride reduction of (**59**) followed by careful treatment with acid (SiO_2, −20 °C, 3 min) gave rise to highly labile β-alkoxy aldehyde (**60**), which was immediately submitted to Wittig reaction. The resulting trimethylsilylpentadecatrienynediol acetonide (**61**) was deprotected smoothly to give a very unstable sample of (**52**) (22% from (**57**)). Although this synthetic route is also effective for the synthesis of (3*E*,6*R*,7*R*)-laurediol (**51**) starting from (2*S*,3*S*)-(−)-tartaric acid, we provided (**51**) more easily from natural laurencin (**1**)[50] (Scheme 8).

Hydrolysis of natural laurencin (**1**) with KOH in EtOH produced (**17**), which was treated with BuLi and TMSCl to afford two separable compounds. The major product (**62**), obtained in 46% yield, was reacted with zinc powder in AcOH and EtOH to give a 20:1 (3*E*/3*Z*)-mixture of (**63**) in 79% yield, which was found to be completely free from the starting compounds ((**17**) and (**62**)) from MS and ^{1}H-NMR (400 MHz) spectra. Compound (**63**) was deprotected to yield quantitatively a 20:1 (3*E*/3*Z*)-mixture of (6*R*,7*R*)-laurediol (**51**), the homogeneity of which was reconfirmed also by MS and ^{1}H-NMR spectra. The compound (**51**) was highly labile and immediately subjected to the enzymatic reaction[50] (Scheme 9).

Compound (**51**) (73.8 mg, 7.9 mM) in DMSO (0.3 ml) was injected into a phosphate buffer (pH 5.5, 50 mM, 40 ml) containing NaBr (7.0 mM), and the mixture was fitted with an Ar balloon and kept at 5 °C. To the mixture was added an aliquot of each solution of H_2O_2 (1.8 mM) and LPO (64 nM) divided into 12 portions during 2 h. The mixture was stirred at 5 °C for 24 h and worked up as usual to lead to production of crude deacetyllaurencin ((**17**), 1.0 mg), an unknown cyclic ether

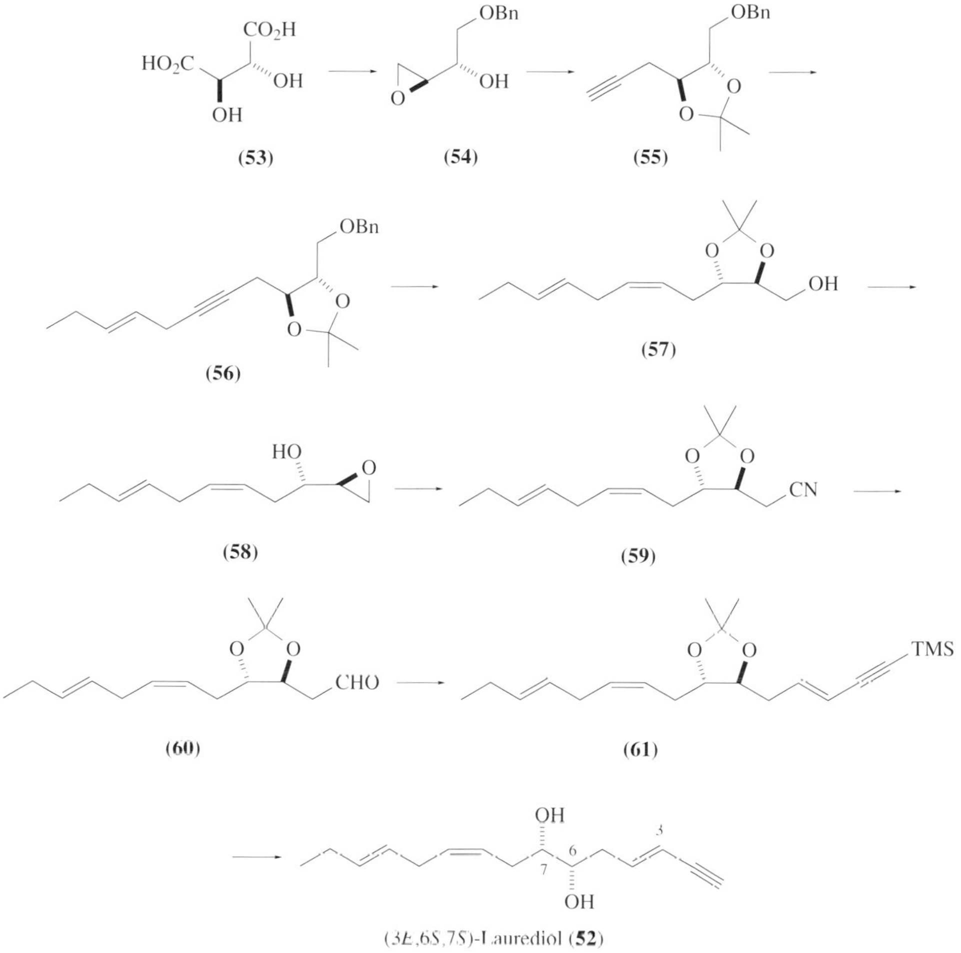

Scheme 7

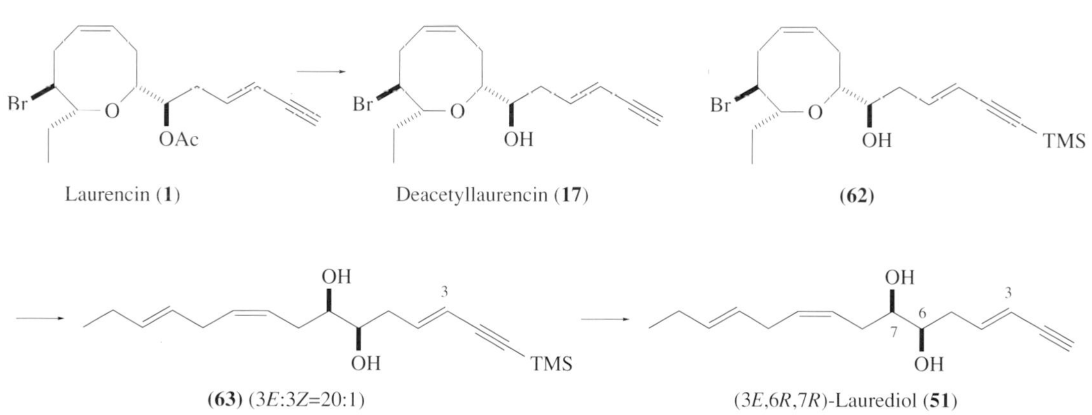

Scheme 8

(0.9 mg), an undetermined mixture of bromohydrins (29.2 mg), and the recovered starting material (38.4 mg). The impure deacetyllaurencin was acetylated under the usual conditions and separated over SiO_2 to afford a pure sample of laurencin (**1**), 0.8 mg, 0.73%). Comparison of ^{1}H-NMR (400 MHz) spectra showed that synthetic laurencin was completely identical with a natural sample (**1**). The results strongly support the hypothesis that laurediols are the real biosynthetic precursors of (**1**) and its related cyclic ether compounds in marine origins. This provides the first example for

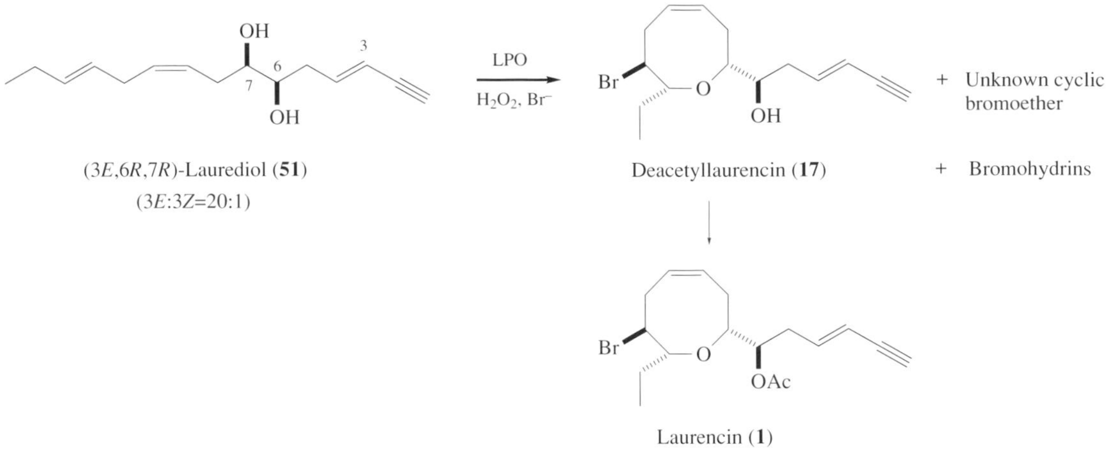

(3*E*,6*R*,7*R*)-Laurediol (**51**) (3*E*:3*Z*=20:1) → LPO, H_2O_2, Br^- → Deacetyllaurencin (**17**) + Unknown cyclic bromoether + Bromohydrins

Deacetyllaurencin (**17**) → Laurencin (**1**)

Scheme 9

clarification of the biosynthetic route to laurencin (**1**). It is to be noted that in this case the bromocyclization proceeded in an *endo*-fashion. This experimental result would strongly suggest that the bromide atom in biosynthetic bromoetherification could be incorporated as a positive bromine, but not as a negative bromine in the algae.

Next, the author's group attempted to check the metabolism of (**17**) with LPO, because the compound still includes the olefin alcohol moiety[22] (Scheme 10). Deacetyllaurencin ((**17**), 205 mg) dissolved in DMSO (2 ml) was added to 200 ml of a phophate buffer (pH 5.5) containing NaBr (0.3 mM). Each solution of LPO (64 nM) and H_2O_2 (0.8 mM) was divided into nine portions and the aliquots were added to the reaction mixture at 10 min intervals in the dark at room temperature. After stirring for 24 h, the mixture was worked up as usual to give laurefucin ((**18**), 13.9 mg), laureoxanyne ((**19**), 8.2 mg), bromohydrins (13.7 mg), and the recovered substrate (124.6 mg). Compounds (**18**) and (**19**) are also natural products isolated from *L. nipponica*. The enzymatic reaction mechanism initiated by LPO strongly suggested the presence of the biosynthetic intermediates ((**64**, (**65**), and (**66**)) as shown in Scheme 10. Although the intermediate (**65**) could not be isolated from the alga, spontaneous conversion of a synthetic sample (**68**) into hexahydrolaurefucin (**69**) in aqueous solvents supported this proposed biosynthetic pathway to laurefucin (**18**). The *E*-isomer of a new metabolite from *L. nipponica* named as notoryne (**67**) having the (6*R*,7*S*)-configuration[54] might be a rearranged product that may be caused by attack of Cl^- at C-7 of (**66**) with inversion of configuration.

1.12.2.3 Laurenan Series

Laureatin (**20**), isolaureatin (**21**), and laurallene (**23**), all possessing (6*S*,7*S*)-configurations, are biogenetically assumed to arise from (3*Z*,6*S*,7*S*)-laurediol (**69**) via the proposed intermediate, prelaureatin (**22**).[24] The author's group were able to isolate (**22**) as a new metabolite from *L. nipponica*.[24] The direct enzymatic cyclization of (3*Z*,6*S*,7*S*)-laurediol (**69**) into both (**22**) and (**20**) in a single-step procedure is described herein.[55]

The enzymatic substrate, (3*Z*,6*S*,7*S*)-laurediol (**69**), was prepared conveniently from natural laureatin (**20**) as shown in Scheme 11.[24] Protection of the terminal acetylene in (**20**) with the TMS group (LDA, THF, −78 °C, 30 min, then TMS-Cl, 1 h) followed by reductive cleavage of (**70**) with Zn powder in AcOH and EtOH gave four products, i.e. a mixture of two oxetanes ((**71**) and (**72**)) in 11% yield, (**73**) (39%), and (**74**) (10%) along with the recovered (**70**) (27%). Removal of the silyl group in (**74**) with TBAF in THF gave a sample of (3*Z*,6*S*,7*S*)-laurediol (**69**) containing 10% of an inseparable (12*Z*)-isomer of (**69**) (Scheme 11).

The substrate (**69**) was subjected immediately to the enzymatic reaction[55] (Equation (1)). To a vigorously stirred solution of (**69**) (24 mg, 3.4 mM) in DMSO (0.3 ml) were added a phosphate buffer (pH 5.5, 50 mM, 30 ml) and NaBr (3 mM, 0.9 ml) (Equation (1)). Two solutions of LPO (64 nM) and H_2O_2 (0.8 mM) were added, in ten portions, to the above mixture every 1 h at 0 °C under Ar atmosphere. After stirring for 24 h, the conventional workup provided 27 mg of a crude mixture, which was separated by repeated column chromatography on SiO_2 and HPLC (Develosil, hexane–

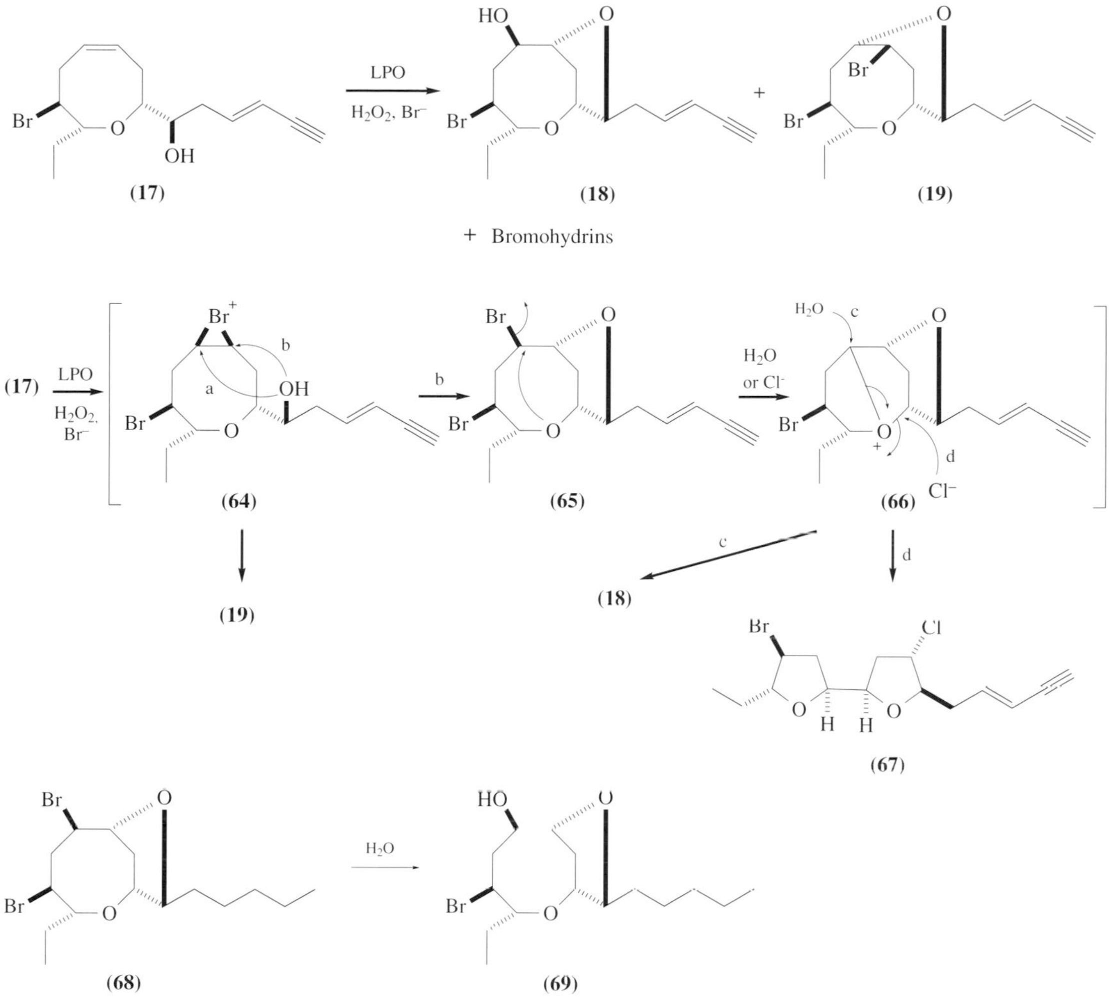

LPO

H_2O_2, Br^-

(17) **(18)** **(19)**

+ Bromohydrins

(64) **(65)** **(66)**

(19) **(18)**

(67)

H_2O

(68) **(69)**

Scheme 10

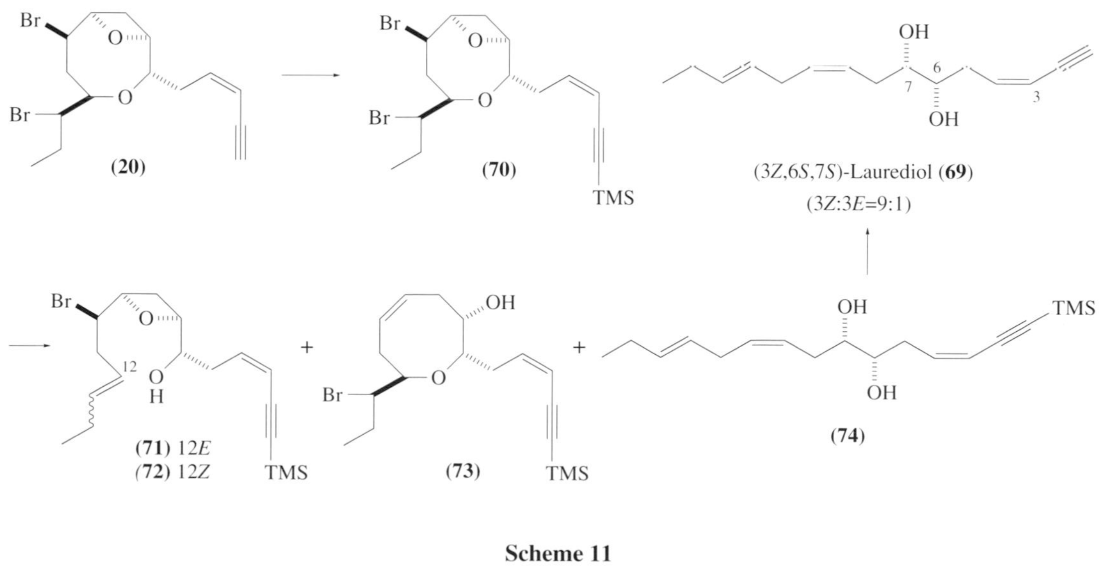

(20) **(70)**

(3*Z*,6*S*,7*S*)-Laurediol (**69**)

(3*Z*:3*E*=9:1)

(**71**) 12*E*

(**72**) 12*Z*

(73) **(74)**

Scheme 11

CH_2Cl_2–MeCN (80:19:1)) to give prelaureatin ((**22**), 1 mg, 3%), laureatin ((**20**), 0.1 mg, 0.3%), two inseparable oxetanes ((**75**) and (**76**), 0.3 mg, 1%), unidentified bromohydrins (3.4 mg, 11%), and the starting substrate ((**69**), 17.5 mg, 72%). Formation of oxetanes ((**75**), (**76**)) does not eliminate the possibility of these compounds as the biosynthetic intermediates for laureatin (**20**) or isolaureatin (**21**), i.e. the biogenetic routes of (**20**) and (**21**) starting from (**69**) via the respective oxetane and oxolane intermediates proposed by Suzuki *et al.*[54] is also to be noted (Scheme 12).

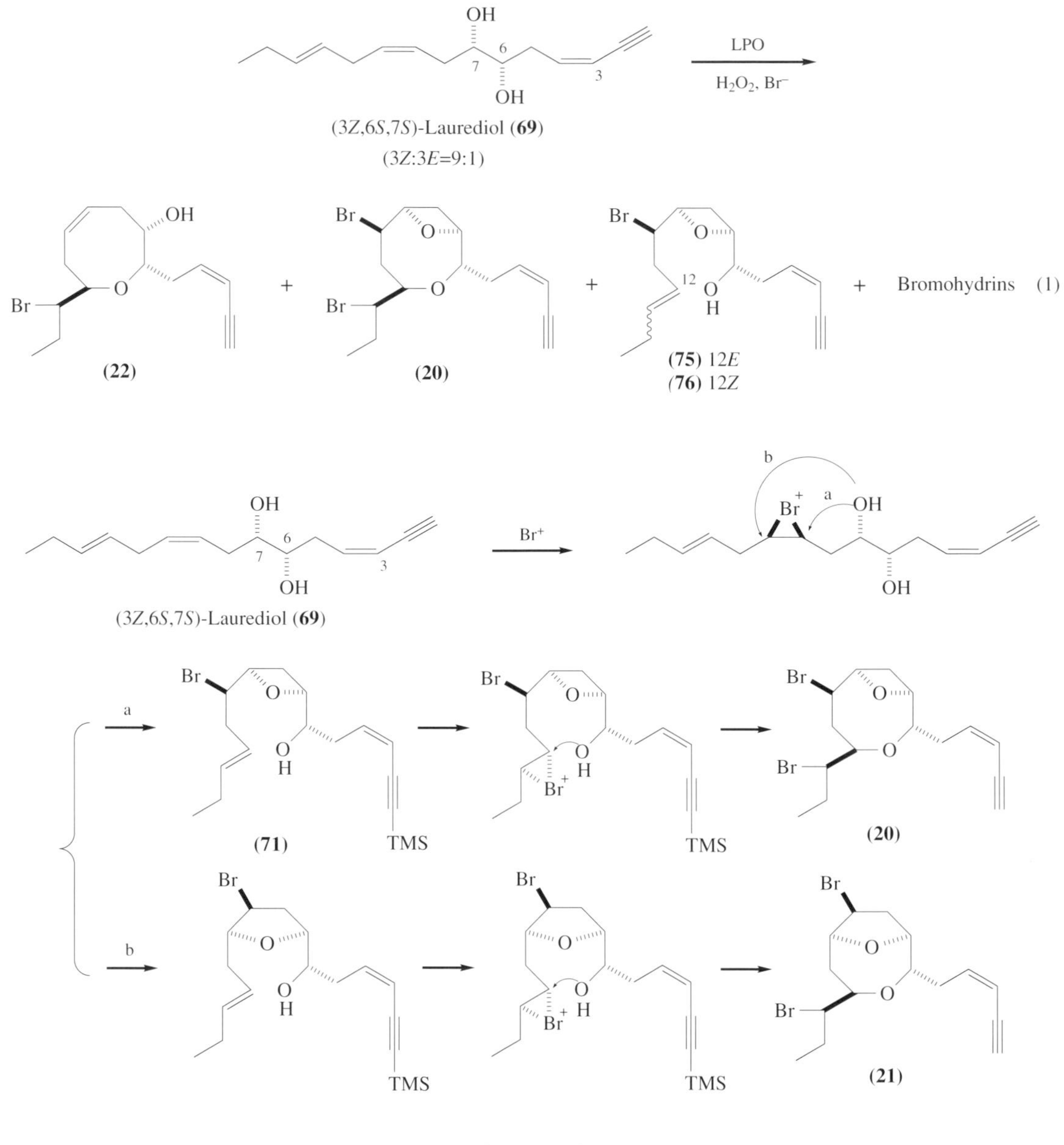

(3*Z*,6*S*,7*S*)-Laurediol (**69**) (3*Z*:3*E*=9:1) —LPO, H_2O_2, Br^-→ (**22**) + (**20**) + (**75**) 12*E* / (**76**) 12*Z* + Bromohydrins (1)

(3*Z*,6*S*,7*S*)-Laurediol (**69**) —Br^+→ … a → (**71**) → … → (**20**); b → … → (**21**)

Scheme 12

[1-^{2}H]-prelaureatin (**22-D**), prepared from (**73**) with TBAF in THF and D_2O at 0 °C for 40 min in 99% yield, was further subjected to the enzymatic reaction[56] (Scheme 13). Compound (**22-D**), when treated with LPO in a 90:1 mixture of H_2O–DMSO adjusting pH 5.5 in the presence of H_2O_2 and NaBr at 22 °C for 24 h, gave rise to [1-^{2}H]-laureatin ((**20-D**), 0.05%), [1-^{2}H]-isolaureatin ((**21-D**), a trace amount), and a bromoallene compound ((**77-D**), 0.07%), along with bromooxolanes ((**78-D**), 0.6% and (**79-D**), 0.9%), three bromohydrins (2.6%), and the recovered starting substrate (79%). The resulting [1-^{2}H]-laureatin (**20-D**) was identified by ^{1}H-NMR spectrum and HPLC analysis. As the bromoallene compound (**77-D**) was slightly different from natural laurallene in ^{1}H-NMR, it was characterized as one isomer of laurallene. In order to clarify the stereochemistry of this compound, the author's group next explored chemically the bromination of (**22-D**) by a brominating reagent. Exposure of (**22-D**) with 2,4,6,6-tetrabromo-2,5-cyclohexadienone (TBCO) in CH_2Cl_2 at 22 °C for 24 h provided two bromoallene products (18% and 11%), which were separated by HPLC. The major component was identified as the aforementioned compound (**77-D**) (18%) and, judging from its levorotation ($[\alpha]_D^{22}$ −148° (c 0.42, $CHCl_3$)), the absolute configuration of the allene was predicted as *R* according to Lowe's method.[57] The stereochemistry at C-4 on the product (**77-D**) was not clarified by nuclear Overhauser effect difference spectroscopy (NOEDS). Therefore, (**77-D**) was converted in two steps (61%: (i) 5% Rh-Al_2O_3, H_2 (20 kgf cm^{-2}), cyclohexane, 40 °C, 4 d; (ii) 10% Pd/C, H_2, EtOH, RT, 17 h) into the monodebromoheptahydro-derivative that was identical with the sample obtained from natural laurallene by the same procedure. Another minor product was found to be in agreement with the [1-^{2}H]-natural laurallene (**23-D**) by spectroscopic data (^{1}H-NMR, optical rotation).

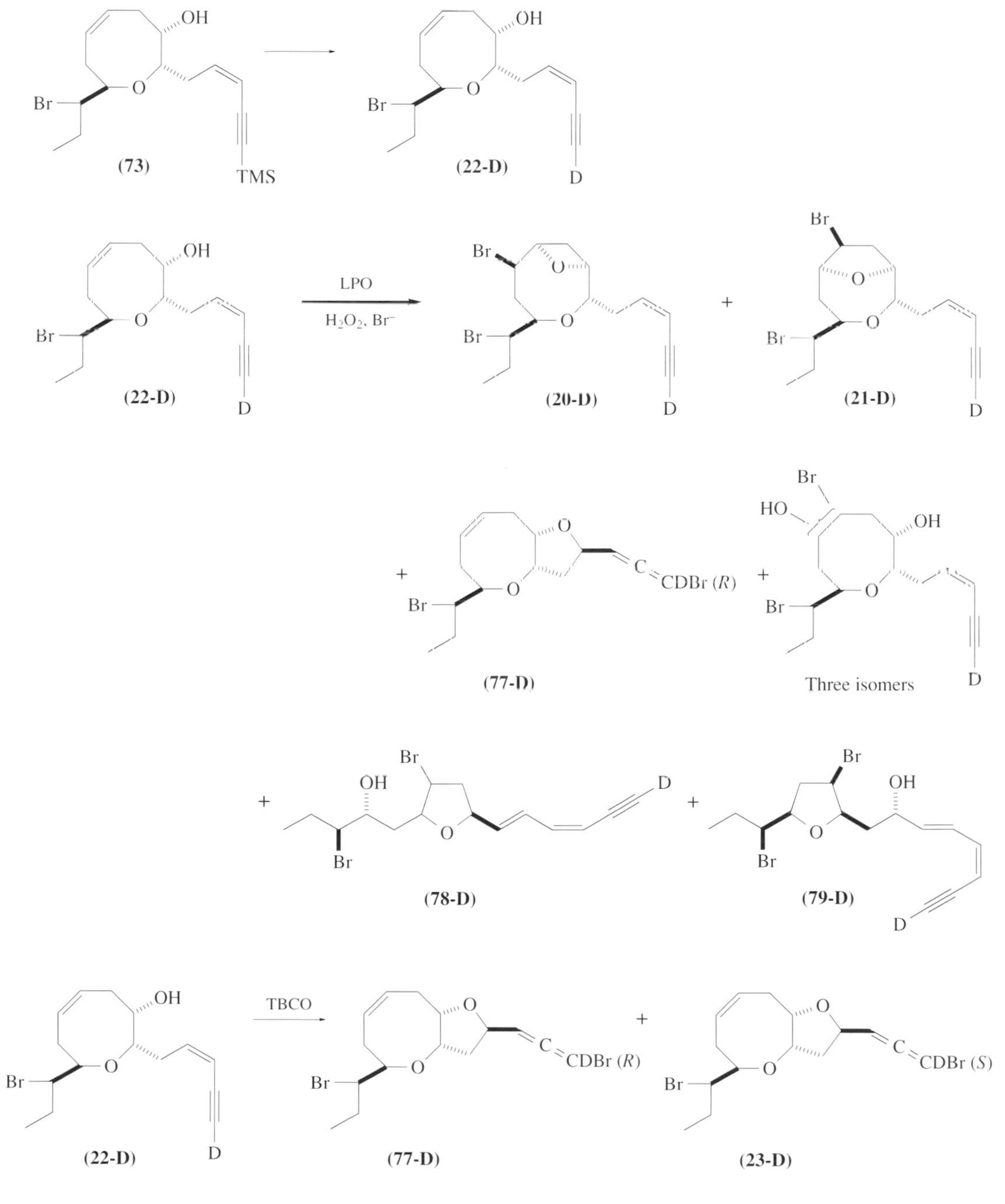

Scheme 13

Thus, the configurations of the vicinal diol units (6*R*,7*R* or 6*S*,7*S*) perfectly control the pathways for either the lauthisan type or the laurenan type. However, the relationship between the geometry of the enyne substituent and the regioselectivity in an intramolecular cyclization remained unclarified. Therefore, the enzymatic cyclization of unnatural (3*E*,6*S*,7*S*)-laurediol (**80**) was carried out,[58] which was prepared from (**70**)[59] (Scheme 14). Compound (**70**), derived from (**20**), was protected with $Co_2(CO)_8$ to give the unstable complex (**81**). When compound (**81**) was treated with CF_3SO_3H in CH_2Cl_2 at 20 °C for 12 h, the isomerization reaction was effected to give exclusively the desired (*E*)-enyne (**82**) accompanied by ring rearrangement of oxetane to oxolane, (cf. Fukuzawa *et al.*[60]), which was decomplexed with $(NH_4)_2Ce(NO_3)_6$ yielding (**83**) as a single product in 70% overall yield from (**70**). Reductive cleavage of (**83**) with zinc powder in AcOH and EtOH at room temperature for 6 h afforded (**84**) in 84% yield, which was detached with TBAF to give (**80**) as a 3:1 mixture of 12*E* and 12*Z* in 96% yield. Compound (**80**) was treated with LPO in a 100:1 mixture of H_2O–DMSO adjusting pH 5.5 in the presence of H_2O_2 and NaBr at 0 °C for 24 h[58] (Equation (2)). The reaction mixture gave rise to an oxetane compound ((**85**), 0.4%), oxolane compound ((**86**), 0.2%),

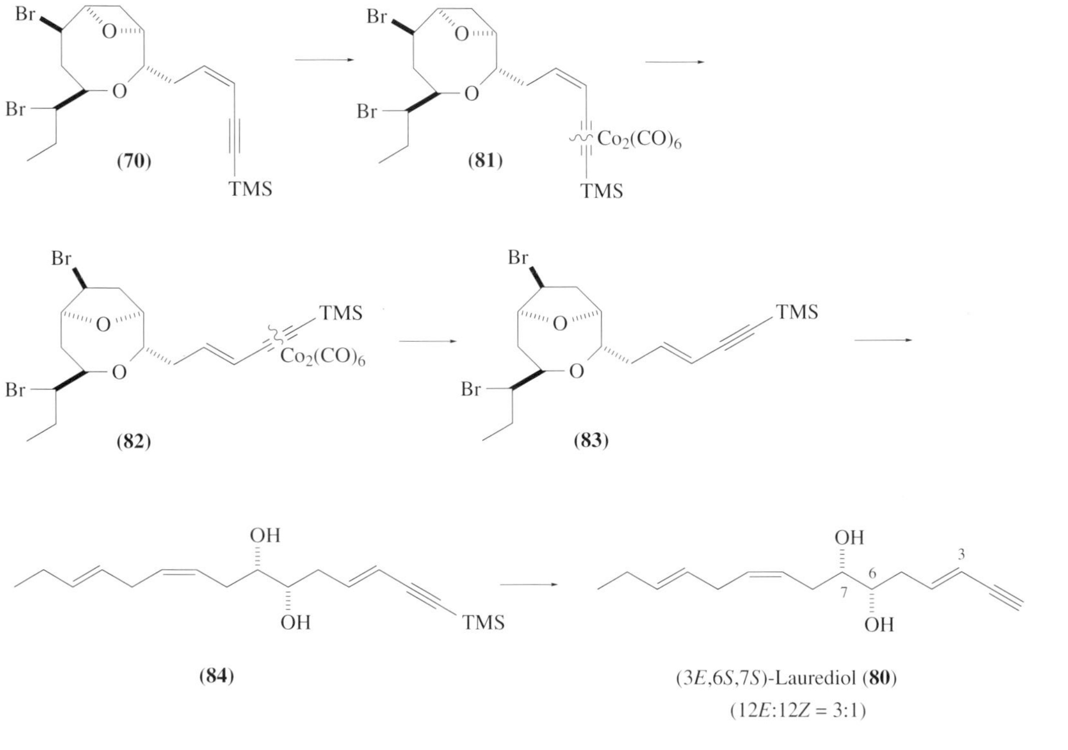

(**70**) → (**81**) → (**82**) → (**83**) → (**84**) → (3*E*,6*S*,7*S*)-Laurediol (**80**) (12*E*:12*Z* = 3:1)

Scheme 14

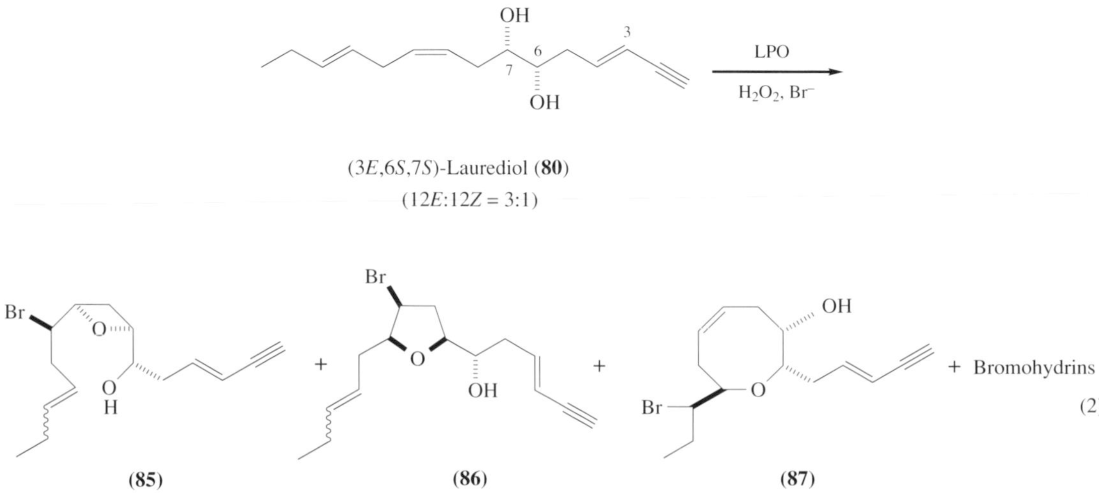

(3*E*,6*S*,7*S*)-Laurediol (**80**) (12*E*:12*Z* = 3:1) —LPO, H_2O_2, Br^-→ (**85**) + (**86**) + (**87**) + Bromohydrins (2)

(3*E*)-prelaureatin ((**87**), 0.05%), the recovered starting material (88%), and a mixture of undetermined bromohydrins (6%). These results reveal that the regioselectivity in the cyclization of laurediols does not depend upon their enyne geometry but their stereochemistry at C-6 and C-7. Since (3*E*)-prelaureatin (**87**) was produced in very low yield (0.05%) comparing to the (3*Z*)-isomer,[55] it was suggested that the steric repulsion of the (*E*)-enyne side-chain in an enzymatic pocket restricted the formation of the eight-membered ring. This might be the reason why the naturally occurring (*E*)-laureatin was a minor product.[61]

In the next stage, attention was focused on the enzymatic reaction of (3*E*)-prelaureatin (**87**). As (*E*)-laureatin was rare in natural products, the author's group had to convert chemically the (*Z*)-enyne in (**20**) into the (*E*)-enyne unit[62] (Scheme 15). Compound (**70**) was repeatedly treated with iodine in benzene to give (**88**). The selective cleavage of the oxetane ring of (**88**) was explored. It was found finally that the metal–halogen exchange with an equimolar amount of BuLi at −78 °C occurred with high selectivity to give (**89**) in 58% yield. The TMS group of (**89**) was then detached smoothly (TBAF, THF, 0 °C, 1.5 h) to yield (**87**) in 90% yield. Next, the bromination reaction of (**87**) with TBCO was carried out.[62] Analogous to the (*Z*)-substrate, compound (**87**) was converted to two bromoallene compounds, one being natural laurallene ((**23**) 23%) and the other, an unnatural isomer ((**77**), 19%). They were also identified by ^{1}H-NMR data, the optical rotations, and HPLC analyses, and the former was also characterized by X-ray crystallography. Consequently, the reactivity of (**87**) was found to be not particularly different from that of the (*Z*)-compound.

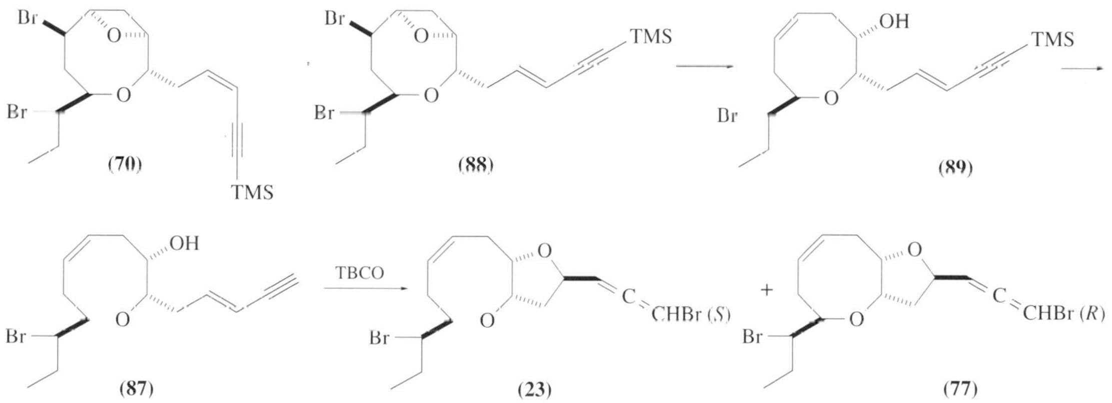

Br O Br O (70) TMS → (88) TMS → OH O Br (89) TMS → (87) OH O Br —TBCO→ (23) O O Br C=CHBr (*S*) + (77) O O Br C=CHBr (*R*)

Scheme 15

The enzymatic reaction of (**87**) was performed with LPO and NaBr in the presence of H_2O_2 at 0 °C for 24 h to give a mixture of a bromoallene compound (0.03%), bromooxolanes ((**91**), 0.7% and (**92**), 0.4%), bromohydrins (7.2%), and the recovered starting material (84%), whereas laureatin-type and isolaureatin-type compounds were not detected[62] (Equation (3)). The bromoallene was found to be in agreement with natural laurallene (**23**) from ^{1}H-NMR and HPLC analysis. These results suggest that laurallene was biogenetically synthesized from (3*E*)-(**87**), and laureatin and isolaureatin were transformed from (3*Z*)-(**22**).

In conclusion, the enzymatic reactions of (*Z*)- and (*E*)-prelaureatins, using LPO and NaBr in the presence of H_2O_2, were performed to afford laureatin and laurallene, respectively. It is revealed that the stereochemistry of the enyne unit plays an important role in the biogenetic production of laureatin or laurallene and product distributions.

1.12.3 PURIFICATION OF BROMOPEROXIDASE

In this section, the partial purification of BPO from the red alga is described.[63] The red alga, *L. nipponica* (3 kg wet weight) collected at the west coast of Oshoro Bay, Hokkaido Island, Japan in

(87) LPO H_2O_2, Br^- (23) CHBr (*S*) + Bromohydrins + (90) + (91) (3)

June 1991, was ground together with crushed dry ice, homogenized in 0.1 M potassium phosphate buffer (pH 5.5, 1.5 l) for 20 min, and filtered through cheesecloth. The filtrate was centrifuged (3000 rpm) at 2 °C for 1 h. The clear supernatant was regarded as the primary enzyme extracts. The extracts were purified further by a modification of the Hager procedure;[64] to the solution was added slowly solid $(NH_4)_2SO_4$ (201 g) with stirring at 0 °C in order to bring the extracts to 25% $(NH_4)_2SO_4$ saturation. The solution, which was kept at 5 °C overnight, was centrifuged at 10 000 rpm for 1 h to afford the supernatant and pellets. The supernatant was brought to 65% saturation by addition of solid $(NH_4)_2SO_4$ (316 g) at 0 °C, and stirred at 5 °C overnight. Centrifugation of the suspension at 9000 rpm gave pellets, which were again dissolved in cold 0.2 M Tris buffer (pH 7.0, 100 ml) and centrifuged at 10 000 rpm at 2 °C for 1 h, yielding the supernatant. This was applied to DEAE-Sephadex column (4 cm × 45 cm) and eluted with 0.2 M Tris buffer (1 l) to remove the pigment fractions at 2 °C overnight, though it was difficult to remove completely the pigments characteristic of red algae. Finally, the active BPO fraction was eluted from the column with 0.2 M–1.0 M Tris buffer (pH 7.0) as a linear gradient (the flow rate was maintained at 7.5 ml/h and 72 fractions consisting of 8.0 ml each were collected). The BPO activity of the fractions thus isolated was measured optically in the decrease of absorbance at 278 nm on reaction with chlorodimedone in the presence of H_2O_2 and KBr in the phosphate buffer (pH 6.8) and the specific activity of the best fraction was 50-fold higher than that of the primary extracts of the red alga. Furthermore, the enzyme was found to contain an iron porphyrin similar to LPO judging from the decrease of the activity by the inhibitors such as KCN or NaN_3.

1.12.4 BIOSYNTHESIS WITH BROMOPEROXIDASE

As a preliminary enzymatic experiment, (6*S*,7*S*,9*E*)-9-pentadecene-6,7-diol (**44**) was employed as the substrate[63] (Equation (4)). A solution of olefin diol (**44**), 27.3 mg, 5.4 mM) in DMSO (0.2 ml) was added into the solution of NaBr (4.8 mM) in a phosphate buffer (pH 5.5, 50 mM, 20 ml). To the mixture was added an aliquot of each solution consisting of 0.1 M H_2O_2 (168 μl, final concentration 0.8 mM) and the BPO (600 μl) in 12 portions during 2 h and the mixture was allowed to react at 23 °C for 24 h. The extracts were subjected to chromatography over SiO_2 and purified by HPLC to provide an oxolane compound ((**47**) 0.5 mg, 1.4%) along with a mixture of bromohydrins (2.0 mg, 5.2%) and the recovered starting material (22.0 mg). Compound (**47**) was identical with an authentic sample in ^{1}H-NMR spectrum. Next, the reaction was carried out under the same conditions as above without BPO and it was found that no reaction occurred. The results reveal that compound (**47**) was produced only in the presence of BPO. Accordingly, the above reaction involving formation of (**47**) can be regarded as the enzymatic reaction.

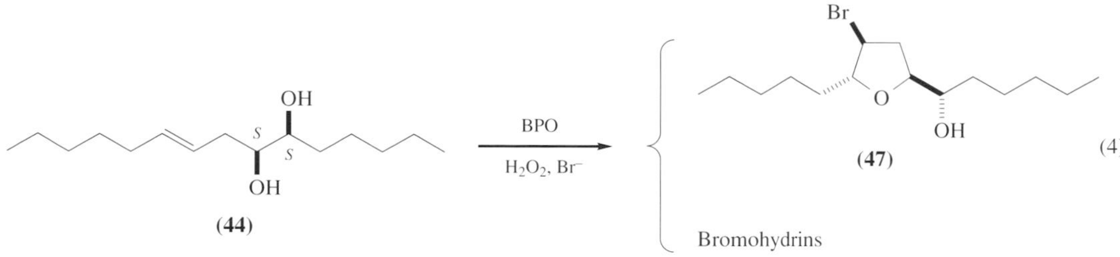

(44) —BPO, H_2O_2, Br^-→ (47) + Bromohydrins (4)

On the basis of the above model experiments, the reaction was extended to use (3*E*,6*R*,7*R*)-laurediol (**51**) as the substrate[63] (Scheme 16). To a solution of (**51**) (101.0 mg, 10.6 mM) and NaBr (10.0 mM) in a 1:100 mixture (40.5 ml) of DMSO and a phosphate buffer (pH 5.5), were injected aliquots of each solution of H_2O_2 (totally 2.0 mM) and the BPO (2.0 ml) divided into 12 portions under an atmosphere of Ar at 5 °C during 2 h. The mixture was stirred at 5 °C for 24 h. The extracts were separated on a SiO_2 column to give rise to deacetyllaurencin ((**17**), 0.02 mg, 0.015%), an unknown cyclic bromoether (0.1 mg, 0.074%), bromohydrins (3.2 mg, 2.24%), and the recovered starting compound ((**51**), 83.2 mg, 82.4%). Deacetyllaurencin (**17**) was identified after acetylation with natural laurencin (**1**) in respect of HPLC (Radial Pak μ-Porasil with hexane–CH_2Cl_2–MeCN (50:49:1)) and ^{1}H-NMR spectrum. These results indicate that BPO is the real enzyme for the direct bromoether cyclization of (**51**) to (**17**). This constitutes the first direct evidence for the plausible biosynthetic route to laurencin skeleton. Furthermore, deacetyllaurencin (**17**) was allowed to react with BPO in the presence of H_2O_2 and NaBr at 20 °C for 24 h to afford laurefucin ((**18**), 1.3%), laureoxanyne ((**19**) 0.8%), bromohydrins (12.6%), and the recovered starting compared ((**17**) 72.2%). Compounds (**18**) and (**19**) were identical with the respective authentic samples as regards the ^{1}H-NMR spectra. These results were analogous to the cases of the enzymatic reactions with LPO.

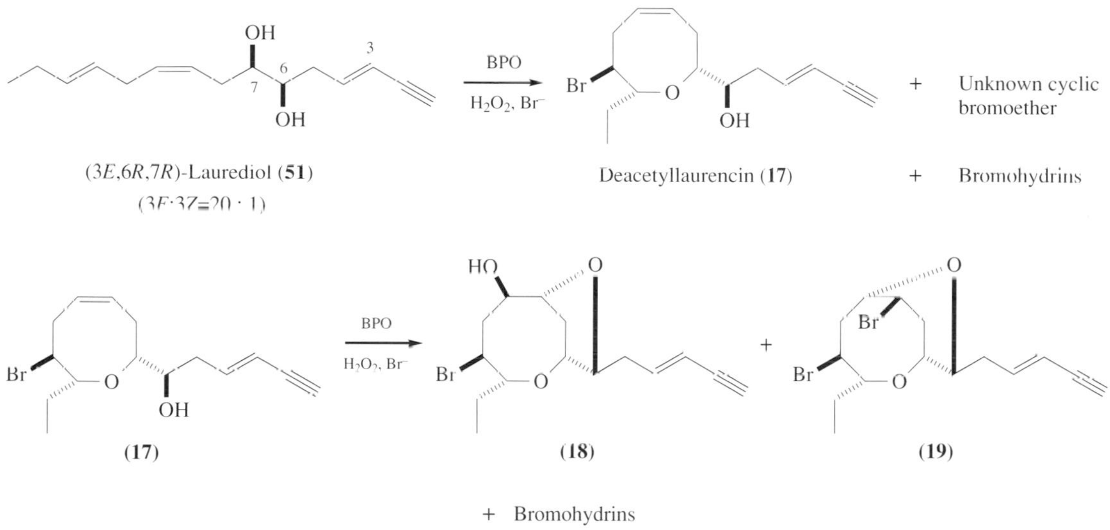

(3*E*,6*R*,7*R*)-Laurediol (**51**) (3*E*:3*Z*=20 : 1) —BPO, H_2O_2, Br^-→ Deacetyllaurencin (**17**) + Unknown cyclic bromoether + Bromohydrins

(**17**) —BPO, H_2O_2, Br^-→ (**18**) + (**19**) + Bromohydrins

Scheme 16

Next, the enzymatic reaction was attempted with (3*Z*,6*S*,7*S*)-laurediol (**69**) as the substrate[63] (Scheme 17). The reaction was carried out in a similar way to the above at 0 °C for 24 h. The reaction afforded prelaureatin ((**22**), a trace amount) as well as a mixture of bromohydrins (8.0%) and the recovered starting compound ((**69**), 82.0%). Although the author's group could not measure any spectral data for the fraction (**22**) for identification, its retention time on HPLC (Develosil 60-3 with hexane–CH_2Cl_2–MeCN (70:20:10)) corresponded exactly to the authentic sample of (**22**). Finally, [1-^{2}H]-prelaureatin (**22-D**) was subjected to the enzymatic reaction (0 °C → 10 °C, 24 h) and yielded [1-^{2}H]-laureatin ((**20-D**), 0.07%), [1-^{2}H]-isolaureatin ((**21-D**), 0.05%), and two bromo-oxolanes ((**78-D**), 0.3%; (**79-D**) 0.9%), along with bromohydrins (5.7%) and the recovered starting compound ((**22-D**), 69.0%). Compounds (**20-D**) and (**21-D**) were identified by comparison with authentic samples by ^{1}H-NMR and the retention times on HPLC (Develosil 60-3 with hexane–

CH_2Cl_2–MeCN (80:19:1)), respectively. The structures of (**78-D**) and (**79-D**) were tentatively elucidated by ^{1}H-NMR spectral data. It should be noted that [1-^{2}H]-bromoallene compound (**77-D**) could not be detected even in a trace amount on HPLC in contrast with the case of LPO (cf., Section 1.12.2.3).

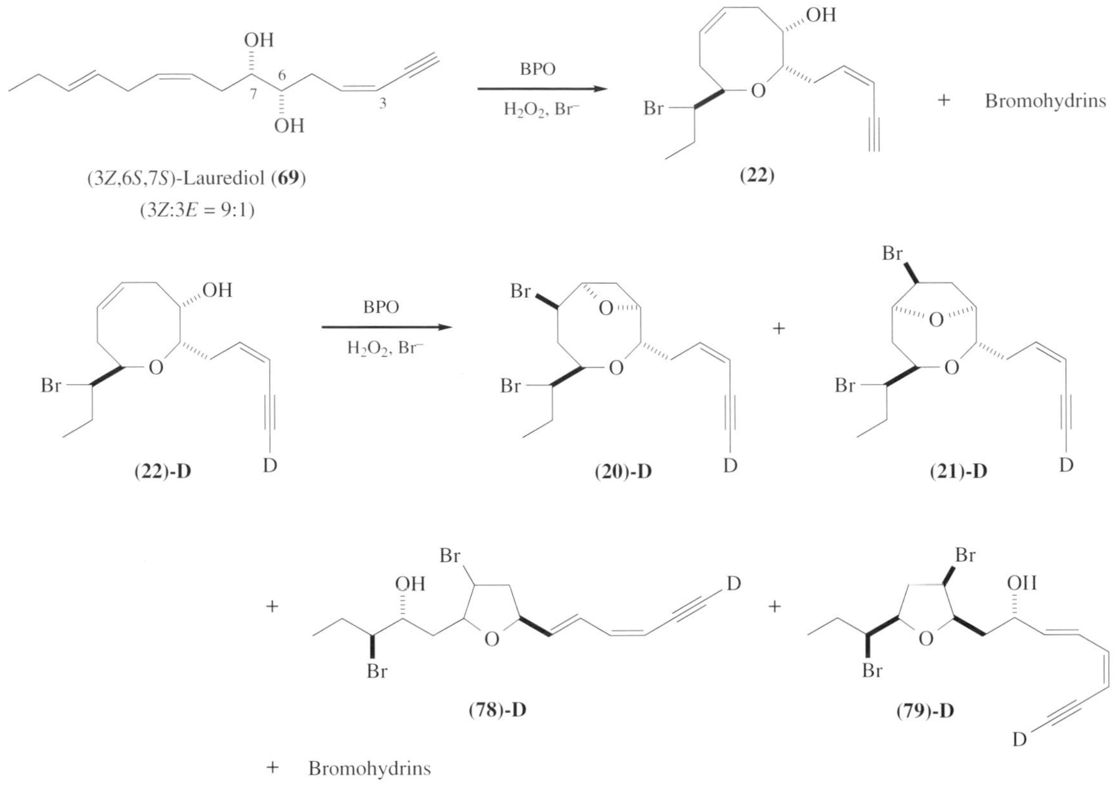

Scheme 17

1.12.5 DISCUSSION

These enzymatic reaction results mentioned above could be summarized in Scheme 18. It is concluded that the same LPO enzyme enabled reaction with the enantiomeric isomers of laurediols ((**51**) and (**80**)) as the substrates in different reaction manners. These enzymatic reactions proceeded stereoselectively and regioselectively without crossing. Such an experimental fact seems to be very rare in nature. While (6*R*,7*R*)-laurediol (**51**) led to formation of laurencin (**1**), the corresponding (6*S*,7*S*)-laurediol (**80**) could not produce the enantiomer of (**1**), but the essentially different bromoether (**87**). Then, how are such different reaction results explained?

The exact whole structure of LPO has been determined from the viewpoint of its lability. Since it is difficult to predict precisely the transition state of this enzymatic reaction, the author's group examined the molecular modeling structure of laurediols.[58] When a substrate enters the pocket in an enzyme and forms an eight-membered ring, the chain should be bent at C-8 to C-11 and the OH group at C-6 or C-7 could approach the olefin–Br^+ complex. As a typical model conformation of lauthisan compounds, laurencin (**1**) was selected which was defined by X-ray crystallography[65,66] (Figure 1). The X-ray analysis revealed that laurencin (**1**) takes a boat-chair conformation and the atom distance between Br atom and the oxygen at C-6, which is independent of cyclization, is 6.8 Å. Based on this conformation, the transition structure close to that of (**1**) for cyclization of (3*E*,6*R*,7*R*)-laurediol (**51**) was predicted. The precursor, (6*R*,7*R*)-laurediol–Br^+ complex, was fixed at C-8 to C-15, and the structure was optimized by MM2. The result revealed that it was represented by the conformer **A**. It shows that the atom length between the carbon at 13-position and the oxygen at C-7 is estimated as 2.0 Å and the angle $\angle$O(C-7)–C-13–Br^+ is 155°. This conformation is

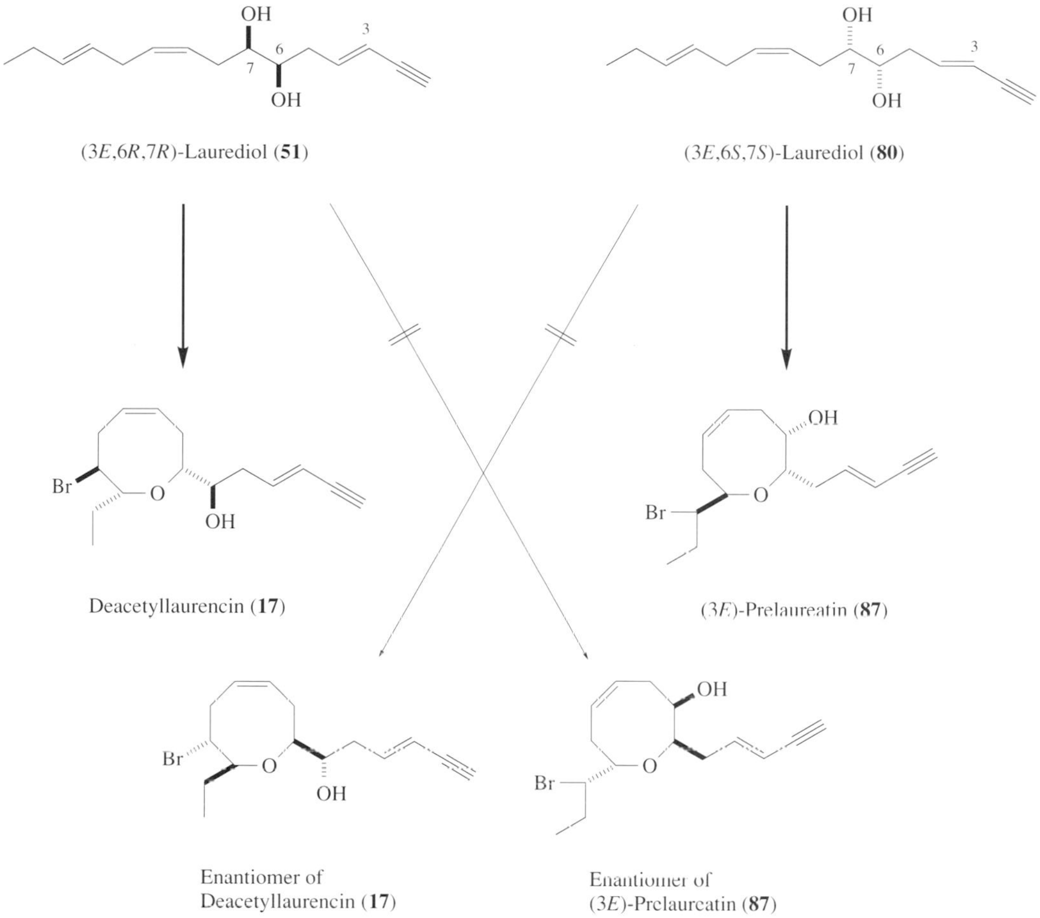

(3*E*,6*R*,7*R*)-Laurediol (**51**) (3*E*,6*S*,7*S*)-Laurediol (**80**)

Deacetyllaurencin (**17**) (3*E*)-Prelaureatin (**87**)

Enantiomer of Deacetyllaurencin (**17**) Enantiomer of (3*E*)-Prelaureatin (**87**)

Scheme 18

considered as one of the most favored transition states leading to formation of (**1**). Similarly, (3*E*,6*S*,7*S*)-laurediol (**80**) was fixed at C-8 to C-15 and the bond length of O(C-7)–C-13 was locked at 2.0 Å. Minimization afforded the conformer **B**, while locking the angle ∠O(C-7)–C-13–Br^+ at 155° gave the conformer **C**. In the structure of **B**, the angle ∠O(C-7)–C-13–Br^+ is only 131° and in the conformer **C**, the bond length of O(C-7)–C-13 is 2.9 Å. In both conformers, the OH group is inadequately located to attack the bromonium ion intramolecularly.

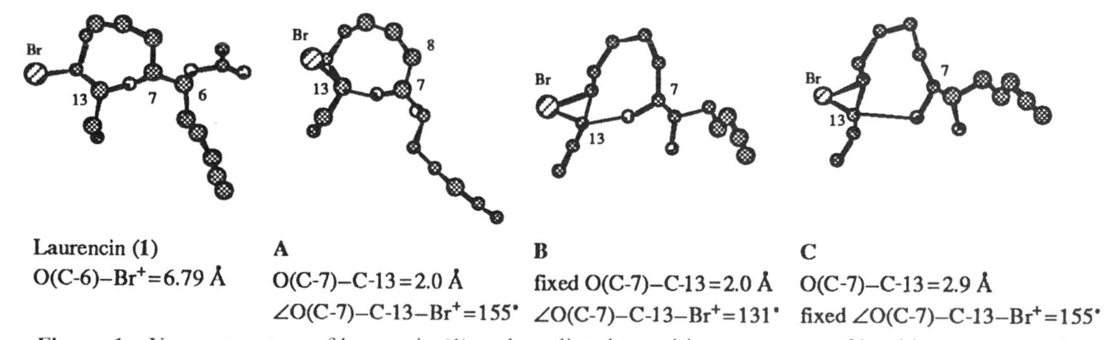

Figure 1 X-ray structure of laurencin (**1**) and predicted transition structures of lauthisan compounds.

On the other hand, in laurenan compounds, the conformation of laurallene (**23**) has been studied by X-ray analysis. The crystals of (**23**) were found to consist of two conformations having a difference of 0.70 kcal mol^{-1}. The more stable conformer (**92**) was selected for the basis of calculation

(Figure 2). The transition structures were also calculated based on another conformation of (**23**) and the minimization afforded similar results to the cases of conformers **D**, **E**, and **F**. The conformer (**92**) reveals that the atom distance between the Br atom and the oxygen at C-7 is 7.0 Å, which is approximately the same as that of (**1**). The fixation of the (3*E*,6*S*,7*S*)-laurediol (**80**), at C-8 to C-15 and minimization gave the conformer **D**. The structure indicated that the bond length of O(C-6)–C-12 is 2.5 Å and the angle ∠O(C-6)–C-12–Br^+ is 151°. Based on this conformation, the *exo*-cyclic transition state of (3*E*,6*R*,7*R*)-laurediol (**51**) was calculated, locking at the bond length or the angle. They are represented by the conformers **E** or **F**, respectively. The conformer **E** was locked at the O(C-6)–C-12 bond length, indicating that the angle ∠O(C-6)–C-12–Br^+ is 135°. The conformer **F**, fixed at the angle ∠O(C-6)–C-12–Br^+ shows that the O(C-6)–C-12 length is 2.9 Å. Both of the transition structures are apparently disfavored to cyclize intramolecularly.

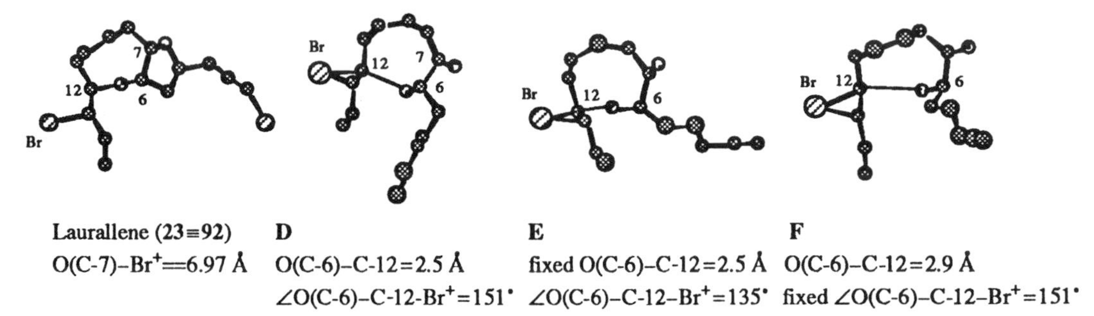

Figure 2 X-ray structure of laurallene (**23**) and predicted transition structures of laurenan compounds.

As a result, when laurediol forms an eight-membered ether ring, it should be bent at C-8 to C-11 and (6*S*,7*S*)-laurediol would take an energetically preferable conformation to produce an *exo*-cyclic ether and (6*R*,7*R*)-isomer to an *endo*-cyclic product. In addition, the atom distances between Br atom and the oxygen atoms at C-6 (laurencin) and C-7 (laurallene) are 6.8 and 7.0 Å, respectively. It is considered that the OH group, free from cyclization, might exert some influence for binding to the enzyme. When this OH group locates ca. 7 Å far from the enzymatic bromocation, the other OH group would attach easily at the carbon to form an eight-membered ether ring.

Scheme 19 summarizes the conclusion that the bromine atom in biosynthetic bromoether cyclization would be incorporated as a positive bromine generated from a negative bromine with H_2O_2.

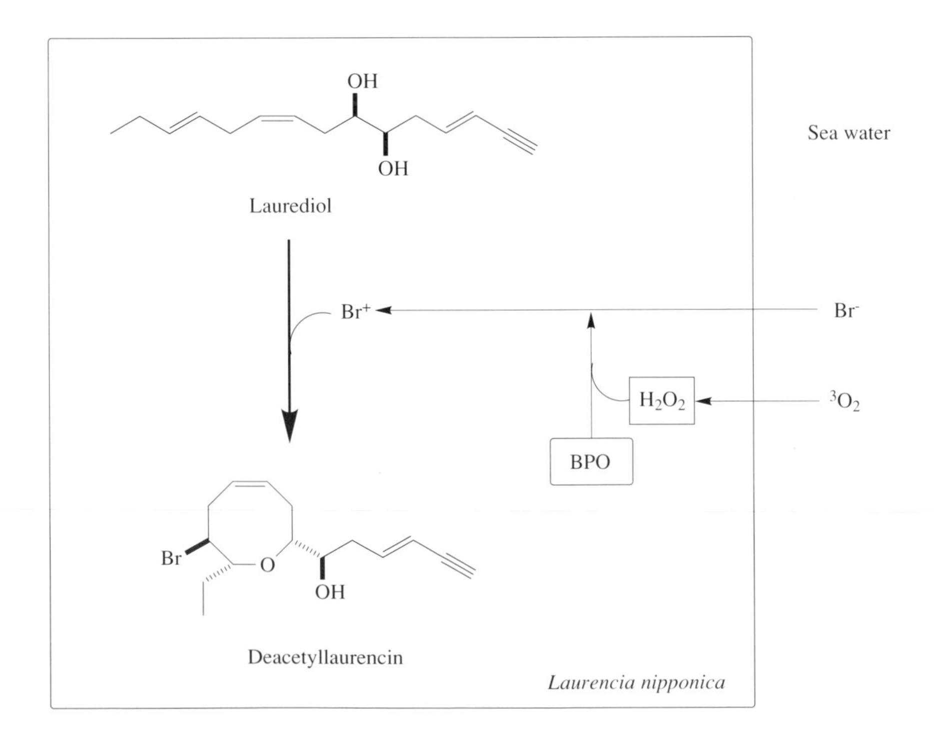

Scheme 19

It has been well known that seaweeds in sea water might always suffer some stress, such as pathogens, changes in temperature, and stream of sea water. It is reasonable to suppose that generation of H_2O_2 would control production of such various brominated cyclic ether compounds from algae in the sea. In 1995, the author's group reported that H_2O_2 is a dynamic substance for triggering the phytoalexin (secondary stress metabolite) production in higher plants, such as potato,[67] sweet potato,[68] kidney bean,[68] and sugar beet.[68] On the basis of these results, the author proposes that these cyclic bromoether compounds might be a kind of stress or abnormal metabolites in the algae as lower plants.

One question which remains is why the bromine atom is preferably introduced into the algae rather than the chlorine atom, regardless of the lower content of the bromide ion which is 1/300 of that of the chloride ion in sea water (the respective concentrations of halide ions are 19 000 mg l^{-1} for Cl^-, 65 mg l^{-1} for Br^-, 1.3 mg l^{-1} for F^-, and 0.06 mg l^{-1} for I^-). It has been well known that, in gaseous state, the electron affinity ($X^- \rightarrow X^o$) and ionization potential ($X^o \rightarrow X^+$) are 77.8 kcal mol^{-1} and 272.4 kcal mol^{-1} for the bromine atom, respectively, while the former value is 83.5 kcal mol^{-1} and the latter 299.0 kcal mol^{-1} for the chlorine atom.[69] It seems to be reasonable to assume that, in view of the comparison of the respective total energies, two-electron oxidation of bromine ($Br^- \rightarrow Br^+$) would proceed more easily than that of chlorine. At the present stage, the author's group has not succeeded in isolation of the chloroperoxidase enzyme from *L. nipponica*. The chlorine atom might be introduced exclusively into algae in an intact form of Cl^- as a nucleophile.

ACKNOWLEDGMENTS

The X-rays of laurencin (**1**) and laurallene (**23**) were kindly measured by Dr. Kazunori Yanagi, Ms. Emiko Fukuyo, and Ms. Misaho Miki, Sumitomo Chemical Co. Ltd., Bioscience Research Laboratory.

1.12.6 REFERENCES

1. T. Irie, M. Suzuki, and T. Masamune, *Tetrahedron Lett.*, 1965, 1091.
2. T. Irie, M. Suzuki, and T. Masamune, *Tetrahedron*, 1968, **24**, 4193.
3. R. E. Moore, in "Marine Natural Products: Chemical and Biological Perspectives," ed. P. J. Scheuer, Academic Press, New York, 1979, vol. 1, chap. 2, p. 43.
4. K. L. Erickson, in "Marine Natural Products: Chemical and Biological Perspectives," ed. P. J. Scheuer, Academic Press, New York, 1983, vol. 5, chap. 4, p. 131.
5. D. J. Faulkner, *Nat. Prod. Rep.*, 1984, **1**, 251.
6. D. J. Faulkner, *Nat. Prod. Rep.*, 1986, **3**, 1.
7. B. M. Howard, W. Fenical, E. V. Arnold, and J. Clardy, *Tetrahedron Lett.*, 1979, 2841.
8. A. G. Gonzalez, J. D. Martin, M. Norte, R. Perez, P. Rivera, J. Z. Ruano, M. L. Rodriguez, J. Fayos, and A. Perales, *Tetrahedron Lett.*, 1983, **24**, 4143.
9. A. Fukuzawa and E. Kurosawa, *Tetrahedron Lett.*, 1980, **21**, 1471.
10. T. Suzuki, K. Koizumi, M. Suzuki, and E. Kurosawa, *Chem. Lett.*, 1983, 1643.
11. T. Suzuki, K. Koizumi, M. Suzuki, and E. Kurosawa, *Chem. Lett.*, 1983, 1639.
12. A. Fukuzawa, Mya Aye, Y. Takaya, H. Fukui, T. Masamune, and A. Murai, *Tetrahedron Lett.*, 1989, **30**, 3665.
13. F. J. McDonald, D. C. Campbell, D. J. Vanderah, F. J. Schmiz, D. M. Washecheck, J. E. Burks, and D. van der Helm, *J. Org. Chem.*, 1975, **40**, 665.
14. D. J. Vanderah and F. J. Schmitz, *J. Org. Chem.*, 1976, **41**, 3480.
15. Y. Gopichand, F. J. Schmitz, J. Shelly, A. Rahman, and D. van der Helm, *J. Org. Chem.*, 1981, **46**, 5192.
16. A. Fukuzawa and T. Masamune, *Tetrahedron Lett.*, 1981, **22**, 4081.
17. E. Kurosawa, A. Fukuzawa, and T. Irie, *Tetrahedron Lett.*, 1973, 4135.
18. E. Kurosawa, A. Fukuzawa, and T. Irie, *Tetrahedron Lett.*, 1972, 2121.
19. A. Fukuzawa, T. Honma, Y. Takasugi, and A. Murai, *Phytochemistry*, 1993, **32**, 1435.
20. A. Fukuzawa, E. Kurosawa, and T. Irie, *Tetrahedron Lett.*, 1972, 3.
21. A. Furusaki, E. Kurosawa, A. Fukuzawa, and T. Irie, *Tetrahedron Lett.*, 1973, 4579.
22. A. Fukuzawa, Mya Aye, M. Nakamura, M. Tamura, and A. Murai, *Tetrahedron Lett.*, 1990, **31**, 4895.
23. T. Irie, M. Izawa, and E. Kurosawa, *Tetrahedron*, 1970, **26**, 851.
24. A. Fukuzawa, Y. Takasugi, and A. Murai, *Tetrahedron Lett.*, 1991, **32**, 5597.
25. A. Fukuzawa and E. Kurosawa, *Tetrahedron Lett.*, 1979, 2797.
26. T. J. King, S. Imre, A. Oztunc, and R. H. Thompson, *Tetrahedron Lett.*, 1979, 1453.
27. B. M. Howard, G. R. Schulte, W. Fenical, B. Solheim, and J. Clardy, *Tetrahedron*, 1980, **36**, 1747.
28. Y. Gopichand, F. J. Schmitz, J. Shelly, A. Rahman, and D. van der Helm, *J. Org. Chem.*, 1981, **46**, 5192.
29. M. Norte, A. G. Gonzalez, F. Cataldo, M. L. Rodriguez, and I. Brito, *Tetrahedron*, 1991, **47**, 9411.
30. K. Kurata, A. Furusaki, K. Suehiro, C. Katayama, and T. Suzuki, *Chem. Lett.*, 1982, 1031.
31. P. J. Cox, S. Imre, S. Islimyeli, and R. H. Thompson, *Tetrahedron Lett.*, 1982, **23**, 579.
32. E. Jones, *Chem. Br.*, 1966, **2**, 6.

33. H. Kigoshi, Y. Shizuri, H. Niwa, and K. Yamada, *Tetrahedron Lett.*, 1981, **22**, 4729.
34. H. Kigoshi, Y. Shizuri, H. Niwa, and K. Yamada., *Tetrahedron Lett.*, 1982, **23**, 1475.
35. A. G. Gonzalez, J. M. Aguiar, J. D. Martin, and M. Norte, *Tetrahedron Lett.*, 1975, 2499.
36. A. G. Gonzalez, J. Darias, A. Diaz, J. D. Fourneron, J. D. Martin, and C. Perez, *Tetrahedron Lett.*, 1976, 3051.
37. E. Kurosawa, in "Kagaku Sosetsu No. 25, Kaiyo Tennenbutsu Kagaku," Chemical Society of Japan, Tokyo, 1979, chap. 4-3, p. 191.
38. R. Kazlauskas, P. T. Murphy, R. J. Wells, J. J. Daly, and W. E. Oberhansli, *Aust. J. Chem.*, 1977, **30**, 2679.
39. S. J. Wratten and D. J. Faulkner, *J. Am. Chem. Soc.*, 1977, **99**, 7367.
40. J. Geigert, S. L. Naidleman, and D. J. Dalietos, *J. Biol. Chem.*, 1983, **258**, 2273.
41. S. Ohtaki, H. Nakagawa, S. Kimura, and I. Yamazaki, *J. Biol. Chem.*, 1981, **256**, 805.
42. S. Ohtaki, H. Nakagawa, M. Nakamura, and I. Yamazaki, *J. Biol. Chem.*, 1982, **257**, 761.
43. S. Ohtaki, H. Nakagawa, M. Nakamura, and I. Yamazaki, *J. Biol. Chem.*, 1982, **257**, 13 398.
44. M. Nakamura, I. Yamazaki, H. Nakagawa, and S. Ohtaki, *J. Biol. Chem.*, 1983, **258**, 3837.
45. M. Nakamura, I. Yamazaki, H. Nakagawa, S. Ohtaki, and N. Ui, *J. Biol. Chem.*, 1984, **259**, 359.
46. S. Ohtaki, H. Nakagawa, S. Nakamura, M. Nakamura, and I. Yamazaki, *J. Biol. Chem.*, 1985, **260**, 441.
47. M. Nakamura, I. Yamazaki, T. Kotani, and S. Ohtaki, *J. Biol. Chem.*, 1985, **260**, 13 546.
48. A. W. Nichol, L. A. Angel, T. Moon, and P. S. Clezy, *Biochem. J.*, 1987, **247**, 147.
49. A. Fukuzawa, Mya Aye, M. Nakamura, M. Tamura, and A. Murai, *Chem. Lett.*, 1990, 1287.
50. A. Fukuzawa, Mya Aye, and A. Murai, *Chem. Lett.*, 1990, 1579.
51. A. Fukuzawa, H. Sate, M. Miyamoto, and T. Masamune, *Tetrahedron Lett.*, 1986, **27**, 2901.
52. E. Hungerbuhler, D. Seebach, and D. Wasmuth, *Angew. Chem.*, 1979, **91**, 1025.
53. F. Johnson and J. A. Panella, *Org. Synth., Coll. Vol. V*, 1973, 614.
54. H. Kikuchi, T. Suzuki, E. Kurosawa, and M. Suzuki, *Bull. Chem. Soc. Jpn.*, 1991, **64**, 1763.
55. A. Fukuzawa, Y. Takasugi, A. Murai, M. Nakamura, and M. Tamura, *Tetrahedron Lett.*, 1992, **33**, 2017.
56. A. Fukuzawa, Y. Takasugi, and A. Murai, unpublished results.
57. G. Lowe, *J. Chem. Soc., Chem. Commun.*, 1965, **17**, 411.
58. J. Ishihara, N. Kanoh, and A. Murai, *Tetrahedron Lett.*, 1995, **36**, 737.
59. J. Ishihara, N. Kanoh, A. Fukuzawa, and A. Murai, *Chem. Lett.*, 1994, 1563.
60. A. Fukuzawa, E. Kurosawa, and T. Irie, *J. Org. Chem.*, 1972, **37**, 680.
61. T. Irie, A. Fukuzawa, M. Izawa, and E. Kurosawa, unpublished results.
62. J. Ishihara, Y. Shimada, and A. Murai, unpublished results.
63. A. Fukuzawa, Mya Aye, Y. Takasugi, M. Nakamura, M. Tamura, and A. Murai, *Chem. Lett.*, 1994, 2307.
64. J. A. Manthey and L. P. Hager, *J. Biol. Chem.*, 1981, **256**, 11232. 49.
65. A. F. Cameron, K. K. Cheung, G. Ferguson, and J. Monteath Robertson, *Chem. Commun.*, 1965, 638
66. A. F. Cameron, K. K. Cheung, G. Ferguson, and J. Monteath Robertson, *J. Chem. Soc. (B)*, 1969, 559.
67. A. Murai, Y. Yoshizawa, T. Toida, M. Sakamoto, T. Monden, and T. Masamune, *Chem. Lett.*, 1995, 171.
68. A. Murai, K. Sato, and T. Hasegawa, *Chem. Lett.*, 1995, 883.
69. D. F. Shriver, P. W. Atkins, and C. H. Langford, "Inorganic Chemistry," Oxford University Press, Oxford, 1990, p. 29.

1.13
Biosynthesis of Lipo-chitin Oligosaccharides: Bacterial Signal Molecules Which Induce Plant Organogenesis

TITA RITSEMA, BEN J. J. LUGTENBERG, and
HERMAN P. SPAINK
Leiden University, The Netherlands

1.13.1 INTRODUCTION

Rhizobium bacteria are able to live in symbiosis with leguminous plants. They elicit the formation of a new organ, the root nodule, by the secretion of lipo-chitin oligosaccharide (LCO) signal molecules. In this chapter the authors describe the importance of LCOs for nodulation and discuss the biosynthesis of LCOs.

Unusual polyunsaturated fatty acids are present in the LCOs of some rhizobial strains, in which they are important as determinants of the host range of nodulation. Rhizobial strains producing LCOs with these unusual fatty acids are a good model system to study the biosynthesis of such fatty acids. The specific use of polyunsaturated fatty acids in LCO assembly is also discussed.

1.13.2 BIOLOGICAL FUNCTION OF LCOs

Farmers have used plants of the legume family, such as clover and lupin, as biological fertilizers for centuries. This is possible, because they accumulate nitrogen in special organs in their roots, which are called nodules. The root nodules are a result of a symbiosis between the plant and bacteria. Bacteria which are able to participate in this symbiosis are called rhizobia. They belong to the genera *Rhizobium*, *Bradyrhizobium*, or *Azorhizobium*. The nodules are beneficial for both plant and bacteria since nutrients are exchanged inside the nodule. Rhizobia fix nitrogen from the air into ammonia, which is used by the plants. This enable legumes to grow on soils with a low nitrogen content. The bacteria are provided with carbon sources by the plant. This is a great advantage for soil bacteria because many soils are very poor in carbon sources.

The formation of a nodule is a very delicate process. Rhizobia cannot nodulate every leguminous plant species, a phenomenon which is known as host-specificity. The host range is called "broad" when many plant genera can be nodulated by one bacteria strain, or "narrow" when one or only a few genera can be nodulated. For example, *Rhizobium* sp. NGR234 can nodulate over 70 different genera of legumes,[1] whereas *Rhizobium leguminosarum* biovar *trifolii* nodulates only *Trifolium* species. The host range is determined at the initiation of nodulation by an exchange of signals between plant and bacterium.

The establishment of the *Rhizobium–Leguminosae* symbiosis is a complex process which can be divided into three stages: nodule initiation, nodule penetration, and nodule maturation. These stages are discussed below. For clarity it is necessary to make a dichotomy between determinate and indeterminate nodules. Determinate nodules are characterized by their appearance as a round nodule in which the meristematic activity is a transient phase. Indeterminate nodules appear as rods, since the nodule keeps on growing as a result of a persistent meristem. Therefore, in indeterminate nodules the different stages of infection can be observed at any time during nodule development.

The type of nodule formed is dependent on the plant species, not on the rhizobial species. Indeterminate nodulation takes place with legumes from temperate regions, for example *Vicia*, *Melilotus*, *Medicago*, *Pisum*, and *Trifolium*. Determinate nodulation is mainly seen on subtropical and tropical legumes, such as *Phaseolus*, *Glycine*, *Macroptilium*, and *Sesbania*, but also on temperate legumes, such as *Lotus* and *Lupinus*.

Nodulation is induced by the exchange of signals between rhizobia and leguminous plants. Plants secrete flavonoids towards which bacteria are chemotactically attracted.[2,3] Rhizobia recognize flavonoids of their host plants.[4–6] In response to these flavonoids, bacterial signals, identified as LCOs, are produced (Figure 1).[7–25] This bacteria signal is needed for nodulation and appears to be a major determinant of host-specificity.[7,8] Each *Rhizobium* produces its own modifications of the basic structure of LCOs and these specific modifications are often required to induce nodule formation on particular host plants.

Upon addition of LCOs, the plant responds with membrane depolarization of root hair cells.[26] In suspension cell cultures alkalinization of the medium is observed.[27] A transient pH shift is also seen in suspension cultures with many different types of elicitors, and therefore this response is not specific for LCOs.[28] On the other hand, membrane depolarization in plants seems a more specific

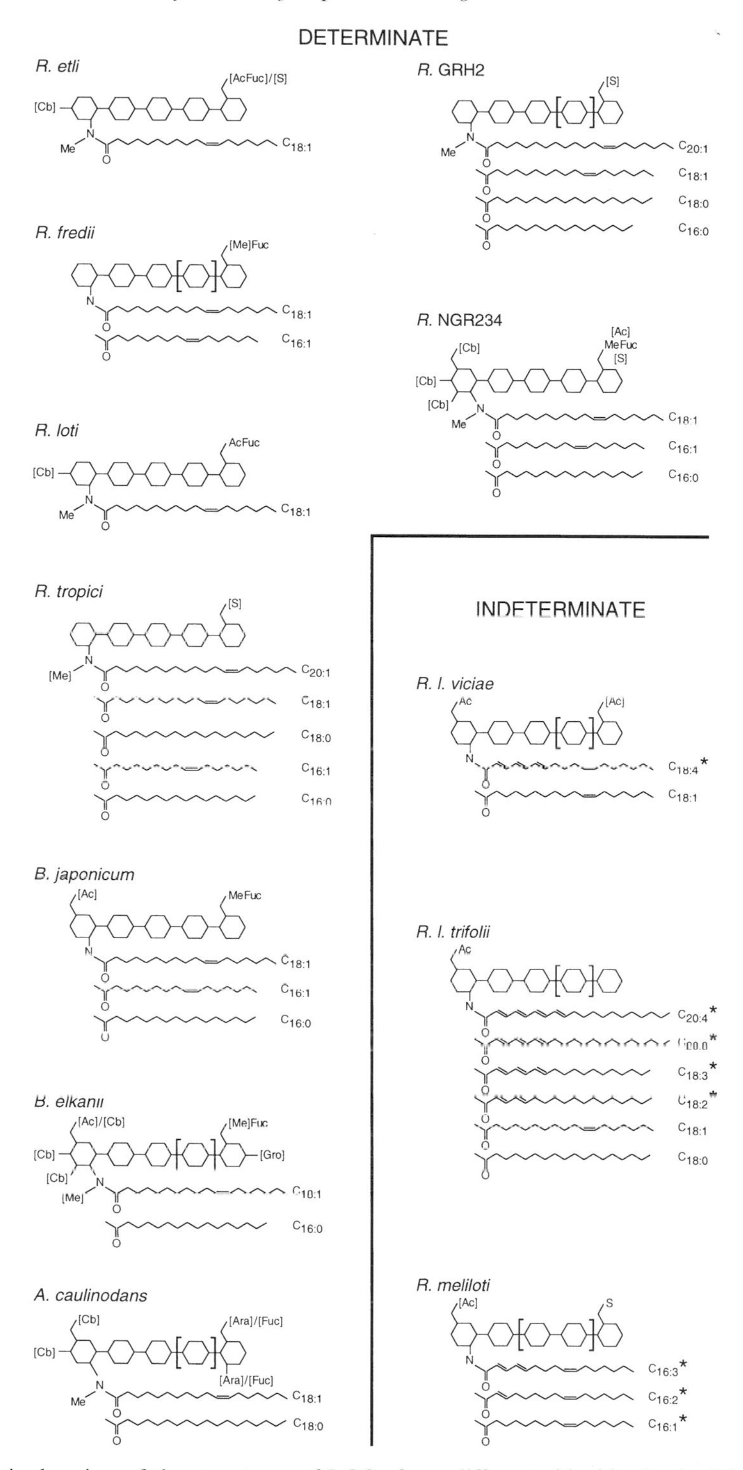

Figure 1 Schematic drawing of the structures of LCOs from different rhizobia. In the "determinate" panel LCOs of rhizobia associated with determinate nodulating plants are shown. In the "indeterminate" panel LCOs of rhizobia associated with indeterminate nodulating plants are shown. *N*-acetyl glucosamine residues are drawn as a hexagon. If a group is placed in brackets, it is not present in all LCOs of that strain. Ara, *O*-arabinose; Ac, *O*-acetyl; Cb, *O*-carbamoyl; Fuc, *O*-fucose; Gro, glycerol; Me, methyl; S, *O*-sulfate.[7–25]

process: it requires LCOs and even some of the modifications are necessary, implying that this is a response with host-specific properties.[29]

After depolarization so-called "spiking" of calcium is observed: calcium concentrations go up and down in a regular fashion. This response seems to be very specific, since it is dependent on the right combination of LCO and host plant.[30] Although the physiological meaning of calcium spiking is not known, it is often reported to be associated with membrane depolarization in animal systems.

Two morphological responses that are observed in early nodulation are root hair deformation[7,31] and the formation of a nodule meristem.[32,33] These responses also occur when LCOs are added to plants in the absence of bacteria, indicating that the LCO signal on its own is sufficient to evoke these morphological responses. Nodule meristems are seen on the root at the position where young root hairs emerge. They are present in the outer or the inner cortex, which is correlated with a determinate or indeterminate nodulation type, respectively.

Plant genes that are induced early on in the nodulation process are called early nodulin genes (*enod*). Some of the *enod* genes are expressed in root hairs (e.g., *enod5* and *enod12A*). Others are expressed in nodule primordia (e.g., *enod12B* and *enod40*).[34–41] These *enod* genes are also expressed when only LCOs are added to the plant roots, showing that the presence of bacteria is not needed. Little is known about the function of *enod* genes. Only for *enod40* has an activity been identified. A peptide encoded by this gene appears to act as a plant growth regulator.[42,43]

In some plants the application of LCOs leads to a response that goes far beyond the formation of a meristem, such as the formation of a complete nodule that does not contain bacteria.[33]

1.13.3 NODULATION GENES

Host-specificity is established in a very early stage of nodule development and is determined by signals that are secreted by both plant and bacterium. Since the 1980s a wealth of information has become available on how host-specificity is determined by the bacterial genes. Much less is known about the plant genes that are involved in nodulation and their host-specific characteristics.

Rhizobia possess specific genes for nodulation that are called *nod*, *nol*, or *noe* genes. Some of these genes are conserved in all rhizobia, whereas others are only present in a limited number of these bacteria. Many of these genes are involved in LCO production. The *nod*, *nol*, or *noe* genes are either encoded by a symbiotic plasmid or by the chromosome. They are present in operons that in most cases are preceded by a specific promoter that is called the *nod* box. The function of some of the proteins encoded by *nod*, *nol*, or *noe* genes is known. These genes can be divided into four groups according to the function of the proteins they encode: transcription factors, synthesis of LCOs, secretion of LCOs, and associated functions.

1.13.3.1 Transcription Factors

NodD acts as a transcription factor for *nod*, *nol*, and *noe* genes. It is present in all rhizobia and recognizes flavonoids (flavonoids are used here to refer to a group of compounds consisting of flavonoids, isoflavonoids, flavones, flavanones, chalcones, and betaines) secreted by the plants.[5,44] NodD binds to a conserved DNA sequence motif that is present in front of most *nod* operons. This consensus sequence is called the *nod* box and typically is 49 base pairs in length, containing 42 base pairs which are strongly conserved in all rhizobial *nod* boxes. Transcription of the *nod* genes starts closely downstream of the *nod* box sequence.[45,46] Plants secrete a distinct mixture of flavonoids which is presumably specifically recognized by NodD. NodD is therefore very important for host-specificity.[47,48] Host-specificity can sometimes be modified by replacing *nodD* by that of a rhizobial strain of a different cross-inoculation group. It has been found that some rhizobia have NodD proteins that recognize many flavonoids, whereas those from other rhizobia react only with a few flavonoids. The *nodD* gene can be present in more than one copy, for instance in *Rhizobium meliloti* where three *nodD*s are found. In *R. meliloti* an analogue of NodD, called SyrM, is involved in *nod* gene induction. It has been found that the different NodDs from one rhizobial strain can have distinct flavonoid specificities. Furthermore, each of the NodDs has a distinct ability to effect transcription of certain operons. As a result, a variation of flavonoid inducers can lead to the production of a different LCO spectrum.[49–51] The multiple *nodD* copies might serve in fine tuning of LCO production.

Some flavonoids can also act as anti-inducers.[52,53] They can probably repress NodD activation or activate *nod* gene repressors. Examples of repressors of *nod* genes are NolR of *R. meliloti*[50,54,55] and NolA of *Bradyrhizobium japonicum*.[56] Flavonoids that activate NodD proteins can repress *nod* gene repressors, as is found for NolR of *R. meliloti*, which is inhibited by the flavonoid luteolin, an activator of NodD1 of this strain.[55] LCOs induce the production of flavonoids in some plants.[57,58] It might be that this response contributes to differential expression of *nod* genes.

B. japonicum contains, in addition to *nodD* genes, the regulatory genes *nodV* and *nodW*. The proteins encoded by these genes are also involved in the transcription of *nod* genes.[59] The NodVW proteins are homologous to two component systems, which consist of a sensor (NodV) and a response-regulator (NodW). The isoflavonoids genistein and daidzein act via NodVW to induce LCO production. It has been reported that *B. japonicum* requires *nodW* for its ability to nodulate *Macroptilium atropurpureum* cv siratro and *Vigna* ssp.[60]

1.13.3.2 Biosynthesis of LCOs

LCOs are oligomers of three to six β-1,4-linked *N*-acetylglucosamine residues in which the *N*-acetyl group of the nonreducing terminal sugar is replaced by a fatty acid.

Most *nod* gene products are involved in the synthesis of LCOs and some of them are essential for LCO production. Some *nod* genes encode enzymes which have counterparts involved in cellular household processes. When these *nod* genes are knocked out, no alteration in LCO structure is detected due to complementation by the household enzyme.

Two types of Nod proteins involved in LCO biosynthesis can be distinguished, those synthesizing the core LCO and those responsible for the presence of strain-specific substituents or modifications of the core LCOs. These substituents and modifications determine the host-specificity of the rhizobia. Numerous investigations have addressed the question of which structural element is needed for nodulation of a certain plant. In general it is found that indeterminate nodulation requires LCOs with polyunsaturated fatty acids and an *O*-acetyl at the nonreducing terminal saccharide residue, whereas for determinate nodulation substituents at the reducing terminal sugar of the LCOs are important (see Figure 1). The biosynthesis of LCOs is discussed in detail in Section 1.13.4.

1.13.3.3 Secretion of LCOs

The NodIJ proteins are involved in secretions of LCOs. They are homologous to ABC (ATP-binding cassette) transporters. NodI has an ATP-binding domain and is associated with the cytosolic membrane[61] and NodJ appears to be an integral membrane protein.[62] The *nodIJ* genes are not essential for secretion of LCOs, since *nodIJ*-minus mutants show a slight delay in secretion which can lead to a slightly delayed nodulation.[63–65] Presumably, chromosomal homologues of the *nodI* and *nodJ* genes exist that can counteract the effect of a *nodIJ* mutation.

The *nodT* gene has been found in *R. leguminosarum* bvs *viciae* and *trifolii*. Based on homology with outer membrane transport proteins, NodT has been suggested to be involved in secretion of LCOs, presumably across the outer membrane.[66] The secretion system could consist of three proteins, NodI, NodJ, and NodT, which act in one complex for the secretion of LCOs. However, a deletion of *nodT* has no effect on secretion of LCOs.[64]

1.13.3.4 Associated Functions

NodO of *R. leguminosarum* bv *viciae* is a secreted protein which is able to form cation-specific pores in lipid-bilayers.[67] This observation led to the idea that NodO forms ion channels in plant membranes. NodO can restore nodulation of a nodulation-minus mutant of *R. leguminosarum* bv *viciae* that does not product LCOs with a polyunsaturated fatty acid.[68] Furthermore, introduction of *nodO* into a similar mutant of *R. leguminosarum* bv *trifolii* leads to the ability of this strain to nodulate *Vicia*, which is normally not a host plant of this strain.[69] It is unclear how NodO can complement for the absence of polyunsaturated fatty acids in LCOs during nodulation on *Vicia*.

1.13.4 BIOSYNTHESIS OF LCOs

1.13.4.1 Core Structure of LCOs

The *nodA*, *nodB*, and *nodC* genes are essential for the synthesis of the core of the LCO molecule (Figure 2). NodC is homologous to chitin synthases. Chitin is a polymer of β-1,4-linked *N*-acetyl glucosamine. It has been shown that NodC synthesizes oligomers of β-1,4-linked *N*-acetyl glucosamine.[70,71] Results indicate that these oligomers are synthesized starting with the reducing sugar residue. Furthermore, these oligomers are not synthesized on a lipid carrier, in contrast to the synthesis of many other oligosaccharides and polysaccharides.[72] *N* NodB is homologous to chitin deacetylases. John *et al.*[73] showed that this protein can hydrolyze the *N*-acetyl ester bond of the nonreducing terminal saccharide residue of chitin oligosaccharides. NodA is needed for the addition of a fatty acyl moiety to the resulting free amino group[74–76] (Figure 2).

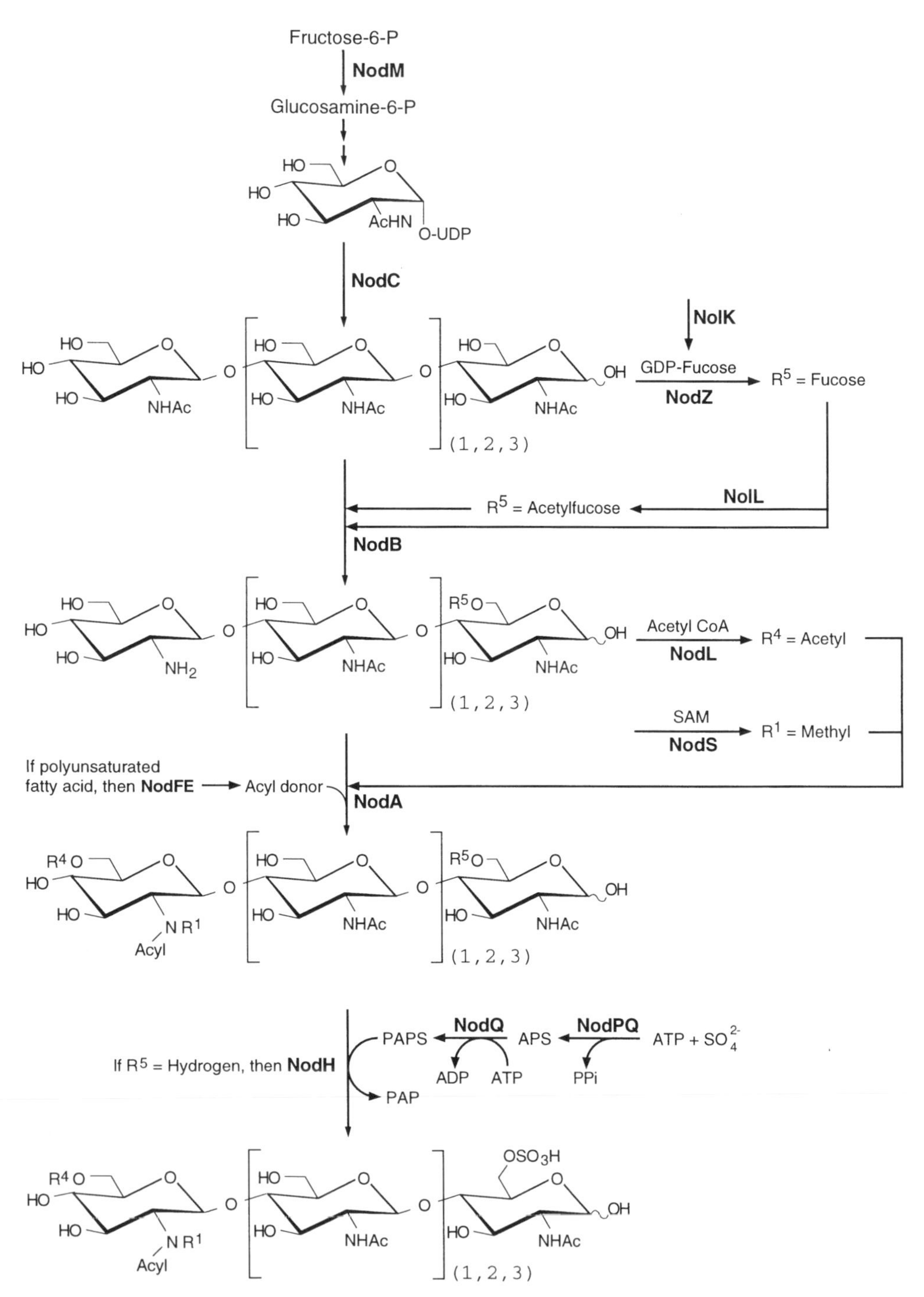

Figure 2 Biosynthesis of a hypothetical LCO. NodC, NodB, and NodA are present in all rhizobia and direct the synthesis of the core of a LCO. SAM, *S*-adenosyl methionine; PAPS, 3′-phosphoadenosine 5′-phosphosulfate.

NodM is homologous to glucosamine synthase, e.g., *glmS* from *Rhizobium*, and is able to complement a GlmS-minus mutant.[77,78] NodN is perhaps an *N*-acetyl glucosamine-1-phosphate uridyl transferase (no homology is found). It is assumed that NodM and NodN are involved in the synthesis of the precursor UDP-*N*-acetyl glucosamine. A mutation in the *nodM* and *nodN* genes seems to have no effect on LCO biosynthesis.[78] UDP-*N*-acetyl glucosamine synthesis in rhizobium is also carried out by other synthases and it seems that NodM and NodN are present to enhance the overall activity.

1.13.4.2 Modifications

Modifications of the core of LCOs determine the host range of nodulation of a rhizobial species. Modifications are found in the reducing terminal saccharide and/or in the nonreducing terminal saccharide. They do not appear in the central *N*-acetyl glucosamine residues of the chitin oligosaccharide (Figure 1).

1.13.4.2.1 Modifications of the reducing terminal saccharide

The LCOs of rhizobia involved in determinate nodulation contain substituents at the reducing terminal *N*-acetyl glucosamine. A sugar moiety can be found coupled to C-6, for example, fucose, methyl-fucose, acetyl-fucose (which can be methylated or sulfated), or arabinose (Figure 1). The results of enzymatic studies show that NodZ adds a fucose moiety to the reducing terminal saccharide.[79] This fucose apparently determines the ability to nodulate *M. atropurpureum* cv. siratro (siratro) since a *nodZ* minus mutant of *B. japonicum* is not able to nodulate siratro.[80] Furthermore, the introduction of *nodZ* into *R. leguminosarum* bv *viciae* leads to the ability to nodulate siratro, *Vigna* ssp., and *Glycine soja*.[79] Determination of K_m values for various substrates of NodZ shows that the preferred substrate is a chitin oligosaccharide, suggesting that NodZ is active after the synthesis of the chitin oligosaccharide by NodC[81] (Figure 2). NolK is probably involved in the synthesis of the GDP-fucose precursor which is needed for transfucosylation by NodZ.[9]

Because NodZ does not seem to determine the transfer of a modified fucose residue, it is assumed that fucose is modified after it is coupled to the reducing *N*-acetyl glucosamine. Candidates for the addition of such moieties are NolL for the addition of an acetyl group[82] (Figure 2) and NoeE for the addition of a sulfate group.[83]

In *Azorhizobium caulinodans* a fucose can also be coupled to C-3 of the reducing saccharide residue. More abundant in this strain, however, is the presence of an arabinose, coupled to either C-6 or C-3. The NoeC and/or NoeD proteins are involved in the addition of the arabinose.[9,10]

Both in *R. meliloti*, which is involved in indeterminate nodulation, and in *R. tropici* and *R.* sp NGR234, which are involved in determinate nodulation, a sulfate group can be present at the C-6 of the reducing terminal *N*-acetyl glucosamine[7,11,12] (Figure 1). The sulfate is required for nodulation of *Medicago* by *R. meliloti* and its presence also prevents nodulation of *Trifolium* and *Vicia*.[13,84–86] The *nod* genes *nodPQH* are responsible for the presence of the sulfate group[87] (Figure 2). NodPQ act as a complex which has ATP sulfurylase and APS kinase activity, resulting in the production of the sulfate precursor 3′-phosphoadenosine 5′-phosphosulfate (PAPS).[88] NodH is responsible for the transfer of the sulfate from PAPS to C-6 of the reducing sugar residue. The proposed *in vivo* acceptor of the sulfate is the LCO since that was found to be the preferred substrate of NodH.[89,90]

In *R. leguminosarum* bv *viciae* strain TOM some LCOs contain an *O*-acetyl group at the C-6 of the reducing terminal saccharide (Figure 1). The presence of the *O*-acetyl group is dependent upon the *nodX* gene. This modification determines nodulation of *R. leguminosarum* bv *viciae* strain TOM on Afghanistan and Iran pea species (*Pisum sativum* cvs. Afghanistan and Iran).[14,91]

In *Bradyrhizobium elkanii* the C-1 of the reducing terminal sugar sometimes carries a glycerol group[15] (Figure 1).

1.13.4.2.2 Modifications of the nonreducing terminal saccharide

An *O*-acetyl at C-6 of the nonreducing terminal saccharide is found in the LCOs of all rhizobia that associate with indeterminate nodulating plants (Figure 1). The *O*-acetyl group is present in the LCOs of *R. leguminosarum* (both biovars), *R. meliloti*, *B. japonicum*, and *B. elkanii*.[8,13,15] It has been shown that the NodL protein is an *O*-acetyl transferase that uses acetyl-CoA as the acetyl donor.[92] The determination of K_m values for different substrates indicates that the *O*-acetyl group is added to the de-*N*-acetylated oligosaccharide backbone before the fatty acid is attached[93] (Figure 2). The presence of an *O*-acetyl group on the nonreducing sugar terminal residue seems to protect the LCOs against chitinases.[94]

A carbamoyl group can be present at the C-3, C-4, or C-6 position of the nonreducing terminal saccharide (Figure 1). It is only found in the LCOs of rhizobia that are associated with determinate nodulating plants. However, not all such bacteria produce a LCO with a carbamoyl group. Carbamoyl groups are found on the LCOs of *R.* sp. NGR234, *Rhizobium loti*, *Rhizobium etli*, *A. caulinodans*, and *B. elkanii*[10,12,15–18] (Figure 1). The *nodU* gene is necessary for the presence of a carbamoyl group at C-6 of the nonreducing terminal saccharide. The presence of a carbamoyl group at the C-3 and/or C-4 position seems not to be determined by NodU.[18] In *B. japonicum* and *Rhizobium fredii* USDA257 the *nodU* gene has also been found;[59,95] however, on the LCOs of these strains no carbamoyl modifications were found[15,19] (Figure 1).

NodS is involved in *N*-methylation of the amino group which results from the activity of NodB, and is active before transacylation by NodA[18,96] (Figure 2). NodS uses *S*-adenosyl methionine (SAM) as the methyl donor[97] (Figure 2). *N*-Methylation is found in the LCOs of most rhizobia that are associated with determinate nodulating plants (Figure 1). The *nodS* gene is needed for nodulation of *Leucaena leucocephala* by *R.* sp. NGR234 and *Rhizobium tropici*.[98,99] A *nodS* gene has also been found in *B. japonicum* and *R. fredii* USDA257, but no methyl group was found on the LCOs of these strains (Figure 1). Krishnan *et al.*[95] reported that *nodS* is expressed at a very low level in *R. fredii* USDA257, probably due to a deletion in the promoter region. A low expression of *nodS* might result in a low expression of *nodU*, since *nodU* is present downstream of *nodS*. The data suggest that the *nodS* and *nodU* genes of *B. japonicum* and *R. fredii* USDA257 are poorly transcribed under the test conditions used.

1.13.5 UNUSUAL FATTY ACIDS IN THE LCOs OF *R. LEGUMINOSARUM*

LCOs contain a fatty acid attached to C-2 via an amide linkage (Figure 1). In most rhizobia a common fatty acid is found. This is either a C_{18} or a C_{16} fatty acid with one or no double bonds. Since these fatty acids are the most abundant fatty acids in rhizobia, they are called common fatty acids. The LCOs of some rhizobia contain a very special polyunsaturated fatty acid. It has *trans*-oriented double bonds conjugated to the carbonyl group.[8] Such fatty acids are found in *R. meliloti* and in both biovars of *R. leguminosarum*, biovar *viciae* and biovar *trifolii*. The difference in host range between biovar *viciae* and biovar *trifolii* is determined by the fatty acids of the LCOs[20] (Figure 1). For the biosynthesis of the polyunsaturated fatty acids the *nod* genes *nodF* and *nodE* are essential.[100] Also *nodG* of *R. meliloti* is proposed to be involved in the biosynthesis of the α,β-unsaturated fatty acids, based on its homology to reductases and dehydrogenases active in fatty acid biosynthesis.[84,101] Mutations in *nodG* do not, however, influence the type of fatty acids in the LCOs.[102] In *R. meliloti* also (ω-1)-hydroxy C_{18} to C_{26} fatty acids have been found in the LCOs.[102] These fatty acids are presumed to be precursors for the C_{26} fatty acid that is present in lipidA of rhizobia.

The LCOs of *R. leguminosarum* bv *viciae* have either a common $C_{18:1}$ (*cis*-11) fatty acid or a $C_{18:4}$ (*trans*-2, *trans*-4, *trans*-6, *cis*-11) fatty acid *N*-linked to a backbone of four or five *N*-acetyl glucosamine residues (Figure 1). LCOs carrying a $C_{18:4}$ fatty acid show both root hair deforming activity and nodule primordium formation on *Vicia* plants, whereas LCOs carrying $C_{18:1}$ fatty acids show no nodule primordium formation.[8]

Some of the LCOs of *R. leguminosarum* bv *trifolii* contain a common $C_{18:1}$ fatty acid. In addition, LCOs from this rhizobium, in contrast to *R. leguminosarum* bv *viciae*, contain several types of polyunsaturated fatty acids. The *cis* double bond is rarely present in the polyunsaturated fatty acids and the length of the fatty acids can vary between 18 and 20 carbon atoms. There are 2, 3, or 4 (in the case of C_{20}) *trans* double bonds conjugated to the carbonyl group (Figure 1). A general feature seems to be that the specific LCOs of *R. leguminosarum* bv *trifolii* are more hydrophobic than the ones of *R. leguminosarum* bv *viciae*[20] (Table 1).

The *nodF* and *nodE* genes are involved in biosynthesis of the *trans*-unsaturated fatty acids. The proteins encoded by these genes are homologous to acyl carrier protein (ACP) and β-keto-acyl synthase (KAS), respectively.[103,104] NodF and NodE are not able to synthesize a fatty acid on their own, since other proteins, encoded by household genes, are also needed in order to synthesize the

Table 1 Chemical properties of α,β-unsaturated fatty acids.

Type of fatty acid	*Mass*	*GC RT*[d] (min)	λ_{max} (nm)	*HPLC RT, LCO*[e] (*IV, Ac*) (min)
$C_{18:1}$(11c)[a]	310	14.3	220	24
$C_{18:1}$(2t)[b]	310	15.1	219	nd
$C_{18:2}$(2t,4t)	308	15.8	260	29
$C_{18:3}$(3t,4t,6t)	306	16.5	303	27
$C_{20:3}$(2t,4t,6t)	334	18.3	303	43
$C_{20:4}$(2t,4t,6t,8t)	nd[c]	nd	330	42
$C_{18:4}$(2t,4t,6t,11c)	304	16.3	303	17
$C_{20:4}$(2t,4t,6t,13c)	332	18.1	303	nd

[a]c, *cis*. [b]t, *trans*. [c]nd, not determined. [d]GC RT, gas chromatography retention times of the ethyl esters of fatty acids prepared by organic synthesis. [e]Fatty acids present in LCO with four saccharide units (IV) and an acetyl modification (AC).

α,β-unsaturated fatty acid. These proteins are probably involved in household fatty acid synthesis (FAS).

1.13.6 α,β-UNSATURATED FATTY ACIDS IN PLANTS

The rare fatty acids in the LCOs of rhizobia are not the only example of this class of fatty acids. In some plants similar α,β-unsaturated fatty acids have been found. In several members of the *Compositae*, for example *Anacyclus pyrethrum*, pellitorine has been found.[105] Pellitorine is identified as an insecticidal compound.[106] It has an *N*-linked C_{10} fatty acid with two *trans* double bonds conjugated to the carbonyl group (Figure 3). The same component is reported in some pepper varieties, for example *Piper nigrum*. In pepper not only is the C_{10} fatty acid found, but also C_8, C_{12}, C_{16}, C_{18}, and C_{20} have been reported, all with two *trans* double bonds (Figure 3).[107] Also two components are found with fatty acids containing, besides the *trans* double bonds, a *cis* double bond. In pepper these components are responsible for the special odor and sharp taste of the peppercorn.

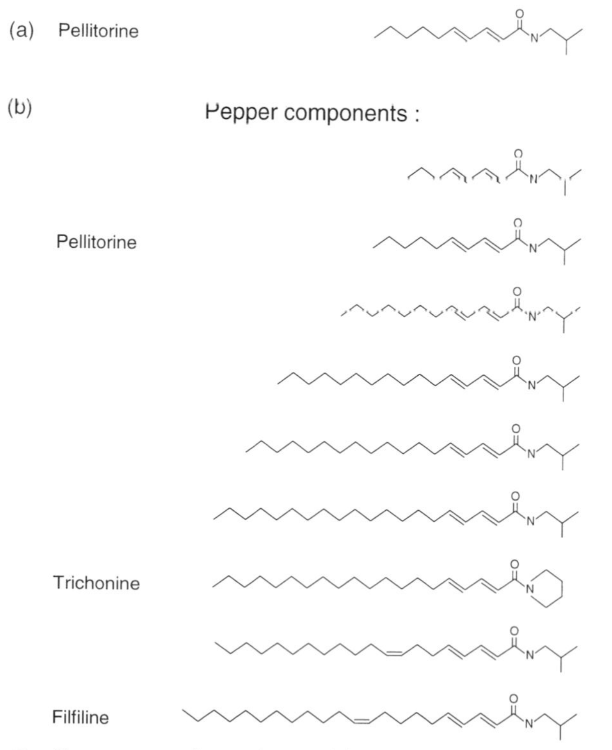

Figure 3 Components from plants with α,β-unsaturated fatty acids.

It is not known what the action of the α,β-unsaturated fatty acids from the LCOs in the plant is and why they are needed for successful nodulation of some plants. However, the above-mentioned fatty acids are known in other plants. As the polyunsaturated fatty acids are only needed for indeterminate nodulation, it is proposed that in indeterminate nodulation an additional prerequisite is required. Geiger *et al.*[108] found that α,β-unsaturated fatty acids in phospholipids are resistant to phospholipase A2 degradation. Maybe those fatty acids also prevent degradation of the LCO. Another possibility is that the rare fatty acids have a role of their own in nodulation, apart from the chitin oligosaccharide backbone of the LCOs.

1.13.7 PROPERTIES OF THE UNUSUAL FATTY ACIDS

The chemical properties of the α,β-unsaturated fatty acids are sometimes surprising (Table 1).[109] The retention time on a gas-chromatograph increases considerably if *trans* double bonds are introduced, unlike *cis* double bonds. For hydrophobicity, as detected by reversed-phase HPLC, *trans* double bonds are of little influence. One *cis* unsaturation gives a more hydrophillic LCO than three *trans* unsaturations (Table 1). The absorption maxima are correlated to the number of double bonds conjugated to the carbonyl group, for instance three *trans* double bonds result in an absorption maximum at 303 nm. It is also striking that an ethyl ester of $C_{20:4}$ (*trans*-2, *trans*-4, *trans*-6, *trans*-8) fatty acid is very unstable in solution, in contrast to the other fatty acids. The $C_{20:4}$ fatty acid is found in the LCOs of *R. leguminosarum* biovar *trifolii*.

1.13.8 BIOSYNTHESIS OF FATTY ACIDS IN BACTERIA

Fatty acid biosynthesis in bacteria is performed by a type II system (for a review see refs.[110,111]). This means that the FAS complex consists of a multienzyme complex in which every enzymatic reaction is performed by an individual protein. Fatty acid synthesis proceeds via cycles of four reactions, performed by four enzymes (Figure 4). Each cycle results in the elongation of the fatty acid with two carbon atoms. The cycles are repeated until the fatty acid reaches its final length. Each cycle consists of (i) condensation with malonate, (ii) keto-reduction, (iii) dehydration, and (iv) enoate-reduction. Condensation leads to the formation of a carbonyl group at carbon number three. This group is removed during the three subsequent steps of the elongation cycle. Keto-reduction yields a hydroxy group, dehydration leads to a double bond in the *trans* orientation, and enoate-reduction saturates this bond. In the synthesis of unsaturated fatty acids, the *trans* double bond that remains after dehydration is transformed into a *cis* double bond.

During synthesis, the fatty acid as well as the malonate extension units are coupled to acyl carrier protein (ACP). ACP needs a prosthetic 4′-phosphopantetheine group in order to be functional as a carrier. This prosthetic group, which is coupled to a serine residue, forms a bridge between the protein and the fatty acid which is coupled to it via a thioester bond. NodF is homologous to ACP. It is, like ACP, a small, very acidic, heat-stable protein and contains the same 4′-phosphopantetheine prosthetic group[112] (see Figure 5(b) for an alignment).

In the first step of fatty acid biosynthesis, the condensation with malonate, one carbon atom is released resulting in an extension of two carbon atoms (Figure 4). The KAS enzymes, which perform this condensation, have a conserved cysteine residue in their active sites. In *Escherichia coli* there are three KAS enzymes, KAS I, KAS II, and KAS III. KAS III is active during the start of fatty acid synthesis, as it forms a C_4 fatty acid. This is the only condensation step in which CoA is involved; malonyl-ACP and acetyl-CoA are condensed to form 3-keto-butyryl-ACP. KAS I and KAS II are active after this initiation reaction. These enzymes have no clear preference for the length of saturated fatty acids, although fatty acids do not become longer than C_{18}. They do, however, show specificity during the elongation of *cis* unsaturated fatty acids. KAS I is essential for the elongation of $C_{10:1}$ fatty acids,[113] whereas KAS II is essential for the elongation of $C_{16:1}$ to $C_{18:1}$.[114]

NodE is homologous to KAS enzymes. It is essential for the synthesis of the polyunsaturated fatty acid present in the LCOs in *R. leguminosarum*. Introduction of the *nodE* of *R. leguminosarum* bv *trifolii* in a *nodE*-minus mutant of *R. leguminosarum* bv *viciae* leads to the appearance of more hydrophobic fatty acids in LCOs. The nodulation behavior of strains in which *nodE* is exchanged between *R. leguminosarum* bvs *trifolii* and *viciae* leads to the conclusion that *nodE* is a major determinant of the difference in host range between *R. leguminosarum* bv *viciae* and *R. leguminosarum* bv *trifolii*.[115,116]

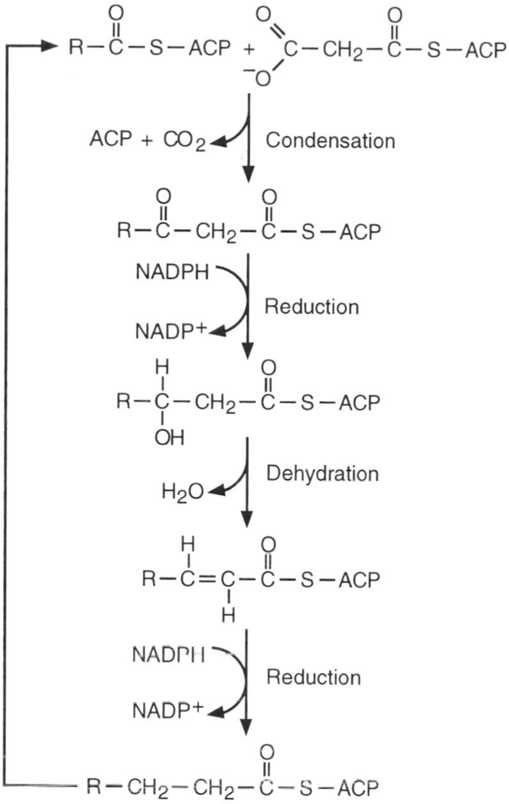

$R{-}C({=}O){-}S{-}ACP + {}^{-}O{-}C({=}O){-}CH_2{-}C({=}O){-}S{-}ACP$

Condensation ($ACP + CO_2$)

$R{-}C({=}O){-}CH_2{-}C({=}O){-}S{-}ACP$

Reduction (NADPH → $NADP^+$)

$R{-}CH(OH){-}CH_2{-}C({=}O){-}S{-}ACP$

Dehydration (H_2O)

$R{-}CH{=}CH{-}C({=}O){-}S{-}ACP$

Reduction (NADPH → $NADP^+$)

$R{-}CH_2{-}CH_2{-}C({=}O){-}S{-}ACP$

Figure 4 Schematic drawing of one elongation cycle in fatty acid biosynthesis. R stands for a fatty acid with a varying length.

1.13.9 STRUCTURE–FUNCTION RELATIONSHIP OF ACYL CARRIER PROTEINS

NodF is homologous to acyl carrier proteins (Figure 5(a)). Some specialized acyl carrier proteins have a high similarity to NodF. These specialized acyl carrier proteins are present in *Streptomyces* strains where they are involved in the biosynthesis of polyketide antibiotics. The biosynthesis of the backbone of polyketides resembles fatty acid biosynthesis. Repetitive cycles of elongation with malonate units leads to a polymer that is used for the synthesis of a polyketide antibiotic. The system of polyketide biosynthesis is highly similar to that of polyunsaturated fatty acid biosynthesis in *Rhizobium*: a specialized acyl carrier protein and a specialized β-keto-acyl-ACP synthase are encoded by genes not involved in household fatty acid biosynthesis. Other genes that are theoretically required for the synthesis of both the polyketide backbone and the fatty acid, for example, the gene encoding malonyl-Coa:ACP transacylase, are not found among the specialized genes and may be recruited from household fatty acid biosynthesis.

NodF is, in contrast to household acyl carrier proteins, not essential for survival of cells. It is therefore a suitable acyl carrier protein for the study of structure–function relationship.

1.13.9.1 The Role of NodF in Fatty Acid Biosynthesis

The authors initiated a study of NodF with the mutagenesis of the active site serine residue, which is situated in a region that is conserved in all known acyl carrier proteins. This serine residue of NodF was changed into a threonine, rendering NodF(S45T). On a native PAGE gel the mutant protein showed the same R_f value as wild-type NodF, indicating that its secondary and tertiary structures are unaltered by the mutation. NodF S45T was, however, not able to accept the prosthetic 4′-phosphopantetheine group and to complement a NodF-minus mutant for nodulation and synthesis of LCOs containing $C_{18:4}$ fatty acids. This indicates that a mutation of the 4′-phosphopantetheine binding serine residue into the very similar amino acid threonine yields a nonfunctional protein.[117]

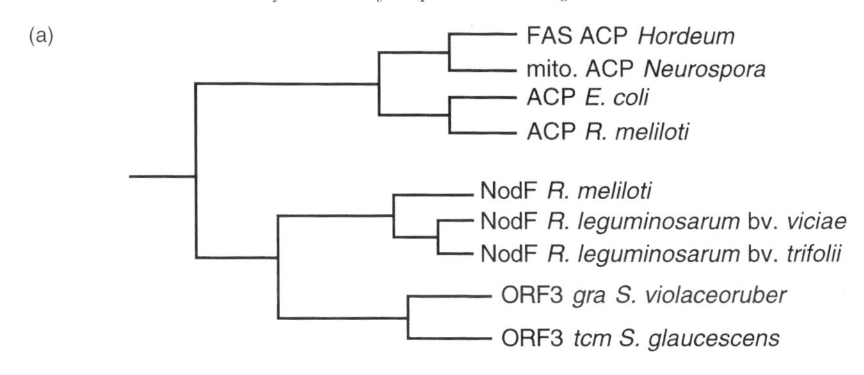

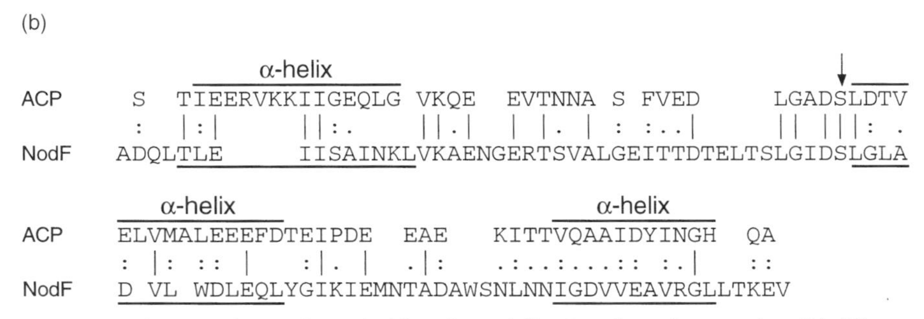

Figure 5 (a) Phylogenic tree of some household and specialized acyl carrier proteins. (b) Alignment of ACP from *E. coli* and NodF from *R. leguminosarum* bv *viciae*. This alignment is based upon the presence of α-helices predicted from NMR data.

ACP of *E. coli* not only functions in fatty acid biosynthesis but is also active in a transglycosylation reaction during synthesis of periplasmic membrane derived oligosaccharides (MDOs). For this activity ACP does not need to contain the 4′-phosphopantetheine prosthetic group.[118,119] NodF cannot replace *E. coli* ACP in a transglycosylation assay.[120] NodF also does not have a similar function in the biosynthesis of the oligosaccharide backbone of LCOs, since a NodF-minus mutant is not defective in LCO biosynthesis.

NodF cannot be replaced by ACP of *E. coli* for nodulation. The activity of key enzymes of fatty acid biosynthesis from *E. coli* was tested in *in vitro* assays using NodF as a substrate. ACP synthase, which adds the prosthetic group, and malonyl-CoA:ACP transferase (MCAT), which couples a malonyl moiety to holo-ACP, are both active with NodF. NodF can also accept long acyl chains, as shown by the coupling of palmitate ($C_{16:0}$) to NodF by acyl-ACP synthetase. It is therefore concluded that NodF functions as an acyl carrier protein.

The condensing enzyme β-keto-acyl synthase III (KAS III) shows no activity with malonyl-NodF as a substrate. KAS III is active in the first condensation step in fatty acid biosynthesis, whereas *trans* double bonds are introduced during the last elongation cycles. These results indicate that NodF is not used in the first condensation step. NodF is apparently only active in the last cycles, in which *trans* double bonds are introduced.

1.13.9.2 NodF Structure

The identity between NodF and ACP of *E. coli* in the primary protein structure is only 25%. Both proteins are small and highly acidic. Ghose *et al.*[121] described an initial NMR study of NodF. NodF has three α-helices which are present in positions in the protein corresponding to those of the three α-helices in ACP of *E. coli.*[122] The helices are present at the N-terminus, the C-terminus and just after the active site serine residue. In Figure 5(b) an alignment of NodF and ACP is shown which is based on the positions of the α-helices in both proteins. Ghose *et al.*[121] did not present structural information other than the α-helices. This gives the impression that outside the α-helices no structural data could be obtained, suggesting that the protein is rather flexible in parts that do not participate in the helix formation. This was also found for ACP.[123]

1.13.9.3 Special Properties of NodF

NodF is essential for the biosynthesis of polyunsaturated fatty acids,[102,117] which explains the requirement of NodF for nodulation of *Vicia* by *R. leguminosarum* bv *viciae*.[117] The nodulation assay is a very sensitive assay to test activity of NodF in $C_{18:4}$ biosynthesis, since a very low production of $C_{18:4}$-containing LCOs is sufficient for nodulation of *Vicia*.

The authors used the nodulation assay to test the ability of an acyl carrier protein from *Streptomyces coelicolor*, which is involved in biosynthesis of the polyketide actinorhodin (*actI*-ORFIII), to function in nodulation. The results show that *actI*-ORFIII gene is not able to complement a *nodF*-minus mutant of *R. leguminosarum* bv *viciae* for nodulation of *Vicia*.[124] This indicates that NodF is indeed a specialized acyl carrier protein with a function that is not shared by other acyl carrier proteins.

Chimeric genes were constructed which contain part of the *nodF* gene and part of the *E. coli acpP* gene (encoding ACP). Functional analysis of the ACP-NodF (ACP aa 1–33 and NodF aa 43–93) and NodF-ACP (NodF aa 1–44 and ACP aa 36–77) proteins encoded by these chimeric genes showed that they are substrates for the fatty acid biosynthetic enzymes holo-ACP synthase, MCAT, and acyl-ACP synthetase. ACP-NodF is able to replace NodF in nodulation. In contrast, NodF-ACP is not able to replace NodF. Since the ability to replace NodF is dependent upon the ability to synthesize polyunsaturated fatty acids, the authors conclude that the C-terminal domain of NodF contains specialized features for recognition by enzymes active in fatty acid biosynthesis.

In the C-terminal part of NodF and ACP two of the three α-helices are encoded.[121] The largest difference between NodF and ACP seems, however, to be the N-terminal helix. This helix is much shorter in NodF than in ACP. The C-terminal parts of NodF and ACP do not contain regions which are clearly different in α-helical structure. There are no indications which amino acids could encode for the specialized protein–protein interaction(s) of NodF with enzymes active in fatty acid biosynthesis.

1.13.10 COMPARISON OF THE FUNCTION OF ACP AND NodF

It was found that some of the key enzymes in fatty acid biosynthesis can use both ACP and NodF, whereas KAS III cannot use NodF. The enzymes that the authors have not tested are responsible for the three enzymatic activities that result in the reduction of the keto-group from C-3 after elongation of the fatty acid. These activities are keto-reduction, dehydration, and enoyl-reduction. A *trans* double bond will remain in the growing acyl chain if the last reduction step in the removal of the keto-group is inhibited. This implies that, in contrast to the keto-reductase and dehydratase, enoyl reductase is not active when a fatty acid is bound to NodF. The escape from enoyl reduction is probably a special feature of NodF which explains why it is essential for biosynthesis of polyunsaturated fatty acids. This leads to the notion that a fatty acid might only be bound to NodF in the last three elongation cycles, which is in agreement with the observation that KAS III does not use malonyl-NodF.

1.13.11 A MODEL FOR THE BIOSYNTHESIS OF POLYUNSATURATED FATTY ACIDS

Rhizobia that contain the specialized acyl carrier protein NodF also contain the specialized condensing enzyme NodE. NodE might be different from the household KAS enzymes in that it can recognize NodF and is able to elongate fatty acids with *trans* double bonds. Condensing enzymes can be restricted in the type of fatty acids that they are able to elongate. KAS I is essential for elongation of $C_{10:1}$ (*cis*-3)[113] and only KAS II is involved in the enlargement of $C_{16:1}$ (*cis*-9) to $C_{18:1}$ (*cis*-11).[114] It has also been shown that exchange of NodE between rhizobia influences the length and number of unsaturations of the fatty acids in the LCOs.[116]

The biosynthesis of the $C_{18:4}$ fatty acid which is found in LCOs of *R. leguminosarum* bv *viciae* could proceed as follows (see Figures 6 and 7). A $C_{12:1}$ (*cis*-5) fatty acid, an intermediate in the synthesis of $C_{18:1}$ fatty acids, is synthesized via common, household, fatty acid synthesis. NodE takes over elongation and condenses $C_{12:1}$-ACP with malonyl-NodE to an acyl-NodF intermediate with 14 C-atoms. After keto-reduction and dehydration by household enzymes, the cycle ends because the acyl-NodF intermediate escapes enoyl-reduction. The product after this elongation

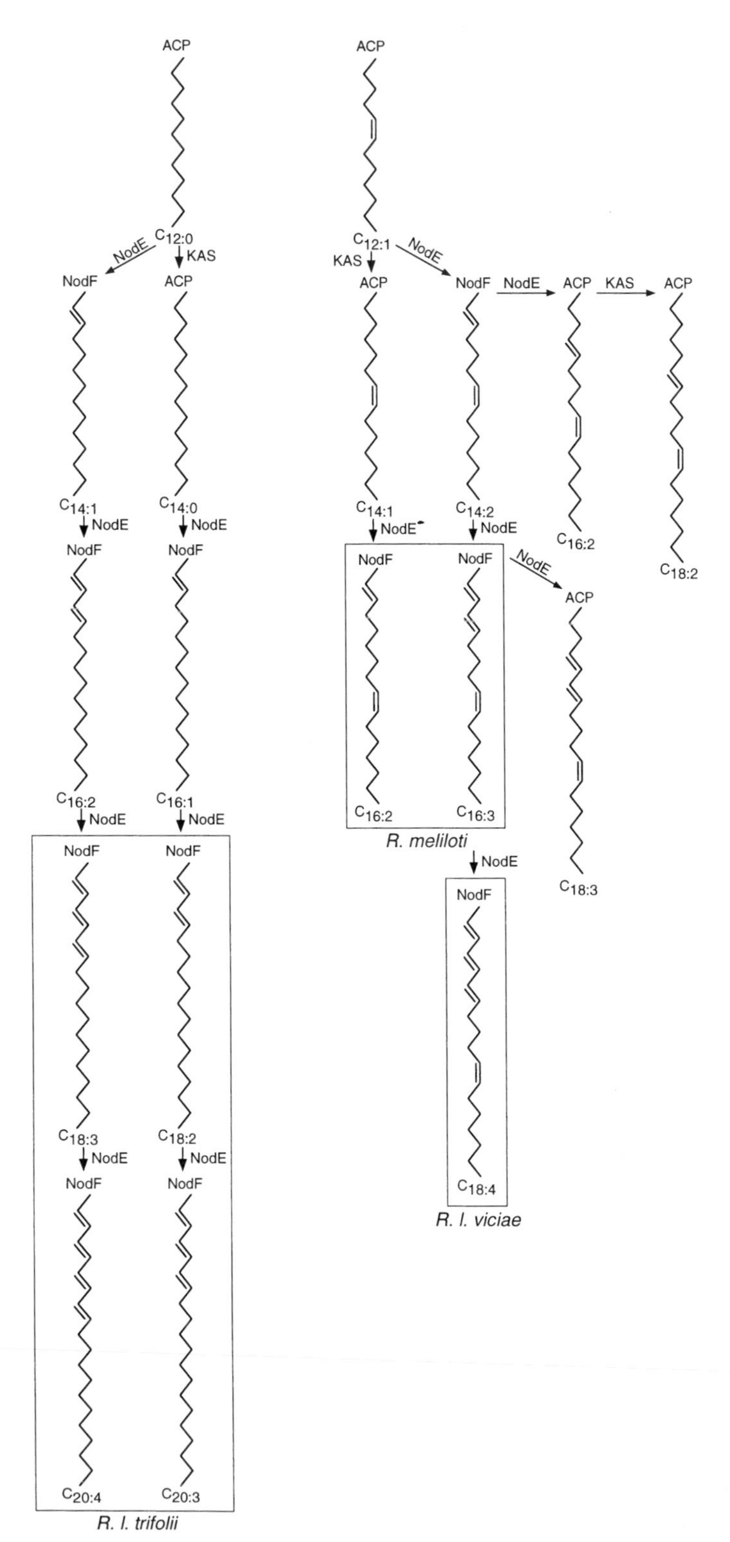

Figure 6 A model for biosynthesis of polyunsaturated fatty acids. Indicated are the fatty acids found in the LCOs of different rhizobia. For each elongation the condensing enzyme responsible is indicated; KAS is used for household β-keto-acyl-ACP synthase.

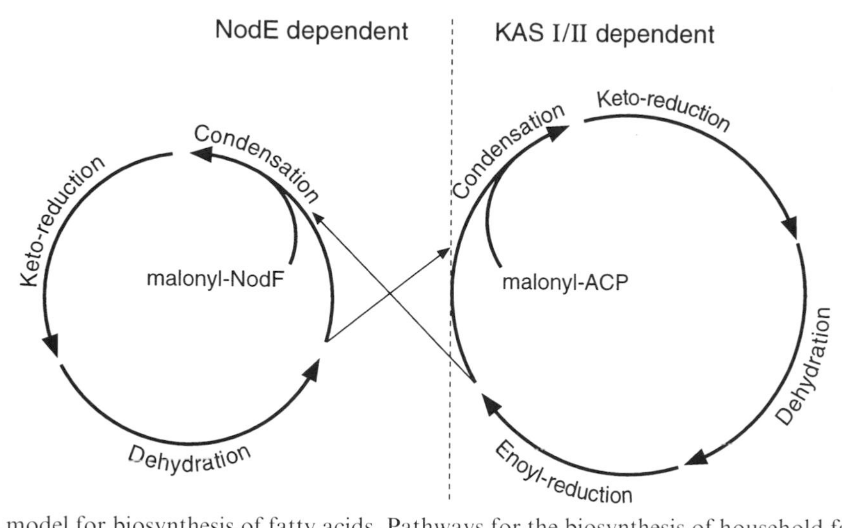

Figure 7 A model for biosynthesis of fatty acids. Pathways for the biosynthesis of household fatty acids and NodE-dependent fatty acids are integrated. NodE controls the transition between household fatty acid synthesis and *trans*-unsaturated fatty acid synthesis, as well as elongation of *trans*-unsaturated fatty acids.

cycle by NodF and NodF is $C_{14:2}$ (*trans*-2, *cis*-7) bound to NodF. NodE is able to elongate this acyl-NodF intermediate using malonyl-NodF, which again escapes enoyl-reduction (resulting in $C_{16:3}$ (*trans*-2, *trans*-4, *cis*-9) bound to NodF). After another elongation cycle the final product of fatty acid biosynthesis is $C_{18:4}$-NodF, which contains three *trans* double bonds.

$C_{18:4}$ fatty acids are also found in the phospholipids of *R. leguminosarum* bv *viciae*.[100] In addition to $C_{18:4}$ fatty acids, the phospholipids also contain $C_{18:2}$ (*trans*-6, *cis*-11) and $C_{18:3}$ (*trans*-4, *trans*-6, *cis*-11) fatty acids (Table 2, Figure 6).[108] The production of these $C_{18:2}$ and $C_{18:3}$ fatty acids can be explained by the hypothesis that NodE can use both malonyl-NodF and malonyl-ACP for elongation. When malonyl-ACP is able to use the resulting acyl-ACP, an intermediate is recognized by enoyl-reductase resulting in a saturated bond (Figure 7). The recognition of both acyl-ACP and acyl-NodF by NodE is in agreement with the idea that $C_{12:1}$-ACP is used as a starter unit for the synthesis of polyunsaturated fatty acids by NodE.

Table 2 Fatty acids in LCOs or phospholipids, whose presence is dependent upon *nodE*.

Rhizobium	*NodE-dependent fatty acids in LCOs*	*NodE-dependent fatty acids in phospholipids*
bv *viciae*	$C_{18:4}$(2,4,6-*t*,11-*c*)[a]	$C_{18:2}$(6-*t*,11-*c*), $C_{18:3}$(4,6-*t*,11-*c*) $C_{18:4}$(2,4,6-*t*,11-*c*)
bv *viciae nodE*$^-$ +*nodE* bv *trifolii*	$C_{18:2}$(2,4-*t*) $C_{18:3}$(2,4,6-*t*)	not tested
bv *trifolii*	$C_{18:2}$(2,4-*t*), $C_{18:3}$(2,4,6-*t*), $C_{20:3}$(2,4,6-*t*), $C_{20:4}$(2,4,6,8-*t*)	not tested
meliloti	$C_{16:1}$(9-*c*), $C_{16:2}$(2-*t*,9-*c*), $C_{16:3}$(2,4-*t*,9-*c*)	not tested
meliloti nodE$^-$ +*nodE* bv *viciae*	$C_{16:1}$, $C_{16:2}$, $C_{16:3}$ $C_{18:2}$, $C_{18:3}$, $C_{18:4}$	not tested

[a] *t*, trans; *c*, *cis*.

In *R. leguminosarum* bv *viciae* no fatty acids with *trans* double bonds other than those described above are found,[20] indicating that NodE of this strain does not use longer or shorter starter units than $C_{12:1}$. The LCOs of *R. leguminosarum* bv *trifolii* contain a large variety of fatty acids, most of which do not have a *cis* unsaturated double bond. The fatty acids that are most abundant are $C_{18:3}$ (*trans*-2, *trans*-4, *trans*-6) and $C_{20:4}$ (*trans*-2, *trans*-4, *trans*-6, *trans*-8), which are synthesized by NodE and NodF from the starter unit $C_{12:0}$. The $C_{18:2}$ (*trans*-2, *trans*-4) and $C_{20:3}$ (*trans*-2, *trans*-4, *trans*-6) fatty acids, which are also found in substantial amounts, are synthesized from the starter unit $C_{14:0}$ (Figure 6).

R. meliloti synthesizes LCOs which contain $C_{16:1}$ (*cis*-9), $C_{16:2}$ (*trans*-2, *cis*-9), or $C_{16:3}$ (*trans*-2, *trans*-4, *cis*-9) fatty acids. For the latter two fatty acids $C_{14:1}$ (*cis*-7) and $C_{12:1}$ (*cis*-5) are probably used as a starter molecule, respectively (Figure 6). It was found that both NodE and NodF are essential for the biosynthesis of the $C_{16:2}$ (*trans*-2, *cis*-9) fatty acid present in LCOs.[102] This indicates that NodF is not used by household KAS enzymes, since condensation of malonyl-NodF with $C_{14:1}$-ACP by KAS enzymes would also lead to the synthesis of $C_{16:2}$ (*trans*-2, *cis*-9) bound to NodF. This is not in agreement with the necessity for NodE in $C_{16:2}$ biosynthesis. Furthermore, since $C_{16:3}$ (*trans*-2, *trans*-4, *cis*-9) fatty acids are only present in a minority of the LCOs of *R. meliloti* and since $C_{16:2}$ (*trans*-2, *cis*-9) has only one *trans* double bond, NodE in *R. meliloti* appears not be important for the elongation of *trans*-unsaturated acyl intermediates. This is in contrast to its role in the synthesis of fatty acids with multiple *trans*-unsaturated double bonds as found in the LCOs of *R. leguminosarum* bvs. *viciae* and *trifolii*.

The presence of $C_{16:1}$ fatty acids in the LCOs of *R. meliloti* is also dependent upon the presence of NodF and NodE.[102] This indicates that in *R. meliloti* part of the $C_{16:2}$ (*trans*-2, *cis*-9) fatty acids bound to NodF are reduced by enoyl reductase to $C_{16:1}$.

1.13.12 FATTY ACID TRANSFER

The specificity of acyl transferases for the nature of the fatty acid has been demonstrated in several cases. For example, during phospholipid biosynthesis unsaturated fatty acids are preferably transferred to the sn-2 position of glycerol. Also the $C_{18:2}$, $C_{18:3}$, and $C_{18:4}$ fatty acids synthesized by NodF and NodE are found in the sn-2 position. It is also known that certain eukaryotic acyl transferases have a specificity towards one particular fatty acid, for example arachidonic acid.

NodA is essential for the transfer of a fatty acid to the chitin oligosaccharide backbone.[75,76] The *nodA* gene is found in all rhizobia and was suggested to be a common nodulation gene. In some rhizobia, however, a polyunsaturated fatty acid is synthesized and is transferred to the oligosaccharide backbone. In *R. leguminosarum* bvs. *trifolii* and *viciae*, the nature of the fatty acid is a major determinant of host-specificity.

The authors investigated the ability of NodAs from strains that do not contain a polyunsaturated fatty acid in their LCOs to transfer polyunsaturated fatty acids. Replacement of *nodA* to *R. leguminosarum* bv *viciae* by *nodA* of a *Bradyrhizobium* species resulted in an inability to nodulate *Vicia*. Further analysis revealed that the *Bradyrhizobium* species is active in LCO biosynthesis, but is unable to direct the transfer of $C_{18:4}$ fatty acids to the chitin oligosaccharide backbone.[125] The results show that this inability of NodA of the *Bradyrhizobium* species leads to an inability to nodulate *Vicia*. It was concluded that *nodA* has host-specific properties and that the original notion that *nodA* is a common protein should be revised. The same conclusion can be drawn from a paper in which it was shown that NodA of *R. meliloti* and NodA of *R. tropici* determine the transfer of different fatty acids in the biosynthesis of LCOs.[126]

It has been found that NodC proteins, like NodA involved in the biosynthesis of the core LCO, also have host-specific properties. It was shown that NodC has an intrinsic preference for the synthesis of pentamers or tetramers.[127] *In vivo* experiments in which *nodC* of *R. meliloti* was introduced into *R. tropici* showed that the length of the oligosaccharide backbone had changed from five to predominantly four sugar residues. This resulted in the ability of *R. tropici* to nodulate *Medicago*.[128]

The authors showed that acyl-ACP is used as a donor for fatty acids in LCO biosynthesis. It is assumed that in the case of polyunsaturated fatty acids, acyl-NodF is the acyl donor. This leads to two possible explanations for the inability of NodA of a *Bradyrhizobium* to synthesize LCOs with $C_{18:4}$ fatty acids. Either NodA of a *Bradyrhizobium* species is not able to recognize the $C_{18:4}$ fatty acid, or it might not recognize the presumed donor acyl-NodF.

In contrast to rhizobial strains that do not contain *nodE* and *nodF*, NodA proteins from strains containing these genes possess the ability to transfer polyunsaturated fatty acids probably donated by NodF. Since also LCOs which contain $C_{18:1}$ fatty acids are found in these strains, it is concluded that the specialized NodAs have not lost the ability to transfer a common $C_{18:1}$ fatty acid donated by ACP.

NodF and NodE of *R. leguminosarum* bv *viciae* are not only involved in the synthesis of $C_{18:4}$ fatty acids, but also of $C_{18:2}$ (*trans*-6, *cis*-11) and $C_{18:3}$ (*trans*-4, *trans*-6, *cis*-11) fatty acids. Even when *nodF* and *nodE* are the only *nod* genes present, the phospholipids contain larger amounts of $C_{18:2}$ and $C_{18:3}$ fatty acids than of $C_{18:4}$ fatty acids, suggesting that $C_{18:2}$ and $C_{18:3}$ fatty acids are produced in higher quantity than $C_{18:4}$ fatty acids.[108] However, in LCOs only $C_{18:4}$ fatty acids are found (Table 2). This indicates that NodA from *R. leguminosarum* bv *viciae* is specific for transfer of $C_{18:4}$ fatty acids. Interesting in this respect is the detection of LCOs containing $C_{18:2}$ (*trans*-2, *trans*-4) and $C_{18:3}$ (*trans*-2, *trans*-4, *trans*-6) fatty acids in a *R. leguminosarum* bv *viciae* strain in which *nodE* is replaced by that of *R. leguminosarum* bv *trifolii*.[116] The *nodA* gene in this strain was from *R. leguminosarum* bv *viciae*. Combining the results of the above-described experiment with the absence of $C_{18:2}$ (*trans*-6, *cis*-11) and $C_{18:3}$ (*trans*-4, *trans*-6, *cis*-11) fatty acids in the LCOs of *R. leguminosarum* bv *viciae*, it is concluded that NodA is not specific for $C_{18:4}$ fatty acids, but for fatty acids containing 18 C-atoms and a *trans*-2 double bond. These are also the fatty acids that are linked to NodF after their synthesis (Figure 6). It is also concluded that NodA of *R. leguminosarum* bv *viciae* is not able to transfer the C_{20} fatty acids that are produced if *nodE* from *R. leguminosarum* bv *trifolii* is present.[20] The appearance of NodE-dependent $C_{16:1}$ (*cis*-9) fatty acids in the LCOs in *R. meliloti* (Table 2) indicates that for a fatty acid the linkage to NodF is advantageous for transfer by NodA. It seems that NodA of *R. leguminosarum* bv *viciae* only transfers C_{18} fatty acids donated by NodF, whereas NodA of *R. leguminosarum* bv *trifolii* can use both C_{18} and C_{20} fatty acids donated by NodF. NodA of a wild-type strain of *R. meliloti* uses only C_{16} fatty acids donated by NodF. However, when NodF and NodE from *R. leguminosarum* bv *viciae* are present, C_{18} fatty acids are also found (Table 2).[102] This is probably due to an overproduction of these fatty acids by NodF and NodE from *R. leguminosarum* bv *viciae*.

The observation that NodAs of all strains are able to transfer $C_{18:1}$ fatty acids donated by ACP indicates that even NodAs which have a specificity for different fatty acids can be forced to accept $C_{18:1}$. This can be explained by the assumption that $C_{18:1}$-ACP is the most abundant acyl-ACP in rhizobia, since the phospholipids of rhizobia contain over 50% $C_{18:1}$.

Expression of *nodABCS* of *A. caulinodans* in *E. coli* leads to the synthesis of LCOs which contain fatty acids ranging from $C_{14:0}$ to $C_{18:0}$ (also odd-numbered species are found) and $C_{18:1}$.[129] This shows that in *E. coli* a range of fatty acids is transferred by NodA that is different from that in *Rhizobium*. This difference may be caused by the fact that in *E. coli* a range of fatty acids is synthesized by household fatty acid biosynthesis that is different from that in *Rhizobium*.

In general, the data suggest that NodA proteins from strains which synthesize polyunsaturated fatty acids have a preference for the transfer of fatty acids of a certain length to the oligosaccharide backbone and that acyl-NodF is used as a donor. If these fatty acids are not available, NodA takes any fatty acid that is provided by household fatty acid biosynthesis using the household acyl-ACP as the preferred donor.

1.13.13 REFERENCES

1. B. Relic, F. Talmont, J. Kopcinska, W. Golinowski, J. C. Promé, and W. J. Broughton, *Mol. Plant–Microbe Interact.*, 1993, **6**, 764.
2. J. M. M. Aguilar, A. M. Ashby, A. J. M. Richards, G. J. Loake, M. D. Watson, and C. H. Shaw, *J. Gen. Microbiol.*, 1988, **134**, 2741.
3. G. Caetano-Anollés and W. D. Bauer, *Planta*, 1988, **175**, 546.
4. N. K. Peters, J. W. Frost, and S. R. Long, *Science*, 1986, **233**, 977.
5. J. W. Redmond, M. Batley, M. A. Djordjevic, R. W. Innes, P. L. Kuempel, and B. G. Rolfe, *Nature (London)*, 1986, **323**, 632.
6. S. A. J. Zaat, A. A. N. van Brussel, T. Tak, E. Pees, and B. J. J. Lugtenberg, *J. Bacteriol.*, 1987, **169**, 3388.
7. P. Lerouge, P. Roche, C. Faucher, F. Maillet, G. Truchet, J. C. Promé, and J. Dénarié, *Nature (London)*, 1990, **344**, 781.
8. H. P. Spaink, D. M. Sheley, A. A. N. van Brussel, J. Glushka, W. S. York, T. Tak, O. Geiger, E. P. Kennedy, V. N. Reinhold, and B. J. J. Lugtenberg, *Nature (London)*, 1991, **354**, 125.
9. P. Mergaert, W. D'Heaze, M. Fernández-López, D. Geelen, K. Goethals, J. C. Promé, M. Van Montagu, and M. Holsters, *Mol. Microbiol.*, 1996, **21**, 409.
10. P. Mergaert, M. Van Montagu, J.-C. Promé, and M. Holsters, *Proc. Natl. Acad. Sci. USA*, 1993, **90**, 1551.
11. R. Poupot, E. Martinez-Romero, and J.-C. Promé, *Biochemistry*, 1993, **32**, 10 430.
12. I. M. López-Lara, J. D. J. van den Berg, J. E. Thomas-Oates, J. Glushka, B. J. J. Lugtenberg, and H. P. Spaink, *Mol. Microbiol.*, 1995, **15**, 627.
13. P. Roche, P. Lerouge, J. C. Promé, C. Faucher, J. Vasse, F. Maillet, S. Camut, F. de Billy, J. Dénarié, and G. Truchet, in "Advances in Molecular Genetics of Plant–Microbe Interactions," eds. H. Hennecke and D. P. S. Verma, Kluwer Academic, Dordrecht, 1991, p. 119.

14. J. L. Firmin, K. E. Wilson, R. W. Carlson, A. E. Davies, and J. A. Downie, *Mol. Microbiol.*, 1993, **10**, 351.
15. R. W. Carlson, J. Sanjuan, R. Bhat, J. Glushka, H. P. Spaink, A. H. M. Wijfjes, A. A. N. van Brussel, T. J. W. Stokkermans, K. Peters, and G. Stacey, *J. Biol. Chem.*, 1993, **268**, 18 372.
16. N. P. J. Price, B. Relic, F. Taimont, A. Lewin, D. Promé, S. G. Pueppke, F. Maillet, J. Dénairé, J. C. Promé, and W. J. Broughton, *Mol. Microbiol.*, 1992, **6**, 3575.
17. L. Cardenas, J. Dominguez, C. Quinto, I. M. López-Lara, B. J. J. Lugtenberg, H. P. Spaink, G. J. Rademaker, J. Haverkamp, and J. E. Thomas-Oates, *Plant Mol. Biol.*, 1995, **29**, 453.
18. S. Jabbouri, R. Fellay, F. Talmont, P. Kamalaprija, U. Burger, B. Relic, J. C. Promé, and W. J. Broughton, *J. Biol. Chem.*, 1995, **270**, 22 968.
19. M. P. Bec-Ferté, H. B. Krishnan, D. Promé, A. Savagnac, S. G. Pueppke, and J. C. Promé, *Biochemistry*, 1994, **33**, 11 782.
20. H. P. Spaink, G. V. Bloemberg, A. A. N. van Brussel, B. J. J. Lugtenberg, K. M. G. M. van der Drift, J. Haverkamp, and J. E. Thomas-Oates, *Mol. Plant–Microbe Interact.*, 1995, **8**, 155.
21. R. Poupot, E. Martinez-Romero, N. Gautier, and J. C. Promé, *J. Biol. Chem.*, 1995, **270**, 6050.
22. J. L. Folch-Mallol, S. Marroqui, C. Sousa, H. Manyani, I. M. López-Lara, *et al.*, *Mol. Plant–Microbe Interact.*, 1996, **9**, 151.
23. J. Sanjuan, R. W. Carlson, H. P. Spaink, U. R. Bhat, W. M. Barbour, J. Glushka, and G. Stacey, *Proc. Natl. Acad. Sci. USA*, 1992, **89**, 8789.
24. I. M. López-Lara, K. M. G. M. van der Drift, A. A. N. van Brussel, J. Haverkamp, B. J. J. Lugtenberg, J. E. Thomas-Oates, and H. P. Spaink, *Plant Mol. Biol.*, 1995, **29**, 465.
25. M. Schultze, B. Quiclet-Sire, E. Kondorosi, H. Virelizier, J. N. Glushka, G. Endre, S. D. Géro, and A. Kondorosi, *Proc. Natl. Acad. Sci. USA*, 1992, **89**, 192.
26. D. W. Ehrhardt, E. M. Atkinson, and S. R. Long, *Science*, 1992, **256**, 998.
27. A. C. Kurkdjian, *Plant Physiol.*, 1995, **107**, 783.
28. C. Staehelin, J. Granado, J. Muller, A. Wiemken, R. B. Mellor, G. Felix, M. Regenass, W. J. Broughton, and T. Boller, *Proc. Natl. Acad. Sci. USA*, 1994, **91**, 2196.
29. H. H. Felle, E. Kondorosi, A. Kondorosi, and M. Schultze, *Plant J.*, 1995, **7**, 939.
30. D. W. Ehrhardt, R. Wais, and S. R. Long, *Cell*, 1996, **85**, 673.
31. P. Y. Yao and J. M. Vincent, *Austr. J. Biol. Sci.*, 1969, **22**, 413.
32. B. G. Rolfe, *BioFactors*, 1988, **1**, 3.
33. G. Truchet, P. Roche, P. Lerouge, J. Vasse, S. Camut, F. de Billy, J.-C. Promé, and J. Dénarié, *Nature* (*London*), 1991, **351**, 670.
34. B. Scheres, C. van de Wiel, A. Zalensky, B. Horvath, H. P. Spaink, *et al.*, *Cell*, 1990, **60**, 281.
35. M. Pichon, E. P. Journet, A. Dedieu, F. de Billy, G. Truchet, and D. G. Barker, *Plant Cell*, 1992, **40**, 1199.
36. H. Kouchi and S. Hata, *Mol. Gen. Genet.*, 1993, **238**, 106.
37. W. C. Yang, P. Katinakis, P. Hendriks, A. Smolders, F. de Vries, J. Spee, A. van Kammen, T. Bisseling, and H. Franssen, *Plant J.*, 1993, **3**, 573.
38. S. Asad, Y. W. Fang, K. L. Wycoff, and A. M. Hirsch, *Protoplasma*, 1994, **183**, 10.
39. P. Bauer, M. D. Crespi, J. Szecsi, L. A. Allison, M. Schultze, P. Ratet, E. Kondorosi, and A. Kondorosi, *Plant Physiol.*, 1994, **105**, 585.
40. I. Vijn, F. Martinez-Abarca, W.-C. Yang, L. das Neves, A. A. N. van Brussel, A. van Kammen, and T. Bisseling, *Plant J.*, 1995, **8**, 111.
41. K. Papadopoulou, A. Roussis, and P. Katinakis, *Plant Mol. Biol.*, 1996, **30**, 403.
42. M. D. Crespi, E. Jurkevitch, M. Poiret, Y. Daubentoncarafa, G. Petrovics, E. Kondorosi, and A. Kondorosi, *EMBO J.*, 1994, **13**, 5099.
43. K. van de Sande, K. Pawlowski, I. Czaja, U. Wieneke, J. Schell, *et al.*, *Science*, 1996, **273**, 370.
44. L. Rossen, C. A. Shearman, A. W. B. Johnston, and J. A. Downie, *EMBO J.*, 1985, **4**, 3369.
45. K. Rostas, E. Kondorosi, B. Horvath, A. Simoncsits, and A. Kondorosi, *Proc. Natl. Acad. Sci. USA*, 1986, **83**, 1757.
46. R. F. Fisher, T. Egelhoff, J. T. Mulligan, and S. R. Long, *Genes Dev.*, 1988, **2**, 282.
47. R. M. Kosslak, R. Bookland, J. Barkei, H. E. Paaren, and E. R. Appelbaum, *Proc. Natl. Acad. Sci. USA*, 1987, **84**, 7428.
48. H. P. Spaink, C. A. Wijffelman, E. Pees, R. J. H. Okker, and B. J. J. Lugtenberg, *Nature* (*London*), 1987, **328**, 337.
49. Z. Györgypal, N. Iyer, and A. Kondorosi, *Mol. Gen. Genet.*, 1988, **212**, 85.
50. E. Kondorosi, J. Gyuris, J. Schmidt, M. John, E. Duda, B. Hoffman, J. Schell, and A. Kondorosi, *EMBO J.*, 1989, **8**, 1331.
51. F. Maillet, F. Debellé, and J. Dénarié, *Mol. Microbiol.*, 1990, **4**, 1975.
52. M. Göttfert, B. Horvath, E. Kondorosi, P. Putnoky, F. Rodriguez-Quinones, and A. Kondorosi, *J. Mol. Biol.*, 1986, **191**, 411.
53. G. L. Bender, M. Nayudu, K. K. Le Strange, and B. G. Rolfe, *Mol. Plant–Microbe Interact.*, 1988, **1**, 259.
54. M. Cren, A. Kondorosi, and E. Kondosori, *J. Bacteriol.*, 1994, **176**, 518.
55. M. Cren, A. Kondorosi, and E. Kondorosi, *Mol. Microbiol.*, 1995, **15**, 733.
56. T. C. Dockendorff, J. Sanjuan, P. Grob, and G. Stacey, *Mol. Plant–Microbe Interact.*, 1994, **7**, 596.
57. A. A. N. van Brussel, K. Recourt, E. Pees, H. P. Spaink, T. Tak, C. A. Wijffelman, J. W. Kijne, and B. J. J. Lugtenberg, *J. Bacteriol.*, 1990, **172**, 5394.
58. K. Recourt, J. Schripsema, J. W. Kijne, A. A. N. van Brussel, and B. J. J. Lugtenberg, *Plant Mol. Biol.*, 1991, **16**, 841.
59. M. Göttfert, P. Grob, and H. Hennecke, *Proc. Natl. Acad. Sci. USA*, 1990, **87**, 2680.
60. J. Sanjuan, P. Grob, M. Gottfert, H. Hennecke, and G. Stacey, *Mol. Plant–Microbe Interact.*, 1994, **7**, 364.
61. H. R. M. Schlaman, R. J. H. Okker, and B. J. J. Lugtenberg, *J. Bacteriol.*, 1990, **172**, 5486.
62. B. P. Surin, J. M. Watson, W. D. O. Hamilton, E. S. A. Economou, and J. A. Downie, *Mol. Microbiol.*, 1990, **4**, 245.
63. H. C. J. Canter Cremers, C. A. Wijffelman, E. Pees, B. G. Rolfe, M. A. Djordjevic, and B. J. J. Lugtenberg, *J. Plant Physiol.*, 1988, **132**, 398.
64. H. P. Spaink, A. H. M. Wijfjes, and B. J. J. Lugtenberg, *J. Bacteriol.*, 1995, **177**, 6276.
65. M. Fernández-López, W. D'Haeze, P. Mergaert, C. Verplancke, J. C. Promé, M. Van Montagu, and M. Holsters, *Mol. Microbiol.*, 1996, **20**, 993.

66. R. Rivilla, J. M. Sutton, and J. A. Downie, *Gene*, 1995, **161**, 27.
67. J. M. Sutton, E. J. A. Lea, and J. A. Downie, *Proc. Natl. Acad. Sci. USA*, 1994, **91**, 9990.
68. J. A. Downie and B. P. Surin, *Mol. Gen. Genet.*, 1990, **222**, 81.
69. E. S. A. Economou, A. E. Davies, A. W. B. Johnston, and J. A. Downie, *Microbiology*, 1994, 2341.
70. R. A. Geremia, P. Mergaert, D. Geelen, M. Van Montagu, and M. Holsters, *Proc. Natl. Acad. Sci. USA*, 1994, **91**, 2669.
71. E. Kamst, K. M. G. M. van der Drift, J. E. Thomas-Oates, B. J. J. Lugtenberg, and H. P. Spaink, *J. Bacteriol.*, 1995, **177**, 6282.
72. E. Kamst, B. J. J. Lugtenberg, and H. P. Spaink, in "Chitin Enzymology," ed. R. A. A. Muzzarelli, Atec Edizioni, Grottammare, 1996, p. 329.
73. M. John, H. Röhrig, J. Schmidt, U. Wieneke, and J. Schell, *Proc. Natl. Acad. Sci. USA*, 1993, **90**, 625.
74. H. P. Spaink, A. H. M. Wijfjes, K. M. G. M. van der Drift, J. Haverkamp, J. E. Thomas-Oates, and B. J. J. Lugtenberg, *Mol. Microbiol.*, 1994, **13**, 821.
75. E. M. Atkinson, M. M. Palcic, O. Hindsgaul, and S. R. Long, *Proc. Natl. Acad. Sci. USA*, 1994, **91**, 8418.
76. H. Röhrig, J. Schmidt, U. Wieneke, E. Kondorosi, I. Barlier, J. Schell, and M. John, *Proc. Natl. Acad. Sci. USA*, 1994, **91**, 3122.
77. N. Baev, G. Endre, G. Petrovics, Z. Banfalvi, and A. Kondorosi, *Mol. Gen. Genet.*, 1991, **228**, 113.
78. C. Marie, M. A. Barny, and J. A. Downie, *Mol. Microbiol.*, 1992, **6**, 843.
79. I. M. López-Lara, L. Blok-Tip, C. Quinto, M. L. Carcia, G. Stacey, G. V. Bloemberg, G. E. M. Lamers, B. J. J. Lugtenberg, J. E. Thomas-Oates, and H. P. Spaink, *Mol. Microbiol.*, 1996, **21**, 397.
80. G. Stacey, S. Luka, J. Sanjuan, Z. Banfalvi, A. J. Nieuwkoop, J. Y. Chun, L. S. Forsberg, and R. Carlson, *J. Bacteriol.*, 1994, **176**, 620.
81. C. Quinto, A. H. M. Wijfjes, G. V. Bloemberg, L. Blok-Tip, I. M. López-Lara, B. J. J. Lugtenberg, J. E. Thomas-Oates, and H. P. Spaink, *Proc. Natl. Acad. Sci. USA*, 1997, **94**, 4336.
82. D. B. Scott, C. A. Young, J. M. Collins-Emerson, E. A. Terzaghi, E. S. Rockman, P. E. Lewis, and C. E. Pankhurst, *Mol. Plant–Microbe Interact.*, 1996, **9**, 187.
83. M. Hanin, S. Jabbouri, V. D. Quesada, C. Freiberg, X. Perret, J. C. Prome, W. J. Broughton, and R. Fellay, *Mol. Microbiol.*, 1997, **24**, 1119.
84. F. Debellé, C. Rosenberg, J. Vasse, F. Maillet, E. Martinez, J. Dénarié, and G. Truchet, *J. Bacteriol.*, 1986, **168**, 1075.
85. B. Horvath, E. Kondorosi, M. John, J. Schmidt, I. Török, Z. Györgypal, I. Barabas, U. Wieneke, J. Schell, and A. Kondorosi, *Cell*, 1986, **46**, 335.
86. C. Faucher, S. Camut, J. Dénarié, and G. Truchet, *Mol. Plant–Microbe Interact.*, 1989, **2**, 291.
87. P. Roche, F. Debellé, F. Maillet, P. Lerouge, C. Faucher, G. Truchet, J. Dénarié, and J. C. Promé, *Cell*, 1991, **67**, 1131.
88. J. Schwedock and S. R. Long, *Nature (London)*, 1990, **348**, 644.
89. D. W. Ehrhardt, E. M. Atkinson, K. F. Faull, D. I. Freedberg, D. P. Sutherlin, R. Armstrong, and S. R. Long, *J. Bacteriol.*, 1995, **177**, 6237.
90. M. Schultze, C. Staehelin, H. Röhrig, M. John, J. Schmidt, E. Kondorosi, J. Schell, and A. Kondorosi, *Proc. Natl. Acad. Sci. USA*, 1995, **92**, 2706.
91. E. O. Davis, I. J. Evans, and A. W. B. Johnston, *Mol. Gen. Genet.*, 1988, **212**, 531.
92. G. V. Bloemberg, J. E. Thomas-Oates, B. J. J. Lugtenberg, and H. P. Spaink, *Mol. Microbiol.*, 1994, **11**, 793.
93. G. V. Bloemberg, R. M. Lagas, S. van Leeuwen, G. A. van der Marel, J. H. van Boom, B. J. J. Lugtenberg, and H. P. Spaink, *Biochemistry*, 1996, **34**, 12 712.
94. C. Staehelin, M. Schultze, E. Kondorosi, and A. Kondorosi, *Plant Physiol.*, 1995, **108**, 1607.
95. H. B. Krishnan, A. Lewin, R. Fellay, W. J. Broughton, and S. G. Pueppke, *Mol. Microbiol.*, 1992, **6**, 3321.
96. D. Geelen, P. Mergaert, R. A. Geremia, S. Goormachtig, M. Van Montagu, and M. Holsters, *Mol. Microbiol.*, 1993, **9**, 145.
97. D. Geelen, B. Leyman, P. Mergaert, K. Klarskov, M. Van Montagu, R. A. Geremia, and M. Holsters, *Mol. Microbiol.*, 1995, **17**, 387.
98. A. Lewin, E. Cervantes, C.-H. Wong, and W. J. Broughton, *Mol. Plant–Microbe Interact.*, 1990, **3**, 317.
99. F. Waelkens, T. Voets, K. Vlassak, J. van der Leyden, and P. van Rhijn, *Mol. Plant–Microbe Interact.*, 1995, **8**, 147.
100. O. Geiger, J. E. Thomas-Oates, J. Glushka, H. P. Spaink, and B. J. J. Lugtenberg, *J. Biol. Chem.*, 1994, **269**, 11 090.
101. A. R. Slabas, D. Chase, I. Nishida, N. Murata, C. Sidebottom, R. Safford, P. S. Sheldon, R. G. Kekwick, D. G. Hardie, and R. W. Mackintosh, *Biochem. J.*, 1992, **283**, 2.
102. N. Demont, F. Debellé, H. Aurelle, J. Dénarié, and J. C. Promé, *J. Biol. Chem.*, 1993, **268**, 20 134.
103. C. A. Shearman, L. Rossen, A. W. B. Johnston, and J. A. Downie, *EMBO J.*, 1986, **5**, 647.
104. M. J. Bibb, S. Biró, H. Motamedi, J. F. Collins, and C. R. Hutchinson, *EMBO J.*, 1989, **8**, 2727.
105. M. Jacobson, *J. Am. Chem. Soc.*, 1949, **71**, 366.
106. R. T. Lalonde, *J. Chem. Ecol.*, 1980, **6**, 35.
107. W. Freist, *Chemie in unserer Zeit, 1991*, **25**, 135.
108. O. Geiger, J. Glushka, B. J. J. Lugtenberg, H. P. Spaink, and J. Thomas-Oates, *Mol. Plant–Microbe Int.*, 1998, **11**, 33.
109. S. H. van Leeuwen, Ph.D. Thesis, Leiden University, 1997.
110. A. J. Fulco, *Prog. Lipid Res.*, 1983, **22**, 133.
111. K. Magnuson, S. Jackowski, C. O. Rock, and J. E. Cronan, Jr., *Microbiol. Rev.*, 1993, **57**, 522.
112. O. Geiger, H. P. Spaink, and E. P. Kennedy, *J. Bacteriol.*, 1991, **173**, 2872.
113. J. E. Cronan, Jr. and C. O. Rock, in "*Escherichia coli* and *Salmonella typhimurium*: Cellular and Molecular Biology," ed. F. C. Neidhardt, American Society for Microbiology, Washington, DC, 1987, p. 474.
114. J. L. Garwin, A. L. Klages, and J. E. Cronan, Jr., *J. Biol. Chem.*, 1980, **255**, 11 949.
115. H. P. Spaink, R. J. H. Okker, C. A. Wijffelman, T. Tak, L. Goosen-deRoo, E. Pees, A. A. N. van Brussel, and B. J. J. Lugtenberg, *J. Bacteriol.*, 1989, **171**, 4045.
116. G. V. Bloemberg, E. Kamst, M. Harteveld, K. M. G. M. van der Drift, J. Haverkamp, J. E. Thomas-Oates, B. J. J. Lugtenberg, and H. P. Spaink, *Mol. Microbiol.*, 1995, **16**, 1123.
117. T. Ritsema, O. Geiger, P. van Dillewijn, B. J. J. Lugtenberg, and H. P. Spaink, *J. Bacteriol.*, 1994, **176**, 7740.
118. H. Therisod, A. C. Weissborn, and E. P. Kennedy, *Proc. Natl. Acad. Sci. USA*, 1986, **83**, 7236.

119. H. Therisod and E. P. Kennedy, *Proc. Natl. Acad. Sci. USA*, 1987, **84**, 8235.
120. A. Weisborn, unpublished results.
121. R. Ghose, O. Geiger, and J. H. Prestegard, *FEBS Lett.*, 1996, **388**, 66.
122. T. A. Holak, S. K. Kearsley, Y. Kim, and J. H. Prestegard, *Biochemistry*, 1988, **27**, 6135.
123. Y. Kim and J. H. Prestegard, *Biochemistry*, 1989, **28**, 8792.
124. T. Ritsema and H. P. Spaink, unpublished results.
125. T. Ritsema, A. H. M. Wijfjes, B. J. J. Lugtenberg, and H. P. Spaink, *Mol. Gen. Genet.*, 1996, **251**, 44.
126. F. Debellé, C. Plazanet, P. Roche, C. Pujol, A. Savagnac, C. Rosenberg, J.-C. Promé, and J. Dénarié, *Mol. Microbiol.*, 1996, **22**, 303.
127. E. Kamst, J. Pilling, L. M. Raamsdonk, B. J. J. Lugtenberg, and H. P. Spaink, *J. Bacteriol.*, 1997, **179**, 2103.
128. F. Debellé, P. Roche, C. Plazanet, F. Maillet, C. Pujol, *et al.*, in "Nitrogen Fixation: Fundamentals and Applications," eds. I. A. Tikhonovich, N. A. Provorov, V. I. Romanov, and W. E. Newton, Kluwer Academic, Dordrecht, 1995, p. 275.
129. P. Mergaert, W. D'Haeze, D. Geelen, D. Promé, M. Van Montagu, R. Geremia, and J.-C. Promé, *J. Biol. Chem.*, 1995, **270**, 29 217.

1.14
Biosynthesis of 6-Methylsalicylic Acid

PETER M. SHOOLINGIN-JORDAN and IAIN D. G. CAMPUZANO
University of Southampton, UK

1.14.1 INTRODUCTION

The biosynthesis of the aromatic ring is accomplished by two very different pathways in biological systems. The shikimate pathway is the major route for aromatic amino acid biosynthesis, the carbon skeleton of shikimate being formed from phosphoenolpyruvate and erythrose 4-phosphate via 3-deoxy-D-*arabino*-heptulosonic acid 7-phosphate[1] (see Chapter 1.24). Shikimate is then transformed in several stages to chorismic acid (Scheme 1) and thence into L-tryptophan or, via prephenate, is converted to L-tyrosine and L-phenylalanine. These amino acids, in addition to their role in protein synthesis, are central building blocks for a variety of secondary metabolites.[1] A related pathway also exists for the biosynthesis of aromatic amines.[2]

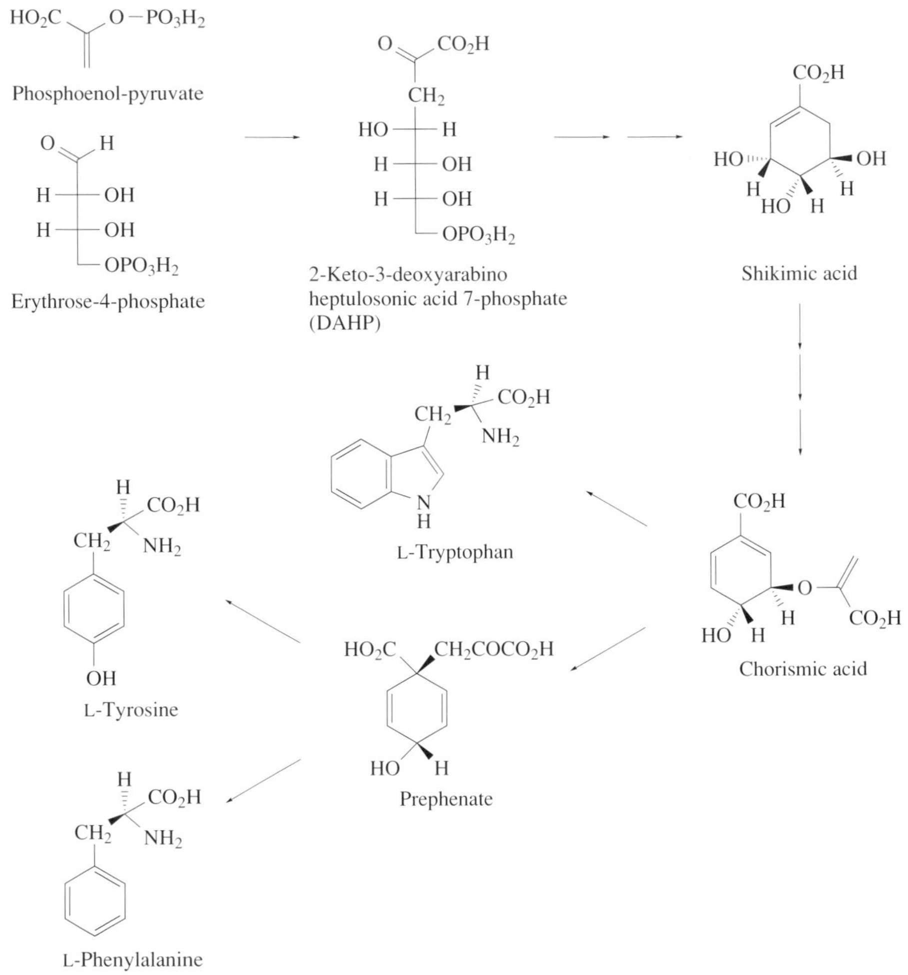

Scheme 1

The other major pathway for the biosynthesis of the aromatic ring involves the participation of polyketide intermediates. Similar to fatty acid synthase, polyketide chain assembly employs the use of acetyl-CoA as a "starter" unit and malonyl-CoA as a chain "extender." However, unlike fatty acid biosynthesis, where the chain is fully reduced during each cycle, in polyketide biosynthesis a variety of permutations are possible, ranging through all the possible intermediates seen in the fatty acid synthase cycle. One of the simplest aromatic compounds, orsellinic acid, is synthesized from a tetraketide intermediate arising from one acetyl-CoA molecule and three malonyl-CoA extender units which, on cylization, dehydration, and enolization results in the aromatic ring product. The closely related 6-methylsalicylic acid is produced similarly, except that a single reduction step with NADPH and dehydration to a *cis*-double bond occurs prior to the condensation with the third malonyl-CoA (Scheme 2). In such polyketide biosynthesis pathways, no free intermediates are formed, all the reactions occurring exclusively through enzyme-bound species.[3,4]

6-Methylsalicylic acid is one of the simplest polyketide-derived metabolites and is the first identifiable product of the enzyme 6-methylsalicylic acid synthase en route to the secondary metabolite, patulin[5] (Scheme 3). Patulin is produced by a range of fungi[6] and is the toxic principle responsible for apple spoilage in cider manufacture. It is of some importance as an antibiotic in veterinary medicine.[7]

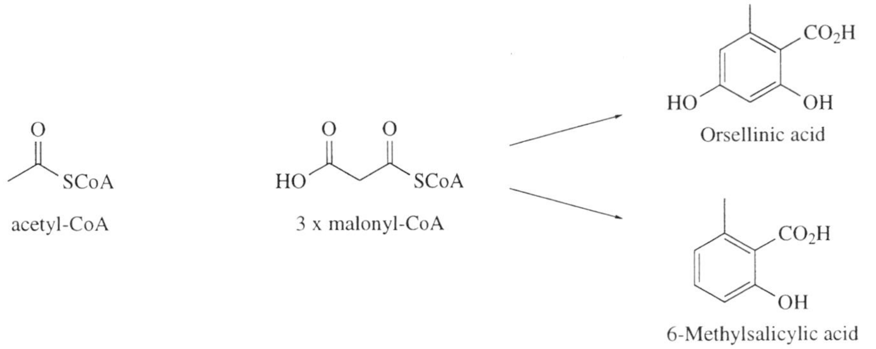

acetyl-CoA

3 x malonyl-CoA

Orsellinic acid

6-Methylsalicylic acid

Scheme 2

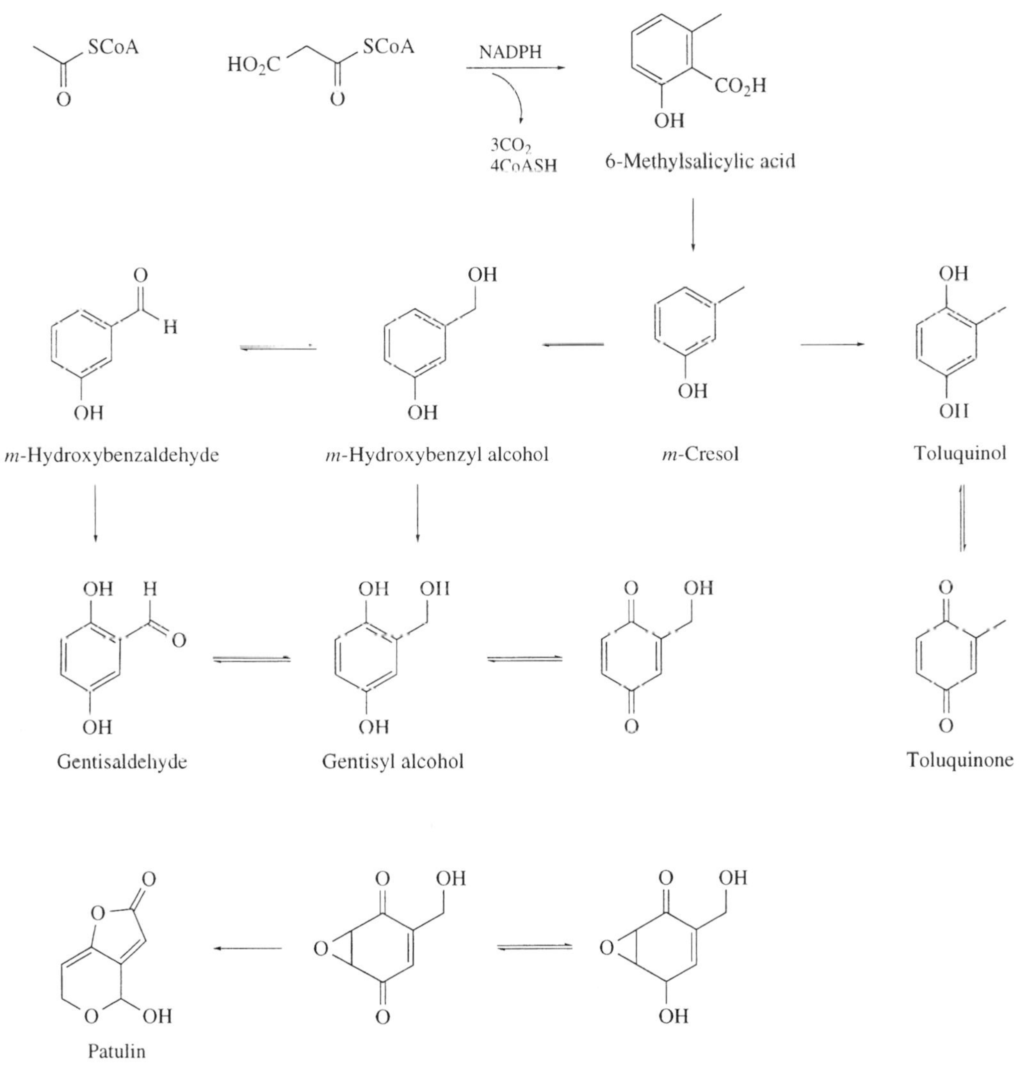

NADPH

$3CO_2$
4CoASH

6-Methylsalicylic acid

m-Hydroxybenzaldehyde

m-Hydroxybenzyl alcohol

m-Cresol

Toluquinol

Gentisaldehyde

Gentisyl alcohol

Toluquinone

Patulin

Scheme 3

1.14.2 FATTY ACID SYNTHASES AS MODELS FOR POLYKETIDE SYNTHASE

The striking similarities in the size, properties, and substrate specificity between vertebrate fatty acid synthases (type I) and polyketide synthases, such as 6-methylsalicylic acid synthase, have been appreciated since the early 1970s.[8] Whereas the equivalent enzymes from bacteria and plants (type

II) are individual enzymes, the eukaryote enzyme families are large multifunctional proteins with subunit M_r values of approximately 200 000. An intermediate class (type I/II) found in yeast, some actinomycetes, and other fungi is composed of two multifunctional proteins.[4] An additional class of enzymes is the modular type of polyketide synthases (type III) such as 6-deoxyerythronolide B synthase from *Saccharopolyspora erythraea* that catalyze the initial stages of erythromycin biosynthesis[9–11] (see Chapter 1.20). These giant enzymes exist as modules, each resembling type I fatty acid synthases but, in contrast to 6-methylsalicylic acid synthase, each module only catalyzes a single reaction cycle. Such polyketide synthases are termed processive rather than iterative enzymes. The way that structural differences in the proteins determine these two types of "programming" strategies (processive or iterative) is one of the most interesting aspects of polyketide synthase investigation.

The similarities between fatty acid synthases and polyketide synthases have been further highlighted by a comparison between the amino acid sequences derived from cDNAs specifying the two enzyme families. These have permitted the alignment of protein components deduced from the nucleotide sequences for rat and yeast fatty acid synthases and 6-methylsalicylic acid synthase[12–14] (Figure 1) confirming the data obtained from enzymology experiments (see below).

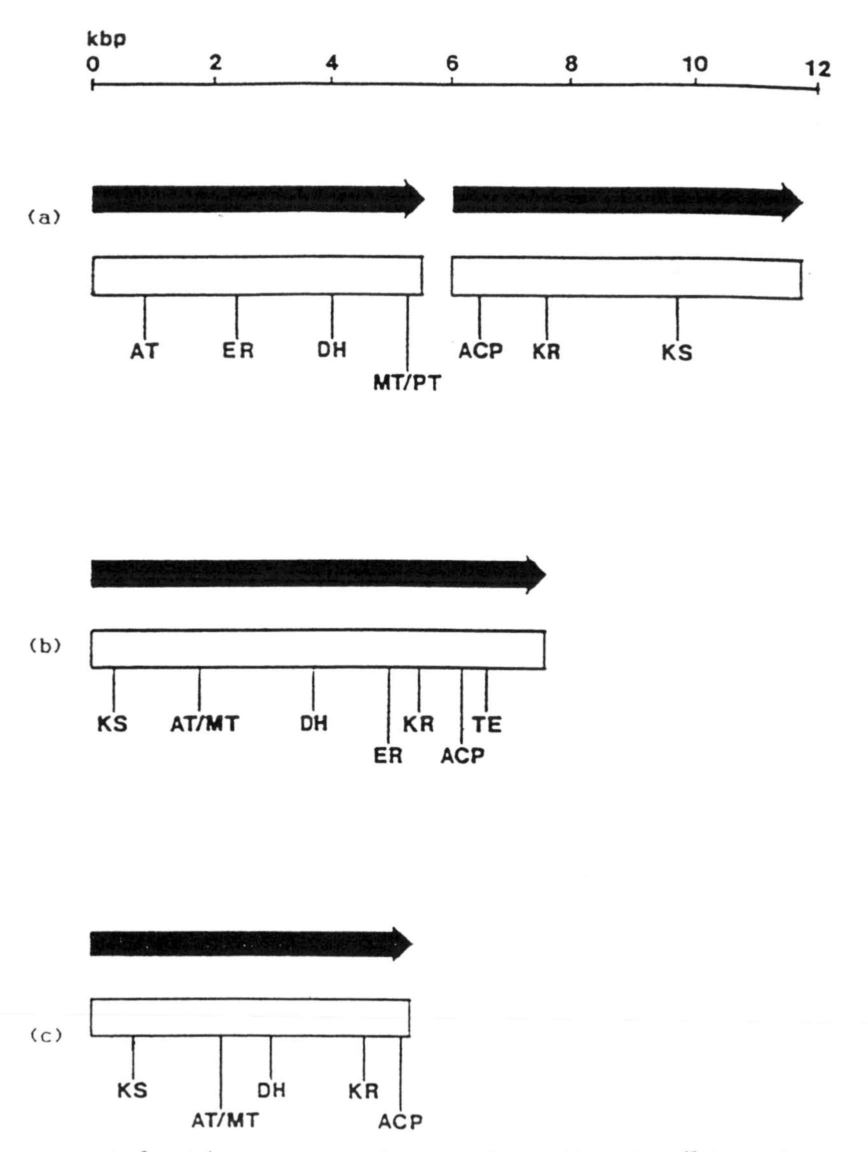

Figure 1 Arrangement of protein components of (a) yeast fatty acid synthase,[12] (b) rat fatty acid synthase,[13] and (c) 6-methylsalicylic acid synthase from *Penicillium patulum*[14] derived from nucleotide sequences.

The well-studied vertebrate fatty acid synthases from rat[15] and chicken[16] have provided a model for 6-methylsalicylic acid synthase. These iterative enzymes utilize one set of catalytic machinery several times during product synthesis. Type I fatty acid synthases contain six enzyme activities:

β-ketoacyl synthase, acyl transferase, β-ketoacyl reductase, β-hydroxyacyl dehydratase, enoyl reductase, thioesterase, and anacyl carrier protein. Of these, the components most essential for polyketide synthases are the β-ketoacyl synthase (condensing enzyme), acyl transferase, the acyl carrier protein (ACP), and an as yet uncharacterized protein component presumed to be important for the structural integrity of the complex. Comparisons between fatty acid synthases and 6-methylsalicylic acid synthase, although of value, need to be exercised with caution since the enzymes have major functional differences that may not be reflected in overall structural identity.

1.14.3 ISOLATION AND PROPERTIES OF 6-METHYLSALICYLIC ACID SYNTHASE

The polyketide synthases are generally less well understood than their fatty acid synthase counterparts, since they are less accessible in large amounts from native sources and suffer from extreme instability during, and following, the isolation process due to proteolytic digestion. The most abundant source of 6-methylsalicylic acid synthase is the mycelia of the ascomycete, *Penicillium patulum* (NRLL 2159A) and it is from this organism that the enzyme has been most studied. The first attempts at isolation were accomplished by a combination of ammonium sulfate precipitation and sucrose density centrifugation leading to enzymes ranging in purity and specific activity.[8,14,17] More recently the enzyme has been purified to homogeneity by PEG (polyethylene glycol) precipitation and DEAE (diethylaminoethyl) sepharose chromatography.[18] The related orsellinic acid synthase has also been purified by a similar method from *Penicillium cyclopium*.[19]

The major difficulty in obtaining the fully active enzyme is due to its instability during isolation. Much of this stems from partial proteolytic digestion during mycelial growth that, while not always affecting the enzyme activity in crude extracts, results in extreme lability during ion exchange chromatography when the structure of the nicked enzyme is further disrupted by the ionic forces of the chromatographic process. This difficulty has been largely overcome by using an inoculum directly from spores which germinate to generate a synchronous culture with cells all approximately of the same age. Cells are then grown in a submerged culture, with vigorous agitation to prevent clumping, over a prescribed time before the 6-methylsalicylic acid levels have reached their optimum levels and before the enzyme has been proteolytically nicked (Figure 2). To minimize proteolytic degradation during purification, cells are disrupted in the presence of the proteolytic inhibitors phenylmethylsulfonylfluoride and benzamidine. Exogenously added thiols and EDTA prevent oxidation of the sensitive cysteine and 4′-phosphopantetheine -SH groups. The purification is carried out using ammonium sulfate precipitation, polyethylene glycol (PEG 6000) fractionation, and DEAE-sepharose ion exchange chromatography.[18] The procedure yields both 6-methylsalicylic acid synthase and fatty acid synthase that can be separated efficiently from one another by hydroxyapatite ion exchange chromatography, the 6-methylsalicylic acid synthase eluting at lower concentrations of phosphate than the fatty acid synthase.

Polyacrylamide gel electrophoresis (PAGE) in the presence of sodium dodecylsulfate has confirmed that the monomer of the enzyme has an M_r value of about 190 000.[18] This is close to the value predicted from the cDNA sequence.[14] It has since been possible to determine the M_r of the 6-methylsalicylic acid synthase subunit directly by mass spectrometry, this technique giving a value of 191 159,[20] close to that predicted from the nucleotide sequence.[14] Gel filtration experiments with purified 6-methylsalicylic acid synthase have established that the enzyme exists as a tetramer.[18] Orsellinic acid synthase has a subunit M_r of 110 k and also appears to exist as a tetramer.

1.14.3.1 Assay of 6-Methylsalicylic Acid Synthase

Because of the presence of fatty acid synthase, which also uses the substrates acetyl-CoA and malonyl-CoA, 6-methylsalicylic acid synthase cannot reliably be assayed by following the oxidation of NADPH except in homogeneous enzyme preparations. The independent assay of 6-methylsalicylic acid synthase was originally achieved by a radiochemical method involving the incorporation of either [^{14}C]acetyl-CoA or [^{14}C]malonyl-CoA into the polyketide product.[8] This proved to be time-consuming during routine purification and was superseded by a novel fluorometric assay in which the fluorescence of the 6-methylsalicylic acid formed is enhanced approximately 30-fold by the inclusion of bovine serum albumin (BSA) in the assay.[21]

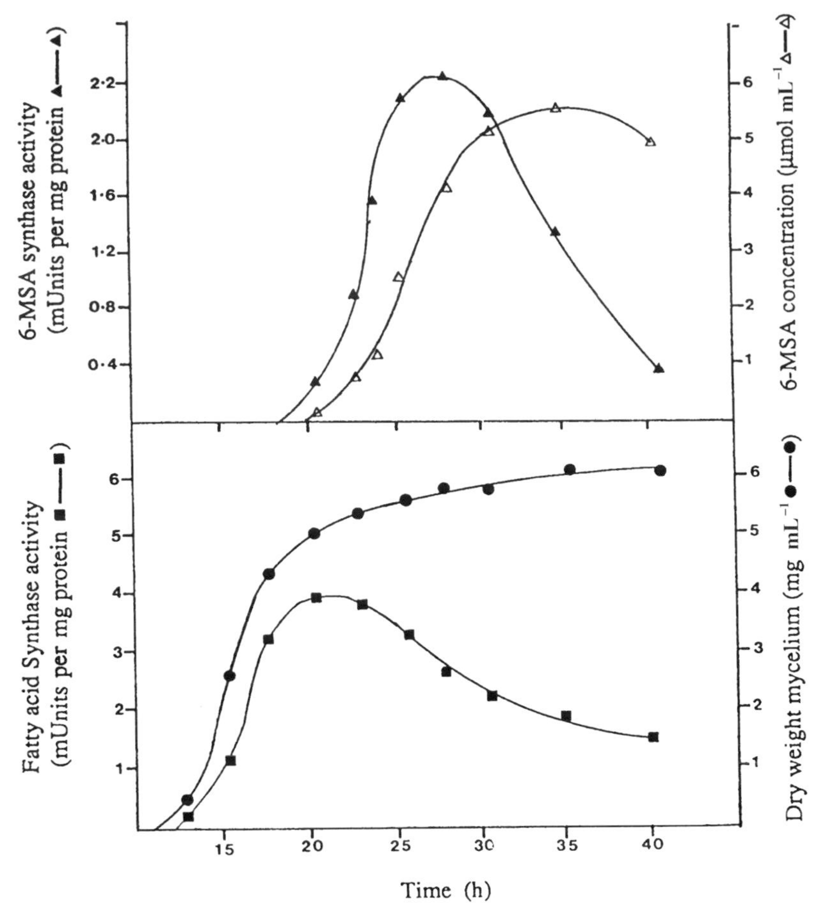

Figure 2 Culture growth characteristics and time courses for the appearance of 6-methylsalicylic acid synthase, 6-methylsalicylic acid, and fatty acid synthase in batch cultures of *Penicillium patulum*.

1.14.3.2 Structure of 6-Methylsalicylic Acid Synthase

In the absence of a complete crystal structure for any polyketide synthase or type I fatty acid synthase, the structure of 6-methylsalicylic acid synthase has been inferred from a range of experimental approaches, many having been carried out with vertebrate fatty acid synthases. These include limited proteolytic analysis, chemical cross-linking, and comparative studies with related enzymes. The nucleotide sequences have also been essential for confirming the precise order of the individual enzyme activities within the multifunctional protein.

1.14.3.2.1 Limited proteolytic digestion

Limited proteolytic digestion studies with 6-methylsalicylic acid synthase from *Penicillium patulum* have been modeled on the earlier experiments with vertebrate fatty acid synthases. Proteolytic digestion of the chicken enzyme[15] (see also Figure 3) resulted in three protein fragments (erroneously called domains), one comprising the β-ketoacyl synthase and acyl transferase, a second containing the two reductases, the dehydratase and the acyl carrier protein, and the third comprising the thioesterase. The M_r values for these three fragments were 127 kDa, 107 kDa and 33 kDa, respectively. The three fragments have been further degraded by proteolysis to the individual components of the multifunctional protein[15,16,22–25]

A similar proteolytic analysis of 6-methylsalicylic acid synthase by limited proteolysis using trypsin and V8 protease has also been employed.[26] Trypsin inactivates 6-methylsalicylic acid synthase rapidly by cleaving initially a C-terminal fragment containing the acyl carrier protein domain. Other proteolytically susceptible sites are found in the β-ketoacyl synthase and β-ketoacyl reductase regions and further trypsin digestion leaves a large proteolytically resistant fragment containing the dehydratase and acyl transferase components. The primary cleavage sites correspond closely to

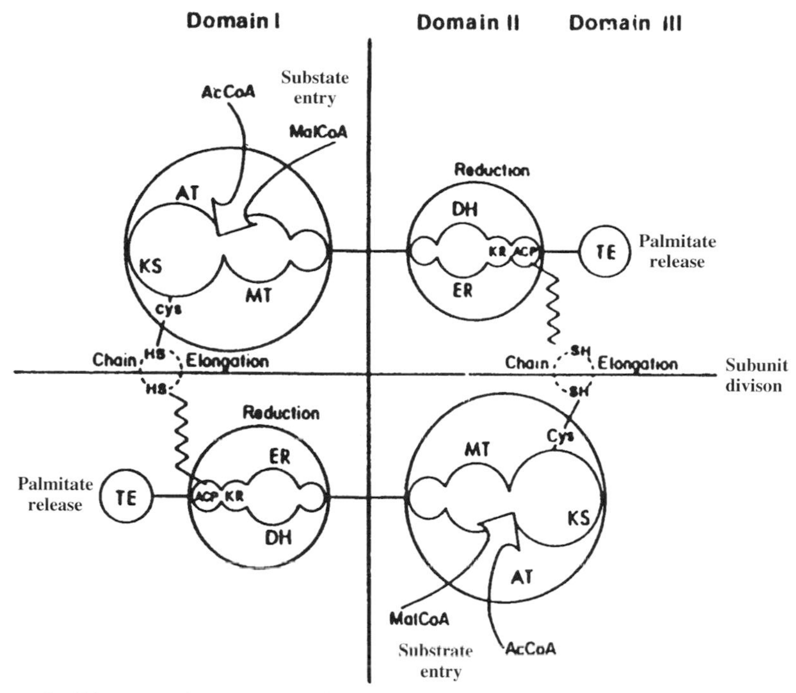

Figure 3 Diagramatic representation of vertebrate fatty acid synthase (after Wakil[15]).

predicted "linker" sites between enzyme moieties (Figure 4) and follow a pattern not only reminiscent of type I fatty acid synthases[15] but also similar to those found in proteolytic studies with the modular polyketide synthase like 6-deoxyerythronolide B synthase.[27] Together with the other similarities in gene/cDNA structure and mechanism of action, it is clear that polyketide synthases and fatty acid synthases arose from a common evolutionary ancestor.

1.14.3.2.2 Protein cross-linking studies

Cross-linking with the bifunctional reagent 1,3-dibromopropan-2-one has provided some of the most valuable insight into the arrangement of subunits within the 6-methylsalicylic acid synthase tetramer.[18] Cross-linking of enzyme, subunit M_r = 190 kDa, yields dimers of M_r = 360 kDa with an additional band at a higher M_r (Figure 5). This additional band is likely to be a singly cross-linked species.[28] Addition of acetyl-CoA or malonyl-CoA protects the enzyme from cross-linking. Acetyl-CoA prevents the reagent from reacting with both the β-ketoacyl synthase and 4′-phosphopantetheine thiols and protects the enzyme from inactivation. On the other hand, malonyl-CoA prevents cross-linking but does not prevent inactivation. The study suggests that acetyl-CoA protects the highly reactive cysteine of the β-ketoacyl synthase moiety and that malonyl-CoA specifically protects the thiol of the 4′-phosphopantetheine group of acyl carrier protein. It may be concluded that dibromopropan-2-one reacts initially with the β-ketoacylsynthase thiol followed by subsequent reaction with the acyl carrier protein. The reactions are summarized in Figure 6.

The results from these experiments closely resemble the findings with type I fatty acid synthases[29] which led to the proposal that two subunits of 6-methylsalicylic acid synthase are also arranged in a "head-to-tail" fashion and that each active site utilizes component activities from two subunits in a functional dimer[18] (Figure 7). This model may need to be modified as a result of elegant complementation experiments using rat fatty acid synthase. In these experiments one of the subunits within the dimer was subjected to mutagenesis and the effects on overall activity were determined.[30] The results suggest that the some of the enzymic components from the same subunit may be able to interact with one another.

This "head-to-tail" model has also been challenged and an alternative "head-to-head" structure has been proposed based on the observation that proteolytically digested 6-deoxyerythronolide B synthase (DEBS) yields dimers of its components.[31] Although it is tempting to relate all polyketide synthases to fatty acid synthases, it is possible that iterative enzymes such as 6-methylsalicylic acid

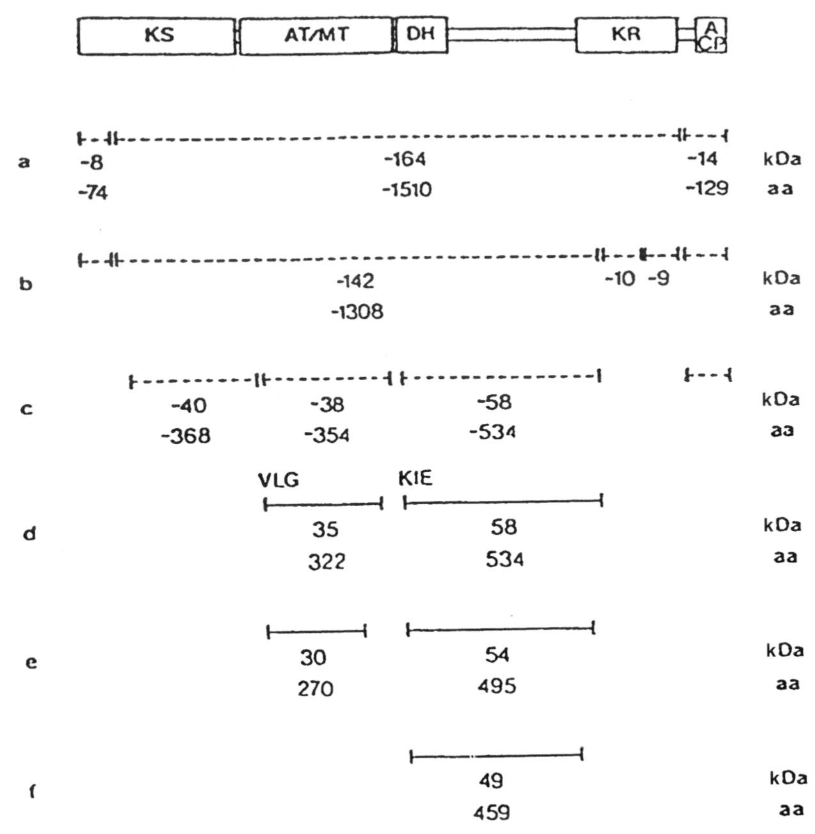

Figure 4 Partial peptide map of 6-methylsalicylic acid synthase digested with trypsin.[26] Peptides formed transiently within 60 s are shown as hatched lines (a, b, and c). The major peptide species commence with the amino acids VLG (35 kDa) and KIE (58 kDa). After 2 min further degradation has occurred (d and e). After 5 min one major peptide remains containing DH and the putative structural protein core (f). After 15 min part of terminus KIE (KS, β-ketoacyl synthase; AT/MT, acyl transferase; DH, dehydratase; KR, β-ketoacyl reductase; ACP, acyl carrier protein). M_r values of peptides are shown with the number of amino acids below.

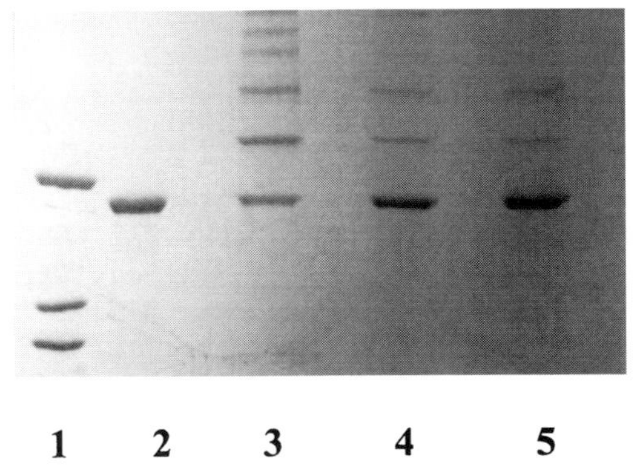

Figure 5 Cross-linking of 6-methylsalicylic acid synthase with the bifunctional reagent 1,3-dibromopropan-2-one. Lane 1, molecular weight markers (myosin, M_r 200 kDa; β-galctosidase, M_r 116 kDa; phosphorylase b, M_r 97 kDa). Lane 2, native 6-methylsalicylic acid synthase. Lane 3, 6-methylsalicylic acid synthase treated with 1,3-dibromopropan-2-one. Lane 4, 6-methylsalicylic acid synthase preincubated with acetyl-CoA and treated with 1,3-dibromopropan-2-one. Lane 5, 6-methylsalicylic acid synthase preincubated with malonyl-CoA and treated with 1,3-dibromopropan-2-one.

synthase and fatty acid synthases share a different topology to the processive enzymes like DEBS and that there are different topological arrangements. Further research is necessary to determine whether a consensus topology is present in all fatty acid synthases and polyketide synthases.

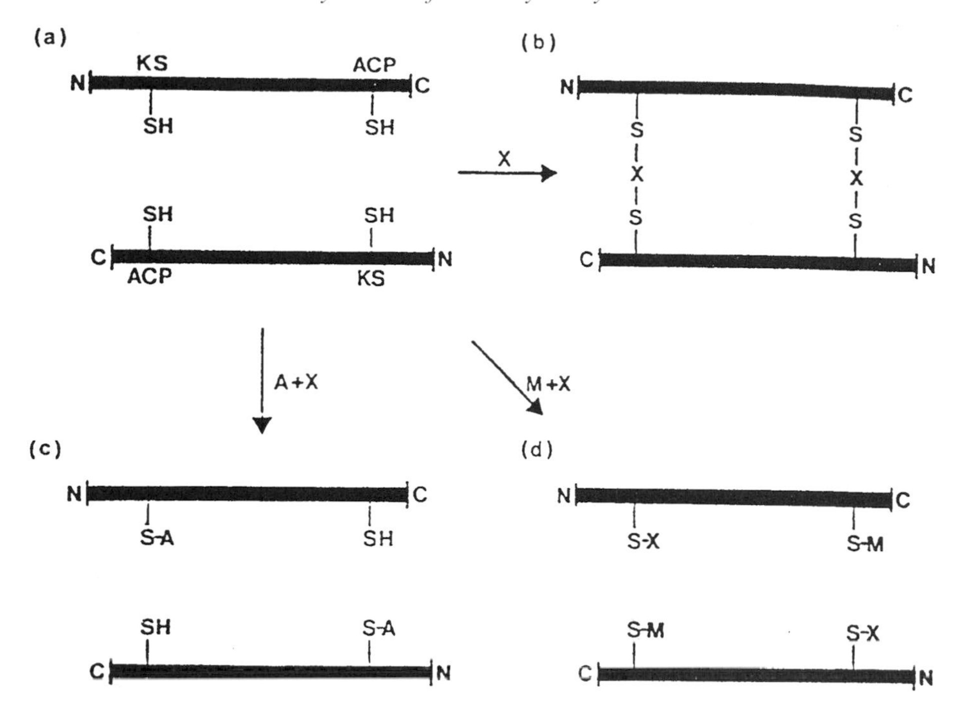

Figure 6 Hypothetical scheme for the cross-linking of 6-methylsalicylic acid synthase with 1,3-dibromopropan-2-one (KS, β-ketoacyl synthase; ACP, acyl carrier protein; A, acetyl-CoA; M, malonyl-CoA and X, 1,3-dibromopropan-2-one). (a) The functional dimer; (b) cross-linked nonactive; (c) non-cross-linked and active; and (d) non-cross-linked and inactive.[18]

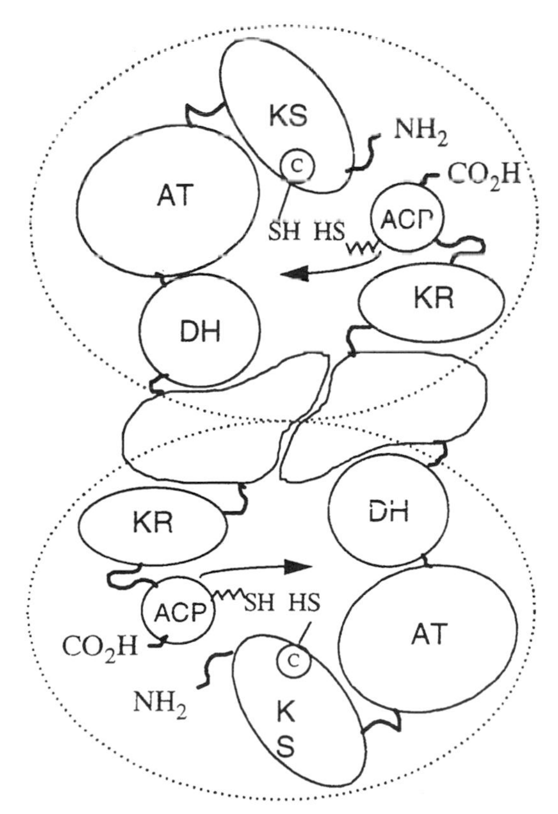

Figure 7 Hypothetical "head-to-tail" arrangement of the functional dimer of 6-methylsalicylic acid synthase.

1.14.4 THE BIOSYNTHETIC PATHWAY FOR 6-METHYLSALICYLIC ACID AND ITS RELATIONSHIP TO FATTY ACID BIOSYNTHESIS

The biosynthesis of fatty acids catalyzed by type I fatty acid synthases is achieved by seven sequential cycles involving one molecule of acetyl-CoA and seven molecules of malonyl-CoA (Scheme 4).

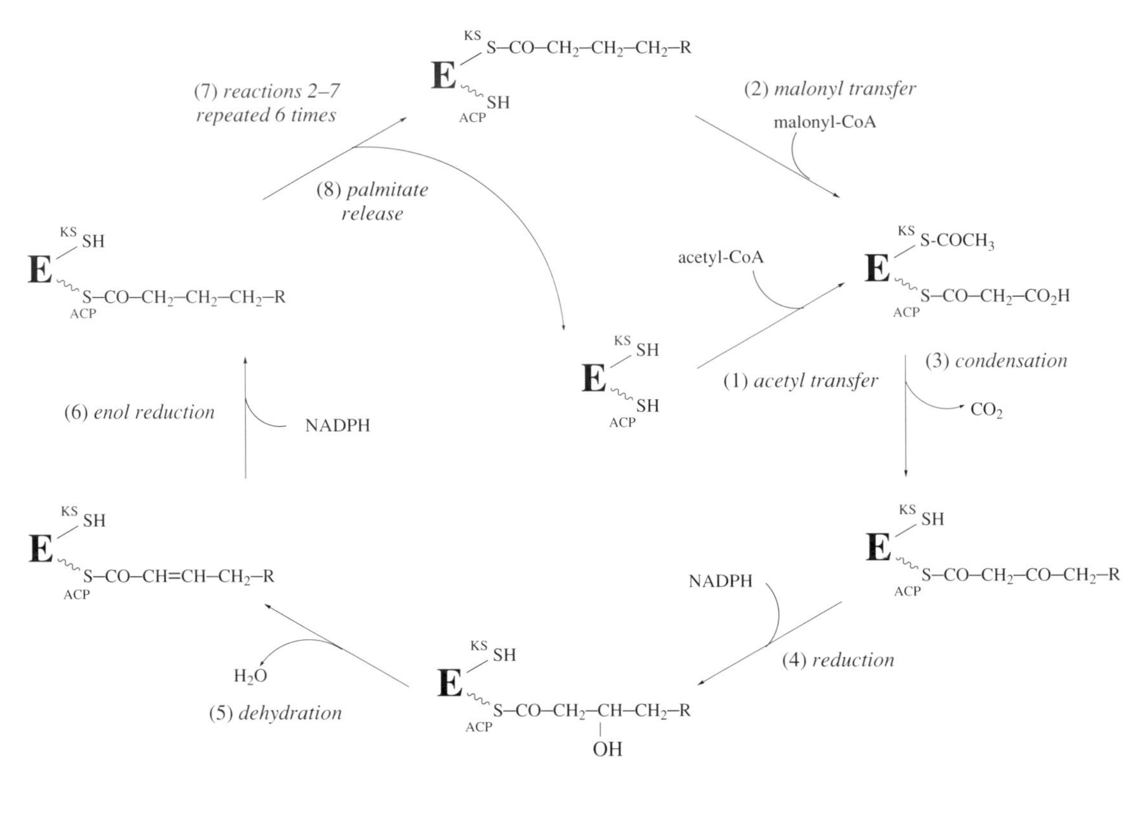

Scheme 4

The first cycle is initiated by the covalent attachment of a "starter unit," acetyl-CoA (or an alternative acyl-CoA), to a reactive cysteine-SH group of β-ketoacyl synthase with the elimination of CoA. The first malonyl "extender" unit is then transferred from malonyl-CoA to the 4′-phosphopantetheine group of ACP and a Claisen condensation occurs resulting in the formation of 3-ketobutyryl-ACP (acetoacetyl-ACP) and the concomitant loss of CO_2.

Whereas in fatty acid biosynthesis the condensation of each malonyl "extender" unit is followed by β-keto-reduction, dehydration, and enoyl-reduction to form a fully reduced methylene chain (Scheme 4), in the reaction catalyzed by 6-methylsalicylic acid synthase (Scheme 5), condensation of the second malonyl extender unit follows immediately to generate an enzyme-bound 3,5-diketohexanoyl–enzyme triketide C-6 intermediate. At this stage the reduction reaction with NADPH leads to the formation of a 3-hydroxy-5-ketohexanoyl–enzyme intermediate. Reduction and dehydration appear to be prerequisites for condensation with the third and final malonyl extender which is then followed by cyclization, dehydration, and enolization. If NADPH is not available then the 3,5-diketohexanoyl enzyme-bound triketide C-6 intermediate cyclizes and is eliminated from the enzyme as triacetic acid lactone (Scheme 6).

The timing of the reduction step at the C-6 polyketide stage has been inferred from inhibition studies with 6-methylsalicylic acid from *Penicillium patulum* in which 3-pentynoyl-*N*-acetyl-cysteamine completely inhibited product formation.[17] This inhibitor was designed to react with any dehydratase activity, the rationale being that the acetylenic bond would form a reactive allene and alkylation of any nearby active site base. The inhibitor was active at a concentration of 10 mM inactivating the enzyme; however, the initial rate of NADPH oxidation was only reduced to 60% of the uninhibited rate. These findings suggested that the inhibitor was acting after the reduction and modifying specifically a reactive histidine at the active site of the dehydratase (Scheme 7).

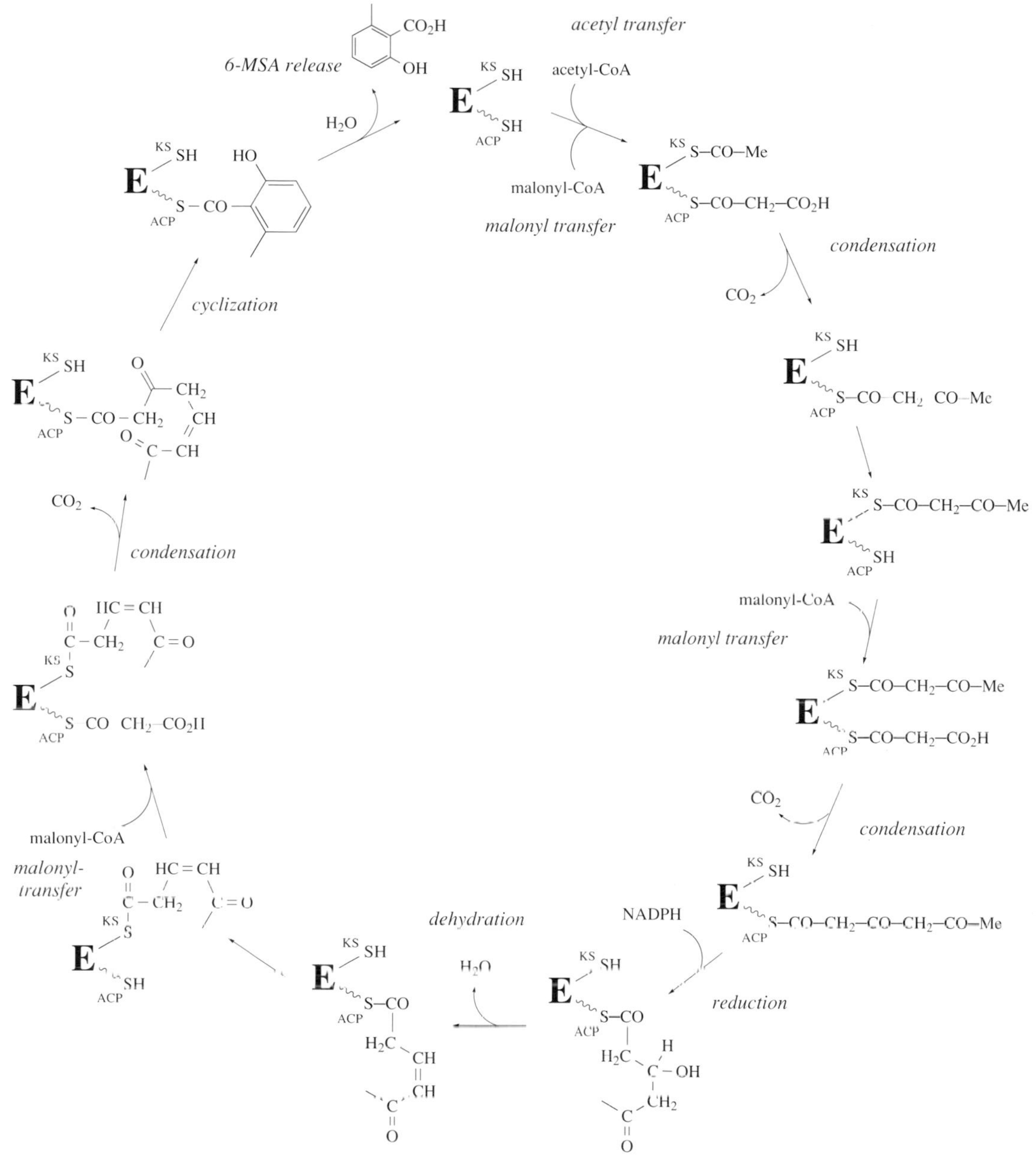

Scheme 5

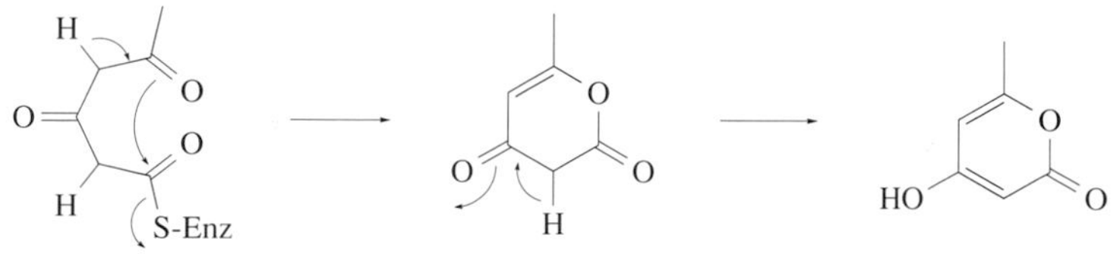

Scheme 6

1.14.5 THE NUCLEOTIDE SEQUENCE OF 6-METHYLSALICYLIC ACID SYNTHASE

One of the most important advances in the understanding of polyketide synthases has come from the identification and sequencing of the oligonucleotides specifying a number of these enzymes. The covalent attachment and inhibition of enzyme

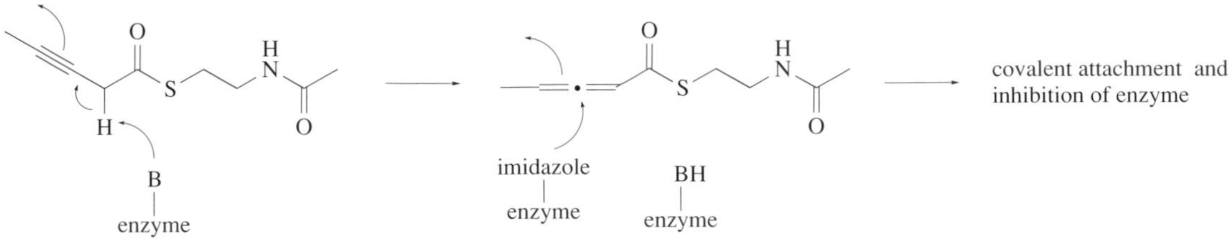

imidazole
enzyme
BH
enzyme
B
enzyme

Scheme 7

deduced amino acid sequences of the proteins have reinforced the claims from enzymic studies that the polyketide synthases are closely related to fatty acid synthases.

In this respect, within the cDNA specifying 6-methylsalicylic acid synthase,[4] six regions are apparent (see Figure 1): encoding β-ketoacyl synthase, malonyl transferase, β-hydroxyacyl dehydratase, β-ketoacyl reductase, and acyl carrier protein. Enoyl reductase is conspicuous by its absence consistent with the incompletely reduced nature of the enzymic product. The absence of a thioesterase could suggest the operation of a mechanism other than hydrolysis (Scheme 8(a)) for release of the 6-methylsalicylic acid product from the enzyme, involving a ketene intermediate (Scheme 8(b)). This could be achieved by the involvement of the adjacent hydroxyl group and hydration of the resulting ketene to the acid.

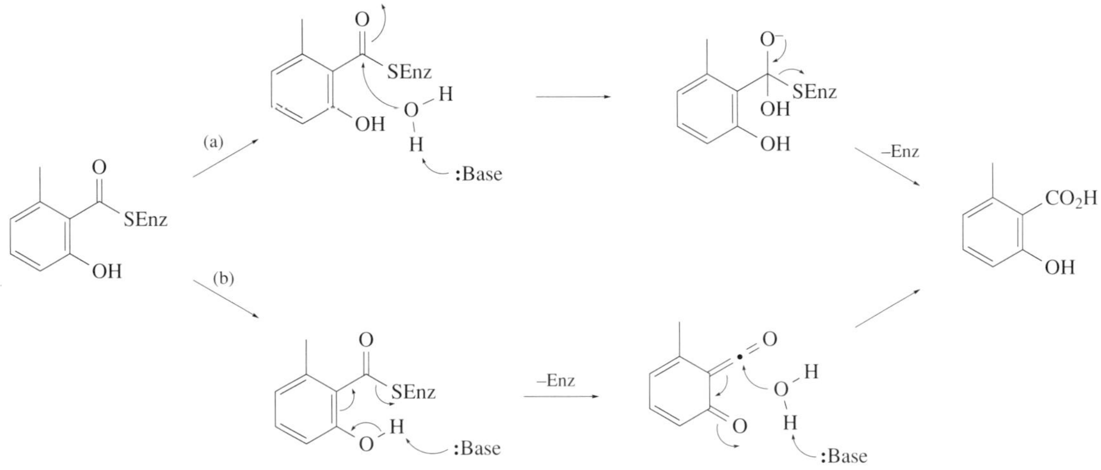

SEnz OH (a) (b) :Base –Enz CO_2H

Scheme 8

Present in the deduced protein sequence of the 6-methylsalicylic acid synthase enzyme is a region of amino acids that does not appear to encode any known protein function and which may play a structural rather than catalytic role. A comparison of the cDNA sequences specifying 6-methylsalicylic acid synthase with the fatty acid synthases from rat and *Penicillium patulum*[14] indicate that 6-methylsalicylic acid synthase is more closely related to the mammalian type I enzymes than to its own fatty acid synthase.[32]

1.14.6 SUBSTRATE SPECIFICITY OF 6-METHYLSALICYLIC ACID SYNTHASE

On incubation of 6-methylsalicylic acid synthase with alternative starter substrate units in the presence of NADPH, the corresponding 6-alkylsalicylic acid derivatives are generated[20] (Table 1). The starter units accepted by the enzyme were acetyl-CoA, acetoacetyl-CoA, propionyl-CoA, butyryl-CoA, β-hydroxybutyryl-CoA, isobutyryl-CoA (C-4), crotonoyl-CoA (unsaturated), valeryl-CoA, hexanoyl-CoA, and heptanoyl-CoA. There proved to be an inverse relationship between the carbon chain length of the starter unit and the level of incorporation into the final product. However, the branched starter units, β-hydroxybutyryl-CoA and isobutyryl-CoA, were not utilized as potential starter units. The results of these studies suggest that the acyl-transferase and keto-synthase components of 6-methylsalicylic acid synthase are only loosely selective, allowing all substrates used to bind to the condensing site and to be subsequently utilized at a relatively high rate for the synthesis of 6-alkylsalicylic acid derivatives.

Table 1 Enzymic synthesis of 6-alkylsalicylic acids and triketide lactones from various acyl-CoA starters by 6-methylsalicylic acid synthase.

Analogue CoA ester	*6-Alkylsalicylic acids*		*Triacetic acid lactones*	
	% of acetyl-CoA	*Observed mass*	*% of acetyl-CoA*	*Observed mass*
Acetate	100.00	150.81	100.00	126.71
Acetoacetate	79.63	150.82	81.96	126.72
Propionate	35.78	164.88	33.14	140.76
Butyrate	22.24	178.92	25.29	154.79
Valerate	16.01	192.98	23.76	168.83
Crotonoate	12.37	177.92	11.76	152.77
Hexanoate	14.82	207.01	15.26	182.82
Heptanoate	4.14	221.12	6.01	196.89
β-Hydroxybutyrate				
Isobutyrate				

Yields of [^{14}C]-labeled 6-alkylsalicylic acid derivatives and [^{14}C]-labeled triacetic acid lactone derivatives produced by 6-methylsalicylic acid synthase using a range of starter substrate analogues in the presence and absence of NADPH, respectively. The amount of 6-alkylsalicylic acid and triacetic acid lactone derivative formed from each analogue was determined by radiochemical assay and related to the observed enzymic rate with acetyl-CoA as starter. Masses were determined on a Micromass Quattro II triple quadrapole electrospray mass spectrometer.

6-Methylsalicylic acid synthase is highly selective for its extender unit, malonyl-CoA. Methylmalonyl-CoA, a substrate for several other polyketide synthases, is not accepted by 6-methylsalicylic acid synthase. The specificity is assigned to the β-ketoacyl synthase enzyme component.

1.14.6.1 Triacetic Acid Lactone Formation

In the absence of NADPH, 6-methylsalicylic acid synthase produces triacetic acid lactone as its major product arising from one molecule of acetyl-CoA and two molecules of malonyl-CoA.[8,17] The lactone is produced at approximately 10% of the rate of 6-methylsalicylic acid synthesis. The addition of increasing amounts of NADPH results in the progressive formation of 6-methylsalicylic acid so that at a concentration of 500 nM NADPH the ratio of 6-methylsalicylic acid:triacetic acid lactone is 5:1. At the K_m concentration of NADPH (12 μM) this rises to 99:1, highlighting the probable importance[18,33,34] of the reduction step (and the dehydratase step) in the programming of subsequent stages in the reaction.

Triketide lactones are also formed by genetically engineered polyketide synthases. For instance, the first two modules of 6-deoxyerythronolide B synthase (DEBS), DEBS 1, with the terminal thioesterase (TE) from module 3 linked to form a DEBS1 + TE system, have been shown to produce a triketide lactone, (2*S*,3*S*,4*S*,5*R*)-2,4-dimethyl-3,5-dihydroxy-*n*-heptanoic acid δ-lactone from a propionyl-CoA starter and two molecules of methylmalonyl-CoA in the presence of NADPH.[35,36] This lactone is also formed in small amounts as a derailment product by the intact DEBS enzyme. In the absence of NADPH, the unreduced C_9-triketide lactone is formed in 20% yield.[37]

As substrates for the formation of triketide lactones, 6-methylsalicylic acid synthase is also able to accept alternative starters to acetyl-CoA with acetoacetyl-CoA incorporation being over 80% of the rate for acetyl-CoA (Table 1). On the other hand, β-hydroxybutyryl-CoA is not accepted as a substrate, although crotonyl-CoA is transformed into the corresponding lactone at 12% of the rate of acetyl-CoA. For saturated starters, increasing chain length decreases the rate of formation of the triketide lactones progressively, from propionyl-CoA at 33% to hepanoyl-CoA at 6% of the acetyl-CoA rate.[20] The enzyme does not recognize branch-chain starters such as isobutyryl-CoA (Table 1).

The engineered DEBS1 + TE protein can also accept alternative starters, yielding a C_8 triketide lactone with acetyl-CoA as a starter and a C_{10} triketide lactone with butyryl-CoA as a starter.[38]

1.14.7 INACTIVATION OF 6-METHYLSALICYLIC ACID BY CERULENIN

The micotoxin cerulenin [(2*S*,3*R*)-2,3-epox-4-oxo-7,10-dodecadienoylamide] produced by *Cephalosporium caerulens*[39] irreversibly inactivates 6-methylsalicylic acid synthase[40] and a range of other polyketide synthases and fatty acid synthases. By the use of [^{3}H]cerulenin the susceptible residue of 6-methylsalicylic acid synthase was shown to be Cys204, a residue with an invariant equivalent in all polyketide and fatty acid synthases.[41] Iodoacetamide was also shown to inactivate 6-methylsalicylic acid synthase by specifically alkylating the same cysteine residue. These findings parallel observations made with type I fatty acid synthases where the analogous cysteine is involved.[15,42]

1.14.8 STEREOCHEMICAL STUDIES ON THE ENZYME MECHANISM

Since 6-methylsalicylic acid is biosynthesized from three molecules of malonyl-CoA, the most valuable mechanistic study is to follow the fate of the methylene hydrogen atoms during the enzyme reaction. Early *in vivo* experiments involving incorporation of deuterated acetate provided the first evidence for the retention of deuterium during the reaction. Thus the labeled acetate produced malonyl-CoA *in vivo* that yielded 6-methylsalicylic acid with deuterium at the 2 and 4 positions[43,44] indicating that the reaction proceeded with some degree of steric control.

The availability of homogeneous 6-methylsalicylic acid synthase allowed these studies to be extended *in vitro* using synthetic chiral (*R*)-[1-^{13}C;2-^{2}H]malonate and (*S*)-[1-^{13}C;2-^{2}H]malonate,[45] each transformed into a pair of malonyl-CoA derivatives by CoA transferase. The chirality of the malonates was checked by transformation into palmitic acid[46–48] relying on the observations from experiments already carried out with tritiated chiral malonyl-CoA derivatives to determine the steric course of fatty acid synthase.[49]

The experimental strategy to incorporate the paired malonyl-CoA derivatives of each chiral malonate dependend on a mass spectrometric analysis of 6-methylsalicylic acid to determine the labeling pattern at the 2 and 4 positions.[50] Both hydrogen atoms from the third malonyl-CoA are lost during the synthesis and play no part in the discussion. The incorporation of chiral (*R*)-[1-^{13}C;2-^{2}H]- and (*S*)-[1-^{13}C;2-^{2}H]malonates (as malonyl-CoA) into 6-methylsalicylic acid are best explained at the putative C_6 intermediate stage at which four steric courses are possible, each of which involves the loss of a different pair of hydrogen atoms from the C-2 and C-4 positions of the triketide intermediate. Initial experiments with acetyl-CoA as a starter partially resolved the problem and indicated that the hydrogen atoms at positions C-3 and C-5 of 6-methylsalicylic acid arise from opposite absolute configurations (H_{Re} and H_{Si} or H_{Si} and H_{Re}) of malonyl-CoA.

These two possibilities were resolved by using acetoacetyl-CoA as a starter, so that the incorporation of hydrogen from the chiral malonates into the C-3 position of 6-methylsalicylic acid could be investigated in isolation. This position arises exclusively from the C-2 of the C-6 triketide intermediate with acetoacetyl-CoA as the starter. Mass spectrometry analysis from experiments with the two samples of chiral malonate indicated conclusively that the hydrogen atom at the 3-position of 6-methylsalicylic acid (□) arises from H_{Re} of malonyl-CoA (Scheme 9). It should be noted that the hydrogen occupies the H_{Si} position in the C-6 intermediate due to the inversion of stereochemistry during the condensation reaction. Since the original experiments with acetyl-CoA as a starter[46] demonstrated that the hydrogens at 3- and 5-positions in 6-methylsalicylic acid arise from opposite orientations of malonyl-CoA, it follows that H_{Re} is eliminated from the 2-position of the C-6 polyketide intermediate (△ in Scheme 9) and that H_{Si} is eliminated from the C-4 position of the C-6 polyketide intermediate. Thus, the hydrogen atoms at C-3 and C-5 of 6-methylsalicylic acid arise from H_{Re} and H_{Si}, respectively, of malonyl-CoA.[47,48]

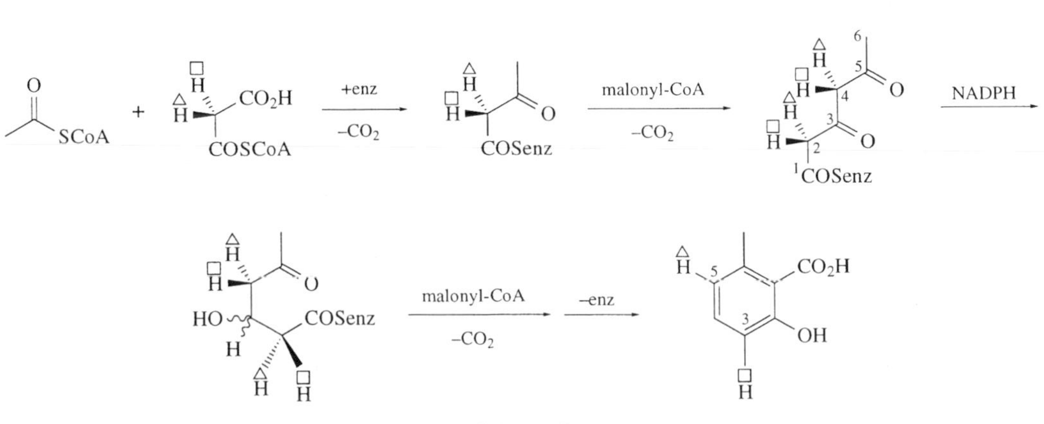

Scheme 9

These findings demonstrate that, as suspected for an enzyme-catalyzed reaction, all the bond forming and breaking events catalyzed by the enzyme occur with strict steric control. These conclusions rely on the not unreasonable assumption that all three Claisen condensations occurring during the mechanism proceed with inversion,[51] consistent with decarboxylation being concerted with C—C bond formation.

1.14.9 MECHANISTIC CONSIDERATIONS FOR THE FORMATION OF 6-METHYLSALICYLIC ACID

Despite the fact that the absolute configuration of the proposed triketide alcohol intermediate is not known, it is productive to speculate on the mechanism of the 6-methylsalicylic acid synthase reaction, taking into account the stereochemical findings. Several mechanisms have been considered, all of which have counterparts in the mechanistic literature. Mechanism A (Scheme 10) is analogous to the reaction catalyzed by β-hydroxyacyl-decanoyl-CoA dehydratase and involves the formation of a *cis*-double bond through initial dehydration followed by allylic rearrangement. This mechanism has been shown to follow a suprafacial process.[52] If a similar mechanistic course is followed in the 6-methylsalicylic acid synthase reaction, then the two hydrogen atoms eliminated are placed on the same face of the molecule and could, in principle, be removed by a single strategically placed enzymic base.

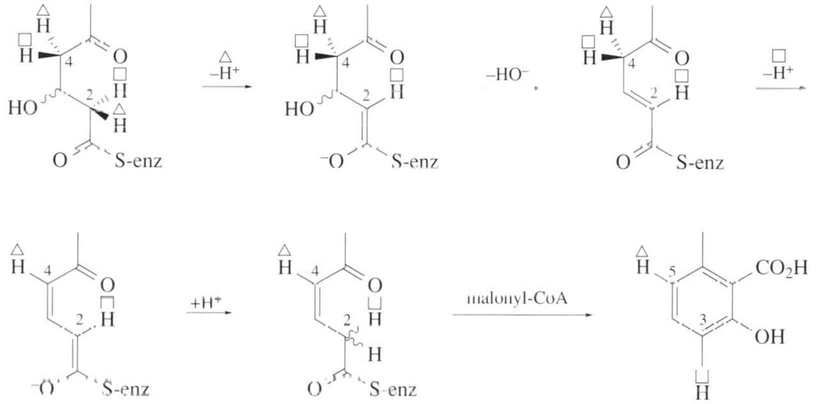

Scheme 10

In mechanism B, the *cis*-double bond could be formed directly by dehydration in an *anti*-elimination process with the loss of H_{Re} (△) from the 2-position. This would require the alcohol at the 3-position to be of *S*-configuration (Scheme 11(a)). Alternatively, dehydration by loss of H_{Si} (□) from the 4-position, as observed, would require an *R*-configuration for the C-3 alcohol if the mechanism followed an *anti*-elimination route (Scheme 11(b)). While a *cis*-elimination is also possible, the precise steric course cannot be determined until the absolute configuration of the alcohol has been established.

An additional possibility, mechanism C (Scheme 12), involves retention of the hydroxyl group until after condensation with the third malonyl-CoA moiety and cyclization. This mechanism has the attraction that a conjugated ketone, produced in the mechanisms shown in Schemes 10 and 11, is not involved, which would make ring closure more facile. The hydrogen atoms would be lost from opposite faces of the ring. This mechanism, however, may not be entirely consistent with the cDNA-derived protein sequence of 6-methylsalicylic acid synthase which indicates the likely presence of a β-hydroxyacyl dehydratase activity.

These findings establish that all bond-forming and bond-breaking events proceed with absolute steric control. It is interesting that the formation of triacetic acid lactone from chiral malonates also appears to occur with stereochemical control,[19] highlighting the fact that elimination of the triketide intermediate from the enzyme as the triacetic acid lactone is also likely to be enzyme catalyzed. The overall mechanism for the formation of 6-methylsalicylic acid, taking into consideration the stereochemical findings from the experiments with chiral maloneates, is shown in Scheme 13, following a route proposed previously.[3]

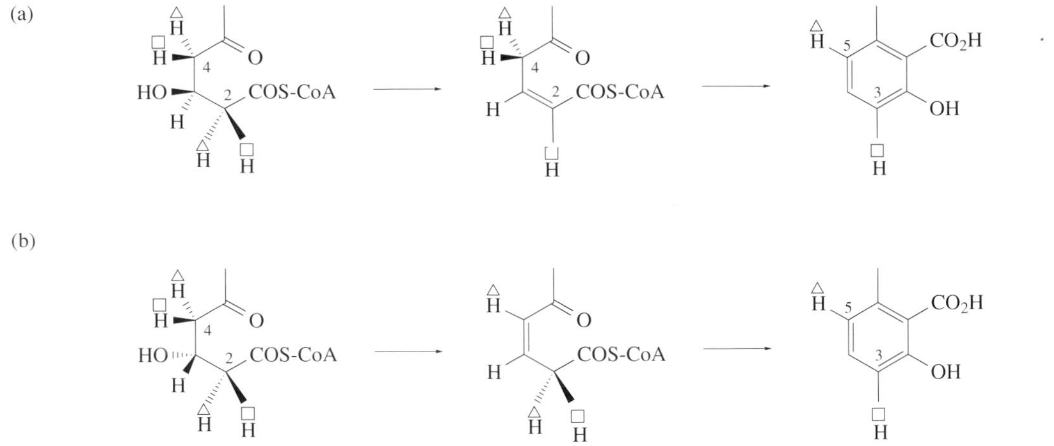

(a)

H H 4 O HO 2 COS-CoA H H H → H H 4 O H 2 COS-CoA H H → H 5 CO_2H 3 OH H

(b)

H H 4 O HO 2 COS-CoA H H H → H O H COS-CoA H H → H 5 CO_2H 3 OH H

Scheme 11

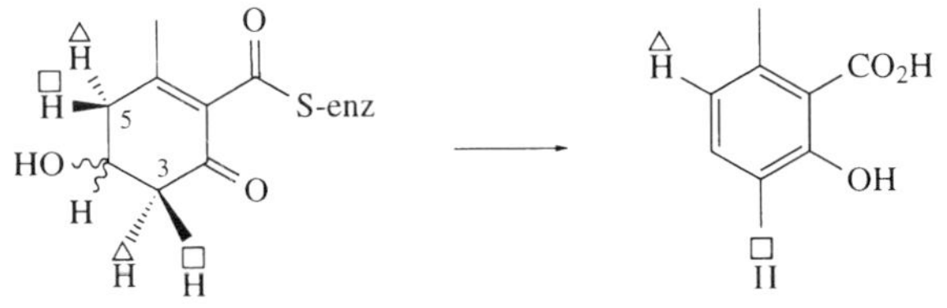

O H H 5 S-enz HO 3 O H H H → H CO_2H OH H

Scheme 12

O SCoA 2 x H H O SCoA O OH → H H O SEnz O O H H → NADPH :Base O H H 2 SEnz O OH H H →

O⁻ SEnz O OH H H –H_2O → O SEnz O H H :Base → O⁻ SEnz O H⁺ Malonyl-CoA →

:Base O O H H SEnz O → HO O SEnz H O :Base –H_2O → O SEnz O H H :Base –Enz → H 5 CO_2H 3 OH H

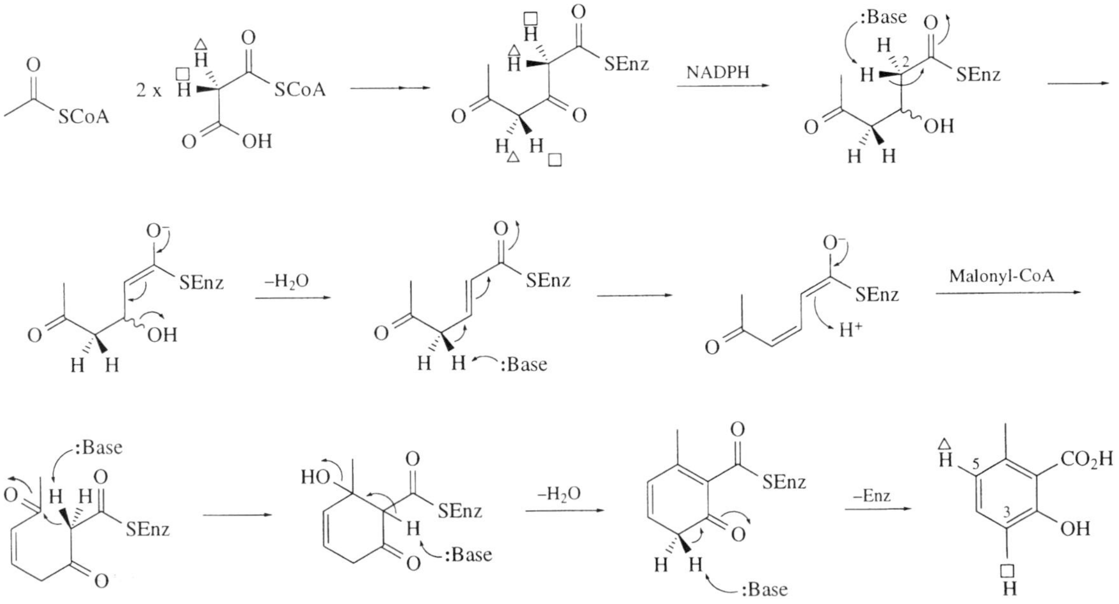

Scheme 13

1.14.10 THE MECHANISM OF ORSELLINIC ACID SYNTHASE

The related enzyme, orsellinic acid synthase, has been isolated from *Penicillium cyclopium*.[19] The reaction requires one molecule of acetyl-CoA and three molecules of malonyl-CoA but differs from that by 6-methylsalicylic acid synthase in that no β-ketoacyl reductase or β-hydroxyacyl dehydratase activities are involved. A study of the steric course of the reaction has also been carried out using the same approach with chiral malonates used for 6-methylsalicylic acid synthase.[53] The results revealed that the hydrogen atoms eliminated from the methylene groups at the 2- and 4-positions of the proposed triketide intermediate are also from opposite absolute orientations in malonyl-CoA (Scheme 14). The results from these experiments were, however, far less clear cut than the results with 6-methylsalicylic acid synthase because of the lability of the malonate-derived hydrogen atoms due to tautomerism and it has not been possible to carry out experiments with acetoacetyl-CoA as starter, as described above for 6-methylsalicylic acid synthase.

SCoA + CO₂H COSCoA —(+enz, –CO₂)→ COSenz —(malonyl-CoA, –CO₂)→ COSenz —(malonyl-CoA, –CO₂)→ COSenz —(–enz)→ CO₂H, HO, OH

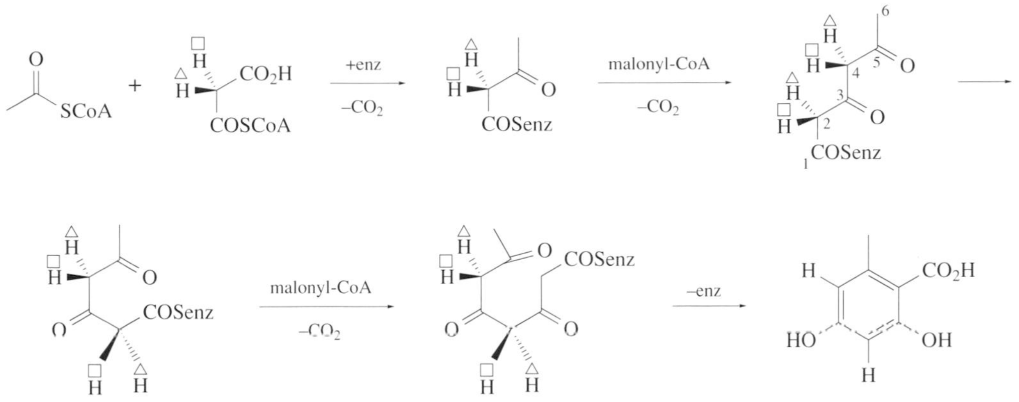

Scheme 14

1.14.11 MALONYL-CoA DECARBOXYLASE ACTIVITY OF 6-METHYLSALICYLIC ACID SYNTHASE

All polyketide synthases, like fatty acid synthases, share a similar condensation reaction mechanism in that a cysteine-bound acyl group acts as an electrophile to facilitate nucleophilic attack by a β-carboxyacyl residue bound to the acyl carrier protein (see Schemes 4 and 5). During the condensation, the latter undergoes decarboxylation to provide the driving force for C—C bond formation. There is a growing awareness, however, that some polyketide synthases may be able to catalyze decarboxylation when the cysteine thiol is unoccupied. For instance, DEBS 1+TE is able to decarboxylate methylmalonyl-CoA to propionate so that even in the absence of the regular propionyl-CoA starter the (2*S*,3*S*,4*S*,5*R*)-2,4-dimethyl-3,5-dihydroxy-*n*-hepanoic acid triketide lactone is formed.[36] Similar activity had been reported with fatty acid synthases from rabbit lactating mammary gland,[54] rat lactating mammary gland,[55] and yeast (Figure 8).[56] The malonyl-CoA decarboxylase activity does not exceed more than 1–2% of the rate of the fatty acid synthase although in the case of the pigeon liver enzyme, rates as high as 25–30% have been reported.[57] Thus, the decarboxylase activity of fatty acid synthases appears to be maximal only when the cysteine thiol is occupied by an acyl group.

The low-level malonyl decarboxylase activity of fatty acid synthases can be greatly increased if the cysteine thiol is modified by an alkylating agent such as iodoacetamide and can be as high as 60% of the rate catalyzed by native fatty acid synthase when both acyl substrates are present.[56] 6-Methylsalicylic acid synthase has also been shown to catalyze malonyl decarboxylation when partially inactivated by iodoacetamide and was able to generate 6-methylsalicylic acid at 20% of the rate of native enzyme.[58] This suggests that the alkylated enzyme is able to generate acetyl-CoA, possibly through the route shown in Scheme 15.

The acetyl-CoA thus generated is then able to act as a starter for the unmodified enzyme molecules still present in the usual way. Enzyme completely inactivated with iodoacetamide still retained decarboxylase activity although even this activity was lost on treatment with excess (1–10 mM) iodoacetamide.

The observations have been rationalized by a model for the active site of the condensing enzyme in which a basic residue is present within hydrogen bonding distance of the cysteine thiol and whose role is to promote acyl transfer from the acyl-4′-phosphopantetheine group as shown in Figure 8. The residue then participates as a base to promote decarboxylation and C—C bond formation. If the cysteine thiol is alkylated by iodoacetamide, the basic group may still be favorably placed to

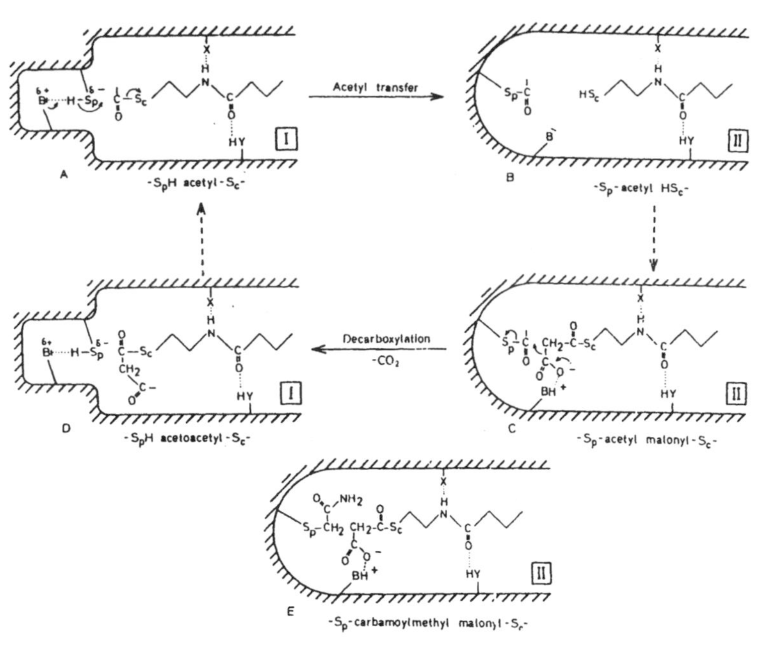

Figure 8 Diagramatic representation of the active site from β-ketoacyl synthase from yeast fatty acid synthase[55] showing a possible route for decarboxylation (A–D) and the effect of alkylation of the active site cysteine with iodoacetamide (E).

$$MSAS_{alkylated} + malonyl\text{-}CoA \longrightarrow malonyl\text{-}MSAS_{alkylated} + CoA$$

$$Malonyl\text{-}MSAS_{alkylated} \longrightarrow acetyl\text{-}MSAS_{alkylated} + CO_2$$

$$Acetyl\text{-}MSAS_{alkylated} + CoA \longrightarrow MSAS_{alkylated} + acetyl\text{-}CoA$$

Scheme 15

catalyze decarboxylation of the malonyl group bound to the 4′-phosphopantetheine arm to form an inactive acetyl–enzyme complex that on reaction with CoA will yield acetyl-CoA.

Malonyl decarboxylase activity has also been observed with native 6-methylsalicylic acid synthase.[20] The activity represents approximately 2% of the normal rate when only malonyl-CoA is used as the acyl substrate. On incubation of the enzyme with a range of starters in the presence of malonyl-CoA, not only were the respective 6-alkylsalicylic acids generated (see Section 1.14.6.1), but a background level of 6-methylsalicylic acid was always formed. It was also observed that the ratio of 6-alkylsalicylic acid:6-methylsalicylic acid was higher when the starter was more inefficient. These data suggest that native 6-methylsalicylic acid synthase possesses malonyl decarboxylase activity.

This has been confirmed by experiments in which [2-^{13}C]malonyl-CoA was incubated with 6-methylsalicylic acid synthase and NADPH in the absence of acetyl-CoA. Four [^{13}C] labels were found in the resulting 6-methylsalicylic acid, one of which was shown to be located in the methyl group of 6-methylsalicylic acid by collision-induced dissociation mass spectrometry (CID MS) (Table 2). As expected, only three [^{13}C] labels were found in 6-methylsalicylic acid derived from a standard incubation with acetyl-CoA and [2-^{13}C]malonyl-CoA (Table 2).

1.14.12 POINTERS TO THE THREE-DIMENSIONAL STRUCTURE OF 6-METHYLSALICYLIC ACID SYNTHASE

In the absence of a three-dimensional structure for 6-methylsalicylic acid synthase, the amino acid residues responsible for catalysis cannot be identified with certainty. Nevertheless, it is tempting to draw analogies from related proteins whose structures have been determined. For instance, the

Table 2 Demonstration of malonyl-CoA decarboxylase activity of 6-methylsalicylic acid synthase by incorporation of [2-^{13}C]malonyl-CoA into 6-methylsalicylic acid.

Structure	*Parent ion*		*Carboxyl loss*		*Methyl loss*	
	Predicted	*Observed*	*Predicted*	*Observed*	*Predicted*	*Observed*
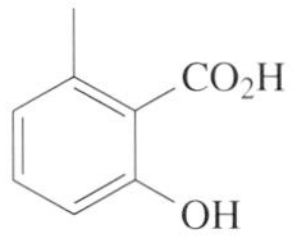 CO_2H, OH	151.14	150.79	107.13	106.79	92.09	91.82
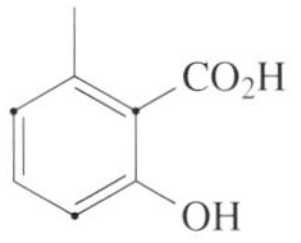 CO_2H, OH	154.14	153.83	110.13	109.85	95.09	94.87
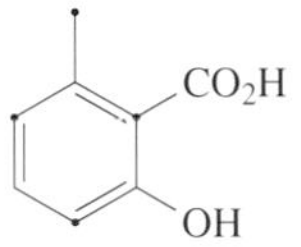 CO_2H, OH	155.14	154.84	111.13	110.84	95.09	94.83

The predicted and observed masses from negative ion CID MS of 6-methylsalicylic acid synthesized from: unlabeled malonyl-CoA; [2-^{13}C]labeled malonyl-CoA in the presence of unlabeled acetyl-CoA; [2-^{13}C]labeled malonyl-CoA alone. The black dots within the structures of 6-methylsalicylic acid represent the position of the ^{13}C labels. Masses were determined on a Micromass Quattro II triple quadrupole electrospray mass spectrometer.

E. coli β-ketoacyl synthase II, encoded by the *fabF* gene, has been crystallized and its X-ray structure determined.[59] The acyl chain appears to bind in a hydrophobic pocket at the base of which is located the reactive Cys165. His303 and His340 are close by with sulfur–nitrogen distances of 4.6 Å and 3.3 Å, respectively, and with His340 well placed to act as a base to assist in the abstraction of a proton from the Cys165 thiol. His303 and His340 may also stabilize the oxygen atom of the tetrahedral intermediate during thioester formation. The most important role for His303 is to act as a base for decarboxylation of the malonyl group. Another invariant residue thought to be important is Lys335 which is situated 7.8 Å from the active site cysteine. The analogous residue to Lys335, in both fatty acid synthase[60] and 6-methylsalicylic acid synthase,[26] is the likely site of inactivation with *o*-phthalaldehyde. In both cases modification with this reagent leads to the formation of a fluorescent thioisoindole ring at the active site thought to involve both this lysine and the catalytic cysteine.[26]

The sequence comparisons between several polyketide synthases and fatty acid synthases indicate two regions containing seven invariant residues within the β-ketoacyl synthase (the numbering relates to 6-methylsalicylic acid synthase), including the histidines and lysines under discussion (Scheme 16).

^{130}EA**H**ATSTPLG**D**PTE**I** ^{165}GS**K**PNIG**H**L

Scheme 16

Although not directly relevant to 6-methylsalicylic acid synthase, the structure of another component of a fatty acid synthase, the enoyl reductase from *Brassica napus*, has also been solved,[61] providing valuable insight into the structure of polyketide synthase modules that catalyze complete reduction cycles like fatty acid synthases.

The only other three-dimensional structure of a protein component of relevance to 6-methylsalicylic acid synthase is the acyl carrier protein from *Streptomyces coelicolor*, purified[62] and solved by ^{1}H NMR.[63] This small protein is made up of four helical regions showing strong structural homology to *E. coli* fatty acid synthase acyl carrier protein.[63] The structure reveals a hydrophobic groove containing basic residues conserved in all polyketide synthase acyl carrier proteins, but absent from fatty acid synthase acyl carrier proteins. It is thought that this groove can harbor and stabilize the growing polyketide chain through enolization of several polyketide chain carbonyl groups.

1.14.13 SUMMARY

This account, although primarily focused on 6-methylsalicylic acid synthase from *Penicillium patulum*, also, unavoidably, covers many aspects of fatty acid synthase structure and mechanism, more details of which are covered in Chapter 1.01 of this volume. There is no doubt that these two classes of enzymes are close cousins, evolving from a common ancestral protein, probably a fatty acid synthase.[64] The exploitation and adaptation of such a successful protein template in evolution to provide the range and diversity of polyketide derived natural products is one of the most fascinating aspects of the field. The dramatic progress in our understanding of enzyme structure and mechanism coming from both X-ray and NMR studies and from the use of molecular biology will in the near future be able to help us relate structure to function, particularly with respect to the factors that define a processive and iterative enzyme reaction mechanism.

1.14.14 REFERENCES

1. M. Luckner, "Secondary Metabolism in Microorganisms, Plants and Animals," Springer, Berlin, 1984, p. 170.
2. C.-G. Kim, T.-W. Yu, C. B. Fryhle, S. Handa, and H. G. Floss, *J. Biol. Chem.*, 1998, 6030.
3. J. D. Bu'lock, "Comprehensive Organic Chemistry," Pergamon, Oxford, 1978, vol. 5, p. 927.
4. D. O'Hagan, in "The Polyketide Metabolites," ed. D. O'Hagan, Ellis Horwood, Chichester, 1991, p. 65.
5. J. F. Martin and A. L. Demain, in "The Filamentous Fungi," eds. J. E. Smith and D. R. Berry, Edward Arnold, London, 1978, vol. 3, p. 426.
6. E. Haslem, "Metabolism and Metabolites," Clarendon, Oxford, 1985, p. 131.
7. P. I. Forester and G. M. Guacher, *Biochemistry*, 1972, **11**, 1102.
8. P. Dimroth, H. Walter, and F. Lynen, *Eur. J. Biochem.*, 1970, **13**, 98.
9. D. J. Bevitt, J. Cortes, S. F. Haydock, and P. F. Leadlay, *Eur. J. Biochem.*, 1992, **204**, 39.
10. J. Cortes, S. F. Haydock, G. A. Roberts, D. J. Bevitt, and P. F. Leadlay, *Nature*, 1990, **348**, 176.
11. S. Donadio and L. Katz, *Gene*, 1992, **111**, 51.
12. C. M. Amy, B. Williams-Ahlf, J. Naggert, and S. Smith, *Biochem. J.*, 1990, **271**, 675.
13. E. Schweizer, B. Kneip, H. Castorph, and U. Holzner, *Eur. J. Biochem.*, 1973, **39**, 353.
14. J. Beck, S. Ripka, A. Siegner, E. Schiltz, and E. Schweizer, *Eur. J. Biochem.*, 1990, **192**, 487.
15. S. Wakil, *Biochemistry*, 1989, **28**, 4523.
16. Y. Tsukamoto and S. Wakil, *J. Biol. Chem.*, 1988, **263**, 16 225.
17. A. I. Scott, I. C. Beadling, N. H. Georgopapadakou, and C. R. Subbarayan, *Bioorg. Chem.*, 1977, **3**, 238.
18. J. B. Spencer and P. M. Jordan, *Biochem. J.*, 1992, **288**, 839.
19. J. B. Spencer and P. M. Jordan, unpublished results.
20. I. D. G. Campuzano, Ph.D. Thesis, University of Southampton, 1998.
21. G. Vogel and F. Lynen, *Methods Enzymol*, 1976, **43**, 520.
22. J. S. Mattick, Y. Tsukamoto, J. Nickless, and S. Wakil, *J. Biol. Chem.*, 1983a, **258**, 15 291.
23. J. S. Mattick, J. Nickless, M. Mizugaki, C. Y. Yang, S. Uchiyama, and S. Wakil, *J. Biol. Chem.*, 1983b, **258**, 15 300.
24. Y. Tsukamoto, H. Wong, H. S. J. Mattick, and S. Wakil, *J. Biol. Chem.*, 1983, **258**, 15 312.
25. H. Wong, J. S. Mattick, and S. Wakil, *J. Biol. Chem.*, 1983, **258**, 15 305.
26. C. Child, Ph.D. Thesis, Queen Mary and Westfield College, University of London, 1994.
27. A. F. A. Marsden, P. Caffrey, J. F. Aparicio, M. S. Loughran, J. Staunton, and P. F. Leadlay, *Science*, 1994, **263**, 378.
28. C. J. Child, J. B. Spencer, P. Bhogal, and P. M. Shoolingin-Jordan, *Biochemistry*, 1996, **35**, 12 267.
29. J. K. Stoops and S. J. Wakil, *J. Biol. Chem.*, 1981, **256**, 5128.
30. J. K. Anil, A. Witkowski, and S. Smith, *Biochemistry*, 1997, **36**, 2316.
31. J. Staunton, P. Caffrey, J. F. Apricio, G. A. Roberts, S. S. Bethel, and P. F. Leadlay, *Nature Struct. Biol.*, 1996, **3**, 188.
32. P. Weisner, J. Beck, K. Beck, S. Ripka, G. Muller, S. Lucke, and E. Schweizer, *Eur. J. Biochem.*, 1988, **177**, 69.
33. G. E. Nixon, J. R. Putz, and J. W. Porter, *J. Biol. Chem.*, 1968, **243**, 5471.
34. M. Yalpani, K. Willecke, and F. Lynen. *Eur. J. Biochem.*, 1969, **8**, 495.
35. J. Cortes, K. E. H. Weismann, G. A. Roberts, M. J. B. Brown, J. Staunton, and P. F. Leadlay, *Science*, 1995, **268**, 1487.
36. R. Pieper, E. Khosla, D. E. Cane, and C. Khosla, *Biochemistry*, 1996, **35**, 2054.
37. R. Pieper, G. Lou, D. E. Cane, and C. Khosla, *J. Am. Chem. Soc.*, 1995, **117**, 11 373.
38. C. M. Kao, L. Katz, and C. Khosla, *Science*, 1994, **265**, 509.
39. H. Funabashi, S. Iwasaki, and S. Okuda, *Tetrahedron Lett.*, 1983, **24**, 2673.
40. H. Ohno, T. Ohno, J. Awaya, and S. Omura, *J. Biochem.*, 1975, **78**, 1149.
41. C. J. Child and P. M. Shoolingin-Jordan, *Biochem. J.*, 1998, **330**, 933.
42. Y. S. Wang, W. X. Tian, and R. Y. Hsu, *J. Biol. Chem.*, 1984, **259**, 13 644.
43. C. Abell, M. J. Garson, F. J. Leeper, and J. Staunton, *J. Chem. Soc., Chem. Commun.*, 1982, 1011.
44. C. Abell and J. Staunton, *J. Chem. Soc., Chem. Commun.*, 1984, 1005.
45. H. G. Floss, M.-D. Tsai, and R. W. Woodward, *Top. Stereochem.*, 1984, **15**, 253.
46. P. M. Jordan, J. B. Spencer, and D. L. Corina, *J. Chem. Soc., Chem. Commun.*, 1986, **314**, 911.

47. J. B. Spencer and P. M. Jordan, *Biochemistry*, 1992, **31**, 9107.
48. P. M. Jordan and J. B. Spencer, *Tetrahedron*, 1991, **47**, 6015.
49. B. Sedgewick, J. W. Conforth, and S. J. French, *J. Chem. Soc., Chem. Commun.*, 1978, 193.
50. R. B. Herbert, "The Biosynthesis of Secondary Metabolites," 2nd edn., Chapman and Hall, London/New York, 1989, p. 50.
51. K. R. Hansen and I. A. Rose, *Acc. Chem. Res.*, 1975, **8**, 1.
52. J. M. Schwab and J. B. Klassen, *J. Am. Chem. Soc.*, 1984, **106**, 7217.
53. J. B. Spencer and P. M. Jordan, *J. Chem. Soc., Chem. Commun.*, 1992, **8**, 646.
54. H. J. M. Hansen, E. M. Carey, and R. Dils, *Biochim. Biophys. Acta*, 1971, **248**, 391.
55. S. Smith and S. Abraham, *J. Biol. Chem.*, 1970, **245**, 3209.
56. G. Kresze, L. Steber, D. Oesterhelt, and F. Lynen, *Eur. J. Biochem.*, 1977, **79**, 191.
57. D. N. Burton, A. G. Haavik, and J. W. Porter, *Arch. Biochem. Biophys.*, 1968, **126**, 141.
58. P. Dimroth, E. Ringlemann, and F. Lynen, *Eur. J. Biochem.*, 1976, **68**, 591.
59. W. Huang, J. Jia, P. Edwards, K. Dehesh, G. Schneider, and Y. Lindquist, *EMBO J.*, 1998, **17**, 1183.
60. J. K. Stoops, S. J. Henry, and S. J. Wakil, *J. Biol. Chem.*, 1983, **258**, 12 482.
61. J. W. Simon, S. Bethel, A. R. Slabas, and D. W. Rice, *Acta Crystallogr., Sect. D., Biol. Cryst.*, 1998, **54**, 427.
62. J. Crosby, D. H. Sherman, M. J. Bibb, W. P. Revill, D. A. Hopwood, and T. J. Simpson, *Biochim. Biophys. Acta*, 1995, **1251**, 32.
63. C. P. Crump, J. Crosby, C. E. Dempsey, J. A. Parkinson, M. Murray, D. A. Hopwood, and T. J. Simpson, *Biochemistry*, 1997, **36**, 6000.
64. D. A. Hopwood and D. H. Sherman, *Annu. Rev. Genet.*, 1990, **24**, 37.

1.15
The Diels–Alder Reaction in Biosynthesis of Polyketide Phytotoxins

AKITAMI ICHIHARA and HIDEAKI OIKAWA
Hokkaido University, Sapporo, Japan

1.15.1 INTRODUCTION

Plant pathogens produce various types of phytotoxin which are significant causal factors in the development of a number of destructive diseases in plants. The main purpose of phytotoxin research is to establish the chemical–ecological relationship between plants and phytopathogenic microorganisms and in some cases the release of toxic molecules by pathogenic microorganisms clearly is the key to understanding their pathogenicity. Discovery of useful plant growth regulators, that is, plant hormones, is also an important area of toxin research. Another important area of phytotoxins (as well as natural products) research is the discovery of new biological functions through biosynthetic studies. Usually the biogenesis of phytotoxins follows established synthetic routes. However, there are unusual compounds whose biogenesis involves uncommon enzymes, for example Diels-Alderase, which catalyzes Diels–Alder reactions. The Diels–Alder reaction is one of the pericyclic reactions and generates a six-membered ring through the 1,4-addition of the double bond of a dienophile to conjugated diene.

Pericyclic reactions are very important in the synthesis of organic compounds, especially natural products. A number of papers dealing with natural product syntheses featuring a pericyclic reaction as the key step have appeared.[1–3] Although the data obtained in most cases have been insufficient, a large number of pericyclic key steps also occur in biological systems. Conversion of (−)-chorismate (**1**) to prephenate (**2**) by chorismate mutase is an example of enzyme-catalyzed sigmatropic rearrangement (Equation (1)). Originally, the enzyme was isolated as a bifunctional enzyme, chorismate mutase–prephenate dehydrogenase, from *Aerobacter aerogenes*.[4,5] The remarkable acceleration ($\sim 10^6$-fold) by the enzymes has been studied extensively and analysis of the active site structures in monofunctional chorismate mutase from *Bacillus subtilis* has led to a general mechanistic hypothesis that the enzymes stabilize the chair-like transition state geometry (**1a**) via a series of electrostatic and hydrogen-bonding interactions.[6–8] Electrocyclic and sigmatropic reactions are well known in vitamin D chemistry.[1] The biosynthesis of racemic endiandric acids A (**3**) and C (**4**) was postulated to involve two successive electrocyclic reactions and one [4 + 2]-cycloaddition[9] and the process was later confirmed by biomimetic total synthesis.[10]

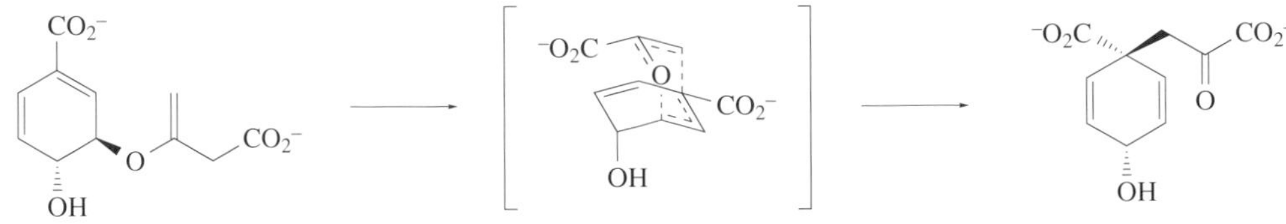

(**1**) (**1a**) (**2**) (1)

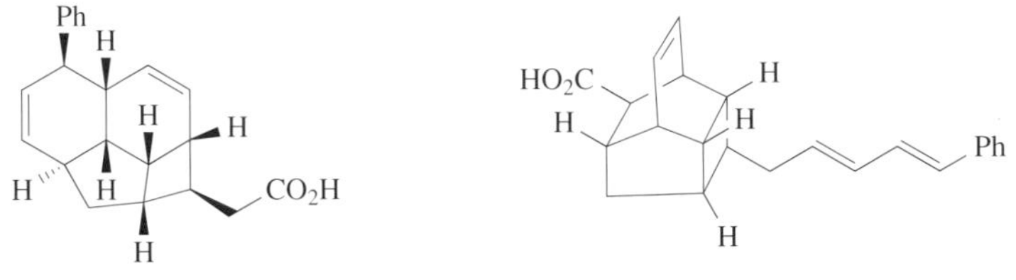

(**3**) (racemic) (**4**) (racemic)

It is not clear whether most potentially Diels–Alder-derived products have been derived through a Diels–Alder reaction or by another process, that is, an ionic reaction. Biosynthetic studies were carried out with several natural cycloadducts and only two examples, solanapyrone and macrophomic acid, were deduced to arise from enzymatic Diels–Alder reactions (see Sections 1.15.3.3 and 1.15.3.4). The reasons why such evidence has not been obtained more generally may be as follows. (i) The Diels–Alder reaction is a symmetry-allowed thermal [4+2]-cycloaddition, which may be unusual in biological processes. (ii) It may be difficult to specify the direct precursor of the Diels–Alder reaction in cases where various possibilities exist. (iii) Synthetic precursors are not readily available. (iv) There is no guarantee that a rather large molecular precursor can be incorporated into organisms. However, rapid progress in synthetic and analytical methodology has made it possible to overcome these difficulties, at least with regard to some molecules and organisms. Therefore, additional enzymes catalyzing the Diels–Alder reaction as well as other pericyclic reactions might be discovered from organisms cited in this article and elsewhere.

On the other hand, several groups have successfully generated catalytic antibodies capable of catalyzing Diels–Alder cycloaddition reactions,[11] although their catalytic activities are not as great as the usual enzymes. In connection with these artificial enzymes, note that Baker's yeast catalyzes the stereoselective Diels–Alder reaction, in which a predominant formation of *exo* isomer is observed.[12] Apart from this biocatalytic reaction, the influence of bovine serum albumin on regio- or enantioselectivity in Diels–Alder reactions has been studied and enantiomeric excess up to 38% is reached in the presence of a catalytic amount of bovine serum albumin.[13]

The structure and biosynthesis (mainly biogenesis) of Diels–Alder-type natural products arbitrarily selected by structural types are described in Section 1.15.2. Major emphasis is placed on biosynthetic studies of phytotoxins in the substrate and enzyme levels, since Diels–Alder enzymes have been found in solanapyrone and macrophomic acid biosynthesis (see Sections 1.15.3.3 and 1.15.3.4) for the first time.

1.15.2 DIELS–ALDER-TYPE NATURAL PRODUCTS

Among a number of Diels–Alder-type natural products,[14–18] some are optically inactive and would be derived from nonenzymatic processes, that is, spontaneous cycloaddition reactions *in vivo* and/or *in vitro*. In another plausible explanation products would be obtained by ionic processes in the physiological conditions of organisms. Only a few examples will be quoted here briefly.

Leaves of *Xanthoxylum procerum* (Rutaceae) yielded two major optically inactive alkaloids, culantraramine (**5**) and culantraraminol (**6**), which were assigned bishordeninyl terpene structures[19] (Scheme 1). Two minor isomeric alkaloids were also found along with hordenine and *N,N*-dimethyltryptamine. Although culantraramine (**5**) could be viewed as a natural self-Diels–Alder adduct of dehydroprenylhordenine (**7**), when this diene was prepared and reacted at room temperature, it yielded instead alternate adduct (**8**) and (**9**) in high yield. Culantraramine (**5**) and culantraraminol (**6**) synthesis was, however, achieved in high yield from (**10**), which was converted to (**5**) and (**6**) by treatment with hydrogen chloride. The particular regio- and stereochemical results observed in the reaction leading to (**5**) to (**6**) can probably best be explained by the stepwise ionic process as depicted in Scheme 1.

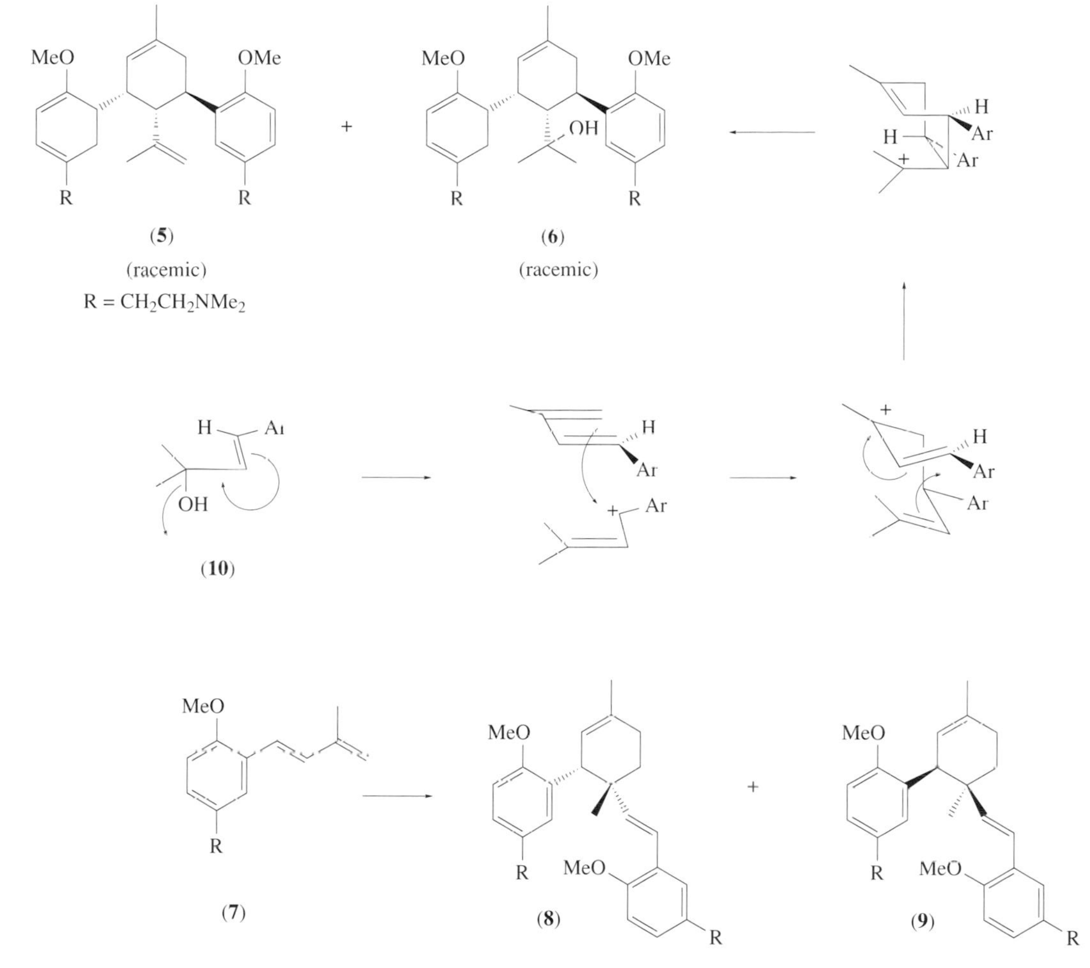

(**5**) (racemic) R = $CH_2CH_2NMe_2$ (**6**) (racemic) (**10**) (**7**) (**8**) (**9**)

Scheme 1

In order to provide (**5**) and (**6**) as the major products, the process would have to take place mainly with retention of configuration of the allylic cation intermediate. Natural products such as cycloadducts (**8**) and (**9**) have been found widely in plants (see Section 1.15.2.1) and arise from a "true" Diels–Alder reaction, while isolates such as (**5**) and (**6**) arise from a cationic process. Another example is yuehchukene (**11**), a novel dimeric indole alkaloid with potent *anti*-implantation activity from the roots of *Murraya paniculata*.[20] Yuehchukene (**11**) appears to be formed via Diels–Alder cyclization of prenylated indoles. As in most of these analogous compounds, the alkaloid is derived by nonenzymatic ionic reaction of the precursor, 3-isoprenylindole. Actually, biomimetic synthesis of (**11**) has been achieved under acidic conditions.[21] The related alkaloids, borreverine (**12**) and isoborreverine (**13**) would be formed by the same ionic mechanisms.[22,23]

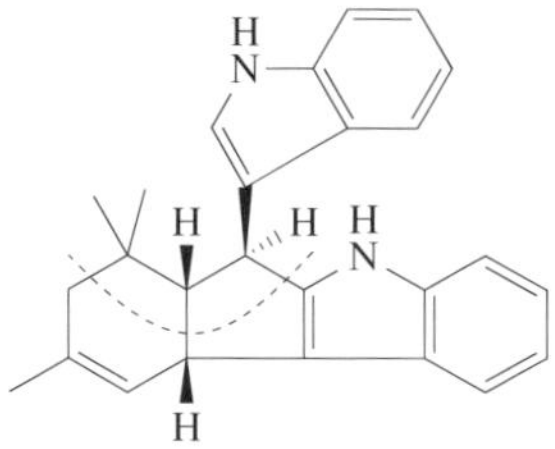

(11) (racemic) (12) (racemic) (13) (racemic)

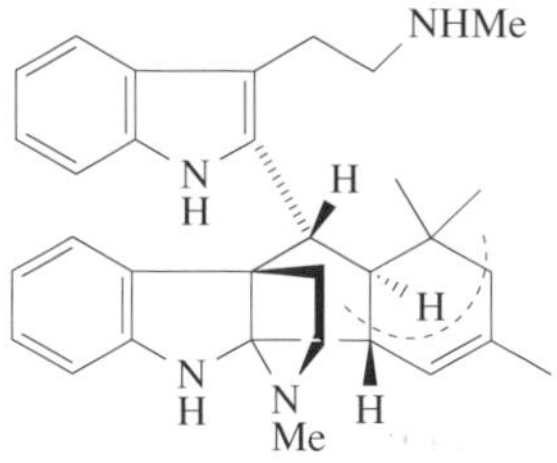

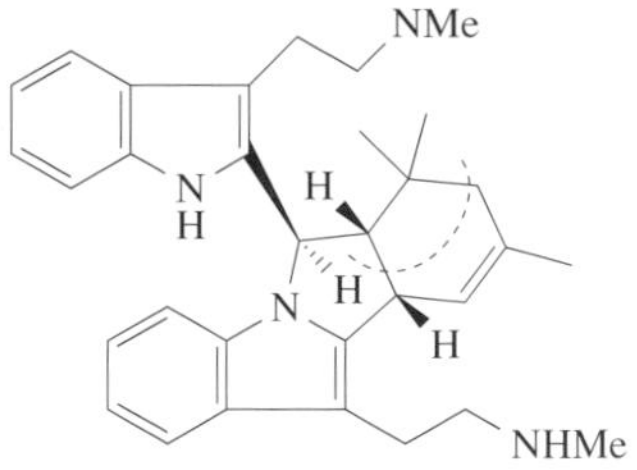

These examples indicate that biogenetic considerations based on structure alone, that is whether or not a Diels–Alder reaction takes place, need to be taken into account from a chemical as well as biosynthetic point of view.

1.15.2.1 Biogenesis of Diels–Alder-type Natural Products

A number of potential Diels–Alder adducts are found in polyketides from various microorganisms. Most of the adducts are optically active and expected to be formed via intramolecular reactions which produce extra five- to 19-membered carbocycles along with cyclohexane rings.

The major class of biological [4+2]-adducts in polyketides is a "1,2-dialkyldecalin polyketide" as shown in Scheme 2. Intramolecular Diels–Alder reaction of the reduced polyketide chain which has a conjugated *E,E*-diene and an *E*-dienophile can explain the stereochemistry found in metabolites of this type. Usually, the dienophile alkene is conjugated with a carbonyl group so that the alkene is activated by the electron-withdrawing group. This empirical rule is supported by the fact that diastereomers corresponding to both *exo*-adduct (PI-201)[24] (**14**) and *endo*-adduct (phomodiol)[25] (**15**) are found in nature. In most cases, *endo*-adducts are predominant and the *Z*-alkene is rarely used as a dienophile. Among the dialkyldecalin polyketides, there are two possible cycloaddition modes involved in the direction of the polyketide chain. Usually, the dienophile is located at the carboxyl terminal and the diene is at the methyl terminal. Opposite cases are only found in ilicicolin H (**16**)[26] and nargenicin A_1.[27] Simple examples of this class produced by fungi and actinomycetes are fusarielin A (**17**),[28] trichoharzin (**18**),[29] rapiculine (**19**),[30] calbistrin A (**20**),[31] versiol,[32] oblongolide,[33] hynapene A,[34] and aldecalmycin.[35] Synthetic studies of this class of metabolites adopted the biogenetic Diels–Alder route and a number of compounds were successfully synthesized.[36–38]

Another feature of this class is the frequent occurrence of the polyketide-bearing acyltetramic acid structure which may be constructed between the polyketide terminals and amino acids. Microbial [4 + 2]-adducts of this type are equisetin (**21**),[39] lydicamycin (**22**),[40] oteromycin (**23**) (deoxytetramate),[41] delaminomycin A,[42] vermisporin,[43] PF1052,[44] and BU4514N.[45] In addition, ilicicolin H (**16**)[26] and fischerin,[46] whose characteristic α-pyridone moiety[47] may be derived via a ring expansion in the corresponding acyltetramic acid, may also be included. The stereochemistry of the decalin system in lydicamycin (**22**) needs comment. To yield the required configuration in its cyclohexene ring, a Diels–Alder reaction between the unusual *Z*-dienophile and *Z,E*-diene, which would give a sterically disfavored transition state, is required. Since elucidation of the configuration is based mainly on NMR data, X-ray analysis or total synthesis is preferable to confirm its stereochemistry.

There is another group represented by chlorothricin (**24**)[48] which possesses the decalin and the cyclohexane ring systems with an acyltetronic acid moiety. Closely related analogues kijanimicins (**25**),[49] tetrocarcins,[50] MM46115,[51] PA-46101 A,[52] and AC6H[53] were also found. Compounds (**24**) and (**25**) have essentially the same configuration on the decalin and the cyclohexene rings. Floss and co-workers proposed that the acylated phosphoenol pyruvate moiety condensed with the methyl terminal triketide moiety in the biosynthesis of (**24**).[54] Alternatively, the cyclohexane ring could be formed between the methylenetetronate and diene at the methyl terminus of dodecaketide via another [4 + 2]-cycloaddition. In the total synthesis of chlorothricin, Yoshii and co-workers adopted this strategy.[55] The isolation of okilactomycin (**26**)[56] and tetronothiodin (**27**)[57] suggests that the cyclohexane ring could be derived from the acyltetramic acid and the terminal diene of the polyketide chain. In addition, occurrence of tetrodecamycin (**28**)[58] implies that the decalin system could be formed via [4+2]-cycloaddition.

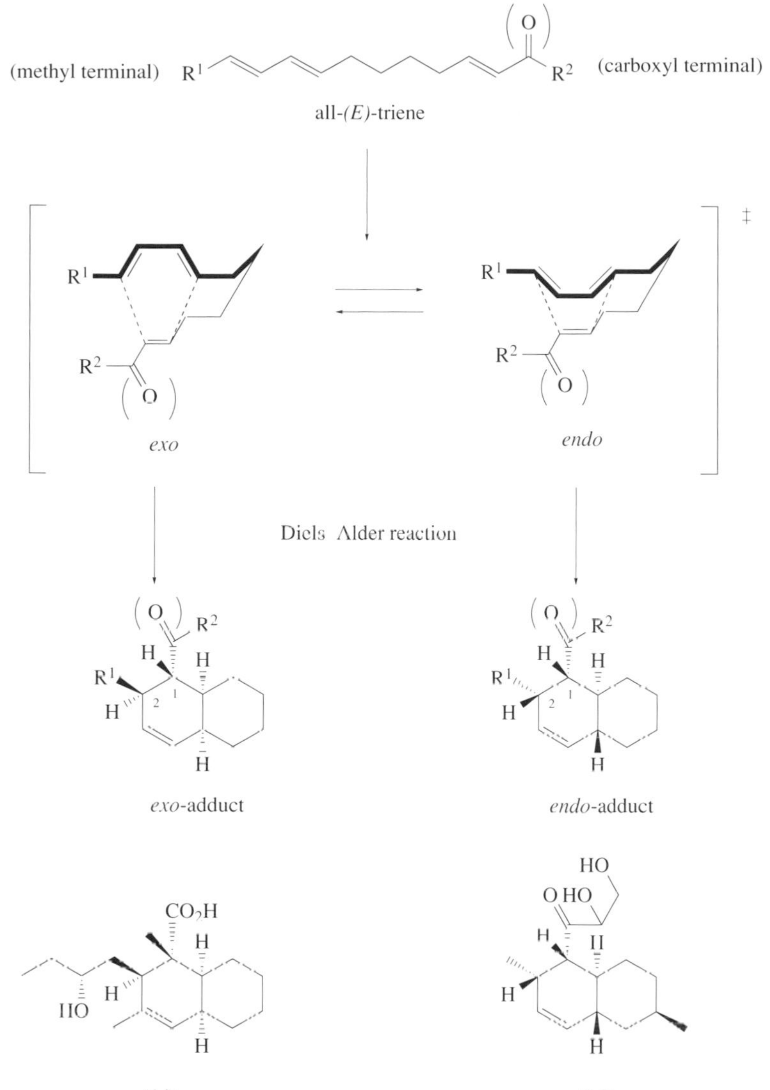

Scheme 2

There are a few decalin polyketides from plant and marine organisms. Himgravin (**29**)[59] from *Galbulimima baccata* is a plausible acetogenin alkaloid which could be considered as an *endo*-adduct. Superstolide A (**30**)[60] is a cytotoxic agent from the marine sponge *Neosiphonia superstes*. Stereochemical analysis of (**30**) implies that it is derived from an unusual *Z*-dienophile.

The second large family of intramolecular biological [4 + 2]-adducts is the macrocyclic polyketides. Among these polyketides, the largest group is the so-called cytochalasins[61] which consist of a characteristic perhydroisoindole skeleton and an 11- to 13-membered carbocycle or 14-membered macrolide. They show structural variations such as, those exhibited by compounds (**31**)–(**34**), which are due to different stages of oxidation, the number of methyl groups flanking the polyketide chain, and the C-2 substituent on the perhydroisoindole skeleton. Cytochalasins are known as mycotoxins and are cytotoxic to mammalian cells. So far, more than 50 natural cytochalasins[62] have been isolated including a phytotoxin, chaetoglobosin O, isolated from *Cylindrocladium floriidanum*.[63]

Based on the stereochemistry of the cyclohexane, it is proposed that *endo*-selective intramolecular Diels–Alder reaction of the polyketide derivatives produces the perhydroisoindole skeleton and the macrocyclic ring. Although this hypothesis is attractive, no experimental evidence is available. Synthetic studies[64] along this line have provided further support for the proposed biogenesis. Biosynthetic studies of one member of this class are described in Section 1.15.3.2.

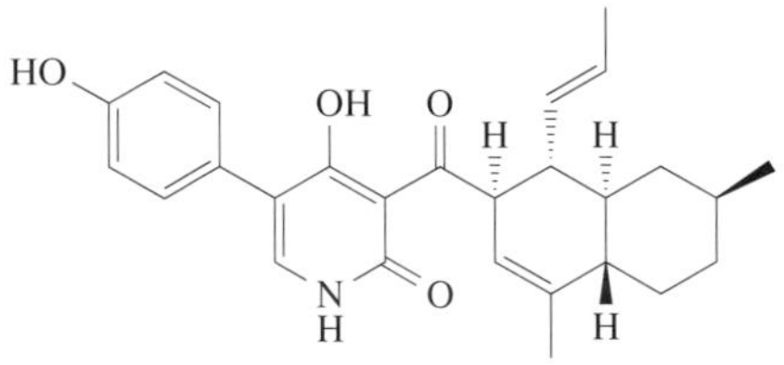

HO OH O H H N H O H (16)

HO H HO H H H O H O (17)

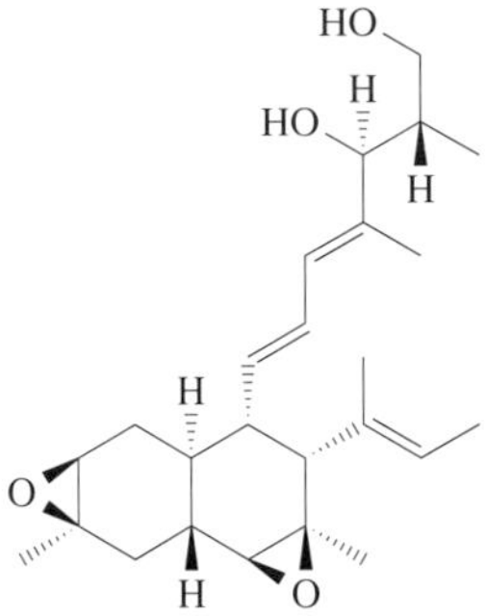

OH O O H O HO$_2$C HO H (18)

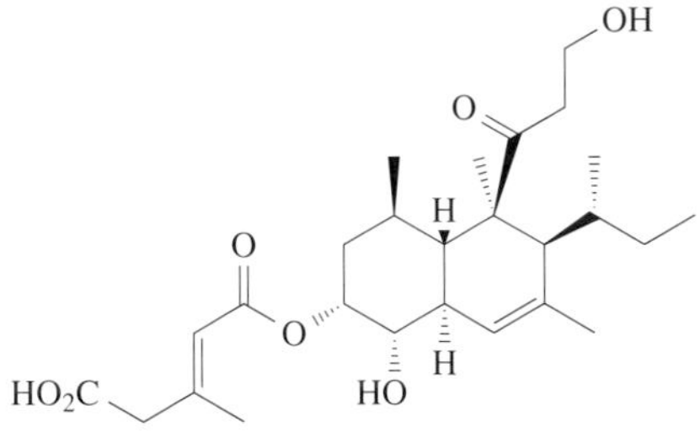

CO$_2$H H O OH (19)

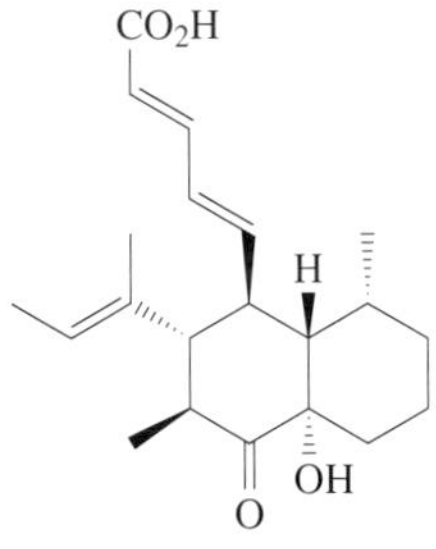

O OH HO O CO$_2$H O H O (20)

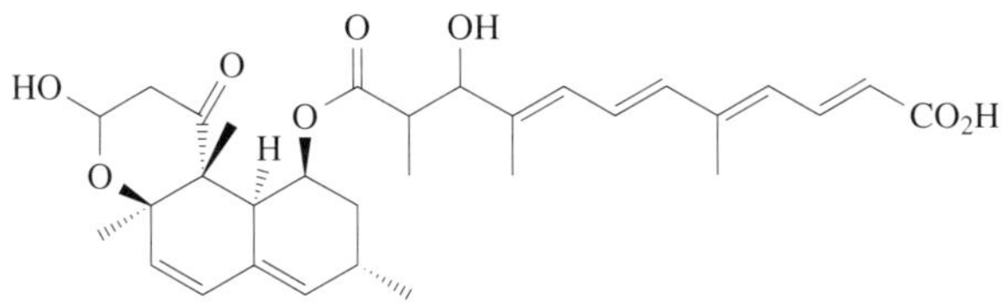

OH OH Me – N O O H H (21)

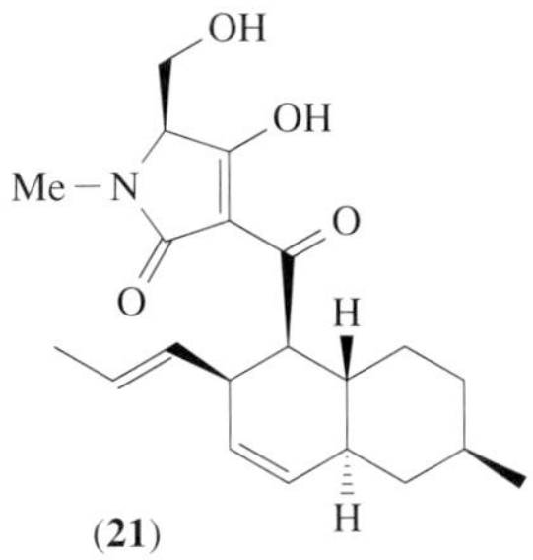

HO NH O O NH$_2$ NH N H H OH OH OH OH OH H OH OH (22)

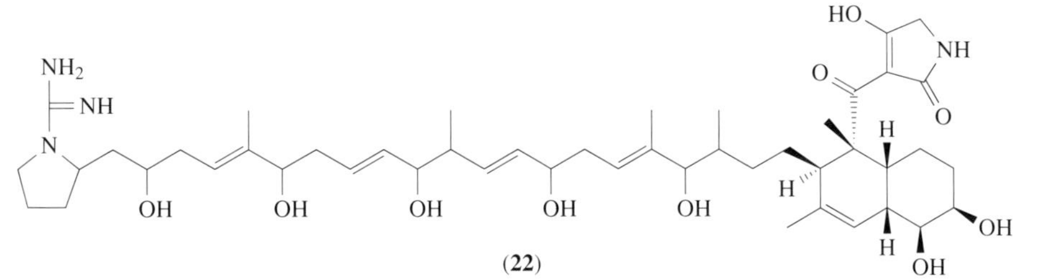

H O O N H OH (23)

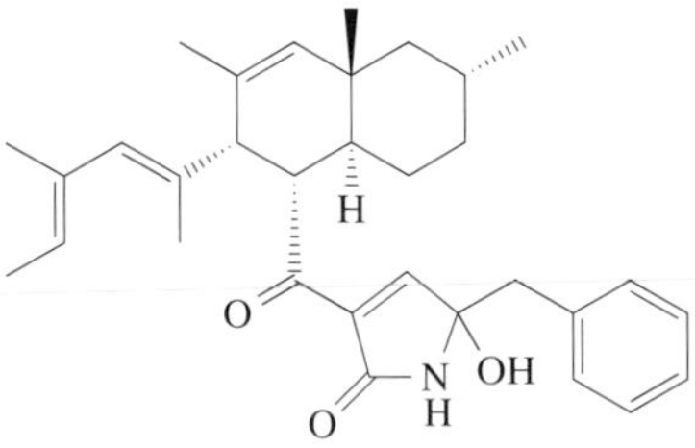

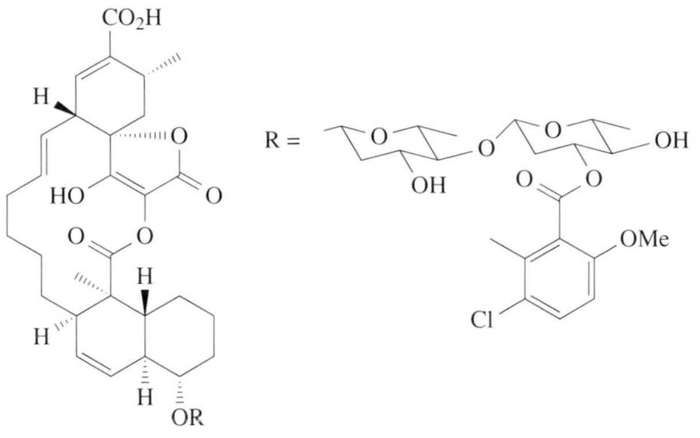

(**24**)

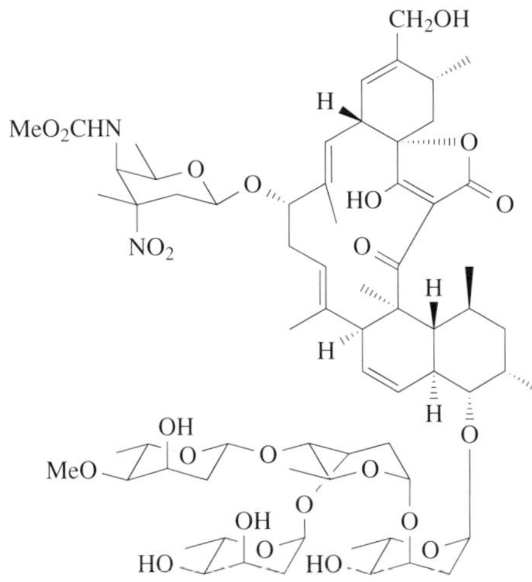

(**25**)

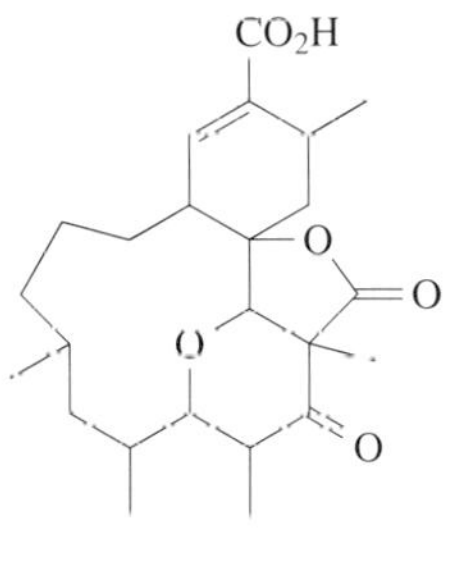

(**26**)

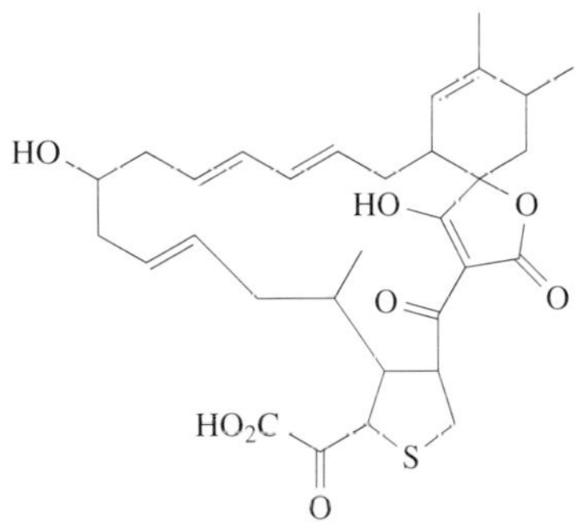

(**27**)

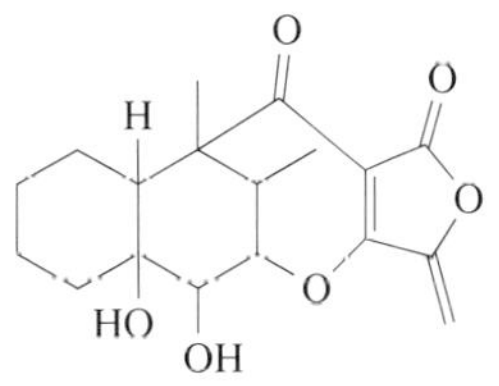

(**28**)

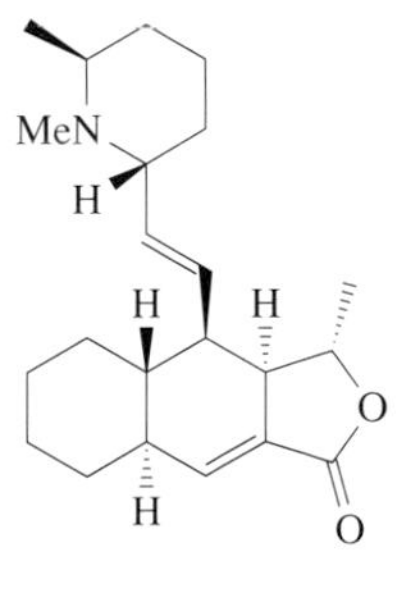

(**29**)

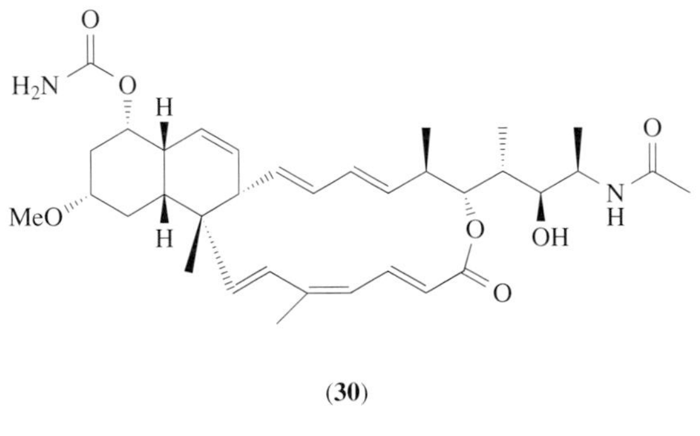

(**30**)

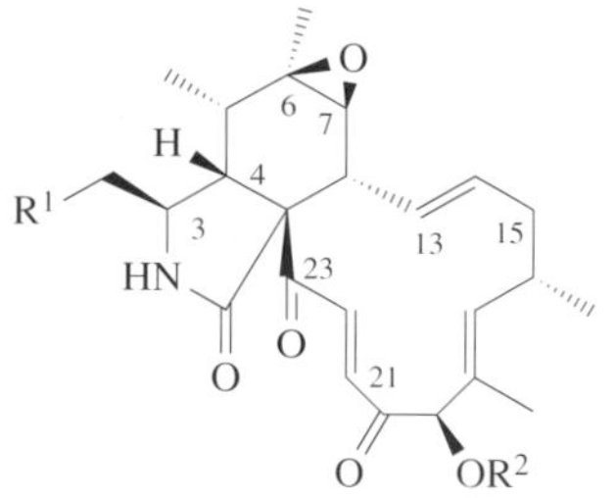

(**31**) R^1 = Ph, R^2 = Ac
(**32**) R^1 = Indolyl, R^2 = H

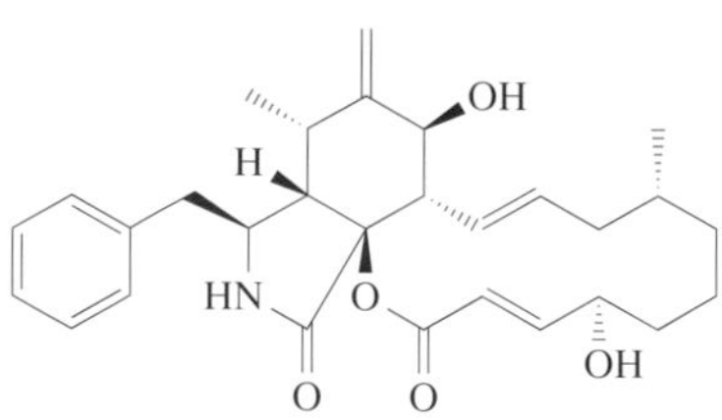

(**33**)

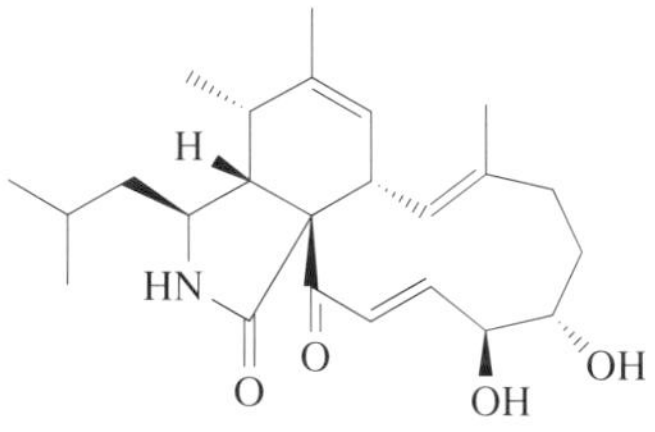

(**34**)

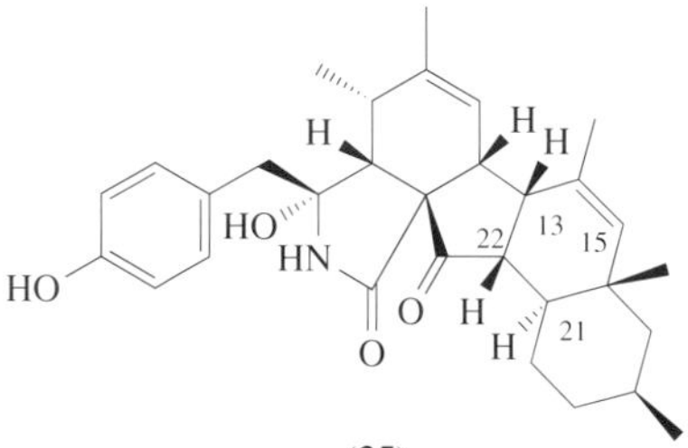

(**35**)

A new cytochalasin, phomopsichalasin (**35**)[65] has been isolated from *Phomopsis* sp., as an antimicrobial agent. In its structure, the macrocycle of other cytochalasins is replaced by the tricyclic system which may be derived from 13,15-diene and C_{21}–C_{23} enone via another Diels–Alder reaction. The marine natural product pulo'upone (**36**)[66] and antibiotics stawamycin (**37**),[67] indanomycin,[68] and A83094A[69] from actinomycetes, belong to this class. Other unique intramolecular [4+2]-adducts are ikarugamycin (**38**)[70] and A83543A (**39**)[71] which have an unusual *as*-hydrindacene skeleton arising from either a single or two polyketide chain(s).

On the basis of the examples described above, various plausible precursors which would give the complex intramolecular [4+2]-adducts have been proposed. The structures of the polyene precursors (**40**) and (**41**) of cytochalasin K (**31**) and oteromycin (**23**), respectively, have common features with respect to chain length and unsaturation mode. A reduced polyketide precursor of asimilar size, such as (**42**), is proposed for the biosynthesis of nargenicin A_1 (see Section 1.15.2.2) and macrolide antibiotics.[72] The formation of different carbocyclic rings could be expected from these linear polyene precursors. The natural diastereomers were not predominant[26] in the total synthesis of natural adducts via intramolecular [4+2]-cycloaddition. Involvement of the enzyme catalyzing the Diels–Alder reaction is strongly suggested in order to explain the formation of the correct ring size and diastereomer.

The final group in the polyketides are the intermolecular [4+2]-adducts. Only three microbial metabolites may be included in this group. Flavoskyrin[73] is a hetero-Diels–Alder adduct of the corresponding dihydroanthraquinones. The antiviral antibiotic quartromicin A1[74] from *Amycolatopsis orientalis* has a unique carbon macrocycle possessing the acyltetroic acid and the diene in an analogous structure to chlorothricin (**24**).

Isoprenoids are another rich source of biological Diels–Alder adducts. Biological Diels–Alder adducts of terpenoids have the following general features: (i) intermolecular [4+2]-adducts are predominant, (ii) most of them are isolated from plants, (iii) α-*exo*-methylene-γ-lactone frequently serves as a dienophile, (iv) *trans*-β-ocimene and myrcene are often found in the diene parts of the adduct and; (v) pyrolysis of the adduct produces both monomers, and the same retro-Diels–Alder reaction is observed in their MS spectra. In several cases, both diene and dienophile components are found in the same plant which produces the [4+2]-adduct. Thus, there are some doubts that the adducts are formed nonenzymatically. Most authors claim that the adducts are not artifacts for the following reasons. (i) Synthesis of the adduct from both components requires much more demanding conditions (temperature and concentration). (ii) The adducts are found in the fresh

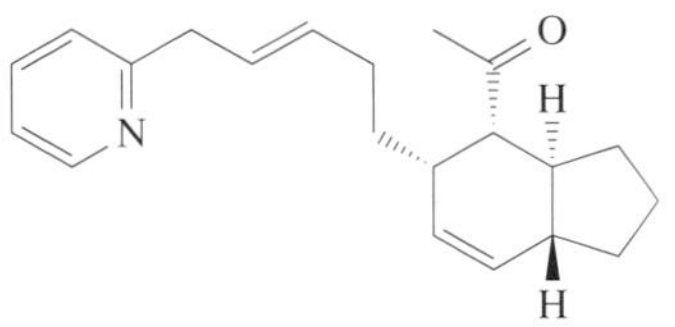

(36)

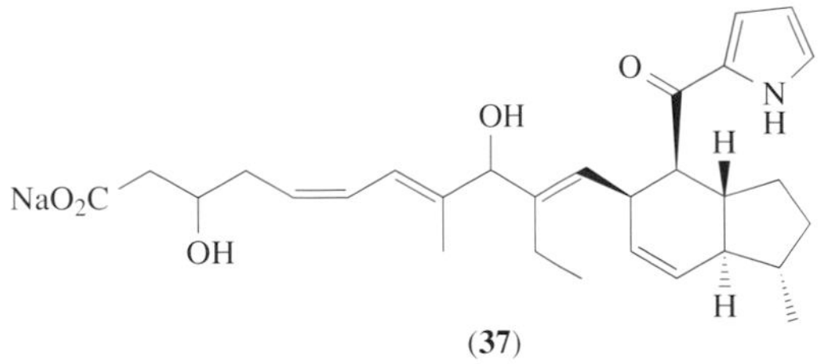

(37)

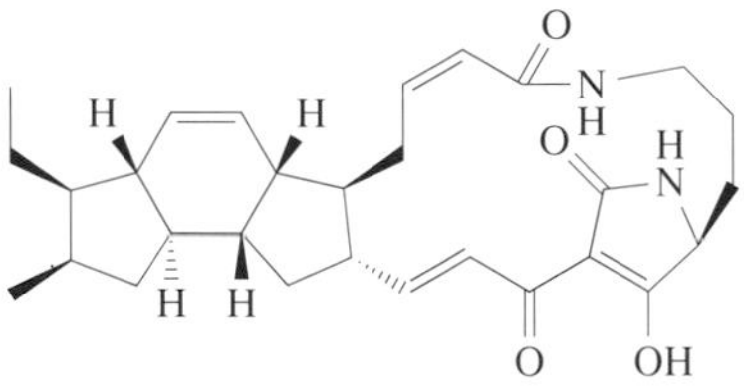

(38)

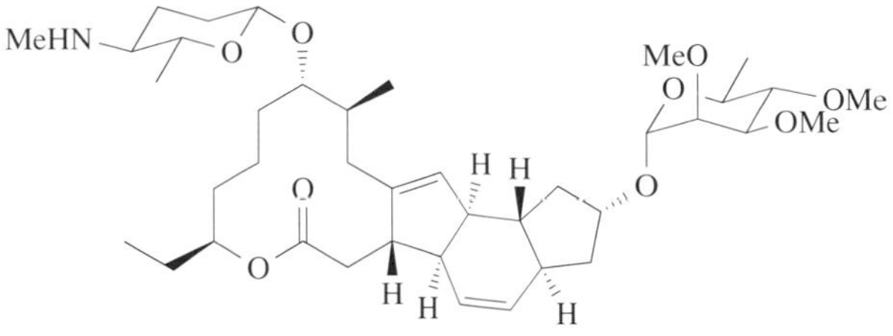

(39)

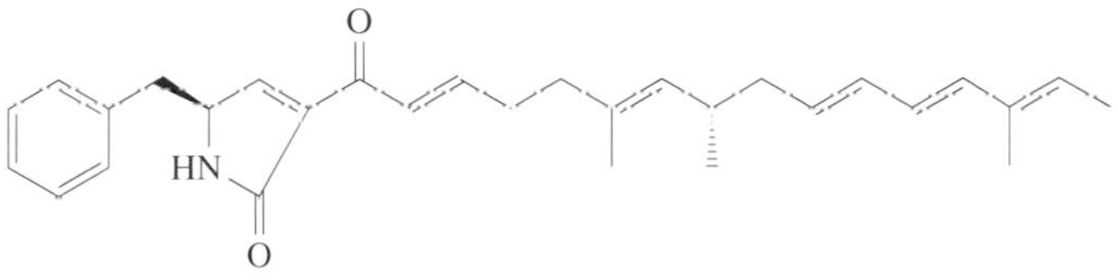

(40)

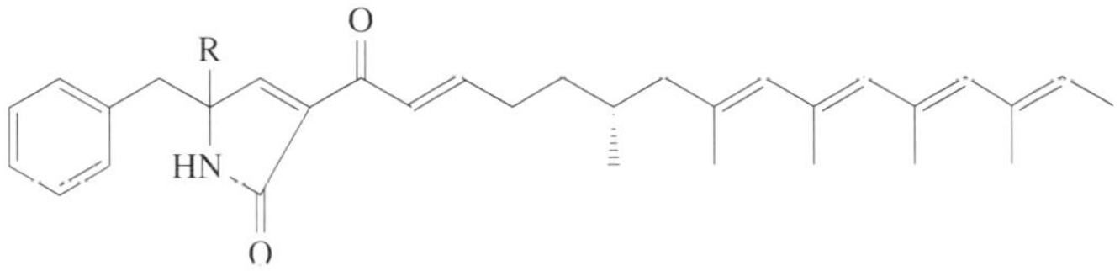

(41)

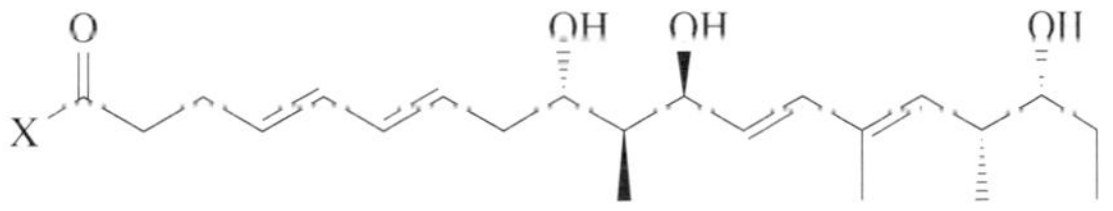

(42)

extract without concentration and purification. (iii) In some cases, the optically active adducts are produced from achiral diene and dienophile.

In various plants, intermolecular [4+2]-adducts using α-*exo*-methylene-γ-lactone are found. The liverwort *Plagiochila moritziana* produces plagiospirolides A (**43**)[75] and E (**44**)[76] which may be derived from the same eudesmanolide and different diene parts. Aestivalin (**45**)[77] from aerial parts of *Gaillardia aestivalis* may be considered as a Diels–Alder adduct between 8-hydroxy-α-phellandrene and a fastigilin analogue. Pungiolide A (**46**)[78] is produced in aerial parts of *Xanthium pungens*, from which the intramolecular adduct xanthipungolide (**49**),[78] probably from the electrocyclic reaction product (**48**) of 2*Z*-8-epixanthatin (**47**), is also isolated. A dimeric guaianolide absinthin (**50**)[79] from *Artemisia absinthium* is considered to be an *endo*-adduct from the closely related monomer artabsin. Ornativolide A (**51**),[80] biennin C (**52**),[81] altenolide,[82] bissesquiterpene lactone,[83] mexicanin F,[84] handelin,[85] and fruticolide[86] are also included in this group.

(**43**)

(**44**)

(**45**)

(**46**)

(**47**) (**48**) (**49**)

(**50**)

(**51**)

(**52**)

(**53**)

(**54**)

(**55**)

Monoterpene myrcene and *trans*-β-ocimene are frequently utilized as dienes. Heliocides H_1 (**53**) and H_4 (**54**)[87] are racemic metabolites from the *Gossypium* species which produces six isomeric congeners. Cycloaddition between *trans*-β-ocimene and 7-*O*-methylhemigossypolone gave both regioisomers in an *endo*-selective manner. Perovscone (**55**)[88] is a constituent of the medicinal plant *Perovskia abrotanoides*. Biogenesis originally proposed construction of the carbon skeleton from the diterpene quinone and the geranylpyrophosphate, but later an alternative [4+2]-cycloaddition route was proposed.[89] Majetich and Zhang succeeded in its chemical synthesis via the latter route.[89] Eudesmanolide adducts (**56**) and (**57**)[90] were isolated from *Artemisia herba-alba* and synthesized via inverse-electron demand Diels–Alder reaction with the corresponding dienone and myrcene.

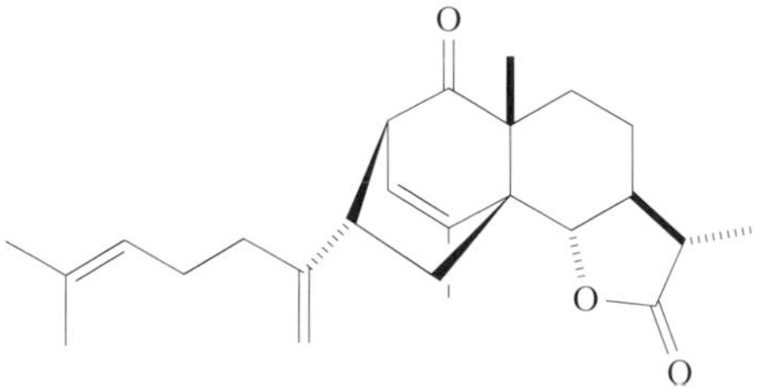

(**56**)

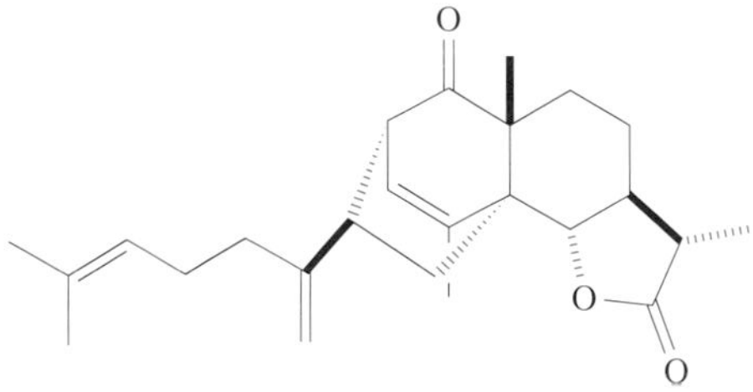

(**57**)

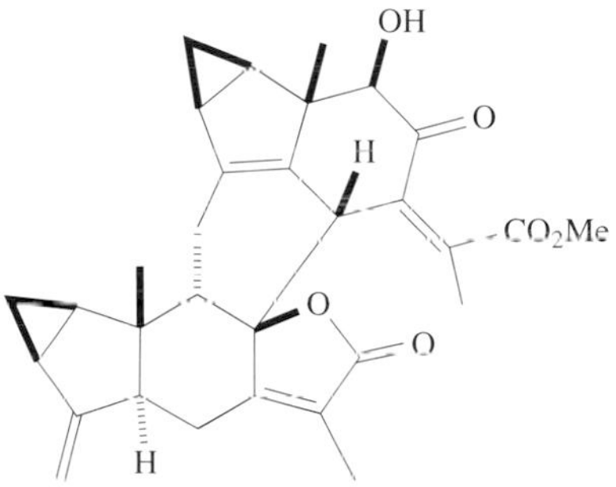

(**58**)

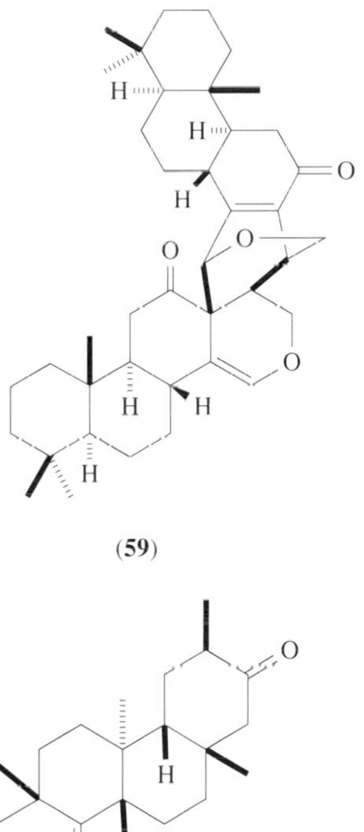

(**59**)

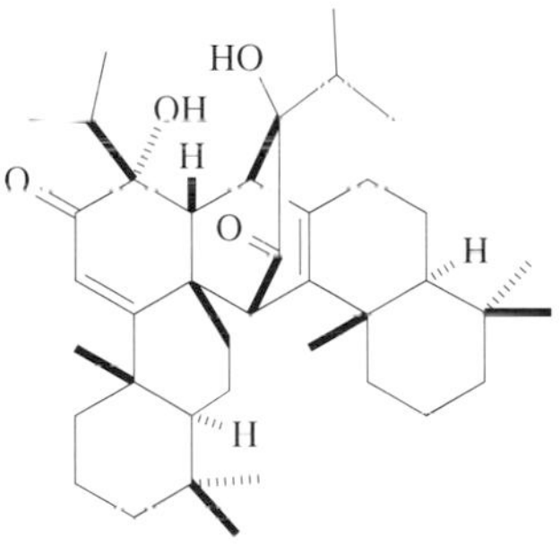

(**60**)

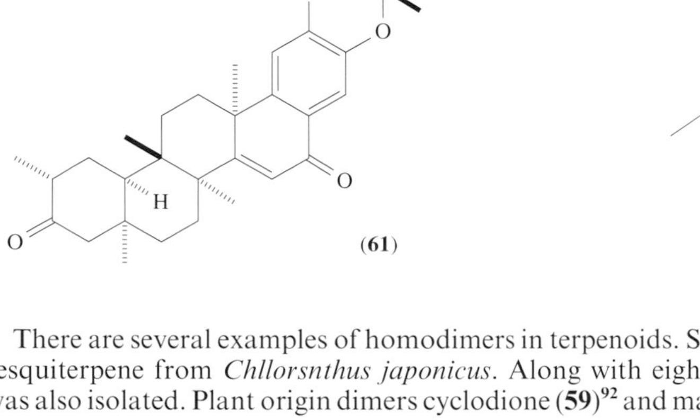

(**61**)

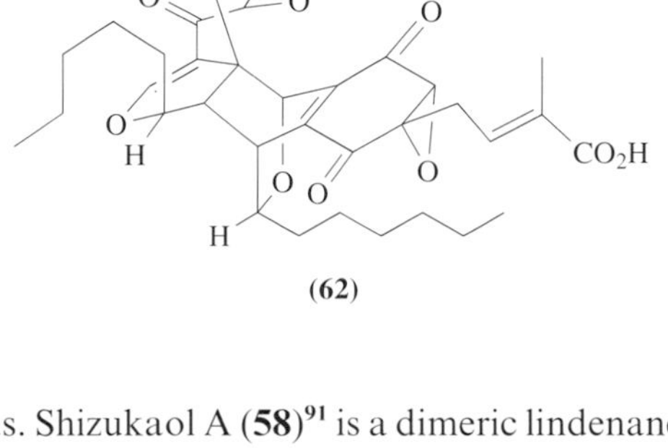

(**62**)

There are several examples of homodimers in terpenoids. Shizukaol A (**58**)[91] is a dimeric lindenane sesquiterpene from *Chllorsnthus japonicus*. Along with eight congeners, a dimeric [6 + 6]-adduct was also isolated. Plant origin dimers cyclodione (**59**)[92] and maytenone (**60**)[93] are dimeric diterpenoids which may be formed by *endo*-selective [4 + 2]-addition. A South American medicinal plant *Maytenus chuchuhuasca* produces a triterpene dimer xuxuarine Aα (**61**)[94] with various adducts which may originate from a combination of structurally related monomers. Cytotoxic agent torreyanic acid (**62**)[95] is a metabolite from the endophytic fungus *Pestalotiopsis microspora*. This complex metabolite may be obtained by Diels–Alder reaction between two epimeric prenylated benzoquinones which probably derived from epoxidation of the racemic precursor. The optically active adduct longithorone (**63**)[96] (Equation (2)) was isolated from the tunicate *Aplydium longithorax*. Its unique dimeric structure suggests that both inter- and intramolecular Diels–Alder reaction of the monomeric quinones would be involved in the formation of its carbon skeleton affording the correct stereochemistry.

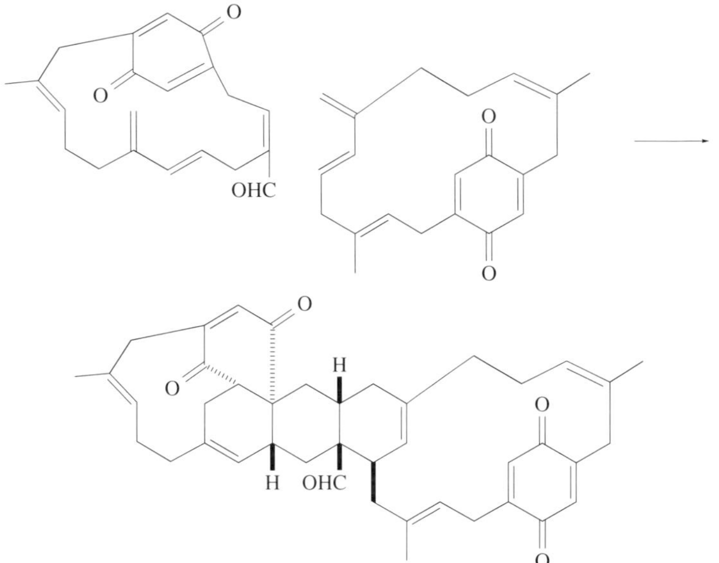

O O OHC O O

O O H H OHC O O (2)

(**63**)

In some cases, the involvement of an intramolecular reaction may be considered in biosynthesis. Andibenin B (**66**)[97] is a highly oxidized meroterpenoid produced by the fungus *Aspergillus variecolor*. On the basis of the biosynthetic study, it was proposed that a hypothetical intermediate (**64**) derived from farnesylpyrophosphate and benzoate provides the adduct (**65**) via the intramolecular inverse-electron demand [4 + 2]-cycloaddition (Equation (3)). Aglycone sordaricin (**70**) from an antibiotic fungal metabolite sordarin[98] is another case where an intramolecular Diels–Alder reaction is proposed in the formation of its carbon skeleton. On the bases of the co-occurrence of congener diterpenes cycloaraneosene (**67**) and 8β-hydroxyanalogue (**68**) in extracts of the fungus *Sordaria araneosa*, it was suggested that ring-opened product (**69**) is recyclized by intramolecular cycloaddition to afford (**70**)[99] (Equation (4)).

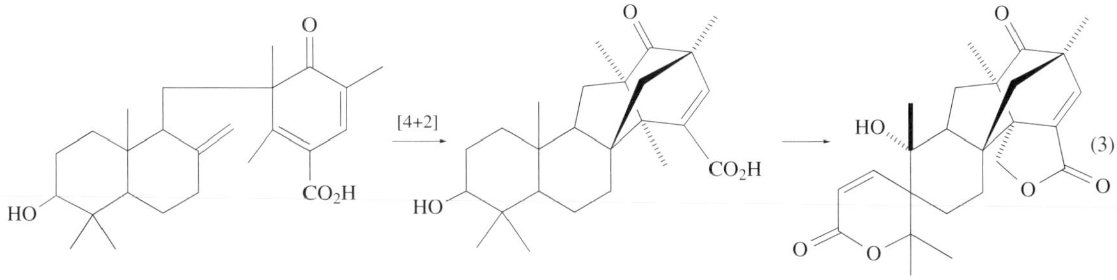

O HO CO_2H [4+2] O HO CO_2H O HO O O O O (3)

(**64**) (**65**) (**66**)

In marine natural products, several unique [4 + 2]-adducts have been reported. Methyl sarcophytoate[100] from soft coral *Sarcophyton glaucum* is an adduct from methyl sarcoate and the corresponding diene. The structurally similar biscembranoids methyl isosartortuoate[101] and methyl neosartortuoate[102] are found in another soft coral. Incinianin,[103] a novel sesterterpene from a marine sponge *Ircinia* sp., is a rare example of an intramolecular [4 + 2]-adduct.

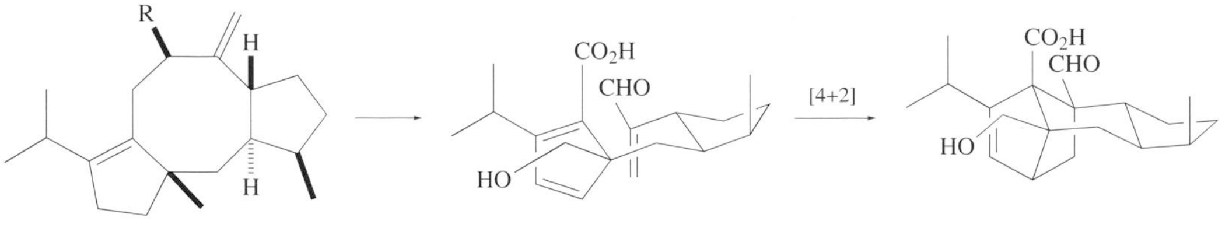

(**67**) R = H
(**68**) R = β-OH
(**69**)
[4+2]
(**70**)
(4)

There are only a few examples of Diels–Alder-type steroids. One of them is bistheonellasterone,[104] a dimeric steroid, isolated together with two new 3-keto-4-methylene steroids, theonellasterone and conicasterone from the Okinawan marine sponge *Theonella swinhoei*. Bistheonellasterone is thought to be biosynthesized through a Diels–Alder cycloaddition of theonellasterone and its Δ^4-isomer.

Two unusual compounds, which are made up of a sterol portion fused at C-5 and C-6 to a polyketide atrovenetin-like part, have been isolated from a fungus that causes Sirococus shoot blight of spruce.[105] Sirosterol is an adduct of ergosterol and atrovenetinone and dehydroazasirosterol is an adduct of 9(11)-dehydroergosterol and an azaatrovenetin. Although a biogenetic pathway uniting ergosterol endoperoxide and an atrovenitin-type molecule has been suggested, another possible route would be cycloaddition of ergosterol and atrovenetinone.

An interesting example of Diels–Alder-type natural products in the lignan field is asatone (**73**) which was isolated from *Asarum teitonense*. Later, two closely related novel neosesquilignans, heterotropatrione (**74**) and isoheterotropatrione (**75**), were isolated from *Heterotropa talai* M.[106,107] Biogenetically, these metabolites arise from oxidation of a phenol (**71**) and successive addition of methanol to produce a dienone (**72**) which dimerizes to (**72**) through a Diels–Alder reaction. Further cycloaddition of (**73**) with (**72**) yielded sesquilignans, heterotropatrione (**74**) and isoheterotropatrione (**75**) (Scheme 3). Having no optical activity, all these lignans would arise by spontaneous cycloaddition of the dienone (**72**) after enzymatic oxidation of the phenol. Actually,

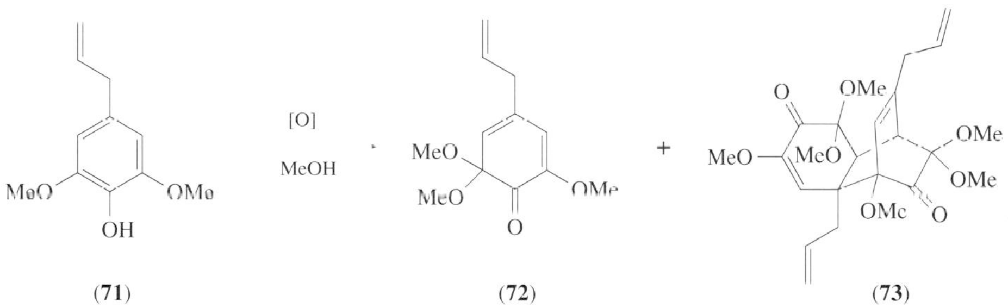

(**71**)
[O]
MeOH
(**72**)
(**73**)
(racemic)
(**74**) 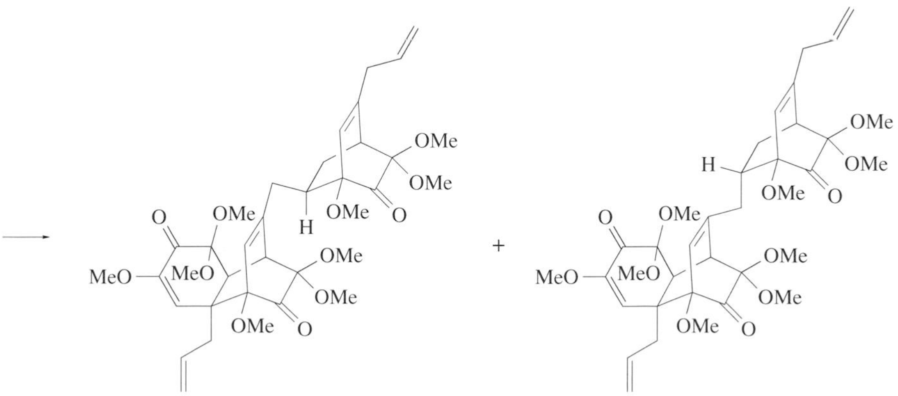
(racemic)
(**75**)
(racemic)

Scheme 3

electrolytic oxidation of the phenol (**71**) in methanol afforded asatone (**73**) and the dienone (**72**) and the latter compound dimerizes spontaneously at room temperature to yield asatone quantitatively.[108–110]

Several naphthalene derivatives of a novel skeletal type have been isolated from the bark of *Brombya platynema* F. Muell (Rutaceae).[111] The metabolites, brombin II (**76**) and brombin III (**77**) could be biogenetically formed via oxidative coupling of two cinnamic acid residues. However, this seems unlikely given the perhydrogenated nature of one of the six-membered rings. An alternative route would involve linkage of a single C_6–C_3 moiety to an acetate chain, a hypothesis supported by the isolation of a structurally related linear chain metabolite. A biosynthetic route leading to a 1-piperonyldodecane intermediate is shown in Scheme 4. The resulting all-*trans*-isomer (**78**) could subsequently undergo two possible Diels–Alder cyclizations to yield two racemic products identical to (**76**) and (**77**) in approximately equal quantities, although (**77**) is further converted to other products.

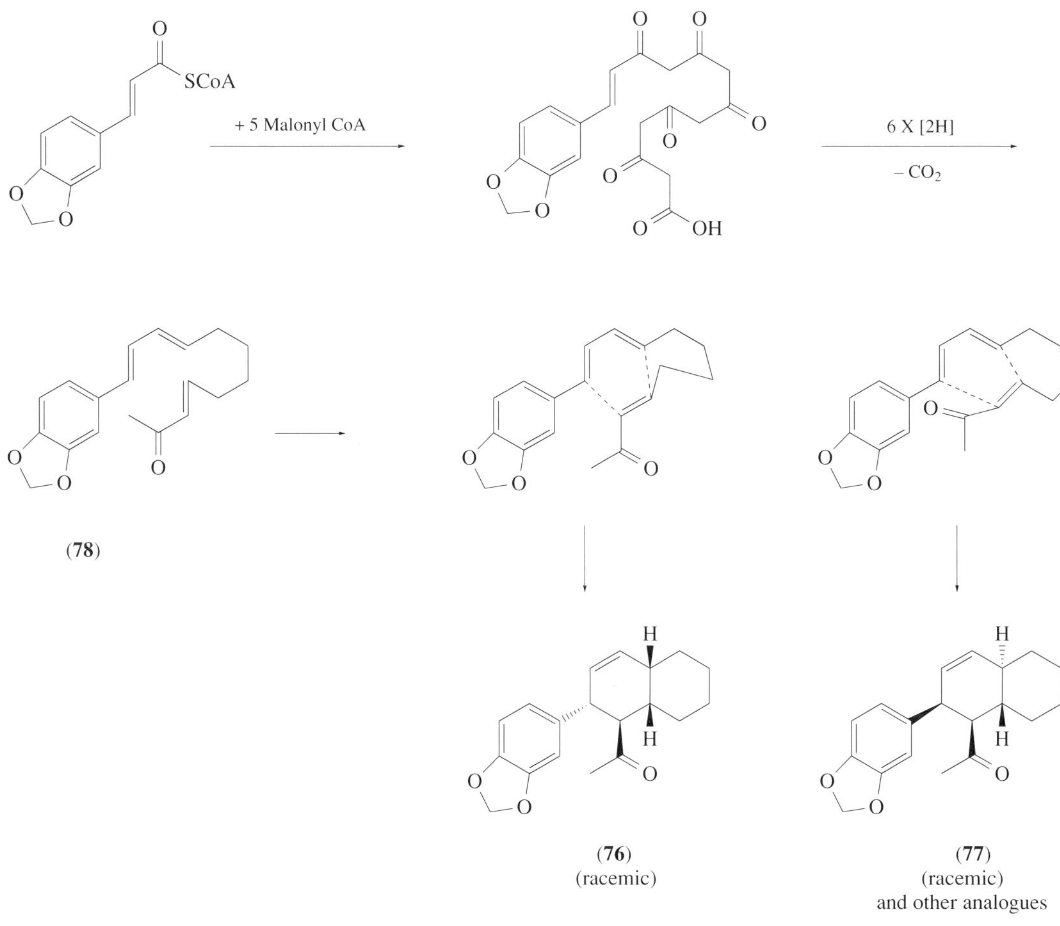

+ 5 Malonyl CoA

6 X [2H]

– CO_2

(**78**)

(**76**)
(racemic)

(**77**)
(racemic)
and other analogues

Scheme 4

After discovery of the first dimeric coumarin system thamnosin (**79**),[112] which contains a cyclohexene ring, several related coumarines, isothamnosins A[113] and B, phebalin (**80**),[114] toddasin (mexolide) (**81**),[115,116] toddacoumalone,[115,117] and microcybin[118] have been found in various plants. Among them, toddacoumalone (**82**) isolated from *Toddalia asiatica* (L.) is the first example of a natural mixed dimer of coumarin (**83**) and quinolone (**84**), each of which was observed in the mass spectrum of (**82**) through retro-Diels–Alder reaction.[117]

A structurally unique group of marine alkaloids, manzamines A (**85**)[118] and B (**86**),[119] has been discovered. In 1992, an elegant biosynthetic pathway involving Diels–Alder reaction of bis-

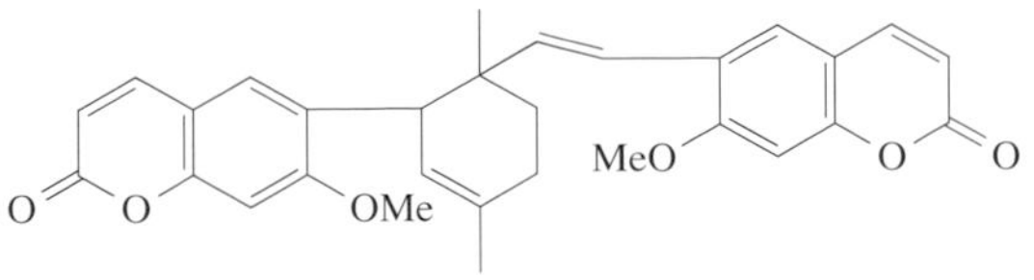

(**79**)
(racemic)

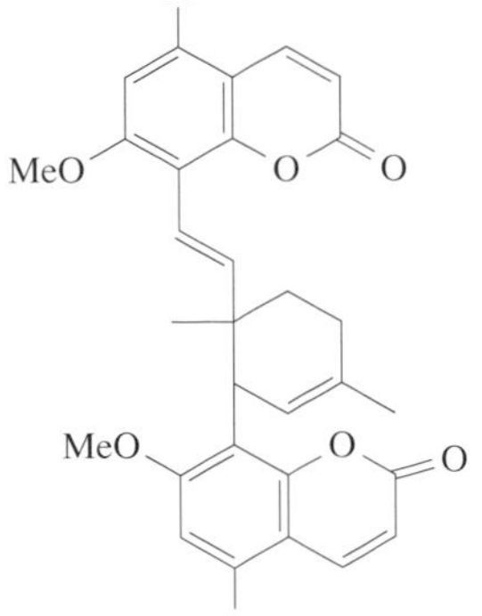

(**80**) R = H
(**81**) (mexolide) R = OMe
(racemic)

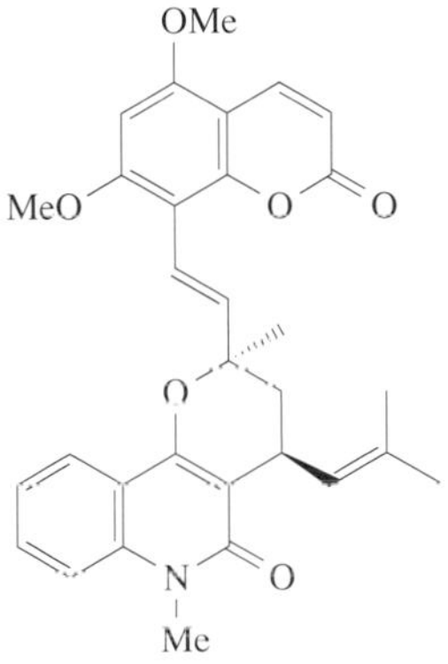

(**82**)
(racemic)

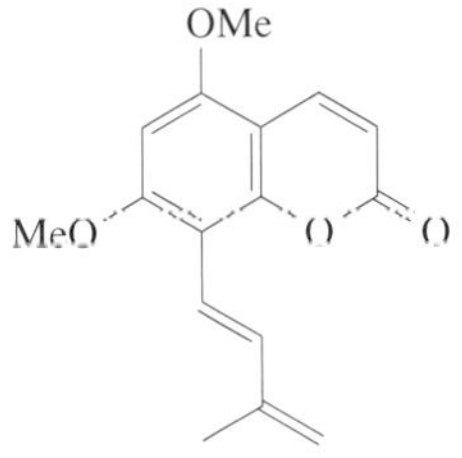

(**83**)

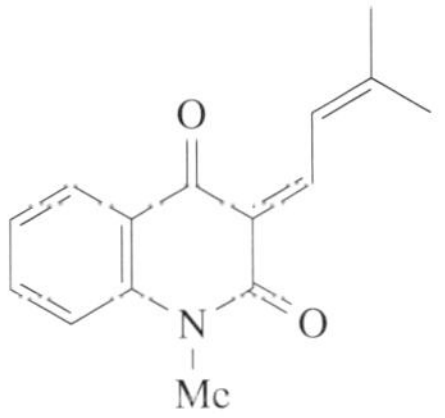

(**84**)

dihydropyridine (**92**) to manzamines was suggested by Baldwin and Whitehead,[120] followed by the discovery of plausible precursors, keramaphidins (**87**), ircinal A (**88**), and ircinal B (**89**).[121,122] In addition, two new manzamine-related alkaloids, ircinol A (**90**) and ircinol B (**91**) have been isolated from the Okinawan marine sponge *Amphimedon* sp.[123] (Scheme 5).

Interestingly, these two alkaloids (**90**) and (**91**) are antipodes of the reduced forms of ircinal A (**88**) and ircinal B (**89**), though these four alkaloids were isolated from the same sponge. Biogenesis including these manzamine-related alkaloids would involve intramolecular Diels–Alder reaction of bisdihydropyridine (**92**) to give a racemic product (**87**) without participation of enzyme. Then each of the racemic keramaphidins (**87**) would be converted by enzymes to ircinals A (**88**) and B (**89**), and ircinols A (**90**) and B (**91**), respectively. Closely related alkaloids, xestocyclamine A[124] and ingenamine,[125] would be formed via a similar biosynthetic pathway.

Novel alkaloids, segoline A (**93**), segoline B (**94**), and isosegoline A (**95**), have been isolated from the Red Sea tunicate *Eudestoma* sp.[126] Since tetracyclic aromatic alkaloids, the cystodytins, were isolated from the Okinawan tunicate *Cystodytes dellechiajel*,[127] the cyclic imide of segolines A and B (**93**) and (**94**) and isosegoline A (**95**) would be formed by intramolecular Diels–Alder reaction of cystodytin B (**96**) involving oxidation at C-12 before or after the cyclization (Equation (5)).

Biologically active extracts of the Caribbian sponge *Agelas conifera* have yielded the diacetate salts of new sceptrin (**97**) and bromopyrroles (**98**).[128–130] These compounds were found to be antiviral and antibacterial and were optically active. The structures of (**97**) and (**98**) suggest [2+2]- and [4+2]-cycloaddition, respectively; however, their optical activity indicates their formation via an enzymatic cycloaddition (Equation (6)).

There are a number of examples which are presumed to arise from a mixed biosynthetic pathway involving the Diels–Alder reaction. Since typical examples according to basic biosynthetic pathways have been described in the previous sections, only two additional reports are reviewed here.

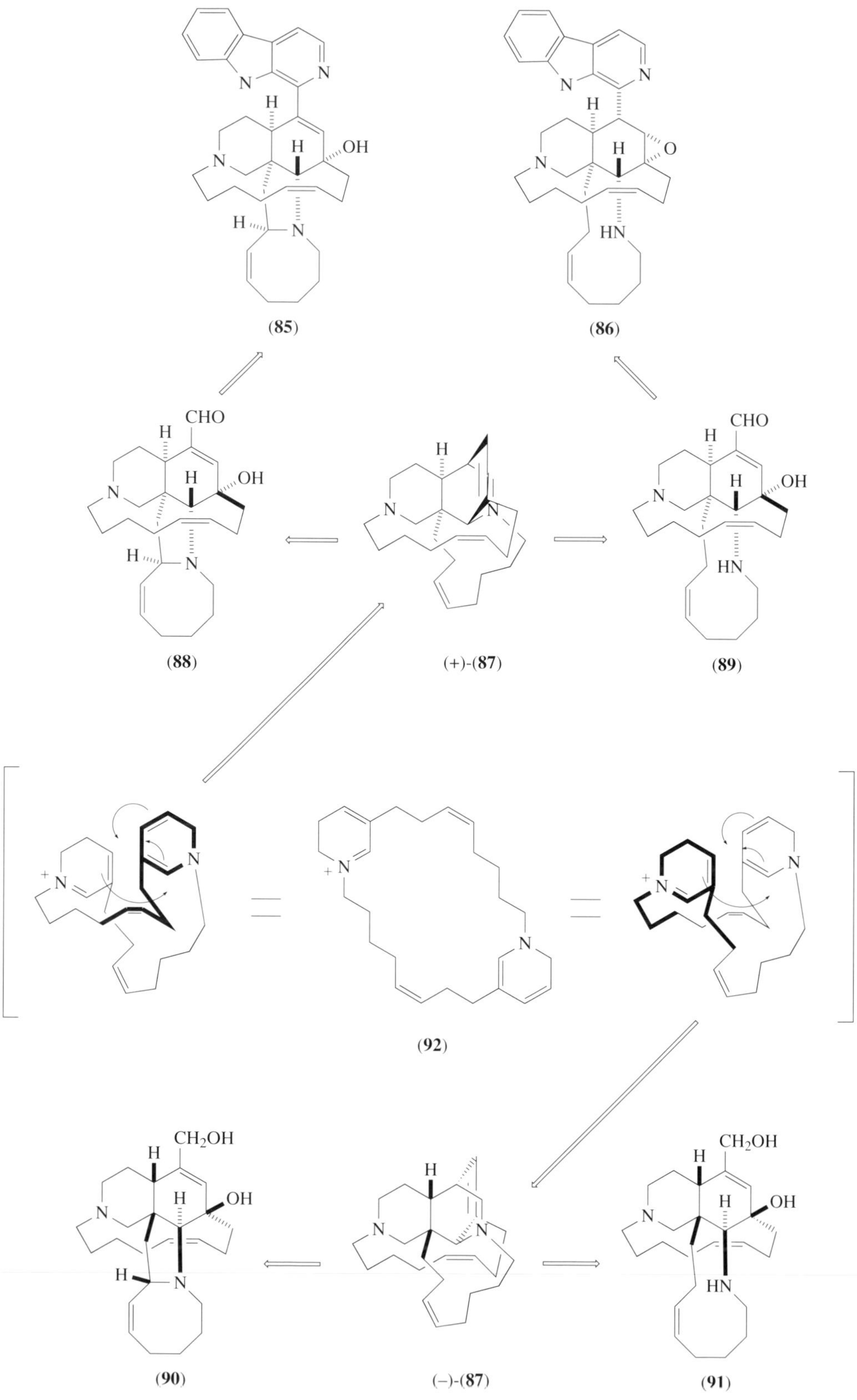

(**85**) (**86**)

(**88**) (+)-(**87**) (**89**)

(**92**)

(**90**) (–)-(**87**) (**91**)

Scheme 5

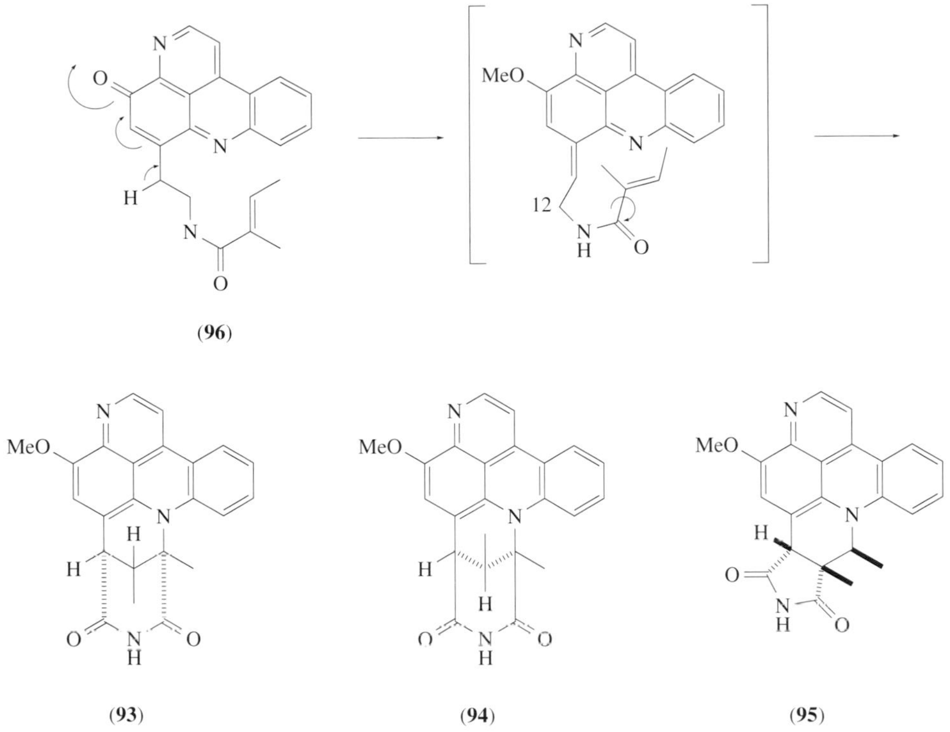

(96)

(93) (94) (95)

(5)

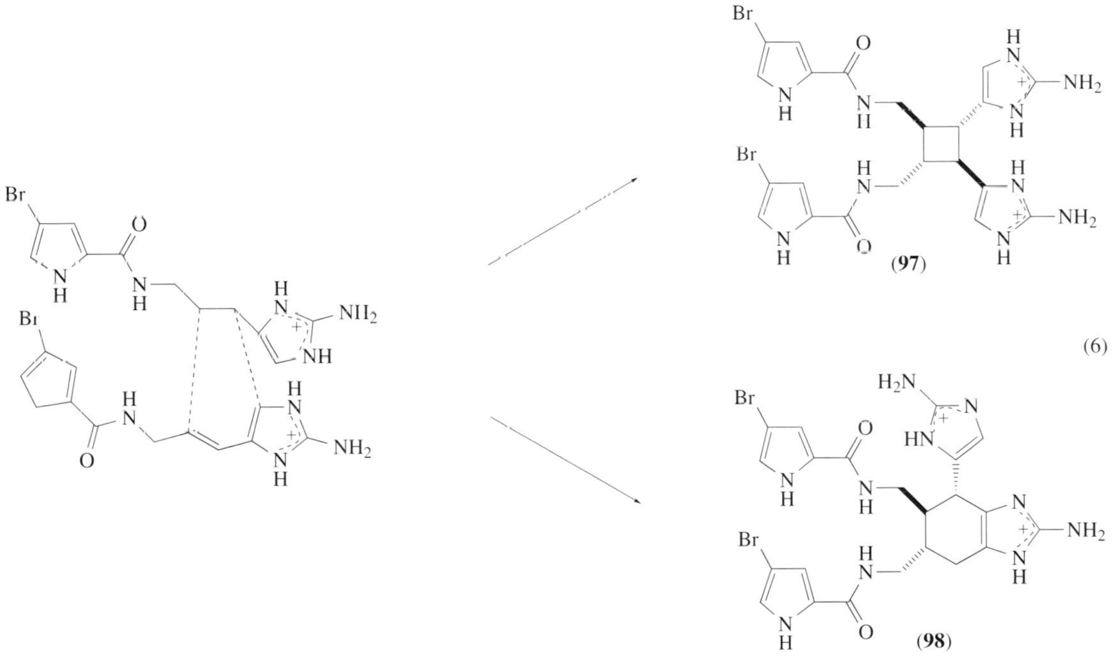

(97)

(6)

(98)

Four new dimeric prenylated quinonolone alkaloids, vepridimerine A (**101**) and vepridimerines B, C, and D,[131] have been isolated from the bark of *Vepris louisii* and *Oricia renieri* (Rutaceae) as a racemic form. Their formation, in this case of (**101**), can be rationalized in terms of the Diels–Alder reaction of veprisin (**99**) and the diene (**100**) followed by addition of a hydroxy group to the residual double bond (Equation (7)).

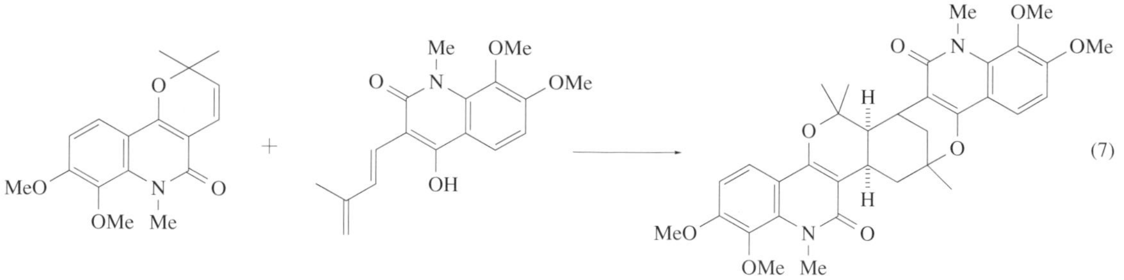

(99) + (100) → (101) (7)

Six novel fatty acid derivatives, manzamenone A (**103**) and mazamenones B, C, D, E, and F possessing a previously unknown skeleton, were isolated from extracts of the sponges of the genus *Plakortis*.[132] These manzamenones could be assumed to have been generated biosynthetically from two fatty acid-derived precursors (**A**) and (**B**) (Scheme 6). Both (**A**) and (**B**) might be derived through condensation of malonate with 4-oxo-2,3-dehydrocarboxylic acid (**C**). A butenolide (**102**), which is an equivalent compound to (**C**), was previously isolated from a Micronesian *Plakortis* sponge. It may be proposed that manzamenones were yielded through an enantioselective intermolecular *endo*-type [4+2]-cycloaddition between (**A**) and (**B**).

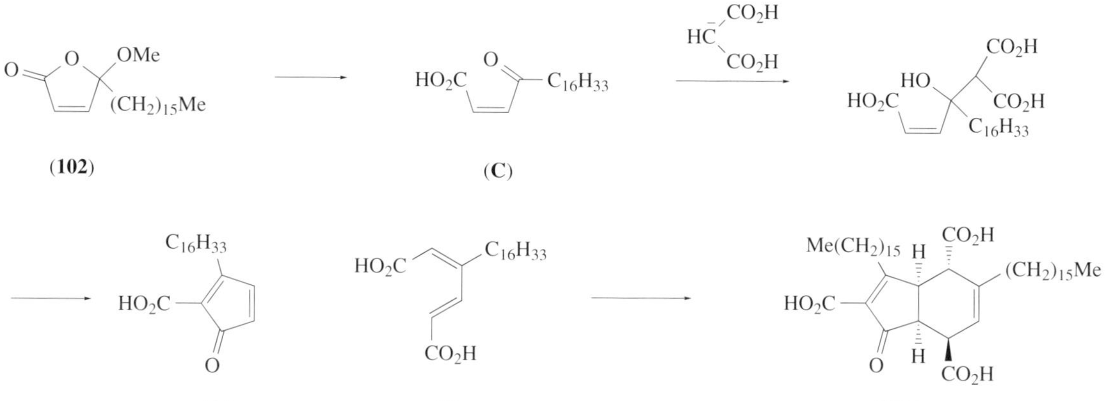

Scheme 6

Other Diels–Alder-type adducts are diclausenan,[133] pummeloquinone,[134] garcilivin A[135] from plant, plakotenin[136] from marine animal antibiotics, ergophilone A,[137] pinnatoxin A,[138] iristectorone A,[139] pittosporumxanthin A1,[140] and procentrolide[141] from marine dinoflagellate.

1.15.2.2 Biosynthetic Studies of Diels–Alder-type Natural Products

There are not many examples of Diels–Alder-type natural products that have been studied biosynthetically.

In the biosynthetic study of nargenicin A_1 (**104**), Cane and Guanglin carried out a number of feeding experiments using di-, tri-, tetra-, and pentaketide precursors[142] (Equation (8)). Intact incorporation of these precursors provided strong support for the idea that small units are directly incorporated into the polyketide chain without further modification and that the *E*-alkene in the pentaketide precursor served as a dienophile. For the hypocholesterolemic agent mevinolin (**105**), Vederas and co-workers established the origin of carbon, hydrogen, and oxygen atoms by experiments incorporating the stable-isotope labeled precursors.[143,144] They also prepared doubly ^{13}C-labeled di-, tri-, tetra-, and hexaketide precursors to prove their intermediacy. The basic idea of this proposal is that the polyketide synthase may catalyze the [4+2]-cycloaddition of the hexaketide (Equation (9)). This is an attractive hypothesis since it can be applied to any decalin polyketide which does not possess a conjugated carbonyl group in its dienophile part. Unfortunately, all their attempts at incorporation were unsuccessful although they did succeed in incorporating similar oligoketides in another system.[145]

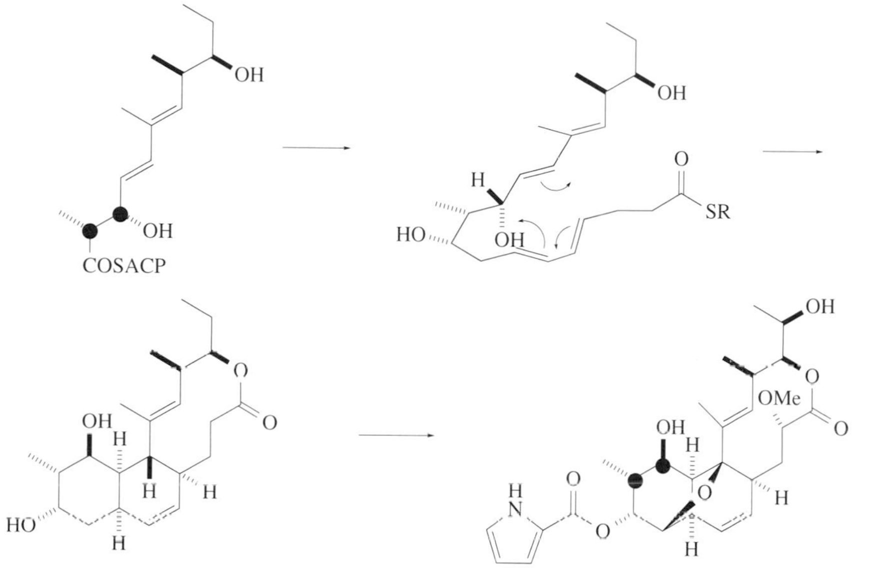

OH OH COSACP OH H HO OH O SR OH O OH H H H O HO H OH O OMe OH H O H N O O H O H (104) ● = ^{13}C (8)

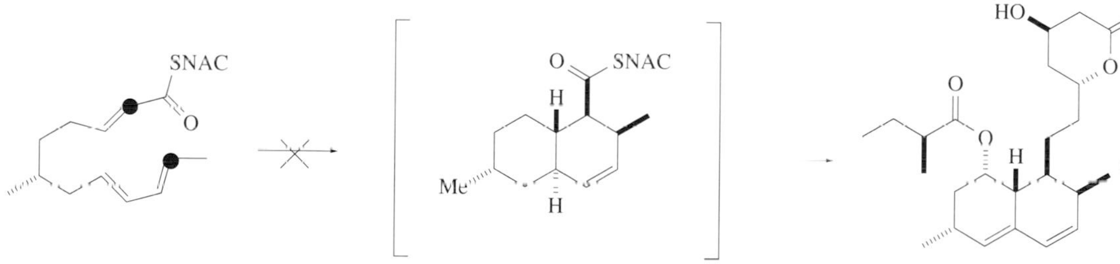

SNAC O SNAC O H Me H HO O O O O H (105) (9) ● = ^{13}C

An interesting biosynthesis involving a biological Diels–Alder reaction has been studied with kuwanon J (**108**), whose congeners were also isolated from *Morus alba* L.[146–148] Experiments feeding 4-methoxychalcone (**106**) into *M. alba* cell cultures yielded optically active cycloadduct (**109**) and two other congeners besides a prenylated product (**107**), indicating that prenylation takes place after aromatization of the cinnamoylpolyketide-derived chalcone skeleton. Metabolites (**109**) and others revealed that the two molecules of precursory chalcone (**106**) were incorporated intact into the optically active products through enzymatic Diels–Alder reaction after dehydrogenation of one of the prenylchalcones[149] (Scheme 7). The structure of optically active artonin (**110**) from an Indonesian moraceous plant was similarly established utilizing the enzyme system of *M. alba* cell cultures.[150] A closely related cyclic compound, brosimonin A, was isolated from a Brazilian moraceous plant *Brosimopsis oblongifolia*. The optically active brosimone A can also be considered to have originated from two enzymatic Diels–Alder cycloadditions between two identical dehydroprenylchalcone units.[151] As similar types of natural product, cassumunarin A (**111**), and cassumunarins B and C were isolated from *Zingiber cassumunar*. Since all these antioxidants are optically active Diels–Alder-type adducts, biosynthetic pathways involving enzymes are deduced for them.[152]

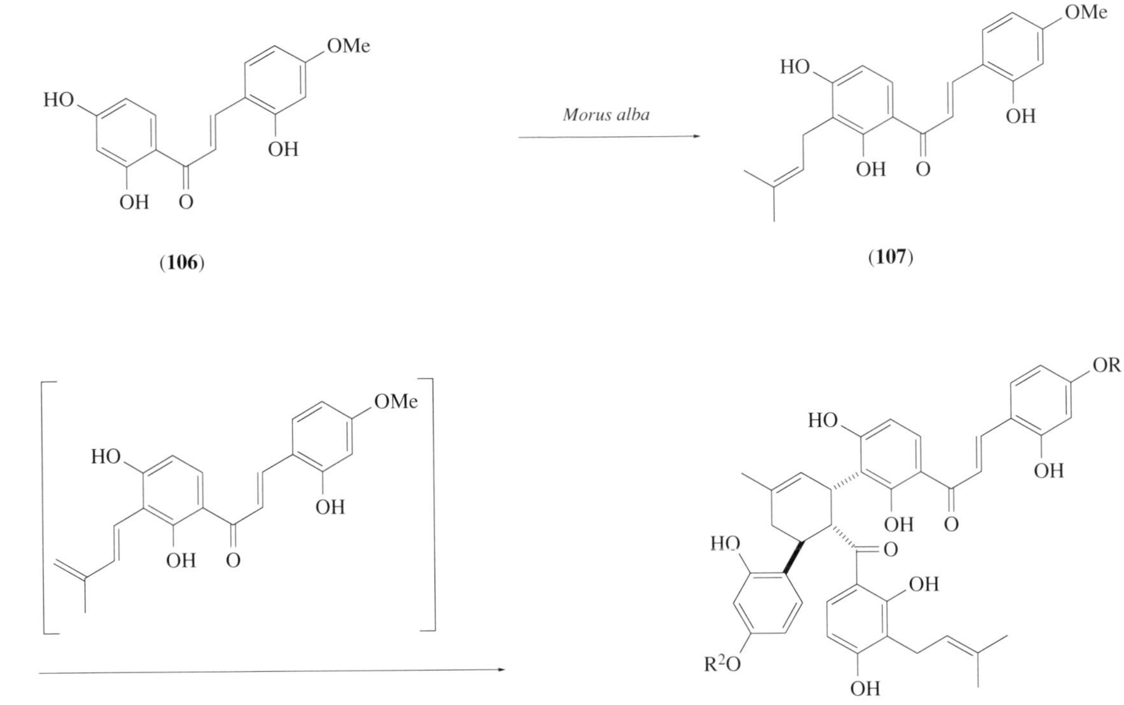

(**106**) *Morus alba* (**107**)

(**108**) $R^1, R^2 = H$
(**109**) $R^1, R^2 = Me$

Scheme 7

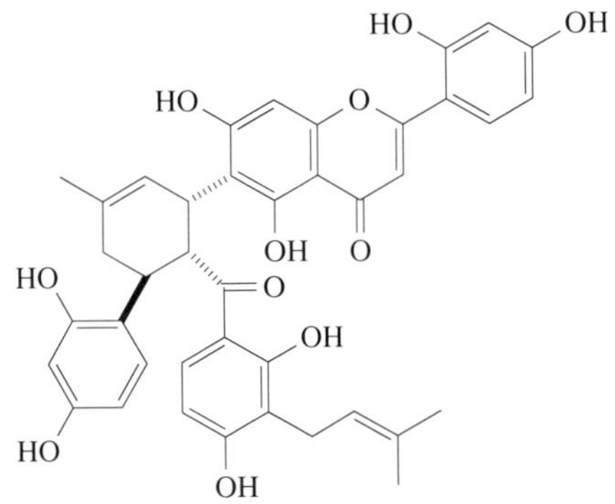

(**110**) 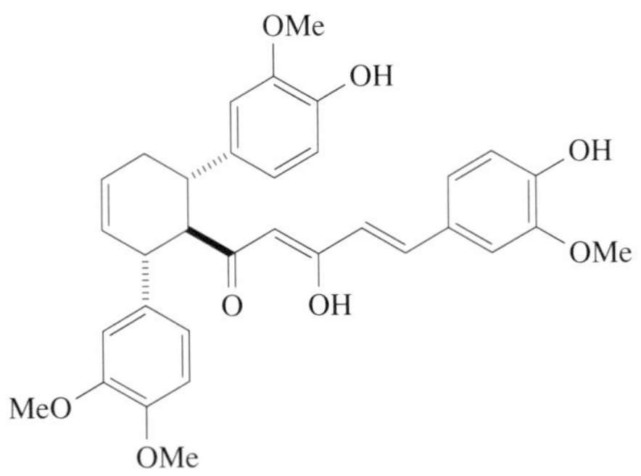 (**111**)

The brevianamides A (**116**) and B (**117**)[153] are the simplest representatives of a curious class of mycotoxins which also includes the paraherquamides[154] and the marcfortines.[155] The incorporation of a precursor in the biosynthesis of brevianamides has provided indirect evidence for biogenesis involving a biological Diels–Alder reaction. Thus, feeding experiments using [8-^{3}H]deoxybrevianamide E (**112**) led to significant incorporation of radioactivity into both (**116**) and (**117**).[156] The results of these feeding experiments suggest a biosynthetic pathway (Scheme 8), which does not involve the intermediate previously presumed. An *R*-selective hydroxylation reaction occurs at the three-position of (**112**) furnishing (**113**). Subsequent pinacol-type rearrangement of (**113**) sets the *R*-absolute stereochemistry (**114**) of the indoxyl, since the 3-hydroxyindolenine (**113**) is the sterically favored product of oxidation, as shown in the autoxidation of (**112**). Oxidation of (**114**) followed by enolization forms the azadiene (**115**). An intramolecular Diels–Alder cyclization from a major rotamer (**115a**) directly leads to (**116**), and a minor rotamer (**115b**) cyclizes to (**117**). It would be interesting to know whether or not the enzyme catalyzing oxidation of (**114**) also participates in the cycloaddition processes.

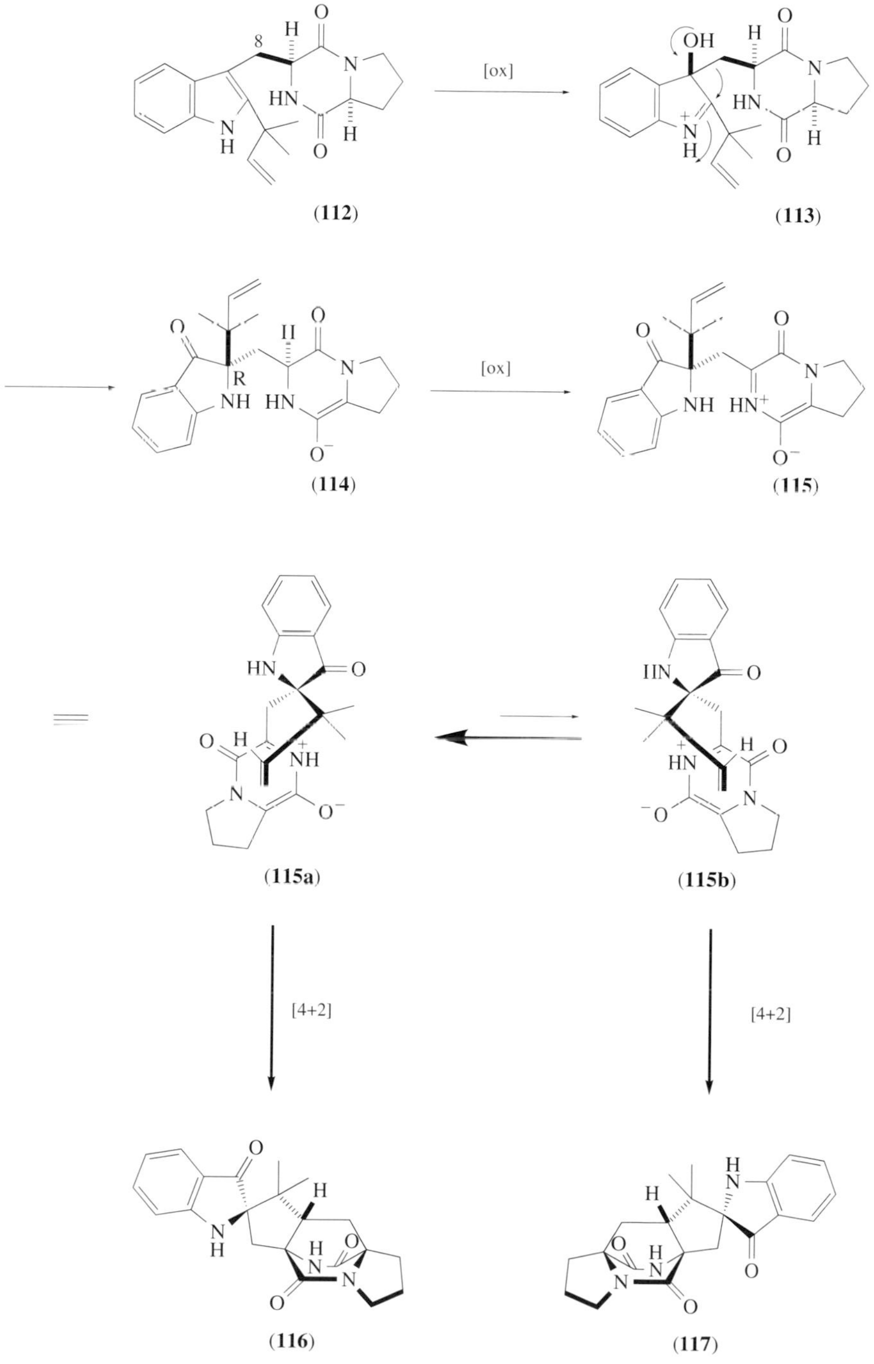

(**112**) [ox] (**113**)

(**114**) [ox] (**115**)

(**115a**) (**115b**)

[4+2] [4+2]

(**116**) (**117**)

Scheme 8

1.15.3 BIOLOGICAL DIELS–ALDER REACTION IN BIOSYNTHESIS OF POLYKETIDE PHYTOTOXINS

In this section, four phytotoxins, betaenone B, chaetoglobosin A, solanapyrone A, and macrophomic acid are selected for biosynthetic studies involving a biological Diels–Alder reaction. These phytotoxins are presumed to have been derived through a polyketide pathway involving intra- and intermolecular Diels–Alder reactions *in vivo*.

1.15.3.1 Betaenones

Betaenones A–F (**118**)–(**123**) are phytotoxic metabolites produced by *Phoma betae*, the causal fungus of leaf spot disease on sugar beet (*Beta vulgaris* L.).[157,158] Among betaenones, betaenone C (**120**) strongly inhibited both protein and RNA synthesis compared with betaenones A and B and exhibited the highest phytotoxic activity causing wilting of the host plant.

Betaenone B (**119**) was obtained as the main product from culture filtrate of the fungus and its structure was deduced from the spectral data of (**119**) and its derivatives. The structure of betaenone A (**118**) was determined by X-ray analyses and the structure of betaenone B (**119**) was elucidated by chemical correlation with betaenone A (**118**). Thus, oxidation of (**119**) with PCC yielded betaenone C (**120**), which was converted to betaenone A (**118**) through aldol condensation. The absolute stereostructure of betaenone B (**119**) was deduced by ORD and c.d. measurements. The same stereostructure was also confirmed by the correlation of (**119**) with betaenone D (**121**), whose absolute configuration was determined by the application of the c.d. exciton chirality method to the benzoate derived from betaenone D (**121**).[159]

At the same time, a phytotoxin, stemphyloxin I (**124**), was isolated from culture filtrate of *Stemphylium botryosum*, the causal fungus of leaf spot disease of tomato.[160] Structurally, stemphyloxin I (**124**) is quite closely related to betaenone C (**120**). Diplodiatoxin (**125**), a mycotoxin, was isolated from maize infected with *Diplodia maydis*, which causes a well-known disease, diplodiosis, among cattle and sheep in southern Africa.[161] Biogenetically, stemphyloxin I (**124**) and diplodiatoxin (**125**) may be derived through intramolecular Diels–Alder reaction of precursor polyketide trienes, similarly to betaenones.

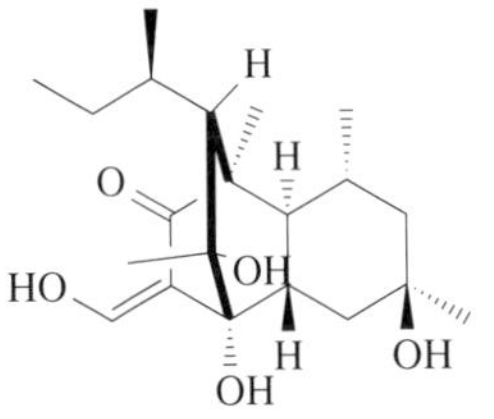

(**118**)

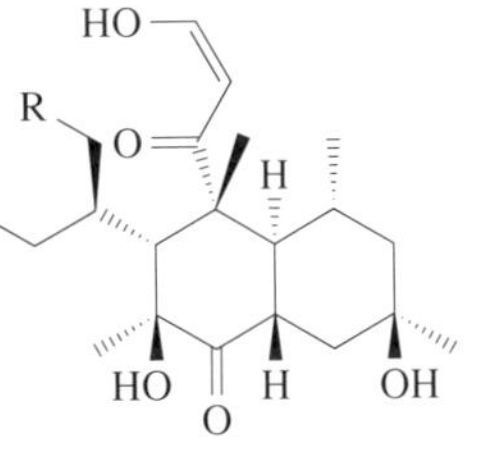

(**120**) R = H
(**124**) R = OH

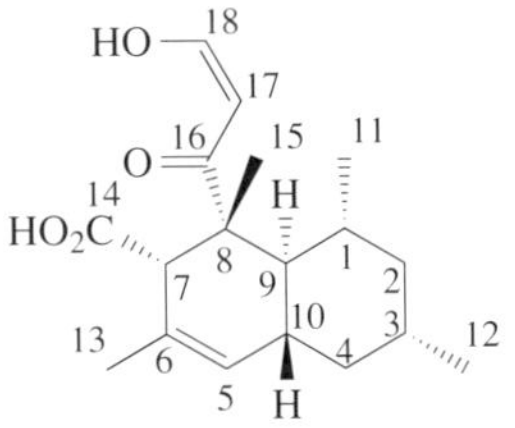

(**125**)

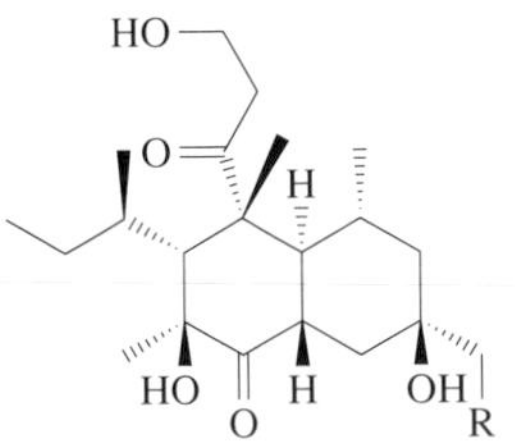

(**119**) R = H
(**121**) R = OH

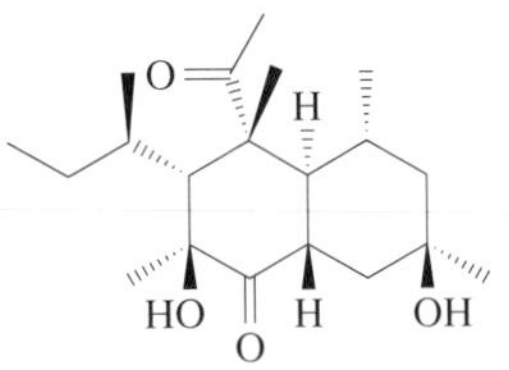

(**122**)

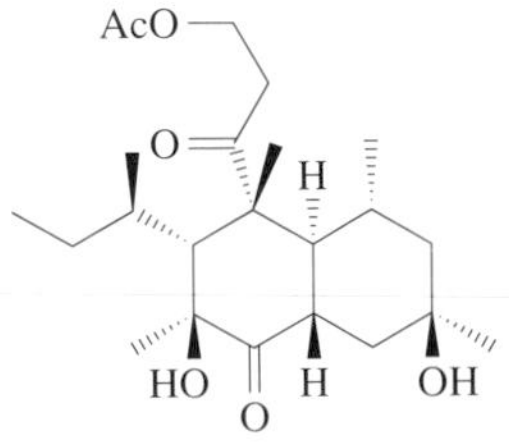

(**123**)

Structurally, betaenone B (**119**) could be regarded as a 1,2-dialkydecalin polyketide, which might be derived from a single polyketide chain (Scheme 2). A number of compounds belonging to this class have been found among microbial secondary metabolites (see Section 1.15.2.1). Intramolecular

Diels–Alder reaction of the reduced polyketide chain which has a conjugated (*E*,*E*)-diene and (*E*)-dienophile can explain the stereochemistry found in metabolites of this type. Along this line, a biogenetic pathway to this family has been proposed by several research groups (see Section 1.15.2.2). In the biosynthesis of polyketide metabolites, the most common six-membered carbocycle formation reaction is an aldol-type condensation like that of aromatic polyketides. Although Diels–Alder reactions using biological systems or catalytic antibody have been found, no enzyme catalyzing biosythetically unusual C—C bond formation has been identified. Thus, the biological Diels–Alder reaction remains a controversial topic. For this reason a biosynthetic study of betaenone B (**119**) was begun.

Before the biosynthetic studies, total synthesis of betaenone C (**120**) was completed through a biomimetic route as shown in Scheme 9.[162] The synthetic strategy involving the intramolecular Diels–Alder reaction of the chiral triene (**126**) is almost the same as that employed in the synthesis of diplodiatoxin (**125**).[163] The triene (**126**) was retrosynthesized into three segments (**a**), (**b**), and (**c**) and segment (**a**) was prepared through three steps starting from (*R*)-2-methylbutanal. Segment (**b**) was synthesized through 11 steps starting from the epoxide, which was prepared by Sharpless oxidation of nerol. The Kocienski–Lythgoe condensation of (**a**) with (**b**) and the Wittig–Horner reaction of the dienals with (**c**) yielded the trienes (**126**), which contain 20% of the *Z*-isomer at C-10. The intramolecular Diels–Alder reaction of the triene (**126**) at 115 °C for 36 h proceeded smoothly to give the single product (**127**). This can also be rationalized in terms of kinetic selectivity in the cycloaddition of the trienes, as in the case of the synthesis of diplodiatoxin. Oxidation of the adduct (**127**) afforded the ketol (**128**). Removal of the protective groups of the ketol (**128**) gave betaenone B (**119**), which was further converted to betaenone C (**120**).

The total syntheses of (+)-diplodiatoxin (**125**) and (−)-betaenone C (**120**) not only confirmed the stereochemistry of these bioactive compounds and probetaenone I,[164] but also provided useful intermediates for biosynthetic studies of (−)-betaenone C (**120**).

Biosynthetic study of betaenones has given indirect evidence for biological Diels–Alder reaction. Conventional incorporation experiments with [1-^{13}C]-, [2-^{13}C]-, and [1,2-$^{13}C_2$]acetate and with [S-^{13}C]-L-methionine have shown that the betaenone skeleton was constructed from an acetate-derived octaketide with five C_1 units derived from methionine as shown in Scheme 10.[165] The origin of oxygen atoms was examined, and the incorporation of [1-^{13}C, $^{18}O_2$]acetate was demonstrated. In the ^{13}C NMR spectrum of the enriched sample, ^{18}O isotope-induced shifts were observed at C-16 and C-18 but not at C-1, where the label derived from acetate was expected.[166] Although an experiment with [$^{18}O_2$]oxygen gas failed to detect any isotope-induced signal, the above data suggest that oxygen atoms at C-1, C-2, and C-8 were derived from molecular oxygen or the medium.

In order to obtain information on late-stage oxidative transformation in betaenone biosynthesis, an experiment was performed with cytochrome P450 inhibitor.[167] Cytochrome P450-dependent hydroxylation is common in the biosynthesis of secondary metabolites.[168–170] This type of oxidation usually takes place in the late stage of biosynthesis and is an important step in introducing biological activity. P450 inhibitor experiments were conducted to study the biosynthesis of various metabolites and were successful in accumulating less oxidized compounds.[171–175] When *P. betae* was treated with the inhibitor, ancymidol (1 mM), a less oxidized precursor named probetaenone I (**129**) was accumulated. The structure of (**129**) was elucidated by extensive NMR analysis[166] and the absolute stereochemistry was confirmed by asymmetric synthesis via a similar route to that of betaenone C.[164]

To prove its intermediacy, isotopically labeled probetaenone I (**129**) was prepared with [*S*-$^{13}CH_3$]methionine and [1-^{14}C]- and [1,2-$^{13}C_2$]acetates in the presence of P450 inhibitors (final concentration, 1–1.8 mM). When a ^{14}C-labeled sample was administered to the fungus, sufficient incorporation (6.02%) into (**119**) was observed. Subsequently, independent feeding experiments were performed with ^{13}C labeled samples. In the ^{13}C NMR spectrum of the compound obtained (**119**), the corresponding signals are enhanced or accompanied by satellite peaks. The high incorporation in these experiments confirms the direct incorporation of ^{13}C labeled (**129**) into (**119**) and excludes the possibility of degradation–reincorporation.[172] The sequence of oxidation at the double bond and C-8 is not clear from these data. However, the retention configuration at C-8 is consistent with other P450 reactions.[173]

The next step was incorporation of the plausible triene precursor (**130**). This and closely related analogues (**131**) and (**132**) (Scheme 10) were synthesized.[18] With the authentic adduct (**133**) in hand, incorporation of the plausible advanced precursor analogue (**131**) in culture was undertaken. However, neither (**133**) nor (**134**) was found in culture extracts and unidentified oxidation products were detected. Then, using a cell-free system of *P. betae*, enzymatic Diels–Alder reaction of the

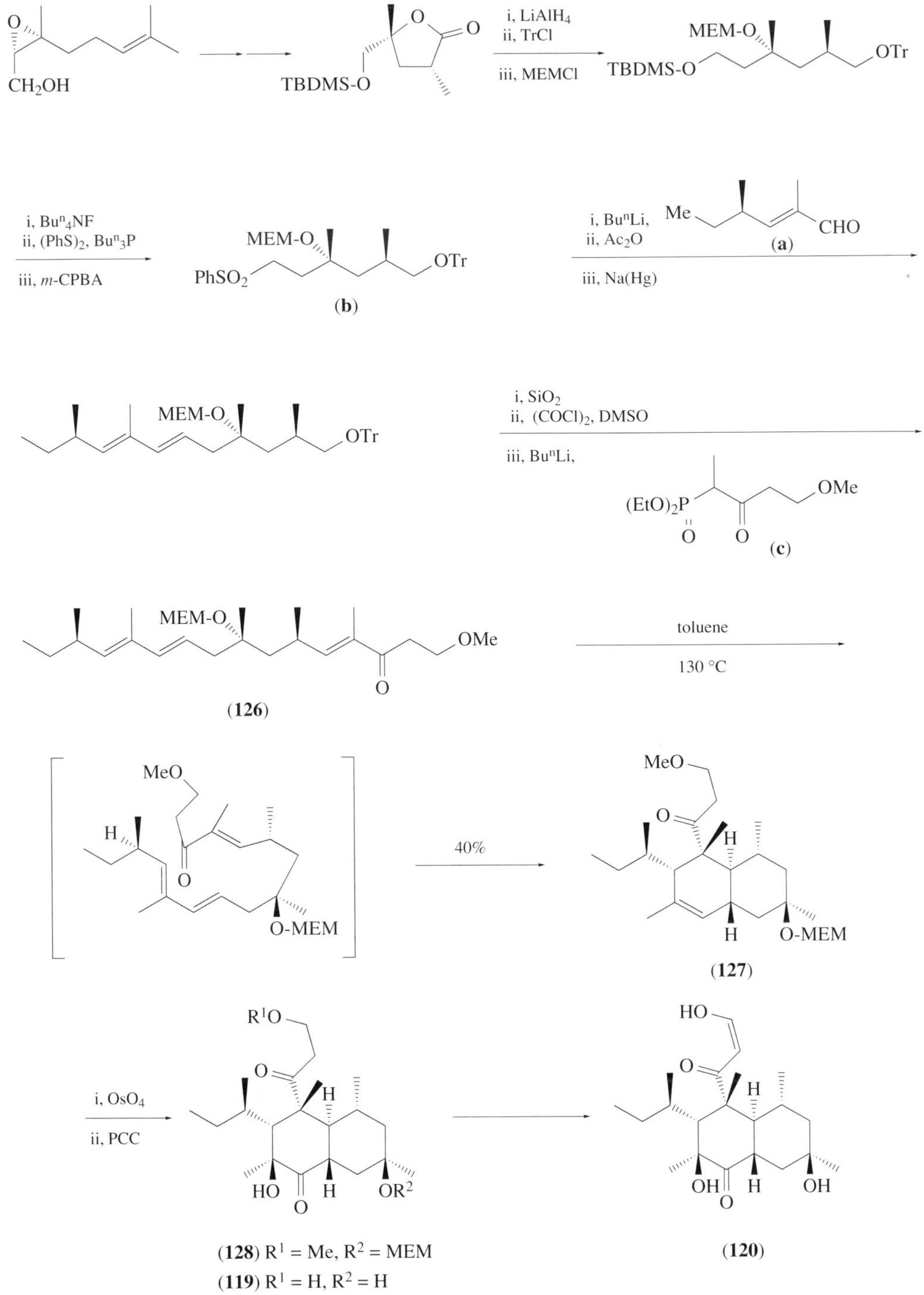

Scheme 9

trienes (**130**) and (**132**) was examined with capillary-column GC detection of the product. However, no enzymatic activity was detected.

Isolation of probetaenone I (**129**) and its incorporation intact into betaenone B (**119**) strongly suggest the involvement of an intramolecular Diels–Alder reaction of a triene such as (**130**) in the biosynthesis of betaenones. Although a feeding experiment with plausible precursor trienes failed, it is believed that the cycloaddition takes place after completion of elongation of the polyketide chain on the basis of the enzymatic Diels–Alder reaction found in the biosynthesis of solanapyrones

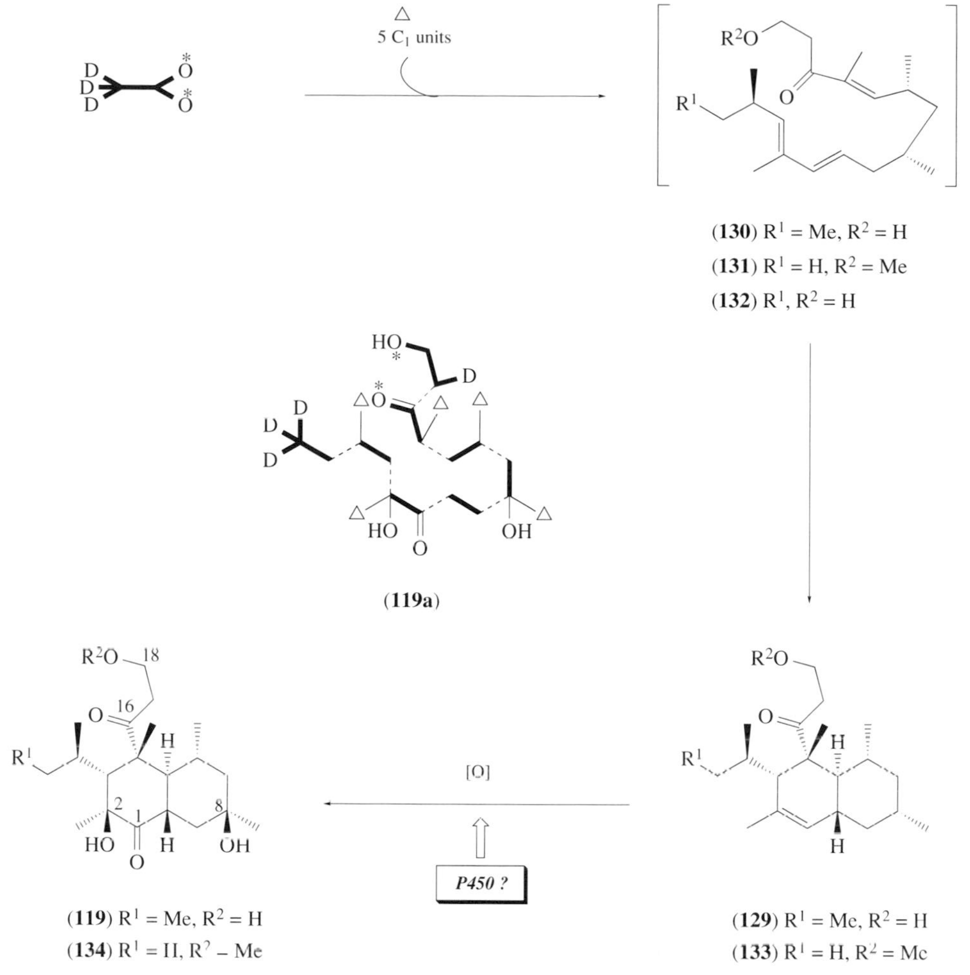

(130) R^1 = Me, R^2 = H
(131) R^1 = H, R^2 = Me
(132) R^1, R^2 = H

(119a)

(119) R^1 = Me, R^2 = H
(134) R^1 = H, R^2 = Me

(129) R^1 = Me, R^2 = H
(133) R^1 = H, R^2 = Me

Scheme 10

(see Section 1.15.3.3) In the case decribed above, the failure of the incorporation can probably be attributed to using the wrong substrate for the Diels–Alder reaction. In the case of enzyme assay, the method of enzyme preparation may not be adequate. At this stage, the possibility cannot be excluded that a compound whose oxidation level at C-18 is different, such as an aldehyde or a thioester, is the real intermediate. Since their efficient synthetic route to the aldehyde and the thioester has been established, further studies should identify the biological Diels–Alder reaction.

1.15.3.2 Chaetoglobosins

Chaetoglobosins (CGs), represented by chaetoglobosin A (**135**), are mycotoxins produced by various fungi, including *Chaetomium* sp. They belong to the group of mycotoxin cytochalasins,[61] which are acutely toxic to mammals and cytotoxic to HeLa cells.[174] Their biological activities, that is, inhibition of cell movement, cytoplasmic division, and induction of multinucleation, make them potentially useful agents for the study of fundamental cellular processes. Chaetoglobosins possess a characteristic 13-member macrocycle and perhydroisoindole skeleton with an indolylmethyl substituent at C-3. So far, 18 natural chaetoglobosins have been isolated.[62,63,175] The variety of their structures originates from the differences in oxidation level, two additional methyl substituents at C-11 and C-12, and five structural types A–E.

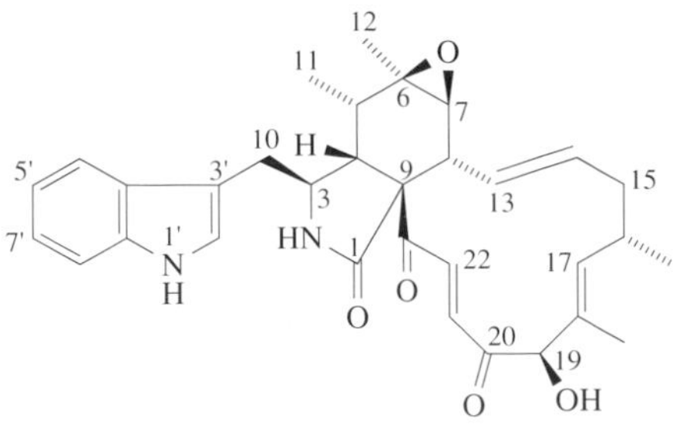

(135)

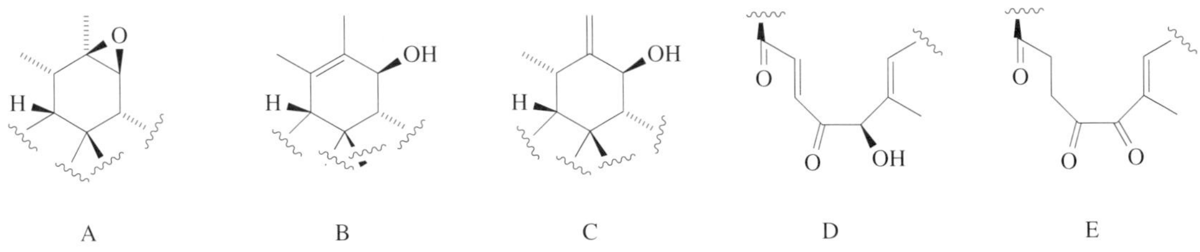

A B C D E

Turner proposed that the unique perhydroisoindole moiety in cytochalasins is derived from a Diels–Alder reaction,[176] based mainly on a consideration of the stereochemistry of the six-membered ring and its incompatibility with normal anion-based cyclization (Equation (10)). This biogenetic pathway could account for the formation of other cytochalasins which were found later. The hypothesis is supported by several circumstantial lines of evidence: (i) The relative stereochemistry at the cyclohexane ring can be explained by *endo*-selective Diels–Alder reaction of the corresponding substrate. (ii) Absolute configurations at the cyclohexane ring are governed by the π-facial selectivity of the corresponding lactam dienophile, whose chiral center is derived from an amino acid. (iii) All-(E)-polyene and reduced acyltetromic acid moieties constructed from a polyketide chain and amino acid are frequently found in natural products.[70,177,178] Although this hypothesis is attractive, no experimental evidence is yet available. Indirect evidence for biological Diels–Alder reaction was obtained in the biosynthetic study of chaetoglobosins.

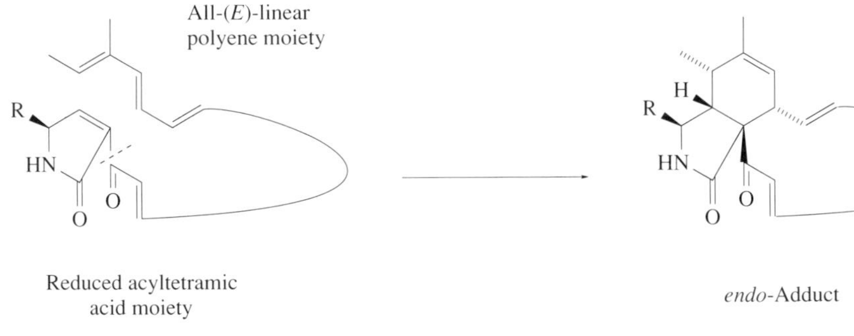

All-(E)-linear polyene moiety

Reduced acyltetramic acid moiety

endo-Adduct

(10)

First, the origin of the oxygen and hydrogen atoms in chaetoglobosin A (**135**) will be described and the late stages of the biosynthetic pathway of (**135**) will be discussed.[179,180] Previous studies have shown that (**135**) is biosynthesized via the coupling of tryptophan with a polyketide derived from 1 mol of starter acetate, 8 mol of malonate, and three C_1 units from methionine.[181,182] Sodium [1-^{13}C,^{18}O$_2$]acetate was administered to a culture of *Chaetomium subaffine*.[179] In the ^{13}C NMR spectrum of the labeled (**135**), isotopically shifted signals were observed at C-1 and C-23. Also, chaetoglobosin A (**135**) obtained by fermentation under [^{18}O$_2$]oxygen gas showed upfield-shifted signals at C-6, C-7, and C-20 in its ^{13}C NMR spectrum. These results clearly show that the oxygen atoms at both C-1 and C-23 are derived from acetate and those at epoxide and carbonyl (C-20) are introduced by oxidation. In these incorporation experiments, the origin of the oxygen atom at C-19 could not be determined due to signal broadening (Equation (11)).

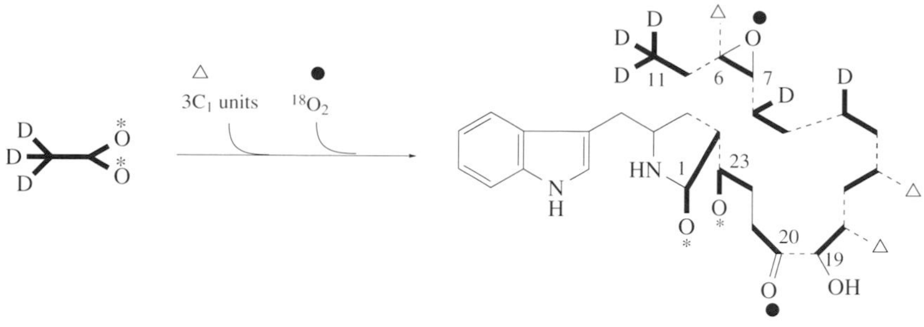

(11)

Analysis of the ^{13}C NMR spectrum of (**135**) obtained by feeding with [1-^{13}C,$^{2}H_3$]acetate indicates the retention of deuterium atoms at C-11, C-8, and C-14.[179] This result was also confirmed from the ^{2}H NMR spectrum, which showed three signals at δ 1.2, 2.1, and 5.3.[179] The low level of enrichment at C-8 and C-14 suggests that significant loss of deuterium occurred during incorporation. Moreover, the presence of deuterium atoms at C-14 excludes the proposed anionic carbon–carbon bond formation,[61] in which the carbonyl group was located at C-14.

From the structure of chaetoglobosins, it was assumed that oxidation occurs at a late stage of the biosynthesis. Although biosynthetic studies of chaetoglobosin building units have been carried out,[181,182] the late biosynthetic route has not been investigated. To obtain information on the sequence of oxidation on chaetoglobosin biosynthesis, a P450 inhibitor experiment (see Section 1.15.3.1) was carried out.

The inhibitor, metyrapone[183] (1 mmol/flask), was added on the fifth day to a culture of *Ch. subaffine*, producing chaetoglobosin A (**135**), and chaetoglobosins C and F[184] as major components. After an additional nine days of fermentation, the mycotoxins were extracted from the mycelia. Four new metabolites, (**136**)–(**139**), named prochaetoglobosins (PCGs) I, II, III, and IV, respectively, were isolated by repeated chromatography.[183] From the molecular formulae of the four components, the accumulated compounds were assumed to be less-oxidized precursors of (**135**). Extensive analysis of NMR data, including H–H-, C–H COSY, and CH–HOHAHA spectra, allowed determination of the planar structures of (**136**)–(**139**). The stereochemistry of these new metabolites was determined by NOESY and NOED experiments and by comparison of NMR data with those of known chaetoglobosins.

The dose effect of the P450 inhibitor, metyrapone, was then investigated.[183] The mycelial extracts treated with metyrapone were subjected to HPLC analysis. The results showed that treatment with inhibitor at a high concentration (1.0 mmol) did not cause a marked decrease in the total amount of CGs and PCGs, but did affect the metabolite pattern. Thus, the sum of the normal CGs (A, C, and F) was dominant in control experiments without inhibitor (98%) but decreased to 63% at 1.0 mmol and decreased concomitantly with an increase in PCGs. Among the accumulated PCGs, more than half was (**136**) and the amount of PCGs increased at 1.0 mmol by 7 to 40 times over that of control experiments without inhibitors. These observations suggest that P450 inhibitors block hydroxylations at C-19 and C-20 and possibly epoxidation as well. The type of inhibitor was also an important factor for the effective accumulation of less-oxidized analogues. Metyrapone was 25 times more effective than S-3307D in the preparation of (**136**)[185] (Scheme 11).

Since treatment of the fungus with a specific inhibitor of monooxygenase cytochrome P450 resulted in accumulation of partially oxidized metabolites (**136**)–(**139**), it is assumed that the plausible intermediate (**136**) is modified by stepwise oxidations, with the introduction of oxygen atoms at three extra sites to produce (**135**). The negative incorporation results mentioned above could be explained if the substrate did not reach the enzyme responsible, owing to poor permeation through the cell membrane, for example. Thus, it is proposed that the biosynthetic pathway of (**135**) is as depicted in Scheme 11.[179] The possibility remains that the oxidative modifications at C-19 and C-20 occur prior to cyclization and then chaetoglobosin J (**140**)[181,182] is formed by cycloaddition. However, the isolation of a range of less oxidized metabolites suggested that oxidation inhibited by a specific P450 inhibitor would occur after cycloaddition. The occurrence of the nonoxidized derivatives, proxiphomin (**141**)[186,187] in cytochalasins and prochaetoglobosin I (**136**) in chaetoglobosins, suggests that these compounds are the first postcyclization intermediates and that a variety of modifications produces a number of analogues.[179]

To obtain the putative intermediate hexaene (**136a**) from accumulated prochaetoglobosin I (**136**) as an isotopically labeled form, the retro-Diels–Alder reaction was examined.[15] The pyrolysis of (**136**) was undertaken in a sealed tube at 180 °C.[179] A less polar compound (**142**) was newly detected

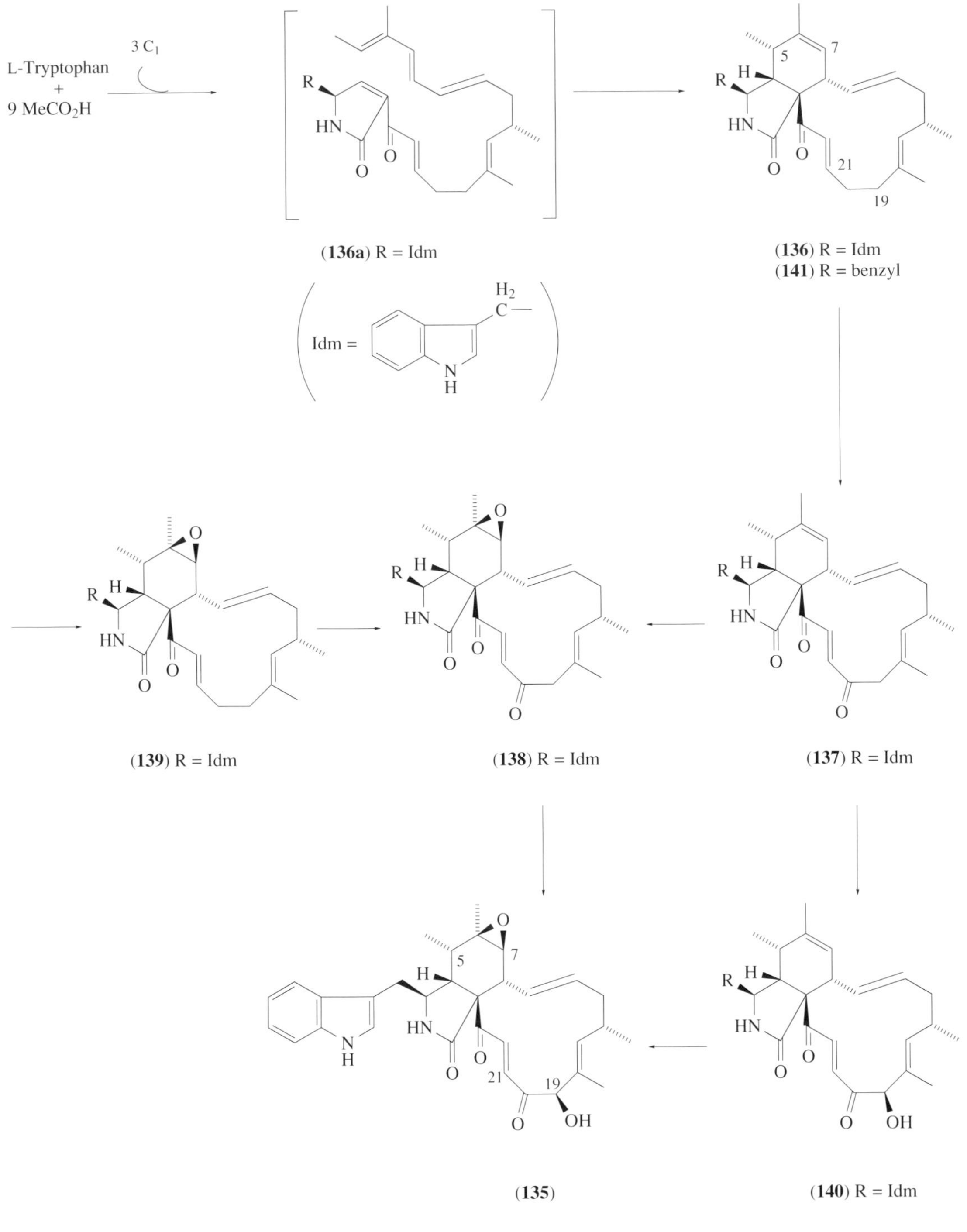

L-Tryptophan + 9 $MeCO_2H$ ($3 C_1$)

(**136a**) R = Idm

(**136**) R = Idm
(**141**) R = benzyl

(**139**) R = Idm

(**138**) R = Idm

(**137**) R = Idm

(**135**)

(**140**) R = Idm

Scheme 11

on TLC after 5 h with a nearly equal amount of the starting material. Prolonged reaction resulted in the degradation of both (**136**) and (**142**). On the basis of detailed NMR analysis and NOE experiments on (**142**), the structure, including stereochemistry, was determined to be as shown in Equation (12).[179] The occurrence of (**142**) demonstrates that the Diels–Alder reaction actually takes place via the hexaene (**136a**) in the absence of a lactam *N*-acyl group, which is believed to be essential to avoid irreversible tautomerization to the enol form of the lactam. Moreover, these results indicate that the *endo*- and *exo*-transition states have only a small energy difference, since the adducts (**136**) and (**142**) are formed in similar amounts. The lack of stereoselectivity in this reaction indicates that the enzyme which is responsible for the [4+2]-cycloaddition should stabilize the *endo*-transition state to afford (**136**) exclusively.[179]

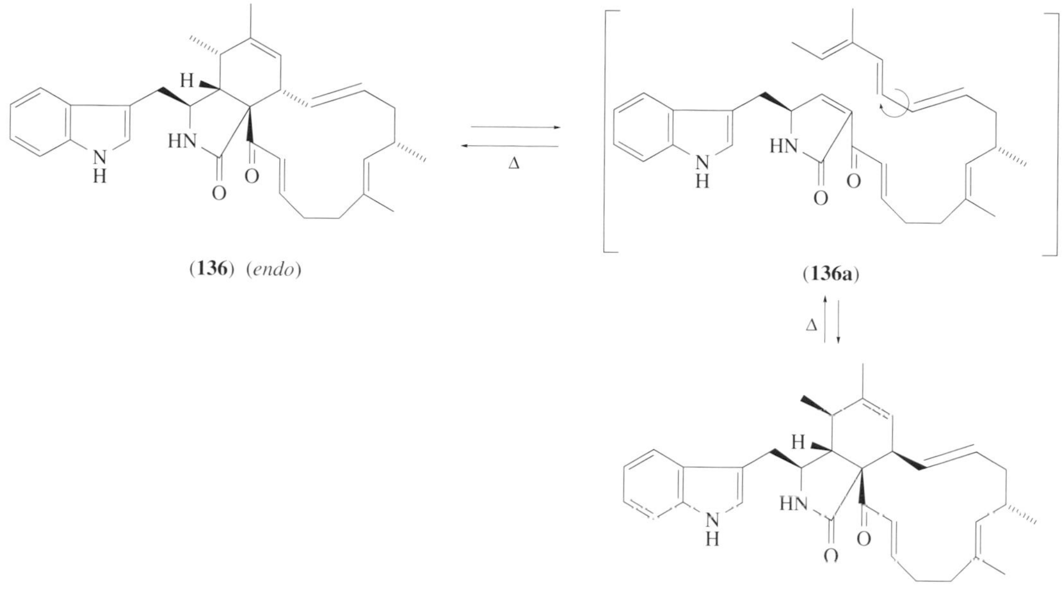

(**136**) (*endo*)

(**136a**)

(12)

(**142**) (*exo*)

In conclusion, a biosynthetic pathway for (**136**) is proposed which involves the cyclization of the hexaene followed by successive oxidations on the basis of the inhibitor experiment and the incorporation data. In addition, the putative biogenetic Diels–Alder reaction has been reproduced chemically. Verification of the involvement of this reaction *in vivo* will require incorporation experiments using a precursor such as (**136a**).

1.15.3.3 Solanapyrones

Alternaria solani, the causal organism of early blight disease of tomato and potato produces several metabolites whose structures have been clarified. It was pointed out that the fungus also secretes two host-specific toxins which induce necrotic symptoms typically associated with the disease. Three phytotoxins were isolated, solanapyrones A (**143**), B (**144**), and C (**145**),[188] and later D (**146**) and E (**147**).[189] Diastereomeric isomers, solanapyrones A and D, and B and E, were obtained in a ratio of 6:1 in the enantiomerically pure state. Solanapyrones A (**143**) and C (**145**) were also isolated from filtrates of stationary cultures of *Ascochyta rabiei*, the causal fungus of chickpea blight.[190] Among the solanapyrones, solanapyrone A (**143**) showed the highest phytotoxicity, inducing a necrotic lesion on the leaf of potato.[188]

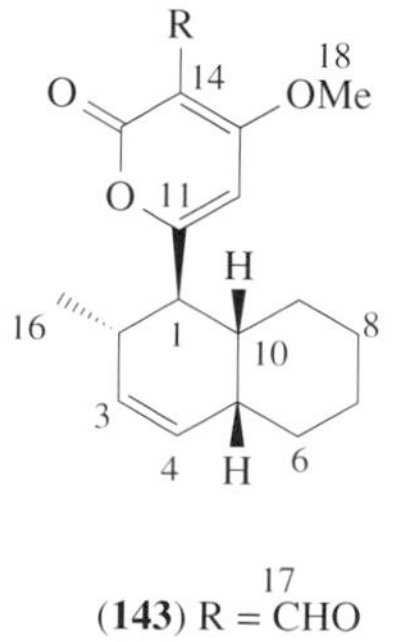

(**143**) R = CHO
(**144**) R = CH_2OH

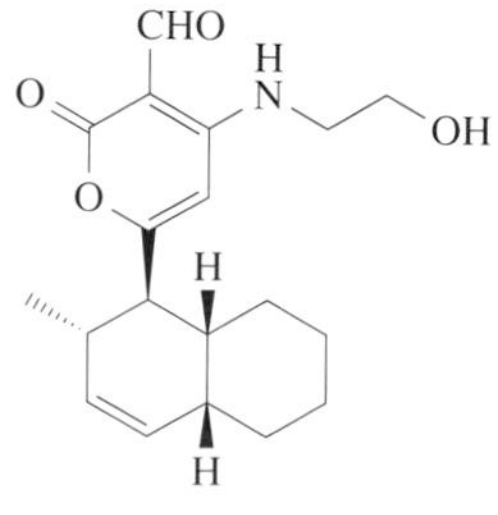

(**145**)

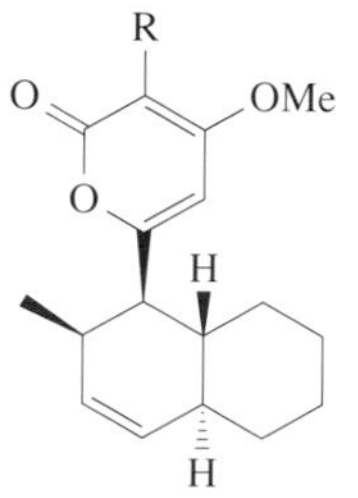

(**146**) R = CHO
(**147**) R = CH_2OH

On the basis of spectroscopic data and chemical reactions, the structures of the solanapyrones have been elucidated.[188,189] The absolute configurations of solanapyrones A (**143**) and D (**146**) were confirmed by the application of the c.d. exciton chirality method to the dibenzoate derivatives.[189,191] All solanapyrone structures are compatible with the biogenetic consideration that these metabolites would be produced through intramolecular Diels–Alder reaction of the precursor, all-*trans* triene, from the polyketide pathway. The relative configurations of solanapyrones A and B, and D and E are the same, and then solanapyrones A and D, and B and E are each in a diastereomeric relationship. This means that the intramolecular Diels–Alder reaction *in vivo* would proceed via *exo* and *endo* transition states to give solanapyrones A, D, and B, E, respectively. Like betaenone B (see Section 1.15.3.1), solanapyrones also belong to the class of 1,2-dialkyldecalin polyketides.

Syntheses of solanapyrones A (**143**) and D (**146**) were attempted based on biogenetic consideration of these phytotoxins.[192] The retro synthesis envisaged intramolecular Diels–Alder reaction of the polyketide triene (**150**), a key intermediate, which is further divided into a pyrone moiety (**148**) and a diene moiety (**149**) (Scheme 12). The moieties (**148**) and (**149**) were prepared from dehydroacetic acid and hexadienyl acetate, respectively. Aldol condensation of the aldehyde (**149**) with the dithioacetal (**148**) gave a dienol, which was further converted to a triene (**150**). The newly formed C_6—C_7 double bond was confirmed to be *trans* from the NMR spectrum. The intramolecular Diels–Alder reaction of (**150**) in toluene at 180 °C for 1 h in a sealed tube yielded a mixture of the adducts (**151**) and (**152**) in a ratio of 1:2. This product ratio depends on the solvents, especially water,[192] and should be useful in differentiating between artificial and enzymatic reactions in a biosynthetic study. In the Diels–Alder reaction, if a mixture of trienes containing small amounts of (*E,E,Z*)-(**150**) and (*E,Z,E*)-(**150**) was used, no product from the *Z*-isomers was obtained. This would be due to the lower activation energy in the transition states from (*E,E,E*)-(**150**) leading to (**151**) and (**152**). Similar kinetic selectivity has also been observed in the synthesis of (+)-diplodiatoxin (**125**).[163] Removal of the protective groups in (**151**) and (**152**) yielded solanapyrones A (**143**) and D (**146**) in a ratio of 3:2.

The first direct evidence for biological Diels–Alder reaction was obtained in biosynthetic studies of solanapyrones. Feeding experiments to elucidate the biosynthetic building blocks were done with [1-^{13}C]-, [2-^{13}C]- and [1,2-$^{13}C_2$]acetate and with [*S*-^{13}C]-L-methionine. The results showed that solanapyrones were biosynthesized from an acetate-derived octaketide with two C_1-units from methionine.[193] The origins of the oxygen and hydrogen atoms were elucidated by studying the incorporation of label from [1-^{13}C, $^{18}O_2$]acetate and [1-$^{13}C_2H_3$]acetate. The results indicated that the oxygen atoms at C-13 and C-15 originate from acetate and the hydrogen atoms at 2-Me, C-3, and C-5 are from acetate, as shown in Equation (13).

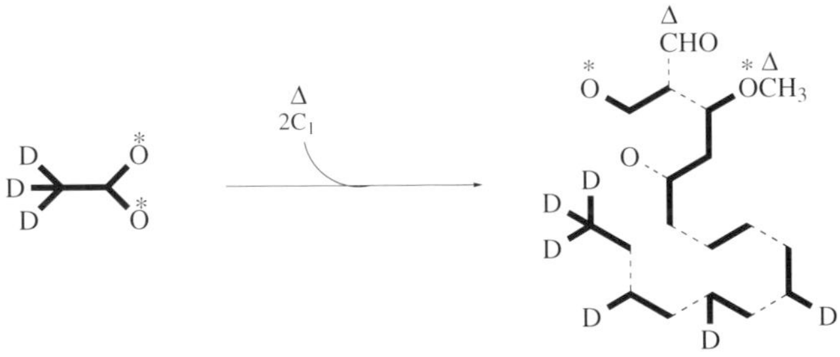

(13)

Since the fungus *A. solani* produces minor plausible Diels–Alder adducts, solanapyrones D (**146**) and E (**147**) as chiral forms,[188,189] the involvement of two biosynthetic pathways A and B were hypothesized (Scheme 13). The difference between these pathways is whether cyclization takes place before or after oxidation of the C_1-unit on pyrone.

The following experiments were performed first to determine the actual pathway. Administration of [*S*-C^2H_3]-L-methionine yielded highly enriched (**144**). The ^{13}C NMR spectrum of this sample displayed a ^{13}C–^{2}H triplet signal (J_{cc} = 22.3 Hz) located at 0.3 ppm upfield of the signal for C-18 at 54.3 ppm. This result clearly indicated that only one deuterium was retained in the hydroxymethyl group of (**144**). Further, [17-C^2H_2]-17-deoxysolanapyrone B (**156**) prepared from (**143**) was fed to a culture of *A. solani*. In the ^{2}H NMR spectrum of (**143**) obtained from the feeding experiment, no signal was observed, indicating no incorporation of deuterium. In combination with the former experiment, these results eliminated pathway B which involves the direct cycloaddition of the triene (**153**) to the decaline system (**156**) and allowed the reaction to be rationalized in terms of pathway A.

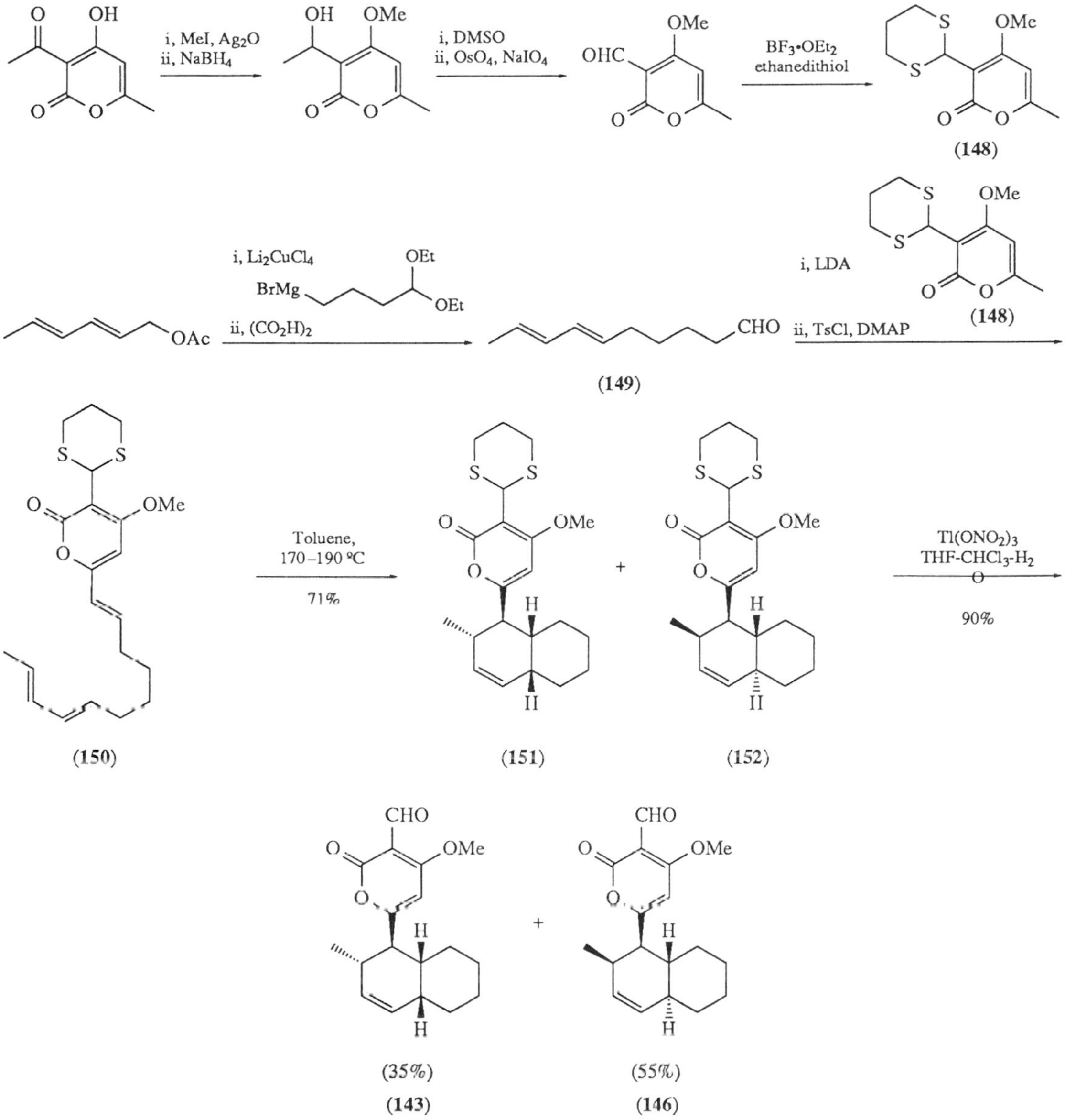

Scheme 12

To establish formation of the decalin system via the [4+2]-cycloaddition route, heavily labeled prosolanapyrones [17-$^{2}H_{2}$, 18-$^{2}H_{3}$]-I (**153**) and [2-^{2}H,3-^{2}H,17-^{2}H,18-$^{2}H_{3}$]-II (**154**) were prepared by similar routes to those described in the synthesis of solanapyrone A (**143**). Solanapyrone A (**143**) derived from [$^{2}H_{5}$]-(**154**) in feeding experiments exhibited two signals due to deuterium at C-17 and C-18 in its ^{2}H NMR spectrum. Furthermore, the product derived from deuterium-labeled (**154**) showed deuterium resonances at C-2, C-3, C-17, and C-18 in its ^{2}H NMR spectrum. The ratio (0.7:1:0.4:3) of integrals essentially reflected that of labeled prosolanapyrone II (**154**). These results demonstrate intact incorporation of intermediates (**153**) and (**154**).[194]

To eliminate the possibility that the labeled prosolanapyrone II (**154**) was cyclized nonenzymatically, the solanapyrone A (**143**) obtained was converted to diastereomers (**157**) and (**158**) in two steps. Since the incorporation of (**154**) was low, nonlabeled *ent*-(**158**) was added to secure complete separation of the diastereomers (**157**) and (**158**). The samples obtained by careful chromatography were examined by ^{2}H-NMR spectroscopy. Enriched signals were found only in compound (**157**) but not (**158**). These data established unambiguously that the biosynthesized product is not racemic and that (**143**) is formed via an enzymatic Diels–Alder reaction.[194] This is the first proof that a Diels–Alder reaction of diene–dienophile precursors operates in biosynthesis.

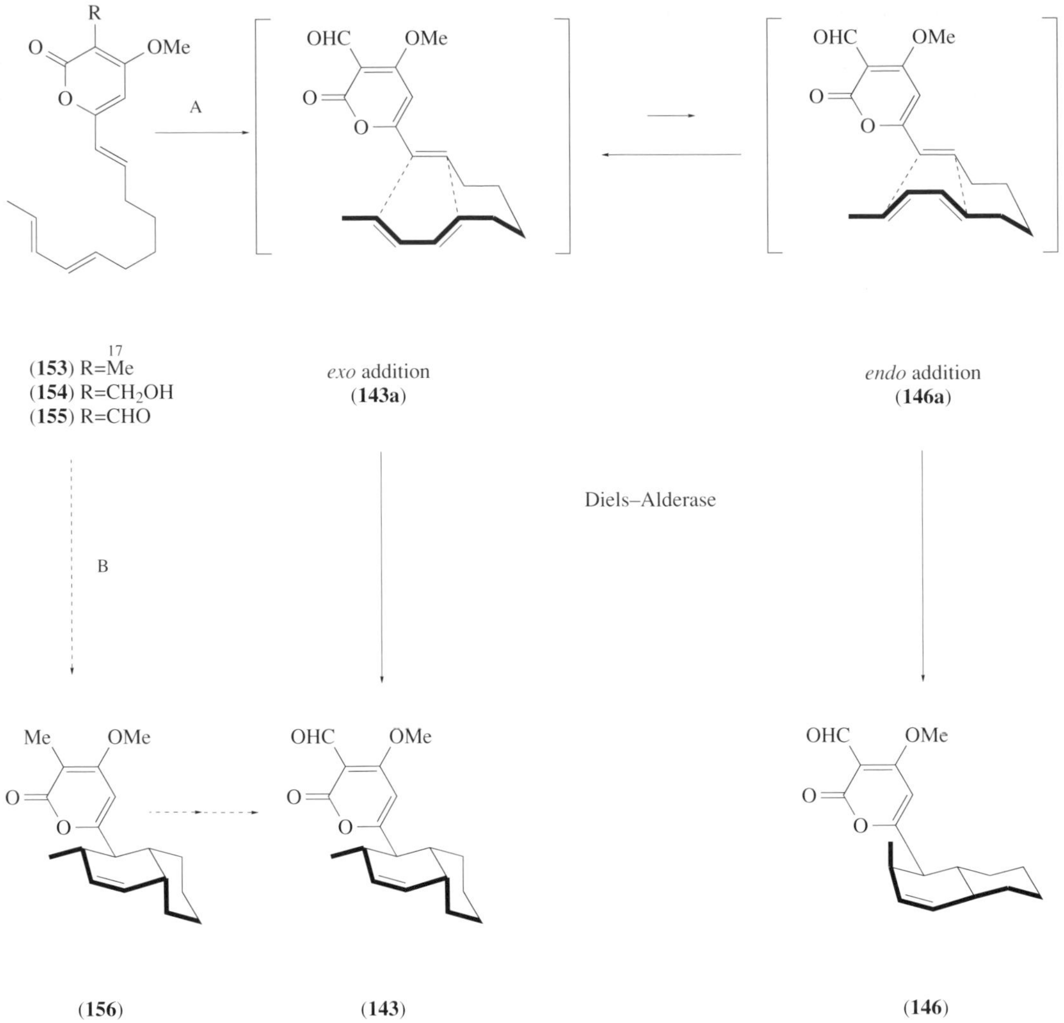

R
OMe
O
O
A
OHC OMe
O
O
OHC OMe
O
O

17
(**153**) R=Me
(**154**) R=CH_2OH
(**155**) R=CHO

exo addition
(**143a**)

endo addition
(**146a**)

Diels–Alderase

B

Me OMe
O
O
OHC OMe
O
O
OHC OMe
O
O

(**156**) (**143**) (**146**)

Scheme 13

OHC OMe
O
O
H
HO
H
RO

MeO CHO
O
O
H
OH
H
OR

(**157**) R = (*R*)-MTPA
ent-(**158**) R = (*S*)-MTPA

(**158**) R = (*R*)-MTPA

Before searching for the enzymatic activity, the reactivities of (**154**) and (**155**) were examined, since it was known that a significant level of cycloaddition occurred in an aqueous solution. Under standard conditions (30 °C, 10 min), 15% of (**155**) was converted to the *exo*-adduct, solanapyrone A (**143**), and the *endo*-adduct, solanapyrone D (**146**), in a 3:97 ratio, with 85% of the substrate unchanged. For Diels–Alder reactions in aqueous medium, similar predominant formation of *endo*-adducts and acceleration of the reaction rate have been reported.[2] On the other hand, no cycloadduct was detected in the case of prosolanapyrone II (**154**) under the same conditions.

After a number of attempts, the enzymatic activity in a cell-free extract of *A. solani* was found. This extract converted (**155**) to (**143**) and (**146**) in an *exo*-selective manner in contrast to *endo*-

selective nonenzymatic cycloaddition.[195] Throughout the study, this criterion was used to differentiate between enzymatic and nonenzymatic reactions. The cell-free extract was partially purified. Incubation (30 °C, 10 min) of (**155**) with this crude enzyme resulted in consumption of 25% of the substrate to give the corresponding amount of the cycloadducts (**143**) and (**146**) in a ratio of 53:47, with the rest of the substrate intact. A control experiment showed that 10% of (**155**) was converted to (**143**) and (**146**) in a ratio of 3:97. These results show that actual enzymatic consumption of (**155**) is 15%, and the *exo*/*endo* ratio for the enzymatic reaction products was calculated as 87:13. At a four times higher protein concentration than normal, the reaction with the enzyme proceeded 4.1 times faster than the nonenzymatic reaction. In order to estimate the optical purity of the products, HPLC analysis with continuous c.d. detection was undertaken.[195] On the basis of the UV and c.d. absorptions of enantiomerically pure natural solanapyrones, the optical purity of (**143**) was calculated as 92 ± 8% *ee* from the negative c.d. absorption at 300 nm.

With this crude enzyme, prosolanapyrone II (**154**) was converted to (**143**) and (**146**) via [4+2]-cycloaddition.[195] Under the same conditions, incubation of (**154**) resulted in consumption of 25% of the substrate to give the cycloadducts (**143**) and (**146**) in a 85:15 ratio in 19% yield, along with 6% of (**155**). The observed *exo*/*endo* ratio was similar to the ratio (**143**:**146** = ~83:17) found in solanapyrones[189] isolated from the culture broth. The reaction products (**143**), (**146**), and (**155**) were characterized by ^{1}H NMR and MS examinations. According to HPLC analysis as described above, the optical purity of the obtained (**143**) was calculated as 99 ± 4% *ee*.

It appeared that a two-step reaction consisting of oxidation and cycloaddition was involved. The sequence of the reactions was unambiguously confirmed as oxidation from (**154**) to (**155**) followed by cycloaddition, based on the following observations: (i) The cycloaddition was completely suppressed under conditions excluding oxygen (argon atmosphere). (ii) A small amount of the intermediate prosolanapyrone III (**155**) was detected in the presence of oxygen. (iii) Neither (**144**) nor (**147**) was detected in the reaction under an ordinary or argon atmosphere. Furthermore, equimolar consumption of molecular oxygen monitored by an oxygen electrode with respect to the total amount of the oxidation products (**143**), (**146**), and (**155**) suggests the conversion of (**154**) to (**143**) is not a dehydrogenation but an oxidation.

Comparing the enzymatic reactions of (**154**) and (**155**), some decrease in *exo*/*endo* ratio and enantioselectivity in the reaction of (**155**) was observed (Scheme 13). This was readily explained by the concomitaint nonenzymatic reaction. Thus, it is speculated that a single bifunctional enzyme which catalyzes a two-step reaction is responsible for producing optically pure solanapyrones with the *exo*/*endo* ratio found in natural solanapyrones. Detection of a small amount of the intermediate (**155**) in the incubation of (**154**) could be explained by leakage from the loosely bound enzyme–substrate complex. Therefore, this enzyme is able most adequately to recognize prosolanapyrone II (**154**).

After studying these enzymatic reactions, further purification of the enzyme was continued. The dialyzed crude extract was loaded onto a DEAE-Sepharose FF column which was washed with potassium phosphate buffer and eluted to a linear gradient. The active fractions were purified successively by a hydroxyapatite column, Superdex-200HR gel filtration column, preparative native PAGE (polyacrylamide gel electrophoresis), and phenyl-Sepharose HR column. The enzyme was purified more than 2000 times compared with the crude extract. Although the enzyme protein was not completely purified, tentative kinetic parameters for this partially purified enzyme were obtained and compared with those of catalytic antibodies catalyzing intermolecular Diels–Alder reaction and chorismate mutase (Table 1). It is easy to understand from this table that the K_m, k_{cat}, and k_{cat}/k_{uncat} values of all these data are superior to those of catalytic antibodies.[196–198]

Table 1 Kinetic parameters of solanapyrone synthase, catalytic antibodies, and chorismate mutase.

Enzymes and antibodies	K_m (μM)	k_{cat} (min^{-1})	k_{cat}/K_m ($\mu M^{-1}\ min^{-1}$)	k_{uncat} (min^{-1})
Solanapyrone synthase	16 (prosolanapyrone II)			
(*endo* + *exo*)	37 (prosolanapyrone III)	32	0.86	0.0062
1E9[196]	2100 (dienophile)	4.3	0.002	
22C8[197]	700 (diene)	0.00317	4.53×10^{-6}	
	7500 (dienophile)		4.23×10^{-7}	
39,AII[198]	1130 (diene)	40.2	0.0356	
	740 (dienophile)		0.0543	
Chorismate mutase[199]	67 (chorismate)	2760	41	0.000744

Substrate specificity was studied next using this enzyme. The results of substrates specificity are shown in Equations (14)–(20), where the hydroxymethyl group of the precursors which have a similar structure to the transition state are easily oxidized to the formyl group, and the oxidation needs some extent of bulkiness of the C_6 alkyl substituent on pyrone ring.

HOH$_2$C OMe → OHC OMe (100%) (14)

HOH$_2$C OMe → OHC OMe (84%) (15)

HOH$_2$C OMe → OHC OMe (5.8%) (16)

HOH$_2$C OMe → OHC OMe (39%) (17)

HOH$_2$C OMe → OHC OMe (~27%) (18)

HOH$_2$C OMe → OHC OMe (only qualitative analysis) (19)

HOH$_2$C OMe → (0%) (20)

1.15.3.4 Macrophomic Acid

In the course of phytotoxin researches, phytotoxins, pyrenocine A (**159**), pyrenocine B (**160**), and pyrenochaetic acids A (**161**), B (**162**), and C (**163**), were isolated from the culture filtrate of *Pyrenochaeta terrestris*, the causal fungus of onion pink root disease.[200,201] After structural determination, syntheses of these phytotoxins were attempted. Although pyrenocines A (**159**) and B (**160**) were easily synthesized, all attempts to synthesize pyrenochaetic acid A (**161**) starting from the closely related benzene derivatives, 2-acetyl-3,5-dimethylphenol and 3-hydroxy-5-methylbenzoic acid have failed. Therefore α-pyrone (**164**) was chosen as a starting material. It was well known that α-pyrone reacts with propiolate to form benzoates evolving carbon dioxide through retro-Diels–Alder reaction.[1] In fact, the Diels–Alder reaction of the pyrone (**164**) with ethyl propiolate at 180–200 °C in a sealed tube afforded two ethyl benzoates, (**165**) and its regioisomer, in a ratio of 2:3 in good yield. From the benzoate (**165**), pyrenochaetic acids A (**161**) and B (**162**) were prepared easily through aldol condensation[202] (Scheme 14).

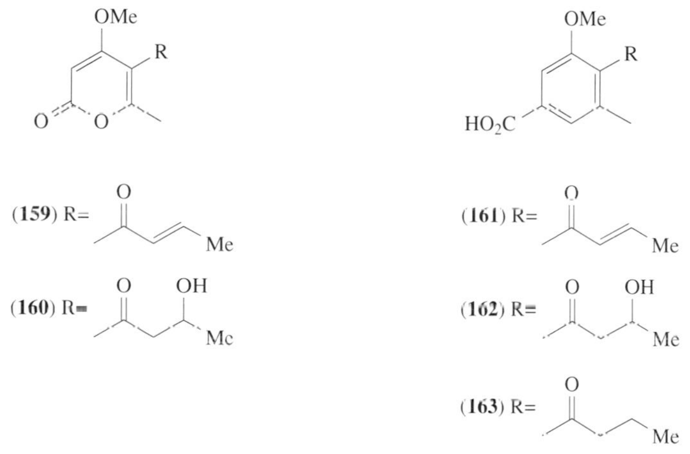

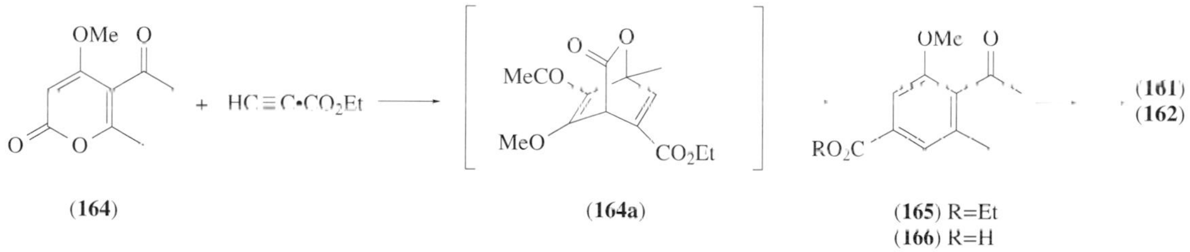

OMe R O O (**159**) R= Me (**160**) R= OH Me

OMe R HO_2C (**161**) R= Me (**162**) R= OH Me (**163**) R= Me

OMe O (**164**) + $HC{\equiv}C{\cdot}CO_2Et$ → [MeCO O O MeO CO_2Et (**164a**)] → OMe O RO_2C (**165**) R=Et (**166**) R=H → (**161**) (**162**)

Scheme 14

After synthetic studies, Yamamoto's group[203] found that *Macrophoma commelinae*, the causal fungus inducing the black rotting of chestnut tree root and fruit rot diseases of some other plants, had an interesting ability to transform 5-acetyl-4-methoxy-6-methyl-2-pyrone (**164**) into 4-acetyl-3-methoxy-5-methylbenzoic acid (**166**) (macrophomic acid) (Equation (21)). These two compounds have a similar relationship to that of pyrenocines and pyrenochaetic acids. After extensive biosynthetic studies, they came to the following conclusions. (i) Each carbon atom from C-3 to C-10 of α-pyrone is retained at the position corresponding to the benzoic acid products, but a carbonyl carbon at C-2 is eliminated as CO_2. (ii) The C-1, C-6, and C-11 carbons of macrophomic acid are derived from another carbon source, probably pyruvate or its equivalent. From these results, they suggested a condensation process for the added 2-pyrone and a catabolic pyruvate (Scheme 15, pathway A). Instead of this mechanism, the authors presumed that macrophomic acid would be derived through a Diels–Alder process (pathway B) similar to the synthesis of pyrenochaetic acid A (**161**), and started to investigate the biosynthesis of macrophomic acid (**166**), in order to answer three questions. (i) What is the real C_3 unit precursor? (ii) How is the reaction mechanism rationalized? (iii) What kind of enzyme is operated in the reaction?

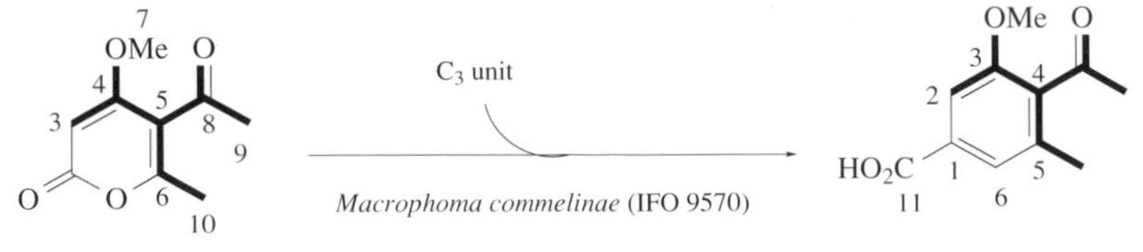

(21)

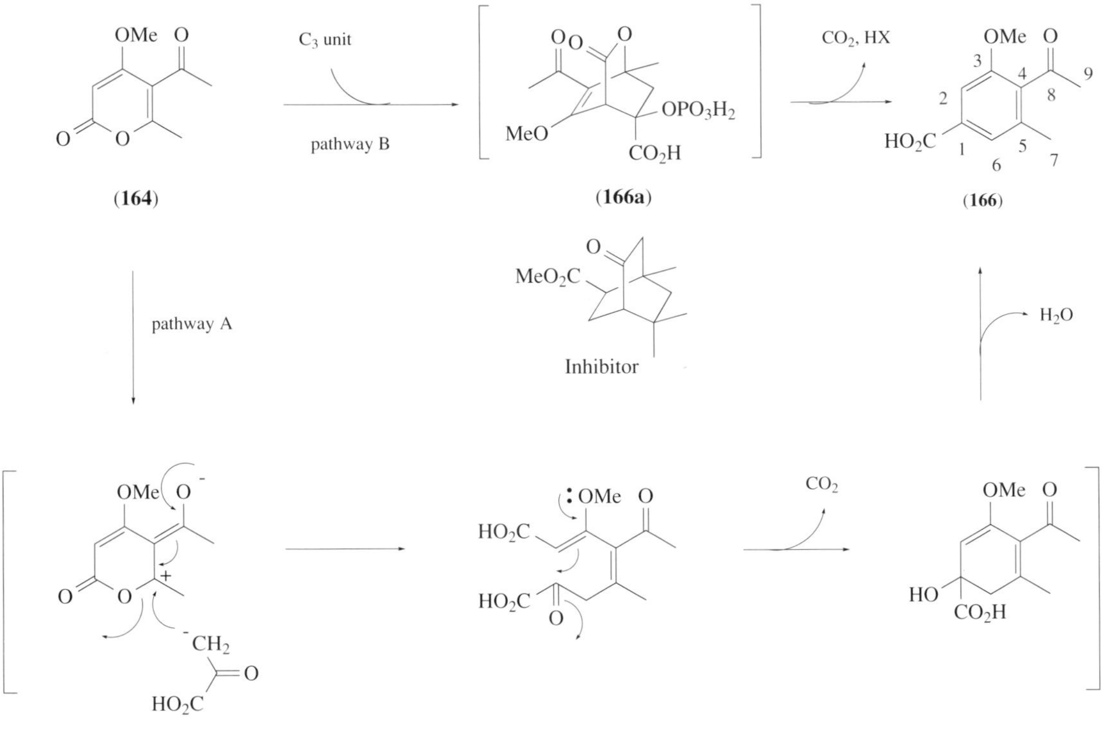

Scheme 15

Feeding experiments of [1-^{13}C]-L-serine and [1-^{13}C]-L-alanine showed high enrichment at C-11 in the ^{13}C NMR spectrum of (**166**). Since pyruvate is also incorporated without the loss of the C-1 carbon, these data indicate that the amino acids and pyruvate are converted to a common intermediate which is incorporated into (**166**). When [*U*-^{13}C]-glycerol was administered, enhancements of the signals due to C-11, C-6, and C-1 were observed.[204,205] The splitting pattern of the C-11 signal suggests the intact incorporation of glycerol. Further administration of (1*RS*,2*S*)- and (1*RS*,2*R*)-[1-^{2}H]glycerols to the fungus revealed that the latter (2*R*)-[^{2}H]glycerol was effectively incorporated at 6-H in (**166**). Feeding experiments with specifically ^{2}H-labeled (1*R*,2*R*)- and (1*S*,2*R*)-[1-^{2}H]glycerols showed that the ^{2}H-label at C-6 in (**166**) was retained at a 5.5 times higher level in the case of (1*R*,2*R*)-glycerol compared with (1*S*,2*R*)-glycerol (Scheme 16).

The selective labeling observed indicated that the C—C bond formation between the pyrone (**164**) and the C$_3$-unit derived from (1*R*,2*R*)-glycerol is stereoselective, and that the pathway involves a C$_3$-unit intermediate which has diastereotopic methylene protons. These feeding experiments and intact incorporation of [*U*-^{13}C]glycerol exclude the possibility that TCA intermediates, pyruvate and alanine are directly converted to (**166**). Although the previous study suggested that phosphoenol pyruvate (PEP) was not incorporated into (**166**), the experimental evidence described above suggests strongly that PEP or its closely related precursors which involve in the glycolytic pathway (Scheme 17), would be a direct substrate in this novel aromatic ring formation and study in the cell free system was initiated.

The crude enzyme was prepared from the mycelium of *M. commelinae* as follows. After preincubation with 1 mM α-pyrone (**164**) to induce the desired enzyme, the mycelia was homogenated

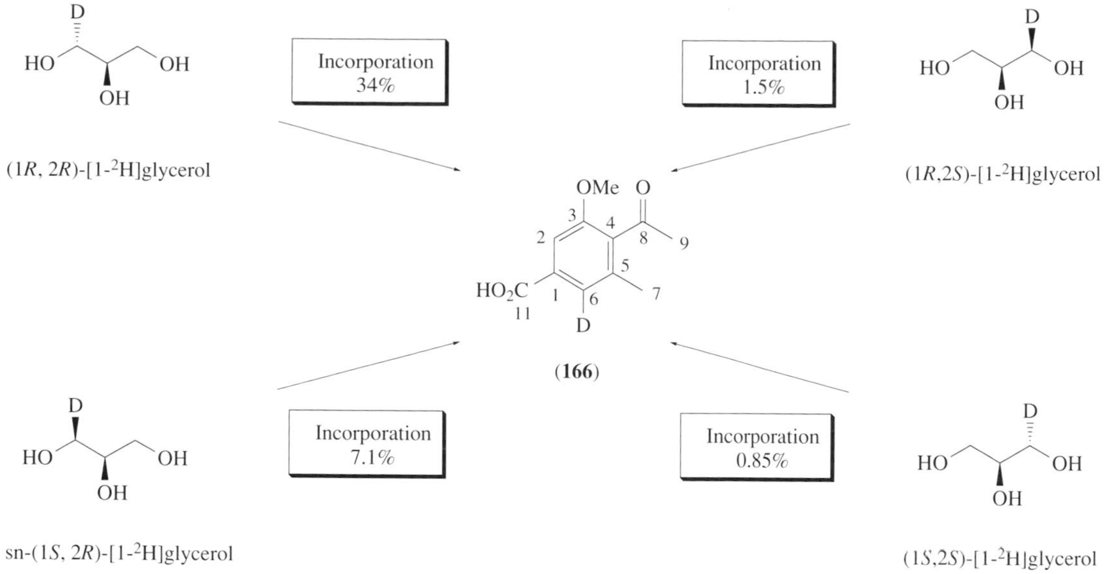

(1*R*, 2*R*)-[1-^{2}H]glycerol — Incorporation 34%

(1*R*,2*S*)-[1-^{2}H]glycerol — Incorporation 1.5%

sn-(1*S*, 2*R*)-[1-^{2}H]glycerol — Incorporation 7.1%

(1*S*,2*S*)-[1-^{2}H]glycerol — Incorporation 0.85%

(**166**)

Scheme 16

glycerol → (2*R*)-2-phosphoglycerate (**167**) → phosphoenolpyruvate → pyruvate → (CO_2) TCA cycle

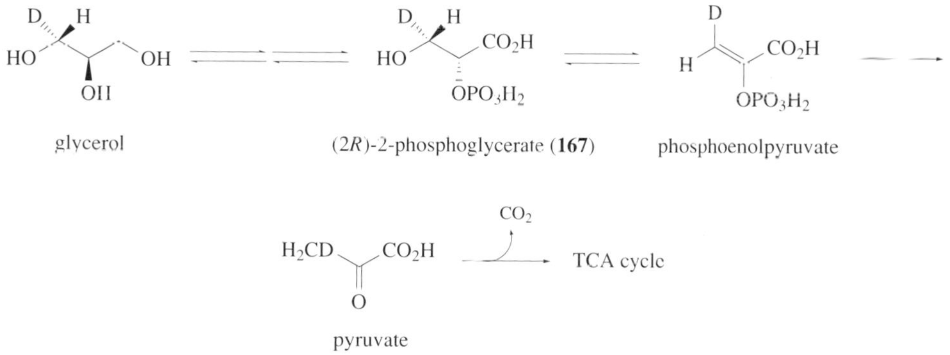

Scheme 17

with PIPES buffer containing KCl and PEG (pH 7.2). The homogenate was disrupted and centrifuged and the supernatant was used as a crude enzyme. By using this crude enzyme, conversion of (**164**) and various plausible C_3 unit precursors were examined. It was found that (2*R*)-3-phosphoglycerate, phosphoenol pyruvate, and pyruvate were all unable to produce macrophomic acid (**166**). However, only (2*R*)-2-phosphoglycerate (**167**) was efficiently converted to macrophomic acid (**166**). Interestingly, addition of PEP did not increase the productivity of (**166**). The results mean that the enzyme only recognizes the substrate (2*R*)-2-phosphoglycerate (**167**).

It was found that the enzyme has the highest activity in the presence of a divalent Mg^{2+} cation, and the activity is inhibited remarkably by addition of fluoride anion (NaF) or enolase inhibitor, 1-hydroxycyclopropane-1-carboxylic acid phosphate (HCP). These findings mean that the specific enzyme catalyzing formation of macrophomic acid (**166**) would be enolase or enolase-like enzyme (Scheme 18).

On the basis of experimental data, instead of the condensation mechanism, an alternative biosynthetic pathway is suggested which involves an intermolecular inverse electron demand Diels–Alder reaction of the α-pyrone (**164**) and dienophile, PEP, which arises from (2*R*)-phosphoglycerate through *anti*-dehydration of the hydroxyl group. The plausible adduct (**166a**) is further converted to (**166**) by successive *syn*-elimination of phosphoric acid and retro-Diels–Alder reaction eliminating carbon dioxide. Compared with Yamamoto's hypothesis, this pathway not only reasonably explains stereospecific formation of (**166**), but also simplifies this novel conversion. In order to obtain evidence for this mechanism, various types of bridged bicyclic compounds were prepared. Among

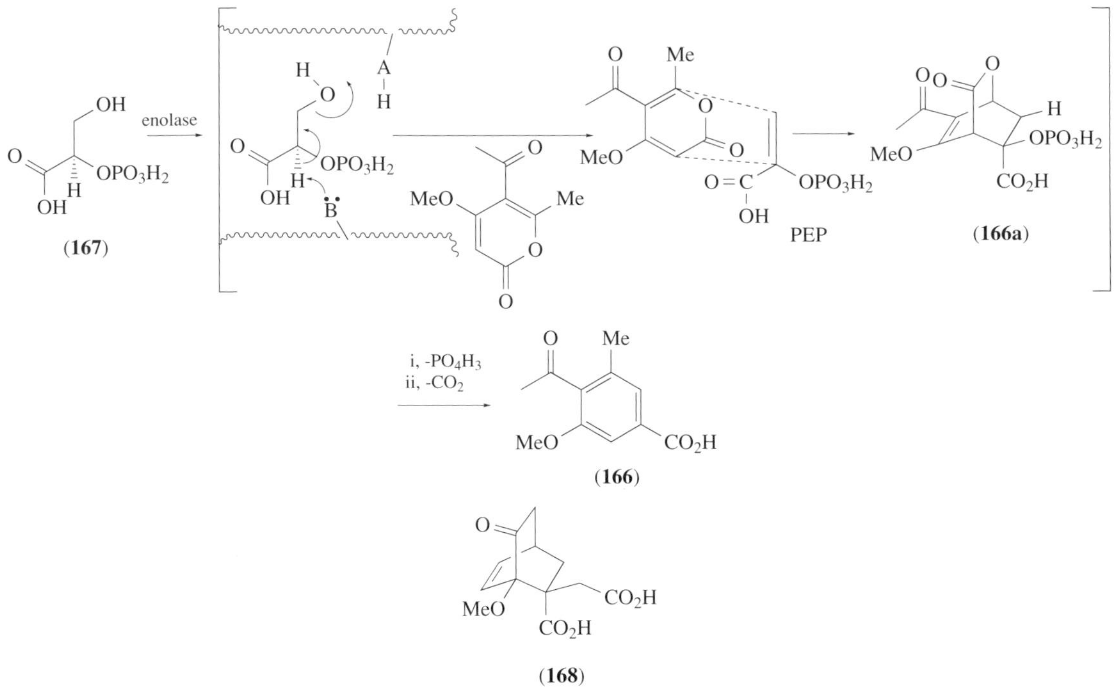

Scheme 18

them, dicarboxylic acid (**168**) inhibited the conversion of (**164**) to (**166**) in 80% in the cell free system at a concentration of one-tenth of the pyrone. This observation provides further support for this biosynthetic pathway, although an anionic condensation process would not be rigorously excluded.

1.15.4 CONCLUSION AND PERSPECTIVES

As described above, we are now able to suggest the existence of two Diels–Alderases, solanapyrone synthase and macrophomic acid synthase, in the biosyntheses of solanapyrones and macrophomic acid, respectively. Interestingly, these enzymes do not catalyze the Diels–Alder reaction directly, but catalyze the preceding reaction and then the Diels–Alder reaction. Thus, solanapyrone synthase catalyzes oxidation of prosolanapyrone II (**154**) and then intramolecular *exo*-cycloaddition, and macrophomic acid synthase catalyzes *anti*-dehydration of (2*R*)-2-phosphoglycerate to PEP, and then intermolecular cycloaddition of α-pyrone (**164**) and PEP. Therefore, these two enzymes are unprecedented bifunctional enzymes. The elucidation of the mechanism of the biological Diels–Alder reaction should provide important information for other biological pericyclic reactions.

Considering factors affecting the rate acceleration of these Diels–Alderases, the proximity effect that holds substrates so as to maximize orbital overlapping of the diene and dienophile in the active site and the activation of the diene and dienophile by hydrogen bonding and electrostatic effects are likely to be important for biological cycloadditions. The acceleration of one of the enzymatic pericyclic reactions chorismate mutase catalyzing the Claisen rearrangement (see Section 1.15.1), is explained by the importance of hydrogen bonding between enzyme residues and the oxygen atom in the cleaving bond in stabilizing the developed negative charge.[4,5]

In discussing natural [4+2]-adducts, distinction should be made between enzymatic and non-enzymatic reactions before searching for the enzyme catalyzing the cycloaddition. When natural cycloadducts are obtained as racemates, it may be concluded that the Diels–Alder reaction involved is not an enzyme-catalyzed reaction. Such possibilities can be noted in the following cases. As mentioned earlier the monomeric alkaloids gave dimers under proper acidic conditions (see Section 1.15.2). This result suggests that an ionic process takes place to give different regioisomers from thermal reaction. In other cases, oxidative coupling between the phenolic radicals producing reactive species such as a cyclohexadienone, which was followed by the spontaneous cyclization shown above, was proposed on the basis of mainly *in vitro* experiments, but experimental evidence has been provided by an incorporation experiment of the acyclic precursor of anigorufone[206] with *Anigozanthos preissii*. Moreover, it is reported that electrochemical oxidation of phenols in SDS (sodium dodecyl sulfate) micelles stabilizes the unstable intermediates and markedly promotes the reactions.[207] These findings imply that the formation of radicals may be responsible for an oxidative enzyme such as a peroxidase, but their coupling and the cyclization proceed without enzyme to give racemic products.

It is well known that the Diels–Alder reaction is greatly accelerated and *endo*-selectivity is enhanced in aqueous media.[208] This is explained mainly by the hydrophobic effect[209] and hydrogen bonding[210] between the dienophile carbonyl and the water molecule. This type of nonenzymatic catalysis could explain the formation of the chiral terpene adducts found in plants as a single diastereomer.

The first examples of Diels–Alderases as observed in the biosynthesis of solanapyrones and macrophomic acid have been described. The authors believe that mechanistic study of the enzymes may give clues for designing an artificial enzyme. If the size and functional groups of the catalytic pocket of the enzyme can be altered without altering other functions, a variety of Diels–Alderases could be prepared by chemical syntheses and genetic engineering.

1.15.5 REFERENCES

1. G. Desimoni, G. Tacconi, A. Barco, and G. P. Pollini, "Natural Products Synthesis Through Pericyclic Reactions," American Chemical Society, Washington, DC, 1983.
2. U. Pindur, G. Lutz, and C. Otto, *Chem. Rev.*, 1993, **93**, 741.
3. S. Laschat, *Angew. Chem., Int. Ed. Engl.*, 1996, **35**, 289.
4. G. L. E. Koch, D. C. Show, and F. Gibson, *Biochim. Biophys. Acta*, 1971, **229**, 795.
5. B. E. Davidson, E. H. Blackburn, and T. A. A. Dopheide, *J. Biol. Chem.*, 1972, **247**, 4441.
6. S. T. Cload, D. R. Liu, R. M. Pastor, and P. G. Schultz, *J. Am. Chem. Soc.*, 1996, **118**, 1787.
7. D. R. Liu, S. T. Cload, R. M. Pastor, and P. G. Schultz, *J. Am. Chem. Soc.*, 1996, **118**, 1789.
8. P. Kast, J. D. Hartgerink, M. Asif-Ullah, and D. Hilvert, *J. Am. Chem. Soc.*, 1996, **118**, 3069.
9. W. M. Bandaranayake, J. E. Banfield, and D. S. C. Black, *J. Chem. Soc., Chem. Commun.*, 1980, 902.
10. K. C. Nicolaou, N. A. Petasis, R. E. Zipkin, and J. Uenishi, *J. Am. Chem. Soc.*, 1982, **104**, 5555.
11. J. T. Yli-Kauhaluoma, J. A. Ashley, C.-H. Lo, L. Tuker, M. M. Wolfe, and K. D. Janda, *J. Am. Chem. Soc.*, 1995, **117**, 7041.
12. K. R. Rao, T. N. Surinivasan, and N. Bhanumathi, *Tetrahedron Lett.*, 1990, **31**, 5959.
13. S. Colonna, A. Manfredi, and R. Annunziata, *Tetrahedron Lett.*, 1988, **29**, 3347.
14. A. C. Bazan, J. M. Edwards, and U. Weiss, *Tetrahedron*, 1978, **34**, 3005.
15. A. Ichihara, *Synthesis*, 1987, 207.
16. A. Ichihara, in "Stereoselective Synthesis (Part C)," ed. A.-u. Rahman, Elsevier, Amsterdam, 1989, vol. 4, p. 579.
17. A. Ichihara, *J. Synth. Org. Chem. (Jpn)*, 1992, **50**, 96.
18. A. Ichihara, H. Oikawa, in "Dynamic Aspects of Natural Products Chemistry; Molecular Biological Approaches," eds. K. Ogura and U. Sankawa, Koudansha Scientific, Tokyo, 1996, p. 119.
19. D. R. Schroeder and F. R. Stermitz, *Tetrahedron*, 1985, **41**, 4309.
20. Y.-C. Kong, K.-F. Cheng, R. C. Cambie, and P. G. Waterman, *J. Chem. Soc., Chem. Commun.*, 1985, 47.
21. K.-F. Cheng, Y. C. Kong, and T.-Y. Chan, *J. Chem. Soc., Chem. Commun.*, 1985, 48.
22. J.-L. Pousset, C. A. A. Chiaroni, and C. Riche, *J. Chem. Soc., Chem. Commun.*, 1977, 261.
23. F. Tillequin, M. Koch, J.-L. Pousset, and A. Cave, *J. Chem. Soc., Chem. Commun.*, 1978, 826.
24. Y. Kishimura, A. Kawasima, T. Kagamizono, M. Yamagishi, K. Matsumoto, Y. Kawasima, and K. Harada, *J. Antibiot.*, 1992, **45**, 892.
25. W. S. Horn, R. E. Schwartz, M. S. J. Simmonds, and W. M. Blaney, *Tetrahedron Lett.*, 1994, **35**, 6037.
26. M. Matsumoto and H. Minato, *Tetrahedron Lett.*, 1976, 3827.
27. W. D. Celmer, G. N. Chmurny, C. E. Moppett, R. S. Ware, P. C. Watts, and E. B. Whipple, *J. Am. Chem. Soc.*, 1980, **102**, 4203.
28. H. Kobayashi, R. Sunaga, K. Furihata, N. Morisaki, and S. Iwasaki, *J. Antibiot.*, 1995, **48**, 42.
29. M. Kobayashi, H. Uehara, K. Matsunami, S. Aoki, and I. Kitagawa, *Tetrahedron Lett.*, 1993, **34**, 7925.
30. K. Nozawa, S. Nakajima, S. Udagawa, and K. Kawai, *J. Chem. Soc., Perkin Trans. 1*, 1991, 537.
31. G. M. Brill, R. H. Chen, R. R. Rasmussen, D. N. Whittern, and J. B. McAlpine, *J. Antibiot.*, 1993, **46**, 39.
32. K. Fukuyama, Y. Katsube, T. Hamasaki, and Y. Hatsuda, *J. Chem. Soc., Perkin Trans. 2*, 1978, 683.
33. M. J. Begley and J. F. Grove, *J. Chem. Soc. Perkin Trans. 1*, 1985, 861.
34. N. Tabata, H. Tomoda, Y. Iwai, and S. Omura, *J. Antibiot.*, 1993, **46**, 1854.
35. R. Sawa, Y. Takahashi, S. Itoh, K. Shimanaka, N. Matsuda, T. Sawa, H. Naganawa, and T. Takeuchi, *J. Antibiot.*, 1992, **45**, 136.
36. J. W. Coe and W. R. Roush, *J. Org. Chem.*, 1989, **54**, 915.
37. W. R. Roush, K. Koyama, M. L. Curtin, and K. J. Moriarty, *J. Am. Chem. Soc.*, 1996, **118**, 7502.
38. M. Hirama and M. Uei, *J. Am. Chem. Soc.*, 1982, **104**, 4251.
39. N. J. Phillips, J. T. Goodwin, A. Fraiman, R. J. Cole, and D. G. Lynn, *J. Am. Chem. Soc.*, 1989, **111**, 8223.
40. Y. Hayakawa, N. Kanamaru, N. Morisaki, and H. Seto, *Tetrahedron Lett.*, 1991, **32**, 213.
41. S. B. Singh, M. A. Goetz, E. T. Jones, G. F. Bills, R. A. Giacobbe, L. Harranz, S. Stevens-Miles, and D. L. Williams, Jr., *J. Org. Chem.*, 1995, **60**, 7040.

42. M. Ueno, T. Someno, R. Sawa, H. Iinuma, Y. Takahashi, H. Naganawa, M. Ishizuka, and T. Takeuchi, *J. Antibiot.*, 1993, **46**, 1020.
43. T. Mikawa, N. Chiba, H. Ogishi, Y. Sato, S. Miyaji, and M. Sezaki, *Chem. Abstr.*, 1990, **113**, 126593.
44. T. Sasaki, M. Takagi, M. Yaguchi, K. Nishiyama, T. Yaguchi, and M. Koyama, *Chem. Abstr.*, 1993, **118**, 211424.
45. S. Toda, S. Yamamoto, O. Tenmyo, T. Tsuno, T. Hasegawa, M. Rosser, M. Oka, Y. Sawada, M. Konishi, T. Oki, and J. Okumura, *J. Antibiot.*, 1993, **46**, 875.
46. H. Fujimoto, M. Ikeda, K. Yamamoto, and M. Yamazaki, *J. Nat. Prod.*, 1993, **56**, 1268.
47. M. Tanabe and S. Urano, *Tetrahedron*, 1983, **39**, 3569.
48. R. Muntwyler and W. Keller-Schierlein, *Helv. Chim. Acta*, 1972, **53**, 2071.
49. A. K. Mallams, M. S. Puar, R. R. Rossman, A. T. McPhail, R. D. Macfarlane, and R. L. Stephens, *J. Chem. Soc., Perkin Trans. 1*, 1983, 1497.
50. N. Hirayama, M. Kasai, K. Shirahata, Y. Ohashi, and Y. Sasada, *Tetrahedron Lett.*, 1980, **21**, 2559.
51. K. Luk and S. A. Readshaw, *J. Chem. Soc., Perkin Trans. 1*, 1991, 1641.
52. M. Matsumoto, Y. Kawamura, Y. Yoshimura, Y. Terui, H. Nakai, T. Yoshida, and J. Shoji, *J. Antibiot.*, 1990, **43**, 739.
53. K. W. Shimotohno, T. Endo, and K. Furihata, *J. Antibiot.*, 1993, **46**, 682.
54. J. J. Lee, J. P. Lee, P. J. Keller, C. E. Cottrell, C.-J. Chang, H. Zähner, and H. G. Floss, *J. Antibiot.*, 1986, **34**, 1123.
55. K. Takeda, Y. Igarashi, K. Okazaki, E. Yoshii, and K. Yamaguchi, *J. Org. Chem.*, 1990, **55**, 3434.
56. H. Imai, H. Kaniwa, T. Tokunaga, S. Fujita, T. Furuya, H. Matsumoto, and M. Shimizu, *J. Antibiot.*, 1987, **40**, 1483.
57. T. Ohtsuka, T. Kudoh, N. Shimma, H. Kotaki, N. Nakamura, Y. Itezono, N. Fujisaki, J. Watanabe, K. Yokose, and H. Seto, *J. Antibiot.*, 1992, **45**, 140.
58. T. Tsuchida, R. Sawa, H. Iinuma, C. Nishida, N. Kinoshita, Y. Takahashi, H. Naganawa, R. Sawa, M. Hamada, and T. Takeuchi, *J. Antibiot.*, 1994, **47**, 386.
59. J. T. Pinhey, E. Ritchie, and W. C. Taylor, *Aust. J. Chem.*, 1961, **14**, 106.
60. M. V. D'Auria, C. Debitus, L. G. Paloma, L. Minale, and A. Zampella, *J. Am. Chem. Soc.*, 1994, **116**, 6658.
61. C. Tamm, in "Cytochalasins-Biochemical and Cell Biological Aspects," ed. S. W. Tanenbaum, Elsevier, Amsterdam, 1978, p. 15.
62. V. Betina, "Bioactive Molecules, Mycotoxins," Elsevier, Amsterdam, 1989, vol. 9, p. 285.
63. A. Ichiahra, K. Katayama, H. Teshima, H. Oikawa, and S. Sakamura, *Biosci. Biotech. Biochem.*, 1995, **60**, 360.
64. E. J. Thomas, *Acc. Chem. Res.*, 1991, **24**, 229 and references cited therein.
65. W. S. Horn, M. S. J. Simmonds, R. E. Schwartz, and W. M. Blaney, *Tetrahedron*, 1995, **51**, 3969.
66. S. J. Coval and P. J. Scheuer, *J. Org. Chem.*, 1985, **50**, 3024.
67. S. Miao, M. R. Anstee, V. Baichwal, and A. Park, *Tetrahedron Lett.*, 1995, **36**, 5699.
68. J. W. Westley, R. H. J. Evans, C.-M. Liu, T. Hermann, and J. F. Blunt, *J. Am. Chem. Soc.*, 1978, **100**, 6784.
69. S. H. Larsen, L. D. Boeck, F. P. Mertz, J. W. Paschal, and J. Occolowitz, *J. Antibiot.*, 1988, **41**, 1170.
70. S. Ito and Y. Hirata, *Bull. Chem. Soc. Jpn.*, 1977, **50**, 1813.
71. H. A. Kirst, K. H. Michel, J. W. Martin, L. C. Creemer, E. H. Chio, R. C. Yao, W. M. Nakatsukasa, L. V. D. Boeck, J. L. Occolowitz, J. W. Paschal, J. B. Deeter, N. D. Jones, and G. D. Thompson, *Tetrahedron Lett.*, 1991, **32**, 4839.
72. D. E. Cane, W. Tan, and W. R. Ott, *J. Am. Chem. Soc.*, 1993, **115**, 527.
73. S. Seo, U. Sankawa, Y. Ogihara, Y. Iitaka, and S. Shibata, *Tetrahedron*, 1973, **29**, 3721.
74. T. Kusumi, A. Ichikawa, H. Kakisawa, M. Tsunakawa, M. Konishi, and T. Oki, *J. Am. Chem. Soc.*, 1991, **113**, 8947.
75. J. Spörle, H. Becker, M. P. Gupta, M. Veith, and V. Huch, *Tetrahedron*, 1989, **45**, 5003.
76. J. Spörle, H. Becker, N. S. Allen, and M. P. Gupta, *Phytochemistry*, 1991, **30**, 3043.
77. W. Herz, K. D. Pethtel, and D. Raulais, *Phytochemistry*, 1991, **30**, 1273.
78. A. A. Ahmed, J. Jakupovic, F. Bohlmann, H. A. Regaila, and A. M. Ahmed, *Phytochemistry*, 1990, **29**, 2211.
79. J. Beauhaire, J. L. Fourrey, and M. Vuilhorgne, *Tetrahedron Lett.*, 1980, **21**, 3191.
80. C. Zdero and F. Bohlmann, *Phytochemistry*, 1989, **28**, 3105.
81. F. Gao, H. Wang, and T. J. Mabry, *Phytochemistry*, 1990, **29**, 3875.
82. A. Ovezdurdyev, N. D. Abdullaev, M. I. Yusupov, and S. Z. Kasymov, *Khim. Prir. Soedin.*, 1987, 667.
83. R. Matsusch and H. Häberlein, *Liebigs Ann. Chem.*, 1987, 455.
84. A. R. Vivar and G. Delgado, *Tetrahedron Lett.*, 1985, **26**, 579.
85. V. A. Tarasov, N. D. Abdullaev, S. Z. Kasymov, G. P. Sidyakin, and M. R. Yagudaev, *Khim. Prir. Soedin.*, 1976, **6**, 745.
86. J. Jakupovic, A. Schuster, F. Bohlmann, and M. O. Dillon, *Phytochemistry*, 1988, **27**, 1113.
87. A. A. Bell, R. D. Stipanovic, D. H. O'Brien, and P. A. Fryxell, *Phytochemistry*, 1978, **17**, 1297.
88. A. Parvez, M. I. Choudhary, F. Akhter, M. Noorwala, F. V. Mohammad, N. M. Hasan, T. Zamir, and V. U. Ahmad, *J. Org. Chem.*, 1992, **57**, 4339.
89. G. Majetich and Y. Zhang, *J. Am. Chem. Soc.*, 1994, **116**, 4979.
90. J. A. Marco, J. F. Sanz, E. Falco, J. Jakupovic, and J. Lex, *Tetrahedron*, 1990, **46**, 7941.
91. J. Kawabata, Y. Fukushi, S. Tahara, and J. Mizutani, *Phytochemistry*, 1990, **29**, 2332.
92. P. Tane, K.-E. Bergquist, M. Tene, B. T. Ngadjui, J. F. Ayafor, and O. Stemer, *Tetrahedron*, 1995, **51**, 11595.
93. C. P. Falshaw and T. J. King, *J. Chem. Soc., Perkin Trans. 1*, 1983, 1749.
94. O. Shirota, H. Morita, K. Takaya, and H. Itokawa, *Tetrahedron*, 1995, **51**, 1107.
95. J. C. Lee, G. A. Strobel, E. Lobkovsky, and J. Clardy, *J. Org. Chem.*, 1996, **61**, 3232.
96. X. Fu, B. Hossain, F. J. Schmitz, and D. Helm, *J. Org. Chem.*, 1997, **62**, 3810.
97. A. J. Bartlett, J. S. Holker, and E. O'Brien, *J. Chem. Soc., Chem. Commun.*, 1981, 1198.
98. D. Hauser and H. P. Sigg, *Helv. Chim. Acta*, 1971, **54**, 1178.
99. N. Kato, S. Kusakabe, X. Wu, M. Kamitamari, and H. Takeshita, *J. Chem. Soc., Chem. Commun.*, 1993, 1002 and references cited therein.
100. T. Kusumi, M. Igari, M. O. Ishitsuka, A. Ichikawa, Y. Itezono, N. Nakayama, and H. Kakisawa, *J. Org. Chem.*, 1990, **55**, 6286.
101. S. Jingyu, L. Kanghou, P. Tangsheng, H. Cun-heng, and J. Clardy, *J. Am. Chem. Soc.*, 1986, **108**, 177.
102. P. A. Leone, B. F. Bowden, A. R. Carroll, J. C. Coll, and G. V. Meehan, *J. Nat. Prod.*, 1993, **56**, 521.

103. W. Hofheinz and P. Schonholzer, *Helv. Chim. Acta*, 1977, **60**, 1367.
104. M. Kobayashi, K. Kawazoe, T. Katori, and I. Kitagawa, *Chem. Pharm. Bull.*, 1992, **40**, 1773.
105. W. A. Ayer and Y.-T. Ma, *Can. J. Chem.*, 1992, **70**, 1905.
106. S. Yamamura, Y. Terada, Y. Chen, M. Hong, H. Hsu, K. Sasaki, and Y. Hirata, *Bull. Chem. Soc. Jpn.*, 1976, **49**, 1940.
107. M. Niwa, Y. Terada, M. Nonoyama, and S. Yamamura, *Tetrahedron Lett.*, 1979, **20**, 813.
108. M. Iguchi, A. Nishiyama, Y. Terada, and S. Yamamura, *Tetrahedron Lett.*, 1977, **18**, 4511.
109. S. Yamamura and M. Niwa, *Chem. Lett.*, 1981, 625.
110. A. Nishiyama, H. Eto, Y. Terada, M. Iguchi, and S. Yamamura, *Chem. Pharm. Bull.*, 1983, **31**, 2820.
111. I. C. Parsons, A. I. Gfay, T. G. Hartlery, and P. G. Waterman, *Phytochemistry*, 1993, **33**, 479.
112. J. P. Kutney, T. Inaba, and D. L. Dreyer, *Tetrahedron*, 1970, **26**, 3171.
113. A. G. Gonzalez, R. J. Cardona, C. E. Diaz, D. H. Lopez, and L. F. Rodriguez, *An. Quim.*, 1977, **73**, 1510.
114. K. L. Brown, A. I. R. Burfitt, R. C. Cambie, D. Hall, and K. P. Mathai, *Aust. J. Chem.*, 1975, **28**, 1327.
115. P. N. Sharma, A. Shoeb, R. S. Kapil, and S. P. Popli, *Phytochemistry*, 1980, **19**, 1258.
116. D. P. Chakraborty, S. Roy, A. Chakraborty, A. K. Mandal, and B. K. Chowdhury, *Tetrahedron*, 1980, **36**, 3563.
117. H. Ishii, J. Kobayashi, and T. Ishikawa, *Tetrahedron Lett.*, 1991, **32**, 6907.
118. R. Sakai, T. Higa, C. W. Jefford, and G. J. Benardinelli, *J. Am. Chem. Soc.*, 1986, **108**, 6404.
119. H. Nakamura, S. Deng, J. Kobayashi, Y. Ohizumi, Y. Tomotake, and T. Matsuzaki, *Tetrahedron Lett.*, 1987, **28**, 621.
120. J. E. Baldwin and R. C. Whitehead, *Tetrahedron Lett.*, 1992, **33**, 2059.
121. J. Kobayashi, M. Tsuda, N. Kawasaki, K. Matsumoto, and T. Adachi, *Tetrahedron Lett.*, 1994, **35**, 4383.
122. K. Kondo, H. Shigemori, Y. Kikuchi, M. Ishibashi, T. Sakai, and J. Kobayashi, *J. Org. Chem.*, 1992, **57**, 2480.
123. M. Tsuda, N. Kawakami, and J. Kobayashi, *Tetrahedron*, 1994, **50**, 7957.
124. J. Rodriguez and P. Crews, *Tetrahedron Lett.*, 1994, **35**, 4719.
125. F. Kong, R. J. Andersen, and T. M. Allen, *Tetrahedron Lett.*, 1994, **35**, 1643.
126. A. Rudi and Y. Kashman, *J. Org. Chem.*, 1989, **54**, 5331.
127. J. Kobayashi, J.-F. Cheng, M. R. Wälchli, H. Nakamura, Y. Hirata, T. Sasaki, and Y. Ohizumi, *J. Org. Chem.*, 1988, **53**, 1800.
128. R. P. Walker, D. H. Faulkner, D. V. Engen, and J. Clardy, *J. Am. Chem Soc.*, 1981, **103**, 6772.
129. P. A. Keifer, R. E. Schwartz, M. E. S. Koker, R. G. Hughes, D. Rittschof, and K. L. J. Rinehart, *J. Org. Chem.*, 1991, **56**, 2956.
130. D. H. Williams and D. J. Faulkner, *Tetrahedron*, 1996, **52**, 5381.
131. T. B. Ngadjui, J. F. Ayafor, B. L. Sandengam, J. D. Connolly, D. S. Rycroft, S. A. Khalid, P. G. Waterman, N. M. D. Brown, M. F. Grundon, and V. N. Ramachandran, *Tetrahedron Lett.*, 1982, **23**, 2041.
132. S. Tsukamoto, S. Takeuchi, M. Ishibashi, and J. Kobayashi, *J. Org. Chem.*, 1992, **57**, 5255.
133. G. S. R. S. Rao, B. Ravindranath, and V. P. S. Kumar, *Phytochemistry*, 1984, **23**, 399.
134. C. Ito, T. Ono, E. Tanaka, Y. Takemura, T. Nakata, H. Uchida, M. Ju-ichi, M. Omura, and H. Furukawa, *Chem. Pharm. Bull.*, 1993, **41**, 205.
135. I. Sordat-Diserens, M. Hamburger, C. Rogers, and K. Hostettmann, *Phytochemistry*, 1992, **31**, 3589.
136. J. Kobayashi, S. Takeuchi, M. Ishibashi, H. Shigemori, and T. Sasaki, *Tetrahedron Lett.*, 1992, **33**, 2579.
137. S. Hyoudo, K. Fujita, O. Kasuya, I. Takahashi, J. Uzawa, and H. Koshino, *Tetrahedron*, 1995, **51**, 6717.
138. T. Cou, O. Kamo, and D. Uemura, *Tetrahedron Lett.*, 1996, **37**, 4023.
139. K. Seki, T. Tomihari, K. Haga, and R. Kaneko, *Phytochemistry*, 1994, **37**, 870.
140. T. Maoka, N. Akimoti, K. Hashimoto, Y. Kuroda, and Y. Fujiwara, in "37th Symposium on the Chemistry of Natural Products, Abstracts, Tokushima, 1995," p. 37.
141. K. Torigoe, M. Murata, T. Yasumoto, and T. Iwashita, *J. Am. Chem Soc.*, 1988, **110**, 7876.
142. D. E. Cane and L. Guanglin, *J. Am. Chem. Soc.*, 1995, **117**, 6633.
143. R. N. Moore, G. Bigam, J. K. Chan, A. M. Hogg, T. T. Nakashima, and J. C. Vederas, *J. Am. Chem. Soc.*, 1985, **107**, 3694.
144. Y. Yoshizawa, D. J. Witter, Y. Liu, and J. C. Vederas, *J. Am. Chem. Soc.*, 1994, **116**, 2693.
145. D. J. Witter and J. C. Vederas, *J. Org. Chem.*, 1996, **61**, 2613.
146. M. Takasugi, S. Nagao, T. Masamune, A. Shirata, and K. Takahashi, *Chem. Lett.*, 1980, 1573.
147. S. Ueda, T. Nomura, and J. Matsumoto, *J. Chem. Pharm. Bull.*, 1982, **30**, 3042.
148. S. Yamamura, Y. Terada, Y. Chen, M. Hong, H. Hsu, K. Sasaki, and Y. Hirata, *Bull. Chem. Soc. Jpn.*, 1976, **49**, 1940.
149. Y. Hano, T. Nomura, and S. Ueda, *J. Chem. Soc., Chem. Commun.*, 1990, 610.
150. Y. Hano, M. Aida, T. Nomura, and S. Ueda, *J. Chem. Soc., Chem. Commun.*, 1992, 1177.
151. I. Messana, F. Ferrari, and M. d. C. M. d. Araujo, *Tetrahedron*, 1988, **44**, 6693.
152. A. Jitoe, T. Masuda, and T. J. Mabry, *Tetrahedron Lett.*, 1994, **35**, 981.
153. A. J. Birch and R. A. Russell, *Tetrahedron*, 1972, **28**, 2999.
154. M. Yamazaki, E. Okuyama, M. Kobayashi, and H. Inoue, *Tetrahedron Lett.*, 1981, **22**, 135.
155. J. Polonsky, M.-A. Merrien, T. Prange, and C. Pascard, *J. Chem. Soc., Chem. Commun.*, 1980, 601.
156. J. F. Sanz-Cervera, T. Glinka, and R. Williams, *J. Am. Chem. Soc.*, 1993, **115**, 347.
157. A. Ichihara, H. Oikawa, K. Hayashi, S. Sakamura, A. Furusaki, and T. Matsumoto, *J. Am. Chem. Soc.*, 1983, **105**, 2907.
158. A. Ichihara, H. Oikawa, M. Hashimoto, S. Sakamura, T. Haraguchi, and H. Nagano, *Agric. Biol. Chem.*, 1983, **47**, 2965.
159. H. Oikawa, A. Ichihara, and S. Sakamura, *Agric. Biol. Chem.*, 1984, **48**, 2603.
160. I. Barash, S. Manulis, Y. Kashman, J. P. Springer, M. H. M. Chen, J. Clardy, and G. A. Strobel, *Science*, 1983, **220**, 1065.
161. P. S. Steyn, P. L. Wessels, C. W. Holzapfel, D. J. Potgieta, and W. K. A. Louw, *Tetrahedron*, 1972, **28**, 4775.
162. A. Ichihara, S. Miki, H. Kawagishi, and S. Sakamura, *Tetrahedron Lett.*, 1989, **30**, 4551.
163. A. Ichihara, H. Kawagishi, N. Tokugawa, and S. Sakamura, *Tetrahedron Lett.*, 1986, **27**, 1347.
164. S. Miki, Y. Sato, H. Tabuchi, H. Oikawa, A. Ichihara, and S. Sakamura, *J. Chem. Soc., Perkin Trans. 1*, 1990, 1228.
165. H. Oikawa, A. Ichihara, and S. Sakamura, *J. Chem. Soc., Chem. Commun.*, 1984, 814.
166. H. Oikawa, T. Yokota, S. Miki, Y. Sato, A. Ichihara, and S. Sakamura, "30th Symposium on the Chemistry of Natural Products, Fukuoka, 1988," p. 371.

167. C. J. Coulson, D. J. King, and A. Wiseman, *Trends Biochem. Sci.*, 1984, **10**, 446.
168. D. W. Nebert, D. R. Nelson, M. J. Coon, R. W. Estabrook, R. Feyereisen, Y. Fujii-Kuriyama, F. J. Gonzalez, F. P. Guengerich, I. C. Gunsalus, E. F. Johnson, J. C. Loper, R. Sato, M. Waterman, and D. J. Waxman, *DNA Cell Biol.*, 1991, **10**, 1.
169. F. P. Guengerich, *J. Biol. Chem.*, 1991, **266**, 10019.
170. J. L. Gaylor, in "Biosynthesis of Isoprenoid Compounds," eds. J. W. Porter and S. L. Spurgeon, Wiley, New York 1981, Vol. 481, p. 1.
171. F. Van Middlesworth, A. E. Desjardins, S. L. Taylor, and R. D. Plattner, *J. Chem. Soc., Chem. Commun.*, 1986, 1156.
172. H. Oikawa, A. Ichihara, and S. Sakamura, *J. Chem. Soc., Chem. Commun.*, 1990, 908.
173. D. E. Stevenson, J. N. Wright, and M. Akhtar, *J. Chem. Soc., Perkin Trans. 1*, 1988, 2043.
174. S. Sekita, K. Yoshihara, S. Natori, S. Udagawa, F. Sakabe, H. Kurata, and M. Umeda, *Chem. Pharm. Bull.*, 1982, **30**, 1609.
175. H. C. Cutler, F. G. Grumley, H. R. Cox, R. J. Cole, J. W. Domer, J. P. Springer, F. M. Latterell, J. M. E. Thean, and A. E. Rossi, *J. Agric. Food Chem.*, 1980, **28**, 139.
176. W. B. Turner, *Postepy Hig. Med. Dosw.*, 1974, **28**, 683.
177. C. E. Stickings and R. J. Townsend, *Biochem. J.*, 1961, **78**, 412.
178. N. J. Phillips, J. T. Goodwin, A. Fraiman, R. J. Cole, and D. G. Lynn, *J. Am. Chem. Soc.*, 1989, **111**, 8223.
179. H. Oikawa, Y. Murakami, and A. Ichihara, *J. Chem. Soc., Perkin Trans. 1*, 1992, 2955.
180. H. Oikawa, Y. Murakami, and A. Ichihara, *Biosci. Biotech. Biochem.*, 1993, **57**, 628.
181. A. Probst and Ch. Tamm, *Helv. Chim. Acta*, 1981, **64**, 2065.
182. S. Sekita, K. Yoshihira, and S. Natori, *Chem. Pharm. Bull.*, 1983, **31**, 490.
183. H. Oikawa, Y. Murakami, and A. Ichihara, *J. Chem. Soc., Perkin Trans. 1*, 1992, 2949.
184. S. Sekita, K. Yoshihira, S. Natori, and H. Kuwano, *Chem. Pharm. Bull.*, 1982, **30**, 1629.
185. H. Oikawa, Y. Murakami, and A. Ichihara, *Tetrahedron Lett.*, 1991, **32**, 1349.
186. M. Binder and Ch. Tamm, *Helv. Chim. Acta*, 1973, **56**, 2387.
187. R. Wyss and Ch. Tamm, *Croat. Chem. Acta*, 1985, **58**, 537.
188. A. Ichihara, H. Tazaki, and S. Sakamura, *Tetrahedron Lett.*, 1983, **24**, 5373.
189. H. Oikawa, T. Yokota, A. Ichihara, and S. Sakamura, *J. Chem. Soc., Chem. Commun.*, 1989, 1284.
190. S. S. Alam, J. M. Bilton, M. Z. Slawin, D. J. Williams, R. N. Sheppard, and R. M. Strange, *Phytochemistry*, 1989, **28**, 2627.
191. A. Ichihara, M. Miki, and S. Sakamura, *Tetrahedron Lett.*, 1985, **26**, 2453.
192. A. Ichihara, M. Miki, H. Tazaki, and S. Sakamura, *Tetrahedron Lett.*, 1987, **28**, 1175.
193. H. Oikawa, T. Yokota, T. Abe, A. Ichihara, S. Sakamura, Y. Yoshizawa, and J. C. Vederas, *J. Chem. Soc., Chem. Commun.*, 1989, 1282.
194. H. Oikawa, Y. Suzuki, A. Naya, K. Katayama, and A, Ichihara, *J. Am. Chem. Soc.*, 1994, **116**, 3605.
195. H. Oikawa, K. Katayama, Y. Suzuki, and A. Ichihara, *J. Chem. Soc., Chem. Commun.*, 1995, 1321.
196. D. Hilvert, K. W. Hill, K. D. Nared, and M.-T. Auditor, *J. Am. Chem. Soc.*, 1989, **111**, 9261.
197. V. E. Gouverneur, K. N. Houk, B. Pascual-Teresa, B. Beno, K. D. Janda, and R. A. Lerner, 1993, **262**, 204.
198. A. C. Braisted and P. G. Schultz, *J. Am. Chem. Soc.*, 1991, **112**, 7430.
199. J. V. Gray, B. Golinelli-Pimpaneau, and J. R. Knowles, *Biochemistry*, 1990, **29**, 376.
200. H. Sato, K. Konoma, S. Sakamura, A. Furusaki, T. Matsumoto, and T. Matsuzaki, *Agric. Biol. Chem.*, 1981, **45**, 795.
201. H. Sato, K. Konoma, and S. Sakamura, *Agric. Biol. Chem.*, 1981, **45**, 1675.
202. A. Ichihara, K. Murakami, and S. Sakamura, *Tetrahedron*, 1987, **43**, 5245.
203. I. Sakurai, H. Miyajima, K. Akiyama, S. Shimizu, and Y. Yamamoto, *Chem. Pharm. Bull.*, 1988, **36**, 2003.
204. H. Oikawa, K. Yagi, K. Watanabe, M. Honma, and A. Ichihara, *J. Chem. Soc., Chem. Commun.*, 1997, 97.
205. K. Yagi, H. Oikawa, and A. Ichihara, *Biosci. Biotech. Biochem.*, 1997, **61**, 1038.
206. D. Hölscher and B. Schneider, *J. Chem. Soc., Chem. Commun.*, 1995, 525.
207. K. Chiba, J. Sonoyama, and M. Tada, *J. Chem. Soc., Chem. Commun.*, 1995, 1381.
208. M. F. Ruiz-López, X. Assfeld, J. García, J. A. Mayoral, and L. Salvatella, *J. Am. Chem. Soc.*, 1993, **115**, 8780.
209. R. Breslow, *Acc. Chem. Res.*, 1991, **24**, 159.
210. D. R. Williams, R. D. Gaston, and I. B. Horton, *Tetrahedron Lett.*, 1985, **26**, 1391.

1.16 Polyketide Biosynthesis in Filamentous Fungi

ISAO FUJII
The University of Tokyo, Japan

1.16.1 INTRODUCTION

Fungi are lower eukaryotic microbes having an important relationship with human beings not only as plant and animal pathogens, but also as major producers of enzymes, amino acids, and biologically active secondary metabolites.

A large number of compounds have been isolated and characterized from fungal origins, as described in books such as the "Fungal Metabolites" series.[1,2] Polyketides are the largest group of fungal metabolites, occurring in the greatest number and variety, especially in the Ascomycetes and the related 'imperfect' fungi such as the well-known *Penicillium* and *Aspergillus*, which are amongst the most prolific sources of polyketide compounds, along with bacteria and plants. Fungal polyketides are mostly aromatic compounds, varying from the simplest tetraketides such as 6-methylsalicylic acid (**1**, 6MSA) and orsellinic acid (**2**) to highly modified aflatoxins (**3**) and (+)-dermalactone (**4**), though some reduced polyketides such as patuolide (**5**),[3] lovastatin (**6**), and related compounds[4] are also produced by fungi. In the late 1990s, the clinical importance of cholesterol-lowering activity of lovastatin (**6**) and its derivatives produced by *Penicillium* and *Aspergillus* fungi has drawn much attention.

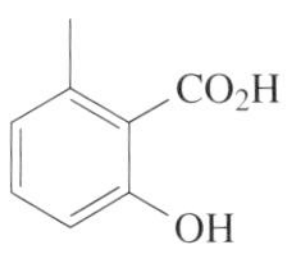

6-Methylsalicylic acid (**1**) 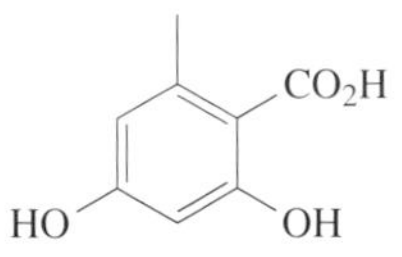 Orsellinic acid (**2**) 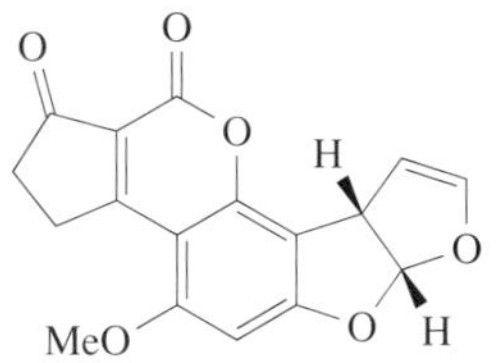 Aflatoxin B_1 (**3**)

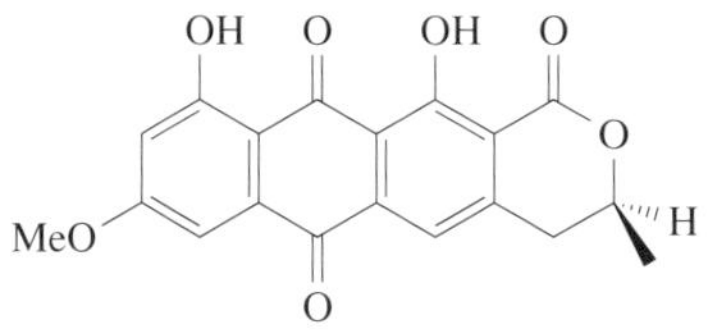

(+)-Dermalactone (**4**) 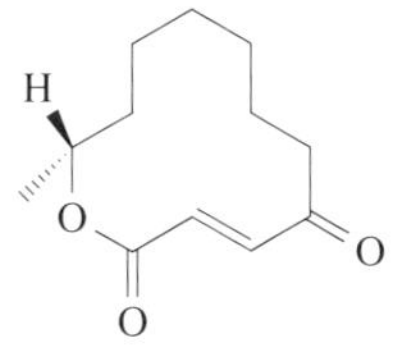 Patuolide A (**5**) 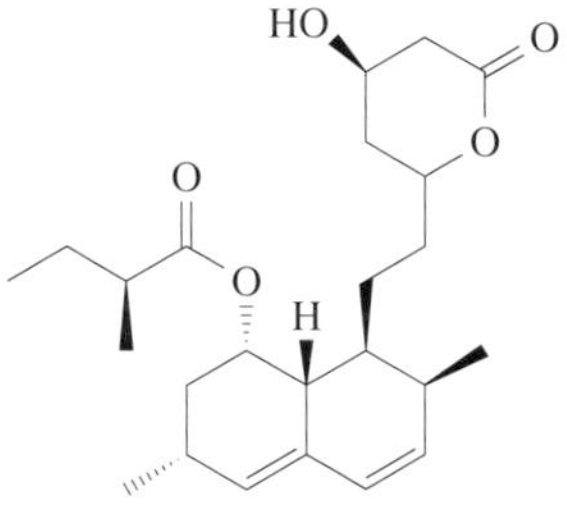 Lovastatin (**6**)

The term "polyketide" was first introduced by Collie[5] from considerations of *in vitro* reactions in which poly-β-ketoacyl compounds afforded aromatic products, notably orsellinic acid (**2**).

6MSA (**1**), a metabolite of *Penicillium patulum*, was the first example experimentally proven to be biosynthesized in the polyketide pathway,[6] which was first proposed by Birch and Donovan as the "polyacetate hypothesis".[7] Since then, many natural products have been identified to be of "polyketide" origin even though some of them were extensively modified structurally by subsequent biosynthetic reactions which obscured their polyketide origin. Most knowledge about polyketide biosynthesis has come from the results of various feeding experiments of isotopically labeled acetates with intact organisms. Now, it has become evident that further understanding in the formation of polyketides, or indeed most families of natural products, must come from studies at cell-free and/or enzyme levels. It is obvious that the use of enzyme systems, if available, not only eliminates the cellular barriers to uptake of exogeneously administered precursors, but also allows us to apply mechanistic enzymology and protein chemistry to investigations on biosynthetic reactions.

Polyketide synthases (PKSs) are enzymes which catalyze the formation and cyclization of the specific "poly β-keto" intermediates to give initial polyketide compounds. 6MSA synthase (MSAS) of *P. patulum* was the first PKS of which enzyme activity was detected *in vitro*.[8] MSAS was later successfully purified and characterized in some detail.[9] However, inherent lability made biochemical studies on other microbial PKSs so difficult that cell-free activities have rarely been detected. Hence the MSAS was the first and sole example of purified microbial PKS for quite some years.

Molecular genetics of bacterial polyketide biosynthesis, especially in Actinomycetes, another rich source of polyketides, opened the way to understanding the structures of bacterial PKS genes, as pioneered by Hopwood and his group.[10–12]

Generally, filamentous fungi have relatively larger genomes, about 10 times as complex as those of bacteria,[13] but their level of complexity, as a class of eukaryotes, is intermediate to unicellular forms such as yeast and truly multicellular forms such as animals and higher plants. Application of molecular genetics to fungal polyketide biosynthesis has been carried out. Cloning of the MSAS gene from *P. patulum* has been the leading accomplishment in this field also.[14] At the time of writing, enzymological and/or molecular genetic investigation have not been carried out extensively on fungal polyketide biosynthesis when compared with progress in the bacterial counterpart. However, some significant progress has been attained in some cases as several individual polyketide synthase genes and gene clusters from fungi have been cloned and analyzed. In this context, studies on fungal polyketide biosynthesis have entered an entirely new era. Detailed analysis of polyketide biosynthesis genes has allowed not only basic mechanistic studies on their biosynthesis, but also practical application of engineered biosynthesis of novel polyketide compounds.[15]

In this chapter, results of polyketide biosynthesis in filamentous fungi mainly at enzyme and molecular genetic levels are described.

1.16.2 FUNGAL POLYKETIDE COMPOUNDS

As evident from their name, polyketide compounds are conveniently classified by the number of C_2-units that constitute their backbone poly-β-ketomethylene chains, such as tetraketide, pentaketide, etc., to denote compounds derived from four, five, etc., C_2-units. Thus, to begin the chapter, it is worthwhile taking an overview of the representative examples of fungal polyketide compounds based on the above classification.

Filamentous fungi most exclusively use acetate as starter units and malonate as extender units for the formation of poly-β-ketomethylene intermediates of their polyketide metabolites, in contrast to higher plants and bacteria, which use, for example, aromatic acid as starter in flavonoid biosynthesis, and propionate starter and methylmalonate as extender units in macrolide biosynthesis, respectively.

Following the formation of carbon skeltons from poly-β-ketomethylene intermediates by cyclization and aromatization, fungi often carry out secondary modification reactions catalyzed by specific enzymes, such as oxidation, alkylation, halogenation, etc. Ring cleavage reactions are key steps in structural rearrangements in fungal polyketide biosynthesis, as is seen in aflatoxin biosynthesis. However, extensive glycosidation is a major secondary modification in bacterial polyketide biosynthesis after PKS reactions. Thus, it is important to characterize reactions involved in the secondary modifications of PKS products.

1.16.3 AROMATIC POLYKETIDE SYNTHASES

Fungal aromatic polyketides vary from single-ring tetraketides such as 6MSA (**1**) and orsellinic acid (**2**) to nonaketide (+)-dermalactone (**4**),[16] for example. Although aromatization is the fundamental characteristic of poly-β-ketomethylene compounds by aldol-type condensation, as first proposed by Collie[5] and later by Birch,[7] regiospecific control of cyclization should be carried out by PKSs for the formation of specific aromatic products. Also, several other key aspects of PKS reactions should be controlled, such as formation and stabilization of the poly-β-ketomethylene intermediates with fixed chain lengths, reduction control and/or other modifications if necessary.

1.16.3.1 Tetraketide Synthases

6MSA (**1**) and orsellinic acid (**2**) are representatives of single aromatic ring tetraketides and direct precursors to fungal metabolites such as gentisic acid (**7**), 4-methoxytoluquinone (**8**), terreic acid (**9**), flavipin (**10**), etc. Also, as discussed later, patulin (**11**) and penicillic acid (**12**) are biosynthesized from 6MSA (**1**) and orsellinic acid (**2**), respectively, via oxidative ring opening reactions.

The enzyme for 6MSA synthesis was detected in the cell-free extract of *P. patulum* as the first example of polyketide synthase from any source.[8] The first purification was reported by Lynen and

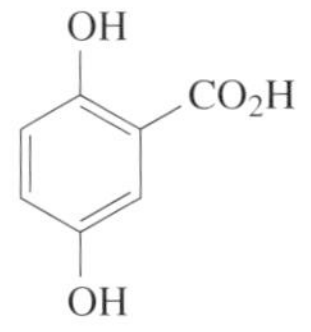

Gentisic acid (**7**) 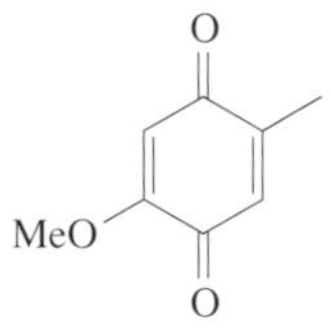 4-Methoxytoluquinone (**8**) 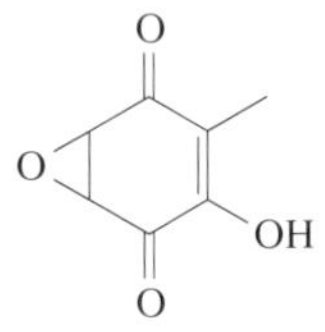 Terreic acid (**9**)

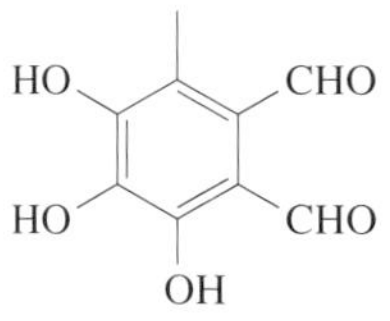

Flavipin (**10**) 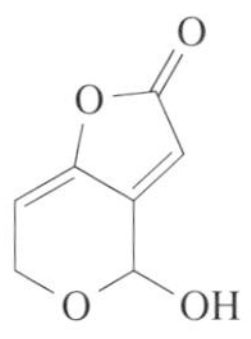 Patulin (**11**) 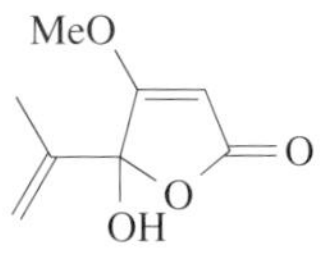 Penicillic acid (**12**)

co-workers in 1970 in a landmark paper.[9] They showed that MSAS contained all of the 11 catalytic activities necessary for the synthesis of 6MSA (**1**) from acetyl- and malonyl-CoA. However, repetition of the enzyme isolation had been hindered by the inherent lability of MSAS. Another successful purification of MSAS was reported much later in 1989 from *Penicillium urticae*.[17] Then purification and characterization of MSAS from *P. patulum* was carried out by other groups in 1990[14] and 1992.[18] MSAS was found to be a 750 kDa homotetramer of a 180 kDa subunit.

In the MSAS reaction, the enzyme catalyzes three cycles of condensations of acyl primer and malonyl extender units. The whole reaction is controlled in the cycle-specific manner. That is, the first cycle proceeds without keto reductase (KR) and dehydratase (DH) reactions, but the second cycle includes KR and DH reactions. In the absence of NADPH, the third condensation cycle does not take place, but gives abortive triacetic lactone (**13**, TAL)[9] (Scheme 1). This cycle-specific control mechanism is still unknown, as is the product-releasing mechanism. In contrast, chalcone synthase catalyzes the condensation to produce chalcone or deoxychalcone depending on the availability of independent reductase protein.[19]

Detection of the synthase activity for orsellinic acid (**2**) (Scheme 2), the simplest aromatic tetraketide, was first reported by Gaucher and Shepherd[20] in 1966. Since then, no purification has been reported although some mechanistic investigations have been carried out using a crude enzyme preparation from *Penicillium cyclopium*.[21] Later, orsellinic acid synthase (OAS) was claimed to be purified from *P. cyclopium*,[22] and a subunit of 130 kDa mentioned, but at the time of writing details have not been reported. The smaller size of the OAS subunit compared to MSAS (180 kDa) may reflect the absence of reductase and dehydratase active sites in the OAS enzyme molecule.

Cloning of the gene of MSAS from *P. patulum*[14] was carried out by screening a cDNA expression library in *E. coli* with an antibody prepared against the purified synthase. The genomic DNA was obtained using the cloned cDNA as a screening probe of a genomic DNA library. Sequencing of the cloned gene revealed a single open reading frame of 5322 bp coding for a 190 731 Da protein of 1774 amino acids. Presence of a 69 bp intron within the N-terminus region was identified by sequence comparison with cDNA and genomic DNA. Although a comparatively low degree of similarity was detected with fungal fatty acid synthase (FAS), a significantly higher sequence similarity was found between MSAS and the rat FAS, especially at their β-ketoacyl synthase (KS), acetyl/malonyl acyltransferase (AT), β-ketoacyl reductase (KR), and acyl carrier protein (ACP) domains in the similar linear organization. However, the deduced amino acid sequence did not contain a thioesterase (TE) domain, which is considered to have a role in chain-length control of the FAS reaction.[23] Partial cDNA and genomic DNA of the MSAS gene were also cloned from *P. urticae*.[24]

Probing with cloned PKS genes proved to be a powerful tool for the screening of related PKS genes, as was exemplified by PKS gene cloning studies in Streptomycetes.[12,25] Southern blot analysis of polyketide-producing fungal genomic DNA was carried out with the *P. patulum* MSAS gene as

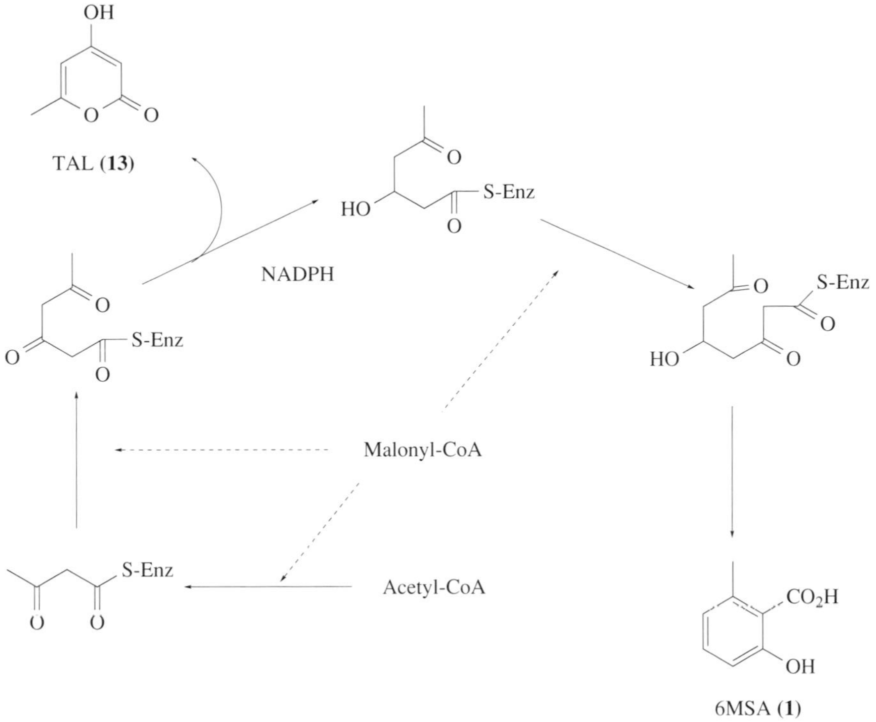

TAL (**13**)

6MSA (**1**)

Scheme 1

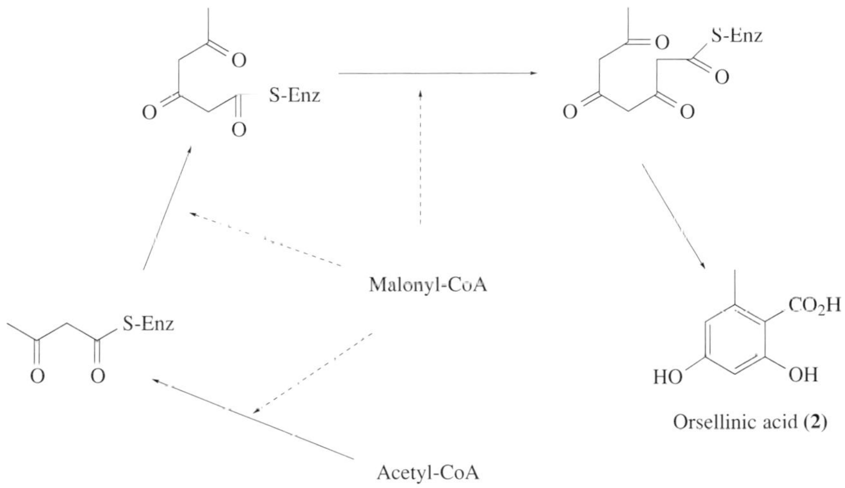

Orsellinic acid (**2**)

Scheme 2

a probe. *Aspergillus terreus* IMI 16043, a strain known to produce (+)-geodin (**14**) as main metabolite, gave a strong hybridizing band. However, no homologous band was detected in the genomic DNA of *P. cyclopium*, which produces orsellinic acid (**2**), indicating much difference between the MSAS and OAS genes.[26]

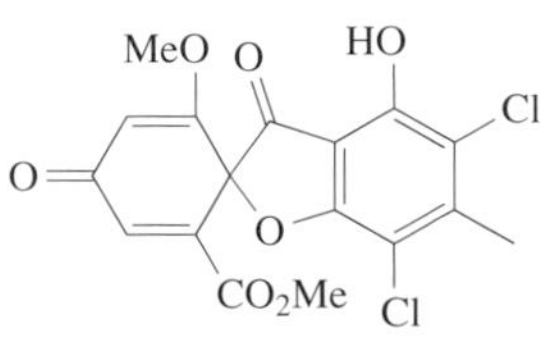

(+)-Geodin (**14**)

From the genomic DNA library of *A. terreus*, a MSAS homologous gene, named *atX*, was cloned and sequenced by Fujii *et al.*[26] Head-to-tail homology with *P. patulum* MSAS was observed. In particular, nearly identical amino acid sequences were found around each KS, AT, KR, and ACP active-site region. Since no transcriptional message of the *atX* gene was observed by Northern blot analysis of all its growth phase in culture, identification of the *atX* gene product was carried out by expression in a heterologous host using fungal expression vector pTAex3.[27] The result, that the *Aspergillus nidulans* transformant with an *atX* expression plasmid produced a significant amount of 6MSA (**1**), confirmed that the *atX* gene codes for MSAS of *A. terreus*. It is not surprising that the MSAS gene is cloned from *A. terreus*, although the gene is not functioning in the strain, since another strain of *A. terreus* is known to produce toluquinones which are derived from 6MSA (**1**). Sequences of probable MSAS genes from other fungal strains were also deposited on DNA database (accession numbers: U31329, U89769) although their functions have not been identified yet. The MSAS gene was also expressed in *Streptomyces coelicolor* to produce 6MSA (**1**) by Khosla and co-workers.[28] These expression systems of fungal PKS genes enable us to dissect fungal PKS reactions using recombinant PKS proteins, and by molecular engineering.

The OAS gene *aviM* has now been cloned, not from fungi but from *Streptomyces viridochromogenes* Tü57, which produces avilamycin.[29] Interestingly, the gene was found to code for multifunctional polypeptide type I PKS as the first example of bacterial aromatic PKS. Its function was identified by expression in heterologous Streptomycete hosts. The deduced size of the PKS protein is about 130 kDa, which accords with the reported subunit size of *P. cyclopium*.[22]

1.16.3.2 Pentaketide Synthases

With an addition of another C_2-unit to tetraketide, pentaketide intermediates could give several folding and cyclization patterns, ((**15**)–(**22**)), as shown in Figure 1.

Although some feeding experiments with ^{13}C-labeled acetate confirmed the pentaketide nature of some aromatic compounds such as scytalone (**23**),[30] only limited biochemical and molecular genetic information has been available on pentaketide synthases.

OH O

HO OH

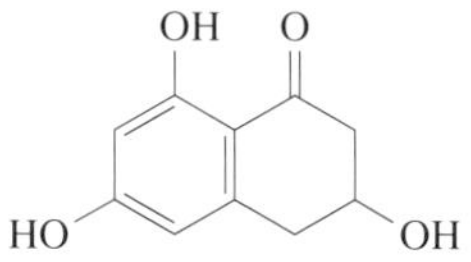

Scytalone (**23**)

Studies on melanin biosynthesis of phytopathogenic fungi identified the pentaketide intermediates of fungal melanin biosynthesis, such as scytalone (**23**), vermelone (**24**), etc. 1,3,6,8-Tetrahydroxynaphthalene (**25**, T4HN) is considered to be an initial product of pentaketide PKS and serves as a biosynthetic precursor to melanins of phytopathogenic fungi.

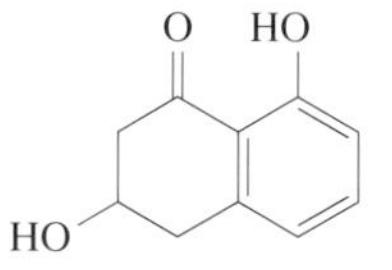

O HO

HO

Vermelone (**24**)

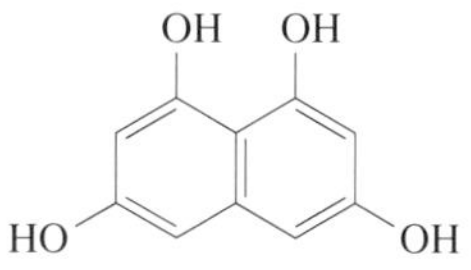

OH OH

HO OH

T4HN (**25**)

Colletotrichum lagenarium, a fungus causing anthracnose of cucumber, was studied for its PKS gene involved in melanin biosynthesis using melanin nonproducing albino mutants. A cosmid library with wild-type strain genomic DNA was constructed and screening to restore melanin formation in albino mutant was carried out. A *Bam*HI DNA fragment of 8.4 kb was identified to be responsible for melanin producing phenotype.[31] Nucleotide sequencing of the cloned gene named *PKS1* revealed the presence of one long open reading frame of which the product possesses homologous active site regions to KS, AT, and ACP of known PKS and FAS.[32] Comparison with other fungal PKS genes indicated the presence of strong homology with the corresponding loci of the *A. nidulans wA* gene,[33] but low homology was observed with the MSAS gene[14] only around the KS region. Thus, the *PKS1* gene was assumed to be a PKS gene for the formation of T4HN (**25**).

To identify the product of PKS coded by the *PKS1* gene of *C. lagenarium*, expression of the *PKS1* gene in heterologous fungus has been carried out by Fujii *et al.*[34] The expression plasmid was

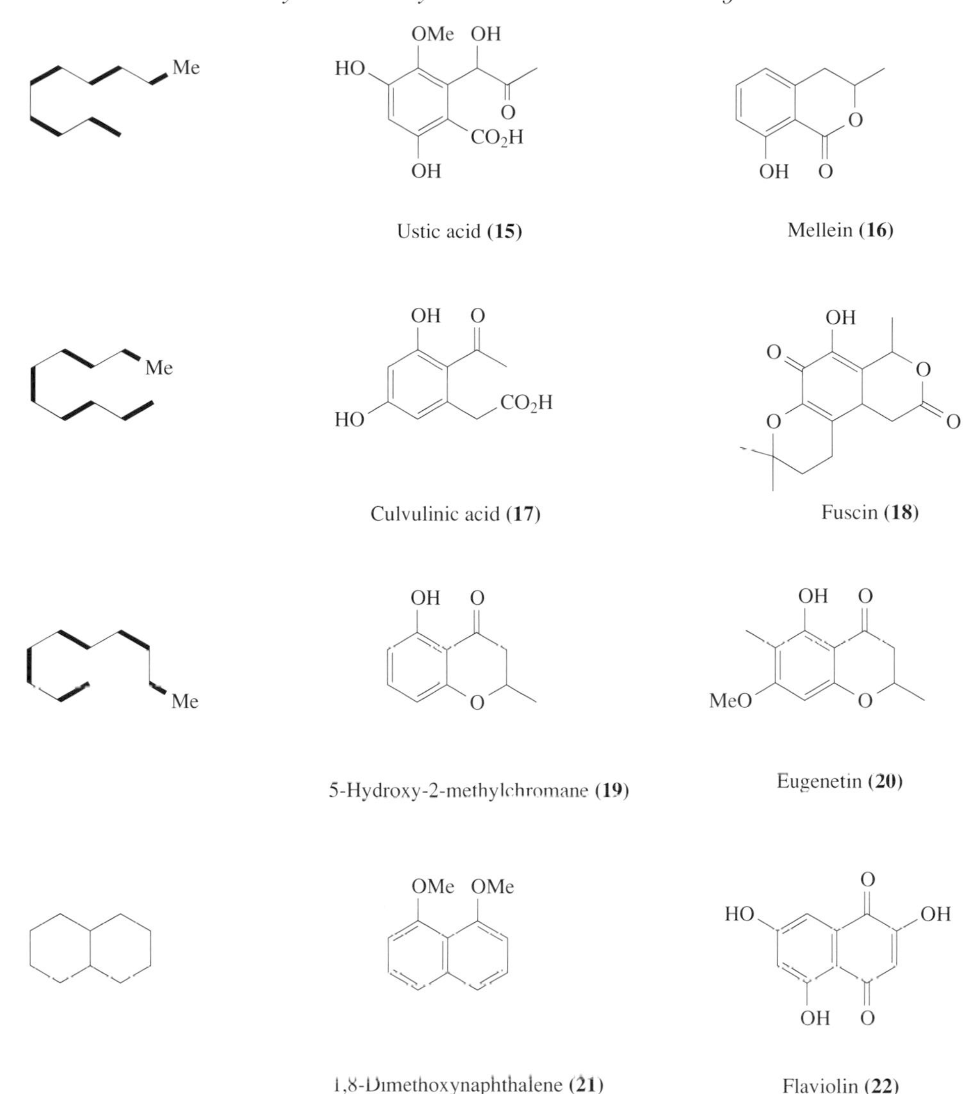

Figure 1 Folding patterns of pentaketides and their representative cyclized products.

constructed with pTAex3, which was successfully used in *atX* gene expression in *A. nidulans*.[26] The transformant of *Aspergillus oryzae* harboring the expression plasmid pTAPSG1 in which the *PKS1* gene was under the control of the α-amylase promoter produced T4HN (**25**) together with a large amount of polymerized dark brown pigments when induced by starch. This result unambiguously identified that the *PKS1* gene codes for a pentaketide T4HN synthase. Several cyclization patterns of a pentaketide are assumed to form a naphthalene skeleton as shown in Figure 2, although a symmetric cyclization was indicated by feeding experiments on scytalone biosynthesis.[30]

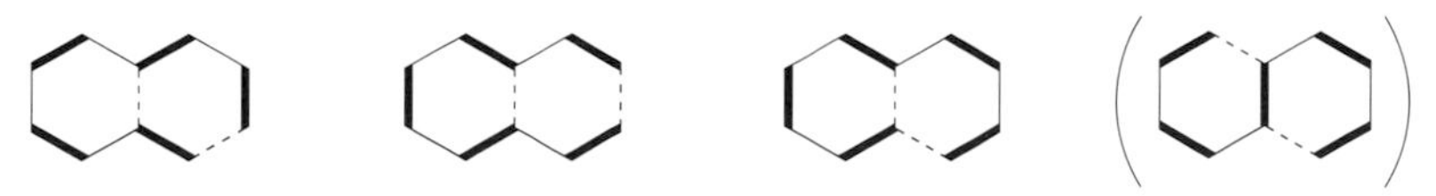

Figure 2 Possible folding patterns for naphthalene cyclization.

Together with T4HN (**25**), 6,8-dihydroxy-3-methylisocoumarin (**26**) was isolated as a minor product of the *A. oryzae*/pTAPSG1 transformant. Production of compound (**26**) indicated a possible folding and cyclization pattern of pentaketide intermediate as shown in Scheme 3. However, isolation of 5-hydroxy-2-methylchromone (**27**) from *Daldinia concentrica* together with 1,8-dihydroxynaphthalene (**28**, DHN)[35] leaves ambiguity in the cyclization pattern of pentaketide.

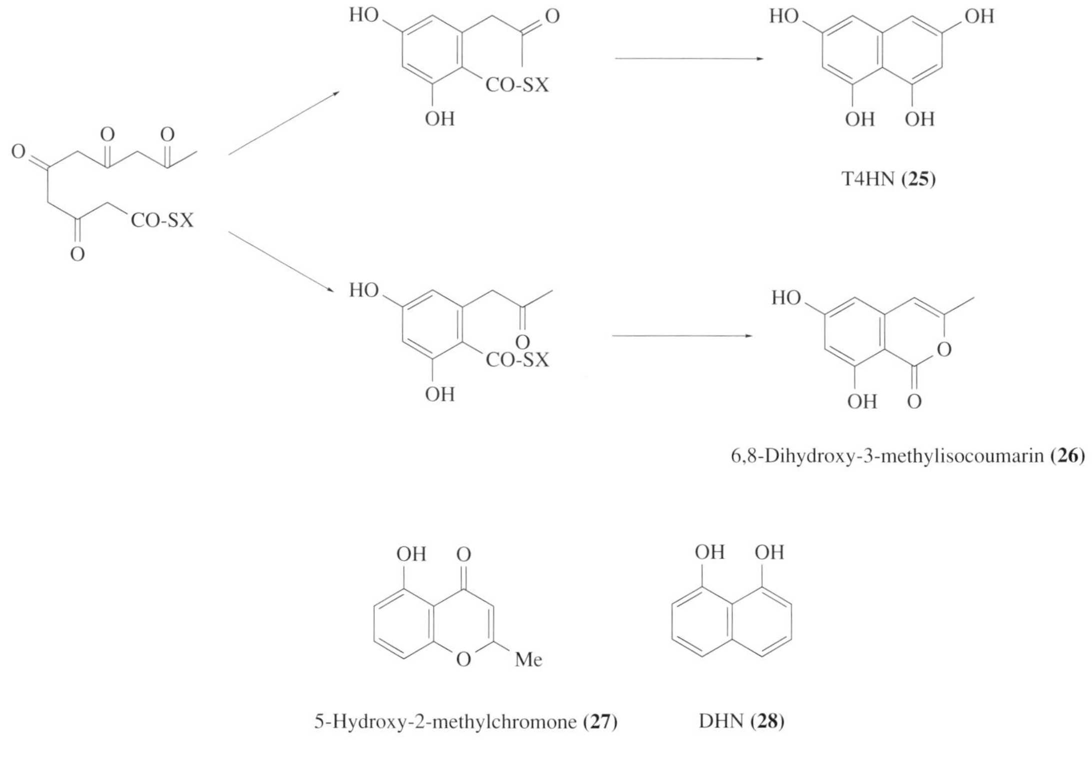

T4HN (**25**)

6,8-Dihydroxy-3-methylisocoumarin (**26**)

5-Hydroxy-2-methylchromone (**27**)

DHN (**28**)

Scheme 3

Mellein, 3-methyl-3,4-dihydro-8-hydroxyisocoumarin (**16**), is also a pentaketide metabolite produced by several fungi.[36] Although fungal PKS or the PKS gene responsible for mellein (**16**) biosynthesis has not been identified at the time of writing, the synthase of related 6-hydroxymellein (**29**, 6HM) has been characterized from a plant source, the carrot (*Daucus carota* L.).[37,38] 6HM synthase is a type I multifunctional PKS[39] induced in carrot cells upon the invasion of pathogenic microorganisms. The enzyme catalyzes the condensation of one acetyl-CoA and four malonyl-CoAs and an NADPH-dependent keto reduction takes place at the triketide intermediate stage to form a dihydroisocoumarin skeleton (Scheme 4).

6HM synthase has been partially purified by Kurosaki *et al.*[40] and estimated to be a homodimer of 130 kDa. Although molecular genetic information is not yet available, they carried out enzymological analysis of the 6HM synthase reaction. When the catalytic reaction of 6HM synthase is carried out in the absence of NADPH, the enzyme is capable of catalyzing the chain elongation reaction by the condensation of acyl-CoAs; however, the third malonyl-CoA molecule is unable to condense to the ketomethylene chain. Therefore, the elongation of the β-ketomethylene chain stops at the triketo stage, and the synthase liberates TAL (**13**) as a derailment product instead of 6HM (**29**)[41] (see Scheme 4). A similar incomplete reaction is also observed when the dimeric synthase is dissociated to monomer subunits in buffers of high ionic strength. The keto reduction of the carbonyl group of the triketomethylene chain does not take place in the monomer enzyme, and TAL (**13**) is liberated as sole product even in the presence of NADPH.[42] It was reported that K_m values for starter acetyl-CoA markedly increase when 6HM synthase is in monomer form or in the absence of NADPH, but the affinity to malonyl-CoA, extender unit, remained unchanged.[41] Thus, the keto-reducing domain of 6HM synthase may play an important role in the entry of the starter unit into the multifunctional enzyme, the earliest step of the reaction. Also, KR and DH participate only once in the second cycle of four condensation cycles of the 6HM synthase reaction. The combination of reiterated and nonreiterated catalytic activities suggests that 6HM synthase precisely determines when and when not to carry out the reduction step by the so-far unknown mechanism.

Although the gene for 6HM synthase has not been cloned at the time of writing, the fact that production of mellein type compounds is rare in plants but not in fungi suggests that horizontal gene transfer might have occurred as assumed in the tripeptide synthase gene of β-lactam biosynthesis.[43]

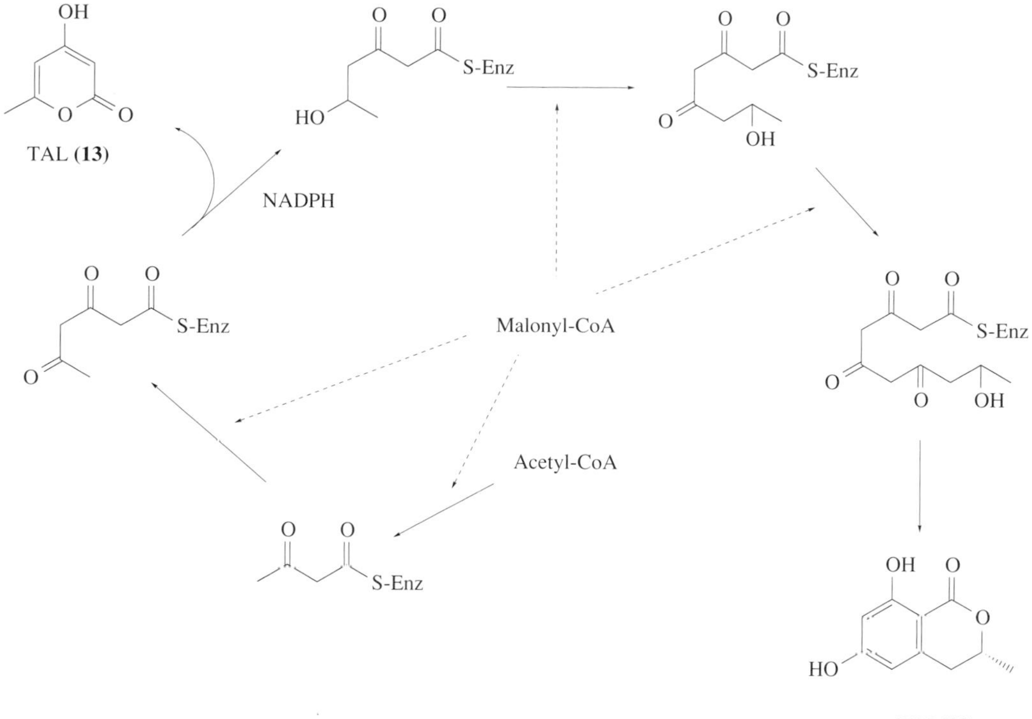

TAL (13)

NADPH

Malonyl-CoA

Acetyl-CoA

6HM (29)

Scheme 4

1.16.3.3 Higher Aromatic Polyketide Synthases

Filamentous fungi produce many other aromatic compounds of higher polyketide origins such as hepta-, octa-, nona- and decaketides. A few such representative types of compounds ((**30**)–(**33**)) are shown in Scheme 5.

Along with elongation of chain-length, variation in folding patterns of poly-β-ketomethylene intermediates increases dramatically to afford structural variety of PKS products. However, quite limited information is available on higher aromatic PKSs in the late 1990s. Following are some examples to show that molecular genetic analyses have been carried out to some degree.

1.16.3.3.1 PKS genes for spore pigment biosynthesis

Aspergillus nidulans wA gene was identified in the study of differentiation of *A. nidulans*.[44] The *wA* mutants produce colorless conidia although the wild type shows green pigmentation in their asexual spores, or conidia. Yellow spores are observed in *yA* mutants and *wA* mutations are epistatic to *yA* mutations.[45] Thus, it is considered that the product of the *wA* gene forms a yellow intermediate, which is subsequently converted to the mature green spore pigment by *yA*-encoded laccase.[46] The structure of the yellow spore pigment intermediate of the related fungus *Aspergillus parasiticus* was determined to be hydroxymethylnaphthopyrone, named parasperone A (**34**),[47] which is considered to be a heptaketide. Also, the ascospore pigment of *A. nidulans* was characterized as a dimeric hydroxylated anthraquinone, ascoquinone A (**35**).[48]

The *wA* gene was cloned by genetic complementation of the *wA* mutation with a cosmid library of wild-type DNA by Timberlake's group.[49] Disruption of the *wA* locus confirmed that the gene is required for synthesis of green pigment present in the walls of mature asexual spores; *wA* disruptants produce colorless (white) conidia. By Northern blot analysis, a 7.5 kb transcript was detected during conidation, beginning when pigmented spores first appeared. The nucleotide sequence indicated that the *wA* gene codes for a 1986 amino acid polypeptide of 217 kDa,[33] and the predicted WA polypeptide showed extensive sequence similarities with bacterial and fungal PKSs, particularly within conserved active sites. The presence of KS, AT, and two ACP motifs was reported in the *wA*

Griseofulvin (**30**)

Rubrofusarin (**31**)

Emodin (**32**)

Sterigma tocystin (**33**)

Scheme 5

Parasperone A (**34**)

Ascoquinone A (**35**)

coded polypeptide. Later, high similarity with the *PKS1* gene of *C. lagenarium*[32] was indicated, such as size and domain organization, although the TE motif was missing in the first report.[33]

Watanabe *et al.*[50] then tried to express the *wA* gene in the heterologous fungus *A. oryzae*. Construction of the expression plasmid was carried out as shown in Figure 3.

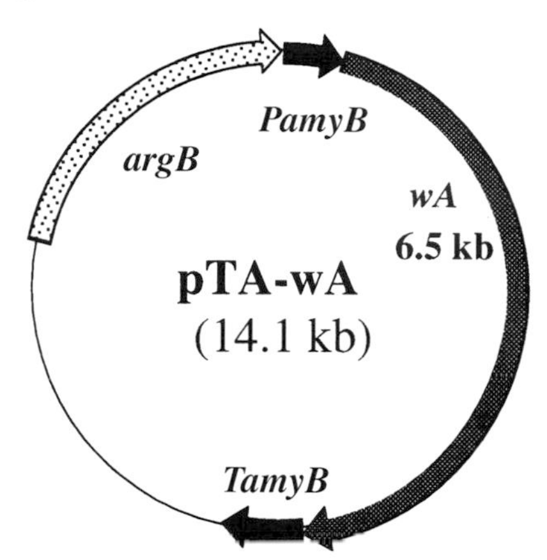

Figure 3 Expression plasmid pTA-wA. Expression plasmid pTAex3 was used to construct WA expression plasmid pTA-wA by inserting the *wA* gene just downstream of α-amylase promoter of *Aspergillus oryzae*. pTAex3 possesses *argB* auxotrophic selection marker in fungi.

The 3′-end of the gene was cut at the *Bam*HI site just below the reported stop codon. Without removing its intervening sequence, the gene was placed under the α-amylase promoter of *A. oryzae* in the fungal expression plasmid pTAex3[77] to construct pTA-wA. The *A. oryzae* transformant harboring this expression plasmid pTA-wA produced compounds which have not been detected in the host *A. oryzae* or even in the *A. nidulans* itself. The newly produced compounds were identified as citreoisocoumarin (**36**) and its derivatives (Figure 4), which are made from heptaketide intermediate with or without reduction in their side chains.

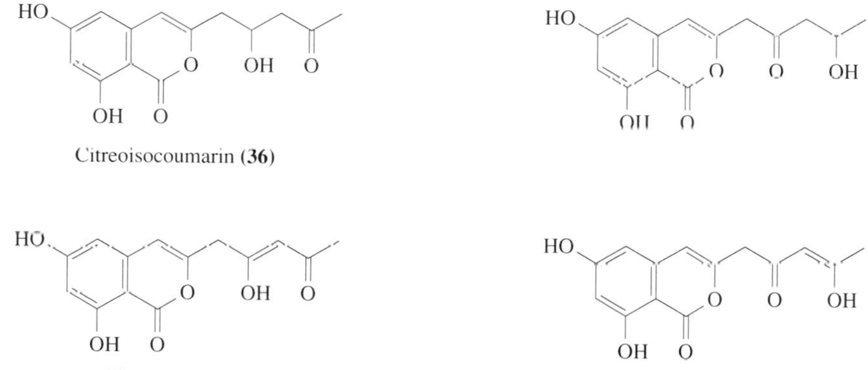

Figure 4 Structure of citreoisocoumarin (**36**) and its derivatives.

Although production of these compounds is directed by the *wA* gene without doubt, there was some ambiguity as to whether the length of polyketomethylene chain and its cyclization occurred correctly in the same way as that in *A. nidulans*, because these compounds do not show significant yellow color.

Re-sequencing of the C-terminal region of the *wA* gene indicated that one base was missing in the reported sequence, just before the predicted stop codon. Thus, the WA polypeptide sequence was corrected and appeared to be 170 amino acids larger (2156 amino acids) than the first predicted polypeptide (1986 amino acids). Comparison with other fungal PKSs indicated the presence of the TE domain in the revised sequence and the identical domain organization to the PKS1 polypeptide of *C. lagenarium*.

The expression plasmid pTA-nwA was constructed to produce a full-length WA polypeptide. The *A. oryzae* transformant with pTA-nwA produced the yellow compound YWA1 (**37**) and its structure was determined by physicochemical analysis. The basic carbon skeleton of naphthopyrone YWA1 (**37**) is identical to the *A. parasiticus* spore pigment parasperone A (**34**)[47] and other related pigments such as fonsecin (**38**) of *Aspergillus fonsecaeus*[51] and rubrofusarin (**31**) of *Fusarium culmorum*[52,53]

(Scheme 6). It seems to be reasonable that the naphthopyrone-type yellow compound YWA1 (**37**) can be converted to green spore pigments by polymerization catalyzed by the *yA* gene product laccase, which has also been cloned by Timberlake's group.[46]

Ruburofusarin (**31**)

YWA1 (**37**)

Citreoisocoumarin (**36**)

Scheme 6

Biosynthetic feeding studies on rubrofusarin (**31**) using sodium [$^{13}C_2$]acetate[54] revealed its polyketide chain folding pattern as shown in Scheme 6. Although similar types of feeding experiments are currently underway for YWA1 (**37**), the same folding pattern of polyketide chain as that of rubrofusarin (**31**) can be assumed for YWA1 (**37**). The folding pattern of citreoisocoumarin (**36**) is a straightforward linear type with one turn. Comparison between the folding patterns of citreoisocoumarin (**36**) and YWA1 (**37**) has given some insight into the cyclization of the heptaketide chain. As shown in Scheme 6, cyclization and aromatization of the first ring proceeds first, in the same way for both compounds. The second ring cyclization occurs in YWA (**37**) formation by Claisen-type condensation, but in the case of citreoisocoumarin (**36**), cyclization by Claisen-type condensation is prohibited and exclusively gives lactone ring formation.

Fonsecin (**38**)

1.16.3.3.2 PKS in aflatoxin biosynthesis

Aflatoxins (**3**) are potent environmental carcinogens produced by *A. parasiticus*, *Aspergillus flavus*, and *Aspergillus nomius*. Because of their direct linkage to carcinomas of the liver and kidney, biosynthesis of aflatoxins (**3**) has been studied intensively by classical genetics including blocked mutant analysis, feeding experiments with labeled acetates, enzymology, and by molecular genetics, as described in detail in Chapter 1.17 of this volume. The biosynthesis pathway, beginning with norsolorinic acid (**39**), is unusually long and complex and proceeds via averufin (**40**), versicolorin

(**41**), and sterigmatocystin (**33**) (Scheme 7). This anthraquinone compound, with a C_6 side chain, is clearly of polyketide origin and derived from the corresponding anthrone.[55]

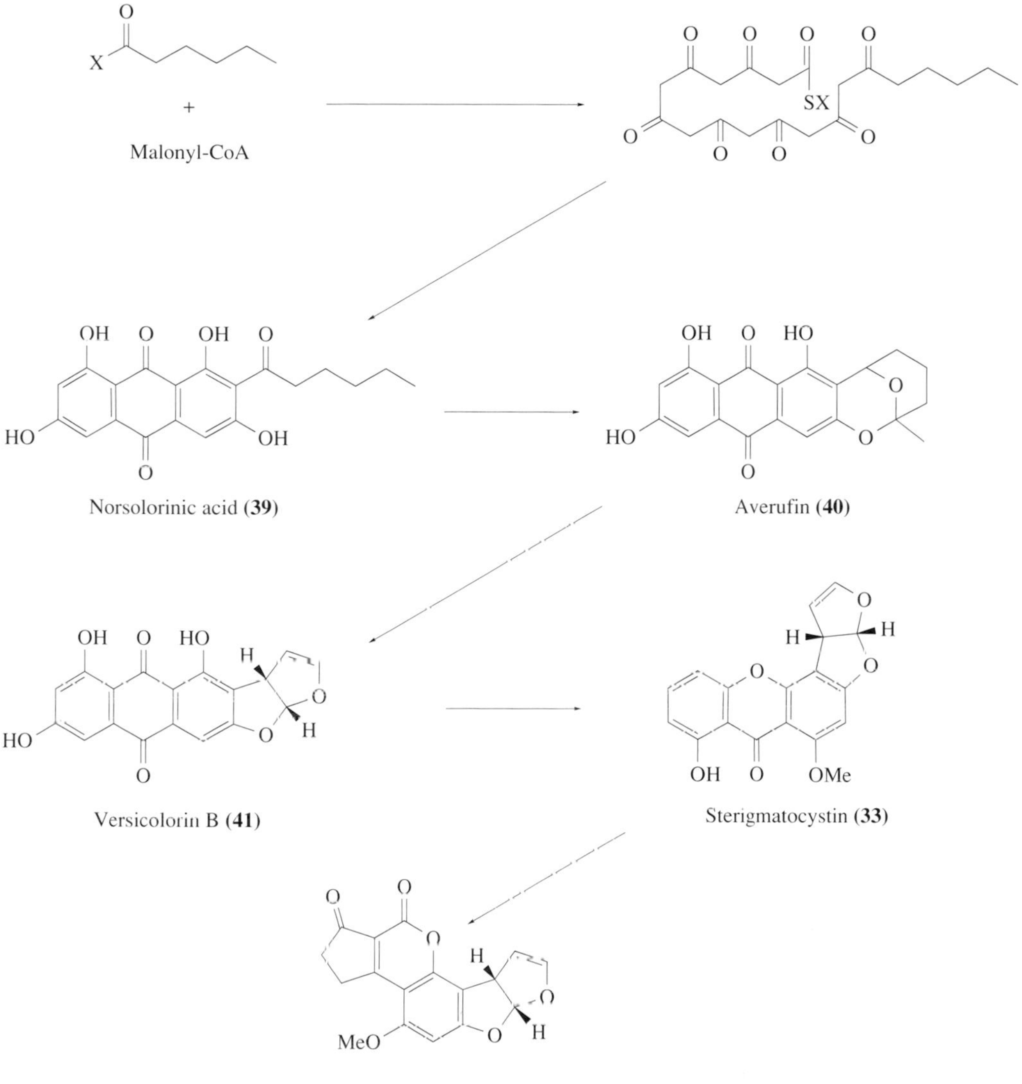

Malonyl-CoA

Norsolorinic acid (**39**)

Averufin (**40**)

Versicolorin B (**41**)

Sterigmatocystin (**33**)

Aflatoxin B_1 (**3**)

Scheme 7

Hexanoate is now identified to be the starter unit of norsolorinic acid (**39**) biosynthesis, by feeding experiments using *N*-acetylcysteamine (NAC) thioesters.[56] Cloning and sequencing analysis of aflatoxin biosynthetic gene clusters from aflatoxigenic fungi[57,58] and that for related sterigmatocystin (**33**) from *A. nidulans*[59] identified the presence of the PKS gene and the specialized fatty acid synthase genes responsible for the formation of the C_6 precursor which serves as a starter unit of the PKS to construct the norsolorinic acid (**39**) carbon skeleton.

Although no direct confirmation of PKS product and/or PKS enzyme activity, sequence alignment shows the presence of domains in the PKS polypeptides in the following order: KS→AT→ACP→TE.[60–62] Also, insertional inactivation of PKS genes gave mutants unable to produce aflatoxin or any biosynthetic intermediates.[60–62] Interestingly, two homologue *pksA* and *stcA* genes from *A. parasiticus*[60,61] and *A. nidulans*,[62] respectively, code for nearly identical products in size (2109 and 2181 amino acids, respectively) and active site arrangements, but the former contains single ACP and the latter tandem ACPs. The specific protein–protein interaction between hexanoate-specialized FAS and PKS is the key feature to be solved for understanding the initial biosynthetic reaction of aflatoxins (**3**) and related compounds in biosynthesis at the enzyme level.[63]

1.16.4 NONAROMATIC POLYKETIDE SYNTHASES

In bacteria, a large number of reduced nonaromatic polyketide compounds is known, as represented by macrolide and polyether antibiotics. However, compounds of this type from fungal sources are limited. Patuolide (**5**), brefeldin (**42**), lovastatin (**6**), and cytochalasins (**43**) are such examples of fungal nonaromatic polyketides. For their biosynthesis, especially PKSs at enzyme and molecular genetic levels, far less information has been available even compared with fungal aromatic PKSs. However, following are two such rare examples of fungal nonaromatic polyketide synthases for which genes have been cloned and analyzed.

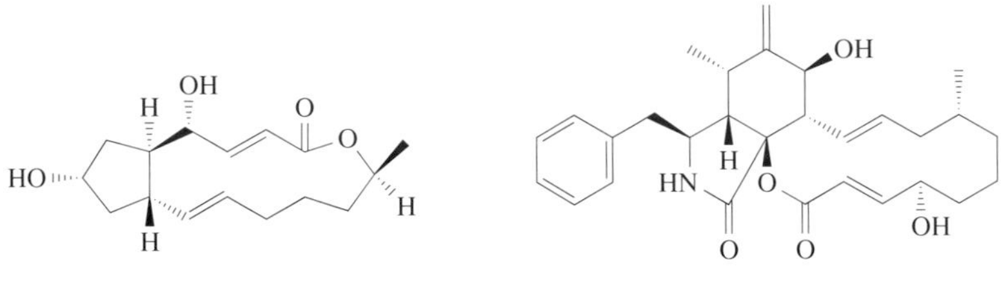

Brefeldin A (**42**)

Cytochalasin B (**43**)

1.16.4.1 Lovastatin Synthase

Lovastatin (**6**)[64] is an important cholesterol biosynthesis inhibitor produced by *A. terreus* ATCC 20542. Compounds closely related are compactin (**44**) from *Monascus ruber*[65] and pravastatin (**45**) from *Penicillium citrinum*.[66]

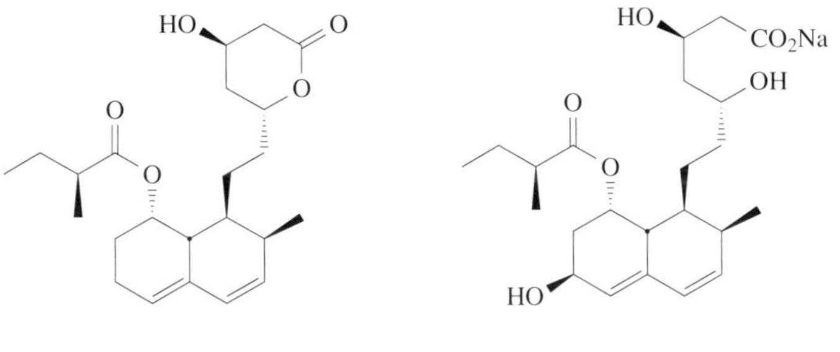

Compactin (**44**)

Pravastatin (**45**)

This class of compounds is known to inhibit cholesterol biosynthesis by inhibiting the rate-limiting step in cellular cholesterol biosynthesis, namely the conversion of hydroxymethylglutaryl-CoA (HMG-CoA) into mevalonate by HMG-CoA reductase. Thus, these HMG-CoA reductase inhibitors show strong cholesterol lowering activity and have been developed for clinical application to hypercholesterolemia.[4]

Lovastatin (**6**) is a reduced complex-type molecule consisting of a conjugated decene ring system joined with a methylbutyryl group by an ester linkage. Feeding experiments revealed that the decene ring moiety of lovastatin (**6**) is derived from nine molecules of acetate units with additional methyl groups at positions 6 and 2′, which are derived from methionine. The side-chain is also composed of two acetate units.[67]

In contrast to aromatic polyketide biosynthesis in fungi, the assembled nonaketide carbon chain is almost fully reduced before the cyclization, which proceeds possibly by a biological Diels–Alder reaction to afford the correct ring system with the right configuration.[68] Also, the presence of a methyl group derived from methionine indicated the involvement of methylation of a polyketomethylene intermediate before cyclization by PKS. A plausible reaction scheme for lovastatin (**6**) biosynthesis is shown in Scheme 8.

The lovastatin PKS gene of *A. terreus* ATCC 20542 has been cloned and identified by the Merck group.[69] They found correlation of the presence of an apparent 269 kDa protein and lovastatin (**6**) production in *A. terreus* and raised antibody against the protein purified by SDS-PAGE, which was then used in the screening of a cDNA library. Subsequent genomic DNA cloning and sequencing

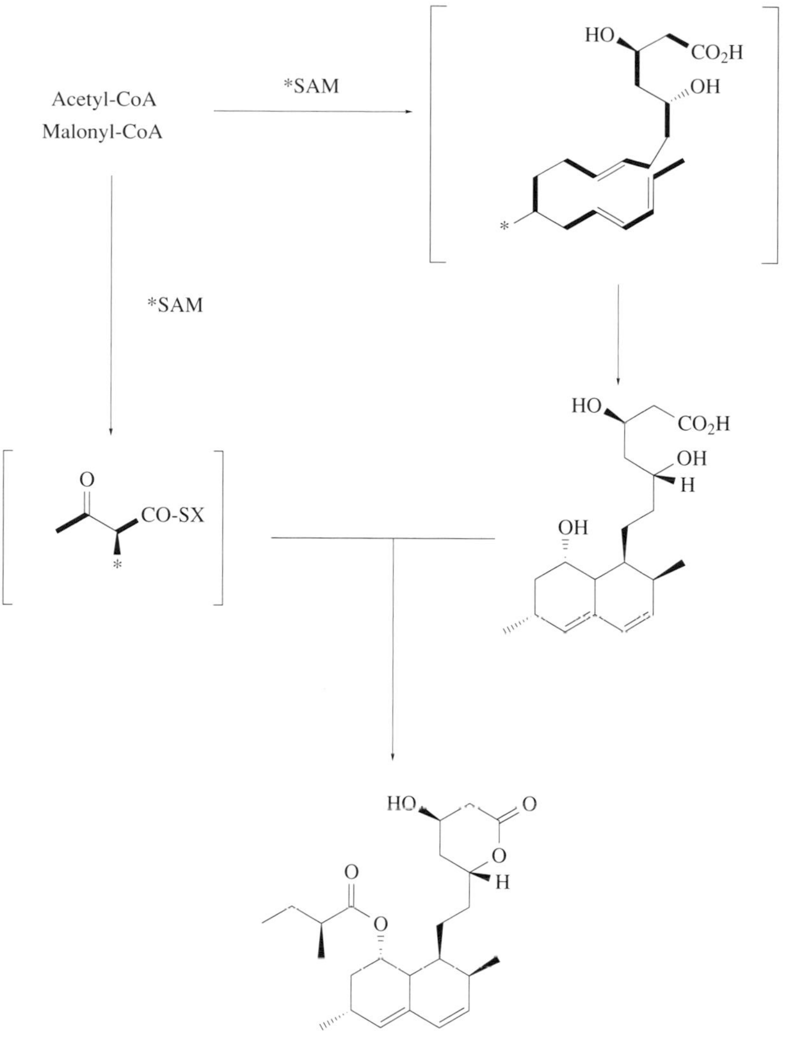

Acetyl-CoA
Malonyl-CoA
*SAM
*SAM
HO
CO_2H
OH
CO-SX
OH
H
Lovastatin (**6**)

Scheme 8

of the cloned 11.5 kb region revealed a 9.6 kb open reading frame with seven short introns coding for a 335 kDa polypeptide of 3038 amino acids.

Comparison of the deduced amino acid sequence with those of known PKSs and FASs resulted in the identification of active site residues and motifs for expected PKS function.

Near the N-terminus is located the KS region, centering around the active site cysteine to which the acyl chain is attached prior to condensation with the malonyl extension unit on ACP-pantetheine-SH. The KS region is most closely related to the rat FAS sequence[70] and exhibited 49% homology over this region compared with 41% to MSAS. The AT functional site with the GXSXG motif[71] is identified in the position next to the KS site from its N-terminus. Alignment with the rat FAS and MSAS also indicated the sites for DH and KR domains. The region between DH and KR shows the best alignment to the rat FAS enoyl reductase (ER), but does not have a strong homology with the GXGXXG motif.[72] In the C-terminus region, the ACP serine was identified, which binds the 4′-phosphopantetheine prosthetic group.

A unique activity domain identified is the methyl transferase (MT) responsible for transfer of the methyl group from *S*-adenosylmethionine (SAM) to the polyketide chain, which is not presented in FASs and other PKSs. Comparison of MTs responsible for the methylation of RNA, DNA, and protein substrate has identified a sequence motif thought to be part of the SAM-binding domain.[73,74]

The linear organization of active sites was observed in the amino acid sequence to be KS→AT→DH→MT→ER→KR→ACP. Interestingly, the MT domain resides in the middle of the PKS.

Coexistence of methyltransferase was found in the *tcmN* gene of the tetracenomycin biosynthesis gene cluster, which codes for cyclase and methyltransferase on a single polypeptide.[75]

Southern blot analysis of genomic DNAs of fungi which produce lovastatin-related compounds, such as *M. ruber*, *P. citrinum*, and *Penicillium brevicompactum*, was carried out with an *A. terreus* lovastatin PKS probe. The results suggested the presence of homologous PKS genes responsible for each lovastatin-related compound's biosynthesis.

1.16.4.2 T-Toxin Synthase

T-toxins (**46**) and related PM-toxins (**47**) are linear long-chain (C_{35} to C_{41}) polyketol compounds isolated from *Cochliobolus heterostrophus* (previously named *Helminthosporium maydis*)[76,77] and *Phyllostica maydis*,[78] respectively. Both are pathogenic fungi, highly virulent toward Texas male sterile maize.

T-toxin (C_{41}) (**46**)

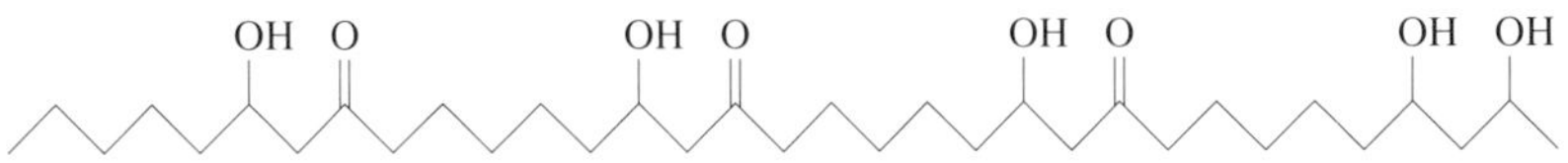

PM-toxin B (C_{33}) (**47**)

The *Tox1* locus of *C. heterostrophus* was defined by Mendelian analysis as a single genetic element that controls production of T-toxin.[79,80] To tag *Tox1*, protoplasts of a $Tox1^+$ (T-toxin producing) strain were transformed with a linearized, nonhomologous plasmid along with an excess of the restriction enzyme used to linearize the plasmid, a procedure called restriction enzyme mediated integration (REMI).[81] Of 1310 transformants recovered, two produced no detectable T-toxin (**46**). In each of these transformants, the Tox^- mutation, mapped at *Tox1*, was tagged with the selection marker residing on the plasmid used in REMI.[82] The DNA recovered from the insertion site of one mutant encodes a 7.6 kb open reading frame (2530 amino acids) that identified a multifunctional PKS-encoding gene (CH-*PKS1*) with six catalytic domains arranged in the following order, starting at the N-terminus: KS→AT→DH→ER→KR→ACP.[83] Thus, the cloned CH-*PKS1* gene is interrupted by four apparent introns and exists in the genome as a single copy surrounded by highly repetitive, A+T-rich DNA. When CH-*PKS1* in race T was inactivated by targeted gene disruption, T-toxin (**46**) production and high virulence were eliminated, indicating that this PKS is required for fungal virulence. Race O strains, which do not produce T-toxin, lack a detectable homologue of CH-*PKS1* gene, suggesting that race T may have acquired the *PKS1* gene by horizontal transfer of DNA rather than by vertical inheritance from an ancestral strain.

The T-toxin (**46**) molecule possesses three of the four functional groups formed in the PKS reaction, that is, keto-, hydroxy-, and alkyl groups (there are no enoyl functions). This structure indicates that at least six enzymatic activities are required for its assembly: KS, AT, and ACP for chain extension and KR, DH, and ER for β-keto group processing. CH-*PKS1* encodes each of these six essential domains and therefore contains all of the information necessary for any of the steps in polyketide construction. Thus, CH-PKS1 enzyme could act by an iterative mechanism, and it alone could account for construction of the T-toxin (**46**) carbon chain.

1.16.5 STRUCTURE OF FUNGAL POLYKETIDE SYNTHASES

As dictated by the convention for FAS,[84] microbial PKSs have been classified into two types, that is type I PKS and type II PKS. Type I PKS is a multifunctional polypeptide enzyme system and possesses several active sites on each polypeptide subunit of the enzyme, while type II PKS consists of several separate, largely monofunctional proteins, which is typical of bacterial aromatic PKSs.

The bacterial type I PKSs consist of polypeptides with modular structures of active sites which work nonreiteratively for synthesis of product compounds, as does the 6-deoxyerythronolide B synthase of *Saccharopolyspora erythrea*.[85]

The fungal PKSs so far known are all classified into the type I system. However, their linear organization of active sites on the enzyme polypeptides are similar to that of the bacterial system; the enzymes use their active sites in the reiterative way, in some cases, partially. Interestingly, some fungal aromatic PKSs such as *A. nidulans* WA,[33] *A. nidulans* STCA,[62] and *C. lagenarium* PKS1[32] possess two ACP motifs in the C-terminal region in a tandem manner. The significance of the duplicated ACPs in fungal aromatic PKSs is still unknown.

Although there have been only two examples of reduced nonaromatic PKS genes cloned from fungi, that is, the lovastatin PKS gene[69] and the T-toxin PKS gene,[83] the fundamental organization of active sites is common in all known fungal PKSs, except that additional KR, DH, and ER regions exist between the AT and ACP domains in reduced complex-type PKSs.

Figure 5 shows schematic comparisons of fungal polyketide synthase active site organizations, including PKSs involved in aflatoxin/sterigmatocystin biosynthesis.

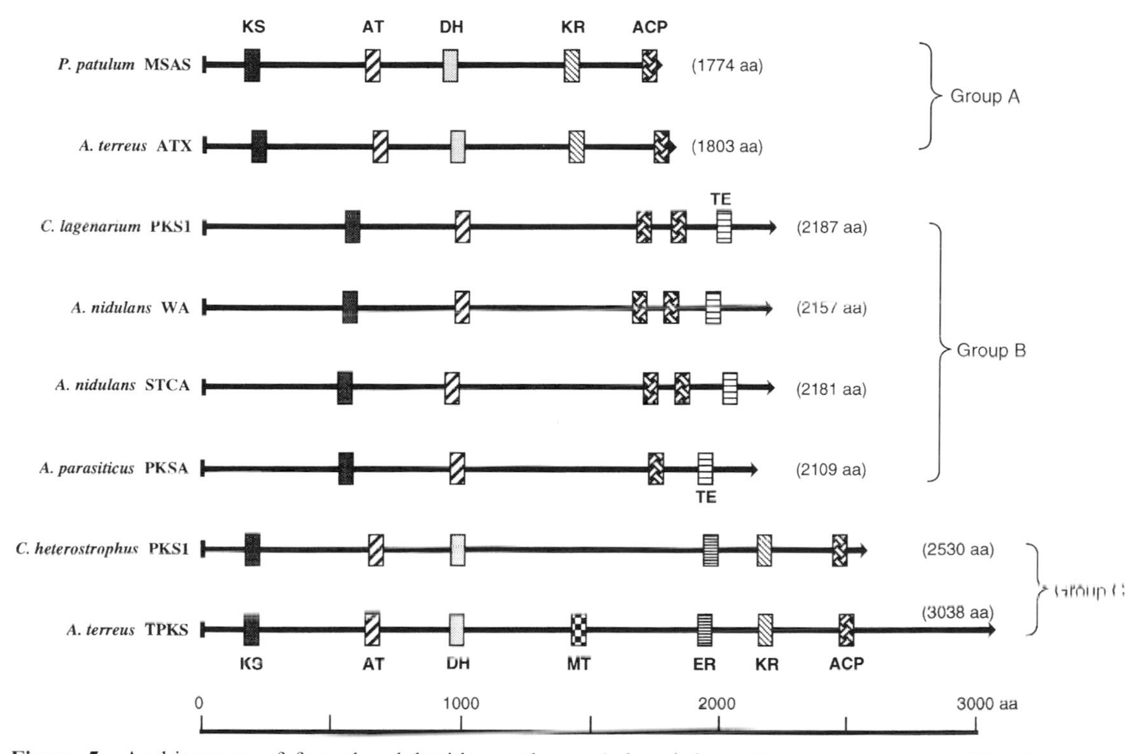

Figure 5 Architecture of fungal polyketide synthases deduced from the gene sequences. KS, β-ketoacyl synthase; AT, acyltransferase; DH, dehydratase, KR, β-keto reductase; ACP, acyl carrier protein; TE, thioesterase. *P. patulum* MSAS, 6-methylsalicylic acid synthase gene (accession number X55776); *A. terreus* ATX, 6-methylsalicylic acid synthase gene (D85860); *C. lagenarium* PKS1, *PKS1* gene for melanin biosynthesis (D83643); *A. nidulans* WA, *wA* gene for spore pigment biosynthesis (X65866); *A. nidulans* STCA, *stcA* gene for stertigmatocystin biosynthesis (L39121); *A. parasiticus* PKSA, *pksA* gene for aflatoxin biosynthesis (L42765); *C. heterostrophus* PKS1, *PKS1* gene for T-toxin biosynthesis (U68040); *A. terreus* TPKS, triol polyketide synthase gene for lovastatin biosynthesis.

The number of fungal PKS genes cloned so far is limited, but they may be classified into three groups: (A) single aromatic ring PKS, (B) multiaromatic ring PKS, and (C) reduced complex-type PKS. Group A consists of the smallest type I PKSs, polypeptides of less than 2000 amino acids. OAS from *S. viridochromogenes*[29] may be classified in this group. Group B PKSs have some different characteristic organizations compared with group A PKSs: the KS region shifts to the C-terminal direction about 340 amino acids and TE and tandem ACPs (typically) exist at the C-terminal region. Their sizes are around 2100 amino acids, which is about 300 amino acids larger than group A PKSs. Group C PKSs are more than 2500 amino acids long and are the largest of reiterative-type PKSs.

Expression experiments have identified the direct products of Cl-PKS1 and WA as T4HN (**25**) and naphthopyrone compound YWA1 (**37**), respectively.[86] The aflatoxin/sterigmatocystin PKS product is assumed to be an anthrone derivative of norsolorinic acid (**39**). Considering their folding pattern from the corresponding polyketomethylene chain, Claisen-type cyclization should occur in the second aromatic ring formation, as shown in Figure 6.

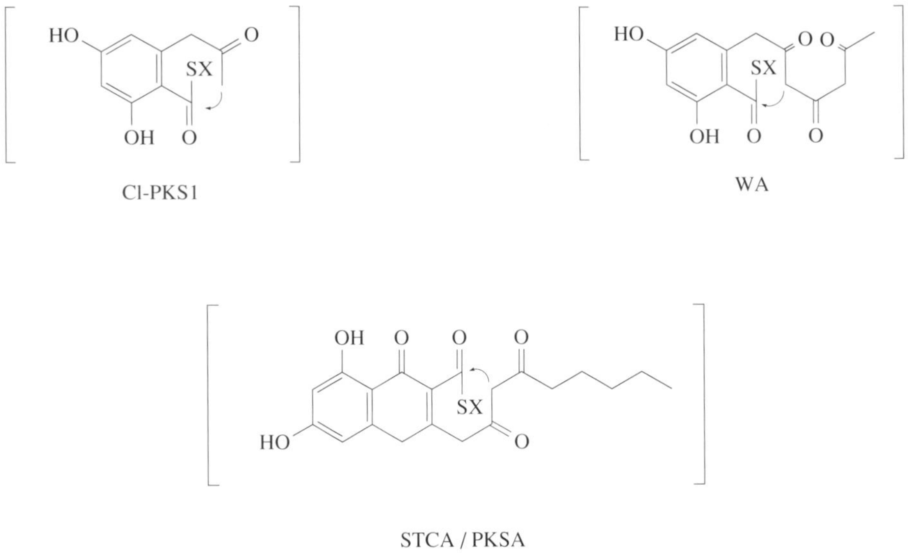

Cl-PKS1

WA

STCA / PKSA

Figure 6 Claisen-type cyclization in group B fungal PKS reactions.

Group A PKSs and most of the bacterial aromatic PKSs catalyze only aldol-type condensation for ring cyclization. Similarity in size and active site location of group B PKSs indicate some key features for Claisen-type condensation reaction, on which the expression of the *A. nidulans wA* gene gave some information. C-terminal modified WA PKS produced citreoisocoumarin (**36**) instead of naphthopyrone compound YWA1 (**37**), suggesting that the C-terminal region might have some role in the Claisen-type condensation. To prove this hypothesis, C-terminal truncated *wA* derivatives were constructed and expressed in *A. oryzae*. Interestingly, deletion of only 30 amino acids from the C-terminus caused the production of isocoumarin instead of YWA1 (**37**), and further truncation over the TE site did not abolish the production of citreoisocoumarin (**36**).[87] Further investigation will reveal how the C-terminus is involved in Claisen-type condensation.

For the construction of specific backbone structures by PKS reaction, PKS should control several key features: (1) how to choose the starter unit—acetyl, propionyl, or other starters, (2) how to select an extension unit, mainly malonyl in fungi, (3) how to control reductions—keto reduction, dehydration, and enoyl reduction, including stereochemistry and positional control, (4) how to control methylation (alkylation) of polyketomethylene chain while on the PKS enzyme, and (5) how to control the folding conformation of the polyketomethylene chain for correct cyclization. In the late 1990s none of these mechanisms has been clarified even on the apparently simple MSAS.

Interesting features about ER and KR motifs in lovastatin PKS have been discussed.[69] In general, ER and KR domains are identified by searching for the GXGXXG/A motif, which is proposed to represent the pyridine nucleotide binding site in many proteins.[72] This structural motif consists of a β-sheet-turn-α-helix, where the glycine-rich region codes for the strong turn signal in the middle. In addition, downstream acidic or basic amino acids are assumed to bind to the phosphate (NADP) or hydroxyl group (NAD) on the 2′ ribose position. Analysis of the structural characteristics using the Chou–Fasman algorithm indicated that this structural motif is conserved in the rat FAS ER and KR domains.[88] The structural predictions of the amino acid sequence of the lovastatin PKS ER and KR, as well as the MSAS KR, show variations of this model. All predicted structures show a β-sheet leading into a turn region, even when amino acid homologies are not strong. Derivation from the structural model may reflect differences in substrate specificity. Thus, it is possible that these structural variations are important in the programming of the PKS, resulting in different levels of reduction of the β-keto group during successive cycles of the biosynthesis of the lovastatin precursor.

1.16.6 BIOSYNTHETIC REACTIONS ACTING ON PKS PRODUCTS

PKS reaction products are initial compounds in the secondary metabolic pathway and, in most cases, subjected to further modification by oxidation, reduction, alkylation, and other types of secondary reactions to afford organism-specific metabolites.

For example, patulin (**11**) and penicillic acid (**12**) are biosynthesized from single aromatic ring tetraketides 6MSA (**1**) and orsellinic acid (**2**), respectively (Scheme 9). Their structural conversion from precursor aromatic compounds has drawn much attention and the biosynthetic schemes for these compounds have been proposed from the results of feeding experiments.[89] Furthermore, some *in vitro* investigations have also been carried out to obtain enzyme activities for the critical ring cleavage reactions. There have been some reports of the successful detection of cell-free activities, but no further biochemical studies have been carried out mainly due to their inherent instability.[89–92]

6MSA (**1**) Patulin (**11**)

Orsellinic acid (**2**) Penicillic acid (**12**)

Scheme 9

Because of the huge structural variety of polyketide compounds, it is difficult to exemplify general principles of secondary modification reactions acting on PKS products, and it is only practicable to provide some representatives. As such examples of enzymologically and/or molecular genetically investigated secondary biosynthetic reactions involved in fungal polyketide biosynthesis, melanin biosynthesis in phytopathogenic fungi and (+)-geodin (**14**) biosynthesis in *A. terreus* IMI 16043 have been chosen and described in the following sections.

1.16.6.1 Enzymes and Genes Involved in Fungal Melanin Biosynthesis

Melanin, a high-molecular-mass black pigment, is biosynthesized by numerous phytopathogenic fungi[93,94] such as *Alternaria alternata*, *Cochliobolus miyabeanus*, *Colletotrichum lagenarium*, *Magnaporthe grisea*, and *Verticillium dahliae*. It has been recognized that melanin is important for the survival and longevity of fungal propagules.[95,96] Analysis of melanin-deficient mutants has shown that fungal melanin is essential for penetration of host leaf tissues by *Colletotrichum lindemuthianum*, *C. lagenarium*, and *M. grisea*, which produce appresoria pigmented with melanin,[97–100] while *A. alternata* produces colorless appresoria. However, *A. alternata* produces melanized conidia,[93,101] while *Magnaporthe* and *Colletotrichum* conidia are colorless. These facts suggest diversity in the regulation and roles of melanin biosynthesis among fungi.

The biosynthetic pathway for melanin has been studied using melanin-deficient mutants which accumulate shunt products and exhibit pigmentation phenotypes.[102] Melanin biosynthesis starts with formation of an assumed pentaketide intermediate, T4HN (**25**). The following steps consist of a series of reduction and dehydration of T4HN (**25**), scytalone (**23**), 1,3,8-trihydroxynaphthalene (**40**, T3HN) and vermelone (**24**), leading to DHN (**28**), polymerization of which yields melanin (Scheme 10).

1.16.6.1.1 Tetrahydroxynaphthalene reductase

By the heterologous expression of *C. lagenarium PKS1* gene in *A. oryzae*, involvement of T4HN (**25**) as the first cyclized PKS product has been confirmed.[103] T4HN (**25**) is then reduced to form

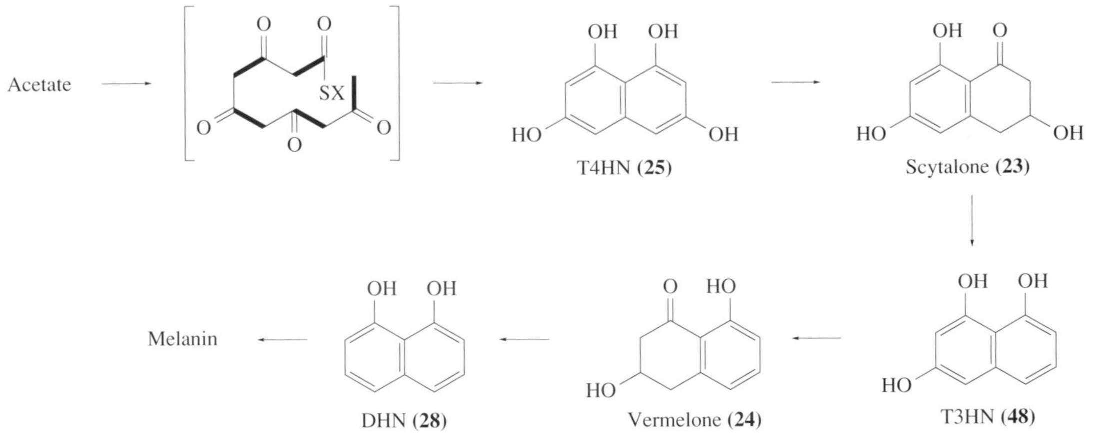

Acetate → [SX] → T4HN (25) → Scytalone (23) → T3HN (48) → Vermelone (24) → DHN (28) → Melanin

Scheme 10

scytalone (**23**). The enzyme responsible for this reduction step was purified from *M. grisea* and its cDNA was cloned.[104]

Owing to the very high sensitivity of T4HN (**25**) to oxidation, spectrophotometric assay in the absence of oxygen was established. The T4HN reductase from *M. grisea* was purified by ammonium sulfate precipitation, and chromatographies on DEAE-Sephacel, AcA 34, and Red-agarose to apparent homogeneity with 32.5% yield and 198-fold purification. The enzyme was found to be a tetramer of identical 30 kDa subunits. A cDNA clone encoding T4HN reductase was then isolated and characterized to code for a 282 amino acid polypeptide of 29.9 kDa. Homology search revealed high similarities with *ver-1*, a gene involved in aflatoxin biosynthesis in *A. parasiticus*.[105] Additional similarities were found with oxidoreductases of the short-chain alcohol dehydrogenase class.[106,107] The β-galactosidase-fusion protein expressed in *E. coli* showed both T4HN and T3HN reductase activities, suggesting that the enzyme is responsible for the two respective reduction steps.

1.16.6.1.2 Scytalone dehydratase

The enzyme converting scytalone (**23**) to T3HN (**48**), known as scytalone dehydratase, was studied at both enzyme and molecular genetic levels. Scytalone dehydratase also catalyzes the dehydration of vermelone (**24**) to form DHN (**28**). Subsequently, DHN (**28**) is polymerized into melanin. This enzyme was first purified from *C. miyabeanus*, the causative agent of leaf spot disease of rice.[108] Enzyme assays were also carried out under anaerobic conditions to prevent unfavorable oxidation of T3HN (**48**). Scytalone dehydratase was purified by ammonium sulfate precipitation and chromatographies on DEAE-Sephadex, Sephadex G-100, DEAE Bio-Gel A, Mono Q columns and finally preparative polyacrylamide gel electrophoresis, with 5.8% recovery and 102-fold purification. SDS-PAGE analysis gave a single band at 23 kDa.

Scytalone dehydratase was also isolated from *M. grisea* and purified to homogeneity by chromatographies on Phenyl Sepharose, DEAE Sephacel, and Mono Q columns.[109] The purified enzyme had no metal or cofactor requirements and was found to be a trimer of three identical single chain subunits. The cDNA clone which codes for 172 amino acid protein was then obtained by screening the cDNA expression library with the specific antibody raised against the purified enzyme. The recombinant enzyme was successfully expressed in *Escherichia coli* and its crystal structure was determined.[109,110]

Using the *M. grisea* cDNA as a probe, the cDNA and genomic DNA encoding scytalone dehydratase were cloned from *C. lagenarium*.[111] Amino acid residues Tyr30, Asp31, Tyr50, His85, Val108, His110, Ser129, and Asn131 involved in the substrate binding of scytalone dehydratase of *M. grisea* were completely conserved in *C. lagenarium*.

1.16.6.1.3 Trihydroxynaphthalene reductase

T3HN (**48**) formed by dehydration of scytalone (**23**) is subjected to another cycle of reduction and dehydration to form DHN (**28**). Mutants deficient in the T3HN (**48**) reduction step were obtained from *A. alternata*, *C. lagenarium* and *M. grisea*.[112–114] The *BRM2* gene of *A. alternata* complemented the Brm2$^-$ mutant, which lacks the T3HN (**48**) reduction step of the wild type phenotype.[112] This *BRM2* gene was then used to clone the T3HN reductase gene of *C. lagenarium*. The cloned gene *THR1* complemented the *C. lagenarium* mutant defect in the conversion of T3HN (**48**) to vermelone (**24**) and the gene disruption of the wild type gave a mutant with the same phenotype to Thr$^-$.[113]

The deduced amino acid sequence of the the *C. lagenarium THR1* reductase gene shows high similarity to that of the T4HN reductase gene of *M. grisea*[104] with 83% identical amino acid residues. Although T4HN reductase can catalyze reduction of both T4HN (**25**) and T3HN (**48**), no evidence has been obtained that the *THR1* gene is involved in both reduction steps in *C. lagenarium* because the Thr$^-$ mutant produces the shunt compound 3,4-dihydro-4,8-dihydroxy-1(2H)naphthalene (**49**, DDN) which is derived from T3HN (**48**) (Scheme 11).[99]

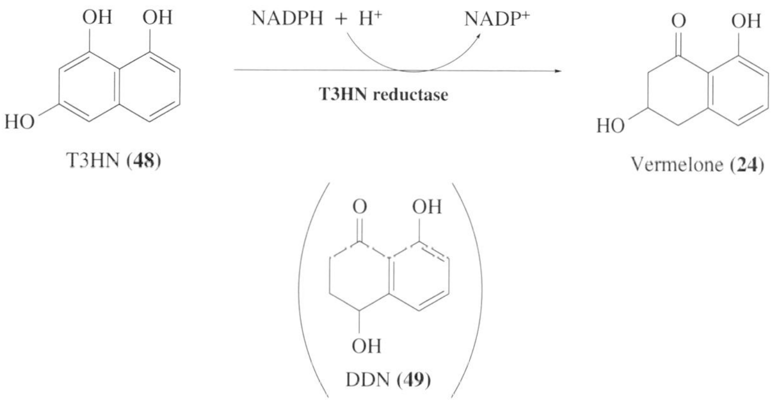

OH OH
NADPH + H$^+$ NADP$^+$
O OH
T3HN reductase
HO
HO
T3HN (**48**)
Vermelone (**24**)
O OH
OH
DDN (**49**)

Scheme 11

1.16.6.1.4 Clustering of melanin biosynthesis genes

In bacteria, clustering of genes for antibiotic biosynthesis has been detected fairly early by genetic analysis.[115] Recent molecular genetic works established this feature unambiguously.[12] In fungi, several examples of clustered biosynthetic genes of secondary metabolites have been observed, for example, trichotecene,[116] and aflatoxins.[58]

Melanin biosynthesis genes cloned from *A. alternata* were found to be a cluster of at least three genes, *ALM*, *BRM1*, and *BRM2*, which code for PKS, scytalone dehydratase, and T3HN reductase, respectively.[112] These genes are located within a genomic region of about 30 kb. The three mRNA species accumulated in cultured mycelia of the wild-type strain synchronously with mycelial melanization. The linkage relationship of melanin biosynthesis genes in other fungi was also analyzed. In *C. heterostrophus* and *C. miyabeanus*, the PKS gene and T3HN reductase gene are closely linked, but the scytalone dehydratase gene is independent of the other two genes.[117,118] In *M. grisea*, no evident linkage of these genes has been observed.[114] Also, the *C. lagenarium* PKS gene *PKS1* was not closely linked to the dehydratase gene.[31] These results demonstrate that the linkage relationship and arrangement of melanin biosynthesis genes are quite different among fungi, although the biosynthetic pathway of melanin is almost identical.

1.16.6.2 Deoxygenation

Hydroxy groups of T4HN (**25**) are removed in the post-PKS reactions catalyzed step-wisely by independent reductase and dehydratases, as mentioned above. A similar type of deoxygenation was shown to be involved in anthraquinonoid biosynthesis.

In the biosynthesis of ergochromes (**50**), fungal pigments produced by *Claviceps purpurea*, the typical polyketide anthraquinone emodin (**32**) and chrysophanol (**51**), 6-deoxyemodin, were shown to be efficient precursors by feeding experiments.[119,120] From the structural consideration of ergochromes (**50**), emodin (**32**) was assumed to be incorporated via chrysophanol (**51**) (Scheme 12).

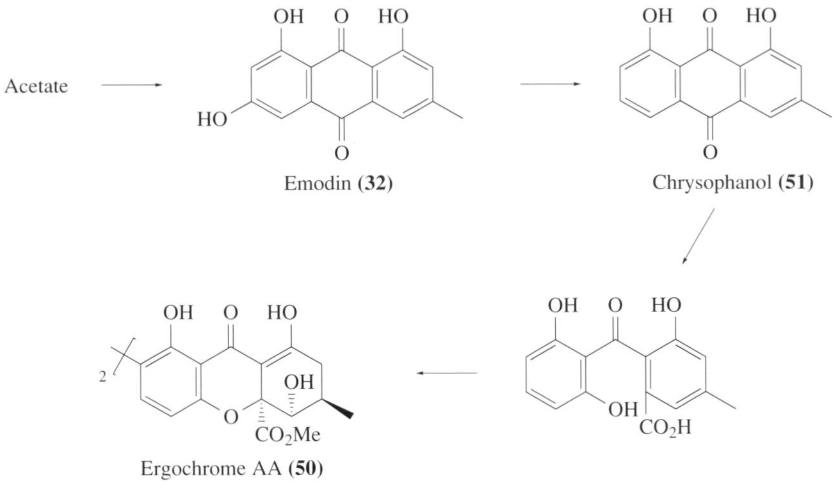

Acetate → Emodin (**32**) → Chrysophanol (**51**) → → Ergochrome AA (**50**)

Scheme 12

Anderson first reported detection of the enzyme activity which converted emodin (**32**) into chrysophanol (**51**) in *Pyrenochaeta terrestris*.[121] This reaction is believed to consist of two steps, reduction of a keto-tautomer of emodin (**32**) to afford dihydroemodin (**52**) and subsequent dehydration. The transient formation of the corresponding keto-tautomer of emodin (**32**) in enzymatic reaction was supported by the incorporation of deuterium from deuterium oxide in an incubation solution into chrysophanol (**51**) carbons adjacent to the deoxygenated phenolic hydroxy group.[122] The exchange of proton with deuterium was also observed in the recovered emodin (**32**) when it was incubated with the enzyme but without NADPH. The overall reaction is regarded as the deoxygenation of emodin (**32**) and the enzyme was thus designated emodin deoxygenase (Scheme 13).

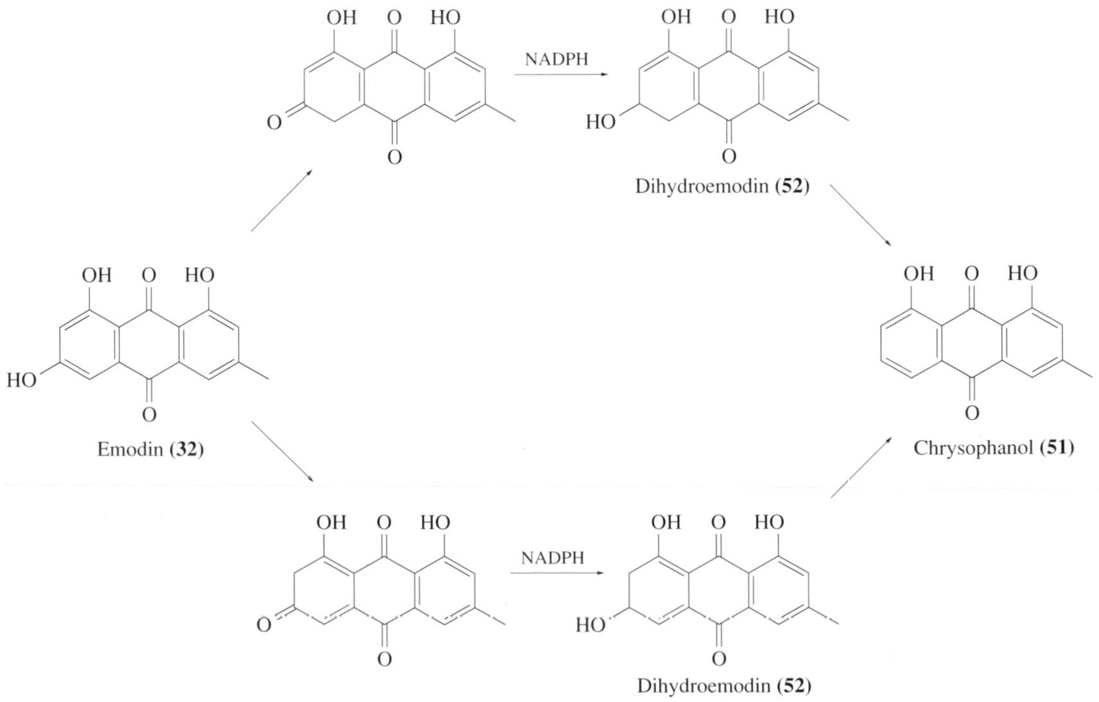

Emodin (**32**) → (NADPH) Dihydroemodin (**52**) → Chrysophanol (**51**)

Scheme 13

Since one of the possible tautomers of dihydroemodin (**52**) is a hypothetical intermediate in the biosynthesis of dimeric anthraquinonoids such as flavoskyrin (**53**) and rugulosin (**54**)[123–125] the enzyme capable of catalyzing reduction of the aromatic ring of emodin (**32**) should have a critical role in the biosynthesis of anthraquinonoids. Anderson *et al.* reported the partial purification and characterization of emodin deoxygenase and particularly emphasized the activation of the enzyme reaction with a low molecular weight cofactor.[126]

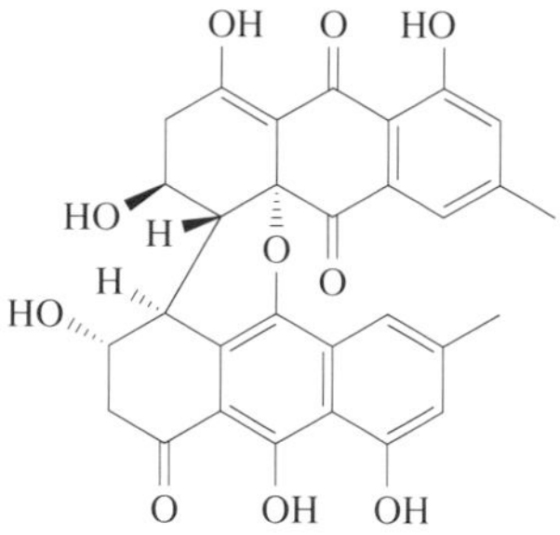

(−)-Flavoskyrin (**53**)

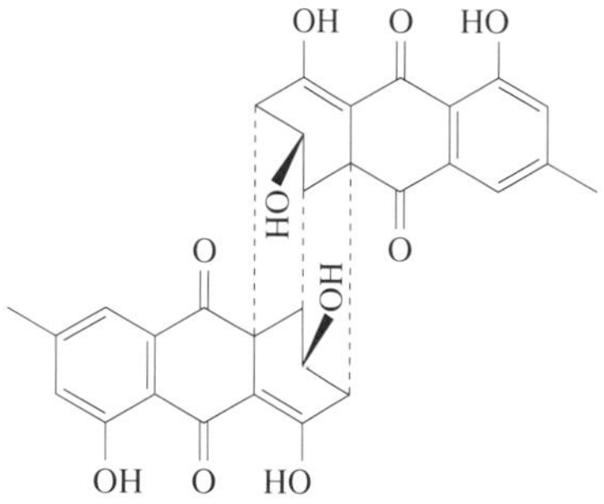

(−)-Rugulosin (**56**)

Fractionation of the crude enzyme preparation of *P. terrestris* was also carried out by Sankawa *et al.*[127] Their results indicated the involvement of reductase and dehydratase in this deoxygenation reaction. One of these may act to generate the keto tautomer of emodin (**32**) to facilitate its reduction into dihydroemodin (**52**).

The enzyme activity of emodin deoxygenase was detected in several *Penicillium* species which produce anthraquinones and related metabolites. Two strains of *Penicillium oxalicum* and *Penicillium islandicum* showed the significant activities of emodin deoxygenase.[127] Thus, emodin deoxygenase is considered to be common among fungi producing 6-deoxy-type anthraquinonoids.

1.16.6.3 Enzymes Involved in (+)-Geodin Biosynthesis

Aspergillus terreus strain IMI 16043 produces (+)-geodin (**14**) as a main metabolite, which is a seco-anthraquinone derived from an octaketide anthraquinone emodin (**32**). Historically, (+)-geodin (**14**) is the first chlorinated compound isolated from fungi[128] and was reported to possess an antibiotic activity.[129] The polyketide nature of (+)-geodin (**14**) was indicated by incorporation experiments using ^{14}C-labeled acetate and malonate[130,131] and the direct incorporation of labeled emodin (**32**)[132] clearly identified it as the seco-anthraquinone class of compounds with modified ring systems. In Scheme 14, the whole biosynthesis of (+)-geodin (**14**) in *A. terreus* is shown via emodinanthrone (**55**), emodin (**32**), questin (**56**), desmethylsulochrin (**57**), sulochrin (**58**), and dihydrogeodin (**59**). This biosynthesis has been extensively studied at enzyme level and most of the enzymes involved have been identified and characterized. Also, some molecular genetic work has been carried out. Thus, as exemplifying anthraquinonoid biosynthesis, (+)-geodin (**14**) biosynthesis in *A. terreus* is described on a step-by-step reaction base although enzymatic activity of PKS forming emodinanthrone (**55**) and its gene has not been identified in *A. terreus* at the time of writing.

1.16.6.3.1 Emodinanthrone oxygenase

The enzyme referred to as emodinanthrone oxygenase incorporated one oxygen atom from molecular oxygen into emodinanthrone (**55**) to generate emodin (**32**) without any requirement for an external electron donor such as NADPH.[133] Thus, the enzyme was formally classified as an internal monooxygenase.[134]

Emodinanthrone oxygenase from *A. terreus* was found to be membrane bound and phospholipids such as phosphatidylglycerol and phosphatidylcholine activated the enzyme after solubilization by nonionic detergent. In spite of the instability and heterogeneity of the solubilized enzyme, emodinanthrone oxygenase was purified to apparent homogeneity by chromatography on DEAE-cellulose, Hydroxyapatite, Q Sepharose, HPLC gel filtration, and Mono Q columns. The purified enzyme showed a single band of 52 kDa on SDS-PAGE.[135]

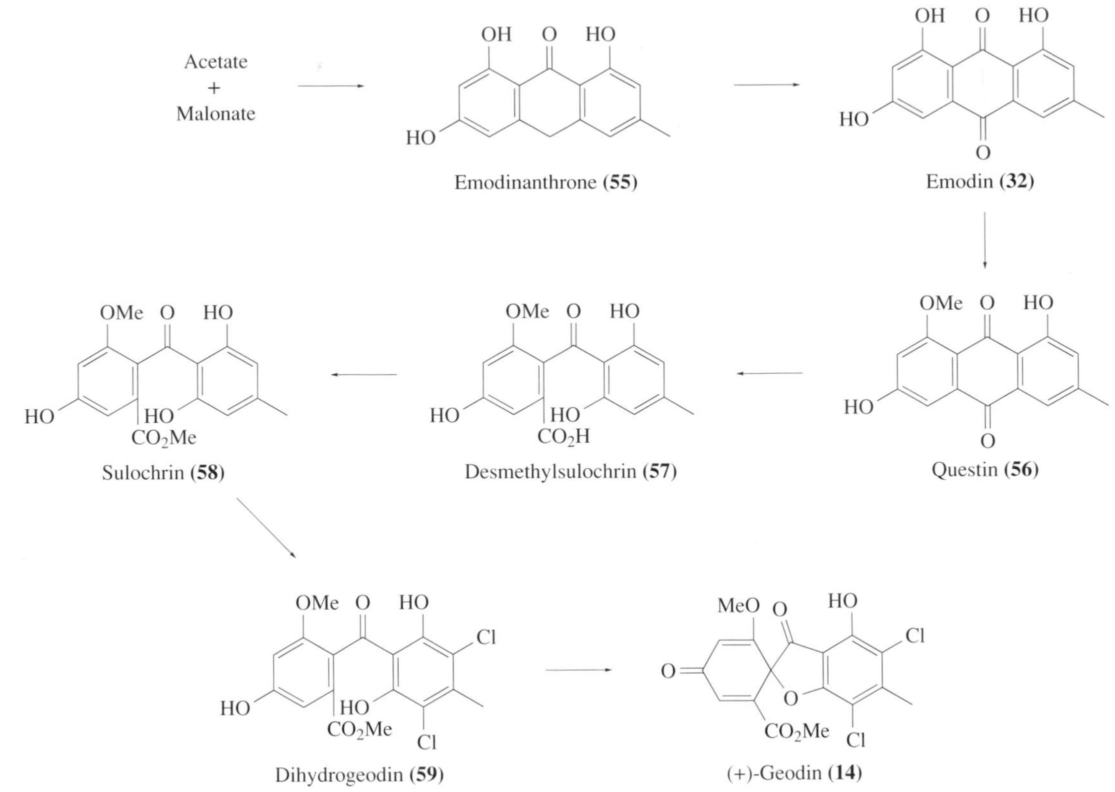

Acetate + Malonate → Emodinanthrone (**55**) → Emodin (**32**) → Questin (**56**) → Desmethylsulochrin (**57**) → Sulochrin (**58**) → Dihydrogeodin (**59**) → (+)-Geodin (**14**)

Scheme 14

Assuming that ferric iron is a prosthetic group of emodinanthrone oxygenase, its reaction mechanism was proposed, based on that of lipoxygenases.[136,137] The hydrogen at C-10 of emodinanthrone (**55**) is very active because of the electron withdrawing effect of the C-9 carbonyl and benzene rings and subsequent stabilization of the resultant radical by delocalization over the benzene rings. The reaction is initiated by the removal of hydrogen at C-10 of emodinanthrone (**55**) to form the emodinanthrone radical with concomitant reduction of the active ferric enzyme to the ferrous state. Molecular oxygen then binds to the emodinanthrone radical–enzyme complex. To the peroxy radical thus formed, one electron transfer from the ferrous iron occurs to give the emodinanthrone peroxy anion. Then, the free active enzyme is regenerated and the emodinanthrone peroxide is formed. The emodinanthrone peroxide is structurally very unstable and decomposes to form emodin (**32**) (Scheme 15).

The same type of enzyme is assumed to be involved in anthracycline biosynthesis. Actually, the tetracenomycin biosynthesis gene *tcmH* coding oxygenase catalyzes the corresponding reaction (Tcm F1 (**60**) to Tcm F2 (**61**), Scheme 16). However, no prosthetic group was found in the purified active enzyme.[138] Also, AknX protein, a product of the *aknX* gene of the aklavinone biosynthesis gene cluster (aklanonic acid anthrone (**62**) to aklanonic acid (**63**), Scheme 16), was expressed in *E. coli* and showed oxygenase activity using emodinanthrone as a substrate.[139] The fungal emodinanthrone oxygenase subunit was estimated to be 52 kDa, but the bacterial oxygenases, TcmH and AknX, were both found to be trimers of about 13 kDa subunits.[138,140] The homologous protein ActVA-Orf6 involved in actinorhodin biosynthesis was reported to catalyze a similar type of oxygenation and was suggested to be a dimer by analytical ultracentrifugation.[141] The reaction mechanism of anthrone oxygenases seems to be quite different between fungi and bacteria.

1.16.6.3.2 Emodin O-methyltransferase

Methylation of the 8-hydroxy group of emodin (**32**) was found to be a prerequisite step for the structural conversion from anthraquinone to benzophenone in the biosynthesis of (+)-geodin (**14**).

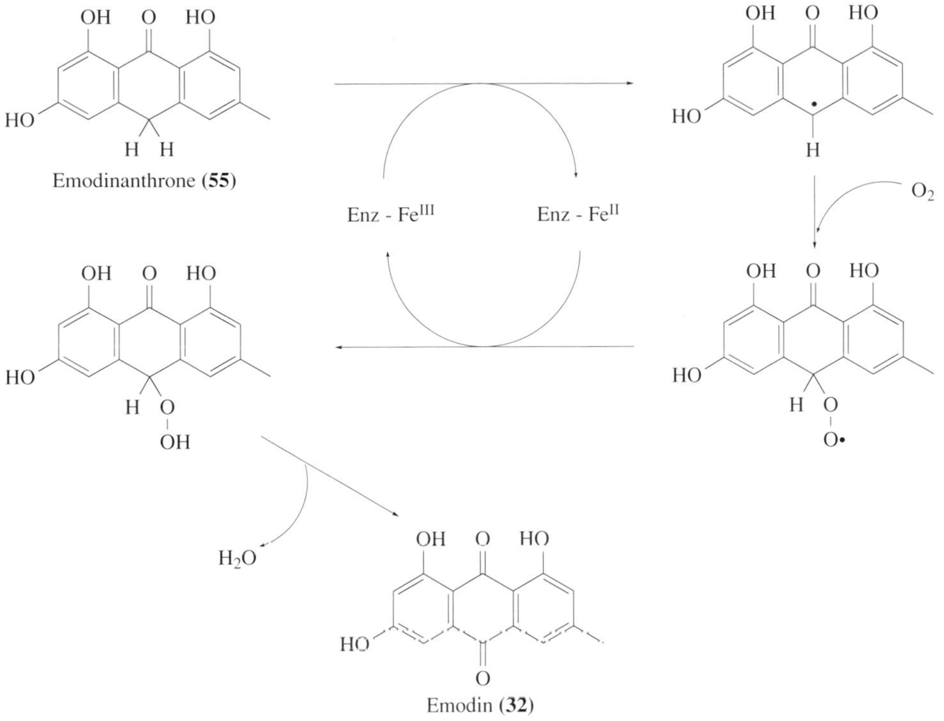

Emodinanthrone (**55**)

Enz - Fe^{III} Enz - Fe^{II}

O_2

H_2O

Emodin (**32**)

Scheme 15

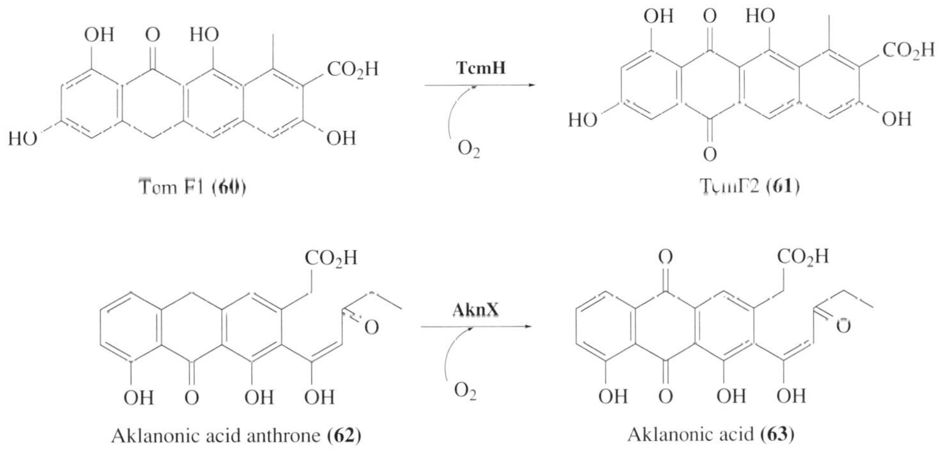

TcmH

O_2

Tcm F1 (**60**)

TcmF2 (**61**)

AknX

O_2

Aklanonic acid anthrone (**62**)

Aklanonic acid (**63**)

Scheme 16

The enzyme activity was found in the cell-free extract of *A. terreus*, which catalyzes methylation of the 8-hydroxy group of emodin (**32**) to form questin (**56**). By chromatography on DEAE-cellulose, Phenyl Sepharose, Q Sepharose, Hydroxyapatite, and CM-cellulose columns, the enzyme was purified to apparent homogeneity.[142] SDS-PAGE and gel filtration analyses indicated that the enzyme is a homohexamer of 53.6 kDa subunits.

1.16.6.3.3 Questin oxygenase

The conversion of anthraquinones to benzophenone or further metabolites was demonstrated by feeding experiments and the involvement of a Baeyer–Villiger type reaction was assumed.[143]

However, attempts to achieve such an oxidative ring cleavage of anthraquinone *in vitro* under chemical Baeyer–Villiger reaction conditions were unsuccessful.[144] A cytochrome P450 model reaction of such *ipso* cleavage of the anthraquinone ring was reported by Hirobe and co-workers.[145]

In vitro enzyme activity of this type of anthraquinone ring cleavage was first identified in the cell-free extract of *A. terreus*.[146] When the 10 000 *g* supernatant of mycelial homogenate was incubated with labeled questin (**56**) and NADPH in phosphate buffer, formation of an acidic compound was observed and the product was identified as benzophenone desmethylsulochrin (**57**).

The enzyme showed an absolute requirement for NADPH and molecular oxygen. Therefore, the enzyme, named questin oxygenase, was classified as a monooxygenase. The enzyme was very unstable but partially stabilized by the presence of polyols and nonionic detergent. Fractionation on a DEAE-cellulose column indicated the involvement of at least two protein components. The results suggested that one of the fractions contained oxygenase, but required additional protein factor(s) to react with questin (**56**). One possible explanation is the participation of some electron transfer component(s) which convey(s) electrons of NADPH to the oxygenase, such as an NADPH-questin oxygenase reductase, like the P450 monooxygenase system. The lactone intermediate formed (**64**) is then hydrolyzed to desmethylsulochrin (**57**), possibly by lactone hydrolase (Scheme 17).

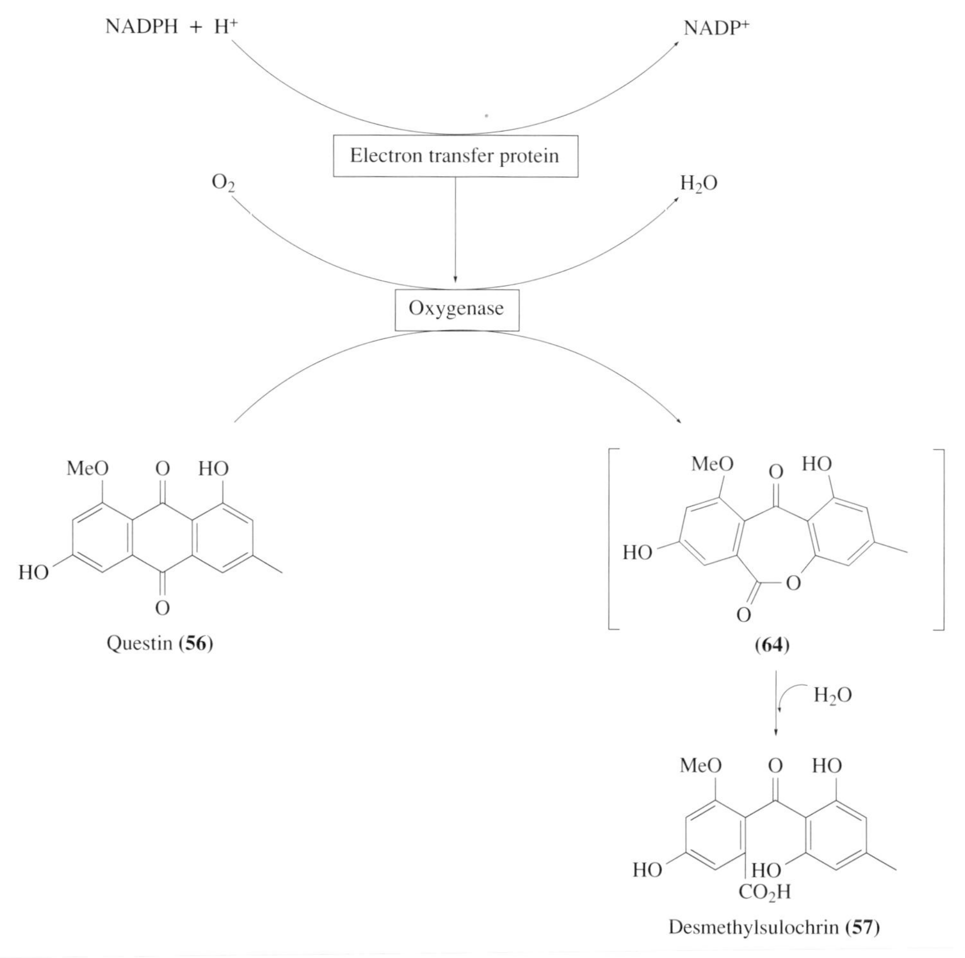

Scheme 17

In the biosynthesis of xanthones, such as tajixanthone (**65**) and shamixanthone (**66**) (Scheme 18), fixation of molecular oxygen at the position of ring cleavage of anthraquinone or anthrone precursor was demonstrated by $^{18}O_2$ feeding experiments.[147] Although no enzymological investigation was reported, the same type of enzymes as questin oxygenase may possibly be involved. Also, the similar type of ring cleavage reaction is considered to be a key reaction in aflatoxin biosynthesis, but such activity has not been detected.

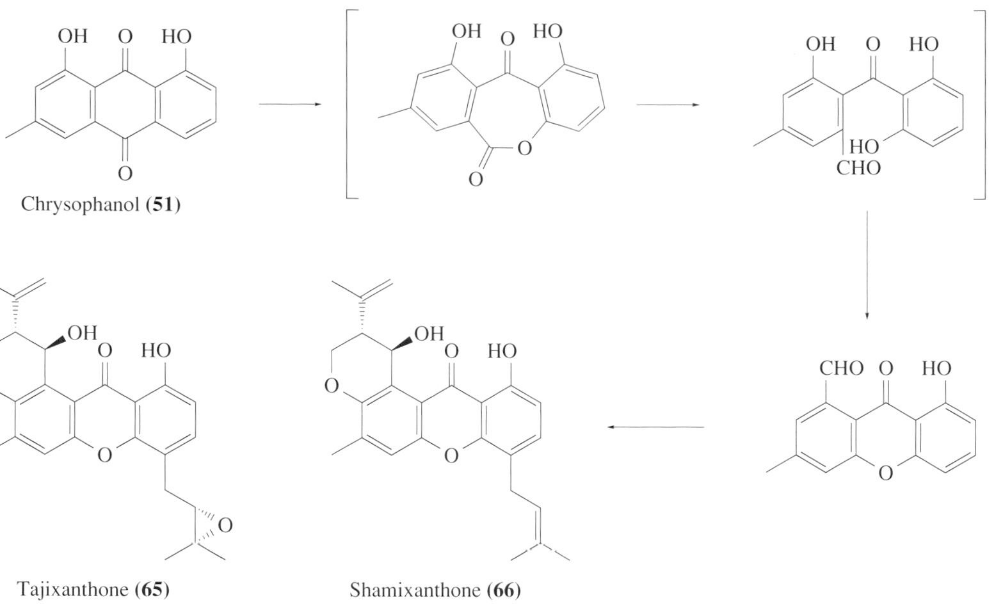

Chrysophanol (**51**)

Tajixanthone (**65**)

Shamixanthone (**66**)

Scheme 18

1.16.6.3.4 Desmethylsulochrin O-methyltransferase

Following the ring cleavage of questin (**56**), the product desmethylsulochrin (**57**) was then converted to sulochrin (**58**) by carboxyl methylation catalyzed by desmethylsulochrin *O*-methyltransferase. This enzyme activity was identified in the cell-free extract of *A. terreus*, but was found to be very unstable. The enzyme showed higher specificity to desmethylsulochrin (**57**) than desmethyldihydrogeodin (**67**).

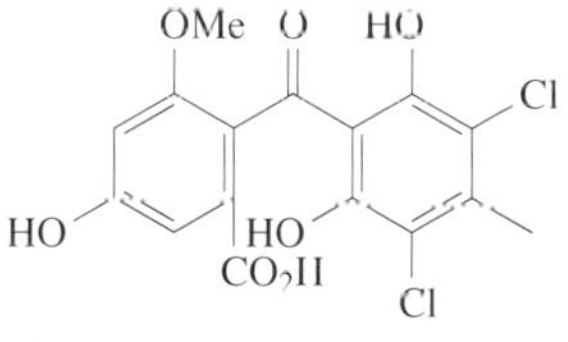

Desmethyldihydrogeodin (**67**)

1.16.6.3.5 Dihydrogeodin oxidase

The last step of (+)-geodin (**14**) biosynthesis is the stereospecific intramolecular phenol oxidative coupling reaction to form the unique spirodienone structure of (+)-geodin (**14**) from benzophenone dihydrogeodin (**59**) catalyzed by dihydrogeodin oxidase (DHGO) (Scheme 19).

It has been recognized that phenol oxidative coupling is one of the most important reactions in the biosynthesis of natural products since Barton and Cohen first proposed that the new C—C or C—O bond could be formed by pairing of radicals of phenolic substrates.[148–151] DHGO catalyzes the regio- and stereospecific phenol oxidative coupling reaction to form (+)-geodin (**14**); it was purified and found to be a homodimer of 76 kDa subunits.[152] The enzyme showed intense blue color with absorption maximum at 600 nm, which suggested it to be a copper protein. The copper content was found to be four atoms per subunit. The EPR spectrum indicated the presence of type-1 and type-2 copper atoms in the enzyme molecule.

Molecular cloning of cDNA and genomic DNA for DHGO was carried out by screening the cDNA and genomic DNA library of *A. terreus*.[153] The sequence of the cloned DHGO genomic DNA and cDNA predicted that the DHGO polypeptide consists of 605 amino acids showing significant homology with multicopper blue proteins such as laccase and ascorbate oxidase. Four potential copper-binding domains were identified in the DHGO polypeptide. The DHGO gene consists of seven exons separated by six short introns. Expression of the DHGO gene in *A. nidulans*

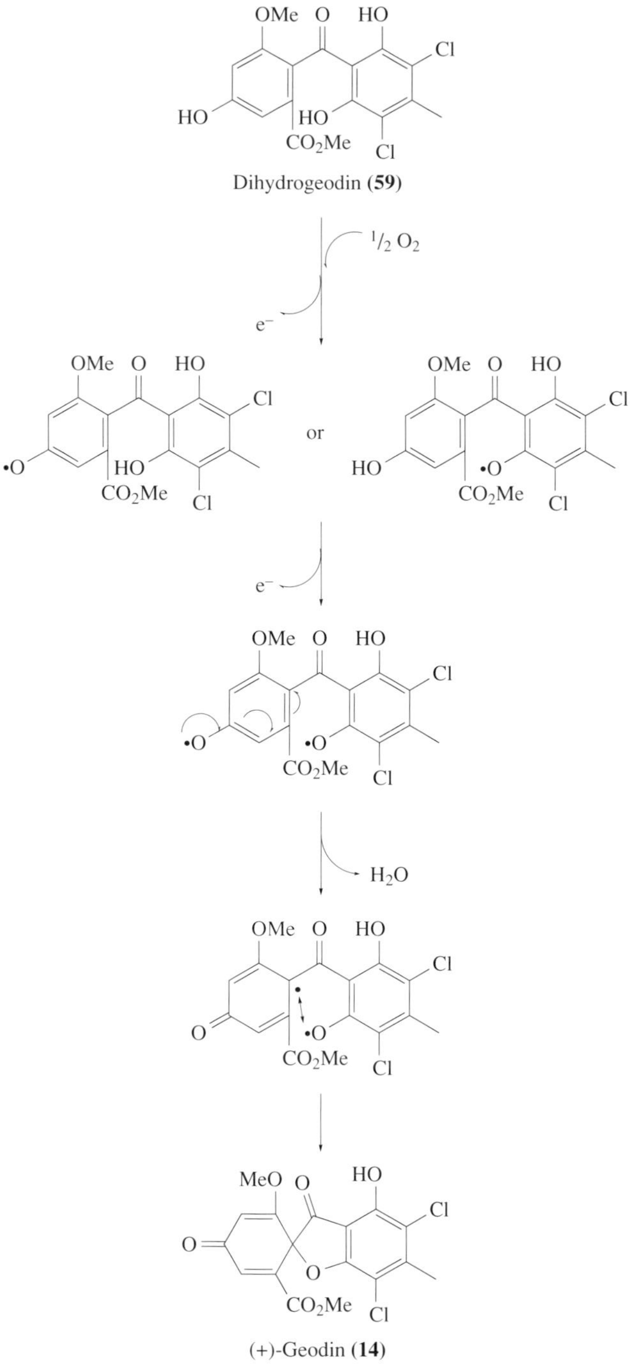

Dihydrogeodin (**59**)

$^1/_2\ O_2$

e^-

or

e^-

H_2O

(+)-Geodin (**14**)

Scheme 19

under the starch-inducible α-amylase promoter using expression plasmid pTAex3[27] as an active enzyme established the functional identity of the gene. Also, introduction of the genomic DNA for DHGO into *Penicillium frequentans* led to the production of DHGO polypeptide, as judged by Western blot analysis.

Similarity between ascorbate oxidase and DHGO—the size of about 600 amino acid subunits, solution structure as homodimer, and nearly identical EPR spectra of complex-type copper proteins—suggested that DHGO might contain four coppers, one type-1, one type-2, and two type-3 coppers per subunit in a relatively similar organization to that of ascorbate oxidase. A catalytic mechanism for ascorbate oxidase has been proposed based on the available kinetic data, the three-dimensional structure, and the associated electron-transfer processes.[154] Thus, the involvement of a similar electron-transfer mechanism was assumed in the phenol oxidative coupling reaction catalyzed by DHGO. That is, type-1 copper is first reduced by one-electron transfer from the substrate dihydrogeodin (**59**), which is oxidized to a free radical. The electron is then transferred from type-1 copper to the type-3 copper pair. After reduction with four equivalents of reductant, the fully reduced enzyme is formed, which is able to bind molecular oxygen into the trinuclear copper center bridging the type-3 copper pair and the type-2 copper. This species must accept protons to release H_2O, while intramolecular C—O coupling of dihydrogeodin diradicals gives (+)-geodin (**14**) under regio- and stereospecific control by DHGO enzyme (Scheme 20).

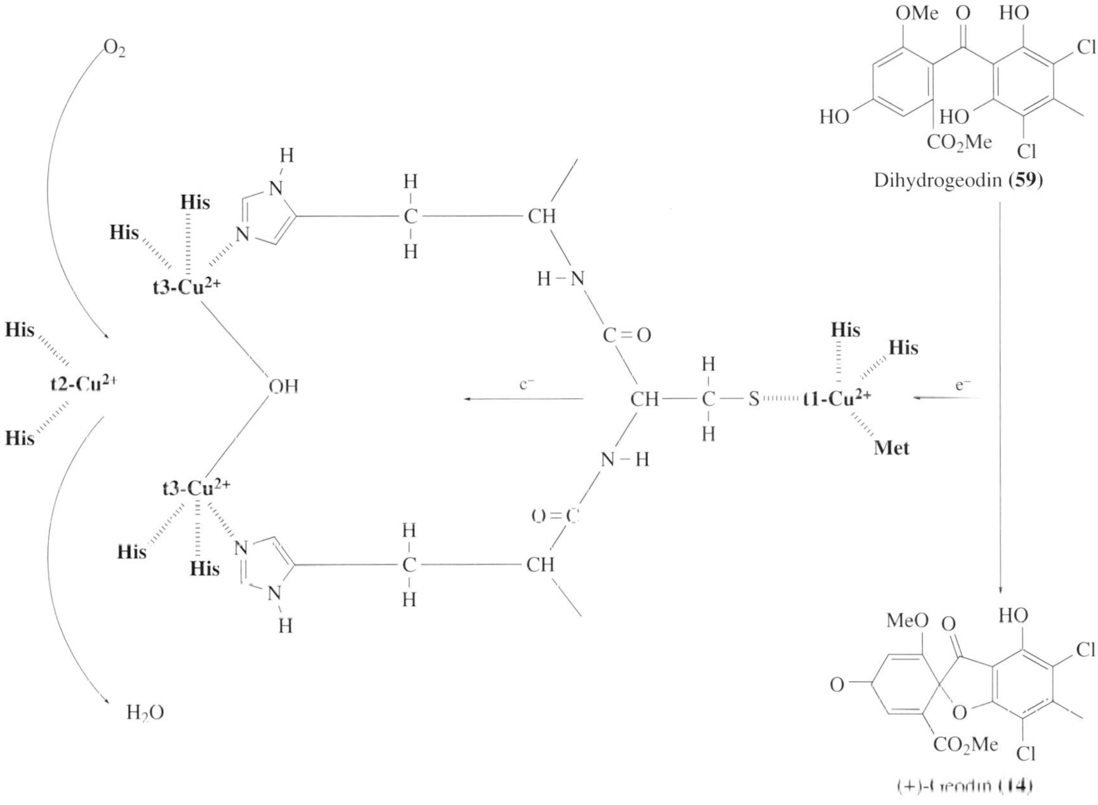

Scheme 20

A similar type of phenol oxidative coupling reaction is involved in the grisan structure formation in (+)- and (−)-bisdechlorogeodin (**68**) from sulochrin (**58**) by sulochrin oxidase from *P. frequentans* and *Oospora sulphrea-ochracea*, respectively[155] (Scheme 21). Purification of the responsible enzymes was also reported.[155,156] In griseofulvin (**30**) biosynthesis, a reaction catalyzed by the same type of phenol oxidative coupling via griseophenone (**69**) and dehydrodemethylgriseofulvin (**70**) was assumed, but there have been no reports on this enzyme activity to our knowledge.

1.16.7 CONCLUDING REMARKS

As exemplified by aflatoxins, fungi have been regarded as mycotoxin producers at least in the sense of secondary metabolites. However, this image has been changing since the advent of lovastatin-related compounds as effective hypercholesterolemia medicines from fungal sources. In this chapter, basic features of fungal polyketide biosynthesis have been described, mainly at the level of enzymology and molecular genetics. Because of rapid progress in molecular genetic techniques, chemists can handle recombinant proteins to carry out mechanistic investigation on enzyme reactions. Thus, it is quite reasonable that biosynthesis of secondary metabolites is now being studied actively at the molecular genetic level. However, progress in fungal secondary metabolism has been relatively slow compared with that of the bacterial counterpart. This might be partly due to limited availability of genetic engineering systems, that is, vectors and transformation methods applicable to filamentous fungi. Self-replicating plasmids practically useful in fungal transformation are not

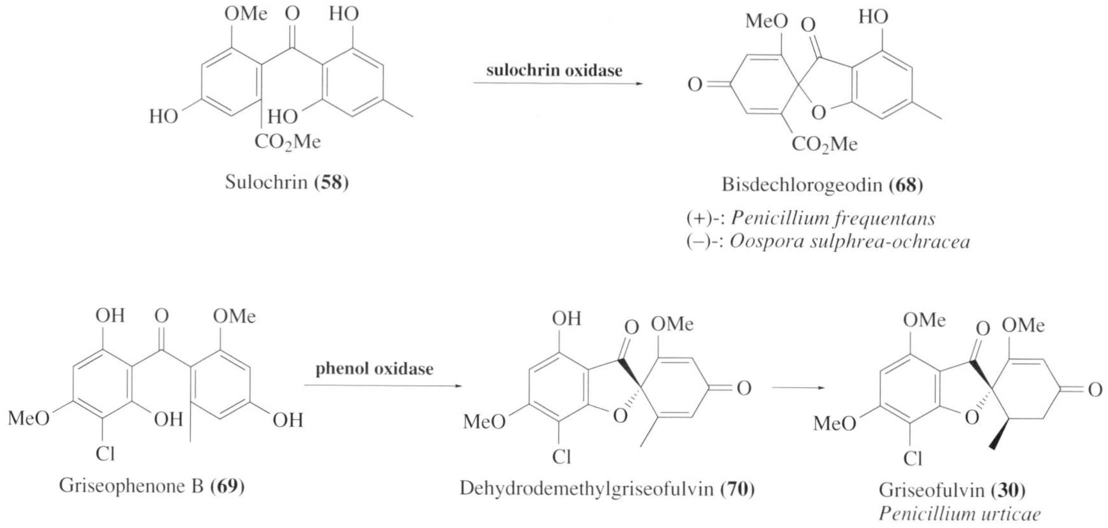

sulochrin oxidase

Sulochrin (**58**) → Bisdechlorogeodin (**68**)

(+)-: *Penicillium frequentans*
(–)-: *Oospora sulphrea-ochracea*

phenol oxidase

Griseophenone B (**69**) → Dehydrodemethylgriseofulvin (**70**) → Griseofulvin (**30**)
Penicillium urticae

Scheme 21

established yet, but some integration vectors and dominant selection markers have been developed. Fungal PKS gene could be expressed in heterologous fungi as active PKS to produce specific polyketide compounds. Thus, it might be possible to carry out the expression of any other fungal biosynthetic genes, which will allow mechanistic investigations of secondary metabolic reactions using recombinant enzymes. Secondary metabolism is a combination of versatile but highly specific enzyme reactions to produce specific compounds. Thus, it will be necessary to study how secondary metabolism enzymes recognize their specific substrates and control their specific reactions, in which fungal systems seem to be much more strict than bacterial systems.

Manipulation of biosynthetic genes is now being recognized to be a promising powerful tool for production of novel compounds. However, simple combinations of biosynthetic genes have limitations due to specificities. Further analyses of PKSs and post-PKS enzymes including their structural features will enable development of designed biosynthetic systems for useful compound production.

ACKNOWLEDGMENTS

The author is very grateful to co-workers Professor Yutaka Ebizuka, Dr. Ke-xue Huang, Messrs. Akira Watanabe, Yuichiro Mori, Yuya Ono, and Hidenori Tada at the University of Tokyo, and Professor Ushio Sankawa of Toyama Medical and Pharmaceutical University. Also, the author thanks Professor Yasuyuki Kubo and Mr. Gento Tsuji of Kyoto Prefectural University, and Dr. Katsuya Gomi of Tohoku University, Japan for their collaboration.

1.16.8 REFERENCES

1. W. B. Turner, "Fungal Metabolites," Academic Press, London, 1971.
2. W. B. Turner and D. C. Aldridge, "Fungal Metabolites II," Academic Press, London, 1983.
3. J. Sekiguchi, H. Kuroda, Y. Yamada, and H. Okada, *Tetrahedron Lett.*, 1985, **26**, 2341.
4. A. Endo and K. Hasumi, *Nat. Prod. Rep.*, 1993, **10**, 541.
5. N. Collie, *J. Chem. Soc.*, 1893, 122.
6. A. J. Birch, R. A. Massy-Westropp, and C. J. Moye, *Aust. J. Chem.*, 1955, **8**, 539.
7. A. J. Birch and F. W. Donovan, *Aust. J. Chem.*, 1953, **6**, 360.
8. F. Lynen and M. Tada, *Angew. Chem.*, 1961, **73**, 513.
9. P. Dimroth, H. Walter, and F. Lynen, *Eur. J. Biochem.*, 1970, **13**, 98.
10. D. A. Hopwood and D. H. Sherman, *Ann. Rev. Genet.*, 1990, **24**, 37.
11. D. A. Hopwood and C. Khosla, in "Secondary Metabolites: Their Function and Evolution, Ciba Foundation Symposium 171, London, 1992," eds. D. J. Chadwick and J. Whelan, Wiley, West Sussex, 1992, p. 88.
12. D. A. Hopwood, *Chem. Rev.*, 1997, **97**, 2465.
13. S. J. Gurr, S. E. Unkles, and J. R. Kinghorn, in "Gene Structure in Eukaryotic Microbes," ed. J. R. Kinghorn, IRL, Oxford, 1987, p. 93

14. J. Beck, S. Ripka, A. Signer, E. Schiltz, and E. Schweizer. *Eur. J. Biochem.*, 1990, **192**, 487.
15. C. R. Hutchinson and I. Fujii, *Ann. Rev. Microbiol.*, 1995, **49**, 201.
16. M. Gill and A. Giménez, *J. Chem. Soc., Perkin Trans. 1*, 1990, 2585.
17. I.-K. Wang and G. M. Gaucher, in "Annual Meeting of the Society of Industrial Microbiology, Seattle, 1989," abstract P-75.
18. J. B. Spencer and P. M. Jordan, *Biochem. J.*, 1992, **288**, 839.
19. R. Welle and H. Grisebach, *FEBS Lett.*, 1988, **236**, 221.
20. G. M. Gaucher and M. G. Shepherd, *Biochem. Biophys. Res. Commun.*, 1968, **32**, 664.
21. E.-R. Woo, I. Fujii, Y. Ebizuka, U. Sankawa, A. Kawaguchi, J. M. Beale, M. Shibuya, U. Mocek, and H. G. Floss, *J. Am. Chem. Soc.*, 1989, **111**, 5498.
22. P. M. Jordan and J. B. Spencer, *Biochem. Soc. Trans.*, 1993, **21**, 222.
23. N. Singh, S. J. Wakil, and J. K. Stoops, *J. Biol. Chem.*, 1984, **259**, 3605.
24. I.-K. Wang, C. Reeves, and G. M. Gaucher, *Can. J. Microbiol.*, 1991, **37**, 86.
25. F. H. Malpartida, S. E. Hallam, H. M. Kieser, H. Motamedi, C. R. Hutchinson, M. J. Butler, D. A. Sugden, M. Warren, C. McKillop, C. R. Bailey, G. O. Humphrey, and D. A. Hopwood, *Nature* (*London*), 1987, **325**, 818.
26. I. Fujii, Y. Ono, H. Tada, K. Gomi, Y. Ebizuka, and U. Sankawa, *Mol. Gen. Genet.*, 1996, **253**, 1.
27. T. Fujii, H. Yamaoka, K. Gomi, K. Kitamoto, and C. Kumagai, *Biosci. Biotech. Biochem.*, 1995, **59**, 1869.
28. D. J. Bedford, E. Schweizer, D. A. Hopwood, and C. Khosla, *J. Bacteriol.*, 1995, **177**, 4544.
29. S. Gaisser, A. Trefzer, S. Stckert, A. Kirshning, and A. Bechthold, *J. Bacteriol.*, 1997, **179**, 6271.
30. U. Sankawa, H. Shimada, T. Sato, T. Kinoshita, and K. Yamasaki, *Chem. Pharm. Bull.*, 1981, **29**, 3586.
31. Y. Kubo, H. Nakamoto, K. Kobayashi, T. Okuno, and I. Furusawa, *Mol. Plant–Microb. Interact.*, 1991, **4**, 440.
32. Y. Takano, Y. Kubo, K. Shimizu, K. Mise, T. Okuno, and I. Furusawa, *Mol. Gen. Genet.*, 1995, **249**, 162.
33. M. E. Mayorga and W. E. Timberlake, *Mol. Gen. Genet.*, 1992, **235**, 205.
34. I. Fujii *et al.*, manuscript in preparation.
35. D. C. Allport and J. D. Bu'Lock, *J. Chem. Soc.*, 1960, 654.
36. J. Blair and G. T. Newbold, *J. Chem. Soc.*, 1955, 2871.
37. F. Kurosaki, Y. Kizawa, and A. Nishi, *Eur. J. Biochem.*, 1989, **185**, 85.
38. F. Kurosaki, M. Itoh, Y. Kizawa, and A. Nishi, *Arch. Biochem. Biophys.*, 1993, **300**, 157.
39. F. Kurosaki, M. Itoh, M. Yamada, and A. Nishi, *FEBS Lett.*, 1991, **288**, 219.
40. F. Kurosaki, *Phytochem.*, 1995, **39**, 515.
41. F. Kurosaki, *Arch. Biochem. Biophys.*, 1996, **328**, 213.
42. F. Kurosaki, *Arch. Biochem. Biophys.*, 1995, **321**, 239.
43. B. J. Weigel, S. G. Burgett, V. J. Chen, P. L. Skatrud, C. A. Frolik, S. W. Queener, and T. D. Ingolia, *J. Bacteriol.*, 1988, **170**, 3817.
44. G. Pontecorvo, J. A. Roper, L. M. Hammons, K. D. Macdonald, and A. W. Bufton, *Adv. Genet.*, 1953, **5**, 141.
45. A. J. Clutterbuck, *J. Gen. Microbiol.*, 1972, **70**, 423.
46. R. Aramayo and W. E. Timberlake, *Nucleic Acids Res.*, 1990, **18**, 341.
47. D. W. Brown, F. M. Hauser, R. Tommasi, S. Corlett, and J. J. Salvo, *Tetrahedron Lett.*, 1993, **34**, 419.
48. D. W. Brown and J. J. Salvo, *Appl. Environ. Microbiol.*, 1994, **60**, 979.
49. M. E. Mayorga and W. E. Timberlake, *Genetics*, 1990, **126**, 73.
50. A. Watanabe, Y. Ono, I. Fujii, U. Sankawa, H. E. Hayorga, W. E. Timberlake, and Y. Ebiz, *Tetrahedron Lett.*, in press.
51. O. L. Galmarini and F. H. Stodola, *J. Org. Chem.*, 1965, **30**, 112.
52. G. H. Stout, D. L. Dreyer, and L. H. Jonson, *Chem. Ind.*, 1961, 289.
53. H. Tanaka and T. Tamura, *Tetrahedron Lett.*, 1961, 151.
54. F. Leeper and J. Stauton, *J. Chem. Soc., Perkin Trans. 1*, 1984, 2919.
55. J. C. Vederas and T. T. Nakashima, *J. Chem. Soc., Chem. Commun.*, 1980, 183.
56. C. A. Townsend, S. B. Christensen, and K. Trautwein, *J. Am. Chem. Soc.*, 1984, **106**, 3868.
57. N, Mahanti, D. Bhatnagar, J. W. Cary, J. Joubran, and J. E. Linz, *Appl. Environ. Microbiol.*, 1996, **62**, 191.
58. J. Yu, P.-K. Chang, J. W. Cary, M. Wright, D. Bhatnagar, T. E. Cleveland, G. A. Payne, and J. E. Linz, *Appl. Environ. Microbiol.*, 1995, **61**, 2365.
59. D. W. Brown, J.-H. Yu, H. S. Kelkar, M. Fernandes, T. C. Nesbitt, N. P. Keller, T. H. Adams, and T. J. Leonards, *Proc. Natl. Acad. Sci. USA*, 1996, **93**, 1418.
60. G. H. Feng and T. J. Leonard, *J. Bacteriol.*, 1995, **177**, 6246.
61. P.-K. Chang, J. W. Cary, J. Yu, D. Bhatnagar, and T. E. Cleveland, *Mol. Gen. Genet.*, 1995, **248**, 270.
62. J.-h. Yu and T. J. Leonard, *J. Bacteriol.*, 1995, **177**, 4792.
63. C. M. H. Watanabe, D. Wilson, J. E. Linz, and C. A. Townsend, *Chem. Biol.*, 1996, **3**, 463.
64. A. W. Alberts, J. Chen, G. Curon, V. Hunt, J. Huff, C. Hoffman, J. Rothrock, M. Lopez, H. Joshua, E. Harris, A. Patchett, R. Monaghan, S. Currie, E. Stapley, G. Albers-Schönberg, O. Hensens, J. Hishfield, K. Hoogsteen, J. Liesch, and J. Springer, *Proc. Natl. Acad. Sci. USA*, 1980, **77**, 3957.
65. A. Endo, *J. Antibiot.*, 1979, **33**, 334.
66. A. Endo, M. Kuroda, and Y. Tsujita, *J. Antibiot.*, 1976, **29**, 1346.
67. R. N. Moore, G. Bigman, J. K. Chan, A. M. Hogg, T. T. Nakashima, and J. C. Vederas, *J. Am. Chem. Soc.*, 1985, **107**, 3694.
68. D. J. Witter and J. C. Vederas, *J. Org. Chem.*, 1996, **61**, 2613.
69. V. A. Vinci, M. J. Conder, P. C. Mcada, C. D. Reeves, J. Rambosek, C. R. Davis, L. E. Hendrickson (Merck & Co., Inc.), *Internat. Publ. No.* WO 95/12661 (1995) (*Chem. Abstr.*, 1995, **123**, 192355).
70. C. M. Amy, A. Wikowski, J. Naggert, B. Williams, Z. Randhawa, and S. Smith, *Proc. Natl. Acad. Sci. USA*, 1989, **86**, 3114.
71. S. J. Wakil, *Biochemistry*, 1989, **28**, 4523.
72. R. K. Wierenga and W. G. J. Hol, *Nature* (*London*), 1983, **302**, 842.
73. G. Wu, H. D. Williams, M. Zamanian, F. Gibson, and R. K. Poole, *J. Gen. Microbiol.*, 1992, **138**, 2101.
74. D. Ingrosso, A. V. Fowler, J. Bleibaum, and S. Clarke, *J. Biol. Chem.*, 1989, **264**, 20 131.

75. R. Summers, E. Wendt-Pienkowski, H. Motamedi, and C. R. Hutchinson, *J. Bacteriol.*, 1992, **174**, 1810.
76. Y. Kono, S. Takeuchi, A. Kawarada, J. M. Daly, and H. W. Knoche, *Tetrahedron Lett.*, 1980, **21**, 1537.
77. Y. Kono and J. M. Daly, *Bioorg. Chem.*, 1979, **8**, 391.
78. Y. Kono, S. J. Danko, Y. Suzuki, S. Takeuchi, and J. M. Daly, *Tetrahedron Lett.*, 1983, **24**, 3803.
79. J. Leach, K. J. Tegtmeier, J. M. Daly, and O. C. Yoder, *Physiol. Plant Pathol.*, 1982, **21**, 327.
80. K. J. Tegtmeier, J. M. Daly, and O. C. Yoder, *Phytopathology*, 1982, **72**, 1492.
81. R. H. Schiest and T. D. Petes, *Proc. Natl. Acad. Sci. USA*, 1991, **88**, 7585.
82. S. Lu, L. Lyngholm, G. Yang, C. Bronson, O. C. Yoder, and B. G. Turgeon, *Proc. Natl. Acad. Sci. USA*, 1994, **91**, 12 649.
83. G. Yang, M. S. Rose, B. G. Turgen, and O. C. Yoder, *Plant Cell*, 1996, **8**, 2139.
84. A. W. Alberts and M. D. Greenspan, in "Fatty Acid Metabolism and Its Regulation," ed. S. Numa, Elsevier, Amsterdam, 1984, vol. 2, p. 29.
85. S. Donadio and L. Katz, *Gene*, 1992, **111**, 51.
86. I. Fujii, A. Watanabe, Y. Mori, and Y. Ebizuka, *Actinomycetol.*, **12**, 1.
87. A. Watanabe *et al.*, manuscript in preparation.
88. A. Witkowski, *Eur. J. Biochem.*, 1991, **198**, 571.
89. L. O. Zamir, in "The Biosynthesis of Mycotoxins," ed. P. S. Steyn, Acdemic Press, New York, 1980, p. 224.
90. K. Axberg and S. Gatenbeck, *FEBS Lett.*, 1975, **54**, 18.
91. H. Iijima, Y. Ebizuka, and U. Sankawa, *Chem. Pharm. Bull.*, 1986, **34**, 3534.
92. J. W. Priest and R. J. Light, *Biochemistry*, 1989, **28**, 9192.
93. M. H. Wheeler, *Trans. Br. Mycol. Soc.*, 1983, **81**, 29.
94. D. H. Ellis and D. A. Griffith, *Can. J. Microbiol.*, 1974, **20**, 1379.
95. A. A. Bell and M. H. Wheeler, *Annu. Rev. Phytopathol.*, 1986, **24**, 411.
96. M. H. Wheeler and A. A. Bell, *Curr. Top. Med. Mycol.*, 1987, **2**, 338.
97. I. Yamaguchi and Y. Kubo, in "Target Sites of Fungicide Action," ed. W. Koeller, CRC Press, London, England, 1992, p. 101.
98. Y. Kubo and I. Furusawa, in "The Fungal Spore and Disease Initiation in Plants and Animals," ed. G. T. Cole and H. C. Hoch, Plenum Publishing, New York, 1991, p. 205.
99. Y. Kubo, K. Suzuki, I. Furusawa, and M. Yamamoto, *Pestic. Biochem. Physiol.*, 1985, **23**, 47.
100. Y. Kubo, I. Suzuki, I. Furusawa, N. Ishida, and M. Yamamoto, *Phytopathology*, 1982, **72**, 498.
101. K. Tanabe, S. Nishimura, and K. Kohmoto, *Ann. Phytopathol. Soc. Jpn.*, 1988, **54**, 54.
102. A. A. Bell, J. E. Puhalia, W. J. Tolmsoff, and R. D. Stipanovic, *Can. J. Microbiol.*, 1976, **22**, 787.
103. I. Fujii *et al.*, manuscript in preparation.
104. A. Vidal-Cros, F. Viviani, G. Laesse, M. Boccara, and M. Gaudry, *Eur. J. Biochem.*, 1994, **219**, 986.
105. C. D. Skory, P. K. Chang, J. Cary, and J. E. Linz, *Appl. Environ. Microbiol.*, 1992, **58**, 3542.
106. B. Presson, M. Krook, and H. Jörnvall, *Eur. J. Biochem.*, 1991, **200**, 537.
107. A. Villaroya, E. Juan, B. Egestd, and H. Jörnvall, *Eur. J. Biochem.*, 1989, **180**, 191.
108. S. Tajima, Y. Kubo, I. Furusawa, and J. Shishiyama, *Exp. Mycol.*, 1989, **13**, 69.
109. T. Lundqvist, P. C. Weber, C. N. Hodge, E. H. Braswell, J. Rice, and J. Pierce, *J. Mol. Biol.*, 1993, **232**, 999.
110. C. N. Hodge and J. Pierce, *Bioorg. Med. Chem. Lett.*, 1993, **3**, 1605.
111. Y. Kubo, Y. Takano, N. Endo, N. Yasuda, S. Tajima, and I. Furusawa, *Appl. Environ. Microbiol.*, 1996, **62**, 4340.
112. N. Kimuara and T. Tsuge, *J. Bacteriol.*, 1993, **175**, 4427.
113. N. S. Perpetua, Y. Kubo, N. Yasuda, Y. Takano, and I. Furusawa, *Mol. Plant–Microbe Interact.*, 1996, **9**, 323.
114. F. G. Chumley and B. Valent, *Mol. Plant–Microbe Interact.*, 1990, **3**, 135.
115. K. F. Chater and C. J. Bruton, *EMBO J.*, 1985, **4**, 1893.
116. T. M. Hohn, S. P. McCormick, and A. E. Desjardins, *Curr. Genet.*, 1993, **24**, 291.
117. Y. Kubo, M. Tsuda, I. Furusawa, and J. Shishiyama, *Exp. Mycol.*, 1989, **13**, 77.
118. C. Tanaka, Y. Kubo, and M. Tsuda, *Mycol. Res.*, 1991, **95**, 45.
119. B. Franck, G. Bringmann, and G. Flohr, *Angew. Chem., Int. Ed. Engl.*, 1980, **19**, 460.
120. B. Franck, F. Hüper, D. Gröger, and D. Erge, *Chem. Ber.*, 1968, **1081**, 1224.
121. J. A. Anderson, *Phytochemistry*, 1986, **25**, 103.
122. J. A. Anderson, B.-K. Lin, H. J. Williams, and A. I. Scott, *J. Am. Chem. Soc.*, 1988, **110**, 1623.
123. N. Takeda, S. Seo, Y. Ogihara, U. Sankawa, Y. Iitaka, I. Kitagawa, and S. Shibata, *Tetrahedron*, 1973, **29**, 3703.
124. D. M. Yang, U. Sankawa, Y. Ebizuka, and S. Shibata, *Tetrahedron*, 1976, **32**, 333.
125. S. Seo, U. Sankawa, Y. Ogihara, Y. Iitaka, and S. Shibata, *Tetrahedron*, 1973, **29**, 3721.
126. J. A. Anderson, B.-K. Lin, and S. S. Wang, *Phytochemistry*, 1990, **29**, 2415.
127. K. Ichinose, J. Kiyono, Y. Ebizuka, and U. Sankawa, *Chem. Pharm. Bull.*, 1993, **41**, 2015.
128. H. Raistrick and G. Smith, *Biochem. J.*, 1936, **30**, 1315.
129. S. Marcus, *Biochem. J.*, 1947, **41**, 349.
130. R. F. Curtis, P. C. Harries, C. H. Hassall, J. D. Levi, and D. M. Phillips, *J. Chem. Soc. C*, 1966, 168.
131. R. F. Curtis, C. H. Hassall, and R. K. Pike, *J. Chem. Soc. C*, 1968, 1807.
132. H. Fujimoto, H. Flash, and B. Frank, *Chem. Ber.*, 1975, **108**, 1224.
133. I. Fujii, Z.-G. Chen, Y. Ebizuka, and U. Sankawa, *Biochem. Internat.*, 1991, **25**, 1043.
134. O. Hayaishi, "Molecular Mechanism of Oxygen Activation", Academic Press, 1974.
135. Z.-G. Chen, I. Fujii, Y. Ebizuka, and U. Sankawa, *Phytochemistry*, 1995, **38**, 299.
136. L. Petersson, S. Slappendel, and J. F. G. Vliegenthart, *Biochim. Biophys. Acta*, 1985, **828**, 81.
137. M. J. Nelson, S. P. Seitz, and R. A. Cowling, *Biochemistry*, 1990, **29**, 6897.
138. B. Shen and C. R. Hutchinson, *Biochemistry*, 1993, **32**, 6656.
139. I. Fujii and Y. Ebizuka, *Chem. Rev.*, 1997, **97**, 2511.
140. J. Y. Chung *et al.*, in preparation.
141. S. G. Kendrew, D. A. Hopwood, and E. N. G. Marsh, *J. Bacteriol*, 1997, **179**, 4305.
142. Z.-G. Chen, I. Fujii, Y. Ebizuka, and U. Sankawa, *Arch. Microbiol.*, 1992, **158**, 29.
143. B. Franck, in "The Biosynthesis of Mycotoxins," ed. P. S. Steyn, Academic Press, New York, 1980, p. 151.

144. B. Franck and B. Berger-Lohr, *Angew. Chem., Int. Ed. Engl.*, 1975, **14**, 818.
145. T. Ohe, T. Mashino, and M. Hirobe, *Drug Metab. Dispos.*, 1997, **25**, 116.
146. I. Fujii, Y. Ebizuka, and U. Sankawa, *J. Biochem.*, 1988, **103**, 878.
147. S. A. Ahmed, E. Bardshir, C. R. McIntyre, and T. Simpson, *Aust. J. Chem.*, 1992, **45**, 249.
148. W. I. Taylor and A. R. Battersby, "Oxidative Coupling of Phenols," Marcel Dekker, New York, 1967.
149. A. I. Scott, *Q. Rev. Chem. Soc.*, 1965, **19**, 1.
150. D. H. R. Barton and T. Cohen, "Festschrift A. Stoll," Birkhauser, Basel, 1957.
151. D. H. R. Barton, *Chemistry in Britain*, 1967, **3**, 330.
152. I. Fujii, H. Iijima, S. Tsukita, Y. Ebizuka, and U. Sankawa, *J. Biochem.*, 1987, **101**, 11.
153. K.-x. Huang, I. Fujii, Y. Ebizuka, K. Gomi, and U. Sankawa, *J. Biol. Chem.*, 1995, **270**, 21 495
154. A. Messerschmidt, R. Ladenstein, R. Huber, M. Bolognesi, L. Avigliano, R. Petruzzelli, A. Rossi, and A. Finaaai-Angró, *J. Mol. Biol.*, 1992, **224**, 179.
155. H. Nordlöv and S. Gatenbeck, *Arch. Microbiol.*, 1982, **131**, 208.
156. K.-x. Huang, Y. Yoshida, K. Mikawa, I. Fujii, Y. Ebizuka, and U. Sankawa, *Biol. Pharm. Bull.*, 1996, **19**, 42.

1.17
Biosynthesis of Aflatoxin

CRAIG A. TOWNSEND and ROBERT E. MINTO
Johns Hopkins University, Baltimore, MD, USA

1.17.1 INTRODUCTION

Aflatoxin B_1 ((**1**), AFB_1) is a potent environmental carcinogen produced by *Aspergillus parasiticus*, *Aspergillus flavus*, and *Aspergillus nomius*, common molds that infect nuts and grains and pose a significant threat to the food supply worldwide.[1–3] The origin of its toxic effects on humans has been extensively investigated and is understood to involve oxidative activation by cytochromes P-450, notably in the liver and kidneys,[4,5] to afford the *exo*-epoxide (**2**) (Scheme 1). The identity and absolute configuration of this species have been inferred from the structure of its covalent adduct with DNA[6] and subsequently have been secured by unambiguous chemical synthesis and its independent reaction with DNA.[7,8] The epoxide (**2**) is short-lived in aqueous solution but has a sufficient lifetime and lipophilicity to intercalate readily into double-stranded DNA selectively forming the N-7 adduct of certain guanine residues.[9,10]

The binding and selective reaction of metabolically activated aflatoxin at guanine residues depends intimately on the absolute configuration of the dihydrobisfuran that uniquely characterizes this family of polyketide metabolites.[11] Detailed ^{1}H NMR analyses of this adduct in short oligonucleotides has shown that the mycotoxin stacks to the 5′-side of the reacting guanine. This

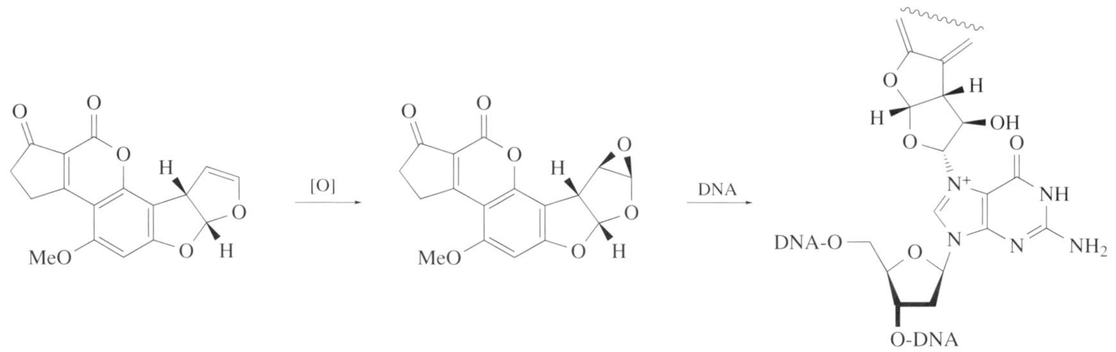

Aflatoxin B_1 (AFB_1) (**1**) (**2**) (**3**)

Scheme 1

orientation presents the *exo*-epoxide (**2**) for reaction with guanine in the major groove to give adduct (**3**) (Scheme 1) in which the cyclopentenone ring hydrogens extend into the minor groove.[6] While the DNA is unwound to accommodate the oxidatively activated aflatoxin, the helix remains right-handed and is slightly perturbed from B-form only in the immediate region of the covalent reaction.[6] A very similar picture has been derived for the corresponding adduct of sterigmatocystin ((**7**), R = H).[12] Thus, reaction of the *exo*-epoxide (**2**) is geometrically matched for covalent reaction with double-stranded DNA to give the *trans*-adduct (**3**). In contrast, the corresponding *endo*-epoxide fails to form an adduct at this or any site in DNA and is essentially nonmutagenic.

A lethal cause and effect relationship between metabolically activated aflatoxin and the incidence of human cancers was linked to a mutational "hot spot" in the p53 gene highly favored for reaction with epoxide (**2**).[13–15] A transversion (G249T) has been observed to occur frequently as a consequence of lesion formation at this site, subsequent depurination, and an error during DNA repair. The p53 protein plans an important role in regulation of the cell cycle. This transversion leads to an amino acid change (Arg→Ser) in the translated protein that becomes defective in its regulatory role and allows unchecked growth of these mutated cells.

Aflatoxin was discovered as the cause of the Turkey X disease, an unidentified disease that resulted in the deaths of tens of thousands of young turkeys on farms in the UK in 1960.[16] These poultry deaths were traced to moldy peanut meal used in their feed. The mold was identified as *Aspergillus flavus*, which produced a number of fluorescent and highly toxic metabolites that were isolated and called collectively aflatoxins (for "*A. flavus* toxin").[17,18] A few reports of structural investigations appeared,[19–21] but the full structures of aflatoxin B_1 (**1**), B_2 (as (**1**) but tetrahydrobisfuran), G_1 (**4**), and G_2 (as (**4**) but tetrahydrobisfuran), the main constituents of *A. flavus*, were provided by Büchi's group[22,23] in a consumate proof of structure relying on classical degradative chemistry and spectroscopic methods. These structures were confirmed shortly thereafter by x-ray crystallography.[24,25] Additional contributions followed from Büchi's laboratory: a total synthesis of racemic (**1**),[26,27] proof of the aflatoxin absolute configuration by stereochemical correlation,[28] and a few years later synthesis of the minor hydroxylation metabolite aflatoxin M_1 (**5**)[29,30] in addition to G_1 (**4**)[29] and an improved synthesis of (**1**).[29] During this period important contributions were made by others in providing the structures of aflatoxin M_1 (**5**) and M_2 (as (**5**) but tetrahydrobisfuran),[31] B_{2a}, G_{2a},[32] and B_3 (**6**),[33] which has also been called parasiticol,[34] and in providing an independent synthesis of (**1**).[35] The isolation and characterization of further metabolites and the progress in the total synthesis of this general class have been thoroughly reviewed by Roberts[36] and Schuda,[37] as have more general accounts of the early interactions of mycologists, toxicologists, and chemists that led to the isolation, structure determination, and biological properties of the aflatoxins.[38–40]

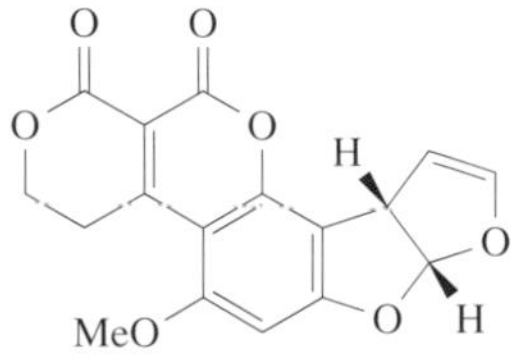

(**4**) 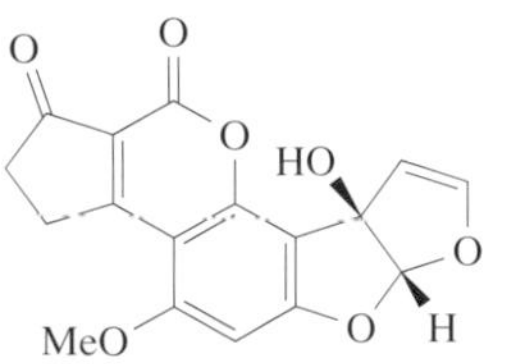 (**5**) 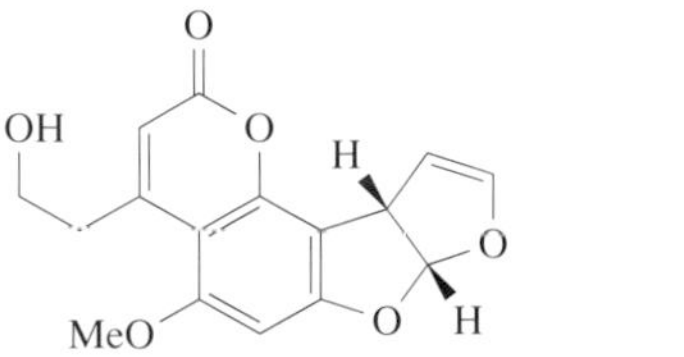 (**6**)

Sterigmatocystin ((**7**), ST) was first isolated in 1954 from *Aspergillus versicolor*, a fungus related to *A. flavus*.[41] The correct structure of (**7**) was proposed in 1962.[42] When the structures of the aflatoxins were published in the following year, the unusual dihydro-tetrahydrobisfuran ring systems common to both skeletal types were immediately recognized and, as will be discussed shortly, influenced biogenetic speculations. The possible precursor relationship of (**7**) to (**1**), or a shared biosynthetic pathway, was further suggested by the isolation of *O*-methylsterigmatocystin ((**12**), OMST) from *A. flavus*.[43]

The earliest biosynthetic experiments were performed by Adye and Mateles[44] using *A. flavus*. Incorporations of radioactivity into (**1**) were reported from [^{14}C]methionine, [^{14}C]acetate, [3-^{14}C]phenylalanine, and [2-^{14}C]tyrosine, but not [2-^{14}C]mevalonate. Utilization of the shikimate-derived aromatic amino acids was later retracted[45] owing to their degradation to acetate and secondary incorporation of radiolabel by way of this simpler precursor.[46] Publication of this finding coincided with the first report from Büchi's laboratory[47,48] of extensive degradative studies to map precisely the location of isotopic labels from [^{14}C]methionine and [1-^{14}C]- and [2-^{14}C]acetate in (**1**). This landmark effort stands as an experimental *tour de force* and remains a high water mark in the application of classical biosynthetic methods. The results of this study are summarized in Scheme 2, in which solid circles and squares indicate individual carbons isolated by degradation or logically implied by difference, and the open circles and squares represent presumed locations of label based on the expected alternating pattern of carboxyl and methyl labels from [1-^{14}C]- and [2-^{14}C]acetate. In sum, these findings demonstrated that (i) AFB_1 was derived entirely from acetate units, except for the *O*-methyl, which was supplied by methionine, (ii) the incorporation of acetate-derived label was virtually equal at all centers, suggesting derivation of the metabolite from a single polyketide chain, and (iii) deep-seated molecular rearrangements must occur in the biosynthesis of aflatoxin to account for the non-alternating pattern of carboxyl and methyl labels and the evident branching from a linear precursor. As is often the case, the experimentally derived distribution of label did not agree with that required by prior biogenetic speculations,[46,49,50] although the scheme of Thomas[51] agreed in part.

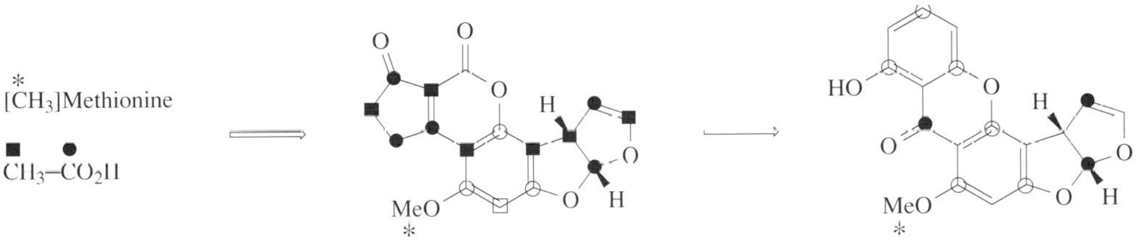

[$\overset{*}{C}H_3$]Methionine

CH_3—CO_2H

Aflatoxin B_1 (AFB_1) (**1**)

Sterigmatocystin (ST) (**7**)

Scheme 2

Soon after the first report of Büchi's biosynthetic experiments in 1968, Holker and Mulheirn[52] published the results of parallel degradation experiments to determine the position of radiolabel in sterigmatocystin ((**7**), ST) derived from cultures of *A. versicolor* administered [1-^{14}C]acetate. Their results are shown schematically in Scheme 2 and, although only three positions were discretely determined (indicated by solid circles), other fragments in the degradation scheme gave specific activities consistent with the labeling pattern shown (open circles). It can be seen that this distribution of label mirrors that obtained for aflatoxin although, curiously, the specific activities of the fragments derived from the bisfuran portion were about 10% lower than those from the rest of the molecule.[52] Given the accuracy of radiochemical methods available at that time, this difference could be dismissed as experimental error, but Holker and Mulheirn[52] suggested that (**7**) might be derived from two polyketide chains. While the interpretation of this experimental result will be seen to be partially correct, the observation itself would have a fundamentally important effect on the course of later biosynthetic investigations.

Viewing the data available in 1968–70, Büchi and co-workers[47,48] advanced a new biogenetic proposal predicated upon the derivation of (**7**) and (**1**) from a single polyketide chain and which accounted for the acetate labeling pattern. Briefly put, it is postulated that a tetracyclic precursor (**8**) (R = OH or H) derived from a C_{18}-polyketide is converted into the *endo*-peroxide (**9**), which opens to (**10**) and rearranges to dihydrobisfuran (**11**) (Scheme 3). Invoking Thomas's oxidative cleavages,[51] as exemplified in the biosynthesis of the ergochromes,[53] (7) can be generated with the correct labeling pattern. Subsequent oxidative cleavage, reduction, cyclization, dehydration, and decarboxylation would be required to transform (7) to (1). No experimental data existed to support such a contention, although the correct acetate labeling pattern would be obtained. The absolute configuration of the dihydrobisfuran in ST was shown to be identical[52] with that established for AFB_1.[28]

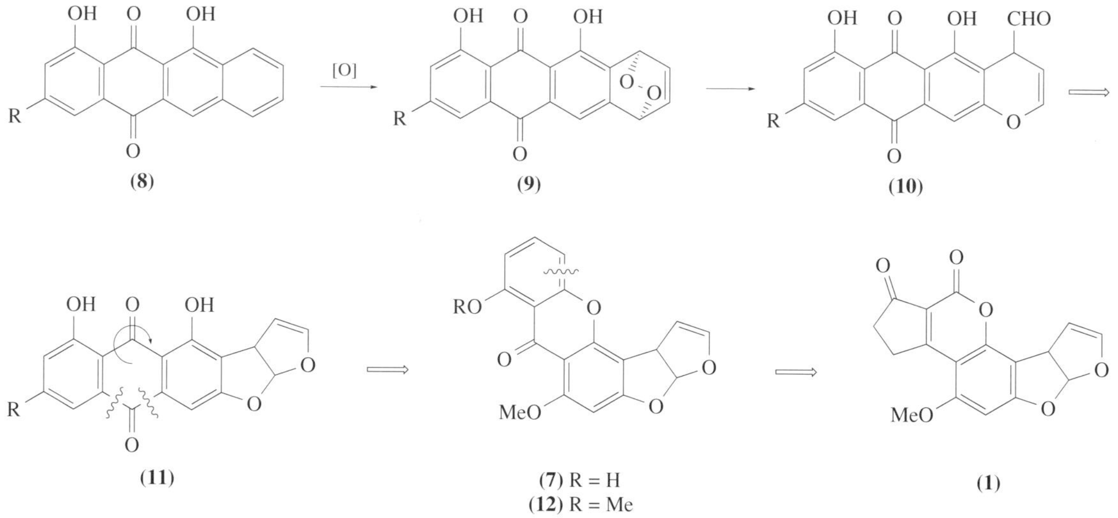

(8) (9) (10)

(11) (7) R = H (12) R = Me (1)

Scheme 3

Since 1970, further development of the aflatoxin biosynthetic problem has progressed through four stages: (i) the generation and use of blocked mutants, (ii) the demonstration of a common polyketide folding pattern among the presumed biosynthetic intermediates and in aflatoxin B_1 itself, (iii) the synthesis of specifically labeled precursors and potential precursors and the demonstration of intact incorporation by some of these into aflatoxin, and (iv) the purification and characterization of individual biosynthetic enzymes and cloning of the genes likely responsible for encoding the biosynthetic pathway. These developments are discussed in the following sections of this chapter.

1.17.2 BLOCKED MUTANTS

The early formation of an aromatic tetracyclic intermediate in tetracycline biosynthesis was a prominent finding that emerged from the detailed mutational analysis of this pathway by McCormick and co-workers.[54,55] While this contemporaneous observation may have influenced Büchi's thinking in the late 1960s with respect to aflatoxin formation, it was a similar turn to classical microbiological methods and the random generation of *A. parasiticus* mutants blocked in aflatoxin production that pointed the way to further progress in understanding the biosynthesis of this mycotoxin.

The B- and G-series aflatoxins are substituted coumarins and, as such, are highly fluorescent. The designations B and G refer, respectively, to their characteristic blue and green fluorescence under UV radiation, readily detectable at picomolar concentrations. The wild-type *A. parasiticus* grows as white colonies initially on agar plates before developing green conidia. Secreted around the colonies are zones of blue-fluorescent AFB_1 (1). Whereas *A. flavus* produces aflatoxins B_1, B_2, G_1, and G_2, *A. parasiticus* affords mainly AFB_1 and some B_2. The latter was chosen by Lee *et al.*[56] and others for extensive mutational experiments and yielded a family of mutants. Spores were collected and mutagenized by UV radiation or *N*-nitrosoguanidine (NTG). On propagation of the treated spores, those survivors showing reduced or undetectable aflatoxin production occasionally accumulated yellow to orange–red pigments readily visible from the obverse side of the agar plates. These pigments were isolated and their structures determined to be 1,3,6,8-tetrahydroxyanthraquinones. Foreshadowing what was to come, Heathcote and Dutton[33] had noted briefly in 1969 the isolation of minor amounts of 1,3,6,8-tetrahydroxyanthraquinone metabolites in *A. flavus*. In rough chronological order the following events took place from 1971 to 1975: mutant NOR-1 (present designation) accumulate was found to norsolorinic acid ((**13**), NA),[56] averufin ((**15**), AVR) was determined from AVR-1 by Donkersloot *et al.*,[57] and versicolorin A ((**17**), VA) was observed from VER-1.[58] Averantin ((**14**), AVN) was accumulated by AVN-1, which was generated by a further mutation of VER-1 (Scheme 4).[59]

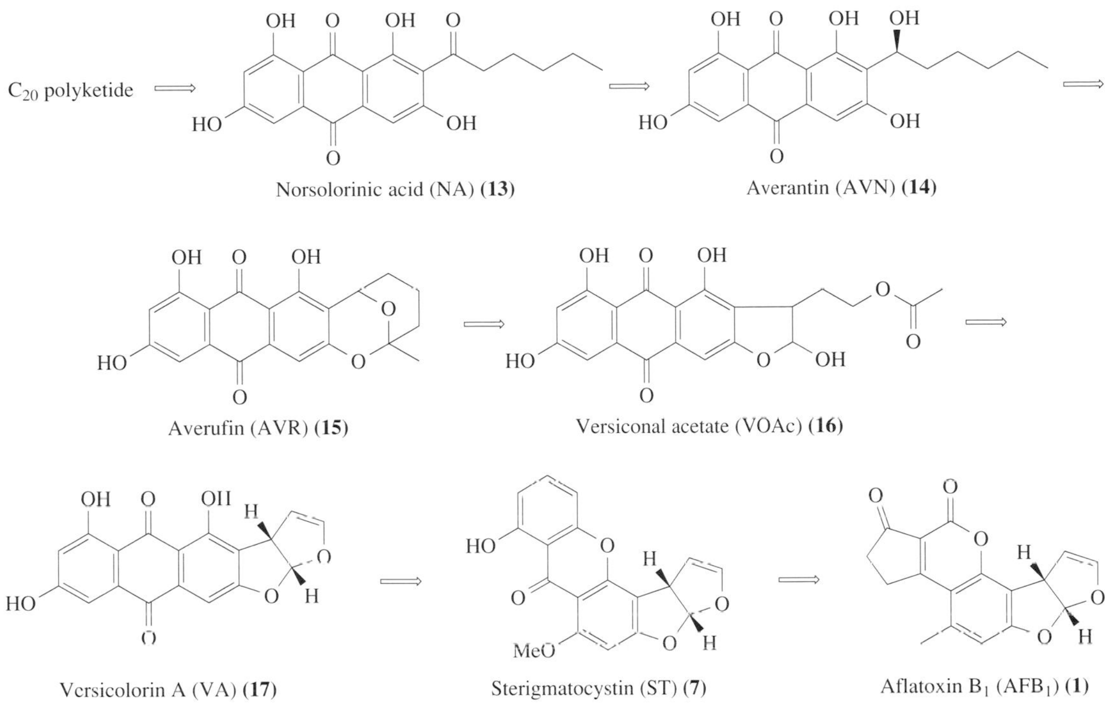

C_{20} polyketide ⟹ Norsolorinic acid (NA) (**13**) ⟹ Averantin (AVN) (**14**) ⟹

Averufin (AVR) (**15**) ⟹ Versiconal acetate (VOAc) (**16**) ⟹

Versicolorin A (VA) (**17**) ⟹ Sterigmatocystin (ST) (**7**) ⟹ Aflatoxin B_1 (AFB_1) (**1**)

Scheme 4

By chance during this period, it was discovered that the insecticide Dichlorvos (dimethyl 2,2-dichlorovinylphosphate) at ppm concentrations inhibited aflatoxin production in *A. flavus* and led to the accumulation of a new orange pigment.[60] The structure of this pigment was eventually correctly formulated as versiconal acetate ((**16**), VOAc).[61]

With this selective chemical inhibitor in hand, four blocked mutants of *A. parasiticus* available and sterigmatocystin (**7**) accessible from *A. versicolor*, biosynthetic experiments could be undertaken. The logic of these was to incorporate [^{14}C]-acetate into the accumulated, presumed intermediate and then (i) demonstrate its conversion into radiolabeled aflatoxin B_1 by wild-type *A. parasiticus* and (ii) establish the biosynthetic order of these putative intermediates by showing that an early precursor would proceed to the blockage point, whereas an intermediate that occurred past a blockage point would be converted by the mutant into labeled aflatoxin B_1. Extensive and interlocking experiments of this kind were conducted, mainly by Bennett and co-workers[59,62] and Hsieh and co-workers[63–65] to support the order of biosynthetic steps shown in Scheme 4.

These incorporation experiments did not establish intact incorporation of radiolabel by chemical degradation and the possibility of randomization of radioisotope could not be strictly eliminated. The point is rendered moot, however, by the fact that the potential intermediates were labeled from [^{14}C]acetate and their degradation to acetate and secondary reincorporation by an alternative route would not necessarily be excluded by systematic degradation as described by Büchi and co-workers.[47,48] Nonetheless, the incorporation efficiencies for these putative intermediates were significantly greater than that of acetate itself and trended higher as the end of the emerging pathway was approached. Overall, these findings lent strong support to the view that a C_{20}-polyketide precursor was responsible for the initiation of aflatoxin biosynthesis, rather than a C_{18} precursor and a tetracyclic intermediate (e.g., (**8**)) as hypothesized earlier.[48] The first half of the pathway involves 1,3,6,8-tetrahydroxyanthraquinone metabolites up to completion of the unique dihydrobisfuran present in versicolorin A ((**17**), VA; Scheme 4). In keeping with the pivotal role of this intermediate, the absolute configuration of its dihydrobisfuran ring system is probably identical[66] with that of (**7**)[52] and (**1**).[28] Similarly, norsolorinic acid (**13**), averufin (**15**), sterigmatocystin (**7**), and

(**17**) have all been shown to be produced in trace quantities in the wild-type *A. parasiticus*, and pulse-labeling experiments[67] with [^{14}C]acetate in the main support the sequence of biosynthetic transformations depicted in Scheme 4. Overall, therefore, the C_6-side chain of (**13**) becomes the dihydrobisfuran of (**17**) with loss of two carbons and is retained through the remainder of the pathway to (**1**). The coumarin nucleus of the latter is derived from the xanthone (**7**), which is in turn derived from the anthraquinone (**17**). The experiments from this phase of the aflatoxin investigation have been thoroughly reviewed.[40,68,69]

1.17.3 COMMON POLYKETIDE FOLDING PATTERN

The advent of practical ^{13}C NMR spectroscopy in the early 1970s revolutionized the study of natural product biosynthesis. Among the first applications of this method was that by Tanabe *et al.*,[70] who analyzed the labeling pattern from [1-^{13}C]- and [2-^{13}C]acetate in sterigmatocystin (**7**). The alternating sequence of enrichments in (**7**) confirmed the polyketide origin of the metabolite, and the chemical shift assignments were in accord with Holker and Mulheirn's earlier observations of the distribution of the corresponding radiolabel.[52] Later the seminal advance of paired ^{13}C labels pioneered by Tanabe[71] (and independently by McInnes and Wright[72]) was illustrated by the incorporation of [1,2-$^{13}C_2$]acetate again into (**7**).[73] The use of paired isotopes, e.g., ^{13}C–^{13}C, ^{2}H–^{13}C, ^{13}C–^{15}N, or ^{13}C–^{18}O, increased the inherent sensitivity of the method for tracer studies and fully evoked the analytical power of NMR spectroscopy in this field. These developments have been reviewed by Vederas.[74]

Unfortunately, at an early stage of these experiments a single transposition of two spectral assignments led to the suggestion by Seto *et al.*[73] that polyketide folding pattern A (Scheme 5) was active in the formation of (**7**). However, a very similar study by Steyn's group in South Africa[75] corrected these assignments to establish that pattern B was active. This single discrepancy underscores the chief hazard in early applications of ^{13}C NMR in biosynthetic studies, namely the difficulty of making assignments among carbons whose chemical shift differences are small. The later application of two-dimensional and long-range coupling methods has largely resolved this problem. Nonetheless, there were further disagreements at this time in the assignments of other metabolites thought to be precursors of aflatoxin that will not be recounted here.

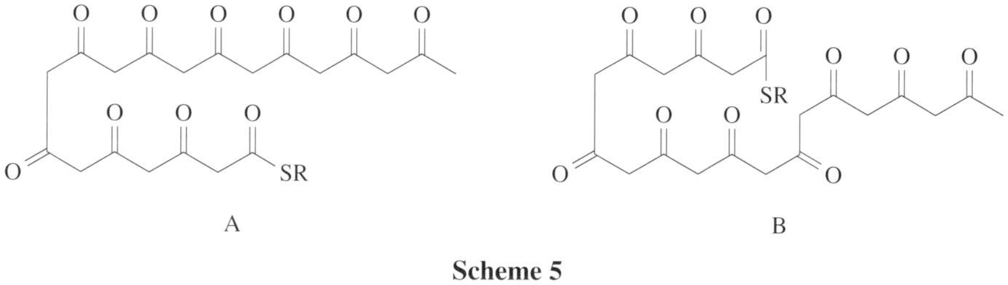

A B

Scheme 5

Through the work largely of the South African group, a consistent pattern emerged that doubly ^{13}C-labeled acetate gave a common folding pattern in norsolorinic acid (**18**),[76] averufin (**19**),[77] versiconal acetate (**20**),[78] versicolorin A (**21**),[66] sterigmatocystin (**22**), and AFB$_1$ (**23**),[79] as summarized in Scheme 6. As had been done radiochemically, each of these compounds was incorporated into (**23**) to give a labeling pattern identical with that from [1,2-$^{13}C_2$]acetate itself. However, as was seen in the corresponding radiochemical experiments, the levels of ^{13}C enrichment in the mycotoxin were higher than would have been seen from [1,2-$^{13}C_2$]acetate alone, suggesting that each had incorporated without prior degradation to acetate. Finally, the observation of three intact acetate units in the central aryl ring of (**23**) was inconsistent with the course of the oxidative rearrangement of a tetracyclic precursor advanced by Büchi and co-workers.[48] It was, however, entirely consistent with derivation from the terminal ring of an anthraquinone progenitor as shown in Scheme 6. The application of doubly ^{13}C-labeled acetate in aflatoxin biosynthesis has been thoroughly reviewed.[80]

Oxygen metabolism in natural product biosynthesis can be monitored in exquisite detail by the paired-isotope NMR method. The heavy isotope of oxygen ^{18}O has no nuclear spin ($I=0$), but when covalently bound to a ^{13}C it results in a slight, but measurable, upfield shift in its ^{13}C NMR resonance frequency whose magnitude reflects the C—O bond order. This fact has been exploited to great effect in studies of polyketide biosynthesis.[74] Among the first applications was that by Vederas and Nakashima,[81] who examined the fate of [1-^{13}C,$^{18}O_2$]acetate on incorporation into averufin (**24**). All

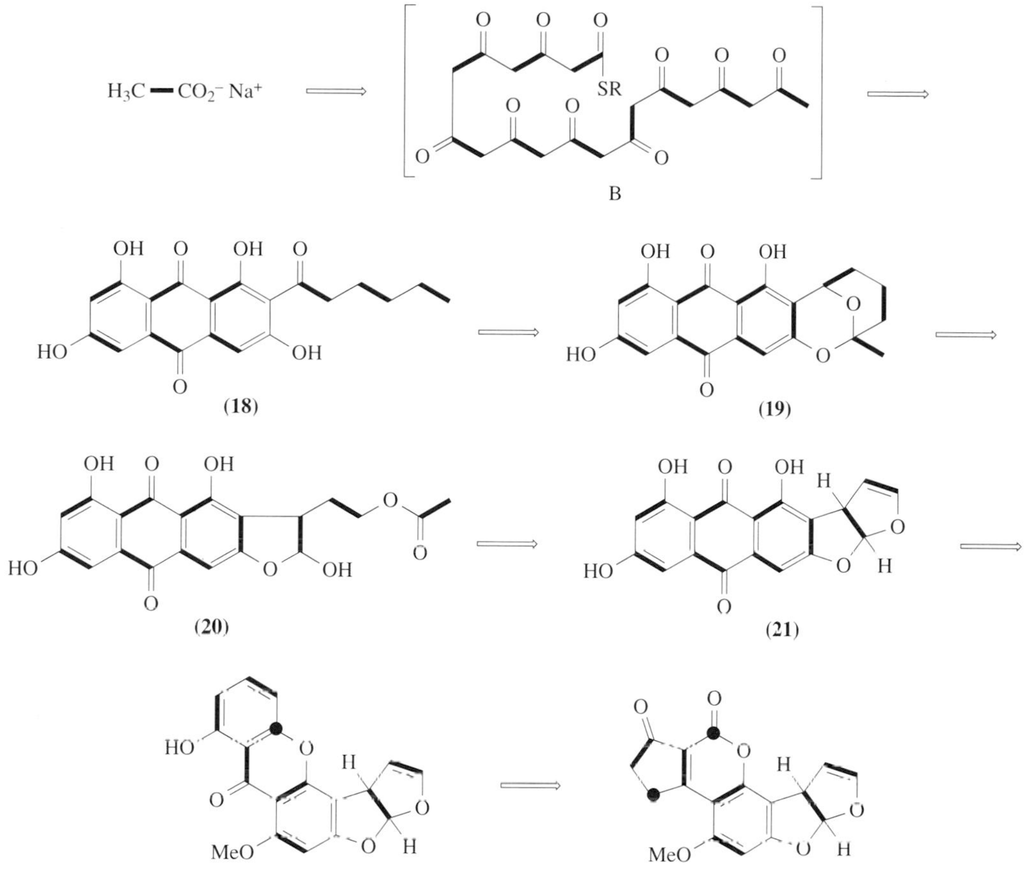

Scheme 6

of the oxygens, except the anthraquinone oxygen at C-10, were demonstrated to be derived from acetate, consistent with the generalized view of polyketide biosynthesis. Incorporation from $^{18}O_2$ at C-10 was as depicted in Scheme 7. Similarly, Sankawa *et al.*[82] have reported the utilization of [1-^{13}C,$^{18}O_2$]acetate by *A. versicolor* in the synthesis of sterigmatocystin (**25**). Polyketide-derived oxygen was observed, as expected, at four locations as illustrated in Scheme 7, but was notably absent in the C-ring methoxyl and in the distal dihydrobisfuran ring. These oxygens were presumably introduced during oxidative rearrangement processes active in the late stages of the biosynthesis, but this point was not directly established experimentally (see below).[82]

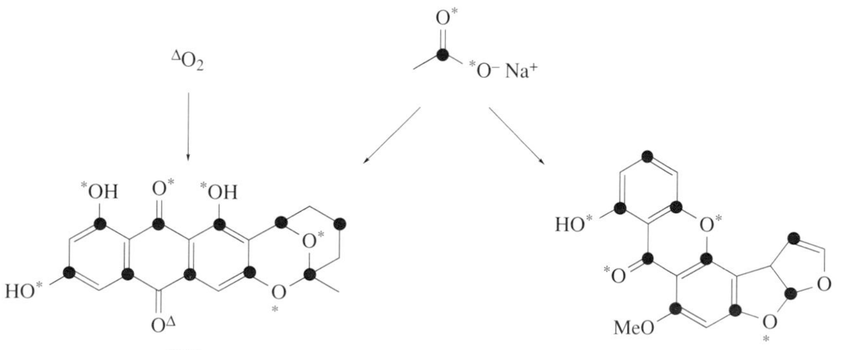

Scheme 7

In summary, therefore, incorporation experiments with [1,2-$^{13}C_2$]acetate into (**18**), (**19**), (**20**), (**21**), (**22**), and (**23**) (Scheme 6) gave a mutually consistent folding pattern from a common polyketide precursor—a precondition of a shared biosynthetic path. In accord with this view, oxygens at carboxyl-derived carbons were shown to have originated similarly from acetate in (**24**) and (**25**) (Scheme 7) with additional oxygen atoms presumably arising from oxidative rearrangement and cleavage processes ($^{\Delta}O_2$). The nature of these more complex processes were examined in specific labeling studies that are considered in the next section.

1.17.4 SYNTHESIS AND TESTING OF POTENTIAL INTERMEDIATES

Further progress in understanding the unusually long and mechanistically diverse pathway to aflatoxin has required molecular probes more discriminating than isotopically labeled acetate. This goal has been achieved in significant measure by the synthesis and testing of specifically labeled intermediates designed to reveal their detailed roles in the biosynthesis. The discussion below is divided into subsections each devoted to the progress made in the individual transformations of the pathway using whole-cell and, more recently, cell-free systems.

1.17.4.1 Norsolorinic Acid

Two observations led to a pivotal experiment to examine the first step of aflatoxin biosynthesis. First the course of dihydrobisfuran formation was seen to begin with norsolorinic acid (**13**), which is reduced to averantin ((**14**), Scheme 4). However, oxidation is required at C-5′, an acetate carboxyl-derived carbon, to a ketone in order to form averufin (**15**). That is, it was striking that the carbonyl from the presumed acetate starter unit would be reduced to a methylene in (**13**) only to be reintroduced a few steps later in the biosynthesis. It is far more often the case, as experiments with [1-^{13}C,$^{18}O_2$]acetate have shown, that oxygens bound to carboxyl-derived carbons typically stem from the progenitor polyketide. For example, even for the highly reduced fungal metabolite brefeldin A (**26**), the lactone oxygen (*O) is derived from the acetate starter unit.[83,84]

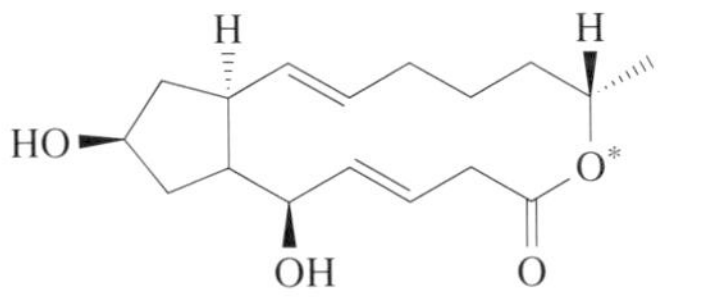

H H HO O* OH O

(**26**)

A possible interpretation of these observations was that hexanoylCoA served as the primer for the putative polyketide synthase (PKS), rather than acetylCoA. The possibility that higher acids than acetate could function as starters had been suggested in 1957 by Birch.[85] Fragmentary circumstantial evidence can be cited in support of this proposal, where in careful radiochemical analyses small but experimentally significant differences had been observed in the specific radioactivities for saturated side-chain carbons (usually lower) vs. nuclear carbons for particular polyketide metabolites.[86–88] Second, the report of Holker and Mulheirn[52] noted in the Introduction is a case in point where the dihydrobisfuran carbons of (**7**) (Scheme 2) were labeled from [^{14}C]acetate at approximately 10% lower specific activity than aryl carbons. While this observation could not be satisfactorily explained at the time, it can be rationalized by the intervention of a hexanoyl starter unit.

Unfortunately, the history of attempted incorporations of long-chain fatty acids, again fragmentary, is far from encouraging, particularly in fungi.[86,89–91] Two successful exceptions can be recalled in plants—butanoate incorporation into margaspidin[92] and octanoate into coniine.[93] These largely negative precedents notwithstanding, [1-^{13}C]hexanoic acid was found to give a surprising 3–4% specific incorporation into averufin ((**27**), Scheme 8) in the appropriate blocked mutant[84] superimposed on the expected background of secondary incorporation from [1-^{13}C]acetate resulting from β-oxidation of the administered primer. The intact incorporation of a C_6-fatty acid was further supported by stereochemical studies in which the absolute configuration of deuterium incorporation from [2-$^2H_3$,^{13}C]acetate in averufin was identical with that of the normal fatty acids of the organism itself,[94] in contrast to polyketide natural products from other fungi.[95,96] Corollary experiments with equimolar amounts of 1-^{13}C-labeled acetate, butanoate, 5-oxohexanoate, and 3-oxooctanoate all failed to give detectable specific incorporation of stable isotope into (**27**). Only secondary incorporation as [1-^{13}C]acetate was observable at a remarkably consistent ca. 0.5% per site irrespective of the labeled substrate tested (Scheme 8).

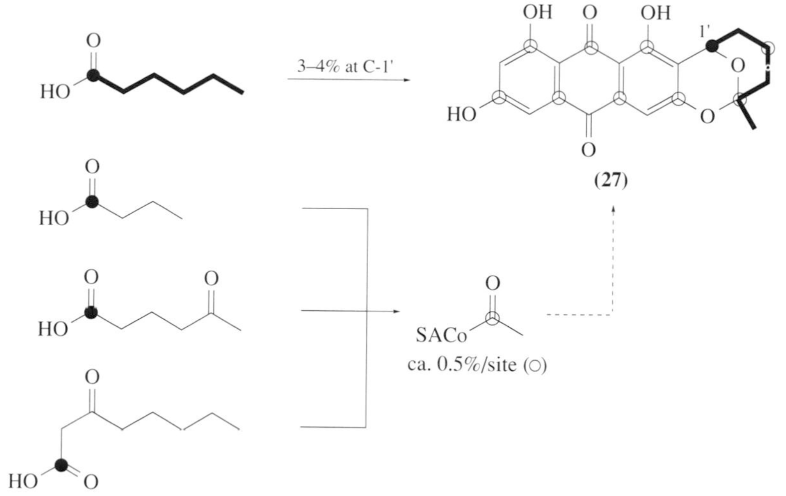

Scheme 8

The signal discovery of hexanoate incorporation into averufin (and, hence, aflatoxin) was soon followed by the incorporation of acetoacetate into nonactin[97] and octanoate into the chain-terminating unit of fungichromin.[98] Successes with microbes, however, were limited. Attempts to demonstrate the intact incorporation of free acids derived from the linking together of two or three ketide units into macrolides, polyethers, and other secondary metabolites were completely fruitless.[99,100] However, the earlier crucial finding of Lynen[101] that *N*-acetylcysteamine (NAC) thioesters could substitute for the corresponding CoA esters to load biosynthetic intermediates on to fatty acid synthase (FAS) led to the testing of potential polyketide intermediates as their NAC thioesters.[99,100] The success of this method in studying polyketide biosynthesis was a breakthrough that has been applied in a number of systems and has secured the central idea of processivity in polyketide chain elongation.[102,103]

The lack of incorporation of butanoate and 3-oxooctanoate (in principle derived in the first C_2-homologation of hexanoate by the putative PKS) was further investigated by synthesis of the corresponding NAC esters. The NAC thioesters of [1-^{13}C]acetate, -butanoate, -hexanoate, and -3-oxooctanoate were administered in equimolar amounts to cell suspensions of *A. parasiticus* blocked at averufin production,[104] as was done before with the free acids.[84,105] The carboxyl-labeled NAC acetate and butanoate each only showed randomized incorporations, again at ca. 0.5% per site (Scheme 9). The apparent absence of intact incorporation from butyrate was re-examined using a more discriminating probe, [2,3-$^{13}C_2$]butanoate. No ^{13}C–^{13}C coupling was observed in the product, clearly indicating no detectable intact incorporation. [1-^{13}C]HexanoylNAC, however, gave an impressive 22% specific incorporation of heavy isotope at C-1′, and intriguingly [1-^{13}C]-3-oxo-octanoylNAC gave a 4–5% incorporation specifically at C-3. The specificity of hexanoylNAC uptake by whole cells of the norsolorinic acid-accumulating mutant of *A. parasiticus* has been probed by Simpson and co-workers.[106] The NAC thioesters of pentanoic acid and 6-fluorohexanoic acid were successfully incorporated to lead to modest yields of modified norsolorinic acids, whereas butanoyl-, heptanoyl-, and octanoylNAC were not utilized under these experimental conditions. The biosynthetic system, therefore, appears fairly selective with respect to primer length.

A considerable weight of evidence had been amassed at this point to favor a hexanoyl unit as the fundamental primer of aflatoxin biosynthesis. This unit could be visualized to be derived either by degradation (β-oxidation) of a longer conventional fatty acid, or by synthesis by a specialized FAS or multifunctional PKS/FAS protein. Were the former true, the thioester oxygen would derive from water, whereas the latter course should result in oxygen arising from acetate/malonate. It was already known from the paired isotope experiments using [1-^{13}C,$^{18}O_2$]acetate that the oxygen at C-1′ in averufin ((**24**), Scheme 7) was derived in the latter sense. To establish this fact unambiguously, [1-^{13}C,$^{18}O_2$]hexanoic acid was prepared and incorporated into averufin ((**28**) Scheme 10) with substantial retention of both isotopes at C-1′.[104]

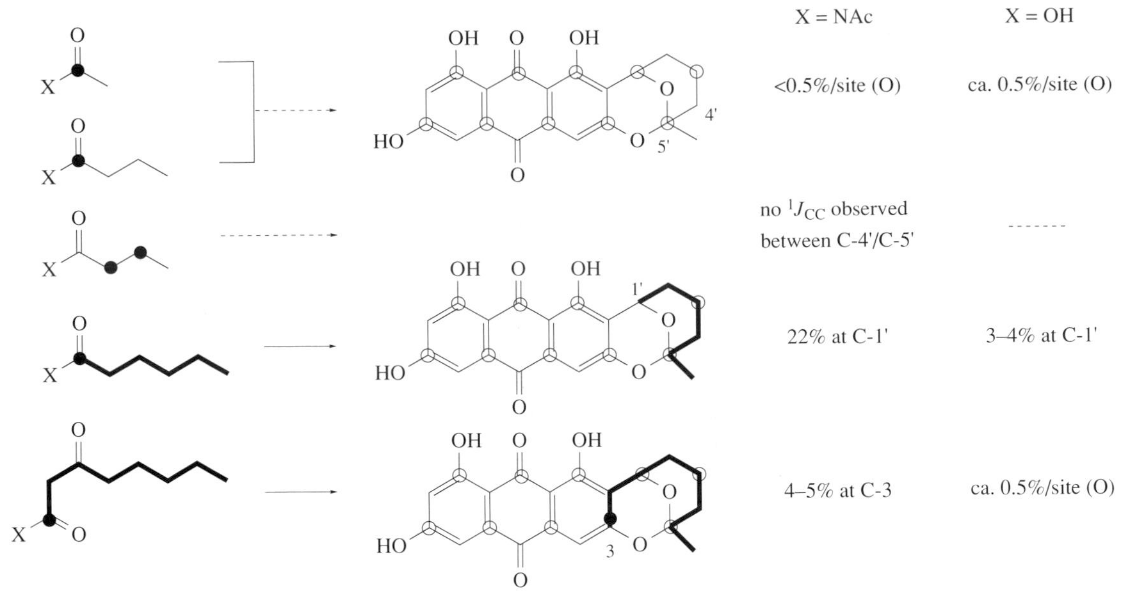

X = NAc	X = OH
<0.5%/site (O)	ca. 0.5%/site (O)
no $^{1}J_{CC}$ observed between C-4'/C-5'	-------
22% at C-1'	3–4% at C-1'
4–5% at C-3	ca. 0.5%/site (O)

Scheme 9

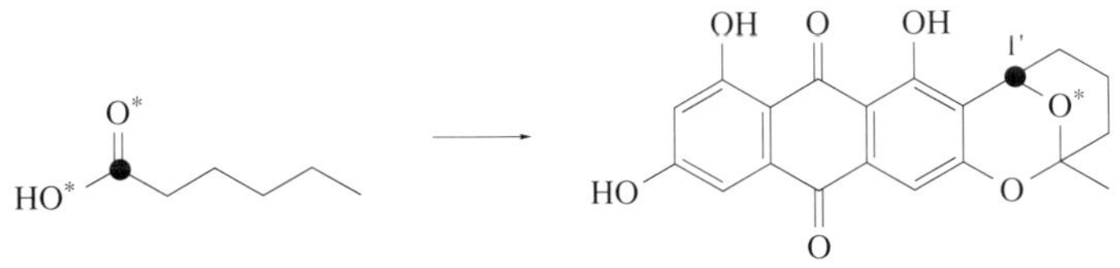

(28)

Scheme 10

Based on these findings, the inescapable conclusion was that either a specialized FAS was present to prepare the unusual hexanoyl primer as its CoA ester or hexanoylACP (Scheme 11, X = –SCoA or –SACP), or this unit is channeled in an enzyme complex with the PKS to lead on to norsolorinic acid (**13**). Alternatively, but indistinguishable based on the labeling experiments above, the FAS and PKS could be visualized as one large polyprotein such that acetylCoA/malonylCoA are reductively homologated to the extent of six carbons followed by further chain elongation nonreductively to a C_{20} precursor. The saturated C_6 intermediate so generated would not be significantly released, e.g., as its CoA ester, but fairly readily exchange with exogenous hexanoylCoA could occur to load the C_6 site for further PKS elaboration with malonylCoA.[104] While the intimate details of these condensation steps and cyclization reactions are not known, they presumably proceed to the anthrone (**29**), which is in turn oxidized (enzyme catalyzed?) to the anthraquinone (**13**).[81] These alternatives will be examined further at the genetic level in the next section.

1.17.4.2 Averantin

The first step in the transformation of the six-carbon benzylic ketone side chain of (**13**) to the bisfuran of the versicolorins and AFB_1 is catalyzed by a dehydrogenase reported from cell-free studies to utilize NADPH or NADH.[107–109] Subsequent further purification revealed a monomeric NADPH-dependent protein.[110,111] with a molecular mass of approximately 40 kDa.[110] However, later investigations of the genetics of aflatoxin biosynthesis have revealed that more than one protein is probably involved in this step. The absolute configuration of the resulting alcohol in the reduction step was shown to be 1′*S* by classical stereochemical correlation. The pentaacetate of natural (–)-averantin (**30**) was oxidized and saponified to give (*S*)-(+)-2-hydroxyheptanoic acid (Scheme 11).[112]

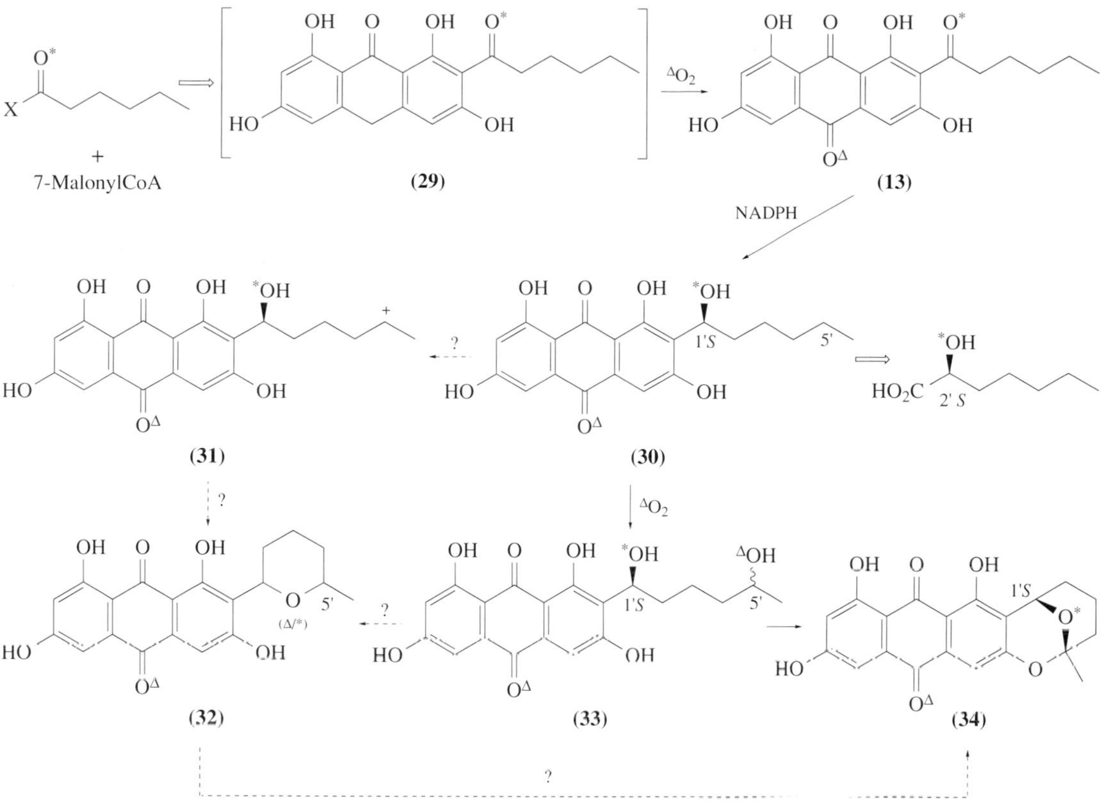

Scheme 11

1.17.4.3 Averufin and the Role of Averufanin

The further reactions of averantin (**30**) to averufin ((**34**), Scheme 11) are surrounded by controversy. Averufanin (**32**) has been isolated from a number of *Aspergillus* species including *A. flavus* and *A. versicolor*.[33,113,114] McCormick *et al.*[115] have reported that [^{14}C]averufanin obtained from the uptake of [1-^{14}C]acetate incorporated radioactivity into AFB_1 in wild-type *A. parasiticus* and into versicolorin A ((**17**), Scheme 4) and (**34**), but not into (**30**) in the appropriate blocked mutants. These experiments suggest, therefore, that (**32**) lies on the biosynthetic pathway between (**30**) and (**34**).

In opposition, however, Yabe and co-workers[109,111] have described a mutant of *A. parasiticus* obtained by UV irradiation, NIAH-26, that produces neither aflatoxin nor known precursors of the mycotoxin. Yet when this mutant was fermented in the presence of (**13**), (**30**), 5′-hydroxyaverantin (**33**) or (**34**), aflatoxins were generated. Moreover, a cytosolic fraction derived from this mutant converted (**30**), interestingly, into a diastereotopic mixture of (1′*S*,5′*S*)- and (1′*S*,5′*R*)-diols (**33**). By comparison, the 1′*R*-enantiomer of (**30**) was inactive under these conditions. Importantly, when (**32**) or averythrin (dehydration product of (**30**) to the *trans* C-1′/C-2′-alkene) were incubated with NIAH-26, no aflatoxins could be detected. Therefore, these findings clearly point to averufanin as a shunt product, probably derived by cyclization of (**31**) or (**33**), while further oxidation to the 5′-ketone would spontaneously and rapidly cyclize to the internal ketal (**34**).

It is difficult to find an argument to accommodate these opposing points of view about the role of averufanin. However, consideration of the absolute configurations of (**30**)[112] and (**34**)[116] reveals that both are 1′*S* as shown in Scheme 11. Moreover, as discussed above, the C-1′ oxygen is derived from the polyketide precursor. That is, the absolute configuration at this center must be set in the reduction of (**13**) to (**30**). In considering the likely course of ether formation in (**32**), loss of the benzylic oxygen (*O) from (**33**) is likely on chemical grounds and would lead to the 1′/5′-oxygen derived presumably from C-5′, i.e., expected from molecular oxygen ($^{\Delta}$O) owing to oxidation at this site. If this were so, (**32**) could not give rise to (**34**) correctly labeled at C-1′. For the polyketide-derived oxygen (*O) to appear in (**32**) as a necessary precondition for its precursor role to (**34**), either the cyclization of (**33**) must occur in the opposite sense with loss of the secondary C-5′ alcohol, or a hypothetical intermediate in the 5′-oxidation of (**30**) itself (e.g., cation (**31**)) is directly cyclized to give (**32**) containing polyketide oxygen at the C-1′/5′ bond (*O). Unfortunately, the source of this oxygen is not known, but the *a priori* greater likelihood of the first of these mechanistic arguments favors the direct intermediacy of the diol (**33**) in averufin biosynthesis. On the other hand, the involvement of averufanin (**32**) could imply a single hydroxylase to first convert averantin (**30**) into the hypothetical intermediate (**31**) or into its 5′-hydroxy derivative (**33**). After closure to the cyclic ether (**32**), the same enzyme could be visualized to oxidize this intermediate to the 5′-hemiketal enroute to averufin (**34**). This route, however, must also accommodate the stereochemical and oxygen labeling constraints noted above.

1.17.4.4 1′-Hydroxyversicolorone

Of the six carbons present in the linear side chain of norsolorinic acid (**13**) or averufin (**15**), four remain in the dihydrobisfuran of versicolorin A (**17**) and AFB_1 ((**1**), Scheme 4). The common polyketide folding pattern summarized in Scheme 6 reveals that three intact acetate units in the former ((**18**) and (**19**)) become two intact units in the bisfuran of the latter ((**21**) and (**23**)). To establish unambiguously the intact incorporation of averufin into aflatoxin (probable, but not strictly proven by earlier experiments) and to identify the paired acetate labels lost in bisfuran formation, three specific labeling experiments were carried out that are represented in combined form in Scheme 12.[117–119] The incorporation of these labeled specimens of averufin ((**35**), Scheme 12) were specific in AFB_1 (**36**) as expected, but additionally, far more detailed insights were afforded as well. The outer intact acetate in averufin ((**19**), Scheme 6), corresponding to C-5′/6′ in (**35**), is lost. In the rearrangement of the linear hydrocarbon side chain of (**35**) to the branched chain of the bisfuran, the migrating C-1′ center changes from the alcohol to the aldehyde oxidation state without loss of the directly bound deuterium label. Finally, by comparing the doubly labeled acetate pattern in averufin ((**19**), Scheme 6) and in AFB_1 (**23**) with that from A-ring labeled (**35**) and (**36**), a complete one-to-one mapping of carbons and identification of those lost was possible.[118]

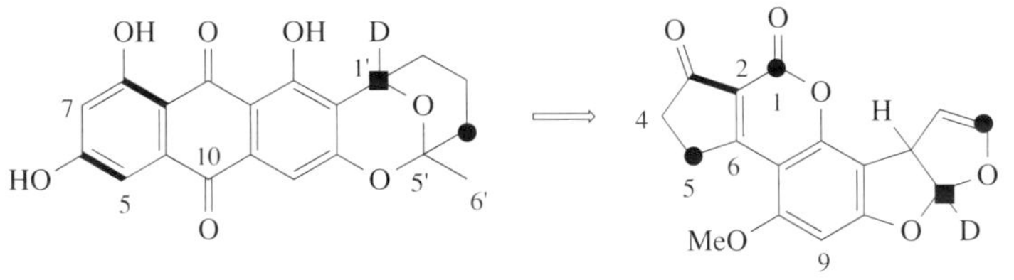

(**35**) (**36**)

(35)		(36)
C-5	→	Lost
C-6	→	C-5
C-7	→	C-4
C-8	→	C-3
C-11	→	C-2
C-9	→	C-6

Scheme 12

The appearance of three intact acetate precursor units in versiconal acetate ((**20**), Scheme 6) suggested that the terminal pair, now recognized as those lost in bisfuran formation, was derived in an internal rearrangement process, such as a Baeyer–Villiger-like oxidation. The issue of intramolecularity in this process was established with a multiply deuteriated sample of averufin (**37**), which was converted in the presence of Dichlorvos into versiconal acetate (**38**) containing heavy isotopes in the proper relative stoichiometry at each labeled site (Scheme 13).[120]

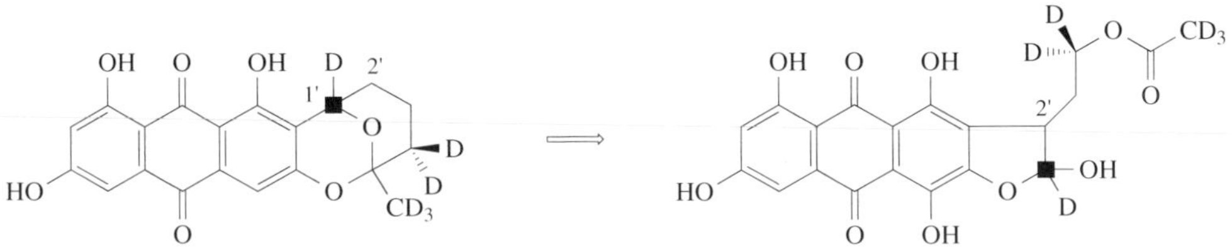

(**37**) (**38**)

Scheme 13

While the acetate in versiconal acetate could have come trivially in a transacylation from acetyl-CoA, a process that would not have been distinguished as such in a double ^{13}C-labeling experiment, the intramolecular nature of the path established by the experiment in Scheme 13 implied the intermediacy of an unknown methyl ketone substrate for Baeyer–Villiger oxidation. This compound, 1′-hydroxyversicolorone (**39**), was isolated from a new mutant of *A. parasiticus*[121] and, when prepared in isotopically labeled form, was demonstrated to give the expected specific labeling pattern in AFB_1 indicative of intact incorporation.[122]

The first furan ring of the bisfuran system, therefore, was formed in the oxidative rearrangement of averufin ((**15**), Scheme 4) to (**39**). The retention of bound deuterium during the change in oxidation state at C-1′ depicted in Scheme 12 is reminiscent of a pinacol rearrangement. The likely precursor of such a rearrangement would be nidurufin (**40**), a known, minor *Aspergillus* metabolite.[114,123] However, when (**40**) and its 2′-epimer, pseudonidurufin (**41**), were administered to mycelial suspensions of *A. parasiticus*, they were excreted as polar, water-soluble conjugates and no incorporation into AFB_1 could be observed.[123]

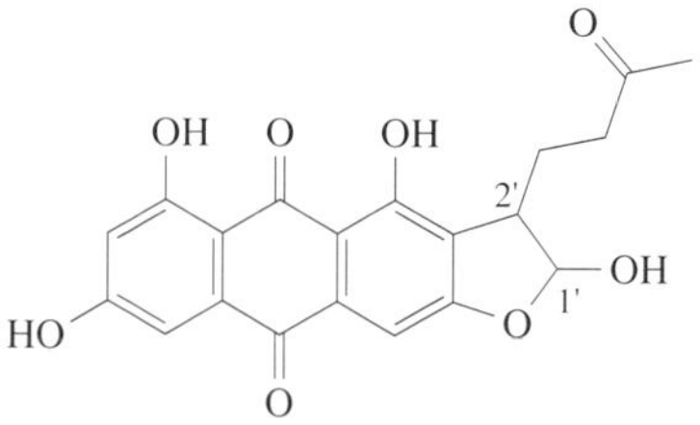

(**39**) 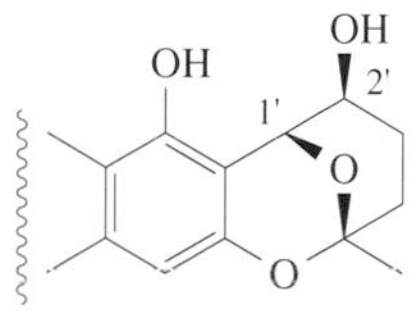 (**40**) 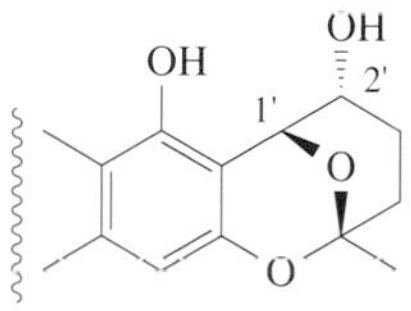 (**41**)

The rearrangement of (**15**) to (**39**) is clearly oxidative, yet the expected intermediates of the required oxidation state, (**40**) or (**41**), were not involved *in vivo*. This curious state of affairs was examined in a further paired isotope experiment as illustrated in Scheme 14. In path A oxidation is invoked at C-2′, but not hydroxylation to (**40**), to generate a reactive intermediate, which can rearrange to (**43**), hydrate/dehydrate to 1′-hydroxyversicolorone (**44**), and, finally, upon Baeyer–Villiger oxidation, yield versiconal acetate (**45**). If oxygen exchange from the methyl ketone (**44**) were slow compared with the rate of oxygen insertion into (**44**), the paired isotopes in [5′-^{13}C,^{18}O]averufin (**42**) would remain directly bound in (**45**). In contrast to this restrictive mechanistic outcome, a number of other possibilities may be represented by path B, which as a group are distinguished by the hydrolytic opening of the averufin ketal and oxidation to an intermediate as (**46**) (X = leaving group). Any such mechanism requires cleavage of the 1′-oxygen–5′-carbon bond with ^{18}O label being rapidly lost from the hemiacetal (**47**). From the magnitude of the isotope effect on the acetyl carbon resonance and its relative intensity, it could be shown that 80% of the ^{18}O (*O) was retained in versiconal acetate specifically labeled as shown in (**45**).[123]

The nature of the intramolecular anthraquinone migration from C-1′ to C-2′ could be analyzed in greater detail knowing the absolute configuration of averufin[116] and the stereochemistry of deuterium incorporation from [2-$^{2}H_3$,^{13}C]acetate at the C-2′ and C-4′ methylenes.[94] These are depicted in (**49**). Knowing also the labeling pattern from deuterated acetate in sterigmatocystin (**51**),[82,124] the labeling pattern shown in 1′-hydroxyversicolorone (**50**) may be deduced. Therefore, in this rearrangement by path A it is the 2′-axial hydrogen (H°) in (**49**) (Scheme 15) that must be specifically lost in the oxidative rearrangement of (**49**) to (**50**).

Radical or cationic intermediates can be visualized to be formed at C-2′ to initiate the *trans*-1,2-diaxial rearrangement of migrating (anthraquinone) and leaving (H°) groups in (**49**). The behavior of these species was probed in model reactions using the 6,8-di-*O*-methylaverufin analogues (**52**) and (**53**) (Scheme 16). The *exo*-mesylate (**52**) (X = $-OSO_2Me$) was found to rearrange under solvolytic conditions to (**54**) whereas the *endo*-isomer (**53**) (X = $-OSO_2Me$) was unreactive.[125] To explore the radical regime, the *exo*- and *endo*-iodides were prepared, (**52**) (X = I) and (**53**) (X = I), respectively. Under a variety of radical-generating conditions, no rearranged products could be obtained. Only reduction (X = H) and small amounts of alcohol (X = OH, from scavenging trace amounts of dissolved oxygen and reduction of the hydroperoxide) were detectable. In contrast, treatment of the *exo*-iodide (**52**) (X = I) with silver trifluoroacetate gave 6,8-di-*O*-methylhydroxyversicolorone (**54**) in good yield. Interestingly, when the more nucleophilic silver acetate was used in this reaction, in addition to (**54**), the intermediate (**43**) was trapped from the face opposite to the migrating anthraquinone to give (**55**).[126]

In conclusion, the biochemical data and the chemical model experiments suggest that oxidation occurs at C-2′ in (**49**) specifically to remove H°. No oxygen is introduced at this carbon to trap the resulting reactive intermediates generated at this carbon to give (**40**), but molecular rearrangement occurs, probably through cationic intermediates represented as path A in Scheme 14, to give (**39**).

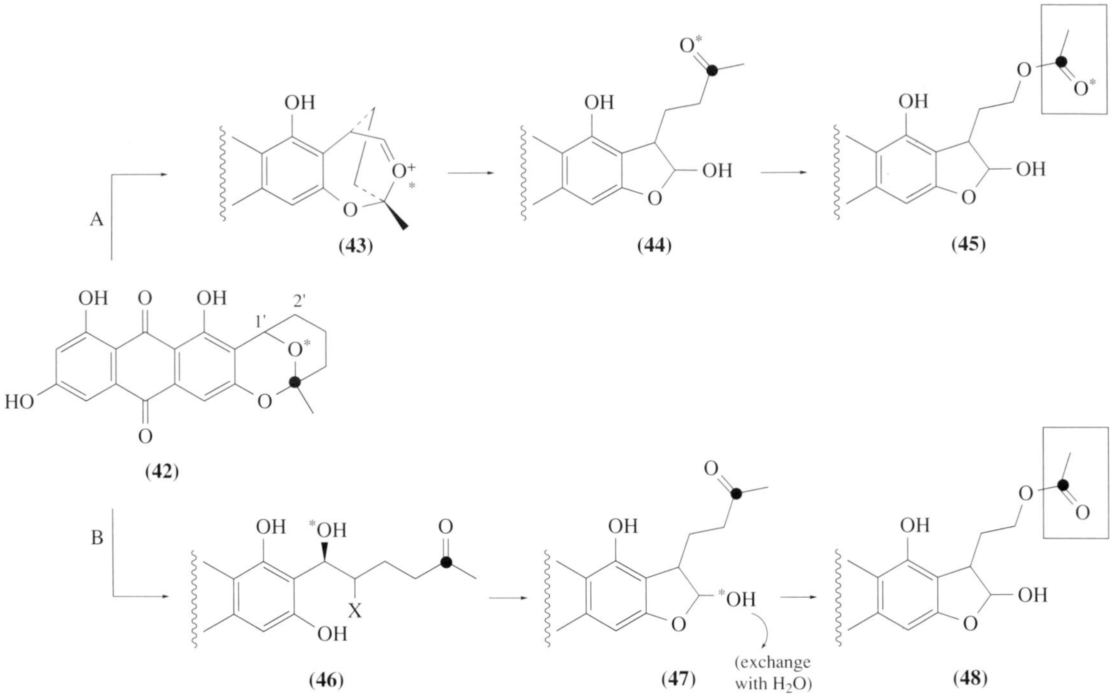

Scheme 14

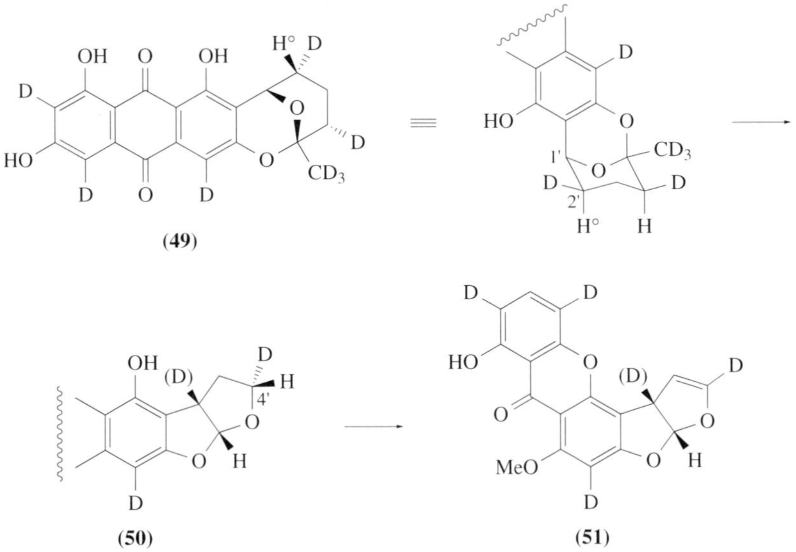

Scheme 15

1.17.4.5 Versiconal Acetate

The insecticide Dichlorvos is a known inactivator of acetylcholine esterases. Wild-type *A. parasiticus* grown in the presence of ppm concentrations of this agent accumulates versiconal acetate.[127] Experiments with multiply deuterated averufin ((**37**), Scheme 13) revealed that the conversion into versiconal acetate (**38**) was strictly intramolecular[120] and support a Baeyer–Villiger-like rearrangement from 1′-hydroxyversicolorone (**39**). Such a process would be predicted to involve incorporation of molecular oxygen in the bridging ester oxygen ($^{\Delta}$O) as shown in (**56**) (Scheme 17). This expectation was shown to be indeed the case[128] to complement derivation of the carbonyl oxygen from the polyketide backbone as summarized in Schemes 11 and 14.

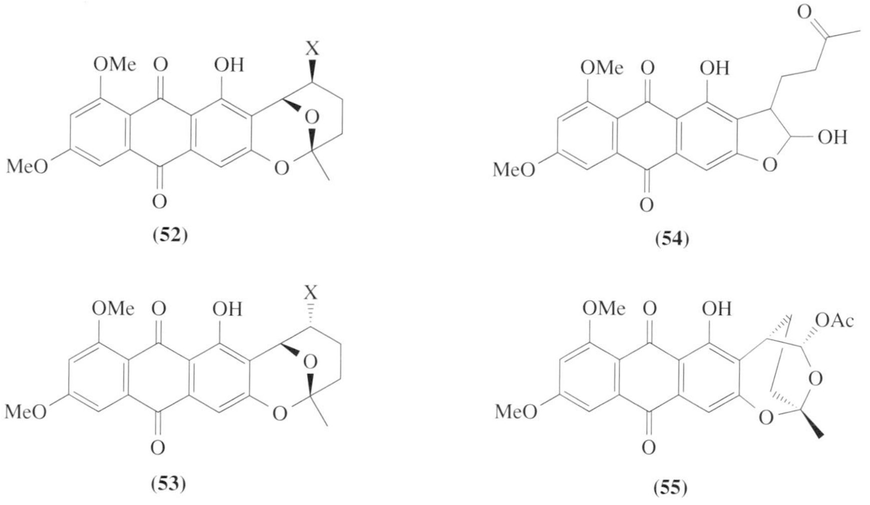

(52) (54)

(53) (55)

Scheme 16

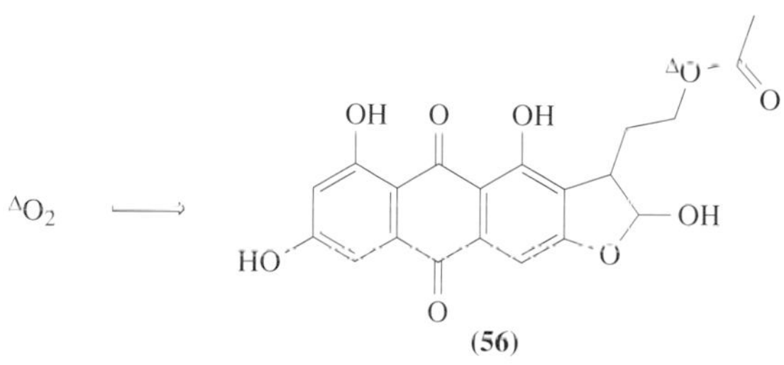

(56)

Scheme 17

1.17.4.6 Versicolorin A and B

In 1980, Wan and Hsieh[129] and Anderson and Dutton[130,131] independently reported cell-free systems from *A. parasiticus* capable of carrying out the conversion of versiconal acetate (**16**) into versicolorin A ((**17**), Scheme 18). Both of these investigators observed that the overall transformation was inhibited by Dichlorvos and reduced in efficiency by anaerobic conditions. Remarkably, when Wan and Hsieh[129] monitored the course of the reaction by autoradiography, at least four apparent intermediates were detected, signaling a complex reaction path. Of several mechanisms considered, and without isolation or characterization of any intermediate, hydrolysis of the ester (**16**) to versiconal (**57**) was proposed. This mechanistically reasonable first step was followed by oxidation to the aldehyde (**58**), which would exist principally in the closed form (**59**), versicolorin A hemiacetal. Dehydration was invoked to yield (**17**).

This process in which the critical dihydrobisfuran was formed was reexamined with synthetic materials of defined structure and led to a different view of the mechanism.[132,133] Preparation of cell-free extracts of *A. parasiticus* similar to those of Wan and Hsieh[129] and Anderson and Dutton[130,131] was found to hydrolyze (**16**) to the primary alcohol (**57**) as proposed by these investigators and consistent with Dichlorvos inhibition of the biosynthetic pathway. However, the product cleanly formed in this cell-free system was not (**17**), but the chromatographically similar tetrahydrobisfuran, versicolorin B (**60**). No other intermediates were detected by analytical HPLC.[132,133]

Faced with this unexpected result, reevaluation of several stereochemical observations led to a new picture of bisfuran formation. First, the absolute configuration of averantin (**30**) and averufin ((**34**), Scheme 11), and the stereochemical course of oxidative rearrangement of the latter to 1′-hydroxyversicolorone ((**44**), path A, Scheme 14) had been predicted earlier[134] to account for the absolute configuration of the bisfuran. However, second, when 1′-hydroxyversicolorone (**39**) and versiconal acetate (**16**) were isolated and characterized, they were found to be racemates.[121,135] It would be expected that (**57**), a rapidly equilibrating mixture of hemiacetals and the open aldehyde,[133] would be similarly racemic. Nonetheless, third, both enantiomers of this intermediate were cyclized

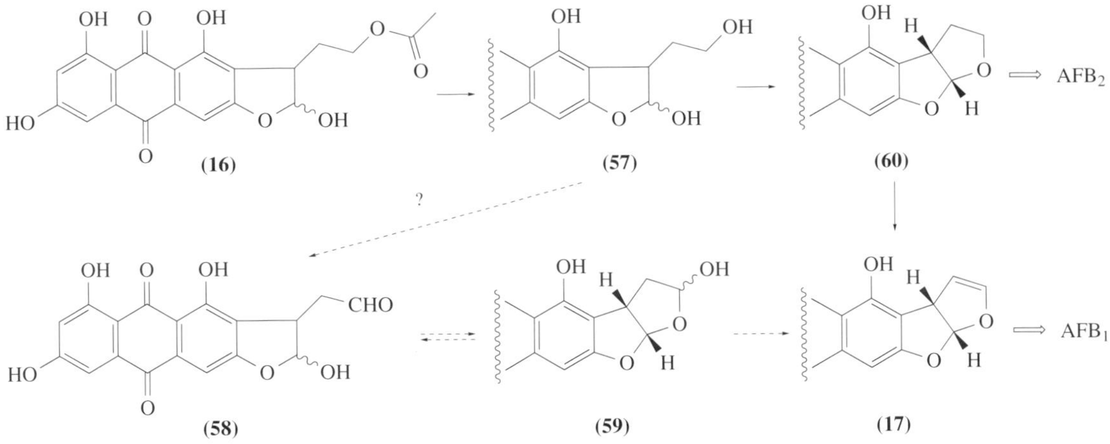

Scheme 18

to optically active (−)-(**60**) (Scheme 18).[132,136] Therefore, it was this cyclizing enzyme, named versicolorin B synthase (VBS), that was responsible for converting (**16**) into a single stereoisomer of (−)-(**17**). This enzyme has been purified[133,137] and its kinetic behavior examined in detail.[138] VBS exists as a dimer of identical 78 kDa subunits, requires no metals or cofactors for activity, and catalyzes the cyclization of (±)-(**57**) to optically pure (−)-(**17**). The inherent rate of racemization of versiconal was determined at neutral pH and found to be fast enough relative to the rate of catalysis to affect minimally, if at all, the flux of the biosynthesis.[138]

When versicolorin A hemiacetal ((**59**), Scheme 18) was incubated in the presence and absence of oxidized or reduced nicotinamide cofactors, no dehydration to versicolorin A (**17**) was observed—only reduction to the tetrahydrobisfuran versicolorin B (**60**).[133,138] In sum, these observations implied that a desaturation step must occur to convert (**60**) into (**17**). The partitioning of labeled samples of (**60**) and (**17**) into AFB_1 and AFB_2 supported this contention.[133,138] Shortly thereafter, Yabe *et al.*[139] described the desaturase, an NADPH-dependent enzyme present in the microsomal fraction of *A. parasiticus*. These facts supported the simplified view of bisfuran formation and the precursor relationship of tetrahydro- to dihydrobisfurans early in the biosynthesis. Similarly, they provided a compelling rationale for the prior observation that natural isolates[140] and mutagenized strains[141] of *A. flavus* product AFB_2 but little or no AFB_1. A defect in the desaturase would account for this observation. These experiments have been reviewed in detail[133] and their conclusions amplified elsewhere.[142]

1.17.4.7 Sterigmatocystin

In contrast to the biosynthetic reactions leading to versicolorin A (**17**), a great deal less is known about the apparently multistep processes by which the anthraquinone (**17**) is converted into the xanthone of demethylsterigmatocystin ((**61**), Scheme 19), a net transformation involving oxidative cleavage, decarboxylation, deoxygenation, and nuclear rearrangement. Two substrate-specific *S*-adenosylmethionine-dependent *O*-methyltransferases have been characterized that carry (**61**) to sterigmatocystin (**7**) and in turn to *O*-methylsterigmatocystin (**12**). Apart from these three species, no further intermediates are known in the conversion of (**17**) into (**61**).

Apparent deoxygenation during the course of aromatic polyketide biosynthesis is a well-known phenomenon. In general, these are thought to occur by "pre-aromatic" reduction/dehydration, as exemplified by 6-methylsalicylic acid.[143] However, exceptions to this general rule are known and perhaps best illustrated by the reduction of emodin (**62**) to chrysophanol (**63**) (Scheme 20) reported at an early stage of ergochrome biosynthesis.[144] Anderson *et al.*[145] unambiguously demonstrated in a cell-free experiment that the reaction was enzyme mediated in the presence of NADPH.

To test directly whether such a sequence of events is active in the aflatoxin pathway to account for the net loss of the anthraquinone C-6 hydroxy group, 6-deoxyversicolorin A (**64**) was prepared

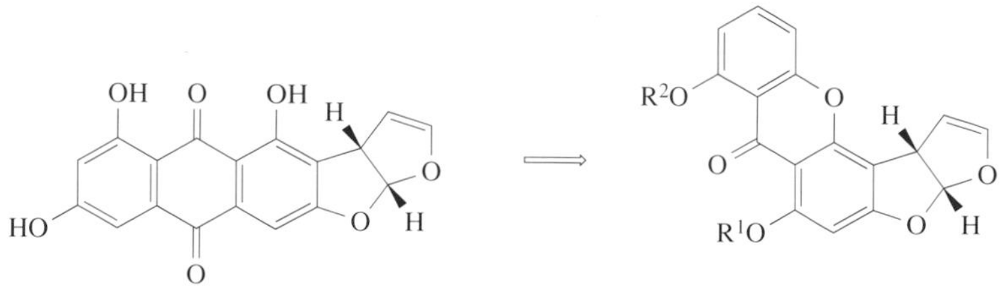

(17)

(61) $R^1 = R^2 = H$
(7) $R^1 = Me$, $R^2 = H$
(12) $R^1 = R^2 = Me$

Scheme 19

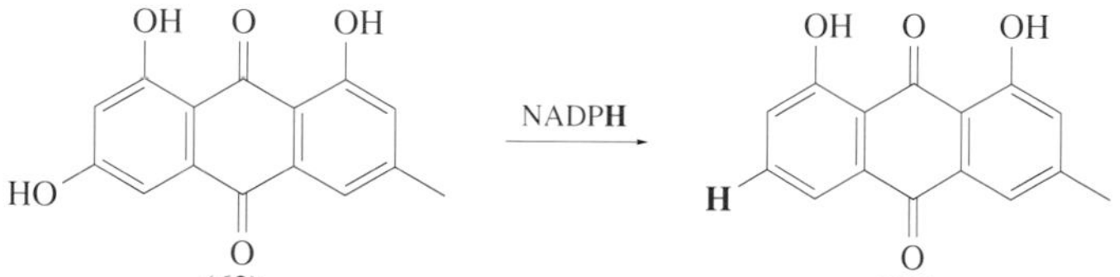

NADPH

(62)

(63)

Scheme 20

in isotopically labeled form. However, although it is a known *Aspergillus* metabolite,[146] (**64**) was not converted into AFB_1 under two distinct experimental protocols. In contrast, (**17**) was efficiently utilized under comparable conditions.[147] It was concluded that reduction is not the first step.

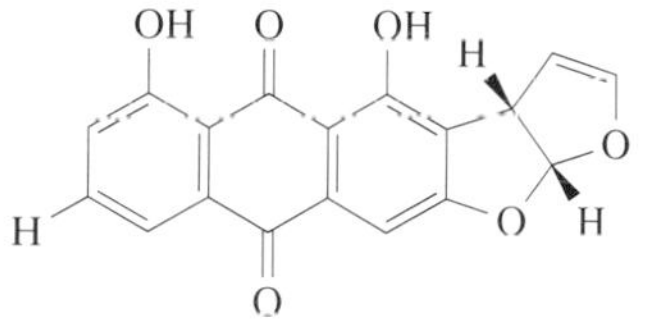

(64)

Set out in Scheme 21 is a proposed mechanism to account for the transformation of (**17**) to (**61**) that is closely related to previous proposals of Sankawa *et al.*[92] and Bennett and Christensen.[68] Consistent with the NADPH dependence seen for this conversion in cell-free systems,[148] Baeyer–Villiger-like cleavage of the anthraquinone is proposed to give (**65**). Heavy isotope is introduced into the C-ring ($^{\Delta}O$) from $^{18}O_2$ in keeping with this putative oxidative change.[149] Opening of this lactone with water and oxidative phenol coupling can be suggested to form the spirodienone (**66**), analogous to griseofulvin or geodin.[69] NADPH-dependent reduction in the A-ring would afford an expectedly very unstable intermediate (**67**), prone to dehydration and rearomatization accompanied by decarboxylation to give (**61**) (one potential sequence of these events is shown in Scheme 21).

1.17.4.8 Aflatoxin B_1

The conversion of *O*-methylsterigmatocystin (**12**) into AFB_1 (**1**) is the second major nuclear rearrangement in the aflatoxin pathway. No intermediates in this process are known, although cell-free experiments have shown a clear NADPH dependence.[148,150] There is evidence for a ca. 200 kDa protein complex that catalyzes this conversion.[142] Further evidence about these late reactions has come from indirect experiments.

The mapping experiments based on ring-A-labeled averufin ((**35**), Scheme 12) allowed the assignment of C-10 in *O*-methylsterigmatocystin ((**69**), R = Me; Scheme 22) as the carbon specifically lost in the net transformation to AFB_1 (**72**). It was shown in a technically demanding series of experiments that C-10 is ultimately lost as CO_2.[151] Knowing the pattern of deuterium incorporation from [2-$^2H_3$,^{13}C]acetate in ST ((**69**), R = H)[82,124] and (**72**),[153] an apparent NIH shift has taken place. The high oxidation state that C-10 eventually attains was rationalized in a monooxygenase step to the proposed aryl epoxide (**70**), which would be expected[153] to rearrange to 10-hydroxy-*O*-methylsterigmatocystin (**71**).[151] Whether further oxidation of this putative intermediate occurs by way of

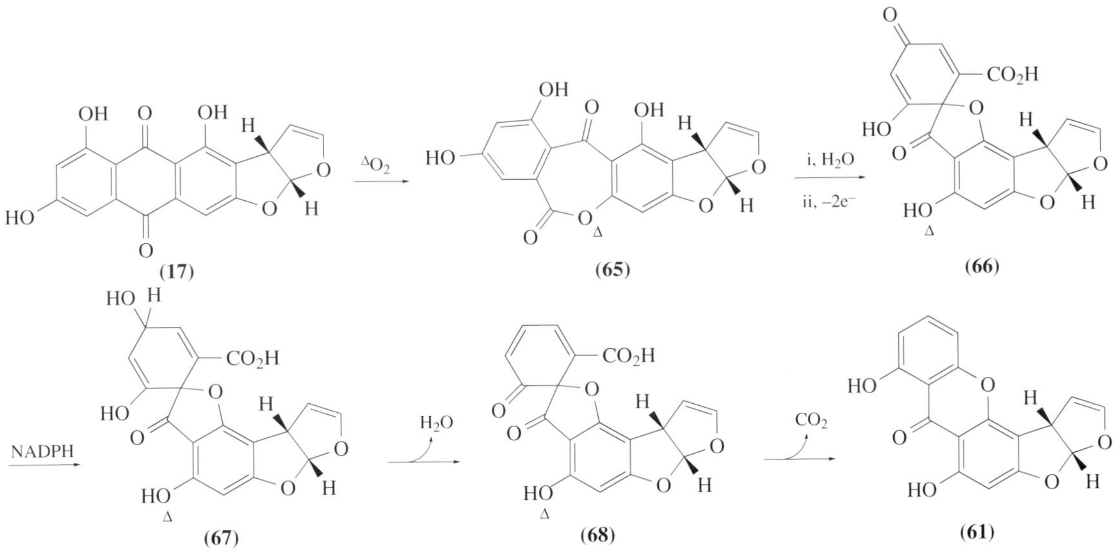

Scheme 21

a monooxygenase (conceivably the same enzyme as catalyzed the first oxidation at this site) or a dioxygenase is not known. However, experiments monitoring the utilization of $^{18}O_2$ in the fermentation clearly demonstrate the incorporation of labeled oxygen notably in the bisfuran and at C-8, as expected, but also at C-1, a mark of the oxidative ring cleavage process. A restricted set of mechanistic possibilities to account for all these data has been proposed.[149]

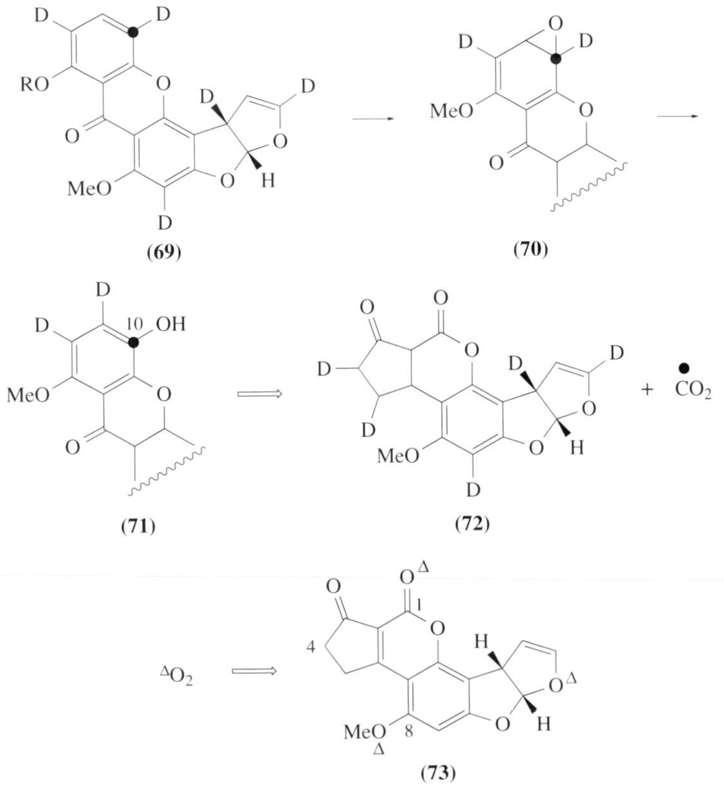

Scheme 22

1.17.5 MOLECULAR BIOLOGICAL APPROACHES TO AFB$_1$/ST BIOSYNTHESIS

Genetic approaches have begun to unravel both structural and regulatory elements of the AFB$_1$/ST biosynthetic pathways. Deletion of key regions of the *Aspergillus* genome through recombinational inactivation is a promising approach for the abolition of carcinogen production, particularly in commercial species such as *A. sojae* which is used in soya sauce fermentations. For agricultural applications a biocompetitive approach in which an aflatoxigenic strain of *Aspergillus* displaces the native organism may be envisioned.[154,155]

Since 1989, three methods have been employed to delineate the molecular biology of AFB$_1$ production: complementation of classically generated mutants, targeted disruption of suspected pathway genes in wild-type species, and reverse genetics to isolate gene sequences from purified biosynthetic enzymes.

1.17.5.1 Complementation

Isolation and characterization of *Aspergillus* mutants to define the AFB$_1$ pathway was originally championed by Lee and co-workers.[56,58] The experimental progression from the biochemically characterized *A. parasiticus* mutants to the operative genetic loci involves the restoration of a mutant to a wild-type phenotype through transformation and genetic recombination. Digestion of the cell wall allows the passage of exogenous DNA into the host cell via polyethylene glycol-mediated cell fusion. The lack of mitotically stable autonomously replicating DNA in *Aspergillus* necessitates stable insertion of the selectable marker-containing construct within the genome (Figure 1). Typical selectable markers confer benomyl (fungicide) resistance[156,157] or nitrate[158,159] or uracil prototrophy.[156,160] The Linz and the Woloshuk/Payne groups were instrumental in developing protoplasting methods for *A. parasiticus* and *A. flavus*, respectively, whose resistance to transformation initially hindered studies of the molecular biology of these organisms.

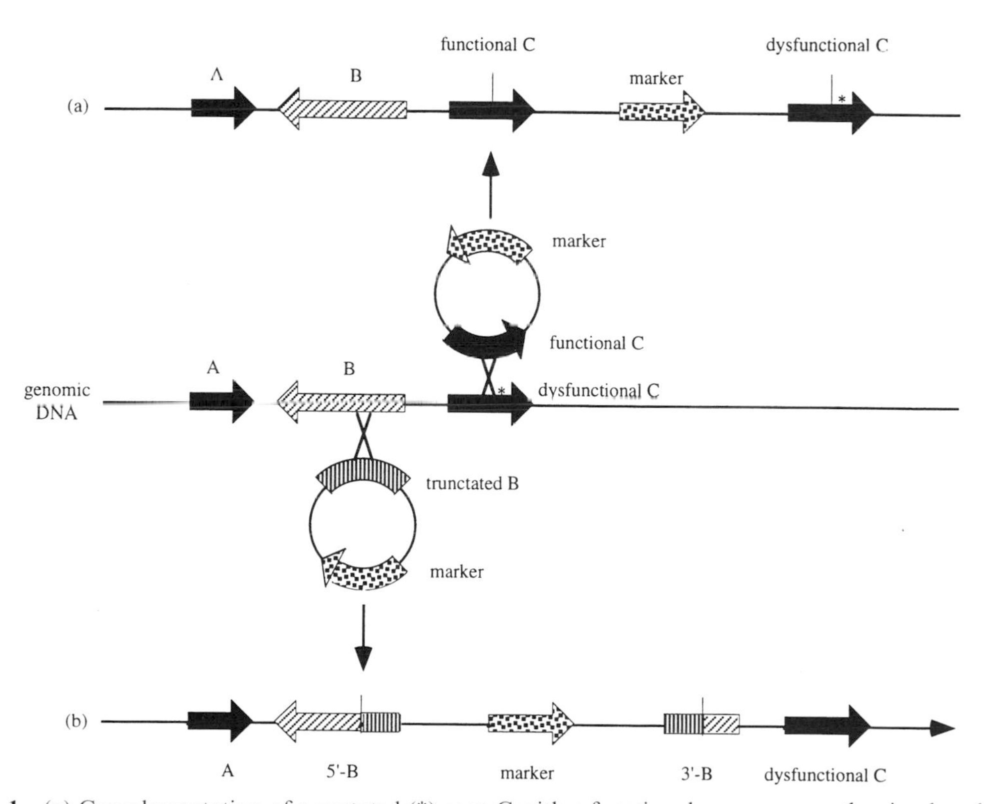

Figure 1 (a) Complementation of a mutated (*) gene C with a functional gene copy on the circular plasmid (exogenous DNA). (b) Recombinational inactivation, or gene disruption, of a functional gene B. Transformation of cells followed by homologous recombination between a 3′,5′-truncated fragment of B on a circular vector and genomic DNA results in two truncated, dysfunctional gene fragments. Insertional junctions are marked by vertical bars. Experimentally, appearance of activity from biosynthetic gene product C or loss of gene B function is observed by the respective processing or accumulation of the respective substrates. Inclusion of a selectable marker facilitates screening of transformed cells.

Single crossover experiments with circular plasmid or cosmid DNA containing a complete wild-type copy of a defective gene installs the vector and yields a functional chromosomal copy of the targeted gene (Figure 1). This method ultimately led to the cloning of the *afl-2* gene from *A. flavus*[161] and *nor-1*,[162] *ver-1*,[163] and *apa-2*[164] from *A. parasiticus*.

The *n*orsolorinic *a*cid *r*elated gene *nar-1*, renamed *nor-1*, was the first gene identified by complementation of an NA-accumulating, nitrate reductase-deficient *A. parasiticus* mutant with a cosmid from a wild-type genomic library.[162] Northern blotting identified a 1.4 kilobase (kb) transcript that hybridized to the *nor-1* fragment, which was further characterized by gene disruption (see below). Installation of this gene sequence removed the established blockage of the enzymatic conversion of (**13**) into (**14**) (Scheme 4).

Conversion of the anthraquinone versicolorin A (**17**) into the xanthone demethylsterigmatocystin ((**61**), Scheme 21) has been postulated to involve a reaction sequence including Baeyer–Villiger cleavage, saponification, reduction, and finally oxidative decarboxylation.[51,68] Transformation of a VA-accumulating strain of *A. parasiticus* with a genomic library from aflatoxigenic *A. parasiticus* yielded a single Afl^+ transformant.[163] Isolation of genomic DNA from the isolated cells followed by digestion with a restriction enzyme and recircularization allowed the recovery, or "marker rescue", in *E. coli*, of the complementing *ver-1* DNA. The predicted 262-amino acid polypeptide, which showed 52% similarity to two NADPH-dependent polyketide ketoreductases[165,166] and a polyhydroxynaphthalene reductase involved in melanin biosynthesis, was transcribed primarily during idiophase.[167]

1.17.5.2 Gene Disruption

Approaches related to complementation may be used to block the biosynthetic pathway. Gene replacement and gene disruption result in the exchange of a segment of genomic region with exogenous DNA or the insertion of a segment of foreign DNA into the genome disrupting the coding region, respectively.

One-step gene replacement of the *nor-1* gene produced blocked strains that accumulated norsolorinic acid ((**13**), Scheme 4) and demonstrated a reduced ability to produce AFB_1.[168] Although leakage was observed in all transformants, homologous recombination was localized to a short-chain alcohol dehydrogenase-like sequence that together with the biochemical results provide strong, direct support for the involvement of a 29 kDa *nor-1* gene product in the biosynthetic pathway.[169] The identification of other reductases competent for the (**13**) to averantin (**14**) transformation has prompted the suggestion that multiple reduction pathways are involved (see below).[170]

The functions of the *stcS* and *stcU* structural genes for sterigmatocystin (**7**) formation in *A. nidulans* have been partially assigned by disruption experiments. Disruption of either gene resulted in the accumulation of versicolorin A (**17**) and nearly complete suppression of (**7**) formation. *stcU*, formerly designated *verA*, possessed an extended 85% amino acid identity with *ver-1* in *A. parasiticus*. This putative NADPH-dependent ketoreductase may be necessary for the reduction of the C-6 phenolic carbon prior to decarboxylation. Genomic sequence analysis of *stcS* (formerly *verB*) indicated that this gene encodes a 505-amino acid P450 monooxygenase which is sufficiently unique to classify it in a new P450 family.[171] The conversion of (**17**) into (**7**) illuminates a primary difficulty in studying the complex transformations within the overall pathway: many intermediates appear to be unstable or enzyme bound. All attempts to determine the functional order of the STC-S and STC-U enzymes failed as neither disruption mutant nor mixed cultures produced visible intermediates or (**7**).[171] Clever approaches to tackling the problems of protein–protein interaction will be essential for the solution of this problem and, for example, characterization of an apparent 200 kDa oxidoreductase complex that carries out the later conversion of *O*-methylsterigmatocystin (**12**) into AFB_1.[142,150]

Gene disruption was similarly employed in the identification of *pksL1* and *stcA* (*pksA*), described later.

1.17.5.3 Reverse Genetics

In reverse genetics, the isolation of a native protein provides an amino acid sequence which is used in the design of molecular probes to the corresponding nucleotide sequence (i.e., in practice, translation to transcription to genome). For structural genes, reverse genetics has the appeal of targeting genes corresponding to proteins identified by an *in vitro* chemical assay. Although the quantity and stability of the aflatoxin-synthesizing enzymes have been obstacles, this method has led to the identification of three pathway genes, *omtA*, *vbs*, and *norA*.

A 40 kDa methyltransferase, OMTA, that catalyzes the conversion of (**7**) into (**12**) (Scheme 19) was isolated in five steps by Bhatnagar and co-workers.[172] A 24 h cDNA library was successfully screened using polyclonal antibodies specific to the purified protein and resulted in the isolation of the 1254 bp *omt-1* coding sequence (subsequently renamed *omtA*). Comparison of native *N*-terminal sequence with the translated 5′-terminal coding sequence indicated a 41-amino acid segment, which precedes the 377-residue protein originally isolated, was absent. The hypothetical function of the hydrophobic leader is to promote membrane association,[172] although other functions may be envisioned. The polyclonal antibodies also recognized an overexpressed OMTA-β-galactosidase fusion protein which was capable of the SAM-mediated conversion of (**7**) to (**12**). Transcripts of *omtA* appeared in *A. parasiticus* cells grown in AFB_1-permissive medium after 24 h.[172] The conservation of the single-copy *O*-methyltransferase genes between *A. parasiticus* and *A. flavus* was evident from the >97% nucleotide identity in the coding sequence and the identical lengths and locations of four introns.[173]

A second heterodimeric cytosolic enzyme composed of 58 and 110 kDa subunits was purified by anion-exchange chromatography.[174,175] Partial characterization of the enzyme demonstrated the requirement for SAM and a K_m of 2.6 μmol l^{-1} for (**7**). Competition experiments indicated a preference for (**7**) over dihydrosterigmatocystin ((**75**), Scheme 23).[142] Antibodies raised against this enzyme weakly cross-reacted with the previously identified 40 kDa species. Disruption of the *A. nidulans stcP* gene, which exhibited only 30% identity to the *omtA* gene, resulted in the accumulation of demethylsterigmatocystin ((**61**), Scheme 19), suggesting its involvement in transfer of the first methyl group.[176]

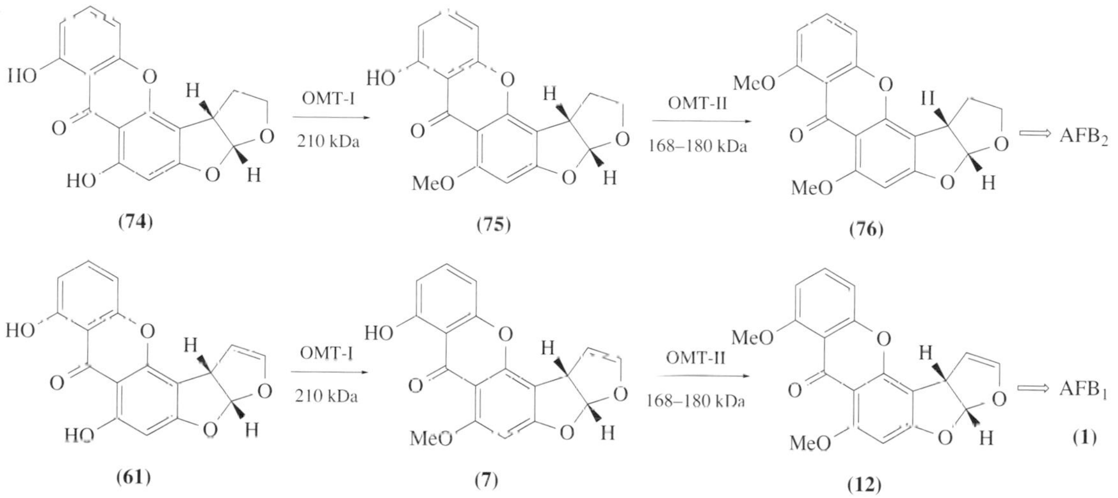

(**74**) —OMT-I, 210 kDa→ (**75**) —OMT-II, 168–180 kDa→ (**76**) ⟹ AFB_2

(**61**) —OMT-I, 210 kDa→ (**7**) —OMT-II, 168–180 kDa→ (**12**) ⟹ AFB_1 (**1**)

Scheme 23

Two distinct methyltransferase activities were also observed by Yabe *et al.*[177] in an AFB_1-permissive medium (Scheme 23). *O*-Methyltransferase I, estimated to be 210 kDa in size by gel filtration chromatography, was proficient in the conversion of dihydrodemethylsterigmatocystin (**74**) and (**61**) into (**75**) and (**7**), respectively. The 180 kDa *O*-methyltransferase II processed dihydrosterigmatocystin and sterigmatocystin to dihydro-*O*-methylsterigmatocystin (**76**) and (**12**), respectively. The type II activity may correspond to the previously reported 168 kDa complex.[177] The two enzyme complexes were further distinguished by the specific *N*-methylmaleimide inhibition of the type I complex. The evidence provided by both groups is consistent with distinct enzymatic activities sequentially converting (**61**), (**7**), and (**12**) into AFB_1 and the tetrahydrobisfuran intermediates into AFB_2.

Creation of the optically active bisfuran ring system required for carcinogenicity occurs at a key branch point in the AFB_1 pathway. Purification of versicolorin B synthase (VBS), which catalyzes the dehydrative cyclization of versiconal (**57**) to versicolorin B ((**60**), Scheme 18), was initially reported in 1992.[133,137] A substantially improved purification of the enzyme was subsequently achieved by McGuire *et al.*[138,178] Homogeneous VBS was isolated by preparative isoelectric focusing followed by Mono-Q anion-exchange chromatography with 53-fold purification.

Native VBS is a homodimeric protein with a molecular weight of 140 kDa.[138] The enzyme stereospecifically catalyzed the cyclization of racemic (**57**) to (−)-(**60**) (≥93% *ee*) with kinetic parameters of $K_m = 2.4\ \mu mol\ L^{-1}$ and a turnover number of $2.5\ s^{-1}$ per VBS subunit for (**57**). The inhibition constant K_p for (**60**) of $11\ \mu mol\ L^{-1}$ indicated that product inhibition is minimal for this enzyme. Racemization of stereochemically labile (**57**) is believed to occur nonenzymatically whereby a single (**57**) enantiomer is turned over by the enzyme to optically pure product (Scheme 18). Chemical modification studies and pH–rate profiles on the native protein suggested the involvement of two carboxylate residues in catalysis.[138] Histidine or the protonation state of (**57**) at the active site appeared to be a controlling factor in the high-pH regime. Compounds tested as potential inhibitors for VBS suggested that the C-ring and side chain were the most important substrate recognition elements.[138]

The *N*-terminal sequence from LysC-endopeptidase-digested VBS was essential for the preparation of partially degenerate oligonucleotide primers.[179] A 750 bp *vbs* gene fragment was PCR-amplified from a 60 h *A. parasiticus* cDNA library, yielding coding sequence which was in agreement with the extended peptide sequence. A PCR-generated internal 615 bp internal fragment was finally used for cDNA and gDNA library hybridization screening. The isolated 1932 bp open reading frame was broken by a single 53 bp intron in the genomic clone. Translation of the *vbs* coding sequence yielded a 643-amino acid, 70.3 kDa protein.

Examination of the protein sequence revealed a 58% similarity to glucose oxidase from *A. niger* and other members of the glucose–methanol–choline (GMC) flavin oxidoreductase family.[179] This observation was unexpected since VBS does not carry out a redox reaction or require flavin for activity. In mandelonitrile lyases, also members of the GMC family, flavin is either not bound or does not actively participate in catalysis. An apparent 23-amino acid gap exists within the FAD binding site of VBS, which may eliminate flavin binding.

Comparison of wild-type VBS before and after PNGase F treatment uncovered extensive *N*-glycosylation of the native protein.[180] Overexpression in *Saccharomyces cerevisiae* was optimized in a vector incorporating a yeast secretion signal which provided VBS, indistinguishable from the native protein, in a 50–100-fold higher yield compared with *Aspergillus*.

Monoclonal antibodies (mAbs) were prepared to a partially purified 43 kDa protein band with norsolorinic acid reductase activity.[170,181] Time-dependent Western analysis of crude *A. parasiticus* extracts using the mAb revealed a single 43 kDa band appearing after 36 h in an aflatoxin-permissive medium.[170] Immunochemical screening yielded a single cDNA clone containing sequence coding for a 388-residue, 43.7 kDa protein. Transcription was concurrent with expression. Both aryl alcohol dehydrogenase (49% amino acid homology to *Phanerochaete chrysosporium*)[182] and NADPH-binding motifs were identified in the translated nucleotide sequence. Multiple reductive pathways leading from norsolorinic acid (**13**) to averantin ((**14**), Scheme 4) have been postulated.[170] Potential *N*-glycosylation sites were apparent in *norA* translations for both *A. parasiticus* and *A. flavus*[170] and absent in *nor-1*, which suggests distinct cellular localizations.

Independently, affinity chromatography resulted in a 138-fold purification of a norsolorinic acid reductase of unreported size involved in the conversion of (**13**) into (**14**).[107,183] A fourth enzyme capable of this reduction has been reported by Chuturgoon and Dutton.[183] This 140 kDa species was isolated by sequential Reactive Green-19 agarose and norsolorinic acid affinity columns. The K_m for (**13**) was $3.45\ \mu mol\ L^{-1}$ and a slight preference for NADPH ($K_m = 103\ \mu mol\ L^{-1}$) over NADH was noted.

Other measurable enzymatic activities, such as the microsomal conversion of *O*-methylsterigmatocystin (**12**) into AFB_1 (apparent K_m of (**12**)$=0.66\ mmol\ L^{-1}$)[150,184] and the demonstrated ability to transform versiconal acetate (**16**) to versicolorin A ((**17**), Scheme 4)[129] and acetate to AFB_1[185] in cell-free extracts may ultimately allow a reverse genetics approach to define other genes required for AFB_1 biosynthesis.

1.17.5.4 Fatty Acid and Polyketide Formation

Specialized fatty acid (FAS) and polyketide synthases (PKS) are required for the initial formation of the hexanoyl starter unit and subsequent malonylCoA extensions to the octaketide backbone ((**29**) in Scheme 11). UV mutagenesis of a norsolorinic acid-accumulating *A. parasiticus* mutant strain yielded a double mutant, UVM8 (*fas-1A*, *nor-1*).[186] Transformation of UVM8 with a plasmid subclone of the NorA genomic DNA cosmid[187] complemented specifically the *fas-1A* locus re-establishing norsolorinic acid (**13**) synthesis. Sequence of the subclone provided a new open reading frame, *fas-1A*, which was transcribed as a single 7.5 kb mRNA. Identification of FAS/PKS genes has been aided by widely conserved sequences within their functional domains. Regions with 58% and 69% amino acid similarity to the malonyl transferase and enoyl reductase palmitoyl domains of yeast fatty acid synthase FAS1 (β-subunit),[188] respectively, and insertional inactivation experiments strongly suggest the involvement of *fas-1A* in hexanoyl primer formation (see below).[186]

Transcriptional analysis of the 35 kb NorA cosmid identified 14 transcripts, eight of which were cotranscribed with *nor-1* and *ver-1*. Three large transcripts (6.5, 7.0, and 7.5 kb) were noted, including *fas-1A*. Partial sequence of the *pksA* gene, corresponding to the 7.0 kb transcript, showed 80% amino acid similarity to the β-ketosynthase and acyl transferase regions of the 9.2 kb *wA* gene responsible for the synthesis of a polyketide conidial pigment in *A. nidulans*.[189] Independent disruption of the *pksA* locus in an *O*-methylsterigmatocystin-accumulating strain of *A. parasiticus* with plasmid pXX, which contained a 2.15 kb 5′3′ truncated segment of *pksA* resulted in an albino strain incapable of producing (**12**).[190] The absence of the enoyl reductase and β-ketoacyl reductase domains necessary for saturated fatty acid biosynthesis indicates that this gene is responsible for the polyketide homologation of the hexanoyl starter unit.[190]

Differential screening, a technique which selectively isolates sequences transcribed during specific growth conditions (i.e., culture time, permissive medium), has also been used for the isolation of *Aspergillus* mRNA enriched in AFB_1 biosynthesis transcripts.[191,192] This method was applied by Feng and Leonard[192] to identify a 6.6-kb mRNA corresponding to the *A. parasiticus pksA* gene (originally denoted *pksL1*) whose transcriptional regulation was linked to growth phase, temperature, and carbon/nitrogen sources in analogous fashion to AFB_1 production.[193] The translated sequence contained four functional domains ordered acyl carrier protein (ACP), β-ketoacyl reductase, acyltransferase, and thioesterase.

Feng and Leonard[193] used a fragment of *nor-1* cDNA to screen an *A. nidulans* genomic cosmid library. Transcriptional mapping of the isolated cosmids identified a 7.2 kb transcript for the *pksA* homolog, *pksST*, and a neighboring gene with similarity to the cytochrome P450 monooxygenase superfamily. The 2181-amino acid *pksST* gene product contained a β-ketoacyl reductase, acyltransferase, two ACP domains, and thioesterase. No dehydrase domain, which may be necessary for anthrone (**29**) formation (Scheme 11), was identified. Both the FAS and PKS genes code for large, multifunctional enzymes (Type I), which are characteristic of most eukaryotes.[194]

Distinct fatty acid synthase genes involved in primary and secondary metabolism have been identified in *A. nidulans*.[195] Long-chain fatty acid synthase subunits *fasA* and *fasB* found on chromosome VIII are necessary for cell growth, whereas *stcJ* and *stcK* are specifically required for the formation of the hexanoyl starter unit. The divergently transcribed *stcJ* and *stcK* genes contain three and five domains, respectively, analogous to the FASα/β genes required for fatty acid biosynthesis in *S. cerevisiae*.[196,197] However, the similarity of chemical function between primary metabolic and the putative hexanoyl-generating FAS is not strongly reflected in the overall 44% and 37% amino acid identity of the *stcJ* and *stcK* sequences to *Penicillium* and *Yarrowia* FAS protein sequences.

A. parasiticus mutants produced by insertional inactivation of *fas-1A* in norsolorinic acid-accumulating strains converted hexanoylNAC into (**13**) with unexpectedly low efficiency, even though the *fas-2A* and *pksA* genes were demonstrated to be intact.[198] Efficient 39% incorporation of hexanoylNAC into NA was verified in control experiments,[198] and agreed with earlier studies.[104] Minimal effects by β-oxidation inhibitors were noted in incorporation experiments with *A. parasiticus fas-1A* disruption mutants.[198] We interpret these data to signify a substantial channeling interaction between the PKS and FAS subunits.[199] In partial support of this idea, the overall yield of (**7**) from *A. nidulans* FAS mutants versus wild-type grown on hexanoic acid-enriched medium is 20-fold lower.[195] Interestingly, *pksA* contains a single ACP domain in contrast to other fungal PKSs, including *pksST*, which all contain two.[189,200,201] The quaternary interactions between the FAS-α/FASβ/PKS subunits, which will be key to unraveling the initial transformations, are not yet understood.

1.17.5.5 Clustering of Pathway Genes

Organization, both the order and the direction of transcription, of clustered genes has regularly been suggested to be important in the efficient, coordinated formation of secondary metabolites. Compact genomic organization is common in prokaryotes; however, it has been only sporadically reported in higher organisms. Examples of eukaryotic structural genes known to be localized include four genes of the *alc* regulon which are clustered on chromosome VII of *A. nidulans*[202] as well as the fungal genes for penicillin,[203] melanin,[204] and trichothecene toxin biosynthesis,[205] nitrate utilization, and quinic acid metabolism.[206,207]

Initially, researchers believed that *Aspergillus* structural genes were distributed throughout the genome, as occurs in most eukaryotes.[208] Papa[209] determined that the mutations *afl*-1 and *afl*-2 were unlinked in *A. flavus* by parasexual analysis. Two other mappable mutations were found in linkage groups II[210] and VIII.[211] Lennox and Davis[212] reported that the AFB_1-linked genes were distributed between six or seven linkage groups. Reproduction by a parasexual cycle, which allows recombination via heterokaryon formation, has presented the major genetic difficulty in studying *A. parasiticus* and *A. flavus* in comparison with the model organism *A. nidulans*.

Later research has shown conclusively that many of the structural genes for aflatoxin biosynthesis are tightly clustered. Despite his earlier work, Papa[213,214] assigned nine mutations to linkage group VII, including *norA*, *afl-1*, and the primary metabolic marker loci *leu* and *arg-7*. The initial direct evidence for physical linkage was provided by Linz and co-workers,[163] who reported that two genes, *nor-1* and *ver-1*, were proximate through the hybridization of restriction fragments to a cosmid. Parasexual analysis of Afl^- mutants in *A. parasiticus* had given inconsistent results concerning the linkage of *nor-1* and *ver-1*.

Restriction mapping of lambda and cosmid clones (including NorA) from *A. parasiticus* and *A. flavus* containing *omtA* provided a 60 kb physical map encompassing nine genes, including *nor-1*, *ver-1*, *aflR*, *pksA*, and *omtA* (Figure 2).[173] The observation that the genes for the entire AFB_1 pathway are ostensibly at a single locus on the chromosome was a major experimental breakthrough enabling researchers to uncover new genes by targeting regions neighboring previously described genes. The genes, *ord-1*, which is a putative cytochrome P450 monooxygenase, and *ord-2*, were identified immediately downstream of *omtA*.[173] Although *ord-1*, *ord-2*, and *omtA* are closely spaced and collinear, no evidence for cotranscription was observed.[187] The right side of the cluster was expanded beyond *omtA* by the mapping of clones neighboring the gene *vbs*.[179] The experimentally determined distance between *vbs* and *omtA* was 3.3 kb, with an intervening cytochrome P450 monooxygenase gene.

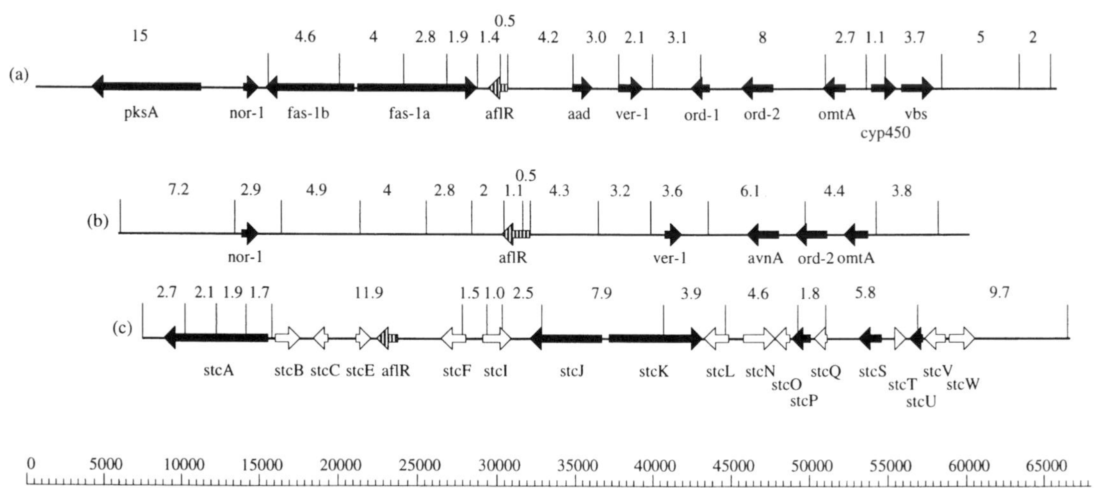

Figure 2 Aflatoxin biosynthetic gene clusters observed in (a) *A. parasiticus* and (b) *A. flavus* and (c) the sequenced sterigmatocystin biosynthetic gene cluster from *A. nidulans*. *Eco*RI restrictions sites are shown by vertical bars and corresponding fragment sizes are listed above the sequence in kilobases (kb). Genes whose function has been experimentally supported are indicated in black arrows. Function based upon homology is denoted by open arrows and regulatory genes are marked by barred arrows. Compounds are shown in Schemes 4, 18, and 23. Gene designations, listed as groups of homologues (former names underlined), are: *aflR*, *apa-2*, *afl-2*, zinc binuclear cluster motif regulatory gene; *omtA*, *omt-1*, *omt-1b*, *O*-methyltransferase (**7** to **12**); *ord-2*, monooxygenase; *stcA*, *pksST*, *pksA*, *pksL1*, polyketide synthase; *stcB*, P450 monooxygenase; *stcC*, peroxidase; *stcE*, *nor-1* ketoreductase (**13** to **14**); *stcF*, *avnA*, *ord-1*, P450 monooxygenase (**15** to **39**); *stcI*, lipase/esterase; *stcJ*, *fas-1A*, fatty acid synthase (β-subunit); *stcK*, *fas-1B*, fatty acid synthase (α-subunit); *stcL*, P450 monooxygenase (**60** to **71**); *stcN*, *vbs*, versicolorin B synthase (**57** to **60**); *stcO*, oxidoreductase; *stcP*, *O*-methyltransferase (**61** to **7**); *stcS*, *apstcS*, *verB*, P450 monooxygenase (**17** to **7**); *stcT*, elongation factor 1γ; *stcU*, *verA*, *ver-1a*, ketoreductase (**17** to **7**); *stcV*, *norA*, *aad*, dehydrogenase/reductase (**13** to **14**); and *stcW*, FAD monooxygenase.

The *afl-1* locus in *A. flavus* possesses a 120 kb deletion including most of the AFB_1 biosynthetic cluster. Complementation with cosmids containing 90 kb of *A. flavus* DNA restores AFB_1 biosynthesis.[215] Importantly, genes involved in the conversion of averufin (**15**) into versiconal acetate (**16**) (*avfl*) and *O*-methylsterigmatocystin ((**12**), Scheme 4) into AFB_1 (named *ordl*, *ord-1* above has been renamed *avnA*) were localized to 7 and 3.3 kb regions, respectively.[216] *stcU* is located at the left end of 2.9 Mb chromosome IV by Southern hybridization to electrophoretically separated *A. nidulans* chromosomes.[217]

Hybridization experiments denoted the presence of a second copy of the *ver-1* gene in wild-type *A. parasiticus* SU-1.[218] The predicted amino acid sequence for the coding regions showed 95% identity between the genes; however, a nonsense mutation was present at codon 87, which strongly suggests that only one copy is functional. The duplicated region extended 12 kb upstream of the renamed functional *ver-1A* and dysfunctional *ver-1B* genes in *A. parasiticus*. It has been observed that for *A. flavus*, which does not contain the duplication, only 40% of the strains produce AFB_1.[218–220] Southern blotting indicated that two copies of the *norA* gene may be present in the *A. parasiticus* genome, although the location of the second copy is unknown.[170] This gene may correspond to *aad*, which is in the duplicated region from *ver-1A* to *aflR*.[173]

The sterigmatocystin pathway is expected to require each cognate enzyme from the entire AFB_1 pathway, excepting at least one protein following (**7**) (Figure 2). A contiguous 60 kb region of the *A. nidulans* genome in the region of *verA* encompassed 24 actively transcribed regions including homologues for *vbs*, *nor-1*, *norA*, *aflR*, *ver-1*, *pksA*, and *fas-1A*.[199] Although only the functions of *stcA*, *stcS*, and *stcU* have been rigorously assigned, other open reading frames code for four P450 monooxygenases, an oxidase, and an FAD monooxygenase gene, which are expected to be involved in many of the cryptic oxidative structural transformations between norsolorinic acid (**13**) and sterigmatocystin (**7**). Notably absent is the cognate gene for *omtA* required for the penultimate conversion in the AFB_1 pathway. The presence of *stcT* and *ord-2* in the clusters, which resemble translation elongation factor γ, remains an enigma.

Although localized physical similarities are present,[173] substantial genetic reorganization between *A. flavus*/*A. parasiticus* and *A. nidulans* is apparent in Figure 2. Furthermore, amino acid identity between the AFB_1-producing species is commonly >90%, whereas in *A. nidulans* often less than 80% similarity is retained. The presence of seemingly closely related biosynthetic pathways in the phylogenically distant organisms *Bipolaris*, *Chaetomium*, *Farrowia*, and *Monocillium* is suggestive of an ancient horizontal gene transfer event.[221]

1.17.5.6 Regulation of Aflatoxin Production

Time-course experiments to monitor the appearance of AFB_1 and the transcripts and the gene products of *nor 1* and *ver 1* during idiophase indicate that partial regulation of AFB_1 biosynthesis occurs at the transcriptional level.[222,223] Similar patterns of transcript accumulation have been reported for each of the previously described AFB_1-related genes; an exception, however, lies with the strongly conserved regulatory factor gene *aflR*, previously named *afl*-2 (*A. flavus*) and *apa-2* (*A. parasiticus*), whose transcript accumulates in a fashion similar to the primary metabolic gene *pyrG* (uracil biosynthesis).[187] The *aflR* genes have been identified in each *Aspergillus* strain examined and exhibited nucleotide sequence homology in excess of 95%.[224] Hybridization of *afl*-2 to similar sequences in *A. parasiticus*, *A. oryzae*, and *A. sojae* has been observed, although the gene is often absent in AFB_1 non-producing strains.

The *A. parasiticus* regulatory gene *aflR*, located 8 kb from *ver-1*, was initially cloned by the identification of Nor-1 and Ver-1 strains which overproduced pathway intermediates (**13**) and (**17**) (Scheme 4), respectively, following transformation with a cosmid library.[164] Transformants resulting from multiple copies of the regulatory gene were shown to have increased enzyme activity for norsolorinic acid and sterigmatocystin metabolism. The 46.7 kDa protein possesses a zinc cluster motif (Cys–Xaa2–Cys–Xaa6–Cys–Xaa6–Cys–Xaa2–Cys–Xaa6–Cys) found in other fungal transcriptional regulatory proteins.[225–228] An overexpressed AFLR fragment weakly bound GAL4-like promoter regions upstream of *aflR* in *A. parasiticus*.[229] This potentially autoregulatory *trans*-acting factor implies that AFB_1 biosynthesis is positively regulated in a similar fashion to sulfur and nitrogen metabolism in *A. nidulans* and *N. crassa*.[230–234] Interestingly, the antisense transcript *aflRas*, which may play a regulatory role, has also been observed.[224] Strong stimulatory and inhibitory effects of ammonium and nitrate ions, respectively, provide a substantive link between nitrogen and AFB_1 regulation.[229,235–237] Other factors which affect AFB_1 production include zinc, cytosolic NADPH/NADP levels, and carbohydrate sources.[222,238–241]

1.17.6 REFERENCES

1. J. W. Dickens, in "Mycotoxins in Human and Animal Health," eds. J. V. Rodricks, C. W. Hesseltine, and M. A. Mehlman, Pathotox, Park Forest South, IL, 1977, p. 99.
2. E. B. Lillehoj and C. W. Hesseltine, in "Mycotoxins in Human and Animal Health," eds. J. V. Rodricks, C. W. Hesseltine, and M. A. Mehlman, Pathotox, Park Forest South, IL, 1977, p. 107.
3. T. W. Kensler and J. D. Groopman, in "Comprehensive Toxicology, Vol. 12: Molecular Mechanisms of Chemical and Radiation Carcinogenesis," eds. I. G. Sipes, C. A. McQueen, and A. J. Gandolfi, Pergamon, Oxford, 1997, p. 201.
4. T. Aoyama, S. Yamano, P. S. Guzelian, and H. V. Gelboin, *Proc. Natl. Acad. Sci. USA*, 1990, **87**, 4790.
5. L. M. Forrester, G. E. Neal, D. J. Judah, M. J. Glancey, and C. R. Wolf, *Proc. Natl. Acad. Sci. USA*, 1990, **87**, 8306.
6. S. Gopalakrishnan, T. M. Harris, and M. P. Stone, *Biochemistry*, 1990, **29**, 10 438.
7. S. W. Baertschi, K. D. Raney, M. P. Stone, and T. M. Harris, *J. Am. Chem. Soc.*, 1988, **110**, 7929.
8. R. S. Iyer, M. W. Voehler, and T. M. Harris, *J. Am. Chem. Soc.*, 1994, **116**, 1603.
9. J. M. Essigmann, R. G. Croy, A. M. Nadzan, W. F. Busby, Jr., V. N. Reinhold, G. Büchi, and G. N. Wogan, *Proc. Natl. Acad. Sci. USA*, 1977, **74**, 1970.
10. E. L. Loechler, M. M. Teeter, and M. D. Whitlow, *J. Biomol. Struc. Dyn.*, 1988, **5**, 1237.
11. R. S. Iyer, M. W. Voehler, and T. M. Harris, *J. Am. Chem. Soc.*, 1994, **116**, 8863.
12. S. Gopalakrishnan, X. Liu, and D. J. Patel, *Biochemistry*, 1992, **31**, 10 790.
13. I. C. Hsu, R. A. Metcalf, T. Sun, J. A. Wesh, N. J. Wang, and C. C. Harris, *Nature (London)*, 1991, **350**, 427.
14. B. Bressac, M. Kew, J. Wands, and M. Ozturk, *Nature (London)*, 1991, **350**, 429.
15. F. Aguilar, S. P. Hussain, and P. Cerutti, *Proc. Natl. Acad. Sci. USA*, 1993, **90**, 8586.
16. K. Sargeant, A. Sheridan, J. O'Kelly, and R. B. A. Carnaghan, *Nature (London)*, 1961, **192**, 1096.
17. A. S. M. van der Zijden, W. A. A. B. Koelensmid, J. Boldingh, C. B. Barrett, W. O. Ord, and J. Philip, *Nature (London)*, 1962, **195**, 1060.
18. B. V. Nesbitt, J. O'Kelly, K. Sargeant, and A. Sheridan, *Nature (London)*, 1962, **195**, 1062.
19. R. D. Hartley, B. F. Nesbitt, and J. O'Kelly, *Nature (London)*, 1963, **195**, 1056.
20. D. A. van Dorp, A. S. M. van der Zuden, R. K. Beerthuis, S. Sparreboom, W. O. Ord, K. De Jong, and R. Kenning, *Recl. Trav. Chim. Pays-Bas*, 1963, **82**, 587.
21. K. J. van der Merwe, L. Fourie, and D. B. Scott, *Chem. Ind. (London)*, 1963, 1660.
22. T. Asao, G. Büchi, M. M. Abdel-Kader, S. B. Chang, E. L. Wick, and G. N. Wogan, *J. Am. Chem. Soc.*, 1963, **85**, 1706.
23. T. Asao, G. Büchi, M. M. Abdel-Kader, S. B. Chang, E. L. Wick, and G. N. Wogan, *J. Am. Chem. Soc.*, 1965, **87**, 882.
24. T. C. van Soest and A. F. Peerdeman, *Acta Crystallogr.*, 1970, **26**, 1940.
25. K. K. Cheung and G. A. Sim, *Nature (London)*, 1964, **201**, 1185.
26. G. Büchi and D. M. Foulkes, *J. Am. Chem. Soc.*, 1966, **88**, 4534.
27. G. Büchi, D. M. Foulkes, M. Kurono, G. F. Mitchell, and R. S. Schneider, *J. Am. Chem. Soc.*, 1967, **89**, 6745.
28. S. Brechbühler, G. Büchi, and G. Milne, *J. Org. Chem.*, 1967, **32**, 2641.
29. G. Büchi and M. Weinreib, *J. Am. Chem. Soc.*, 1971, **93**, 746.
30. G. Büchi, M. A. Francisco, J. M. Liesch, and P. F. Schuda, *J. Am. Chem. Soc.*, 1981, **103**, 3497.
31. C. W. Holtzapfel, P. S. Steyn, and I. F. H. Purchase, *Tetrahedron Lett.*, 1966, 2799.
32. M. F. Dutton and J. G. Heathcote, *Chem. Ind. (London)*, 1968, 418.
33. J. G. Heathcote and M. F. Dutton, *Tetrahedron*, 1969, **25**, 1497.
34. R. D. Stubblefield, O. L. Shotwell, G. M. Shannon, D. Weisleder, and W. K. Rohwedder, *J. Agric. Food Chem.*, 1970, **18**, 391.
35. J. C. Roberts, A. H. Sheppard, J. A. Knight, and P. Roffey, *J. Chem. Soc. C*, 1968, 22.
36. J. C. Roberts, *Fortschr. Chem. Org. Naturst.*, 1974, **31**, 119.
37. P. F. Schuda, *Top. Curr. Chem.*, 1980, **91**, 75.
38. A. Ciegler and E. B. Lillehoj, *Adv. Appl. Microbiol.*, 1968, **10**, 155.
39. L. A. Goldblatt, "Aflatoxin: Scientific Background, Control and Implications," Academic Press, New York, 1969.
40. J. G. Heathcote and J. R. Hibbert, "Aflatoxins: Chemical and Biological Aspects," Elsevier North-Holland, New York, 1978, p. 151.
41. Y. Hatsuda and S. Kuyama, *J. Agric. Chem. Soc. Jpn.*, 1954, **28**, 989.
42. E. Bullock, J. C. Roberts, and J. C. Underwood, *J. Chem. Soc.*, 1962, 4179.
43. H. J. Burkhardt and J. Forgacs, *Tetrahedron*, 1968, **24**, 717.
44. J. Adye and R. I. Mateles, *Biochim. Biophys. Acta*, 1964, **86**, 418.
45. J. A. Donkersloot, D. P. H. Hsieh, and R. I. Mateles, *J. Am. Chem. Soc.*, 1968, **90**, 5020.
46. J. G. Heathcote, M. F. Dutton, and J. R. Hibbert, *Chem. Ind. (London)*, 1973, 1027.
47. M. Biollaz, G. Büchi, and G. Milne, *J. Am. Chem. Soc.*, 1968, **90**, 5017.
48. M. Biollaz, G. Büchi, and G. Milne, *J. Am. Chem. Soc.*, 1970, **92**, 1035.
49. D. P. Moody, *Nature (London)*, 1964, **202**, 188.
50. J. S. E. Holker and J. G. Underwood, *Chem. Ind. (London)*, 1964, 1865.
51. R. Thomas, in "Biogenesis of Antibiotic Substances," eds. Z. Vanek and Z. Hostalek, Academic Press, New York, 1965, p. 155.
52. J. S. E. Holker and L. J. Mulheirn, *J. Chem. Soc., Chem. Commun.*, 1968, 1576.
53. B. Franck and H. Flasch, *Fortschr. Chem. Org. Naturst.*, 1973, **30**, 151.
54. J. R. D. McCormick and E. R. Jensen, *J. Am. Chem. Soc.*, 1968, **90**, 7126.
55. J. R. D. McCormick, E. R. Jensen, N. H. Arnold, H. S. Corey, U. H. Joachim, S. Johnson, P. A. Miller, and N. O. Sjolander, *J. Am. Chem. Soc.*, 1968, **90**, 7127.
56. L. S. Lee, J. W. Bennett, L. A. Goldblatt, and R. E. Lundin, *J. Am. Oil Chem. Soc.*, 1971, **48**, 93.
57. J. A. Donkersloot, R. I. Mateles, and S. S. Yang, *Biochem. Biophys. Res. Commun.*, 1972, **47**, 1051.
58. L. S. Lee, J. W. Bennett, A. F. Cucullu, and J. B. Stanley, *J. Agric. Food Chem.*, 1975, **23**, 1132.
59. J. W. Bennett, L. S. Lee, S. M. Shoss, and G. H. Boudreaux, *Appl. Environ. Microbiol.*, 1980, **39**, 835.

60. D. P. H. Hsieh, *J. Agric. Food Chem.*, 1973, **21**, 468.
61. R. H. Cox, F. Churchill, R. J. Cole, and J. W. Dorner, *J. Am. Chem. Soc.*, 1977, **99**, 3159.
62. L. S. Lee, J. W. Bennett, A. F. Cucullu, and R. L. Ory, *J. Agric. Food Chem.*, 1976, **24**, 1167.
63. M. T. Lin, D. P. H. Hsieh, R. C. Yao, and J. A. Donkersloot, *Biochemistry*, 1973, **12**, 5167.
64. D. P. H. Hsieh, M. T. Lin, R. C. Yao, and R. Singh, *J. Agric. Food Chem.*, 1976, **24**, 1170.
65. R. Singh and D. P. H. Hsieh, *Arch. Biochem. Biophys.*, 1977, **178**, 285.
66. C. P. Gorst-Allman, P. S. Steyn, P. L. Wessels, and D. B. Scott, *J. Chem. Soc., Perkin Trans. 1*, 1978, 961.
67. L. O. Zamir and K. D. Hufford, *Appl. Environ. Microbiol.*, 1981, **42**, 168.
68. J. W. Bennett and S. B. Christensen, *Adv. Appl. Microbiol.*, 1983, **29**, 53.
69. W. B. Turner and D. C. Aldridge, "Fungal Metabolites II," Academic Press, London, 1983, p. 121.
70. M. Tanabe, T. Hamasaki, and H. Seto, *J. Chem. Soc., Chem. Commun.*, 1970, 1539.
71. M. Tanabe, in "Specialist Periodical Reports: Biosynthesis," ed. T. A. Geissman, Chemical Society, London, 1975, vol. 3, p. 247.
72. A. G. McInnes and J. L. C. Wright, *Acc. Chem. Res.*, 1975, **8**, 313.
73. H. Seto, L. W. Cary, and M. Tanabe, *Tetrahedron Lett.*, 1974, 4491.
74. J. C. Vederas, *Nat. Prod. Rep.*, 1987, **4**, 277.
75. K. G. R. Pachler, P. S. Steyn, R. Vleggaar, and P. L. Wessels, *J. Chem. Soc., Chem. Commun.*, 1975, 355.
76. P. S. Steyn, R. Vleggaar, and P. L. Wessels, *S. Afr. J. Chem.*, 1980, **34**, 12.
77. C. P. Gorst-Allman, K. G. R. Pachler, P. S. Steyn, P. L. Wessels, and D. B. Scott, *J. Chem. Soc., Perkin Trans. 1*, 1977, 2181.
78. P. S. Steyn, R. Vleggaar, P. L. Wessels, and D. B. Scott, *J. Chem. Soc., Perkins Trans. 1*, 1979, 460.
79. K. G. R. Pachler, P. S. Steyn, R. Vleggaar, P. L. Wessels, and D. B. Scott, *J. Chem. Soc., Perkin Trans. 1*, 1976, 1182.
80. P. S. Steyn, R. Vleggaar, and P. L. Wessels, in "The Biosynthesis of Mycotoxins. A Study in Secondary Metabolism," ed. P. S. Steyn, Academic Press, New York, 1980, p. 105.
81. J. C. Vederas and T. T. Nakashima, *J. Chem. Soc., Chem. Commun.*, 1980, 183.
82. U. Sankawa, H. Shimada, T. Kobayashi, Y. Ebizuka, Y. Yamamoto, H. Noguchi, and H. Seto, *Heterocycles*, 1982, 1053.
83. C. T. Mabuni, L. Garlaschelli, R. A. Ellison, and C. R. Hutchinson, *J. Am. Chem. Soc.*, 1979, **101**, 707.
84. C. A. Townsend and S. B. Christensen, *Tetrahedron*, 1983, **21**, 3575.
85. A. J. Birch, *Fortschr. Chem. Org. Naturst.*, 1957, **14**, 187.
86. S. W. Tannenbaum and S. Nakajima, *Biochemistry*, 1969, **11**, 4626.
87. J. R. Hadfield, J. S. E. Holker, and D. N. Stanway, *J. Chem. Soc. C*, 1967, 751.
88. J. L. Gellerman, W. H. Anderson, and H. Schlenk, *Lipids*, 1974, **9**, 722.
89. A. J. Birch, *Proc. Chem. Soc.*, 1962, 3.
90. J. F. Grove, *J. Chem. Soc. C*, 1970, 1860.
91. J. H. Birkinshaw and A. Gowlland, *Biochem. J.*, 1962, **84**, 342.
92. P. G. Gordon, A. Penttila, and H. M. Fales, *J. Am. Chem. Soc.*, 1968, **90**, 1376.
93. E. Leete, *J. Am. Chem. Soc.*, 1970, **92**, 3835.
94. C. A. Townsend, S. W. Brobst, S. E. Ramer, and J. C. Vederas, *J. Am. Chem. Soc.*, 1988, **110**, 318.
95. B. J. Rawlings, P. B. Reese, S. E. Ramer, and J. C. Vederas, *J. Am. Chem. Soc.*, 1989, **111**, 3382.
96. K. Arai, B. J. Rawlings, Y. Yoshizawa, and J. C. Vederas, *J. Am. Chem. Soc.*, 1989, **111**, 3391.
97. C. A. Clark and J. A. Robinson, *J. Chem. Soc., Chem. Commun.*, 1985, 1568.
98. H. Noguchi, P. H. Harrison, K. Arai, T. T. Nakashima, L. A. Trimble, and J. C. Vederas, *J. Am. Chem. Soc.*, 1988, **110**, 2938.
99. S. Yue, J. S. Duncan, Y. Yamamoto, and C. R. Hutchinson, *J. Am. Chem. Soc.*, 1987, **109**, 1253.
100. D. E. Cane and C. C. Yang, *J. Am. Chem. Soc.*, 1987, **109**, 1255.
101. F. Lynen, *Fed. Proc. Fed. Am. Soc. Exp. Biol.*, 1961, **20**, 941.
102. D. O'Hagan, *Nat. Prod. Rep.*, 1991, **8**, 447.
103. D. O'Hagan, "The Polyketide Metabolites," Ellis Horwood, New York, 1991, p. 65.
104. S. W. Brobst and C. A. Townsend, *Can. J. Chem.*, 1994, **72**, 200.
105. C. A. Townsend, S. B. Christensen, and K. Trautwein, *J. Am. Chem. Soc.*, 1984, **106**, 3868.
106. D. S. J. McKeown, C. McNicholas, T. J. Simpson, and N. J. Willett, *J. Chem. Soc., Chem. Commun.*, 1996, 301.
107. A. A. Chuturgoon, M. F. Dutton, and R. K. Berry, *Biochem. Biophys. Res. Commun.*, 1990, **166**, 38.
108. A. A. Chuturgoon and M. F. Dutton, *Mycopathologia*, 1991, **113**, 41.
109. K. Yabe, Y. Nakamura, H. Nakajima, Y. Ando, and T. Hamasaki, *Appl. Environ. Microbiol.*, 1991, **57**, 1340.
110. D. Bhatnagar and T. E. Cleveland, *FASEB J.*, 1990, **4**, A2164.
111. K. Yabe, Y. Matsuyama, Y. Ando, H. Nakajima, and T. Hamasaki, *Appl. Environ. Microbiol.*, 1993, **59**, 2486.
112. C. A. Townsend and S. B. Christensen, *Tetrahedron Lett.*, 1986, 887.
113. J. S. E. Holker, S. A. Kagal, L. J. Milheirn, and P. M. White, *J. Chem. Soc., Chem. Commun.*, 1966, 911.
114. P. J. Aucamp and C. W. Holzapfel, *J. S. Afr. Chem. Inst.*, 1970, **23**, 40.
115. S. P. McCormick, D. Bhatnagar, and L. S. Lee, *Appl. Environ. Microbiol.*, 1987, **53**, 14.
116. M. Koreeda, B. Hulin, M. Yoshihara, C. A. Townsend, and S. Christensen, *J. Org. Chem.*, 1985, **50**, 5426.
117. C. A. Townsend, S. B. Christensen, and S. G. Davis, *J. Am. Chem. Soc.*, 1982, **104**, 6152.
118. C. A. Townsend and S. G. Davis, *J. Chem. Soc., Chem. Commun.*, 1983, 1420.
119. T. J. Simpson, A. E. de Jesus, P. S. Steyn, and R. Vleggaar, *J. Chem. Soc., Chem. Commun.*, 1982, 631.
120. C. A. Townsend, S. B. Christensen, and S. G. Davis, *J. Am. Chem. Soc.*, 1982, **104**, 6154.
121. C. A. Townsend, K. A. Plavcan, K. Pal, S. W. Brobst, M. S. Irish, E. W. Ely, Jr., and J. W. Bennett, *J. Org. Chem.*, 1988, **53**, 2472.
122. C. A. Townsend, P. R. O. Whittamore, and S. W. Brobst, *J. Chem. Soc., Chem. Commun.*, 1988, 726.
123. C. A. Townsend and S. B. Christensen, *J. Am. Chem. Soc.*, 1985, **107**, 270.
124. T. J. Simpson and D. J. Stenzel, *J. Chem. Soc., Chem. Commun.*, 1982, 890.
125. C. A. Townsend, S. G. Davis, M. Koreeda, and B. Hulin, *J. Org. Chem.*, 1985, **50**, 5428.
126. C. A. Townsend, Y. Isomura, S. G. Davis, and J. A. Hodge, *Tetrahedron*, 1989, **45**, 2263.

127. R. C. Yao and D. P. H. Hsieh, *Appl. Microbiol.*, 1974, **28**, 52.
128. S. M. McGuire and C. A. Townsend, *Bioorg. Med. Chem. Lett.*, 1993, **3**, 653.
129. N. C. Wan and D. P. H. Hsieh, *Appl. Environ. Microbiol.*, 1980, **39**, 109.
130. M. S. Anderson and M. F. Dutton, *Appl. Environ. Microbiol.*, 1980, **40**, 706.
131. M. F. Dutton and M. S. Anderson, *Appl. Environ. Microbiol.*, 1982, **43**, 548.
132. S. M. McGuire, S. W. Brobst, T. L. Graybill, K. Pal, and C. A. Townsend, *J. Am. Chem. Soc.*, 1989, **111**, 8308.
133. C. A. Townsend, S. M. McGuire, S. W. Brobst, T. L. Graybill, K. Pal, and C. E. Barry, III, in "Secondary-Metabolite Biosynthesis and Metabolism," ed. R. J. Petroski, Plenum, New York, 1992, vol. 44, p. 141.
134. C. A. Townsend, *Pure Appl. Chem.*, 1986, **58**, 227.
135. P. S. Steyn, R. Vleggaar, P. L. Wessels, R. J. Cole, and D. B. Scott, *J. Chem. Soc., Perkin Trans. 1*, 1979, 451.
136. K. Yabe and T. Hamasaki, *Appl. Environ. Microbiol.*, 1993, **59**, 2493.
137. B.-K. Lin and J. A. Anderson, *Arch. Biochem. Biophys.*, 1992, **293**, 67.
138. S. M. McGuire, J. C. Silva, E. G. Casillas, and C. A. Townsend, *Biochemistry*, 1996, **35**, 11 470.
139. K. Yabe, Y. Ando, and T. Hamasaki, *Agric. Biol. Chem.*, 1991, **55**, 1907.
140. H. W. Schroeder and W. W. Carlton, *Appl. Microbiol.*, 1973, **25**, 146.
141. K. E. Papa, *Appl. Environ. Microbiol.*, 1977, **33**, 206.
142. D. Bhatnager, T. E. Cleveland, and D. G. I. Kingston, *Biochemistry*, 1991, **30**, 4343.
143. P. Dimroth, H. Walter, and F. Lynen, *Eur. J. Biochem.*, 1970, **13**, 98.
144. B. Franck, in "The Biosynthesis of Mycotoxins: A Study in Secondary Metabolism," ed. P. S. Steyn, Academic Press, New York, 1980, p. 157.
145. J. A. Anderson, B.-K. Lin, H. J. Williams, and A. I. Scott, *J. Am. Chem. Soc.*, 1988, **110**, 1623.
146. G. C. Elsworthy, J. S. E. Holker, J. B. McKeown, J. B. Robinson, and L. J. Mulheirn, *J. Chem. Soc., Chem. Commun.*, 1970, 1069.
147. T. L. Graybill, K. Pal, S. M. McGuire, S. W. Brobst, and C. A. Townsend, *J. Am. Chem. Soc.*, 1989, **111**, 8306.
148. R. Singh and D. P. H. Hsieh, *Appl. Environ. Microbiol.*, 1976, **31**, 743.
149. C. M. H. Watanabe and C. A. Townsend, *J. Org. Chem.*, 1996, **61**, 1990.
150. T. E. Cleveland and D. Bhatnager, *Can. J. Microbiol.*, 1987, **33**, 1108.
151. M. Chatterjee and C. A. Townsend, *J. Org. Chem.*, 1994, **59**, 4424.
152. T. J. Simpson, A. E. deJesus, P. S. Steyn, and R. J. Vleggaar, *J. Chem. Soc., Chem. Commun.*, 1983, 338.
153. G. Guroff, J. W. Daly, D. W. Jerina, J. Renson, B. Witkop, and S. Udenfriend, *Science*, 1967, **157**, 1524.
154. F. Trail, N. Mahanti, and J. Linz, *Microbiology*, 1995, **141**, 755.
155. E. Robens, "A Perspective on Aflatoxin in Field Crops and Animal Food Products in the United States (ARS-83)," US Department of Agriculture, Washington, DC, 1990.
156. C. Woloshuk, E. Seip, G. Payne, and C. Adkins, *Appl. Environ. Microbiol.*, 1989, **55**, 86.
157. G. A. Payne and C. P. Woloshuk, *Mycopathologia*, 1989, **107**, 139.
158. J. S. Horng, P.-K. Chang, J. J. Petska, and J. E. Linz, *Mol. Gen. Genet.*, 1990, **224**, 294.
159. T. S. Wu and J. E. Linz, *Appl. Environ. Microbiol.*, 1993, **59**, 2998.
160. C. D. Skory, J. S. Horng, J. J. Pestka, and J. E. Linz, *Appl. Environ. Microbiol.*, 1990, **56**, 3315.
161. G. A. Payne, J. Nystrom, D. Bhatnagar, T. E. Cleveland, and C. P. Woloshuk, *Appl. Environ. Microbiol.*, 1993, **59**, 156.
162. P. K. Chang, C. D. Skory, and J. E. Linz, *Curr. Genet.*, 1992, **21**, 231.
163. G. D. Skory, P. K. Chang, J. Cary, and J. E. Linz, *Appl. Environ. Microbiol.*, 1992, **58**, 3527.
164. P. K. Chang, J. W. Cary, D. Bhatnagar, T. E. Cleveland, J. W. Bennett, J. E. Linz, C. P. Woloshuk, and G. A. Payne, *Appl. Environ. Microbiol.*, 1993, **59**, 3273.
165. M. Delledonne, R. Porcari, and C. Fogher, *Nucleic Acids Res.*, 1990, **18**, 6435.
166. J. J. Heilmann, H. J. Maegert, and H. G. Gassen, *Eur. J. Biochem.*, 1988, **174**, 485.
167. A. Vidal-Cros, F. Viviani, G. Labesse, M. Boccara, and M. Gaudry, *Eur. J. Biochem.*, 1994, **219**, 985.
168. F. Trail, P.-K. Chang, J. Cary, and J. E. Linz, *Appl. Environ. Microbiol.*, 1994, **60**, 4078.
169. B. Persson, M. Krook, and H. Joernvall, *Eur. J. Biochem.*, 1991, **200**, 537.
170. J. W. Cary, M. Wright, D. Bhatnagar, R. Lee, and F. S. Chu, *Appl. Environ. Microbiol.*, 1996, **62**, 360.
171. N. P. Keller, S. Segner, D. Bhatnagar, and T. H. Adams, *Appl. Environ. Microbiol.*, 1995, **61**, 3628.
172. J. Yu, J. W. Cary, D. Bhatnagar, T. E. Cleveland, N. P. Keller, and F. S. Chu, *Appl. Environ. Microbiol.*, 1993, **59**, 3564.
173. J. Yu, P.-K. Chang, J. W. Cary, M. Wright, D. Bhatnagar, T. E. Cleveland, G. A. Payne, and J. E. Linz, *Appl. Environ. Microbiol.*, 1995, **61**, 2365.
174. D. Bhatnagar, A. H. Ullah, and T. E. Cleveland, *Prep. Biochem.*, 1988, **18**, 321.
175. N. P. Keller, H. C. Dischinger, D. Bhatnagar, T. E. Cleveland, and A. H. J. Ullah, *Appl. Environ. Microbiol.*, 1993, **59**, 479.
176. H. S. Kelkar, N. P. Keller, and T. H. Adams, *Appl. Environ. Microbiol.*, 1996, **62**, 4296.
177. K. Yabe, Y. Ando, J. Hashimoto, and T. Hamasaki, *Appl. Environ. Microbiol.*, 1989, **55**, 2172.
178. S. M. McGuire, Ph.D. Thesis, Johns Hopkins University, 1992.
179. J. C. Silva, R. E. Minto, C. E. Barry, III, K. A. Holland, and C. A. Townsend, *J. Biol. Chem.*, 1996, **23**, 13 600.
180. J. C. Silva and C. A. Townsend, *J. Biol. Chem.*, 1997, **272**, 804.
181. R. C. Lee, J. W. Cary, D. Bhatnagar, and F. S. Chu, *Food Agric. Immunol.*, 1995, **7**, 21.
182. A. Muheim, R. Waldner, D. Sanglard, J. Reiser, H. E. Schoemaker, and M. S. A. Leisola, *Eur. J. Biochem.*, 1991, **195**, 369.
183. A. A. Chuturgoon and M. F. Dutton, *Prep. Biochem.*, 1991, **21**, 125.
184. T. E. Cleveland, A. R. Lax, L. S. Lee, and D. Bhatnagar, *Appl. Environ. Microbiol.*, 1987, **53**, 1711.
185. M. S. Anderson and M. F. Dutton, *Experientia*, 1979, **35**, 21.
186. N. Mahanti, D. Bhatnagar, J. W. Cary, J. Joubran, and J. E. Linz, *Appl. Environ. Microbiol.*, 1996, **62**, 191.
187. F. Trail, N. Mahanti, M. Rarick, R. Mehigh, S.-H. Liang, R. Zhou, and J. E. Linz, *Appl. Environ. Microbiol.*, 1995, **61**, 2665.
188. H. Köttig, G. Rottner, K.-F. Beck, M. Schweizer, and E. Schweizer, *Mol. Gen. Genet.*, 1991, **226**, 310.

189. M. E. Mayorga and W. E. Timberlake, *Genetics*, 1990, **126**, 73.
190. P.-K. Chang, J. W. Cary, J. Yu, D. Bhatnagar, and T. E. Cleveland, *Mol. Gen. Genet.*, 1995, **248**, 270.
191. C. P. Woloshuk and G. A. Payne, *Appl. Environ. Microbiol.*, 1994, **60**, 670.
192. G. H. Feng, F. S. Chu, and T. J. Leonard, *Appl. Environ. Microbiol.*, 1992, **58**, 455.
193. J.-H. Yu and T. J. Leonard, *J. Bacteriol.*, 1995, **117**, 4792.
194. D. A. Hopwood and D. H. Sherman, *Annu. Rev. Genet.*, 1990, **24**, 37.
195. D. W. Brown, T. H. Adams, and N. P. Keller, *Proc. Natl. Acad. Sci. USA*, 1996, **93**, 14 873.
196. M. Schweizer, L. M. Roberts, H. J. Holtke, K. Takabayashi, E. Hollerer, B. Hoffmann, G. Muller, H. Kottig, and E. Schweizer, *Mol. Gen. Genet.*, 1986, **203**, 479.
197. A. H. Mohamed, S. S. Chirala, N. H. Mody, W. Huang, and S. J. Wakil, *J. Biol. Chem.*, 1988, **263**, 12 315.
198. C. M. H. Watanabe, D. Wilson, J. E. Linz, and C. A. Townsend, *Chem. Biol.*, 1996, **3**, 463.
199. D. W. Brown, J.-H. Yu, H. S. Kelkar, M. Fernandes, T. C. Nesbitt, N. P. Keller, T. H. Adams, and T. J. Leonard, *Proc. Natl. Acad. Sci. USA*, 1996, **93**, 1418.
200. C. Scotti, M. Piatti, A. Cuzzoni, P. Perani, A. Tognoni, G. Grandi, A. Galizzi, and A. M. Albertini, *Gene*, 1993, **130**, 65.
201. Y. Takano, Y. Kubo, K. Shimizu, K. Mise, T. Okuno, and I. Furusawa, *Mol. Gen. Genet.*, 1995, **249**, 162.
202. B. Felenbok and H. M. Sealy-Lewis, in "Aspergillus: 50 Years On," eds. S. D. Martinelli and J. R. Kinghorn, Elsevier, Amsterdam, *Progress in Industrial Microbiology*, 1994, vol. 29, p. 141.
203. D. J. Smith, K. R. Burnham, J. H. Bull, J. E. Hodgson, J. M. Ward, P. Browne, J. Brown, B. Barton, J. E. Alison, and G. Turner, *EMBO J.*, 1990, **9**, 741.
204. N. Kimura and T. Tsuge, *J. Bacteriol.*, 1993, **175**, 4427.
205. T. M. Hohn, S. P. McCormick, and A. E. Desjardins, *Curr. Genet.*, 1993, **24**, 291.
206. J. R. Kinghorn and S. E. Unkles, in "Aspergillus: 50 Years On," eds. S. D. Martinelli and J. R. Kinghorn, Elsevier, Amsterdam, *Progress in Industrial Microbiology*, 1994, vol. 29, p. 181.
207. A. R. Hawkins, H. K. Lamb, and C. F. Roberts, in "Aspergillus: 50 Years On," eds. S. D. Martinelli and J. R. Kinghorn, Elsevier, Amsterdam, *Progress in Industrial Microbiology*, 1994, vol. 29, p. 195.
208. S. J. Gurr, S. E. Unkles, and J. R. Kinghorn, in "Gene Structure in Eukaryotic Microbes," ed. J. R. Kinghorn, IRL, Oxford, 1990, p. 93.
209. K. E. Papa, *Mycologia*, 1978, **70**, 766.
210. K. E. Papa, *Genet. Res.*, 1979, **34**, 1.
211. K. E. Papa, *Mycologia*, 1977, **68**, 159.
212. J. E. Lennox and C. K. Davis, *Exp. Mycol.*, 1983, **7**, 192.
213. K. E. Papa, J. Gen. Microbiol., 1982, **128**, 1345.
214. K. E. Papa, *Can. J. Microbiol.*, 1984, **30**, 68.
215. R. Prieto, G. L. Yousibova, and C. P. Woloshuk, *Appl. Environ. Microbiol.*, 1996, **62**, 3567.
216. C. P. Woloshuk, G. L. Yousibova, J. A. Rollins, D. Bhatnagar, and G. A. Payne, *Appl. Environ. Microbiol.*, 1995, **61**, 3019.
217. N. P. Keller, N. J. Kantz, and T. H. Adams, *Appl. Environ. Microbiol.*, 1994, **60**, 1444.
218. S.-H. Liang, C. D. Skory, and J. E. Linz, *Appl. Environ. Microbiol.*, 1996, **62**, 4568.
219. T. E. Cleveland and D. Bhatnagar, in "Mycotoxins, Cancer, and Health," eds. G. A. Bray and D. H. Ryan, Louisiana State University Press, Baton Rouge, LA, 1991, p. 270.
220. J. W. Bennett and K. E. Papa, *Adv. Plant Pathol.*, 1988, **6**, 265.
221. R. J. Cole and R. H. Cox, "Handbook of Toxic Fungal Metabolites," Academic Press, New York, 1981, p. 1.
222. C. D. Skory, P.-K. Chang, and J. E. Linz, *Appl. Environ. Microbiol.*, 1993, **59**, 1642.
223. T. E. Cleveland and D. Bhatnagar, *Can. J. Microbiol.*, 1990, **36**, 1.
224. C. P. Woloshuk, K. R. Foutz, J. F. Brewer, D. Bhatnagar, T. E. Cleveland, and G. A. Payne, *Appl. Environ. Microbiol.*, 1994, **60**, 2408.
225. J. A. Baum, R. Geever, and N. H. Giles, *Mol. Cell. Biol.*, 1987, **7**, 1256.
226. M. A. Davis and M. J. Hines, in "More Gene Manipulations in Fungi," eds. J. W. Bennett and L. L. Lasure, Academic Press, London, 1991, p. 151.
227. Y.-H. Fu and G. A. Marzluf, *Mol. Cell. Biol.*, 1990, **10**, 1056.
228. C. Scazzocchio, in "*Aspergillus*: Biology and Industrial Applications," eds. J. W. Bennett and M. A. Klich, Butterworth–Heinemann, London, 1992, p. 43.
229. P. K. Chang, K. C. Ehrlich, J. Yu, D. Bhatnagar, and T. E. Cleveland, *Appl. Environ. Microbiol.*, 1995, **61**, 2732.
230. G. A. Marzluf, *Annu. Rev. Microbiol.*, 1993, **47**, 31.
231. G. A. Marzluf, *Adv. Genet.*, 1994, **31**, 187.
232. G. F. Yuan, Y.-H. Fu, and G. A. Marzluf, *Mol. Cell. Biol.*, 1991, **11**.
233. G. Burger, J. Strauss, C. Scazzocchio, and B. F. Lang, *Mol. Cell. Biol.*, 1991, **11**, 5746.
234. G. A. Marzluf and R. L. Metzenberg, *J. Mol. Biol.*, 1968, **33**, 423.
235. G. W. Niehaus and W. Jiang, *Mycopathologia*, 1989, **107**, 131.
236. J. W. Bennett, P. L. Rubin, L. S. Lee, and P. N. Chen, *Mycopathologia*, 1979, **69**, 161.
237. T. Kachholz and A. L. Demain, *J. Nat. Prod.*, 1983, **46**, 499.
238. R. L. Buchanan and D. F. Lewis, *Appl. Environ. Microbiol.*, 1984, **48**, 306.
239. R. L. Buchanan, S. B. Jones, W. V. Gerasimowicz, L. L. Zaika, H. G. Stahl, and L. A. Ocker, *Appl. Environ. Microbiol.*, 1987, **53**, 1224.
240. S. K. Gupta, K. K. Maggon, and T. A. Venkitasubramanian, *Appl. Environ. Microbiol.*, 1976, **32**, 753.
241. S. K. Gupta, K. K. Maggon, and T. A. Venkitasubramanian, *J. Gen. Microbiol.*, 1977, **99**, 43.

1.18
Structure, Function, and Engineering of Bacterial Aromatic Polyketide Synthases

MAIA RICHARDSON and CHAITAN KHOSLA
Stanford University, CA, USA

1.18.1 INTRODUCTION

1.18.1.1 Bacterial Aromatic Polyketides

Polyketides were named in the 1890s to refer to a structurally diverse group of natural products that contained many carbonyls and alcohols, generally separated by methylene carbons.[1] They are synthesized by a series of decarboxylative condensation reactions between small carboxylic acids and malonate using enzyme complexes homologous to fatty acid synthases, called polyketide synthases (PKSs). Many polyketides are biologically active as antibiotics, anticancer agents, and immunosuppressants, and are pharmacologically important products. There are several general reviews of PKSs.[2–8]

Polyketides and the enzymes that make them have many similarities at the genetic, biosynthetic, and product structural levels; however, there are some differences that allow division into subcategories. There are significant differences between kingdoms and one natural division is bacteria, fungi, and plants. Much research has focused on the two types of bacterial PKSs. Modular PKSs have large, multifunctional proteins which use each enzyme once to synthesize macrolide or polyether polyketides.[4] This chapter will focus on bacterial aromatic PKSs.

Although bacteria produce a broad range of aromatic natural products, the term "bacterial aromatic polyketides" generally refers to polycyclic molecules of bacterial origin that are derived from multiple decarboxylative condensations of malonyl units leading to the formation of relatively long chain polyketide intermediates (>14 carbon atoms). In each case this highly reactive intermediate is further processed through a unique set of intramolecular cyclization, elimination, oxido-reduction, and group transfer reactions to generate a highly functionalized natural product consisting of 3–6 fused and/or disconnected six (and occasionally five)-membered carbocyclic and/or heterocyclic rings. These compounds are produced in soil bacteria of the Actinomycetes class, being particularly plentiful in the *Streptomyces* genus.[4]

1.18.1.2 Bacterial Aromatic Polyketide Synthases

The defining feature of a bacterial aromatic polyketide is a pathway intermediate (or, in some cases, even the final product itself) that is synthesized by a multiprotein assembly called a bacterial aromatic PKS. Other terms including Type II PKSs[3] and iterative PKSs[9] have also been used in the literature to describe this family of multifunctional catalysts. Bacterial aromatic PKSs can be defined as complexes of several monofunctional (and occasionally bifunctional) proteins that convert specific carboxylic acid primers (activated as CoA thioesters) and malonyl-CoA derived extenders into polycyclic aromatic intermediates. Below it will become clear that, since the physical existence of these protein complexes has yet to be proven, the above definition of an aromatic PKS is arbitrary. However, given the highly reactive nature of nascent polyketone chains, the well-established precedent for the existence of covalent intermediates in fatty acid biosynthesis (a mechanistically and evolutionally related process),[10,11] and the available insights from genetic and protein chemical analysis of aromatic PKSs (see below), it is reasonable to expect that the proteins involved in chain building, modification, and early-stage cyclization and dehydration reactions must associate with each other to form a multienzyme complex in a manner that precludes diffusion of intermediates into the surrounding medium. Therefore, although neither the precise composition nor the exact product of a bacterial aromatic PKS can be unambiguously identified, the above definition should provide a basis for further structural and mechanistic studies of the PKSs and also the post-PKS enzymes in bacterial aromatic polyketide pathways.

Whereas a majority of bacterial aromatic PKSs studied thus far use acetate units as primers, alternative primer units such as propionate (in daunomycin biosynthesis)[12] and malonate or malonamide (in oxytetracycline biosynthesis)[13] are also used. Further examination of the structures of other actinomycete-derived aromatic polyketides suggests that their PKSs may also be capable of using other primers, including butyrate (frenolicin),[14,15] isovalerate (espicufolin),[16] and isoprene-derived carboxylic acids.[17] Unlike the modular PKSs, however, bacterial aromatic PKSs invariably derive their extended units from malonyl-CoA. In cases where methyl branches are introduced into the polyketide skeleton, this most likely occurs through the action of *S*-adenosylmethionine-dependent methyltransferases, as in the case of oxytetracycline biosynthesis.[18]

Bacterial aromatic PKSs comprise a relatively small set of enzymatic activities. The "minimal PKS," which is essential for the biosynthesis of a polyketide chain, consists of a *β*-ketoacyl synthase or ketosynthase (KS), a chain length factor (CLF), an acyl carrier protein (ACP),[19] and most probably a malonyl-CoA:ACP malonyltransferase or acyltransferase (AT). Other activities may also be required, especially in cases where primer units are derived from acyl-CoA species other than acetyl-CoA. The minimal PKS is responsible for iteratively catalyzing decarboxylative condensation reactions leading to the formation of a polyketone intermediate. Whereas the exact stage at which the first cyclization occurs has not yet been established, proposed models favor synthesis of the complete carbon chain backbone before cyclization (see below). In the related fatty acid synthase, the growing polyketide chain is attached to the KS via a thioester linkage at an active-site cysteine, and the KS also contains the active sites for decarboxylation and condensation.[20–22] An AT generally does not appear as a separate gene in bacterial aromatic PKS gene clusters. An acyltransferase-like motif has been reported to exist at the *C*-terminal end of all KSs from bacterial aromatic PKSs; however, its function remains to be established and mutational analysis suggests that it is non-essential for polyketide biosynthesis.[23,24] Alternatively, it is possible that an AT encoded elsewhere in the genome may be required for PKS activity.[25] The chain length factor shares considerable sequence similarity to the KS but lacks the active-site cysteine residue; it appears to be involved in determining the polyketide chain length (see below).[19,26] The ACP is a small protein with a covalently bound 4′-phosphopantethine cofactor; malonyl extender units are transferred to the terminal thiol of this cofactor in preparation for each condensation reaction.[27] The genes encoding the KS and ACP are homologous to their fatty acid synthase counterparts.

In addition to the minimal PKS, bacterial aromatic PKSs can also include additional enzymatic activities, including ketoreductase (KR), aromatase (ARO), and cyclase (CYC) activities. These activities are also encoded as monofunctional or bifunctional proteins. Whereas KRs show some sequence similarity to certain other NAD(P)H-dependent dehydrogenases,[28] AROs and CYCs appear to be evolutionally unrelated to any other known family of enzymes.[23] As discussed above, the assignment of these proteins as PKS subunits (as opposed to independent post-PKS enzymes) is tentative. Possibly together with the minimal PKS, these enzymes convert the reactive polyketone intermediate into a relatively stable polycyclic aromatic compound through a series of temporally and/or regiochemically controlled reduction, cyclization, and dehydration steps.

The biosynthetic pathways of most naturally occurring bacterial aromatic polyketides also include post-PKS enzymatic reactions. For example, *S*-adenosylmethionine-dependent methyltransferases add methyl groups to carbons or oxygens, whereas glycosyltransferases modify core polyketide intermediates with a variety of sugar moieties. NAD(P)H-dependent oxidoreductases and cytochrome P-450 dependent oxygenases modulate the oxidation level of individual carbon atoms. Indeed, post-PKS pathways display impressive enzymological diversity; many of these transformations are reviewed elsewhere in this series. The remainder of this chapter will focus on bacterial aromatic PKSs themselves.

1.18.2 GENES ENCODING BACTERIAL AROMATIC PKSs

1.18.2.1 Background and Overview of Methodologies

Thus far, the genes encoding several bacterial aromatic PKSs have been cloned and sequenced. A monumental discovery that has facilitated such progress was that all genes exclusively involved in the biosynthesis of a given natural product tend to be clustered in the bacterial genome.[29,30] Indeed, since these biosynthetic gene clusters also include resistance and transcriptional regulatory genes, a variety of genetic strategies have been developed over the years to take advantage of nature's benevolence in order to clone complete biosynthetic gene clusters of interest. Frequently used strategies include complementation of blocked mutants, transfer of partial or complete pathways in a surrogate host, homology-based gene isolation, identification of resistance gene(s) through selection in a heterologous host, and reverse genetics based on limited amino acid sequence of a purified pathway enzyme. It should be noted, however, that considerable variability exists within this technological repertoire with respect to degree of difficulty and likelihood of success. For example, whereas homology-based cloning is relatively straightforward and has the important advantage of not requiring any genetic tools in the producer host organism, it may lead to the cloning of a wrong gene cluster, since many actinomycetes carry more than one polyketide gene cluster within their chromosomes. In contrast, although heterologous expression of an entire gene cluster in a surrogate host provides definitive evidence for having cloned all the genes of interest, it is risky since it requires functional expression of all genes from their natural promoters, and technically challenging since it requires the construction of large plasmids and their introduction into the host.[29,31–33]

Within the entire biosynthetic gene cluster, the genes encoding bacterial aromatic PKSs are usually (but not always) contiguous. More significant from the viewpoint of genetic analysis is the fact that the genes encoding individual PKS subunits share a high degree of sequence similarity with their counterparts in other gene clusters. To date, 16 bacterial aromatic PKS gene clusters have been completely or partially sequenced. These include PKSs involved in the biosynthesis of structurally diverse natural products including benzoisochromanequinones, anthracyclines, tetracyclines, angucylines, and spore pigments. The genetic content of these gene clusters is summarized below and in Figure 1.

1.18.2.2 Benzoisochromanequinones

Benzoisochromanequinones, such as actinorhodin, granaticin, frenolicin, nanaomycin, kalafungin, and griseusin, have the core structure shown in (**1**)–(**6**) (here and in later structures, unless noted otherwise, R refers to sugar groups attached to the polyketide).

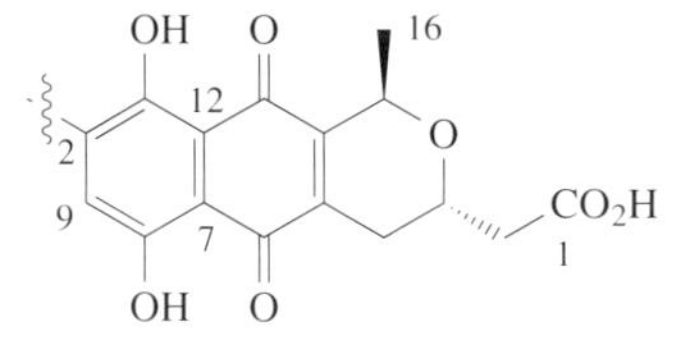

(**1**) 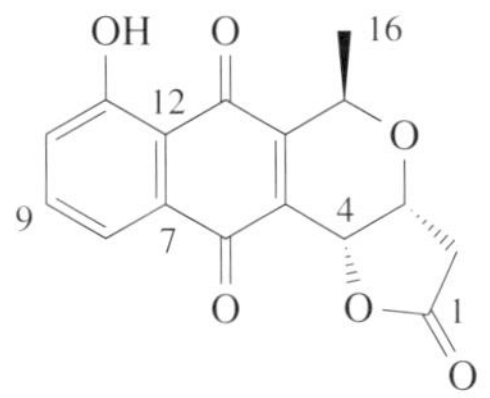 (**2**) 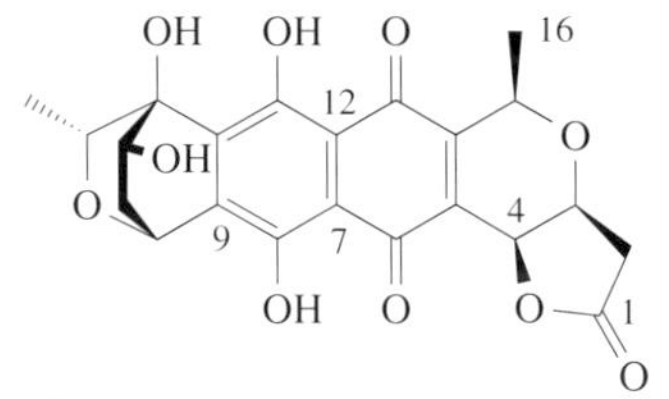 (**3**)

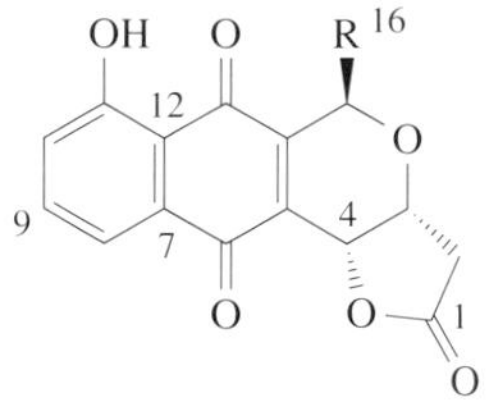

(**4**) R= Me
(**5**) R = Pr

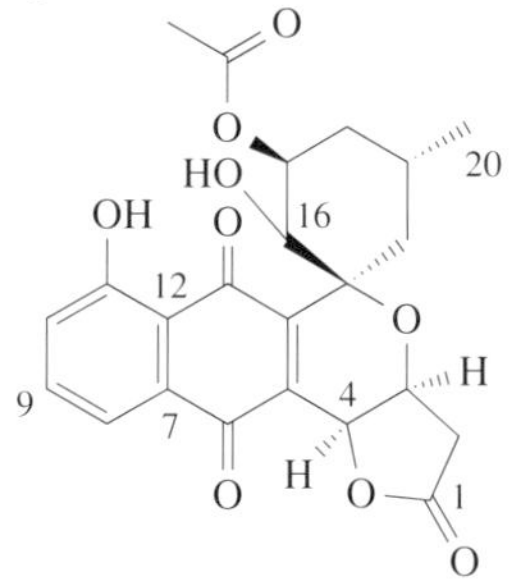

(**6**)

1.18.2.2.1 Actinorhodin

Actinorhodin (**1**) is a blue pigment synthesized by *S. coelicolor* A3(2), whose polyketide backbone must undergo two regiospecific reductions, two intramolecular aldol condensations, hemiketalization, aromatization of two rings, oxidation to form a quinone, hydroxylation, and dimerization. The entire *act* gene cluster (which includes ca. 22 genes and stretches over ca. 26 kilobases (kb)) has been cloned[29] and sequenced,[23,28,34–36] and a variety of randomly generated and targeted mutants (including biosynthetic, regulatory, and resistance gene mutations) have been isolated within this chromosomal locus.[37,38] Indeed, since the actinorhodin biosynthetic pathway was the first actinomycete-derived natural product pathway to be studied at a genetic level, a dis-

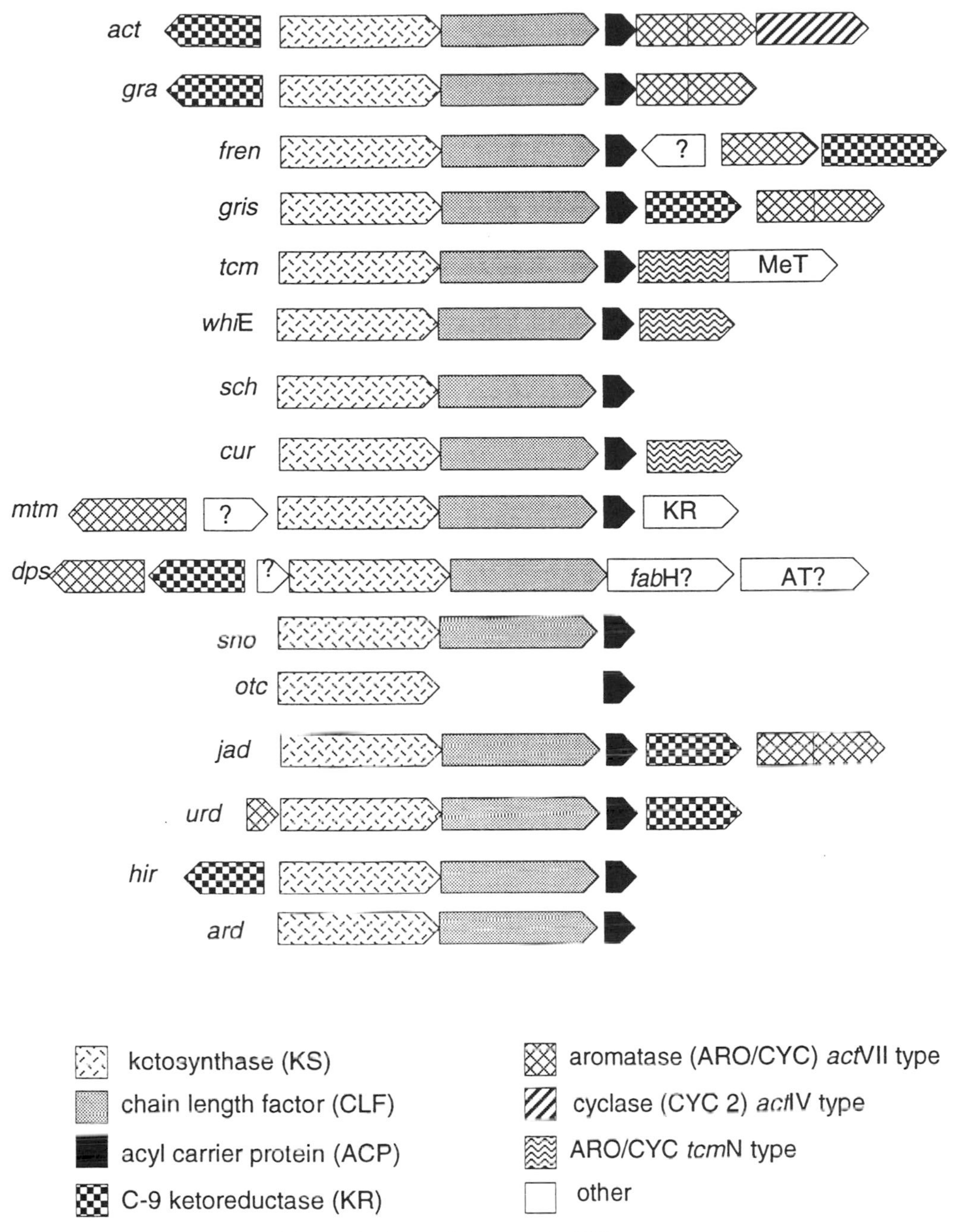

Figure 1 Bacterial aromatic PKS genes.

proportionately large fraction of the genetic toolbox for dissecting microbial natural products has been developed in the context of this model system.

The *act* PKS includes the minimal PKS components (KS, CLF, ACP) which together synthesize the octaketide backbone, a C-9 ketoreductase (KR), a didomain aromatase/cyclase (ARO/CYC) which is required for the formation of the first aromatic ring, and a second ring cyclase (CYC2) (see Scheme 1). Additionally, based on the above PKS definition, the C-3 enoyl reductase and the third ring cyclase/dehydratase (if one exists) may also associate with the PKS complex. The KS and CLF genes appear translationally coupled in this and most other aromatic PKS gene clusters. Although this putative translational coupling has been shown not to be essential for functional gene expression, the detection of almost universal translational coupling led to the hypothesis that these two proteins probably associate in a stoichiometric ratio.[23,26,39] In the *act* gene cluster, the ACP, ARO/CYC, and CYC2 genes are also translationally coupled, and lie immediately downstream of the KS and CLF genes, whereas the KR gene runs in the opposite direction (as shown in Figure 1).[23,28] The *act* ARO/CYC appears to be a didomain protein, whose *N*-terminal half is similar in sequence to its own *C*-terminal half.[40]

The *act* KS and ACP genes share a high degree of sequence similarity to their fatty acid synthase counterparts. As indicated above, the CLF is homologous to the KS, but lacks the conserved cysteine residue of the KS. The C-9 KR was found to share the greatest similarity with a bacterial ribitol dehydrogenase.[28] Neither the ARO/CYC nor the CYC2 possess any mechanistically meaningful sequence similarity with any non-PKS proteins in the database.[23]

In addition to the above PKS genes, the *act* gene cluster also encodes several post-PKS genes including an oxidoreductase which may play a part in either dimerization or hydroxylation reactions,[41] four genes which may catalyze the two late-stage hydroxylation reactions,[36] a C-3 dehydrogenase, and two putative enoyl reductases.[35] The exact functions of most of these genes remain unclear.

1.18.2.2.2 ***Kalafungin***

Kalafungin (**2**) is produced as a natural product by *S. tanashiensis*, and is also formed as shunt product in actinorhodin mutants defective in the oxidoreductase referred to above. The kalafungin gene cluster from *S. tanashiensis* has been cloned. Although the genes have not been sequenced, Southern hybridization studies have revealed an architecture similar to, but different and distinct from, that of the actinorhodin gene cluster.[42]

1.18.2.2.3 ***Granaticin***

Granaticin (**3**), another benzoisochromanequinone produced by *S. violaceoruber* Tü22, is structurally very similar to actinorhodin. Two key structural differences can be seen. First, the stereochemistry around the oxygen bridge of the pyran ring in granaticin is opposite to that of actinorhodin. Second, the presumed tricyclic intermediate undergoes *C*-glycosylation instead of dimerization. Thus, in addition to a similar complement of genes as in actinorhodin, the *gra* gene cluster must also encode for the biosynthesis and transfer of the appropriate deoxysugar group.

The *gra* PKS genes have been cloned and partially sequenced. They include the minimal PKS genes, a KR gene, and an ARO/CYC gene. This portion of the gene cluster is closely related to its *act* counterpart, both in terms of its overall architecture and at the level of DNA sequence. A second KR gene was also identified, although its function remains to be established.[39,43] Further sequencing of this gene cluster has led to the identification of several genes involved in deoxysugar biosynthesis.[44]

1.18.2.2.4 ***Frenolicin***

Six PKS genes from the frenolicin B (**4**) producer *S. roseofulvus* have been cloned and sequenced to reveal the *fren* minimal PKS gene set, a KR, an ARO/CYC, and a gene of unknown function. In addition, partial sequence of the *fren* CYC2, which is homologous to its *act* counterpart and lies immediately downstream of the ARO/CYC gene, is also available. None of the minimal PKS genes are translationally coupled.[40]

It should be noted that unambiguous proof for the fact that the *fren* gene cluster controls frenolicin biosynthesis remains to be provided. However, the assignment seems plausible, since *S. roseofulvus* produces a second closely related benzoisochromanequinone antibiotic, nanaomycin (**5**), and the *fren* PKS has been shown to produce both 16- and 18-carbon polyketides. Thus, since only one aromatic PKS gene cluster was identified via homology based cloning, it is likely that the *fren* gene cluster produces both frenolicin (**4**) and nanaomycin (**5**). If so, then the exact origin of the C-4 unit of frenolicin represents an important unsolved problem. It could either be derived through condensation and complete β-ketoreduction of two acetate units, or (more likely) it might result from direct incorporation of a butyryl primer unit.[14,40]

1.18.2.2.5 Griseusin

Griseusin (**6**) is a 20-carbon benzoisochromanequinone produced by *S. griseus*. The genes encoding the KS, CLF, ACP, KR, and ARO/CYC of the *gris* PKS have been cloned and sequenced.[45] It should be noted that, based on limited sequence analysis, the region immediately downstream of the ARO/CYC gene lacks a CYC2 homologue. Thus, either the gene is located elsewhere in the gene cluster, or the second ring cyclization is catalyzed by an enzyme that is evolutionally unrelated to the *act* and *fren* subunits. Another as yet unidentified gene of interest from the *gris* cluster should be the C-17 ketoreductase. In particular, the timing of this reaction in the overall pathway may indicate whether it is a PKS-associated subunit or a post-PKS enzyme.

1.18.2.3 Unreduced Aromatic Polyketides

A group of bacterial aromatic natural products, (**7**)–(**9**), including tetracenomycin, elloramycin, several actinomycete spore pigments, and possibly even the polyketide precursors of the aureolic acid group of antitumor compounds, share a feature that is not observed in other natural products of this class: a completely unreduced backbone that undergoes a C-9/C-14 initial cyclization instead of the more commonly observed C-7/C-12 cyclization pattern. These structural relationships are reflected in the content and architecture of their PKS gene clusters, as described below.

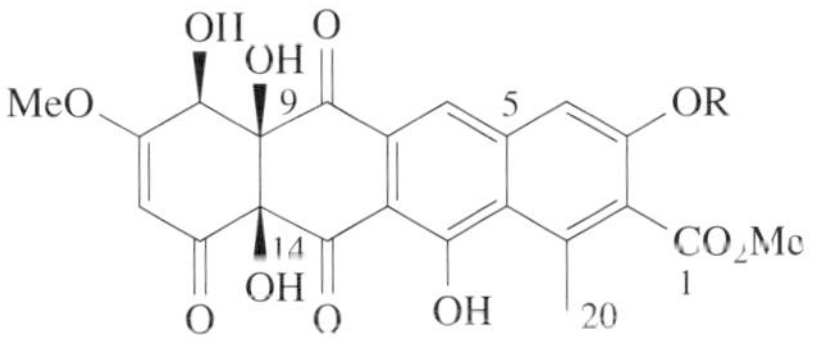

OH O OH MeO 9 5 OR 14 CO_2Me 1 OH O O OH 20

(**7**)

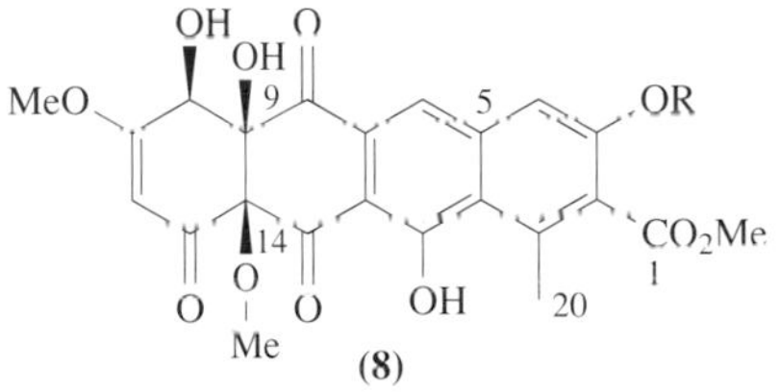

OH O OH MeO 9 5 OR 14 CO_2Me 1 O O O OH 20 Me

(**8**)

OH OH O 14 18 OR O OH RO 9 5 2 19 20 H OMe OH

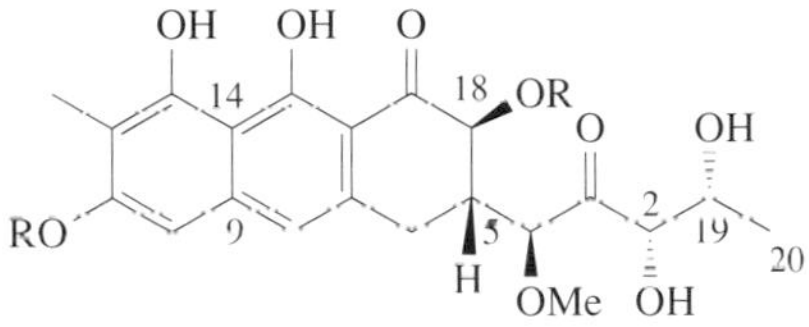

(**9**)

1.18.2.3.1 Tetracenomycin

The tetracenomycins, such as tetracenomycin C (7), from *S. glaucescens* are decaketides. The *tcm* PKS genes were among the first few aromatic PKSs to be cloned and sequenced, and subsequently the entire gene cluster has been sequenced. The PKS, which is encoded as part of a large operon, includes a KS, a CLF, an ACP, and *tcm*N, a member of the ARO/CYC family of proteins. The *N*-terminal domain of *tcm*N is homologous to the *N*-terminal half of its benzoisochromanequinone counterparts; however, it lacks their *C*-terminal domain. Instead it is fused to a post-PKS methyltransferase domain at its *C*-terminal end (Figure 1). Since the *N*-terminal domain of *tcm*N is fully functional when expressed by itself, it can be thought of as a representative of a separate subfamily of (monodomain) ARO/CYCs.[46] This notion is strengthened upon an examination of the spore pigment PKS gene clusters (see below). Together, the KS, CLF, ACP, and ARO/CYC may be sufficient for the *in vivo* synthesis of TcmF2, the first isolable intermediate in the tetracenomycin pathway[46] (although heterologous expression of these four proteins in *S. coelicolor* leads to the synthesis of a slightly different product (see below)).[47]

In addition to the above PKS subunits, the *tcm* gene cluster also encodes a final ring cyclase (*tcm*I), two oxygenases (*tcm*G and *tcm*H), three methyltransferases (*tcm*O, *tcm*P, and the *C*-terminal domain of *tcm*N), and *tcm*J, which has homologues encoded within other unreduced polyketide gene clusters but does not appear to be necessary for tetracenomycin biosynthesis, although it may increase the amount of product made.[46,48–50]

1.18.2.3.2 Elloramycin

Elloramycin (**8**) is structurally and genetically similar to tetracenomycin. Using a *tcm* probe, a cosmid clone with the *elm* gene cluster was isolated from *S. olivaceus*. It complemented *tcm* and *elm* mutants, had an *elm* resistance gene, and could direct production of an unglycosylated *elm* biosynthetic intermediate in a heterologous host. The gene cluster has not been sequenced.[51]

1.18.2.3.3 **Streptomyces** *spore pigments*

Many *Streptomyces* produce pigmented spores. None of the structures of these spore pigments has been elucidated. Indeed, this is a technically demanding task, since these pigments remain spore bound and the production of the spores cannot be scaled up via liquid fermentations. (*Streptomycetes* do not typically produce spores in submerged cultures.)[52] Thus, the discovery that many (if not all) of these spore pigments are polyketides was originally made through sequence analysis of genomic DNA fragments that complemented mutants with altered spore color. The genes encoding spore pigment PKSs have been cloned and sequenced from at least two strains, *S. coelicolor* A3(2) and *S. halstedii*. Additionally, as described below, it is very likely that the *cur* PKS gene cluster from *S. curacoi* is also involved in spore pigment production. In each case architectural and functional homology has been observed between these gene clusters and the *tcm* gene cluster.

The *whi*E locus from *S. coelicolor* encodes a minimal PKS and a monodomain ARO/CYC protein resembling the *N*-terminal domains of *tcm*N and the (didomain) benzoisochromanequinone ARO/CYCs. The presence of a monodomain ARO/CYC instead of a didomain one, together with the absence of a ketoreductase homologue within this PKS, suggests that spore pigments are derived from completely unreduced polyketide backbones.[53] The minimal *whi*E PKS has been found to synthesize undecaketides and dodecaketides, suggesting that this PKS has relaxed chain length specificity (as in the case of the *fren* PKS).[54] Furthermore, the *whi*E ARO/CYC can exclusively direct the formation of a C-9/C-14 cyclized intermediate, as in the case of tetracenomycin.[55,56] Taken together, one can conclude that the *S. coelicolor* spore pigment is an unreduced dodecaketide that undergoes a C-9/C-14 cyclization. Additional genes have been sequenced within the *whi*E locus, including putative *tcm*J and *tcm*I cyclase homologues, a late-step putative hydroxylase, and protein involved in targeting or binding the pigment in the correct location.[57]

The spore pigment PKS genes (*sch*) from *S. halstedii* were cloned using an *act* probe. This clone could complement a *whi*E mutant; furthermore, deletion of this region in the genome of *S. halstedii* led to loss of spore pigment production.[58] The content and organization of this gene cluster are very similar to those of the *whi*E gene cluster. Furthermore, the *sch* PKS (KS, CLF, ACP, ARO/CYC) also makes an unreduced dodecaketide with a C-9/C-14 cyclization pattern.[54,55] Hence it appears that many, if not all, *Streptomycete* spore pigments are derived from structurally identical polyketide intermediates and that interstrain variations in spore colors are most likely due to differences in post-PKS steps, such as hydroxylation. This hypothesis is strengthened by two other findings.[57,59] First, it has been reported that 12 out of an arbitrary set of 21 *Streptomyces* species contained a homologue of *whi*E.[59] Second, another *whi*E homologue (with a very similar gene content and architecture), cloned from the curamycin producer *S. curacoi*,[60] has been shown to direct production of an unreduced dodecaketide that undergoes C-9/C-14 cyclization.[56]

1.18.2.3.4 Mithramycin

The mithramycin (*mtm*) PKS gene cluster from *S. argillaceus* was isolated using *act* PKS and *gris* sugar biosynthesis probes. Although mithramycin (**9**) appears to be structurally unrelated to tetracenomycin or elloramycin, heterologous expression of the minimal *mtm* PKS resulted in exclusive production of a decaketide product. Together with the results of isotopic labeling studies on mithramycin, it was concluded that the mithramycin aglycone is derived from a tetracenomycin F1-like intermediate that subsequently undergoes decarboxylation, oxidative ring cleavage between C-18 and C-19, and other post-PKS modification reactions. Consistent with this notion, the polycyclic portion of the mithramycin aglycone shows no evidence of ketoreduction.[61] Remarkably, however, the genes discovered thus far within this gene cluster are more characteristic of a reduced than an unreduced polyketide pathway. Specifically, two KRs and a didomain ARO/CYC have been identified, although it should be noted that one KR is most similar to the *gra* ORF6 gene product (a putative *gra* KR that is not required for C-9 ketoreduction in mithramycin (**9**) biosynthesis; see above) and *dnr*H, which is a gene in doxonubicin biosynthesis and also thought to catalyze a post-PKS reaction.[33,62] Hence further functional analysis of these potential PKS subunits may reveal a link between the *act* and the *tcm* types of aromatic PKSs.

1.18.2.4 Reduced Anthracycline and Tetracycline Polyketides

This class of bacterial aromatic polyketides includes several clinically important natural products such as oxytetracycline and daunorubicin/doxorubicin, **(10)**–**(14)**. Since the early-stage intermediates generated by these PKSs are very similar to their benzoisochromanequinone counterparts, not surprisingly the PKS genes also reflect this similarity.

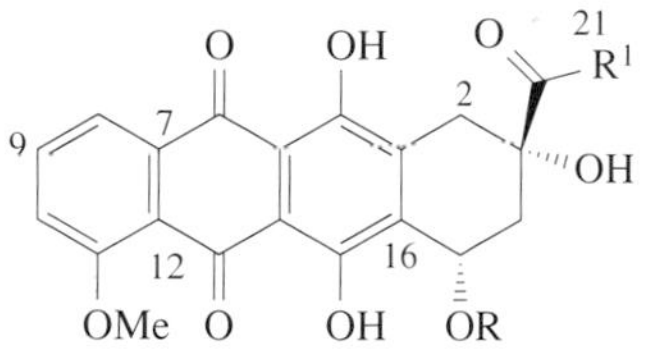

(10) R^1= Me
(11) R^1= CH_2OH

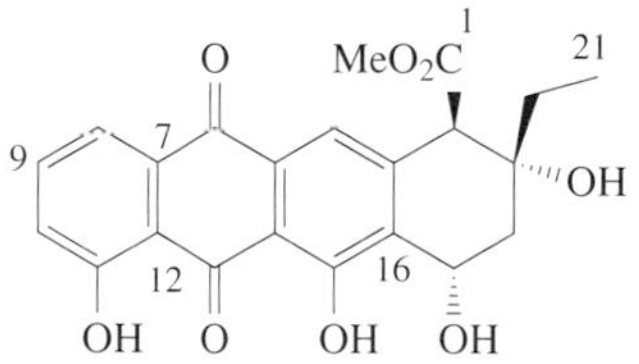

(12)

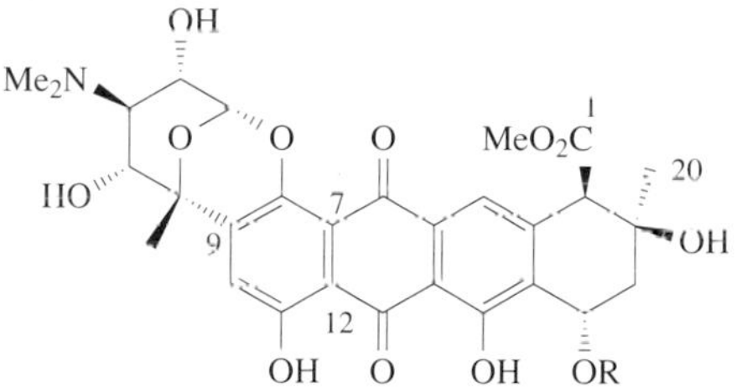

(13)

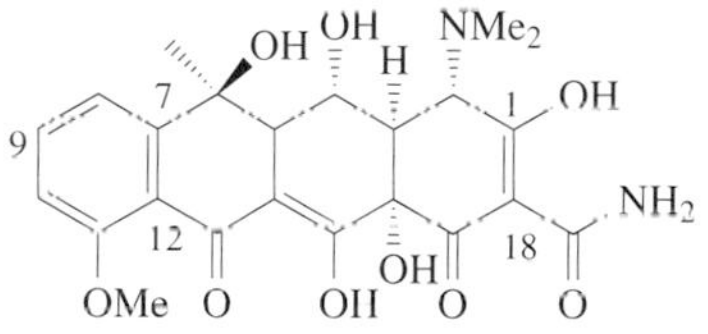

(14)

1.18.2.4.1 Daunomycin and doxorubicin

Daunomycin **(10)** and doxorubicin **(11)** are very closely related anticancer agents. Large portions (and possibly all) of the daunomycin/doxorubicin gene cluster from *S. peuceticus* and the daunomycin gene cluster from *Streptomyces* sp. C5 have been cloned and sequenced, and have been found to be very similar to each other at the DNA sequence level. Since biosynthesis of these polyketides starts with a propionate primer that is extended using nine malonyl units, genetic analysis of these systems was expected to reveal insights into the mechanisms by which aromatic PKSs can be primed by carboxylic acids other than acetate. The architecture of both these gene clusters is unusual in that the ACP gene lies ca. 10 kb away from the KS and CLF genes. Other putative PKS genes in these clusters include three KRs that share homology to the *act* KR family of proteins (although the degree of similarity differs), a didomain ARO/CYC, a putative acyltransferase, and a gene that resembles *fab*H, which encodes the third fatty acid synthase KS in *E. coli*. This KS is responsible for priming fatty acid biosynthesis through the formation of acetoacetyl-ACP. No homologues of any other characterized cyclases have been found; it therefore remains a mystery as to which proteins control the regiochemistry of the second, third, and final cyclization reactions. The presence of additional AT and KS genes in these two gene clusters led to the hypothesis that the encoded proteins were responsible for introducing a propionate primer into the polyketide backbone through the formation of 3-ketopentanoyl-ACP, which would then be extended further by the usual KS.[33,63] However, available data suggest that these genes are not essential for the biosynthesis of a C-21 backbone.[33] Hence it is unlikely that either the AT or the *fab*H homologue controls the specificity of these PKSs for propionate primers.

In addition to the above PKS genes, the daunomycin cluster from *Streptomyces* sp. strain C5 also revealed putative genes for two methyltransferases (based on mutant complementation and *in vitro* activity), a glycosyltransferase, a monooxygenase (based on homology to *tcm*H), a final ring cyclase (based on mutant complementation and heterologous expression), and a putative carboxymethyl esterase (based on homology).[63–65]

The gene for the biosynthesis of the daunomycin intermediate aklavinone (**12**) from *Streptomyces galilaeus* has also been cloned[66] and partially sequenced.[104]

1.18.2.4.2 Nogalamycin

A close relative of doxorubicin, nogalamycin (**13**) from *S. nogalater* is made from acetate and nine malonates. Its nine cloned *sno* genes encode the minimal PKS, a KR, a methylase, an oxygenase, a methyltransferase, and a regulatory gene. No ARO/CYC homologues have been found. The gene functions have been confirmed by mutant complementation.[67]

1.18.2.4.3 Tetracyclines

The tetracycline backbone has an unusual malonamyl end group, although it remains to be proved whether the primer unit for tetracycline biosynthesis is malonate (which is further modified via aminotransfer) or malonamide. The complete oxytetracycline (**14**) gene cluster (*otc*) from *S. rimosus* has been cloned and heterologously expressed, and homologues of the *act* minimal PKS and *act* KR were identified within this gene cluster via Southern hybridization.[32,68] Sequence analysis has revealed a minimal PKS gene set.[69] Post-PKS genes have been identified from blocked mutants and mapped by chromosomal linkage.[70] Further analysis of this gene cluster should yield valuable new insights into PKS as well as post-PKS enzyme function.

1.18.2.5 Angucyclines

The family of angucyclines are characterized by a bent fused ring structure, such as in jadomycin (**15**), urdamycin (**16**), tetrangulol (**17**), and PD116740 (**18**). Although these natural products are closely related in structure to linear anthracyclines, as is described below, several interesting features make the PKSs responsible for their biosynthesis interesting targets in themselves.

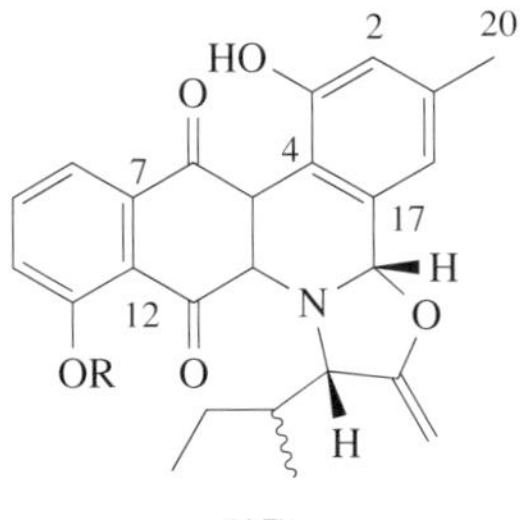

HO 2 20 O 4 7 17 H N O 12 OR O H

(15)

O 2 20 O RO 4 OH 7 17 OH R 12 OR O

(16)

HO 2 20 O 4 9 7 17 12 15 OH O

(17)

HO 2 20 O OH 4 7 17 OH 12 MeO O OH

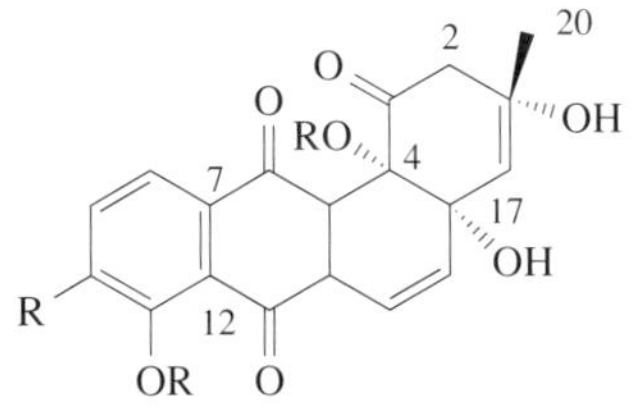

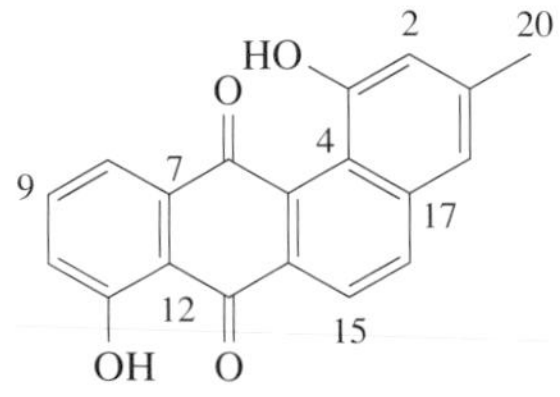

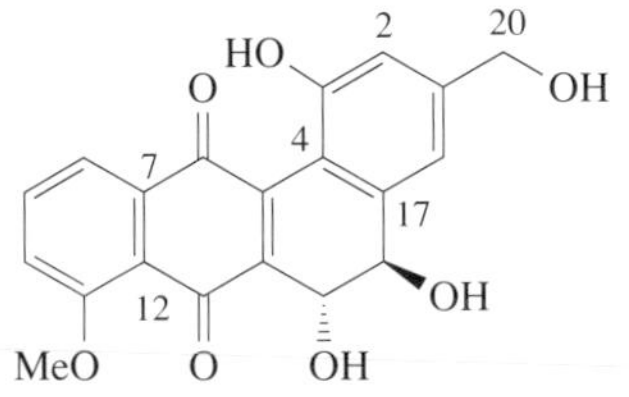

(18)

1.18.2.5.1 Jadomycin

The jadomycin B (**15**) (*jad*) biosynthetic genes from *S. venezuelae* have been cloned and partially sequenced to reveal a minimal PKS gene set, a KR gene, and a didomain ARO/CYC gene. These enzymes are responsible for the biosynthesis, reduction, and C-7/C-12 cyclization of a decaketide backbone. No information is available regarding the proteins responsible for the formation of the second and third rings or the insertion of isoleucine.[71]

1.18.2.5.2 Urdamycin

The urdamycin (**16**) (*urd*) PKS genes from *S. fradiae* were cloned and sequenced, revealing a minimal PKS gene set, a KR gene, a homologue of the *tcm*I final ring cyclase, and a gene encoding an oxygenase. Again, no functional insights into candidate angucyclases are available.[72]

1.18.2.5.3 Tetrangulol and PD116740

The complete gene clusters for these two natural products, (**17**) and (**18**), from *S. rimosus* and *S.* WP4669, respectively, have been cloned, as demonstrated by heterologous expression in *S. lividans*. No sequence information is available. In addition to yielding insights into the biochemical basis for the angular cyclization of nascent polyketide chains, further analysis of the PD116740 gene cluster should also lead to the identification of a C-15 ketoreductase, whose action is also believed to precede cyclization, as in the case of the C-9 ketoreductases.[73]

1.18.2.6 Other Aromatic PKS Gene Clusters

In addition to the above PKS gene clusters, a few others have also been cloned using homology-based approaches. In each case it is now reasonably certain that the cloned PKS gene clusters are not involved in the biosynthesis of the natural product that was the original target of the study. However, the actual products of these gene clusters (if any) are unknown.

A PKS gene cluster designated *mon* from *S. cinnamonensis* has been cloned and sequenced. It contains the KS, CLF, ACP, KR, and ARO/CYC genes, all transcribed in the same direction. *S. cinnamonensis* produces the polyether polyketide monensin, but the cloned gene cluster is not involved in monensin biosynthesis since it did not complement monensin mutants and its deletion did not affect monensin production. Its deletion also did not affect spore pigment production, and so its function remains unknown.[74] However, heterologous expression of this minimal PKS in *S. coelicolor* led to the production of a decaketide (unpublished results). Therefore, the natural product is likely to possess a 20-carbon backbone with a C-9 ketoreduction. Interestingly, the minimal PKS encoded by this gene cluster shares greater sequence similarity with the *jad* and *urd* PKSs than with the other PKSs. Perhaps this might suggest that the natural product of this gene cluster is an angucycline.

Using an *act* probe, a minimal PKS gene set, a KR gene, and two other genes were cloned and sequenced from *Saccharopolyspora hirsuta*. One unknown protein resembles a biotin carboxyl carrier protein and the other is homologous to the genes of unknown function, *fren* ORFX and *act*VI ORFA. The other four genes are in the same order as the *act* cluster and they were able to complement an *act* mutant.[75]

Kibdelosporangium aridum produces ardacin, a glycopeptide antibiotic which contains two acetate-derived *p*-hydroxyphenylglycine rings. A gene cluster (*ard*) was cloned using the *act* probe, and its sequence revealed a bacterial aromatic minimal PKS and *tcm*I cyclase homologue. Judging from this homology, the *ard* cluster probably is not involved in the synthesis of ardacin but may encode the genes for a spore pigment or another polyketide.[76]

1.18.3 GENETIC CONSTRUCTION AND CHEMICAL ANALYSIS OF RECOMBINANT PKSs

1.18.3.1 Background and Methodological Overview

In the case of a few well studied bacterial aromatic polyketide gene clusters (e.g., the *act*, *tcm*, and *otc* loci) classical genetic analysis had been used to isolate several nonproducing mutants.[37,70,77] Analysis of these mutants served two useful purposes. First, by determining the structure of the resulting intermediates or shunt products, one could make some deductions about the order of reactions in the overall biosynthetic pathway. Second, since some of these mutants mapped in putative PKS genes, they provided a convenient starting point for the construction and analysis of recombinant PKSs through mutant complementation. Specifically, mutants blocked in the *act* KS, CLF, ACP, KR, and ARO/CYC,[43,78] the *tcm* KS, CLF, and ARO/CYC,[79] and the *dau* KR gene(s)[80] were used as hosts for complementation with heterologous PKS subunits.

As increasingly powerful techniques for targeted gene disruption/replacement were developed in *Streptomyces* species, they were used to replace specific PKS genes in chromosomally occurring gene clusters with homologues from other gene clusters. The advantages of this approach were that they allowed for regulated coexpression of the entire PKS gene set in a "natural-like" manner and avoided problems due to plasmid instability or recombinogenicity.[38]

In 1993, a new host–vector system was developed for expression of natural and recombinant polyketide gene clusters in *S. coelicolor*. The entire *act* gene cluster, except *act*VI ORFA, has been deleted from *S. coelicolor* CH999; this permitted the construction of a low copy shuttle vector containing the relevant expression signals from the *act* gene cluster, together with useful cloning sites.[19] The development of this host–vector system has facilitated the construction and analysis of numerous recombinant bacterial aromatic and modular PKSs, and even fungal PKSs.[14,81,82] Other heterologous hosts such as *S. lividans*[32,46,83,84] and *S. parvulus*[85] have also been used for polyketide production.

1.18.3.2 Deletion Analysis of PKS Gene Sets

Although the construction and analysis of blocked mutant strains threw valuable light on the sequence of reactions within the overall biosynthetic pathways of some well-studied bacterial aromatic polyketides (e.g., actinorhodin and tetracenomycin), they were of limited utility in dissecting the function of aromatic PKSs themselves. This could have been due either to the extreme lability of the shunt products or to the fact that many PKS mutants led to the appearance of a null phenotype. The advent of molecular genetics in the context of these biosynthetic pathways facilitated the development of more powerful approaches for deletion analysis of naturally occurring PKS gene sets.

1.18.3.2.1 Actinorhodin

Seven classes of blocked mutants were found in the overall *act* pathway; in biosynthetic sequence they were designated as classes I, III, VII, IV, VI, Va, and Vb.[37,86] However, only two of the shunt products identified within these mutants yielded insights into *act* PKS function. The *act*VII mutant made mutactin (**21**), a shunt product that had the correct chain length, a C-9 ketoreduction, and the regiochemically correct C-7/C-12 intramolecular aldol condensation, but with an aliphatic first ring and incorrectly cyclized second ring.[87] The *act*VI mutant made aloesaponarin II (**24**), a shunt product that was indistinguishable from the actinorhodin polyketide precursor with respect to its first two rings.[80] Thus, the *act*VII gene was thought to be a dehydratase that abstracted a water molecule from the prearomatic first ring in addition to a second ring cyclase, and the *act*VI genes were implicated in the formation of the pyran ring of actinorhodin.

DNA sequence analysis subsequently revealed that the *act*I locus included the minimal PKS gene set, the *act*III locus encoded the KR, the *act*VII locus encoded the ARO/CYC, and the *act*IV locus encoded the CYC2. These new designations were based on analysis of subsets of these six genes in *S. coelicolor* CH999. Specifically, in the presence of the minimal PKS alone, a mixture of SEK4 (**19**) and SEK4b (**20**) was produced.[88] The minimal PKS together with the KR produced a full-length shunt product, mutactin (**21**), suggesting that ketoreduction occurs after the complete chain is synthesized.[89] If the ARO/CYC was also included, SEK34 (**22**) was produced,[90] and if the CYC2 gene was also coexpressed, DMAC (**23**) and aloesaponarin II (**24**) were synthesized.[19] Removal of an upstream PKS subunit resulted in the loss of downstream subunit functions. Hence the absence of any minimal PKS component resulted in complete loss of polyketide production. Absence of the KR led to SEK4 and SEK4b production even in the presence of the ARO/CYC and CYC2,[89] whereas absence of ARO/CYC activity resulted in mutactin production even in the presence of

CYC2.[19] However, in the absence of the KR, the ARO/CYC influenced the relative distribution of SEK4 and SEK4b.[88,89] Based on these results, the model for the *act* PKS pathway was revised, and individual PKS subunits were assigned the functions shown in Scheme 1. It should be noted that the above analysis could not unambiguously pinpoint the relative positions of the C-9 ketoreduction and the first aldol condensation. However, as described below, indirect evidence argues in favor of a model where reduction precedes condensation.

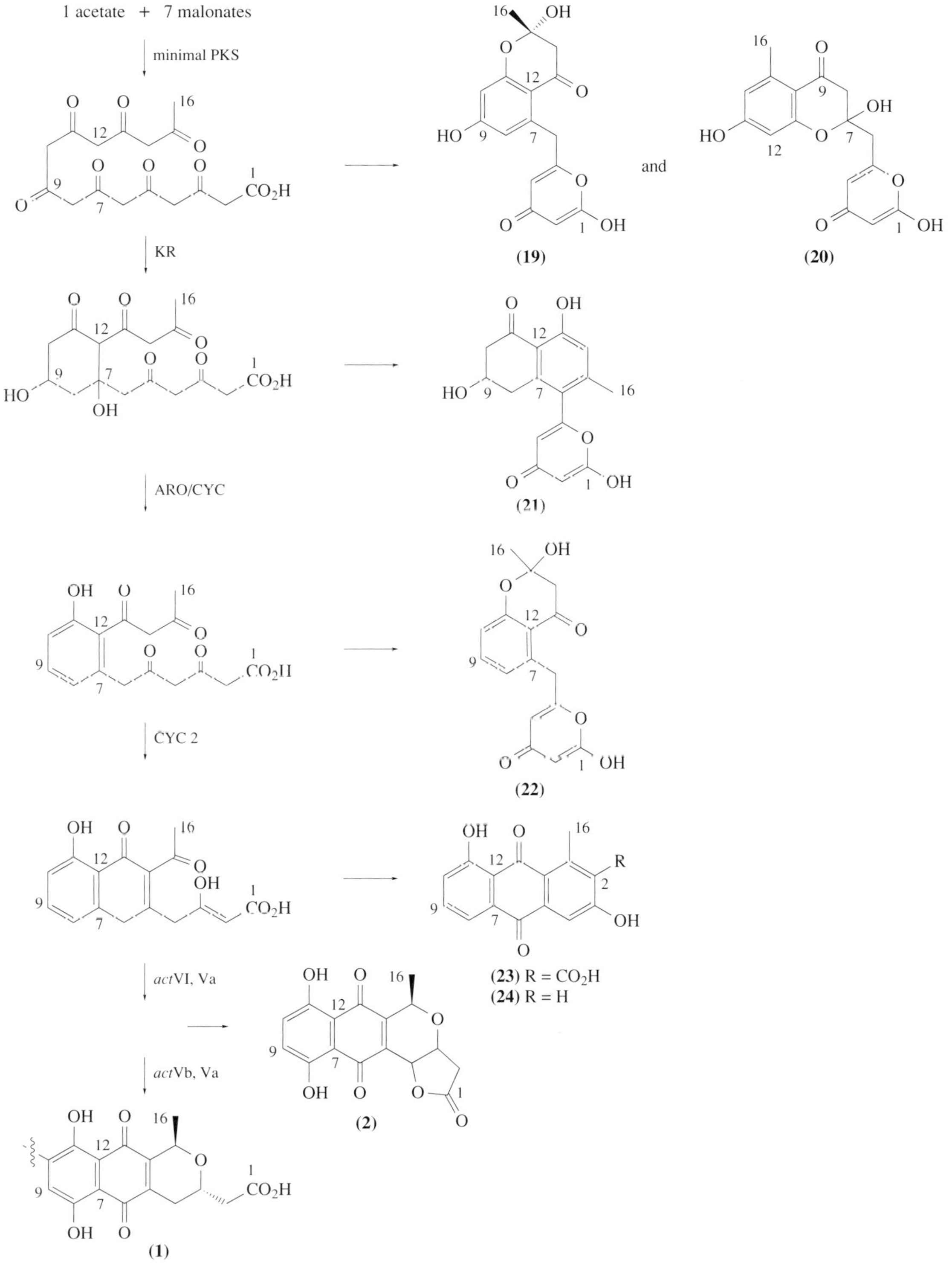

1 acetate + 7 malonates

minimal PKS

KR

ARO/CYC

CYC 2

*act*VI, Va

*act*Vb, Va

(19) (20) (21) (22)

(23) R = CO_2H
(24) R = H

(2) (1)

Scheme 1

1.18.3.2.2 Tetracenomycin

Studies of the *tcm* pathway (Scheme 2) followed a similar course of events as in the case of the *act* pathway. Analysis of a series of classical mutants shed much light on the reaction sequence in the latter part of the pathway, but was of limited use for the analysis of the PKS itself since the earliest intermediates were tetracyclic compounds.[77] Expression of the *tcm* minimal PKS gene set in CH999 produced SEK15 (**25**) and SEK15b (**26**), 20-carbon compounds with alternate first-ring cyclizations.[91] Expression of the *tcm* minimal PKS together with the *tcm*N ARO/CYC resulted in the formation of tetracenomycin F2 (**27**) in *S. glaucescens* and *S. livians*[46] and RM80 (**28**) in *S. coelicolor* CH999.[47] This, coupled with the fact that purified *tcm*I had been shown to be necessary and sufficient for conversion of tetracenomycin F2 into tetracenomycin F1 (**29**),[92] led to the notion that the *tcm* PKS comprised the minimal PKS and the monodomain ARO/CYC.

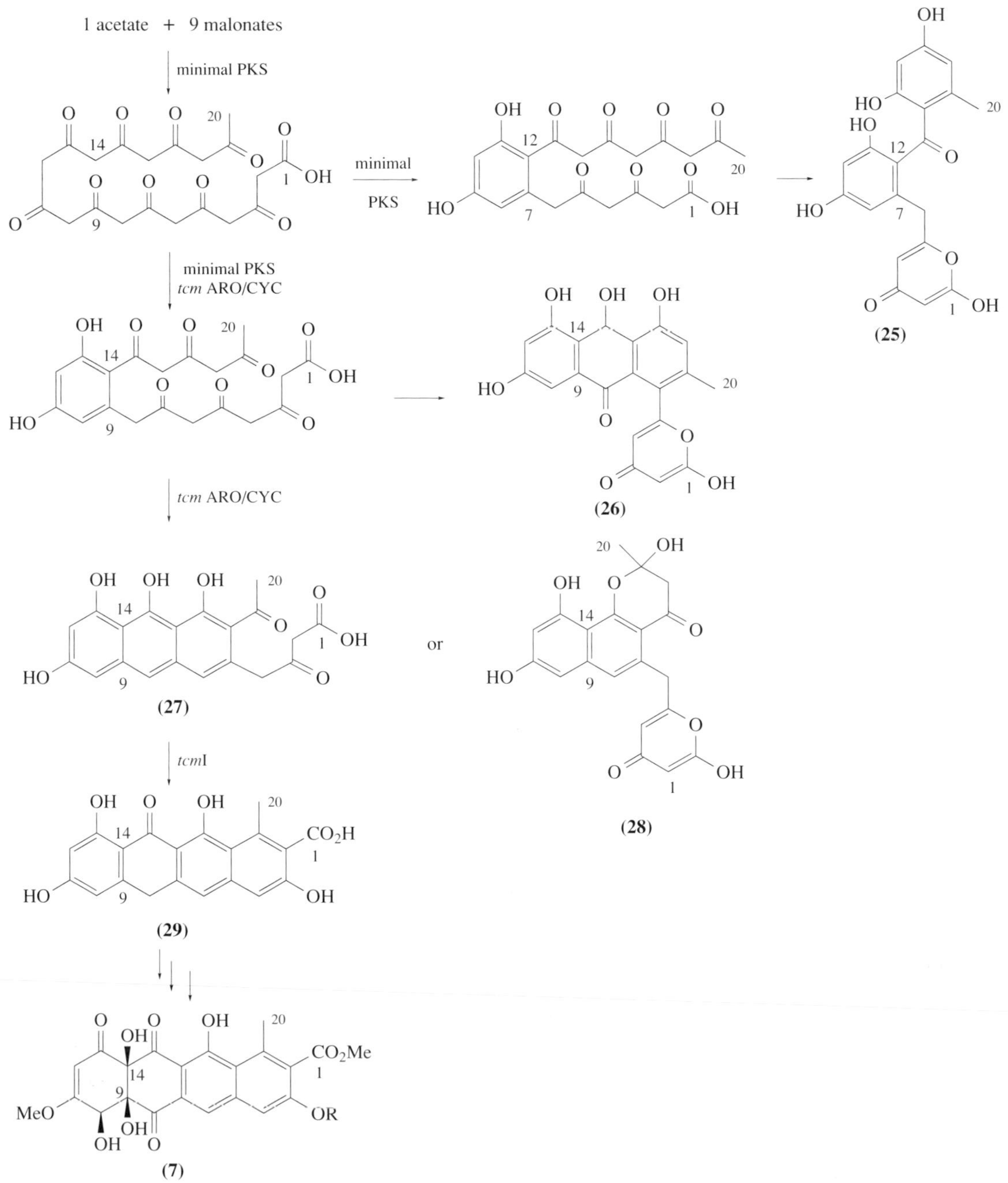

Scheme 2

1.18.3.2.3 Doxorubicin/daunomycin

Expression of some of the *dps* genes from *S. peucetius* in *S. lividans* led to the formation of aklanonic acid (**30**), the earliest intermediate of the daunorubicin pathway (Scheme 3). The genes used were the minimal PKS, two KR, and ARO/CYC. Production of this intermediate was independent of whether the *fab*H homologue and putative AT genes were included or not.[33]

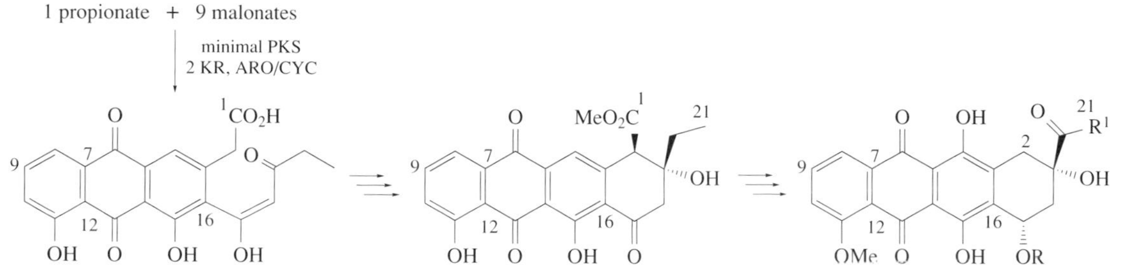

1 propionate + 9 malonates

minimal PKS
2 KR, ARO/CYC

(**30**) (**12**) (**10**) R^1 = Me
(**11**) R^1 = CH_2OH

Scheme 3

Several *dau* mutants were isolated from *S.* sp. strain C5 and the structures of some latter-stage intermediates were determined. In biosynthetic order, *dau*A is the minimal PKS, *dau*C is a methyltransferase, *dau*D is a cyclase, *dau*E is a reductase, *dau*F is a hydroxylase, and *dau*H is one or more later steps. Interestingly, the *dau*F reacts with the product of *dau*D, so a *dau*E/*dau*F double mutant produces a compound that is different from the product of either of the single mutants.[93] Cell-free extracts of *dau*C–*dau*F were able to convert aklanonic acid into the same compounds as found *in vivo*.[94]

1.18.3.2.4 The whi*E spore pigment pathway*

Deletion analysis of the *whi*E locus revealed that ORFIII–V, which encode the minimal PKS, were essential for any kind of pigment production.[57] By itself, the *whi*E minimal PKS in *S. coelicolor* CH999 produced 22- and 24-carbon compounds with C-7/C-12 first-ring cyclizations (**31** and **32**). Expressing the ARO/CYC gene with the minimal PKS gene set yields only a 24-carbon compound that has undergone C-9/C-14 and C-7/C-16 cyclizations (**33**), as shown in Scheme 4. From this it appears that the *whi*E PKS is similar to *tcm* PKS with respect to many of its molecular recognition properties.[54,56]

1.18.3.3 Combinatorial Analysis of PKS Gene Sets

Whereas deletion analysis of PKS genes facilitated the assignment of function to individual subunits, it stopped short of providing insights into their catalytic specificity. In contrast, the construction and analysis of "hybrid" PKSs has proved to be an effective way to dissect the molecular recognition features of PKS subunits.

1.18.3.3.1 The minimal PKS

Not only is the minimal PKS the primary determinant of polyketide chain length, it also influences the regiospecificity of other reactions in the overall PKS catalytic cycle, presumably by virtue of its ability to bind to the highly reactive polyketone backbone and/or release it in a temporally controlled manner. These molecular recognition features have been explored through several combinatorial experiments. Direct evidence supporting the notion that the minimal PKS (i.e., the KS, CLF, and ACP) alone was sufficient to dictate chain length specificity came from analysis of the *act* and *tcm* minimal PKSs in a host lacking other PKS functions. In both cases the major products were octaketides and decaketides, respectively.[19] Occasionally however, certain minimal PKSs such as the *whi*E minimal PKS (see above) show relaxed chain-length specificity within a window of one

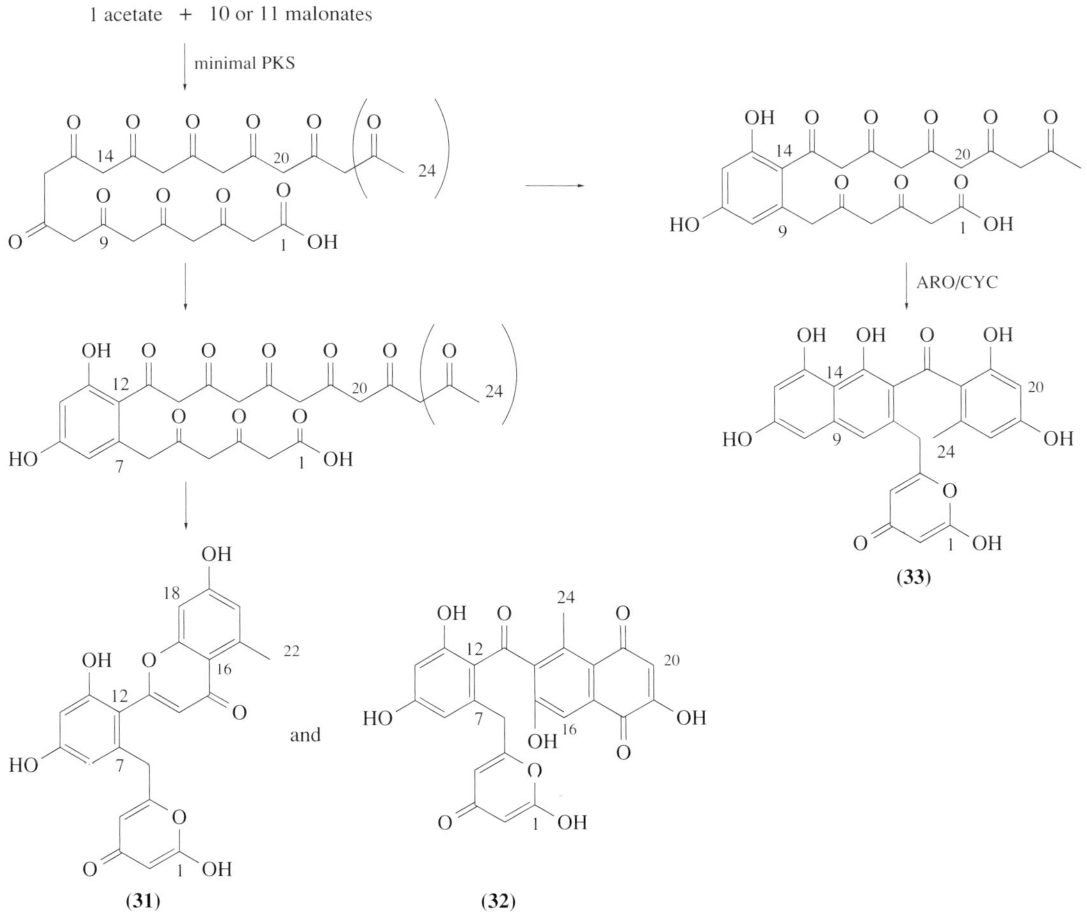

1 acetate + 10 or 11 malonates

minimal PKS

ARO/CYC

(31) and **(32)**

(33)

Scheme 4

acetate unit.[56] Furthermore, as discussed below, auxiliary subunits such as KRs and ARO/CYCs can also induce minimal PKSs to synthesize polyketide chains of variable lengths (±one acetate unit).

Within the minimal PKS, the CLF plays a crucial role in determining chain length, possibly together with the KS. In *S. coelicolor* CH999, a hybrid PKS containing the *tcm* KS and the *act* CLF produced a 16-carbon polyketide; in contrast, the reverse experiment leads to the creation of an inactive PKS and therefore precludes determination of whether the CLF is sufficient for dictating chain length specificity. KSs and CLFs from systems with the same chain length specificity, such as the *act* and *gra* PKSs, appeared to be interchangeable.[19] However, a different result was obtained in a complementation experiment in *S. glaucescens*. Whereas the *act* KS was still nonfunctional with *tcm* CLF, the *act* CLF complemented *tcm* CLF mutants without changing the product length.[79]

The ACPs of *act*, *gra*, *otc*, *tcm*, and *fren* were found to be interchangeable apparently without changing the nature of the product.[95] Indeed, the *act* ACP can be replaced by even the ACP of a heterologous fatty acid synthase without affecting chain-length specificity, although this greatly decreases the yield of product.[38,96]

In addition to determining chain-length specificity, the minimal PKS may influence the regiospecificity of cyclization by predisposing the backbone to undergo an intramolecular aldol condensation between the C-7 carbonyl and the C-12 methylene (Scheme 1). It should be noted, however, that although C-7/C-12 cyclized products tend to be relatively abundant, cyclization control is not absolute in the absence of other auxiliary subunits such as KRs and ARO/CYCs. Thus, in addition to SEK4 (**19**), the *act* minimal PKS alone also synthesizes the unusually cyclized octaketide SEK4b (**20**) in comparable amounts.[89] Likewise, by itself the *tcm* minimal PKS synthesizes SEK15 (**25**) and SEK15b (**26**) in similar amounts, which have undergone C-7/C-12 and C-9/C-14 cyclizations, respectively. Once it is formed, this (natural) C-9/C-14 cyclized intermediate will also undergo a "natural" C-7/C-16 second cyclization to generate a naphthalenic moiety in the presence of the minimal PKS alone.[91] Given the difficulty of synthesizing authentic full-length intermediates predicted for these pathways, it is unclear as yet whether the observed cyclization propensities of minimal PKSs simply reflect the intrinsic reactivity of nascent polyketide chains, or whether enzyme–substrate interactions are required in order to observe this limited cyclization regiospecificity.

1.18.3.3.2 Ketoreductases

The substrate specificity of only the *act* KR has been actively investigated. This KR can catalyze a C-9 reduction on nascent polyketide chains ranging from 16 to 24 carbon atoms (and possibly others), including those generated by the *gra*, *fren*, *tcm*, *mtm*, *mon*, *otc*, *whi*E, and *sch* minimal PKSs.[19,56,89] Since ketoreduction occurs at a constant distance relative to the carboxyl end of the nascent chain in all these cases, it has been speculated that the KR probably acts after the entire chain is synthesized.[80] Although no direct evidence is available for this hypothesis, it is reinforced by the fact that full-length products are synthesized even in the absence of the KR. Interestingly, several PKSs with relaxed chain-length specificity also show variability with respect to the regiospecificity of ketoreduction. Thus, the *fren* minimal PKS and the *act* KR together produce 16- and 18-carbon chains that undergo C-9 reduction in addition to 16-carbon chains that undergo C-7 reduction.[14] Likewise, the *whi*E minimal PKS together with the *act* KR produces 22- and 24-carbon backbones undergoing C-9 reduction and 24-carbon backbones undergoing C-11 reduction.[56] Relaxed chain-length specificity and ketoreduction specificity are also observed in the case of the *tcm* PKS in the presence of certain types of mutant ARO/CYC proteins.[97] Assuming that this overlapping relaxation in chain length and reduction specificity arises from the same structural feature(s), these results argue in favor of direct association of one or more minimal PKS subunits with the KR.

Although the *dau* KR that is responsible for C-9 ketoreduction of daunomycin has not been unambiguously identified, it can recognize and regiospecifically reduce the nascent chain product of the *act* minimal PKS.[80]

1.18.3.3.3 Aromatases and cyclases

As described above, the *act* VII locus and its homologues encode a family of proteins designated ARO/CYC on account of their ability to influence the cyclization and/or the aromatization of the first ring. Unreduced rings can aromatize spontaneously, but reduced rings need an enzyme. In the event of ketoreduction, the regiochemistry of the first cyclization is constant relative to the ketone that undergoes reduction, even in situations where ketoreduction regiospecificity is altered (see above). This suggests that ketoreduction precedes cyclization, and that the didomain ARO/CYC family of proteins do not play a role in controlling cyclization specificity. (However, in the absence of ketoreduction, didomain ARO/CYCs have been reported to influence the relative distribution of alternatively cyclized products.)[89]

With the possible exception of the mithramycin PKS (see above), didomain ARO/CYCs have only been found in gene clusters that control the biosynthesis of C-9 reduced polyketides. They can be exchanged between different PKSs but they exhibit an interesting hierarchy with respect to chain-length specificity. An aromatase derived from a given natural product pathway is only active on polyketide backbones of the same length or shorter. Thus the 16-carbon *act* ARO/CYC can aromatize octaketides but not longer substrates, while 20-carbon *gris* works with 16-, 18-, and 20-carbon substrates.[98]

The primary role of monodomain ARO/CYC proteins appears to be controlling the regiospecificity of the first cyclization in the case of unreduced backbones. For example, the *tcm* minimal PKS by itself can generate C-7/C-12 and C-9/C-14 cyclized products; however, in the presence of the *tcm* ARO/CYC it exclusively produces C-9/C-14 cyclized products.[47] A similar influence on cyclization specificity by the *tcm* ARO/CYC is also observed with 16- and 18-carbon backbones, indicating relaxed chain-length specificity in this subunit.[47,99] Likewise, the *whi*E monodomain

ARO/CYC can also catalyze C-9/C-14 cyclizations of 20-carbon backbones in addition to its natural 24-carbon substrate.[55] However, in the presence of a KR, both the *tcm* and *whi*E proteins are inactive.[47,55] Interestingly, when monodomain ARO/CYCs are coexpressed with minimal PKSs showing relaxed chain-length specificity, they result in a restriction of chain-length specificity in favor of the longer backbone.[99] For example, the *fren* minimal PKS and *tcm* ARO/CYC produce only a nonaketide backbone, whereas the *whi*E minimal PKS and *whi*E ARO/CYC lead to the production of a dodecaketide.[55,99] Again, this interplay between chain length specificity and aromatization argues in favor of the PKS being a complex of physically associated proteins.

Only the *act* CYC2 and its *fren* counterpart have been definitively identified as second-ring cyclases. Both cyclases can only recognize C-9 reduced substrates that have an aromatic first ring. The *act* and *fren* CYC2 proteins can be interchanged, suggesting at least some tolerance for chains of alternative lengths.[98]

1.18.3.4 Domain Analysis

An important feature of bacterial aromatic PKSs is the existence of a high degree of sequence similarity between subunits with similar function but different specificity. In the absence of direct tertiary and quaternary structural information, one could potentially take advantage of this homology to gain insights into the mechanisms by which PKS subunits control the structure of their final product. A step in this direction was taken through the construction and analysis of a series of deletion and hybrid ARO/CYC mutants. By themselves, neither the *N*- nor the *C*-domain of didomain ARO/CYC proteins was found to be active. However, when coexpressed as separate proteins, they yielded aromatase activity, suggesting that the two domains fold independently but act in concert to catalyze first-ring aromatization. Hybrid *act*/*gris* ARO/CYCs revealed that chain-length specificity in this class of proteins lies in the *N*-domain, but that the *C*-domain of the *gris* ARO/CYC can induce the *tcm* minimal PKS to produce an 18-carbon product in addition to the expected 20-carbon product. These results bode well for the use of protein fusions to study aromatic PKSs, and also demonstrate the flexibility of the PKS enzymes to create unexpected compounds when expressed in new combinations.[97]

1.18.4 BIOCHEMICAL ANALYSIS OF BACTERIAL AROMATIC PKSs

1.18.4.1 Background and Methodological Overview

In the past, a key missing component in the study of bacterial aromatic PKSs was the availability of cell-free systems suitable for the purification and mechanistic analysis of these multiprotein catalysts. The development of cell-free systems was crucially influenced by the availability of plasmid-based expression systems for polyketide biosynthetic genes. All reported *in vitro* systems for bacterial aromatic PKSs have been taken in the form of relatively crude cell extracts, hence the notion of the bacterial aromatic PKSs as a physical multiprotein complex remains to be formally demonstrated.

1.18.4.2 Tetracenomycin

The *tcm* cell-free system was the first *in vitro* system to be developed for a bacterial PKS. Using crude extracts containing the *tcm* minimal PKS, *tcm*J, and the *tcm* ARO/CYC that were overproduced in a nonproducing strain, Tcm F2 was synthesized from acetate and malonate. Confirmation of the Tcm F2 structure was obtained by converting it into Tcm F1 using purified *tcm*I protein. Only acetyl-CoA and malonyl-CoA (which presumably can be decarboxylated by the PKS to yield an acetyl primer) can be used as sources of starter units. Without the *tcm* ARO/CYC there was no production of Tcm F2, and the ACP could not be exchanged with one from another PKS. This deviates from the *in vivo* results and may be due to more stringent structural considerations in the *in vitro* complex. Antibodies against these proteins and chemical inhibitors of fatty acid synthase prevent polyketide synthesis.[50] In 1996 the *tcm*N protein was purified to homogeneity.[100]

1.18.4.3 Actinorhodin

A cell-free system has also been developed with the *act* PKS. From malonyl-CoA, the "minimal PKS" made SEK4 and 4b, and with the KR, ARO/CYC, CYC, and NADPH, the crude cell extract made DMAC. This replicated the *in vivo* results described above. Adequate quantities of the reporter molecules were made to confirm their structures by HPLC–UV analysis, conversion into an expected derivative by an exogenously added enzyme, and NMR. Synthesis was inhibited by fatty acid synthase inhibitors, *N*-ethylmaleimide and cerulenin.[101] Further purification of the *act* and *tcm* minimal PKSs has demonstrated that the KS and CLF proteins coelute over a series of chromatographic columns, suggesting the existence of a physical heteromeric complex of proteins.[102]

1.18.5 ENGINEERING OF BACTERIAL AROMATIC PKSs

1.18.5.1 Rational Design

Unlike modular PKSs, where a unique active site is used for each reaction within the overall catalytic cycle, the active sites in bacterial aromatic PKSs are used iteratively. Therefore, until recently the feasibility of predictive manipulation of product structure by genetically modifying active sites was not conceptually straightforward in the latter case. However, much has been learned about bacterial aromatic PKS function and specificity in recent years; in turn, this has motivated a few successful attempts at the rational design of novel aromatic polyketides. By choosing a minimal PKS together with an appropriate subset of KR, ARO/CYC, and CYC2 subunits, one can set out to make a polyketide of a given length, reduction, aromatization, and cyclization pattern. For example, the *tcm* minimal PKS, the *act* KR, and a *gris* ARO/CYC were used to make a reduced, aromatized 20-carbon polyketide, the *fren* minimal PKS, *act* KR, *fren* ARO/CYC and *act* CYC2 were used to make a reduced and aromatized nonaketide anthraquinone,[98] and the *fren* minimal PKS was used with the *tcm* ARO/CYC to make an unreduced nonaketide with a C-9/C-14 ring.[99] In each case the salient structural features of the reporter polyketide product were predictable *a priori* from a defined set of design heuristics.

1.18.5.2 Combinatorial Biosynthesis

An important motivating factor for the above studies has been the potential for using genetically manipulated PKSs for the generation of molecular diversity. With the increasing repertoire of cloned and sequenced genes coupled with improving tools for their combinatorial manipulation, the library of "unnatural" natural products of this class keeps growing. For example, based on LC–MS–UV analysis it is estimated that the library of (mostly novel) aromatic polyketides generated by the combinatorial manipulation of the *act*, *fren*, *gris*, *tcm*, and *whi*E gene clusters alone exceeds 200 compounds.[103] In the future, the availability of minimal PKSs with novel chain-length specificities and possibly even primer unit specificity, ketoreductases with novel regiospecificities, and aromatases and cyclases with novel cyclization regiospecificity will undoubtedly add to the combinatorial biosynthesis toolbox.

1.18.5.3 Protein Engineering

A cursory examination of the structures of nature's aromatic polyketides suggests bounds to the size and structural diversity of a polyketide library that could be generated via combinatorial biosynthesis. Whereas it remains to be seen whether the protein engineer can expand this repertoire of molecular recognition features, as described in Section 1.18.3.4, early results suggest that mutagenesis of naturally occurring PKS subunits have the potential of generating new allelic forms with novel catalytic functions and/or specificities.

1.18.6 CONCLUSIONS AND FUTURE DIRECTIONS

From the above discussion, it is hoped that it is clear that, from several viewpoints, bacterial aromatic PKSs have been forerunners in the development of a new genetics-led paradigm for natural products science and technology. The very notion that natural product biosynthetic genes are clustered in the genomes of microorganisms was first demonstrated in the context of actinorhodin, a bacterial aromatic polyketide. The actinorhodin gene cluster was also the first complete natural product gene cluster to be cloned. The notion of homology based cloning of natural product gene clusters was initially developed using actinorhodin gene probes. DNA sequencing of the actinorhodin, granaticin, and tetracenomycin PKS loci led to the discovery that bacterial aromatic polyketide biosynthesis was catalyzed by families of homologous multigene systems. The notion that this modularity could be systematically exploited by a mix-and-match approach to generate novel "unnatural" natural products was initially demonstrated using bacterial aromatic PKSs as model systems. Since then, the power of combinatorial biosynthesis using bacterial aromatic PKSs has been harnessed through a variety of loss-of-function, change-of-specificity, gain-of-function, and indeed even protein-engineering strategies, leading to the emergence of "design rules" for rationally guided polyketide design. In the course of all these discoveries, a variety of generally applicable concepts and technologies have been developed that have subsequently been applied to other classes of PKSs, including the modular PKSs and even the fungal PKSs. Whereas future efforts will likely extend this application to other classes of nontemplate-derived natural products, including the nonribosomal peptides, the glycosides, and the isoprenoids, studies on bacterial aromatic PKSs themselves will undoubtedly continue to increase our understanding and exploitation of the structure and function of this remarkable class of multienzyme systems.

ACKNOWLEDGMENTS

We thank David Hopwood for comments on the manuscript, and T. W. Yu and C. Carreras for sharing unpublished results. Research on bacterial aromatic PKSs in the authors' laboratory has been supported in part by the National Science Foundation, the David and Lucile Packard Foundation, Merck Research Laboratories, Procter and Gamble, and Sankyo Co.

1.18.7 REFERENCES

1. N. Collie, *J. Chem. Soc.*, 1893, 122.
2. D. A. Hopwood and C. Khosla, *Ciba Found. Symp.*, 1992, **171**, 88.
3. D. A. Hopwood and D. H. Sherman, *Annu. Rev. Genet.*, 1990, **24**, 37.
4. D. O. O'Hagan, "The Polyketide Metabolites," Ellis Horwood, Chichester, 1991.
5. D. O'Hagan, *Nat. Prod. Rep.*, 1992, **9**, 447.
6. D. O'Hagan, *Nat. Prod. Rep.*, 1993, **10**, 593.
7. L. Katz and S. Donadio, *Annu. Rev. Microbiol.*, 1993, **47**, 875.
8. C. R. Hutchinson and I. Fujii, *Annu. Rev. Microbiol.*, 1995, **49**, 201.
9. C. J. Tsoi and C. Khosla, *Chem. Biol.*, 1995, **2**, 355.
10. S. J. Wakil, *Biochemistry*, 1989, **28**, 4523.
11. S. Smith, *FASEB J.*, 1994, **8**, 1248.
12. R. C. Paulick, M. L. Casey, and H. W. Whitlock, *J. Am. Chem. Soc.*, 1976, **98**, 3370.
13. R. Thomas and D. J. Williams, *J. Chem. Soc., Chem. Commun.*, 1983, 677.
14. R. McDaniel, S. Ebert-Khosla, D. A. Hopwood, and C. Khosla, *J. Am. Chem. Soc.*, 1993, **115**, 11 671.
15. K. Tsuzuki, Y. Iwai, and S. Omura, *J. Antibiot.*, 1986, **39**, 1343.
16. J.-S. Kim, K. Shin-Ya, J. Eishima, K. Furihata, and H. Seto, *J. Antibiot.*, 1996, **49**, 947.
17. H. Seto, H. Wanatabe, and K. Furihata, *Tetrahedron Lett.*, 1996, **37**, 7979.
18. J. R. D. McCormick and E. R. Jensen, *J. Am. Chem. Soc.*, 1968, **90**, 7126.
19. R. McDaniel, S. Ebert-Khosla, D. A. Hopwood, and C. Khosla, *Science*, 1993, **262**, 1546.
20. M. J. S. Dewar and K. M. Dieter, *Biochemistry*, 1988, **27**, 3302.
21. K. Magnuson, S. Jackowski, C. O. Rock, and J. E. J. Cronan, *Microbiol. Rev.*, 1993, **57**, 522.
22. M. Siggaard-Andersen, *Protein Sequences Data Anal.*, 1993, **5**, 325.
23. M. A. Fernandez-Moreno, E. Martinez, L. Boto, D. A. Hopwood, and F. Malpartida, *J. Biol. Chem.*, 1992, **267**, 19 278.
24. G. Meurer and C. R. Hutchinson, *J. Bacteriol.*, 1994, **177**, 477.
25. W. P. Revill, M. J. Bibb, and D. A. Hopwood, *J. Bacteriol.*, 1995, **177**, 3946.
26. M. J. Bibb, S. Biro, H. Motamedi, J. F. Collins, and C. R. Hutchinson, *EMBO J.*, 1989, **8**, 2727.
27. P. R. Vagelos, in "The Enzymes," ed. P. D. Boyer, Academic Press, New York, 3rd edn., vol. 8, p. 155.
28. S. E. Hallam, F. Malpartida, and D. A. Hopwood, *Gene*, 1988, **74**, 305.
29. F. Malpartida and D. A. Hopwood, *Nature (London)*, 1984, **309**, 462.

30. D. A. Hopwood, *Curr. Opin. Biotechnol.*, 1993, **4**, 531.
31. H. Motamedi and C. R. Hutchinson, *Proc. Natl. Acad. Sci. USA*, 1987, **84**, 4445.
32. C. Binnie, M. Warren, and M. J. Butler, *J. Bacteriol.*, 1989, **171**, 887.
33. A. Grimm, K. Madduri, A. Ali, and C. R. Hutchinson, *Gene*, 1994, **151**, 1.
34. M. A. C. Fernandez-Moreno, J. L. Caballero, D. A. Hopwood, and F. Malpartida, *Cell*, 1991, **66**, 769.
35. M. A. C. Fernandez-Moreno, E. Martinez, J. L. Caballero, K. Ichinose, D. A. Hopwood, and F. Malpartida, *J. Biol. Chem.*, 1994, **269**, 24 854.
36. J. L. Caballero, E. Martinez, F. Malpartida, and D. A. Hopwood, *Mol. Gen. Genet.*, 1991, **230**, 401.
37. B. A. M. Rudd and D. A. Hopwood, *J. Gen. Microbiol.*, 1979, **114**, 35.
38. C. Khosla, S. Ebert-Khosla, and D. A. Hopwood, *Mol. Microbiol.*, 1992, **6**, 3237.
39. D. H. Sherman, F. Malpartida, M. J. Bibb, H. M. Kieser, M. J. Bibb, and D. A. Hopwood, *EMBO J.*, 1989, **8**, 2717.
40. M. J. Bibb, D. H. Sherman, S. Omura, and D. A. Hopwood, *Gene*, 1994, **141**, 31–39.
41. S. G. Kendrew, S. E. Harding, D. A. Hopwood, and E. N. G. Marsh, *J. Biol. Chem.*, 1995, **270**, 17 339.
42. S. Kakinuma, H. Ikeda, and S. Omura, *Tetrahedron*, 1991, **47**, 6059.
43. D. H. Sherman, E.-S. Kim, M. J. Bibb, and D. A. Hopwood, *J. Bacteriol.*, 1992, **174**, 6184.
44. A. Bechthold, J. K. Sohng, T. M. Smith, X. Chu, and H. G. Floss, *Mol. Gen. Genet.*, 1995, **248**, 610.
45. T.-W. Yu, M. J. Bibb, P. Revill, and D. A. Hopwood, *J. Bacteriol.*, 1994, **176**, 2627.
46. R. G. Summers, E. Wendt-Pienkowski, H. Motamedi, and C. R. Hutchinson, *J. Bacteriol.*, 1992, **174**, 1810.
47. R. McDaniel, C. R. Hutchinson, and C. Khosla, *J. Am. Chem. Soc.*, 1995, **117**, 6805.
48. R. G. Summers, E. Wendt-Pienkowski, H. Motamedi, and C. R. Hutchinson, *J. Bacteriol.*, 1993, **175**, 7571.
49. H. Decker, H. Motamedi, and C. R. Hutchinson, *J. Bacteriol.*, 1993, **175**, 3876.
50. B. Shen and C. R. Hutchinson, *Science*, 1993, **262**, 1535.
51. H. Decker, J. Rohr, H. Motamedi, H. Zahner, and C. R. Hutchinson, *Gene*, 1995, **166**, 121.
52. K. E. Kendrick and J. C. Ensign, *J. Bacteriol.*, 1983, **155**, 357.
53. N. K. Davis and K. F. Chater, *Mol. Microbiol.*, 1990, **4**, 1679.
54. T.-W. Yu, Ph.D. Thesis, University of East Anglia, 1995.
55. M. A. Alvarez, H. Fu, C. Khosla, D. A. Hopwood, and J. E. Bailey, *Nature Biotechnol.*, 1996, **14**, 335.
56. T.-W. Yu, unpublished results, 1996.
57. T.-W. Yu and D. A. Hopwood, *Microbiology*, 1995, **141**, 2779.
58. G. Blanco, A. Pereda, C. Mendez, and J. A. Salas, *Gene*, 1992, **112**, 59.
59. G. Blanco, P. Brian, A. Pereda, C. Mendez, J. A. Salas, and K. F. Chater, *Gene*, 1993, **130**, 107.
60. S. Bergh and M. Uhlen, *Gene*, 1992, **117**, 131.
61. G. Blanco, H. Fu, C. Mendez, C. Khosla, and J. A. Salas, *Chem. Biol.*, 1996, **3**, 193.
62. F. Lombo, G. Blanco, E. Fernandez, C. Mendez, and J. A. Salas, *Gene*, 1996, **172**, 87.
63. J. Ye, M. L. Dickens, R. Plater, Y. Li, J. Lawrence, and W. R. Strohl, *J. Bacteriol.*, 1994, **176**, 6270.
64. M. L. Dickens, J. Ye, and W. R. Strohl, *J. Bacteriol.*, 1995, **177**, 536.
65. W. R. Strohl, J. Ye, and M. L. Dickens, *Biotekhnologiya*, 1995, **7–8**, 45.
66. N. Tsukamoto, I. Fujii, Y. Ebizuka, and U. Sankawa, *J. Antibiot.*, 1992, **45**, 1286.
67. K. Ylihonko, J. Tuikkanen, S. Jussila, L. Cong, and P. Mantsala, *Mol. Gen. Genet.*, 1996, **251**, 113.
68. F. Malpartida, S. E. Hallam, H. M. Kieser, H. Motamedi, C. R. Hutchinson, M. J. Butler, D. A. Sugden, M. Warren, C. McKillop, C. R. Bailey, G. O. Humphreys, and D. A. Hopwood, *Nature (London)*, 1987, **325**, 818.
69. E.-S. Kim, M. J. Bibb, M. J. Butler, D. A. Hopwood, and D. H. Sherman, *Gene*, 1994, **141**, 141.
70. P. M. Rhodes, N. Winskill, E. J. Friend, and M. Warren, *J. Gen. Microbiol.*, 1981, **124**, 329.
71. L. Han, K. Yang, E. Ramalingam, R. H. Mosher, and L. C. Vining, *Microbiology*, 1994, **140**, 3379.
72. H. Decker and S. Haag, *J. Bacteriol.*, 1995, **177**, 6126.
73. S.-T. Hong, J. R. Carney, and S. J. Gould, *J. Bacteriol.*, 1997, **179**, 470.
74. T. J. Arrowsmith, F. Malpartida, D. H. Sherman, A. Birch, D. A. Hopwood, and J. A. Robinson, *Mol. Gen. Genet.*, 1992, **234**, 254.
75. C. LeGouill, D. Desmarais, and C. V. Dery, *Mol. Gen. Genet.*, 1992, **240**, 146.
76. M. Piecq, P. Dehottay, A. Biot, and J. Dusart, *DNA Sequence*, 1994, **4**, 219.
77. H. Motamedi, E. Wendt-Pienkowski, and C. R. Hutchinson, *J. Bacteriol.*, 1986, **167**, 575.
78. D. H. Sherman, M. J. Bibb, T. J. Simpson, D. Johnson, F. Malpartida, M. Fernandez-Moreno, E. Martinez, C. R. Hutchinson, and D. A. Hopwood, *Tetrahedron*, 1991, **47**, 6029.
79. B. Shen, R. G. Summers, E. Wendt-Pienkowski, and C. R. Hutchinson, *J. Am. Chem. Soc.*, 1995, **117**, 6811.
80. P. L. Bartel, C.-B. Zhu, J. S. Lampel, D. C. Dosh, N. C. Connors, W. R. Strohl, J. M. Beale Jr., and H. G. Floss, *J. Bacteriol.*, 1990, **172**, 4816.
81. C. M. Kao, L. Katz, and C. Khosla, *Science*, 1994, **265**, 509.
82. D. J. Bedford, E. Schweizer, D. A. Hopwood, and C. Khosla, *J. Bacteriol.*, 1995, **177**, 4544.
83. S. L. Otten, K. J. Stuzman-Engwall, and C. R. Hutchinson, *J. Bacteriol.*, 1990, **172**, 3427.
84. S. Horinouchi and T. Beppu, *J. Bacteriol.*, 1985, **162**, 406.
85. E.-S. Kim, K. D. Cramer, A. L. Shreve, and D. H. Sherman, *J. Bacteriol.*, 1995, **177**, 1202.
86. S. P. Cole, B. A. M. Rudd, D. A. Hopwood, C.-J. Chang, and H. G. Floss, *J. Antibiot.*, 1987, **40**, 340.
87. H. L. Zhang, X. G. He, A. Adefarati, J. Gallucci, S. P. Cole, J. M. Beale, P. J. Keller, C. J. Chang, and H. G. Floss, *J. Org. Chem.*, 1990, **55**, 1682.
88. H. Fu, D. A. Hopwood, and C. Khosla, *Chem. Biol.*, 1994, **1**, 205.
89. H. Fu, S. Ebert-Khosla, D. A. Hopwood, and C. Khosla, *J. Am. Chem. Soc.*, 1994, **116**, 4166.
90. R. McDaniel, S. Ebert-Khosla, D. A. Hopwood, and C. Khosla, *J. Am. Chem. Soc.*, 1994, **116**, 10 855.
91. R. McDaniel, S. Ebert-Khosla, H. Fu, D. A. Hopwood, and C. Khosla, *Proc. Natl. Acad. Sci. USA*, 1994, **91**, 11 542.
92. B. Shen and C. R. Hutchinson, *Biochemistry*, 1993, **32**, 1149.
93. P. L. Bartel, N. C. Connors, and W. R. Strohl, *J. Gen. Microbiol.*, 1990, **136**, 1877.
94. N. C. Connors, P. L. Bartel, and W. R. Strohl, *J. Gen. Microbiol.*, 1990, **136**, 1887.
95. C. Khosla, R. McDaniel, S. Ebert-Khosla, R. Torres, D. H. Sherman, M. J. Bibb, and D. A. Hopwood, *J. Bacteriol.*, 1993, **175**, 2197.

96. W. P. Revill, M. J. Bibb, and D. A. Hopwood, *J. Bacteriol.*, 1996, **178**, 5660.
97. R. J. X. Zawada and C. Khosla, *J. Biol. Chem.*, 1997, **272**, 16 184.
98. R. McDaniel, S. Ebert-Khosla, D. A. Hopwood, and C. Khosla, *Nature (London)*, 1995, **375**, 549.
99. P. J. Kramer, R. J. X. Zawada, R. McDaniels, C. R. Hutchinson, D. A. Hopwood, and C. Khosla, *J. Am. Chem. Soc.*, 1997, **119**, 635.
100. B. Shen and C. R. Hutchinson, *Proc. Natl. Acad. Sci. USA*, 1996, **93**, 6600.
101. C. W. Carreras, R. Pieper, and C. Khosla, *J. Am. Chem. Soc.*, 1996, **118**, 5158.
102. C. Carreras, unpublished results, 1996.
103. G. Ashley, unpublished results, 1996.
104. N. Tsukamoto, J. Fujii, Y. Ebizuk, and U. Sankawa, *J. Bacteriol.*, 1994, **176**, 2473.

1.19
Biosynthesis of Erythromycin and Related Macrolides

JAMES STAUNTON
University of Cambridge, UK

and

BARRIE WILKINSON
GlaxoWellcome, Stevenage, UK

1.19.1 INTRODUCTION TO POLYKETIDE STRUCTURES

The main focus of this chapter will be the erythromycin family of macrolide antibiotics, represented here by erythromycin A (**1**), the structural variant which is widely used as an antibiotic in human medicine. This has a large lactone ring (hence the title macrolide) containing 14 atoms. Many structural variants have been isolated from natural sources having different patterns of hydroxylation and glycosylation around the macrolide core. For the most part these are either biosynthetic intermediates on the biosynthetic pathway leading to erythromycin A or degradation products of it.

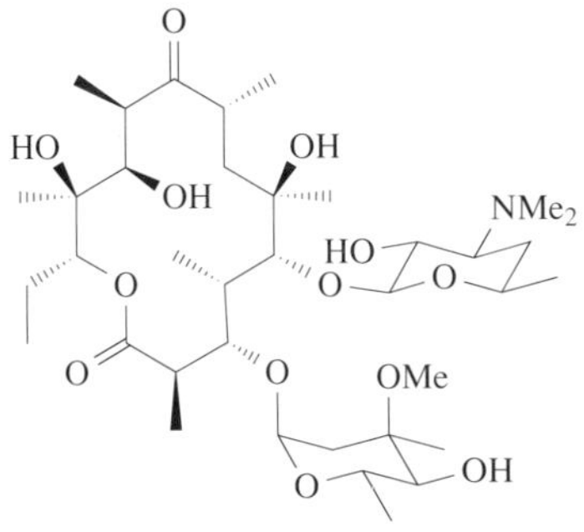

(**1**) Erythromycin A

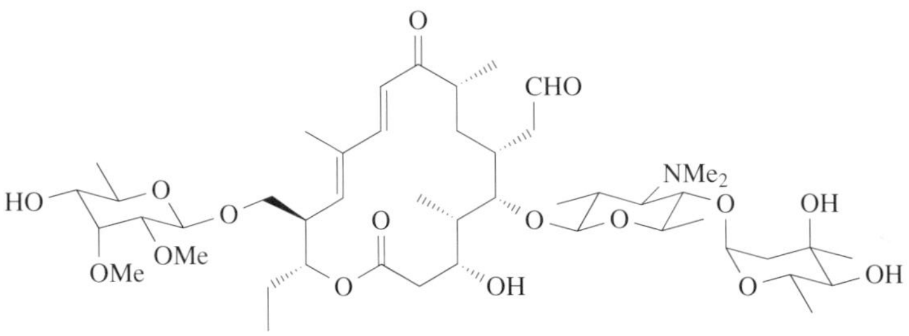

(**2**) Tylosin

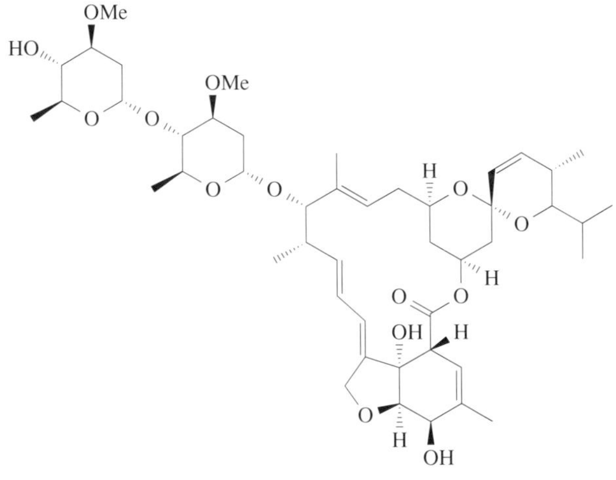

(**3**) Avermectin B1b

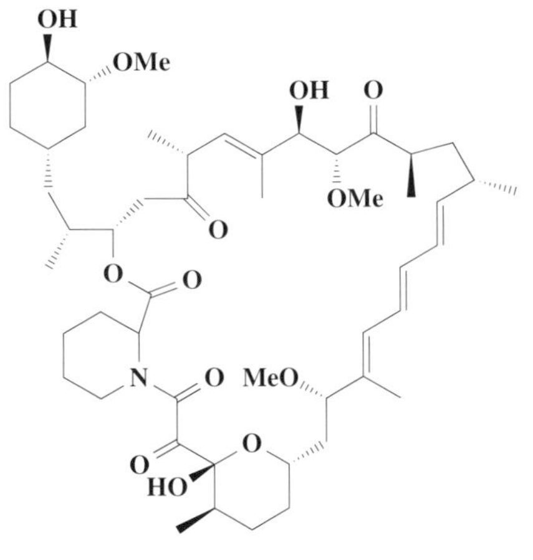

(**4**) Rapamycin

Other families of macrolides are known containing 12- or 16-membered macrolide rings.[1] A representative of the latter family is tylosin (**2**), a commercially important antibacterial agent used in animal medicine. Note that as in the case of erythromycin A, the macrolide core is heavily substituted and is rich in chiral centers. Careful comparison will also reveal the strong structural homology between the right-hand portions of these two compounds; not only are the comparable structural elements similar, but there is also a strong affinity at the stereochemical level.[2] It is possible that the tylosin and erythromycin biosynthetic pathways may have evolved from a common primitive system. One of the aims of this chapter is to consider the genetics of these biosynthetic pathways in this light.

Two other macrolides are introduced here to show the wider diversity of structure which has been identified. Avermectin B1b (**3**) contains a 16-membered macrolide core. The structure shows no structural homology to erythromycin A. If the two biosynthetic pathways are related in terms of evolution, the point of divergence must have occurred very early. Even so, as we shall see, the genes for these pathways, and the resulting enzymes, can be made by hybridization, leading to the production of new natural product structures.

Finally, rapamycin (**4**), an important immunosuppressant, diverges even further from erythromycin A in having a 29-membered macrolide ring. More significantly, the macrolide ring contains a heteroatom in what is clearly an amino acid-derived portion of the molecule. No obvious structural homology with erythromycin A is apparent in the rest of the macrolide ring. Once again, therefore, it can be assumed that any evolutionary relationship is very distant but the biosynthetic genes for the two structures have been hybridized, leading to the creation of novel natural products.

This "forced evolution" of novel structures in the polyketide family is one of the major growth areas in biosynthetic study. The field has attracted the attention of industrialists as well as academics because it has enormous potential for the creation of new pharmaceutical lead compounds and therapeutic agents in the battle against disease.

It will be apparent that the mode of biosynthesis rather than homology of structure is the feature that links polyketide metabolites. They are all biosynthesized by the repeated addition of small building blocks, usually comprising two or three carbons, to form a linear chain. This chain can have a series of keto groups on alternate carbons—hence the term polyketide. Alternatively, one or more of the keto groups may be reduced. Three levels of reduction of the keto group are observed: hydroxyl, double bond, or methylene. The carbon frameworks of the macrolide rings in (**1**)–(**4**) are formed by this type of biosynthetic process. How this takes place and how it can be altered is the subject of this chapter.

1.19.2 THE ERYTHROMYCINS

1.19.2.1 Structures of the Erythromycins

Erythromycin A was isolated in 1952 from *Saccharopolyspora erythraea*.[3] The structure was elucidated by classical methods[4] and confirmed by X-ray crystallography.[5] Other members of the family occur as biosynthetic intermediates and so will be presented in that context later. Erythromycin A is an important antibiotic widely used in clinical medicine against infections caused by Gram-positive bacteria, and it is the main treatment for many pulmonary infections such as Legionnaire's disease. In recent years, semisynthetic derivatives produced by chemical modification have been found to be even more effective. For example, in azithromycin (**5**), the lactone ring is expanded by insertion of a nitrogen atom; this is achieved by Beckmann rearrangement of an oxime generated at the C-9 carbonyl group. Clarithromycin (**6**) has the 6-hydroxy group methylated.

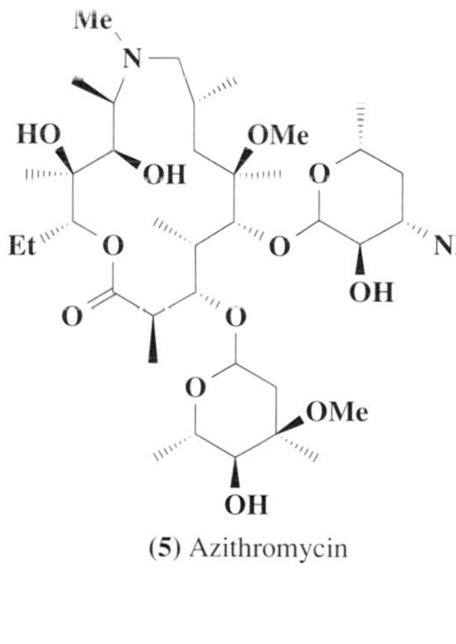

(**5**) Azithromycin

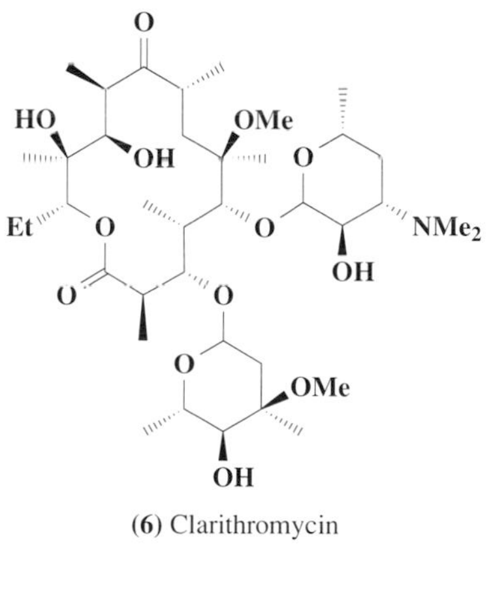

(**6**) Clarithromycin

1.19.2.2 Overview of the Erythromycin Biosynthetic Pathway

The biosynthesis of erythromycin takes place in two phases (Scheme 1). First, a set of key enzymes collectively known as the polyketide synthases (PKSs) assembles the polyketide chain by sequential condensation of one unit of propionyl-CoA with six units of methylmalonyl-CoA. The linear chain is then cyclized to give the first macrolide intermediate 6-deoxyerythronolide B (**7**).[6,7] Generation of the core polyketide structure by PKS is the characteristic polyketide part of the pathway.

Subsequently, in the second phase, 6-deoxyerythronolide B is elaborated by a series of "tailoring" enzymes which carry out regiospecific hydroxylations, glycosylations, and a methylation (of an added sugar residue) to give finally erythromycin A. These later elaboration steps are essential for producing active antibiotics.

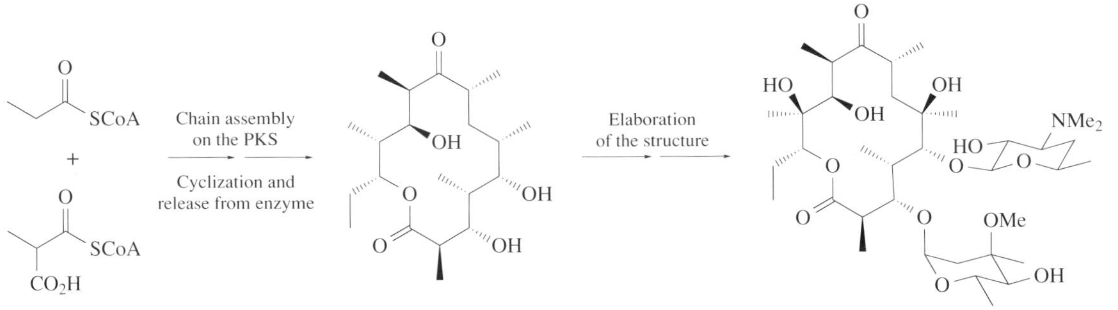

Chain assembly on the PKS → Cyclization and release from enzyme

Elaboration of the structure →

(6) **(7)** 6-Deoxyerythronolide B **(1)** Erythromycin A

Scheme 1

1.19.2.3 The Elaboration Steps of the Pathway

In a search for new antibiotics or to improve the yields of the existing compounds, the producing organism has been subjected to intensive mutation studies in the hope of achieving useful modifications to the biosynthetic machinery. Most of the information concerning the late stages of the biosynthetic pathway came as a spin-off from these studies. When a late step on the pathway is blocked by lack of the requisite enzyme, the previous intermediate may accumulate by the resulting blocked mutant in sufficient quantities for it to be isolated and identified. Mutants blocked in each step in the pathway from 6-deoxyerythronolide B (7) to erythromycin A (**1**) have been produced, allowing the late intermediates to be identified[8] (Scheme 2).

Initially, erythronolide B (**8**) is produced by hydroxylation at C-6.[9] Incorporation of radioactive (7)[10] and isotope dilution experiments[11] confirmed that it was a true intermediate. Next, L-mycarose is attached to the C-3 hydroxyl group of (**8**) by a TDP-mycarose glycosyltransferase,[12,13] followed by addition of the amino sugar D-desosamine to the C-5 hydroxyl of (**9**) by the action of the enzyme TDP-desosamine glycosyltransferase.[13,14] The resulting intermediate, erythromycin D (**10**), is the first to contain an amino sugar and to show antibacterial activity. At this stage there is a branch in the biosynthetic pathway. Either C-12 hydroxylation takes place with retention of configuration[15] to produce erythromycin C (**11**), or *O*-methylation of the C-3″ hydroxyl of the mycarose moiety with SAM, catalyzed by an *O*-methyltransferase, produces erythromycin B (**12**). Finally erythromycin A (**1**) can be generated either by *O*-methylation of (**11**) catalyzed by an *O*-methyltransferase and SAM, or by C-12 hydroxylation of (**12**). A single *O*-methyltransferase can serve on both branches of the pathway.[16] Similarly, a single hydroxylase has been implicated.[17] Since this has a very strong selectivity for erythromycin C in preference to erythromycin B, it is likely that the pathway via erythromycin C is the one with greatest flux.

A speculative biosynthetic pathway starting from glucose has been proposed for the sugar residues based on genetic analysis.[18,19]

1.19.2.4 Assembly of the Macrolide Core—Classical Precursor Studies

Corcoran and co-workers fed labeled precursors to *S. erythraea* and provided the first evidence that the macrolide core is derived from one propionyl-CoA and six methylmalonyl CoA units (Scheme 3).[20] Subsequently, Cane and co-workers[21] fed [^{18}O]propionate and demonstrated that all the oxygens attached to carboxyl-derived carbons in the macrolide core were retained from the propionate precursor and were not derived from molecular oxygen or water.

In studies using blocked mutants, no intermediates preceding 6-deoxyerythronolide B were isolated, implying that these intermediates are either too unstable to accumulate or are enzyme-bound.[8] In an important pioneering experiment, Cane and Yang investigated a possible diketide

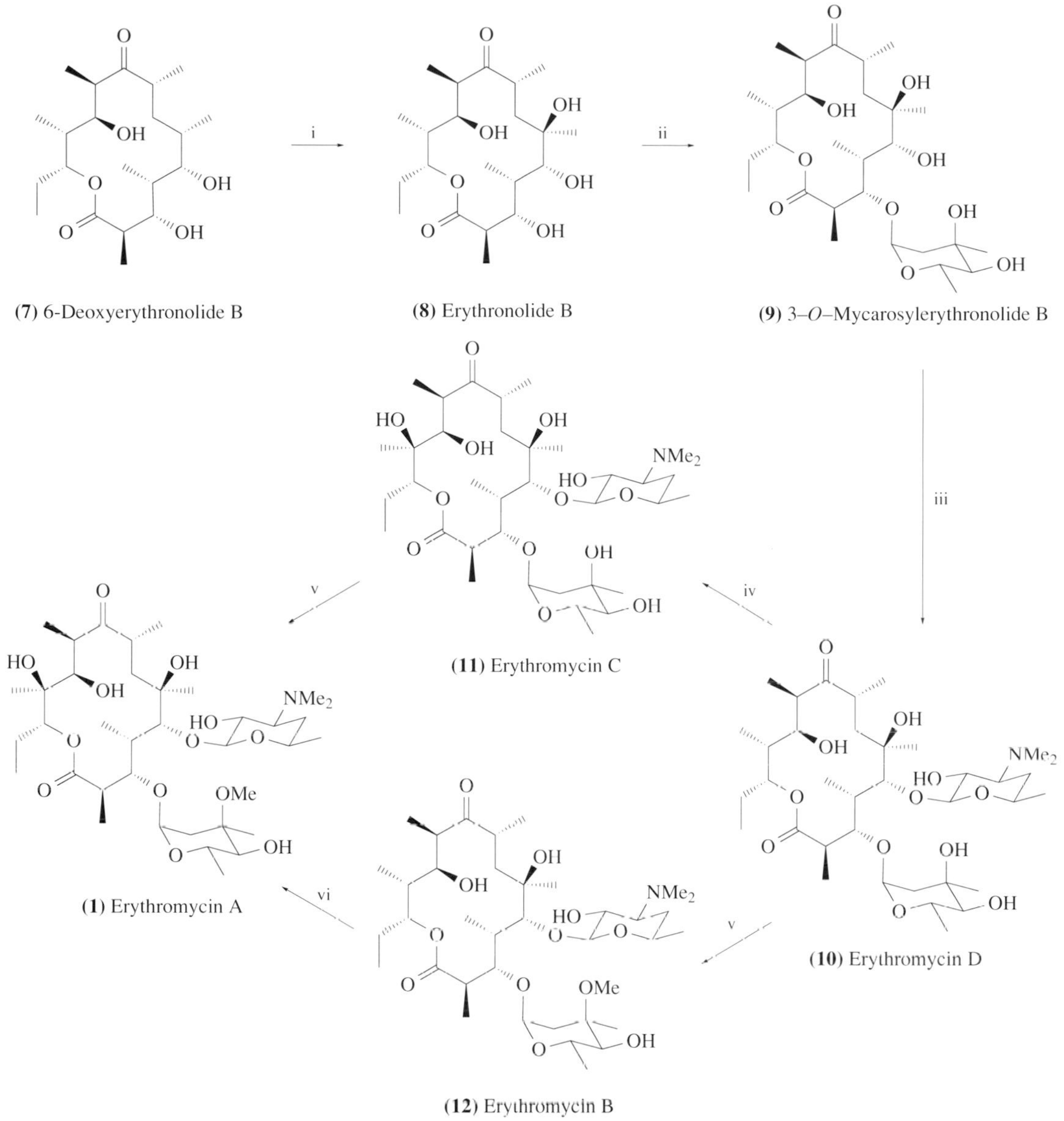

(7) 6-Deoxyerythronolide B

(8) Erythronolide B

(9) 3–O–Mycarosylerythronolide B

(11) Erythromycin C

(1) Erythromycin A

(10) Erythromycin D

(12) Erythromycin B

i, C 6 erythronolide hydroxylase; ii, TDP–mycarose glycosyltransferase;
iii, TDP desosamine glycosyltransferase; iv, C-12 hydroxylase; v, O-methyltransferase

Scheme 2

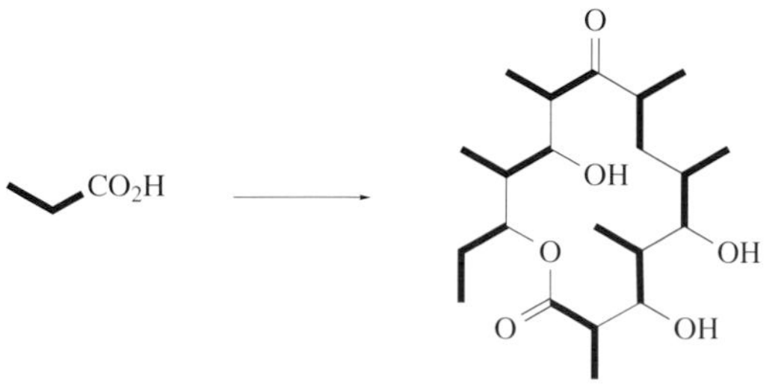

(7) 6-Deoxyerythronolide B

Scheme 3

intermediate in which the keto group has been reduced to hydroxyl.[22] The free acid showed random incorporation into erythromycin B (**12**), indicating that the diketide had been degraded to propionate before being used in the biosynthesis. However, the *N*-acetylcysteamine (NAC) thioester analogue (**13**) was incorporated intact with retention of the coupling between the two simultaneously labeled sites of ^{13}C enrichment (Scheme 4). The possibility that (**13**) was oxidized to the β-keto thioester before incorporation was eliminated by feeding the analogous deuterium labeled diketide (**14**)[23] which resulted in erythromycin doubly labeled with ^{13}C and deuterium at the expected sites.

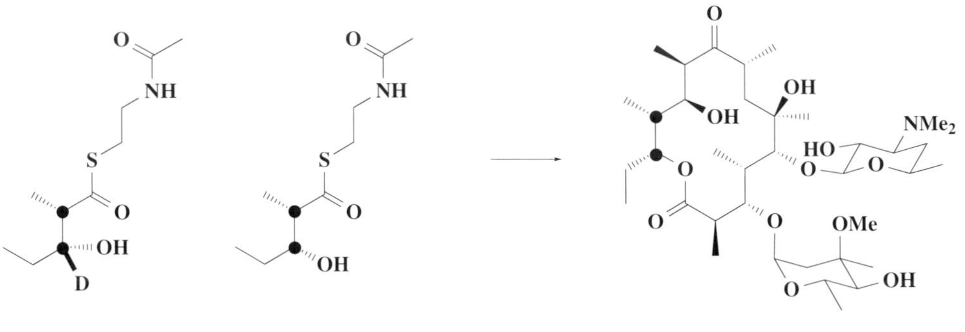

(**14**) (**13**) (**12**) Erythromycin B

● = ^{13}C label

Scheme 4

The *N*-acetylcysteamine (NAC) thioester was employed because it exhibits a high degree of structural homology with the thiol terminus of coenzyme A and the 4′-phosphopantetheine group of the active acyl carrier protein. From these results it can be concluded that the diketide ketoester intermediate produced by the first chain extension is reductively processed to the carbinol prior to the addition of the second chain extension unit. Similar evidence for a processive mechanism has been obtained for other polyketides including tylosin,[24] methymycin,[25] and avermectin.[26]

1.19.2.5 Relationship Between Polyketide and Fatty Acid Biosynthesis

It may come as a surprise to find a discussion of fatty acid biosynthesis in the middle of an account of erythromycin biosynthesis, but it was long suspected that the two processes may be closely related. This has been confirmed and so a short digression is appropriate at this point.[27,28] Investigations using isolated fatty acid synthases (FAS) have established that formation of a fatty acid is initiated by condensation of a starter unit (commonly acetate) with an extender unit (malonate) (Scheme 5).[27,28] The resulting β-ketoester is then fully processed (reduced, dehydrated, and reduced again) to give an elongated saturated fatty chain. A second chain extension cycle then takes place with the condensation of a new extender unit, followed by complete reductive processing of the new keto group. To carry out this repetitive sequence of operations, the fatty acid synthase requires a characteristic set of catalytic activities, with each activity being responsible for one step of the cycle of reactions, i.e., acyltransferase (AT), β-ketoacyl synthase (KS), and acyl carrier protein (ACP) for chain elongation; β-ketoreductase (KR), dehydratase (DH), and enoyl reductase (ER) for processing of the β-keto group to a methylene; and a thioesterase (TE) for hydrolytic release of the full-length chain. Remarkably, a FAS can carry out all the successive chain extension cycles using only a single set of these enzymes. Control over chain length is achieved by the substrate specificity of the thioesterase which in contrast to the other enzymes has a tightly defined substrate specificity in terms of chain length.

Note that the intermediates remain covalently bound to the various proteins throughout. Initially, the starter acyl group is transferred from the pool of appropriate acyl CoA to form a thioester link to a cysteine residue located in the active site of the KS. The malonyl unit is similarly transferred from its pool of CoA ester to the phosphopantetheine arm of the ACP. The arm is long and flexible and so can insert the attached malonyl unit into the active site of the KS for carbon–carbon bond formation. The extended chain remains bound to the ACP until the methylene group is produced. If further extension is required, the chain is transferred back to the KS for the next cycle. Once the

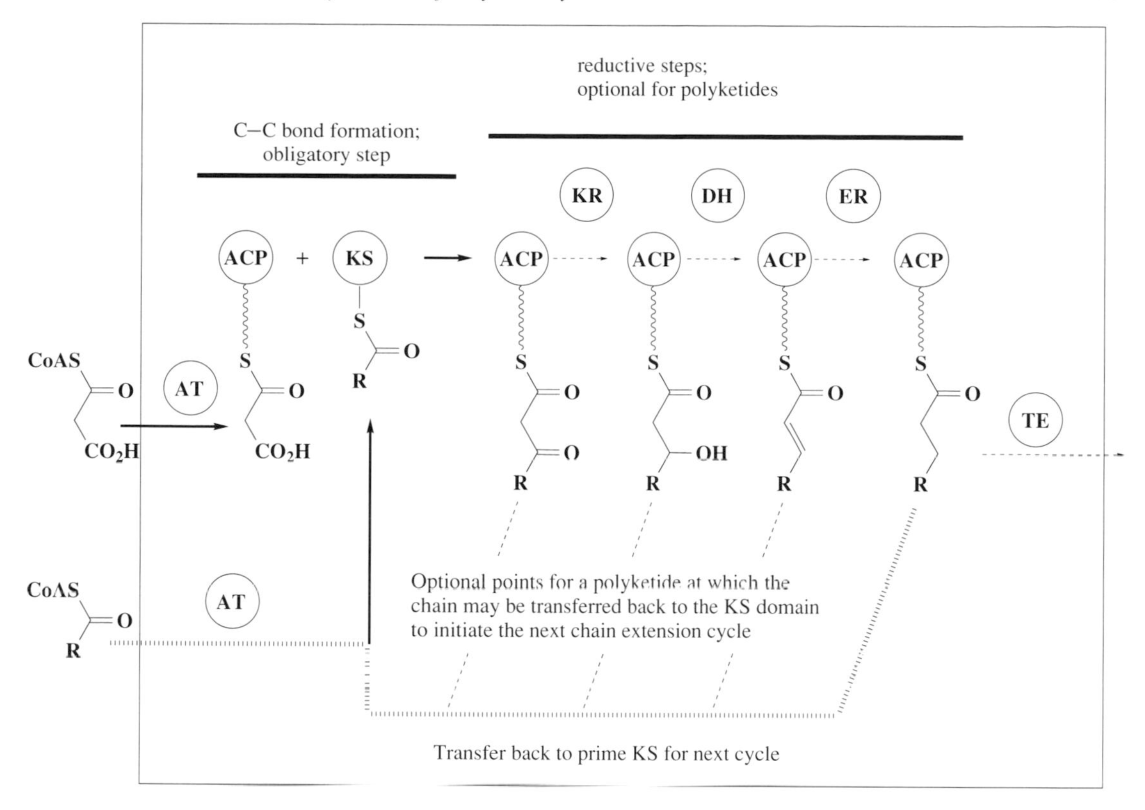

ACP = acylcarrier protein; KS = ketosynthase; KR = ketoreductase;
DH = dehydratase; ER = enoyl reductase; AT = acyltransferase; TE = thioesterease

Scheme 5

chain has reached the required length, it is cleared from the thioester by the thioesterase (TE), usually by hydrolysis, to give a free carboxylic acid.

The constituent proteins of an FAS can be organized in different ways depending on the organism. In bacteria, the individual activities (domains) are freely dissociable and can be isolated separately.[29] In animals and yeast, the domains are bound together by peptide links to form large highly organized nondissociable multidomain proteins.[27,28] The multidomain proteins are designated "Type I" and the dissociable synthases are designated "Type II." The same nomenclature has since been adopted in the PKS field.

1.19.2.6 Sequencing of the Genes Coding for the Erythromycin PKS

Early attempts over many years to isolate the PKS enzymes associated with assembly of the macrolide core of erythromycin proved unsuccessful. The first clue to the type of domains and their organization in the erythromycin PKS came therefore with the sequencing of the corresponding genes. These were located in the vicinity of the gene coding for erythromycin resistance *ermE*.

On both sides of *ermE* there are regions of DNA with open reading frames, thought on the basis of sequence homology with known enzymes to encode for non-PKS proteins responsible for the late stages of the biosynthesis, from 6-deoxyerythronolide B (**7**) to erythromycin A (**1**). Many of these assignments have been confirmed by targeted disruption of these genes to give mutants from which late intermediates have been isolated (see above). A map showing these regions of the genome is presented in Figure 1.

Sequencing further away from the resistance gene, the Leadlay group in Cambridge[30] and Katz and co-workers[31] at the Abbott Laboratories jointly discovered three exceptionally large open reading frames, each containing about 10 kbp of DNA designated *eryAI*, *AII*, and *AIII*. Each of these genes codes for a giant (>3000 amino acids) multifunctional protein. It was proposed therefore that the erythromycin PKS consists of three giant proteins called DEBS 1, DEBS 2, and DEBS 3,

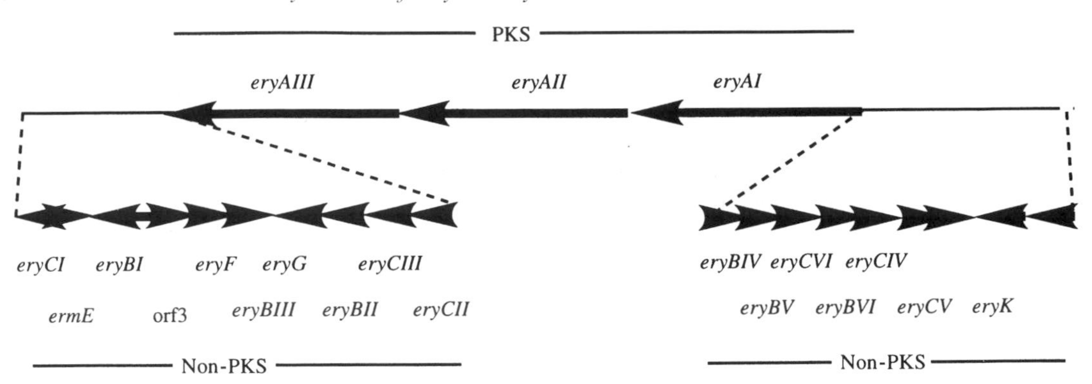

Figure 1 Map of the *S. erythraea* genome containing genes associated with the late stages of the erythromycin biosynthetic pathway.

respectively.[30–34] In the primary sequence of the putative proteins, it was possible to identify regions with motifs showing strong sequence homology with the active site motifs of proteins involved in fatty acid synthases. It was this homology that encouraged the investigators in their belief that they had discovered the genes coding for the elusive PKS proteins.

In the linear representations of the primary sequence of the proteins shown in Figure 2, sections thought to be associated with specific catalytic activities are indicated by coded blocks. These catalytic centers are termed *domains*. Each domain is thought to be folded to form a localized globular structure with a catalytic role dictated by its specific active site. The various regions between the domains which are unmarked in this diagram are thought to play a passive but vital structural role in maintaining the various domains in the correct topology for cooperation in the overall catalytic process. Most of these structural regions consist of short sections of amino acids rich in alanine, proline, and charged residues, up to 30 residues long, which are thought to serve as linkers between the globular domains. There are also intriguing, much longer sections of primary sequence in all three proteins preceding the KR domains which cannot be associated with any specific catalytic activity.

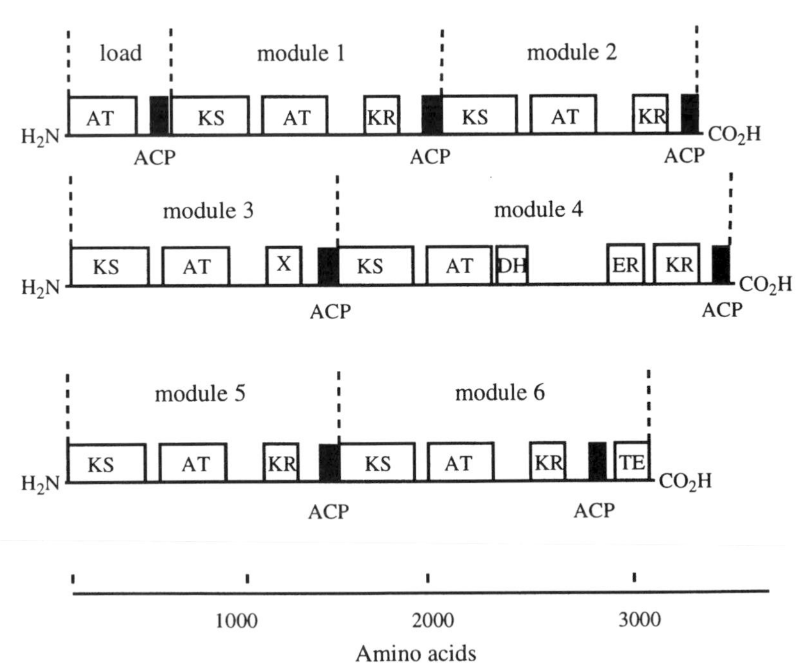

Figure 2 Predicted domain organization of the DEBS proteins. Ketoacyl synthase (KS); acetyl transferase (AT); dehydratase (DH); enoyl reductase (ER); keto reductase (KR); acyl carrier protein (ACP); thioesterase (TE). Each domain is represented by a box with coded shading whose length is proportional to the size of the domain; (KR) indicates an inactive KR domain. The ruler indicates the residue number within the primary structure of the constituent proteins. Linker regions are shown in proportion.

1.19.2.7 Organization and Function of the Erythromycin PKS

To help explain the synthetic implications of this organization of the domains, a different representation of the PKS is shown in Scheme 6, in which domains are represented by circles and the linker regions and structural residues are ignored. In all there are six ketosynthase (KS) domains, which allows one for each chain extension cycle. Following each KS there is an AT domain, then a variable set of domains associated with keto group modification, and finally an ACP. The set of domains starting with each KS and finishing with the next ACP is defined as a "module." Each module carries out one chain extension cycle. This organization has given rise to the term "modular" PKS.

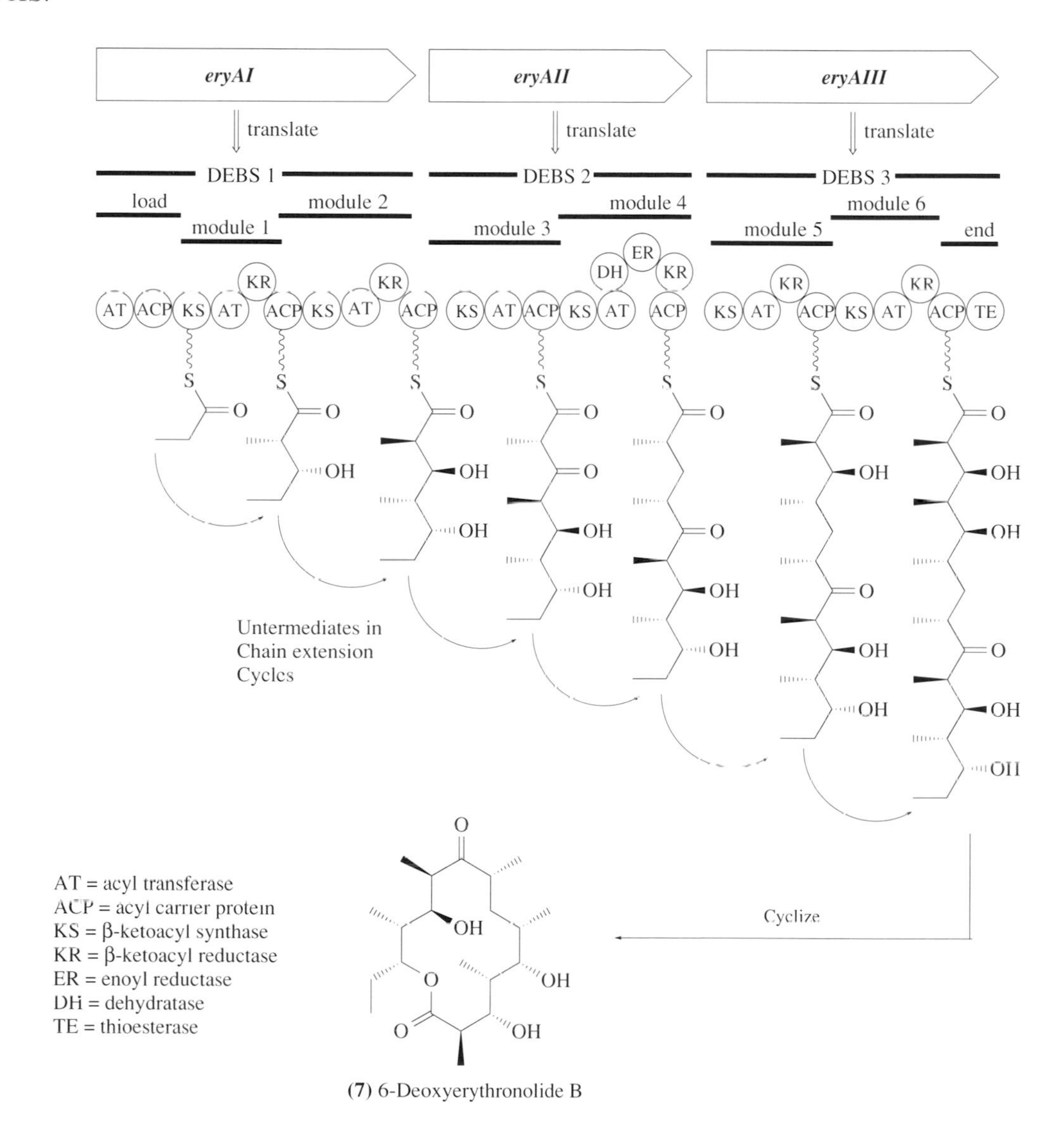

(7) 6-Deoxyerythronolide B

Scheme 6

Closer inspection of the modules allows some predictions to be made concerning the nature of the chain extension unit expected to be formed in the developing polyketide chain. The three essential domains KS, AT, and ACP cooperate to form the new carbon–carbon bond resulting in a keto ester intermediate. The variable set of domains positioned between the AT and ACP (raised as a loop above the line of essential domains) carries out modification of the keto group before the next chain extension sequence. Module 3 has no extra domain and so the keto group should survive unchanged; modules 1, 2, 5, and 6 have a lone KR domain which would be expected to result in reduction of the keto group to a hydroxyl group; in module 4 there is a full set of FAS-type reductive domains (DH + ER + KR) and so complete reduction to produce a methylene would be expected.

Therefore, if the modules operate in the sequence in which they are positioned, the polyketide product would be predicted to have the following sequence of functional groups on alternate carbons, starting with the first chain extension unit: hydroxyl, hydroxyl, keto, methylene, hydroxyl, hydroxyl, and finally an acyl group. Examination of the uncyclized form of the PKS product, 6-deoxyerythronolide B (**7**), shows that this has precisely this substitution pattern. Therefore, it is proposed that the three DEBS proteins function in succession as a molecular assembly line which is programmed by the variable sets of reductive domains to produce the specific polyketide chain.

There is an ambiguity in this analysis because DEBS 1 and DEBS 3 produce comparable structures. Reliable differentiation and role assignment is possible, however, based on the extra domains not associated with chain extension modules. At the start of DEBS 1 there is a set of two domains, AT plus ACP, which would be appropriate to load the starter unit; propionate, from propionyl CoA to the active site of KS1; this is called the "loading" module and it identifies DEBS 1 with the start of the production line. Conversely, at the terminus of DEBS 3 there is a thioesterase (TE) domain which is appropriate for release of the completed chain at the end of the assembly line. In fatty acid biosynthesis, the product is not cyclized but in the case of DEBS it is thought that the TE catalyzes release of the chain by macrolactone formation.

According to this modular analysis, each protein catalyzes two cycles of chain extension. The term "*cassette*" has been proposed for the giant proteins.[34] All three cassettes in the erythromycin cluster are bimodular, but in other systems, such as the rapamycin[35] and tylosin[36] PKSs, the size of a cassette can vary from one to six chain extension modules. The three cassettes cooperate in some way to form an extraordinarily complex molecular assembly line. The biosynthetic intermediates remain PKS-bound throughout the whole synthetic sequence via *thioester* links. A challenging feature of this organization is the mechanism which controls the ordering of the cassettes in the assembly line so that the transfer of the growing chain from one cassette to the next is correctly controlled.

The correspondence between the domain composition of successive modules and the structure of each newly added C_3 unit in successive intermediates is persuasive but it does not prove that the modules function in the order indicated. The protein could fold in many different ways to bring sets of noncontiguous domains together to form functional modules. It was a high priority to establish that modules are indeed made up of adjacent sets of domains as indicated, and also that the individual modules and domains are used in the order shown.

The first strong supporting evidence came from Katz and co-workers who carried out pioneering gene disruption experiments. In 1991 they disrupted the β-ketoacyl reductase (KR) in module 5 and from the resulting mutant they isolated a partly processed erythromycin analogue (**15**), in which a keto group has survived at the predicted position of the macrolide in place of the normal hydroxyl group.[37] The presumed product of the PKS is therefore the 5-keto analogue (**16**) of 6-deoxyerythronolide B (Scheme 7). More recently the enoyl reductase (ER) in module 4 was disrupted and an analogue (**17**) of erythromycin was isolated containing a double bond at C-6/C-7 (Scheme 8).[38] It can be concluded that the disrupted keto reductase operates in the fifth cycle of chain extension and that the disrupted enoyl reductase operates in the fourth cycle as predicted by the simple successional arrangement of modules in Scheme 6.

These two experiments also had important implications for genetic engineering of polyketide biosynthesis to produce novel metabolites. The formation of the two structural analogues of the polyketide chain and the fact that they are correctly lactonized shows that the structure of the nascent chain may not play a critical role in polyketide biosynthesis, and at least some altered polyketides can be substrates for further cycles of chain extension. This implies that the correct transfer of the growing chain from one synthase unit to the next may reside more in the specific juxtaposition of the various domains rather than in the conventional substrate specificity of a particular synthase domain for the structure of the approaching chain. It is interesting and encouraging that the product released from the PKS in both experiments was at least partly processed towards an analogue of erythromycin A, showing that the elaboration enzymes also possess a useful degree of relaxation in their substrate specificity. Further examples of genetic engineering to produce novel metabolites are presented in Section 1.19.2.9.

1.19.2.8 Isolation and Structural Studies of the Erythromycin PKS

All three DEBS proteins have been overexpressed and successfully purified to homogeneity.[39] Gel filtration indicated that all were dimeric under native conditions, and a DEBS multienzyme has

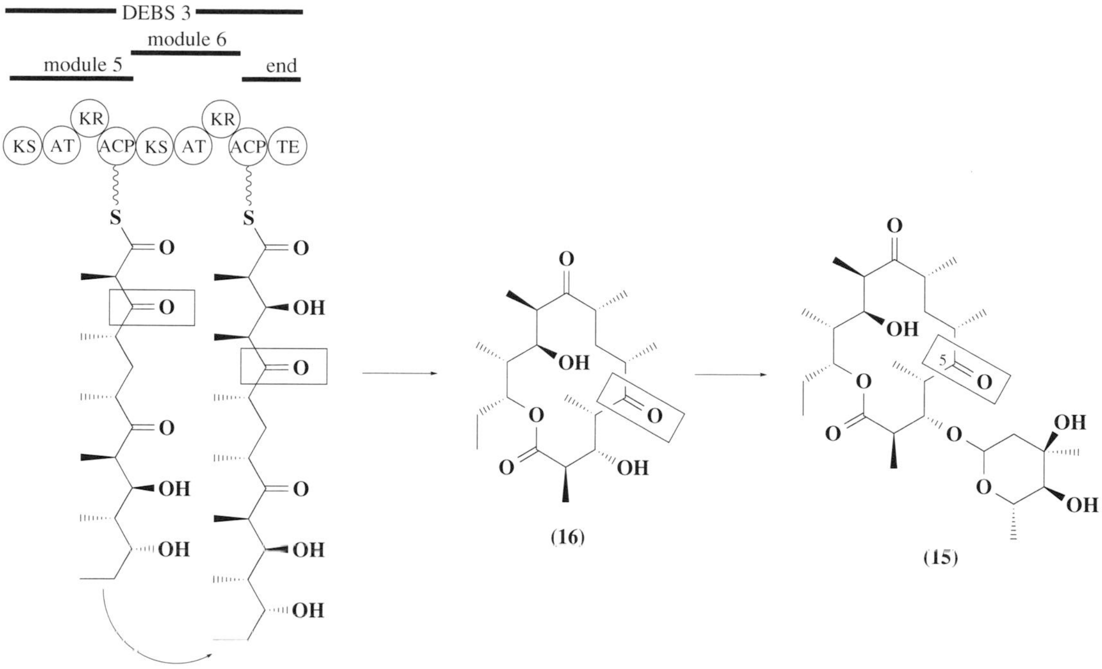

(16) **(15)**

Scheme 7

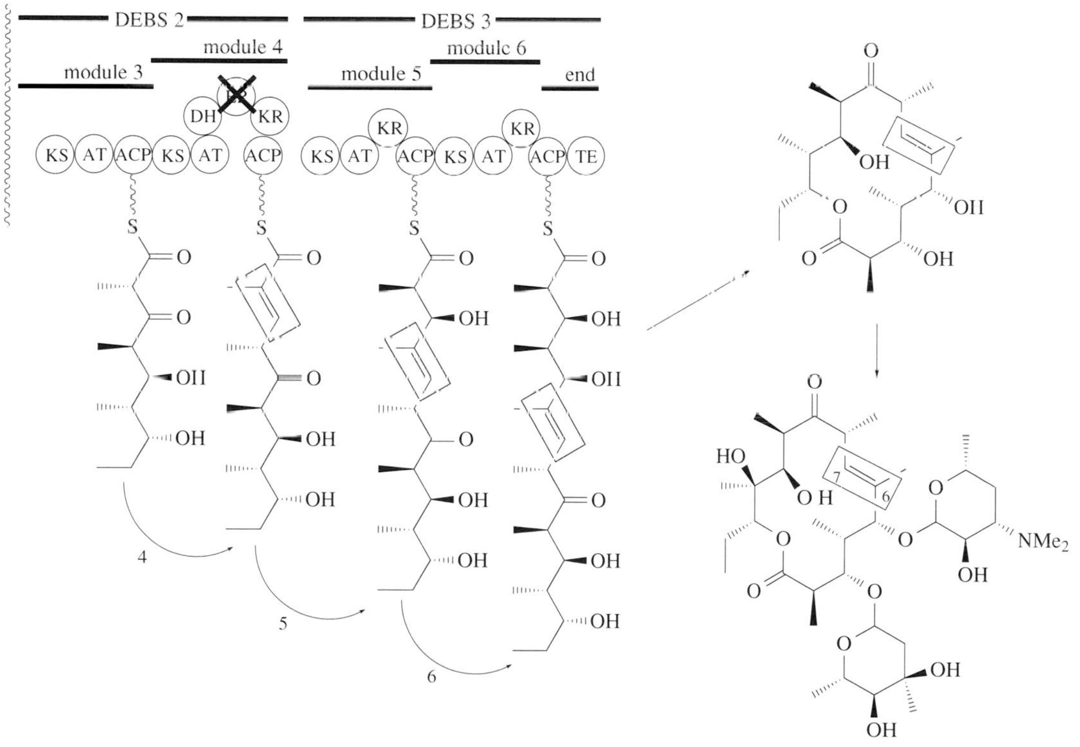

(17)

Scheme 8

since been confirmed to be exclusively homodimeric under the conditions of analytical ultracentrifugation.[40]

The first information concerning the structure of the DEBS homodimers came from limited proteolysis studies.[40–42] For each DEBS protein the cleavage pattern was highly specific and reproducible under given conditions. Sets of fragments were obtained containing one or more intact domains produced by cleavages in the linker regions predicted on the basis of sequence alignments. The results are summarized for DEBS 1 in Figure 3.

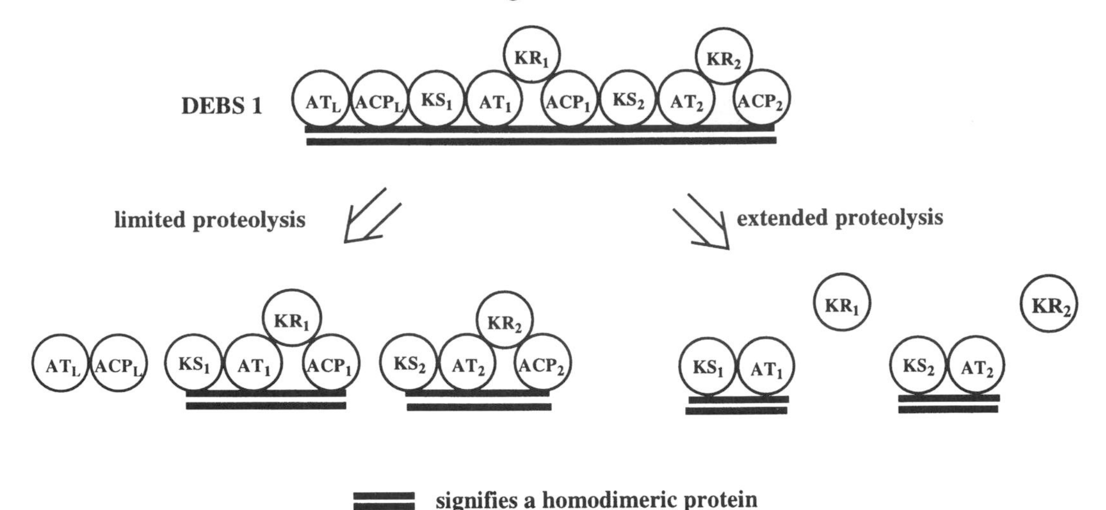

Figure 3 Pattern of fragments generated by controlled proteolysis of DEBS 1.

Significantly, initial cleavage occurred between the two chain extension modules housed in each multienzyme.[40–42] Other initial cleavages split the N-terminal loading didomain from DEBS 1 and a didomain consisting of the C-terminal offloading thioesterase from DEBS 3 together with its attached ACP domain.[39,40] Under prolonged conditions of reaction, more extensive proteolysis of DEBS 1, DEBS 2, and DEBS 3 occurred to give smaller fragments.[40–42] For example, the six KS domains remained attached to the adjacent AT domains giving rise to didomain fragments, whereas the four KR domains were cleared from their partners and were isolated as individual fragments.

In further studies it was found that certain fragments of the proteins existed as homodimers, whereas others were monomeric. Thus, any fragments which retained the KS–AT didomain were homodimeric. In contrast, all the isolated enzymes associated with keto group reduction existed as monomers. On this basis it was argued that each KS–AT pair is probably in contact with the equivalent residues in the associated chain in the original intact DEBS proteins. This can be achieved by lining up two protein chains side by side as in Figure 4. In one of the current models of the structure, the two chains are then twisted to form a double helix as shown (sense of twist is arbitrarily chosen).[40]

The helical twist is introduced to explain other experimental data. Of particular significance is the discovery that the KS domain of one chain can be cross-linked with the downstream ACP of the partner chain. The twist brings these two domains into the required proximity. This arrangement means that the developing polyketide molecule passes from one chain to the other within each chain extension cycle. This result has been independently demonstrated.[43]

The optional reductive domains in each module needed for ketone group modification (DH, ER, and KR) form loops that protrude out sideways from the central core, while maintaining the active sites of all reductive domains within the range of the phosphopantetheine group on the adjacent ACP in the *same* chain. There is therefore no contact between reductive domains of complementary chains, which is consistent with the monomeric nature of these domains in the isolated state (see Figure 3). The architecture of the homodimers is, of course, crucial to understanding how these fascinating synthases operate. The helical model will no doubt be subjected to further investigation by X-ray crystallography, but for the moment it provides a useful model with good predictive power and it is the only model consistent with all the available data.

1.19.2.9 Genetic Engineering of the Erythromycin PKS

The pioneering studies by the Katz group have already been discussed in support of the proposed modular structure of the PKS (Section 1.19.2.7). A common characteristic of these genetic engineering experiments was the essentially destructive character of the mutations. The activity of individual domains could be destroyed without destroying the overall capacity of the PKS to make macrolide structures.

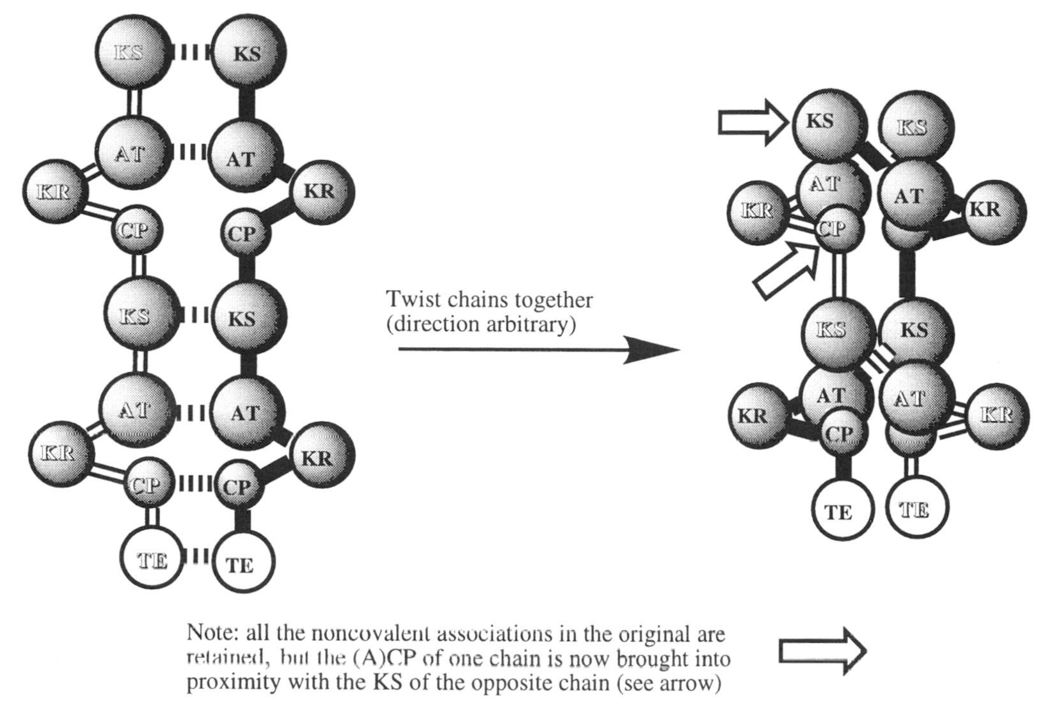

Figure 4 Proposed folding of DEBS 3 to give the "double helical" model.

The erythromycin PKS has been genetically engineered in many different ways. Much of the early work relied upon an expression plasmid pRM5 developed by Hopwood and Khosla for the expression of genes in *Streptomyces coelicolor* CH999;[44] for this mutant the genes which encode actinhorodin biosynthesis have been removed. In initial experiments it was demonstrated that the whole of the erythromycin PKS could be transferred to *S. coelicolor* CH999 to give a mutant with the capacity to produce the unelaborated macrolide product, 6-deoxyerythronolide B (**7**), as well as the analogue with an acetate starter unit.[45] This highly efficient vehicle for genetic engineering has been widely used for the production of modified polyketide products.

The first significant breakthrough in this area involved the repositioning of a domain within the PKS to show that the repositioned domain could carry out its normal type of reaction in its new context and on a foreign substrate. This entailed adding a copy of the thioesterase (TE) domain to the terminus of DEBS 1.[46] The aim was first to prevent further chain extension by blocking the docking of the C-terminus of DEBS 1 onto the N-terminal end of DEBS 2 (this was the predicted consequence of this change according to the "helical" model for the PKS structure). It was further reasoned that the TE domain would play an *active* role in the engineered protein by catalyzing the release of the triketide intermediate from module 2 as a δ-lactone.

A mutant of *S. erythraea* containing the engineered protein (called DEBS 1-TE) (Scheme 9) did produce the anticipated δ-lactone (**18**) at levels comparable to the wild-type organism. As expected, erythromycin production was shut down completely. A vital control experiment was then carried out to establish the second point—that the relocated domain was acting as an *active* agent of chain release rather than just a passive block to further chain extension. A further mutant was made in which an inactivated copy of the TE was placed at the end of DEBS 1. The inactivated TE differed from the normal enzyme only in the replacement of the key serine residue at the active site by an alanine; this conservative change should not alter the folding of the protein significantly. This control mutant did not produce erythromycin showing that the block to further transfer of the chain still existed. The δ-lactone (**18**) was produced but in much reduced yield, demonstrating that the TE in the first mutant was playing an active role in chain release.

Further examples of this strategy have since been reported. The first of these experiments mirrored the design of DEBS 1-TE in that it involved placing the thioesterase at the end of DEBS 1. The construct was engineered in a different way, however, to yield a protein with a different primary structure, so it is appropriate that it was given a slightly different name: DEBS 1+TE.[47] For this experiment no control experiment was performed, although using the *S. coelicolor* CH999 system

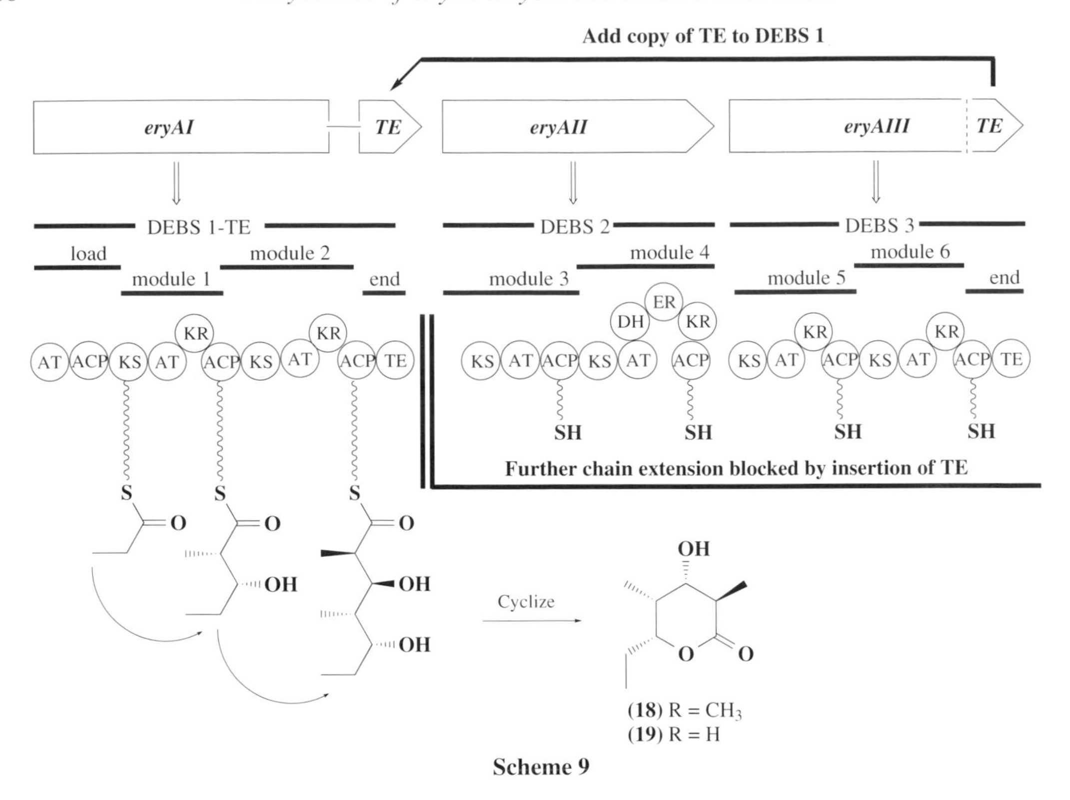

Scheme 9

employed, it was shown that expression of DEBS 1 alone produces (**18**), albeit at reduced levels compared to DEBS 1 + TE.[48] Interestingly, the two similar constructs DEBS 1-TE and DEBS 1 + TE gave significantly different results when expressed in *S. coelicolor* CH999; DEBS 1-TE but not DEBS 1 + TE produced the δ-lactone analogue (**19**) derived from an acetate starter unit as well as (**18**).[49]

The TE has since been relocated to the terminus of modules 5[47] and 3.[50] Products consistent with truncation of chain extension at the expected stages were obtained from both mutants (Scheme 10). The truncation at module 5 was especially significant because it led to release of the hexaketide intermediate as a 12-membered ring macrolide (**20**). Truncation after module 3 caused the formation of two tetraketide products, (**21**) and (**22**); the second is possibly formed from a ketoacid by decarboxylation after release from the enzyme.

A more versatile strategy for generating novel products is the transfer of domains between foreign PKS clusters. This would amount in effect to "spare part surgery" or "mix and match" swapping of structural residues between different natural products. In the first demonstration of the concept, sections of two different type I PKS systems (rapamycin and erythromycin) were hybridized to produce a hybrid type I PKS (Scheme 11).[51] The experimental model was DEBS 1-TE and the experiment involved replacing the AT of DEBS 1-module 1, by the AT derived from module 2 of the rapamycin PKS (which will be described later). The transplanted AT specifies a malonate unit as chain extender rather than methylmalonate. δ-Lactones (**23**) and (**24**) were produced lacking a methyl group at C-4, but otherwise the normal triketide lactone structure was maintained. The hybrid PKS therefore produced the predicted hybrid product.

As described above, the transplanted AT domain from the rapamycin PKS is responsible for the incorporation of a malonate, rather than a methylmalonate derived unit, thereby giving rise to a variation in alkyl branching along the polyketide chain. Detailed sequence analysis has revealed that the sequences for the AT domains in the rapamycin PKS can be categorized into two groups depending upon whether the predicted substrate is malonyl- or methylmalonyl-CoA.[52]

In further AT domain-swapping experiments, modified erythromycin analogues have been generated in *S. erythraea* mutants by utilizing the AT domain from module 14 of the rapamycin PKS "RapAT," and also sequence data for other PKS-like clusters in *S. hygroscopicus* "HygAT" and *S. venezuelae* "VenAT" (the pikromycin producer).[53] Based on the known function and sequence alignments, these three AT domains were believed to be specific for malonyl-CoA. The replacements

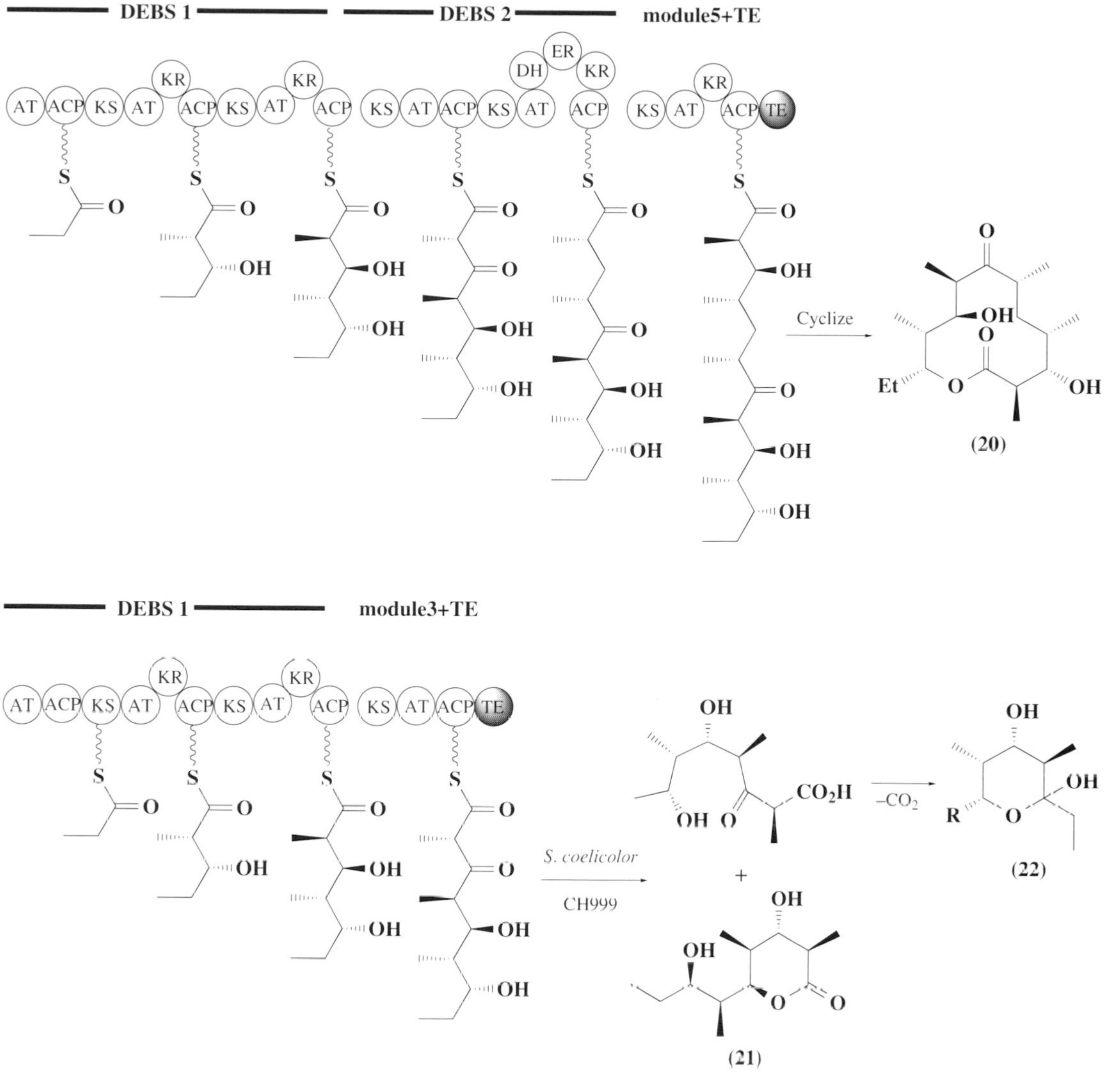

Scheme 10

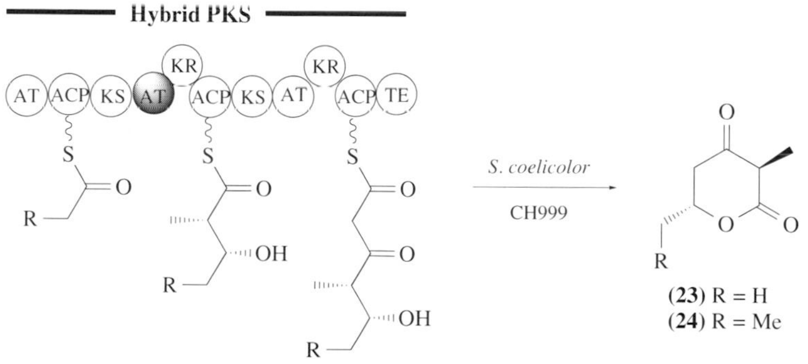

Scheme 11

were made in modules 1 and 2 of DEBS 1 and led respectively to 12-desmethylerythromycin B (**25**) and 10-desmethylerythromycin B (**26**) and A (**27**) analogues (Scheme 12). It is not surprising that hydroxylation at C-12 is completely removed for the first of these experiments, as the hydroxylase acting at C-12, eryK, is known to be rather substrate specific.

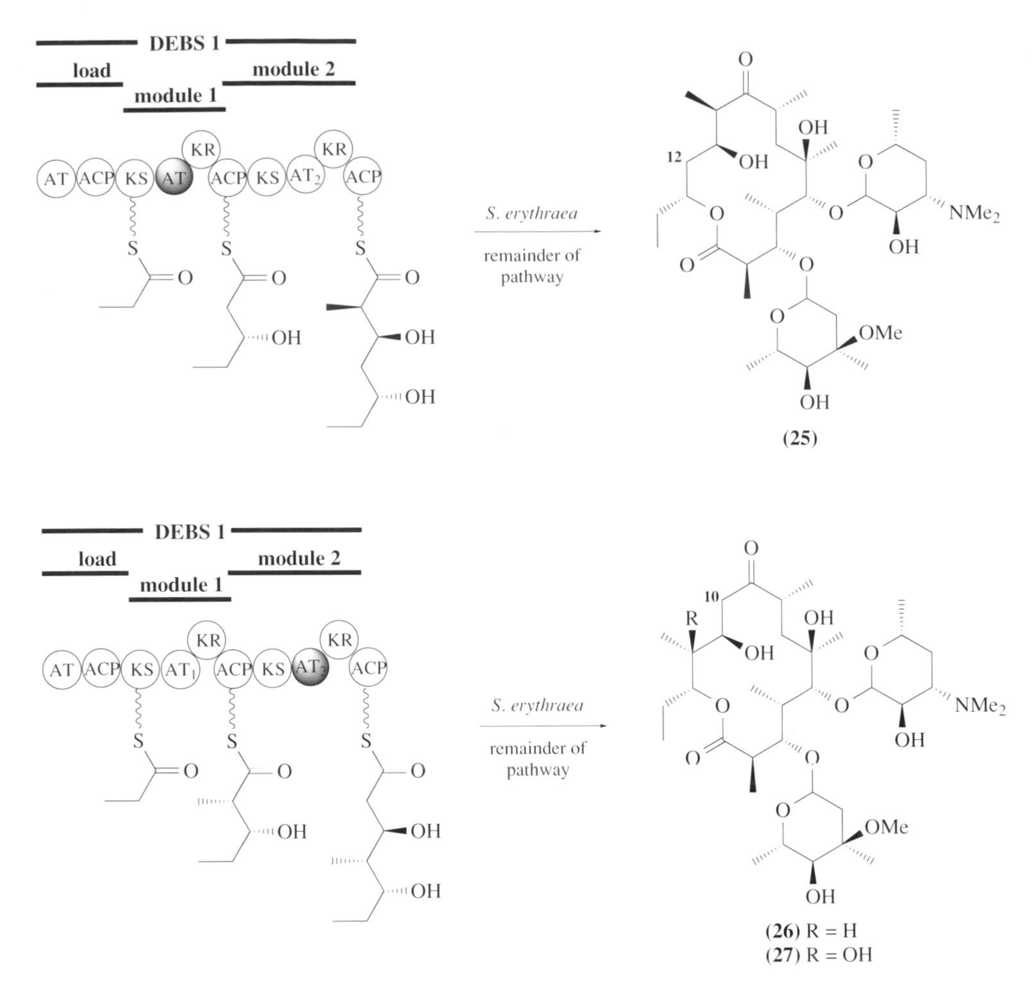

Scheme 12

The three heterologous AT domains employed gave rise to reduced levels of polyketides compared to the wild-type organism. For exchange in module 1 of DEBS 1, the HygAT was the most successful at approximately 50% of wild-type levels, with the RapAT and VenAT levels at approximately 25% of the HygAT level. The use of VenAT in module 2 was not successful, and the HygAT and RapAT levels of production for exchange in module 2 were only approximately 20% of that observed for exchange in module 1. The decreasing trend of efficiency became absolute when attempts to exchange these three heterologous domains in any of the further downstream modules (3–6) was attempted. The reduction of product levels for successful domain exchanges is not considered to be a consequence of precursor levels for either translation of the PKSs or production of the polyketides,[54] and it appears therefore that the variation in levels of polyketide production is a reflection of the modified structure of these heterologous PKSs.

In contrast to the experiments discussed above, the production of 2-desmethyl erythromycins has been reported for the complementation of a nonproducing *S. erythraea* mutant with a genomic library derived from *S. antibioticus*;[55] however, the biochemical nature of this result is unknown. More recently the AT domain from the rapamycin PKS module 2 has been used to replace the AT of module 6 in DEBS 3.[56] Expression of this construct in *S. coelicolor* CH999 produced 2-desmethyl-6-deoxyerythronolide B as predicted. It is therefore clear that some AT domains will function in downstream modules of the erythromycin PKS. The availability of a large pool or "toolbox" of PKS genes with which to perform rational engineering is of great importance.

A more ambitious approach to increasing the diversity within polyketide structures is to exchange a series of catalytic domains within a PKS. An example of such an experiment came when the propionate selective loading module (AT-ACP didomain) of the erythromycin PKS in DEBS 1-TE was replaced with that from the avermectin PKS.[56] Previous studies had shown that this loading module, which naturally recruits branched chain starter acids, can also accept more than 40 exogenous carboxylic acids under the correct conditions (this will be described in more detail in Section 1.19.4). The products of the fermentation of this mutant strain included two novel lactones (**28**) and (**29**) as well as the normal products (**18**) and (**19**) (Scheme 13). The novel products had starter acyl groups characteristic of the avermectins and therefore can reasonably be viewed as hybrid molecules incorporating elements of both the avermectin and erythromycin structures. In the subsequent experiment, the avermectin loading module was placed at the start of the entire erythromycin PKS in *S. erythraea*.[57] As anticipated, new erythromycins were identified from this mutant organism, which once again contained starter acids characteristic of the avermectin structures (Scheme 14). Through the use of LCMS-MS, LCMS-MS coupled with accurate mass determination, and NMR analysis, it was clear that A, B, and D analogues generated during the post-PKS steps had accumulated. In total nine erythromycins were observed in the fermentation extract, six of which were novel. The broad specificity of the avermectin loading module is well documented[58,59] (see Section 1.19.4) and it is anticipated that this approach will allow the facile production of many more novel erythromycin analogues through feeding exogenous carboxylic acids.

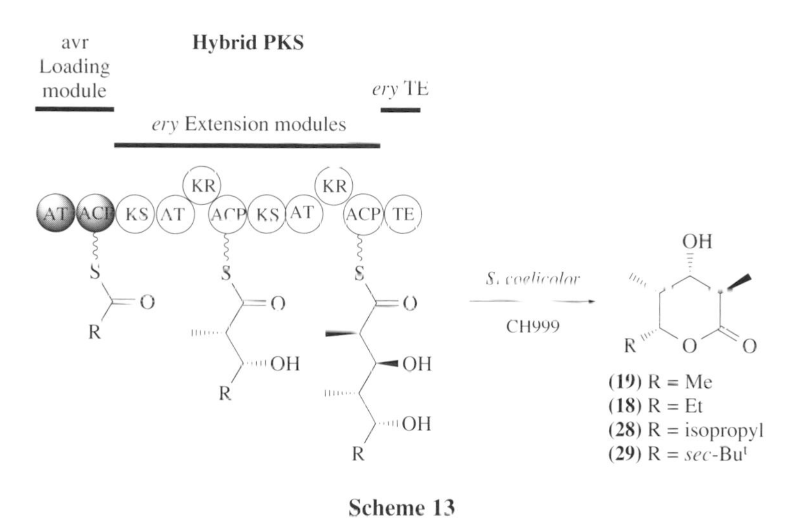

Scheme 13

It is further anticipated that this approach may be applied to other PKSs, or that other loading modules can be swapped to generate specific new products. Indeed, in an earlier example the acetate specific loading module of the platenolide PKS in *S. ambofaciens* was exchanged with that of the tylactone PKS from *S. fradiae* to generate a new macrolide derived from a propionate starter unit.[36] An interesting feature of the erythromycin PKS which is worth noting is the ability for an engineered PKS to produce erythromycin A even when the loading module didomain has been removed, albeit at reduced levels compared to the wild type.[60] It would therefore appear that the KS1 domain of DEBS 1 can be acylated directly by propionyl-CoA.

In order to produce modified oxidation states at the β-keto derived carbons of the erythromycin structure, the existing catalytic domains must be altered in some manner. As well as specific domain inactivations as demonstrated previously, possible methods which can be utilized include removal, addition, or exchange of the optional KR, DH, and ER domains. In the first experiment of this type the KR domain of module 2 of DEBS 1+TE was replaced with that from module 3 of DEBS 2 which is nonfunctional in its normal context.[61] Upon expression in *S. coelicolor* CH999 novel δ-lactone products were produced in which the keto group at C-3 generated by the second condensation step survived and with the normal erythromycin pattern at all other carbons (Scheme 15). While a successful outcome required the new PKS to be functional when a catalytic domain was relocated, there was also the need for later domains, in this case the TE domain, to accept a new structural feature.

The exchange of *functional* domains came when the KR2 domain of DEBS 1-module 2 was successfully swapped for the reductive domains DH4-KR4 derived from module 4 of the rapamycin PKS.[62] Expression of this new DEBS 1 hybrid and module 3+TE as a trimodular PKS in *S. coelicolor* CH999 resulted in two new products (**30**) and (**31**), derived from acetate and propionate starter units, in which an alkene moiety replaced the normal module 2 hydroxyl functionality (Scheme 16). The isolated compounds were observed to have undergone decarboxylation, consistent

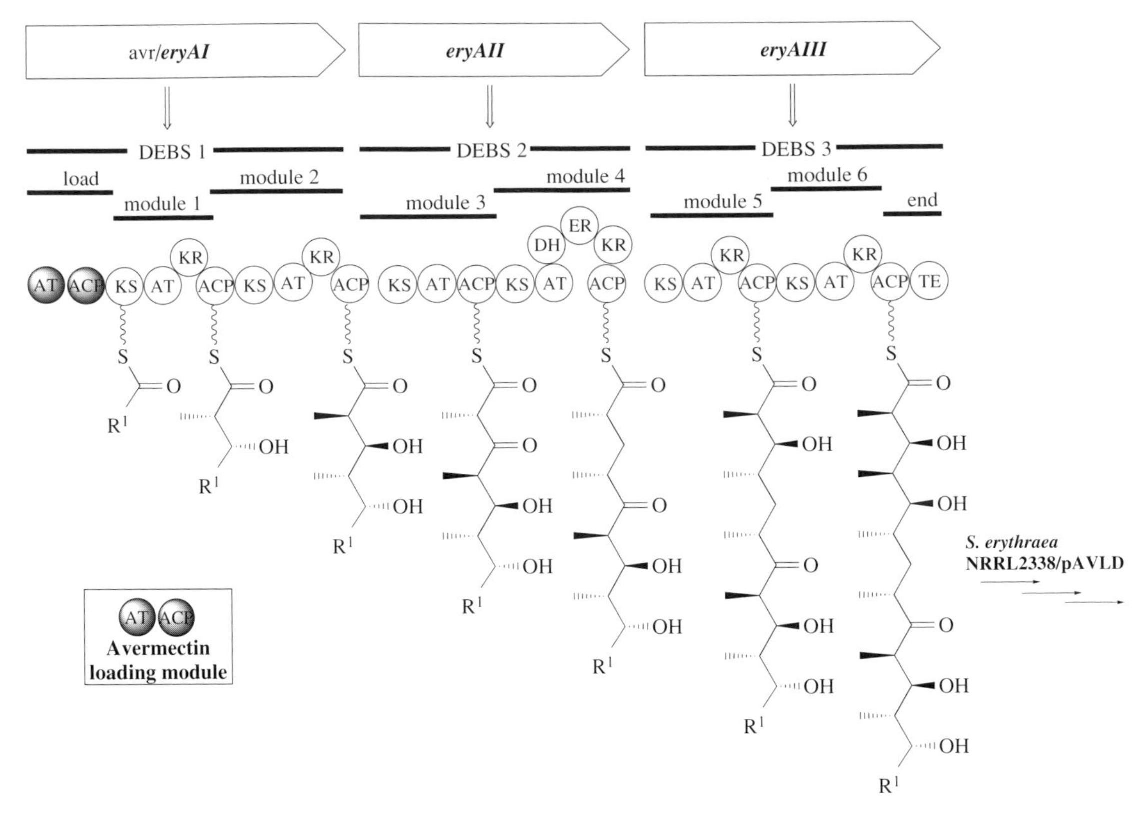

R^1 = *sec*-Bu, R^2 = OH, R^3 = Me; (A)
R^1 = *sec*-Bu, R^2 = H, R^3 = Me; (B)
R^1 = *sec*-Bu, R^2 = H, R^3 = H; (D)

R^1 = iso-Pr, R^2 = OH, R^3 = Me (A)
R^1 = iso-Pr, R^2 = H, R^3 = Me; (B)
R^1 = iso-Pr, R^2 = H, R^3 = H; (D)

R^1 = Et, R^2 = OH, R^3 = Me; (*ery A*)
R^1 = Et, R^2 = H, R^3 = Me; (*ery B*)
R^1= Et, R^2 = H, R^3 = H; (*ery D*)

Scheme 14

Engineered PKS

AT ACP KS AT KR ACP KS AT KR ACP TE

S. coelicolor
CH999

R = H, Me

Scheme 15

with the production of β-keto acid PKS products.[50] This work was extended further by demonstration that the KR2 domain of DEBS 1 could be successfully swapped for the DH1-ER1-KR1 tridomain of module 1 of the rapamycin PKS[63] (Scheme 17). In this example loss of the C-5 hydroxyl group (as a consequence of the tridomain substitution) promotes the formation of an unusual eight-membered lactone ring (**32**), a result which again demonstrates the remarkable substrate tolerance of the erythromycin TE domain; reduction of the carbonyl group at C-3 is presumed to be a consequence of an additional metabolic enzyme. Also described was a result which indicates that not all such tridomain swaps will be successful. When the experiment was performed using the DH4-ER4-KR4 tridomain of DEBS 2-module 4 (from the erythromycin PKS), only the linear alkene product as exemplified by Scheme 16 was obtained. Once again it is clear that having simply the correct juxtaposition of domains is not the only requirement for successful catalytic behavior.

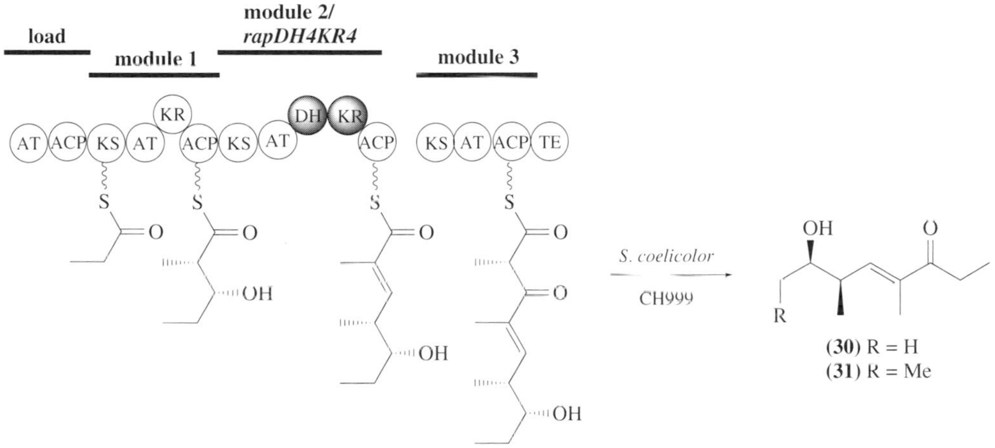

Scheme 16

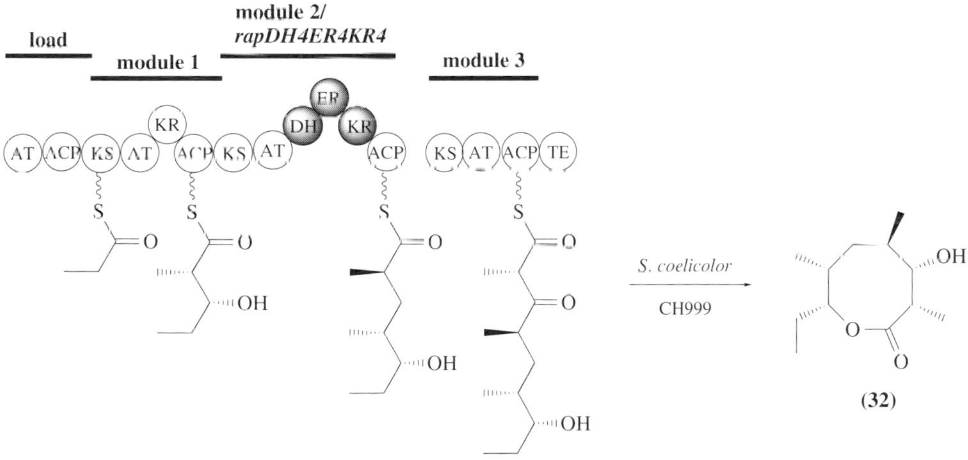

Scheme 17

The ability to fuse PKS proteins has been demonstrated for the DEBS 1 and 2 proteins. The model was the DEBS 1+module 3+TE system and the trimodular PKS was engineered in two forms.[64] First, a fusion was made between the stop codon of DEBS 1 and the start codon of DEBS 2; in this case all 117 amino acids between ACP2 and KS3 were retained. In the second experiment these 117 amino acids were exchanged for the 25 amino acid "linker region" which exists between modules 5 and 6 of DEBS 3. Expression in *S. coelicolor* CH999 led to the production of tetraketide products (**21**) and (**22**) from both constructs in yields comparable to that of the noncovalently linked trimodular system,[50] indicating no loss of catalytic potential. It would appear therefore that the amino and carboxy termini of these two proteins are not required for chain transfer or catalytic activity, but are most likely required for the correct intermolecular "docking" of DEBS 1 and DEBS 2 in the natural context.

Finally, the ability to successfully swap ketoreductase domains and therefore alter the stereochemistry of hydroxyl groups in the erythromycin backbone has been described.[65] Once again DEBS 1+module 3+TE was the model vehicle employed. Exchange of the KR domain in module 2 for that of module 5 of DEBS 3 (a reductase domain which in its native environment generates the same stereochemistry as the KR2 of DEBS 1) gave the expected products which were identical to those of DEBS 1+module 3+TE (see e.g., Scheme 10). However, when KR2 of this system was replaced with the KR domains from either module 2 or module 4 of the rapamycin PKS, a triketide lactone product (**33**) was obtained in which the stereochemistry at C-3 was opposite to that normally exhibited by the DEBS 1-TE system, but was in keeping with that observed for KR2 of rapamycin in its native environment (Scheme 18). It would therefore appear that the transplanted KR domains are capable of accepting structurally diverse β-ketothioesters and of reducing them in a stereospecific manner in keeping with that observed in the native environment. The stereochemistry of rapamycin KR4 is normally hidden by a subsequent dehydration but is revealed in this experiment. The reason for isolation of aborted triketide products rather than the expected tetraketides is not clear but may be a consequence of specificity for KS3 of DEBS 2 for a particular hydroxyl stereochemistry in the processing thioester.

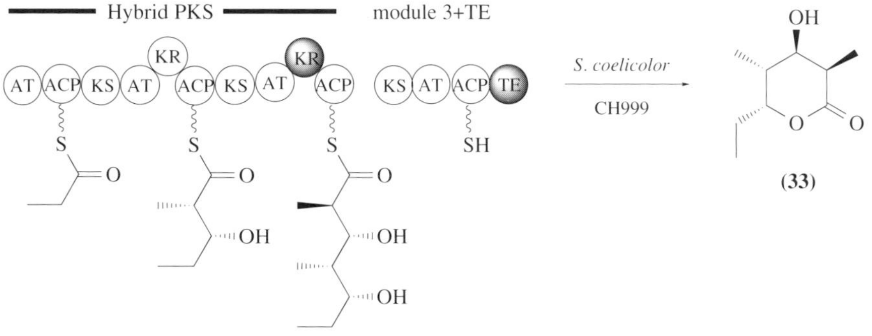

Scheme 18

Subsequently, both the KR domains of DEBS 1 were shown to catalyze the stereospecific transfer of the 4-*pro*-S hydride from the NADPH cofactor, although the thioester products of these domains contain opposite stereochemistries.[66] This finding is consistent with the current data for the stereochemistry of hydride transfer in fatty acid biosynthesis. The mechanisms which determine the stereochemistry of alkyl groups along the polyketide backbone will be discussed in the following section.

The experiments described within this section clearly demonstrate the ability to generate hybrid PKSs with the capacity for producing erythromycin analogues which have been altered in practically every conceivable fashion. These examples, however, remain bounded by the ability to perform modifications generated upon a specific macrolide backbone. What remains unclear still is whether individual modules or proteins from different systems can be assembled in a stepwise fashion to produce a totally novel polyketide backbone of our own design, or randomly to generate statistical libraries. This remains the distant goal of truly combinatorial polyketide biosynthesis.

1.19.2.10 *In Vitro* Experiments with DEBS Domains and Proteins

In some of the early structural studies of the isolated DEBS proteins, it was demonstrated that specific domains were catalytically active. For example, cleavage of DEBS 1 with trypsin gave the N-terminal didomain together with two other fragments, comprising modules 1 and 2, respectively. It was found that the N-terminal didomain was specifically radiolabeled with [^{14}C]propionyl-CoA, after incubation, providing the first evidence for its proposed role as the "loading module" for the propionate starter unit. In contrast, modules 1 and 2 were specifically acylated with [^{14}C]methylmalonyl-CoA, indicating that the other two acyltransferases were enzymatically active after proteolysis.[41,42] Using intact DEBS proteins, it was discovered that all the AT domains of the chain extension modules specifically catalyzed the hydrolysis of (S)-methylmalonyl-CoA but were inert to the (*R*)-isomer. This was the first evidence that the (*S*)-isomer of methylmalonyl-CoA serves as building block for both chain extension modules of DEBS 1 (see later).[42]

In 1995 the engineered, self-sufficient model PKS, DEBS 1-TE, was isolated in pure form.[67] When incubated with the appropriate building blocks, propionyl-CoA and methylmalonyl-CoA, and a reducing agent, specifically NADPH, the protein catalyzed the formation of the δ-lactone triketide product (**18**) (Scheme 19). This was the first demonstration of product formation *in vitro* by any modular PKS. Given the high state of purity, it was established that the PKS could carry out its normal function without the assistance of external proteins. The PKS showed a more relaxed degree of substrate specificity in its loading domain than was apparent *in vivo*. *n*-Butyryl- and isobutyryl-CoA esters were acceptable as starter acyl groups *in vitro* to give novel products (**34**) and (**35**) as well as the acetyl and propionyl starter acyl groups found to be effective *in vivo* (see Scheme 19).

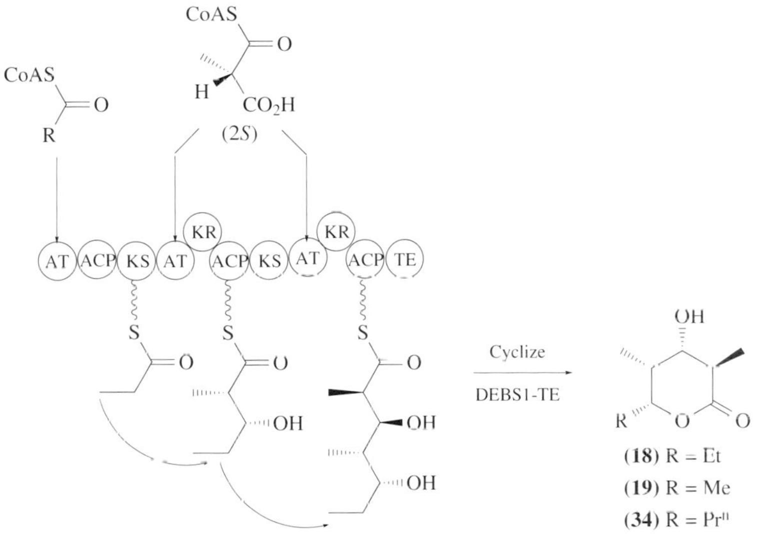

(**18**) R = Et
(**19**) R = Me
(**34**) R = Pr^n
(**35**) R = Pr^i

Scheme 19

Similar results have come from studies of the closely related DEBS 1+TE system *in vitro*.[68] When DEBS 1+TE was incubated with appropriate precursors in the absence of the reducing agent NADPH, the keto (**36**) and the pyrone (**37**) analogues of the triketide lactone were obtained (Scheme 20).[68,69] This latter result resembles the similar formation of a pyrone product by the animal FAS and the 6-methylsalicylic acid PKS under similar conditions.[70] Clearly, the β-ketoester intermediates can be transferred from one module to the next without further processing, even in modules which are geared to the production of a hydroxyester intermediate.

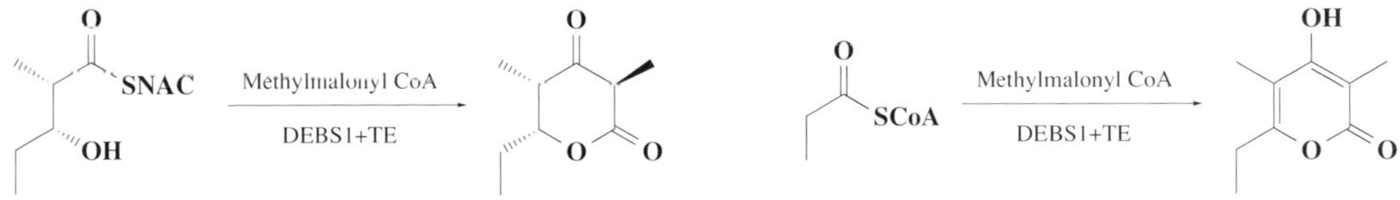

(**36**) (**37**)

Scheme 20

Other significant *in vitro* studies include the demonstration that a cell-free extract of all three DEBS proteins expressed heterologously in *S. coelicolor* was capable of synthesising 6-deoxyerythronolide B (**7**), albeit with rather low efficiency.[71] The engineered protein module 3+TE has been purified to homogeneity.[72] In collaboration with a pure extract of DEBS 1 this produced the tetraketide products (**21**) and (**22**) (see Scheme 10). It may be possible to use one of these systems to catch the two cassettes in contact as the developing chain is passed from one to the other by treatment with appropriate cross-linking reagents. Unfortunately, attempts so far have not met with success and the nature of cassette docking in multicassette PKS systems remains a mystery.

With the availability of the pure isolated proteins DEBS 1-TE, it became possible to carry out rigorous systematic investigations into the mechanisms of stereochemical control. First, it was shown that only the (2*S*)-isomer of the chain extension building block 2-methylmalonate would serve as a precursor[67] for both modules 1 and 2. This meant that the mechanism of chain extension in these modules had to be different to give the two different stereochemistries in the sites of methyl branching of the triketide product. The nature of the divergence was determined to be an epimerization step in module 1 which was detected by use of deuterium labeling.[73] In both modules, therefore, it is thought that the two condensation steps catalyzed by the KS domains take place with inversion of configuration to give the (2*R*)-isomer of the newly formed keto ester (Scheme 21). In module 1 this is epimerized to the (2*S*)-isomer but in module 2 the absolute configuration is preserved. The differing configurations at the two hydroxyl groups of the triketide are established by the differing stereospecificity of hydride addition catalyzed by the two KR domains. The first KR domain also plays the crucial role of selecting the (2*S*)-isomer of the methyl group as its substrate while rejecting the initially formed (2*R*)-isomer. The control of stereochemistry, therefore, relies on different forms of cooperation between the KS and KR domains in each. This complexity needs to be considered carefully in attempts to design altered synthases with different stereochemistry of chain extension processes.

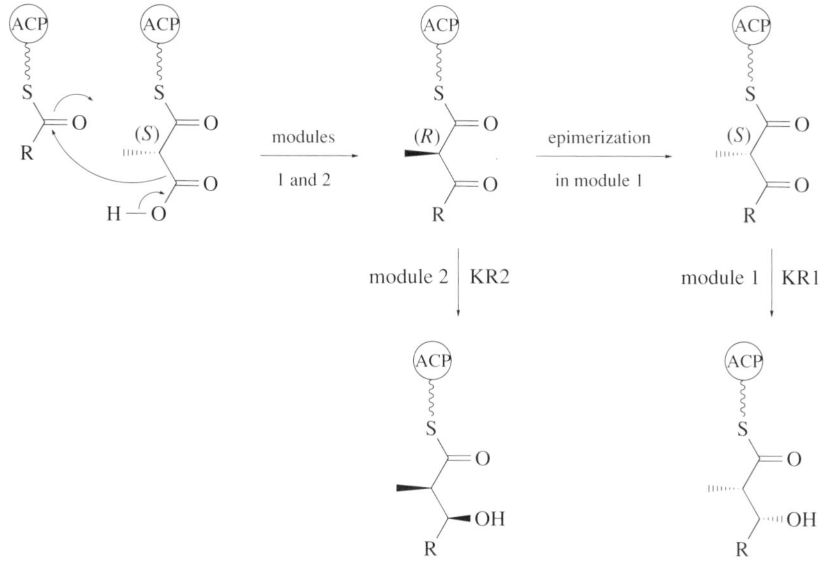

ACP S O R ACP S (S) O O H—O modules 1 and 2 ACP S (R) O O R epimerization in module 1 ACP S (S) O O R

module 2 KR2 module 1 KR1

ACP S O OH R ACP S O OH R

Scheme 21

The substrate specificity of the thioesterase domain is of crucial concern in the quest for a rational basis for genetic engineering of PKSs to produce novel products: it would be futile to engineer a PKS system to produce a particular novel product if there is no effective mechanism for its release from the final ACP.

To facilitate a study of the substrate specificity of the TE, the didomain from the end of DEBS 3 was overexpressed in *E. coli*.[74] The protein was shown by electrospray mass spectrometry to have the correct molecular weight for the didomain with an *apo*-ACP in which the phosphopantetheine group had not been added. The planned experiments concerned only the TE domain so this deficiency did not matter. First, it was shown that the protein was able to bind *p*-methylphenylsulfonyl fluoride (PMSF), a standard inhibitor of chymotrypsin-like enzymes, to the hydroxyl of a specific serine residue located in the putative active site of the enzyme. On this basis it was suggested that the release of the completed polyketide chain involved its initial transfer to the hydroxyl group of this active serine residue (Scheme 22). The resulting oxygen ester is then cleaved by attack of a suitable nucleophile to release the product. In forming the macrolide ring, the nucleophile would be the appropriate hydroxyl group at C-13 of the polyketide chain.

In experiments designed to assess its substrate specificity, the thioesterase was challenged with synthetic acylesters derivatized as *p*-nitrophenyl and *N*-acetylcysteamine (NAC) esters (exemplified in Scheme 23). A wide range of structurally varied substrates with both leaving groups was cleaved.[75,76] By electrospray mass spectrometry it was shown that acyl enzyme intermediates were formed as predicted. The resulting acyl enzyme intermediates could be cleaved by either water to

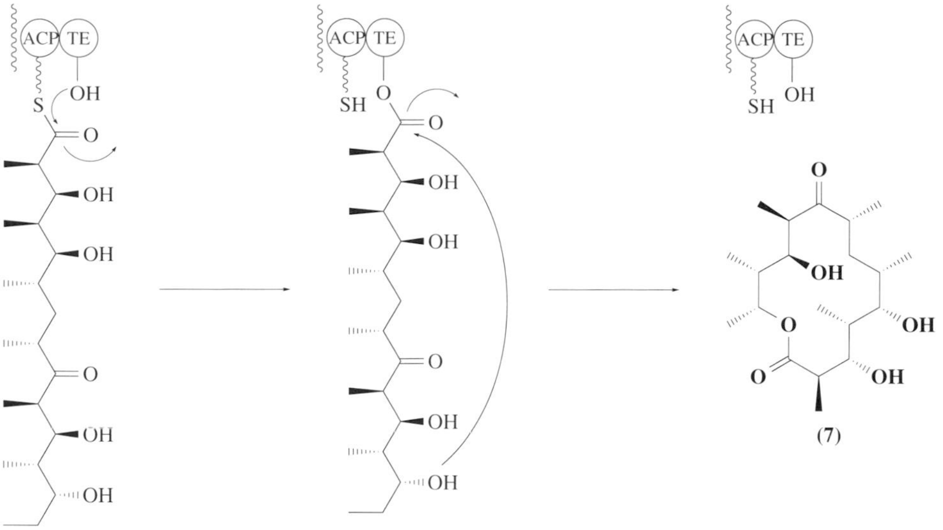

Scheme 22

give the corresponding acid or by an added alcohol such as ethanol to give an ester. None of the substrates tested gave any macrocyclic lactone even when there was a hydroxyl group elsewhere in the chain at a suitable distance from the acyl terminus. The work established that the TE has alternative mechanisms for product release (hydrolysis or ester formation) when lactonization is not possible. These studies demonstrated that the TE domain has a very relaxed substrate specificity which makes it suitable as a mechanism of release for widely varying novel polyketides produced by genetic engineering of upstream chain assembly modules.

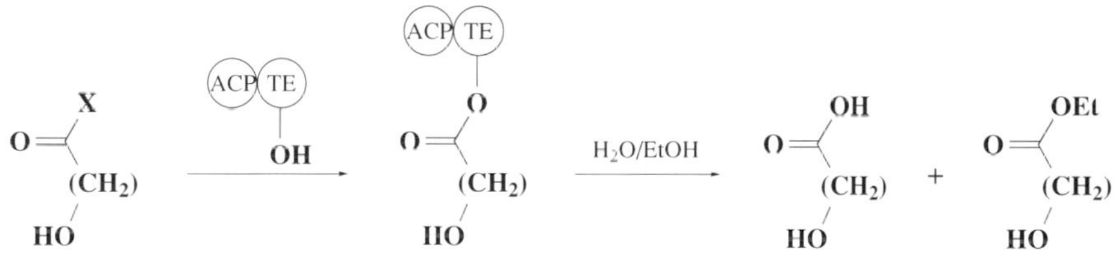

X = *p*-nitrophenyl or SNAC

1.0 : 1.5

Scheme 23

1.19.2.11 Mutasynthesis of Novel Analogues

An alternative method for producing novel erythromycin analogues involves modification of the biosynthetic machinery in order to generate a nonproducing mutant of *S. erythraea*. When supplied with the appropriate alternative substrates, erythromycin biosynthesis is re-established to yield products which display the structural elements of the exogenously supplied precursors. This approach is termed mutasynthesis.

The erythromycin PKS has been inactivated by a point mutation within the ketosynthase active site of DEBS 1.[77] This mutation of the active site cysteine to an alanine residue precludes the normal chain extension process from operating, but leaves the remainder of the PKS functional. By feeding synthetic di- and triketide NAC thioesters to this $KS1^-$ mutant expressed in *S. coelicolor* CH999, several new analogues of 6-dEB were biosynthesized (Scheme 24). Incorporation of the diketide (**38**) gave rise to a new compound (**39**) bearing an aromatic, benzyl moiety at C-13; incorporation of this compound would appear to demonstrate that structures radically different to the natural starter acid derived ethyl side chain can process through the later PKS modules. More interestingly, when a triketide intermediate of the tylosin pathway[24] is fed to this system a novel 16-membered macrolide (**40**) was produced, a result once again showing the broad substrate tolerance of chain extension and also the TE domain. Of note in this experiment is the ability of the KS2 domain to load a triketide rather than a diketide analogue of a chain elongation intermediate. After isolation of the novel 6-dEB analogues and feeding to a ΔDEBS mutant of *S. erythraea*, bioconversion as far as the erythromycin D analogues was observed (Scheme 24). Attempts to bioconvert the 16-membered aglycone (**40**) in an analogous manner were unsuccessful.

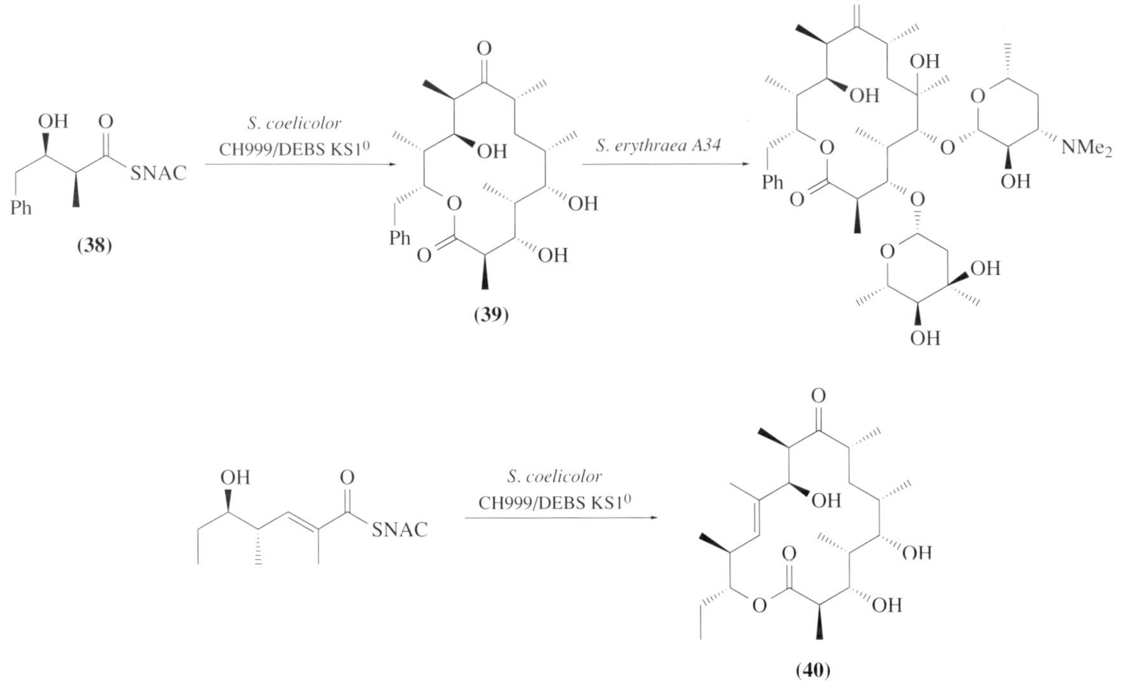

Scheme 24

It is interesting to compare the mutasynthesis approach described in this section with that of loading module exchange described in Section 1.19.2.7. In the latter approach a single organism is capable of converting a commercially available starting material (a simple carboxylic acid) through to a fully processed analogue of erythromycin. In the diketide approach there is a requirement not only for several steps of synthetic chemistry to generate the diketide precursors, but as set up further work is also involved in isolation and biotransformation steps to provide further processing. There are of course limitations to the substrate tolerance of both the avermectin loading module towards some unnatural starter acids,[58,59] and of the KS2 domain of the erythromycin PKS towards certain diketide analogues.[78] It may be that some erythromycin analogues are accessible by one approach only and others are exclusively produced by the alternative. Finally, in order to scale up production levels, it may well prove much easier by using the loading module strategy as was found in the work with the avermectin system which is described in Section 1.19.4.

1.19.2.12 Summary

The erythromycin PKS is typical of a large number of type I systems. It has provided the experimental basis for the development of an impressive range of genetic engineering strategies which should be widely applicable. The most pressing need now is for a more precise definition of the structural architecture of the type I PKSs. The current model based on a "double helical" dimer is probably conceptually correct but it cannot provide guidance on the distances between cooperating domains which are an essential condition for effective cooperation. We also have no idea how rigid the folded structure is.

1.19.3 RAPAMYCIN, FK506, AND FK520

The structurally related compounds rapamycin (**4**), FK506 (**41**), and FK520 (**42**) belong to the family of polyketide polyenes and demonstrate antitumor, antifungal, and immunosuppressant activities. This latter immunosuppressant activity has generated great interest due to applications in the therapeutic area of organ transplant surgery, and a detailed model for their mode of action has been developed.[79–81] Due to the similarity of their structures and the parallel biosynthetic studies, rapamycin will be the focus of this section and cross-reference will be made where appropriate.

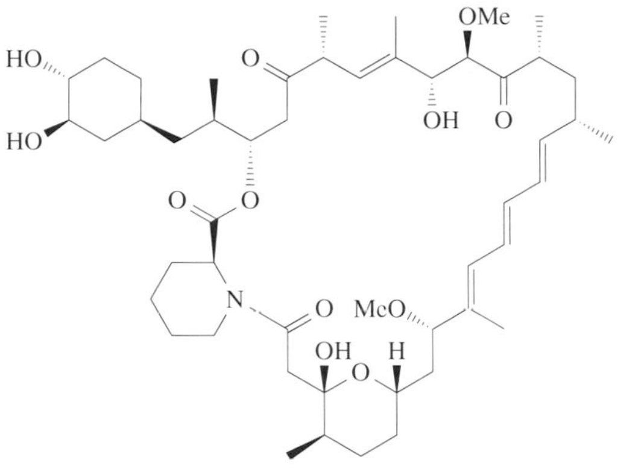

(4)

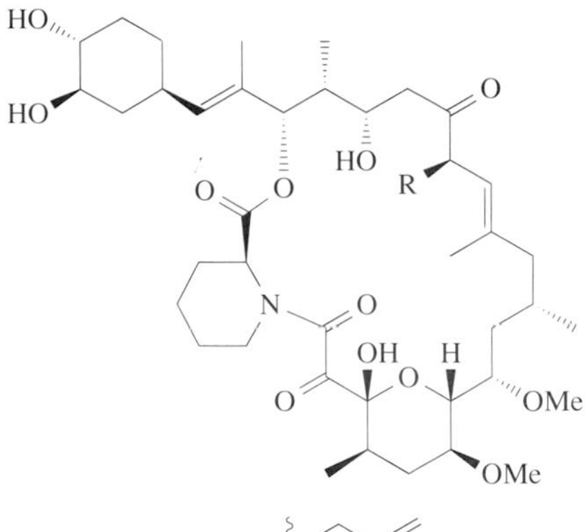

(41) R = allyl
(42) R = Et

1.19.3.1 Precursor Studies

Rapamycin obtained from *Streptomyces hygroscopicus* subsp. *hygroscopicus* is biosynthesized from seven acetate and seven propionate units, with *O*-methyl groups derived from L-methionine[82] (Scheme 25). Competitive incorporation studies with radiolabeled precursors demonstrated that the pipecolate ring is derived from lysine via free pipecolate.[83] The *trans*-dihydroxycyclohexane carboxylic acid (DHCHC) ring is derived from shikimate,[84] and studies have shown that free DHCHC can act as a precursor of the cyclohexyl unit.[85] Related experiments with *Streptomyces hygroscopicus* subsp. *yakushimaensis* have elaborated some of the steps involved in the formation of the cyclohexyl starter unit for FK520.[86] The early steps closely resemble the biosynthesis of CHC in *S. collinus*. Shikimic acid undergoes a 1,4-conjugate elimination of the C-3 hydroxyl and C-6 hydrogen followed by reduction, double bond isomerization, and finally reduction to the saturated DHCHC.

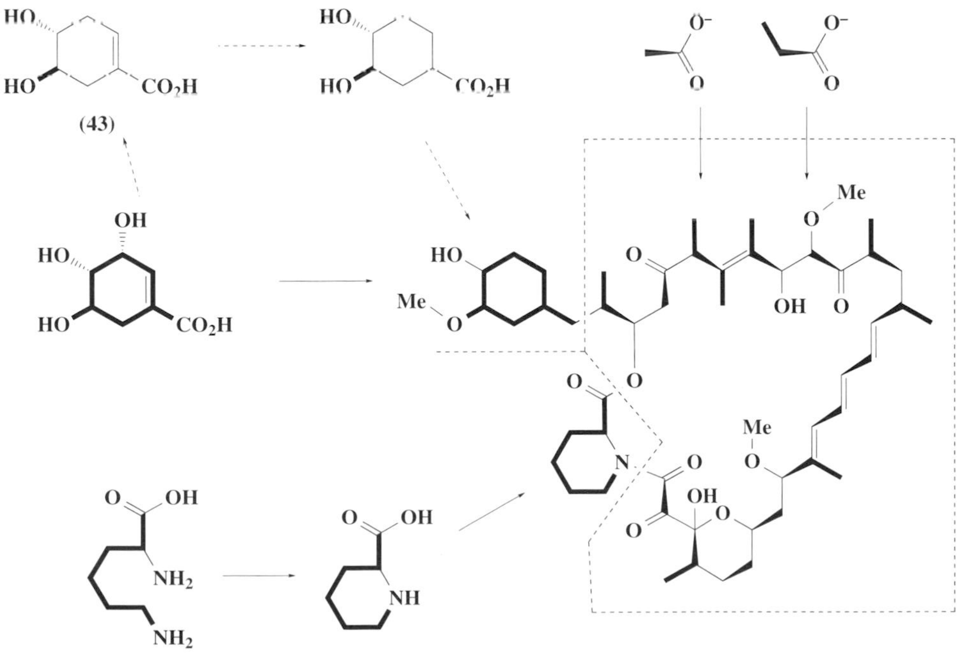

(43)

Scheme 25

1.19.3.2 Genetics of the Rapamycin Biosynthetic Cluster

The entire biosynthetic gene cluster for rapamycin biosynthesis has been sequenced and published.[35] The PKS genes were identified by hybridization with DNA from the PKS genes for erythromycin biosynthesis. Sequencing beyond the PKS region identified other genes that are predicted to be involved in the late stages and regulation of rapamycin biosynthesis.[87,88] The sequence and organization of a gene encoding a four-module PKS protein which is responsible in part for the biosynthesis of FK506 in *Streptomyces tsukabaensis* has also been published.[89]

As expected, the PKS for rapamycin showed a type I organization strongly reminiscent of the erythromycin PKS, with catalytic activities arranged in modules (Scheme 26) and with sets of modules housed in turn in three multimodular cassettes designated RAPS 1, RAPS 2, and RAPS 3. RAPS 1 contains modules 1–4, RAPS 2 modules 5–10, and RAPS 3 modules 11–14. The domain structure of the rapamycin PKS may not correspond in every detail to the pattern expected from the proposed structure for the PKS product, however. In modules 3 and 6, there appear to be potentially active KR and DH domains which are not required; module 3 also contains a potentially active but functionally redundant ER domain. It is possible that the active sites of these extra domains have been inactivated in a way that is not apparent from the primary sequence, and that the now redundant protein residues have still to be edited out by the random processes of evolution. There is also a chance that all these domains are indeed active and that the true rapamycin PKS product is more fully reduced than that shown. Extra post-PKS reoxidations would then be required to reintroduce the oxygen functionality at the relevant sites in the final structure.

The loading module comprises three domains. The first (CL) shows homology to ATP-dependent carboxylic acid-CoA ligases, the second is a putative enoyl reductase (ER), and the third an ACP. The probable sequence of operations starts with the enoic acid (**43**) derived from shikimic acid which is reduced by the ER domain (see Scheme 26). The first domain will activate the carboxylic acid to an active acyl derivative ready for transfer to the thiol residue of the ACP. The final saturated product will end up attached to the ACP as a thioester derivative ready for transfer to the KS domain of the first chain extension module. The timing of the reduction in this sequence of operations cannot be predicted.

Chain termination in rapamycin biosynthesis is not performed by a thioesterase as in the erythromycin PKS, but is effected by a specialized multidomain protein coded by the gene *rapP*.[87] This gene has strong similarity to genes involved in nonribosomal peptide biosynthesis[90] and the corresponding protein is believed to catalyze the formation of the ester and amide bonds to pipecolic acid (Scheme 27). The final product of the PKS is coupled to the nitrogen of the pipecolate unit, and the carboxyl group of this unit is esterified with the first free hydroxyl at the remote end of the polyketide chain to yield the first enzyme free intermediate. The order of these steps is not known. It is this key multidomain protein, designated the pipecolate incorporating enzyme (PIE), that provides the pivotal link between polyketide and polypeptide biosynthesis. Isolation and characterization studies have demonstrated that this enzyme is monomeric.[91]

An enzyme has been isolated from the FK520 producer which is believed to be the key one responsible for inserting pipecolic acid into the macrocycle.[92] It is reported to be dimeric and activates pipecolic acid and several structural analogues in an ATP-dependent reaction to give an enzyme-bound amino-acyl adenylate. There is evidence that this then reacts to form a thioester linkage to the enzyme. This mechanism of activation is the same as that found in the nonribosomal biosynthesis of peptide natural products such as gramicidin.[90]

1.19.3.3 Mutasynthesis of Rapamycin Analogues

Knowledge of the mechanism by which rapamycin is biosynthesized has precipitated a mutasynthesis approach for the production of analogues in which the pipecolate moiety is replaced with prolyl derivatives. The *rapL* gene of the rapamycin biosynthetic cluster has been assigned as a putative lysine cyclodeaminase responsible for the provision of pipecolic acid during rapamycin biosynthesis.[87] A frameshift mutation of this gene made by utilizing phage-mediated gene deletion gave rise to a mutant which did not produce significant levels of rapamycin unless supplemented with exogenous pipicolic acid.[93] By addition of exogenous analogues of proline in place of the pipecolic acid, analogues of rapamycin were generated which derived from incorporation of these alternative amino acids into the polyketide backbone (Scheme 28). New molecules were isolated which had been processed through fully to rapamycin analogues, as well as 26-demethoxy compounds.

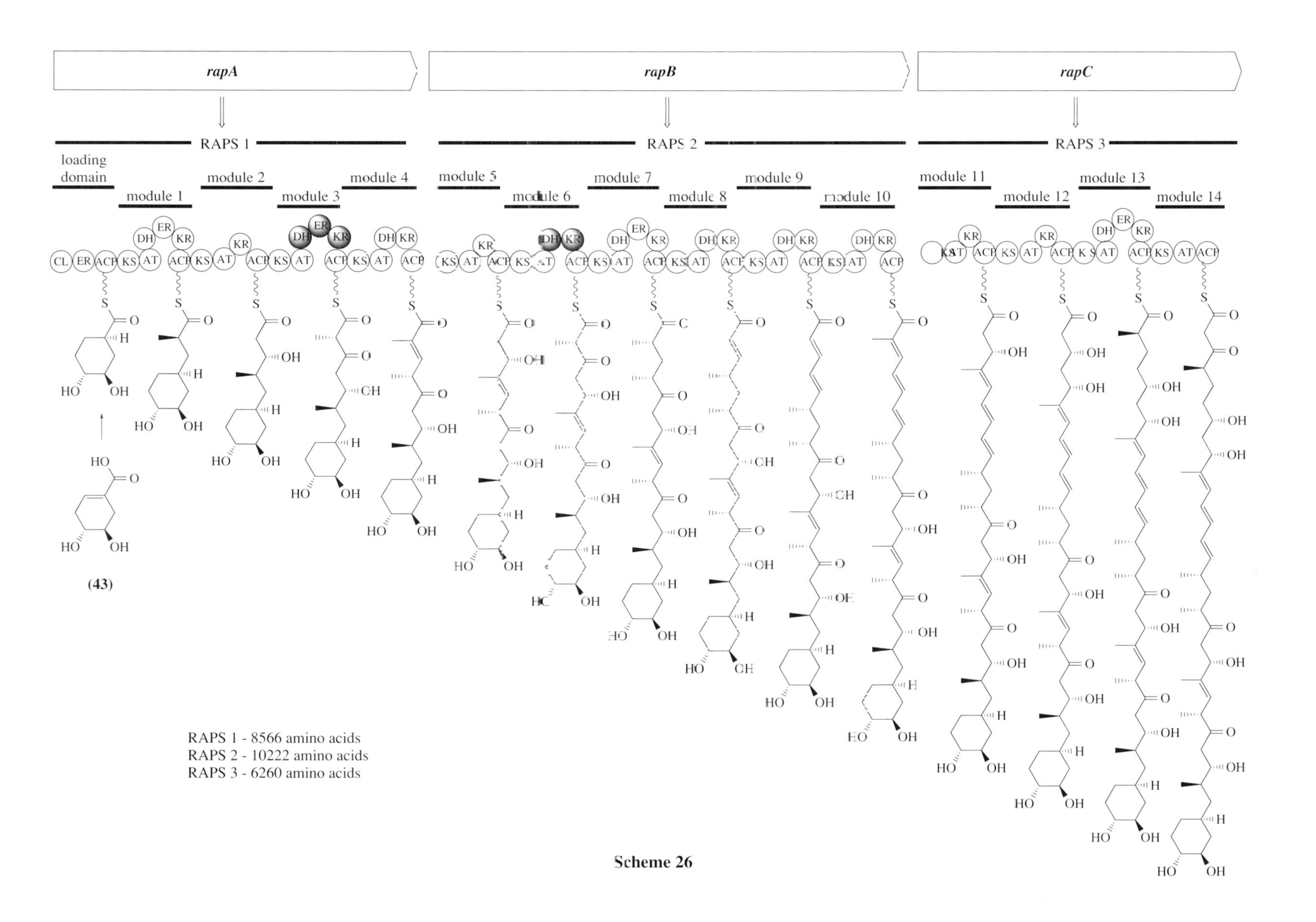
rapA
rapB
rapC
RAPS 1
RAPS 2
RAPS 3
loading domain
module 1
module 2
module 3
module 4
module 5
module 6
module 7
module 8
module 9
module 10
module 11
module 12
module 13
module 14
(43)
RAPS 1 - 8566 amino acids
RAPS 2 - 10222 amino acids
RAPS 3 - 6260 amino acids

Scheme 26

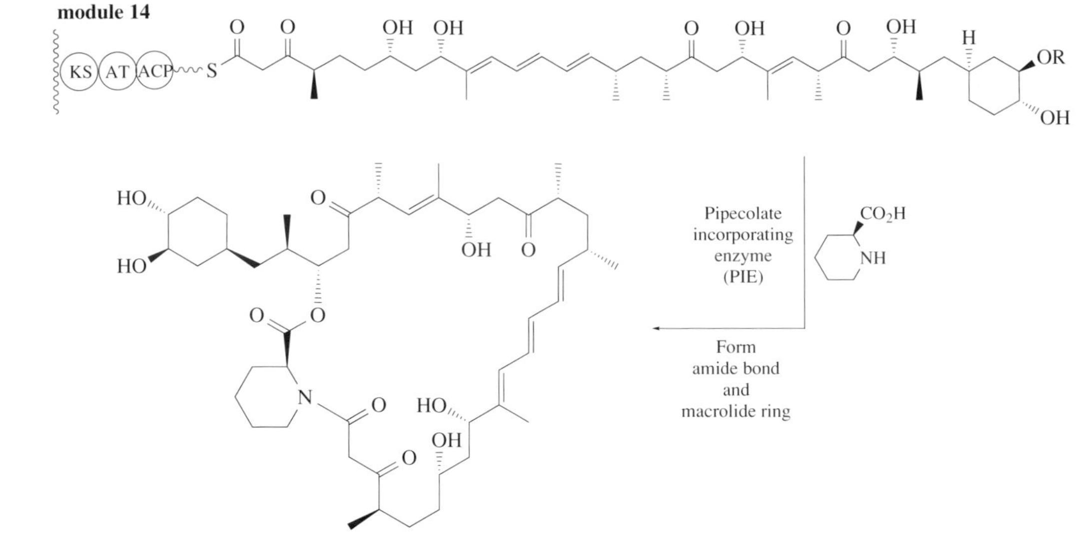

Scheme 27

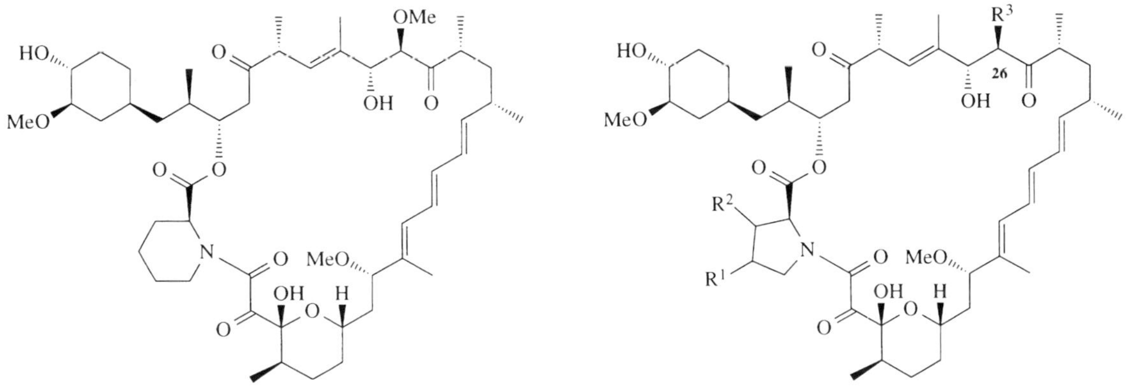

Rapamycin analogues

R^1 = H, R^2 = H, R^3 = OMe

R^1 = OH, R^2 = H, R^3 = OMe (*cis* and *trans*)

R^1 = OH, R^2 = H, R^3 = H (*cis* and *trans*)

R^1 = H, R^2 = OH, R^3 = OMe (*cis*)

R^1 = H, R^2 = OH, R^3 = H (*cis*)

Scheme 28

Immunoassay of the new rapamycin analogues demonstrated significantly reduced activity compared to the parent compound, although in light of the mechanism by which rapamycin exerts its immunosuppressant function[79] one can speculate that these engineered polyketides will most probably be important examples of complex structures capable of mediating highly specific protein–protein interactions, thereby finding utility in fields other than that of immunotherapy.

1.19.4 THE AVERMECTINS

The naturally occurring avermectins are a family of eight pentacyclic structures containing a 16-membered macrolide ring which are produced by *Streptomyces avermitilis* (Figure 5).[94] These eight compounds differ in structure due to the variability of the PKS starter unit at C-25, the *O*-methylation pattern at C-5, and the hydration–dehydration pattern at C-22/C-23. Other closely related natural products include the milbemycins and the nemadectins. The avermectins exhibit potent broad-spectrum antiparasitic activity through their action as agonists of the γ-amino butyric acid-insensitive chloride channel.[95] The semisynthetic Ivermectin® (hydrogenated avermectin B1) displays even greater efficacy.[94,96,97]

Avermectin	R^1	R^2	X–Y
A1a	Me	Et	CH=CH
A1b	Me	Me	CH=CH
A2a	Me	Et	CH_2–CHOH
A2b	Me	Me	CH_2–CHOH
B1a	H	Et	CH=CH
B1b	H	Me	CH=CH
B2a	H	Et	CH_2–CHOH
B2b	H	Me	CH_2–CHOH
(29)	H	Et	CH_2–CH_2

Figure 5 Structures of the avermectins.

1.19.4.1 Precursor Studies

Feeding studies have shown the avermectin aglycone to be derived from seven acetate and five propionate units.[98] The isobutyrate and 2-methylbutyrate starter units derive from catabolism of L-valine and L-isoleucine, respectively.[99] [^{18}O]Acetate and propionate feedings have shown that all the oxygen atoms other than those attached to C-6 and C-25 are derived from their carboxylate precursors.[98] The C-25 oxygen is most probably derived from the starter unit and that of the furan from molecular oxygen. The three methoxy groups derive from L-methionine and the oleandrose units from glucose.[99,100] An impressive and detailed review of all aspects of avermectin biosynthesis is available.[101]

1.19.4.2 Genetics of the Avermectin Biosynthetic Pathway

At the genetic level, a 95 kb gene cluster has been cloned and demonstrated to be responsible for avermectin biosynthesis (Figure 6).[102] Sequencing of the cluster revealed the structural organization for a type I PKS. A central 70 kb fragment contains two convergent transcripts, *aveAI* and *AII*, which appear to encode multifunctional proteins with six modules present in each transcript. Further genes responsible for post-PKS modifications and regulation both flank and occur between the two PKS transcripts.

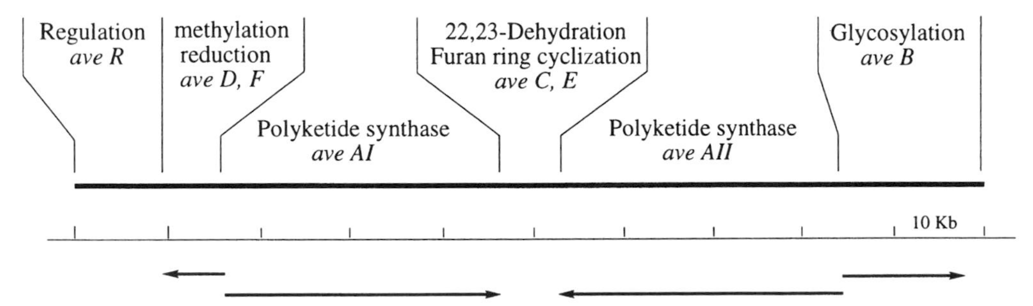

Figure 6 Organization of the genes for avermectin biosynthesis.

The structure (but not the sequence) of the two *aveA* transcripts has become available.[101] Each consists of two co-directional ORFs. *AveAI* encodes for two and four modular PKSs, the first of which, PKS-1, begins with a loading didomain which is followed by two condensation modules; the adjacent PKS-2 then appears to carry the next four modules required for avermectin biosynthesis. The transcript *aveAII* encodes for two, three-module PKSs. Following the direction of transcription we see PKS-3 and PKS-4. A thioesterase domain is present at the end of PKS-3, indicating that the order in which these proteins cooperate. PKS-1 contains modules 1–2, PKS-2 contains modules 3–6, PKS-4 contains modules 7–9, and PKS-3 contains modules 10–12.

Analysis of the domain structure is in keeping with that expected for the presumed avermectin PKS aglycone product, even where inactive domains are observed. One of the unique features of avermectin biosynthesis lies in the unsaturation at C-22/C-2323. As both saturated and unsaturated versions at this position are obtained from the same fermentation broth, one may consider the possibility that the dehydration step to yield the 1-series of compounds occurs post-PKS. Feeding studies of 2-series avermectins to *S. avermitilis* strains blocked in the PKS, however, fails to produce any 1-series compounds, indicating that dehydration occurs at the PKS stage.[103] Analysis of PKS-1 has indicated a functional dehydratase domain in the appropriate module responsible for the C-22/C-23 positions,[101] and it is assumed therefore that this domain is only partially active, allowing a mixture of 1- and 2-series functionality to pass onto the adjacent PKS-2. The complete activity of this domain would appear to be somewhat in question due to a small number of mismatched amino acid residues when compared to consensus dehydratase sequences for the active site region.

A final interesting point concerning the genetics of this system is the presence of a monofunctional type-II ketoreductase responsible for reduction of the C-5 keto group. The gene *aveD* responsible for this KR protein has been deactivated and appears to be transcriptionally linked to *aveF* which interestingly, is responsible for *O*-methylation of the reduced C-5 product.[104]

1.19.4.3 Mutasynthesis of Avermectin Analogues

The range of naturally occurring avermectins has been extended by utilizing a mutasynthesis approach.[58,59] The *S. avermitilis* blocked mutant (bkd$^-$) employed lacked a functional branched chain 2-oxoacid dehydrogenase activity. This activity is essential for conversion of the 2-oxo products of valine and isoleucine catabolism to isobutyrate and (2*S*)-2-methylbutyrate, respectively, the starter units of avermectin biosynthesis.[105] Feeding of the natural starter acids restored the production of natural avermectins, whereas more than 40 "unnatural" starter acids led to many novel avermectin analogues modified at C-25.[58,59] Doramectin (**44**), an extremely effective anti-parasitic treatment now marketed worldwide as Dectomax®,[106] was isolated after feeding cyclohexane carboxylic acid (**45**) (Scheme 29). This broad substrate tolerance of the loading module is transferable to the hybrid avermectin–erythromycin PKS described in Section 1.19.2.11.[107]

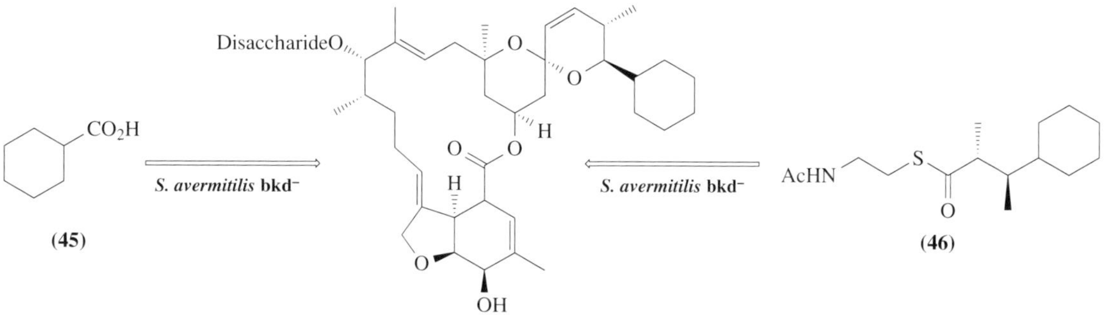

Scheme 29

In a related experiment the NAC thioester analogue (**46**) of the diketide biosynthetic intermediate was also incorporated into (**44**) using this bkd$^-$ mutant strain (see Scheme 29).[26] This was the first example of a mutasynthesis approach towards novel macrolide analogues by incorporation of diketide NAC thioesters. The diketide feeding approach was, however, much less efficient than the feeding of simple carboxylic acids, even though the mutational method employed in this study left the avermectin PKS apparently unaltered and thereby (presumably) fully functional.

1.19.5 TYLOSIN, SPIRAMYCIN, AND NIDDAMYCIN

Tylosin (**2**) from *Streptomyces fradiae*, spiramycin (**47**) from *S. ambofaciens*, and niddamycin (**48**) from *S. caelestis* are members of a large family of 16-membered macrolide antibiotics commonly utilized in veterinary medicine.[1]

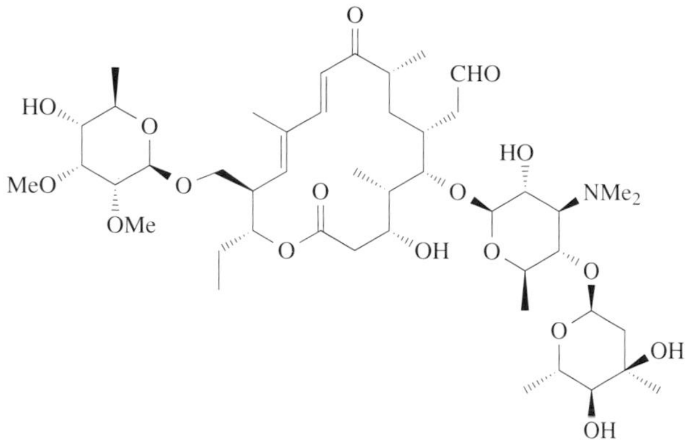

(**2**) Tylosin

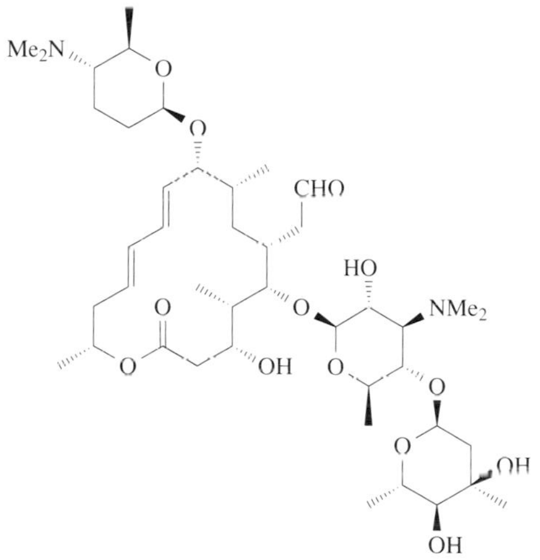

(**47**) Spiramycin

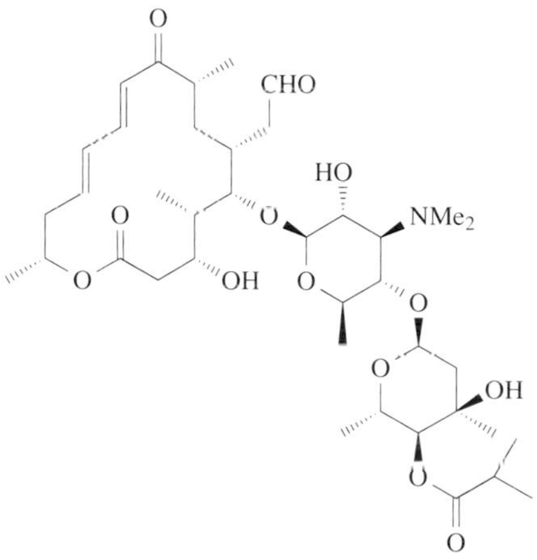

(**48**) Niddamycin

1.19.5.1 Precursor Studies

The aglycone core of these molecules is built up from a series of acetate and propionate extender units as one would expect, and also involves the addition of butyrate derived extender units which give rise to ethyl branches at C-6 of each macrolide ring.[108] Interestingly, the incorporation of hydroxylated 2-carbon extender units, possibly in the form of coenzyme-A esters of hydroxy-malonate, has been proposed to occur in module-6 of the spiramycin[109] and niddamycin[108] PKSs. The various post-PKS tailoring steps for these systems are closely related, and have been described in detail for tylosin biosynthesis.[110,111]

The processive mechanism for tylosin biosynthesis was first demonstrated by the pioneering contributions of Hutchinson and co-workers who demonstrated the incorporation of NAC thioesters of di- and triketide intermediates into the tylosin aglycone tylactone (**49**) (Scheme 30).[24] Further supporting evidence came from the isolation of partially elaborated intermediates of chain elongation in mutants of *S. fradiae*.[112]

1.19.5.2 Genetics of the Biosynthetic Pathways

The gene clusters responsible for the biosynthesis of each of these metabolites have been cloned and sequenced;[36,108,109,113] the respective PKS sequences have been deposited electronically and in

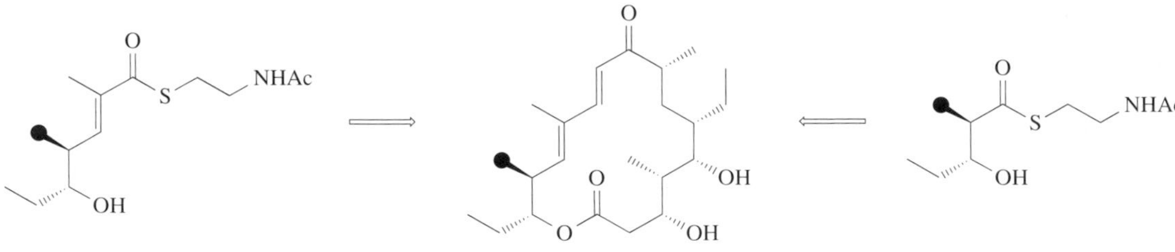

(49)

Scheme 30

patent form. The organization of these systems is strongly conserved, but differs markedly from that of the erythromycin or avermectin PKSs (Figure 7). The genes encoding the PKS proteins are transcribed colinearly in each system, with five proteins encoding, respectively, 2-, 1-, 2-, 1-, and 1-modules in each case. In each system the final protein contains a terminal TE domain. The first protein in each case begins with a loading module, although this again differs from the AT-ACP didomain of the erythromycin system in that an inactive KS domain precedes the loading module AT. These KS domains have been termed KS^Q by virtue of an amino acid substitution (C to Q) present in the active site motif. The sequence of domains is identical in all cases and matches closely that expected from the aglycone structures of the three polyketides. Interestingly in each case an apparently inactive KR domain is present in module 4; the structure of the third protein in all these systems is therefore strongly conserved and is reminiscent of that exhibited by *eryAII*.

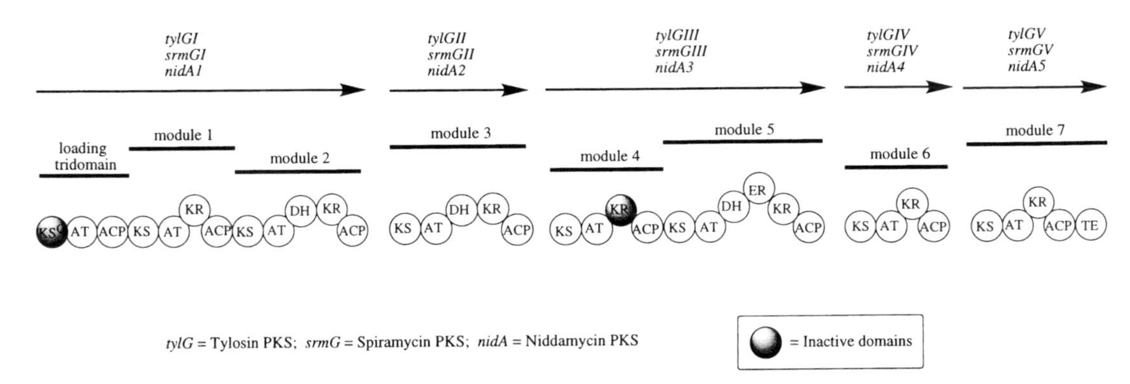

Figure 7 Organization of the genes and proteins for the PKSs of tylosin, spiramycin, and niddamycin.

A hybrid PKS has been constructed which allowed specific incorporation of the tylactone starter acid (propionate) in place of the platenolide (**50**) starter acid (acetate) by generating a hybrid PKS.[36] In this hybrid the tylactone loading tridomain (KS^Q-AT-ACP) was used to replace the equivalent spiramycin loading tridomain (see Figure 7). The hybrid PKS protein *tylGI*/*srmGI* was expressed in a mutant of *S. ambofaciens* which is blocked in sugar additions and from which the chromosomal *srmGI* gene had been exised. Interestingly the new product 10-methylplatenolide I (**51**) was produced at levels 3–4 times greater than levels of platenolide I production in an equivalent *S. ambofaciens* mutant blocked in sugar addition.

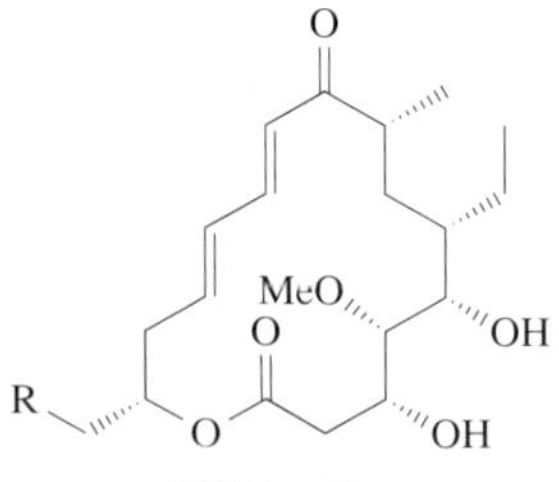

(**50**) R = H
(**51**) R = Me

1.19.6 RIFAMYCIN

Rifamycin B (**52**) from *Amycolatopsis mediterranei* is a member of the ansamycin family of aliphatic polyketides which also includes ansatrienin A (**53**) among others. A common structural feature of this family of metabolites is the "ansa" bridge system of mixed (acetate/propionate) polyketide origin which is initiated from a "mC_7N" PKS starter unit. The presence of a (biosynthetically unique) mC_7N unit is recognized by a six-membered carbocyclic ring carrying the extra carbon and nitrogen atoms in a *meta* arrangement. This structure is generally quinonoid or benzenoid in nature, and found either with minimal further modification, as in (**53**), or as part of a naphthalenic structure, as in (**52**).

(**52**)

(**53**)

(**54**)

1.19.6.1 Precursor Studies

Evidence for the processive mechanism of rifamycin B biosynthesis was first observed through the accumulation of a "tetraketide" chain elongation intermediate P8/1-OG (**54**) from a mutant of *A. mediterranei*.[114] This evidence and the demonstration from feeding experiments that the carbons on either side of the ether link in the ansa bridge derive from the same propionate unit[115] indicated that this ether moiety is formed after the initial biosynthesis of a fully extended polyketide chain.

The specific and proximate precursor of the mC_7N unit in ansamycin polyketides is 3-amino-5-hydroxybenzoic acid (AHBA) (**55**).[116,117] Through the use of elegant biochemical experiments, the biosynthesis of AHBA has been described by Floss and co-workers from the initial branch point of the shikimic acid pathway prior to 3-deoxy-D-*arabino*-heptulosonic acid 7-phosphate (DAHP).[118] The pathway is shown in Scheme 31 and was delineated by feedings of the proposed AHBA precursors, in labeled forms, to cell-free extracts of both the rifamycin B producer *A. mediterranei* S699 and also to the ansatrienin A producer *S. collinus* Tü1892. In these experiments each of the compounds (**56**)–(**60**) was converted into AHBA with generally increasing efficiency. Most importantly, the shikimate pathway compound DAHP cannot replace phosphoenolpyruvate (**56**) and erythrose 4-phosphate (**57**), or aminoDAHP (**58**) as the precursor of AHBA (**55**). More recently the gene for AHBA synthase has been cloned and the protein overexpressed.[119] The enzyme displays an absolute requirement for the co-factor pyridoxal phosphate and a mechanism has been proposed for the conversion of aminoDHS (**60**) to AHBA (**55**).

1.19.6.2 Genetics of the Rifamycin Biosynthetic Pathway

In two independent studies the gene cluster for rifamycin biosynthesis has been identified.[120,121] In both of the these the genes responsible for the rifamycin PKS, *rifA-E*, were described. These five

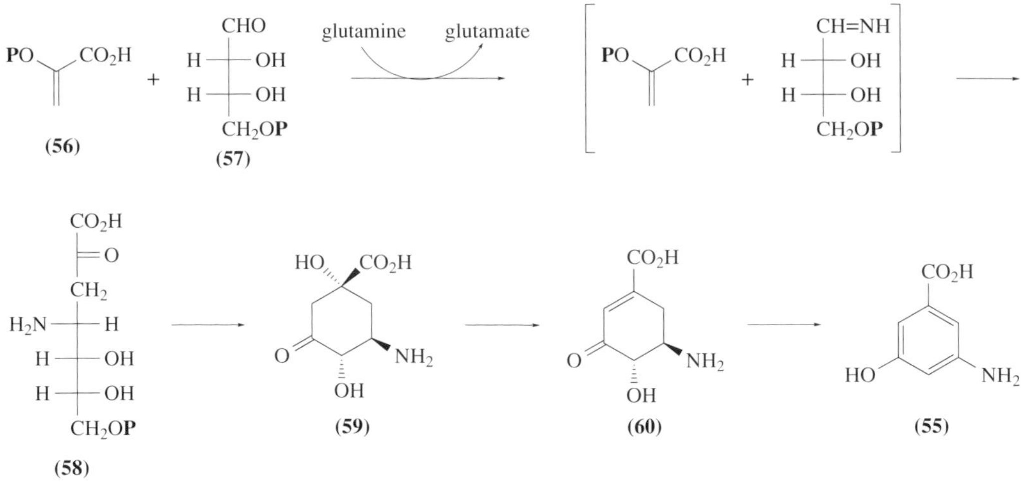

Scheme 31

ORFS are transcribed unidirectionally and encode for 10 type I PKS modules. The order of these modules also follows the order in which the polyketide chain is assembled; *rifA* encodes for modules 1–3, *rifB* encodes for modules 4–6, *rifC* encodes for module 7, *rifD* encodes for module 8, and *rifE* encodes for modules 9 and 10. The loading module for AHBA at the beginning of *rifA* comprises an acyl-CoA ligase and an ACP domain which has a similar organization to that observed for the rapamycin PKS.[35] At the end of *rifE* no thioesterase is present, but in keeping with the structure of rifamycin an aryl-amide synthase encoding gene *rifF* is present following the last PKS protein; this is again similar to the rapamycin cluster. It is assumed that the *rifF* product catalyzes closure of the macrolactam ring by condensation of the aryl-amine of the AHBA starter unit and the final PKS linked undecaketide thioester.

The former report[120] also fully describes the rifamycin cluster including genes responsible for biosynthesis of the AHBA starter unit, modification of the polyketide derived framework, and rifamycin resistance and export. A subcluster of some seven or eight genes (*RifG*-1 and *K-O*) immediately downstream of the *rifA-F* genes appear to encode activities responsible for the biosynthesis of AHBA. AHBA synthase is encoded by *rifK*.[119] Strikingly absent from this cluster is a gene with homology to either type I or type II dehydroquinate (DHQ) dehydratases. An activity corresponding to an aminoDHQ dehydratase should be present based on previous biochemical studies, and a gene *rifJ* located 30 kb downstream of *rifK* appears to encode this protein. The absolute requirement of *rifJ* for AHBA production was verified by insertional inactivation.

1.19.6.3 Mutasynthesis of Rifamycin Analogues

Mutasynthesis type experiments have been carried out with a rifamycin nonproducing strain of *A. mediterranei* which is blocked in AHBA synthesis.[122] Feeding of AHBA (**55**) demonstrated the resumption of rifamycin production. The use of AHBA analogues 3-hydroxybenzoic acid (**61**) and 3,5-dihydroxybenzoic acid (**62**) led to the production of new compounds (**63**) and (**64**) analogous to the tetraketide shunt metabolite P8/1-OG (**54**) described earlier (Scheme 32). The interpretation of these results was that early enzyme(s) in the polyketide pathway have relatively low specificity towards alternative primer units and initial PKS bound extension intermediates, and that some later enzyme(s) is dependent on the presence of an aromatic amino group for correct function.

1.19.7 NARGENICIN A

Nargenicin A (**65**) isolated from *Nocardia argentinesis* is a nonaketide containing an intriguing cyclohexeneyl moiety adjacent to a *cis*-fused octalin system, and is known to be active against *Staphylococcus aureus*.[123,124] It has been proposed that this system may arise via an intramolecular Diels–Alder reaction.

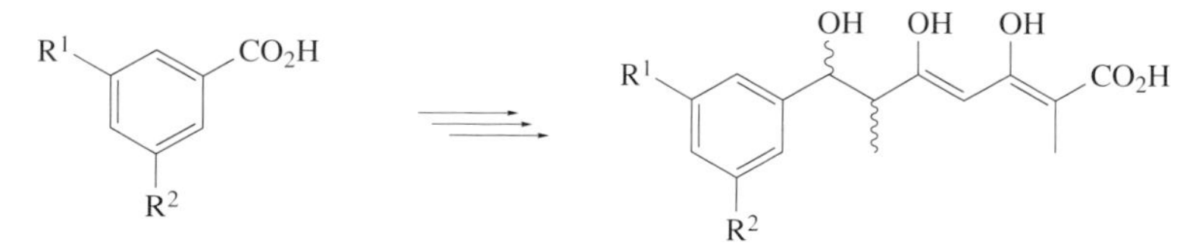

(55) $R^1 = NH_2$, $R^2 = OH$
(61) $R^1 = OH$, $R^2 = H$
(62) $R^1 = R^2 = OH$

(54) $R^1 = NH_2$, $R^2 = OH$
(63) $R^1 = OH$, $R^2 = H$
(64) $R^1 = R^2 = OH$

Scheme 32

The polyketide skeleton of nargenicin A is derived from five acetate and four propionate units.[125,126] Further incorporation experiments with [^{18}O,^{13}C]-labeled acetate and propionate demonstrated that the oxygen atoms attached to C-1, C-9, C-11, and C-17 derive from the carboxylate precursors.[125] Similarly, the use of [^{18}O]O_2 established that the oxygen atoms attached to C-2 and C-18 as well as the bridging ether atom were derived from molecular oxygen.[127] These results rule out the possibility that the *cis*-fused octalin system is formed via any plausible epoxy–alkene cyclization mechanisms.

Further experiments with [^{13}C]-labeled NAC thioesters of di-, tri-, tetra-, and pentaketide intermediates demonstrated regiospecific incorporations into nargenicin (**65**) (Scheme 33) which are consistent with an intramolecular Diels–Alder cyclization of a linear nonaketide intermediate (Scheme 34).[128,129]

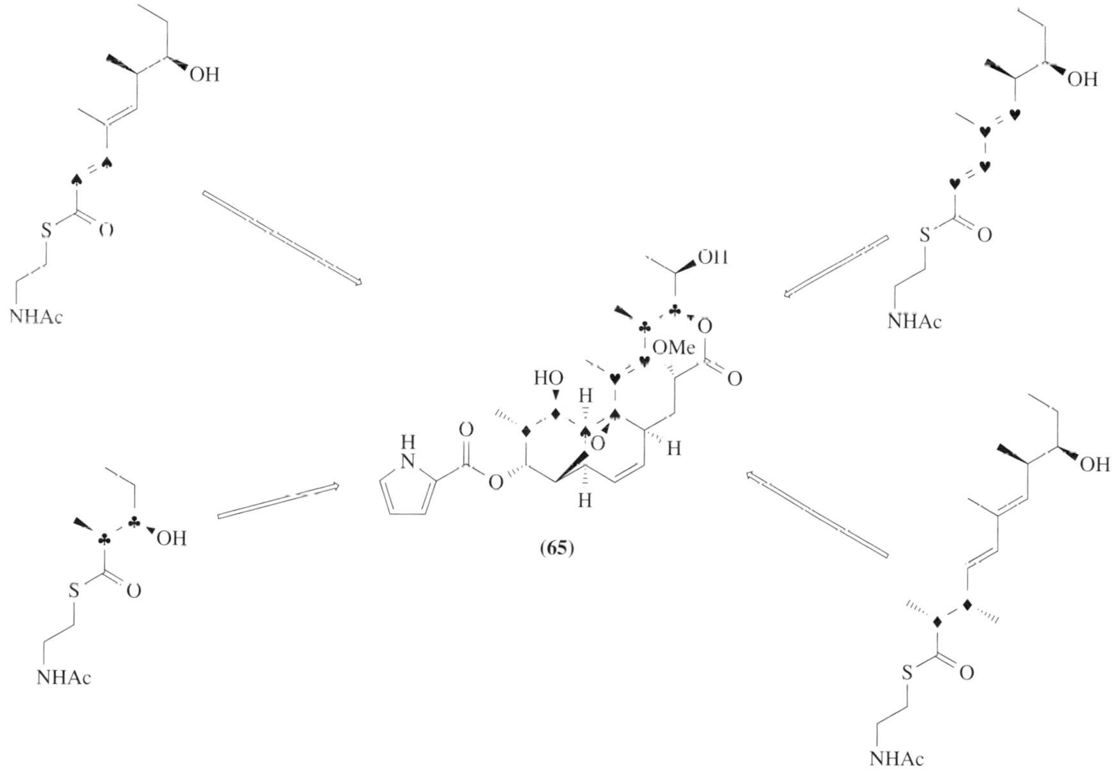

(**65**)

Scheme 33

1.19.8 CONCLUSIONS

A striking feature of this review of aliphatic polyketide biosynthesis is the large amount of information which is known concerning the genetics and enzymology of the various pathways. All this knowledge has been gained since 1990.

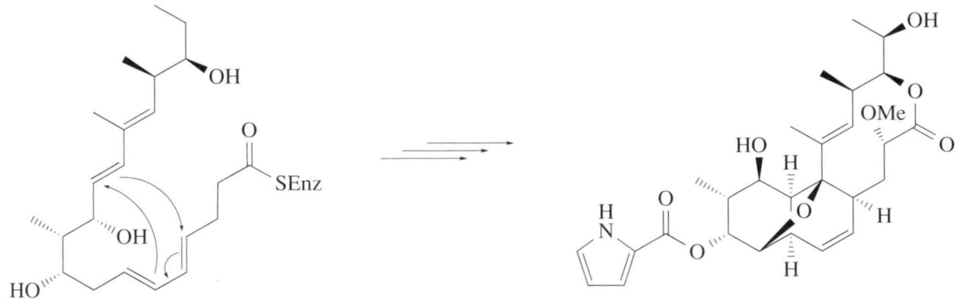

(65)

Scheme 34

The classical approach used over the previous 30 years in which isotopically labeled precursors are fed to intact organisms continues to play an important role. The knowledge gained from such experiments is essential to understanding the significance of the gene sequences.

This field has a very exciting future as advances in genetic engineering lead to the production of novel PKSs which produce novel metabolites. The results of such studies will have a profound impact not just on research into biosynthetic pathways, but also on synthetic chemistry and pharmaceutical chemistry.

1.19.9 REFERENCES

1. S. Omura, in "*Macrolide Antibiotics: Chemistry, Biology and Practice*," ed. S. Omura, Academic Press, New York, 1984.
2. W. D. Celmer, *J. Am. Chem. Soc.*, 1965, **87**, 1801.
3. J. M. McGuire, R. L. Bunch, R. C. Anderson, H. E. Boaz, E. H. Flynn, M. Powell, and J. W. Smith, *Antibiot. Chemother.*, 1952, **2**, 281.
4. P. F. Wiley, K. Gerzon, E. H. Flynn, M. V. Sigal, Jr., O. Weaver, U. C. Quarck, R. R. Chauvette, and R. Monahan, *J. Am. Chem. Soc.*, 1957, **79**, 6062.
5. D. R. Harris, S. G. McGeachin, and H. H. Mills, *Tetrahedron Lett.*, 1965, **6**, 679.
6. J. R. Martin and W. Rosenbrook, *Biochemistry*, 1967, **6**, 435.
7. J. W. Corcoran and A. M. Vygantas, *Biochemistry*, 1982, **21**, 263.
8. J. M. Weber, C. K. Wierman, and C. R. Hutchinson, *J. Bacteriol.*, 1985, **164**, 425.
9. J. W. Corcoran, in "*Antibiotics Volume IV: Biosynthesis*," ed. J. W. Corcoran, Springer, New York, 1981, p. 132
10. A. M. Vygantas and J. W. Corcoran, *Fed. Proc.*, 1974, **33**, 1233.
11. P. P. Hung, C. L. Marks, and P. L. Tardrew, *J. Biol. Chem.*, 1965, **240**, 1322.
12. J. R. Martin, T. J. Perun, and R. L. Girolami, *Biochemistry*, 1966, **5**, 2852.
13. J. M. Weber, J. O. Leung, G. T. Maine, R. H. B. Potenz, T. J. Paulus, and J. P. DeWitt, *J. Bacteriol.*, 1990, **172**, 2372.
14. J. Majer, J. R. Martin, R. S. Egan, and J. W. Corcoran, *J. Am. Chem. Soc.*, 1977, **99**, 1620.
15. J. W. Corcoran and A. M. Vygantas, *Fed. Proc.*, 1977, **36**, 663.
16. J. W. Corcoran, in "*Methods in Enzymology XLIII*," ed. J. H. Hash, Academic Press, London, 1975, p. 487.
17. R. H. Lambalot, D. E. Cane, J. J. Aparicio, and L. Katz, *Biochemistry*, 1995, **34**, 1858.
18. L. Katz and S. Donadio, in "*Genetics and Biochemistry of Antibiotic Production*," eds. L. C. Vining and C. Stoddard, Butterworth-Heinmann, Boston, MA, 1995, p. 223.
19. H.-W. Liu and J. S. Thorson, *Annu. Rev. Microbiol.*, 1994, **48**, 223.
20. T. Kaneda, J. C. Butte, S. B. Taubman, and J. W. Corcoran, *J. Biol. Chem.*, 1962, **237**, 322.
21. D. E. Cane, H. Hasler, and T. Liang, *J. Am. Chem. Soc.*, 1981, **103**, 5960.
22. D. E. Cane and J. C-.C. Yang, *J. Am. Chem. Soc.*, 1987, **109**, 1255.
23. D. E. Cane, P. C. Prabhakaran, W. Tan, and W. R. Ott, *Tetrahedron Lett.*, 1991, **32**, 5457.
24. S. Yue, J. S. Duncan, Y. Yamamoto, and C. R. Hutchinson, *J. Am. Chem. Soc.*, 1987, **109**, 1253.
25. D. E. Cane, R. H. Lambalot, P. C. Prabhakaran, and W. R. Ott, *J. Am. Chem. Soc.*, 1993, **115**, 522.
26. C. J. Dutton, A. M. Hooper, P. F. Leadlay, and J. Staunton, *Tetrahedron Lett.*, 1994, **35**, 327.
27. S. J. Wakil, *Biochemistry*, 1989, **28**, 4523.
28. S. Smith, *FASEB*, 1994, **8**, 1248.
29. A. J. Fulco, *Prog. Lipid Res.*, 1983, **22**, 133.
30. J. Cortés, S. F. Haydock, G. A. Roberts, D. J. Bevitt, and P. F. Leadlay, *Nature*, 1990, **348**, 176.
31. S. Donadio, M. J. Staver, J. B. McAlpine, S. J. Swanson, and L. Katz, *Science*, 1991, **252**, 675.
32. S. Donadio and L. Katz, *Gene*, 1992, **111**, 51.
33. D. J. Bevitt, J. Cortes, S. F. Haydock, and P. F. Leadlay, *Eur. J. Biochem.*, 1992, **204**, 39.
34. J. Staunton, *Angew. Chem., Int. Ed. Engl.*, 1991, **30**, 1302.
35. T. Schwecke, J. F. Aparicio, I. Molnár, A. König, L. E. Khaw, S. F. Haydock, M. Oliynyk, P. Caffrey, J. Cortés, J. B. Lester, G. A. Böhm, J. Staunton, and P. F. Leadlay, *Proc. Natl. Acad. Sci. USA*, 1995, **92**, 7839.
36. S. Kuhstoss, M. Huber, J. R. Turner, J. W. Pashal, and R. N. Rao, *Gene*, 1996, **183**, 231.

37. S. Donadio, M. J. Staver, J. B. McAlpine, S. J. Swanson, and L. Katz, *Science*, 1991, **252**, 675.
38. S. Donadio, J. B. McAlpine, P. L. Sheldon, M. Jackson, and L. Katz, *Proc. Natl. Acad. Sci. USA*, 1993, **90**, 7119.
39. P. Caffrey, D. J. Bevitt, J. Staunton, and P. F. Leadlay, *FEBS Lett.*, 1992, **304**, 225.
40. J. Staunton, P. Caffrey, J. F. Aparicio, G. A. Roberts, S. S. Bethell, and P. F. Leadlay, *Nature Struct. Biol.*, 1996, **3**, 188.
41. J. F. Aparicio, P. Caffrey, A. F. A. Marsden, J. Staunton, and P. F. Leadlay, *J. Biol. Chem.*, 1994, **269**, 8524.
42. A. F. A. Marsden, P. Caffrey, J. F. Aparicio, M. S. Loughran, J. Staunton, and P. F. Leadlay, *Science*, 1994, **263**, 378.
43. C. M. Kao, R. Pieper, D. E. Cane, and C. Khosla, *Biochemistry*, 1996, **35**, 12 363.
44. R. McDaniel, S. Ebert-Khosla, D. A. Hopwood, and C. Khosla, *Science*, 1993, **262**, 1546.
45. C. M. Kao, L. Katz, and C. Khosla, *Science*, 1994, **265**, 509.
46. J. Cortés, K. E. H. Wiesmann, G. A. Roberts, M. J. B. Brown, J. Staunton, and P. F. Leadlay, *Science*, 1995, **268**, 1487.
47. C. M. Kao, G. Luo, L. Katz, D. E. Cane, and C. Khosla, *J. Am. Chem. Soc.*, 1995, **117**, 9105.
48. C. M. Kao, G. Luo, L. Katz, D. E. Cane, and C. Khosla, *J. Am. Chem. Soc.*, 1994, **116**, 11 612.
49. M. J. B. Brown, J. Cortés, A. L. Cutter, P. F. Leadlay, and J. Staunton, *J. Chem. Soc., Chem. Commun.*, 1995, 1517.
50. C. M. Kao, G. Luo, L. Katz, D. E. Cane, and C. Khosla, *J. Am. Chem. Soc.*, 1996, **118**, 9184.
51. M. Oliynyk, M. J. B. Brown, J. Cortés, J. Staunton, and P. F. Leadlay, *Chem. Biol.*, 1996, **3**, 833.
52. S. F. Haydock, J. F. Aparicio, I. Molnár, T. Schwecke, L. E. Khaw, A. König, A. F. A. Marsden, I. S. Galloway, J. Staunton, and P. F. Leadlay, *FEBS Lett.*, 1995, **374**, 246.
53. X. Ruan, A. Pereda, D. L. Stassi, D. Zeidner, R. G. Summers, M. Jackson, A. Shivakumar, S. Kakavas, M. J. Staver, S. Donadio, and L. Katz, *J. Bacteriol.*, 1997, **179**, 6416.
54. J. B. McApline, J. S. Tuan, D. P. Brown, K. D. Grebner, D. N. Whittern, A. Buko, and L. Katz, *J. Antibiot.*, 1987, **40**, 1115.
55. L. Liu, A. Thamchaipenet, H. Fu, M. Betlach, and G. Ashley, *J. Am. Chem. Soc.*, 1997, **119**, 10 553.
56. P. F. Leadlay, J. Staunton, A. F. A. Marsden, B. Wilkinson, N. J. Dunster, J. Cortés, M. Oliynyk, U. Hanefeld and M. J. B. Brown, in "*Industrial Micro-Organisms: Basic and Applied Molecular Genetics: Developments in Industrial Microbiology GMBIM*," eds. R. H. Baltz, G. D. Hegeman, and P. L. Skatrud, American Society for Microbiology, Washington, DC, 1996, vol. **34**, p. 33.
57. A. F. A. Marsden, B. Wilkinson, J. Cortés, N. J. Dunster, J. Staunton, and P. F. Leadlay, *Science*, 1998, **279**, 199.
58. C. J. Dutton, S. P. Gibson, A. C. Goudie, K. S. Holdom, M. S. Pacey, J. C. Ruddock, J. D. Bu'Lock, and M. K. Richards, *J. Antibiot.*, 1991, **44**, 357.
59. E. W. Hafner, B. W. Holley, K. S. Holdom, S. E. Lee, R. G. Wax, D. Beck, H. A. I. McArthur, and W. C. Wernau, *J. Antibiot.*, 1991, **44**, 349.
60. A. Pereda, R. G. Summers, D. L. Stassi, X. Ruan, and L. Katz, *Microbiology*, 1998, **144**, 543.
61. D. Bedford, J. R. Jacobsen, G. Luo, D. E. Cane, and C. Khosla, *Chem. Biol.*, 1996, **3**, 827.
62. R. McDaniel, C. M. Kao, H. Fu, P. Hevezi, C. Gustafsson, M. Betlach, G. Ashley, D. E. Cane, and C. Khosla, *J. Am. Chem. Soc.*, 1997, **119**, 4309.
63. C. M. Kao, M. McPherson, R. N. McDaniel, H. Fu, D. E. Cane, and C. Khosla, *J. Am. Chem. Soc.*, 1997, **119**, 11339.
64. R. McDaniel, C. M. Kao, S. J. Hwang, and C. Khosla, *Chem. Biol.*, 1997, **4**, 667.
65. C. M. Kao, M. McPherson, R. N. McDaniel, H. Fu, D. E. Cane, and C. Khosla, *J. Am. Chem. Soc.*, 1998, **120**, 2478.
66. M. McPherson, C. Khosla, and D. E. Cane, *J. Am. Chem. Soc.*, 1998, **120**, 3267.
67. K. E. H. Wiesmann, J. Cortes, M. J. B. Brown, A. L. Cutter, J. Staunton, and P. F. Leadlay, *Chem. Biol.*, 1995, **2**, 583
68. R. Pieper, G. Luo, D. E. Cane, and C. Khosla, *J. Am. Chem. Soc.*, 1995, **117**, 11 373.
69. G. Luo, R. Pieper, A. Rosa, C. Khosla, and D. E. Cane, *Bioorg. Med. Chem. Lett.*, 1996, **4**, 995.
70. D. O'Hagan, in "*The Polyketide Metabolites*," Ellis Horwood, Chichester, 1991, p. 70
71. R. Pieper, G. Luo, D. E. Cane, and C. Khosla, *Nature*, 1995, **378**, 263.
72. R. Pieper, R. S. Gokhale, G. Luo, D. E. Cane, and C. Khosla, *Biochemistry*, 1997, **36**, 1846.
73. K. J. Weissman, M. Timoney, M. Bycroft, P. Grice, U. Hanefeld, J. Staunton, and P. F. Leadlay, *Biochemistry*, 1997, **36**, 13 849.
74. P. Caffrey, B. Green, L. C. Packman, B. J. Rawlings, J. Staunton, and P. F. Leadlay, *Eur. J. Biochem.*, 1991, **195**, 823.
75. R. Aggarwal, P. Caffrey, P. F. Leadlay, C. J. Smith, and J. Staunton, *J. Chem. Soc., Chem. Commun.*, 1995, 1519.
76. K. J. Weissman, C. J. Smith, U. Hanefeld, R. Aggarwal, M. Bycroft, J. Staunton, and P. F. Leadlay, *Angew. Chem., Int. Ed. Engl.*, 1998, **37**, 1437.
77. J. R. Jacobsen, C. R. Hutchinson, D. E. Cane, and C. Khosla, *Science*, 1997, **277**, 367.
78. J.-A. Chuck, M. McPherson, H. Huang, J. R. Jacobsen, C. Khosla, and D. E. Cane, *Chem. Biol.*, 1997, **4**, 757.
79. S. L. Schreiber and G. R. Crabtree, *Immunol. Today*, 1992, **13**, 136.
80. J. Kunz and M. N. Hall, *Trends Biochem. Sci.*, 1993, **18**, 334.
81. F. J. Dumont and Q. Su, *Life Sci.*, 1995, **58**, 373.
82. N. L. Paiva, A. L. Demain, and M. F. Roberts, *J. Nat. Prod.*, 1991, **54**, 167.
83. N. L. Paiva, A. L. Demain, and M. F. Roberts, *Enzyme Microb. Technol.*, 1993, **15**, 581.
84. N. L. Paiva, M. F. Roberts, and A. L. Demain, *J. Ind. Microbiol.*, 1993, **12**, 423.
85. P. A. S. Lowden, G. A. Böhm, J. Staunton, and P. F. Leadlay, *Angew. Chem., Int. Ed. Engl.*, 1996, **35**, 2249.
86. K. W. Wallace, K. A. Reynolds, K. Koch, H. A. I. McArthur, M. S. Brown, R. G. Wax, and B. S. Moore, *J. Am. Chem. Soc.*, 1994, **116**, 11 600.
87. I. Molnár, J. F. Aparicio, S. F. Haydock, L. E. Khaw, T. Schwecke, A. König, J. Staunton, and P. F. Leadlay, *Gene*, 1996, **169**, 1.
88. J. F. Aparicio, I. Molnár, T. Schwecke, A. König, S. F. Haydock, L. E. Khaw, J. Staunton, and P. F. Leadlay, *Gene*, 1996, **169**, 9.
89. H. Motamedi, S.-J. Cai, A. Shafiee, and K. O. Elliston, *Eur. J. Biochem.*, 1997, **244**, 74.
90. H. KleinKauf and H. Von Döhren, *Eur. J. Biochem.*, 1996, **236**, 335.
91. A. König, T. Schwecke, I. Molnár, G. A. Böhm, P. A. S. Lowden, J. Staunton, and P. F. Leadlay, *Eur. J. Biochem.*, 1997, **247**, 526.
92. J. B. Nielsen, M.-J. Hsu, K. M. Byrne, and L. Kaplan, *Biochemistry*, 1991, **30**, 5789.

93. L. E. Khaw, G. A. Böhm, S. Metcalfe, J. Staunton, and P. F. Leadlay, *J. Bacteriol.*, 1998, **180**, 809.
94. R. W. Burg, B. M. Miller, E. E. Baker, J. Birnbaum, S. A. Currie, R. Hartman, Y.-L. Kong, R. L. Monaghan, G. Olson, I. Putter, J. B. Tunac, H. Wallick, E. O. Stapley, R. Oiwa, and S. Omura, *Antimicrob. Agents Chemother.*, 1979, **15**, 361.
95. D. F. Cully, D. K. Vassilatis, K. K. Liu, P. S. Paress, L. H. T. Van der Ploeg, J. M. Schaeffer, and J. P. Arena, *Nature*, 1994, **371**, 707.
96. J. C. Chabala, H. Mrozik, R. L. Tolman, P. Eskola, A. Lusi, L. H. Peterson, M. F. Woods, and M. H. Fisher, *J. Med. Chem.*, 1980, **29**, 1134.
97. J. Remme, R. H. A. Baker, G. DeSole, K. Y. Dadzie, J. F. Walsh, M. A. Adams, E. S. Alley, and H. S. K. Avissey, *Trop. Med. Parasitol.*, 1989, **40**, 367.
98. D. E. Cane, T.-C. Liang, L. K. Kaplan, M. K. Nallin, M. D. Schulman, O. D. Hensens, A. W. Douglas, and G. Albers-Schönberg, *J. Am. Chem. Soc.*, 1983, **105**, 4110.
99. S. T. Chen, O. D. Hensens, and M. D. Schulman, in "*Ivermectin and Abamectin*," ed. W. C. Campbell, Springer, New York, 1989, p. 55.
100. S. T. Chen, B. H. Arison, V. P. Gullo, and E. S. J. Inamine, *Ind. Microbiol.*, 1989, **4**, 231.
101. H. Ikeda and S. Omura, *Chem. Rev.*, 1997, **97**, 2591.
102. D. J. MacNeil, J. L. Occi, K. M. Gewain, T. MacNeil, P. H. Gibbons, C. L. Ruby, and S. J. Danis, *Gene*, 1992, **115**, 119.
103. S. T. Chen and E. S. Inamine, *Arch. Biochem. Biophys.*, 1989, **270**, 521.
104. H. Ikeda, Y. Takada, C.-H. Pang, K. Matsuzaki, H. Tanaka, and S. Omura, *J. Antibiot.*, 1995, **48**, 95.
105. E. W. Hafner, B. W. Holley, K. S. Holdom, S. E. Lee, R. G. Wax, D. Beck, H. A. I. McArthur, and W. C. Wernau, *J. Antibiot.*, 1991, **44**, 349.
106. H. A. I. McArthur, in "*Proceedings from the* 1997 *Biotechnology of Microbial Products* (*BMP*) *Conference, Williamsburg*," eds. C. R. Hutchinson and J. McAlpine, Society for Industrial Microbiology, Fairfax, VA, 1997, p. 43.
107. H. A. I. McArthur, personnal communication.
108. S. J. Kakava, L. Katz, and D. Stassi, *J. Bacteriol.*, 1997, **179**, 7515.
109. S. G. Burgett, S. A. Khustoss, R. N. Rao, M. A. Richardson, and P. R. Rosteck, Jr., *Eur. Pat.* EP 0791 656 A2 (1997).
110. R. H. Baltz, E. T. Seno, J. Stonesifer, and G. M. Wild, *J. Antibiot.*, 1983, **36**, 131.
111. E. T. Seno, and R. H. Baltz, *Antimicrob. Agents Chemother.*, 1982, **27**, 758.
112. M. L. B. Huber, J. W. Paschal, J. P. Leeds, H. A. Hirst, J. A. Wind, F. D. Miller, and R. J. Turner, *Antimicrob. Agents Chemother.*, 1990, **34**, 1535.
113. B. S. Dehoff, S. A. Khustoss, P. R. Rosteck, Jr., and K. L. Sutton, *Eur. Pat.* EP 0 791 655 A2 (1997).
114. O. Ghisalba, H. Fuhrer, W. J. Richter, and S. Moss, *J. Antibiot.*, 1981, **34**, 58.
115. R. J. White, E. Martinelli, G. G. Gallo, G. Lancini, and P. Beynon, *Nature*, 1973, **243**, 273.
116. J. J. Kibby, I. A. McDonald, and R. W. Rickards, *J. Chem. Soc., Chem. Commun.*, 1980, 768.
117. O. Ghisalba and J. Nüesch, *J. Antibiot.*, 1981, **34**, 64.
118. C.-G. Kim, A. Kirschning, P. Bergon, P. Zhou, E. Su, B. Sauerbrei, S. Ning, Y. Ahn, M. Breuer, E. Leistner, and H. G. Floss, *J. Am. Chem. Soc.*, 1996, **118**, 7486.
119. C-.G. Kim, T-.W. Yu, C. B. Fryle, S. Handa, and H. G. Floss, *J. Bacteriol.*, 1998, **273**, 6030.
120. P. R. August, L. Tang, Y. J. Yoon, S. Ning, R. Muller, T-.W. Yu, M. Taylor, D. Hoffmann, C-.G. Kim, X. Zhang, C. R. Hutchinson, and H. G. Floss, *Chem. Biol.*, 1998, **5**, 69.
121. T. Schupp, C. Toupet, N. Engel, and S. Goff, *FEMS Microbiol. Lett.*, 1998, **159**, 201.
122. D. Hunziker, T-.W. Yu, C. R. Hutchinson, H. G. Floss, and C. Khosla, *J. Am. Chem. Soc.*, 1998, **120**, 1092.
123. W. D. Celmer, G. N. Chmurny, C. E. Moppet, R. S. Ware, P. C. Watts, and E. B. Wipple, *J. Am. Chem. Soc.*, 1980, **102**, 4203.
124. W. D. Celmer, W. P. Cullen, C. E. Moppett, M. T. Jefferson, L. H. Huang, R. Shibakawa, and J. Tone (Pfizer Inc.), *US Pat.* 4 148 883 (1979).
125. D. E. Cane and C. Yang, *J. Am. Chem. Soc.*, 1984, **106**, 784.
126. W. C. Snyder and K. L. Rinehart, *J. Am. Chem. Soc.*, 1984, **106**, 787.
127. D. E. Cane and C. Yang, *J. Antibiot.*, 1985, **38**, 423.
128. D. E. Cane, W. T. Tan, and W. R. Ott, *J. Am. Chem. Soc.*, 1993, **115**, 527.
129. D. E. Cane and G. Luo, *J. Am. Chem. Soc.*, 1995, **117**, 6633.

1.20
Cyclosporin: The Biosynthetic Path to a Lipopeptide

HANS VON DÖHREN and HORST KLEINKAUF
Technische Universität Berlin, Germany

1.20.1 INTRODUCTION

Cyclosporins are cycloundecapeptides produced by a variety of fungi. Their cyclic structure with the unusual amino acid (4*R*)-4-[(*E*)-2-butenyl]-4-methyl-L-threonine (Bmt), the nonprotein constituents D-alanine and α-aminobutyrate, and the seven *N*-methylated peptide bonds underlines their nonribosomal origin. The first example was originally discovered by its selective antifungal properties, and proved to be pharmacologically useful not only as an immunosuppressant in transplantation medicine, but to hold promising activity against viral and parasitic infections and psoriasis.[1,2] The ecological function of cyclosporins may well be control of fungal populations, since in 1997 screening has demonstrated that cyclosporin can control postharvest mold diseases.[3] As in the immunosuppressive action, fungal control is presumably exerted by the transcriptional inhibition of 1,3-glucan synthase leading to a distorted morphology.[4–6] Such a mode of action implies protective features in the producer strains, which may include altered targets and export functions. Cyclosporin A does not affect cells of the producer *Tolypocladium inflatum*, while protoplasts are impaired.[7] Attention here will be focused on biosynthesis of the peptide, which is fairly well understood at the molecular level. The Bmt precursor, 3(*R*)-hydroxy-4(*R*)-methyl-6(*E*)-octenoate, is provided by a polyketide synthase. The assembly of cyclosporin from the respective amino acids including *N*-methylations is carried out by one protein molecule integrating 40 sequential reactions, cyclosporin synthetase. This synthetase is the most complex enzyme studied thus far, and has been used to produce a variety of structural analogues of cyclosporin *in vitro*. Similar enzymes are involved in the formation of various peptides including penicillins, gramicidins, and siderophores, as well as in the synthesis of mixed structural types like chromopeptides and peptidolactones. They also mediate the formation of polyketides containing aminoacyl structures such as rapamycin.[8]

1.20.2 DISCOVERY AND PRODUCTION STRAINS

Cyclosporins were first described as new metabolites (Figure 1) isolated from fungi imperfecti in soil samples from Wisconsin (*Cylindrocarpum lucidum* Booth) and the Hardanger Vidda (*Tolypocladium inflatum* Gams, originally classified as *Trichoderma polysporum* Link ex. Pers.).[9] Since growth in submerged culture was achieved in the latter strain, it was selected for large-scale production by fermentation. Cyclosporins are not released but are retained in the mycelium.[9] Other producers like *Stachybotrys chartarum*,[10] strains of *Neocosmospora*,[6] or *Acremonium luzulae*[3] have been found to excrete up to 50% of the cyclosporins produced, and the complete medium is extracted for recovery. The fungal extracts displayed a narrow antifungal spectrum, e.g., against *Aspergillus niger* and *Neurospora crassa*, but became important drug leads soon after their discovery owing to their immunosuppressive effects and low toxicity.[11] The first screening results disclosed the most important biological effects of these preparations (containing mainly cyclosporin A): effective immunosuppression, and no general inhibition of cell proliferation or impairment of kidney function at very high doses.[12] Further biological testing faced difficulties, as intraperitoneal and oral treatments of animals led to only marginal or no immunosuppressive effects. Soon it became evident that the addition of Tween 80 had improved bioavailability, and this opened up the route to further galenical and clinical work.[13]

The antifungal spectrum has since been investigated in more detail, and found to be broader for example, the phytopathogenic fungi such as *Botrytis cinerea*[3] or *Alternaria kikuchina* are targets of cyclosporins.[10] So far cyclosporins have been found in a number of fungi not only of soil origin,[14] but also those growing on plants, and since cyclosporins display insecticidal activity,[15] insects presumably represent another source. A selection of producers is shown in Table 1.

1.20.3 ENVIRONMENTAL IMPLICATIONS

A strain of *S. chartarum*, No. 19 392, was discovered by an assay for antifungal activity where morphological alterations of mycelia were detected.[6] The strain was found to excrete the cyclosporin analogue FR901459. Speculation on a possible ecological role of cyclosporins should also consider additional metabolites produced by the relevant strains. An isolate of *A. luzulae* (Fuckel) W. Gams from strawberry fruits also reported as *Gliomastix luzulae* (Fuckel) Mason ex Hughes, shows a strong antagonistic activity against a large number of phytopathogenic fungi.[3] This strain is known to produce cerevisterol, ergosterol peroxide, and some diterpenes such as virescenosides (β-D-altropyranosides of virescenols).[16–19] A spontaneous albino mutant produced the sesquiterpene ascochlorine.[20] Antifungal action and antagonistic activities in this case have been related mainly to the presence of cyclosporin C.[3] However, other compounds may act synergistically, and equally importantly, lytic enzymes may play a significant role.[21] Strains of *Neocosmospora* are known to produce the antifungal nonaketide monordcn as well as cyclosporin A and C.[8] *Neocosmospora vasinfecta* E.F. Smith var. *africana* (von Arx) Cannon et Hawksworth NHL 2298 is a well-known phytopathogen which causes root and fruit rot, and seedling damping-off in the Malvaceae, Leguminosae, Piperaceae, and Cucurbitaceae.[22] Besides these presumably antifungal compounds, plant growth regulators of the α-pyrone-type have been reported from this strain.[22] Ergocornine C, a carboxysterol antifungal, has been isolated from a mutant strain of *T. inflatum*.[23] All these metabolites may well be involved in the complex interactions required for the colonization of plant surfaces, either in plant–fungus or in plant–microbe interactions or competing with other microbes.

Insecticidal effects of cyclosporin have also been reported,[15] and it is well established that producers may act as insect pathogens. In the analysis of the insecticidal effects, an additional metabolite, tolypin, has been detected,[24] but the production of extracellular lipase and chitinase did not correlate

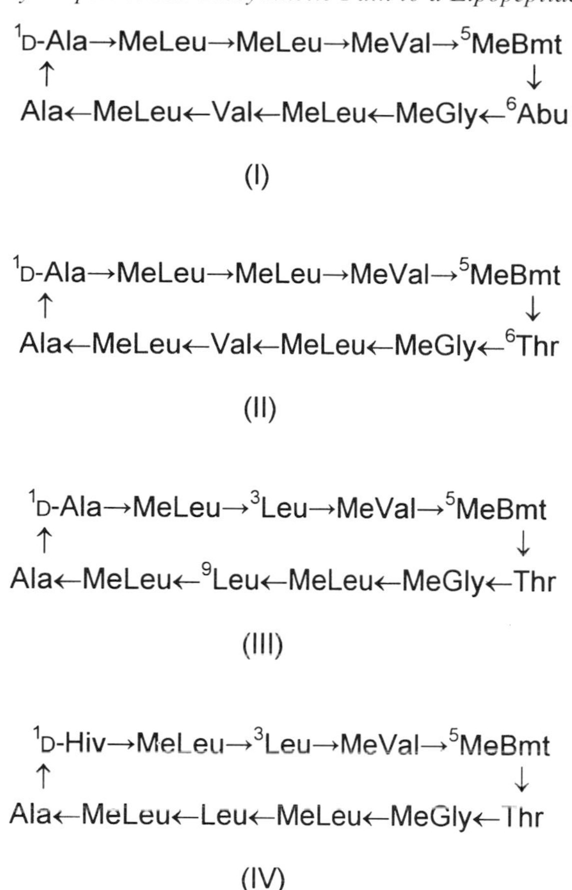

Figure 1 Structures of cyclosporins A (I) and C (II), the analogue FR 901459[9] (III), and the peptidolactone analogue SDZ 214-103 (IV). Differing from work dealing with chemical synthesis, residues are numbered according to the biosynthetic reaction sequence. The highly variable amino acid at position 6 distinguishes cyclosporins A and C. Note the positional exchange of ^{9}valine by leucine and the loss of *N*-methylation at position 3 in the newly described analogue, and the replacement of D-alanine in initiating position 1 by D-hydroxyisovalerate in the peptidolactone.

with cyclosporin production in shake flask cultures.[25] Such growth conditions, however, do not resemble the complex invasion processes of plant or animal tissues. Strains of *Tolypocladium* sp. have been used in mosquito control,[26,27] and laboratory trials on black fly and house fly have been conducted.[28,29] Indeed, *T. cylindrosporum*, *T. tundrense*, and *T. terricola* conidia have been shown to infect larvae of these species, and cyclosporin A has been identified as a predominant metabolite of the spore surface layer.[30] Both spore extracts and tolypin were efficient in killing the larvae of, for example, *Aedes aegyptii*.[31] If the efficiency of biological control trials could be improved, a cyclosporin producer would certainly be of interest.

1.20.4 PRODUCTION LEVELS AND STRUCTURAL ANALOGUES OBTAINED FROM CULTURES

The enzyme systems forming precursors and directing the assembly of complex peptides are subject to control mechanisms at several levels. A significant factor is the *in situ* availability of substrates at the cellular level. It is known that in filamentous fungi, compartmentation of enzyme systems plays an essential role in metabolite production.[32,33] No data are yet available on cyclosporin producers. Effective precursor concentrations, which may be altered by external feeding or metabolic engineering of the respective pathways, compete at the respective amino acid binding sites of the synthetase. In this primary selection process, aminoacyl adenylates are formed and stabilized as enzyme-attached intermediates.[34] The architecture of the respective adenylate-forming domains determines an individual substrate profile, which may be altered, for example, by point mutations

Table 1 Selected cyclosporin producer strains.

Strain	Volumetric production[a] CycA	CycC	Remarks	Ref.
Tolypocladium inflatum NRRL8044	100^{L}	30^{L}		56
	500^{L}		UV mutant	57
	300^{SF}			48
	230^{L}		Carrageenan-immobilized	48
Tolypocladium inflatum ATCC 34291, local isolate	199^{SF}		White strain	25
	318^{S}		Epichlorohydrin mutant	55
	205^{L}			152
Tolypocladium cyclindrosporum				
260	79^{SF}			25
285	89^{SF}			25
286	71^{SF}			25
287	113^{SF}			25
Tolypocladium niveum UAMH 2472	200^{L}		Isolate from NRRL8044	39
Acremonium luzulae		110^{SF}		3
Beauverai bassiana	135^{SF}			3
Fusarium solanii ATCC 46829	39^{SF}			3
Neocosmospora vasinfecta				
ATCC 24402	17^{S}			10
IFO 8966		1.7^{S}		10
IFO 31377	0.78^{S}			10
Neocosmospora boninensis NHL 2919	19^{S}			10
Stachybotrys chartarum No. 19392	148^{L}		FR 901459	6

[a]Volumetric production in mg L^{-1}: L, liquid culture; SF, shake flask; S, solid culture.

in individual strains.[35,36] Not all adenylates end the catalytic cycle by aminoacylating enzyme thiols of carrier proteins allowing peptide bond formation. Some intermediates may be lost by internal cyclizations or hydrolysis before completing the reaction sequence.[37] Therefore, several selection principles operate in the control of the final product(s). Although the selectivity and fidelity of the nonribosomal system are generally considered to be less efficient compared with the ribosomal system regarding its proofreading controls, certain sites in nonribosomally produced peptides are remarkably invariable. It has been suggested that such sites are essential for interactions with natural target molecules, and thus are connected with a target-oriented evolution of biosynthetic systems.[38]

Cyclosporin analogues obtained by fermentation of *T. inflatum* and surface cultures of *T. terricola* are listed in Table 2. Positions 5 and 6 are especially variable, in contrast to the conserved positions 1, 2, 3, 4, 7, 10, and 11. The numbering of residues corresponds to the biosynthetic template and the sequence of incorporation. A different numbering system is used by the chemists working on cyclosporin, where Bmt as position 1 is assigned.

1.20.4.1 Strain Improvement

Some attempts have been made to correlate morphological and physiological traits with cyclosporin production. Not all strains of *T. niveum* produce the peptide.[39,40] White, red, orange, and brown variants have been isolated, and a red variant showed increased productivity.[41] The cyclosporin production phase has been tentatively correlated with high glycolytic activities,[42] lowering of polyphosphate content, and pyrophosphatase and polyphosphatase activities.[43] An alternative cyanide-resistant respiration pathway was found to be low in producer strains and suppressed during production.[44] Random mutagenesis with UV or epichlorohydrin yielded increases in production (Table 1).

1.20.4.2 Fermentation Improvement

The manipulation of fermentation parameters has shifted the product yields into the g L^{-1} range for *Tolypocladium niveum*. Such factors include medium selection, phosphate and carbon source

Table 2 Cyclosporins produced by *Tolypocladium niveum* in different nutrient broths.

Peptide	*1*	*2*	*3*	*4*	*5*	*6*	*7*	*8*	*9*	*10*	*11*
CyA	D-Ala	MeLeu	MeLeu	MeVal	MeBmt	Abu	Sar	MeLeu	Val	MeLeu	Ala
CyB	D-Ala	MeLeu	MeLeu	MeVal	MeBmt	**Ala**	Sar	MeLeu	Val	MeLeu	Ala
CyC	D-Ala	MeLeu	MeLeu	MeVal	MeBmt	**Thr**	Sar	MeLeu	Val	MeLeu	Ala
CyD	D-Ala	MeLeu	MeLeu	MeVal	MeBmt	**Val**	Sar	MeLeu	Val	MeLeu	Ala
CyE	D-Ala	MeLeu	MeLeu	**Val**	MeBmt	Abu	Sar	MeLeu	Val	MeLeu	Ala
CyF	D-Ala	MeLeu	MeLeu	MeVal	**MedBmt**	**Nva**	Sar	MeLeu	Val	MeLeu	Ala
CyG	D-Ala	MeLeu	MeLeu	MeVal	MeBmt	Abu	Sar	MeLeu	Val	MeLeu	Ala
CyH	D-Ala	MeLeu	MeLeu	**D-MeVal**	MeBmt	Abu	Sar	MeLeu	Val	MeLeu	Ala
CyI	D-Ala	MeLeu	**Leu**	MeVal	MeBmt	**Val**	Sar	MeLeu	Val	MeLeu	Ala
CyK	D-Ala	MeLeu	MeLeu	MeVal	**MedBmt**	**Val**	Sar	MeLeu	Val	MeLeu	Ala
CyL	D-Ala	MeLeu	MeLeu	MeVal	**Bmt**	Abu	Sar	MeLeu	Val	MeLeu	Ala
CyM	D-Ala	MeLeu	MeLeu	MeVal	MeBmt	**Nva**	Sar	MeLeu	**Nva**	MeLeu	Ala
CyN	D-Ala	MeLeu	**Leu**	MeVal	MeBmt	**Nva**	Sar	MeLeu	Val	MeLeu	Ala
CyO	D-Ala	MeLeu	MeLeu	MeVal	**MeLeu**	**Nva**	Sar	MeLeu	Val	MeLeu	Ala
CyP	D-Ala	MeLeu	MeLeu	MeVal	**Bmt**	**Thr**	Sar	MeLeu	Val	MeLeu	Ala
CyQ	D-Ala	MeLeu	MeLeu	MeVal	MeBmt	Abu	Sar	**Val**	Val	**MeLeu**	Ala
CyR	D-Ala	MeLeu	**Leu?**	MeVal	MeBmt	Abu	Sar	MeLeu	Val	**Leu?**	Ala
CyS	D-Ala	MeLeu	MeLeu	MeVal	MeBmt	**Thr**	Sar	**Val**	Val	MeLeu	Ala
CyU	D-Ala	MeLeu	**Leu**	MeVal	MeBmt	Abu	Sar	MeLeu	Val	**Leu**	Ala
CyV	D-Ala	MeLeu	MeLeu	MeVal	MeBmt	Abu	Sar	MeLeu	Val	MeLeu	**Abu**
CyW	D-Ala	MeLeu	MeLeu	**Val**	MeBmt	Thr	Sar	MeLeu	Val	MeLeu	Ala
CyX	D-Ala	**Leu**	MeLeu	MeVal	MeBmt	**Nva**	Sar	MeLeu	Val	MeLeu	Ala
CyY	D-Ala	MeLeu	MeLeu	MeVal	MeBmt	**Nva**	Sar	MeLeu	Val	**Leu**	Ala
CyZ	D-Ala	MeLeu	MeLeu	MeVal	MeAOA	Abu	Sar	MeLeu	Val	MeLeu	Ala
Cy26	D-Ala	MeLeu	MeLeu	MeVal	MeBmt	**Nva**	Sar	MeLeu	**Leu**	MeLeu	Ala
Cy27	D-Ala	MeLeu	MeLeu	MeVal	**Bmt**	**Val**	Sar	MeLeu	Val	MeLeu	Ala
Cy28	D-Ala	MeLeu	MeLeu	MeVal	**MeLeu**	Abu	Sar	MeLeu	Val	MeLeu	Ala
Cy29	D-Ala	MeLeu	MeLeu	MeVal	MeBmt	Abu	Sar	**MeIle**	Val	MeLeu	Ala
Cy30	D-Ala	MeLeu	MeLeu	MeVal	**MeLeu**	**Val**	Sar	MeLeu	Val	MeLeu	Ala
Cy31	D-Ala	MeLeu	MeLeu	MeVal	MeBmt	Abu	Sar	**Ile**	Val	MeLeu	Ala
Cy32	D-Ala	MeLeu	MeLeu	MeVal	MeBmt	Abu	Gly	MeLeu	Val	MeLeu	Ala

concentration,[45] oxygen and redox control, and control of the state of mycelia (see e.g., references 46,47). In addition, attempts have been described to employ carrageenan-entrapped mycelia, and comparable results to shake flask conditions have been achieved.[48] Cells immobilized in the pores of celite beads were used successfully in high-density fermentation, continuously producing free cyclosporin A-containing cells.[49] The productivity was about 6–10 times higher than in batch procedures. Spores were found to be entrapped *in situ*, generating new immobilized cells. Precursor feeding in the case of valine was only successful during the late exponential growth phase,[50] indicating a preferential channeling to secondary metabolism.

With reference to the molecular background of genes involved in the biosynthetic cluster, so far no manipulations are known for cyclosporin producers. As has been largely evaluated from fermentations in the penicillin field, the manipulation of single factors such as the level of key enzymes in biosynthesis will likely lead to improvements in conventional g-range producers.[51]

Current high-yield strains of fungal β-lactam producers derived in continuous selection procedures over decades have not been analyzed in sufficient detail. It is now agreed that amplification of the peptide synthetase cluster linked to two modification genes observed in these strains is mainly responsible for metabolite overproduction.[52,53] Molecular genetic attempts to generate over-producers by defined alterations are just beginning.[51] The productivity of strains varies considerably depending on the growth conditions applied (Table 1). Cyclosporin production may be enhanced severalfold by mutagenesis[54,55] or precursor feeding.[37] The main products observed are cyclosporin A and C (Figure 1). Supplying the cyclosporin A component aminobutyrate suppresses cyclosporin C formation and increases the yield of cyclosporin A 2.5-fold.[56] Feeding of the cyclosporin C component threonine increases the yield from 30 to 275 mg L^{-1}, increasing at the same time the cyclosporin A level from 100 to almost 400 mg L^{-1}. Strains like *N. vasinfecta* IFO 8966 produce only cyclosporin C, half of which is excreted (12.7 mg L^{-1}).[10] In cyclosporin A producers of the same genus, only 15–25% of the peptide produced is excreted. Likewise, *A. luzulae* produces only cyclosporin C, and even feeding of alanine or valine does not lead to the respective analogues cyclosporin B and D (Table 2). Cyclosporin C is recovered from this strain by medium extraction.[3] The molecular mechanisms involved in excretion have not been studied to date.

1.20.4.3 Directed Biosynthesis

Since exogenously added amino acids may dramatically affect the precursor pool and thus the yield and composition of the peptides produced, feeding may be applied to (i) improve the level of minor components and (ii) generate new unnatural products.[56,57] The addition of D,L-α-aminobutyrate (8 g L^{-1}) had no effect on the cell mass but increased the amount of cyclosporin A from 100 mg L^{-1} to 249 mg L^{-1}, suppressing at the same time the formation of cyclosporin C. The analogue, D,L-3-aminobutyrate suppressed cyclosporin biosynthesis, while D,L-4-aminobutyrate had no effect.[57] If L-threonine is fed to promote cyclosporin C formation, the C level rises from 30 mg L^{-1} to 275 mg L^{-1}; however, cyclosporin A increases at the same time from 100 mg L^{-1} to 396 mg L^{-1}. Similarly, feeding of L-alanine or L-valine dramatically increases the levels of cyclosporin B and D, respectively, but A and C are also affected.[56] A minor natural component, cyclosporin M, carries the unusual amino acid norvaline in position 6.[58] Feeding of norvaline leads to yields of 237 mg mL^{-1}, while cyclosporin A formation decreases from 100 mg mL^{-1} to 23 mg mL^{-1}. No other cyclosporins are detected.[56] To understand these impressive feeding data, a detailed analysis of amino acid pools needs to be carried out, since uptake and metabolization need to be evaluated.

The attempt to replace the initiating D-alanine with D-threonine gave an unexpected result. Culture feeding led to *N*-methyl-^{4}isoleucyl-cyclosporin.[59] Presumably the D-threonine is metabolized to L-*allo*-threonine by the nonspecific alanine racemase (see below), and not being a substrate for cyclosporin synthetase, it may enter the isoleucine pathway. Alternatively, as known from the bacterium *Serratia marcescens*, a specific D-threonine dehydratase may be induced which directly converts D-threonine to L-isoleucine. Site 4 of cyclosporin synthetase, which activates leucine, differs from the other leucine-activating sites by its low level of discrimination between leucine and isoleucine. The initial ratio of these is 5 : 1 in the medium employed, and the postulated shift in this ratio leads to a cyclosporin analogue devoid of immunosuppressive activity. However, it retains affinity to cyclophilin, but the complex no longer binds to calcineurin.[60] Physiological effects for the producer have not been followed, but interestingly the analogue has potential anti-HIV1-activity. *In vitro* studies have shown that the D-alanine activation site does accept D-alanine, D-serine, and related derivatives, but not D-threonine.[61]

Consequently, replacement of D-alanine with D-serine proceeds as expected, and combined feeding of position 6 analogues valine, threonine, and norvaline leads to the respective analogues in good yields of 20–100 mg L^{-1}.[56] The position 6 analogue allylglycine, however, is substantially reduced to norvaline, and thus both ^{6}allylglycine-cyclosporin A and cyclosporin G are the main products. The selectivity of specific sites can be illustrated with L-β-cyclohexylalanine, a position 5 analogue. Here ^{5}*N*-methylcyclohexylalanine-cyclosporin A is obtained, while the formation of cyclosporins A, B, and C is completely inhibited. On the other hand, D,L-threo-β-cyclohexylserine is not recognized by the synthetase and does not affect the formation of products. Suppression of incorporation of an available natural precursor indicates a superior substrate or suppression of precursor formation.

Compounds obtained by directed biosynthesis *in vivo* are summarized in Figure 2. The general precursor pool has been analyzed by Senn *et al.*[62] employing C-13 labeled glucose as the only carbon source. The isotope patterns in cyclosporin A agreed well with the known biosynthetic routes. The only pool estimates available concern the valine distribution between medium and cells by Lee and Agathos.[63] The intracellular concentrations do not necessarily reflect the availability of a precursor to the enzyme system, since the enzyme system may be localized in a compartment. The modeling carried out in this connection provided valuable simulations for optimal timing and level of external valine addition,[63,64] yielding almost 0.5 g L^{-1} of cyclosporin A with *T. inflatum* ATCC 34921. However, external feeding of valine also increases the intracellular threonine pool and leads not only to an increase of cyclosporin C (with Thr in position 6) but also to the formation of cyclosporin D with valine in the variable 6-position.[56] These products may not be intended, and additional manipulations of metabolic pathways would be required to direct the integration of externally added compound. If cyclosporin A were to be produced exclusively, these pathways would have to include engineering of the amino acid selection site of the peptide synthetase, a task yet to be demonstrated. Alternatively, product selection could be achieved by selective secretion mechanisms.

1.20.5 BIOSYNTHESIS

1.20.5.1 Genetics

Genetic approaches for identifying sets of genes and verification of their role in product formation, for example by gene disruption, are only partially available for the cyclosporin system. It is evident

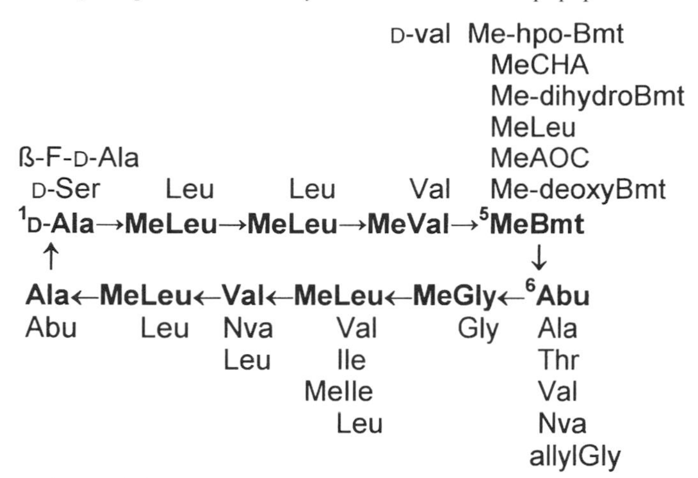

Figure 2 Cyclosporins isolated from fermentation broth including feeding studies. Abu, aminobutyrate; allylGly, allylglycine; AOC, aminooctanoic acid; CHA, cyclohexylalanine; D, D-configuration; Me, *N*-methyl. All compounds are single-substituted analogues with the exception of cyclosporin F (^{1}deoxy-MeBmt^{2}Abu), cyclosporin K (^{1}deoxy-MeBmt^{2}Val), cyclosporin M (^{2,5}Nva), cyclosporin O (^{1}MeLeu^{2}Nva), ^{2}Nva^{5}Leu-cyclosporin, and ^{1}MeLeu^{2}Val-cyclosporin. Demethylated cyclosporins are lacking an *N*-methyl in one position only, with the exception of cyclosporin R (positions 6 and 10). D-Valine in position 4 (cyclosporin H) is an artefact and formed by rearrangement during isolation.[43] Me-hpo-Bmt presumably represents another artefact due to the oxidation of Bmt to 3-hydroxy-7-hydroperoxy-4-methyl-amino-5*E*-octenoic acid.[142]

from the various producers analyzed so far that sets of genes for cyclosporin biosynthesis are present in a variety of fungi. Structural alterations and variations in the product profile point to minor alterations in the genes involved. Strains of *Fusarium* producing sets of analogues of the cyclohexadepsipeptide enniatin have been shown to contain variants of enniatin synthetase differing in substrate binding and kinetic properties.[65] The reverse genetics approach has been employed to obtain sequence data, and to identify the cyclosporin synthetase.

1.20.5.1.1 Isolation of the cyclosporin synthetase gene

Characterization of cyclosporin synthetase was only successful when high producer mutants *T. niveum* was used, suggesting the presence of higher enzyme levels exerted, for example, by gene dosage, relaxed regulation at the transcriptional level, or a reduced level of protein degradation. The multienzyme was purified from lyophilized mycelia by ammonium sulfate precipitation and gel filtration.[66] To obtain protein sequence data, samples were digested with trypsin or the endoproteinases Lys-C or Glu-C, and the separated fragments N-terminally sequenced. In addition, two functional domains (see Section 1.20.5.2.3(iii)) were specifically labeled by aminoacylation with L-alanine in the presence of ATP, and by *S*-adenosylmethionine employing photoactivation, respectively.[66] One of the 20 internal sequences obtained was used to design an oligonucleotide mixture to screen a genomic library of *T. niveum* ATCC 34921.[67] Regions of interest were selected by Northern hybridization. The enormous size of the mRNA involved permitted only the detection of a heterogeneous RNA population with a size above 9.5 kb. The respective clones were sequenced and assembled to a continuous stretch of nearly 47 kb containing an intron-free reading frame of 45 823 bp termed *simA*.[68] The ATG start codon was identified by comparing fungal promoter consensus sequences and the codon potential of the upstream region.[66] Identification of the synthetase gene is inferred by correlating the positions of sequence elements defining the biosynthetic reactions of adenylate formation, aminoacylation, *N*-methylation, and condensation, respectively. (see Section 1.20.5.2.3(iii)). All peptide sequences determined including the affinity labeled fragments were present; in particular, the alanyl label matched the predicted position (Figure 3).

Compared to other peptide synthetase genes, *simA* is the largest currently known, and like other fungal genes fully integrates the functions of amino acid activation and condensation (Table 3). On the contrary, most prokaryotic peptide synthetase genes are only partially integrated, requiring the interactions of two or more multienzymes. The limit of integration may reside at the 19 amino acid-size of peptaibols like alamethicin, currently the largest fungal peptides known of nonribosomal origin. Strains of Pseudomonads are known to produce 25 amino acid-peptides,[75] while the largest

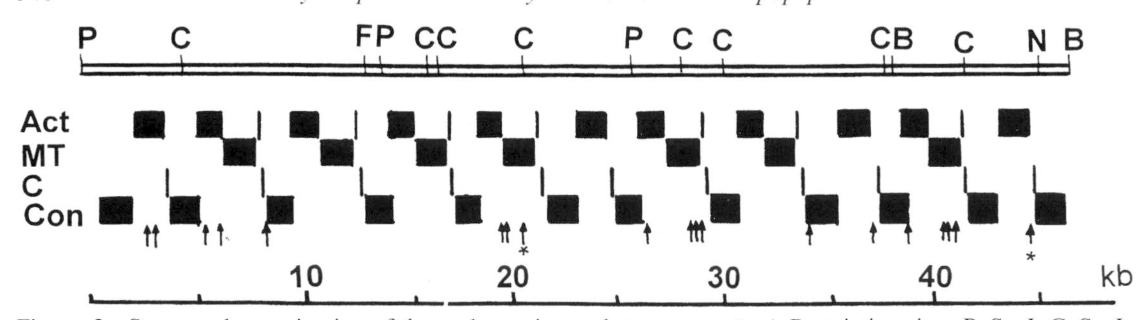

Figure 3 Structural organization of the cyclosporin synthetase gene *simA*. Restriction sites: P, SpeI; C, ScaI; F, SfiI; B, XbaI; N, NotI; Act, adenylate-forming domains (note that *N*-methyltransferase domains are inserted into adenylate domains); MT, *N*-methyltransferase domains; C, carrier domains (4-phosphopantetheine binding consensus site); Con, condensation domains. Arrows indicate positions of proteinase cleavage sites, identified by N-terminal sequencing of the respective peptide fragments (see text); the arrows marked * correspond to the affinity-labeled methyltransferase and the alanine binding site.

peptide characterized so far, polytheonamide, which is presumably made in the nonribosomal mode, contains 48 amino acids.[76] The actual producer of this peptide isolated from the sponge *Theonella swinhoei* remains to be identified. Prokaryotic systems are generally encoded by polycistronic loci, a structural feature not found in eukaryotes. Thus the complete integration of modules seems the preferred solution in these complex pathways. The absence of introns seems surprising; however, the destruxin synthetase gene in *Metarrhizium anisopliae* does contain introns.[71]

Table 3 Current state of research on fungal peptide synthetases.[1,8,66–74]

Peptide	*Organism*	*Structural type*	*Gene cloned*	*Enzymology*
ACV	*Aspergillus nidulans*	P-3	+	+
	Penicillium chrysogenum		+	(+)
	Acremonium chrysogenum		+	+
Ergotpeptides	*Claviceps purpurea*	R-P-3-M	+	+
Alamethicin	*Trichoderma viride*	P-19-M	−	+
Cyclopeptin	*Penicillium cyclopium*	C-2	−	+
HC toxin	*Cochliobolus carbonum*	C-4	+	+
Tentoxin	*Alternaria alternata*	C-4	−	+
Ferrichrome	*Aspergillus quadricinctus*	C-6	(+)	+
Echinocandin	*Aspergillus nidulans*	R-C-6	−	(+)
Cyclosporin	*Tolypocladium niveum*	C-11	+	+
Destruxin	*Metarhizium anisopliae*	L-6	+	(+)
R106	*Aureobasidium pullulans*	L-9	−	(+)
SDZ90-215	*Septoria* sp.	L-10	−	+
SDZ214-103	*Cylindrotrichum oligosporum*	L-11	(+)	+
Enniatin	*Fusarium oxysporum*	D-6	+	+
	F. scirpi			
	F. sambucinum			
	F. lateritium			
Beauvericin	*Beauveria bassiana*	D-6	−	+
PF1022	*Mycelia sterilia*	D-8	−	(+)

P, peptide; C, cyclopeptide; L, lactone; D, depsipeptide; R, acyl; M, modified. The structural types are defined by the number of amino, imino, or hydroxy acids in the precursor chain. The ring sizes of cyclic structures are indicated in the number following C, L, or D, defining the type of ring closure. ACV = δ-(L-α-aminoadipoyl)-L-cysteinyl-D-valine, HC = *Helmintosporium carbonum*.

1.20.5.1.2 The cyclophilins

The successful application of cyclosporin A in transplantation to prevent allograft rejection and in autoimmune disease treatment have promoted numerous investigations on the mode of action in human cells. The search for cyclosporin acceptors led to cyclophilins, a family of highly conserved low molecular weight proteins. Cyclophilin A was isolated as a cytosolic protein binding cyclosporin A with high affinity ($K_d = 10^{-8}$ M^{-1}).[77,78] It soon became established that cyclophilins have prolyl–peptidyl-*cis–trans* isomerase activity,[79,80] and a protein folding step was thus considered essential in the immunosuppressive action interfering with T-lymphocyte activation. Cyclosporin itself has a *cis*-peptide bond between $MeLeu^2$ and $MeLeu^3$ in the crystal state and in chloroform solution.[81] In the complexed state all amide bonds are *trans*.[82,83] More detailed studies revealed that the cyclosporin–cyclophilin complex forms a ternary complex with calcineurins, inhibiting their phosphatase activity.[84,85] Their pharmacological relevance has been substantiated by the correlation between calcineurin activity and IL-2 production in T-cells, which is dose-dependently inhibited by cyclosporin A.[86] In addition, a number of other cyclosporin A-binding proteins have been described and have been discussed in relation to binding proteins and targets of the polyketide-type immunosuppressors rapamycin and immunomycin (FK 506).[87–90] However, neither the natural ligands for cyclophilins nor the targets of calcineurin action have been identified to date. Unwanted side effects such as renal dysfunction and nephrotoxicity, and emerging applications in the treatment of HIV infections, malaria, or interference with drug resistance leave open large areas of research with cyclosporins as potential tools.

The antifungal action of cyclosporins, which is manifested in distorted mycelial growth, might also be traced to their interference with the transcriptional regulation of certain genes. The unraveling of regulatory properties of the cyclosporin–cyclophilin complex upon binding calcineurin, for example in the yeast 1,3-β-glucan synthase system,[91,92] extends the suggested complexity of intrinsic functions of the peptide.

Cyclophilins have been identified in various sources, including the filamentous fungi *N. crassa* and the cyclosporin producer *T. niveum*.[93] Zocher *et al.*[65] isolated a 17 kDa peptidyl–prolyl-*cis–trans* isomerase from *T. niveum* DSM 915 that was sensitive to cyclosporin. The authors proposed that the strain must possess a self-protection system to survive cyclosporin production. Cyclosporin A resistant mutants of calcineurin are indeed known in mammalian cells.[94]

The presumed corresponding cyclophilin gene has been cloned by Weber and Leitner probing a cDNA library in λgt10 with a fragment of the *N. crassa* cyclophilin gene.[95] The respective genomic DNA fragment was obtained from a λEMBL3 library revealing an 890 bp gene (cptA) containing three introns of 220, 57, and 60 bp, respectively. The corresponding mRNA encodes a protein of 19 569 Da with an 80.2% similarity to the *N. crassa* cyclophilin. The respective promoter region has been used to construct a transformation vector utilizing hygromycin resistance.[95] A similar gene has been sequenced from *T. niveum* by Hornbogen and Zocher,[96] with a 956 bp reading frame and four introns of 76, 221, 58, and 61 bp size. Possible endogenous functions of cyclophilins in the cyclosporin producers remain to be studied.

1.20.5.1.3 Manipulations of the cyclosporin synthetase gene

To perform DNA manipulations in *T. niveum*, Weber and Leitner inserted the cyclophilin promoter region into an *N. crassa* transformation vector.[95] This construct employs bacterial hygromycin resistance as a selectable marker, and contains the hygromycin phosphotransferase gene together with the transcriptional terminator of the *Aspergillus nidulans trpC* gene[97] (Figure 4). Protoplasts obtained in the presence of sorbitol and Ca^{2+} were isolated by centrifugation and transformed in the presence of polyethyleneglycol. Resistant colonies were selected on hygromycin-containing medium (600 μg mL^{-1}) after 7–20 days at 25 °C. The transformation frequencies (transformants/μg plasmid DNA) varied considerably. By employing 5×10^6 protoplasts (0.5 mL volume), a frequency of 10^2 was achieved using 0.2 μg of DNA. With 5 μg of DNA, frequencies ranged from 4 to 34, and were not altered if linearized DNA was used.

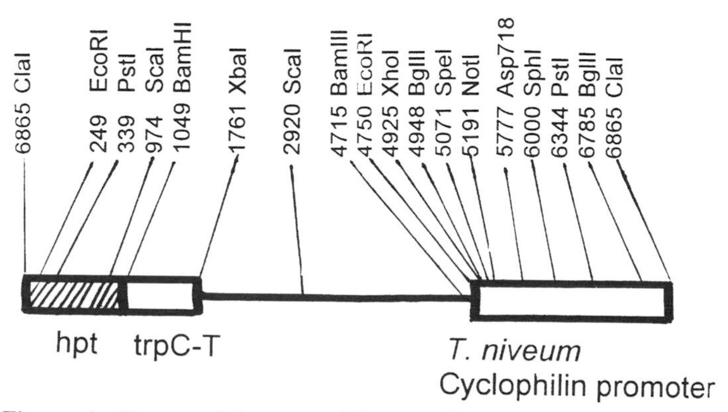

Figure 4 Structural features of the transformation vector pSIM10.

Disruption of the cyclosporin synthetase gene (*simA*) was achieved by double crossovers of derivatives of pSIM10 (Figure 5). These derivatives contained fragments close to the 3′-end of *simA* with sizes of 3.6 kb (pSIM11), 0.8 kb (pSIM12), and 2.1 kb (pSIM13). Fragment sizes correspond roughly to the nonproducer yield, since insertion frequency is favored for longer stretches. The predicted integration sites were confirmed by restriction analysis by Southern hybridization. Non-predicted fragments in some transformants were ascribed to gene rearrangements or additional integration sites and could be reduced by further purification of the gene fragments used.

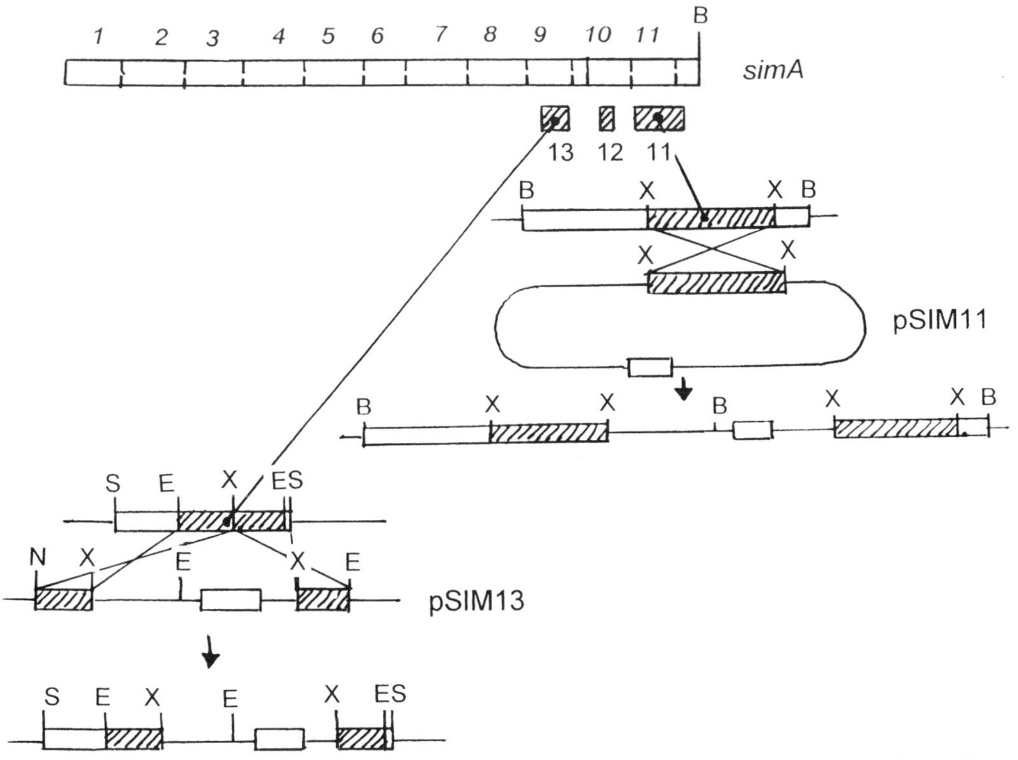

Figure 5 Gene disruption strategy used for identification of the *simA* gene. Top: positions of *simA* fragments inserted into the vector pSIM10. Middle: DNA structure formed in a single recombinational event with pSIM11. Bottom: DNA structure formed in the double recombinational event with pSIM13.

The high transformation frequencies (8% for the 0.8 kb fragment, 30% for the 2.1 kb fragment, and 60% for the 3.6 kb fragment) imply that only one copy of the *simA* gene is present in *T. niveum*. In contrast, the peptide synthetase gene encoding the cyclotetrapeptide HC toxin resides as two copies in one chromosome of the corn pathogen *Helminthosporium carbonum*.[69] In industrial high-producer strains of β-lactam antibiotics, biosynthetic clusters have been found to be amplified to more than 20 copies.[52,53]

The generated cyclosporin nonproducers should still have transcribed most of the peptide synthetase gene, since fragments of the late region had been used for disruption. Thus, accumulation of the precursor Bmt had been expected, but could not be confirmed, even when high producers were transformed. A nonproducer mutant of *T. niveum* Cyb156 isolated before was found to accumulate Bmt, but showed reduced sporulation and reverted to producers at a high frequency.[98] In contrast, pSIM11-transformands displayed normal physiology and growth. The results indicate that even if the biosynthetic genes are expressed coordinately, the functionality of individual components is part of a regulatory system. Truncated forms of the cyclosporin synthetase may be subject to degradation, or the product itself could act, by maintaining the pathway in a functional state.

Characterization of the cyclosporin nonproducer *T. niveum* YP582 has shown that only the cyclodipeptide cyclo-D-Ala-MeLeu is formed and released.[99,100] The synthetase seemed to be expressed in full size. At the enzymatic level, no defects in activation and aminoacylation have been found, but to ascertain, for example, a loss of leucine activation, e.g., in the third module, is a difficult task, since four leucine-activating modules are present. Such an analysis can only be achieved by point mutation search or sequencing of the mutant gene.

The transformation system now permits module and fragment exchange experiments in *simA* to design variants of the synthetase favoring the production of cyclosporins present only as by-products. The feasibility of such strategies as well as the difficulties involved have already been demonstrated in the surfactin system in *Bacillus subtilis*.[101,102]

1.20.5.1.4 Producer identification

Analysis of strains of *Tolypocladium* employing sequence analysis of 28S-rRNA has been used to identify *T. cylindrosporum* and *T. extinguens*, which apparently do not belong to the same genus.[103] Differential hybridization analysis of lambda clones of the cyclosporin producer *T. inflatum* ATCC 34921 with total fungal DNA and rDNA probes led to the discovery of a repeated DNA sequence. This sequence appears to be strain specific and has been termed a CPA (cyclosporin production associated) element.[104] Such elements belong to the hAT transposon family of class II transposons, which are present in plants and animals. The 4097 bp transposon of *T. inflatum* has been termed *restless*, and hybridization experiments indicate the presence of 15 copies in the genome.[105]

1.20.5.2 Enzyme Characterization

Two of the immediate precursors of cyclosporins are supplied by associated pathways: the unusual amino acid Bmt ((4*R*)-4[(*E*)-2-butenyl]-4-methyl-L-threonine) and D-alanine. The genetic linkage of the respective enzymes to the synthetase gene has not yet been established. The polyketide synthase forming the precursor of Bmt, the alanine racemase, and cyclosporin synthetase have been largely characterized at the enzyme level.

1.20.5.2.1 Formation of Bmt

Although cyclosporin devoid of Bmt has been detected (Me^1Leu-cyclosporin G = cyclosporin O)[58] or produced *in vitro* (see below), the low immunosuppressive and antifungal activity has so far not justified efforts to uncouple Bmt from the pathway. Bmt is crucial for binding to cyclophilin,[2,89] and the entire side chain is essential for immunosuppressive activity.[106] Feeding experiments employing C-13 labeled acetate and methionine have revealed the polyketide route of the precursor of this uncommon amino acid.[62,107] Short-term feeding of L-[^{14}C methyl]methionine apparently labeled only the *N*-methyl function of Bmt in cyclosporin A,[108] however, this has not been verified by degradation. Well-established concepts of polyketide biosynthesis propose two principal phases: (i) formation of the polyketide backbone from four acetate units, followed by reduction, dehydration, and methylation steps; and (ii) the transformation process introducing the amino group.

(i) The polyketide synthase

In an impressive study, Offenzeller *et al.*[109] have demonstrated *in vivo* and *in vitro* formation of the Bmt precursor acid. *In vivo* studies with [1-^{13}C,$^{18}O_2$]acetate have shown oxygen retention in the 3-hydroxyl group, suggesting 3(*R*)-hydroxy-4(*R*)-methyl-6(*E*)-octenoate as key intermediate (Figure 6). A partially purified enzyme system from *T. niveum* was indeed shown to produce this intermediate from acetyl-CoA, malonyl-CoA, and *S*-adenosyl-methionine as a CoA-thioester. Active enzyme fractions were prepared from lyophilized mycelia by grinding in liquid nitrogen, extraction in a glycerol-containing buffer, high-speed centrifugation, ammonium sulfate fractionation followed by ion exchange chromatography and gel filtration. The chromatographic steps efficiently removed the fatty acid synthase from the respective polyketide synthase fraction.

Product formation dependent on malonyl-CoA and *S*-adenosyl-methionine was analyzed by HPLC employing fluorescence detection with 4-bromomethyl-6,7-dimethoxy-coumarin. With the ion exchange fraction, three products have been identified: 3(*R*)-hydroxy-4(*R*)-methyl-6(*E*)-octenoic acid, 4(*R*)-methyl-(*E*,*E*)-2,6-octadienoic acid, and 4(*R*)-methyl-6(*E*)-octenoic acid. Thus, apparently, the key intermediate 3(*R*)-hydroxy-4(*R*)-methyl-6(*E*)-octenoate has been processed by unknown enzyme activities deviating from the *in vivo* pathway. These modifications could be largely eliminated by introducing a further gel chromatographic step. The results were confirmed by TLC analysis of [C^{14}]-*S*-adenosyl-methionine-labeled products. The CoA-thioester state of the polyketide inter-

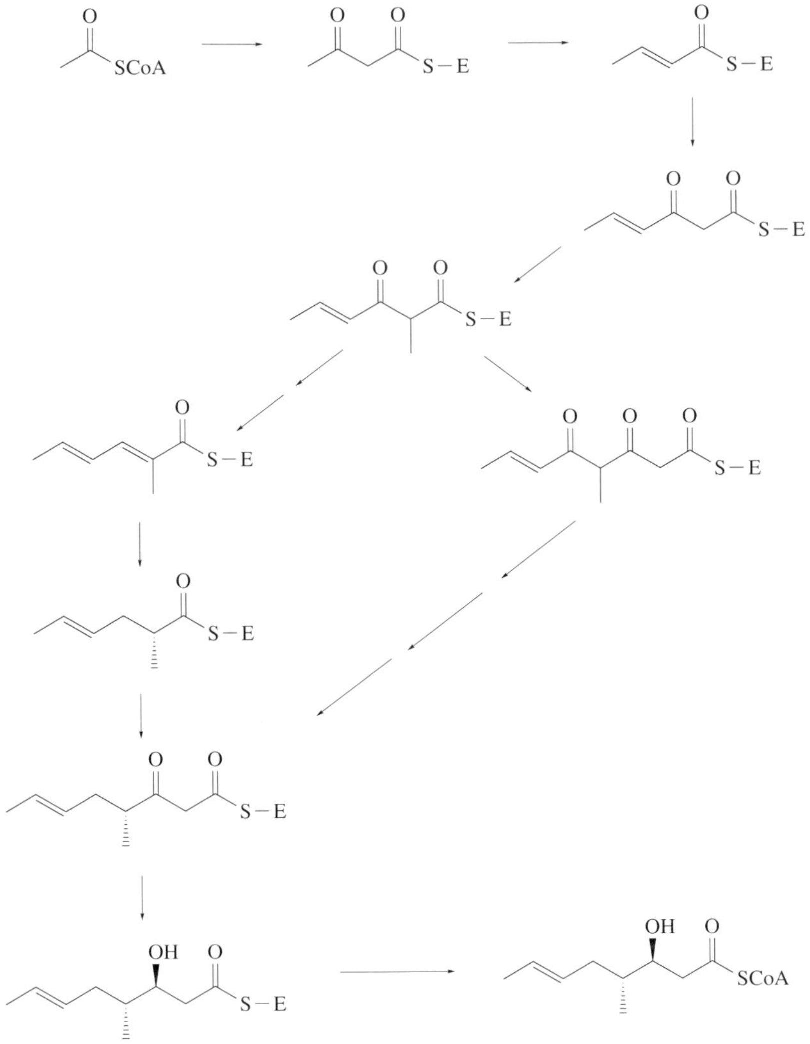

Figure 6 Polyketide pathway to Bmt: the basic assembly reaction leading to 3(*R*)-hydroxy-4(*R*)-methyl-6(*E*)-octenoyl-CoA. The intermediates remain covalently attached at the polyketide synthase as thioesters (indicated as S–E).

mediate was concluded from the required heating under alkaline conditions for extraction, and the molecular size estimate by ultrafiltration of <3000 Da to exclude protein attachment.

The polyketide synthase showed optimal activity at substrate concentrations of 200 μM acetyl-CoA, 150 μM malonyl-CoA, and 200 μM *S*-adenosyl-methionine, at pH 7 in phosphate buffer at 35 °C. The *in situ* emerging CoA (about 5 μM) seemed to be sufficient to saturate the enzyme system, since additions had no effect. Addition of excess dithiothreitol (DTT) to the 2 mM present in the buffer was inhibitory (5 mM DTT 50%).

The polyketide pathway was further dissected in a following study.[110] All acyl compounds either released as coenzyme A thioesters or all carboxylic acids released by alkaline treatment of the polyketide synthase (presumably consisting of a single polypeptide chain) were identified by HPLC. The processive mechanism of the first elongation cycle has been analyzed with respect to the starter unit. Acetyl-CoA (100%), crotonyl-CoA (32%), and acetoacetyl-CoA (65%) were incorporated and correctly processed with malonyl-CoA to 3(*R*)-hydroxy-4(*R*)-methyl-6(*E*)-octenoate, or in the case of butyryl-CoA (38%) to its saturated analogue. The high incorporation rate of butyryl-CoA and the specific methylation of all compounds indicates exclusive involvement of the polyketide synthase in this process.

To determine the timing of the methylation reaction, hexanoyl-CoA was introduced as starter molecule, but failed to be methylated, although it was elongated to finally yield octanoyl-CoA. 3-Oxo-hexanoyl-CoA was chosen in place of the unstable 3-oxo-4(*E*)-hexenoyl-CoA, and the saturated analogue of the end product was obtained in 20–30% yield compared to the synthesis starting with acetyl-CoA. In the absence of malonyl-CoA and NADPH, methylation to 2-methyl-3-oxohexanoyl-CoA was achieved in good yield. Reduction in the presence of NADPH to hexanoyl-CoA was thus shown to interfere with the processing *in vitro*. Attempts to reduce these side reactions by the consecutive addition of substrates and cofactor failed. Chase experiments demonstrated that in the case of substrate depletion, the polyketide synthase releases the respective intermediates either as coenzyme A thioesters or, in the case of a C-8 backbone with 3,5-dioxo groups, as a lactone shunt product (Figure 6).

The last intermediate to be unambiguously introduced in the pathway is 3-oxo-4(*E*)-hexenoic acid. Methylation definitely prevents transfer to the synthase and elongation. So this region of the multienzyme is inaccessible for externally supplied CoA thioesters.

Transformation to Bmt presumably by C-2 hydroxylation, oxidation, and transamination (Figure 7), is currently under investigation. Since Bmt is activated as aminoacyl adenylate, a respective thioesterase can be assumed to act during or after generation of the amino acid from the precursor-CoA derivative. So far aminoacyl-CoA derivatives have not been observed in the cellular context, and are considered to be fairly unstable.

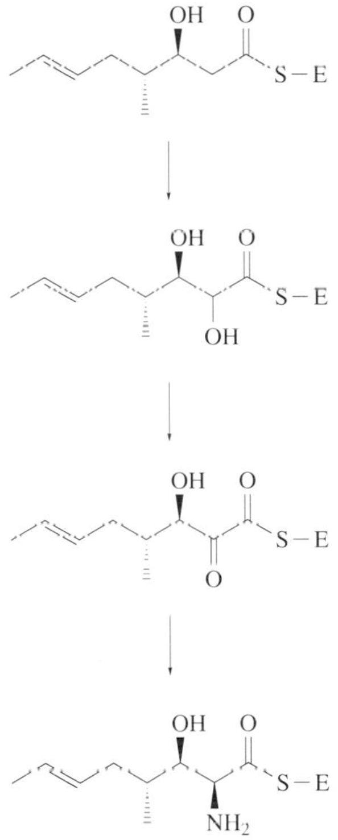

Figure 7 Transformation of the polyketide precursor 3(*R*)-hydroxy-4(*R*)-methyl-6(*E*)-octenoyl-CoA to Bmt. Presumably, free Bmt is the direct precursor accepted by cyclosporin synthetase.

1.20.5.2.2 Alanine racemase

First attempts to achieve cell-free biosynthesis of cyclosporin led to cyclo-D-alanyl-*N*-methyl-leucine and established the direct activation of D-alanine.[111] The characterization and purification of alanine racemase was carried out from the high-producer *T. niveum* 7939/45 using cyclosporin synthesis in absence of D-alanine as an assay.[112,113] The enzyme is a trimer or tetramer of a 37 kDa peptide, as judged by gel filtration, and constitutes about 0.01% of the total protein. The wild-type strain contained about 30% of the enzyme activity, while it was absent in the SDZ 214-103 producer *Cylindrosporum oligospermum*, in which D-alanine is replaced by D-hydroxyisovalerate (Figure 1).

Pyridoxal phosphate has been detected as the exclusive cofactor; and kinetic constants were K_m(L-Ala) 38 mM, K_m(D-Ala) 2 mM, V_{max}(D-L) 8 nM min^{-1}, and V_{max}(L-D) 126 nM min^{-1}. This activity is sufficient to fully account for the observed rate of cyclosporin formation. The potential substrates L-alanine L-serine, L-2-aminobutyrate and L-leucine were isomerized with reduced rates of 23%,

15%, and 13%. Both D-serine and D-2-aminobutyrate may replace D-alanine in biosynthesis, while D-leucine has not been assayed *in vitro*.

If transcriptional and translational rates of alanine racemase and cyclosporin synthetase were similar, and thus equimolar amounts of both proteins were formed, we would observe a 50-fold higher relative content of the synthetase due to its size of 1.7 MDa corresponding to about 0.5% of the total protein. So although the racemase is observed to have low abundance, both enzymes could well be coexpressed with similar transcriptional and translational rates.

Isomers of L-amino acids in peptides are generated by at least three different mechanisms. Direct incorporation of the D-isomer has been verified, apart from the case of cyclosporin, only for D-alanine in the HC toxin from *H. carbonum*[114] and for D-hydroxyproline in viridogrisein in *Streptomyces viridogriseus*.[115] HC toxin contains D-proline, D-alanine, and L-alanine. Only one epimerizing site has been localized in the HC toxin synthetase gene, which has been tentatively ascribed to the proline site, since activation of D-alanine but not D-proline has been detected. Biosynthesis of viridogrisein requires D-hydroxyproline, which is provided by a two-step transformation of L-proline. D-proline is also accepted as a substrate if the hydroxylating activity is suppressed.

In-chain epimerization seems to be the main path to D-residues, and evidence for this mechanism has been obtained in the case of actinomycin, penicillin, and bacitracin biosynthesis.[8,34] The respective multienzymes contain an epimerization domain, which closely resembles the condensation domain structure.[116] A similar reaction is found in peptides of ribosomal origin,[117] although it is catalyzed by a serine-proteinase structural type of enzyme.[118,119] The direct activation of an L-amino acid that is epimerized at the aminoacyl stage, as found in gramicidin S and tyrocidin biosynthesis,[8,120] so far represents a unique case. These two multienzymes are not stereospecific in their adenylate formation, and accept both isomers.

1.20.5.2.3 Cyclosporin synthetase

(*i*) In vitro *systems*

Reconstitution of the biosynthetic system forming cyclosporin from amino acid precursors was attempted by employing procedures developed earlier for fungal peptide synthetases, in particular the extraction of lyophilized mycelia with glycerol-containing buffers. The first experiments were conducted using a nitrosoguanidine mutagenized strain of *T. niveum*, and a 540-fold purified protein fraction was obtained by gel filtration.[111] This high molecular weight fraction catalyzed adenylate formation of the constituent amino acids, including D-alanine and Bmt, as detected by the amino acid dependent ATP–PPi exchange reaction. In addition, thioester attachment of valine and leucine as well as their *N*-methylated forms could be demonstrated by acid precipitation of enzyme–substrate complexes and performic acid release of intermediates. The fraction produced cyclo-D-alanyl-*N*-methyl-leucine, which was later shown to be the initiation point of biosynthesis. This piperazinedione is usually not detected in producer strains, but is found in the mycelium of a nonproducer mutant, *T. niveum* YP 582.[99] Soon afterwards Billich and Zocher achieved the total *in vitro* synthesis of cyclosporins, employing high glycerol concentrations in the initial extraction step, thus stabilizing the multienzyme fraction.[66]

(ii) Characterization of the synthetase

Protein chemical characterization of cyclosporin synthetase as a single polypeptide chain was carried out by Lawen and Zocher with a sample purified to homogeneity from the high-producer strain *T. niveum* 7939/45. A protein of very high molecular mass was correlated with biosynthetic activity using denaturating polyacrylamide gel electrophoresis.[121] The presence of methyltransferase activity was established by photoaffinity labeling with *S*-adenosylmethionine. The synthetase also cross-reacted with monoclonal antibodies raised against the related fungal enniatin synthetase, and was shown to contain 4′-phosphopantetheine. The lack of marker proteins delayed correct size estimations considerably. Thus, first estimates employing denaturing gel electrophoresis led to a significant underestimate of 600 kDa. Molecular cloning and sequence determination of δ-L-α-aminoadipyl-L-cysteinyl-D-valine synthetases[122] and gramicidin-(*S*)-synthetase 2[123,124] provided reliable sizes for extrapolation, especially when the modular construction of peptide synthetases was employed to predict molecular weights. A size of 1.54 MDa was proposed for cyclosporin synthetase, and 1.46 MDa for SDZ 214-103 synthetase, lacking one methyltransferase function.[125] Cesium chloride density gradient centrifugation gave an estimate of 1.4 MDa, suffering, however, from variations of the sedimentation coefficients due to denaturation and the unknown shape of the synthetase. First attempts to obtain electron micrographs were unsuccessful, apparently due to the collapse of structure during fixation procedures. The cloning and sequence determination of the cyclodepsipeptide enniatin synthetase, which contains a methyltransferase function,[126] later permitted excellent estimations of the size, which were perfectly verified by analysis of the cyclosporin synthetase gene by Weber *et al.* as 1.69 MDa.[67] Thus, cyclosporin synthetase still presents the largest known enzyme polypeptide, combining 40 catalytic functions.

(iii) Correlation of gene and enzyme structures

The modular construction of peptide synthetases at the gene level and their corresponding domains at the protein level have been reviewed.[8,34,127] Minimal peptide synthetase modules contain activation, carrier, and condensation domains. Activating adenylate-forming domains are associated with an acyl carrier domain, which structurally resembles acyl carrier proteins.[128] This pair of domains catalyzing amino acid activation and aminoacylation of the carrier protein-attached 4′-phosphopantetheine thiol group correspond well with the number of amino acids introduced into each peptide structure. Adenylate domains may contain an *N*-methyltransferase insert, as has been shown first for enniatin synthetase.[126] This transferase function catalyzes *N*-methylation of the aminoacyl-thioester via *S*-adenosyl-methionine. Condensation occurs at elongation domains, which have to be accessible from adjacent carrier proteins.[120] Epimerization domains are similar in structure, but contain specific sequence motifs.[116,129,130] These domains are missing if a D-amino acid is directly activated. Finally, domains with structural similarities to thioesterases have been detected, and these are thought to be involved in hydrolysis (generation of free peptide) or cyclization reactions. Some of these data are summarized in Figure 8.

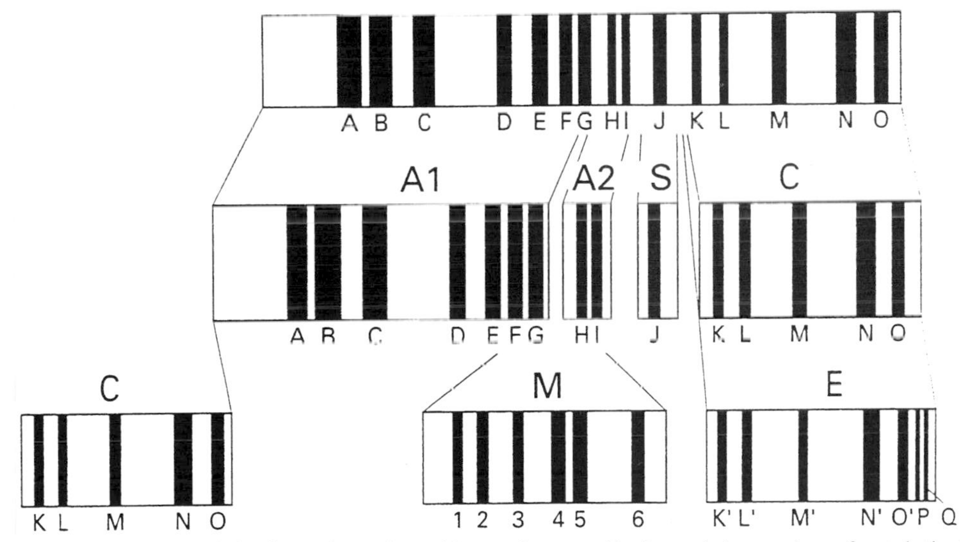

Figure 8 Biosynthetic module dissection of peptide synthetases. Each module consists of catalytic domains and subdomains. Domains can be identified by highly conserved amino acid sequences (indicated as bars). A, adenylate forming activation domain, composed of two subdomains A1 and A2; M, *N*-methyltransferase domains; S, carrier domain containing the cofactor 4′-phosphopantetheine with the active thiol of the terminal cysteamine; C, condensation domain; E, epimerization domain.

The cyclosporin synthetase gene consists of 11 modules each for activation, elongation, and carrier proteins, and seven *N*-methyltransferase modules, inserted in the second, third, fourth, sixth, seventh, and ninth adenylate modules. A schematic view correlating structure and function is given in Figure 9. Cloning of the 45.8 kb gene, which has no introns, was facilitated by an oligonucleotide corresponding to a peptide sequence of the last domain. Additional 19 N-terminal sequences derived from endoproteolytically generated fragments of the synthetase have been traced in the derived amino acid sequence. Two sequences have been assigned to alanine binding and *N*-methyltransferase, respectively, and correlated with the C-terminal domain and the insert within the sixth domain. The functional interpretation as depicted in Figure 9 shows 11 adenylate-forming and aminoacylation domains, with the seven *N*-methyltransferase domains inserted in the adenylation domains colinear with the cyclosporin amino acid sequence. Each aminoacylation domain is modified with 4′-phosphopantetheine by a holo-enzyme synthase,[131,132] still unidentified in filamentous fungi, utilizing CoA.

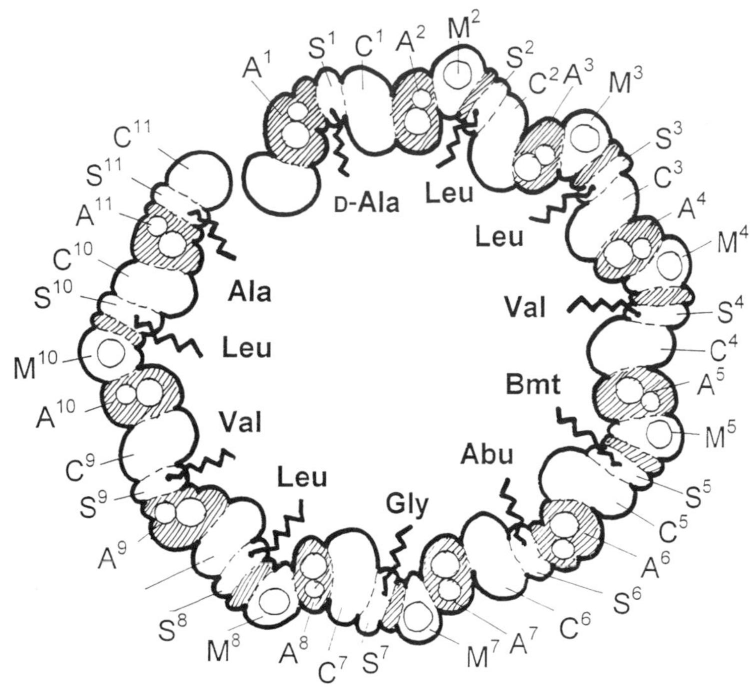

Figure 9 Schematic view of the organization of cyclosporin synthetase. Domains are numbered according to the sequence within the *simA* gene. The actual shape of the multienzyme is expected to be quite globular, as indicated by sedimentation data.

The module organization in the case of cyclosporin remains unsolved. While peptide-forming systems generally contain an arrangement of activation domain–carrier domain–condensation domain, the structure of cyclosporin synthetase has been interpreted differently. The N-terminal condensation domain is unexpected, and generally found only if acyl-coenzyme A compounds initiate synthesis, or aminoacyl or peptidyl intermediates are transferred from another peptide synthetase. The gene structure seems to support a module organization of condensation domain–activation domain–carrier domain.[2,67] However, the C-terminal domain also resembles condensation domain features, so the question has not yet been solved.

(iv) Functional analysis

In the functional analysis of the multienzyme, the presence of four leucine and two valine binding sites does not permit their separate study. So far adenylate formation has been investigated with the amino acid dependent ATP-[P^{32}]pyrophosphate exchange reaction.[100,121] Aminoacylation has been demonstrated with labeled amino acids or, if not available, using labeled *S*-adenosylmethionine to detect the *N*-methylated amino acids. Evidence for the involvement of thiol groups of either the cofactor cysteamine or cysteines can be derived from modification with 4-chloromercuribenzoate or *N*-ethylmaleimide, or by reversible blocking with 2,2′-dithiopyridine.[100]

To prove the predicted sequence of reactions (Scheme 1), labeled intermediate peptides were generated, the thioesters cleaved with performic acid, and analyzed by two-dimensional TLC. Only D-alanine was confirmed as an N-terminal amino acid, and among several peptides D-Ala-MeLeu, D-Ala-MeLeu-MeLeu, D-Ala-MeLeu-MeLeu-MeVal, and the nonapeptide intermediate D-Ala-MeLeu-MeLeu-MeVal-MeBmt-Abu-MeGly-MeLeu were confirmed by cochromatography. These data support the proposed scheme; however, as yet unidentified peptides may originate from side reactions. One such reaction is the formation of cyclodipeptides in incomplete reaction mixtures, as observed by Glinski *et al.*[133] Upon incubation of cyclosporin synthetase with L-alanine, L-leucine,

E[1] + DAla+ ATP → **E1**(DAlaAMP) → 1DAla-S-**E1**
↓
E2 + Leu+ ATP → **E2**(LeuAMP) → Leu-S-**E2** + SAM → MeLeu-S-**E2**
↓
1DAla-MeLeu-S-**E2**
↓
E3 + Leu+ ATP → **E3**(LeuAMP) → Leu-S-**E3** + SAM → MeLeu-S-**E3**
↓
1DAla-MeLeu-MeLeu-S-**E**
↓
E4 + Val+ ATP → **E4**(ValAMP) → Val-S-**E4** + SAM → MeVal-S-**E4**
↓
1DAla-MeLeu-MeLeu-MeVa
↓
E5 + Bmt+ ATP → **E5**(BmtAMP) → Bmt-S-**E5** + SAM → MeBmt-S-**E5**
↓
1DAla-MeLeu-MeLeu-MeVal-MeBmt
↓
E6 + Abu+ ATP → **E6**(AbuAMP) → Abu-S-**E6**
↓
1DAla-MeLeu-MeLeu-MeVal-MeBmt-Ab
↓
E7 + Gly+ ATP → **E7**(GlyAMP) → Gly-S-**E7** + SAM → MeGly-S-**E7**
↓
1DAla-MeLeu-MeLeu-MeVal-MeBmt-Abu-MeGl
↓
E8 + Leu+ ATP → **E8**(LeuAMP) → Leu-S-**E8** +SAM → MeLeu-S-**E8**
↓
1DAla-MeLeu-MeLeu-MeVal-MeBmt-Abu-MeGly-MeLeu-
↓
E9 + Val+ ATP → **E9**(ValAMP) → Val-S-**E9**
↓
1DAla-MeLeu-MeLeu-MeVal-MeBmt-Abu-MeGly-MeLeu-V
↓
E10 + Leu+ ATP → **E10**(LeuAMP) → Leu-S-**E10** +SAM → MeLeu-S-**E10**
↓
1DAla-MeLeu-MeLeu-MeVal-MeBmt-Abu-MeGly-MeLeu-Val-MeLeu
↓
E11 + Ala+ ATP → **E11**(AlaAMP) → Ala-S-**E11**
↓
1DAla-MeLeu-MeLeu-MeVal-MeBmt-Abu-MeGly-MeLeu-Val-MeLeu-Ala S
↓
cyclization

Scheme 1

and *S*-adenosylmethionine the authors observed the formation of cyclo-L-Ala-MeLeu. Apart from the estimate of cyclosporin formation at a rate of 16 nkatal mg^{-1},[121] no enzyme kinetic data are available.

(v) Synthesis of cyclosporin analogues

So far a total of 32 cyclosporin analogues isolated from fermentation broths have been described[54,56,134–142] and their positional variations are shown in Figure 2. Although 20 positional changes have been compiled, not all of the predictable 72 000 analogues can of course be traced. Obviously only one or at most two variations are tolerated by the enzyme system. Thus, only one compound has been found devoid of more than one *N*-methylation. The rate of synthesis decreases about 10-fold if unmethylated intermediates have to be processed in enniatin synthesis.[86] Similarly, if analogue substrates are poorly accepted, the overall rate will decrease.[143] Dramatic variations in the yield of analogues can be achieved by external feeding. Thus, feeding of L-norvaline doubled the peptide yield and decreased the cyclosporin A content to 9% compared with 91% of the ^{2}Nva-analogue cyclosporin G.[58] This analogue has similar immunosuppressive properties to cyclosporin A, but reduced nephrotoxicity.[144] The *in vitro* approach of cyclosporin synthesis has the advantages that transport and metabolization of precursors do not have to be considered. Thus, several analogues have been synthesized which were not available by feeding experiments,[145] including the D-aminobutyrate replacement of D-alanine. The ring-extended $^8\beta$-alanine cyclosporin has also been produced, which is surprisingly a side product of D-serine fed cultures.[146] Enzymatically synthesized cyclosporin analogues are compiled in Figure 10. Most of the compounds, however, have been only tentatively identified by TLC or HPLC, without actual verification of their structures. The *in vitro* approach uses a synthetase fraction, the precursor amino acids, ATP for activation, and *S*-adenosylmethionine for *N*-methylation. If one position is omitted, and the respective amino acid analogue supplied, a new compound indicates the formation of an analogue. This poses problems if substrate specificities are close. Feeding of norvaline targeted at the variable position 2 also replaces the valines in positions 5 and 11 to some extent. Since position 1 may also be replaced by leucine, substitutions upon feeding Bmt analogues are not necessarily restricted to this site, but could also replace other positions. Substrate specificities of individual site vary and these differences, rather than precursor levels, may account for the composition of peptide mixtures isolated from various strains. This has been demonstrated by Pieper *et al.*[142] for enniatin synthetase from different strains of *Fusarium*. A similar study compared the substrate specificities of cyclosporin synthetase and the similar SDZ 214-103 synthetase.[61,147] It was demonstrated, although not quantified, that substrate discrimination varied in these related multienzymes (Figure 11). The starting activation domain (position 8) of both systems is strictly specific for either D-hydroxy- or D-amino acids.

(vi) Preparative enzymatic synthesis of cyclosporins

Although this procedure is elegant and convincing in the production of, at most, milligram quantities, which is sufficient for many activity tests, it still needs improvement. The synthesis of 8-β-Ala-cyclosporin A, for example, leads to significant by-products including cyclosporins A and V or glycine in positions 7 and 8, which are unexpected from the substrates supplied. The enzyme system may contain intermediates which are completed during synthesis, and in case of unfavorable products could even dominate. Yields depend not only on stabilities of the multienzyme, ATP, and *S*-adenosylmethionine in the system, but more critically on the acceptance of precursors. Data obtained on the gramicidin *S*-system indicate that rates of synthesis with analogues are additive.[148] This means if an analogue is incorporated with 10% efficiency, and a second analogue is added having a likewise reduced rate of 10%, the overall rate for the disubstituted peptide will only be 1%. In addition, product inhibition may be severe and cause low yields. Many synthetic approaches use an excess of ATP limiting the yields by either a labeled amino acid or labeled *S*-adenosylmethionine, as in most protocols producing cyclosporin analogues. This new technique is still in need of vast improvement and development, including enzyme stabilization by chaperones, ATP regeneration, and product removal to avoid product inhibition.

1.20.6 FUTURE PROSPECTS

A large number of cyclosporin analogues have been prepared by chemical and biotechnical means. Some of the new compounds, like cyclosporin G (^{2}Nva), which are only minor fermentation products, show an improved pharmaceutical profile with reduced nephrotoxicity.[144] Others, like ^{4}MeIle-cyclosporin A, an unexpected feeding derivative, have reduced immunosuppressive properties, but are active against HIV1.[149] New fields of application for cyclosporins may emerge in antiviral therapy (Herpes[150]) or psoriasis[151] and again require the evaluation of analogues.

By comparison, chemical synthesis currently is unsuited for large-scale production. The biotechnical production of certain analogues, if not available by feeding strategies, remains an area of development. Equally promising is the screening for peptide analogue producers with slight alterations, like SDZ 214-103 or FR901459. Such systems show subsite specificities differing from the parent compound, and the respective peptide synthetases are a valuable source of amino acid introducing functional units. Such units may be used as modules in the reconstruction of altered peptide synthetase genes. Reconstruction of genes and reintroduction into defined nonproducers are available techniques for many filamentous fungi. The main result of cyclosporin research has been the emergence of new application fields besides the presently dominant transplantation areas. Divergent targets are obvious for many cyclic peptides and this should be a concern in new evaluations. Especially promising is the study of endogenous binding proteins and receptors. The binding of cyclosporins to cyclophilins and their complex to calcineurin may differ in the producer from nonproducers or yeasts. Inhibition of transcription of β-1,3-glucan synthase may be related to the antifungal spectrum of cyclosporin. Similar targets may thus operate in seemingly unrelated fields. However, producers might have altered targets and the selection of analogues for specific targets mimics the evolution of these versatile effectors. The question as to whether such an evolved

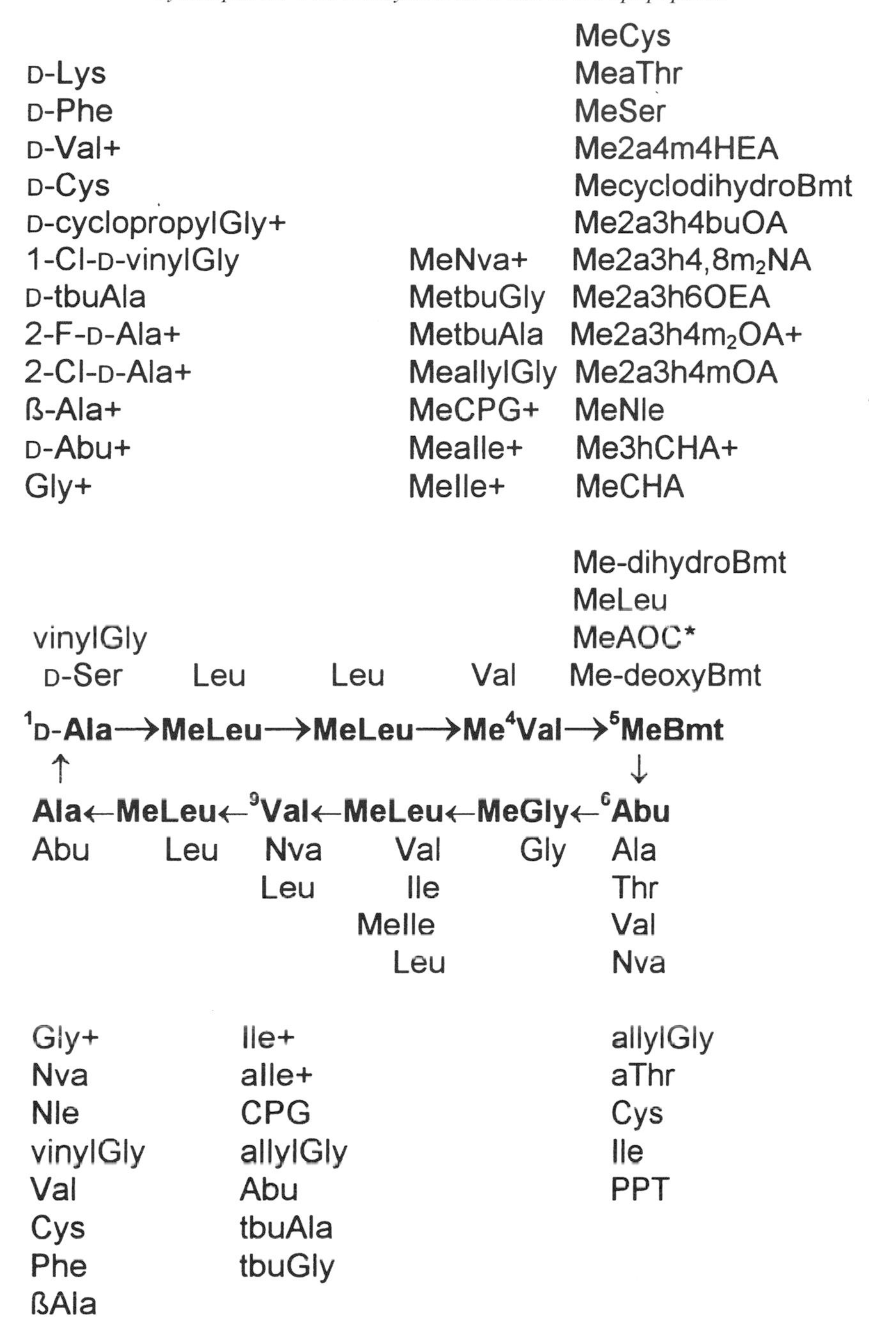

Figure 10 Cyclosporins synthesized *in vitro*. Changed positions are indicated, and generally single replacements have been reported, except for substitutions at positions 5 and 11 showing double replacement, or even triple replacement for 2, 5, and 11 in the case of, e.g., Nva or allylglycine. Compounds directly placed at the cyclosporin A structure have also been isolated from fermentations and were available as reference compounds. All other compounds have been described by chromatographic evidence or additional mass spectra (+). Abu, aminobutyrate; allylGly, allylglycine; 2a3h4buOH, 2-amino-3-hydroxy-4-butyloctanoic acid; 2a3h4,8m2NA, 2-amino-3-hydroxy-4,8-dimethylnonanoic acid; 2a3h6OEA, 2-amino-3-hydroxyoct-6-enoic acid; 2a3h4m2OA, 2-amino-3-hydroxy-4,4-dimethylocanoic acid; 2a4m4HEA, 2-amino-4-methylhex-4-enoic acid; AOC, aminooctanoic acid; CHA, cyclohexylalanine; 2-Cl-DAla, 2-chloro-D-alanine; CPG, cyclopropylglycine; cyclopropylGly, cyclopropylglycine; D, D-configuration; 2-F-DAla, 2-fluoro-D-alanine; 3hCHA, 3-hydroxycyclohexylalanine; Me, *N*-methyl; Nle, norleucine; PPT, phosphinothricine; tbuAla, *t*-butylalanine; tbuGly, *t*-butylglycine.

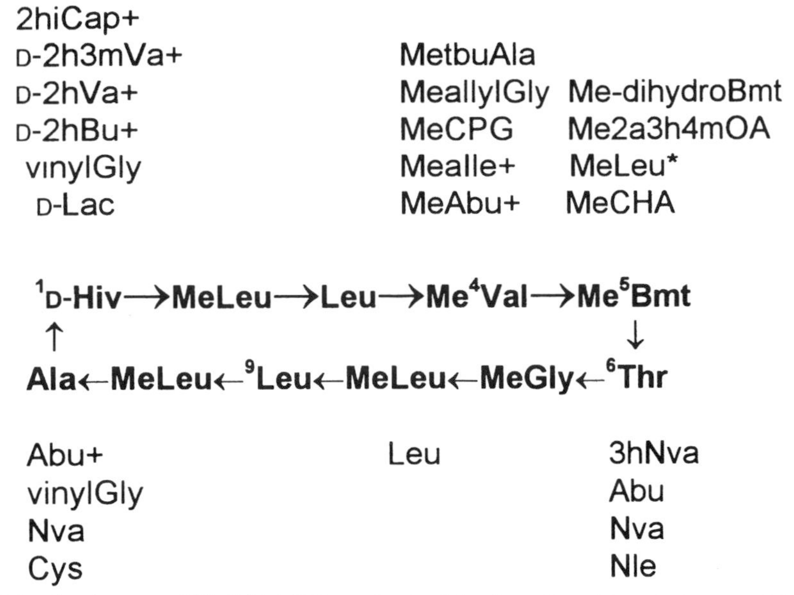

Figure 11 Analogues of SDZ 214-103 synthesized *in vitro*. For abbreviations see Figure 10.

structure like cyclosporin could be improved, or if this structure is already the most effective, should be considered in such a perspective. Obviously in different fungi, altered peptides of the cyclosporin type have evolved, and the intended targets have no relation to their formation. In fact, we can only speculate on the evolutionary pressure. Improvements can thus be envisaged.

1.20.7 REFERENCES

1. H. Kleinkauf and H. von Döhren, in "Fungal Biotechnology," ed. T. Anke, Chapman & Hall, Weinheim, 1997, p. 147.
2. J. Kallen, V. Mikol, V. F. J. Quesniaux, M. D. Walkinshaw, E. Schneider-Scherzer, K. Schörgendörfer, G. Weber, and H. G. Fliri, in "Products of Secondary Metabolism," eds. H. Kleinkauf and H. von Döhren, "Biotechnology," 2nd ed., eds. H. Rehm and L. G. Reed, Verlag Chemie, Weinheim, 1997, vol. VII, p. 535.
3. M. Moussaif, P. Jacques, P. Schaarwachter, H. Budzikiewicz, and P. Thonart, *Appl. Environ. Microbiol.*, 1997, **63**, 1739.
4. N. H. Georgopapdakou and J. S. Tkacz, *Trends Microbiol.*, 1995, **3**, 98.
5. J. F. Borel, in "Cyclosporin A," ed. D. J. G. White, Elsevier Biomedical Press, Amsterdam, 1982, p. 5.
6. K. Sakamoto, E. Tsuji, M. Miyauchi, T. Nakanishi, M. Yamashita, N. Shigematsu, T. Tada, S. Izumi, and M. Okuhara, *J. Antibiot.*, 1993, **46**, 1788.
7. IuO. Sazukin, T. V. Salova, and V. P. Ivanov, *Antibiot. Khimioter.*, 1997, **42**, 3.
8. H. Kleinkauf and H. von Döhren, *Eur. J. Biochem.*, 1996, **236**, 335.
9. M. M. Dreyfuss, E. Härn, H. Hofmann, H. Kobel, W. Pache, and T. Tscherter, *Eur. J. Appl. Microbiol.*, 1976, **3**, 125.
10. H. Nakajima, T. Hamasaki, K. Tanaka, Y. Kimura, S. Udagawa, and Y. Horie, *Agric. Biol. Chem.*, 1989, **53**, 2291.
11. J. F. Borel, *Prog. Allergy*, 1986, **38**, 9.
12. H. Stähelin, *Prog. Allergy*, 1986, **38**, 19.
13. J. F. Borel and H. Stähelin, *Sandorama*, 1983, **11**, 5.
14. M. M. Dreyfuss, *Sydowia*, 1986, **39**, 22.
15. J. Weiser and V. Matha, *J. Invertebr. Pathol.*, 1986, **51**, 92.
16. N. C. Bellavita, P. Ceccherelli, and M. Raffaele, *Eur. J. Biochem.*, 1970, **15**, 356.
17. J. Berdy, A. Aszalos, M. Bostian, and K. L. McNitt, in "Handbook of Antibiotic Compounds," eds. J. Berdy, A. Aszalos, M. Bostian, and K. L. McNitt, CRC Press, Boca Raton, FL, 1981, vol. VI, p. 93.
18. J. Polonsky, Z. Baskevitch, N. C. Bellavita, P. Ceccherelli, B. L. Buckwalter, and E. Wenkert, *J. Am. Chem. Soc.*, 1972, **94**, 4369.
19. W. B. Turner and D. C. Aldridge, "Fungal Metabolites II," Academic Press, London, 1983.
20. N. C. Bellavita, P. Ceccherelli, R. Fringuelli, and M. Raffaele, *Experientia*, 1975, **14**, 807.
21. M. Schirmbock, M. Lorito, Y. L. Wang, C. K. Hayes, I. Arison-Atac, F. Scala, G. E. Harman, and C. P. Kubicek, *Appl. Environ. Microbiol.*, 1994, **60**, 4364.
22. H. Nakajima, T. Hamasaki, K. Nishimura, T. Kondo, Y. Kimura, S. Udagawa, and S. Sato, *Agric. Biol. Chem.*, 1988, **52**, 1621.
23. U. Gräfe, W. Ihn, B. Schlegel, G. Höfle, H. Augustiniak, and P. Sandor, *Pharmazie*, 1991, **46**, 613.
24. J. Weiser and V. Matha, *J. Invertebr. Pathol.*, 1986, **51**, 94.
25. T. H. Aarnio and S. N. Agathos, *Biotechnol. Lett.*, 1989, **11**, 759.

26. J. M. Gardner and J. S. Pillai, *Mycopathologia*, 1987, **97**, 83.
27. M. Ravallec, A. Vey, and G. Riba, *J. Invertebr. Pathol.*, 1989, **53**, 7.
28. M. P. Nadeau and J. L. Boisvert, *J. Am. Mosq. Control Assoc.*, 1994, **10**, 487.
29. G. Barson, N. Renn, and A. F. Bywater, *J. Invertebr. Pathol.*, 1994, **64**, 107.
30. V. Matha, A. Jegorov, J. Weiser, and J. S. Pillai, *Cytobios*, 1992, **69**, 163.
31. V. Matha, J. Weiser, and J. Olejnicek, *Folia Parasitol. (Prague)*, 1988, **35**, 379.
32. W. H. Müller, T. P. van der Krift, A. J. Krouwer, H. A. Wosten, L. H. van der Voort, E. B. Smaal, and A. J. Verkleij, *EMBO J.*, 1991, **10**, 489.
33. T. Lendenfeld, D. Ghali, M. Wolschek, E. M. Kubicek-Pranz, and C. P. Kubicek, *J. Biol. Chem.*, 1993, **268**, 665.
34. H. von Döhren, U. Keller, J. Vater, and R. Zocher, *Chem. Rev.*, 1997, **97**, 2675.
35. E. Conti, T. Stachelhaus, and M. A. Marahiel, *EMBO J.*, 1997, **16**, 4174.
36. H. Husi, K. Schörgendörfer, G. Stempfer, P. Taylor, and M. D. Walkinshaw, *FEBS Lett.*, 1997, **414**, 532.
37. H. Kleinkauf and H. von Döhren, *Acta Biochim. Pol.*, 1997, **44**, 839.
38. H. Kleinkauf and H. von Döhren, *Antonie van Leeuwenhoek*, 1994, **67**, 229.
39. C. E. Isaac, A. Jones, and M. A. Pickard, *Antimicrob. Agents Chemother.*, 1990, **34**, 121.
40. G. F. Gauze, L. P. Terekhova, T. S. Maksimova, M. G. Brazhnikova, and G. B. Fedorova, *Antibiotiki*, 1983, **28**, 243.
41. T. H. Aarnio and S. N. Agathos, *Appl. Microbiol. Biotechnol.*, 1990, **33**, 435.
42. I. V. Sotnikova, G. N. Telesnina, I. N. Krakhmaleva, IuO. Sazykin, and S. M. Navashin, *Antibiot. Khimioter.*, 1991, **36**, 10.
43. I. V. Sotnikova, G. N. Telesnina, I. N. Krakhmaleva, IuO. Sazykin, and S. M. Navashin, *Antibiot. Khimioter.*, 1991, **35**, 9.
44. I. V. Sotnikova, G. N. Telesnina, R. A. Zviagil'skaia, L. P. Ivanitskaia, M. V. Bibikova, A. M. Rybakova, IuO. Sazykin, and S. M. Navashin, *Antibiot. Khimioter.*, 1990, **35**, 3.
45. D. X. Zhao, M. Bcran, J. Kozova, and Z. Rehacek, *Folia Microbiol. (Prague)*, 1991, **36**, 549.
46. S. N. Agathos, C. Madhosingh, J. W. Marshall, and J. Lee, *Ann. NY Acad. Sci.*, 1987, **506**, 657.
47. S. N. Agathos, J. W. Marshall, C. Moraiti, R. Parekh, and C. Madhosingh, *J. Ind. Microbiol.*, 1986, **1**, 39.
48. B. C. Foster, R. T. Coutts, F. M. Pasutto, and J. B. Dossetor, *Biotechnol. Lett.*, 1983, **5**, 693.
49. T. H. Lee, G. T. Chun, and Y. K. Chang, *Biotechnol. Prog.*, 1997, **13**, 546.
50. G. T. Chun and S. N. Agathos, *J. Biotechnol.*, 1993, **27**, 283.
51. P. L. Skatrud, T. Schwecke, H. van Liempt, and M. B. Tobin, in "Products of Secondary Metabolism," eds. H. Kleinkauf and H. von Döhren, "Biotechnology," 2nd ed., eds. H. Rehm, A. Pühler, P. Stadler, and L. G. Reed, Verlag Chemie, Weinheim, 1997, vol. VII, p. 247.
52. F. Fierro, J. L. Barredo, B. Diez, S. Guttierrez, F. J. Fernandez, and J. F. Martin, *Proc. Natl. Acad. Sci. USA*, 1995, **92**, 6200.
53. D. J. Smith, J. H. Bull, J. Edwards, and G. Turner, *Mol. Gen. Genet.*, 1989, **216**, 492.
54. R. Traber, H. Hofmann, and H. Kobel, *J. Antibiot.*, 1989, **42**, 591.
55. S. N. Agathos and R. Parekh, *J. Biotechnol.*, 1990, **13**, 73.
56. H. Kobel and R. Traber, *Eur. J. Appl. Microbiol. Biotechnol.*, 1986, **14**, 237.
57. R. Traber, H. Hofmann, and H. Kobel, *J. Antibiot.*, 1989, **42**, 591.
58. R. Traber, H. Hofmann, H.-R. Loosli, M. Ponelle, and A. von Wartburg, *Helv. Chim. Acta*, 1987, **70**, 13.
59. R. Traber, H. Kobel, H.-R. Loosli, H. Senn, B. Rosenwirth, and A. Lawen, *Antiviral Chem. Chemother.*, 1994, **5**, 331.
60. A. Billich, F. Hammerschnmid, P. Reichl, R. Wenger, G. Zenke, V. Quesniaux, and B. Rosenwirth, *J. Virol.*, 1995, **69**, 2451.
61. A. Lawen and R. Traber, *J. Biol. Chem.*, 1993, **268**, 20452.
62. H. Senn, C. Weber, H. Kobel, and R. Traber, *Eur. J. Biochem.*, 1991, **199**, 653.
63. S. N. Agathos and J. Lee, *Biotechnol. Prog.*, 1993, **9**, 54.
64. J. Lee and S. N. Agathos, *Biotechnol. Lett.*, 1991, **34**, 513.
65. R. Zocher, U. Keller, C. Lee, and K. Hoffmann, *J. Antibiot.*, 1992, **45**, 265.
66. A. Billich and R. Zocher, *J. Biol. Chem.*, 1987, **262**, 17258.
67. G. Weber, K. Schörgendörfer, E. Schneider-Scherzer, and E. Leitner, *Curr. Genet.*, 1994, **26**, 120.
68. Sandoz Ltd., *Eur. Pat. Appl.* 0578616, EMBL accession No. Z28383.
69. J. H. Ahn and J. D. Walton, *Plant Cell*, 1996, **8**, 887.
70. A. N. Nikolskaya, D. G. Panaccione, and J. D. Walton, *Gene*, 1995, **165**, 207.
71. A. M. Bailey, M. J. Kershaw, B. A. Hunt, I. C. Paterson, A. K. Charnley, S. E. Reynolds, and J. M. Clarkson, *Gene*, 1996, **173**, 195.
72. K. D. Siegmund, H. J. Plattner, and H. Diekmann, *Biochim. Biophys. Acta*, 1991, **1076**, 123.
73. C. Lee and A. Lawen, *Biochem. Mol. Biol. Int.*, 1993, **31**, 797.
74. H. Mohr and H. Kleinkauf, *Biochim. Biophys. Acta*, 1978, **526**, 375.
75. P. Lavermicocca, N. Sante Iacobellis, M. Simmaco, and A. Graniti, *Physiol. Mol. Plant Pathol.*, 1997, **50**, 129.
76. T. Hamada, T. Sugawaka, S. Masunagi, and N. Fusetani, *Tetrahedron Lett.*, 1994, **35**, 719.
77. R. Handschumacher, M. Harding, J. Rice, and R. Drugge, *Science*, 1986, **226**, 544.
78. M. Harding, R. Handschumacher, and D. Speicher, *J. Biol. Chem.*, 1986, **261**, 8547.
79. N. Takahashi, T. Hayano, and M. Suzuki, *Nature*, 1989, **337**, 473.
80. G. Fischer, B. Wittmann-Liebold, U. Lang, T. Kiefhaber, and F. X. Schmid, *Nature*, 1989, **337**, 476.
81. H. R. Loosli, H. Kessler, H. Oschkinat, H. Weber, and T. Petcher, *Helv. Chim. Acta*, 1985, **68**, 682.
82. C. Weber, G. Wider, B. von Freyberg, R. Traber, W. Braun, H. Widmer, and K. Wüthrich, *Biochemistry*, 1991, **30**, 6563.
83. P. Neri, G. Gemmecker, L. D. Zydowsky, C. T. Walsh, and S. W. Fesik, *FEBS Lett.*, 1991, **290**, 195.
84. J. Friedman and I. Weismann, *Cell*, 1991, **66**, 799.
85. J. Liu, J. D. Farmer, U. S. Lane, J. Friedman, I. Weismann, and S. L. Schreiber, *Cell*, 1991, **66**, 807.
86. D. A. Fruman, P. E. Mather, S. J. Burakoff, and B. E. Bierer, *Eur. J. Immunol.*, 1992, **22**, 2513.

87. A. Ruhlmann and A. Nordheim, *Immunobiology*, 1997, **198**, 192.
88. A. R. Marcks, *Physiol. Rev.*, 1996, **76**, 631.
89. P. Taylor, H. Husi, G. Kontopidis, and M. D. Walkinshaw, *Prog. Biophys. Mol. Biol.*, 1997, **67**, 155.
90. T. Hunter, *Cell*, 1998, **92**, 141.
91. C. M. Douglas, F. Foor, J. A. Marrinan, N. Morin, J. B. Nielsen, A. M. Dahl, P. Mazur, W. Baginsky, W. Li, M. el-Sherbeini *et al.*, Proc. Natl. Acad. Sci. USA, 1994, **91**, 12 907.
92. P. Mazur, N. Morin, W. Baginsky, M. el-Sherbeini, J. A. Clemas, J. B. Nielsen, and F. Foor, *Mol. Cell Biol.*, 1995, **15**, 5671.
93. J. Kunz and M. N. Hall, *Trends Biochem. Sci.*, 1993, **18**, 334.
94. D. Zhu, M. E. Cardenas, and J. Heitman, *Mol. Pharmacol.*, 1996, **50**, 506.
95. G. Weber and E. Leitner, *Curr. Genet.*, 1994, **26**, 461.
96. T. Hornbogen and R. Zocher, *Biochem. Mol. Biol. Int.*, 1995, **36**, 169.
97. C. Staben, B. Jensen, M. Singer, J. Pollock, M. Schechtman, J. Kinsey, and E. Selker, *Fungal Genet. Newsletter*, 1989, **36**, 79.
98. J. J. Sanglier, R. Traber, R. H. Buck, H. Hofmann, and H. Kobel, *J. Antibiot.*, 1990, **43**, 707.
99. J. Dittmann, A. Lawen, R. Zocher, and H. Kleinkauf, *Biol. Chem. Hoppe-Seyler*, 1990, **371**, 829.
100. J. Dittmann, R. M. Wenger, H. Kleinkauf, and R. Zocher, *J. Biol. Chem.*, 1994, **269**, 2841.
101. T. Stachelhaus, A. Schneider, and M. A. Marahiel, *Science*, 1995, **269**, 69.
102. A. Schneider, T. Stachelhaus, and M. A. Marahiel, *Mol. Gen. Genet.*, 1998, **257**, 308.
103. M. S. Rakotonirainy, M. Dutewrtre, Y. Brygoo, and G. Riba, *J. Invertebr. Pathol.*, 1991, **57**, 17.
104. F. Kempken, C. Schreiner, K. Schörgendörfer, and U. Kück, *Exp. Mycol.*, 1995, **19**, 305.
105. F. Kempken and U. Kück, *Mol. Cell Biol.*, 1996, **16**, 6563.
106. N. Sadeg, C. Pham-Huy, P. Rucay, S. Righezi, O. Halle-Pannenko, J.-R. Claude, H. Bismuth, and H.-T. Duc, *Immunopharm. Immunotoxicol.*, 1993, **15**, 163.
107. H. Kobel, H.-R. Loosli, and R. Voges, *Experientia*, 1983, **39**, 873.
108. R. Zocher, N. Madry, H. Peeters, and H. Kleinkauf, *Phytochemistry*, 1984, **23**, 549.
109. M. Offenzeller, Z. Su, G. Santer, H. Moser, R. Traber, K. Memmert, and E. Schneider-Scherzer, *J. Biol. Chem.*, 1993, **268**, 26 167.
110. M. Offenzeller, G. Santer, K. Totschnig, Z. Su, H. Moser, R. Traber, and E. Schneider-Scherzer, *Biochemistry*, 1996, **35**, 8401.
111. R. Zocher, T. Nihira, E. Paul, N. Madry, H. Peeters, and H. Kleinkauf, *Biochemistry*, 1986, **25**, 550.
112. K. Hoffmann, E. Schneider-Scherzer, H. Kleinkauf, and R. Zocher, *J. Biol. Chem.*, 1994, **269**, 12 710.
113. H. P. Kocher, E. Schneider-Scherzer, K. Schörgendörfer, and G. Weber, *Int. Pat. Publ.* 1994, No. WO94/25606.
114. J. S. Scott-Craig, D. G. Panaccione, J. A. Pocard, and J. D. Walton, *J. Biol. Chem.*, 1992, **267**, 26 044.
115. Y. Okumura, in "Biochemistry of Peptide Antibiotics," eds. H. Kleinkauf and H. von Döhren, de Gruyter, Berlin, 1990, p. 365.
116. V. de Crécy-Lagard, P. Marlière, and W. Saurin, *C.R. Acad. Sci. Paris*, 1995, **318**, 927.
117. G. Kreil, *Annu. Rev. Biochem.*, 1997, **66**, 337.
118. S. D. Heck, W. S. Faraci, P. R. Kelbaugh, N. A. Saccomano, P. F. Tadeio, and R. A. Volkmann, *Proc. Natl. Acad. Sci. USA*, 1995, **93**, 4036.
119. Y. Shikata, T. Watanabe, T. Teramoto, A. Inoue, Y. Kawakami, Y. Nishizawa, K. Katayama, and M. Kuwada, *J. Biol. Chem.*, 1995, **270**, 16 719.
120. T. Stachelhaus, H. D. Mootz, V. Bergendahl, and M. A. Marahiel, *J. Biol. Chem.*, 1998, **273**, 22 773.
121. A. Lawen and R. Zocher, *J. Biol. Chem.*, 1990, **265**, 11 355.
122. Y. Aharonowitz, H. Bergmeyer, J. M. Cantoral, G. Cohen, A. L. Demain, U. Fink, J. Kinghorn, H. Kleinkauf, A. MacCabe, H. Palissa, E. Pfeifer, T. Schwecke, H. van Liempt, H. von Döhren, S. Wolfe, and J. Zhang, *Biotechnology*, 1993, **11**, 807.
123. K. Turgay, M. Krause, and M. A. Marahiel, *Mol. Microbiol.*, 1992, **6**, 529.
124. F. Saito, K. Hori, M. Kanda, T. Kurotsu, and Y. Saito, *J. Biochem.*, 1994, **116**, 357.
125. B. Schmidt, D. Riesner, A. Lawen, and H. Kleinkauf, *FEBS Lett.*, 1992, **307**, 355.
126. A. Haese, M. Schubert, M. Herrmann, and R. Zocher, *Mol. Microbiol.*, 1993, **7**, 905.
127. M. A. Marahiel, T. Stachelhaus, and H. Mootz, *Chem. Rev.*, 1997, **97**, 2651.
128. R. Dieckmann, Y.-O. Lee, H. van Liempt, H. von Döhren, and H. Kleinkauf, *FEBS Lett.*, 1995, **357**, 212.
129. E. Pfeifer, M. Pavela-Vrancic, H. von Döhren, and H. Kleinkauf, *Biochemistry*, 1995, **34**, 7450.
130. T. Stein and J. Vater, *Amino Acids*, 1996, **10**, 201.
131. C. T. Walsh, A. M. Gehring, P. H. Weinreb, L. E. N. Quadri, and R. S. Flugel, *Curr. Opin. Chem. Biol.*, 1997, **1**, 309.
132. L. E. N. Quadri, P. H. Weinreb, M. Lei, M. M. Nakano, P. Zuber, and C. T. Walsh, *Biochemistry*, 1998, **37**, 1585.
133. M. Glinski, M. Walther, and R. Zocher, *Abstracts Symp. Enzymology of Biosynthesis of Natural Products*, Berlin Technical University, 1996, abstract 56.
134. R. Traber, H.-R. Loosli, H. Hofmann, M. Kuhn, and A. von Wartburg, *Helv. Chim. Acta*, 1982, **65**, 1655.
135. K. Sakamoto, E. Tsuji, and M. Miyauchi, *J. Antibiot.*, 1993, **46**, 1788.
136. R. Wenger, *Transplant. Proc.*, 1986, **18**, 213.
137. A. von Wartburg and R. Traber, *Prog. Allergy*, 1986, **38**, 28.
138. H. G. Fliri and R. Wenger, in "Peptide Antibiotics," eds. H. Kleinkauf and H. von Döhren, de Gruyter, Berlin, 1990, p. 246.
139. V. Havlícek, A. Jegorov, P. Sedmera, and M. Ryska, *Org. Mass Spectrom.*, 1993, **28**, 1440.
140. V. Havlícek, A. Jegorov, P. Sedmera, W. Wagner-Redeker, and M. Ryska, *Org. Mass Spectrom.*, 1995, **30**, 940.
141. A. Jevorov, V. Matha, P. Sedmera, V. Havlicek, J. Stuchlik, P. Seidl, and P. Simek, *Phytochemistry*, 1994, **38**, 403.
142. P. Sedmera, V. Havlicek, A. Jegorov, and A. L. Segre, *Tetrahedron Lett.*, 1995, **36**, 6953.
143. R. Pieper, H. Kleinkauf, and R. Zocher, *J. Antibiot.*, 1992, **45**, 1273.
144. P. C. Hiestand, H. C. Gunn, J. M. Gale, B. Ryffel, and J. F. Borel, *Immunology*, 1985, **55**, 249.
145. A. Lawen, R. Traber, D. Geyl, and R. Zocher, *J. Antibiot.*, 1989, **42**, 1283.

146. A. Lawen, R. Traber, R. Reuille, and M. Ponelle, *Biochem. J.*, 1994, **300**, 395.
147. H. Kleinkauf and H. von Döhren, in "Trends in Antibiotic Research," Japan Antibiotics Research Association, Tokyo, 1982, p. 220.
148. A. Lawen, R. Traber, and D. Geyl, *J. Biol. Chem.*, 1991, **266**, 15 567.
149. S. R. Bartz, E. Hohenwalter, M.-K. Hu, D. H. Rich, and M. Malkovsky, *Proc. Natl. Acad. Sci. USA*, 1995, **92**, 5381.
150. A. Vahlne, P.-A. Larsson, J. Ahlmén, B. Svennerholm, J. S. Gronowitz, and S. Olofsson, *J. Arch. Virol.*, 1992, **122**, 61.
151. H. Zachariae and T. S. Olsen, *Clin. Nephrol.*, 1995, **43**, 154.
152. K. Balakrishnan and A. Pandey, *Folia Microbiol. (Prague)*, 1996, **41**, 401.

1.21
Biosynthesis of Enediyne Antibiotics

SHIGO IWASAKI
The University of Tokyo, Japan

1.21.1 NEW CLASS OF ANTIBIOTICS

The enediyne antibiotics are extremely potent antitumor agents with a unique molecular structure. This new class of antibiotics, having the interesting bicyclo[7.3.1]enediyne substructure, includes calicheamycin (CAL; (**1**)),[1,2] esperamycin (ESP; (**2**)),[3,4] and dynemycin A (DNM-A; (**3**)).[5,6] The chromophores of neocarzinostatin (NCS; (**4**))[7] and related compounds,[8] such as maduropeptin (**5**), C-1027 (**6**), and kedarcidin (**7**), having bicyclo[7.3.0]enediyne cores, have also been classified in this family. At room temperature in the presence of DNA, the core system of an enediyne antibiotic undergoes a remarkable reaction yielding diradicals on sp^2 carbon that cause DNA strand breakage. During studies of the structures of calicheamycin γ^1 and esperamycin A_1, it became apparent that enediynes could be triggered to aromatize via cleavage of the trisulfide with formation of a diradical intermediate as the biologically active species. From this observation it was also speculated that neocarzinostatin chromophore A might undergo a similar aromatization process,[9] since it had been known that the neocartinostatin DNA-damaging mechanism involved free radicals when activated by thiol addition.

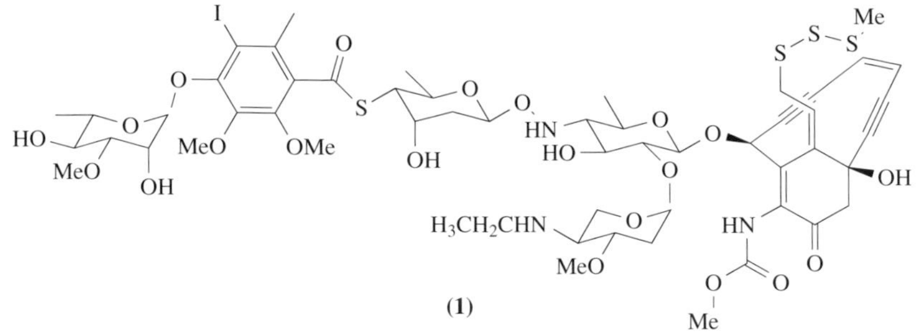

(1)

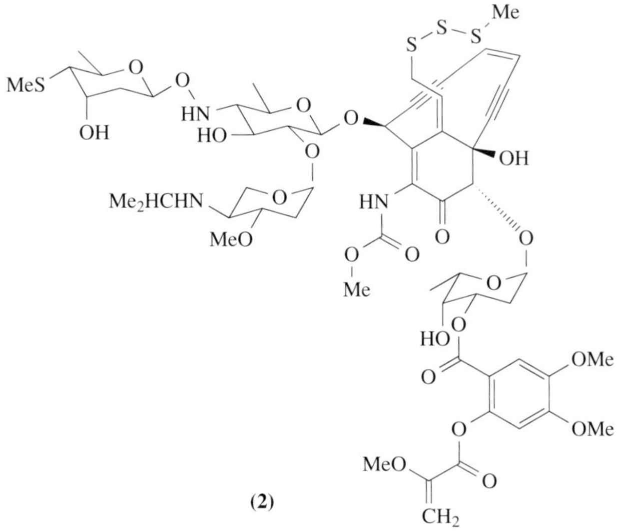

(2)

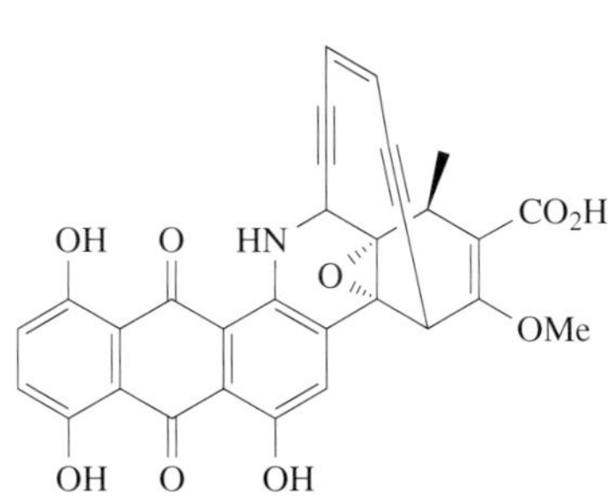

(3)

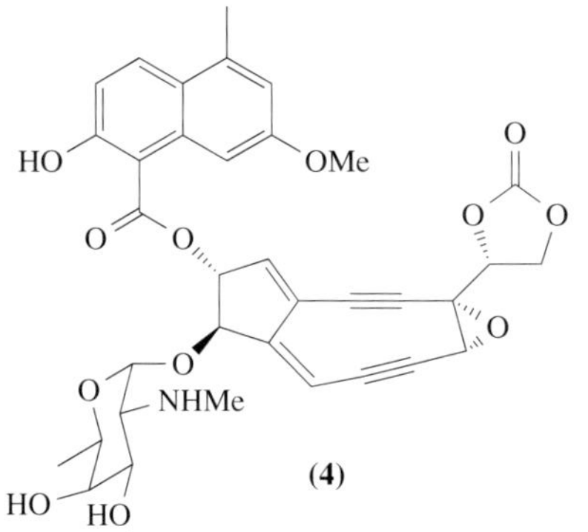

(4)

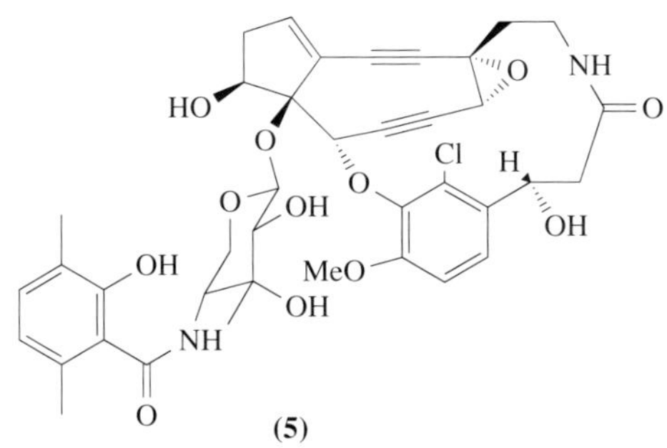

(5)

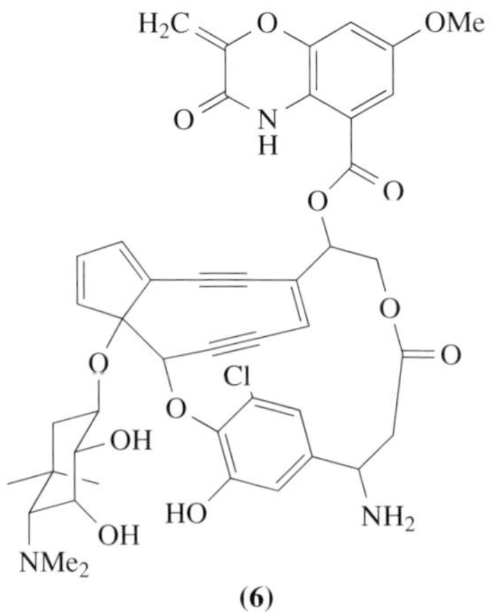

(6)

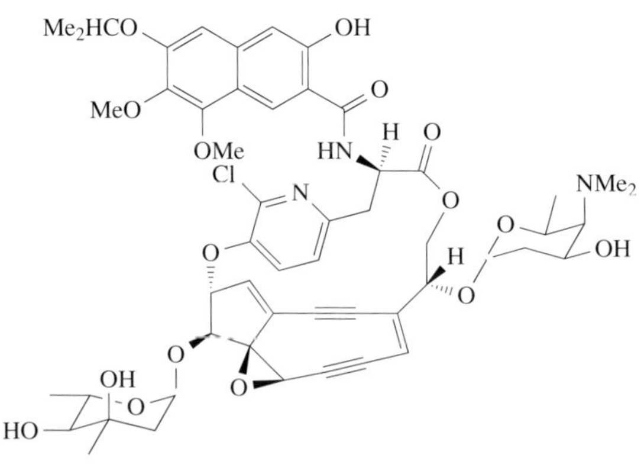

(7)

A few years after the structural papers on calicheamycin and esperamycin appeared, a new enediyne-type antibiotic, dynemycin A, was reported. This new, extremely potent antitumor antibiotic had an enediyne unit associated with a hydroxyanthraquinone chromophore. Remarkable DNA scission by dynemycin A was induced in the presence of reducing agents[10] and the reaction mechanism involving an arene diradical intermediate triggered by reduction of the hydroxyanthraquinone moiety was chemically proved.[11]

Not much work has been done on the biosynthesis of the enediyne class of compounds. There are only two published reports in this area: on the biosynthesis of the neocarzinostatin chromophore by Hensens *et al.*[12] and of dynemycin A by Iwasaki's group.[13] The study on the biosynthesis of ESP A_1 (**2**) is described in a monograph by Lam and Veitch.[14]

1.21.2 BIOSYNTHESIS OF DYNEMYCIN A

1.21.2.1 Introduction

Dynemycin A (DNM-A; (**3**)) is a potent antibacterial and antitumor antibiotic isolated from *Micromonospora chersina* M956-1 strain[5,6] and has a striking hybrid structure combining the characteristics of both the enediyne and anthracycline classes of antibiotics (Figure 1). The compound is the third member of a series of antibiotics that have a bicyclo[7.3.1]enediyne substructure, which may be related biosynthetically to the cores of esperamycin and calicheamycin. The neocarzinostatin chromophore A, having a bicyclo[7.3.0]dienediyne core has also been classified in this family, by analogy of the structures and of the mode of action. A study on the biosynthesis of DNM-A (**3**) was carried out by Iwasaki's group,[13] and this section deals with the full assignment of the ^{13}C NMR signals of (**3**) and the incorporation of various ^{13}C- and ^{15}N-labeled precursors into (**3**). A possible biosynthetic scheme of (**3**) is proposed.

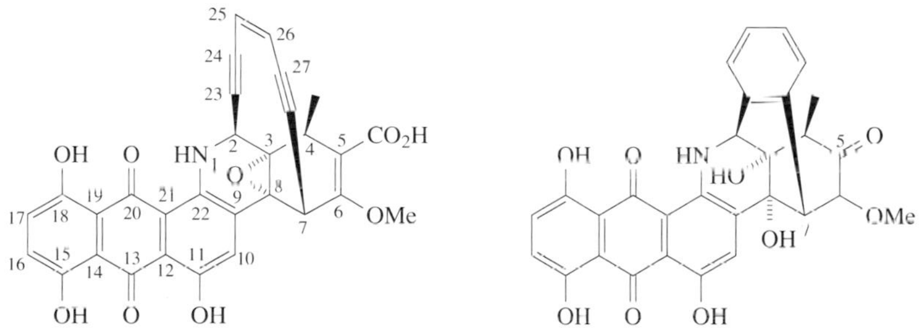

Dynemicin A (**3**) Dynemicin M (**8**)

Figure 1 Structure of dynemycin A and M with numbering.

1.21.2.2 ^{13}C NMR Assignments

^{13}C NMR signals of DNM-A (**3**) measured in DMSO-d_6 had previously been partially assigned,[5] though many of the assignments were regarded as interchangeable. Unambiguous assignments of the ^{13}C NMR signals of (**3**) were first established by combining ^{1}H-^{13}C COZY, INEPT, and heteronuclear multiple-bond [^{1}H—^{13}C] correlation spectroscopy (HMBC), and also by analyzing the incorporation patterns of [1-^{13}C]-, [2-^{13}C]-, and [1,2-$^{13}C_2$] acetates into (**3**). Although the assignments of each of the four pairs of signals due to C-13 and C-20, C-14 and C-19, C-15 and C-18, and C-16 and C-17 were interchangeable on the basis of the NMR techniques, the incorporation patterns of labeled acetates allowed to differentiate these signals. The ^{13}C NMR data and assignments thus obtained are summarized in Table 1.

Table 1 ^{13}C NMR assignments of [1-^{13}C]- and [2-^{13}C]acetate-labeled DNM-A and $^{1}J_{cc}$ of [1,2-$^{13}C_2$]acetate-labeled DNM-A (in DMSO-d_6).

Carbon	δ^a (ppm)	$^{1}J_{CH}$ (Hz)	$^{1}J_{cc}$ (Hz)
2	43.9 d (1)	154	68.4[b]
3	70.2 s (2)		31.4[c]
4-Me$_3$	18.5 q (2)	129	33.2
4	35.5 d (1)	131	33.2
5-CO$_2$H	167.4 s (2)		
5	114.6 s (2)		86.0
6	153.3 s (1)		86.0
6-OMe	57.8 q (M, G)	146	
7	31.4 d (2)	143	66.6[d]
8	63.2 s (1)		31.4[c]
9	135.2 s (1)		65.7
10	127.4 d (2)	163	65.7
11	156.2 s (1)		64.7
12	113.1 s (2)		64.7
13	189.0 s (1)		57.4
14	113.1 s (2)		57.4
15	156.0 s (1)		64.8
16	127.0 d (2)	166	64.8
17	129.1 d (1)	165	64.7
18	155.7 s (2)		64.7
19	113.8 s (1)		55.5
20	186.6 s (2)		55.5
21	110.8 s (1)		66.5
22	142.8 s (2)		66.5
23	98.0 s (2)		68.4[b]
24	89.6 s (1)		88.7
25	124.4 d (2)	176	88.7
26	124.2 d (1)	176	88.7
27	88.9 s (2)		88.8
28	99.4 s (1)		66.6[d]

[a]CH coupling multiplicity: s, singlet; d, doublet; q, quartet. In parentheses, ^{13}C enrichment from: 1, [1-^{13}C]acetate; 2, [2-^{13}C]acetate; G, [2-^{13}C]glycin; M, [methyl-^{13}C]methionine. [b-d]Mutual $^{1}J_{cc}$ coupling was observed.

1.21.2.3 Incorporation of ^{13}C- and ^{15}N-Labeled Precursors

In order to establish the origins of carbon atoms in (**3**), incorporation experiments with a variety of ^{13}C- and ^{15}N-labeled precursors were carried out with shaken cultures of *M. chersina* M956-1. Addition of Diaion HP-20, a nonionic highly porous resin, to the culture medium significantly increased the yield of (**3**) to 10 mg L^{-1}. The resin was considered to adsorb excreted (**3**), greatly reducing the contact of (**3**) with the producing organism, which is itself sensitive to this antibiotic.

The ^{13}C-labeling patterns after incorporation of [1-^{13}C]-, [2-^{13}C]-, and [1,2-$^{13}C_2$]acetates, L-[methyl-^{13}C]methionine and [2-^{13}C, ^{15}N]glycine were determined from the respective ^{13}C NMR spectra and are shown in Figure 2. The incorporation pattern of [1,2-$^{13}C_2$]acetate was confirmed by matching of $^{1}J_{cc}$ values, as shown in Table 1. The carboxyl group attached to C-5 was found to be derived from C-2 of an acetate unit. Addition of [1,2,3-^{13}C]malonate to the culture of *M. chersina* induced overall ^{13}C-enrichment in (**3**), but no ^{13}C—^{13}C coupling was observed between C-5 and the carboxyl carbon, indicating that intact incorporation of a C_3 malonate unit at this point does not occur. Propionate and succinate were not specifically incorporated into (**3**). Feeding of [^{15}N,2-^{13}C]glycine enhanced the intensity only of the *O*-methyl signal, which should be caused by one-carbon transfer from glycine to methionine, and neither ^{13}C-enrichment at C-22 nor N-1—C-22 coupling was observed. This indicates that glycine is not the precursor of the C-21—C-22—N unit. But, instead, [^{15}N]ammonium sulfate was incorporated into N-1, as verified by observation of N-1—H-1 coupling.

The incorporation results summarized in Figure 2 demonstrated a remarkable feature in the labeling pattern at the two sets of vicinal carbons: both C-5 and the carboxyl carbon are derived from C-2 of acetate and both C-8 and C-9 are derived from C-1 of acetate. The result indicated that the polyketide sequence of (**3**) should be biogenetically disconnected at these bonds.

The experiment with doubly labeled [1,2-$^{13}C_2$]acetate to establish the carbon–carbon connectivities strongly suggested that the respective C_{14} bicyclo[7.3.1]enediyne and anthraquinone moieties are

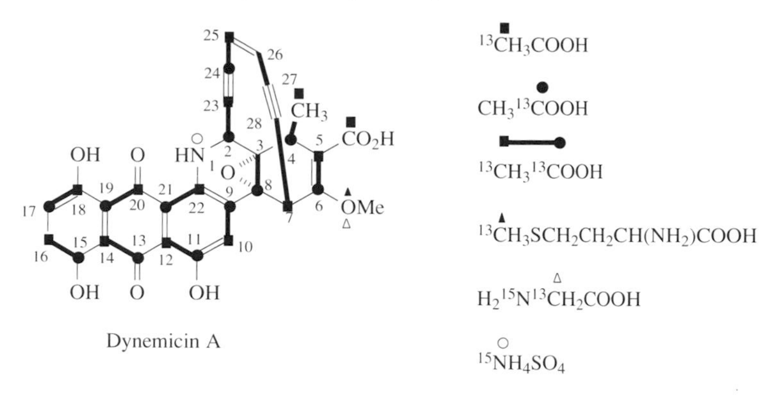

Figure 2 Incorporation patterns of ^{13}C- and ^{15}N-labeled precursors.

biosynthesized separately as two different heptaketide chains derived from seven head-to-tail coupled acetate units and should be connected at a later stage.

From inspection of the established carbon–carbon connectivities in (**3**), a possible precursor of the bicyclo[7.3.1]enediyne substructure is assumed to be a C_{14} chain, such as (**11**) or (**12**) (Scheme 1), or their biogenetic analogue, which can be connected with the anthraquinone moiety through a Friedel–Crafts type of acylation reaction to form the C-8—C-9 bond and can also cyclize to the bicyclo structure through two condensation reactions to form the C-3—C-4 and C-7—C-8 bonds. Introduction of an additional acetate unit at C-5 followed by oxidative degradation to form the carboxyl group and *O*-methylation should occur at some point. Biosynthesis of anthraquinones from heptaketide precursors is unknown. Though a heptacarbonyl acid (**9**) is tentatively drawn as a possible precursor, the location of the starter acetate unit in the anthraquinone moiety is not defined. Moreover, an enediyne chain such as (**11**) is also conceivable as a precursor of the anthraquinone moiety (as depicted in Scheme 1).

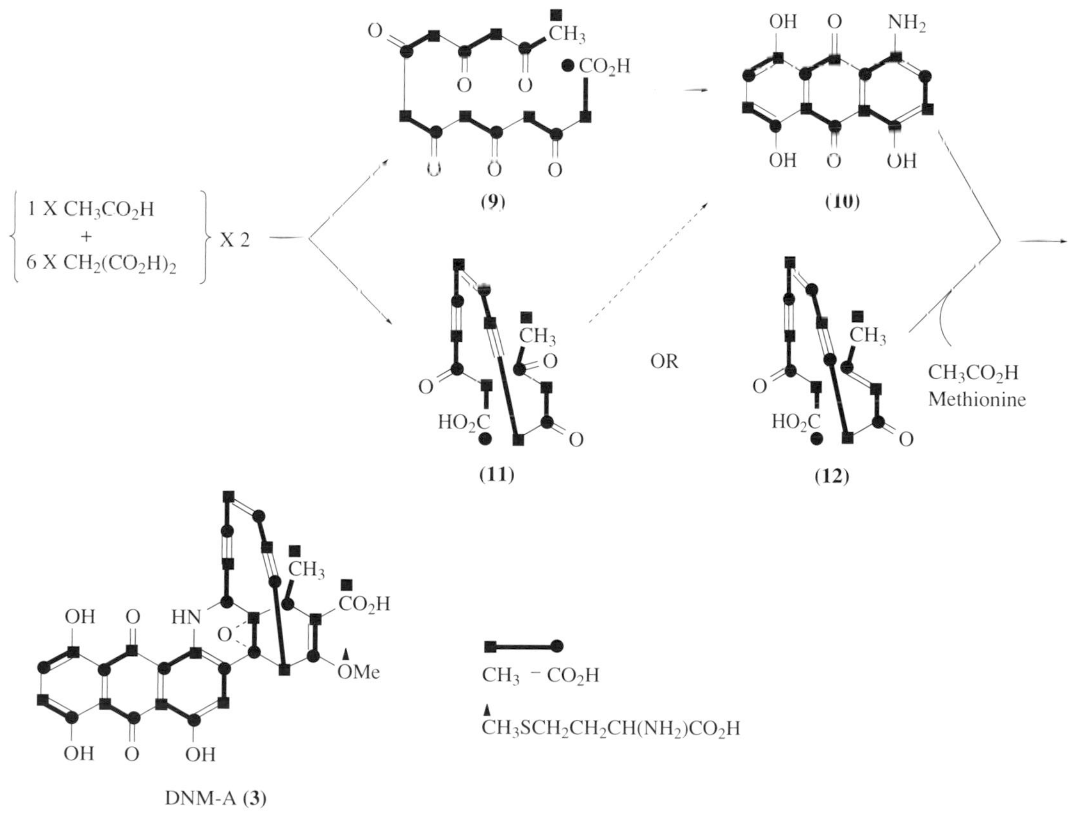

Scheme 1

Tokiwa *et al.*[13] suggested that, in the light of the above data on DNM-A (**3**), the biosynthesis of the C_{15} enediyne skeleton of the ESP/CAL class of antibiotics may be explained similarly to that of DNM-A (shown in Scheme 2), with an octaketide (**13**) or its biogenetic analogue as the common precursor. A formal loss of a two-carbon unit from the carboxylate end (path b) leads to (**11**), and loss of one carbon from either end (path a or c) results in the ESP/CAL bicyclic core, (**14a**), (**14b**), (**14c**), or (**14d**). The four possible pathways leading to (**14a–d**) can be easily differentiated by analyzing incorporation patterns of 1,2-^{13}C doubly labeled acetate into the ESP/CAL core. As discussed below, the Bristol-Myers Squibb group, indeed, proposed that path c leading to (**14d**) should be the scheme for the biosynthesis of esperamycin A_1. The precursors such as (**11**) and (**13**) could be derived from the oleate-crepenynate pathway, as discussed for NCS Chrom A biosynthesis and further for the biosynthesis of the C_{15} enediyne core of ESP/CAL. This pathway, however, has so far been shown to operate only in higher plants and fungi.

The stereochemistry of (**11**) and (**14**) in Scheme 2 accords with the established absolute stereochemistry of ESP/CAL cores and with the conventional structure of DNM-A. A report on a model study for the dynemycin-DNA complex predicted that this enantiomer (2*S*,7*R*) of DNM-A should be the correct absolute stereochemical structure.[15] Several compounds such as dynemycins M[6] (**8**), O, P, and Q,[16] which have an oxo group at C-5 (type 2) in place of the carboxyl group in DNM-A (type 1), were isolated from the culture broth of *M. chersina* M956-1. The biogenetic relationship between these compounds is intriguing with respect to the functional group at C-5. Type 2 compounds may be formed from type 1 compounds by oxidative decarboxylation, but it is also possible that the latter compounds are directly derived from some heptaketide precursor(s) without introduction of an extra acetate unit.

1.21.2.4 Summary

In summary, the bicyclic enediyne skeleton and anthraquinone moiety were shown to be biosynthesized from two separate heptaketides consisting of seven head-to-tail coupled acetate units. Inorganic nitrogen ($^{15}NH_4SO_4$) was incorporated into N-1. A scheme was proposed for the biosynthesis of the bicyclic core in DNM-A via a heptaketide precursor such as (**5**) or (**6**); such a scheme can be extended to the biosynthesis of the ESP/CAL enediyne core, as illustrated in Scheme 2. The heptaketide (**9**) is tentatively suggested as a possible precursor of the anthraquinone moiety, although a C_{14} enediyne chain such as (**11**) could also be the precursor. In this regard, it would be informative to determine the starter acetate unit of the heptaketide composing the anthraquinone moiety, which has not been accomplished.

1.21.3 BIOSYNTHESIS OF ESPERAMYCIN A_1

1.21.3.1 Introduction

Esperamycin A_1 (ESP A_1; (**2**)), one of the most potent antitumor antibiotics, was isolated from cultures of *Actinomadura verrucosospora*.[3] The producing organism was isolated from a soil sample collected at Pto Esperanza, Misiones, Argentina. The isolation and the elucidation of the structure of ESP A_1 and its related compounds have been reported (Figure 3).[3,4] The absolute configurations of each sugar and of the bicyclic core have also been established. ESP A_1 (**2**) consists of a bicyclic core to which are attached a trisaccharide and a substituted 2-deoxy-L-fucose, with an aromatic chromophore attached to the sugar 3 position. The individual sugars of the trisaccharide are novel sugars and contain an unusual hydroxylamino sugar linked to a thiomethyl sugar via an *O*-glycosidic linkage at the 4 position. The hydroxylamino sugar is further attached to an isopropyl sugar at the 2 position. The bicyclic core contains the very unusual enediyne, an allylic trisulfide, and a bridgehead enone.

The results described here on the biosynthesis of ESP A_1 are recorded in a monograph by Lam and Veitch of a Bristol-Myers Squibb research group.[14]

1.21.3.2 Radioactive Precursor Labeling Study

The major problem in elucidating the biosynthesis of ESP A_1 is due to its low production in fermentation. The estimated yield of ESP A_1 from the original strain SA-24868 was about

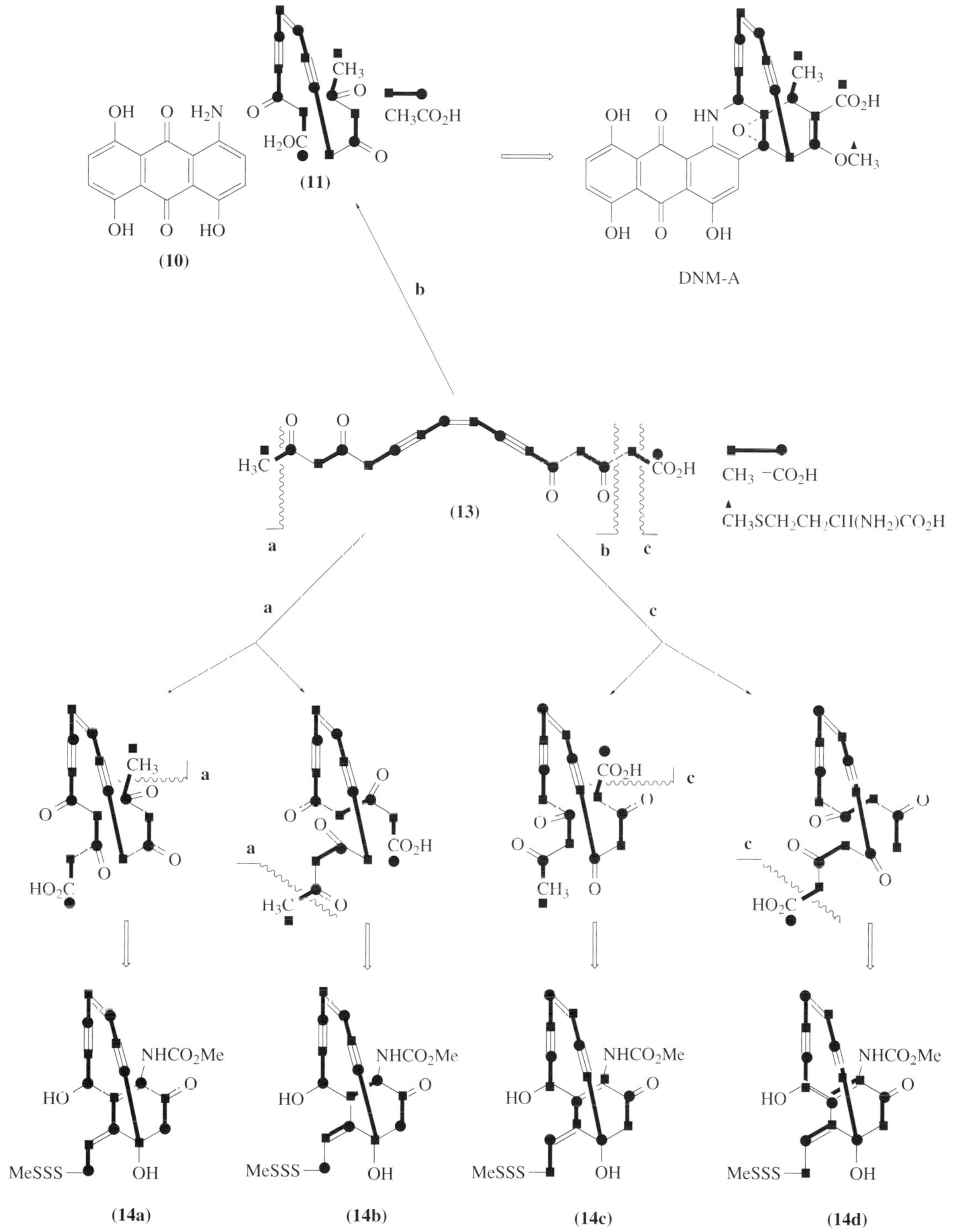

Scheme 2

0.05 μg ml^{-1}. Through extensive media development and strain-improvement studies, the yield was increased to 5 μg ml^{-1}. Table 2 summarizes the results obtained from the incorporation studies with ^{14}C-labeled and ^{35}S-labeled radioactive precursors.

The incorporation values of the radiolabeled precursors into ESP A_1 were 0.0037 to 0.11% (Table 2), which appeared to be very small. However, the low incorporation rates were due to the low yield of ESP A_1, and the specific activities calculated for ESP A_1 ranged from 16 200 to

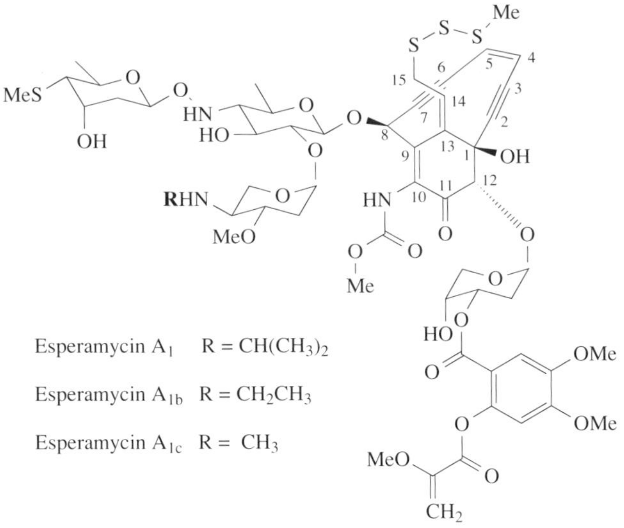

Esperamycin A_1 R = $CH(CH_3)_2$

Esperamycin A_{1b} R = CH_2CH_3

Esperamycin A_{1c} R = CH_3

Figure 3 Structures of esperamycins A_1, A_{1b}, and A_{1c}.

Table 2 Incorporation of radiolabeled precursors into esperamycin A_1. Radiolabeled precursors were added to the culture at day 3 of the fermentation.

Radiolabeled precursor	*Amount of radioactivity added* (dpm)	*Radioactivity incorporated into esperamycin* (dpm)	*% Incorporation*
Sodium [1-^{14}C]acetate	2.22×10^8	8 100	0.0037
Sodium [1-^{14}C]pyruvate	5.55×10^7	19 300	0.035
L-[Methyl-^{14}C]methionine	2.22×10^8	236 900	0.11
D-[U-^{14}C]Glucose	2.22×10^8	9 200	0.0041
L-[^{35}S]Methionine	2.22×10^8	210 400	0.095
L-[^{35}S]Cysteine	2.22×10^8	41 000	0.019
Sodium [^{35}S]sulfate	2.22×10^8	74 600	0.034

473 800 dpm mg^{-1}, which are acceptable incorporation rates for the biosynthetic precursors, considering the radioactivity (0.25–1 $\mu Ci\ ml^{-1}$) fed into the cultures. The result, therefore, suggested that these radiolabeled compounds were the possible precursors in ESP A_1 biosynthesis. To determine the incorporation sites of these precursors in ESP A_1 molecule, precursors labeled with stable isotopes were used in the feeding experiments.

1.21.3.3 Biosynthesis of the Enediyne Core

The incorporation rate of sodium [1-^{14}C]acetate into ESP A_1 was very low (0.0037%). The result did not decisively indicate that acetate is the precursor of ESP A_1, and, moreover, difficulty was experienced in obtaining a ^{13}C-labeled sample for NMR analysis. Therefore, Lam *et al.*[14] chose, at first, a method to determine the effect of cerulenin on the ESP A_1 production. Cerulenin is a fungal antibiotic that inhibits the condensing enzyme involved in fatty acid and polyketide biosynthesis.[17] If the biosynthesis of ESP A_1 is derived from head-to-tail condensation of acetate units, i.e., via the polyketide pathway, cerulenin should inhibit its production. At all concentrations tested, cerulenin did not affect the growth of the organism or the pH of the fermentation. Significant inhibition of ESP A_1 production (61.5%) was observed when 0.25 mM cerulenin was added to the culture. At 1 mM cerulenin concentration, no ESP A_1 could be detected in the culture. This indicated that ESP A_1 is biosynthesized, at least in part, by the polyketide pathway and that its enediyne moiety should also be derived from acetate units.

Later, a hyperproducer mutant, MU-5019 strain, was isolated. The yields of ESP A_1 by this mutant strain were, at about 25–30 $\mu g\ ml^{-1}$, five- to sixfold higher than the production by the parent strain. This improvement in ESP A_1 production enabled feeding experiments on a 5- to 10-liter scale to obtain ^{13}C-enriched ESP A_1 satisfactory for NMR studies. Usually, 0.2% sodium [1-^{13}C]acetate, sodium [2-^{13}C]acetate, and sodium [1,2-$^{13}C_2$]acetate each were added to the cultures, and 25–30 mg of pure samples were obtained by a 10-liter fermentation.

Since ESP A_1 is very soluble in chloroform but has limited solubility in methanol, $CDCl_3$ solution has to be used in NMR analysis. However, the signals of carbons 8, 9, and 10 of the enediyne moiety were broadened in $CDCl_3$ and the signal due to carbon 1 of the enediyne moiety overlapped with the $CDCl_3$ signal. In order to obtain accurate integration of the signals for the above carbons, the ^{13}C-enriched ESP A_1 was first converted to diacetyl derivative and CD_3OD was used for NMR analysis. Sharp signals for all 15 carbons of the enediyne core of diacetyl ESP A_1 were obtained and could be assigned as shown in Table 3.

Table 3 ^{13}C NMR assignments of [1-^{13}C]- and [2-^{13}C]acetate-labeled diacetyl-esperamycin A_1 and J_{cc} of [1,2-$^{13}C_2$]acetate-labeled diacetyl-esperamycin.

Carbon	δ (ppm)	*[1-^{13}C]acetate* (relative intensity)	*[2-^{13}C]acetate* (relative intensity)	*[1,2-$^{13}C_2$]acetate*	
				$J^{13}C_1$-$J^{13}C_2$	*Satellites*
1	77.7	4.5	1	79.9	45.1
2	99.9	1	3.0	80.2	186
3	84.2	4.2	1	89.4	187
4	126.8	1	3.8	90.3	72.2
5	123.5	4.1	1	89.7	72.1
6	88.9	1	3.5	89.8	
7	98.1	4.2	1	74.4	165
8	72.5	1	4.7	74.0	
9	148.8	4.3	1	76.7	52.2
10	134.4	1	3.0	77.0	
11	194.8	2.0	1	45.2	Broad
12	84.3	1	3.3	45.5	Obscured
13	136.7	1	4.7	80.7	45.0
14	131.8	5.8	1	80.5	45.4
15	41.0	1	6.0		45.2

Figure 4 summarizes the ^{13}C-labeling pattern into ESP A_1 fed with respective ^{13}C-acetate precursors. Feeding with sodium [1-^{13}C]acetate enriched carbons 1, 3, 5, 7, 9, 11, and 14 of the enediyne ring (Figure 4(a)). By addition of sodium [2-^{13}C]acetate, carbons 2, 4, 6, 8, 10, 12, 13, and 15 were enriched (Figure 4(b)). The intensities of these enriched signals were 2.0- to 5.8-fold greater relative to those of the control (ESP A_1 obtained without addition of sodium [^{13}C]acetate).

In order to determine the biosynthetic connectivity of the carbon units in the enediyne skeleton, doubly labeled sodium [1,2-$^{13}C_2$]acetate was fed to the esperamycin-producing culture. The resulting incorporation pattern is shown in Figure 4(c), which was confirmed by matching of J_{cc} values, as shown in Table 3. This data showed labeling of seven coupled pairs and an uncoupled carbon at C-15 derived from C-2 of an acetate.

There are only four possible folding patterns of an octaketide such as (**11**), leading to (**14a–d**) with one carbon loss from either end (Scheme 2), for a linear C_{15} unit of esperamycin's enediyne skeleton, as discussed previously. The incorporation pattern of the acetate units shown in Figure 4(c) agree only with the path leading to (**14d**). The labeling patterns of the diyne moieties of ESP A_1 and DNM-A are also the same in that the two carbons comprising the respective yne moieties are derived from separate acetate units.

1.21.3.4 Biosynthetic Origin of Sulfur

ESP A_1 contains four sulfur atoms, a thiomethyl group of a sugar moiety, and an allylic trisulfide moiety attached to the bicyclic core. The biosynthetic process of the allylic trisulfide is very intriguing, because such a functional group is unique in natural products and this function has an important role in highly efficient DNA strand scission by ESP A_1. In feeding experiments using $Na_2{}^{35}SO_4$, L-[^{35}S]cysteine, and L-[^{35}S]methionine, sulfur incorporation was very low (Table 2). In attempts to increase the incorporation of ^{35}S-precursors, an experiment using a medium with sodium sulfate as the sole sulfur source gave no ESP A_1, but instead gave ESP A_{1b} and ESP A_{1c} (Figure 3). Incorporation of ^{35}S into ESP A_{1c} in this experiment was 1.7%.

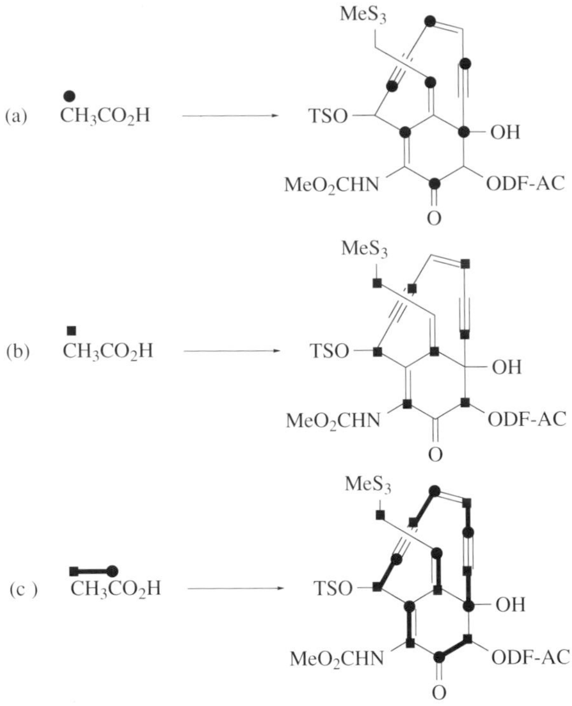

(a) CH_3CO_2H → TSO, MeS$_3$, OH, MeO$_2$CHN, ODF-AC, O

(b) CH_3CO_2H → TSO, MeS$_3$, OH, MeO$_2$CHN, ODF-AC, O

(c) CH_3CO_2H → TSO, MeS$_3$, OH, MeO$_2$CHN, ODF-AC, O

Figure 4 ^{13}C incorporation pattern of esperamycin A_1 from cultures fed with (a) [1-^{13}C]acetate, (b) [2-^{13}C]acetate, and (c) [1,2-$^{13}C_2$]acetate.

Under these fermentation conditions, Lam *et al.*[14] used $Na_2{}^{34}SO_4$ as a source of the stable isotope of sulfur and analyzed the incorporation of ^{34}S into ESP A_{1c} by high-resolution fast atom bombardment mass spectrometry (HRFAB-MS). ^{34}S-Esp A_{1c} thus obtained showed a molecular weight of 1304 (HRFAB-MS: $[M + H]^+$ 1305.3743, calculated for $C_{57}H_{77}N_4O_{22}{}^{34}S_4$ 1305.3743). This molecular weight is 8 Da greater than the native ESP A_{1c} suggesting the incorporation of four atoms of ^{34}S from the inorganic source. The labeling of each sulfur atom was confirmed by comparisons of the substructure fragments in HRFAB-MS for the native and the labeled compounds. No explanation for the ^{35}S incorporation from L-[^{35}S]cysteine and L-[^{35}S]methionine was given.

1.21.3.5 Biosynthetic Origin of the Methyl Groups

Methylation of hydroxyl and amino groups in bacteria normally involves L-methionine. Table 1 shows that the most efficient precursor of ESP A_1 is L-[methyl-^{14}C]methionine. The incorporation rate of L-[methyl-^{14}C]methionine was 30-fold higher than that of sodium [1-^{14}C]acetate, which is frequently observed in feeding experiments with bacteria and fungi.

Addition of L-methionine to the culture medium, at a concentration as low as 0.02%, inhibited the production of ESP A_1 by 36.6%. At a 0.1% L-methionine concentration, no ESP A_1 could be detected in the fermentation, even though there was no effect on growth of the organism and the pH of the culture. A similar result has been reported by Gairola and Hurley[18] in anthramycin production by *Streptomyces refuineus*, in which they reported that addition of L-methionine at 0.03% concentration inhibited the biosynthesis of anthramycin by 56%. Although the reason remained unknown, it was suggested that a higher concentration of L-methionine in the culture medium should be avoided. Consequently, L-[methyl-^{13}C]methionine was added stepwise to the culture on two separate days up to 0.05% of the total amount.

^{13}C NMR analysis showed that ^{13}C of the seven *O*- and *S*-methyl groups were evidently enriched by L-[methyl-^{13}C]methionine, and that the three carbons of the isopropyl group of an aminosugar were also slightly enriched. The possible incorporation of the L-methionine methyl into the isopropyl carbons was also suggested by blocked mutants studies. The blocked mutants, DG-111-10-6 and DG-108-9-3, produced no ESP A_1 (isopropylamino) but did produce ESP A_{1b} (ethylamino) and

ESP A_{1c} (methylamino), respectively, as the major esperamycin. These facts may support the possibility of a sequential biosynthesis of the alkylamino groups by L-methionine.

1.21.3.6 Summary

Esperamycins are composed of (i) a unique enediyne core with a methyl trisulfide group, (ii) four sugar moieties attached to the enediyne core, and (iii) an aromatic aglycone moiety. This study on the biosynthesis of esperamycin was focused, as a matter of course, on the C_{15} enediyne core. The skeleton was proved to be derived from head-to-tail condensation of eight acetate units followed by the loss of the terminal carboxyl carbon. The sulfur atoms of the interesting methyl trisulfide group were derived from inorganic sulfur. Biogenesis of the anthranilate moiety was not determined, and a scheme from L-tryptophan is proposed according to the tomaymycin and anthramycin biosynthesis. Five *O*-methyl and two *S*-methyl carbons, and possibly the three carbons of isopropyl amino group on an aminosugar were derived from the methyl group of L-methionine.

1.21.4 BIOSYNTHESIS OF NEOCARZINOSTATIN CHROMOPHORE

1.21.4.1 Introduction

Neocarzinostatin (NCS, (**4**)) is the first member of a family of chromoprotein antitumor antibiotics obtained from culture filtrates of *Streptomyces*. The drug causes DNA strand breakage *in vivo* and *in vitro* in a reaction greatly stimulated by a sulfhydryl compound. All biological activity resides in a methanol-extractable nonprotein chromophore that is tightly and specifically bound to an apoprotein ($M = 11\ 000$). The instability of neocarzinostatin chromophore A (NCS Chrom A) has made its structural investigation quite difficult. The light-sensitive chromophore, when unprotected by the apoprotein, is extremely unstable in aqueous solution, especially upon concentration, and treatment with a variety of reagents results in decomposition.

Edo *et al.*[7] reported the complete structure determination of the epoxy bicyclo-[7.3.0]dodecadienediyne ring system of NCS Chrom A, based primarily on NMR investigation of a stable chlorohydrin derivative, incorporating the previous findings by Hensens *et al.*[12] and proposing for the C_{12} substructure the unprecedented bicyclo[7.3.0]dodecadienediyne ring system. The absolute stereochemistry of the molecule, as in (**4**), has since been determined.

This section deals with the study of the biosynthetic origin of the carbon skeletons of the naphthoic, cyclic carbonate, and epoxy bicyclo[7.3.0]dodecadienediyne ring systems of neocarzinostatin.

1.21.4.2 Incorporation of ^{13}C-Labeled Acetate

To produce biosynthetically ^{13}C-enriched samples for determination of biosynthesis of NCS Chrom A, four incorporation experiments with singly and doubly ^{13}C-enriched acetate were carried out with shake culture of *Streptomyces carzinostaticus* (ATCC #15944 F-42). High-purity NCS Chrom A was obtained through a multistep isolation procedure.

The incorporation results indicated that six carbons of the C_{12} naphthoic acid and seven carbons of the C_{14} bicyclic diyne/cyclic carbonate ring systems are enriched by [1-^{13}C]- and [2-^{13}C]-acetate, respectively (see Table 4), with no labeling observed in the methoxyl, *N*-methyl, carbonate, and sugar carbons. This suggests that both ring systems are derived from six and seven acetate units, respectively. Incorporation experiments with [methyl-^{3}H]methionine and [^{14}C]sodium bicarbonate provide evidence that the *N*-methyl of the fucosamine and the *O*-methyl of the naphthoic acid moieties derive from methionine and the cyclic carbonate carbon is carbonate derived. The naphthoic acid moiety therefore appears to consist of a single polyketide chain of six intact acetate units which can fold in only one way. Thus, it can be envisaged that the C_{14} substructure is derived from a single heptaketide chain of seven head-to-tail coupled acetate units, by either pathway a or b (Scheme 3).

To establish the carbon–carbon connectivities of the C_{14} substructure of NCS Chrom A, experiments with doubly labeled [1,2-$^{13}C_2$]acetate and mixed-labeled [1 + 2-^{13}C]acetate were conducted. In the case of the [1 + 2-^{13}C]acetate incorporation, many C—H bond couplings were evident for protonated carbons (Table 4). With [1,2-$^{13}C_2$]acetate-enriched NCS Chrom A, C—C couplings of

Table 4 ^{13}C NMR assignments and $^1J_{cc}$ of [1+2-^{13}C]- and [1,2-$^{13}C_2$]acetate labeled NCS Chrom A.

	[1+2-^{13}C]acetate[a]		*[1,2-$^{13}C_2$]acetate*[b]		
Carbon	δ (ppm)[c]	$^1J_{cc}$ (Hz)[d]	δ (ppm)[c]	$^1J_{^{13}C-^1H}$ (Hz)	$^1J_{cc}$ (Hz)
C-1″	108.7 s (2)		105.9 s (2)		obsc
C-1″—CO_2	172.4 s (1)		173.3 s (1)		75
C-2″	162.5 s (1)		165.8 s (1)		64
C-3″	116.5 d (2)	62	116.9 d (2)	166	64
C-4″	132.9 d (1)	62	134.5 d (1)	159	56
C-4a″	124.4 s (2)		124.5 s (2)		~55
C-5″	138.6 s (1)	60	138.6 s (1)		~46
C-5″—Me	19.9 q (2)		~20.0 q (2)		obsc
C-6″	118.2 d (2)	63	118.1 d (2)	160	64.5
C-7″	161.2 s (1)		161.1 s (1)		64.5
C-7″—OMe	55.8 q		55.7 q		
C-8″	103.4 d (2)	69	104.8 d (2)	161.5	61
C-8a″	135.4 s (1)		135.5 s (1)		61
C-1	130.4 s (1)		130.2 s (1)		~72
C-2	87.6 s (2)		87.7 s (2)		188
C-3	97.9 s (1)		97.7 s (1)		188
C-4	63.8 s (2)		63.9 s (2)		57
C-5	55.2 d (1)		55.1 d (1)	198.5	s (103)[d]
C-6	100.0 s (2)		99.8 s (2)		186
C-7	90.8 s (1)		90.8 s (1)		186
C-8	106.8 d (2)	91	106.6 d (2)	172	81
C-9	161.1 s (1)		160.8 s (1)		81
C-10	82.0 d (2)	48	81.3 d (2)	152	43
C-11	82.8 d (1)	~45	83.0 d (1)	161	43
C-12	140.0 d (2)	45	140.0 d (2)	179.5	72
C-13	76.4 d (1)	32	75.8 d (1)	163	57 (32)[c]
C-14	68.2 dd (2)	32	67.9 dd (2)	156, 160	s (32)[d]
C-C=O	~160.0 s		155.9 s		
C-1′	95.9 d		95.7 d		
C-2′	59.5 d		59.3 d		
C-2′—NMe	33.5 q		33.5 q		
C-3′	68.6 d		68.1 d		
C-4′	72.6 d		72.7 d		
C-5′	69.2 d		69.1 d		
C-5′—Me	16.5 q		16.4 q		

[a]At ~0 °C in CD_3OD/CD_3COOD (9:1). [b]At ~5 °C in CD_3COOD. [c]Those carbons enriched by [1-^{13}C]acetate (1) and [2-^{13}C]acetate (2), respectively. Abbreviations: s, singlet; d, doublet; q, quartet; obsc, obscured. [d] $^1J_{cc}$(Hz) values of satellites.

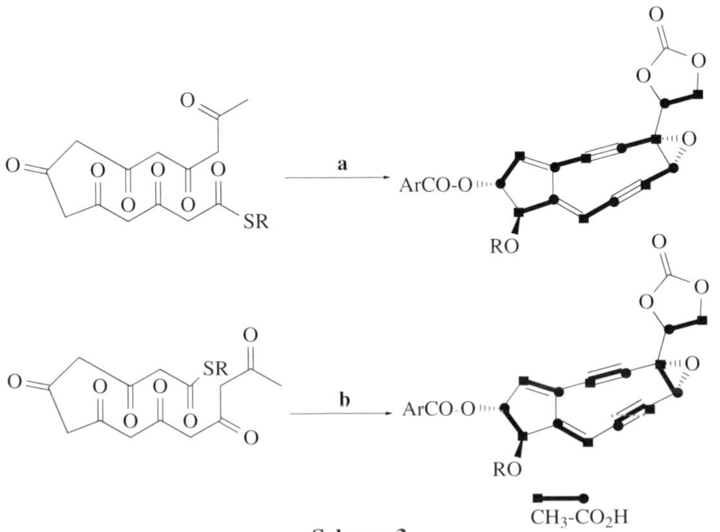

Scheme 3

protonated carbon were clearly visible, whereas many of the quaternary carbons were weak, and it allowed, nevertheless, to identify all acetate pairs. The matching of $^{1}J_{cc}$ values confirmed the polyketide origin of the naphthoic acid ring, as depicted in Figure 5, but the carbon signals for the epoxide C_5 and cyclic carbonate C_{14} carbons were observed as singlets and found to be inconsistent with the labeling patterns a and b for the C_{14} substructure in Scheme 3. Loss of methyl and carboxylate carbons from either end of an octaketide chain appears to be implicated.

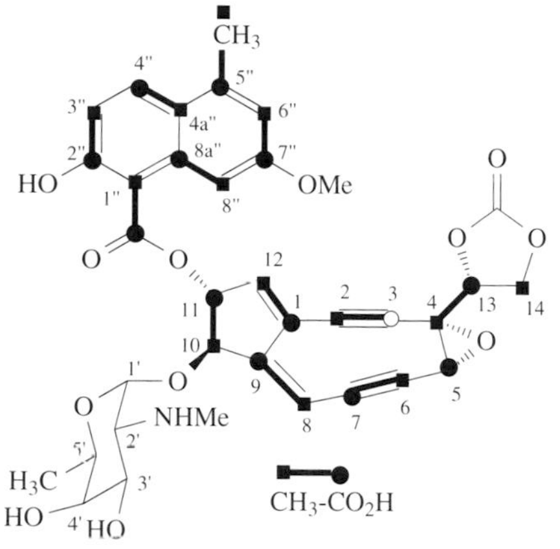

Figure 5 ^{13}C incorporation pattern of neocarzinostatin chromophore from a culture fed with [1,2-$^{13}C_2$]acetate.

Hensens *et al.*[12] discussed this unexpected result in the following way, according to the biosynthesis of polyacetylenes derived from oleic acid via the well-accepted oleate–crepenynate pathway, in which shortening of the C_{18} fatty acid is common by loss of carbon from either end. Although their results are consistent with the degradation of two carbons from an octaketide, they favored a linear C_{18} polyketide precursor of acetate units coupled in head-to-tail fashion for the C_{14} substructure of NCS Chrom A, and extended the idea to the ESP/CAL class of antibiotics, as illustrated in Scheme 4 (the numbering in Scheme 4 for both skeletons reflects the origin of the carbons from crepenynate).

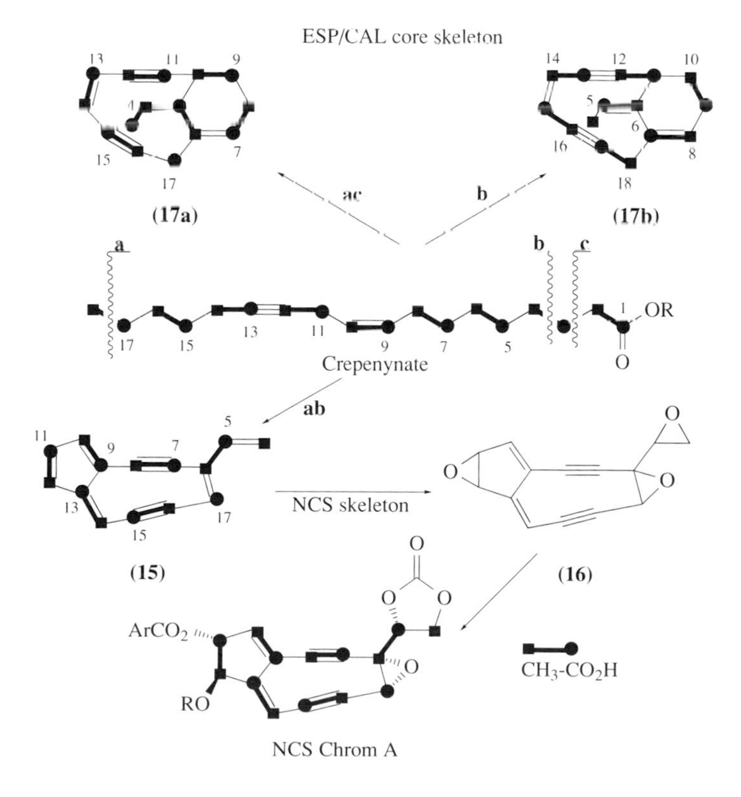

Scheme 4

This scheme would suggest that all oxygens attached at C-4, C-5, C-10, C-11, C-13, and C-14 are introduced by oxidation at a late stage of the biosynthetic pathway of a long deoxygenated precursor. The presence of epoxides in polyacetylenes is not uncommon, whereas the presence of a cyclic carbonate moiety has only been demonstrated in a C_{16} triyne carboxylic acid isolated from *Actinomycetes* and *Microbispora* species. A linear chain oleate-crepenynate pathway was postulated for NCS Chrom A (path ab) and for ESP/CAL endiyne core (path ac or b), as illustrated in Scheme 4. The scheme differs from that proposed by Schreiber and Kiessling[19] in that a dual polyketide pathway would have to be invoked to account for the postulated common precursor to both NCS Chrom A and the ESP/CAL class of antibiotics. In particular, they hypothesized that C-3 of (**17a**) (or C-4 of (**17b**)) is derived from the C_1 pool, whereas Hensens *et al.*[12] suggested its origin as C-3 or C-4 of crepenynate.

1.21.4.3 Summary

In summary, this work demonstrates that the C_{12} and C_{14} ring skeletons of the chromophore are formed by a polyketide pathway. The C_{12} naphthoic acid ring is derived from six intact acetate units, linked in head-to-tail fashion, whereas the novel C_{14} dienediyne ring skeleton incorporates six intact acetates and two terminal acetate units which undergo C—C bond cleavage. It was proposed that the C_{14} chain is derived from degradation of oleate via the now well-accepted oleate–crepenynate pathway for polyacetylenes rather than by *de novo* synthesis from acetate, and that the C_{15} enediyne skeleton of ESP/CAL class antibiotics can similarly be derived via the oleate–crepenynate pathway.

1.21.5 COMPREHENSIVE ANALYSIS OF ENEDIYNE CORE BIOSYNTHESIS

Biosynthetic studies of enediyne antibiotics have been reported for dynemycin A,[13] esperamycins,[14] and neocarzinostatin chromophores.[12] In the biosynthesis of these antibiotics, interest was focused on the construction of their bicyclodiynene cores. Incorporation patterns of ^{13}C-labeled acetates into respective compounds were analyzed to establish the carbon–carbon connectivities, especially of their enediyne cores.

Dynemycin A has a striking hybrid structure of an unusual bicyclo[7.3.1]-1,5-diyn-3-ene and anthracyclin, the former unit being similar to the core portions of esperamycins and calicheamycins. The labeling pattern of dynemycin A after incorporation of various ^{13}C-labeled acetates, L-methionine and glycine, and of [^{15}N]ammonium sulfate is shown in Figure 2. The result established the carbon–carbon connectivities and suggested that the respective C_{14} bicyclo[7.3.1]enediyne and anthraquinone moieties are biosynthesized separately from two separate heptaketide chains, and connected at a later stage (Scheme 1). In this scheme, a heptacarbonyl acid (**9**) is tentatively drawn as the possible precursor of the anthraquinone (**10**) but an enediyne chain (**11**) is also a conceivable precursor, as shown by the dotted arrow in Scheme 1.

The scheme proposed for the biosynthesis of the bicyclic core of DNM-A via heptaketide precursor (**11**) was extended to the biosynthesis of the C_{15} enediyne skeleton of the ESP/CAL class of antibiotics, involving an octaketide (**13**) which has the same partial structure as the enediyne chain (**11**) (Scheme 2). Lam *et al.*[14] examined the incorporation pattern of ^{13}C-labeled acetates into the bicyclo-enediyne core of ESP A_1, and proved that the labeling pattern was consistent with that drawn for (**14d**) (Scheme 2), as shown in Figure 4(c). Neocarzinostatin is the first of the so-called enediyne antibiotics. Biosynthesis of its novel C_{14} dienediyne ring system was reported as early as 1989.[12] It was envisaged that the substructure was derived from a single heptaketide chain of seven head-to-tail coupled acetate units, but an incorporation pattern of doubly labeled [1,2-$^{13}C_2$]acetate was inconsistent with this expectation and the carbon signals of C_5 and C_{14} appeared without ^{13}C–^{13}C coupling, suggesting loss of methyl and carboxylate carbons from either end of an octaketide chain.

Hensens *et al.*[12] proposed that the C_{14} chain of NCS Chrom A is derived from degradation of oleate via the oleate–crepenynate pathway for polyacetylenes rather than by *de novo* synthesis from acetate. They further postulated that the C_{15} enediyne cores of ESP/CAL can similarly be derived via the oleate–crepenynate pathway (Scheme 4). This pathway, however, has been shown to operate only in higher plants and fungi.

Lam *et al.*[14] do not agree with this biosynthetic scheme of ESP A_1, based on their observation that ESP A_1 production from cultures supplemented with cerulenin and sodium oleate was inconsistent with the above hypothesis. Addition of sodium oleate (0.1–0.5%) did not reverse the inhibitory effect of cerulenin on ESP A_1 production by *A. verrucosospora* SA-25262.

It should be noted that the incorporation patterns of [1,2-$^{13}C_2$]acetate into DNM-A and ESP A_1 and that into NCS Chrom A evidently differ in the labeling patterns of diynes. Namely, in DNM-A and ESP A_1 the two carbons composing the respective yne moieties derive from separate acetate units, whereas the corresponding carbons in NCS Chrom A derive from the same acetate units. It appears that the biosynthetic pathway of NCS Chrom A is somewhat different from that of the bicyclodiynene substructure in the DNM-A and ESP/CAL class of antibiotics.

1.21.6 REFERENCES

1. M. D. Lee, T. S. Dunne, M. M. Siegel, C. C. Chang, G. O. Morton, and D. B. Borders, *J. Am. Chem. Soc.*, 1987, **109**, 3464.
2. M. D. Lee, T. S. Dunne, C. C. Chang, G. A. Ellestad, M. M. Siegel, G. O. Morton, W. J. McGahren, and D. B. Borders, *J. Am. Chem. Soc.*, 1987, **109**, 3466.
3. J. Golik, J. Clardy, G. Dubay, G. Groenewold, H. Kawaguchi, M. Konishi, B. Krishnan, H. Ohkuma, K. Saitoh, and T. W. Doyle, *J. Am. Chem. Soc.*, 1987, **109**, 3461.
4. J. Golik, G. Dubay, G. Groenewold, H. Kawaguchi, M. Konishi, B. Krishnan, H. Ohkuma, K. Saitoh, and T. W. Doyle, *J. Am. Chem. Soc.*, 1987, **109**, 3462.
5. M. Konishi, H. Ohkuma, T. Tsuno, H. Kamei, T. Miyaki, T. Oki, H. Kawaguchi, G. D. VanDuyne, and J. Clardy, *J. Antibiot.*, 1989, **42**, 1449.
6. M. Konishi, H. Ohkuma, K. Matsumoto, T. Tsuno, T. Oki, H. Kawaguchi, G. D. VanDuyne, and J. Clardy, *J. Am. Chem. Soc.*, 1990, **112**, 3715.
7. K. Edo, M. Mizugaki, Y. Koide, H. Seto, K. Furihata, N. Ohtake, and N. Ishida, *Tetrahedron Lett.*, 1985, **26**, 331.
8. T. W. Doyle and D. B. Borders, in "Enediyne Antibiotics as Antitumor Agents," eds. D. B. Borders and T. W. Doyle, Dekker, New York, 1995, p. 1.
9. A. G. Myers, *Tetrahedron Lett.*, 1987, **28**, 4493.
10. T. Kusakabe, M. Uesugi, and Y. Sugiura, *Biochemistry*, 1995, **34**, 9944.
11. M. Miyoshi, N. Morisaki, Y. Tokiwa, H. Kobayashi, S. Iwasaki, M. Konishi, and T. Oki, *Tetrahedron Lett.*, 1991, **42**, 6007.
12. O. D. Hensens, J.-L. Giner, and I. H. Goldberg, *J. Am. Chem. Soc.*, 1989, **111**, 3295.
13. Y. Tokiwa, M. Miyoshi-Saitoh, H. Kobayashi, R. Sunaga, M. Konishi, T. Oki, and S. Iwasaki, *J. Am. Chem. Soc.*, 1992, **114**, 4107.
14. K. S. Lam and J. A. Veitch, in "Enediyne Antibiotics as Antitumor Agents," eds. D. B. Borders and T. W. Doyle, Dekker, New York, 1995, p. 217.
15. D. R. Langley, T. W. Doyle, and D. L. Beveridge, *J. Am. Chem. Soc.*, 1991, **113**, 4395.
16. M. Miyoshi-Saitoh, N. Morisaki, Y. Tokiwa, S. Iwasaki, M. Konishi, and T. Oki, *J. Antibiot.*, 1991, **44**, 1037.
17. S. Ohmura, *Bacteriol. Rev.*, 1976, **40**, 681.
18. C. Gairola and L. H. Hurley, *Eur. J. Appl. Microbiol.*, 1976, **2**, 95.
19. S. L. Schreiber and L. L. Kiessling, *J. Am. Chem. Soc.*, 1988, **110**, 631.

1.22
Enzymology and Molecular Biology of the Shikimate Pathway

CHRIS ABELL
University of Cambridge, UK

1.22.1 INTRODUCTION

The shikimate pathway is the biosynthetic pathway from erythrose 4-phosphate (**1**) and phosphoenol pyruvate (PEP, **2**) to chorismate (**9**) (Scheme 1), the precursor of the aromatic amino acids phenylalanine (**11**), tyrosine (**12**), and tryptophan (**16**), and other important aromatic compounds such as anthranilate (**15**), *para*-hydroxybenzoate (**13**), and *para*-aminobenzoate (**14**) (Scheme 2). Each step on the shikimate pathway will be discussed in turn, with the emphasis being on understanding the mechanism of the reactions. Many mechanistic studies have been dependent on advances in molecular biology and protein overexpression. These are described briefly, primarily in relation to the *E. coli* enzymes. Details of reactions after chorismate, the pathway in other organisms, and regulation of the pathway have been reviewed elsewhere.[1]

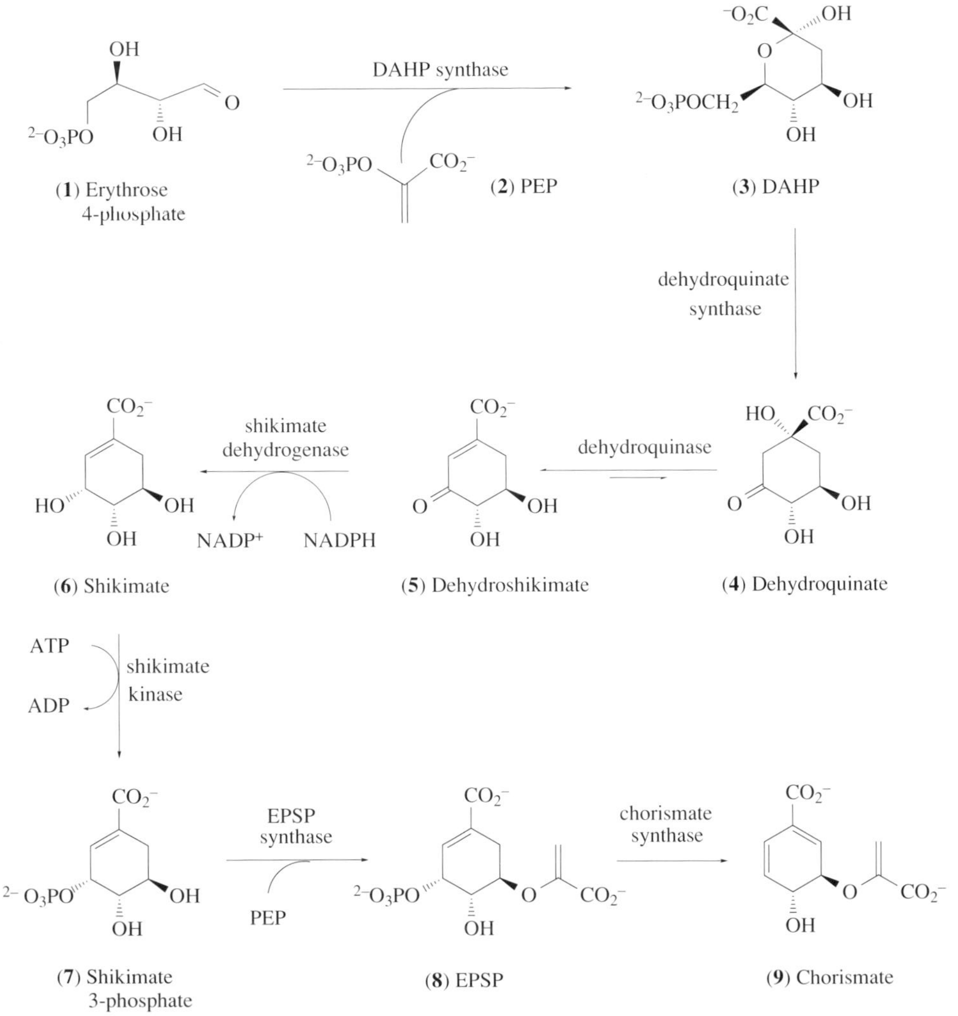

(**1**) Erythrose 4-phosphate — DAHP synthase, (**2**) PEP → (**3**) DAHP — dehydroquinate synthase → (**4**) Dehydroquinate — dehydroquinase → (**5**) Dehydroshikimate — shikimate dehydrogenase, NADPH → NADP+ → (**6**) Shikimate — shikimate kinase, ATP → ADP → (**7**) Shikimate 3-phosphate — EPSP synthase, PEP → (**8**) EPSP — chorismate synthase → (**9**) Chorismate

Scheme 1

The shikimate pathway is named after shikimic acid (**6**). This compound was isolated from *Illicium religiosum*, shikiminoki (star anise tree), in 1886.[2,3] The shikimate pathway is present in microorganisms, fungi, and especially in plants (where it can account for 20–30% of the carbon flux), but not in mammals. The pathway is the target of the very successful broad-spectrum herbicide glyphosate, which inhibits EPSP (5-enolpyruvyl-shikimate 3-phosphate) synthase (Scheme 1).[4]

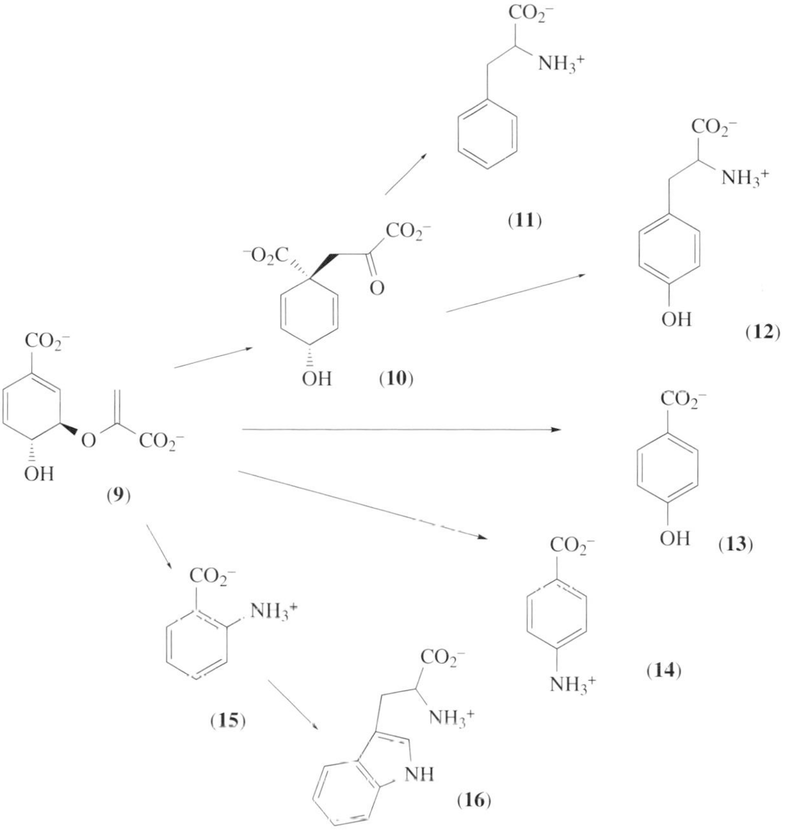

Scheme 2

1.22.1.1 Enzyme Organization

The seven enzymes required to catalyze the conversion of erythrose 4-phosphate and PEP to chorismate are organized differently in different organisms (Figure 1). In *E. coli* the seven reactions are catalyzed by separate enzymes which are independently transcribed. In *Bacillus subtilis* the genes are more clustered than in *E. coli* and the *aro* and *trp* genes may be part of a supra-operon.[5]

In *Saccharomyces cerevisiae*, *Euglena gracilis*, and fungi such as *Aspergillus nidulans* and *Neurospora crassa*, the five catalytic activities required to convert DAHP (3-deoxy-D-*arabino*-heptulosonate 7-phosphate) into EPSP are catalyzed by a pentafunctional polypeptide, referred to as the *arom* complex. In *N. crassa* the *arom* complex is a homodimer of subunit M_r 165 000. In higher plants dehydroquinase and shikimate dehydrogenase activities are present as a bifunctional polypeptide. In the pea this enzyme is monomeric and has a molecular weight which is close to the sum of the molecular weights of the two corresponding *E. coli* enzymes.[6] In plants the aromatic amino acids are synthesized by the shikimate pathway in the chloroplast.[7] However, there is evidence for a separate cytosolic pathway which may be involved in secondary metabolite biosynthesis.[8]

1.22.2 DAHP SYNTHASE

The first enzyme on the shikimate pathway is DAHP synthase (7-phospho-2-dehydro-3-deoxy-heptonate aldolase, EC 4.1.2.15). It catalyzes the reaction between erythrose 4-phosphate (**1**), produced by the pentose phosphate pathway, and PEP (**2**) to form DAHP (**3**) (Scheme 3).

There are three isozymes of DAHP synthase in *E. coli*. They are each subject to feedback inhibition by one of the three aromatic amino acid products of the pathway. Carbon-13 NMR spectroscopic studies of whole cells of *E. coli* have been used to show that feedback inhibition of the DAHP synthase isozymes is the main mechanism for controlling the flux through the shikimate pathway.[9] The three isozymes are designated DAHP synthase(tyr), DAHP synthase(phe), and DAHP synthase(trp) and are encoded by the genes *aroF*, *aroG*, and *aroH*, respectively. All three genes have

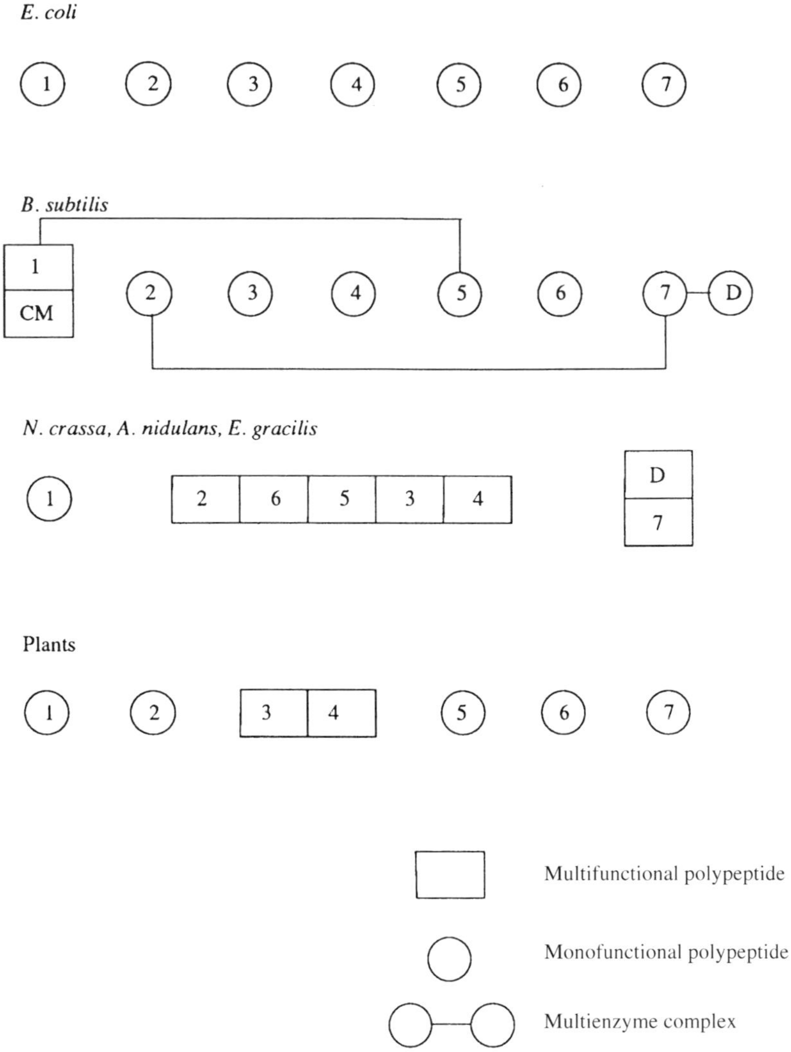

E. coli

1 2 3 4 5 6 7

B. subtilis

1 / CM 2 3 4 5 6 7—D

N. crassa, A. nidulans, E. gracilis

1 | 2 | 6 | 5 | 3 | 4 | D / 7

Plants

1 2 | 3 | 4 | 5 6 7

Multifunctional polypeptide

Monofunctional polypeptide

Multienzyme complex

Figure 1 Organization of enzymes catalyzing the conversion of erythrose 4-phosphate and PEP to chorismate in different organisms. 1, DAHP synthase; 2, dehydroquinate synthase; 3, dehydroquinase; 4, shikimate dehydrogenase; 5, shikimate kinase; 6, EPSP synthase; 7, chorismate synthase; D, diaphorase; CM, chorismate mutase.

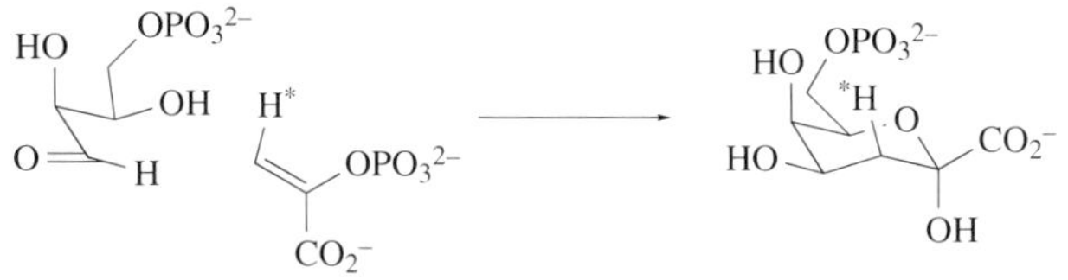

HO OPO_3^{2-} OH H* O H OPO_3^{2-} CO_2^- → HO OPO_3^{2-} *H O CO_2^- HO OH

(**1**) Erythrose 4-phosphate (**2**) PEP (**3**) DAHP

Scheme 3

been cloned and overexpressed.[10–12] The proteins share 41% identity in their amino acid sequences, suggesting a common evolutionary origin.

DAHP synthase(phe) is a tetramer of subunit M_r 37 995. DAHP synthase(tyr) and DAHP synthase(trp) are both dimeric.[13] DAHP synthase(phe) and DAHP synthase(tyr) account for 80% and 20% of the DAHP synthase activity in wild type *E. coli*, with DAHP synthase(trp) contributing less than 1%.[14]

The isozyme distribution for DAHP synthase is different in different organisms,[15] for example, in *B. subtilis*, where the DAHP synthase and chorismate mutase activities are associated, the feedback inhibition is by chorismate and prephenate (**10**). The complexity of isozyme distribution and inhibition of DAHP synthase is a complex subject with more allosteric regulatory mechanisms known for it than any other protein.[1]

Genes encoding DAHP synthase have been isolated from a number of plants including potato,[16,17] tobacco,[18] and tomato.[19] Each plant has been shown to possess at least two DAHP synthase isozymes, products of the *shkA* and *shkB* genes.[20] Comparison of the plant and bacterial sequences shows low sequence identity. However, expression of the potato *shkA* gene has been shown to complement an *E. coli* mutant lacking DAHP synthase.[21]

1.22.2.1 Mechanistic Studies on DAHP Synthase

The mechanism of DAHP synthase formally involves attack of C-3 of PEP (**2**) on the aldehyde of erythrose 4-phosphate (**1**) (Scheme 3). A mechanism whereby PEP serves as an enolate equivalent and the phosphate group of PEP is lost by P—O bond cleavage can be discounted as it has been firmly established that the reaction involves C—O rather than P—O bond cleavage.[22,23] Labeling studies using (*Z*)- and (*E*)-[3-^{3}H]PEP established that H_Z of PEP becomes H_S at C-3, i.e., the reaction involves the attack of the *si* face of PEP on the *re* face of the aldehyde.[24,25] This observation effectively rules out any mechanism that involves the transient formation of a methyl group at C-3 of PEP. As will be described later, this distinguishes the DAHP synthase reaction (and that of KDO8P synthase) from EPSP synthase and UDP–GlcNAc enolpyruvyl transferase.

Some of the key observations relating to the reaction mechanism, and the irreversibility of the reaction are:

(i) DAHP synthase does not catalyze the exchange of ^{32}P from inorganic phosphate into PEP,[22,26]

(ii) [U-^{14}C]erythrose 4-phosphate does not exchange into DAHP, in the presence or absence of phosphate,[26]

(iii) phosphate is not released from DAHP synthase until erythrose 4-phosphate binds,[26] and

(iv) product inhibition studies are consistent with an ordered sequential mechanism in which PEP binds first and phosphate is the first product to leave.[27]

Several mechanisms have been proposed, including the one shown in Scheme 4.[28] This involves an initial conjugate addition of an active site sulfhydryl group to PEP to form the intermediate (**17**). This undergoes a sulfur migration with a concomitant 1,2-phosphate shift, followed by β-elimination of the phosphate to form an enzyme-bound thioenolpyruvate which in turn condenses with erythrose 4-phosphate. Although there is some precedent for the key rearrangement step, this mechanism seems unlikely and does not fit observed kinetics, unless the loss of phosphate occurs after binding of erythrose 4-phosphate. More plausible, if somewhat more prosaic, mechanisms are shown in Scheme 5. Here attack of C-3 of PEP on the aldehyde transiently generates an oxonium species which is either attacked by water (mechanism **a**) or the C-3 hydroxyl of erythrose 4-phosphate (mechanism **b**). Both mechanisms result in formation of an intermediate, (**18**) or (**19**), from which phosphate is expelled followed by capture of the resulting oxonium ion by either the C-6 hydroxyl group (mechanism **a**) or water (mechanism **b**).

1.22.2.2 Metal Dependence of DAHP Synthase

DAHP synthase is a metalloprotein, but there are ambiguities about the nature and role of the metal. Early studies reported sensitivity of the enzyme to the presence of metal ions and metal ion chelators,[27,29,30] and pointed to cobalt being the required metal ion. *E. coli* DAHP synthase(phe) was reported to contain one equivalent of Fe^{2+} per tetramer,[31] whereas a detailed study of DAHP synthase(tyr) suggested it was a copper metalloenzyme.[32] The native enzyme was found to contain 0.5 moles of Cu^{2+} per mole of enzyme subunit. The *apo* enzyme had 6% of the activity of the native enzyme and was reactivated by Cu^{2+} with a stoichiometry of one copper per enzyme subunit. The Zn^{2+} ion also resulted in reactivation, but Fe^{2+} did not.

In 1991 the activation parameters of all three *E. coli* isozymes with a number of divalent metal ions were reported.[33] The native isozymes were found to contain 0.2–0.3 equivalents of iron per monomer. Treatment of the three isozymes with EDTA produced *apo* enzymes of very low

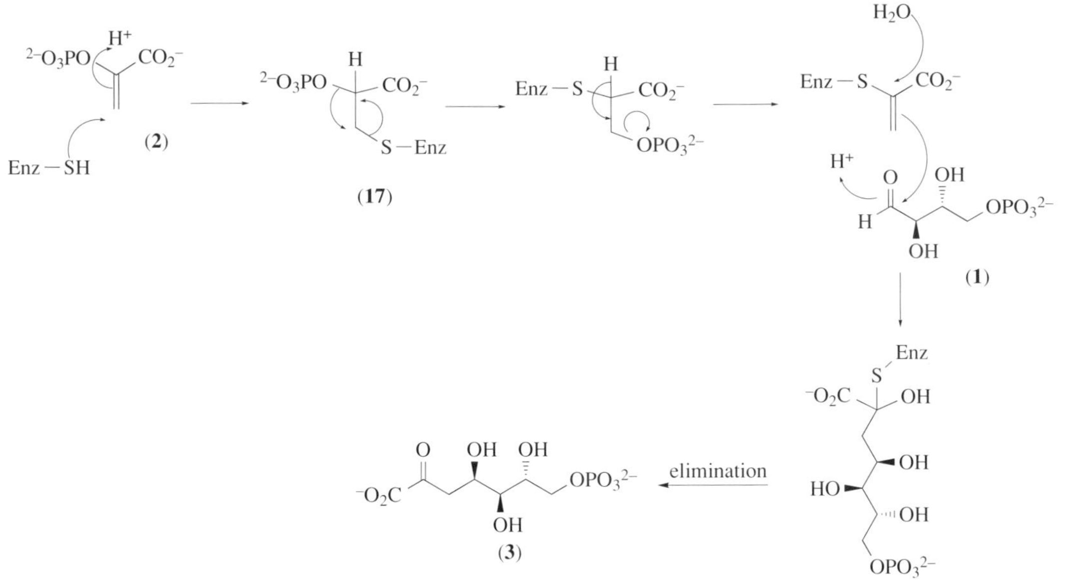

Scheme 4

activity. Reactivation with a variety of metal ions, gave the following order of activity: $Mn^{2+} > Cd^{2+}$, $Fe^{2+} > Co^{2+} > Ni^{2+}$, Cu^{2+}, $Zn^{2+} \gg Ca^{2+}$.

Mn^{2+}-activated enzymes show the highest activity. However, enzyme-bound Mn^{2+} is susceptible to displacement by other metal ions. In contrast, the enzymes complexed to Fe^{2+}, Co^{2+}, and Zn^{2+} form more stable metal–enzyme complexes. It is proposed that Fe^{2+} and perhaps Zn^{2+} are the preferred metal cofactors *in vivo*, based on their presence before removal by chelation, their affinity for the enzyme, and their high bioavailability.

It has been shown that there is cooperativity between metal and erythrose 4-phosphate for binding to the DAHP synthase(phe) enzyme and a dependency of K_m^{E4P} on the identity of the metal ion. For example, the K_m for the Mn^{2+} enzyme is 170 μM compared with 16 μM of the Zn^{2+} enzyme. No corresponding effect on K_m^{PEP} was observed.[33]

1.22.2.3 Active Site Studies on DAHP Synthase

Studies on the residues at the active site of DAHP synthase have focused on the evidence for a thiol.[13] For example, *E. coli* DAHP synthase(phe) was modified with 5,5′-dithiobis(2-nitrobenzoate) (DNTB).[34] In the absence of other ligands, DNTB initially modified two cysteines with loss of 90% of activity. This inactivation was slowed by PEP and Mn^{2+}.

There are only two cysteines conserved in the known bacterial and fungal DAHP synthase sequences: Cys61 and Cys328. Mutagenesis of Cys328 gave catalytically active protein, whereas all four Cys61 mutants were inactive. For these mutants Mn^{2+} did not protect from DNTB modification, suggesting that Cys61 is an essential residue for metal binding. Furthermore the 350 nm absorption band produced when Cu^{2+} is bound to the wild-type DAHP synthase(phe), is absent in the Cys61 mutants, but present in the Cys328 mutants.[34] Cys61 is part of a conserved Cys-X-X-His motif in the *E. coli* isozymes but this motif is not conserved in plant sequences. However, there is a Cys-X-X-His motif elsewhere in these sequences.[20]

1.22.2.4 Substrate and Substrate Analogues for DAHP Synthase

Erythrose 4-phosphate (**1**) can be prepared by oxidation of glucose 6-phosphate by lead tetraacetate.[35] It dimerizes readily to form three major dimeric species. At low concentrations the dimers dissociate to form the monomer, which exists mainly as the hydrate in solution.[36] Although erythrose 4-phosphate is generally present as the monomer at concentrations used in assays of DAHP synthase

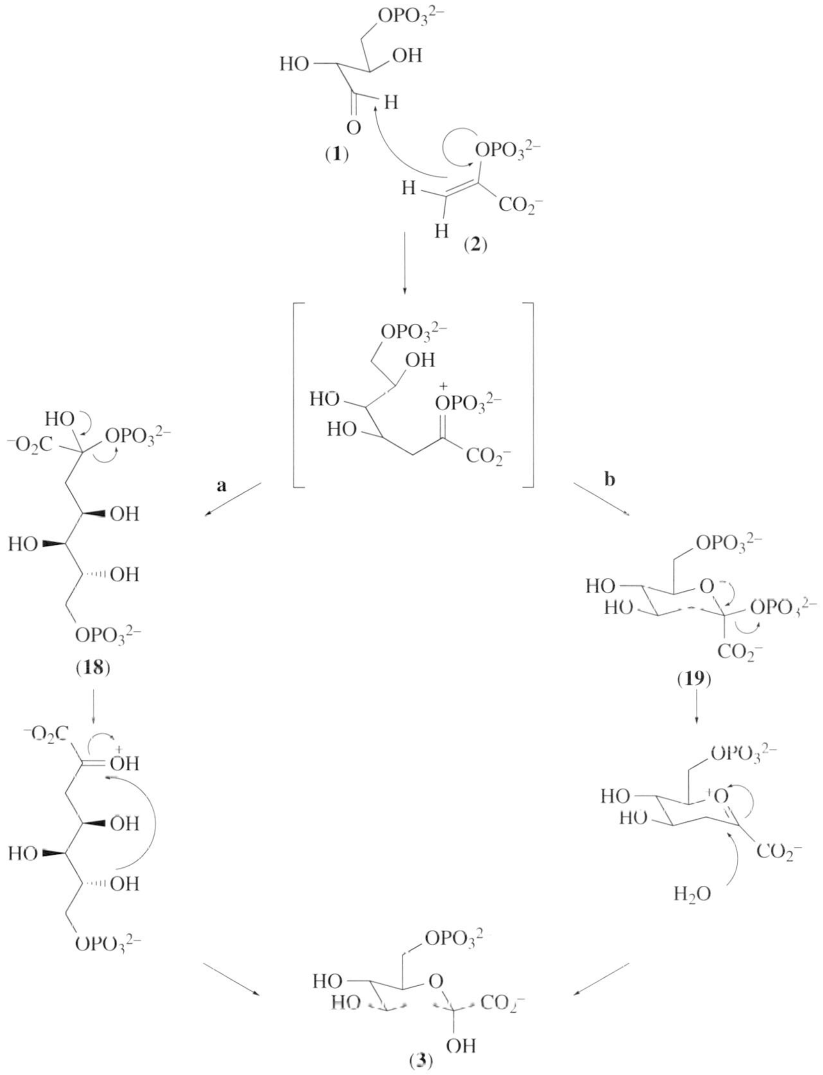

Scheme 5

activity, the rate of dissociation of the dimers is slow and care has to be taken to make sure the system has time to equilibrate. Otherwise a burst of DAHP synthase activity can be observed as the erythrose 4-phosphate monomer is consumed, followed by a slower rate which is limited by the dissociation of the dimers.[37]

The phosphonate (**20**) and the homophosphonate (**21**) analogues of erythrose 4-phosphate are substrates for *E. coli* DAHP synthase(tyr) with V_{max} values of approximately 30% and 5% of erythrose 4-phosphate, respectively.[38] Both isomers of 3-fluoroPEP (**22**) and (**23**) have been shown to be substrates for DAHP synthase,[39] whereas 3-bromopyruvate (**24**) has been found to be an irreversible inhibitor.[34,40] The inactivation is competitive with respect to PEP and the presence of metal ions protects against inactivation.

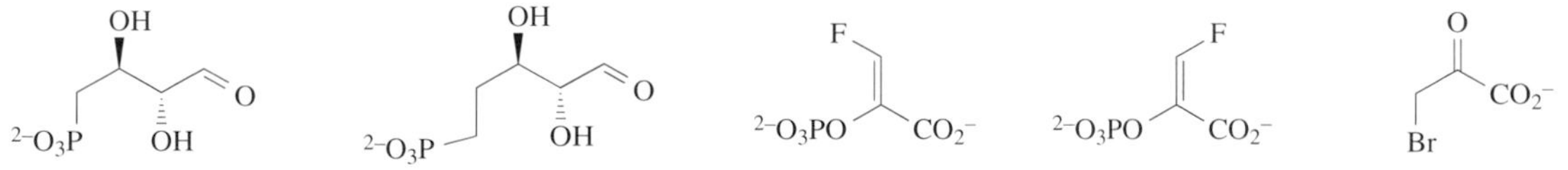

(**20**) (**21**) (**22**) (**23**) (**24**)

1.22.2.5 Mechanistic Studies on KDO8P Synthase

KDO8P synthase catalyzes the reaction between PEP and arabinose 5-phosphate (**25**) to form 3-deoxy-D-*manno*-octulosonate 8-phosphate (KDO8P, **26**) (Scheme 6). This reaction is analogous to the reaction catalyzed by DAHP synthase and several mechanistic similarities have been established: the reaction involves C—O bond cleavage of PEP;[41] the *si* face of PEP attacks the *re* face of the aldehyde;[42,43] and the reaction has an ordered sequential mechanism.[44]

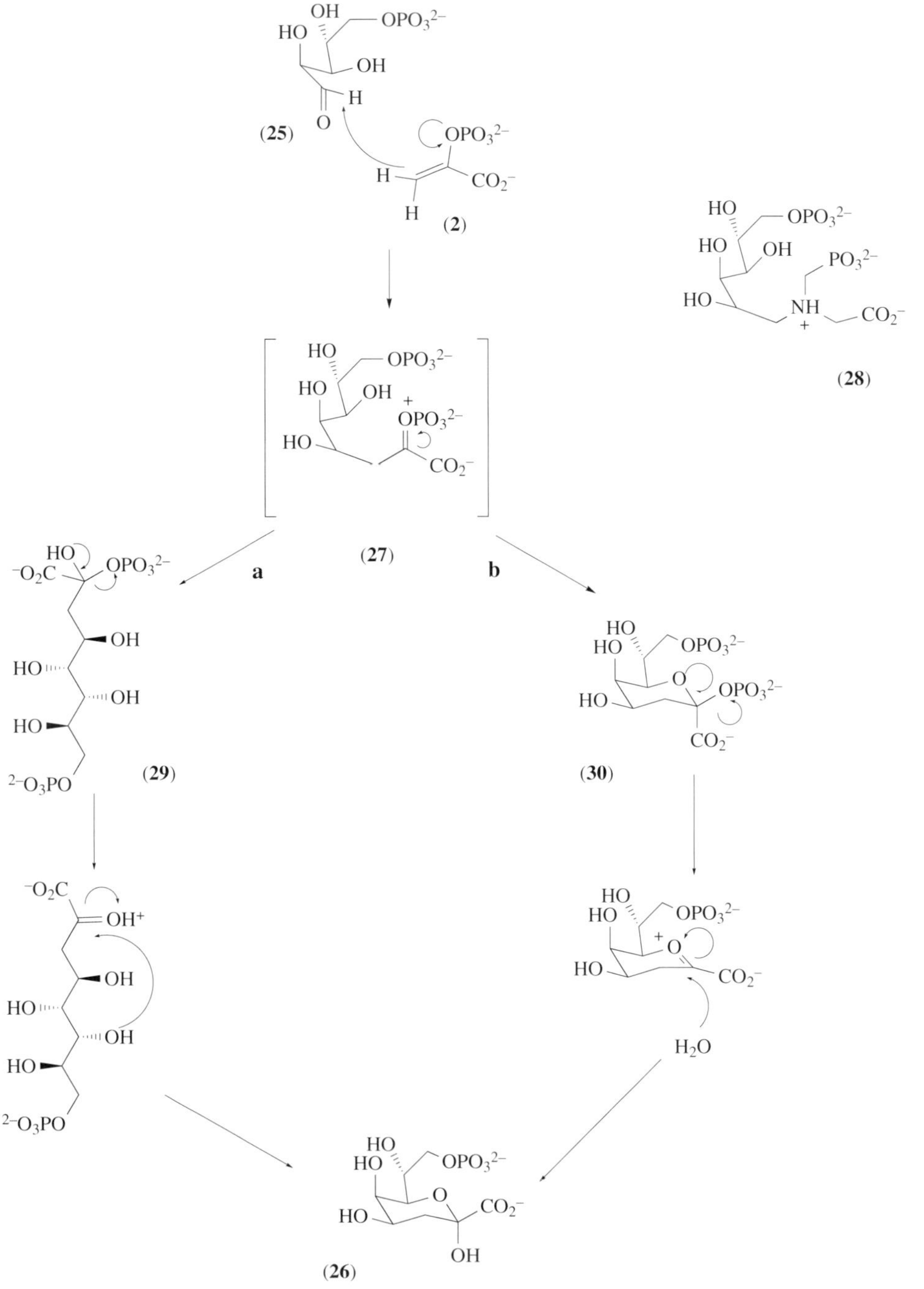

Scheme 6

Mechanisms similar to those proposed in Scheme 5 for DAHP synthase are shown for KDO8P synthase in Scheme 6. After nucleophilic attack on arabinose 5-phosphate by PEP, a cationic intermediate (**27**) may form. Compound (**28**) has been designed to mimic this cation and is a potent inhibitor of the enzyme with a K_i of 3 μM.[45] The cationic species (**27**) can either be quenched by attack of water (path **a**) to form the acyclic intermediate (**29**) or intramolecularly by the C-5 alcohol (path **b**) to form the cyclic intermediate (**30**). A number of lines of evidence suggested path **b**.[46] These included the observation that 3-deoxyarabinose 5-phosphate shows no substrate or inhibitory activity and that the acyclic product analogue (**31**) is not an inhibitor.[47] More persuasive was the finding that the phosphonate analogue (**32**) of the proposed cyclic intermediate (**30**) was an inhibitor with a K_i of 5 μM.[48]

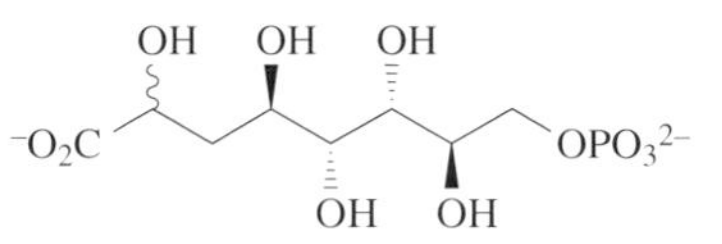

(**31**) 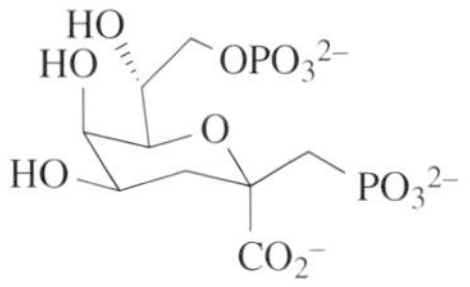 (**32**)

However, (**30**) has been synthesized and shown to be neither a substrate nor a potent inhibitor.[49] Pre-steady-state experiments which looked for the presence of (**30**) in the enzyme reaction also gave negative results—and additionally did not detect any enzyme-bound intermediate.[50] These results have prompted a major reevaluation of the mechanism in favor of path **a**, via the acyclic intermediate (**29**), albeit that there is not direct evidence for this mechanism.

In considering these results with KDO8P synthase, it must be borne in mind that unlike DAHP synthase, KDO8P synthase is not a metalloenzyme.[51]

1.22.3 DEHYDROQUINATE SYNTHASE

Dehydroquinate synthase (EC 4.6.1.3) catalyzes the conversion of DAHP (**3**) into the carbocyclic dehydroquinate (**4**) (Scheme 7). This transformation is mechanistically complex and involves an oxidation and reduction reaction, a phosphate cleavage, and an intramolecular aldol condensation. The complexity of this chemistry and the relative smallness of the enzyme (a monomer of 362 amino acids, requiring one NAD^+ and one divalent metal ion for catalysis),[52] have prompted rigorous enquiry into how the transformations are catalyzed.

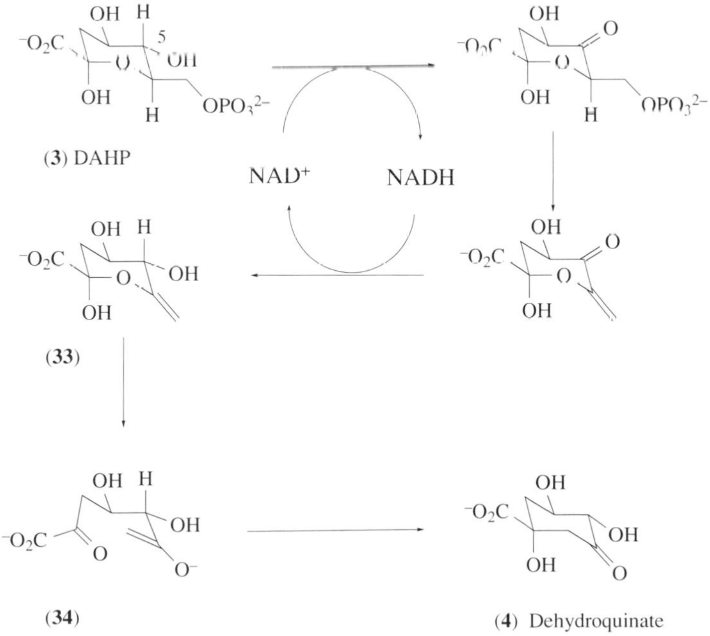

(**3**) DAHP

NAD^+ NADH

(**33**)

(**34**) (**4**) Dehydroquinate

Scheme 7

Dehydroquinate synthase was purified 9000-fold from *E. coli* K12. Subcloning of the *aroB* gene behind a tac promoter gave *E. coli* transformants that produced 2×10^4 times more enzyme.[53] The homogeneous enzyme is a monomeric protein of M_r 3.9×10^4 that contains one tightly bound Co^{2+}

and one NAD^+.[54] The dissociation of NAD^+ from the holoenzyme is slow but is accelerated 40-fold under turnover conditions with saturating substrate. This dissociation allows nonfunctional nicotinamide analogues to bind to the enzyme if these are present as impurities in the NAD^+ used. This causes an exponential loss of enzyme activity. This is an important practical consideration when using this enzyme. The *apo* enzyme, prepared by incubating the enzyme with EDTA, is stable, but catalytically inactive. Incubation with Co^{2+} and NAD^+ fully restores the enzyme activity, whereas Zn^{2+} gives enzyme with 53% the activity of the Co^{2+} enzyme. It is thought likely that Zn^{2+} is the functioning metal *in vivo*. The apparent K_m for DAHP is 4 μM.[55]

The dehydroquinate synthase functional domain from the pentafunctional *arom* complex in *A. nidulans* has been overproduced in *E. coli*[56] and purified to homogeneity. The apparent k_{cat} is 8 s^{-1}, and the apparent K_m values for NAD^+ and DAHP are 3 μM and 2.2 μM, respectively. The monofunctional domain is inactivated by metal-chelating agents. Reactivation with Zn^{2+} gave the highest k_{cat}/K_m. Both substrate and phosphate protect against diethyl pyrocarbonate inactivation.[57]

1.22.3.1 Mechanistic Studies on Dehydroquinate Synthase

The overall transformation to form dehydroquinate from DAHP does not involve a change in oxidation state, and so the requirement for a nicotinamide cofactor was surprising. A mechanism that utilizes NAD^+ was proposed whereby the C-5 hydroxyl is oxidized to a ketone. This makes the C-6 hydrogen more acidic and facilitates the loss of phosphate in a stepwise β-elimination reaction.[58,59] Reduction of the C-5 ketone by the enzyme-bound NADH then regenerates the C-5 hydroxyl group with unchanged stereochemistry and forms the enolpyranose (**33**), a key intermediate in the overall transformation. Ring opening of the enolpyranose directly forms an enolate (**34**) which attacks back on the C-2 ketone to form dehydroquinate (**4**).

Indirect evidence for the oxidation of the C-5 hydroxyl to the ketone comes from the observation of a k_H/k_T tritium isotope effect of 1.7 on the reaction when [1-^{14}C,5-^{3}H]DAHP is the substrate.[59] The formation of NADH in the catalytic cycle cannot be detected from its absorption at 340 nm with DAHP under steady-state conditions. However, when the carbahomophosphonate analogue (**35**) was incubated with the enzyme approximately 85% of the enzyme-bound NAD^+ was converted to NADH (Scheme 8).[52] This experiment exploited the more favorable oxidation potential of a secondary alcohol with fewer oxygen substituents on adjacent carbons.

NAD^+ NADH

(**35**)

Scheme 8

The cyclic 2-deoxy substrate analogue (**36**) has been used to great effect to study the first part of the dehydroquinate synthase reaction. It behaves as a substrate for the initial part of the transformation, eliminating phosphate and forming the 2-deoxy analogue of the enol pyranose (**37**), albeit at a slower rate (2%) than the overall enzymatic transformation with the natural substrate (Scheme 9).[60,61] By incubating the enzyme with (**36**) labeled stereospecifically with deuterium at the *proS* position on C-7, (**37**) was isolated labeled in the (*E*) position (Scheme 9). This is the labeling expected if the elimination of the phosphate is stereospecific and *syn*.

The *syn* elimination of phosphate is likely to proceed through an enolate intermediate by a stepwise E1CB mechanism. In order to probe this mechanism, a series of substrate analogues were incubated with the enzyme in D_2O to see if there was exchange of the C-6 proton. Such exchange, at a rate about 0.1% k_{cat}, was observed for the carbahomophosphonate (**35**) but not the phosphonate (**38**) and the *cis*-vinyl homophosphonate (**39**) but not the *trans*-vinyl homophosphonate (**40**). This is despite the fact that (**38**) is the most potent competitive inhibitor of dehydroquinate synthase known, and is readily oxidized at C-5 when bound.[55] Consideration of these results led to the proposal that the phosphate dianion could be the base involved in deprotonation. At physiological pH it would be expected to be a strong base. Furthermore, it was argued that such intramolecular deprotonation would overcome some of the difficulties of accessing such a hindered proton.

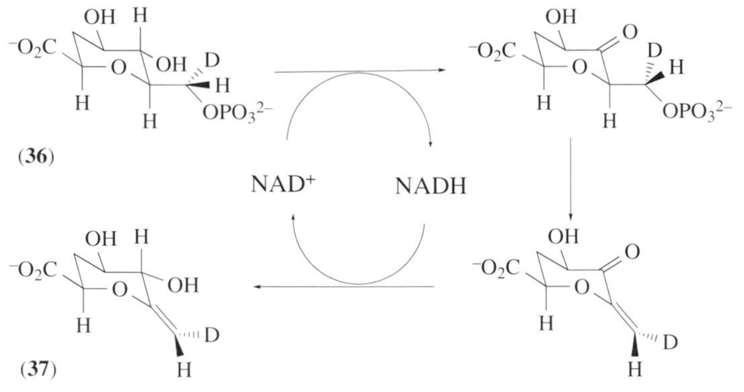

Scheme 9

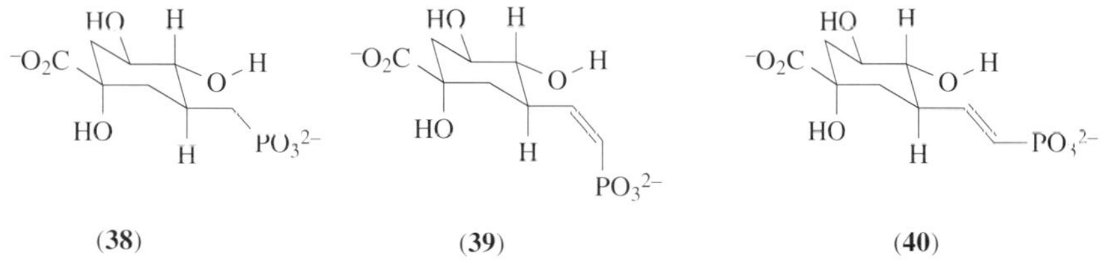

(38) (39) (40)

From the stereochemistry of the phosphate elimination and the stereochemistry of the overall transformation in the conversion of DAHP to dehydroquinate, shown previously to involve inversion of stereochemistry at C-7,[62,63] the stereochemistry of the intramolecular aldol reaction can be deduced. It was proposed that this reaction proceeds through a chair transition state (**41**) (Scheme 10).[60] Although the stereochemical results preclude boat transition state (**42**) and chair transition state (**43**), they do not exclude the alternative boat transition state (**44**).

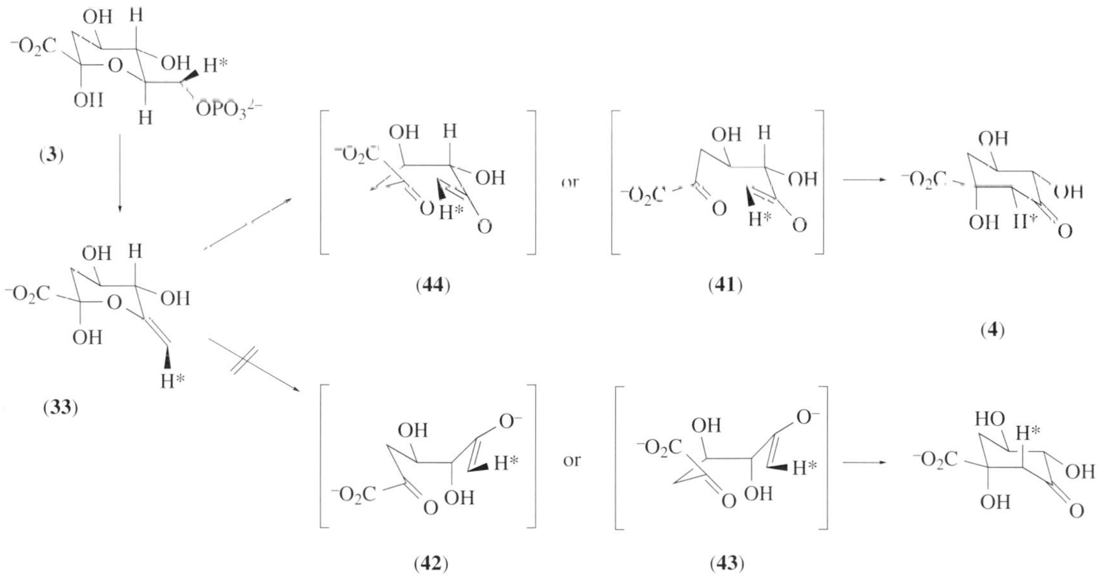

Scheme 10

The enolpyranose intermediate was synthesized with a photolabile *o*-nitrobenzyl protecting group on the C-2 hydroxyl (**45**). Release of the protecting group was reported to yield exclusively dehydroquinate.[64] Furthermore, the conversion was shown to proceed with the same stereochemistry as the enzymatic reaction. This led to the proposal that the enolpyranose (**33**) was the product of dehydroquinate synthase, and the suggestion that it was unlikely that an enzyme would evolve to catalyze a reaction that occurs rapidly in its absence.[64] This proposal was subsequently revised when the 1-*epi*-dehydroquinate (**46**) was synthesized and shown to be produced in small amounts (2–4%) in the nonenzymatic cyclization of the enolpyranose, which was consequently not entirely specific (Scheme 11).[65] By contrast the enzymatic reaction is entirely specific. Addition of excess enzyme did not affect the ratio of (**46**) to (**4**) formed, suggesting that the rate of spontaneous rearrangement of the enol pyranose was faster than uptake by the enzyme. From this result it is necessary to assume that the enzyme is still playing some role in directing the correct cyclization.

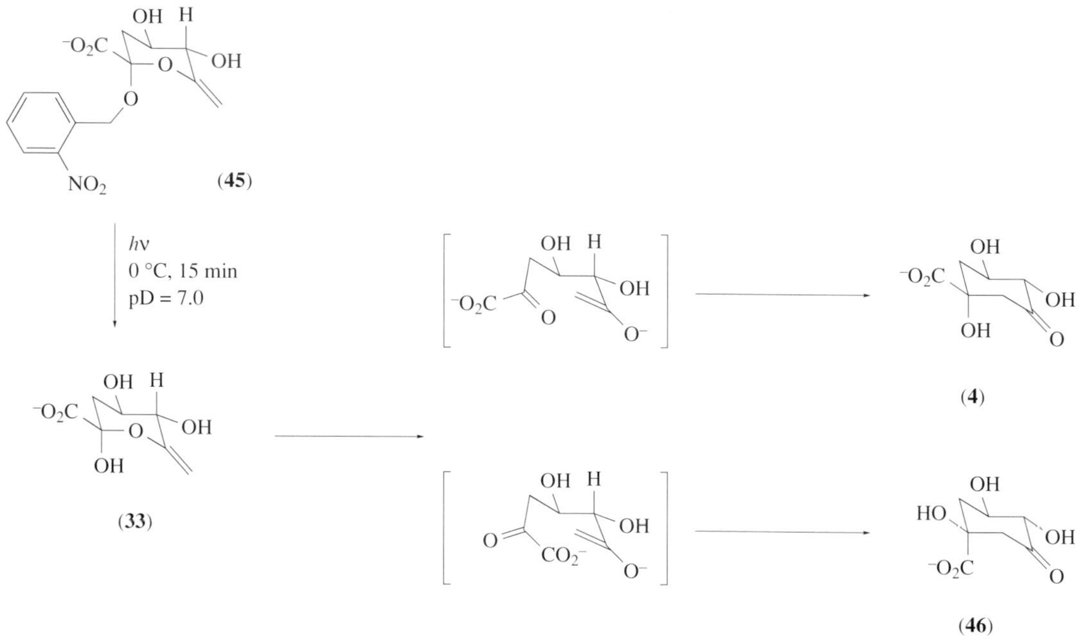

(**45**) (**33**) (**4**) (**46**)

Scheme 11

There is a twist in the tail in this story. It has been found that whereas (3*R*)-3-fluoroDAHP (**47**) is smoothly converted by dehydroquinate synthase into (6*R*)-6-fluorodehydroquinate (**48**), (3*S*)-3-fluorodehydroquinate (**49**) is converted much more slowly into a 2:1 mixture of (6*S*)-6-fluoro-dehydroquinate (**50**) and (6*S*)-6-fluoro-1-*epi*-dehydroquinate (**51**) (Scheme 12). It is proposed that the axial fluorine stabilizes the enol pyranose (**52**) long enough for it to dissociate from the enzyme. It then cyclizes nonenzymatically in solution.[66]

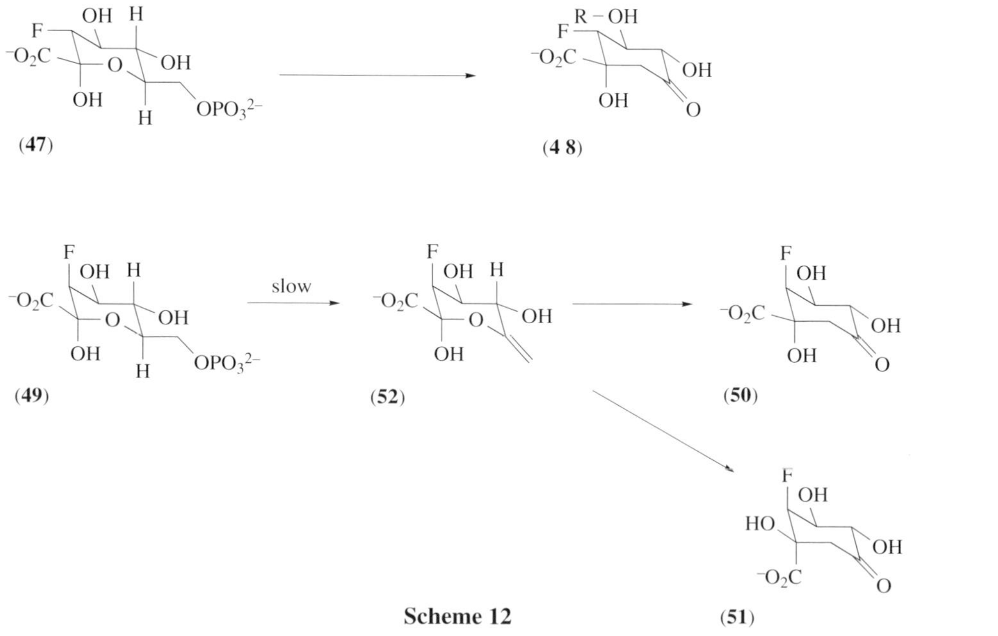

(**47**) (**48**) (**49**) (**52**) (**50**) (**51**)

Scheme 12

1.22.3.2 Substrate, Substrate Analogues, and Inhibitors of Dehydroquinate Synthase

DAHP exists almost exclusively as the cyclic pyranose.[67] There are several synthetic routes to DAHP, most notably from 2-deoxyglucose.[68] This synthesis compares favorably with isolation of DAHP from auxotrophic strains of *E. coli* lacking dehydroquinate synthase activity.[68] DAHP synthase, hexokinase, pyruvate kinase, and transketolase were all immobilized and used together in a multienzyme reactor to form DAHP from fructose.[69]

A series of cyclic analogues of DAHP have been tested as inhibitors of *E. coli* dehydroquinase.[70] Of note is the observation that the homophosphonate (**53**) shows no inhibition, whereas the corresponding phosphonate (**54**) is a reasonable inhibitor. A similar trend is seen for the carbocyclic homophosphonate (**35**) and the carbocyclic phosphonate (**38**), where the latter is a potent inhibitor.[55] Somewhat surprisingly, the 2-deoxy-compounds (**55**) and (**56**) are less good inhibitors than their C-1 epimers (**57**) and (**58**). Compounds (**53**), (**54**), (**56**), and (**58**) have also been tested against the purified enzyme from *Pisum sativum*.[71] The inhibition pattern generally mirrored that seen with the *E. coli* enzyme. Furthermore, when (**54**) was applied to several postemergent plants, 3-deoxy-D-*arabino*-heptulosonate accumulation and effects on plant growth were observed. These results offer the exciting possibility of herbicidal activity by inhibition of enzymes on the shikimate pathway other than EPSP synthase. Synthetic approaches and the use of immobilized enzymes or whole cells to make (**54**) have been compared.[69]

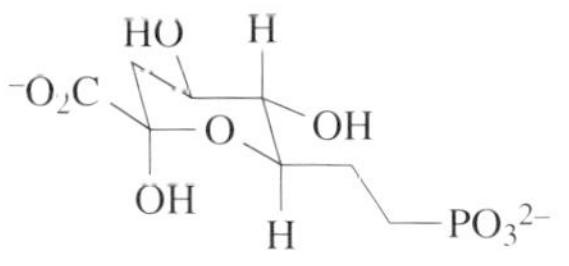

(**53**) 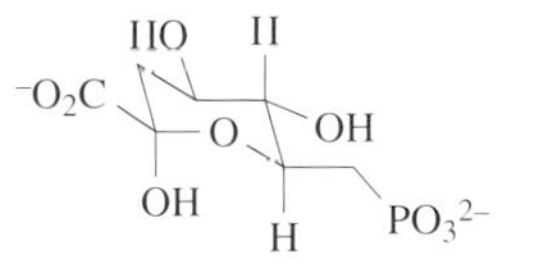 (**54**) 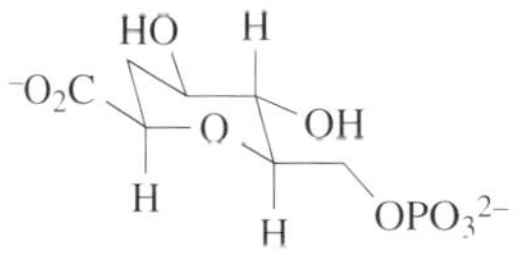 (**55**)

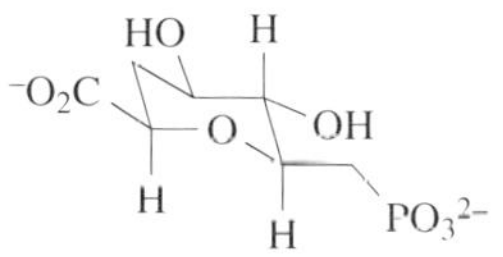

(**56**) 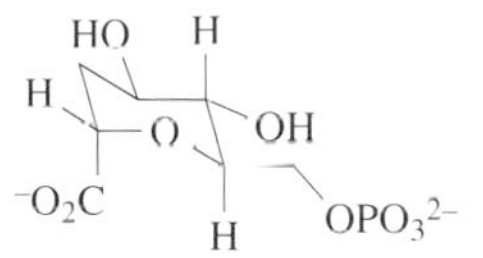 (**57**) 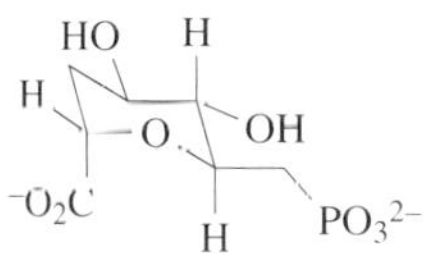 (**58**)

1.22.4 DEHYDROQUINASE

Dehydroquinase (3-dehydroquinate dehydratase, EC 4.2.1.10) catalyzes the reversible dehydration of dehydroquinate (**4**) to form dehydroshikimate (**5**). This reaction is a step in both the biosynthetic shikimate pathway to chorismate (**9**) and the catabolic quinate pathway, the first steps of which involve the conversion of quinate (**59**) to protocatechuate (**60**) (Scheme 13). The quinate pathway is an inducible pathway in fungi that enables them to use quinate as a food source. Unlike the shikimate pathway it is not operating in cells all the time. Two distinct classes of dehydroquinase are now recognized, based on amino acid sequences and biophysical criteria.[72] These are referred to as type I and type II dehydroquinases. Most of the literature relates to type I enzymes, which have only been reported as part of the shikimate pathway. The type II enzymes function in both the quinate and shikimate pathways.

Type I and type II dehydroquinases share no similarity at the DNA and protein sequence level,[73] although type II dehydroquinase has a limited degree of sequence similarity with dehydroquinate synthase.[74] This observation, coupled with the isolation of type II dehydroquinases functioning in the biosynthetic pathways of *S. coelicolor* and *Mycobacterium tuberculosis*, has led to the suggestion that the type I and type II dehydroquinases both arose from enzymes that had evolved independently as part of the shikimate biosynthetic pathways in different prokaryotes.[74]

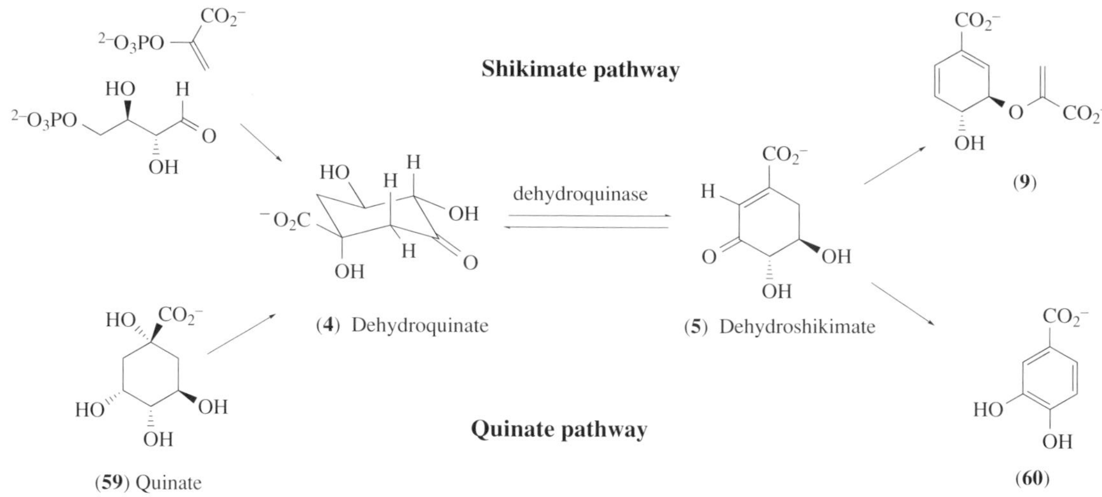

Shikimate pathway

dehydroquinase

(4) Dehydroquinate

(5) Dehydroshikimate

(9)

(59) Quinate

(60)

Quinate pathway

Scheme 13

1.22.4.1 Type I Dehydroquinase

Type I dehydroquinases have been isolated in three different polypeptide forms. The bacterial enzymes from *E. coli*[75] and *Salmonella typhi*[76,77] are monofunctional. A bifunctional form with shikimate dehydrogenase has been identified in the plants *P. sativum*[6,78] and *Physcomitrella patens*.[79] The sequence of the dehydroquinase domain of the *P. sativum* bifunctional dehydroquinase–shikimate dehydrogenase has been determined.[80] In *N. crassa*,[81] *A. nidulans*,[82] *E. gracilis*,[83] and *S. cerevisiae*[84] dehydroquinase is part of the pentafunctional *arom* protein. The dehydroquinase domain of the *arom* protein has been separately overproduced and purified from *A. nidulans*.[85] The amino acid sequences of dehydroquinases from all these organisms show a high degree of similarity.[76]

The *E. coli* dehydroquinase has been purified to homogeneity from wild type,[86] and in milligram quantities from an overproducing strain.[75,87] The native enzyme is dimeric, consisting of subunits of 252 amino acids and molecular weight of 27 466.[88] The K_m for dehydroquinate is 16 μM.[72] This compares with the values of 5 μM and 18 μM for the *N. crassa* and *S. typhi* enzymes, respectively.[72,77] The *E. coli* and *S. typhi* enzymes were crystallized in 1992.[89] The enzyme from *S. typhi* is very similar to the *E. coli* enzyme, sharing 69% sequence identity at the amino acid level. Catalytically active residues identified in *E. coli* enzyme are conserved in *S. typhi*.[76]

1.22.4.1.1 Mechanistic studies on type I dehydroquinase

The enzymatic conversion of 3-dehydroquinate (**4**) to dehydroshikimate (**5**) is reversible, with an equilibrium constant of 15 in favor of dehydroshikimate.[90] The reverse reaction was shown to proceed with *syn* addition of water across the double bond of dehydroshikimate (**5**).[91] The forward reaction was subsequently studied using partially deuteriated (2*S*)-[2-^{2}H]- and (2*R*)-[2-^{2}H]dehydroquinate and the *syn* stereochemistry of the elimination confirmed.[92,93]

The enzymatic reaction proceeds via an imine (Schiff base) mechanism between the ε-NH_2 of an active-site lysine, and the C-3 ketone of dehydroquinate (Scheme 14). The key diagnostic experiment was the addition of sodium borohydride to dehydroquinase in the presence of an equilibrium mixture of (**4**) and (**5**), resulting in irreversible inactivation of the enzyme.[94] The active site lysine in the *E. coli* enzyme has been identified as Lys-170,[88] by using Na[^{3}H]BH_4 in the inactivation experiment, followed by proteolytic degradation and sequencing. Lys-170 is conserved in all type I dehydroquinases, but not in type II dehydroquinases.[74] Replacement of Lys-170 by an alanine residue gives a mutant protein which is still able to bind substrate and product but is 10^6-fold less catalytically active.[95]

The imine intermediate (**61**) (Scheme 14) has been detected by electrospray mass spectrometry.[96] The electrospray spectrum of enzyme inactivated by addition of sodium borohydride in the presence

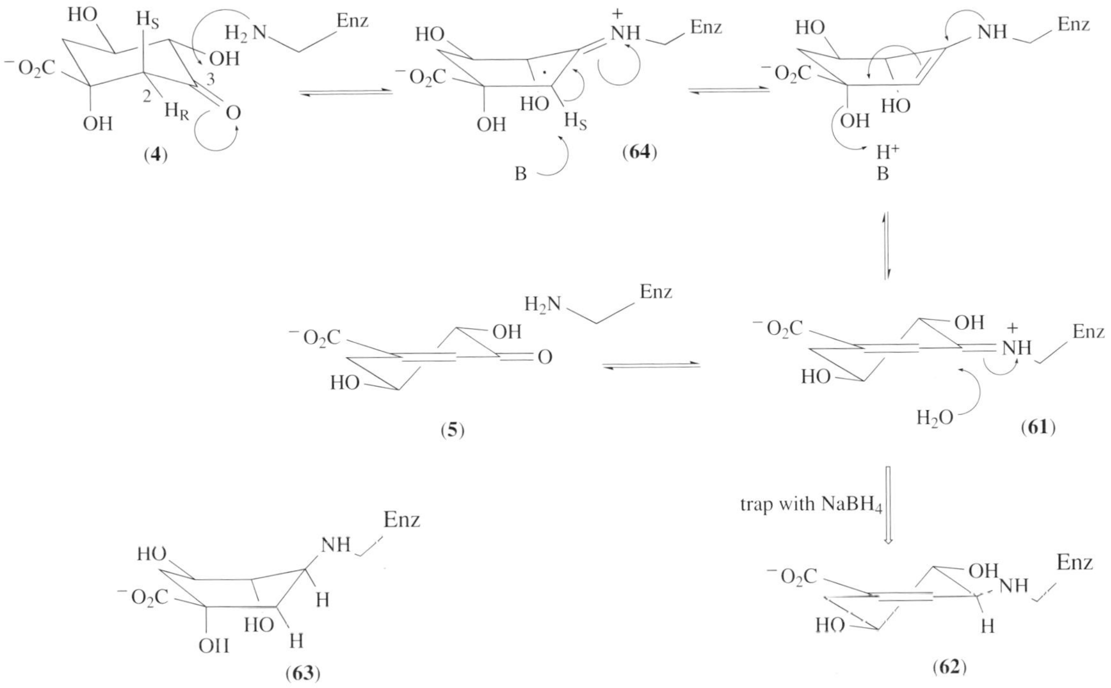

Scheme 14

of an equilibrium mixture of (4) and (5) showed only (62), the reduced form of the product-imine (61) (Scheme 14). There is no evidence for (63), the reduced form of imine (64). No unmodified enzyme was detected, showing that both active sites in the dimer are used and the enzyme does not exhibit half of sites reactivity. Borohydride reduction of the imine intermediate on the enzyme increases the conformational stability of the protein. The melting temperature for the protein increases by 40 °C and, as a result, the concentration of unfolded protein at room temperature is decreased by over three orders of magnitude.[97] The dimeric quaternary structure is also stabilized.[98]

In the mechanism shown in Scheme 14, initial imine formation is followed by abstraction of the C-2 *pro-R* proton by the active site base. The imine acts as an electron sink to stabilize the carbanion thus formed, before the expulsion of the hydroxyl group with protonation by an acidic residue in the active site. In solution, dehydroquinate adopts a chair conformation. In this conformation the axial *pro-S* proton at C-2 is the more acidic, having the greater overlap with the π orbital of the carbonyl group. Under nonenzymatic conditions this proton is lost,[62] presumably by an E1CB reaction. In order to activate the *pro-R* proton towards abstraction it has been suggested that the enzyme must enforce a conformational change on the imine-bound substrate to bring the *pro-R* proton coplanar with the π orbital of the imine, and thereby increase its acidity. Mechanisms have been proposed in which the imine adopts either a boat conformation,[62] or a twist-boat conformation.[99]

Using (2*R*)-[2-^{3}H]dehydroquinate, a primary kinetic tritium isotope effect of 2.3 was observed for the dehydroquinase reaction.[59] A secondary kinetic isotope effect of 1.19 for the conversion of (2*S*)-[2-^{3}H]dehydroquinate to [2-^{3}H]dehydroshikimate was also reported. A primary deuterium kinetic isotope effect on V_{max}/K_m of 2.1 has subsequently been measured.[100]

The active site base has been shown to be a histidine.[86] Studies of the variation of V_{max} with pH showed that for maximal activity there is a requirement for deprotonation of a group with a pK_a of 6.1, and also that diethylpyrocarbonate irreversibly inhibited the enzyme. Inactivation and peptide mapping studies have identified the histidine as His-143,[101] which is in a highly conserved region of the amino acid sequence. A His143Ala mutant was prepared and shown to be 10^6-fold less active.[95] A significant fraction of this mutant was isolated from overexpressing cells strain having the product already bound as the imine (61). It appears that even without this key catalytic residue (4) binds and is slowly converted to (61), but that the breakdown of this imine is stalled. Active-site labeling studies with iodoacetate (K_i = 30 mM) labeled two of the 11 methionines, Met23 and Met205.[102] However, a Met205Leu mutant dehydroquinase was shown to have a very similar K_m and k_{cat}.[95]

Type I dehydroquinase is also irreversibly inactivated by 3-chloroacetylcyclohexanone (**65**), the 1,2-epoxy substrate analogue (**66**), and the photoaffinity label (**67**).[103]

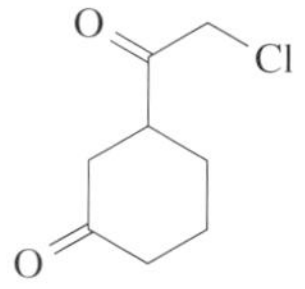

(**65**) 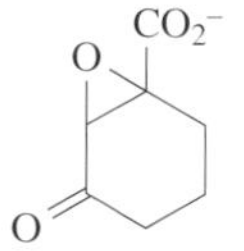 (**66**) 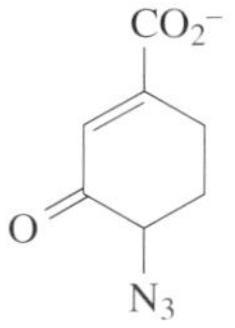 (**67**)

1.22.4.2 Type II Dehydroquinase

The type II enzymes are found on both the shikimate and quinate pathways. Quinate pathway type II dehydroquinases have been characterized from *A. nidulans*[72,104] and *N. crassa*.[85,105–107] Type II enzymes that are on the shikimate pathway have been described from *M. tuberculosis*,[74,108] *Helicobacter pylori*,[109] and *S. coelicolor*.[110] The type II dehydroquinase from *A. methanolica* operates on both pathways.[111] Type II dehydroquinases show significant sequence identity with each other, but not with the type I enzymes. The catabolic dehydroquinases from *A. nidulans* and *N. crassa* form part of the *qut* and *qa* gene clusters, respectively. The genes encoding the dehydroquinases from *M. tuberculosis* and *A. methanolica* are genetically linked to the corresponding genes encoding dehydroquinate synthase, but their transcription results in the synthesis of two separate proteins.[74,111]

Type II dehydroquinases are reported to be dodecameric with subunit molecular weights between 1.2×10^4 (*A. methanolica*) and 1.85×10^4 (*N. crassa*). Preliminary crystallization data on *M. tuberculosis* type II dehydroquinase have been reported.[112] Analysis of native and refolded enzyme by electron microscopy showed that the *A. nidulans* enzyme adopts a ring-like structure similar to that of glutamine synthase, suggesting an arrangement of two hexameric rings stacked on top of one another.[113] The type II dehydroquinases are thermally stable, with none losing any activity on heating at 70 °C for 10 minutes.

The reported K_m values for type II dehydroquinases range from 9 μM for the *M. tuberculosis* enzyme[100] to 650 μM for the enzyme from *S. coelicolor*.[110] In general, dehydroquinases on the inducible quinate pathway have relatively high K_m values, whereas biosynthetic dehydroquinases, active in the constitutive shikimate pathway, tend to have relatively low K_m values. The dual function enzyme from *A. methanolica* has an intermediate value.[111] The *A. nidulans* enzyme has been shown to be competitively inhibited by phosphate and bicarbonate, whilst no enzymatic activity is detected in the presence of citrate.[72] However, unlike the type I enzymes, the type II dehydroquinases are not inhibited by chloride or acetate ions.[110] Consequently, routine assays of type II dehydroquinases are carried out in Tris–HCl or Tris–OAc buffer, whilst the type I enzymes are assayed in phosphate buffer.

1.22.4.2.1 Mechanistic studies on type II dehydroquinase

Type II dehydroquinases are not inactivated by treatment with sodium borohydride or cyanoborohydride in the presence of an equilibrium mixture of substrate and product.[72] Furthermore, there is no conserved lysine in the type II sequences.[74] These two results effectively rule out the kind of mechanism used by the type I enzyme involving an imine intermediate. The mechanism of the type II dehydroquinase also differs from that of the type I enzyme in a significant, if cryptic way. The reaction proceeds with *anti* stereochemistry, involving the loss of the more acidic *proS* hydrogen from C-2.[114,115] This is the same stereochemical course as is observed in the acid- or base-catalyzed conversion of (**4**) to (**5**).[62]

Diethyl pyrocarbonate irreversibly inhibits the dehydroquinase from *A. nidulans*, suggesting the presence of a histidine at the active site.[72] Reactive arginine residues which are essential for catalytic activity have been identified in the *S. coelicolor* and *A. nidulans* enzymes by treatment with phenylglyoxal, followed by proteolytic digestion and analysis using electrospray mass spectrometry.[113,116,117] These studies identified a specific reactive arginine residue, Arg19 of the *A. nidulans* enzyme and Arg23 in the *S. coelicolor* enzyme. This arginine is in a conserved structural motif that might reflect a common substrate binding fold shared by type I and type II enzymes. Replacement of Arg23 in *S. coelicolor* dehydroquinase using site-directed mutagenesis by lysine, glutamine, and alanine residues led to mutant enzymes which were very much less active.[117]

A study of the effect of pH on the V_{max} of the *A. nidulans*[72] enzyme showed a maximal value at pH 8.5–9.0. However, subsequent studies of the pH dependence of both V_{max} and K_m showed they both increased with pH up to at least pH 10. It was proposed that these data were consistent with an arginine at the active site (Scheme 15). As the arginine becomes deprotonated it would bind the substrate less well and so increase K_m. However, the loss of an electrostatic interaction with the active site base (presumed to be a histidine) could increase its basicity and so lead to an increase in k_{cat}.[100] The results of these mechanistic studies are consistent with a mechanism proceeding by an E1CB mechanism through an enolate intermediate (**68**) (Scheme 15). Arginine residues may also be involved in stabilizing the negative charge on the enolate in the transition state. It is known that the type II dehydroquinase does not employ a metal for this stabilization.[113]

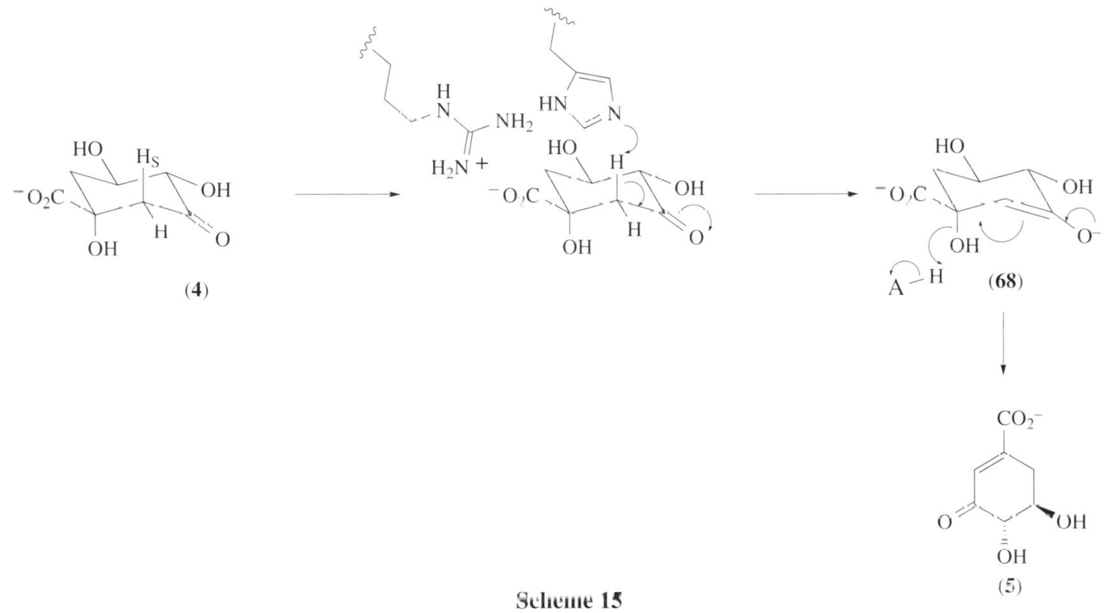

Scheme 15

What little is known about the substrate specificity of type II dehydroquinases reveals subtle differences from the type I enzymes.[118] 5-Deoxydehydroquinate (**69**) and 4,5-dideoxy dehydroquinate (**70**) are equally poor substrates for the *M. tuberculosis* type II enzyme as measured by a drop of 10^5-fold in their k_{cat}/K_m compared to dehydroquinate. In comparison, (**69**) is a reasonable substrate for the type I enzyme, the drop in k_{cat}/K_m being only 100-fold, but (**70**) is not a substrate at all.

1.22.5 THE QUINATE PATHWAY

The quinate pathway is an inducible catabolic pathway which enables fungi to utilize quinate (**59**) as a carbon source via the β-oxoadipate pathway.[119] Quinate is converted to protocatechuate (**60**) via dehydroquinate (**4**) and dehydroshikimate (**5**) in reactions catalyzed by a quinate dehydrogenase, dehydroquinase, and dehydroshikimate dehydratase (Scheme 13). Although the dehydration of dehydroquinate to dehydroshikimate is identical to the reaction in the shikimate pathway, it has been shown that there are two distinct dehydroquinases in *A. nidulans* and *N. crassa*. The gene for the shikimate pathway dehydroquinase is in the *arom* gene cluster in *N. crassa* and *A. nidulans*, whilst that for the quinate pathway enzyme is part of the *qa* gene cluster in *N. crassa*[119] and the *qut* gene cluster in *A. nidulans*.[120]

The catabolic *N. crassa* quinate dehydrogenase is monomeric with a molecular weight of approximately 40 kDa and is reported to show comparable activities with quinate and shikimate.[121] The gene has been sequenced[122] and shown to have an aspartate in the final position of the nucleotide binding domain, whereas *E. coli* shikimate dehydrogenase has a threonine in this position. This may account for quinate dehydrogenase's specificity for NAD^+ rather than $NADP^+$. The corresponding quinate dehydrogenase from *A. nidulans* has been overexpressed.[85] It is bifunctional and shows significant sequence similarity with the shikimate dehydrogenase in the *arom* protein.[123] The dehydroshikimate dehydratase has been constitutively overexpressed in *A. nidulans*,[124] and more recently overproduced in *E. coli*. The purified enzyme was shown to have a K_m of 530 μM for (**5**) and a requirement for bivalent metal cations such as Mg^{2+}, Mn^{2+}, or Zn^{2+}.[125]

In *N. crassa* the tightly linked *qa* gene cluster is made up from seven genes in a 17.5 kb region of DNA. The transcriptional map of this region is shown in Figure 2. Quinate dehydrogenase is encoded by *qa-3*, dehydroquinase by *qa-2*, and dehydroshikimate dehydratase by *qa-4*. There are two regulatory genes, *qa-1F* and *qa-1S*. The *qa-Y* gene shares sequence identity with the *A. nidulans* *qutD* gene which encodes a permease involved in uptake of quinate.[123,126]

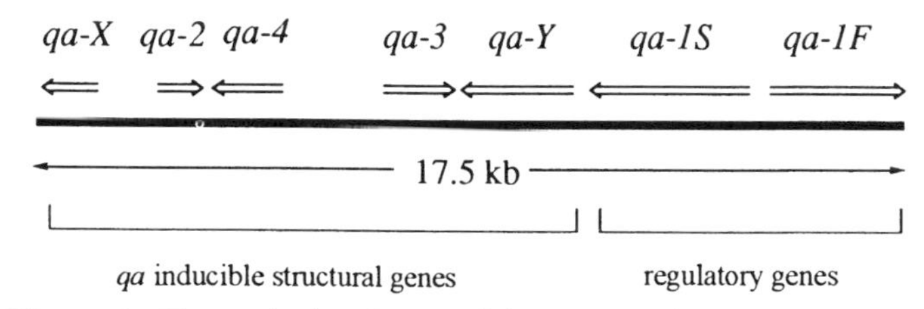

Figure 2 Transcriptional map of the *qa* gene cluster in *N. crassa*.

The regulatory genes encode repressor (*qa-1S*) and activator (*qa-1F*) proteins of 101 kDa and 89 kDa, respectively. These two proteins regulate the utilization of quinate as a carbon source. The repressor protein controls the expression of the *qa-1F* and consequently of the *qa* structural genes. Addition of an inducer releases the inhibition of the activator gene expression by the repressor protein, whereupon the activator initiates its own synthesis as well as transcription of all the other *qa* genes. The activator also stimulates production of the repressor protein, ensuring that sufficient repressor is present to turn off the system when the concentration of quinate falls.[119]

The amino acid sequence of the *qa-1S* repressor protein shows a high level of sequence similarity to the three C-terminal domains of the *N. crassa arom* complex, which in turn resemble the corresponding monofunctional *E. coli* enzymes: shikimate kinase, dehydroquinase, and shikimate dehydrogenase.[127] Although the repressor protein can bind quinate, dehydroquinate, or dehydroshikimate as an inducer, the three domains lack key residues associated with catalysis. This is most obvious for the dehydroquinase which has an arginine in place of the active site lysine.

In *A. nidulans* the six genes (three structural and three regulatory) required to catabolize quinate to protocatechuate are linked in a tight *qut* gene cluster on chromosome VIII.[120] The genes are organized in the order: *qutC* (dehydroshikimate dehydratase), *qutD* (permease), *qutB* (quinate dehydrogenase), *qutE* (dehydroquinase), *qutA* (activator), and *qutR* (repressor). A seventh gene *qutG*, which has a quinate-inducible message but no known function, has also been isolated.[128] As in *N. crassa* the repressor protein QutR (encoded by *qutR*) controls transcription of the *qut* genes by interacting with the activator gene or protein, and shows sequence similarity with the three C-terminal domains of the *A. nidulans arom* polypeptide.[129] The dehydroquinase-like domain of the QutR protein and the dehydroquinase domain of the *arom* protein were purified in bulk and shown to have virtually superimposable c.d. spectra.[130]

The story does not end there. The QutA transcriptional activator protein (encoded by the *qutA* gene) shows sequence similarity with the two N-terminal domains of the *arom* protein.[85,131] Assuming the sequence similarity corresponds to structural similarity, the QutA and QutR proteins may recognize each other in the same way that the corresponding domains of the *arom* complex interact. It is proposed that the recruitment of such preformed enzymatically active domains to a regulatory role, in which they lose their catalytic activity but retain their binding sites, may represent a general mechanism for the evolution of pathway-specific regulator proteins.

1.22.5.1 The Flux Between the Shikimate and Quinate Pathways

In *N. crassa* and *A. nidulans*, there are two types of dehydroquinase, the catabolic type II enzyme and the biosynthetic type I enzyme which is part of the *arom* complex. It was thought that the *arom* protein kept the pools of dehydroquinate (**4**) and dehydroshikimate (**5**) separate in the quinate and shikimate pathways by channeling intermediates;[132] this would enable their concentrations to be kept low and so not trigger induction of the catabolic pathway.[133]

However, this idea is contradicted by experiments with a double mutant of *A. nidulans* (*qutR*^{c}; *qutE*),[134] which had a mutation in the repressor gene (*qutR*) leading to constitutive expression of the quinate pathway enzymes, and a mutation in the *qutE* gene which abolished the catabolic dehydroquinase activity. This mutant could not synthesize enough dehydroshikimate to survive on quinate as the sole carbon source. However, when the *arom* protein was overproduced *in vivo*, the double mutant grew on quinate. This suggests that there was sufficient leakage of (**4**) and (**5**) from the shikimate pathway to allow enough flow through the *qut* pathway to metabolize the quinate carbon source. It was subsequently shown that overproduction of dehydroshikimate dehydratase in the absence of quinate resulted in flux of dehydroshikimate from the shikimate pathway into the quinate pathway.[124] This in turn led to the organism attempting to compensate by increasing the concentration of the *arom* protein. The isolation of a dual function dehydroquinase from *A. methanolica* is further evidence for the existence of an interaction between the two pathways, in the form of a common pool of the metabolites dehydroquinate and dehydroshikimate.[111]

In *N. crassa* flux through the biosynthetic pathway is favored by the lower K_m of the biosynthetic dehydroquinase for dehydroquinate (**4**) compared to the catabolic enzyme,[72] and the lower K_m of the biosynthetic shikimate dehydrogenase (30 μM) than the catabolic dehydroshikimate dehydratase (K_m = 590 μM).[105]

1.22.6 SHIKIMATE DEHYDROGENASE

Shikimate dehydrogenase catalyzes the reversible reduction of dehydroshikimate (**5**) to shikimate (**6**), utilizing NADPH as the reducing nicotinamide cofactor (Scheme 16).[135–137] Like dehydroquinase, shikimate dehydrogenase occurs either as a monofunctional protein, a bifunctional protein, or as part of the *arom* pentafunctional multifunctional enzyme.

(**5**) Dehydroshikimate NADPH (**6**) Shikimate NADP$^+$

Scheme 16

E. coli shikimate dehydrogenase (EC 1.1.1.25) has been purified to homogeneity,[137] and the *aroE* gene has been cloned, overexpressed, and sequenced.[138] The derived sequence of 272 amino acids has a predicted molecular weight of 29 380. In plants the enzyme is part of a shikimate dehydrogenase/dehydroquinase bifunctional polypeptide. These have been purified from the moss *P. patens*[79] and pea seedlings (*P. sativum*).[6] The shikimate dehydrogenase domains of *A. nidulans*[133] and *S. cerevisiae arom* gene have been sequenced.[84]

The reduction of dehydroshikimate to shikimate involves hydride transfer from the A-side of NADPH (Scheme 16).[139,140] The kinetics of the *P. sativum* enzyme are consistent with an ordered BiBi mechanism, in which the nicotinamide cofactor binds first and leaves last.[139] The enzyme can be assayed at pH 7.0 in the forward direction or around pH 10 in the reverse direction.[137] At pH 7.0 the equilibrium constant for the reduction is only 28.[135] The pH–rate profile shows a dependence upon groups of pK_a 8.6 and 9.4.[141]

Very little is known about the substrate specificity of shikimate dehydrogenase. Studies with 5-deoxydehydroshikimate (**71**) and 4,5-dideoxydehydroshikimate (**72**) highlighted the importance of the 4-hydroxyl group in binding.[142] (6*R*)- and (6*S*)-6-Fluorodehydroshikimates, (**73**) and (**74**), have been reported to be converted slowly into the corresponding 6-fluoroshikimates, although little characterization of either starting material or product was presented.[143] A series of derivatives of 1,6-dihydroxy-2-oxoisonicotinic acid have been shown to be modest inhibitors of the pea enzyme but did not show herbicidal activity.[144]

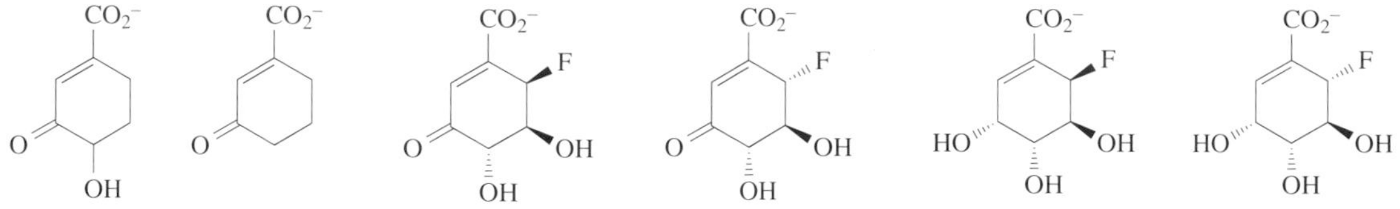

(**71**) (**72**) (**73**) (**74**) (**75**) (**76**)

1.22.7 SHIKIMATE KINASE

Shikimate kinase (EC 2.7.1.71) catalyzes the transfer of the γ-phosphate of ATP onto the C-3 hydroxy of shikimate to form shikimate 3-phosphate (**7**) and ADP (Scheme 17). Two isoenzymes of shikimate kinase are known for *E. coli*[145] and *S. typhi*.[146] *E. coli* shikimate kinase II gene *aroL* has been cloned and overexpressed.[147] It is a monomeric enzyme of 173 amino acids with a calculated M_r of 18 937. The amino acid sequence includes a region homologous with other kinases and ATP-requiring enzymes. The apparent K_m for shikimate is 200 μM at 5 mM ATP, and 160 μM for ATP at 1 mM shikimate.[148] The *E. coli* gene encoding shikimate kinase I has 34% amino acid identity to shikimate kinase II in a 97 amino-acid region.[149] The K_m of shikimate kinase I is about 100-fold higher than that for shikimate kinase II.[148]

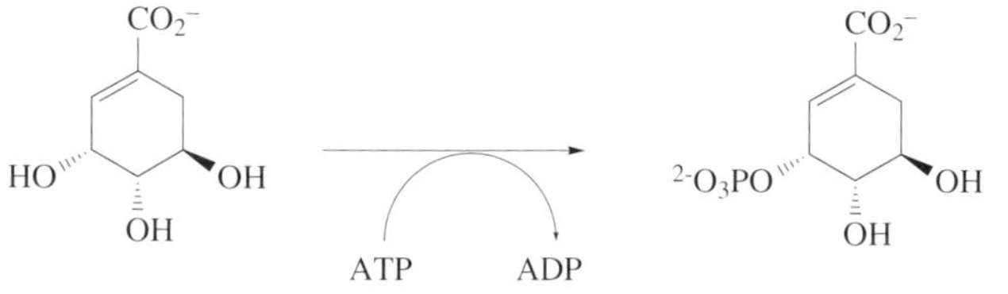

ATP ADP

(**6**) Shikimate

(**7**) Shikimate 3-phosphate

Scheme 17

In *B. subtilis* a single shikimate kinase is reported as a component of a trifunctional multienzyme complex along with DAHP synthase and chorismate mutase.[150] The kinase protein, which is active only in the complex, has been purified to homogeneity and has an M_r of around 10^4.[151] The gene for this enzyme has been sequenced.[152] The derived amino acid sequence shows 40% sequence identity with shikimate kinase I from *E. coli* and 26% sequence identity with shikimate kinase II.

Shikimate kinase has also been purified to near homogeneity from spinach chloroplasts and was found to be a single 31 kDa polypeptide.[153] The apparent K_m for ATP in the absence of stabilizing proteins and thiol compounds was found to be 1.8 mM. This value decreased to 0.3 mM when they were present. Shikimate kinases have been purified from mung bean[154] and sorghum.[155] The nucleotide sequence of an *Erwinia chrysanthemi* gene encoding shikimate kinase has been reported.[156] The shikimate kinase activity of the pentafuntional *arom* enzymes of *A. nidulans*,[133] *S. cerevisiae*,[84] and *N. crassa*[157] have been reported. The *N. crassa* enzyme has a k_{cat} of 18 s^{-1}, K_m for shikimate is 100 μM and for ATP is 320 μM.

Shikimate kinase can be assayed by measuring the conversion of [^{14}C]shikimate to [^{14}C]shikimate 3-phosphate,[148] or by coupling the release of ADP to the pyruvate kinase and lactate dehydrogenase reactions.[147] The enzyme mechanism has not been studied in detail and the stereochemistry of phosphoryl transfer is unknown, but is likely to proceed by an in-line displacement mechanism.

Almost nothing is known about the substrate specificity of shikimate kinase. (6*R*)- and (6*S*)-6-Fluoroshikimates, (**75**) and (**76**), have been shown to be good substrates for shikimate kinase II.[158] 4,5-Dideoxyshikimate, 5-deoxyshikimate, and 5-aminoshikimate have also been phosphorylated by this enzyme.[159]

1.22.8 5-ENOLPYRUVYL-SHIKIMATE 3-PHOSPHATE SYNTHASE

5-Enolpyruvyl-shikimate 3-phosphate synthase (EPSP synthase, EC 2.5.1.19) catalyzes the reversible transfer of the enolpyruvyl group of PEP (**2**) onto shikimate 3-phosphate (**7**) to give 5-enolpyruvyl-shikimate 3-phosphate (EPSP, **8**) and phosphate (Scheme 18). This enzyme is the target of the successful herbicide glyphosate. Both the enzyme mechanism and its mode of inhibition by glyphosate have been studied in detail. There has been a corresponding interest in the protein chemistry and molecular biology.

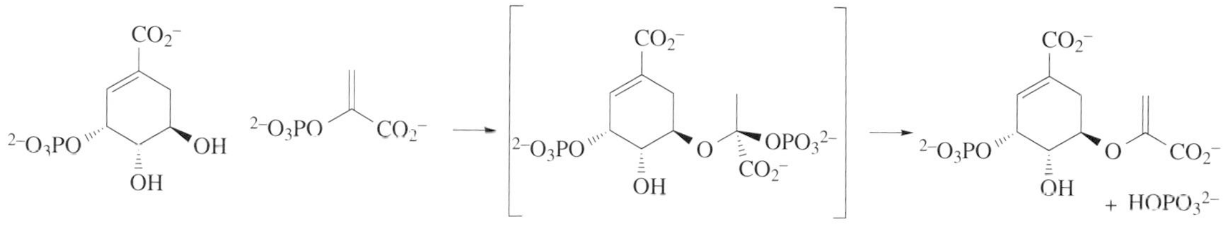

(**7**) Shikimate 3-phosphate (**2**) PEP (**78**) (**8**) EPSP

Scheme 18

EPSP synthases have been purified from many sources including *E. coli*. The enzyme was shown to be monomeric.[160] The *E. coli aroA* gene encodes a protein of M_r 46 112.[161] It is in an operon with the *SerA* gene,[162] in contrast to *Staphylococcus aureus* where the *aroA* gene is in an operon with the *aroC* and *aroB* genes.[163] The gene encoding EPSP synthase has been cloned from *Campylobacter jejeuni*,[164] and from the gram-negative pathogen *Dichelobacter nodosus*.[165]

The 3-dehydroquinate synthase–EPSP synthase bifunctional protein of the *arom* multifunctional enzyme from *A. nidulans* has been overexpressed. EPSP synthase activity is only observed when the EPSP synthase domain is covalently attached to the dehydroquinate synthase domain.[166] In *E. gracilis* there are two forms of EPSP synthase. One is a domain of the cytosolic 165 kDa multi-functional *arom* protein, the other is a monofunctional 59 kDa protein, found in the chloroplast.[167] EPSP synthase has been purified from several plant sources, including pea seedlings[168] and cultured maize cells (*Zea mays*) [169] Two isoforms were purified from maize, both located in the plastids. One isoform was present throughout the culture growth cycle, whereas the amount of the other increased in exponentially growing cells then declined in late logarithmic phase. The amino acid sequences of the plant enzymes show some sequence identity with the bacterial and fungal enzymes.[170]

The three-dimensional structure of the *E. coli* EPSP synthase has been determined to 3 Å by X-ray crystallography. The structure consists of two domains, each of which comprises three units of two parallel helices and a four-stranded sheet. The domains are related by an approximate dyad. The active site is thought to be located near the interdomain crossover segments. The structure helps explain how glyphosate tolerance can be conferred by specific sequence alterations.[171] The structure of the mechanistically related enzyme UDP–*N*-acetylglucosamine enolpyruvyl transferase has been reported at 1.8 Å resolution, complexed with UDP–*N*-acetylglucosamine (UDP–GlcNAc) and fosfomycin. The structure consists of two domains with the active site located between them. The overall protein architecture is similar to that of EPSP synthase.[172]

1.22.8.1 Mechanistic Studies on EPSP Synthase

The mechanism of the EPSP synthase reaction is well understood as a result of numerous studies over many years. The mechanism is similar to the first committed step in peptidoglycan synthesis, the formation of UDP–*N*-acetylenolpyruvylglucosamine, catalyzed by UDP–GlcNAc enolpyruvyl transferase (Scheme 18).[173,174]

The kinetics of the EPSP synthase reaction have been studied extensively. A steady-state kinetics study showed that the substrate binding is ordered with shikimate 3-phosphate (**7**) binding first, followed by PEP (**2**). The k_{cat} values measured were 60 s^{-1} for the forward reaction and of 10 s^{-1} for the reverse reaction.[175] It was subsequently reported that the enzyme exhibits a random kinetic mechanism.[176,177] The reaction is reversible with an equilibrium constant of 180.[178] Pre-steady-state kinetics have been used to build up a complete kinetic description of the reaction.[178] Equilibrium binding studies were used to determine the dissociation constants for shikimate 3-phosphate (7 μM) and EPSP (1 μM).[179]

One of the earliest mechanistic studies on EPSP synthase looked at the fate of [^{18}O]PEP labeled in the bridging oxygen. When this was incubated with enzyme and shikimate 3-phosphate, the label was found only in the released phosphate.[180] This is consistent with a mechanism involving C—O rather than P—O bond cleavage in PEP. A similar observation was made with DAHP synthase.[23] However, unlike DAHP synthase, exchange of the methylene protons in PEP with solvent was observed. These results led to the proposal of an addition–elimination mechanism proceeding through the tetrahedral intermediate (**78**), with transient formation of a methyl group (Scheme 18).[180] The protonation and deprotonation steps were shown to proceed with a primary kinetic isotope effect.[181,182] Furthermore, double labeling of the methylene hydrogens of PEP with deuterium and tritium showed that the addition and elimination steps occur with opposite stereochemistry (i.e., *anti*/*syn* or *syn*/*anti*). This result is that expected if the reaction involves only a single base in the enzyme (Scheme 19). If the addition proceeds with *anti* stereochemistry, the base is well placed to participate in a *syn* elimination (path **a**). For an *anti* elimination it is necessary to either rotate part of the intermediate through 120° (path **b**) or involve a second base on the enzyme (path **c**).

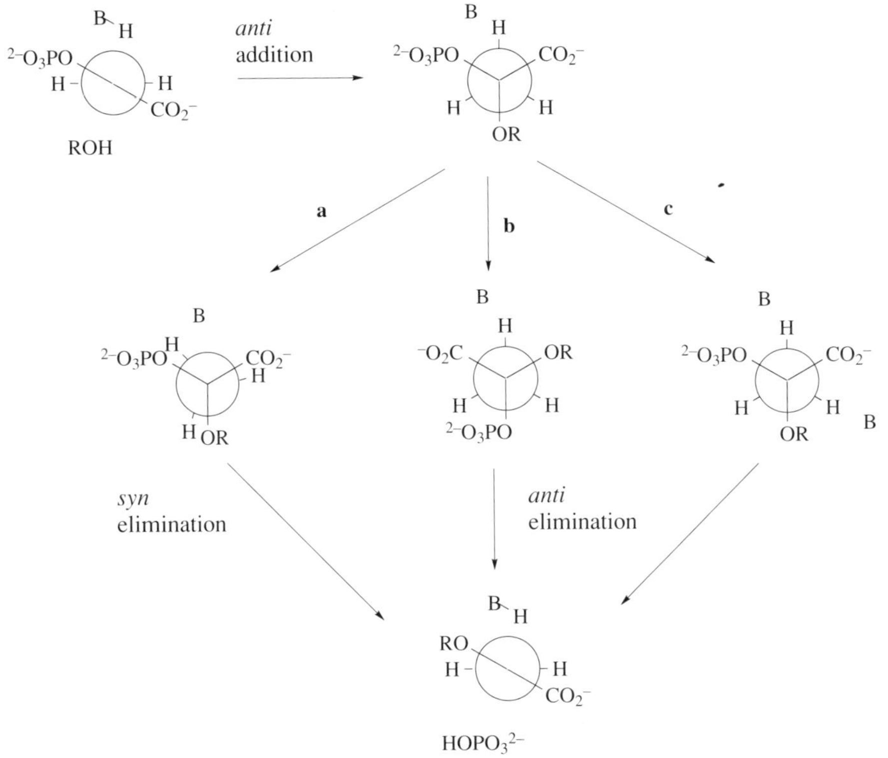

Scheme 19

Substrate trapping experiments, following the partitioning of [^{14}C]shikimate 3-phosphate on the enzyme between conversion to EPSP or dissociation from the enzyme in a single turnover experiment, showed that substrate dissociation is very fast (4500 s^{-1}).[179] These experiments also provided the first evidence for the intermediate (**78**) in the reaction. This was subsequently isolated by quenching the enzyme reaction with 100% triethylamine. The intermediate has been characterized by ^{1}H, ^{13}C, and ^{31}P NMR spectroscopy,[183] and shown to be both kinetically and chemically competent.[179,183]

Additional insight about the tetrahedral intermediate was inferred from studies with (*Z*)- and (*E*)-3-fluoroPEP, (**22**) and (**23**). Both stereoisomers of 3-fluoroPEP act as pseudosubstrates and react with shikimate 3-phosphate in the presence of EPSP synthase to give a fluorinated tetrahedral intermediate (**79**). This did not eliminate to give 9-fluoroEPSP (**80**) (Scheme 20).[184,185] The fluorinated intermediate (**79**) has high affinity for the enzyme (K_d estimated to be 600 pM), and so incubation of EPSP synthase with 3-fluoroPEP and shikimate 3-phosphate results in time-dependent inactivation of the enzyme.

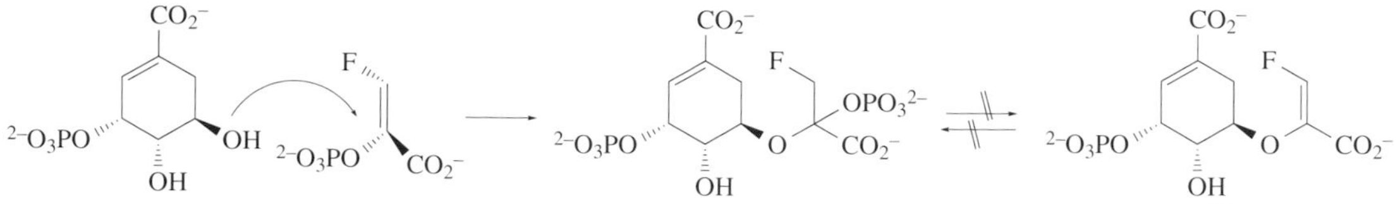

Scheme 20

PEP undergoes enzyme-catalyzed exchange of the methylene protons in D_2O in the presence of 4,5-dideoxyshikimate 3-phosphate (**81**).[186] This may occur via a protonated oxycarbenium species such as (**82**) (Scheme 21). It is possible that similar oxocarbenium reactivity may be involved in both the addition and elimination steps, and could explain why the presence of a fluorine on the methyl group of (**79**) prevents it being converted to (**80**). This would suggest that the transition state for elimination involves more extensive oxycarbenium character than the addition reaction. The observation that (*Z*)-9-fluoroEPSP (**80**) is not a substrate or pseudosubstrate for this enzyme in the reverse direction is consistent with this conclusion.[187]

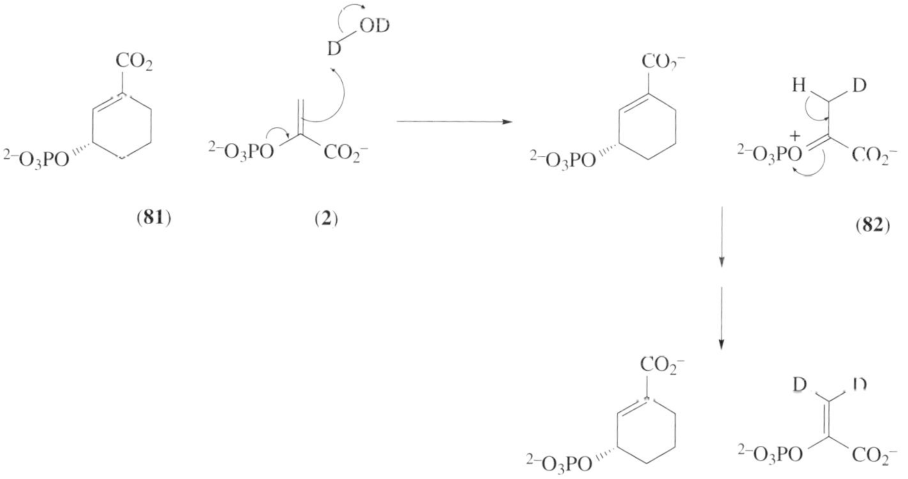

Scheme 21

The use of 3-substituted PEP analogues as substrates for UDP–GlcNAc enolpyruvyl transferase provided further evidence that the addition and elimination steps in this reaction, and by analogy in the EPSP synthase reaction, proceed with the opposite stereochemistry.[188] Using 3-fluoroPEPs it was shown that the proton was added to the *re* face of 3-fluoroPEP (equivalent to the *si* face of PEP) for both enolpyruvyl transferases.[184] Other PEP-utilizing enzymes also catalyze addition to the *si* face of PEP, suggesting evolutionary conservation of the PEP binding site.

A phosphonate analogue of intermediate (**78**) was synthesized and the (*R*)-diastereoisomer (**83**) shown to be a potent competitive inhibitor of *Petunia hybrida* EPSP synthase (apparent $K_i = 15$ nM against EPSP, and 110 nM against phosphate).[189] A K_i of 5 nM was subsequently reported, compared with 1.1 μM for the (*S*)-phosphonate (**84**).[190] The analogue of the tetrahedral intermediate containing two fluorines in the methyl group is a slightly better inhibitor.[190] The apparent K_i of the (*R*)-phosphate (**85**) is 4 nM, whereas the (*S*)-phosphate (**86**) has an apparent K_i of 75 nM (note (*R*) and (*S*) for these compounds are opposite to (**83**) and (**84**) due to the priority of fluorine). The best inhibitors have dissociation constants approaching the dissociation constant for the tetrahedral intermediate, which has been variously estimated to be 0.05 nM[191] and 0.25–2.5 nM.[192]

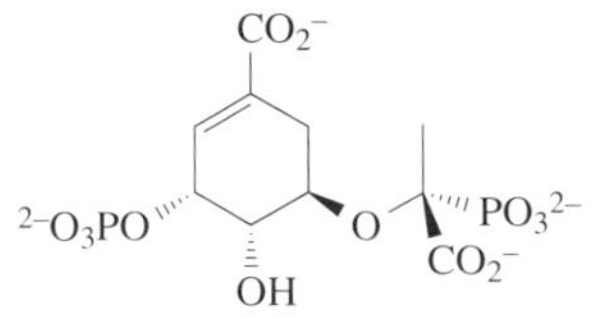

(83) 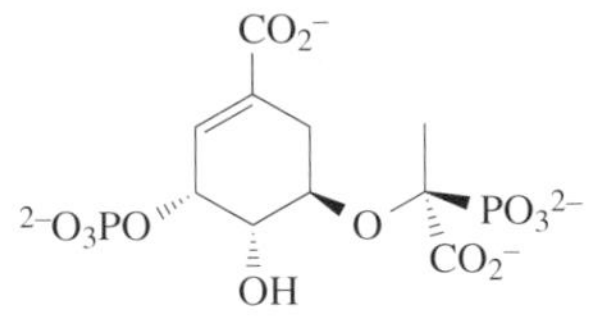 (84)

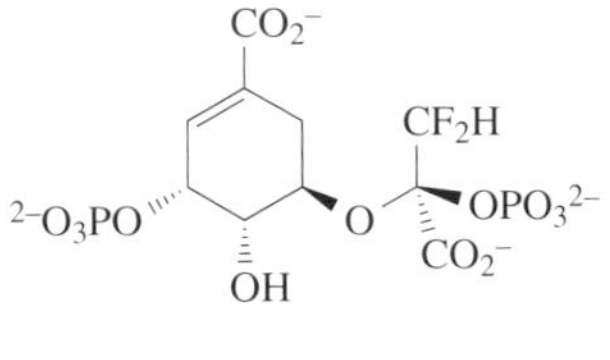

(85) 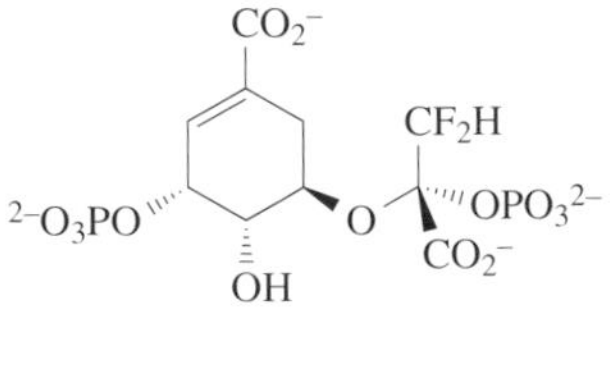 (86) 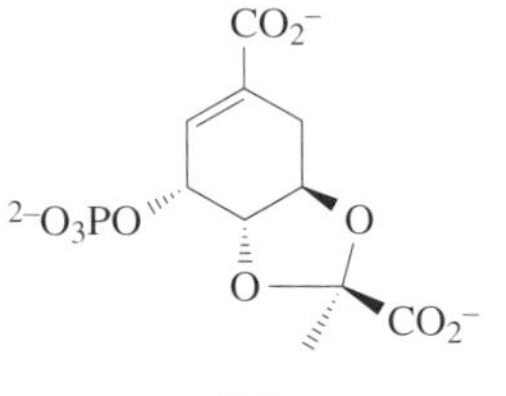 (87)

It is tempting to use the relative inhibition data for the pairs of inhibitors (**83**) and (**84**), and (**85**) and (**86**) to deduce the configuration at the tetrahedral center of the intermediate (**78**). In this respect, the two sets of results are contradictory. An alternative way to deduce the stereochemistry of the tetrahedral center in (**78**) is from the configuration of the ketal (**87**), isolated as a by-product from the normal enzymatic reaction.[193] If it is assumed that the reaction to form this from (**78**) involves an inversion of stereochemistry, it implies that the tetrahedral intermediate on the enzyme had the (*S*)-configuration at the tetrahedral center. This is the configuration predicted by the higher affinity of the (*R*)-phosphate (**85**) than the (*S*)-phosphate, but opposite to the initial prediction[189] based on the higher affinity of the (*R*)-phosphonate (**83**) than the (*S*)-phosphonate.

The EPSP synthase reaction has been studied using several different NMR techniques. Two-dimensional transfer NOE (nuclear Overhauser effect) experiments were used to show that both shikimate 3-phosphate and EPSP synthase retain their half-chair conformation with the phosphate group axial when bound to the enzyme.[194] Rapid quench freeze methods have been used to prepare enzymes for time-resolved solid state NMR spectroscopy. These studies showed the appearance of a peak due to the tetrahedral intermediate being formed on a millisecond timescale.[195,196]

1.22.8.2 Mutagenesis of EPSP Synthase

Extensive chemical modification studies on EPSP synthase have identified several residues at the active site, including: lysine (Lys22 in *E. coli*, Lys23 in *P. hybrida*),[197] arginine (Arg28 in *P. hybrida*),[198] glutamate (Glu418 in *E. coli*),[197] and two cysteines (Cys208 and Cys385 in *E. coli*).[199]

Site-directed mutagenesis of Lys23 in the *Petunia hybrida* enzyme to Ala23 gave inactive enzyme, whereas the Arg23 mutant was still active. It was suggested that Lys23 is part of an anionic binding site, together with Arg28.[78] This site could be responsible for binding of either of the carboxylates or phosphates. The Gly96Ala mutant has a 5000-fold increased apparent K_i for glyphosate. Substitution of this glycine with serine abolishes EPSP synthase activity but results in a novel EPSP hydrolase activity which converts EPSP to shikimate 3-phosphate and pyruvate. The region around this glycine is critical for the interaction of the phosphate moiety of phosphoenolpyruvate with EPSP synthase.[200] Replacement of either Lys19 or Arg24 of the *B. subtilis* enzyme by glutamate or aspartic acid leads to inactive enzyme which has a reduced affinity for shikimate 3-phosphate.[201] Replacement of His385 by asparagine in the *E. coli* enzyme gives an enzyme with a much reduced activity.[202]

1.22.8.3 Inhibition of EPSP Synthase by Glyphosate

EPSP synthase is strongly inhibited by phosphonomethylglycine (glyphosate, **88**) with a K_i of about 1 μM. Glyphosate forms a ternary complex with the enzyme and shikimate 3-phosphate (**7**). This complex inhibits enzyme activity and is thought to be responsible for its herbicidal activity. Glyphosate is the active ingredient in the herbicides Roundup and Tumbleweed.[4]

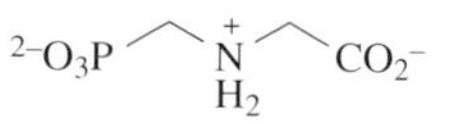

(88) 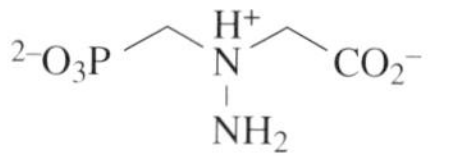 (89)

It has been proposed that glyphosate acts as a transition state analogue for a putative PEP oxycarbenium ion (such as (**82**) formed transiently during the reaction).[186,203] Kinetic experiments have shown that glyphosate acts as an uncompetitive inhibitor with respect to shikimate 3-phosphate, and a competitive inhibitor with respect to PEP, with an apparent K_i of 0.2–0.9 μM.[175,179] Glyphosate is not a ground state analogue of PEP, and does not inhibit other PEP-utilizing enzymes.[170]

The ternary complex of glyphosate, shikimate 3-phosphate, and EPSP synthase has been studied using ^{31}P, ^{15}N, and ^{13}C NMR spectroscopy. These studies revealed the ionization state of glyphosate.[204] Rotational-echo, double-resonance ^{31}P NMR spectroscopy was used to show the proximity of the phosphate to the labeled carbon of [1-^{13}C]glyphosate (internuclear distance 7.2 Å). An intramolecular ^{31}P–^{13}C distance of 5.6 Å was measured between the phosphonate and the labeled carbon of glyphosate, indicating that the glyphosate is fully extended when bound to the enzyme. This is not the conformation expected if it were acting as a transition state analogue.[205] Similar experiments have been used to measure other distances between glyphosate and shikimate 3-phosphate,[206] and to identify protein side chains near these compounds (three lysine, four arginines, and a histidine).[207] Titration calorimetry data show that the formation of the ternary complex is enthalpy driven but must offset a substantial negative entropy term.[208] Stronger synergy in binding is seen between shikimate 3-phosphate and glyphosate than shikimate 3-phosphate and PEP. No glyphosate binding was detected when shikimate 3-phosphate was replaced with 5-deoxyshikimate 3-phosphate.[159]

Almost any alterations in the structure of glyphosate result in loss of potency as an inhibitor, except *N*-aminoglyphosate (**89**) which shows comparable activity.[209] The inhibitor (**90**) incorporates features of both shikimate 3-phosphate and glyphosate. If this structure were a good representation of these molecules at the active site, it might be expected to be a better inhibitor than glyphosate. However, on assaying the reaction in the reverse direction it showed surprisingly weak competitive inhibition with EPSP (apparent $K_i = 7.4$ μM), and mixed inhibition against phosphate (apparent $K_i = 13$ μM).[210] These results were interpreted as showing that the inhibitor binds well into the shikimate 3-phosphate site, but that there is incomplete overlap with the phosphate site. The binding of (**90**) into the shikimate 3-phosphate site has been confirmed by ^{31}P NMR spectroscopic studies and been shown to be entropy driven.[211]

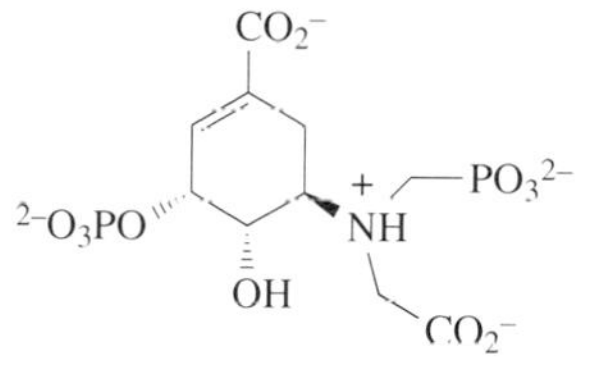

(90) 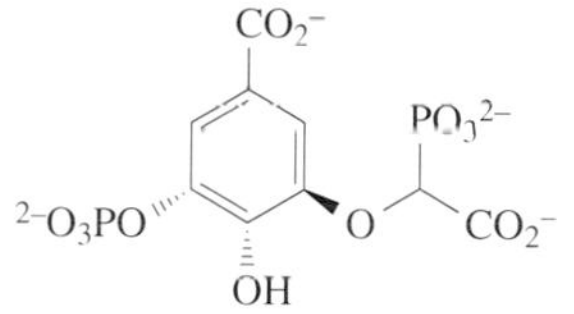 (91) 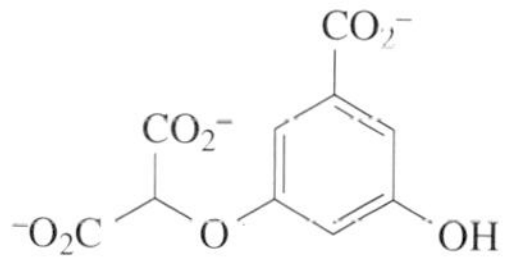 (92)

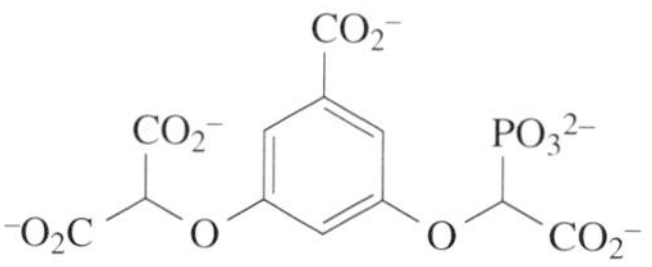

(93) 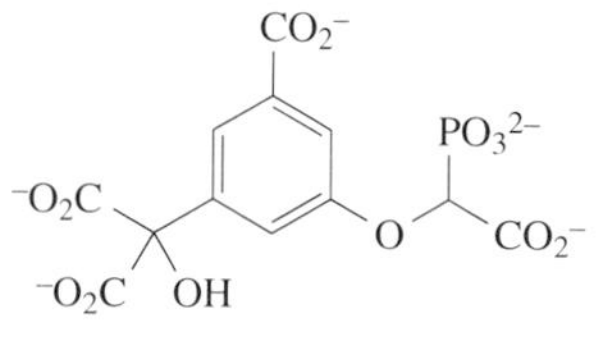 (94) 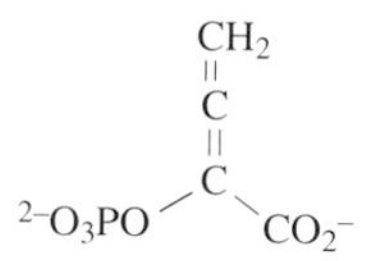 (95)

A detailed study using rapid gel filtration experiments showed that not only can glyphosate and EPSP synthase form the expected ternary complex with shikimate 3-phosphate, they can also form a ternary complex with EPSP. The K_d for glyphosate in this complex is 56 μM, compared with a K_d of 12 mM with EPSP synthase alone.[208] Glyphosate is an uncompetitive inhibitor versus EPSP, and a mixed inhibitor versus phosphate.[212] Changes in the fluorescence spectra of EPSP synthase show that glyphosate induces an additional conformational change which is not observed when only EPSP is bound. These results are not consistent with glyphosate inhibiting EPSP synthase by acting as a transition state analogue. Furthermore, because glyphosate exhibits mixed inhibition with respect to phosphate, it implies that the quaternary complex [enzyme.EPSP.glyphosate.phosphate] can form. This is not consistent with glyphosate binding site in the main active site.[212] It has been proposed that the inhibition data on glyphosate can be rationalized if it is acting as an adventitious allosteric inhibitor which causes a conformational change that stops PEP binding at the active site.[213]

Some organisms are tolerant to glyphosate. This tolerance can be due to changes in the interaction of the EPSP synthase with glyphosate. For example, EPSP synthase purified from the glyphosate-tolerant cyanobacterium *Anabaena variabilis* has an elevated K_i for glyphosate.[214] In species of *Pseudomonas*, tolerance is due to a specific single amino acid change.[215] Alternatively, tolerance to glyphosate can result from changes in the level of expression of EPSP synthase. When cells of plastid-free *E. gracilis* were grown in the presence of increasing amounts of glyphosate a corresponding overexpression of the *arom* complex was observed.[216] Glyphosate-tolerant cell cultures of *Corydalis sempervirens* were shown to have 10-fold higher levels of mRNA and 30–40-fold higher levels of EPSP synthase. The higher enzyme levels were ascribed to its stabilization by glyphosate.[217]

Glyphosate-tolerant plants have been generated by introduction of a gene for a glyphosate tolerant EPSP synthase, for example a mutant *aroA* gene was introduced into poplar using *Agrobacterium*-mediated transformation.[218] Similarly, a glyphosate tolerant soybean line has been generated by insertion of a bacterial EPSP synthase.[219] Such studies have shown that the degree of glyphosate tolerance depends upon, *inter alia*, the tissue specificity of expression.[220] One important reason for the continued success of glyphosate as a herbicide is considered to be the limited of evolution of weed resistance.[221]

1.22.8.4 Other Inhibitors of EPSP Synthase

The most potent inhibitors of EPSP synthase are the analogues of the tetrahedral intermediate (**78**). There are, however, many other inhibitors which are substrate or product analogues. Inhibition studies have shown that the ring in shikimate 3-phosphate can be replaced by a benzene ring,[222] or even a pyrrole.[223] The aromatic analogue (**91**) of the tetrahedral phosphonate inhibitors (**83**) and (**84**) has a K_i of 0.16 μM. The phosphate group has been shown to contribute over 8 kcal mol^{-1} in binding energy in shikimate 3-phosphate.[176,224] Inhibitors have been synthesized where this group is replaced by a malonate ether[222,224] or a hydroxymalonate.[225] Compound (**92**) is a substrate for EPSP synthase,[224] whereas (**93**) is an inhibitor with an apparent K_i of 1.3 μM.[222] The corresponding hydroxymalonate (**94**) has an apparent K_i of 0.57 μM.

The enolpyruvyl group present in EPSP has been replaced by several groups including an oxamic acid,[226] a glycolate group, a malonate ether,[227] and a phosphonoacetoxy group.[222] There appears to be less scope for structural variation in the PEP site, although both carboxyallenyl phosphate (**95**) and (*Z*)-3-fluoroPEP (**22**) are good inhibitors.[228]

1.22.9 THE *AROM* PENTAFUNCTIONAL PROTEIN

In fungi, yeast and *Euglena* the *arom* pentafunctional enzyme catalyzes the five steps on the shikimate pathway from DAHP to EPSP.[83,84,157,229] The *A. nidulans* enzyme has been shown to comprise of two independently folding regions, the N-terminal region including dehydroquinate synthase and EPSP synthase and the C-terminal region including shikimate kinase, dehydroquinase, and shikimate dehydrogenase.[229] The five enzymatic activities of the *arom* protein from *S. cerevisiae* have been shown to be in the same ratio as they are in crude cell extracts.[230]

The *ARO1* gene encoding the *arom* complex in *S. cerevisiae* has been cloned and overexpressed.[230] There have been extensive studies on the *aromA* gene in *A. nidulans*.[73,82,133] It has been overexpressed more than 120-fold in an *A. nidulans* mutant such that the *arom* protein comprised over 5% of total protein in crude cell extracts.[108]

Fragments of the *A. nidulans arom* protein have been expressed. For example, the dehydroquinate synthase[56,57] and dehydroquinase[56,85] domains of the *A. nidulans arom* protein have been subcloned and expressed in *E. coli*. A bifunctional protein containing the dehydroquinate synthase and EPSP synthase domains have been expressed in *E. coli*. The EPSP synthase domain was only enzymatically active when covalently attached to the dehydroquinate synthase domain.[166] The two C-terminal domains of the *arom* protein specifying the dehydroquinase and shikimate dehydrogenase have been overproduced in *A. nidulans*, but could not be overexpressed in *E. coli*.[166]

These experiments with domains from the *arom* protein show that it cannot be thought of as five independent units covalently linked together like beads on a string, but rather that interdomain interactions can play an important role in enzyme stability and activity.

1.22.10 CHORISMATE SYNTHASE

The last step on the shikimate pathway is the elimination of a hydrogen atom from C-6 and the 3-phosphate group of EPSP to form chorismate (Scheme 22). Chorismate synthase has been purified from microbial, fungal, and plant sources. These studies show that there are two types of enzyme, one monofunctional, the other with an associated flavin reductase activity.

(**8**) EPSP

(**96**) $FMNH_2$

chorismate synthase

(**9**) Chorismate

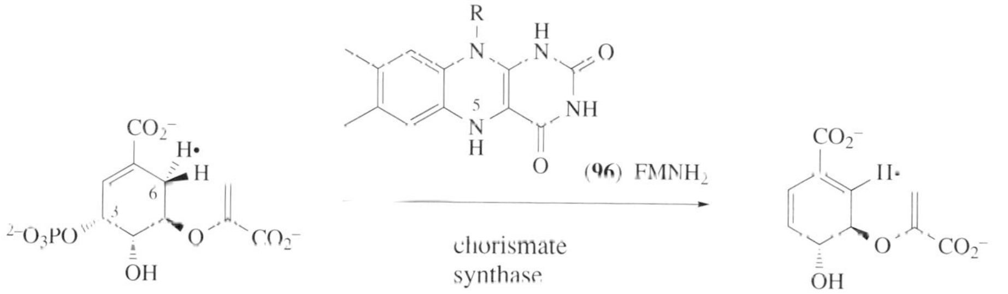

Scheme 22

The chorismate synthase found in most microorganisms and plants is monofunctional. All these enzymes show homology to the *E. coli* enzyme, which has a M_r of 39 138 and is tetrameric.[231] The *aroC* gene has been cloned from several microbial sources including *E. coli*,[231,232] *S. typhi*,[231] and *S. aureus*.[233] The chorismate synthase gene has also been cloned from cyanobacteria[234] and plant sources.[235]

The chorismate synthases from *N. crassa* and *S. cerevisiae* are bifunctional, also having flavin reductase (diaphorase) activity.[232,236] These enzymes are larger than the monofunctional proteins (46.4 kDa per subunit for the tetrameric *N. crassa* enzyme), and so it was initially assumed that the flavin reductase activity would be associated with an extra domain. Both *N. crassa* and *S. cerevisiae* genes have now been cloned and show a 79% sequence identity.[237,238] The *S. cerevisiae* enzyme also shares 47% identity with the *E. coli* sequence. Very surprisingly, however, excising parts of the sequence from the *N. crassa* enzyme that are not found in the monofunctional enzymes does not lead to a loss in flavin reductase activity.

1.22.10.1 Mechanistic Studies on Chorismate Synthase

Mechanistic studies have been concentrated on the *E. coli* and *N. crassa* enzymes, both of which have an absolute requirement for the reduced flavin mononucleotide, $FMNH_2$ (**96**). The lack of flavin reductase activity in the *E. coli* enzyme means that to assay the enzyme, a reduced flavin must be generated *in situ* (e.g., from FMN and dithionite) and the assay must be performed under anaerobic conditions.[239] Assaying the *N. crassa* enzyme is more straightforward and simply requires addition of NADPH. The reaction can then either be followed by UV spectrophotometry by monitoring production of chorismate at 275 nm,[240] or fluorimetrically by adding excess anthranilate synthase to convert the chorismate to anthranilate.[157]

The most intriguing aspect of the reaction catalyzed by chorismate synthase is the absolute requirement for $FMNH_2$. The requirement for a reduced flavin appears to have been largely ignored in early studies on the reaction mechanism, probably because the overall transformation does not involve any change in oxidation state, and consequently the cofactor is not stoichiometrically consumed during the course of the reaction. There are other flavin-dependent enzymes that catalyze reactions that do not involve a redox change, for example, mandelonitrile lyase,[241] and glyoxylate carboligase.[242] However, these enzymes use oxidized FAD. Furthermore, whereas substitution with 5-deazaFAD does not alter V_{max} for the glyoxylate carboligase reaction,[242] substitution with 5-deaza$FMNH_2$ abolishes the activity of both the *E. coli* and the *N. crassa* chorismate synthase.[243,244]

The reduced cofactor is either sequestered from solution (e.g., for the *E. coli* enzyme) or generated on the enzyme by reduction of an oxidized flavin using a reduced nicotinamide cofactor (as for the *N. crassa* enzyme). The *E. coli* enzyme binds oxidized FMN with a K_d of 30 μM.[245] Upon formation of ternary complex with EPSP, the K_d for oxidized FMN drops to 20 nM. The corresponding value for the ternary complex with chorismate is 0.54 μM. The redox potential of the oxidized/reduced FMN couple is 95 mV more positive than for free FMN, indicating 1660 times tighter binding of the reduced form, a stabilization of 4.4 kcal mol^{-1}. This corresponds to a K_d for reduced FMN of 18 nM. Spectroscopic data show that the reduced FMN is bound as the monoanion.

Reduction of oxidized FMN in the presence of chorismate synthase and (6*R*)-6-fluoroEPSP (**97**), EPSP, or chorismate results in varying amounts of the neutral flavin semiquinone (100%, 40%, and 14%, respectively).[245] This suggest that *in vivo* reduction of FMN must occur prior to binding of EPSP, consistent with an ordered binding mechanism for reduced cofactor and substrate.

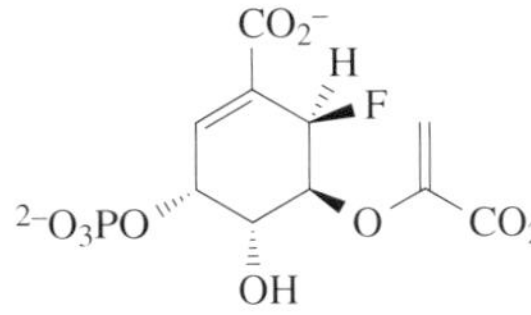

CO_2^- H F $^{2-}O_3PO$ O CO_2^- OH

(**97**)

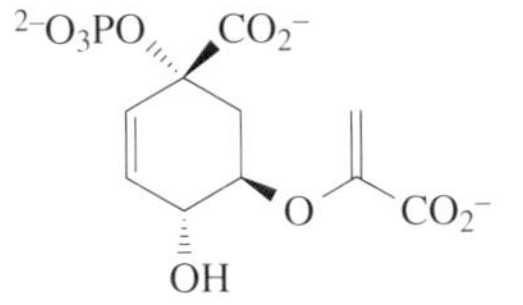

$^{2-}O_3PO$ CO_2^- O CO_2^- OH

(**98**)

The involvement of the flavin in the catalytic cycle of *E. coli* chorismate synthase was confirmed by detection of a flavin intermediate using pre-steady-state kinetics.[246] A modified FMN species was detected which formed very rapidly ($k_{obs} = 160\ s^{-1}$) and decayed with a first-order rate constant of 30 s^{-1}. It was suggested that the difference spectrum of the intermediate resembled that of a C_{4a} flavin adduct or a charge transfer complex. A kinetic model for the reaction has been proposed which suggests that this flavin intermediate accounts for almost all the flavin under turnover conditions.[247] The spectral changes are consistent with the monoanionic reduced flavin being protonated and its binding site becoming more apolar when EPSP binds.

In considering the bond breaking events at C-3 and C-6 of EPSP, it should be noted that the hydrogen at C-6 is relatively unactivated, and that the extended elimination of phosphate across a double bond proceeds with *anti* stereochemistry.[24,248,249] This stereochemistry is not favored for concerted $E_{2'}$ reactions.[250,251] Such considerations seem to have been the stimulus for some of the early proposals about the mechanism. The different mechanistic proposals are summarized below. Although none of them now appears correct, each stimulated research.

(i) To avoid a disfavored concerted 1,4-*anti* elimination, an initial 1,3-suprafacial rearrangement of the phosphate group to form isoEPSP (**98**) was proposed.[28] IsoEPSP was subsequently synthesized and shown not to be a substrate for the enzyme, although it binds with an affinity similar to that of EPSP itself ($K_i = 8.7$ μM, $K_{m(EPSP)} = 2.7$ μM).[252]

(ii) A mechanism was proposed involving initial attack of an enzymic nucleophile to effect an S_N2' displacement of phosphate giving a covalently bound intermediate (**99**) that can undergo *anti*-1,2-elimination to give chorismate (Scheme 23).[24] There is no evidence for this nucleophile, and incubation with (6*R*)-6-fluoroEPSP (**97**) did not result in irreversible inactivation,[158] as might be expected if the intermediate analogue (**100**) were formed.

(iii) A radical mechanism has been proposed along the lines shown in Scheme 24.[253] Abstraction of a hydrogen atom from C-6 of EPSP gives an allylic radical (**101**) which, on heterolytic cleavage of the phosphate group, forms a radical cation which is converted to chorismate upon electron transfer. This mechanism would seem to be supported by the observation of a flavin semiquinone radical on incubation of *E. coli* chorismate synthase with the substrate analogue (6*R*)-6-fluoro-EPSP,[254] and a model study which showed that similar radical 1,4-eliminations can occur in substituted cyclohexene systems.[255] However, the proposed mechanism is unlikely to be correct as it involves an *oxidized* flavin.

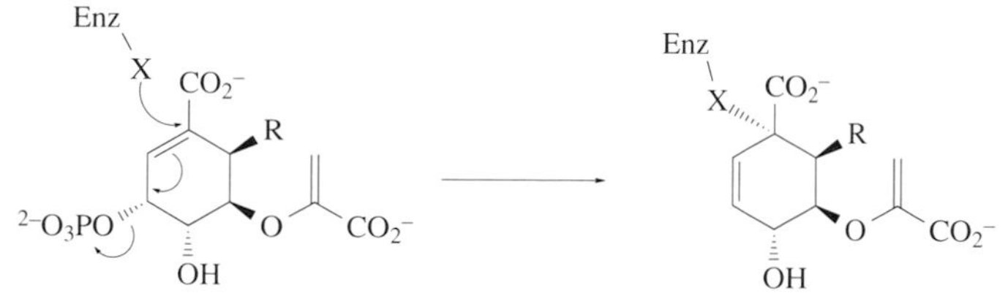

(**8**) R = H
(**97**) R = F

(**99**) R = H
(**100**) R = F

Scheme 23

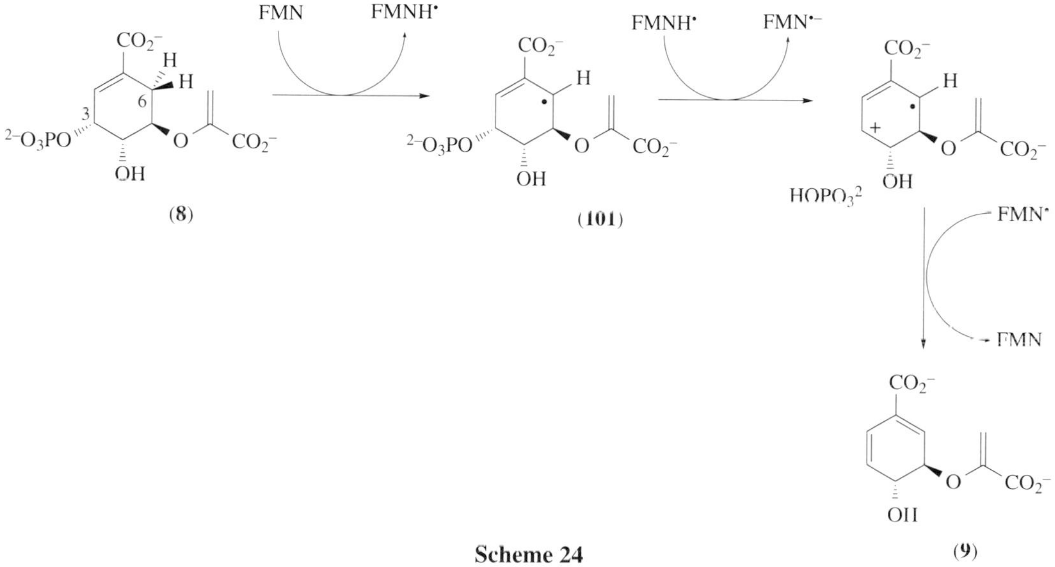

(**8**) (**101**) (**9**)

Scheme 24

Reduction of a disulphide by the $FMNH_2$ to give a thiyl radical could provide a species capable of abstracting the C-6 hydrogen. However, all four cysteinyl thiol groups in *E. coli* chorismate synthase were quantitatively detected using 5,5′-dithiobis(2-nitrobenzoic acid).[246] In addition there are no conserved cysteine residues among the known chorismate synthase sequences.

The kinetics of release of phosphate from the *E. coli* enzyme have been studied by pre-steady-state kinetics using a rapid quench technique.[256] No burst or lag in phosphate release was detected, suggesting that phosphate release is concomitant with the rate-determining step. An isotope effect of 1.13 is observed with the *E. coli* enzyme when (6*R*)-[6-^{2}H]EPSP[257] is used as the substrate.[258] This may be a small primary effect on C—H bond cleavage at C-6 or may be a large vinylogous β-effect on C(3)—O cleavage. In a related study of the *N. crassa* enzyme, a primary deuterium isotope effect of 2.7 on V_{max} and of 1.6 on $^{D}(V/K)$ were measured with (6*R*)-[6-^{2}H]EPSP.[259] By isotopic substitution at C-3 a secondary tritium kinetic isotope effect at C-3 of 1.05 was also determined using the *N. crassa* enzyme.[260]

The studies so far still do not give an unambiguous picture of the mechanism, although a stepwise mechanism involving initial cleavage of the C—O bond now appears more probable. There is a potentially interesting parallel to be drawn between chorismate synthase and dehydratases found in anaerobic bacteria which ferment amino acids.[261] These enzymes catalyze a reaction which involves removal of an unactivated hydrogen and contain redox active centers. For example, the (*R*)-lactyl-CoA dehydratase from *Clostridium propionicum* contains a reduced FMN, riboflavin, and iron–sulfur clusters and is extremely oxygen sensitive. Possible mechanisms for chorismate synthase based on the mechanistic proposals for dehydratases are shown in Scheme 25.[261] The first step involves electron donation from the strongly reducing flavin monoanion to EPSP to form the radical anion (**102**). This then undergoes heterolytic loss of phosphate to form the radical (**103**). From here two possible mechanisms are shown. Either (**a**) homolytic removal of the C-6 hydrogen by the flavin radical anion, or (**b**) a second electron donation from the flavin radical anion to form the anion (**104**), which then transfers hydride back to the FMN.

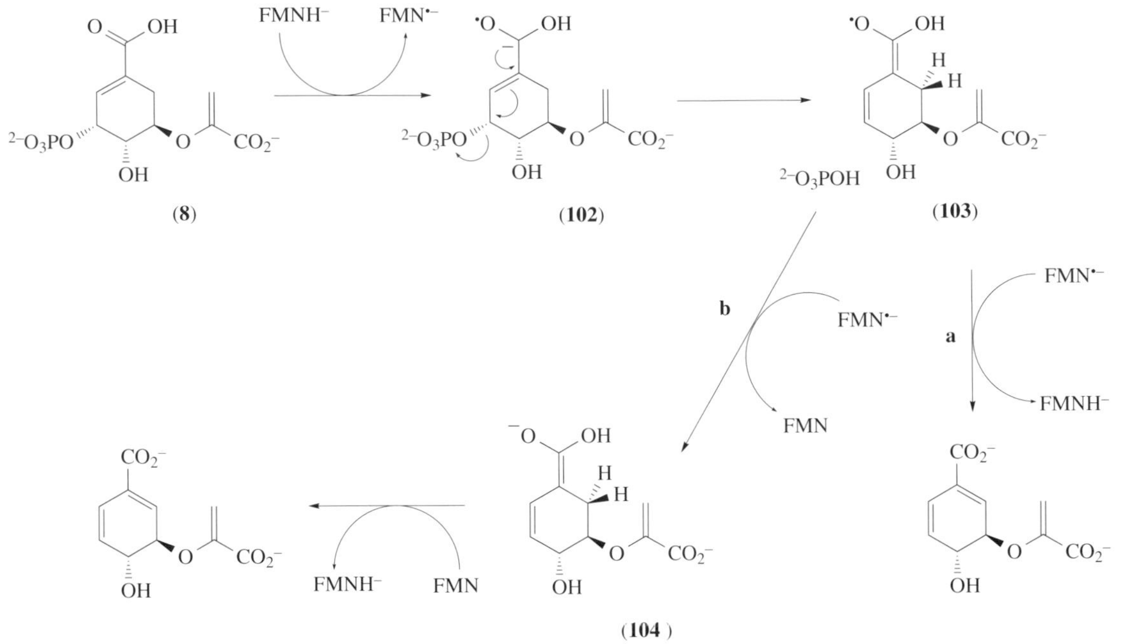

Scheme 25

After formation of chorismate, chorismate and phosphate dissociate. The rate for phosphate dissociation is 63 s^{-1}.[247] The flavin intermediate then decays at a rate of 52 s^{-1}. The overall k_{cat} for the reaction is 29 s^{-1}.

The (6*R*)- and (6*S*)-6-fluoroEPSPs, (**97**) and (**105**), have been synthesized and shown to be potent competitive inhibitors of the *N. crassa* enzyme. The (6*S*)-6-fluoroEPSP has a K_i of 0.2 μM, making it the most potent inhibitor of this enzyme described.[158] (6*S*)-6-FluoroEPSP (**105**) was reported not to be a substrate for *N. crassa* chorismate synthase (detection limit 0.1% of V_{max}).[158] It has, however, been shown to be converted to 6-fluorochorismate (**106**) by *E. coli* chorismate synthase at a rate 280 times slower than turnover of EPSP.[262] This behavior could be due to fluorine substitution destabilizing the buildup of positive charge at C-3. It is thought that the antibiotic properties of (6*S*)-6-fluoroshikimate (**76**)[263] are due to it being converted into 6-fluorochorismate (**106**) *in vivo* (Scheme 26). This then inhibits PABA synthase by an unknown mechanism, leading to death of the microorganism.

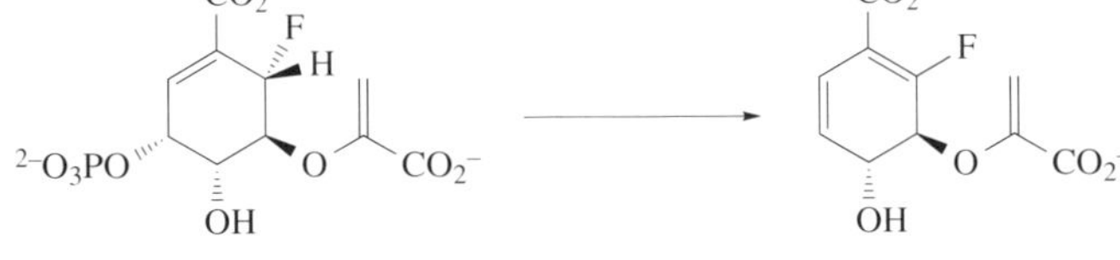

Scheme 26

1.22.11 FINAL COMMENTS

Many aspects of the shikimate pathway are sufficiently well understood for us to exploit that knowledge. Examples of this include the use of purified enzymes in multistep biotransformations to prepare fluorinated substrated analogues,[264] and the introduction of plasmid-borne pathway genes into a blocked *E. coli* mutant individually and in various combinations.[265] Using this approach it was possible to adjust levels of pathway intermediates seen directly by ^{1}H NMR spectroscopy of culture supernatants, and so show that dehydroquinate synthase, shikimate kinase, EPSP synthase, and chorismate synthase were all rate-limiting enzymes. In addition, a previously unknown feedback loop was identified involving inhibition of shikimate dehydrogenase by shikimic acid.

Over the last 100 years the shikimate pathway has been a focus of research for chemists, biochemists, geneticists, and microbiologists. More recently it has come to the attention of the structural biologists and their results will take the study of this pathway one stage further on. At the same time, the shikimate pathway has assumed commercial importance as the target for a major herbicide. It remains to be seen whether our very sophisticated knowledge of this pathway will enable us to manipulate it in ways to target other organisms for which this pathway is vital.

ACKNOWLEDGMENTS

I am grateful to Dr. Finian Leeper for proof-reading this review. I acknowledge the contribution of Professor John R. Coggins and the students who have worked with me on the shikimate pathway for teaching me what I know about this area.

1.22.12 REFERENCES

1. R. Bentley, *CRC Crit. Rev. Biochem.*, 1990, **25**, 307.
2. J. F. Eykmann, *Recl. Trav. Chim.*, 1885, **4**, 32.
3. J. F. Eykmann, *Chem. Ber.*, 1891, **24**, 1278.
4. H. C. Steinrucken and N. Amhrein, *Biochem. Biophys. Res. Commun.*, 1980, **94**, 1207.
5. H. Zalkin and D. J. Ebbole, *J. Biol. Chem.*, 1988, **263**, 1595.
6. D. M. Mousdale, M. S. Campbell, and J. R. Coggins, *Phytochemistry*, 1987, **26**, 2665.
7. D. M. Mousdale and J. R. Coggins, *Planta*, 1985, **163**, 241.
8. P. F. Morris, R.-L. Doong, and R. A. Jensen, *Plant Physiol.*, 1989, **89**, 10.
9. T. Ogino, C. Garner, J. L. Markley, and K. M. Herrmann, *Proc. Natl. Acad. Sci. USA*, 1982, **79**, 5828.
10. J. Shultz, M. A. Hermodson, C. C. Garner, and K. M. Herrmann, *J. Biol. Chem.*, 1984, **259**, 9655.
11. W. D. Davies and B. E. Davidson, *Nucleic Acids Res.*, 1982, **10**, 4045.
12. J. M. Ray, C. Yanofsky, and R. Bauerle, *J. Bacteriol.*, 1988, **170**, 5500.
13. J. M. Ray and R. Bauerle, *J. Bacteriol.*, 1991, **173**, 1894.
14. D. E. Tribe, H. Camakaris, and J. Pittard, *J. Bacteriol.*, 1976, **127**, 1085.
15. C. H. Doy, *Rev. Pure Appl. Chem.*, 1968, **18**, 41.
16. W. E. Dyer, L. M. Weaver, J. Zhao, D. N. Kuhn, S. C. Weller, and K. M. Herrmann, *J. Biol. Chem.*, 1990, **265**, 1608.
17. J. M. Zhao and K. M. Herrmann, *Plant Physiol.*, 1992, **100**, 1075.
18. Y. X. Wang, K. M. Herrmann, S. C. Weller, and P. B. Goldsbrough, *Plant Physiol.*, 1991, **97**, 847.
19. J. Gorlach, A. Beck, J. M. Henstrand, A. K. Handa, K. M. Herrmann, J. Schmid, and N. Amrhein, *Plant Mol. Biol.*, 1993, **23**, 697.
20. K. M. Herrmann, *Plant Physiol.*, 1995, **107**, 7.
21. L. M. Weaver, J. Pinto, and K. M. Herrmann, *Bioorg. Med. Chem. Lett.*, 1993, **3**, 1421.
22. H. Nagano and H. Zalkin, *Arch. Biochem. Biophys.*, 1970, **138**, 58.
23. A. B. DeLeo and D. B. Sprinson, *Biochem. Biophys. Res. Commun.*, 1968, **32**, 873.
24. D. K. Onderka and H. G. Floss, *J. Am. Chem. Soc.*, 1969, **91**, 5894.
25. H. G. Floss, D. K. Onderka, and M. Carroll, *J. Biol. Chem.*, 1972, **247**, 736.
26. A. B. DeLeo, J. Dayan, and D. B. Sprinson, *J. Biol. Chem.*, 1973, **218**, 2344.
27. R. Schoner and K. M. Herrmann, *J. Biol. Chem.*, 1976, **251**, 5440.
28. B. Ganem, *Tetrahedron*, 1978, **34**, 3353.
29. R. J. McCandliss, M. D. Poling, and K. M. Herrmann, *J. Biol. Chem.*, 1978, **253**, 4259.
30. M. Staub and G. Denes, *Biochim. Biophys. Acta*, 1969, **178**, 599.
31. R. J. McCandliss and K. M. Herrmann, *Proc. Natl. Acad. Sci. USA*, 1978, **75**, 4810.
32. T. Baasov and J. R. Knowles, *J. Bacteriol.*, 1989, **171**, 6155.
33. C. M. Stephens and R. Bauerle, *J. Biol. Chem.*, 1991, **266**, 20 810.
34. C. M. Stephens and R. Bauerle, *J. Biol. Chem.*, 1992, **267**, 5762.
35. F. J. Simpson, A. S. Perlin, and A. S. Sieben, *Methods Enzymol.*, 1966, **9**, 35.
36. C. C. Duke and J. K. MacLeod, *Carbohydr. Res.*, 1981, **95**, 1.
37. F. Stuart and I. S. Hunter, *Biochim. Biophys. Acta*, 1993, **1161**, 209.
38. P. Le Marechal, C. Froussios, M. Level, and R. Azerad, *Biochem. Biophys. Res. Commun.*, 1980, **92**, 1097.
39. P. F. Pilch and R. L. Somerville, *Biochemistry*, 1976, **15**, 5315.
40. M. Staub and G. Denes, *Biochim. Biophys. Acta*, 1969, **178**, 588.
41. L. Hedstrom and R. Abeles, *Biochem. Biophys. Res. Commun.*, 1988, **157**, 816.
42. A. Kohen, R. Berkovich, V. Belakhov, and T. Baasov, *Bioorg. Med. Chem. Lett.*, 1993, **3**, 1577.
43. G. D. Dotson, P. Nanjappan, M. D. Reily, and R. W. Woodard, *Biochemistry*, 1993, **32**, 12 392.
44. A. Kohen, A. Jakob, and T. Baasov, *Eur. J. Biochem.*, 1992, **208**, 443.
45. S. C. Du, H. Tsipori, and T. Baasov, *Bioorg. Med. Chem. Lett.*, 1997, **7**, 2469.
46. T. Baasov, S. Shefferdeenoor, A. Kohen, A. Jakob, and V. Belakhov, *Eur. J. Biochem.*, 1993, **217**, 991.
47. P. H. Ray, J. E. Kelsey, E. C. Bigham, C. D. Benedict, and T. A. Miller, *ACS Symp. Ser.*, 1983, **231**, 141.

48. S. Shefferdeenoor, V. Belakhov, and T. Baasov, *Bioorg. Med. Chem. Lett.*, 1993, **3**, 1583.
49. F. W. Dsouza, Y. Benenson, and T. Baasov, *Bioorg. Med. Chem. Lett.*, 1997, **7**, 2457.
50. P. H. Liang, A. Kohen, T. Baasov, and K. S. Anderson, *Bioorg. Med. Chem. Lett.*, 1997, **7**, 2463.
51. P. H. Ray, *J. Bacteriol.*, 1980, **141**, 635.
52. S. L. Bender, T. Widlanski, and J. R. Knowles, *Biochemistry*, 1989, **28**, 7560.
53. J. W. Frost, J. L. Bender, J. T. Kadonaga, and J. R. Knowles, *Biochemistry*, 1984, **23**, 4470.
54. S. L. Bender, S. Mehdi, and J. R. Knowles, *Biochemistry*, 1989, **28**, 7555.
55. T. Widlanski, S. L. Bender, and J. R. Knowles, *J. Am. Chem. Soc.*, 1989, **111**, 2299.
56. J. van den Hombergh, J. D. Moore, I. G. Charles, and A. R. Hawkins, *Biochem. J.*, 1992, **284**, 861.
57. J. D. Moore, J. R. Coggins, R. Virden, and A. R. Hawkins, *Biochem. J.*, 1994, **301**, 297.
58. P. R. Srinivasan, J. Rothschild, and D. B. Sprinson, *J. Biol. Chem.*, 1963, **238**, 3176.
59. S. L. Rotenberg and D. B. Sprinson, *J. Biol. Chem.*, 1978, **253**, 2210.
60. T. Widlanski, S. L. Bender, and J. R. Knowles, *Biochemistry*, 1989, **28**, 7572.
61. T. S. Widlanski, S. L. Bender, and J. R. Knowles, *J. Am. Chem. Soc.*, 1987, **109**, 1873.
62. M. J. Turner, B. W. Smith, and E. Haslam, *J. Chem. Soc., Perkin Trans.*, 1975, **1**, 52.
63. S. L. Rotenberg and D. B. Sprinson, *Proc. Natl. Acad. Sci. USA*, 1970, **67**, 1669.
64. P. A. Bartlett and K. Satake, *J. Am. Chem. Soc.*, 1988, **110**, 1628.
65. P. A. Bartlett, K. L. McLaren, and M. A. Marx, *J. Org. Chem.*, 1994, **59**, 2082.
66. E. J. Parker, J. R. Coggins, and C. Abell, *J. Org. Chem.*, 1997, **62**, 8582.
67. J. M. Lambert, M. R. Boocock, and J. R. Coggins, *Biochem. J.*, 1985, **226**, 817.
68. J. W. Frost and J. R. Knowles, *Biochemistry*, 1984, **23**, 4465.
69. L. M. Reimer, D. L. Conley, D. L. Pompliano, and J. W. Frost, *J. Am. Chem. Soc.*, 1986, **108**, 8010.
70. S. Myrvold, L. M. Reimer, D. L. Pompliano, and J. W. Frost, *J. Am. Chem. Soc.*, 1989, **111**, 1861.
71. D. L. Pompliano, L. M. Reimer, S. Myrvold, and J. W. Frost, *J. Am. Chem. Soc.*, 1989, **111**, 1866.
72. C. Kleanthous, R. Deka, K. Davis, S. M. Kelly, A. Cooper, S. E. Harding, N. C. Price, A. R. Hawkins, and J. R. Coggins, *Biochem. J.*, 1992, **282**, 687.
73. A. R. Hawkins, *Curr. Genet.*, 1987, **11**, 491.
74. T. Garbe, S. Servos, A. Hawkins, G. Dimitriadis, D. Young, G. Dougan, and I. Charles, *Mol. Gen. Genet.*, 1991, **228**, 385.
75. K. Duncan, S. Chaudhuri, M. S. Campbell, and J. R. Coggins, *Biochem. J.*, 1986, **238**, 475.
76. S. Servos, S. Chatfield, D. Hone, M. Levine, G. Dimitriadis, D. Pickard, G. Dougan, N. Fairweather, and I. Charles, *J. Gen. Microbiol.*, 1991, **137**, 147.
77. J. D. Moore, A. R. Hawkins, I. G. Charles, R. Deka, J. R. Coggins, A. Cooper, S. M. Kelly, and N. C. Price, *Biochem. J.*, 1993, **295**, 277.
78. Q. K. Huynh, S. C. Bauer, G. S. Bild, G. M. Kishore, and J. R. Borgmeyer, *J. Biol. Chem.*, 1988, **263**, 11 636.
79. L. D. Polley, *Biochim. Biophys. Acta*, 1978, **526**, 259.
80. R. K. Deka, I. A. Anton, B. Dunbar, and J. R. Coggins, *FEBS Lett.*, 1994, **349**, 397.
81. J. Lumsden and J. R. Coggins, *Biochem. J.*, 1977, **161**, 599.
82. I. G. Charles, J. W. Keyte, W. J. Brammar, and A. R. Hawkins, *Nucleic Acids Res.*, 1985, **13**, 8119.
83. V. B. Patel and N. H. Giles, *Biochim. Biophys. Acta*, 1979, **567**, 24.
84. K. Duncan, R. M. Edwards, and J. R. Coggins, *Biochem. J.*, 1987, **246**, 375.
85. A. R. Hawkins, J. D. Moore, and A. M. Adeokun, *Biochem. J.*, 1993, **296**, 451.
86. S. Chaudhuri, J. M. Lambert, L. A. Mccoll, and J. R. Coggins, *Biochem. J.*, 1986, **239**, 699.
87. S. Chaudhuri, K. Duncan, and J. R. Coggins, *Methods Enzymol.*, 1987, **142**, 320.
88. S. Chaudhuri, K. Duncan, L. D. Graham, and J. R. Coggins, *Biochem. J.*, 1991, **275**, 1.
89. C. W. G. Boys, S. M. Bury, L. Sawyer, J. D. Moore, I. G. Charles, A. R. Hawkins, R. Deka, C. Kleanthous, and J. R. Coggins, *J. Mol. Biol.*, 1992, **227**, 352.
90. S. Mitsuhashi and B. D. Davis, *Biochim. Biophys. Acta*, 1954, **15**, 268.
91. K. R. Hanson and I. A. Rose, *Proc. Natl. Acad. Sci. USA*, 1963, **50**, 981.
92. B. W. Smith, M. J. Turner, and E. Haslam, *J. Chem. Soc., Chem. Commun.*, 1970, 842.
93. E. Haslam, M. J. Turner, D. Sargent, and R. S. Thompson, *J. Chem. Soc. C*, 1971, 1489.
94. J. R. Butler, W. L. Alworth, and M. J. Nugent, *J. Am. Chem. Soc.*, 1974, **96**, 1617.
95. A. P. Leech, R. James, J. R. Coggins, and C. Kleanthous, *J. Biol. Chem.*, 1995, **270**, 25 827.
96. A. Schneier, C. Kleanthous, R. Deka, J. R. Coggins, and C. Abell, *J. Am. Chem. Soc.*, 1991, **113**, 9416.
97. C. Kleanthous, M. Reilly, A. Cooper, S. Kelly, N. C. Price, and J. R. Coggins, *J. Biol. Chem.*, 1991, **266**, 10 893.
98. A. Reilly, P. Morgan, K. Davis, S. M. Kelly, J. Greene, A. J. Rowe, S. E. Harding, N. C. Price, J. R. Coggins, and C. Kleanthous, *J. Biol. Chem.*, 1994, **269**, 5523.
99. A. D. N. Vaz, J. R. Butler, and M. J. Nugent, *J. Am. Chem. Soc.*, 1975, **97**, 5914.
100. J. M. Harris, C. Gonzalezbello, C. Kleanthous, A. R. Hawkins, J. R. Coggins, and C. Abell, *Biochem. J.*, 1996, **319**, 333.
101. R. K. Deka, C. Kleanthous, and J. R. Coggins, *J. Biol. Chem.*, 1992, **267**, 22 237.
102. C. Kleanthous and J. R. Coggins, *J. Biol. Chem.*, 1990, **265**, 10 935.
103. T. D. H. Bugg, C. Abell, and J. R. Coggins, *Tetrahedron Lett.*, 1988, **29**, 6783.
104. A. J. F. Da Silva, H. Whittington, J. Clements, C. Roberts, and A. R. Hawkins, *Biochem. J.*, 1986, **240**, 481.
105. P. Strøman, W. R. Reinert, and N. H. Giles, *J. Biol. Chem.*, 1978, **253**, 4593.
106. A. R. Hawkins, W. R. Reinert, and N. H. Giles, *Biochem. J.*, 1982, **203**, 769.
107. J. A. Hautala, J. W. Jacobson, M. E. Case, and N. H. Giles, *J. Biol. Chem.*, 1975, **250**, 6008.
108. J. D. Moore, H. K. Lamb, T. Garbe, S. Servos, G. Dougan, I. G. Charles, and A. R. Hawkins, *Biochem. J.*, 1992, **287**, 173.
109. J. R. Bottomley, C. L. Clayton, P. A. Chalk, and C. Kleanthous, *Biochem. J.*, 1996, **319**, 559.
110. P. J. White, J. Young, I. S. Hunter, H. G. Nimmo, and J. R. Coggins, *Biochem. J.*, 1990, **265**, 735.
111. G. J. W. Euverink, G. I. Hessels, J. W. Vrijbloed, J. R. Coggins, and L. Dijkhuizen, *J. Gen. Microbiol.*, 1992, **138**, 2449.
112. D. G. Gourley, J. R. Coggins, N. W. Isaacs, J. D. Moore, I. G. Charles, and A. R. Hawkins, *J. Mol. Biol.*, 1994, **241**, 488.

113. J. R. Bottomley, A. R. Hawkins, and C. Kleanthous, *Biochem. J.*, 1996, **319**, 269.
114. A. Shneier, J. Harris, C. Kleanthous, J. R. Coggins, A. R. Hawkins, and C. Abell, *Bioorg. Med. Chem. Lett.*, 1993, **3**, 1399.
115. J. Harris, C. Kleanthous, J. R. Coggins, A. R. Hawkins, and C. Abell, *J. Chem. Soc., Chem. Commun.*, 1993, 1080.
116. T. Krell, A. R. Pitt, and J. R. Coggins, *FEBS Lett.*, 1995, **360**, 93.
117. T. Krell, M. J. Horsburgh, A. Cooper, S. M. Kelly, and J. R. Coggins, *J. Biol. Chem.*, 1996, **271**, 24492.
118. J. M. Harris, W. J. Watkins, A. R. Hawkins, J. R. Coggins, and C. Abell, *J. Chem. Soc., Perkins Trans. 1*, 1996, 2371.
119. N. H. Giles, M. E. Case, J. A. Baum, R. F. Geever, L. Huiet, V. B. Patel, and B. M. Tyler, *Microbiol. Rev.*, 1985, **49**, 338.
120. A. R. Hawkins, N. H. Giles, and J. R. Kinghorn, *Biochem. Genet.*, 1982, **20**, 271.
121. J. L. Barea and N. H. Giles, *Biochim. Biophys. Acta*, 1978, **524**, 1.
122. N. K. Alton, F. Buxton, V. Patel, N. H. Giles, and D. Vapnek, *Proc. Natl. Acad. Sci. USA*, 1982, **79**, 1955.
123. A. R. Hawkins, H. K. Lamb, M. Smith, J. W. Keyte, and C. F. Roberts, *Mol. Gen. Genet.*, 1988, **214**, 224.
124. H. K. Lamb, J. van den Hombergh, G. H. Newton, J. D. Moore, C. F. Roberts, and A. R. Hawkins, *Biochem. J.*, 1992, **284**, 181.
125. K. A. Wheeler, H. K. Lamb, and A. R. Hawkins, *Biochem. J.*, 1996, **315**, 195.
126. S. Grant, C. F. Roberts, H. Lamb, M. Stout, and A. R. Hawkins, *J. Gen. Microbiol.*, 1988, **134**, 347.
127. I. A. Anton, K. Duncan, and J. R. Coggins, *J. Mol. Biol.*, 1987, **197**, 367.
128. R. K. Beri, S. Grant, C. F. Roberts, M. Smith, and A. R. Hawkins, *Biochem. J.*, 1990, **265**, 337.
129. A. R. Hawkins, H. K. Lamb, and C. F. Roberts, *Gene*, 1992, **110**, 109.
130. H. K. Lamb, J. D. Moore, J. H. Lakey, L. J. Levett, K. A. Wheeler, H. Lago, J. R. Coggins, and A. R. Hawkins, *Biochem. J.*, 1996, **313**, 941.
131. I. Levesley, G. H. Newton, H. K. Lamb, E. van Schothorst, R. W. M. Dalgleish, A. C. R. Samson, C. F. Roberts, and A. R. Hawkins, *Microbiology*, 1996, **142**, 1909.
132. N. H. Giles, *Am. Nat.*, 1978, **112**, 641.
133. I. G. Charles, J. W. Keyte, W. J. Brammar, M. Smith, and A. R. Hawkins, *Nucleic Acids Res.*, 1986, **14**, 2201.
134. H. K. Lamb, C. R. Bagshaw, and A. R. Hawkins, *Mol. Gen. Genet.*, 1991, **227**, 187.
135. H. Yaniv and C. Gilvarg, *J. Biol. Chem.*, 1955, 787.
136. D. Balinsky and D. D. Davies, *Biochem. J.*, 1961, **80**, 292.
137. S. Chaudhuri and J. R. Coggins, *Biochem. J.*, 1985, **226**, 217.
138. I. A. Anton and J. R. Coggins, *Biochem. J.*, 1988, **249**, 319.
139. J. R. Dowsett, J. R. Corbett, B. Middleton, and P. K. Tubbs, *Biochem. J.*, 1971, **123**, 23.
140. P. Dansette and R. Azerad, *Biochimie*, 1974, **56**, 751.
141. A. W. Dennis and D. Balinsky, *Int. J. Biochem.*, 1972, **3**, 93.
142. T. D. H. Bugg, C. Abell, and J. R. Coggins, *Tetrahedron Lett.*, 1988, **29**, 6779.
143. P. Le Marechal, C. Froussios, and R. Azerad, *Biochimie*, 1986, **68**, 1211.
144. A. C. Baillie, J. R. Corbett, J. R. Dowsett, and P. McCloskey, *Pestic. Sci.*, 1972, **3**, 113.
145. R. Defeyter, *Methods Enzymol.*, 1987, **142**, 355.
146. M. B. Berlin and N. H. Giles, *J. Bacteriol.*, 1969, **99**, 222.
147. G. Millar, A Lewendon, M. G. Hunter, and J. R. Coggins, *Biochem. J.*, 1986, **237**, 427.
148. R. C. Defeyter and J. Pittard, *J. Bacteriol.*, 1986, **165**, 331.
149. A. Lobner-Olesen and M. G. Marinus, *J. Bacteriol.*, 1992, **174**, 525.
150. W. M. Nakatsukasa and E. W. Nester, *J. Biol. Chem.*, 1972, **247**, 5972.
151. L. Huang, A. L. Montoya, and E. W. Nester, *J. Biol. Chem.*, 1975, **250**, 7675.
152. A. Nakane, K. I. Ogawa, K. Nakamura, and K. Yamane, *J. Ferment. Bioeng.*, 1994, **77**, 312.
153. C. L. Schmidt, H. J. Danneel, G. Schultz, and B. B. Buchanan, *Plant Physiol.*, 1990, **93**, 758.
154. T. Koshiba, *Biochim. Biophys. Acta*, 1978, **522**, 10.
155. J. R. Bowen and T. Kosuge, *Plant Physiol.*, 1979, **64**, 382.
156. N. P. Minton, P. J. Whitehead, T. Atkinson, and H. J. Gilbert, *Nucleic Acids Res.*, 1989, **17**, 1769.
157. J. R. Coggins, M. R. Boocock, S. Chaudhuri, J. M. Lambert, J. Lumsden, G. A. Nimmo, and D. D. S. Smith, *Methods Enzymol.*, 1987, **142**, 325.
158. S. Balasubramanian, G. M. Davies, J. R. Coggins, and C. Abell, *J. Am. Chem. Soc.*, 1991, **113**, 8945.
159. P. D. Pansegrau, K. S. Anderson, T. S. Widlanski, J. E. Ream, R. D. Sammons, J. A. Sikorski, and J. R. Knowles, *Tetrahedron Lett.*, 1991, **32**, 2589.
160. A. Lewendon and J. R. Coggins, *Biochem. J.*, 1983, **213**, 187.
161. K. Duncan, A. Lewendon, and J. R. Coggins, *FEBS Lett.*, 1984, **170**, 59.
162. J. Duncan and J. R. Coggins, *Biochem. J.*, 1986, **234**, 49.
163. C. O'Connell, P. A. Pattee, and T. J. Foster, *J. Gen. Microbiol.*, 1993, **139**, 1449.
164. M. Wosten, H. J. Dubbink, and B. A. M. Vanderzeijst, *Gene*, 1996, **181**, 109.
165. R. A. Alm, B. P. Dalrymple, and J. S. Mattick, *Gene*, 1994, **145**, 97.
166. J. D. Moore and A. R. Hawkins, *Mol. Gen. Genet.*, 1993, **240**, 92.
167. C. Reinbothe, B. Ortel, B. Parthier, and S. Reinbothe, *Mol. Gen. Genet.*, 1994, **245**, 616.
168. D. M. Mousdale and J. R. Coggins, *Planta*, 1984, **160**, 78.
169. G. Forlani, B. Parisi, and E. Nielsen, *Plant Physiol.*, 1994, **105**, 1107.
170. G. M. Kishore and D. M. Shah, *Annu. Rev. Biochem.*, 1988, **57**, 627.
171. W. C. Stallings, S. S. Abdelmeguid, L. W. Lim, H. S. Shieh, H. E. Dayringer, N. K. Leimgruber, R. A. Stegeman, K. S. Anderson, J. A. Sikorski, S. R. Padgette, and G. M. Kishore, *Proc. Natl. Acad. Sci. USA*, 1991, **88**, 5046.
172. T. Skarzynski, A. Mistry, A. Wonacott, S. E. Hutchinson, V. A. Kelly, and K. Duncan, *Structure*, 1996, **4**, 1465.
173. K. G. Gunetileke and R. A. Anwar, *J. Biol. Chem.*, 1968, **243**, 5770.
174. E. D. Brown, E. I. Vivas, C. T. Walsh, and R. Kolter, *J. Bacteriol.*, 1995, **177**, 4194.
175. M. R. Boocock and J. R. Coggins, *FEBS Lett.*, 1983, **154**, 127.
176. K. J. Gruys, M. C. Walker, and J. A. Sikorski, *Biochemistry*, 1992, **31**, 5534.
177. K. J. Gruys, M. R. Marzabadi, P. D. Pansegrau, and J. A. Sikorski, *Arch. Biochem. Biophys.*, 1993, **304**, 345.

178. K. S. Anderson and K. A. Johnson, *Chem. Rev.*, 1990, **90**, 1131.
179. K. S. Anderson, J. A. Sikorski, and K. A. Johnson, *Biochemistry*, 1988, **27**, 7395.
180. W. E. Bondinell, J. Vnek, P. F. Knowles, M. Sprecher, and D. B. Sprinson, *J. Biol. Chem.*, 1971, **246**, 6191.
181. C. E. Grimshaw, S. G. Sogo, and J. R. Knowles, *J. Biol. Chem.*, 1982, **257**, 596.
182. Y. Asano, J. J. Lee, T. L. Shieh, F. Spreafico, C. Kowal, and H. G. Floss, *J. Am. Chem. Soc.*, 1985, **107**, 4314.
183. K. S. Anderson, J. A. Sikorski, A. Benesi, and K. A. Johnson, *J. Am. Chem. Soc.*, 1988, **110**, 6577.
184. D. H. Kim, G. W. Tuckerkellogg, W. J. Lees, and C. T. Walsh, *Biochemistry*, 1996, **35**, 5435.
185. M. C. Walker, C. R. Jones, R. L. Somerville, and J. A. Sikorski, *J. Am. Chem. Soc.*, 1992, **114**, 7601.
186. D. Anton, L. Hedstrom, S. Fish, and R. Abeles, *Biochemistry*, 1983, **22**, 5903.
187. C. T. Seto and P. A. Bartlett, *J. Org. Chem.*, 1994, **59**, 7130.
188. W. J. Lees and C. T. Walsh, *J. Am. Chem. Soc.*, 1995, **117**, 7329.
189. D. G. Alberg and P. A. Bartlett, *J. Am. Chem. Soc.*, 1989, **111**, 2337.
190. D. G. Alberg, C. T. Lauhon, R. Nyfeler, A. Fassler, and P. A. Bartlett, *J. Am. Chem. Soc.*, 1992, **114**, 3535.
191. K. A. Anderson and K. A. Johnson, *J. Biol. Chem.*, 1990, **265**, 5567.
192. W. W. Cleland, *Biochemistry*, 1990, **29**, 3194.
193. G. C. Leo, J. A. Sikorski, and R. D. Sammons, *J. Am. Chem. Soc.*, 1990, **112**, 1653.
194. G. C. Leo, S. Castellino, R. D. Sammons, and J. A. Sikorski, *Bioorg. Med. Chem. Lett.*, 1992, **2**, 151.
195. J. N. S. Evans, R. J. Appleyard, and W. A. Shuttleworth, *J. Am. Chem. Soc.*, 1993, **115**, 1588.
196. R. J. Appleyard, W. A. Shuttleworth, and J. N. S. Evans, *Biochemistry*, 1994, **33**, 6812.
197. Q. K. Huynh, G. M. Kishore, and G. S. Bild, *J. Biol. Chem.*, 1988, **263**, 735.
198. S. R. Padgette, C. E. Smith, Q. K. Huynh, and G. M. Kishore, *Arch. Biochem. Biophys.*, 1988, **266**, 254.
199. S. R. Padgette, Q. K. Huynh, S. Aykent, R. D. Sammons, J. A. Sikorski, and G. M. Kishore, *J. Biol. Chem.*, 1988, **263**, 1798.
200. S. R. Padgette, D. B. Re, C. S. Gasser, D. A. Eichholtz, R. B. Frazier, C. M. Hironaka, E. B. Levine, D. M. Shah, R. T. Fraley, and G. M. Kishore, *J. Biol. Chem.*, 1991, **266**, 22 364.
201. A. Selvapandiyan, S. Ahmad, K. Majumder, N. Arora, and R. K. Bhatnagar, *Biochem. Mol. Biol. Int.*, 1996, **40**, 603.
202. W. A. Shuttleworth and J. N. S. Evans, *Arch. Biochem. Biophys.*, 1996, **334**, 37.
203. H. C. Steinrucken and N. Amerhein, *Eur. J. Biochem.*, 1984, **143**, 351.
204. S. Castellino, G. C. Leo, R. D. Sammons, and J. A. Sikorski, *Biochemistry*, 1989, **28**, 3856.
205. A. M. Christensen and J. Schaefer, *Biochemistry*, 1993, **32**, 2868.
206. L. M. McDowell, C. A. Klug, D. D. Beusen, and J. Schaefer, *Biochemistry*, 1996, **35**, 5395.
207. L. M. McDowell, A. Schmidt, E. R. Cohen, D. R. Studelska, and J. Schaefer, *J. Mol. Biol.*, 1996, **256**, 160.
208. J. E. Ream, H. K. Yuen, R. B. Frazier, and J. A. Sikorski, *Biochemistry*, 1992, **31**, 5528.
209. W. S. Knowles, K. S. Anderson, S. S. Andrew, D. P. Phillion, J. E. Ream, K. A. Johnson, and J. A. Sikorski, *Bioorg. Med. Chem. Lett.*, 1993, **3**, 2863.
210. M. R. Marzabadi, J. L. Font, K. J. Gruys, P. D. Pansegrau, and J. A. Sikorski, *Bioorg. Med. Chem. Lett.*, 1992, **2**, 1435.
211. M. R. Marzabadi, K. J. Gruys, P. D. Pansegrau, M. C. Walker, H. K. Yuen, and J. A. Sikorski, *Biochemistry*, 1996, **35**, 4199.
212. R. D. Sammons, K. J. Gruys, K. S. Anderson, K. A. Johnson, and J. A. Sikorski, *Biochemistry*, 1995, **34**, 6433.
213. J. A. Sikorski and K. J. Gruys, *Acc. Chem. Res.*, 1997, **30**, 2.
214. H. A. Powell, N. W. Kerby, P. Rowell, D. M. Mousdale, and J. R. Coggins, *Planta*, 1992, **188**, 484.
215. D. M. Stalker, W. R. Hiatt, and L. Comai, *J. Biol. Chem.*, 1985, **260**, 4724.
216. S. Reinbothe, B. Ortel, and B. Parthier, *Mol. Gen. Genet.*, 1993, **239**, 416.
217. H. Hollanderczytko, I. Sommer, and N. Amrhein, *Plant Mol. Biol.*, 1992, **20**, 1029.
218. R. A. Donahue, T. D. Davis, C. H. Michler, D. E. Riemenschneider, D. R. Carter, P. E. Marquardt, N. Sankhla, D. Sankhla, B. E. Haissig, and J. G. Isebrands, *Can. J. Forest Res.*, 1994, **24**, 2377.
219. S. R. Padgette, K. H. Kolacz, X. Delannay, D. B. Re, B. J. Lavallee, C. N. Tinius, W. K. Rhodes, Y. I. Otero, G. F. Barry, D. A. Eichholtz, V. M. Peschke, D. L. Nida, N. B. Taylor, and G. M. Kishore, *Crop Sci.*, 1995, **35**, 1451.
220. G. M. Kishore, S. R. Padgette, and R. T. Fraley, *Weed Technol.*, 1992, **6**, 626.
221. L. D. Bradshaw, S. R. Padgette, S. L. Kimball, and B. H. Wells, *Weed Technol.*, 1997, **11**, 189.
222. M. J. Miller, D. G. Cleary, J. E. Ream, K. R. Snyder, and J. A. Sikorski, *Bioorg. Med. Chem.*, 1995, **3**, 1685.
223. M. L. Peterson, S. D. Corey, J. L. Font, M. C. Walker, and J. A. Sikorski, *Bioorg. Med. Chem. Lett.*, 1996, **6**, 2853.
224. M. J. Miller, K. S. Anderson, D. S. Braccolino, D. G. Cleary, K. J. Gruys, C. Y. Han, K. C. Lin, P. D. Pansegrau, J. E. Ream, R. D. Sammons, and J. A. Sikorski, *Bioorg. Med. Chem. Lett.*, 1993, **3**, 1435.
225. A. Shah, J. L. Font, M. J. Miller, J. E. Ream, M. C. Walker, and J. A. Sikorski, *Bioorg. Med. Chem. Lett.*, 1997, **5**, 323.
226. S. D. Corey, P. D. Pansegrau, M. C. Walker, and J. A. Sikorski, *Bioorg. Med. Chem. Lett.*, 1993, **3**, 2857.
227. M. J. Miller, D. S. Braccolino, D. G. Cleary, J. E. Ream, M. C. Walker, and J. A. Sikorski, *Bioorg. Med. Chem. Lett.*, 1994, **4**, 2605.
228. M. C. Walker, J. E. Ream, R. D. Sammons, E. W. Logusch, M. H. Oleary, R. L. Somerville, and J. A. Sikorski, *Bioorg. Med. Chem. Lett.*, 1991, **1**, 683.
229. A. R. Hawkins and M. Smith, *Eur. J. Biochem.*, 1991, **196**, 717.
230. L. D. Graham, F. M. Gillies, and J. R. Coggins, *Biochim. Biophys. Acta*, 1993, **1216**, 417.
231. I. G. Charles, H. K. Lamb, D. Pickard, G. Dougan, and A. R. Hawkins, *J. Gen. Microbiol.*, 1990, **136**, 353.
232. P. J. White, G. Millar, and J. R. Coggins, *Biochem. J.*, 1988, **251**, 313.
233. M. J. Horsburgh, T. J. Foster, P. T. Barth, and J. R. Coggins, *Microbiology*, 1996, **142**, 2943.
234. J. Schmidt, M. Bubunenko, and A. R. Subramanian, *J. Biol. Chem.*, 1993, **268**, 27 447.
235. A. Schaller, J. Schmid, U. Leibinger, and N. Amrhein, *J. Biol. Chem.*, 1991, **266**, 21 434.
236. J. M. Henstrand, A. Schaller, M. Braun, N. Amrhein, and J. Schmid, *Mol. Microbiol.*, 1996, **22**, 859.
237. J. M. Henstrand, N. Amrhein, and J. Schmid, *J. Biol. Chem.*, 1995, **270**, 20 447.
238. D. G. L. Jones, U. Reusser, and G. H. Braus, *Mol. Microbiol.*, 1991, **5**, 2143.
239. P. J. White, D. M. Mousdale, and J. R. Coggins, *Biochem. Soc. Trans.*, 1987, **15**, 144.

240. F. Gibson, *Methods Enzymol.*, 1970, **17**, 362.
241. P.-H. Xu, B. K. Singh, and E. E. Conn, *Arch. Biochem. Biophys.*, 1986, **250**, 322.
242. T. M. Cromartie and C. Walsh, *J. Biol. Chem.*, 1976, **251**, 329.
243. S. Bornemann, J. R. Coggins, D. J. Lowe, and R. N. F. Thorneley, "Perspectives in Protein Engineering," Mayflower Worldwide, Oxford, UK, 1995, p. 134.
244. C. T. Lauhon and P. A. Bartlett, *Biochemistry*, 1994, **33**, 14 100.
245. P. Macheroux, J. Petersen, S. Bornemann, D. J. Lowe, and R. N. F. Thorneley, *Biochemistry*, 1996, **35**, 1643.
246. M. N. Ramjee, J. R. Coggins, T. R. Hawkes, D. J. Lowe, and R. N. F. Thorneley, *J. Am. Chem. Soc.*, 1991, **113**, 8566.
247. S. Bornemann, D. J. Lowe, and R. N. F. Thorneley, *Biochemistry*, 1996, **35**, 9907.
248. R. K. Hill and G. R. Newkome, *J. Am. Chem. Soc.*, 1969, **91**, 5893.
249. D. K. Onderka and H. G. Floss, *J. Biol. Chem.*, 1972, **247**, 736.
250. K. Fukui, *Tetrahedron Lett.*, 1965, 2427.
251. N. T. Anh, *J. Chem. Soc., Chem. Commun.*, 1968, 1089.
252. P. A. Bartlett, U. Maitra, and P. M. Chouinard, *J. Am. Chem. Soc.*, 1986, **108**, 8068.
253. P. A. Bartlett, K. L. McLaren, D. G. Alberg, A. Fassler, R. Nyfeler, C. T. Lauhon, and C. B. Grissom, Proc. Soc. Chem. Ind. Pesticides Group Meeting, *BCPC Monogr.*, 1989, **42**, 155.
254. M. N. Ramjee, S. Balasubramanian, C. Abell, J. R. Coggins, G. M. Davies, T. R. Hawkes, D. J. Lowe, and R. N. F. Thorneley, *J. Am. Chem. Soc.*, 1992, **114**, 3151.
255. B. Giese and N. G. Almstead, *Tetrahedron Lett.*, 1994, **35**, 1677.
256. T. R. Hawkes, T. Lewis, J. R. Coggins, D. M. Mousdale, D. J. Lowe, and R. N. F. Thorneley, *Biochem. J.*, 1990, **265**, 899.
257. S. Balasubramanian and C. Abell, *Tetrahedron Lett.*, 1991, **32**, 963.
258. S. Bornemann, S. Balasubramanian, J. R. Coggins, C. Abell, D. J. Lowe, and R. N. F. Thorneley, *Biochem. J.*, 1995, **305**, 707.
259. S. Balasubramanian, C. Abell, and J. R. Coggins, *J. Am. Chem. Soc.*, 1990, **112**, 8581.
260. S. Balasubramanian, J. R. Coggins, and C. Abell, *Biochemistry*, 1995, **34**, 341.
261. W. Buckel and R. Keese, *Angew. Chem., Int. Ed. Engl.*, 1995, **34**, 1502.
262. S. Bornemann, M. K. Ramjee, S. Balasubramanian, C. Abell, J. R. Coggins, D. J. Lowe, and R. N. F. Thorneley, *J. Biol. Chem.*, 1995, **270**, 22 811.
263. D. A. Jude, D. C. D. Ewart, J. L. Thain, G. M. Davies, and W. W. Nichols, *Biochim. Biophys. Acta*, 1996, **1279**, 125.
264. P. J. Duggan, E. Parker, J. Coggins, and C. Abell, *Bioorg. Med. Chem. Lett.*, 1995, **5**, 2347.
265. K. A. Dell and J. W. Frost, *J. Am. Chem. Soc.*, 1993, **115**, 11 581.

1.23
The Role of Isochorismic Acid in Bacterial and Plant Metabolism

ECKHARD LEISTNER
Rheinische Friedrich-Wilhelms-Universität Bonn, Germany

1.23.1 DISCOVERY OF ISOCHORISMIC ACID

The shikimic acid pathway is responsible for the production of vitamins and aromatic amino acids such as phenylalanine, tyrosine, and tryptophan.[1] It is also a source of precursors which are converted to an array of natural products. The sequence of reactions leading to aromatic amino acids branch at chorismic acid (**1**), a central intermediate of the shikimic acid pathway. Chorismic acid is also converted to its structural isomer isochorismic acid (**2**) (Scheme 1).

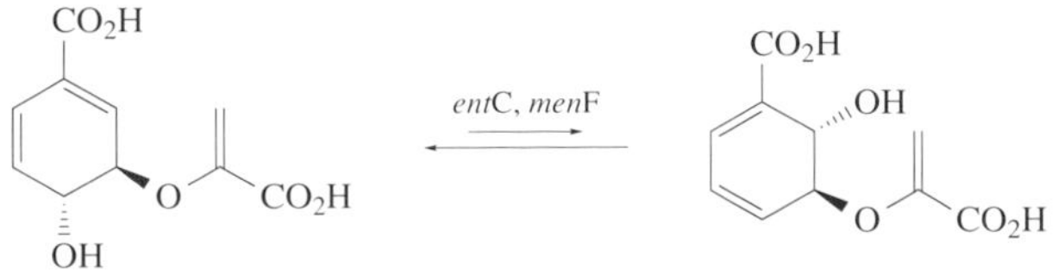

CO_2H OH O CO_2H (**1**) — *ent*C, *men*F → CO_2H OH O CO_2H (**2**)

Scheme 1

Isochorismic acid (**2**) is the less celebrated of the two isomers, yet it is an important branch point from which primary and secondary metabolites of bacterial and plant origin are formed.[2] Isochorismic acid (**2**) was discovered after incubating an enzyme extract from *Klebsiella pneumoniae* 62-1 (*Aerobacter aerogenes*, *Enterobacter aerogenes*) with chorismic acid (**1**). While a crude extract converted chorismic acid (**1**) to 2,3-dihydroxybenzoic acid in the presence of NAD, an enriched enzyme fraction obtained after chromatography on diethylaminoethyl cellulose did not form the benzoate derivative, but accumulated an intermediate in this conversion, which was isolated by electrophoresis and ion exchange chromatography. Thermal decomposition of this compound gave *m*-hydroxybenzoic, salicylic, pyruvic, and *m*-carboxyphenylpyruvic acid. These degradation studies and spectroscopic data (NMR, UV) disclosed the structure.[3] The absolute configuration was determined after ozonolysis which gave tartaric acid.[4] The data showed that the stereochemistry of chorismic acid (**1**) was retained during its conversion to isochorismic acid (**2**). Thus, the structure is *trans*-3-((1-carboxyethenyl)oxy)-2-hydroxy-4,6-cyclohexadiene-1-carboxylic acid with *S* configuration at C-2 and C-3.

1.23.2 BIOSYNTHESIS OF ISOCHORISMIC ACID

1.23.2.1 Tracer Studies

Several different mechanisms have been discussed to account for the origin of oxygen at C-2 of isochorismic acid (**2**) in its formation from chorismic acid (**1**). In principle, the oxygen of the C-2 hydroxy in isochorismic acid (**2**) may originate from molecular oxygen or water. Moreover, it was possible that biosynthesis of isochorismic acid (**2**) involved an intramolecular shift of the hydroxy group from C-4 of chorismic acid (**1**) to the C-2 position.

Three different groups addressed this question.[5–7] It was found in each case that ^{18}O from $H_2{}^{18}O$ was incorporated into isochorismic acid (**2**) with concomitant elimination of the C-4 hydroxy group of chorismic acid (**1**) and a shift of the double bonds. The resulting isochorismic acid (**2**) was analyzed for its ^{18}O content, either by mass spectrometry[7] or by mass spectrometry after enzymic conversion of the labeled isochorismic (**2**) to 2,3-dihydroxybenzoic acid.[5] It was also observed that incorporation of ^{18}O into the C-2 hydroxy group resulted in an additional ($^{13}C—^{18}O$) isotope-shifted resonance of C-2 upfield of the normal ($^{13}C—^{16}O$) resonance.[6] All experiments consistently indicated that the reaction is formally a double S_N2' reaction involving water and that a pericyclic reaction is excluded. A reaction sequence which accounts for these observations was published by Walsh *et al.*[5]

1.23.2.2 Genes Encoding Isochorismate Hydroxymutase Enzymes

In *Escherichia coli* and *Bacillus subtilis*, isochorismic acid (**2**) is a precursor of menaquinones (i.e., vitamin K_2) (**3**) and siderophores such as enterobactin (**4**) (*E. coli*) or 2,3-dihydroxybenzoic acid (*B. subtilis*). The latter two compounds are involved in iron sequestration. Two distinct genes have been cloned and sequenced which are responsible for the interconversion of chorismic (**1**) and isochorismic acid (**2**) in *B. subtilis*[8] and in *E. coli*.[9,10] Both genes are clustered with genes which specify enzymes for menaquinone (**3**) or enterobactin (**4**) biosynthesis. While the former isochorismate hydroxymutase gene has been designated *men*F, the latter was named *ent*C. The *ent* genes are grouped at ~13.5 min of the *E. coli* genome.[11] The cluster (*entC*, *entE*, *entB*, *entA*) is followed by an unknown open reading frame which is called p15 or *entX*.[5]

(3)

(4)

entC exhibits extensive homology with *trpE* and *pabB*, genes encoding like *entC* chorismate utilizing enzymes. *trpE* and *pabB* are involved in anthranilic or *p*-aminobenzoic acid (**5**) biosynthesis, respectively. Homologous regions are particularly obvious in the carboxy termini of the proteins. The amino acid sequences are 19% (Ent–PabB), 22.3% (PabB–TrpE) and 22.3% (EntC–TrpE) homologous. It is possible that the three genes share a common ancestor.[12]

The contiguous *ent* genes beginning with *entC* and extending to p15 are cotranscribed from a control region ("iron box") upstream of *entC*. The ferric uptake regulatory protein (Fur) acts as a repressor binding to the iron box in the presence of ferrous iron.[13] The ATG start codon of *entC* and the SD ribosomal binding site sequence are only four base pairs separate. This is at the short extreme of all such distances known in *E. coli* (normal SD position is ca. 8 bases upstream of ATG) and may indicate that *entC* expression may be limited at the translational level.[5]

Another isochorismate hydroxymutase gene (*menF*) has been discovered.[9,10] The gene is associated with genes *menC*, *menE*, and *menD* involved in menaquinone biosynthesis. The genes form a cluster at 48.5 min.[14] The *men*F gene is responsible for the synthesis of isochorismic acid (**2**) which is exclusively channeled into menaquinone (**3**) synthesis. This is evident from mutant studies: selective disruption of *menF* resulted in 95% loss of menaquinone whereas a reduced enterobactin synthesis was not observed. An *E. coli* mutant with a disrupted *entC*, however, does not synthesize enterobactin (**4**) but produces menaquinones.[9] Thus, as opposed to our previous assumption,[15] both isochorismate hydroxymutase genes play distinct roles in enterobactin (**4**) and menaquinone (**3**) synthesis in *E. coli*; a *menF* gene has also been detected in *B. subtilis*.[8]

1.23.2.3 Enzyme Studies

The enzyme isochorismate synthase (systematic name isochorismate hydroxymutase, E.C. 5.4.99.6) was first detected in *E. coli* strains but is much more active in the mutant *K. pneumoniae* 62-1. Pure enzyme was available after the *entC* gene had been cloned, sequenced, and overexpressed.[16] Overexpression was much improved by changing the distance from the promoter and the distance from the SD sequence to the ATG start codon. The enzyme was purified to homogeneity with a yield of 11 mg enzyme from 3 liters of recombinant cultured *E. coli* cells. The enzyme is an active monomer with a native molecular weight of 42.9 kDa. The enzyme catalyzed the isomerization of chorismic acid (**1**) and isochorismic acid (**2**) in both directions with a K_m for chorismate (**1**) (14 μM) slightly higher than for isochorismate (**2**) (5 μM). Thus, the binding affinity for isochorismate is higher with the equilibrium lying toward the side of chorismate (**1**). The enzyme activity depends on the presence of Mg^{2+}. An isochorismate hydroxymutase which like EntC may be involved in the biosynthesis of 2,3-dihydroxybenzoic acid has also been detected in cultured cells of *Catharanthus roseus*.[17] Biosynthesis of the phenolic acid and isochorismate hydroxymutase activity are both elicited by addition to the cell culture of an elicitor preparation obtained from *Pythium aphanidermatum*.

The gene product of *menF*, the second isochorismate hydroxymutase gene in *E. coli* has not yet been characterized. Unlike EntC, however, it is involved in menaquinone (**3**) biosynthesis[9] and it is evident that *menF* is regulated in a different way from that which is known for the regulation of *entC* activity by Fur (see above). The sequence upstream of *menF* in *E. coli* exerts a stimulating effect on isochorismic acid synthesis when placed onto an overexpression plasmid together with *menF*.[18]

Flavobacterium K_{3-15} is another organism which has been used as a source for isochorismate hydroxymutase.[19,20] This organism is particularly interesting because it is an overproducer of menaquinones (**3**) and does not seem to produce enterobactin (**4**). The isochorismate hydroxymutase was purified 750-fold and characterized. The most striking difference between this and the *E. coli* Ent C enzyme is that the K_m values are significantly higher for both substrates (chorismic acid 350 μM, isochorismic acid 254 μM).

The isochorismate hydroxymutase enzyme was immobilized on different solid supports.[20] Stable preparations were obtained on CNBr activated Sepharose and alkylamine glass. The thermal stability of the enzyme was much improved over the native dissolved enzyme. On the other hand, after immobilization a reduced enzyme activity was observed.

An isochorismate hydroxymutase has also been isolated from a *Galium mollugo* cell culture.[21] This culture produces large amounts of anthraquinones which are known to be derived from isochorismic acid (**2**).[22] The enzyme can be greatly stimulated by methyljasmonate as an elicitor. Many attempts to purify the enzyme to homogeneity, however, met with little success. The enzyme turned out to be very labile. In spite of this drawback a 572-fold purification was accomplished.

The K_m values of the enzyme exceeded those found in *Flavobacterium*, indicating again that this may be essential to secure aromatic amino acid and *p*-hydroxybenzoic acid formation in a branched pathway where different enzymes compete for the same substrate (chorismic acid (**1**)).

1.23.3 METABOLISM OF ISOCHORISMIC ACID IN BACTERIA

1.23.3.1 Chorismic Acid, Isochorismic Acid, and *p*-Aminobenzoic Acid Biosynthesis

p-Aminobenzoic acid (**5**) is a structural unit of tetrahydrofolic acid and of heptaene antibiotics such as candicidin and perimycin. The latter are produced by *Streptomyces griseus* (IMRU 3570, *Streptomyces coelicolor*) or *Streptomyces aminophilus*, respectively. Three proteins, PabA, PabB, and PabC, are involved in the biosynthesis of *p*-aminobenzoic acid (**5**) (Scheme 2). The amino acid is synthesized in two steps catalyzed by 4-amino-4-deoxychorismate (ADC) synthase (PabA and PabB) and a lyase (PabC) which eliminates the pyruvyl moiety with concomitant aromatization.[23] The substrate for the ADC synthase was proposed to be either chorismic (**1**) or isochorismic acid (**2**). This question was solved in favor of chorismic acid being converted to 4-amino-4-deoxychorismate (**6**) in a net 1,1-replacement of NH_2 for OH.[24,25]

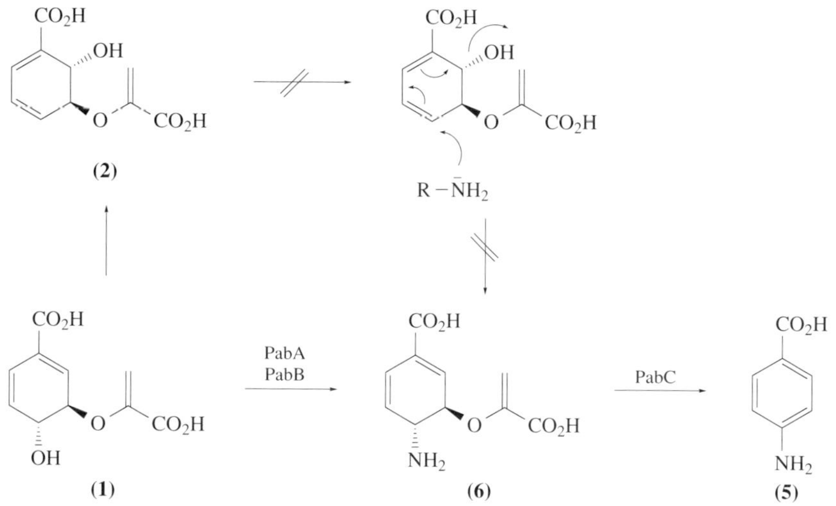

CO_2H OH O CO_2H (**2**)

CO_2H OH O CO_2H R – $\bar{N}H_2$

CO_2H O CO_2H OH (**1**) PabA PabB CO_2H O CO_2H NH_2 (**6**) PabC CO_2H NH_2 (**5**)

Scheme 2

Enzyme extracts from *K. pneumoniae* 62-1, *S. aminophilus*, and *S. coelicolor* were used to investigate the biosynthesis of *p*-aminobenzoic acid (**5**). The enzyme preparations from *Klebsiella pneumoniae* 62-1 and *Streptomyces aminophilus* contained both *p*-aminobenzoate synthase activity and isochorismate synthase activity, and were able to convert both chorismic (**1**) and isochorismic acid (**2**) to *p*-aminobenzoic acid (**5**). The apparent K_m for chorismic acid (**1**) was, however, significantly lower than that for isochorismic acid (**2**), while the V_{max} was identical for both substrates. Isochorismate synthase activity was not detectable in enzyme preparations from *S. coelicolor* and *p*-aminobenzoic acid synthesis took place from chorismic acid (**1**) only. It was concluded that isochorismic acid (**2**) is not an obligatory intermediate in *p*-aminobenzoic acid biosynthesis from chorismic acid (**1**) and that *p*-aminobenzoic acid (**5**) is directly derived from chorismic acid (**1**).[24,25]

1.23.3.2 Isochorismic Acid and Salicylic Acid Biosynthesis

Salicylic acid (**7**) (Scheme 3) plays an important role in the induction of plant resistance to pathogens, the so-called systematic acquired resistance (SAR).[26] Salicylic acid (**7**) is produced by the monooxygenase catalyzed oxidation of benzoic acid (**8**). It is likely that an arene oxide:oxepin

(**9**) functions as an intermediate in this reaction.[27] The origin of the benzoic acid in this reaction is unknown but could be derived from cinnamic acid coenzyme A ester by β-oxidation.

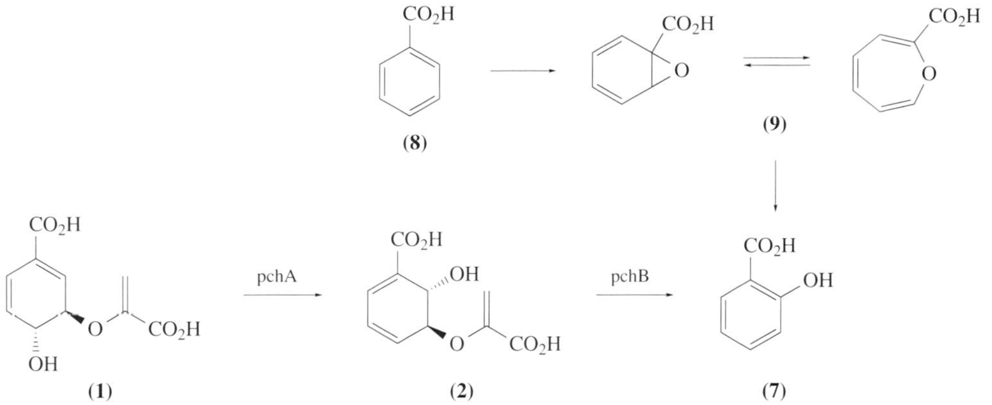

CO_2H (**8**) → CO_2H O ⇌ CO_2H O (**9**) ↓ CO_2H OH (**7**)

CO_2H O CO_2H OH (**1**) —pchA→ CO_2H OH O CO_2H (**2**) —pchB→ (**7**)

Scheme 3

Salicylic acid (**7**) is a compound which is also involved in iron transport in bacteria.[28] Salicylate may form a 3:1 complex with Fe^{III} and act as a compound involved in iron sequestration. Alternatively, salicylic acid (**7**) may be converted to mycobactins. These are compounds with a growth-promoting activity which are involved in iron transport.[29] The bacterial salicylic acid moiety of mycobactins is derived from a metabolite of the shikimate pathway. Experiments with crude protein extracts indicate that this metabolite is isochorismic acid (**2**). The biosynthesis of salicylic acid (**7**) is under investigation in *Pseudomonas aeruginosa*.[30] Genetic studies indicate that two genes are involved in salicylic acid biosynthesis and that one gene (*pchA*) very likely specifies an isochorismate hydroxymutase gene. This gene is genetically linked to an upstream open reading frame (*pchB*) which overlaps the putative ATG start codon of *pchA*. Mutation of either *pchA* or *pchB* results in the inability to synthesize salicylic acid (**7**) and pyochelin. *pchA* is expressed only when *pchB* is transcribed and translated simultaneously. Isochorismate hydroxymutase activity was not observed after expression of *pchA* alone. It was speculated that when *pchB* is separated from *pchA* the corresponding PchA protein folds incorrectly. PchA and PchB are thought to form a stable complex which catalyzes two subsequent steps in the conversion of chorismic (**1**) to isochorismic acid (**2**) (PchA) and of isochorismic acid (**2**) to salicylic acid (**7**) (PchB). Thus, PchA would be an isochorismate hydroxymutase, whereas PchB would be an enzyme with a dual function in which an isochorismate pyruvate lyase activity and a dehydrogenase activity catalyze the conversion of isochorismic (**2**) to salicylic acid (**7**). An enzyme complex reminiscent of the anthranilate and *p*-aminobenzoate synthase would thus be functioning in salicylic acid (**7**) biosynthesis. Indeed, *pchA* is a gene with a significant homology to genes *trpE* and *pabB* which are involved in chorismic acid metabolism and anthranilic and *p*-aminobenzoic acid biosynthesis, respectively.

1.23.3.3 Isochorismic Acid and Menaquinone Biosynthesis

1.23.3.3.1 Biosynthesis of o*-succinylbenzoic acid*

o-Succinylbenzoic acid (OSB; (**10**)) (Scheme 4) is the first aromatic intermediate in the biosynthesis of menaquinones[31] (**3**) from isochorismic acid (**2**). The compound was proposed to be a precursor of menaquinones (**3**) on theoretical grounds.[32] Incorporation studies with radio-labeled OSB showed indeed that this material was specifically converted to menaquinones (**3**) and biosynthetically related quinones from higher plants. The observation that *E. coli* mutants blocked in menaquinone biosynthesis accumulate OSB (**10**) in the culture broth showed for the first time that OSB (**10**) is a naturally occurring compound.[33] The same technique revealed the presence of 1,4-dihydroxy-2-naphthoic acid (**11**) in the supernatant of cultured mutants and provided strong evidence in favor of a pathway leading via OSB (**10**) and naphthoic acid (**11**) to menaquinones (**3**).

One early observation, however, was difficult to reconcile with later data: a mutant (*entC*$^-$) blocked in isochorismate synthesis produced normal levels of menaquinones (**3**). This led to the assumption that chorismic acid (**1**) is the branch point for menaquinone biosynthesis. It was later

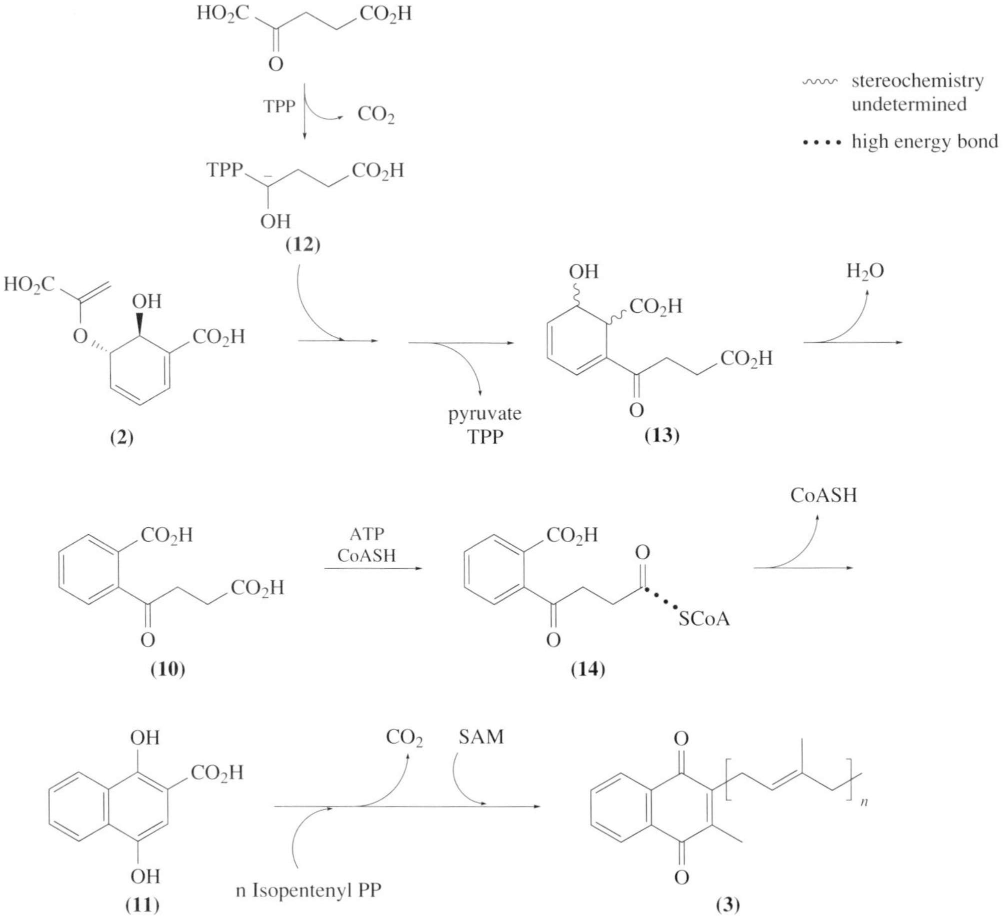

Scheme 4

demonstrated, however, that the branch point is indeed isochorismic acid (**2**).[34,35] The fact that a mutant blocked in the *entC* gene is able to produce menaquinones (**3**) can now be explained by the finding that there are two genes for isochorismate synthesis, *entC* and *menF*, both of which specify an isochorismate hydroxymutase.[9,10]

OSB (**10**) is a simple and stable aromatic compound which is derived by a new and hitherto unprecedented aromatization process. Enzyme extracts and partially purified enzyme preparations convert isochorismic acid (**2**) and 2-oxoglutaric acid to OSB (**10**) in the presence of thiamine pyrophosphate and a divalent metal ion.[1,31]

The succinyl moiety of OSB (**10**) is generated from C-2 to C-5 of 2-oxoglutarate with simultaneous loss of C_1 in the form of CO_2. The reaction resembles the initial steps catalyzed by the 2-oxoglutarate dehydrogenase. It was therefore unclear whether the active succinic semialdehyde thiamine pyrophosphate adduct (**12**) was generated by the 2-oxoglutarate dehydrogenase and subsequently channeled into both succinyl CoA and OSB (**10**) synthesis, or whether both enzyme systems are distinct entities, each of which is capable of catalyzing the decarboxylation of 2-oxoglutarate in the presence of thiamine pyrophosphate. It was eventually shown by different techniques that both systems can be easily separated.[34–37] The finding that both systems are distinct is also evident from the fact that the 2-oxoglutarate dehydrogenase of the citric acid cycle is only active under aerobic growth conditions, while menaquinones (**3**) are essential under anaerobic growth conditions in *E. coli*.

Attempts to purify the *o*-succinylbenzoate synthase system met with little success. It turned out that the enzyme complex (molecular weight ~195.5 kDa) disintegrated into a polypeptide of ~66.5 kDa and another portion which remained undetected. The 66.5 kDa polypeptide generated succinsemialdehyde from 2-oxoglutarate. The former was identified after conversion to its hydrazone. Evidence for the presence of the succinsemialdehyde thiamine pyrophosphate adduct (**12**) was also obtained.[34,35]

Studies with mutants have shown that there is another intermediate in the conversion of isochorismic acid (**2**) to OSB (**10**). This compound was suggested to be 2-succinyl-6-hydroxy-2,4-cyclohexadiene-1-carboxylate (SHCHC; (**13**)).[38] The gene that encodes the catalyzing enzyme system was named *menD*.[39] The gene that is responsible for the conversion of SHCHC (**13**) to OSB (**10**) is *menC*.[40]

1.23.3.3.2 Metabolism of OSB

The cell-free conversion of OSB (**10**) to 1,4-dihydroxy-2-naphthoic acid (**11**) in an enriched enzyme solution from *E. coli* was first reported by Bryant and Bentley.[41] The enzyme system which was active in the presence of OSB (**10**), ATP, Mg^{2+}, and coenzyme A was termed naphthoate synthetase. The enzyme was thought to have a molecular weight of 45 kDa. Separation of the crude enzyme into two fractions by protamine sulfate precipitations showed that the activity was separable into two fractions which carried out the reactions only in combination. This indicated that there are two steps involved in the overall reaction from OSB (**10**) to 1,4-dihydroxy-2-naphthoic acid (**11**). This was confirmed by mutant studies. Moreover, the occurrence of an intermediate OSB coenzyme A ester (**14**) was postulated. This coenzyme A ester was found and was shown to be a mono-coenzyme A ester.[42] There were two options as to which of the carboxyl groups of OSB (**10**) would be activated, either the "aromatic" or the "aliphatic" carboxyl group. Chemical and enzymic synthesis of the coenzyme A ester[43] in question and their comparison showed that contrary to expectations the "aliphatic" carboxyl group of OSB (**10**) was activated during the conversion of OSB (**10**) to 1,4-dihydroxy-2-naphthoic acid (**11**).[44] The structure of the intermediate activated OSB coenzyme A ester (**14**) was confirmed by Arigoni's group.[45] The ligase activating OSB (**10**) was purified 134-fold from *Mycobacterium phlei* and was shown to be an enzyme of low substrate specificity. Coenzyme A and dephosphocoenzyme A, various nucleotides and analogues of OSB such as *o*-malonylbenzoic acid or benzoylpropionic acid were accepted by the enzyme. The molecular weight of the *M. phlei* enzyme was found to be 28 kDa by gel filtration.[46] This contrasts with the molecular weight of the corresponding ligase from *E. coli* which is encoded by *menE* and has a deduced molecular weight of 50.2 kDa.[47] Stereochemical aspects of the conversion of OSB coenzyme A ester (**14**) to 1,4-dihydroxynaphthoic acid (**11**) has also been investigated. The succinyl side chain of OSB (**10**) was enantiotopically labeled using a series of enzyme reactions of known stereochemistry and OSB activated as coenzyme A ester. The chirally labeled samples were converted into 1,4-dihydroxy-2-naphthoic acid (**11**) in the presence of naphthoate synthase extracted from menaquinone-producing bacteria. The results showed that the enzyme catalyzes a stereospecific process. The enzyme exchanges the hydrogen at the 2-H_R position of the coenzyme A ester much faster than the hydrogen at the 2-H_S position. This indicated that during the ring closure a carbanion is formed in a fast and reversible reaction (Scheme 5). These observations are only possible because the ring closure reaction proper which leads from the carbanion to 1,4-dihydroxy-2-naphthoic acid (**11**) is a rate limiting step in which the 2-H_S and the 3-H_S hydrogens are eliminated, whereas the 3-H_R hydrogen is retained and gives the proton at position 3 of 1,4-dihydroxy-2-naphthoic acid[48] (**11**), (Scheme 5).

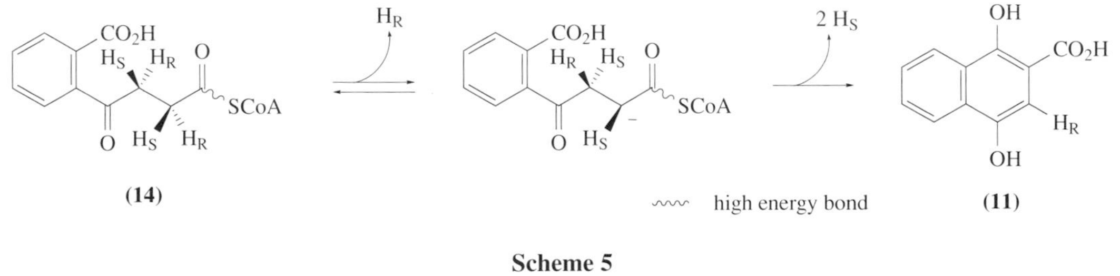

(**14**) ~~~ high energy bond (**11**)

Scheme 5

The gene (*men*B) encoding the naphthoate synthase in *E. coli* has been identified and sequenced.[49]

1.23.3.4 Biosynthesis of Enterobactin and 2,3-Dihydroxybenzoic Acid

Iron is essential for the growth of virtually all organisms. At neutral pH, however, iron forms insoluble salts. As a result, many organisms have evolved siderophores which are essential for iron acquisition. One such siderophore is 2,3-dihydroxybenzoic acid and enterobactin (enterochelin) (**4**) is another. The latter siderophore consists of three serine residues linked together via ester bonds. The amino groups of the three serine residues form peptide bonds with three 2,3-dihydroxybenzoic acid molecules. Early cotransduction experiments showed that three of the genes (*entA*, *entB*, *entC*) involved in enterobactin biosynthesis map near 14 min of the *E. coli* chromosome.[50] The functions assigned to these genes were isochorismate synthase (isochorismate hydroxymutase, E.C. 5.4.99.6) (*entC*), 2,3-dihydro-2,3-dihydroxybenzoate synthase (isochorismatase, E.C. 3.3.2.1) (*entB*) and 2,3-dihydro-2,3-dihydroxybenzoate dehydrogenase (E.C. 1.3.1.28) (*entA*) (Scheme 6). The last gene and enzyme (*entA* and its product) are responsible for the formation of 2,3-dihydroxybenzoic acid. The reaction catalyzed by EntA was investigated with substrate analogues. It was demonstrated that conversion of 2,3-dihydro-2,3-dihydroxybenzoic acid (**15**) to 2,3-dihydroxybenzoic acid (**16**) is inhibited by oxidation at C-3 rather than C-2. The hydride ion abstracted from the substrate is transferred to the *si* face of the enzyme-bound NAD.[51]

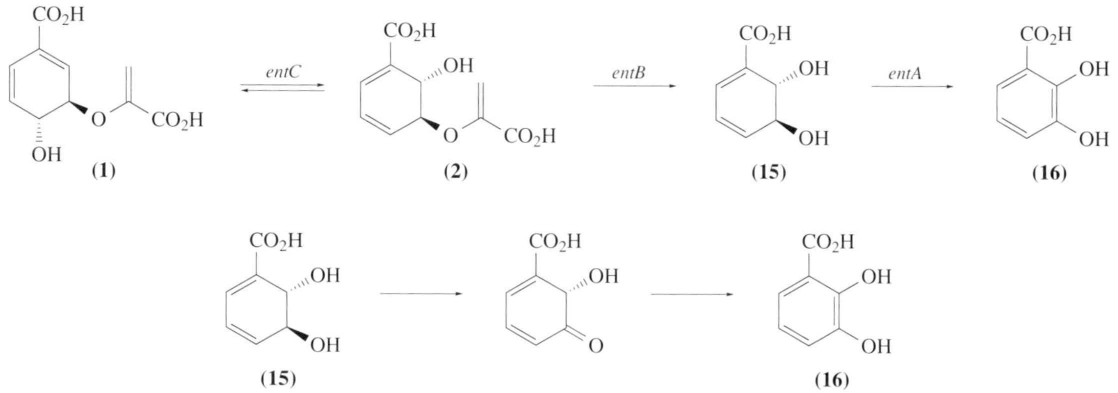

Scheme 6

The sequence of genes in this operon has been established as *entC*, *entE*, *entB*, and *entA* followed by p15[52] (also called *entX*). The function of the latter open reading frame is unknown, while *entE* is involved in the activation of 2,3-dihydroxybenzoic acid (**16**). The last steps in enterobactin (**4**) synthesis are carried out by an enterobactin-synthetase, a multienzyme complex believed to be composed of EntD, EntE, EntF, and EntG.[53] Surprisingly, there is no separate *entG* gene. EntG activity is encoded by the 3′-terminus of the *entB* gene. The precise function of EntG is unknown, albeit essential for enterobactin (**4**) biosynthesis. The *entF* gene specifies a protein with 160 000 Da. The enzyme contains a 4′-phosphopantetheine cofactor. This, and the fact that a 267 amino acid stretch—roughly in the middle of EntF—is 38% homologous with the amino terminus of tyrocidine synthase, suggests a reaction mechanism reminiscent of nonribosomal peptide synthases. Indeed, partially purified EntF catalyzes an ATP-[^{32}P]PPi exchange as would be expected for such an enzyme system. A combined incubation of polypeptides EntF, EntB, and EntE with serine, 2,3-dihydroxybenzoic acid (**16**), and ATP, however, did not give benzoyl serine.[54]

In the presence of iron the *entCEBA*(*X*) operon is negatively regulated by the ferric uptake regulatory protein Fur.[13]

The biochemical and molecular biological investigations of the enterobactin biosynthesis has been reviewed.[5]

1.23.3.5 Biosynthesis of Gabaculin and Sarubicin A

In the course of studies on enzyme regulators from microorganisms, Mishima and co-workers have discovered a new 7-carbon amino acid from *Streptomyces toyoycaensis* which inhibits γ-aminobutyric acid transferase.[55] The compound was named gabaculine (**17**). Structure elucidation showed that this compound has an unusual *meta* disposition of a carboxyl- and an amino group. It was proposed that gabaculine (**17**) may be derived from isochorismic acid (**2**) after reduction, incorporation of nitrogen by S_N2' displacement of the enolpyruvate, and dehydration.[56] Any experimental evidence concerning this hypothetical pathway is missing.

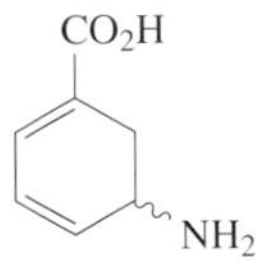

(17)

The possible participation of isochorismic acid (**2**) in sarubicin A (**18**) biosynthesis is also speculative. The assumption that isochorismic acid (**2**) would give rise to sarubicin A is based on the observation that labeled 6-hydroxyanthranilic acid was converted to sarubicin A by cultures of *Streptomyces helicus* and that the aromatic amino acid would most reasonably be derived from the shikimate pathway via isochorismic acid (**2**).[57]

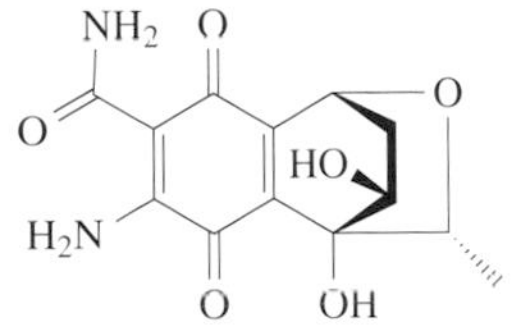

(18)

1.23.4 METABOLISM OF ISOCHORISMIC ACID IN HIGHER PLANTS

1.23.4.1 Biosynthesis of Phylloquinone

Phylloquinone and menaquinones (**3**) share a 2-methyl-1,4-naphthoquinone chromophore which is linked to either an isoprenoid side chain of varying length (menaquinone, vitamin K_2) or a phythyl side chain (phylloquinone, vitamin K_1). While menaquinones occur in bacteria, phylloquinone is a constituent of higher plants. Both compounds are derived from OSB (**10**).[58–60]

The OSB synthase catalyzing the conversion of isochorismic acid (**2**) and α-oxoglutaric acid to OSB (**10**) in the presence of thiamine pyrophosphate and Mn^{2+} has been detected in cells of a photoautotrophic green suspension culture of *Morinda lucida* (*Rubiaceae*).[59,60] An enzyme system catalyzing the same reaction was detected in protein extracts of the unicellular green alga *Euglena gracilis*.[61] The activity of the OSB synthase is linked to the presence of the intact photosynthetic apparatus. In spite of this, the enzyme is not located within the chloroplast but is cytosolic. No activity was found in an aplastidic mutant of the alga.

Tentative evidence indicated that OSB (**10**) is activated and metabolized via its coenzyme A ester to 1,4-dihydroxy-2-naphthoic acid (**11**), as has been observed in bacteria (see above). Incubation of fractionated chloroplasts indicated that the enzymes are associated with the envelope of the chloroplasts. Phytylation of 1,4-dihydroxy-2-naphthoic acid (**11**) occurs also in chloroplast envelope membranes and the terminal methylation reaction of phylloquinone biosynthesis in thylakoid membranes requires the addition of soluble stroma protein.[62]

1.23.4.2 Biosynthesis of Naphthoquinones and Anthraquinones

Feeding experiments with isotopically labeled precursors revealed that most plant naphthoquinones and anthraquinones are derived either from the polyketide or the isochorismic–OSB pathway. These results have been reviewed.[31]

While early experiments were carried out with intact plants it was later shown that quinones belong to a group of natural products which are abundantly present in plant cell suspension cultures.[22,63] Factors that influence quinone production are hormones, light, nutritional factors, and heterotrophic or autotrophic growth conditions.

It is therefore possible and tempting to study the enzymes involved in quinone biosynthesis and their regulation.

The isochorismate hydroxymutase has been detected in cell cultures derived from *Catharanthus roseus*[17] or *G. mollugo*[21] plants. The enzyme responds in both cases to elicitors. In spite of an inherent lability of the enzyme it has been purified 572-fold from cultured cells of *G. mollugo*. Glycine betaine in the incubation mixture increased the enzyme activity when present at a 0.5 M concentration. The K_m values were rather high for both the forward (807 μM) and reverse reaction (675 μM) and the K_m value for isochorismic acid was lower, when compared with chorismic acid (**1**). This has also been observed with the isochorismate hydroxymutase from other sources (*E. coli*, *Flavobacterium*, see above). The enzymic conversion in protein extracts of isochorismic acid (**2**) to OSB (**10**) has also been detected. There seem to be at least two intermediates involved in this reaction sequence. One intermediate is SHCHC (**13**); however, the other intermediate remained unidentified.[64]

In the conversion of OSB (**10**) to anthraquinones, again the acid undergoes activation at the "aliphatic" carboxyl group.[65] The activating ligase is an enzyme with a high substrate specificity when compared to the corresponding enzyme from *M. phlei*. The mechanism of activation, however, is much different from what is known from the bacterial enzyme. The latter releases AMP from ATP during the activation, whereas the former generates ADP. Isoelectric focusing of the ligase from *G. mollugo* gave two separate isoenzyme activities. It was speculated that one activity represented a phosphorylated ligase, whereas the other one was assumed to be the nonphosphorylated species.

The biosynthesis of the alkaloid shihunine (**19**) in the orchid *Dendrobium pierardii*[66] is interesting in the present context. ^{14}C- and ^{13}C-labeled OSB (**10**) was specifically converted to shihunine (**19**). The labeling pattern of this alkaloid was consistent with a pathway in which the "aliphatic" carboxyl group of OSB (**10**) was reduced to an aldehyde with subsequent amination, methylation, and eventual cyclization of the side chain to give shihunine (**19**) (Scheme 7). Reduction of the carboxyl group would be initiated by an activating ligase in the presence of coenzyme A, ATP, and Mg^{2+}. This is interesting in the light of a discussion on the site of activation of OSB (**10**) at either the "aromatic" or "aliphatic" carboxyl group.[67] All available data on shihunine (**19**), menaquinone (**3**), and anthraquinone biosynthesis support the view that the "aliphatic" carboxyl group is activated. The coenzyme A ester has not been tested as an intermediate in anthraquinone biosynthesis.

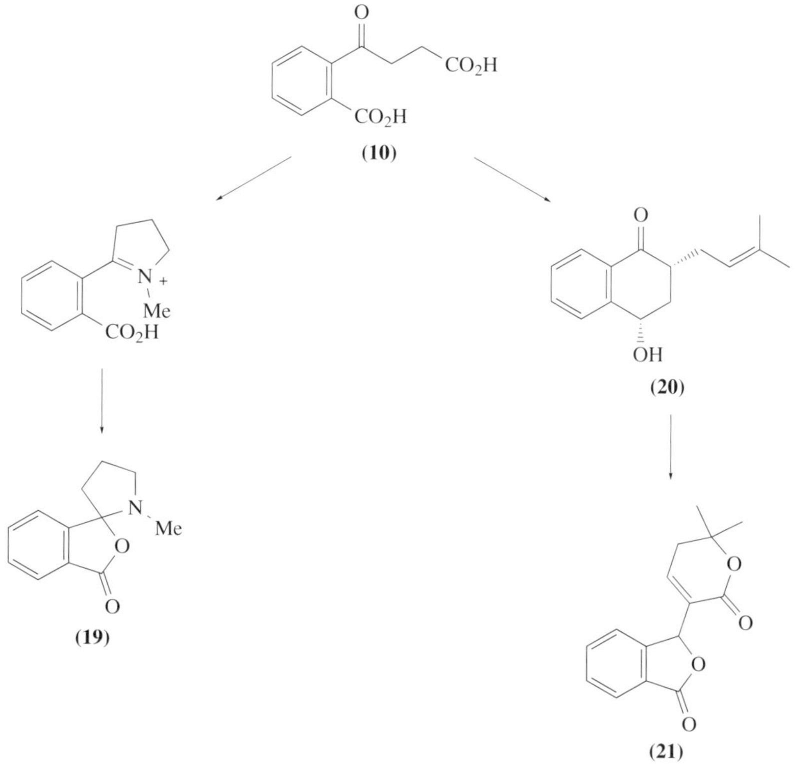

Scheme 7

Tracer studies[31] carried out on *Catalpa ovata* cell cultures indicate, however, that as opposed to menaquinone biosynthesis the ring closure reaction leads to nonaromatic compounds like catalponol (**20**) (or its diketone catalponone). This metabolite is a precursor of catalpalactone.[31,68]

1.23.4.3 Biosynthesis of *m*-Carboxy Amino Acids

The Claisen rearrangement of chorismic (**1**) to prephenic acid is the only known sigmatropic reaction in primary metabolism. Visual analysis of some natural products including compounds with a terpenoid moiety leads to the suggestion that Claisen rearrangements may also be involved in the biosynthesis of secondary metabolites. Feeding experiments have shown that *m*-carboxyphenylalanine (**22**), a natural product occurring in plants belonging to the *Resedaceae* and *Brassicaceae*, is derived from the shikimate pathway (Scheme 8). The reaction sequence may lead from shikimic and isochorismic acid (**2**) via a Claisen rearrangement to isoprephenic acid (**23**). Dehydration and amination may give *m*-carboxyphenylalanine[69] (**22**) and *m*-carboxytyrosine (**24**). A thermal conversion of isochorismic to isoprephenic acid (**23**) has also been observed *in vitro*.[5]

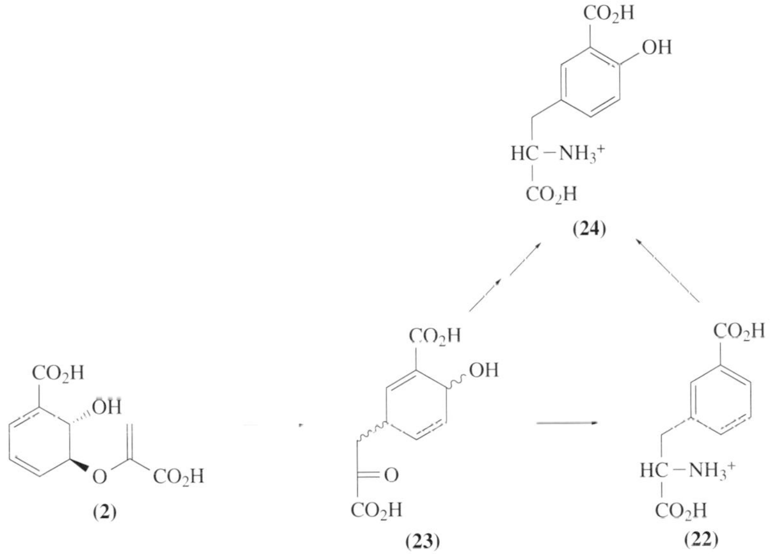

CO_2H OH $HC-NH_3^+$ CO_2H (**24**)

CO_2H OH O CO_2H (**2**) CO_2H OH O CO_2H (**23**) CO_2H $HC-NH_3^+$ CO_2H (**22**)

Scheme 8

Surprisingly, an enzyme system that catalyzes the conversion of isochorismic (**2**) to isoprephenic acid (**23**) has been postulated to be present in a cell suspension culture of *Nicotiana sylvestris*. The presence of *m*-carboxyamino acids in this plant or in the corresponding cell culture was not reported.[70]

1.23.5 PREPARATION BY METABOLIC PATHWAY ENGINEERING OF METABOLITES DERIVED FROM CHORISMIC ACID

Investigation of the metabolism of chorismic acid (**1**) has been greatly facilitated by the availability of large amounts of this compound. *K. pneumoniae* 62-1 (ATCC 25306), a mutant blocked in aromatic amino acid biosynthesis, excretes up to 800 mg $liter^{-1}$ chorismic acid (**1**) into the culture broth.

This strain was used by the author's group to prepare metabolites of chorismic acid (**1**) by a recombinant *K. pneumoniae* 62-1 strain harboring plasmid with genes involved in chorismic acid (**1**) metabolism.[71,72] Transformation of the mutant with a plasmid carrying isochorismate hydroxy-mutase genes *menF* or *entC* results in the appearance of a mixture of chorismic acid (**1**) and isochorismic acid (**2**) in the culture broth. The gene *entC* was more efficient in the production of isochorismic acid (**2**). Both isomers were isolated from the broth by ion exchange and reversed-phase chromatography. When *entB* (Scheme 6) was introduced into the mutant, (−)-*trans*-(3*R*,4*R*)-3,4-dihydroxycyclohexa-1,5-dienecarboxylic acid appeared in the broth at a concentration of 200 mg $liter^{-1}$. Similarly, (+)-*trans*-(2*S*,3*S*)-2,3-dihydroxycyclohexa-4,6-dienecarboxylic acid (**15**) was excreted at a concentration of 200 mg $liter^{-1}$ broth when both *entC* and *entB* were introduced into the mutant. The genes were placed behind the lac promoter. The data show that *K. pneumoniae* 62-1 can be employed in the production of cyclohexadiene carboxylic acid *trans* diols.[72] A bio-technological process has been used to produce *cis* diols. These diols play an important role in enantioselective syntheses. *Trans* diols could be used for a similar purpose. *K. pneumoniae* 62-1 can also be used in the preparation of specifically isotope-labeled metabolites. The only carbon source in the medium is glucose and specifically labeled glucose will give specifically labeled metabolites, as exemplified by *p*-hydroxybenzoic acid which is excreted into the culture broth when an over-expressed *ubiC* gene is present.[73]

1.23.6 CONCLUSIONS

The metabolic steps discussed in this chapter are summarized in Scheme 9. The scheme shows that a multitude of metabolites are derived from isochorismic acid (**2**) and that these metabolites are part of primary as well as secondary metabolism. The experimental evidence in support of this scheme is limited in some cases. In other cases, however, there is no evidence at all. No experiments have been carried out on the biosynthesis of gabaculine (**17**) or arene oxides. However, the proposal for a metabolic relationship is plausible.[56] The biosynthesis of catalponol (**20**), catalpalactone (**21**),[31] sarubicin A (**18**), and the orchid alkaloids (**19**) is based on isotope labeling studies, whereas enzyme studies were carried out on the biosynthesis of menaquinones (**3**), phylloquinone, anthraquinones,

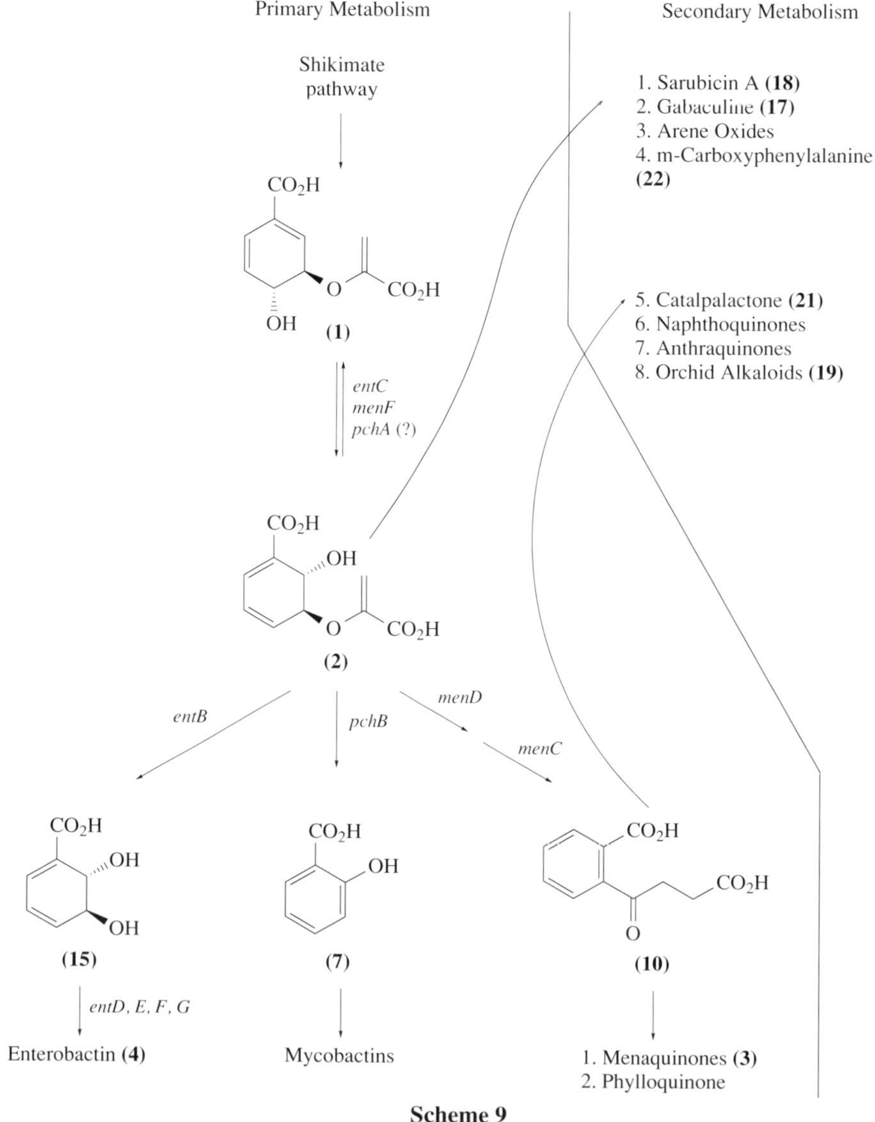

Scheme 9

and enterobactin (**4**). The steps leading to enterobactin (**4**), salicylic acid (**7**), and menaquinones (**3**), however, are also based on genetic evidence.

In conclusion, the data summarized here show that isochorismic acid (**2**) is a metabolic branch point which is at least as important as chorismic acid (**1**).

ACKNOWLEDGEMENTS

The author's work cited herein has been supported by the Deutsche Forschungsgemeinschaft and Fonds der Chemischen Industrie.

1.23.7 REFERENCES

1. R. Bentley, *Crit. Rev. Biochem. Mol. Biol.*, 1990, **25**, 307.
2. A. Kaiser and E. Leistner, *World J. Microbiol. Biotechnol.*, 1992, **8**, 92.
3. I. G. Young, T. J. Batterham, and F. Gibson, *Biochim. Biophys. Acta*, 1969, **177**, 389.
4. I. G. Young and F. Gibson, *Biochim. Biophys. Acta*, 1969, **177**, 348.
5. Ch. T. Walsh, J. Liu, F. Rusnak, and M. Sakaitani, *Chem. Rev.*, 1990, **90**, 1105.
6. S. J. Gould and R. L. Eisenberg, *Tetrahedron*, 1991, **47**, 5979.
7. L. O. Zamir, K. A. Devor, R. A. Jensen, R. Tiberio, F. Sauriol, and O. Mamer, *Can. Microbiol.*, 1991, **37**, 276.
8. B. M. Rowland and H. S. Taber, *J. Bacteriol.*, 1996, **178**, 854.
9. R. Müller, C. Dahm, G. Schulte, and E. Leistner, *FEBS Lett.*, 1996, **378**, 131.
10. R. Daruwala, O. Kwon, R. Meganathan, and M. E. S. Hudspeth, *FEMS Microbiol. Lett.*, 1996, **140**, 159.
11. C. L. Pickett, L. Hayes, and C. F. Earhart, *FEMS Microbiol. Lett.*, 1984, **24**, 77.
12. M. F. Elkins and C. F. Earhart, *FEMS Microbiol. Lett.*, 1988, **56**, 35.
13. T. J. Brickman, B. A. Ozenberger, and M. A. McIntosh, *J. Mol. Biol.*, 1990, **212**, 669.
14. D. J. Shaw, E. C. Robinson, R. Meganathan, R. Bentley, and J. R. Guest, *FEMS Microbiol. Lett.*, 1983, **17**, 63.
15. A. Kaiser and E. Leistner, *Arch. Biochem. Biophys.*, 1990, **276**, 1.
16. J. Liu, G. A. Berchtold, and C. T. Walsh, *Biochemistry*, 1990, **29**, 1417.
17. P. R. H. Moreno, R. van der Heijden, and R. Verpoorte, *Plant Cell Rep.*, 1994, **14**, 188.
18. R. Müller, C. Dahm, and E. Leistner, unpublished results.
19. P. M. Schaaf, L. E. Heide, and E. Leistner, *J. Nat. Prod.*, 1993, **56**, 1294.
20. P. M. Schaaf, L. E. Heide, and E. Leistner, *J. Nat. Prod.*, 1993, **56**, 1304.
21. C. Leduc, I. Birgel, R. Muller, and E. Leistner, *Planta*, 1997, **202**, 206.
22. E. Leistner, in "Biotechnology in Agriculture and Forestry," ed. Y. P. S. Bajaj, Springer Verlag, Berlin, 1995, vol. 33, p. 296.
23. V. K. Viswanathan, J. M. Green, and B. P. Nichols, *J. Bacteriol.*, 1995, **177**, 5918.
24. M. Johanni, P. Hofmann, and E. Leistner, *Arch. Biochem. Biophys.*, 1989, **271**, 495.
25. C.-Y. P. Teng, B. Ganem, S. Z. Doktor, B. P. Nichols, R. K. Bhatnagar, and L. C. Vining, *J. Am. Chem. Soc.*, 1985, **107**, 5008.
26. J. León, N. Yalpani, I. Raskin, and M. A. Lawton, *Plant Physiol.*, 1993, **103**, 323.
27. W. A. Ayer and E. R. Cruz, *J. Nat. Prod.*, 1995, **58**, 622.
28. R. G. Ankenbauer and C. D. Cox, *J. Bacteriol.*, 1988, **170**, 5364.
29. A. T. Hudson and R. Bentley, *Biochemistry*, 1970, **9**, 3984.
30. L. Serino, C. Reimmann, H. Baur, M. Beyeler, P. Visca, and D. Haas, *Mol. Gen. Genet.*, 1995, **249**, 217.
31. H. Inouye and E. Leistner, in "The Chemistry of Quinonoid Compounds," eds. S. Patai and Z. Rappoport, Wiley, Chichester, 1988, vol. II, p. 1293.
32. P. Dansette and R. Azerad, *Biochem. Biophys. Res. Commun.*, 1970, **40**, 1090.
33. I. G. Young, *Biochemistry*, 1975, **14**, 399.
34. A. Weische, M. Johanni, and E. Leistner, *Archiv. Biochem. Biophys.*, 1987, **256**, 212.
35. A. Weische, W. Garvert, and E. Leistner, *Archiv. Biochem. Biophys.*, 1987, **256**, 223.
36. M. G. Marley, R. Meganathan, and R. Bentley, *Biochemistry*, 1986, **25**, 1304.
37. C. Palaniappan, H. Taber, and R. Meganathan, *J. Bacteriol.*, 1994, **176**, 2648.
38. G. T. Emmons, I. M. Campbell, and R. Bentley, *Biochem. Biophys. Res. Commun.*, 1985, **131**, 956.
39. C. Palaniappan, V. Sharma, M. E. S. Hudspeth, and R. Meganathan, *J. Bacteriol.*, 1992, **174**, 8111.
40. V. Sharma, R. Meganathan, and M. E. S. Hudspeth, *J. Bacteriol.*, 1993, **175**, 4917.
41. R. W. Bryant Jr. and R. Bentley, *Biochemistry*, 1976, **15**, 4792.
42. L. Heide, S. Arendt, and E. Leistner, *J. Biol. Chem.*, 1982, **257**, 7396.
43. R. Kolkmann and E. Leistner, *Z. Naturforsch., Sect. C Biosci.*, 1987, **42c**, 542.
44. R. Kolkmann and E. Leistner, *Z. Naturforsch., Sect. C Biosci.*, 1987, **42c**, 1207.
45. I. Hubacek, Ph.D. Thesis, University of Zürich, 1991.
46. H. J. Sieweke and E. Leistner, *Z. Naturforsch., Sect. C Biosci.*, 1991, **46c**, 585.
47. V. Sharma, M. E. S. Hudspeth, and R. Meganathan, *Gene*, 1996, **168**, 43.
48. U. Igbavboa and E. Leistner, *Eur. J. Biochem.*, 1990, **192**, 441.
49. V. Sharma, K. Suvarna, R. Meganathan, and M. E. S. Hudspeth, *J. Bacteriol.*, 1992, **174**, 5057.
50. I. G. Young, L. Langman, R. K. J. Luke, and F. Gibson, *J. Bacteriol.*, 1971, **106**, 51.
51. M. Sakaitani, F. Rusnak, N. R. Quinn, C. Tu, T. B. Frigo, G. A. Berchtold, and C. T. Walsh, *Biochemistry*, 1990, **29**, 6789.

52. M. Schrodt Nahlik, T. J. Brickman, B. A. Ozenberger, and M. A. McIntosh, *J. Bacteriol.*, 1989, **171**, 784.
53. J. F. Staab and C. F. Earhart, *J. Bacteriol.*, 1990, **172**, 6403.
54. F. Rusnak, M. Sakaitani, D. Drueckhammer, J. Reichert, and C. T. Walsh, *Biochemistry*, 1991, **30**, 2916.
55. K. Kobayashi, S. Miyazawa, A. Terahara, H. Mishima, and H. Kurihara, *Tetrahedron Lett.*, 1976, **7**, 537.
56. B. Ganem, *Tetrahedron*, 1978, **34**, 3353.
57. L. R. Hillis and S. J. Gould, *J. Am. Chem. Soc.*, 1985, **107**, 4593.
58. K. G. Hutson and D. R. Threlfall, *Phytochemistry*, 1980, **19**, 535.
59. U. Igbavboa, H. J. Sieweke, E. Leistner, I. Röwer, W. Hüsemann, and W. Barz, *Planta*, 1985, **166**, 537.
60. M. Simantiras and E. Leistner, *Z. Naturforsch., Sect. C, Biosci.*, 1991, **46c**, 364.
61. J. W. Seeger Jr. and R. Bentley, *Phytochemistry*, 1991, **30**, 3585.
62. S. Kaiping, J. Soll, and G. Schultz, *Phytochemistry*, 1984, **23**, 89.
63. M. H. Zenk, H. El-Shagi, and U. Schulte, *Planta Med.*, Supplement, 1975, 79.
64. M. Simantiras and E. Leistner, *Z. Naturforsch., Sect. C, Biosci.*, 1991, **46c**, 364.
65. H.-J. Sieweke and E. Leistner, *Phytochemistry*, 1992, **31**, 2329.
66. E. Leete and G. B. Bodem, *J. Am. Chem. Soc.*, 1976, **98**, 6321.
67. E. Leistner, in "Recent Advances in Phytochemistry," ed. E. E. Conn, Plenum Press, New York, 1986, vol. 20, p. 243.
68. K. Inoue, S. Ueda, Y. Shiobara, and H. Inouye, *Tetrahedron Lett.*, 1976, **21**, 1795.
69. P. O. Larsen, D. K. Onderka, and H. G. Floss, *J. Chem. Soc., Chem. Commun.*, 1972, **14**, 842.
70. L. O. Zamir, A. Nikolakakis, C. A. Bonner, and R. Y. Jensen, *Bioorg. Med. Chem. Lett.*, 1993, **3**, 1441.
71. K. Schmidt and E. Leistner, *Biotechnol. Bioeng.*, 1995, **45**, 285.
72. R. Müller, M. Breuer, A. Wagener, K. Schmidt, and E. Leistner, *Microbiology*, 1996, **142**, 1005.
73. R. Müller, A. Wagener, K. Schmidt, and E. Leistner, *Appl. Microbiol. Biotechnol.*, 1995, **43**, 985.

1.24
Biosynthesis of Coumarins

ULRICH MATERN, PATRICIA LÜER, and DIETER KREUSCH
Philipps-Universität Marburg, Germany

1.24.1 INTRODUCTION

1.24.1.1 Definition and Survey

Coumarins are classified by their 2*H*-1-benzopyran-2-one core structure (**1**) and are distinguished from other benzopyranones such as the isocoumarins (**2**)[1] essentially for biosynthetic reasons, although coumarins and isocoumarins may accumulate coincidentally in the same tissues.[2] Coumarins are widely distributed in higher plants and a few examples showing noteworthy bioactivity, such as novobiocin,[3] have also been isolated from fungal and bacterial sources. Following the comprehensive treatise of Murray *et al.* in 1982,[1] a vast array of further reports has been published on the taxonomic distribution of familiar coumarins as well as on the synthesis and isolation of new derivatives. An update of chemical structures covering the period up to 1989 was published in 1991[4] and the ongoing interest in these compounds is best documented by the numerous reports that have appeared since then. This chapter is dedicated exclusively to coumarins from higher plants and focuses primarily on research concerning their biosynthesis and physiological regulation. Several constitutive plant coumarins have been isolated as glycosidic conjugates, for example, from *Ammi majus*.[5] Glycosidation generally appears to affect the subcellular distribution of phenolic metabolites rather than the biosynthesis of coumarins and will not be considered in detail. Among the dicotyledoneous plants, the *Apiaceae*, *Rutaceae*, and *Moraceae* are particularly rich sources of coumarins.[1,4] Several members of these plant families are used as spices and vegetables in human nutrition or for medicinal purposes and, accordingly, multiple studies have addressed the beneficial effects and potential hazards of the respective coumarin metabolites as well as of the related synthetic derivatives. The toxic potential of linear furanocoumarins (psoralens), which is much less pronounced in the angular furanocoumarin series, has in fact been a major driving force behind research, and the capabilities of detoxification in herbivores feeding on plants such as celery have been intensively studied.[6]

(1) Coumarin

(2) 6-Methoxymellein (Isocoumarin)

1.24.1.2 Medicinal and Biotechnological Applications

The bioactivities of dicoumarol derivatives (e.g. (**3**)) and of phototoxic psoralens (**4**–**7**) are common knowledge and several of these compounds are used in anticoagulant and antipsoriatic therapy, respectively. However, coumarins show further effects of medicinal value which justify and nourish the current committed research. In particular the platelet and lipoxygenase inhibitory activities[7,8] or the mechanistically related antiinflammatory activity[9] as well as the inhibition of DNA gyrase[10] or topoisomerase,[11] which may be linked to the anti-HIV activity,[12] the antitumor activity,[13] the inhibitory effect on T-cell activation,[14] and the inhibition of superoxide generation in activated neutrophiles[15] were significant in pharmacological testing and are under investigation. Among the active coumarins, the calanolides from the rainforest tree *Calophyllum lanigerum*, indigenous to Borneo, deserve particular mention; these compounds inhibited *in vitro* the replication and cytopathicity of HIV-1, but not HIV-2, and were active even against azidothymidine- (AZT) and pyridinone-resistant strains.[16] Calanolide A (**8**) was identified as an inhibitor of HIV-1-specific reverse transcriptase and represented a novel anti-HIV chemotype for drug development. Although the initially inappropriate taxonomic classification caused some confusion,[17] the active principle has meanwhile also been isolated from *Calophyllum teysmannii*[18] and *C. inophyllum*,[19] and the synthesis of calanolide A has been accomplished in several laboratories.[20] The fluorescence of substituted coumarins and the reactivity of 4-hydroxycoumarins have furthermore been exploited in diverse applications such as the labeling of transcription factors for the kinetic evaluation of dimer association[21] or the synthesis of L-proline conjugated chirality reagents.[22]

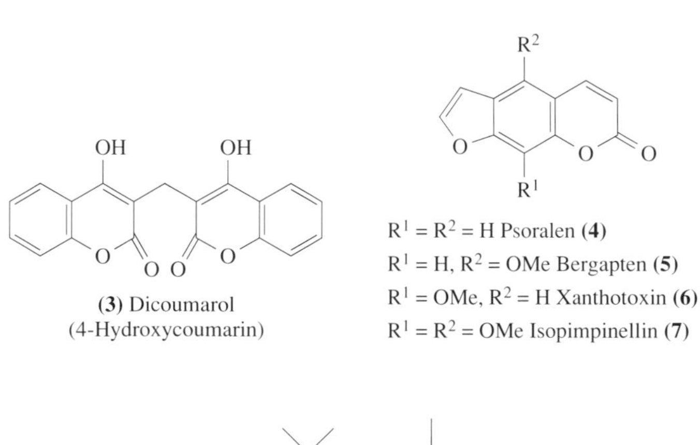

(3) Dicoumarol (4-Hydroxycoumarin)

$R^1 = R^2 = H$ Psoralen (**4**)
$R^1 = H$, $R^2 = OMe$ Bergapten (**5**)
$R^1 = OMe$, $R^2 = H$ Xanthotoxin (**6**)
$R^1 = R^2 = OMe$ Isopimpinellin (**7**)

(8) Calanolide A

1.24.1.3 Ontogenetic Pattern and Physiological Significance

Healthy plants often accumulate considerable amounts of coumarins in oil tubes of the fruit and in the seed coats, as has been reported for wild parsnip, *Pastinaca sativa*,[23] but the genetic control of seed chemistry is barely understood.[23] A similar analysis of *Angelica archangelica* revealed high levels of coumarins in the seeds with low levels in the fruit tissues.[24] Nevertheless, coumarins were also found in the green tissues. For example, clear seasonal trends were reported for the petiole and leaf tissue of *Apium graveolens* where the bergapten (**5**) level increased during development and declined only at later stages of maturity.[25] This trend was basically supported by experiments with parsley plants, where older leaves appeared to be a richer source of coumarin-specific *O*-methyltransferases.[26] Many plants excrete their coumarins to the leaf surface, and this was monitored by a selective experimental approach which involved the extraction of leaf surface coumarins;[27] with this procedure, seasonal changes of surface levels of xanthotoxin (**6**), psoralen (**4**), and bergapten (**5**) were confirmed for *Heracleum lanatum*[27] and other umbelliferous plants,[28] which increased until mid-May and decreased thereafter until maturity. A similar analysis of mature *Ruta graveolens* plants revealed furthermore that the proportion of bergapten in comparison to psoralen and xanthotoxin increased during senescence.[28] Coumarins, and in particular furanocoumarins, are known to inhibit root tip growth and seem to induce membrane disturbances,[29] and their excretion on seed surfaces might be a means to delay germination. Coumarins are leached from the roots of some plants, such as wild *Avena*, into the soil,[30] where they provide a defense against hostile microorganisms. Alternatively, coumarins in the soil might play a specific role in bacterial root symbiosis as was reported for umbelliferone (**9**).[31] The excretion of coumarins to the leaf surface is likely to serve other allelopathic functions.[32]

$R^1 = R^2 = H$ Umbelliferone (**9**)
$R^1 = H$, $R^2 = OH$ Aesculetin (**10**)
$R^1 = H$, $R^2 = OMe$ Scopoletin (**11**)

1.24.1.4 Biosynthetic Classification

Coumarins and isocoumarins differ greatly in terms of biogenesis. Whereas isocoumarins are synthesized by polyketide synthases,[1,33,34] coumarins result from the cyclization of cinnamic acids.[1] Furthermore, plant phenylcoumarins and 4-hydroxycoumarins appear to fall into the same category as the isocoumarins, e.g., the formation of 4-hydroxy-5-methylcoumarin in the course of gerberacoumarin (**12**) biosynthesis has been shown to proceed via the polyketide pathway in *Gerbera jamesoni* similar to the fungal synthesis of a 5-methylcoumarin.[35] It is likely that bacteria also use the polyketide route for the synthesis of 4-hydroxycoumarins, accounting for the lack of the general phenylpropanoid pathway. This aspect is particularly relevant in the context of ^{18}O-labeling studies on novobiocin biosynthesis,[1] which revealed that the ring-oxygen of the 4-hydroxycoumarin nucleus was derived from a carboxyl group and suggested the lactonization of a suitable hydroxylated precursor acid. It is obvious, therefore, that this result cannot be considered as a precedent for the cyclization reaction of plant coumarins. Analogous to the isocoumarins, phenyl- and 4-hydroxycoumarins are beyond the scope of this review.

(**12**) Gerberacoumarin

1.24.2 MODULATION OF COUMARIN ACCUMULATION

Several environmental factors have been found to influence the coumarin content of plants. Treatment with heavy metals or irradiation with short wavelength UV light induced the accumulation of scopoletin (**11**) and ayapin (**13**) in sunflower.[36] Experimental exposure of celery plants to acid fog (pH 2.0) for 4 h, which was initiated to simulate the conditions near urban centers in California, increased the coumarin content of leaves and petioles approximately fivefold as measured at 120 h post treatment.[37] Spraying *Ruta graveolens* plants with sulfuric acid at pH 2.4 or saturated sodium chloride solutions decreased the total concentration of furanocoumarins but increased the relative percentage on the surface of the leaves.[38] A single ozone treatment (200 nL L^{-1}) for 10 h acted as a cross-inducer of flavonoid glucoside and furanocoumarin biosynthesis in parsley plants,[39] and in other instances the accumulation of furanocoumarins was induced by mechanical wounding and herbivore attack of plants,[40] infection by pathogens,[1,41] or by treatment with airborne methyl jasmonate, which is a "broad spectrum plant activator".[42] Furthermore, scopoletin was reported as a phytoalexin from taxonomically diverse plants such as sweet potato,[43] tobacco,[44] carrot,[2] sunflower,[45] *Citrus*,[46] *Hevea*,[47] and cotton[48] and the accumulation of coumarins may thus be regarded as a very general defense response. Nevertheless, the induction of celery leaves with jasmonic acid or analogues of amino acid conjugates of jasmonate revealed a peculiar effect on the proportion of furanocoumarins sequestered to the surface.[32] Whereas the ratio of bergapten (**5**) to xanthotoxin (**6**) within the leaf remained at 1.3 : 1.0, xanthotoxin dominated in the surface lipids at a ratio of 0.8 : 1.0 and indicated that the export is not a simple diffusive translocation.[32] These results are reminiscent of the data reported for *Ruta graveolens*[28] and fit the observation that jasmonate conceivably induces rapid senescence.

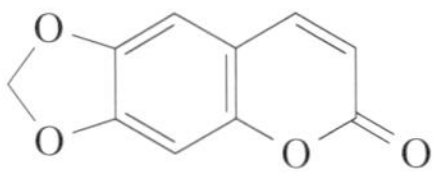

(**13**) Ayapin

Cell cultures of umbelliferous plants proved particularly suitable for regulatory studies and the induction of liner furanocoumarin phytoalexins was first demonstrated in parsley cultures treated with fungal elicitors.[49] Most of the coumarins were secreted into the culture fluid under the conditions of elicitation. Numerous studies have followed on the regulation and molecular biology of the phenylpropanoid pathways in such cell cultures[50,51] and these aspects will be elaborated in more detail below. In case of *Ammi majus* cell cultures, the effects of growth media nutrients or of known intermediates in the biosynthetic pathway on coumarin accumulation were investigated and 3% sucrose or the addition of fairly high concentrations of L-phenylalanine were found most effective.[52] Besides, preincubations with methyl jasmonate[53] or 2,6-dichloroisonicotinic and 5-chlorosalicyclic acid,[54] respectively, conditioned the cells for a greatly enhanced accumulation of coumarins on subsequent challenge with fungal elicitor. 2,6-Dichloroisonicotinic acid is available commercially as an activator of the plant systemic acquired resistance response; its mode of action appears to be more subtle than that of methyl jasmonate, generating much less active oxygen species and causing no phenotypic change of the cells.[54] Data suggest that plant activators like 2,6-dichloroisonicotinic acid induce particularly the late enzymes of the phenylpropanoid pathways, catalyzing reactions beyond that of phenylalanine ammonia-lyase (PAL), while PAL is not induced in the plant cells.[55] This raises the possibility that the enzyme system mediating the cyclization of the benzopyranone moiety might also be selectively induced by plant activator chemicals. Another interesting facet of the fungal-induced furanocoumarin synthesis is the concomitant *in vivo* inhibition of phytosterol biosynthesis, which was observed in *Ammi majus*[56] as well as in *Petroselinum crispum* cells[57] and suggested the specific inhibition of one of the enzymes on the mevalonic acid to dimethylallyl diphosphate pathway span of the cytosolic microsomal pathway of the terpenoid biosynthesis.[56] This effect was indirectly supported by the observation that various furanocoumarins inhibited the production of trichothecenes in *Fusarium culmorum*.[58] However, the induction of *Petroselinum crispum* with fungal elicitor was shown to cause pleiotropic effects on gene expression and included the repression of cell cycle-related genes.[59]

1.24.3 BIOSYNTHESIS

1.24.3.1 Coumarin and Umbelliferone

Most intermediates in the formation of coumarins in higher plants were identified by classical precursor feeding studies,[1,52] and the overall pattern of biosynthesis was proposed from such experiments. The biosynthesis of coumarins proceeds from *trans*-cinnamic acid,[1] which is generated from primary metabolites through the shikimate and general phenylpropanoid pathways. It has become clear, however, that an in depth mechanistic description of the pathway can only be elaborated from supplementary, thorough *in vitro* studies. The enzymology of the shikimate pathway has been reviewed in detail,[60,61] (see also Chapter 1.22), and the general phenylpropanoid pathway is outlined in Chapter 1.26. Precise knowledge of the subcellular topology of these pathways is particularly desirable with respect to the subsequent formation of the benzopyranone moiety. Several reports published since the late 1980s seem to support the view that the entire shikimate pathway has to be assigned exclusively to the plastid compartment, which would also favor the plastids for the synthesis of coumarins. Subsequent data,[60] however, have called this idea into question, and it appears reasonable at present to leave the case of compartmentation of the coumarin-committed pathway open, pending further experimental evidence.

Cinnamate (**14**) may be hydroxylated to 4-coumarate (**15**) by a cytochome P450-dependent monooxygenase (Scheme 1), which was initially cloned from *Helianthus tuberosus* and later functionally characterized after expression in an optimized yeast system.[62] This enzyme has meanwhile been cloned also from *Phaseolus aureus*, *Medicago sativa*, and *Pisum sativum*.[63] Formally, cinnamate and 4-coumarate were considered the immediate precursors of coumarin (**1**) and umbelliferone (**9**), respectively, biosyntheses which require the 2-hydroxylation of the aromatic ring and/or the cyclization reaction (Scheme 1).[1,50,52] Both the lactonization of 2-hydroxycinnamic acids and the direct cyclization of coumaric acid via a spirodiene intermediate appeared feasible and have been outlined.[50,64] In the case of cinnamate the 2′-hydroxylation is likely to precede the cyclization to coumarin, since in *Melilotus* the 2′-*O*-glucoside of *trans*-2′-hydroxycinnamic acid is delivered through the tonoplast and stored in the vacuole predominantly in the *cis*-configuration.[65] Nevertheless, early reports on the enzymatic *ortho*-hydroxylation of cinnamic and 4-coumaric acids *in vitro* could not be repeated and the discrepancies have been discussed elsewhere.[50,64] However, in an entirely different context the *ortho*-hydroxylation of benzoate to salicylate has been reported and this enzyme was identified as a soluble cytochrome P450-dependent monooxygenase.[66] The benzoate 2-hydroxylase might be considered as a precedent enzyme for cinnamate 2′-hydroxylation, but this assumption definitely requires further experimental verification. The gap in our knowledge of coumarin biosynthesis therefore concerns both the mechanism of cyclization, including the prerequisite *cis*- to *trans*-isomerization of cinnamic acids, as well as the exact chemical nature of the substrates, since the respective CoA-esters cannot be ruled out at the present time. Provided that the formation of the benzoypyranone moiety proceeds in two steps (i.e., hydroxylation and subsequent cyclization) the CoA-activation of the acids might increase the rate of lactonization which can also proceed spontaneously.[1]

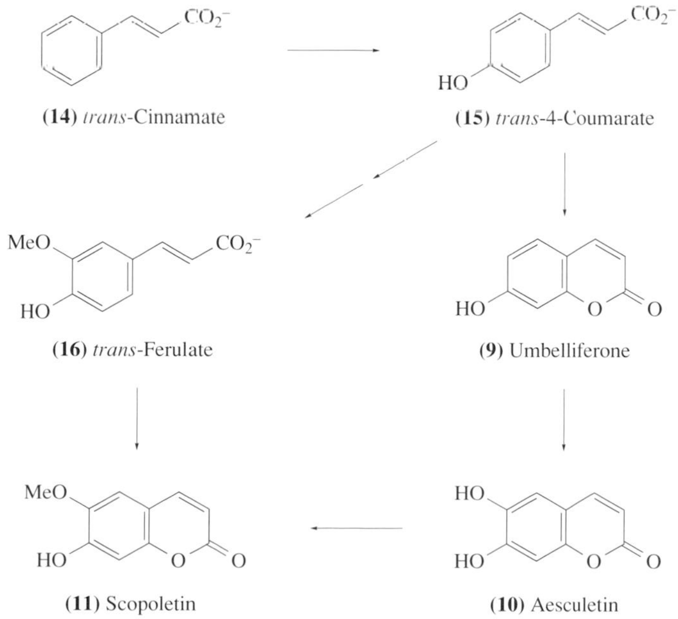

(14) *trans*-Cinnamate

(15) *trans*-4-Coumarate

(16) *trans*-Ferulate

(9) Umbelliferone

(11) Scopoletin

(10) Aesculetin

Scheme 1

1.24.3.2 Aesculetin and Scopoletin

Coumarin itself and the hydroxylated, alkoxylated, or alkylated derivatives were regarded as simple coumarins in contrast to, for example, the furanocoumarins.[52] The biosynthetic path of simple coumarins, e.g., aesculetin (**10**) and scopoletin (**11**), had to be traced on the basis of *in vivo* precursor feeding studies,[1,52,61,64] owing to the fact that the benzopyranone cyclization had not been reliably accomplished *in vitro*. Most of these experiments were carried out on plant tissues which constitutively produce coumarins, yielding fairly low rates of precursor incorporation and the results were summarized previously.[50,52,64] The studies revealed inconsistent results concerning the sequence of cyclization and hydroxylation in different plants. In *Cichorium intybus* aesculetin was proposed to be formed by hydroxylation of umbelliferone (**9**),[67] whereas enzyme activities from other plants readily catalyzed the synthesis from caffeic acid *in vitro*.[68,69] Scopoletin (**11**) was proposed to be formed directly from ferulic acid (**16**) in tobacco mosaic virus (TMV) infected tobacco, although tobacco expresses an *O*-methyltransferase activity methylating aesculetin to scopoletin.[1,44,50] However, feeding studies in *Daphne mezereum*[70] and *Agathosma puberula*[71] corroborated the role of aesculetin as the intermediate between umbelliferone and scopoletin, and the pathway in tobacco was considered exceptional.[52,71] An investigation of scopoletin biosynthesis in elicited sunflower plants again yielded ambiguous results, and here also the methylation of aesculetin to scopoletin was catalyzed *in vitro*.[36]

1.24.3.3 Prenylcoumarins

The prenylations of umbelliferone (**9**) in the 6- or 8-position yield demethylsuberosin (**17**) and osthenol (**18**), respectively, and give access to the branch pathways to linear or angular furano- and pyranocoumarins, which are predominantly found in the *Umbelliferae* (Scheme 2). Pyranocoumarins have been reported as major metabolites of, for example, *Petroselinum crispum*[72] and *Ammi visnaga*.[73] Prenylated coumarins have furthermore been isolated from various plants and, in particular, from *Rutaceae* species.[74] Following the first report by Dhillon and Brown on the *in vitro* prenylation of umbelliferone with fractions from *Ruta graveolens* plastids in the presence of Mn^{2+} ions,[1,52] several dimethylallyl transferase activities prenylating umbelliferone[50,64,75] and other aromatic substrates such as flavonoids,[76,77] acridones,[78] and 4-hydroxybenzoate[79] or forming geranyl and farnesyl diphosphate for lower terpenoid biosynthesis[80,81] have been described from plants. The enzyme activities were always associated with the microsomal fraction and assigned in soybean and French bean to the envelope membrane of plastids.[76] Although only the 6-*C*- and 7-*O*-prenylation of umbelliferone have been described *in vitro*,[64,75] it is to be expected that the prenylation at the 8-position (Scheme 2) in the course of angular furano- and pyranocoumarin biosyntheses is catalyzed by a homologous membrane-bound transferase. This 8-*C*-prenyltransferase must clearly be a separate enzyme entity.

Despite the numerous prenylated phenolic plant metabolites that have been isolated particularly in the field of flavonoids,[76] neither one of the corresponding prenyltransferases has been purified and thoroughly studied. This is in sharp contrast to fungal and mammalian farnesyl- and geranylgeranyltransferases,[82–84] which fulfil important regulatory functions in cellular homoeostasis, or to the polyprenyl diphosphate synthases from plant[85,86] and bacterial sources.[87] The stereochemical mode of action of bacterial polyprenyl diphosphate synthases has been investigated and several such enzymes have been cloned;[87] highly conserved sequence domains of these genes might be helpful in the cloning of plant dimethylallyl transferases.

1.24.3.4 Dihydrofuranocoumarins and Cytochrome P450 Catalysis

The oxidative cyclization reactions of 6- and 8-prenylated umbelliferone were proposed to yield the dihydrofuranocoumarins (+)-marmesin (**19**) and (−)-columbianetin (**20**), respectively, which are the immediate precursors of linear and angular furanocoumarins[1] (Scheme 2). The cyclization of demethylsuberosin to (+)-marmesin (Scheme 2) was accomplished *in vitro* in the presence of NADPH using microsomal fractions of umbelliferous plant cells (*Petroselinum crispum*, *Ammi majus*, *Arracacia xanthorhiza*) that had been elicited with fungal elicitor.[50,64] Inhibition studies identified the "marmesin synthase" as a cytochrome P450-dependent monooxygenase,[88] which did not release any intermediates during the reaction.[50,64] Following an earlier proposal,[1,52] it was nevertheless suggested that the epoxidation of the side-chain double bond as a hypothetical intermediate initiated the reaction.[50] The transfer of oxygen to alkenes by P450 enzymes typically results

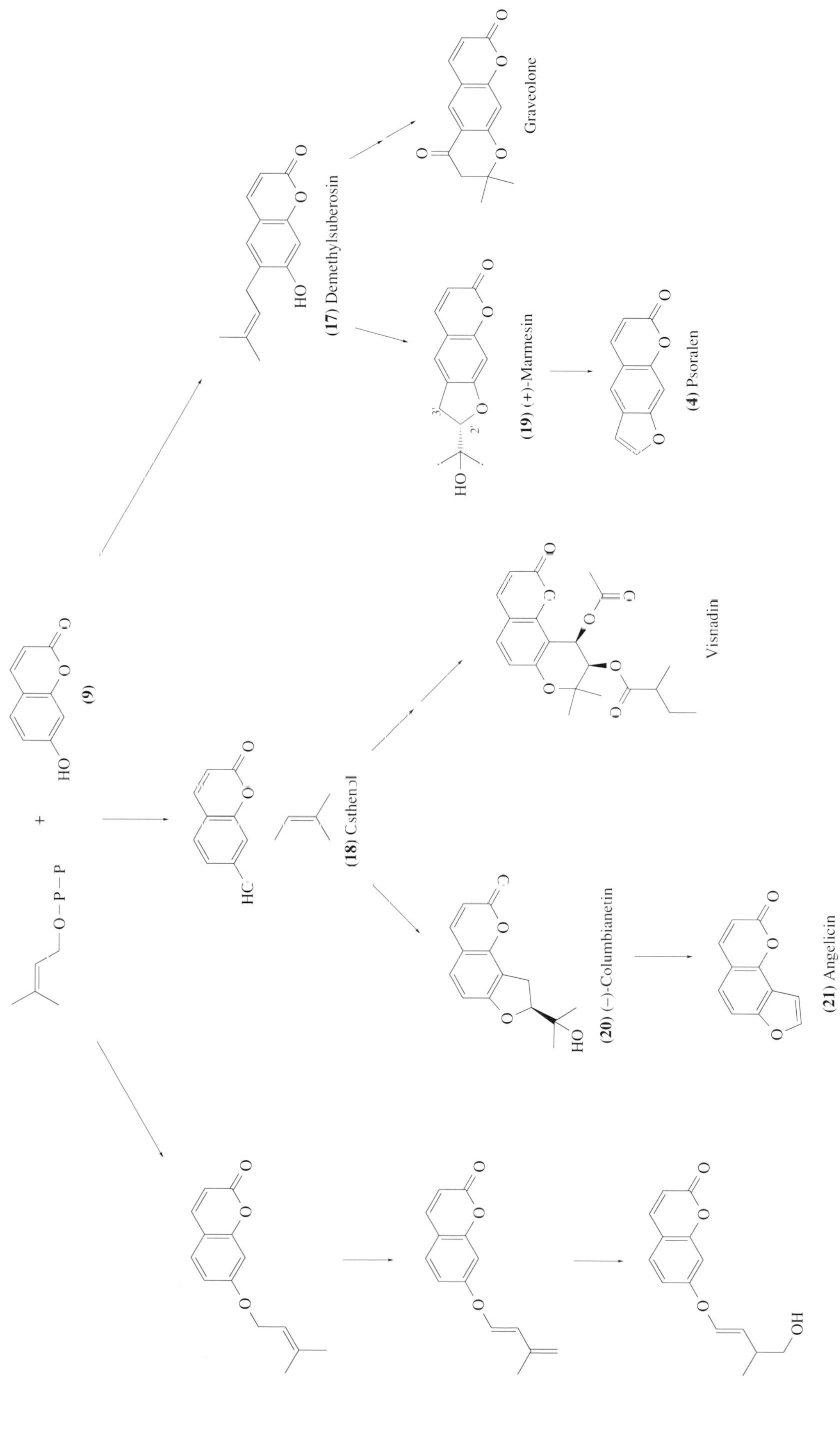

(9)

(17) Demethylsuberosin

Graveolone

(19) (+)-Marmesin

(4) Psoralen

(18) Osthenol

Visnadin

(20) (–)-Columbianetin

(21) Angelicin

Scheme 2

in epoxidation rather than hydroxylation reactions[89] and the reaction is formally considered as an insertion of an "oxen".[90] Furthermore, the product often deactivates the enzyme by concomitant alkylation of the prosthetic heme group.[91] In the case of marmesin synthase, the proper spatial orientation of the 7-hydroxy group of demethylsuberosin (**17**) during catalysis is probably responsible for delocalization of the electron density from the double bond and favors the cyclization reaction. Conceivably, the formation of (−)-columbianetin from osthenol (Scheme 2) follows an analogous route and is also catalyzed by a P450 monooxygenase, but the experimental support for this assumption is still lacking.

Cytochrome P450 monooxygenases appear to play an outstanding role in the overall biosynthesis of furanocoumarins and some basic mechanistic parameters will be recalled briefly for a better understanding of the coumarin-specific pathway. P450 monooxygenases catalyze the reductive activation of molecular oxygen to yield "active oxygen" species, capable of supporting the free radical homolytic cleavage of C—H or C—C bonds. Several steps of the activation process comprise one-electron transfers, that is, movement of either an electron or a hydrogen atom.[90,91] The enzymes contain Fe–protoporphyrin IX and, in the resting state, the ferric prosthetic group is held in a hexacoordinated position by a cysteinyl thiolate as a fifth ligand and coordination of a water molecule provides a sixth ligand (Figure 1). Binding of substrate displaces the water molecule and the iron spin state changes from a low spin to a high spin, pentacoordinated state. Furthermore, the reduction of the prosthetic group to a high spin, ferrous state by a single-electron transfer from cytochrome P450–NADPH reductase is facilitated in the substrate-bound P450, which then readily binds molecular oxygen to form a ferrous dioxygen complex. The structural information on the short-lived radicals beyond the ferrous dioxygen complex is rather limited and must be inferred from model systems.[90,91] Two steps lead to the "reactive oxygen species" as the eventual catalysts. A redox reaction in the complex produces a ferrisuperoxide ion, $[Fe^{III}—O—O]^{\cdot}$, which is most probably converted to the ferrihydroperoxy derivative, $[Fe^{III}—O—OH]$ (Figure 1), by transfer of another single electron (from NADPH–cytochrome P450 reductase or ferrous cytochrome b_5) and a proton or abstraction of a free radical hydrogen atom (from the substrate). The stoichiometric generation of H_2O_2 from NADPH and O_2 by liver microsomes in the absence of substrate was taken as evidence for the existence of the ferric-hydroperoxy species, which cannot be observed directly. $[Fe^{III}—O—OH]$ thus formed is itself a strong nucleophile and was proposed to act as the reactive intermediate in various P450 reactions, e.g., nitric oxide synthase or acyl-carbon cleavage reaction in the conversion of pregnenolone to the 16-en steroid products.[90–92] The instantaneous heterolytic cleavage of the oxygen–oxygen bond in the ferric-hydroperoxy species, which is probably assisted by the cysteinyl thiolate ligand, yields the reactive oxo-derivative as well as water. The oxo-derivative (oxoferryl or iron-oxo radical) may be represented in several resonance forms, for example, $Fe^{V}{=}O$, $[Fe^{IV}{=}O]^{+\bullet}$ or $[Fe^{IV}—O]^{\bullet}$. $[Fe^{IV}{=}O]^{+\bullet}$ and the oxo-Fe^{IV} porphyrin radical $[Fe^{IV}—O]^{\bullet}$ are electrophilic and behave like an alkoxy radical, respectively; these radicals are capable of abstracting hydrogen from a C—H bond, e.g., in the hydroxylation cycle, to produce $Fe^{IV}—OH$ and a substrate carbon radical,[90,91] which immediately recombine ("oxygen rebound" in the hydroxylation pathway) to yield an alcohol. This recombination event must take place very rapidly in a reaction cage to account for the stereospecificity of the hydroxylation. Both the nucleophilic ferri-hydroperoxy species and the electrophilic oxo-Fe^{IV} radical species might participate in the reactions catalyzed by P450 monooxygenases during the course of furanocoumarin biosynthesis.

Besides the hydroxylation and C—C bond cleavage reactions P450 enzymes are capable of desaturating aliphatic substrates. It is noteworthy in the context of dihydrofuranocoumarin formation that P450 enzyme activities of the microsomes isolated from elicited *Ammi majus* cells metabolized 7-*O*-prenylumbelliferone (Scheme 2), which cannot cyclize to a dihydrofurane, by $\Delta^{1'}$-desaturation to yield butenyl ethers rather than epoxidation of the prenyl residue.[93] Desaturation reactions brought about by mammalian P450 monooxygenases[90,91] or by the mechanistically analogous non-heme-iron oxygenases of plants[94] can proceed without the intermediacy of an alcohol and instead require the successive abstraction of two hydrogen atoms. Accordingly, the iron-bound hydroxy radical formed initially by abstraction of the first hydrogen atom from the substrate functions here as an oxidant and abstracts the second hydrogen atom to give rise to a water molecule rather than to recombine with the substrate radical as suggested for the P450 hydroxylations.[90,91,94] The underlying mechanism has been outlined in detail elsewhere.[91,94]

The reaction mechanism of epoxidation of alkenic double bonds by P450 monooxygenases has not yet been solved, although a concerted process has been ruled out already.[90] Nevertheless, the model mechanisms proposed for the primary event in the interaction of the P450 oxo-derivative with aliphatic double bonds[91] provide a reasonable basis for the cyclization of demethylsuberosin to (+)-marmesin without the formation of an intermediate epoxide. The double-bond π system is

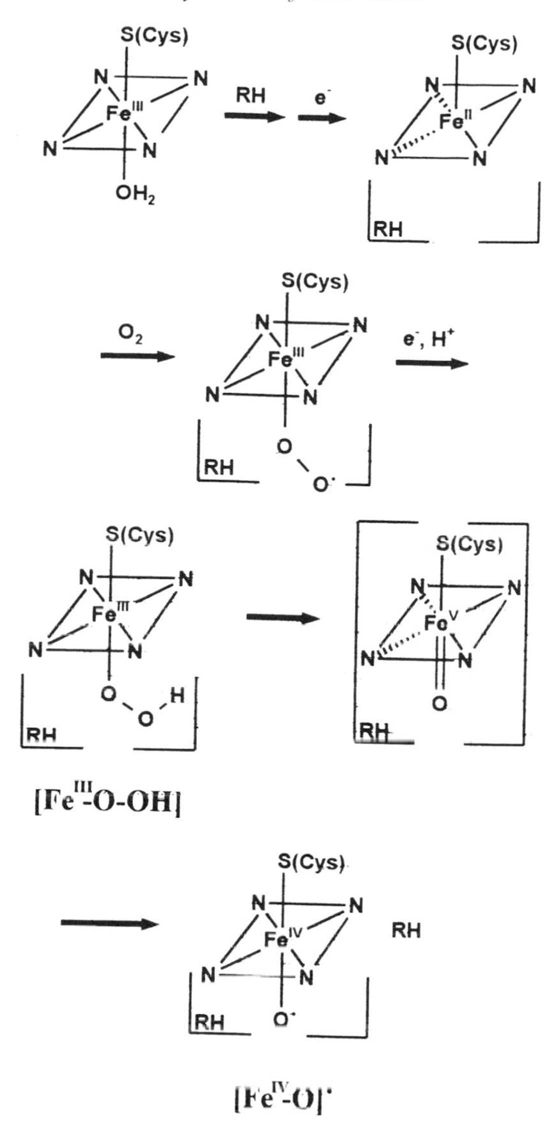

Figure 1 The sequence of events involved in the activation of cytochrome P450 leading to the ferric-hydroperoxy species and to the iron-oxo radical. The enzyme contains a low-spin iron in the resting state, which is converted to high-spin iron upon substrate binding.

predominantly suited to undergo electrophilic additions and the electrophilic rather than the radical character of the oxo-derivative might be emphasized. Under these premises, the addition in 3′ of demethylsuberosin (**17**) produces a 2′-cationic intermediate which would support the spontaneous cyclization to the dihydrofuran coincidentally with the formation of the 3′-tertiary alcohol (Scheme 2). It is experimentally difficult to distinguish this type of reaction mechanism from the pathway involving an intermediate epoxide, since in either case the dihydrofuran-ring oxygen must stem entirely from the umbelliferone substrate and the tertiary marmesin alcohol originates from molecular oxygen. Provided that the substrate specificity of marmesin synthase resides entirely in the prenyl chain, the same enzyme species might also catalyze the cyclization of osthenol to (−)-columbianetin (Scheme 2). However, this reaction has not yet been studied *in vitro*.

1.24.3.5 Psoralen

The conversion of (+)-marmesin (**19**) to psoralen (**4**) involves a C—C bond cleavage, which formally releases a C_3-fragment in the form of acetone and the concomitant 1′-desaturation of the dihydrofuran moiety[1] (one-step process). The system of carbon numbering (Scheme 2) was chosen for convenience, as in most of the relevant literature, and does not conform to the standard nomenclature. The release of acetone in a plant biosynthetic pathway is rather unusual and the scheme dates back to an earlier proposal by Birch *et al.*[95] which included the generation of a 3′-carbenium ion with subsequent 1,3-elimination. However, this proposal has not received experimental confirmation. An alternative mechanism might involve the stepwise oxidation and removal of, for example, one carbon of the isopropyloxy side chain in the form of an aldehyde or acid followed by removal of the residual acetyl group (two-step process). Precedent enzymes for the two-step process and C—C bond cleavage can be commonly found in the P450-type steroid metabolism.[90,91] The psoralen synthase was identified as a cytochrome P450 monooxygenase.[50,64] Furthermore, inhibitor studies clearly revealed that the psoralen synthase reaction was catalyzed by a P450 entity different from marmesin synthase.[64] These results are in accordance with the pertinent literature suggesting that plant P450s involved in biosynthetic pathways show narrow substrate specificities in contrast to the detoxifying P450 enzymes.[63,89] The high regio- and stereoselectivity of such enzymes is thought to be imposed by structural constraints within the active site rather than by the nature of the catalytic mechanism.[90]

The common mode of action of P450 monooxygenases, in which the oxo-iron radical species promotes the initial abstraction of hydrogen atoms to yield a carbon radical, is difficult to reconcile with the direct generation of a carbenium ion as anticipated by Birch *et al.*[95] By analogy to the chemical elimination mechanism, the formation of psoralen by psoralen synthase was therefore assumed to involve the 3′-hydroxylation of (+)-marmesin as a typical P450-catalyzed oxygen-rebound process and followed by base-catalyzed *anti*-elimination.[50] A different point of view was put forward by Hakamatsuka *et al.*[96] in the context of rearrangement reactions, suggesting the initial homolytic abstraction of one of the 3′-hydrogens of (+)-marmesin, followed by disproportionation of the primary radical to psoralen concomitant with the release of the isopropyloxy side-chain radical which eventually recombines with the hydroxy radical of the P450-iron-hydroxy (oxygen-rebound process) to yield acetone and water. Neither of these two mechanisms had been sufficiently founded on experimental evidence. Therefore, the reaction mechanism of psoralen synthase was revisited in a fruitful collaboration with Boland and Stanjek.[97] A set of enzyme assays was conducted with microsomal fractions from induced *Ammi majus* cells and employing stereospecifically deuterated (±)-marmesin (**19**) or (±)-2′-acetyl-2′,3′-dihydropsoralen (**22**) (Schemes 3 and 4). The latter compound was included as a pseudosubstrate in an effort eventually to distinguish the one-step from the two-step process of catalysis. The microsomes converted both the substrate and pseudosubstrate to psoralen. Furthermore, the experiments clearly revealed that the reaction proceeds exclusively by *syn*-elimination. The side-chain release of acetate from (±)-2′-acetyl-2′,3′-dihydropsoralen (**22**) (Scheme 4) was reminiscent of the reaction catalyzed by 17α-hydroxylase-17,20-lyase involved in the formation of Δ^{16}-steroid from pregnenolone.[90,92] By analogy, psoralen synthase might therefore catalyze the initial addition of the nucleophilic ferric-hydroperoxy species to the side-chain carbonyl, forming a peroxyhemiketal, followed by decomposition of the adduct to yield acetate and a carbon radical, which then loses a hydrogen atom from the neighboring carbon with coincident desaturation (Scheme 4). In the course of this reaction, one atom of molecular oxygen must be incorporated into the acetate released.[90,92] In contrast to psoralen synthase, the authors reported a *trans*-scission process for the 17α-hydroxylase-17,20-lyase and claimed a radical rather than a concerted mechanism as a first example in C—C bond cleavage by P450 enzymes. This *anti*-elimination reaction, however, must be regarded as an exception to the rule, and the type of elimination is primarily governed by the spatial configuration of the active site rather than the catalytic mechanism. Additional model reactions nevertheless suggested that psoralen synthase operates via a different mode. Tetraphenyl-21*H*,23*H*-porphin-Fe^{III}-complex activated with iodosobenzene also converted deuterated (±)-marmesin or (±)-2′-acetyl-2′,3′-dihydropsoralen to psoralen, but under these conditions both *syn*- and *anti*-elimination was observed. More importantly, however, this activated porphin-Fe^{III} model complex does not deliver the ferrihydroperoxy species and the elimination reactions thus are more likely to have followed the common elimination mechanism which is initiated by the oxo-iron radical species.

In order to obtain unequivocal proof of the type of carbonyl compound released upon the psoralen synthase reaction *O*-(2,3,4,5,6-pentafluorobenzyl)hydroxylamine[98] was used as a trapping reagent in further assays conducted with appropriately deuterated (±)-marmesin. These experiments, in fact, revealed the stoichiometric release of acetone and psoralen from (+)-marmesin and provided the first solid evidence for the mechanistic mode of action of psoralen synthase (Scheme 3). The results predicted that the *syn*-elimination catalyzed by psoralen synthase is not necessarily the consequence of a concerted elimination mechanism. More likely, the oxo-Fe^{IV}-porphyrin radical $[Fe^{IV}—O]^{\bullet}$, acting like an alkoxy radical, abstracts the hydrogen atom from carbon-3′ of

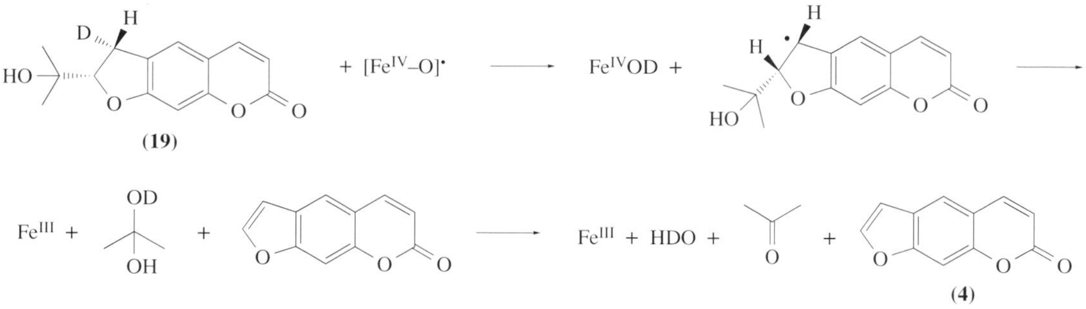

Scheme 3

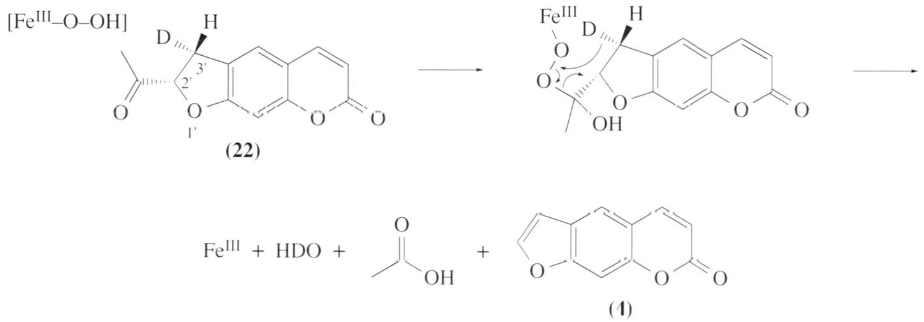

Scheme 4

(+)-marmesin in *syn*-orientation to the isopropyloxy side chain (Scheme 3) and instead of entering an oxygen-rebound process with the carbon radical, the reactive intermediate loses the side-chain isopropyl radical in a disproportionation, which coincidentally recombines with the hydroxy radical (oxygen-rebound) furnishing psoralen, acetone, and water. A primary kinetic isotopic effect ($k_{H\text{-}3'}/k_{D\text{-}3'}$) of about 4 was observed suggesting that the abstraction of the 3′-hydrogen atom is rate-limiting and supporting experimentally the proposed mechanistic sequence. The experiments *stricto sensu* do not rule out an intermediate (+)-3′-hydroxymarmesin. However, such an intermediate was not released during the enzyme assays, a hypothetical hydroxymarmesin would be likely to desaturate by *trans*-elimination, and precedent desaturation reactions by other P450 monooxygenases suggest the direct radical abstraction.[90,91] Overall, this again emphasizes the impact of the particular steric constraints of plant P450s on the course of the reaction, which is also the basis of their narrow substrate specificities. Nevertheless, as mentioned for marmesin synthase the eventual involvement of psoralen synthase in the formation of the angular series of furanocoumarins, i.e., the conversion of (−)-columbianetin (**20**) to angelicin (**21**), remains to be tested. Assays conducted with synthetic (±)-columbianetin and *Ammi majus* microsomes revealed that psoralen synthase does not catalyze the formation of angelicin.[99] Angelicin synthase, therefore, represents a separate enzyme entity which probably acts by a mechanism analogous to that of psoralen synthase.

1.24.3.6 Oxygenated Psoralens

The hydroxylation of psoralen in the 5- and/or 8-position formally yields bergaptol (**23**) or xanthoxol (**24**) and 5,8-dihydroxypsoralen (**25**), respectively. The latter compound is a fairly labile hydroquinone which has nevertheless been employed in precursor feeding studies concerned with the biosynthesis of isopimpinellin (**7**) (Scheme 5) in *Ruta graveolens* shoots,[1,52] although an alternative route via 5-hydroxyxanthotoxin (**26**) may be envisage.[52] These studies suggested that 5,8-dihydroxypsoralen is the major precursor of isopimpinellin,[52] which underlines the metabolic function of the hydroquinone. The sequence of hydroxylations leading to 5,8-dihydroxypsoralen via bergaptol or xanthotoxol has not yet been solved, but a psoralen 5-monooxygenase activity converting psoralen to bergaptol was identified in the microsomes from elicited *Ammi majus* cells.[100] Also this enzyme was identified as a P450 monooxygenase and inhibition studies again pointed to an enzyme different from marmesin and psoralen synthases.[88] The reaction probably proceeds analogously to cinnamate and many other P450 hydroxylases by the oxygen-rebound process.[90,91] Bergaptol was the only product observed in these assays, suggesting that the formation of 8-hydroxypsoralen (xanthoxol) and 5,8-dihydroxypsoralen requires an additional one or two enzyme(s).

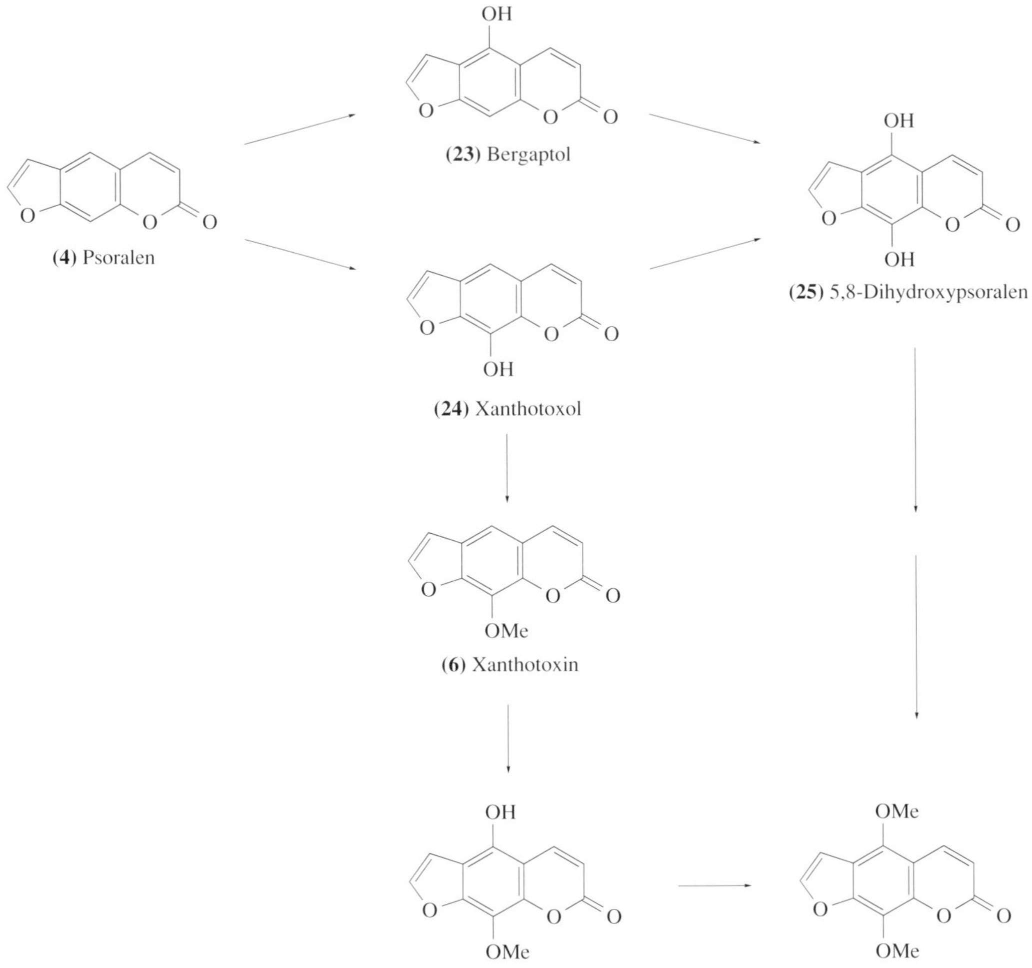

(**4**) Psoralen

(**23**) Bergaptol

(**24**) Xanthotoxol

(**25**) 5,8-Dihydroxypsoralen

(**6**) Xanthotoxin

(**26**) 5-Hydroxyxanthotoxin

(**7**) Isopimpinellin

Scheme 5

The hydroxylated furanocoumarins may be further processed by *O*-alkylation and the methoxylated psoralens bergapten, xanthotoxin, and isopimpinellin (Scheme 5) accumulate in many plants as the final products of the pathway. Two *O*-methyltransferases catalyzing the methylation of bergaptol or xanthotoxol to bergapten (BMT) and xanthotoxin (XMT), respectively, were distinguished and extensively purified by affinity chromatography from *Ruta graveolens* and later from *Petroselinum crispum*.[1,52] The substrate specificities of the methyltransferases were considered helpful in clarifying the sequential order of biosynthetic reactions in the course of isopimpinellin biosynthesis. Precursor feeding studies with *Ruta graveolens* cultures had suggested that both xanthotoxin and bergapten were converted to isopimpinellin, although the route from xanthotoxin via 5-hydroxyxanthotoxin was preferred.[1,52] In *Petroselinum crispum* the XMT activity was shown to methylate exclusively xanthotoxol (8-hydroxyxanthotoxol), whereas the BMT catalyzed the 5-*O*-methylation of bergaptol (5-hydroxypsoralen) as well as the 5- and 8-*O*-methylations of 5,8-dihydroxypsoralen[101] (Scheme 5). Furthermore, 5-hydroxyxanthotoxin was methylated at a significantly higher rate than bergaptol and the authors considered another unidentified methyltransferase for the methylation of 8-hydroxybergapten in the course of isopimpinellin biosynthesis. Taken together, the methyltransferase activities did not shed much light on the late steps of isopimpinellin formation and the outcome is remeniscent of the difficulties encountered in studies on the formation of simple coumarins where the rather broad substrate specificities of *O*-methyltransferases caused ambiguous results.[36] The subcellular topology of *Petroselinum crispum* BMT in the epithelial cells of oil ducts[102] and the expression of the coding gene have been intensively studied in response to wounding or fungal infection.[103] The induction of BMT gene expression upon fungal elicitation was characterized as rather late in comparison to other genes of the inducible phenylpropanoid pathways,[103] which correlates with the catalytic function in the course of furanocoumarin biosynthesis.

1.24.4 CONCLUSIONS

Most of the biosynthetic pathway leading from cinnamic acid to alkylated furanocoumarins such as isopimpinellin has been unravelled by precursor feeding and *in vitro* studies. Psoralen synthase catalyzes a pivotal reaction in this pathway and the reaction mechanism has been elucidated and shown to differ from that previously proposed by Birch *et al.*[95] Cell culture systems inducible for the accumulation of coumarins provided a major breakthrough for the investigation of the enzymology, and these systems will certainly be helpful in tackling the still controversial course of *ortho*-hydroxylation and/or cyclization reactions of cinnamate and 4-coumarate as well as the classification of the enzyme introducing the 8-hydroxyl function in furanocoumarins. It is striking that 5–6 steps (cinnamate 4-hydroxylation, *ortho*-hydroxylation, formation of marmesin, psoralen, bergaptol and xanthotoxol/dihydroxypsoralen) of the total of 10 steps on the path to isopimpinellin in *Ammi majus* are catalyzed by cytochrome-P450-dependent monooxygenases, which carry out hydroxylation, desaturation, and cyclization reactions. Furthermore, the conversion of *O*-prenylumbelliferone to the corresponding butenyl ethers in *Ammi majus* is most likely to be brought about by P450 monooxygenases, which underlines the importance of this class of enzymes for overall coumarin accumulation. Coumarins accumulate in various plants in response to biotic or abiotic stressors, and it is conceivable that the expression of the committed enzymes must follow a coordinated pattern. One coumarin-specific cDNA (BMT) has been cloned[103] and the cloning of genes coding for the respective P450s will provide insight into their common regulatory sequences as well as into the basis of differential substrate specificities.

ACKNOWLEDGMENTS

The work cited from the authors' laboratory was financially supported by the Deutsche Forschungsgemeinschaft and Fonds der Chemischen Industrie, which is gratefully acknowledged. We thank S. Brown (Peterborough, Ontario, Canada) and W. Boland (Institut für Organische Chemie und Biochemie, Universität Bonn, Germany) for helpful suggestions and critical reading of the manuscript.

1.24.5 REFERENCES

1. R. D. H. Murray, J. Méndez, and S. A. Brown, "The Natural Coumarins. Occurrence, Chemistry and Biochemistry," Wiley, New York, 1982.
2. D. T. Coxon, R. F. Curtis, K. R. Price, and G. Levett, *Phytochemistry*, 1973, **12,** 1881.
3. A. M. Feigin, E. V. Aronov, J. H. Teeter, and J. G. Brand, *Biochim. Biophys. Acta*, 1995, **1234,** 43.
4. R. D. H. Murray, *Prog. Chem. Org. Nat. Prod.*, 1991, **58,** 83–316.
5. M. H. A. Elgamal, N. M. M. Shalaby, H. Duddeck, and M. Hiegemann, *Phytochemistry*, 1993, **34,** 819.
6. M. R. Berenbaum and A. R. Zangerl, *Recent Adv. Phytochem.*, 1996, **30,** 1.
7. H. C. Huang, M. W. Lai, H. R. Wang, Y. L. Chung, L. M. Hsieh, and C. C. Chen, *Eur. J. Pharmacol.*, 1993, **237,** 39.
8. Y. L. Chen, T. C. Wang, K. H. Lee, and C. C. Tzeng, *Helv. Chim. Acta*, 1996, **79,** 651.
9. A. M. Silván, M. J. Abad, P. Bermejo, M. Sollhuber, and A. Villar, *J. Nat. Prod.*, 1996, **59,** 1183.
10. P. Yogey, J. Lee, J. Kordel, E. Vivas, P. Warner, D. Jebaratnam, and R. Kolter, *Proc. Natl. Acad. Sci. USA*, 1994, **91,** 4519.
11. H. Peng and K. J. Marians, *J. Biol. Chem.*, 1993, **268,** 24481.
12. L. Huang, Y. Kashiwada, L. M. Cosentino, S. Fan, C.-H. Chen, A. T. McPhail, T. Fujioka, K. Mihashi, and K. H. Lee, *J. Med. Chem.*, 1994, **37,** 3947.
13. A. Mizuno, M. Takata, Y. Okada, T. Okuyama, H. Nishino, A. Nishino, J. Takayasu, and A. Iwashima, *Planta Med.*, 1994, **60,** 333.
14. S. Von Kruedener, W. Schneider, and E. F. Elstner, *Arzneim. Forsch.*, 1995, **45,** 169.
15. M. Paya, M. L. Ferrandiz, F. Miralles, C. Montesinos, A. Ubeda, and M. J. Alcaraz, *Arzneim. Forsch.*, 1993, **43,** 655.
16. Y. Kashman, K. R. Gustafson, R. W. Fuller, J. H. Cardellina, J. B. McMahon, M. J. Currens, R. W. Buckheit, S. H. Hughes, G. M. Cragg, and M. R. Boyd, *J. Med. Chem.*, 1992, **35,** 2735.

17. J. H. Cardellina, K. R. Gustafson, J. A. Beutler, T. C. McKee, Y. F. Hallock, R. W. Fuller, and M. R. Boyd, in "Human Medicinal Agents from Plants," ACS Symposium Series 534, eds. A. D. Kinghorn and M. F. Baladrin, American Chemical Society, Washington, DC, 1993, p. 218.
18. K. R. Gustafson, H. R. Bokesch, R. W. Fuller, J. H. Cardellina, M. R. Kadushin, D. D. Soejarto, and M. R. Boyd, *Tetrahedron Lett.*, 1994, **35,** 5821.
19. T. C. McKnee, J. H. Cardellina, G. B. Dreyer, and M. R. Boyd, *J. Nat. Prod.*, 1995, **58,** 916.
20. K. S. Rehder, M. K. Hristovakazmierski, and J. A. Kepler, *J. Labelled Compd. Radiopharm.*, 1996, **8,** 1077.
21. H. Wendt, C. Berger, A. Baici, R. M. Thomas, and H. R. Bosshard, *Biochemistry*, 1995, **34,** 4097.
22. K. Nagasawa, Y. Yamashita, S. Katoh, K. Ito, and K. Wade, *Chem. Pharm. Bull.*, 1995, **43,** 344.
23. A. R. Zangerl, M. R. Berenbaum, and E. Levine, *J. Hered.*, 1989, **80,** 404.
24. A. M. Zobel and S. A. Brown, *Environ. Exp. Bot.*, 1991, **31,** 447.
25. J. T. Trumble, J. G. Millar, D. E. Ott, and W. C. Carson, *J. Agric. Food Chem.*, 1992, **40,** 1501.
26. R. Lois and K. Hahlbrock, *Z. Naturforsch.*, 1992, **47c,** 90.
27. A. M. Zobel and S. A. Brown, *J. Chem. Ecol.*, 1990, **16,** 1623.
28. A. M. Zobel and S. A. Brown, *J. Chem. Ecol.*, 1991, **17,** 1801.
29. E. Kupidlowska, M. Kowalec, G. Sulkowski, and A. M. Zobel, *Ann. Bot.*, 1994, **73,** 525.
30. F. J. Perez and J. Ormeno Nunez, *Phytochemistry*, 1991, **30,** 2199.
31. J. R. Rao and J. E. Cooper, *Mol. Plant-Microbe Interact.*, 1995, **8,** 855.
32. V. Stanjek, C. Herhaus, U. Ritgen, W. Boland, and E. Städler, *Helv. Chim. Acta*, 1997, in press.
33. F. Kurosaki, *Phytochemistry*, 1994, **37,** 727.
34. F. Kurosaki, *Phytochemistry*, 1995, **39,** 515.
35. T. Inoue, T. Toyonaga, S. Nagumo, and M. Nagai, *Phytochemistry*, 1989, **28,** 2329.
36. M.-C. Gutierrez, A. Parry, M. Tena, J. Jorrin, and R. Edwards, *Phytochemistry*, 1995, **38,** 1185.
37. W. Dercks, J. Trumble, and C. Winter, *J. Chem. Ecol.*, 1990, **16,** 443.
38. A. M. Zobel, S. A. Brown, and J. E. Nighswander, *Ann. Bot.*, 1991, **67,** 213.
39. H. Eckey-Kaltenbach, D. Ernst, W. Heller, and H. Sandermann Jr., *Plant Physiol.*, 1994, **104,** 67.
40. A. R. Zangerl and M. R. Berenbaum, *Ecology*, 1990, **71,** 1933.
41. C. Johnson, D. R. Brannon, and J. Kuc, *Phytochemistry*, 1973, **12,** 2961.
42. M. Miksch and W. Boland, *Experientia*, 1996, **52,** 739.
43. T. Minamikawa, T. Akazawa, and I. Uritani, *Plant Physiol.*, 1963, **8,** 493.
44. B. Fritig, L. Hirth, and G. Ourisson, *Phytochemistry*, 1970, **9,** 1963.
45. B. Tal and D. J. Robeson, *Phytochemistry*, 1986, **25,** 77.
46. I. A. Dubery, *Phytochemistry*, 1990, **29,** 2107.
47. A. Giesemann, B. Biehl, and R. Lieberei, *J. Phytopathol.*, 1986, **117,** 373.
48. H. J. Zeringue Jr., *Phytochemistry*, 1984, **23,** 2501.
49. K. G. Tietjen, D. Hunkler, and U. Matern, *Eur. J. Biochem.*, 1983, **131,** 401.
50. U. Matern, H. Strasser, H. Wendorff, and D. Hamerski, in "Cell Culture and Somatic Cell Genetics of Plants," eds. F. Constabel and I. K. Vasil, Academic Press, New York, 1988, vol. 5, p.3.
51. K. Hahlbrock and D. Scheel, *Annu. Rev. Plant Physiol. Plant Mol. Biol.*, 1989, **40,** 347.
52. S. A. Brown, in "Chemistry and Biochemistry of Organic Natural Products," ed. A. Zobel, 1998, manuscript in preparation.
53. H. Kauss, K. Krause, and W. Jeblick, *Biochem. Biophys. Res. Commun.*, 1992, **189,** 304.
54. H. Kauss, W. Jeblick, J. Ziegler, and W. Krabler, *Plant Physiol.*, 1994, **105,** 89.
55. G. Busam, K. T. Junghanns, R. E. Kneusel, H.-H. Kassemeyer, and U. Matern, *Plant Physiol.*, 1997, **115**, 1039.
56. D. C. Fulton, P. A. Kroon, U. Matern, D. R. Trelfall, and I. M. Whitehead, *Phytochemistry*, 1993, **34,** 139.
57. C. Haudenschild and M.-A. Hartmann, *Phytochemistry*, 1995, **40,** 1117.
58. A. R. Hesketh, L. Gledhill, D. C. Marsh, B. W. Bycroft, P. M. Dewick, and J. Gilbert, *Phytochemistry*, 1991, **30,** 2237.
59. E. Logemann, S.-C. Wu, J. Schröder, E. Schmelzer, I. E. Somssich, and K. Hahlbrock, *Plant J.*, 1995, **8,** 865.
60. E. Haslam, *Prog. Chem. Org. Nat. Prod.*, 1996, **69,** 157.
61. D. Strack, in "Plant Biochemistry," eds. J. B. Harborne and P. M. Dey, Academic Press, New York, 1997, p. 387.
62. M. A. Pierrel, Y. Batard, M. Kazmaier, C. Mignotte-Vieux, F. Durst, and D. Werck-Reichhart, *Eur. J. Biochem.*, 1994, **224,** 835.
63. M. A. Schuler, *Crit. Rev. Plant Sci.*, 1996, **15,** 235.
64. U. Matern, *Planta Med.*, 1991, **57,** S15.
65. P. Rataboul, G. Alibert, T. Boller, and A. M. Boudet, *Biochim. Biophys. Acta*, 1985, **816,** 25.
66. H. Lee, J. Leon, and I. Raskin, *Proc. Natl. Acad. Sci. USA*, 1995, **92,** 4076.
67. S. A. Brown, *Can. J. Biochem. Cell Biol.*, 1985, **63,** 292.
68. M. Sato, *Phytochemistry*, 1967, **5,** 1363.
69. R. E. Kneusel, Diploma Thesis, Universität Freiburg, 1987.
70. S. A. Brown, *Z. Naturforsch.*, 1986, **41c,** 247.
71. S. A. Brown, R. E. March, D. E. A. Rivett, and H. J. Thompson, *Phytochemistry*, 1988, **27,** 391.
72. R. C. Beier, G. W. Ivie, and E. H. Oertli, *Phytochemistry*, 1994, **36,** 869.
73. H. W. Rauwald, O. Brehm, and K.-P. Odenthal, *Planta Med.*, 1994, **60,** 101.
74. T. Kinoshita, J.-B. Wu, and F.-C. Ho, *Phytochemistry*, 1996, **43,** 125.
75. D. Hamerski, D. Schmitt, and U. Matern, *Phytochemistry*, 1990, **29,** 1131.
76. D. Barron and R. K. Ibrahim, *Phytochemistry*, 1996, **43,** 921.
77. H. Yamamoto, J. Kimata, M. Senda, and K. Inoue, *Phytochemistry*, 1997, **44,** 23.
78. W. Maier, A. Baumert, B. Schumann, H. Furukawa, and D. Gröger, *Phytochemistry*, 1993, **32,** 691.
79. R. Boehm, S.-M. Li, M. Melzer, and L. Heide, *Phytochemistry*, 1997, **44,** 419.
80. D. V. Banthorpe, S. A. Branch, V. C. O. Njar, M. G. Osborne, and D. G. Watson, *Phytochemistry*, 1986, **25,** 629.
81. S. W. Zito, V. Srivastava, and E. Adebayo-Olojo, *Planta Med.*, 1991, **57,** 425.
82. C. A. Omer and J. B. Gibbs, *Mol. Microbiol.*, 1994, **11,** 219.
83. R. Roskoski Jr., P. Ritchie, and L. G. Gahn, *Anal. Biochem.*, 1994, **222,** 275.

84. Y. Q. Mu, C. A. Omer, and R. A. Gibbs, *J. Am. Chem. Soc.*, 1996, **118,** 1817.
85. T. Koyama, Y. Kokubun, and K. Ogura, *Phytochemistry*, 1988, **27,** 2005.
86. K. Cornish, *Eur. J. Biochem.*, 1993, **218,** 267.
87. K. Ogura and U. Sankawa, "Dynamic Aspects of Natural Products Chemistry—Molecular Biological Approaches," Gordon and Breach, Tokyo, 1996.
88. U. Matern, H. Wendorff, D. Hamerski, A. E. Pakusch, and R. E. Kneusel, *Bull. Liaison Groupe Polyphenols*, 1988, **14,** 173.
89. G. P. Bolwell, K. Bozak, and A. Zimmerlin, *Phytochemistry*, 1994, **37,** 1491.
90. M. Akhtar and J. N. Wright, *Nat. Prod. Rep.*, 1991, 527.
91. B. A. Halkier, *Phytochemistry*, 1996, **43,** 1.
92. M. Akhtar, D. Corina, S. Miller, A. Z. Shyadehi, and J. N. Wright, *Biochemistry*, 1994, **33,** 4410.
93. D. Hamerski, R. C. Beier, R. E. Kneusel, U. Matern, and K. Himmelspach, *Phytochemistry*, 1990, **29,** 1137.
94. L. Britsch, *Arch. Biochem. Biophys.*, 1990, **282,** 152.
95. A. J. Birch, M. Maung, and A. Pelter, *Aust. J. Chem.*, 1969, **22,** 1923.
96. T. Hakamatsuka, M. F. Hashim, Y. Ebizuka, and U. Sankawa, *Tetrahedron*, 1991, **47,** 5969.
97. V. Stanjek, M. Miksch, P. Lüer, U. Matern, and W. Boland, *Angew. Chemie*, 1998, in press.
98. X. P. Luo, *Anal. Biochem.*, 1995, **228,** 294.
99. W. Boland, personal communication.
100. D. Hamerski and U. Matern, *FEBS Lett.*, 1988, **239,** 263.
101. K. D. Hauffe, K. Hahlbrock, and D. Scheel, *Z. Naturforsch.*, 1986, **41c,** 228.
102. S. C. Wu and K. Hahlbrock, *Z. Naturforsch.*, 1992, **47c,** 591.
103. K. Hahlbrock, D. Scheel, E. Logemann, T. Nürnberger, M. Parniske, S. Reinold, W. R. Sacks, and E. Schmelzer, *Proc. Natl. Acad. Sci. USA*, 1995, **92,** 4150.

1.25
Lignans: Biosynthesis and Function

NORMAN G. LEWIS and LAURENCE B. DAVIN
Washington State University, Pullman, WA, USA

1.25.1 INTRODUCTION

The lignans are a very common, structurally diverse, group of plant natural products of phenylpropanoid origin.[1] They display important physiological functions *in planta*, particularly in plant defense,[2–5] and are most efficacious in human nutrition and medicine, given their extensive health protective and curative properties.[6–9]

This chapter primarily describes the intricate biochemical pathways established in lignan biosynthesis, particularly as regards phenylpropanoid coupling. It must be emphasized at the outset, however, that the biochemical outcome of enzymatic coupling was previously widely held to result only in formation of randomly linked racemic products. This hypothesis, originally adopted for lignin biopolymer formation (but see Chapter 3.18), could not explain the observed optical activities of the vast majority of naturally occurring lignans (discussed in Section 1.25.5). In contrast, research has established that most, if not all, phenylpropanoid coupling reactions catalyzed by purified plant proteins and enzymes *in vitro* can be either regio- or stereoselectively controlled, as are their subsequent stereospecific metabolic conversions. This contribution therefore attempts to summarize the progress made in this area, as well as in providing an assessment of the rapidly growing importance of lignans in plant growth, development, and survival, and in human usage.

1.25.2 DEFINITION AND NOMENCLATURE

Around the turn of the nineteenth century, a number of plant phenolic substances from various species were isolated and given trivial names, prior to their chemical structures being determined. One of these, guaiaretic acid, from guaiacum resin obtained from *Guaiacum officinale* heartwood, was later shown to contain the skeletal formula (**1**) by Schroeter *et al.*[10] It was subsequently proposed[11] that it and related compounds represented a unique class of dimeric phenylpropanoid substances linked exclusively through 8—8′ bonds, e.g., (**2**). In order to provide a system for their classification, Haworth[11] introduced the term "lignane" (later shortened to lignan) to define these substances; they were considered to result from regiospecifically linking two "cinnamyl" (C_6C_3) molecules to give compounds such as guaiaretic acid (**1**) and pinoresinols (**3a,b**). This initial classification, unfortunately, failed to account for either other dimeric lignan skeletal types that were present in many plant species and tissues or much larger molecules (oligomeric lignans) that could also exist. A derivative term, neolignan, was introduced to account for the other coupling modes, e.g., megaphone (**4**) (8—1′ linked), dehydrodiconiferyl alcohols (**5a,b**) (8—5′ linked) and *erythro*/*threo* guaiacylglycerol 8—*O*—4′ coniferyl alcohol ethers (**6a,b**) (8—*O*—4′ linked),[12–14] but this was later modified to encompass only presumed allylphenol-derived coupling products, such as the lignans

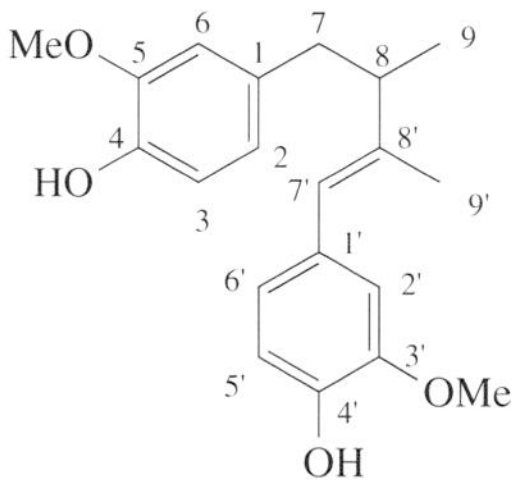

(**1**) Guaiaretic acid
$[\alpha]_D = -94°$
(Guaiacum officinale)

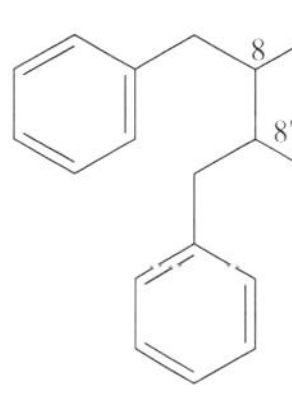

(**2**)

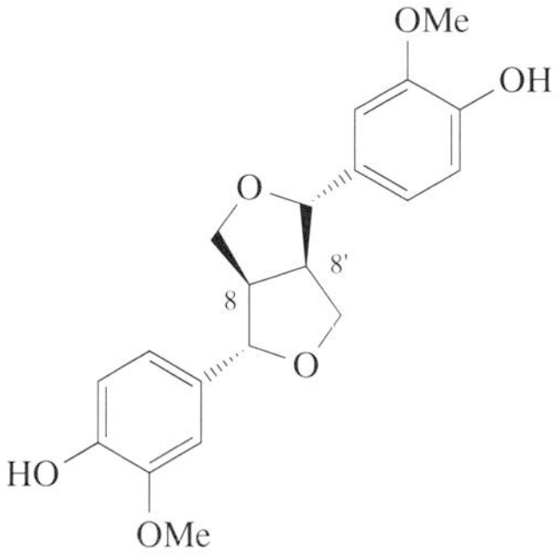

(**3a**) (+)-Pinoresinol
$[\alpha]^{22}_{D} = +61.6°$ ($CHCl_3$)
(Forsythia europaea)

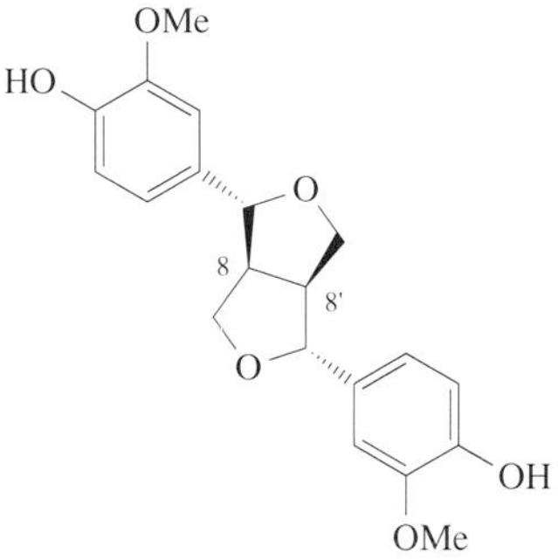

(**3b**) (−)-Pinoresinol
$[\alpha]^{25}_{D} = -34.7°$ ($CHCl_3$, $c = 0.91$)
(Daphne tangutica)

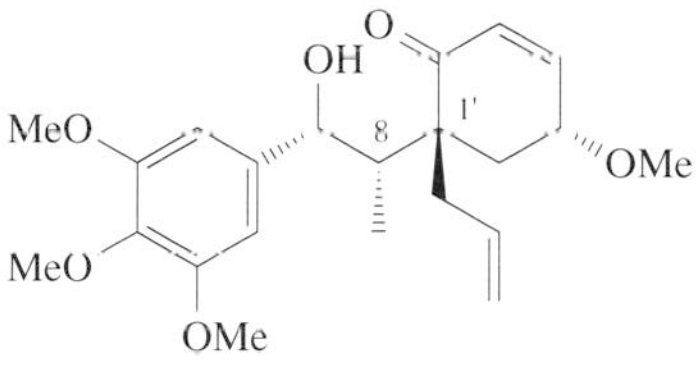

(**4**) (−)-Megaphone
$[\alpha]^{27}_{D} = -23.0°$ (EtOH, $c = 0.15$)
(Aniba megaphylla)

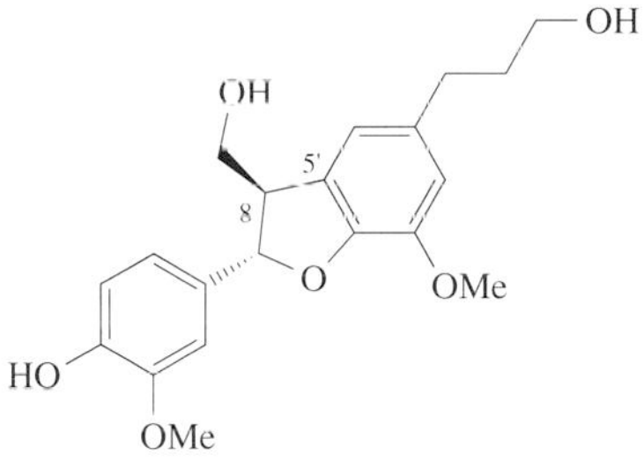

(**5a**,b) (±)-Dehydrodiconiferyl alcohols

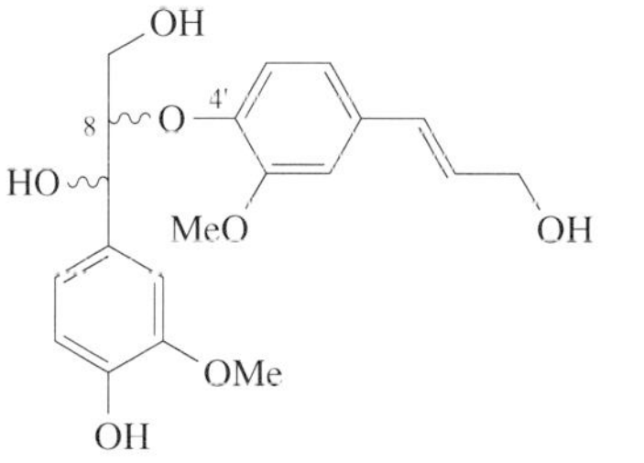

(**6a**,b) (±)-*erythro*/*threo* Guaiacylglycerol 8–*O*–4′ coniferyl alcohol ethers

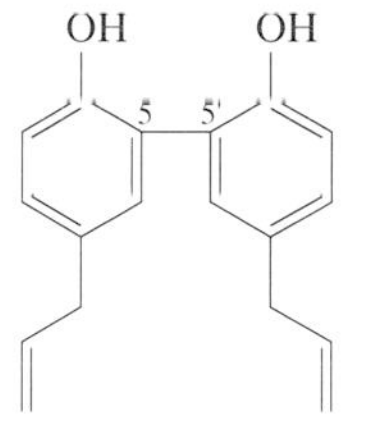

(**7**) Magnolol
(Magnolia virginiana)

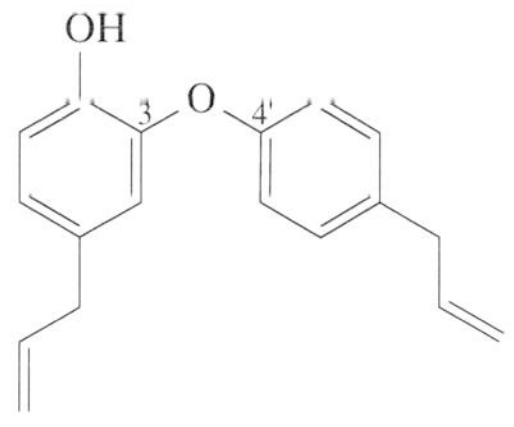

(**8**) Isomagnolol
(Sassafras randaiense)

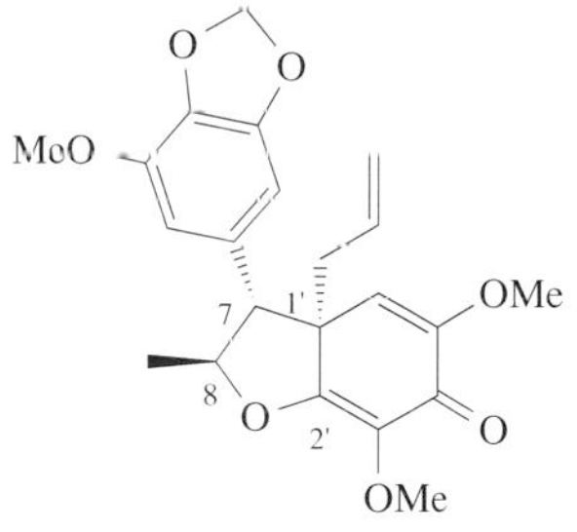

(**9**) Chrysophyllon 1A
(Licaria chrysophylla)

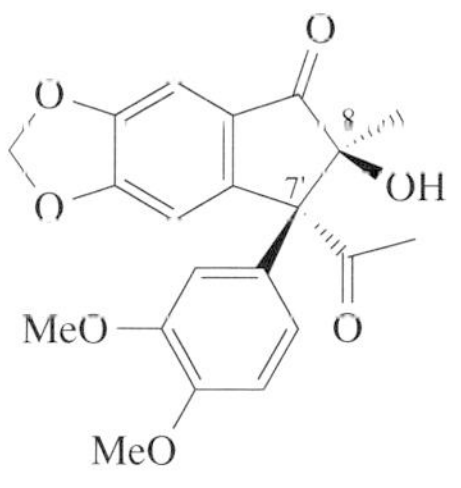

(**10**)
(Virola sebifera)

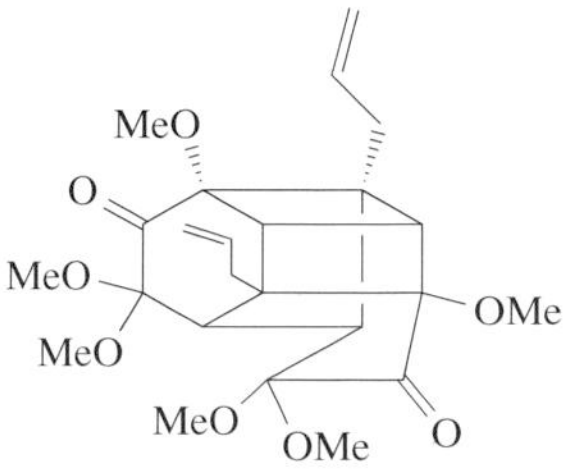

(**11**) Isoasatone
$[\alpha]^{20}_{D} = 0°$
(Asarum taitonense)

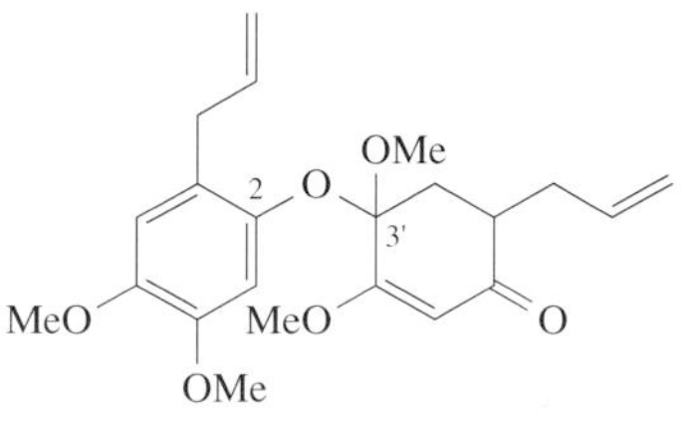

(**12**) Lancilin
(Aniba lancifolia)

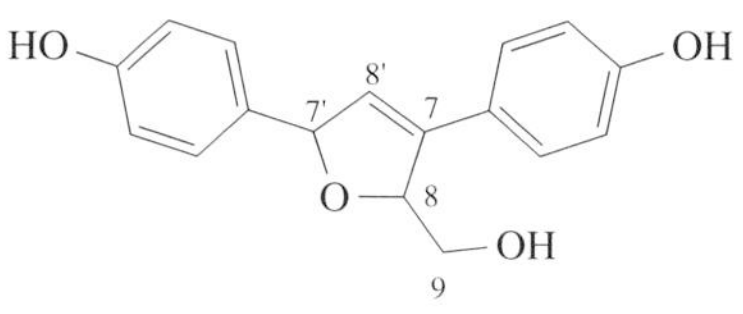

(**13**) (−)-Cryptoresinol
$[\alpha]^{25}_{D} = -170.4°$ (MeOH, $c = 1.25$)
(Cryptomeria japonica)

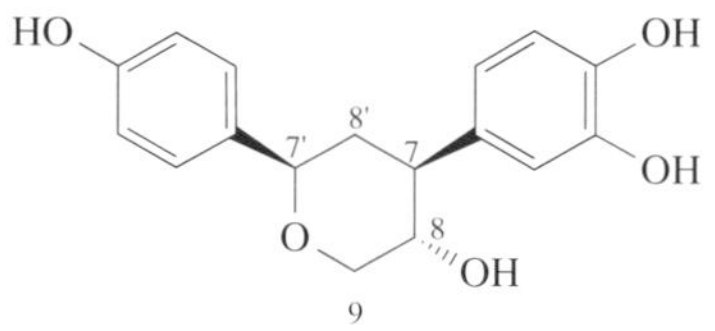

(14) (–)-Hydroxysugiresinol
$[\alpha]_D^{25} = -19°$ (EtOH, c 1)
(Sequoia sempervirens)

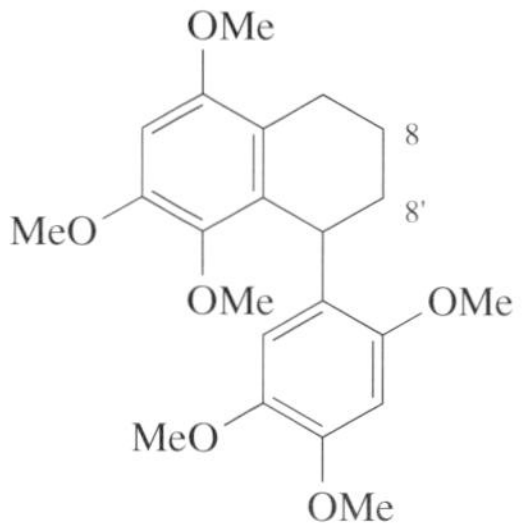

(15) Pachypostaudin A
$[\alpha]_D^{24} = 0°$ ($CHCl_3$, $c = 0.8$)
(Pachypodanthium staudtii)

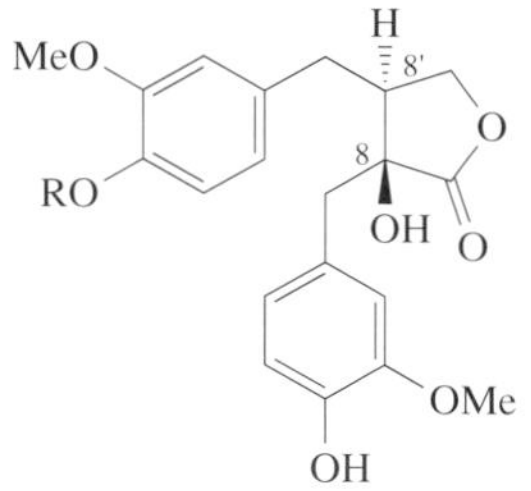

(16) (–)-Nortrachelogenin, R = H
$[\alpha]_D^{17} = -16.8°$ (EtOH, $c = 0.178$)
(17) (–)-Trachelogenin, R = Me
$[\alpha]_D^{23} = -43.3°$ (EtOH, $c = 0.25$)
(Trachelospermum asiaticum)

(**4**) and (**7**)–(**12**).[15] As a further complication, the term norlignan[16] was adopted to depict lignan-like metabolites which lacked either a carbon at the C-9 and/or C-9′ positions, e.g., cryptoresinol (**13**),[17] hydroxysugiresinol (**14**),[18] and pachypostaudin A (**15**),[19] or a methyl group on the aromatic methoxyl, e.g., nortrachelogenin (**16**)[20,21] versus trachelogenin (**17**).[20,22]

It is now well established that the products of phenylpropanoid coupling have a range of structural motifs[8,14,15,23,24] and molecular sizes,[25–30] rather than being restricted to 8—8′ linked moieties as previously contemplated. Moreover, since there appear to be only a relatively small number of distinct skeletal forms, the term lignan can be conveniently used to encompass all skeletal types, provided that the precise linkage type is stipulated, e.g., 8—8′, 8—1′, 8—5′, 8—*O*—4′, 5—5′, 3—*O*—4′, 7—1′, 8—7′, 1—5′, and 2—*O*—3′.

Of the various lignan types known, however, those that are 8—8′ linked appear to be the most widespread in nature, based on current chemotaxonomic data;[23] they can also be further subdivided into the substituted furofurans, tetrahydrofurans, dibenzylbutanes, dibenzylbutyrolactones, aryltetrahydronaphthalenes, arylnaphthalenes, dibenzocyclooctadienes, etc. (e.g., Structures (**18**)–(**25**)). Although not lignans proper, there are also related natural products of mixed metabolic origin, e.g., so-called flavonolignans[31–33] which are described in Section 1.25.11.

Finally, given that lignans can also exist in oligomeric form,[25–29] it is necessary at this point to introduce terminology to distinguish them from the corresponding lignin biopolymers. In this regard, since upper limits of molecular size have not been established for the oligomeric lignans, perhaps the easiest distinction lies with their physiological roles/functions: the lignin biopolymers have cell wall structural roles, whereas oligomeric lignans are nonstructural components (discussed in Section 1.25.13.6).

1.25.3 EVOLUTION OF THE LIGNAN PATHWAY

The evolutionary adaptation of plants to land was accompanied by massive elaboration of the phenylpropanoid pathway.[23] The salient features are summarized in Figure 1, and the reader is referred to Chapter 3.18 for a comprehensive description of each enzymatic step in the pathway. Nevertheless, assimilated carbon is directed to provide a variety of plant phenolics, such as the lignans and lignins, suberins and flavonoids, with the latter two metabolic classes being formed via merging both phenylpropanoid and acetate pathways. Accordingly, the transition of plants from an aquatic to a terrestrial environment was greatly facilitated via formation of lignans (protective/defense functions), lignins (structural support), suberins (formation of protective barriers preventing excessive water loss), and flavonoids (flower petal pigments, signaling molecules, UV-B screening, and so forth).[23] Together, these metabolites account for more than 30% of the carbon in vascular plants; however, note that lignans and lignins have never been convincingly demonstrated as present in aquatic plants, such as algae.

Furofuran

(18) (+)-Sesamin
$[\alpha]_D^{20} = +68.36°$ ($CHCl_3$, c = 24.45)
(*Sesamum indicum*)

Tetrahydrofuran

(19a) (+)-Lariciresinol
$[\alpha]_D^{25} = +18°$ (Me_2CO, c = 1.0)
(*Araucaria angustifolia*)

(19b) (–)-Lariciresinol
$[\alpha]_D^{25} = -12.3°$ (Me_2CO, c = 0.94)
(*Daphne tangutica*)

Dibenzylbutane

(20a) (–)-Secoisolariciresinol
$[\alpha]_D^{25} = -35.6°$ (Me_2CO, c = 1.07)
(*Podocarpus spicatus*)

(20b) (+)-Secoisolariciresinol
(*Linum usitatissimum*)

Dibenzylbutyrolactone

(21a) (–)-Matairesinol, R = H
$[\alpha]_D^{18} = -48.6°$ (Me_2CO, c = 2.409)
(*Podocarpus spicatus*)

(21b) (+)-Matairesinol, R = H
$[\alpha]_D^{20} = +37.0°$ (MeOH, c = 1.0)
(*Selaginella doederleinii*)

(22a) (–)-Arctigenin, R = Me
$[\alpha]_D^{23} = -34.6°$ (EtOH)
(*Forsythia intermedia*)

(22b) (+)-Arctigenin, R = Me
$[\alpha]_D^{23} = +28.05°$ (EtOH, c = 1.23)
(*Wikstroemia indica*)

Aryltetrahydronaphthalene

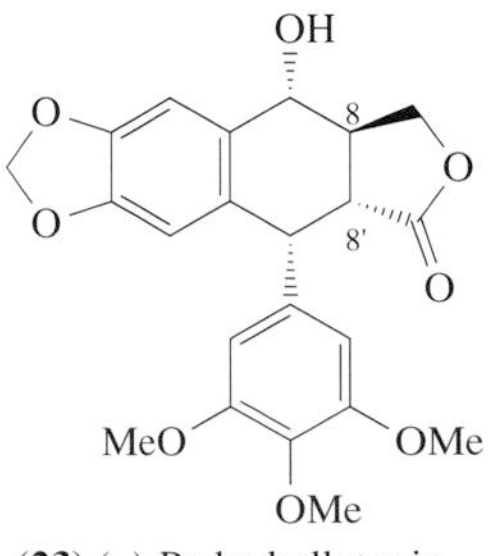

(23) (–)-Podophyllotoxin
$[\alpha]_D = -109°$ (EtOH)
(*Podophyllum hexandrum*)

Arylnaphthalene

(24) Justicidin E
(*Justicia procumbens*)

Dibenzocyclooctadiene

(25) (–)-Steganacin
$[\alpha]_D = -122.6°$ ($CHCl_3$, c = 1.02)
(*Steganotaenia araliacea*)

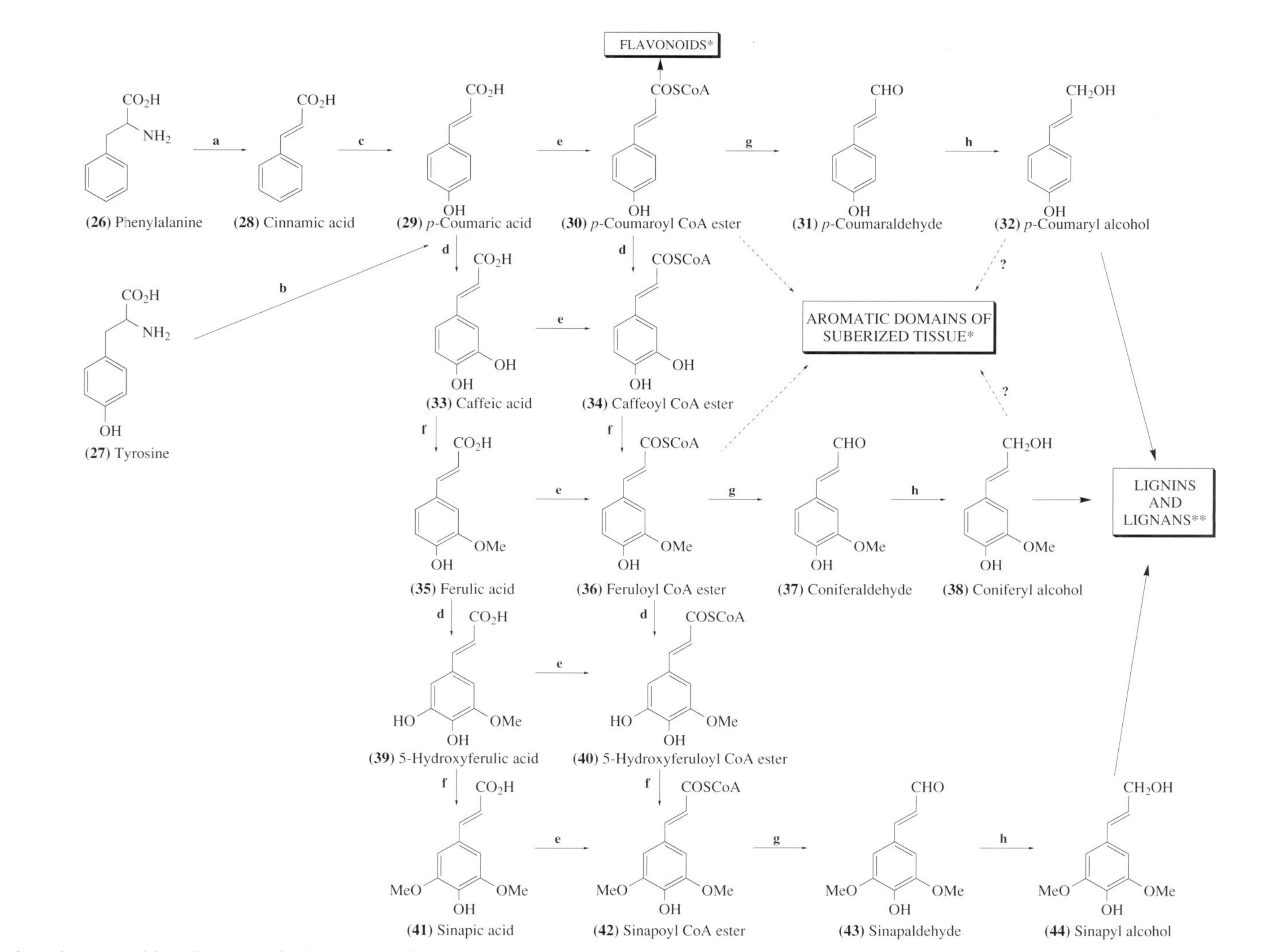

Figure 1 The phenylpropanoid pathway and selected metabolic branchpoints. (**a**) Phenylalanine ammonia-lyase, (**b**) tyrosine ammonia-lyase, (**c**) 4-cinnamate hydroxylase, (**d**) hydroxylases, (**e**) CoA ligases, (**f**) *O*-methyltransferases, (**g**) cinnamoyl-CoA reductase, and (**h**) cinnamyl alcohol dehydrogenase. *: plus acetate pathway for both flavonoids and suberins, **: main sources of lignan skeleta.

1.25.4 OCCURRENCE

Lignans are present in a large number of vascular plant species, ranging from "primitive" hornworts, liverworts, and ferns to the woody gymnosperms and woody/herbaceous angiosperms. They have been found in all plant parts including (woody) stems, rhizomes, roots, seeds, oils, exuded resins, flowers, fruits, leaves, and bark tissues.[8,23,24,34] However, their amounts can vary extensively between tissues and species, and in some instances their deposition can be massive. For example, in western red cedar (*Thuja plicata*) heartwood, dimeric lignans and higher oligomers can constitute up to 20% of the dry weight of the plant tissue,[35] where both (oligomeric) lignans and lignin biopolymers coexist in the same (heartwood) tissue.[36] In terms of localization of lignans *in planta*, evidence points to different sites depending upon the tissue/species involved. For example, in western red cedar heartwood, they appear to be mainly deposited as nonstructural infusions (secreted from specialized cells such as the ray parenchyma) into the prelignified sapwood,[28] whereas in flax (*Linum usitatissimum*) they seem to be covalently bound to carbohydrate moieties in the seed. On the other hand, in *Linum flavum*, lignans apparently accumulate in vacuoles (discussed in Section 1.25.13.6).[37]

The lignans are a structurally very diverse class of natural products, with more than several thousand distinct structures known at the dimer level alone.[8,14–16,38–42] Interestingly, they are typically found in optically active form, although the particular antipode can vary with plant species. For example, (+)-pinoresinol (**3a**) is present in *Forsythia europaea*,[43] whereas its (−)-antipode (**3b**) occurs in *Daphne tangutica*,[44] and (−)-arctigenin (**22a**) has been isolated from *Forsythia intermedia*,[45] with the (+)-form (**22b**) being found in *Wikstroemia indica*.[46] A discussion of optical activity in lignans and radical coupling mechanisms is given in Section 1.25.5.

1.25.4.1 Lignans in "Early" Land Plants

Based on DNA sequence analyses and neontological/paleontological data,[47–49] it is considered that bryophyte-like plants, which include the liverworts (Hepaticae), hornworts (Anthocerotae), and mosses (Musci), were part of the terrestrial flora that emerged some 20–40 million years or so prior to the "pretracheophytes" and vascular plants. The earliest documented examples of presumed bryophyte-like fossilized remains come from both lower Devonian (~408 million years ago) and early Silurian (~438 million years ago) records: the early Silurian spore fossils possess cell wall ultrastructures reminiscent of those present in extant liverworts,[50] whereas lower Devonian fossils show some similarities to modern thaloid hepatics (liverworts).[51] Moreover, gene sequence analyses of the large subunit of ribulose 1,5-bisphosphate carboxylase/oxygenase (rbcL) in the chloroplast, has suggested that liverworts are at the base of the embryophyte (embryo containing plants) lineage and that the hornworts represent the closest lineage to vascular plants,[52] i.e., the liverworts emerged first, then the mosses, and finally the hornworts, before the first tracheophytes.

Lignans are found in both liverworts and hornworts, although none have been reported in mosses. Consideration of their structures is of interest, as it may give useful insight into how the lignan pathway evolved. Thus, lignans present in liverworts (*Pellia epiphylla* (**45**)–(**48**),[53,54] *Jamesoniella autumnalis* (**49**)–(**54**),[55] and *Scapania undulata* (**54**)[56]) contain optically active 8—8′ linked lignans which provisionally appear to be caffeic acid (**33**) derived. They can also be ligated to shikimic acid moieties (**55**) as shown in (**47**),[53] or contain pendant aryl groups that have undergone fission and recyclization to afford lactones as in (**49**)–(**52**) and (**54**).[54] On the other hand, egonol-2-methylbutanoate (**56**) and the optically active lignan, (−)-licarin A (**57**), present in the liverworts, *Riccardia multifida* subsp. *decrescens*[57,58] and *Jackiella javanica*,[59,60] respectively, appear to be derived from 8—5′ coupling of allylphenols, eugenol (**58**)/isoeugenol (**59**) and *p*-hydroxyarylpropene (**60**). Of these, egonol-2-methylbutanoate (**56**) in *R. multifida* has two noteworthy features: namely, the apparent absence of a C-9 carbon and the introduction of a methylenedioxy group. Hitherto, methylenedioxy bridge formation was thought to occur in the lignan pathway with the advent of the gymnosperms.[23]

The only report of a lignan present in hornworts (*Megaceros flagellaris*, *Notothylas temperata*, and *Phaeoceros laevis*)[61,62] is that of (+)-megacerotonic acid. Two different structures (**61**) and (**62**) were reported by the same authors for this metabolite in two different publications in the same year, but without any cross-referencing or explanation given to account for the different skeletal representations.[61,62] In a subsequent total chemical synthesis study by Brown *et al.*,[63] the structure of megacerotonic acid was unequivocally established to be compound (**61**).

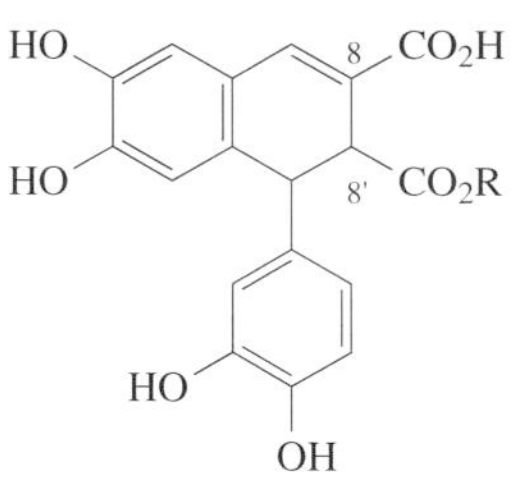

(45) R = H, $[\alpha]_D^{20} = -130.77°$ (MeOH)
(46) R = Me, $[\alpha]_D^{20} = -112.11°$ (MeOH)
(Pellia epiphylla)

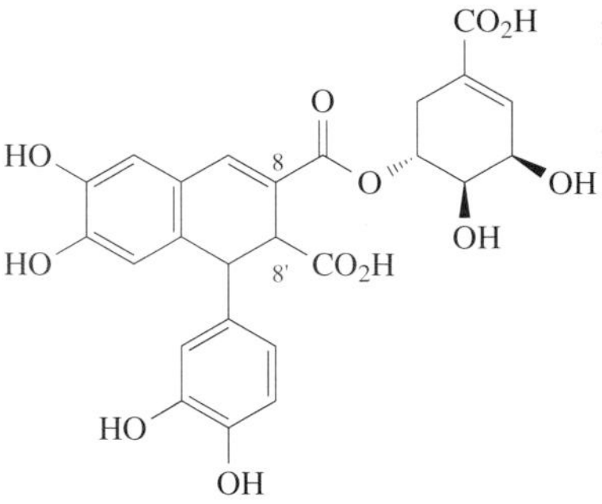

(47) $[\alpha]_D^{20} = -122.19°$ (MeOH)
(Pellia epiphylla)

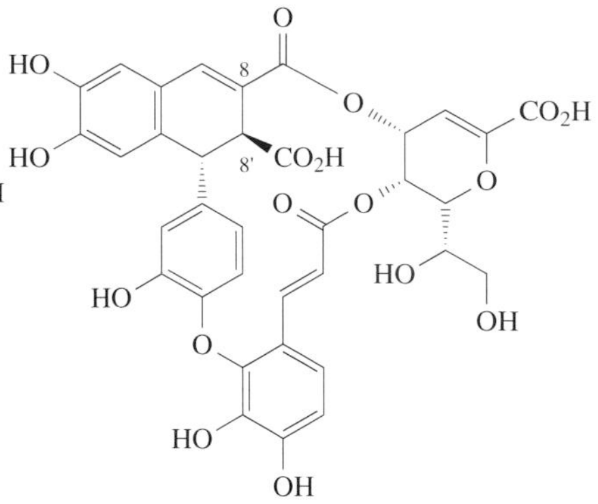

(48) Pelliatin $[\alpha]_D^{20} = -106.68°$ (MeOH)
(Pellia epiphylla)

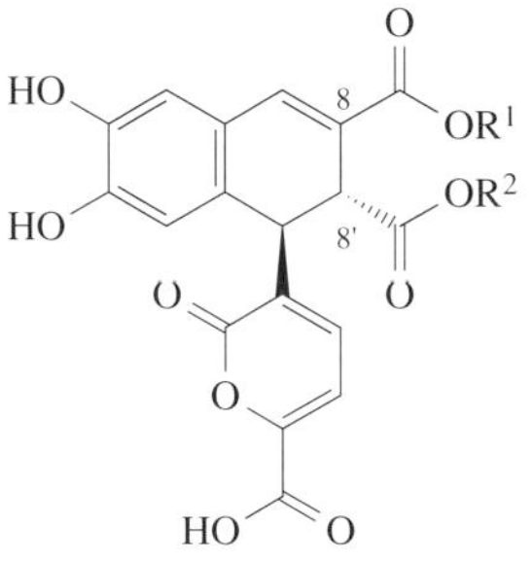

(49) $R^1 = R^2 = H$, $[\alpha]_D^{20} = +54.88°$ (MeOH, *c* = 0.30)
(50) $R^1 = Me$, $R^2 = H$, $[\alpha]_D^{20} = +33.49°$ (MeOH, *c* = 1.70)
(51) $R^1 = H$, $R^2 = Me$, $[\alpha]_D^{20} = +50.48°$ (MeOH, *c* = 0.95)
(Jamesoniella autumnalis)

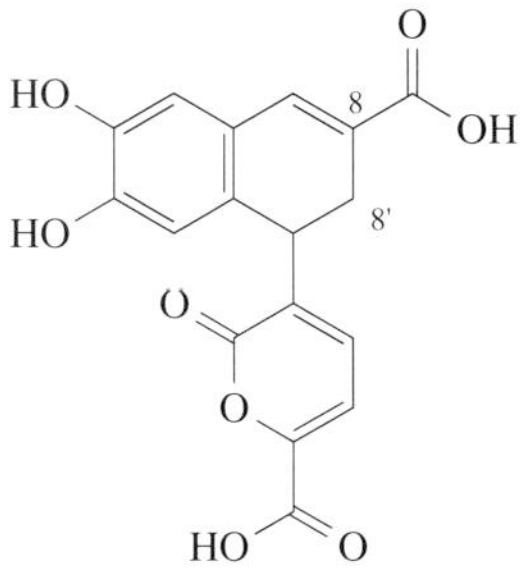

(52) $[\alpha]_D^{20} = +156.52°$ (MeOH, *c* = 0.08)
(Jamesoniella autumnalis)

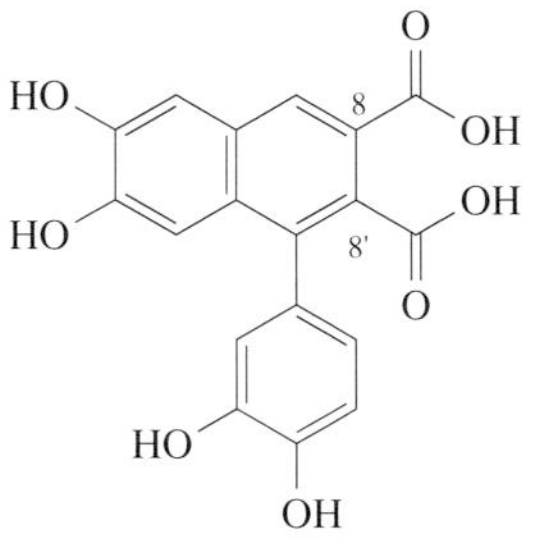

(53)
(Jamesoniella autumnalis)

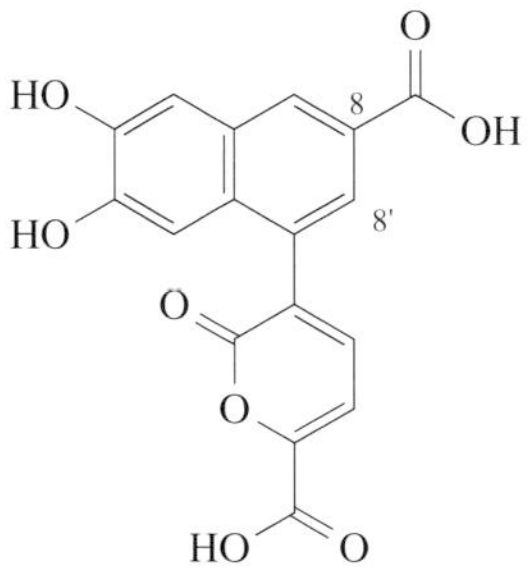

(54) Scapaniapyrone A
(Jamesoniella autumnalis)
(Scapania undulata)

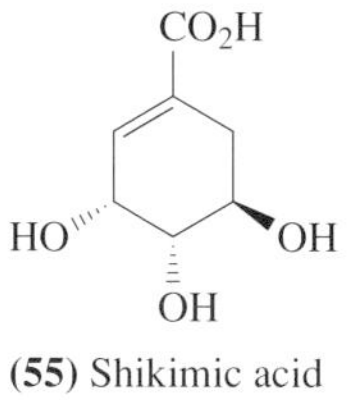

(55) Shikimic acid

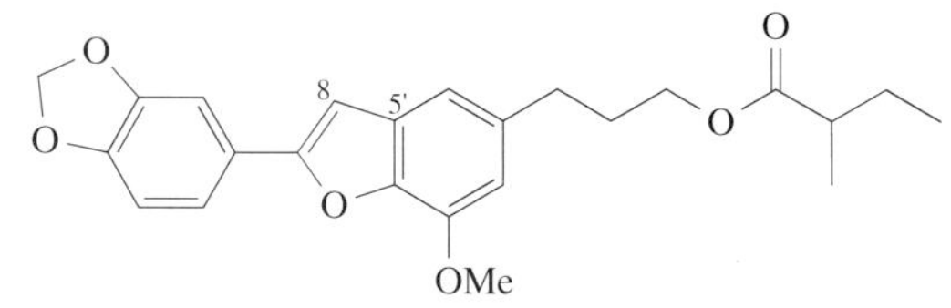

(56) Egonol 2-methylbutanoate
(Riccardia multifida, subsp. *decrescens)*

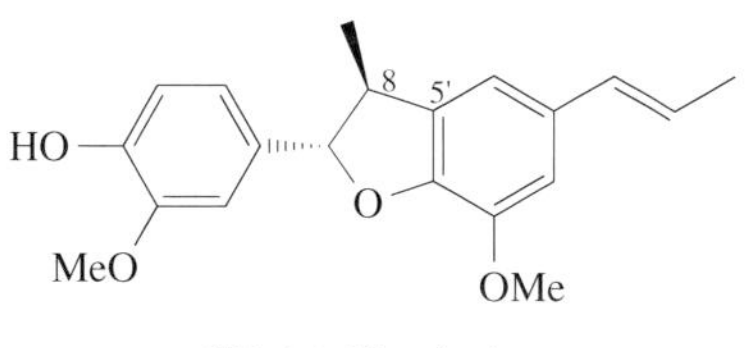

(57) (–)-Licarin A
$[\alpha]_D = -43°$ (c 9.0)
(Jackiella javanica)

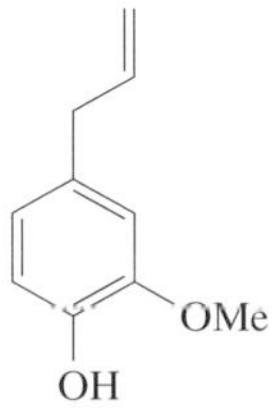

(58) Eugenol

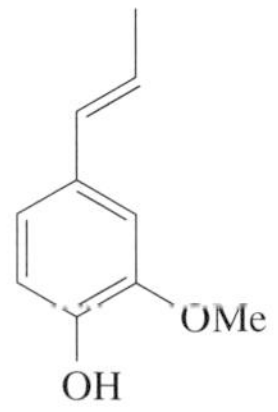

(59) Isoeugenol

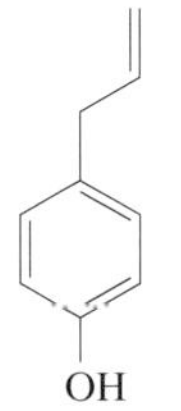

(60) *p*-Hydroxyarylpropene

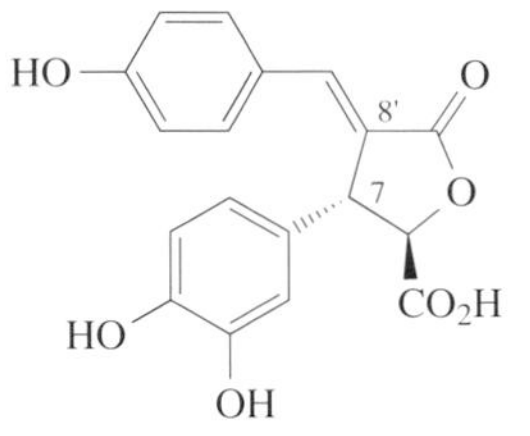

(61) 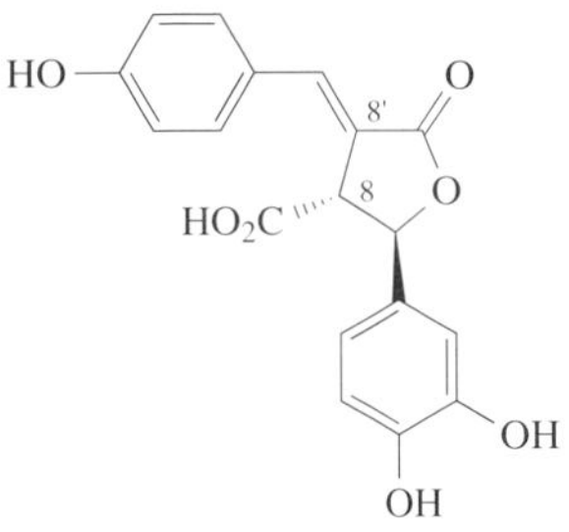 **(62)**

(+)-Megacerotonic acid
$[\alpha]_D = +233.0°$ (5% AcOH, $c = 1.66$)
(Megaceros flagellaris)

Lignans are also found in the ferns (Pteridophytes), and include the presumed caffeic acid (**33**) derived 8—2′ linked lignans, (−)-blechnic acid (**63**) and its shikimate derivative, (−)-brainic acid (**64**) from *Blechnum orientale*, *Struthiopteris amabilis*, *Struthiopteris niponica*, *Woodwardia orientalis*, *Woodwardia prolifera*, and *Brainea insignis* (Blechnaceae).[64] Additionally, (−)-lirioresinol A [=(−)-epi-syringaresinol] (**65**), (−)-lirioresinol B [=(−)-syringaresinol] (**66b**), (+)-wikstromol [=(+)-nortrachelogenin] (**67**), (−)-nortracheloside (**68**), and (+)-matairesinol (**21b**) are present in *Selaginella doederleinii* Hieron. (Selaginellaceae), a small perennial pteridophyte found in south and southwestern China at low altitude.[65]

The remaining lignans known to be present in the fern *Pteris vittata* (Pteridaceae), are the 8—5′ and 8—8′ linked glucosides of (−)-dihydrodehydrodiconiferyl alcohol (**69**) and (+)-lariciresinol (**70**).[66] Initially, lignan (**69**) was claimed to have the *cis* configuration at positions 7 and 8; however, subsequent studies by Wallis and co-workers established it to be *trans* (**71**).[67] Other than the lignan dimers, no higher oligomers have been reported in either liverworts, hornworts, or ferns.

1.25.4.2 Lignans in Gymnosperms and Angiosperms (General Features)

In contrast to the relatively few examples of lignans in early land plants, the gymnosperms were accompanied by a massive increase in lignan structures, particularly representatives of the optically active 8—8′ linked tetrahydrofurans, dibenzylbutanes, dibenzylbutyrolactones, aryltetrahydronaphthalenes, and arylnaphthalenes.[23] The vast majority reported thus far are *E*-coniferyl alcohol (**38**)-derived dimers; however, they can also exist in trimeric and higher oligomeric forms.[25–29,36] Other lignan types, e.g., 8—5′ and 8—*O*—4′ linked, are present in varying amounts in the gymnosperms.

The evolutionary transition to the angiosperms also witnessed a massive increase in formation of (oligomeric) lignan structural types. While the most widespread are again the optically active 8—8′ linked, other coupling modes, affording 8—1′, 8—5′, 5—5′, 7—1′, 8—7′, 1—5′, 8—*O*—4′, 2—*O*—3′, and 3—*O*—4′ coupled products also appeared—particularly in the Magnoliiflorae.[23] By contrast, relatively few lignans of any type have been reported in the monocotyledons. The few described primarily consist of cyclobutane dimers, such as the dihydroxytruxillic acids (**72**) and (**73**) from *Setaria anceps* cv *Nandi* (Poaceae, Commeliniflorae)[68,69] and acoradin (**74**) found in *Acorus calamus* (Araceae, Ariflorae).[70]

However, rather than attempting to distinguish between the various lignans on the basis of plant family or origin, it is perhaps more instructive to discuss the distinct structural types in terms of known or perceived biosynthetic pathways, as understood at the enzyme, protein, and molecular levels. A discussion of optical activity is, however, first required.

1.25.5 OPTICAL ACTIVITY OF LIGNAN SKELETAL TYPES AND LIMITATIONS TO THE FREE RADICAL RANDOM COUPLING HYPOTHESIS

No description of lignan structure would be complete without first discussing the changing paradigm for control of enzymatic free-radical coupling of phenylpropanoids. Initially, based on the study of presumed lignin biopolymer formation, a random free-radical coupling process was favored as being the only operative mechanism.[71,72] However, this earlier concept has been dramatically revised,[73,74] and it is important to discuss the basis for this changing view.

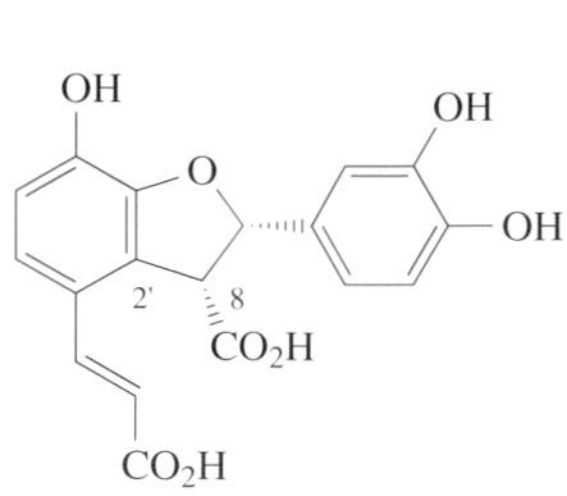

(63) (–)-Blechnic acid
$[\alpha]_D^{23} = -28°$ (MeOH, $c = 1.0$)
(Blechnum orientale)

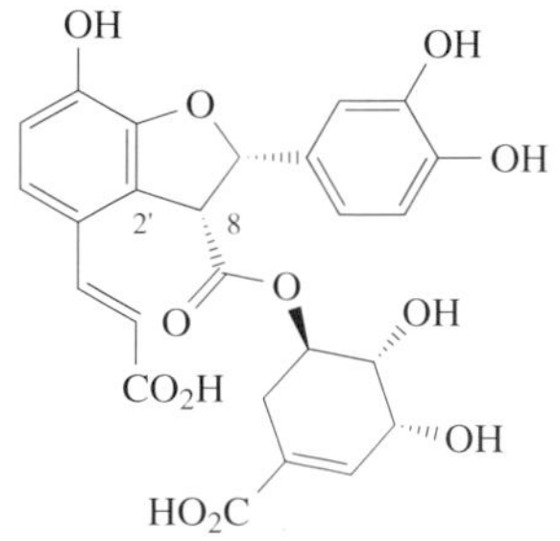

(64) (–)-Brainic acid
$[\alpha]_D^{24} = -42°$ (MeOH, $c = 1.0$)
(Blechnum orientale)

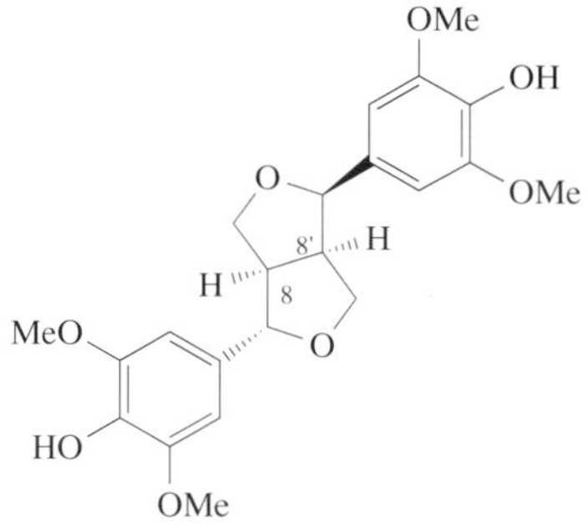

(65) (–)-Lirioresinol
= (–)-Episyringaresinol
(Selaginella doederleinii)

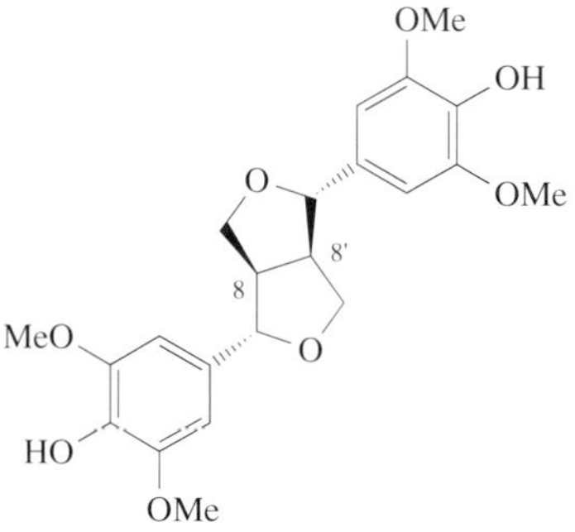

(66a) (+)-Lirioresinol B
= (+)-Syringaresinol

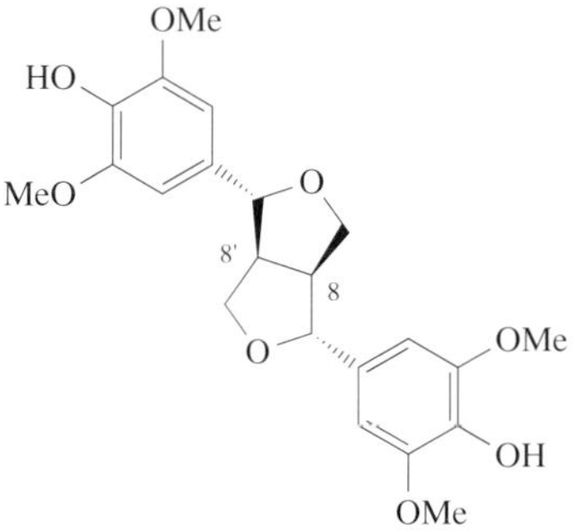

(66b) (–)-Lirioresinol B
= (–)-Syringaresinol
(Selaginella doederleinii)

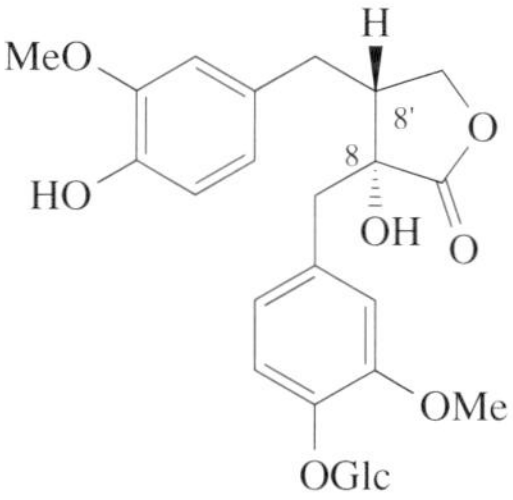

(67) (+)-Wikstromol
= (+)-Nortrachelogenin
(Selaginella doederleinii)

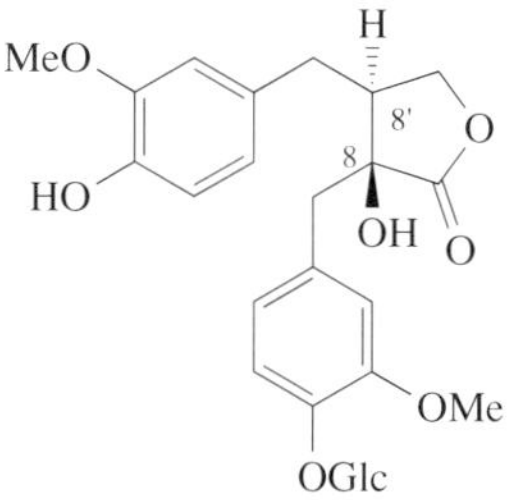

(68) (–)-Nortracheloside
(Selaginella doederleinii)

In the original random free-radical coupling hypothesis, substrate monomers, such as the monolignols, *p*-coumaryl (**32**), coniferyl (**38**), and sinapyl (**44**) alcohols, were envisaged to undergo single-electron oxidations (oxidase-catalyzed) to afford the corresponding free-radical species, where the only enzymatic requirement was generation of the free-radical intermediates. Under such conditions, at least *in vitro*, random coupling can then occur at several sites on the molecule to initially afford racemic lignan dimers, and Figure 2 illustrates this random coupling hypothesis using *E*-coniferyl alcohol (**38**) as an example: the major coupling products formed are the three racemic 8—5′, 8—8′, and 8—*O*—4′ linked lignan dimers (**5**), (**3**), and (**6**), with the 8—5′ dimer predominating in coupling frequency *in vitro*.

According to this view, there was believed to be no further requirement for additional enzyme/protein involvement in determining the outcome of lignin biopolymer assembly, other than reoxidation of the dimeric species. Therefore, unlike any other biopolymer, lignin formation was thought to be satisfactorily duplicated by the random encounter of its free-radical precursors *in vitro*. However, this random assembly mechanism, in fact, never gave a biopolymer duplicating lignin structure (see Chapter 3.18); nor could it explain the observed optical activity of numerous lignans

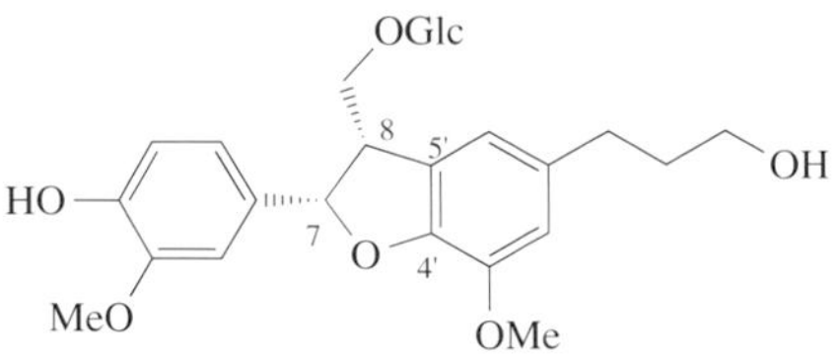

(69) *cis*-Dihydrodehydrodiconiferyl alcohol-9–*O*–β-D-glucoside (incorrectly assigned)

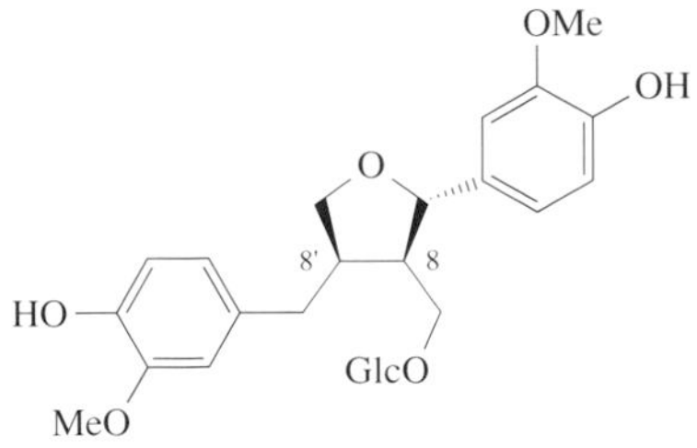

(70) Lariciresinol-9–*O*–β-D-glucoside
$[\alpha]_D^{27} = -39.77°$ (Me_2CO, $c = 0.88$)
$[\alpha]_D^{20}$ of aglycone = $+15.7°$ (Me_2CO, c 0.35)
(Pteris vittata)

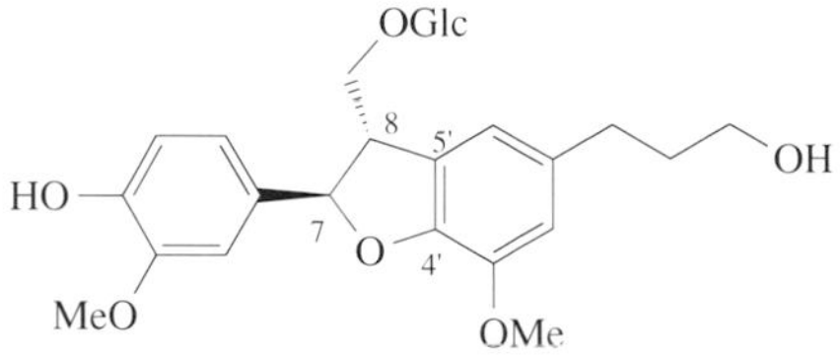

(71) *trans*-Dihydrodehydrodiconiferyl alcohol-9–*O*–β-D-glucoside
$[\alpha]_D^{22} = -23.6°$ (MeOH, c = 0.98)
$[\alpha]_D^{22}$ of aglycone = $+8.5°$ (Me_2CO, $c = 0.96$)
(Pteris vittata)

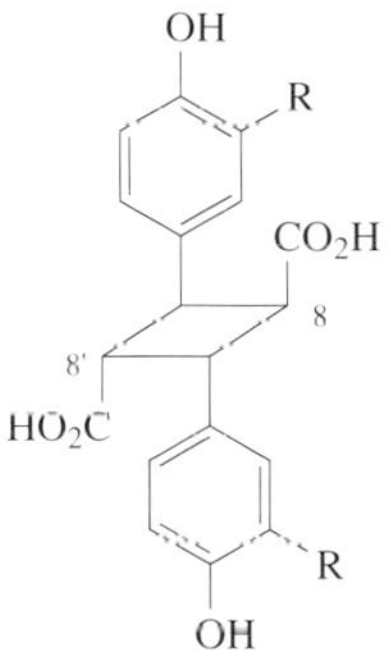

(72) Dihydroxytruxillic acid
R = H or OMe
(Setaria anceps)

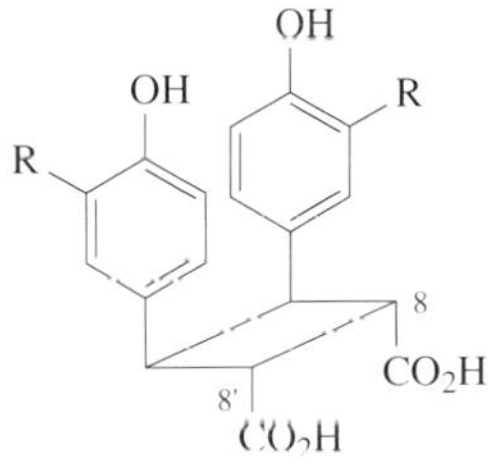

(73) Dihydroxytruxinic acid
R = H or OMe
(Setaria anceps)

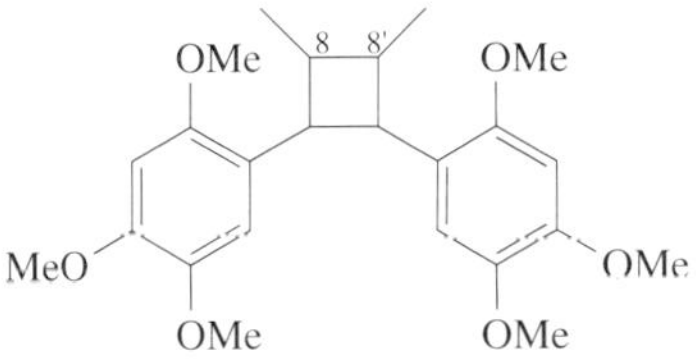

(74) Acoradin
$[\alpha]_D = 0°$ ($c = 1.1$)
(Acorus calamus)

occurring in a wide variety of plant species. Indeed, lignan optical activity could presumably only result via one of three ways: (i) where stereoselective phenylpropanoid coupling occurs, thereby explicitly controlling both the regio- and stereochemistries; (ii) where racemic coupling occurs, but where one of the enantiomeric forms is selectively metabolized; and/or (iii) where both stereoselective and random coupling can occur, but to different extents in different tissues and cell types.

Each of these three possibilities had to be considered since, in many cases, the optical rotation, $[\alpha]_D$, value for a particular lignan can differ substantially between plant species: For example, Table 1 shows the variation in reported optical rotations for syringaresinols **(66)** obtained from different plants.[44,75–85] As can be seen, these range from being optically active [(+)- or (−)-], to being nearly racemic. These observations, however, only serve to remind us that an $[\alpha]_D$ determination does not reveal the enantiomeric purity of isolated lignans, or, indeed, provide any insight as to whether the enantiomer(s) originate from the same subcellular compartment, cell type, or tissue.

The application of chiral HPLC methodologies, suitable for the facile separation of various lignan optical antipodes, has significantly helped in the study of this class of natural products, i.e., whether for examination of enantiomeric purity and/or outcome of phenylpropanoid coupling.[86–88] The two major chiral stationary phases currently employed for separation of (+)- and (−)-lignan antipodes are either cellulose carbamate coated on silica gel (Chiralcel columns, Daicel), or a macrocyclic

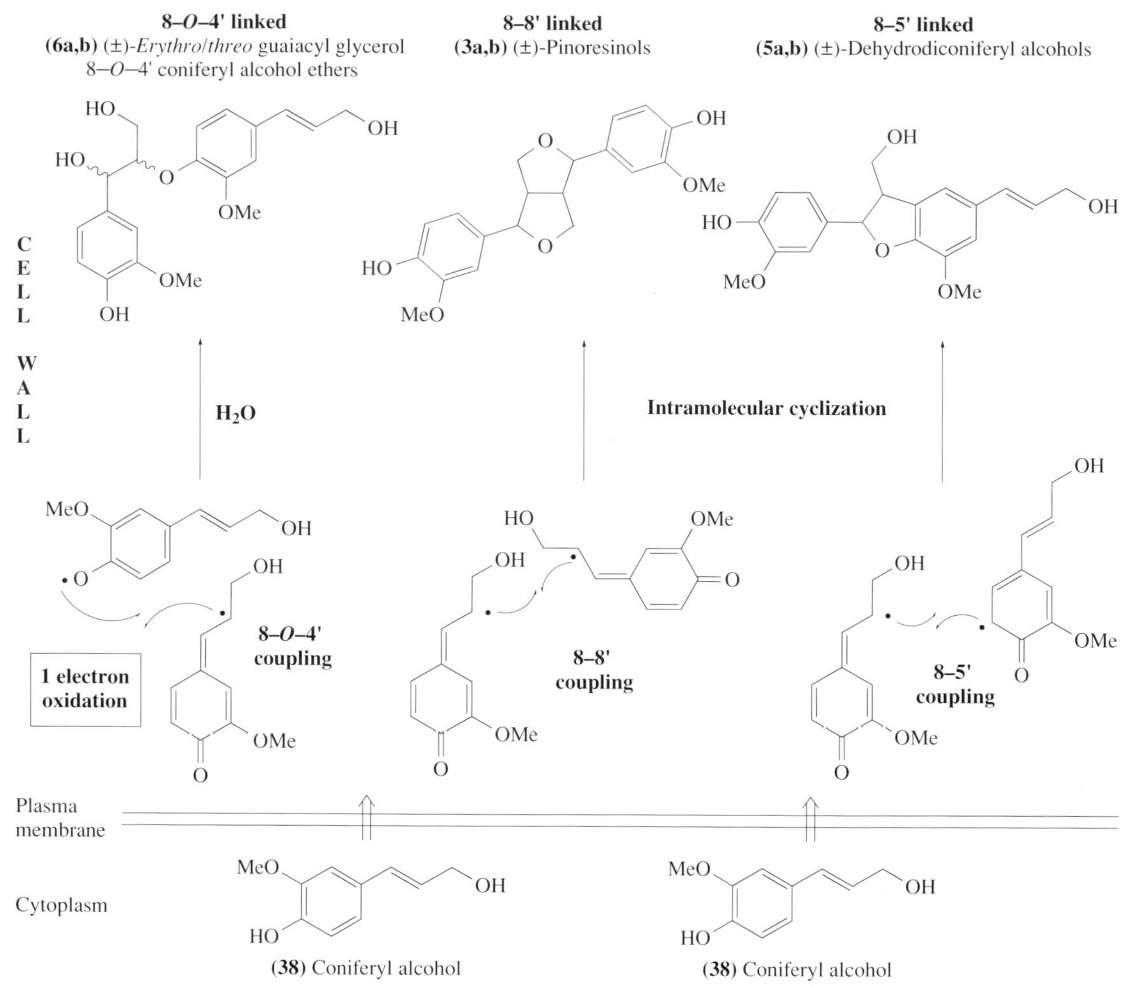

Figure 2 Illustration of the monolignol random coupling hypothesis leading to lignin as according to Freudenberg.[71,72]

Table 1 Specific rotations and enantiomeric compositions of syringaresinol (**66**) isolated from different species.[76] (Enantiomeric composition was determined following chiral column chromatography.)

Species	$[\alpha]_D$ *reported*	*Ref.*	*Syringaresinol* (%)	
			(+)-(**66a**)	(−)-(**66b**)
Aspidosperma marcgravianum	−34.8	77		
Holocantha emoryi	−32.5	78		
Xanthoxylum ailanthoides	−9.6	79	38	62
Daphne tangutica	−2.1	44	48	52
Xanthoxylum inerme	0	80	48	52
Stellera chamaejasme	+3.0	81		
Liriodendron tulipifera	+19.0	82		
Hedyotis lawsoniae	+23.0	83		
Eucommia ulmoides	+44.0	84	88	12
Liriodendron tulipifera	+48.9	85		

glycopeptide covalently bound to silica (5 μm), such as Vancomycin (Chirobiotic V column, Advanced Separation Technologies). The basis for their separations is the reversible formation of transient diastereomeric complexes formed between each enantiomer and the chiral stationary phase.[76]

Conditions have thus been developed for separation of (+)- and (−)-enantiomers of various lignans obtained, for example, by chemical syntheses, such as the pinoresinols (**3a,b**), secoisolariciresinols (**20a,b**), syringaresinols (**66a,b**), and dehydrodiconiferyl alcohols (**5a,b**) (see Figures 3(a) to 3(d)). Further, chiral HPLC analysis of certain lignans from selected plant species, e.g., *F. intermedia*, has revealed that, in many instances, lignans, such as pinoresinol (**3**) and secoisolariciresinol (**20**), occur in optically pure form (Figures 3(e) and 3(f)). On the other hand, with the syringaresinols (**66a,b**) isolated from different sources, it was observed (Figure 4 and Table 1) that both antipodes could either occur in enantiomeric excess (e.g., in *Xanthoxylum ailanthoides* and *Eucommia ulmoides*; Figures 4(b) and 4(a)) or were in nearly racemic amount (e.g., *D. tangutica* and *Xanthoxylum inerme*; Figures 4(c) and 4(d)). Some convincing scientific explanation was thus required for the noted, but often differing, optical activities in the lignans.

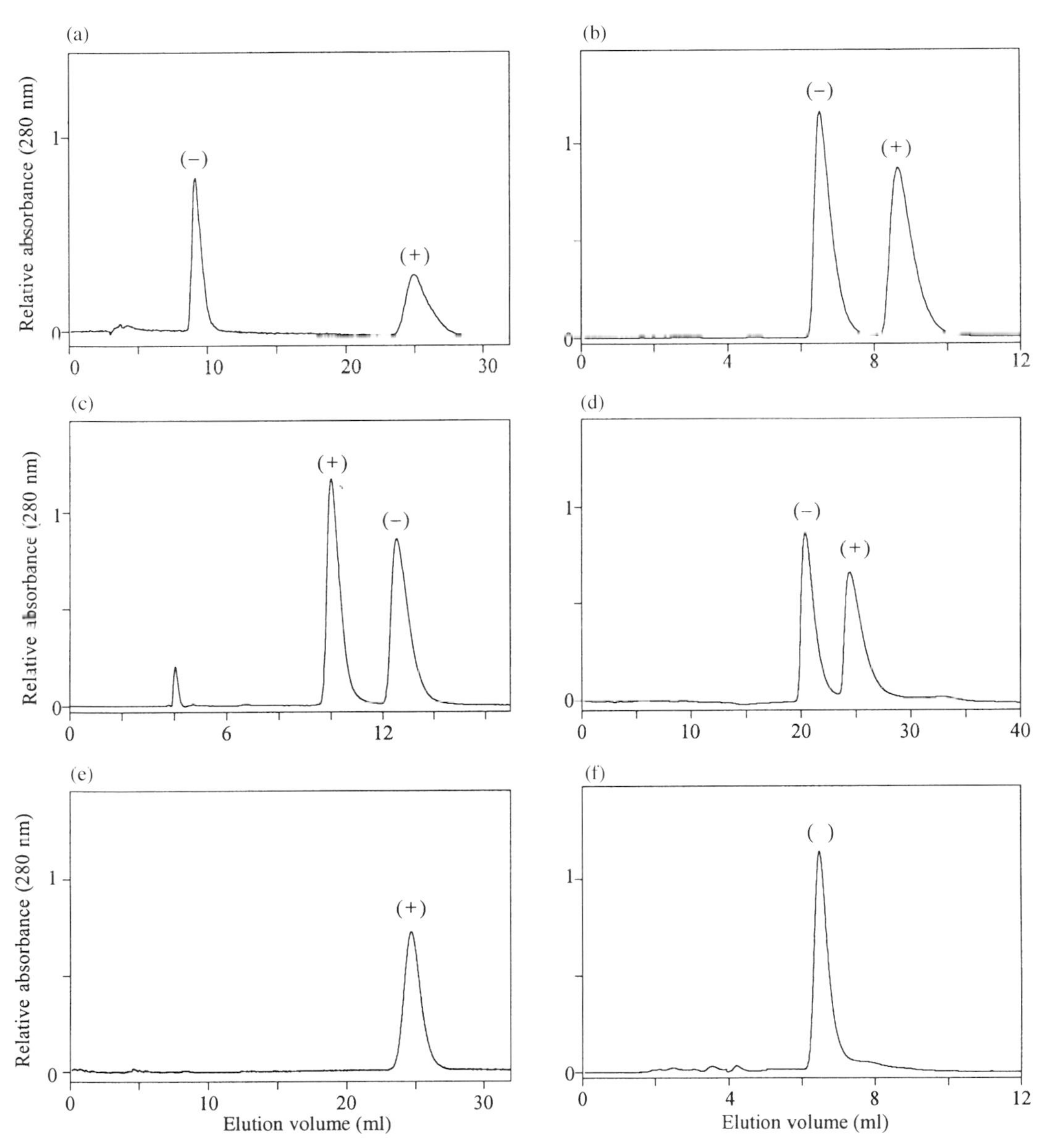

Figure 3 Chiral HPLC separation of selected lignans. (**a**) (±)-Pinoresinols (**3a,b**), (**b**) (±)-secoisolariciresinols (**20a,b**), (**c**) (±)-syringaresinols (**66a,b**), (**d**) (±)-dehydrodiconiferyl alcohols (**5a,b**) as well as (**e**) (+)-pinoresinol (**3a**), and (**f**) (−)-secoisolariciresinol (**20a**) isolated from *Forsythia intermedia*. (±)-Pinoresinols (**3a,b**), (±)-syringaresinols (**66a,b**), and (±)-secoisolariciresinols (**20a,b**) were individually resolved using a Chiralcel OD column (Daicel),[86,88] whereas resolution of (±)-dehydrodiconiferyl alcohols (**5a,b**) employed a Chirobiotic V column (Advanced Separation Technologies) with MeOH-NH_4NO_3 (1:9) as solvent system (flow rate: 1 ml min^{-1}).

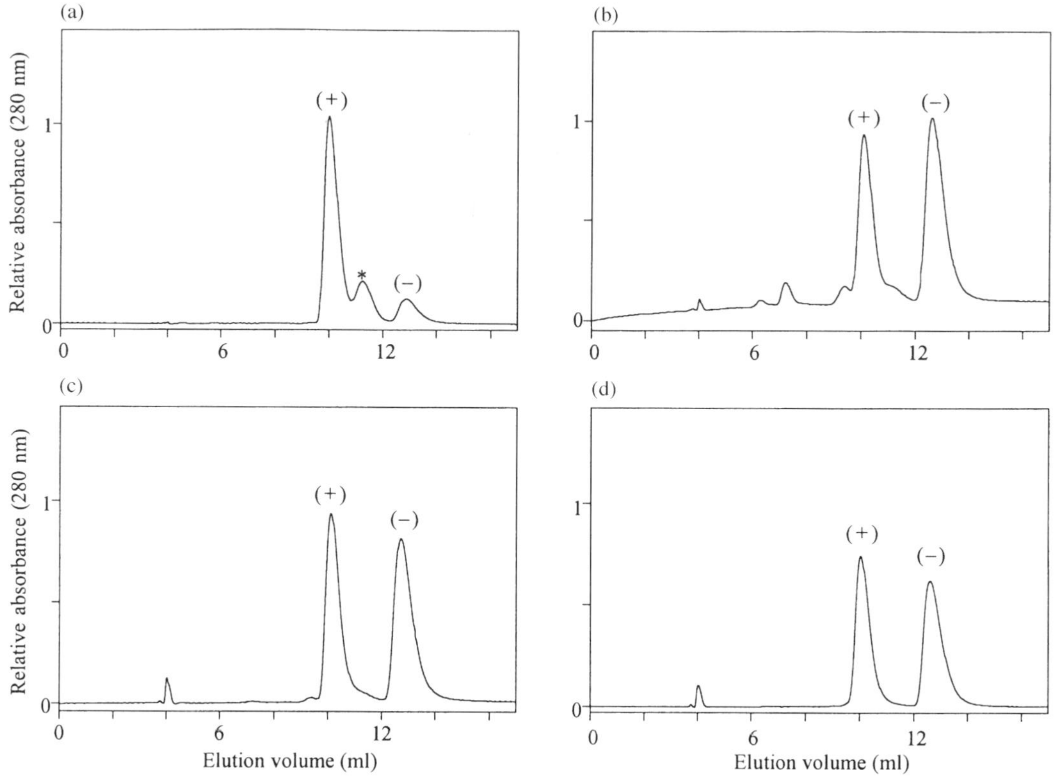

Figure 4 Chiral HPLC separation of syringaresinols (**66a,b**) isolated from different plant species. (**a**) *Eucommia ulmoides*, (**b**) *Xanthoxylum ailanthoides*, (**c**) *Daphne tangutica*, and (**d**) *Xanthoxylum inerme*.[76] For elution conditions, see Figure 3 legend (* = contaminant).

1.25.6 8—8′ STEREOSELECTIVE COUPLING: DIRIGENT PROTEINS AND *E*-CONIFERYL ALCOHOL RADICALS

1.25.6.1 Dirigent Proteins Stipulate Stereoselective Outcome of *E*-Coniferyl Alcohol Radical Coupling in Pinoresinol Formation

Since lignan optical activity cannot readily be explained by random coupling, it was essential to delineate how monolignol coupling control might be achieved *in planta*. Initial investigations used *Forsythia* species, given that it is an abundant source of the optically pure 8—8′ linked lignans, (+)-pinoresinol (**3a**) and (−)-matairesinol (**21a**): it was discovered that *Forsythia* stem residues, following removal of readily soluble proteins, were capable of preferentially stereoselectively converting *E*-coniferyl alcohol (**38**) into the 8—8′ linked (+)-pinoresinol (**3a**) in ~60% enantiomeric excess.[86] In contrast, all previous bimolecular phenoxy radical coupling processes, whether engendered biochemically[89] or chemically,[71] only afforded the racemic lignans, (±)-dehydrodiconiferyl alcohols (**5a,b**), (±)-pinoresinols (**3a,b**), and (±)-*erythro*/*threo* guaiacylglycerol 8—*O*—4′ coniferyl alcohol ethers (**6a,b**) as previously discussed.

(+)-Pinoresinol (**3a**) was ultimately demonstrated to be formed *in vitro* when two distinct proteins were added together; this resulted in conferring the requisite stereoselectivity to the coupling of two *E*-coniferyl alcohol (**38**) molecules. The first protein was of circa 78 kDa molecular size, as determined by both gel permeation chromatography (Sepharose S200, Pharmacia) and analytical ultracentrifugation. SDS-PAGE analysis, on the other hand, gave a single band of 26–27 kDa suggesting that the native protein existed as a trimer. This 78 kDa protein lacked any (oxidative) catalytic capacity by itself, and was unable to directly engender formation of (+)-pinoresinol (**3a**) from *E*-coniferyl alcohol (**38**). The second protein exhibited a typical plant laccase EPR spectrum, but did not catalyze stereoselective coupling. It did, however, oxidatively convert *E*-coniferyl alcohol (**38**) into the well-known racemic lignan products *in vitro*, i.e. (**5a,b**), (**3a,b**), and (**6a,b**) in a ratio of circa 4:2:1 (Figure 5(a)).

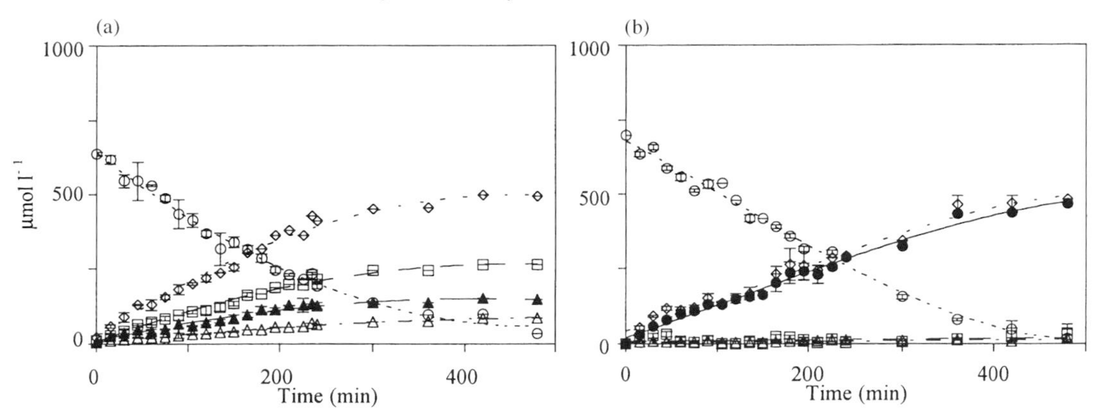

Figure 5 Time courses for *E*-coniferyl alcohol (**38**) depletion and formation of corresponding lignans during incubation in the presence of (**a**) an oxidase, and (**b**) dirigent protein and an oxidase.[87] ○, Coniferyl alcohol (calculated as dimer equivalents); □, (±)-dehydrodiconiferyl alcohols (**5a,b**); ●, (+)-pinoresinol (**3a**); ▲, (±)-pinoresinols (**3a,b**); △, (±)-*erythro*/*threo* guaiacylglycerol 8—*O*—4′ coniferyl alcohols (**6a,b**); and ◇, total of all lignans (after Lewis *et al.*[87]).

When both proteins were combined together in judicious amounts, however, essentially only conversion of *E*-coniferyl alcohol (**38**) into (+)-pinoresinol (**3a**) was observed. This is illustrated in Figure 5(b) which shows the effect of both proteins on stereoselective coupling, and the term dirigent protein[87] (Latin, *dirigere*: to guide or align) was introduced to describe the perceived unique function of the 78 kDa protein (discussed in Section 1.25.6.5). Significantly, stereoselective coupling occurred regardless of whether a one-electron oxidase (e.g., laccase) or a one-electron oxidant (such as FMN) was used in conjunction with the dirigent protein (see Figures 6(b) and 6(d)). In both cases, dirigent

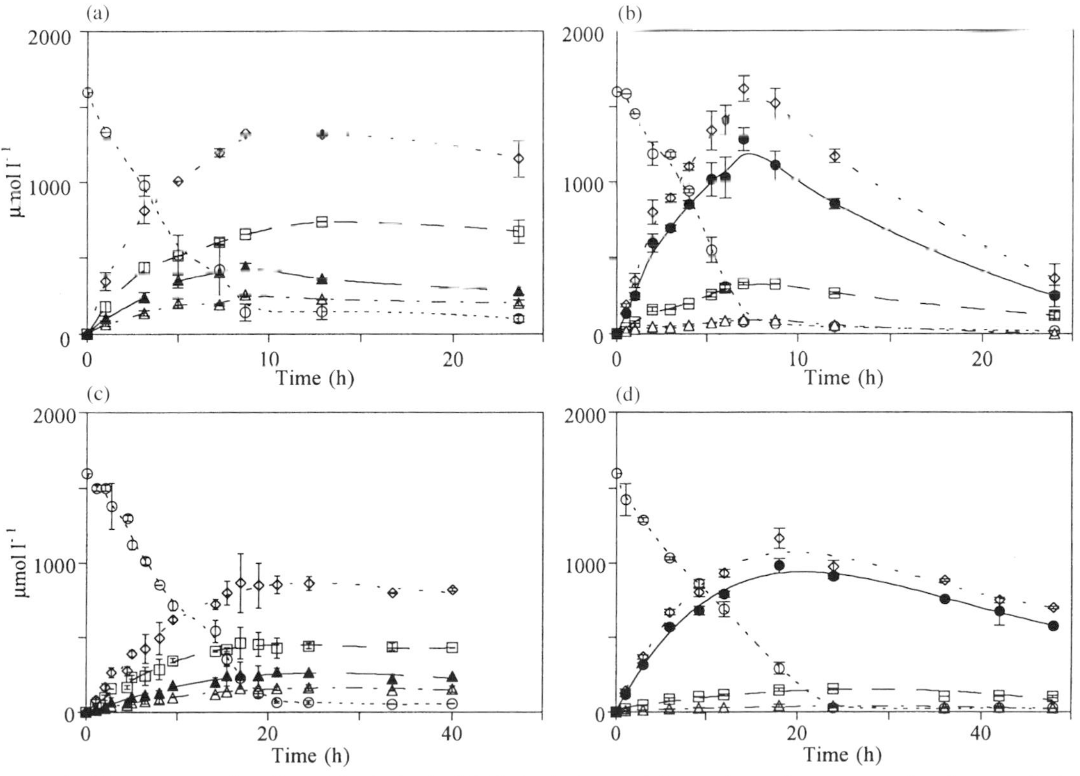

Figure 6 Time courses for *E*-coniferyl alcohol (**38**) depletion and formation of corresponding lignans during incubation in the presence of (**a**) laccase, (**b**) dirigent protein and laccase, (**c**) FMN, and (**d**) FMN and dirigent protein. See Figure 5 for key (after Lewis *et al.*[87]).

protein addition had little, if any, effect on the rate of *E*-coniferyl alcohol (**38**) depletion (see Figures 6(a) to (d)). Moreover, stereoselectivity was only observed with *E*-coniferyl alcohol (**38**) as a substrate, but not when either *E*-*p*-coumaryl (**32**) or *E*-sinapyl (**44**) alcohol was used.

1.25.6.2 Cloning of the Gene Encoding the Dirigent Protein and Recombinant Protein Expression in Heterologous Systems

The cDNA of two *Forsythia* dirigent protein genes (psd-Fi1 and psd-Fi2; Figures 7(a) and 7(b)) were obtained by a polymerase chain reaction guided strategy, and sequence analyses revealed the presence of secretory signal peptides, potential *N*-glycosylation and serine/tyrosine phosphorylation sites.[74,90] Each cDNA encoded a protein subunit of only circa 18 kDa, after taking into account the secretory signaling sequences, indicating that the presumed 26 kDa subunit was extensively glycosylated.

The authenticity of the dirigent protein clone(s) was proven by obtaining recombinant dirigent proteins: these were heterologously expressed in both *Spodoptera frugiperda* (fall army worm) Sf9 cells[74] using a baculovirus based expression system,[91] and in *Drosophila melanogaster* cells cotransfected with the DES (drosophila expression system) vector (Invitrogen);[92] both expression systems were chosen over *Escherichia coli* since they carry out glycosylation/post-translational modifications. As illustrated in Figure 8, the purified recombinant proteins were functionally capable of conferring the requisite stereoselectivity to [9-^{3}H]coniferyl alcohol (**38**) coupling, provided that single-electron oxidative capacity (e.g., laccase) was supplied, thereby establishing the gene sequence(s) to be correct.[74,90] These findings were also confirmed using deuterated [9-$^{2}H_2$, OC$^{2}H_3$]coniferyl alcohol (**38**) as substrate.

1.25.6.3 Sequence Homology Comparisons

The *Forsythia* dirigent protein sequence has no significant level of homology to any other protein of known function when using the BLAST/BLAST-Beauty database search tool.[93,94] It did, however, show significant identity (~64%) with a "pea disease resistance response gene 206-d" of unknown function, which is induced in conjunction with isoflavone reductase.[95] Low sequence homology levels were also noted with portions of three other plant genes, of no known function, i.e., *Arabidopsis thaliana*, wheat (*Triticum aestivum*), and barley (*Hordeum vulgare*). The *Arabidopsis* gene possessing limited sequence homology was the "hypothetical protein" random BAC clone (40% similarity over 82% of the sequence, Genbank accession number SF000657), whereas that from wheat (47% similarity over 56% of the sequence, Genbank accession number U32427) is associated with systemic acquired resistance and is induced by benzothiadiazole.[96] Finally, the barley gene encoded a "putative 32.7 kDa jasmonate-induced" gene (Genbank accession number U43497), with some 50% similarity over 35% of the sequence of the dirigent protein. No meaningful comparisons could be made with other gene sequences outside of the plant kingdom, and it can, therefore, be concluded that the dirigent protein genes are unique. Current information does not, however, give any incisive insight into how these gene(s) may have evolved, or as to what its progenitor function may have been.

1.25.6.4 Comparable Systems

Since various 8—8′ linked lignans are present in many plant species, isolation and characterization of dirigent protein analogues and homologues was of interest. Accordingly, dirigent protein involve-

---

Figure 7 Complete sequence of *Forsythia intermedia* dirigent protein cDNA (**a**) psd-Fi1, and (**b**) psd-Fi2. The signal peptide cleavage sites are indicated by an arrow (before Arg-25 and His-25 in psd-Fi1 and psd-Fi2, respectively, in the native proteins, and before Thr-22 in the recombinant psd-Fi1 proteins). Potential *N*-glycosylation sites (Asn-52, Asn-65, Asn-122, and Asn-140 in psd-Fi1; Asn-51, Asn-64, Asn-121, and Asn-139 in psd-Fi2) and serine (Ser-123 in psd-Fi1, Ser-28 in psd-Fi2) and tyrosine phosphorylation (Tyr-183 in psd-Fi1, Tyr-182 in psd-Fi2) sites are indicated by underlining. The stop codons are indicated by an asterisk (after Lewis *et al.*[74]).

(a)

```
     ATTTCGGCACGAGATTAAACCAAACATGGTTTCTAAAACACAAATTGTAGCTCTTTTCCT
                                 M  V  S  K  T  Q  I  V  A  L  F  L

 36  TTGCTTCCTCACTTCCACCTCTTCCGCCACCTACGGCCGCAAGCCACGCCCTCGCCGGCC
 13   C  F  L  T  S  T  S  S  A  T  Y  G  R  K  P  R  P  R  R  P
                                ⇑        ⇑
 96  CTGCAAAGAATTGGTGTTCTATTTCCACGACGTACTTTTCAAAGGAAATAATTACCACAA
 33   C  K  E  L  V  F  Y  F  H  D  V  L  F  K  G  N  N  Y  H  N

156  TGCCACTTCCGCCATAGTCGGGTCCCCCCAATGGGGCAACAAGACTGCCATGGCCGTGCC
 53   A  T  S  A  I  V  G  S  P  Q  W  G  N  K  T  A  M  A  V  P

216  ATTCAATTATGGTGACCTAGTTGTGTTCGACGATCCCATTACCTTAGACAACAATCTGCA
 73   F  N  Y  G  D  L  V  V  F  D  D  P  I  T  L  D  N  N  L  H

276  TTCACCCCCAGTGGGTCGGGCGCAAGGGATGTACTTCTATGATCAAAAAAATACATACAA
 93   S  P  P  V  G  R  A  Q  G  M  Y  F  Y  D  Q  K  N  T  Y  N

336  TGCTTGGCTAGGGTTCTCATTTTTGTTCAATTCAACTAAGTATGTTGGAACCTTGAACTT
113   A  W  L  G  F  S  F  L  F  N  S  T  K  Y  V  G  T  L  N  F

396  TGCTGGGGCTGATCCATTGTTGAACAAGACTAGAGACATATCAGTCATTGGTGGAACTGG
133   A  G  A  D  P  L  L  N  K  T  R  D  I  S  V  I  G  G  T  G

456  TGACTTTTTCATGGCGAGAGGGGTTGCCACTTTGATGACCGATGCCTTTGAAGGGGATGT
153   D  F  F  M  A  R  G  V  A  T  L  M  T  D  A  F  E  G  D  V

516  GTATTTCCGCCTTCGTGTCGATATTAATTTGTATGAATGTTGGTAAACAATTTAGCCGTA
173   Y  F  R  L  R  V  D  I  N  L  Y  E  C  W  *

576  TATATATATATATATGGCTATACATATTTCATAGAATCCAGATTTGCTGTTTCAAATGTG
636  TGTTTCTTTAGTTGTGCCACCAATAAAAAAATGTACACATTATTTAATAAATATAATTAT
696  TTAATGTGTTCATTTTTGAAGTTAAATTTAAGTTGTATTTATTTGATTATGTATAAATTC
756  TCTATTAGTAAAATAGTCAAAGTGACACATATTCAAGACGACATATGTAACTTTATTTCA
816  TATCTTCAACAAGTTCAATAATGTCATATATATTGTACTATTGAAAAAAAAAAAAAAAAA
```

(b)

```
     AATTCGGCACGAGGAAAAATGGCAGCTAAAACACAAACCACAGCCCTTTTCCTCTGCCTC
                        M  A  A  K  T  Q  T  T  A  L  F  L  C  L

 43  CTCATCTGCATCTCCGCCCTCTACCCCCACAAAACCAGGTCTCGACGCCCCTGTAAAGAG
 15  L  I  C  I  S  A  V  Y  G  H  K  T  R  S  R  R  P  C  K  E
                                ⇑
103  CTCGTTTTCTTCTTCCACGACATCCTCTACCTAGGATACAATAGAAACAATGCCACCGCT
 35  L  V  F  F  F  H  D  I  L  Y  L  G  Y  N  R  N  N  A  T  A

163  GTCATAGTAGCCTCTCCTCAATGGGGAAACAAGACTGCCATGGCTAAACCTTTCAATTTT
 55  V  I  V  A  S  P  Q  W  G  N  K  T  A  M  A  K  P  F  N  F

223  GGTGATTTGGTTGTGTTTGATGATCCCATTACCTTAGACAACAACCTGCATTCTCCTCCG
 75  G  D  L  V  V  F  D  D  P  I  T  L  D  N  N  L  H  S  P  P

283  GTCGGCCGGGCTCAGGGAACTTATTTCTACGATCAATGGAGTATTTATGGTGCATGGCTT
 95  V  G  R  A  Q  G  T  Y  F  Y  D  Q  W  S  I  Y  G  A  W  L

343  GGATTTTCATTTTTGTTCAATTCTACTGATTATGTTGGAACTCTAAATTTTGCTGGAGCT
115  G  F  S  F  L  F  N  S  T  D  Y  V  G  T  L  N  F  A  G  A

403  GATCCATTGATTAACAAAACTAGGGACATTTCAGTAATTGGAGGAACTGGTGATTTTTTC
135  D  P  L  I  N  K  T  R  D  I  S  V  I  G  G  T  G  D  F  F

463  ATGGCTAGAGGGGTAGCCACTGTGTCGACCGATGCTTTTGAAGGGGATGTTTATTTCAGG
155  M  A  R  G  V  A  T  V  S  T  D  A  F  E  G  D  V  Y  F  R

523  CTTCGTGTTGATATTAGGTTGTATGAGTGTTGGTAAATTTACCTTATTTTTCCATTTTCT
175  L  R  V  D  I  R  L  Y  E  C  W  *

583  TGAGTTTGACTCGGATTTGACTAATAATGTCTTCTGTAATCCTTGTTTTTGATCAATTTG
643  TGGCGATTTTATCAATTAGTGATTGTTTGGTTCATATTTTAATCTGTTAAAAAAAATTGT
703  GGTCAAAAGCCAATAACCACAACCGTAGGGAGTTTTTTCCGTTAAGGGGAAAAAAAAGTA
763  TGTCCATGTGTTACTACGTTTTCAATTTCATTCAAAATTTGCTTTTCAATCATCTTCTTC
823  AAAAAAAAAAAAAAAAAA
```

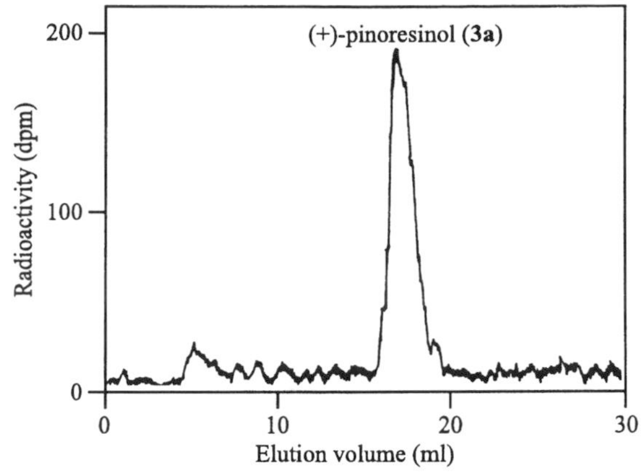

Figure 8 Chiral HPLC analysis of [9,9′-^{3}H]-pinoresinol (**3**) formed by incubation of recombinant dirigent protein with [9-^{3}H]coniferyl alcohol (**38**) in the presence of laccase as oxidant.[74] For elution conditions and column employed see Figure 3 legend.

ment in (+)-pinoresinol formation was detected in sesame (*Sesamum indicum*), whereas in flax (*Linum usitatissimum*) the corresponding dirigent protein stipulated formation of the (−)-antipode[97] i.e., the proteins from different plant sources can have different stereoselectivities to give (+)- and (−)-pinoresinols (**3a**) and (**3b**), respectively.

The *Forsythia* dirigent protein cDNA, psd-Fi1, was used to probe the cDNA libraries of two gymnosperms, western red cedar (*T. plicata*) and western hemlock (*Tsuga heterophylla*), which accumulate 8—8′ linked lignans, plicatic acid (**75**) and (α)-conidendrin (**76**),[30] respectively, in their heartwood. Two dirigent protein-like cDNAs were isolated from *T. plicata* (psd-Th1 and psd-Th2) and eight from *T. heterophylla* (psd-Tp1 to psd-Tp8).[74,90] Using a PCR-guided approach on DNA extracted from different plant species, dirigent protein genes were also identified in Manchurian ash (*Fraxinus mandschurica*, Salicaceae: psd-Fm1 and psd-Fm2) and aspen (*Populus tremuloides*, Oleaceae: psd-Pop1 and psd-Pop2). These have a very high degree of similarity/identity[74] (from 91.9% and 85.4% for psd-Fm1 and psd-Fm2 versus psd-Fi1 to 66.1% and 61.2% for psd-Tp1 versus psd-Fi1), as shown in Table 2.

Table 2 Similarity/identity between the different dirigent protein-like clones isolated and the dirigent protein clone psd-Fi1 from *Forsythia intermedia*.

		With signal peptide	
Plant species	*cDNA*	*Similarity (%)*	*Identity (%)*
Forsythia intermedia	psd-Fi2	87.6	81.6
Thuja plicata	psd-Tp1	66.1	61.2
	psd-Tp2	77.4	61.8
	psd-Tp3	78.5	61.8
	psd-Tp4	83.3	67.4
	psd-Tp5	78.4	60.0
	psd-Tp6	76.3	57.5
	psd-Tp7	79.0	59.1
	psd-Tp8	79.0	59.1
Tsuga heterophylla	psd-Th1	69.9	60.8
	psd-Th2	68.5	61.4
Fraxinus mandschurica	psd-Fm1	91.9	85.4
	psd-Fm2	91.9	85.4
Populus tremuloides	psd-Pop1	77.2	66.7
	psd-Pop2	77.2	66.7

The high similarity/identity between different dirigent protein clones isolated and detected indicates that the dirigent proteins may be ubiquitous throughout the plant kingdom, although it

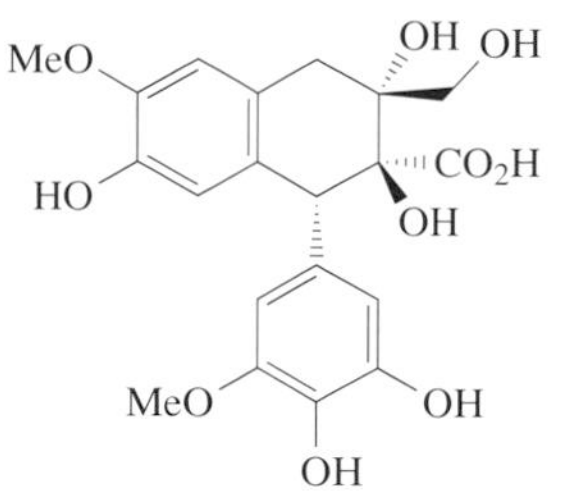

(75) Plicatic acid

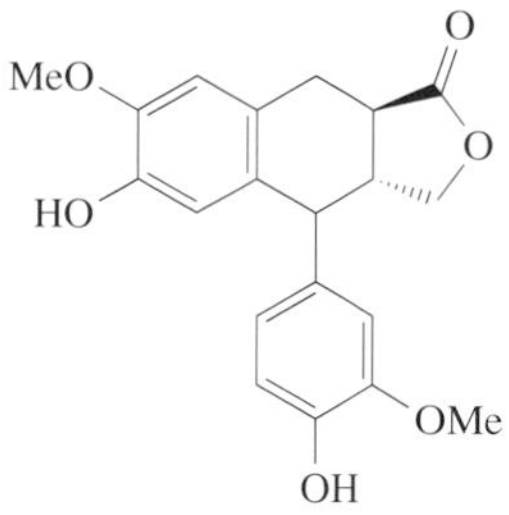

(76) (–)-α-Conidendrin

remains to be established whether they stipulate only 8—8′ coupling or other modes (e.g., 8—1′, etc.) as well (see Sections 1.25.8 and 1.25.9).

1.25.6.5 Perceived Biochemical Mechanism of Action

Free-radical coupling reactions, as catalyzed by nonspecific (per)oxidases, cannot control either the regio- or stereochemistries of product formation, when more than one potential coupling site is on the substrate molecule. On the other hand, nature extensively utilizes free-radical coupling processes, with circa 30–40% of all organic carbon being linked together in this way, e.g., lignans, lignins, suberins, melanins, insect cuticles, etc. Given the extensive deployment of such coupling reactions *in vivo*, it could therefore be anticipated that nature had some means to control or stipulate the outcome of phenoxy free-radical coupling.

The discovery of the *Forsythia* dirigent protein, stipulating both the regio- and stereochemical fate of *E*-coniferyl alcohol (**38**) coupling during (+)-pinoresinol (**3a**) formation, thus gave a new perspective as to how control of free-radical reactions occurred *in vivo*; this was further supported by analysis of its unique 18 kDa gene sequence which revealed no counterpart elsewhere. It is believed that the dirigent protein effectuating (+)-pinoresinol (**3a**) formation requires glycosylation of the 18 kDa subunit to give the 26 kDa glycoprotein which then forms the functional ~78 kDa dirigent protein trimer.

There are three distinct biochemical mechanisms[87] that can be envisaged as operative, with each discussed in terms of relative likelihood based on data interpretation. The first is generation of free-radicals from *E*-coniferyl alcohol (**38**), by action of nonspecific oxidase(s)/oxidant(s), with the free radicals then binding to the dirigent protein prior to stereoselective coupling. In this case, the free-radical species are orientated on the 78 kDa protein prior to coupling, hence the term dirigent (Latin, *dirigere*: to guide or align). Alternative possibilities are that *E*-coniferyl alcohol (**38**) molecules are themselves bound to the dirigent protein, and appropriately orientated to give (+)-pinoresinol (**3a**) following one-electron oxidation. It is anticipated that this could only occur when either the substrate phenolic groups were exposed so that they could readily be oxidized by an oxidase or oxidant, or via an electron transfer mechanism between the oxidase/oxidant and an electron acceptor site or sites on the dirigent protein.

Lines of evidence, however, suggest "capture" of phenoxy radical intermediates by the dirigent protein. This is because both the rates of substrate depletion and product formation were largely unaffected by the presence of the dirigent protein. This is consistent with a free-radical capture mechanism which would only affect coupling specificity when single-electron oxidation of coniferyl alcohol (**38**) is rate-determining. The electron transfer mechanism is ruled out, on the basis that no new ultraviolet-visible chromophores were observed in either the presence or absence of an auxiliary oxidase or oxidant, under oxidizing conditions. Preliminary kinetic data were also in agreement with free-radical capture based on the formal Michaelis constant (K_m) and maximum velocity (V_{max}) values characterizing the conversion of *E*-coniferyl alcohol (**38**) into (+)-pinoresinol (**3a**), with the dirigent protein alone and in the presence of the various oxidases or oxidants.[87] With a free-radical capture process, the Michaelis–Menten parameters obtained will only represent formal rather than true values, given that the highest free-energy intermediate state during the conversion of *E*-coniferyl alcohol (**38**) into (+)-pinoresinol (**3a**) is unknown and the relation between the concentration of substrate and that of the corresponding intermediate free-radical in open solution has not been delineated.

With these qualifications in mind, formal K_m and V_{max} values for the dirigent protein preparation were estimated in the presence and absence of the oxidase, laccase and the oxidant, FMN. The preliminary kinetic parameters so obtained were in harmony with the finding that the dirigent protein does not substantially affect the rate of *E*-coniferyl alcohol (**38**) depletion in the presence of the oxidase/oxidant, and are thus in accord with the working hypothesis that the dirigent protein functions by capturing free-radical intermediates which then undergo stereoselective coupling. Accordingly, it is of considerable importance to establish the site(s) on the dirigent protein that involve substrate binding, together with the number of monomeric forms binding to the 18 kDa subunit and to the preformed 78 kDa protein.

1.25.7 PINORESINOL METABOLISM AND ASSOCIATED METABOLIC PROCESSES

The 8—8′ linked lignan, pinoresinol (**3**), is a central intermediate in lignan metabolism, which, depending upon the plant species, can be converted into a rather large number of natural products of important plant physiological and pharmacological functions (see Sections 1.25.13 and 1.25.14). Delineation of the various biochemical pathways associated with its metabolism has been carried out using sesame (*S. indicum*), *Magnolia kobus*, *Forsythia* species, flax (*L. usitatissimum*), western red cedar (*T. plicata*), western hemlock (*T. heterophylla*), *Linum flavum*, and *Podophyllum* species. Together, these plants contain various furanofuran, furano, dibenzylbutane, dibenzylbutyrolactone, and aryltetrahydronaphthalene lignans, and, therefore, provide the opportunity to study the formation of distinct 8—8′ linked lignan skeletal forms.

1.25.7.1 *Sesamum indicum*: (+)-Piperitol, (+)-Sesamin, and (+)-Sesamolinol Synthases

Sesame (*S. indicum*) seeds are rich in furanofuran lignans, of which (+)-sesamin (**18**) and the unusual oxygenated derivative, (+)-sesamolin (**77**) are the most abundant (Figure 9). They differ from (+)-pinoresinol (**3a**) by methylenedioxy bridge formation and, in the latter case, by an oxygen insertion between the furanofuran group and the aryl moiety. Their physiological roles appear to be as protective antioxidants, thereby helping stabilize sesame seed oil from the rapid onset of rancidity.[4,98,99]

Sesame seed pod development is a gradual maturation process, being initiated some 14 days or so after flowering in the oldest tissues nearest the stem base. New seed pods are then continually formed over a six-week period, and Figure 10 shows a maturing *S. indicum* plant with its seed pods at different maturation stages. Plants at this stage were used for tracer experiments, employing both radio- and stable isotopically labeled lignan precursors, where it was found that sesamin (**18**) and sesamolin (**77**) were both formed *de novo* in maturing seed.[100,101] The possible biochemical routes to both lignans are shown in Figure 9.

When racemic (±)-[3,3′-O^{14}CH$_3$]pinoresinols (**3a,b**) were administered to intact sesame seeds taken from pods from eight-week-old plants, it was found that only the (+)-antipode (**3a**) was metabolized into (+)-piperitol (**78**), (+)-sesamin (**18**), and (+)-sesamolin (**77**); the corresponding (−)-enantiomer (**3b**) was not utilized.[101] It was also found that the relative efficacy of metabolism of (+)-[3,3′-O^{14}CH$_3$] pinoresinol (**3a**) into each of these lignans varied with the stage of maturation of the developing seed pod. Incorporation into sesamolin (**77**) was highest at the early stages of seed pod maturation, whereas conversion into sesamin (**18**) and piperitol (**78**) was greatest at the later stages.

Confirmation of these radiochemical observations was attained by administering (±)-[9,9′-^{2}H$_2$, 3,3′-OC^{2}H$_3$]pinoresinols (**3a,b**) to sesame seed. The enzymatically synthesized (+)-[9,9′-^{2}H$_2$, 3-OC^{2}H$_2$, 3′-OC^{2}H$_3$]piperitol (**78**) so obtained displayed a molecular ion [M$^+$ +9] at *m/z* 365, corresponding to the presence of nine deuterium atoms.[101] This established intact incorporation of the precursor during methylenedioxy bridge formation, in a conversion catalyzed by (+)-piperitol synthase.

Microsomal preparations from this seed pod maturation stage were examined as a potential source of (+)-piperitol synthase. These were incubated with (±)-[3,3′-O^{14}CH$_3$]pinoresinols (**3a,b**) but only gave radiolabeled [3-O^{14}CH$_2$, 3′-O^{14}CH$_3$]piperitol (**78**) when NADPH (1 mM) was added.[101] The conversion was completely enantiospecific, since only the radiolabeled (+)-antipode was formed (Figure 11). Confirmation of the radiochemical findings was achieved using isotopically labeled

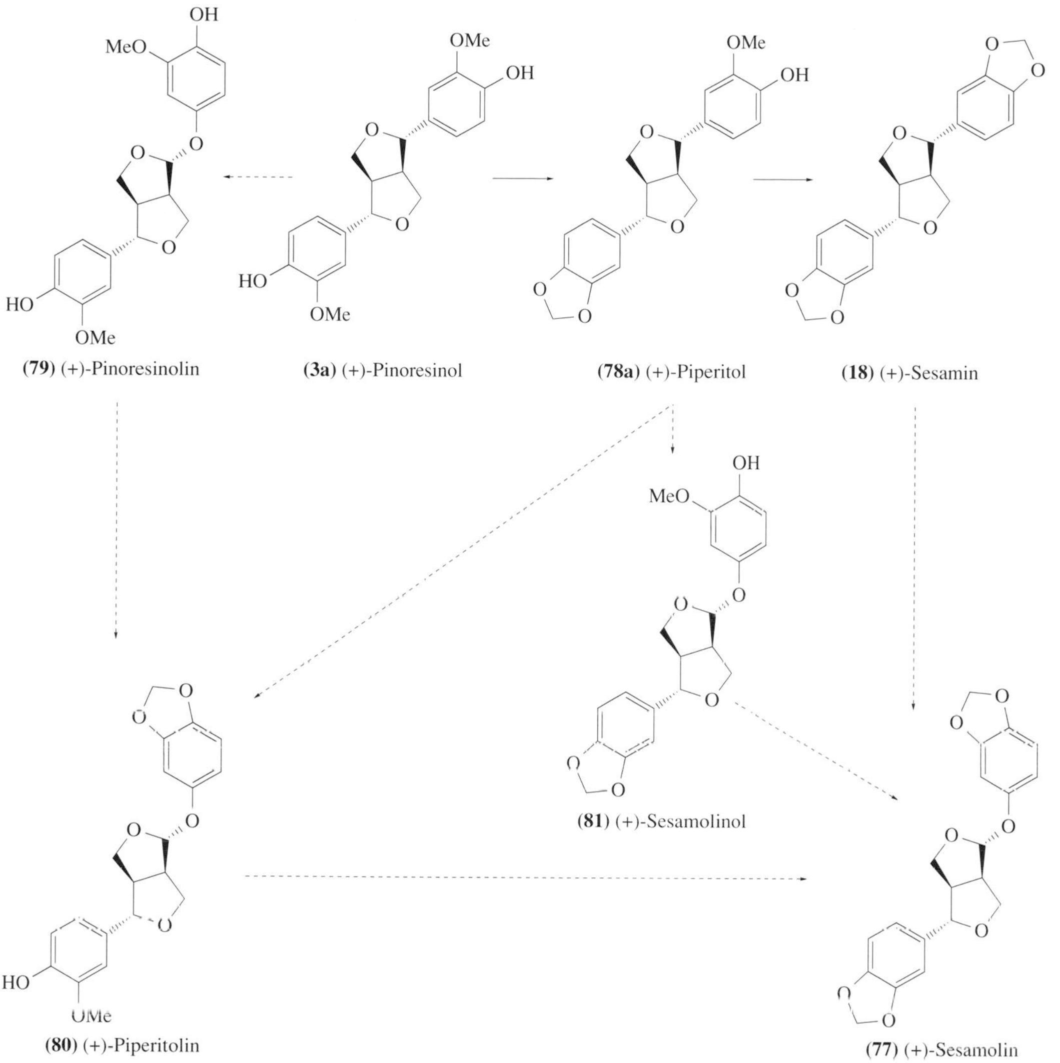

(79) (+)-Pinoresinolin (3a) (+)-Pinoresinol (78a) (+)-Piperitol (18) (+)-Sesamin

(81) (+)-Sesamolinol

(80) (+)-Piperitolin (77) (+)-Sesamolin

Figure 9 Possible biosynthetic pathways to (+)-sesamolin (**77**) from (+)-pinoresinol (**3a**) in *S. indicum*. (**78b** would be the opposite enantiomer.)

(±)-[9,9′-$^2H_2$, 3,3′-OC^2H_3]pinoresinols (**3a,b**) as substrates, where the enzymatically generated (+)-piperitol (**78**) displayed the expected molecular ion [M^+ +9] at *m/z* 365 as before.[101]

(+)-Piperitol synthase is an O_2-requiring, NADPH-dependent, cytochrome P450 enzyme, with temperature and pH maxima of 40 °C and 7.5, respectively. Cytochrome P450 inhibitors, clotrimeizol (300 μM), miconazol (300 μM), and cytochrome c (170 μM), completely inhibited the enzyme, with tropolone and metyrapone being less efficient (Table 3). It is also strongly inhibited by carbon monoxide (90% inhibition in a CO/O_2 (9:1) atmosphere).[101] These results are consistent with previous studies of benzophenanthridine alkaloid[102,103] and isoflavonoid[104] biosynthetic pathways, where methylenedioxy bridge generation during the formation of such natural products was catalyzed by NADPH-dependent, cytochrome P450 monooxygenase(s).

Interestingly, the *S. indicum* microsomal preparation only effectively converted (+)-pinoresinol (**3a**) into (+)-piperitol (**78**), but not into (+)-sesamin (**18**). Incubation with (+)-piperitol (**78**) did, however, give (+)-sesamin (**18**).[105] Accordingly, there may be two distinct cytochrome P450s involved in (+)-piperitol (**78**) and (+)-sesamin (**18**) formation, respectively. The enzymology associated with oxygen insertion between the aryl group and furanofuran to give sesamolin (**77**) also awaits delineation. It is perhaps significant, however, that ketone (**82**) was isolated from *S. indicum* seeds,[106] and studies will establish if it is a pathway intermediate to (+)-sesamolin (**77**) via rearrangement.

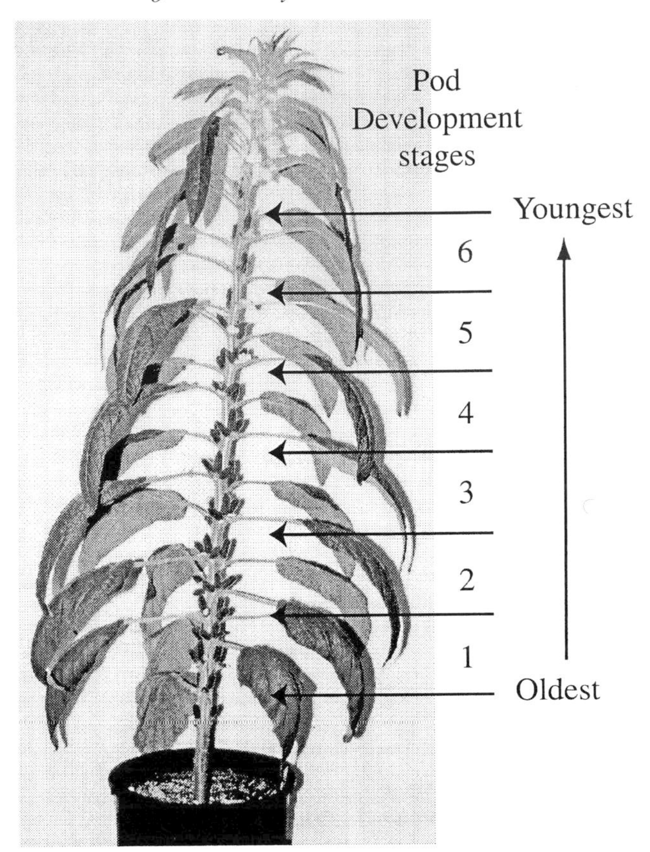

Figure 10 Sesame (*S. indicum*) plant showing pods at different maturation stages.

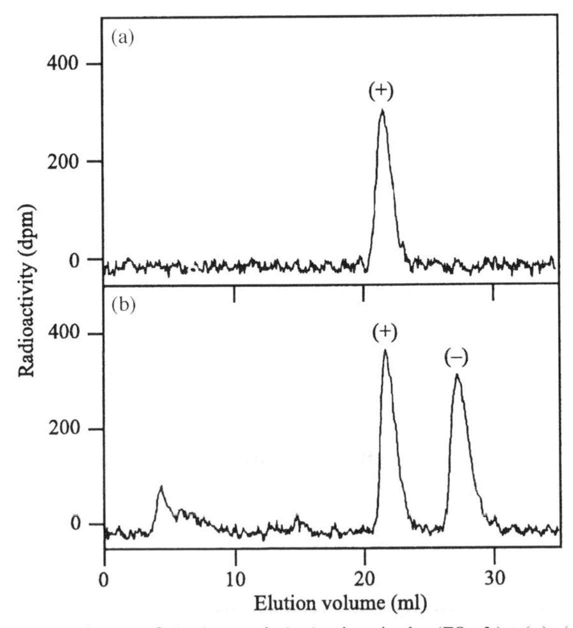

Figure 11 Chiral HPLC separations of (+)- and (−)-piperitols (**78a,b**): (a) (+)-[3-$O^{14}CH_2$, 3′-$O^{14}CH_3$]-Piperitol (**78a**) obtained after incubation of (±)-[3,3′-$O^{14}CH_3$]pinoresinols (**3a,b**) with a *S. indicum* microsomal preparation and (b) (+)- and (−)-antipodes of synthetic piperitols (**78a**) and (**78b**) (after Jiao *et al.*[101]).

Table 3 Effect of various cytochrome P450 enzyme inhibitors on the activity of microsomal bound (+)-piperitol synthase from *S. indicum*. (Standard assay conditions were employed, where 100% activity = 1.46 pkat for (+)-piperitol synthase.)[101]

Cytochrome P450 inhibitor	*Inhibitor concentration* (μM)	*Inhibition* (%)
Clotrimeizol	300	100.0
	35	100.0
Miconazol	300	100.0
	35	71.2
Cytochrome c	170	100.0
	17	52.4
Tropolone	300	65.5
	35	54.4
Metyrapone	300	46.4
	35	26.8

(82) Ketone

1.25.7.2 *Magnolia kobus*: Pinoresinol and Pinoresinol Monomethyl Ether *O*-Methyltransferase(s)

Magnolia kobus var. *borealis* is a member of the Magnoliaceae, which accumulates furanofuran and tetrahydrofuran lignans in its leaves, e.g., (+)-eudesmin (**83**), (+)-magnolin (**84**), (+)-yangambin (**85**), (+)-kobusin (**86**), and (+)-kobusinols A (**87**) and B (**88**).[107] Its lignans primarily differ from pinoresinol (**3**) by the degree of methoxylation and/or methylenedioxy bridge formation on the aromatic rings, as well as reductive modifications (discussed in Section 1.25.7.3.1). Thus, the methylation patterns (i.e., at C-4/C-4′) for these lignans have a very different regiospecificity to that observed for monolignol formation, which only occurs at the *meta* (C-3) position (see Figure 1 for comparison). Miyauchi and Ozawa have carried out studies to begin to characterize the *M. kobus* lignan *O*-methyltransferases using crude cell-free extracts from leaves (Scheme 1).[108] These were individually incubated with both racemic (±)-pinoresinols (**3a,b**) and (±)-pinoresinol monomethyl ethers (**89**), in the presence of *S*-adenosyl-L-[methyl-^{14}C]methionine, in order to ascertain whether the *O*-methyltransferases were enantiospecific or not. With (±)-pinoresinols (**3a,b**) as substrates, both (+)- and (−)-[4-O$^{14}CH_3$]pinoresinol monomethyl ethers (**89**) were formed, where the (+)-enantiomer predominated (3:1) over the (−)-form; dimethylation affording (+)-eudesmin (**83**), however, was not observed. On the other hand, with (±)-pinoresinol monomethyl ethers (**89**) as substrates, both (+)- and (−)-eudesmins (**83**) were obtained with formation of the (+)-enantiomer again being favored (~2:1).

Thus, the crude *O*-methyltransferase preparation catalyzing formation of eudesmin (**83**) *in vitro* from pinoresinol (**3**) is either not stereospecific, or more likely contains various *O*-methyltransferases which differ in their specificities towards each enantiomeric form. These possibilities will only be distinguished when the purified *O*-methyltransferases are obtained. As discussed below (in Section 1.25.7.3.3), this lack of being fully enantiospecific is consistent with previous observations made, using *Forsythia* crude cell-free preparations, in the study of *O*-methylation of matairesinol (**21**) to give arctigenin (**22**).[109]

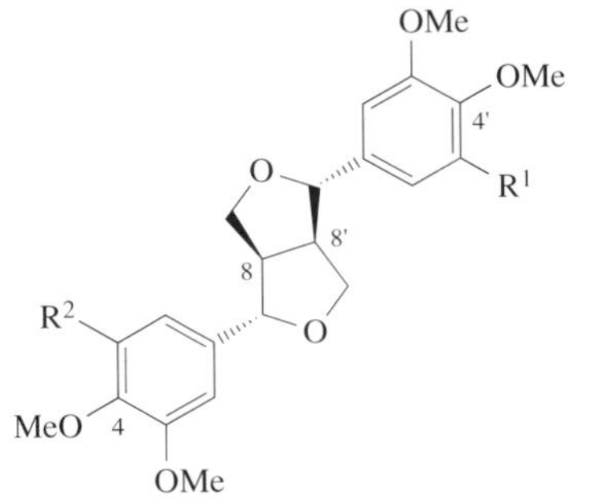

(**83**) R^1 = R^2 = H, Eudesmin, $[\alpha]_D^{25}$= +64.0° (MeOH, c = 0.01)
(**84**) R^1 = H, R^2 = OMe, Magnolin, $[\alpha]_D^{25}$= +55.9° (MeOH, c = 0.35)
(**85**) R^1 = R^2 = OMe, Yangambin, $[\alpha]_D^{25}$= +63.49° (MeOH, c = 0.03)

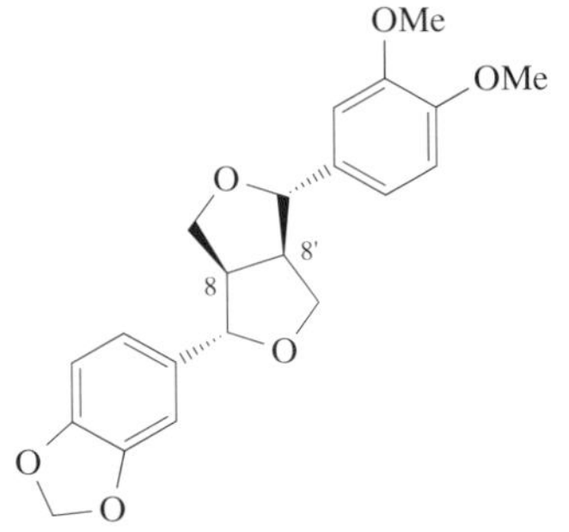

(**86**) Kobusin, $[\alpha]_D^{25}$= +48.0° (MeOH, c = 0.31)

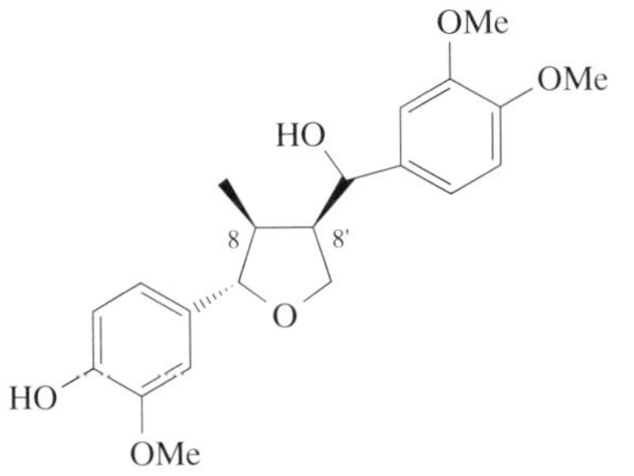

(**87**) Kobusinol A, $[\alpha]_D^{25}$= +98.2° (MeOH, c = 0.28)

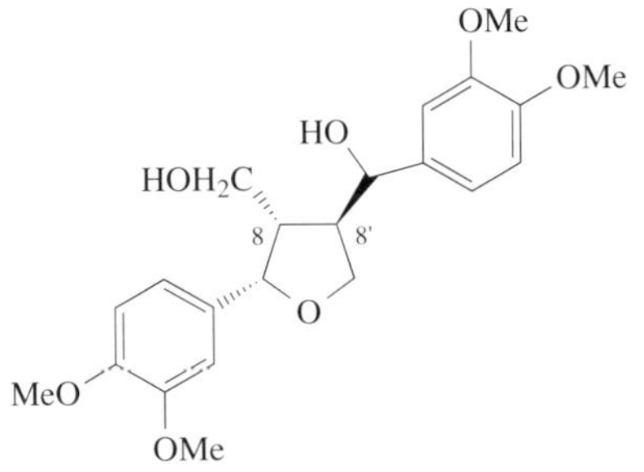

(**88**) Kobusinol B, $[\alpha]_D^{25}$= +16.4° (MeOH, c = 0.22)

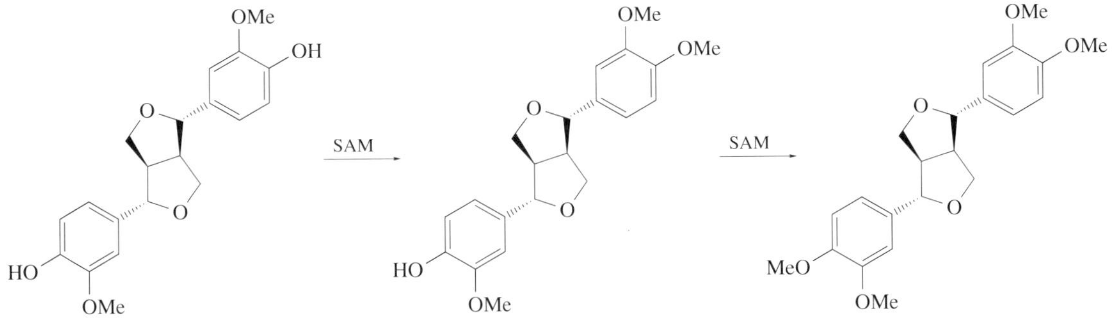

(**3**) Pinoresinol — SAM → (**89**) Pinoresinol monomethyl ether — SAM → (**83**) Eudesmin

Scheme 1

1.25.7.3 *Forsythia intermedia* and *Forsythia suspensa*

Forsythia species accumulate various 8—8′ linked lignans in differing amounts: (+)-pinoresinol (**3a**), (+)-phillygenin (**90**), and (+)-phyllirin (**91**) are present in *F. suspensa*, whereas *F. viridissima* accumulates (−)-matairesinol (**21a**), (−)-arctigenin (**22a**), and (−)-arctiin (**92**). The hybrid, *F. intermedia* (*F. suspensa* × *F. viridissima*), on the other hand, contains all of these lignans.[45]

The lignan biosynthetic pathway defined in the *Forsythia* plant family is summarized in Scheme 2, being initiated with the dirigent protein mediated stereoselective coupling of *E*-coniferyl alcohol (**38**) to give (+)-pinoresinol (**3a**) (Section 1.25.6). This is sequentially reduced to afford (+)-lariciresinol (**19a**) and (−)-secoisolariciresinol (**20a**), respectively, with stereospecific dehydrogenation of the latter giving (−)-matairesinol (**21a**), which is subsequently converted into (−)-arctigenin (**22a**) and (−)-arctiin (**92**). The enzymology associated with (+)-pinoresinol (**3a**) metabolism in *F. intermedia* is described below.

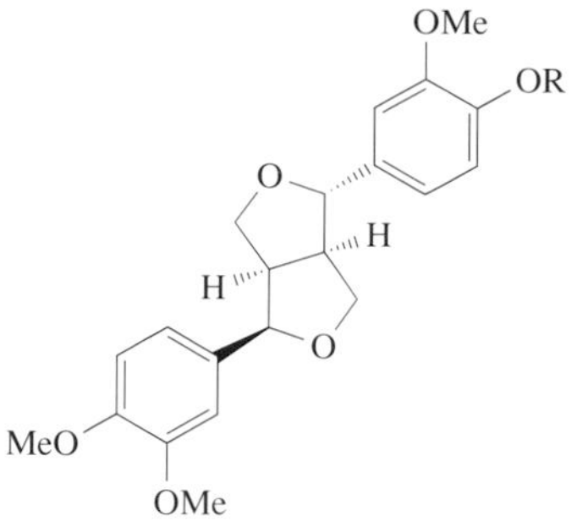

(**90**) R = H, (+)-Phillygenin
(**91**) R = Glc, (+)-Phillyrin

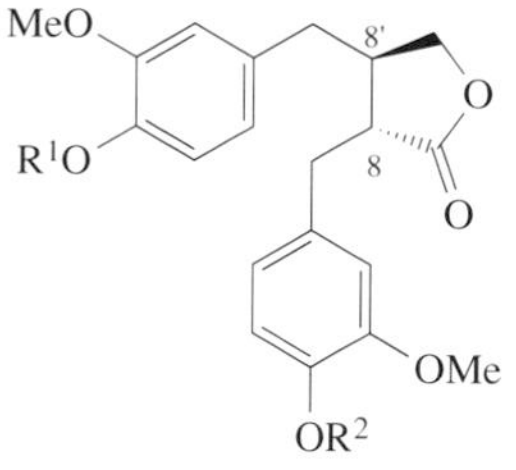

(**21a**) $R^1 = R^2 = H$, (–)-Matairesinol
(**22a**) $R^1 = Me$, $R^2 = H$, (–)-Arctigenin
(**92**) $R^1 = Me$, $R^2 = Glc$, (–)-Arctiin

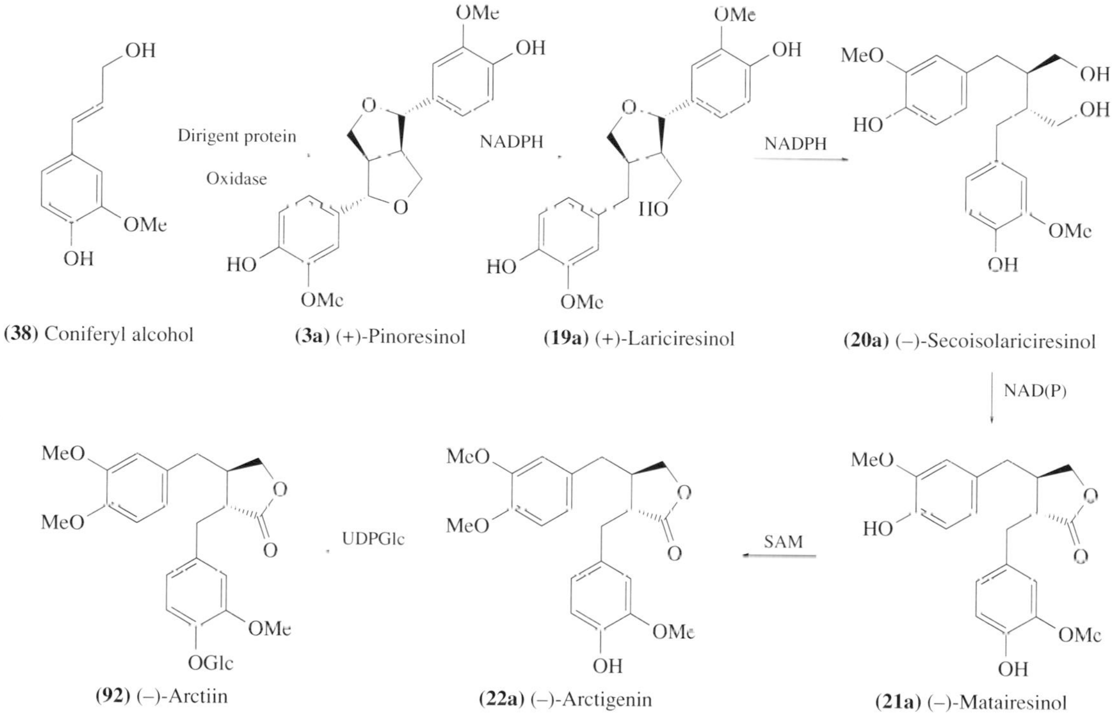

(**38**) Coniferyl alcohol — Dirigent protein, Oxidase → (**3a**) (+)-Pinoresinol — NADPH → (**19a**) (+)-Lariciresinol — NADPH → (**20a**) (–)-Secoisolariciresinol — NAD(P) → (**21a**) (–)-Matairesinol — SAM → (**22a**) (–)-Arctigenin — UDPGlc → (**92**) (–)-Arctiin

Scheme 2

1.25.7.3.1 (+)-Pinoresinol/(+)-lariciresinol reductase

The first enzyme identified in the lignan pathway was the bifunctional, NADPH-dependent, (+)-pinoresinol/(+)-lariciresinol reductase.[88,110,111] This circa 35 kDa enzyme was purified (> 3000-fold) to apparent homogeneity from *F. intermedia*,[111] and shown to catalyze the enantiospecific conversion of (+)-pinoresinol (**3a**) into (+)-lariciresinol (**19a**), and (+)-lariciresinol (**19a**) into (–)-secoisolariciresinol (**20a**); the corresponding antipodes (**3b**) and (**19b**) did not serve as substrates.

It is a type A reductase, as established using [4*R*-^{3}H] and [4*S*-^{3}H]NADPH as cofactors, since only the pro-*R* hydrogen on the nicotinamide ring of NADPH was abstracted and transferred to the lignan product.[88] As shown in Scheme 3, this hydride transfer from NADPH during furanofuran lignan reduction could occur in any one of three ways, i.e., via direct hydride attack onto the furano ring or the regenerated quinone methide, resulting in either retention or inversion of apparent configuration at C-7/C-7′. Alternatively, hydride attack could occur at either face and result in racemization. That an "inversion" of configuration mechanism occurred was established by examination of the products obtained following individual incubation of (±)-[7,7′-$^2H_2$]pinoresinols (**3a,b**) and [7,7′-$^2H_3$]lariciresinols (**19a,b**) with (+)-pinoresinol/(+)-lariciresinol reductase, in the presence of NADPH (1.6 mM): ^{1}H NMR spectroscopic analyses (Figure 12) of the enzymatically generated lariciresinol (**19**) and secoisolariciresinol (**20**) revealed that hydride (deuteride) transfer occurred in

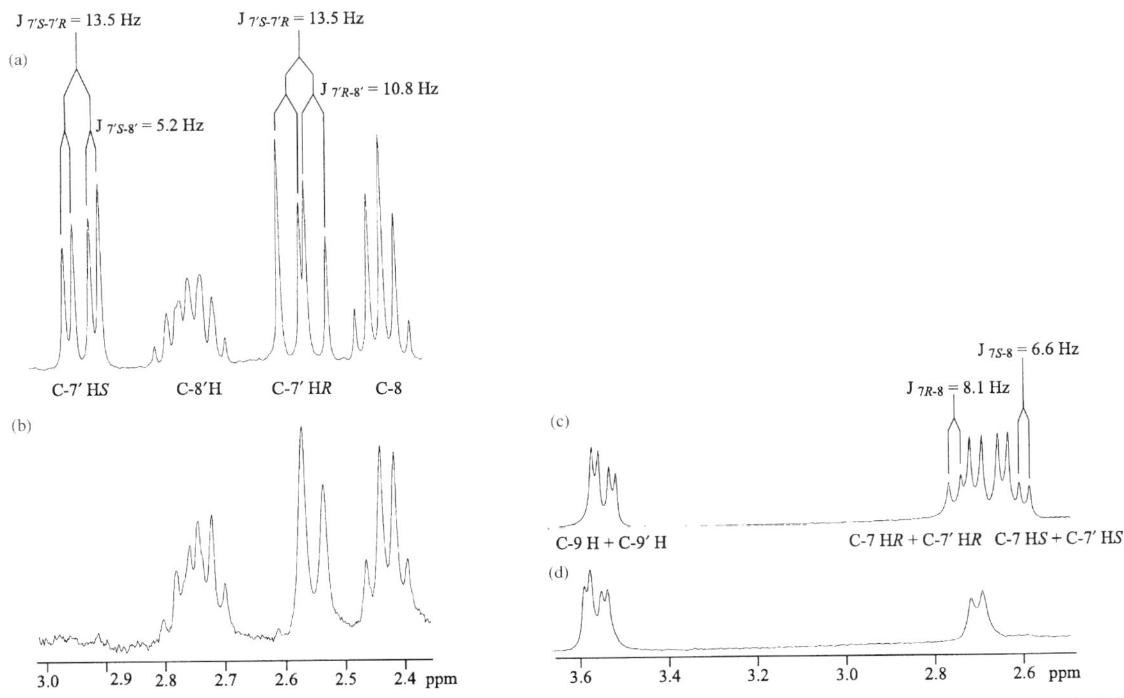

Figure 12 Partial ^{1}H NMR spectra of lariciresinol (**19**) showing spectral regions for C-7′, C-8′, and C-8 proton resonances ((a) and (b)) and secoisolariciresinol (**20**) showing spectral regions for C-7 and C-9 proton resonances ((c) and (d)). (a) Synthetic (±)-lariciresinols (**19a,b**), (b) enzymatically synthesized (+)-[7,7′*S*-$^2H_2$]lariciresinol (**19a**) obtained following incubation with (±)-[7,7′-$^2H_2$]pinoresinols (**3a,b**), (c) synthetic (±)-secoisolariciresinols (**20a,b**), and (d) enzymatically synthesized (−)-[7,7′-$^2H_3$]secoisolariciresinol (**20a**) obtained following incubation with (±)-[7,7′*S*-$^2H_3$]lariciresinols (**19a,b**).[88]

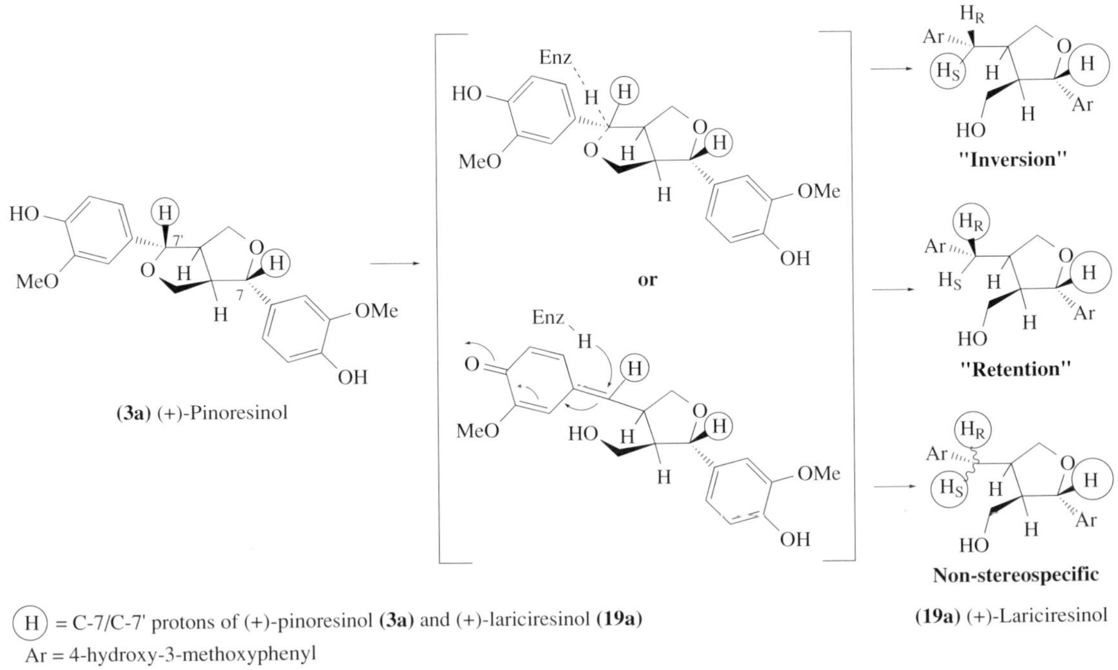

(**3a**) (+)-Pinoresinol

"Inversion"

or

"Retention"

Non-stereospecific

(**19a**) (+)-Lariciresinol

(H) = C-7/C-7′ protons of (+)-pinoresinol (**3a**) and (+)-lariciresinol (**19a**)
Ar = 4-hydroxy-3-methoxyphenyl

Scheme 3

a completely stereospecific manner, with the incoming hydride (deuteride) taking up the pro-*R* position to give (+)-[7,7′*S*-$^2H_2$]lariciresinol (**19a**) and (−)-[7,7′*S*-$^2H_3$]secoisolariciresinol (**20a**), respectively, i.e., with "inversion" of configuration at C-7 and C-7′.

The nucleotide sequence for the gene encoding (+)-pinoresinol/(+)-lariciresinol reductase has a single open reading frame encoding a polypeptide of 312 amino acids (Figure 13), giving a calculated molecular mass of 34.9 kDa.[111] The authenticity of its sequence was established via assay of functionally recombinant (+)-pinoresinol/(+)-lariciresinol reductase, as either its β-galactosidase fusion protein[111] or the "native" protein in a pSBETa[112] vector system expressed in *E. coli*: both recombinant proteins catalyzed the enantiospecific conversion of (+)-pinoresinol (**3a**) into (+)-lariciresinol (**19a**), and the latter into (−)-secoisolariciresinol (**20a**): the corresponding antipodes were not used as substrates (Figure 14). Sequence analysis of the cDNA also revealed that the NADPH-binding domain was situated close to the N-terminus, consisting of three conserved glycine (as GXGXXG, with X representing any residue) and four conserved hydrophobic residues (underlined in Figure 13); these are required for correct packaging for domain formation.[113] The sequence also revealed considerable homology (63.5% to 61.6% similarity and 44.4% to 41.3% identity) to isoflavone reductases (see Table 4), which catalyze the analogous reduction of α,β unsaturated ketones during isoflavonoid formation.[111] For example, *Pisum sativum* isoflavone reductase converts 2′-hydroxy-pseudobaptigenin (**93**) into (3*R*)-sophorol (**94**), a precursor of the phytoalexin, (+)-pisatin (**95**) (Scheme 4). Moreover, given that pinoresinol-derived lignans appear to be found in even the most primitive extant plants (e.g., ferns, gymnosperms), whereas the isoflavonoids may be more restricted (e.g., to the Leguminosae), it is tempting to speculate that pinoresinol/lariciresinol reductase is the evolutionary forerunner of the isoflavonoid reductases.[34] (+)-Pinoresinol/(+)-lariciresinol reductase also contains five conserved possible phosphorylation sites, including Thr-302 (casein kinase II-type protein phosphorylation site) suggesting that its activity might be regulated by protein kinase cascades;[34,111] a somewhat analogous situation also holds for isoflavonoid reductase(s).

In agreement with these findings, (+)-pinoresinol/(+)-lariciresinol reductase has also been detected in cell-free extracts from *Forsythia koreana* by Umezawa *et al.*[114]

Table 4 Homology between (+)-pinoresinol/(+)-lariciresinol reductase from *Forsythia intermedia* and isoflavone reductases from three legumes.[111]

Plant species	*Similarity (%)*	*Identity (%)*
Cicer arietinum	63.5	44.4
Medicago sativa	62.6	42.0
Pisum sativum	61.6	41.3

1.25.7.3.2 (−)-Secoisolariciresinol dehydrogenase

This NAD(P)$^+$-requiring enzyme catalyzes the enantiospecific conversion of (−)-secoisolariciresinol (**20a**) into (−)-matairesinol (**21a**).[115,116] It has been purified (> 6000-fold) to apparent homogeneity from *F. intermedia* stem tissue, and is an ~32 kDa protein, as estimated by SDS-PAGE analysis.[117,118] Interestingly, the presumed lactol (**96**) intermediate was not detected in any of

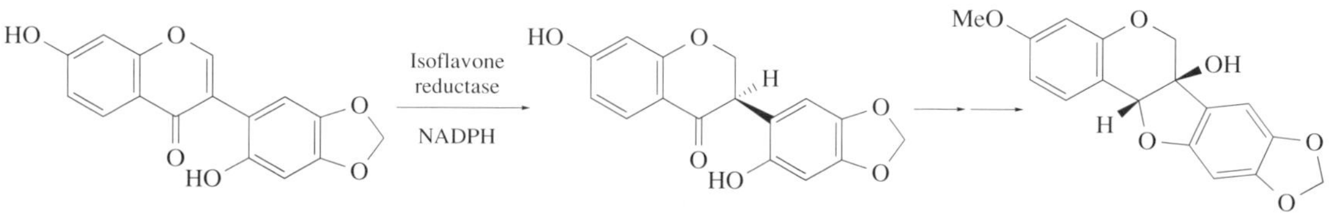

Isoflavone reductase

NADPH

(**93**) 2-Hydroxypseudobaptigenin

(**94**) (3*R*)-Sophorol

(**95**) (+)-Pisatin

Scheme 4

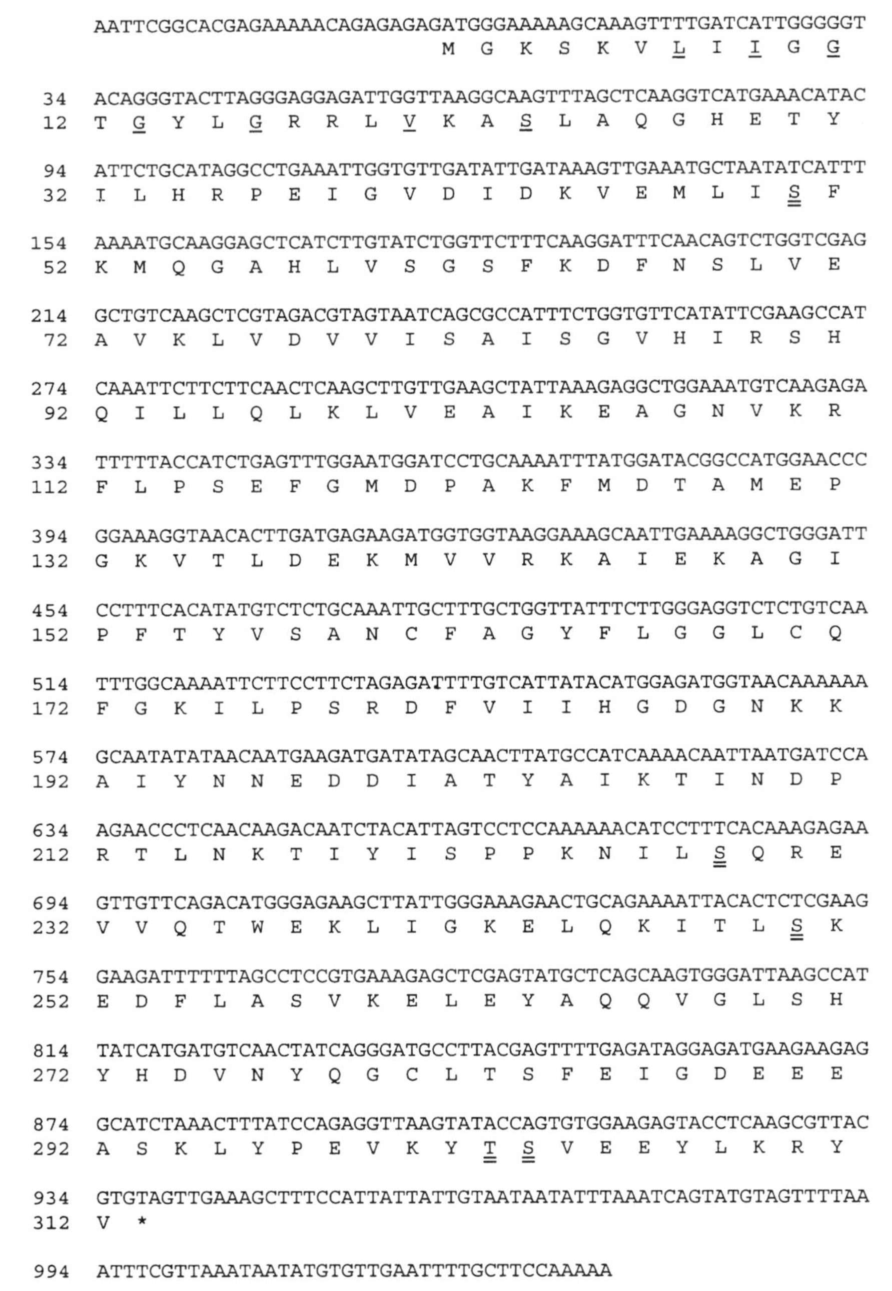

```
    AATTCGGCACGAGAAAAACAGAGAGAGATGGGAAAAAGCAAAGTTTTGATCATTGGGGGT
                                  M  G  K  S  K  V  L  I  I  G  G

 34 ACAGGGTACTTAGGGAGGAGATTGGTTAAGGCAAGTTTAGCTCAAGGTCATGAAACATAC
 12 T  G  Y  L  G  R  R  L  V  K  A  S  L  A  Q  G  H  E  T  Y

 94 ATTCTGCATAGGCCTGAAATTGGTGTTGATATTGATAAAGTTGAAATGCTAATATCATTT
 32 I  L  H  R  P  E  I  G  V  D  I  D  K  V  E  M  L  I  S  F

154 AAAATGCAAGGAGCTCATCTTGTATCTGGTTCTTTCAAGGATTTCAACAGTCTGGTCGAG
 52 K  M  Q  G  A  H  L  V  S  G  S  F  K  D  F  N  S  L  V  E

214 GCTGTCAAGCTCGTAGACGTAGTAATCAGCGCCATTTCTGGTGTTCATATTCGAAGCCAT
 72 A  V  K  L  V  D  V  V  I  S  A  I  S  G  V  H  I  R  S  H

274 CAAATTCTTCTTCAACTCAAGCTTGTTGAAGCTATTAAAGAGGCTGGAAATGTCAAGAGA
 92 Q  I  L  L  Q  L  K  L  V  E  A  I  K  E  A  G  N  V  K  R

334 TTTTTACCATCTGAGTTTGGAATGGATCCTGCAAAATTTATGGATACGGCCATGGAACCC
112 F  L  P  S  E  F  G  M  D  P  A  K  F  M  D  T  A  M  E  P

394 GGAAAGGTAACACTTGATGAGAAGATGGTGGTAAGGAAAGCAATTGAAAAGGCTGGGATT
132 G  K  V  T  L  D  E  K  M  V  V  R  K  A  I  E  K  A  G  I

454 CCTTTCACATATGTCTCTGCAAATTGCTTTGCTGGTTATTTCTTGGGAGGTCTCTGTCAA
152 P  F  T  Y  V  S  A  N  C  F  A  G  Y  F  L  G  G  L  C  Q

514 TTTGGCAAAATTCTTCCTTCTAGAGATTTTGTCATTATACATGGAGATGGTAACAAAAAA
172 F  G  K  I  L  P  S  R  D  F  V  I  I  H  G  D  G  N  K  K

574 GCAATATATAACAATGAAGATGATATAGCAACTTATGCCATCAAAACAATTAATGATCCA
192 A  I  Y  N  N  E  D  D  I  A  T  Y  A  I  K  T  I  N  D  P

634 AGAACCCTCAACAAGACAATCTACATTAGTCCTCCAAAAAACATCCTTTCACAAAGAGAA
212 R  T  L  N  K  T  I  Y  I  S  P  P  K  N  I  L  S  Q  R  E

694 GTTGTTCAGACATGGGAGAAGCTTATTGGGAAAGAACTGCAGAAAATTACACTCTCGAAG
232 V  V  Q  T  W  E  K  L  I  G  K  E  L  Q  K  I  T  L  S  K

754 GAAGATTTTTTAGCCTCCGTGAAAGAGCTCGAGTATGCTCAGCAAGTGGGATTAAGCCAT
252 E  D  F  L  A  S  V  K  E  L  E  Y  A  Q  Q  V  G  L  S  H

814 TATCATGATGTCAACTATCAGGGATGCCTTACGAGTTTTGAGATAGGAGATGAAGAAGAG
272 Y  H  D  V  N  Y  Q  G  C  L  T  S  F  E  I  G  D  E  E  E

874 GCATCTAAACTTTATCCAGAGGTTAAGTATACCAGTGTGGAAGAGTACCTCAAGCGTTAC
292 A  S  K  L  Y  P  E  V  K  Y  T  S  V  E  E  Y  L  K  R  Y

934 GTGTAGTTGAAAGCTTTCCATTATTATTGTAATAATATTTAAATCAGTATGTAGTTTTAA
312 V  *

994 ATTTCGTTAAATAATATGTGTTGAATTTTGCTTCCAAAAA
```

Figure 13 Complete sequence of *F. intermedia* (+)-pinoresinol/(+)-lariciresinol reductase cDNA plr-Fi1. The NADPH binding domain conserved residues are underlined. The five possible phosphorylation sites that are conserved among (+)-pinoresinol/(+)-lariciresinol reductase and the characterized isoflavone reductases are double underlined. The stop codon is indicated by an asterisk.[111]

the enzyme assays using native protein (Scheme 5). The genes encoding (−)-secoisolariciresinol dehydrogenase were obtained using a PCR-guided strategy, coupled with the screening of the *F. intermedia* cDNA library. Two of the five cDNA clones obtained, SMDEHY631 and DEHY130, were subsequently expressed in *E. coli*, and both gave proteins capable of enantiospecifically converting (−)-secoisolariciresinol (**20a**) into (−)-matairesinol (**21a**). Both had single open-reading frames encoding polypeptides of 272 amino acids, giving calculated molecular masses of 29 kDa, and isoelectric points of ~6.1. However, in contrast to the native dehydrogenase, the corresponding lactol (**96**) in each instance was observed as an intermediate, during conversion of (−)-secoisolariciresinol (**20a**) into (−)-matairesinol (**21a**) (Scheme 5).

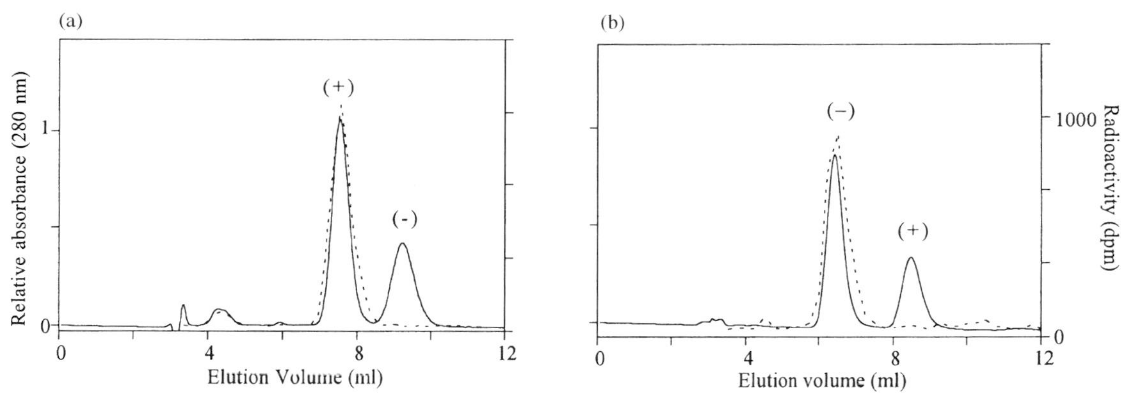

Figure 14 Chiral column HPLC analysis of lignans (a) lariciresinol (**19**) and (b) secoisolariciresinol (**20**). (±)-Pinoresinols (**3a,b**) were incubated with recombinant (+)-pinoresinol/(+)-lariciresinol reductase in the presence of [4*R*-^{3}H]NADPH, with unlabeled (±)-lariciresinols (**19a,b**) and (±)-secoisolariciresinols (**20a,b**) added as radiochemical carriers. The solid line represents the UV absorbance trace (280 nm), whereas the dashed line shows that radioactivity is only incorporated into (+)-lariciresinol (**19a**) and (−)-secoisolariciresinol (**20a**), respectively.[111] For elution conditions see Figure 3.

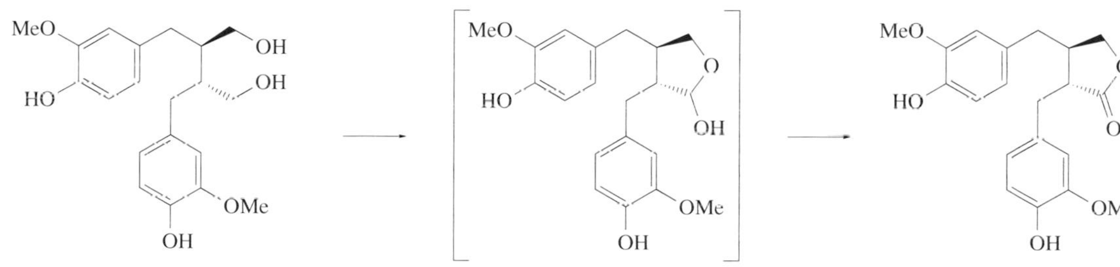

(20a) (–)-Secoisolariciresinol **(96)** (–)-Lactol **(21a)** ( ) Matairesinol

Scheme 5

1.25.7.3.3 *Matairesinol* O-*methyltransferase*

Cell-free extracts of *F. intermedia* were examined in order to attempt to establish whether *O* methylation of matairesinol (**21**) giving arctigenin (**22**) occurred in an enantio- and regiospecific manner.[109] Incubation with (±)-[Ar-^{3}H]matairesinols (**21a,b**), in the presence of *S*-adenosyl-L-methionine, gave both radiolabeled arctigenin (**22**) and isoarctigenin (**97**) (Scheme 6), with the (−)-enantiomeric form being preferentially obtained in both cases (2:1 and 2.5:1, respectively), i.e., *O*-methylation was neither regio- nor stereospecific. As for eudesmin (**83**) biosynthesis, purification of the actual *O*-methyltransferase involved in catalyzing the formation of (−)-arctigenin (**22a**) is required in order to determine whether the *O*-methyltransferase is truly enantio- and regiospecific.

SAM SAM

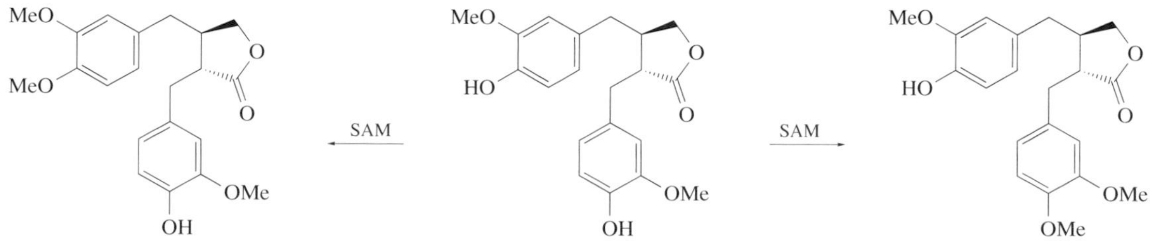

(22) Arctigenin **(21)** Matairesinol **(97)** Isoarctigenin

Scheme 6

1.25.7.4 *Linum usitatissimum*: (−)-Pinoresinol/(−)-Lariciresinol Reductase and (+)-Secoisolariciresinol Glucosyltransferase(s)

Flaxseed (*L. usitatissimum*, Linaceae) has been used for several millennia for medicinal purposes. There is considerable renewed interest in it because of the high levels (3–3.5%) of secoisolariciresinol diglucoside (**98**), which has an important role in the diet due to its protection against onset of breast and prostate cancers[119,120] (see Section 1.25.14), and can be solubilized from flaxseed under very basic conditions (e.g., 0.3 M sodium methoxide).[121] Hydrolysis of secoisolariciresinol diglucoside (**98**) gives the aglycone (**20**), the chiral analysis (Chiralcel OD, Daicel) of which revealed that essentially only the (+)-antipode (**20b**) was present.[97]

The proposed biochemical pathway to secoisolariciresinol diglucoside (**98**) is shown in Scheme 7. Preliminary experiments[97] have indicated that stereoselective coupling of *E*-coniferyl alcohol (**38**) occurs to give (−)-pinoresinol (**3b**). Based on results from enzyme assays, using cell-free extracts of developing seeds, this is then reduced to first give (−)-lariciresinol (**19b**) and then (+)-secoisolariciresinol (**20b**). Screening procedures are under way to obtain the corresponding gene(s) for both the dirigent protein and reductase(s) from this source.

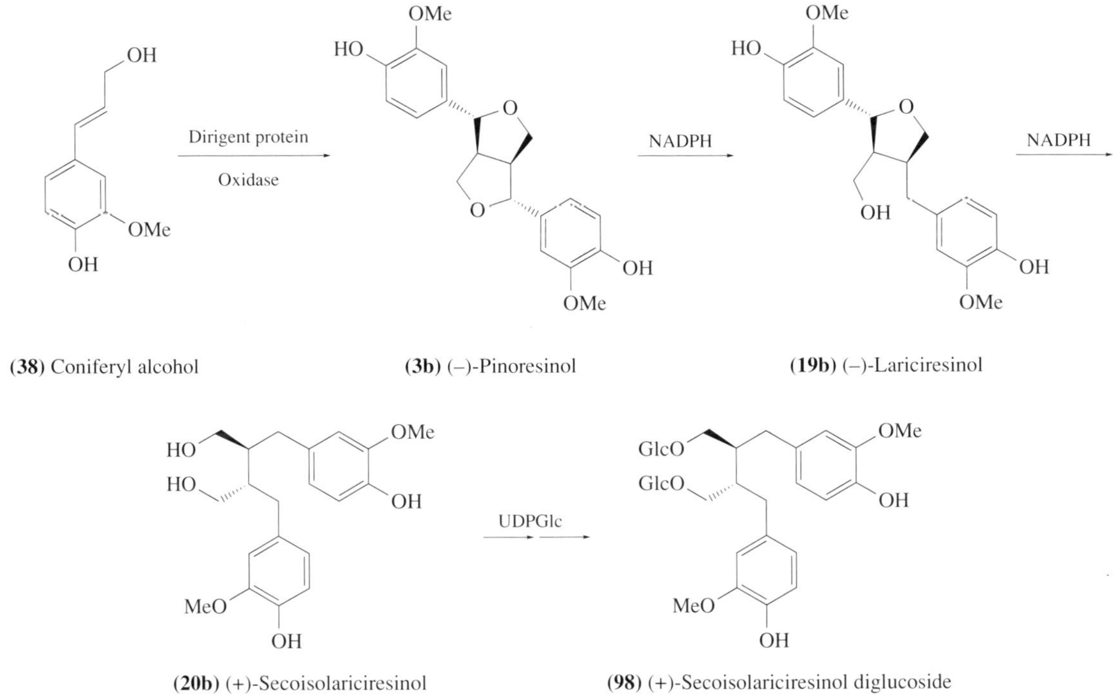

(**38**) Coniferyl alcohol — Dirigent protein / Oxidase → (**3b**) (−)-Pinoresinol — NADPH → (**19b**) (−)-Lariciresinol — NADPH →

(**20b**) (+)-Secoisolariciresinol — UDPGlc → (**98**) (+)-Secoisolariciresinol diglucoside

Scheme 7

Preliminary assays, with partially purified secoisolariciresinol glucosyltransferase, in the presence of UDP[1-^{3}H]glucose and (±)-secoisolariciresinols (**20a,b**), have revealed that maximum activity of the glucosyltransferase occurs during the second week of seed development.

1.25.7.5 *Thuja plicata* and *Tsuga heterophylla*: Pinoresinol/Lariciresinol Reductases and Other Enzymatic Conversions

Western red cedar (*Thuja plicata*) and western hemlock (*Tsuga heterophylla*) are two long-living gymnosperms, with the former having life spans in excess of 3000 years. Of the two species, cedar heartwood is particularly valued for its color, durability, and texture, this in part being due to the massive deposition (~20% of dry weight) of heartwood (oligomeric) lignans, derived from plicatic acid (**75**) and its congeners.[35] In a somewhat analogous manner, western hemlock accumulates various lignans, such as 7′-hydroxymatairesinol (**99**) (0.3% in sapwood) and α-conidendrin (**76**) (0.05% in sapwood and 0.15 to 0.2% in heartwood).[122] Based on their respective chemical structures, and the results from studies in the authors' laboratory, it was rationalized that both plicatic acid (**75**) and α-conidendrin (**76**) were pinoresinol-derived (Scheme 8). Plicatic acid (**75**) could result from matairesinol (**21**) via ring closure (aryltetrahydronaphthalene ring formation), hydroxylations (at both aromatic and aliphatic positions), and lactone ring cleavage, whereas α-conidendrin (**76**) might result from regiospecific hydroxylation at the 7′-position of matairesinol (**21**) to give 7′-hydroxymatairesinol (**99**), with subsequent ring closure. It was important to establish, therefore, whether the corresponding pinoresinol/lariciresinol reductases were present in these plant systems in addition to the dirigent proteins (see Section 1.25.6.4).

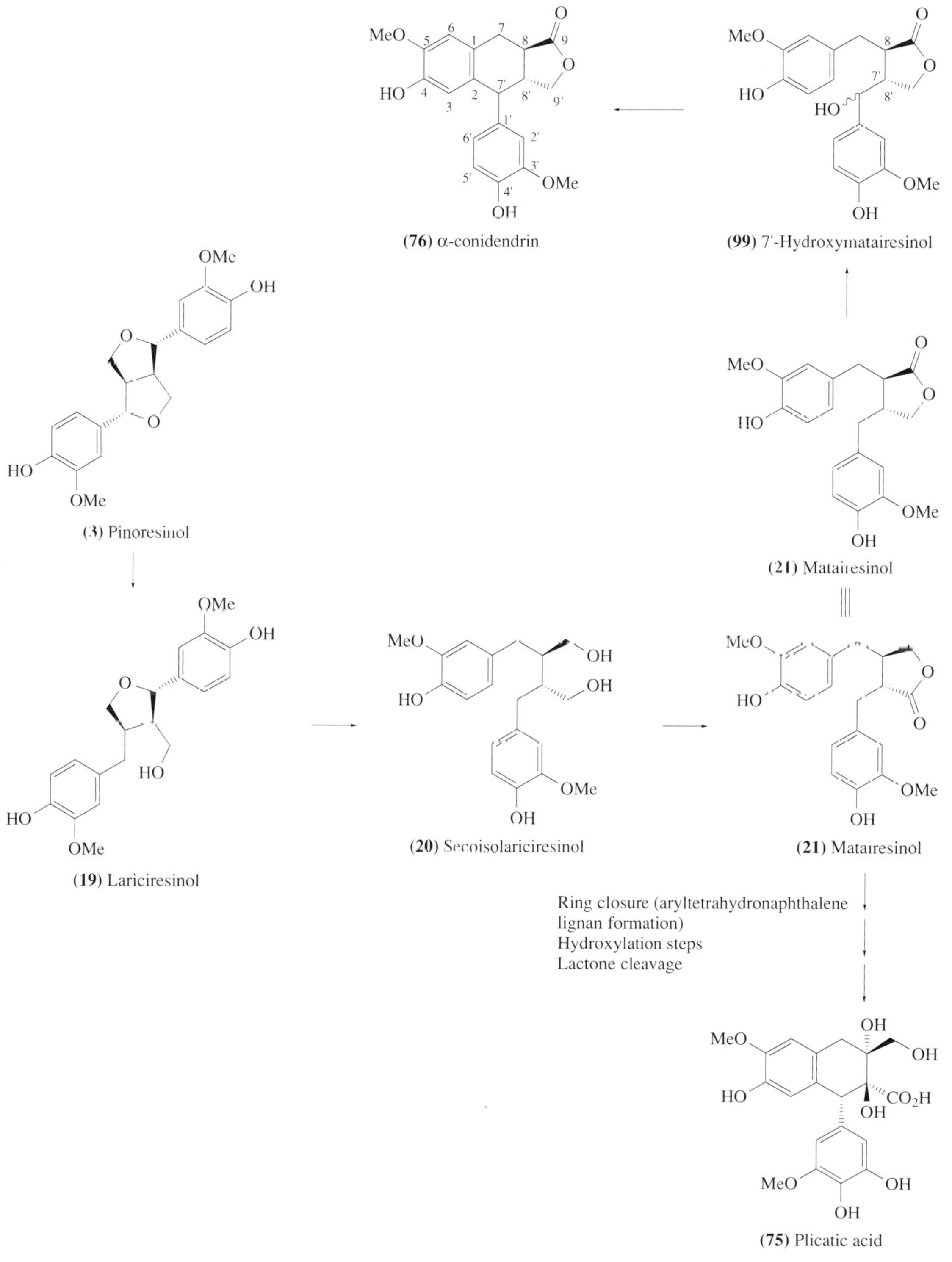

(**76**) α-conidendrin

(**99**) 7′-Hydroxymatairesinol

(**3**) Pinoresinol

(**21**) Matairesinol

(**19**) Lariciresinol

(**20**) Secoisolariciresinol

(**21**) Matairesinol

Ring closure (aryltetrahydronaphthalene lignan formation)
Hydroxylation steps
Lactone cleavage

(**75**) Plicatic acid

Scheme 8

Thus, screening of a western red cedar cDNA library, via a PCR-guided strategy, gave two pinoresinol/lariciresinol reductase cDNAs with ~58% identity and ~70% similarity to plr-Fi1. Their recombinant proteins, when expressed in *E. coli*, displayed distinct pinoresinol/lariciresinol reductase activities.[30,123,124] The first, plr-Tp1, catalyzed the NADPH-dependent conversion of (−)-pinoresinol (**3b**) into (−)-lariciresinol (**19b**), this being subsequently reduced to (+)-secoisolariciresinol (**20b**) (see Scheme 9), with the enantiospecificity of the conversion being determined by chiral HPLC analysis of the enzymatic products so obtained (Figure 15(a)). The second, plr-Tp2, isolated from the same plant source, converted (+)-pinoresinol (**3a**) into (+)-lariciresinol (**19a**), and then into (−)-secoisolariciresinol (**20a**) (see Scheme 10 and Figure 15(b)). That is, western red cedar has genes encoding two reductive pathways with differing enantiospecificities. Although the biological significance of both pathways needs to be established, this study reveals that the pathway to plicatic acid (**75**) indeed occurs via involvement of both dirigent proteins and pinoresinol/lariciresinol reductase(s).

(3b) (−)-Pinoresinol —NADPH, plr-Tp1→ **(19b)** (−)-Lariciresinol —NADPH, plr-Tp1→ **(20b)** (+)-Secoisolariciresinol

Scheme 9

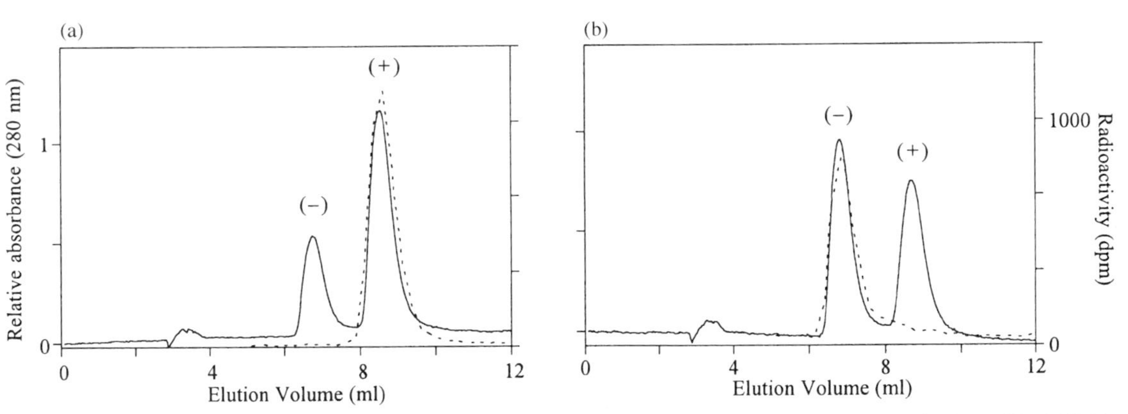

Figure 15 Chiral column HPLC analysis of the secoisolariciresinol (**20**) obtained after incubation of (±)-[3,3′-$O^{14}CH_3$]pinoresinols (**3a,b**) with recombinant pinoresinol/lariciresinol reductase, in the presence of NADPH. (a) plr-Tp1, and (b) plr-Tp2. Unlabeled (±)-secoisolariciresinols (**20a,b**) were added as radiochemical carriers. The solid line represents the absorbance trace at 280nm, whereas the dashed line shows the radioactivity incorporated into the corresponding secoisol ariciresinol (**20**) product.

Further characterization of the recombinant reductases, plr-Tp1 and plr-Tp2, resulted in the discovery that both differ substantially from their *Forsythia* counterpart: plr-Tp1 preferentially converts (−)-pinoresinol (**3b**) into (−)-lariciresinol (**19b**), but can also slowly reduce (+)-pinoresinol (**3a**) to give (+)-lariciresinol (**19a**) although the latter is not further metabolized. The converse situation was observed for plr-Tp2. It preferentially converts (+)-pinoresinol (**3a**) into (+)-lariciresinol (**19a**), but can also slowly reduce (−)-pinoresinol (**3b**) into (−)-lariciresinol (**19b**). As for plr-Tp1, plr-Tp2 is fully enantiospecific for lariciresinol (**19**) reduction, since it only utilizes (**19a**) as a substrate and not its (−)-enantiomer (**19b**).[30,124]

In a comparable manner, a *T. heterophylla* cDNA library was screened, this resulting in two

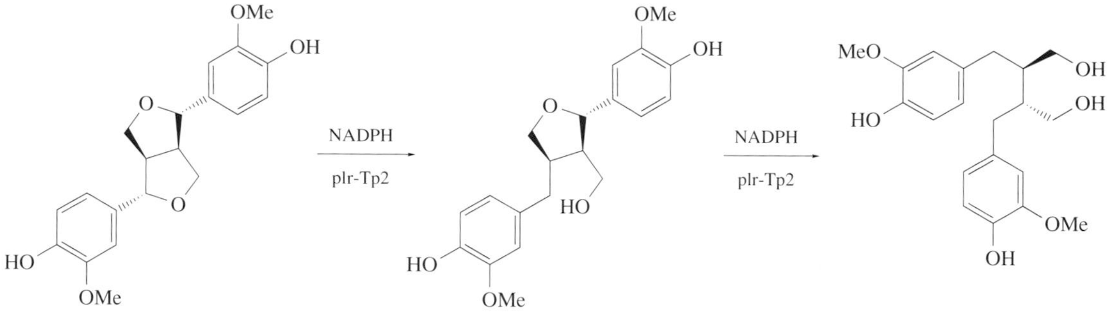

NADPH plr-Tp2 → NADPH plr-Tp2 →

(**3a**) (+)-Pinoresinol (**19a**) (+)-Lariciresinol (**20a**) (–)-Secoisolariciresinol

Scheme 10

putative cDNAs encoding pinoresinol/lariciresinol reductase with ~73% similarity and ~62% identity.

Taken together, these findings may help explain the results of Umezawa and Shimada where the formation of secoisolariciresinol (**20**), catalyzed by cell-free extracts from *Arctium lappa* petioles, was investigated.[125] In that study, coniferyl alcohol (**38**) was incubated with the cell-free preparation in the presence of H_2O_2 and NADPH. Under these conditions, nonspecific peroxidase catalyzed coupling occurs to give the racemic lignans, (±)-(**5a,b**), (±)-(**6a,b**), and (+)-(**3a,b**), with the latter then undergoing NADPH-dependent reduction to afford secoisolariciresinol (**20**), whose (+)-enantiomer (**20b**) was in enantiomeric excess (e.e. 20%). Additionally, (+)-secoisolariciresinol (**20b**) was isolated from MeOH extracts of *A. lappa* petioles in 78% enantiomeric excess.[125] These data are consistent with both (+)- and (−)-pinoresinol/lariciresinol reductases being present in this species, as previously described for western red cedar.

1.25.7.6 *Linum flavum* and *Podophyllum hexandrum*: Podophyllotoxin and its Pinoresinol Precursor

(−)-Podophyllotoxin (**23**) accumulates up to 4% of the dry weight in *P. hexandrum* (Podophyllaceae) rhizomes[126] and (−)-5-methoxypodophyllotoxin (**100**) up to 3.5% of the dry weight in *L. flavum* (Linaceae) roots.[127] Podophyllotoxin (**23**), as its etoposide (**101**) and teniposide (**102**) derivatives, is used in treatment of various cancers (see Section 1.25.14.2). Because of the limited supply of *Podophyllum* rhizomes, due to their intensive collection in the wild, there is a growing interest in defining and exploiting the podophyllotoxin (**23**) biosynthetic pathway. This is in part because studies, using both cell suspension and root tissue cultures, have not been very successful in producing significantly elevated levels of either podophyllotoxin (**23**) or (−)-5-methoxypodophyllotoxin (**100**).[127–132] Accordingly, both defining the pathway and establishing what factors affect its induction seem to be worthwhile goals in preparation for future biotechnological manipulations.

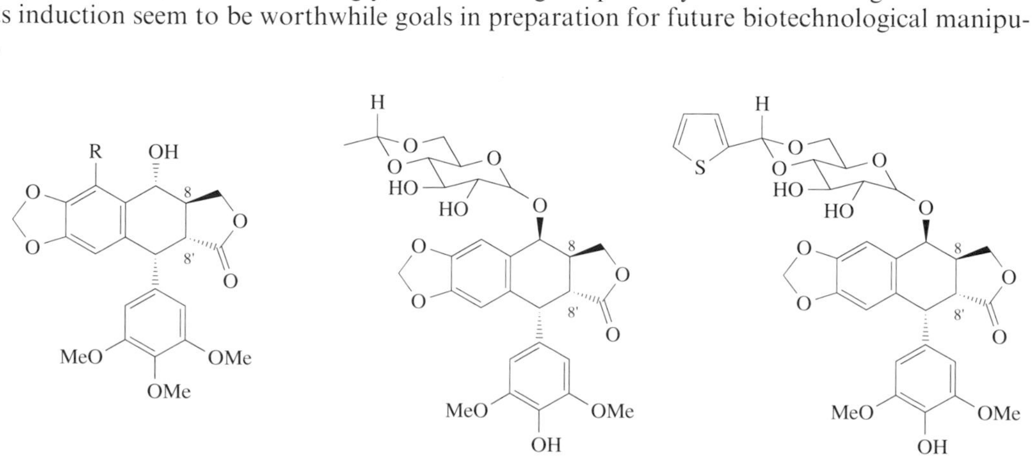

(**23**) R = H, (–)-Podophyllotoxin
(**100**) R = OMe, (–)-5-Methoxypodophyllotoxin

(**101**) Etoposide

(**102**) Teniposide

Preliminary investigations by Broomhead *et al.* suggested that (−)-matairesinol (**21a**) might be a precursor of podophyllotoxin (**23**).[133] Accordingly, it was important to establish whether the biochemical pathway to matairesinol (**21**) was also present in *L. flavum*. Thus, cell-free extracts from *L. flavum* roots were incubated with (±)-pinoresinols (**3a,b**) (20 mM) in the presence of NADPH (20 mM), where it was established that both (+)-lariciresinol (**19a**) and (−)-secoisolariciresinol (**20a**) were formed.[134] Furthermore, a partially purified secoisolariciresinol dehydrogenase preparation was also obtained which, when incubated with (±)-[Ar-^{2}H]secoisolariciresinols (**20a,b**) and NAD, only gave (−)-[Ar-^{2}H]matairesinol (**21a**). The enzymatically formed (−)-[Ar-^{2}H]matairesinol (**21a**) had a cluster of ions centered at *m/z* 360 (Figure 16(a)) relative to that of (±)-[Ar-^{2}H]secoisolariciresinol (**20a,b**) cluster of ions at *m/z* 364 (Figure 16(b)), thereby establishing the intact enzymatic conversion, and thus verification of the pathway to (−)-matairesinol (**21a**) in this species.

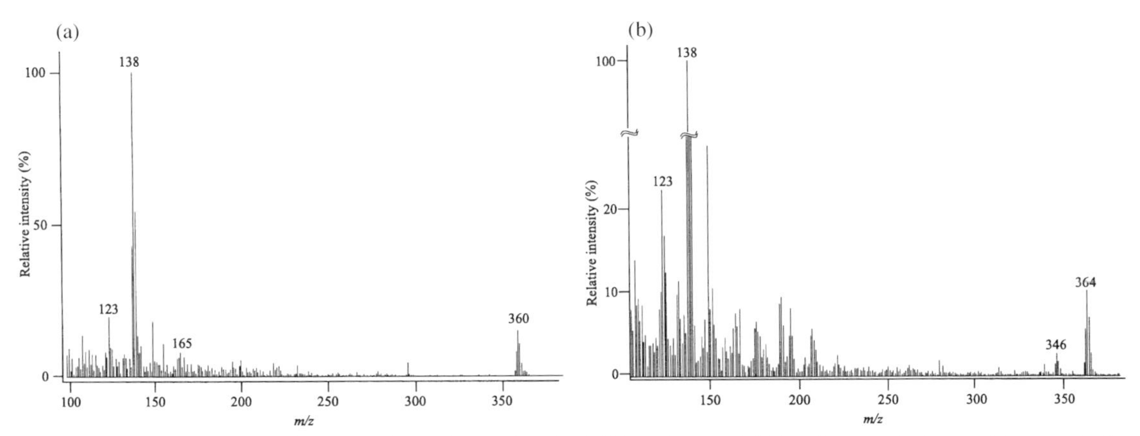

Figure 16 Mass spectra of (a) (−)-[Ar-^{2}H]matairesinol (**21a**) obtained following incubation with a partially purified secoisolariciresinol dehydrogenase preparation of *Linum flavum* and (±)-[Ar-^{2}H]secoisolariciresinols (**20a,b**) in the presence of NAD, (b) (±)-[Ar-^{2}H]secoisolariciresinol (**20a,b**) substrates.

Thus, in summary, pinoresinol (**3**) formation and metabolism have been shown to occur in *Forsythia* species, *Linum usitatissimum*, *Thuja plicata*, *Tsuga heterophylla*, *Linum flavum*, and *Podophyllum peltatum*, thereby establishing its pivotal role in 8—8′ linked lignan biosynthesis.

1.25.8 ARE DIRIGENT PROTEIN HOMOLOGUES INVOLVED IN OTHER 8—8′ PHENOXY RADICAL COUPLING PROCESSES?

In addition to coniferyl alcohol (**38**) derived 8—8′ linked lignans, there are optically active 8—8′ linked lignans which provisionally appear to result from coupling of other monomeric precursors, such as *E*-*p*-coumaryl alcohol (**32**), *E*-sinapyl alcohol (**44**), various hydroxycinnamic acids/allylphenols, and their derivatives. Although each of their biochemical pathways await delineation, the occurrence of these lignans suggests the presence of dirigent proteins with differing substrate specificities. This, in turn, is consistent with the existence of an entire new class of proteins, and this possibility is discussed below using just a few examples for illustrative purposes only.

1.25.8.1 Ligballinol (*p*-Coumarylresinol) and Related Structures

Based on the chemical structures of ligballinol (= *p*-coumarylresinol) (**103**), termilignan (**104**), anolignan B (**105**), and thannilignan (**106**), it can be proposed that all are either *p*-coumaryl alcohol (**32**) or *p*-hydroxyarylpropene (**60**) derived. Of these, ligballinol (*p*-coumarylresinol) (**103**) was first isolated in optically active form from squirting cucumber (*Ecballium elaterium*, Cucurbitaceae) fruits,[135] but also accumulates in enantiomeric excess in red bean (*Vigna angularis*, Leguminosae) cell suspension cultures treated with actinomycin D.[136] Interestingly, although both isolates have very distinct $[\alpha]_D$ values suggesting different enantiomeric compositions, the chiral HPLC analyses of the ligballinol (*p*-coumarylresinol) (**103**) preparations have not been reported. Termilignan

(**104**), anolignan B (**105**), and the optically active thannilignan (**106**), from *Terminalia bellerica* (Combretaceae),[137] can also be proposed to partly result from regiospecific/stereoselective *E*-*p*-coumaryl alcohol (**32**) coupling, although they might also derive from the corresponding allylphenol (**60**). Whether such coupling is under the control of dirigent-like proteins awaits delineation, as does identification of the nature of the precursors involved.

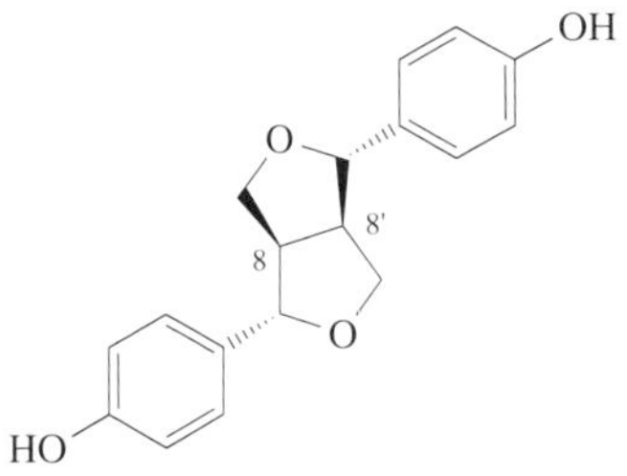

(**103**) Ligballinol = *p*-coumarylresinol
(*Ecballium elaterium*), $[\alpha]_D = -7.1°$ (MeOH, $c = 1.61$)
(*Vigna angularis*), $[\alpha]_D^{28} = +25.3°$ (MeOH, $c = 0.06$)

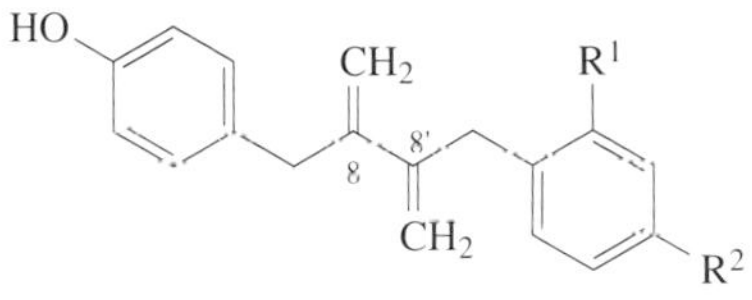

(**104**) R^1 = OH, R^2 = OMe, Termilignan
(**105**) R^1 = H, R^2 = OH, Anolignan B
(*Terminalia bellerica*)

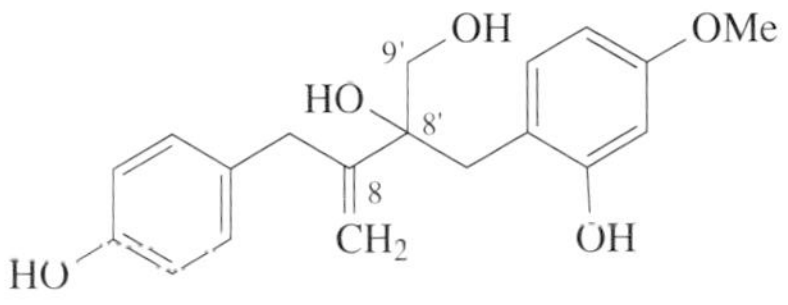

(**106**) (–)-Thannilignan
$[\alpha]_D^{24} = -63.9°$ ($CHCl_3$, $c = 0.3$)
(*Terminalia bellerica*)

1.25.8.2 Syringaresinol and Medioresinol

As summarized in Table 1, syringaresinol (**66**) is found in different species, with predictably different degrees of enantiomeric purity depending upon the plant species involved. Its formation could result from either dirigent protein mediated coupling of sinapyl alcohol (**44**), albeit with more than one mode of stereoselective coupling to afford both (+)- and (–)-enantiomers (**66a**) and (**66b**), respectively, or perhaps less likely, it could result from modification of preformed pinoresinol (**3**). As an added consideration, optically active medioresinol (**107**) present in *Eucommia ulmoides* (Eucommiaceae),[138] might result from either heterogeneous coupling of *E*-coniferyl (**38**) and sinapyl (**44**) alcohols, or be formed through pinoresinol (**3**) modification (Scheme 11). Accordingly, the precise enzymology involved in their formation awaits full clarification, including that of the role of dirigent proteins.

1.25.8.3 *Pellia* Liverwort Lignans

Lignans present in liverworts,[53–60] such as *Pellia epiphylla*, appear to be formed via stereoselective coupling of *E*-caffeic acid (**33**) molecules, as shown in Scheme 12. Thus, following stereoselective coupling, the regenerated diphenol on ring A can then participate in nucleophilic attack onto the quinone methide of ring B to generate the resulting aryldihydronaphthalene derivative (**45**). Although the enantiomeric composition of these lignans has yet to be described, their pronounced optical rotations suggest involvement of dirigent proteins.

1.25.8.4 Lignanamides

Various plant species, particularly in, but not restricted to, the Solanaceae accumulate *p*-coumaroyl and feruloyl tyramine derivatives (**108**) and (**109**), with these being considered to be incorporated (at least in part) into the aromatic component of the suberin biopolymer.[139–141] Feruloyl (**109**), sinapoyl (**110**), and caffeoyl (**111**) tyramine derivatives, and the optically active aryldihydronaphthalene lignanamides (**112**) and (**113**), have also been isolated from *Porcelia macrocarpa*

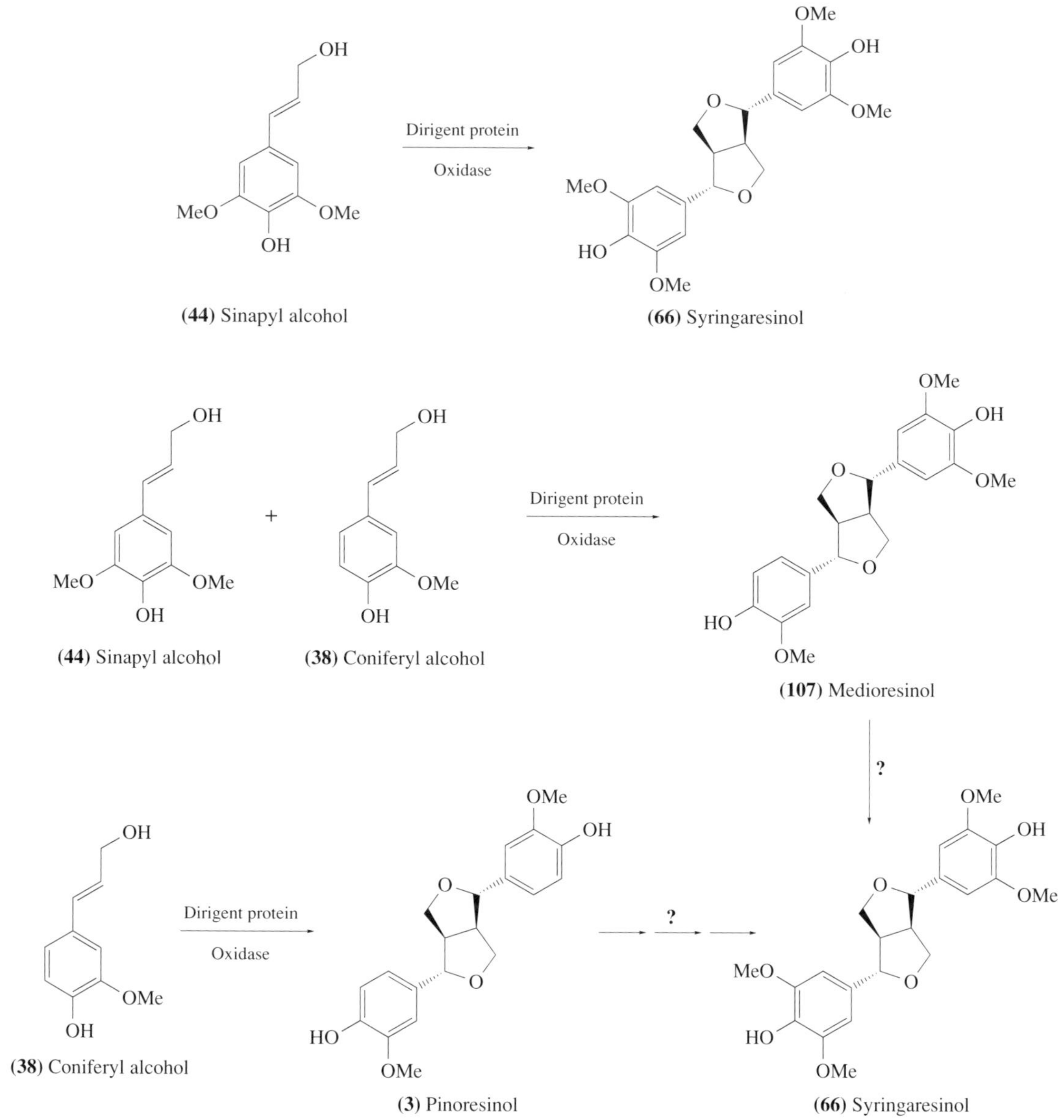

Dirigent protein
Oxidase

(44) Sinapyl alcohol

(66) Syringaresinol

(44) Sinapyl alcohol + (38) Coniferyl alcohol

Dirigent protein
Oxidase

(107) Medioresinol

?

(38) Coniferyl alcohol

Dirigent protein
Oxidase

(3) Pinoresinol

?

(66) Syringaresinol

Scheme 11

(Annonaceae) branch tissue.[142] It is, accordingly, again tempting to speculate that the lignanamides result from dirigent protein mediated stereoselective coupling (cf. the liverwort lignans).

Fruits of *Xylopia aethiopica* (Annonaceae) also contain optically active 8—8′ linked (−)-cannabisin B (**114**) and (−)-cannabisin D (**115**), together with smaller amounts of the racemic lignans, (±)-grossamide (**116**) and (±)-demethylgrossamide (**117**).[143] Perhaps significantly, the small amounts of racemic products (**116**) and (**117**) co-occur with larger amounts of the corresponding "monomeric" tyramine derivatives (**108**), (**109**), and (**111**). Indeed, based on the methodologies employed for their isolation (plant tissue grinding over long periods, lengthy extractions, numerous chromatographic steps), formation of the racemic dimers could result from nonspecific coupling (artifact formation) during isolation. This is because during their isolation disruption of the plant material will result in the tyramine derivatives (**108**), (**109**), and (**111**) coming into contact with nonspecific oxidases/oxidants. Surprisingly, few studies seem to consider the possibility that isolated (racemic) phenolic dimers (and higher oligomers) may be artifacts, in spite of the readily oxidizable nature of such compounds.

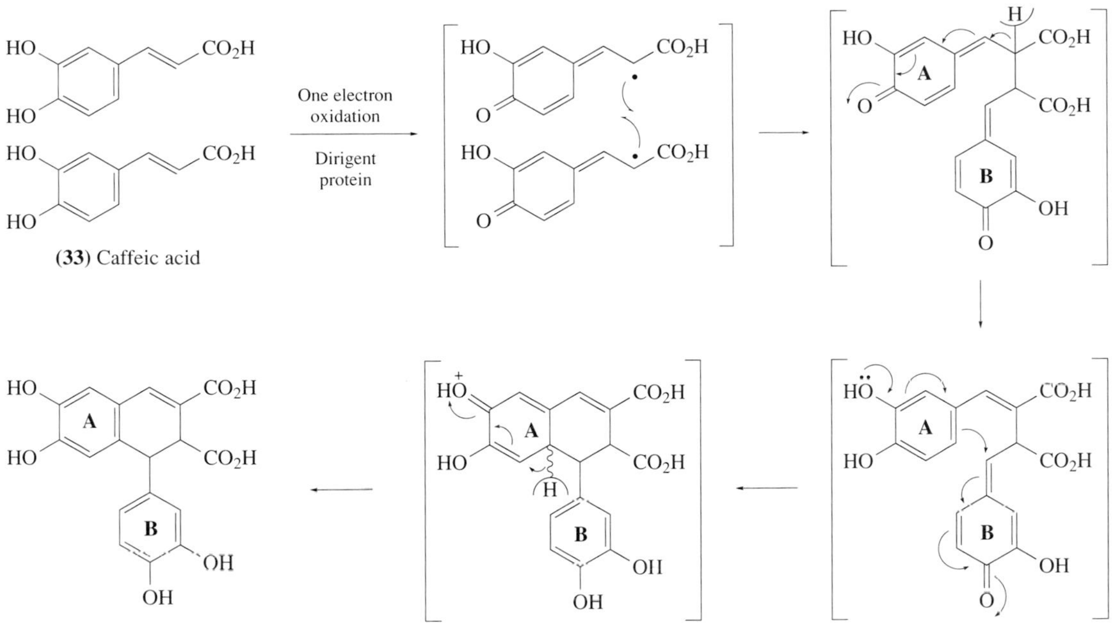

HO HO CO_2H HO HO CO_2H

(33) Caffeic acid

One electron oxidation

Dirigent protein

HO O CO_2H HO O CO_2H

HO A O H CO_2H CO_2H B OH O

HÖ A HO CO_2H CO_2H B OH O

HO A HO H CO_2H CO_2H B OH OH

HO A HO CO_2H CO_2H B OH OH

(45)

$[\alpha]_D^{20} = -130.77°$ (MeOH)

(*Pellia epiphylla*)

Scheme 12

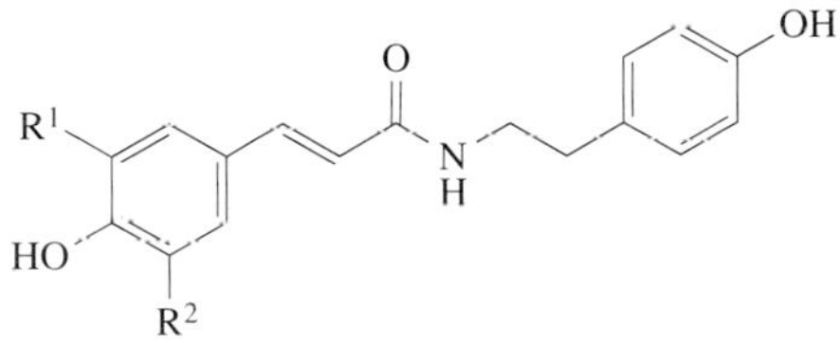

R^1 HO R^2 O N H OH

(108) $R^1 = R^2 = H$, *p*-Coumaroyl tyramine
(109) $R^1 = OMe$, $R^2 = H$, Feruloyl tyramine
(110) $R^1 = R^2 = OMe$, Sinapoyl tyramine
(111) $R^1 = OH$, $R^2 = H$, Caffeoyl tyramine

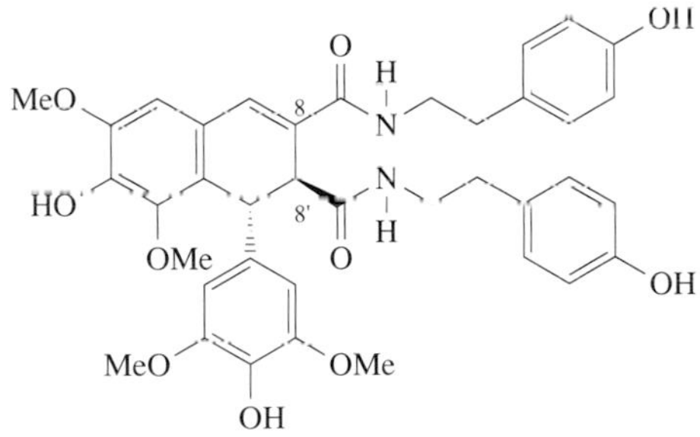

MeO HO OMe 8 8' O H N N H O OH OH MeO OMe OH

(112)

$[\alpha]_D^{25} = -20°$ (MeOH, $c = 0.062$)

(*Porcelia macrocarpa*)

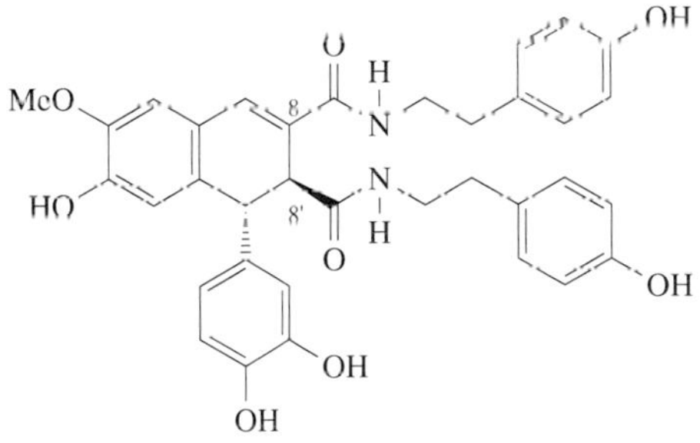

MeO HO 8 8' O H N N H O OH OH OH OH

(113)

$[\alpha]_D^{25} = -12°$ (MeOH, $c = 0.085$)

(*Porcelia macrocarpa*)

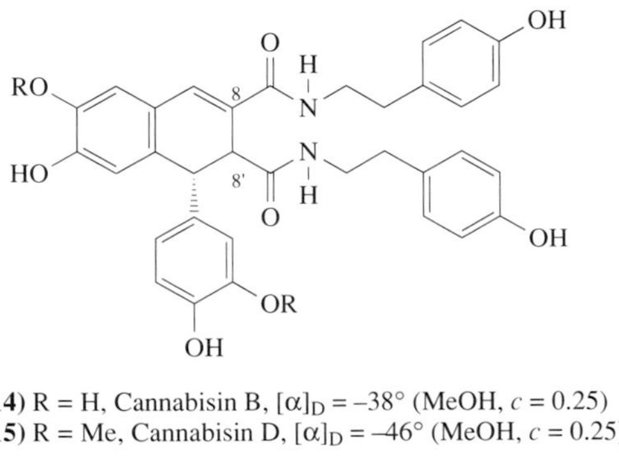

RO HO 8 8' O H N N H O OH OH OR OH

(114) R = H, Cannabisin B, $[\alpha]_D = -38°$ (MeOH, $c = 0.25$)
(115) R = Me, Cannabisin D, $[\alpha]_D = -46°$ (MeOH, $c = 0.25$)

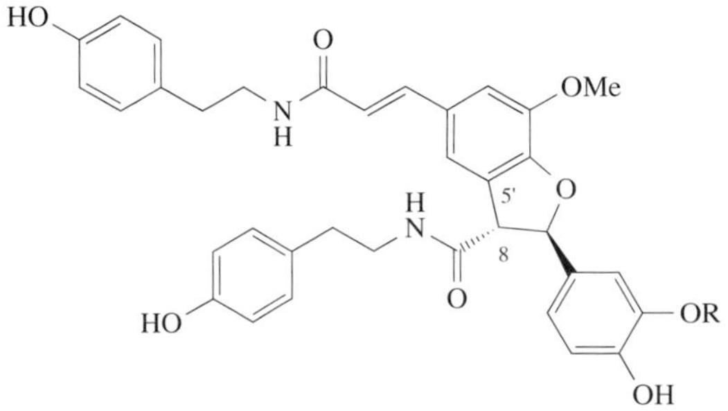

HO O N H OMe 5' O H N 8 O HO OR OH

(116) R = Me, (±)-Grossamide
(117) R = H, (±)-Demethylgrossamide

Cannabis sativa (Cannabidaceae) fruits[144–146] also reputedly contain racemic 8—8′ ((**114**), (**115**), (**118**)), 8—*O*—4′ ((**119**), (**120**)), and 8—5′ ((**116**), (**117**)) linked lignan amides, together with their "monomeric" precursors (**108**), (**109**), and (**111**). At a first glance, the occurrence of these dimers would appear to favor random coupling mechanisms *in vivo*. However, such reports of racemic products being present in plant tissues must be viewed with some caution, since once again the isolation conditions employed were very harsh, and it cannot be ruled out that these "randomly linked" dimers are instead just isolation artifacts.

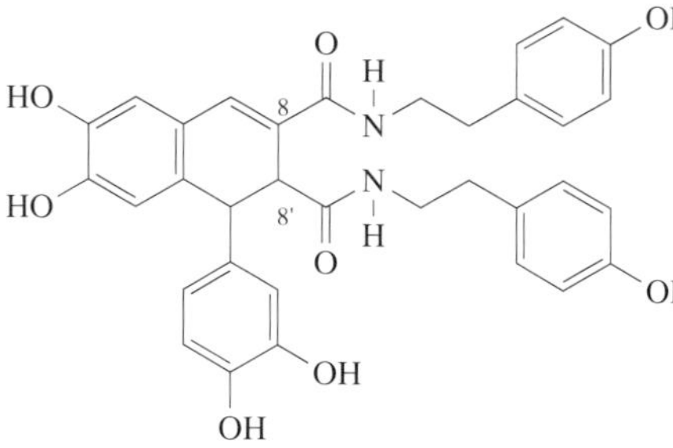

(**118**) Cannabisin A

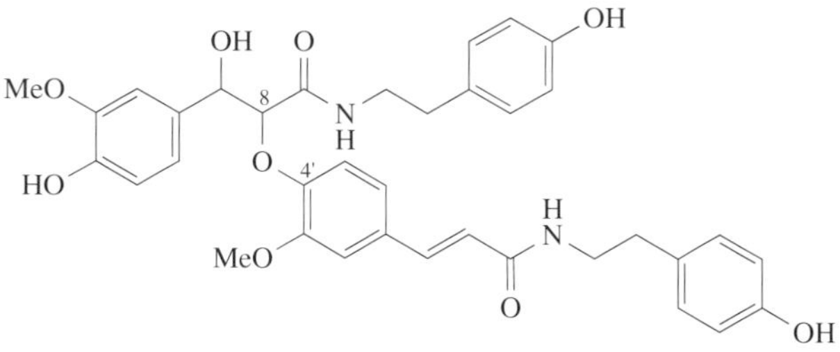

(**119**) Cannabisin E

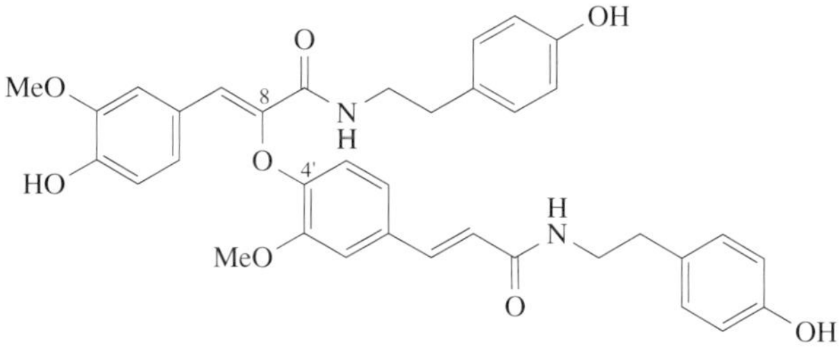

(**120**) Cannabisin F

Artifact formation may also account for the occurrence of racemic grossamide (**116**) in bell pepper (*Capsicum annuum* var. *grossum*, Solanaceae) roots.[147] In this case, homogenized roots (21 kg) were extracted for 336 h with 70% EtOH (160 L), prior to any chromatographic purification, i.e., under conditions which would certainly favor nonspecific coupling and artifact formation from the monomeric precursors present in the tissue.

1.25.8.5 Guaiaretic Acid, Steganacin, and Gomisin A

Certain plant species contain relatively unusual types of optically active 8—8′, as well as 8—8′, 2—2′ linked lignans that appear to be derived via either allylphenol and/or monolignol coupling. Examples include the 8—8′ linked (−)-guaiaretic acid (**1**) from *Guaiacum officinale*,[10] (−)-steganacin (**25**) from *Steganotaenia araliacea* (Umbelliferae),[148] and (+)-gomisin A (**121**) from the fruits of *Schizandra chinensis* (Schizandraceae).[149,150] The formation of these optically active lignans is again of considerable interest: (−)-guaiaretic acid (**1**) can be envisaged to result via stereoselective coupling of (iso)eugenol (**58**/**59**) (cf. the mechanism shown in Scheme 12), in a process presumably mediated by the corresponding dirigent protein.

(−)-Steganacin (**25**), on the other hand, could result from dirigent protein mediated coupling to give (+)-pinoresinol (**3a**), this being metabolized into (−)-matairesinol (**21a**) or a substituted butyrolactone equivalent (Scheme 13). A most interesting biochemical conversion then follows which stipulates 2—2′ intramolecular coupling: this may involve a cytochrome P450 catalyzed process, in a manner comparable to diphenol intramolecular coupling processes leading to the alkaloids (*R*,*S*)-berbamunine, (*R*,*S*)-norberbamunine, and (*R*,*R*)-guattegaumerine from (*S*)-coclaurine, (*R*)-*N*-methylcoclaurine, and (*S*)-*N*-methylcoclaurine, respectively, in *Berberis stolonifera* plant cell cultures.[151]

(+)-Gomisin A (**121**) appears to be derived from dirigent protein mediated stereoselective coupling of (iso)eugenol (**58**/**59**) (Scheme 14). If correct, this would initially afford the bis-quinone methide which could then be acted upon by the corresponding reductase to generate the putative

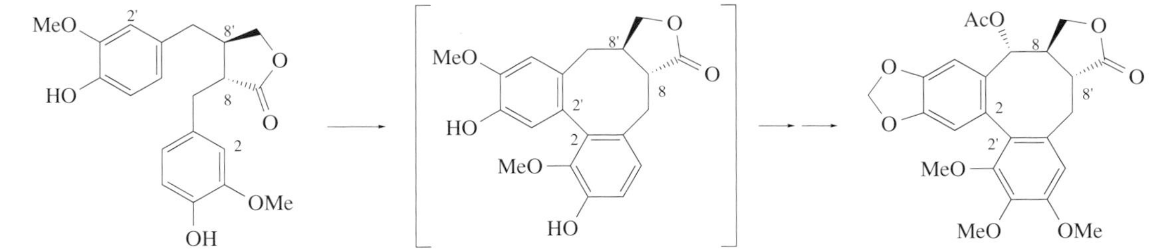

(21) Matairesinol or a substituted butyrolactone equivalent

2–2' intramolecular coupling

(25) (–)-Steganacin
[α]$_D$ = –122.6° ($CHCl_3$, c 1.02)
(*Steganotaenia araliacea*)

Scheme 13

dibenzylbutane (**122**). The latter (or some equivalent) could then undergo 2—2′ coupling, as well as various hydroxylation, methylation, and methylenedioxy bridge formation steps, to ultimately afford (+)-gomisin A (**121**).

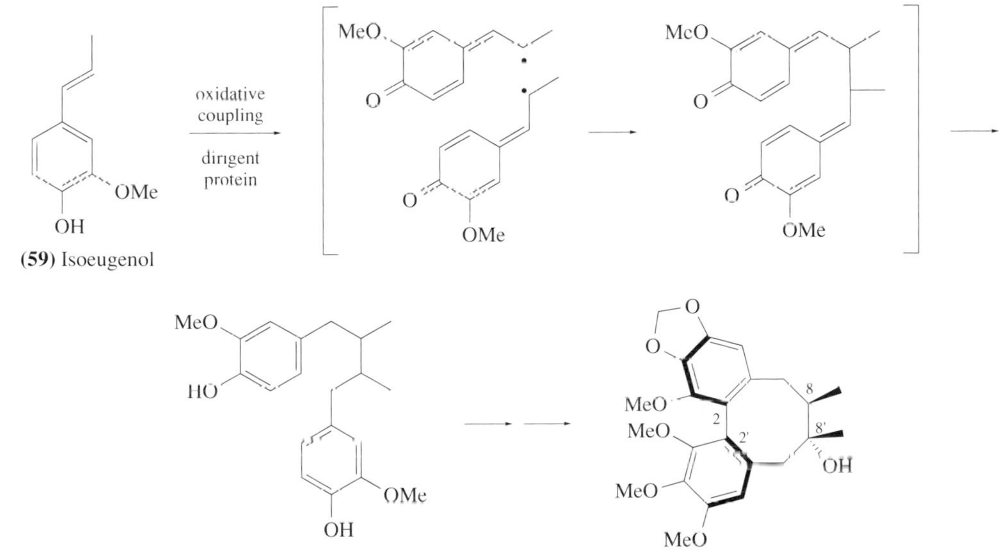

oxidative coupling

dirigent protein

(59) Isoeugenol

(122)

(121) (+)-Gomisin A
[α]$_D$ = +67.9° ($CHCl_3$)
(*Schizandra chinensis*)

Scheme 14

1.25.9 MISCELLANEOUS COUPLING MODES: ARE DIRIGENT PROTEINS ALSO INVOLVED?

Consideration of the chemical structures and optical rotations of many other isolated lignans strongly suggests involvement of dirigent proteins stipulating distinct coupling modes. This is illustrated with only three examples, namely the 8—2′ linked (−)-blechnic acid (**63**) from the fern *B. orientale*,[64] the 8—5′ linked (+)-denudatin B (**123**) from *Magnolia denudata* (Magnoliaceae)[152] and the 8—1′ linked (−)-megaphone (**4**) from *Aniba megaphylla* (Lauraceae).[153]

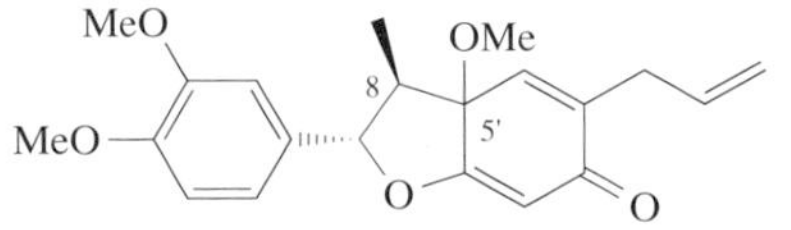

(123) (+)-Denudatin B
[α]$_D$ = +82.7° (MeOH, *c* = 2.67)
(*Magnolia denudata*)

Blechnic acid (**63**), which reputedly co-occurs with its epimeric form, 7-epiblechnic acid (**124**), and other conjugates in *B. orientale*, provisionally appears to result from stereoselective coupling of two caffeic acid (**33**) molecules. Its formation can be envisaged to occur through the involvement of a dirigent protein-mediated process in a manner analogous to that for (+)-pinoresinol (**3a**) formation in *Forsythia* species, with the proposed biosynthetic scheme shown in Scheme 15. In a somewhat analogous manner, formation of both (+)-denudatin B (**123**) and (−)-megaphone (**4**) can be envisaged to occur via stereoselective 8—5′ and 8—1′ coupling of two (iso)eugenol (**58**/**59**) derived compounds, followed by skeletal modifications as needed. Accordingly, these and many other examples again strongly suggest that a class of dirigent proteins exists with each engendering distinct coupling modes.

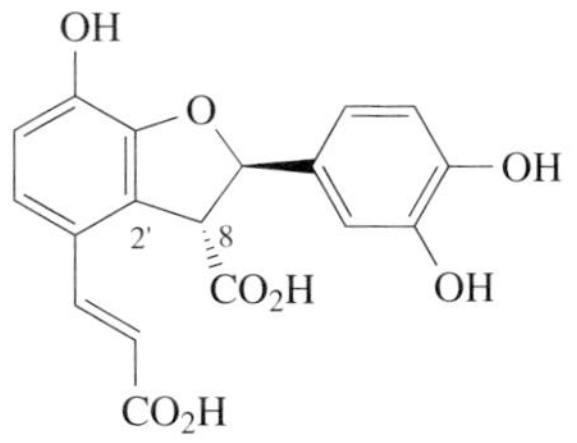

(**124**) (–)-7-Epiblechnic acid
$[\alpha]_D^{23} = -145°$ (MeOH, c = 1.0)
(*Blechnum orientale*)

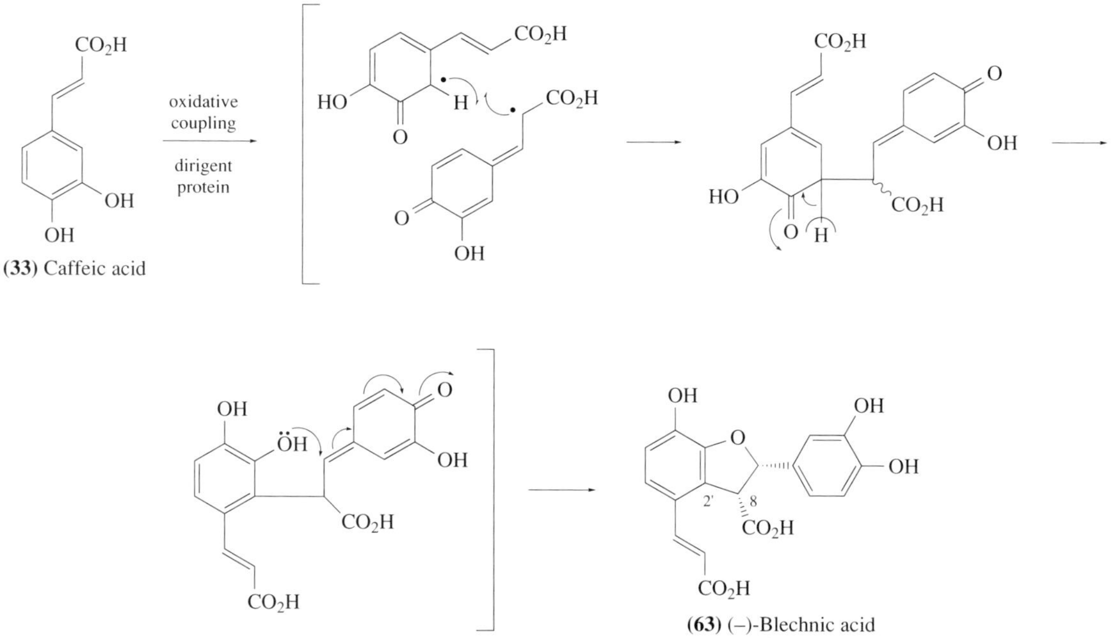

(**33**) Caffeic acid

oxidative coupling

dirigent protein

(**63**) (–)-Blechnic acid
or (**64**) epimer

Scheme 15

1.25.10 8—5′ AND 8—*O*—4′ COUPLING OF MONOLIGNOLS AND ALLYLPHENOLS AND THEIR ASSOCIATED METABOLIC PROCESSES

A number of plant species in the pteridophytes, gymnosperms, and angiosperms contain 8—5′ and 8—*O*—4′ linked lignans, in addition to their 8—8′ constituents. These lignans can, depending upon the plant species from which they are isolated, exist in either optically pure or near racemic form. Of these, the most frequently reported are the dehydrodiconiferyl alcohols (**5a**) and (**5b**), the guaiacylglycerol 8—*O*—4′ coniferyl alcohol ethers (**6**), and derivatives thereof. Another example of an 8—5′ linked lignan is that of the presumed allylphenol-derived lignan, licarin A (**57**).[59,154] The proposed main metabolic role of lignans (**5**) and (**6**) *in vivo* are as intermediates in lignin formation[71] (but see Chapter 3.18), whereas other derivatives, such as dehydrodiconiferyl alcohol glucosides (**125a,b**), are reputed to be cell wall components functioning as cytokinins[155–157] (but see Section 1.25.13.5).

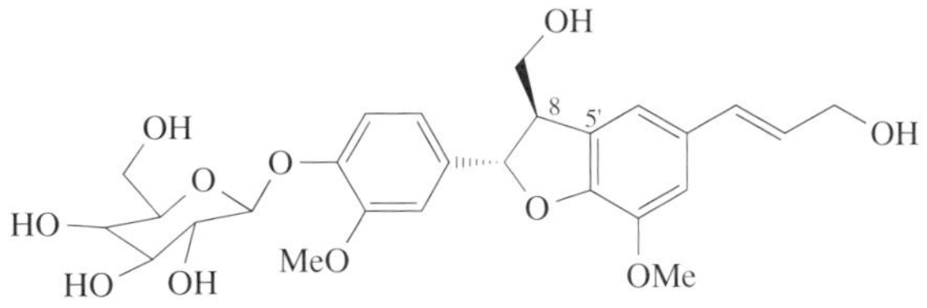

(**125a,b**)

Prior to discussing the biochemical processes known to involve the common 8—5′/8—*O*—4′ coupled products, a brief discussion of their relative configurations and optical activities is required. In a study by Hirai *et al.*,[158] each enantiomer of dehydrodiconiferyl alcohol (**5a,b**) was separated and subsequently degraded by chemical means to give methylsuccinic acid (**126**) via aromatic ring fission (Scheme 16). Following analysis of the configurations of the resulting methylsuccinic acids (**126a**) and (**126b**) at carbon 2 (corresponding to C-8 of lignan (**5**)), it was deduced that the (+)- and (−)-forms of dehydrodiconiferyl alcohol (**5a**) and (**5b**) had 7*S*, 8*R* and 7*R*, 8*S* configurations, respectively. Interestingly, the enantiomers of dehydrodiconiferyl alcohol (**5a**) and (**5b**) have $[\alpha]_D^{25}$ values of $+36.3°$ (MeOH, $c = 0.168$) and $-44.6°$ (MeOH, $c = 0.186$),[159] whereas the synthetic 7′—8′ dihydro derivatives (**127a**) and (**127b**) (Scheme 17) have lower $[\alpha]_D^{25}$ values, i.e., $+5.1°$ (MeOH, $c = 0.196$) and $-9.18°$ (MeOH, $c = 0.244$), respectively. On the other hand, phenols (**128a**) and (**128b**), which have only a single chiral center due to 7—*O*—4′ ether reductive cleavage, have $[\alpha]_D^{25}$ values of $+25.84°$ (MeOH, $c = 0.209$) and $-29.35°$ (MeOH, $c = 0.184$).

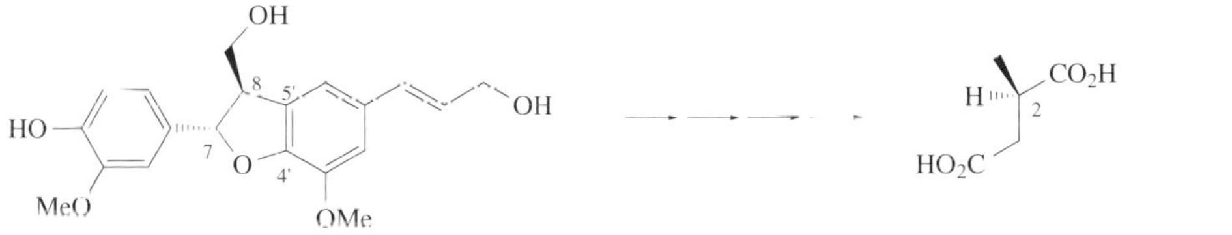

(**5a**) (7*S*, 8*R*)-(+)-Dehydrodiconiferyl alcohol

(**126a**) (*R*)-(+)-Methylsuccinic acid

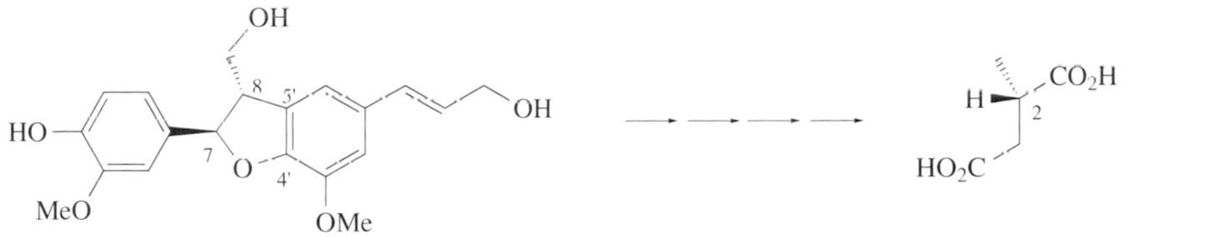

(**5b**) (7*R*, 8*S*)-(−)-Dehydrodiconiferyl alcohol

(**126b**) (*S*)-(−)-Methylsuccinic acid

Scheme 16

While most ^1H and ^{13}C NMR spectroscopic studies of 8—5′ linked lignans have established a *trans* configuration for the constituents attached to carbons 7 and 8,[160–162] there are a number of studies which concluded that various 8—5′ linked lignans were in a *cis* configuration.[66,163–165] Detailed analyses by Wallis and co-workers, however, established that only the *trans* isomers were natural products, and that the proposed *cis* configurations resulted from misinterpretation of spectroscopic data.[67]

With this background regarding optical activities and configurations, naturally occurring 8—5′ linked lignans seem to be present in different plants with differing degrees of enantiomeric purity, based on their observed optical rotations, e.g., (−)-licarin A (**57**) from the liverwort *Jackiella javanica*,[59] which has an $[\alpha]_D = -43°$ ($c = 9.0$), is presumed to be optically pure, whereas the same compound (**57**) in *Eupomatia laurina* is apparently racemic.[166] Additionally, the aglycone (**127**) of *trans*-dihydrodehydrodiconiferyl alcohol glucoside (**71**), $[\alpha]_D^{21} = -8.5°$ ($c = 0.96$, acetone), from the fern, *P. vittata*,[66] is also apparently in significant enantiomeric excess, based on comparison with the optical rotation data obtained for the synthetic analogue (**127b**) described above. Optical

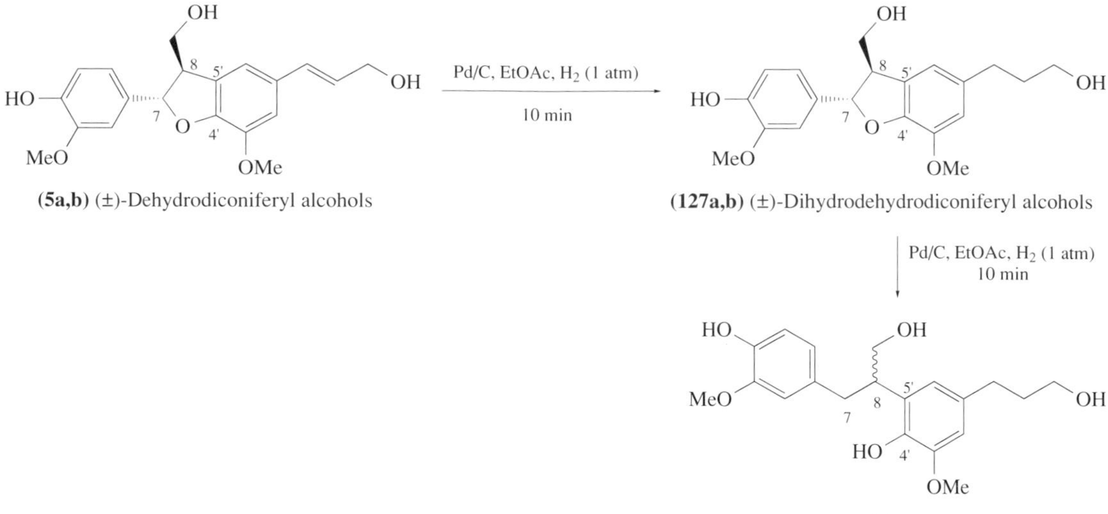

Pd/C, EtOAc, H_2 (1 atm)
10 min

(5a,b) (±)-Dehydrodiconiferyl alcohols

(127a,b) (±)-Dihydrodehydrodiconiferyl alcohols

Pd/C, EtOAc, H_2 (1 atm)
10 min

(128a,b)

Scheme 17

activities for numerous other 8—5′ linked lignans have also been described in the gymnosperms and angiosperms. In general, however, these have a wide range of optical rotations.

Finally, the 8—*O*—4′ linked lignans (**6**), which exist *in vivo* in both *erythro* and *threo* form, have diastereomers that are readily separable by conventional chromatographic methodologies.[87] Although the chiral separation of each form has not been reported, enantiomeric separation was achieved with synthetic analogues, i.e., *erythro* and *threo* forms of lignan (**129**).[16]

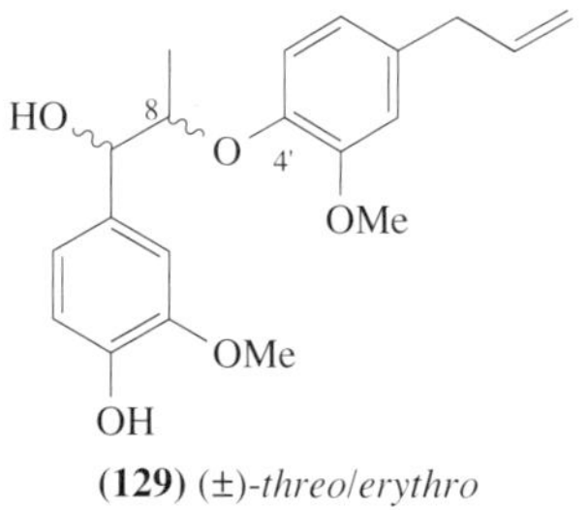

(129) (±)-*threo*/*erythro*

1.25.10.1 Formation and Metabolism of 8—5′ and 8—*O*—4′ Linked Lignans

As for the 8—8′ linked lignans, the fact that many 8—5′ and 8—*O*—4′ lignans are either enantiomerically pure or in enantiomeric excess requires an explicit biochemical explanation, in terms of whether the coupling steps are stereoselective, and/or if subsequent metabolic conversions are enantiospecific.

However, since no systematic study has been reported, it is premature to discuss in any detail how 8—*O*—4′ and 8—5′ coupling may be either regio- or stereoselectively controlled, and whether dirigent protein mediation is involved. On the other hand, the optical activities noted for both (−)-licarin A (**57**) from *J. javanica*[59] and the (−)-aglycone (**127b**) from deglycosylation of (−)-*trans*-dihydrodehydrodiconiferyl alcohol glucoside (**71**) found in *P. vittata*[66] suggest that stereoselective coupling of (iso)eugenol (**58**/**59**) and coniferyl alcohol (**38**), respectively, occurs—at least in these species—but this needs to be established at the protein/enzyme level.

Significant progress has been made though in delineating postcoupling metabolic conversions of dehydrodiconiferyl alcohol (**5**) and guaiacylglycerol 8—*O*—4′ coniferyl alcohol ether (**6**) in loblolly pine (*Pinus taeda*) cell suspension cultures.[159,168–170] Not unexpectedly, based on chemical characterization of previously isolated lignans in the Pinaceae,[168–170] the most common modifications are the regiospecific reduction of the 7′—8′ allylic double bond of (**5**) and (**6**), and demethylation at carbon 3′ (see compounds (**127**) and (**130**)–(**132**)). Additional regiospecific transformations also occur, which involve 7—*O*—4′ reduction (**128**), and in other species, acylation of the 9 and 9′ aliphatic hydroxyl groups with acetate, *p*-coumaroyl, and feruloyl moieties (**133**)–(**136**).

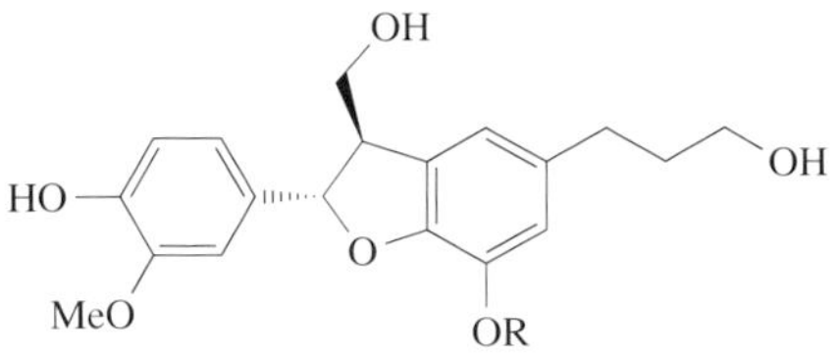

(**127a**) R = Me, Dihydrodehydrodiconiferyl alcohol
$[\alpha]_D^{25}$ = +5.1° (MeOH, c = 0.6)
(**130**) R = H, Cedrusin, $[\alpha]_D^{25}$ = +3.2° (MeOH, *c* = 1.0)
(*Pinus massonian*) *a*

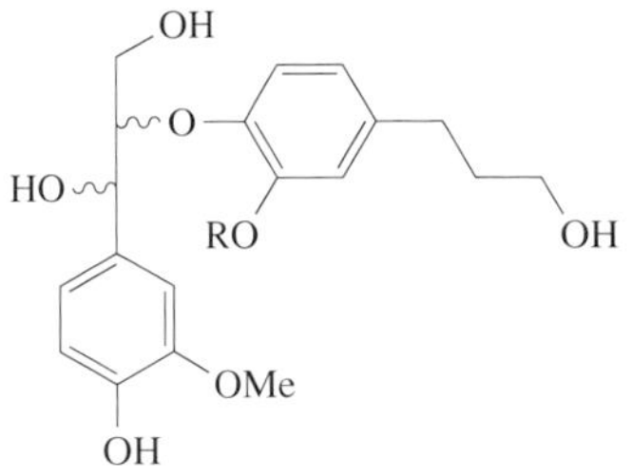

(**131**) R = H
(**132**) R = Me
(*Pinus massonian*) *a*

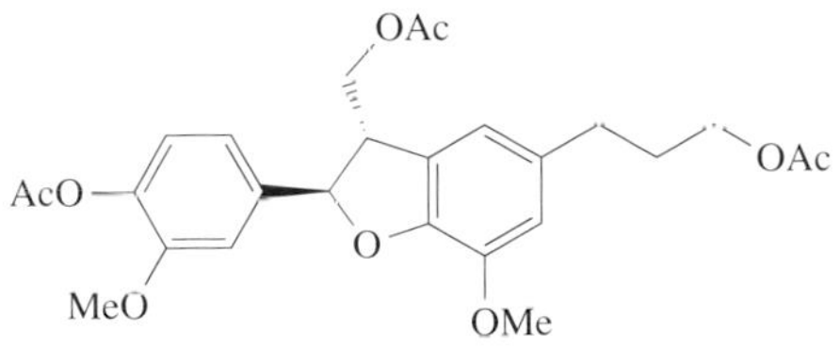

(**133**) *Tran s*Dihydrodehydrodiconiferyl alcohol triacetate
$[\alpha]_D^{25}$ = −66° ($CHCl_3$, *c* = 0.8)
(*Cryptomeria japonic*) *a*

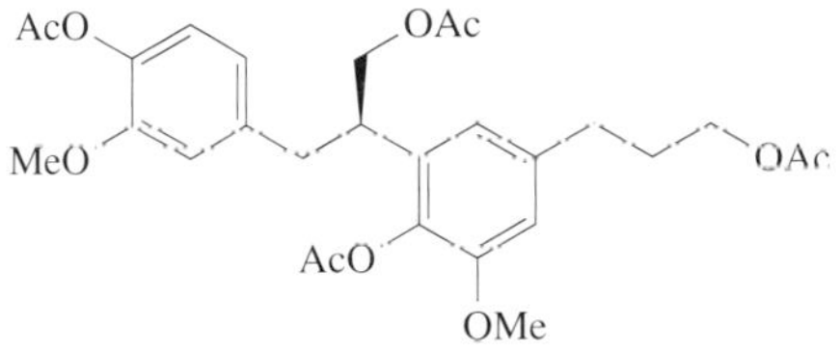

(**134**)
$[\alpha]_D^{25}$ = −2.5° ($CHCl_3$ *c* = 2.0)
(*Cryptomeria japonic*) *a*

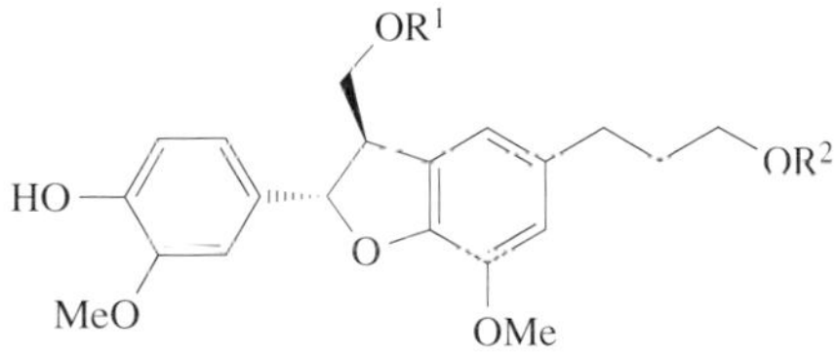

(**135**) $R^1 = R^2$ = *p*-coumaroyl, $[\alpha]_D^{28}$ = +2.0° (MeOH, *c* = 0.5)
(**136**) R^1 = *p*-coumaroyl, R^2 = feruloyl, $[\alpha]_D^{28}$ = +2.3° (MeOH, *c* = 0.5)
(*Corylus sieboldian*) *a*

1.25.10.1.1 Phenylcoumaran 7—O—4′ ring reduction

The 8—5′ linked lignans, such as dehydrodiconiferyl alcohol (**5**) and licarin A (**57**), both contain phenylcoumaran ring structures, which result from nucleophilic attack of the phenol group of ring A onto the quinone methide, as shown in Scheme 18. In certain gymnosperms, such as *Cryptomeria japonica*, there are other 8—5′ linked lignans whose structures might be considered to result from reduction of the 7—*O*—4′ benzylic ether of the phenylcoumaran ring. In *C. japonica*, these metabolites e.g., (**134**) have low optical rotations ($[\alpha]_D^{25} = -2.5°$, $c = 2.0$, $CHCl_3$)[171] suggesting that reductive cleavage might only be regio- rather than enantiospecific. Such reductions are mechanistically analogous to those for pinoresinol/lariciresinol and isoflavonoid reductases.

In this regard, there have been a number of reports of so-called isoflavone reductase homologues, which show considerable sequence homology to both isoflavone and pinoresinol/lariciresinol reductases, e.g., from *A. thaliana*, *Nicotiana tabacum*, *Solanum tuberosum*, *Zea mays*, and *Lupinus albus* (summarized by Dinkova-Kostova *et al.*[111]). None of these homologues have, however, any known catalytic function.

During screening of a *P. taeda* cDNA library, one such homologue was obtained,[159] and Figure 17 shows its gene sequence. The corresponding recombinant protein was subsequently expressed in *E. coli*, but was unable to reduce (±)-pinoresinols (**3a,b**). On the other hand, when incubated with

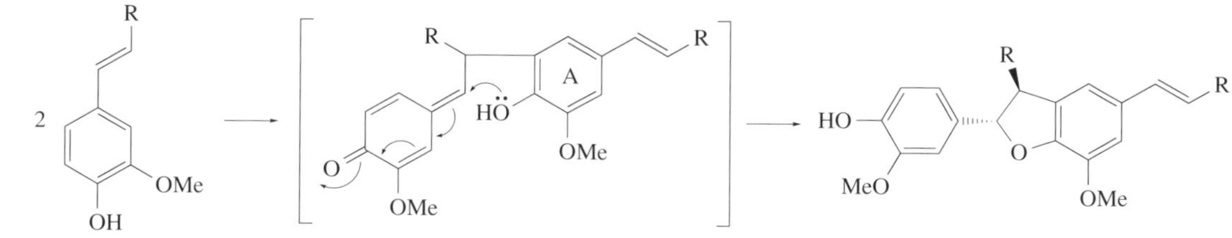

(38) R = CH_2OH, Coniferyl alcohol
(59) R = Me, Isoeugenol

(5) R = CH_2OH, Dehydrodiconiferyl alcohol
(57) R = Me, Licarin A

Scheme 18

(±)-dehydrodiconiferyl alcohols (**5a,b**), it catalyzed 7—*O*—4′ reduction of both enantiomers to give the products (**128a,b**),[172] albeit at low specific activity compared with pinoresinol/lariciresinol reductase. As for pinoresinol/lariciresinol and isoflavone reductases, it is a type A reductase, since only the 4 pro-*R*, but not the 4 pro-*S*, hydride is abstracted from NADPH during reductive cleavage. In contrast to the enantiospecific *Forsythia* pinoresinol/lariciresinol reductase, the recombinant 7—*O*—4′ reductase was capable of effectively reducing both (+)- and (−)-antipodes (**5a**) and (**5b**).

```
      GGCACGAGGTTTCAGGGCCGACATGGGAAGCAGGAGCAGGATACTCCTAATTGGCGCAACAGGATACATTGGTCGCCATGTTGCCAAGGC   68
   1                         M  G  S  R  S  R  I  L  L  I  G  A  T  G  Y  I  G  R  H  V  A  K  A     23
  69  TAGCCTTGATCTCGGCCATCCCACCTTCCTTCTGGTTAGAGAGTCCACTGCTTCTTCTAATTCTGAGAAAGCCCAGCTCCTGGAATCCTT   158
  24   S  L  D  L  G  H  P  T  F  L  L  V  R  E  S  T  A  S  S  N  S  E  K  A  Q  L  L  E  S  F     53
 159  CAAGGCCTCTGGTGCTAATATAGTCCATGGATCCATAGATGATCATGCAAGCCTTGTGGAGGCAGTGAAGAATGTGGATGTAGTAATCTC   248
  54   K  A  S  G  A  N  I  V  H  G  S  I  D  D  H  A  S  L  V  E  A  V  K  N  V  D  V  V  I  S     83
 249  CACAGTTGGATCACTACAGATAGAGAGCCAGGTCAATATTATCAAGGCTATTAAAGAAATTGGAACCGTCAAGAGGTTTTTCCATCTGA   338
  84   T  V  G  S  L  Q  I  E  S  Q  V  N  I  I  K  A  I  K  E  I  G  T  V  K  R  F  F  P  S  E    113
 339  GTTCGGGAATGATGTTGATAACGTCCATGCAGTGGAGCCTGCAAAGAATGTGTTTGAGGTGAAAGCCAAAGTCCGTAGGGCAATCGAAGC   428
 114   F  G  N  D  V  D  N  V  H  A  V  E  P  A  K  N  V  F  E  V  K  A  K  V  R  R  A  I  E  A    143
 429  AGAGGGTATTCCTTATACATACGTCTCTAGCAACTGTTTTGCAGGGTATTTCCTGCGAAGCCTCGCACAGGCTGGCCTAACAGCTCCTCC   518
 144   E  G  I  P  Y  T  Y  V  S  S  N  C  F  A  G  Y  F  L  R  S  L  A  Q  A  G  L  T  A  P  P    173
 519  AAGAGATAAAGTTGTCATTCTTGGAGATGGAAATGCCAGAGTTGTCTTTGTAAAAGAGGAAGACATTGGAACATTTACAATCAAGGCAGT   608
 174   R  D  K  V  V  I  L  G  D  G  N  A  R  V  V  F  V  K  E  E  D  I  G  T  F  T  I  K  A  V    203
 609  GGACGACCCCAGAACGTTGAACAAGACCTTATACTTGAGGCTTCCTGCCAATACTCTGTCTTTAAATGAGCTTGTGGCTCTCTGGGAGAA   698
 204   D  D  P  R  T  L  N  K  T  L  Y  L  R  L  P  A  N  T  L  S  L  N  E  L  V  A  L  W  E  K    233
 699  GAAGATTGATAAGACTCTGGAGAAGGCCTACGTGCCCGAGGAGGAGGTTCTTAAATTAATCGCAGATACACCATTCCCAGCTAATATTAG   788
 234   K  I  D  K  T  L  E  K  A  Y  V  P  E  E  E  V  L  K  L  I  A  D  T  P  F  P  A  N  I  S    263
 789  CATAGCAATTAGTCATTCTATCTTCGTGAAAGGAGATCAAACAAATTTTGAAATTGGACCTGCTGGTGTAGAGGCTAGTCAGCTGTACCC   878
 264   I  A  I  S  H  S  I  F  V  K  G  D  Q  T  N  F  E  I  G  P  A  G  V  E  A  S  Q  L  Y  P    293
 879  TGATGTGAAATACACCACCGTCGACGAGTACCTGAGCAATTTTGTGTGAACTATGCAACATTCTTCAACGAACCTAATAGATTAAATAGT   968
 294   D  V  K  Y  T  T  V  D  E  Y  L  S  N  F  V  *                                               309
 969  GGTTTCTATGAAAGTTTATATAAGGCATTTTGCCAAATTGTTGGTTATCGCCTTTCATAGATATTCGACAAAATAGAACCTCCAAAATAG   1058
1059  AACCTCCTGTTTAGTATATTATGATAAGCCAGGCTAGGTAGATCCTTCACACATTTCTGTTCTGTAAGAAGCAAACATATAAAATATCCG   1148
1149  TTTTTTGTTTTTTGATTAAAAAAAAAAAAAAAAAACTCGAGGGGGGCCCGTACCCAAT                                  1205
```

Figure 17 Complete sequence of *Pinus taeda* 7—*O*—4′ -reductase cDNA, plrh-Pt.

Thus, it appears there is a family of reductases in phenylpropanoid metabolism, which catalyze the reduction of isoflavonoids (e.g., hydroxypseudobaptigenin (**93**)), 8—8′ linked lignans (e.g., pinoresinol (**3**) and lariciresinol (**19**)), and 8—5′ linked phenylcoumarans (e.g., dehydrodiconiferyl alcohol (**5**)). The latter, however, differs from the others by the ability to effectively reduce both (+)- and (−)-antipodes (**5a**) and (**5b**) of the substrate, i.e., it is regio- rather than enantiospecific.

1.25.10.1.2 Allylic 7—8′ bond reduction

A common structural modification of 8—5′ and 8—*O*—4′ lignans *in planta*, including in the Pinaceae such as *Pinus taeda* cell suspension cultures, is that of the so-called "dihydro" derivatives, whose structures are based mainly on dihydrodehydrodiconiferyl alcohol (**127**)[160–162,164,165,168,171–185] and guaiacylglycerol 8—*O*—4′ dihydroconiferyl alcohol ethers (**132**).[165,172,178–180] However, in addition to these substances, other "dihydro" compounds frequently co-occur, including dihydro-*p*-coumaryl (**137**)[178] and dihydroconiferyl (**138**)[162,165,173,176,178] alcohols, as well as various dihydrocinnamic acids e.g., (**139**) and (**140**).[173,178]

(**137**) $R^1 = CH_2OH$, $R^2 = H$, Dihydro *p*-coumaryl alcohol
(**138**) $R^1 = CH_2OH$, $R^2 = OMe$, Dihydroconiferyl alcohol
(**139**) $R^1 = CO_2H$, $R^2 = H$, Dihydro *p*-coumaric acid
(**140**) $R^1 = CO_2H$, $R^2 = OMe$, Dihydroferulic acid

P. taeda suspension cultures have been used to define the biochemical processes involved in allylic double bond reductions of dehydrodiconiferyl alcohols (**5a,b**), where using partially purified reductase preparations, it was established that NADPH (but not NADH) was required as a cofactor.[169] Individual incubation of (**5a,b**) with [4*R*-^{3}H] and [4*S*-^{3}H]NADPH, respectively, revealed that only the 4 pro-*R* hydride was transferred to the reduced product (**127**), with the 7′—8′ allylic bond reductase being capable of utilizing both (+)- and (−)-antipodes (**5a**) and (**5b**) as substrates. It needs to be determined, however, whether the reductase reduces the allylic double bond directly, or whether some (oxidized) intermediate is involved. The only other known allylic bond reductase is that of enoyl acyl carrier protein reductase in fatty acid biosynthesis,[186,187] which requires the double bond to be in conjugation with a carbonyl group. The purified dehydrodiconiferyl alcohol 7′—8′ double bond reductase is thus required, in order to mechanistically determine how this reduction is achieved. Additionally, it needs to be established as to whether the same enzyme catalyzes formation of all of the various "dihydro" natural products albeit with different substrate specificities, or if a family of these reductases exists with each catalyzing distinct transformations.

1.25.10.2 Regiospecific *O*-Demethylation at Carbon 3′ and Monosaccharide Functionalization

In *P. taeda* cell suspension cultures, (±)-dehydrodiconiferyl alcohols (**5**) can also undergo further metabolism involving regiospecific demethylation at C-3′. As shown in Figure 18, when (±)-[9,9′-$^3H_2$]dehydrodiconiferyl alcohols (**5**) were incubated with *P. taeda* cell suspension cultures for different time intervals, both enantiomers were either rapidly demethylated to give demethyldehydrodiconiferyl alcohol (**141**), or reduced to afford dihydrodehydrodiconiferyl alcohols

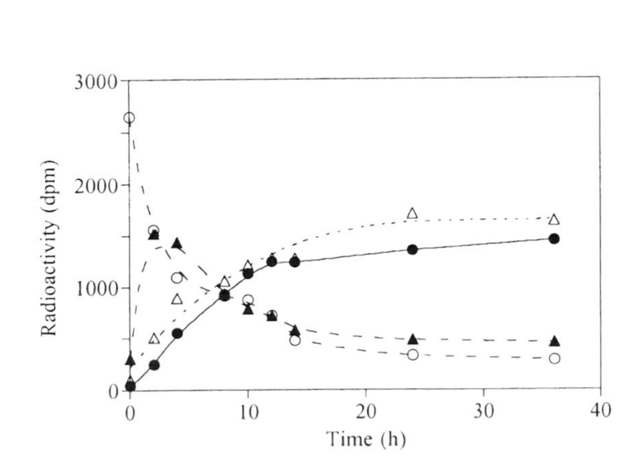

(**5a,b**) R=Me, (±)-Dehydrodiconiferyl alcohols
(**141a,b**) R=H, (±)-Demethyldehydrodiconiferyl alcohols

(**127a,b**) R=Me, (±)-Dihydrodehydrodiconiferyl alcohols
(**130a,b**) R=H, (±)-Cedrusins

Figure 18 Time course of metabolism of (±)-dehydrodiconiferyl alcohols (**5a,b**) by *Pinus taeda* cell suspension cultures. ○, (±)-dehydrodiconiferyl alcohols (**5a,b**); ▲, (±)-dihydrodehydrodiconiferyl alcohols (**127a,b**); ●, (±)-demethyldehydrodiconiferyl alcohols (**141a,b**); △, (±)-cedrusins (**130a,b**).

(**127a,b**).[170] In turn, both metabolites (**127a,b**) were finally converted into (±)-demethyldihydrodehydrodiconiferyl alcohols = (±)-cedrusins (**130a,b**);[170] the enzymology involved in this demethylation awaits full characterization.

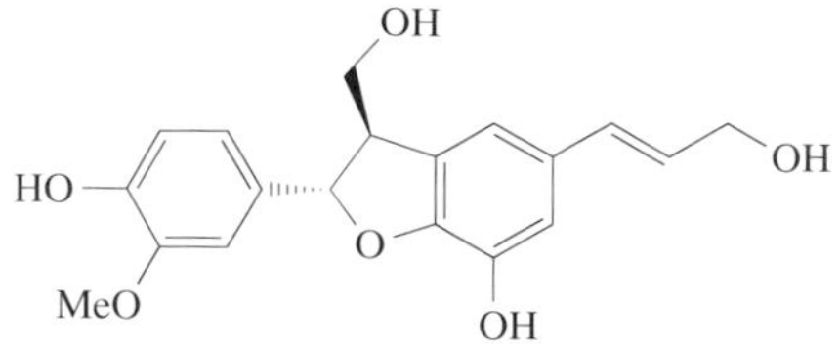

(**141**)
(*Pinus taeda*)

Additionally, these and related derivatives, whether from dihydrodehydrodiconiferyl alcohol (**127**), dehydrodiconiferyl alcohol (**5**), and guaiacylglycerol 8—*O*—4′ (dihydro)coniferyl alcohol ethers (**6**) and (**132**) are frequently found conjugated to monosaccharides. They are typically attached to the C-4 phenolic hydroxyl groups with either glucose or rhamnose, e.g., (**142**) and (**143**) in *Picea abies*[172] and *Pinus silvestris*,[165,182] and even if the demethylated 3′ hydroxyl group is in its free phenolic form. In other species, such as *Clematis stans*,[177] the glucose moiety is attached to the 3′ hydroxyl position of cedrusin (**130**) (see (**144**)), but not to the 4-hydroxy functionality, whereas in *Licaria chrysophylla* (Lauraceae),[162] it is attached to C-9 of dihydrodehydrodiconiferyl alcohol (**127**) (see (**145**)). The significance, if any, of these different monosaccharide attachments is unknown.

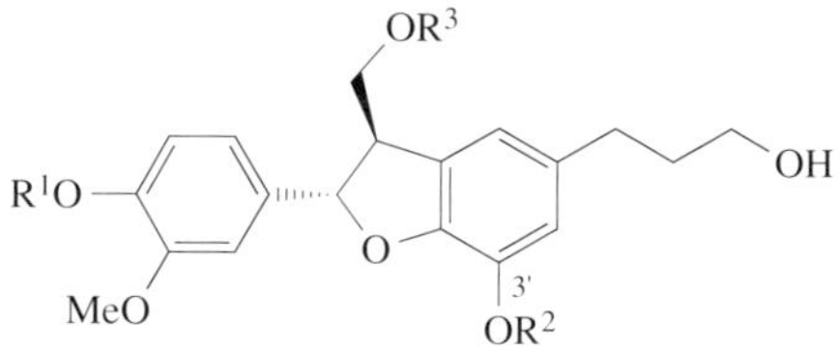

(**142**) R^1 = Glc, R^2 = R^3 = H
(**143**) R^1 = Rha, R^2 = R^3 = H
(*Pinus silvestris*)
(**144**) R^1 = H, R^2 = Glc, R^3 = H, Clemastanin A
(*Clematis stans*)
(**145**) R^1 = R^2 = H, R^3 = Glc
(*Licaria chrysophylla*)

1.25.10.3 Acylation

A few lignans have also been reported in species such as *C. japonica* and *Corylus sieboldiana* which contain acyl, *p*-coumaroyl, or feruloyl moieties conjugated to the lignan skeleton (**133**)–(**136**).[171,174] These have been noted for both 8—5′ (**133**)–(**136**) and 8—8′ (**146**)[171] linked lignans, but nothing has been established about how these *trans*-esterifications and coupling reactions occur.

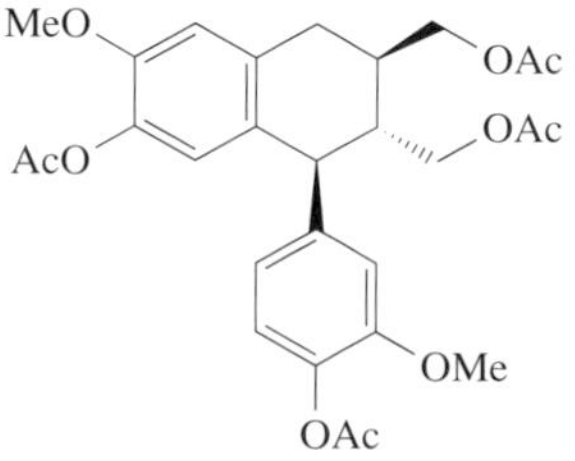

(**146**)
(*Cryptomeria japonica*)

1.25.11 MIXED DIMERS CONTAINING MONOLIGNOLS AND RELATED MONOMERS

In addition to the *bona fide* lignan skeletal types described throughout this chapter, there are other natural products trivially described as being flavonolignans, coumarinolignans, stilbenolignans, etc. These appear to result either from coupling of monolignols, or in some cases, lignans, with other phenolic substances, such as flavonoids. These include, for example, the flavonolignan, sinaiticin (**147**) from *Verbascum sinaiticum* (Scrophulariaceae) leaves[188] which appears to result from regio/stereospecific coupling of luteolin (**148**) with *p*-coumaryl alcohol (**32**). Indeed, given that the only optical center present in the molecule results from monolignol attachment, how this coupling is controlled needs to be defined. Related substances such as the optically active hydnowightin (**149**) are also present in seeds of *Hydnocarpus wightiana* (Flacourtiaceae),[32] again suggesting enzymatic control of coupling in both a regio- and stereospecific manner. More complex flavonolignans also exist, such as pseudotsuganol (**150**), which consists of pinoresinol (**3**) linked to a dihydroquercetin (**151**) moiety,[31,189] this being found in the outer bark of Douglas fir (*Pseudotsuga menziesii*) in optically active form ($[\alpha]_D = +20.2°$ ($c = 0.25$, MeOH)).

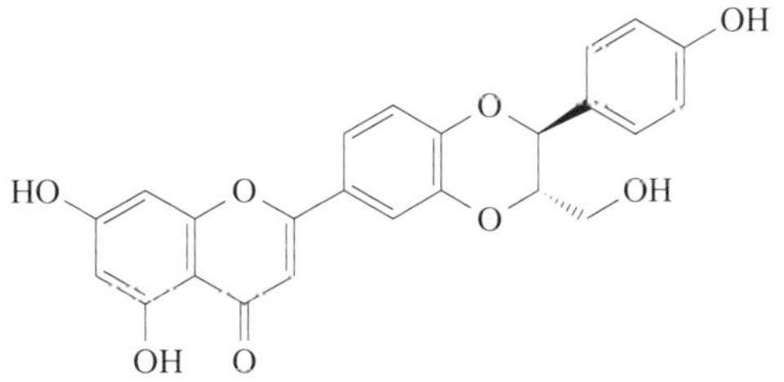

(**147**) Sinaiticin
$[\alpha]_D = -15.2°$ (MeOH, $c = 0.0033$)
(*Verbascum sinaiticum*)

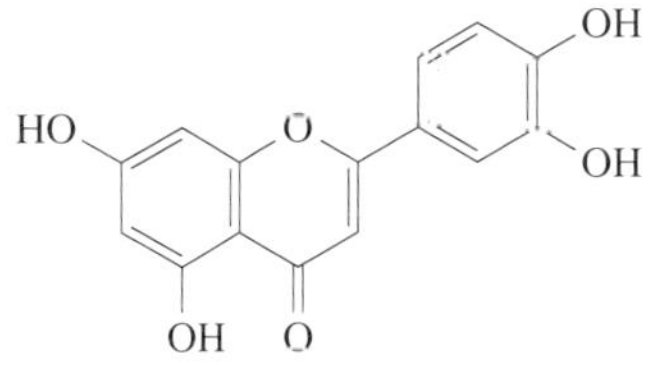

(**148**) Luteolin

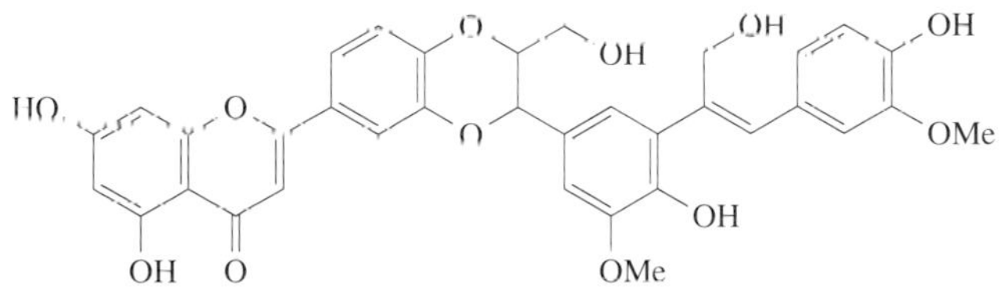

(**149**) Hydnowightin
$[\alpha]_D = +40°$ (MeOH, $c = 0.55$)
(*Hydnocarpus wightiana*)

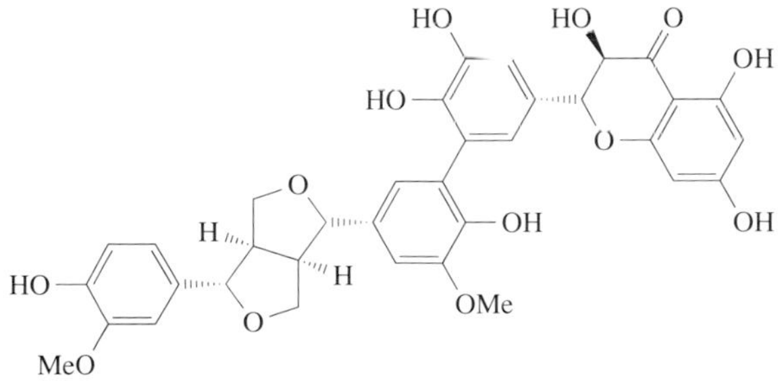

(**150**) Pseudotsuganol
$[\alpha]_D = +20.2°$ (MeOH, $c = 0.25$)
(*Pseudotsuga menziesii*)

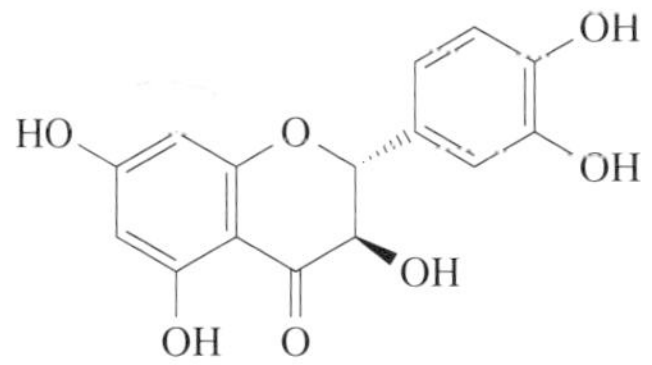

(**151**) Dihydroquercetin
(*Pseudotsuga menziesii*)

Another interesting group of mixed dimers are the lignanamides, the jacpaniculines (**152**) and (**153**) from the fruits of *Jacquemontia paniculata* (Convolvulaceae).[190] These presumably result from coupling of *E*-coniferyl alcohol (**38**) with feruloyl tyramine (**109**), although nothing can be concluded about how coupling might be carried out, since the enantiomeric purity of these metabolites has not been described. On the other hand, the stilbenolignans, such as maackoline (**154**) from *Maackia amurensis* (Leguminosae) heartwood,[191] are apparently racemic suggesting nonspecific coupling of the presumed sinapyl alcohol (**44**) and stilbene precursors.

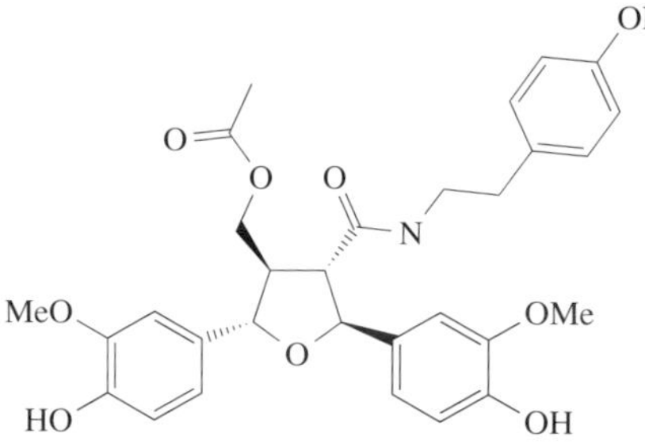

OH OH O O O N MeO OMe O HO OH

(152) Jacpaniculine
(*Jacquemontia paniculata*)

OH O O O N MeO OMe O HO OH

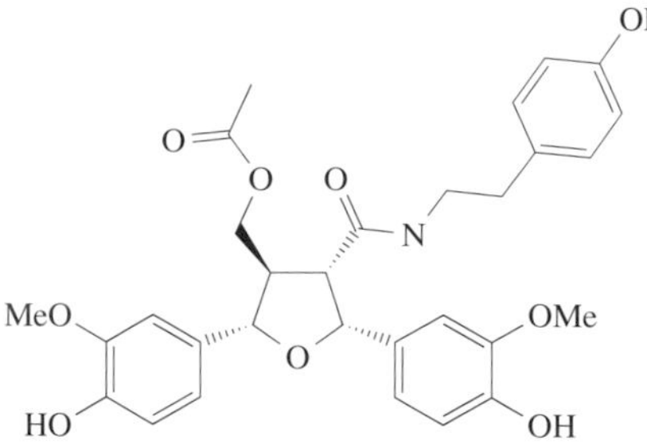

(153) Isojacpaniculine
(*Jacquemontia paniculata*)

OMe HO OH OH OMe H H O HO OH

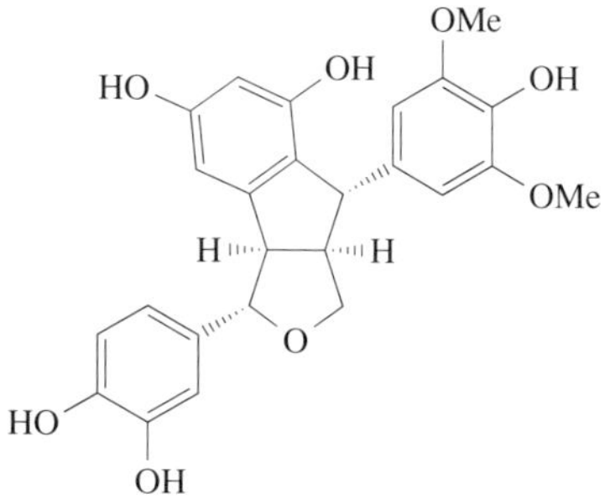

(154) Maackoline
$[\alpha]_D = 0$
(*Maackia amurensis*)

Other "mixed dimers" include (+)-megacerotonic acid (**61**) from *M. flagellaris*,[61–63] (−)-cryptoresinol (**13**) from *C. japonica* (Pinaceae),[17] and the diarylheptanoids, alnusdiol (**155**) and maximowicziol A (**156**) from *Betula maximowicziana* (Betulaceae) heartwood.[192] (+)-Megacerotonic acid (**61**) co-occurs with rosmarinic acid (**157**), and its formation (Scheme 19) may result from oxidative coupling as shown, i.e., where generation of the transient biradical species, derived from (**158**) gives, following ring closure and rearomatization, the 7—8′ linked lignan (**61**). A comparable mechanism may account for the formation of the lignan, (−)-cryptoresinol (**13**). The diarylheptanoids (**155**) and (**156**), by contrast, appear to result from oxidative coupling of two *p*-coumaryl alcohol (**32**) molecules via 3—3′ or 3—*O*—4′ bonds. Interestingly, these metabolites contain an additional carbon attached to the 9/9′ carbons. This suggests that the linear diarylheptanoid moiety is first formed, followed by phenoxy radical coupling; in the latter case, intramolecular coupling may result from action of a cytochrome P-450 type oxidase, such as previously noted for (*R*,*S*)-berbamunine, (*R*,*S*)-norberbamunine, and (*R*,*R*)-guattegaumerine.[151]

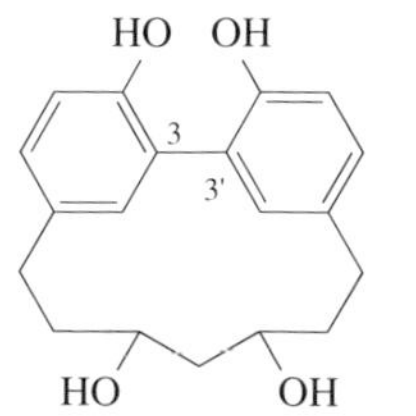

HO OH 3 3' HO OH

(155) (−)-(a*S*, 9*S*, 9′*S*)-Alnusdiol
$[\alpha]_D^{23} = -46.0°$ (MeOH, $c = 0.85$)
(*Betula maximowicziana*)

OH 3 O 4' OH OH

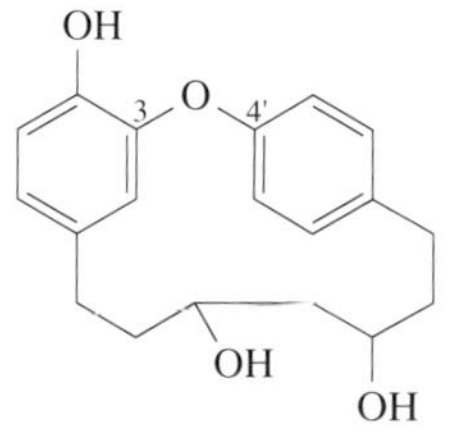

(156) (−)-rel-(*p*R, 9*S*, 9′*S*)-Maximowicziol A
$[\alpha]_D^{23} = -86.3°$ (MeOH, $c = 0.66$)
(*Betula maximowicziana*)

Finally, a quite abundant group of "mixed" dimers are the so-called cyclobutane lignans found in the Poaceae (Commeliniflorae) (**72**) and (**73**)[68,69] and the Ariflorae (**74**).[70] These are presumed to be primarily formed via photodimerization of either juxtaposed hydroxycinnamic acids (e.g., (**29**) and (**35**)), aldehydes (e.g., (**31**) and (**37**)), alcohols (e.g., (**32**) and (**38**)), or even allylphenols (e.g., (**58**)–(**60**)). These molecules are typically attached to the various cell wall fractions of grasses and grains. They can apparently be formed by either head-to-head, head-to-tail, or tail-to-tail coupling; all are believed to be optically inactive.

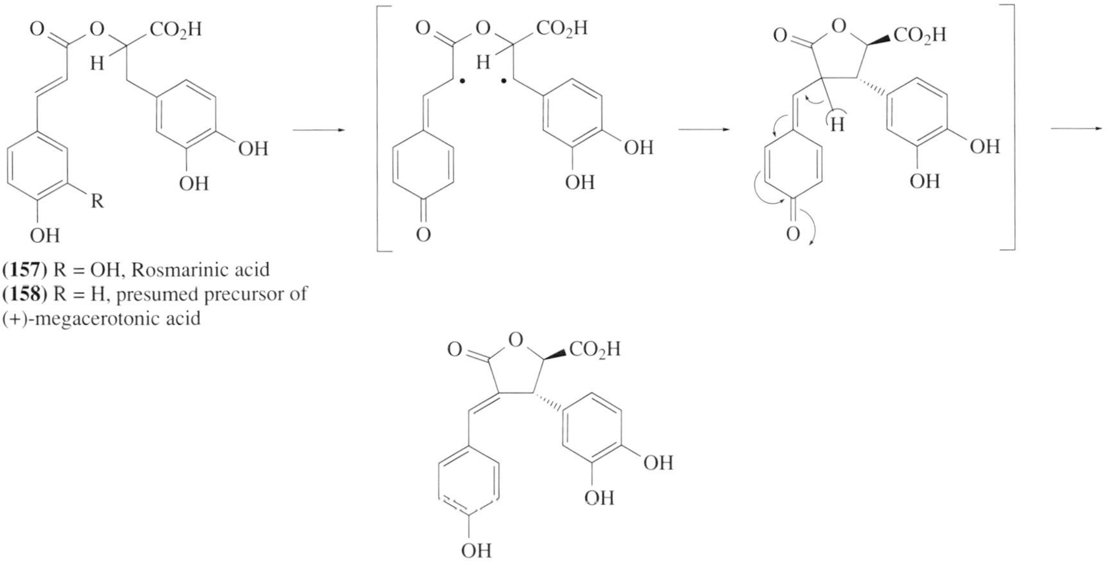

(157) R = OH, Rosmarinic acid
(158) R = H, presumed precursor of (+)-megacerotonic acid

(61)

Scheme 19

1.25.12 LIGNANS AND SESQUILIGNANS: WHAT IS THE RELATIONSHIP TO LIGNIN FORMATION?

The view has long been held,[72] without rigorous scientific proof, that monolignol (glycoside)s are transported from the cytoplasm into lignifying cell walls where they undergo sequential random coupling to give biopolymeric lignins via transient (oligomeric) lignan formation. However, this view must be tempered by the fact that "random" coupling of monolignols *in vitro* never gave an adequate representation of the natural lignin biopolymer(s), in terms of, for example, frequencies of inter-unit linkages[73] (see Chapter 3.18). In contrast, lignification proper has been proposed to occur via end-wise polymerization of the monolignols at discrete points in the cell wall, this being envisaged to occur along a template viewed to consist of arrays of dirigent protein sites[74] or some proteinaceous equivalent, with the initial lignin strand then replicating via a template mechanism[193–195] (see Chapter 3.18 and work by Lewis *et al.*[73]). Such a process would preclude the formation of transient dimeric and higher oligomeric forms undergoing random coupling, but would explain all the known features of lignin proper *in situ*.

On the other hand, the original random coupling hypothesis was supported in part from analyses of acetone-water extracts from homogenized sapwood and heartwood tissues. Such preparations were mainly obtained from the Pinaceae, and were thought at the time to contain "native or Brauns lignins."[196] Curiously, while they shared some structural similarities with lignin biopolymers (e.g., being coniferyl alcohol (**38**)-derived), they were never proven: (i) to be present in lignifying cell walls, (ii) to have any structural roles, and (iii) to be formed via direct monomer polymerization.

Today, these "Brauns and native lignins" appear only to be nonstructural, nonlignin oligomeric sesquilignans, which depending upon the plant system involved can apparently exist in a variety of different molecular sizes (see Section 1.25.13.6). Their roles appear to be primarily in defense. For example, the roots of the herbaceous plant *Phryma leptostachya* (Phrymaceae) contain a series of insecticidal oligolignans, such as haedoxan A (**159**) and its congeners,[197] whose formation is clearly under stereospecific enzymatic control (Scheme 20). It contains a modified 8—8′ furanofuran skeleton, with an additional C_6C_3 moiety linked through the 7″—8″ positions as noted for flavonolignans such as sinaiticin (**147**). Presumably it is formed via stereospecific radical–radical coupling of coniferyl alcohol (**38**) to the preformed dimer (**160**) or some equivalent thereof, which then undergoes further metabolism to give haedoxan A (**159**). Such substances clearly would be unable to undergo further conversion to give lignin biopolymer(s) proper, or even to be incorporated into the same via a template-replication process.

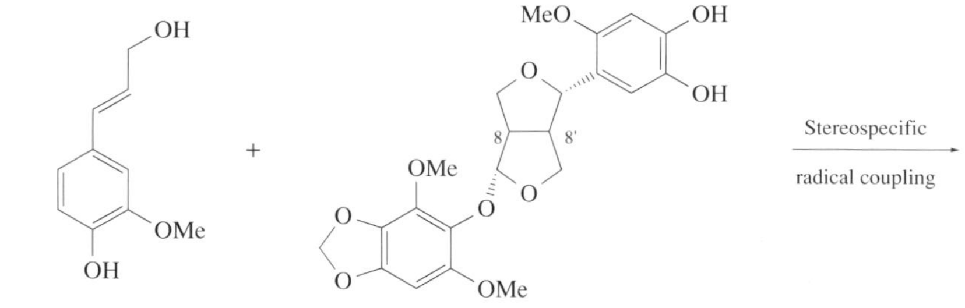

(38) Coniferyl alcohol + **(160)** Dimer

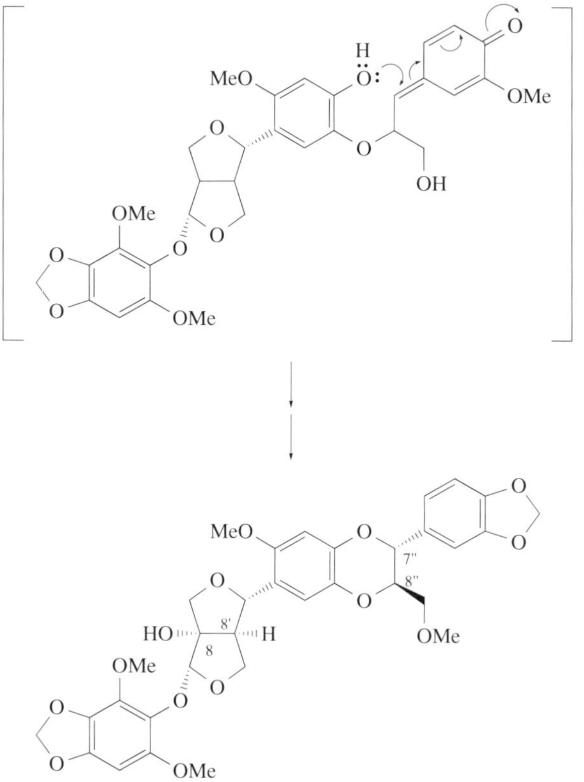

Stereospecific radical coupling

(159) Haedoxan A

$[\alpha]_D^{27} = +125°$ ($EtOH-CH_2Cl_2$, $c = 0.32$)

(*Phryma leptostachya*)

Scheme 20

Other sesquilignans frequently consist of monolignols conjugated to optically active lignan dimers, such as medioresinol (**107**), secoisolariciresinol (**20**), matairesinol (**21**), dihydrodehydrodiconiferyl alcohol (**127**), and olivil (**161**). These include substances such as hedyotol C (**162**) from *E. ulmoides*[198] bark tissue and *Hedyotis lawsoniae* (Rubiaceae) leaves,[83] which can be viewed to result from coupling of medioresinol (**107**) with coniferyl alcohol (**38**). Another example is dihydrobuddlenol B (**163**) from *Prunus jamasakura* (Rosaceae) bark,[199] which consists of a 5-methoxydihydrodehydrodiconiferyl (**164**) moiety attached to a coniferyl alcohol (**38**) residue via an 8—*O*—4′ linkage. Others include 7′-hydroxylappaol E (**165**) from hemlock (*T. heterophylla*) sapwood,[200] consisting of 7′-hydroxymatairesinol (**99**) linked to coniferyl alcohol (**38**) via an 8—*O*—4′ bond, as well as the lappaols (**166**) and (**167**) from the roots of *Arctium lappa* (Compositae)[26,201] and the cerberalignans (**168**)–(**170**) from *Cerbera manghas* and *Cerbera odollam* (Apocynaceae) stem tissue.[202–204] The lappaols (**166**) and (**167**) are primarily matairesinol (**21**)-derived lignans linked via 8—5′ bonds to coniferyl alcohol (**38**) residues, whereas the cerberalignans (**168**)–(**170**) are mainly

(161) (–)-Olivil
$[\alpha]_D^{26} = -48.5°$ (MeOH, c = 1.5)
(*Cerbera manghas, C. odollam*)

(162) Hedyotol C
$[\alpha]_D^{27} = +30.7°$ (MeOH, c = 0.3)
(*Eucommia ulmoides*)
(*Hedyotis lawsoniae*)

(163) Dihydrobuddlenol B
$[\alpha]_D^{29} = -28.2°$ (MeOH, c = 0.43)
(*Prunus jamasakura*)

(164) 5-Methoxydihydrodehydrodiconiferyl alcohol

(165) (–)-7'-Hydroxylappaol E
$[\alpha]_D^{25} = -3.7°$ (MeOH)
(*Tsuga heterophylla*)

(166) Lappaol A
$[\alpha]_D^{20} = -17.4°$ (MeOH, c = 1.0)
(*Arctium lappa*)

(167) Lappaol C
$[\alpha]_D^{20} = -55°$ (MeOH, c = 1.0)
(*Arctium lappa*)

(168) Cerberalignan A
$[\alpha]_D^{20} = -76°$ (MeOH, c = 0.55)
(*Cerbera manghas, C. odollam*)

(169) Cerberalignan B
$[\alpha]_D^{20} = -75.4°$ (MeOH, c = 0.35)
(*Cerbera manghas, C. odollam*)

(170) Cerberalignan D
$[\alpha]_D^{20} = -50.6°$ (MeOH, c = 0.98)
(*Cerbera manghas, C. odollam*)

(−)-olivil (**161**)-derived molecules linked head-to-head, tail-to-tail, or head-to-tail. As before, however, isolation procedures, such as for the cerberalignans, were extremely lengthy and the various lignans isolated were present together with much larger amounts of the presumed precursors, e.g., olivil (**160**), again raising the possibility of artifact formation.

Perhaps one of the most unusual sesquilignan structures proposed is that of herpepentol (**171**), described as present in methanol extracts of grains from *Herpetospermum caudigerum* Wall. (Cucurbitaceae).[205] Based on its preliminary characterization by FAB MS and ^{1}H NMR spectroscopy, a coniferyl alcohol (**38**) derived 8—5′, 8—5′, 8—5′, 8—8′ linked pentamer was proposed. Although such a substance could only result from endwise coupling (cf. lignin), a more definitive study is, however, necessary to establish that its proposed structure is correct.

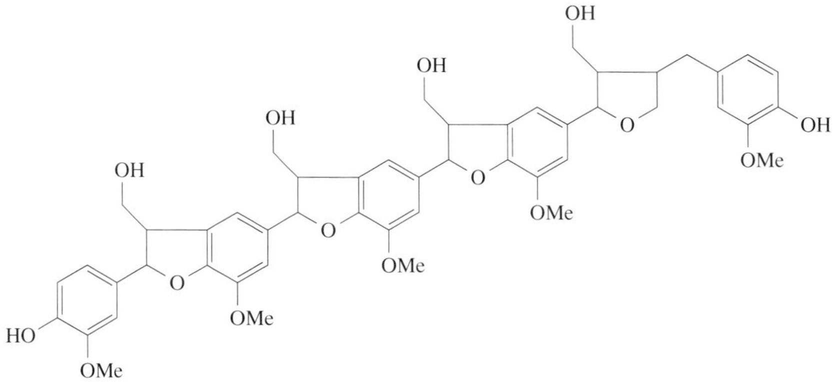

OH OH OH OH OH OH HO O O O O OMe OMe OMe OMe OMe

(**171**) Herpepentol
(*Herpetospermum caudigerum* Wall.)

Thus, the preponderance of available evidence reveals that (sesqui)lignans have structures which cannot be derived directly via simple free-radical coupling of monolignols, and therefore lead to lignin formation. Moreover, the structural modifications typically encountered (e.g., methylenedioxy bridge formation, allylic double bond reduction, etc.) give substances which (bio)chemically cannot readily undergo further coupling to form high molecular weight lignin biopolymers. Accordingly, there is no convincing evidence linking (sesqui)lignan formation to that of the structural lignin biopolymers, and both pathways must be viewed as being biochemically, configurationally, temporally, and spatially distinct (discussed in Section 1.25.13.6).

1.25.13 PHYSIOLOGICAL ROLES *IN PLANTA*

Although identification of the physiological roles of (oligomeric) lignans is still very much in its infancy, significant progress in defining some of their functions has been made. The best documented roles appear to be primarily defense related, namely, antioxidant, biocidal, feeding deterrent, and allelopathic properties. As for other systems, and although this aspect is seldom examined, a specific biological activity can be associated with a particular enantiomeric form. Using a pharmacological example to illustrate this point, (−)-trachelogenin (**17**) inhibits the replication of HIV-1 *in vitro*, whereas the corresponding (+)-antipode is much less effective.[7] Another significant role of lignans in certain species is in heartwood formation, since they affect color, durability, quality, and texture of the resulting wood: these substances have occasionally been erroneously described as "abnormal or secondary lignins." Other putative roles proposed for the (oligomeric) lignans, include that of cytokinins[155–157] and as intermediates in lignification,[206] although there are significant counterarguments to both proposed functions (see Sections 1.25.13.6 and 1.25.12).

1.25.13.1 Antioxidant Properties

Nordihydroguaiaretic acid (**172**), a major constituent of the resinous exudate of the creosote bush (*Larrea tridentata*),[207] is one of the most powerful antioxidants known,[208–210] whereas others such as sesamolin (**77**), sesamin (**18**), and sesamolinol (**81**) from sesame (*S. indicum*) seeds display less potent but still striking antioxidant properties.[4,98,99] Indeed, it is perhaps no coincidence that several oil seed bearing plants, such as sesame and flax, contain high lignan levels, which accordingly help stabilize lipid (oil) components against oxidative degradation and onset of rancidity.

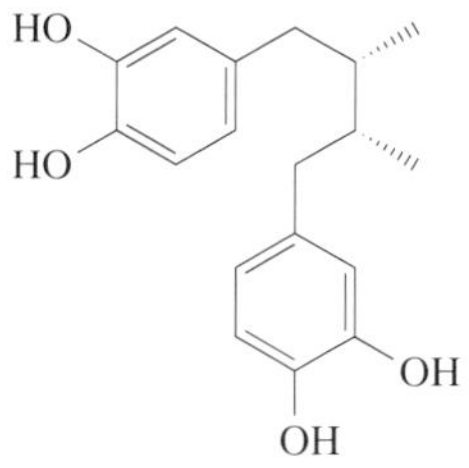

(**172**) Nordihydroguaiaretic acid

Because of these properties, several studies have been directed toward defining the precise modes of action of lignans *in planta* and in human applications. Thus, nordihydroguaiaretic acid (NDGA) (**172**) competitively inhibits soybean meal lipoxidase-catalyzed oxidation of sodium linoleate,[210,211] an important antioxidant property given that lipoxidase is directly involved in the autooxidation of unsaturated fatty acids during vegetable and seed oil manufacture. Indeed, NDGA (**172**) was a common antioxidant in various foodstuffs until 1972, when its use was discontinued following indications that it had toxic effects on the kidneys.[209] It is used instead in nonfood applications, such as in stabilizing polymers, rubber, perfumery oils, and photographic formulations.

1.25.13.2 Antifungal and Antimicrobial Effects

Many lignans have antifungal and antimicrobial properties. For example, termilignan (**104**) and (−)-thannilignan (**106**), from the popular Indian traditional medicinal plant, *T. bellerica* (Combretaceae), have potent activities against the fungus *Penicillium expansum*: the minimum amounts of each lignan required to inhibit fungal growth (1 and 2 μg respectively), compared favorably with conventional treatment levels with nystatin (0.5 μg).[137]

Additionally, formation of (−)-matairesinol (**21a**) and related metabolites, e.g., 7′-hydroxymatairesinol (**99**) and α conidendrin (**76**), is induced in *Picea abies* upon infection by *Fomes annosus*,[212,213] which, in turn, limits further fungal growth.[212–214] This observation may partly help explain the massive deposition of lignans in the heartwood of western red cedar (*T. plicata*)[35] which, in conjunction with tropolones, helps confer protection to this plant species thereby enabling life spans in excess of 3000 years to be reached.

Other lignans reputed to have antifungal properties include representatives from *Podophyllum hexandrum*[215] and the katsura tree (*Cercidiphyllum japonicum*, Cercidiphyllaceae):[216] 4′-*O*-demethyldehydropodophyllotoxin (**173**), and picropodophyllone (**174**) from *P. hexandrum* reportedly have antifungal activities against *Epidermophyton floccosum*, *Curvularia lunata*, *Nigrospora oryzae*, *Microsporum canis*, *Allescheria boydii*, and *Pleurotus ostreatus*, although no quantitative data were given,[215] and magnolol (**7**) from *C. japonicum* accumulates in twig cortical tissue in response to *Fusarium solani* f. sp. *mori* invasion.[216] Magnolol (**7**) is also effective against *Aspergillus niger* and *Tricophyton mentagrophytes* with minimum inhibitory concentrations (MIC) of 30 and 2.5 $\mu g\ ml^{-1}$, respectively, which compares well with the control using amphotericin B (MIC = 30 and 15 $\mu g\ ml^{-1}$).[217]

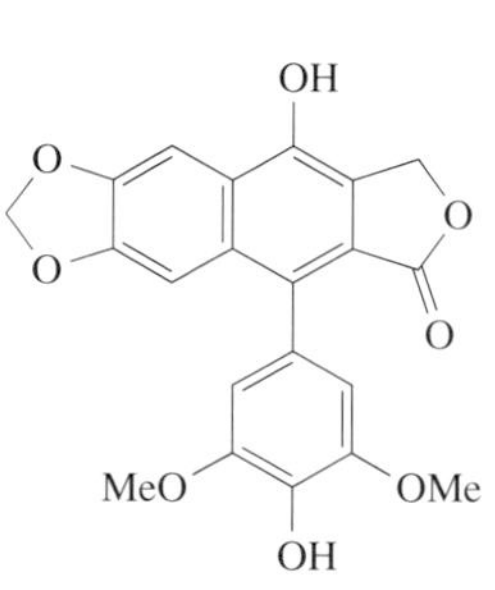

(**173**) 4′-*O*-Demethyldehydropodophyllotoxin

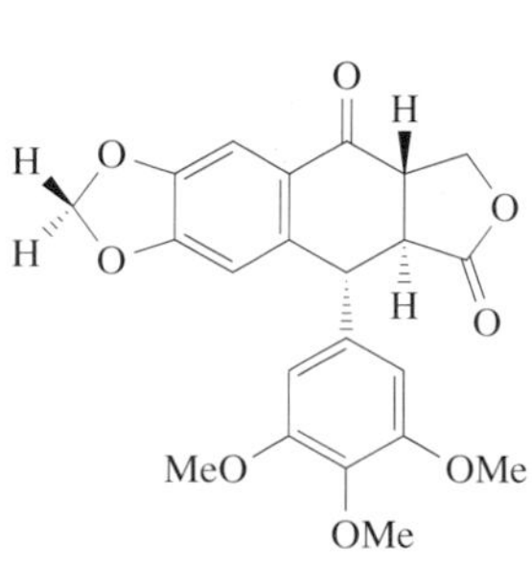

(**174**) Picropodophyllone

There have also been a limited number of studies examining both regio- and enantiospecific conversions of lignans in response to exposure to various fungi. For example, (±)-eudesmins (**83**), when incubated with *Aspergillus niger*, are converted into both pinoresinol monomethyl ether (**89**) ($[\alpha]_D^{29} = -12.8°$, enantiomeric excess 39.3%) and pinoresinol (**3**) ($[\alpha]_D^{29} = -57.6°$, enantiomeric excess 100%), where the (−)-antipode was more rapidly demethylated than its (+)-counterpart.[218] Why this organism preferentially metabolizes one particular enantiomer may be of significance in plant–fungus interactions. This may also be particularly important in lignin biodegradation studies, which most commonly use lignans rather than lignins for their assays. Unfortunately, none of the lignin-degradation studies have examined whether conversions with lignans are enantiospecific.

A number of other lignans are also regiospecifically demethylated at the *para* position by *A. niger*, notably (+)-magnolin (**84**), (+)-epimagnolin A (**175**), (+)-veraguensin (**176**), (+)-galbelgin (**177**), and galgravin (**178**), but not (+)-yangambin (**85**).[219–221] This fungus can also de-ethylate diethyl pinoresinol (**179**) and its monoethyl derivative (**180**) to give pinoresinol (**3**), but has no effect on dipropyl (**181**) and dibutyl (**182**) analogues.[222]

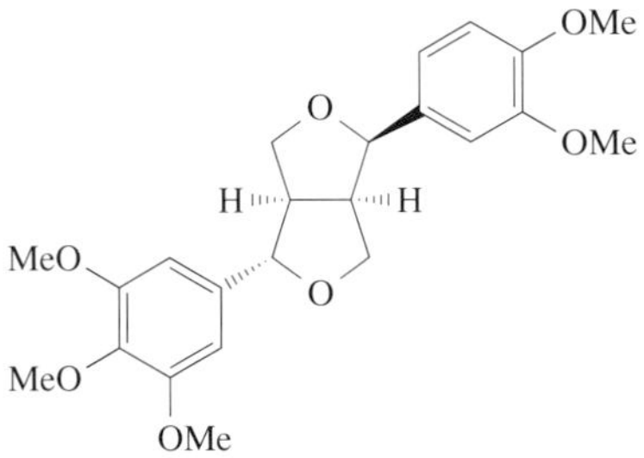

(**175**) (+)-Epimagnolin

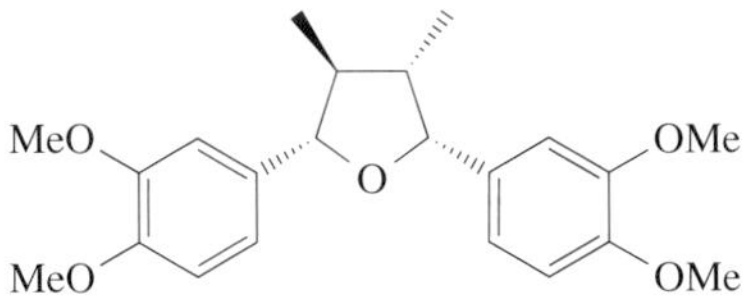

(**176**) (+)-Veraguensin

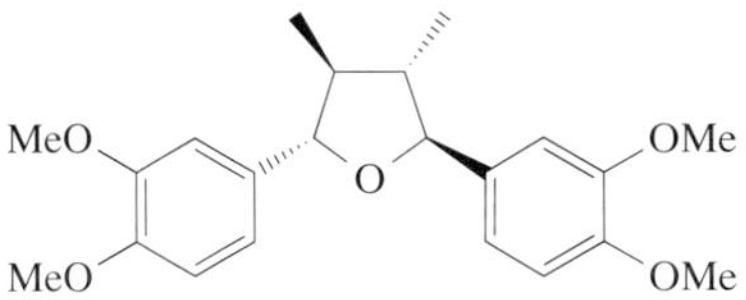

(**177**) (+)-Galbelgin

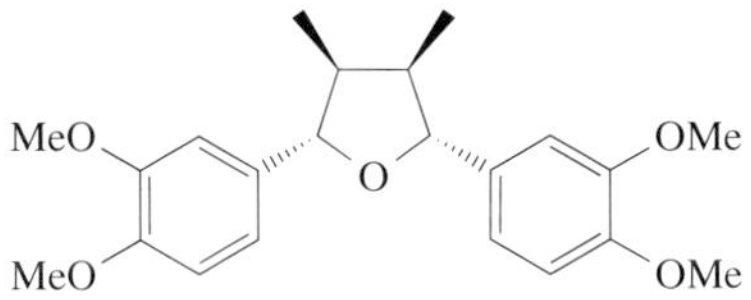

(**178**) Galgravin

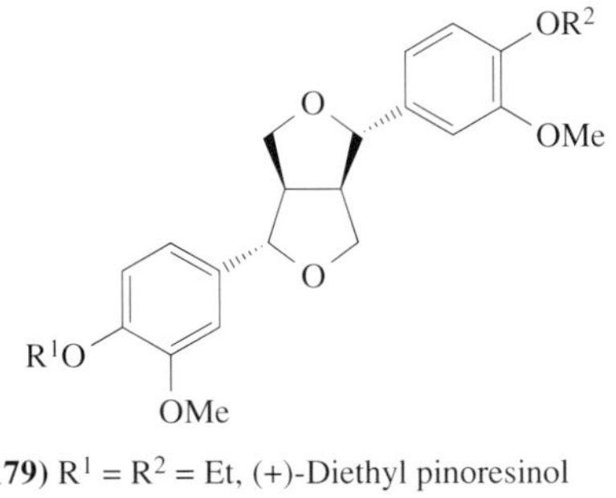

(**179**) $R^1 = R^2 =$ Et, (+)-Diethyl pinoresinol
(**180**) $R^1 =$ Et, $R^2 =$ H, (+)-Monoethyl pinoresinol
(**181**) $R^1 = R^2 = n$Pr, (+)-Dipropyl pinoresinol
(**182**) $R^1 = R^2 = n$Bu, (+)-Dibutyl pinoresinol

Interestingly, and while the significance is again unknown, *Fusarium solani* enantiospecifically converts Δ8′-hydroxy-3,3′dimethoxy-7-oxo-8-*O*-4′-lignan (**183**)—when incubated in the presence of the ketone (**184**)—into the *threo*/*erythro* lignans (**129**) in a ratio of 2:3 (Scheme 21). For both *threo* and *erythro* forms, only one enantiomer was formed, revealing that the reductive step was fully enantiospecific.[167]

Lignans also have antibacterial properties: magnolol (**7**) and honokiol (**185**) inhibit *Staphylococcus aureus*, *Bacillus subtilis*, and *Mycobacterium smegmatis* bacterial growth with a MIC of 5–10 μg ml^{-1}, which is comparable to or better than the activity of streptomycin sulfate (MIC = 10, 10, and 2.5 μg ml^{-1}, respectively)[217], and nordihydroguaiaretic acid (**172**) is effective against salmonella, penicillium, *M. pyrogenes*, and *Saccharomyces cerevisiae*.[209] Extracts from the aril tissue of *Myristica*

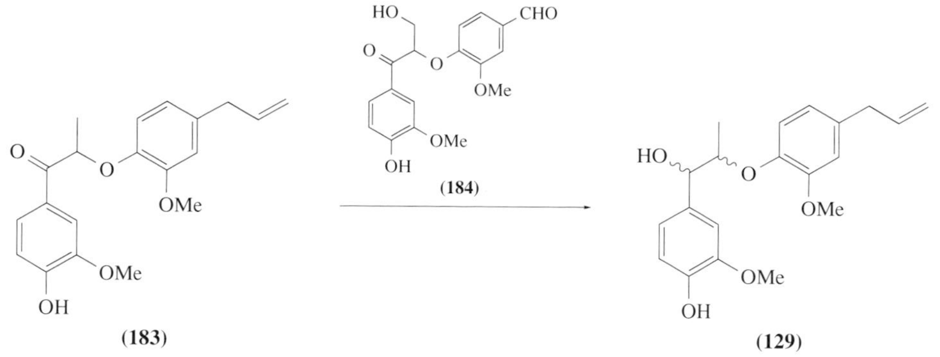

Scheme 21

fragrans (mace) are also used in Sri Lanka for dental caries prevention (antiplaque formation), with dehydrodiisoeugenol (licarin A) (**57**) and 5′-methoxydehydrodiisoeugenol (**186**) being characterized as the major antibacterial principles. Both inhibit the growth of *Streptococcus mutans* at concentrations of 12.5 μg ml^{-1}.[223]

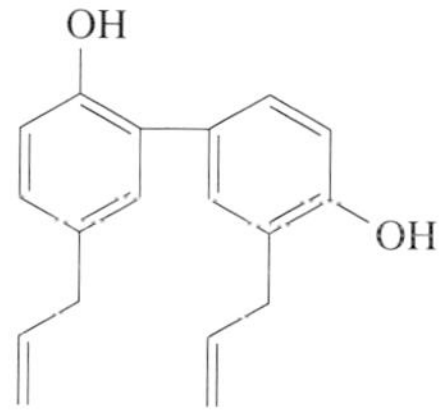

(**185**) Honokiol

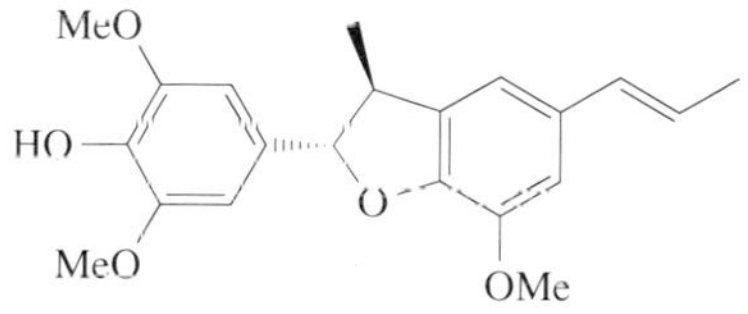

(**186**) 5-Methoxydehydrodiisoeugenol

1.25.13.3 Insecticides, Nematocides, Antifeedants, and Poisons

(+)-Haedoxan A (**159**), isolated from the roots of the herbaceous perennial plant, *Phryma leptostachya*, is perhaps the best known insecticidal lignan.[197,224–226] In combination with piperonyl butoxide (a synergist), it has excellent insecticidal activity, when administered orally to several lepidopterous insect larvae and houseflies, e.g., *Musca domestica*, $LD_{50} = 0.25$ ng per fly,[225] this being comparable to that of commercial synthetic pyrethroids. Its physiological effect results in muscle relaxation, feeding cessation, general paralysis, and death, thereby causing similar effects to the insect neurotoxins, nereistoxin, ryanodine, and reserpine.[197] Another insecticidal lignan is the 8,5′-linked, licarin B (**187**), isolated from *Myristica fragrans*, which is effective against silkworm (*Bombyx mori*) fourth instar larvae (at 300 ppm in the diet), with death occurring 3 days or so after treatment.[227] Additionally, the lignans (**188**), (**189**), and magnolol (**7**), isolated from *Magnolia virginiana* (Magnoliaceae), cause 100% mortality to mosquito (*Aedes aegypti*) larvae at concentrations of ~10 ppm within 2 hours, this being comparable to valinomicin treatment as a control.[228]

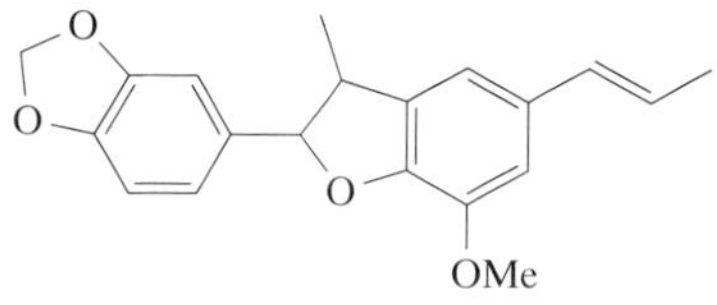

(**187**) Licarin B

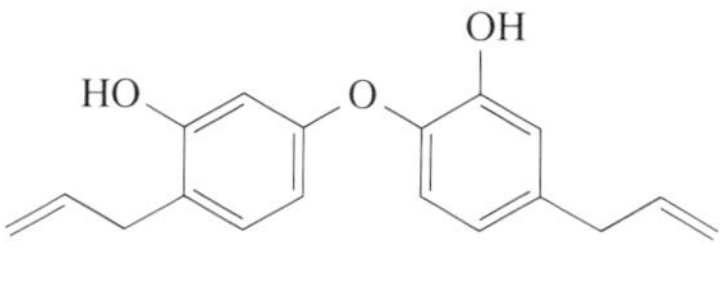

(**188**)

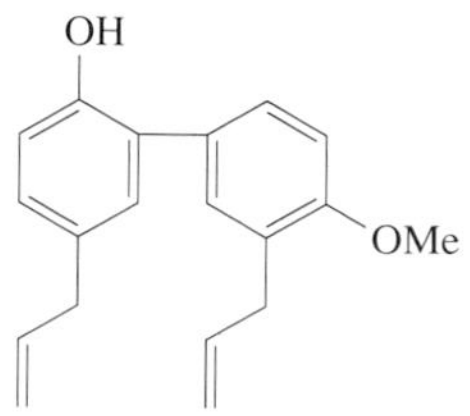

(**189**)

Growth inhibitory properties of lignans have also been described: (+)-epimagnolin A (**175**), isolated from the flower buds of *Magnolia fargesii*, inhibits the growth of *D. melanogaster* larvae[229] at concentrations greater than 1 mg ml^{-1}, whereas licarin A (**57**), (−)-machilusin (**190**), and lignans

(**191**) and (**192**) from the leaves of *Machilus japonica* function by inhibiting *Spodoptera litura* larval growth when added to their diets ($EC_{50} = 0.20$, 0.19, 0.13, and 0.24% w/w, respectively).[230] (+)-Sesamin (**18**) and (+)-sesamolin (**77**) from *S. indicum* also synergistically act with natural juvenile hormone to prevent metamorphosis in the milkweed bug (*Oncopeltus fasciatus*) at amounts of 10 and 1 μg, respectively.[231] Moreover, (+)-sesamin (**18**) and *epi*-sesamin (**193**) function with pyrethrum insecticides, by inhibiting oxidative degradation (cytochrome P450 oxygenase system) in the gut of the ingesting organism.[5,232]

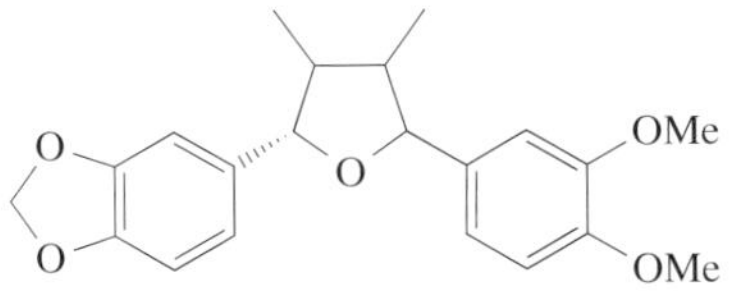

(**190**) (–)-Machilusin

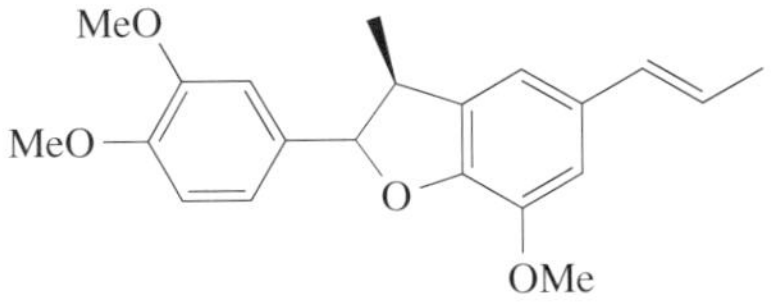

(**191**)

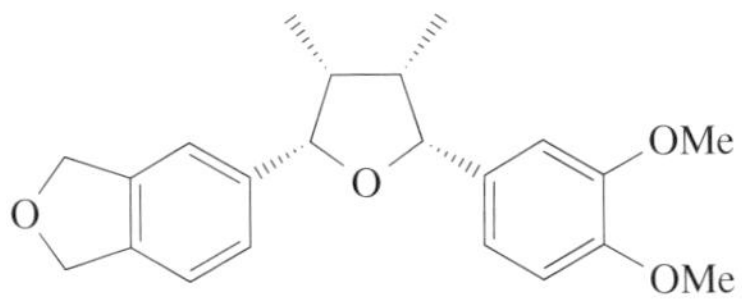

(**192**)

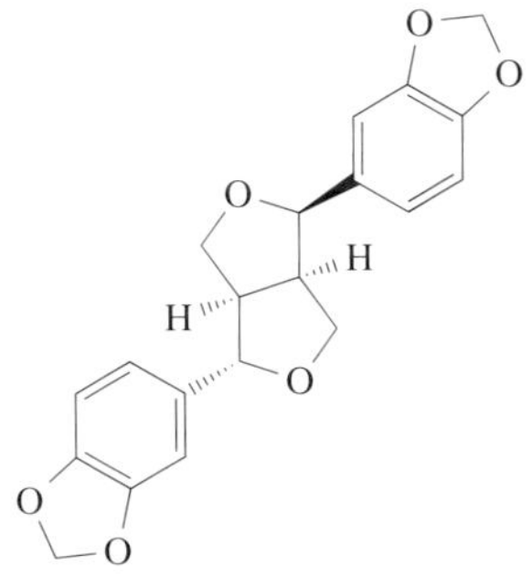

(**193**) *Epi*-sesamin

Lignans confer protection against nematodes.[233,234] For example, (−)-matairesinol (**21a**) and (−)-bursehernin (**194**), at concentrations of ~50 μg ml^{-1}, inhibit the hatching of potato cyst nematodes, *Globodera pallida* and *G. rostochiensis*, by 55% and 70%, respectively, when compared with controls using $ZnSO_4$ and ethanol: the hatching inhibitory dose (16.42 μg ml^{-1}) for bursehernin (**194**) reduced hatching by 50% over a 2-week period. The presence of the methylenedioxy bridge seems to play an important role, since when replaced with either a methoxyl, one hydroxyl, or two hydroxyl groups, the inhibitory activity was greatly reduced. Interestingly, lariciresinol 9-*O*-*β*-D-glucoside (**70**), isolated from the roots and stolons of potatoes, accumulates in response to infection with *G. rostochiensis*,[235] perhaps also suggesting a nematocidal role.

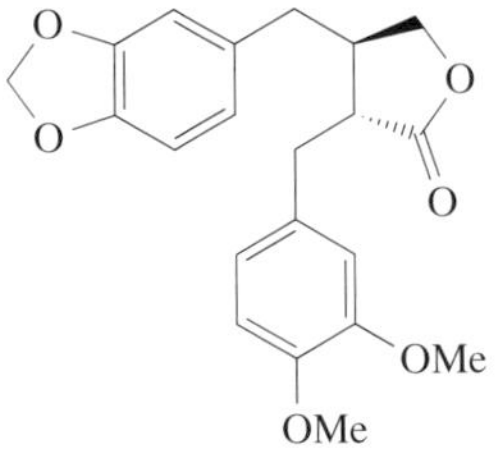

(**194**) Bursehernin

Several lignans display antifeedant and/or toxicity effects. (−)-Yatein (**195**) has antifeedant properties when added to the diets of adult granary weevil beetles (*Sitophilus granarius*) and confused flour beetle (*Tribolium confusum*) with an estimated total coefficient of deterrency of 189 and 158, respectively, where values > 200 indicate very good feeding-deterrent activity.[236] (+)-Eudesmin (**83**)/(+)-*epi*-eudesmin (**196**) from *Parabenzoin praecox*, and (−)-piperenone (**197**) from *Piper futo-kadsura*, have antifeedant activities (90–100%) against *S. litura* larvae when provided at concentrations of 0.05%, 1.0%, and 0.005% in the diet.[237,238] Justicidins A (**198**) and B (**199**) from *Justicia hayatai* var. *decumbens*, on the other hand, show strong toxicity against *Oryzias latipes* at levels comparable to that of rotenone and 10 times higher than that of pentachlorophenol. *J. hayatai* has been used for many centuries as a fish-poison by the natives of the Pescadores (Pung Fu islands) of Taiwan.[239]

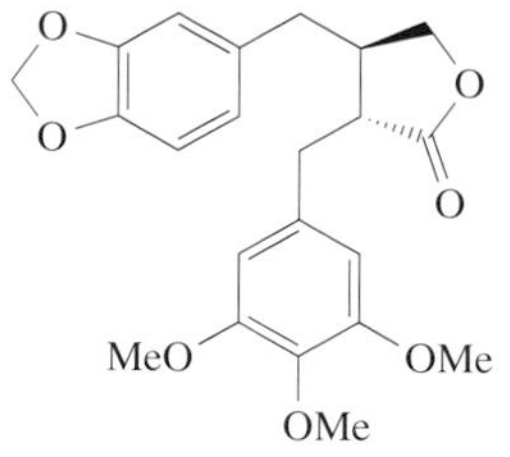

(195) Yatein

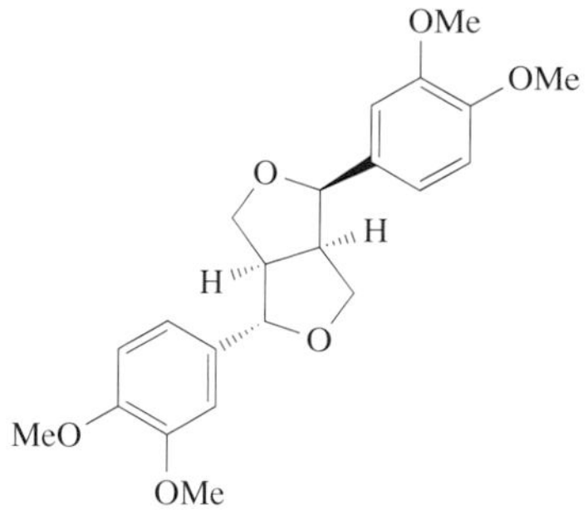

(196) *Epi*-eudesmin

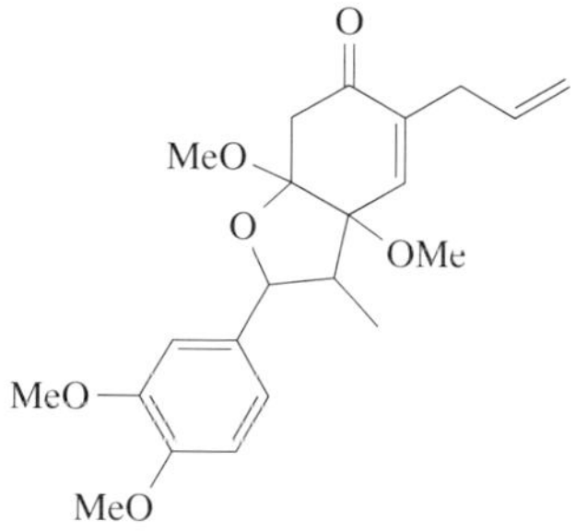

(197) Piperenone

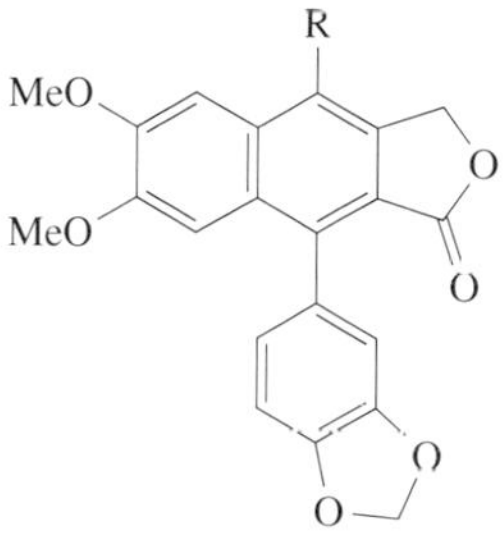

(198) R = OMe, Justicidin A
(199) R = H, Justicidin B

The lignanamides, (±)-grossamide (**116**), (±)-demethylgrossamide (**117**), (−)-cannabisin B (**114**), and (−)-cannabisin D (**115**), isolated from *X. aethiopica* also display antifeedant properties at 5000 ppm against subterranean termite (*Reticulitermes speratus*) workers: index values of 1.91, 29.49, 7.10, and 12.93, respectively, were obtained, where < 20 indicates significant feeding-deterrent activity.[143]

1.25.13.4 Allelopathy

Various lignans have powerful allelopathic properties. For example, nordihydroguaiaretic acid (**172**), when supplied at concentrations of ~20 μg l^{-1}, is able to dramatically reduce the seedling root growth of barnyard grass, green foxtail, perennial ryegrass, annual ryegrass, red millet, lambsquarter, lettuce, and alfalfa, as well as the hypocotyl growth of lettuce and green foxtail.[240] Additionally, the monoepoxylignanolide (**200**) of *Aegilops ovata* is reputedly a unique germination inhibitor of *Lactuca sativa* (lettuce) achenes, this effect being greater in the light than under darkness.[241] Arctiin (**92**) from *Arctium lappa* inhibits the germination of 11 out of 12 different plant species tested at concentrations of ~5 μg μl^{-1}, and the levels of arctiin (**92**) parallel that of the annual rhythm of germination.[242] The furofurans, fargesin (**201**) and sesamin (**18**), are germination inhibitors of peanut (*Arachis hypogaea*) seeds (lipid-storing seeds), but not of rice (*Oryza sativa*) seeds (carbohydrate-storing seeds),[243] where the investigators suggested that this effect might be on processes/enzymes controlling lipid metabolism. Interestingly, in the same studies, eudesmin (**83**), which lacks a methylenedioxy group, was much less active, suggesting that the methylenedioxy group is needed for biological activity.

1.25.13.5 Cytokinin-like Activities

Certain lignans have been implicated to function as cytokinins during plant growth and development, although they are only effective at very high concentrations. For example, a cytokinin-like

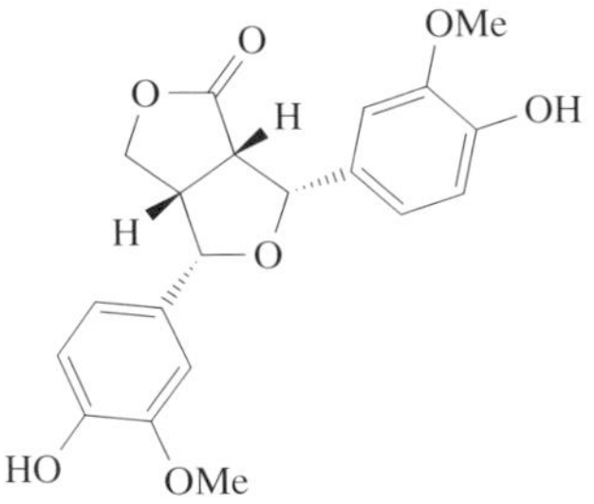

(200)

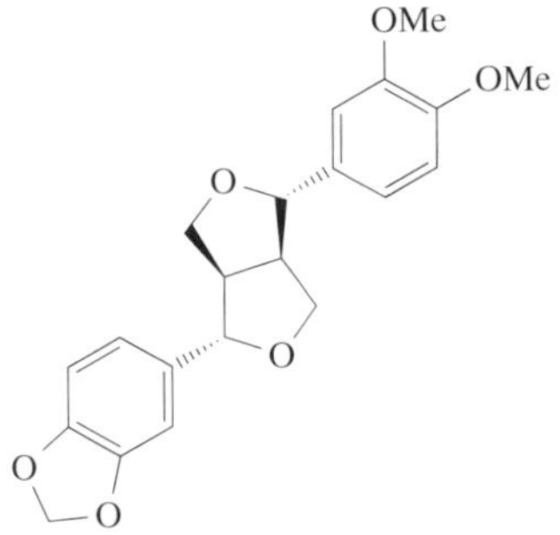

(201) Fargesin

function has been proposed for ($\pm$)-dehydrodiconiferyl alcohol 4-*O*-glucosides (**125a,b**), since they can stimulate cell division of tobacco (*N. tabacum*) cells and replace cytokinin in pith and callus cultures:[155–157] both (+)- and (−)-dehydrodiconiferyl alcohol 4-*O*-glucosides (**125a**) and (**125b**) stimulated pith growth, at concentrations of $\sim 10\ \mu M$, in a manner comparable to that of the cytokinin, zeatin riboside ($0.1\ \mu M$). On the other hand, they apparently did not stimulate shoot formation from leaf explants as normally observed for cytokinins.[155] These investigators also proposed that the glucosides were being mobilized from their roles as cell-wall components, even though this could not be the case since they were extracted from the callus cultures by methanol/water extraction. That is, although their subcellular origins were not determined, they could not have been cell-wall constituents based on this solubilization property. Moreover, these data must be viewed as only a tentative indication of any cytokinin role *in vivo*, and more extensive studies are required to prove that this is indeed a true function.

1.25.13.6 Constitutive and Inducible (Oligomeric) Lignan Deposition and Nonstructural Infusions, "Abnormal" and "Stress" Lignins

The physiological roles of lignans discussed thus far are those of antioxidant, biocidal, allelopathic, and antifeedant agents, as well as a putative role as cytokinins. These properties, in turn, lead to the question of what are the factors controlling their induction/constitutive formation *in planta*? However, with the appropriate gene[87,90,111] and promoter sequences (data not shown) in hand, together with their proteins and antibodies, meaningful experiments can be undertaken to define how their formation is regulated, and the temporal and spatial nature of expression of the biosynthetic pathways involved.

What is known about sites of lignan accumulation is both rudimentary and variable, with evidence from different studies perhaps pointing to different locations. One study suggests that 5-methoxypodophyllotoxin (**100**) is present in the vacuolar compartments of *Linum album* cell suspension cultures, this being based on crude subcellular (organelle) fractionation studies.[37] That it is supposedly present in the vacuoles might also suggest a similar location for the constitutively formed insecticidal lignan, haedoxan A (**159**), in the roots of *P. leptostachya*, and the cytotoxin, podophyllotoxin (**23**), in the rhizomes of *P. peltatum*/*P. hexandrum*. In seed tissues, such as sesame and flax, nothing is definitively known about their lignan subcellular locations, a situation which also holds for lignans in flowers, fruits, leaves, and bark.

Lignan deposition in sapwood and heartwood tissues of certain woody plants has been a matter of particular interest, given the long-standing confusion surrounding the nature of various heartwood metabolites and whether they are lignins, lignans, or oligomeric lignans (see Section 1.25.12); this, in turn, has led to the use of lax terminology to describe such constituents, e.g., as "abnormal and secondary" lignins.

Heartwood formation itself is initiated, at some undetermined point, in the center (pith) region of mature prelignified secondary xylem wood. What initiates or induces its formation is unknown, although the metabolic composition (lignans, flavonoids, alkaloids, etc.) can vary extensively with the species. Nevertheless, it is this phenomenon that provides the various woody types, with these differing in terms of color, quality, durability, and rot resistance, e.g., the black color of ebony wood and the reddish-brown color of western red cedar are due to their distinctive heartwood metabolites, whereas the whitish-yellow of spruce results from the near absence of any heartwood metabolites.

Heartwood-forming substances are considered to be initially formed and released from specialized ray parenchyma cells, with these substances further infusing into neighboring tracheids and/or fibers/vessels (Figure 19). As proposed by Chattaway nearly 50 years ago,[244,245] they are exuded through parenchyma cells via pit apertures into the lumen of adjacent dead, prelignified, cells and then diffuse into neighboring, prelignified cells to ultimately afford the heartwood tissue. In agreement with this contention, Hergert's analysis of western red cedar and western hemlock constituents also led to the conclusion that heartwood-forming substances accumulated in ray parenchyma cells, prior to becoming insoluble infusions/deposits in tracheid cells.[29] Indeed, for these reasons, Hergert cautioned that analytical results obtained from the analysis of wood samples must specify the physiological conditions and/or tissue, e.g., whether sapwood, heartwood, compression wood, diseased wood, etc. Unfortunately, most investigations treat woody tissue as if it were homogeneous.

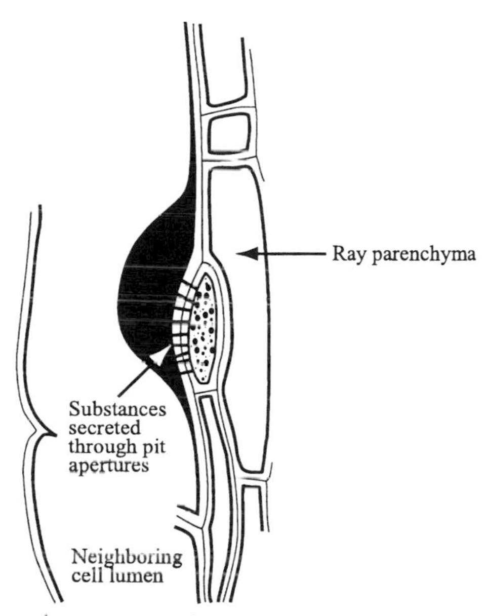

Figure 19 Secretion of heartwood constituents by ray parenchyma cells into the lumen of neighboring cells appears to occur through pit apertures. After Chattaway.[244]

Heartwood metabolites are frequently, but erroneously, also described as "extractives," based on the fact that a portion can be removed by aqueous/organic solvent treatment. It is seldom acknowledged, however, that only a proportion of these metabolites is solubilized, with the remaining often requiring harsher conditions, e.g., as commonly employed for lignin dissolution. To a lesser extent, comparable substances to those present in heartwood can also be formed in sapwood, in response to biological challenges, such as insect attack.

Unfortunately, imperfect characterization of such heartwood/sapwood metabolites, particularly when lignan derived, has led to their descriptions as "abnormal" lignins, "secondary" lignins, and "Brauns native" lignins. These substances are, however, not lignins, but instead constitute (see Section 1.25.12) a fundamentally distinct biochemical class of nonstructural, nonlignin metabolites which can be produced in a range of sizes and form insoluble deposits during heartwood formation, i.e., they differ from lignins in terms of temporal and spatial deposition, configurations of the molecules, postcoupling modifications, and physiological function. In western red cedar, for example, the lignan-derived components can range from monomers, such as plicatic acid (**75**) and plicatin (**202**) to oligomers (MW ~ 10 000), with a portion only being removed under conditions required for lignin removal.[27–29,36] Another example is western hemlock heartwood, the lumen of whose tracheid cells can contain either 7′-hydroxymatairesinol (**99**) or α-conidendrin (**76**).[246] This heterogeneity is particularly interesting, since it suggests that certain parenchyma cells may be involved in formation of specific metabolites, regardless of the fact that complex mixtures may ultimately result in the developed heartwood via the infusion process previously discussed.[244,245]

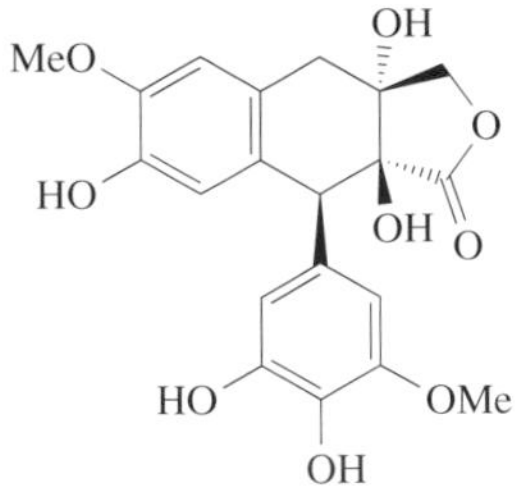

(202) (–)-Plicatin

As described in Section 1.25.12, other nonlignin, nonstructural metabolites are that of the "Brauns native" lignins[196] and "abnormal" lignins[247] present in both Pinaceae heartwood and sapwood (although typically by 10-fold less in the latter). These substances, e.g., in loblolly pine (*Pinus taeda*), are primarily dehydrodiconiferyl alcohol (**5a,b**) derived, and result from lignan modifications via allylic double bond reduction, demethylation, and phenylcoumaran ring opening.[171–173] As before, they are released into the prelignified sapwood via specialized cells, and have chemical structures which preclude them from being able to undergo polymerization to give lignin biopolymers (see Section 1.25.12 and Chapter 3.18).

The term, "stress" lignins, has also been introduced to describe inducible "lignin-like" responses, but again with no explicit biochemical explanation as to what it actually meant. For example, *P. taeda* cell suspension cultures (and those of other plants in the Pinaceae) can be induced, at high sucrose concentrations, to form a lignin-like "extracellular" precipitate.[168] Detailed analysis of these substances revealed, however, that this inducible response resulted from lignan coupling as for the "abnormal" lignins, and as such they more closely resemble constituents formed during heartwood deposition which are formed by a distinct biochemical pathway to that of the lignins.

This, however, again underscores the necessity to use molecular probes to comprehensively distinguish, both temporally and spatially, between lignin and (oligomeric) lignan formation and deposition.

1.25.14 ROLES IN HUMAN NUTRITION/HEALTH PROTECTION AND DISEASE TREATMENT

Lignans have long held considerable importance in medicine, human health, and nutrition, and a brief description of some of their most significant applications is summarized below.

1.25.14.1 Nutrition/Health and Protection against Onset of Breast and Prostate Cancers: Secoisolariciresinol, Matairesinol, and Sesamin

Dietary lignans, e.g., secoisolariciresinol (**20**) and matairesinol (**21**), have significant roles in conferring health protection, particularly against the onset of breast and prostate cancers. During digestion, they are metabolized into the "mammalian" lignans, enterodiol (**203**) and enterolactone (**204**), these being first detected in the urine of female rats and humans.[248,249] Interestingly, excretion of these lignans had a cyclic pattern during the menstrual cycle, which reached a maximum during the luteal phase. It has also been shown with rats that the levels of enterodiol (**203**) increase substantially when flaxseed was provided, due to metabolism of secoisolariciresinol diglucoside (**98**).[250]

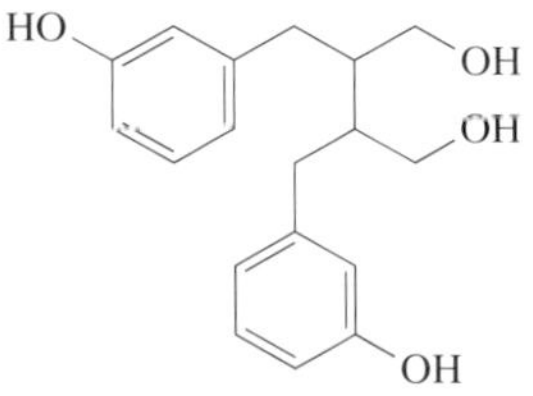

(203) Enterodiol

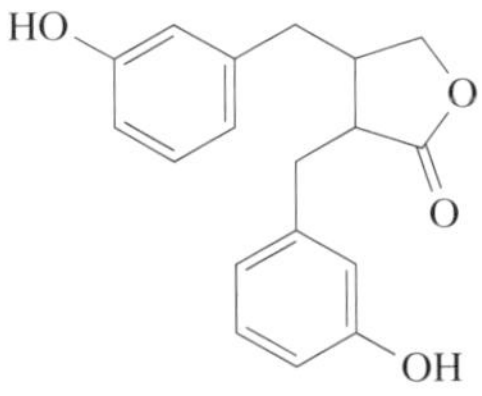

(204) Enterolactone

Recognition of the health-protection effects of dietary lignans began with observations of significant metabolic profile differences in the urinary excretions from individuals at low risk for breast/prostate cancers, and those at high risk or who had contracted these cancers. Low breast cancer risk Finnish women, for example, had high levels of mammalian lignan-derived metabolites, enterodiol (**203**) and enterolactone (**204**), in their body fluids (urine, plasma, and bile), whereas those at high risk did not (< 5% of total).[251–253] This was subsequently recognized to be a form of chemoprotection which was ultimately correlated with a vegetarian-like diet (grains, fibers, seeds, and berries) rich in the lignans, secoisolariciresinol (**20**) and matairesinol (**21**). Their subsequent conversion into the protective "mammalian" lignans, enterodiol (**203**) and enterolactone (**204**), occurs in the gut[254,255] via loss of the hydroxyl functionalities at C-4 and C-4′ and demethylation at C-3 and C-3′. Scheme 22 shows a possible biochemical pathway.

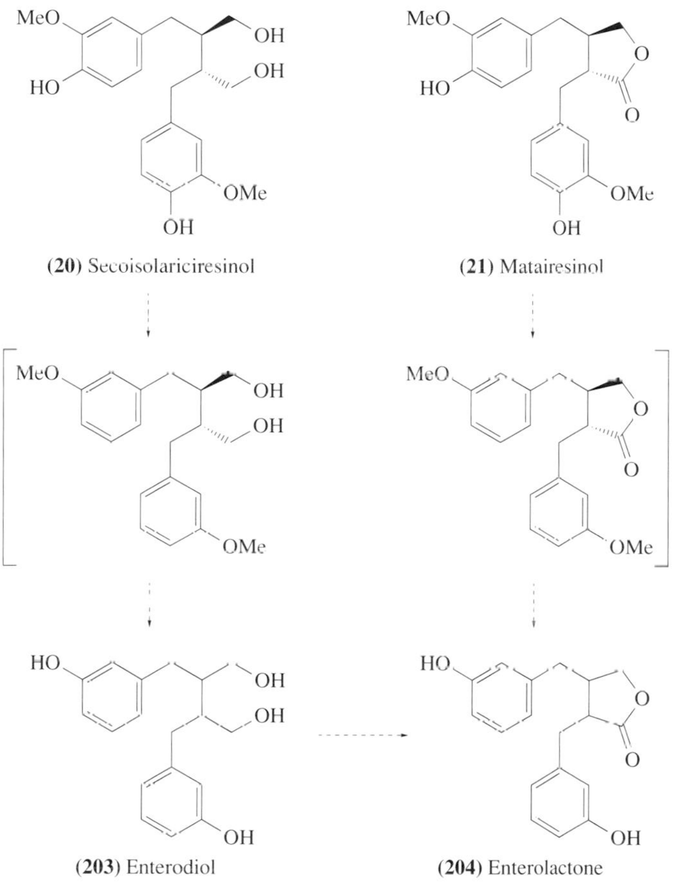

(**20**) Secoisolariciresinol

(**21**) Matairesinol

(**203**) Enterodiol

(**204**) Enterolactone

Scheme 22

Dietary lignans (as well as isoflavonoids) impart chemoprotective effects due to their antioxidant,[4,98,99] weak oestrogenic/antioestrogenic,[256–258] anti-aromatase,[259,260] and anticarcinogenic/antitumor[261–265] properties, thereby protecting against the initiation of various sex hormone-induced cancers. Their importance can be illustrated using as an example the sex hormone binding globulin (SHBG). This binds circa 50% of circulating testosterone in men, and 80% of the oestrogen in women[258] and thus the availability of sex hormone to target cells is greatly affected by changes in both its concentration and/or binding properties. Postmenopausal women excreting large amounts of mammalian lignans have higher levels of SHBG than omnivores or breast cancer patients,[265] and other studies have shown that mammalian lignans and isoflavonoids interact with SHBG in a dose-dependent manner, with enterolactone (**204**) > equol (**205**) > genistein (**206**) for displacing estradiol (**207**), and equol (**205**) > enterolactone (**204**) or enterodiol (**203**) > genistein (**206**) for testosterone (**208**) displacement.[258,266] These effects occur at levels (5–50 μM) which correspond well with the concentrations of plant-derived dietary diphenols in body fluids, following ingestion of vegetarian and soy diets, thereby suggesting that sex-hormone binding is modulated by their presence.

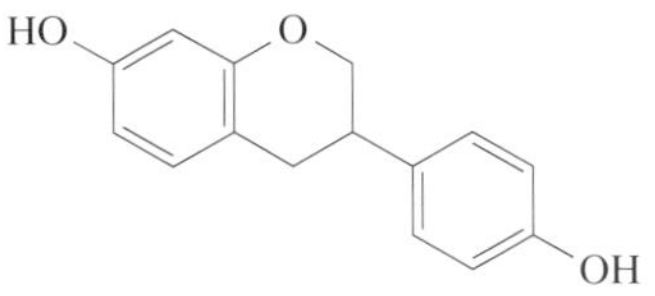

HO O OH

(**205**) Equol

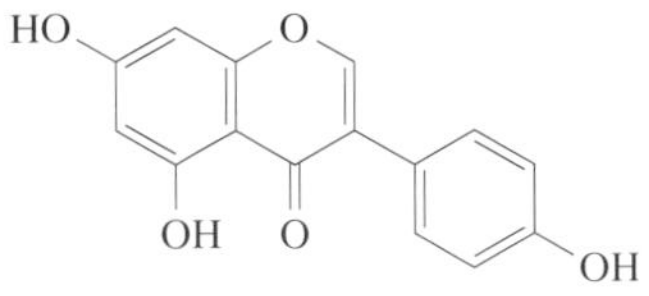

HO O OH O OH

(**206**) Genistein

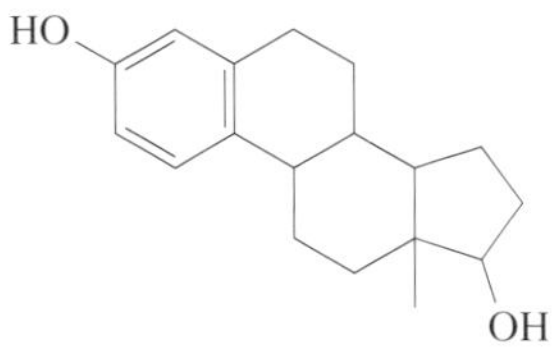

HO OH

(**207**) Estradiol

OH H H H O

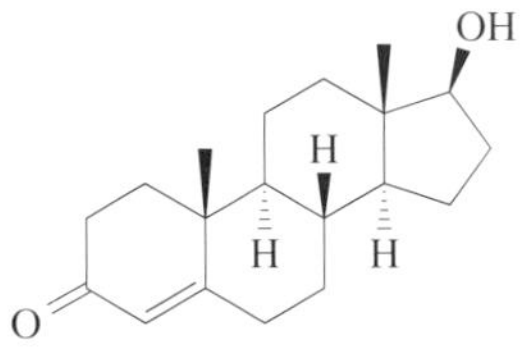

(**208**) Testosterone

There is also chemoprotection against the onset of tumors. Sprague–Dawley female rats, pretreated with the (pro)carcinogen dimethylbenzanthracene, had reduced incidence rates (37% reduction) and tumor masses (46% reduction in mammary tissues)[261] when rats were administered a diet containing secoisolariciresinol diglucoside (**98**), and the number of valid putative preneoplastic markers for colon carcinogenesis in male rats also declined.[263] Treatment with flaxseed or the flax lignan, secoisolariciresinol diglucoside (**98**), also significantly reduced epithelial cell proliferation, as well as the number of aberrant crypts and aberrant crypt foci, viewed to be early indicators of colon cancer risk.[263] In this regard, it is noteworthy that both Japanese and Caucasian men have comparable numbers of precancerous colonic aberrations; however, in Japanese men, somehow these do not become cancerous to the same extent, an observation which is viewed as a dietary consequence.[267] Such dietary differences may also help explain the difficulties in inducing carcinogenesis in primates, even when procarcinogens are added to the diet. For example, administering oestrogens with benzpyrene/dibenzanthracene to rhesus monkeys did not result in carcinogenesis, even after 8 years.[264] This effect is presumed to be due to the fact that primates have diets which result in the massive accumulation and excretion of lignan (as well as isoflavone) diphenolics.

Flaxseed is the richest source of mammalian lignan precursors, containing levels 75–800 higher than any other plant food, and is being widely investigated for its cancer protective effects. Incubation of its most abundant lignan, (+)-secoisolariciresinol diglucoside (**98**), with cultured human fecal microflora *in vitro*, suggested the metabolic pathway (Scheme 22) for its conversion into the mammalian lignans: that is, intestinal bacteria hydrolyzed the sugar moiety to release secoisolariciresinol (**20**), this being presumably followed by dehydroxylation and demethylation to give the mammalian lignan enterodiol (**203**), which was oxidized to enterolactone (**204**). As indicated in Scheme 22, enterolactone (**204**) is also considered to result from catabolism of the plant lignan, matairesinol (**21**).[268] Once formed, both mammalian lignans undergo enterohepatic circulation,[269] where a good correlation exists with their presence and the reduced incidence rates of hormone-related cancers.

Chemoprotection can also be correlated with antioxidant abilities, e.g., of secoisolariciresinol diglucoside (**98**) as a radical scavenger.[270] For example, hydrogen peroxide, when subjected to photolysis under ultraviolet light in the presence of salicylic acid, is involved in the formation of the $OH^{\cdot}$ adduct products, 2,3-dihydroxybenzoic acid (DHBA) and 2,5-DHBA. When secoisolariciresinol diglucoside (**98**) was present, however, a concentration-dependent decrease in the formation of 2,3-DHBA and 2,5-DHBA occurred due to scavenging of $OH^{\cdot}$ radicals by secoisolariciresinol diglucoside (**98**). On a somewhat related topic, dietary sesamin (**18**) also has the effect *in vivo* of elevating levels of the antioxidant, γ-tocopherol, in rat plasma and liver, leading to the suggestion that a sesame-rich diet increases availability of antioxidants (vitamin E) in the body. This, in turn, decreases the risk of a number of diseases directly related to free-radical formation, such as more rapid onset of aging;[271] sesame lignans can also cause an increase in vitamin E activity in rats fed a low α-tocopherol diet.[272]

1.25.14.2 Antitumor Properties: Podophyllotoxin and other 8—8′ Lignans

There are essentially only a handful of plant natural products used in medicine today in cancer treatment, of which one is podophyllotoxin (**23**) from *Podophyllum peltatum* and *Podophyllum hexandrum*. Its pharmacological usage dates back many centuries, when may apple (*P. hexandrum*) alcoholic extracts, obtained from rhizomes and roots, were employed first as a poison and later, in smaller doses, for treatment of various pathological conditions.[273] The cytotoxic effect of these extracts was subsequently found to be due to podophyllotoxin (**23**), which led to its use as an antitumor agent against various malignancies. It was later shown that podophyllotoxin (**23**) was readily taken up by the cells due to its small size and hydrophobicity, with tubulin binding (at a different site to that occupied by the *Vinca rosea* alkaloids) and microtubular assembly inhibition arresting cells in mitosis;[274,275] this occurred in a manner more rapid and reversible than colchicine. Its action leads to cytoskeletal arrest of cell division and ultimately cell death.

In spite of its antitumor promise, the clinical applications of direct administration of podophyllotoxin (**23**) were greatly compromised by severe (gastrointestinal) toxicity experienced by those under treatment. Accordingly, a significant effort was launched to identify means whereby the drug could be delivered with reduced cytotoxicity. This led to the development, and subsequent widespread application, of its semisynthetic derivatives, etoposide (**101**) and teniposide (**102**),[276,277] which are used (alone or in conjunction with other drugs, e.g., cisplatin) for treatment of Hodgkin's lymphomas, non-Hodgkin's lymphomas, small cell lung cancers, and acute leukemias.[278–280] The semisynthetic derivatives (**101**) and (**102**), however, were found to differ in their mechanism of action. The added sugar moieties prevented tubulin interactions from occurring, and thus microtubular assembly was not inhibited. The antitumor action was, instead, a consequence of an ability to form stable tertiary complexes with topoisomerase II and its substrate DNA leading to formation of numerous double-stranded DNA breaks. In turn, this results in large DNA fractures, thereby arresting cells in their life cycle at the G_2 phase, and ultimately causing cell death.[281] At present, etoposide phosphate (**209**), the phosphorylated form of etoposide (**101**), is undergoing clinical tests, since its application may be more convenient due to its increased water solubility.[282,283]

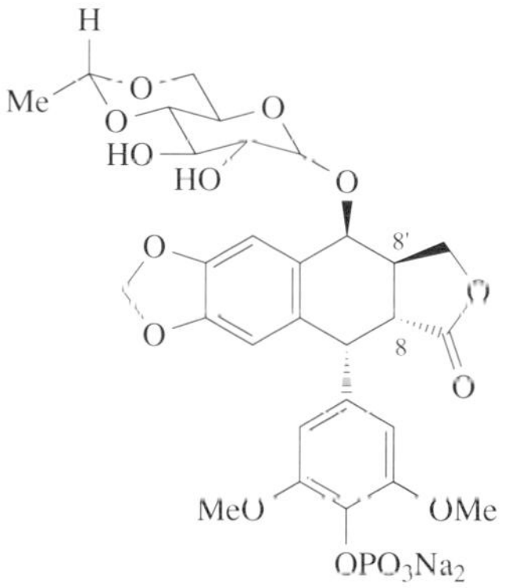

(**209**) Etoposide phosphate

Other lignans have promising anticancer properties: (−)-steganacin (**25**) and (−)-steganangin (**210**), isolated from *S. araliacea* stem bark and wood, exhibit antileukemia activities both against the *in vivo* murine P-388 lymphocytic leukemia test system and *in vitro* against cells derived from a human carcinoma of the nasopharynx cell culture.[148] It is thought that this antimitotic activity is through an effect on spindle microtubules, as for podophyllotoxin (**23**), with the chirality about the pivotal biphenyl bond and the orientation of the lactone carbonyl being essential for antitumor activity.[284] *Epi*-steganangin (**211**) and steganoate B (**212**) also have cytotoxic properties against 11 different human tumor cell lines,[285] and (−)-burseran (**213**) from *Bursera microphylla* (Burseraceae) displays antitumor properties against human epidermoid carcinoma of the nasopharynx cell culture.[286,287] (+)-Wikstromol (**67**), from *Wikstroemia foetida* var. *oahuensis* Gray (Thymeleaceae) is also active against the P-388 lymphocytic leukemia test system.[288]

Finally, phyllanthin (**214**) and hypophyllanthin (**215**), from *Phyllanthus amarus* Schum. & Thonn. (Euphorbiaceae), enhance cytotoxic responses mediated by vinblastine in the multidrug-resistant KB cell line. However, alone they had no significant cytotoxic activity with a large number of mammalian cells.[289]

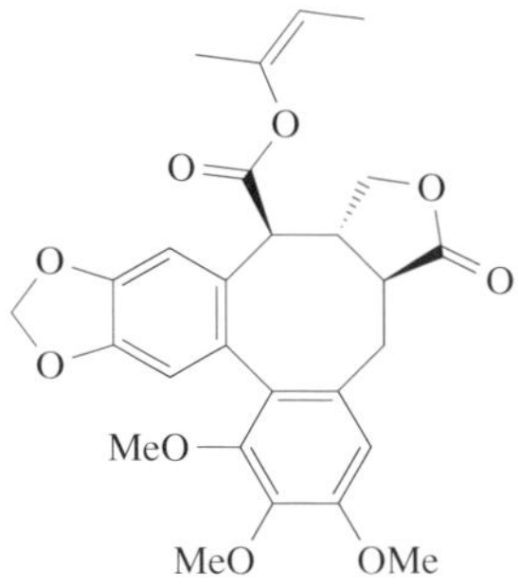

(**210**) (–)-Steganangin 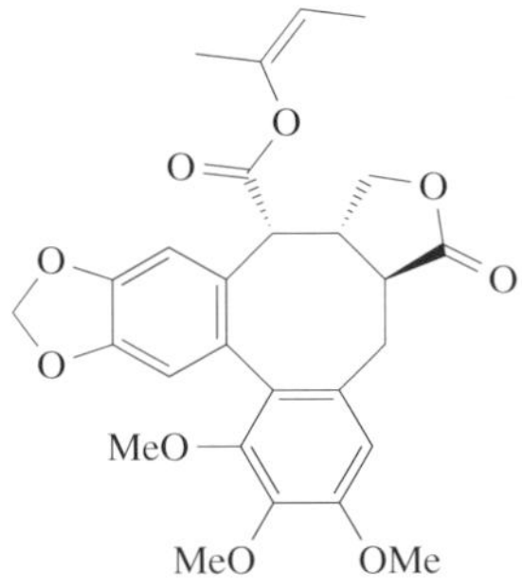 (**211**) (–)-Episteganangin 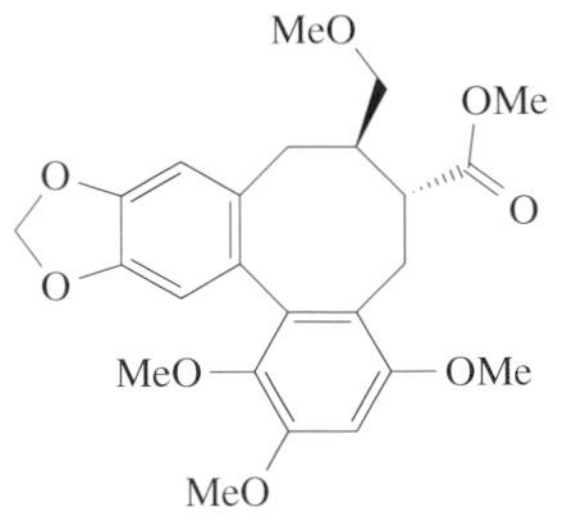 (**212**) (–)-Steganoate B 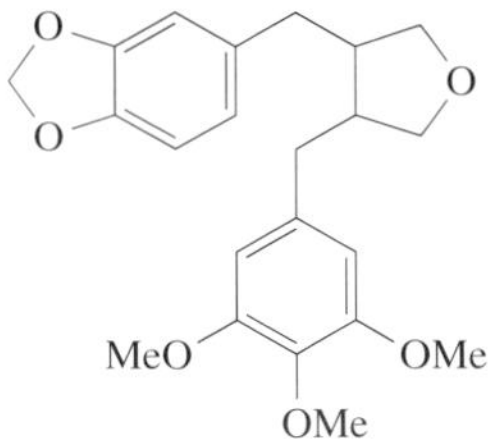 (**213**) Burseran

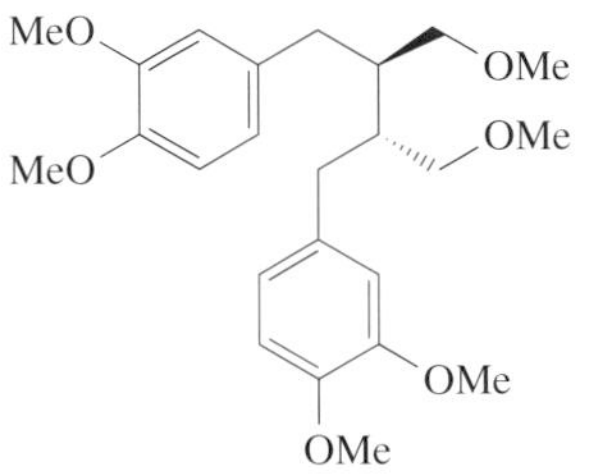

(**214**) Phyllanthin 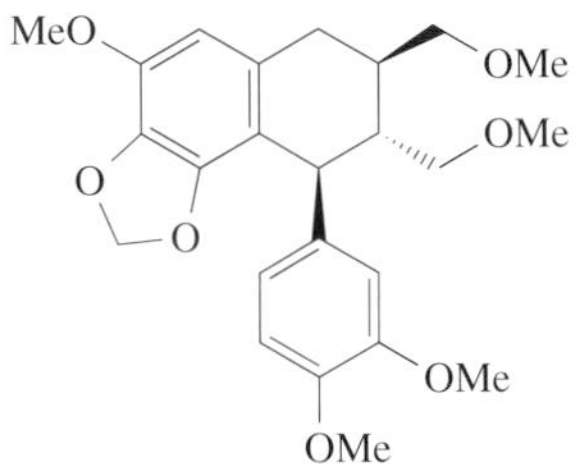 (**215**) Hypophyllanthin

1.25.14.3 Hepatotoxic Preventive Effects

Schizandra chinensis fruit is an important component of various traditional Asian medicines. Its "kita-gomisi" extract is used as an antitussive and tonic,[149] whereas the "Sheng Mai San" formulation is employed in the treatment of coronary heart disease treatment.[290] The fruits have also been used in Japan and East Asia for the treatment of elevated serum aminotransferase activity in acute hepatitis.[291]

Schizandra fruit contains significant levels of the 8—8′,2-2′ linked lignans, such as gomisin A (**121**).[292] This lignan appears to have an excellent ability in protecting the liver from a variety of liver-damaging agents, such as the hepatotoxic compounds, CCl_4, galactosamine, and lipopolysaccharides.[291,293,294] This protection has been correlated with several enzymatic processes, e.g., inhibition of leukotriene biosynthesis preventing arachidonic acid release,[295] and prevention of acetaminophen-induced liver injury in rats by inhibiting lipid peroxidation.[291] Prevention or limitation of acetaminophen-caused intoxication by gomisin A (**121**) is considered to be due to the reduction of aminotransferase activity in serum and suppression of lipoperoxide accumulation in liver.[291] Liver regeneration, following partial hepatectomy, is also stimulated by gomisin A (**121**), and this is thought to occur via stimulation of ornithine decarboxylase activity leading to putrescine and spermidine accumulation (polyamines play an important biochemical role in liver regeneration)[296] as well as DNA and RNA biosynthesis.[297] Gomisin A (**121**) also has an ability to inhibit 3′-methyl-4-dimethylamino-azobenzene induced liver carcinogenesis,[298,299] as well as limiting muscular damage induced by excessive exercise.[300]

1.25.14.4 Antiviral Properties

Many lignans have been demonstrated to exhibit quite potent antiviral properties: podophyllotoxin (**23**) and (−)-α-peltatin (**216**) prevent development of murine cytomegalovirus plaques in mouse 3T3-L1 cells, as demonstrated *in vitro* by reduction of plaque numbers by ~50% at a concentration of 10 ng ml^{-1}.[301] Rhinacanthin E (**217**) and rhinacanthin F (**218**), from the medicinal plant *Rhinacanthus nasutus* (Acanthaceae), also have antiviral activities against the influenza type A virus. Using the hemadsorption inhibition assay, these lignans have EC_{50} values of 1.7 and < 0.94 μg ml^{-1}, whereas with a cytopathic effect assay dosage levels were 7.4 and 3.1 μg ml^{-1}, respectively. For both assays, control values with amantadine and ribavirin were 0.054 and 3.7 μg ml^{-1}, respectively.[302]

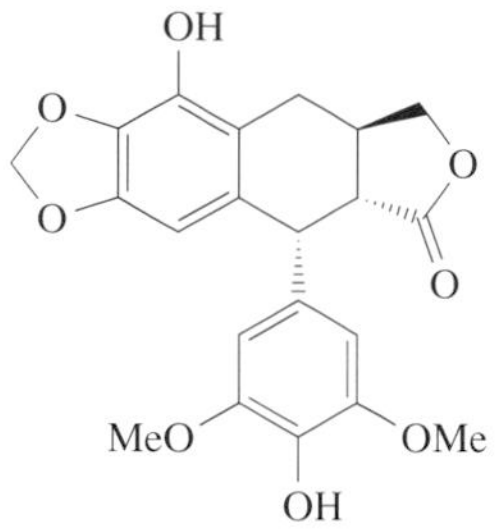

(216) (–)-α-Peltatin 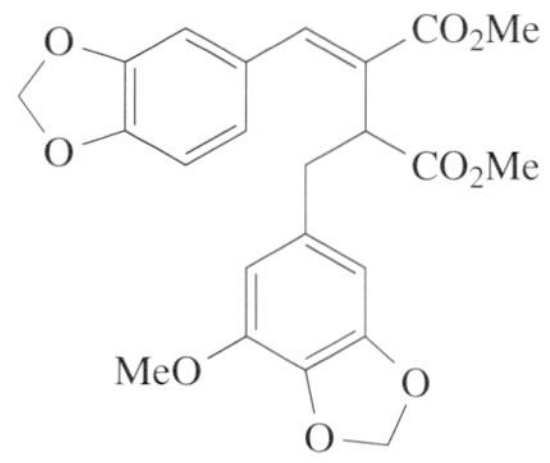 **(217)** Rhinacanthin E 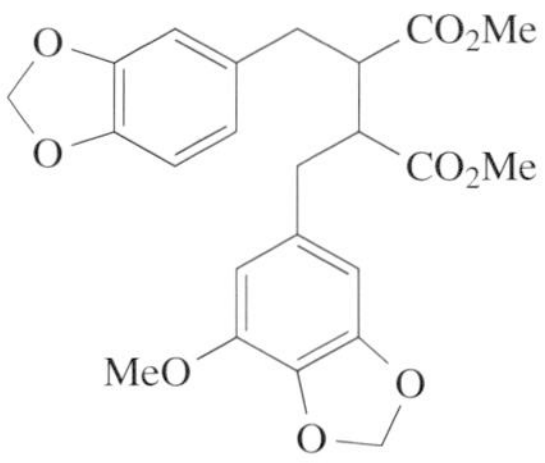 **(218)** Rhinacanthin F

A number of lignans inhibit replication of the human immunodeficiency virus (HIV), albeit with different modes of action. (−)-Arctigenin (**22a**) and (−)-trachelogenin (**17**) inhibit HIV-1 replication in infected human cell systems,[7] with (−)-arctigenin (**22a**) suppressing integration of proviral DNA into the cellular DNA genome,[303] whereas it was inactive with purified HIV-1 integrase.[304] 3,3′-Demethylarctigenin (**219**) (a catechol analogue), on the other hand, gave a strong inhibition of HIV-1 integrase.[304]

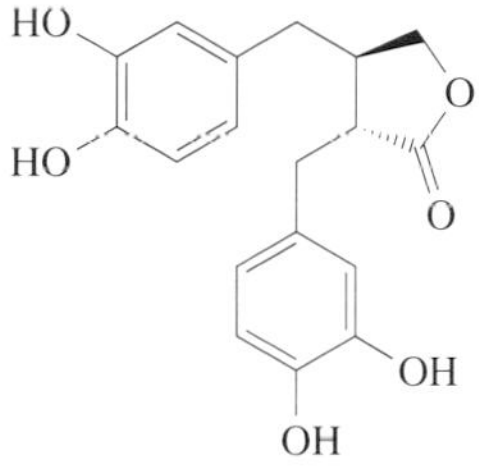

(219) 3,3′-Demethylarctigenin

Antiviral effects have also been noted for constituents from the creosote bush,[6,305] a plant widely used in traditional medicine among the indigenous people of America for digestive disorders, rheumatism, venereal diseases, and sores. One of its lignans, 3′-*O*-methylnordihydroguaiaretic acid (**220**),[6] inhibits HIV Tat-regulated transactivation *in vivo* (EC_{50} = 25 μM), and it is thought that this occurs by interrupting not only the life cycle of wild-type HIV, but also of reverse transcriptase or protease mutant viruses.[305] Other tetrahydronaphthalene lignan analogues have been tested as potent inhibitors of HIV-1; of these, the most effective is compound (**221**), which has an ED_{50} of 0.8 μM with an IC_{50} of 58 μM. It functions as a noncompetitive inhibitor of HIV-1 reverse transcriptase with respect to both template-primer and substrate (dGTP),[306] as does phyllamycin B (**222**) and retrojusticidin B (**223**) from *Phyllanthus myrtifolius* (Euphorbiaceae; IC_{50} = 3.5 and 5.5 μM, respectively) with respect to template primer and triphosphate substrate.[307]

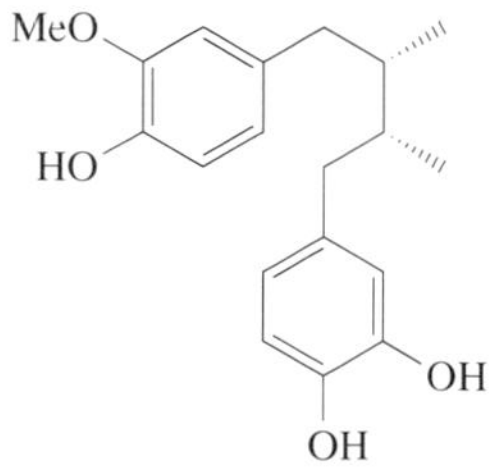

(220) 3′-*O*-Methylnordihydro-guaiaretic acid 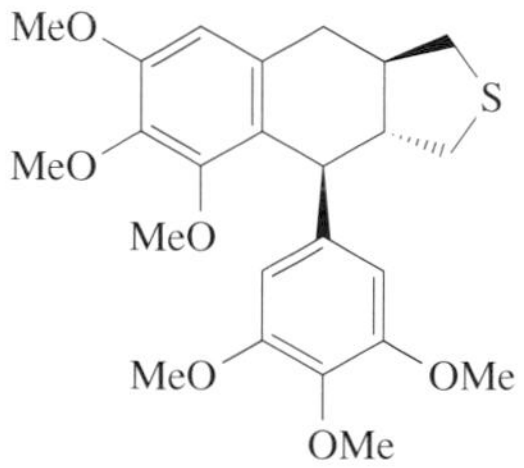 **(221)** 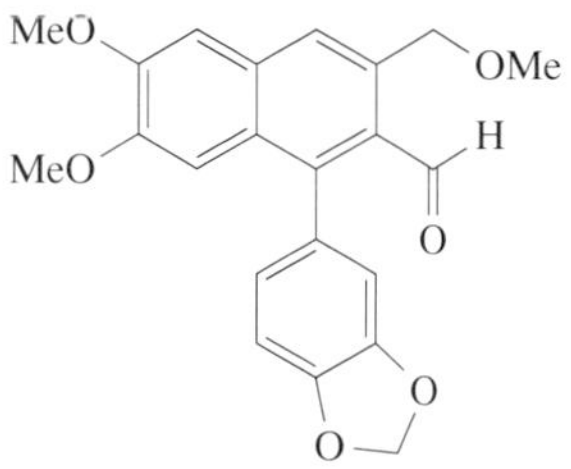 **(222)** Phyllamycin B 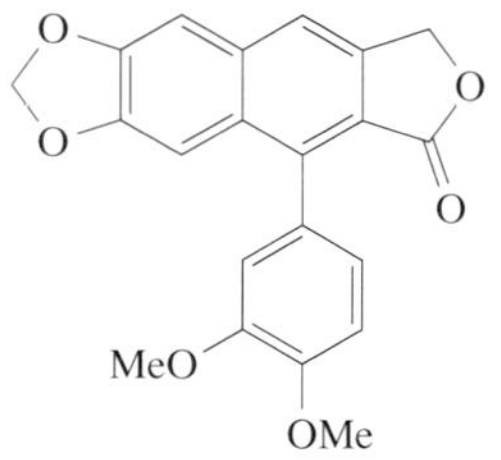 **(223)** Retrojusticidin B

Of the *Schizandra* lignans, examined for anti-HIV activity, (−)-gomisin J (**224a**) displays beneficial effects, whereas gomisins A (**121**), D (**225**), E (**226**), and N (**227**), deoxyschizandrin (**228**), and (+)-gomisin J (**224b**) do not.[308] Interestingly, the synthetic bromine analogue (**229**) of (−)-gomisin J was 33-fold more effective than (−)-gomisin J (**224a**) itself. Studies of its action suggested it to be both a noncompetitive inhibitor of HIV-1 reverse transcriptase, and a mixed (noncompetitive and uncompetitive) inhibitor with respect to the primer-template.[308] The bromo derivative (**229**) was also effective against 3′-azido-3′-deoxythymidine (AZT) resistant HIV-1, as well as synergistically acting with AZT. Finally, anolignan A (**230**) and anolignan B (**105**), from *Anogeissus acuminata* (Combretaceae), synergistically inhibit HIV-1 reverse transcriptase,[309] whereas interiotherin A (**231**) and schisantherin D (**232**), isolated from *Kadsura interior* (Schizandraceae), inhibit HIV replication with EC_{50} values of 6.1 and 1.0 μM, respectively.[310]

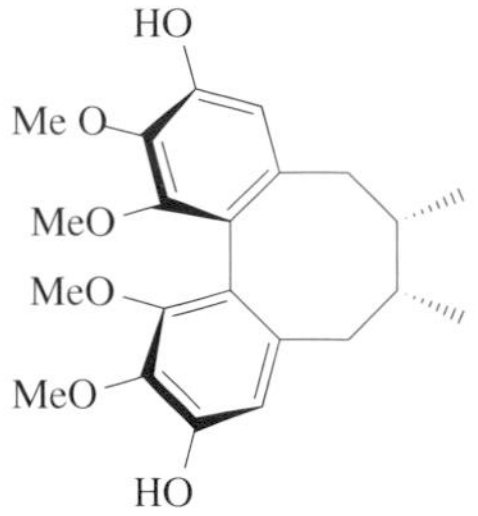

(**224a**) (–)-Gomisin-J 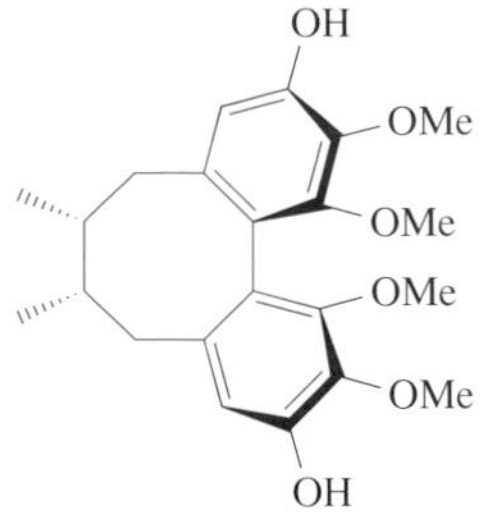 (**224b**) (+)-Gomisin-J 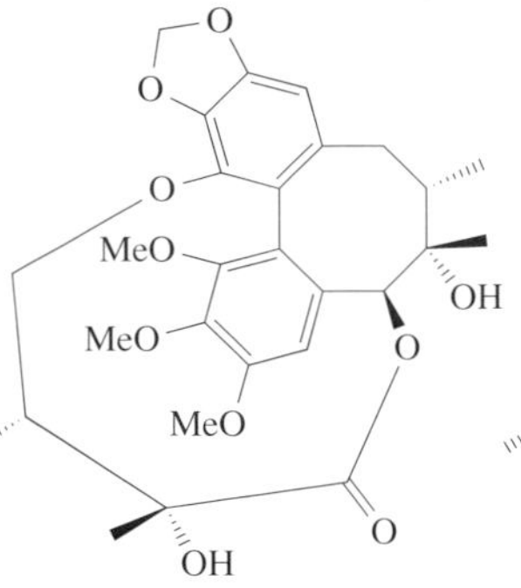 (**225**) Gomisin-D 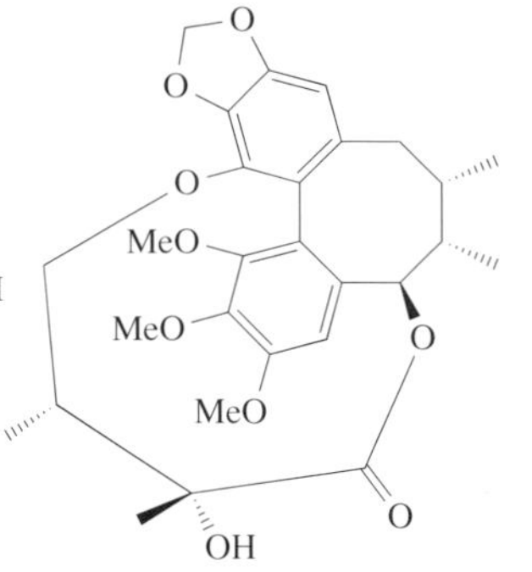 (**226**) Gomisin-E

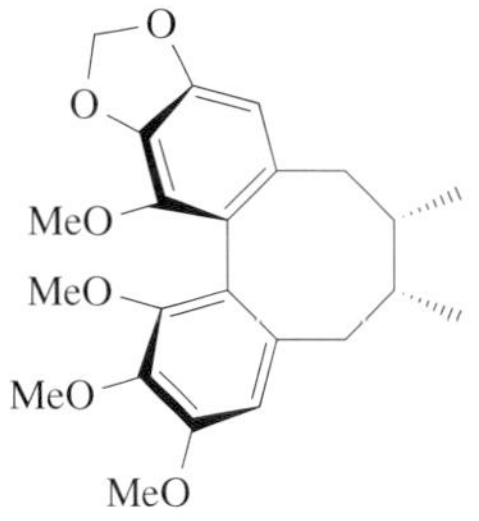

(**227**) Gomisin-N 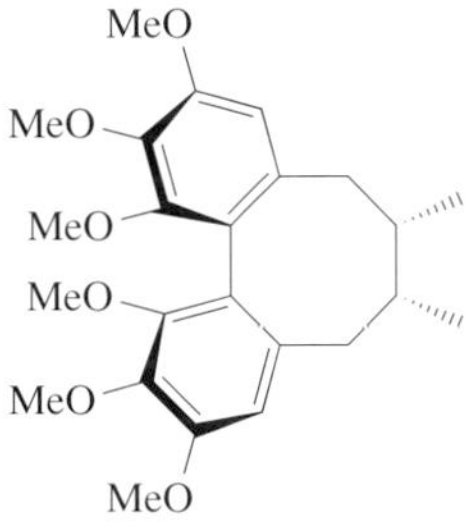 (**228**) Deoxyschizandrin 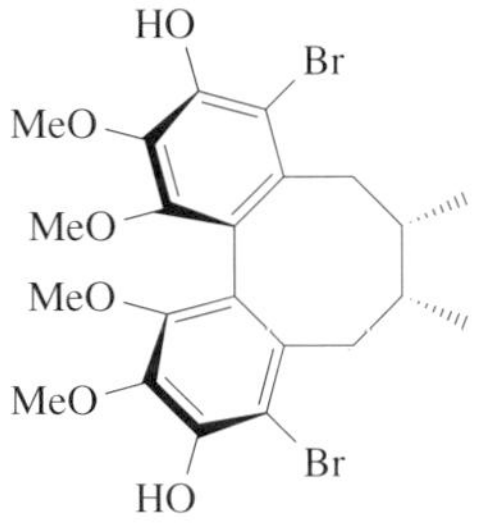 (**229**) 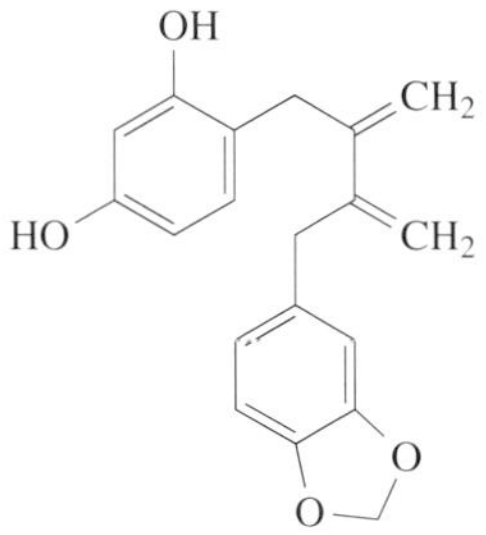 (**230**) Anolignan A

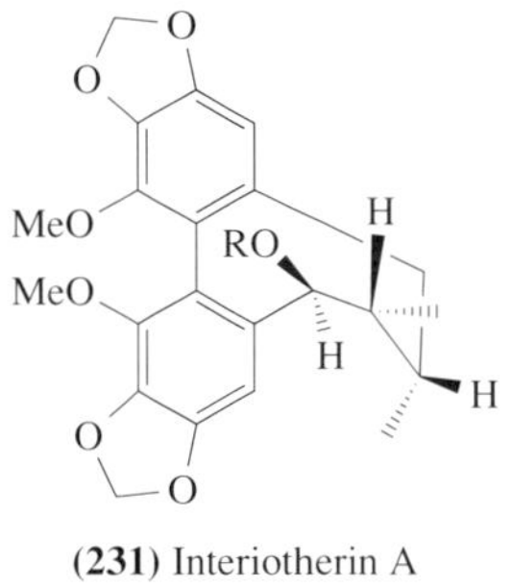

(**231**) Interiotherin A 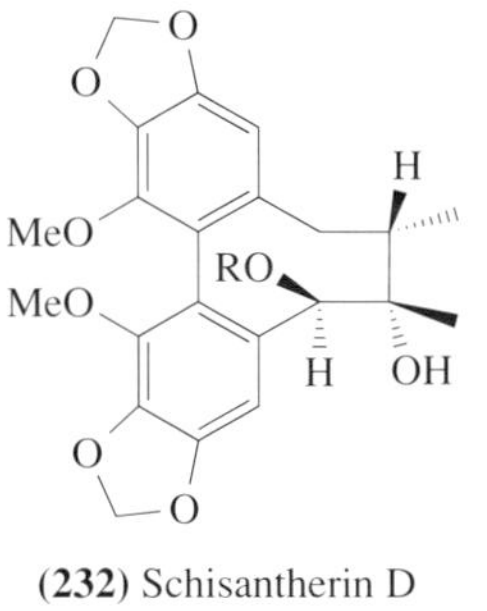 (**232**) Schisantherin D

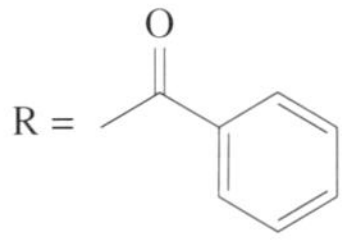

1.25.14.5 Miscellaneous Health Benefits: Anti-inflammatory, Antiasthmatic, and Antidepressant Effects

Kadsurenone (**233**), from *P. futokadsura* (Piperaceae), is a platelet-activating factor, a potent mediator of inflammation, and an asthma antagonist.[311] Such effects have also been ascribed to fargesin (**201**)/eudesmin (**83**) from *Magnolia biondii* (Magnoliaceae),[312] as well as yangambin A (**85**) from *Ocotea duckei* (Lauraceae)[313] and neojusticin A (**234**), justicidin B (**199**), taiwanin E (**235**), and its methyl ether (**236**) from *Justicia procumbens* (Acanthaceae).[314]

Potential antiasthmatic agents are evaluated on their abilities to inhibit cyclic nucleotide phosphodiesterase, this being responsible for hydrolyzing cAMP and cGMP into their respective 5′-mononucleotides. That is, increases in cellular levels of cAMP and cGMP have been implicated in relaxation of the airway smooth muscle, given that elevated levels of cAMP prevent the activation of pro-inflammatory cells. Lignans and norlignans with demonstrable cAMP phosphodiesterase

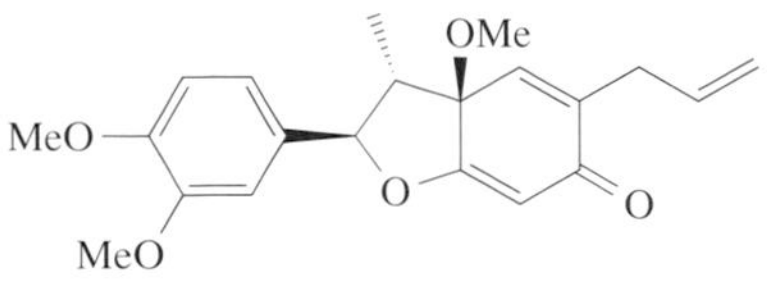

(233) Kadsurenone

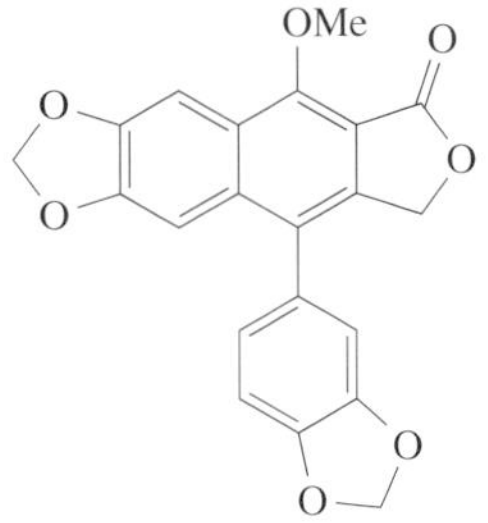

(234) Neojusticin A

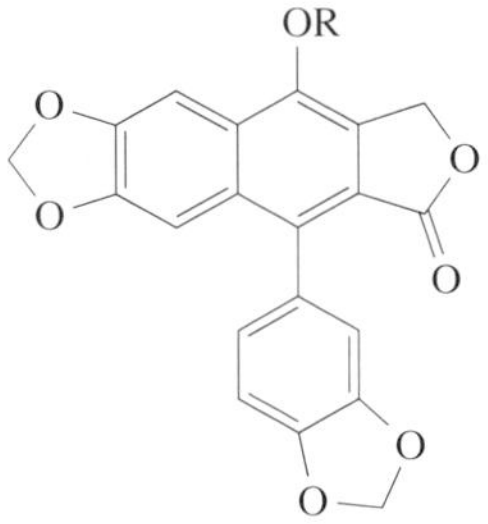

(235) R = H, Taiwanin E
(236) R = Me, Taiwanin E methyl ether

inhibitory properties include: (+)-pinoresinol (**3a**)/(−)-matairesinol (**21a**) from *Forsythia* species,[315] *cis*-hinokiresinol (**237**)/oxy-*cis*-hinokiresinol (**238**) from *Anemarrhena asphodeloides*,[316] and (+)-syringaresinol-di-*O*-*β*-D-glucopyranoside (**239**)/(+)-hydroxypinoresinol 4′,4″-di-*O*-*β*-D-glucopyranoside (**240**) from *Eucommia ulmoides* bark.[317] Additionally, arylnaphthalene analogues have been synthesized and tested as cyclic nucleotide phosphodiesterase IV inhibitors: compound (**241**) inhibits phosphodiesterase (IC_{50} = 0.057 μM) and displays antispasmodic activities; it is 8-fold more active than rolipram (ED_{50} = 2.3 mg kg^{-1} versus 19 mg kg^{-1} i.v.) in the guinea pig antigen-induced bronchoconstriction model. However, it was 8-fold less active than rolipran in a histamine-induced bronchospasmotic assay (ED_{50} = 0.08 mg kg^{-1} versus 0.01 mg kg^{-1} i.v.).[318]

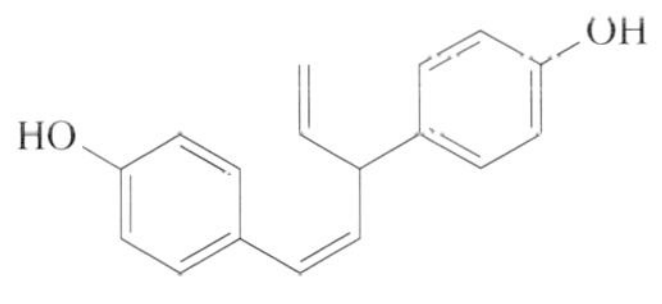

(237) *cis*-Hinokiresinol

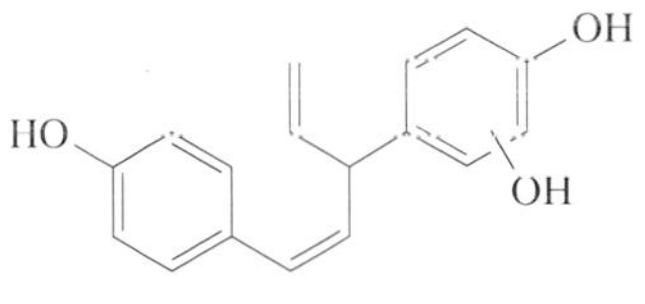

(238) Oxy-*cis*-hinokiresinol

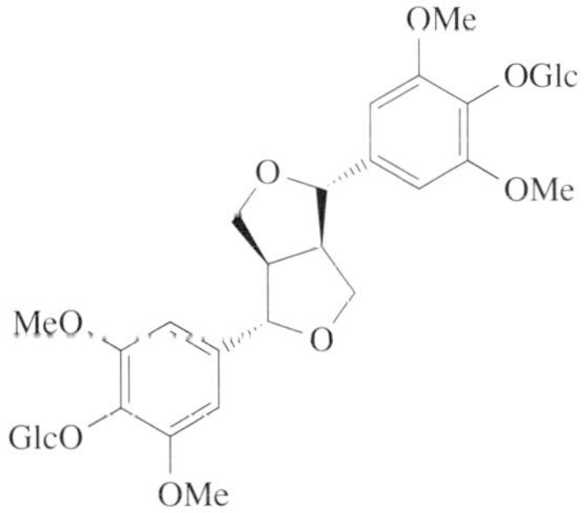

(239) (+)-Syringaresinol di-*O*-β-D-glucopyranoside

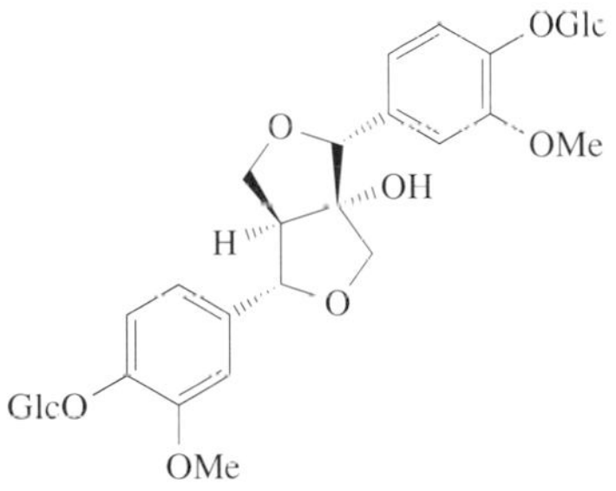

(240) (+)-Hydroxypinoresinol 4,4'-di-*O*-β-D-glucopyranoside

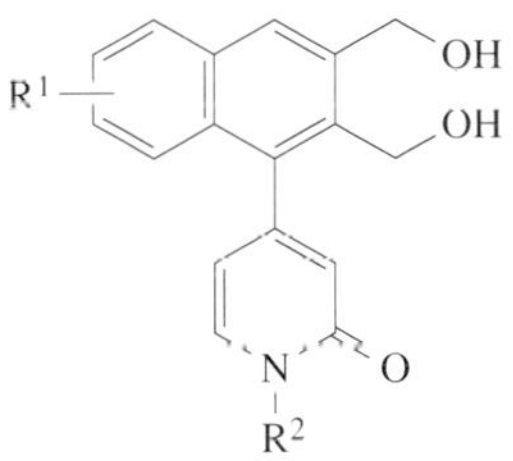

(241)
R^1 = 6,7-$(OEt)_2$
R^2 = *N*-(2-methoxyethyl)

Magnoshinin (**242**) and magnosalin (**243**), isolated from *Magnolia salicifolia* buds, also display anti-inflammatory effects comparable to hydrocortisone acetate.[319,320] Diphyllin acetyl apioside (**244**) and tuberculatin (**245**) are active against inflammation induced by 12-*O*-tetradecanoylphorbol acetate (TPA): the 50% inhibitory doses for acute TPA inflammation were 0.27 and 1.23 μmol/ear for (**244**) and (**245**), respectively; the former is a more potent inhibitor than indomethacin.[321]

Lignans also exhibit antidepressant activities, e.g., prostalidins A, B, and C (**246**)–(**248**), from *Justicia prostata* (Acanthaceae), a plant native to the Western Himalayas.[322]

Finally, the cardiovascular effects of lignans are very significant: Siberian ginseng (*Acanthopomax senticosus*), which is widely used in Asia, has the effect of helping sustain cardiovascular activity during prolonged exercise;[323] this has been attributed to the lignan, (+)-syringaresinol di-*O*-*β*-D-glucoside (**239**).[324]

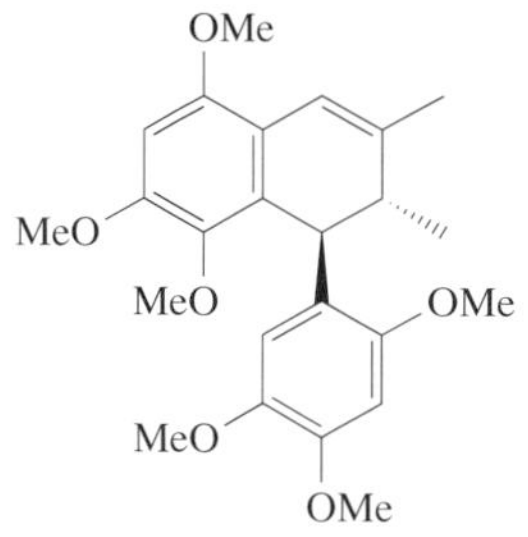

(242) Magnoshinin

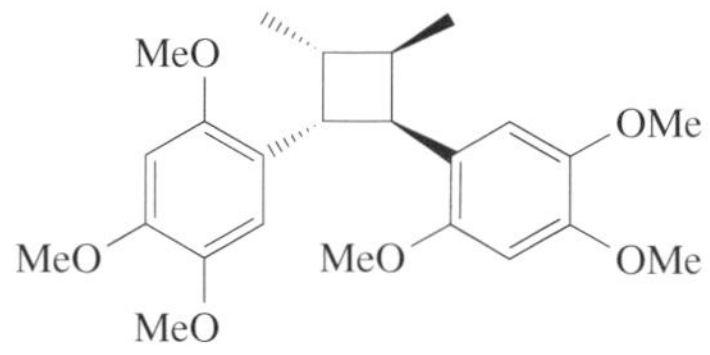

(243) Magnosalin

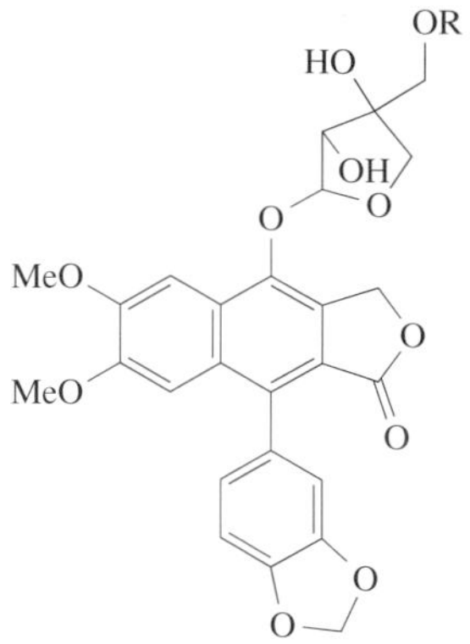

(244) R = Ac, Diphyllin acetyl apioside
(245) R = H, Tuberculatin

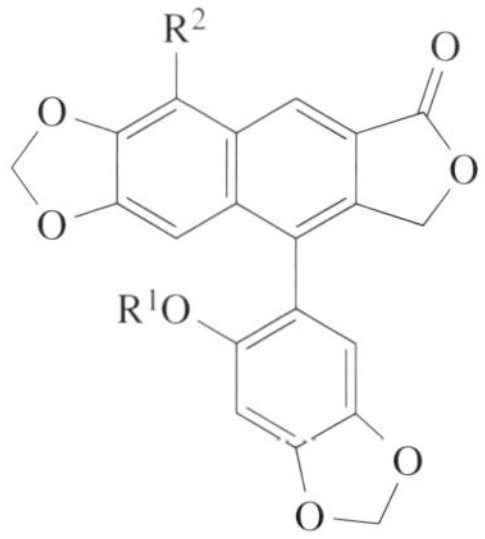

(246) R^1 = H, R^2 = OMe, Prostalidin A
(247) R^1 = Me, R^2 = OMe, Prostalidin B
(248) R^1 = R^2 = H, Prostalidin C

1.25.15 CONCLUDING REMARKS

The foregoing discourse has described the knowledge gained in delineating how stereoselective and regiospecific control of coupling occurs *in planta*. The discovery of dirigent protein mediated phenolic coupling, leading to both (+)- and (−)-pinoresinols (**3a**) and (**3b**), depending upon the species, strongly suggests the involvement of related proteins which stipulate distinct coupling modes. Indeed, all research studies reveal that steps associated with both coupling and subsequent metabolic conversions are, as for all other natural products, under full biochemical control.

Re-examination of various claims for "abnormal" lignins, "secondary" lignins and related substances has revealed that they are nonlignin, nonstructural, oligomeric lignan infusions being secreted into neighboring prelignified cells (such as in heartwood) via specialized cells. Additionally, because of their susceptibility to further oxidation during isolation, it cannot be ruled out that some of the sesquilignans are not, in fact, artifacts of the isolation process.

Additional work will lead to further clarification of the (oligomeric) lignan and lignin forming processes, which must now be viewed as being fully distinct, in terms of biochemical processes involved in their formation, and their structural configurations, as well as in their temporal and spatial deposition *in vivo*.

Finally, the importance of the various lignan skeleta in both plant physiology (particularly defense) and in human nutrition and medicine continues to grow, as the properties of this massive class of natural products continue to be discovered.

ACKNOWLEDGMENTS

The authors thank the United States Department of Energy (DE-FG0397ER20259), the National Science Foundation (MCB09631980), the National Aeronautics and Space Administration

(NAG100164), the United States Department of Agriculture (9603622), McIntire-Stennis, the Arthur M. and Kate Eisig Tode Foundation, and the Lewis B. and Dorothy Cullman and G. Thomas Hargrove Center for Land Plant Adaptation Studies for generous support of this study.

1.25.16 REFERENCES

1. L. B. Davin and N. G. Lewis, in "Recent Advances in Phytochemistry," eds. H. A. Stafford and R. K. Ibrahim, Plenum, New York, 1992, vol. 26, p. 325.
2. H. Belmares, A. Barrera, E. Castillo, L. F. Ramos, F. Hernandez, and V. Hernandez, *Ind. Eng. Chem. Prod. Res. Dev.*, 1979, **18**, 220.
3. D. P. Figgitt, S. P. Denyer, P. M. Dewick, D. E. Jackson, and P. Williams, *Biochem. Biophys. Res. Commun.*, 1989, **160**, 257.
4. Y. Fukuda, T. Osawa, M. Namiki, and T. Ozaki, *Agric. Biol. Chem.*, 1985, **49**, 301.
5. H. L. Haller, F. B. LaForge, and W. N. Sullivan, *J. Org. Chem.*, 1942, **7**, 185.
6. J. N. Gnabre, Y. Ito, Y. Ma, and R. C. Huang, *J. Chromatogr. A*, 1996, **719**, 353.
7. H. C. Schröder, H. Merz, R. Steffen, W. E. G. Müller, P. S. Sarin, S. Trumm, J. Schulz, and E. Eich, *Z. Naturforsch.*, 1990, **45c**, 1215.
8. D. C. Ayres and J. D. Loike (eds.), "Chemistry and Pharmacology of Natural Products. Lignans: Chemical, Biological and Clinical Properties," Cambridge University Press, Cambridge, 1990.
9. K. Griffiths, H. Adlercreutz, P. Boyle, L. Denis, R. I. Nicholson, and M. S. Morton (eds.), "Nutrition and Cancer," Isis Medical Media, Oxford, 1996.
10. G. Schroeter, L. Lichtenstadt, and D. Irineu, *Chem. Ber.*, 1918, **51**, 1587.
11. R. D. Haworth, *Annu. Rept. Prog. Chem.*, 1937, **33**, 266.
12. O. R. Gottlieb, *Phytochemistry*, 1972, **11**, 1537.
13. O. R. Gottlieb, *Rev. Latinoamer. Quim.*, 1974, **5**, 1.
14. O. R. Gottlieb, *Progr. Chem. Org. Nat. Prod.*, 1978, **35**, 1.
15. O. R. Gottlieb and M. Yoshida, in "Natural Products of Woody Plants—Chemicals Extraneous to the Lignocellulosic Cell Wall," eds. J. W. Rowe and C. H. Kirk, Springer-Verlag, Berlin, 1989, p. 439.
16. D. A. Whiting, *Nat. Prod. Rep.*, 1987, **4**, 499.
17. K. Takahashi, M. Yasue, and K. Ogiyama, *Phytochemistry*, 1988, **27**, 1550.
18. R. Riffer and A. B. Anderson, *Phytochemistry*, 1967, **6**, 1557.
19. B. T. Ngadjui, D. Lontsi, J. F. Ayafor, and B. L. Sondengam, *Phytochemistry*, 1989, **28**, 231.
20. S. Nishibe, S. Hisada, and I. Inagaki, *Phytochemistry*, 1971, **10**, 2231.
21. S. Nishibe, S. Hisada, and I. Inagaki, *Chem. Pharm. Bull.*, 1973, **21**, 1108.
22. I. Inagaki, S. Hisada, and S. Nishibe, *Chem. Pharm. Bull.*, 1972, **20**, 2710.
23. N. G. Lewis and L. B. Davin, in "Isopentenoids and Other Natural Products: Evolution and Function," ed. W. D. Nes, ACS Symposium Series, Washington, DC, 1994, vol. 562, p. 202.
24. N. G. Lewis, M. J. Kato, N. Lopes, and L. B. Davin, in "Chemistry of the Amazon. Biodiversity, Natural Products, and Environmental Issues," eds. P. R. Seidl, O. R. Gottlieb, and M. A. C. Kaplan, ACS Symposium Series, Washington, DC, 1995, vol. 588, p. 135.
25. A. F. Barrero, A. Haïdour, M. M. Dorado, and J. M. Cuerva, *Phytochemistry*, 1996, **41**, 605.
26. A. Ichihara, Y. Numata, S. Kanai, and S. Sakamura, *Agric. Biol. Chem.*, 1977, **41**, 1813.
27. H. L. Hergert, *J. Org. Chem.*, 1960, **25**, 405.
28. H. L. Hergert, in "Lignins—Occurrence, Formation, Structure and Reactions," eds. K. V. Sarkanen and C. H. Ludwig, Wiley-Interscience, New York, 1971, p. 267.
29. H. L. Hergert, in "Cellulose Chemistry and Technology," ed. J. C. Arthur, Jr., American Chemical Society, Washington, DC, 1977, vol. 48, p. 227.
30. D. R. Gang, M. Fujita, L. B. Davin, and N. G. Lewis, in "Lignin and Lignan Biosynthesis," eds. N. G. Lewis and S. Sarkanen, ACS Symposium Series, Washington, DC, 1998, vol. 697, p. 389.
31. L. Y. Foo and J. Karchesy, *J. Chem. Soc., Chem. Commun.*, 1989, 217.
32. D. K. Sharma, K. R. Ranganathan, M. R. Parthasarathy, B. Bhushan, and T. R. Seshadri, *Planta Medica*, 1979, **37**, 79.
33. M. L. Cardona, B. Garcia, J. R. Pedro, and J. F. Sinisterra, *Phytochemistry*, 1990, **29**, 629.
34. D. R. Gang, A. T. Dinkova-Kostova, L. B. Davin, and N. G. Lewis, in "Phytochemical Pest Control Agents," eds. P. A. Hedin, R. M. Hollingworth, E. P. Masler, J. Miyamoto, and D. G. Thompson, American Chemical Society, Washington, DC, 1997, vol. 658, p. 58.
35. H. MacLean and J. A. F. Gardner, *For. Prod. J.*, 1956, **6**, 510.
36. R. P. Beatson, W. Wang, C. I. Johansson, and J. N. Saddler, in "7th International Conference on Biotechnology in the Pulp and Paper Industry," Technical Section, Canadian Pulp and Paper Association, 1998, p. 211.
37. A. Henges, M. Petersen, and A. W. Alfermann, in "Abstracts, Botanikertagung, Düsseldorf, 1996," Deutsche Botanische Gesellschaft Vereinigung für Angewandte Botanik, 1996, p. 253.
38. D. A. Whiting, *Nat. Prod. Rep.*, 1985, **2**, 191.
39. D. A. Whiting, *Nat. Prod. Rep.*, 1990, **7**, 349.
40. R. S. Ward, *Nat. Prod. Rep.*, 1993, **10**, 1.
41. R. S. Ward, *Nat. Prod. Rep.*, 1995, **12**, 183.
42. R. S. Ward, *Nat. Prod. Rep.*, 1997, **14**, 43.
43. S. Kitagawa, S. Nishibe, R. Benecke, and H. Thieme, *Chem. Pharm. Bull.*, 1988, **36**, 3667.
44. L.-G. Zhuang, O. Seligmann, K. Jurcic, and H. Wagner, *Planta Medica*, 1982, **45**, 172.
45. S. Nishibe, A. Sakushima, S. Kitagawa, B. Klimek, R. Benecke, and H. Thieme, *Shoyakugaku Zasshi*, 1988, **42**, 324.
46. H. Suzuki, K.-H. Lee, M. Haruna, T. Iida, K. Ito, and H.-C. Huang, *Phytochemistry*, 1982, **21**, 1824.

47. L. W. Wilcox, P. A. Fuerst, and G. L. Floyd, *Amer. J. Bot.*, 1993, **80**, 1028.
48. R. Hiesel, B. Combettes, and A. Brennicke, *Proc. Natl. Acad. Sci. USA*, 1994, **91**, 629.
49. L. E. Graham, *J. Plant Res.*, 1996, **109**, 241.
50. W. A. Taylor, *Nature*, 1995, **373**, 391.
51. D. Edwards, J. G. Duckett, and J. B. Richardson, *Nature*, 1995, **374**, 635.
52. L. A. Lewis, B. D. Mishler, and R. Vilgalys, *Mol. Phyl. Evol.*, 1997, **7**, 377.
53. F. Cullmann, K.-P. Adam, and H. Becker, *Phytochemistry*, 1993, **34**, 831.
54. F. Cullmann, K.-P. Adam, J. Zapp, and H. Becker, *Phytochemistry*, 1996, **41**, 611.
55. H. Tazaki, K.-P. Adam, and H. Becker, *Phytochemistry*, 1995, **40**, 1671.
56. R. Mues, S. Huneck, J. D. Connolly, and D. S. Rycroft, *Tetrahedron Lett.*, 1988, **29**, 6793.
57. T. Yoshida, M. Toyota, and Y. Asakawa, *J. Nat. Prod.*, 1997, **60**, 145.
58. Y. Asakawa, *Heterocycles*, 1997, **46**, 795.
59. F. Nagashima, M. Toyota, and Y. Asakawa, *Phytochemistry*, 1990, **29**, 2169.
60. Y. Asakawa (ed.), "Progress in the Chemistry of Organic Natural Products," Springer-Verlag, Wien, 1995, vol. 65.
61. R. Takeda, J. Hasegawa, and M. Shinozaki, *Tetrahedron Lett.*, 1990, **31**, 4159.
62. R. Takeda, J. Hasegawa, and K. Sinozaki, in "Bryophytes: Their Chemistry and Chemical Taxonomy," eds. H. D. Zinsmeister and R. Mues, Clarendon Press, Oxford, 1990, p. 201.
63. E. Brown, R. Dhal, and N. Papin, *Tetrahedron*, 1995, **51**, 13 061.
64. H. Wada, T. Kido, N. Tanaka, T. Murakami, Y. Saiki, and C.-M. Chen, *Chem. Pharm. Bull.*, 1992, **40**, 2099.
65. R. C. Lin, A. L. Skaltsounis, E. Seguin, F. Tillequin, and M. Koch, *Planta Medica*, 1994, **60**, 168.
66. T. Satake, T. Murakami, Y. Saiki, and C.-M. Chen, *Chem. Pharm. Bull.*, 1978, **26**, 1619.
67. S. Li, T. Iliefski, K. Lundquist, and A. F. A. Wallis, *Phytochemistry*, 1997, **46**, 929.
68. R. D. Hartley and C. W. Ford, in "Plant Cell Wall Polymers. Biogenesis and Biodegradation," eds. N. G. Lewis and M. G. Paice, ACS Symposium Series, Washington, DC, 1989, vol. 399, p. 137.
69. C. W. Ford and R. D. Hartley, *J. Sci. Food Agric.*, 1990, **50**, 29.
70. A. Patra and A. K. Mitra, *Indian J. Chem.*, 1979, **17B**, 412.
71. K. Freudenberg, *Science*, 1965, **148**, 595.
72. K. Freudenberg, in "Constitution and Biosynthesis of Lignin," eds. K. Freudenberg and A. C. Neish, Springer-Verlag, New York, 1968, p. 47.
73. N. G. Lewis and L. B. Davin, in "Lignin and Lignan Biosynthesis," eds. N. G. Lewis and S. Sarkanen, ACS Symposium Series, Washington, DC, 1998, vol. 697, p. 334.
74. D. R. Gang, M. A. Costa, M. Fujita, A. T. Dinkova-Kostova, H. B. Wang, V. Burlat, W. Martin, S. Sarkanen, L. B. Davin, and N. G. Lewis, 1998, submitted for publication.
75. H. Yamaguchi, F. Nakatsubo, Y. Katsura, and K. Murakami, *Holzforschung*, 1990, **44**, 381.
76. N. G. Lewis and L. B. Davin, in "Plant Polyphenols," eds. R. W. Hemingway and P. E. Laks, Plenum, New York, 1992, p. 73.
77. R. R. Arndt, S. H. Brown, N. C. Ling, P. Roller, C. Djerassi, J. M. Ferreira, F. B. Gilbert, E. C. Miranda, S. E. Flores, A. P. Duarte, and E. P. Carrazzoni, *Phytochemistry*, 1967, **6**, 1653.
78. W. Stöcklin, L. B. De Silva, and T. A. Geissman, *Phytochemistry*, 1969, **8**, 1565.
79. H. Ishii, T. Ishikawa, M. Mihara, and M. Akaike, *Yakugaku Zasshi*, 1983, **103**, 279.
80. H. Ishii, H. Ohida, and J. Haginiwa, *Yakugaku Zasshi*, 1972, **92**, 118.
81. H. Tatematsu, M. Kurokawa, M. Niwa, and Y. Hirata, *Chem. Pharm. Bull.*, 1984, **32**, 1612.
82. H. Fujimoto and T. Higuchi, *Mokuzai Gakkaishi*, 1977, **23**, 405.
83. T. Kikuchi, S. Matsuda, S. Kadota, and T. Tai, *Chem. Pharm. Bull.*, 1985, **33**, 1444.
84. T. Deyama, *Chem. Pharm. Bull.*, 1983, **31**, 2993.
85. E. E. Dickey, *J. Org. Chem.*, 1958, **23**, 179.
86. L. B. Davin, D. L. Bedgar, T. Katayama, and N. G. Lewis, *Phytochemistry*, 1992, **31**, 3869.
87. L. B. Davin, H.-B. Wang, A. L. Crowell, D. L. Bedgar, D. M. Martin, S. Sarkanen, and N. G. Lewis, *Science*, 1997, **275**, 362.
88. A. Chu, A. Dinkova, L. B. Davin, D. L. Bedgar, and N. G. Lewis, *J. Biol. Chem.*, 1993, **268**, 27 026.
89. J. Iqbal, B. Bhatia, and N. K. Nayyar, *Chem. Rev.*, 1994, **94**, 519.
90. N. G. Lewis, L. B. Davin, A. T. Dinkova-Kostova, M. Fujita, D. R. Gang, and S. Sarkanen, Patent: "Recombinant Pinoresinol/Lariciresinol Reductase, Recombinant Dirigent Protein, and Methods of Use", 1997, p. 146. Chemical Abstract XX129:38116.
91. D. R. O'Reilly, L. K. Miller, and V. A. Luckow (eds.), "Baculovirus Expression Vectors: A Laboratory Manual," Oxford University Press, New York, 1994.
92. A. van der Straten, H. Johansen, M. Rosenberg, and R. W. Sweet, *Curr. Methods Mol. Cell. Biol.*, 1989, **1**, 1.
93. S. F. Altschul, W. Gish, W. Miller, E. W. Myers, and D. J. Lipman, *J. Mol. Biol.*, 1990, **215**, 403.
94. K. C. Worley, B. A. Wiese, and R. F. Smith, *Genome Res.*, 1995, **5**, 173.
95. B. Fritensky, D. Horovitz, and L. A. Hadwiger, *Plant Mol. Biol.*, 1988, **11**, 713.
96. J. Görlach, S. Volrath, G. Knauf-Beiter, G. Hengy, U. Beckhove, K.-H. Kogel, M. Oostendorp, T. Staub, E. Ward, H. Kessmann, and J. Ryals, *Plant Cell*, 1996, **8**, 629.
97. J. D. Ford, L. B. Davin, and N. G. Lewis, in "Plant Polyphenols 2: Chemistry and Biology," eds. G. G. Gross, R. W. Hemingway, and T. Yoshida, Plenum, New York, 1999.
98. Y. Fukuda, M. Nagata, T. Osawa, and M. Namiki, *J. Amer. Oil Chem. Soc.*, 1986, **63**, 1027.
99. T. Osawa, M. Nagata, M. Namiki, and Y. Fukuda, *Agric. Biol. Chem.*, 1985, **49**, 3351.
100. M. J. Kato, A. Chu, L. B. Davin, and N. G. Lewis, *Phytochemistry*, 1998, **47**, 583.
101. Y. Jiao, L. B. Davin, and N. G. Lewis, *Phytochemistry*, 1998, **49**, 387.
102. W. Bauer and M. H. Zenk, *Phytochemistry*, 1991, **30**, 2953.
103. M. Rueffer and M. H. Zenk, *Phytochemistry*, 1994, **36**, 1219.
104. S. Clemens and W. Barz, *Phytochemistry*, 1996, **41**, 457.
105. Y. Jiao, unpublished results, 1998.
106. P. A. Marchand, M. J. Kato, and N. G. Lewis, *J. Nat. Prod.*, 1997, **60**, 1189.

107. Y.-G. Kim, S. Ozawa, Y. Sano, and T. Sasaya, *Res. Bull. Hokkaido University Forests*, 1996, **53**, 1.
108. T. Miyauchi and S. Ozawa, *Phytochemistry*, 1998, **47**, 665.
109. S. Ozawa, L. B. Davin, and N. G. Lewis, *Phytochemistry*, 1993, **32**, 643.
110. T. Katayama, L. B. Davin, A. Chu, and N. G. Lewis, *Phytochemistry*, 1993, **33**, 581.
111. A. T. Dinkova-Kostova, D. R. Gang, L. B. Davin, D. L. Bedgar, A. Chu, and N. G. Lewis, *J. Biol. Chem.*, 1996, **271**, 29 473.
112. P. M. Schenk, S. Baumann, R. Mattes, and H.-H. Steinbiß, *Biotechniques*, 1995, **19**, 196.
113. C. Branden and J. Tooze (eds.), "Introduction to Protein Structure," Garland Publishing, New York, 1991.
114. T. Umezawa, H. Kuroda, T. Isohata, T. Higuchi, and M. Shimada, *Biosci. Biotech. Biochem.*, 1994, **58**, 230.
115. T. Umezawa, L. B. Davin, E. Yamamoto, D. G. I. Kingston, and N. G. Lewis, *J. Chem. Soc., Chem. Commun.*, 1990, 1405.
116. T. Umezawa, L. B. Davin, and N. G. Lewis, *J. Biol. Chem.*, 1991, **266**, 10 210.
117. Z.-Q. Xia, M. A. Costa, L. B. Davin, and N. G. Lewis, 1999, submitted for publication.
118. N. G. Lewis, M. A. Costa, L. B. Davin, and Z.-Q. Xia, Patent Application: "Recombinant Secoisolariciresinol Dehydrogenase, and Methods of Use," 1998, p. 40.
119. H. Adlercreutz, K. Höckerstedt, C. Bannwart, E. Hämäläinen, T. Fotsis, and S. Bloigu, *Progress in Cancer: Research and Therapy*, 1988, **35**, 409.
120. H. Adlercreutz, in "Natural Antioxidants and Food Quality in Atherosclerosis and Cancer Prevention," eds. J. T. Kumpulainen and J. K. Salonen, Royal Society of Chemistry, Cambridge, 1996, p. 349.
121. J. E. Bakke and H. J. Klosterman, *Proc. No. Dakota Acad. Sci.*, 1956, **10**, 18.
122. O. Goldschmid and H. L. Hergert, *Tappi*, 1961, **44**, 858.
123. L. B. Davin, D. R. Gang, M. Fujita, A. M. Anterola, and N. G. Lewis, in "Proc. 9th Internat. Symp. Wood Pulp. Chem.," 1997, p. H3.
124. M. Fujita, D. R. Gang, L. B. Davin, and N. G. Lewis, *J. Biol. Chem.*, 1999, **274**.
125. T. Umezawa and M. Shimada, *Biosci. Biotech. Biochem.*, 1996, **60**, 736
126. D. E. Jackson and P. M. Dewick, *Phytochemistry*, 1984, **23**, 1147.
127. W. van Uden, N. Pras, and H. J. Woerdenbag, in "Biotechnology in Agriculture and Forestry," ed. Y. P. S. Bajaj, Springer-Verlag, Berlin, 1994, vol. 26, p. 219.
128. H. J. Wichers, G. G. Versluis-De Haan, J. W. Marsman, and M. P. Harkes, *Phytochemistry*, 1991, **30**, 3601.
129. W. van Uden, N. Pras, J. F. Visser, and T. M. Malingré, *Plant Cell Rep.*, 1989, **8**, 165.
130. W. van Uden, N. Pras, and T. M. Malingré, *Plant Cell, Tiss. Org. Cult.*, 1990, **23**, 217.
131. W. van Uden, N. Pras, E. M. Vossebeld, J. N. M. Mol, and T. M. Malingré, *Plant Cell, Tiss. Org. Cult.*, 1990, **20**, 81.
132. W. van Uden, *Pharm. World. Sci.*, 1993, **15**, 41.
133. A. J. Broomhead, M. M. A. Rahman, P. M. Dewick, D. E. Jackson, and J. A. Lucas, *Phytochemistry*, 1991, **30**, 1489.
134. Z.-Q. Xia, L. B. Davin, and N. G. Lewis, *Phytochemistry*, 1999, submitted for publication.
135. M. M. Rao and D. Lavie, *Tetrahedron*, 1974, **30**, 3309.
136. M. Kobayashi and Y. Ohta, *Phytochemistry*, 1983, **22**, 1257.
137. R. Valsaraj, P. Pushpangadan, U. W. Smitt, A. Adsersen, S. B. Christensen, A. Sittie, U. Nyman, C. Nielsen, and C. E. Olsen, *J. Nat. Prod.*, 1997, **60**, 739.
138. T. Deyama, T. Ikawa, and S. Nishibe, *Chem. Pharm. Bull.*, 1985, **33**, 3651.
139. M. A. Bernards, M. L. Lopez, J. Zajicek, and N. G. Lewis, *J. Biol. Chem.*, 1995, **270**, 7382.
140. M. A. Bernards and N. G. Lewis, *Polyphénols Actualités*, 1996, **14**, 4.
141. M. A. Bernards and N. G. Lewis, *Phytochemistry*, 1998, **47**, 915.
142. M. H. Chaves and N. F. Roque, *Phytochemistry*, 1997, **46**, 879.
143. L. Lajide, P. Escoubas, and J. Mizutani, *Phytochemistry*, 1995, **40**, 1105.
144. I. Sakakibara, T. Katsuhara, Y. Ikeya, K. Hayashi, and H. Mitsuhashi, *Phytochemistry*, 1991, **30**, 3013
145. I. Sakakibara, Y. Ikeya, K. Hayashi, and H. Mitsuhashi, *Phytochemistry*, 1992, **31**, 3219.
146. I. Sakakibara, Y. Ikeya, K. Hayashi, M. Okada, and M. Maruno, *Phytochemistry*, 1995, **38**, 1003.
147. T. Yoshihara, K. Yamaguchi, S. Takamatsu, and S. Sakamura, *Agric. Biol. Chem.*, 1981, **45**, 2593.
148. S. M. Kupchan, R. W. Britton, M. F. Ziegler, C. J. Gilmore, R. J. Restivo, and R. F. Bryan, *J. Am Chem. Soc.*, 1973, **95**, 1335.
149. H. Taguchi and Y. Ikeya, *Chem. Pharm. Bull.*, 1975, **23**, 3296.
150. H. Taguchi and Y. Ikeya, *Chem. Pharm. Bull.*, 1977, **25**, 364.
151. R. Stadler and M. H. Zenk, *J. Biol. Chem.*, 1993, **268**, 823.
152. T. Iida, K. Ichino, and K. Ito, *Phytochemistry*, 1982, **21**, 2939.
153. S. M. Kupchan, K. L. Stevens, E. A. Rohlfing, B. R. Sickles, A. T. Sneden, R. W. Miller, and R. F. Bryan, *J. Org. Chem.*, 1978, **43**, 586.
154. C. J. Aiba, R. G. C. Corrêa, and O. R. Gottlieb, *Phytochemistry*, 1973, **12**, 1163.
155. A. N. Binns, R. H. Chen, H. N. Wood, and D. G. Lynn, *Proc. Natl. Acad. Sci. USA*, 1987, **84**, 980.
156. D. G. Lynn, R. H. Chen, K. S. Manning, and H. N. Wood, *Proc. Natl. Acad. Sci.*, 1987, **84**, 615.
157. J. D. Orr and D. G. Lynn, *Plant Physiol.*, 1992, **98**, 343.
158. N. Hirai, M. Okamoto, H. Udagawa, M. Yamamuro, M. Kato, and K. Koshimizu, *Biosci. Biotech. Biochem.*, 1994, **58**, 1679.
159. D. R. Gang, H. Kasahara, Z.-Q. Xia, K. Vander Mijnsbrugge, W. Boerjan, H. van Montagu, L. B. Davin, and N. G. Lewis, *J. Biol. Chem.*, 1999, **274**.
160. B. Singh, P. K. Agrawal, and R. S. Thakur, *J. Nat. Prod.*, 1989, **52**, 48.
161. Y. Fukuyama, M. Nakahara, H. Minami, and M. Kodama, *Chem. Pharm. Bull.*, 1996, **44**, 1418.
162. M. S. Da Silva, J. M. Barbosa-Filho, M. Yoshida, and O. R. Gottlieb, *Phytochemistry*, 1989, **28**, 3477.
163. O. Salama, R. K. Chaudhuri, and O. Sticher, *Phytochemistry*, 1981, **20**, 2603.
164. E. Smite, H. Pan, and L. N. Lundgren, *Phytochemistry*, 1995, **40**, 341.
165. H. Pan and L. N. Lundgren, *Phytochemistry*, 1996, **42**, 1185.
166. R. W. Read and W. C. Taylor, *Aust. J. Chem.*, 1979, **32**, 2317.
167. T. Katayama, in "Lignin and Lignan Biosynthesis," eds. N. G. Lewis and S. Sarkanen, ACS Symposium Series, Washington, DC, 1998, vol. 697, p. 362.

168. M. Nose, M. A. Bernards, M. Furlan, J. Zajicek, T. L. Eberhardt, and N. G. Lewis, *Phytochemistry*, 1995, **39**, 71.
169. Y. Jiao, D. L. Bedgar, Z.-Q. Xia, L. B. Davin, and N. G. Lewis, 1999 submitted for publication.
170. P. Jäger-Vottero, Y. Jiao, L. B. Davin, and N. G. Lewis, 1999, submitted for publication.
171. W.-C. Su, J.-M. Fang, and Y.-S. Cheng, *Phytochemistry*, 1995, **40**, 563.
172. H. Pan and L. N. Lundgren, *Phytochemistry*, 1995, **39**, 1423.
173. T. Suga, S. Ohta, K. Munesada, N. Ide, M. Kurokawa, M. Shimizu, and E. Ohta, *Phytochemistry*, 1993, **33**, 1395.
174. N. Watanabe, T. Sasaya, and S. Ozawa, *Mokuzai Gakkaishi*, 1992, **38**, 796.
175. P. K. Agrawal, S. K. Agarwal, and R. P. Rastogi, *Phytochemistry*, 1980, **19**, 1260.
176. T. Deyama, T. Ikawa, S. Kitagawa, and S. Nishibe, *Chem. Pharm. Bull.*, 1987, **35**, 1785.
177. H. Kizu, H. Shimana, and T. Tomimori, *Chem. Pharm. Bull*, 1995, **43**, 2187.
178. C. Kraus and G. Spiteller, *Phytochemistry*, 1997, **44**, 59.
179. L. N. Lundgren, T. Popoff, and O. Theander, *Phytochemistry*, 1981, **20**, 1967.
180. L. N. Lundgren, Z. Shen, and O. Theander, *Acta Chem. Scand.*, 1985, **B39**, 241.
181. K. Miki and T. Sasaya, *Mokuzai Gakkaishi*, 1979, **25**, 437.
182. T. Popoff and O. Theander, *Phytochemistry*, 1975, **14**, 2065.
183. D. Strack, J. Heilemann, V. Wray, and H. Dirks, *Phytochemistry*, 1989, **28**, 2071.
184. R. X. Tan, J. Jakupovic, and Z. J. Jia, *Planta Medica*, 1990, **56**, 475.
185. F. Kawamura, H. Ohashi, S. Kawai, F. Teratani, and Y. Kai, *Mokuzai Gakkaishi*, 1996, **42**, 301.
186. A. R. Slabas, C. M. Sidebottom, A. Hellyer, R. M. J. Kessell, and M. P. Tombs, *Biochim. Biophys. Acta*, 1986, **877**, 271.
187. I. R. Cottingham, A. J. Austin, and A. R. Slabas, *Biochim. Biophys. Acta*, 1989, **995**, 273.
188. M. S. A. Afifi, M. M. Ahmed, J. M. Pezzuto, and A. D. Kinghorn, *Phytochemistry*, 1993, **34**, 839.
189. J. C. S. Malan, J. Chen, L. Y. Foo, and J. J. Karchesy, in "Plant Polyphenols," eds. R. W. Hemingway and P. E. Laks, Plenum, New York, 1992, p. 411.
190. A. Henrici, M. Kaloga, and E. Eich, *Phytochemistry*, 1994, **37**, 1637.
191. N. I. Kulesh, V. A. Denisenko, and O. B. Maksimov, *Phytochemistry*, 1995, **40**, 1001.
192. F. Hanawa, M. Shiro, and Y. Hayashi, *Phytochemistry*, 1997, **45**, 589.
193. S.-Y. Guan, J. Hlynár, and S. Sarkanen, *Phytochemistry*, 1997, **45**, 911.
194. S. Sarkanen, in "Lignin and Lignan Biosynthesis," eds. N. G. Lewis and S. Sarkanen, ACS Symposium Series, Washington, DC, vol. 697, p. 194
195. N. G. Lewis, L. B. Davin, and S. Sarkanen, in "Lignin and Lignan Biosynthesis", eds. N. G. Lewis and S. Sarkanen, ACS Symposium Series, Washington, DC, vol. 697, p. 1.
196. F. E. Brauns, *J. Am. Chem. Soc.*, 1939, **61**, 2120.
197. E. Taniguchi, K. Imamura, F. Ishibashi, T. Matsui, and A. Nishio, *Agric. Biol. Chem.*, 1989, **53**, 631.
198. T. Deyama, T. Ikawa, S. Kitagawa, and S. Nishibe, *Chem. Pharm. Bull.*, 1986, **34**, 4933.
199. K. Yoshinari, N. Shimazaki, Y. Sashida, and Y. Mimaki, *Phytochemistry*, 1990, **29**, 1675.
200. F. Kawamura, S. Kawai, and H. Ohashi, *Phytochemistry*, 1997, **44**, 1351.
201. A. Ichihara, K. Oda, Y. Numata, and S. Sakamura, *Tetrahedron Lett.*, 1976, **44**, 3961.
202. F. Abe, T. Yamauchi, and A. S. C. Wan, *Phytochemistry*, 1989, **28**, 3473.
203. F. Abe, T. Yamauchi, and A. S. C. Wan, *Phytochemistry*, 1988, **27**, 3627.
204. F. Abe, T. Yamauchi, and A. S. C. Wan, *Chem. Pharm. Bull.*, 1988, **36**, 795.
205. M. Kaouadji and J. Favre-Bonvin, *Tetrahedron Lett.*, 1984, **25**, 5137.
206. M. M. A. Rahman, P. M. Dewick, D. E. Jackson, and J. A. Lucas, *Phytochemistry*, 1990, **29**, 1841.
207. C. W. Waller and O. Gisvold, *J. Am. Pharm. Assoc.*, 1945, **34**, 78.
208. W. O. Lundberg, H. O. Halvorson, and G. O. Burr, *Oil and Soap*, 1944, **21**, 33.
209. E. P. Oliveto, *Chem. Ind.*, 1972, 677.
210. K. Yasumoto, A. Yamamoto, and H. Mitsuda, *Agric. Biol. Chem.*, 1970, **34**, 1162.
211. A. L. Tappel, W. O. Lundberg, and P. D. Boyer, *Arch. Biochem. Biophys.*, 1953, **42**, 293.
212. L. Shain and W. E. Hillis, *Phytopathology*, 1971, **61**, 841.
213. L. Shain, *Phytopathology*, 1971, **61**, 301.
214. P. Rudman, *Holzforschung*, 1965, **19**, 57.
215. Atta-ur-Rahman, M. Ashraf, M. I. Choudhary, Habib-ur-Rehman, and M. H. Kazmi, *Phytochemistry*, 1995, **40**, 427.
216. M. Takasugi and N. Katui, *Phytochemistry*, 1986, **25**, 2751.
217. A. M. Clark, F. S. El-Feraly, and W.-S. Li, *J. Pharm. Sci.*, 1981, **70**, 951.
218. H. Kasahara, M. Miyazawa, and H. Kameoka, *Phytochemistry*, 1997, **44**, 1479.
219. M. Miyazawa, H. Kasahara, and H. Kameoka, *Phytochemistry*, 1993, **34**, 1501.
220. M. Miyazawa, H. Kasahara, and H. Kameoka, *Phytochemistry*, 1994, **35**, 1191.
221. H. Kasahara, M. Miyazawa, and H. Kameoka, *Phytochemistry*, 1996, **43**, 111.
222. H. Kasahara, M. Miyazawa, and H. Kameoka, *Nat. Prod. Lett.*, 1997, **9**, 277.
223. M. Hattori, S. Hada, A. Watahiki, H. Ihara, Y.-Z. Shu, N. Kakiuchi, T. Mizuno, and T. Namba, *Chem. Pharm. Bull.*, 1986, **34**, 3885.
224. S. Yamauchi, F. Ishibashi, and E. Taniguchi, *Biosci. Biotech. Biochem.*, 1992, **56**, 1760.
225. S. Yamauchi and E. Taniguchi, *Biosci. Biotech. Biochem.*, 1992, **56**, 1744.
226. S. Yamauchi and E. Taniguchi, *Biosci. Biotech. Biochem.*, 1992, **56**, 1751.
227. A. Isogai, S. Murakoshi, A. Suzuki, and S. Tamura, *Agric. Biol. Chem.*, 1973, **37**, 889.
228. J. K. Nitao, M. G. Nair, D. L. Thorogood, K. S. Johnson, and J. M. Scriber, *Phytochemistry*, 1991, **30**, 2193.
229. M. Miyazawa, Y. Ishikawa, H. Kasahara, J.-I. Yamanaka, and H. Kameoka, *Phytochemistry*, 1994, **35**, 611.
230. A. González-Coloma, P. Escoubas, J. Mizutani, and L. Lajide, *Phytochemistry*, 1994, **35**, 607.
231. W. S. Bowers, *Science*, 1968, **161**, 895.
232. H. L. Haller, E. R. McGovran, L. D. Goodhue, and W. N. Sullivan, *J. Org. Chem.* 1942, **7**, 183.
233. J. A. González, A. Estevez-Braun, R. Estevez-Reyes, and A. G. Ravelo, *J. Chem. Ecol.*, 1994, **20**, 517.
234. J. A. González, A. Estévez-Braun, R. Estévez-Reyes, I. L. Bazzocchi, L. Moujir, I. A. Jimenez, A. G. Ravelo, and A. G. González, *Experientia*, 1995, **51**, 35.

235. T. Yoshihara, K. Yamaguchi, and S. Sakamura, *Agric. Biol. Chem.*, 1982, **46**, 853.
236. J. Harmatha and J. Nawrot, *Biochem. Syst. Ecol.*, 1984, **12**, 95.
237. K. Matsui and K. Munakata, *Tetrahedron Lett.*, 1975, **24**, 1905.
238. K. Matsui, K. Wada, and K. Munakata, *Agric. Biol. Chem.*, 1976, **40**, 1045.
239. K. Munakata, S. Marumo, K. Ohta, and Y.-L. Chen, *Tetrahedron Lett.*, 1965, **47**, 4167.
240. S. D. Elakovich and K. L. Stevens, *J. Chem. Ecol.*, 1985, **11**, 27.
241. D. Lavie, E. C. Levy, A. Cohen, M. Evenari, and M. Y. Guttermann, *Nature*, 1974, **249**, 388.
242. M. Szabó and A. Garay, *Acta Botanica Academiae Scientiarum Hungaricae*, 1970, **16**, 207.
243. P. V. Bhiravamurty, R. Das Kanakala, E. Venkata Rao, and K. V. Sastry, *Curr. Sci.*, 1979, **48**, 949.
244. M. M. Chattaway, *Aust. J. Sci. Res. B*, 1949, **2**, 227.
245. M. M. Chattaway, *Aust. For.*, 1952, **16**, 25.
246. R. L. Krahmer, R. W. Hemingway, and W. E. Hillis, *Wood Sci. Technol.*, 1970, **4**, 122.
247. J. Ralph, J. J. MacKay, R. D. Hatfield, D. M. O'Malley, R. W. Whetten, and R. R. Sederoff, *Science*, 1997, **277**, 235.
248. K. D. R. Setchell, A. M. Lawson, F. L. Mitchell, H. Adlercreutz, D. N. Kirk, and M. Axelson, *Nature*, 1980, **287**, 740.
249. S. R. Stitch, J. K. Toumba, M. B. Groen, C. W. Funke, J. Leemhuis, J. Vink, and G. F. Woods, *Nature*, 1980, **287**, 738.
250. M. Axelson, J. Sjövall, B. E. Gustafsson, and K. D. R. Setchell, *Nature*, 1982, **298**, 659.
251. H. Adlercreutz, in "Nutrition, Toxicity, and Cancer," ed. I. R. Rowland, CRC Press, Boca Raton, FL, 1991, p. 137.
252. H. Adlercreutz, T. Fotsis, J. Lampe, K. Wähälä, T. Mäkelä, G. Brunow, and T. Hase, *Scand. J. Clin. Lab. Invest.*, 1993, **53**, 5.
253. H. Adlercreutz, J. van der Wildt, J. Kinzel, H. Attalla, K. Wähälä, T. Mäkelä, T. Hase, and T. Fotsis, *J. Steroid Biochem. Molec. Biol.*, 1995, **52**, 97.
254. M. Axelson and K. D. R. Setchell, *FEBS Lett.*, 1981, **123**, 337.
255. K. D. R. Setchell, A. M. Lawson, S. P. Borriello, R. Harkness, H. Gordon, D. M. L. Morgan, D. N. Kirk, H. Adlercreutz, L. C. Anderson, and M. Axelson, *Lancet*, 1981, 4.
256. M. S. Kurzer, J. W. Lampe, M. C. Martini, and H. Adlercreutz, *Cancer Epidemiol., Biomarkers and Prev.*, 1995, **4**, 353.
257. S. I. Mäkelä, L. H. Pylkkänen, R. S. S. Santti, and H. Adlercreutz, *J. Nutr.*, 1995, **125**, 437.
258. M. E. Martin, M. Haourigui, C. Pelissero, C. Benassayag, and E. A. Nunez, *Life Sci.*, 1996, **58**, 429.
259. H. Adlercreutz, C. Bannwart, K. Wähälä, T. Mäkelä, G. Brunow, T. Hase, P. J. Arosemena, J. T. Kellis, Jr., and L. E. Vickery, *J. Steroid Biochem. Molec. Biol.*, 1993, **44**, 147.
260. C. Wang, T. Mäkelä, T. Hase, H. Adlercreutz, and M. S. Kurzer, *J. Steroid Biochem. Molec. Biol.*, 1994, **50**, 205.
261. L. U. Thompson, M. M. Seidl, S. E. Rickard, L. J. Orcheson, and H. H. S. Fong, *Nutr. Cancer*, 1996, **26**, 159.
262. L. U. Thompson, S. E. Rickard, L. J. Orcheson, and M. M. Seidl, *Carcinogenesis*, 1996, **17**, 1373.
263. M. Jenab and L. U. Thompson, *Carcinogenesis*, 1996, **17**, 1343.
264. P. I. Musey, H. Adlercreutz, K. G. Gould, D. C. Collins, T. Fotsis, C. Bannwart, T. Mäkelä, K. Wähälä, G. Brunow, and T. Hase, *Life Sci.*, 1995, **57**, 655.
265. H. Adlercreutz, Y. Mousavi, J. Clark, K. Höckerstedt, E. Hämäläinen, K. Wähälä, T. Mäkelä, and T. Hase, *J. Steroid Biochem. Molec. Biol.*, 1992, **41**, 331.
266. M. Schöttner, G. Spiteller, and D. Gansser, *J. Nat. Prod.*, 1998, **61**, 119.
267. H. Adlercreutz, H. Honjo, A. Higashi, T. Fotsis, E. Hämäläinen, T. Hasegawa, and H. Okada, *Am. J. Clin. Nutr.*, 1991, **54**, 1093.
268. S. P. Borriello, K. D. R. Setchell, M. Axelson, and A. M. Lawson, *J. Appl. Bacteriol.*, 1985, **58**, 37.
269. K. D. R. Setchell, A. M. Lawson, S. P. Borriello, H. Adlercreutz, and M. Axelson, *Falk Symp.*, 1982, **31 (Colonic Carcinog.)**, 93.
270. K. Prasad, *Mol. Cell. Biochem.*, 1997, **168**, 117.
271. K. Yamashita, Y. Nohara, K. Katayama, and M. Namiki, *J. Nutr.*, 1992, **122**, 2440.
272. K. Yamashita, Y. Iizuka, T. Imai, and M. Namiki, *Lipids*, 1995, **30**, 1019.
273. M. G. Kelly and J. L. Hartwell, *J. Nat. Cancer Inst.*, 1954, **14**, 967.
274. J. D. Loike and S. B. Horwitz, *Biochemistry*, 1976, **15**, 5435.
275. M. B. Chabner (ed.), "Pharmacological Principles of Cancer Treatment," W. B. Saunders, Philadelphia, PA, 1982.
276. H. Stähelin, *Eur. J. Cancer*, 1970, **6**, 303.
277. H. Stähelin, *Eur. J. Cancer*, 1973, **9**, 215.
278. P. J. O'Dwyer, B. Leyland-Jones, M. T. Alonso, S. Marsoni, and R. E. Wittes, *N. Engl. J. Med.*, 1985, **312**, 692.
279. S. D. Williams, R. Birch, and L. H. Einhorn, *N. Engl. J. Med.*, 1987, **316**, 1433.
280. R. C. Young, *Semin. Oncol.*, 1992, **19 (Suppl. 13)**, 19.
281. G. L. Chen, L. Yang, T. C. Rowe, B. D. Halligan, K. M. Tewey, and L. F. Liu, *J. Biol. Chem.*, 1984, **259**, 13 560.
282. P. D. Senter, M. G. Saulnier, G. J. Schreiber, D. L. Hirschberg, J. P. Brown, I. Hellström, and K. E. Hellström, *Proc. Natl. Acad. Sci. USA*, 1988, **85**, 4842.
283. D. J. Brooks, N. R. Srinivas, D. S. Alberts, T. Thomas, L. M. Igwemzie, L. M. McKinney, J. Randolph, L. Schacter, S. Kaul, and R. H. Barbhaiya, *Anti-Cancer Drugs*, 1995, **6**, 637.
284. K. Tomioka, T. Ishiguro, H. Mizuguchi, N. Komeshima, K. Koga, S. Tsukagoshi, T. Tsuruo, T. Tashiro, S. Tanida, and T. Kishi, *J. Med. Chem.*, 1991, **34**, 54.
285. D. B. M. Wickramaratne, T. Pengsuparp, W. Mar, H.-B. Chai, T. E. Chagwedera, C. W. W. Beecher, N. R. Farnsworth, A. D. Kinghorn, J. M. Pezzuto, and G. A. Cordell, *J. Nat. Prod.*, 1993, **56**, 2083.
286. J. R. Cole, E. Bianchi, and E. R. Trumbull, *J. Pharm. Sci.*, 1969, **58**, 175.
287. K. Tomioka, T. Ishiguro, and K. Koga, *Chem. Pharm. Bull.*, 1985, **33**, 4333.
288. S. J. Torrance, J. J. Hoffmann, and J. R. Cole, *J. Pharm. Sci.*, 1979, **68**, 664.
289. A. Somanabandhu, S. Nitayangkura, C. Mahidol, S. Ruchirawat, K. Likhitwitayawuid, H.-L. Shieh, H. Chai, J. M. Pezzuto, and G. A. Cordell, *J. Nat. Prod.*, 1993, **56**, 233.
290. P. C. Li, D. H. F. Mak, M. K. T. Poon, S. P. Ip, and K. M. Ko, *Phytomedicine*, 1996, **3**, 217.
291. S. Yamada, Y. Murawaki, and H. Kawasaki, *Biochemical Pharmacology*, 1993, **46**, 1081.
292. Y. Ikeya, H. Taguchi, I. Yosioka, and H. Kobayashi, *Chem. Pharm. Bull.*, 1979, **27**, 1383.
293. S. P. Ip, D. H. F. Mak, P. C. Li, M. K. T. Poon, and K. M. Ko, *Pharmacology and Toxicology*, 1996, **78**, 413.

294. K. M. Ko, S. P. Ip, M. K. T. Poon, S. S. Wu, C. T. Che, K. H. Ng, and Y. C. Kong, *Planta Medica*, 1995, **61**, 134.
295. Y. Ohkura, Y. Mizoguchi, S. Morisawa, S. Takeda, M. Aburada, and E. Hosoya, *Japan. J. Pharmacol.*, 1990, **52**, 331.
296. S. Kubo, I. Matsui-Yuasa, S. Otani, S. Morisawa, H. Kinoshita, and K. Sakai, *J. Surg. Res.*, 1986, **41**, 401.
297. S. Kubo, Y. Ohkura, Y. Mizoguchi, I. Matsui-Yuasa, S. Otani, S. Morisawa, H. Kinoshita, S. Takeda, M. Aburada, and E. Hosoya, *Planta Medica*, 1992, **58**, 489.
298. M. Nomura, M. Nakachiyama, T. Hida, Y. Ohtaki, K. Sudo, T. Aizawa, M. Aburada, and K.-I. Miyamoto, *Cancer Lett.*, 1994, **76**, 11.
299. M. Nomura, Y. Ohtaki, T. Hida, T. Aizawa, H. Wakita, and K.-I. Miyamoto, *Anticancer Res.*, 1994, **14**, 1967.
300. K. M. Ko, D. H. F. Mak, P. C. Li, M. K. T. Poon, and S. P. Ip, *Phytotherapy Research*, 1996, **10**, 450.
301. W. D. MacRae, J. B. Hudson, and G. H. N. Towers, *Planta Medica*, 1989, **55**, 531.
302. M. K. Kernan, A. Sendl, J. L. Chen, S. D. Jolad, P. Blanc, J. T. Murphy, C. A. Stoddart, W. Nanakorn, M. J. Balick, and E. J. Rozhon, *J. Nat. Prod.*, 1997, **60**, 635.
303. K. Pfeifer, H. Merz, R. Steffen, W. E. G. Mueller, S. Trumm, J. Schulz, E. Eich, and H. C. Schroeder, *J. Pharm. Med.*, 1992, **2**, 75.
304. E. Eich, H. Pertz, M. Kaloga, J. Schulz, M. R. Fesen, A. Mazumder, and Y. Pommier, *J. Med. Chem.*, 1996, **39**, 86.
305. J. N. Gnabre, J. N. Brady, D. J. Clanton, Y. Ito, J. Dittmer, R. B. Bates, and R. C. C. Huang, *Proc. Natl. Acad. Sci. USA*, 1995, **92**, 11 239.
306. H. Hara, T. Fujihashi, T. Sakata, A. Kaji, and H. Kaji, *AIDS Res. Hum. Retroviruses*, 1997, **13**, 695.
307. C.-W. Chang, M.-T. Lin, S.-S. Lee, K. C. S. Chen Liu, F.-L. Hsu, and J.-Y. Lin, *Antiviral Res.*, 1995, **27**, 367.
308. T. Fujihashi, H. Hara, T. Sakata, K. Mori, H. Higuchi, A. Tanaka, H. Kaji, and A. Kaji, *Antimicrob. Agents Chemother.*, 1995, **39**, 2000.
309. A. M. Rimando, J. M. Pezzuto, N. R. Farnsworth, T. Santisuk, V. Reutrakul, and K. Kawanishi, *J. Nat. Prod.*, 1994, **57**, 896.
310. D.-F. Chen, S.-X. Zhang, K. Chen, B.-N. Zhou, P. Wang, L. M. Cosentino, and K.-H. Lee, *J. Nat. Prod.*, 1996, **59**, 1066.
311. T. Y. Shen, *Lipids*, 1991, **26**, 1154.
312. J. X. Pan, O. D. Hensens, D. L. Zink, M. N. Chang, and S.-B. Hwang, *Phytochemistry*, 1987, **26**, 1377.
313. H. C. Castro-Faria-Neto, C. V. Araújo, S. Moreira, P. T. Bozza, G. Thomas, J. M. Barbosa-Filho, R. S. B. Cordeiro, and E. V. Tibiriçá, *Planta Medica*, 1995, **61**, 106.
314. C.-C. Chen, W.-C. Hsin, F.-N. Ko, Y.-L. Huang, J.-C. Ou, and C.-M. Teng, *J. Nat. Prod.*, 1996, **59**, 1149.
315. T. Nikaido, T. Ohmoto, T. Kinoshita, U. Sankawa, S. Nishibe, and S. Hisada, *Chem. Pharm. Bull.*, 1981, **29**, 3586.
316. T. Nikaido, T. Ohmoto, H. Noguchi, T. Kinoshita, H. Saitoh, and U. Sankawa, *Planta Medica*, 1981, **43**, 18.
317. T. Deyama, S. Nishibe, S. Kitagawa, Y. Ogihara, T. Takeda, T. Ohmoto, T. Nikaido, and U. Sankawa, *Chem. Pharm. Bull.*, 1988, **36**, 435.
318. T. Iwasaki, K. Kondo, T. Kuroda, Y. Moritani, S. Yamagata, M. Sugiura, H. Kikkawa, O. Kaminuma, and K. Ikezawa, *J. Med. Chem.*, 1996, **39**, 2696.
319. M. Kimura, J. Suzuki, T. Yamada, M. Yoshizaki, T. Kikuchi, S. Kadota, and S. Matsuda, *Planta Medica*, 1985, **51**, 291.
320. S. Kadota, K. Tsubuno, K. Makino, M. Takeshita, and T. Kikuchi, *Tetrahedron Lett.* 1987, **28**, 2857.
321. J. M. Prieto, M. C. Recio, R. M. Giner, S. Máñez, A. Massmanian, P. G. Waterman, and J. L. Ríos, *Z. Naturforsch.*, 1996, **51c**, 618.
322. S. Ghosal, S. Banerjee, and A. W. Frahm, *Chem. Ind.*, 1979, 854.
323. N. R. Farnsworth, A. D. Kinghorn, D. D. Soejarto, and D. P. Waller, in "Economic and Medicinal Plant Research," eds. H. Wagner, H. Hikino, and N. R. Farnsworth, Academic Press, London, 1985, vol. 1, p. 155.
324. S. Nishibe, H. Kinoshita, H. Takeda, and G. Okano, *Chem. Pharm. Bull.*, 1990, **38**, 1763.

1.26
Biosynthesis of Flavonoids

GERT FORKMANN
Technische Universität München, Freising, Germany

and

WERNER HELLER
GSF Forschungszentrum für Umwelt und Gesundheit, Oberschleissheim, Germany

1.26.1 INTRODUCTION

Flavonoids represent an important class of natural products occurring in all vascular plants, and also in some mosses. In a single species, dozens of different flavonoids may be present. To date, more than 4000 flavonoid compounds have been isolated and identified.[1] Flavonoids exhibit a wide range of functions in biochemistry, physiology, and ecology of plants, for example in the coloration of flower petals,[2] the fertility and germination of pollen, and the activation of bacterial nodulation genes, which are involved in the nitrogen fixation process,[3–5] in warding off pathogenic microorganisms (phytoalexins),[6] and in protection against ultraviolet light (UV-B screening pigments).[7–9] Furthermore, flavonoids can act as plant growth regulators, enzyme inhibitors, insect antifeedants as well as oviposition stimulants for some butterflies.[10] They can also function as antioxidants, and are discussed as potential anticancer agents in humans.[11]

1.26.2 FLAVONOID STRUCTURE AND MAJOR FLAVONOID CLASSES

Early work in structure elucidation of anthocyanins, flavones, and flavonols revealed that flavonoids have a basic $C_6—C_3—C_6$ skeleton structure in common, consisting of two aromatic rings (A and B), and a heterocyclic ring (C) containing one oxygen atom (Figure 1 (**1**)).

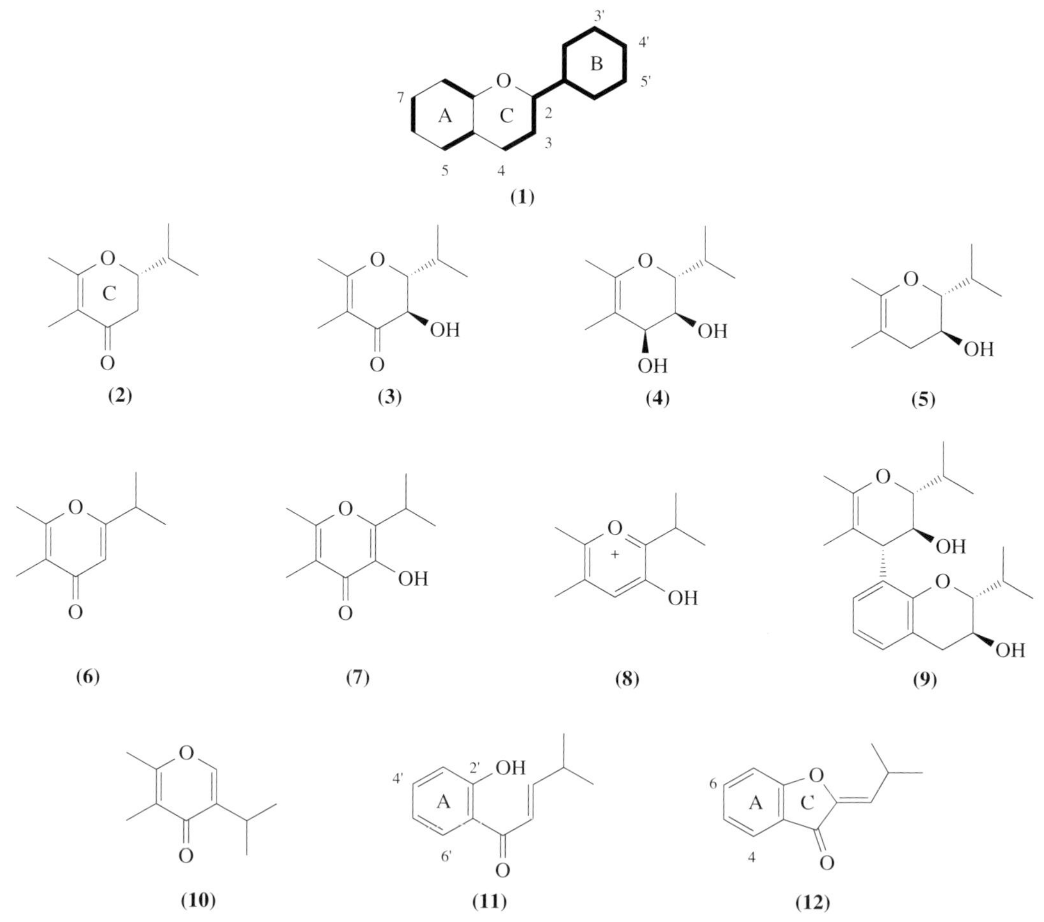

Figure 1 Elements of flavonoid structures. The basic flavonoid structure (**1**) is shown with biosynthetic building blocks marked by thick bonds. Only ring C is presented for the main flavonoid classes (**2**)–(**9**), while ring A is additionally given for chalcone (**11**) and aurone (**12**), because of their divergent ring numbering.

Robinson postulated as early as 1936 that this structure might be biogenetically derived from a C_6—C_3 (ring B and carbon atoms 2, 3 and 4) and a C_6 unit (ring A).[12] In 1953, Birch and Donovan proposed the polyacetate hypothesis for the biosynthesis of phloroglucinol and resorcinol derivatives, which also includes the flavonoids.[13] These authors suggested that the compounds are formed from a hydroxycinnamic acid and three acetate units to give a tri-oxo acid intermediate, which would then cyclize to a chalcone. Tracer experiments with labeled precursors basically confirmed Birch's hypothesis.[14] Shikimate was shown to contribute to ring B, and then phenylalanine and cinnamate to both rings B and C. Ring A was found to be formally derived from acetate units by head-to-tail condensation. A period of extensive tracer studies followed, where the importance of chalcones as the key intermediates of flavonoid formation was demonstrated, and the biogenetic relationships between the various flavonoid classes were established.

According to the oxidation level of the central heterocyclic ring C, flavonoids are grouped into nearly a dozen structural classes, the major ones being shown in Figure 1. The first flavonoid and central compound formed is a flavanone (**2**). Hydroxylation in position 3 leads to dihydroflavonols (**3**), reduction of the carbonyl group in position 4 to flavan-3,4-diols (**4**), and further reduction at the same position to flavan-3-ols (**5**). Formation of a double bond between positions 2 and 3 of flavanones and dihydroflavonols leads to flavones (**6**) and flavonols (**7**), respectively. Anthocyanins (**8**) possess a conjugated system of double bonds throughout the molecule giving rise to the typical red or blue color characteristic for these compounds. Proanthocyanidins (**9**) are condensation products of flavan-3,4-diols with a flavan-3-ol starter unit. Isoflavones (**10**) are distinct from other flavonoid classes by having the B-ring attached to position 3 of the heterocyclic ring C, instead of position 2. Chalcones (**11**) and aurones (**12**) (note the different numbering in Figure 1) lack the typical flavonoid structure, but they are biosynthetically closely related to flavonoids. Chalcones are furthermore precursors of many diarylpropanoid metabolites. Studies on flavonoid structures have been reviewed by Grisebach,[14,15] Heller,[16] and Stafford.[17]

1.26.3 ELUCIDATION AND GENERAL OVERVIEW OF THE FLAVONOID PATHWAY

In addition to the tracer experiments, genetic studies have greatly contributed to the understanding of the biogenetic sequence of the different flavonoid classes. Since Mendel derived his laws of inheritance from the results of experiments crossing red- and white-flowering peas, changes in the readily observable coloring of flowers have been widely applied in genetic studies. In the course of structure elucidation of anthocyanins and other flavonoids, single genes controlling the visible color changes were successfully correlated with either presence or absence of particular flavonoid metabolites. Efficient methods for flavonoid isolation and identification including chromatographic and spectroscopic techniques enabled the rapid analysis of the genetic control of flavonoid expression within a broad range of plant species. More recently, structure, function, regulation, and interaction of flavonoid genes have been studied at the molecular level. This work, which also includes the expression of flavonoid genes in transgenic plants, has resulted in the rapid accumulation of molecular genetic data. The more classical genetic studies on flavonoids have been extensively reviewed earlier.[18–21]

The discovery of phenylalanine ammonia-lyase activity by Koukol and Conn in 1961 marked the starting point of the enzymology of phenylpropanoids, and thus also of flavonoid biosynthesis.[22] In the early 1970s, plant cell suspension cultures, in particular of *Petroselinum crispum*, were a valuable source for the isolation and characterization of flavonoid enzymes. More recently, the wide range of chemicogenetic information available from flowering plants has been exploited for the biochemical analysis of flavonoid formation. Flowers of defined genotypes have turned out to be very useful in experiments of feeding with potential precursors and also as an enzyme source. This allowed, for the first time, the unequivocal correlation of single genes with particular enzymes, and clearly confirmed the *in vivo* function of an enzyme activity measured *in vitro*.[23]

Since the mid-1970s, rapid and substantial progress has been made in elucidating the pathways to the precursors of flavonoids, the individual steps to the flavonoid classes, flavonoid modifications, and regulation of the respective enzyme steps. At present, the essential reactions are known, although a few gaps still exist. The latter include even the key reaction to one of the best-known flavonoid classes, the anthocyanins, as well as the formation of epicatechin and the proanthocyanidins. Information is also lacking on the enzymology of some minor-flavonoid-related compounds such as aurones and dihydrochalcones.

The pathways to the preflavonoid precursors 4-coumaroyl-CoA and malonyl-CoA, and the reaction steps leading to the various flavonoid classes, are outlined in Scheme 1. The enzymes involved are summarized in Table 1.

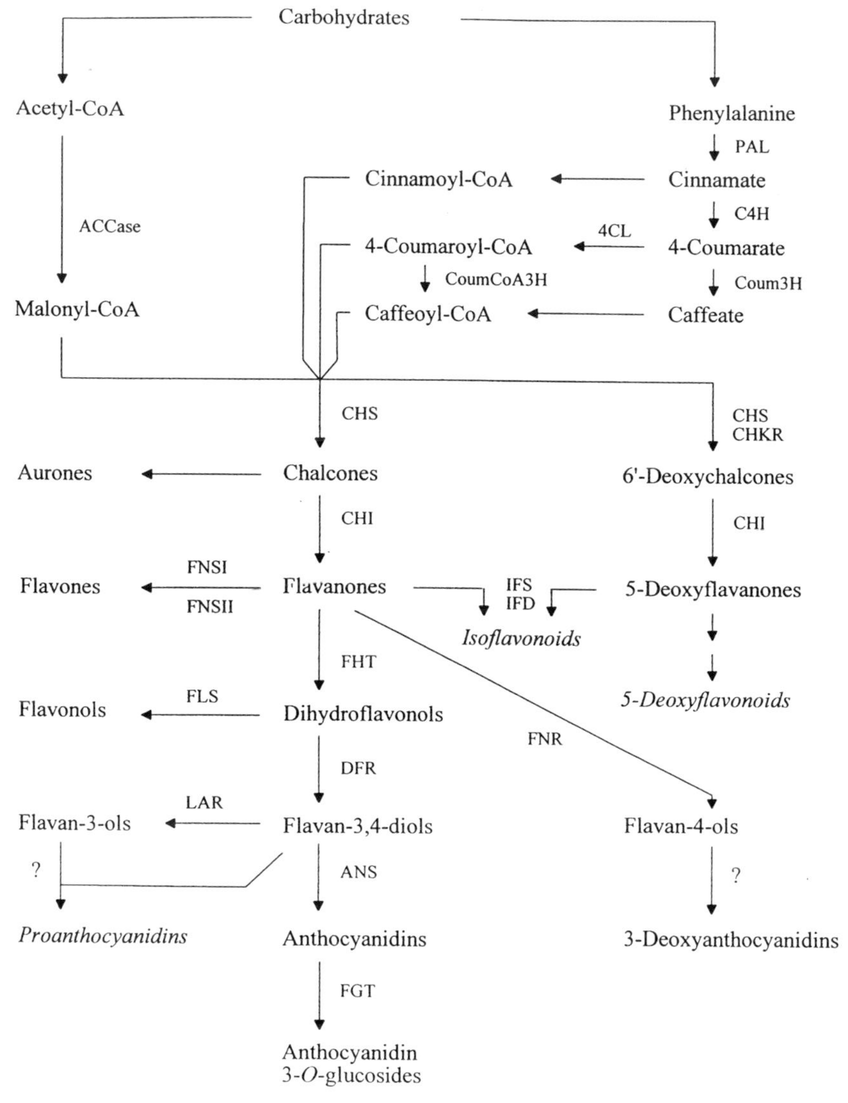

Scheme 1

The preflavonoid precursors are both derived from carbohydrate metabolism. Malonyl-CoA is directly synthesized from acetyl-CoA and CO_2 by means of acetyl-CoA:carboxyligase (ACCase). The formation of 4-coumaroyl-CoA and related hydroxycinnamic acid esters is more complex. It involves the shikimate/arogenate pathway, which leads to the aromatic amino acid phenylalanine, the starting point of the general phenylpropanoid pathway. The following three steps to the second precursor, 4-coumaroyl-CoA, are catalyzed by the enzymes phenylalanine ammonia-lyase (PAL), cinnamate 4-hydroxylase (C4H), and 4-coumarate:CoA ligase (4CL). 4-Coumaroyl-CoA may be hydroxylated in position 3 to give caffeoyl-CoA, which can also serve as substrate in chalcone synthesis in some plant species. 4-Coumarate can similarly be hydroxylated to give caffeate, which may then be transformed to its CoA ester by action of the ligase.

The key reaction in flavonoid biosynthesis is the stepwise condensation of three molecules of malonyl-CoA with a suitably hydroxycinnamic CoA ester, in most cases 4-coumaroyl-CoA, giving 2′,4′,6′-oxygenated (phloroglucinol-type) C_{15} chalcones. This reaction is catalyzed by chalcone synthase (CHS). It confirmed the results of the earlier tracer experiments that the basic flavonoid structure originates from a hydroxycinnamic acid and three acetate units. 6′-Deoxychalcones (resorcinol-type) are likewise synthesized from malonyl-CoA and a suitably substituted cinnamic acid CoA ester by CHS, but this transformation additionally requires the presence of an NADPH-

Table 1 List of enzymes leading to nonflavonoid precursors and various flavonoid classes.

Enzyme	*Acronym*	*EC number*
Nonflavonoid precursors		
Acetyl-CoA carboxyligase	ACCase	6.4.1.2
Phenylalanine ammonia-lyase	PAL	4.3.1.5
Cinnamate 4-hydroxylase	C4H	1.14.13.11
4-Coumarate:CoA ligase	4CL	6.2.1.12
4-Coumaroyl-CoA 3-hydroxylase	CoumCoA3H	
4-Coumarate 3-hydroxylase	Coum3H	
Flavonoid classes		
Chalcone synthase	CHS	2.3.1.74
Chalcone polyketide reductase	CHKR	
Chalcone isomerase	CHI	5.5.1.6
2-Hydroxyisoflavanone synthase	IFS	
2-Hydroxyisoflavanone dehydratase	IFD	
Flavone synthase I	FNS I	
Flavone synthase II	FNS II	
Flavanone 4-reductase	FNR	1.1.1.234
Flavanone 3-hydroxylase	FHT	1.14.11.9
Flavonol synthase	FLS	
Dihydroflavonol 4-reductase	DFR	1.1.1.219
Leucoanthocyanidin 4-reductase	LAR	
Anthocyanidin synthase	ANS	
Flavonoid (anthocyanidin/flavonol) 3-*O*-Glucosyltransferase	FGT	2.4.1.91

dependent chalcone ketide reductase (CHKR). Both 6′-hydroxy- and 6′-deoxychalcones are the precursors for aurones and other diarylpropanoids, but the respective enzymes catalyzing these reactions are still unknown.

The first flavonoid is provided by the stereospecific conversion of the chalcone to the respective flavanone with 2*S*-configuration by the action of chalcone isomerase (CHI). Two types of chalcone isomerases are known: one type can isomerize both 6′-hydroxy- and 6′-deoxychalcones to 5-hydroxy- and 5-deoxyflavanones, respectively, the other exclusively transforms 6′-hydroxychalcones.

The flavanone is one of the main branch points in the flavonoid pathway. Flavanones are the direct intermediates in the formation of isoflavones, flavones, dihydroflavonols and flavan-4-ols. The 5-deoxyflavanones are important intermediates in the isoflavone/pterocarpan pathway. The enzymatic conversion of 5-deoxyflavanones into other 5-deoxyflavonoids has not been demonstrated so far.

The first step in isoflavone formation is catalyzed by 2-hydroxyisoflavanone synthase (IFS), an NADPH-dependent cytochrome P450 mixed-function monooxygenase. Subsequent action of a dehydratase (IFD) leads to the respective isoflavone. The synthase accepts both 5-hydroxy- and 5-deoxyflavanones as substrates. The biosynthesis of isoflavones is described in detail in Chapter 1.28 of this volume and will not be further discussed here.

Introduction of a double bond between C-2 and C-3 of flavanones leads to the abundant class of flavones. Two types of enzymes can catalyze this reaction: flavone synthase I (FNS I), a 2-oxoglutarate-dependent dioxygenase, and flavone synthase II (FNS II), an NADPH-dependent cytochrome P450 species which accomplishes dehydrogenation.

Hydroxylation of flavanones in position 3 leads to dihydroflavonols. This step is catalyzed by flavanone 3-hydroxylase (FHT), another 2-oxoglutarate-dependent dioxygenase. Dihydroflavonols are the substrates for flavonol and flavan-3,4-diol synthesis, the latter being the direct biosynthetic intermediate to catechin, proanthocyanidins, and anthocyanidins. Flavonols are formed from dihydroflavonols by the introduction of a double bond between C-2 and C-3. Flavonol synthase (FLS), the enzyme catalyzing this step, is a third 2-oxoglutarate-dependent dioxygenase in the flavonoid pathway.

Reduction of the carbonyl group of dihydroflavonols in position 4 leads to flavan-3,4-diols, also called leucoanthocyanidins. This reaction is catalyzed by dihydroflavonol 4-reductase (DFR), which uses NADPH as the reducing cofactor. Flavan-3,4-diols are the immediate precursors for the synthesis of catechins and proanthocyanidins. Catechins are formed by further reduction of flavan-3,4-diols in position 4 by leucoanthocyanidin (flavan-3,4-*cis*-diol) reductase (LAR) in the presence of NADPH. Enzymes responsible for proanthocyanidin synthesis are as yet unknown. Proanthocyanidins are assumed to be formed by condensation of flavan-3,4-diol units with catechin or epicatechin as the starter unit.

Although there is clear evidence from genetic studies as well as from precursor experiments that flavan-3,4-diols are the direct precursors of anthocyanidins, *in vitro* conversion of leucoanthocyanidins to anthocyanidins has not yet been achieved. Results of molecular biological work suggest that a further 2-oxoglutarate- or ascorbate-dependent dioxygenase maybe involved in this step. This enzyme has tentatively been named anthocyanidin synthase (ANS). It is not yet known whether ANS also catalyzes the conversion of flavan-4-ols to the respective 3-deoxyanthocyanidins.

Since anthocyanidins with a free hydroxyl group in position 3 are not stable under physiological pH conditions, and therefore have never been observed in plant tissues, glycosylation of this hydroxyl group is supposed to be an obligatory step in the biosynthetic reaction sequence. A glycosyltransferase that has been proved to be responsible for this step is the long-known UDP-glucose:flavonoid 3-*O*-glucosyltransferase (FGT), giving an anthocyanidin 3-*O*-glucoside as the first stable product of the pathway.

The basic 4′,7-di- and 4′,5,7-trihydroxylated flavonoid structure can be extensively modified by further hydroxylation of both aromatic rings (A and B), methylation, glycosylation, acylation, prenylation, *C*-glycosylation, sulfation, and other reactions. Numerous enzymes catalyzing these modifications have been described. As a rule, these enzymes are highly specific for the position in the molecule, and in some cases even to the extent that they use only flavonoids with a specific substitution pattern. These modification steps finally result in the immense diversity of flavonoid metabolites observed in nature.

1.26.4 FORMATION OF FLAVONOID PRECURSORS AND FLAVONOID CLASSES

The enzymes involved in the supply of flavonoid precursors and the reactions to the various flavonoid classes are described here in more detail. In addition, the essential genetic and molecular biological data will be presented. The extensive enzymatic, genetic, and molecular biological work has been subject of several reviews.[17,24–30] For a detailed presentation of earlier data, the reader is referred to the reviews by Hahlbrock and Grisebach,[31] Ebel and Hahlbrock,[32] Grisebach,[33] and Hahlbrock.[34]

1.26.4.1 Steps to the Major Precursors

1.26.4.1.1 Acetyl-CoA carboxyligase (ACCase)

Carboxylation of acetyl-CoA is catalyzed by ACCase in the presence of adenosine 5′-triphosphate (ATP), and Mg^{2+} as a cofactor. This reaction has been studied extensively in various organisms in relation to fatty acid biosynthesis.[35] Data for the plant enzyme have been summarized by Harwood.[36,37] The first ACCase related to a secondary metabolic pathway was isolated from UV-B irradiated parsley cell cultures. It has been shown that UV-B treatment concomitantly induced the complete range of enzymes required for flavonoid formation. This indicates the role of UV-B radiation in this biosynthetic pathway.[8] ACCase has been purified to apparent homogeneity, and extensive kinetic studies have been performed. The native enzyme protein had a molecular weight of 420 000, and two subunits were identified having a molecular weight between 200 000 and 240 000.[38] Similar results have been obtained for ACCase from various other plant systems, and application of both protease inhibitors and rapid purification procedures have been developed in order to obtain the undegraded protein.[17,36] Further evidence indicated that isoenzymes may exist in different organs and cell compartments, and a cytosolic form seems to be involved in the biosynthesis of secondary metabolites, for example flavonoids and anthocyanins.[37,39] Besides its role as one of the substrates in the chalcone synthase reaction, the first committed step of the flavonoid pathway, malonyl-CoA may also be the acyl donor in the malonylation of sugar residues of flavonoid glycosides (see Section 1.26.5.5.2).

1.26.4.1.2 The shikimate/arogenate pathway

The second biosynthetic precursor of flavonoids, cinnamic acid, is derived from the aromatic amino acid phenylalanine, which is formed via the shikimate/arogenate pathway. Because this pathway is described in full detail in Chapter 1.22 of this volume, the reaction sequence is only briefly outlined. Condensation of erythrose 4-phosphate and phosphoenolpyruvate gives 3-deoxy-D-*arabino*-heptulosonate 7-phosphate (DAHP synthase), which is cyclized to dehydroquinate (3-dehydroquinate synthase), transformed to dehydroshikimate by the elimination of water (3-dehydroquinate dehydratase), and reduced to shikimate (shikimate NADP oxidoreductase). Shikimate is the central intermediate to a large range of secondary metabolites in plants and microorganisms. Phosphorylation of shikimate in position 3 (shikimate kinase) and enolpyruvylation in position 5 (5-*enol*pyruvylshikimate 3-phosphate [EPSP] synthase) is followed by the elimination of the phosphate residue to give chorismate (chorismate synthase). Intramolecular shift of the *enol*pyruvyl residue then leads to prephenate (chorismate mutase), which is transformed to L-phenylalanine in two steps either via phenlypyruvate (prephenate dehydratase, phenylpyruvate aminotransferase) or via L-arogenate (prephenate aminotransferase, L-arogenate dehydratase).

1.26.4.1.3 Phenylalanine ammonia-lyase (PAL)

(i) Biochemistry

PAL is the first committed enzyme of the cinnamate-related secondary metabolism in plants. It catalyzes the *trans*-elimination of ammonia from L-phenylalanine, selectively removing the pro-*S* hydrogen in position 3. It has been found that a dehydroalanine residue in the active center is involved in the elimination mechanism; no cofactor is required for the reaction.[40] PAL activity was first measured in protein extracts from *Sorghum bicolor* by Koukol and Conn in 1961.[22] It was later detected in many other plant species and even some fungi, and the enzyme protein has been purified to homogeneity from various sources.[30,41–43] Molecular weights between 320 000 and 340 000 were found for the native enzyme, depending on the organism studied as well as the purification method used. Subunit molecular weights ranged from 53 000 up to 82 000, mostly depending on the rapidity of the isolation procedure, and degradation during preparation has variously been observed. A one-step immunoaffinity technique has been developed for the rapid isolation of PAL from anthers of *Tulipa*.[44] Affinity chromatography on L-aminooxy-(*p*-hydroxyphenyl)propionic acid matrices and chromatofocusing allowed the separation of four PAL isoforms from *Phaseolus* with different p*I* values and kinetic properties.[45,46] *Petroselinum crispum* PAL transgenically expressed in *E. coli* has now also been purified and its kinetic properties analyzed.[47] Properties of PAL including enzyme kinetic data and enzyme inhibition have variously been summarized.[30,41,48]

(ii) Genetics and molecular biology

Mutants completely lacking PAL activity have so far not been observed. It can be expected that such mutants of PAL and of other essential enzymes of the general phenylpropanoid metabolism may be lethal, not allowing a natural development of the plant even at the embryonic stage. Molecular analysis has been performed on genomic and/or cDNA clones of PAL isolated from many plant species.[26,49,50] The regulation of PAL genes has been extensively studied in relation with the various phenylpropanoid pathways,[49,51,52] but also specifically as part of flavonoid biosynthesis.[53,54] PAL genes were found to comprise small multigene families, which are differentially expressed during plant development and in response to biotic and abiotic stresses. The presence of multiple copies of PAL genes might explain why mutants lacking PAL activity have not yet been detected.

1.26.4.1.4 Cinnamate 4-hydroxylase (C4H)

(i) Biochemistry

trans-4-Coumarate is the second important intermediate on the way to phenylpropanoid metabolites. It is formed by hydroxylation of *trans*-cinnamate in the presence of molecular oxygen, and NADPH as a reducing cofactor. The reaction is catalyzed by a membrane-bound cytochrome P450-dependent monooxygenase that co-acts with a cytochrome P450 (cytochrome c) reductase (CPR).[55–57] Cinnamate hydroxylation was first unequivocally demonstrated with endoplasmic membranes from *Pisum sativum* in 1967 by Russell and Conn.[58] Studies on the reaction mechanism using mixtures of 4-^{3}H- and ^{14}C-labeled cinnamate revealed that 50% of the ^{3}H label reappeared in position 3 of the product, 4-coumarate. This mechanism, that has become known as NIH shift,[59] involves an unstable 3,4-epoxy intermediate which isomerizes to a methylene ketone by shifting the hydrogen from position 4 to position 3. Tautomerization of the structure by random removal of either of the two methylene hydrogen atoms in position 3, thus losing half of the label, leads to formation of the phenolic end product, 4-coumarate. Carbon monoxide characteristically inhibits the reaction by forming an enzyme complex with a typical absorption band at 450 nm. Irradiation at 450 nm restores the activity. Cytochrome c as well as a series of complex nitrogen-containing heterocyclic compounds, recognized as typical cytochrome P450 inhibitors, for example ketoconazole and tetcyclacis, also inhibit C4H activity.[60,61] 1-Aminobenzotriazole has been found to be a suicide inhibitor for the enzyme.[62] C4H was first purified to homogeneity from *Helianthus tuberosus* tissue[63] and *Glycine max* cell cultures.[64] Presence of detergents, such as Triton X-114 or Chaps, was required during purification of the protein. Antibodies were raised from both homogenous proteins, and they were used in Western blot analyses of the denatured enzyme proteins from both plant species, revealing molecular weights of 57 000 and 58 000, respectively. Details on the biochemistry of C4H have been summarized by Werck-Reichhart.[55]

(ii) Genetics and molecular biology

For the same reason mentioned for PAL, mutants, which are devoid of C4H activity, have not been observed so far. Sequence information on C4H genes were first reported in 1993, initiating very rapid further work. Thus, up to now, genomic and/or cDNA clones have been isolated and characterized from at least 11 plant species including *Helianthus tuberosus*, *Medicago sativa* and *Phaseolus aureus*[55] as well as *Arabidopsis thaliana*,[65] *Catharantus roseus*,[66] *Glycine max*,[67] *Glycyrrhiza echinata*,[68] *Petroselinum crispum*,[53] *Populus tremuloides*,[69] *Populus kitakamiensis*,[70] and *Zinnia elegans*.[71] Comparison of amino acid sequences deduced from C4H clones showed that the C4H protein is highly conserved among widely divergent plant species (between 83% and 98%), including the common motifs of the oxygen and heme binding regions typical for cytochrome P450 enzymes. Within the P450 families, the clones were classified as CYP 73 sequences, and detection of many more representatives can be expected in the near future.[57] The identity of the encoded gene product with C4H was confirmed by expression of clones in yeast or as translation fusion with cytochrome P450 reductase in *E. coli*, respectively. From both organisms the expressed enzymes were successfully isolated, and their activities confirmed.[66,72]

1.26.4.1.5 Hydroxycinnamate:CoA ligase (4CL)

(i) Biochemistry

Coenzyme A ligases activate hydroxycinnamic acids for further enzymatic transformations at the side chain, and also at the aromatic ring in some specific cases. The ligase reaction strictly depends on Mg^{2+} as a cofactor and ATP as the activating cosubstrate. ATP cannot be substituted by any of the other nucleoside triphosphates. There is good evidence that an acyl-AMP intermediate is formed in the first step, which then reacts with CoASH to give the CoA ester.[32] 4-Coumarate, and in some plants presumably cinnamate, is the main substrate for the CoA ligases involved in the biosynthesis of flavonoids, and of many other low molecular weight phenylpropanoids, as well as in guaiacyl lignin formation. Owing to this specificity, this ligase type is called 4-coumarate:CoA ligase (4CL). A separate type of ligase showed high specificity towards sinapate (3,5-dimethoxy-4-hydroxycinnamate), and is specifically involved in the formation of sinapic acid-derived metabolites including syringyl lignins.[29,32]

A 4CL related to flavonoid biosynthesis was first characterized and partially purified from light-exposed *Petroselinum* cell suspension cultures.[73] The transient irradiation-dependent increase of activity correlated with the accumulation of flavonoid metabolites, demonstrating the significance of the enzyme in flavonoid formation. Two isoenzymes have frequently been separated, which differ primarily in the presence or absence of activity towards sinapate.[29,32] Only one form has been identified and isolated from young xyleme tissue of *Picea abies*,[74] but two forms were partially purified from cell cultures of *Glycine max*,[75] and three from *Populus* x *euramericana*, which also differed in their tissue distribution.[29,76] More recently, substrate specificities of ligase preparations from elicited *Cephalocereus senilis* cell suspension cultures[77] and *Populus tremuloides* xylem tissue have been studied.[78] Caffeate was the best substrate for the *C. senilis* enzyme, and activities of 70–80% for other hydroxycinnamates were observed, while the enzyme from *P. tremuloides* showed highest activity with 4-coumarate, and 40–70% with the other substrates. Sinapate was not transformed by either of the preparations. Early data on hydroxycinnamate:CoA ligases have been reviewed.[29,31,32] Standard protocols were applied for purification of the enzyme protein from *Glycine max*,[75] cell cultures of *Petroselinum crispum*,[73] and xylem tissue of *Picea abies*.[74] The enzyme occurs as a monomer of a molecular weight between 40 000 and 50 000. The metabolic grid from 4-coumarate to sinapate derivatives has since been discussed extensively in relation to lignin biosynthesis.[79]

(ii) Genetics and molecular biology

Up to now, no mutant lacking 4CL activity has been observed, indicating that 4CL also belongs to the essential enzymes of the general phenylpropanoid pathway. Genomic and/or cDNA clones have been isolated and characterized from at least eight plant species, including *Oryza sativa*, *Petroselinum crispum*, and *Solanum tuberosum*[26] as well as *Arabidopsis thaliana*,[80] *Glycine max*,[81] *Lithospermum erythrorhizon*,[82] *Nicotiana tabacum*[83,84] and *Pinus taeda*.[85,86] Comparison of amino acid sequences deduced from the respective clones revealed a homology between 52% and 93%. As a rule, 4CL genes comprise small gene families, which might explain the lack of respective mutants.

1.26.4.2 Individual Steps to Flavonoid Classes

1.26.4.2.1 Chalcone synthase (CHS)

(i) Biochemistry

CHS activity was first demonstrated in cell cultures of *Petroselinum crispum*.[87] The enzyme catalyzes the sequential condensation of three acetate units from malonyl-CoA (**13**) with a suitable hydroxycinnamic acid CoA ester (**14**)–(**16**) giving chalcones (**18**)–(**20**) as the central C_{15} intermediates (Scheme 2) from which all flavonoids originate. Compound (**17**) has not been tested so far as a substrate for the condensation reaction to (**21**). Because (**15**) is the main physiological substrate for CHS, ring B of chalcone is primarily hydroxylated in position 4 (**19**), and the respective flavonoids therefore in position 4′ (see Scheme 3). The generation of other substitution patterns of ring B (Scheme 2) is discussed in more detail in Section 1.26.5.2.1. CHS has no cofactor requirements, as is typical for key enzymes. The biochemistry of CHS is described in detail in Chapter 1.27 of this volume and will therefore not be further discussed here.

(ii) Genetics and molecular biology

Genetic control of CHS activity has been demonstrated in some plant species, including *Antirrhinum majus*, *Arabidopsis thaliana*, *Lycopersicon esculentum*, *Matthiola incana*, and *Zea mays*.[26] As a rule, recessive mutants completely lack CHS activity, leading to white flower or pollen color.

Genomic and/or cDNA clones have been isolated and characterized from about one hundred different plant species. The expression of the *chs* genes, particularly in response to biotic and abiotic stresses and to plant development, as well as the structure and function of the CHS proteins, have been studied extensively. For a detailed description the reader is referred to Chapter 1.27 of this volume.

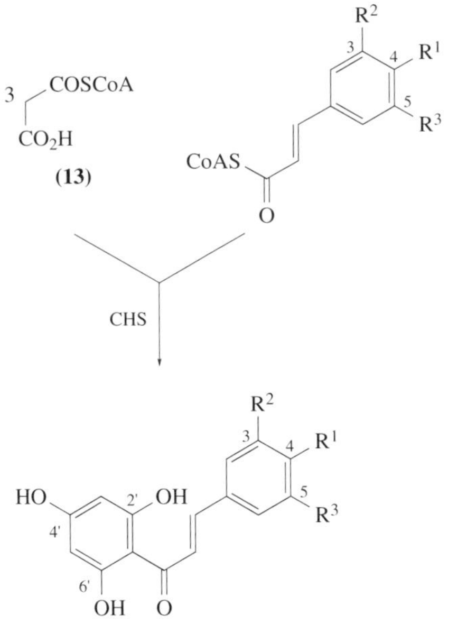

3 COSCoA
CO$_2$H
(13)
CoAS
O
CHS
HO OH
OH O

	R^1	*R^2*	*R^3*	
(14)	H	H	H	Cinnamoyl-CoA
(15)	OH	H	H	4-Coumaroyl-CoA
(16)	OH	OH	H	Caffeoyl-CoA
(17)	OH	OH	OH	3,4,5-Trihydroxy-cinnamoyl-CoA
(18)	H	H	H	Pinocembrin chalcone
(19)	OH	H	H	Naringenin chalcone
(20)	OH	OH	H	Eriodictyol chalcone
(21)	OH	OH	OH	2',3,4,4'5,6'-Hexa-hydroxychalcone

Scheme 2

1.26.4.2.2 Chalcone isomerase (CHI)

(i) Biochemistry

Chalcone isomerase was the first enzyme of the flavonoid pathway detected.[88] It catalyzes the stereospecific cyclization of chalcones to (2*S*)-flavanones, which were found to be the exclusive substrates for the reactions to other flavonoid classes (Scheme 3). In the case of the 2′,4′,6′-trihydroxylation pattern (phloroglucinol-type; see Scheme 2, (**18**)–(**21**)), the cyclization reaction can also proceed spontaneously under physiological conditions, but giving a racemic mixture of the respective 2*S*- and 2*R*-flavanones. This chemical reaction may also occur *in vivo*, and it allows the formation of moderate amounts of flavonoids even if CHI activity is absent. This can easily been seen, when anthocyanin is formed on a background of large concentrations of the yellow-colored 2′,4,4′,6′-tetrahydroxychalcone 2′-glucoside (isosalipurposide) in CHI-deficient flower petals (see below). The chemical isomerization is considerably reduced in the presence of elevated protein concentrations.[89] There is good evidence that generally high activities of CHI are present in flavonoid-forming plant tissues that allow the virtually exclusive formation of the 2*S*-stereomeric flavanones required for further enzymatic transformations. In contrast to chalcones of the phloroglucinol type, chalcones with a 2′,4′-dihydroxylation pattern (resorcinol-type; Scheme 3, (**26**)–(**29**)) are chemically stable under physiological conditions, and therefore strictly require the presence of CHI for isomerization.[90] CHI activity has been demonstrated in cell-free extracts from many plant species. With regard to substrate specificity, two types of CHI enzymes can be classified (Scheme 3), and their occurrence is closely related to the ring-A substitution pattern of chalcones and flavonoids present in the particular plant. In plants containing 5,7-dihydroxyflavonoids, the CHI enzyme only isomerizes phloroglucinol-type chalcones,[29,77,91] whereas in plants with both 7-hydroxy- and 5,7-dihydroxyflavonoids, the enzyme accepts chalcones of the phloroglucinol- as well as the resorcinol-type.[29,92] Detailed studies on substrate specificity of CHI in crude extracts from five cactus species as well as from *Petunia* flowers and *Petroselinum* cell cultures revealed that the enzymes of these plant species also accept (**18**) with an unsubstituted B ring besides (**19**) as substrate.[77] The CHI enzymes investigated so far did not transform chalcone glucosides. The isoflavonoid derivatives coumestrol and kievitone (2′,4′,5,7-tetrahydroxy-8-(3,3-dimethylallyl)isoflavone) were found to act as competitive inhibitors for CHI from *Phaseolus vulgaris*, exhibiting K_i values of 2.5 and 9.2 μM, respectively.[93] For CHI from developing fruit tissue of *Citrus sinensis*, competitive inhibition was observed with (**23**), (**44**) (see Scheme 6), and morin, exhibiting K_i values of 180, 45, and 30 μM, respectively.[91]

The enzymatic ring closure is an overall *syn*-addition to the *E*-double bond, whereas the chemical reaction is an *anti*-addition. At first, a mechanism including participation of a histidyl[94] or an amino

(18), **(19)**, **(20)**, **(21)**

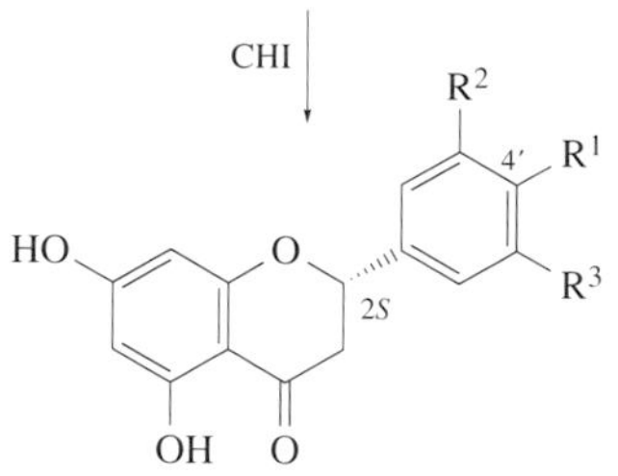

CHI

	R^1	R^2	R^3	
(22)	H	H	H	Pinocembrin
(23)	OH	H	H	Naringenin
(24)	OH	OH	H	Eriodictyol
(25)	OH	OH	OH	3',4',5,5',7-Pentahydroxyflavanone
(26)	H	H	H	2',4'-Dihydroxychalcone
(27)	OH	H	H	Isoliquiritigenin
(28)	OH	OH	H	Butein
(29)	OH	OH	OH	2',3,4,4',5-Penta hydroxychalcone
(30)	H	H	H	7-Hydroxyflavanone
(31)	OH	H	H	Liquiritigenin
(32)	OH	OH	H	Butin
(33)	OH	OH	OH	3',4',5',7-Tetrahydroxyflavanone

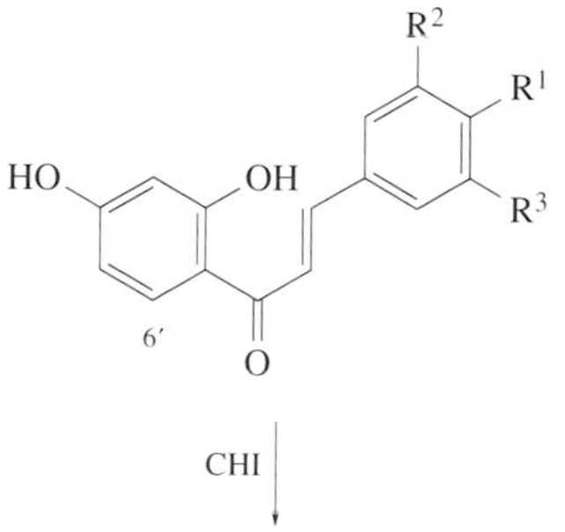

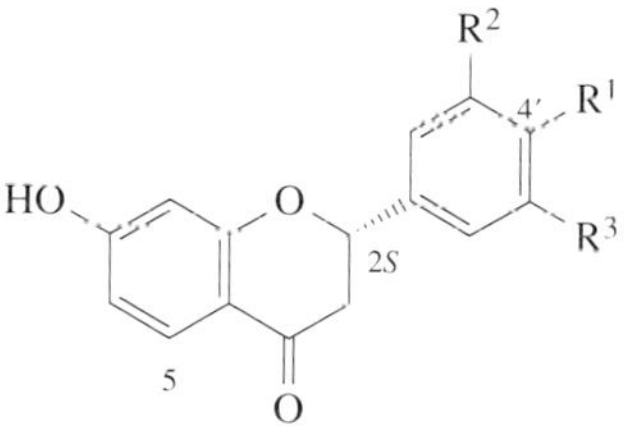

CHI

Scheme 3

group[93] was suggested. Chemical modification of the purified CHI using diethylpyrocarbonate, however, revealed that the five histidine residues of the enzyme protein are not essential for catalysis.[95] Chemical modification of the single cystein residue of the enzyme protein purified from *Glycine max* by mercurials or tetrathionate, however, abolished the reaction completely, indicating the importance of this amino acid in the catalytic process.[96] This coincides well with the Michael type addition of thiol rather than amino functions.[97] CHI has been partially purified or purified to apparent homogeneity from flowers and callus cultures of *Petunia hybrida*, and from cell cultures of *Glycine max*, *Medicago sativa*, and *Phaseolus vulgaris*,[90,92,98] as well as from developing fruit tissue of *Citrus sinensis*.[91] Specific antisera were raised against the CHI protein from *Petunia* flowers and *Phaseolus* cell cultures.[98,99] CHI enzymes are monomers with molecular weights between 24 000 (*Glycine*) and 29 000 (*Petunia*). Apparent p*I* value ranges of 5.0 (*Phaseolus*), 5.7 (*Glycine*), 4.7–5.0 (*Petunia* corolla), and 5.3 (*Petunia* pollen) have been determined. Isoelectric focusing of CHI of *Citrus sinensis* revealed four protein bands with p*I* values of 6.55, 6.38, 6.35, and 6.04. Two isoforms of CHI were separated from *Cephalocereus senilis* by chromatofocusing, showing p*I* values of 5.0 and 4.8.[77]

(ii) Genetics and molecular biology

Experiments with defined genotypes of several plants proved the importance of CHI in flavonoid biosynthesis. The genetic control of CHI activity has been demonstrated in flowers of *Callistephus chinensis* and *Dianthus caryophyllus* as well as pollen of *Petunia hybrida*.[26] Strong anthocyanin formation is observed in pollen or flowers of genotypes with the respective wild-type allele, whereas in recessive mutants, **(19)** or its 2′-glucoside (isosalipurposide) are accumulated, leading to a yellow coloration of the pollen or flowers, respectively. The accumulation of chalcones is clearly due to lack of a functional CHI. Spontaneous chemical isomerization of **(19)** in those mutants which lack

CHI activity provides some racemic flavanone. This allows the formation of moderate amounts of anthocyanin from the 2*S* epimeric form on a usually yellow background. Pure yellow flowers are only produced when a second block in the pathway to anthocyanins is present. A CHI mutant influencing seed coat coloration has been identified in *Arabidopsis thaliana*.[100]

Genomic and/or cDNA clones of CHI have been isolated and characterized from at least 12 plant species including *Antirrhinum majus*, *Arabidopsis thaliana*, *Petunia hybrida*, and *Phaseolus vulgaris*[26] as well as *Callistephus chinensis*,[101] *Dianthus caryophyllus*,[102] *Malus* sp.,[103] *Medicago sativa*,[104] *Pisum sativum*,[105] *Pueraria lobata*,[106] *Vitis vinifera*,[107] and *Zea mays*.[108] Comparison of amino acid sequences deduced from the respective clones revealed a homology for the CHI enzymes of between 44% and 87%. However, the molecular basis of the different specificities of CHI enzymes towards the substitution pattern of ring A of chalcone is still unknown.

CHI gene expression was analyzed in more detail in *Petunia* and *Phaseolus*. Promoter studies showed that the *chi-A* gene of *Petunia* is regulated by a tandem promoter, which independently controls CHI activities in corollas and anthers, respectively.[109,110] In *Phaseolus*, induction of CHI gene expression correlated well with pathogen infection as well as mechanical wounding.[111]

1.26.4.2.3 Flavone synthases (FNS I and FNS II)

(i) Biochemistry

In vitro conversion of flavanones to flavones was first demonstrated with cell-free extracts from very young primary leaves of *Petroselinum crispum*.[112] The reaction has been characterized in more detail in flower extracts of *Antirrhinum majus*[113] and in enzyme preparations from *P. crispum* cell suspension cultures.[114] These studies revealed that two different enzyme systems exist, which introduce the double bond between C-2 and C-3 of flavanones. In *Petroselinum*, a soluble 2-oxoglutarate-dependent dioxygenase, FNS I, catalyzed this reaction, whereas an NADPH-dependent microsomal enzyme activity, FNS II, was responsible in *Antirrhinum* flowers (Scheme 4). Both FNS I and FNS II catalyze the reaction from **(23)** to **(35)**, and **(24)** to **(36)**, whereas **(25)** was a poor substrate. (2*R*)-**(23)** and **(39)** (see Scheme 5) were not converted, and **(22)** has not been tested as substrate so far.

(22), **(23)**, **(24)**, **(25)**

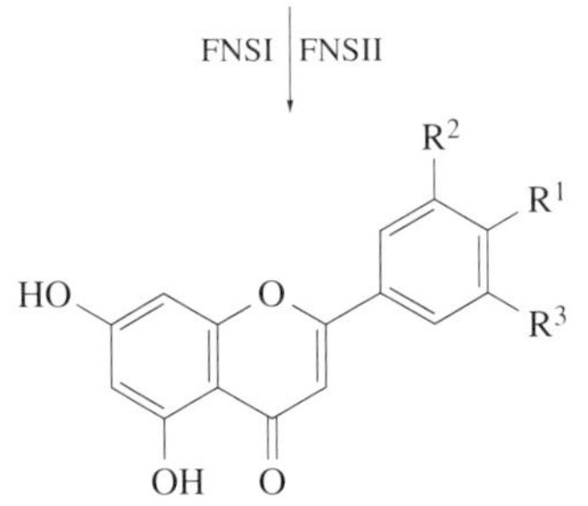

FNSI | FNSII

HO, O, OH, O, R^1, R^2, R^3

	R^1	R^2	R^3	
(34)	H	H	H	Chrysin
(35)	OH	H	H	Apigenin
(36)	OH	OH	H	Luteolin
(37)	OH	OH	OH	Tricetin

Scheme 4

FNS I is apparently not very widespread, since it has only been demonstrated in some members of the plant family Apiaceae. In contrast, FNS II has been observed in many plant species, for example *Chrysanthemum morifolium*, *Columnea hybrida*, *Dahlia variabilis*, *Gerbera hybrida*, *Sinningia cardinalis*, *Streptocarpus hybridus*, *Verbena hybrida*, and *Zinnia elegans*.[29,30] The distribution of FNS I and FNS II in the plant kingdom might therefore be of taxonomic significance and might also be important in considering the evolution of flavonoid biosynthesis.

FNS II activity has also been observed in osmotically stressed cell cultures of *Glycine max*, from which it was partially characterized.[115] The FNS II reaction has an absolute requirement for NADPH and molecular oxygen. Inhibition by cytochrome c and by typical cytochrome P450 inhibitors, such as ancymidol, ketoconazole, and tetcyclacis, and a number of other properties including light reversible inhibition by carbon monoxide, identified FNS II as a cytochrome P450-dependent monooxygenase.[115,116]

FNS I has been purified to apparent homogeneity from *Petroselinum crispum* cell suspension cultures, and specific antisera have been raised.[117] The native enzyme has a molecular weight of

48 000 and consists of two subunits with a molecular weight between 24 000 and 25 000. A p*I* of 4.8 was determined. FNS I requires 2-oxoglutarate and ferrous iron as cofactors, and the presence of molecular oxygen. Ascorbate stimulates the enzyme activity. The cofactor requirement classifies FNS I as a 2-oxoglutarate-dependent dioxygenase.[118] 2,4-Pyridinedicarboxylate and 2*S*-naringenin 7-*O*-glucoside were found to be competitive inhibitors with respect to 2-oxoglutarate and (**23**), respectively. Furthermore, the enzyme activity was strongly inhibited by Cu^{2+} and Zn^{2+}.

The catalytic mechanisms of double-bond formation by both FNS I and FNS II are still unclear. The first postulate, that a 2-hydroxyflavanone may be an intermediate which is subsequently converted to the flavone via enzymatic elimination of water,[113,114] could not be confirmed. Tests with purified FNS I unequivocally proved that chemically prepared 2-hydroxynaringenin did not serve as substrate in flavone formation. Moreover, there was no competition of 2-hydroxynaringenin for the flavanone substrate. It is now assumed that introduction of the double bond between C-2 and C-3 is achieved by sequential abstraction of two vicinal hydrogen atoms in a radical-type mechanism, but unequivocal proof is still lacking.[117]

(ii) Genetics and molecular biology

Genetic control of flavone formation was first demonstrated with flowers of *Gerbera hybrida*.[119] In this plant, flavones are exclusively found in the presence of the dominant allele of the gene *fns*. Respective enzymic studies on defined genotypes have revealed that a clear correlation also exists between the occurrence of flavones and the presence of FNS II activity.[120] This proves, for the first time, that FNS II is unequivocally responsible for flavone formation at least in *Gerbera hybrida*. A respective correlation for FNS I is still lacking. Up to now, neither the gene encoding FNS I nor the one encoding FNS II have been isolated.

1.26.4.2.4 Flavanone 3-hydroxylase (FHT)

(i) Biochemistry

Enzyme studies with flower extracts of *Matthiola incana* provided clear evidence for the hydroxylation in position 3 of flavanones to give dihydroflavonols.[121] The reaction is catalyzed by FHT, which was classified as a 2-oxoglutarate-dependent dioxygenase according to its requirement of 2-oxoglutarate and ferrous iron in the enzyme assay (Scheme 5). Ascorbate stimulates the enzyme activity. FHT catalyzes the stereospecific 3β-hydroxylation of (**23**) and (**24**) to the respective (2*R*,3*R*)-dihydroflavonols (**39**) and (**40**). The (2*R*)-flavanone enantiomers were not transformed. Substrate (**25**) was found to be hydroxylated at a considerably rate to (**41**) by FHT from *Verbena*, but not by the enzyme from *Petunia* flowers.[122,123] Flavanones with an unsubstituted B-ring (e.g., (**22**)) and 5 deoxyflavanones (e.g., (**31**)) or flavanone glycosides have not been tested as substrates as yet.

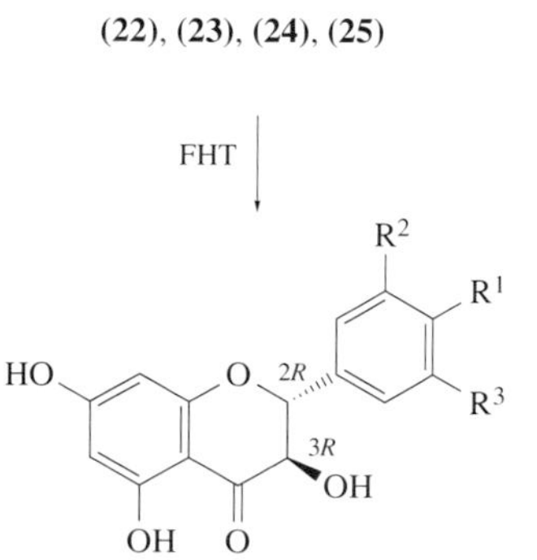

(22), **(23)**, **(24)**, **(25)**

FHT ↓

HO, O, 2*R*, 3*R*, OH, OH, O, R^1, R^2, R^3

	R^1	R^2	R^3	
(38)	H	H	H	Pinobanksin
(39)	OH	H	H	Dihydrokaempferol (Aromadendrin)
(40)	OH	OH	H	Dihydroqercetin (Taxifolin)
(41)	OH	OH	OH	Dihydromyricetin

Scheme 5

FHT activity has been demonstrated in protein extracts from flowers of a wide variety of plant species including *Antirrhinum majus*, *Dahlia variabilis*, *Dianthus caryophyllus*, *Petunia hybrida*, *Streptocarpus hybridus*, *Verbena hybrida*, *Zinnia elegans*, as well as from *Hordeum* grains, hypocotyls of *Lycopersicon esulentum*, seedlings of *Sinapis alba*, anthers of *Tulipa*, and various tissues of *Zea mays*.[29,30]

In spite of a rapid proteolytic degradation in crude protein extracts, accompanied by a corresponding decline of enzyme activity, FHT was successfully purified to apparent homogeneity from young flower buds of *Petunia*, and specific antisera have been raised.[124,125] The purified protein turned out to be a degradation product with a molecular weight of 34 000 and a p*I* of 4.8. Upon improving enzyme extraction procedure, immunoblot analyses revealed that the undegraded FHT protein has a molecular weight between 41 000 and 42 000. Similar values were found with FHT immunoblots of several other plants including *Antirrhinum majus*, *Dahlia variabilis*, *Dianthus caryophyllus*, and *Zinnia elegans*.[126] FHT can occur as a dimer with a molecular weight between 70 000 and 75 000.[124,125]

(ii) Genetics and molecular biology

Genetic control of FHT activity has been demonstrated in most of the plant species mentioned above.[26] As a rule, recessive mutants lacking FHT activity accumulate flavanones and exhibit a white flower color as a consequence. Northern blot analyses revealed the absence of FHT-specific mRNA in respective mutants of *Petunia hybrida* and *Hordeum vulgare*.[127,128] Furthermore, mutants with considerably reduced FHT activity have been characterized in *Petunia* and *Dianthus caryophyllus*. In *Dianthus*, this reduction is caused by a post-translational process rather than inactivation of the promoter.[129]

Genomic and/or cDNA clones of FHT have been isolated and characterized from at least 16 plant species including *Antirrhinum majus*,[130] *Arabidopsis thaliana*,[131] *Bromheadia finlaysoniana*,[132] *Callistephus chinensis*,[133] *Dianthus caryophyllus*,[129,133] *Hordeum vulgare*,[128] *Ipomoea purpurea*,[134] *Malus* sp.,[135] *Matthiola incana*,[133] *Medicago sativa*,[136] *Perilla frutescens*,[137] *Persea americana*,[138] *Petunia hybrida*,[127] *Rosa hybrida*,[139] *Vitis vinifera*,[107] and *Zea mays*.[140] Comparison of amino acid sequences deduced from FHT clones revealed that the FHT protein is highly conserved among widely divergent plant species (between 68% and 95% homology). A strict genetic conservation of 14 amino acids, in particular, of one aspartate and of three histidines was found. These histidines are possibly involved in the iron binding site of 2-oxoglutarate-dependent dioxygenases.[133] Heterologous expression of the FHT cDNA clone of *Petunia* in bacteria resulted in the production of large quantities of active FHT protein, which exceeded the activity found in plant extracts by two orders of magnitude.[127]

1.26.4.2.5 Flavonol synthase (FLS)

(i) Biochemistry

FLS activity was first observed and characterized in protein preparations from *Petroselinum crispum* cell suspension cultures.[114] The enzyme catalyzes the introduction of a double bond between C-2 and C-3 of (2*R*,3*R*)-dihydroflavonols giving flavonols (Scheme 6). Like FNS I and FHT, FLS belongs to the 2-oxoglutarate-dependent dioxygenases, according to the cofactors required, 2-oxoglutarate and ferrous iron. Compounds (**39**) and (**40**) were substrates for the conversion to (**43**) and (**44**), respectively. Compound (**39**) was consistently the better of the two substrates. Compound (**41**) was only tested with enzyme preparations from *Petunia* flowers, where it was a poor substrate, however. Compound (**38**) has not yet been tested as a substrate for the conversion to (**42**).

FLS activity has also been demonstrated in flower extracts of *Dianthus caryophyllus*, *Gerbera hybrida*, *Matthiola incana*, *Petunia hybrida*, and in anthers of *Tulipa* sp.[29,30] Clear changes in FLS activity during flower development have been observed in *Dianthus*, *Matthiola*, and *Petunia*, which correlate with flavonol accumulation. FLS activity is high in as yet uncolored flower buds and declines rapidly as anthocyanin formation starts. This may circumvent a strong competition for the dihydroflavonols as the common intermediates of anthocyanin and flavonol formation. Both FLS and FNS I belong to the same enzyme class, and both catalyze the introduction of a double bond between C-2 and C-3 of their respective substrates. It can therefore be assumed that flavonol formation may similarly be achieved by direct abstraction of the two vicinal hydrogen atoms in positions C-2 and C-3, as postulated for flavone formation, rather than by 2-hydroxylation and subsequent dehydration.

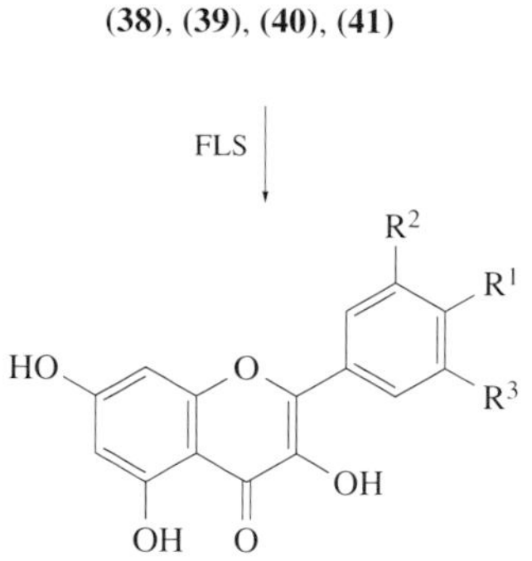

(38), (39), (40), (41)

FLS

	R^1	R^2	R^3	
(42)	H	H	H	Galangin
(43)	OH	H	H	Kaempferol
(44)	OH	OH	H	Quercetin
(45)	OH	OH	OH	Myricetin

Scheme 6

(ii) Genetics and molecular biology

Genetic control of FLS activity has so far only been demonstrated in defined lines of *Petunia hybrida*. In flowers of a recessive mutant, the flavonol content is greatly reduced in comparison to the respective wild-type line. In agreement with this observation, high FLS activity is present in enzyme extracts from wild-type flowers but only low activity is found in preparations from flowers of the mutant line.[141] Flavonol mutants have been reported for *Cyclamen persicum*[142,143] and *Primula praenitens*[144] at the metabolite level, but these lines are not yet characterized enzymatically.

Genomic and/or cDNA clones of FLS have been isolated from at least four plant species including *Arabidopsis thaliana*,[145] *Matthiola incana* (J. Henkel, and G. Forkmann, unpublished),[146] *Petunia hybrida*,[147] and *Solanum tuberosum*.[148] Comparison of amino acid sequences deduced from the respective clones revealed a homology of the FLS enzymes between 52% and 86%. The FLS clone from *Petunia* has been characterized in more detail. FLS gene expression during bud and flower development corresponded to the earlier studies on FLS enzyme activity (see above). Heterologous expression of an FLS cDNA clone in yeast resulted in an active FLS protein. Antisense expression of the cloned FLS in *Petunia* strongly reduced flavonol synthesis in flower corollas.[147]

1.26.4.2.6 Dihydroflavonol 4-reductase (DFR)/Flavanone 4-reductase (FNR)

(i) Biochemistry

First evidence for the enzymic conversion of dihydroflavonols to flavan-3,4-diols was obtained with enzyme preparations from cell suspension cultures of *Pseudotsuga menziesii*, where DFR is involved in catechin and proanthocyanidin formation.[149] With NADPH as cofactor, DFR catalyzes the stereospecific reduction of (**40**) and (**41**) to the respective (2*R*,3*S*,4*S*)-flavan-2,3-*trans*-3,4-*cis*-diols, (**48**)[150] and (**49**),[151] respectively (Scheme 7). Further examples of DFR activity related to catechin and proanthocyanidin synthesis were reported from cell suspension cultures of *Cryptomeria japonica* and *Gingko biloba* as well as from maturing grains of *Hordeum vulgare*.[29,30]

The important role of DFR in anthocyanin biosynthesis was proved by supplementation experiments with flavan-3,4-diols and enzymic studies on flowers of *Matthiola incana*.[152,153] In this plant, DFR activity was measured with (**39**), (**40**), and (**41**) as substrates. The presence of DFR activity in a number of other anthocyanin-producing plant species further confirmed the involvement of DFR in anthocyanin biosynthesis. These studies included flowers of *Callistephus chinensis*, *Dahlia variabilis*, *Dianthus caryophyllus*, *Nicotiana alata*, and *Petunia hybrida*, leaves of *Hedera helix*, hypocotyls of *Lycopersicon esculentum*, seedlings of *Sinapis alba*, and various tissues of *Zea mays*[29,30] as well as lignocellulosic tissues of different conifer species[154] and cell suspension cultures of *Vitis vinifera*.[155]

DFR has been purified to apparent homogeneity and characterized in detail from young flower buds of *Dahlia variabilis*.[156] A molecular weight of 41 000 was estimated. The enzyme requires NADPH as cofactor and catalyzes the transfer of the pro-*S* hydrogen of NADPH. The reaction was also observed with NADH, but the rate was only about 20% of that found with NADPH. The DFR enzymes from *Matthiola incana* and *Hordeum vulgare*, on the other hand, exhibited reaction rates up to 90% with NADH as cofactor.[153,157]

(38), (39), (40), (41)

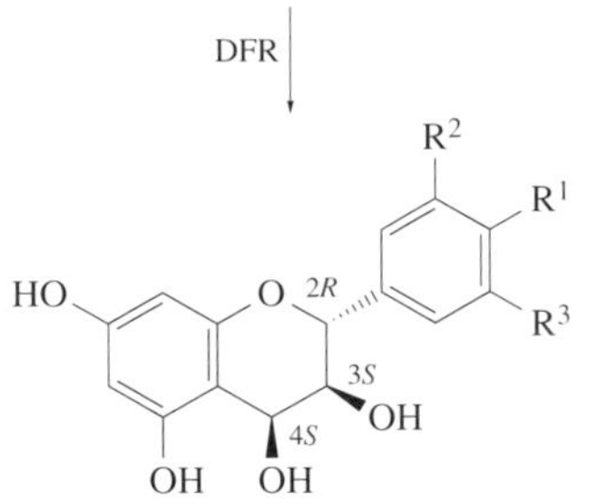

DFR

	R^1	R^2	R^3	
(46)	H	H	H	5,7-Dihydroxy-flavan-3,4-diol
(47)	OH	H	H	Leucopelargonidin
(48)	OH	OH	H	Leucocyanidin
(49)	OH	OH	OH	Leucodelphinidin

(22), (23), (24), (25)

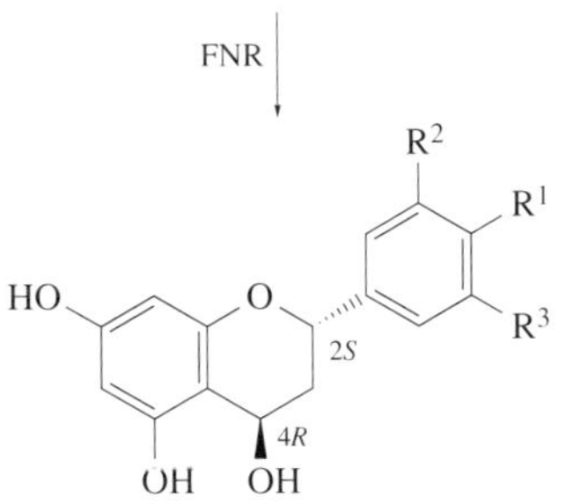

FNR

	R^1	R^2	R^3	
(50)	H	H	H	5,7-Dihydroxy-flavan-4-ol
(51)	OH	H	H	Apiforol
(52)	OH	OH	H	Luteoforol
(53)	OH	OH	OH	3',4',5,5',7-Penta-hydroxyflavan-4-ol

Scheme 7

It is noteworthy that the DFR enzymes known so far exhibit striking substrate specificities (Scheme 7). The enzymes from *Lycopersicon esculentum* seedlings and flowers of *Nicotiana alata* and *Petunia hybrida* do not accept (**39**) at all, but reduce (**40**) reasonably well and, most efficiently, (**41**). Consequently, these plants contain anthocyanins based on (**61**) and (**62**) (see Scheme 9) whereas derivatives of (**60**) are lacking completely.[158] Although the DFR enzymes from *Callistephus chinensis* and *Dianthus caryophyllus* use (**39**) as a substrate, this compound was converted to an appreciably lower extent than (**40**) and (**41**). On the other hand, the DFR enzymes from *Dahlia variabilis*, *Dianthus caryophyllus*, and *Matthiola incana* accept (**41**) as a substrate *in vitro*, although anthocyanins based on (**62**) are not formed in the flowers of these plant species. Compound (**38**) has not yet been tested as substrate for the DFR enzymes. The preference of DFR enzymes for dihydroflavonols with a higher degree of oxygenation in the B ring, and the preference of FLS enzymes for (**39**) as substrate, together with the sequential expression of FLS and DFR during flower development may explain the observation that often relatively high amounts of derivatives based on (**43**) are even formed in anthocyanin-containing flowers.

An NADPH-dependent reduction of the carbonyl group of (2*S*)-flavanones was first demonstrated in enzyme extracts from flowers of *Sinningia cardinalis* producing the rare 3-deoxyanthocyanidins. Reduction of (**23**) and (**24**) leads to the respective (2*S*,4*R*)-*trans*-flavan-4-ols, (**51**) and (**52**), respectively (Scheme 7), which are the immediate precursors for the 3-deoxyanthocyanidins.[159] Other (2*S*)-flavanones have not been tested as substrates so far. In agreement with the presence of 3-deoxyanthocyanidins, FNR activity was also demonstrated in flower extracts of *Columnea hybrida*[160] and in various tissues of *Zea mays*.[161] In all cases, a remarkably low pH optimum around pH 6.0 was found for the reaction.

DFR and FNR act on structurally closely related substrates, i.e., dihydroflavonols (3-hydroxyflavanones) and flavanones (Scheme 7). The assumption that the two reactions may be catalyzed by one and the same enzyme is supported by the observation that enzyme preparations from *Columnea hybrida*, *Sinningia cardinalis*, and *Zea mays* catalyze the transformation of both dihydroflavonols and flavanones. Purified DFR from *Dahlia variabilis* also catalyzes the reduction of both substrates at pH 6.0, although neither 3-deoxyanthocyanidins nor flavan-4-ols have been detected so far in *Dahlia* flowers.[156] On the other hand, DFR preparations from *Dianthus caryophyllus* and *Matthiola incana* did not accept flavanones as substrates for the conversion to flavan-4-ols.[162]

(ii) Genetics and molecular biology

Genetic control of DFR activity by structural genes has been demonstrated in at least six plant species, for example *Callistephus chinensis*, *Dianthus caryophyllus*, *Hordeum vulgare*, *Lycopersicon esculentum*, *Petunia hybrida*, and *Zea mays*.[26]

DFR cDNA and/or genomic clones have been isolated from a number of plant species, and the temporal and spatial expression of the respective genes have been characterized. These studies include *Antirrhinum majus*, *Arabidopsis thaliana*, *Gerbera hybrida*, *Hordeum vulgare*, *Petunia hybrida*, and *Zea mays*[26] as well as *Callistephus chinensis*,[163,164] *Dianthus caryophyllus*,[163,165] *Gentiana triflora*,[166] *Ipomoea purpurea*,[167] *Matthiola incana*,[163], *Forsythia* x *intermedia*,[168] *Lycopersicon esculentum*,[169] *Medicago sativa*,[136] *Oryza sativa*,[170] *Perilla frutescens*,[171] *Rosa hybrida*,[172] and *Vitis vinifera*.[107] Comparison of the deduced amino acid sequences revealed a homology of the DFR enzymes between 57% and 87%. Sense expression of the *Zea mays* or *Gerbera hybrida* DFR gene in a suitable line of *Petunia* resulted in the generation of the pelargonidin-type plants with orange-red flowers that do not naturally exist.[173–175] Antisense expression or co-suppression using heterologous or homologous DFR clones strongly lowered the formation of flavan-3,4-diols, therefore leading to reduction or even suppression of anthocyanin and/or catechin and proanthocyanidin formation.[176,177] Introduction of a genomic DFR clone from *Lycopersicon* or *Hordeum* into respective DFR mutant lines resulted in complementation of the mutation.[178,179] Whereas numerous studies have been performed on DFR, there is no unequivocal information available on the genetics and molecular biology of the FNR reaction so far.

1.26.4.2.7 Leucoanthocyanidin 4-reductase (LAR) and proanthocyanidin formation

(i) Biochemistry

LAR catalyzes the NADPH-dependent reduction of (2*R*,3*S*,4*S*) flavan-2,3-*trans*-3,4-*cis*-diols, for example (**48**) and (**49**), to the respective (2*R*,3*S*)-flavan-2,3-*trans*-3-ols, (**56**) and (**57**) (Scheme 8). *In vitro* formation of (**56**) from (**48**) was first demonstrated with enzyme preparations from cell suspension cultures of *Pseudotsuga menziesii*.[150] LAR activity was also found in enzyme extracts from cell suspension cultures of *Cryptomeria japonica*,[180] suspension and callus cultures of *Gingko biloba*,[151] maturing grains of *Hordeum vulgare*,[157] young leaves of *Onobrychis viciifolia*,[181,182] as well as from various Fabaceae species.[183] As a rule, LAR activity was measured in combination with DFR starting from dihydroflavonols as substrates in a two-step reduction reaction.

(46), **(47)**, **(48)**, **(49)**

LAR

HO, OH, O, 2*R*, 3*S*, OH, R^1, R^2, R^3

	R^1	R^2	R^3	
(54)	H	H	H	not observed
(55)	OH	H	H	Afzelechin
(56)	OH	OH	H	Catechin
(57)	OH	OH	OH	Gallocatechin

HO, OH, O, 2*R*, 3*R*, OH, R^1, R^2, R^3

(58)	OH	OH	H	Epicatechin
(59)	OH	OH	OH	Epigallocatechin

Scheme 8

In vitro reduction of (**46**) to (**54**) and (**47**) to (**55**) has not been reported so far. There is also no information available on the pathway to the flavan-2,3-*cis*-3-ols, (**58**) and (**59**). Furthermore, the postulated condensing enzyme in the synthesis of oligomeric proanthocyanidins from flavan-3,4-diols and flavan-3-ols has not yet been identified. Possible mechanisms for the reaction to (**58**) or (**59**) as well as to proanthocyanidins have been discussed in reviews by Stafford.[17,184,185]

(ii) Genetics and molecular biology

In *Hordeum vulgare*, two genes have been described, which might control LAR activity and the putative flavanol condensing enzyme, respectively.[156,186,187] To date, molecular cloning of genes involved in catechin or proanthocyanidin formation has not been performed.

1.26.4.2.8 Anthocyanidin synthase (ANS)

(i) Biochemistry

Clear evidence that leucoanthocyanidins are precursors of anthocyanin biosynthesis has been provided by supplementation experiments with flavan-3,4-diols on flowers of DFR mutants of *Callistephus chinensis*, *Dendranthema grandiflora*, *Dianthus caryophyllus*, *Matthiola incana*, and *Petunia hybrida*. Supplementation of acyanic flowers with (**47**), (**48**), or (**49**) initiated formation of anthocyanins based on (**60**), (**61**), or (**62**) (Scheme 9).[152,158,162,188] The important role of leucoanthocyanidins as precursors in anthocyanin biosynthesis was further supported by the incorporation of the ^{3}H label of [4-^{3}H]-(**47**) into the derivatives of (**60**).[152]

(47), **(48)**, **(49)**

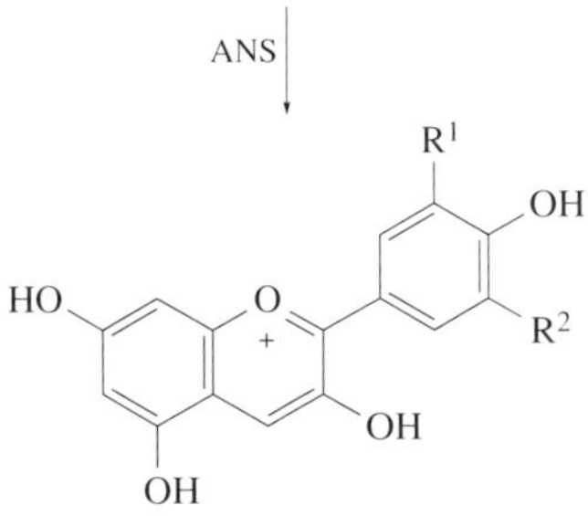

ANS

	R^1	R^2	
(60)	H	H	Pelargonidin
(61)	OH	H	Cyanidin
(62)	OH	OH	Dephinidin

Scheme 9

From the deduced amino acid sequence of the cloned genes concerning this step (see below) a 2-oxoglutarate- or ascorbate-dependent dioxygenase is involved in this reaction sequence. However, no attempts to demonstrate the *in vitro* transformation of leucoanthocyanidins to anthocyanidins using a variety of different cofactor combinations have been successful so far. The assays have additionally been performed in the presence of FGT activity (see Section 1.26.4.2.9), that would lead to stabilization of the anthocyanidin or a possibly as yet unknown intermediate by glucosylation of the hydroxyl group in position 3. Assuming that a 2-oxoglutarate-dependent dioxygenase is involved, the conversion of leucoanthocyanidins to anthocyanidins may proceed by a similar mechanism proposed for flavone and flavonol formation.[30] Introduction of a double bond between C-2 and C-3 of a flavan-3,4-diol can be expected, which would lead to a 2-flaven-3,4-diol. The enzyme catalyzing this step was tentatively named anthocyanidin synthase (ANS). Spontaneous isomerization would lead to the thermodynamically more stable 3-flaven-2,3-diol.[189] 3-*O*-Glucosylation would stabilize either of the two pseudobase products.

(ii) Genetics and molecular biology

ANS cDNA or genomic clones have been isolated and characterized from *Antirrhinum majus*,[130] *Arabidopsis thaliana*,[190] *Dianthus caryophyllus*,[191] *Callistephus chinensis*, *Matthiola incana* and *Rosa hybrida*,[163] *Petunia hybrida*,[192] *Vitis vinifera*,[107] and *Zea mays*,[193] and mutants of structural or regulatory genes concerning the ANS reaction have been investigated in these plants. Clones concerning the ANS reaction have also been isolated from *Forsythia* x *intermedia*,[194] *Malus* sp.,[195] and *Oryza sativa*.[196] Sequence comparisons on the amino acid level indicated that the clones isolated so far code for one and the same type of enzyme with a homology between 46% and 87%. The striking sequence homology to 2-oxoglutarate-dependent dioxygenases, such as FHT[133,197] and FLS, suggest that this enzyme also belongs to this class of proteins. However, demonstration of ANS activity in protein extracts of plant tissue or by heterologous expression of an ANS clone in bacteria has not been successful as yet.

1.26.4.2.9 UDP-Glucose: flavonoid 3-O-glucosyltransferase (FGT)

(i) Biochemistry

FGT catalyzes the transfer of glucose from UDP-glucose to the hydroxyl group in position 3 of flavonols **(42)**–**(45)** and anthocyanidins **(60)**–**(62)**, leading to **(63)**–**(66)** and **(67)**–**(69)**, respectively (Scheme 10). Anthocyanidins, which bear a free hydroxyl group in this position, are unstable under physiological pH conditions, and have not been observed in nature. FGT is therefore regarded as an indispensable enzyme of the main biosynthetic pathway to anthocyanins rather than a modifying enzyme. FGT activity was first demonstrated with protein extracts from pollen of *Zea mays*.[198,199] Further studies were performed with protein preparations from seedlings of *Brassica oleracea* and cell suspension cultures of *Haplopappus gracilis*, as well as from flowers of *Matthiola incana*, *Petunia hybrida*, and *Silene dioica*. Furthermore, the enzyme has been purified from seedlings of *Brassica oleracea*, cell cultures of *Daucus carota*, petals of *Hippeastrum* sp., and needles of *Picea abies*,[29,30,32] as well as from *Vigna mungo* seedlings.[200] Further biochemical properties of FGT are summarized with those of other glycosyltransferases (see Section 1.26.5.3).

(42), **(43)**, **(44)**, **(45)**

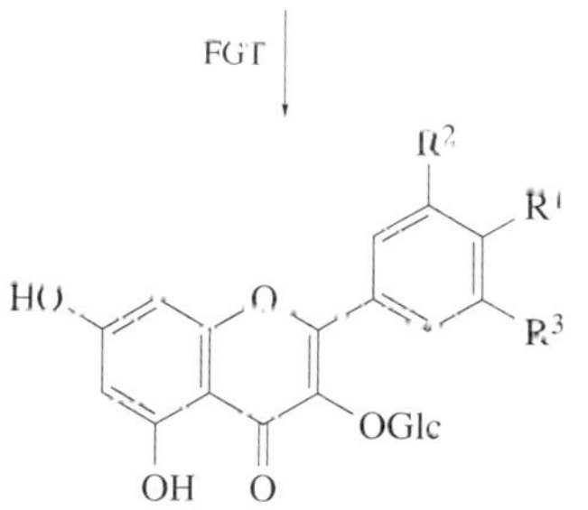

FGT

	R^1	R^2	R^3	
(63)	H	H	H	Galangin 3-*O*-glc
(64)	OH	H	H	Kaempferol 3-*O*-glc
(65)	OH	OH	H	Quercetin 3-*O*-glc
(66)	OH	OH	OH	Myricetin 3-*O*-glc

(60), **(61)**, **(62)**

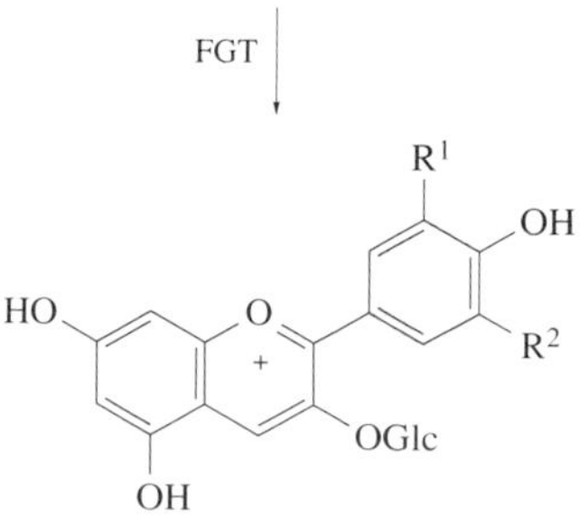

FGT

(67)	H	H	Pelargonidin 3-*O*-glc
(68)	OH	H	Cyanidin 3-*O*-glc
(69)	OH	OH	Delphinidin 3-*O*-glc

Scheme 10

(ii) Genetics and molecular biology

Genetic control of FGT activity by structural or regulatory genes has been reported in *Matthiola incana*, *Petunia hybrida*, and *Zea mays*.[29] It is noteworthy that recessive mutants interfering with late steps (DFR, ANS) in the anthocyanin biosynthesis also show considerably reduced FGT activity. This observation might indicate that either the respective genes exert a regulatory effect on the expression of other enzymes involved in the pathway or the "late" enzymes of the biosynthetic sequence are closely associated in a functional complex.[17,201,202]

Genomic or cDNA clones of FGT have been isolated from *Antirrhinum majus*, *Hordeum vulgare*, and *Zea mays*[26] as well as *Gentiana triflora*,[166] *Perilla frutescens*,[203] and *Vitis vinifera*.[107] Comparison of the amino acid sequences deduced from the respective clones showed a quite low homology ranging from 34% to 76%. Introduction of the cDNA clone from *Antirrhinum majus* into *Eustoma grandiflorum* resulted in the production of a range of novel anthocyanins in the petals.[204]

1.26.4.3 Removal of Hydroxy Functions: 6′-Deoxychalcone Formation

6′-Deoxychalcones are the central intermediates to the large group of 5-deoxyflavonoids, mainly represented by isoflavonoid and pterocarpanoid metabolites (see Chapter 1.28 of this volume), but also of various flavonoids. It has long been a matter of debate at which level and by which mechanism the oxygen function in position 6′ of the chalcone would be removed. Early precursor studies with [1,2-$^{13}C_2$] doubly labeled acetate had shown that incorporation into the A ring occurred exclusively in one direction, indicating that oxygen had to be eliminated prior to chalcone formation.[205] Removal of a hydroxyl group from the phloroglucinol-type chalcone would have led to formal randomization of the direction of incorporation of acetate units into ring A. It was therefore postulated that reduction had to occur with the enzyme-bound polyketide intermediate after condensation of at least two molecules of (**13**) to a suitable hydroxycinnamoyl-CoA substrate. It was only in 1988, when Ayabe *et al.* added rather high concentrations of NADPH to CHS assays with crude extracts from protoplasts and cell cultures of *Glycyrrhiza echinata*, which contained CHI activity, that formation of (**27**) from (**13**) and (**15**) was observed.[206,207] The reaction was later confirmed with protein extracts from cell suspension cultures of *Glycine max*[208] and *Pueraria lobata*.[209] However, purified CHS proteins from any plant source studied only produced phloroglucinol-type chalcones, even in the presence of high concentrations of NADPH, indicating that an additional factor was required for 6′-deoxychalcone formation. The hypothesis that this factor might be a second enzyme with reductase function was first proved by Welle and Grisebach[208] and later by Harano *et al.*,[210] who successfully supplemented the assays of CHS from *Glycine max* and *Pueraria lobata*, respectively, with suitable protein fractions from the CHS purification protocol. It became obvious that an NADPH-dependent reductase had to co-act with CHS at the level of a chalcone ketide to give 6′-deoxychalcones. The enzyme was therefore named chalcone ketide reductase (CHKR).

(i) *Biochemistry*

CHKR has been purified to apparent homogeneity from cell suspension cultures of soybean, and specific antisera against the protein were raised.[208,211] A molecular weight of 34 000 and a p*I* of 6.3 were found. CHKR was shown to be a monomer. The functional coaction with CHS, which appears to be a dimer *in vitro*, is as yet unknown. Inhibition of the reduction step by higher salt concentrations indicates that an ionic interaction may occur. Besides (**15**), (**16**) was also accepted as substrate in CHS/CHKR assays, leading to the formation of (**28**) (see Scheme 2), and both substrates were used at comparable rates. It is noteworthy that 6′-deoxychalcone formation also proceeds in combined assays with purified CHKR of *Glycine max* and a heterologous CHS protein from, for example, *Petroselinum crispum*, despite the fact that the latter plant does not contain 5-deoxyflavonoids naturally. Purified CHKR did not catalyze chalcone formation in the absence of CHS, indicating that both enzymes are simultaneously required for the reaction.

(ii) *Genetics and molecular biology*

CHKR appears to be the key enzyme in the formation of pterocarpan phytoalexins in *Glycine max* and other members of the Fabaceae. A mutant lacking CHKR activity has not yet been reported.

CHKR cDNA clones have been isolated from *Glycine max* and were further characterized.[212] The deduced amino acid sequence revealed a marked similarity with other oxido-reductases. Moreover, a leucine zipper motif was found, which could be of functional importance. Heterologous expression of a CHKR clone in bacteria not only resulted in the production of a highly active protein, but this also allowed the isolation of milligram amounts of pure CHKR protein for further characterization.[212,213] CHKR cDNA clones have also been isolated from *Glycyrrhiza echinata*,[214] *Glycyrrhiza glabra*,[215] and *Medicago sativa*.[216,217] Comparison of amino acid sequences deduced from the respective clones revealed a high homology of the CHKR enzymes, between 84% and 98%.

1.26.5 MODIFICATION REACTIONS

Flavonoid aglycones with a simple substitution pattern are the substrates for the enzymes catalyzing the main steps to the various flavonoid classes. The immense diversity of flavonoids found in nature is due to the fact that these simple flavonoids can be extensively modified by a number of different reactions. Modification by hydroxylation, methylation, glycosylation, acylation, and prenylation occurs within virtually all flavonoid classes. However, some other modifications, such as *C*-glycosylation and sulfation, are restricted to a few classes as known so far. Hydroxylation of both aromatic rings (A and B) and methylation of hydroxyl groups leads to the formation of the different aglycones within each flavonoid class. Glycosylation of the hydroxyl groups of flavonoid aglycones increases not only water solubility but also provides new substrates for further glycosylation leading to di- and even oligoglycosides, and for acylation with both aliphatic or aromatic acids. Various enzymes have been described catalyzing hydroxylation, methylation, glycosylation, acylation, and other reactions on flavonoids. The principal reactions and positions of modification are summarized in Figure 2.

1.26.5.1 A-Ring Hydroxylation Pattern

1.26.5.1.1 Flavonol 6- or 8-hydroxylation

In some flavonoid classes such as flavones, flavonols, and anthocyanins, additional hydroxy groups in the A-ring are found in positions 6 or 8, and these compounds are correlated with particular flower colors.[218] An enzyme activity catalyzing hydroxylation of the flavonol (**44**) in position 6 to quercetagetin (Figure 2, A), has been demonstrated in protein extracts from flowers of *Tagetes patula*.[219] The enzyme activity was found to be localized in the microsomal fraction and the reaction required NADPH and molecular oxygen. The presence of this enzyme activity is in good agreement with the natural occurrence of quercetagetin 7-*O*-glucoside in the flowers of this plant. Besides (**44**), (**43**) was found to be hydroxylated to give 6-hydroxykaempferol. Inhibition of the reaction by typical cytochrome P450 inhibitors, such as ketoconazole and tetcyclacis, indicated that the enzyme is another cytochrome P450 mixed-function monooxygenase in the flavonoid pathway, which was tentatively named flavonol 6-hydroxylase.

Another enzyme activity catalyzing hydroxylation of the flavonol (**44**) in position 8 to gossypetin in the presence of NADPH and molecular oxygen has been observed in microsomal fractions prepared from protein extracts of flowers of *Chrysanthemum segetum*, which contain gossypetin 7-*O*-glucoside as the main flower pigment (Figure 2, B).[219]

1.26.5.2 B-Ring Hydroxylation Pattern

1.26.5.2.1 Metabolic grid of B-ring hydroxylation

Based on chemicogenetic results and tracer experiments, two basically different hypotheses concerning the determination of the oxygenation pattern of ring B of flavonoids were disputed for many years: (i) genetically controlled selection and incorporation of a properly substituted cinnamic acid derivative during synthesis of the C_{15} skeleton,[220] and (ii) hydroxylation and methylation of the B ring at the C_{15} stage by specific enzymes.[15] The first hypothesis would mean that 4-coumarate is the precursor for the 4′-hydroxylated, caffeate for the 3′,4′-dihydroxylated, and 3,4,5-trihydroxycinnamate for 3′,4′,5′-trihydroxylated flavonoids. Methoxylated cinnamic acids, such as ferulate or sinapate, would also be the precursors for flavonoids with a B ring bearing the respective substitution pattern (see Figure 3). In the second case, 4-coumarate would be the most plausible precursor

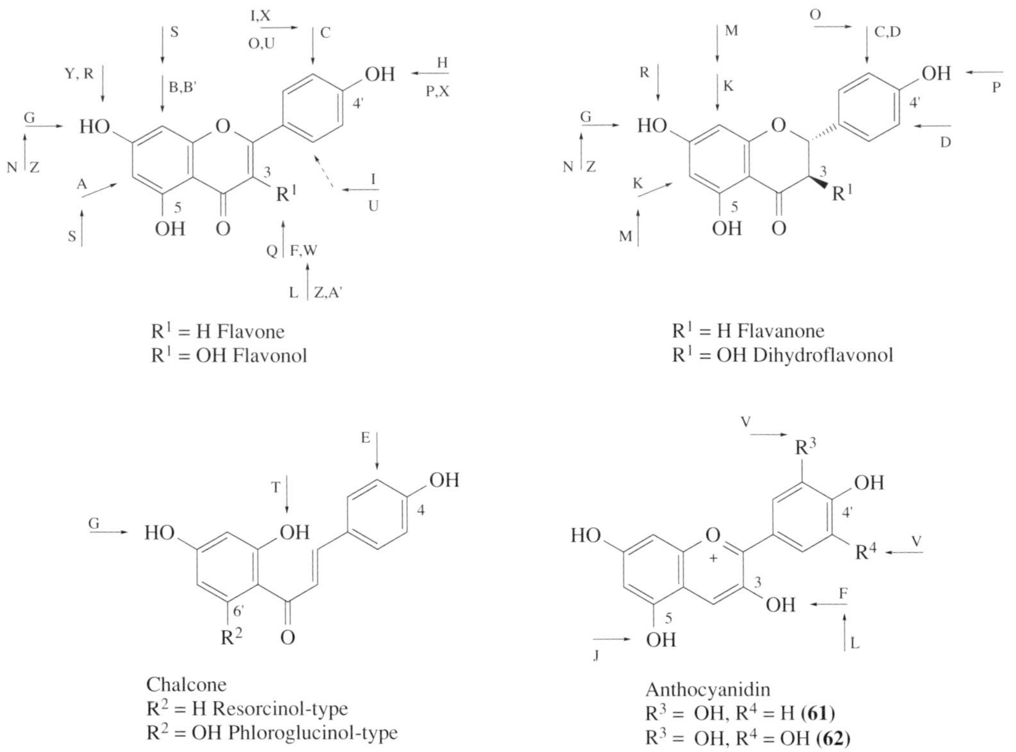

R^1 = H Flavone
R^1 = OH Flavonol

R^1 = H Flavanone
R^1 = OH Dihydroflavonol

Chalcone
R^2 = H Resorcinol-type
R^2 = OH Phloroglucinol-type

Anthocyanidin
R^3 = OH, R^4 = H **(61)**
R^3 = OH, R^4 = OH **(62)**

Figure 2 Positions of frequent modification reactions with selected flavonoid classes. Arrows indicate positions of modification. The broken arrow indicates hydroxylation prior to the reaction indicated. Enzymes involved are indicated with capital letters as follows: (A) flavonol 6-hydroxylase, (B) flavonol 8-hydroxylase, (C) flavonoid 3′-hydroxylase (F3′H), (D) flavonoid 3′,5′-hydroxylase (F3′5′H), (E) chalcone 3-hydroxylase, (F) flavonoid 3-*O*-glycosyltransferases (FGT), (G) flavonoid 7-*O*-glycosyltransferases, (H) flavonoid 4′-*O*-glucosyltransferases, (I) flavonol 2′- or 5′-*O*-glucosyltransferases, (J) anthocyanin 5-*O*-glycosyltransferases, (K) 2-hydroxyflavanone 6-(8-)-*C*-glycosyltransferase, (L) flavonoid 3-*O*-glycoside glycosyl-*O*-glycosyltransferases, (M) flavone 6- or 8-*C*-glycoside glycosyl-*O*-glycosyltransferases, (N) flavonoid 7-*O*-glycoside glycosyl-*O*-glycosyltransferases, (O) flavonoid 3′-*O*-methyltransferase, (P) flavonoid 4′-*O*-methyltransferases, (Q) flavonoid 3-*O*-methyltransferases, (R) flavonoid 7-*O*-methyltransferases, (S) flavonol 6- or 8-*O*-methyltransferases, (T) chalcone 2′-*O*-methyltransferases, (U) flavonol 2′- or 5′-*O*-glucoside 5′- or 2′-*O*-methyltransferases, (V) anthocyanin 3′,5′-*O*-methyltransferases, (W) flavonol 3-*O*-sulfotransferase, (X) flavonol 3′- or 4′-*O*-sulfotransferase, (Y) flavonol 7-*O*-sulfotransferase, (Z) flavonoid 3- or 7-*O*-glycoside (aliphatic) acyltransferases, (A′) flavonoid 3- or 7-*O*-glycoside (aromatic) acyltransferases, (B′) flavonol 8-*C*-prenyltransferase.

for all flavonoids, and flavonoid-specific hydroxylases and methlytransferases would catalyze the introduction of additional B ring substituents. Extensive information is available supporting the second hypothesis, which is outlined by bold arrows in Figure 3.

(i) Biochemistry

All CHS enzymes tested so far showed appreciably higher product formation with **(15)** than with **(16)** (see Scheme 2), in particular, when both substrates were present in the same assay.[29,30] Thus, generation of the 3′,4′-hydroxylation pattern by incorporation of **(16)** instead of **(15)** seems to be of minor importance or aberrant biosynthesis occurred in the presence of excess of unnatural substrate. Evidence for an *in vivo* incorporation of **(16)** besides **(15)** by CHS has been provided from flowers of *Silene dioica*[221,222] and *Verbena hybrida*[223] as well as from anthers of *Tulipa* cv. Apeldoorn and petals of *Cosmos sulphureus*, where even feruloyl-CoA was incorporated *in vitro*.[224] Compound **(17)** has not been tested so far as a substrate for CHS, but generation of the 3′,4′,5′-hydroxylation pattern by incorporation of **(17)** is highly unexpected, since 3,4,5-trihydroxycinnamate has never occurred in nature. With regard to the naturally occurring flavonoids without any hydroxy groups in the B ring, for example in *Pinus sylvestris*, it is noteworthy that protein preparations from

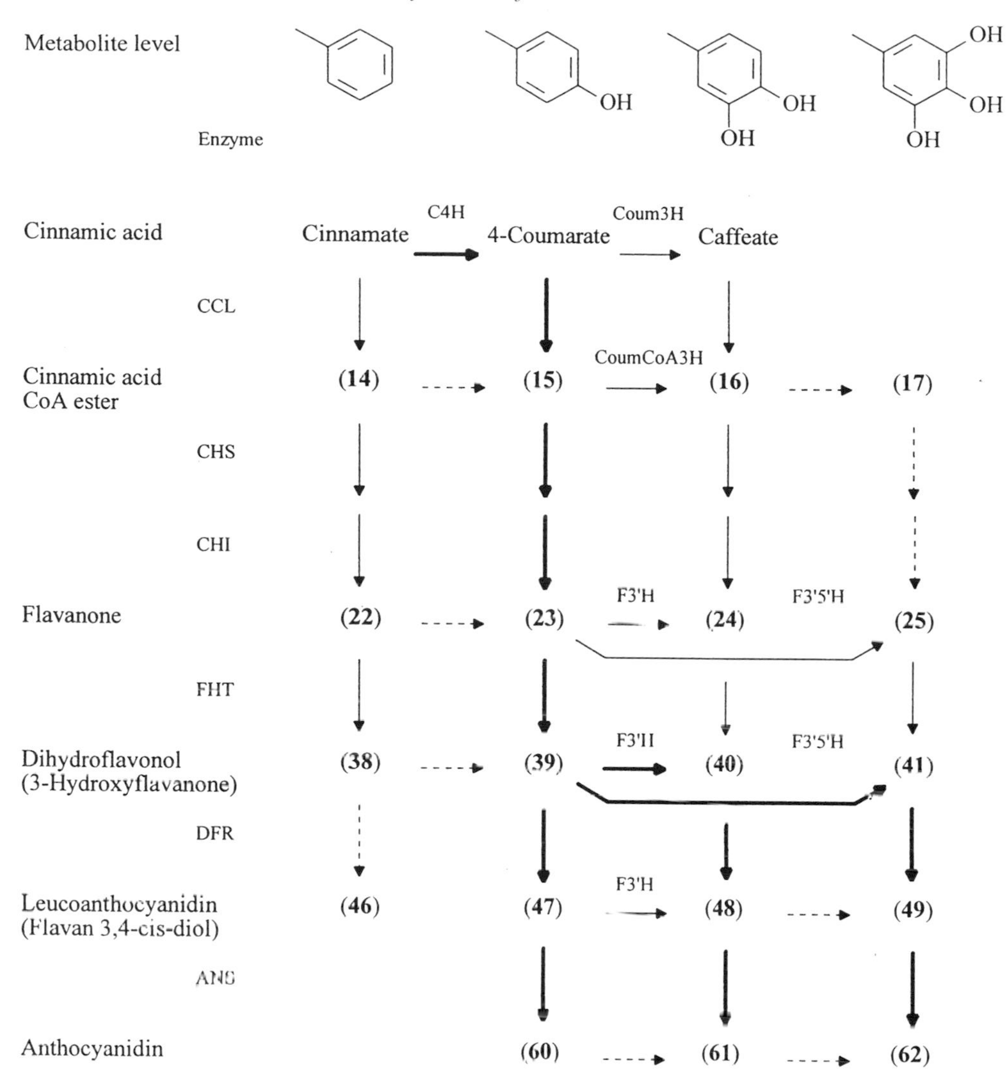

Figure 3 Metabolic grid of B-ring oxygenation of flavonoids and anthocyanins.

seedlings of this plant, which contain both CHS and CHI activity, accepted (**15**) as well as (**14**), giving (**23**) and (**22**), respectively (see Schemes 2 and 3).[225,226]

In addition to the substrate specificity of CHS, the identification of specific flavonoid B-ring hydroxylases catalyzing introduction of hydroxyl groups in positions 3′ and 5′, and the observation of specific methyltransferases for these groups, strongly indicated that the B-ring substitution pattern is determined at the C_{15} rather than the hydroxycinnamate level. Flavonoid 3′-hydroxylase (F3′H) activity (Figure 2, C) was first demonstrated in microsomal fractions prepared from cell cultures of *Haplopappus gracilis*.[227] Enzyme activity was later detected in microsomal preparations from flowers of *Antirrhinum majus*, *Columnea hybrida*, *Dahlia variabilis*, *Dianthus caryophyllus*, *Gerbera hybrida*, *Matthiola incana*, *Petunia hybrida*, *Sinningia cardinalis*, and *Streptocarpus hybridus* as well as from *Petroselinum crispum* cell suspension cultures, *Zea mays* seedlings, and developing grains of *Hordeum vulgare*.[29,30] F3′H from *Petroselinum crispum* has been studied in more detail.[228] The enzyme hydroxylates the flavanone (**23**), the dihydroflavonal (**39**), the flavone (**33**), and the flavonol (**43**) in position 3′ to give (**24**), (**40**), (**36**), and (**44**), respectively. However, the reaction was not observed with the flavan-3,4-diol (**47**), or the anthocyanidin (**60**), but F3′H from *Petunia* may use (**47**) as a substrate.[229] The activity of F3′H was found in the microsomal fraction, and the reaction required molecular oxygen and NADPH. Inhibition was observed with $NADP^+$, cytochrome c, and carbon monoxide,[228] as well as with typical cytochrome P450 inhibitors, such as ketoconazole and tetcyclacis.[116] These results classified F3′H as a cytochrome P450 mixed-function monooxygenase.

Flavonoid 3′,5′-hydroxylase (F3′5′H) activity (Figure 2, D) was first observed with microsomal preparations from flowers of *Verbena hybrida*.[230] Enzyme activity has also been found in similar fractions prepared from flowers of *Callistephus chinensis*, *Lathyrus odoratus*, and *Petunia hybrida*.[231,232] F3′5′H catalyzes the hydroxylation of (**23**) and (**39**) in both positions 3′ and 5′ to give (**25**) and (**41**), respectively. The 3′,4′-hydroxylated flavonoids (**24**) and (**40**) are also substrates for this enzyme, leading to the reaction products mentioned. Other flavonoid substrates such as flavones and flavonols have not been tested so far. Similar to F3′H, the F3′5′H is localized in the microsomal fraction and requires molecular oxygen and NADPH as cofactor. Studies on *Petunia* revealed that F3′5′H activity is inhibited by carbon monoxide, cytochrome c, and tetcyclacis, classifying F3′5′H as a cytochrome P450 mixed-function monooxygenase. Moreover, a polyclonal antibody that inhibits higher plant NADPH-cytochrome P450 reductase also inhibited F3′5′H.[232]

While B-ring hydroxylation in position 2′ has been measured with isoflavonoids,[233,234] this reaction has not yet been demonstrated with flavonoids. 2′-Hydroxyflavones and -flavonols are of special interest in relation to their chelating properties and their contribution to yellow flower coloration.

(ii) Genetics and molecular biology

The genes that govern the B-ring hydroxylation pattern have been shown to control F3′H and F3′5′H, respectively, in a range of plant species, for example *Antirrhinum majus*, *Callistephus chinensis*, *Dianthus caryophyllus*, *Hordeum vulgare*, *Matthiola incana*, *Petunia hybrida*, and *Verbena hybrida*,[26] whereas no correlation has ever been observed between these genes and the substrate specificity of CHS from any plant species studied. The latter is implied by the "cinnamic acid starter hypothesis".[220]

F3′5′H cDNA clones have been isolated and characterized from at least four different plant species including *Eustoma grandiflorum*,[235] *Gentiana triflora*,[166] *Petunia hybrida*,[236,237] and *Solanum melongena*.[238] Comparison of amino acid sequences deduced from the respective cDNA clones revealed a high homology (between 64% and 98%) including the common motifs of the oxygen and heme binding regions typical for cytochrome P450 enzymes. F3′5′H clones were classified as CYP 75 sequences.[57] For *Petunia*, the identity of the encoded gene product with F3′5′H was confirmed by complementation of an appropriate *Petunia* mutant and by expression of a full-sized cDNA clone in yeast from which enzyme preparations with F3′5′H activity were successfully isolated.[236] Heterologous expression of a F3′5′H clone in *Dianthus caryophyllus* resulted in the generation of bluish flower color due to the presence of delphinidin derivatives, which are not synthesized naturally in *Dianthus*. In context with the molecular work on F3′5′H, isolation of F3′H clones from *Petunia* has also been reported.[236,239]

1.26.5.2.2 Chalcone 3-hydroxylase

(i) Biochemistry

The yellow flower color of several cultivars of ornamental plants is based on the presence of 6′-hydroxychalcones (phloroglucinol-type, e.g., in *Antirrhinum majus*, *Helichrysum bracteatum*) or 6′-deoxychalcones (resorcinol-type, e.g., in *Dahlia variabilis*, *Coreopsis grandiflora*, and other species of the Asteraceae), with hydroxylated B ring. The results of metabolite studies on *Antirrhinum* and *Helichrysum* showed that the 3′,4′- or 3′,4′,5′-hydroxylation pattern of the aurones and respective hydroxylated chalcones, did not correlate with B-ring hydroxylation pattern of the flavonoids concomitantly present. Thus, F3′H and F3′5′H, which are governing flavonoid B-ring hydroxylation, are most probably not involved in the hydroxylation of the chalcone/aurone B-ring in these plants.[21]

In vitro hydroxylation of chalcone in position 3 (Figure 2, E) was first demonstrated with microsomal fractions prepared from flowers of *Dahlia variabilis* and *Coreopsis grandiflora*.[240] The enzyme catalyzes the hydroxylation of the 6′-deoxychalcone (**27**) to (**28**). The reaction requires NADPH and molecular oxygen. Inhibition of enzyme activity was observed with $NADP^+$, cytochrome c, and typical cytochrome P450 inhibitors, such as tetcyclacis and ketoconazole. These results indicate that chalcone 3-hydroxylase also belongs to the group of the cytochrome P450 mixed-function monooxygenases, as already demonstrated for the flavonoid B-ring hydroxylases F3′H and F3′5′H. Microsomal preparations of *Matthiola incana* and *Dianthus caryophyllus* flowers containing F3′H activity did not hydroxylate (**27**) to (**28**), indicating that the chalcone hydroxylating enzyme activity in *Dahlia* and *Coreopsis* flower extracts might be an enzyme separate from F3′H. The enzyme was therefore tentatively addressed as chalcone 3-hydroxylase.

1.26.5.3 Glycosylation Reactions

(i) Biochemistry

Flavonoids and anthocyanidins most frequently occur in plant tissues as glycosides substituted with various sugars, mostly glucose, but also galactose, rhamnose, and other sugars. While anthocyanidins oxygenated in position 3 naturally occur as their 3-*O*-glycosides owing to the low stability of the aglycone under physiological pH conditions, flavonoid aglycones may interfere unspecifically with cellular membrane functions based on their lipophilicity. Glycosylation of aglycones or further glycosylation of glycosides is therefore an important reaction in flavonoid pathways. It has also been shown that this modification not only enhances water solubility of the metabolites, but may serve, in combination with additional acylation of the carbohydrate moiety, as a recognition signal for transmembrane transport and vacuolar storage.[30] Glycosylation at the flavonoid structure not only occurs with the phenolic hydroxyl groups in various positions, but also upon carbon atoms of ring A,[241] and rarely on the aliphatic hydroxyl group in position 3 of dihydroflavonols[3] and flavan-3-ols.[242]

UDP-Glucose: flavonoid 3-*O*-glucosyltransferase (FGT; Figure 2, F) is assumed to be a key enzyme of anthocyanin biosynthesis, and its plant sources as well as the genetic and molecular biology of the enzyme have been discussed (see Section 1.26.4.2.9). The enzyme is a soluble protein, and it has commonly been characterized, with flavonols as the phenolic substrate, and UDP-glucose or UDP-galactose as the donor of the sugar residue. The enzyme from *Daucus carota* accepted UDP-galactose with comparable activity to UDP-glucose *in vitro*, while only 3-*O*-galactoside derivatives occurred *in vivo*.[243] A 3-*O*-galactosyltransferase has been detected in *Vigna mungo*, which was different from the 3-*O*-glucosyltransferase concomitantly detected in this plant.[200] Interestingly, the enzyme from *Picea abies* accepted TDP-glucose as an alternative glucosyl donor besides UDP-glucose.[244] FGT from *Petunia* pollen was strongly associated with membrane fractions, indicated by its exclusive dependence on the presence of a detergent in the assay *in vitro*.[245] Whole pollen preparations glucosylated position 3 of various flavonal aglycones having a free 4′-hydroxy group without addition of UDP-glucose. However, the nucleotide sugar had to be added when disintegrated pollen was used. Flavonol 7-*O*-glycosides were also transformed under these conditions, and reaction rates of up to threefold compared to the respective aglycones. A 3-*O*-glucuronosyltransferase from cell cultures of *Anethum graveolens* has been characterized with (**44**) and UDP-glucuronic acid as substrates.[246] A flavonol 3-*O*-xylosyltransferase has been partially purified from *Euonimus alatus* f. *ciliato-denatus* leaves. This enzyme exhibited a broad specificity for flavonol aglycones, and it also accepted dihydroflavonols to some extent.[247] Enzyme purification of a 3-*O*-glucosyltransferase activity resulted in two distinct isoenzymes.[248] Glycosylation in position 7 of flavonoids (Figure 2, G) is also frequently observed. Three separate flavanone 7-*O*-glucosyltransferases have been described from *Citrus paradisi* that differ in their specificities for the flavanone substrates naringenin and hesperetin.[249] A similar enzyme from *Citrus limon* has been partially purified,[250] which showed high specificity for hesperetin, but was also active with some flavones and flavonols. Since more than one product was formed from flavonols, it can be assumed that at least two separate glycosylating activities are present. Another protein preparation with a similar broad substrate specificity has been obtained from ripening strawberry fruits,[251] but product analysis with flavonols as substrates did not indicate enzyme heterogeneity. A flavonol 7-*O*-glucosyltransferase has also been purified from *Euonymus alatus f. ciliato-dentatus*.[248] Two distinct glucosyl- and galactosyltransferases for position 7 of the *C*-glucosyl flavone, vitexin (6-*C*-glucosyl derivative of (**35**)), have been demonstrated from petals and green tissues of *Silene latifolia* (syn. *S. pratensis*)[29,252,253] and the glucosyltransferase has later been purified taking advantage of a very specific interaction between the enzyme and phenyl-Sepharose.[254] A 7-*O*-glucuronosyltransferase for flavones has been detected in primary leaves of *Secale cereale*, which accepts various UDP-sugars besides the main substrate UDP-glucuronic acid *in vitro*.[255] A flavonoid 7-*O*-glucosyltransferase has been characterized from *Chrysanthemum segetum* flower petals, which prefers flavonol substrates with a high number of hydroxyl groups, particularly in ring A, exhibiting highest reaction rates with gossypetin.[256] The respective position 4′ of 6′-deoxychalcones was glucosylated by a transferase from *Dahlia variabilis*[257] and *Coreopsis grandiflora*.[258] The glucosyltransferase from the latter plant has been studied in more detail including 6-hydroxylated 4-deoxyaurones as substrates. Developmental characteristics and further properties of the enzyme activities indicated that only one protein species might be responsible for both reactions.[258]

A 4′-*O*-glucosyltransferase using various flavone and flavonol aglycones as substrates (Figure 2, H) has been prepared from *Allium cepa* bulb scales. A glucuronosyltransferase for the same position of luteolin 7-*O*-diglucuronoside has been measured in *Secale cereale*. Two flavonoid B-ring-specific glucosyltransferases have been characterized from *Chrysosplenium americanum*, which transferred glucose specifically into positions 2′ and 5′, respectively, of highly substituted flavonol substrates (Figure 2, I).[29,30] Separation of the two enzyme proteins, which behaved very similarly during purification, was only successful with affinity chromatography on Reactive Brown 10 agarose.[259]

Position 5 is often glucosylated in anthocyanins (Figure 2, J), and the reaction usually requires previous acylation of the glycosyl residue in position 3 of the substrate. Such 5-*O*-glucosyltransferases have been characterized with protein extracts from flower petals of *Matthiola incana* and *Petunia hybrida*.[29] On the other hand, acylation of anthocyanidin 3-rhamnosyl(1 → 6)glucosides does not seem to be a prerequisite for consecutive 5-*O*-glucosylation in *Silene dioica*, but cyanidin 3-*O*-glucoside was no more substrate.[260]

Enzymic glucosylation of flavonoid A-ring carbon positions 6 or 8 (Figure 2, K) has, up to now, only once been described. 2-Hydroxynaringenin was found to be a substrate for a *C*-glucosyltransferase from *Fagopyrum esculentum*. UDP-galactose, UDP-xylose, as well as ADP-glucose could substitute for UDP-glucose, which is apparently the natural substrate.[261,262] Subsequent elimination of water between positions 2 and 3 of the products leads to the respective flavones, vitexin (8-*C*-glucosylapigenin) and isovitexin (6-*C*-glucosylapigenin).

Enzymes for further glycosylation of sugar residues of flavonoid glycosides have variously been described. Glucosyltransferases that produce di- and triglucosides from flavonol 3-*O*-glucosides (Figure 2, L) were measured with flower extracts from *Pisum sativum*, and rhamnosyl- as well as xylosyltransferases for flavonol 3-*O*-glucosides, galactosides, and diglycosides have been determined in *Tulipa* anthers. Further enzymes were a xylosyltransferase for anthocyanidin 3-*O*-glucosides and their 4-coumaroyl derivatives in flower petals of *Matthiola incana* and a rhamnosyltransferase for position 6″ of anthocyanidin 3-*O*-mono- and 3,5-di-*O*-glucosides. A xylosyltransferase for position 2″ of vitexin (Figure 2, M) from *Silene latifolia* (syn. *S. pratensis*) as well as a glucuronosyltransferase for position 2″ of flavone 7-*O*-glucuronosides from *Secale cereale* and a rhamnosyltransferase for position 2″ of the flavanone hesperetin 7-*O*-glucoside from young leaves of *Citrus* spp. have been described (Figure 2, N).[29,30] While these glycosyltransferases are soluble proteins, a membrane-bound 2″-*O*-galactosyltransferase for flavonol 3-*O*-glucosides has been isolated from *Petunia* pollen.[245] Addition of the donor substrate, UDP-galactose, was necessary when a disintegrated pollen fraction, but not whole pollen, preparations were used.

Glycosyltransferases are frequently inhibited by higher concentrations of the UDP sugars, as well as by UDP. Mercurials, Zn^{2+} and Cu^{2+}, usually strongly inhibit the reaction, but reversion of this inhibition is achieved with thiol reagents. Some of the enzymes were also inhibited by Fe^{2+}, Mn^{2+}, and Co^{2+}. Accordingly, EDTA addition often increased the enzymes' activities *in vitro*. Other inhibitors were iodoacetic acid and its amide, and various flavonoid aglycones, in particular the flavonoid substrates. Molecular weights ranged between 30 000 and 60 000, and two subunits of half of the size of the native enzyme have been observed. The pH optima determined ranged between 6 and 9.5, but were most frequently between 7 and 8,[29,30] and may depend on the specific substrate studied.[256] Values for p*I* are somewhat lower than the respective pH optima of the reactions, ranging between 4 and 6.

(ii) Genetics and molecular biology

For *Silene dioica* and *S. latifolia* (syn. *S. pratensis*) extensive genetic biochemical information is available for the various glycosylation reactions with isovitexin (6-*C*-glucosylapigenin) as substrate.[26] Moreover, genes governing glycosylation of anthocyanins have been correlated with the respective glycosyltransferase activities. Examples have been reported for *Callistephus chinensis*, *Matthiola incana*, and *Silene dioica*,[26] but none of these genes have ben isolated so far. In *Petunia hybrida*, however, the gene encoding UDP-rhamnose:anthocyanidin 3-*O*-glucoside rhamnosyltransferase has been isolated, and its expression has been analyzed.[263]

1.26.5.4 Methylation Reactions

(i) Biochemistry

Methylation is another widespread modification of flavonoids. While methyl ethers occur with virtually all aromatic hydroxyl groups of most flavonoid classes, and many *O*-methyltransferases have been described, *C*-methyl derivatives are rather scarce, and *C*-methyltransferases are unknown so far. *S*-Adenosylmethionine (SAM) is the exclusive methyl donor for the methyltransferases characterized up to now.[29,30]

Three different classes of flavonoid methyltransferases may be differentiated based on the type of substrate: group (i) enzymes exhibit rather broad specificity, group (ii) enzymes catalyze a specific reaction in a sequence of methylation reactions, and group (iii) enzymes catalyze a specific late step in a complex biosynthetic pathway, for example on complex glycosides. Flavonoid 3′-*O*-methyltransferase (Figure 2, O) is a frequently observed representative of group (i). The enzyme has been characterized from *Glycine max* and *Petroselinum crispum* cell cultures, *Nicotiana tabacum* leaves and cell cultures, as well as from *Tulipa* anthers. In addition to aglycones, i.e., (**24**), (**36**), and (**44**), the 7-*O*-glucosides were transformed with even better efficiency.[32] The respective enzyme from *Lotus corniculatus* flowers was tested with gossypetin, (**36**), (**44**), and (**45**).[264,265] Activity with the flavanone (**24**) besides various flavonols was observed with a protein preparation from *Zea mays* vegetative tissue[266] or pollen.[267] Another 3′-*O*-methyltransferase from *Silene latifolia* (syn. *S. alba*) leaves had good activity with the *C*-glucosylflavone iso-orientin and its 2″-*O*-rhamnoside.[268] A 4′-*O*-methyltransferase (Figure 2, P) was measured from *Robinia pseudacacia* shoots using (**23**) and (**35**).[269] Position-7-specific methyltransferases (Figure 2, R) for (**35**) and its 8-*C*-glycosyl derivative vitexin 2″-*O*-rhamnoside have been identified in *Avena sativa* leaves.[270] A similar enzyme from leaves of *Prunus* x *yedoensis* displayed a rather broad substrate specificity, transforming the flavanone (**23**), the dihydroflavonol (**39**), and the flavonols (**43**), (**44**), and (**65**) in addition to the isoflavones sophoricoside, genistein, and biochanin A with high reaction rates.[271] Methylation of a 6- and 8-hydroxy group in flavonols with varying numbers of methoxyl substituents (Figure 2, S) has been demonstrated in protein extracts from *Chrysosplenium americanum* shoots.[272,273] A chalcone 2′-*O*-methyltransferase (Figure 2, T) that acts on several 6′-deoxychalcones, and a separate methyl transferase that is specific for the retrochalcone, licodione, have been described from elicitor-induced *Medicago sativa* cell cultures and roots.[274,275] *Glycyrrhiza echinata* cell cultures also contained licodione-specific enzyme, but not chalcone 2′-*O*-methyltransferase under the same conditions.[275]

Sequential methylation of phenolic hydroxy groups in polyhydroxylated substrates (group (ii) enzymes) has been extensively studied with flavonols in spinach leaf tissue,[276] apple cell cultures,[277] and *Chrysosplenium americanum* tissue.[272,273] It was found that methylation in position 3 (Figure 2, Q) was always the first step of the sequence, followed by position 7 (Figure 2, R), and then position 3′ or 4′ (Figure 2, O, P).

Enzymes of group (iii) have been described as part of the flavonol glycoside pathway of *Chrysosplenium americanum*, and have also been studied extensively in relation to anthocyanin formation in *Petunia hybrida* flowers. In *Chrysosplenium*, derivatives of (**44**) and quercetagetin are further hydroxylated in position 6′ at a high methylation stage, then glucosylated in either of the positions 2′ or 5′ by specific glucosyltransferases (Figure 2, I), and finally methylated at the respective free hydroxyl group (Figure 2, U). Two distinct methyltransferases have been described for the two positions, and the proteins could be separated using the chromatofocusing technique.[278] Methylation in position 3′ of cyanidin and positions 3′ and 5′ of delphinidin 3-*O*-(4-coumaroyl)-rutinosido-5-*O*-glucosides (Figure 2, V) were the last steps in anthocyanin formation in flower petals of *Petunia*. Four different enzyme species have been separated by ion exchange chromatography, two proteins with specificity for position 3′ and two separate ones for positions 3′ and 5′.[279]

Most methyltransferases of the flavonoid pathway needed Mg^{2+} for full activity. This divalent metal ion could be substituted by Mn^{2+} with some of the enzymes, and more rarely by Co^{2+}. Accordingly EDTA as a strong complexing agent inhibited these transferases. Other divalent transition metal ions such as Cu^{2+}, Zn^{2+}, and with some of the enzymes also Co^{2+}, inhibited the reaction. The activities were generally inhibited by mercurials, but were reactivated by adding thiol reagents such as DTE, DTT, or 2-mercaptoethanol. 2-Iodoacetic acid derivatives and *N*-ethylmaleimide were also inhibitors, further indicating the involvement of a thiol group in the enzyme reaction. *S*-Adenosylhomocysteine (SAH), the product of SAM, was a strong inhibitor for all methyltransferases. Flavonoid substrates and methylated products inhibited some of the enzymes. pH optima of the reactions ranged between 7 and 9. p*I* Values were between 4.0 and 5.8 and differences in p*I* of otherwise closely related enzyme isoforms allowed the separation of these enzyme species by electrophoretic techniques or by chromatofocusing. The molecular weights of the enzyme proteins ranged between 43 000 and 66 000. Subunit molecular weights of 43 000 were usually determined, indicating monomeric rather than dimeric structures of the native proteins.[29,30] Photo-affinity labeling with *S*-adenosyl-L (*methyl*-^{3}H)methionine has been used to identify further the subunit of isoliquiritigenin 2′-*O*-methyltransferase from *Medicago sativa*.[274]

(ii) Genetics and molecular biology

Genetic control of anthocyanin 3′- and 3′,5′-*O*-methyltransferase activity, respectively, has been demonstrated in *Petunia hybrida*.[280,281] Cloning of the respective genes is underway.[282] A cDNA clone of the chalcone 2′-*O*-methyltransferase has been isolated from roots of *Medicago sativa*[283] and a cDNA clone of flavonol 3′-*O*-methyltransferase from *Chrysosplenium americanum*.[284]

1.26.5.5 Acylation Reactions

Basically, two groups of acylation reactions may be differentiated: (i) acylation of flavonoid aglycone with an inorganic acid, for example sulphuric acid, and (ii) acylation of flavonoid glycosides at their sugar moieties with aliphatic or aromatic organic acids.

1.26.5.5.1 Sulfuric ester formation

(i) Biochemistry

Natural occurrence of flavonoid sulfate esters was reported as early as 1937, but it was only in the 1970s that their common occurrence in plants became obvious.[285–287] A decade later, the first sulfotransferase specific for flavonoids in *Flaveria choraefolia* was discovered.[286] Flavonol 3-*O*-sulfotransferase (Figure 2, W), the first enzyme in the biosynthesis of oligosulfated flavonols, was characterized with protein extracts from *F. choraefolia* using 3′-phosphoadenosine 5′-phospho[^{35}S]sulfate (PAPS) as the acyl donor.[288,289] While 3-*O*-monosubstituted products were predominantly formed in *F. choraefolia* extracts using (**43**), products up to tetrasulfated have been obtained with extracts from *F. bidentis* and (**44**) as the substrate. Protein purification with *F. choraefolia* led to the separation of three different enzymes with specificities for position 3 of various flavonols, and positions 3′ and 4′ (Figure 2, X) of quercetin 3-*O*-sulfate, respectively. The specificity of flavonol 3-*O*-sulfotransferase for the flavonol substrate was in a decreasing range: rhamnetin ~ isorhamnetin > (**44**) ~ patuletin ~ (**43**) > ombuin ~ tamarixetin. Quercetin 3-sulfate was the best substrate for both B-ring-specific sulfotransferases compared to the respective patuletin, tamarixetin, kaempferol, and isorhamnetin derivatives. Two isoforms of a 7-*O*-sulfotransferase (Figure 2, Y) have been partially purified by affinity chromatography on 3′-phosphoadenosine 5′-phosphate agarose and chromatofocusing from *F. bidentis*.[286] Both isoforms acted equally well on quercetin 3,3′- and 3,4′-disulfates as well as on isorhamnetin 3-sulfate, but not on quercetin 3-sulfate. The occurrence of quercetin 3,7-disulfate in this plant therefore suggested the presence of a sulfatase specifically hydrolyzing the B-ring sulfate esters. The physicochemical properties of these sulfotransferases are very similar. Molecular weights for active proteins of 30–35 000, and sometimes of 60 000, were obtained in gel filtration, indicating that these enzymes may also occur as dimers. No requirement for metal ions and thiol protection was found, and equally low K_m values in the submicromolar region were determined. Distinct differences were observed with the pH optima for the reaction *in vitro* ranging between 6.5 and 8.5, and p*I* values between 4.5 and 6.5, which enabled the separation of the different enzyme species by chromatofocusing.[290] These enzymes are structurally very closely related, which is reflected by a strong cross-reactivity of antibodies for the 3- and 4′-specific enzymes.[291]

(ii) Genetics and molecular biology

cDNA clones of three position-specific flavonol sulfotransferases have been isolated from terminal buds of *Flaveria chloraefolia* and *F. bidentis*, respectively, and the expression of respective genes in response to auxins and sulfated products have been studied.[291–293]

1.26.5.5.2 Carboxylic ester formation

(i) Biochemistry

Acyltransferases for aliphatic and aromatic organic acids are known for different specific positions of the sugar moiety of flavonoid and anthocyanin glycosides. There is good evidence that acylation may be a signal for the import of flavonoid metabolites into the vacuole, for example malonylation of flavonoid glycosides in parsley, and sinapoylation of anthocyanins in *Daucus carota*. Acylation of anthocyanins with hydroxycinnamic acids may also involve internal molecular stacking, which prevents addition of water to the positively charged anthocyanin structure, thus stabilizing the molecule, which is also part of a copigmentation mechanism.[2] Coumaroylated flavonol glucosides in leaves of Mediterranean *Quercus* sp. possess antimicrobial activity, and their occurrence in leaf surface structures indicated a possible function in plant defense.[287] High concentrations of flavonol 3-*O*-glucosides doubly acylated with hydroxycinnamic acids were identified in leaf epidermal cell layers of several trees, for example *Pinus sylvestris*,[9] *Picea abies*,[294] and some *Quercus ilex*.[295] These compounds accumulated upon UV-B irradiation of the plants, and their high molar absorbance in the UV-B region (280–315 nm) underline their possible role as a radiation screen.[9] Specific acylation of intermediates of the anthocyanin pathway, which was a prerequisite for 5-*O*-glucosylation and B-ring *O*-methylation, further stressed the importance of this type of modification.[279]

Malonyltransferases for flavonoid 7-*O*- and flavonol 3-*O*-glucoside derivatives (Figure 2, Z) have been characterized in protein extracts from irradiated parsley cell suspension cultures.[296,297] A flavonoid 7-*O*-glucoside-specific enzyme has been extensively characterized, and acylation in position 6 of the glucose substituent was proved by NMR analysis of the enzyme product of cosmosein (apigenin 7-*O*-glucoside).[298] It has also been shown that malonylation acts as a signal for vacuolar transport of flavonoid glycosides.[299] A similar enzyme has been purified from *Cicer arietinum* cell cultures, which preferentially malonylated the 7-*O*-glucosides of the isoflavones formononetin and biochanin A, but also accepted a chalcone 4′-*O*-glucoside.[300] A malonyltransferase for anthocyanidin 3-*O*-glucosides has been found in extracts from flower petals of *Callistephus chinensis*. This enzyme furthermore transformed cyanidin-3,5-bis-*O*-glucoside, but not cyanidin 3-*O*-xylosylglucoside. Malonyl-CoA, which was the most efficient acyl donor, could be substituted by methylmalonyl-, succinyl- and glutaryl-CoA, respectively.[301] Similar malonyltransferases that acylate position 6″ of anthocyanidin 3-*O*-mono- but not 3,5-bis-*O*-glucosides have been described from flower petals of *Dendranthema morifolium* cultivars,[302] *Centaurea cyanus*,[303] and from leaves of *Lactuca sativa*.[304] While the enzyme preparation from *L. sativa* exhibited activity with succinyl-CoA of about 10% compared to malonyl-CoA, the enzyme preparation from *C.cyanus* showed comparable activities with both succinyl- and malonyl-CoA, reflecting the acylation pattern of the natural flower pigments. The *C. cyanus* enzyme was therefore named a succinyl- rather than a malonyltransferase. A 6″-*O*-acetyltransferase with a substrate specificity similar to that of the last three malonyltransferases has been demonstrated in *Zinnia elegans* flower petals.[305] Another malonyltransferase with a rather broad substrate specificity was characterized from *Ajuga reptans* cell cultures. It preferred anthocyanidin 3-*O*-coumaroylglucoside-5-*O*-glucosides and 3-*O*-glucosides, but also transformed anthocyanidin 3,5-bis-*O*-glucosides, 3-*O*-sophoroside-5-*O*-glucosides, 3-*O*-diglycosides, and cyanidin 3,3′-bis-*O*-glucoside, although less effectively, and occurred even in acyanic cells.[306]

The first transferase for aromatic acids has been demonstrated in *Pisum sativum*, where it catalyzed the last step of kaempferol 3-*O*-*p*-coumaroyltriglucoside formation (Figure 2, A′).[307] A similar activity has been shown in flower extracts of *Silene dioica*,[308] which transferred the acyl residue of 4-coumaroyl-CoA, and caffeoyl-CoA to some extent, to the rhamnose position 4 of anthocyanidin 3-*O*-rutinosides and 3-*O*-rutinoside-5-*O*-glucosides. Another acyltransferase from flowers of defined lines of *Matthiola incana* exhibited the same specificity towards the two CoA esters, and used anthocyanidin 3-*O*-glucosides and 3-*O*-xylosylglucosides, but not 3,5-bis-*O*-glucosylated derivatives as substrates.[309] A hydroxycinnamoyltransferase from *Ajuga reptans* cell cultures acylated cyanidin

3-*O*-sophoroside, anthocyanidin 3,5-bis-*O*-glucosides and 3-*O*-sophoroside-5-*O*-glucosides as well as delphinidin 3-*O*-glucoside.[306]

While the acyltransferases mentioned so far exclusively used CoA esters as donor substrates, a second type of transferase has been demonstrated in protein extracts from anthocyanin-forming *Daucus carota* cell cultures, which exclusively transfers the acyl moiety from hydroxycinnamoyl-1-*O*-glucosides to position 6 of the glucose of cyanidin 3-*O*-(6″-*O*-glucosido-2″-*O*-xylosido)galactoside.[310] This acylation was an indispensable step for the anthocyanin being transported into the vacuole, as has been shown with vacuolar preparations from these cell cultures.

(ii) Genetics and molecular biology

Genetic control of transferases catalyzing acylation of definite anthocyanidin glycosides with hydroxycinnamic acids has been observed in *Silene dioica*[308] and *Matthiola incana*,[309] but cloning of respective genes has not yet been reported.

1.26.5.6 Prenylation Reactions

C-Prenylated metabolites are known from several flavonoid classes. This modification has commonly been studied with phytoalexins of the isoflavonoid- and pterocarpan type. Prenylation in *ortho* positions and subsequent cyclization to furanoid and pyranoid products enhanced the antimicrobial properties of the compounds. While several prenyltransferases have been described for isoflavonoid structures, only one flavonoid-specific prenyltransferase, kaempferol 8-dimethylallyltransferase (Figure 2, B′), has so far been detected.[311] The enzyme was membrane-bound, required divalent cations, preferably Mg^{2+}, but also Mn^{2+}, and to a lower extent Zn^{2+} and Co^{2+}. Virtually no activity was observed in the presence of Ca^{2+} and Cu^{2+}. There was a broad pH optimum between pH 7.5–11.0. The activity was strictly dependent on dimethylallyl diphosphate as prenyl donor. Compounds (**35**), (**36**), and (**44**) were prenyl acceptors besides (**43**), the best substrate, whereas glycosides of (**43**), (**23**), and genistein were not prenylated.

1.26.5.7 Glutathione Transfer Reaction

1.26.5.7.1 Biochemistry

Glutathionetransferase activity measured with 2,4-dinitrochlorobenzene has been associated with the presence or absence of a specific vacuolar anthocyanin-derived metabolite in tissues of the two *Zea mays* genotypes *Bronze-2* (wild-type) and *bronze-2* (mutant line).[312] Using ^{35}S-labeled glutathione *in vivo*, formation of radiolabeled cysteinyl metabolites of anthocyanins has been demonstrated.[312] Glutathione *S*-transferases are enzymes that have become mainly known by their potential to detoxify anthropogenic chemicals released into the ecosystem (xenobiotics) by covalently linking glutathione (GSH) to the substrate, forming water-soluble glutathione *S*-conjugates.[313,314] Glutathionylation allows recognition and entry of the molecule into the vacuole by means of a glutathione pump in the vacuolar membrane.[315] The glutathione-containing products are then processed in the vacuoles, and are finally stored as *S*-cysteinyl metabolites.

1.28.5.7.2 Genetics and molecular biology

The Bronze-2 gene encoding a glutathione *S*-transferase in *Zea mays* has been cloned.[316,317] Expression of the respective cDNA in *Arabidopsis thaliana* and in bacteria resulted in a significant increase in glutathione *S*-transferase activity in both organisms.[312]

1.26.6 PERSPECTIVES

To date, the essential reactions leading to the various flavonoid classes and to many of the structural modifications have been elucidated. While earlier work was mainly performed on the characterization of the biochemical properties of the enzyme proteins, recent studies increasingly concentrate on the molecular characterization and expression of the respective structural and regulatory genes. Besides phylogenetic analyses, the molecular information provides the means for comparisons of specific enzymes and enzyme classes at the DNA and protein levels. In particular, these techniques allow the production of large quantities of active enzyme proteins using suitable organisms. The heterologously expressed proteins can favorably be used for further biochemical analysis of the enzymes, preparation of specific antibodies, and especially for the enzymatic synthesis of stereospecifically uniform radiolabeled flavonoid metabolites in profuse amounts. The ready availability of such compounds will further the characterization of the as yet unknown steps of flavonoid biosynthesis. These mainly include the last reactions in anthocyanidin synthesis, the formation of proanthocyanidins and epicatechin, various steps in the 5-deoxyflavonoid pathway, as well as reactions to the minor flavonoid-related compounds such as dihydrochalcones and aurones.

1.26.7 REFERENCES

1. J. B. Harborne, "Introduction to Ecological Biochemistry," 4th edn., Academic Press, London, 1993, p. 318.
2. R. Brouillard and O. Dangles, in "The Flavonoids: Advances in Research Since 1986," ed. J. B. Harborne, Chapman & Hall, London, 1993, p. 565.
3. B. A. Bohm, in "The Flavonoids: Advances in Research Since 1986," ed. J. B. Harborne, Chapman & Hall, London, 1993, p. 387.
4. R. E. Koes, F. Quattrocchio, and J. N. M. Mol, *BioEssays*, 1994, **16**, 123.
5. B. W. Shirley, *Trends Plant Sci.*, 1996, **1**, 377.
6. C. J. Smith, *New Phytol.*, 1996, **132**, 1.
7. M. Tevini, in "UV-B Radiation and Ozone Depletion. Effects on Humans, Animals, Plants, Microorganisms, and Materials," ed. M. Tevini, Lewis, Boca Raton, FL, 1993, p. 125.
8. C. J. Beggs and E. Wellmann, in "Photomorphogenesis in Plants," 2nd edn., Kluwer, Dordrecht, The Netherlands, 1994, p. 733.
9. J.-P. Schnitzler, T. P. Jungblut, W. Heller, M. Köfferlein, P. Hutzler, U. Heinzmann, W. Schmelzer, D. Ernst, C. Langebartels, and H. Sandermann, *New Phytol.*, 1996, **132**, 247.
10. J. B. Harborne and R. J. Grayer, in "The Flavonoids: Advances in Research Since 1986," ed. J. B. Harborne, Chapman & Hall, London, 1993, p. 589.
11. E. Middleton, and C. Kandaswami, in "The Flavonoids: Advances in Research Since 1986," ed. J. B. Harborne, Chapman & Hall, London, 1993, p. 619.
12. R. Robinson, *Nature*, 1936, **137**, 172.
13. A. J. Birch, and F. W. Donovan, *Aust. J. Chem.*, 1953, **6**, 360.
14. H. Grisebach, *Planta Med.*, 1962, **10**, 385.
15. H. Grisebach, in "Recent Advances in Phytochemistry," eds. T. J. Mabry, R. E. Alston, and V. C. Runeckles, Appleton-Century-Crofts, New York, 1968, vol. 1, p. 379.
16. W. Heller, in "Plant Flavonoids in Biology and Medicine: Biochemical, Pharmacological, and Structure Activity Relationships," eds. V. Cody, E. Middleton, and J. B. Harborne, Liss, New York, 1986, p. 25.
17. H. A. Stafford, "Flavonoid Metabolism," CRC Press, Boca Raton, FL, 1990, p. 298.
18. C. D. Paris, W. J. Haney, and G. B. Wilson, "A Survey of the Interactions of Genes for Flower Color," ed. Department of Horticulture, Botany and Plant Pathology, Michigan State University, East Lansing, MI, 1960, Technical Bulletin No. 281.
19. R. E. Alston, in "Biochemistry of Phenolic Compounds," ed. J. B. Harborne, Academic Press, London, 1964, p. 171.
20. J. B. Harborne, in "The Chemistry of Flavonoid Compounds," ed. T. A. Geissman, Pergamon, Oxford, 1962, p. 593.
21. J. B. Harborne (ed.), "Comparative Biochemistry of the Flavonoids," Academic Press, London, 1967.
22. J. Koukol and E. E. Conn, *J. Biol. Chem.*, 1961, **236**, 2692.
23. W. Seyffert, *Biol. Zentralbl.*, 1982, **101**, 465.
24. M. N. Zaprometov, in "Advances in Cell Culture," ed. K. Maramorosch, Academic Press, New York, 1989, vol. 7, p. 201.
25. H. K. Dooner, T. P. Robbins, and R. A. Jorgensen, *Annu. Rev. Genet.*, 1991, **25**, 173.
26. G. Forkmann, in "The Flavonoids: Advances in Research Since 1986," ed. J. B. Harborne, Chapman & Hall, London, 1993, p. 537.
27. A. G. M. Gerats and C. Martin, in "Recent Advances in Phytochemistry," eds. H. A. Stafford and R. K. Ibrahim, Plenum Press, New York, 1992, vol. 26, p. 165.
28. C. Martin and A. G. M. Gerats, in "The Molecular Biology of Flowering," ed. B. Jordan, CAB International, Wallingford, Oxford, 1993, p. 219.
29. W. Heller and G. Forkmann, in "The Flavonoids: Advances in Research Since 1980," ed. J. B. Harborne, Chapman & Hall, London, 1988, p. 399.
30. W. Heller and G. Forkmann, in "The Flavonoids: Advances in Research since 1986," ed. J. B. Harborne, Chapman & Hall, London, 1993, p. 499.
31. K. Hahlbrock and H. Grisebach, in "The Flavonoids," eds. J. B. Harborne, T. J. Mabry, and H. Mabry, Chapman & Hall, London, 1975, p. 866.
32. J. Ebel and K. Hahlbrock, in "The Flavonoids—Advances in Research," eds. J. B. Harborne and T. J. Mabry, Chapman & Hall, London, 1982, p. 641.

33. H. Grisebach, in "Recent Advances in Phytochemistry," eds. T. Swain, J. B. Harborne, and C. Van Sumere, Plenum Press, New York, 1979, vol. 12, p. 221.
34. K. Hahlbrock, in "The Biochemistry of Plants," eds. P. K. Stumpf and E. E. Conn, Academic Press, New York, 1981, vol. 7, p. 425.
35. J. M. Lowenstein, in "Methods in Enzymology," eds. S. P. Colowick and N. O. Kaplan, Academic Press, New York, 1981, vol. 71, p. 5.
36. J. L. Harwood, *Annu. Rev. Plant Physiol. Plant Mol. Biol.*, 1988, **39**, 101.
37. J. L. Harwood, *Biochim. Biophys. Acta*, 1996, **1301**, 7.
38. B. Egin-Bühler, and J. Ebel, *Eur. J. Biochem.*, 1983, **133**, 335.
39. Y. Sasaki, T. Konishi, and Y. Nagano, *Plant Physiol.*, 1995, **108**, 445.
40. B. Schuster and J. Retey, *Proc. Natl. Acad. Sci. USA*, 1995, **18**, 8433.
41. K. R. Hanson and E. A. Havir, in "The Biochemistry of Plants," eds. P. K. Stumpf and E. E. Conn, Academic Press, New York, 1981, vol. 7, p. 577.
42. G. B. D'Cunha, V. Satyanarayan, and P. M. Nair, *Phytochemistry*, 1996, **42**, 17.
43. S. H. Kim, J. W. Kronstad, and B. E. Ellis, *Phytochemistry*, 1996, **43**, 351.
44. B. Kehrel and R. Wiermann, *Planta*, 1985, **163**, 183.
45. G. P. Bolwell, J. N. Bell, C. L. Cramer, W. Schuch, C. J. Lamb, and R. A. Dixon, *Eur. J. Biochem.*, 1985, **149**, 411.
46. R. A. Dixon, G. P. Bowell, R. L. Sunley, D. K. Lawrence, and I. G. Bridges, *Phytochemistry*, 1987, **26**, 659.
47. C. Appert, E. Logemann, K. Hahlbrock, J. Schmid, and N. Amrhein, *Eur. J. Biochem.*, 1994, **225**, 491.
48. D. H. Jones, *Phytochemistry*, 1984, **23**, 1349.
49. K. Hahlbrock and D. Scheel, *Annu. Rev. Plant Physiol. Plant Mol. Biol.*, 1989, **40**, 347.
50. J. L. Dangl, in "Genes Involved in Plant Defense," eds. T. Boller and F. Meins, Jr., Springer, Wien, 1992, Plant Gene Research, vol. 8, p. 303.
51. T. Fukasawa-Akada, S.-D. Kung, and J. C. Watson, *Plant Mol. Biol.*, 1996, **30**, 711.
52. S. V. N. Prasad, M. Thungapathra, V. Mohindra, and K. C. Upadhyaya, *J. Genet.*, 1996, **74**, 111.
53. E. Logemann, M. Parniske, and K. Hahlbrock, *Proc. Natl. Acad. Sci. USA*, 1995, **92**, 5905.
54. C. E. Lister, J. E. Lancaster, and J. R. L. Walker, *J. Am. Soc. Horticult. Sci.*, 1996, **121**, 281.
55. D. Werck-Reichhart, *Drug Metab. Drug Interact.*, 1995, **12**, 221.
56. F. Durst and D. P. O'Keefe, *Drug Metab. Drug Interact.*, 1995, **12**, 171.
57. F. Durst and D. R. Nelson, *Drug Metab. Drug Interact.*, 1995, **12**, 189.
58. D. W. Russell and E. E. Conn, *Arch. Biochem. Biophys.*, 1967, **122**, 256.
59. D. J. Reed, J. Vimmerstedt, D. M. Jerina, and J. W. Daly, *Arch. Biochem. Biophys.*, 1973, **154**, 642.
60. F. J. Schwinn, *Pestic. Sci.*, 1984, **15**, 40.
61. C. J. Coulson, D. J. King, and A. Wiseman, *Trends Biochem. Sci.*, 1984, **9**, 446.
62. D. Reichhart, A. Simon, and F. Durst, *Arch. Biochem. Biophys.*, 1982, **216**, 522.
63. B. Gabriac, D. Werck-Reichhart, H. Teutsch, and F. Durst, *Arch. Biochem. Biophys.*, 1991, **288**, 302.
64. G. Kochs, D. Werck-Reichhart, and H. Grisebach, *Arch. Biochem. Biophys.*, 1992, **293**, 187.
65. D. A. Bell-Lelong, J. C. Cusumano, K. Meyer, and C. Chapple, *Plant Physiol.*, 1997, **113**, 729.
66. M. Hotze, G. Schröder, and J. Schröder, *FEBS Lett.*, 1995, **374**, 345.
67. C. R. Schopfer and J. Ebel, GenBank, 1995, Accession No. X92437.
68. T. Akashi, T. Aoki, T. Takahashi, N. Kameya, I. Nakamura, and S. Ayabe, GenBank, 1997, Accession No. D87520.
69. L. Ge and V. L. Chiang, GenBank, 1996, Accession No. U47293.
70. S. Kawai, A. Mori, T. Shiokawa, S. Kajita, Y. Katayama, and N. Morohoshi, *Biosci. Biotechnol. Biochem.*, 1995, **60**, 1586.
71. Z. Ye and J. E. Varner, *Plant Sci.*, 1996, **121**, 133.
72. P. Urban, D. Werck-Reichhart, H. G. Teutsch, F. Durst, S. Regnier, M. Kazmaier, and D. Pompon, *Eur. J. Biochem.*, 1994, **222**, 843.
73. K.-H. Knobloch and K. Hahlbrock, *Arch. Biochem. Biophys.*, 1977, **184**, 237.
74. T. Lüderitz, G. Schatz, and H. Grisebach, *Eur. J. Biochem.*, 1982, **123**, 583.
75. K.-H. Knobloch and K. Hahlbrock, *Eur. J. Biochem.*, 1975, **52**, 311.
76. C. Grand, A. Boudet, and A. M. Boudet, *Planta*, 1983, **158**, 225.
77. Q. Liu, M. S. Bonness, M. Liu, E. Seradge, R. A. Dixon, and T. J. Mabry, *Arch. Biochem. Biophys.*, 1995, **321**, 397.
78. H. Meng and W. H. Campbell, *Phytochemistry*, 1997, **44**, 605.
79. A. M. Boudet, D. B. Goffner, and J. Grima-Pettenati, *C. R. Acad. Sci. Paris, Sci. Vie*, 1996, **319**, 317.
80. D. Lee, M. Ellard, L. A. Wanner, K. R. Davis, and C. J. Douglas, *Plant Mol. Biol.*, 1995, **28**, 871.
81. A. Uhlmann and J. Ebel, *Plant Physiol.*, 1993, **102**, 1147.
82. K. Yazaki, K. Inushima, M. Kataoka, and M. Tabata, *Phytochemistry*, 1995, **38**, 1127.
83. D. Lee and C. J. Douglas, *Plant Physiol.*, 1996, **112**, 193.
84. Y. Katayama, S. Kawai, N. Morohoshi, and S. Kajita, GenBank, 1997, Accession No. D43773.
85. K. S. Voo, R. W. Whetten, D. M. O'Malley, and R. R. Sederoff, *Plant Physiol.*, 1995, **108**, 85.
86. X.-H. Zang and V. L. Chiang, *Plant Physiol.*, 1997, **113**, 65.
87. F. Kreuzaler and K. Hahlbrock, *FEBS Lett.*, 1972, **28**, 69.
88. E. Moustafa and E. Wong, *Phytochemistry*, 1967, **6**, 625.
89. J. N. M. Mol, M. P. Robbins, R. A. Dixon, and E. Veltkamp, *Phytochemistry*, 1985, **24**, 2267.
90. R. A. Bednar and J. R. Hadcock, *J. Biol. Chem.*, 1988, **263**, 9582.
91. S. D. Fouché and I. A. Dubery, *Phytochemistry*, 1994, **37**, 127.
92. R. A. Dixon, E. R. Blyden, M. P. Robbins, A. J. van Tunen, and J. N. Mol, *Phytochemistry*, 1988, **27**, 2801.
93. R. A. Dixon, P. M. Dey, and I. M. Whitehead, *Biochim. Biophys. Acta*, 1982, **715**, 25.
94. M. J. Boland and E. Wong, *Bioorg. Chem.*, 1979, **8**, 1.
95. R. A. Bednar and A. J. Adeniran, *Arch. Biochem. Biophys.*, 1990, **282**, 393.
96. R. A. Bednar, W. B. Fried, Y. W. Lock, and B. Parmanik, *J. Biol. Chem.*, 1989, **264**, 14 272.
97. E. Kuss, in "Glutathione: Chemical, Biochemical, and Medical Aspects," eds. D. Dolphin, R. Poulson, and O. Avramovic, Wiley, New York, 1989, Part B, p. 511.

98. A. J. van Tunen and J. N. M. Mol, *Arch. Biochem. Biophys.*, 1987, **257**, 85.
99. M. P. Robbins and R. A. Dixon, *Eur. J. Biochem.*, 1984, **145**, 195.
100. B. W. Shirley, S. Henley, and H. M. Goodman, *Plant Cell*, 1992, **4**, 333.
101. J. Henkel, M. Wassenegger, H. Sommer, and G. Forkmann, GenBank, 1995, Accession No. Z67980.
102. J. Henkel, B. Ruhnau-Brich, J. Dedio, M. Wassenegger, H. Sommer, and G. Forkmann, GenBank, 1995, Accession No. Z67989.
103. E. Podivinsky, J. M. Bradley, and K. M. Davis, *Plant Mol. Biol.*, 1993, **21**, 737.
104. H. I. McKhann and A. M. Hirsch, *Plant Mol. Biol.*, 1994, **24**, 767.
105. A. J. Wood and E. Davies, *Plant Physiol.*, 1994, **104**, 1465.
106. Y. Terai, I. Fujii, S. H. Byun, O. Nakajima, T. Hakamatsuka, Y. Ebizuka, and U. Sankawa, *Protein Expr. Purif.*, 1996, **8**, 183.
107. F. Sparvoli, C. Martin, A. Scienza, G. Gavazzi, and C. Tonelli, *Plant Mol. Biol.*, 1994, **24**, 743.
108. E. Grotewold and T. Peterson, *Mol. Gen. Genet.*, 1994, **242**, 1.
109. A. J. van Tunen, L. A. Mur, G. S. Brouns, J.-D. Rienstra, R. E. Koes, and J. N. M. Mol, *Plant Cell*, 1990, **2**, 393.
110. A. J. van Tunen, L. A. Mur, K. Recourt, A. G. M. Gerats, and J. N. M. Mol, *Plant Cell*, 1991, **3**, 39.
111. M. C. Mehdy and C. J. Lamb, *EMBO J.*, 1987, **6**, 1527.
112. A. Sutter, J. Poulton, and H. Grisebach, *Arch. Biochem. Biophys.*, 1975, **170**, 547.
113. G. Stotz and G. Forkmann, *Z. Naturforsch.*, 1981, **36c**, 737.
114. L. Britsch, W. Heller, and H. Grisebach, *Z. Naturforsch.*, 1981, **36c**, 742.
115. G. Kochs and H. Grisebach, *Z. Naturforsch.*, 1986, **42c**, 343.
116. K. Stich, R. Ebermann, and G. Forkmann, *Phyton (Austria)*, 1988, **28**, 237.
117. L. Britsch, *Arch. Biochem. Biophys.*, 1990, **282**, 152.
118. M. T. Abbot and S. Udenfriend, in "Molecular Mechanism of Oxygen Activation," ed. O. Hayaishi, Academic Press, New York, 1974, p. 187.
119. A. Tyrach and W. Horn, *Plant Breeding*, 1997, **116**, 377.
120. S. Martens and G. Forkmann, in "Polyphenols Communications 96," eds. J. Vercauteren, C. Chèze, M. C. Dumon, and J. F. Weber, Groupe Polyphenols, Bordeaux, 1996, vol. 2, p. 545.
121. G. Forkmann, W. Heller, and H. Grisebach, *Z. Naturforsch.*, 1980, **35c**, 691.
122. G. Forkmann and G. Stotz, *Planta*, 1984, **161**, 261.
123. S. Froemel, P. de Vlaming, G. Stotz, H. Wiering, G. Forkmann, and A. W. Schram, *Theor. Appl. Genet.*, 1985, **70**, 561.
124. L. Britsch and H. Grisebach, *Eur. J. Biochem.*, 1986, **156**, 569.
125. L. Britsch, *Arch. Biochem. Biophys.*, 1990, **276**, 348.
126. I. Koch, PhD Thesis, Universität Tübingen, 1992.
127. L. Britsch, B. Ruhnau-Brich, and G. Forkmann, *J. Biol. Chem.*, 1992, **267**, 5380.
128. M. Meldgaard, *Theor. Appl. Genet.*, 1992, **83**, 695.
129. J. Dedio, H. Saedler, and G. Forkmann, *Theor. Appl. Genet.*, 1995, **90**, 611.
130. C. Martin, A. Prescott, S. Machay, J. Bartlett and E. Vrijlandt, *Plant J.*, 1991, **1**, 37.
131. M. K. Pelletier and B. W. Shirley, *Plant Physiol.*, 1996, **111**, 339.
132. S. H. Lim, GenBank, 1995, Accession No. X89199.
133. L. Britsch, J. Dedio, H. Saedler, and G. Forkmann, *Eur. J. Biochem.*, 1993, **217**, 745.
134. M. D. Rausher, P. L. Triffin, and R. E. Miller, GenBank, 1997, Accession No. U74081.
135. K. M. Davies, *Plant Physiol.*, 1993, **103**, 1015.
136. B. Charrier, C. Coronado, A. Kondorosi, and P. Ratet, *Plant Mol. Biol.*, 1995, **29**, 773.
137. Z. Gong, M. Yamazaki, M. Sugiyama, M. Kobayashi, and K. Saito, GenBank, 1997, Accession No. AB002816.
138. D. Prusky, GenBank, 1995, Accession No. U23066.
139. J. Dedio, PhD Thesis, Universität Köln, 1993.
140. G. B. Deboo, M. C. Albertsen, and L. P. Taylor, *Plant J.*, 1995, **7**, 703.
141. G. Forkmann, P. de Vlaming, R. Spribille, H. Wiering, and A. W. Schram, *Z. Naturforsch.*, 1986, **41c**, 179.
142. W. Seyffert, *Züchter*, 1955, **70**, 117.
143. W. Seyffert, *Z. Vererbungsl.*, 1955, **87**, 311.
144. R. Scott-Moncrieff, *J. Genet.*, 1936, **32**, 117.
145. M. K. Pelletier, J. R. Murrell, and B. W. Shirley, GenBank, 1996, Accession No. U72631.
146. J. Henkel and G. Forkmann, unpublished results.
147. T. A. Holton, F. Brugliera, and Y. Tanaka, *Plant J.*, 1993, **4**, 1003.
148. G. J. van Eldik, R. K. Ruiter, W. H. Reijnen, M. M. A. Van Herpen, J. A. M. Schrauwen, and G. J. Wullems, *Plant J.*, 1997, **11**, 105.
149. H. A. Stafford and H. H. Lester, *Plant Physiol.*, 1982, **70**, 695.
150. H. A. Stafford and H. H. Lester, *Plant Physiol.*, 1984, **76**, 184.
151. H. A. Stafford and H. H. Lester, *Plant Physiol.*, 1985, **78**, 791.
152. W. Heller, L. Britsch, G. Forkmann, and H. Grisebach, *Planta*, 1985, **163**, 191.
153. W. Heller, G. Forkmann, L. Britsch, and H. Grisebach, *Planta*, 1985, **165**, 284.
154. V. Dellus, W. Heller, H. Sandermann, and A. Scalbert, *Phytochemistry*, 1997, **45**, 1415.
155. F. Dedaldechamp, C. Uhel, and J.-J. Macheix, *Phytochemistry*, 1995, **40**, 1357.
156. D. Fischer, K. Stich, L. Britsch, and H. Grisebach, *Arch. Biochem. Biophys.*, 1988, **264**, 40.
157. K. N. Kristiansen, *Carlsberg Res. Commun.*, 1986, **51**, 51.
158. G. Forkmann, in "The Genetics of Flavonoids," eds. D. E. Styles, G. A. Gavazzi, and M. L. Racchi, Edizioni Unicopli, Milano, 1989, p. 50.
159. K. Stich and G. Forkmann, *Phytochemistry*, 1988, **27**, 785.
160. K. Stich and G. Forkmann, *Z. Naturforsch.*, 1988, **43**, 311.
161. K. Stich and G. Forkmann, unpublished results.
162. K. Stich, T. Eidenberger, F. Wurst, and G. Forkmann, *Planta*, 1992, **187**, 103.
163. B.-W. Min, Thesis, Technische Universität München, 1994.
164. B. Min, H. Sommer, and G. Forkmann, GenBank, 1995, Accession No. Z67981.

165. B. Min, H. Sommer, and G. Forkmann, GenBank, 1995, Accession No. Z97983.
166. Y. Tanaka, *Plant Cell Physiol.*, 1996, **37**, 711.
167. P. L. Triffin, M. D. Rausher, and R. E. Miller, GenBank, 1997, Accession No. U90432.
168. C. Rosati, A. Cadic, M. Duron, J. P. Renoú, and P. Simoneau, *Plant Mol. Biol.*, 1997, **35**, 303.
169. M. Bongue-Bartelsman, S. D. O'Neill, Y. Tong, and J. I. Yoder, *Gene*, 1994, **138**, 153.
170. V. S. Reddy, B. E. Scheffler, U. Wienand, and A. R. Reddy, GenBank, 1996, Accession No. Y07956.
171. Z. Gong, M. Yamazaki, M. Sugiyama, M. Kobayashi, K. Saito, and Y. Tanaka, GenBank, 1997, Accession No. AB0022817.
172. Y. Tanaka, Y. Fukui, M. Fukuchi-Mizutani, T. A. Holton, E. Higgins, and T. Kusumi, *Plant Cell Physiol.*, 1995, **36**, 1023.
173. P. Meyer, I. Heidmann, G. Forkmann, and H. Saedler, *Nature*, 1987, **330**, 677.
174. Y. Helariutta, P. Elomaa, M. Kotilainen, P. Seppänen, and T. H. Teeri, *Plant Mol. Biol.*, 1993, **22**, 183.
175. P. Elomaa, Y. Helariutta, R. J. Griesbach, M. Kotilainen, P. Seppänen, and T. H. Teeri, *Mol. Gen. Genet.*, 1995, **248**, 649.
176. A. R. van der Krol, L. A. Mur, M. Beld, J. N. M. Mol, and A. R. Stuitje, *Plant Cell*, 1990, **2**, 291.
177. T. R. Carron, M. P. Robbins, and P. Morris, *Theor. Appl. Genet.*, 1994, **87**, 1006.
178. J. I. Yoder, F. Belzile, Y. Tong, and A. Goldsbrough, *Euphytica*, 1994, **79**, 163.
179. X. Wang, O. Olsen, and S. Knudsen, *Hereditas* (*Lund*), 1993, **119**, 67.
180. N. Ishikura, H. Murakami, and Y. Fujii, *Plant Cell*, 1988, **29**, 795.
181. G. J. Tanner and K. N. Kristiansen, *Anal. Biochem.*, 1993, **209**, 274.
182. S. Singh, J. McCallum, M. Y. Gruber, G. H. N. Towers, A. D. Miur, and B. A. Gohm, *Phytochemistry*, 1997, **44**, 425.
183. B. Skadhauge, M. Y. Gruber, K. K. Thomsen, and D. von Wettstein, *Am. J. Bot.*, 1997, **84**, 494.
184. H. A. Stafford, *Phytochemistry*, 1988, **27**, 1.
185. H. A. Stafford, in "Chemistry and Significance of Condensed Tannins," eds. R. W. Hemingway and J. J. Karchesy, Plenum Press, New York, 1989, p. 47.
186. B. Jende-Strid, in "Barley Genetics VI," ed. L. Munck, Muncksgaard, Copenhagen, 1991, p. 504.
187. B. Jende-Strid, *Hereditas*, 1991, **119**, 187.
188. K. E. Schwinn, K. R. Markham, and N. K. Given, *Phytochemistry*, 1994, **35**, 145.
189. R. Brouillard and J. Lang, *Can. J. Chem.*, 1990, **68**, 755.
190. M. K. Pelletier and B. W. Shirley, GenBank, 1996, Accession No. U70478.
191. J. Henkel and G. Forkmann, GenBank, 1997, Accession No. U82432.
192. D. Weiss, A. H. van der Luit, J. T. M. Kroon, J. N. M. Mol, and J. M. Kooter, *Plant Mol. Biol.*, 1993, **22**, 893.
193. A. Menssen, S. Höhmann, W. Martin, P. S. Schnable, P. A. Peterson, H. Saedler, and A. Gierl, *EMBO J.*, 1990, **9**, 3051.
194. C. Rosati, GenBank, 1997, Accession No. Y12489.
195. K. M. Davies, *Plant Physiol.*, 1993, **103**, 1015.
196. V. S. Reddy, A. R. Reddy, U. Wienand, and B. E. Scheffler, GenBank, 1997, Accession No. Y07955.
197. A. Menssen, H. Saedler, and A. Gierl, *Coop. Newslett.*, 1991, **65**, 50.
198. R. L. Larson, and E. H. Coe, in "Proceedings of the 12th International Congress on Genetics," Tokyo, 1968, p. 131.
199. R. L. Larson and C. M. Lonergan, *Cereals Res. Commun.* (*Hungary*), 1973, **1**, 13.
200. N. Ishikura and M. Mato, *Plant Cell Physiol.*, 1993, **34**, 329.
201. G. Hrazdina and G. J. Wagner, *Arch. Biochem. Biophys.*, 1985, **237**, 88.
202. G. Hrazdina and R. A. Jensen, *Annu. Rev. Plant Physiol. Plant Mol. Biol.*, 1992, **43**, 241.
203. Z. Gong, M. Yamazaki, M. Sugiyama, M. Kobayashi, K. Saito, and Y. Tanaka, GenBank, 1997, Accession No. AB002818.
204. K. R. Markham, *Phytochemistry*, 1996, **42**, 1035.
205. P. M. Dewick, in "The Flavonoids Advances in Research since 1980," ed. J. B. Harborne, Chapman & Hall, London, 1988, p. 125.
206. S. Ayabe, A. Udagawa, and T. Furuya, *Plant Cell Rep.*, 1988, **7**, 35.
207. S. Ayabe, A. Udagawa, and T. Furuya, *Arch. Biochem. Biophys.*, 1988, **261**, 458.
208. R. Welle and H. Grisebach, *FEBS Lett.*, 1988, **236**, 221.
209. T. Hakamatsuka, H. Noguchi, Y. Ebizuka, and U. Sankawa, *Chem. Pharm. Bull.*, 1988, **36**, 4225.
210. K. Harano, N. Okada, T. Furuno, T. Takahashi, S. Ayabe, and R. Welle, *Plant Cell Rep.*, 1993, **12**, 66.
211. R. Welle and H. Grisebach, *Arch. Biochem. Biophys.*, 1989, **272**, 97.
212. R. Welle, G. Schröder, E. Schiltz, H. Grisebach, and J. Schröder, *Eur. J. Biochem.*, 1991, **196**, 423.
213. R. Welle and J. Schröder, *Arch. Biochem. Biophys.*, 1992, **293**, 377.
214. T. Akashi, T. Furuno, K. Futami, M. Honda, T. Takahashi, R. Welle, and S. Ayabe, *Plant Physiol.*, 1996, **111**, 347.
215. H. Hayashi, K. Murayama, N. Hiraoka, and Y. Ikeshiro, *Plant Physiol.*, 1996, **112**, 864.
216. C. Sallaud, J. el-Turk, L. Bigarre, H. Sevin, R. Welle, and R. Esnault, *Plant Physiol.*, 1995, **108**, 869.
217. G. M. Ballance and R. A. Dixon, *Plant Physiol.*, 1995, **107**, 1027.
218. J. B. Harborne, in "Chemistry and Biochemistry of Plant Pigments," ed. T. W. Goodwin, Academic Press, London, 1976, vol. 1, p. 737.
219. H. Halbwirth, F. Wurst, G. Forkmann, and K. Stich, in "Polyphenols Communications 98," Groupe Polyphenols, Lille 1–4 September, 1998.
220. D. Hess, "Biochemische Genetik," Springer, Berlin, 1968.
221. J. Kamsteeg, J. van Brederode, and G. van Nigtevecht, *Phytochemistry*, 1980, **19**, 1459.
222. J. Kamsteeg, J. van Brederode, P. M. Verschuren, and G. van Nigtevecht, *Z. Pflanzenphysiol.*, 1981, **102**, 435.
223. G. Stotz, R. Spribille, and G. Forkmann, *J. Plant Physiol.*, 1984, **116**, 173.
224. R. Sütfeld and R. Wiermann, *Z. Naturforsch.*, 1981, **36c**, 30.
225. D. Rosemann, W. Heller, and H. Sandermann, *Plant Physiol.*, 1991, **97**, 1280.
226. J. Fliegmann, G. Schröder, S. Schanz, L. Britsch, and J. Schröder, *Plant Mol. Biol.*, 1992, **18**, 489.
227. H. Fritsch and H. Grisebach, *Phytochemistry*, 1975, **14**, 2437.
228. M.-L. Hagmann, W. Heller, and H. Grisebach, *Eur. J. Biochem.*, 1983, **134**, 547.

229. K. E. Schwinn, *Polyphénols Actualités*, 1994, No. 11, 58.
230. G. Stotz and G. Forkmann, *Z. Naturforsch.*, 1982, **37c**, 19.
231. G. Stotz, Thesis, Universität Tübingen, 1983.
232. J. G. T. Menting, R. K. Scopes, and T. W. Stevenson, *Plant Physiol.*, 1994, **106**, 633.
233. G. Kochs and H. Grisebach, *Eur. J. Biochem.*, 1986, **155**, 311.
234. W. Hinderer, U. Flentje, and W. Barz, *FEBS Lett.*, 1987, **214**, 101.
235. K. M. Nielsen and E. Podivinsky, GenBank, 1997, Accession No. U72654.
236. T. A. Holton, F. Brugliera, D. R. Lester, Y. Tanaka, C. D. Hyland, J. G. T. Menting, C.-Y. Lu, E. Farcy, T. W. Stevenson, and E. C. Cornish, *Nature*, 1993, **366**, 276.
237. T. Toguri, M. Azuma, and T. Ohtani, *Plant Sci.*, 1993, **94**, 119.
238. T. Toguri, N. Umemoto, O. Kobayashi, and T. Ohtani, *Plant Mol. Biol.*, 1993, **23**, 933.
239. T. A. Holton, E. C. Cornish, and Y. Tanaka, 1993, Int. Patent Publ. No. WO93/20206: Int. Appl. No. PCT/AU93/00127.
240. G. Wimmer, H. Halbwirth, F. Wurst, G. Forkmann, and K. Stich, *Phytochemistry*, 1998, **47**, 1013.
241. M. Jay, in "The Flavonoids: Advances in Research since 1986," ed. J. B. Harborne, Chapman & Hall, London, 1993, p. 57.
242. L. J. Porter, in "The Flavonoids: Advances in Research since 1986," ed. J. B. Harborne, Chapman & Hall, London, 1993, p. 23.
243. W. E. Gläßgen, and H. U. Seitz, *Planta*, 1992, **186**, 582.
244. J. Heilemann and D. Strack, *Phytochemistry*, 1991, **30**, 1773.
245. T. Vogt and L. P. Taylor, *Plant Physiol.*, 1995, **108**, 903.
246. B. Möhle, W. Heller, and E. Wellmann, *Phytochemistry*, 1985, **24**, 465.
247. N. Ishikura and Z.-Q. Yang, *Z. Naturforsch.*, 1991, **46c**, 1003.
248. N. Ishikura and Z.-Q. Yang, *Phytochemistry*, 1994, **36**, 1139.
249. C. A. McIntosh, L. Latchinian, and R. L. Mansell, *Arch. Biochem. Biophys.*, 1990, **282**, 50.
250. M. A. Berhow and D. Smolensky, *Plant Sci.*, 1995, **112**, 139.
251. G. W. Cheng, D. A. Melencik, and P. J. Breen, *Phytochemistry*, 1994, **35**, 1435.
252. J. van Brederode and J. M. Steyns, *Z. Naturforsch.*, 1983, **38c**, 549.
253. J. M. Steyns and J. van Brederode, *Biochem. Gent.*, 1986, **24**, 349.
254. P. Vellekoop, L. Lugones, and J. van Brederode, *FEBS Lett.*, 1993, **330**, 36.
255. M. Schulz and G. Weissenböck, *Phytochemistry*, 1988, **27**, 1261.
256. K. Stich, H. Halbwirth, F. Wurst, and G. Forkmann, *Z. Naturforsch.*, 1997, **52**, 153.
257. K. Stich, H. Halbwirth, F. Wurst, and G. Forkmann, *Z. Naturforsch.*, 1994, **49c**, 737.
258. H. Halbwirth, G. Wimmer, F. Wurst, G. Forkmann, and K. Stich, *Plant Sci.*, 1997, **122**, 125.
259. L. Latchinian-Sadek and R. K. Ibrahim, *Arch. Biochem. Biophys.*, 1991, **289**, 230.
260. J. Kamsteeg, J. van Brederode, and G. van Nigtevecht, *Biochem. Genet.*, 1978, **16**, 1059.
261. F. Kerscher and G. Franz, *Z. Naturforsch.*, 1987, **42c**, 519.
262. F. Kerscher and G. Franz, *J. Plant Physiol.*, 1988, **132**, 110.
263. F. Brugliera, T. A. Holton, T. W. Stevenson, E. Farcy, C.-Y. Lu, and E. C. Cornish, *Plant J.*, 1994, **5**, 81.
264. M. Jay, V. de Luca, and R. K. Ibrahim, *Z. Naturforsch.*, 1982, **38c**, 413.
265. M. Jay, V. de Luca, and R. K. Ibrahim, *Eur. J. Biochem.*, 1985, **153**, 321.
266. R. L. Larson, *Biochem. Physiol. Pflanzen*, 1989, **184**, 453.
267. R. B. Tobias and R. L. Larson, *Biochem. Physiol. Pflanzen*, 1991, **187**, 243.
268. J. van Brederode, R. Kamps-Heinsbroek, and O. Mastenbroek, *Z. Pflanzenphysiol.*, 1982, **106**, 43.
269. G. Kuroki and J. E. Poulton, *Z. Naturforsch.*, 1981, **36c**, 916.
270. W. Knogge and G. Weissenböck, *Eur. J. Biochem.*, 1984, **140**, 113.
271. N. Ishikura, S. Nakamura, M. Mato, and K. Yamamoto, *Bot. Mag. Tokio*, 1992, **105**, 83.
272. R. K. Ibrahim, V. de Luca, H. Khouri, L. Latchinian, L. Brisson, and P. M. Charest, *Phytochemistry*, 1987, **26**, 1237.
273. R. K. Ibrahim, L. Latchinian, and L. Brisson, in "Plant Cell Wall Polymers. Biogenesis and Biodegradation," eds. N. G. Lewis and M. G. Paice, American Chemical Society, Washington, DC, 1989, ACS Symp. Ser., vol. 399, p. 122.
274. C. A. Maxwell, R. Edwards, and R. A. Dixon, *Arch. Biochem. Biophys.*, 1992, **293**, 158.
275. M. Ichimura, T. Furuno, T. Takahashi, R. A. Dixon, and S. Ayabe, *Phytochemistry*, 1997, **44**, 991.
276. K. Thresh and R. K. Ibrahim, *Z. Naturforsch.*, 1985, **40c**, 331.
277. J.-J. Macheix and R. K. Ibrahim, *Biochem. Physiol. Pflanzen*, 1984, **179**, 659.
278. H. E. Khouri and R. K. Ibrahim, *J. Chromatogr.*, 1987, **407**, 291.
279. L. M. V. Jonsson, M. E. G. Aarsman, J. E. Poulton, and A. W. Schram, *Planta*, 1984, **160**, 174.
280. L. M. V. Jonsson, P. de Vlaming, H. Wiering, M. E. G. Aarsman, and A. W. Schram, *Theor. Appl. Genet.*, 1983, **66**, 349.
281. L. M. V. Jonsson, M. E. G. Aarsman, P. de Vlaming, and A. W. Schram, *Theor. Appl. Genet.*, 1984, **68**, 459.
282. J. Mol, personal communication.
283. C. Maxwell, M. Harrison, and R. A. Dixon, *Plant J.*, 1993, **4**, 971.
284. A. Gauthier and R. K. Ibrahim, in "Polyphenols Communications 96," eds. J. Vercauteren, C. Chèze, M. C. Dumon, and J. F. Weber, Groupe Polyphenols, Bordeaux, 1996, vol. 2, p. 491.
285. J. B. Harborne, in "Progress in Phytochemistry," ed. L. Reinhold, Interscience, London, 1977, vol. 4, 189.
286. D. Barron, L. Varin, R. K. Ibrahim, J. B. Harborne, and C. A. Williams, *Phytochemistry*, 1988, **27**, 2375.
287. C. A. Williams and J. B. Harborne, in "The Flavonoids: Advances in Research since 1986," ed. J. B. Harborne, Chapman & Hall, 1993, p. 337.
288. L. Varin, D. Barron, and R. K. Ibrahim, *Phytochemistry*, 1987, **26**, 135.
289. L. Varin and R. K. Ibrahim, *J. Biol. Chem.*, 1992, **267**, 1858.
290. L. Varin and R. K. Ibrahim, *Plant Physiol.*, 1991, **95**, 1254.
291. L. Varin, V. de Luca, R. K. Ibrahim, and N. Brisson, *Proc. Natl. Acad. Sci. USA*, 1992, **89**, 1286.
292. S. Ananvoranich, L. Varin, P. Gulick, and R. K. Ibrahim, *Plant Physiol.*, 1994, **106**, 485.
293. R. K. Ibrahim, S. Ananvoranich, L. Varin, and P. J. Gulick, 1994, in "Polyphenols 94," eds. R. Brouillard, M. Jay, and A. Scalbert, INRA, Paris, 1995, Les Colloques No. 69, p. 79.

294. J.-P. Schnitzler, personal communication.
295. H. Skaltsa, E. Verykokidou, C. Harvala, G. Karabourniotis, and Y. Manetas, *Phytochemistry*, 1994, **37**, 987.
296. U. Matern, J. R. M. Potts, and K. Hahlbrock, *Arch. Biochem. Biophys.*, 1981, **208**, 233.
297. U. Matern, C. Feser, and D. Hammer, *Arch. Biochem. Biophys.*, 1983, **226**, 206.
298. U. Matern, W. Heller, and K. Himmelspach, *Eur. J. Biochem.*, 1983, **133**, 439.
299. U. Matern, C. Reichenbach, and W. Heller, *Planta*, 1986, **167**, 183.
300. J. Köster, R. Bussmann and W. Barz, *Arch. Biochem. Biophys.*, 1984, **234**, 513.
301. M. Teusch and G. Forkmann, *Phytochemistry*, 1987, **26**, 2181.
302. I. Ino, H. Nishiyama, and M.-A. Yamaguchi, *Phytochemistry*, 1993, **32**, 1425.
303. M.-A. Yamaguchi, T. Maki, T. Ohishi, and I. Ino, *Phytochemistry*, 1995, **39**, 311.
304. M.-A. Yamaguchi, S. Kawanobu, T. Maki, and I. Ino, *Phytochemistry*, 1996, **42**, 661.
305. I. Ino and M.-A. Yamaguchi, *Phytochemistry*, 1993, **33**, 1415.
306. A. Callebaut, N. Terahara, and M. Decleire, *Plant Sci.*, 1996, **118**, 109.
307. M. H. Saylor and R. L. Mansell, *Z. Naturforsch.*, 1977, **32c**, 765.
308. J. Kamsteeg, J. van Brederode, C. H. Hommels, and G. van Nigtevecht, *Biochem. Physiol. Pflanzen*, 1980, **175**, 403.
309. M. Teusch, G. Forkmann, and W. Seyffert, *Phytochemistry*, 1987, **26**, 991.
310. W. E. Gläßgen and H. U. Seitz, *Planta*, 1992, **186**, 582.
311. H. Yamamoto, J. Kimata, M. Senda, and K. Inoue, *Phytochemistry*, 1997, **44**, 23.
312. K. A. Marrs, M. R. Alfenito, A. M. Lloyd, and V. Walbot, *Nature*, 1995, **375**, 397.
313. C. B. Pickett and A. Y. H. A. Lu, *Rev. Biochem.*, 1989, **58**, 743.
314. T. Ishikawa, *Trends Biochem. Sci.*, 1992, **17**, 463.
315. E. Martinoia, E. Grill, R. Tommasini, K. Kreuz, and N. Amrhein, *Nature*, 1993, **364**, 247.
316. M. McLaughlin and V. Walbot, *Genetics*, 1987, **117**, 771.
317. N. Theres, T. Scheele, and P. Starlinger, *Mol. Gen. Genet.*, 1987, **209**, 193.

1.27 The Chalcone/Stilbene Synthase-type Family of Condensing Enzymes

JOACHIM SCHRÖDER
Universität Freiburg, Germany

1.27.1 INTRODUCTION

Chalcone synthases (CHSs) are plant-specific polyketide synthases that appear to be ubiquitous in higher plants. They utilize CoA-esters from the phenylpropanoid pathway and malonyl-CoA to synthesize the chalcones that are the starting material for the biosynthesis of a large number of biologically important substances; their roles include flower colors, UV protection, defense against pathogens (phytoalexins), interaction with microorganisms, and fertility (in some plants). These and related topics have been the subject of several reviews,[1–9] and they are discussed in Chapters 1.26 and 1.28 of this volume. CHS is the first committed step in these pathways.

In contrast to CHS derivatives, stilbenes and stilbene derivatives are relatively rare in higher plants. They occur in widely unrelated plant families, and in some cases only a few species of a family are able to synthesize stilbenes.[10] They have significant roles in the resistance of wood against microbial degradation, and in other parts of the plants they are considered to act as phytoalexins in general stress response. The enzyme stilbene synthase (STS) in transgenic plants has also been used to introduce the capacity to produce new phytoalexins in species that otherwise do not synthesize stilbenes.[11,12] A review by Gorham[13] covers most of the information on stilbenes and their functions. The stilbene backbone is synthesized by STS, which is structurally and functionally closely related to CHS. The two enzymes are polyketide synthases (PKS) by the definition that such enzymes catalyze the linking of acyl-CoA units by repetitive condensations associated with decarboxylation. The complex reactions are performed by relatively small homodimeric proteins (subunits 40–45 kDa), and in most cases (see Section 1.27.6, except for carbonyl reduction by polyketide reductase) there is no modification of reaction intermediates by additional enzymes. CHS and STS represent a line of evolution that is separate from the other PKS (including fatty acid synthases; FAS). Those PKS and their complexity have been described in several reviews.[14–18]

This chapter summarizes the available information on CHS and STS. Dominant topics include the contribution of molecular techniques to the understanding of reaction mechanisms and evolution, and the emerging evidence that CHS and STS are only the most well-known members of a superfamily of related proteins which function in widely different pathways. This development requires a clear definition of the enzyme activities. In the context of this review, CHS and STS are defined as enzymes that use phenylpropanoid starter molecules and three condensation reactions to synthesize chalcones and stilbenes, respectively. All other related enzymes are described as CHS/STS-type proteins.

1.27.2 CHS AND STS

1.27.2.1 Reactions

The principle of the reaction is shown in Scheme 1. Both enzymes use a starter CoA-ester (**1**) from the phenylpropanoid pathway (see Table 1 for typical examples) and perform three sequential condensation reactions with C_2 units from decarboxylated malonyl-CoA (**2**), to form a linear tetraketide intermediate (**3**) which is folded to form a new aromatic ring system. The intermediates cannot be demonstrated directly because of their instability. The scheme is consistent with all experimental data, and in particular with the structure of CHS by-products originating from only one or two condensation reactions (see Section 1.27.2.2).

The reactions of CHS and STS are identical up to the tetraketide stage. The difference is in the formation of the new aromatic ring systems: the tetraketide must be folded differently to connect different carbon atoms during the ring closure to form either the chalcone (**4**) or the stilbene (**5**), and the mechanisms are formally different (Claisen and aldol condensation). All STSs analyzed *in vitro* remove the terminal carboxy group of the tetraketide (**3**), and it is not known whether this occurs before or after the ring closure. The existence of stilbenoids retaining the carboxy group[13] indicates that the decarboxylation is not a necessary consequence or an essential part of the STS-type ring closure.

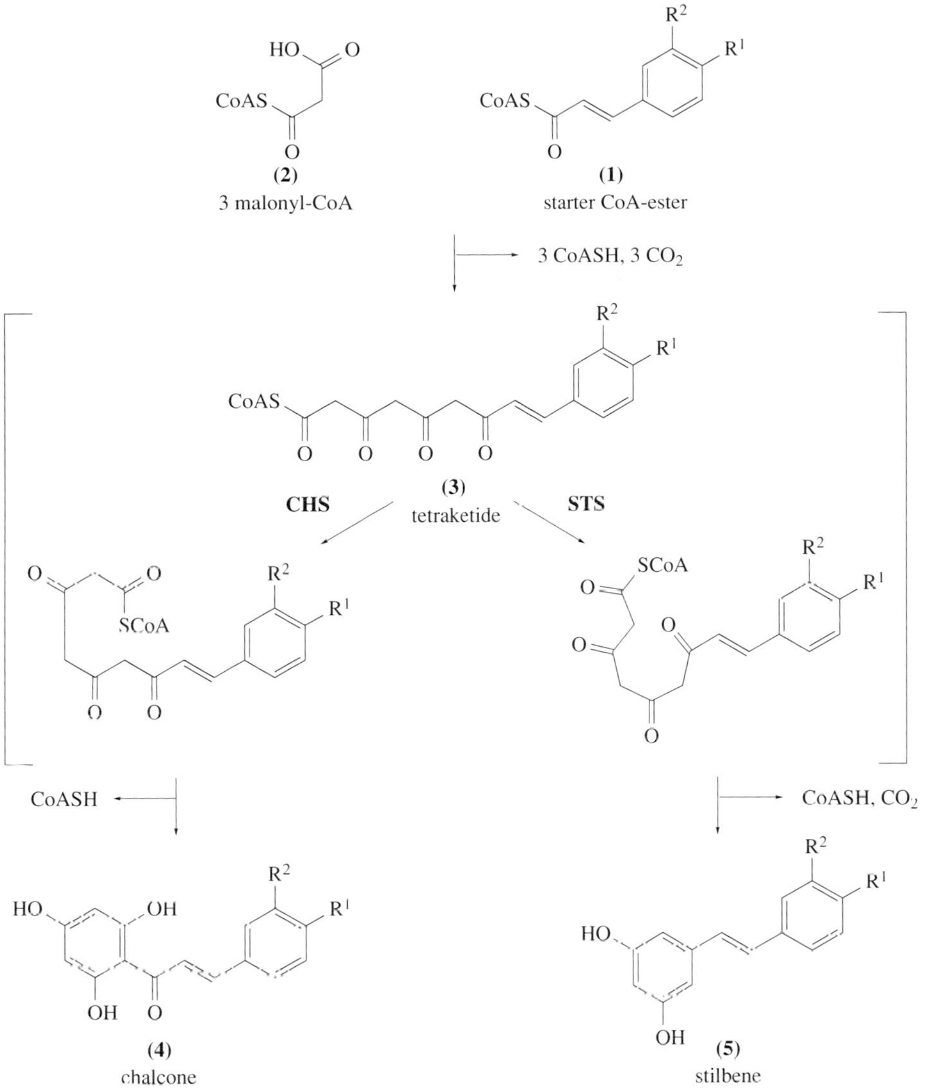

Scheme 1

Table 1 Typical phenylpropanoid-CoA starter substrates of CHSs and STSs, and trivial names of the products. R^1 and R^2 refer to Scheme 1.

R^1	R^2	*Substrate*	*Product*	
			Chalcone	*Stilbene*
-OH	-H	4-coumaroyl-CoA	naringenin	resveratrol
-H	-H	cinnamoyl-CoA	pinocembrin	pinosylvin
-OH	-OH	caffeoyl-CoA	eriodictyol	piceatannol
-H	-H	dihydrocinnamoyl-CoA[a]	dihydropinocembrin	dihydropinosylvin

[a] Reduced double bond in the propenoyl moiety.

The enzymes are either CHS or STS, and intermediate forms possessing both high CHS and STS activities are not known. Such forms were also not observed during attempts to convert a CHS into a STS by site-directed mutagenesis,[19] or in attempts to produce hybrids between CHS and STS.[20,21] This indicates an either/or switch mechanism in the CHS- and STS-type ring folding. Data suggest, however, that the mutual exclusion is probably not perfect (at least *in vitro*), because a few percent of the CHS products are stilbenes, and likewise STS may synthesize a very low percentage of the chalcone.[22] These small amounts are not detected in routine assays and cross-reactions between CHS and STS were only confirmed by dilution analysis after extensive purification with HPLC.

CHS and STS activities are usually measured by using [2-^{14}C]malonyl-CoA as substrate, followed by ethyl acetate extraction of the products, and quantification after TLC analysis. The ethyl acetate extracts may be counted directly,[23] but care should be taken that no other radioactive, extractable products are formed in the incubations (e.g. by-products). A nonradioactive assay with HPLC analysis of the products has also been described.[24]

1.27.2.2 CHS Properties

The first demonstration of CHS activity *in vitro* was reported in 1972 with extracts from parsley (*Petroselinum crispum*) cell suspension cultures.[25] This work was extended in a series of papers describing the properties of the enzyme from parsley and more than 30 other plants (reviewed by Heller and Forkmann[8] and Martin[26]). Many of these studies also investigated the induction kinetics of CHSs under various conditions, but the turnover of the protein has rarely been investigated. A study with parsley cell cultures indicated that the activity disappeared faster than the protein detectable in immunoreactions. The inactive protein revealed no difference in the size of the subunits, suggesting that the inactivation was not simply the result of proteolytic degradation.[27]

Initially, the CHS product was identified as the isomeric flavanone which *in vivo* is the result of the chalcone/flavanone isomerase (CHI) activity on the chalcone. Later experiments with the parsley[28] and tulip[29] enzymes showed that the apparent *in vitro* product was caused either by the presence of very stable CHI or by the rapid and nonenzymatic isomerization of the chalcone to the flavanone[30] at the pH of the assay (pH 8). The isomerization is pH-dependent, and about 50% conversion is observed even at pH 6 and in very short incubations. The complication of quantifying two products can be avoided in routine incubations. At the end of the incubation, it is sufficient to raise the pH to 9 for 10 min to obtain the flavanone as the only product. The K_m for malonyl-CoA is in the range of 30 μM, and most of the enzymes prefer 4-coumaroyl-CoA as starter unit (K_m usually below 10 μM). Other phenylpropanoid starters (e.g., cinnamoyl-CoA, caffeoyl-CoA) are also accepted *in vitro*, but usually at lower efficiency. The possibility that CHSs in some plants synthesize eriodictyol *in vivo* from caffeoyl-CoA has been suggested.[8] Data with two enzymes cloned from *Hordeum vulgare* indicate that CHS2 prefers caffeoyl-CoA rather than 4-coumaroyl-CoA,[31] and this appears to be the first direct evidence that 4-coumaroyl-CoA is not the only physiological substrate in some plants.

The characterization of the proteins showed that the enzymes are dimers of identical subunits (41–44 kDa). Several properties indicated similarities with the condensing enzymes of fatty acid biosynthesis (e.g., CO_2 exchange at the malonate moiety,[32] inhibition by cerulenin[33]), but there is no evidence for an acyl carrier protein or a 4′-phosphopantetheine arm being involved in the reactions.[34] Two other points emerged from the experiments *in vitro*: (a) the substrate specificity is not confined to phenylpropanoid starters, because the enzyme also accepted alkoxy CoA-esters,[35] and (b) CHS can synthesize products that are the result of the release of intermediates after one (benzalacetone, aryldihydropyrone) or two condensation reactions (styrylpyrones).[33,36–38] Later data with improved enzyme preparation techniques and assays indicated that these products are largely a consequence of nonoptimal preparation and assay conditions.[39] The results are nevertheless interesting because they demonstrate the flexibility of CHS with respect to various starters, and more recent data indicate the existence of CHS-related enzymes that can perform these reactions in the biosynthesis of natural products (see Section 1.27.4).

1.27.2.3 STS Properties

The first reaction *in vitro* was demonstrated in 1978 in crude extracts from rhizomes of rhubarb (*Rheum rhaponticum*), with 4-coumaroyl-CoA as starter CoA-ester and resveratrol as product.[40] STS activities were also demonstrated in extracts from Scots pine (*Pinus sylvestris*),[41] groundnut (*Arachis hypogaea*),[42–44] and *Vitis* species.[45] In some cases, a substantial part of the activity was detected in membrane fractions,[40,41] and this corresponds to observations with CHSs[46] and the proposal of pathway channeling.[46–48] The elucidation of the biosynthetic reactions by feeding studies and the early enzymatic work have been reviewed.[49]

The characterization of the enzymes showed that they accepted various CoA-esters from the phenylpropanoid pathway, but, as with CHS, the normal precursor for the natural products occurring in the plants was preferred.[50–53] The purified proteins were homodimers (90–95 kDa) with subunits of 43–47 kDa.[51–55] This work also showed that STS has no significant CHS activity, and in extracts from cultured cells of *Picea excelsa* it was shown that the two proteins can be separated physically.[51] The pH optima and the K_m for the preferred starter CoA-ester and malonyl-CoA[52] were in the same range as determined for CHS. In contrast to CHS, by-products originating from only one or two condensation reactions have not been described.

A special subgroup of STSs are the enzymes characterized from *Dioscorea* species[56] and several orchids (e.g., *Epipactis palustris*,[57] *Bletilla striata*,[58] and *Phalaenopsis* sp.[59,60]). The proteins can be distinguished from other STSs by their substrate preference: all of them distinctly prefer phenylpropanoyl-CoA (double bond in the coumaroyl moiety reduced) to all other investigated starter substrates, in particular to 4-coumaroyl-CoA. The products are formally bibenzyl derivatives, and therefore the enzymes are often called bibenzyl synthases. Otherwise there are no basic differences from the other STSs (e.g., subunit size, native enzyme as dimer, K_m with starter CoA-ester and malonyl-CoA). The cDNA sequence of the *Phalaenopsis* sp. protein[60] revealed extensive similarities to the other STSs.

1.27.2.4 What do we Learn from Sequences?

1.27.2.4.1 Overall relationships

The first sequences were described in 1983 for CHS (*Petroselinum crispum*)[61] and in 1988 for STS (*Arachis hypogaea*).[62] In October 1996, the databases contained more than 100 CHS entries from at least 40 plants, and STS entries from five plants. Figure 1 presents a protein tree reflecting the overall relationship of most of the CHSs and the STSs available as DNA sequences. The tree was rooted to a consensus derived from 18 CHSs that have been shown to be functional (see below), and very closely related isoforms from the same plant were summarized in a single subbranch. The tree was not designed to reflect the actual relationships between the plant families.

The tree shows that the CHSs from the same plant family are often in a distinct main branch (e.g., Leguminosae). Within that branch, isoforms from a single plant may be on different subbranches that also contain CHSs from other plants (e.g., *Medicago sativa*), or a specific isoform may be clearly separated from other CHSs of the same plant (e.g., *Glycine* max CHS7). CHSs from the same plant may even be found in very distant branches (e.g. *Daucus carota*, 81% identity between the two proteins). The significance is not clear in any of these cases. It may reflect functional differences, but that has not been investigated.

The STSs are closely related to CHSs, as predicted from the similarities in the reaction. In view of the functional differences it is an interesting question whether the STSs are on a branch that is separate from the main body of the CHSs. This is clearly not the case, and this basic result is also found after using other program options in the development of the tree, or after rooting the tree to other sequences. In all cases the STSs (marked with ▮) from different plants show close homology with CHSs of the same plants or families rather than with other STSs. Often they cluster with CHSs from the same species (e.g., in *Pinus*) or the same family (STS *Arachis hypogaea* with CHSs from Leguminosae; the CHS from *A. hypogaea* has not been described). The STS and CHS from *Vitis*, however, are on widely separated branches, and no comment on the STS from *Phalaenopsis* is possible because the CHS from this or a closely related plant has not been cloned. The distance between the various STSs reflects that the proteins from different species share less than 70% identity to each other.

The close functional relationship between STS and CHS, the finding that STSs cluster with CHSs rather than with each other, the presence of stilbenes in just a few plants of a given family, and the fact that these families are often only distantly related led to the proposal that the present-day STSs evolved from CHSs independently several times.[19] All CHS and STS genes contain an intron at the same position; it splits a cysteine in a highly conserved location (position 65 in the protein, Figure 2). The strict conservation also argues for a close relation. There is no evidence supporting a hypothesis that the intron separates domain functions.

Figure 1 Protein relationship tree. Marked sequences: ▮, STS; ▮▮, not CHS or STS, or unknown function. The sequences were aligned with CLUSTAL V,[63] and the tree was developed with the program TREECON[64] using the inbuilt matrix for amino acid sequences and the neighbor-joining method.[65] The root is a consensus derived from 18 functionally identified CHSs.

1.27.2.4.2 Conserved motifs and differences between CHS and STS

Crystal structures are not available so far from any of the proteins. One aspect of the wealth of sequences is the possibility that it might allow the establishment of consensus sequences that highlight motifs that are conserved in all condensing enzymes,[66] conserved in all CHSs and STSs,[67] or different in CHSs and STSs.[26,68] These considerations are of interest because the functional differences between CHSs and STSs should be reflected in some way in the primary sequence. A more detailed comparison could also detect small systematic differences that do not show up in the overall comparison that the relationship tree reflects.

It should be noted that the attempt to correlate primary sequence and functional significance imposes an important constraint: the analysis must be restricted to proteins that have been demonstrated to be functional, either by genetic evidence or by the demonstration of enzyme activity after heterologous expression. This is no problem with STSs because at least one enzyme from each of the five species was tested for function after heterologous expression; these experiments in fact were crucial for the identification as STS. The application of this criterion to CHSs, however, eliminates about 80% of the sequences for either one of the following reasons.

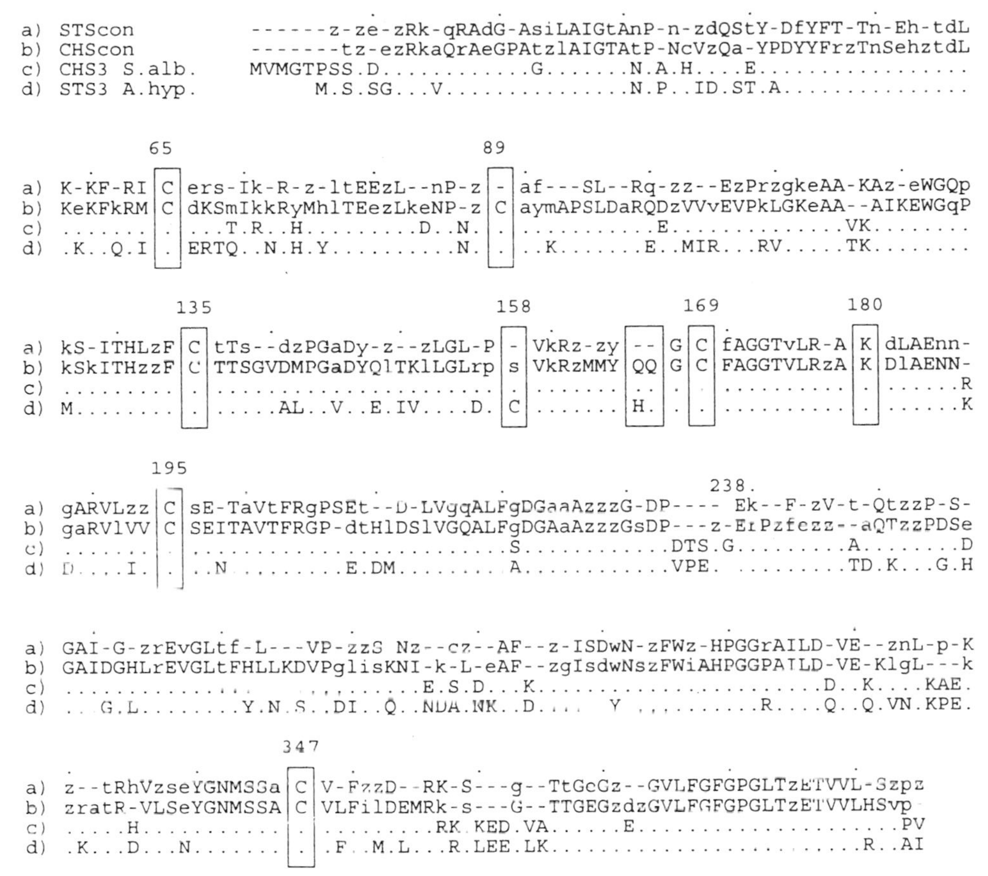

Figure 2 Consensus sequences and some conserved residues: (a) STS (from five different plants), (b) CHS (from 18 CHSs demonstrated to be functional). The alignment also shows as examples (c) the CHS3 from *Sinapis alba* and (d) the STS3 from *Arachis hypogaea*; the dots indicate identity with the CHS consensus. The numbering follows the example of the first published CHS (parsley)[61] to obtain common reference positions for proteins that are shorter at the N-terminal. Large letters, conserved; small letters, predominant; z, hydrophobic (I, L, V, M); -, variable residues. Boxed: conserved Cys, CHS Gln-Gln-motif directly left of the active site Cys (Cys169), and contact sites between subunits (position 158 and 180). The gap at position 238 indicates that this residue is absent in most CHSs and all STSs. Cys65 is the amino acid split by an intron in all known CHS and STS genes, and Cys169 is the active site of the condensing reactions.

Only a few plants have been demonstrated to contain only one CHS gene (e.g., *Petroselinum crispum*,[69] *Antirrhinum majus*,[67,70] and *Arabidopsis thaliana*[71–73]). Most plants contain several CHS genes (gene families), and in many cases the analysis was restricted to genomic or cDNA sequences, or the extent of the family was not fully investigated. Often it is not known whether all of the predicted CHS isoenzymes are expressed and functional, and heterologous expression and functional assays were usually not performed. *Petunia hybrida* represents a well-investigated example of CHS genes that are expressed or predict unusual proteins.[74–76]

Many of the proteins predicted to be CHS are identified by sequence similarity only, and a few examples may highlight that this can be misleading. The acridone synthase (ACS, see Section 1.27.4.1.1) is not a CHS, but shares ≥65% identity with typical CHSs. It might easily be described as CHS if only the sequences were available. Another case is *Gerbera hybrida*.[77] Heterologous expression identified CHS1 and CHS3 as typical CHS, but CHS2 (74% identity with CHS1) should not be considered as CHS because it showed no activity with any of the phenylpropanoid starter CoA-esters. It produced a detectable product only with benzoyl-CoA, and the *in vivo* substrate and the physiological role of CHS2 remain to be clarified. Two CHS-type cDNAs from *Pinus strobus* are a third example. The analysis after heterologous expression showed that only CHS1 had CHS activity, while CHS2 was inactive with any of the phenylpropanoid starter CoA-esters, although the predicted protein had 87.6% identity with the CHS1 from the same plant.[78] The relationship tree shows that all of these proteins (Figure 1, marked with ▮▮) are no more distant from the CHS consensus than the STSs, and in particular not more distant than the *Ipomoea* sequences designated as CHSs by sequence similarity.[79] These examples show that the assignment of function solely based on sequence similarities should be avoided, and the conclusion is supported by other cases (see Somssich *et al.*[80] for an interesting example). Some of the sequences designated as CHS may encode other functions than serving the flavonoid pathway (see Section 1.27.4).

Figure 2 shows a consensus derived for STSs (five different plants) and a consensus from 18 CHSs proven to synthesize the chalcone intermediate in the flavonoid pathway. The two sequences are easily aligned without significant gaps, and both share several highly conserved regions. There are actually very few systematic differences between CHSs and STSs, and these are distributed throughout the protein.

The restriction applied to the selection of the sequences most likely excludes a large number of CHSs that simply lack the functional verification. The inclusion of these proteins remarkably increases the number of variable positions in the CHS consensus (not shown), and the differences to the STSs become close to oblique, i.e., it seems not possible to predict from the primary sequence whether the protein is a CHS, a STS, or another enzyme sharing the same type of condensing reactions (see Section 1.27.4). The presence or absence of certain secondary plant products in a given plant may provide some clues for new sequences, but it should be realized that only a few plants have been thoroughly analyzed. Stilbenes, for example, have been described in plants where they were previously not suspected (e.g., in monocots).[81–84]

1.27.3 FUNCTION AND STRUCTURE: ANALYSIS OF MUTANT PROTEINS

Several studies have addressed the importance of specific amino acids, either for the activity of the enzymes in general or with respect to the functional differences between CHS and STS. Many of these investigations employed site-directed mutagenesis and heterologous expression of the mutant proteins in *E. coli*.

1.27.3.1 Active Site of the Condensing Reaction

The condensing reactions of CHS and STS are very similar to those in other polyketide and fatty acid synthases, but no acyl carrier protein is involved. With those enzymes, the starter residue must be bound to a Cys-SH group of the condensing enzyme prior to the condensation. One would therefore predict that CHS and STS contain a Cys that is essential because it serves the same function. They could also contain a second essential Cys that accepts the malonyl moiety and plays the role of the acyl carrier protein.

This was investigated with the CHS from *Sinapis alba* and the STS from *Arachis hypogaea*. The two proteins contain six conserved Cys (boxed in Figure 2; note that Cys89 was later shown not to be absolutely conserved).[60,85,86] The functional importance was investigated by mutating all of them individually to Ser and/or Ala, and the activity was tested after protein expression in *E. coli*. The results[87] showed with both proteins that only the mutation of Cys169 led to a complete loss of enzyme activity. None of the five other Cys was essential, although most of the mutants had reduced activities. In two cases (Cys135 and Cys195), CHS and STS showed a differential reaction to the mutation (the loss of activity was much higher with STS than with CHS), and it is tempting to speculate that this may reflect some aspect of the product folding to different ring systems. Based on the similarities of the CHS and STS reactions with other polyketide synthases, the data suggest that Cys169 is the active site of the condensing reaction that covalently binds the starter residue prior to the condensation. This conclusion was confirmed by biochemical experiments analyzing the binding site for the starter residue.[88] The absence of a second essential Cys indicates that malonyl-CoA is used directly in the reaction. Scheme 2 presents a model of the reactions derived from these and previous investigations.[32–34,36–38]

1.27.3.2 Environment of the Active Site

The environment of the active site Cys is highly conserved in CHSs and STSs, except for the positions minus 2 and 3. All CHSs contain Gln-Gln (Figure 2), while the STSs revealed either His-Gln or Gln-His. This suggested that the motif may participate in the product specificity, i.e., formation of chalcone or stilbene. Site-directed mutagenesis of the Gln-Gln in the CHS from *Pinus*

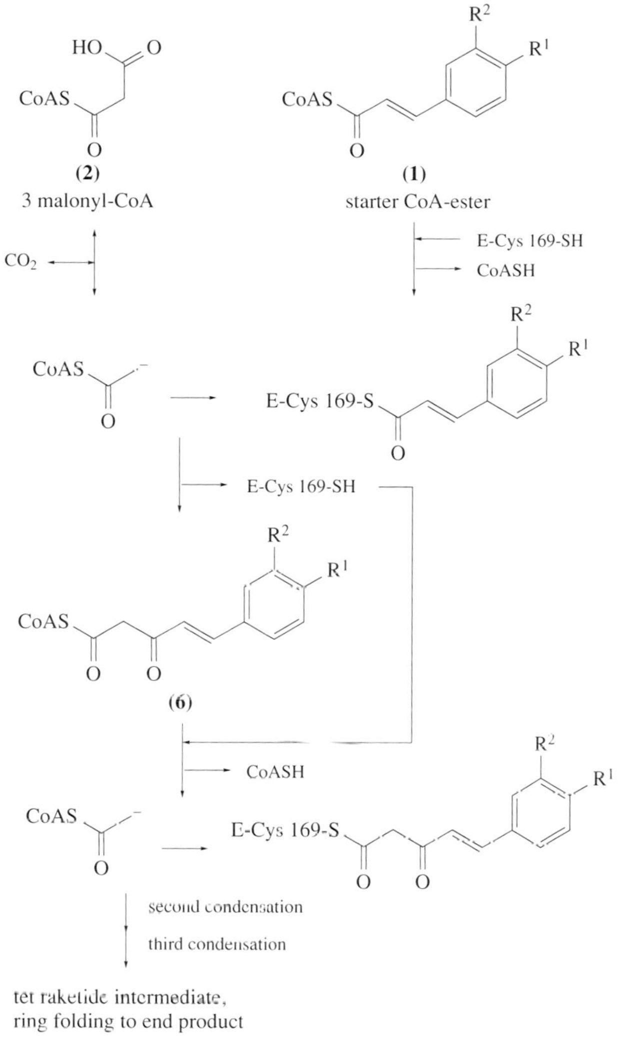

(2) 3 malonyl-CoA

(1) starter CoA-ester

CO_2

E-Cys 169-SH

CoASH

E-Cys 169-S

E-Cys 169-SH

(6)

CoASH

E-Cys 169-S

second condensation

third condensation

tetraketide intermediate, ring folding to end product

Scheme 2

sylvestris to either the His-Gln or Gln-His motifs of STS, and in the STS from the same plant from Gln-His to His-Gln or the CHS-type Gln-Gln, however, only led to a general reduction of activity, and not to a change in the type of ring folding. The mutagenesis of the *Arachis hypogaea* STS produced some unexpected effects. The type of the reaction (stilbene product) was not changed, but the simple inversion of His-Gln to Gln-His almost completely abolished the enzyme activity. The change to the CHS-type Gln-Gln led to an alteration of the substrate preference from 4-coumaroyl-CoA to dihydrocinnamoyl-CoA, converting the resveratrol forming STS into a bibenzyl synthase.[89] The later published bibenzyl synthase sequence from *Phalaenopsis* sp. indeed predicts the Gln-Gln motif.[60] It may therefore be possible that it participates in influencing the substrate preference in some STSs. The only other tested example (STS from *Pinus sylvestris*) did not reveal a comparable change.

1.27.3.3 Subunit Interactions

CHSs and STSs are dimers, and cross-linking studies[20] showed that contact sites between the subunits were at position 158 and most likely at position 180, i.e., flanking the active site Cys in position 169 (see Figure 2). This suggested the possibility that the active sites are close together in the dimer and that the subunits alternate in performing the three sequential condensing reactions rather than operating independently. One of the consequences of that model would be that heterodimers with only one active site could only perform the first condensation reaction, possibly leading to the release of products that had been identified previously as by-products of CHS reactions (benzalacetone, arylpyrone). In particular the benzalacetone would be of interest because an enzyme synthesizing this product has been described (see Section 1.27.4.2.1).

The model was tested, with CHS as well as with STS, by coexpression of two inactive mutants in *E. coli*: one of them was a protein inactivated by mutagenesis of the active site (Cys169), and the other protein was inactive because of other changes.[20] Under these conditions enzyme activity could be obtained only by functional complementation of the inactive subunits, and the active enzyme had to be a heterodimer with only one active site. Enzyme activity was obtained with CHS as well as with STS, and both heterodimers synthesized the end-products (chalcone or stilbene). This indicated that each subunit can perform all three condensation reactions. The simplest explanation for the successful complementation of two inactive subunits to give active enzyme is that the subunits cooperate in other ways in the formation of the products.

These and other experiments also addressed the question of whether it is possible to obtain proteins with both CHS and STS activity, but the results were negative.[19,20,90]

1.27.3.4 Evolution *In Vitro*: Conversion of CHS to STS by Mutagenesis

The relationship tree (Figure 1) suggested that STSs evolved from CHSs, and that possibly a relatively small number of amino acid exchanges in key positions may be sufficient to produce a protein with STS activity. This was investigated with a hybrid between the CHS from *Sinapis alba* (N-terminal 107 residues) and the STS from *Arachis hypogaea* (rest of the fusion protein).[19] The hybrid was completely inactive, and the approach attempted to restore STS activity by changing amino acids typical for the CHS consensus towards the residues in the STS. The results showed that exchanges of three residues in the CHS part were sufficient to create a protein with low STS activity (Gln100 to Glu, Val103 to Met, Val105 to Arg; see sequences in Figure 2). Disappointingly, many additional mutations towards STS and their various combinations failed to improve the activity, and in many cases it was lost entirely. Finally, it was discovered that only one additional exchange (Gly23 to Thr) was necessary to increase the activity to 25% of the parent STS. The high activity of that protein was entirely unexpected because the importance of Gly23 was not recognizable from consensus considerations; most CHSs contain Thr in position 23, like the STS from *Arachis hypogaea*, and a Gly appears to be present only in the CHSs from Brassicaceae. These results stress that consensus observations are not necessarily the only points to be considered. It is also noteworthy that the *de novo* created STS could be called a bibenzyl synthase because it preferred dihydrocinnamoyl-CoA to 4-coumaroyl-CoA, in contrast to the parent enzyme.

Another important point emerged from these experiments.[19] Most of the inactive mutant proteins (including the original CHS/STS hybrid) were found to form insoluble aggregates, but not soluble dimers. This suggests that the proteins were incapable of correct dimerization, and that the monomeric subunits are unstable and denature very quickly, aggregating to inactive complexes. The conclusion is supported by other experiments showing that the dissociation of active dimers into monomers leads to irreversible denaturation of the subunits. Taken together, these and the findings with the heterodimers (see Section 1.27.3.3) suggest that correct dimerization is a prerequisite for active enzyme. This indicates that the conversion of the inactive hybrid to a protein with STS activity involves both structural and functional elements, and that it is not yet possible to distinguish between these two aspects.

1.27.3.5 Other Functionally Important Amino Acids

The sensitivity of the proteins to seemingly minor changes in widely scattered locations of the primary sequence was also demonstrated in other cases. One example is the CHS from *Antirrhinum majus*.[91] The imprecise excision of a transposable element led to a seemingly minor alteration at the position corresponding to Asn50 in the CHS consensus (Figure 2). An additional Asn or Ile reduced anthocyanin formation to 45% (*niv*-630, pale red) and 3% (*niv*-631, very pale red), respectively. Immunoblots for CHS revealed no significant differences between wild-type and mutants, indicating that the slight protein modifications were responsible for the remarkable reduction of anthocyanin synthesis.

In *Arabidopsis thaliana*, a chemically induced CHS mutant which had a single mutation Gly to Ser at position 268 in the CHS consensus showed the same mRNA levels as in the wild-type, but the synthesis of anthocyanins was completely abolished.[73] Gly268 is conserved in all CHSs and STSs, but its role in the reaction is not well understood.

A final example is of two STSs from *Pinus strobus* that were obtained as cDNAs from stress-induced seedlings.[92] One of the proteins possessed normal STS activity after heterologous expression, whereas the other had only 3–5% of that activity, a lowered pH-optimum (pH 6), and synthesized with cinnamoyl-CoA a second unknown product. The proteins were different in only five amino acids, and site-directed mutagenesis demonstrated that a single change from Arg to His (position 313 in Figure 2) was responsible for all of the unusual properties of the enzyme.[92] The physiological significance of this STS is unknown.

1.27.4 A SUPERFAMILY OF CHS/STS-TYPE ENZYMES?

This section summarizes the emerging evidence suggesting that CHSs and STSs are simply the most well-studied members of a superfamily of related proteins that use the same type of condensing reaction, but with different starter CoA-esters or with programming for only one or two C_2-condensation reactions. Much of the data are from the mid-1990s, and the acridone synthase is the first example where the relationship of the protein to CHSs and STSs has been established by sequence analysis. This needs to be done with the other enzymes. The examples with three condensation reactions are all of the CHS-type ring folding. The differentiation between CHS- and STS-type enzyme is not possible with the enzymes programmed for only one or two C_2-condensations, because the distinction defines differences occurring after the third condensation reaction (Scheme 1).

1.27.4.1 Three Condensation Reactions with Nonphenylpropanoid Starters

1.27.4.1.1 Acridone synthase (ACS)

Acridone alkaloids are only found in some genera of the Rutaceae family, and about 100 of these substances are known, with a remarkable variety in structure. The enzymatic formation of the acridone backbone was investigated with cell cultures from *Ruta graveolens*. Crude extracts showed that the acridone was synthesized from *N*-methylanthraniloyl-CoA (**7**) and three malonyl-CoA,[93,94] and the new aromatic ring system was obviously formed by a CHS-type ring folding (Scheme 3). The product detected *in vitro* (1,3-dihydroxy-*N*-methylacridone) (**8**) is the result of a second ring closure, and it is not known whether this is an intrinsic activity of the enzyme or a nonenzymatic process. The purified enzyme[95] has K_m of about 11 μM and 32 μM for the starter CoA-ester and malonyl-CoA, respectively, a size of about 69 kDa in gel permeation chromatography, and subunits of about 40 kDa that cross-react with CHS antisera. The microsequences from seven peptides revealed a high degree of similarity to CHSs. The cDNA sequence confirmed the overall similarity to CHSs and STSs, and the function as ACS was established by heterologous expression.[96]

ACS is not a CHS because it uses an *N*-methylanthraniloyl-CoA as starter, and the product belongs to an entirely different pathway. It belongs to the CHS/STS-type superfamily because the protein shares ≥65% overall identity with most CHSs and STSs. The relationship tree indicates that it is not clearly separated from the other sequences designated as CHSs (Figure 1). As in the comparison of CHSs with STSs, it is difficult to identify motifs that could explain the different starter specificities. Without the functional identification, the ACS would probably have been labeled as CHS (stilbenes have not been described in *Ruta graveolens*), and this confirms the notion that sequence similarities even to ≥65% and immunoreactions of the proteins with known antisera are not sufficient for functional assignments.

1.27.4.1.2 Benzophenone synthase (BPS)

Xanthones are a group of natural products with interesting pharmaceutical properties,[97–101] and the majority occur in two plant families (Gentianaceae and Hypericaceae).[97,102,103] Feeding experiments indicated that the xanthone ring system was synthesized from shikimate-derived pre-

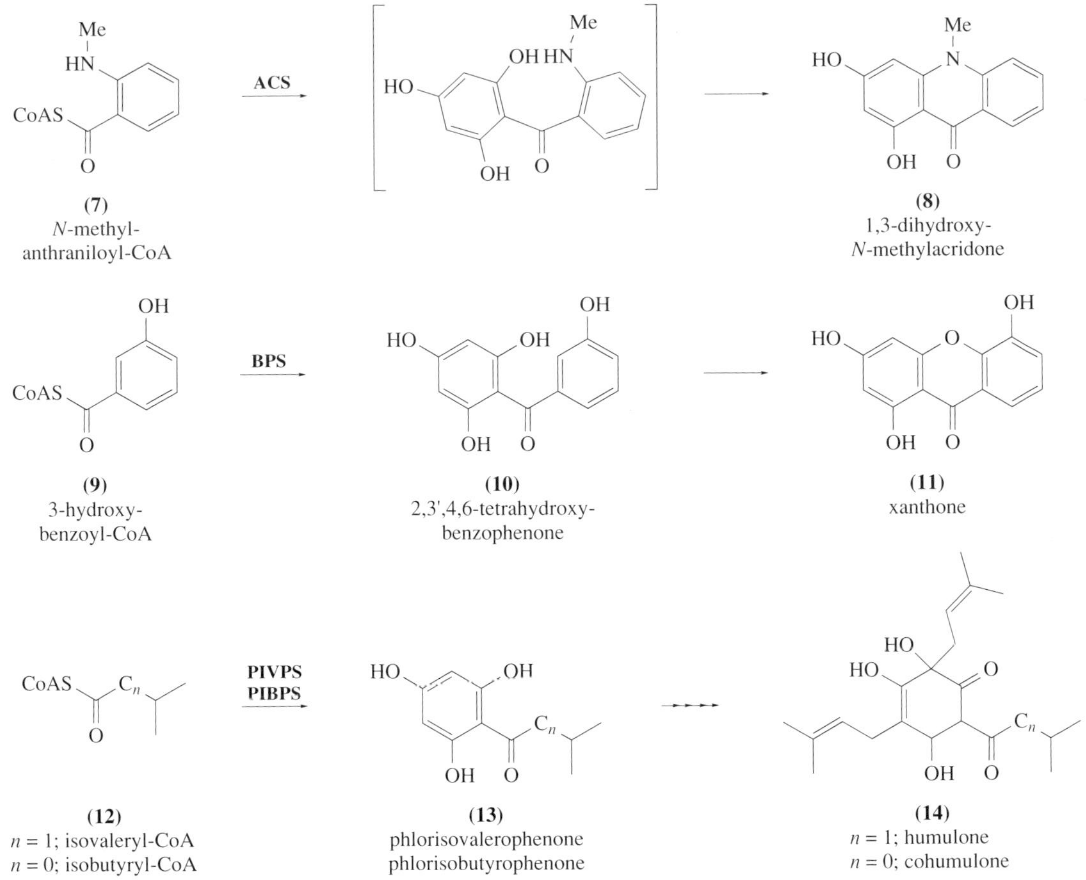

(7) N-methyl-anthraniloyl-CoA → ACS → (8) 1,3-dihydroxy-N-methylacridone

(9) 3-hydroxy-benzoyl-CoA → BPS → (10) 2,3',4,6-tetrahydroxy-benzophenone → (11) xanthone

(12) $n = 1$; isovaleryl-CoA, $n = 0$; isobutyryl-CoA → PIVPS PIBPS → (13) phlorisovalerophenone phlorisobutyrophenone → (14) $n = 1$; humulone, $n = 0$; cohumulone

Scheme 3

cursors and via the acetate/malonate route (see Sultanbawa[102] for review). Cell cultures of *Centaurium erythraea* and *Centaurium littorale* have been used to investigate the xanthones and the regulation of their formation.[104,105] The proposed biosynthetic pathway[97,105] suggested the condensation of a benzoyl-CoA derivative with three C_2 units derived from malonyl-CoA to form the benzophenone, the precursor of the xanthones (Scheme 3). The reaction to the benzophenone (**10**) was confirmed *in vitro* with a partially purified protein preparation incubated with 3-hydroxybenzoyl-CoA (**9**) and malonyl-CoA;[106] the second ring closure to the xanthone (**11**) apparently requires an additional enzyme. The BPS also accepted benzoyl-CoA as starter (44% efficiency), but was completely inactive with 2-hydroxybenzoyl-CoA and 4-hydroxybenzoyl-CoA, and that indicated a clear specificity for only one position of the hydroxy group. The ring closure shows that the BPS is a CHS-type enzyme using a benzoyl-CoA instead of phenylpropanoid-CoA derivative as starter. The actual relationship of the protein with CHSs needs to be established. Significant differences are to be expected because typical CHSs do not accept 3-hydroxybenzyol-CoA as starter.[107]

1.27.4.1.3 Phloroisovalerophenone and phloroisobutyrophenone synthase (PIVPS and PIBPS)

The ripe cones of hop (*Humulus lupulus*, Cannabaceae) contain up to 20% of bitter acids,[108] and Scheme 3 shows as examples humulone and cohumulone (**14**) which are converted during the brewing process to their isoforms which are important for the flavor and taste of beer. It had been proposed[109] that the first aromatic intermediates in the pathway were produced via prenylated long-chain acids. Based on the detection of PIVP and PIBP (**13**) as intermediates,[110,111] Verpoorte's group suggested instead that the formation of the aromatic ring could precede all of the prenylation steps. Previous experiments had shown that CHS from parsley accepted alkoxy CoA-esters as starter substrates,[35] and therefore it seemed possible that PIVP and PIBP were synthesized via a CHS-type reaction, using isovaleryl-CoA and isobutyryl-CoA (**12**) as starters and three malonyl-CoA (Scheme 3).

The hypothesis was confirmed by demonstrating the predicted enzyme activities in crude extracts.[111] Immunoblots with an antiserum against parsley CHS and denatured proteins revealed a band of ~45 kDa, and this is the size expected for CHSs and CHS-related proteins. CHS activity was also present (starter 4-coumaroyl-CoA), but the activity with isovaleryl-CoA was always higher than with 4-coumaroyl-CoA at each developmental stage, and CHS and PIVPS/PIBPS followed different kinetics during the development from flower buds to ripe cones. The presence of different enzymes needs to be confirmed by the separation of CHS from the other activities, and ideally by cloning and heterologous expression of the single proteins. This seems essential because CHSs cloned from plants not producing these particular secondary products do accept isovaleryl- and isobutyryl-CoA as substrates.[112]

1.27.4.2 Programming for One or Two Condensation Reactions

1.27.4.2.1 One condensation reaction: benzalacetone synthase (BAS)

The characteristic aroma of raspberries is caused by 4-hydroxyphenylbutan-2-one (pHPB, "raspberry ketone"),[113] and this compound or its glycosides[114] have also been found in *Pinus contorta*,[115,116] *Rhizoma rhei*,[117] *Vaccinium oxycoccus* (European cranberry),[118] *Hippophae rhamnoides* (sea buckthorn),[119] and *Scutellaria rivularis*.[120] The biosynthesis involves two enzymes; these were investigated in raspberry fruits and tissue cultures.[121] The first enzyme (named BAS,[122] benzalacetone synthase) uses a phenylpropanoid starter CoA-ester (**1**), performs a one step condensation reaction with malonyl CoA, and liberates a decarboxylated product (**16**). The second enzyme reduces the double bond with NADPH to form the aroma component (**17**) (Scheme 4). BAS was characterized with a preparation from raspberry fruits that was enriched 172-fold for BAS and 14-fold for CHS activity.[122] Stained gels revealed a single band; the native and denatured proteins were 83 kDa and 41.5 kDa, respectively. The BAS activity had K_m of 3 µM and 1 µM for 4-coumaroyl-CoA and malonyl-CoA, respectively. All of these values are typical for CHS-type enzymes, except for the very high affinity to malonyl-CoA.

The partial copurification of BAS and CHS did not allow a definite conclusion as to whether the CHS activity represented a contamination, or whether the BAS is an enzyme that can perform either one or three condensation reactions. This question is important because previous experiments showed that purified CHS can produce benzalacetone as a by-product.[37,38] Several lines of evidence argue that BAS and CHS are different enzymes: (a) feruloyl-CoA was an excellent substrate for BAS (surprisingly, threefold better than 4-coumaroyl-CoA), but not for the CHS activity (6%, when compared with 4-coumaroyl-CoA); (b) BAS and CHS revealed different responses to 2-thioethanol and ethylene glycol *in vitro*; (c) BAS and CHS were differentially inactivated by antiserum against the purified protein; (d) BAS in tissue cultures was strongly induced after treatment with yeast extract, but CHS was not. The unambiguous answer to the question of whether there are one or two enzymes requires the cloning and heterologous expression of BAS.

1.27.4.2.2 Two condensation reactions: styrylpyrone synthase (SPS)

Styrylpyrones are common constituents in fungi, mainly in Basidiomycetes,[123–125] but they also occur in Pteridophytes,[126–128] and in Angiosperm families.[129–131] Feeding experiments[132–134] suggested that the styrylpyrones are synthesized from the same precursors as used by CHS, but with only two condensation reactions which are followed by a ring closure of the triketide intermediates (**15**) to the styrylpyrones (**18**) (Scheme 4).

Equisetum arvense is an interesting plant system because it shows a developmental switch:[126,127,135,136] gametophytes and rhizomes accumulate styrylpyrones as major phenolic constituents, but no flavonoids, while green sprouts contain various flavonoid glycosides, but no styrylpyrones. This made extracts from gametophytes a suitable system to investigate SPS without interference from CHS. The data with partially purified SPS activities[136] showed that the enzyme used 4-coumaroyl-CoA and caffeoyl-CoA to synthesize with malonyl-CoA the corresponding styrylpyrones (**18**) (bisnoryangonin and hispidin). It is not clear which starter is used preferentially *in vivo*. The apparent K_m was >210 µM in both cases, and 80 µM for malonyl-CoA. These values

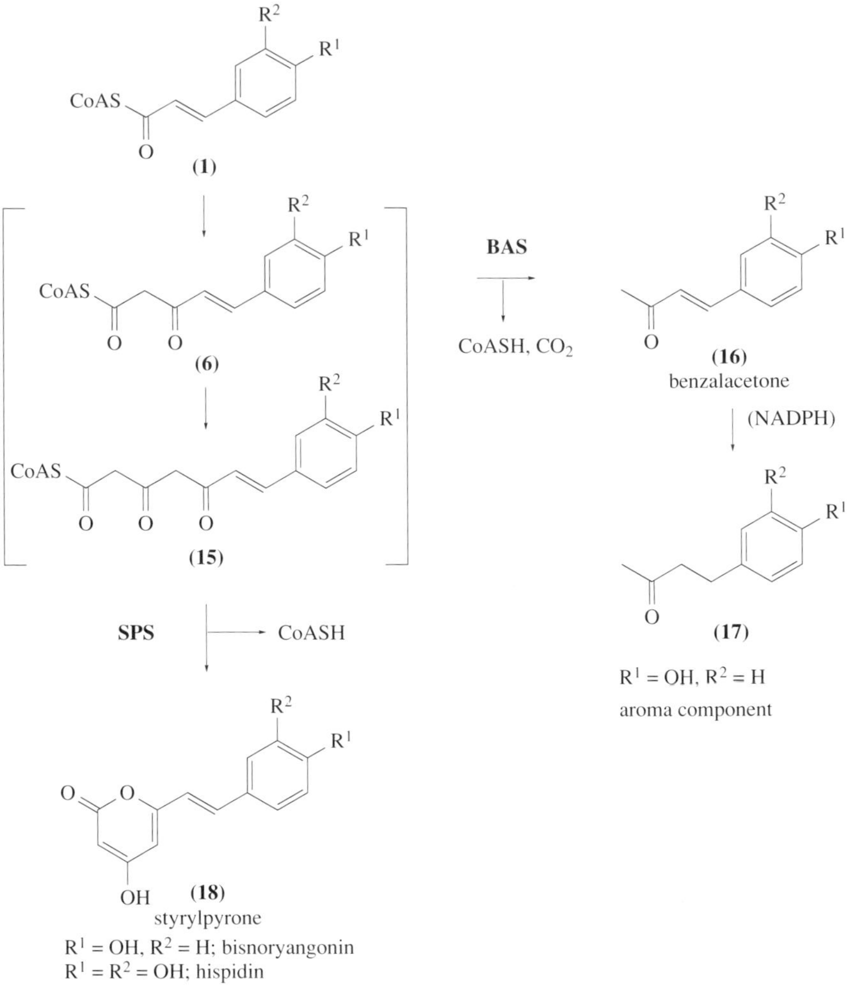

CoAS (1)

CoAS (6)

CoAS (15)

BAS

$CoASH, CO_2$

(16) benzalacetone

(NADPH)

(17)

$R^1 = OH, R^2 = H$

aroma component

SPS CoASH

(18) styrylpyrone

$R^1 = OH, R^2 = H$; bisnoryangonin

$R^1 = R^2 = OH$; hispidin

Scheme 4

appear high when compared with CHSs, STSs, or BAS, in particular for the starter units. It is an attractive speculation that the SPS expressed in the gametophytes represents an evolutionarily older form of the present-day CHSs.[136] In this context it should be repeated that styrylpyrones have been found as by-products of CHS reactions.[35–38]

1.27.5 EVOLUTION: HOW OLD ARE PROTEINS OF THE CHS/STS-TYPE?

The reactions of CHS and STS share many similarities with the condensing activities in the biosynthesis of fatty acids and other polyketides, and it was proposed that certain basic amino acids required for the stabilization of reaction intermediates are conserved in all of these enzymes.[66] It seems reasonable to assume that CHSs and STSs evolved from an ancestral condensing enzyme, like the other PKS. A long line of evolution separate from the other PKS is indicated by a very distant relationship, e.g., the use of CoA-esters rather than of ACP-derivatives, and the absence of easily detectable overall sequence similarities; even the environment of the active site Cys does not reveal more than a superficial similarity.[87] One of the attractive scenarios is that a change in the starter specificity, the capacity for more than one chain elongation (two: styrylpyrones, see Section 1.27.4.2.2; three: chalcones, stilbenes, and other products), and the concomitant development of structures for the stabilization of the unstable intermediates were steps in the evolution. The BAS and SPS are formally representatives of the steps to CHS/STS-type activities. However, the present-day enzymes are not necessarily ancient because it is not excluded that they represent "reprogrammed" enzymes. In this context it is interesting that proteins with significant similarities to

CHS/STS and possibly performing CHS-type reactions have been found in bacteria. Many *Pseudomonas fluorescens* strains synthesize 2,4-diacetylphloroglucinol, an antibiotic of considerable interest because of its ability to suppress root and seedling diseases of crop plants caused by soilborne pathogens.[137–142] The biosynthetic genes are in a 6.5 kbp gene cluster[143] (accession no. U41818). One of the open reading frames (*phlD*) encodes a protein of about 25% similarity with several CHSs, and it contains a cysteine in a position equivalent to the active site of the condensing reaction in CHS. The other open reading frames reveal no obvious similarities with other condensing enzymes or acyl carrier proteins. It is possible that the protein encoded in *phlD* synthesizes mono-acetylphloroglucinol (the first product of the pathway) from acetyl-CoA and three malonyl-CoA with a CHS-type ring folding of the tetraketide intermediate. If this could be demonstrated *in vitro*, it would establish the existence of a CHS-type reaction with three condensations already in bacteria, indicating that CHS/STS-related proteins and the reaction type may be much older than previously suspected (see below). Another interesting case is a small gene cluster in *Streptomyces griseus* (accession no. D45916) that has been shown to confer the production of a red-brown pigment.[144] The combination of the open reading frames 1 and 2 encodes proteins with 45% identity to the *Pseudomonas fluorescens* CHS-like protein and about 25% identity to several CHSs; these two open reading frames are sufficient for pigment production in the foreign host *E. coli*. The structure of the pigment was not identified, and therefore speculation on a CHS/STS-like reaction is not yet possible. An open reading frame (*bcsA*) with comparable similarities to CHSs is also present in *Bacillus subtilis* (accession no. L77246), but nothing seems to be known about the function.

It is thought that CHSs (chalcone as the product) first appeared in Charophyceae or in simple Bryophytes,[9,145–148] but there seems to be no direct information on these enzymes. Even less is known about STSs. A large variety of stilbenoids are known in liverworts,[13] but the enzymology of the proteins synthesizing the stilbene backbone has not been explored. Information on CHSs in those plants is similarly scant and does not extend beyond the demonstration of enzyme activity in crude extracts from the liverwort *Marchantia polymorpha*.[149] It has been proposed that the present-day STSs in higher plants evolved from CHSs several times independently (see 1.27.2.4.1). The lack of information on enzymes or sequences in plants thought to be more primitive actually does not exclude the possibility that stilbenes were at one time ubiquitous, that they played roles later take over by other molecules (e.g., hormones, see Gorham[13] for a review), and that the STSs were progressively lost in most species.

1.27.6 MODIFICATION OF REACTION INTERMEDIATES

1.27.6.1 CHS: Polyketide Reductase (PKR)

The reactions of CHS and STS do not involve a modification of the enzyme-bound intermediates. However, chalcone derivatives reduced at a specific position in the *de novo* synthesized aromatic ring are widespread in nature,[150] and precursor studies indicated that the reduction has to occur prior to the formation of the chalcone, most likely at the level of a polyketide intermediate.[151] The puzzle was solved after the demonstration that a reduced chalcone (6′-deoxychalcone) (**20**) could be synthesized in crude extracts from a phenylpropanoid starter (**1**) and malonyl-CoA in the presence of NADPH,[152,153] if the assay conditions are adjusted to pH 6. Further analysis showed that the reduction was performed by an NADPH-dependent, monomeric reductase (35 kDa) that interacted with high affinity with CHS,[154] reducing a specific keto group of the tri- or tetraketide intermediate (**19**). The hydroxy group presumably is lost by removal of water during the aromatizing ring closure to the 6′-deoxychalcone (**20**) (Scheme 5). The enzyme was named PKR (polyketide reductase),[8] and the use of the name chalcone reductase should be discouraged because it is misleading.

The PKR from soybean was purified and characterized;[155] the first cDNAs were obtained from the same plant,[156] and the protein was functionally overexpressed in *E. coli*.[157] Experiments with CHS and the PKR individually expressed in *E. coli* showed that no other plant factor is necessary for the reaction.[20] Later on, closely related sequences were described from *Medicago sativa*,[158] *Sesbania rostrata*,[159] *Pueraria lobata*,[160] and *Glycyrrhiza echinata*.[161] In all cases the proteins are encoded in small gene families, and the PKRs from the different species are very similar. Interestingly, they show no significant similarity with enzymes involved in the reduction of intermediates in the biosynthesis of fatty acids or other polyketides, but rather with various aldo/keto-reductases, mostly in carbohydrate metabolism (30–39% identity).[156] The PKR was apparently acquired from other pathways during evolution of the flavonoid biosynthesis.

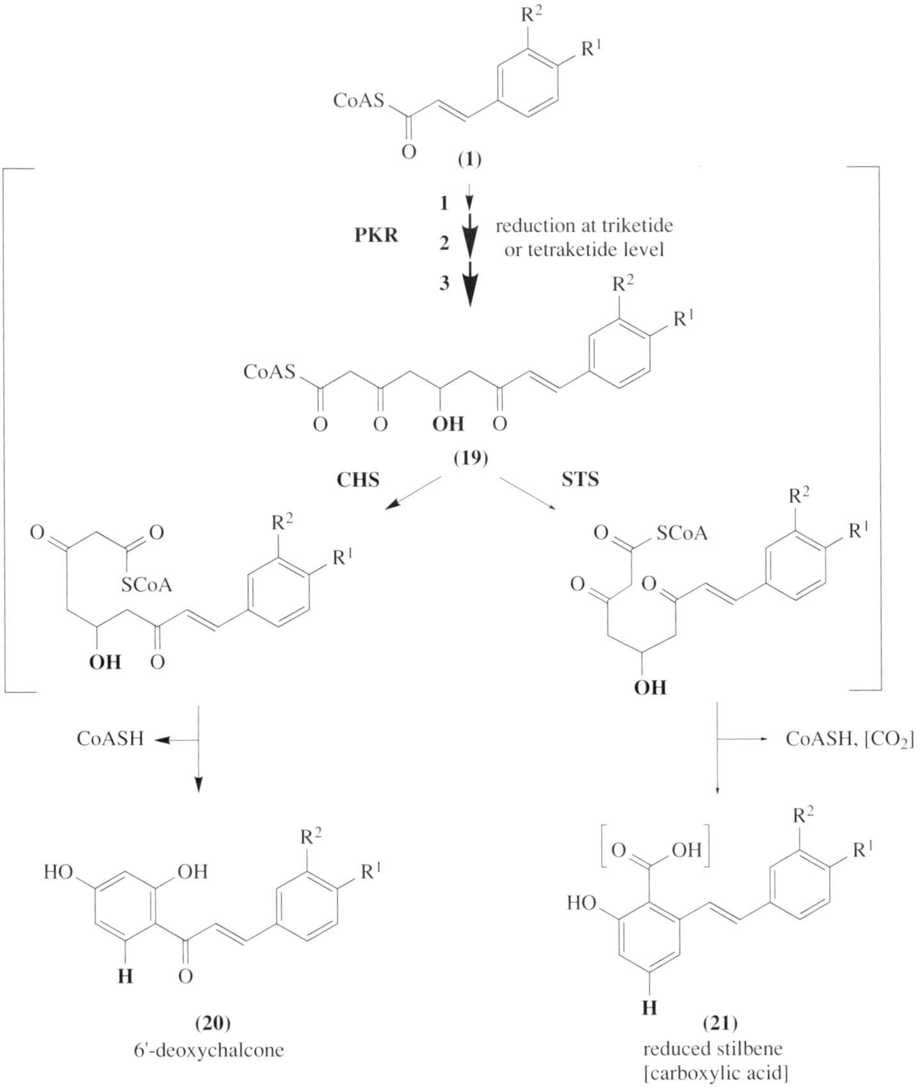

CoAS (1) PKR 1 2 3 reduction at triketide or tetraketide level (19) CHS STS CoASH CoASH, $[CO_2]$

(20)
6'-deoxychalcone

(21)
reduced stilbene
[carboxylic acid]

Scheme 5

Although the basics of the reaction are understood, there are several aspects that remain to be explored. It is not known how the monomeric PKR interacts with the dimeric CHS (two active sites for the condensing reactions), and whether the reduction step occurs at the tri- or the tetraketide intermediate level. It is also remarkable that the reaction *in vitro* always leads to both the reduced **(20)** and the nonreduced product **(4)** (6′-deoxychalcone and chalcone, respectively), and at most a 1 : 1 ratio was achieved even with a large excess of the PKR.[155] One of the possible explanations could have been that CHS preparations from plant cells contain several isoenzymes (soybean, for example, contains at least seven CHS genes),[162–164] and that only some of them interact with the PKR. However, there is no evidence for specificity on the side of the CHSs. The PKR from soybean interacted with CHSs from all other plants tested,[155] regardless of whether these synthesize 6′-deoxychalcone derivatives *in vivo* or not. A mixture of 6′-deoxychalcone and chalcone was also obtained with both soybean PKR and *Sinapis alba* CHS expressed as single enzymes in *E. coli*.[20] The PKR is the key enzyme directing the metabolite flow into the biosynthesis of isoflavonoids, and it is difficult to imagine that the channeling does not discriminate between the need for 6′-deoxychalcone or chalcone. It seems therefore likely that the *in vitro* conditions do not reflect all aspects of the *in vivo* situation. One of the likely explanations is that the enzymes of the pathway are associated to a complex *in vivo* and thus channel the metabolite flow in a highly organized manner.[46–48]

1.27.6.2 STS: A Proposal for the Biosynthesis of Reduced Stilbenes

In view of the close relationship between CHSs and STSs, it is an intriguing question whether a comparable PKR exists in the biosynthesis of stilbenes. Various plants indeed contain stilbenes and derivatives that lack one or even both of the hydroxy groups that originate from the aromatization of the *de novo* synthesized ring system, and liverworts are a particularly rich source.[13] The loss of the hydroxy groups could be explained by reductase activities co-acting with the STSs like the PKR with CHSs, and Scheme 5 outlines a scheme for an example with the reduction at the same keto group of an intermediate as in the known CHS/PKR interaction (**19**), but followed by an STS-type ring closure (**21**). Interestingly, several of these reduced stilbenes (e.g., hydrangeic acid in *Hydrangea macrophylla*) retain the carboxy group that is removed during the biosynthesis of the standard stilbene backbone, and both carboxylated (e.g., lunularic acid) and decarboxylated forms (e.g., lunularin) have been described in liverworts. However, neither the biosynthesis of stilbenecarboxylic acids nor the formation of reduced stilbenes has been demonstrated *in vitro*, although many of these substances have interesting biological activities.[13]

1.27.7 PERSPECTIVES

The advancement of molecular biology techniques in plant science brought a large amount of information on CHS/STS genes and proteins. It is expected that a combination of these techniques with heterologous expression and enzymology will make important contributions to the identification and understanding of the precise functions of proteins that have been cloned already and will be cloned in the future. These will be of particular importance in view of the emerging evidence that sequences with a high degree of similarity are not necessarily CHS or STS, but might encode proteins with functions in quite different pathways.

One of the most intriguing developments is indeed the finding that CHS and STS are probably only the well-investigated examples of a superfamily of related proteins with widely varying substrate specificities and programming for one, two, or three condensation reactions. Although the ACS represents the only thoroughly documented case, there are good chances that the other examples described here will confirm the concept.

The concept also allows testable predictions as to the properties of key enzymes involved in the biosynthesis of other natural products where feeding experiments indicate that condensing reactions are or may be involved. One of the prime examples is probably the biosynthesis of the aucuparin backbone (a biphenyl) because it is readily explained by an STS-type reaction with a benzoyl-CoA derivative, i.e., three condensations followed by STS-type ring closure, including removal of the terminal carboxy group (reviewed by Sultanbawa).[102] A substance retaining the carboxy group has also been described (hermonionic acid), suggesting a similar situation as with the stilbenecarboxylic acids. Benzophenone derivatives (xanthones) co-occur with aucuparins in some plants, suggesting that BPS (CHS-type, Scheme 3) and the postulated aucuparin synthase (STS-type) use the same or very similar starter substrates.

The BAS (see Section 1.27.4.2.1) is of particularly interest in the context of the proposed superfamily, because it is a CHS/STS-type activity that performs only one condensation reaction (Scheme 4). Feeding studies suggest that the biosynthesis of a fairly large number of natural products could involve a diketide intermediate (**6**) as a first reaction product that is further processed in various ways. A few examples are discussed below.

An interesting group of substances are the psilotins[165–167] found in the Psilotaceae which are considered to be closely related to ferns. They are synthesized from phenylpropanoid precursors, and it was proposed[168] that the biosynthesis of the aglycone involves the condensation of a starter CoA-ester with one C_2 unit from malonyl-CoA, followed by a reduction step, water removal, and ring closure to an arylpyrone (**22**) (Scheme 6). Compounds probably synthesized via this route are also known in higher plants.[169] It is an attractive hypothesis that the postulated arylpyrone synthase is a CHS/STS-type enzyme closely related to BAS.

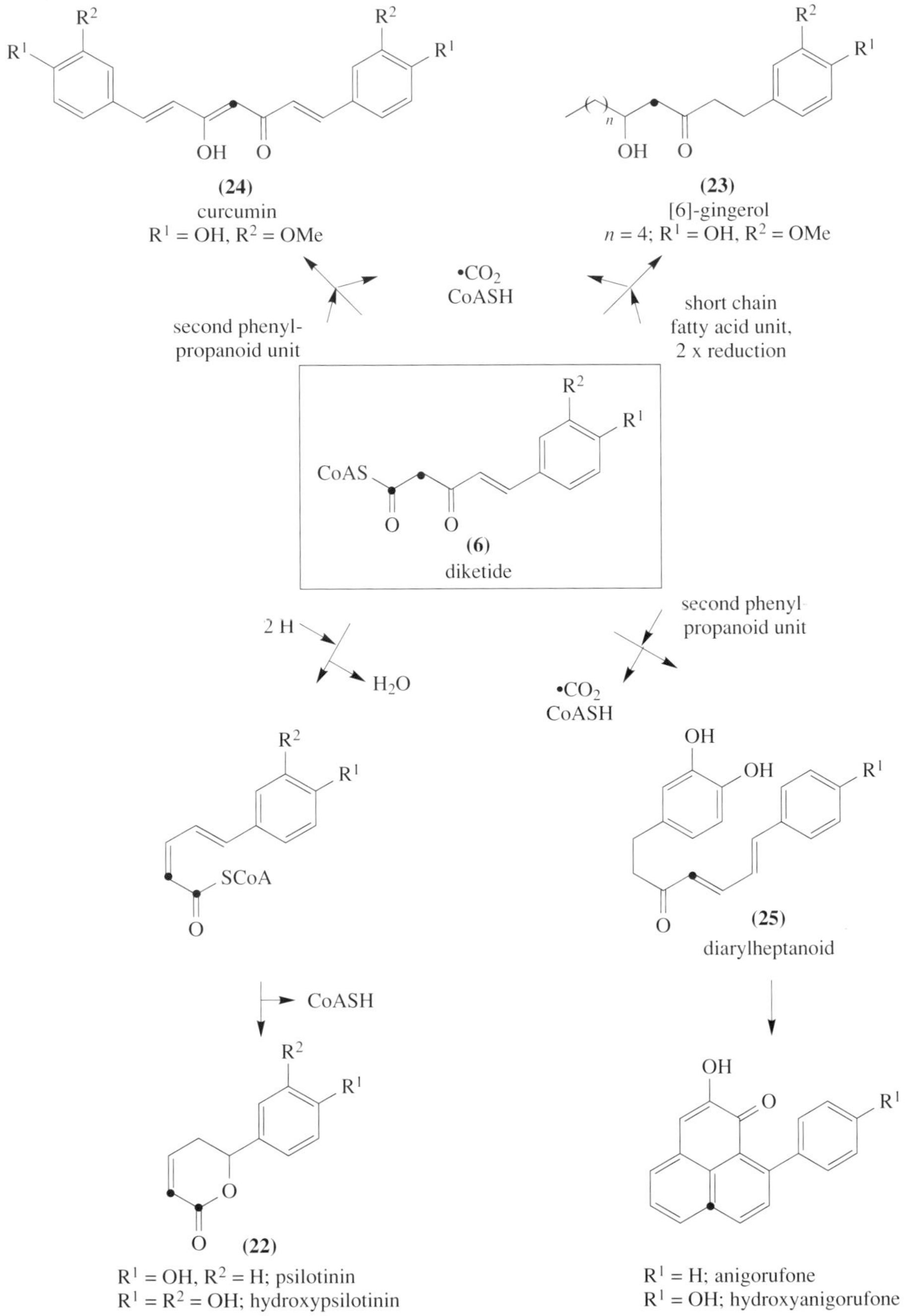

(24)
curcumin
R^1 = OH, R^2 = OMe

(23)
[6]-gingerol
n = 4; R^1 = OH, R^2 = OMe

second phenyl-propanoid unit

•CO_2
CoASH

short chain fatty acid unit, 2 x reduction

CoAS
(6)
diketide

2 H
H_2O

second phenyl propanoid unit
•CO_2
CoASH

SCoA

(25)
diarylheptanoid

CoASH

(22)
R^1 = OH, R^2 = H; psilotinin
R^1 = R^2 = OH; hydroxypsilotinin

R^1 = H; anigorufone
R^1 = OH; hydroxyanigorufone

Scheme 6

The biosynthesis of a large number of natural products could involve a diketide (**22**) from which the terminal carboxy group is removed. Feeding studies with [6]-gingerol (**23**), the pungent principle of ginger (rhizomes of *Zingiber officinale*), indicate that the biosynthesis uses ferulic acid, malonate, and hexanoate, and that it includes two reduction steps.[170,171] The scheme in Scheme 6 proposes a phenylpropanoid diketide intermediate; this seems more likely than the corresponding diketide intermediate from hexanoic acid, since the short-chain fatty acid is variable in gingerols, but not the phenylpropanoid moiety.[171] Replacing the fatty acid residue by a phenylpropanoid unit and omission of the reductions leads to curcumin (**24**) (Scheme 6), the main pigment of turmeric (*Curcuma longa*). Similar C_6-C_7-C_6 structures (diarylheptanoids) (**25**) and the biosynthetically related phenylphenalenones are known from several plants,[172–178] including the genus *Anigozanthos*[179] (e.g., anigorufone and hydroxyanigorufone (**26**)) (Scheme 6). Experiments with root cultures of *Anigozanthos preissii* (Kangaroo paw) showed that the phenylphenalenones are synthesized from two phenylpropanoid precursors via a diarylheptanoid intermediate,[180] and that the central one-carbon unit (marked with a dot in Scheme 6) originates from acetate.[181] These data provided the first experimental evidence for an earlier hypothesis,[182,183] and they are consistent with the possibility that a diketide intermediate (**6**) is involved in the biosynthesis.

The enzymology of these reactions and pathways has hardly been explored. The concept of the CHS/STS-type superfamily of related proteins might provide interesting clues for approaching these questions, and that includes direct strategies on the molecular level. Related enzymes of this type may be involved in more reactions than previously suspected, and the combination of biochemical and molecular techniques has the potential to contribute significantly to understanding the biosynthesis of many biologically interesting natural products. In several of the cases discussed above one has to postulate that additional proteins are involved in the formation of the end-products, and it will be interesting to see whether the enzymes interact in a similarly close manner as demonstrated for the CHS/PKR co-action.

1.27.8 APPENDIX: SEQUENCE ACCESSION NUMBERS

In brackets: D = gene, i = size of intron, R = RNA, p = partial sequence.

1.27.8.1 CHS

Antirrhinum majus: X03710 (D, $i = 109$); *Arabidopsis thaliana*: M20308 (D, $i = 86$); S80554 (mutant, D); *Betula pendula*: X77513 (D, p); *Brassica napus*: (A1, A2, B1, B2) (R);[184] X70976 (R, p); X70977 (R,p), and [185] (R, p); *Callistephus chinensis*: Z67988 (R); *Camellia sinensis*: D26593 (R); D26594 (R); D26595 (R); *Dahlia* sp.: X91341 (D, p); X91342 (D, p); *Daucus carota*: D16256 (CHS9; R); D16255 (CHS2; D); AJ006780 (CHS17; R); AJ006779 (CHS84; R); *Dianthus caryophyllus*: Z67982 (R); *Fragaria* x *ananassa*: U19942 (R, p); *Gerbera hybrida*: Z38096 (CHS1; R); Z38097 (CHS2; R); Z38098 (CHS3, R); X91340 (D, $i = 485$); *Gerbera* sp.: X91339 (D, p, $i = 1638$); *Glycine* max: S46989 (R, p); L07647 (CHS4; D, $i = 121$); L03352 (CHS6, D, $i = 617$); M98871 (CHS7A; D, $i = 445$); X53958 (CHS3; D, $i = 122$); X54644 (CHS1; D, $i = 122$); X65636 (CHS2; D, $i = 137$); X52097 (D, $i = 121$); *Hordeum vulgare*: X58339 (CHS1; D, $i = 1649$); Y09233 (CHS2; R); U43494 (R, p); *Ipomoea cordatotriloba*: U15941 (D, p, $i = 119$); U15942 (D, p, $i = 103$); *Ipomoea nil*: U15943 (D, p, $i = 105$); U15944 (D, p, $i = 89$); *Ipomoea platensis*: U15945 (R, p); *Ipomoea purpurea*: U15946 (D, $i = 82$); U15947 (D, $i = 112$); U15948 (D, p, $i = 90$); U15949 (D, p, $i = 117$); *Ipomoea trifida*: U15950 (D, p, $i = 118$); U15951 (D, p, $i = 125$); *Ipomoea triloba*: U15952 (D, p, $i = 84$); U15953 (D, p, $i = 84$); *Juglans* sp. (*J. nigra* x *J. regia*): X94706 (CHS2; R); X94995 (CHS1; R); *Leibnitzia*: X91343 (D, p); *Lycopersicon esculentum*: X55194 (CHS1; R); X55195 (CHS2; R); *Magnolia liliflora*:[67] (R); *Malus* sp.: X68977 (R, p); *Matthiola incana*: X17577 (R); *Medicago sativa*: L02901 (R); L02902 (R); L02903 (R); L02904 (R); L02905 (R); U01021 (R); U01018 (R); U01019 (R); U01020 (R); X68106 (R); X68107 (R); *Onoseris sagittatis* X91344 (D, p); *Oryza sativa*: X89859 (R); X91811 (R); D29697 (R, p); *Petroselinum crispum*: V01538 (R); M35516 (D, p); *Petunia hybrida*: X04080 (V30, R); S80857 (CHSA; R); X14599 (CHSJ; R); X14591 CHSA; D, $i = 1347$); X14592 (CHSB; D, $i = 3777$); X14593 (CHSD; D, $i = 685$); X14594 (CHSF; D, $i = 564$); X14595 (CHSG; D, $i = 2439$); X14596 (CHSH; D, p); X14597 (CHSJ; D, $i = 729$); X14598 (CHSL; D, p); *Phaseolus vulgaris*: X06411 (R); K02953 (R; p); *Pinus strobus*: AJ004800 (CHS1; R); AJ002156 (CHS2; R); *Pinus sylvestris*: X60754 (D, $i = 109$); *Pisum sativum*: D10661 (CHS1; D, $i = 88$); D10662 (CHS2; D, $i = 88$); X80007 (CHSAB; D, $i = 110$); X63333 (CHS1; R); X63334 (CHS2; R); X63335 (CHS3; R); *Pueraria lobata*: D10223 (R); D63855 (D, p); *Ranunculus acer*:[67] (R); *Rosa hybrida*:[186] (R); *Secale cereale*: X92548 (D, $i = 94$); X92547 (R); *Sinapis alba*: X16437 (CHSG; D, $i = 523$); X14314 (CHS3; R); X14315 (CHS1; R); *Solanum tuberosum*: U47738 (R); U47739 (R); U47740 (R); *Sorghum bicolor*: U51569 (D, p); *Taraxacum* sp.: X91345 (D, p); *Trifolium subterraneum*: L24515 (CHS3; D, $i = 116$); L24516 (CHS4; D, $i = 140$); L24517 (CHS5; D, $i = 109$); M91193 (CHS1; D, $i = 106$); M91194 (CHS2; D, $i = 182$); M91195 (CHS6; D, $i = 98$); *Vigna unguiculata*: X74821 (R); *Vitis vinifera*: X75969 (R); *Zea mays*: X60205 (C2, D, $i = 1524$); X60204 (WHP, D, $i = 2157$).

1.27.8.2 STS

Arachis hypogaea: X62300 (STS2, R, p); L00952–L00954 (STS3, D, $i = 355$); X62298 (STS1A, D, p, $i = 369$); X62299 (STS1B, D, p); A00769–70 (D); *Phalaenopsis* sp.: X79903 (R); X79904 (R); *Pinus sylvestris*: X60753 (D, $i = 567$); A24145 (R, p); S50350 (R); *Pinus strobus*: Z46914 (STS1, R); Z46915 (STS2, R); *Vitis vinifera*: S63221 (R); S63225 (R); S63227 (R); X76892 (R).

1.27.8.3 CHS-type (ACS)

Ruta graveolens acridone synthase: Z34088 (R).

ACKNOWLEDGMENTS

I would like to thank all of the colleagues who provided unpublished data; and I am grateful to U. Sankawa, B. Schneider, and M. Veit for supplying literature on the occurrence and biosynthesis of natural products.

1.27.9 REFERENCES

1. R. A. Dixon, C. J. Lamb, S. Masoud, V. J. H. Sewalt, and N. L. Paiva, *Gene*, 1996, **179**, 61.
2. B. W. Shirley, *Trends Plant Sci.*, 1996, **1**, 377.
3. R. A. Dixon, M. J. Harrison, and N. L. Paiva, *Physiol. Plant.*, 1995, **93**, 385.
4. R. A. Dixon and N. L. Paiva, *Plant Cell*, 1995, **7**, 1085.
5. T. A. Holton and E. C. Cornish, *Plant Cell*, 1995, **7**, 1071.
6. J.-P. Biolley, M. Jay, and G. Forkmann, *Phytochemistry*, 1994, **36**, 1189.
7. G. Forkmann, in "The Flavonoids. Advances in Research since 1986," ed. J. B. Harborne, Chapman & Hall, London, 1994, p. 537.
8. W. Heller and G. Forkmann, in "The Flavonoids. Advances in Research since 1986," ed. J. B. Harborne, Chapman & Hall, London, 1994, p. 499.
9. R. E. Koes, F. Quattrocchio, and J. N. M. Mol, *BioEssays*, 1994, **16**, 123.
10. J. L. Ingham, *Z. Naturforsch.*, 1990, **45c**, 829.
11. R. Hain, B. Bieseler, H. Kindl, G. Schröder, and R. Stöcker, *Plant Mol. Biol.*, 1990, **15**, 325.
12. R. Hain, H.-J. Reif, E. Krause, R. Langebartels, H. Kindl, B. Vornam, W. Wiese, E. Schmelzer, P. H. Schreier, R. Stöcker, and K. Stenzel, *Nature (London)*, 1993, **361**, 153.
13. J. Gorham, "The Biochemistry of the Stilbenoids," Chapman & Hall, London, 1995.
14. D. Hopwood and D. H. Sherman, *Annu. Rev. Genet.*, 1990, **24**, 37.
15. L. Katz and S. Donadio, *Annu. Rev. Microbiol.*, 1993, **47**, 875.
16. C. R. Hutchinson and I. Fujii, *Annu. Rev. Microbiol.*, 1995, **49**, 201.
17. J. Ohlrogge and J. Browse, *Plant Cell*, 1995, **7**, 957.
18. R. McDaniel, S. Ebert-Khosla, D. Hopwood, and C. Khosla, *Nature (London)*, 1995, **375**, 549.
19. S. Tropf, T. Lanz, S. A. Rensing, J. Schröder, and G. Schröder, *J. Mol. Evol.*, 1994, **38**, 610.
20. S. Tropf, B. Kärcher, G. Schröder, and J. Schröder, *J. Biol. Chem.*, 1995, **270**, 7922.
21. J. Schröder, unpublished results.
22. U. Sankawa, personal communication.
23. J. Schröder, W. Heller, and K. Hahlbrock, *Plant Sci. Lett.*, 1979, **14**, 281.
24. K. W. M. Zuurbier, S.-Y. Fung, J. J. C. Scheffer, and R. Verpoorte, *Phytochemistry*, 1993, **34**, 1225.
25. F. Kreuzaler and K. Hahlbrock, *FEBS Lett.*, 1972, **28**, 69.
26. C. R. Martin, in "International Review of Cytology: A Survey of Cell Biology," eds. K. Jeon and J. Jarvik, Academic Press, New York, 1993, vol. 147, p. 233.
27. J. Schröder and E. Schäfer, *Arch. Biochem. Biophys.*, 1980, **203**, 800.
28. W. Heller and K. Hahlbrock, *Arch. Biochem. Biophys.*, 1980, **200**, 617.
29. R. Sütfeld and R. Wiermann, *Arch. Biochem. Biophys.*, 1980, **201**, 64.
30. J. N. M. Mol, M. P. Robbins, R. A. Dixon, and E. Veltkamp, *Phytochemistry*, 1985, **24**, 2267.
31. A. B. Christensen, P. L. Gregersen, J. Schröder, and D. B. Collinge, *Plant. Mol. Biol.*, 1998, **37**, 849.
32. F. Kreuzaler, R. J. Light, and K. Hahlbrock, *FEBS Lett.*, 1978, **94**, 175.
33. F. Kreuzaler and K. Hahlbrock, *Eur. J. Biochem.*, 1975, **56**, 205.
34. F. Kreuzaler, H. Ragg, W. Heller, R. Tesch, I. Witt, D. Hammer, and K. Hahlbrock, *Eur. J. Biochem.*, 1979, **99**, 89.
35. R. Schüz, W. Heller, and K. Hahlbrock, J. Biol. Chem., 1983, **258**, 6730.
36. F. Kreuzaler and K. Hahlbrock, *Arch. Biochem. Biophys.*, 1975, **169**, 84.
37. G. Hrazdina, F. Kreuzaler, K. Hahlbrock, and H. Grisebach, *Arch. Biochem. Biophys.*, 1976, **175**, 392.
38. N. A. M. Saleh, H. Fritsch, F. Kreuzaler, and H. Grisebach, *Phytochemistry*, 1978, **17**, 183.
39. L. Britsch and H. Grisebach, *Phytochemistry*, 1985, **24**, 1975.
40. N. Rupprich and H. Kindl, *Hoppe-Seyler's Z. Physiol. Chem.*, 1978, **359**, 165.
41. A. Schöppner and H. Kindl, *FEBS Lett.*, 1979, **108**, 349.
42. C.-H. Rolfs, K.-H. Fritzemeier, and H. Kindl, *Plant Cell Rep.*, 1981, **1**, 83.

43. K.-H. Fritzemeier, C.-H. Rolfs, J. Pfau, and H. Kindl, *Planta*, 1983, **159**, 25.
44. C.-H. Rolfs, H. Schön, M. Steffens, and H. Kindl, *Planta*, 1987, **172**, 238.
45. K.-H. Fritzemeier and H. Kindl, *Planta*, 1981, **151**, 48.
46. G. Hrazdina, A. M. Zobel, and H. C. Hoch, *Proc. Natl. Acad. Sci. USA*, 1987, **84**, 8966.
47. G. Hrazdina and G. Wagner, *Arch. Biochem. Biophys.*, 1985, **237**, 88.
48. G. Hrazdina and R. A. Jensen, *Annu. Rev. Plant Physiol. Plant Mol. Biol.*, 1992, **43**, 241.
49. H. Kindl, in "Biosynthesis and Biodegradation of Wood Components," ed. T. Higuchi, Academic Press, New York, 1985, p. 349.
50. N. Rupprich, H. Hildebrand, and H. Kindl, *Arch. Biochem. Biophys.*, 1980, **200**, 72.
51. C.-H. Rolfs and H. Kindl, *Plant Physiol.*, 1984, **75**, 489.
52. A. Schöppner and H. Kindl, *J. Biol. Chem.*, 1984, **259**, 6806.
53. R. Gehlert, A. Schöppner, and H. Kindl, *Mol. Plant-Microbe Interact.*, 1990, **3**, 444.
54. B. Vornam, H. Schön, and H. Kindl, *Plant Mol. Biol.*, 1988, **10**, 235.
55. F. M. Liswidowati, F. Hohmann, B. Schwer, and H. Kindl, *Planta*, 1991, **183**, 307.
56. K.-H. Fritzemeier, H. Kindl, and E. Schlösser, *Z. Naturforsch.*, 1984, **39c**, 217.
57. R. Gehlert and H. Kindl, *Phytochemistry*, 1991, **30**, 457.
58. T. Reinecke and H. Kindl, *Phytochemistry*, 1994, **35**, 63.
59. T. Reinecke and H. Kindl, *Mol. Plant-Microbe Interact.*, 1994, **7**, 449.
60. R. Preisig-Müller, P. Gnau, and H. Kindl, *Arch. Biochem. Biophys.*, 1995, **317**, 201.
61. U. Reimold, M. Kroeger, F. Kreuzaler, and K. Hahlbrock, *EMBO J.*, 1983, **2**, 1801.
62. G. Schröder, J. W. S. Brown, and J. Schröder, *Eur. J. Biochem.*, 1988, **172**, 161.
63. D. G. Higgins, A. J. Bleasby, and R. Fuchs, *Cabios*, 1992, **8**, 189.
64. Y. Van de Peer and R. De Wachter, *Comput. Appl. Biosci.*, 1993, **9**, 177.
65. N. Saitou and M. Nei, *Mol. Biol. Evol.*, 1987, **4**, 406.
66. M. Siggaard-Andersen, *Protein Sequences Data Anal.*, 1993, **5**, 325.
67. U. Niesbach-Klösgen, E. Barzen, J. Bernhardt, W. Rohde, Z. Schwarz-Sommer, H. J. Reif, U. Wienand, and H. Saedler, *J. Mol. Evol.*, 1987, **26**, 213.
68. J. Schröder and G. Schröder, *Z. Naturforsch.*, 1990, **45c**, 1.
69. A. Herrmann, W. Schulz, and K. Hahlbrock, *Mol. Gen. Genet.*, 1988, **212**, 93.
70. H. Sommer and H. Saedler, *Mol. Gen. Genet.*, 1986, **202**, 429.
71. R. L. Feinbaum and F. M. Ausubel, *Mol. Cell. Biol.*, 1988, **8**, 1985.
72. I. E. Burbulis, M. Iacobucci, and B. W. Shirley, *Plant Cell*, 1996, **8**, 1013.
73. B. W. Shirley, W. L. Kubasek, G. Storz, E. Bruggemann, M. Koorneef, F. M. Ausubel, and H. M. Goodman, *Plant J.*, 1995, **8**, 659.
74. H. J. Reif, U. Niesbach, B. Deumling, and H. Saedler, *Mol. Gen. Genet.*, 1985, **199**, 208.
75. R. E. Koes, C. E. Spelt, J. N. M. Mol, and A. G. M. Gerats, *Plant Mol. Biol.*, 1987, **10**, 159.
76. R. E. Koes, C. E. Spelt, P. J. M. van den Elzen, and J. N. M. Mol, *Gene*, 1989, **81**, 245.
77. Y. Helariutta, P. Elomaa, M. Kotilainen, R. J. Griesbach, J. Schröder, and T. H. Teeri, *Plant Mol. Biol.*, 1995, **28**, 47.
78. J. Schröder, S. Raiber, T. Berger, A. Schmidt, J. Schmidt, A. M. Soares-Sello, E. Bardshiri, D. Strack, T. J. Simpson, M. Veit, and G. Schröder, *Biochemistry*, 1998, **37**, 8417.
79. M. L. Durbin, G. H. Learn, G. A. Huttley, and M. T. Clegg, *Proc. Natl. Acad. Sci. USA*, 1995, **92**, 3338.
80. I. E. Somssich, P. Wernert, S. Kiedrowski, and K. Hahlbrock, *Proc. Natl. Acad. Sci. USA*, 1996, **93**, 14 199.
81. R. G. Powell, M. R. TePaske, R. D. Plattner, J. F. White, and S. L. Clement, *Phytochemistry*, 1994, **35**, 335.
82. A. C. Casabuono and A. B. Pomilio, *Phytochemistry*, 1994, **35**, 479.
83. R. J. Grayer and J. B. Harborne, *Phytochemistry*, 1994, **37**, 19.
84. D. Hölscher and B. Schneider, *Phytochemistry*, 1996, **43**, 471.
85. F. Melchior and H. Kindl, *Arch. Biochem. Biophys.*, 1991, **288**, 552.
86. F. Sparvoli, C. Martin, A. Scienza, G. Gavazzi, and C. Tonelli, *Plant Mol. Biol.*, 1994, **24**, 743.
87. T. Lanz, S. Tropf, F.-J. Marner, J. Schröder, and G. Schröder, *J. Biol. Chem.*, 1991, **266**, 9971.
88. T. Simpson, personal communication.
89. G. Schröder and J. Schröder, *J. Biol. Chem.*, 1992, **267**, 20 558.
90. J. Schröder, unpublished results.
91. D. Luo, E. S. Coen, S. Doyle, and R. Carpenter, *Plant J.*, 1991, **1**, 59.
92. S. Raiber, G. Schröder, and J. Schröder, *FEBS Lett.*, 1995, **361**, 299.
93. A. Baumert, A. Porzel, J. Schmidt, and D. Gröger, *Z. Naturforsch.*, 1992, **47c**, 365.
94. W. Maier, A. Baumert, B. Schumann, H. Furukawa, and D. Gröger, *Phytochemistry*, 1993, **32**, 691.
95. A. Baumert, W. Maier, D. Gröger, and R. Deutzmann, *Z. Naturforsch.*, 1994, **49c**, 26.
96. K. T. Junghanns, R. E. Kneusel, A. Baumert, W. Maier, D. Gröger, and U. Matern, *Plant Mol. Biol.*, 1995, **27**, 681.
97. G. J. Bennett and H. H. Lee, *Phytochemistry*, 1989, **28**, 967.
98. W. Jinsart, D. Buddhasukh, and G. M. Polya, *Phytochemistry*, 1992, **31**, 3711.
99. C. N. Lin, S. S. Liou, F. N. Ko, and C. M. Teng, *J. Pharm. Sci.*, 1993, **82**, 11.
100. H. Minami, M. Kinoshita, Y. Fukuyama, M. Kodama, T. Yoshizawa, M. Sugiura, K. Nakagawa, and H. Tago, *Phytochemistry*, 1994, **36**, 501.
101. L. Rocha, A. Marston, M. A. C. Kaplan, H. Stoeckli-Evans, U. Thull, B. Testa, and K. Hostettmann, *Phytochemistry*, 1994, **36**, 1381.
102. M. U. S. Sultanbawa, *Tetrahedron*, 1980, **36**, 1465.
103. K. Hostettmann and M. Hostettmann, in "Methods in Plant Biochemistry, Vol. 1, Plant Phenolics," ed. J. B. Harborne, Academic Press, London, 1989, p. 493.
104. L. Beerhues and U. Berger, *Phytochemistry*, 1994, **35**, 1227.
105. L. Beerhues and U. Berger, *Planta*, 1995, **197**, 608.
106. L. Beerhues, *FEBS Lett.*, 1996, **383**, 264.
107. J. Schröder, unpublished results.
108. M. Verzele, *J. Inst. Brew. London*, 1986, **92**, 32.

109. F. Drawert and J. Beier, *Phytochemistry*, 1976, **15**, 1695.
110. S.-Y. Fung, J. Brussee, R. A. M. Van der Hoeven, W. M. A. Niessen, J. J. C. Scheffer, and R. Verpoorte, *J. Nat. Prod.*, 1994, **57**, 452.
111. K. W. M. Zuurbier, S.-Y. Fung, J. J. C. Scheffer, and R. Verpoorte, *Phytochemistry*, 1995, **38**, 77.
112. K. W. M. Zuurbier, J. Leser, T. Berger, A. J. P. Hofte, G. Schröder, R. Verpoorte, and J. Schröder, *Phytochemistry*, 1998, in press.
113. M. Larsen, L. Poll, O. Callesen, and M. Lewis, *Acta Agric. Scand.*, 1991, **41**, 447.
114. A. Pabst, D. Barron, J. Adda, and P. Schreier, *Phytochemistry*, 1990, **29**, 3853.
115. L. Bauer, A. J. Birch, and A. J. Ryan, *Aust. J. Chem.*, 1955, **8**, 534.
116. R. Higuchi and D. M. X. Donnelly, *Phytochemistry*, 1977, **16**, 1587.
117. T. Murakami and K. Tanaka, *Tetrahedron Lett.*, 1972, **29**, 2965.
118. T. Hirvi, E. Honkanen, and T. Pyysalo, *Z. Lebensm. Unters. Forsch.*, 1981, **172**, 365.
119. T. Hirvi and E. Honkanen, *Z. Lebensm. Unters. Forsch.*, 1984, **179**, 387.
120. Y. L. Lin and C. J. Chou, *Chem. Abstr.*, 1985, **102**, 92 951.
121. W. Borejsza-Wysocki and G. Hrazdina, *Phytochemistry*, 1994, **35**, 623.
122. W. Borejsza-Wysocki and G. Hrazdina, *Plant Physiol.*, 1996, **110**, 791.
123. J.-L. Fiasson, *Biochem. Syst. Ecol.*, 1982, **10**, 289.
124. M. Gill and W. Steglich, *Prog. Chem. Org. Nat. Prod.*, 1987, **51**, 86.
125. G. H. N. Towers, C. P. Vance, and A. M. D. Nambudiri, *Recent Adv. Phytochem.*, 1974, **8**, 81.
126. M. Veit, H. Geiger, V. Wray, A. Abou-Mandour, W. Rozdzinski, L. Witte, D. Strack, and F.-C. Czygan, *Phytochemistry*, 1993, **32**, 1029.
127. M. Veit, C. Beckert, C. Höhne, K. Bauer, and H. Geiger, *Phytochemistry*, 1995, **38**, 881.
128. C.-B. Cui, Y. Tezuka, H. Yamashita, T. Kikuchi, H. Nakano, T. Tamaoki, and J.-H. Park, *Chem. Pharm. Bull.* (*Tokyo*), 1992, **40**, 1711.
129. C. M. A. M. Rezende, M. V. Von Bülow, O. R. Gottlieb, S. L. V. Pinho, and A. Da Rocha, *Phytochemistry*, 1971, **10**, 3167.
130. R. Hänsel, A. Pelter, J. Schulz, and C. Hille, *Chem. Ber.*, 1976, **109**, 1617.
131. R. M. Smith, *Phytochemistry*, 1983, **22**, 1055.
132. P. W. Perrin and G. H. N. Towers, *Phytochemistry*, 1973, **12**, 589.
133. G. M. Hatfield and L. R. Brady, *Lloydia*, 1973, **36**, 59.
134. C.-K. Wat and G. H. N. Towers, in "Biochemistry of Plant Phenolics", eds. T. Swain, J. B. Harborne, and C. F. Van Sumere, Plenum, New York, 1979, p. 371.
135. M. Veit, H. Geiger, B. Kast, C. Beckert, C. Horn, K. R. Markham, H. Wong, and F.-C. Czygan, *Phytochemistry*, 1995, **39**, 915.
136. C. Beckert, C. Horn, J.-P. Schnitzler, A. Lehning, W. Heller, and M. Veit, *Phytochemistry*, 1997, **44**, 275.
137. M. Tada, T. Takakuwa, M. Nagai, and T. Yoshii, *Agric. Biol. Chem.*, 1990, **54**, 3061.
138. C. Keel, U. Schnider, M. Maurhofer, C. Voisard, J. Lavilee, U. Burger, P. Wirthner, D. Haas, and G. Défago, *Mol. Plant-Microbe Interact.*, 1992, **5**, 4.
139. P. Shanahan, D. J. O'Sullivan, P. Simpson, J. D. Glennon, and F. O'Gara, *Appl. Environ. Microbiol.*, 1992, **58**, 353.
140. G. Defago, *Plant Pathol.*, 1993, **42**, 311.
141. L. A. Harrison, L. Letendre, P. Kovacevich, E. A. Pierson, and D. M. Weller, *Soil Biol. Biochem.*, 1993, **25**, 215.
142. B. Nowak-Thompson, S. Gould, J. Kraus, and J. Loper, *Can. J. Microbiol.*, 1994, **40**, 1064.
143. M. G. Bangera and L. S. Thomashow, *Mol. Plant-Microbe Interact.*, 1996, **9**, 83.
144. K. Ueda, K.-M. Kim, T. Beppu, and S. Horinouchi, *J. Antibiot.*, 1995, **48**, 638.
145. K. Kubitzki, *J. Plant Physiol.*, 1987, **131**, 17.
146. K. R. Markham, in "The Flavonoids," ed. J. B. Harborne, Chapman & Hall, London, 1988, p. 427.
147. K. R. Markham, in "Bryophytes: Their Chemistry and Chemical Taxonomy," eds. H. D. Zinsmeister and R. Mues, Clarendon, Oxford, 1990, p. 143.
148. H. A. Stafford, *Plant Physiol.*, 1991, **96**, 680.
149. S. Fischer, U. Böttcher, S. Reuber, S. Anhalt, and G. Weissenböck, *Phytochemistry*, 1995, **39**, 1007.
150. P. M. Dewick, in "The Flavonoids: Advances in Research since 1986," ed. J. B. Harborne, Chapman & Hall, London, 1994, p. 117.
151. P. M. Dewick, in "The Flavonoids: Advances in Research since 1980," ed. J. B. Harborne, Chapman & Hall, London, 1988, p. 125.
152. T. Hakamatsuka, H. Noguchi, Y. Ebizuka, and U. Sankawa, *Chem. Pharm. Bull* (*Tokyo*), 1988, **36**, 4225.
153. S.-I. Ayabe, A. Udagawa, and T. Furuya, *Arch. Biochem. Biophys.*, 1988, **261**, 458.
154. H. Grisebach, L. Edelmann, D. Fischer, G. Kochs, and R. Welle, in "Signal Molecules in Plants and Plant–Microbe Interactions," ed. B. J. J. Lugtenberg, Springer Verlag, Berlin, 1989, vol. H36, p. 57.
155. R. Welle and H. Grisebach, *FEBS Lett.*, 1988, **236**, 221.
156. R. Welle, G. Schröder, E. Schiltz, H. Grisebach, and J. Schröder, *Eur. J. Biochem.*, 1991, **196**, 423.
157. R. Welle and J. Schröder, *Arch. Biochem. Biophys.*, 1992, **293**, 377.
158. C. Sallaud, J. El-Turk, L. Bigarré, H. Sevin, R. Welle, and R. Esnault, *Plant Physiol.*, 1995, **108**, 869.
159. S. Goormachtig, M. Valerio-Lepiniec, K. Szczyglowski, M. Van Montagu, M. Holsters, and F. J. De Bruijn, *Mol. Plant-Microbe Interact.*, 1995, **8**, 816.
160. U. Sankawa, T. Hakamatsuka, T. Shinkai, M. Yoshida, H.-H. Park, and Y. Ebizuka, in "Current Issues in Plant Molecular and Cellular Biology," eds. M. Terzi, R. Cella, and A. Falavigna, Kluwer, Dordrecht, 1995, p. 595.
161. T. Akashi, T. Furuno, K. Futami, M. Honda, T. Takahashi, R. Welle, and S.-I. Ayabe, *Plant Physiol. Plant Gene Register*, 1996, PGR96-023.
162. D. Grab, R. Loyal, and J. Ebel, *Arch. Biochem. Biophys.*, 1985, **243**, 523.
163. R. Wingender, H. Röhrig, C. Höricke, D. Wing, and J. Schell, *Mol. Gen. Genet.*, 1989, **218**, 315.
164. S. Akada and S. K. Dube, *Plant Mol. Biol.*, 1995, **29**, 189.
165. A. G. McInnes, S. Yoshida, and G. H. N. Towers, *Tetrahedron*, 1965, **21**, 2939.
166. F. Balza, A. D. Muir, and G. H. N. Towers, *Phytochemistry*, 1985, **24**, 529.

167. S. Takahashi, F. Nakamura, N. Sahashi, T. Ohmoto, U. Mizushima, U. Sankawa, and G. H. N. Towers, *Biochem. Syst. Ecol.*, 1990, **18**, 11.
168. E. Leete, A. Muir, and G. H. N. Towers, *Tetrahedron Lett.*, 1982, **23**, 2635.
169. K. Ishiguro, S. Nagata, H. Fukumoto, M. Yamaki, K. Isoi, and Y. Yamagata, *Phytochemistry*, 1994, **37**, 283.
170. I. Macleod and D. A. Whiting, *J. Chem. Soc., Chem. Commun.*, 1979, 1152.
171. P. Denniff, I. Macleod, and D. A. Whiting, *J. Chem. Soc., Perkin Trans. 1*, 1980, 2637.
172. R. G. Cooke and J. M. Edwards, *Fortsch. Chem. Org. Naturstoffe*, 1980, **40**, 153.
173. T. Inoue, N. Kenmochi, N. Furukawa and M. Fujita, *Phytochemistry*, 1987 **26**, 1409.
174. M. D. Greca, R. Lanzetta, A. Molinaro, P. Monaco, and L. Previtera, *Bioorg. Med. Chem. Lett.*, 1992, **2**, 311.
175. M. D. Greca, A. Molinaro, P. Monaco, and L. Previtera, *Tetrahedron*, 1992, **48**, 3971.
176. J. G. Luis, F. Echeverri, W. Quinones, I. Brito, M. Lopez, F. Torres, G. Cardona, Z. Aguiar, C. Pelaez, and M. Rojas, *J. Org. Chem.*, 1993, **58**, 4306.
177. N. Hirai, H. Ishida, and K. Koshimizu, *Phytochemistry*, 1994, **37**, 383.
178. J. G. Luis, W. Q. Fletcher, F. Echeverri, T. Abad, M. P. Kishi, and A. Perales, *Nat. Prod. Lett.*, 1995, **6**, 23.
179. R. G. Cooke and R. L. Thomas, *Aust. J. Chem.*, 1975, **28**, 1053.
180. D. Hölscher and B. Schneider, *J. Chem. Soc., Chem. Commun.*, 1995, 525.
181. D. Hölscher and B. Schneider, *Nat. Prod. Lett.*, 1995, **7**, 177.
182. R. Thomas, *Biochem. J.*, 1961, **78**, 807.
183. R. Thomas, *Pure Appl. Chem.*, 1973, **34**, 515.
184. B. Ylstra, Ph.D. Thesis, Vrije Universiteit, 1995.
185. J. B. Shen and F. C. Hsu, *Mol. Gen. Genet.* 1992, **234**, 379.
186. B.-W. Min, "Klonierung flavonoidspezifischer Gene aus cDNA-Bibliotheken verschiedener Blütenpflanzen und die Charakterisierung ihrer Expression in einem genetisch definierten Pflanzenmaterial", Ph.D. Thesis, Technische Universität München, 1994.

1.28
Isoflavonoids: Biochemistry, Molecular Biology, and Biological Functions

RICHARD A. DIXON
Samuel Roberts Noble Foundation, Ardmore, OK, USA

1.28.1 INTRODUCTION: CHEMICAL CLASSES AND BIOLOGICAL OCCURRENCE OF ISOFLAVONOIDS

The flavonoids represent one of the major classes of phenylpropanoid-derived compounds. The 15-carbon (C_6—C_3—C_6) backbone of the flavonoids can be arranged as a 1,3-diphenylpropane skeleton (flavonoid nucleus) (**1**) or as a 1,2-diphenylpropane skeleton (isoflavonoid nucleus) (**2**). More than 4000 different 1,3-diphenylpropane flavonoid derivatives have been characterized from terrestrial plants, in which such flavonoids are almost ubiquitous. In contrast, the isoflavonoids are restricted primarily to leguminous plants, although they occur rarely in other families such as the Apocynaceae, Meliaceae, Pinaceae, Polygalaceae, Compositae, and Myristicaceae.[1] The occurrence of an ester of the isoflavone genistein (**3**) has been reported in *Cotoneaster* (Rosaceae).[2]

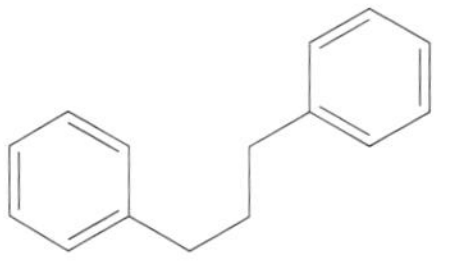

(**1**) 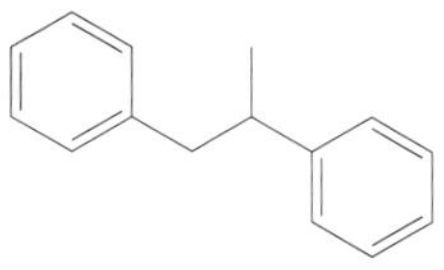 (**2**)

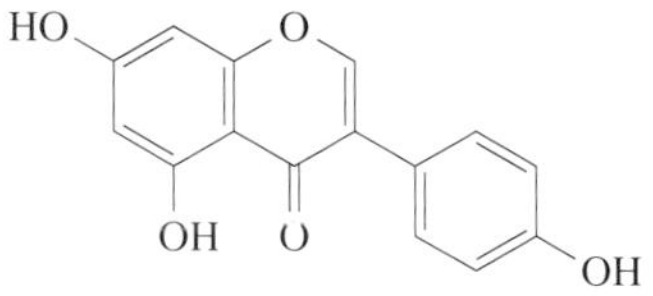

HO O OH O OH

(**3**)

In a review, Tahara and Ibrahim[1] described the classification of 870 naturally occurring isoflavonoid aglycones into nine major classes based on their skeletal modifications. These classes, and their proposed biosynthetic interrelationships, are outlined in Scheme 1, which also indicates the numbering systems used for the different classes of isoflavonoids. Of these classes, the isoflavones and pterocarpans are the most abundant, with 334 and 152 different structures, respectively, having been described as of September 1994.

The limited taxonomic distribution of the isoflavonoids is linked to the occurrence of the enzyme isoflavone synthase, which catalyzes the aryl migration reaction (a two-step process involving hydroxylation/aryl migration followed by dehydration) that leads to the formation of an isoflavone from a flavanone (see below). Many of the subsequent ring modifications that occur in isoflavonoids (e.g., *O*-methylation, isoprenylation, methylenedioxy bridge formation) are also common in the flavonoids *per se*, and are catalyzed by highly regiospecific enzymes with tight substrate (e.g., isoflavonoid class) specificity.

Within a particular species, several different isoflavonoids usually occur. These may be of different classes, and with a variety of substitution patterns. Furthermore, some may be formed constitutively in various plant organs and tissues as part of the plant's developmental program, whereas others are synthesized *de novo* in response to biotic and abiotic stress. Compounds that are synthesized constitutively in a limited range of tissues may accumulate in most tissues of the plant if that tissue is microbially infected. Members of several of the isoflavonoid classes are commonly found in the bark or heartwood of tropical leguminous trees, where they may act as protective compounds. The stress-inducible isoflavonoids with antimicrobial activity (phytoalexins) have been the most studied of the isoflavonoids with respect to biosynthesis, and discussion of these compounds will form the basis of much of the present chapter. Lima bean (*Phaseolus lunatus*) exhibits one of the most varied inducible isoflavonoid responses, with a report of 25 different compounds formed in response to abiotic elicitation.[3]

Compounds (**4**)–(**16**) illustrate the diversity of isoflavonoids that have been reported from the forage legume alfalfa (*Medicago sativa* L.). Glycosides and malonyl glycosides of the isoflavones daidzein (**4**) and formononetin (**5**), and of the pterocarpan medicarpin (**6**), occur constitutively, primarily in root tissue, whereas medicarpin aglycone and the isoflavans sativan (**7**) and vestitol (**8**) are the predominant phytoalexins. The remainder of the compounds shown are minor isoflavonoids that have been identified from large-scale purifications.[4–7] Note that the major ring substituent in alfalfa isoflavonoids is the methoxyl group. In contrast, the major phytoalexins from bean (phase-

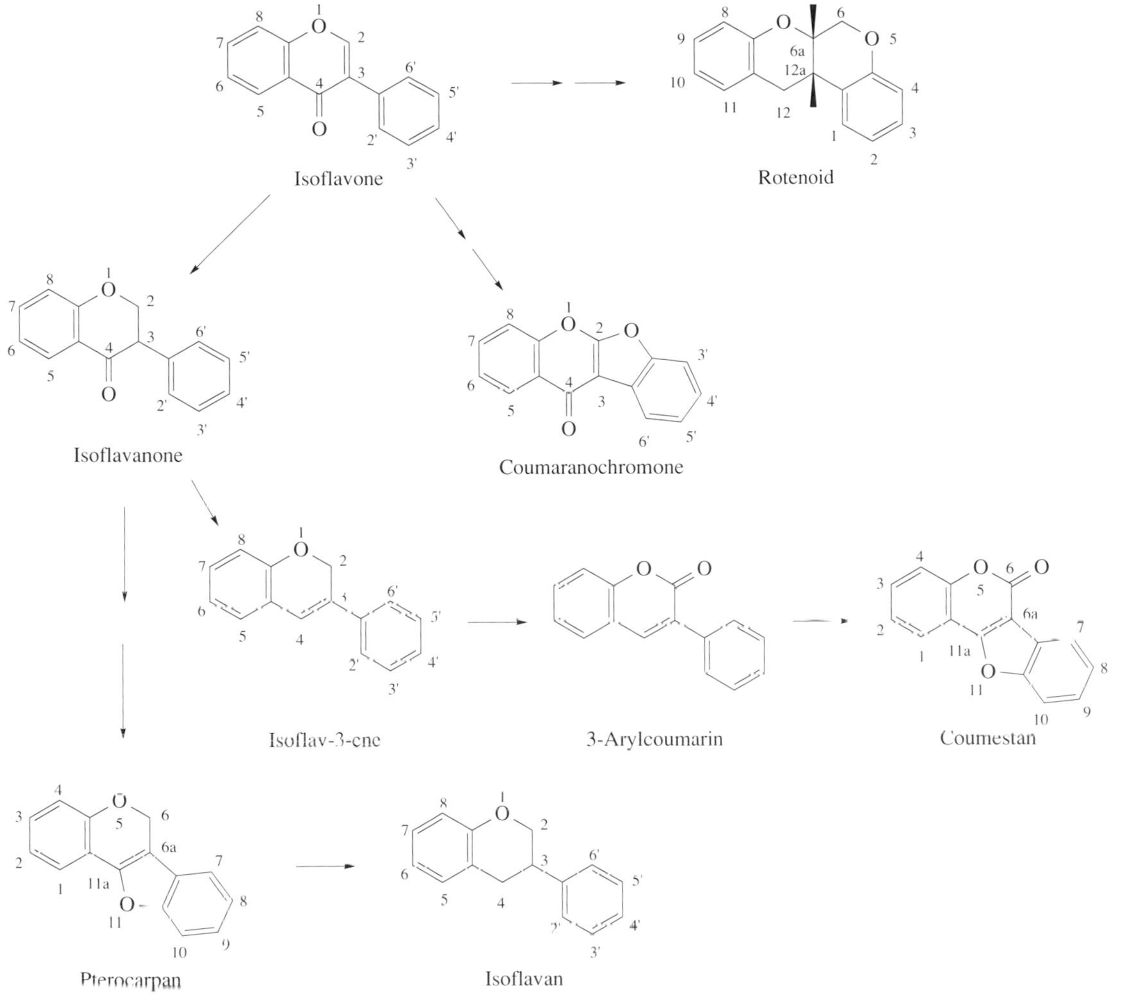

Isoflavone → → Rotenoid

Isoflavanone   Coumaranochromone

Isoflav-3-ene → 3-Arylcoumarin → Coumestan

Pterocarpan → Isoflavan

Scheme 1

ollin (**17**)) and soybean (the glyceollins I, II and III (**18–20**, respectively)) bear prenyl substituents, whereas (+)-pisatin (**21**) from peas, the first phytoalexin to be structurally characterized, bears methylenedioxy, methoxy, and 6a-hydroxyl substituents. 6a-Hydroxylation of pterocarpans is a common substitution, also occurring in the glyceollins and in the phytoalexins of red clover,[8] among others. Hydroxylation at the 6a-position is also employed by fungi as an early step in the detoxification of pterocarpan phytoalexins.[9]

Pea (*Pisum sativum*) tissues exposed to the biotic elicitor copper chloride have been reported to accumulate, in addition to (**21**), several isoflavones including afrormosin (**22**), (+)-2-hydroxypisatin (**23**), (−)-pisatin (**24**), and the pterocarpene anhydropisatin (**25**).[10–12] The minor compound (**23**) has less antifungal activity than (**21**), but, unlike (**21**), suppresses alfalfa seed germination,[12] suggesting that it might have allelochemical activity.

In addition to the commonly occurring isoflavonoids with the typical substituents shown in Scheme 1 and (**4**)–(**16**), more complex isoflavonoids have also been described. Some of these are shown in (**26**)–(**31**). They include rare heteroatom-containing 4′-aminoisoflavonoids from the root bark of *Piscidia erythrina* (reviewed in reference 1) (**26, 27**), the lupinols (coumaranochroman-4-ones) from white lupin (reviewed in reference 1) (**28**), a range of isoflavonoid oligomers (isoflavan–isoflavan such as vestitol-(4→5′)-vestitol (**29**) from heartwood of *Dalbergia odorifera*, isoflavan–isoflavone, isoflavan–flavanone, isoflavan–flavone, isoflavone–flavone, isoflavan–flavene, isoflavone–cinnamyl alcohol, isoflavone–stilbene, and isoflavan–chalcone) (reviewed

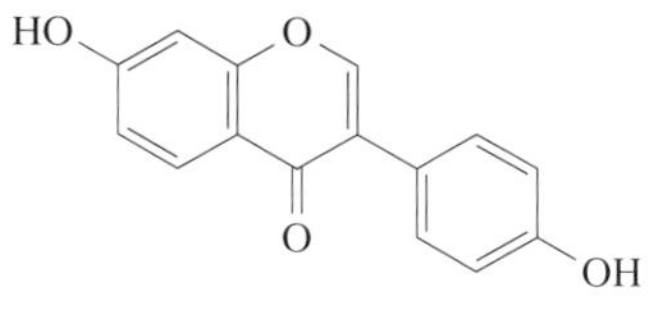

(4) Daidzein

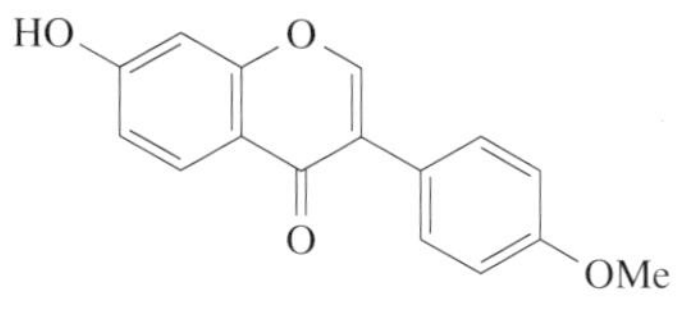

(5) Formononetin

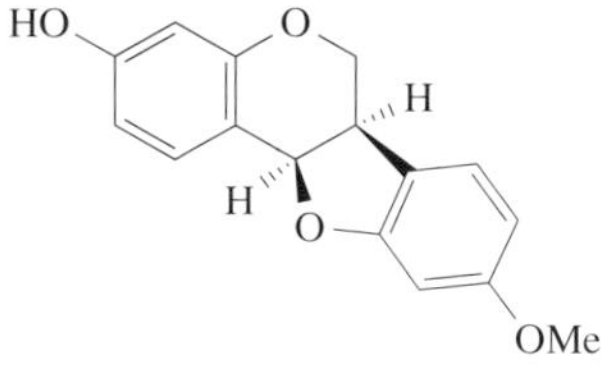

(6) (–)-Medicarpin

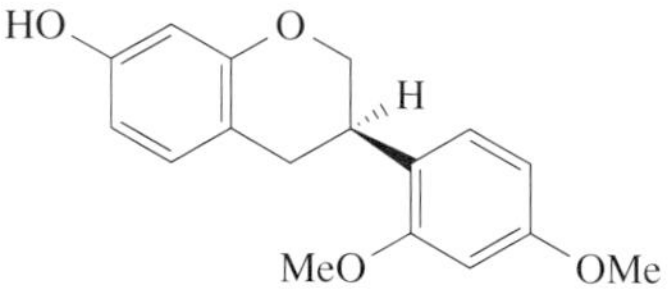

(7) (–)-Sativan

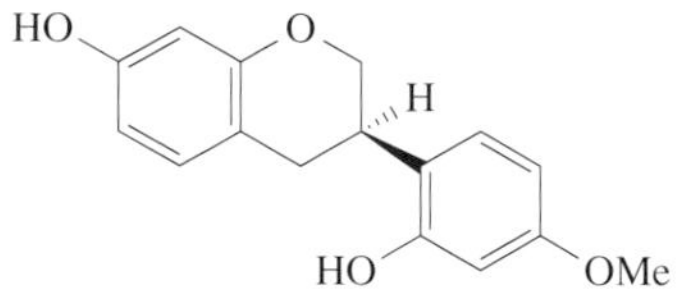

(8) (–)-Vestitol

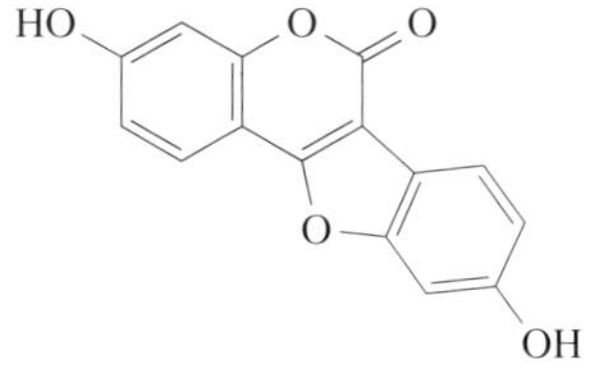

(9) Coumestrol

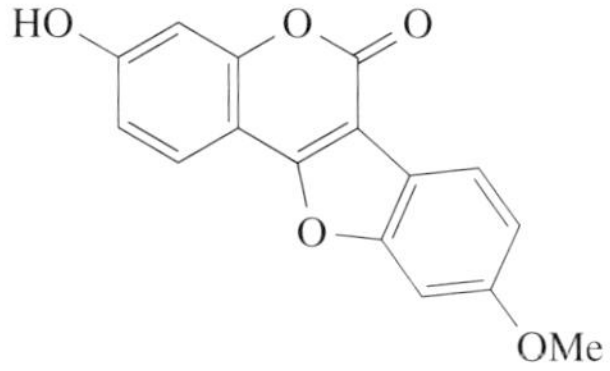

(10) 9-*O*-Methyl coumestrol

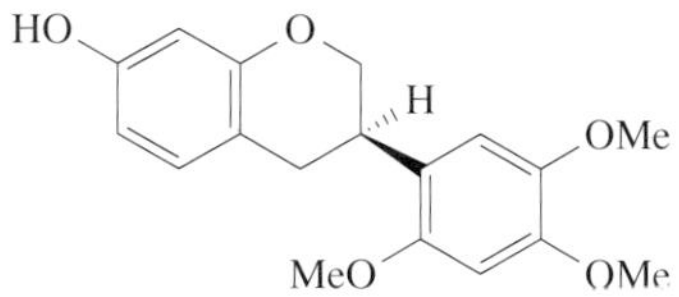

(11) (–)-5'-Methoxysativan

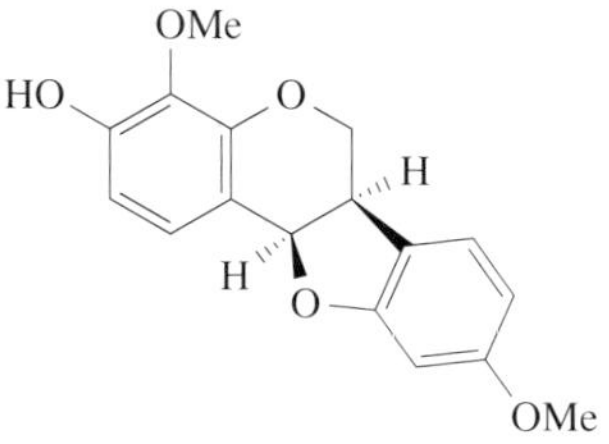

(12) (–)-4-Methoxymedicarpin

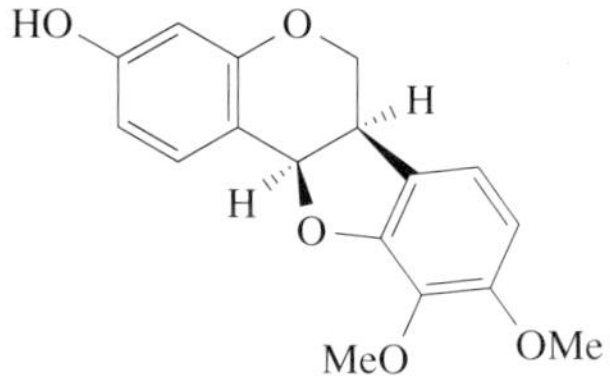

(13) 10-Methoxymedicarpin

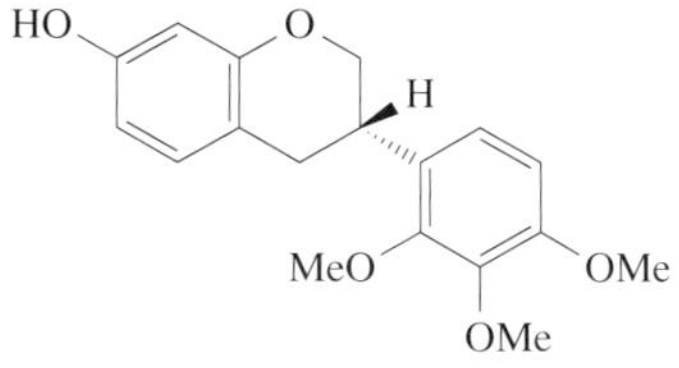

(14) (+)-7-Hydroxy-2',3',4'-trimethoxyisoflavan

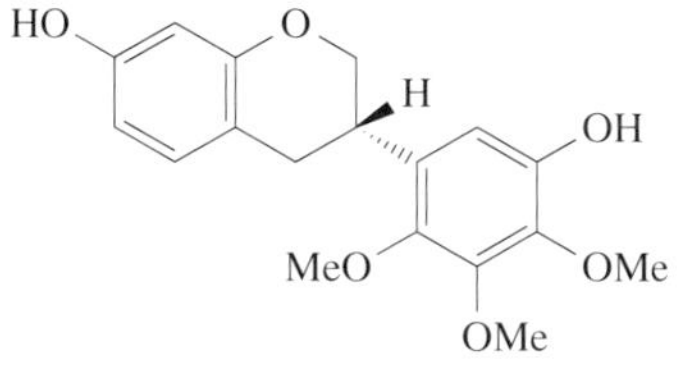

(15) (+)-7,5'-Dihydroxy-2',3',4'-trimethoxyisoflavan

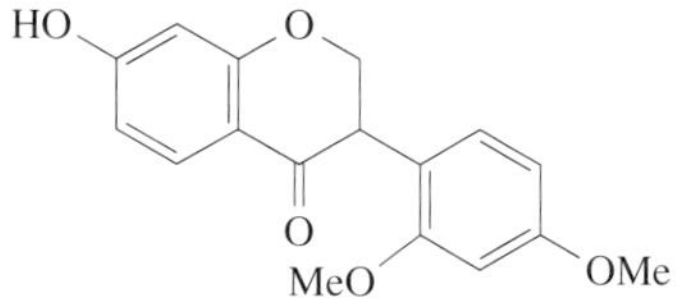

(16) Sativanone

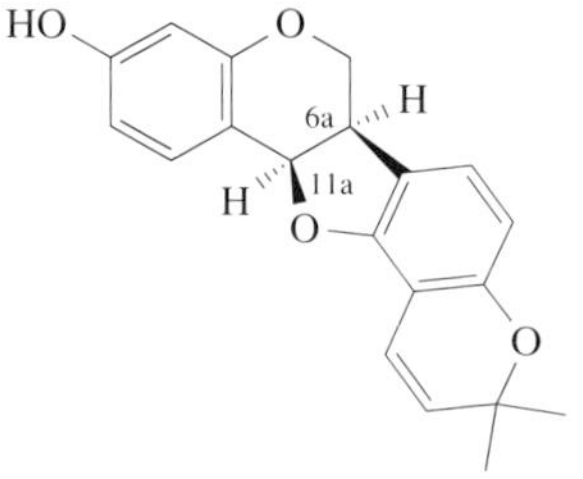

(17)

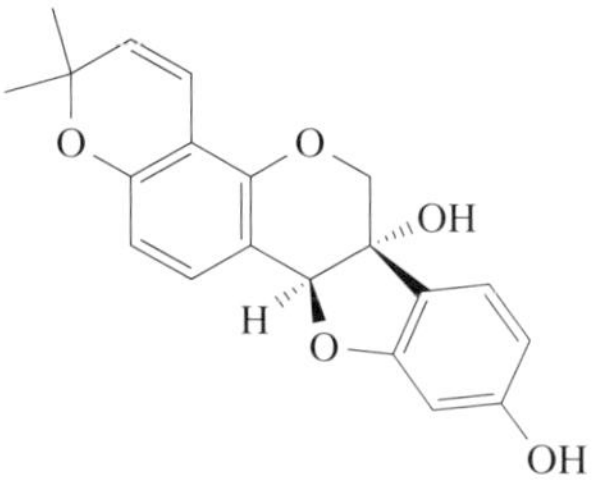

(18)

(19)

(20)

(21) (+)-Pisatin

(22) Afrormosin

(23) (+)-2-Hydroxypisatin

(24) (−)-Pisatin

(25) Anhydropisatin

by Dewick[13]), and the complex santarubin dyes (**30**) that occur along with substituted isoflavenes (**31**) in the red heartwood of the West African tree *Baphia nitida*.[14]

Isoflavonoids have been widely used as taxonomic markers within the Leguminosae. Typical examples include comparative studies among the genera *Medicago* and *Trigonella*.[6,15]

Many excellent reviews have been written on the structure and occurrence of isoflavonoids. The reader is referred to papers by Wong,[16] Dewick,[13,17] Ingham,[18] and Tahara and Ibrahim,[1] and the references cited therein, for detailed information on this topic.

1.28.2 BIOLOGICAL ACTIVITIES OF ISOFLAVONOIDS

1.28.2.1 Overview

The biological activities of isoflavonoids range from properties that suggest important functions in the plant's interaction with its environment to pharmacological properties in animal cells that may or may not reflect corresponding functions/activities in the plant. It is not the purpose of this chapter to review all these biological properties. Many of them are listed in Tables 1 and 2, and only the better studied aspects reviewed in the following section. The multiple roles of isoflavonoids in the relations of plants with their environment have been addressed elsewhere.[39]

1.28.2.2 Role of Isoflavonoids in Plant–Microbial Pathogen Interactions

Isoflavonoids have been ascribed key roles in plant–pathogen interactions because many have strong antimicrobial activity. Antimicrobial isoflavonoids fall into two functional classes, the pre-formed "phytoanticipins" and the inducible "phytoalexins".[40] Examples of the former class include the prenylated isoflavones of lupin, which are synthesized in various organs of the plant during seedling development.[41] Examples of the latter include several pterocarpans, the biosynthesis of which has been studied in detail, particularly with respect to the phytoalexin response of bean, alfalfa, pea, and soybean (**6, 17–21**).

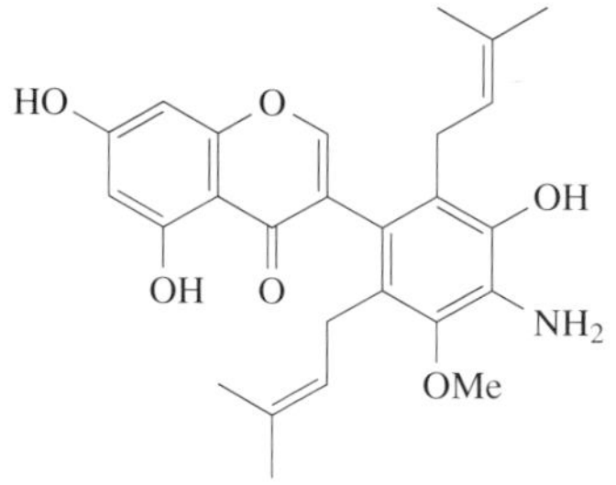

(26) Piscerythramine

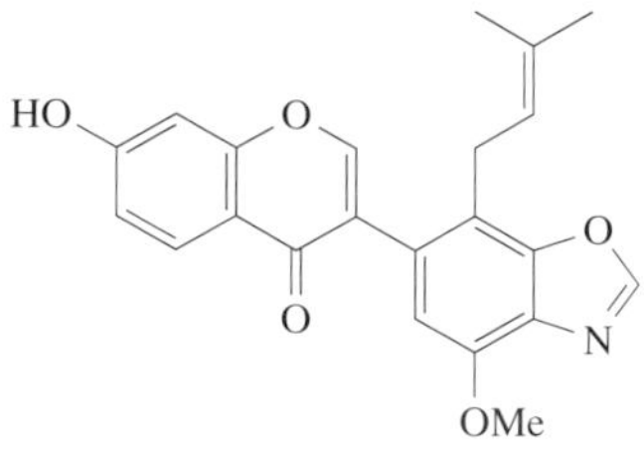

(27) Piscerythoxazole

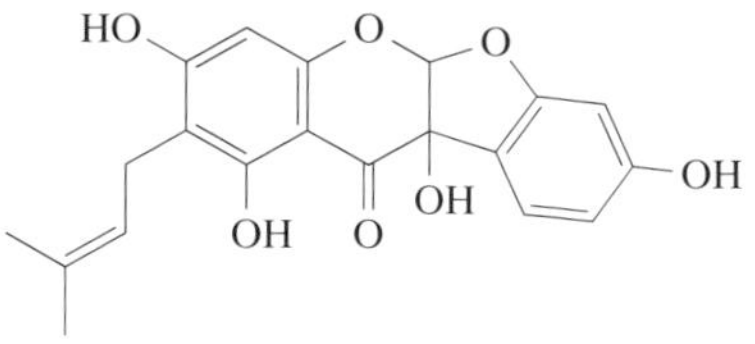

(28) Lupinol C

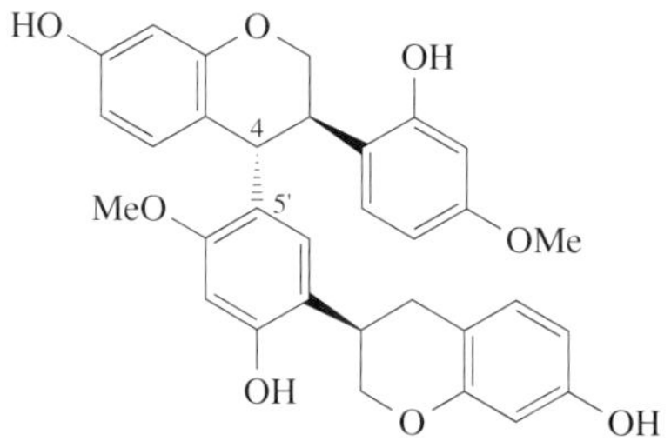

(29) Vestitol-(4→5') vestitol

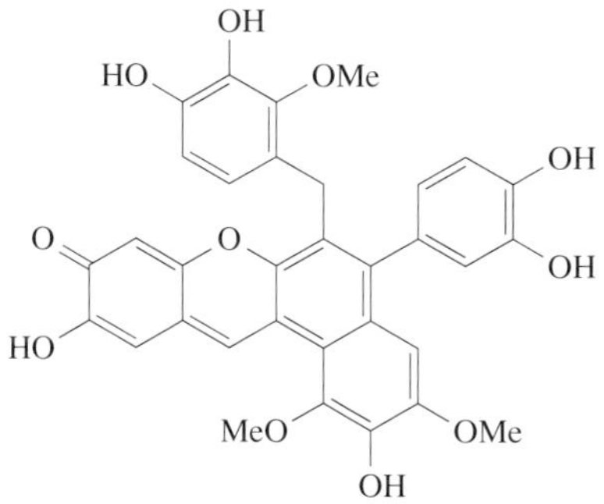

(30) Santarubin C

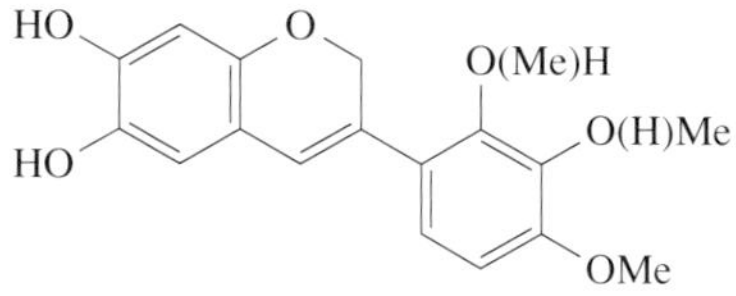

(31)

Table 1 Biological activities of isoflavonoids: activities with functional implications for the plant.

Biological activity	*Examples*	*Reference*
Preformed antimicrobial	prenylated isoflavones	see Section 1.28.2.2
Phytoalexin	isoflavans, pterocarpans	see Section 1.28.2.2
Inducer of fungal pathogen spore germination	pisatin (**21**)	see Section 1.28.2.2
Nodulation gene inducer	isoflavones	see Section 1.28.2.4
VAM interaction inducer	various	see Section 1.28.2.3
Nematocidal	glyceollin (**18**)	Huang and Barker[19]
Antiinsect	rotenoids	Fukami and Nakajima[20]
Allelochemical	medicarpin (**6**)	Miller *et al.*[21]
Phytotoxic	phaseollin (**17**), glyceollin (**18**)	Skipp *et al.*[22]; Glazener and VanEtten[23]; Giannini *et al.*[24]
IAA oxidase modulator	lupin isoflavones	Ferrer *et al.*[25]
Control of cell division	sayanedine	Bailey and Francis[26]
Iron chelator	2-(3′,5′-dihydroxyphenyl)-5,6-dihydroxybenzofuran	Masaoka *et al.*[27]
Various pharmacological effects on animals and humans	genistein (**3**), daidzein (**4**), coumestrol (**9**)	see Table 2

There is a vast literature on the isoflavonoid phytoalexins of the Leguminosae. In view of all this research activity, it is surprising that major questions still exist concerning structure–activity relationships and the exact role of these compounds as determinants of disease resistance. Because many plant pathogens have the ability to metabolize, and therefore detoxify, isoflavonoid compounds, structure–activity relations are highly dependent on the fungi used in the bioassays.

Table 2 Biological activities of isoflavonoids: pharmacological activities.

Pharmacological activity	*Compounds*[a]	*Reference*
Mitochondrial ADH inhibitor	daidzin	Keung *et al.*[28]
Antiulcer	G, F, 3′-methoxydaidzein	Takai *et al.*[29]
Antiarthritis	Pseudobaptigenin	Malhotra *et al.*[30]
Estrogenic, proestrogenic	G, F, C	see Section 1.28.2.5
Estrogen receptor binding	C, G, D, F, BA	see Section 1.28.2.5
Antiangiogenic	G	Fotsis *et al.*[31]
Antioxidant	G, many others	Wang and Murphy[32]
Anticancer	G	See Section 1.28.2.5
Protein tyrosine kinase inhibitor (e.g., EGF receptor)	G	Akiyama *et al.*[33]
Protein histidine kinase inhibitor	G	Huang *et al.*[34]
Prostaglandin synthesis inhibitor	isoflavans, isoflavenes	Goda *et al.*[35]
DNA synthesis/cell cycle arrest	G	Takano *et al.*[36]
Topoisomerase inhibitor	G	Okura *et al.*[37]
P_i-purinergic receptor antagonist	G	Okajima *et al.*[38]

[a]G, genistein (**3**); F, formononetin (**5**); 3′-OH-F, 3′-hydroxyformononetin; C, coumestrol (**9**); D, daidzein (**4**); BA, biochanin A (**33**).

A comparison of the effects of a series of isoflavones, isoflavanones, pterocarpans, and isoflavans on the growth of *Aspergillus niger* and *Cladosporium cucumerinum* suggested that lipophilicity and the presence of at least one unsubstitued phenolic hydroxyl group correlated with fungitoxicity.[42] Other studies have suggested that a skewed, aplanar ring structure is essential for high activity of isoflavans, although this has been seriously questioned, and alternative suggestions made that specific combinations of hydrophobic (methoxy) and hydroxyl groups are important, but that no absolute generalizations are possible.[43–45] In the case of the pterocarpan phytoalexins, the stereochemistry associated with the 6a and 11a chiral centers (compare (**17**) and (**21**)) plays an important role in determining antifungal activity, because plant pathogenic fungi are often able to degrade the isomer produced by their host plant but may be highly sensitive to the opposite isomer.[46]

The phytoalexin "hypothesis" is based for the most part on indirect determinations of causality, with major reliance on correlative data. Thus, isoflavonoid compounds have been shown to accumulate in infected plant cells to levels shown to be antimicrobial *in vitro*. The temporal, spatial, and quantitative aspects of accumulation are consistent with a role for these compounds in disease resistance.[47–50] However, few studies have directly tested this hypothesis. Inhibition of the synthesis of (**18**) by application of an inhibitor of L-phenylalanine ammonia-lyase (PAL) to soybean seedlings breaks resistance to *Phytophthora megasperma* f. sp. *glycinea*,[51] but this could be due to pleiotropic effects related to down-regulation of the phenylpropanoid pathway as a whole; for example, salicylic acid, a product of the early part of the phenylpropanoid pathway after the PAL reaction, is known to be important for expression of disease resistance.[52] Isolates of *Nectria hematococca* with reduced ability to degrade the pea phytoalexin (**21**) have reduced virulence on pea, suggesting that (**21**) is indeed a factor in the disease resistance response.[53] Ultimately, it will be necessary to produce mutant plants lacking only the isoflavonoid phytoalexins in order to test rigorously the roles of these compounds in plant–microbe interactions. Mutants of *Arabidopsis* lacking the indole phytoalexin camalexin were not less resistant to incompatible bacteria, but did show increased disease symptoms following infection with a compatible race, suggesting that this phytoalexin may play a role in disease symptom limitation rather than in determination of resistance *per se*.[54]

Antifungal isoflavonoids are often as phytotoxic as they are fungitoxic. Compound (**18**) causes proton leakage from *Phytophthora* plasma membrane vesicles and from red beet and soybean tonoplast vesicles.[24] Treatment of bean suspension cells with 30 $\mu g\ mL^{-1}$ of (**17**) resulted in inhibition of respiration within 2 min, and subsequent death of most of the cell population.[22] Pretreatment with lower concentrations of (**17**) does not induce tolerance to higher concentrations, although bean cells do have mechanisms for degrading exogenously added (**17**).[23] These results indicate a requirement for sequestration of bioactive isoflavonoids, be they phytoalexins or phytoanticipins, away from the sensitive molecular sites of the host. In this respect, an immunolocalization study indicated the presence of a diprenylated isoflavone, 2′-hydroxylupalbigenin (**32**) in secondary walls and pericycle cells of lupin roots, and provided evidence for compartmentation in the wall mediated via membrane vesicles.[55]

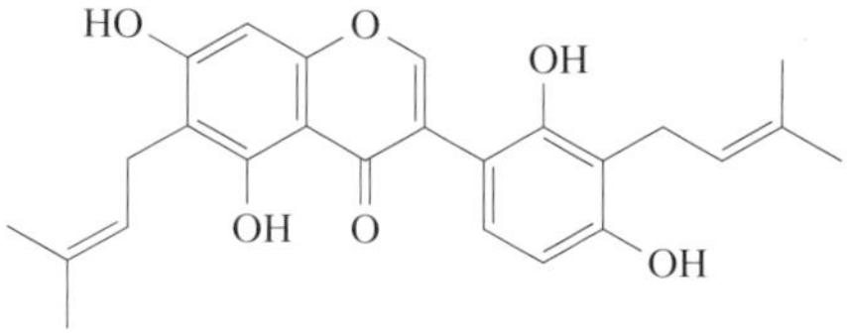

(32)

Isoflavonoids can act as stimulatory, as well as inhibitory, factors in interactions of legumes with fungi. Daidzein (**4**) and genistein (**3**), important components of soybean root exudates, act at very low concentrations (10 nmol L^{-1}) as chemoattractants for zoospores of *Phytophthora sojae*, and also induce encystment and germination of the zoospores.[56] The isoflavones appear to be inactive with nonpathogens of soybean. Likewise, biochanin A (**33**), (**3**), and several pterocarpan phytoalexins including (**6**) and (**21**) stimulate spore germination of *Fusarium solani* forma speciales pathogenic on pea or bean at a concentration of 10 μmol L^{-1}).[57]

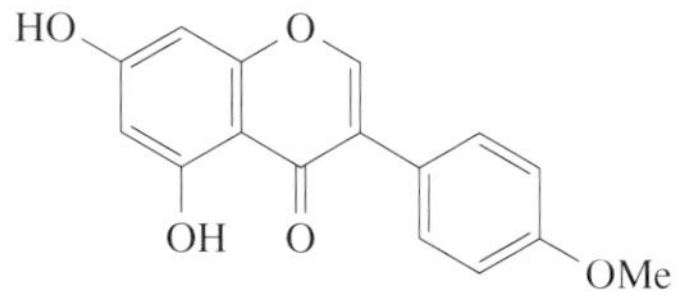

(33)

1.28.2.3 Role of Isoflavonoids in Mycorrhizal Interactions

There has been much debate on the possible role of isoflavonoids during the establishment of the symbiotic vesicular arbuscular mycorrhizal (VAM) association of fungi of the species *Glomus* with legume roots. At low concentrations (2–5 μmol L^{-1}), daidzein (**4**) increases the percentage germination of *Glomus* spores by ~35%,[58] and coumestrol (**9**), (**4**), and (**3**) have small but significant stimulatory effects on the degree of mycorrhizal colonization of soybean.[59] It has been suggested that one effect of isoflavonoids on the soybean mycorrhizal symbiosis could be via induction of nodulation (Nod) factors (see below) from cocolonizing *Rhizobia*, since Nod factors have also been shown to stimulate fungal colonization.[59] Once mycorrhizal fungi begin to colonize the host root, an initial increase in isoflavonoid levels is rapidly suppressed.[60,61] The decrease in isoflavonoid levels correlates with a reduction in transcripts of enzymes specific for the later stages of isoflavonoid synthesis throughout the root cortex, although the root cells harboring the fungal arbuscules contain elevated transcript levels for enzymes of the central phenylpropanoid pathway and flavonoid branch pathway, indicating tight and differential control of flavonoid and isoflavonoid synthesis during the establishment and maintenance of this mutually beneficial association.[60,62]

1.28.2.4 Role of Isoflavonoids in the *Rhizobium*–Legume Symbiosis

The establishment of nitrogen-fixing nodules in leguminous plants is initiated by the recognition by the *Rhizobium* bacteria of compounds released in root and seed exudates. Recognition of these compounds (*nod* gene inducers) by the bacterial *NodD* gene products leads to transcription of a set of genes in the bacteria (*nod* genes) which encode biosynthetic enzymes for the formation of substituted lipochitooligosaccharide signal molecules (Nod factors) that in turn induce root hair curling and the cortical cell divisions that characterize the early development of the nodule. Flavonoid and isoflavonoid compounds can play critical roles in these processes. Alfalfa and red clover root exudates contain flavones that potently activate *nod* gene expression,[63,64] whereas the major *nod* gene inducers in the soybean–*Bradyrhizobium* symbiosis are the isoflavones daidzein (**4**) and genistein (**3**).[65] Reduced synthesis of (**3**) in roots at suboptimal temperatures may represent a limitation to Rhizobial colonization.[66]

Root exudates from alfalfa plants inoculated with *Rhizobium meliloti* contain the pterocarpan medicarpin (**6**) and its glucoside, as well as formononetin 7-*O*-glucoside-6″-*O*-malonate (FGM) (**34**). Levels of (**34**) are increased when plants are grown under low nitrogen conditions.[67] Formononetin (**5**) and its 7-*O*-glucoside do not possess *nod* gene-inducing activity for the alfalfa symbiont. Surprisingly, however, (**34**) can induce *Rhizobium nod* genes through interactions with both the NodD1 and NodD2 recognition proteins.[68] The Nod factors synthesized as a result of *nod* gene activation are active on alfalfa roots at concentrations of around 10^{-9} mol L^{-1}. At higher concentrations (10^{-6} mol L^{-1}), they have been shown to induce genes of the isoflavonoid biosynthetic pathway in microcallus cultures,[69] but it is not clear whether this represents a physiologically relevant defensive response by the plant.

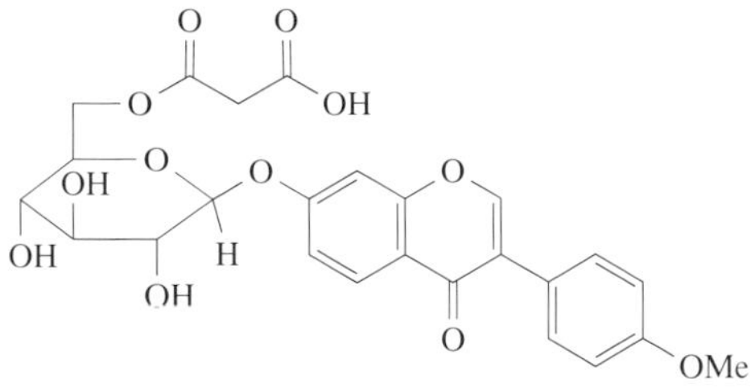

(**34**)

Following inoculation of bean (*Phaseolus vulgaris*) roots with *Rhizobium leguminosarum*, coumestrol (**9**), (**4**), (**3**), and genistin (**35**) are released in the root exudate. Of these compounds, (**9**) and (**4**) activate transcription under control of the *R. leguminosarum nodD*1 gene.[70]

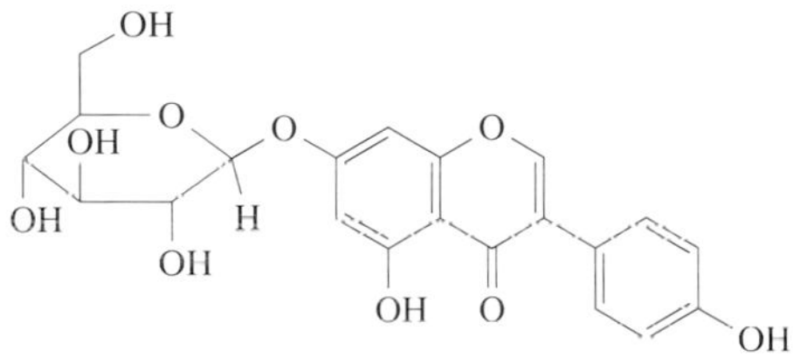

(**35**)

Compounds (**3**), (**4**), and (**9**) produced by soybean are inactive as chemoattractants for *Bradyrhizobium japonicum*, whereas hydroxycinnamic acid precursors are strong attractants.[71] The chalcone precursor of (**4**), isoliquiritigenin (**36**), is an order of magnitude more potent than (**4**) as a *nod* gene inducer, but is not a chemoattractant.[72]

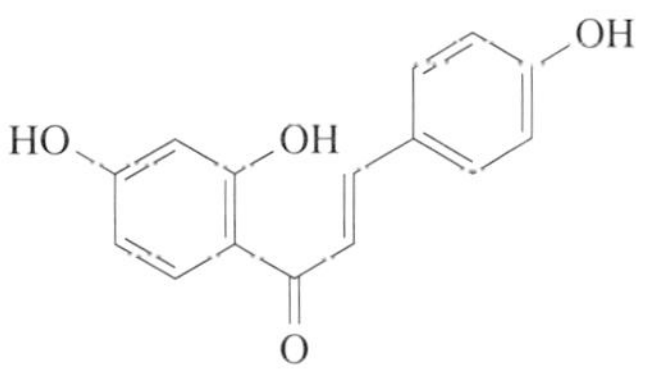

(**36**)

It has been suggested that internal isoflavonoids may also play a role in later stages of nodulation in soybean, because their levels are elevated in hypernodulating mutants, and nitrogen application reduces isoflavonoid levels in parallel with decreased nodule number, weight, and nitrogenase activity.[73] Increased nodule numbers in a hypernodulating mutant are not observed until ~9 d after initial bacterial inoculation.[74] Reciprocal grafting experiments between a wild-type and a hypernodulating soybean mutant have shown that root isoflavonoid levels are controlled by the shoot.[74] Treatment of soybean roots with abscisic acid (ABA) leads to reduction in both nodulation and isoflavonoid levels, although a comparison of wild-type, hyper-, and hyponodulating mutants did not reveal significant differences in endogenous ABA levels, suggesting that ABA is not the factor produced by the shoots to autoregulate nodulation.

Levels of glyceollin I (**18**) increase by a factor of 50 in soybean root exudates following inoculation with *B. japonicum*, and increases are also observed in the levels of the *nod* gene inducers (**4**), (**3**), and (**9**).[75,76] The induction of (**18**) does not require bacterial penetration, as it can be mimicked with heat-killed cells.[75] The levels of (**18**) obtained are, however, significantly lower than those observed in the response of soybean roots to an incompatible race of the fungal pathogen *Phytophthora megasperma* f. sp. *glycinea*. Levels of (**4**), (**3**), and (**9**) are not elevated in root exudates following inoculation with mutant *B. japonicum* that cannot produce Nod factors, and pure nonsulfated Nod factor (as produced by *B. japonicum*) itself can induce these isoflavonoids.[76]

B. japonicum is sensitive to (**18**), but can tolerate it following adaptation to low concentrations. This resistance, which does not involve detoxification or degradation, can also be induced by (**4**) and (**3**), and is not dependent on NodD, the protein that binds isoflavonoids to induce *nod* genes.[77] Therefore, isoflavonoids act at different sites to induce glyceollin tolerance and *nod* gene activity. Induced glyceollin tolerance may be important for survival of *B. japonicum* in the rhizosphere.

It appears that isoflavonoids can induce rhizobial genes in addition to those involved in production of Nod factors.[78] Further studies are necessary to confirm whether isoflavonoid turnover and isoflavonoid-induced gene expression are important features of the regulatory cross-talk between host and symbiont.

1.28.2.5 Effects of Isoflavonoids on Animal and Human Health

It has been known for many years that dietary isoflavonoids can exert estrogenic effects in animals. For example, reports as early as 1946 documented the occurrence of infertility in sheep resulting from grazing on clover rich in the isoflavone formononetin (**5**),[79] and breeding programs have been devised to select for low isoflavone lines of subterranean clover.[80] It has been suggested that California quails might control their natural populations during periods of low food supply by feeding on legumes rich in daidzein (**4**), genistein (**3**), and/or coumestrol (**9**).[81]

Compounds (**4**), (**3**), and equol (**37**) (a major metabolite of dietary isoflavonoids formed by the gastrointestinal flora) share structural features with the potent estrogen estradiol-17β (**38**), particularly the phenolic ring and the distance (11.5 Å) between the two hydroxyl groups, features that determine ability to bind estrogen receptors. Isoflavonoids can thus exert both estrogenic and antiestrogenic activity, the latter by competing for receptor binding by (**38**). However, isoflavonoids and their gastrointestinal metabolites have relatively weak estrogenic activity.[82,83] Compounds (**37**) and (**3**) are active in displacing bound estrogen and testosterone from human sex steroid binding protein,[84] suggesting that phytoestrogens might also affect clearance rates of androgens and estrogens and thus the availability of the hormones to target cells.

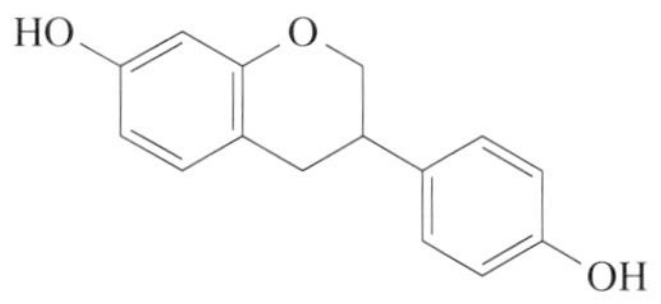

(**37**) 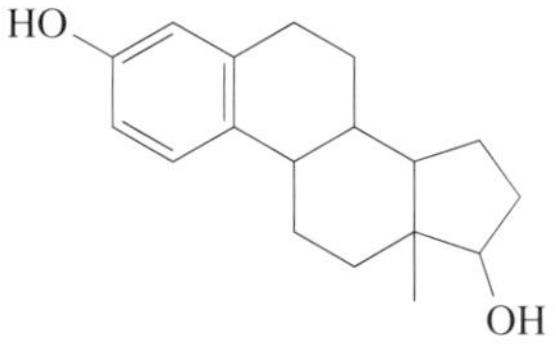 (**38**)

Compound (**5**) has been shown to stimulate mammary gland proliferation and to increase estrogen receptor and plasma prolactin levels in mice.[83] Although these effects mimic the action of (**38**), (**5**) is 15 000 times less potent than (**38**) for binding to murine mammary estrogen receptors.[83] Hence its major biological activity is probably as an estrogen agonist, although its concentration in the "normal Western" diet is probably too low to have any physiological effects. However, in humans eating a soy protein-rich diet, isoflavonoids may be present in the urine at very high levels. Thus, human adults given a diet containing 40 g of soy protein per day secreted 5.3 mg of (**37**) per day, compared with only 2–27 μg of the principal urinary estrogen, estrone glucuronide, released during the follicular phase of the menstrual cycle.[82] This is a 100-fold increase in urinary (**37**) above that observed in adults who consume very little soy products in their diet.

The major interest in dietary isoflavonoid phytoestrogens is because of the significant correlations demonstrated between a soy-rich diet and reduced incidence of breast cancer or mortality from prostate cancer. The incidence of breast cancer is 5–8-fold lower among women from Japan who consume a traditional diet than among women in the United States or Europe who consume a diet higher in animal fat but with very little soy products.[85] A detailed epidemiological study of Singapore Chinese women (420 healthy controls and 200 with histologically confirmed breast cancer), indicated that soy consumption was directly correlated with reduced risk of cancer.[86] The offspring of oriental women who have emigrated to the United States have the same risk of breast cancer as American women if they adopt the Western diet, suggesting that oriental women are not simply genetically predisposed against breast cancer.

Feeding rats a diet containing powdered soybean chips strongly reduced mammary tumor formation in response to the direct carcinogen *N*-methyl-*N*-nitrosourea, with no effect on estrus cycling.[85] The soybean diet resulted in elevated levels of hepatic PAPS:sulfotransferase activity, which may be involved in deactivating metabolically activated carcinogens. It was concluded that the effects of the soybean diet could be the result of estrogenic substances acting as inhibitors of estrogen action or, because similar reduction in tumor formation was observed in response to the procarcinogen 7,12-dimethylbenz[*a*]anthracene, as inducers of hepatic metabolism.

Urinary excretion of (**4**), (**3**), and (**37**) was shown to be at least 10-fold higher in a population of farm workers from Japan compared with Americans or Europeans, and it was suggested that the isoflavonoids found in soy products might be the agents responsible for reduced cancer risk.[87] In fermented soybean foods, the isoflavonoids are usually present as the aglycones, whereas the β-glycosides predominate in nonfermented products.[88] A 1 g amount of powdered soybean chips contains nearly 800 μg of (**4**) and >500 μg of (**3**), whereas 1 g of soya protein has ~ 150 μg of (**4**) and 250 μg of (**3**).[85] Structural similarities have been noted between soybean isoflavones and tamoxifen (**39**), an antiestrogen which has been clinically tested as a chemopreventive agent in women with high risk of breast cancer.[88]

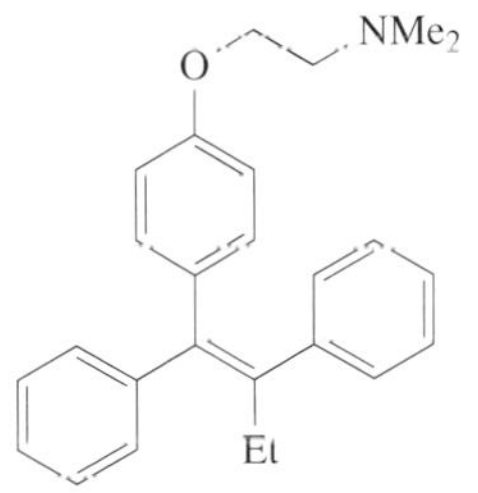

NMe$_2$ O Et

(**39**)

When administered neonatally, (**3**) effectively protects against chemically induced mammary tumors in rats.[89] The effects include increased latency, reduced tumor incidence and multiplicity, and more rapid maturation of undifferentiated end buds to differentiated lobules. Although no clinical trials have been reported documenting effects of controlled dietary supplementation with (**3**) on breast cancer incidence in humans, it has been shown that a high soy diet containing up to 45 mg of isoflavones per day causes changes in the menstrual cycle that may help reduce cancer risk.[90] It has been suggested that the high levels of isoflavones in breast milk of humans consuming a high soy diet may provide the infant with protection against cancer later in life.[91]

Compounds (**3**) and (**33**) inhibit the growth of human stomach cancer cell lines *in vitro*, apparently by stimulating a signal transduction pathway leading to apoptosis.[92] When these cancer cells were transplanted into mice, (**33**), but not (**3**), significantly inhibited tumor growth.

Compound (**3**) can affect a number of molecular processes, one or more of which may be associated with its pharmacological effects. Thus, in addition to showing estrogenic properties in receptor binding and whole tissue studies, (**3**) is an inhibitor of several enzymes, including DNA topoisomerase and tyrosine protein kinase,[33] and also exhibits antioxidant properties and cell cycle arrest activity. Kinase inhibition is generally regarded as being specific for tyrosine kinases such as epidermal growth factor receptor, pp60$^{v\text{-}src}$ and pp110$^{gag\text{-}fes}$, although at higher concentrations (**3**) also inhibits protein histidine kinase.[34] Compound (**3**) blocks EGF-mediated tyrosine phosphorylation *in vivo* in human epidermal carcinoma cells.[33] However, as (**3**) does not block epidermal growth factor phosphorylation *in vivo* at a concentration that reduces mammary tumor formation, it is unlikely that its chemopreventive activity is a result of its activity as a protein kinase inhibitor.[93] Nevertheless, when specifically targeted to the B-cell-specific receptor CD-19 by conjugation to a monoclonal antibody, (**3**) selectively inhibited CD-19-associated tyrosine kinase activities, resulting in death of human B-cell precursor leukemia cells.[94] Other isoflavones such as (**4**) do not inhibit tyrosine kinase activity, and are therefore used as controls in pharmacological experiments utilizing (**3**).

Compound (**3**) potently, specifically, and directly inhibits glucose and dehydroascorbate uptake by the mammalian facilitative hexose transporter GLUT 1.[95] It also inhibits fast sodium channels in human uterine leiomyosarcoma cells, a process that is also inhibited by (**4**) and is therefore independent of tyrosine kinase activity.[96] Unlike other isoflavonoids, (**3**) only exerts toxicity at concentrations greatly in excess of those at which it first exerts its biological effects, making it an important subject for future studies on cancer chemoprevention. Further information on the clinical effects of isoflavonoid phytoestrogens can be found in reviews by Adlercreutz *et al.*,[97] Messina *et al.*,[98] Knight and Eden,[99] and Wiseman.[100]

1.28.3 BIOSYNTHESIS OF ISOFLAVONOIDS

1.28.3.1 Experimental Systems for the Study of Isoflavonoid Biosynthesis

As described above, isoflavonoids may be either constitutively synthesized under the control of developmental programs, and/or induced in response to environmental cues such as pathogen infection. Tissues in which isoflavonoids are made constitutively have been less popular as model systems for biosynthetic studies than have inducible systems. Exceptions include cases where the constitutive isoflavonoids have particular substitutions of interest, such as the prenylated isoflavonoids of *Lupinus* and several other species,[1] or where a very wide range of isoflavonoids occurs in plants with medicinal value, such as in Kudzu vine (*Pueraria lobata*).[101]

Elicitor-treated plants and cell suspension cultures have been widely used for studies of isoflavonoid biosynthesis, enzymology, and molecular biology. The first studies to define the basic pathways leading to pterocarpanoid compounds utilized seedlings exposed to $CuCl_2$ through the roots.[102,103] This toxic abiotic elicitor induces the synthesis of most of the isoflavonoids encountered in infected tissues. Improved exposure of cells to both elicitor and labeled precursors, and simplicity of metabolite extraction due to lack of interfering chlorophyll, are features that have made elicitor-treated cell suspension cultures the most popular system for isoflavonoid biosynthetic studies.[104] Such cultures also provide large amounts of elicited material for enzyme purification. Heavy metal ions are often poor elicitors in cell suspension systems owing to their overall toxicity to the culture. Preferred elicitors are preparations, usually containing glycans and glycoproteins, obtained from the cell walls of phytopathogenic fungi or from yeast. The nature and properties of such biotic elicitors have been extensively reviewed.[105,106] The tripeptide glutathione has been used as a cheap, convenient, and biologically reproducible elicitor with some cell and organ culture systems, although it is inactive with others.[107,108]

The response of cell cultures to exogenously applied elicitors is dependent upon the growth stage of the culture, the exact nature of the culture medium (particularly the plant growth regulators and their levels), and the nature and concentration of elicitor applied. The systems for which these various parameters have been most studied are bean, soybean, alfalfa, and chickpea cell suspension cultures. Table 3 outlines the features of a range of cell and organ culture systems that have provided important information on isoflavonoid biosynthesis and its control. The references cited point the reader to the conditions for establishment and use of the culture system.

Although cell culture systems have obvious advantages for basic biochemical and molecular studies, they lack the spatial organization of the intact plant. Isoflavonoid accumulation may vary quantitatively and qualitatively depending on the distance from the applied stimulus. Such distinct proximal and distal cell responses have been best characterized in soybean using an elicitor-treated cotyledon system.[113,114]

In the intact plant, the rate and extent of production of isoflavonoid phytoalexins in response to microbial pathogens depend on the genotype of the host and the particular race of the pathogen.[115] This so-called race-specific or gene-for-gene resistance, which is often determined by single complementary genes in both host and pathogen, is usually not easy to mimic in a cell culture system, as most elicitors that have been isolated do not share the specificity for a particular host genotype that is characteristic of the organism from which they were isolated. A soybean cell culture system has been developed in which gene-for-gene resistance is expressed in response to the bacterial pathogen *Pseudomonas syringae* pv *glycinea*,[116,117] and such systems will be of value for future studies on the molecular genetic control of isoflavonoid synthesis in response to specific host resistance gene-mediated signaling pathways.

1.28.3.2 Role of the Central Phenylpropanoid and Acetate–Polymalonate Pathways in Isoflavonoid Synthesis

The B-ring of the C_{15} skeleton of flavonoids and isoflavonoids originates from the phenylpropane unit of 4-coumaroyl-CoA (**40**), whereas the A-ring is derived from head-to-tail condensation of three molecules of malonyl CoA (**41**), derived from acetyl-CoA (Scheme 2). In elicitor-treated cell

Table 3 Plant cell and organ culture systems used for the study of isoflavonoid biosynthesis.

Species	*Culture*	*Elicitor*	*Compounds*	*Ref.*
Cicer arietinum (chickpea)	cell suspension cultures (resistant and susceptible to *Ascochyta rabei*)	from *A. rabei*; yeast elicitor	pterocarpans	109
Glycine max (soybean)	cell suspension cultures	from *Phytophthora megasperma* f. sp. *glycinea*	prenylated pterocarpans	110
Lotus corniculatus (birdsfoot trefoil)	*Agrobacterium rhizogenes*—transformed hairy root cultures	from *Rhyncosporium orthosporum*; glutathione	isoflavans	107
Lupinus polyphyllus, *Lupinus hartwegii*	*Agrobacterium rhizogenes*—transformed hairy root and suspension cultures	not used	isoflavone glycosides	111
Medicago sativa (alfalfa)	cell suspension cultures	from *Colletotrichum lindemuthianum*; yeast elicitor	pterocarpans	112
Phaseolus vulgaris (French bean)	cell suspension cultures	from *Colletotrichum lindemuthianum*; yeast elicitor	prenylated pterocarpans	104
Pueraria lobata (Kudzu vine)	cell suspension cultures	yeast elicitor	isoflavones and isoflavone dimers	101

cultures, the enzymes of the core phenylpropanoid pathway, L-phenylalanine ammonia-lyase (PAL), cinnamate 4-hydroxylase (C4H), and 4-coumarate:CoA ligase (4CL), as well as the acetyl-CoA carboxylase (ACCase), are coinduced with the later enzymes specific for isoflavonoid synthesis.[118,119] Indeed, elicitor-treated cell cultures provided the biological material for cloning of the cDNAs encoding these four elicitor-induced enzymes,[119–122] and many papers have described the induction of PAL and 4CL at the enzyme activity, protein, and transcript levels in relation to isoflavonoid synthesis (reviewed by Dixon and Harrison[123]).

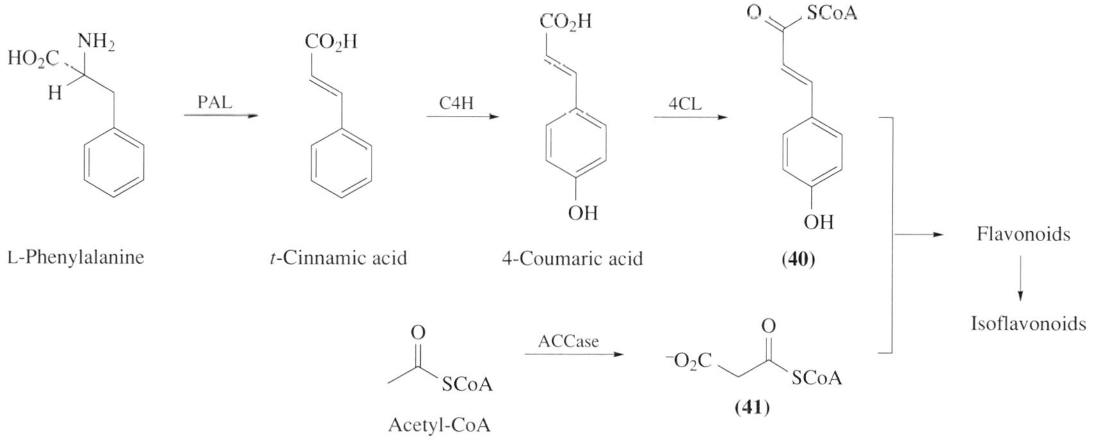

L-Phenylalanine —PAL→ *t*-Cinnamic acid —C4H→ 4-Coumaric acid —4CL→ **(40)** → Flavonoids → Isoflavonoids

Acetyl-CoA —ACCase→ **(41)**

Scheme 2

PAL is encoded by a family of at least three genes in most species studied. In bean cell cultures, elicitation leads to the preferential appearance of the PAL isoenzymes with the highest affinity (lowest K_m value) for phenylalanine, prior to the accumulation of phaseollin (**17**).[124] Whether this reflects a mechanism for the specific channeling of phenylalanine into isoflavonoid synthesis, or whether it simply increases the overall flux into the phenylpropanoid pathway, remains to be determined. It is also not clear whether specific isoforms of 4CL play a regulatory role in isoflavonoid synthesis.

Studies on transgenic tobacco plants that overexpress or underexpress PAL have demonstrated that, under normal conditions, PAL is the rate-limiting step for the synthesis of hydroxycinnamic acid esters such as chlorogenic acid, but not for the flavonoid rutin.[125,126] It is therefore likely that the rate-determining steps for isoflavonoid synthesis in legumes are downstream of the enzymes of the core phenylpropanoid pathway, although the enzymes may have to be induced to accommodate the increased flux following elicitation. The acetyl-CoA carboxylase cDNA cloned from elicited alfalfa cell cultures encodes a cytoplasmic form of the enzyme that could be involved in both fatty acid elongation and flavonoid/isoflavonoid synthesis.[119] Demonstration of whether the carboxylase activity is rate limiting for flavonoid/isoflavonoid synthesis will require the generation of transgenic plants with altered activity levels of this enzyme.

1.28.3.3 Reactions of Isoflavonoid Biosynthesis as Determined by Radiotracer Experiments

A series of pioneering experiments by Dewick and collaborators between 1978 and 1983 helped define the sequence of individual reactions leading to the formation of isoflavones, isoflavans, and pterocarpans.[102,103,127–131] These studies measured incorporation of radiolabeled precursors in $CuCl_2$-treated seedlings of alfalfa, bean, pea, and red clover. The conclusions were essentially supported by results of parallel studies in which many of the proposed biosynthetic intermediates were isolated and characterized.[132,133]

The pathways as they are currently understood are shown in Scheme 3, which outlines several clear principles for the elaboration of isoflavonoid structures. First, the isoflavone daidzein (**4**) is the first isoflavonoid product, having been formed by the isoflavone synthase reaction. It should be noted that genistein (**3**) is the corresponding precursor for a series of isoflavonoids that retain the 5-hydroxyl group that originates as the 6′-hydroxyl of 2′,4,4′,6′-tetrahydroxychalcone (naringenin chalcone) (**42**), the normal product of the chalcone synthase reaction that is formed in the absence of chalcone reductase (see below). Pterocarpans and isoflavans are then formed following 2′-hydroxylation of the isoflavone and reduction to the corresponding isoflavanone. Some substitutions of the B-ring, such as *O*-methylation, further hydroxylation, and methylenedioxy ring formation, occur prior to reduction to 2′-hydroxyisoflavanone.[102,103,113,114]

Further substitutions such as prenylation or 6a-hydroxylation occur after ring closure to yield the pterocarpan nucleus.[127,129] The final reaction of pisatin (**21**) biosynthesis is the A-ring *O*-methylation of (+)-6a-hydroxymaackiain (**43**).[129]

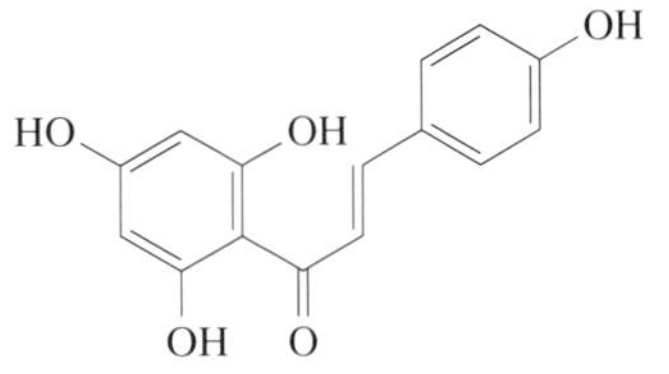

OH
HO OH
OH O
(42)

The above labeling experiments demonstrated that medicarpin (**6**) and vestitol (**8**) were interconvertible in alfalfa, and this was explained on the basis of a common carbonium ion intermediate (**44**), formed from an isoflavanol (**45**), that could be involved in the synthesis of pterocarpans, isoflavans, and possibly also coumestans[103] (Scheme 4). On the basis of radiotracer experiments, coumestans were proposed to arise via isoflav-3-enes (**46**) and 3-arylcoumarins (**47**) [131] (Scheme 4). One apparently contradictory result to arise from these studies was the unexpected lack of incorporation of (**4**) into (**6**) (via formononetin (**5**)) in alfalfa, although this compound was a good precursor of coumestrol (**9**). The possible mechanism of isoflavone 4′-*O*-methylation is discussed in detail in Section 1.28.3.10.1.

The isoflavanones and pterocarpans have one and two chiral centers, respectively, and the enzymes involved in their biosynthesis exhibit strict stereochemical requirements. In most species that have been described, the pterocarpans are of the (−)-configuration, although (+)-pisatin (**21**) occurs as the major phytoalexin in pea. Labeling studies with enatiomeric precursors indicated the preferential incorporation of (+)-(6a*S*, 11a*S*)-maackiain (**52**) over (−)-(6a*R*, 11a*R*)-maackiain (**59**) into (+)-

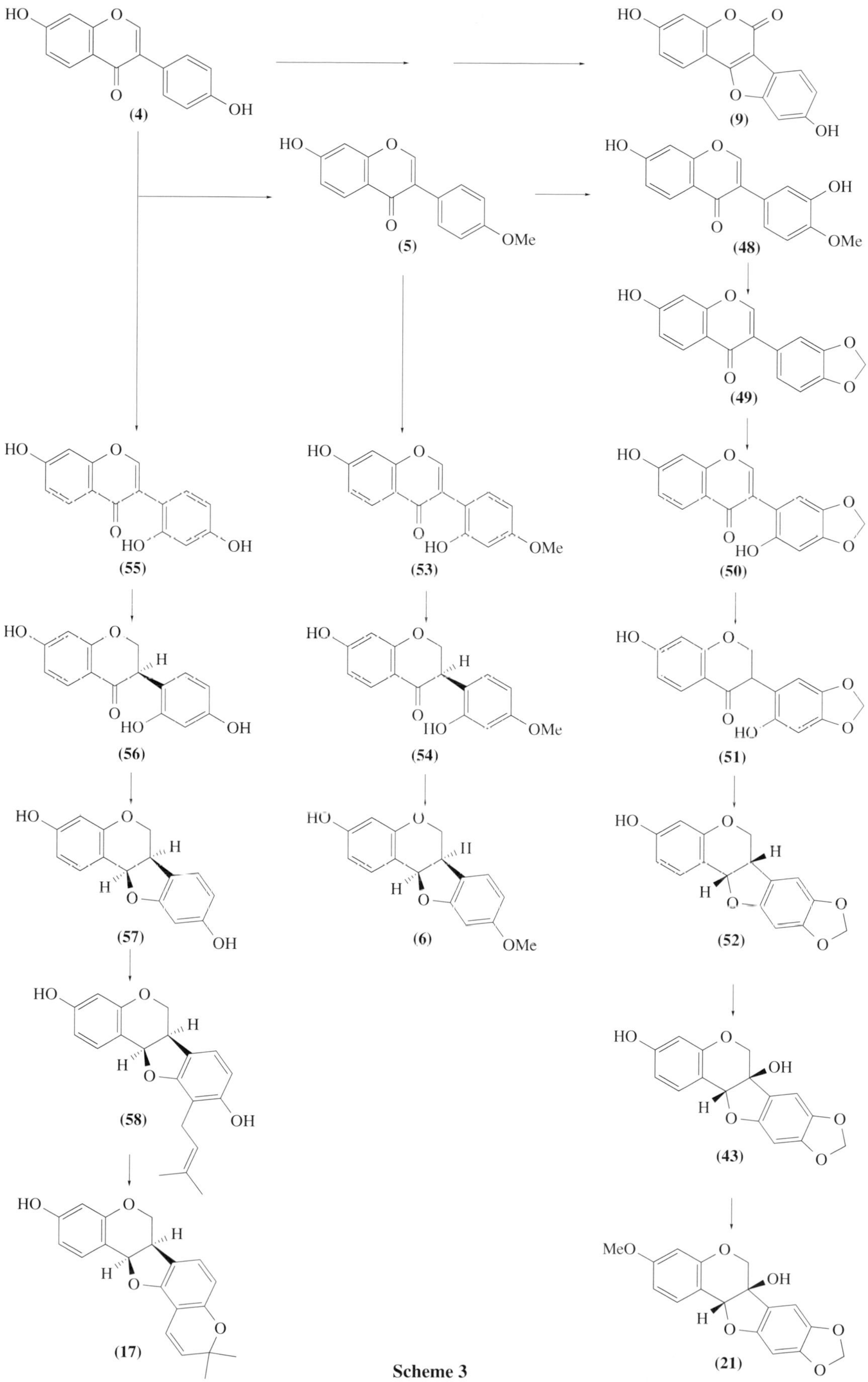

HO O (4) OH

HO O O (9) OH

HO O (5) OMe

HO O OH (48) OMe

HO O O O (49)

HO O HO OH (55)

HO O HO OMe (53)

HO O HO O O (50)

HO O H HO OH (56)

HO O H HO OMe (54)

HO O HO O O (51)

HO O H H O (57) OH

HO O H H O (6) OMe

HO O H H O O O (52)

HO O H H O (58) OH

HO O OH H O O O (43)

HO O H H O O (17)

MeO O OH H O O O (21)

Scheme 3

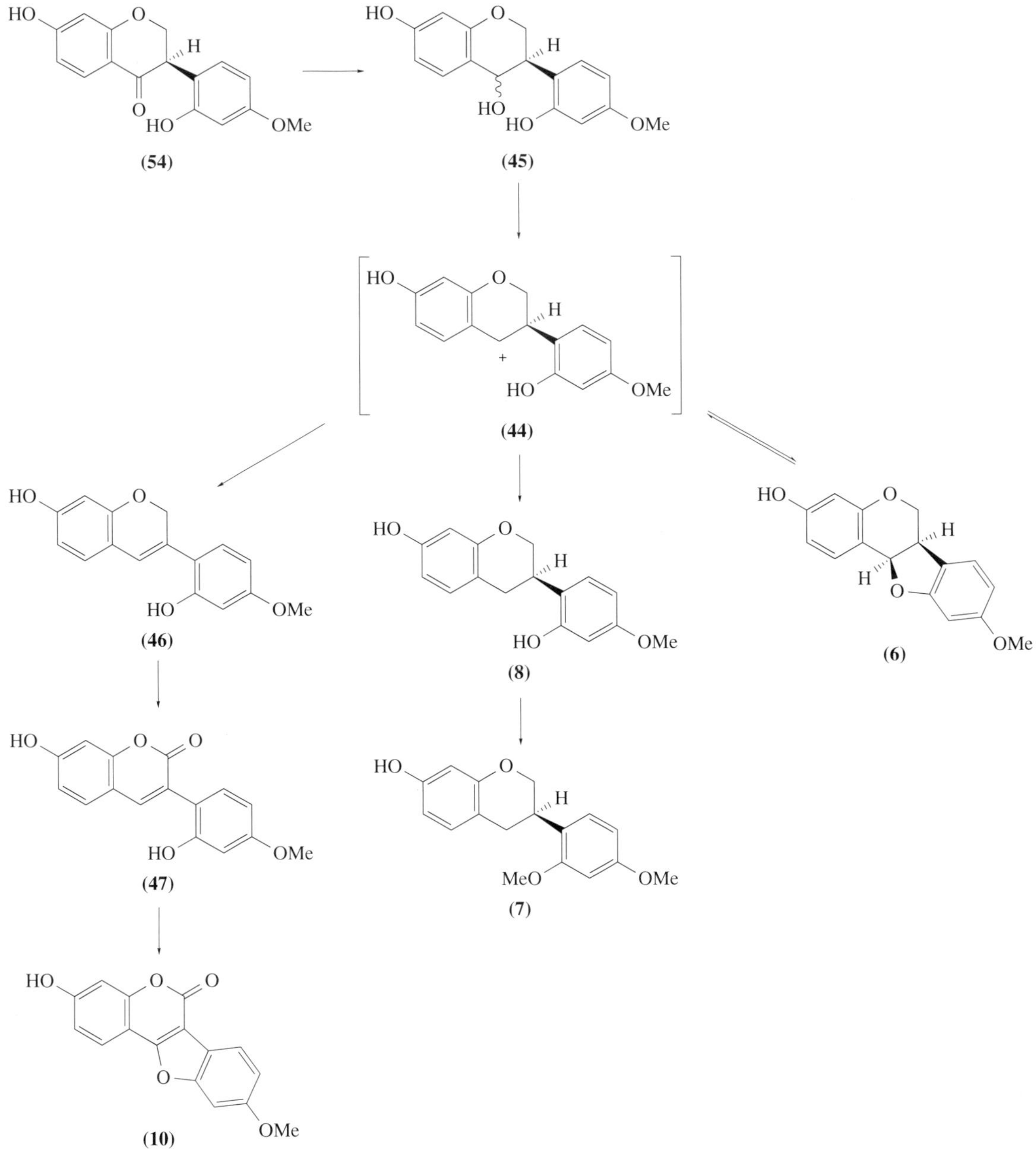

HO O H O HO OMe (54)

HO O H HO HO OMe (45)

HO O H + HO OMe (44)

HO O HO OMe (46)

HO O H HO OMe (8)

HO O H H O OMe (6)

HO O O HO OMe (47)

HO O H MeO OMe (7)

HO O O O OMe (10)

Scheme 4

(6a*R*, 11a*R*)-pisatin (**21**) (note that assignment of *S* and *R* configuration is changed following hydroxylation of the 6a position), establishing that the 6a-hydroxylation of pterocarpans occurs with retention of configuration.[128] This would be predicted because inversion would also require additional inversion at C-11a as pterocarpans have a Z-fused ring system. A full understanding of the chemical and enzymological basis of pterocarpan stereochemistry is of significant importance for attempts to improve phytoalexin efficiency through genetic engineering, because plant pathogens are often unable to metabolize the opposite stereoisomer to that produced in their host plant(s).

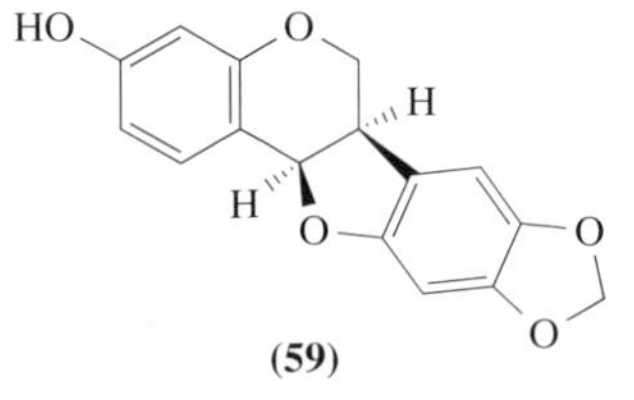

HO O H H O O O (59)

1.28.3.4 Chalcone Synthase and Chalcone Reduction

The first C_{15} precursor of the isoflavonoids is the chalcone derived from the head-to-tail condensation of 4-coumaroyl-CoA (**40**) and three molecules of malonyl CoA (**41**) catalyzed by the enzyme chalcone synthase (CHS). CHS is a dimeric polyketide synthase, subunit $M_r \approx 42\,000$, which catalyzes the addition, condensation, and cyclization reactions leading to the formation of 2′,4,4′,6′-tetrahydroxychalcone (naringenin chalcone) (**42**) (Scheme 5). CHS has been purified and characterized, and its genes cloned, from many plant species.[134–138] Further discussions here will be limited to aspects of its action and expression specifically related to its participation in the synthesis of isoflavonoids.

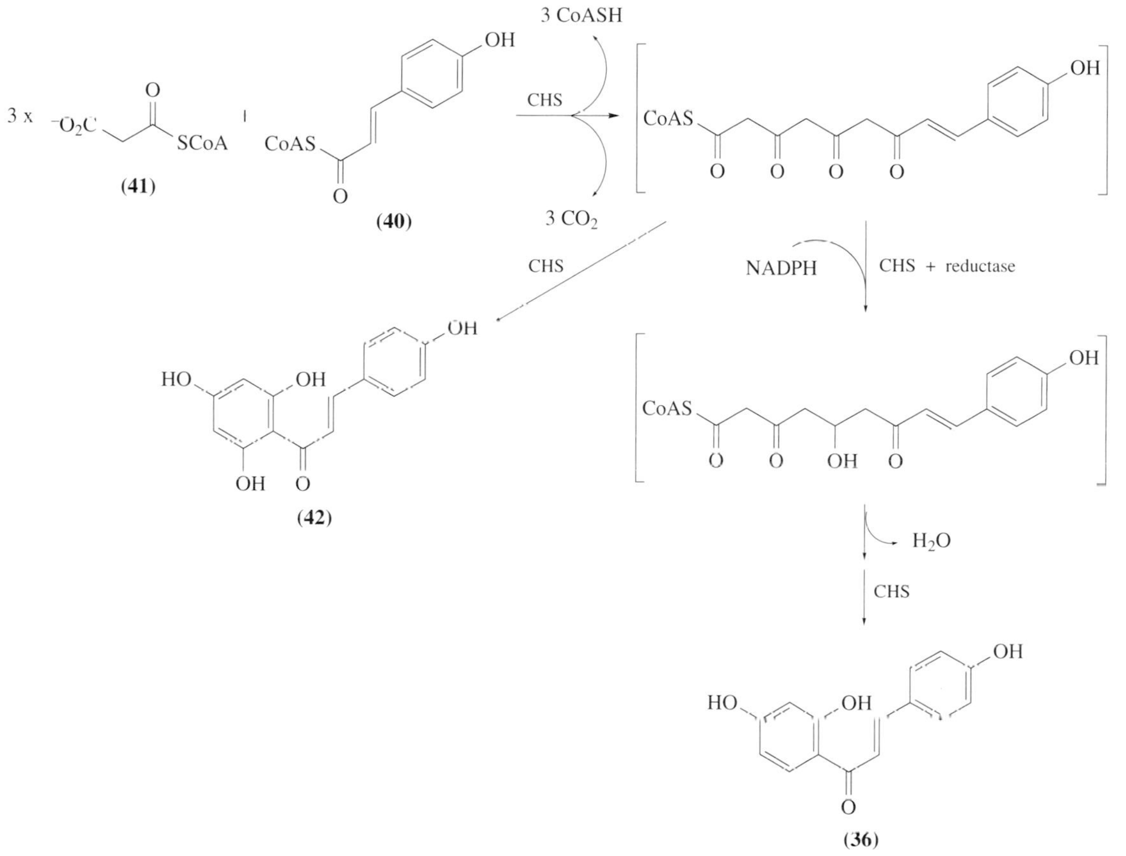

Scheme 5

The genetic model plant *Arabidopsis thaliana* contains a single *CHS* gene,[139] which is clearly sufficient for the basic functions of plant growth and development. However, in most legume species, CHS is encoded by multigene families, consisting of 6–8 members in green bean (*P. vulgaris*),[140] at least seven in soybean,[135] at least seven in pea,[136] at least four in subterranean clover,[137] six or seven in *Pueraria lobata*,[141] and more than seven in alfalfa.[138] Gene family members are often tightly clustered in the genome,[135,137,140] suggesting that they have arisen from fairly recent gene duplication events. It has been suggested that the multiple forms of CHS in legumes may have evolved to serve particular specializations of the flavonoid pathway, for production of isoflavonoid phytoalexins and flavonoid/isoflavonoid/chalcone nodulation gene inducers. However, there is currently no direct evidence in support of this hypothesis. In alfalfa, at least five different members of the *CHS* gene family are constitutively expressed in roots and root nodules, but not in the aerial parts of the plant. However, these family members are expressed in leaves, at the onset of the isoflavonoid phytoalexin defense response, following exposure to pathogens or elicitors.[138] The CHS proteins encoded by the different gene family members are generally very similar in primary sequence, and it is not known if they possess different kinetic properties or are differentially localized in the cell.

Induction of CHS at the level of activity, protein, transcript levels, translatable mRNA activity, or transcription rate has been demonstrated in cells of many legume species in relation to elicitation of isoflavonoids.[138,142–145] Considerable attention has been paid to the regulatory mechanisms whereby *CHS* genes are activated in response to developmental and environmental cues, and promoter elements and their cognate transcription factors involved in the switching on of expression of the gene during the isoflavonoid phytoalexin response have been identified.[146]

Many isoflavonoids lack the 5-hydroxyl group (6′-hydroxyl, chalcone numbering), and are derived from 2′,4,4′-trihydroxychalcone (**36**) rather than from (**42**). The 5-deoxyisoflavonoids are particularly prevalent in legume roots, and the pterocarpan phytoalexins are invariably of this class. ^{13}C-labeling studies indicated that the 5-hydroxyl group was lost prior to the cyclization of the A-ring of the chalcone,[147] presumably at the polyketide stage. After many unsuccessful attempts to demonstrate the reaction *in vitro*, it was shown that a crude extract from elicited cell cultures of *Glycyrrhiza echinata* could produce (**36**) and its corresponding flavanone liquiritigenin (**60**), in addition to naringenin (**61**), from (**40**) and (**41**) in the presence of high concentrations of NADPH.[148] Compound (**36**) was produced first, and then converted to (**60**) by chalcone isomerase present in the preparation. The activity was described as 6′-deoxychalcone synthase, and was also demonstrated in *G. echinata* protoplasts.[148]

The mechanism of 6′-deoxychalcone formation became apparent when it was shown that purified soybean CHS required the presence of a separate protein, given the trivial name "chalcone reductase" (CHR), for NADPH-dependent formation of (**36**).[149] The reductase was purified to apparent homogeneity, and was shown to be a monomer, of M_r 34 000, that catalyzed the transfer of the pro-*R*-hydrogen of [4-^{3}H]NADPH to the polyketide bound to CHS, with resultant loss of the hydroxyl function as water (Scheme 5). The enzyme had a pH optimum of 6.0, a K_m for NADPH of 17 μmol L^{-1}, exhibited approximately 90% maximum activity at a molar ratio (CHS:reductase) of 2:1, and could coact with CHS from parsley, a species that does not synthesize 6′-deoxychalcone derivatives.[149] This latter point suggests that the multiple forms of CHS found in legumes are unlikely to be involved differentially in the formation of 6′-deoxy and 6′-hydroxychalcones.

Antibodies were raised against the soybean reductase[150] and cDNA clones were obtained.[151] CHR is encoded by a small gene family in soybean[151] and alfalfa,[152–155] and has also been cloned from *Pueraria lobata* and *Glycyrrhiza echinata*;[101,155] the gene does not appear to be present in species such as carrot and parsley that do not accumulate 5-deoxyisoflavonoids.[151] CHR can be functionally expressed in *Escherichia coli*[151] and recombinant enzyme can be obtained in milligram quantities from this source.[156] The enzyme possesses a leucine zipper domain, but it is not known if this is involved in interactions with CHS. Although the enzyme is a polyketide reductase, it does not share significant sequence identity to the reductases of fatty acid synthesis; rather, it is related to a mammalian aldose reductase and prostaglandin synthase, and to 2,5-diketo-D-gluconic acid reductase from *Corynebacterium*.[151] It is still not clear why coaction of CHR with CHS never results in more than 50% formation of the 6′-deoxychalcone. This is the case with the enzyme if purified from plant sources or if produced in *E. coli*. Interaction of recombinant CHR with a CHS heterodimer containing a single active site produced no significant difference in 6′-deoxy to 6′-hydroxyl product ratio from that observed with wild-type CHS, indicating that the production of both chalcones cannot result from the presence of two functionally distinct active sites (i.e., one coupled to CHR and one not).[157]

Studies using CHR antibodies and cDNA probes have demonstrated closely coordinated induction of CHR with CHS at the protein, mRNA activity or transcription rate levels in elicited cell suspension cultures of soybean,[150] *G. echinata*,[158] and alfalfa.[159]

1.28.3.5 Chalcone Isomerase. Formation of the Immediate Precursor for Isoflavone Formation

At alkaline pH, naringenin chalcone (**42**) isomerizes spontaneously to the corresponding flavanone, naringenin, yielding a racemic mixture of (+)- and (−)-forms (**61**,**62**) (Scheme 6). This reaction occurs less readily with (**36**). The substrate for aryl migration to isoflavone is the (−)-(2*S*)-flavanone (see below), and it is formed *in planta* from the corresponding chalcone by the activity of chalcone isomerase (CHI). CHI has been purified from many sources,[160–162] and cDNAs and genomic clones have been characterized.[163–168] The enzyme is generally present at significantly higher activity levels than CHS, but is nevertheless induced by elicitor treatment in various legume cell suspension cultures at the onset of isoflavonoid accumulation.[169,170]

CHI catalyzes a net intramolecular *cis*-addition to the chalcone double bond (Scheme 6), and various models have been proposed to describe the stereochemical course of the reaction.[171,172] The size of the enzyme appears to vary depending on its source. In legumes such as bean and alfalfa, it has an M_r of ~28 000 and, in contrast to CHS, appears to exist as a single form.[173] The CHIs from

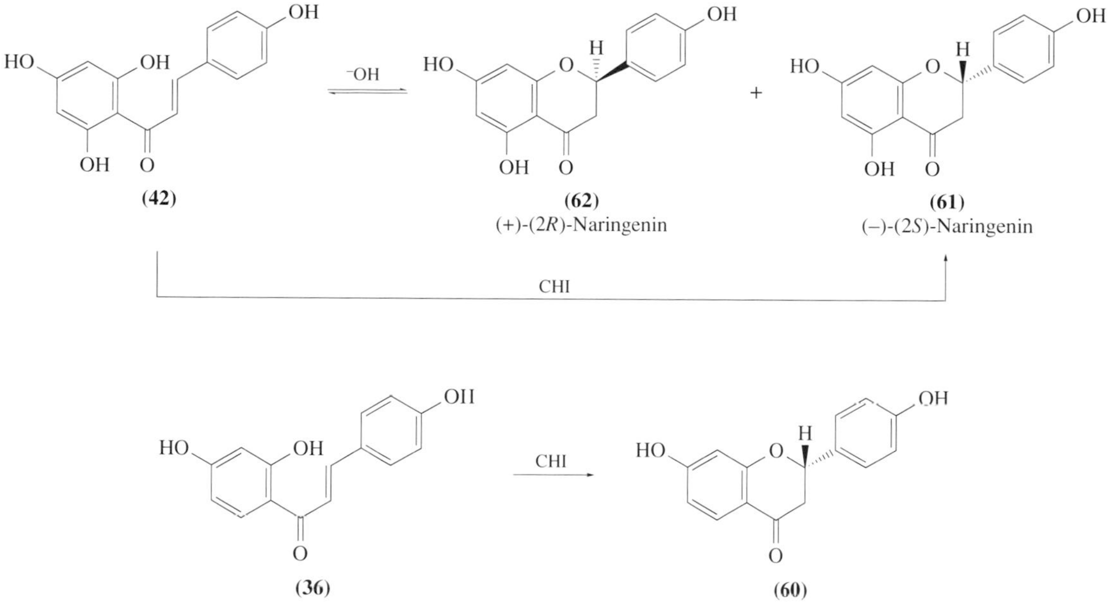

(42) ⇌ **(62)** (+)-(2*R*)-Naringenin + **(61)** (−)-(2*S*)-Naringenin

CHI

(36) → CHI → **(60)**

Scheme 6

legumes catalyze the isomerization of both (**36**) and (**42**), whereas the former is not utilized by the enzyme from parsley, *Petunia*, *Dianthus*, or *Callistephus*, plants that do not make the 5-deoxy class of flavonoids/isoflavonoids.[173]

1.28.3.6 "Isoflavone Synthase." The First Committed Step of the Isoflavonoid Pathway

The chalcone/flavanone pair represents the branch point for the elaboration of the various flavonoid and isoflavonoid secondary metabolites found in the plant kingdom. Various *in vivo* labeling experiments prior to 1984 (reviewed by Dixon *et al.*[174]) had demonstrated that chalcone/flavanone was incorporated into isoflavonoids. For example, Grisebach and Brandner[175] showed that 2′,4,4′-trihydroxychalcone 4-*O*-glucoside (**63**) was converted to daidzein (**4**) in chickpea, although at very low levels, and (2*S*)-naringenin (**61**) was stereoselectively converted to biochanin A (**33**), a finding that was interpreted as indicating that the flavanone rather than its nonoptically active isomeric chalcone was the substrate for the proposed enzyme-catalyzed B-ring aryl migration. Such a reaction was assumed to be oxidative, leading to the isoflavone as the first isoflavonoid product. At the same time, putative mechanisms for such a reaction were proposed, based either on theoretical considerations or on results of direct chemical oxidations of flavonoid compounds. Such mechanisms included epoxidation and/or the formation of a spirodienone intermediate with aryl migration associated with *O*-methylation at the 4′-position (Scheme 7).

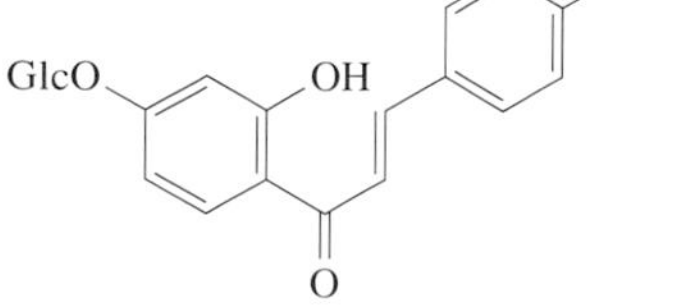

(63)

In 1984, Hagmann and Grisebach[176] provided the first evidence for the enzymatic conversion of flavanone to isoflavone (the "isoflavone synthase" (IFS) reaction) in a cell-free system. They demonstrated that microsomes from elicitor-treated soybean cell suspension cultures could catalyze the conversion of (**61**) to genistein (**3**) or of (2-*S*)-liquiritigenin (**60**) to (**4**) in the presence of NADPH. The crude microsomal enzyme preparation, which was stable at $-70\,^{\circ}C$ but had a half-life of only 10 min at room temperature, was absolutely dependent on NADPH and molecular oxygen. It was

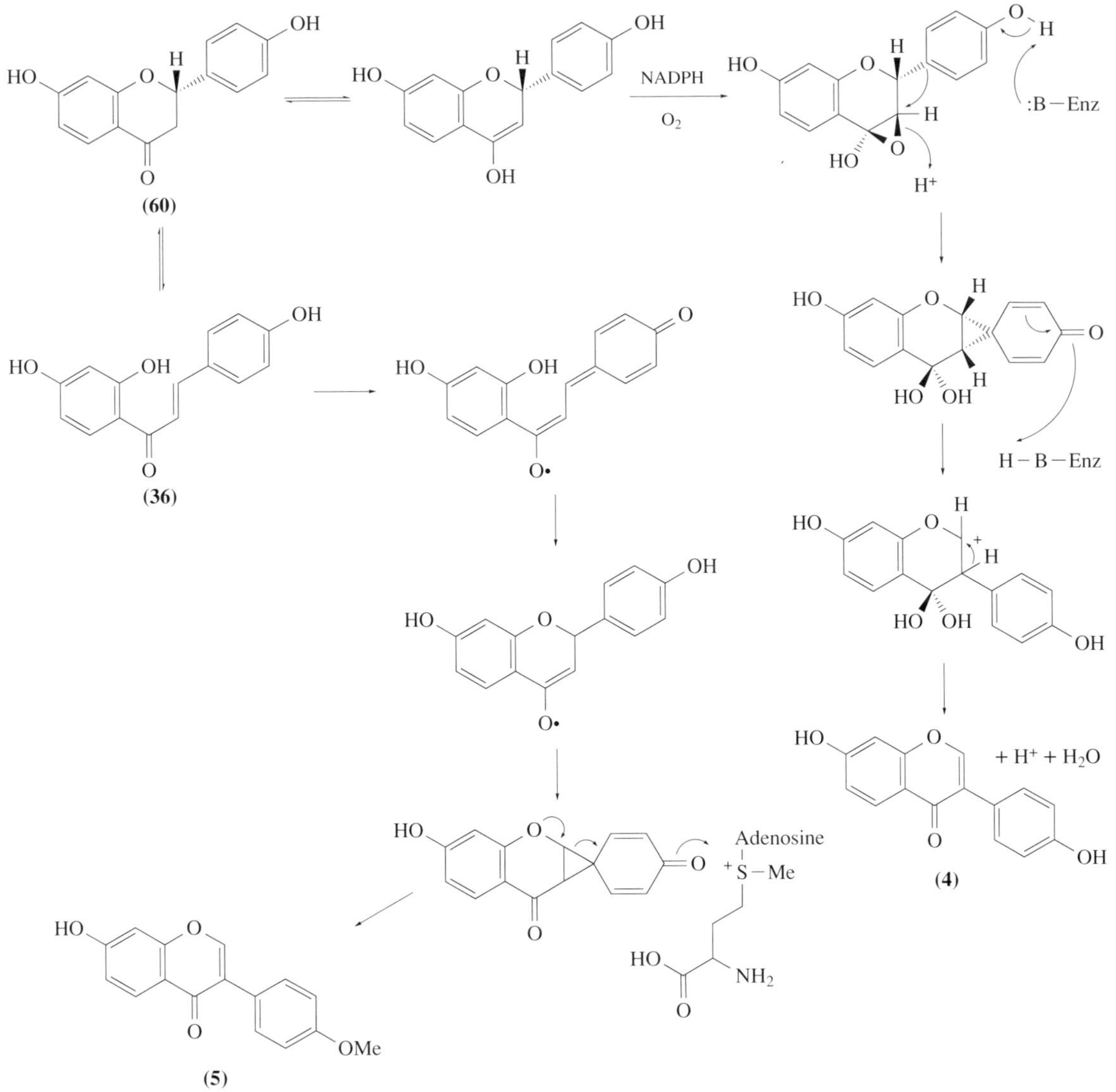

Scheme 7

subsequently shown[177] that the reaction proceeded in two steps. Naringenin was converted in a cytochrome P450-catalyzed reaction requiring NADPH and O_2 to the corresponding 2-hydroxy-isoflavanone (**64**). This relatively unstable compound then underwent dehydration to yield (**3**) (Scheme 8). The dehydration reaction appeared to be catalyzed by an activity present predominantly in the cytoplasmic supernatant, although it was not possible to remove all this activity from the microsomes. Compound (**64**) can spontaneously convert to (**3**), for example in MeOH at room temperature. Kinetic analysis indicated that (**64**) is formed prior to (**3**), consistent with its being an intermediate.

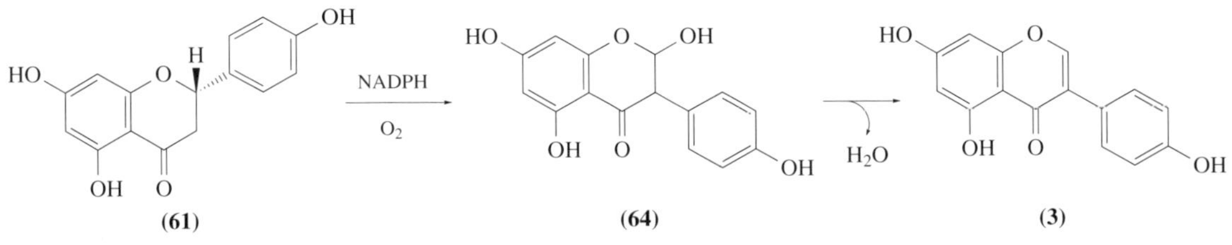

Scheme 8

Involvement of cytochrome P450 in the 2-hydroxyisoflavanone synthase reaction was confirmed by inhibition by CO, replacing O_2 with N_2, and a range of known P450 inhibitors of which ancymidol was the most effective. The enzyme comigrated with the endoplasmic reticulum markers cinnamate 4-hydroxylase and cytochrome b5 reductase on Percoll gradients. The properties of the crude microsomal IFS are shown in Table 4. The enzyme is stereoselective, and (2*R*)-naringenin (**62**) is not a substrate.

Table 4 Properties of isoflavone synthase activity in microsomes from elicited cell suspensions of *Glycine max* and *Pueraria lobata*.

Property	*G. max*	*P. lobata*
Specific activity:		
Unelicited	not reported	76 nkat kg^{-1}
Elicited	302 nkat kg^{-1}	1164 nkat kg^{-1}
Optimum pH	8.0–8.6	not reported
K_m:		
naringenin (**61**)	8.7 μmol L^{-1}	20 μmol L^{-1}
liquiritigenin (**60**)	not reported	6.9 μmol L^{-1}
NADPH	39 μmol L^{-1}	not reported
$t_{1/2}$ at 4 °C	200 min	60 h

Source: Kochs and Grisebach[177] and Hakamatsuka *et al.*[178]

The pioneering work of Grisebach's group did not unequivocally address the origin of the 2-hydroxyl group (i.e., from molecular oxygen or water). Indeed, their model for the reaction involved the initial formation of a diol at position C-4 with subsequent addition of a hydroxyl to the carbocation formed at position 2 (Equation (1)). The origin of the 2-hydroxyl group was determined from studies on the isoflavone synthase present in microsomes from elicited cell cultures of *Pueraria lobata*, some properties of which are summarized in Table 4. Carefully washed microsomes produced predominantly (**65**) from (**60**), whereas (**4**) was the only product in cruder microsome preparations. As with the soybean system, a soluble enzyme was shown to catalyze the dehydration of (**65**) to (**4**).[179] ^{18}O from $^{18}O_2$ was incorporated into the 2-hydroxyl group, resulting in a 2-hydroxyisoflavanone with the molecular ion shifted by two mass units, whereas there was no corresponding shift in the molecular ion of (**4**), consistent with the subsequent dehydration reaction. Furthermore, use of [4-^{18}O]-(**60**) as substrate demonstrated that there was no exchange of the carbonyl oxygen, a finding which disproves the earlier suggestion of the formation of a diol at position C-4. The currently accepted model for the reaction pathway of "isoflavone synthase" (Scheme 9)[180] therefore involves P450-catalyzed hydroxylation coupled to aryl migration, a reaction with mechanistic similarities to the well-described proton migration mechanism of some P450 reactions. Similarities between the mechanism of IFS and other reactions such as ring condensation of *ent*-7-hydroxykaurenoic acid to GA_{12} aldehyde in gibberellin biosynthesis, formation of the furan ring in furanocoumarin synthesis, and sterol demethylation have been discussed.[180]

(**64**) (1)

Dual-labeling experiments with [^{14}C]chalcone and [^{3}H]flavanone confirmed that the flavanone and not the chalcone was the substrate for the *Pueraria* IFS.[178] This confirms the role of chalcone isomerase as a key enzyme of isoflavonoid synthesis. The *Pueraria* IFS prefers the 5-deoxyflavanone (**60**) to (**61**), and this is reflected by the chalcone isomerase from cell cultures of this species being active against (**36**) but not (**42**).[178]

There have been no reports on the purification to homogeneity, or the molecular cloning, of either of the two enzymes of the IFS complex. The flavanone 2-hydroxylase cytochrome P450 from *Pueraria* has been solubilized with Triton X-100 and partially purified by DEAE-Sepharose chromatography; the enzymatic reaction could be reconstituted by addition of NADPH cytochrome P450 reductase that separated from the hydroxylase on the ion-exchange column.[177] The 2-hydroxy-

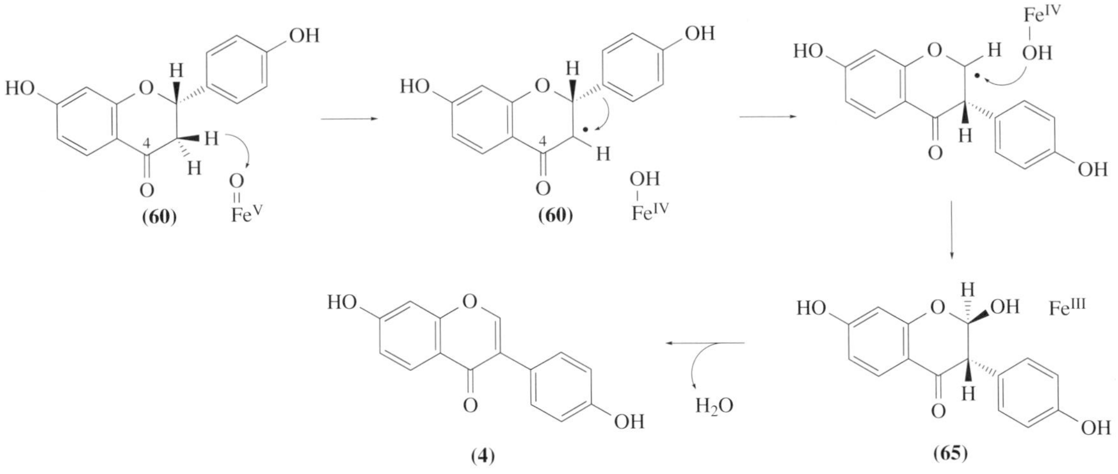

Scheme 9

isoflavanone dehydratase has been purified from elicitor-treated *P. lobata* cells, and is a soluble monomeric enzyme of subunit M_r 38 000.[101] It is not clear whether this enzyme physically associates with the P450 hydroxylase catalyzing the aryl migration.

Flavanone is a potential substrate for more than one type of hydroxylation reaction at the 2-position. Thus, elicitor-treated cell cultures of alfalfa and *Glycyrrhiza echinata* have been shown to accumulate the dibenzoylmethane licodione (**66**).[181,182] Licodione synthase is, by classical criteria, a cytochrome P450, the activity of which is induced by yeast elicitor in *Glycyrrhiza* cells.[182] The reaction it catalyzes involves 2-hydroxylation of flavanone followed by hemiacetal opening, and may have mechanistic similarities to the flavone synthase II enzyme previously characterized from soybean.[183] A comparison of the reactions catalyzed by IFS, licodione synthase, and flavone synthase II (leading to the formation of 7,4′-dihydroxyflavone (**67**)) is shown in Scheme 10.

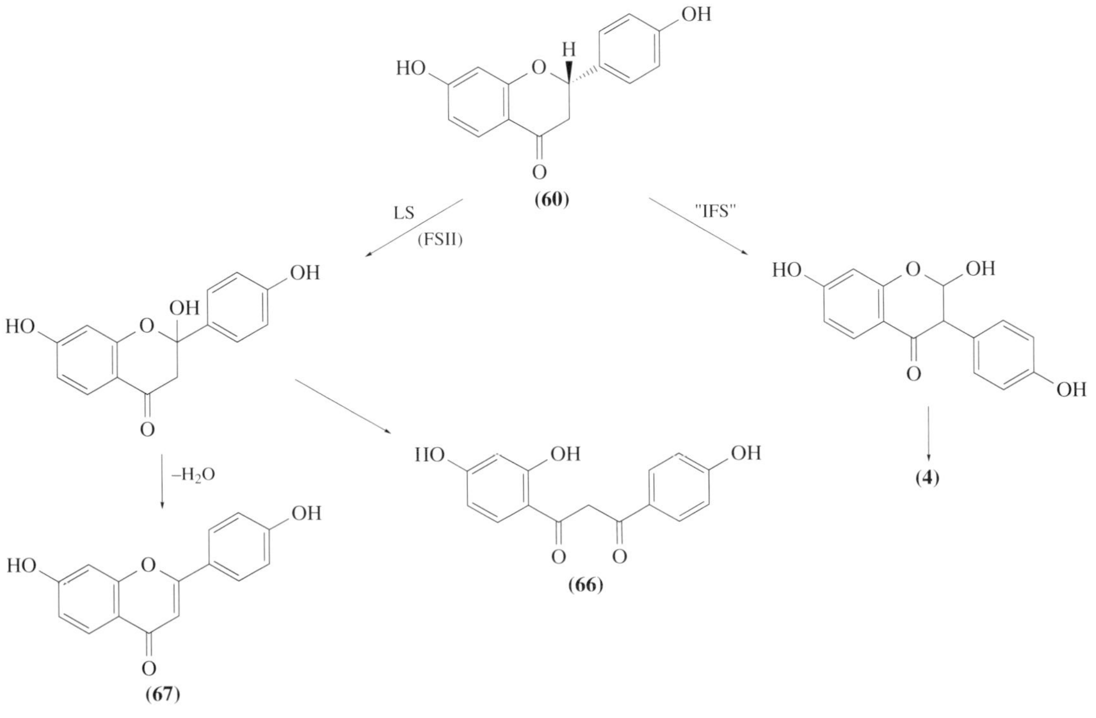

Scheme 10

1.28.3.7 2′- and 3′-Hydroxylation of Isoflavones

In addition to the formation of 2-hydroxyflavanone (**64**) and genistein (**3**), soybean microsomes incubated with naringenin (**61**) also produced small amounts of 2′-hydroxygenistein (**68**), indicating the presence of an isoflavone 2′-hydroxylase activity.[177] 2′-Hydroxylation is a prerequisite for subsequent reduction and ring closure in the formation of pterocarpans, and elicitor-induced increases in microsomal isoflavone 2′-hydroxylase activities have been described in cell cultures of alfalfa[184] and chickpea,[185–187] associated with phytoalexin accumulation.

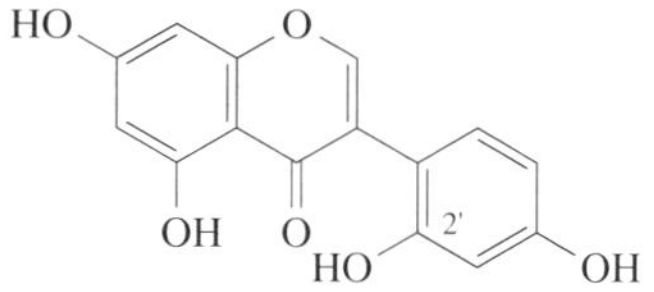

(68)

Microsomes isolated from yeast elicitor-treated chickpea cell cultures catalyzed the formation of 2′-hydroxyformononetin (**53**) and calycosin (3′-hydroxyformononetin) (**69**) from formononetin (**5**) (K_m values 3.3 and 11.0 μmol L^{-1}, respectively), and 2′-hydroxybiochanin A (**70**) and pratensin (3′-hydroxybiochanin A) (**71**) from biochanin A (**33**) (K_m values 13.0 and 12.5 μmol L^{-1}, respectively) (Scheme 11). No 2′,3′-dihydroxylated products were formed. No activity was observed with daidzein (**4**) or (**3**) as substrates, suggesting that, at least in chickpea, 4′-*O*-methylation is a prerequisite for 2′-hydroxylation. Furthermore, no activity was observed with formononetin-7-ethyl ether (**72**), (**53**), vestitione (**54**), or medicarpin (**6**). The 2′-hydroxylation reactions were shown to be catalyzed by a cytochrome P450 enzyme system on the basis of inhibition with cytochrome *c*, juglone, or flushing with nitrogen.[185,187]

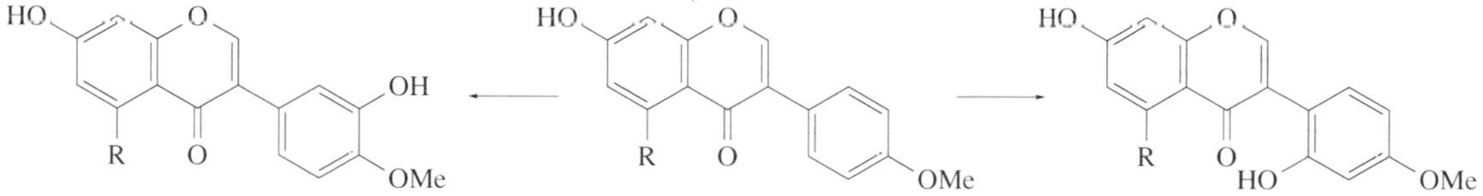

(69) R = H
(71) R = OH

(5) R = H
(33) R = OH

(53) R = H
(70) R = OH

Scheme 11

Several pieces of indirect evidence point to there being more than one enzyme involved in the 2′- and 3′-hydroxylation of chickpea isoflavones. Thus, the optimum pH for 2′-hydroxylation is 7.4, whereas that for 3′-hydroxylation is 8.0.[187] The cytochrome P450 inhibitors BAS 110 and BAS 111 have differential effects on 2′- and 3′-hydroxylase activities. Finally, the induction kinetics of the 2′- and 3′-hydroxylase activities are different in both cell cultures and roots.[186,187] It is still not clear whether one or two enzymes catalyze the 2′-hydroxylation of (**5**) and (**33**).

It is interesting to note that, unlike the situation in chickpea, 2′-hydroxylase activity in soybean, a species that does not appear to accumulate 4′-methoxyisoflavonoids, does not require a 4′-methoxylated isoflavone as substrate.[177] Compound (**5**) is a substrate for 2′-hydroxylation in alfalfa, but no 3′-hydroxylase activity appears to be present in microsomes from elicited alfalfa cells.[184] The 3′-hydroxylation reaction observed in chickpea microsomes is presumably involved in the synthesis of (−)-maackiain (**59**), which has a 3′,4′-methylenedioxy substituent. It is probable that 3′-hydroxylation and formation of the methylenedioxy bridge occur prior to 2′-hydroxylation during the biosynthesis of (**59**) in chickpea[185] (see Section 1.28.3.10.2).

Comparative studies of the activities of a range of enzymes involved in the formation of medicarpin in elicited chickpea cell cultures from lines resistant or susceptible to the fungal pathogen *Ascochyta rabiei* have indicated that the increased production of (**6**) in the resistant line is most likely determined by its high activity of formononetin 2′-hydroxylase.[186]

1.28.3.8 Conversion of Isoflavone to Isoflavanone

Reduction at the C-2 atom of the heterocyclic ring of isoflavones leads to the formation of isoflavanones, which are obligatory intermediates in the formation of pterocarpans.[130,188] *In vivo* labeling and early enzymological studies confirmed that 2′-hydroxylation is a prerequisite for enzymatic reduction of isoflavone to isoflavanone. A crude enzyme preparation from yeast elicitor-treated chickpea cell suspension cultures was shown to catalyze the NADPH-dependent reduction of 2′-hydroxyformononetin (**53**) to vestitione (**54**); the only other substrate was the methylenedioxy-substituted isoflavone 2′-hydroxypseudobaptigenin (**73**), which is converted to (−)-sophorol (**74**) (Scheme 12).[189] This activity was rapidly and strongly induced in response to elicitors in a chickpea cell line that accumulated medicarpin, but was only weakly induced in a line that produced little phytoalexin and was susceptible to the fungus *Ascochyta rabiei*.[109,189]

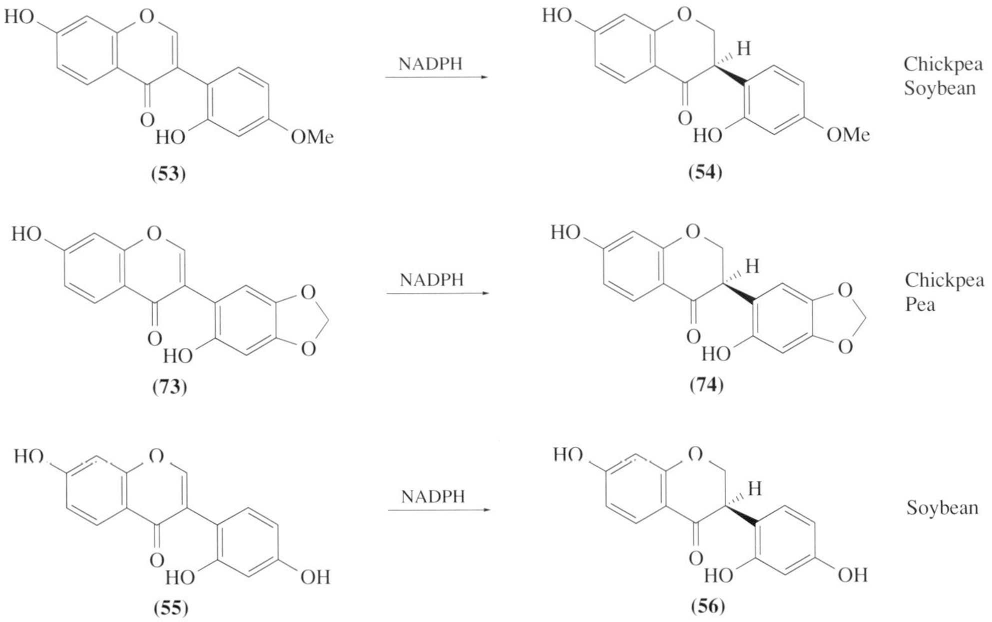

(53) →NADPH→ (54) Chickpea Soybean

(73) →NADPH→ (74) Chickpea Pea

(55) →NADPH→ (56) Soybean

Scheme 12

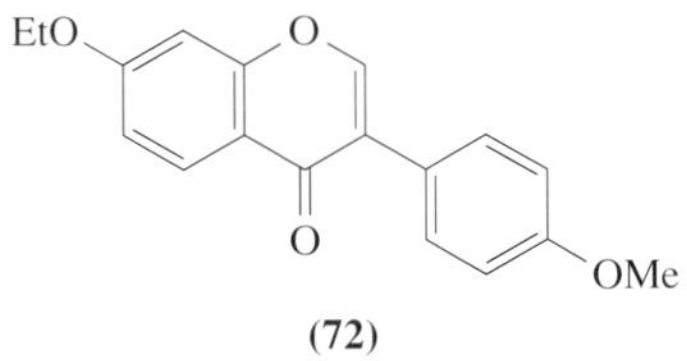

(72)

The chickpea isoflavone reductase (IFR) has been purified to homogeneity and a cDNA clone obtained.[190] The highly purified enzyme is a mononer of M_r 36 000, and has K_m values of 6, 6, and 20 μmol L^{-1} for (**53**), (**73**), and NADPH, respectively.[190] Substrate specificity studies confirmed absolute requirements for the 2′-hydroxyl group and either a 4′-methoxy or 4′,5′-methylenedioxy substitution on the B-ring. This contrasts with the enzyme from soybean, a species that does not produce 4′-*O*-methylated isoflavonoids. Soybean IFR has been purified to homogeneity from elicitor-treated cell cultures. Although having an absolute requirement for the 2′-hydroxyl substitution pattern, it can convert 2′-hydroxydaidzein (**55**) to 2′-hydroxydihydrodaidzein (**56**) (K_m 50 μmol L^{-1}) and also (**53**) to (**54**) (K_m 60 μmol L^{-1}.)[191]

Extracts from $CuCl_2$-elicited pea seedlings, which accumulate the methylenedioxy-substituted pterocarpans (+)-maackiain (**52**) and (+)-pisatin (**21**), can catalyze the reduction of (**73**) to (**74**). This activity is strongly induced on elicitation in parallel with that of a 6a-hydroxymaackiain-3-*O*-methyltransferase.[192] It would appear, from the substrate specificities of the IFRs from chickpea, soybean, and pea, that reduction of isoflavone to isoflavanone generally occurs after B-ring methylation in species that make B-ring methoxy isoflavanone-derived compounds.

Antibodies to the soybean IFR protein recognize IFR and two additional bands of slightly higher M_r on Western blots of crude soybean protein extracts.[191] Likewise, antibodies against a pea IFR recognize multiple bands, one constitutive and two induced, on Western blots of elicited alfalfa cell cultures,[193] suggesting that plants may contain multiple IFR-like proteins (see Section 1.28.6).

Reduction of the C-2 atom of isoflavones generates a chiral center. The CD spectrum of the product formed from (**55**) by the purified soybean IFR confirmed the 3*R* stereochemistry of the isoflavanone.[191] Studies with stereospecifically tritiated NADPH suggested *trans*-addition from H_A of NADPH to the C-2 atom of (**53**) by the chickpea IFR.[194] This has been confirmed using recombinant alfalfa IFR expressed in *E. coli* to catalyze the formation of (3*R*)-vestitone (**54**) from (**53**).[193] However, the CD spectrum of (**74**) produced from (**73**) by recombinant pea IFR expressed in *E. coli* also revealed the 3*R* stereochemistry,[195] which was unexpected because the final pterocarpan products in pea, (+)-maackiain (**52**) and (+)-pisatin (**21**), have the opposite stereochemistry at their two chiral centers (6a and 11a) compared with the (−)-medicarpin (**6**), (−)-maackiain (**59**), or (−)-3,9-dihydroxypterocarpan (**57**) produced in alfalfa, chickpea, or soybean. This problem is discussed further in Section 1.28.3.9.

Amino acid sequence data indicate a high degree of similarity between the IFRs cloned from chickpea,[190] alfalfa,[193] and pea.[195] Alfalfa IFR is 92% identical with pea IFR at the overall amino acid level, and shares 62% identity in its *N*-terminal region to *Antirrhinum* dihydroflavanol reductase. IFR is encoded by a single gene in pea and alfalfa, and its transcripts are highly induced in response to elicitors.[193,195] This induction has been shown, by nuclear transcript run on analyses, to be the result of increased *de novo* transcription rather than effects on RNA stability.[159,196]

Northern blot analysis has indicated that IFR transcripts in alfalfa are most abundant in roots and nodules, consistent with the constitutive accumulation of isoflavanoid malonyl glycosides in these organs.[193] This same pattern of expression is observed for the alfalfa IFR promoter driving expression of the β-glucuronidase (GUS) reporter gene in transgenic alfalfa, although unexpected ectopic expression in a range of tissues was observed following transformation of tobacco with IFR–GUS fusions.[196]

1.28.3.9 Isoflavanone Reductase and the Synthesis of Pterocarpans

Pterocarpans contain a fused furan ring structure that arises from ring closure between the C-4-carbonyl and C-2′-positions of 2′-hydroxyisoflavanones. The mechanism of this reaction has been elucidated following the characterization of the enzymatic system involved in this late stage of isoflavonoid phytoalexin synthesis.

The first report of the *in vitro* formation of pterocarpans was the demonstration of the NADPH-dependent conversion of vestitone (**54**) to (6a*R*, 11a*R*)-medicarpin (**6**) catalyzed by a soluble enzyme extract from yeast elicitor treated chickpea cell suspensions.[197] The greater incorporation into (**6**) of optically active (−)-(**54**) than the racemic mixture indicated a stereochemical preference for the 3*R* optical isomer of (**54**) by the enzyme system. The enzyme was partially purified through ammonium sulfate precipitation and ion-exchange chromatography, and appeared to elute as a single activity, optimum pH 6.0, with K_m values for (**54**) and NADPH of 17 μmol L^{-1} and 40 μmol L^{-1} respectively.[197] Essentially similar observations were made in the case of the enzyme system from elicited soybean cell suspensions,[198] which was purified only 7.3-fold by a five-step procedure including ion-exchange chromatography, blue Sepharose (to which the enzyme did not bind), and gel filtration (revealing an M_r of 29 000). This enzyme converted (3*R*)-2′-hydroxydihydrodaidzein (**56**) to (6a*R*, 11a*R*)-3,9-dihydroxypterocarpan (**57**), with an optimum pH of 6.0 and K_m values for the isoflavanone and NADPH of 75 μmol L^{-1} and 45 μmol L^{-1}, respectively. Compound (**54**) was converted to (**6**) at approximately half of the rate for the conversion of (**56**) to its corresponding pterocarpan. In contrast, the chickpea enzyme has an absolute requirement for the presence of the 4′-methoxy group of (**54**), and (**56**) is therefore not a substrate (Scheme 13).

Both of the above reports suggested that pterocarpan formation was catalyzed by a single enzyme, which was termed "pterocarpan synthase," and this activity, with 2′-hydroxyisoflavanone as substrate, is induced by elicitor treatment when measured in crude extracts from chickpea, soybean, and alfalfa cell suspension cultures.[193,197,198] Bless and Barz[197] indicated the possibility that the reaction might proceed through an isoflavan-4-ol intermediate (**75**) (Scheme 13). Confirmation of this came from studies on the "pterocarpan synthase" from elicited alfalfa cell suspension cultures.[199] Attempted purification on red agarose indicated that an enzyme activity consuming (**54**) was bound to the column, but, on elution, this activity did not form medicarpin (**6**). Rather, an intermediate was formed which could be converted to (**6**) by a second enzyme present in the flow-through fraction from the red agarose affinity column. The intermediate compound was shown to be 7,2′-dihydroxy-4′-methoxyisoflavanol (DMI) (**75**) when (**54**) was used as substrate. The "pterocarpan synthase" reaction was therefore catalyzed by two enzymes, a reductase (in the case

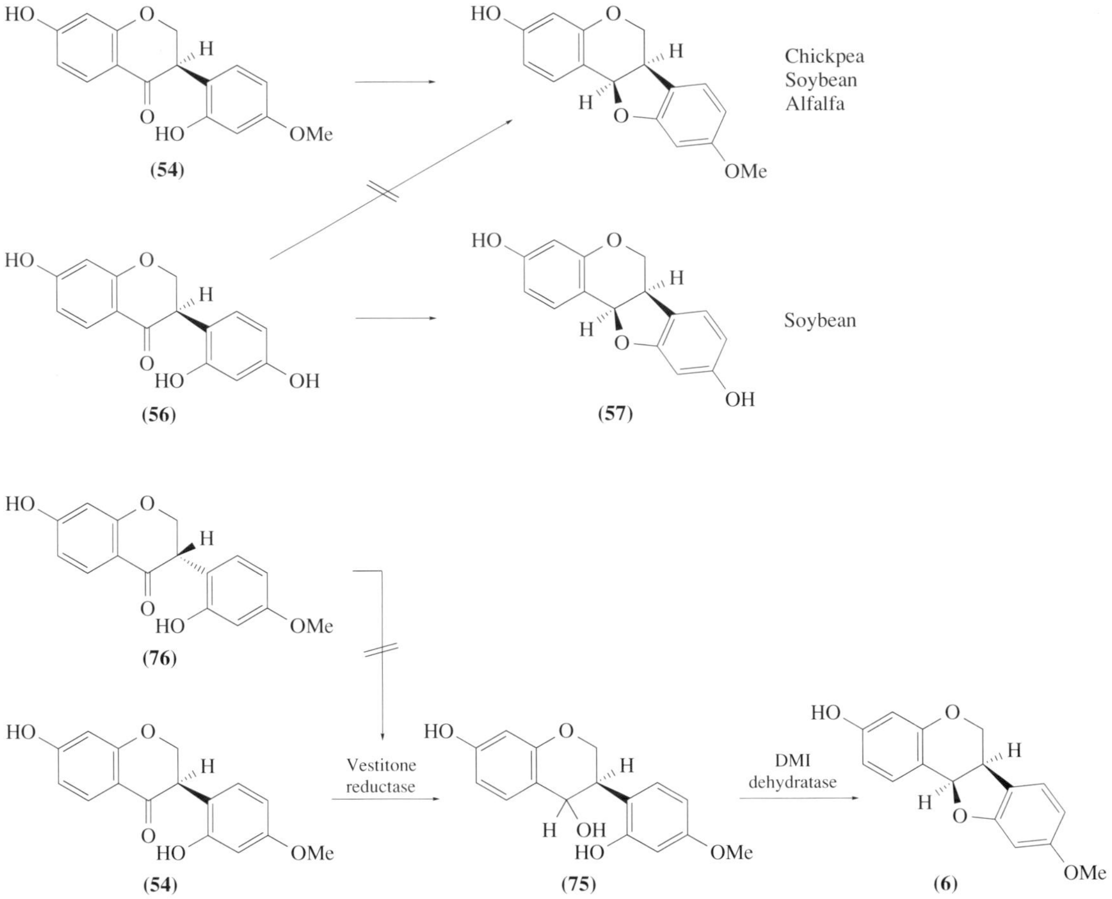

Scheme 13

of alfalfa a vestitone reductase) that converts the 2′-hydroxyisoflavanone to its corresponding isoflavanol, and a dehydratase that catalyzes the final ring closure. Alfalfa vestitone reductase and DMI dehydratase were both extensively purified.[199] The reductase is a monomeric enzyme of subunit M_r 38 000, optimum pH 6.0, with a K_m value for (**54**) of 40 μmol L^{-1}. The activity is inhibited by concentrations of (**54**) above 50 μmol L^{-1}. The enzyme was specific for (3*R*)-(**54**), and (3*S*)-vestitone (**76**) did not inhibit the enzyme. DMI dehydratase has a native M_r of 38 000, optimum pH 6.0, and a K_m value for (**75**) of 5 μmol L^{-1}. It produces only (6a*R*, 11a*R*)-medicarpin (**6**) from (3*R*)-vestitone (**54**).[200] It appears to be a very hydrophobic enzyme, and can form a physical association with vestitone reductase *in vitro* at low salt concentrations.[200] The dehydratase is coinduced with the reductase in elicitor-treated alfalfa cells.[200]

It is very likely that the "pterocarpan synthases" from chickpea and soybean also consist of a separate reductase and dehydratase. This would explain why multi-step purification protocols only gave very restricted purifications from these sources,[197,198] the reductase and dehydratase will partially copurify on size-exclusion chromatography, and a small amount of dehydratase may then fractionate with the reductase on other matrices due to protein–protein interactions. It will be interesting to determine whether the reductase and dehydratase are physically associated *in vivo*.

1.28.3.10 Substitution of the Isoflavonoid Nucleus

1.28.3.10.1 O-Methylation

The isoflavonoid phytoalexins of several species, including alfalfa and chickpea, are methylated at the 4′-position of the B-ring (Figure 1). However, in spite of many metabolic and enzymatic studies, the nature of the enzymatic step resulting in this methylation is still unclear.

Radiolabeled precursor feeding experiments with elicited alfalfa seedlings indicated that, although 2′,4,4′-trihydroxychalcone (**36**) and formononetin (**5**) were good precursors of medicarpin (**6**), daidzein (**4**) (the presumed substrate of the 4′-*O*-methyltransferase) was not incorporated.[130] These results were originally interpreted as indicating a requirement for methylation of the B-ring during the aryl migration reaction catalyzed by isoflavone synthase[130] (see Scheme 7). This now seems unlikely in view of the demonstration that the aryl migration catalyzed by the 2-hydroxyisoflavanone synthase described in Section 1.28.3.6 can occur in the absence of methylation in species in which the 4′-hydroxyl group is either free (e.g., in soybean)[177] or methylated (e.g., in alfalfa).[184] Our knowledge of the substrate specificities of the enzymes preceding and following (**4**) in the isoflavonoid pathway points to (**4**) as the substrate for 4′-*O*-methylation. Furthermore, a mutant of subterranean clover (*Trifolium subterraneum*), which produced greatly reduced levels of (**5**) and biochanin A (**33**), accumulated high levels of (**4**), suggesting that (**4**) is the immediate precursor of (**5**)[201] (Table 5). Normally, free (**4**) is present at very low levels in subterranean clover (and in many other legumes), although genistein (**3**) can accumulate to appreciable levels (Table 5).

Table 5 Levels of isoflavonoids in two varieties and one mutant line (A258) of subterranean clover.

Isoflavone	*Clare*	*Geraldton*	*Geraldton A258*
Genistein (**3**)	242	14.6	60.5
Biochanin A (**33**)	4.1	23.4	0.4
Pratensin (**71**)	0	1.6	trace
Daidzein (**4**)	0	0.2	76.5
Formononetin (**5**)	4.0	42.7	4.0

Source: Wong and Francis[201,202].

The contradiction between the labeling studies described above (which should be reevaluated in a cell culture system which is more optimal for precursor uptake) and the enzymological and genetic studies pointing to (**4**) as a substrate for 4′-*O*-methylation is compounded by attempts to demonstrate the enzymatic basis for the origin of the 4′-methoxy group of isoflavones. It would be expected that a simple isoflavone 4′-*O*-methyltransferase reaction would be involved in the conversion of (**4**) to (**5**), or of (**3**) to (**33**) (Scheme 14). However, in a study of isoflavone 4′-*O*-methylation in chickpea cell cultures, an isoflavone 7-*O*-methyltransferase activity, which methylated the A ring of (**3**) to yield prunetin (5,4′-dihydroxy-7-methoxyisoflavone) (**77**) was described[176] (Scheme 17). This enzyme activity had been initially described as a 4′-*O*-methyltransferase occurring as a dimer of M_r 110 000, with an optimum pH of 9.0 and a K_m for (**4**) of 80 μmol L^{-1}.[203]

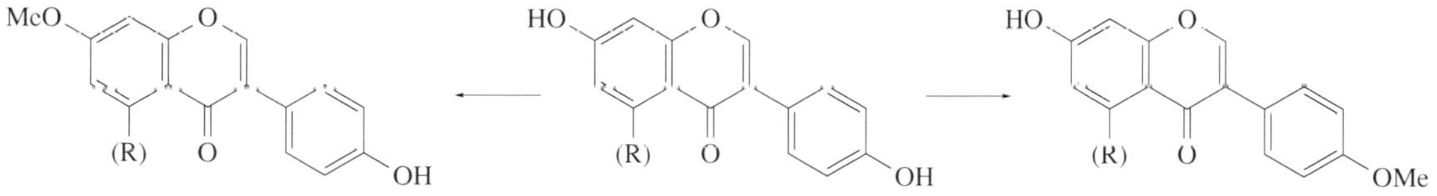

(**78**) R = H
(**77**) R = OH

(**4**) R = H
(**3**) R = OH

(**5**) R = H
(**33**) R = OH

Scheme 14

Fungal infection of jackbean (*Canavalia ensiformis*) callus led to a 3–4-fold increase in the extractable activities of enzymes that could methylate (**4**) and (**3**), and it was reported that the products cochromatographed with the 4′-methoxy derivatives.[204] However, treatment of alfalfa cell suspension cells with yeast elicitor results in a massive induction of isoflavone 7-*O*-methyltransferase activity,[205] which methylates the A-ring of (**4**) to produce isoformononetin (4′-hydroxy-7-methoxy-isoflavone) (**78**)[205] (Scheme 14), a rare naturally occurring compound which is unlikely to be involved in the formation of medicarpin (**6**). The enzyme, which is monomeric, was purified by SDS-PAGE to a single band of M_r 41 000 that could be photoaffinity labeled with [^{3}H]-*S*-adenosyl-L-methionine,[205] although the preparation was contaminated with high levels of caffeic acid 3-*O*-methyltransferase (COMT) activity. Partially purified alfalfa isoflavone 7-OMT had an optimum pH of 8.5, a K_m value of 20 μmol L^{-1} for (**4**), and exhibited a very low level of 4′-*O*-methyltransferase activity resulting in the formation of (**5**).[205] It has not proven possible to purify this 4′-OMT activity further.[206] The extremely low level of daidzein 4′-*O*-methyltransferase activity in elicited alfalfa cultures contrasts with the strongly increased extractable activity of the isoflavone 7-*O*-methyltransferase in parallel with all the other known enzymes in the pathway leading to (**6**).[205,207]

The author's research group has developed a substrate-based affinity chromatographic system to purify the 41 kDa isoflavone 7-OMT to homogeneity.[206] Four internal peptide sequences were obtained from the purified protein, one of which had high (72%) sequence identity with a region of a catechol-*O*-methyltransferase from barley. All four internal peptides had about 55% amino acid sequence identity with four regions of 6a-hydroxymaackiain 3-*O*-methyltransferase from *Pisum sativum* (see below), but had no sequence identity with the alfalfa COMT or chalcone 2′-*O*-methyltransferase (ChalOMT) genes previously cloned. The purified isoflavone *O*-methyltransferase had substrate specificity toward isoflavones with a free 7-hydroxyl group, and could also methylate the 5-hydroxyl group of (**3**). It was inactive against (**5**).

It is proposed that the enzyme with isoflavone 7-OMT activity *in vitro* may methylate the 4′-position *in vivo*. The unexpected precursor feeding results in alfalfa can be explained if the OMT is in a "metabolic compartment" or "channel," and its association with the enzymes producing its substrate or removing its product could account for the different product specificities observed *in vivo* and *in vitro*. The isoflavone synthase and 2′-hydroxylase are both microsomal cytochrome P450s, with which the 4′-OMT could be physically associated. Thus, only (**4**) formed *in situ* by microsomal isoflavone synthase, but not exogenously supplied (**4**), might act as substrate for the OMT (Figure 1). This hypothesis can be tested by molecular genetic strategies. To this end, full-length cDNA clones encoding the isoflavone OMT have been obtained. These can be transformed into plant cells that normally produce isoflavonoids with a free 4′-hydroxyl group, such as green bean or soybean, and the effects on metabolites determined. At the same time, testing is continuing to determine whether the isoflavone OMT can be used to identify other interacting proteins, using the yeast two-hybrid system for cloning genes based on physical interactions between their products.[208]

The 5-hydroxyl group of isoflavones is energetically the most difficult to methylate owing to its chelation to the carbonyl oxygen of the heterocyclic ring. However, yellow lupin roots accumulate a range of 5-methoxyisoflavones (e.g., 5-*O*-methylgenistein (**79**) and 5-*O*-methylderrone (**80**)) based on (**3**) and its 8-prenyl derivative (**81**). Khouri *et al.*[209] have reported the 810-fold purification of an isoflavone 5-*O*-methyltransferase from this source. The enzyme is a monomer of subunit M_r 55 000, with a pH optimum of 7.0 and a K_m value for (**3**) of 1 μmol L^{-1}. Its substrates, in order of decreasing activity, are 8-prenyl-2′-hydroxygenistein (**82**), 2′-hydroxygenistein (**68**), (**3**), and (**81**). This suggests that methylation of the 5-position can occur at several stages during the biosynthesis of the lupin isoflavonoids.

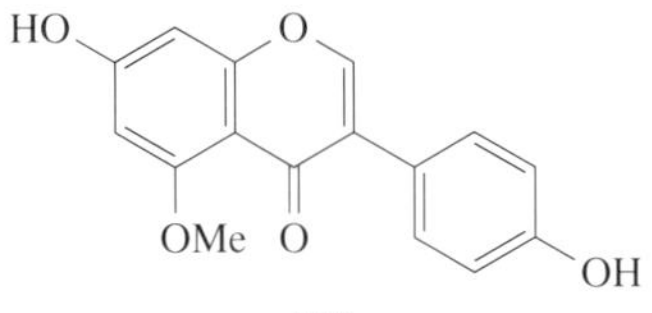

(**79**) 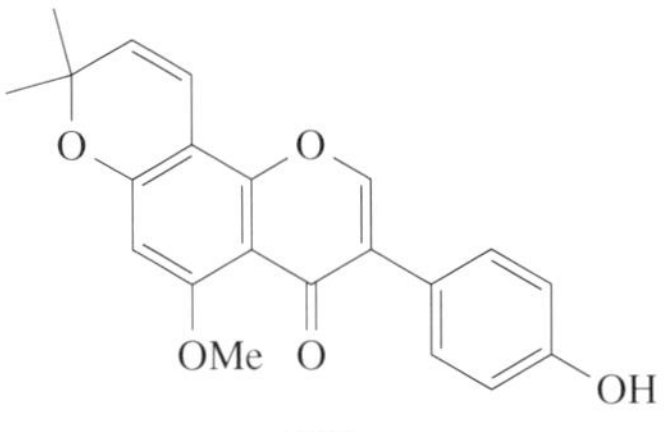 (**80**)

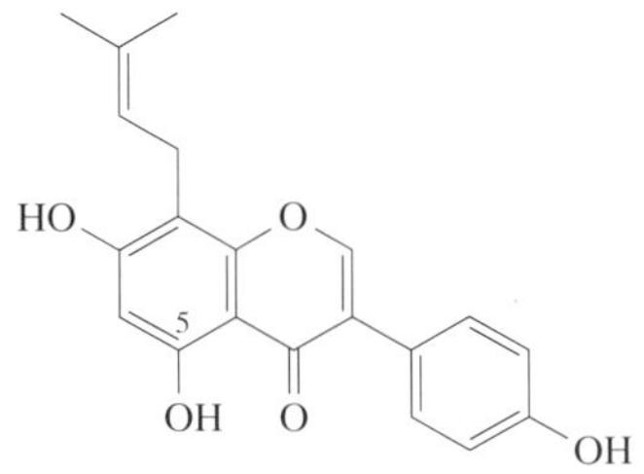

(**81**) 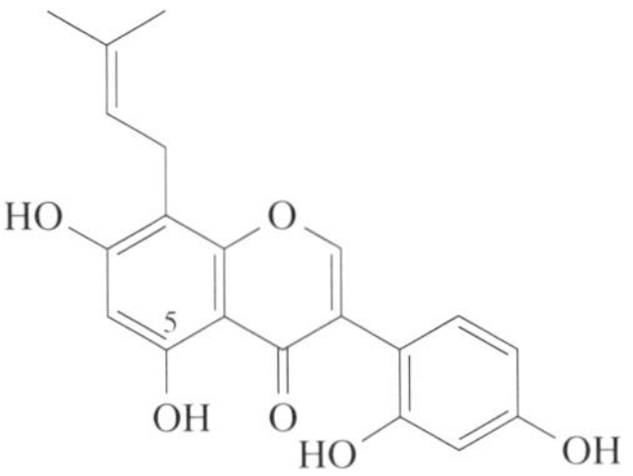 (**82**)

The final step in the biosynthesis of the pea phytoalexin (+)-pisatin (**21**) is the *O*-methylation of the 3-position of the pterocarpan (+)-6a-hydroxymaackiain (**43**) (Equation (2)). The 3-position of a pterocarpan is equivalent to the 7-position of the isoflavone nucleus. Preisig *et al.*[210] have purified an enzyme from $CuCl_2$-treated pea seedlings that catalyzes this methylation reaction. The monomeric enzyme has a subunit M_r of 43 000, exists as two isoforms of p*I* 5.2 and 4.9, and has an optimum

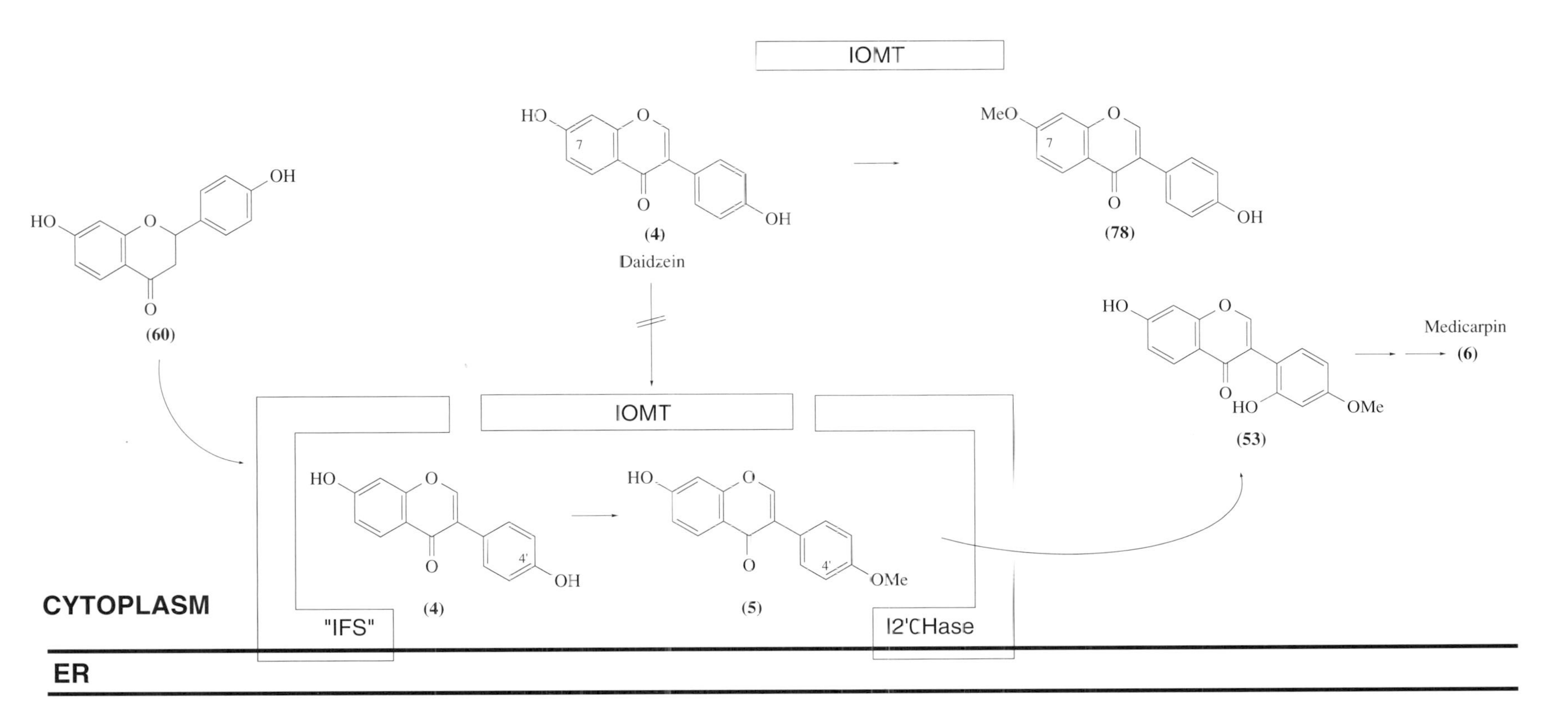

Figure 1 Potential metabolic channeling involving isoflavone *O*-methyltransferase. Association of the enzyme with endoplasmic reticulum (ER)-associated cytochrome P450 systems that catalyze isoflavone formation and 2′-hydroxylation is proposed to alter the product specificity of the enzyme, resulting in synthesis of the B-ring methylated isoflavone formononetin (**5**) rather than the A-ring methylated isoformononetin (**78**) which is the major product of the enzyme *in vitro*.

pH of 7.9 and a K_m value for (**43**) of 2.3 μmol L^{-1}. The enzyme has highest activity against (**43**), and low activity against (−)-6a-hydroxymaackiain (**83**), (+)-maackiain (**52**), and (+)-medicarpin (**84**). Antibodies were raised against the purified enzyme and used to demonstrate induction of enzyme protein and translatable mRNA activity in response to elicitation.[211] A cDNA encoding this OMT has been cloned;[212] it shows 51% amino acid sequence identity with the isoflavone 7-OMT from alfalfa.[213]

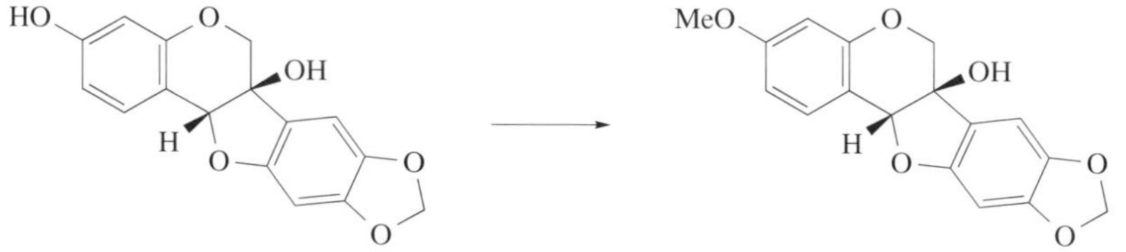

(**43**) → (**21**) (2)

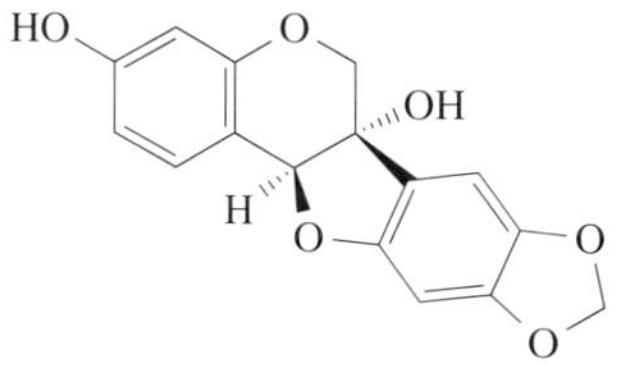

(**83**) 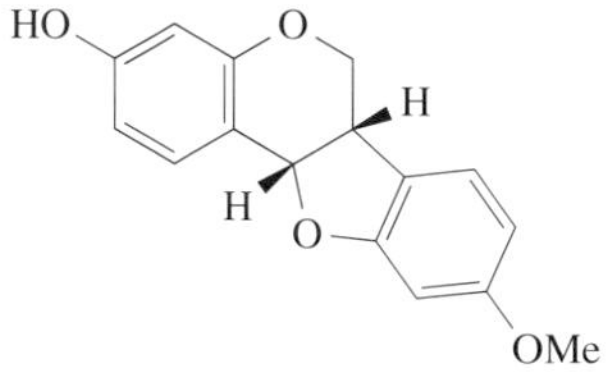 (**84**)

1.28.3.10.2 Formation of methylenedioxy rings

Methylenedioxy bridge functions occur in a number of isoflavonoids, the best known of which are the pterocarpan phytoalexins maackiain (**52**) and pisatin (**21**). Studies with elicitor-treated chickpea cell suspensions have indicated that maackiain synthesis proceeds via cytochrome P450-mediated 3′-hydroxylation of formononetin (**5**) to yield calycosin (**69**) (see above), followed by methylenedioxy bridge formation to yield pseudobaptigenin (**85**) (Equation (3)). A parallel series of reactions occur with the 5-hydroxylated isoflavone pratensin (**71**), although further metabolism of the methylenedioxy-substituted product leads to a substituted isoflavanone rather than to a pterocarpan. Chickpea microsomes convert (**69**) and (**71**) to (**85**) and 5′-hydroxypseudobaptigenin (**86**), respectively, in a reaction requiring oxygen and NADPH that shows all the classical characteristics of a cytochrome P450-catalyzed reaction.[214] This enzyme system has apparent K_m values for (**69**) and NADPH of 2 μmol L^{-1} and 70 mol L^{-1}, respectively, and is strongly induced, from a zero background level, in response to elicitor.

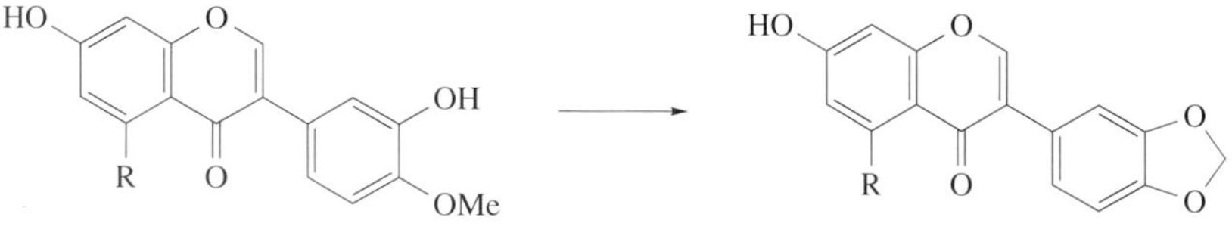

(**69**) R = H
(**71**) R = OH
→
(**85**) R = H
(**86**) R = OH
(3)

1.28.3.10.3 6a-Hydroxylation of Pterocarpans

The 6a position is hydroxylated in several pterocarpan phytoalexins such as pisatin (**21**) from peas and the glyceollins (**18–20**) from soybean. Fungal degradation of pterocarpans can also involve hydroxylation of the 6a position (see Section 1.28.4.2). The mechanism of 6a-hydroxylation of the glyceollin precursor 3,9-dihydroxypterocarpan (**57**) in soybean has been clearly established (Scheme 15). The cytochrome P450 enzyme catalyzing the formation of 3,6a,9-trihydroxypterocarpan (**87**) was solubilized from soybean microsomal membranes utilizing 1% Chaps detergent,[215] although this detergent strongly inhibits the hydroxylase at concentrations below the critical micellar concentration. In spite of low recovery, the enzyme was purified 765-fold to yield a major component of subunit M_r 55 000, the activity of which could be reconstituted with purified soybean NADPH:

cytochrome P450 reductase in the presence of added lipid, with dilauroylphosphatidylcholine being the most effective. The pH optimum of the enzyme was 7.0. The enzyme could be resolved from the more abundant cinnamate 4-hydroxylase cytochrome P450, and this was in fact the first direct demonstration of the presence of distinct molecular species of P450s with different catalytic activities in plants. Soybean pterocarpan 6a-hydroxylase activity was not detectable in unelicited cells.

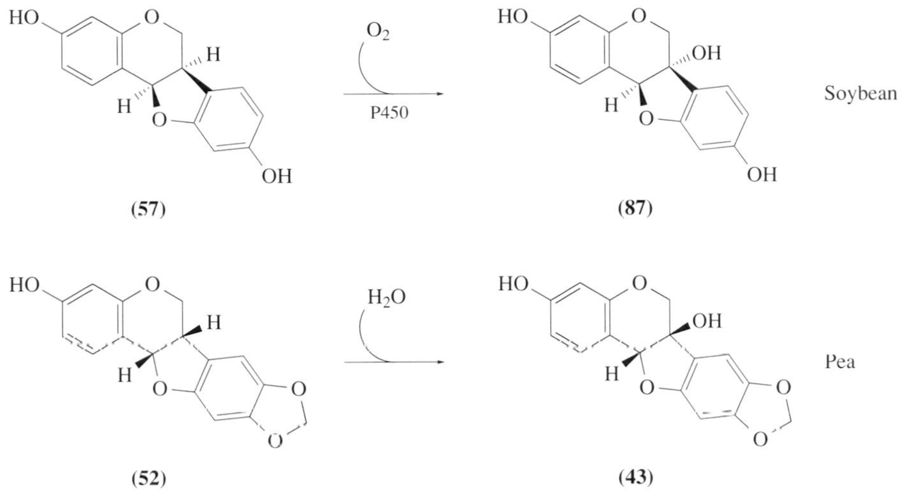

(57) $\xrightarrow{O_2,\ P450}$ (87) Soybean

(52) $\xrightarrow{H_2O}$ (43) Pea

Scheme 15

In soybean, the pterocarpan precursor for 6a-hydroxylation is of the (−)-6a*R*,11a*R* stereochemistry, whereas in peas the maackiain (**52**) that is hydroxylated in the penultimate step of pisatin (**21**) biosynthesis is of the opposite stereochemistry. ^{18}O-labeling experiments with $CuCl_2$-treated pea seedlings led to the surprising conclusion that the 6a-hydroxyl group of (**52**) destined for biosynthesis of (**21**) is derived from water rather than from molecular oxygen[216] (Scheme 15). This contrasts with the fact that fungal degradation of maackiain proceeds via 6a-hydroxylation utilizing molecular oxygen although, in this case, the maackiain that the fungus degrades in this manner is the (−)-stereoisomer.[216]

It is not clear at present whether these different mechanisms for 6a hydroxylation of (6a*R*, 11a*R*)- and (6a*S*, 11a*S*)-pterocarpans are associated with an as yet unidentified pathway for the biosynthesis of the (+)-pterocarpans. The finding that the pea IFR produces the 3*R* steroisomer of sophorol (**74**)[195] raises questions about the mechanism of (+)-pterocarpan formation. The problem is how (**74**) is converted to (**21**). (+)-6a-Hydroxymaackiain (**52**) is clearly the immediate substrate for the methyltransferase that is the final step in pisatin biosynthesis.[210] Production of the (+)-pterocarpan from the (−)-isoflavanone could occur either via epimerization at the isoflavanone level, although there is at present no enzymological evidence to support this, or at the level of insertion of the 6a-hydroxyl group, which could conceivably involve the addition of water across a pterocarpene double bond, with subsequent inversion of stereochemistry. Arguments in favor of these two alternatives have been discussed in detail elsewhere[195] and the resolution of this problem requires more detailed enzymological studies.

1.28.3.10.4 Isoprenylation

At least half of the isoflavonoids that have been characterized from leguminous plants have side attachments derived from the 3,3-dimethylallyl (prenyl) group,[1] and several of these compounds have already been listed (**17–20, 26–28, 32, 80–82**). Further examples of prenylated isoflavonoids are given ((**88**)–(**92**)), from which the variability in position and complexity of prenylation, and also the subsequent metabolic modifications of the prenyl side chain, can be appreciated. Most of the enzymological work on isoflavonoid prenylation has been performed in relation to the elicitor/infection-induced prenylated pterocarpans in bean and soybean, or the constitutively expressed prenylated isoflavones in white lupin.

[^{14}C]Mevalonic acid is incorporated into the glyceollins (**18–20**) of soybean, and a cell-free preparation from elicited soybean cotyledons was shown to catalyze prenyl group addition from

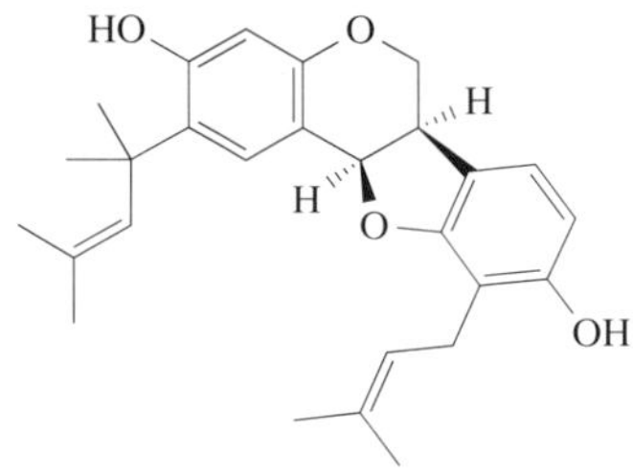

(**88**) Striatine (*Mundulea striata*)

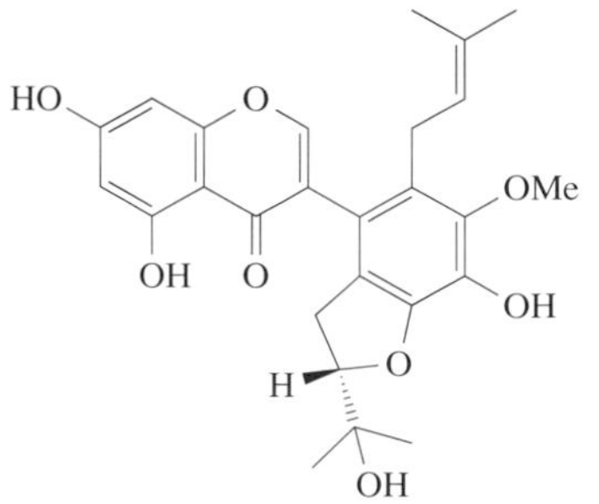

(**89**) Erythbigenol A (*Piscidia erythrina*)

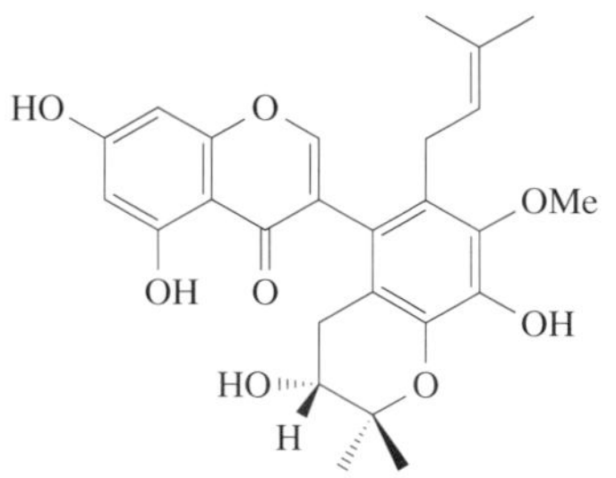

(**90**) Erythbigenone A (*Piscidia erythrina*)

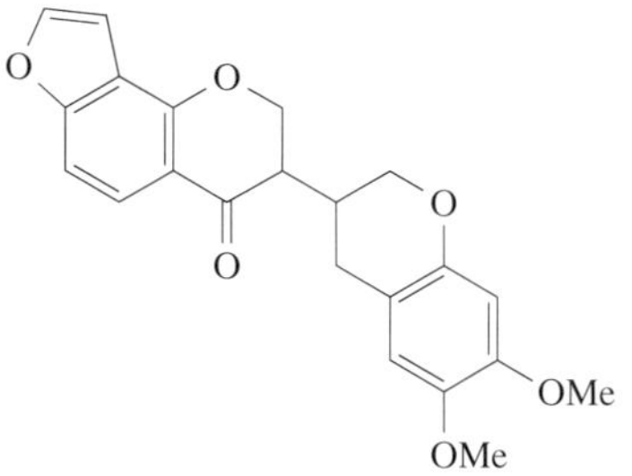

(**91**) Elliptone

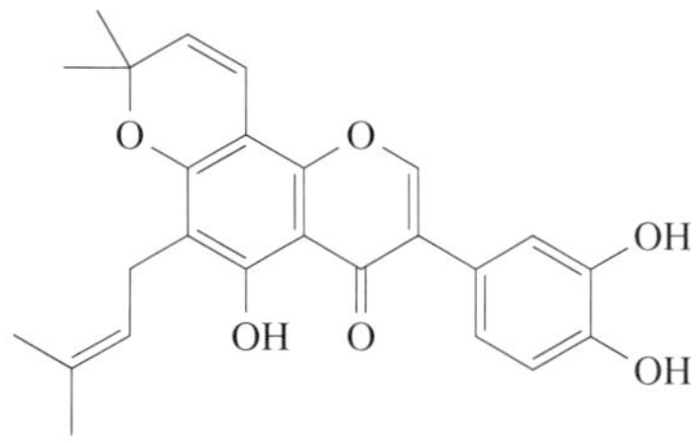

(**92**) Pomiferin

dimethylallyl pyrophosphate to the C-2 (major product (**93**)) and C-4 (minor product (**94**)) positions of 3,6a,9-trihydroxypterocarpan (**87**), as shown in Scheme 16.[217] The enzymatic activity appeared to be localized in a particulate fraction. Similarly, a microsomal fraction from yeast extract elicited bean (*Phaseolus vulgaris*) cell suspension cultures catalyzed the prenylation of 3,9-dihydroxy-pterocarpan (**57**) to 3,9-dihydroxy-10-dimethylallyl-pterocarpan (phaseollidin) (**58**). The K_m values were 1.5 μmol L^{-1} for DMAPP and 1.4 μmol L^{-1} (assuming that only the 6a*R*, 11a*R* enantiomer can act as substrate) for (**57**), and the enzyme was strongly product inhibited.[218] Sucrose- and Percoll-gradient centrifugation studies revealed that the pterocarpan prenyltransferases from both soybean and bean were localized to the chloroplast inner envelope membrane,[219] implying movement of isoflavonoid precursors among different cellular compartments.

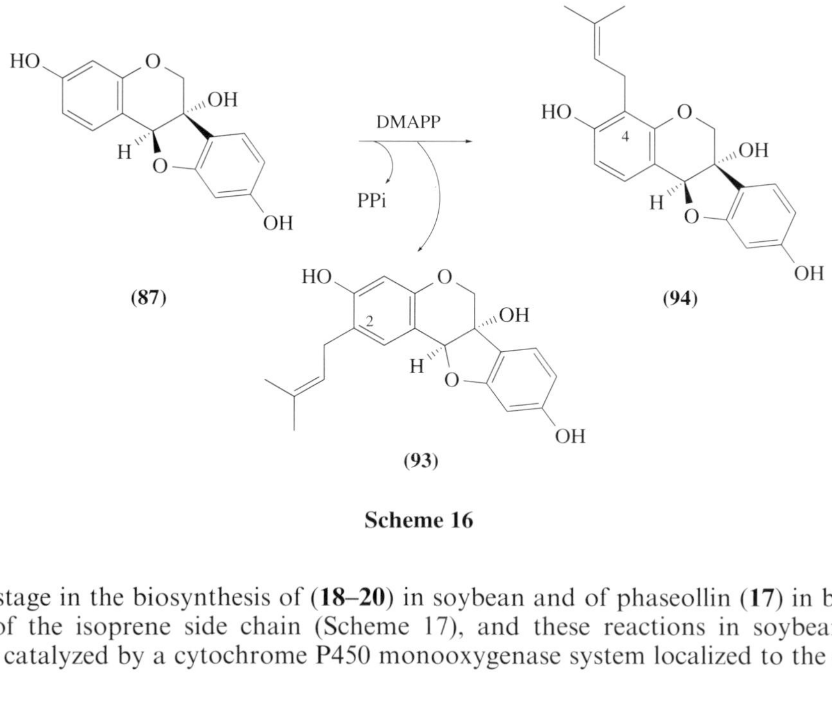

(**87**) → DMAPP, PPi → (**93**) + (**94**)

Scheme 16

The final stage in the biosynthesis of (**18–20**) in soybean and of phaseollin (**17**) in bean involves cyclization of the isoprene side chain (Scheme 17), and these reactions in soybean have been shown to be catalyzed by a cytochrome P450 monooxygenase system localized to the endoplasmic reticulum.[220] As with the soybean prenyltransferase, cyclase activity is absent from unelicited cells and is strongly induced by exposure to elicitors from yeast or the fungal pathogen *Phytophthora megasperma* f. sp. *glycinea*.[220] The involvement of a membrane-bound cytochrome P450 in the cyclization of the prenyl group in the soybean pterocarpans contrasts with the demonstration of a soluble non-heme iron protein from *Tephrosia vogelii* that catalyzes oxidative ring closure of rot-2′-enoic acid (**95**) in the formation of the rotenoid deguelin (**96**)[221] (Equation (4)).

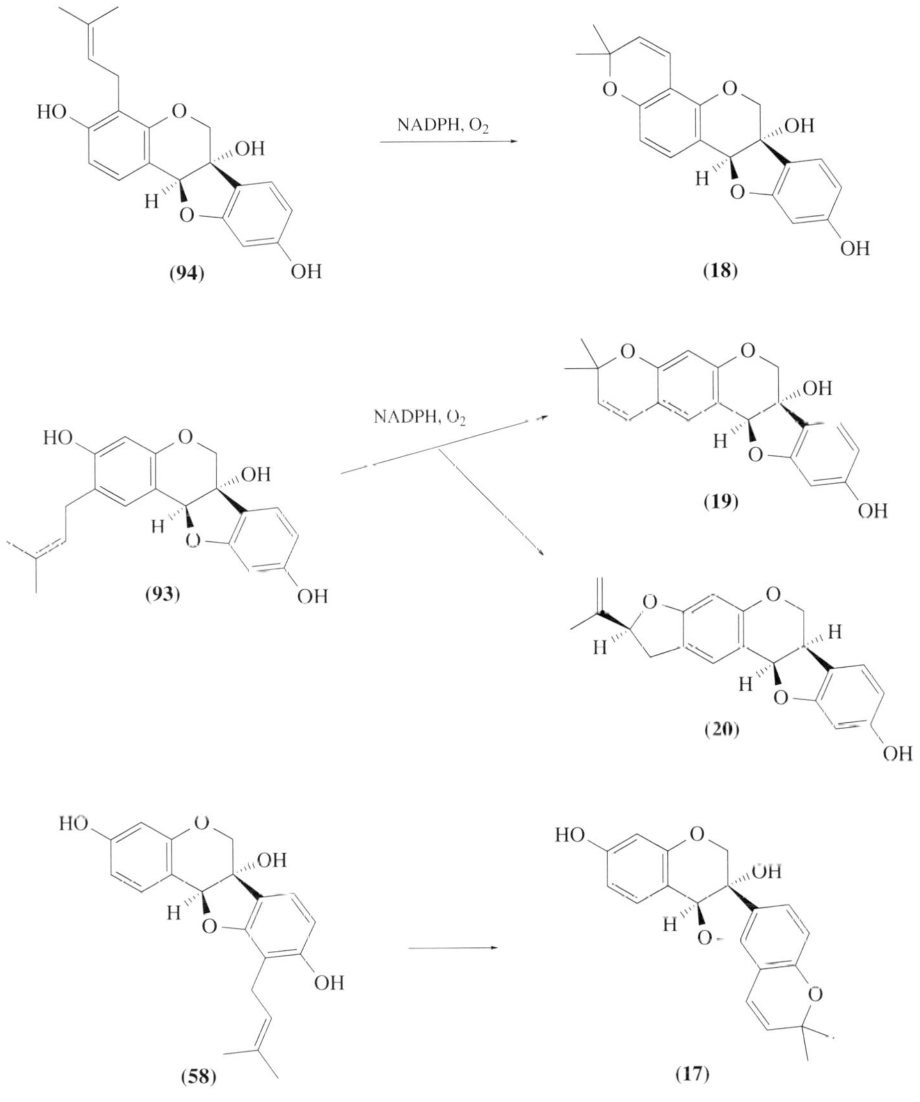

HO O OH H O OH (**94**) — NADPH, O_2 → O O OH H O OH (**18**)

HO O OH H O OH (**93**) — NADPH, O_2 → O O OH H O OH (**19**); O O H H H O OH (**20**)

HO O OH H O OH (**58**) → HO O OH H O O (**17**)

Scheme 17

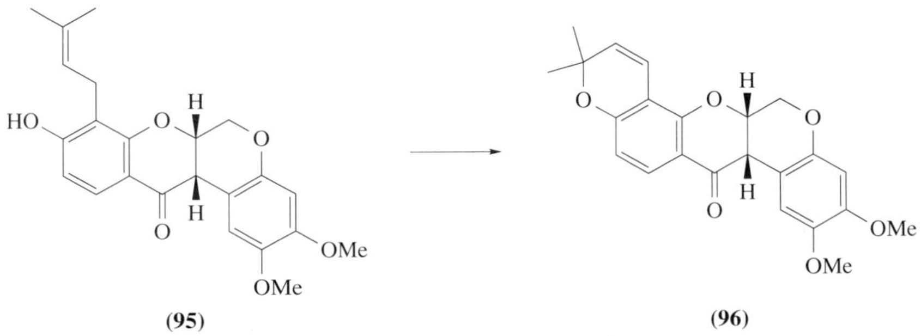

HO O H O H O O OMe OMe (**95**) → O O H O H O O OMe OMe (**96**) (4)

White lupin leaves, hypocotyls, roots, root nodules, and cell suspension cultures accumulate a range of prenylated isoflavones, and lupin microsomal enzyme systems have been described that catalyze the prenylation of genistein (**3**) and 2′-hydroxygenistein (**68**), from which these compounds are derived.[222,223] The prenyltransferase activity is constitutively expressed. The activity from radicles catalyzed prenylation of the two isoflavones at the 6-, 8-, and 3′-positions. Because different ratios of products were obtained with the enzyme preparation from different sources, and after different detergent solubilization steps, it was proposed that a number of distinct position-specific prenyltransferases may be present in lupin.[223]

Prenylation often increases the antifungal and anti-insect properties of isoflavonoids, and the value of introducing novel prenylated isoflavonoids into plants via genetic engineering was first proposed in 1987.[218] However, the isolation of the required prenyltransferase genes has been hampered by the difficulties inherent in the effective solubilization of the enzyme from the inner chloroplast membrane; although this has been achieved, recoveries are low.[223,224] It is to be hoped that molecular approaches such as differential library screening or PCR-based differential display will lead directly to the cloning of isoflavonoid-specific prenyltransferases, and that functional expression studies with different related sequences will then answer the outstanding questions concerning the molecular basis for substrate and position specificity.

1.28.3.10.5 Formation of isoflavone dimers

Dimeric isoflavonoids appear to be relatively rare in nature, with limited reports of the isolation and structural elucidation of such compounds as isoflavone–isoflavan, isoflavan–isoflavan, and isoflavone–isoflavone dimers.[1,225] Elicitation of cell cultures of *Pueraria lobata* results in the appearance of small quantities of kudzuisoflavones A and B (**97, 98**), dimeric isoflavones coupled through the B-rings.[225] As the latter is obtained as a racemic mixture, it is likely that these molecules arise via radical coupling, and may be viewed as artificial by-products of the action of peroxidase on daidzein (**4**).[225]

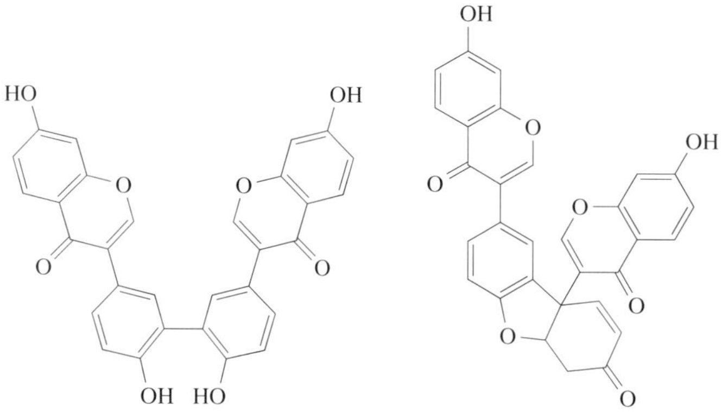

(97) **(98)**

1.28.3.10.6 Formation and vacuolar storage and efflux of isoflavone glycosides

Isoflavonoids often occur constitutively as their glycosides and malonyl glycosides, and these compounds have generally been considered as "storage forms." The most studied systems for isoflavonoid glycoside metabolism are white lupin roots and cell cultures, soybean seeds and seedlings, and chickpea cell suspension cultures. Features of the accumulation and metabolism of these compounds differ somewhat in the different species.

In soybean seed hypocotyls, the 7-*O*-glucosides, 7-*O*-glucoside-6″-*O*-malonates, and 7-*O*-glucoside-6″-*O*-acetates of the isoflavones daidzein (**4**), genistein (**3**), and glycitein (**99**) occur, and have all been shown to increase during seed development in the pod, to maximum levels between 45 and 60 days after flowering.[226] Three days after germination, the metabolism of the young leaf shifts from isoflavonoid to flavonoid accumulation,[227] although low levels of isoflavone conjugates remain.[228] Cotyledons maintain high levels of conjugates of (**4**) and (**3**), but these same compounds decrease dramatically in the hypocotyl hook between 2 d and 4 d post-germination.[227] In soybean roots, conjugates of (**4**) predominate throughout development. A similar pattern of developmental distribution of isoflavonoids is observed in alfalfa, where formononetin 7-*O*-glucoside-6″-*O*-malonate (**34**) accumulates in roots, although in this case along with the malonyl glucoside of the pterocarpan medicarpin (**6**).[229]

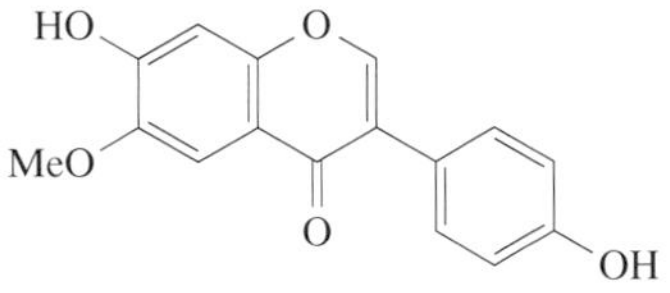

(**99**)

Infection of soybean with *Phytophthora sojae* leads to dramatic changes in isoflavone glycoside profiles and distribution. Thus, in leaves, the pterocarpan glyceollin (**18**) accumulates to high levels only in the hypersensitive lesion formed in a resistant interaction, whereas the glucosides and malonyl glucosides of (**4**), (**3**), and (**99**) accumulate in a broad area around the lesion.[228] In cotyledons, the already large constitutive pools of isoflavone glycosides are rapidly mobilized in the incompatible (resistant) interaction with *P. sojae*, and, in the case of (**4**), the aglycone may be utilized for phytoalexin (glyceollin) synthesis.[230]

Hairy root and untransformed cell suspension cultures of white lupin (*Lupinus albus*) contain a range of mono- and diglucosides of (**3**), 2′-hydroxygenistein (**68**), and their 6- or 3′-prenyl derivatives[231–233] (Scheme 18; source references 111, 232 and 233). The same compounds are found in young plantlets, but unusually high levels are found constitutively in the hairy root and cell suspension cultures.[231,232] Although significant attention has been paid to the enzymology of isoflavone prenylation in lupin, nothing is known of the specificity of the enzyme system(s) required for 7-, 2′-, or 4′-*O*-glucosylation.

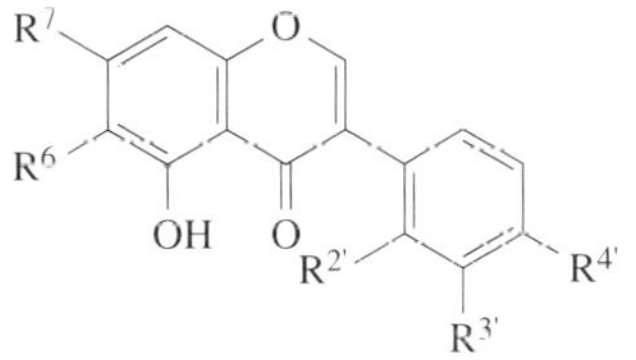

R^6	R^7	$R^{2'}$	$R^{3'}$	$R^{4'}$
H	OGlc	OGlc	H	OH
H	OGlc	H	H	OGlc
H	OGlc	OH	H	OH
H	OGlc	H	H	OH
H	OGlc Mal	OH	H	OH
H	OGlc Mal	H	H	OH
H	OGlc	OGlc	Pre	OH
Pre	OGlc	OGlc	H	OH
H	OGlc Mal	OGlc	Pre	OH
H	OH	H	H	OGlc
Pre	OGlc	OH	H	OH
H	OGlc	OH	Pre	OH
H	OH	OH	Pre	OGlc

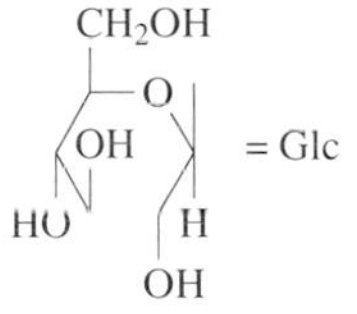

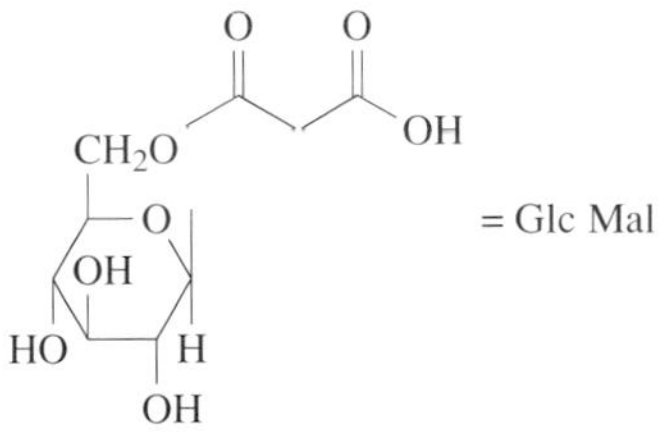

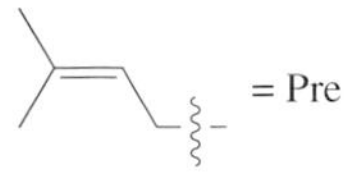

= Glc

= Glc Mal

= Pre

Scheme 18

In cell suspension cultures of *Pueraria lobata*, (**4**) and (**3**) exist as their 6″-*O*-malonyl glucosides (7-*O*- and 8-*C*-glucosides for (**4**), 7-*O*-glucosides for (**3**)) (Scheme 19; source reference 234). Levels of the malonylglucosides rapidly decrease following exposure to yeast and fungal elicitors, or to hydrogen peroxide.[235,236] In elicitor-treated cells, the conjugates then reaccumulate, along with isoflavone aglycones and their dimers. The rapid decrease in conjugate levels, which correlates with accumulation of the derived aglycone into insoluble cell wall material, is unaffected by the protein synthesis inhibitor cycloheximide which does, however, prevent isoflavone reaccumulation, suggesting that metabolism of the conjugates utilizes preexisting enzymatic machinery.[235] A range of exogenously applied flavonoids and isoflavonoids were converted to the corresponding 7-*O*-glucoside-6″-*O*-malonates by the *Pueraria* cells. However, following elicitation, only the isoflavonoids were remobilized.[235] In alfalfa seedlings, elicitation with $CuCl_2$ leads to rapid mobilization of isoflavone conjugate stores, with resultant accumulation of (**6**) and formononetin (**5**).[237]

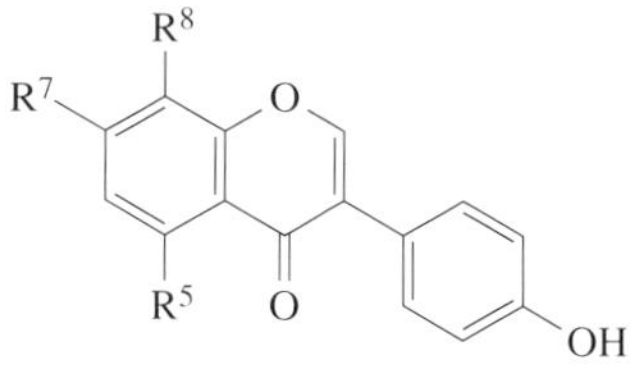

R^5	R^7	R^8
H	OGlc	H
OH	OGlc	H
H	OH	Glc
OH	OH	Glc
H	OGlc Mal	H
OH	OGlc Mal	H
H	OH	Glc Mal
OH	OH	Glc Mal

Scheme 19

Chickpea cell suspension cultures have provided the most detailed information on the physiology and biosynthesis/catabolism of isoflavonoid glycosides. These cultures contain the malonyl glycosides of biochanin A (**33**), (**5**), cicerin (**100**), homoferreirin (**101**), (**6**), and maackiain (**59**), all of which are localized in the vacuole.[238] Treatment of the cells with an elicitor preparation from *Ascochyta rabiei* leads to accumulation of (**6**), synthesized *de novo* from L-phenylalanine. However, if the cells are elicited in the presence of the potent and specific phenylalanine ammonia-lyase (PAL) inhibitor L-α-aminooxy-β-phenylpropionic acid (AOPP), the vacuolar pool of (**34**) is mobilized as a precursor for medicarpin synthesis, indicating metabolic cross-talk between the central phenylpropanoid pathway and the pathway of isoflavone glycoside catabolism.[239] Furthermore, treatment of cells with low levels of yeast extract elicitor leads to acumulation of the malonyl glucosides of (**6**) and (**59**), whereas at high elicitor doses the pterocarpan aglycones accumulate, partially as a result of formation from mobilized conjugates.[240] Likewise, (**6**) is in part formed from mobilized (**34**), and from medicarpin malonyl glucoside, in elicitor-treated alfalfa cell suspenions.[241] The elicitor-mediated vacuolar efflux of isoflavone conjugates in chickpea is blocked by *trans*-cinnamic acid, the product of the PAL reaction,[242] providing a potential mechanism for how the mobilization machinery can "sense" the flux through the phenylpropanoid pathway.

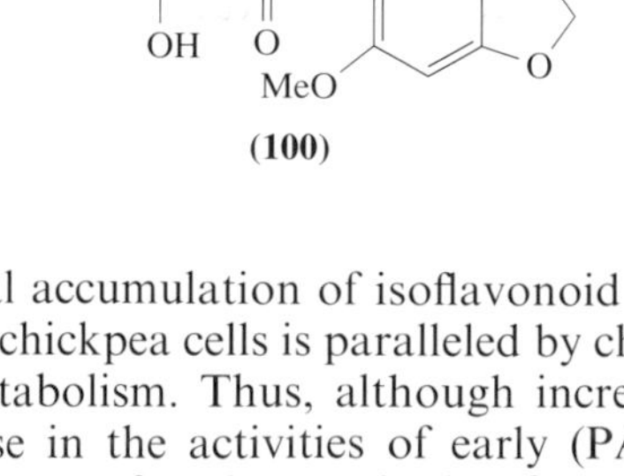

(100) 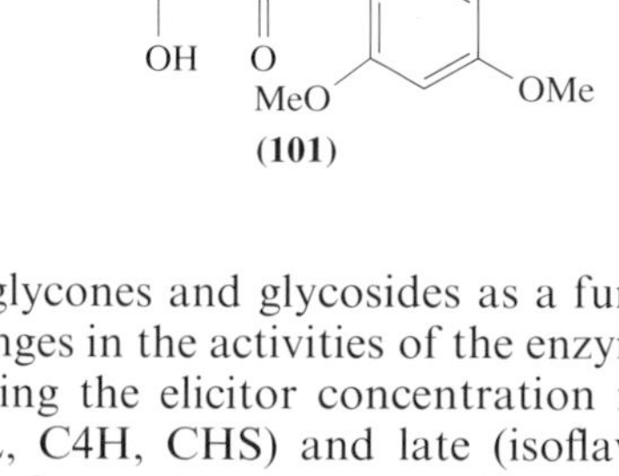 **(101)**

The differential accumulation of isoflavonoid aglycones and glycosides as a function of elicitor concentration in chickpea cells is paralleled by changes in the activities of the enzymes of conjugate synthesis and catabolism. Thus, although increasing the elicitor concentration results in a proportional increase in the activities of early (PAL, C4H, CHS) and late (isoflavone 2′- and 3′-hydroxylase) enzymes for the synthesis of the isoflavonoid aglycones, the catabolic isoflavone malonylesterase and glucosidase activities remain constant and then increase at the highest elicitor concentrations, whereas the glucosyltransferase and malonyltransferase of conjugate formation are highest at low elicitor concentrations, and then decrease.[240]

The isoflavone glucosyltransferase has been purified 120-fold from chickpea roots.[243] It is a cytoplasmic enzyme[238] of M_r 50 000, which preferentially glycosylates the 7-position of the 4′-*O*-methylated isoflavones (**5**) (K_m 24 μmol L^{-1}) and (**33**) (K_m 12 μmol L^{-1}). Compounds (**3**) and (**4**) are poor substrates, and 6,7-dihydroxy-4′-methoxyisoflavone (texasin) (**102**) is not glucosylated. A 47 kDa glucosyltransferase which acts on the A-ring hydroxyl groups of coumestrol (**9**) (K_m 57 μmol L^{-1}) and (**6**) (K_m 24 μmol L^{-1}) has been partially purified from alfalfa cell suspension cultures.[244] Little is known concerning the enzymology of isoflavone *C*-glycoside formation, as occurs in *Pueraria lobata*.

HO HO O O OMe

(102)

The isoflavone glucoside malonyl transferase also appears to be a cytoplasmic enzyme in chickpea.[238] It has an M_r of 112 000 and specifically malonylates the 6-position of the glucose residues of isoflavone 7-*O*-glucosides.[245] Its substrates, in order of decreasing activity, are the glucosides of (**33**), (**5**), (**3**), pratensin (**71**), (**4**), and (**59**). The enzyme does not act on 4′-*O*-glucosides.

The first stage of isoflavone conjugate mobilization appears to be catalyzed by an isoflavone glucoside malonylesterase that is vacuolar-localized in chickpea.[238] The enzyme has been purified over 700-fold from chickpea roots,[247] and possesses some unusual properties. The subunit M_r is ~32 kDa, although the holoenzyme M_r appears to be in excess of 2×10^6. The enzyme has very little activity against standard non-specific esterase substrates, and is unaffected by standard esterase inhibitors. It has a high affinity for biochanin A 7-*O*-glucoside-6″-*O*-malonate (K_m 4.4 μmol L^{-1}).

As most β-glucosidases lack absolute specificity for any particular substrate, it is often difficult to know their true metabolic functions. Chickpea roots contain three isoforms of β-glucosidase that are only found in the isoflavonoid-containing tissues of the plant.[247] The enzymes are dimers of M_r ≈130 000, and have low K_m (20–40 μmol L^{-1}) and high V_{max} values for the 7-*O*-β-glucosides of (**5**) and (**33**), although apigenin 7-*O*-β-glucoside (**103**) is an equally good substrate. In contrast to the malonylesterase, isoflavone β-glucosidase activity appears to be primarily cytoplasmic.[238]

OH GlcO O OH O

(103)

Unconjugated medicarpin (**6**) is taken up into isolated vacuoles of mung bean (*Vigna radiata*) at a low rate in the presence or absence of MgATP. However, following incubation of [^{3}H]medicarpin and glutathione with a maize glutathione *S*-transferase preparation, the resultant [^{3}H]medicarpin–glutathione conjugate(s) is taken up eight times faster in the presence than in the absence of MgATP.[248] Uptake of medicarpin–glutathione is not significantly inhibited by the protonophore gramicidin-D, but is strongly inhibited by vanadate and the alternative transport substrate *S*-(2,4-dinitrophenyl)glutathione.[248] These results demonstrate the operation of the high-affinity, high-capacity, glutathione conjugate (GS-X) pump, previously shown to be involved in the vacuolar uptake of xenobiotics[249] and anthocyanin,[250] in the vacuolar transport of an isoflavonoid phytoalexin.

Because glycosylated anthocyanin is a substrate for the GS-X pump[250] the question remains as to whether medicarpin conjugates, in addition to the aglycone, are also substrates for the vacuolar GS-X pump. It is therefore important to clarify whether glycosylation and malonylation of isoflavonoids occur in the cytoplasm, the vacuole, or both, to determine how glutathione is removed from conjugates, and, most importantly, to understand what determines whether a vacuolar compound will be permanently or temporarily sequestered. Answers to these questions with respect to isoflavonoids will provide clues as to how the host plant integrates the balance between attack and defense during the determinative stages of pathogen infection. It is interesting to speculate that one function for the GST(s) induced following the hypersensitive response to avirulent fungal pathogens[116,251] could be in facilitating the vacuolar storage of antimicrobial isoflavonoids in the healthy cells around the hypersensitive lesion.

1.28.3.11 Biosynthesis of Coumestans

An early hypothesis proposed that coumestrol (**9**) might be formed from daidzein (**4**) by hydroxylation at the 2-position, followed by tautomerization and ring closure, as shown in Scheme 20. Subsequent ^{14}C-labeling studies revealed good incorporation of (**4**), 2′,4′,7-trihydroxyisoflavone (**55**) and dihydrodaidzein (**104**) into (**9**) in mung bean seedlings, leading to the proposal that the pathway operated via a metabolic grid leading to a pterocarp-6a-ene intermediate (**105**) (Scheme 21).[252,253] In $CuCl_2$-treated alfalfa seedlings, in which accumulation of 9-*O*-methylcoumestrol (**10**) was induced with little effect on the level of (**9**), (**4**) was incorporated into (**9**) but not into (**10**), although formononetin (**5**) was incorporated into the latter.[130] This labelling pattern reflects that discussed in Section 1.28.3.10.1 in relation to the paradox concerning the origin of the 4′-*O*-methyl group in isoflavones destined for pterocarpan synthesis.

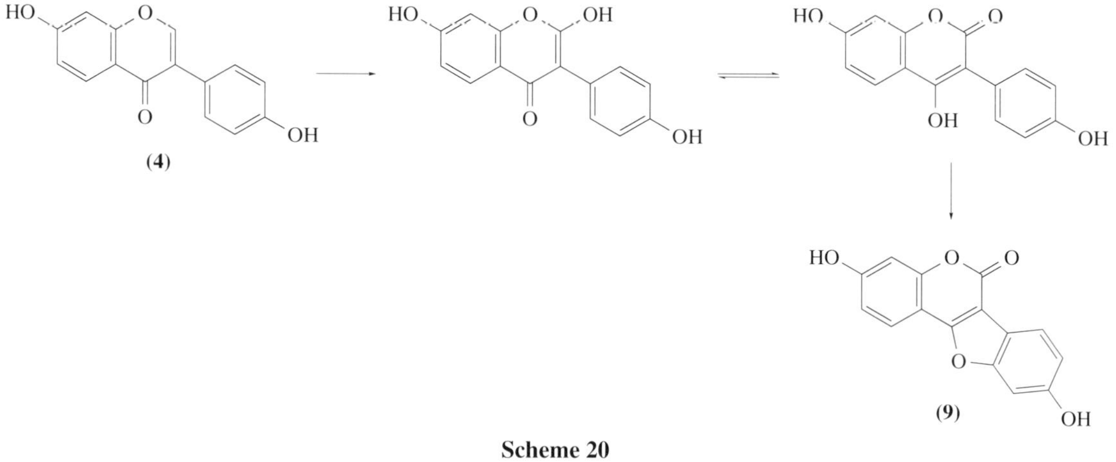

HO O O OH (4) HO O OH O OH HO O O OH OH HO O O O (9) OH

Scheme 20

7,2′-Dihydroxy-4′-methoxyisoflav-3-ene (**46**) and 7,2′-dihydroxy-4′-methoxyisoflav-3-ene-2-one (**47**) were excellent precursors of (**10**) in elicited alfalfa seedlings, leading to the proposal of a pathway to coumestans involving the intermediacy of isoflav-3-enes and 3-arylcoumarins (Scheme 4).[131] To the best of the author's knowledge, the enzymology of these reactions remains to be worked out.

1.28.3.12 Biosynthesis of Rotenoids

The rotenoids are characterized by their four-ring chromanochromanone structure. Biosynthetically, the carbon skeleton can be envisaged as arising from an isoflavanone with an extra carbon atom that could arise initially as a methoxyl group at the 2′-position (isoflavanone numbering), yielding the position 6 carbon of rotenoids. This basic scheme is supported by the results of radiolabeled precursor studies in seedlings of *Amorpha fruticosa*, which accumulate rotenone (**106**) and amorphigenin (**107**).[254] In the case of (**107**), it appears that formononetin (**5**) is first hydroxylated at the 3′-position, and methylation of this position may then occur prior to substitution of the 2′-position and subsequent closure of ring B via conjugate addition of a methoxyl radical (Scheme 20).

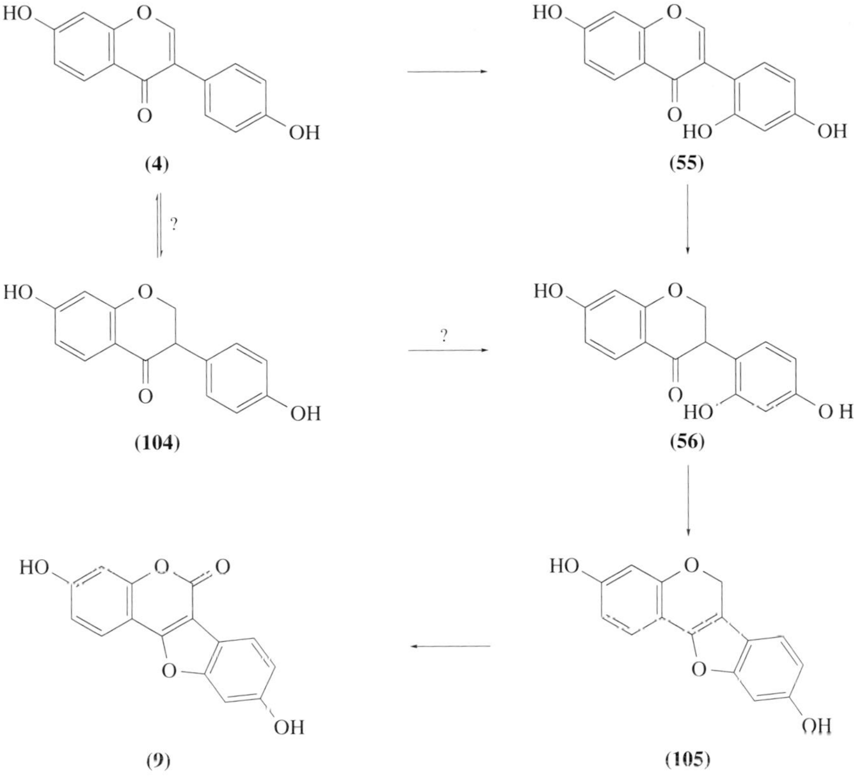

(4) (55) (104) (56) (9) (105)

Scheme 21

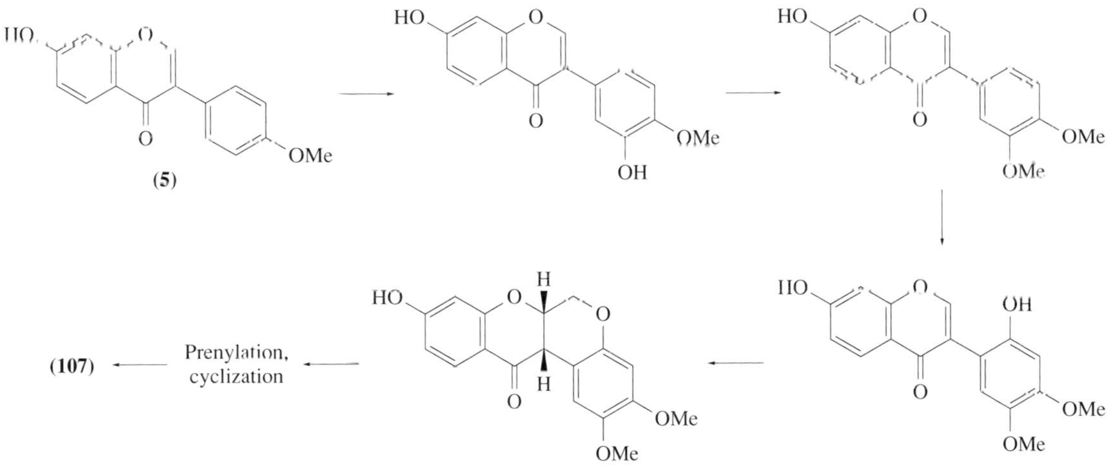

(5) (107) Prenylation, cyclization

Scheme 22

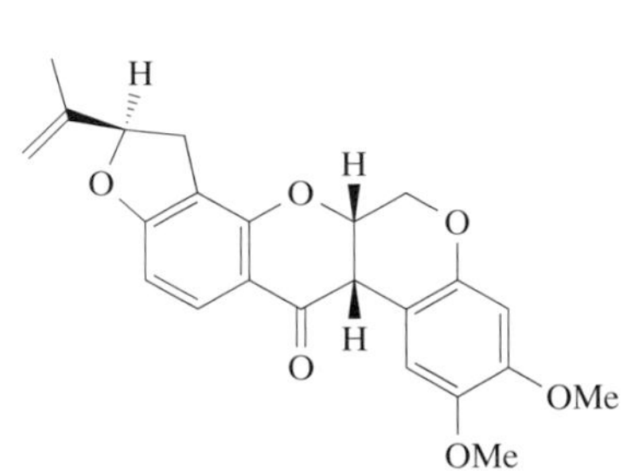

(106) 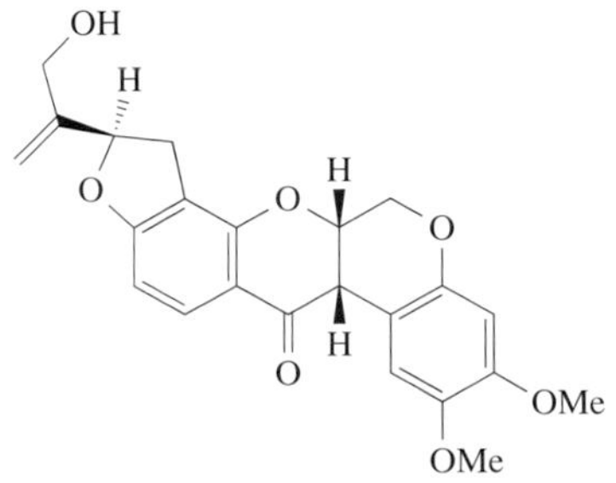 (107)

Prenylation and subsequent prenyl cyclization are late steps in rotenoid biosynthesis. Seeds, plants, and plant cell cultures of the West African tropical plant *Tephrosia vogelii* have been used to study the formation of the chromen ring that occurs in the rotenoid deguelin (**96**).[221,255] A soluble enzyme, deguelin cyclase, was isolated and partially purified. The cyclase catalyzes the direct formation of (**96**) from rot-2′-enoic acid (**95**) (Equation (4)) without the appearance of a hydroxylated intermediate.[221] The enzyme, which is inhibited by metal chelators, utilizes molecular oxygen, requires no cofactors, and has a K_m value for (**95**) of 4.6 μmol L^{-1} and an M_r of ~39 000. It also catalyzes the conversion of sumatrolic acid (**108**) to α-toxicarol (**109**), but does not convert (**95**) to (**106**). The stereochemistry of the reaction has been determined[256] and a reaction mechanism proposed that parallels a suggested mechanism for the formation of (**106**) from (**95**).[256] Deguelin cyclase is clearly not a cytochrome P450, and is therefore distinct from the enzyme(s) catalyzing prenyl to chromen transformations in the pterocarpans.[220]

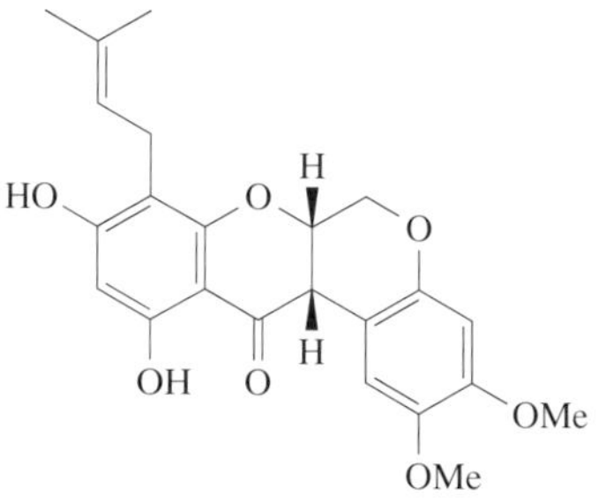

(108) 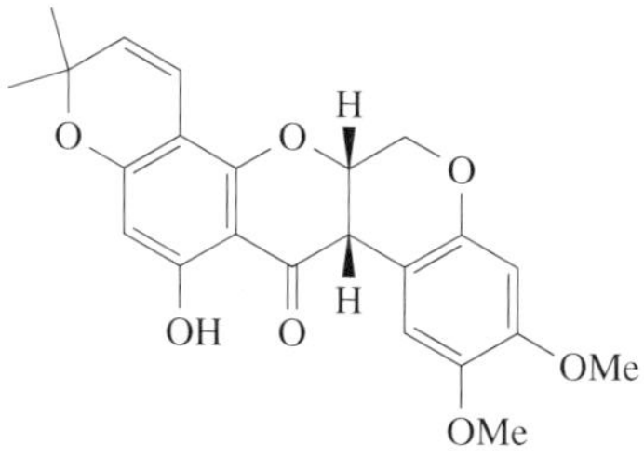 **(109)**

1.28.4 CATABOLISM OF ISOFLAVONOIDS

1.28.4.1 Metabolism by Plant Tissues

Isoflavonoids may not be end products of plant metabolism. In addition to demonstrating their mobilization from vacuolar stores and subsequent metabolism (often to more highly modified isoflavonoid derivatives, see above), some studies have documented metabolism of endogenously applied isoflavonoids by plant tissue. However, the presence of contaminating microorganisms can seriously compromise the interpretation of such experiments. For example, studies with chickpea and mungbean seedlings indicated half lives for exogenously added daidzein (**4**), formononetin (**5**) or coumestrol (**9**) of ~50 h. However, repeating these experiments with sterile mung bean seedlings revealed little appreciable metabolism of (**5**) (95% recovery after 24 h), although [^{14}C]-(**4**) was rapidly metabolized (8.5% recovery) with label incorporated into most cellular/chemical fractions, including the cell wall.[257]

The interconversions of medicarpin (**6**) and its corresponding isoflavan vestitol (**8**) in alfalfa and red clover[102,103] have been described above. Ring opening of a pterocarpan to yield the corresponding isoflavan (**110**) has also been reported when phaseollin (**17**) is fed to bean cell suspension cultures,[258] and this is accompanied by the opening of the ring formed from the cyclized prenyl side chain. Compound (**17**) is also converted to (**110**) by the fungal pathogen *Septoria nodorum*.[259]

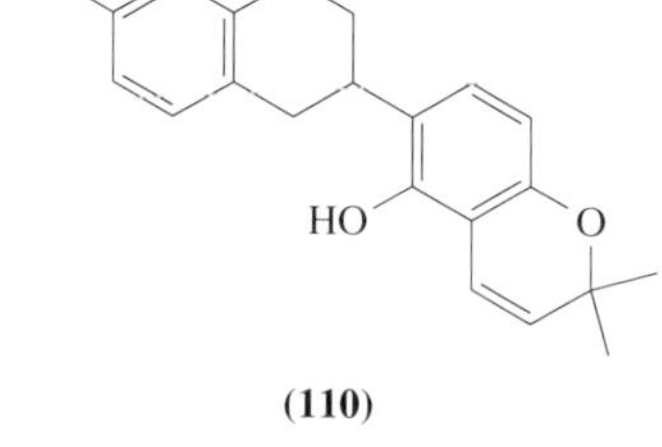

(110)

The role of isoflavonoid degradation as a factor in the elicitor- and pathogen-induced accumulation of isoflavonoid phytoalexins received considerable attention when it was proposed that elicitation by abiotic elicitors or incompatible races of pathogens was associated with strongly inhibited phytoalexin degradation (assessed using exogenously applied radiolabeled phytoalexin), whereas an increased biosynthetic rate was the major factor determining phytoalexin levels in response to biotic elicitors.[260,261] These conclusions were challenged when it was demonstrated, using $^{14}CO_2$ labeling *in vivo*, that the half-lives of glyceollin (**18**) and its trihydroxypterocarpan precursor (**87**) were long, ~100 h and ~38 h, respectively.[262] Apparently, the metabolic fates of exogenously applied and endogenously synthesized glyceollin are different. Studies of isoflavonoid turnover have subsequently been eclipsed by the vast body of work on the induced biosynthesis of these compounds, and more studies are needed to determine the biological half-lives and metabolic fates *in planta* of biologically active isoflavonoids.

1.28.4.2 Metabolism by Plant Pathogenic Fungi

There is a considerable body of literature on the metabolism of isoflavonoids by phytopathogenic fungi. A driving force for much of this work is the observation that metabolism of isoflavonoid phytoalexins, leading to their detoxification, is a mechanism by which successful pathogens may overcome the resistance response of their host.[9] The following discussion outlines metabolic strategies for phytoalexin detoxification for a limited selection of compounds, concentrating on pathways for which molecular genetic and/or enzymological data are available. This topic has been reviewed in more detail elsewhere.[263]

The simple isoflavone biochanin A (**33**) is degraded by *Nectria haematocca* to 3,4-dihydroxyphenylacetic acid (**111**) (Scheme 23),[264] and similar metabolic products have been obtained with other related fungi. The pathway involves sequential fission of the A-ring and the heterocyclic ring.

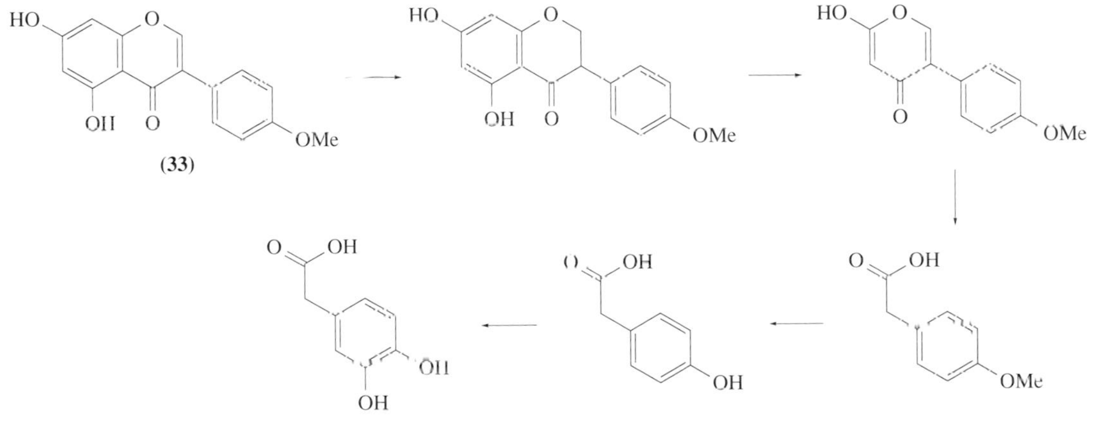

Scheme 23

The pea phytoalexin pisatin (**21**) is detoxified by *N. haematococca* by demethylation at position 3 (Equation (5)).[265] The enzyme that catalyzes this reaction, pisatin demethylase (PDA), is a cytochrome P450 that has been studied extensively at both the biochemical[266] and genetic[9,267] levels. Several different *PDA* genes are present in the *N. haematocca* genome, and their expression levels appear to confer different levels of demethylating activity.[268] The *PDA*6 gene is localized on a small meiotically unstable chromosome that is dispensable for normal growth.[269] The *PDA*1 and *PDA*9 genes contain an upstream sequence that is important for the induction of the genes by (**21**),[270,271] and this element is absent from *PDA*6. The sequences responsible for induction of *PDA*1 by (**21**), and the transcription factors that bind to these elements, have been studied in detail.[272–275]

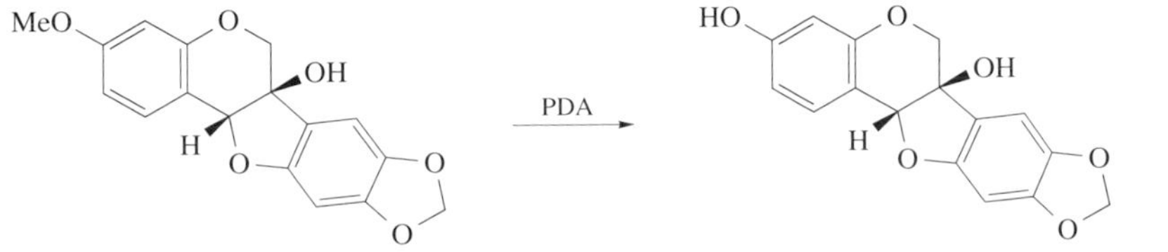

(5)

Ascochyta rabiei converts the chickpea pterocarpan phytoalexin (−)-maackiain (**59**) to a range of catabolites via reduction to 2′-hydroxyisoflavan (**112**) or oxidation to a 1a-hydroxypterocarp-1,4-dien-3-one (**113**)[276] (Scheme 24). The latter reaction has also been observed during the metabolism of phaseollin (**17**) by *Fusarium solani*.[277] A flavoprotein monooxygenase catalyzing 1a-hydroxylation of (**59**) in the presence of NAD(P)H, FAD, and molecular oxygen, and a pterocarpan reductase catalyzing the conversion of pterocarpan to 2′-hydroxyisoflavan (**112**) in the presence of NADPH, have been purified and characterized from *A. rabiei*[278,279] in which they are constitutively expressed. Three genes involved in maackiain detoxification by *N. haematococca*, which occurs via 1a- or 6a-hydroxylation, have been characterized. The *MAK*1 gene which, like *PDA*6, is located on a dispensable chromosome, appears, from nucleotide-derived amino acid sequence data, to encode a soluble flavoprotein monooxygenase,[280] similar to the pterocarpan hydroxylase described from *A. rabiei*.[278]

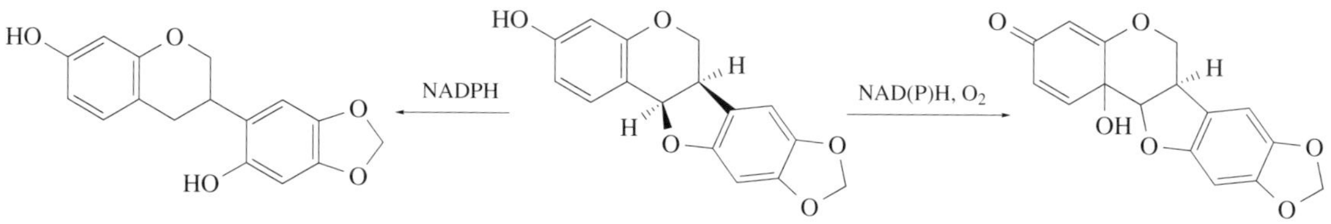

HO O NADPH HO O H H O O O NAD(P)H, O_2 O O H OH O O O HO O O

(**112**) (**59**) (**113**)

Scheme 24

Fusarium solani f. sp. *phaseoli* detoxifies the bean phytoalexins kievitone (**114**) and phaseollidin (**58**) by hydration of the prenyl side chain[281,282] (Scheme 25). The enzyme kievitone hydratase has been purified from *F. solani*, and its gene cloned.[283,284] The protein is extensively glycosylated, and is encoded by a single locus in *F. solani*.

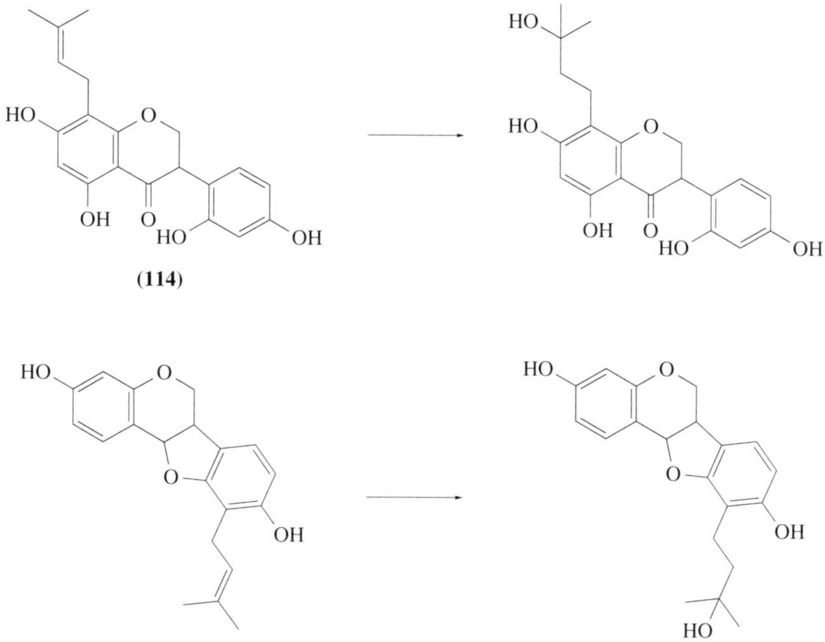

HO O OH O HO OH HO HO O OH O HO OH

(**114**)

HO O O OH HO O O OH HO

(**58**)

Scheme 25

1.28.4.3 Metabolism by Bacteria

Isoflavonoid *nod* gene inducers can be metabolized by *Rhizobia* to a number of simpler compounds. Thus, genistein (**3**) is converted to naringenin chalcone (**42**), which is then converted to phloroglucinol (**115**), phloroglucinol carboxylic acid (**116**), 4-coumarate (**117**), *p*-hydroxybenzoic acid (**118**), and phenylacetic acid (**119**) (Scheme 26).[285] The degradative reactions generally involve C-ring fission, and coumestrol (**9**) is produced as a C-ring metabolite of daidzein (**4**).[286] Some of the metabolic products are themselves effective *nod* gene inhibitors.

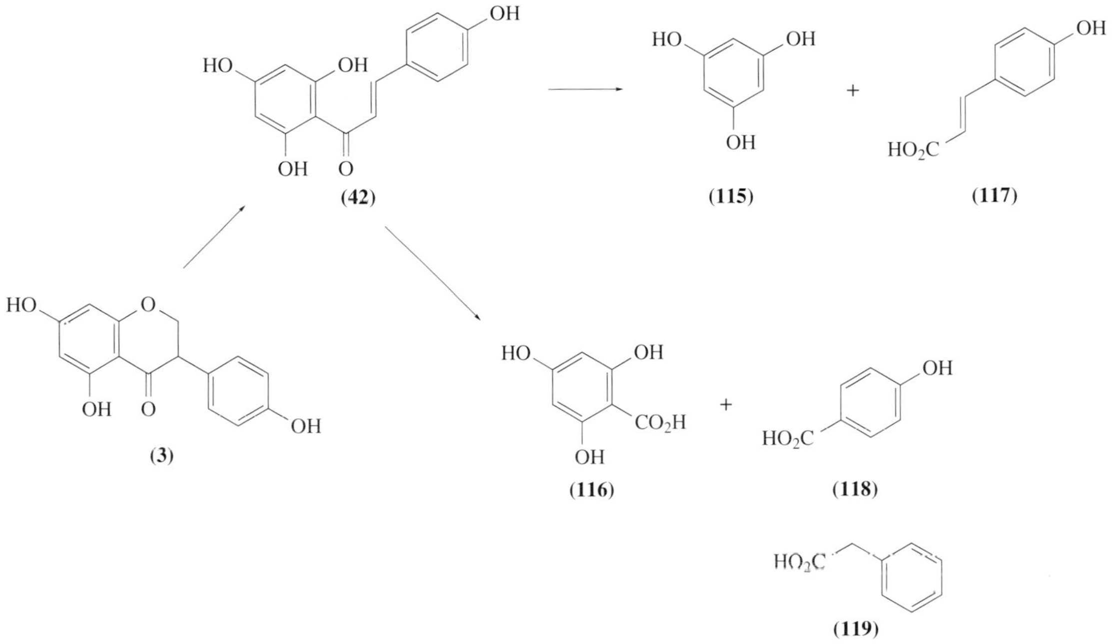

Scheme 26

Interest has been shown in the bacterial metabolism of soybean isoflavones in relation to the occurrence of isoflavone metabolites in tempe, a fermented soybean food from Indonesia. One important metabolite is 6,7,4′-trihydroxyisoflavone (**120**), which has powerful antioxidant activity. *Brevibacterium epidermidis* and *Micrococcus luteus* demethylate the soybean isoflavone glycitein (**99**) to (**120**), whereas *Microbacterium arborescens* was shown to convert (**4**) to (**99**) via (**120**) (Scheme 27).[287]

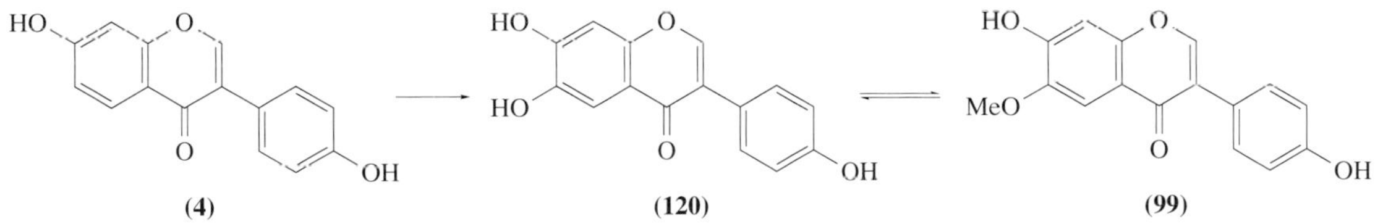

Scheme 27

1.28.4.4 Metabolism by Animals

The conversion of dietary soybean isoflavones to urinary equol (**37**) has already been referred to. This is catalyzed by gastrointestinal flora. The proposed pathway[97] is summarized in Scheme 28.

Studies have been performed on the metabolism of the prenylated lupin isoflavone luteone (**121**) in a rat-liver homogenate.[288] All the interconversions involved side-chain oxidation reactions, most of which were distinct from those observed following incubation of 7-methylluteone (**122**) with *Botrytis cinerea*[289,290] (Scheme 29). The enzymology of the side chain oxidation in rat liver homogenates has yet to be determined. In *B. cinerea*, formation of the epoxide is catalyzed by a microsomal monooxygenase that is inducible by prenylated isoflavones.[290]

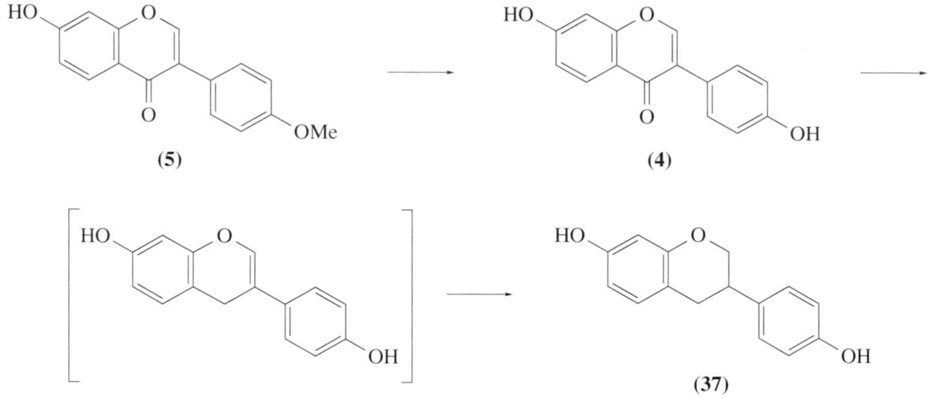

Scheme 28

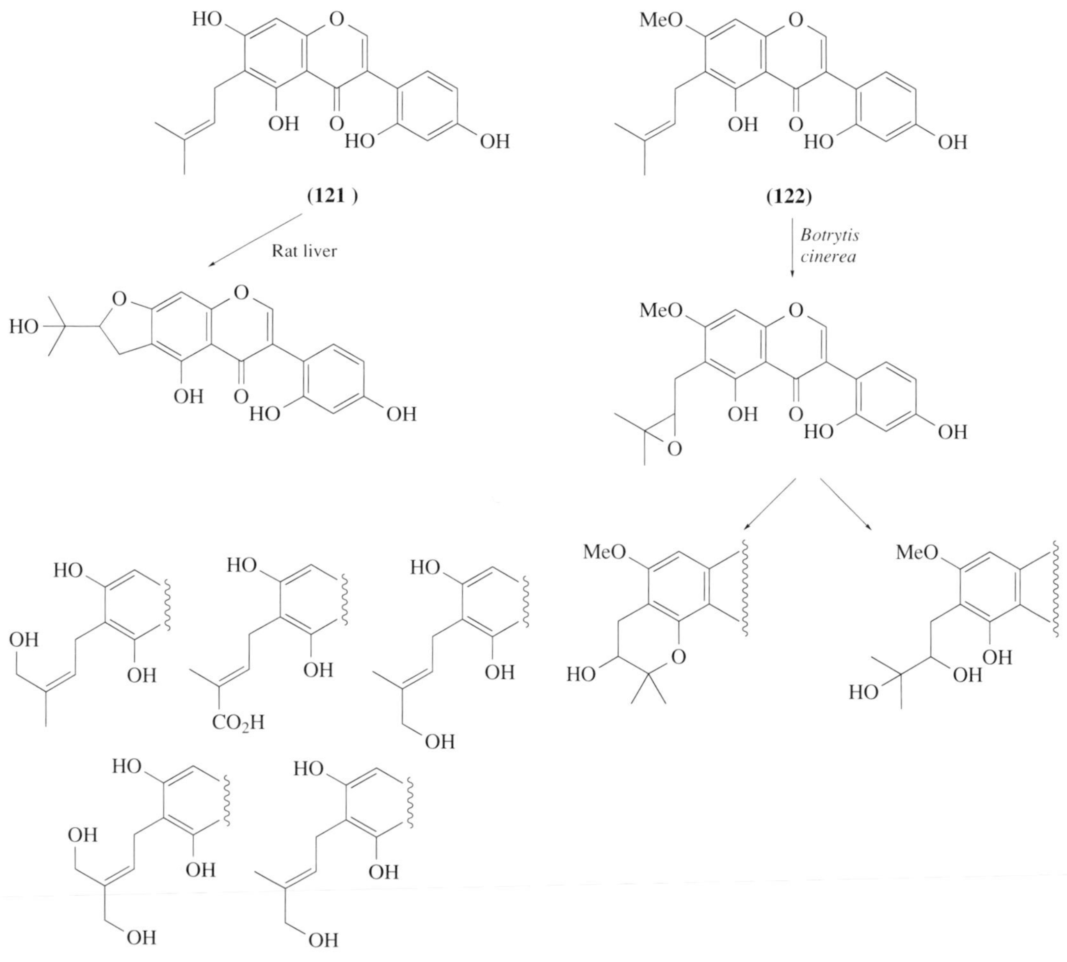

Scheme 29

1.28.5 INTEGRATED CONTROL OF ISOFLAVONOID BIOSYNTHESIS

In elicited plant cells, the accumulation of isoflavonoids is usually accompanied by coordinated increases in the extractable activities of all the biosynthetic enzymes of the pathway. These increases are primarily the result of increased transcription of the genes encoding the various biosynthetic enzymes,[146,159] and the evidence for increased gene transcription has already been described above in relation to those enzymes for which molecular probes are available.

Two major questions are now attracting increased attention: what are the signal transduction pathways linking elicitor perception at the cell surface to increased transcription of phytoalexin biosynthetic genes, and how are these integrated within the total program of induced defense responses? A related question is, how many different transcriptional activators are necessary to orchestrate the complete pathway response? To address these questions, it is necessary to isolate the transcriptional regulators responsible for the elicitation response. Because of the ease of selection for mutations affecting the synthesis of colored flavonoid derivatives, genetic approaches have been successfully used for the cloning of transcriptional regulators of the anthocyanin pathway.[291] Such a strategy is much less easy for colorless isoflavonoids, and the genetic intractability of many of the legume species used for isoflavonoid research is a further problem. The development of model genetic systems in legumes such as *Medicago truncatula*[292] provides hope that it will soon be possible to use mutation followed by positional cloning to isolate genes involved in the regulation of isoflavonoid synthesis.

The alternative approach is to use a combination of molecular and biochemical techniques to identify the *cis*-elements in isoflavonoid pathway gene promoters that confer infection- or elicitor-inducibility,[293,294] to use this information to generate probes for the isolation of the transcription factors that bind to these sequences, and then to search for molecules that might interact with these factors to modulate their activity. The author's group have performed such a series of experiments using an elicitor-inducible *CHS* gene promoter as the starting point for walking back up the signal transduction pathway for elicitor modulation of isoflavonoid synthesis.[146,295] At least two distinct classes of DNA-binding protein interact with the elicitor-response transducing element in the bean *chs*15 promoter, and their activity is regulated by a highly poised phosphorylation/dephosporylation cascade.[296] Several reviews give detailed background on the status of our understanding of defense gene signal transduction pathways.[298,299]

Transcription of the *IFR* gene is slightly delayed compared with that of *PAL* or *CHS* genes in elicited alfalfa cell suspensions, consistent with its responding to different transcriptional regulators.[159] This idea is confirmed by the results of *in vivo* (functional) and *in vitro* analyses of the alfalfa isoflavone reductase promoter,[297] which appears to be regulated by transcription factor(s) recognizing sequences not involved in CHS regulation.

1.28.6 EVOLUTION OF ISOFLAVONOID PHYTOALEXIN BIOSYNTHETIC PATHWAYS

Genes encoding enzymes of plant natural product biosynthesis have most likely arisen from duplication and subsequent mutation of genes encoding enzymes of primary metabolism. Natural selection will act to preserve such altered gene sequences if they confer a selective advantage, i.e., if some new functionality is produced. It has been argued that the original function leading to selection for flavonoid pigments was their ability to modulate internal growth regulator levels (e.g., by stimulating or inhibiting IAA oxidation), because this could be manifest at relatively low levels of product present in the cytoplasm.[300] UV-filtering vacuolar flavonoid pigments would have evolved later.

In view of their increased complexity and limited taxonomic distribution compared with 1,3-diphenylpropane flavonoids, it is clear that the isoflavonoids evolved more recently than the UV-protective flavonoids. The selective advantage of the first isoflavonoids could have been as antimicrobial agents. Thus, microbial infection would be a strong selection pressure for somatic mutations occurring even in a single branch or shoot that could give rise to protected tissue, as argued by Walbot.[301] However, simple isoflavonoids generally have low antimicrobial activity, and it is unlikely that the position-specific modifying enzymes that generate highly bioactive isoflavonoids would have arisen in parallel with the isoflavone synthase system. Other functions for the earliest isoflavonoids must therefore be sought, and the evidence outlined below suggests that antioxidant activity is a likely possibility.

Many flavonoids, and also lignans and hydroxycinnamic acid derivatives, have powerful anti-oxidant activity,[302] and can scavenge superoxide anion,[303] or help remove apoplastic hydrogen peroxide by acting as cosubstrates for ascorbate-dependent peroxidases.[304] These active oxygen species are produced during the oxidative burst, a key component of plant disease resistance responses that has many common features with mammalian neutrophil activation,[294] and is probably therefore of ancient evolutionary origin. Using a screen that selects for plant cDNA clones that can confer oxidative stress tolerance when expressed in yeast, an Arabidopsis NADPH reductase was characterized that has striking amino acid sequence similarity to alfalfa IFR,[305] and further IFR homologues have been reported from the Arabidopsis expressed sequence tag (EST) project.[306] A similar IFR-like protein is induced in maize by treaments that affect redox balance by reducing cellular glutathione levels.[307] Neither the Arabidopsis nor maize proteins exhibit IFR activity, and neither plant has been reported to make isoflavonoids; the proteins appear to belong to a new class of oxidoreductases that may function in a thiol-independent response to oxidative stress under conditions of reduced glutathione shortage.

Other enzymes with very strong amino acid sequence identity to IFR have been identified in tobacco floral, stem, and root tissue,[308,309] but are clearly not involved in isoflavonoid synthesis. The gene encoding the IFR-like protein that is expressed in tobacco flowers and stems has several introns that are in the identical position to the introns in the alfalfa IFR gene,[196] indicating close evolutionary relatedness. The functions of the tobacco IFR-related proteins remain to be determined.

ACKNOWLEDGMENTS

I would like to thank Drs. Shin-ichi Ayabe (Nihon University, Kanagawa), Jochen Berlin (GBF, Braunschweig), Jim Cooper (Queen's University, Belfast), Yutaka Ebizuka (University of Tokyo), Robert Esnault and Pascal Ratet (CNRS, Gif-sur-Yvette), Adrian Franke (Cancer Research Center of Hawaii), Terrence Graham (Ohio State University), James Harper (USDA, Urbana), Mark Robbins (IGER, Aberystwyth), David Straney (University of Maryland), Satoshi Tahara (Hokkaido University, Sapporo), and Dietrich Werner (Phillipps-Universität, Marburg), for providing reprints and preprints of unpublished work for inclusion in this chapter. I also thank Dr. Maria Harrison for critical review of the manuscript and Cuc Ly for assistance with the artwork. Work from the author's laboratory was supported by the Samuel Roberts Noble Foundation.

1.28.7 REFERENCES

1. S. Tahara and R. K. Ibrahim, *Phytochemistry*, 1995, **38**, 1073.
2. E. Palme, A. R. Bilia, and I. Morelli, *Phytochemistry*, 1996, **42**, 903.
3. M. J. O'Neill, S. A. Adesanya, M. F. Roberts, and I. R. Pantry, *Phytochemistry*, 1986, **25**, 1315.
4. R. W. Miller and G. F. Spencer, *J. Nat. Prod.*, 1989, **52**, 634.
5. G. F. Spencer, B. E. Jones, R. D. Plattner, D. E. Barnekow, L. S. Brinen, and J. Clardy, *Phytochemistry*, 1991, **30**, 4147.
6. J. L. Ingham, *Biochem. Syst. Ecol.*, 1979, **7**, 29.
7. A. F. Olah and R. T. Sherwood, *Phytopathology*, 1971, **61**, 65.
8. J. N. Bilton, J. R. Debnam, and I. M. Smith, *Phytochemistry*, 1976, **15**, 1411.
9. H. D. VanEtten, D. E. Matthews, and P. S. Matthews, *Annu. Rev. Phytopathol.*, 1989, **27**, 143.
10. S. W. Banks and P. M. Dewick, *Phytochemistry*, 1982, **21**, 1605.
11. R. E. Carlson and D. H. Dolphin, *Phytochemistry*, 1981, **20**, 2281.
12. A. Kobayashi, K. Akiyama, and K. Kawazu, *Phytochemistry*, 1993, **32**, 77.
13. P. M. Dewick, in "The Flavonoids: Advances in Research Since 1986," ed. J. B. Harborne, Chapman and Hall, London, 1993, p. 117.
14. A. Arnone, L. Camarda, L. Merlini, G. Nasini, and D. A. H. Taylor, *Phytochemistry*, 1981, **20**, 799.
15. N. A. M. Saleh, L. Boulos, S. I. El-Negoumy, and M. F. Abdalla, *Biochem. Syst. Ecol.*, 1982, **10**, 33.
16. E. Wong, in "The Flavonoids. Part 2," eds. J. B. Harborne, T. J. Mabry, and H. Mabry, Academic Press, New York, 1975, p. 743.
17. P. M. Dewick, in "The Flavonoids: Advances in Research Since 1980," ed. J. B. Harborne, Chapman and Hall, London, 1988, vol. 5, p. 125.
18. J. L. Ingham, in "Phytoalexins," eds. J. A. Bailey and J. W. Mansfield, Halstead Press, New York, 1982, p. 21.
19. J. S. Huang and K. R. Barker, *Plant Physiol.*, 1991, **96**, 1302.
20. H. Fukami and M. Nakajima, in "Naturally Occurring Insecticides," eds. M. Jacobson and D. G. Crosby, Dekker, New York, 1971, p. 71.
21. R. W. Miller, R. Kleiman, R. G. Powell, and A. R. Putnam, *J. Nat. Prod.*, 1988, **51**, 328.
22. R. A. Skipp, C. Selby, and J. A. Bailey, *Physiol. Plant Pathol.*, 1977, **10**, 221.
23. J. A. Glazener and H. D. VanEtten, *Phytopathology*, 1978, **68**, 111.
24. J. L. Giannini, J. S. Halvorson, and G. O. Spessard, *Phytochemistry*, 1991, **30**, 3233.
25. M. A. Ferrer, M. A. Pedreño, R. Muñoz, and A. R. Barceló, *Phytochemistry*, 1992, **31**, 3681.
26. E. T. Bailey and C. M. Francis, *Aust. J. Agric. Res.*, 1971, **22**, 731.
27. Y. Masaoka, M. Kojima, S. Sugihara, T. Yoshihara, M. Koshino, and A. Ichihara, *Plant Soil*, 1993, **155/156**, 75.
28. W. M. Keung, O. Lazo, L. Kunze, and B. L. Vallee, *Proc. Natl. Acad. Sci. USA*, 1995, **92**, 8990.
29. M. Takai, H. Yamaguchi, T. Saitoh, and S. Shibata, *Chem. Pharm. Bull.*, 1972, **20**, 2488.
30. A. Malhotra, V. V. S. Murti, and T. R. Seshadri, *Curr. Sci.*, 1967, **36**, 484.

31. T. Fotsis, M. Pepper, H. Adlercreutz, G. Fleischmann, T. Hase, R. Montesano, and L. Schweigerer, *Proc. Natl. Acad. Sci. USA*, 1993, **90**, 2690.
32. H. J. Wang and P. A. Murphy, *J. Agric. Food Chem.*, 1994, **42**, 1666.
33. T. Akiyama, J. Ishida, S. Nakagawa, H. Ogawara, S. Watanabe, N. Itoh, M. Shibuya, and Y. Fukami, *J. Biol. Chem.*, 1987, **262**, 5592.
34. J. Huang, M. Nasr, Y. Kim, and H. R. Matthews, *J. Biol. Chem.*, 1992, **267**, 15511.
35. Y. Goda, F. Kiuchi, M. Shibuya, and U. Sankawa, *Chem. Pharm. Bull.*, 1992, **40**, 2452.
36. T. Takano, K. Takada, H. Tada, S. Nishiyama, and N. Amino, *Biochem. Biophys. Res. Commun.*, 1993, **190**, 801.
37. A. Okura, H. Arakawa, H. Oka, T. Yoshinari, and Y. Monden, *Biochem. Biophys. Res. Commun.*, 1988, **157**, 183.
38. F. Okajima, M. Akbar, M. A. Majid, K. Sho, H. Tomura, and Y. Kondo, *Biochem. Biophys. Res. Commun.*, 1994, **203**, 1488.
39. F. Dakora and D. Phillips, *Physiol. Mol. Plant Pathol.*, 1996, **49**, 1.
40. H. VanEtten, J. W. Mansfield, J. A. Bailey, and E. E. Farmer, *Plant Cell*, 1994, **6**, 1191.
41. J. L. Ingham, S. Tahara, and J. B. Harborne, *Z. Naturforsch., Teil C*, 1983, **38**, 194.
42. S. A. Adesanya, M. J. O'Neill, and M. F. Roberts, *Physiol. Mol. Plant Pathol.*, 1986, **29**, 95.
43. R. P. Krämer, H. Hindorf, H. Chandra, J. Kallage, and F. Zilliken, *Phytochemistry*, 1984, **23**, 2203.
44. M. Weidenbörner, H. Hindorf, H. J. Chandra, P. Tsotsonos, and H. Egge, *Phytochemistry*, 1989, **28**, 3317.
45. M. Weidenbörner, H. Hindorf, H. Chandra Jha, P. Tsotsonos, and H. Egge, *Phytochemistry*, 1990, **29**, 801.
46. L. M. Delserone, D. E. Matthews, and H. D. VanEtten, *Phytochemistry*, 1992, **31**, 3813.
47. J. E. Rahe, *Can. J. Bot.*, 1973, **51**, 2423.
48. L. A. Hadwiger and D. M. Webster, *Phytopathology*, 1984, **74**, 1312.
49. M. Long, P. Barton-Willis, B. J. Staskawicz, D. Dahlbeck, and N. T. Keen, *Phytopathology*, 1985, **75**, 235.
50. M. K. Bhattacharyya and E. W. B. Ward, *Physiol. Mol. Plant Pathol.*, 1987, **31**, 387.
51. P. Moesta and H. Grisebach, *Physiol. Plant Pathol.*, 1982, **21**, 65.
52. I. Raskin, *Plant Physiol.*, 1992, **99**, 799.
53. H. C. Kistler and H. D. Vanetten, *J. Gen. Microbiol.*, 1984, **130**, 2605.
54. J. Glazebrook and F. M. Ausubel, *Proc. Natl. Acad. Sci. USA*, 1994, **91**, 8955.
55. J. Grandmaison and R. Ibrahim, *J. Exp. Bot.*, 1995, **46**, 231.
56. P. F. Morris and E. W. B. Ward, *Physiol. Mol. Plant Pathol.*, 1992, **40**, 17.
57. Y. Ruan, V. Kotraiah, and D. C. Straney, *Mol. Plant–Microbe Interact.*, 1995, **8**, 929.
58. R. Kape, K. Wex, M. Parniske, E. Görge, A. Wetzel, and D. Werner, *J. Plant Physiol.*, 1992, **141**, 54.
59. Z.-P. Xie, C. Staehelin, H. Vierheilig, A. Wiemken, S. Jabbouri, W. J. Broughton, R. Vögeli-Lange, and T. Boller, *Plant Physiol.*, 1995, **108**, 1519.
60. M. J. Harrison and R. A. Dixon, *Mol. Plant–Microbe Interact.*, 1993, **6**, 643.
61. H. Volpin, D. A. Phillips, Y. Okon, and Y. Kapulnik, *Plant Physiol.*, 1995, **108**, 1449.
62. M. J. Harrison and R. A. Dixon, *Plant J.*, 1994, **6**, 9.
63. N. K. Peters, J. W. Frost, and S. R. Long, *Science*, 1986, **233**, 977.
64. J. W. Redmond, M. Batley, M. A. Djordjevic, R. W. Innes, P. L. Kuempel, and B. G. Rolfe, *Nature (London)*, 1986, **323**, 632.
65. R. M. Kosslak, R. Bookland, J. Barkei, H. E. Paaren, and E. R. Applebaum, *Proc. Natl. Acad. Sci. USA*, 1987, **84**, 7428.
66. F. Zhang and D. L. Smith, *J. Exp. Bot.*, 1996, **47**, 785.
67. C. Coronado, J. A. S. Zuanazzi, C. Sallaud, J.-C. Quirion, R. Esnault, H.-P. Husson, A. Kondorosi, and P. Ratet, *Plant Physiol.*, 1995, **108**, 533.
68. F. Dakora, C. Joseph, and D. Phillips, *Plant Physiol.*, 1993, **101**, 819.
69. A. Savoure', Z. Magyar, M. Pierre, S. Brown, M. Schultze, D. Dudits, A. Kondorosi, and E. Kondorosi, *EMBO J.*, 1994, **13**, 1093.
70. F. D. Dakora, C. M. Joseph, and D. A. Phillips, *Mol. Plant–Microbe Interact.*, 1993, **6**, 665.
71. R. Kape, M. Parniske, and D. Werner, *Appl. Environ. Microbiol.*, 1991, **57**, 316.
72. R. Kape, M. Parniske, S. Brandt, and D. Werner, *Appl. Environ. Microbiol.*, 1992, **58**, 1705.
73. M. J. Cho and J. E. Harper, *Plant Physiol.*, 1991, **95**, 435.
74. M. Cho and J. Harper, *Plant Physiol.*, 1991, **96**, 1277.
75. P. Schmidt, M. Parinske, and D. Werner, *Bot. Acta*, 1992, **105**, 18.
76. P. E. Schmidt, W. J. Broughton, and D. Werner, *Mol. Plant–Microbe Interact.*, 1994, **7**, 384.
77. M. Parniske, B. Ahlborn, and D. Werner, *J. Bacteriol.*, 1991, **173**, 3432.
78. J. E. Cooper, J. R. Rao, L. De Cooman, T. M. Corry, A. J. Bjourson, H. L. Steele, W. J. Broughton, and D. Werner, in "Biological Fixation of Nitrogen for Ecology and Sustainable Agriculture," eds. A. Legocki, H. Bothe, and A. Puhler, NATO ASI Series G, Vol. 39, Springer, Berlin, 1997, p. 115.
79. D. A. Shutt, *Endeavour*, 1976, **75**, 110.
80. R. R. Gildersleeve, G. R. Smith, I. J. Pemberton, and C. L. Gilbert, *Crop Sci.*, 1991, **31**, 889.
81. A. S. Leopold, M. Erwin, J. Oh, and B. Browning, *Science*, 1976, **191**, 98.
82. K. D. R. Setchell, S. P. Borriello, P. Hulme, D. N. Kirk, and M. Axelson, *Am. J. Clin. Nutr.*, 1984, **40**, 569.
83. W. Wang, Y. Tanaka, Z. Han, and C. M. Higuchi, *Nutr. Cancer*, 1995, **23**, 131.
84. M. E. Martin, M. Haourigui, C. Pelissero, C. Benassayag, and E. A. Nunez, *Life Sci.*, 1996, **58**, 429.
85. S. Barnes, C. Grubbs, K. D. R. Setchell, and J. Carlson, in "Mutagens and Carcinogens in the Diet," ed. M. W. Pariza, Wiley-Liss, New York, 1990, p. 239.
86. H. P. Lee, L. Gourley, S. W. Duffy, J. Esteve, J. Lee, and N. E. Day, *Lancet*, 1991, **337**, 1197.
87. H. Adlercreutz, H. Honjo, A. Higashi, T. Fotsis, E. Hämäläinen, T. Hasegawa, and H. Okada, *Am. J. Clin. Nutr.*, 1991, **54**, 1093.
88. L. Coward, N. C. Barnes, K. D. R. Setchell, and S. Barnes, *J. Agric. Food Chem.*, 1993, **41**, 1961.
89. C. A. Lamartiniere, J. Moore, M. Holland, and S. Barnes, *Proc. Soc. Exp. Biol. Med.*, 1995, **208**, 120.
90. A. Cassidy, S. Bingham, and K. D. Setchell, *Am. J. Clin. Nutr.*, 1994, **60**, 333.
91. A. A. Franke, W. Wang, and C. Y. Shi, *Proc. Soc. Exp. Biol. Med.*, 1998, **217**, 263.

92. K. Yanagihara, A. Ito, T. Toge, and M. Numoto, *Cancer Res.*, 1993, **53**, 5815.
93. S. Barnes, in "American Chemical Society Symposium on Flavonoids and Related Compounds, 1996, Orlando, Florida," American Chemical Society, Washington DC, in press.
94. F. M. Uckun, W. E. Evans, C. J. Forsyth, K. G. Waddick, L. T. Ahlgren, L. M. Chelstrom, A. Burkhardt, J. Bolen, and D. E. Myers, *Science*, 1995, **267**, 886.
95. J. C. Vera, A. M. Reyes, J. G. Carcamo, F. V. Velasquez, C. I. Rivas, R. H. Zhang, P. Strobel, R. Iribarren, H. I. Scher, J. C. Slebe, and D. W. Golde, *J. Biol. Chem.*, 1996, **271**, 8719.
96. M. Kusaka and N. Sperelakis, *Biochim. Biophys. Acta*, 1996, **1278**, 1.
97. H. Adlercreutz, K. Hockerstedt, S. Bloigu, E. Hamalainen, T. Fotsis, and A. Ollus, *J. Steroid Biochem.*, 1987, **27**, 1135.
98. M. J. Messina, V. Persky, K. D. R. Setchell, and S. Barnes, *Nutr. Cancer*, 1994, **21**, 113.
99. D. C. Knight and J. A. Eden, *Obstet. Gynecol.*, 1996, **87**, 897.
100. H. Wiseman, *Biochem. Soc. Trans.*, 1996, **24**, 795.
101. T. Hakamatsuka, Y. Ebizuka, and U. Sankawa, in "Biotechnology in Agriculture and Forestry," ed. Y. P. S. Bajai, Springer, Berlin, 1994, vol. 28, p. 336.
102. P. M. Dewick and D. Ward, *Phytochemistry*, 1978, **17**, 1751.
103. P. M. Dewick and M. Martin, *Phytochemistry*, 1979, **18**, 591.
104. R. A. Dixon, in "Tissue Culture Methods for Plant Pathologists," eds. D. S. Ingram and J. P. Helgeson, Blackwell, Oxford, 1980, p. 185.
105. P. Albersheim, M. McNeil, A. G. Darvill, B. S. Valent, M. G. Hahn, B. K. Robertsen, and P. Aman, *Recent Adv. Phytochem.*, 1981, **15**, 37.
106. R. A. Dixon, *Biol. Rev.*, 1986, **61**, 239.
107. M. P. Robbins, J. Hartnoll, and P. Morris, *Plant Cell Rep.*, 1991, **10**, 59.
108. R. Edwards, J. W. Blount, and R. A. Dixon, *Planta*, 1991, **184**, 403.
109. H. Keßmann, S. Daniel, and W. Barz, *Z. Naturforsch., Teil C*, 1988, **43**, 529.
110. J. Ebel, A. R. Ayers, and P. Albersheim, *Plant Physiol.*, 1976, **57**, 775.
111. J. Berlin, L. Fecker, C. Rügenhagen, C. Sator, D. Strack, L. Witte, and V. Wray, *Z. Naturforsch., Teil C*, 1991, **46**, 725.
112. K. Dalkin, R. Edwards, B. Edington, and R. A. Dixon, *Plant Physiol.*, 1990, **92**, 440.
113. T. L. Graham and M. Y. Graham, *Mol. Plant–Microbe Interact.*, 1991, **4**, 60.
114. T. L. Graham and M. Y. Graham, *Plant Physiol.*, 1996, **110**, 1123.
115. J. N. Bell, R. A. Dixon, J. A. Bailey, P. M. Rowell, and C. J. Lamb, *Proc. Natl. Acad. Sci. USA*, 1984, **81**, 3384.
116. A. Levine, R. Tenhaken, R. A. Dixon, and C. J. Lamb, *Cell*, 1994, **79**, 583.
117. K. Shirasu, H. Nakajima, V. K. Rajasekhar, R. A. Dixon, and C. J. Lamb, *Plant Cell*, 1997, **9**, 261.
118. R. A. Dixon and D. S. Bendall, *Physiol. Plant Pathol.*, 1978, **13**, 295.
119. B. S. Shorrosh, R. A. Dixon, and J. B. Ohlrogge, *Proc. Natl. Acad. Sci. USA*, 1994, **91**, 4323.
120. C. L. Cramer, K. Edwards, M. Dron, X. Liang, S. L. Dildine, G. P. Bolwell, R. A. Dixon, C. J. Lamb, and W. Schuch, *Plant Mol. Biol.*, 1989, **12**, 367.
121. C. Douglas, H. Hoffmann, W. Schulz, and K. Hahlbrock, *EMBO J.*, 1987, **6**, 1189.
122. T. Fahrendorf and R. A. Dixon, *Arch. Biochem. Biophys.*, 1993, **305**, 509.
123. R. A. Dixon and M. J. Harrison, *Adv. Genet.*, 1990, **28**, 165.
124. G. P. Bolwell, J. N. Bell, C. L. Cramer, W. Schuch, C. J. Lamb, and R. A. Dixon, *Eur. J. Biochem.*, 1985, **149**, 411.
125. N. J. Bate, J. Orr, W. Ni, A. Meroni, T. Nadler-Hassar, P. W. Doerner, R. A. Dixon, C. J. Lamb, and Y. Elkind, *Proc. Natl. Acad. Sci. USA*, 1994, **91**, 7608.
126. P. A. Howles, N. L. Paiva, V. J. H. Sewalt, N. L. Elkind, Y. Bate, C. J. Lamb, and R. A. Dixon, *Plant Physiol.*, 1996, **112**, 1617.
127. P. M. Dewick and M. J. Steele, *Phytochemistry*, 1982, **21**, 1599.
128. S. W. Banks and P. M. Dewick, *Phytochemistry*, 1983, **22**, 1591.
129. S. W. Banks and P. M. Dewick, *Phytochemistry*, 1982, **21**, 2235.
130. P. M. Dewick and M. Martin, *Phytochemistry*, 1979, **18**, 597.
131. M. Martin and P. M. Dewick, *Phytochemistry*, 1980, **19**, 2341.
132. M. D. Woodward, *Physiol. Plant Pathol.*, 1980, **17**, 17.
133. M. D. Woodward, *Physiol. Plant Pathol.*, 1981, **18**, 33.
134. U. Niesbach-Klösgen, E. Barzen, J. Bernhardt, W. Rohde, Z. Schwarz-Sommer, H. J. Reif, U. Wienand, and H. Saedler, *J. Mol. Evol.*, 1987, **26**, 213.
135. S. Akada and S. K. Dube, *Plant Mol. Biol.*, 1995, **29**, 189.
136. C. An, Y. Ichinose, T. Yamada, Y. Tanaka, T. Shiraishi, and H. Oku, *Plant Mol. Biol.*, 1993, **21**, 789.
137. T. Arioli, P. A. Howles, J. J. Weinman, and B. G. Rolfe, *Gene*, 1994, **138**, 79.
138. H. Junghans, K. Dalkin, and R. A. Dixon, *Plant Mol. Biol.*, 1993, **22**, 239.
139. R. L. Feinbaum and F. M. Ausubel, *Mol. Cell. Biol.*, 1992, **8**, 1985.
140. T. B. Ryder, S. A. Hedrick, J. N. Bell, X. Liang, S. D. Clouse, and C. J. Lamb, *Mol. Gen. Genet.*, 1987, **210**, 219.
141. O. Nakajima, M. Shibuya, T. Hakamatsuka, H. Noguchi, Y. Ebizuka, and U. Sankawa, *Biol. Pharmacol. Bull.*, 1996, **19**, 71.
142. T. B. Ryder, C. L. Cramer, J. N. Bell, M. P. Robbins, R. A. Dixon, and C. J. Lamb, *Proc. Natl. Acad. Sci. USA*, 1984, **81**, 5724.
143. S. Daniel and W. Barz, *Planta*, 1990, **182**, 279.
144. S. Dhawale, G. Souciet, and D. N. Kuhn, *Plant Physiol.*, 1989, **91**, 911.
145. S. Ayabe, A. Udagawa, and T. Furuya, *Plant Cell Rep.*, 1988, **7**, 35.
146. R. A. Dixon, M. J. Harrison, and N. L. Paiva, *Physiol. Plant.*, 1995, **93**, 385.
147. P. M. Dewick, M. J. Steele, R. A. Dixon, and I. M. Whitehead, *Z. Naturforsch., Teil C*, 1982, **37**, 363.
148. S. I. Ayabe, A. Udagawa, and T. Furuya, *Arch. Biochem. Biophys.*, 1988, **261**, 458.
149. R. Welle and H. Grisebach, *FEBS Lett.*, 1988, **236**, 221.
150. R. Welle and H. Grisebach, *Arch. Biochem. Biophys.*, 1989, **272**, 97.
151. R. Welle, G. Schröder, E. Schiltz, H. Grisebach, and J. Schröder, *Eur. J. Biochem.*, 1991, **196**, 423.

152. G. M. Ballance and R. A. Dixon, *Plant Physiol.*, 1994, **107**, 1027.
153. C. Sallaud, J. El-Turk, L. Bigarré, H. Sevin, R. Welle, and R. Esnault, *Plant Physiol.*, 1995, **108**, 869.
154. C. Sallaud, J. Elturk, C. Breda, D. Buffard, I. Dekozak, R. Esnault, and A. Kondorosi, *Plant Sci.*, 1995, **109**, 179.
155. T. Akashi, T. Furuno, K. Futami, M. Honda, T. Takahashi, R. Welle, and S. Ayabe, *Plant Physiol.*, 1996, **111**, 347.
156. R. Welle and J. Schröder, *Arch. Biochem. Biophys.*, 1992, **293**, 377.
157. S. Tropf, B. Kärcher, G. Schröder, and J. Schröder, *J. Biol. Chem.*, 1995, **270**, 7922.
158. K. Harano, N. Okada, T. Furuno, T. Takahashi, S. Ayabe, and R. Welle, *Plant Cell Rep.*, 1993, **12**, 66.
159. W. Ni, T. Fahrendorf, G. M. Ballance, C. J. Lamb, and R. A. Dixon, *Plant Mol. Biol.*, 1996, **30**, 427.
160. R. A. Dixon, P. M. Dey, and I. M. Whitehead, *Biochim. Biophys. Acta*, 1982, **715**, 25.
161. M. J. Boland and E. Wong, *Eur. J. Biochem.*, 1975, **50**, 383.
162. R. A. Bednar and J. R. Hadcock, *J. Biol. Chem.*, 1988, **263**, 9582.
163. M. C. Mehdy and C. J. Lamb, *EMBO J.*, 1987, **6**, 1527.
164. H. I. McKhann and A. M. Hirsch, *Plant Mol. Biol.*, 1994, **24**, 767.
165. E. R. Blyden, P. W. Doerner, C. J. Lamb, and R. A. Dixon, *Plant Mol. Biol.*, 1991, **16**, 167.
166. A. J. Wood and E. Davies, *Plant Physiol.*, 1994, **104**.
167. E. Grotewold and T. Peterson, *Mol. Gen. Genet.*, 1994, **242**, 1.
168. A. J. van Tunen, R. E. Koes, C. E. Spelt, A. R. van der Krol, A. R. Stuitje, and J. N. M. Mol, *EMBO J.*, 1988, **7**, 1257.
169. M. P. Robbins and R. A. Dixon, *Eur. J. Biochem.*, 1984, **145**, 195.
170. K. Dalkin, J. Jorrin, and R. A. Dixon, *Physiol. Mol. Plant Pathol.*, 1990, **37**, 293.
171. K. Hahlbrock, H. Zilg, and H. Grisebach, *Eur. J. Biochem.*, 1970, **15**, 13.
172. M. J. Boland and E. Wong, *Bioorg. Chem.*, 1979, **8**, 1.
173. R. A. Dixon, E. R. Blyden, M. P. Robbins, A. J. van Tunen, and J. N. M. Mol, *Phytochemistry*, 1988, **27**, 2801.
174. R. A. Dixon, P. M. Dey, and C. J. Lamb, *Adv. Enzymol. Relat. Areas Mol. Biol.*, 1983, **55**, 1.
175. H. Grisebach and G. Brandner, *Biochim. Biophys. Acta*, 1962, **60**, 51.
176. M. Hagmann and H. Grisebach, *FEBS Lett.*, 1984, **175**, 199.
177. G. Kochs and H. Grisebach, *FEBS Lett.*, 1986, **155**, 311.
178. T. Hakamatsuka, H. Noguchi, Y. Ebizuka, and U. Sankawa, *Chem. Pharm. Bull.*, 1990, **7**, 1942.
179. M. F. Hashim, T. Hakamatsuka, Y. Ebizuka, and U. Sankawa, *FEBS Lett.*, 1990, **271**, 219.
180. T. Hakamatsuka, M. F. Hashim, Y. Ebizuka, and U. Sankawa, *Tetrahedron*, 1991, **47**, 5969.
181. Y. Kirikae, M. Sakurai, T. Furuno, T. Takahashi, and S. I. Ayabe, *Biosci. Biotechnol. Biochem.*, 1993, **57**, 1353.
182. K. Otani, T. Takahashi, T. Furuya, and S. I. Ayabe, *Plant Physiol.*, 1994, **105**, 1427.
183. G. Kochs and H. Grisebach, *Z. Naturforsch., Teil C*, 1987, **42**, 343.
184. H. Kessmann, A. D. Choudhary, and R. A. Dixon, *Plant Cell Rep.*, 1990, **9**, 38.
185. W. Hinderer, U. Flentje, and W. Barz, *FEBS Lett.*, 1987, **214**, 101.
186. W. Gunia, W. Hinderer, U. Wittkampf, and W. Barz, *Z. Naturforsch., Teil C*, 1991, **46**, 58.
187. S. Clemens, W. Hinderer, U. Wittkampf, and W. Barz, *Phytochemistry*, 1993, **32**, 653.
188. P. M. Dewick and S. W. Banks, *Planta Med.*, 1980, **39**, 287.
189. K. Tiemann, W. Hinderer, and W. Barz, *FEBS Lett.*, 1987, **213**, 324.
190. K. Tiemann, D. Inzé, V. Montagu, and W. Barz, *Eur. J. Biochem.*, 1991, **200**, 751.
191. D. Fischer, C. Ebenau-Jehle, and H. Grisebach, *Arch. Biochem. Biophys.*, 1990, **276**, 390.
192. C. L. Preisig, J. N. Bell, Y. Sun, G. Hrazdina, D. E. Matthews, and H. D. VanEtten, *Plant Physiol.*, 1990, **94**, 1444.
193. N. L. Paiva, R. Edwards, Y. Sun, G. Hrazdina, and R. A. Dixon, *Plant Mol. Biol.*, 1991, **17**, 653.
194. D. Schlieper, K. Tiemann, and W. Barz, *Phytochemistry*, 1990, **29**, 1519.
195. N. L. Paiva, Y. Sun, R. A. Dixon, H. D. VanEtten, and G. Hrazdina, *Arch. Biochem. Biophys.*, 1994, **312**, 501.
196. A. Oommen, R. A. Dixon, and N. L. Paiva, *Plant Cell*, 1994, **6**, 1789.
197. W. Bless and W. Barz, *FEBS Lett.*, 1988, **235**, 47.
198. D. Fischer, C. Ebenau-Jehle, and H. Grisebach, *Phytochemistry*, 1990, **29**, 2879.
199. L. Guo, R. A. Dixon, and N. L. Paiva, *J. Biol. Chem.*, 1994, **269**, 22372.
200. L. Guo, R. A. Dixon, and N. L. Paiva, *FEBS Lett.*, 1994, **356**, 221.
201. E. Wong and C. M. Francis, *Phytochemistry*, 1968, **7**, 2131.
202. E. Wong and C. M. Francis, *Phytochemistry*, 1968, **7**, 2139.
203. H. Wengenmayer, J. Ebel, and H. Grisebach, *Eur. J. Biochem.*, 1974, **50**, 135.
204. D. L. Gustine and R. T. Sherwood, *Plant Physiol.*, 1978, **61**, 226.
205. R. Edwards and R. A. Dixon, *Phytochemistry*, 1991, **30**, 2597.
206. X.-Z. He and R. A. Dixon, *Arch. Biochem. Biophys.*, 1996, **336**, 121.
207. R. A. Dixon, A. D. Choudhary, K. Dalkin, R. Edwards, T. Fahrendorf, G. Gowri, M. J. Harrison, C. J. Lamb, G. J. Loake, C. A. Maxwell, J. Orr, and N. L. Paiva, in "Phenolic Metabolism in Plants," eds. H. A. Stafford and R. K. Ibrahim, Plenum Press, New York, 1992, p. 91.
208. S. Fields and O.-K. Song, *Nature (London)*, 1989, **340**, 245.
209. H. E. Khouri, S. Tahara, and R. K. Ibrahim, *Arch. Biochem. Biophys.*, 1988, **262**, 592.
210. C. L. Preisig, D. E. Matthews, and H. D. VanEtten, *Plant Physiol.*, 1989, **91**, 559.
211. C. L. Preisig, H. D. VanEtten, and R. A. Moreau, *Arch. Biochem. Biophys.*, 1991, **290**, 468.
212. H. D. VanEtten, personal communication.
213. X.-Z. He and R. A. Dixon, unpublished results.
214. S. Clemens and W. Barz, *Phytochemistry*, 1996, **41**, 457.
215. G. Kochs and H. Grisebach, *Arch. Biochem. Biophys.*, 1989, **273**, 543.
216. D. E. Matthews, E. J. Weiner, P. S. Matthews, and H. D. VanEtten, *Plant Physiol.*, 1987, **83**, 365.
217. U. Zähringer, J. Ebel, L. J. Mulheirn, R. L. Lyne, and H. Grisebach, *FEBS Lett.*, 1979, **101**, 90.
218. D. R. Biggs, R. Welle, F. R. Visser, and H. Grisebach, *FEBS Lett.*, 1987, **220**, 223.
219. D. R. Biggs, R. Welle, and H. Grisebach, *Planta*, 1990, **181**, 244.
220. R. Welle and H. Grisebach, *Arch. Biochem. Biophys.*, 1988, **263**, 191.
221. L. Crombie, J. T. Rossiter, N. Van Bruggen, and D. A. Whiting, *Phytochemistry*, 1992, **31**, 451.
222. G. Schröder, U. Zähringer, W. Heller, J. Ebel, and H. Grisebach, *Arch. Biochem. Biophys.*, 1979, **194**, 635.

223. P. LaFlamme, H. Khouri, P. Gulick, and R. Ibrahim, *Phytochemistry*, 1993, **34**, 147.
224. R. Welle and H. Grisebach, *Phytochemistry*, 1991, **30**, 479.
225. T. Hakamatsuka, K. Shinkai, H. Noguchi, Y. Ebizuka, and U. Sankawa, *Z. Naturforsch., Teil C*, 1992, **47**, 177.
226. S. Kudou, Y. Fleury, D. Welti, D. Magnolato, T. Uchida, K. Kitamura, and K. Okubo, *Agric. Biol. Chem.*, 1991, **55**, 2227.
227. T. L. Graham, *Plant Physiol.*, 1991, **95**, 594.
228. P. F. Morris, M. E. Savard, and E. W. B. Ward, *Physiol. Mol. Plant Pathol.*, 1991, **39**, 229.
229. S. A. Tiller, A. D. Parry, and R. Edwards, *Physiol. Plant.*, 1994, **91**, 27.
230. T. L. Graham, J. E. Kim, and M. Y. Graham, *Mol. Plant–Microbe Interact.*, 1990, **3**, 157.
231. J. Berlin, L. Fecker, C. Rugenhagen, C. Sator, D. Strack, L. Witte, and V. Wray, *Z. Naturforsch., Teil C*, 1991, **46**, 725.
232. D. Hallard, E. Bleichert, H. Gagnon, S. Tahara, and R. Ibrahim, *Z. Naturforsch., Teil C*, 1992, **47**, 346.
233. Y. Shibuya, S. Tahara, Y. Kimura, and J. Mizutani, *Z. Naturforsch., Teil C*, 1991, **46**, 513.
234. H. H. Park, T. Hakamatsuka, H. Noguchi, U. Sankawa, and Y. Ebizuka, *Chem. Pharm. Bull.*, 1992, **40**, 1978.
235. H. H. Park, T. Hakamatsuka, U. Sankawa, and Y. Ebizuka, *Phytochemistry*, 1995, **38**, 373.
236. H. H. Park, T. Hakamatsuka, U. Sankawa, and Y. Ebizuka, *Z. Naturforsch., Teil C*, 1995, **50**, 824.
237. A. D. Parry, S. A. Tiller, and R. Edwards, *Plant Physiol.*, 1994, **106**, 195.
238. U. Mackenbrock, R. Vogelsang, and W. Barz, *Z. Naturforsch., Teil C*, 1992, **47**, 815.
239. U. Mackenbrock and W. Barz, *Z. Naturforsch., Teil C*, 1991, **46**, 43.
240. U. Mackenbrock, W. Gunia, and W. Barz, *J. Plant Physiol.*, 1993, **142**, 385.
241. H. Kessmann, R. Edwards, P. Geno, and R. A. Dixon, *Plant Physiol.*, 1990, **94**, 227.
242. W. Barz and U. Mackenbrock, *Plant Cell Tissue Organ Cult.*, 1994, **38**, 199.
243. J. Köster and W. Barz, *Arch. Biochem. Biophys.*, 1981, **212**, 98.
244. A. D. Parry and R. Edwards, *Phytochemistry*, 1994, **37**, 655.
245. J. Koester, R. Bussmann, and W. Barz, *Arch. Biochem. Biophys.*, 1984, **234**, 513.
246. W. Hinderer, J. Köster, and W. Barz, *Arch. Biochem. Biophys.*, 1986, **248**, 570.
247. W. Hösel and W. Barz, *Eur. J. Biochem.*, 1975, **57**, 607.
248. Z.-S. Li, M. Alfenito, P. Rea, V. Walbot, and R. A. Dixon, *Phytochemistry*, in press.
249. E. Martinoia, E. Grill, R. Tommasini, K. Kreuz, and N. Amrhein, *Nature (London)*, 1993, **364**, 247.
250. K. A. Marrs, M. R. Alfenito, A. M. Lloyd, and V. Walbot, *Nature (London)*, 1995, **375**, 397.
251. K. A. Marrs, *Annu. Rev. Plant Physiol. Plant Mol. Biol.*, 1996, **47**, 127.
252. P. M. Dewick, W. Barz, and H. Grisebach, *Phytochemistry*, 1970, **9**, 775.
253. J. Berlin, P. M. Dewick, W. Barz, and H. Grisebach, *Phytochemistry*, 1972, **11**, 1689.
254. P. Bhandari, L. Crombie, P. Daniels, I. Holden, N. Van Bruggen, and D. A. Whitting, *J. Chem. Soc., Perkin Trans. 1*, 1992, 839.
255. N. Lambert, M.-F. Trouslot, C. Nef-Campa, and H. Chrestin, *Phytochemistry*, 1993, **34**, 1515.
256. P. Bhandari, L. Crombie, M. F. Harper, J. T. Rossiter, M. Sanders, and D. A. Whiting, *J. Chem. Soc., Perkin Trans. 1*, 1992, 1685.
257. W. Barz, C. Adamek, and J. Berlin, *Phytochemistry*, 1970, **9**, 1735.
258. J. A. Hargreaves and C. Selby, *Phytochemistry*, 1978, **17**, 1099.
259. V. J. Higgins, S. A., and M. C. Heath, *Phytopathology*, 1974, **64**, 105.
260. M. Yoshikawa, *Nature (London)*, 1978, **275**, 546.
261. M. Yoshikawa, K. Yamauchi, and H. Masago, *Physiol. Plant Pathol.*, 1979, **14**, 157.
262. P. Moesta and H. Grisebach, *Nature (London)*, 1980, **286**, 710.
263. H. D. VanEtten, D. E. Matthews, and D. A. Smith, in "Phytoalexins," eds. J. A. Bailey and J. W. Mansfield, Blackie, Glasgow, 1982, p. 180.
264. U. Willeke, K. M. Weltring, W. Barz, and H. D. VanEtten, *Phytochemistry*, 1983, **22**, 1539.
265. H. D. VanEtten, S. G. Pueppke, and T. C. Kelsey, *Phytochemistry*, 1975, **14**, 1103.
266. A. E. Desjardins, D. E. Matthews, and H. D. VanEtten, *Plant Physiol.*, 1984, **75**, 611.
267. K. M. Weltring, B. G. Turgeon, O. C. Yoder, and H. D. VanEtten, *Gene*, 1988, **68**, 335.
268. K. Hirschi and H. Van Etten, *Mol. Plant–Microbe Interact.*, 1996, **9**, 483.
269. V. P. Miao, S. F. Covert, and H. D. VanEtten, *Science*, 1991, **254**, 1773.
270. C. Reimmann and H. D. VanEtten, *Gene*, 1994, **146**, 221.
271. A. P. Maloney and H. D. VanEtten, *Mol. Gen. Genet.*, 1994, **243**, 506.
272. D. C. Straney and H. D. VanEtten, *Mol. Plant–Microbe Interact.*, 1994, **7**, 256.
273. Y. Ruan and D. C. Straney, *Curr. Genet.*, 1994, **27**, 46.
274. Y. Ruan and D. C. Straney, *Mol. Gen. Genet.*, 1996, **250**, 29.
275. J. He, Y. Ruan, and D. Straney, *Mol. Plant–Microbe Interact.*, 1996, **9**, 171.
276. B. Höhl, N. Arnemann, L. Schwenen, D. Stöckl, G. Bringmann, J. Jansen, and W. Barz, *Z. Naturforsch., Teil C*, 1989, **44**, 771.
277. J. van den Heuvel, H. D. VanEtten, and J. W. Serum, *Phytochemistry*, 1974, **13**, 1129.
278. R. Tenhaken, H. C. Salmen, and W. Barz, *Arch. Microbiol.*, 1991, **155**, 353.
279. B. Höhl and W. Barz, *Z. Naturforsch., Teil C*, 1987, **42**, 897.
280. S. F. Covert, J. Enkerli, V. P. W. Miao, and H. D. VanEtten, *Mol. Gen. Genet.*, 1996, **251**, 397.
281. P. J. Kuhn, D. A. Smith, and D. F. Ewing, *Phytochemistry*, 1977, **16**, 296.
282. D. A. Smith, P. J. Kuhn, J. A. Bailey, and R. S. Burden, *Phytochemistry*, 1980, **19**, 1673.
283. C. S. Turbek, D. Li, G. H. Choi, C. L. Schardl, and D. A. Smith, *Phytochemistry*, 1990, **29**, 2841.
284. D. X. Li, K. R. Chung, D. A. Smith, and C. L. Schardl, *Mol. Plant–Microbe Interact.*, 1995, **8**, 388.
285. J. Rao and J. Cooper, *J. Bacteriol.*, 1994, **176**, 5409.
286. J. R. Rao and J. E. Cooper, *Mol. Plant–Microbe Interact.*, 1995, **8**, 855.
287. K. Klus, G. Börger-Papendorf, and W. Barz, *Phytochemistry*, 1993, **34**, 979.
288. K. Sugawara, S. Tahara, and J. Mizutani, *Agric. Biol. Chem.*, 1991, **55**, 1799.
289. S. Tahara, F. Saitoh, and J. Mizutani, *Z. Naturforsch., Teil C*, 1993, **48**, 16.

290. M. Tanaka, J. Mizutani, and S. Tahara, *Biosci. Biotechnol. Biochem.*, 1996, **60**, 171.
291. T. A. Holton and E. C. Cornish, *Plant Cell*, 1995, **7**, 1071.
292. V. Bénaben, G. Duc, V. Lefebvre, and T. Huguet, *Plant Physiol.*, 1995, **107**, 53.
293. W. P. Lindsay, C. J. Lamb, and R. A. Dixon, *Trends Microbiol.*, 1993, **1**, 181.
294. C. Lamb and R. A. Dixon, *Annu. Rev. Plant Physiol. Plant Mol. Biol.*, 1997, **48**, 251.
295. L. M. Yu, C. J. Lamb, and R. A. Dixon, *Plant J.*, 1993, **3**, 805.
296. W. Dröge-Laser, A. Kaiser, W. P. Lindsay, B. A. Halkier, G. J. Loake, P. Doerner, R. A. Dixon, and C. Lamb, *EMBO J.*, 1997, **16**, 726.
297. B. Miao and N. L. Paiva, personal communication.
298. C. J. Lamb, *Cell*, 1994, **76**, 419.
299. R. A. Dixon, M. J. Harrison, and C. J. Lamb, *Annu. Rev. Phytopathol.*, 1994, **32**, 479.
300. H. A. Stafford, *Plant Physiol.*, 1991, **96**, 680.
301. V. Walbot, *Trends Plant Sci.*, 1996, **1**, 27.
302. R. A. Larson, *Phytochemistry*, 1988, **27**, 969.
303. H. Yamasaki, H. Uefusi, and Y. Sakihama, *Arch. Biochem. Biophys.*, 1996, **332**, 183.
304. H. Mehlhorn, M. Lelandais, H. G. Korth, and C. H. Foyer, *FEBS Lett.*, 1996, **378**, 203.
305. E. Babiychuk, S. Kushnir, E. Bellesboix, M. Vanmontagu, and D. Inze, *J. Biol. Chem.*, 1995, **270**, 26224.
306. R. Cooke, M. Raynal, M. Laudie, F. Grellet, M. Delseny, P. Morris, D. Guerrier, J. Giraudat, F. Quigley, G. Clabault, Y. Li, R. Mache, M. Krivitzky, I. Gy, M. Kreis, A. Lecharny, Y. Parmentier, J. Marbach, J. Fleck, B. Clement, G. Phillips, C. Herve, C. Bardet, D. Tremousaygue, B. Lescure, C. Lacomme, D. Roby, M. Jourjon, P. Chabrier, J. Charpenteau, T. Desprez, J. Amselem, H. Chiapello, and H. Hofte, *Plant J.*, 1996, **9**, 101.
307. S. Petrucco, A. Bolchi, C. Foroni, R. Percudani, G. L. Rossi, and S. Ottonello, *Plant Cell*, 1996, **8**, 69.
308. G. N. Drews, T. P. Beals, A. Q. Bui, and R. B. Goldberg, *Plant Cell*, 1992, **4**, 1383.
309. N. Hibi, S. Higashiguchi, T. Hashimoto, and Y. Yamada, *Plant Cell*, 1994, **6**, 723.

1.29
Biosynthesis of Sulfur-containing Natural Products

RONALD J. PARRY
Rice University, Houston, TX, USA

1.29.1 INTRODUCTION

Sulfur-containing compounds are present in all living organisms. These compounds can be divided into two groups: the primary metabolites that are essential for life, and the nonessential, secondary metabolites whose function is often obscure. The family of known sulfur-containing compounds including both primary and secondary metabolites exhibits an astonishing degree of structural variety.[1] For this reason, the current understanding of the biosynthesis of sulfur compounds is still quite limited. This review will focus upon those examples where significant progress has been achieved in unraveling the biosynthetic pathways. The coverage includes both primary and secondary metabolites, but it is limited to compounds of low molecular mass.

1.29.2 ENZYME COFACTORS

1.29.2.1 Biotin

The vitamin biotin (**5**) is an essential enzyme cofactor that is involved in a number of enzymatic carboxylation reactions.[2] The biosynthetic pathway for biotin has been investigated in both plants and microorganisms. The major steps of the biotin biosynthetic pathway have been elucidated in *Escherichia coli*[3] and *Bacillus sphaericus*.[4] These are shown in Scheme 1. The pathway begins with pimeloyl-CoA (**1**) and proceeds via 8-amino7-oxope-largonic acid (**2**) and 7,8-diaminopelargonic acid (**3**) to dethiobiotin (**4**). The final step in the pathway is the insertion of sulfur into dethiobiotin to give biotin (**5**). The designations of the *E. coli* and *B. sphaericus* genes coding for each of these stages in the pathway are also shown in the Scheme.

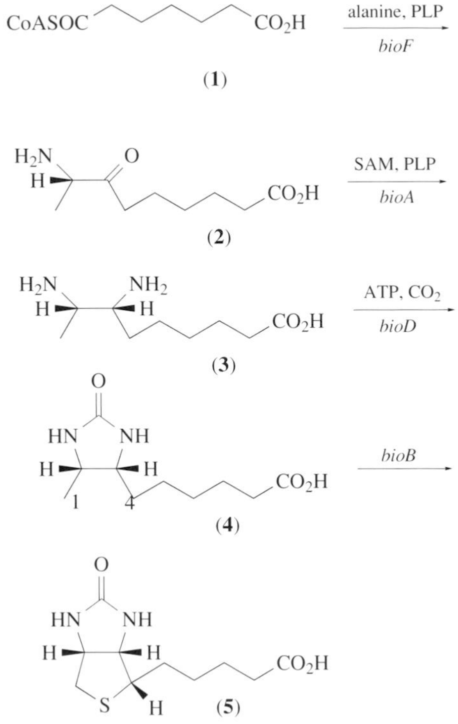

CoASOC CO$_2$H (**1**) —alanine, PLP / *bioF*→

H$_2$N H O CO$_2$H (**2**) —SAM, PLP / *bioA*→

H$_2$N H NH$_2$ H CO$_2$H (**3**) —ATP, CO$_2$ / *bioD*→

O HN NH H H 1 4 CO$_2$H (**4**) —*bioB*→

O HN NH H H CO$_2$H S H (**5**)

Scheme 1

Relatively little is known about the origin of the pimeloyl-CoA required for the first step in the pathway. In *E. coli*, two genes, *bioC* and *bioH*, encode enzymes that appear to be involved, and it has been known that free pimelic acid is not a biotin precursor in *E. coli*. Administration of [1-^{13}C]-, [2-^{13}C]-, and [1,2-$^{13}C_2$]acetate to an *E. coli* strain carrying the biotin operon on a multicopy plasmid yielded biotin whose labeling pattern was consistent with the biosynthesis of pimelate via a modified fatty acid pathway, and it was suggested that pimelate may be formed by extension of a malonate starter unit (Scheme 2).[5] However, attempts to verify this hypothesis by administration of either [1-^{13}C] malonate or [^{13}C]CO$_2$ gave inconclusive results. If pimelate is indeed formed via malonate in *E. coli*, this presents an interesting quandary as to where the first biotin came from, since biotin is required for the carboxylation of acetyl-CoA. Biotin would therefore be needed to catalyze its own biosynthesis. A similar puzzle arises in thiamine biosynthesis (see below). In contrast to *E. coli*, *B. sphaericus* can utilize free pimelic acid for biotin biosynthesis and the *bioW* gene of this organism has been shown to encode a pimeloyl-CoA synthase.[6]

The enzyme catalyzing the formation of 8-amino-7-oxopelargonic acid (**2**) from pimeloyl-CoA and alanine has been purified from *B. sphaericus*, overexpressed in *E. coli*, and detailed mechanistic studies carried out.[7] Because the enzyme requires pyridoxal phosphate as a cofactor, two plausible mechanisms can be envisioned for the overall transformation. The first mechanism would involve formation of a pyridoxal phosphate stabilized carbanion by abstraction of the C-2 hydrogen of alanine followed by acylation and decarboxylation. The second mechanism would proceed by

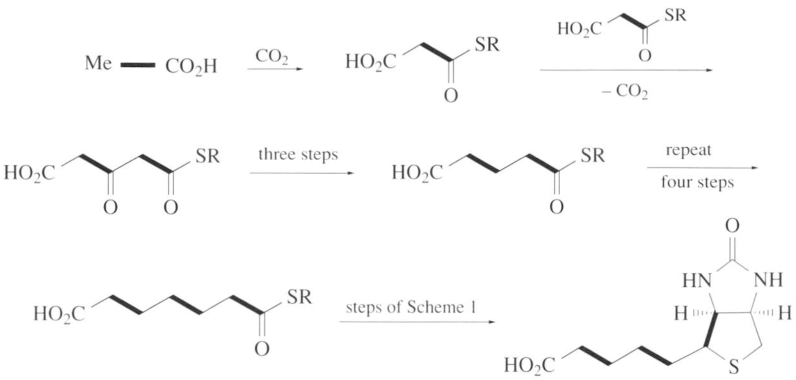

Me—CO_2H —CO_2→ HO_2C–CH$_2$–C(=O)–SR —[HO_2CCH$_2$C(=O)SR, $-CO_2$]→

HO_2C–C(=O)–C(=O)–SR —three steps→ HO_2C–C(=O)–SR —repeat four steps→

HO_2C–C(=O)–SR —steps of Scheme 1→ biotin (HN, NH, O, H, H, S, HO_2C)

Scheme 2

formation of the carbanion by decarboxylation of the pyridoxal phosphate adduct of alanine followed by acylation. A decision between these two mechanisms was reached by examining the fate of the C-2 hydrogen of alanine during the course of the reaction. Proton NMR spectra of (**2**) formed from L-[2-$^2H_1$]alanine in H_2O or L-alanine in D_2O revealed that the C-2 hydrogen atom of alanine is removed during the reaction and that the C-8 hydrogen atom of the product is derived from the solvent. In addition, 8-amino-7-oxopelargonate synthase will catalyze exchange of the C-2 hydrogen atom of L-alanine with solvent in the absence of pimeloyl-CoA. The exchange is stereospecific and proceeds with retention of configuration. Finally, the use of L-[2-$^2H_1$]alanine as a substrate gave a primary deuterium isotope effect ($^DV = 1.3$). All of these results are consistent with the first mechanism in which the formation and acylation of a carbanion precedes the decarboxylation step. The same mechanism has been shown to be followed by 5-aminolevulinate synthase, an enzyme that catalyzes the pyridoxal phosphate-dependent decarboxylative acylation of glycine by succinyl-CoA.[7]

The *E. coli* enzyme catalyzing the conversion of 8-amino-7-oxopelargonic acid to 7,8-diaminopelargonic acid (**3**) was isolated and characterized in the mid-1970s by Eisenberg and coworkers.[8,9] The enzyme requires pyridoxal phosphate and is unusual in that it utilizes *S*-adenosylmethionine as the amino donor. Kinetic studies indicate that the enzyme uses a bi-bi ping-pong mechanism. Incorporation of the methionine nitrogen atom of *S*-adenosylmethionine into 8-amino-7-oxopelargonic acid was subsequently verified by isotopic labeling.[10]

The enzyme catalyzing the conversion of (**3**) to dethiobiotin (**4**) was first purified by Krell and Eisenberg who showed that the preferred substrates for the reaction are 7,8-diaminopelargonic acid, ATP, Mg^{2+}, and CO_2 and that the products are dethiobiotin, ADP, and inorganic phosphate.[11] This allowed two reaction paths to be formulated for the conversion (Scheme 3). The reaction could proceed by carboxylation of either the C-7 or C-8 amino group of (**3**) to give carbamic acid (**6**) or carbamic acid (**7**). Phosphorylation of the carboxyl group of one of these carbamates would then be followed by cyclization and loss of P_i. A more detailed mechanistic picture of this enzymatic reaction has been provided by subsequent investigations. Gibson *et al.*[12] and Huang *et al.*[13] have shown that solutions of (**3**) at neutral pH react readily with CO_2 to form a mixture of the two carbamates (**6**) and (**7**). However, the results of additional experiments clearly indicated that only the carbamate (**6**) is an intermediate in the conversion of (**3**) to (**4**). Some of the evidence in support of this conclusion is as follows. (i) Incubation of (**3**) with [^{13}C]CO_2 gave a mixture of labeled carbamates (**6**) and (**7**) which could be observed by ^1H and ^{13}C NMR. In the presence of excess dethiobiotin synthetase, only one ^{13}C NMR signal was observed for enzyme-bound carbamate, a result indicating that only one of the two carbamates binds to the enzyme. By using [7-^{15}N]-(**3**) and [^{13}C]CO_2, it was shown that the enzyme binds carbamate (**6**) since the ^{13}C signal of the bound carbamate was split by ^{15}N. (ii) Pulse-chase experiments indicated that the binary complex of (**6**) and dethiobiotin synthetase becomes kinetically committed upon addition of ATP. (iii) The two carbamate analogues (**8**) and (**9**) were synthesized and their behavior with dethiobiotin synthetase was examined (Scheme 4). Compound (**8**), which is an analogue of (**6**), cyclized to the corresponding lactam in the presence of dethiobiotin synthetase and ATP with concomitant formation of ADP and P_i. In contrast, the isomeric analogue (**9**) was a very poor substrate. Finally, the crystal structures of six complexes of dethiobiotin synthetase with a variety of substrates, substrate analogues, and

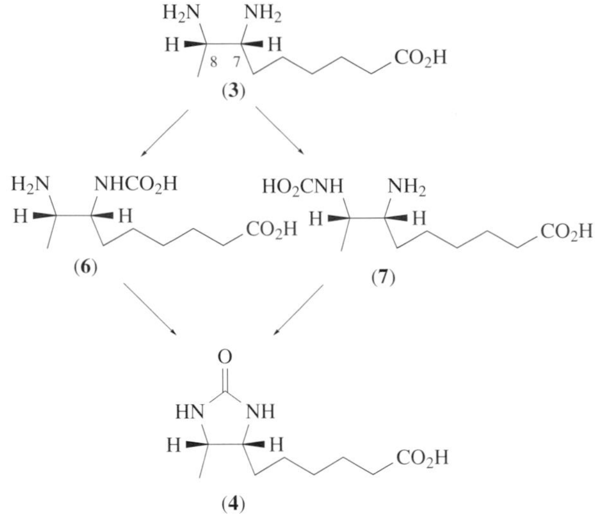

Scheme 3

products were determined at high resolution. These structure determinations revealed the location of the active site and identified protein residues involved in binding of (**3**) and ATP. This information is summarized in Figure 1 which displays the interactions of Mn^{2+}, the ATP analogue AMPPCP, and (**6**) with amino acid residues at the active site of dethiobiotin synthetase. All of these observations are in conflict with chemical trapping experiments, which indicated that the carbamate (**7**) is the intermediate involved in the conversion of (**3**) to dethiobiotin.[14] The reasons for this discrepancy are unclear, and the weight of the evidence obviously favors the intermediacy of the carbamate (**6**).

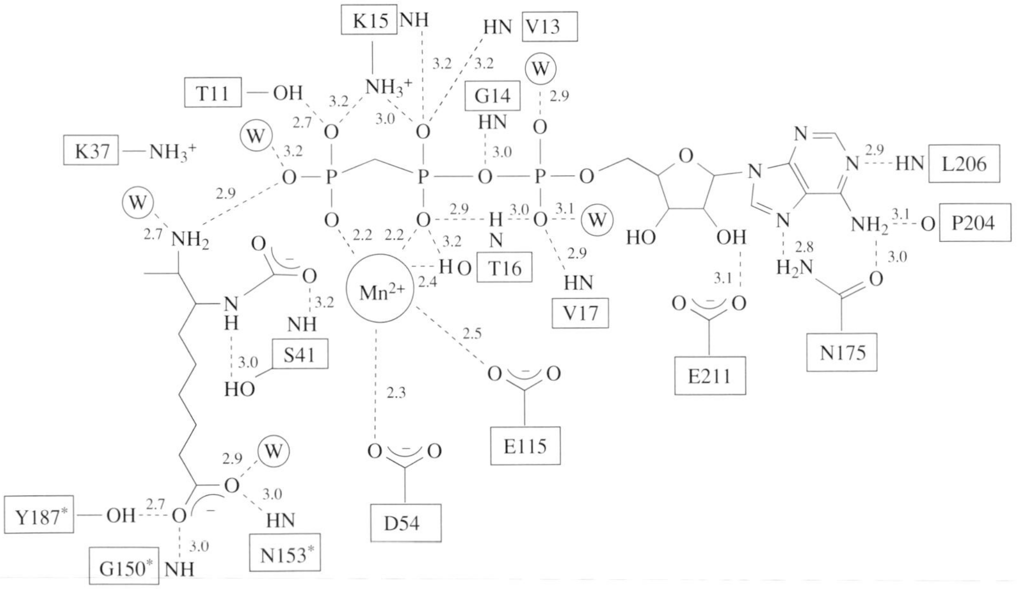

Figure 1 Schematic view of the interactions of AMPCP and carbamate (**6**) with enzymatic residues and the metal ion at the active site of dethiobiotin synthetase. Hydrogen bonds are indicated by dashed lines. Reproduced by permission of the American Chemical Society from *Biochemistry*, 1995, **34**, 10 985.

The mechanism of the ring-closure of the carbamate (**6**) to dethiobiotin has also been examined.[15] Since only one equivalent of ATP is required for the reaction, and ADP and P_i are the products, two mechanisms can be formulated for the cyclization reaction (Scheme 5). One mechanism (a) would proceed via the formation of a carbamoyl phosphate, while the other (b) would involve the formation of a mixed anhydride between the carbamate carboxyl group and ADP. The two mechanisms were distinguished using γ-[$^{18}O_3$]-$\gamma\beta$-[^{18}O]ATP as a substrate and examining the distribution of ^{18}O in the products formed in the dethiobiotin synthetase reaction. If the reaction proceeds by

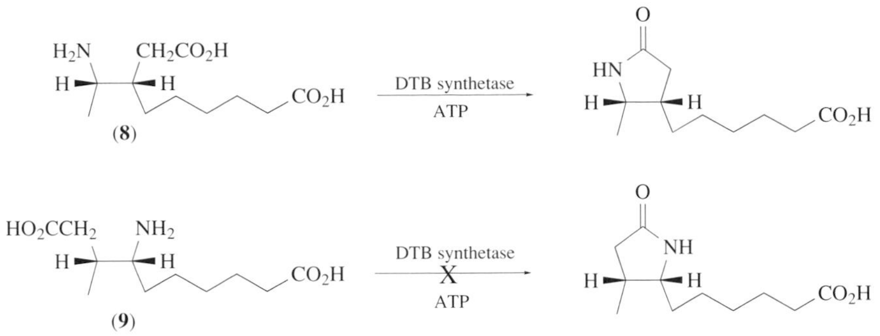

Scheme 4

mechanism (a), then [$^{18}O_3$]phosphate and *β*-[^{18}O]ADP would be produced, while mechanism (b) would yield [$^{18}O_4$]phosphate and unlabeled ADP. Examination of the ^{31}P NMR spectrum of the products formed from the ^{18}O-labeled ATP demonstrated that the cyclization reaction proceeds via mechanism (a).

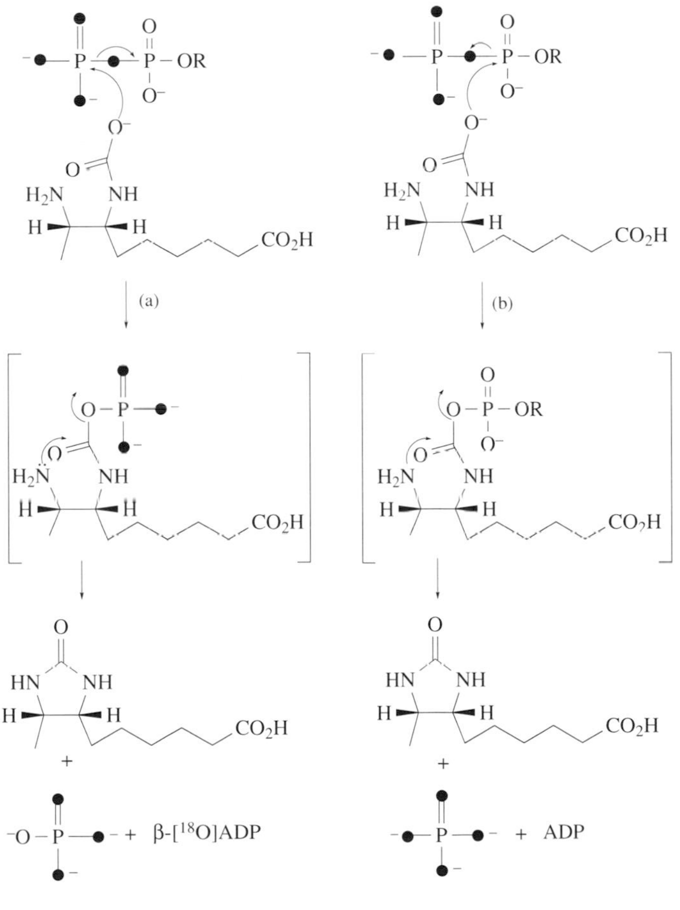

Scheme 5

The conversion of dethiobiotin (**4**) into biotin (**5**), which is catalyzed by biotin synthase, is the most novel step in biotin biosynthesis since it involves the formal insertion of sulfur at two saturated carbon atoms. The early investigations of the mechanism of this conversion were reviewed by the author in 1983.[16] What follows is a summary and an update. Some insight into the nature of this reaction was obtained by an examination of the incorporation of specifically tritiated forms of dethiobiotin into biotin by *Aspergillus niger*. These experiments demonstrated that only two hydrogen atoms are lost from dethiobiotin as a consequence of this transformation: one from C-1 and one from C-4.[16] This finding suggests that unsaturated intermediates are not likely to be involved in the sulfur insertion process. Marquet *et al.* reported similar results for experiments conducted with *E. coli*, indicating that the mechanism of the conversion of dethiobiotin to biotin is likely to be the same in both organisms.[16] By using dethiobiotin that was stereospecifically tritiated at C-4, Parry *et al.* were also able to determine that, in *A. niger*, sulfur is inserted into the C-4 position of dethiobiotin with overall retention of configuration.[16] Stereochemical information with respect to the introduction of sulfur at C-1 of dethiobiotin was obtained by Arigoni *et al.* Experiments with dethiobiotin bearing a chiral methyl group at C-1 revealed that the insertion of sulfur into this position occurs with racemization.[17] These observations suggest that the sulfur introduction mechanism probably does not involve hydroxylated intermediates, since biological hydroxylations usually proceed with retention of configuration. Hydroxylation, hydroxyl group activation, and displacement by a sulfur nucleophile would therefore be expected to lead to sulfur introduction with overall inversion of configuration. Additional evidence against hydroxylated intermediates was obtained by studies in which 1-hydroxydethiobiotin, the two epimers of 4-hydroxydethiobiotin, and (4*R*)-1,4-dihydroxydethiobiotin were synthesized and evaluated for their ability to support growth of an auxotrophic *E. coli* mutant whose biotin biosynthetic pathway is blocked before dethiobiotin. Although transport studies indicated that all of the compounds could enter the cells, none of the compounds was able to support the growth of the mutant.[18]

Since two carbon–sulfur bonds are formed during the conversion of dethiobiotin to biotin, it is likely that the reaction proceeds through one or more intermediates that possess only a single carbon–sulfur bond. This question has been examined by two groups. The most definitive results were obtained from experiments with resting cells of *B. sphaericus*.[19] The reduced form of the disulfide (**10**) derived from 1-mercaptodethiobiotin and the two epimeric 4-mercaptodethiobiotins (**11**) and (**12**) were incubated with resting *B. sphaericus* cells and the amount of biotin produced from each was quantitated by a microbiological assay. The (4*R*)-4-mercapto compound (**11**), which has the opposite configuration at C-4 to that of biotin, produced a very small amount of biotin, whereas reduced (**10**) gave a significantly higher quantity of the cofactor. No excess biotin was produced from the (4*S*)-mercapto isomer (**12**). Additional evidence for the intermediacy of reduced (**10**) was obtained by incorporation experiments. The ^{35}S-labeled disulfide (+)-(**10**) was synthesized and incubated with *B. sphaericus* in the presence of DTT. The biotin that was produced in this experiment retained radioactivity on purification. A second experiment with reduced (+)-1-[$^{2}H_2$]-(**10**) gave more conclusive results. CI mass spectra of the methyl ester of the biotin produced from this compound by resting *B. sphaericus* cells exhibited molecular MH^+ peaks at *m/z* 259 and 261 (intensities 10:90) which corresponded exactly to the deuterium content of the precursor. A similar experiment was carried out with reduced ^{34}S-labeled (+)-(**10**) and a recombinant strain of *B. sphaericus* that overproduces biotin synthase. In this case, mass spectrometric analysis showed that ca. 80% of the ^{34}S was retained in the biotin. Independent experiments have shown that the reduced form of racemic (**10**) can support the growth of both an *E. coli bioA* mutant and an *E. coli* mutant that lacks the biotin operon but contains the biotin synthase gene encoded on a plasmid.[20] All of this information is consistent with the intermediacy of 1-mercaptodethiobiotin or a closely related compound in the biotin synthase reaction.

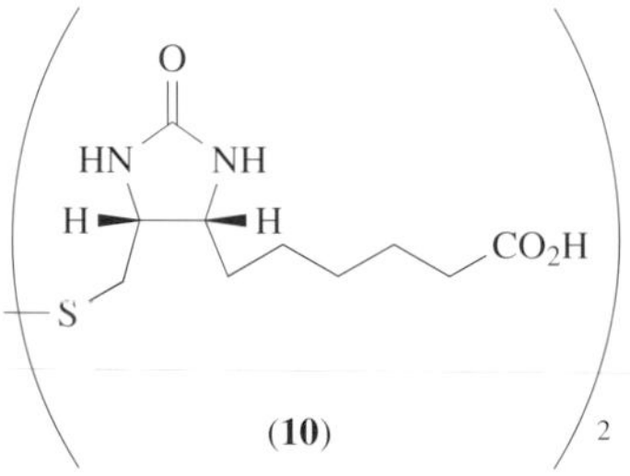

(**10**)

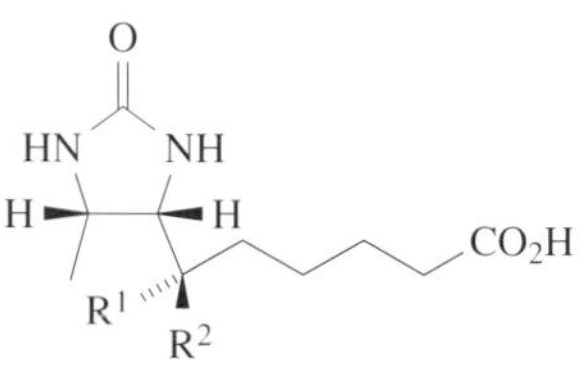

(**11**) R^1 = H, R^2 = SH
(**12**) R^1 = SH, R^2 = H

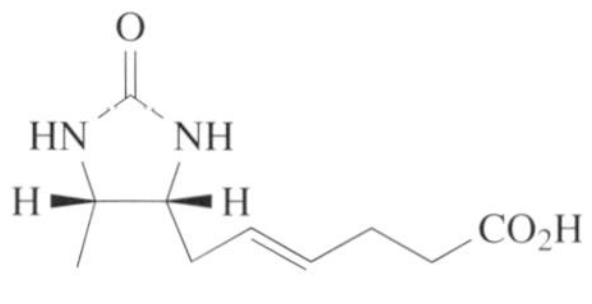

(**13**)

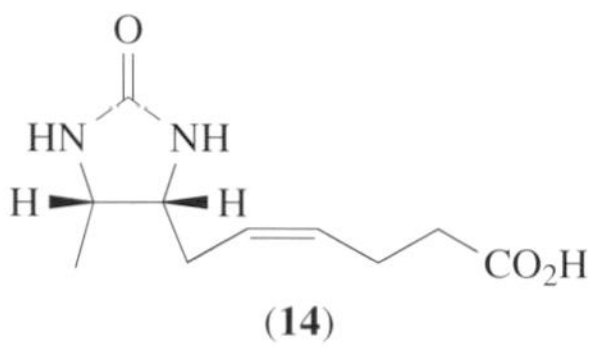

(**14**)

Experiments with growing cells of an *E. coli bioA* mutant gave less easily interpretable results.[19] Both (4*R*)-4-mercaptodethiobiotin (**11**) and (**10**) exhibited growth-promoting activity with this mutant. However, the ^{35}S-labeled forms of both of these compounds were converted into biotin with loss of the labels. Partial deuterium loss also accompanied the conversion of 1-[$^2H_2$]-labeled (**10**) into biotin by the growing *E. coli* cells. Similar results were obtained when 1-[$^2H_2$]-labeled (**10**) was administered to growing *B. sphaericus* cells. It appears that growing cells of both species have a complex desulfurizing pathway that competes with the more direct pathway observed in resting cells.

Investigations of the enzymology of the dethiobiotin to biotin conversion have yielded some important insights, but major questions remain concerning the mechanism of the carbon–sulfur bond-forming reactions involved in this unusual transformation. Biotin synthase has been purified to homogeneity from both *B. sphaericus*[21] and *E. coli*.[22] The enzymes have subunit molecular masses of 38 and 39 kDa, respectively, and the native *E. coli* enzyme exists as a homodimer. Both enzymes contain one [2Fe-2S] cluster per subunit. Six cysteine residues are conserved in the predicted amino acid sequences of both biotin synthases, and resonance Raman studies of the *E. coli* enzyme indicate that there are four cysteinyl ligands to the [2Fe-2S] clusters of this protein. Three of the conserved cysteines are in a Cys-X-X-X-Cys-X-X-Cys motif that is also found in the lipoic acid synthase protein (LipA, see below). A defined mixture of components that will support the reaction catalyzed by the *E. coli* enzyme contains biotin synthase, dethiobiotin, flavodoxin,[23] flavodoxin reductase, NADPH-, *S*-adenosylmethionine, Fe^{2+} or Fe^{3+}, fructose-1,6-biphosphate, cysteine, and dithiothreitol.[24] However, there are four lines of evidence which suggest that this mixture does not contain all of the important factors required for the *in vivo* biotin synthase reaction. First, in the defined reaction mixture, a maximum of 2 mol of biotin is produced per mol of biotin synthase monomer. Second, the requirement for fructose-1,6-biphosphate can be satisfied by an unknown component in the low molecular weight fraction of crude *E. coli* extracts. Third, it has been found that a labile, low-molecular-mass product of the 7,8-diaminopelargonic acid aminotransferase reaction stimulates the rate of biotin formation and increases the amount of biotin produced by biotin synthase in the defined reaction mixture. Fourth, in the defined reaction mixture, neither [^{35}S]cysteine nor [^{35}S]*S*-adenosylmethionine gives rise to radioactive biotin.[24] The biotin synthase from *B. sphaericus* appears to be similar to the *E. coli* enzyme. It requires NADPH, *S*-adenosylmethionine, and unidentified components from a crude extract of *B. sphaericus*.[21] The unidentified components and NADPH could be replaced by photoreduced deazaflavin. However, the *in vitro* system containing biotin synthase, *S* adenosylmethionine, and reduced deazaflavin produced only about 0.05 0.1 mol of biotin per mole of enzyme. This system also did not incorporate radioactivity from [^{35}S]cysteine into biotin.[21] The critical question of the identity of the sulfur donor in the biotin synthase reaction has yet to be answered.

The mechanism of the sulfur introduction reactions catalyzed by biotin synthase remains elusive. The fact that dethiobiotin bearing a chiral methyl group at C-1 is converted into biotin with complete racemization suggests that radical intermediates may be involved. Additional support for this hypothesis is provided by the fact that biotin synthase requires flavodoxin, flavodoxin reductase, NADPH, and *S*-adenosylmethionine. The same combination of cofactors and proteins is required by anaerobic ribonucleotide reductase and by pyruvate formate lyase.[25–29] The reactions catalyzed by the latter two enzymes involve radical mechanisms and proceed by transfer of an electron from reduced flavodoxin to *S*-adenosylmethionine which then dismutates into methionine and the 5′-deoxyadenosyl radical. The 5′-deoxyadenosyl radical is then involved in the formation of an enzyme-based glycyl radical. However, it should be noted that the glycyl radicals formed during the activation of anaerobic ribonucleotide reductase and pyruvate formate lyase are very sensitive to molecular oxygen, while the biotin synthase reaction functions either anaerobically or aerobically.[24] This lack of oxygen sensitivity may indicate that glycyl radicals are not formed in the biotin synthase reaction. Both the *E. coli* and *B. sphaericus* biotin synthases also lack the consensus sequence RVXGY which is the site of the glycyl radical in both anaerobic ribonucleotide reductase and pyruvate formate lyase.[25]

The only other evidence bearing on the possible intermediacy of radical species in the biotin synthase reaction comes from preliminary studies with the substrate analogs (*E*)- and (*Z*)-5,6-dehydrodethiobiotin, (**13**) and (**14**).[30] The rationale for the investigation of these compounds stems from the possibility that a radical may be generated at C-4 of dethiobiotin as part of the sulfur introduction process, and that the radical produced at C-4 of (**13**) or (**14**) would be allylic. The formation of this delocalized radical might lead to novel reaction products. This type of approach has been successfully employed with isopenicillin N synthase (see below). Another possible outcome would be the formation of a covalent adduct between the delocalized radical and the enzyme. In the event, incubation of ^{14}C-labeled (**13**) and (**14**) with purified biotin synthase, *S*-adenosylmethionine, and photoreduced deazaflavin led to some covalent labeling of the protein. The extent of labeling was small (1% for the *E* isomer, 0.6% for the *Z* isomer), but it will be recalled that this system only produces about 0.05–0.1 mol of biotin per mole of enzyme. While the results are intriguing, it appears unlikely that they can be analyzed until a more active system for the *in vitro* synthesis of biotin from dethiobiotin has been devised.

1.29.2.2 Lipoic Acid

R-(+)-Lipoic acid (6,8-thioctic acid) (**16**) (Equation (1)) is a widely distributed coenzyme that is found in multienzyme complexes that catalyze the oxidative decarboxylation of α-keto acids[31–33] and in the glycine cleavage system.[34] Early investigations of lipoic acid biosynthesis have been the subject of a previous review.[16] In this review, that information will be summarized and updated. The first insight into the biosynthesis of lipoic acid was obtained by Reed, who also played a key role in the isolation and structural determination of the coenzyme.[16] In 1964, Reed reported that octanoic acid (**15**) served as a specific precursor of lipoic acid in *E. coli*.[16] This observation was confirmed by Parry *et al.* who then examined the mechanism of the conversion of octanoic acid into lipoic acid by using specifically tritiated forms of octanoic acid.[16] The results of these studies showed that sulfur is introduced into octanoic acid with the loss of only two hydrogen atoms, one from C-6 and one from C-8. By using octanoic acid that was stereospecifically tritiated at C-6, it was also shown that sulfur is introduced at C-6 of octanoic acid with overall inversion of configuration. These results were confirmed by White using precursors labeled with deuterium.[16] Additional stereochemical information was provided by Arigoni,[35] who found that octanoic acid bearing a chiral methyl group at C-8 was converted into lipoic acid with racemization of stereochemistry. White also examined the potential intermediacy of hydroxylated forms of octanoic acid in the sulfur introduction process by means of precursor incorporation experiments with deuterated forms of 6-hydroxyoctanoic acid, 8-hydroxyoctanoic acid, and 6,8-dihydroxyoctanoic acid.[16] None of these compounds appeared to be efficiently incorporated into lipoic acid (<0.5%). Uptake studies indicated that the lack of incorporation of 8-hydroxyoctanoic acid could not be due to poor transport of this compound into the cells. On the other hand, experiments with deuterated forms of 8-thiooctanoic acid and 6-thiooctanoic acid indicated that both of these compounds could be converted into lipoic acid, with the 8-thio acid being the more efficient precursor (19% vs. 2%). All of these results suggest that there is probably a close relationship between the mechanism of sulfur introduction associated with biotin biosynthesis and that associated with lipoic acid biosynthesis.

8 6 1 CO_2H (**15**) → S–S H CO_2H (**16**) (1)

Further support for such a relationship has been provided by investigations of the genetics of lipoic acid biosynthesis in *E. coli*. Two genes, *lipA* and *lipB*, have been clearly identified as being involved in lipoic acid biosynthesis or metabolism.[36–39] The amino acid sequence of the *lipA* gene exhibits some similarity to that of biotin synthase. The central region of the protein (residues 141–201) makes up the largest stretch of similarity and is 36% identical with the biotin synthase from *E. coli*. The protein also contains the same cysteine triad Cys-X-X-X-Cys-X-X-Cys motif that is found in biotin synthase (see above). This suggests that the LipA protein may contain an iron–sulfur cluster. The phenotypes of *lipA* mutants suggests that the LipA protein is probably involved in the introduction of both sulfur atoms into octanoic acid. The *E. coli* lipoic acid auxotrophs W1485-*lip2* and JRG33-*lip9* have been shown to contain mutations in *lipA*.[38] Both mutations consist of a single G/C to A/T substitution resulting in the conversion of Ser307 to Phe in the *lip2* mutant and Glu195 to Lys in the *lip9* mutant. Neither octanoate or 6-thiooctanoate will satisfy the lipoic acid requirement of these two mutants, but the *lip2* mutant allele can use 8-thiooctanoate. On the other hand, it has been reported that a Tn1000 *lipA* mutant (*lipA*150::Tn1000dKn) can utilize both 6-thio and 8-thiooctanoate to make lipoic acid.[39] The *lipB* gene has been shown to code for a lipoic acid ligase that attaches the cofactor to the α-ketoglutarate and glycine cleavage enzyme complexes.[40] Curiously, *E. coli* has been found to contain a second gene called *lplA* that also codes for lipoic acid ligase. Current evidence suggests the LplA protein utilizes ATP to activate free lipoic acid as a lipoyl-AMP derivative, while the LipB protein may be involved in the transfer of lipoate from a covalently bound form of lipoate that is produced from octanoic acid.[40]

1.29.2.3 Coenzyme M

The methanogenic bacteria are anaerobic organisms that convert CO_2 to methane via a sequence of reactions that utilizes several unusual cofactors. One of these cofactors is coenzyme M (**17**), whose structure was determined in 1974 by Taylor and Wolfe.[41] The final stages in the reduction of CO_2 to methane involve the transfer of a CO_2-derived methyl group from a methylcobamide-containing protein to coenzyme M to give *S*-methyl coenzyme M (MeCoM, Scheme 6). This is followed by a reaction that utilizes *N*-(7-mercaptoheptanoyl)threonine phosphate ((**18**), component B), *S*-methyl coenzyme M, and a Ni corphin (coenzyme F_{430})-containing methyl reductase to produce methane and a mixed disulfide between coenzyme M and *N*-(7-mercaptoheptanoyl)threonine phosphate. The last step of the reaction involves the reduction of the mixed disulfide to (**17**) and (**18**) by a heterodisulfide reductase (Scheme 6).[42]

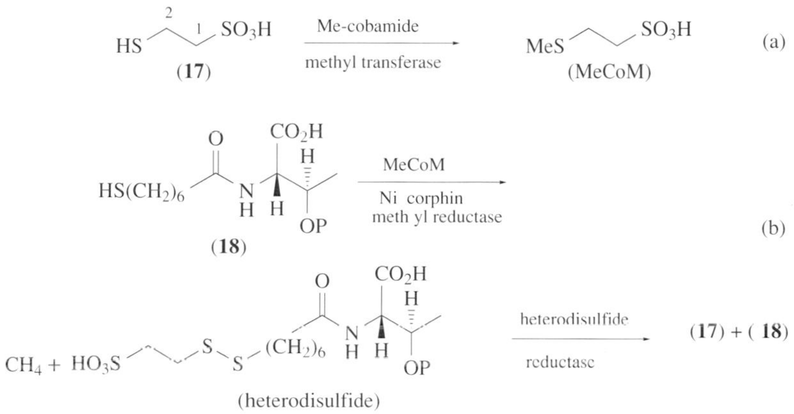

HS–CH2–CH2–SO3H (**17**) —Me-cobamide / methyl transferase→ MeS–CH2–CH2–SO3H (MeCoM) (a)

$HS(CH_2)_6$–CO–NH–CH(CO_2H)–CH(OP)–CH3 (**18**) —MeCoM / Ni corphin / meth yl reductase→ CH_4 + HO_3S–CH2–CH2–S–S–$(CH_2)_6$–CO–NH–CH(CO_2H)–CH(OP)–CH3 (heterodisulfide) —heterodisulfide / reductase→ (**17**) + (**18**) (b)

Scheme 6

The biosynthesis of coenzyme M has been investigated by White.[43–45] Using mass spectrometric analysis, it was found that growth of three different strains of methanogenic bacteria (*Methanobacterium formicicum*, *Methanosarcina strain* TM-1, and rumen isolate 10-16B) in the presence of [$^2H_3$]acetate led to the formation of coenzyme M with up to two deuterium atoms present at C-1. The extent of labeling was the same as that calculated for the phosphoenolpyruvate in the cells. Using strain 10-16B, [1,2-$^{13}C_2$]acetate was shown to be incorporated into coenzyme M as a unit. Additional experiments revealed that DL-[3-$^2H_2$]sulfolactic acid and $H^{34}SO_3^-$ also serve as coenzyme M precursors in strain 10-16B (3.2% and 3.1% incorporation, respectively), while labeled forms of sulfate, cysteic acid, sulfoacetic acid, taurine, and isethionate (2-hydroxyethanesulfonic acid) do not. On the basis of these results, the biosynthetic pathway shown in Scheme 7 was proposed. It is postulated that phosphoenolpyruvate (**19**) reacts with bisulfite anion via conjugate addition to produce sulfolactate which is then oxidized to sulfopyruvate (**20**). The next stage of the pathway is suggested to proceed by decarboxylation of sulfopyruvate to sulfoacetaldehyde (**21**) followed by reaction of the latter with L-cysteine to give the thiazolidine derivative (**22**). The final stages of the pathway are postulated to involve reduction of the thiazolidine (**22**) to *S*-(2-sulfoethyl)cysteine (**23**) which is then converted to coenzyme M and pyruvate by a transformation that presumably requires pyridoxal phosphate. Additional evidence is available that supports several stages of this pathway. Partially purified cell-free extracts of *Methanobacterium formicicum* were found to produce coenzyme M when incubated with phosphoenolpyruvate, bisulfite, and cysteine. When pyruvate was substituted for phosphoenolpyruvate, no coenzyme M was produced. When extracts incubated with phosphoenolpyruvate, bisulfite, and cysteine were analyzed for the presence of sulfonic acids by GC-MS, three of the proposed intermediates in coenzyme M biosynthesis, sulfolactic acid, sulfopyruvic acid, and sulfoacetaldehyde, were identified. Incubation of the cell-free extracts with sulfopyruvate in the presence or absence of cysteine also produced coenzyme M, sulfolactate, and sulfoacetaldehyde. This clearly suggests that sulfopyruvate lies on the biosynthetic pathway to the coenzyme. Incubation of a cell-free extract of *M. formicicum* with [2-$^2H_2$]sulfoacetaldehyde and L-cysteine under a hydrogen atmosphere led to the formation of coenzyme M in which 78% of the cofactor retained two deuterium atoms. The mass spectral fragmentation pattern demonstrated that the deuterium label was present at C-1 of coenzyme M, as expected. When a similar experiment was carried out with [2-$^2H_2$]sulfoacetaldehyde and L-[^{34}S]cysteine, the thiol group of the resulting coenzyme M contained 90 atom % ^{34}S. Incubation of [*ethylene*-$^2H_4$]*S*-(2-sulfoethyl)-L-cysteine with the cell-free extracts led to the isolation of coenzyme M with 88% of the molecules containing four deuterium atoms. All of these results support the biosynthetic pathway shown in Scheme 7. However, no direct evidence is available for the postulated intermediacy of the thiazolidine (**22**).

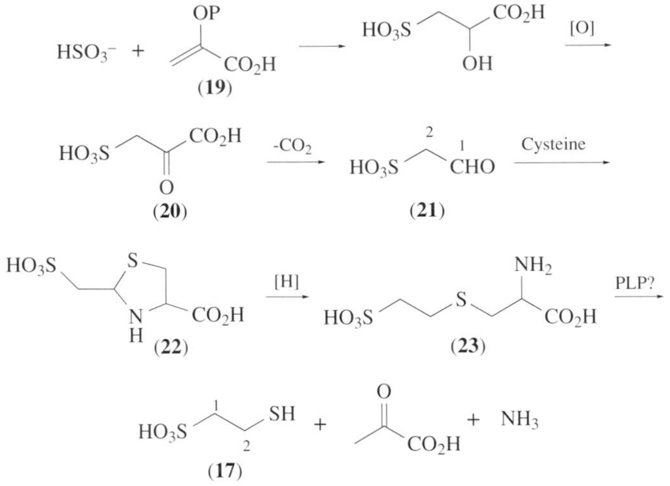

Scheme 7

The mechanism of formation of the bisulfite required for coenzyme M biosynthesis has not been clearly defined. The only sulfur sources in the growth media for methanogenic bacteria are sulfide and sulfate. Since sulfate does not support the growth of these bacteria[46] and is not incorporated into the coenzyme (see above), it appears that bisulfite must be formed by the oxidation of sulfide. It has been suggested that this oxidation could be carried out by a P_{590} enzyme that has been isolated from *Methanosarcina barkeri* and shown to possess sulfite reductase activity.

1.29.2.4 *N*-(7-Thioheptanoyl)threonine Phosphate

N-(7-Thioheptanoyl)threonine phosphate ((**18**), component B) is the second sulfur-containing cofactor that is required by methanogenic bacteria for the conversion of CO_2 into methane (see above and Scheme 6). The biosynthesis of this compound has been investigated by White[47,48] by means of incorporation experiments with ^2H- and ^{13}C-labeled precursors in rumen isolate 10-16B and *Methanococcus volta*. The extent and position of the incorporated isotopic labels were determined by mass spectrometry of suitable derivatives. [2-$^2H_3$]Acetate was found to label four separate positions of 7-thioheptanoic acid (**31**). One deuterium label was equally distributed between C-2 and C-3 of the 7-thioheptanoic acid moiety, while the remaining three isotopic labels were at C-4 to C-6. The degree of incorporation of deuterium at C-2 and C-3 was approximately the same as for the glutamic acid and proline in the cells, suggesting the intermediacy of α-ketoglutarate (**26**) in the biosynthesis of 7-thioheptanoic acid. Administration of [1,2-$^{13}C_2$]acetate led to the incorporation of an intact acetate unit at C-2 and C-3 of (**31**), while single labels were incorporated at C-5, C-6, and C-7. These observations are also consistent with the intermediacy of α-ketoglutarate, and can be explained by the conversion of acetate into α-ketoglutarate via pyruvate, oxaloacetate, and succinate using the incomplete, reverse TCA cycle that is known to occur in methanogens.[49] The α-ketoglutarate would then be elongated to α-ketosuberate (**29**) via α-ketoadipate (**27**) and α-ketopimelate (**28**) (Scheme 8). However, in order to explain the deuterium labeling pattern from [2-$^2H_3$]acetate it is necessary to assume that the dehydration and rehydration steps in the chain elongation processes proceed with internal return of a deuterium atom. Additional evidence in support of the pathway shown in Scheme 8 was obtained in two ways. First, when [2,3-$^2H_4$]succinate was administered, it was incorporated into (**31**) with retention of all four deuterium atoms. This result can also be explained by the conversion of succinyl-CoA into α-ketoglutarate by the incomplete, reverse TCA cycle.[49] Second, a GC-MS analysis of the distribution of α-ketodicarboxylic acids in various bacteria showed that while α-ketoglutarate and α-ketoadipate are widely distributed, α-ketopimelate and α-ketosuberate are found only in the methanogenic archaebacteria.[50] The final stages of the pathway to (**31**) were postulated to involve decarboxylation of (**29**) to 7-oxoheptanoic acid (**30**) followed by conversion of (**30**) into (**31**) by a process analogous to the final stages of coenzyme M biosynthesis (Scheme 8). The conversion of thioheptanoic acid into 7-thioheptanoylthreonine phosphate (**18**) was studied in *Methanococcus volta* and *Methanosarcina thermophila*.[48] Growth of these bacteria in the presence of [7-$^2H_2$]-7-thioheptanoic acid, [3,4-$^2H_4$-threonine]-*N*-(7-thioheptanoyl)threonine, [7-$^2H_2$]-*N*-(7-thioheptanoyl)threonine, or DL-[3,4-$^2H_4$]threonine led to the formation of (**18**) bearing the same number of deuterium atoms as the precursor. Cell-free extracts of the same species of methanogens were shown to carry out the ATP-dependent phosphorylation of *N*-(7-thioheptanoyl)threonine to (**18**). These results indicate that the biosynthesis of (**18**) from 7-thioheptanoic acid is likely to proceed via the formation of *N*-(7-thioheptanoyl)threonine which is then phosphorylated to give the cofactor (Scheme 8).

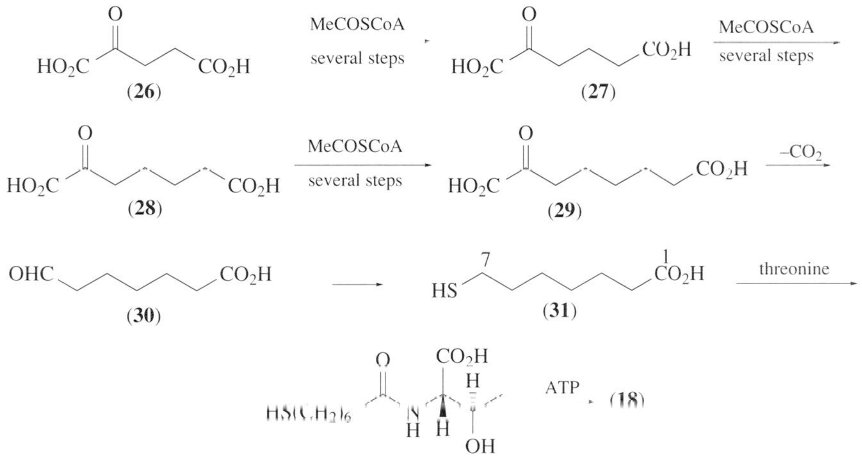

HO$_2$C–CO–CH$_2$CH$_2$–CO$_2$H (**26**) → [MeCOSCoA, several steps] → (**27**) → [MeCOSCoA, several steps] → (**28**) → [MeCOSCoA, several steps] → (**29**) → [$-CO_2$] → OHC(CH$_2$)$_5$CO$_2$H (**30**) → HS(CH$_2$)$_6$CO$_2$H (**31**) → [threonine] → HS(CH$_2$)$_6$CO–NH–threonine → [ATP] → (**18**)

Scheme 8

1.29.2.5 The Pterin Molybdenum Cofactors

The pterin molybdenum cofactors have been the subject of two reviews.[51,52] Consequently, this review of the biosynthetic pathway will be somewhat abbreviated. The molybdenum cofactors are all derived from the dihydropterin molybdopterin (**32**). The active forms of molybdopterin consist of the molybdenum cofactor itself (**33**) and four dinucleotide forms of the molybdenum cofactor containing GMP, CMP, AMP, or IMP residues ((**34**)–(**37**)) (Scheme 9). Enzymes containing the molybdenum cofactors are found in animals, plants, bacteria and archaebacteria. The reactions catalyzed by the enzymes involve either hydroxylations or reductive deoxygenations. Examples of the former include sulfite oxidase and xanthine oxidase, while the latter include nitrate reductase and dimethyl sulfoxide reductase.

The current state of knowledge with respect to the biosynthesis of the molybdenum cofactors is outlined in Scheme 9. This Scheme also indicates the *E. coli* genes that are known to be associated with various stages of the pathway. Investigations of the biosynthesis of the molybdenum cofactors have been hampered by the instability of these compounds. The key to the elucidation of the structure of the cofactors has been the isolation of a series of stable derivatives. Early studies showed that oxidation of the cofactor with I_2 gave Form A (**38**), while air oxidation of the cofactor gave Form B (**39**). Additional insight was obtained by the isolation of the stable bis(carboxamidomethyl) derivative (**40**) after treatment of chicken liver sulfite oxidase and milk xanthine oxidase with sodium dodecyl sulfate in the presence of iodoacetamide. A clue to the nature of the biosynthetic pathway was obtained by the investigation of chlorate-resistant *E. coli* mutants. Extracts of an *E. coli chlN* mutant (now known as *moeB*) were found to contain an unstable precursor (precursor Z) that could be converted to molybdopterin by an *E. coli chlAl*(*moaA*) mutant. Precursor Z could be oxidized by air or iodine to a stable substance which was called compound Z (**41**) and assigned the structure shown. Additional studies on the nature of precursor Z subsequently established that it is a dihydro derivative of compound Z. A possible structure for precursor Z is shown in Scheme 9, but the position of the double-bond in ring B has not be determined. It should also be noted that the bis(carboxamidomethyl) derivative (**40**) isolated from mammalian sources is a tetrahydropterin, a

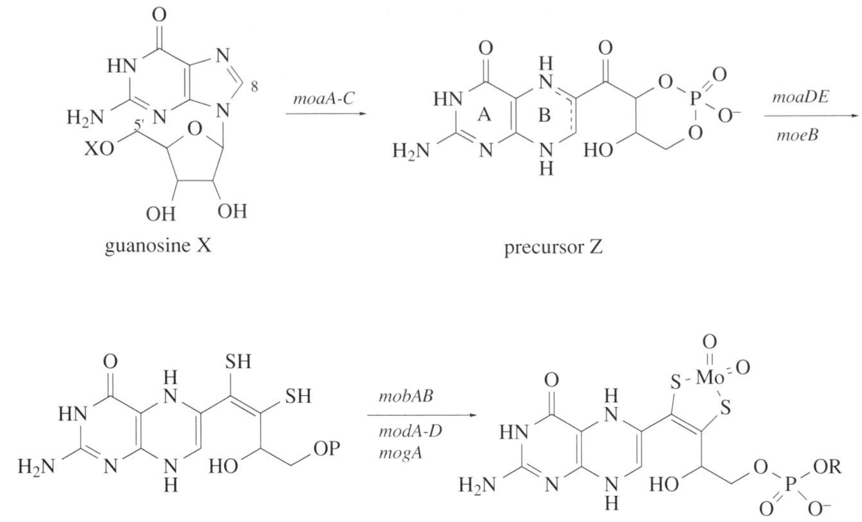

moaA-C *moaDE* *moeB*

guanosine X

precursor Z

mobAB *modA-D* *mogA*

(**32**)

(**33**) R = H
(**34**) R = GMP
(**35**) R = CMP
(**36**) R = AMP
(**37**) R = IMP

Scheme 9

fact which suggests that the oxidation level of the pterin ring may differ depending on the source of the cofactor.

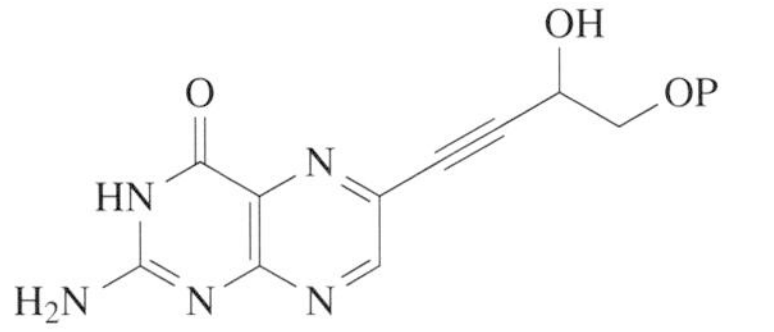

(**38**) 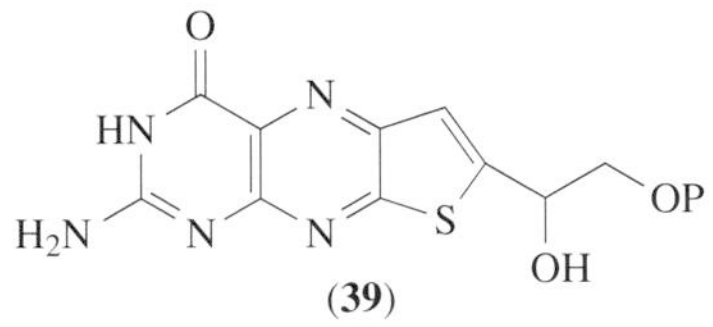 (**39**)

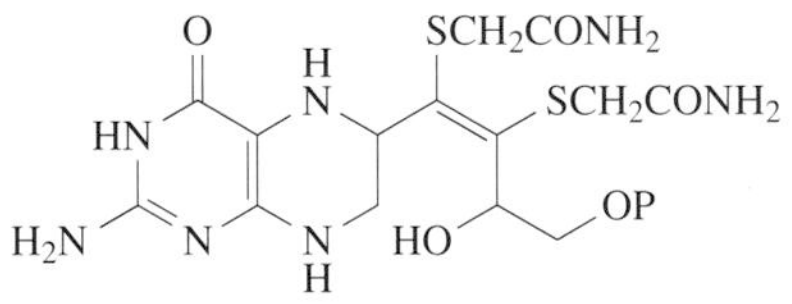

(**40**) 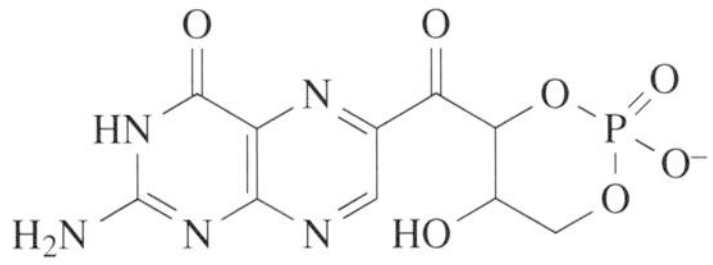 (**41**)

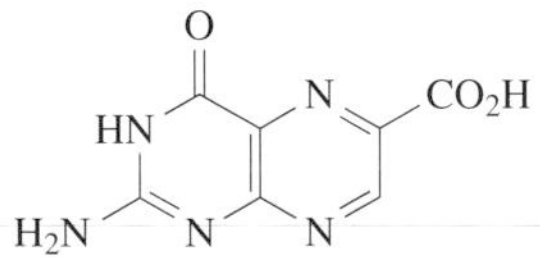

(**42**) 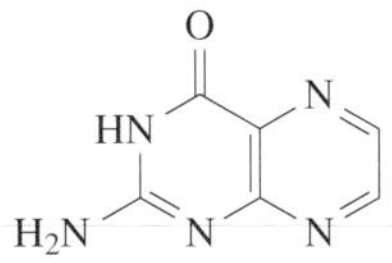 (**43**)

The conversion of precursor Z to molybdopterin has been found to require a two-subunit protein called the converting factor. The larger subunit has a molecular mass of 16.8 kDa and is encoded by the *moaE* gene, while the smaller subunit has a molecular mass of 8.5 kDa and is encoded by the *moaD* gene. In an *in vitro* system containing excess precursor Z and one equivalent of the converting factor, only one equivalent of molybdopterin is produced and it remains bound to the large subunit of the converting factor. The absence of catalytic turnover is consistent with the hypothesis that the sulfur introduced into precursor Z during its conversion to molybdopterin is derived from the converting factor itself. Preliminary evidence suggests that the smaller of the two subunits of converting factor may be the sulfur donor. An additional gene, *moeB*, has been identified that appears to play a role in the reactivation of MoaD by transfer of labile sulfur. The MoeB protein has been shown to contain zinc. MoeB exhibits significant sequence similarities to a number of other proteins. These include the ThiF protein which appears to be required for the formation of the thiazole ring of thiamine (see below), and the HesA protein which may be involved in the assembly of an iron-sulfur protein in *Anabaena* spp.

The early stages in the biosynthesis of the molybdopterin portion of the molybdenum cofactors have been investigated by means of precursor incorporation experiments. Since other pteridines are known to be derived from GTP, radioactively labeled forms of guanosine were evaluated as precursors of molybdopterin in an *E. coli moeB* mutant that accumulates precursor Z. The labeling pattern in precursor Z was determined by its oxidation to compound Z (**41**) followed by further degradation to pterin-6-carboxylic acid (**42**) and pterin (**43**). Administration of [U-^{14}C]guanosine yielded radioactive (**41**) which was degraded to show that ca. 87% of the radioactivity was present in (**42**), while 74% remained in (**43**). When [8-^{14}C]guanosine was the precursor, compound Z was also radioactive and 98.5% of the radiolabel was shown to be present in the carboxyl group of (**42**) by degradation. This result was unanticipated since the known pathways for pteridine biosynthesis led to the loss of C-8 of GTP. Additional experiments with [8-^{3}H]guanosine and [5′,8-^{3}H]guanosine revealed that the tritium label is lost from C-8 of guanosine as the result of its conversion to precursor Z, while the C-5′ tritium label is retained at C-3′, C-4′ or C-5′ of this intermediate. These results indicate that a novel rearrangement has taken place during the formation of precursor Z. The information that is currently available is insufficient to provide a clear picture of how this rearrangement might take place. A formal pathway consistent with the labeling results would begin with the hydrolytic opening of a guanosine phosphate (**44**) to the Amadori rearrangement product (**45**) (Scheme 10). This compound could then undergo cleavage between C-2′ and C-3' with insertion of the C-8 carbon atom to yield the diketo compound (**46**). Cyclization of (**46**) would then give precursor Z (Scheme 10).

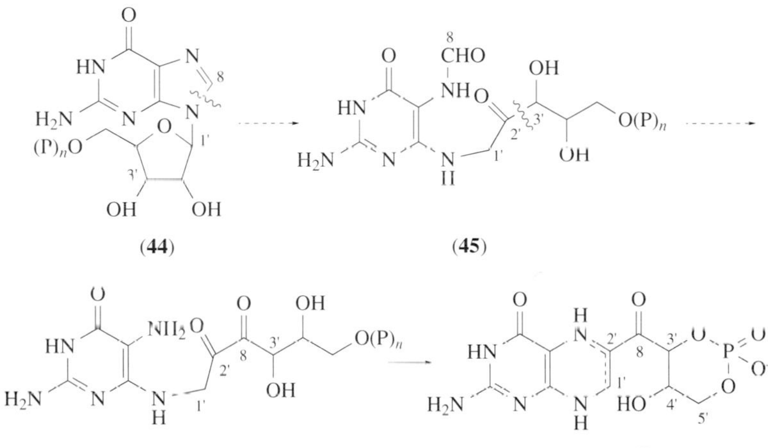

(**44**) (**45**)

(**46**) precursor Z

Scheme 10

1.29.2.6 Thiamine

Thiamine (vitamin B_1) (**49**) (Scheme 11) is biosynthesized by both prokaryotes and eukaryotes, and serves as an essential cofactor for enzymatic reactions that catalyze carbon–carbon bond-forming or bond-breaking reactions alpha to a carbonyl group.[2] Since the biosynthesis of thiamine has been the subject of a thorough review by White and Spenser,[53] a more concise survey of thiamine biosynthesis will be provided here.

The late stages of thiamine biosynthesis, which involve the linking of the pyrimidine subunit (**47**) to the thiazole subunit (**48**), are relatively well understood. Scheme 11 summarizes the current state of knowledge with respect to this portion of the biosynthetic pathway and it also indicates the genes known to encode enzymes that catalyze the various steps (*E. coli* genes unless otherwise indicated). The steps leading to thiamine pyrophosphate ((**50**), co-carboxylase) are also shown.

In contrast to the late stages of thiamine biosynthesis, a great deal less is known about the

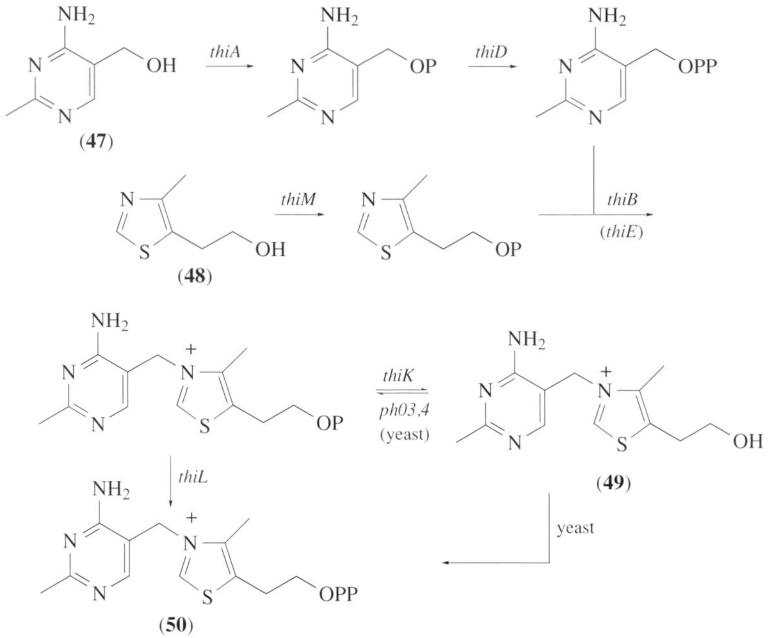

Scheme 11

biosynthesis of the pyrimidine and thiazole subunits. As pointed out by White and Spenser,[53] progress in understanding this aspect of thiamine biosynthesis was hampered by an unwillingness to recognize the possibility that different routes to thiamine might be employed by different organisms. However, it is now firmly established that the biosynthesis of both the pyrimidine and thiazole subunits exhibits species variation.

From extensive investigations of the biosynthesis of the pyrimidine subunit in *E. coli* and *Salmonella typhimurium*, the picture shown in Scheme 12 has emerged. The biosynthesis proceeds via 5-aminoimidazole riboside (**54**), which is also an intermediate in purine biosynthesis. As in purine biosynthesis, the heterocyclic ring of (**54**) is derived from formate (**51**), glycine (**52**), and the amide nitrogen of glutamine (**53**) in the manner indicated by the labeling pattern in Scheme 12(a). The final disposition of these labels in (**47**) is shown in Scheme 12(b). In this transformation, the connection between C-1 and C-2 of glycine is broken by the insertion of two carbon atoms (C-5, C-5′). Remarkably, C-5 and C-5′ of (**47**) have been shown to be derived from C-4′ and C-5′ of the ribose ring of (**54**) (Scheme 12(c)). More remarkable still, the methyl group of (**47**) has been found to be derived from C-2′ of the ribose moiety of (**54**).

It has been discovered that *S. typhimurium* mutants blocked in *purF*, a gene encoding the first enzyme in the purine biosynthetic pathway, are able to grow without thiamine under certain conditions. The evidence currently available suggests that a second route to 5-aminoimidazole riboside (**54**) is present in this organism. In eukaryotes, the biosynthesis of the pyrimidine subunit follows a completely different course.[53] Formate labels C-4 of this subunit, and (**54**) does not appear to play a role in its biosynthesis.

The building blocks for the thiazole subunit (**48**) in *E. coli* and *S. typhimurium* have been established to be L-tyrosine (**55**) and 1-deoxy-D-xylulose (**57**) (Scheme 13). Only C-2 and the amino group of tyrosine are incorporated into (**48**), and the remaining portion of the tyrosine molecule is probably released as *p*-hydroxybenzyl alcohol (**56**) (Scheme 13(a)). The amino group and C-2 of tyrosine are probably incorporated as an intact unit. Glycine does not serve as a precursor of the thiazole nucleus of thiamine in *S. typhimurium*. 1-Deoxy-D-xylulose appears to be the source of the remaining carbon atoms of the thiazole subunit (Scheme 13(b)). The 1-deoxy-D-xylulose is apparently formed from the condensation of pyruvate with D-glyceraldehyde, a reaction that can be catalyzed by pyruvate dehydrogenase. However, this process requires thiamine pyrophosphate so that thiamine would be required to catalyze its own biosynthesis. A similar conundrum appears in biotin biosynthesis (see above). Nothing appears to be known about the nature of the intermediates formed from (**55**) and (**57**) on the way to the thiazole subunit. However, two closely related thiazoles, 5-(1,2-dihydroxyethyl)-4-methylthiazole and 5-(2-hydroxyethyl)-4-methylthiazole-2-carboxylic acid, have been isolated from *E. coli* mutants. [*carboxyl*-^{14}C]Tyrosine has been shown to label the

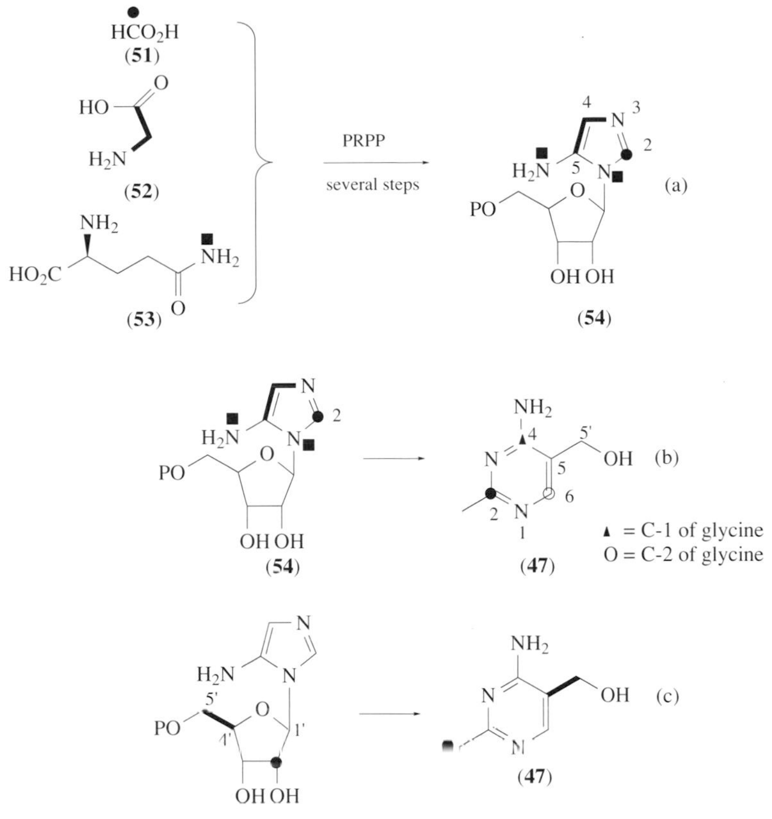

Scheme 12

latter compound, which could serve as a precursor of (**48**) via decarboxylation. Neither compound has been evaluated as an intermediate, however. In yeast, the thiazole subunit appears to be derived from glycine (**52**) and a D-pentulose (**58**) (Scheme 13(c)).

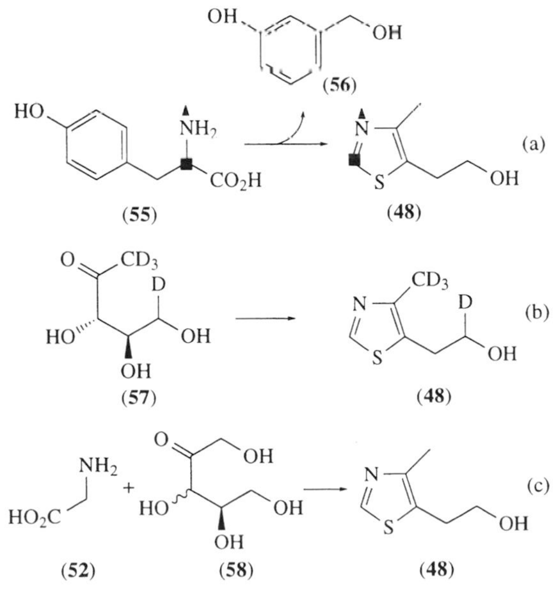

Scheme 13

While *thi*⁻ *E. coli* mutants have been known for a number of years, the use of genetic approaches to elucidate the thiamine biosynthetic pathway has proved to be difficult. A group of five genes (*thiCEFGH*) clustered at 90 min on the *E. coli* chromosome and identified with the *thi*$^-$ phenotype has been analyzed. The *thiC* gene is known to be required for the synthesis of the pyrimidine subunit (**47**), while the *thiE*, *thiF*, *thiG*, and *thiH* genes have been shown to be necessary for the synthesis of the thiazole subunit (**48**). Expression of the sequenced genes gave polypeptides of 70, 21, 27, 34, and 43 kDa from *thiC*, *thiE*, *thiF*, *thiG*, and *thiH*, respectively. 1-Deoxy-D-xylulose (**57**) did not satisfy the thiamine requirement of *thiE-thiH* mutants, indicating that the products of these genes are involved in steps to the thiazole subunit that lie beyond (**57**). The amino acid sequence of the ThiF protein derived from the *thiF* gene shares substantial sequence similarity with MoeB (ChlN), a protein believed to be involved in the introduction of sulfur during molybdopterin biosynthesis (see above). Unfortunately, the amino acid sequences derived from the *thiC*, *thiE*, *thiG*, and *thiH* genes did not show significant similarity to other sequences in the databases. As in the case of biotin, lipoic acid, and molybdopterin, the mechanism of the sulfur introduction process associated with thiamine biosynthesis is presently unknown. Clearly, many major questions remain to be answered with respect to the pathways used for assembly of the pyrimidine and thiazole rings of thiamine.

1.29.3 ANTIBIOTICS

1.29.3.1 Penicillins and Cephalosporins

The clinical importance of the penicillins and the cephalosporins has led to extensive investigations of the biosynthesis of these β-lactam antibiotics. The progress in this field was summarized by Baldwin in 1988 and 1990.[54,55] Consequently, this review will emphasize more recent developments. The structure, organization, regulation, and evolution of the penicillin and cephalosporin biosynthetic genes were reviewed in 1992 by Aharonowitz *et al.*[56] The interested reader is referred to the work of these authors for a detailed discussion of this aspect of penicillin and cephalosporin biosynthesis (see also Volume 4, Chapter 11).

The major stages in the biosynthesis of the penicillins and cephalosporins are shown in Schemes 14 and 15. The pathway begins with the ACV synthetase-catalyzed condensation of L-aminoadipic acid (**59**), L-cysteine (**60**), and L-valine (**61**) to give the ACV tripeptide (**62**). The incorporation of L-valine into (**62**) proceeds with epimerization at the α-carbon atoms of this amino acid. The second step is the cyclization of (**62**) to isopenicillin N (**63**) catalyzed by IPN synthase. The third step involves epimerization at the α-carbon of the α-aminoadipoyl moiety of (**63**) to give penicillin N (**64**). Lastly, the thiazolidine ring of (**64**) undergoes expansion to give deacetoxycephalosporin C (**65**) followed by hydroxylation to produce deacetyl cephalosporin C (**66**). In eukaryotes, these two steps are catalyzed by a single enzyme (DAOC/DAC synthase), while in prokaryotes the expandase and hydroxylase activities are present as separate enzymes (DAOC synthase and DAC synthase). Schemes 14 and 15 also indicate the designations that have been assigned to the genes coding for each of these stages in the pathway.

ACV synthetase has now been isolated and purified from three organisms: *Aspergillus nidulans*, *Streptomyces clavuligerus*, and *Cephalosporium acremonium* (syn. *Acremonium chrysogenum*).[57] Genes coding for ACV synthetase have been cloned from the prokaryotes *Lysobacter lactamgenus*, *Flavobacterium* sp., *Nocardia lactamdurans*, and *S. clavuligerus*, and from the eukaryotes *A. nidulans*, *Pencillium chrysogenum*, and *C. acremonium*. The *Lysobacter*, *Nocardia*, *Aspergillus*, *Penicillium*, and *Cephalosporium* genes have been fully sequenced.[57] These genes code for enzymes with molecular masses in the range of ca 404–424 kDa. The sequences of the genes coding for ACV synthetase reveal that this enzyme is a member of the family of nonribosomal peptide synthetases.[58] This family of enzymes has been proposed to function by a "thiol template" mechanism that involves the activation of the carboxyl groups of the substrate amino acids by reaction with ATP to yield aminoacyl adenylates and inorganic pyrophosphate. The aminoacyl groups are then transferred to an enzyme-bound thiol. The thiol template mechanism also postulates the translocation of the enzyme-bound amino acid thioesters and the peptide intermediates via enzyme-bound 4′-phosphopantetheine moieties. A detailed analysis has been carried out of the enzymes that are involved in the biosynthesis of the cyclo-decapeptide gramicidin S [cyclo-(D-Phe-L-Pro-L-Val-L-Orn-L-Leu)$_2$]. This analysis has revealed a highly conserved and ordered modular structure.[59] Each module, which occupies about 1000 amino acids, contains domains for the activation of a specific amino and for the formation of a thioester with a 4′-phosphopantetheine moiety that is also present in the same module.[60,61] A region of gramicidin S synthetase 1 has also been identified that is responsible for the epimerization of L-phenylalanine to D-phenylalanine.[60]

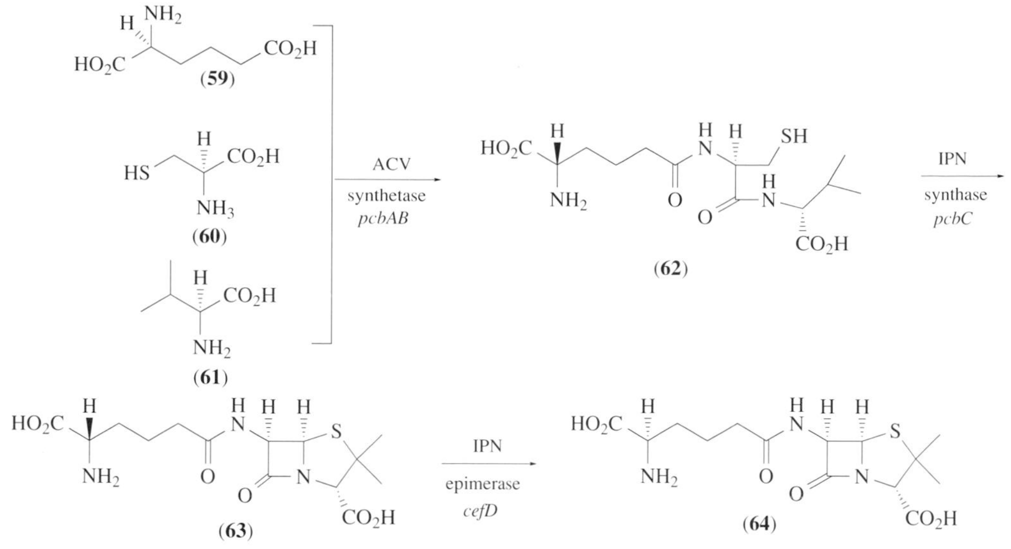

Scheme 14

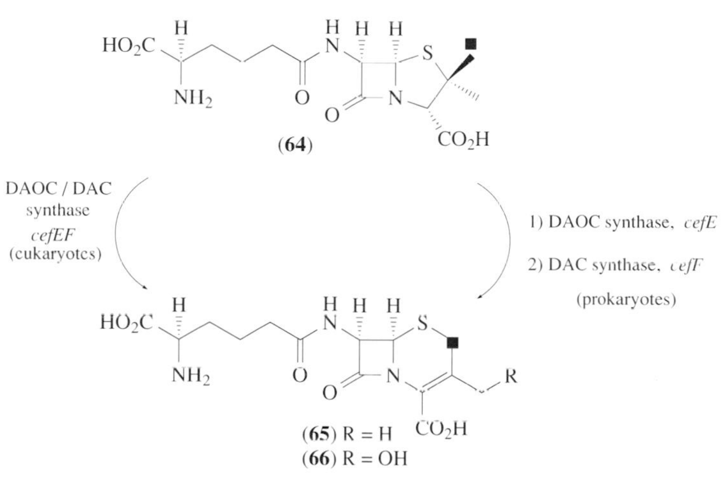

Scheme 15

The enzymatic properties of the purified forms of ACV synthetase are also consistent with a thiol template mechanism. A characteristic feature of the nonribosomal peptide synthetases is their ability to catalyze the exchange of labeled inorganic pyrophosphate into ATP in the presence of the substrate amino acids. The ACV synthetases from *A. nidulans* and *C. acremonium* have been shown to catalyze the exchange of inorganic pyrophosphate into ATP in the presence of all three substrate amino acids. In the case of ACV synthetase from *S. clavuligerus*, the enzyme is able to catalyze pyrophosphate exchange in the presence of L-cysteine and L-valine, but not L-α-aminoadipic acid.[57,62] This may due to a slow rate for the reverse of the aminoadipic acid adenylation reaction. D-Valine is not a substrate for the ACV synthetases, nor does it induce the ATP-pyrophosphate exchange reaction. Incubation of L-[^{2}H]valine with purified ACV synthetase from *C. acremonium* and *S. clavuligerus* led to ACV tripeptide in which essentially complete loss of deuterium has occurred from the α-position of the valinyl residue.[63] Incubations in deuterium oxide/water with the *C. acremonium* enzyme produced ACV tripeptide with significant incorporation of deuterium into the valine moiety. These results are consistent with the postulate that a single multifunctional enzyme is responsible for formation of the peptide bonds of ACV and the epimerization of the valine residue. The cofactor 4′-phosphopantetheine has been detected in the *C. acremonium* and *S. clavuligerus* ACV synthetases, but its presence in the *A. nidulans* enzyme has not yet been reported. The number of 4′-phosphopantetheine moieties attached to each ACV molecule has not been conclusively determined for any of the enzymes.

Substrate specificity studies have shown that the ACV synthetase from *C. acremonium* will accept L-*S*-carboxymethylcysteine in place of α-aminoadipate, both L-vinylglycine and L-allylglycine in place of cysteine, and L-alloisoleucine in place of valine. Tripeptide products could be isolated when each of these substrate analogues was used.[62] In the case of the *S. clavuligerus* enzyme, L-*S*-carboxymethylcysteine could be substituted for α-aminoadipate, and L-homocysteine for cysteine. A number of amino acids including L-alloisoleucine, L-α-aminobutyric acid, L-allylglycine, and L-norvaline could be substituted for L-valine.[64] Studies of the substrate specificity of the *C. acremonium* synthetase have also provided some new insights into the order of peptide bond formation and the timing of the epimerization reaction.[65,66] Replacement of cysteine with L-*O*-methylserine in preparative-scale incubations led to the isolation of both L-*O*-methylserinyl-L-valine and L-*O*-methylserinyl-D-valine dipeptides. Yields of both diastereomers were significantly decreased when L-α-aminoadipate was absent from the incubations, suggesting that the binding of aminoadipate is required for the full activity of the enzyme. When incubations were carried out with L-glutamate, L-cysteine, and L-valine, L-cysteinyl-D-valine was formed, but no tripeptide product was produced. These observations indicate that the assembly of ACV by ACV synthetase may proceed by initial formation of a peptide bond between L-cysteine and L-valine followed by formation of a peptide bond between L-α-aminoadipate and L-cysteinyl-D-valine. The results further suggest that the epimerization of valine takes place at the dipeptide stage.

The cyclization of the ACV tripeptide to isopenicillin N is catalyzed by isopenicillin N synthase. This enzyme has been purified from *C. acremonium*, *Flavobacterium* sp., and *Nocardia* (*Streptomyces*) *lactamdurans*.[67–69] The isopenicillin N structural genes have been cloned from a variety of β-lactam producing organisms, including *C. acremonium*, *Penicillium chrysogenum*, *A. nidulans*, *N. lactamdurans*, and several species of *Streptomyces*.[56] The *P. chrysogenum*, *A. nidulans*, *C. acremonium*, and *S. clavuligerus* enzymes have been overexpressed in *E. coli*.[70–74] All of the isopenicillin N synthases have molecular masses in the range of 30–40 kDa. The enzymes are non-heme iron-containing proteins which require the presence of Fe^{2+}, oxygen, and ascorbate in order to catalyze the cyclization of the ACV tripeptide to isopenicillin N. The reaction involves removal of four hydrogen atoms from the substrate and is accompanied by the reduction of one molecule of oxygen to water.[54] Early incorporation studies indicated that both the C—N bond and C—S bond of isopenicillin N are formed with overall retention of configuration. These results were subsequently confirmed using stereospecifically labeled forms of ACV and cell-free systems.[54] Once the *C. acremonium* and *P. chrysogenum* enzymes became available, it was discovered that isopenicillin N synthase exhibits a relatively broad substrate specificity.[54,75] This lack of rigid specificity allowed important insights into the mechanism of isopenicillin N synthase to be gained by detailed studies of the products formed from analogues of the natural substrate.[54,55,76–78] These studies provided evidence that the reaction proceeds by initial formation of the β-lactam ring and that the subsequent formation of the C—S bond involves abstraction of a hydrogen atom from C-3 of the valinyl residue of the enzyme-bound intermediate by a ferryl(IV) species to produce a radical intermediate.

Detailed spectroscopic investigations using Mossbauer, NMR, EPR, EXAFS, and ESEEM have been carried out to examine the nature of the iron-binding site of the isopenicillin N synthase from *C. acremonium*.[79–83] The results of all these studies are consistent with a six-coordinate Fe^{2+} site containing two or three endogenous histidine ligands, an endogenous aspartate ligand, and a water ligand. The studies also indicate that the binding of ACV to the synthase is accompanied by coordination of the thiol group of ACV to the iron center. The culmination of the structural studies has been the determination of the crystal structure of the isopenicillin N synthase from *A. nidulans* complexed with manganese.[84] The secondary structure of the enzyme is composed of 10 α-helices and 16 β-strands. Eight of the β-strands are folded to give a jelly-roll motif, but this motif is unlike those encountered in other proteins since the jelly roll is not completely enclosed, and the active site resides within the β-barrel. The active site structure, with manganese substituting for ferrous ion, exhibits a distorted octahedral geometry with four protein ligands (His 214, Asp 216, His 270, Gln 330) and two water molecules coordinated to the metal atom. The two water atoms occupy coordination sites that are directed into a hydrophobic cavity within the protein. It is believed that ACV and dioxygen bind to the coordination sites occupied by water molecules and Gln 330. Location of the active site within a predominantly hydrophobic cavity may reflect the fact that the reaction proceeds via highly reactive intermediates. On the basis of the crystal structure and previous studies of the mechanism of isopenicillin N synthase, the reaction pathway shown in Scheme 16 has been proposed for the cyclization.

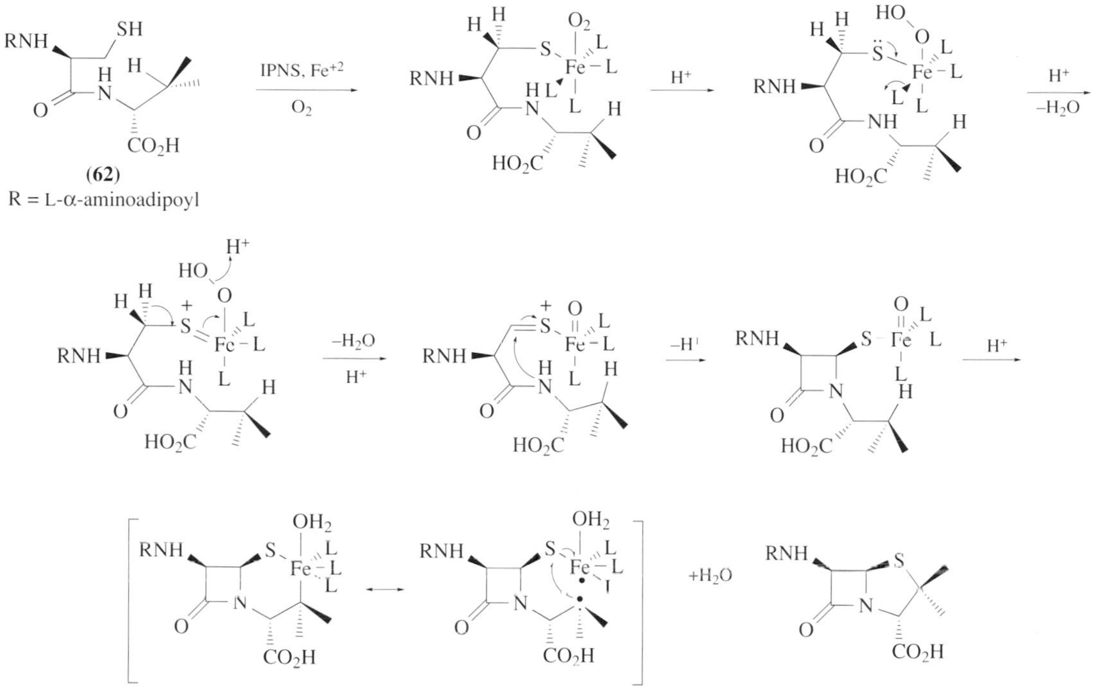

(62)
R = L-α-aminoadipoyl

Scheme 16

The enzyme isopenicillin N epimerase has been purified from *N. lactamdurans* and *S. clavuligerus*.[85,86] The *Nocardia* enzyme is monomeric with a molecular mass between 47 and 50 kDa. The *Streptomyces* enzyme is also monomeric and has a molecular mass of 59 kDa. Both enzymes utilize pyridoxal phosphate as a cofactor. The *Streptomyces* enzyme has been shown to contain one mole of covalently bound pyridoxal phosphate per mole of protein. This enzyme catalyzes the epimerization of both the L-α-aminoadipyl sidechain of isopenicillin N and the D-α-aminoadipyl sidechain of penicillin N. At equilibrium, the enzyme produces a 1:1 mixture of the two epimers.

Enzymatic activity catalyzing the expansion of penicillin N to the cephalosporin ring system was first reported in cell-free extracts of *C. acremonium*.[87,88] These investigations also established that the reaction requires ferrous iron, molecular oxygen, 2-oxoglutarate, and DTT. Efforts to purify the expandase and hydroxylase activities then revealed that the enzymology differed in prokaryotic and eukaryotic organisms (Scheme 15). In the case of the fungus *C. acremonium*, purification led to a single monomeric expandase-hydroxylase with a molecular mass of ca. 40 kDa.[54] However, purification of expandase and hydroxylase activities from *S. clavuligerus* led to two distinct proteins.[89–91] The bacterial expandase and the hydroxylase are both monomeric, with M_r values of 35 000 and 35 000–38 000, respectively. Ferrous iron, molecular oxygen, and 2-oxoglutarate are required by both enzymes for catalysis. A similar expandase protein has also been purified from *N. lactamdurans*.[92] The genes coding for the expandase-hydroxylase of *C. acremonium* and the expandase and hydroxylase of *S. clavuligerus* have been cloned.[56,93] Comparison of the deduced amino acid sequence for the bifunctional fungal enzyme with those for the *S. clavuligerus* expandase and hydroxylase indicate a high degree (55–60%) of similarity. The bacterial expandase and hydroxylase also exhibit a high similarity to each other. In each case, the regions of high similarity are evenly spread throughout the sequences. The apparent close relationship between the bacterial expandase and hydroxylase has been confirmed by the discovery that the expandase exhibits a slight hydroxylation activity for deacetoxycephalosporin C (**65**), while the hydroxylase displays a weak expandase activity for penicillin N (**64**).[91] The expandase-hydroxylase from *C. acremonium* and the expandase of *S. clavuligerus* have both been overexpressed in soluble form in *E. coli*.[94,95]

Investigations of the mechanism of the expandase and hydroxylase reactions have revealed a number of interesting features. *In vivo* labeling experiments with *C. acremonium* using DL-(3*R*) and L-(3*S*)-[4-^{13}C]valine have shown that the β-methyl group of penicillin N (**64**) is incorporated into the dihydrothiazine ring system of the cephalosporins ((**65**), (**66**)) (Scheme 15). *In vitro* experiments with the *S. clavuligerus* expandase demonstrated that the same stereochemistry is followed by this enzyme.[96] Incorporation experiments with DL-(3*R*,4*R*) and (3*R*,4*S*)-[4-^{2}H$_1$,^{3}H]valine in *C. acremonium* also revealed that the ring-expansion leads to complete racemization of the chirality of the β-methyl group.[54] Similar experiments with DL-(3*S*,4*S*) and (3*S*,4*R*)-[4-^{2}H$_1$,^{3}H]valine demonstrated that the hydroxylation of deacetoxycephalosporin C (**65**) to deacetylcephalosporin C (**66**) proceeds with retention of configuration.[97] The hydroxyl group of deacetylcephalosporin C was also shown to incorporate label from ^{18}O-labeled molecular oxygen.[98] A careful examination of the products formed from penicillin N (**64**) (X = H) by the action of the *C. acremonium* expandase-hydroxylase revealed that a small amount of the 3β-hydroxycepham (**67**) (X = H) is formed in addition to the expected products (**65**) and (**66**) (ratio of (**65**):(**66**):(**67**) = 40:20:1) (Scheme 17(a)).[99] Incubation of [3-^{2}H$_1$]penicillin N (**64**) (X = D) with the same enzyme led to an increase in the amount of (**67**) (X = D) (ratio of (**65**):(**66**):(**67**) = 40:35:25). This result was rationalized as being due to the influence of a kinetic isotope effect on the partitioning of an intermediate in the pathway (Scheme 18). Additional experiments demonstrated that the oxygen of the 3β-hydroxyl group of (**67**) incorporated label from ^{18}O-labeled molecular oxygen and that the 2β-methyl group of (**64**) was incorporated into the dihydrothiazine ring system of (**67**) as expected. Other studies showed that neither the 3β-methylenehydroxy penam (**68**) nor the β-sulfoxide (**69**) were utilized as substrates by the expandase-hydroxylase (Scheme 17(b)),[54] but that the 3-exomethylene cepham (**70**) (X = H) was converted to (**66**) without the apparent intermediacy of (**65**) (Scheme 17(c)).[54] The formation of (**66**) from (**70**) can be rationalized as resulting from the addition of a ferryl iron species to the double bond of (**70**).[100] When deuterated (**70**) (X = D) was used as the substrate, the epoxide (**71**) was formed in addition to (**66**).[100] This result can be explained as the result of a kinetic isotope effect on the partitioning of an enzyme-bound intermediate. Label from ^{18}O-labeled molecular oxygen is also incorporated into the epoxide moiety of (**71**).[101]

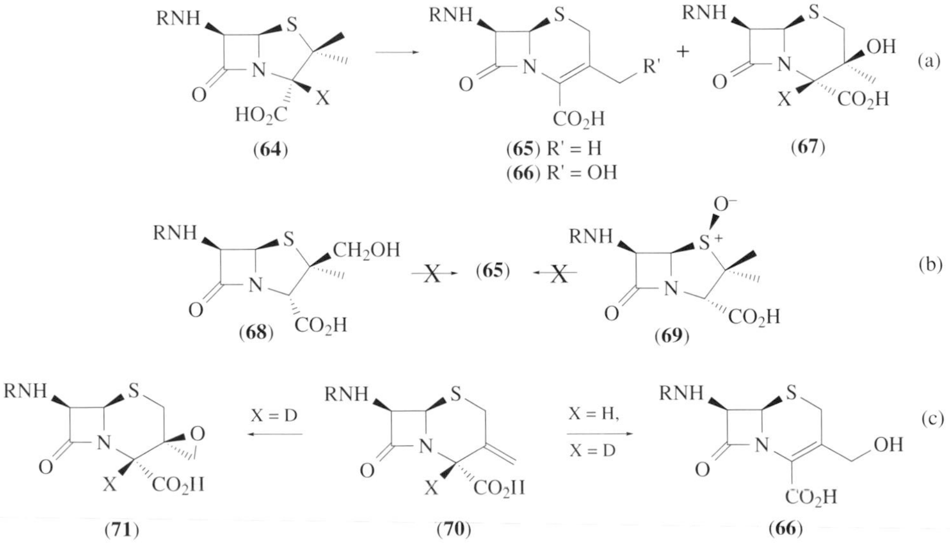

Scheme 17

From these and other studies,[95] it appears that the ring expansion and hydroxylation reactions catalyzed by the *C. acremonium* expandase-hydroxylase and the distinct expandase and hydroxylase of *S. clavuligerus* involve the intermediacy of a ferryl iron species. This is consistent with the fact that the mechanism of 2-oxoglutarate dependent iron-containing dioxygenases is believed to proceed by an oxidative decarboxylation of 2-oxoglutarate (**72**) in the presence of molecular oxygen to generate a ferryl, iron-oxo (**73**) moiety and succinate (**74**).[102] In this transformation, one atom of oxygen resides in (**73**) and the other is found in the carboxyl group of (**74**) (Scheme 18(a)). The

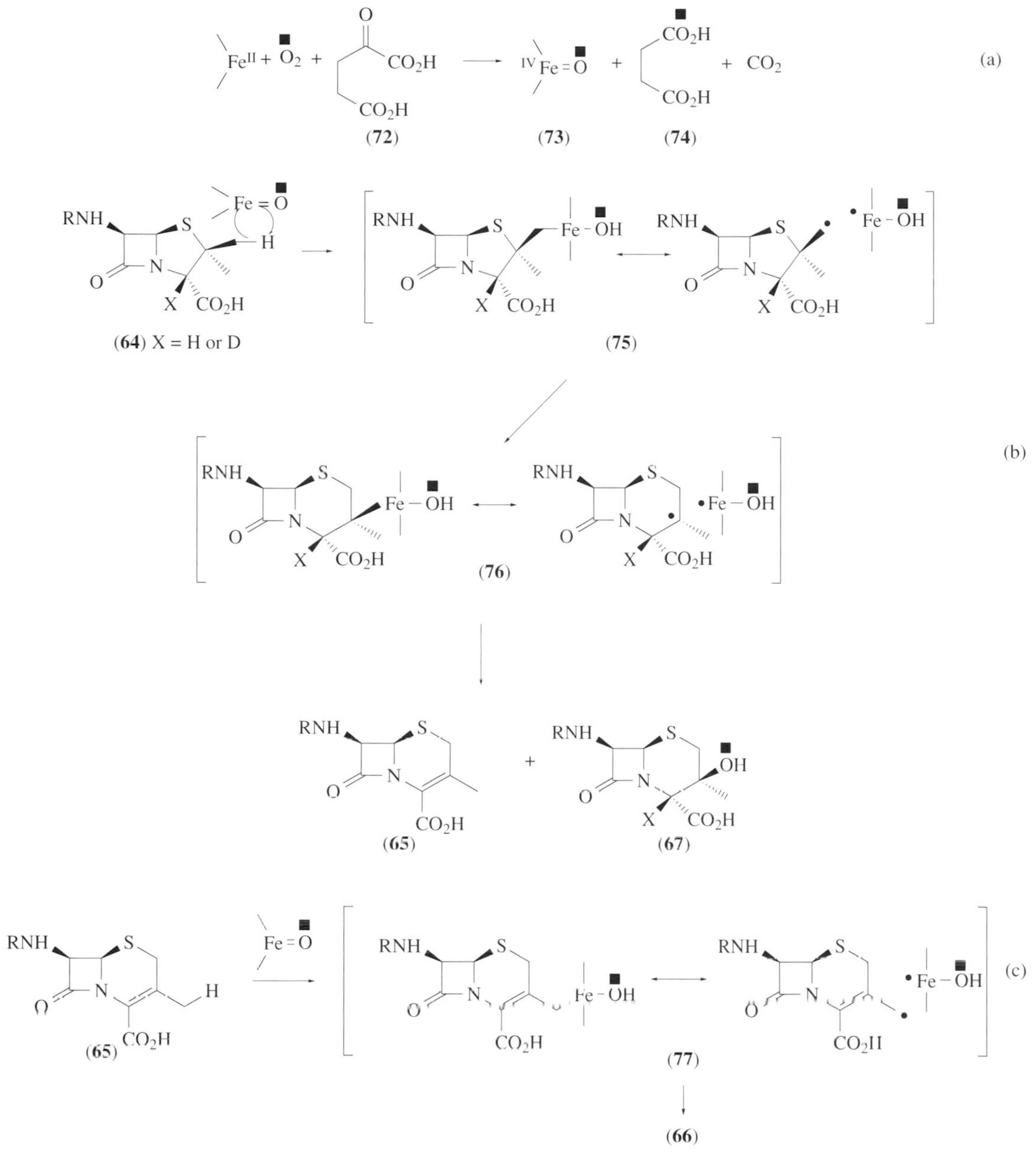

Scheme 18

ferryl species (**73**) is believed to be responsible for the oxidation of the substrate and the donation of an oxygen atom in the case of an oxygenation reaction. Scheme 18(b) outlines a plausible mechanism for the ring-expansion of penicillin N to deacetoxycephalosporin C using ferryl iron chemistry. This scheme also illustrates how a ring-expanded intermediate (**76**) derived from [3-$^{2}H_{1}$]penicillin N could give rise to a higher proportion of the 3β-hydroxycepham (**67**). Scheme 18(c) provides a mechanistic rationalization for the hydroxylation of deacetoxycephalosporin C to deacetylcephalosporin C using ferryl iron chemistry. These schemes can also accommodate the discovery that the deacetylcephalosporin C and 3β-hydroxycepham (**67**) formed by the *C. acremonium* enzyme incorporate label from ^{18}O-labeled water, if it is assumed that the hydroxyl group bound to iron in species (**75**), (**76**), and (**77**) can undergo exchange with water.[101]

1.29.3.2 Sparsomycin

Sparsomycin (**78**) (Scheme 19) is a unique antibiotic produced by *Streptomyces sparsogenes*[103] and *Streptomyces cuspidosporus*.[104] It exhibits antibiotic activity against both gram-negative and gram-positive bacteria and it displays potent antitumor activity against KB human epidermoid carcinoma cells *in vitro*.[105] The biological activity of sparsomycin is the result of its ability to inhibit the peptide bond-forming step of protein biosynthesis. The mechanism of action of sparsomycin is not fully understood, but it appears to involve the interaction of sparsomycin with the ribosomal RNA component of the peptidyl transferase.[106,107]

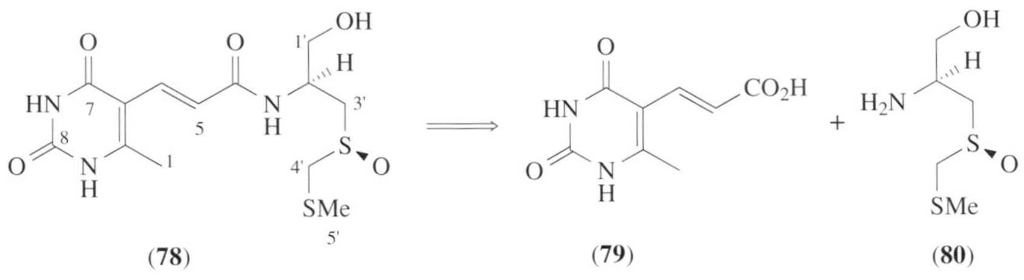

Scheme 19

The sparsomycin molecule is composed of two unusual building blocks, the uracil acrylic acid moiety (**79**) and the mono-oxodithioacetal moiety (**80**) (Scheme 19). The biosynthesis of each of these moieties has been investigated by means of precursor incorporation experiments.[108,109] Initially, it was hypothesized that (**79**) would be derived from L-tryptophan (**81**) via *N*-formyl kynurenine (**82**) and/or *N*-formyl anthranilic acid (**83**) (Scheme 20). The hypothesis that tryptophan would be a specific precursor of (**79**) was then proven by experiments with DL-[2-^{13}C]-and DL-[5-^{2}H$_1$]tryptophan. These two forms of tryptophan labeled C-8 and C-5 of the uracil moiety, respectively, with enrichments of 22% and 6%. However, additional experiments with [5-^{2}H$_1$]- and [*formyl*-^{13}C]-*N*-formyl kynurenine revealed that only the deuterium labeled from these precursors was incorporated into (**78**). Similarly, [*formyl*-^{13}C,5-^{2}H$_1$]- and [*carboxyl*-^{13}C,3,5-^{2}H$_2$]*N*-formyl anthranilic acid were incorporated into sparsomycin with deuterium labeling at C-5 of the antibiotic, but no ^{13}C-labeling at C-8 or C-7. These results are consistent with the degradation of (**82**) and (**83**) to anthranilic acid (**84**) which is known to be a precursor of tryptophan and to be converted into tryptophan with loss of its carboxyl group. An alternative hypothesis which can account for the specific incorporation of tryptophan into the uracil moiety of sparsomycin and is consistent with the results described above is shown in Scheme 21. This hypothesis postulates that ring A of tryptophan is cleaved between C-6 and C-7 prior to the cleavage of ring B between C-2 and C-3. The obligatory loss of the tryptophan sidechain could occur at an early stage in the pathway by the action of tryptophanase to give indole, or it could occur later, possibly at the stage indicated in the Scheme. Hydroxylation of ring A of tryptophan or indole at C-6 or C-7 prior to ring cleavage was ruled out by precursor incorporation experiments with 6-hydroxy- and 7-hydroxytryptophan and the corresponding indoles.[108]

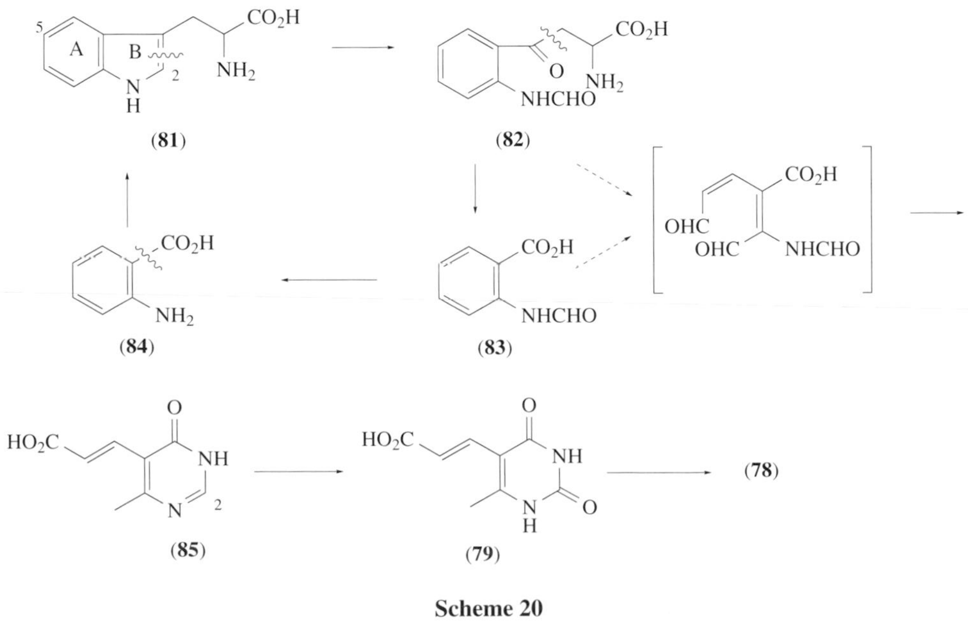

Scheme 20

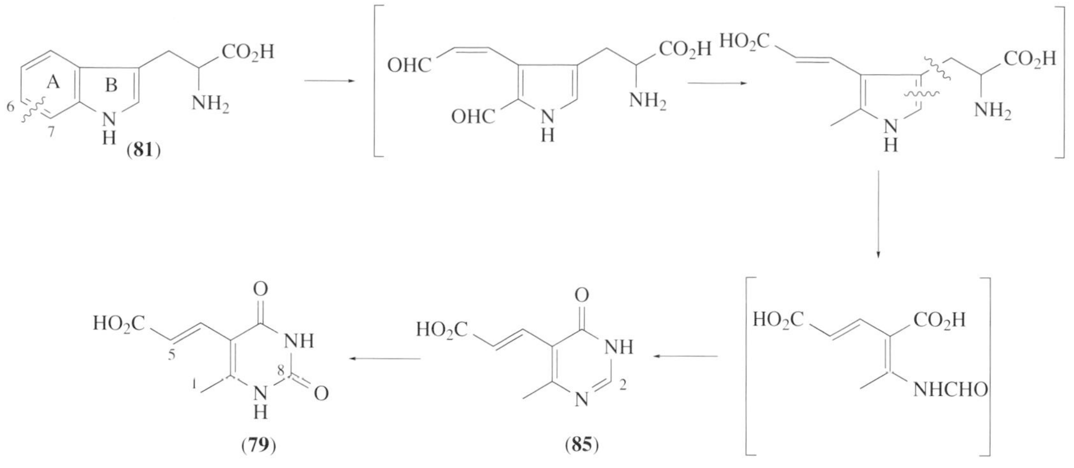

Scheme 21

Some insight into the mechanism of formation of the uracil ring of (**79**) was gained when the resemblance between the heterocyclic ring of xanthosine and (**79**) was noted. On the basis of the analogy provided by the known conversion of inosine 5′-monophosphate to xanthosine 5′-monophosphate by inosine-5′-monophosphate dehydrogenase, it was postulated that the pyrimidine acrylic acid (**85**) would be the immediate precursor of (**79**). This was first proven by synthesis of 2-^{13}C-labeled (**85**) and administration of the labeled compound to cultures of *S. sparsogenes*. The sparsomycin produced in this experiment was specifically labeled with ^{13}C at C-8, as predicted (7% enrichment).[108] Additional evidence for the intermediacy of (**85**) in sparsomycin biosynthesis was obtained from cell-free studies.[109] These investigations eventually led to the isolation and purification to homogeneity of an enzyme that catalyzes the conversion of (**85**) to (**79**).[110] The enzyme is monomeric with a molecular mass of 87 kDa. It requires NAD^+ and monovalent cations for activity, and is irreversibly inhibited by 6-chloropurine. The properties of the enzyme indicate that it is related to inosine-5′-monophosphate dehydrogenase.

The structure of the mono-oxodithioacetal moiety (**80**) suggests that L-methionine and L-cysteine are likely to be precursors. Administration[108] of L-[*methyl*-^{13}C]methionine to *S. sparsogenes* yielded sparsomycin (**78**) that was labeled at both C-4′ and C-5′ (2.7% enrichment at C-4′,6.7% enrichment at C-5′), thereby demonstrating that both of these carbon atoms are derived from the methyl group of methionine. The fact that the C-methyl group (C-1) of sparsomycin (**78**) was not labeled by this precursor was also noteworthy. Administration[108] of DL-[3-^{13}C]cysteine produced antibiotic that exhibited enrichment at C-3′, as anticipated, but some label from the precursor also appeared at C-4′. Labeling at the latter carbon atom might arise by conversion of the labeled cysteine to [3-^{13}C]serine which could then label the C_1 pool. Administration[108] of DL-[1-^{13}C]serine labeled sparsomycin at C-1′ of the mono-oxodithioacetal moiety, a result that can be rationalized by the conversion of serine into cysteine. Taken together, the incorporation experiments with cysteine and serine imply that C-1′ to C-3′ of (**80**) are derived from L-cysteine.

The next phase of the investigation of the biosynthesis of the moiety (**80**) examined the timing of the reduction of the cysteine carboxyl group versus the formation of the dithioacetal group.[108] Since the absolute configuration at C-2′ of sparsomycin corresponds to that of D-cysteine, the timing of the epimerization at the C-2 position was also investigated. Administration of both L- and D-[4-^{13}C]-*S*-(methylthiomethyl)cysteine (**86**) (Scheme 22) yielded antibiotic that was highly enriched at C-4′ (84% from L-(**86**), 72% from D-(**86**)). Two conclusions could be drawn from these results. First, the high levels of enrichment achieved with both precursors suggest that the epimerization of the asymmetric center present at C-2 can occur at a late stage. Second, the results suggest that the dithioacetal moiety is formed prior to the reduction of the cysteine carboxyl group. Additional information on the timing of the carboxyl group reduction was obtained by incorporation experiments with L- and D-[4-^{13}C]-*S*-(methylthiomethyl)cysteinol (**87**). Both compounds were found to label sparsomycin at the expected position (C-4′), but the incorporation figures were lower than those observed with the L and D forms of the amino acid (**86**) (L, 1.3%; D, 6.7%). The possibility that incorporation of L- and D-(**87**) was occurring by oxidation back to L- and D-(**86**) was ruled out by incorporation experiments with 1-[$^2H_2$]-labeled L- and D-(**87**). A deuterium NMR analysis demonstrated that both forms of deuterated (**87**) labeled sparsomycin at C-1′. The specific incorporation of the L- and D-forms of (**87**) into sparsomycin suggests that the reduction of the cysteine-derived carboxyl group precedes sulfoxide formation and coupling to the uracil moiety (**79**).

The results of the preceding incorporation experiments support the hypothesis that cysteine is converted to sparsomycin via the intermediacy of L and/or D-*S*-(methylthiomethyl)cysteine (**86**). A crucial question with respect to sparsomycin biosynthesis therefore concerns the mechanism of the conversion of cysteine into *S*-(methylthiomethyl)cysteine(**86**). Scheme 22 summarizes the most plausible routes that can be envisioned for this transformation. One pathway would proceed by methylation of cysteine with *S*-adenosylmethionine (SAM) serving as the methyl donor to produce *S*-methylcysteine (**88**). *S*-Methylcysteine could then be transformed into *S*-(methylthiomethyl)-cysteine (**86**) in one of two ways. The first would be via the direct insertion of sulfur into the *S*-methyl group of (**88**) to produce the unstable dithio hemiacetal (**89**), whose subsequent *S*-methylation would then generate (**86**). The postulated sulfur insertion reaction would presumably bear some relationship to the sulfur insertion processes encountered in the biosynthesis of biotin, lipoic acid, and the penicillins and cephalosporins (see above). An alternative way to convert *S*-methylcysteine to (**86**) would proceed by hydroxylation of the *S*-methyl group of (**88**) to yield the monothio-hemiacetal (**90**). Reaction of (**90**) with hydrosulfide and SAM or with methanethiol would then lead to the formation of (**86**). Another route to (**86**) could proceed by reaction of (**90**) with cysteine to produce djenkolic acid (**91**), an amino acid that occurs in the plants *Pithecolobium lobatum* and *Albizzia lophanta*. The djenkolic acid could then undergo fragmentation in a pyridoxal phosphate (PLP) mediated process to form (**89**) which would then be converted to (**86**). Still another route that can be imagined for the conversion of cysteine into *S*-(methylthiomethyl)cysteine would begin with the reaction of cysteine with methylene-tetrahydrofolate to produce the monothio hemiacetal (**90**). Once formed, (**90**) could then be transformed into (**86**) in one of the ways already outlined.

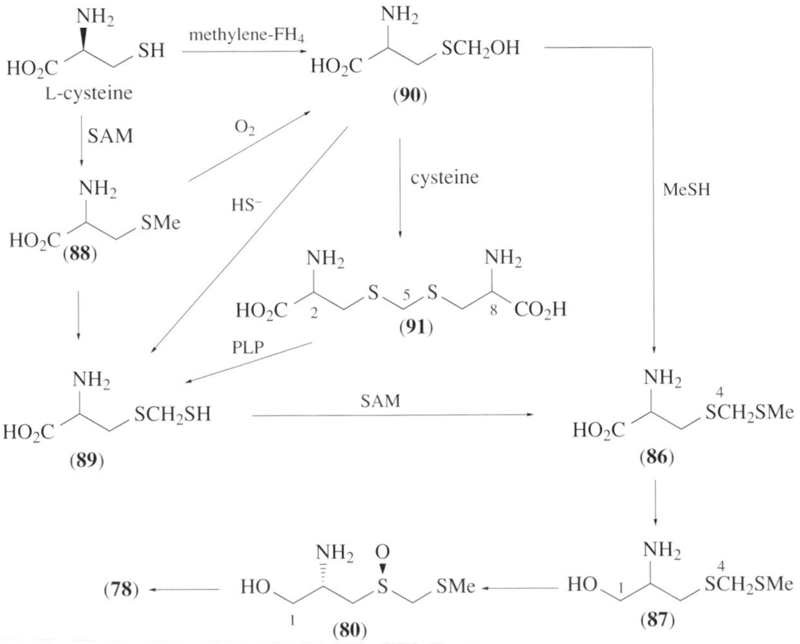

Scheme 22

A number of precursor incorporation experiments were carried out to evaluate the alternative pathways to (**86**) outlined in Scheme 22.[108] Administration of both L- and D-[*methyl*-^{13}C]-*S*-methyl-cysteine led to labeling of sparsomycin at C-4′. The D-form of the precursor also produced some ^{13}C-labeling of the C-5′ position of the antibiotic, although the enrichment was lower at C-5′ than at C-4′ (1.1% vs. 0.6%). This result suggested that some demethylation of the precursor might be occurring *in vivo*. This supposition was confirmed by means of an incorporation experiment with [*methyl*-^{14}C,3-^3H]-*S*-methylcysteine. The sparsomycin derived from this precursor exhibited an eight-fold increase in its tritium to carbon ratio compared to that of the precursor. This observation raised concerns about the intact incorporation of *S*-methylcysteine into sparsomycin. These concerns were addressed by an isotope dilution experiment. L-[U-^{14}C]Cystine was administered to the *S. sparsogenes* fermentation and the mycelium harvested after 10 hours. The washed mycelium was sonicated in the presence of unlabeled L-*S*-methylcysteine (**88**) as carrier, and the carrier was then reisolated and purified to constant radioactivity. The final incorporation figure (0.02%) suggested that L-*S*-methylcysteine (**88**) is present in *S. sparsogenes*.

The route to (**86**) that proceeds by the reaction of cysteine with methylene-tetrahydrofolate to give the hemi thioacetal (**90**) was examined by means of an incorporation experiment with [2-^{13}C]glycine, since C-2 of glycine can serve as a direct source of the C_1 unit of methylene-tetrahydrofolate. Administration of [2-^{13}C]glycine led to sparsomycin exhibiting a higher enrichment at C-5′ than at C-4′ (2.2% vs. 0.5%). Since administration of L-[*methyl*-^{13}C]methionine also leads to a higher degree of enrichment at C-5′ than at C-4′ (see above), the labeling pattern observed with [2-^{13}C]glycine suggests that the label from glycine is first incorporated into the *S*-methyl group of SAM before serving as the source of the two C_1 units in the mono-oxodithioacetal moiety of sparsomycin.

The next experiment examined the possible intermediacy of djenkolic acid (**91**) in sparsomycin biosynthesis. A precursor incorporation experiment with (2*S*,8*S*)-[5-^{13}C]-djenkolic acid yielded antibiotic that did not exhibit any ^{13}C enrichment.

Finally, the potential role of methanethiol in the biosynthesis of sparsomycin was investigated.[111] In order to simplify handling of this volatile compound, [*methyl*-^{13}C]-*S*-methyl-*n*-thiolhexanoate was synthesized with the presumption that the thiol ester would be hydrolyzed *in vivo*. Administration of this compound to *S. sparsogenes* yielded sparsomycin that exhibited ^{13}C enrichment only at C-4′ (1.2%) and not at C-5′, whereas the pathway shown in Scheme 22 predicts labeling at C-5′. The C-4′ labeling encountered in the experiment may be most easily explained by the *in vivo* formation of labeled *S*-methylcysteine from *O*-acetylserine and methanethiol.

Since the weight of the evidence seemed to favor the intermediacy of *S*-methylcysteine in the formation of *S*-(methylthiomethyl)cysteine, experiments were conducted to determine the number of hydrogen atoms removed from the *S*-methyl group of *S*-methylcysteine as the result of formation of the dithioacetal group. The first of these experiments involved the administration of D-[*methyl*-^{13}C,*methyl*-$^{2}H_3$]-*S*-methylcysteine. Unexpectedly, no labeling of the sparsomycin derived from the precursor was detectable by either ^{13}C- or ^{2}H-NMR spectroscopy. A second experiment with L-[*methyl*-^{13}C,*methyl*-$^{2}H_3$]methionine was therefore carried out. In this instance, no labeling at C-4′ was encountered, but examination of the ^{13}C NMR spectrum of the sparsomycin while simultaneously decoupling both protons and deuterons confirmed the presence of three deuterium atoms at C-5′. The failure of *S*-methylcysteine and methionine bearing a trideuterated *S*-methyl group to label the C-4′ position of sparsomycin could be due to the operation of a significant kinetic isotope effect during the functionalization of the *S*-methyl group of *S*-methylcysteine. The apparent magnitude of such an isotope effect could be increased in the case of *S*-methylcysteine if the demethylation of this precursor does not involve a deuterium isotope effect.

To summarize, the investigations of the biosynthesis of the mono-oxodithioacetal moiety (**80**) of sparsomycin appear to support a pathway that proceeds from L-cysteine via *S*-methylcysteine (**88**), *S*-(methylthiomethyl)cysteine (**86**), and *S*-(methylthiomethyl)cysteinol (**87**) (Scheme 22). The *S*-methyl group of (**80**) (C-5′) is clearly derived from the *S*-methyl group of methionine, but the origin of the sulfur atom in the *S*-methyl group of (**80**) is unknown, as is the mechanism of introduction of this sulfur atom.

1.29.3.3 Thiotropocin

Thiotropocin (**92**) (Scheme 23) is an antibiotic of unusual structure isolated from the fermentation broth of *Pseudomonas* sp. CB-104.[112,113] The compound exhibits antibacterial, antifungal, and antiprotozoal activities *in vitro* with the greatest biological activity being exhibited at low pH. Thiotropocin also causes morphological changes in both *E. coli* and *Proteus mirabilis*. Biosynthetic investigations of thiotropocin have been carried out by Cane and coworkers.[114]

The initial investigations of thiotropocin biosynthesis were predicated upon the hypothesis that the antibiotic would be of polyketide origin. However, attempts to observe incorporation of labeled acetate and labeled 6-methylsalicylic acid were unsuccessful. Consequently, a precursor incorporation experiment was carried out with [U-$^{13}C_6$]glucose in order to determine the nature of the

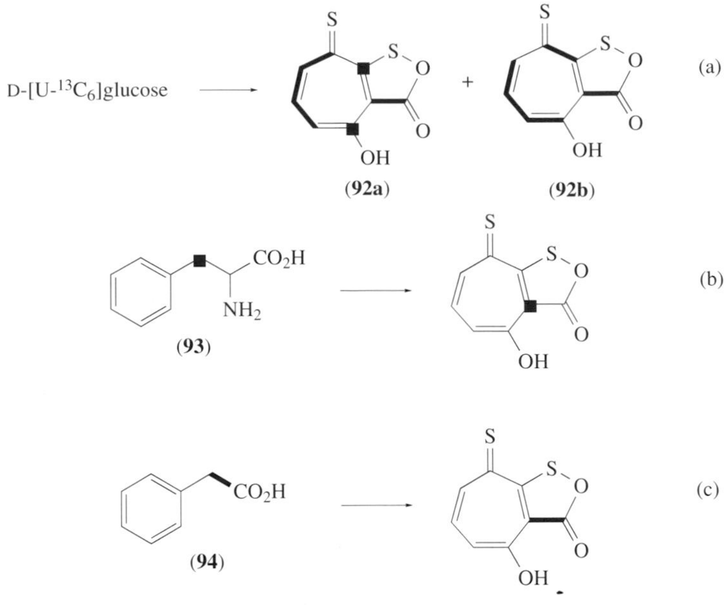

Scheme 23

primary building blocks for the antibiotic on the basis of the observed pattern of carbon-13 connectivity. The ^{13}C NMR spectrum of the thiotropocin produced from [U-$^{13}C_6$]glucose was analyzed as its *p*-bromobenzyl thioether. The analysis revealed the presence of two isotopomerically labeled species, (**92a**) and (**92b**) (Scheme 23(a)). Additional precursor incorporation experiments with [3-^{13}C]phenylalanine (**93**) and [1,2-$^{13}C_2$]phenylacetic acid (**94**) produced thiotropocin labeled as shown in Scheme 23(b) and (c). The observed labeling patterns are consistent with the intermediacy of a symmetrical intermediate derived from the shikimate pathway. The results observed with [U-$^{13}C_6$]glucose can be explained by conversion of the labeled hexose into uniformly labeled pyruvate and erythrose-4-phosphate, followed by formation of shikimic acid (**95**) with the labeling pattern shown in Scheme 24(a). Reaction of the labeled shikimic acid with uniformly labeled phosphoenol pyruvate (PEP) would then produce chorismic acid (**96**) labeled as shown, and the chorismic acid could then be transformed via prephenic acid into labeled phenylacetic acid (**98**) via phenylpyruvic acid (**97**). The conversion of the labeled phenylacetic acid into thiotropocin can be envisioned as occurring via an oxidative ring expansion that would have an equal probability of either cleaving the bond between the two and four carbon units in the aromatic ring of (**98**) or of cleaving the two carbon unit itself (Scheme 24(b) and (c)). The origin of the sulfur atoms of thiotropocin and the mechanisms(s) involved in the introduction of these sulfur atoms are unknown. Given the possible intermediacy of a 4-hydroxytropolone-3-carboxylic acid in the biosynthetic pathway, it has been suggested that the introduction of both sulfur atoms may involve nucleophilic forms of sulfur.

1.29.3.4 Microcin B17

The *E. coli* peptide antibiotic microcin B17 (**101**) (Scheme 25) is a member of a class of DNA gyrase inhibitors that is distinct from coumarin or quinolone drugs. Genetic investigations of the microcin B17 operon have revealed the presence of seven plasmid-encoded open reading frames and assigned tentative functions to each of them.[115] The *mcbA* gene encodes a 69 amino acid polypeptide called pre-microcin B17 (**99**). Three genes, *mcbB*, *-C*, and *-D* have been shown to be required for the conversion of (**99**) into pro-microcin B17 (**100**). A chromosomally encoded protease then removes the first 26 amino acids from pro-microcin B17 to produce the active antibiotic. Of the three remaining genes, *mcbE* and *mcbF* encode proteins that transport microcin B17 out of the cell, while *mcbG* confers microcin B17 resistance on the producing organism.

The roles assigned to the *mcbB*, *-C*, and *-D* genes have been confirmed by purification and characterization of the microcin B17 synthase complex.[116] The purified synthase exhibited a Michaelis constant (K_m) of 2.3 μM for a substrate consisting of the first 46 amino acids of (**99**) (McbA(1–46)). The k_{cat} was 0.2 min^{-1} based upon the formation of both of the heterocyclic rings

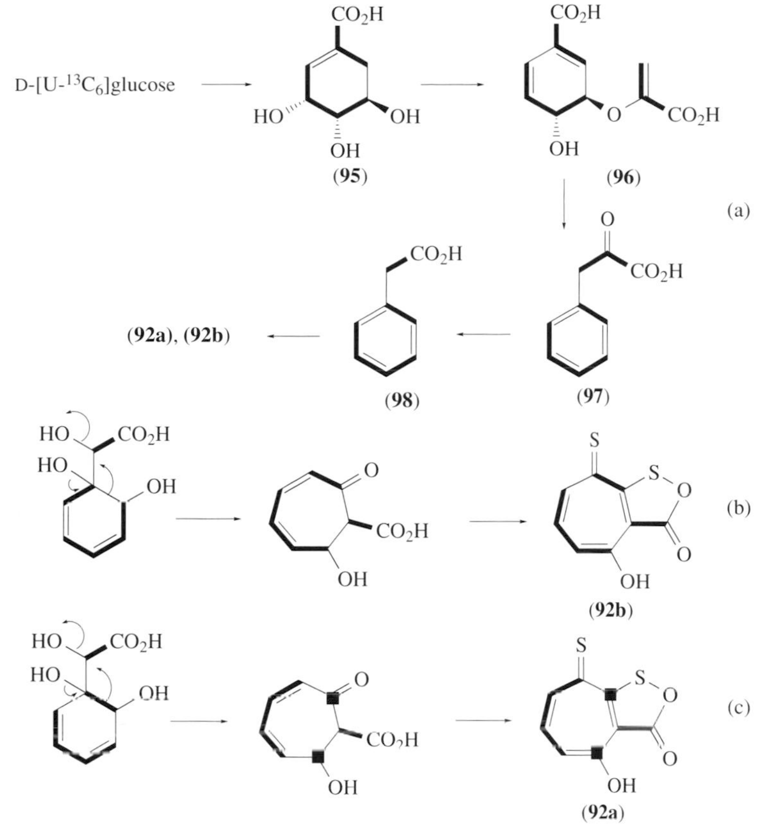

D-[U-$^{13}C_6$]glucose → CO_2H HO OH OH (**95**) → CO_2H O CO_2H OH (**96**)

↓ (a)

O CO_2H (**97**) → CO_2H (**98**) → (**92a**), (**92b**)

HO CO_2H HO OH → O CO_2H OH → S S O O OH (**92b**) (b)

HO CO_2H HO OH → O CO_2H OH → S S O O OH (**92a**) (c)

Scheme 24

MELKASEFGVVLSVDALKLSRQSPLG[26] [27]VGIGGGGGGG[36] GGGSCGGQGG[46] GCGGCSNGCS[56] GGNGGSGGSG[66] SHI

(**99**)

↓

M[1]------- G[38] –N(H) O N S N (O) GGQGG–N(H) S N (O) G–N(H) S N (O) SN—

—N(H) S N O N (O) GGNG–N(H) O N (O) G–N(H) O N (O) GSHI

(**100**)

↓ protease

[27]VGIGGGGGGGGG–N(H) O N S N (O) GGQGG–N(H) S N (O) G–N(H) S N (O) SN—

—N(H) S N O N (O) GGNG–N(H) O N (O) G–N(H) O N (O) GSHI

(**101**)

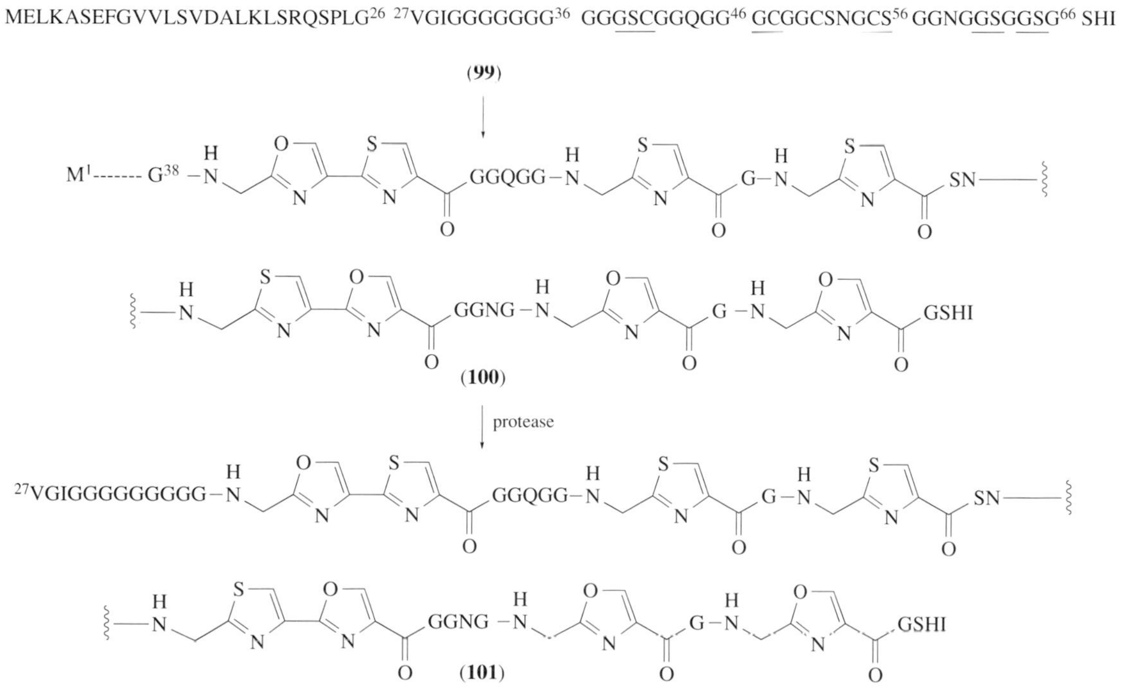

Scheme 25

that can be produced from this 46 amino acid substrate. SDS-PAGE analysis of the purified synthase revealed the presence of three protein bands with apparent molecular masses of 68 kDa, 45 kDa, and 31 kDa. N-Terminal sequencing and protein immunoblot analyses indicated that the 31 kDa band contained two proteins, McbB (33 kDa) and McbC (31 kDa), and that the 45 kDa band corresponded to McbD. The protein band at 68 kDa was identified as HrpG, a member of the Hsp90 heat shock protein family. Since the molecular mass of the purified microcin B17 synthase complex determined by gel filtration was ca. 100 kDa, the ratio of the protein components in the complex appears to be 1:1:1 (McbB, 33 kDa; McbC, 31 kDa; McbD, 45 kDa). While the pre-microcin B17 fragment McbA(1–46) was efficiently processed by the purified microcin B17 synthase complex, the fragment McbA(27–69) which contains the 43 amino acids that reside in the mature antibiotic did not serve as a substrate for the synthase. Furthermore, this fragment and a peptide consisting of residues 27 through 46 (McbA(27–46)) did not inhibit the processing of McbA(1–46) at 80 μM concentration. On the other hand, the fragment McbA(1–26), which consists of the leader sequence that is cleaved to convert pro-microcin B17 to microcin B17, inhibited the modification of Mcb(1–46) with an IC_{50} value of 2 μM. These observations suggest that the leader sequence plays an important role in the recognition of substrate by the microcin B17 synthase complex.

Investigations of the cofactor requirements for the synthase complex revealed that ATP is needed (K_m = 89 μM) and that it is converted into ADP and inorganic phosphate by the complex. The conversion of a cysteine or serine residue to a thiazole or oxazole ring requires a two-electron oxidation. It appears that this role may be fulfilled by McbC since this has been found to be a flavoprotein which contains a stoichiometric amount of noncovalently bound FMN. The terminal electron acceptor in the oxidation process is likely to be molecular oxygen since degassing of the enzymatic reaction mixture led to a reduced rate of product formation. The roles played by McbB and McbD in the cyclization reaction are unknown at present. The mechanism of the cyclization reactions leading to the oxazole and thiazole rings (Scheme 26) can be envisioned as occurring in three steps: an initial cyclization, a loss of water (Y = H) or phosphate (Y = P), and an FMN dependent dehydrogenation. Nothing is currently known regarding the order and timing of the these steps with respect to the creation of the eight heterocyclic rings that are present in microcin B17.

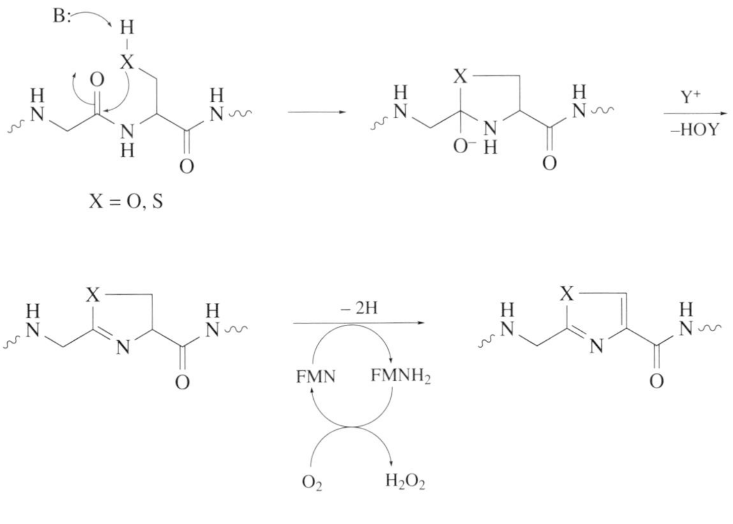

X = O, S

Y^+, –HOY

– 2H, FMN, $FMNH_2$, O_2, H_2O_2

Scheme 26

1.29.3.5 Nosiheptide and Thiostrepton

The genetics and biosynthesis of the antibiotics nosiheptide (**101**) and thiostrepton (**102**) (Figures 2 and 3) have been the subject of review.[117] Consequently, a more concise summary will be provided here. Nosiheptide is produced by *Streptomyces actuosus*, while thiostrepton is produced by *Streptomyces laurentii* and *S. azureus*. They belong to a small family of sulfur-rich antibiotics that are highly modified peptides. Other members of the family include the siomycins, thiopeptins, micrococcins, and thiocillins,[117] as well as berninamycin[118] and antibiotic A10255.[119] The compounds exhibit close structural similarities, and possess a common mode of action that leads to the inhibition of protein biosynthesis in Gram-positive bacteria. The mechanism of action appears to involve binding of the antibiotics to the complex of 23S rRNA and the ribosomal protein L11 which then inhibits the GTP-dependent elongation factors EF and Tu. The self-resistance gene of the producing organisms has been shown to code for a methylase that specifically methylates the 2′ position of adenosine 1067 in the 23S rRNA thereby blocking the binding of the antibiotics.[117]

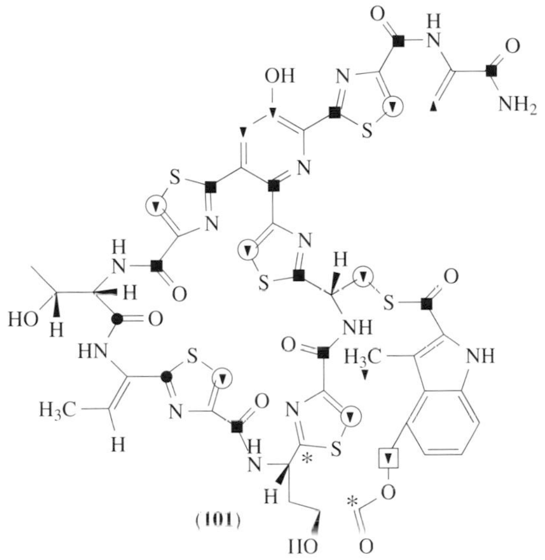

(101)

DL-[1-^{13}C]serine, ■ DL-[3-^{13}C]cysteine, ●
L-[3-^{13}C]serine, ▼ L-[*methyl*-^{13}C]methionine, ○
L-[1,5-$^{13}C_2$]glutamate, * DL-[1-^{13}C]threonine, □

Figure 2 Incorporation of labeled amino acids into nosiheptide. Reproduced by permission of the American Chemical Society from *J. Am. Chem. Soc.*, 1993, **115**, 7557.

An extensive series of incorporation experiments has been carried out with ^{13}C-labeled precursors to determine the primary metabolic building blocks used in the biosynthesis of nosiheptide and thiostrepton. The results of these experiments are summarized in Figures 2 and 3. These experiments demonstrated that the dehydroalanine and dehydrobutyrine residues arise from serine and threonine, respectively, that the thiazole rings are generated from cysteine and the carboxyl group of the adjacent N-terminal amino acid residue, that the 3,4-dihydroxyisoleucine residue of thiostrepton is derived from isoleucine, and that the 4-hydroxyglutamate residue of nosiheptide is derived from glutamate. More intriguing findings from these experiments were the discoveries that the indolic acid moiety of nosiheptide and the quinaldic acid moiety of thiostrepton are derived from tryptophan, and that the tetrahydropyridine ring of thiostrepton and the pyridine ring of nosiheptide are both formed from a tail-to-tail linkage of two molecules of serine plus the carboxyl group of an adjacent cysteine.[117,120,121]

Investigations of the mechanism of formation of the pyridine and tetrahydropyridine rings of nosiheptide and thiostrepton were carried out using L-[1,2-$^{13}C_2$]- and [2,3-$^{13}C_2$]serine as well as L-[3-^{13}C,3-$^{2}H_2$]- and L-(3*S*)-[3-^{13}C,3-$^{2}H_1$]serine. These experiments demonstrated that serine is incorporated as intact three carbon units into these rings and that the 3-*pro-S* hydrogen atom of serine is eliminated during formation of the pyridine ring of nosiheptide. The experiments with deuterated forms of serine also showed that both hydrogen atoms at C-4 of the tetrahydropyridine ring of thiostrepton are derived from C-3 of serine, while the β-hydrogen atom at C-3 of the tetrahydropyridine moiety is derived from the 3-*pro-R* hydrogen atom of serine. A hetero-Diels-Alder reaction between two dehydroalanine residues was postulated to account for the formation of the heterocyclic ring.

The precursor incorporation experiments with [^{13}C,^{2}H]-labeled serine residues also disclosed three other interesting features of nosiheptide and thiostrepton biosynthesis. The first is that the elimination of water to generate the dehydroalanine moieties of the antibiotics follows *anti* geometry. The second is that the formation of all of the thiazole rings proceeds with the stereospecific loss of the 3-*pro-S* hydrogen atoms of serine and retention of the 3-*pro-R* hydrogen atom. Finally, the

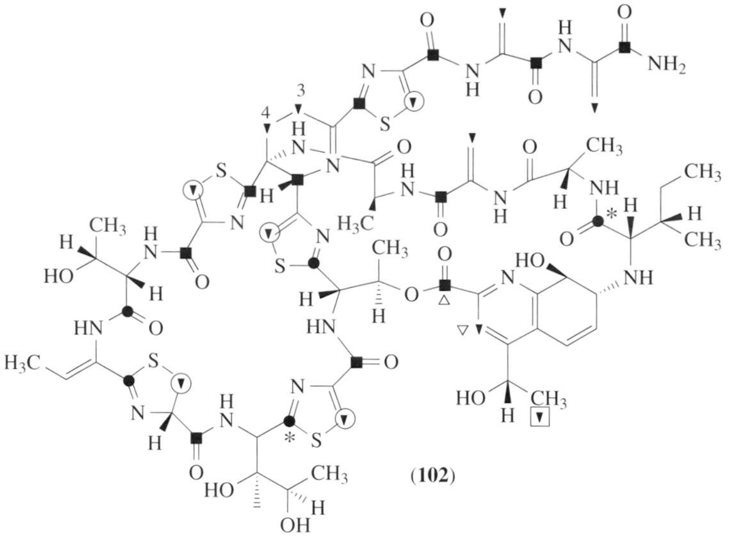

(102)

DL-[1-^{13}C]serine, ■ L-[*methyl*-^{13}C]methionine, □ L-[1'-^{13}C]tryptophan, △

DL-[3-^{13}C]serine, ▼ DL-[1-^{13}C]threonine, ● DL-2-methyl-[3'-^{13}C]tryptophan, ▽

DL-[3-^{13}C]cysteine, ○ DL-[1-^{13}C]isoleucine, *

Figure 3 Incorporation of labeled amino acids into thiostrepton. Reproduced by permission of the American Chemical Society from *J. Am. Chem. Soc.*, 1993, **115**, 7992.

single dihydrothiazole ring in thiostrepton retained both of the C-3 hydrogen atoms of serine. The last observation suggests that formation of the dihydrothiazole ring proceeds in a manner that is identical to that suggested for the first step of thiazole ring formation in microcin B17 (see Scheme 26) rather than proceeding via a thioaldehyde intermediate.

Investigations of the mechanism of formation of the indolic acid moiety of nosiheptide (**103**) (Scheme 27) have clearly demonstrated that this residue is derived from tryptophan (**81**), although the precise details of this interesting conversion remain unclear. An experiment with tryptophan doubly labeled with ^{13}C at C-2 of the indole ring and in the tryptophan carboxyl group established that the carboxyl group of the indolic acid moiety is derived by intramolecular transfer of the tryptophan carboxyl group to the C-2 position. An experiment with DL-[3′-^{13}C,3′-$^{2}H_2$]tryptophan showed that the C-3 methyl group of (**103**) is derived from C-3′ of tryptophan and retains both of the C-3′ hydrogen atoms of the amino acid. The hydroxymethyl group at C-4 of (**103**) was found to be derived from methionine. Various potential intermediates lying between tryptophan and (**103**) were evaluated by incorporation experiments. The highest degree of incorporation (65% enrichment) was exhibited by 3-methylindole-2-carboxylic acid (**104**), while 3,4-[4′-^{3}H]dimethylindole-2-carboxylic acid (**105**) gave an incorporation of 6.8%. Degradation of the nosiheptide obtained from the latter precursor showed that the labeling was specific (C-4′). 4-[4′-^{3}H]Hydroxymethyl-3-methylindole-2-carboxylic acid (**103**) was a relatively poor precursor (1.2% incorporation). On the basis of these and other experiments, the pathway from tryptophan to (**103**) shown in Scheme 27 was proposed. An enzyme that catalyzes the formation of an acyl adenylate from (**105**) and ATP has been detected in partially purified extracts of *S. actuosus*.[122]

Precursor incorporation experiments have demonstrated that the quinaldic acid moiety of thiostrepton (**106**) is derived from tryptophan and the *S*-methyl group of methionine (Scheme 28). DL-[1′-^{13}C]Tryptophan was found to label the carboxyl group of (**106**) (8.5% enrichment), while L-[*methyl*-^{13}C]methionine was incorporated into the methyl group of the hydroxyethyl sidechain (22.9% enrichment). Insight into the mechanism of formation of the quinoline ring system was obtained by means of an incorporation experiment with (*S*)-[1′,2′-^{13}C, *indole*-^{15}N]tryptophan. The quinaldic acid moiety derived from this precursor exhibited ^{13}C enrichment at the carboxyl group and at C-2, with ^{13}C–^{13}C coupling between the two adjacent labels. In addition, both ^{13}C signals exhibited coupling to ^{15}N, a result consistent with cleavage of the indole nucleus between the N1—C2 bond rather than between the N1—C7a bond (Scheme 28). The timing of the introduction of the

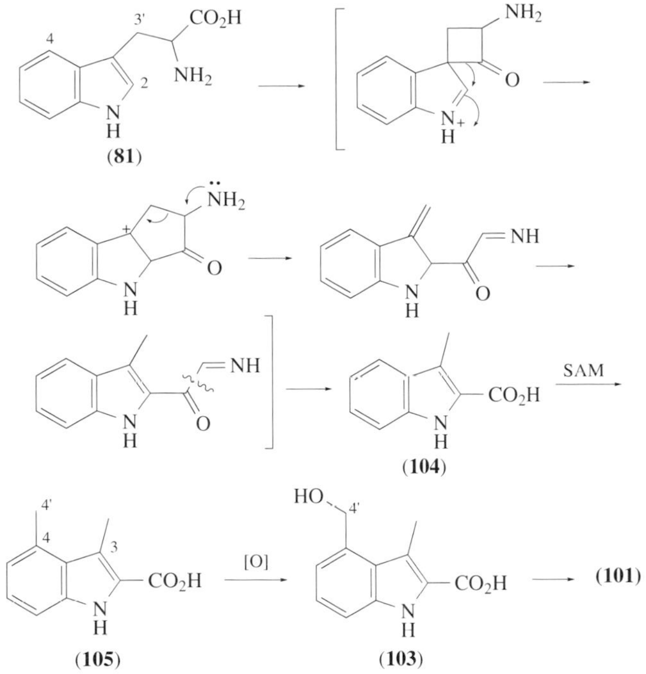

Scheme 27

methyl group during the biosynthesis of the quinaldic acid moiety was investigated by administration of DL-[3′-^{13}C]-2-methyltryptophan to *S. laurentii*. This precursor labeled C-3 of (**106**) with 40% enrichment, implying that methylation of tryptophan occurs prior to the opening of the indole ring. Additional evidence for the role of 2-methyltryptophan in thiostrepton biosynthesis was obtained by detection of this compound in butanol extracts of *S. laurentii* by GC-MS analysis. Administration of the methyl esters of [*carboxy*-^{13}C]-4-acetylquinoline-2-carboxylic acid (**107**) and (*RS*)-[*carboxy*-^{13}C]-4-(1-hydroxyethyl)quinoline-2-carboxylic acid (**108**) led to the formation of thiostrepton specifically labeled in the carboxyl group of the quinaldic acid moiety (ca. 29% and 81% enrichment, respectively).[123] On the basis of these results, the pathway for the conversion of tryptophan into (**106**) shown in Scheme 28 was formulated. An enzyme that catalyzes the activation of the carboxyl group of (**108**) by formation of an acyl adenylate from (**108**) and ATP has also been detected in cell-free extracts of *S. laurentii* and partially purified.[123]

An investigation of the steric course of the tryptophan methylation reaction leading to (**106**) using samples of methionine bearing chiral *S*-methyl groups gave a surprising result: the methylation at C-2 of tryptophan was found to proceed with overall retention of configuration. Extensive investigations of the stereochemistry of SAM-dependent methyltransferase reactions have shown that the majority of these reactions proceed with inversion of configuration through a transition-state that is presumed to be S_N2-like in character.[124] The simplest explanation for the observed stereochemistry of the tryptophan methylation process would be a mechanism that involves an even number of methyl transfers. A cell-free extract that catalyzes the conversion of L-tryptophan to L-2-methyltryptophan has been prepared from *S. laurentii*.[125] Unfortunately, the enzymatic activity proved to be unstable.

In most of the thiopeptide antibiotics, the carboxy terminus of the peptide chain exists as a carboxamide group. Experiments with ^{15}N-labeled precursors indicated that the terminal amide nitrogen atom of nosiheptide is likely to be derived from serine, although a possible derivation from glycine could not be completely ruled out. This suggests that the peptide precursors of nosiheptide may contain one or more additional serine residues at the carboxy terminus which are cleaved off by a reaction that resembles the peptidylglycine α-amidating monooxygenase/α-hydroxyglycine amidating dealkylase system.[126] A similar situation may obtain with thiostrepton, although proof is lacking due to the inconclusive results obtained from experiments with ^{15}N labeled precursors.

The amino acid building blocks for the thiopeptide antibiotics berninamycin and A10255 have also been established by precursor incorporation experiments.[118,119]

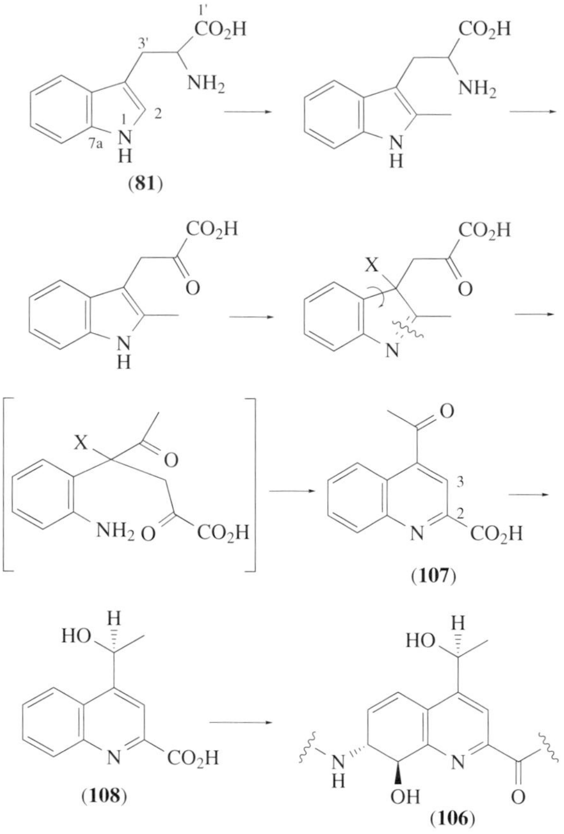

(81)

(107)

(108)

(106)

Scheme 28

1.29.4 MISCELLANEOUS

1.29.4.1 Caldariellaquinone

Caldariellaquinone (**111**) (Scheme 29) was first isolated from the extremely thermophilic and acidophilic bacterium *Caldariella acidophila* in 1977.[127] Subsequently, the compound was also found to be a major component of the quinone fraction of *Sulfolobus* spp.[128–130] In *Sulfolobus* spp., caldariellaquinone may play a role in electron transport reactions since menaquinone and ubiquinone do not occur in these bacteria.

The biosynthesis of the isoprenoid sidechain of caldariellaquinone has been examined by administration of [1-^{13}C]- and [2-^{13}C]acetate to *Caldariella acidophila*.[127] The label from these precursors was incorporated solely into the sidechain of caldariellaquinone with a labeling pattern consistent with the derivation of the sidechain from mevalonic acid. Administration of DL-[2-^{13}C]tyrosine and L-[*aromatic*-^{13}C$_6$]tyrosine to *Sulfolobus acidocaldarius* gave caldariellaquinone whose labeling pattern, as analyzed by mass spectrometry, indicated that the benzothiophene ring system of (**111**) is derived from tyrosine (**81**).[131] Additional information on the mode of tyrosine incorporation was obtained by administration of DL-[3-^{2}H$_2$]tyrosine to *S. solfataricus*. Mass spectrometric analysis of the caldariellaquinone revealed that one deuterium atom was incorporated into the benzothiophene ring system, presumably at C-3.[131]

The stereospecificity of hydrogen loss from C-3 of tyrosine was determined by precursor incorporation experiments with L-(3*S*)-[2-^{2}H$_1$,3-^{2}H$_1$]- and L-(3*R*)-[2-^{2}H$_1$,3-^{2}H$_1$]tyrosine and *S. acidocaldarius*. Mass spectral analysis of the caldariellaquinone derived from these two forms of labeled tyrosine revealed that only the 3-*pro-S* hydrogen atom of tyrosine is incorporated.[132] *S. acidocaldarius* was also able to incorporate deuterium efficiently into caldariellaquinone from D-(3*S*)-[3-^{2}H$_1$]tyrosine, a result that is consistent with the intermediacy of *p*-hydroxyphenylpyruvic acid (**109**) (Scheme 29) in caldariellaquinone biosynthesis. Administration of [U-^{2}H$_5$]homogentisic acid (**110**) to

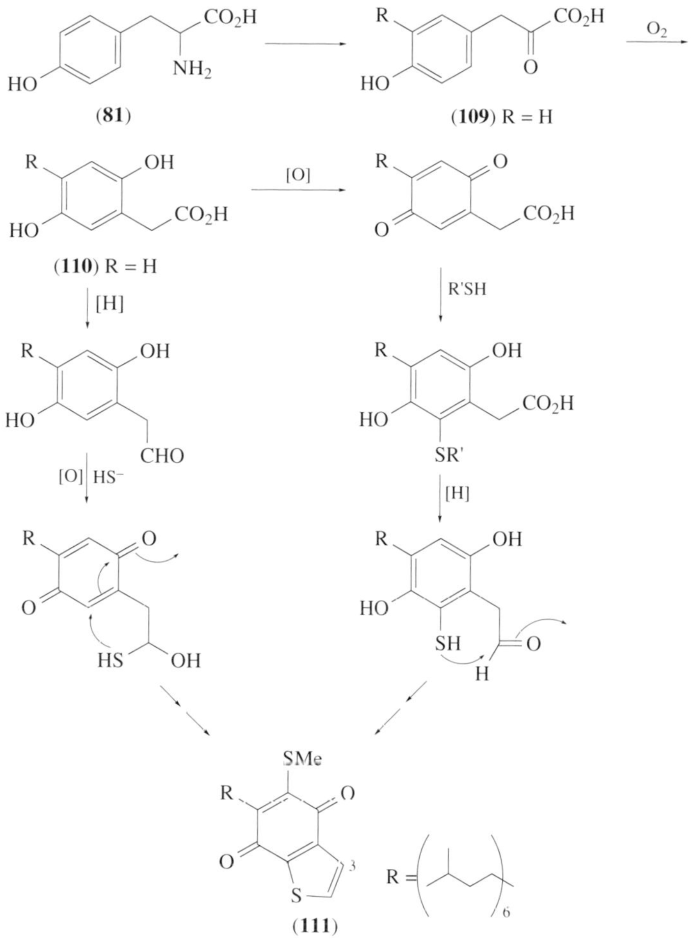

Scheme 29

S. acidocaldarius failed to yield labeled (**111**), suggesting that this compound is not an intermediate on the biosynthetic pathway. Similarly, DL-[3-$^2H_2$]-2′,5′-dihydroxyphenylalanine did not label caldariellaquinone when administered to *S. acidocaldarius*.

The origin of the methylthio group of (**111**) was examined by precursor incorporation experiments with labeled forms of methionine.[133] Administration of L-[*methyl*-$^2H_3$]methionine to *S. acidocaldarius* produced caldariellaquinone that appeared to contain a trideuterated methyl group, as judged by mass spectrometric analysis. The use of L-[^{35}S]methionine gave radioactive caldariellaquinone, but the relative extent of incorporation of the ^{35}S label into the methylthio group versus the benzothiophene ring was not determined. Administration of L-[^{34}S, *methyl*-$^2H_3$]methionine produced (**111**) that exhibited a mixture of labeling patterns. Approximately 21.5% of the molecules were labeled with ^{34}S, 12.6% were labeled with a CD_3 group, and 7.4% were labeled with both ^{34}S and a trideuterated methyl group. These observations suggest that the methylthio group of (**111**) is not derived by the intact incorporation of a methylthio moiety. However, an examination of the isotopic distribution in the cellular methionine revealed that a significant amount of demethylation and remethylation of the labeled methionine had occurred: 10.4% of the molecules contained ^{34}S only, while 21.2% contained both ^{34}S and a CD_3 group. This finding introduces some uncertainty into the interpretation of the labeling pattern in the caldariellaquinone derived from L-[^{34}S,*methyl*-$^2H_3$]methionine.

Since the foregoing results did not reveal the timing of the introduction of either the prenyl sidechain or the methylthio group of caldariellaquinone, a large number of pathways can be envisioned for the biosynthesis of this compound. Some of the pathways are outlined in Scheme 29.

1.29.4.2 Glucosinolates

The glucosinolates are a unique class of thioglucosides restricted to certain families of dicotyledenous plants including the *Brassicaceae*, *Resedaceae*, *Capparidaceae*, and *Moringaceae*. Their presence in all members of the *Brassicaceae* is of considerable economic significance since many members of this family are cultivated as a source of vegetables, condiments, fodder and forage. Although nearly 100 glucosinolates are known, they all contain the same structural unit (**112**) and differ primarily in the nature of the R group.[134,135,136] The structure proposed for the glucosinolate moiety has been confirmed by an X-ray analysis of allyl glucosinolate (sinigrin).[137] The glucosinolates were originally known as mustard oil glycosides since glucosinolate-containing plants contain an endogenous thioglucoside glucohydrolase called myrosinase whose action causes the formation of volatile isothiocyanates (**113**), thiocyanates (**114**), and nitriles (**115**) (Scheme 30).

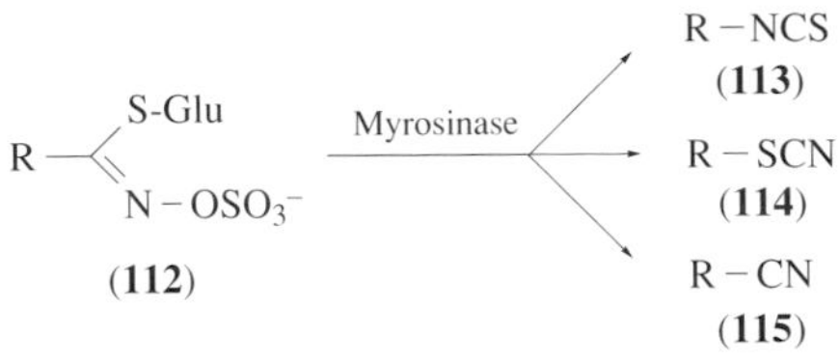

R–C(S-Glu)=N–OSO_3^- (**112**) —Myrosinase→ R–NCS (**113**); R–SCN (**114**); R–CN (**115**)

Scheme 30

In vivo biosynthetic investigations of the glucosinolates[138] using seedlings or excised tissues have revealed a general pathway (Scheme 31) that proceeds from an α-amino acid to an *N*-hydroxy-α-amino acid (**116**), which then undergoes oxidative decarboxylation to give the corresponding oxime (**117**). The oxime is next converted into a thiohydroxamate (**118**), which is glucosylated to yield the desulfoglucosinolate (**119**). The final step in the biosynthesis of the glucosinolate moiety appears to involve sulfation of the desulfoglucosinolate.

R–CH(NH_2)–CO_2H → R–CH(NHOH)–CO_2H (**116**) → R–CH=NOH (**117**) → R–C(SH)=NOH (**118**) → R–C(S-Glu)=NOH (**119**) → R–C(S-Glu)=N–OSO_3^- (**112**)

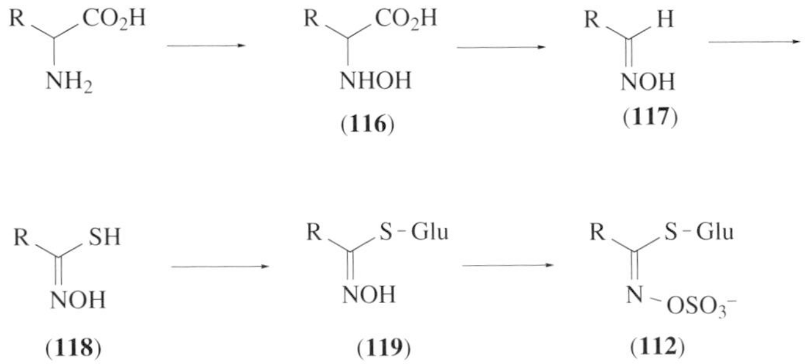

Scheme 31

Although the pathway shown in Scheme 31 has been defined for a number of years, information concerning the enzymology of glucosinolate biosynthesis has only become available since the mid-1990s. A microsomal enzyme system that catalyzes the conversion of L-tyrosine to *p*-hydroxyphenylacetaldehyde oxime has been isolated from jasmonic acid-induced seedlings of *Sinapis alba*.[139] The formation of oxime was strictly dependent on NADPH, and the reaction was inhibited by CO. This inhibition was photoreversible, a result which indicates that a cytochrome P450 system is involved in the oxidation. Several cytochrome P450 inhibitors also caused inhibition of the reaction, providing further support for the involvement of a P450 system in the oxidation. A similar P450 enzyme system has been reported to be involved in the biosynthesis of the cyanogenic glucoside dhurrin in *Sorghum bicolor*.[140] The reaction catalyzed by the *Sorghum* enzyme appears to proceed by two successive *N*-hydroxylations of L-tyrosine to give *N*-hydroxy-L-tyrosine, followed by *N*,*N*-dihydroxytyrosine, which then undergoes dehydration and decarboxylation to yield *p*-hydroxyphenylacetaldehyde oxime. Two microsomal aldoxime-forming enzyme systems have also been reported to be present in the leaves of *Brassica napus* cv Bienvenu (oilseed rape).[141] These enzymes catalyze the NADPH-dependent conversion of homophenylalanine and dihomomethionine to their respective aldoximes. Neither enzyme appears to have the characteristics of a cytochrome P450 system.

No information is available on the enzymology of the conversion of oximes into thiohydroxamates. Biosynthetic experiments indicate that both cysteine and methionine can serve as sulfur donors for glucosinolates,[136] but the nature of the immediate sulfur donor is unknown. Incorporation experiments do not support the intermediacy of a 1-thio sugar.[136]

More is known about the glucosylation step of glucosinolate biosynthesis. A UDP-glucose:thiohydroxamate glucosyltransferase has been purified to near homogeneity from *Brassica napus* L. seedlings.[142] The enzyme is monomeric with a molecular mass of 46 kDa. However, it shows multiple isoforms between pH 4.6 and 4.3 on isoelectric focusing. The glucosyltransferase exhibits a high degree of specificity with respect to the thiohydroxamic acid functional group, but shows little specificity for the sidechains associated with the thiohydroxamates. The enzyme was isolated and purified using the thiohydroxamate of phenylacetic acid ((**118**), R = benzyl) (Scheme 31) as a substrate even though *B. napus* plants do not produce benzyl glucosinolate. The *Brassica* glucosyltransferase requires the presence of sulfhydryl compounds for activity and is strongly inhibited by Cu^{2+} and Zn^{2+}. A similar enzyme has been partially purified from inflorescences of *Arabidopsis thaliana*[143] using the thiohydroxamate of phenylacetic acid as a substrate. This enzyme exhibited a native molecular mass of ca. 57.8 kDa. Its activity was stimulated by thiols and inhibited by Ni^{2+}, Co^{2+}, Zn^{2+} and Cu^{2+}. The substrate specificity was not examined.

Some information is also available on the enzymology of the final step in glucosinolate biosynthesis. An enzyme catalyzing the sulfation of desulfobenzylglucosinolate ((**119**), R = benzyl) (Scheme 31) has been partially purified from *Lepidium sativum* L. seedlings.[144] The enzyme utilizes 3′-phosphoadenosine 5′-phosphosulfate as the sulfate donor and has a pH optimum of 9.0. The activity of the enzyme is stimulated by thiols and by Mg^{2+} and Mn^{2+}. It is inhibited by sulfhydryl directed reagents and by Zn^{2+}. The enzyme also converts desulfoallylglucosinolate to allyl glucosinolate (sinigrin), but it is unable to utilize hydroxylated phenylpropanoids and flavonoids as substrates. An enzyme with similar properties has been partially purified from cell cultures of *Brassica juncea* (cv Cutlass).[145] In addition to sulfation activity, this enzyme displays thiohydroxamate glucosyltransferase activity at all stages of the purification. Since the copurifying enzymatic activities exhibit different pH and temperature stabilities, it appears that the dual activity may be due to the existence of an enzyme complex rather than to a single bifunctional protein.

A notable feature of the glucosinolates is the occurrence of homologous compounds whose side chains differ only in the number of methylene groups.[134] For example, glucosinolates with side chains that begin with a methylthio or methylsulfinyl group occur as a series of compounds with differing chain length (Figure 4).[146] Biosynthetic experiments indicate that these compounds are formed from the homologous series of amino acids. The homologous amino acids arise from α-keto acids by a chain elongation reaction sequence that is analogous to the one involved in the biosynthesis of leucine from valine.[146,147] A homologous series of glucosinolates whose sidechains begin with a vinyl group has also been isolated.[134] The simplest member of this series, allyl glucosinolate (sinigrin) (**123**) has been shown by precursor incorporation experiments in *Armoracia lapathifolia* Gilib. (horseradish) to be derived from homomethionine (**120**) via [3-(methylthio)propyl]glucosinolate (**121**) (Scheme 32).[148,149] [3-(Methylsulfinyl)propyl]glucosinolate (**122**) was also found to be a highly efficient precursor (Scheme 32). The conversion of (**121**) and (**122**) into (**123**) is an unusual biochemical reaction for which at least three possible mechanisms can be envisioned (Scheme 33). One mechanism would proceed by the β-elimination of methanethiol (pathway (a)). A more likely mechanism would involve the conversion of (**121**) into a sulfonium salt followed by β-elimination of a sulfide (pathway (b)). Lastly, a third mechanism (pathway (c)) would proceed by an enzyme-catalyzed pericyclic elimination of methanesulfenic acid from (**122**). In principle, the pericyclic pathway might be distinguished from the two alternative pathways since the pericyclic elimination must proceed by a *syn* elimination process. Precursor incorporation experiments to examine the question of the stereochemistry of the elimination reaction were carried out in *A. lapathifolia* using forms of homomethionine that were stereospecifically tritiated at C-4 or stereospecifically deuterated at C-5. The results of these experiments indicated that the 4-*pro-S* hydrogen atom is lost from homomethionine as a result of its conversion to allyl glucosinolate and that the conversion of homomethionine to allyl glucosinolate proceeds via an *anti* elimination process.[150]

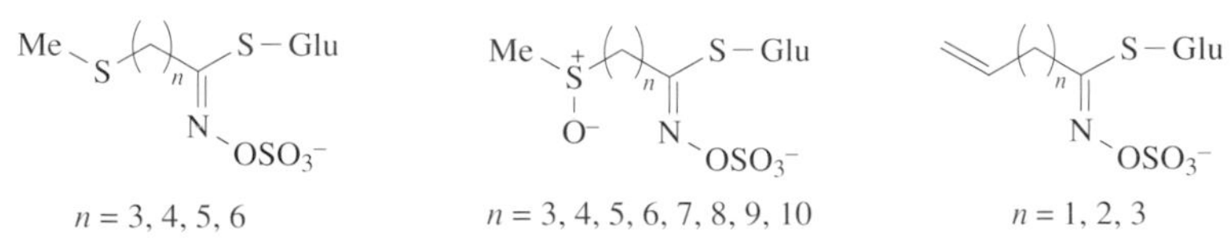

Figure 4 Some homologous series of glucosinolates.

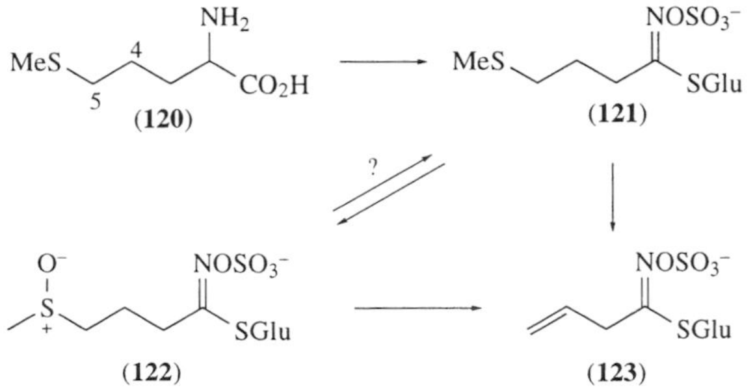

Scheme 32

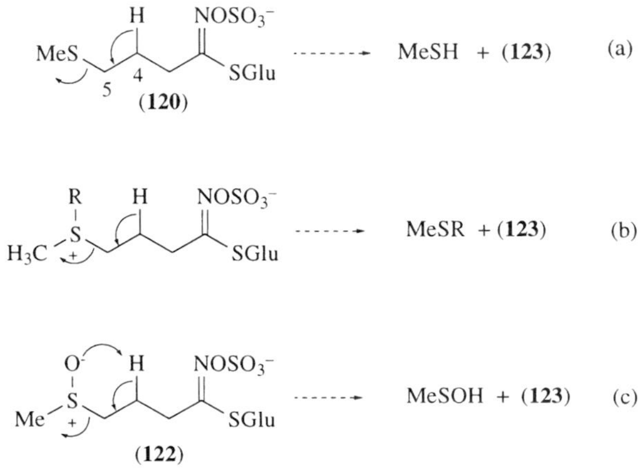

Scheme 33

1.29.5 CONCLUSION

It has surely become apparent to the reader of this review that a great deal remains to be learned concerning the biosynthesis of naturally occurring sulfur compounds. Two major areas of ignorance concern the nature of the sulfur donors and mechanisms of the sulfur introduction reactions. With respect to the latter, it appears that some patterns are beginning to emerge. The mechanism of the sulfur introduction reactions associated with biotin and lipoic acid biosynthesis may be related to each other and to the mechanism of carbon–sulfur bond formation associated with the biosynthesis of the penicillins and cephalosporins. The available evidence also suggests that the mechanism of the sulfur introduction processes associated with thiamine and molybdopterin biosynthesis are probably related to one another. It is likely that a number of additional mechanisms exist for the introduction of sulfur into biomolecules. For example, previous work from the author's laboratory[151] has shown that the sulfur introduction process associated with the biosynthesis of 1,2-dithiolane-4-carboxylic acid (asparagusic acid) takes place by the 1,4-addition of cysteine to an α,β-unsaturated acid. Because of the great structural diversity exhibited by naturally occurring sulfur compounds, it appears likely that investigations of the biosynthesis of this group of natural products will continue to provide challenges and surprises for many years to come.

ACKNOWLEDGMENT

The author would like to thank S. Jiralerspong for a critical reading of the manuscript.

1.29.6 REFERENCES

1. A Kjaer, *Pure Appl. Chem.*, 1977, **49**, 137.
2. J. Kyte, "Mechanism in Protein Chemistry," Garland, New York, 1995.

3. E. DeMoll, in "*Escherichia coli* and *Salmonella*. Cellular and Molecular Biology," ed. F. C. Neidhardt, ASM Press, Washington, DC, 1996, vol. 1, p. 704.
4. J. B. Perkins and J. G. Pero, in "*Bacillus subtilis* and Other Gram-positive Bacteria. Biochemistry, Physiology, and Molecular Genetics," eds. A. L. Sonenshien and J. A. Hoch, American Society for Microbiology, Washington, DC, 1993, p. 319.
5. I. Sanyal, S. L. Lee, and D. H. Flint, *J. Am. Chem. Soc.*, 1994, **116**, 2637.
6. O. Ploux, P. Soularue, A. Marquet, R. Gloeckler, and Y. Lemoine, *Biochem. J.*, 1992, **287**, 685.
7. O. Ploux and A. Marquet, *Eur. J. Biochem.*, 1996, **236**, 301.
8. G. L. Stoner and M. A. Eisenberg, *J. Biol. Chem.*, 1975, **250**, 4029.
9. G. L. Stoner and M. A. Eisenberg, *J. Biol. Chem.*, 1975, **250**, 4037.
10. E. DeMoll, R. H. White, and W. Shive, *Biochemistry*, 1984, **23**, 558.
11. K. Krell and M. A. Eisenberg, *J. Biol. Chem.*, 1970, **245**, 6558.
12. K. J. Gibson, G. H. Lorimer, A. R. Rendina, W. S. Taylor, G. Cohen, A. A. Gatenby, W. G. Payne, D. C. Roe, B. A. Lockett, A. Nudelman, D. Marcovici, A. Nachum, B. A. Wexler, E. L. Marsilii, I. M. Turner, L. D. Howe, C. E. Kalbach, and H. J. Chi, *Biochemistry*, 1995, **34**, 10 976.
13. W. Huang, J. Jia, K. J. Gibson, W. S. Taylor, A. R. Rendina, G. Schneider, and Y. Lindqvist, *Biochemistry*, 1995, **34**, 10 985.
14. R. L. Baxter, A. J. Ramsey, L. A. McIver, and H. C. Baxter, *J. Chem. Soc., Chem. Commun.*, 1994, 559.
15. R. L. Baxter and H. C. Baxter, *J. Chem. Soc., Chem. Commun.*, 1994, 759.
16. R. J. Parry, *Tetrahedron*, 1983, **39**, 1215.
17. F. B. Marti, Thesis ETH 7236, Eidgenössischen Technischen Hochschule Zurich, 1983.
18. F. Frappier, G. Guillerm, G. Salib, and A. Marquet, *Biochem. Biophys. Res. Commun.*, 1979, **91**, 521.
19. A. Marquet, F. Frappier, G. Guillerm, M. Azoulay, D. Florentin, and J. C. Tabet, *J. Am. Chem. Soc.*, 1993, **115**, 2139.
20. R. L. Baxter, D. J. Camp, A. Coutts, and N. Shaw, *J. Chem. Soc., Perkin Trans. 1*, 1992, 255.
21. A. Mejean, B. T. S. Bui, D. Florentin, O. Ploux, Y. Izumi, and A. Marquet, *Biochem. Biophys. Res. Commun.*, 1995, **217**, 1231.
22. I. Sanyal, G. Cohen, and D. H. Flint, *Biochemistry*, 1994, **33**, 3625.
23. O. Ifuku, N. Koga, S. Haze, J. Kishimoto, and Y. Wachi, *Eur. J. Biochem.*, 1994, **224**, 173.
24. I. Sanyal, K. J. Gibson, and D. H. Flint, *Arch. Biochem. Biophys*, 1996, **326**, 48.
25. X. Sun, J. Harder, K. Maria, H. Jornvall, B.-M. Sjoberg, and P. Reichard, *Proc. Natl. Acad. Sci. USA*, 1993, **90**, 577.
26. E. Mullicz, M. Fontecave, J. Gaillard, and P. Reichard, *J. Biol. Chem.*, 1993, **268**, 2296.
27. X. Sun, R. Eliasson, E. Pontis, J. Andersson, G. Buist, B.-M. Sjöberg, and P. Reichard, *J. Biol. Chem.*, 1995, **270**, 2443.
28. A. F. V. Wagner, M. Frey, F. A. Neugebauer, W. Schäfer, and J. Knappe, *Proc. Natl. Acad. Sci. USA*, 1992, **89**, 996.
29. K. K. Wong, B. W. Murray, S. A. Lewisch, M. K. Baxter, T. W. Ridky, L. Ulissi-DeMario, and J. W. Kozarich, *Biochemistry*, 1993, **32**, 14 102.
30. E. Jestin, F. Moreau, D. Florentin, and A. Marquet, *Bioorg. Med. Chem.*, 1996, **4**, 1065.
31. L. J. Reed, in "Comprehensive Biochemistry," eds. M. Florkin and E. Stotz, Elsevier, New York, 1966, vol. 14, p. 99.
32. L. J. Reed, *Acc. Chem. Res.*, 1974, **7**, 40.
33. U. Schmidt, P. Graffen, K. Atland, and H. W. Goedde, *Adv. Enzymol.*, 1969, **32**, 423.
34. K. Fujiwara, K. Okamura-Ikeda, and Y. Motokawa, *J. Biol. Chem.*, 1986, **261**, 8836.
35. D. Arigoni, personal communication.
36. T. J. V. Boom, K. Reed, and J. J. E. Cronan, *J. Bacteriol.*, 1991, **173**, 6411.
37. M. A. Hayden, I. Huang, D. E. Bussiere, and G. W. Ashley, *J. Biol. Chem.*, 1992, **267**, 9512.
38. M. A. Hayden, I. Y. Huang, G. Iliopoulos, M. Orozco, and G. W. Ashley, *Biochemistry*, 1993, **32**, 3778.
39. K. E. Reed and J. E. Cronan, *J. Bacteriol.*, 1993, **175**, 1325.
40. T. W. Morris, K. E. Reed, and J. E. Cronan, *J. Bacteriol.*, 1995, **177**, 1.
41. C. D. Taylor and R. S. Wolfe, *J. Biol. Chem.*, 1974, **249**, 4879.
42. J. G. Ferry (ed.), "Methanogenesis (Ecology, Physiology, Biochemistry, Genetics)," Chapman & Hall, New York, 1993.
43. R. H. White, *Biochemistry*, 1985, **24**, 6487.
44. R. H. White, *Biochemistry*, 1986, **25**, 5304.
45. R. H. White, *Biochemistry*, 1988, **27**, 7458.
46. L. Bhatnagar, M. Henriquet, J. G. Zeikus, and J.-P. Aubert, *FEMS Microbiol. Lett.*, 1984, **22**, 155.
47. R. H. White, *Biochemistry*, 1989, **28**, 860.
48. R. H. White, *Biochemistry*, 1994, **33**, 7077.
49. W. L. Jones, J. D. P. Nagle, and W. B. Whitman, *Microbiol. Rev.*, 1987, **51**, 135.
50. R. H. White, *Arch. Biochem. Biophys.*, 1989, **270**, 691.
51. K. V. Rajagopalan and J. L. Johnson, *J. Biol. Chem.*, 1992, **267**, 10 199.
52. K. V. Rajagopalan, in "*Escherichia coli* and *Salmonella*. Cellular and Molecular Biology," ed. F. C. Neidhardt, ASM Press, Washington, DC, 1996, vol. 1, p. 674.
53. R. L. White and I. D. Spenser, in "*Escherichia coli* and *Salmonella*. Cellular and Molecular Biology," ed. F. C. Neidhardt, ASM Press, Washington, DC, 1996, vol. 1, p. 680.
54. J. E. Baldwin and E. Abraham, *Nat. Prod. Rep.*, 1988, 129.
55. J. E. Baldwin and M. Bradley, *Chem. Rev.*, 1990, **90**, 1079.
56. Y. Aharonowitz, G. Cohen, and J. F. Martin, *Annu. Rev. Microbiol.*, 1992, **46**, 461.
57. Y. Aharonowitz, J. Bergmeyer, J. S. Cantoral, G. Cohen, A. L. Demain, U. Fink, J. Kinghorn, H. Kleinkauf, A. MacCabe, H. Palissa, E. Pfiefer, T. Schwecke, H. Liempt, H. v. Döhren, S. Wolfe, and J. Zhang, *Bio/technology*, 1993, **11**, 807.
58. H. Kleinkauf and H. v. Döhren, *Eur. J. Biochem.*, 1990, **192**, 1.
59. K. Turgay, M. Krause, and M. A. Marahiel, *Mol. Microbiol.*, 1992, **6**, 529.
60. T. Stachelhaus, and M. A. Marahiel, *J. Biol. Chem.*, 1995, **270**, 6163.
61. T. Stein, J. Vater, V. Kruft, A. Otto, B. Wittmann-Liebold, P. Franke, M. Panico, R. McDowell, and H. R. Morris, *J. Biol. Chem.*, 1996, **271**, 15 428.

62. J. E. Baldwin, C. Y. Shiau, M. F. Byford, and C. J. Schofield, *Biochem. J.*, 1994, **301**, 367.
63. J. E. Baldwin, M. F. Byford, R. A. Field, C. Y. Shiau, W. J. Sobey, and C. J. Schofield, *Tetrahedron*, 1993, **49**, 3221.
64. J. Zhang, S. Wolfe, and A. L. Demain, *Biochem. J.*, 1992, **283**, 691.
65. C. Y. Shiau, J. E. Baldwin, M. F. Byford, W. J. Sobey, and C. J. Schofield, *FEBS Lett.*, 1995, **358**, 97.
66. C. Y. Shiau, J. E. Baldwin, M. F. Byford, and C. J. Schofield, *FEBS Lett.*, 1995, **373**, 303.
67. C.-P. Pang, B. Chakravarti, R. M. Adlington, H.-H. Ting, R. L. White, G. S. Jayatilake, J. E. Baldwin, and E. P. Abraham, *Biochem. J.*, 1984, **222**, 789.
68. J. M. Castro, P. Liras, L. Liaz, J. Cortes, and J. F. Martin, *J. Gen. Microbiol.*, 1988, **134**, 133.
69. H. Palissa, H. v. Döhren, H. Kleinkauf, H. Ting, and J. E. Baldwin, *J. Bacteriol.*, 1989, **171**, 5720.
70. S. M. Samson, R. Belagaje, D. T. Blankenship, J. L. Chapman, P. L. Skatrud, R. M. VanFrank, E. P. Abraham, J. E. Baldwin, S. W. Queener, and T. D. Ingolia, *Nature* (*London*), 1985, **318**, 191.
71. L. G. Carr, P. L. Skatrud, R. M. VanFrank, E. Abraham, J. E. Baldwin, S. W. Queener, and T. D. Ingolia, *Gene*, 1986, **48**, 257.
72. A Kriauciunas, C. A. Frolik, T. C. Hassell, P. L. Skatrud, M. G. Johnson, N. L. Holbrook, and V. J. Chen, *J. Biol. Chem.*, 1991, **266**, 11 779.
73. J. E. Baldwin, J. M. Blackburn, J. D. Sutherland, and M. C. Wright, *Tetrahedron*, 1991, **47**, 5991.
74. M. Duraijai and S. E. Jensen, *J. Ind. Microbiol.*, 1996, **16**, 197.
75. G. W. Huffman, P. D. Gesellchen, J. R. Turner, R. B. Rothenberger, H. E. Osborne, F. D. Miller, J. L. Chapman, and S. W. Queener, *J. Med. Chem.*, 1992, **35**, 1897.
76. J. E. Baldwin, G. P. Lynch, and C. J. Schofield, *Tetrahedron*, 1992, **48**, 9085.
77. J. M. Blackburn, J. D. Sutherland, and J. E. Baldwin, *Biochemistry*, 1995, **34**, 7548.
78. J. E. Baldwin, R. M. Adlington, D. G. Marquess, A. R. Pitt, M. J. Porter, and A. T. Russell, *Tetrahedron*, 1996, **52**, 2537.
79. V. J. Chen, A. M. Orville, M. R. Harpel, C. A. Frolik, K. K. Surerus, E. Minck, and J. D. Lipscomb, *J. Biol. Chem.*, 1989, **264**, 21 677.
80. F. Jiang, J. Peisach, L.-J. Ming, J. L. Que, and V. J. Chen, *Biochemistry*, 1991, **30**, 11 437.
81. L.-J. Ming, J. L. Que, A. Kriauciunas, C. A. Frolik, and V. J. Chen, *Biochemistry*, 1991, **30**, 11 653.
82. R. A. Scott, S. Wang, M. K. Eidsness, A. Kriauciunas, C. A. Frolik, and V. J. Chen, *Biochemistry*, 1992, **31**, 4596.
83. A. M. Orville, V. J. Chen, A. Kriauciunas, M. R. Harpel, B. G. Fox, E. Munck, and J. D. Lipscomb, *Biochemistry*, 1992, **31**, 4602.
84. P. L. Roach, I. J. Clifton, V. Fulop, K. Harlos, G. J. Barton, J. Hajdu, I. Andersson, C. J. Schofield, and J. E. Baldwin, *Nature* (*London*), 1995, **375**, 700.
85. S. Usui and C.-A. Yu, *Biochem. Biophys. Acta.*, 1989, **999**, 78.
86. L. Laiz, P. Liras, J. M. Castro, and J. F. Martin, *J. Gen. Microbiol.*, 1990, **136**, 663.
87. M. Kohsaka and A. L. Demain, *Biochem. Biophys. Res. Commun.*, 1976, **70**, 465.
88. M. Yoshida, T. Konomi, M. Kohsaka, J. E. Baldwin, S. Herchen, P. Singh, N. A. Hunt, and A. L. Demain, *Proc. Natl. Acad. Sci. USA*, 1978, **75**, 6253.
89. S. E. Jensen, D. W. S. Westlake, and S. Wolfe, *J. Antiobiot.*, 1985, **38**, 263.
90. J. E. Dotzlaf and W.-K. Yeh, *J. Biol. Chem.*, 1989, **264**, 10 219.
91. B. J. Baker, J. E. Dotzlaf, and W.-K. Yeh, *J. Biol. Chem.*, 1991, **266**, 5087.
92. J. Cortes, J. F. Martin, J. M. Castro, L. Laiz, and P. Liras, *J. Gen. Microbiol.*, 1987, **133**, 3165.
93. W. K. Yeh, S. K. Ghag, and S. W. Queener, *Ann. N. Y. Acad. Sci.*, 1992, **672**, 396.
94. J. E. Baldwin, J. M. Blackburn, R. J. Heath, and J. D. Sutherland, *Bioorg. Med. Chem. Lett.*, 1992, **2**, 663.
95. N. Morgan, I. A. C. Periera, I. A. Andersson, R. M. Adlington, J. E. Baldwin, S. C. J. Cole, N. P. Crouch, and J. D. Sutherland, *Bioorg. Med. Chem. Lett.*, 1994, **4**, 1595.
96. J. E. Baldwin, R. M. Adlington, N. P. Crouch, L. C. Mellor, N. Morgan, A. M. Smith, and J. D. Sutherland, *Tetrahedron*, 1995, **51**, 4089.
97. C. A. Townsend and E. B. Barrabee, *J. Chem. Soc.*, *Chem. Commun.*, 1984, 1586.
98. J. E. Baldwin, R. M. Adlington, N. P. Crouch, and C. J. Schofield, *Tetrahedron*, 1988, **4**, 643.
99. J. E. Baldwin, R. M. Adlington, N. P. Crouch, C. J. Schofield, N. J. Turner, and R. T. Aplin, *Tetrahedron*, 1991, **47**, 9881.
100. J. E. Baldwin, R. M. Adlington, N. P. Crouch, and I. A. C. Pereira, *Tetrahedron*, 1993, **49**, 4907.
101. J. E. Baldwin, R. M. Adlington, N. P. Crouch, and I. A. C. Pereira, *Tetrahedron*, 1993, **49**, 7499.
102. B. Siegel, *Bioorg. Chem.*, 1979, **8**, 219.
103. A. D. Argoudelis and R. R. Herr, *Antimicrob. Agents Chemother.*, 1962, 780.
104. E. Higashide, T. Hasegawa, M. Shibata, K. Mizuno, and H. Akaike, *Takeda Kenkyusho Nempo*, 1966, **25**, 1.
105. S. P. Owen, A. Dietz, and G. W. Camiener, *Antimicrob. Agents Chemother.*, 1962, 772.
106. E. Lazaro, C. Rodriguez-Fonseca, B. Porse, D. Urena, R. A. Garrett, and J. P. G. Ballesta, *J. Mol. Biol.*, 1996, **261**, 231.
107. G. T. Tan, A. DeBlasio, and A. S. Mankin, *J. Mol. Biol.*, 1996, **261**, 222.
108. R. J. Parry, Y. Li, and E. E. Gomez, *J. Am. Chem. Soc.*, 1992, **114**, 5946.
109. R. J. Parry, J. C. Hoyt, and Y. Li, *Tetrahedron Lett.*, 1994, **35**, 7497.
110. R. J. Parry and J. C. Hoyt, *J. Bacteriol.*, 1997, **179**, 1385.
111. R. J. Parry and Y. Li, unpublished observations.
112. K. Kintaka, H. Ono, S. Tsubotani, S. Harada, and H. Okazaki, *J. Antibiot.*, 1984, **37**, 1294.
113. S. Tsubotani, Y. Wada, K. Kamiya, H. Okazaki, and S. Harada, *Tetrahedron Lett.*, 1984, **25**, 419.
114. D. E. Cane, Z. Wu, and J. E. Vanepp, *J. Am. Chem. Soc.*, 1992, **114**, 8479.
115. P. Yorgey, J. Davagnino, and R. Kolter, *Mol. Microbiol.*, 1993, **9**, 897.
116. Y.-M. Li, J. C. Milne, L. T. Madison, R. Kolter, and C. T. Walsh, *Science*, 1996, **274**, 1188.
117. W. R. Strohl and H. G. Floss, in "Genetics and Biochemistry of Antibiotic Production," eds. L. C. Vining and C. Stuttard, Butterworth-Heinemann, Newton, MA, 1995, p. 223.
118. R. C. M. Lau and K. L. Rinehart, *J. Am. Chem. Soc.*, 1995, **117**, 7606.
119. M. E. Favret, J. W. Paschal, T. K. Elzey, and L. D. Boeck, *J. Antibiot.*, 1992, **45**, 1499.

120. U. Mocek, Z. P. Zeng, D. O'Hagan, P. Zhou, L. D. G. Fan, J. M. Beale, and H. G. Floss, *J. Am. Chem. Soc.*, 1993, **115**, 7992.
121. U. Mocek, A. R. Knaggs, R. Tsuchiya, T. Nguyen, J. M. Beale, and H. G. Floss, *J. Am. Chem. Soc.*, 1993, **115**, 7557.
122. T. M. Smith, N. D. Priestley, A. R. Knaggs, T. Nguyen, and H. G. Floss, *Chem. Commun.*, 1993, 1612.
123. N. D. Priestley, T. M. Smith, P. R. Shipley, and H. G. Floss, *Bioorg. Med. Chem.*, 1996, **4**, 1135.
124. H. G. Floss and S. Lee, *Acc. Chem. Res.*, 1993, **26**, 116.
125. T. Frenzel, P. Zhou, and H. G. Floss, *Arch. Biochem. Biophys.*, 1990, **278**, 35.
126. A. G. Katapodis, D. Ping, and S. W. May, *Biochemistry*, 1990, **29**, 6115.
127. M. D. Rosa, S. D. Rosa, A. Gambacorta, L. Minale, R. H. Thomson, and R. D. Worthington, *J. Chem. Soc., Perkin Trans. 1*, 1977, 653.
128. M. D. Collins and T. A. Langworthy, *Syst. Appl. Microbiol.*, 1983, **4**, 295.
129. V. Lanzotti, A. Trincone, M. Gambacorta, M. D. Rosa, and E. Breitmaier, *Eur. J. Biochem.*, 1986, **160**, 37.
130. S. Thurl, W. Witke, I. Buhrow, and W. Schafer, *Biol. Chem. Hoppe-Seyler*, 1986, **367**, 191.
131. D. Zhou and R. H. White, *J. Bacteriol.*, 1989, **171**, 6610.
132. D. Zhou and R. H. White, *J. Chem. Soc., Perkin Trans. 1*, 1991, 1335.
133. D. Zhou and R. H. White, *J. Chem. Soc., Perkin Trans. 1*, 1990, 2346.
134. A. Kjaer in "Progress in the Chemistry of Organic Natural Products," ed. L. Zechmeister, Springer-Verlag, Vienna, 1960, p. 122.
135. A. Kjaer in "Chemistry in Botanical Classification," Nobel Symposium 25, Academic Press, New York, 1974, p. 229.
136. J. E. Poulton and B. C. Møller in "Methods in Plant Biochemistry," ed. P. J. Lea, Academic Press, New York, 1993, vol. 9, p. 209.
137. R. E. Marsh and J. Waser, *Acta Crystallogr. Sect. B*, 1970, **26**, 1030.
138. P. O. Larsen in "The Biochemistry of Plants," ed. E. E. Conn, Academic Press, New York, 1981, vol. 7, p. 502.
139. L. Du, J. Lykkesfeldt, C. E. Olsen, and B. A. Halkier, *Proc. Natl. Acad. Sci. USA*, 1995, **92**, 12 505.
140. O. Sibbesen, B. Koch, B. A. Halkier, and B. L. Møller, *J. Biol. Chem.*, 1995, **270**, 3506.
141. R. N. Bennett, A. J. Hick, G. W. Dawson, and R. M. Wallsgrove, *Plant Physiol.*, 1995, **109**, 299.
142. D. W. Reed, L. Davin, J. C. Jain, V. Deluca, L. Nelson, and E. W. Underhill, *Arch. Biochem. Biophys.*, 1993, **305**, 526.
143. L. Guo and J. E. Poulton, *Phytochemistry*, 1994, **36**, 1133.
144. T. M. Glendening and J. E. Poulton, *Plant Physiol.*, 1988, **86**, 319.
145. J. C. Jain, J. W. D. GrootWassink, A. D. Kolenovsky, and E. W. Underhill, *Phytochemistry*, **29**, 1425.
146. M. D. Chisolm, *Phytochemistry*, 1972, **11**, 197.
147. G. W. Dawson, A. J. Hick, R. N. Bennett, A. Donald, J. A. Pickett, and R. M. Wallsgrove, *J. Biol. Chem.*, 1993, **268**, 27 154.
148. M. Matsuo and M. Yamazaki, *Chem. Pharm. Bull.*, 1968, **16**, 1034.
149. M. D. Chisolm and M. Matsuo, *Phytochemistry*, 1972, **11**, 203.
150. R. J. Parry and M. V. Naidu, *J. Am. Chem. Soc.*, 1982, **104**, 3217.
151. R. J. Parry, A. E. Mizusawa, I. C. Chiu, M. V. Naidu, and M. Ricciardone, *J. Am. Chem. Soc.*, 1985, **107**, 2512.

1.30
Biosynthesis of the Natural C—P Compounds, Bialaphos and Fosfomycin

HARUO SETO
The University of Tokyo, Japan

1.30.1 INTRODUCTION

Several antibiotics and physiologically active substances containing unique C—P bond(s) have been isolated mainly as metabolites of *Streptomyces* (Figure 1). These include bialaphos (BA),[1] which is identical to phosphinothricylalanylalanine[2] and related to phosalacine,[3] fosfomycin (FM),[4] FR-33289,[5] plumbemycin,[6] fosmidomycin (FR-31564),[7] fosfazinomycin,[8] and K-26.[9] More recently, a new member of this group, phosphonothrixin, has been added as a herbicide produced by *Saccharothrix* sp.[10] and its stereochemistry determined by chemical synthesis.[11] Since the discovery of the first natural C—P compound, 2-aminoethylphosphonic acid (AEP),[12] which was produced by *Tetrahymena pyriformis*, the biosyntheses of these compounds, in particular the formation mechanism of the C—P bond, have attracted considerable attention.[13] Among these metabolites, BA and FM are being used mainly in Japan as a herbicide and an antibacterial antibiotic, respectively, and therefore, their biosynthetic studies are considered to be important to improve the production yield of these metabolites by fermentation. FM is commercially prepared in a racemic form by chemical synthesis. Improvement of the production yield of FM by several times is expected to enable commercial production of this antibiotic by fermentation.

In this chapter, the detailed biosynthetic pathways of these two metabolites will be explained based mainly on the results obtained in the author's laboratory.

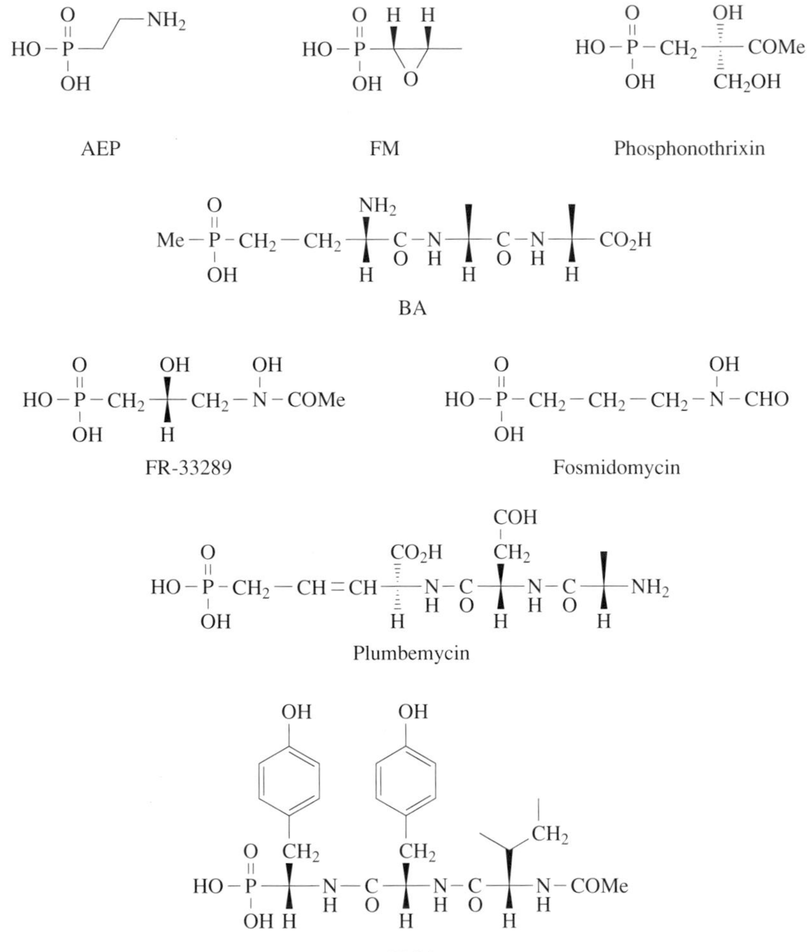

AEP FM Phosphonothrixin

BA

FR-33289 Fosmidomycin

Plumbemycin

K-26

Figure 1 Representative natural C—P compounds.

1.30.2 BIOSYNTHESIS OF BIALAPHOS

BA is a herbicide produced by *Streptomyces hygroscopicus* consisting of two L-alanine residues and an unusual amino acid, phosphinothricin (PT)[1,2] (see Scheme 1). Its analogue, phosalacine[3] showed the same biological activity. PT is an analogue of glutamate which contains a methylated phosphino group in the position of the ω-carboxyl group of glutamate. This unique structure of BA poses profound questions as to the mechanism of formation of the C—P—C bond. Therefore, interest in the BA biosynthetic pathway has focused mostly on the reaction mechanisms and enzymes needed for the formation of C—P bonds. Extensive studies on the biosynthesis of BA prove that PT is synthesized via a pathway (see Scheme 4, later) consisting of at least 14 steps (presumably more than 20 steps) which were revealed by chemical and biochemical methods, especially by the use of a series of well characterized mutant strains of *S. hygroscopicus* defective for various reaction steps.

1.30.2.1 Origin of the Carbon Skeleton

Since BA has a unique methylated phosphinic function, initial studies were directed to reveal the origin of the PT carbon skeleton[14] (Scheme 1). The structure of PT suggested that it might be derived from a common α-amino acid having a relevant carbon skeleton such as methionine and glutamic acid. However, these expected candidates were not incorporated into the PT skeleton, except for the methyl group of methionine.

In contrast, [1,2-$^{13}C_2$]acetate was incorporated selectively into C-1 and C-2 positions of PT. The reported biosynthetic pathway of related compounds having phosphonate groups, including AEP,[15] FM,[16] and FR-33289[17] (Figure 1), in which the phosphonate residue and its adjacent two-carbon unit originated from phosphoenolpyruvate (PEP), strongly suggested that the remaining —CH_2—CH_2—P unit in the PT molecule derived from the same precursor. In agreement with this expectation, the ^{13}C NMR spectrum of BA labeled with [U-$^{13}C_6$]glucose which was used in place of PEP, showed *inter alia* a ^{13}C—^{13}C coupling between C-4 and C-3. This incorporation pattern of the labeled glucose strongly indicated the involvement of PEP for the formation of the two-carbon units adjacent to a phosphinic acid moiety of BA as were reported for C—P compounds such as AEP,[15] FM,[16] and FR-33289.[17] The formation of the C—P bond is believed to proceed via intramolecular rearrangement of PEP to phosphonopyruvate (PnPy)[14] (see Scheme 2, later). The C—P bond formation mechanism of PT, however, turned out to involve more complicated reactions (see Section 1.30.2.3.3). The origin of the *P*-methyl group was shown by the selective incorporation of the CD_3 residue of [CD_3]methionine into C-5 of PT. These results suggested the origins of PT carbon residues as shown in Scheme 1.[14]

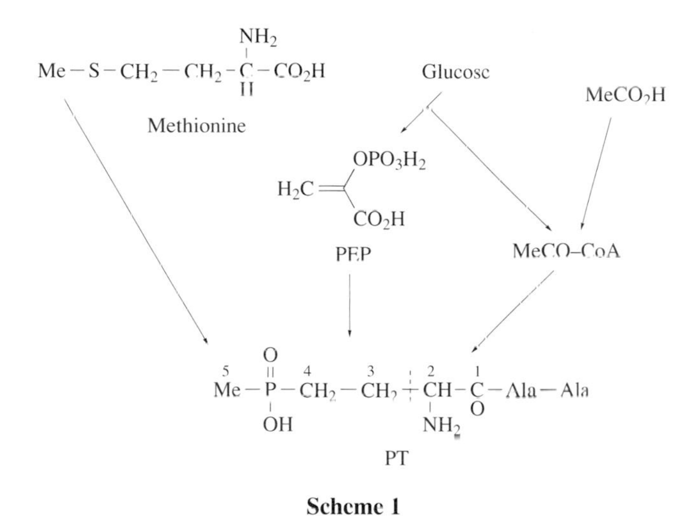

Scheme 1

1.30.2.2 Identification of Biosynthetic Intermediates

At the beginning of this biosynthetic work, it was known that Co^{2+} played an important role in improvement of the production yield of BA.[18] In order to clarify the effect of this ion on the biosynthesis of BA, *S. hygroscopicus* was cultivated in the absence of Co^{2+}. Direct analysis of the fermentation broth[19] by ^{31}P NMR revealed the presence of two new signals characteristic of —C—P(=O)OH structures[20] which were never observed with samples prepared under the normal fermentation conditions.

NMR spectral analysis of the compounds showing these two signals defined their structures as demethylbialaphos (DMBA), and demethylphosphinothricin (DMPT)[19] (Figure 2). The unique structural feature of these two compounds is the nonmethylated phosphinate residue, not found in any other natural compounds.

Both the demethyl compounds can be quantitatively converted to BA by incubation with washed mycelia of a mutant of *S. hygroscopicus* blocked at an early step of the biosynthetic pathway. This result clearly shows that the reduction of phosphonic acid to phosphinic acid is a prerequisite for *P*-methylation[19] and that the methylation takes place at the later or final stage of the biosynthetic pathway.

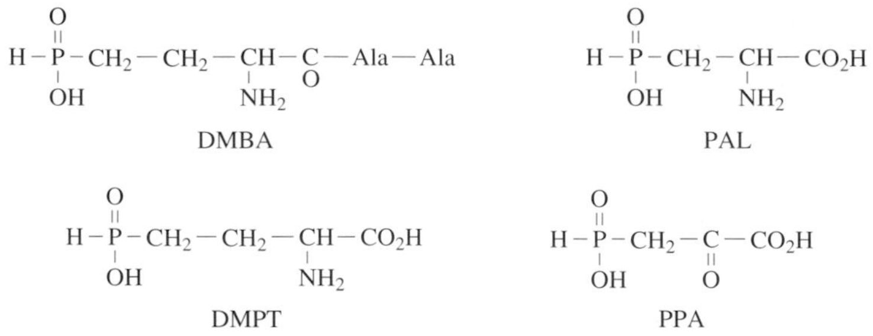

$H{-}P(=O)(OH){-}CH_2{-}CH_2{-}CH(NH_2){-}C(=O){-}Ala{-}Ala$

DMBA

$H{-}P(=O)(OH){-}CH_2{-}CH(NH_2){-}CO_2H$

PAL

$H{-}P(=O)(OH){-}CH_2{-}CH_2{-}CH(NH_2){-}CO_2H$

DMPT

$H{-}P(=O)(OH){-}CH_2{-}C(=O){-}CO_2H$

PPA

Figure 2 Biosynthetic intermediates of BA with a P—H bond.

DMPT and DMBA also accumulated in the fermentation broth of vitamin B_{12} auxotrophic mutants derived from *S. hygroscopicus*. These results indicate that *P*-methylation is mediated by an enzyme which uses vitamin B_{12} as a cofactor (see Section 1.30.2.3.4). *N*-Acetyl derivatives of these two metabolites were also accumulated by a mutant NP45,[21] which could not catalyze the *P*-methylation reaction (see Section 1.30.2.3.4).

In addition to these two biosynthetic intermediates, analysis of the fermentation broth of a mutant identified a key intermediate to reveal the formation mechanism of the phosphinate function in BA. A blocked mutant of the BA producer (NP44) accumulated a derivative with a nonmethylated phosphinate residue, phosphinoalanine (PAL) which is regarded as a biological equivalent of phosphinopyruvate (PPA)[22] (Figure 2). PAL can be converted to BA by several mutants which are blocked at earlier steps of the biosynthetic pathway. Since this reaction proceeds quantitatively, PPA is concluded to be a biosynthetic intermediate which accumulates as a more stable amino acid, PAL. PPA is a reduced form of PnPy (Scheme 2, double vertical lines show blocked points of mutants) which was believed to be the earliest precursor for the biosynthesis of all C—P compounds.[13]

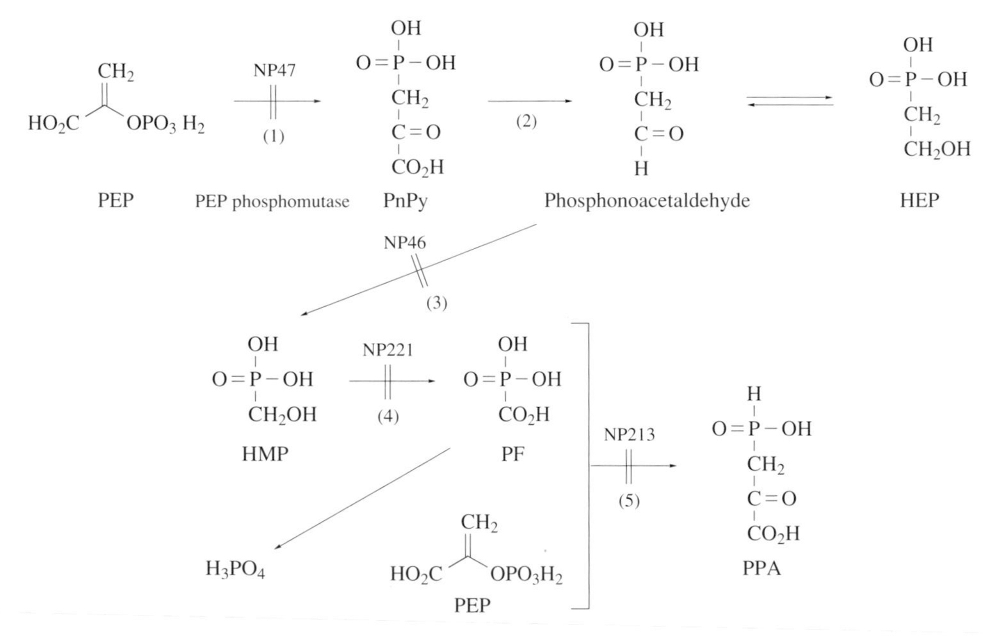

Scheme 2

1.30.2.3 Formation Reactions of C—P Bonds

1.30.2.3.1 Formation of the first C—P bond catalyzed by PEP phosphomutase

Labeling experiments indicate that the C—P bonds of BA,[15] FM,[16] and FR-33289[17] are formed by intramolecular rearrangement of PEP to PnPy (step 1 in Scheme 2) via the mechanism proposed for the formation of AEP in *T. pyriformis*.[15] Curiously, however, extensive attempts to detect enzyme activities which catalyze C—P bond formation in cell-free systems of various organisms were uniformly unsuccessful. The reason for this failure was clarified by a finding that the equilibrium between PEP and PnPy favors the formation of the phosphate ester (in the ratio of more than 500:1).[23–25] This result was quite unexpected, since it had been generally believed that the C—P bond was far more stable than the C—O—P bond[13] and that the forward reaction should proceed irreversibly. In fact, vigorous acid pretreatment to convert usual phosphate esters to phosphoric acid was employed as a procedure for quantitative analysis of C—P compounds.

This reverse reaction forming PEP was used for purification of the enzyme PEP phosphomutase from *T. pyriformis*,[23] *S. hygroscopicus*,[24] and *Pseudomonas gladioli*[26] as well as for screening of C—P compound producing organisms.[27] However, the failure to demonstrate the forward reaction in C—P compound producing organisms left open the question as to whether the enzyme which catalyzed the reverse reaction was truly involved in BA biosynthesis. Two blocked mutants defective in step 1 were independently isolated to address this question.[24] One of them, NP47 was obtained by conventional *N*-methyl-*N'*-nitro-*N*-nitrosoguanidine (NTG) treatment of a parent strain and the other, E26 was prepared by a gene replacement technique developed by Anzai *et al.*[28] Both mutants could not form the C—P bond or catalyze the formation of PEP from PnPy. This suggested that these reactions were one and the same and that PEP phosphomutase was the first step in the BA biosynthetic pathway.[24]

Since the equilibrium of the PEP phosphomutase reaction lies extremely toward the formation of PEP, the reaction product, which is assumed to be PnPy, must be removed from the reaction system to catalyze the forward reaction. By exploiting structural similarity between PnPy and oxaloacetic acid, PnPy could be detected by converting to 3-phosphonolactic acid using malate dehydrogenase. When ^{14}C-labeled PEP was incubated with partially purified PEP phosphomutase and NADH/malate dehydrogenase, the formation of ^{14}C-labeled 3-phosphonolactic acid was detected by autoradiography of a TLC plate.[24] This result clearly showed that C—P bond formation and the reverse reaction are catalyzed by the same enzyme.

Since this reaction is associated with the biosynthesis of many C—P compounds, the enzymatic activity to convert PEP to PnPy was analyzed in extracts of *Streptomyces wedmorensis*, *Streptomyces rubellomurinus*, *Streptomyces plumbeus*, and *Actinomycetes* K-26, which produced FM,[4] FR-33289,[5] plumbemycin,[6] and K-26,[9] respectively. Among them, FR-33289- and FM-producing organisms showed the expected activity.[24] Therefore, PEP phosphomutase seems to be common to C—P compound producing organisms.

1.30.2.3.2 Formation of phosphonoformic acid, a substrate for the phosphinic acid formation reaction

The reduction mechanism of phosphonic acid (H_2O_3P—C—) to phosphinic acid (H_2O_2P—C—) was clarified by the isolation of three key metabolites, 2-hydroxyethylphosphonic acid (HEP), hydroxymethylphosphonic acid (HMP), and phosphonoformic acid (PF) from cultures of blocked mutants NP46, NP221, and NP213, respectively[29] (Scheme 2). A blocked mutant NP47 could transform all three of these compounds to the final product.[29] It is interesting to note that syntheses of these three C—P compounds had been reported before their isolation from a natural source.[30–32]

HEP, which was obtained from culture broth of a blocked mutant NP46, is presumably in equilibrium with phosphonoacetaldehyde (PnAA), which is believed to be the true biosynthetic intermediate of BA. Oxidation of the next intermediate, HMP, generates PF which is spontaneously decarboxylated to give phosphorous acid under acidic condition.[33] This chemical feature of PF suggested that HMP plays an important role in the biosynthesis of BA.

The reaction mechanism that converts PnAA to HMP remains unknown; one plausible mechanism may be formation of a formate ester of HMP by Baeyer–Villiger oxidation of PnAA. Transformation experiments using NP47 and this candidate compound were inconclusive, since the ester was spontaneously cleaved to HMP under the experimental conditions. So far, mutants blocked at step 2 could not be obtained with the BA-producing organism, but the corresponding mutant was obtained with FM-producing organism which enabled the identification of the step 2 gene of BA (see Section 1.30.3.1). Since two carbon atoms of the skeleton of PnPy are removed in the process of the formation of HMP, PF must react with a second molecule of PEP to generate the two-carbon unit adjacent to the phosphorus atom in the PT moiety (Scheme 1). The detailed reaction mechanism is explained in the next section.

1.30.2.3.3 Formation of the second C—P bond by carboxyphosphonoenolpyruvate phosphonomutase

Based on the results explained above, PF was assumed to react with PEP to give PPA with a C_3 skeleton. Direct formation of PPA by the condensation between these two compounds seemed to be unlikely and involvement of unidentified intermediate(s) such as carboxylated PEP (carboxyphosphonoenolpyruvate—CPEP) was postulated (Scheme 3). However, since only one mutant (NP213) defective in the step 5 reaction was available, construction of additional mutants with a block at step 5 was attempted by a gene disruption technique developed by Anzai *et al.*[28]

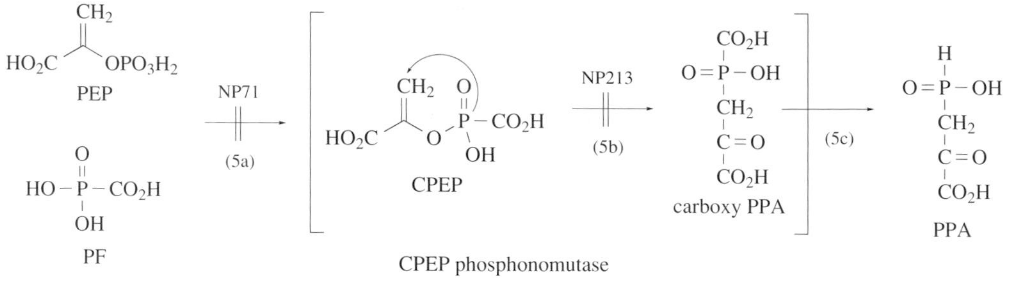

PEP + PF —(5a, blocked in NP71)→ [CPEP —(5b, blocked in NP213)→ carboxy PPA] —(5c)→ PPA

CPEP phosphonomutase

Scheme 3

At this stage of the biosynthetic studies of BA, most of the BA biosynthetic genes had been cloned from *S. hygroscopicus* and proved that most, if not all, of them were clustered in about 35 kb of the genome[34,35] (Figure 3). A region near the step 5 gene contained no known BA biosynthetic genes. A restriction site (*Sac*I) located just upstream of the gene encoding the step 5 enzyme was mutated *in vitro* and introduced into the *S. hygroscopicus* chromosome to give a new mutant NP71, which was unable to produce BA, but complementary to NP213.[36]

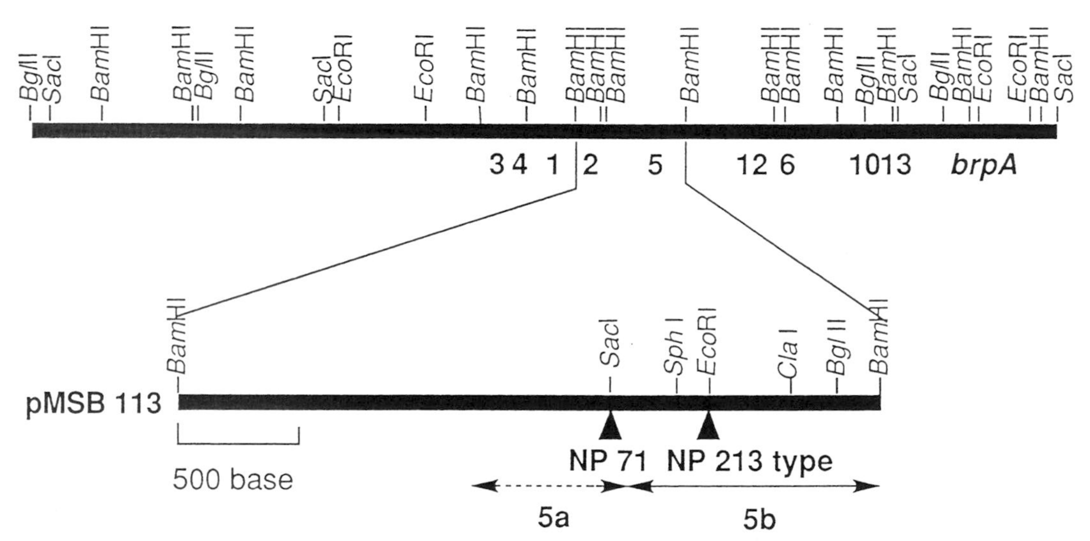

Figure 3 The BA biosynthetic gene cluster of *S. hygroscopicus*. Numbers of ORFs correspond to the BA biosynthetic steps.

Although these two mutants could not utilize PF which was accumulated by NP213, they could convert PPA to BA. The fact that they produced BA in their mixed cultures demonstrated that NP71 and NP213 were defective for different enzymes and that the step 5 reaction consists of at least two reactions, as assumed in Scheme 3.

Mycelia of NP213 could form BA upon incubation with cell-free extracts of NP71. However, NP71 mycelia could not produce BA with cell-free extracts of NP213. These results could be reasonably interpreted by assuming that NP213 secreted an unknown and hitherto overlooked intermediate and that the cell-free extract of NP71 contained an enzyme absent in NP213 which catalyzed the conversion of the unknown intermediate to the next biosynthetic precursor of BA. This enzyme, CPEP phosphonomutase, was purified to homogeneity[36] (32 kDa, monomeric protein) and shown to carry out a reaction similar to step 1. Introduction of a mutation to an *Eco*RI site (Figure 3) by the same technique gave a mutant with a phenotype similar to that of NP213.[28]

As mentioned above, NP213 mycelia could not produce BA in the presence of the cell-free extracts of NP71. Addition of the broth filtrate of NP213 treated with CPEP mutase to the mycelium of NP213, however, allowed BA biosynthesis. This result suggests that NP213 produces a hitherto unidentified intermediate which is a substrate of CPEP phosphonomutase. This compound was isolated from the fermentation broth of NP213 and its structure was determined to be a carboxylated derivative of PEP (CPEP).[37] The structure of CPEP (Scheme 3), a hybrid of PF and PEP, suggested that it was formed by transesterification between these two metabolites. Since NP71 could produce BA from CPEP, which was accumulated by NP213, NP71 was concluded to be defective in the formation of CPEP from PF and PEP (Scheme 3, step 5a). Later, CPEP was prepared by chemical synthesis by Knowles and co-workers.[38]

Since the hypothetical intermediate, carboxylated PPA (Scheme 3) could not be detected in the reaction mixture, decarboxylation should occur spontaneously with intramolecular rearrangement. The above results proved that step 5 consisted of two reactions 5a (transesterification) and 5b (rearrangement and decarboxylation).[39] Failure to recognize the presence of CPEP in the fermentation broth of NP213 was caused by the decomposition of CPEP to PF during isolation procedure; this change could not be detected because both compounds were transformed to BA by the assay method employed.

1.30.2.3.4 ***P-methylation, the third C—P bond formation reaction***

As described above, mutant NP45 accumulated *N*-acetyl demethyl compounds, *N*-acetyl DMPT and *N*-acetyl DMBA. This finding suggested that these two metabolites might be substrates of *P*-methylation enzyme.

Among various methyl group donors tested, including *S*-adenosylmethionine, betaine, and 5-methylfolate, a cell-free extract containing methyltransferase utilized methylcobalamin as the best methyl donor.[40] These results were supported by incorporation of the radioactivity into *N*-acetyl PT when [$^{14}CH_3$]methylcobalamin was incubated with a crude enzyme system and *N*-acetyl DMPT.[40]

Among plausible substrates for the *P*-methylation reaction, only *N*-acetylated derivatives, *N*-acetyl DMPT, and *N*-acetyl DMBA,[21] were methylated by the crude enzyme system. Failure to utilize free amino acids, DMPT and DMBA, suggested that the *N*-acetyl group is an essential structural requirement for the substrates of *P*-methylation.[40] This conclusion is supported by the observation that *N*-acetyl DMPT and *N*-acetyl PT were accumulated in the fermentation broth of NP45 which is unable to catalyze *P*-methylation.[21]

The Co^{2+} requirement for the improved production of BA[18] (see above) is rationalized by the essential role of the metal for the biosynthesis of methylcobalamin.

Since the phosphorus atom in the phosphinic acid function is linked directly to negatively charged oxygen atoms, it is most unlikely that *P*-methylation takes place via electrophilic attack of methyl cation to the phosphorus as in the case of usual methylation reactions. Therefore, involvement of methylcobalamin, which can generate methyl anions, in *P*-methylation seems to be quite reasonable. Similar mechanisms may be operative in the *C*-methylation reactions of fortimicin A[41] and FM[42] requiring the participation of methylcobalamin or Co^{2+} (see Section 1.30.3.1).

1.30.2.4 Chain Elongation Reaction, Transformation of the C_3 Intermediate to the C_4 Intermediate

As in the analysis of the aforementioned step 1 to step 5, blocked mutants were expected to be powerful tools for studying the reaction mechanisms of the next steps. All attempts to prepare desirable mutants, however, were unsuccessful. However, use of enzyme inhibitors gave fruitful results to reveal the transformation mechanism of a C_3 intermediate such as PPA to a C_4 intermediate such as DMPT. In view of the structural similarity between the —PH(=O)OH and —C(=O)OH functions, PPA and DMPT can be regarded as analogues of oxaloacetic acid and glutamic acid, respectively. Thus, the enzyme system converting PPA to DMPT may be assumed to be identical or closely related to that of the tricarboxylic acid (TCA) cycle.

In accord with this assumption, addition of monofluoroacetic acid (MFA), a strong inhibitor of aconitase, to the culture of a BA-producing organism inhibited BA biosynthesis and resulted in accumulation of a new C—P compound. The new metabolite was isolated and identified as (*R*)-phosphinomethylmalic acid (PMM), which is an analogue of citric acid (Scheme 4, double vertical lines show blocked points of mutants).[43,44] Feeding of this compound to cultures of *S. hygroscopicus* stimulated BA production.

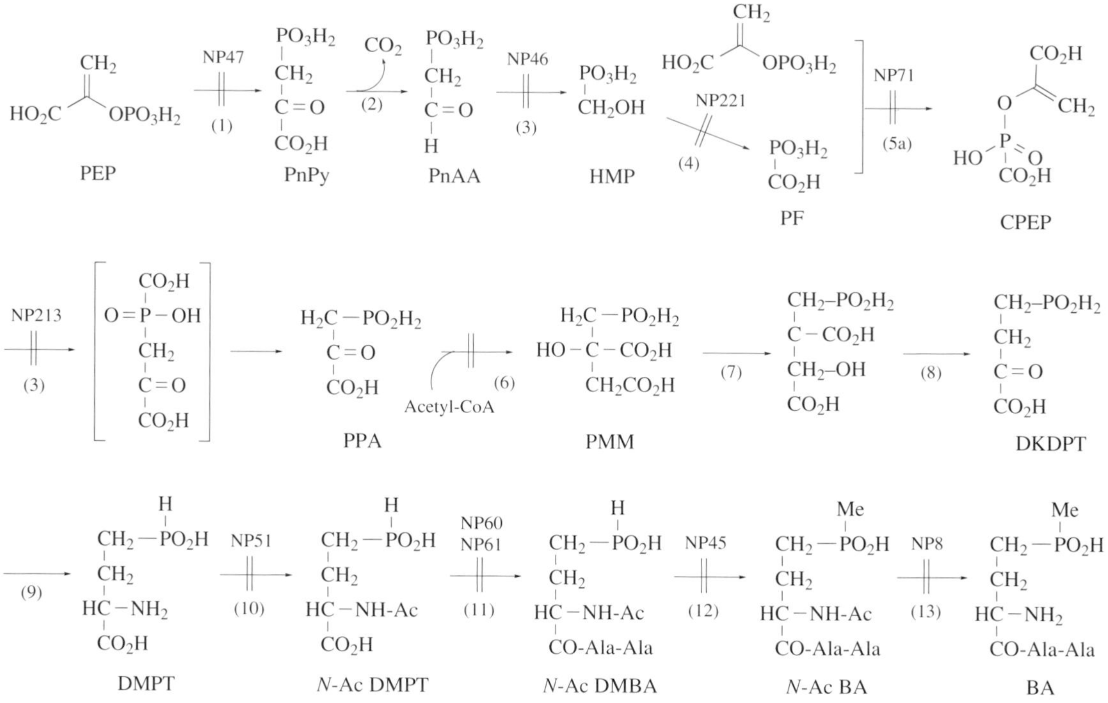

PEP —NP47 (1)→ PnPy —CO_2 (2)→ PnAA —NP46 (3)→ HMP; PEP —NP221 (4)→ PF —NP71 (5a)→ CPEP

—NP213 (3)→ [CO_2H–P(=O)(OH)–CH_2–C(=O)–CO_2H] → PPA —Acetyl-CoA (6)→ PMM —(7)→ CH_2–PO_2H_2 / C–CO_2H / CH_2–OH / CO_2H —(8)→ DKDPT

—(9)→ DMPT —NP51 (10)→ *N*-Ac DMPT —NP60, NP61 (11)→ *N*-Ac DMBA —NP45 (12)→ *N*-Ac BA —NP8 (13)→ BA

Scheme 4

MFA inhibited conversion of PAL (a biological equivalent of PPA with a C_3 skeleton), while it did not affect the transformation of DMPT and PT (with a C_4 skeleton) to BA, thus revealing selective inhibition of the carbon skeleton elongation step. These experimental results and the structure of PMM clearly indicate that the conversion of PPA to PMM is either catalyzed by the ubiquitous bacterial citrate synthase of primary metabolism or a closely related enzyme.

In view of the structural similarity of PMM and PPA with citric acid and oxaloacetic acid, respectively, the condensation reaction to form PMM was considered analogous to that catalyzed by citrate synthase. Therefore, PMM synthase[45] which formed (*R*)-type PMM may be closely related to (*R*)-citrate synthase isolated from a few obligate anaerobic bacteria.[46,47] In fact, PMM synthase isolated from *S. hygroscopicus* and (*R*)-citrate synthase of anaerobic bacteria showed very similar biochemical properties; PMM synthase was strongly inhibited by *p*-chloromercuribenzoate and iodoacetamide,[45] as was (*R*)-citrate synthase from *Clostridium acidi-urici*.[47] In addition, PMM synthase and (*R*)-citrate synthase showed similar divalent metal ion requirement[45] such as Mn^{2+} or Co^{2+}. However, (*S*)-citrate synthases from the BA-producing organism,[48] animal tissues, yeast and *Escherichia coli*[46] possessed quite different enzymatic properties from (*R*)-citrate synthase. These similar properties suggested a common evolutionary origin of the two enzymes; unfortunately, however, the amino acid sequence of (*R*)-citrate synthase of *C. acidi-urici* is not presently available. The amino acid sequence of PMM synthase is clearly different from that of citrate synthase purified from the same organism.[48]

Since the stereochemistry of PMM is (*R*), the conversion from PMM to deaminoketodemethylphosphinothricin (DKDPT), which are structurally related to citric acid and α-ketoglutaric acid, respectively, might be carried out by enzymes of the TCA cycle. In fact, PMM could be converted to DMPT by extracts of *Brevibacterium lactofermentum*, a glutamic acid producer.[49] Thus, the mechanism to transform PAL or PPA with a C_3 skeleton to DMPT with a C_4 skeleton can be explained by involvement of a new kind of (*R*)-citrate synthase and TCA cycle enzymes which are assumed to produce phosphinic acid analogues of *cis*-aconitic acid, isocitric acid and oxalosuccinic acid as biosynthetic intermediates. The failure to obtain mutants blocked in these steps may be rationalized by the involvement of the TCA cycle in these transformations.

Wohlleben *et al.* identified *acnP* encoding an aconitase-like gene, whose gene product catalyzed the isomerization of PMM, from another BA-producing organism, *Streptomyces viridochromogenes*.

The gene product AcnP was highly similar to the *E. coli* aconitase AcnA and to other members of the aconitase family.[50] This finding may indicate that the BA-producing organisms possess the specific enzyme system required for the chain elongation reaction.

1.30.2.5 Other Reactions

Conversion of PMM to DMPT by *B. lactofermentum* strongly suggested that the final transamination reaction to form the amino acid be catalyzed by ubiquitous enzyme(s). Experiments using three common enzymes, glutamate oxaloacetate transaminase (GOT), glutamate pyruvate transaminase (GPT), and glutamate dehydrogenase (GDH) revealed that they utilized DKDPT as a substrate four to five times less efficiently compared to their normal substrates.

The chemical modifications formally required to convert DMPT to BA are only peptide bond formation and *P*-methylation. Analysis of the metabolites accumulated by four different blocked mutants (NP81,[12] NP45, NP51, and NP60[35]) indicated the sequence of these final reactions and involvement of unexpected *N*-acetylated intermediates.

NP8 accumulated *N*-acetyl derivatives of BA which were converted to BA by other mutants blocked earlier in the pathway (NP51 and NP45).[21] Accumulation of these metabolites revealed the important role of *N*-acetyl derivatives in the biosynthesis of BA. NP8 lacks the enzyme system which removes the acetyl group from *N*-acetyl BA (step 13 in Scheme 4).

NP51 which accumulated DMPT could transform *N*-acetyl DMPT and *N*-acetyl PT to BA.[21] Unlike NP8, NP51 lacked the ability to form the *N*-acetyl derivative of DMPT in the presence of acetyl CoA and, therefore, its defect is related to the formation of *N*-acetyl derivatives (step 10 in Scheme 4). Similar results were obtained when PT was utilized as a substrate. These results suggest that *N*-acetyl DMPT is a biosynthetic intermediate of BA.

NP45 cannot convert *N*-acetyl DMPT and *N*-acetyl DMBA to BA, while it can transform *P*-methyl derivatives such as PT and *N*-acetyl PT to BA. This mutant was proved to be defective in *P*-methylation enzyme (step 12 in Scheme 4) by employing a cell-free system as described above.

NP60 accumulated *N*-acetyl DMPT. The properties of NP60 revealed by detailed analysis of the products it accumulated and transformation experiments using plausible intermediates proved that NP60 cannot catalyze peptide bond formation[35] (step 11 in Scheme 4).

The biological features of these nonproducing mutants and the structures of the metabolites they accumulated indicated that the biosynthetic steps of BA following the formation of DMPT are as summarized in Scheme 4.

1.30.2.6 Molecular Cloning of Bialaphos Biosynthetic Genes

Mutagenesis of the parent strain with NTG yielded BA nonproducing mutants that accumulated various kinds of biosynthetic intermediates of BA. Complementation of these mutants revealed that BA biosynthetic genes including the self-resistance gene (*bar* = step 10) and the regulatory gene (*brp*A)[34] were clustered in about 35 kb DNA segment on the chromosome (Figure 3). In addition to these genes, three regions coding for alanylation reaction(s) were found, one between *brp*A and step 13 gene and two at the far left of step 3 gene in Figure 3.

Although all conversion steps are not covered by these mutants, this region must contain most of the genes needed for BA biosynthesis. Regions of the gene cluster which do not correspond to any of nonproducing mutants may contain undefined genes related to BA biosynthesis. Using a technique developed by Anzai *et al.*[28] (see earlier), new mutants that could not be isolated by conventional methods were generated (NP71 defective in step 5a[36] (see above), NP60, NP61, and NP62 defective in the alanylation step).[35]

The BA resistance gene, *bar*, was first isolated by shotgun cloning in the BA-sensitive host, *Streptomyces lividans*.[51] In addition to conferring antibiotic resistance, its gene product (an acetyltransferase) catalyzes one of the biosynthetic steps, conversion of DMPT to *N*-acetyl DMPT[52] (step 10 in Scheme 4).

The regulatory gene of BA biosynthesis, *brp*A, was also cloned by the complementation of a pleiotropic mutant, NP57, defective in at least six steps of BA biosynthesis.[53] Since NP57 lacked mRNAs for these six steps, *brp*A could turn on BA biosynthetic genes by activating their transcription.

Thompson and co-workers determined the nucleotide sequence of the 5.0 kb region next to *brp*A, and found five open reading frames (ORFs).[54] The amino acid sequences deduced from their nucleotide sequences predicted their functions, *N*-acetyl hydrolase, thioesterase, and a transport protein. Among these biosynthetic genes, genes coded for C—P bond-forming enzymes were precisely analyzed and sequenced (steps 1, 5b, and 12 in Figure 3).

1.30.2.6.1 ***Phosphoenolpyruvate phosphomutase***

Analysis of mutant NP47 defective in the first step of BA biosynthesis and site-specific mutagenesis revealed the position of the PEP phosphomutase gene on the cluster.[24,34] The nucleotide sequence of the 1.3 kb fragment revealed an ORF of 314 codons.[55] When this fragment was cloned onto pIJ680 and the resulting plasmid introduced into *S. lividans*, high expression of PEP phosphomutase was detected. It was thus indicated that PEP phosphomutase gene was expressed under the control of the *aph* gene promoter of pIJ680 in *S. lividans*. The Harvard group reported the nucleotide sequence of PEP phosphomutase of *T. pyriformis* producing AEP.[56] The deduced amino acid sequence of PEP phosphomutase of *S. hygroscopicus* showed significant homology to that of *T. pyriformis* indicating that this enzyme is common to the organisms producing C—P compounds. As mentioned in Section 1.30.3.2, PEP phosphomutase of FM-producing *S. wedmorensis* also showed homology to those of *T. pyriformis* and *S. hygroscopicus*.[57]

1.30.2.6.2 ***Carboxyphosphonoenolpyruvate phosphonomutase***

Analysis of mutant NP213 and NP71 described above indicated that CPEP phosphonomutase was encoded in the 1.2 kb fragment containing the mutational point of NP213[36] (5b in Figure 3). By the nucleotide sequence of the fragment, an ORF of 296 codons was identified and the *N*-terminal 30 amino acid sequence deduced from the nucleotide sequence was in complete agreement with that of purified CPEP phosphomutase.[58] As in the case of the PEP phosphomutase gene, introduction of CPEP phosphonomutase gene, whose expression is controlled by the *aph* gene promoter, into *S. lividans* resulted in the production of the corresponding enzyme at an almost equivalent level to that of the BA high-producing strain of *S. hygroscopicus*.[58]

CPEP phosphonomutase showed approximately 26% homology to PEP phosphomutase of *S. hygroscopicus*.[55] Since CPEP phosphonomutase catalyzes a reaction analogous to PEP phosphomutase, i.e., the intramolecular rearrangement of a phosphate ester to form C—P bond, this similarity seems to be quite reasonable. CPEP synthase that catalyzes transesterification between PEP and PF to generate CPEP (Scheme 3, step 5a), was shown to be encoded on the region next to CPEP phosphonomutase. The deduced amino acid sequence of this protein showed homology to some enolases.[59]

1.30.2.6.3 **P-*methyltransferase***

As described above, an enzymatic activity catalyzing direct *P*-methylation of phosphonic acid was detected in a cell extract of *S. hygroscopicus*.[40] In addition, a 2.5 kb *Bcl*I-*Bam*HI fragment, which restored BA productivity of *P*-methylation-deficient mutants, was identified in the BA biosynthetic gene cluster.[34] This fragment cloned onto pAK114 conferred the activity to carry out *P*-methylation of the phosphinic acid, DMPT, to *S. lividans*.[40] An ORF according to the *P*-methyltransferase was detected in the nucleotide sequence of the fragment.[60] The deduced polypeptide showed homology to magnesium-protoporphyrin IX monomethyl ester oxidative cyclase[61] and the step 3 enzyme in FM biosynthesis which catalyzes the methylation of the aldehyde carbon of PnAA (see Section 1.30.3.1).[43,57] The similarity between *P*-methyltransferase and the step 3 enzyme of FM means that the reaction mechanisms of these two methylations are similar; both of them are assumed to utilize methylcobalamin as the methyl donor. However, the homology between *P*-methyltransferase and magnesium-protoporphyrin IX monomethyl ester oxidative cyclase presumably results from the structural similarity between methylcobalamin and protoporphyrin.

1.30.3 BIOSYNTHESIS OF FOSFOMYCIN

FM was discovered in 1968 as a metabolite of several *Streptomyces* species[4] and later was proved to be produced by taxonomically quite different microorganisms such as *Pseudomonas syringae*[62] and *Pseudomonas viridiflava*.[63] This compound, acting as a PEP analogue, irreversibly inhibits PEP UDP-*N*-acetylglucosamine-3-*O*-enolpyruvyltransferase (enolpyruvyl transferase), an enzyme catalyzing the first step of peptideglycan biosynthesis.[64] Since FM shows negligible toxicity in humans and exhibits broad antibacterial spectrum covering many pathogenic and opportunistic Gram-positive and Gram-negative bacteria, it is widely used in chemotherapy, especially against *E. coli* O 157 in Japan.

The structure of FM attracted considerable interest in its biosynthesis, because despite its simple structure, FM possesses unique functional groups such as an epoxide and a direct C—P bond (Scheme 5, double vertical lines indicate blocked points of mutants); both of which are important for its antimicrobial activity.

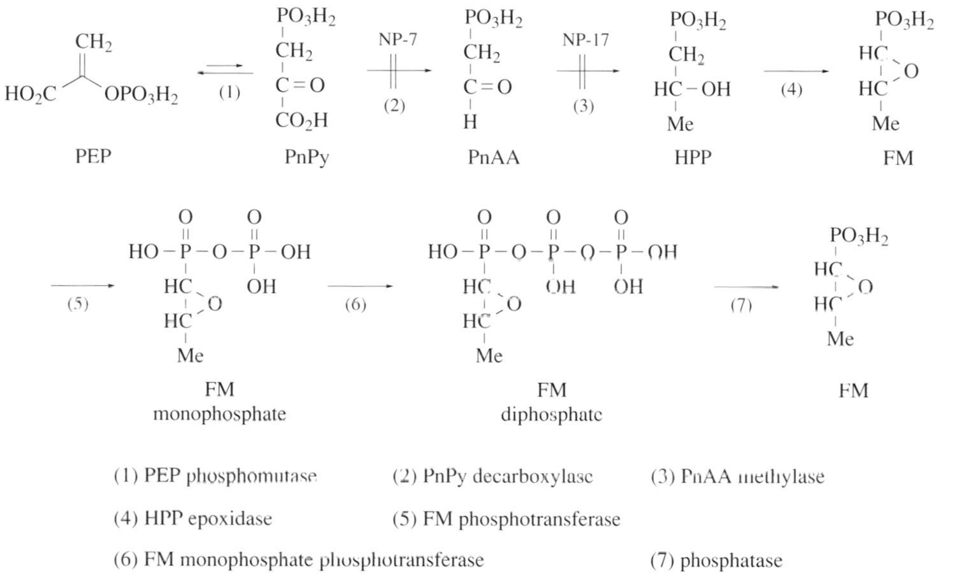

PEP →(1) PnPy →(2, NP-7) PnAA →(3, NP-17) HPP →(4) FM →(5) FM monophosphate →(6) FM diphosphate →(7) FM

(1) PEP phosphomutase (2) PnPy decarboxylase (3) PnAA methylase

(4) HPP epoxidase (5) FM phosphotransferase

(6) FM monophosphate phosphotransferase (7) phosphatase

Scheme 5

1.30.3.1 Biosynthetic Pathway of Fosfomycin and Intermediates

The first biosynthetic experiment on FM carried out using ^{14}C-labeled glucose and [$^{14}CH_3$]methionine proved that the carbon skeleton of FM was synthesized by condensation of a two-carbon unit originated from PEP and a methyl group of methionine.[16] Although this result suggested that the early steps of FM biosynthesis might be similar to those proposed for AEP, the detailed mechanism of C—P bond formation of FM remained unknown for more than 10 years.

As described above, the first reaction of the biosynthesis of BA is the intramolecular rearrangement of PEP to form the C—P bond of PnPy catalyzed by PEP phosphomutase.[24] This reaction is believed to be common to the biosyntheses of all C—P compounds including FM.[13] In support of this hypothesis, Imai *et al.* reported that HEP and AEP, early intermediates of BA biosynthesis, were converted to FM by a blocked mutant of the FM-producing organism, *Streptomyces wedmorensis* NP7.[65] This finding strongly suggested that the initial FM biosynthetic steps related to the C—P bond formation are identical to those of BA.

In agreement with this assumption, PEP phosphomutase activity, which catalyzes the step 1 reaction of the BA biosynthetic pathway, was detected in a cell-free extract of an FM high producer, *S. wedmorensis* 144-91, though at a much lower level[66] than that of the BA-producing organism. This result confirmed that the C—P bond of FM is also formed by PEP phosphomutase reaction (step 1 in Scheme 5) in a similar manner as for BA[24]; the enzymatic properties of *S. wedmorensis* PEP phosphomutase, however, seemed to be different from those of *S. hygroscopicus*.[66]

Since the equilibrium of PEP phosphomutase greatly favors the cleavage of the C—P bond[23,24] (see earlier), the reaction product (PnPy) must be removed from the reaction system by the next enzyme (step 2 reaction) which presumably catalyzes decarboxylation of PnPy[67] (step 2 in Scheme 5). This reaction mechanism, however, remained unknown, because step 2 mutants could not be obtained with the BA-producing organism (see above).

To analyze the step 2 reaction in detail, the rigorous characterization of mutant NP7, the only strain of the FM-producing organism available at that time, was necessary. For reasons that were not known, all attempts to prepare other kinds of FM-blocked mutants were totally unsucessful. Since NP7 could convert AEP and HEP to FM presumably via PnAA,[65] its mutational point was assumed to be at the step 1 or step 2 reaction (Scheme 5). However, the intrinsic nature of the step 1 reaction favoring the catalysis of PEP formation prevented the characterization of the nature of NP7, because mutants blocked at step 1 or step 2 would show the same phenotypic properties. The mutational point of NP7 was clarified to be at the step 2 reaction, i.e., decarboxylation of PnPy, by the experimental result that despite the successful expression of the PEP phosphomutase gene of *S. hygroscopicus* in *S. wedmorensis* NP7, FM productivity could not be restored in this mutant.[66]

In order to reveal the mechanism leading from HEP to FM, the conversion to FM of two putative intermediates *cis*-propenylphosphonic acid (PPOH) and 2-hydroxypropylphosphonic acid (HPP),[68] was investigated using mutant NP7 as a converter (see Scheme 5). Although epoxidation of PPOH seemed to be the most likely route to FM, attempts to convert PPOH to FM using this mutant were totally unsuccessful. However, NP7 surprisingly produced FM upon the addition of HPP.[68] In agreement with these results, Hammerschmidt *et al.* observed the incorporation of ^{18}O in the alcohol function of ^{18}O-labeled HEP into the FM molecule but not of $^{18}O_2$ gas by another FM-producing organism, *Streptomyces fradiae*.[69,70] It was thus concluded that the epoxide ring of FM was formed by dehydrogenation of HPP (step 4 in Scheme 5) and not by the addition of molecular oxygen to the double bond of PPOH. This type of unique epoxidation had only been reported for the biosynthesis of scopolamine[71] (see Section 1.30.3.2). They also studied the stereochemical course of the epoxidation reaction in detail; use of stereospecifically labeled intermediates proved the retention of *pro*-(*S*)-hydrogen at C-1 of HEP during its transformation to FM.[72,73]

The remaining problem in the biosynthesis of FM, introduction of a methyl group to PnAA (step 3 in Scheme 5) was investigated by the use of another mutant of *S. wedmorensis* named A16 defective in the biosynthesis of methylcobalamin.[42] Since the mutant could convert HPP to FM, incorporation of the methyl group into the FM molecule was suggested to take place during a stage between a putative C_2 biosynthetic intermediate, PnAA (or its equivalent, HEP) and HPP with the involvement of methylcobalamin. The confirmative evidence for this mechanism was obtained by incorporation of [$^{14}CH_3$]methylcobalamin into FM by the mutant A16.[42]

As described above, methylcobalamin was also shown to serve as the direct methyl donor for methylation of the phosphorus atom of the phosphinic acid function in a BA biosynthetic intermediate using a cell-free system.[40] Since the carbonyl carbon of PnAA is positively charged as is the phosphorus atom in the phosphinic acid function, it is reasonably concluded that PnAA is methylated by a nucleophilic attack of the methyl anion derived from methylcobalamin to generate HPP.

Based on the results described above, an FM biosynthetic pathway consisting of four steps is summarized as shown in Scheme 5; C—P bond formation (step 1), decarboxylation of PnPy (step 2), methylation of PnAA (step 3), and epoxide formation (dehydrogenation of HPP, step 4).

1.30.3.2 Molecular Cloning of Fosfomycin Biosynthetic Genes

As described above, mutant NP7 was shown to be defective in the step 2 reaction of FM biosynthesis.[66] By utilizing this mutant for cloning experiments, a 5.9 kb DNA fragment (Figure 4, pFBG21) that restored FM productivity to this mutant[74] was obtained. A complementation assay using a BA nonproducing mutant (E26) deficient in the PEP phosphomutase reaction revealed that the 5.9 kb fragment also contained step 1 gene (*fom1* in Figure 4).

For the cloning of the other biosynthetic genes of FM biosynthesis, several FM nonproducing mutants were prepared by treating *S. wedmorensis* 144-91 with NTG, and their transformation efficiency was checked by using plasmid pIJ702. However, for unknown reasons, none of these mutants could be transformed, but mutant NP7 proved to be a good host for gene cloning.[74] Thus, protoplasts of strain NP7 were treated with NTG and resultant mutants which could not convert AEP or HEP to FM were screened. Among 5000 colonies, one such mutant, NP17 was isolated

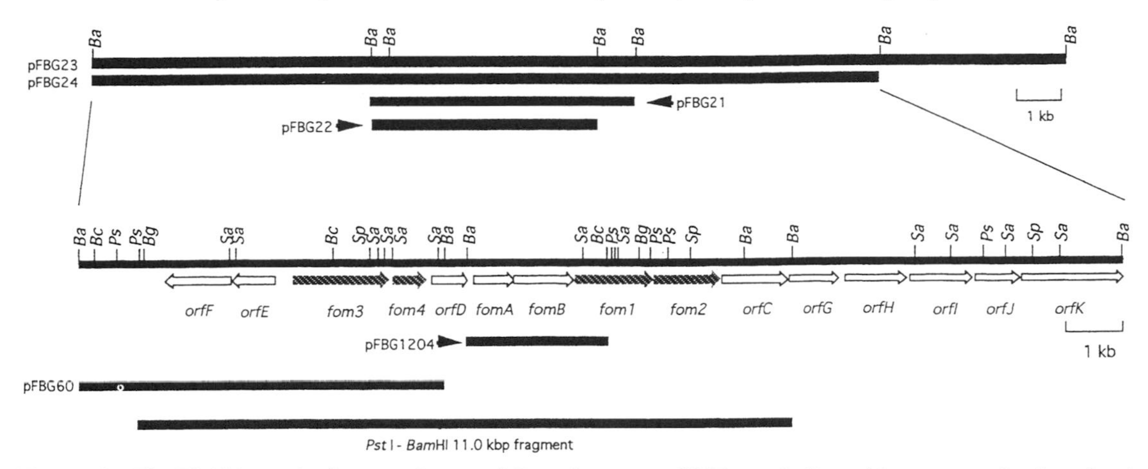

Figure 4 The FM biosynthetic gene cluster of *S. wedmorensis*. ORFs are indicated by arrows. *fom1* to *fom4* under the map correspond to the fosfomycin biosynthetic steps.

which could transform HPP to FM.[57] Thus, NP17 was defective for both the step 3 reaction and the step 2 reaction (original NP7 mutation) and was easily transformed with pIJ702 or pIJ922.

In order to reveal the organization of FM biosynthetic genes, ~20 kb DNA fragments (Figure 4, pFBG23 and pFGB24) which contained the previously cloned 5.9 kb area were cloned and analyzed. Since this fragment could restore FM productivity to NP17 defective in step 3 conversion, step 3 gene (*fom3*) should exist in this region. Subcloning experiments defined a 6.0 kb fragment (Figure 4, pFBG60) for *fom3*.[57]

Since step 4 mutants of *S. wedmorensis* could not be isolated, *S. lividans* was used as the host for step 4 gene (*fom4*) cloning. The above-mentioned 20 kb fragment containing *fom1* to *fom3* was digested with *Bam*HI, ligated to pIJ702 and used to transform *S. lividans*. The conversion of HPP to FM by the transformants enabled the identification of *fom4* (epoxidase gene) on the chromosomal region next to *fom3*.[57] Subcloning experiments suggested that *fom4* may have its own promoter which functions in *S. lividans*.

Figure 4 shows the FM biosynthetic gene cluster of *S. wedmorensis*. Nucleotide sequencing of the 11.0 kb *Pst*I–*Bam*HI fragment identified 10 ORFs.[57] Subcloning experiments showed that *fom1* to *fom4* coded for step 1 to step 4 enzymes, respectively.

The molecular mass of the deduced protein coded in *fom1* was 50 300 Da. As explained above, the PEP phosphomutase genes had also been cloned from BA-producing *S. hygroscopicus*[58] and *T. pyriformis*, a producer of AEP.[56] The deduced amino acid sequence of *fom1* showed significant identity to both PEP phosphomutases (33.2% to that of *S. hygroscopicus* and 34.1% to that of *T. pyriformis*), suggesting conservation of this enzyme in all C—P compound-producing organisms. *fom1* also showed homology to CPEP phosphonomutase,[55] another C—P bond-forming enzyme that catalyzes an intramolecular rearrangement of CPEP to PPA, a reaction reminiscent of the rearrangement of PEP by PEP phosphomutase in the BA biosynthetic pathway (Scheme 4).

Complementation experiments revealed[72] that *fom2* was the step 2 gene of FM biosynthesis. The predicted protein had a molecular mass of 39 700 Da. It showed no significant similarity to any other sequences in the SWISS-PROT database. As mentioned above, all attempts to prepare a step 2 mutant of BA were unsuccessful. Use of NP7, however, enabled this problem to be overcome; the step 2 gene of *S. hygroscopicus* (BA producer) could be cloned which encodes the enzyme-catalyzing decarboxylation of PnPy[67] (PnPy decarboxylase). Approximately 17% identity was observed between *fom2* of *S. wedmorensis* and PnPy decarboxylase of *S. hygroscopicus*.[75] This considerably low similarity suggests that the primary structures of PnPy decarboxylases are not conserved as compared to those of PEP phosphomutases among the C—P compound-producing organisms.

Similar to the *P*-methylation enzyme in BA biosynthesis, the direct methyl donor for methylation of PnAA to form HPP was assumed to be methylcobalamin.[43] A search of the SWISS-PROT protein sequence database indicated a 24.8% amino acid similarity between the deduced amino acid sequences of *fom3* and magnesium-protoporphyrin monomethyl ester oxidative cyclase.[61] This result may be rationalized by the structural similarity between protoporphyrin and methylcobalamin. In addition, *fom3* protein exhibited 19.0% identity to *P*-methyl transferase involved in BA biosynthesis.[60] The methyl group of methylcobalamin was directly incorporated into the *P*-methyl group of BA by this enzyme.[40] Therefore, there is reasonable similarity among these three proteins.

Although blocked mutants defective in the epoxide formation were not isolated, success of the epoxidase gene expression in *S. lividans* enabled this step to be studied in more detail. The very well known mechanism of epoxide formation in natural products is the addition of molecular oxygen to the double bond. Surprisingly, the epoxide of FM is generated by the dehydrogenation of a secondary alcohol.[70] Only one precedent of such a reaction is found: the biosynthesis of scopolamine, which is formed from 6α-hydroxyhyoscyamine by epoxidation catalyzed by 6α-hydroxylase.[73,76] Unlike epoxidase in FM biosynthesis, this enzyme can catalyze the epoxidation in two different ways, i.e., by the addition of oxygen to the double bond and by dehydrogenation of a secondary alcohol. As expected from those different enzymatic properties, there was no similarity between the amino acid sequences of hyoscyamine 6α-hydroxylase[77] and the *fom4* protein. Success in the expression of this enzyme in *E. coli* and purification of the recombinant enzyme enabled the determination of the *N*-terminus 15 amino acids arrangement that turned out to be identical with that deduced from the nucleotide sequence of *fom4*. The detailed reaction mechanism of this unique enzyme remains to be clarified, but preliminary results imply the requirement of Fe^{2+} for the reaction.[78]

In addition to *fom1* to *fom4*, 11 *orfs* (*orfA* to *orfK*) were found in the sequenced fragment (Figure 4). A database search revealed that the deduced products of *orfC* and *orfF* were similar to yeast alcohol dehydrogenase[79] (28% similarity) and a putative amino acid racemase (24.8% similarity),[80] respectively. *orfA*, *orfB*, *orfD*, and *orfE* showed no significant similarity to any proteins in the database. At present *orfE* and *orfF* are assumed to be unrelated to the biosynthesis of FM.

Since *orfA* and *orfB* were located in the center of the FM gene cluster, they were considered to play important roles in the biosynthesis of FM. Transformation of *E. coli* with these two genes conferred high-level resistance to the host against FM and the inactivated FM products were identified as FM monophosphate and FM diphosphate (Scheme 5).[81] These two metabolites were easily reactivated to FM by the action of alkaline phosphatase. Detailed experiments proved that *orfA* and *orfB* encode FM monophosphate phosphotransferase and FM diphosphate phosphotransferase, respectively. Thus, the functions of these two genes are reasonably assumed to be related to self-resistance of the FM-producing organism. This reversible inactivation mechanism is in sharp contrast to the irreversible inactivation of FM by clinically isolated FM-resistant pathogens which cleave the epoxide ring of FM by addition of glutathione.[82]

orfI, *orfJ*, and *orfK* showed similarity to *phnD*, *phnC*, and *phnE* genes of *E. coli* whose functions were related to incorporation into the cells of outer cellular phosphonic acid.[83] Therefore, it may be reasonable to assume that the roles of *orfI*, *orfJ*, and *orfK* are related to transportation of FM across the cell membrane. These findings clearly show that the biosynthesis of FM is considerably more complicated than expected. The functions of *orfC* to *orfF* in FM biosynthesis which remain unknown will be revealed by gene disruption analysis.

1.30.4 CONCLUSION

The experimental results described above revealed the detailed formation mechanisms of two representative members of the natural C—P compounds, BA and FM. The total biosynthetic pathway of BA consisting of at least 14 steps (Scheme 4) is briefly summarized as follows: (i) first formation of the C—P bond to give a phosphonic acid derivative (PnPy), (ii) reduction of the phosphonic acid through removal of the carbon skeleton from PnPy, (iii) second C—P bond formation by reaction with PEP to give PPA, (iv) chain elongation through a TCA cycle-related mechanism, (v) third C—P bond formation by *P*-methylation, and (vi) peptide bond formation. Thus, the biosynthetic pathway of BA is far more complicated than one would assume from the structure of BA.

However, the biosynthetic pathway of FM is simpler than that of BA and formally consists of four steps: (i) formation of the C—P bond to give PnPy via intramolecular rearrangement of PEP, (ii) decarboxylation of PnPy to form PnAA, (iii) addition of a methyl group to PnAA by the use of methylcobalamin, and (iv) epoxide formation by dehydrogenation of a secondary alcohol. In addition, this pathway contains three steps which are apparently related to the self-resistance mechanism of the producing organism, i.e., monophosphorylation and diphosphorylation followed by removal of the diphosphate residue. The first two steps (steps 1 and 2) are believed to be common to the biosynthesis of all natural C—P compounds.

The information on the C—P bond formation mechanisms obtained through these works has contributed to the improvement of the production yield of BA, and will do so for the production of FM, and is expected to be useful for understanding the biosynthetic mechanisms of other C—P compounds.

1.30.5 REFERENCES

1. Y. Ogawa, T. Tsuruoka, S. Inoue, and T. Niida, *Sci. Reports of Meiji Seika Kaisha*, 1973, **13**, 42.
2. E. Bayer, K. H. Gugel, K. Hagele, H. Hagenmaier, S. Jessipow, W. A. Konig, and H. Zahner, *Helv. Chim. Acta*, 1972, **55**, 224.
3. S. Omura, K. Hinotozawa, N. Imamura, and M. Murata, *J. Antibiot.*, 1984, **37**, 939.
4. D. Hendlin, E. O. Stapley, M. Jackson, H. Wallick, A. K. Miller, F. J. Wolf, T. W. Miller, L. Chaiet, F. M. Kahan, E. L. Foltz, H. B. Woodruff, J. M. Mata, S. Hernandez, and S. Mochales, *Science*, 1969, **166**, 122.
5. Y. Kuroda, M. Okuhara, T. Goto, M. Okamoto, H. Terano, M. Kohsaka, H. Aoki, and H. Imanaka, *J. Antibiot.*, 1980, **33**, 29.
6. B. K. Park, A. Hirota, and H. Sakai, *Agric. Biol. Chem.*, 1977, **41**, 573.
7. Y. Kuroda, M. Okuhara, T. Goto, M. Okamoto, H. Terano, M. Kohsaka, H. Aoki, and H. Imanaka, *J. Antibiot.*, 1980, **33**, 29.
8. T. Ogita, S. Gunji, Y. Fukazawa, A. Terahara, T. Kinoshita, H. Nagai, and T. Beppu, *Tetrahedron Lett.*, 1983, **24**, 2283.
9. M. Kasai, M. Yoshida, N. Hirayama, and K. Shirahata, in "Symposium Papers, The 27th Symposium on the Chemistry of Natural Products," Hiroshima University, Hiroshima, Japan, 1985, p. 577.
10. E. Takahashi, T. Kimura, K. Nakamura, M. Arahira, and M. Iida, *J. Antibiot.*, 1995, **48**, 1124.
11. K. Nakamura and S. Yamamura, *Tetrahedron Lett.*, 1997, **38**, 437.
12. M. Horiguchi and M. Kandatsu, *Nature*, 1959, **184**, 901.
13. T. Hori, M. Horiguchi, and A. Hayashi, "Biochemistry of Natural C—P Compounds," Japanese Association for Research on the Biosynthesis of C—P Compounds, Maruzen, Kyoto, 1984.
14. H. Seto, S. Imai, T. Tsuruoka, A. Satoh, M. Kojima, S. Inouye, T. Sasaki, and N. Otake, *J. Antibiot.*, 1982, **35**, 1719.
15. M. Horiguchi, J. S. Kittredge, and E. Roberts, *Biochim. Biophys. Acta*, 1968, **165**, 164.
16. O. Rogers and J. Birnbaum, *Antimicrob. Agents Chemother.*, 1974, **5**, 121.
17. M. Okuhara, Ph.D. Thesis, University of Tokyo, 1989.
18. H. Takebe, S. Imai, H. Ogawa, A. Satoh, and H. Tanaka, *J. Ferment. Bioeng.*, 1989, **67**, 226.
19. H. Seto, T. Sasaki, S. Imai, T. Tsuruoka, H. Ogawa, A. Satoh, S. Inouye, T. Niida, and N. Otake, *J. Antibiot.*, 1983, **36**, 96.
20. M. M. Crutchfield, C. H. Dungan, J. H. Letcher, V. Mark, and J. R. Van Wazer, "^{31}P Nuclear Magnetic Resonance," Wiley Interscience, Chichester, 1967.
21. S. Imai, H. Seto, T. Sasaki, T. Tsuruoka, H. Ogawa, A. Satoh, S. Inoue, T. Niida, and N. Otake, *J. Antibiot.*, 1985, **38**, 687.
22. H. Seto, S. Imai, T. Tsuruoka, H. Ogawa, A. Satoh, T. Sasaki, and N. Otake, *Biochem. Biophys. Res. Commun.*, 1983, **111**, 1008.
23. H. M. Seidel, S. Freeman, H. Seto, and J. R. Knowles, *Nature*, 1988, **335**, 457.
24. T. Hidaka, M. Mori, S. Imai, O. Hara, K. Nagaoka, and H. Seto, *J. Antibiot.*, 1989, **42**, 491.
25. E. Bowman, M. McQueney, R. J. Barry, and D. Dunaway-Mariano, *J. Am. Chem. Soc.*, 1988, **110**, 5575.
26. H. Nakashita, A. Shimazu, T. Hidaka, and H. Seto, *J. Bacteriol.*, 1992, **174**, 6857.
27. H. Nakashita and H. Seto, *Agric. Biol. Chem.*, 1991, **55**, 2825.
28. H. Anzai, Y. Kumada, O. Hara, T. Murakami, R. Itoh, E. Takano, S. Imai, A. Satoh, and K. Nagaoka, *J. Antibiot.*, 1988, **41**, 226.
29. S. Imai, H. Seto, T. Sasaki, T. Tsuruoka, H. Ogawa, A. Satoh, S. Inoue, T. Niida, and N. Otake, *J. Antibiot.*, 1984, **37**, 1505.
30. I. L. Knunyants and R. N. Sterlin, *Compt. Rend. Acad. Sci. URSS*, 1947, **58**, 49.
31. H. J. Page, *J. Chem. Soc.*, 1912, **101**, 423.
32. E. Helgstrand, B. Eriksson, N. G. Johansson, B. Lannero, A. Larsson, A. Misiorny, J. O. Noren, B. Sjoberg, K. Stenberg, G. Stening, S. Stridh, B. Oberg, K. Alenius, and L. Philipson, *Science*, 1978, **201**, 819.
33. S. Warren and M. R. Williams, *J. Chem. Soc.* (*B*), 1971, 618.
34. T. Murakami, H. Anzai, S. Imai, A. Satoh, K. Nagaoka, and C. J. Thompson, *Mol. Gen. Genet.*, 1986, **205**, 42.
35. O. Hara, H. Anzai, S. Imai, Y. Kumada, T. Murakami, R. Itoh, E. Takano, A. Satoh, and K. Nagaoka, *J. Antibiot.*, 1988, **41**, 538.
36. T. Hidaka, S. Imai, O. Hara, H. Anzai, T. Murakami, K. Nagaoka, and H. Seto, *J. Bacteriol.*, 1990, **172**, 3066.
37. T. Hidaka, S. Imai, and H. Seto, *J. Am. Chem. Soc.*, 1989, **111**, 8012.
38. S. Freeman, S. J. Pollack, and J. R. Knowles, *J. Am. Chem. Soc.*, 1992, **114**, 377.
39. T. Hidaka, O. Hara, S. Imai, H. Anzai, T. Murakami, K. Nagaoka, and H. Seto, *Agric. Biol. Chem.*, 1990, **54**, 2121.
40. K. Kamigiri, T. Hidaka, S. Imai, K. Murakami, and H. Seto, *J. Antibiot*, 1992, **45**, 781.
41. S. Okumura, T. Deguchi, and H. Marumo, *J. Antibiot.*, 1981, **34**, 1360.
42. T. Kuzuyama, T. Hidaka, K. Kamigiri, S. Imai, and H. Seto, *J. Antibiot*, 1992, **45**, 1812.
43. H. Seto, S. Imai, T. Sasaki, K. Shimotohno, T. Tsuruoka, H. Ogawa, A. Satoh, S. Inoue, T. Niida, and N. Otake, *J. Antibiot.*, 1984, **37**, 1509.
44. K. Shimotohno, H. Seto, N. Otake, S. Imai, and A. Satoh, *J. Antibiot.*, 1986, **39**, 1356.
45. K. Shimotohno, H. Seto, and N. Otake, *J. Antibiot.*, 1988, **41**, 1057.
46. L. B. Spector, in "The Enzymes," ed. P. D. Boyer, Academic Press, New York, 1972, vol. 7, p. 357.
47. G. Gottschalk and H. A. Barker, *Biochemistry*, 1966, **5**, 1125.
48. K. Shimotohno, S. Imai, T. Murakami, and H. Seto, *Agric. Biol. Chem.*, 1990, **54**, 463.

49. H. Seto and K. Shimotohno, unpublished results.
50. D. Schwartz, J. Recktenwald, S. Kasper, G. Kienzlen, and W. Wohlleben, "Symposium Papers, Japan–UK Joint Study on Molecular Genetics of *Streptomyces*," John Innes Research Centre, Norwich, UK, 1997, p. 21.
51. C. J. Thompson, N. R. Movva, R. Tizard, R. Crameri, J. E. Davies, M. Lauwereys, and J. Botterman, *EMBO J.*, 1987, **6**, 2519.
52. Y. Kumada, H. Anzai, E. Takano, T. Murakami, O. Hara, R. Itoh, S. Imai, A. Satoh, and K. Nagaoka, *J. Antibiot.*, 1988, **41**, 1938.
53. H. Anzai, T. Murakami, S. Imai, A. Satoh, K. Nagaoka, and C. J. Thompson, *J. Bacteriol.*, 1987, **169**, 3482.
54. A. Raibaud, M. Zalacain, T. G. Holt, R. Tizard, and C. J. Thompson, *J. Bacteriol.*, 1991, **173**, 4454.
55. T. Hidaka, M. Hidaka, and H. Seto, *J. Antibiot.*, 1992, **45**, 1977.
56. H. M. Seidel, D. L. Pompliano, and J. R. Knowles, *Biochemistry*, 1992, **31**, 2598.
57. T. Hidaka, M. Goda, T. Kuzuyama, N. Takei, M. Hidaka, and H. Seto, *Mol. Gen. Genet.*, 1995, **249**, 274.
58. T. Hidaka, M. Hidaka, T. Uozumi, and H. Seto, *Mol. Gen. Genet.*, 1992, **233**, 476.
59. S.-H. Lee, T. Hidaka, H. Nakashita, and H. Seto, *Gene*, 1995, **153**, 143.
60. T. Hidaka, M. Hidaka, T. Kuzuyama, and H. Seto, *Gene*, 1995, **160**, 149.
61. D. H. Burke, M. Alberti, G. A. Armstrong, and J. E. Heast, EMBL, GenBank and DDBJ data banks, accession number P26168.
62. J. Shoji, T. Kato, H. Hinoo, T. Hattori, K. Hirooka, K. Matsumoto, T. Tanimoto, and E. Kondo, *J. Antibiot.*, 1986, **39**, 1011.
63. N. Katayama, S. Thubotani, Y. Nozaki, S. Harada, and H. Ono, *J. Antibiot.*, 1990, **43**, 238.
64. F. M. Kahan, J. S. Kahan, P. J. Cassidy, and H. Kropp, *Ann. NY Acad. Sci.*, 1974, **235**, 364.
65. S. Imai, H. Seto, H. Ogawa, A. Satoh, and N. Otake, *Agric. Biol. Chem.*, 1985, **49**, 873.
66. T. Hidaka, H. Iwakura, S. Imai, and H. Seto, *J. Antibiot.*, 1992, **45**, 1008.
67. H. Nakashita, K. Watanabe, T. Hidaka, O. Hara, and H. Seto, *J. Antibiot.*, 1997, **50**, 212.
68. H. Seto, T. Hidaka, T. Kuzuyama, S. Shibahara, T. Usui, O. Sakanaka, and S. Imai, *J. Antibiot.*, 1991, **44**, 1286.
69. F. Hammerschmidt, G. Bovermann, and K. Bayer, *Liebigs Ann. Chem.*, 1990, 1055.
70. F. Hammerschmidt, *J. Chem. Soc. Perkin Trans. 1* 1991, 1993.
71. T. Hashimoto and Y. Yamada, *Eur. J. Biochem.*, 1987, **164**, 277.
72. F. Hammerschmidt and H. Kahlig, *J. Org. Chem.*, 1991, **56**, 2364.
73. F. Hammerschmidt, *Monatshefte für Chemie*, 1991, **122**, 389.
74. T. Kuzuyama, T. Hidaka, S. Imai, and H. Seto, *J. Antibiot.*, 1993, **46**, 1478.
75. T. Hidaka, T. Kuzuyama, and H. Seto, *Actinomycetologica*, 1994, **8**, 41.
76. T. Hashimoto, A. Hayashi, Y. Amano, J. Kohno, H. Iwanari, S. Usuda, and Y. Yamada, *J. Biol. Chem.*, 1991, **266**, 4648.
77. J. Matsuda, S. Okabe, T. Hashimoto, and Y. Yamada, *J. Biol. Chem.*, 1991, **266**, 9460.
78. H. Seto, unpublished results.
79. M. Peretz and Y. Burstein, *Biochemistry*, 1989, **28**, 6549.
80. D. L. Popham and P. Setlow, *J. Bacteriol.*, 1993, **175**, 2917.
81. T. Kuzuyama, S. Kobayashi, K. Ohara, T. Hidaka, and H. Seto, *J. Antibiot.*, 1996, **49**, 502.
82. J. E. Suarez and M. C. Mendoza, *Antimicrob. Agents Chemother.*, 1991, **35**, 791.
83. W. W. Metcalf and B. L. Wanner, *J. Bacteriol.*, 1993, **175**, 3430.

1.31 Biosynthesis and Degradation of Cyanogenic Glycosides

MONICA A. HUGHES
University of Newcastle upon Tyne, UK

1.31.1 INTRODUCTION

1.31.1.1 Cyanogenesis

The term cyanogenesis describes the release of hydrogen cyanide (HCN) from damaged plant (and some insect[1]) tissue. Although HCN can occur in small quantities in all plant tissues, for example cyanide is a product of ethene biosynthesis,[2] in cyanogenic species large quantities of HCN are produced only following tissue disruption. Cyanogenesis was first described in plants in 1803 and has now been reported in at least 2600 species, within 130 different families.[3,4] In approximately 475 of these species, the source of HCN has been identified[4] and shown to result from the enzymatic degradation of cyanogenic glycosides. In most species this degradation involves hydrolysis by one or more β-glycosidases, followed by enzymatic breakdown of the aglycone (cyanohydrin) to a carbonyl compound and HCN by an α-hydroxynitrile lyase.

The last general review of cyanogenesis was published in 1990[5] and since that time there have been major advances in our understanding of the biosynthesis of cyanoglycosides[6,7] and the crystal structure of two of the enzymes responsible for the degradation of cyanogenic glycosides has been solved.[8,9]

1.31.1.2 Phylogenic Distribution of Cyanogenesis in Plants

Angiosperm families that are noted for cyanogenesis are Rosaceae (150 species), Leguminosae (125 species), Gramineae (100 species), Araceae (50 species), Euphorbiaceae (50 species), Compositae (50 species), and Passifloraceae (30 species).[10] In addition, some gymnosperms (for example, *Taxus baccata* L.[11]) and ferns (for example, *Davallia trichomanoides*[12]) are cyanogenic. In many plant species only a single cyanogenic glycoside has been reported, however, in an increasing number of plants, more than one compound has been found. In barley (*Hordeum vulgare* L.), for example, five cyanogenic glycosides have been identified in the leaf epidermal cells.[13] In barley, these compounds differ in the structure of the aglycone but in other species cyanogenic glycosides with different sugar residues are found. For example, both monoglucosides and diglucosides are found in *Prunus serotina* Ehrh. (black cherry)[14] and in *Linum usitatissimum* L. (flax).[15] Cyanogenesis is of limited use in angiosperm phylogeny studies because it occurs in both primitive and advanced groups and because the occurrence is erratic in most families and even within some genera (*Trifolium*).[16] An interesting situation is found in the genus *Acacia*, where cyclic cyanogenic glucosides are found in Australian species whereas aliphatic groups are found in African, Asian, and American species.[16]

Some plant species are polymorphic for the cyanogenic phenotype, that is both cyanogenic and acyanogenic plants may occur in the same species. The most extensive studies of the cyanogenic polymorphism have been carried out in the herbage legume, *Trifolium repens* L. (white clover) (reviewed by Hughes[17]) and in *Lotus corniculatus* L. (birds's foot trefoil).[18] The cyanogenic polymorphism in white clover is controlled by alleles of two independently segregating loci (*Li* and *Ac*). It shows diploid inheritance and only plants which contain a functional, dominant allele of both loci are cyanogenic. In two papers published in 1954, Daday[19,20] demonstrated a clear association between the frequency of the cyanogenic morph in natural populations of white clover and the mean January isotherm, such that populations at higher altitudes and higher latitudes have lower frequencies of cyanogenic plants. This association has been confirmed by other workers and a survey of the US white clover germplasm collection has shown that accessions from low altitudes and from sites with a high winter temperature, lower summer precipitation, spring sunshine, and snow cover, have higher frequencies of cyanogenic plants.[21]

The cyanogenic polymorphism in white clover is thought to be maintained by selection for the acyanogenic morph by increased frost damage in cyanogenic plants and a balancing selection for the cyanogenic morph, caused by increased predation of acyanogenic plants by small predators (see Section 1.31.1.3).[17] Quantitative variation in levels of cyanogenesis has also been documented and, for example, variation in levels of cyanogenic glucoside in mature plants of the tropical species *Turnera ulmifolia* L., collected from different locations in Jamaica, has been shown to have a genetic basis,[22] however, the selective agents have not been firmly identified.

Variation in levels of cyanogenesis has also been reported for single plants, depending upon both external environmental factors and development. In white clover cyanogenic glucoside synthesis is influenced by temperature[23] and, in cassava (*Manihot esculenta* Crantz) leaves, diurnal variation in cyanogenic glucoside levels has been reported.[24] Patterns of changes in cyanogenic glucoside content during development vary between species. In *Hevea* species, the seeds contain high levels of cyanogenic glucosides and as the seeds germinate levels in the endosperm fall but levels in the embryo/plantlet increase[25] (see Section 1.31.1.4). In cassava, which is also a member of the Euphorbiaceae and produces the same cyanogenic glucosides as *Hevea* (linamarin (**1**) and lotaustralin (**2**), see Section 1.31.2), the seeds contain virtually no cyanogenic glucoside. Both cyanogenic glucosides, and the enzymes responsible for cyanogenic glucoside degradation during cyanogenesis, are synthesized rapidly *de novo* during germination in cassava (Figure 1).

1.31.1.3 Cyanogenesis in Plant–Animal and Plant–Microbe Interactions

The toxicity of hydrogen cyanide to insects is well known and cyanogenesis is widely regarded as a defense mechanism which has evolved because it protects plants from predation by small herbivores;

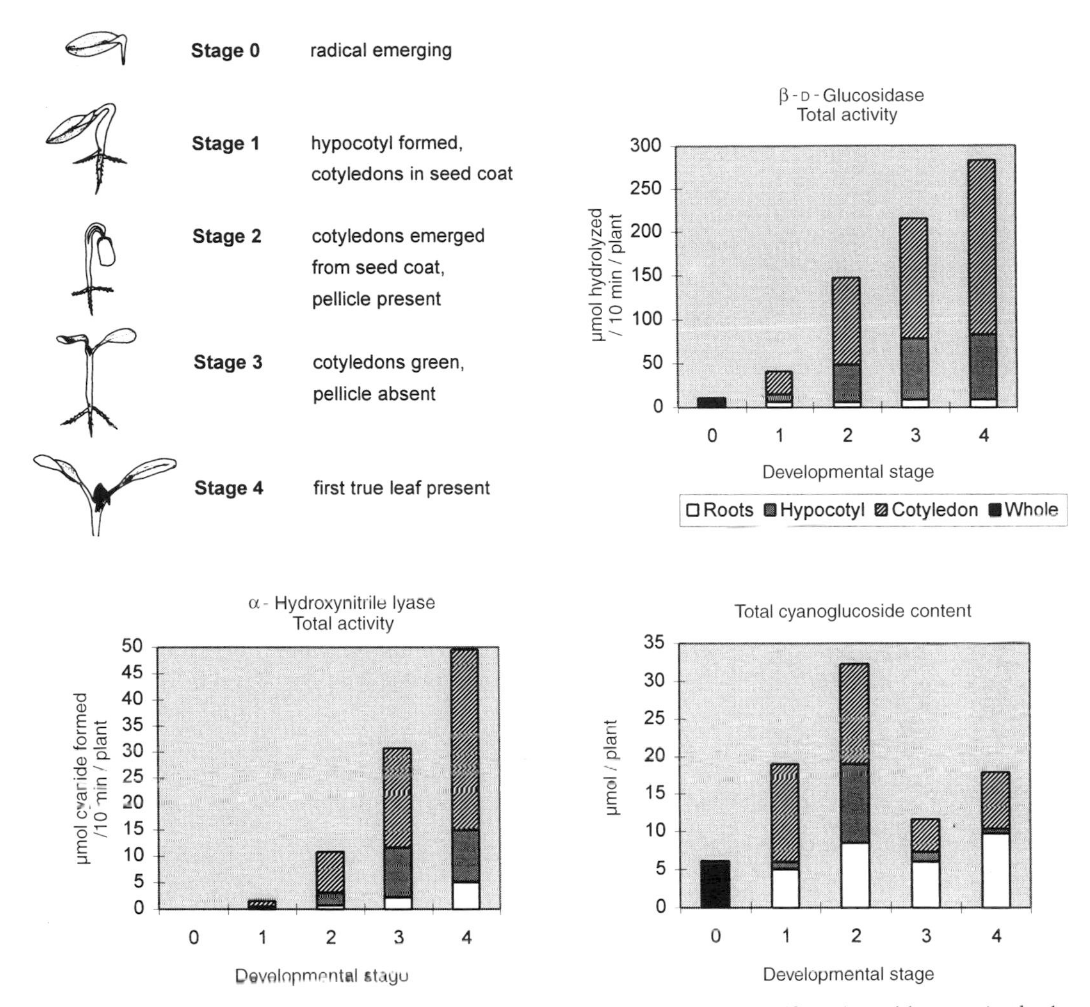

Figure 1 Production of cyanogenic glucosides and the degrading enzymes (β-D-glucosidase and α-hydroxynitrile lyase) in different organs during germination of *Manihot esculenta* Crantz (cassava) seeds.

however, the interaction of cyanogenic host plant and herbivore is complicated in many cases by co-evolution. In species which are polymorphic for cyanogenesis, there is abundant evidence from both natural habitats and experimental data that selective grazing of the acyanogenic form occurs;[17,18,26] however, there is considerable variation between species of herbivores in selective feeding behaviour. In plant species such as sorghum (*Sorghum bicolor* L. Moench.) and cassava, which are not polymorphic for cyanogenesis, there is quantitative variation in the levels of cyanide produced. Studies of herbivore grazing in these species have compared feeding/damage in cultivars with different levels of cyanide production (cyanogenic potential). Cassava is a crop which evolved in South America and although it appears to be resistant to many pests there are several specialized pests, such as the hornworm (*Erinnyis ello*) and the green mite (*Mononychellus tanajoa*), which have co-evolved with cassava and show no preference between plants that are highly cyanogenic or not.[27] The burrowing bug *Cyrtomenus bergi*, however, appears to be a recent pest of the cassava crop in Colombia and laboratory and field experiments show that root damage from this pest is reduced in those cassava cultivars with a high cyanogenic potential.[27]

A number of herbivorous insect species are also cyanogenic[1] and these may feed upon cyanogenic plants. The larvae of *Acraea horta* (Lepidoptera) feed upon the leaves of the cyanogenic species *Kiggelaria africana* L., and the cyanogenic glucoside (gynocardin (**17**)) which is taken up by the larvae, is sequestered by the insect.[28] However, although some *Zygaena* species eat birds' foot trefoil, *de novo* synthesis of cyanogenic glucosides (linamarin (**1**) and lotaustralin (**2**)) has been shown to occur in this genus of Lepidoptera.[1,29] *T. repens* L. (white clover) is one of the primary host plants of the sulfur butterfly (*Colias erate poliographys*) and cyanogenic glucosides have been shown to serve as synergistic oviposition stimulants for this insect, suggesting that they play a positive role in host selection.[30]

Many plant pathogens have the ability to detoxify the HCN produced from cyanogenic glycosides during cyanogenesis. In only a small number of examples do high levels of cyanogenesis correlate with resistance to pathogens. Lehman *et al.*[31] report that infection of white clover by *Sclerotina trifoliorum* is reduced in the highly cyanogenic cultivar, Arau. However, Lieberei *et al.*[32] have shown that cyanogenesis inhibits active pathogen defence in plants and *Microcyclus ulei*, which causes blight of *Hevea brasiliensis* Muell. Arg. (rubber tree), is not only tolerant of HCN but grows better in an HCN-containing atmosphere.[33] This means that weakly cyanogenic plants may generally show more resistance to the pathogen than highly cyanogenic plants.

1.31.1.4 Cyanogenesis and Metabolism

Although it is widely accepted that the cyanogenic system is a plant mechanism for protection against herbivores, two other roles have been suggested, namely that cyanogenic glycosides are either (i) waste products or (ii) intermediates in nitrogen metabolism.[18]

The hypothesis that cyanogenic glycosides are nitrogenous waste compounds is difficult to defend. Nitrogen is often limiting for plant growth and cyanogenic species are generally not limited to ecological habitats with nitrogen-rich soil. Further, other examples of nitrogenous waste products are not known in plants.

The presence of cyanogenic and acyanogenic individuals in polymorphic species, such as white clover, argues against a role in primary metabolism, in these species at least. Although cyanogenic glycosides do not have an essential role in primary metabolism and despite the general observation that they are stable compounds stored in cellular compartments that lack degrading β-glycosidases, a number of examples exist where turnover of cyanogenic glycosides has been reported. It has been suggested that cyanogenic diglucosides are metabolites of cyanogenic monoglucosides, which can be translocated within the plant because they are resistant to the abundant monoglucosidase enzyme. In seeds of *H. brasiliensis* Muell. Arg., the cyanogenic monoglucoside linamarin (**1**) accumulates in the endosperm. After the onset of germination, the levels of this glucoside in the endosperm decrease, with a concomitant increase in the level of the diglucoside linustatin (**4**) in endosperm exudates.[34] It is proposed that during germination the stored monoglucoside linamarin (**1**) is glycosylated to the diglucoside linustatin (**4**). This makes it resistant to the abundant apoplastic monoglucosidase so that it can be transported from the endosperm to the growing seedling, where it is cleaved by a diglucosidase to produce HCN. Negligible amounts of gaseous HCN are produced because cyanide is reassimilated into noncyanogenic compounds.[34] Detoxification of HCN to asparagine by β-cyanoalanine synthase produced in developing *Hevea* seedlings allows the HCN to reenter general metabolic pools.[34] The monoglucoside linamarin (**1**), produced in the developing plantlet is synthesized *de novo* from the precursor amino acid, valine (see Sections 1.31.2.2 and 1.31.3).

The diglucoside linustatin (**4**) has also been isolated from flax,[35] which also produces cyanogenic seeds, suggesting that this mechanism may be general in plants. In fact, small amounts of dhurrin 6′-glucoside (the diglucoside produced by further glycosylation of the monoglucoside dhurrin (**14**) have been identified in guttation droplets of *S. bicolor* (L.) Moench. seedlings,[36] although this species does not have cyanogenic seeds. Turnover of dhurrin (**14**) in green sorghum seedlings has also been demonstrated using an inhibitor of tyrosine (the precursor amino acid of dhurrin (**14**), see Section 1.31.2.5) biosynthesis and radiolabeled tyrosine in *in vivo* feeding experiments.[37]

1.31.1.5 Compartmentation

Consistent with a role in plant defense, cyanogenic glucosides are stored and separated from the catabolic enzymes in the intact plant by compartmentation at either tissue or subcellular levels. Information about compartmentalization is not available for many cyanogenic species but it is clear, from those which have been studied, that the details of compartmentation differ between species.

In the leaves of sorghum seedlings, the cyanogenic glucoside is sequestered within the vacuoles of epidermal cells, whereas the two degrading enzymes, β-glucosidase and α-hydroxynitrile lyase, are present almost exclusively in the underlying mesophyll cells, within the chloroplasts and cytosol respectively.[38,39] Large-scale hydrolysis of the cyanogenic glucoside can therefore only occur following tissue disruption, such as during herbivore attack.

Cassava is a member of the Euphorbiaceae and contains a network of latex vessels which run throughout the plant. The number of vessels in different organs and in different tissues within organs varies. These vessels are, for example, abundant in young leaf spongy mesophyll tissue but relatively rare in parenchyma of the swollen roots. White *et al.*[40] have demonstrated the presence of the cyanogenic glucoside, linamarin (**1**), in cassava leaf vacuoles. The first degrading enzyme, a cyanogenic β-glucosidase with the trivial name linamarase (see Section 1.31.4.1), is primarily located in the latex vessels.[41,42] The exact location of the α-hydroxynitrile lyase is not known but the structure of the protein (having no signal sequence or organelle retention signals) suggests that it is cytosolic.[43] In white clover, which produces the same cyanogenic glucosides as cassava but is a legume and therefore possesses no latex vessel system, the cyanogenic β-glucosidase (see Section 1.31.4.1) has been shown to be apoplastic, possibly present in cell walls.[44] The techniques which are commonly used to demonstrate an apoplastic location for proteins[45] are difficult to interpret in a species with latex vessels containing latex under pressure, and a number of reports of the apoplastic location of the cassava linamarase have to be interpreted with caution, particularly since a latex control enzyme such as chitinase was not included in the experiments.[46]

Cyanogenesis in black cherry (*P. serotina*) has been extensively studied by Poulton's group.[47–51] The kernels of black cherry seeds contain large quantities of the cyanogenic diglucoside (*R*)-amygdalin (**12**) and three catabolic enzymes: the diglucosidase amygdalin hydrolase; the monoglucosidase, prunasin hydrolase; and an α-hydroxynitrile lyase, (*R*)-(+)-mandelonitrile lyase. These enzymes first appear in the seeds about 6 weeks after flowering. The two β-glucosidases are restricted to protein bodies in the procambium, whereas the hydroxynitrile lyase occurs primarily in protein bodies in the cotyledonary parenchyma cells, which is also the location of the cyanogenic diglucoside, amygdalin (**12**). Thus, in black cherry, cyanogenesis in intact tissues of the developing seed is prevented by segregation of the first degrading enzyme, amygdalin hydrolase, and amygdalin (**12**) in different tissues.

1.31.2 CHEMICAL NATURE OF CYANOGENIC GLYCOSIDES

1.31.2.1 Precursor Amino Acids and Nicotinic Acid

Cyanogenic glycosides are of intermediate polarity, being water-soluble compounds which are typically *O*-β-glycosides of α-hydroxynitriles (cyanohydrins), and are themselves relatively nontoxic to most organisms. All of the 57 known higher plant cyanogenic glycosides are probably derived from the five hydrophobic L-amino acids, valine, isoleucine, leucine, phenylalanine and tyrosine, the nonprotein amino acid (2-cyclopentenyl)glycine, and nicotinic acid.[4,36] Glucose is the sugar directly attached to the hydroxy of the cyanohydrin. In addition to the optically active centres of the sugars, the carbon of the cyanohydrin which is attached to the sugar, the nitrile group, and the hydrogen are also usually chiral. Thus (*R*)- and (*S*)-epimers are known for most series and in some plants both epimers can co-occur, for example, (*R*)-prunasin (**10**) and (*S*)-sambunigrin (**11**) in *Acacia* species.[52] The structures of cyanogenic glycosides and related compounds have been reviewed by Seigler,[4] where all of their structures can be found.

1.31.2.2 Cyanogenic Glycosides Derived from Valine and Isoleucine

The cyanogenic glucosides linamarin (**1**) and lotaustralin (**2**), which are derived from the amino acids valine and isoleucine, respectively, commonly co-occur, although the proportion of each may vary. Thus, in cassava 95% of the total cyanogenic glucoside is linamarin (**1**),[53] whereas in white clover they are more or less equally abundant.[54] These cyanogenic glycosides are recorded in more plant species than any of the other cyanogenic glycosides. The (*S*)-epimer of lotaustralin, (*S*)-epilotaustralin (**3**), is difficult to distinguish from (*R*)-lotaustralin (**2**) and most reports do not specify which epimer is produced. The diglucosides of linamarin (**1**) and lotaustralin (**2**) (linustatin (**4**) and neolinustatin (**5**) respectively) have been found at low levels in flax,[35] *Hevea*,[34] and cassava,[55] all of which produce linamarin (**1**) and lotaustralin (**2**).

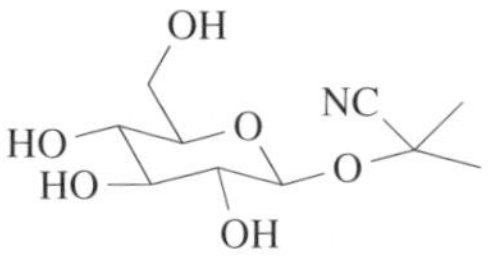

Linamarin (**1**) 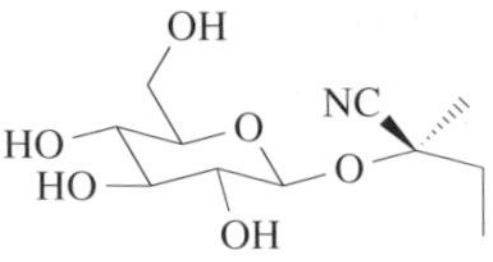   (*R*)-Lotaustralin (**2**) 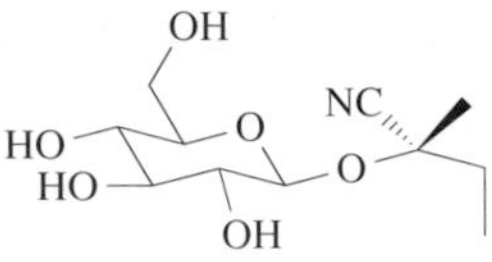   (*S*)-Epilotaustralin (**3**)

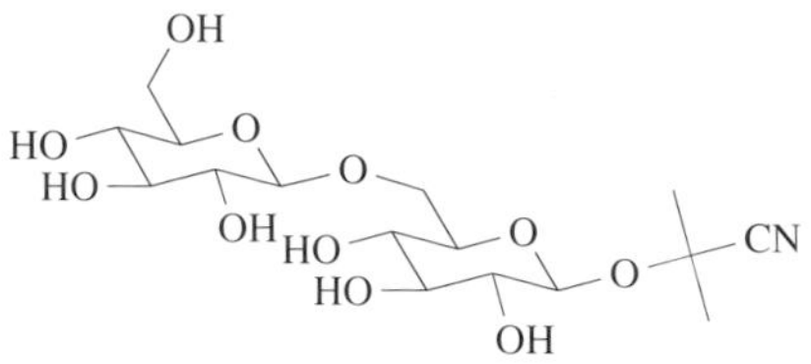

Linustatin (**4**) 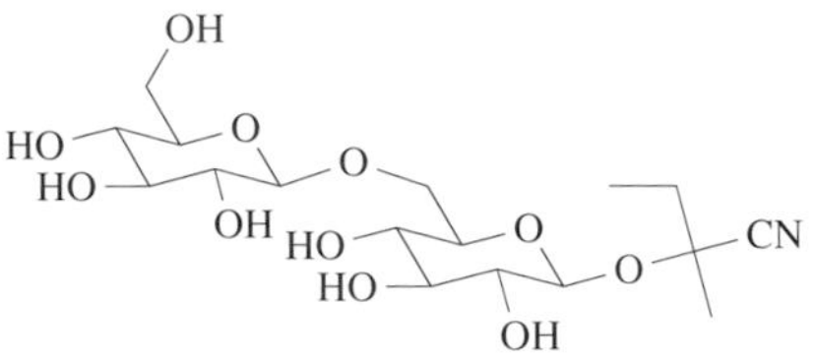   Neolinustatin (**5**)

1.31.2.3 Cyanogenic Glycosides Derived from Leucine

Ten cyanogenic glycosides are known which have L-leucine as the precursor. Four of these are illustrated ((**6**) to (**9**)). (*R*)-Epiheterodendrin (**9**) is produced by germinating barley. Its breakdown during fermentation of malted barley is a problem because the HCN released can react with ethanol to produce ethylcarbamate, an established carcinogen.[56]

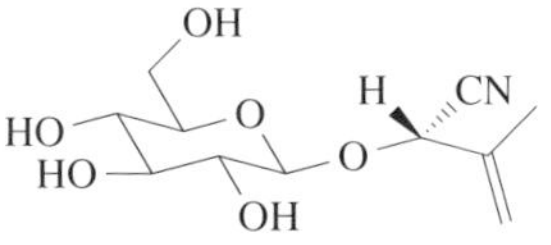

(*S*)-Proacacipetalin (**6**) 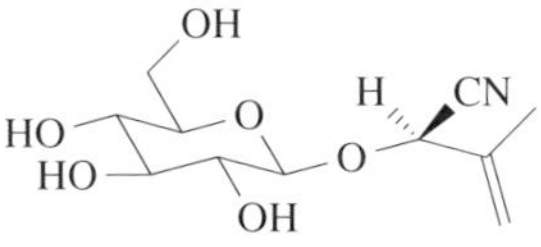   (*R*)-Epiproacacipetalin (**7**)

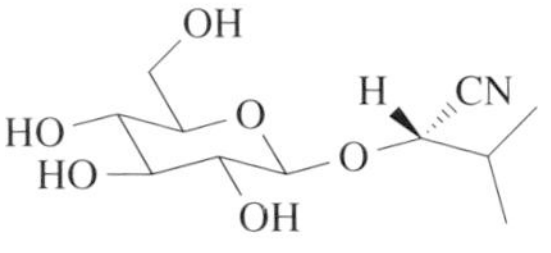

(*S*)-Heterodendrin (**8**) 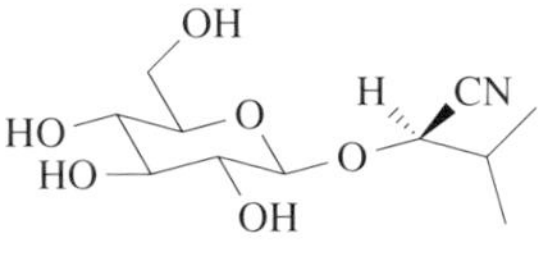   (*R*)-Epiheterodendrin (**9**)

1.31.2.4 Cyanogenic Glycosides Derived from Phenylalanine

One of the best-known cyanogenic glycosides, (*R*)-amygdalin (**12**) is a diglucoside derived from phenylalanine. This compound is commonly found in seeds of members of the Rosaceae, such as black cherry, almonds, peaches, and apricots.[4] There are six monoglucosides known to be derived from phenylalanine, including the monoglucoside equivalent of amygdalin, (*R*)-prunasin (**10**), and the *S*-epimer of (*R*)-prunasin, (*S*)-sambunigrin (**11**).

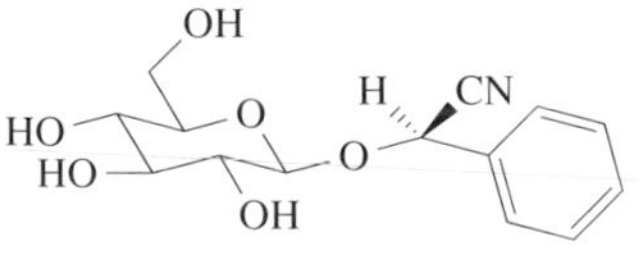

(*R*)-Prunasin (**10**) 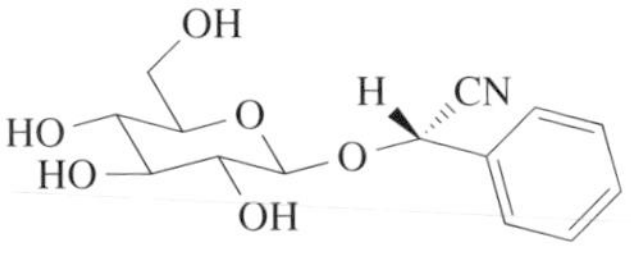   (*S*)-Sambunigrin (**11**)

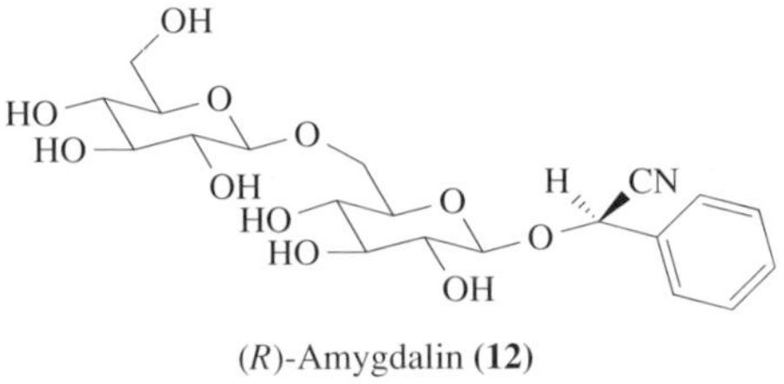

(*R*)-Amygdalin (**12**)

1.31.2.5 Cyanogenic Glycosides Derived from Tyrosine

Seven cyanogenic glycosides are known which have tyrosine as a precursor. This series also includes (*R*)- and (*S*)-epimers of monoglucosides and equivalent diglucosides.[4,36] The structure of (*R*)-taxiphyllin (**13**) and (*S*)-dhurrin (**14**) are shown. The biosynthesis of dhurrin (**14**) in etiolated sorghum seedlings is the best-understood biosynthetic system (see Section 1.31.3.1).

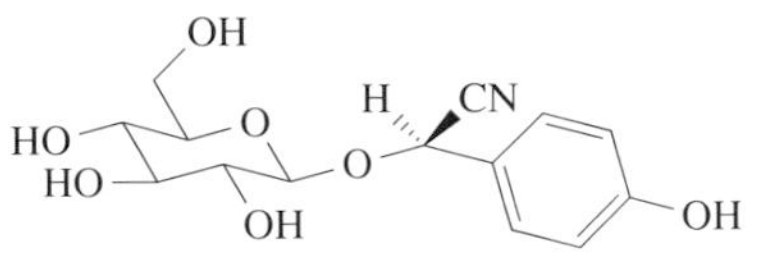

(*R*)-Taxiphyllin (**13**)

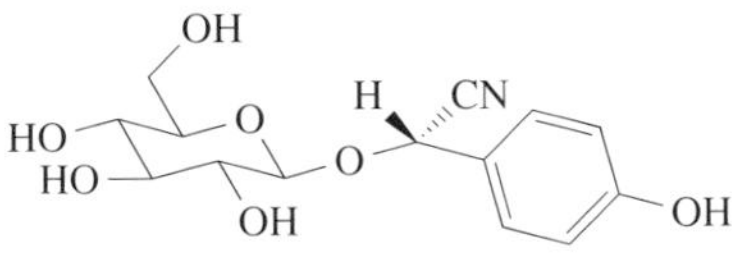

(*S*)-Dhurrin (**14**)

1.31.2.6 Cyanogenic Glycosides Derived from the Non-protein Amino Acid (2-Cyclopentenyl)glycine and Nicotinic Acid

A group of 14 cyanogenic glycosides has been identified, probably derived from (2-cyclopentenyl)glycine, three of which are illustrated (**15**, **16**, **17**). As with the other series, the group includes *R*- and *S*-epimers and both mono- and diglucosides. They are commonly found in Passifloraceae.[4] The compound acalyphin (**18**) from a member of the Euphorbiaceae (*Acalypha indica* L.) appears to be derived from nicotinic acid.[57]

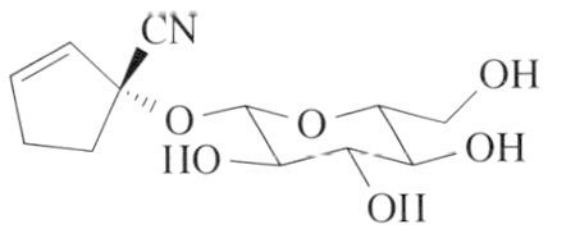

Deidaclin (**15**)

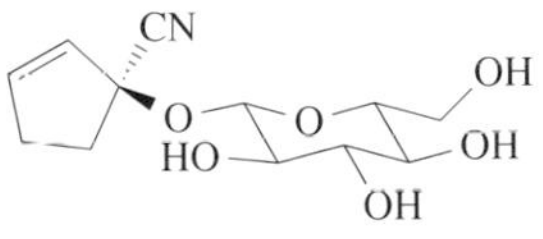

Tetraphyllin A (**16**)

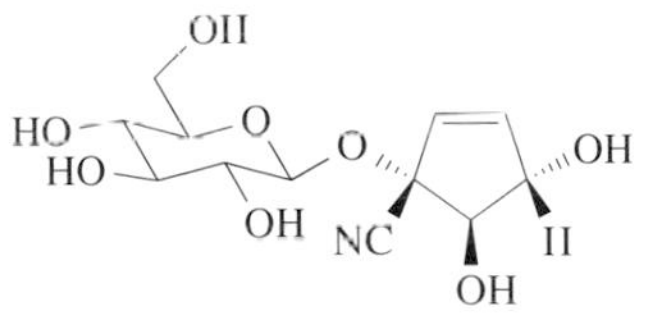

Gynocardin (**17**)

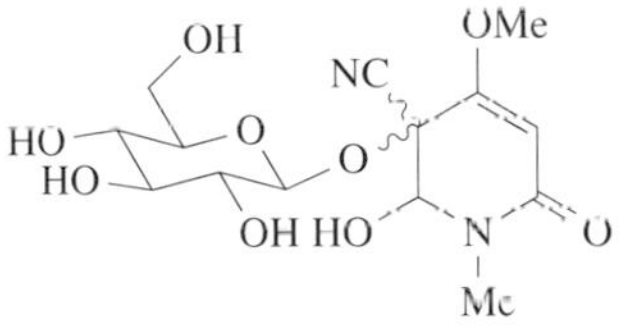

Acalyphin (**18**)

1.31.3 BIOSYNTHESIS OF CYANOGENIC GLYCOSIDES

1.31.3.1 Cyanohydrin (α-Hydroxynitrile) Biosynthesis

The biosynthetic pathway of cyanogenic glycosides in higher plants is considered to be closely related to the pathways producing glucosinolates, organic nitro- compounds and possibly nitrile glycosides. In outline, a membrane-bound enzyme system converts a precursor amino acid to an α-hydroxynitrile via an oxime. The α-hydroxynitrile is then glucosylated by a soluble UDP-glucose glucosyltransferase. Scheme 1 shows the biosynthetic pathway for the cyanogenic glucoside, dhurrin, in *Sorghum bicolor* (L.) Moench. The reactions catalyzed by cytochrome $P450_{TYR}$ are boxed.[67] *In vivo* feeding of labeled amino acid precursors to plants which are actively synthesizing cyanogenic glycoside commonly results in extremely efficient labeling of the cyanogenic glycosides, reflecting the very large quantities of these compounds which can accumulate in some tissues.

Early biosynthetic studies demonstrated the direct incorporation of the C_{β}—C_{α}—N moiety of the precursor amino acid into the cyanogenic glycoside, indicating that all of the intermediates contain it.[58] The carboxyl carbon of the amino acid is lost, the α-carbon bearing the amine group is oxidized to the level of nitrile, and the β-carbon is oxygenated to yield a hydroxy group, which bears the glucose of the glucoside. All early studies found negligible accumulation of pathway intermediates

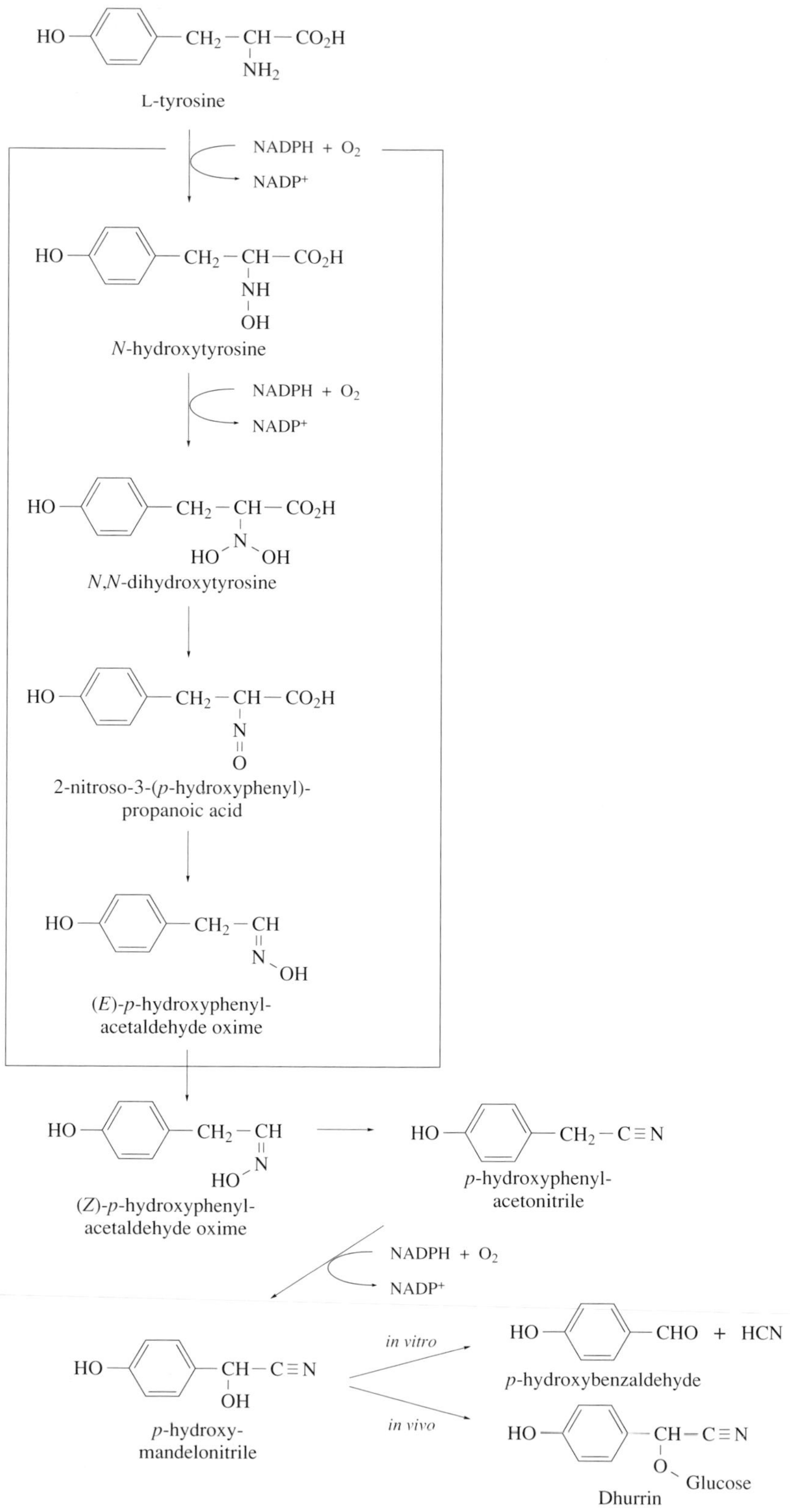

HO–C6H4–CH_2–CH(NH2)–CO_2H

L-tyrosine

NADPH + O_2 → $NADP^+$

HO–C6H4–CH_2–CH(NH–OH)–CO_2H

N-hydroxytyrosine

NADPH + O_2 → $NADP^+$

HO–C6H4–CH_2–CH(N(OH)2)–CO_2H

N,*N*-dihydroxytyrosine

HO–C6H4–CH_2–CH(N=O)–CO_2H

2-nitroso-3-(*p*-hydroxyphenyl)-propanoic acid

HO–C6H4–CH_2–CH=N–OH

(*E*)-*p*-hydroxyphenyl-acetaldehyde oxime

HO–C6H4–CH_2–CH=N–OH

(*Z*)-*p*-hydroxyphenyl-acetaldehyde oxime

→ HO–C6H4–CH_2–C≡N

p-hydroxyphenyl-acetonitrile

NADPH + O_2 → $NADP^+$

HO–C6H4–CH(OH)–C≡N

p-hydroxy-mandelonitrile

in vitro → HO–C6H4–CHO + HCN

p-hydroxybenzaldehyde

in vivo → HO–C6H4–CH(O–Glucose)–C≡N

Dhurrin

Scheme 1

and the pathway was therefore referred to as 'channeled'. To date, the most detailed studies on cyanogenic glycoside biosynthesis have been carried out for dhurrin (**14**) in sorghum. Based on the similarity of the biosynthetic reactions (when studied) in other cyanogenic plants, it is believed that the information obtained in sorghum can be extrapolated to other plants.

In vitro, biosynthetic studies were made possible by the isolation of a biosynthetically active microsomal preparation from etiolated seedlings of sorghum, which was capable of converting L-tyrosine into (*S*)-*p*-hydroxymandelonitrile, the α-hydroxynitrile (cyanohydrin) precursor of dhurrin (**14**).[59] Thus this preparation can carry out all except the final glycosylation step in the dhurrin biosynthetic pathway (Scheme 1). It has been demonstrated that the sorghum microsomal preparation can produce and metabolize the intermediates *N*-hydroxytyrosine, 2-nitroso-3-(*p*-hydroxyphenyl)propanoic acid, (*E*)- and (*Z*)-*p*-hydroxyphenylacetaldehyde oxime, *p*-hydroxyphenylacetonitrile and *p*-hydroxymandelonitrile, which are shown in the biosynthetic pathway in Scheme 1.[59–64] The compound *N*,*N*-dihydroxytyrosine is very labile and has not been isolated. The only intermediate in this pathway that freely exchanges with exogenously supplied material is (*Z*)-*p*-hydroxyphenylacetaldehyde oxime.

Stoichiometric measurements of oxygen consumption and biosynthetic activity have shown that two molecules of oxygen are consumed in two consecutive *N*-hydroxylation reactions in the conversion of L-tyrosine to *p*-hydroxyphenylacetaldehyde oxime and one oxygen in the C-hydroxylation converting *p*-hydroxyphenylacetonitrile to *p*-hydroxymandelonitrile.[64]

The L-tyrosine *N*-hydroxylase is inhibited by carbon monoxide and this inhibition is reversed by 450 nm light. This demonstrates that the enzyme is a cytochrome P450 dependent monooxygenase.[65] A sorghum seedling heme-thiolate enzyme, cytochrome $P450_{TYR}$, which catalyzes the conversion of L-tyrosine to *p*-hydroxyphenylacetaldehyde oxime, has been isolated, purified, and characterized.[6,7] Cytochrome P450 dependent monooxygenase reactions are dependent on small electron transport chains, where reducing equivalents from NADPH are transferred via a flavin-containing oxidoreductase to the terminal cytochrome P450. When a reconstituted complex containing purified cytochrome $P450_{TYR}$, NADPH-cytochrome P450 oxidoreductase, and L-α-dilauroylphosphatidylcholine is fed L-tyrosine in the presence of NADPH, *p*-hydroxyphenylacetaldehyde oxime accumulates.[7] A cDNA clone encoding the sorghum cytochrome $P450_{TYR}$ has been isolated and expressed in *Escherichia coli*.[66,67] The purified *E. coli* recombinant protein also catalyzes the conversion of L-tyrosine to *p*-hydroxyphenylacetaldehyde oxime in reconstitution experiments, using sorghum NADPH–cytochrome P450 reductase.[66] The surprising biosynthetic properties of this cytochrome P450 raise questions about the nature of the intermediates between amino acid and oxime, shown in Scheme 1. Could some of the compounds detected in earlier experiments represent artificially generated stable forms of transition states?

Conversion of the *p*-hydroxyphenylacetaldehyde oxime to *p*-hydroxymandelonitrile in the presence of oxygen, by isolated sorghum microsomes, involves a C-hydroxylation reaction of *p*-hydroxyphenylacetonitrile. This reaction also shows inhibition by carbon monoxide that is reversed by 450 nm light, characteristic of cytochrome P450;[65] in addition, the reaction is inhibited by antibodies to NADPH-cytochrome P450 oxidase. It was not possible to dissect the cofactor requirements for the conversion of *p*-hydroxyphenylacetaldehyde oxime to *p*-hydroxymandelonitrile into two separate reactions and the intermediate *p*-hydroxyphenylacetonitrile does not accumulate in the reaction mixture.[65] Purification and characterization of the protein(s) responsible for oxime metabolism in sorghum have not been reported.

Microsomal preparations, which can carry out metabolism of valine and isoleucine to produce linamarin (**1**) and lotaustralin (**2**), respectively, have also been isolated from white clover,[68] flax,[69] and cassava.[70] Further it has been shown that the metabolism of valine to linamarin and isoleucine to lotaustralin is carried out by the same proteins.[71] In white clover, plants possessing only non-functional *ac* alleles are unable to synthesize either linamarin or lotaustralin.[72] *In vivo* and *in vitro* labeling experiments have shown that *ac ac* plants have at least two steps in the conversion of amino acids to α-hydroxynitrile missing from the microsomal preparations. Thus microsomes from *ac ac* plants are (i) unable to produce the oxime intermediate and (ii) unable to convert fed oxime to α-hydroxynitrile, whilst *Ac Ac* microsomes can carry out both steps.[68] Microsomes have also been isolated from *Triglochin maritima* L. which produce the cyanohydrin of taxiphyllin (**13**) from tyrosine.[73] It is a general feature of these biosynthetic studies that the oxime is the only intermediate which can be easily detected or used in the cyanohydrin product, either when fed *in vivo* or incorporated into the microsome reaction mix. It is tempting to speculate that in all higher plant species, the metabolism of amino acid to cyanohydrin involves just two cytochrome P450 enzymes with the oxime being the only true intermediate.

1.31.3.2 UDP-glucose Glucosyltransferase

The final step in the biosynthesis of cyanogenic glycosides is glucosylation of the α-hydroxylnitrile (see Scheme 1). UDP-glucose glucosyltransferase enzymes, which are capable of glucosylating the respective α-hydroxylnitrile, have been partially purified from black cherry,[74] *T. maritima* L.,[75] flax,[76] sorghum,[77] and cassava.[78] All of the enzymes behave as soluble enzymes and are not found associated with the microsomal preparations which synthesize the α-hydroxynitrile. The soluble nature of these glucosyltransferases is perhaps surprising given (i) the 'channeled' nature of α-hydroxynitrile biosynthesis by membrane-bound (microsomal) enzymes, (ii) the instability of the α-hydroxynitriles, and (iii) the localization of the cyanogenic glycosides in vacuoles.

Glycosylation of a number of secondary plant compounds, including flavonoids and steroidal alkaloids, occurs at the end of their biosynthetic pathway. The most common sugar is glucose and these reactions are also catalyzed by a UDP-glucose glucosyltransferase to produce stable water-soluble compounds which are often transported into the vacuole. Given the reported specificity of these enzymes and the large number of potential substrates, a wide range of different glucosyltransferases may be expected to occur within a single plant species.[79] Further, considerable deduced amino acid sequence homology exists between those plant glucosyltransferases which have been cloned.[80] These factors have contributed to the difficulty of purifying and characterizing the cyanogenic glycoside UDP-glucose glucosyltransferase and detailed information about these proteins does not exist.

1.31.4 DEGRADATION OF CYANOGENIC GLYCOSIDES

1.31.4.1 β-Glycosidases

Scheme 2 shows the degradation of linamarin (**1**) and lotaustralin (**2**) by β-D-glucosidase and α-hydroxynitrile lyase. The first step in the degradation of cyanogenic glycosides is hydrolysis by one or more β-glycosidases. In most cyanogenic plant species one or more β-glycosidases are produced which have pronounced specificity for their endogenous cyanogenic glycoside(s);[81] however, examples of enzymes with very broad specificity exist (e.g., flax[82]). All of the cyanogenic β-glycosidases tested will hydrolyze the synthetic substrates, *p*-nitrophenyl-β-D-glucoside and *p*-nitrophenyl-β-D-galactoside, and are therefore not entirely specific for either the aglycone or the sugar moiety of the substrate. The cyanogenic β-glycosidases that have been investigated are all glycoproteins with a subunit molecular mass of $55–65 \times 10^3$, isoelectric points between pH 4.0–5.5, and acidic pH optima (pH 4.0–6.2).[5]

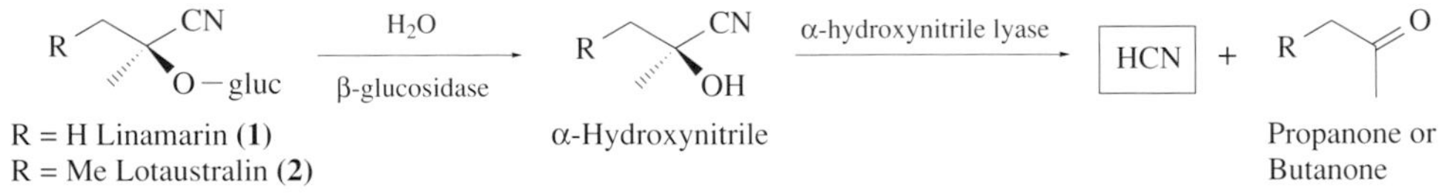

R = H Linamarin (**1**)
R = Me Lotaustralin (**2**)
H_2O / β-glucosidase → α-Hydroxynitrile → α-hydroxynitrile lyase → HCN + Propanone or Butanone

Scheme 2

Hydrolysis of cyanogenic disaccharides may be either 'simultaneous', such as linustatin(**4**) in *H. brasiliensis* Muell. Arg.[83] and vicianin in *Vicia augustifolia* L.,[84] where hydrolysis yields a disaccharide plus aglycone. Alternatively hydrolysis can be sequential, where two hydrolytic reactions are catalyzed by two separate β-glycosidases. The best documented example of sequential hydrolysis is amygdalin (**12**) hydrolysis in black cherry.[85] Amygdalin hydrolase degrades the cyanogenic diglucoside, amygdalin (**12**), to produce glucose and the cyanogenic monoglucoside, prunasin(**10**). Prunasin is subsequently hydrolyzed by prunasin hydrolase to produce glucose and the α-hydroxynitrile, mandelonitrile. Four isozymes of amygdalin hydrolase and three isozymes of prunasin hydrolase have been purified from black cherry.[85]

The cyanogenic β-glucosidase responsible for hydrolysis of the monoglucosides, linamarin (**1**) and lotaustralin (**2**), has the trivial name, linamarase. This enzyme has been cloned as cDNA from white clover[86] and from cassava.[87] Classification of these enzymes on the basis of amino acid sequence similarity[88,89] places them in Family 1 of the glycosyl hydrolases. These are known as retaining glycosidases due to retention of the configuration of the anomeric centre of the substrate during hydrolysis in a double displacement mechanism[90] (Scheme 3). Stereoselective substitution in this position, by water or another nucleophile, is supported by a catalytic dyad composed of an acid catalyst residue acting in the departure of the aglycone from the substrate. A nucleophile group (generally a carboxylate) stabilizes the oxocarbonium ion-like transition state and a proposed glucosyl-enzyme intermediate from an axial direction.

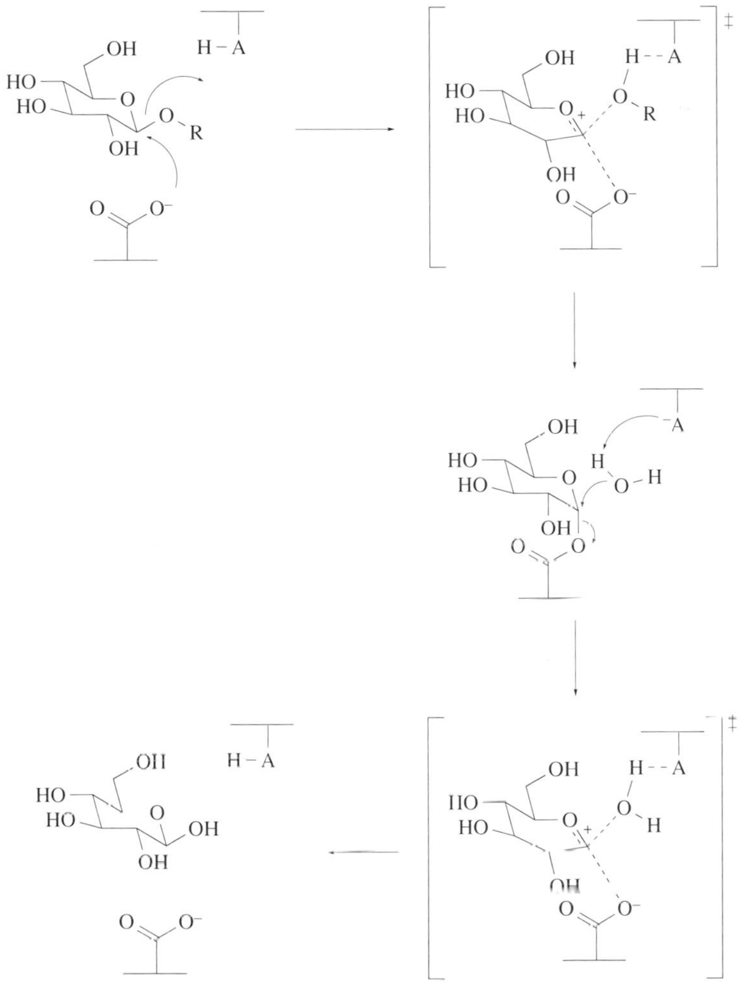

Scheme 3

The cyanogenic β-glucosidases from white clover and cassava have the closest homology to the Family 1 β-glucosidase from *Agrobacterium* spp.[91] and in this protein Glu-358 has been identified as the catalytic nucleophile.[92] This glutamate lies within the highly conserved peptide I/VTENG, which is also present in the two cyanogenic β-glucosidases.

The acid catalyst group (Glu-198) was identified in the cassava cyanogenic β-glucosidase by affinity labeling with the inhibitor, *N*-bromoacetyl-β-D-glucosylamine.[91] This amino acid (Glu-198) also lies within a highly conserved peptide (NEP). The crystal structure of the white clover cyanogenic β-glucosidase has been solved at 2.15 Å resolution.[9] The overall fold of the molecule is an $(\alpha/\beta)_8$ barrel, a structure found in a number of other glycosyl hydrolases, with all of the residues located in a single domain. Residues Glu-183 (in NEP) and Glu-397 (in I/VTENG) are highly conserved and are predicted to be the acid catalyst (proton donor) and the nucleophile catalyst (for stabilization of the glycosylium cation-like transition state), respectively. These roles are consistent with the molecular environments of these two residues in the white clover enzyme. The pocket itself is typical of a sugar-binding site as it contains a number of charged, aromatic and polar groups. Molecular modeling has shown that the active site of the protein encoded by the cassava cyanogenic β-glucosidase cDNA has high structural homology to the white clover protein.[93]

Table 1 Major plant cyanoglucosides and associated α-hydroxynitrile lyases.

Plant	*Precursor amino acids*	*Cyanoglucosides*	*α-Hydroxynitrile lyase*	*Properties of α-hydroxynitrile lyase*	*Catalyzed synthesis of cyanohydrins*	*Ref.*
Black cherry (*Prunus serotina*)	phenylalanine	(*R*)-prunasin[a]	(*R*)-mandelonitrile lyase	60 000 MW, monomer, FAD,[e] PMSF[e] inhib., glycoprotein	(*R*)-aromatic	96
Phlebodium aureum (fern)	phenylalanine	(*R*)-vicianin[b]	(*R*)-mandelonitrile lyase	20 000 MW, homomultimer, no FAD, no PMSF inhib., no CHO[e]	(*R*)-aromatic	100
Ximenia americana	phenylalanine	(*S*)-sambunigrin	(*S*)-mandelonitrile lyase	38 000 MW, monomer, no FAD, PMSF inhib. NS, glycoprotein	NS[e]	101
Sorghum bicolor	tyrosine	(*S*)-dhurrin	(*S*)-*p*-hydroxymandelonitrile lyase	33 000 MW + 18 000 MW heterotetramer, no FAD, DFP[e] inhib., glycoprotein	(*S*)-aromatic	97
Cassava (*Manihot esculenta*)	valine	linamarin[c]	α-hydroxynitrile lyase (acetone cyanohydrin lyase)[d]	29 000 MW, homotrimer, no FAD, PMSF inhib., no CHO	(*S*)-aliphatic	43
Flax (*Linum usitatissimum*)	valine	linamarin[c]	α-hydroxynitrile lyase (acetone cyanohydrin lyase)[d]	42 000 MW, dimer, no FAD, PMSF inhib. NS, no CHO	(*R*)-aliphatic	102
Rubber (*Hevea brasiliensis*)	valine	linamarin[c]	α-hydroxynitrile lyase (acetone cyanohydrin lyase)	30 000 MW, no FAD, no CHO	(*S*)-aliphatic	98

[a] Seeds of *Prunus* species accumulate the diglucoside (*R*)-amygdalin, which is produced by further glycosylation of (*R*)-prunasin. [b] Diglycoside (glucose-arabinose). [c] The same biosynthetic enzymes produce linamarin from valine and (*R*)-lotaustralin from isoleucine. [d] There is no evidence for a separate acetone cyanohydrin and (*R*)-butanone cyanohydrin lyase in cassava or in flax. [e] FAD, flavin prosthetic group; CHO, oligosaccharide; PMSF, phenylmethanesulfonyl fluoride; DFP, diisopropylfluorophosphate; NS, not studied.

The cyanogenic β-glucosidase is encoded by a multigene family in cassava. All of the genes contain 12 introns and the sequence variation between them reflects conservation of those amino acid residues which are important in the structure and function of the protein.[93] The promoter of one of these cassava genes has been analyzed using reporter gene expression,[93] and this analysis indicates that the gene encodes a root-specific cyanogenic β-glucosidase.

In white clover, which is polymorphic for cyanogenesis, the locus *Li* controls the presence of cyanogenic β-glucosidase activity in plants, such that plants homozygous for the nonfunctional allele, *ac*, have no enzyme activity.[17] These plants have been shown to produce no cyanogenic β-glucosidase transcript (mRNA).[94] These observations suggest that there is only one gene encoding the cyanogenic β-glucosidase in white clover.

1.31.4.2 α-Hydroxynitrile Lyases

A number of α-hydroxynitrile lyases from cyanogenic plant species have been characterized, largely because there is considerable interest in the use of α-hydroxynitrile lyases as biocatalysts for the synthesis of optically active α-hydroxynitriles, which are important building blocks in the fine chemical and pharmaceutical industries.[95]

Table 1 summarizes the known properties of α-hydroxynitrile lyase enzymes from seven plant species. From this list the genes for α-hydroxynitrile lyase from black cherry,[96] sorghum,[97] cassava,[43] and *Hevea*[98] have been cloned as cDNA. It is clear from Table 1 and comparison of the cDNA deduced amino acid sequences that, unlike the β-glucosidases, there is little structural similarity among the α-hydroxynitrile lyases from different taxonomic groups. The only two enzymes which show sequence homology are those from cassava and *Hevea*, which are both members of the Euphorbiaceae. This structural heterogeneity is confirmed by serological cross-reactivity of hydroxynitrile lyases. Thus, for example, antibodies raised against the cassava enzyme, cross react with the *Hevea* proteins but not with the other enzymes, including the α-hydroxynitrile lyase from flax (Linaceae), which has the same substrate.[99] These enzymes are therefore a good example of convergent evolution.

Although the deduced amino acid sequences of the cloned *Prunus* enzyme[96] and the cassava/*Hevea*[43,98] enzyme show no significant homology with other proteins in the databanks, the α-hydroxynitrile lyase from sorghum,[97] shows high sequence homology with a wheat serine carboxypeptidase. In addition, the sorghum enzyme contains the carboxypeptidase catalytic triad, Ser, Asp, His and is inhibited by serine/cysteine modifying agents. Despite the lack of sequence homology, cassava α-hydroxynitrile lyase also contains this catalytic triad and is similarly inhibited by the serine protease inhibitor, phenylmethanesulfonyl fluoride (PMSF).[103] Site-directed mutagenesis of Ser-80, which is part of a typical serine protease G-X-S-X-G/A consensus motif, in this cassava enzyme confirms that it is essential for enzyme activity.[104]

The crystal structure of the *H. brasiliensis* Muell. Arg, α-hydroxynitrile lyase has been determined at 1.9 Å resolution.[8] It belongs to the α/β hydrolase superfamily, with an active site containing the catalytic triad Ser-80, Asp-207, His-207 deeply buried within the protein and connected to the surface by a narrow tunnel. By analogy to other α/β hydrolases, the reaction catalyzed by the *Hevea* and cassava α-hydroxynitrile lyase involves a tetrahedral hemiketal or hemiacetal intermediate, formed by nucleophilic attack of Ser-80 on the substrate, stabilized by the oxyanion hole.

1.31.5 REFERENCES

1. A. Nahrstedt, in "Cyanide Compounds in Biology," CIBA Found. Symp. 140, eds. D. Evered and S. Harnett, Wiley, Chichester, 1988, p. 1312.
2. G. Peiser, T.-T. Wang, N. E. Hoffman, S. F. Yang, H.-W. Lui, and C. T. Walsh, *Proc. Natl. Acad. Sci. USA*, 1984, **81**, 3059.
3. E. E. Conn, in "The Biochemistry of Plants. A Comprehensive Treatise: Secondary Plant Products," eds. P. K. Stumpf and E. E. Conn, Academic Press, New York, 1981, vol. 7, p. 479.
4. D. S. Seigler, in "Herbivores: Their Interactions with Secondary Plant Metabolites—the Chemical Participants," eds. G. A. Rosenthal and M. R. Berenbaum, Academic Press, New York, 1991, vol. 1, p. 35.
5. J. E. Poulton, *Plant Physiol.*, 1990, **94**, 401.
6. O. Sibbesen, B. M. Koch, B. A. Halkier, and B. L. Møller, *Proc. Natl. Acad. Sci. USA*, 1994, **91**, 9740.
7. O. Sibbesen, B. M. Koch, B. A. Halkier, and B. L. Møller, *J. Biol. Chem.*, 1995, **270**, 3506.
8. U. G. Wagner, M. Hasslacher, H. Griengl, H. Schwab, and C. Kratky, *Structure*, 1996, **4**, 811.
9. T. Barrett, C. G. Suresh, S. P. Tolley, E. J. Dodson, and M. A. Hughes, *Structure*, 1995, **3**, 951.
10. E. E. Conn, *Annu. Rev. Plant Physiol.*, 1980, **31**, 433.

11. D. S. Seigler, in "Progress in Phytochemistry," eds. L. Reinhold, J. B. Harborne, and T. Swain, Springer Verlag, Berlin, 1977, vol. 4, p. 83.
12. P. A. Lizotte and J. E. Poulton, *Plant Physiol.*, 1988, **86**, 322.
13. H. Pourmohseni, W. D. Ibenthal, R. Machinek, G. Remberg, and V. Wray, *Phytochemistry*, 1993, **33**, 295.
14. E. Swain, C.-P. Li, and J. E. Poulton, *Plant Physiol.*, 1992, **98**, 1423.
15. C. R. Smith, D. Weisleder, and R. W. Miller, *J. Org. Chem.*, 1980, **45**, 507.
16. S. G. Saupe, in "Phytochemistry and Angiosperm Phylogeny," eds. D. A. Young and D. S. Seigler, Praeger, New York, 1981, p. 80.
17. M. A. Hughes, *Heredity*, 1991, **66**, 105.
18. D. A. Jones, in"Cyanide Compounds in Biology," CIBA Found. Symp. 140, eds. D. Evered and S. Harnett, Wiley, Chichester, 1988, p. 151.
19. H. Daday, *Heredity*, 1954, **8**, 61.
20. H. Daday, *Heredity*, 1954, **8**, 377.
21. G. A. Pederson, T. E. Fairbrother, and S. L. Greene, *Crop Sci.*, 1996, **36**, 427.
22. P. J. Schappert and J. S. Shore, *Heredity*, 1995, **74**, 392.
23. D. B. Collinge and M. A. Hughes, *J. Exp. Bot.*, 1982, **33**, 154.
24. P. N. Okolie and B. N. Obasi, *Phytochemistry*, 1993, **33**, 775.
25. D. Selmar, R. Lieberei, N. Junqueira, and B. Biehl, *Phytochemistry*, 1991, **30**, 2135.
26. I. Shreiner, D. Nafus, and D. Pimentel, *Ecol. Entomol.*, 1984, **9**, 69.
27. A. C. Bellotti and L. Riis, in "Acta Horticulturae Number 375: International Workshop on Cassava Safety," eds. M. Bokanga, A. J. A. Essers, N. Poulter, H. Rosling, and O. Tewe, WOCAS/ISIS/ISATRC, Wageningen, Netherlands, 1994, p. 141.
28. D. Raubenheimer, M.Sc. Thesis, University of Capetown, 1987.
29. G. Holzkamp and A. Nahrstedt, *Insect Biochem. Mol. Biol.*, 1994, **24**, 161.
30. K. Honda, W. Nishii, and N. Hayashi, *J. Chem. Ecol.*, 1997, **23**, 323.
31. J. Lehman, E. Meister, A. Gutzwiller, F. Jans, J. P. Charles, and J. Blum, *Rev. Suisse Agric.*, 1991, **23**, 107.
32. R. Lieberei, R. Fock, and B. Biehl, *Angew. Bot.*, 1996, **70**, 230.
33. R. Lieberei, *J. Phytopathol.*, 1988, **122**, 54.
34. D. Selmar, R. Lieberei, and B. Biehl, *Plant Physiol.*, 1988, **86**, 711.
35. C. R. Smith Jr, D. Weisleder, R. W. Miller, I. S. Palmer, and O. E. Olsen, *J. Org. Chem.*, 1980, **45**, 507.
36. D. Selmar, Z. Irandoost, and V. Wray, *Phytochemistry*, 1996, **43**, 569.
37. S. R. A. Adewusi, *Plant Physiol.*, 1990, **94**, 1219.
38. M. Kojima, J. E. Poulton, S. S. Thayer, and E. E. Conn, *Plant Physiol.*, 1979, **63**, 1022.
39. S. S. Thayer and E. E. Conn, *Plant Physiol.*, 1981, **67**, 617.
40. W. L. B. White, J. M. McMahon, and R. T. Sayre, in "Acta Horticulturae Number 375: International Workshop on Cassava Safety," eds. M. Bokanga, A. J. A. Essers, N. Poulter, H. Rosling, and O. Tewe, WOCAS/ISIS/ISATRC, Wageningen, Netherlands, 1994, p. 69.
41. A. Pancoro and M. A. Hughes, *Plant J.*, 1992, **2**, 821.
42. M. Elias, B. Nambisan, and P. R. Sudhakaran, *Arch. Biochem. Biophys.*, 1997, **341**, 222.
43. J. Hughes, F. J. P. De C. Carvalho, and M. A. Hughes, *Arch. Biochem. Biophys.*, 1994, **311**, 496.
44. P. Kakes, *Planta*, 1985, **166**, 156.
45. M. Frehner and E. E. Conn, *Plant Physiol.*, 1987, **84**, 1296.
46. C. Gruhnert, B. Biehl, and D. Selmar, *Planta*, 1994, **195**, 36.
47. H.-C. Wu and J. E. Poulton, *Plant Physiol.*, 1991, **96**, 1329.
48. E. Swain, C. P. Li, and J. E. Poulton, *Plant Physiol.*, 1992, **98**, 1423.
49. E. Swain and J. E. Poulton, *Plant Physiol.*, 1994, **106**, 437.
50. J. E. Poulton and C. P. Li, *Plant Physiol.*, 1994, **104**, 29.
51. L. Zheng and J. E. Poulton, *Plant Physiol.*, 1995, **109**, 31.
52. B. R. Maslin, E. E. Conn, and J. E. Dunn, *Phytochemistry*, 1985, **24**, 961.
53. J. H. Cock, in "Cassava: New Potential for a Neglected Crop," Westfield Press, London, 1985, p. 191.
54. D. B. Collinge and M. A. Hughes, *Plant Sci. Lett.*, 1984, **34**, 119.
55. J. Lykkesfeldt and B. L. Møller, *Acta Chem. Scand.*, 1994, **48**, 178.
56. R. Cook, N. McCaig, J. M. B. McMillan, and W. B. Lumsden, *J. Inst. Brew.*, 1990, **96**, 233.
57. A. Nahrstedt, in "Biologically Active Natural Products," eds. K. Hostettmann and P. J. Lea, Clarendon Press, Oxford, 1987, p. 213.
58. J. Koukol, P. Miljanich, and E. E. Conn, *J. Biol. Chem.*, 1962, **237**, 3223.
59. I. J. McFarlane, E. M. Lees, and E. E. Conn, *J. Biol. Chem.*, 1975, **250**, 4708.
60. M. Shimada and E. E. Conn, *Arch. Biochem. Biophys.*, 1977, **180**, 199.
61. B. L. Møller and E. E. Conn, *J. Biol. Chem.*, 1979, **254**, 8575.
62. B. A. Halkier, C. E. Olsen, and B. L. Møller, *J. Biol. Chem.*, 1989, **264**, 19 487.
63. B. A. Halkier and B. L. Møller, *J. Biol. Chem.*, 1990, **265**, 21 114.
64. B. A. Halkier, J. Lykkesfeldt, and B. L. Møller, *Proc. Natl Acad. Sci. USA*, 1991, **88**, 487.
65. B. A. Halkier and B. L. Møller, *Plant Physiol.*, 1991, **96**, 10.
66. B. A. Halkier, H. L. Nielsen, B. M. Koch, and B. L. Møller, *Arch. Biochem. Biophys.*, 1995, **322**, 369.
67. B. M. Koch, O. Sibbesen, B. A., Halkier, I. Svendsen, and B. L. Møller, *Arch. Biochem. Biophys.*, 1995, **323**, 177.
68. D. B. Collinge and M. A. Hughes, *Arch. Biochem. Biophys.*, 1982, **218**, 38.
69. A. J. Cutler and E. E. Conn, *Arch. Biochem. Biophys.*, 1981, **212**, 468.
70. B. M. Koch, V. S. Nielsen, B. A. Halkier, C. E. Olsen, and B. L. Møller, *Arch. Biochem. Biophys.*, 1992, **292**, 141.
71. D. B. Collinge and M. A. Hughes, *Plant Sci. Lett.*, 1984, **34**, 119.
72. M. A. Hughes and E. E. Conn, *Phytochemistry*, 1976, **15**, 687.
73. W. Hösel and A. Nahrstedt, *Arch. Biochem. Biophys.*, 1980, **203**, 753.
74. J. E. Poulton and S.-I. Shin, *Z. Naturforsch.*, 1983, **38c**, 369.
75. W. Hösel and O. Schiel, *Arch. Biochem. Biophys.*, 1984, **229**, 177.

76. K. Hahlbrock and E. E. Conn, *J. Biol. Chem.*, 1970, **245**, 917.
77. P. F. Reay and E. E. Conn, *J. Biol. Chem.*, 1974, **249**, 5826.
78. H. Mederacke, D. Selmar, and B. Biehl, *Angew. Bot.*, 1995, **69**, 119.
79. G. Hrazdina and G. J. Wagner, *Annu. Proc. Phytochem. Soc. Europe*, 1985, **25**, 120.
80. J. Hughes and M. A. Hughes, *DNA Sequence*, 1994, **5**, 41.
81. E. E. Conn, in "β-Glucosidases: Biochemistry and Molecular Biology," ed. A. Esen, ACS Symposium Series 533, American Chemical Society, Washington DC, 1993, p. 15.
82. T. W.-M. Fan and E. E. Conn, *Arch. Biochem. Biophys.*, 1985, **243**, 361.
83. D. Selmar, Ph.D. Thesis, University of Braunschweig, 1986.
84. T. Kasai, M. Kisimoto, and S. Kawamura, *Kagawa Diagaku Nagakubu Hokoku*, 1981, **32**, 111.
85. C. P. Li, E. Swain, and J. E. Poulton, *Plant Physiol.*, 1992, **100**, 282.
86. E. Oxtoby, M. A. Dunn, A. Pancoro, and M. A. Hughes, *Plant Mol. Biol.*, 1991, **17**, 209.
87. M. A. Hughes, K. Brown, A. Pancoro, B. S. Murray, E. Oxtoby, and J. Hughes, *Arch. Biochem. Biophys.*, 1992, **295**, 273.
88. B. Henrissat, *Biochem. J.*, 1991, **280**, 309.
89. A. Rojas, L. Arola, and A. Romeu, *Biochem. Mol. Biol. Int.*, 1995, **35**, 1223.
90. D. Trimbur, R. A. J. Warren, and S. G. Withers, in "β-Glucosidases: Biochemistry and Molecular Biology," ed. A. Esen, ACS Symposium Series 533, American Chemical Society, Washington DC, 1993, p. 42.
91. Z. Kerseztessy, L. Kiss, and M. A. Hughes, *Arch. Biochem. Biophys.*, 1994, **315**, 323.
92. S. G. Withers, R. A. J. Warren, I. P. Street, K. Rupitz, J. B. Kempton, and R. Aebersold, *J. Am. Chem. Soc.*, 1990, **112**, 5887.
93. S. Liddle, Z. Keresztessy, J. Hughes, and M. A. Hughes, *Afr. J. Root Tuber Crops*, 1997, **2**, 158.
94. M. A. Hughes and M. A. Dunn, *Plant Mol. Biol.*, 1982, **1**, 169.
95. H. Wajant and F. Effenberger, *Biol. Chem.*, 1996, **377**, 611.
96. I.-P. Cheng and J. E. Poulton, *Plant Cell Physiol.*, 1993, **34**, 1139.
97. H. Wajant, K.-W. Mundry, and K. Pfizenmaier, *Plant Mol. Biol.*, 1994, **26**, 735.
98. M. Hasslaeher, M. Schall, M. Hayn, H. Griengl, S. D. Kohlwein, and H. Schwab, *J. Biol. Chem.*, 1996, **271**, 5884.
99. H. Wajant, S. Förster, H. Böttinger, F. Effenberger, and K. Pfizenmaier, *Plant Sci.*, 1995, **108**, 1.
100. H. Wajant, S. Förster, D. Selmar, F. Effenberger, and K. Pfizenmaier, *Plant Physiol.*, 1995, **109**, 1231.
101. G. W. Kuroki and E. E. Conn, *Proc. Natl Acad. Sci. USA*, 1986, **86**, 6978.
102. L.-L. Xu, B. K. Singh, and E. E. Conn, *Arch. Biochem. Biophys.*, 1988, **266**, 256.
103. J. Hughes, J. H. Lakey, and M. A. Hughes, *Biotechnol. Bioeng.*, 1997, **53**, 332.
104. H. Wajant and K. Pfizenmaier, *J. Biol. Chem.*, 1996, **271**, 25 830.

Author Index

This Author Index comprises an alphabetical listing of the names of the authors cited in the text and the references listed at the end of each chapter in this volume.

Each entry consists of the author's name, followed by a list of numbers, for example

Templeton, J. L., 366, 385[233] (350, 366), 387[370] (363)

For each name, the page numbers for the citation in the reference list are given, followed by the reference number in superscript and the page number(s) in parentheses of where that reference is cited in the text. Where a name is referred to in text only, the page number of the citation appears with no superscript number. References cited in both the text and in the tables are included.

Although much effort has gone into eliminating inaccuracies resulting from the use of different combinations of initials by the same author, the use by some journals of only one initial, and different spellings of the same name as a result of the transliteration processes, the accuracy of some entries may have been affected by these factors.

Subject Index

PHILIP AND LESLEY ASLETT
Marlborough, Wiltshire, UK

Every effort has been made to index as comprehensively as possible, and to standardize the terms used in the index in line with the IUPAC Recommendations. In view of the diverse nature of the terminology employed by the different authors, the reader is advised to search for related entries under the appropriate headings.

The index entries are presented in letter-by-letter alphabetical sequence. Compounds are normally indexed under the parent compound name, with the substituent component separated by a comma of inversion. An entry with a prefix/locant is filed after the same entry without any attachments, and in alphanumerical sequence. For example, 'diazepines', '1,4-diazepines', and '2,3-dihydro-1,4-diazepines' will be filed as:-

diazepines
1,4-diazepines
1,4-diazepines, 2,3-dihydro-

The Index is arranged in set-out style, with a maximum of three levels of heading. Location references refer to volume number (in bold) and page number (separated by a comma); major coverage of a subject is indicated by bold, elided page numbers; for example;

triterpene cyclases, **299–320**
  amino acids, 315

See cross-references direct the user to the preferred term; for example,

olefins *see* alkenes

See also cross-references provide the user with guideposts to terms of related interest, from the broader term to the narrower term, and appear at the end of the main heading to which they refer, for example,

thiones
  see also thioketones